Glencoe

ADVANCED Mathematical Concepts

Precalculus with Applications

New York, New York Columbus, Ohio Woodland Hills, California Peoria, Illinois

Visit the Glencoe Mathematics Internet Site for
Advanced Mathematical Concepts: Precalculus with Applications at

www.amc.glencoe.com

You'll find:

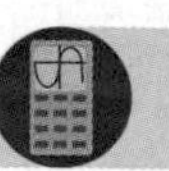

- Research Helps
- Data Updates
- Career Data
- Graphing Calculator Programs
- Review Activities
- Test Practice

links to Web sites relevant to Unit Projects, Career Choices features, exercises, and much more.

Glencoe/McGraw-Hill

A Division of The **McGraw·Hill** *Companies*

Send all inquiries to:
Glencoe/McGraw-Hill
8787 Orion Place
Columbus, OH 43240

ISBN: 0-02-834175-9 (Student Edition)
0-02-834176-7 (Teacher's Wraparound Edition)

7 8 9 10 11 12 071/055 10 09 08 07 06 05 04 03

AUTHORS

Counterclockwise from lower center: Berchie Holliday, Gilbert Cuevas, John Carter, Daniel Marks, Melissa McClure

Berchie Holliday
Mathematics Teacher
Northwest Local School District
Cincinnati, Ohio

Gilbert J. Cuevas
Professor of Mathematics Education
University of Miami
Miami, Florida

Melissa S. McClure
Mathematics Consultant
Teaching for Tomorrow
Weatherford, Texas

John A. Carter
Mathematics Department Chairperson
Community High School
West Chicago, Illinois

Daniel Marks
Associate Professor of Mathematics
Auburn University Montgomery
Montgomery, Alabama

ACADEMIC CONSULTANTS AND REVIEWERS

Each of the Academic Consultants read all 15 chapters, while each Teacher Reviewer read at least two chapters. The Consultants and Reviewers gave suggestions for improving the Student Edition and the Teacher's Wraparound Edition.

Academic Consultants

Glenn Gould
Mathematics Teacher/Educational Consultant
Liberty High School
Bealeton, Virginia

Dr. Roland Minton
Co-author of *McGraw-Hill Calculus*
Professor of Mathematics
Roanoke College
Salem, Virginia

Barbara Nunn
Secondary Mathematics Curriculum Specialist
Broward County Schools
Fort Lauderdale, Florida

Margene Ryberg
Mathematics Teacher/Department Chairperson
Northwest High School
Cincinnati, Ohio

Dr. Robert Smith
Co-author of *McGraw-Hill Calculus*
Professor of Mathematics and Department Chair
Millersville University of Pennsylvania
Millersville, Pennsylvania

Dr. Paul Zitzewitz
Professor of Physics
University of Michigan, Dearborn
Dearborn, Michigan

Teacher Reviewers

Debra K. Anderson
Mathematics Teacher
Renton High School
Renton, Washington

Sue Ellen Baker
Mathematics Teacher
Hoover High School
Hoover, Alabama

Melody Boring
Mathematics Coordinator, Grades 7–12
St. Joseph School District
St. Joseph, Missouri

Mary S. Burkholder
Mathematics Department Chairperson
Chambersburg Area Senior High School
Chambersburg, Pennsylvania

Rod Camden
Mathematics Department Chairperson
E. C. Glass High School
Lynchburg, Virginia

Robyn W. Carlin
La SIP Site Coordinator
East Baton Rouge Parish Schools
Baton Rouge, Louisiana

Mark Daniels
Mathematics Department Chairperson
McNeil High School
Austin, Texas

Dr. Nestor Diaz
Mathematics and Computer Science Department Chairperson
Coral Gables Senior High School
Coral Gables, Florida

James R. Ebbert
Mathematics Teacher
Deland High School
Deland, Florida

Everett J. Gaston
Mathematics Department Chairperson
Bullard High School
Fresno, California

Nancy Donlon Grigassy
Mathematics Department Chairperson
Stephen F. Austin High School
Sugar Land, Texas

Rebecca M. Gummerson
District Mathematics Curriculum Coordinator
Burley High School
Burley, Idaho

Glenn Hoit
Mathematics Department Co-Chair
Air Academy High School
United States Air Force Academy
Colorado

Paul Kunes
Mathematics Instructor
Papillion LaVista High School
Papillion, Nebraska

Carol Marsteiner
Mathematics Department Chairperson
Lyons High School
Lyons, New York

Gerald E. Martau
Retired Mathematics Instructor
Lakewood City Schools
Lakewood, Ohio

Albert H. Mauthe, Ed.D.
Supervisor of Mathematics
Norristown Area School District
Norristown, Pennsylvania

Kathleen McKinley
Mathematics Teacher/Consultant
School District of Philadelphia
Philadelphia, Pennsylvania

Movelle Murdock
Mathematics Department Chairperson
Alexander High School
Douglasville, Georgia

Marilyn J. Parker
Technology Consultant
Clark County School District
Las Vegas, Nevada

Ron Phillis
Mathematics Department Chairperson
Troy High School
Troy, Ohio

H. Russell Pittman
Mathematics Department Chairperson
Abington Senior High School
Abington, Pennsylvania

Kenneth Rockwell
Mathematics Teacher
Theodore High School
Theodore, Alabama

Paul H. Sanderson
Mathematics Teacher
Hickman High School
Columbia, Missouri

Sandy Schoff
Mathematics Coordinator
Anchorage School District
Anchorage, Alaska

Sharon Schueler
Mathematics Department Chairperson
Roosevelt High School
Sioux Falls, South Dakota

Vicki D. Shull
Mathematics Teacher
Goose Creek High School
Goose Creek, South Carolina

Amie Sketel
Mathematics Educator
Union Local High School
Belmont, Ohio

Roberta L. Stewart
Mathematics Department Chairperson
University City High School
University City, Missouri

Joyce Svabek
Mathematics Department Chairperson
East Lake High School
Tarpon Springs, Florida

David Thiel
Mathematics Department Chairperson
Green Valley High School
Henderson, Nevada

Dickie Thomasson
Mathematics Teacher
Jacksonville High School
Jacksonville, Arkansas

Bobby G. Tyus
Teacher
Mt. Spokane High School
Spokane, Washington

Cynthia D. Westermann
Mathematics Teacher
Amelia High School
Batavia, Ohio

Rick Westfall
Mathematics Teacher
Dunbar High School-HSEP
Fort Worth, Texas

UNIT 1 TABLE OF CONTENTS

RELATIONS, FUNCTIONS, AND GRAPHS

Unit 1 *interNET* Projects

TELECOMMUNICATION
Introduction, 3
Projects, 61, 123, 201, 271

Matrices, 97

Chapter 1 LINEAR RELATIONS AND FUNCTIONS 4

Chapter 2 SYSTEMS OF LINEAR EQUATIONS AND INEQUALITIES 66

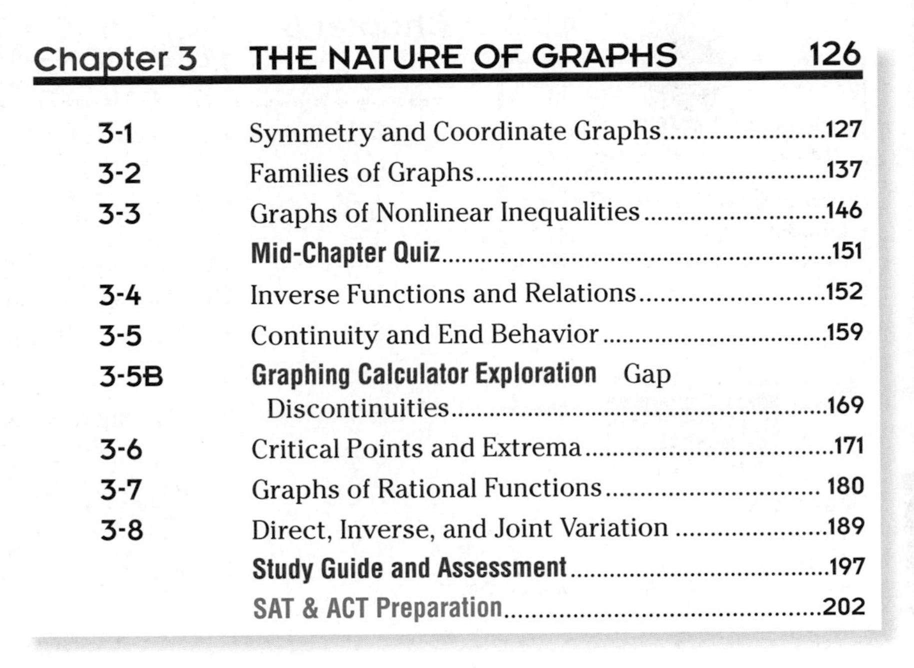

CAREER CHOICES

*inter*NET CONNECTION

GRAPHING CALCULATOR EXPLORATIONS

Chapter 3 THE NATURE OF GRAPHS 126

Chapter 4 POLYNOMIAL AND RATIONAL FUNCTIONS 204

TRIGONOMETRY

Unit 2 *interNET* Projects

THE CYBERCLASSROOM
Introduction, 275
Projects, 339, 417, 481, 547

HISTORY of MATHEMATICS

Angles, 319
Functions, 367
Trigonometry, 462
Angle Measure, 534

Chapter 5 THE TRIGONOMETRIC FUNCTIONS 276

Chapter 6 GRAPHS OF TRIGONOMETRIC FUNCTIONS 342

CAREER CHOICES

interNET CONNECTION

GRAPHING CALCULATOR EXPLORATIONS

Chapter 7 TRIGONOMETRIC IDENTITIES AND EQUATIONS 420

Chapter 8 VECTORS AND PARAMETRIC EQUATIONS 484

UNIT 3 ADVANCED FUNCTIONS AND GRAPHING

Unit 3 interNET Projects

SPACE—THE FINAL FRONTIER

Introduction, 551
Projects, 611, 691, 753

Chapter 9 POLAR COORDINATES AND COMPLEX NUMBERS 552

Chapter 10 CONICS 614

CAREER CHOICES

*inter*NET CONNECTION

GRAPHING CALCULATOR EXPLORATIONS

Chapter 11 EXPONENTIAL AND LOGARITHMIC FUNCTIONS 694

Lesson 10-7, page 675

UNIT

DISCRETE MATHEMATICS

Unit 4 *inter*NET Projects

THE UNITED STATES CENSUS BUREAU

CAREER CHOICES

*inter*NET CONNECTION

Chapter 12 SEQUENCES AND SERIES 758

Chapter 13 COMBINATORICS AND PROBABILITY 836

Lesson 12-5, page 798

Chapter 14 STATISTICS AND DATA ANALYSIS 888

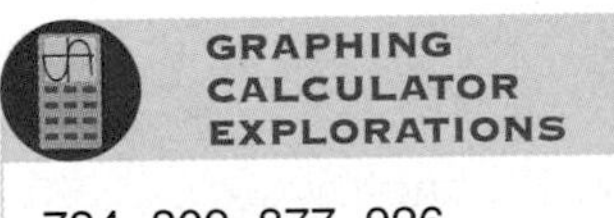

UNIT 5 CALCULUS

Chapter 15 INTRODUCTION TO CALCULUS 940

Unit 5 interNET Project

DISEASES

CAREER CHOICES

HISTORY of MATHEMATICS

interNET CONNECTION

GRAPHING CALCULATOR EXPLORATIONS

Student Handbook

USING A GRAPHING CALCULATOR

***inter*NET CONNECTION**

To find keystroke instructions for each Graphing Calculator Exploration, visit **www.amc. glencoe.com**

In ***Glencoe Advanced Mathematical Concepts,*** the graphing calculator is an important tool in exploring precalculus topics. A graphing calculator is actually a specialized hand-held computer. **Graphing Calculator Explorations** throughout this text offer opportunties for you to use a graphing calculator to explore the mathematical topics you are studying. You will also see graphing calculator screens actually generated by a graphing calculator. If you do not recognize the keys, commands, or screens shown in this test, you may have a different model of calculator. Refer to your owner's manual to verify the proper commands and keystrokes to use.

Graphing Calculator Appendix

At the back of this text, you will find a **Student Handbook.** The **Graphing Calculator Appendix** begins on page A2. This appendix introduces you to some of the common functions of the graphing calculator most widely used in high school classrooms. It is divided into nine sections:

1. Introduction to the Graphing Calculator
2. Graphing Functions
3. Analyzing Functions
4. Graphing Inequalities
5. Matrices
6. Graphing Trigonometric Functions
7. Graphing Special Functions
8. Parametric and Polar Equations
9. Statistics and Statistical Graphs

In each section, you will find keystroke instructions for specific examples that demonstrate some of the basic operations and most common menus available on the calculator. Whenever you have specific questions about commands or procedures, you should refer to your owner's manual for instruction.

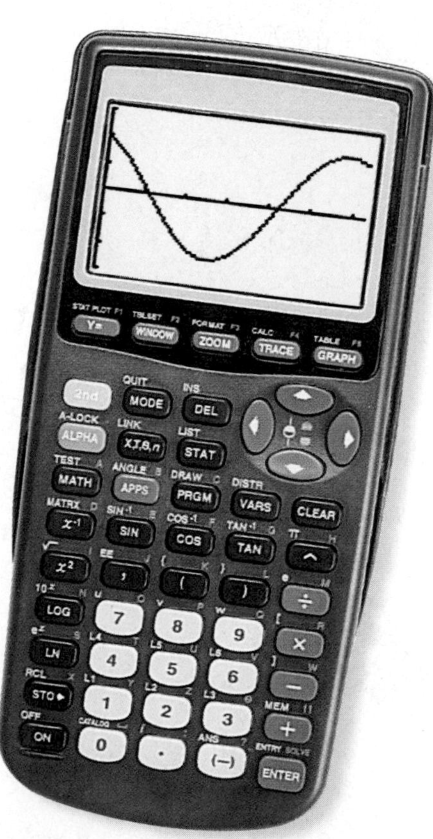

SAT and ACT Preparation

It is important to note that graphing calculators are permitted when taking the ACT or SAT and are required for the AP Calculus test. Through your experiences with the graphing calculator in studying ***Glencoe Advanced Mathematical Concepts,*** you should become comfortable with its operation and appreciate the power of this tiny computer.

UNIT 1

Relations, Functions, and Graphs

Throughout this text, you will see that many real-world phenomena can be modeled by special relations called functions that can be written as equations or graphed. As you work through Unit 1, you will study some of the tools used for mathematical modeling.

Unit 1 interNET Projects

TELECOMMUNICATION

In today's world, there are various forms of communication, some that boggle the mind with their speed and capabilities. In this project, you will use the Internet to help you gather information for investigating various aspects of modern communication. At the end of each chapter, you will work on completing the Unit 1 Internet Project. Here are the topics for each chapter.

CHAPTER 1
(page 61)

Is Anybody Listening? Everyday that you watch television, you are bombarded by various telephone service commercials offering you the best deal for your dollar.
Math Connection: How could you use the Internet and graph data to help determine the best deal for you?

CHAPTER 2
(page 123)

You've Got Mail! The number of homes connected to the Internet and e-mail is on the rise. Use the Internet to find out more information about the types of e-mail and Internet service providers available and their costs.
Math Connection: Use your data and a system of equations to determine if any one product is better for you.

CHAPTER 3
(page 201)

Sorry, You Are Out of Range for Your Telephone Service ... Does your family have a cell phone? Is its use limited to a small geographical area? How expensive is it? Use the Internet to analyze various offers for cellular phone service.
Math Connection: Use graphs to describe the cost of each type of service. Include initial start-up fees or equipment cost, beginning service offers, and actual service fees.

CHAPTER 4
(page 271)

The Pen is Mightier Than the Sword! Does anyone write letters by hand anymore? Maybe fewer people are writing by pen, but most people use computers to write letters, reports, and books. Use the Internet to discover various types of word processing, graphics, spreadsheet, and presentation software that would help you prepare your Unit 1 presentation.
Math Connection: Create graphs using computer software to include in your presentation.

interNET CONNECTION For more information on the Unit Project, visit: **www.amc.glencoe.com**

LINEAR RELATIONS AND FUNCTIONS

CHAPTER OBJECTIVES

- **Determine whether a given relation is a function and perform operations with functions.** *(Lessons 1-1, 1-2)*
- **Evaluate and find zeros of linear functions using functional notation.** *(Lesson 1-1, 1-3)*
- **Graph and write functions and inequalities.** *(Lessons 1-3, 1-4, 1-7, 1-8)*
- **Write equations of parallel and perpendicular lines.** *(Lesson 1-5)*
- **Model data using scatter plots and write prediction equations.** *(Lesson 1-6)*

1-1 Relations and Functions

OBJECTIVES
- Determine whether a given relation is a function.
- Identify the domain and range of a relation or function.
- Evaluate functions.

METEOROLOGY Have you ever wished that you could change the weather? One of the technologies used in weather management is cloud seeding. In cloud seeding, microscopic particles are released in a cloud to bring about rainfall. The data in the table show the number of acre-feet of rain from pairs of similar unseeded and seeded clouds.

An acre-foot is a unit of volume equivalent to one foot of water covering an area of one acre. An acre-foot contains 43,560 cubic feet or about 27,154 gallons.

Acre-Feet of Rain	
Unseeded Clouds	**Seeded Clouds**
1.0	4.1
4.9	17.5
4.9	7.7
11.5	31.4
17.3	32.7
21.7	40.6
24.4	92.4
26.1	115.3
26.3	118.3
28.6	119.0

Source: Wadsworth International Group

We can write the values in the table as a set of ordered pairs. A pairing of elements of one set with elements of a second set is called a **relation.** The first element of an ordered pair is the *abscissa.* The set of abscissas is called the **domain** of the relation. The second element of an ordered pair is the *ordinate.* The set of ordinates is called the **range** of the relation. *Sets D and R are often used to represent domain and range.*

Relation, Domain, and Range

A relation is a set of ordered pairs. The domain is the set of all abscissas of the ordered pairs. The range is the set of all ordinates of the ordered pairs.

Example 1

METEOROLOGY State the relation of the rain data above as a set of ordered pairs. Also state the domain and range of the relation.

Relation: {(28.6, 119.0), (26.3, 118.3), (26.1, 115.3), (24.4, 92.4), (21.7, 40.6), (17.3, 32.7), (11.5, 31.4), (4.9, 17.5), (4.9, 7.7), (1.0, 4.1)}

Domain: {1.0, 4.9, 11.5, 17.3, 21.7, 24.4, 26.1, 26.3, 28.6}

Range: {4.1, 7.7, 31.4, 17.5, 32.7, 40.6, 92.4, 115.3, 118.3, 119.0}

There are multiple representations for each relation. You have seen that a relation can be expressed as a set of ordered pairs. Those ordered pairs can also be expressed as a table of values. The ordered pairs can be graphed for a pictorial representation of the relation. Some relations can also be described by a rule or equation relating the first and second coordinates of each ordered pair.

Example 2 **The domain of a relation is all positive integers less than 6. The range y of the relation is 3 less x, where x is a member of the domain. Write the relation as a table of values and as an equation. Then graph the relation.**

Table:

x	y
1	2
2	1
3	0
4	−1
5	−2

Graph:

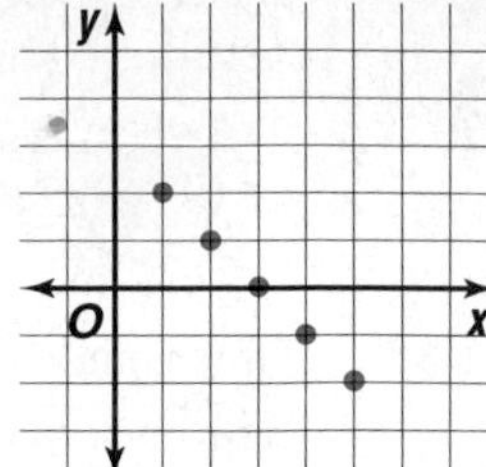

Equation: $y = 3 - x$

You can use the graph of a relation to determine its domain and range.

Example 3 **State the domain and range of each relation.**

a.

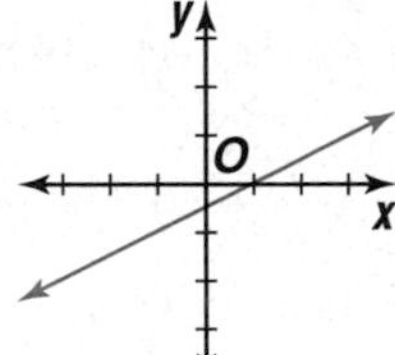

It appears from the graph that all real numbers are included in the domain and range of the relation.

b.

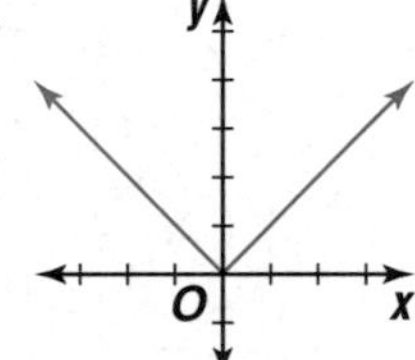

It appears from the graph that all real numbers are included in the domain. The range includes the non-negative real numbers.

The relations in Example 3 are a special type of relation called a **function.**

Function	A function is a relation in which each element of the domain is paired with exactly one element in the range.

Example 4 **State the domain and range of each relation. Then state whether the relation is a function.**

a. {(−3, 0), (4, −2), (2, −6)}

The domain is {−3, 2, 4}, and the range is {−6, −2, 0}. Each element of the domain is paired with exactly one element of the range, so this relation is a function.

b. {(4, −2), (4, 2), (9, −3), (−9, 3)}

For this relation, the domain is {−9, 4, 9}, and the range is {−3, −2, 2, 3}. In the domain, 4 is paired with two elements of the range, −2 and 2. Therefore, this relation is *not* a function.

An alternate definition of a function is a set of ordered pairs in which no two pairs have the same first element. This definition can be applied when a relation is represented by a graph. If every vertical line drawn on the graph of a relation passes through no more than one point of the graph, then the relation is a function. This is called the **vertical line test.**

a relation that is a function	a relation that is not a function
y, O, x	y, O, x

Example 5 **Determine if the graph of each relation represents a function. Explain.**

a. 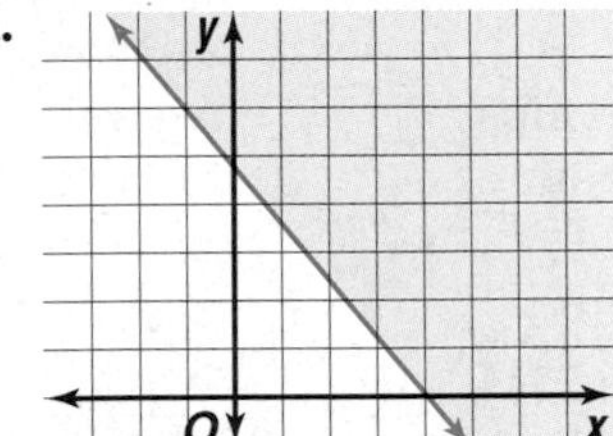

No, the graph does not represent a function. A vertical line at $x = 1$ would pass through infinitely many points.

b. 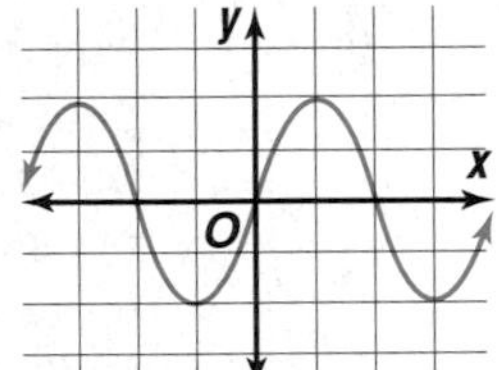

Every element of the domain is paired with exactly one element of the range. Thus, the graph represents a function.

x is called the independent variable, and y is called the dependent variable.

Any letter may be used to denote a function. In **function notation,** the symbol $f(x)$ is read "f of x" and should be interpreted as the value of the function f at x. Similarly, $h(t)$ is the value of function h at t. The expression $y = f(x)$ indicates that for each element in the domain that replaces x, the function assigns one and only one replacement for y. The ordered pairs of a function can be written in the form (x, y) or $(x, f(x))$.

Every function can be evaluated for each value in its domain. For example, to find $f(-4)$ if $f(x) = 3x^3 - 7x^2 - 2x$, evaluate the expression $3x^3 - 7x^2 - 2x$ for $x = -4$.

Example 6 **Evaluate each function for the given value.**

a. $f(-4)$ if $f(x) = 3x^3 - 7x^2 - 2x$

$$f(-4) = 3(-4)^3 - 7(-4)^2 - 2(-4)$$
$$= -192 - 112 - (-8) \text{ or } -296$$

b. $g(9)$ if $g(x) = |6x - 77|$

$$g(9) = |6(9) - 77|$$
$$= |-23| \text{ or } 23$$

Functions can also be evaluated for another variable or an expression.

Example 7 **Evaluate each function for the given value.**

a. $h(a)$ **if** $h(x) = 3x^7 - 10x^4 + 3x - 11$

$$h(a) = 3(a)^7 - 10(a)^4 + 3(a) - 11 \quad x = a$$
$$= 3a^7 - 10a^4 + 3a - 11$$

b. $j(c - 5)$ **if** $j(x) = x^2 - 7x + 4$

$$j(c - 5) = (c - 5)^2 - 7(c - 5) + 4 \quad x = c - 5$$
$$= c^2 - 10c + 25 - 7c + 35 + 4$$
$$= c^2 - 17c + 64$$

When you are given the equation of a function but the domain is not specified, the domain is all real numbers for which the corresponding values in the range are also real numbers.

Example 8 **State the domain of each function.**

a. $f(x) = \dfrac{x^3 + 5x}{x^2 - 4x}$

Any value that makes the denominator equal to zero must be excluded from the domain of f since division by zero is undefined. To determine the excluded values, let $x^2 - 4x = 0$ and solve.

$$x^2 - 4x = 0$$
$$x(x - 4) = 0$$
$$x = 0 \text{ or } x = 4$$

Therefore, the domain includes all real numbers except 0 and 4.

b. $g(x) = \dfrac{1}{\sqrt{x - 4}}$

Any value that makes the radicand negative must be excluded from the domain of g since the square root of a negative number is not a real number. Also, the denominator cannot be zero. Let $x - 4 \le 0$ and solve for the excluded values.

$$x - 4 \le 0$$
$$x \le 4$$

The domain excludes numbers less than or equal to 4. The domain is written as $\{x \mid x > 4\}$, which is read "the set of all x such that x is greater than 4."

CHECK FOR UNDERSTANDING

Communicating Mathematics

Read and study the lesson to answer each question.

1. **Represent** the relation {(−4, 2), (6, 1), (0, 5), (8, −4), (2, 2), (−4, 0)} in two other ways.
2. **Draw** the graph of a relation that is not a function.
3. **Describe** how to use the vertical line test to determine whether the graph at the right represents a function.

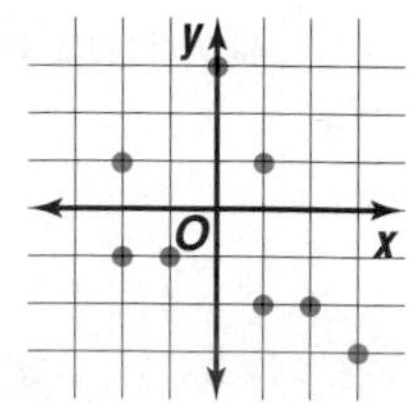

4. **You Decide** Keisha says that all functions are relations but not all relations are functions. Kevin says that all relations are functions but not all functions are relations. Who is correct and why?

Guided Practice

5. The domain of a relation is all positive integers less than 8. The range y of the relation is x less 4, where x is a member of the domain. Write the relation as a table of values and as an equation. Then graph the relation.

State each relation as a set of ordered pairs. Then state the domain and range.

6.

x	y
−3	4
0	0
3	−4
6	−8

7.

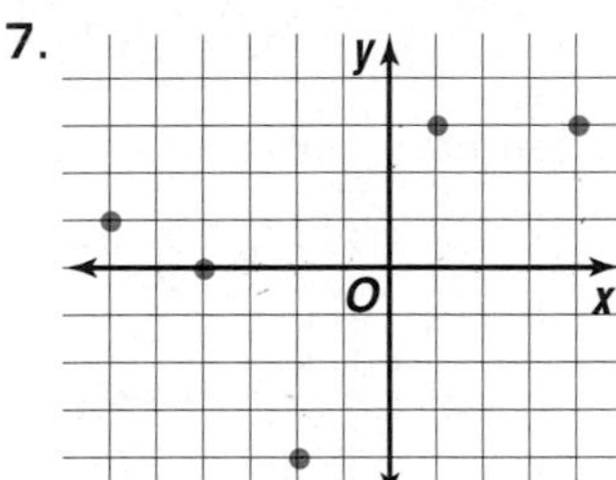

Given that x is an integer, state the relation representing each equation by making a table of values. Then graph the ordered pairs of the relation.

8. $y = 3x + 5$ and $-4 \leq x \leq 4$

9. $y = -5$ and $1 \leq x \leq 8$

State the domain and range of each relation. Then state whether the relation is a function. Write *yes* or *no*. Explain.

10. $\{(1, 2), (2, 4), (-3, -6), (0, 0)\}$

11. $\{(6, -2), (3, 4), (6, -6), (-3, 0)\}$

12. Study the graph at the right.
    a. State the domain and range of the relation.
    b. State whether the graph represents a function. Explain.

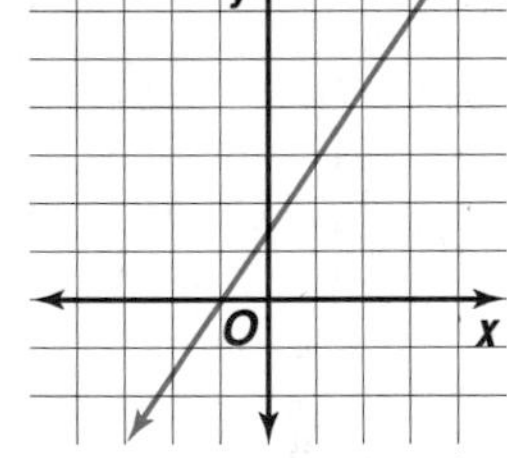

Evaluate each function for the given value.

13. $f(-3)$ if $f(x) = 4x^3 + x^2 - 5x$

14. $g(m + 1)$ if $g(x) = 2x^2 - 4x + 2$

15. State the domain of $f(x) = \sqrt{x + 1}$.

16. **Sports** The table shows the heights and weights of members of the Los Angeles Lakers basketball team during a certain year.
    a. State the relation of the data as a set of ordered pairs. Also state the domain and range of the relation.
    b. Graph the relation.
    c. Determine whether the relation is a function.

Los Angeles Lakers

Height (in.)	Weight (lb)
83	240
81	220
82	245
78	200
83	255
73	200
80	215
77	210
78	190
73	180
86	300
77	220
82	260

Source: Preview Sports

EXERCISES

Practice

Write each relation as a table of values and as an equation. Graph the relation.

17. the domain is all positive integers less than 10, the range is 3 times x, where x is a member of the domain

18. the domain is all negative integers greater than -7, the range is x less 5, where x is a member of the domain

19. the domain is all integers greater than -5 and less than or equal to 4, the range is 8 more than x, where x is a member of the domain

State each relation as a set of ordered pairs. Then state the domain and range.

20.

x	y
−5	−5
−3	−3
−1	−1
1	1

21.

x	y
−10	0
−5	0
0	0
5	0

22.

x	y
4	0
5	1
8	0
13	1

23.
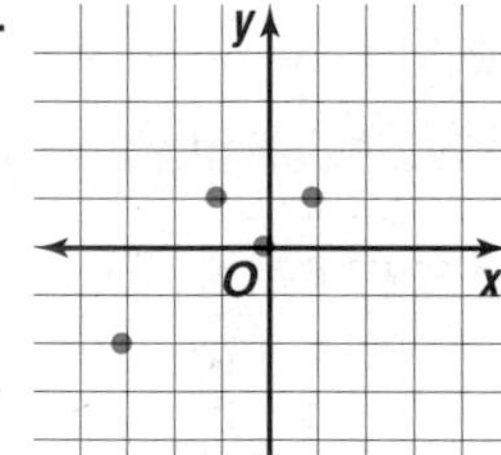

24.
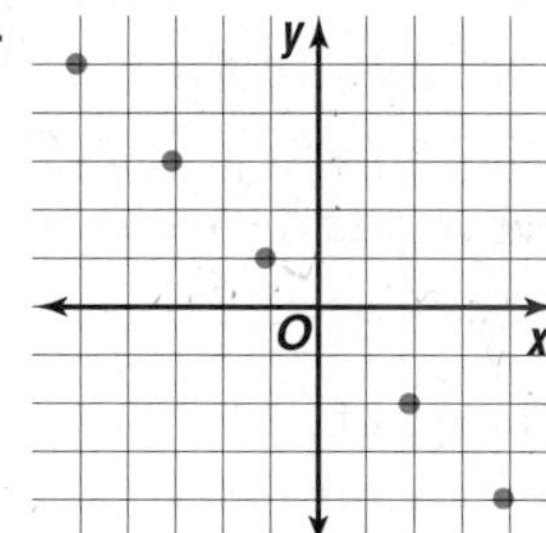

25.
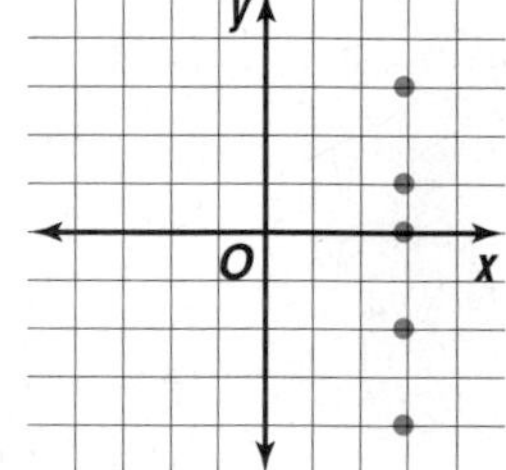

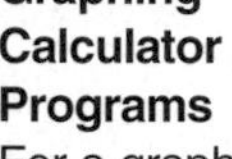

***inter*NET CONNECTION**

Graphing Calculator Programs
For a graphing calculator program that plots points in a relation, visit **www.amc.glencoe.com**

Given that x is an integer, state the relation representing each equation by making a table of values. Then graph the ordered pairs of the relation.

26. $y = x - 5$ and $-4 \leq x \leq 1$

27. $y = -x$ and $1 \leq x < 7$

28. $y = |x|$ and $-5 \leq x \leq 1$

29. $y = 3x - 3$ and $0 < x < 6$

30. $y^2 = x - 2$ and $x = 11$

31. $|2y| = x$ and $x = 4$

State the domain and range of each relation. Then state whether the relation is a function. Write *yes* or *no*. Explain.

32. $\{(4, 4), (5, 4), (6, 4)\}$

33. $\{(1, -2), (1, 4), (1, -6), (1, 0)\}$

34. $\{(4, -2), (4, 2), (1, -1), (1, 1), (0, 0)\}$

35. $\{(0, 0), (2, 2), (2, -2), (5, 8), (5, -8)\}$

36. $\{(-1.1, -2), (-0.4, -1), (-0.1, -1)\}$

37. $\{(2, -3), (9, 0), (8, -3), (-9, 8)\}$

For each graph, state the domain and range of the relation. Then explain whether the graph represents a function.

38.
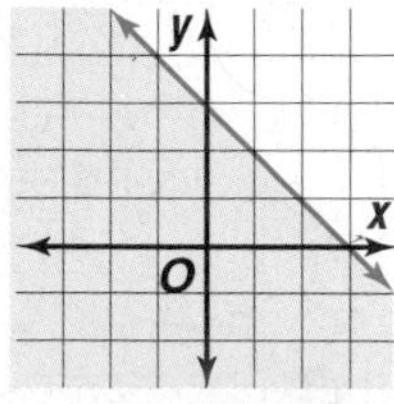

39.
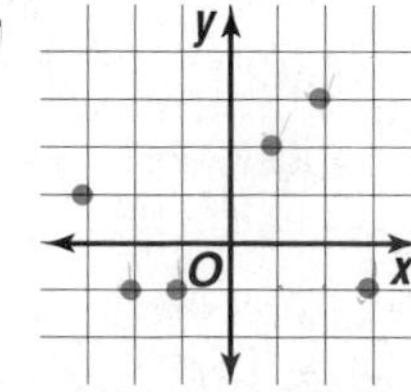

40.
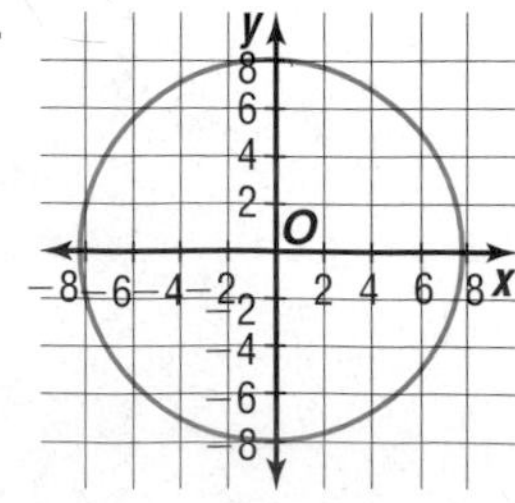

Evaluate each function for the given value.

41. $f(3)$ if $f(x) = 2x + 3$

42. $g(-2)$ if $g(x) = 5x^2 + 3x - 2$

43. $h(0.5)$ if $h(x) = \frac{1}{x}$

44. $j(2a)$ if $j(x) = 1 - 4x^3$

45. $f(n - 1)$ if $f(x) = 2x^2 - x + 9$

46. $g(b^2 + 1)$ if $g(x) = \frac{3 - x}{5 + x}$

47. Find $f(5m)$ if $f(x) = |x^2 - 13|$.

State the domain of each function.

48. $f(x) = \frac{3x}{x^2 - 5}$

49. $g(x) = \sqrt{x^2 - 9}$

50. $h(x) = \frac{x + 2}{\sqrt{x^2 - 7}}$

Graphing Calculator

51. You can use the table feature of a graphing calculator to find the domain of a function. Enter the function into the Y= list. Then observe the y-values in the table. An error indicates that an x-value is excluded from the domain. Determine the domain of each function.

a. $f(x) = \frac{3}{x - 1}$

b. $g(x) = \frac{3 - x}{5 + x}$

c. $h(x) = \frac{x^2 - 12}{x^2 - 4}$

Applications and Problem Solving

52. **Education** The table shows the number of students who applied and the number of students attending selected universities.

a. State the relation of the data as a set of ordered pairs. Also state the domain and range of the relation.

b. Graph the relation.

c. Determine whether the relation is a function. Explain.

University	Number Applied	Number Attending
Auburn University	9244	3166
University of California, Davis	18,584	3697
University of Illinois-Champaign-Urbana	18,140	5805
University of Maryland	16,182	3999
State University of New York – Stony Brook	13,589	2136
The Ohio State University	18,912	5950
Texas A&M University	13,877	6233

Source: *Newsweek, "How to get into college, 1998"*

53. **Critical Thinking** If $f(2m + 1) = 24m^3 + 36m^2 + 26m$, what is $f(x)$? (*Hint:* Begin by solving $x = 2m + 1$ for m.)

54. **Aviation** The temperature of the atmosphere decreases about 5°F for every 1000 feet that an airplane ascends. Thus, if the ground-level temperature is 95°F, the temperature can be found using the function $t(d) = 95 - 0.005d$, where $t(d)$ is the temperature at a height of d feet. Find the temperature outside of an airplane at each height.

a. 500 ft
b. 750 ft
c. 1000 ft
d. 5000 ft
e. 30,000 ft

55. **Geography** A global positioning system, GPS, uses satellites to allow a user to determine his or her position on Earth. The system depends on satellite signals that are reflected to and from a hand-held transmitter. The time that the signal takes to reflect is used to determine the transmitter's position. Radio waves travel through air at a speed of 299,792,458 meters per second. Thus, the function $d(t) = 299{,}792{,}458t$ relates the time t in seconds to the distance traveled $d(t)$ in meters.

a. Find the distance a sound wave will travel in 0.05, 0.2, 1.4, and 5.9 seconds.

b. If a signal from a GPS satellite is received at a transmitter in 0.08 seconds, how far from the transmitter is the satellite?

Extra Practice See p. A26.

56. Critical Thinking $P(x)$ is a function for which $P(1) = 1$, $P(2) = 2$, $P(3) = 3$, and $P(x + 1) = \dfrac{P(x - 2)\,P(x - 1) + 1}{P(x)}$ for $x \geq 3$. Find the value of $P(6)$.

57. SAT Practice Quantitative Comparison

A if the quantity in Column A is greater
B if the quantity in Column B is greater
C if the two quantities are equal
D if the relationship cannot be determined from the information given

Column A	Column B
$3^2 + 4^2$	7^2

CAREER CHOICES

Veterinary Medicine

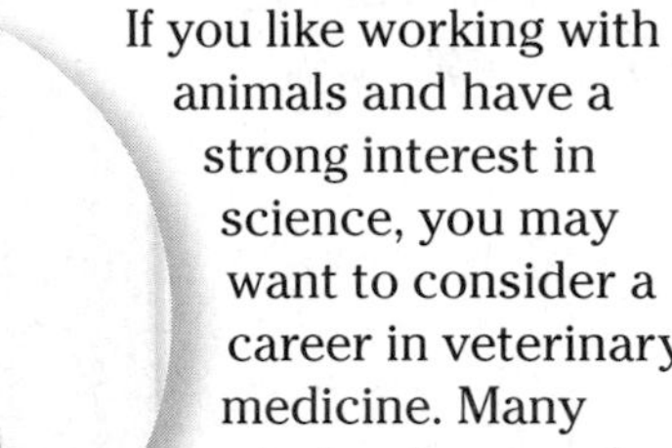

If you like working with animals and have a strong interest in science, you may want to consider a career in veterinary medicine. Many veterinarians work with small animals, such as pets, maintaining their good health and treating illnesses and injuries. Some veterinarians work with large animals, such as farm animals, to ensure the health of animals that we depend upon for food. Still other veterinarians work to control diseases in wildlife.

Duties of veterinarians can include administering medications to the animals, performing surgeries, instructing people in the care of animals, and researching genetics, prevention of disease, and better animal nutrition.

Many veterinarians work in private practice, but jobs are also available in industry and governmental agencies.

Career Overview

Degree Preferred:
D.V.M. (doctor of veterinary medicine) consisting of six years of college

Related Courses:
biology, chemistry, mathematics

Outlook:
number of jobs expected to increase through 2006

Spending on Pet Health Care

Source: American Veterinary Medical Association

***inter*NET CONNECTION** For more information on careers in veterinary medicine, visit: **www.amc.glencoe.com**

1-2 Composition of Functions

OBJECTIVES

- Perform operations with functions.
- Find composite functions.
- Iterate functions using real numbers.

BUSINESS Each year, thousands of people visit Yellowstone National Park in Wyoming. Audiotapes for visitors include interviews with early settlers and information about the geology, wildlife, and activities of the park. The revenue $r(x)$ from the sale of x tapes is $r(x) = 9.5x$. Suppose that the function for the cost of manufacturing x tapes is $c(x) = 0.8x + 1940$. What function could be used to find the profit on x tapes? *This problem will be solved in Example 2.*

To solve the profit problem, you can subtract the cost function $c(x)$ from the revenue function $r(x)$. If you have two functions, you can form new functions by adding, subtracting, multiplying, or dividing the functions.

GRAPHING CALCULATOR EXPLORATION

Use a graphing calculator to explore the sum of two functions.

- Enter the functions $f(x) = 2x - 1$ and $f(x) = 3x + 2$ as **Y1** and **Y2**, respectively.
- Enter **Y1** + **Y2** as the function for **Y3**. To enter **Y1** and **Y2**, press VARS, then select **Y-VARS.** Then choose the equation name from the menu.
- Use **TABLE** to compare the function values for **Y1**, **Y2**, and **Y3**.

TRY THESE

Use the functions $f(x) = 2x - 1$ and $f(x) = 3x + 2$ as Y1 and Y2. Use TABLE to observe the results for each definition of Y3.

1. Y3 = Y1 − Y2

2. Y3 = Y1 · Y2

3. Y3 = Y1 ÷ Y2

WHAT DO YOU THINK?

4. Repeat the activity using functions $f(x) = x^2 - 1$ and $f(x) = 5 - x$ as **Y1** and **Y2**, respectively. What do you observe?

5. Make conjectures about the functions that are the sum, difference, product, and quotient of two functions.

The Graphing Calculator Exploration leads us to the following definitions of operations with functions.

Operations with Functions

Sum: $(f + g)(x) = f(x) + g(x)$

Difference: $(f - g)(x) = f(x) - g(x)$

Product: $(f \cdot g)(x) = f(x) \cdot g(x)$

Quotient: $\left(\frac{f}{g}\right)(x) = \frac{f(x)}{g(x)},\ g(x) \neq 0$

For each new function, the domain consists of those values of x common to the domains of f and g. The domain of the quotient function is further restricted by excluding any values that make the denominator, $g(x)$, zero.

Example 1 **Given $f(x) = 3x^2 - 4$ and $g(x) = 4x + 5$, find each function.**

a. $(f + g)(x)$

$$(f + g)(x) = f(x) + g(x) = 3x^2 - 4 + 4x + 5 = 3x^2 + 4x + 1$$

b. $(f - g)(x)$

$$(f - g)(x) = f(x) - g(x) = 3x^2 - 4 - (4x + 5) = 3x^2 - 4x - 9$$

c. $(f \cdot g)(x)$

$$(f \cdot g)(x) = f(x) \cdot g(x) = (3x^2 - 4)(4x + 5) = 12x^3 + 15x^2 - 16x - 20$$

d. $\left(\frac{f}{g}\right)(x)$

$$\left(\frac{f}{g}\right)(x) = \frac{f(x)}{g(x)} = \frac{3x^2 - 4}{4x + 5},\ x \neq -\frac{5}{4}$$

You can use the difference of two functions to solve the application problem presented at the beginning of the lesson.

Example 2 **BUSINESS** **Refer to the application at the beginning of the lesson.**

a. Write the profit function.

b. Find the profit on 500, 1000, and 5000 tapes.

a. Profit is revenue minus cost. Thus, the profit function $p(x)$ is $p(x) = r(x) - c(x)$.

The revenue function is $r(x) = 9.5x$. The cost function is $c(x) = 0.8x + 1940$.

$$p(x) = r(x) - c(x) = 9.5x - (0.8x + 1940) = 8.7x - 1940$$

b. To find the profit on 500, 1000, and 5000 tapes, evaluate $p(500)$, $p(1000)$, and $p(5000)$.

$p(500) = 8.7(500) - 1940$ or 2410

$p(1000) = 8.7(1000) - 1940$ or 6760

$p(5000) = 8.7(5000) - 1940$ or 41,560

The profit on 500, 1000, and 5000 tapes is \$2410, \$6760, and \$41,560, respectively. *Check by finding the revenue and the cost for each number of tapes and subtracting to find profit.*

Functions can also be combined by using **composition.** In a composition, a function is performed, and then a second function is performed on the result of the first function. You can think of composition in terms of manufacturing a product. For example, fiber is first made into cloth. Then the cloth is made into a garment.

In composition, a function g maps the elements in set R to those in set S. Another function f maps the elements in set S to those in set T. Thus, the range of function g is the same as the domain of function f. A diagram is shown below.

R	S
x	$g(x) = \frac{1}{4}x$
4	1
8	2
12	3

S	T
x	$f(x) = 6 - 2x$
1	4
2	2
3	0

domain of g(x)   *The range of g(x) is the domain of f(x).*   *range of f(x)*

The function formed by composing two functions f and g is called the **composite** of f and g. It is denoted by $f \circ g$, which is read as "*f composition g*" or "*f of g*."

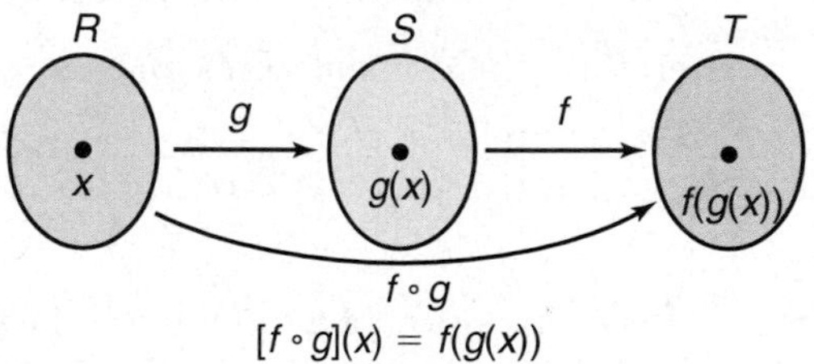

Composition of Functions

Given functions f and g, the composite function $f \circ g$ can be described by the following equation.

$$[f \circ g](x) = f[g(x)]$$

The domain of $f \circ g$ includes all of the elements x in the domain of g for which $g(x)$ is in the domain of f.

Example 3 **Find $[f \circ g](x)$ and $[g \circ f](x)$ for $f(x) = 2x^2 - 3x + 8$ and $g(x) = 5x - 6$.**

$[f \circ g](x) = f(g(x))$

$= f(5x - 6)$ *Substitute $5x - 6$ for $g(x)$.*

$= 2(5x - 6)^2 - 3(5x - 6) + 8$ *Substitute $5x - 6$ for x in $f(x)$.*

$= 2(25x^2 - 60x + 36) - 15x + 18 + 8$

$= 50x^2 - 135x + 98$

$[g \circ f](x) = g(f(x))$

$= g(2x^2 - 3x + 8)$ *Substitute $2x^2 - 3x + 8$ for $f(x)$.*

$= 5(2x^2 - 3x + 8) - 6$ *Substitute $2x^2 - 3x + 8$ for x in $g(x)$.*

$= 10x^2 - 15x + 34$

The domain of a composed function $[f \circ g](x)$ is determined by the domains of both $f(x)$ and $g(x)$.

Example 4 **State the domain of $[f \circ g](x)$ for $f(x) = \sqrt{x - 4}$ and $g(x) = \frac{1}{x^2}$.**

$f(x) = \sqrt{x - 4}$ *Domain: $x \geq 4$*

$g(x) = \frac{1}{x^2}$ *Domain: $x \neq 0$*

If $g(x)$ is undefined for a given value of x, then that value is excluded from the domain of $[f \circ g](x)$. Thus, 0 is excluded from the domain of $[f \circ g](x)$.

The domain of $f(x)$ is $x \geq 4$. So for x to be in the domain of $[f \circ g](x)$, it must be true that $g(x) \geq 4$.

$g(x) \geq 4$

$\frac{1}{x^2} \geq 4$ $\quad g(x) = \frac{1}{x^2}$

$1 \geq 4x^2$ *Multiply each side by x^2.*

$\frac{1}{4} \geq x^2$ *Divide each side by 4.*

$\frac{1}{2} \geq |x|$ *Take the square root of each side.*

$-\frac{1}{2} \leq x \leq \frac{1}{2}$ *Rewrite the inequality.*

Therefore, the domain of $[f \circ g](x)$ is $-\frac{1}{2} \leq x \leq \frac{1}{2}, x \neq 0$.

The composition of a function and itself is called iteration. Each output of an iterated function is called an iterate. To iterate a function $f(x)$, find the function value $f(x_0)$, of the initial value x_0. The value $f(x_0)$ is the first iterate, x_1. The second iterate is the value of the function performed on the output; that is, $f(f(x_0))$ or $f(x_1)$. Each iterate is represented by x_n, where n is the iterate number. For example, the third iterate is x_3.

Example 5 **Find the first three iterates, x_1, x_2, and x_3, of the function $f(x) = 2x - 3$ for an initial value of $x_0 = 1$.**

To obtain the first iterate, find the value of the function for $x_0 = 1$.

$x_1 = f(x_0) = f(1)$

$= 2(1) - 3$ or -1

To obtain the second iterate, x_2, substitute the function value for the first iterate, x_1, for x.

$x_2 = f(x_1) = f(-1)$

$= 2(-1) - 3$ or -5

Now find the third iterate, x_3, by substituting x_2 for x.

$x_3 = f(x_2) = f(-5)$

$= 2(-5) - 3$ or -13

Thus, the first three iterates of the function $f(x) = 2x - 3$ for an initial value of $x_0 = 1$ are -1, -5, and -13.

CHECK FOR UNDERSTANDING

Communicating Mathematics

Read and study the lesson to answer each question.

1. **Write** two functions $f(x)$ and $g(x)$ for which $(f \cdot g)(x) = 2x^2 + 11x - 6$. Tell how you determined $f(x)$ and $g(x)$.
2. **Explain** how iteration is related to composition of functions.
3. **Determine** whether $[f \circ g](x)$ is always equal to $[g \circ f](x)$ for two functions $f(x)$ and $g(x)$. Explain your answer and include examples or counterexamples.
4. *Math Journal* Write an explanation of function composition. Include an everyday example of two composed functions and an example of a realworld problem that you would solve using composed functions.

Guided Practice

5. Given $f(x) = 3x^2 + 4x - 5$ and $g(x) = 2x + 9$, find $f(x) + g(x)$, $f(x) - g(x)$, $f(x) \cdot g(x)$, and $\left(\frac{f}{g}\right)(x)$.

Find $[f \circ g](x)$ and $[g \circ f](x)$ for each $f(x)$ and $g(x)$.

6. $f(x) = 2x + 5$
   $g(x) = 3 + x$
7. $f(x) = 2x - 3$
   $g(x) = x^2 - 2x$
8. State the domain of $[f \circ g](x)$ for $f(x) = \frac{1}{(x-1)^2}$ and $g(x) = x + 3$.
9. Find the first three iterates of the function $f(x) = 2x + 1$ using the initial value $x_0 = 2$.
10. **Measurement** In 1954, the Tenth General Conference on Weights and Measures adopted the kelvin K as the basic unit for measuring temperature for all international weights and measures. While the kelvin is the standard unit, degrees Fahrenheit and degrees Celsius are still in common use in the United States. The function $C(F) = \frac{5}{9}(F - 32)$ relates Celsius temperatures and Fahrenheit temperatures. The function $K(C) = C + 273.15$ relates Celsius temperatures and Kelvin temperatures.
    a. Use composition of functions to write a function to relate degrees Fahrenheit and kelvins.
    b. Write the temperatures $-40°$F, $-12°$F, $0°$F, $32°$F, and $212°$F in kelvins.

EXERCISES

Practice

Find $f(x) + g(x)$, $f(x) - g(x)$, $f(x) \cdot g(x)$, and $\left(\frac{f}{g}\right)(x)$ for each $f(x)$ and $g(x)$.

•11. $f(x) = x^2 - 2x$
   $g(x) = x + 9$

12. $f(x) = \frac{x}{x+1}$
   $g(x) = x^2 - 1$

•13. $f(x) = \frac{3}{x-7}$
   $g(x) = x^2 + 5x$

14. If $f(x) = x + 3$ and $g(x) = \frac{2x}{x-5}$, find $f(x) + g(x)$, $f(x) - g(x)$, $f(x) \cdot g(x)$, and $\left(\frac{f}{g}\right)(x)$.

Find $[f \circ g](x)$ and $[g \circ f](x)$ for each $f(x)$ and $g(x)$.

15. $f(x) = x^2 - 9$
 $g(x) = x + 4$

16. $f(x) = \frac{1}{2}x - 7$
 $g(x) = x + 6$

17. $f(x) = x - 4$
 $g(x) = 3x^2$

18. $f(x) = x^2 - 1$
 $g(x) = 5x^2$

19. $f(x) = 2x$
 $g(x) = x^3 + x^2 + 1$

20. $f(x) = 1 + x$
 $g(x) = x^2 + 5x + 6$

21. What are $[f \circ g](x)$ and $[g \circ f](x)$ for $f(x) = x + 1$ and $g(x) = \frac{1}{x-1}$?

State the domain of $[f \circ g](x)$ for each $f(x)$ and $g(x)$.

22. $f(x) = 5x$
 $g(x) = x^3$

23. $f(x) = \frac{1}{x}$
 $g(x) = 7 - x$

24. $f(x) = \sqrt{x - 2}$
 $g(x) = \frac{1}{4x}$

Find the first three iterates of each function using the given initial value.

25. $f(x) = 9 - x; x_0 = 2$

26. $f(x) = x^2 + 1; x_0 = 1$

27. $f(x) = x(3 - x); x_0 = 1$

Applications and Problem Solving

28. **Retail** Sara Sung is shopping and finds several items that are on sale at 25% off the original price. The items that she wishes to buy are a sweater originally at \$43.98, a pair of jeans for \$38.59, and a blouse for \$31.99. She has \$100 that her grandmother gave her for her birthday. If the sales tax in San Mateo, California, where she lives is 8.25%, does Sara have enough money for all three items? Explain.

29. **Critical Thinking** Suppose the graphs of functions $f(x)$ and $g(x)$ are lines. Must it be true that the graph of $[f \circ g](x)$ is a line? Justify your answer.

30. **Physics** When a heavy box is being pushed on the floor, there are two different forces acting on the movement of the box. There is the force of the person pushing the box and the force of friction. If W is work in joules, F is force in newtons, and d is displacement of the box in meters, $W_p = F_p d$ describes the work of the person, and $W_f = F_f d$ describes the work created by friction. The increase in kinetic energy necessary to move the box is the difference between the work done by the person W_p and the work done by friction W_f.
 a. Write a function in simplest form for net work.
 b. Determine the net work expended when a person pushes a box 50 meters with a force of 95 newtons and friction exerts a force of 55 newtons.

31. **Finance** A sales representative for a cosmetics supplier is paid an annual salary plus a bonus of 3% of her sales *over* \$275,000. Let $f(x) = x - 275{,}000$ and $h(x) = 0.03x$.
 a. If x is greater than \$275,000, is her bonus represented by $f[h(x)]$ or by $h[f(x)]$? Explain.
 b. Find her bonus if her sales for the year are \$400,000.

32. **Critical Thinking** Find $f\left(\frac{1}{2}\right)$ if $[f \circ g](x) = \frac{x^4 + x^2}{1 + x^2}$ and $g(x) = 1 - x^2$.

33. International Business Value-added tax, VAT, is a tax charged on goods and services in European countries. Many European countries offer refunds of some VAT to non-resident based businesses. VAT is included in a price that is quoted. That is, if an item is marked as costing \$10, that price includes the VAT.

a. Suppose an American company has operations in The Netherlands, where the VAT is 17.5%. Write a function for the VAT amount paid $v(p)$ if p represents the price including the VAT.

b. In The Netherlands, foreign businesses are entitled to a refund of 84% of the VAT on automobile rentals. Write a function for the refund an American company could expect $r(v)$ if v represents the VAT amount.

c. Write a function for the refund expected on an automobile rental $r(p)$ if the price including VAT is p.

d. Find the refunds due on automobile rental prices of \$423.18, \$225.64, and \$797.05.

Mixed Review

34. Finance The formula for the simple interest earned on an investment is $I = prt$, where I is the interest earned, p is the principal, r is the interest rate, and t is the time in years. Assume that \$5000 is invested at an annual interest rate of 8% and that interest is added to the principal at the end of each year. *(Lesson 1-1)*

a. Find the amount of interest that will be earned each year for five years.

b. State the domain and range of the relation.

c. Is this relation a function? Why or why not?

35. State the relation in the table as a set of ordered pairs. Then state the domain and range of the relation. *(Lesson 1-1)*

x	y
−1	8
0	4
2	−6
5	−9

36. What are the domain and the range of the relation {(1, 5), (2, 6), (3, 7), (4, 8)}? Is the relation a function? Explain. *(Lesson 1-1)*

37. Find $g(-4)$ if $g(x) = \dfrac{x^3 + 5}{4x}$. *(Lesson 1-1)*

38. Given that x is an integer, state the relation representing $y = |-3x|$ and $-2 \leq x \leq 3$ by making a table of values. Then graph the ordered pairs of the relation. *(Lesson 1-1)*

39. SAT/ACT Practice Find $f(n - 1)$ if $f(x) = 2x^2 - x + 9$.

A $2n^2 - n + 9$

B $2n^2 - n + 8$

C $2n^2 - 5n + 12$

D 9

E $2n^2 + 4n + 8$

Extra Practice See p. A26.

Graphing Linear Equations

OBJECTIVES
- Graph linear equations.
- Find the x- and y-intercepts of a line.
- Find the slope of a line through two points.
- Find zeros of linear functions.

AGRICULTURE American farmers produce enough food and fiber to meet the needs of our nation and to export huge quantities to countries around the world. In addition to raising grain, cotton and other fibers, fruits, or vegetables, farmers also work on dairy farms, poultry farms, horticultural specialty farms that grow ornamental plants and nursery products, and aquaculture farms that raise fish and shellfish. In 1900, the percent of American workers who were farmers was 37.5%. In 1994, that percent had dropped to just 2.5%. What was the average rate of decline? *This problem will be solved in Example 2.*

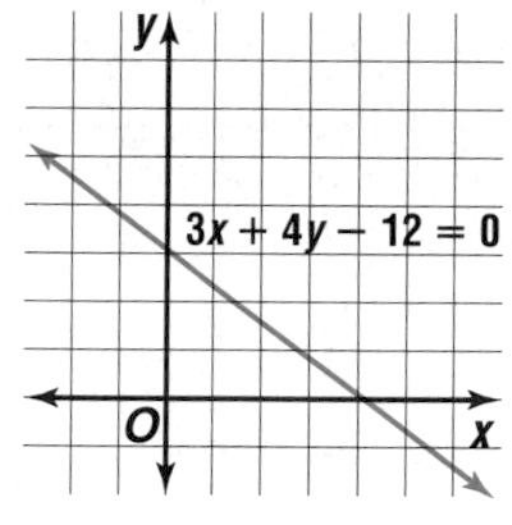

The problem above can be solved by using a linear equation. A **linear equation** has the form $Ax + By + C = 0$, where A and B are not both zero. Its graph is a straight line. The graph of the equation $3x + 4y - 12 = 0$ is shown.

The solutions of a linear equation are the ordered pairs for the points on its graph. An ordered pair corresponds to a point in the coordinate plane. Since two points determine a line, only two points are needed to graph a linear equation. Often the two points that are easiest to find are the ***x*-intercept** and the ***y*-intercept.** The x-intercept is the point where the line crosses the x-axis, and the y-intercept is the point where the graph crosses the y-axis. In the graph above, the x-intercept is at $(4, 0)$, and the y-intercept is at $(0, 3)$. *Usually, the individual coordinates 4 and 3 called the x- and y-intercepts.*

Example 1 **Graph $3x - y - 2 = 0$ using the x-and y-intercepts.**

Substitute 0 for y to find the x-intercept. Then substitute 0 for x to find the y-intercept.

***x*-intercept**	***y*-intercept**
$3x - y - 2 = 0$	$3x - y - 2 = 0$
$3x - (0) - 2 = 0$	$3(0) - y - 2 = 0$
$3x - 2 = 0$	$-y - 2 = 0$
$3x = 2$	$-y = 2$
$x = \frac{2}{3}$	$y = -2$

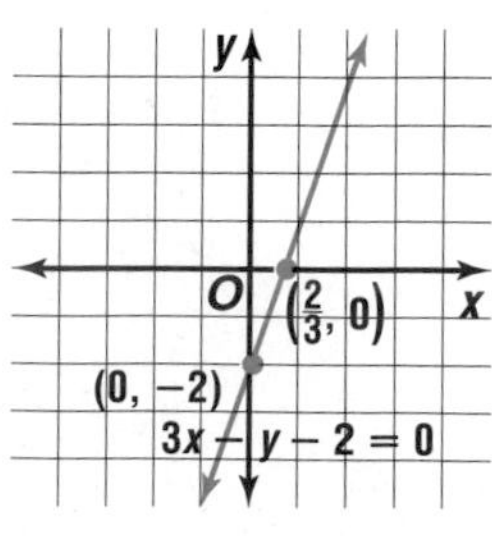

The line crosses the x-axis at $\left(\frac{2}{3}, 0\right)$ and the y-axis at $(0, -2)$. Graph the intercepts and draw the line.

The **slope** of a nonvertical line is the ratio of the change in the ordinates of the points to the corresponding change in the abscissas. The slope of a line is a constant.

Slope

The slope, m, of the line through (x_1, y_1) and (x_2, y_2) is given by the following equation, if $x_1 \neq x_2$.

$$m = \frac{y_2 - y_1}{x_2 - x_1}$$

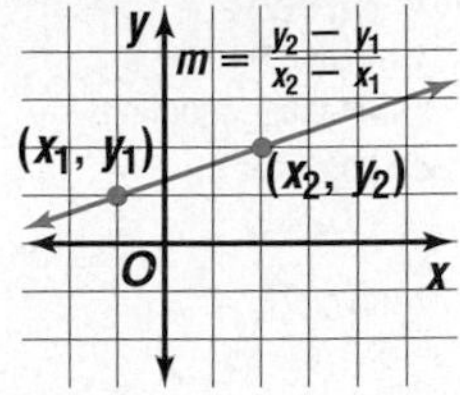

The slope of a line can be interpreted as the rate of change in the y-coordinates for each 1-unit increase in the x-coordinates.

Example 2 **AGRICULTURE** **Refer to the application at the beginning of the lesson. What was the average rate of decline in the percent of American workers who were farmers?**

The average rate of change is the slope of the line containing the points at (1900, 37.5) and (1994, 2.5). Find the slope of this line.

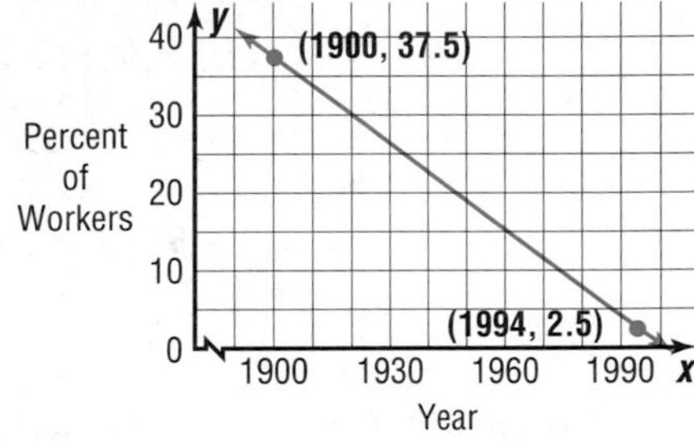

$$m = \frac{y_2 - y_1}{x_2 - x_1}$$

$$= \frac{2.5 - 37.5}{1994 - 1900} \quad \textit{Let } x_1 = 1900,\ y_1 = 37.5,\ x_2 = 1994, \textit{ and } y_2 = 2.5.$$

$$= \frac{-35}{94} \text{ or about } -0.37$$

On average, the number of American workers who were farmers decreased about 0.37% each year from 1900 to 1994.

A linear equation in the form $Ax + By = C$ where A is positive is written in **standard form.** You can also write a linear equation in **slope-intercept form.** Slope-intercept form is $y = mx + b$, where m is the slope and b is the y-intercept of the line. You can graph an equation in slope-intercept form by graphing the y-intercept and then finding a second point on the line using the slope.

Slope-Intercept Form

If a line has slope m and y-intercept b, the slope-intercept form of the equation of the line can be written as follows.

$$y = mx + b$$

Example 3 **Graph each equation using the y-intercept and the slope.**

a. $y = \frac{3}{4}x - 2$

The y-intercept is -2. Graph $(0, -2)$.
Use the slope to graph a second point.
Connect the points to graph the line.

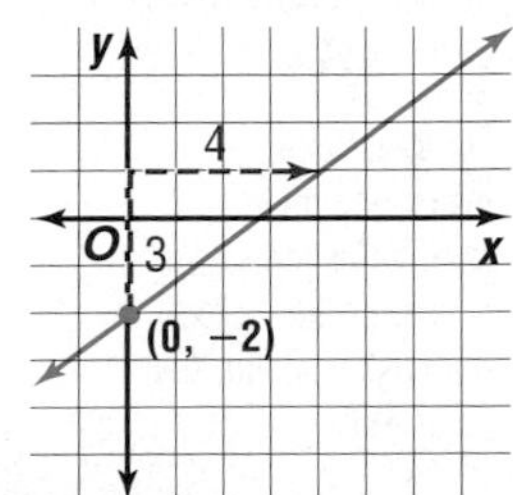

Graphing Calculator Appendix

For keystroke instruction on how to graph linear equations, see page A5.

b. $2x + y = 5$

Rewrite the equation in slope-intercept form.

$2x + y = 5 \rightarrow y = -2x + 5$

The y-intercept is 5. Graph (0, 5). Then use the slope to graph a second point. Connect the points to graph the line.

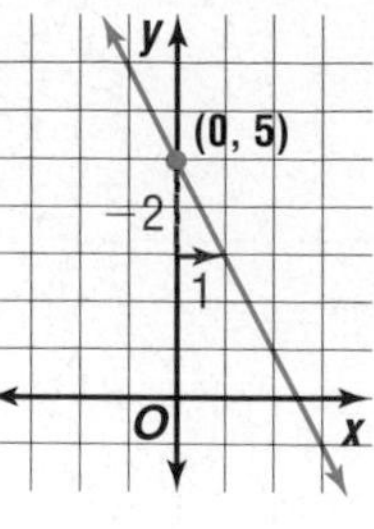

There are four different types of slope for a line. The table below shows a graph with each type of slope.

A line with undefined slope is sometimes described as having "no slope."

Types of Slope

positive slope	negative slope	0 slope	undefined slope
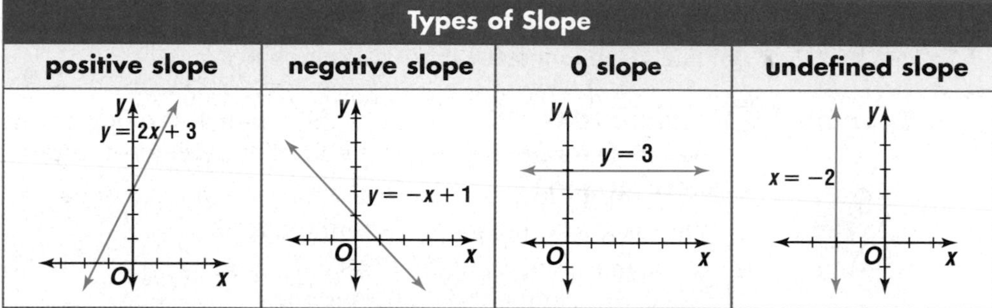			

Notice from the graphs that not all linear equations represent functions. A linear function is defined as follows. *When is a linear equation not a function?*

Linear Functions	A linear function is defined by $f(x) = mx + b$, where m and b are real numbers.

Values of x for which $f(x) = 0$ are called **zeros of the function *f*.** For a linear function, the zeros can be found by solving the equation $mx + b = 0$. If $m \neq 0$, then $-\frac{b}{m}$ is the only zero of the function. The zeros of a function are the x-intercepts. Thus, for a linear function, the x-intercept has coordinates $\left(-\frac{b}{m}, 0\right)$.

In the case where $m = 0$, we have $f(x) = b$. This function is called a **constant function** and its graph is a horizontal line. The constant function $f(x) = b$ has no zeros when $b \neq 0$ or every value of x is a zero if $b = 0$.

Example  **Find the zero of each function. Then graph the function.**

Graphing Calculator Appendix

For keystroke instruction on how to find the zeros of a linear function using the **CALC** menu, see page A11.

a. $f(x) = 5x + 4$

To find the zeros of $f(x)$, set $f(x)$ equal to 0 and solve for x.

$5x + 4 = 0 \Rightarrow x = -\frac{4}{5}$

$-\frac{4}{5}$ is a zero of the function. So the coordinates of one point on the graph are $\left(-\frac{4}{5}, 0\right)$. Find the coordinates of a second point. When $x = 0$, $f(x) = 5(0) + 4$, or 4. Thus, the coordinates of a second point are (0, 4).

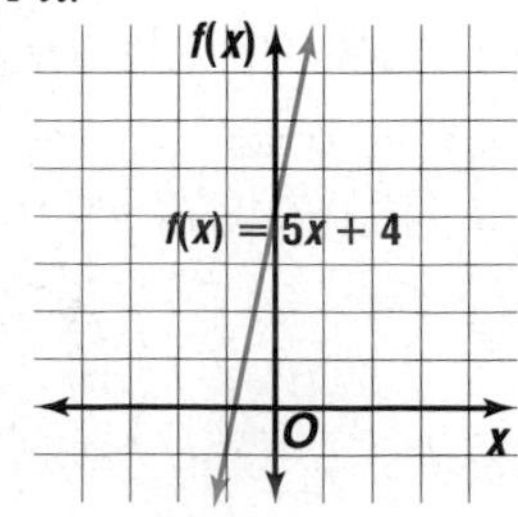

b. $f(x) = -2$

Since $m = 0$ and $b = -2$, this function has no x-intercept, and therefore no zeros. The graph of the function is a horizontal line 2 units below the x-axis.

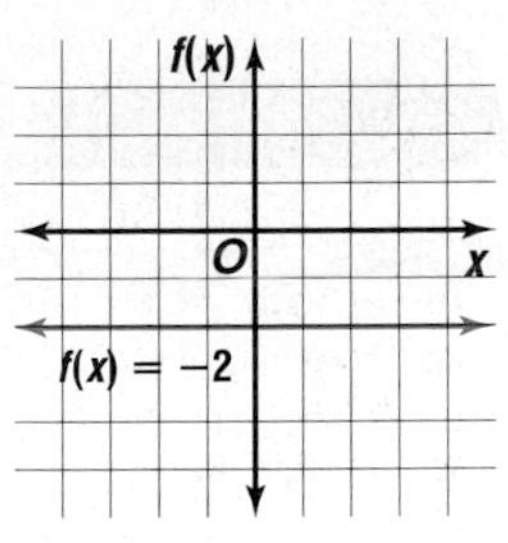

CHECK FOR UNDERSTANDING

Communicating Mathematics

Read and study the lesson to answer each question.

1. **Explain** the significance of m and b in $y = mx + b$.
2. **Name** the zero of the function whose graph is shown at the right. Explain how you found the zero.
3. **Describe** the process you would use to graph a line with a y-intercept of 2 and a slope of -4.
4. **Compare and contrast** the graphs of $y = 5x + 8$ and $y = -5x + 8$.

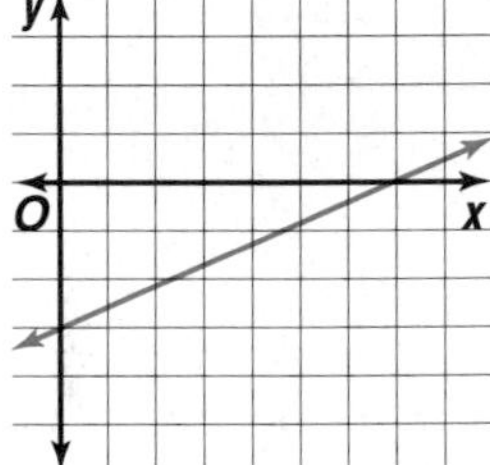

Guided Practice

Graph each equation using the *x*- and *y*-intercepts.

5. $3x - 4y + 2 = 0$
6. $x + 2y - 5 = 0$

Graph each equation using the *y*-intercept and the slope.

7. $y = x + 7$
8. $y = 5$

Find the zero of each function. If no zero exists, write *none.* Then graph the function.

9. $f(x) = \frac{1}{2}x + 6$
10. $f(x) = 19$

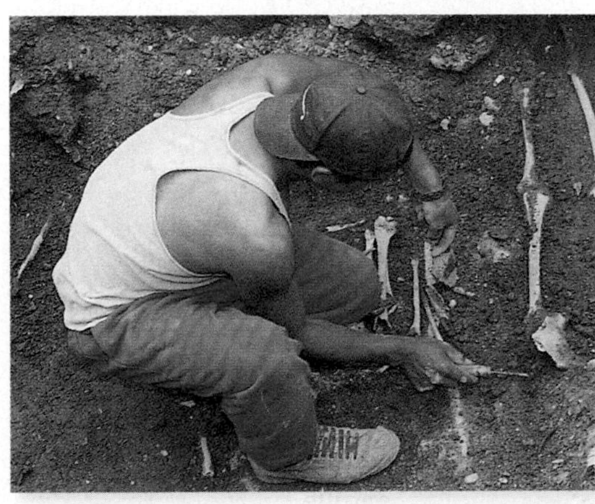

11. **Archaeology** Archaeologists use bones and other artifacts found at historical sites to study a culture. One analysis they perform is to use a function to determine the height of the person from a tibia bone. Typically a man whose tibia is 38.500 centimeters long is 173 centimeters tall. A man with a 44.125-centimeter tibia is 188 centimeters tall.
    a. Write two ordered pairs that represent the function.
    b. Determine the slope of the line through the two points.
    c. Explain the meaning of the slope in this context.

EXERCISES

Practice

Graph each equation.

12. $y = 4x - 9$
13. $y = 3$
14. $2x - 3y + 15 = 0$
15. $x - 4 = 0$
16. $y = 6x - 1$
17. $y = 5 - 2x$
18. $y + 8 = 0$
19. $2x + y = 0$
20. $y = \frac{2}{3}x - 4$
21. $y = 25x + 150$
22. $2x + 5y = 8$
23. $3x - y = 7$

Find the zero of each function. If no zero exists, write *none.* Then graph the function.

24. $f(x) = 9x + 5$
25. $f(x) = 4x - 12$
26. $f(x) = 3x + 1$
27. $f(x) = 14x$
28. $f(x) = 12$
29. $f(x) = 5x - 8$

30. Find the zero for the function $f(x) = 5x - 2$.
31. Graph $y = -\frac{3}{2}x + 3$. What is the zero of the function $f(x) = -\frac{3}{2}x + 3$?
32. Write a linear function that has no zero. Then write a linear function that has infinitely many zeros.

Applications and Problem Solving

33. **Electronics** The voltage V in volts produced by a battery is a linear function of the current i in amperes drawn from it. The opposite of the slope of the line represents the battery's effective resistance R in ohms. For a certain battery, $V = 12.0$ when $i = 1.0$ and $V = 8.4$ when $i = 10.0$.
    a. What is the effective resistance of the battery?
    b. Find the voltage that the battery would produce when the current is 25.0 amperes.
34. **Critical Thinking** A line passes through $A(3, 7)$ and $B(-4, 9)$. Find the value of a if $C(a, 1)$ is on the line.
35. **Chemistry** According to Charles' Law, the pressure P in pascals of a fixed volume of a gas is linearly related to the temperature T in degrees Celsius. In an experiment, it was found that when $T = 40$, $P = 90$ and when $T = 80$, $P = 100$.
    a. What is the slope of the line containing these points?
    b. Explain the meaning of the slope in this context.
    c. Graph the function.
36. **Critical Thinking** The product of the slopes of two non-vertical perpendicular lines is always -1. Is it possible for two perpendicular lines to both have positive slope? Explain.
37. **Accounting** A business's capital costs are expenses for things that last more than one year and lose value or wear out over time. Examples include equipment, buildings, and patents. The value of these items declines, or depreciates over time. One way to calculate depreciation is the straight-line method, using the value and the estimated life of the asset. Suppose $v(t) = 10{,}440 - 290t$ describes the value $v(t)$ of a piece of software after t months.
    a. Find the zero of the function. What does the zero represent?
    b. Find the slope of the function. What does the slope represent?
    c. Graph the function.

38. **Critical Thinking** How is the slope of a linear function related to the number of zeros for the function?

39. **Economics** Economists call the relationship between a nation's disposable income and personal consumption expenditures the marginal propensity to consume or MPC. An MPC of 0.7 means that for each $1 increase in disposable income, consumption increases $0.70. That is, 70% of each additional dollar earned is spent and 30% is saved.

    a. Suppose a nation's disposable income, x, and personal consumption expenditures, y, are shown in the table at the right. Find the MPC.

x (billions of dollars)	y (billions of dollars)
56	50
76	67.2

    b. If disposable income were to increase $1805 in a year, how many additional dollars would the average family spend?

    c. The marginal propensity to save, MPS, is 1 − MPC. Find the MPS.

    d. If disposable income were to increase $1805 in a year, how many additional dollars would the average family save?

Mixed Review

40. Given $f(x) = 2x$ and $g(x) = x^2 - 4$, find $(f + g)(x)$ and $(f - g)(x)$. *(Lesson 1-2)*

41. **Business** Computer Depot offers a 12% discount on computers sold Labor Day weekend. There is also a $100 rebate available. *(Lesson 1-2)*

    a. Write a function for the price after the discount $d(p)$ if p represents the original price of a computer.

    b. Write a function for the price after the rebate $r(d)$ if d represents the discounted price.

    c. Use composition of functions to write a function to relate the selling price to the original price of a computer.

    d. Find the selling prices of computers with original prices of $799.99, $999.99, and $1499.99.

42. Find $[f \circ g](-3)$ and $[g \circ f](-3)$ if $f(x) = x^2 - 4x + 5$ and $g(x) = x - 2$. *(Lesson 1-2)*

43. Given $f(x) = 4 + 6x - x^3$, find $f(9)$. *(Lesson 1-1)*

44. Determine whether the graph at the right represents a function. Explain. *(Lesson 1-1)*

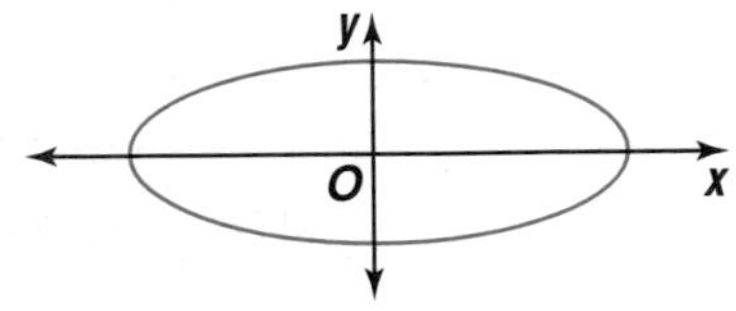

45. Given that x is an integer, state the relation representing $y = 11 - x$ and $3 \leq x \leq 0$ by listing a set of ordered pairs. Then state whether the relation is a function. *(Lesson 1-1)*

46. **SAT/ACT Practice** What is the sum of four integers whose average is 15?

    A 3.75

    B 15

    C 30

    D 60

    E cannot be determined

Extra Practice See p. A26.

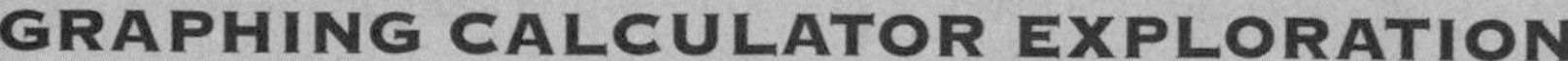

GRAPHING CALCULATOR EXPLORATION

1-3B Analyzing Families of Linear Graphs

An Extension of Lesson 1-3

OBJECTIVE
- Investigate the effect of changing the value of m or b in $y = mx + b$.

A **family of graphs** is a group of graphs that displays one or more similar characteristics. For linear functions, there are two types of families of graphs. Using the slope-intercept form of the equation, one family is characterized by having the same slope m in $y = mx + b$. The other type of family has the same y-intercept b in $y = mx + b$.

You can investigate families of linear graphs by graphing several equations on the same graphing calculator screen.

Example **1** **Graph $y = 3x - 5$, $y = 3x - 1$, $y = 3x$, and $y = 3x + 6$. Describe the similarities and differences among the graphs.**

Graph all of the equations on the same screen. Use the viewing window, $[-9.4, 9.4]$ by $[-6.2, 6.2]$.

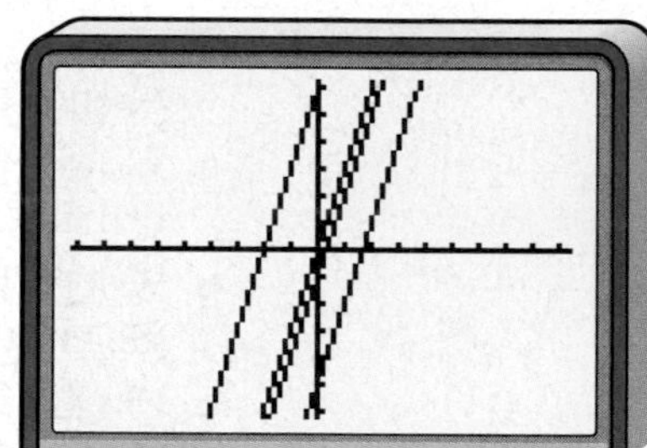

[−9.4, 9.4] scl:1 by [−6.2, 6.2] scl:1

Notice that the graphs appear to be parallel lines with the same positive slope. They are in the family of lines that have the slope 3.

The slope of each line is the same, but the lines have different y-intercepts. Each of the other three lines are the graph of $y = 3x$ shifted either up or down.

equation	slope	y-intercept	relationship to graph of $y = 3x$
$y = 3x - 5$	3	−5	shifted 5 units down
$y = 3x - 1$	3	−1	shifted 1 unit down
$y = 3x$	3	0	same
$y = 3x + 6$	3	6	shifted 6 units up

TRY THESE

1. Graph $y = 4x - 2$, $y = 2x - 2$, $y = -2$, $y = -x - 2$, and $y = -6x - 2$ on the same graphing calculator screen. Describe how the graphs are similar and different.

WHAT DO YOU THINK?

2. Use the results of the Example and Exercise 1 to predict what the graph of $y = 3x - 2$ will look like.
3. Write a paragraph explaining the effect of different values of m and b on the graph of $y = mx + b$. Include sketches to illustrate your explanation.

1-4 Writing Linear Equations

OBJECTIVE
- Write linear equations.

ECONOMICS Each year, the U.S. Department of Commerce publishes its *Survey of Current Business*. Included in the report is the average personal income of U.S. workers.

Years since 1980	Average Personal Income ($)
0	9916
5	13,895
10	18,477
11	19,100
12	19,802
13	20,810
14	21,846
15	23,233
16	24,457
17	25,660

Personal income is one indicator of the health of the U.S. economy. How could you use the data on average personal income for 1980 to 1997 to predict the average personal income in 2010? *This problem will be solved in Example 3.*

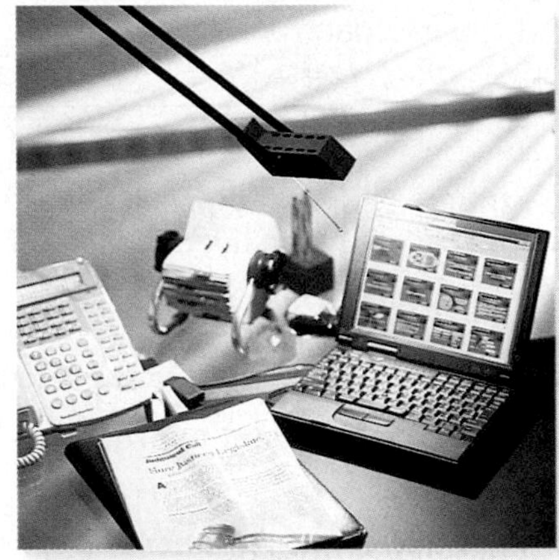

A mathematical **model** may be an equation used to approximate a real-world set of data. Often when you work with real-world data, you know information about a line without knowing its equation. You can use characteristics of the graph of the data to write an equation for a line. This equation is a model of the data. Writing an equation of a line may be done in a variety of ways depending upon the information you are given. If one point and the slope of a line are known, the slope-intercept form can be used to write the equation.

Example 1 **Write an equation in slope-intercept form for each line described.**

a. a slope of $-\frac{3}{4}$ and a y-intercept of 7

Substitute $-\frac{3}{4}$ for m and 7 for b in the general slope-intercept form.

$y = mx + b \rightarrow y = -\frac{3}{4}x + 7$

The slope-intercept form of the equation of the line is $y = -\frac{3}{4}x + 7$.

b. a slope of −6 and passes through the point at (1, −3)

Substitute the slope and coordinates of the point in the general slope-intercept form of a linear equation. Then solve for b.

$y = mx + b$

$-3 = -6(1) + b$ *Substitute −3 for y, 1 for x, and −6 for m.*

$3 = b$ *Add 6 to each side of the equation.*

The y-intercept is 3. Thus, the equation for the line is $y = -6x + 3$.

Example 2  **BUSINESS Alvin Hawkins is opening a home-based business. He determined that he will need $6000 to buy a computer and supplies to start. He expects expenses for each following month to be $700. Write an equation that models the total expense y after x months.**

The initial cost is the y-intercept of the graph. Because the total expense rises $700 each month, the slope is 700.

$y = mx + b$

$y = 700x + 6000$ *Substitute 700 for m and 6000 for b.*

The total expense can be modeled by $y = 700x + 6000$.

When you know the slope and a point on a line, you can also write an equation for the line in **point-slope form.** Using the definition of slope for points (x, y) and (x_1, y_1), if $\frac{y - y_1}{x - x_1} = m$, then $y - y_1 = m(x - x_1)$.

Point-Slope Form

If the point with coordinates (x_1, y_1) lies on a line having slope m, the point-slope form of the equation of the line can be written as follows.

$$y - y_1 = m(x - x_1)$$

If you know the coordinates of two points on a line, you can find the slope of the line. Then the equation of the line can be written using either the slope-intercept or the point-slope form.

Example 3  **ECONOMICS Refer to the application at the beginning of the lesson.**

a. Find a linear equation that can be used as a model to predict the average personal income for any year.

b. Assume that the rate of growth of personal income remains constant over time and use the equation to predict the average personal income for individuals in the year 2010.

c. Evaluate the prediction.

a. Graph the data. Then select two points to represent the data set and draw a line that might approximate the data. Suppose we chose (0, 9916) and (17, 25,660). Use the coordinates of those points to find the slope of the line you drew.

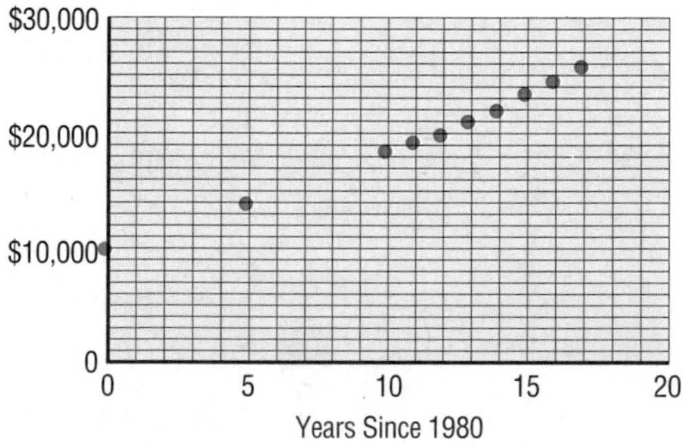

$\$30,000$ / $\$20,000$ / $\$10,000$ / 0 (vertical axis); 0, 5, 10, 15, 20 (horizontal axis); Years Since 1980

$$m = \frac{y_2 - y_1}{x_2 - x_1}$$

$$= \frac{25{,}660 - 9916}{17 - 0} \quad x_1 = 0,\ y_1 = 9916,\ x_2 = 17,\ y_2 = 25{,}660$$

≈ 926 *Thus for each 1-year increase, average personal income increases $926.*

Use point-slope form.

$y - y_1 = m(x - x_1)$

$y - 9916 = 926(x - 0)$ *Substitute 0 for x_1, 9916 for y_1, and 926 for m.*

$y = 926x + 9916$

The slope-intercept form of the model equation is $y = 926x + 9916$.

b. Evaluate the equation for $x = 2010$ to predict the average personal income for that year. The years since 1980 will be 2010 − 1980 or 30. So $x = 30$.

$y = 926x + 9916$

$y = 926(30) + 9916$ *Substitute 30 for x.*

$y = 37{,}696$

The predicted average personal income is about $37,696 for the year 2010.

c. Most of the actual data points are close to the graph of the model equation. Thus, the equation and the prediction are probably reliable.

CHECK FOR UNDERSTANDING

Communicating Mathematics

Read and study the lesson to answer each question.

1. **List** all the different sets of information that are sufficient to write the equation of a line.
2. **Demonstrate** two different ways to find the equation of the line with a slope of $\frac{1}{4}$ passing through the point at (3, −4).
3. **Explain** what 55 and 49 represent in the equation $c = 55h + 49$, which represents the cost c of a plumber's service call lasting h hours.
4. **Write** an equation for the line whose graph is shown at the right.

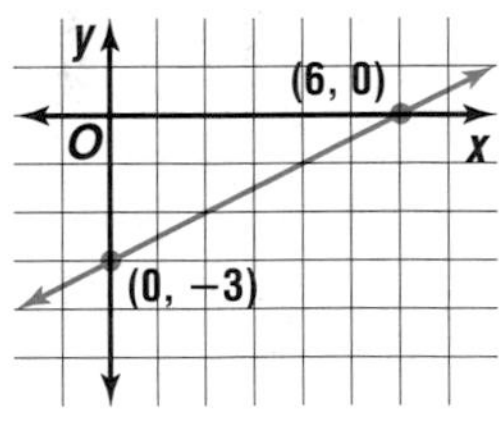

5. *Math Journal* Write a sentence or two to describe when it is easier to use the point-slope form to write the equation of a line and when it is easier to use the slope-intercept form.

Guided Practice

Write an equation in slope-intercept form for each line described.

6. slope $= -\frac{1}{4}$, y-intercept $= -10$
7. slope = 4, passes through (3, 2)
8. passes through (5, 2) and (7, 9)
9. horizontal and passes through (−9, 2)
10. **Botany** Do you feel like every time you cut the grass it needs to be cut again right away? Be grateful you aren't cutting the Bermuda grass that grows in Africa and Asia. It can grow at a rate of 5.9 inches per day! Suppose you cut a Bermuda grass plant to a length of 2 inches.
    a. Write an equation that models the length of the plant y after x days.
    b. If you didn't cut it again, how long would the plant be in one week?
    c. Can this rate of growth be maintained indefinitely? Explain.

EXERCISES

Practice

Write an equation in slope-intercept form for each line described.

11. slope = 5, y-intercept = -2
12. slope = 8, passes through $(-7, 5)$
13. slope = $-\frac{3}{4}$, y-intercept = 0
14. slope = -12, y-intercept = $\frac{1}{2}$
15. passes through $A(4, 5)$, slope = 6
16. no slope and passes through $(12, -9)$
17. passes through $A(1, 5)$ and $B(-8, 9)$
18. x-intercept = -8, y-intercept = 5
19. passes through $A(8, 1)$ and $B(-3, 1)$
20. vertical and passes through $(-4, -2)$
21. the y-axis
22. slope = 0.25, x-intercept = 24

23. Line ℓ passes through $A(-2, -4)$ and has a slope of $-\frac{1}{2}$. What is the standard form of the equation for line ℓ?

24. Line m passes through $C(-2, 0)$ and $D(1, -3)$. Write the equation of line m in standard form.

Applications and Problem Solving

25. **Sports** Skiers, hikers, and climbers often experience altitude sickness as they reach elevations of 8000 feet and more. A good rule of thumb for the amount of time that it takes to become acclimated to high elevations is 2 weeks for the first 7000 feet. After that, it will take 1 week more for each additional 2000 feet of altitude.
    a. Write an equation for the time t to acclimate to an altitude of f feet.
    b. Mt. Whitney in California is the highest peak in the contiguous 48 states. It is located in Eastern Sierra Nevada, on the border between Sequoia National Park and Inyo National Forest. About how many weeks would it take a person to acclimate to Mt. Whitney's elevation of 14,494 feet?

26. **Critical Thinking** Write an expression for the slope of a line whose equation is $Ax + By + C = 0$.

27. **Transportation** The mileage in miles per gallon (mpg) for city and highway driving of several 1999 models are given in the chart.

Model	City (mpg)	Highway (mpg)
A	24	32
B	20	29
C	20	29
D	20	28
E	23	30
F	24	30
G	27	37
H	22	28

    a. Find a linear equation that can be used to find a car's highway mileage based on its city mileage.
    b. Model J's city mileage is 19 mpg. Use your equation to predict its highway mileage.
    c. Highway mileage for Model J is 26 mpg. How well did your equation predict the mileage? Explain.

28. **Economics** Research the average personal income for the current year.

a. Find the value that the equation in Example 2 predicts.

b. Is the average personal income equal to the prediction? Explain any difference.

29. **Critical Thinking** Determine whether the points at (5, 9), (−3, 3), and (1, 6) are collinear. Justify your answer.

Mixed Review

30. Graph $3x - 2y - 5 = 0$. *(Lesson 1-3)*

31. **Business** In 1995, retail sales of apparel in the United States were \$70,583 billion. Apparel sales were \$82,805 billion in 1997. *(Lesson 1-3)*

a. Assuming a linear relationship, find the average annual rate of increase.

b. Explain how the rate is related to the graph of the line.

32. If $f(x) = x^3$ and $g(x) = 3x$, find $g[f(-2)]$. *(Lesson 1-2)*

33. Find $(f \cdot g)(x)$ and $\left(\frac{f}{g}\right)(x)$ for $f(x) = x^3$ and $g(x) = x^2 - 3x + 7$. *(Lesson 1-2)*

34. Given that x is an integer, state the relation representing $y = x^2$ and $-4 \leq x \leq -2$ by listing a set of ordered pairs. Then state whether this relation is a function. *(Lesson 1-1)*

35. **SAT/ACT Practice** If $xy = 1$, then x is the reciprocal of y. Which of the following is the arithmetic mean of x and y?

A $\frac{y^2 + 1}{2y}$ B $\frac{y + 1}{2y}$ C $\frac{y^2 + 2}{2y}$

D $\frac{y^2 + 1}{y}$ E $\frac{x^2 + 1}{y}$

MID-CHAPTER QUIZ

1. What are the domain and the range of the relation {(−2, −3), (−2, 3), (4, 7), (2, −8), (4, 3)}? Is the relation a function? Explain. (Lesson 1-1)

2. Find $f(4)$ for $f(x) = 7 - x^2$. (Lesson 1-1)

3. If $g(x) = \frac{3}{x - 1}$, what is $g(n + 2)$? (Lesson 1-1)

4. **Retail** Amparo bought a jacket with a gift certificate she received as a birthday present. The jacket was marked 33% off, and the sales tax in her area is 5.5%. If she paid \$45.95 for the jacket, use composition of functions to determine the original price of the jacket. (Lesson 1-2)

5. If $f(x) = \frac{1}{x - 1}$ and $g(x) = x + 1$, find $[f \circ g](x)$ and $[g \circ f](x)$. (Lesson 1-2)

Graph each equation. (Lesson 1-3)

6. $2x - 4y = 8$

7. $3x = 2y$

8. Find the zero of $f(x) = 5x - 3$. (Lesson 1-3)

9. Points $A(2, 5)$ and $B(7, 8)$ lie on line ℓ. What is the standard form of the equation of line ℓ? (Lesson 1-4)

10. **Demographics** In July 1990, the population of Georgia was 6,506,416. By July 1997, the population had grown to 7,486,242. (Lesson 1-4)

a. If x represents the year and y represents the population, find the average annual rate of increase of the population.

b. Write an equation to model the population change.

Extra Practice See p. A26.

Writing Equations of Parallel and Perpendicular Lines

OBJECTIVE

- Write equations of parallel and perpendicular lines.

E-COMMERCE Have you ever made a purchase over the Internet? Electronic commerce, or e-commerce, has changed the way Americans do business. In recent years, hundreds of companies that have no stores outside of the Internet have opened.

Suppose you own shares in two Internet stocks, Bookseller.com and WebFinder. One day these stocks open at \$94.50 and \$133.60 per share, respectively. The closing prices that day were \$103.95 and \$146.96, respectively. If your shares in these companies were valued at \$5347.30 at the beginning of the day, is it possible that the shares were worth \$5882.03 at closing? *This problem will be solved in Example 2.*

This problem can be solved by determining whether the graphs of the equations that describe the situation are parallel or coincide. Two lines that are in the same plane and have no points in common are **parallel lines.** The slopes of two nonvertical parallel lines are equal. The graphs of two equations that represent the same line are said to **coincide.**

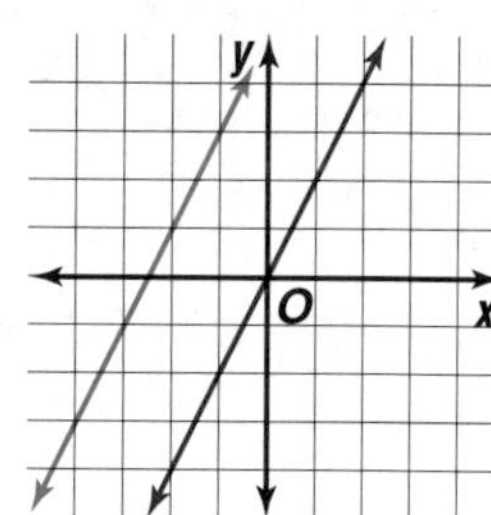

Parallel Lines Two nonvertical lines in a plane are parallel if and only if their slopes are equal and they have no points in common. Two vertical lines are always parallel.

We can use slopes and y-intercepts to determine whether lines are parallel.

Example 1 **Determine whether the graphs of each pair of equations are *parallel, coinciding,* or *neither.***

a. $3x - 4y = 12$
$9x - 12y = 72$

Write each equation in slope-intercept form.

$3x - 4y = 12$ $\qquad$ $9x - 12y = 72$

$y = \frac{3}{4}x - 3$ $\qquad$ $y = \frac{3}{4}x - 6$

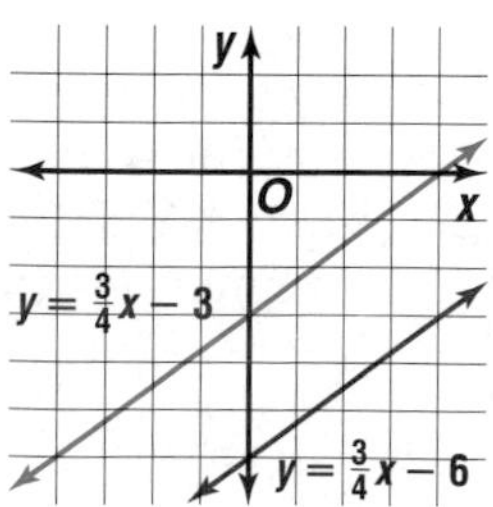

The lines have the same slope and different y-intercepts, so they are parallel. The graphs confirm the solution.

b. $\mathbf{15x + 12y = 36}$
$\mathbf{5x + 4y = 12}$

Write each equation in slope-intercept form.

$15x + 12y = 36$ $\qquad$ $5x + 4y = 12$

$y = -\frac{5}{4}x + 3$ $\qquad$ $y = -\frac{5}{4}x + 3$

The slopes are the same, and the y-intercepts are the same. Therefore, the lines have all points in common. The lines coincide. *Check the solution by graphing.*

You can use linear equations to determine whether real-world situations are possible.

Example 2

FINANCE **Refer to the application at the beginning of the lesson. Is it possible that your shares were worth $5882.03 at closing? Explain.**

Let x represent the number of shares of Bookseller.com and y represent the number of shares of WebFinder. Then the value of the shares at opening is $94.50x + 133.60y = 5347.30$. The value of the shares at closing is modeled by $103.95x + 146.96y = 5882.03$.

Write each equation in slope-intercept form.

$94.50x + 133.60y = 5347.30$	$103.95x + 146.96y = 5882.03$
$y = \frac{945}{1336}x + \frac{53,473}{1336}$	$y = -\frac{945}{1336}x + \frac{53,473}{1336}$

Since these equations are the same, their graphs coincide. As a result, any ordered pair that is a solution for the equation for the opening value is also a solution for the equation for the closing value. Therefore, the value of the shares could have been $5882.03 at closing.

In Lesson 1-3, you learned that any linear equation can be written in standard form. The slope of a line can be obtained directly from the standard form of the equation if B is not 0. Solve the equation for y.

$$Ax + By + C = 0$$
$$By = -Ax - C$$
$$y = -\frac{A}{B}x - \frac{C}{B}, \quad B \neq 0$$

↑ *slope* ↑ *y-intercept*

So the slope m is $-\frac{A}{B}$, and the y-intercept b *is* $-\frac{C}{B}$.

Example 3 **Write the standard form of the equation of the line that passes through the point at (4, −7) and is parallel to the graph of $2x - 5y + 8 = 0$.**

Any line parallel to the graph of $2x - 5y + 8 = 0$ will have the same slope. So, find the slope of the graph of $2x - 5y + 8 = 0$.

$$m = -\frac{A}{B}$$
$$= -\frac{2}{(-5)} \text{ or } \frac{2}{5}$$

Use point-slope form to write the equation of the line.

$$y - y_1 = m(x - x_1)$$
$$y - (-7) = \frac{2}{5}(x - 4) \quad \textit{Substitute 4 for } x_1,\ -7 \textit{ for } y_1, \textit{ and } \frac{2}{5} \textit{ for } m.$$
$$y + 7 = \frac{2}{5}x - \frac{8}{5}$$
$$5y + 35 = 2x - 8 \quad \textit{Multiply each side by 5.}$$
$$2x - 5y - 43 = 0 \quad \textit{Write in standard form.}$$

There is also a special relationship between the slopes of perpendicular lines.

Perpendicular Lines

Two nonvertical lines in a plane are perpendicular if and only if their slopes are opposite reciprocals.

A horizontal and a vertical line are always perpendicular.

You can also use the point-slope form to write the equation of a line that passes through a given point and is perpendicular to a given line.

Example 4 **Write the standard form of the equation of the line that passes through the point at (−6, −1) and is perpendicular to the graph of $4x + 3y - 7 = 0$.**

The line with equation $4x + 3y - 7 = 0$ has a slope of $-\frac{A}{B} = -\frac{4}{3}$. Therefore, the slope of a line perpendicular must be $\frac{3}{4}$.

$$y - y_1 = m(x - x_1)$$
$$y - (-1) = \frac{3}{4}[x - (-6)] \quad \textit{Substitute } -6 \textit{ for } x_1,\ -1 \textit{ for } y_1, \textit{ and } \frac{3}{4} \textit{ for } m.$$
$$y + 1 = \frac{3}{4}x + \frac{9}{2}$$
$$4y + 4 = 3x + 18 \quad \textit{Multiply each side by 4.}$$
$$3x - 4y + 14 = 0 \quad \textit{Write in standard form.}$$

You can use the properties of parallel and perpendicular lines to write linear equations to solve geometric problems.

Example 5 **GEOMETRY** **Determine the equation of the perpendicular bisector of the line segment with endpoints $S(3, 4)$ and $T(11, 18)$.**

Recall that the coordinates of the midpoint of a line segment are the averages of the coordinates of the two endpoints. Let S be (x_1, y_1) and T be (x_2, y_2). Calculate the coordinates of the midpoint.

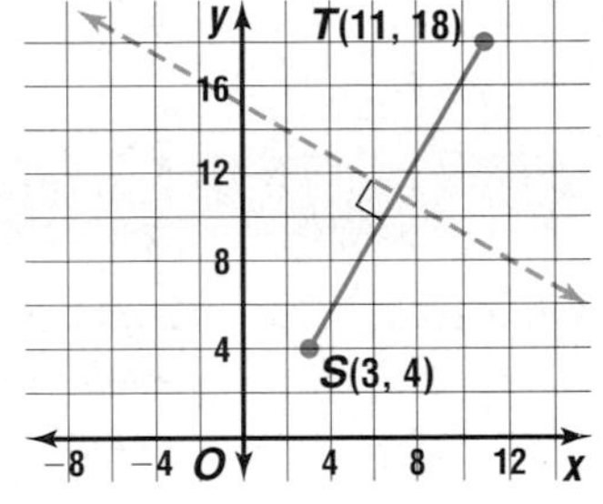

$$\left(\frac{x_1 + x_2}{2}, \frac{y_1 + y_2}{2}\right) = \left(\frac{3 + 11}{2}, \frac{4 + 18}{2}\right)$$
$$= (7, 11)$$

The slope of $\overline{ST}$ is $\frac{18 - 4}{11 - 3}$ or $\frac{7}{4}$.

The slope of the perpendicular bisector of $\overline{ST}$ is $-\frac{4}{7}$. The perpendicular bisector of $\overline{ST}$ passes through the midpoint of $\overline{ST}$, (7, 11).

$y - y_1 = m(x - x_1)$ *Point-slope form*

$y - 11 = -\frac{4}{7}(x - 7)$ *Substitute 7 for x_1, 11 for y_1, and $\frac{-4}{7}$ for m.*

$7y - 77 = -4x + 28$ *Multiply each side by 7.*

$4x + 7y - 105 = 0$ *Write in standard form.*

CHECK FOR UNDERSTANDING

Communicating Mathematics

Read and study the lesson to answer each question.

1. **Describe** how you would tell that two lines are parallel or coincide by looking at the equations of the lines in standard form.
2. **Explain** why vertical lines are a special case in the definition of parallel lines.
3. **Determine** the slope of a line that is parallel to the graph of $4x + 3y + 19 = 0$ and the slope of a line that is perpendicular to it.
4. **Write** the slope of a line that is perpendicular to a line that has undefined slope. Explain.

Guided Practice

Determine whether the graphs of each pair of equations are *parallel, coinciding, perpendicular,* or *none of these.*

5. $y = 5x - 5$
   $y = -5x + 2$
6. $y = -6x - 2$
   $y = \frac{1}{6}x - 8$
7. $y = x - 6$
   $x - y + 8 = 0$
8. $y = 2x - 8$
   $4x - 2y - 16 = 0$

9. Write the standard form of the equation of the line that passes through $A(5, 9)$ and is parallel to the graph of $y = 5x - 9$.
10. Write the standard form of the equation of the line that passes through $B(-10, -5)$ and is perpendicular to the graph of $6x - 5y = 24$.

11. Geometry A quadrilateral is a parallelogram if both pairs of its opposite sides are parallel. A parallelogram is a rectangle if its adjacent sides are perpendicular. Use these definitions to determine if the *EFGH* is a *parallelogram*, a *rectangle*, or *neither*.

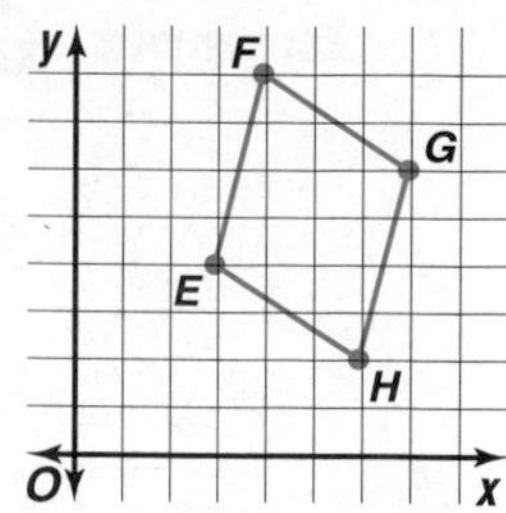

EXERCISES

Practice

Determine whether the graphs of each pair of equations are *parallel, coinciding, perpendicular,* or *none of these.*

12. $y = 5x - 18$
$2x + 10y + 10 = 0$

13. $y - 7x + 5 = 0$
$y - 7x - 9 = 0$

14. $y = \frac{1}{3}x + 11$
$y = 3x - 9$

15. $y = -3$
$x = 6$

16. $y = 4x - 3$
$4.8x - 1.2y = 3.6$

17. $4x - 6y = 11$
$3x + 2y = 9$

18. $y = 3x - 2$
$3x + y = 2$

19. $5x + 9y = 14$
$y = -\frac{5}{9}x + \frac{14}{9}$

20. $y + 4x - 2 = 0$
$y + 4x + 1 = 0$

21. Are the graphs of $y = 3x - 2$ and $y = -3x + 2$ *parallel, coinciding, perpendicular,* or *none of these*? Explain.

Write the standard form of the equation of the line that is parallel to the graph of the given equation and passes through the point with the given coordinates.

22. $y = 2x + 10$; $(0, -8)$ **23.** $4x - 9y = -23$; $(12, -15)$ **24.** $y = -9$; $(4, -11)$

Write the standard form of the equation of the line that is perpendicular to the graph of the given equation and passes through the point with the given coordinates.

25. $y = 5x + 12$; $(0, -3)$ **26.** $6x - y = 3$; $(7, -2)$ **27.** $x = 12$; $(6, -13)$

28. The equation of line ℓ is $5y - 4x = 10$. Write the standard form of the equation of the line that fits each description.
a. parallel to ℓ and passes through the point at $(-15, 8)$
b. perpendicular to ℓ and passes through the point at $(-15, 8)$

29. The equation of line m is $8x - 14y + 3 = 0$.
a. For what value of k is the graph of $kx - 7y + 10 = 0$ parallel to line m?
b. What is k if the graphs of m and $kx - 7y + 10 = 0$ are perpendicular?

Applications and Problem Solving

30. Critical Thinking Write equations of two lines that satisfy each description.
a. perpendicular and one is vertical
b. parallel and neither has a y-intercept

31. Geometry An altitude of a triangle is a segment that passes through one vertex and is perpendicular to the opposite side. Find the standard form of the equation of the line containing each altitude of $\triangle ABC$.

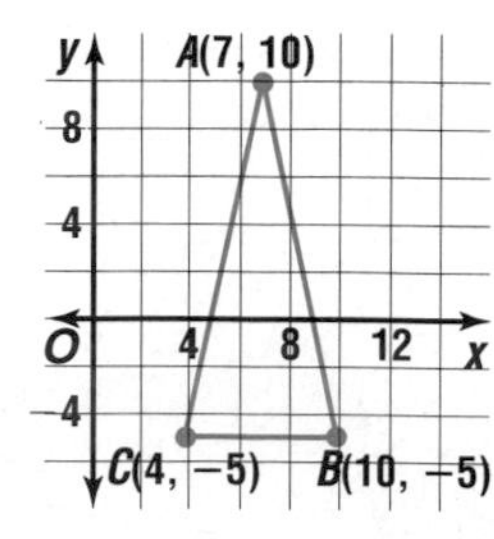

32. **Critical Thinking** The equations $y = m_1x + b_1$ and $y = m_2x + b_2$ represent parallel lines if $m_1 = m_2$ and $b_1 \neq b_2$. Show that they have no point in common. (*Hint:* Assume that there is a point in common and show that the assumption leads to a contradiction.)

33. **Business** The Seattle Mariners played their first game at their new baseball stadium on July 15, 1999. The stadium features Internet kiosks, a four-story scoreboard, a retractable roof, and dozens of espresso vendors. Suppose a vendor sells 216 regular espressos and 162 large espressos for a total of $783 at a Monday night game.
   a. On Thursday, 248 regular espressos and 186 large espressos were sold. Is it possible that the vendor made $914 that day? Explain.
   b. On Saturday, 344 regular espressos and 258 large espressos were sold. Is it possible that the vendor made $1247 that day? Explain.

34. **Economics** The table shows the closing value of a stock index for one week in February, 1999.
   a. Using the day as the x-value and the closing value as the y-value, write equations in slope-intercept form for the lines that represent each value change.
   b. What would indicate that the rate of change for two pair of days was the same? Was the rate of change the same for any of the days shown?
   c. Use each equation to predict the closing value for the next business day. The actual closing value was 1241.87. Did any equation correctly predict this value? Explain.

Stock Index February, 1999

Day	Closing value
8	1243.77
9	1216.14
10	1223.55
11	1254.04
12	1230.13

Mixed Review

35. Write the slope-intercept form of the equation of the line through the point at (1, 5) that has a slope of -2. *(Lesson 1-4)*

36. **Business** Knights Screen Printers makes special-order T-shirts. Recently, Knights received two orders for a shirt designed for a symposium. The first order was for 40 T-shirts at a cost of $295, and the second order was for 80 T-shirts at a cost of $565. Each order included a standard shipping and handling charge. *(Lesson 1-4)*
   a. Write a linear equation that models the situation.
   b. What is the cost per T-shirt?
   c. What is the standard shipping and handling charge?

37. Graph $3x - 2y - 6 = 0$. *(Lesson 1-3)*

38. Find $[g \circ h](x)$ if $g(x) = x - 1$ and $h(x) = x^2$. *(Lesson 1-2)*

39. Write an example of a relation that is not a function. Tell why it is not a function. *(Lesson 1-1)*

40. **SAT Practice** **Grid-In** If $2x + y = 12$ and $x + 2y = -6$, what is the value of $2x + 2y$?

Extra Practice See p. A27.

1-6

Modeling Real-World Data with Linear Functions

OBJECTIVES

- Draw and analyze scatter plots.
- Write a prediction equation and draw best-fit lines.
- Use a graphing calculator to compute correlation coefficients to determine goodness of fit.
- Solve problems using prediction equation models.

Education The cost of attending college is steadily increasing. However, it can be a good investment since on average, the higher your level of education, the greater your earning potential. The chart shows the average tuition and fees for a full-time resident student at a public four-year college. Estimate the average college cost in the academic year beginning in 2006 if tuition and fees continue at this rate. *This problem will be solved in Example 1.*

Academic Year	Tuition and Fees
1990–1991	2159
1991–1992	2410
1992–1993	2349
1993–1994	2537
1994–1995	2681
1995–1996	2811
1996–1997	2975
1997–1998	3111
1998–1999	3243

Source: The College Board and National Center for Educational Statistics

As you look at the college tuition costs, it is difficult to visualize how quickly the costs are increasing. When real-life data is collected, the data graphed usually does not form a perfectly straight line. However, the graph may approximate a linear relationship. When this is the case, a **best-fit line** can be drawn, and a **prediction equation** that models the data can be determined. Study the **scatter plots** below.

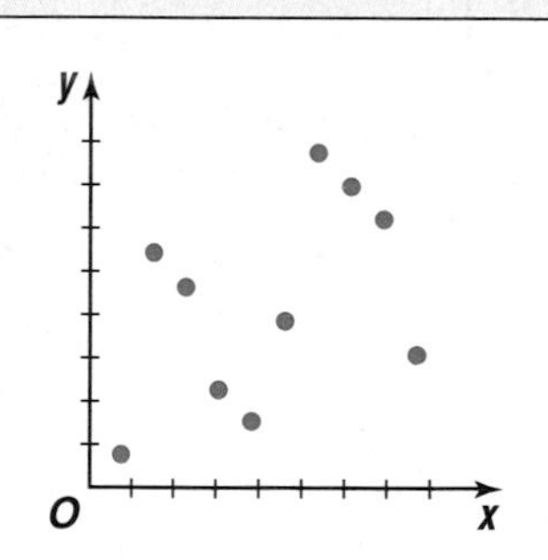

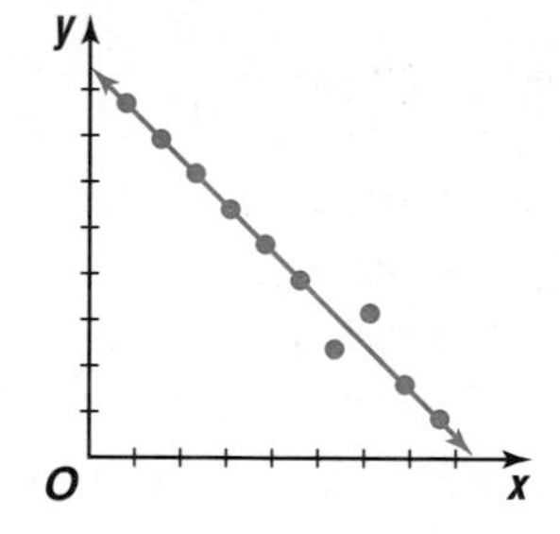

Linear Relationship		No Pattern
This scatter plot suggests a linear relationship. Notice that many of the points lie on a line, with the rest very close to it. Since the line has a positive slope, these data have a positive relationship.	This scatter plot also implies a linear relationship. However, the slope of the line suggested by the data is negative.	The points in this scatter plot are very dispersed and do not appear to form a linear pattern.

A prediction equation can be determined using a process similar to determining the equation of a line using two points. The process is dependent upon your judgment. You decide which two points on the line are used to find the slope and intercept. Your prediction equation may be different from someone else's. A prediction equation is used when a rough estimate is sufficient.

Example 1 **EDUCATION** **Refer to the application at the beginning of the lesson. Predict the average college cost in the academic year beginning in 2006.**

Graph the data. Use the starting year as the independent variable and the tuition and fees as the dependent variable.

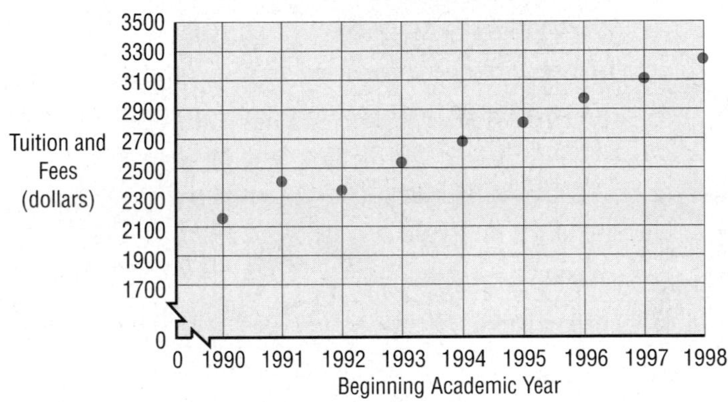

Select two points that appear to represent the data. We chose (1992, 2349) and (1997, 3111). Determine the slope of the line.

$m = \frac{y_2 - y_1}{x_2 - x_1}$ *Definition of slope*

$= \frac{3111 - 2349}{1997 - 1992}$ $(x_1, y_1) = (1992, 2349), (x_2, y_2) = (1997, 3111)$

$= \frac{762}{5}$ or 152.4

Now use one of the ordered pairs, such as (1992, 2349), and the slope in the point-slope form of the equation.

$y - y_1 = m(x - x_1)$ *Point-slope form of an equation*

$y - 2349 = 152.4(x - 1992)$ $(x_1, y_1) = (1992, 2349)$, *and* $m = 152.4$

$y = 152.4x - 301{,}231.8$

Thus, a prediction equation is $y = 152.4x - 301{,}231.8$. Substitute 2006 for x to estimate the average tuition and fees for the year 2006.

$y = 152.4x - 301{,}231.8$

$y = 152.4(2006) - 301{,}231.8$

$y = 4482.6$

According to this prediction equation, the average tuition and fees will be $4482.60 in the academic year beginning in 2006. *Use a different pair of points to find another prediction equation. How does it compare with this one?*

Data that are linear in nature will have varying degrees of **goodness of fit** to the lines of fit. Various formulas are often used to find a **correlation coefficient** that describes the nature of the data. The more closely the data fit a line, the closer the correlation coefficient r approaches 1 or -1. Positive correlation coefficients are associated with linear data having positive slopes, and negative correlation coefficients are associated with negative slopes. Thus, the more linear the data, the more closely the correlation coefficient approaches 1 or -1.

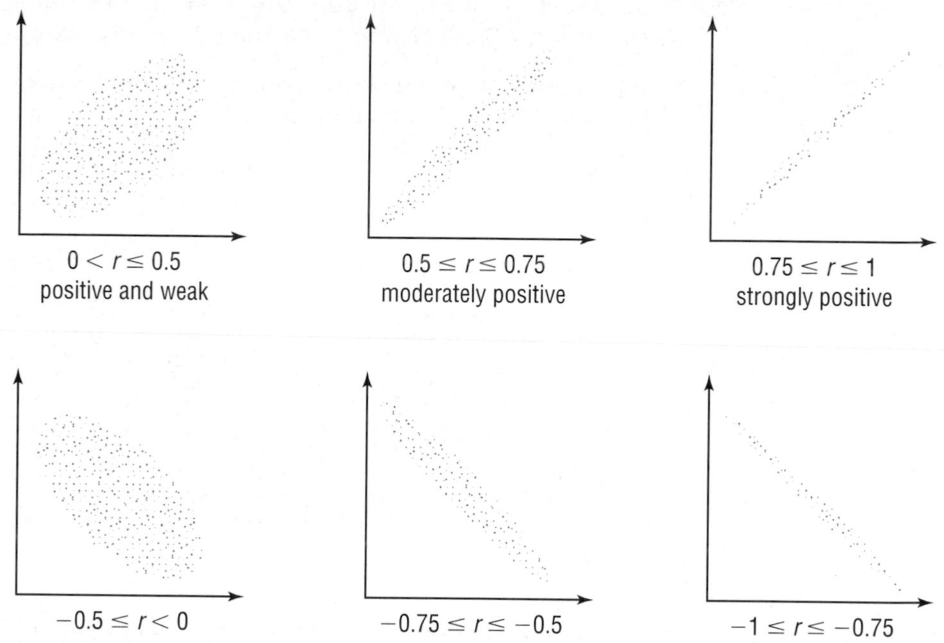

Statisticians normally use precise procedures, often relying on computers to determine correlation coefficients. The graphing calculator uses the **Pearson product-moment correlation,** which is represented by r. When using these methods, the best fit-line is often called a **regression line.**

Example 2 

NUTRITION The table contains the fat grams and Calories in various fast-food chicken sandwiches.

a. Use a graphing calculator to find the equation of the regression line and the Pearson product-moment correlation.

b. Use the equation to predict the number of Calories in a chicken sandwich that has 20 grams of fat.

Chicken Sandwich (cooking method)	Fat (grams)	Calories
A (breaded)	28	536
B (grilled)	20	430
C (chicken salad)	33	680
D (broiled)	29	550
E (breaded)	43	710
F (grilled)	12	390
G (breaded)	9	300
H (chicken salad)	5	320
I (breaded)	26	530
J (breaded)	18	440
K (grilled)	8	310

Graphing Calculator Appendix

For keystroke instruction on how to enter data, draw a scatter plot, and find a regression equation, see pages A22-A25.

a. Enter the data for fat grams in list L1 and the data for Calories in list L2. Draw a scatter plot relating the fat grams, x, and the Calories, y.

Then use the linear regression statistics to find the equation of the regression line and the correlation coefficient.

The Pearson product-moment correlation is about 0.98. The correlation between grams of fat and Calories is strongly positive. Because of the strong relationship, the equation of the regression line can be used to make predictions.

b. When rounding to the nearest tenth, the equation of the regression line is $y = 11.6x + 228.3$. Thus, there are about $y = 11.6(20) + 228.3$ or 460.3 Calories in a chicken sandwich with 20 grams of fat.

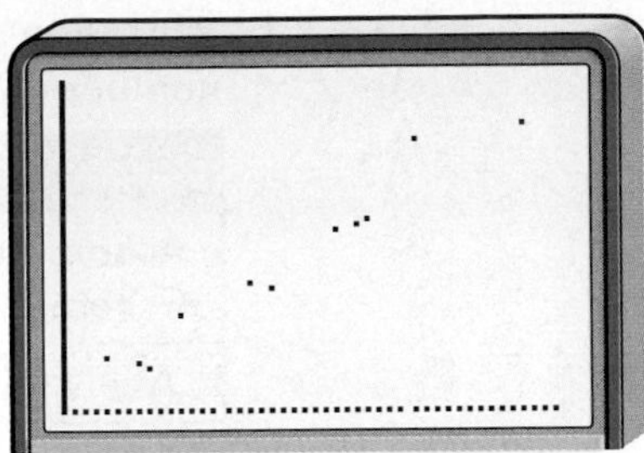

[0, 45] scl: 1 by [250, 750] scl: 50

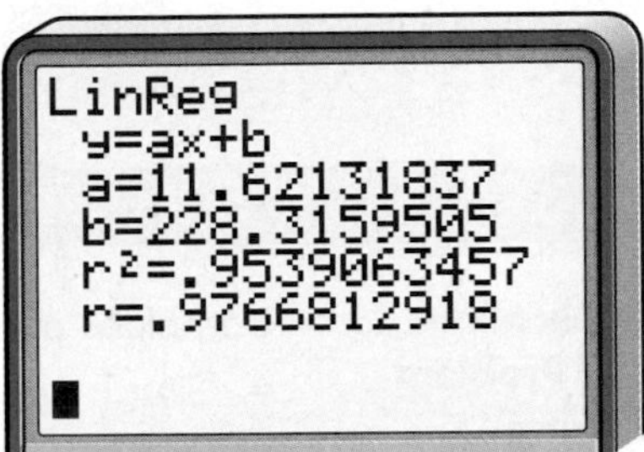

It should be noted that even when there is a large correlation coefficient, you cannot assume that there is a "cause and effect" relationship between the two related variables.

CHECK FOR UNDERSTANDING

Communicating Mathematics

Read and study the lesson to answer each question.

1. **Explain** what the slope in a best-fit line represents.
2. **Describe** three different methods for finding a best-fit line for a set of data.
3. **Write** about a set of real-world data that you think would show a negative correlation.

Guided Practice

Complete parts a–d for each set of data given in Exercises 4 and 5.

a. Graph the data on a scatter plot.
b. Use two ordered pairs to write the equation of a best-fit line.
c. Use a graphing calculator to find an equation of the regression line for the data. What is the correlation coefficient?
d. If the equation of the regression line shows a moderate or strong relationship, predict the missing value. Explain whether you think the prediction is reliable.

4. **Economics** The table shows the average amount that an American spent on durable goods in several years.

Personal Consumption Expenditures for Durable Goods

Year	1990	1991	1992	1993	1994	1995	1996	1997	2010
Personal Consumption ($)	1910	1800	1881	2083	2266	2305	2389	2461	?

Source: U.S. Dept. of Commerce

5. **Education** Do you share a computer at school? The table shows the average number of students per computer in public schools in the United States.

Students per Computer

Academic Year	1983–1984	1984–1985	1985–1986	1986–1987	1987–1988	1988–1989	1989–1990	1990–1991
Average	125	75	50	37	32	25	22	20

Academic Year	1991–1992	1992–1993	1993–1994	1994–1995	1995–1996	1996–1997	?
Average	18	16	14	10.5	10	7.8	1

Source: QED's Technology in Public Schools

EXERCISES

Applications and Problem Solving

Complete parts a–d for each set of data given in Exercises 6–11.

a. Graph the data on a scatter plot.

b. Use two ordered pairs to write the equation of a best-fit line.

c. Use a graphing calculator to find an equation of the regression line for the data. What is the correlation coefficient?

d. If the equation of the regression line shows a moderate or strong relationship, predict the missing value. Explain whether you think the prediction is reliable.

6. **Sports** The table shows the number of years coaching and the number of wins as of the end of the 1999 season for selected professional football coaches.

NFL Coach	Years	Wins
Don Shula	33	347
George Halas	40	324
Tom Landry	29	270
Curly Lambeau	33	229
Chuck Noll	23	209
Chuck Knox	22	193
Dan Reeves	19	177
Paul Brown	21	170
Bud Grant	18	168
Steve Owen	23	153
Marv Levy	17	?

Source: *World Almanac*

7. **Economics** Per capita personal income is the average personal income for a nation. The table shows the per capita personal income for the United States for several years.

Year	1990	1991	1992	1993	1994	1995	1996	1997	2005
Personal Income ($)	18,477	19,100	19,802	20,810	21,846	23,233	24,457	25,660	?

Source: U.S. Dept. of Commerce

8. Transportation Do you think the weight of a car is related to its fuel economy? The table shows the weight in hundreds of pounds and the average miles per gallon for selected 1999 cars.

Weight (100 pounds)	17.5	20.0	22.5	22.5	22.5	25.0	27.5	35.0	45.0
Fuel Economy (mpg)	65.4	49.0	59.2	41.1	38.9	40.7	46.9	27.7	?

Source: U.S. Environmental Protection Agency

9. Botany Acorns were one of the most important foods of the Native Americans. They pulverized the acorns, extracted the bitter taste, and then cooked them in various ways. The table shows the size of acorns and the geographic area covered by different species of oak.

Acorn size (cm^3)	0.3	0.9	1.1	2.0	3.4	4.8	8.1	10.5	17.1
Range (100 km^2)	233	7985	10,161	17,042	7900	3978	28,389	7646	?

Source: *Journal of Biogeography*

10. Employment Women have changed their role in American society in recent decades. The table shows the percent of working women who hold managerial or professional jobs.

Percent of Working Women in Managerial or Professional Occupations										
Year	1986	1988	1990	1992	1993	1994	1995	1996	1997	2008
Percent	23.7	25.2	26.2	27.4	28.3	28.7	29.4	30.3	30.8	?

Source: U.S. Dept. of Labor

11. Demographics The world's population is growing at a rapid rate. The table shows the number of millions of people on Earth at different years.

World Population							
Year	1	1650	1850	1930	1975	1998	2010
Population (millions)	200	500	1000	2000	4000	5900	?

Source: *World Almanac*

12. Critical Thinking Different correlation coefficients are acceptable for different situations. For each situation, give a specific example and explain your reasoning.

a. When would a correlation coefficient of less than 0.99 be considered unsatisfactory?

b. When would a correlation coefficient of 0.6 be considered good?

c. When would a strong negative correlation coefficient be desirable?

13. **Critical Thinking** The table shows the median salaries of American men and women for several years. According to the data, will the women's median salary ever be equal to the men's? If so, predict the year. Explain.

Median Salary ($)					
Year	**Men's**	**Women's**	**Year**	**Men's**	**Women's**
1985	16,311	7217	1991	20,469	10,476
1986	17,114	7610	1992	20,455	10,714
1987	17,786	8295	1993	21,102	11,046
1988	18,908	8884	1994	21,720	11,466
1989	19,893	9624	1995	22,562	12,130
1990	20,293	10,070	1996	23,834	12,815

Source: U.S. Bureau of the Census

Mixed Review

14. **Business** During the month of January, Fransworth Computer Center sold 24 computers of a certain model and 40 companion printers. The total sales on these two items for the month of January was $38,736. In February, they sold 30 of the computers and 50 printers. *(Lesson 1-5)*

    a. Assuming the prices stayed constant during the months of January and February, is it possible that their February sales could have totaled $51,470 on these two items? Explain.

    b. Assuming the prices stayed constant during the months of January and February, is it possible that their February sales could have totaled $48,420 on these two items? Explain.

15. Line ℓ passes through $A(-3, -4)$ and has a slope of -6. What is the standard form of the equation for line ℓ? *(Lesson 1-4)*

16. **Economics** The equation $y = 0.82x + 24$, where $x \geq 0$, models a relationship between a nation's disposable income, x in billions of dollars, and personal consumption expenditures, y in billions of dollars. Economists call this type of equation a *consumption function.* *(Lesson 1-3)*

    a. Graph the consumption function.

    b. Name the y-intercept.

    c. Explain the significance of the y-intercept and the slope.

17. Find $[f \circ g](x)$ and $[g \circ f](x)$ if $f(x) = x^3$ and $g(x) = x + 1$. *(Lesson 1-2)*

18. Determine if the relation $\{(2, 4), (4, 2), (-2, 4), (-4, 2)\}$ is a function. Explain. *(Lesson 1-1)*

19. **SAT/ACT Practice** Choose the equation that is represented by the graph.

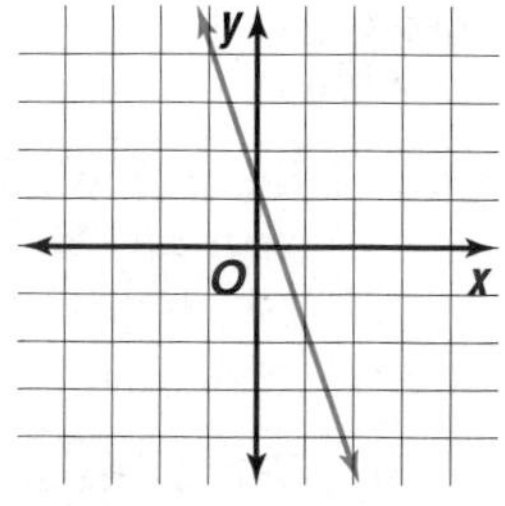

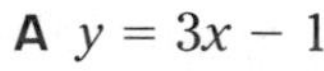

A $y = 3x - 1$

B $y = \frac{1}{3}x - 1$

C $y = 1 - 3x$

D $y = 1 - \frac{1}{3}x$

E none of these

Extra Practice See p. A27.

Piecewise Functions

OBJECTIVE

- Identify and graph piecewise functions including greatest integer, step, and absolute value functions.

ACCOUNTING The Internal Revenue Service estimates that taxpayers who itemize deductions and report interest and capital gains will need an average of almost 24 hours to prepare their returns. The amount that a single taxpayer owes depends upon his or her income. The table shows the tax brackets for different levels of income for a certain year.

Single Individual Income Tax	
Limits of Taxable Income	**Tax Bracket**
\$0 to \$25,350	15%
\$25,351 to \$61,400	28%
\$61,401 to \$128,100	31%
\$128,101 to \$278,450	36%
over \$278,450	39.6%

Source: *World Almanac*

A problem related to this will be solved in Example 3.

The tax table defines a special function called a **piecewise function.** For piecewise functions, different equations are used for different intervals of the domain. The graph below shows a piecewise function that models the number of miles from home as a function of time in minutes. Notice that the graph consists of several line segments, each of which is a part of a linear function.

Brittany traveled at a rate of 30 mph for 8 minutes. She stopped at a stoplight for 2 minutes. Then for 4 minutes she traveled 15 mph through the school zone. She sat at the school for 3 minutes while her brother got out of the car. Then she traveled home at 25 mph.

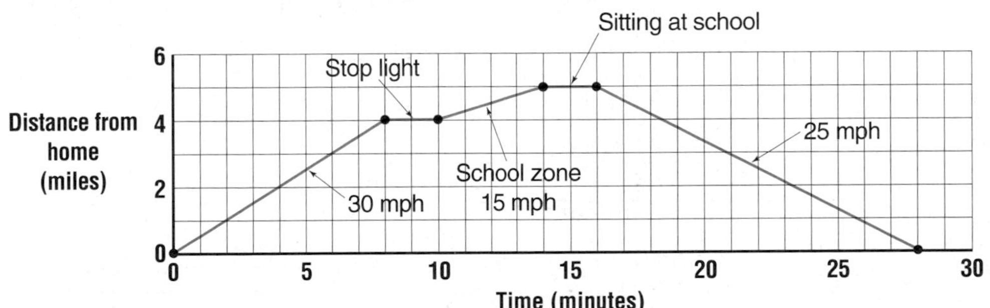

When graphing piecewise functions, the partial graphs over various intervals do not necessarily connect. The definition of the function on the intervals determines if the graph parts connect.

Example 1 Graph $f(x) = \begin{cases} 1 \text{ if } x \leq -2 \\ 2 + x \text{ if } -2 < x \leq 3. \\ 2x \text{ if } x > 3 \end{cases}$

First, graph the constant function $f(x) = 1$ for $x \leq -2$. This graph is part of a horizontal line. Because the point at $(-2, 1)$ is included in the graph, draw a closed circle at that point.

Second, graph the function $f(x) = 2 + x$ for $-2 < x \leq 3$. Because $x = -2$ is not included in this part of the domain, draw an open circle at $(-2, 0)$. $x = 3$ is included in the domain, so draw a closed circle at $(3, 5)$ since for $f(x) = 2 + x$, $f(3) = 5$.

Third, graph the line $y = 2x$ for $x > 3$. Draw an open circle at $(3, 6)$ since for $f(x) = 2x$, $f(3) = 6$.

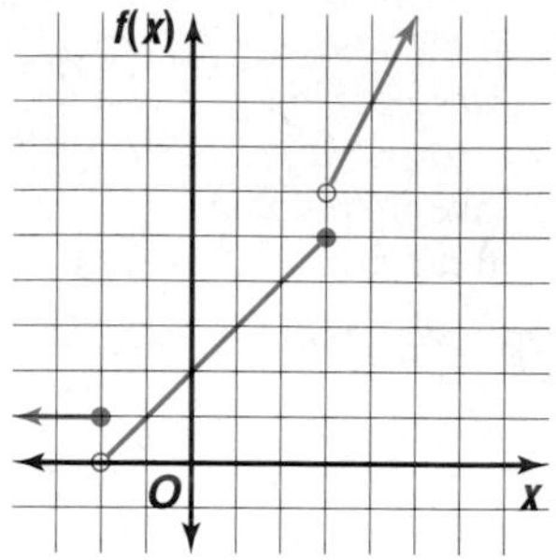

Graphing Calculator Tip

On a graphing calculator, **int(X)** indicates the greatest integer function.

A piecewise function where the graph looks like a set of stairs is called a **step function.** In a step function, there are breaks in the graph of the function. You cannot trace the graph of a step function without lifting your pencil. One type of step function is the **greatest integer function.** The symbol $[\![x]\!]$ means *the greatest integer not greater than x.* This does not mean to round or truncate the number. For example, $[\![8.9]\!] = 8$ because 8 is the greatest integer not greater than 8.9. Similarly, $[\![-3.9]\!] = -4$ because -3 is greater than -3.9. The greatest integer function is given by $f(x) = [\![x]\!]$.

Example 2 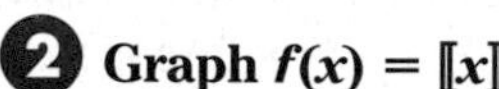 Graph $f(x) = [\![x]\!]$.

Make a table of values. The domain values will be intervals for which the greatest integer function will be evaluated.

x	f(x)
$-3 \leq x < -2$	-3
$-2 \leq x < -1$	-2
$-1 \leq x < 0$	-1
$0 \leq x < 1$	0
$1 \leq x < 2$	1
$2 \leq x < 3$	2
$3 \leq x < 4$	3
$4 \leq x < 5$	4

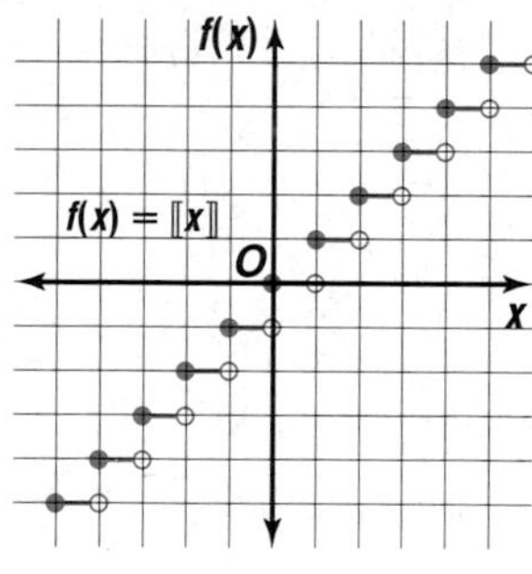

Notice that the domain for this greatest integer function is all real numbers and the range is integers.

The graphs of step functions are often used to model real-world problems such as fees for cellular telephones and the cost of shipping an item of a given weight.

Example 3 Refer to the application at the beginning of the lesson.

Real World Application

a. Graph the tax brackets for the different incomes.

b. What is the tax bracket for a person who makes \$70,000?

a.

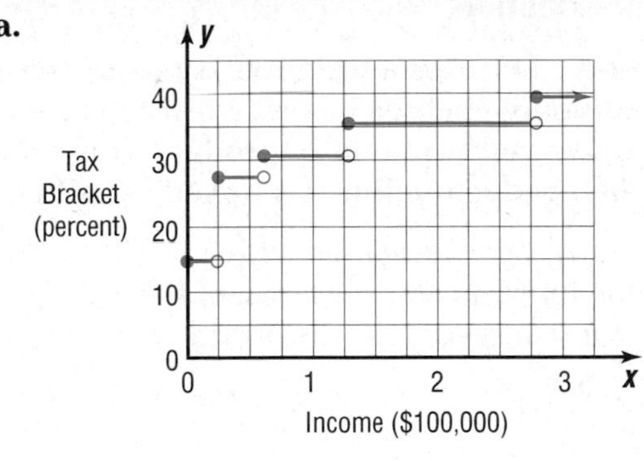

b. \$70,000 falls in the interval \$61,401 to \$128,100. Thus, the tax bracket for \$70,000 is 31%.

The **absolute value function** is another piecewise function. Consider $f(x) = |x|$. The absolute value of a number is always nonnegative. The table lists a specific domain and resulting range values for the absolute value function. Using these points, a graph of the absolute value function can be constructed. Notice that the domain of the graph includes all real numbers. However, the range includes only nonnegative real numbers.

table

$f(x) = \|x\|$	
x	**f(x)**
−3	3
−2.4	2.4
0	0
0.7	0.7
2	2
3.4	3.4

graph

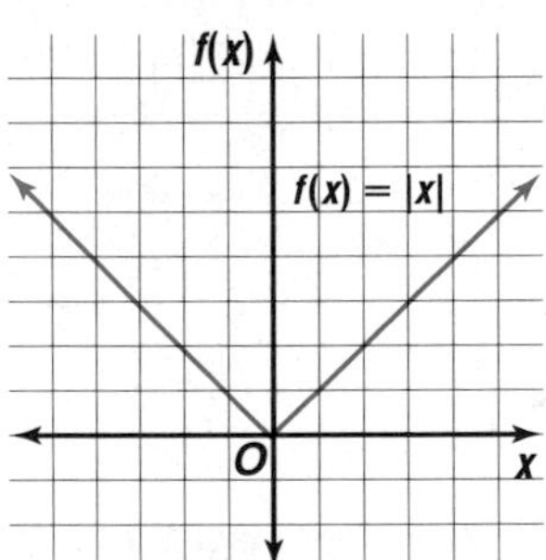

piecewise function

$$f(x) = \begin{cases} -x \text{ if } x < 0 \\ x \text{ if } x \geq 0 \end{cases}$$

Example  **Graph $f(x) = 2|x| - 6$.**

Use a table of values to determine points on the graph.

x	2|x| − 6	(x, f(x))
−6	2|−6| − 6 = 6	(−6, 6)
−3	2|−3| − 6 = 0	(−3, 0)
−1.5	2|−1.5| − 6 = −3	(−1.5, −3)
0	2|0| − 6 = −6	(0, −6)
1	2|1| − 6 = −4	(1, −4)
2	2|2| − 6 = −2	(2, −2)

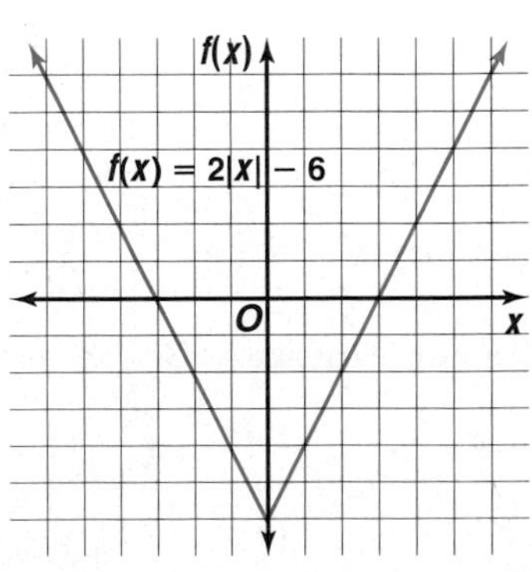

Many real-world situations can be modeled by a piecewise function.

Example 5 **Identify the type of function that models each situation. Then write a function for the situation.**

a. Manufacturing **The stated weight of a box of rice is 6.9 ounces. The company randomly chooses boxes to test to see whether their equipment is dispensing the right amount of product. If the discrepancy is more than 0.2 ounce, the production line is stopped for adjustments.**

The situation can be represented with an absolute value function. Let w represent the weight and $d(w)$ represent the discrepancy. Then $d(w) = |6.9 - w|$.

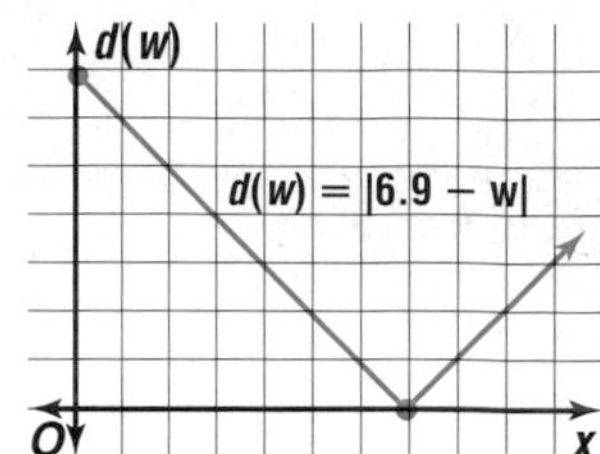

b. Business **On a certain telephone rate plan, the price of a cellular telephone call is 35¢ per minute or fraction thereof.**

This can be described by a greatest integer function.

Let m represent the number of minutes of the call and $c(m)$ represent the cost in cents.

$$c(m) = \begin{cases} 35m \text{ if } [\![m]\!] = m \\ 35[\![m + 1]\!] \text{ if } [\![m]\!] < m \end{cases}$$

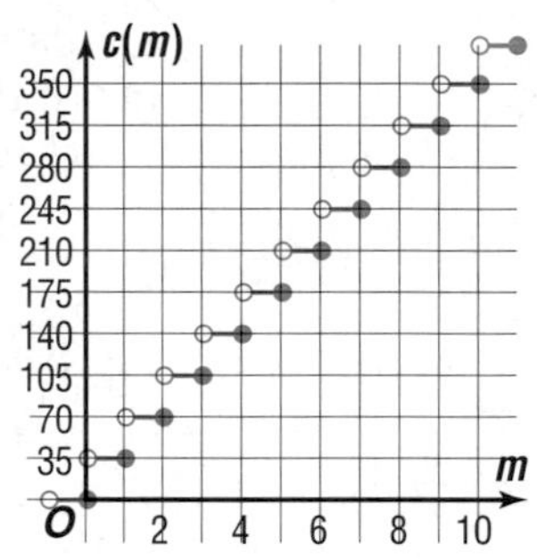

CHECK FOR UNDERSTANDING

Communicating Mathematics

Read and study the lesson to answer each question.

1. **Write** $f(x) = |x|$ as a piecewise function.
2. **State** the domain and range of the function $f(x) = 2[\![x]\!]$.
3. **Write** the function that is represented by the graph.
4. **You Decide** Misae says that a step graph does not represent a function because the graph is not connected. Alex says that it does represent a function because there is only one y for every x. Who is correct and why?

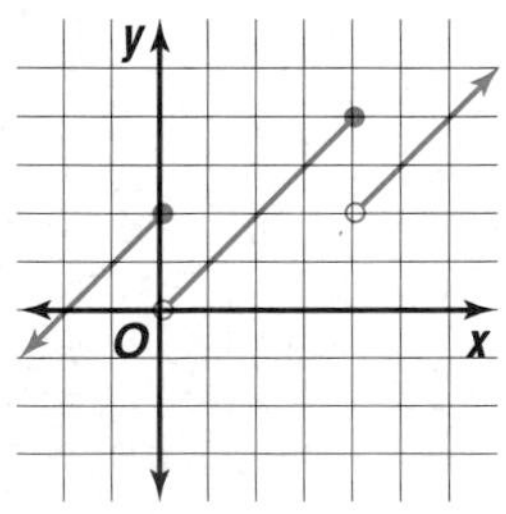

Guided Practice

Graph each function.

5. $f(x) = \begin{cases} 2x \text{ if } 0 \leq x \leq 4 \\ 8 \text{ if } 4 < x \leq 7 \end{cases}$

6. $f(x) = \begin{cases} 6 \text{ if } x \leq -6 \\ |x| \text{ if } -6 < x < 6 \\ 6 \text{ if } x > 6 \end{cases}$

7. $f(x) = -[\![x]\!]$

8. $f(x) = |x - 3|$

9. **Business** Identify the type of function that models the labor cost for repairing a computer if the charge is \$50 per hour or fraction thereof. Then write and graph a function for the situation.

10. **Consumerism** Guillermo Lujan is flying from Denver to Dallas for a convention. He can park his car in the Denver airport long-term parking lot at the terminal or in the shuttle parking facility closeby. In the long-term lot, it costs \$1.00 per hour or any part of an hour with a maximum charge of \$6.00 per day. In shuttle facility, he has to pay \$4.00 for each day or part of a day. Which parking lot is less expensive if Mr. Lujan returns after 2 days and 3 hours?

EXERCISES

Practice

Graph each function.

11. $f(x) = \begin{cases} 2x + 1 \text{ if } x < 0 \\ 2x - 1 \text{ if } x \geq 0 \end{cases}$

12. $g(x) = |x - 5|$

13. $h(x) = [\![x]\!] + 2$

14. $g(x) = |2x + 3|$

15. $f(x) = [\![x - 1]\!]$

16. $h(x) = \begin{cases} 3 \text{ if } -1 \leq x \leq 1 \\ 4 \text{ if } 1 < x \leq 4 \\ x \text{ if } x > 4 \end{cases}$

17. $g(x) = 2|x - 3|$

18. $f(x) = [\![-3x]\!]$

19. $h(x) = \begin{cases} x + 3 \text{ if } x \leq 0 \\ 3 - x \text{ if } 1 < x \leq 3 \\ 3x \text{ if } x > 3 \end{cases}$

20. $f(x) = \begin{cases} -2x \text{ if } x < 1 \\ 3 \text{ if } x = 1 \\ 4x \text{ if } x > 1 \end{cases}$

21. $j(x) = \dfrac{2}{[\![x]\!]}$

22. $g(x) = |9 - 3|x||$

Identify the type of function that models each situation. Then write and graph a function for the situation.

23. **Tourism** The table shows the charge for renting a bicycle from a rental shop on Cumberland Island, Georgia, for different amounts of time.

Island Rentals	
Time	**Price**
$\frac{1}{2}$ hour	\$6
1 hour	\$10
2 hours	\$16
Daily	\$24

24. **Postage** The cost of mailing a letter is \$0.33 for the first ounce and \$0.22 for each additional ounce or portion thereof.

25. **Manufacturing** A can of coffee is supposed to contain one pound of coffee. How does the actual weight of the coffee in the can compare to 1 pound?

Applications and Problem Solving

26. **Retail Sales** The table shows the shipping charges that apply to orders placed in a catalog.
    a. What type of function is described?
    b. Write the shipping charges as a function of the value of the order.
    c. Graph the function.

Shipping to or within the United States	
Value of Order	**Shipping, Packing, and Handling Charge**
\$0.00–25.00	\$3.50
\$25.01–75.00	\$5.95
\$75.01–125.00	\$7.95
\$125.01 and up	\$9.95

27. **Critical Thinking** Describe the values of x and y which are solutions to $[\![x]\!] = [\![y]\!]$.

28. **Engineering** The degree day is used to measure the demand for heating or cooling. In the United States, 65°F is considered the desirable temperature for the inside of a building. The number of degree days recorded on a given date is equal to the difference between 65 and the mean temperature for that date. If the mean temperature is above 65°F, cooling degree days are recorded. Heating degree days are recorded if the mean temperature is below 65°F.
    a. What type of function can be used to model degree days?
    b. Write a function to model the number of degree days $d(t)$ for a mean temperature of t°F.
    c. Graph the function.
    d. The mean temperature is the mean of the high and low temperatures for a day. How many degree days are recorded for a day with a high of temperature of 63°F and a low temperature of 28°F? Are they heating degree days or cooling degree days?

29. **Accounting** The income tax brackets for the District of Columbia are listed in the tax table.

Income	**Tax Bracket**
up to \$10,000	6%
more than \$10,000, but no more than \$20,000	8%
more than \$20,000	9.5%

    a. What type of function is described by the tax rates?
    b. Write the function if x is income and $t(x)$ is the tax rate.
    c. Graph the tax brackets for different taxable incomes.
    d. Alicia Davis lives in the District of Columbia. In which tax bracket is Ms. Davis if she made \$36,000 last year?

30. **Critical Thinking** For $f(x) = [\![x]\!]$ and $g(x) = |x|$, are $[f \circ g](x)$ and $[g \circ f](x)$ equivalent? Justify your answer.

Mixed Review

31. **Transportation** The table shows the percent of workers in different cities who use public transportation to get to work. *(Lesson 1-6)*

a. Graph the data on a scatter plot.

b. Use two ordered pairs to write the equation of a best-fit line.

c. Use a graphing calculator to find an equation for the regression line for the data. What is the correlation value?

d. If the equation of the regression line shows a moderate or strong relationship, predict the percent of workers using public transportation in Baltimore, Maryland. Is the prediction reliable? Explain.

City	Workers 16 years and older	Percent who use Public Transportation
New York, NY	3,183,088	53.4
Los Angeles, CA	1,629,096	10.5
Chicago, IL	1,181,677	29.7
Houston, TX	772,957	6.5
Philadelphia, PA	640,577	28.7
San Diego, CA	560,913	4.2
Dallas, TX	500,566	6.7
Phoenix, AZ	473,966	3.3
San Jose, CA	400,932	3.5
San Antonio, TX	395,551	4.9
San Francisco, CA	382,309	33.5
Indianapolis, IN	362,777	3.3
Detroit, MI	325,054	10.7
Jacksonville, FL	312,958	2.7
Baltimore, MD	307,679	22.0

Source: U.S. Bureau of the Census

32. Write the standard form of the equation of the line that passes through the point at (4, 2) and is parallel to the line whose equation is $y = 2x - 4$. *(Lesson 1-5)*

33. **Sports** During a basketball game, the two highest-scoring players scored 29 and 15 points and played 39 and 32 minutes, respectively. *(Lesson 1-3)*

a. Write an ordered pair of the form (minutes played, points scored) to represent each player.

b. Find the slope of the line containing both points.

c. What does the slope of the line represent?

34. **Business** For a company, the revenue $r(x)$ in dollars, from selling x items is $r(x) = 400x - 0.2x^2$. The cost for making and selling x items is $c(x) = 0.1x + 200$. Write the profit function $p(x) = (r - c)(x)$. *(Lesson 1-2)*

35. **Retail** Winston bought a sweater that was on sale 25% off. The original price of the sweater was \$59.99. If sales tax in Winston's area is 6.5%, how much did the sweater cost including sale tax? *(Lesson 1-2)*

36. State the domain and range of the relation $\{(0, 2), (4, -2), (9, 3), (-7, 11), (-2, 0)\}$. Is the relation a function? Explain. *(Lesson 1-1)*

37. **SAT Practice** **Quantitative Comparison**

A if the quantity in Column A is greater

B if the quantity in Column B is greater

C if the two quantities are equal

D if there is not enough information to determine the relationship

Column A	Column B
5×6^{12}	6×5^{12}

Extra Practice See p. A27.

1-8 Graphing Linear Inequalities

OBJECTIVE
- Graph linear inequalities.

NUTRITION Arctic explorers need endurance and strength. They move sleds weighing more than 1100 pounds each for as much as 12 hours a day. For that reason, Will Steger and members of his exploration team each burn 4000 to 6000 Calories daily!

An *endurance diet* can provide the energy and nutrients necessary for peak performance in the Arctic. An endurance diet has a balance of fat and carbohydrates and protein. Fat is a concentrated energy source that supplies nine calories per gram. Carbohydrates and protein provide four calories per gram and are a quick source of energy. What are some of the combinations of carbohydrates and protein and fat that supply the needed energy for the Arctic explorers? *This problem will be solved in Example 2.*

This situation can be described using a **linear inequality.** A linear inequality is not a function. However, you can use the graphs of linear functions to help you graph linear inequalities.

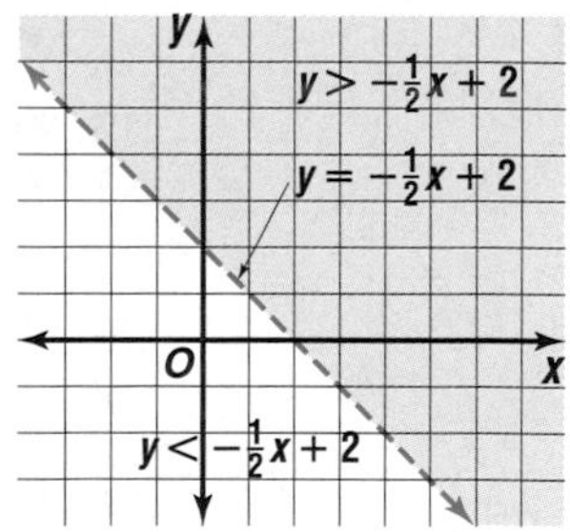

The graph of $y = -\frac{1}{2}x + 2$ separates the coordinate plane into two regions, called **half planes.** The line described by $y = -\frac{1}{2}x + 2$ is called the **boundary** of each region. If the boundary is part of a graph, it is drawn as a solid line. A boundary that is not part of the graph is drawn as a dashed line. The graph of $y > -\frac{1}{2}x + 2$ is the region above the line. The graph of $y < -\frac{1}{2}x + 2$ is the region below the line.

When graphing an inequality, you can determine which half plane to shade by testing a point on either side of the boundary in the original inequality. If it is not on the boundary, the origin (0, 0) is often an easy point to test. If the inequality statement is true for your test point, then shade the half plane that contains the test point. If the inequality statement is false for your test point, then shade the half plane that does not contain the test point.

Example 1 **Graph each inequality.**

a. $x > 3$

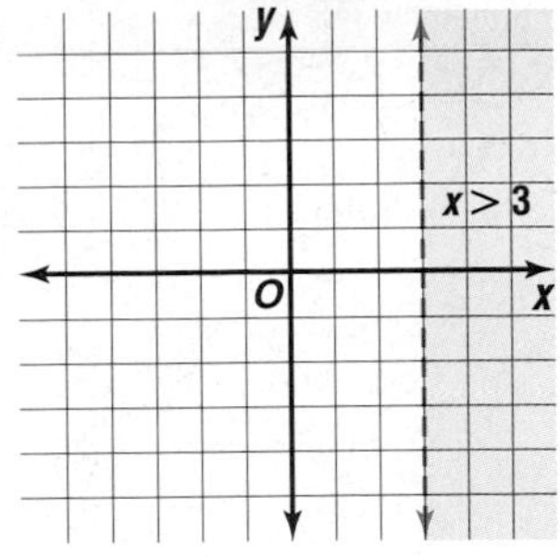

The boundary is not included in the graph. So the vertical line $x = 3$ should be a dashed line.

Testing (0, 0) in the inequality yields a false inequality, $0 > 3$. So shade the half plane that does not include (0, 0).

b. $x - 2y - 5 \leq 0$

$$x - 2y - 5 \leq 0$$
$$-2y \leq -x + 5$$
$$y \geq \frac{1}{2}x - \frac{5}{2}$$

Reverse the inequality when you divide or multiply by a negative.

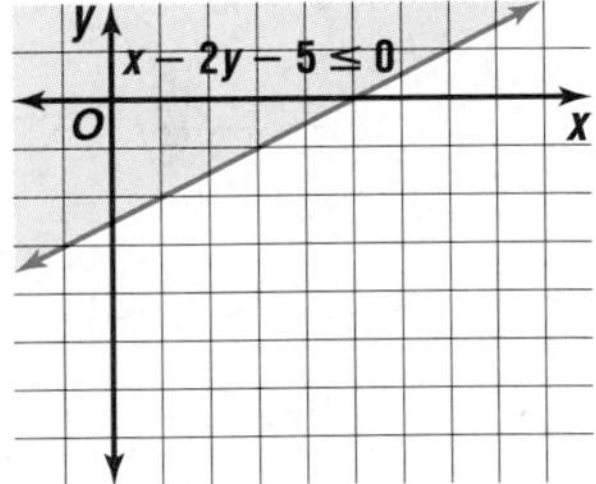

The graph does include the boundary. So the line is solid.

Testing (0, 0) in the inequality yields a true inequality, so shade the half plane that includes (0, 0).

c. $y > |x - 2|$

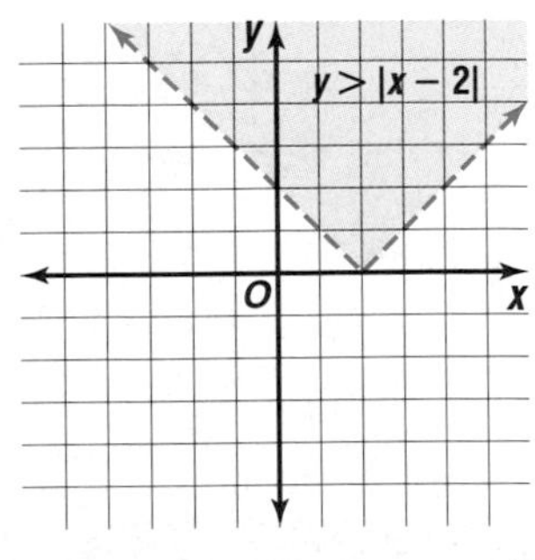

Graph the equation $y = |x - 2|$ with a dashed boundary.

Testing (0, 0) yields the false inequality $0 > 2$, so shade the region that does not include (0, 0).

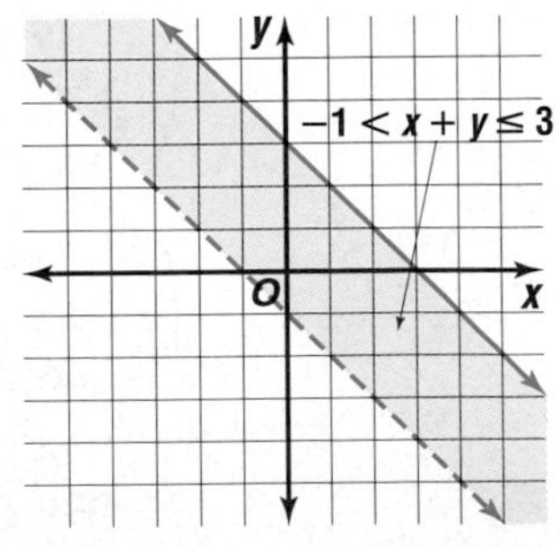

You can also graph relations such as $-1 < x + y \leq 3$. The graph of this relation is the intersection of the graph of $-1 < x + y$ and the graph of $x + y \leq 3$. Notice that the boundaries $x + y = 3$ and $x + y = -1$ are parallel lines. The boundary $x + y = 3$ is part of the graph, but $x + y = -1$ is not.

Example **2** **NUTRITION** **Refer to the application at the beginning of the lesson.**

a. Draw a graph that models the combinations of grams of fat and carbohydrates and protein that the arctic team diet may include to satisfy their daily caloric needs.

Let x represent the number of grams of fat and y represent the number of grams of carbohydrates and protein. The team needs at least 4000, but no more than 6000, Calories each day. Write an inequality.

$4000 \leq 9x + 4y \leq 6000$

You can write this compound inequality as two inequalities, $4000 \leq 9x + 4y$ and $9x + 4y \leq 6000$. Solve each part for y.

$$4000 \leq 9x + 4y \quad \text{and} \quad 9x + 4y \leq 6000$$

$$4000 - 9x \leq 4y \qquad\qquad 4y \leq 6000 - 9x$$

$$1000 - \frac{9}{4}x \leq y \qquad\qquad y \leq 1500 - \frac{9}{4}x$$

Graph each boundary line and shade the appropriate region. The graph of the compound inequality is the area in which the shading overlaps.

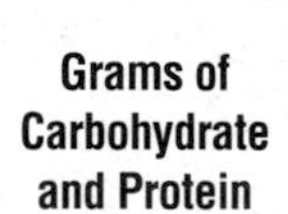

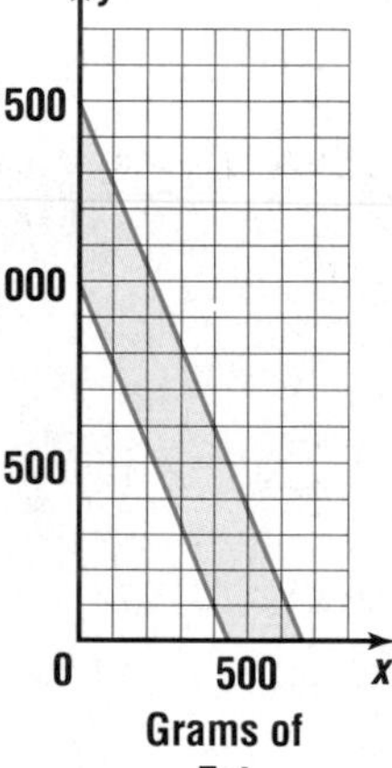

b. Name three combinations of fat or carbohydrates and protein that meet the Calorie requirements.

Any point in the shaded region or on the boundary lines meets the requirements. Three possible combinations are (100, 775), (200, 800), and (300, 825). These ordered pairs represent 100 grams of fat and 775 grams of carbohydrate and protein, 200 grams of fat and 800 grams of carbohydrate and protein, and 300 grams of fat and 825 grams of carbohydrate and protein.

CHECK FOR UNDERSTANDING

Communicating Mathematics

Read and study the lesson to answer each question.

1. **Write** the inequality whose graph is shown.
2. **Describe** the process you would use to graph $-3 \leq 2x + y < 7$.
3. **Explain** why you can use a test point to determine which region or regions of the graph of an inequality should be shaded.

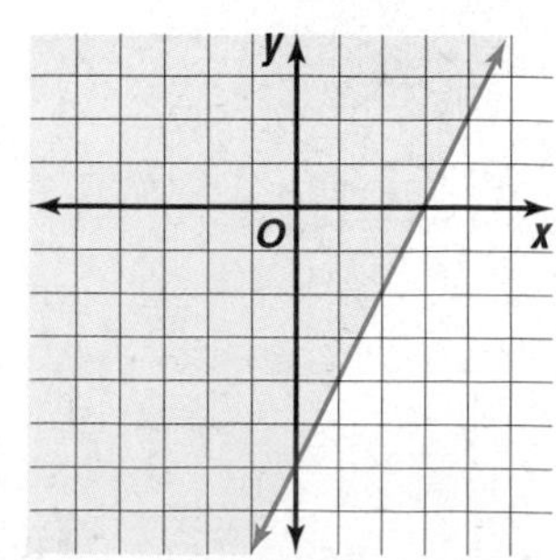

Guided Practice Graph each inequality.

4. $x + y < 4$
5. $3x - y \leq 6$
6. $7 < x + y \leq 9$
7. $y < |x + 3|$

8. **Business** Nancy Stone has a small company and has negotiated a special rate for rental cars when she and other employees take business trips. The maximum charge is \$45.00 per day plus \$0.40 per mile. Discounts apply when renting for longer periods of time or during off-peak seasons.
   a. Write a linear inequality that models the total cost of the daily rental $c(m)$ as a function of the total miles driven, m.
   b. Graph the inequality.
   c. Name three combinations of miles and total cost that satisfy the inequality.

EXERCISES

Practice Graph each inequality.

9. $y < 3$
10. $x - y > -5$
11. $2x + 4y \geq 7$
12. $-y < 2x + 1$
13. $2x - 5y + 19 \leq 0$
14. $-4 \leq x - y \leq 5$
15. $y \geq |x|$
16. $-2 \leq x + 2y \leq 4$
17. $y > |x| + 4$
18. $y < |2x + 3|$
19. $-8 \leq 2x + y < 6$
20. $y - 1 > |x + 3|$
21. Graph the region that satisfies $x \geq 0$ and $y \geq 0$.
22. Graph $2 < |x| \leq 8$.

Applications and Problem Solving

23. **Manufacturing** Many manufacturers use inequalities to solve production problems such as determining how much of each product should be assigned to each machine. Suppose one bakery oven at a cookie manufacturer is being used to bake chocolate cookies and vanilla cookies. A batch of chocolate cookies bakes in 8 minutes, and a batch of vanilla cookies bakes in 10 minutes.
   a. Let x represent the number of batches of chocolate cookies and y represent the number of batches of vanilla cookies. Write a linear inequality for the number of batches of each type of cookie that could be baked in one oven in an 8-hour shift.
   b. Graph the inequality.
   c. Name three combinations of batches of chocolate cookies and vanilla cookies that satisfy the inequality.
   d. Often manufacturers' problems involve as many as 150 products, 218 facilities, 10 plants, and 127 customer zones. Research how problems like this are solved.

24. **Critical Thinking** Graph $|y| \geq x$.

25. **Critical Thinking** Suppose $xy > 0$.

a. Describe the points whose coordinates are solutions to the inequality.

b. Demonstrate that for points whose coordinates are solutions to the inequality, the equation $|x + y| = |x| + |y|$ holds true.

26. **Engineering Mechanics** The production cost of a job depends in part on the accuracy required. On most sheet metal jobs, an accuracy of 1, 2, or 0.1 mils is required. A mil is $\frac{1}{1000}$ inch. This means that a dimension must be less than $\frac{1}{1000}$, $\frac{2}{1000}$, or $\frac{1}{10{,}000}$ inch larger or smaller than the blueprint states. Industrial jobs often require a higher degree of accuracy.

a. Write inequalities that models the possible dimensions of a part that is supposed to be 8 inches by $4\frac{1}{4}$ inches if the accuracy required is 2 mils.

b. Graph the region that shows the satisfactory dimensions for the part.

27. **Exercise** The American College of Sports Medicine recommends that healthy adults exercise at a target level of 60% to 90% of their maximum heart rate. You can estimate your maximum heart rate by subtracting your age from 220.

a. Write a compound inequality that models age, a, and target heart rate, r.

b. Graph the inequality.

Mixed Review

28. **Business** Gatsby's Automotive Shop charges $55 per hour or any fraction of an hour for labor. *(Lesson 1-7)*

a. What type of function is described?

b. Write the labor charge as a function of the time.

c. Graph the function.

29. The equation of line ℓ is $3x - y = 10$. *(Lesson 1-5)*

a. What is the standard form of the equation of the line that is parallel to ℓ and passes through the point at $(0, -2)$?

b. Write the standard form of the equation of the line that is perpendicular to ℓ and passes through the point at $(0, -2)$.

30. Write the slope-intercept form of the equation of the line through (1, 4) and (5, 7). *(Lesson 1-4)*

31. **Temperature** The temperature in Indianapolis on January 30 was 23°F at 12:00 A.M. and 48°F at 4:00 P.M. *(Lesson 1-3)*

a. Write two ordered pairs of the form (hours since midnight, temperature) for this date. What is the slope of the line containing these points?

b. What does the slope of the line represent?

32. **SAT/ACT Practice** Which expression is equivalent to $\frac{9^5 - 9^4}{8}$?

A $\frac{1}{8}$ B $\frac{9}{8}$ C $\frac{9^3}{8}$

D $\frac{9^9}{8}$ E 9^4

Extra Practice See p. A27.

VOCABULARY

abscissa (p. 5)
absolute value function (p.47)
boundary (p. 52)
coinciding lines (p. 32)
composite (p. 15)
composition of functions (pp. 14-15)
constant function (p. 22)
domain (p. 5)
family of graphs (p. 26)
function (p. 6)
function notation (p. 7)
greatest integer function (p. 46)
half plane (p. 52)
iterate (p. 16)
iteration (p. 16)
linear equation (p. 20)
linear function (p. 22)
linear inequality (p. 52)
ordinate (p. 5)
parallel lines (p. 32)
perpendicular lines (p. 34)
piecewise function (p. 45)
point-slope form (p. 28)
range (p. 5)
relation (p. 5)
slope (pp. 20-21)
slope-intercept form (p. 21)
standard form (p. 21)
step function (p. 46)
vertical line test (p. 7)
x-intercept (p. 20)
y-intercept (p. 20)
zero of a function (p. 22)

Modeling

best-fit line (p. 38)
correlation coefficient (p. 40)
goodness of fit (p. 40)
model (p. 27)
Pearson-product moment correlation (p. 40)
prediction equation (p. 38)
regression line (p. 40)
scatter plot (p. 38)

UNDERSTANDING AND USING THE VOCABULARY

Choose the letter of the term that best matches each statement or phrase.

1. for the function f, a value of x for which $f(x) = 0$
2. a pairing of elements of one set with elements of a second set
3. has the form $Ax + By + C = 0$, where A is positive and A and B are not both zero
4. $y - y_1 = m(x - x_1)$, where (x_1, y_1) lies on a line having slope m
5. $y = mx + b$, where m is the slope of the line and b is the y-intercept
6. a relation in which each element of the domain is paired with exactly one element of the range
7. the set of all abscissas of the ordered pairs of a relation
8. the set of all ordinates of the ordered pairs of a relation
9. a group of graphs that displays one or more similar characteristics
10. lie in the same plane and have no points in common

a. function
b. parallel lines
c. zero of a function
d. linear equation
e. family of graphs
f. relation
g. point-slope form
h. domain
i. slope-intercept form
j. range

For additional review and practice for each lesson, visit: **www.amc.glencoe.com** 

SKILLS AND CONCEPTS

OBJECTIVES AND EXAMPLES

Lesson 1-1 Evaluate a function.

- Find $f(-2)$ if $f(x) = 3x^2 - 2x + 4$.
  Evaluate the expression $3x^2 - 2x + 4$ for $x = -2$.

$$\begin{aligned} f(-2) &= 3(-2)^2 - 2(-2) + 4 \\ &= 12 + 4 + 4 \\ &= 20 \end{aligned}$$

REVIEW EXERCISES

Evaluate each function for the given value.

11. $f(4)$ if $f(x) = 5x - 10$
12. $g(2)$ if $g(x) = 7 - x^2$
13. $f(-3)$ if $g(x) = 4x^2 - 4x + 9$
14. $h(0.2)$ if $h(x) = 6 - 2x^3$
15. $g\left(\frac{1}{3}\right)$ if $g(x) = \frac{2}{5x}$
16. $k(4c)$ if $k(x) = x^2 + 2x - 4$
17. Find $f(m + 1)$ if $f(x) = |x^2 + 3x|$.

Lesson 1-2 Perform operations with functions.

- Given $f(x) = 4x + 2$ and $g(x) = x^2 - 2x$, find $(f + g)(x)$ and $(f \cdot g)(x)$.

$$\begin{aligned} (f + g)(x) &= f(x) + g(x) \\ &= 4x + 2 + x^2 - 2x \\ &= x^2 + 2x + 2 \end{aligned}$$

$$\begin{aligned} (f \cdot g)(x) &= f(x) \cdot g(x) \\ &= (4x + 2)(x^2 - 2x) \\ &= 4x^3 - 6x^2 - 4x \end{aligned}$$

Find $(f + g)(x)$, $(f - g)(x)$, $(f \cdot g)(x)$, and $\left(\frac{f}{g}\right)(x)$ for each $f(x)$ and $g(x)$.

18. $f(x) = 6x - 4$
    $g(x) = 2$
19. $f(x) = x^2 + 4x$
    $g(x) = x - 2$
20. $f(x) = 4 - x^2$
    $g(x) = 3x$
21. $f(x) = x^2 + 7x + 12$
    $g(x) = x + 4$
22. $f(x) = x^2 - 1$
    $g(x) = x + 1$
23. $f(x) = x^2 - 4x$
    $g(x) = \frac{4}{x - 4}$

Lesson 1-2 Find composite functions.

- Given $f(x) = 2x^2 + 4x$ and $g(x) = 2x - 1$, find $[f \circ g](x)$ and $[g \circ f](x)$.

$$\begin{aligned} [f \circ g](x) &= f(g(x)) \\ &= f(2x - 1) \\ &= 2(2x - 1)^2 + 4(2x + 1) \\ &= 2(4x^2 - 4x + 1) + 8x + 4 \\ &= 8x^2 + 6 \end{aligned}$$

$$\begin{aligned} [g \circ f](x) &= g(f(x)) \\ &= g(2x^2 + 4x) \\ &= 2(2x^2 + 4x) - 1 \\ &= 4x^2 + 8x - 1 \end{aligned}$$

Find $[f \circ g](x)$ and $[g \circ f](x)$ for each $f(x)$ and $g(x)$.

24. $f(x) = x^2 - 4$
    $g(x) = 2x$
25. $f(x) = 0.5x + 5$
    $g(x) = 3x^2$
26. $f(x) = 2x^2 + 6$
    $g(x) = 3x$
27. $f(x) = 6 + x$
    $g(x) = x^2 - x + 1$
28. $f(x) = x^2 - 5$
    $g(x) = x + 1$
29. $f(x) = 3 - x$
    $g(x) = 2x^2 + 10$
30. State the domain of $[f \circ g](x)$ for $f(x) = \sqrt{x - 16}$ and $g(x) = 5 - x$.

OBJECTIVES AND EXAMPLES

Lesson 1-3 Graph linear equations.

- Graph $f(x) = 4x - 3$.

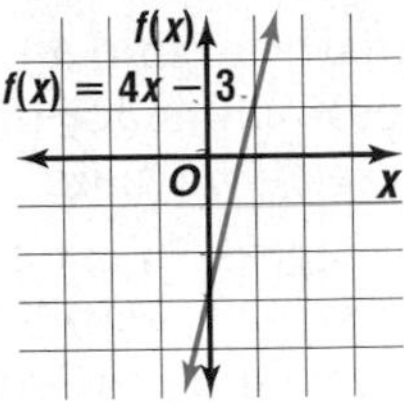

REVIEW EXERCISES

Graph each equation.

31. $y = 3x + 6$

32. $y = 8 - 5x$

33. $y - 15 = 0$

34. $0 = 2x - y - 7$

35. $y = 2x$

36. $y = -8x - 2$

37. $7x + 2y = -5$

38. $y = \frac{1}{4}x - 6$

Lesson 1-4 Write linear equations using the slope-intercept, point-slope, and standard forms of the equation.

- Write the slope-intercept form of the equation of the line that has a slope of 24 and passes through the point at (1, 2).

$y = mx + b$ *Slope-intercept form*
$2 = -4(1) + b$ *$y = 2$, $x = 1$, $m = -4$*
$6 = b$ *Solve for b.*

The equation for the line is $y = -4x + 6$.

Write an equation in slope-intercept form for each line described.

39. slope = 2, y-intercept = -3

40. slope = -1, y-intercept = 1

41. slope $= \frac{1}{2}$, passes through the point at $(-5, 2)$

42. passes through $A(-4, 2)$ and $B(2, 5)$

43. x-intercept = 1, y-intercept = -4

44. horizontal and passes through the point at $(3, -1)$

45. the x-axis

46. slope = 0.1, x-intercept = 1

Lesson 1-5 Write equations of parallel and perpendicular lines.

- Write the standard form of the equation of the line that is parallel to the graph of $y = 2x - 3$ and passes through the point at $(1, -1)$.

$y - y_1 = m(x - x_1)$ *Point-slope form*
$y - (-1) = 2(x - 1)$ *$y_1 = -1, m = 2, x = 1$*
$2x - y - 3 = 0$

Write the standard form of the equation of the line that is perpendicular to the graph of $y = 2x - 3$ and passes through the point at $(6, -1)$.

$y - y_1 = m(x - x_1)$ *$y_1 = -1$,*
$y - (-1) = -\frac{1}{2}(x - 6)$ *$m = -\frac{1}{2}$, $x = 6$*
$x + 2y - 2 = 0$

Write the standard form of the equation of the line that is parallel to the graph of the given equation and passes through the point with the given coordinates.

47. $y = x + 1$; (1, 1)

48. $y = \frac{1}{3}x - 2$; $(-1, 6)$

49. $2x + y = 1$; $(-3, 2)$

Write the standard form of the equation of the line that is perpendicular to the graph of the given equation and passes through the point with the given coordinates.

50. $y = -2x + \frac{1}{4}$; $(4, -8)$

51. $4x - 2y + 2 = 0$; (1, 4)

52. $x = -8$; $(4, -6)$

OBJECTIVES AND EXAMPLES

Lesson 1-6 Draw and analyze scatter plots.

- *This scatter plot implies a linear relationship. Since data closely fits a line with a positive slope, the scatter plot shows a strong, positive correlation.*

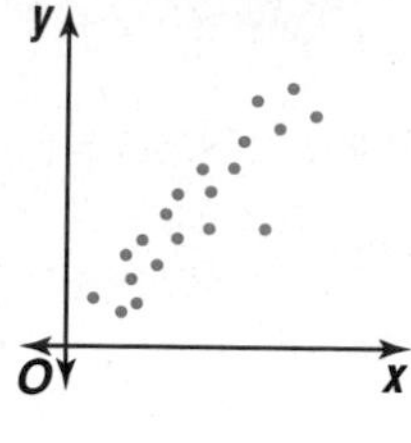

This scatter plot implies a linear relationship with a negative slope.

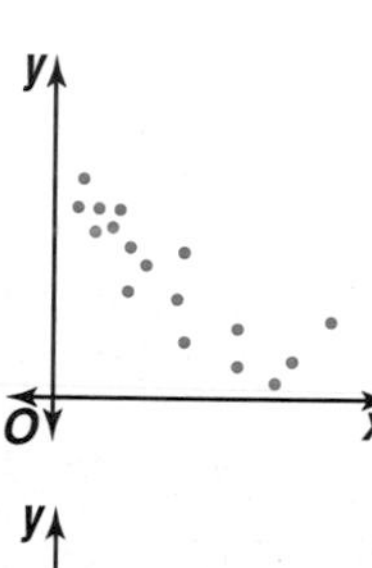

The points in this scatter plot are dispersed and do not form a linear pattern.

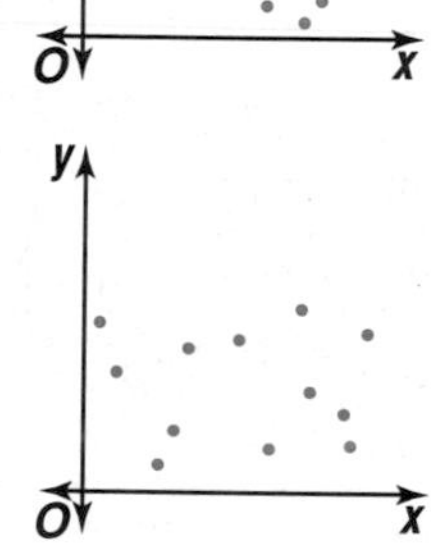

REVIEW EXERCISES

53. a. Graph the data below on a scatter plot.

b. Use two ordered pairs to write the equation of a best-fit line.

c. Use a graphing calculator to find an equation of the regression line for the data. What is the correlation value?

d. If the equation of the regression line shows a moderate or strong relationship, predict the number of visitors in 2005. Explain whether you think the prediction is reliable.

Overseas Visitors to the United States (thousands)

Year	1987	1988	1989	1990	1991
Visitors	10,434	12,763	12,184	12,252	12,003
Year	1992	1993	1994	1995	1996
Visitors	11,819	12,024	12,542	12,933	12,909

Source: U.S Dept. of Commerce

Lesson 1-7 Identify and graph piecewise functions including greatest integer, step, and absolute value functions.

- Graph $f(x) = |3x - 2|$.

This is an absolute value function. Use a table of values to find points to graph.

x	(x, f(x))
0	(0, 2)
1	(1, 1)
2	(2, 4)
3	(3, 7)
4	(4, 10)

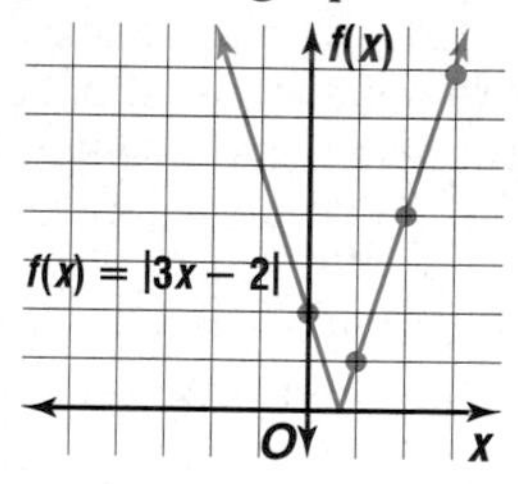

Graph each function.

54. $f(x) = \begin{cases} x \text{ if } 0 \leq x \leq 5 \\ 2 \text{ if } 5 < x \leq 8 \end{cases}$

55. $h(x) = \begin{cases} -1 \text{ if } -2 \leq x \leq 0 \\ -3x \text{ if } 0 < x \leq 2 \\ 2x \text{ if } 2 < x \leq 4 \end{cases}$

56. $f(x) = [\![x]\!] + 1$

57. $g(x) = |4x|$

58. $k(x) = 2|x| + 2$

Lesson 1-8 Graph linear inequalities.

- Graph the inequality $2x - y < 4$.

$2x - y < 4$
$y > 2x - 4$

The boundary is dashed. Testing (0, 0) yields a true inequality, so shade the region that includes (0, 0).

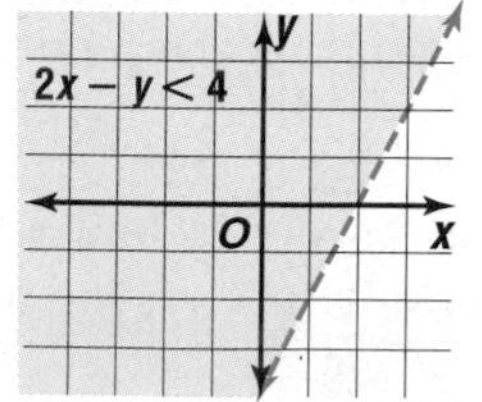

Graph each inequality.

59. $y > 4$

60. $x \leq 5$

61. $x + y \leq 1$

62. $2y - x < 4$

63. $y \leq |x|$

64. $y - 3x > 2$

65. $y > |x| - 2$

66. $y < |x - 2|$

APPLICATIONS AND PROBLEM SOLVING

67. **Aviation** A jet plane start from rest on a runway. It accelerates uniformly at a rate of 20 m/s^2. The equation for computing the distance traveled is $d = \frac{1}{2}at^2$. *(Lesson 1-1)*
   a. Find the distance traveled at the end of each second for 5 seconds.
   b. Is this relation a function? Explain.

68. **Finance** In 1994, outstanding consumer credit held by commercial banks was about \$463 billion. By 1996, this amount had grown to about \$529 billion. *(Lesson 1-4)*
   a. If x represents the year and y represents the amount of credit, find the average annual increase in the amount of outstanding consumer credit.
   b. Write an equation to model the annual change in credit.

69. **Recreation** Juan wants to know the relationship between the number of hours students spend watching TV each week and the number of hours students spend reading each week. A sample of 10 students reveals the following data.

Watching TV	Reading
20	8.5
32	3.0
42	1.0
12	4.0
5	14.0
28	4.5
33	7.0
18	12.0
30	3.0
25	3.0

Find the equation of a regression line for the data. Then make a statement about how representative the line is of the data. *(Lesson 1-6)*

ALTERNATIVE ASSESSMENT

OPEN-ENDED ASSESSMENT

1. If $[f \circ g](x) = 4x^2 - 4$, find $f(x)$ and $g(x)$. Explain why your answer is correct.
2. Suppose two distinct lines have the same x-intercept.
   a. Can the lines be parallel? Explain your answer.
   b. Can the lines be perpendicular? Explain your answer.
3. Write a piecewise function whose graph is the same as each function. The function should not involve absolute value.
   a. $y = x + |4 - x|$
   b. $y = 2x + |x + 1|$

Additional Assessment See p. A56 for Chapter 1 Practice Test.

Unit 1 interNET Project

WORLD WIDE WEB

TELECOMMUNICATION

Is Anybody Listening?

- Research several telephone long-distance services. Write and graph equations to compare the monthly fee and the rate per minute for each service.
- Which service would best meet your needs? Write a paragraph to explain your choice. Use the graphs to support your choice.

PORTFOLIO

Select one of the functions you graphed in this chapter. Write about a real-world situation this type of function can be used to model. Explain what the function shows about the situation that is difficult to show by other means.

Quantitative Comparison and Grid-In Questions

TEST-TAKING TIP
When you take the SAT, do the first 10 QC questions, and then do the first five Grid-Ins. Then with any remaining time, work on the remaining problems in that section.

At the end of each chapter in this textbook, you will find practice for the SAT and ACT tests. Each group of 10 questions contains eight multiple-choice questions, one quantitative comparison question, and one grid-in question.

QUANTITATIVE COMPARISON

One of the math sections on the SAT contains 15 *Quantitative Comparison* (QC) questions. As the name implies, these questions ask you to compare two quantities.

The fifteen QC questions are ordered by difficulty.

Level	Easy	Medium	Difficult
Problem Number	1–5	6–10	11–15

The instructions for these questions are different from the rest of the SAT. *Memorize* the instructions before you take the test.

There are four possible answers—

A if the quantity in Column A is greater;

B if the quantity in Column B is greater;

C if the two quantities are equal;

D if the relationship cannot be determined from the information given

Choice **A** means that the quantity in Column A is *always* greater than the other quantity—all the time, no matter what. Choice **B** means that the quantity in Column B is *always* greater. Choice **C** means that the two quantities are *always* equal —under all possible circumstances.

Choice **D** means that choices **A**, **B**, and **C** are all false; it means that which quantity is greater *depends* on what value is used for a variable or how a figure is drawn. Remember **D** stands for *cannot determine* or *depends*.

General strategies, like the following, can be helpful.

Arithmetic

If the question contains no variables, only choices **A, B,** or **C** are possible. Do not choose **D** for arithmetic questions.

In this example, simplify the quantities.

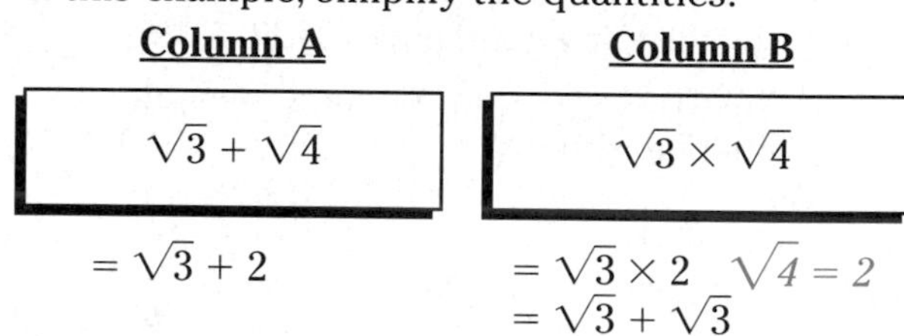

Compare the two quantities using $\sqrt{3} \approx 1.7$.

$$1.7 + 2 > 1.7 + 1.7.$$

The answer is choice **A.**

Algebra

$$x > 0$$

Column A	**Column B**
$x^2 + 1$	$x^3 - 1$

Don't be fooled into thinking that the quantity in Column B is "obviously" greater. Carefully read and use the given information.

Use the substitution strategy and look for a pattern. Choose only positive values since you are given that $x > 0$.

Record your results. Circle the greater quantity.

x	1	2	$\frac{1}{2}$	$\frac{3}{2}$
A	(2)	5	($\frac{5}{4}$)	($3\frac{1}{4}$)
B	0	(7)	$-\frac{7}{8}$	$2\frac{3}{8}$

In some cases, quantity A is greater, and in others, quantity B is greater. The correct answer is choice **D.**

Geometry

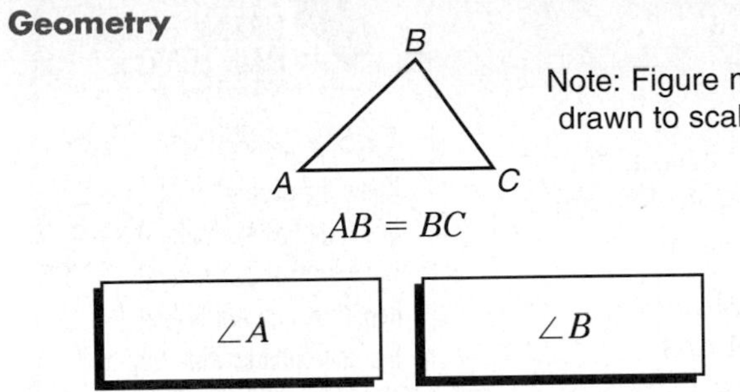

Never determine your answer by simply looking at a figure in an SAT or ACT test. $\angle B$ *looks* greater in this figure. Since $AB = BC$, $\angle A = \angle C$. If $\angle B = 40°$, then $\angle A = 70°$, but if $\angle B = 120°$, then $\angle A = 30°$. The answer is **D**.

GRID-IN

The same math section on the SAT that contains the quantitative comparison questions also includes ten problems in which you must mark your answer on a grid printed on the answer sheet. These are called *Student Produced Response* questions (or Grid-Ins), because you must create the answer yourself, not just choose from five possible answers.

These questions are *not* more difficult than the multiple-choice questions, but you'll want to be extra careful when you fill in your answers on the grid, so that you don't make careless errors. Grid-in questions are arranged in order of difficulty.

Level	Easy	Medium	Difficult
Problem Number	16–18	19–22	23–25

The instructions for using the grid are printed in the SAT test booklet. *Memorize* these instructions before you take the test.

The grid contains a row of four boxes at the top, two rows of ovals with decimal and fraction symbols, and four columns of numbered ovals.

After you solve the problem, always write your answer in the boxes at the top of the grid.

Start with the left column. Write one numeral, decimal point, or fraction line in each box. Shade the oval in each column that corresponds to the numeral or symbol written in the box. Only the shaded ovals will be scored, so work carefully. Don't make any extra marks on the grid.

Suppose the answer is $\frac{2}{3}$ or 0.666 … . You can record the answer as a fraction or a decimal. For the fraction, write $\frac{2}{3}$. For a decimal answer, you must enter the most accurate value that will fit the grid. That is, you must enter as many decimal place digits as space allows. An entry of .66 would not be acceptable.

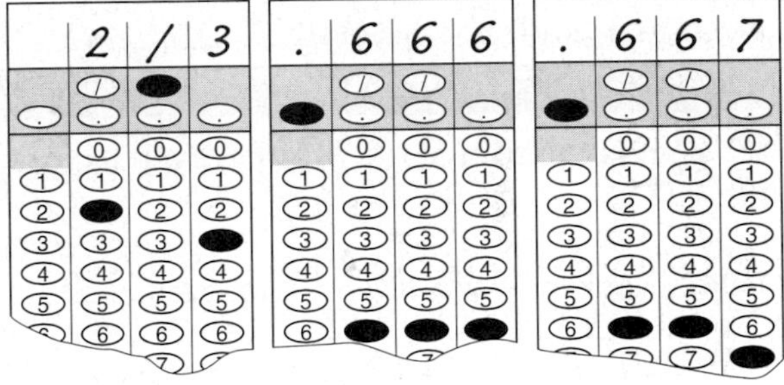

There is no 0 in bubble column 1. This means that you do *not* enter a zero to the left of the decimal point. For example, enter .25 and not 0.25.

Here are some other helpful hints for successfully completing grid-in questions.

- You don't have to write fractions in lowest terms. Any equivalent fraction that fits the grid is counted as correct. If your fraction does not fit (like 15/25), then either write it in lowest terms or change it to a decimal before you grid it.
- There is no negative symbol. Grid-in answers are never negative, so if you get a negative answer, you've made an error.
- If a problem has more than one correct answer, enter just one of the answers.
- Do not grid mixed numbers. Change the mixed number to an equivalent fraction or decimal. If you enter 11/2 for $1\frac{1}{2}$, it will be read as $\frac{11}{2}$. Enter it as 3/2 or 1.5.

Chapter 1 SAT & ACT Preparation

Arithmetic Problems

TEST-TAKING TIP

Know the properties of zero and one. For example, 0 is even, neither positive nor negative, and not prime. 1 is the only integer with only one divisor. 1 is not prime.

All SAT and ACT tests contain arithmetic problems. Some are easy and some are difficult. You'll need to understand and apply the following concepts.

odd and even	factors	divisibility
positive, negative	integers	fractions
scientific notation	exponents	roots
prime numbers	decimals	inequalities

Several concepts are often combined in a single problem.

SAT EXAMPLE

1. What is the sum of the positive even factors of 12?

HINT Look for words like *positive, even,* and *factor.*

Solution First find all the factors of 12.

1 2 3 4 6 12

Re-read the question. It asks for the sum of *even* factors. Circle the factors that are even numbers.

1 (2) 3 (4) (6) (12)

Now add these even factors to find the sum.

$2 + 4 + 6 + 12 = 24$ The answer is 24.

This is a grid-in problem. Record your answer on the grid.

2	4		

ACT EXAMPLE

2. $(-2)^3 + (3)^{-2} + \frac{8}{9}$

A -7 B $-1\frac{7}{9}$ C $\frac{8}{9}$

D $1\frac{7}{9}$ E 12

HINT Analyze what the – (negative) symbol represents each time it is used.

Solution Use the properties of exponents to simplify each term.

$(-2)^3 = (-2)(-2)(-2)$ or -8

$(3)^{-2} = \frac{1}{3^2}$ or $\frac{1}{9}$

Add the terms.

$$(-2)^3 + (3)^{-2} + \frac{8}{9} = -8 + \frac{1}{9} + \frac{8}{9}$$
$$= -8 + 1 \text{ or } -7$$

The answer is choice **A.**

Always look at the answer choices before you start to calculate. In this problem, three (incorrect) answer choices include fractions with denominators of 9. This may be a clue that your calculations may involve ninths.

Never assume that because three answer choices involve ninths and two are integers, that the correct answer is more likely to involve ninths. Also don't conclude that because the expression contains a fraction that the answer will necessarily have a fraction in it.

After you work each problem, record your answer on the answer sheet provided or on a piece of paper.

Multiple Choice

1. Which of the following expresses the prime factorization of 54?

 A 9×6

 B $3 \times 3 \times 6$

 C $3 \times 3 \times 2$

 D $3 \times 3 \times 3 \times 2$

 E 5.4×10

2. If 8 and 12 each divide K without a remainder, what is the value of K?

 A 16

 B 24

 C 48

 D 96

 E It cannot be determined from the information given.

3. After $\dfrac{4\frac{1}{3}}{2\frac{3}{5}}$ has been simplified to a single fraction in lowest terms, what is the denominator?

 A 2 B 3 C 5

 D 9 E 13

4. For a class play, student tickets cost \$2 and adult tickets cost \$5. A total of 30 tickets were sold. If the total sales must exceed \$90, what is the minimum number of adult tickets that must be sold?

 A 7 B 8 C 9

 D 10 E 11

5. $-|-7| - |-5| - 3|-4| = ?$

 A -24 B -11 C 0

 D 13 E 24

6. $(-4)^2 + (2)^{-4} + \frac{3}{4}$

 A $16\frac{13}{16}$ B $16\frac{3}{4}$ C $-15\frac{7}{32}$

 D $15\frac{7}{32}$ E 16

7. Kerri subscribed to four publications that cost \$12.90, \$16.00, \$18.00, and \$21.90 per year. If she made an initial down payment of one half of the total amount and paid the rest in 4 equal monthly payments, how much was each of the 4 monthly payments?

 A \$8.60

 B \$9.20

 C \$9.45

 D \$17.20

 E \$34.40

8. $\sqrt{64 + 36} = ?$

 A 10 B 14 C 28

 D 48 E 100

9. **Quantitative Comparison**

 A if the quantity in Column A is greater

 B if the quantity in Column B is greater

 C if the two quantities are equal

 D if the relationship cannot be determined from the information given

Column A	Column B
The number of distinct prime factors of 60.	The number of distinct prime factors of 30.

10. **Grid-In** There are 24 fish in an aquarium. If $\frac{1}{8}$ of them are tetras and $\frac{2}{3}$ of the remaining fish are guppies, how many guppies are in the aquarium?

interNET CONNECTION **SAT/ACT Practice** For additional test practice questions, visit: **www.amc.glencoe.com**

SYSTEMS OF LINEAR EQUATIONS AND INEQUALITIES

CHAPTER OBJECTIVES

- **Solve systems of equations and inequalities.** *(Lessons 2-1, 2-2, 2-6)*
- **Define matrices.** *(Lesson 2-3)*
- **Add, subtract, and multiply matrices.** *(Lesson 2-3)*
- **Use matrices to model transformations.** *(Lesson 2-4)*
- **Find determinants and inverses of matrices.** *(Lesson 2-5)*
- **Use linear programming to solve problems.** *(Lesson 2-7)*

Solving Systems of Equations in Two Variables

OBJECTIVES

- Solve systems of equations graphically.
- Solve systems of equations algebraically.

CONSUMER CHOICES Madison is thinking about leasing a car for two years. The dealership says that they will lease her the car she has chosen for \$326 per month with only \$200 down. However, if she pays \$1600 down, the lease payment drops to \$226 per month. What is the break-even point when comparing these lease options? Which 2-year lease should she choose if the down payment is not a problem? *This problem will be solved in Example 4.*

The *break-even point* is the point in time at which Madison has paid the same total amount on each lease. After finding that point, you can more easily determine which of these arrangements would be a better deal. The break-even point can be determined by solving a system of equations.

A **system of equations** is a set of two or more equations. To "solve" a system of equations means to find values for the variables in the equations, which make all the equations true at the same time. One way to solve a system of equations is by graphing. The intersection of the graphs represents the point at which the equations have the same x-value and the same y-value. Thus, this ordered pair represents the solution common to both equations. This ordered pair is called the **solution** to the system of equations.

Graphing Calculator Tip

You can estimate the solution to a system of equations by using the **TRACE** function on your graphing calculator.

Example 1 **Solve the system of equations by graphing.**

$$\begin{aligned} 3x - 2y &= -6 \\ x + y &= -2 \end{aligned}$$

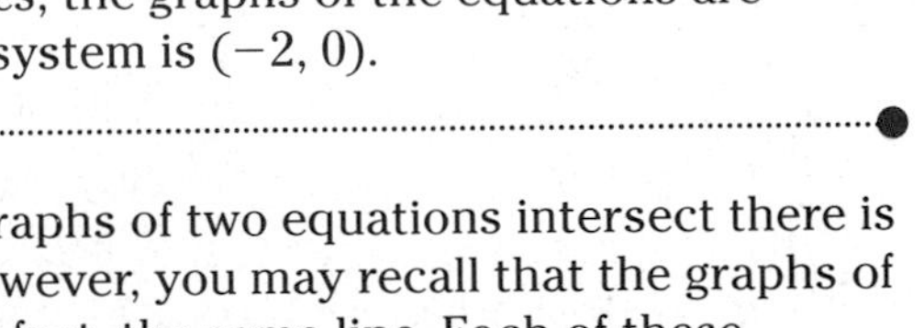

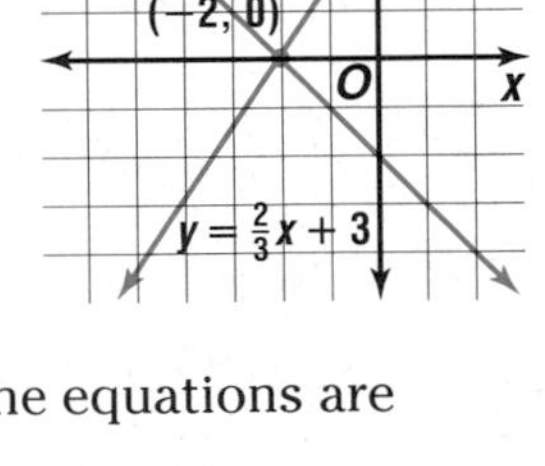

First rewrite each equation of the system in slope-intercept form by solving for y.

$$\begin{aligned} 3x - 2y &= -6 \\ x + y &= -2 \end{aligned} \quad \textbf{becomes} \quad \begin{aligned} y &= \frac{3}{2}x + 3 \\ y &= -x - 2 \end{aligned}$$

Since the two lines have different slopes, the graphs of the equations are intersecting lines. The solution to the system is $(-2, 0)$.

As you saw in Example 1, when the graphs of two equations intersect there is a solution to the system of equations. However, you may recall that the graphs of two equations may be parallel lines or, in fact, the same line. Each of these situations has a different type of system of linear equations.

A **consistent** system of equations has at least one solution. If there is exactly one solution, the system is **independent.** If there are infinitely many solutions, the system is **dependent.** If there is no solution, the system is **inconsistent.** By rewriting each equation of a system in slope-intercept form, you can more easily determine the type of system you have and what type of solution to expect.

The chart below summarizes the characteristics of these types of systems.

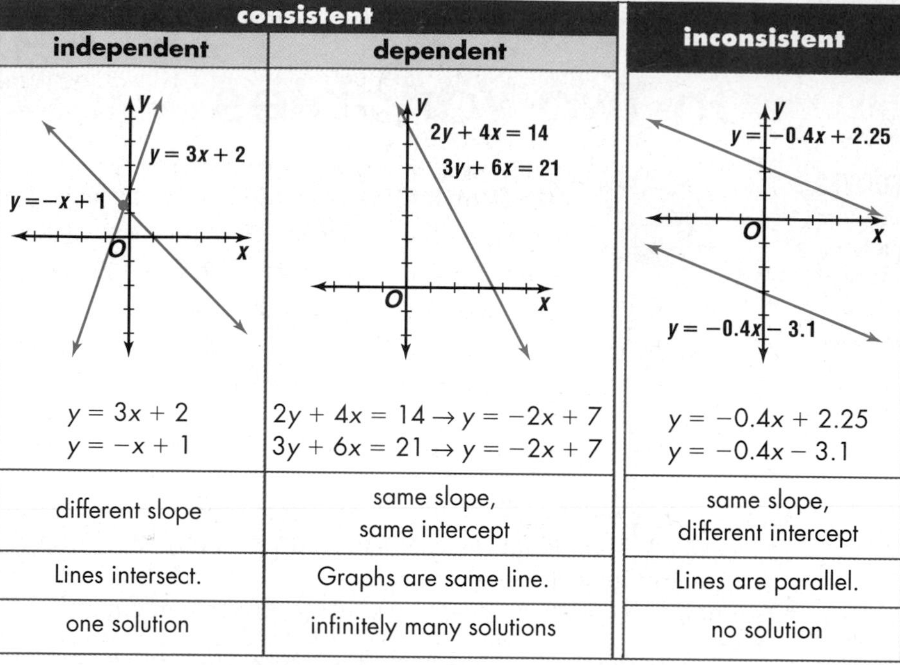

consistent		inconsistent
independent	dependent	
$y = 3x + 2$ $y = -x + 1$	$2y + 4x = 14 \rightarrow y = -2x + 7$ $3y + 6x = 21 \rightarrow y = -2x + 7$	$y = -0.4x + 2.25$ $y = -0.4x - 3.1$
different slope	same slope, same intercept	same slope, different intercept
Lines intersect.	Graphs are same line.	Lines are parallel.
one solution	infinitely many solutions	no solution

When graphs result in lines that are the same line, we say the lines coincide.

Often, graphing a system of equations is not the best method of finding its solution. This is especially true when the solution to the system contains non-integer values. Systems of linear equations can also be solved algebraically. Two common ways of solving systems algebraically are the **elimination method** and the **substitution method.** In some cases, one method may be easier to use than the other.

Example 2 **Use the elimination method to solve the system of equations.**

$$1.5x + 2y = 20$$
$$2.5x - 5y = -25$$

One way to solve this system is to multiply both sides of the first equation by 5, multiply both sides of the second equation by 2, and add the two equations to eliminate y. Then solve the resulting equation.

$$5(1.5x + 2y) = 5(20) \quad \Rightarrow \quad 7.5x + 10y = 100$$
$$2(2.5x - 5y) = 2(-25) \quad \Rightarrow \quad 5x - 10y = -50$$
$$12.5x = 50$$
$$x = 4$$

Now substitute 4 for x in either of the *original* equations.

$$1.5x + 2y = 20$$
$$1.5(4) + 2y = 20 \quad x = 4$$
$$2y = 14$$
$$y = 7$$

The solution is (4, 7). *Check it by substituting into $2.5x - 5y = -25$. If the coordinates make both equations true, then the solution is correct*

If one of the equations contains a variable with a coefficient of 1, the system can often be solved more easily by using the substitution method.

Example **3** **Use the substitution method to solve the system of equations.**

$$2x + 3y = 8$$
$$x - y = 2$$

You can solve the second equation for either y or x. If you solve for x, the result is $x = y + 2$. Then substitute $y + 2$ for x in the first equation.

$$2x + 3y = 8$$
$$2(y + 2) + 3y = 8 \quad x = y + 2$$
$$5y = 4$$
$$y = \frac{4}{5}$$

Now substitute $\frac{4}{5}$ for y in either of the original equations, and solve for x.

$$x - y = 2$$
$$x - \frac{4}{5} = 2 \quad y = \frac{4}{5}$$
$$x = \frac{14}{5}$$

The solution is $\left(\frac{14}{5}, \frac{4}{5}\right)$.

GRAPHING CALCULATOR EXPLORATION

You can use a graphing calculator to find the solution to an independent system of equations.

- Graph the equations on the same screen.
- Use the **CALC** menu and select **5:intersect** to determine the coordinates of the point of intersection of the two graphs.

TRY THESE

Find the solution to each system.

1. $y = 500x - 20$
   $y = -20x + 500$
2. $3x - 4y = 320$
   $5x + 2y = 340$

WHAT DO YOU THINK?

3. How accurate are solutions found on the calculator?
4. What type of system do the equations $5x - 7y = 70$ and $10x - 14y = 120$ form? What happens when you try to find the intersection point on the calculator?
5. Graph a system of dependent equations. Find the intersection point. Use the **TRACE** function to move the cursor and find the intersection point again. What pattern do you observe?

You can use a system of equations to solve real-world problems. Choose the best method for solving the system of equations that models the situation.

Example **4** CONSUMER CHOICES **Refer to the application at the beginning of the lesson.**

a. What is the break-even point in the two lease plans that Madison is considering?

b. If Madison keeps the lease for 24 months, which lease should she choose?

a. First, write an equation to represent the amount she will pay with each plan. Let C represent the total cost and m the number of months she has had the lease.

Lease 1 ($200 down with monthly payment of $326): $C = 326m + 200$

Lease 2 ($1600 down with monthly payment of $226): $C = 226m + 1600$

Now, solve the system of equations. Since both equations contain C, we can substitute the value of C from one equation into the other.

(continued on the next page)

$$C = 326m + 200$$

$$226m + 1600 = 326m + 200 \quad C = 226m + 1600$$

$$1400 = 100m$$

$$14 = m$$

With the fourteenth monthly payment, she reaches the break-even point.

b. The graph of the equations shows that after that point, Lease 1 is more expensive for the 2-year lease. So, Madison should probably choose Lease 2.

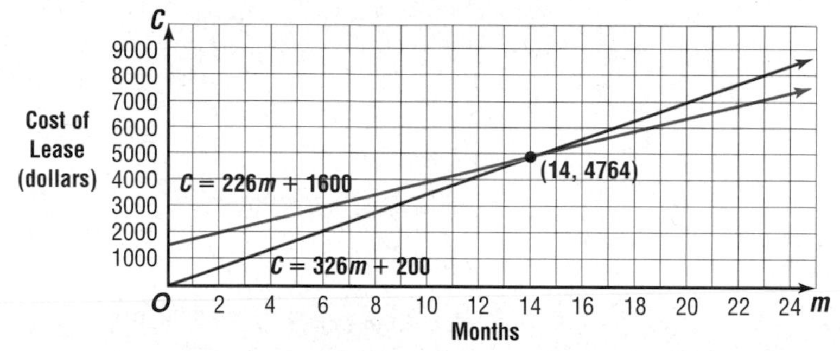

CHECK FOR UNDERSTANDING

Communicating Mathematics

Read and study the lesson to answer each question.

1. **Write** a system of equations in which it is easier to use the substitution method to solve the system rather than the elimination method. Explain your choice.
2. Refer to the application at the beginning of the lesson. **Explain** what factors Madison might consider before making a decision on which lease to select.
3. *Math Journal* **Write** a description of the three different possibilities that may occur when graphing a system of two linear equations. Include examples and solutions that occur with each possibility.

Guided Practice

4. State whether the system $2y + 3x = 6$ and $4y = 16 - 6x$ is *consistent and independent, consistent and dependent,* or *inconsistent.* Explain your reasoning.

Solve each system of equations by graphing.

5. $y = 5x - 2$
   $y = -2x + 5$
6. $x - y = 2$
   $2x = 2y + 10$

Solve each system of equations algebraically.

7. $7x + y = 9$
   $5x - y = 15$
8. $3x + 4y = -1$
   $6x - 2y = 3$
9. $\frac{1}{3}x - \frac{3}{2}y = -4$
   $5x - 4y = 14$

10. **Sales** HomePride manufactures solid oak racks for displaying baseball equipment and karate belts. They usually sell six times as many baseball racks as karate-belt racks. The net profit is \$3 from each baseball rack and \$5 from each karate-belt rack. If the company wants a total profit of \$46,000, how many of each type of rack should they sell?

EXERCISES

Practice

State whether each system is *consistent and independent, consistent and dependent,* or *inconsistent.*

11. $x + 3y = 18$
$-x + 2y = 7$

12. $y = 0.5x$
$2y = x + 4$

13. $-35x + 40y = 55$
$7x = 8y - 11$

Solve each system of equations by graphing.

14. $x = 5$
$4x + 5y = 20$

15. $y = -3$
$2x = 8$

16. $x + y = -2$
$3x - y = 10$

17. $x + 3y = 0$
$2x + 6y = 5$

18. $y = x - 2$
$x - 2y = 4$

19. $3x - 2y = -6$
$x = 12 - 4y$

20. Determine what type of solution you would expect from the system of equations $3x - 8y = 10$ and $16x - 32y = 75$ without graphing the system. Explain how you determined your answer.

Solve each system of equations algebraically.

21. $5x - y = 16$
$2x + 3y = 3$

22. $3x - 5y = -8$
$x + 2y = 1$

23. $y = 6 - x$
$x = 4.5 + y$

24. $2x + 3y = 3$
$12x - 15y = -4$

25. $-3x + 10y = 5$
$2x + 7y = 24$

26. $x = 2y - 8$
$2x - y = -7$

27. $2x + 5y = 4$
$3x + 6y = 5$

28. $\frac{3}{5}x - \frac{1}{6}y = 1$
$\frac{1}{5}x + \frac{5}{6}y = 11$

29. $4x + 5y = -8$
$3x - 7y = 10$

30. Find the solution to the system of equations $3x - y = -9$ and $4x - 2y = -8$.

31. Explain which method seems most efficient to solve the system of equations $a - b = 0$ and $3a + 2b = -15$. Then solve the system.

Applications and Problem Solving

32. Sports Spartan Stadium at San Jose State University in California has a seating capacity of about 30,000. A newspaper article states that the Spartans get four times as many tickets as the visiting team. Suppose S represents the number of tickets for the Spartans and V represents the number of tickets for the visiting team's fans.

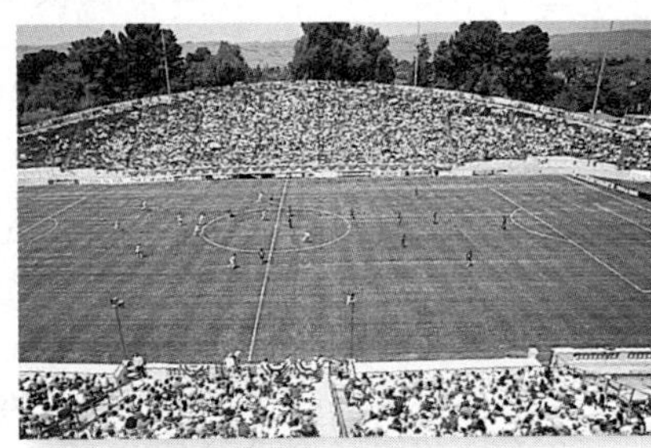

a. Which system could be used by a newspaper reader to determine how many tickets each team gets?

A $4S + 4V = 30{,}000$
$S = 4V$

B $S - 4V = 0$
$S + V = 30{,}000$

C $S + V = 30{,}000$
$V - 4S = 0$

b. Solve the system to find how many tickets each team gets.

33. Geometry Two triangles have the same perimeter of 20 units. One triangle is an isosceles triangle. The other triangle has a side 6 units long. Its other two sides are the same lengths as the base and leg of the isosceles triangle.

a. What are the dimensions of each triangle?

b. What type of triangle is the second triangle?

34. **Critical Thinking** The solution to a system of two linear equations is (4, −3). One equation has a slope of 4. The slope of the other line is the negative reciprocal of the slope of the first. Find the system of equations.

35. **Business** The first Earth Day was observed on April 22, 1970. Since then, the week of April 22 has been Earth Week, a time for showing support for environmental causes. Fans Café is offering a reduced refill rate for soft drinks during Earth Week for anyone purchasing a Fans mug. The mug costs $2.95 filled with 16 ounces of soft drink. The refill price is 50¢. A 16-ounce drink in a disposable cup costs $0.85.
    a. What is the approximate break-even point for buying the mug and refills in comparison to buying soft drinks in disposable cups?
    b. What does this mean? Which offer do you think is best?
    c. How would your decision change if the refillable mug offer was extended for a year?

36. **Critical Thinking** Determine what must be true of a, b, c, d, e, and f for the system $ax + by = c$ and $dx + ey = f$ to fit each description.
    a. consistent and independent
    b. consistent and dependent
    c. inconsistent

37. **Incentive Plans** As an incentive plan, a company stated that employees who worked for four years with the company would receive $516 and a laptop computer. Mr. Rodriquez worked for the company for 3.5 years. The company pro-rated the incentive plan, and he still received the laptop computer, but only $264. What was the value of the laptop computer?

38. **Ticket Sales** In November 1994, the first live concert on the Internet by a major rock'n'roll band was broadcast. Most fans stand in lines for hours to get tickets for concerts. Suppose you are in line for tickets. There are 200 more people ahead of you than behind you in line. The whole line is three times the number of people behind you. How many people are in line for concert tickets?

Mixed Review

39. Graph $-2x + 7 \geq y$. *(Lesson 1-8)*

40. Graph $f(x) = 2|x| - 3$. *(Lesson 1-7)*.

41. Write an equation of the line parallel to the graph of $y = 2x + 5$ that passes through the point at (0, 6). *(Lesson 1-5)*

42. **Manufacturing** The graph shows the operational expenses for a bicycle shop during its first four years of business. How much was the startup cost of the business? *(Lesson 1-3)*

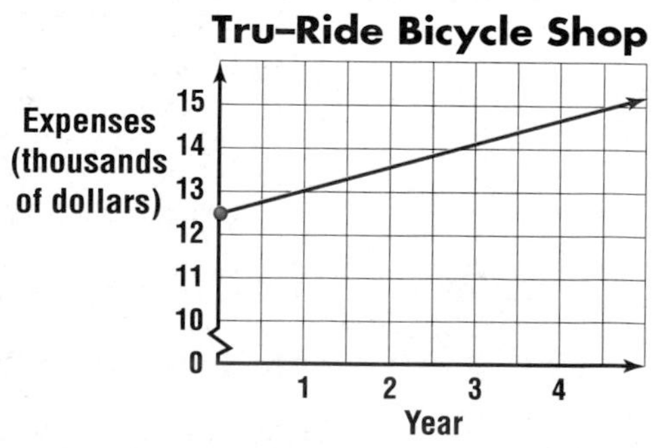

43. Find $[f \circ g](x)$ if $f(x) = 3x - 5$ and $g(x) = x + 2$. *(Lesson 1-2)*

44. State the domain and range of the relation {(18, −3), (18, 3)}. Is this relation a function? Explain. *(Lesson 1-1)*

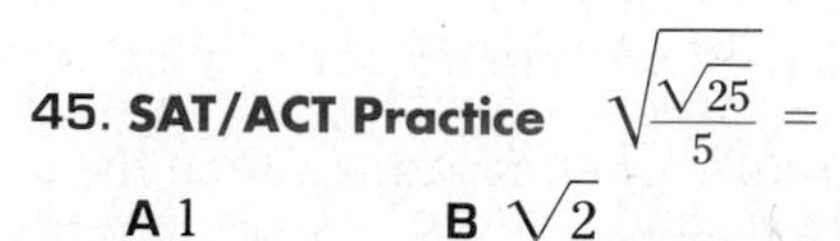

45. **SAT/ACT Practice** 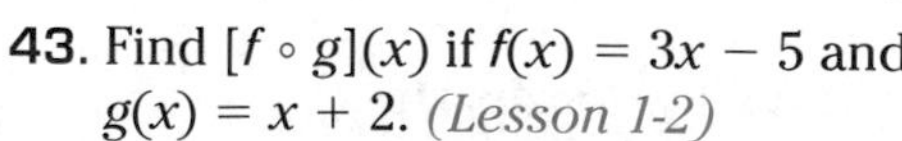 $\sqrt{\dfrac{\sqrt{25}}{5}} =$

A 1 **B** $\sqrt{2}$ **C** 2 **D** 5 **E** $5\sqrt{2}$

Extra Practice See p. A28.

Solving Systems of Equations in Three Variables

OBJECTIVE

- Solve systems of equations involving three variables algebraically.

SPORTS In 1998, Cynthia Cooper of the WNBA Houston Comets basketball team was named Team Sportswoman of the Year by the Women's Sports Foundation. Cooper scored 680 points in the 1998 season by hitting 413 of her 1-point, 2-point, and 3-point attempts. She made 40% of her 160 3-point field goal attempts. How many 1-, 2-, and 3-point baskets did Ms. Cooper complete? *This problem will be solved in Example 3.*

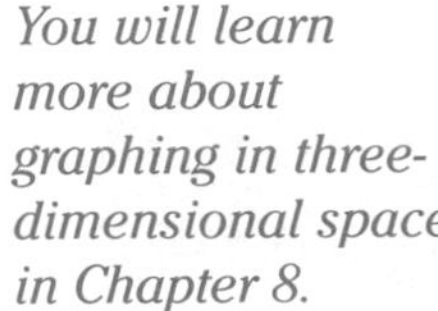

You will learn more about graphing in three-dimensional space in Chapter 8.

This situation can be described by a system of three equations in three variables. You used graphing to solve a system of equations in two variables. For a system of equations in three variables, the graph of each equation is a plane in space rather than a line. The three planes can appear in various configurations. This makes solving a system of equations in three variables by graphing rather difficult. However, the pictorial representations provide information about the types of solutions that are possible. Some of them are shown below.

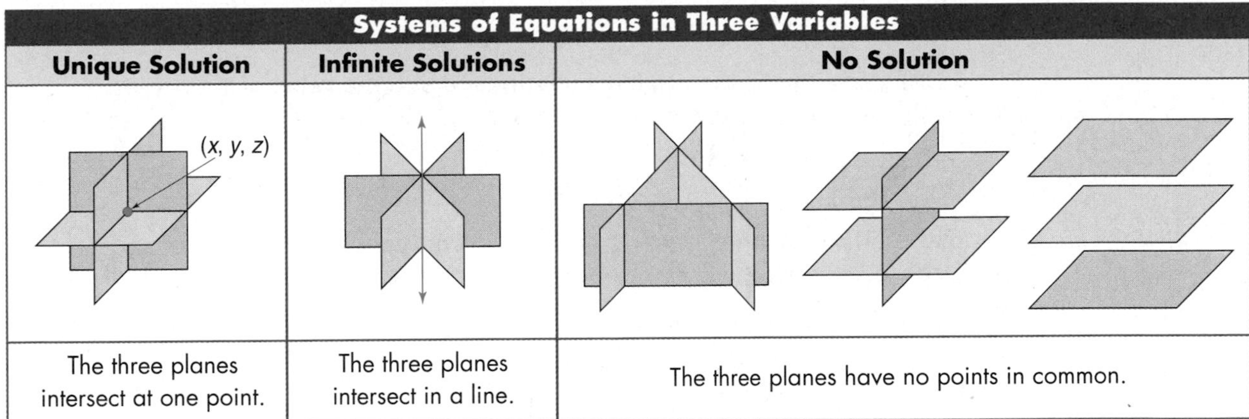

Systems of Equations in Three Variables

Unique Solution	Infinite Solutions	No Solution
(x, y, z)		
The three planes intersect at one point.	The three planes intersect in a line.	The three planes have no points in common.

You can solve systems of three equations more efficiently than graphing by using the same algebraic techniques that you used to solve systems of two equations.

Example 1 **Solve the system of equations by elimination.**

$$\begin{aligned} x - 2y + z &= 15 \\ 2x + 3y - 3z &= 1 \\ 4x + 10y - 5z &= -3 \end{aligned}$$

One way to solve a system of three equations is to choose pairs of equations and then eliminate one of the variables. Because the coefficient of x is 1 in the first equation, it is a good choice for eliminating x from the second and third equations.

(continued on the next page)

To eliminate x using the first and second equations, multiply each side of the first equation by -2.

$$-2(x - 2y + z) = -2(15)$$
$$-2x + 4y - 2z = -30$$

Then add that result to the second equation.

$$\begin{array}{r} -2x + 4y - 2z = -30 \\ 2x + 3y - 3z = 1 \\ \hline 7y - 5z = -29 \end{array}$$

To eliminate x using the first and third equations, multiply each side of the first equation by -4.

$$-4(x - 2y + z) = -4(15)$$
$$-4x + 8y - 4z = -60$$

Then add that result to the third equation .

$$\begin{array}{r} -4x + 8y - 4z = -60 \\ 4x + 10y - 5z = -3 \\ \hline 18y - 9z = -63 \end{array}$$

Now you have two linear equations in two variables. Solve this system. Eliminate z by multiplying each side of the first equation by -9 and each side of the second equation by 5. Then add the two equations.

$$\begin{array}{r} -9(7y - 5z) = -9(-29) \\ 5(18y - 9z) = 5(-63) \\ \hline \end{array} \quad \Rightarrow \quad \begin{array}{r} -63y + 45z = 261 \\ 90y - 45z = -315 \\ \hline 27y = -54 \end{array}$$

$y = -2$ The value of y is -2.

By substituting the value of y into one of the equations in two variables, we can solve for the value of z.

$$7y - 5z = -29$$
$$7(-2) - 5z = -29 \quad y = -2$$
$$z = 3$$

The value of z is 3.

Finally, use one of the original equations to find the value of x.

$$x - 2y + z = 15$$
$$x - 2(-2) + 3 = 15 \quad y = -2, z = 3$$
$$x = 8$$

The value of x is 8.

The solution is $x = 8$, $y = -2$, and $z = 3$. This can be written as the **ordered triple** $(8, -2, 3)$. *Check by substituting the values into each of the original equations.*

The substitution method of solving systems of equations also works with systems of three equations.

Example 2 **Solve the system of equations by substitution.**

$$\mathbf{4x = -8z}$$
$$\mathbf{3x - 2y + z = 0}$$
$$\mathbf{-2x + y - z = -1}$$

You can easily solve the first equation for x.

$$4x = -8z$$
$$x = -2z \quad \textit{Divide each side by 4.}$$

Then substitute $-2z$ for x in each of the other two equations. Simplify each equation.

$$3x - 2y + z = 0$$
$$3(-2z) - 2y + z = 0 \quad x = -2z$$
$$-2y - 5z = 0$$

$$-2x + y - z = -1$$
$$-2(-2z) + y - z = -1 \quad x = -2z$$
$$y + 3z = -1$$

Solve $y + 3z = -1$ for y. $y + 3z = -1$
$y = -1 - 3z$ *Subtract 3z from each side.*

Substitute $-1 - 3z$ for y in $-2y - 5z = 0$. Simplify.

$$-2y - 5z = 0$$
$$-2(-1 - 3z) - 5z = 0 \quad y = -1 - 3z$$
$$z = -2$$

Now, find the values of y and x. Use $y = -1 - 3z$ and $x = -2z$. Replace z with -2.

$y = -1 - 3z$
$y = -1 - 3(-2)$ *z = −2*
$y = 5$

$x = -2z$
$x = -2(-2)$ *z = −2*
$x = 4$

The solution is $x = 4$, $y = 5$, *and* $z = -2$. *Check each value in the original system.*

Many real-world situations can be represented by systems of three equations.

Example 3 

SPORTS Refer to the application at the beginning of the lesson. Find the number of 1point free throws, 2-point field goals, and 3-point field goals Cynthia Cooper scored in the 1998 season.

Write a system of equations. Define the variables as follows.

x = the number of 1-point free throws

y = the number of 2-point field goals

z = the number of 3-point field goals

The system is:

$x + 2y + 3z = 680$ *total number of points*

$x + y + z = 413$ *total number of baskets*

$\frac{z}{160} = 0.40$ *percent completion*

The third equation is a simple linear equation. Solve for z.

$\frac{z}{160} = 0.40$, so $z = 160(0.40)$ or 64.

Now substitute 64 for z to make a system of two equations.

$$x + 2y + 3z = 680$$
$$x + 2y + 3(64) = 680 \quad z = 64$$
$$x + 2y = 488$$

$$x + y + z = 413$$
$$x + y + 64 = 413 \quad z = 64$$
$$x + y = 349$$

Solve $x + y = 349$ for x. Then substitute that value for x in $x + 2y = 488$ and solve for y.

$$x + y = 349$$
$$x = 349 - y$$

$$x + 2y = 488$$
$$(349 - y) + 2y = 488 \quad x = 349 - y$$
$$y = 139$$

(continued on the next page)

Solve for x.
$x = 349 - y$
$x = 349 - 139 \quad y = 139$
$x = 210$

In 1998, Ms. Cooper made 210 1-point free throws, 139 2-point field goals, and 64 3-point field goals. *Check your answer in the original problem.*

CHECK FOR UNDERSTANDING

Communicating Mathematics

Read and study the lesson to answer each question.

1. **Compare and contrast** solving a system of three equations to solving a system of two equations.
2. **Describe** what you think would happen if two of the three equations in a system were consistent and dependent. Give an example.
3. **Write** an explanation of how to solve a system of three equations using the elimination method.

Guided Practice

Solve each system of equations.

4. $4x + 2y + z = 7$
   $2x + 2y - 4z = -4$
   $x + 3y - 2z = -8$

5. $x - y - z = 7$
   $-x + 2y - 3z = -12$
   $3x - 2y + 7z = 30$

6. $2x - 2y + 3z = 6$
   $2x - 3y + 7z = -1$
   $4x - 3y + 2z = 0$

7. **Physics** The height of an object that is thrown upward with a constant acceleration of a feet per second per second is given by the equation $s = \frac{1}{2}at^2 + v_0 t + s_0$. The height is s feet, t represents the time in seconds, v_0 is the initial velocity in feet per second, and s_0 is the initial height in feet. Find the acceleration, the initial velocity, and the initial height if the height at 1 second is 75 feet, the height at 2.5 seconds is 75 feet, and the height at 4 seconds is 3 feet.

EXERCISES

Practice

Solve each system of equations.

8. $x + 2y + 3z = 5$
   $3x + 2y - 2z = -13$
   $5x + 3y - z = -11$

9. $7x + 5y + z = 0$
   $-x + 3y + 2z = 16$
   $x - 6y - z = -18$

10. $x - 3z = 7$
    $2x + y - 2z = 11$
    $-x - 2y + 9z = 13$

11. $3x - 5y + z = 9$
    $x - 3y - 2z = -8$
    $5x - 6y + 3z = 15$

12. $8x - z = 4$
    $y + z = 5$
    $11x + y = 15$

13. $4x - 3y + 2z = 12$
    $x + y - z = 3$
    $-2x - 2y + 2z = 5$

14. $36x - 15y + 50z = -10$
    $2x + 25y = 40$
    $54x - 5y + 30z = -160$

15. $-x - 3y + z = 54$
    $4x + 2y - 3z = -32$
    $2y + 8z = 78$

16. $1.8x - z = 0.7$
    $1.2y + z = -0.7$
    $1.5x - 3y = 3$

17. If possible, find the solution of $y = x + 2z$, $z = -1 - 2x$, and $x = y - 14$.

18. What is the solution of $\frac{1}{8}x - \frac{2}{3}y + \frac{5}{6}z = -8$, $\frac{3}{4}x + \frac{1}{6}y - \frac{1}{3}z = -12$, and $\frac{3}{16}x - \frac{5}{8}y - \frac{7}{12}z = -25$? If there is no solution, write *impossible*.

Applications and Problem Solving

19. **Finance** Ana Colón asks her broker to divide her 401K investment of \$2000 among the International Fund, the Fixed Assets Fund, and company stock. She decides that her investment in the International Fund should be twice her investment in company stock. During the first quarter, the International Fund earns 4.5%, the Fixed Assets Fund earns 2.6%, and the company stock falls 0.2%. At the end of the first quarter, Ms. Colón receives a statement indicating a return of \$58 on her investment. How did the broker divide Ms. Colón's initial investment?

20. **Critical Thinking** Write a system of equations that fits each description.
    a. The system has a solution of $x = -5, y = 9, z = 11$.
    b. There is no solution to the system.
    c. The system has an infinite number of solutions.

21. **Physics** Each year the Punkin' Chunkin' contest is held in Lewes, Delaware. The object of the contest is to propel an 8- to 10-pound pumpkin as far as possible. Steve Young of Hopewell, Illinois, set the 1998 record of 4026.32 feet. Suppose you build a machine that fires the pumpkin so that it is at a height of 124 feet after 1 second, the height at 3 seconds is 272 feet, and the height at 8 seconds is 82 feet. Refer to the formula in Exercise 7 to find the acceleration, the initial velocity, and the initial height of the pumpkin.

22. **Critical Thinking** Suppose you are using elimination to solve a system of equations.
    a. How do you know that a system has no solution?
    b. How do you know when it has an infinite number of solutions?

23. **Number Theory** Find all of the ordered triples (x, y, z) such that when any one of the numbers is added to the product of the other two, the result is 2.

Mixed Review

24. Solve the system of equations, $3x + 4y = 375$ and $5x + 2y = 345$. *(Lesson 2-1)*

25. Graph $y \leq -\frac{1}{3}x + 2$. *(Lesson 1-8)*

26. Show that points with coordinates $(-1, 3)$, $(3, 6)$, $(6, 2)$, and $(2, -1)$ are the vertices of a square. *(Lesson 1-5)*

27. **Manufacturing** It costs ABC Corporation \$3000 to produce 20 of a particular model of color television and \$5000 to produce 60 of that model. *(Lesson 1-4)*
    a. Write an equation to represent the cost function.
    b. Determine the fixed cost and variable cost per unit.
    c. Sketch the graph of the cost function.

28. **SAT/ACT Practice** In the figure, the area of square $OXYZ$ is 2. What is the area of the circle?

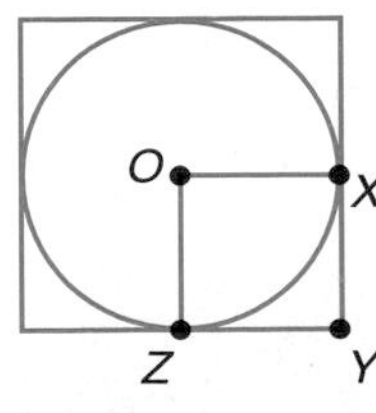

A $\frac{\pi}{4}$ **B** $\pi\sqrt{2}$ **C** 2π

D 4π **E** 8π

Extra Practice See p. A28.

Modeling Real-World Data with Matrices

OBJECTIVES
- Model data using matrices.
- Add, subtract, and multiply matrices.

Real World Application

TRAVEL Did you ever go on a vacation and realize that you forgot to pack something that you needed? Sometimes purchasing those items while traveling can be expensive. The average cost of some items bought in various cities is given below.

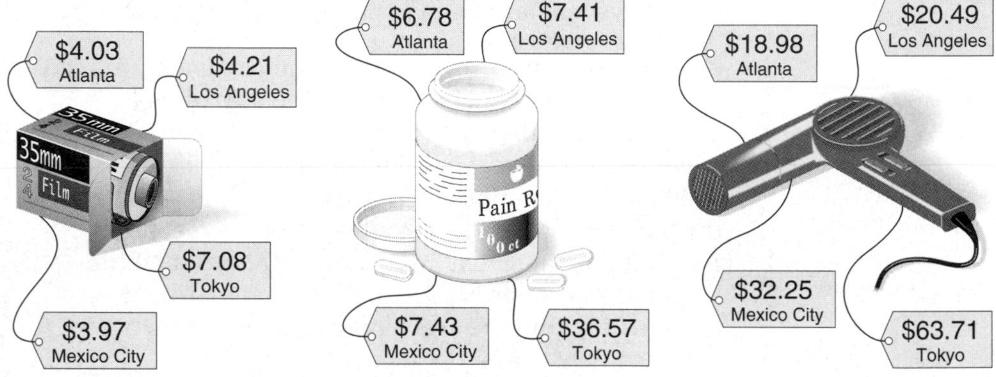

Source: Runzheimer International

Data like these can be displayed or modeled using a matrix. *A problem related to this will be solved in Example 1.*

The plural of matrix is matrices.

A **matrix** is a rectangular array of terms called **elements.** The elements of a matrix are arranged in rows and columns and are usually enclosed by brackets. A matrix with m rows and n columns is an $\boldsymbol{m \times n}$ **matrix** (read "m by n"). The **dimensions** of the matrix are m and n. Matrices can have various dimensions and can contain any type of numbers or other information.

2 × 2 matrix

$$\begin{bmatrix} -\frac{3}{5} & \frac{1}{2} \\ 3 & -\frac{3}{4} \end{bmatrix}$$

2 × 5 matrix

$$\begin{bmatrix} 0.2 & 3.4 & -1.1 & 2.5 & 6.7 \\ 3.4 & -3.4 & -22 & 0.5 & 7.2 \end{bmatrix}$$

3 × 1 matrix

$$\begin{bmatrix} -2 \\ 3 \\ 11 \end{bmatrix}$$

The element 3 is in row 2, column 1.

Special names are given to certain matrices. A matrix that has only one row is called a **row matrix,** and a matrix that has only one column is called a **column matrix.** A **square matrix** has the same number of rows as columns. Sometimes square matrices are called matrices of $\boldsymbol{n}$**th order,** where n is the number of rows and columns. The elements of an $m \times n$ matrix can be represented using double subscript notation; that is, a_{24} would name the element in the second row and fourth column.

$$\begin{bmatrix} a_{11} & a_{12} & a_{13} & \cdots & a_{1n} \\ a_{21} & a_{22} & a_{23} & \cdots & a_{2n} \\ a_{31} & a_{32} & a_{33} & \cdots & a_{3n} \\ \vdots & \vdots & \vdots & \vdots & \vdots \\ a_{m1} & a_{m2} & a_{m3} & \cdots & a_{mn} \end{bmatrix}$$

a_{ij} is the element in the ith row and the jth column.

Example **1** **TRAVEL** **Refer to the application at the beginning of the lesson.**

a. Use a matrix to represent the data.

b. Use a symbol to represent the price of pain reliever in Mexico City.

a. To represent data using a matrix, choose which category will be represented by the columns and which will be represented by the rows. Let's use the columns to represent the prices in each city and the rows to represent the prices of each item. Then write each data piece as you would if you were placing the data in a table.

	Atlanta	Los Angeles	Mexico City	Tokyo
film (24 exp.)	\$4.03	\$4.21	\$3.97	\$7.08
pain reliever (100 ct)	\$6.78	\$7.41	\$7.43	\$36.57
blow dryer	\$18.98	\$20.49	\$32.25	\$63.71

Notice that the category names appear outside of the matrix.

b. The price of pain reliever in Mexico City is found in the row 2, column 3 of the matrix. This element is represented by the symbol a_{23}.

Just as with numbers or algebraic expressions, matrices are equal under certain conditions.

Equal Matrices Two matrices are equal if and only if they have the same dimensions and are identical, element by element.

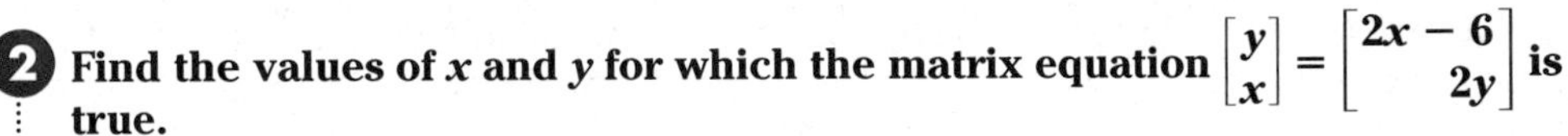

Example  **Find the values of x and y for which the matrix equation $\begin{bmatrix} y \\ x \end{bmatrix} = \begin{bmatrix} 2x - 6 \\ 2y \end{bmatrix}$ is true.**

Since the corresponding elements are equal, we can express the equality of the matrices as two equations.

$y = 2x - 6$
$x = 2y$

Solve the system of equations by using substitution.

$y = 2x - 6$
$y = 2(2y) - 6$ *Substitute 2y for x.*
$y = 2$ *Solve for y.*

$x = 2(2)$ *Substitute 2 for y in the second equation to find x.*
$x = 4$

The matrices are equal if $x = 4$ and $y = 2$. *Check by substituting into the matrices.*

Matrices are usually named using capital letters. The sum of two matrices, $A + B$, exists only if the two matrices have the same dimensions. The ijth element of $A + B$ is $a_{ij} + b_{ij}$.

Addition of Matrices The sum of two $m \times n$ matrices is an $m \times n$ matrix in which the elements are the sum of the corresponding elements of the given matrices.

Example 3 **Find $A + B$ if $A = \begin{bmatrix} -2 & 0 & 1 \\ 0 & 5 & -8 \end{bmatrix}$ and $B = \begin{bmatrix} -6 & 7 & -1 \\ 4 & -3 & 10 \end{bmatrix}$.**

Graphing Calculator Appendix

For keystroke instruction on entering matrices and performing operations on them, see pages A16-A17.

$$A + B = \begin{bmatrix} -2 + (-6) & 0 + 7 & 1 + (-1) \\ 0 + 4 & 5 + (-3) & -8 + 10 \end{bmatrix}$$
$$= \begin{bmatrix} -8 & 7 & 0 \\ 4 & 2 & 2 \end{bmatrix}$$

You know that 0 is the additive identity for real numbers because $a + 0 = a$. Matrices also have additive identities. For every matrix A, another matrix can be found so that their sum is A. For example,

$$\text{if } A = \begin{bmatrix} a_{11} & a_{12} \\ a_{21} & a_{22} \end{bmatrix}, \text{ then } \begin{bmatrix} a_{11} & a_{12} \\ a_{21} & a_{22} \end{bmatrix} + \begin{bmatrix} 0 & 0 \\ 0 & 0 \end{bmatrix} = \begin{bmatrix} a_{11} & a_{12} \\ a_{21} & a_{22} \end{bmatrix}.$$

The matrix $\begin{bmatrix} 0 & 0 \\ 0 & 0 \end{bmatrix}$ is called a **zero matrix.** The $m \times n$ zero matrix is the **additive identity matrix** for any $m \times n$ matrix.

You also know that for any number a, there is a number $-a$, called the additive inverse of a, such that $a + (-a) = 0$. Matrices also have additive inverses. If $A = \begin{bmatrix} a_{11} & a_{12} \\ a_{21} & a_{22} \end{bmatrix}$, then the matrix that must be added to A in order to have a sum of a zero matrix is $\begin{bmatrix} -a_{11} & -a_{12} \\ -a_{21} & -a_{22} \end{bmatrix}$ or $-A$. Therefore, $-A$ is the additive inverse of A. The additive inverse is used when you subtract matrices.

Subtraction of Matrices	The difference $A - B$ of two $m \times n$ matrices is equal to the sum $A + (-B)$, where $-B$ represents the additive inverse of B.

Example 4 **Find $C - D$ if $C = \begin{bmatrix} 9 & 4 \\ -1 & 3 \\ 0 & -4 \end{bmatrix}$ and $D = \begin{bmatrix} 8 & -2 \\ -6 & 1 \\ 5 & -5 \end{bmatrix}$.**

$$C - D = C + (-D)$$
$$= \begin{bmatrix} 9 & 4 \\ -1 & 3 \\ 0 & -4 \end{bmatrix} + \begin{bmatrix} -8 & 2 \\ 6 & -1 \\ -5 & 5 \end{bmatrix}$$
$$= \begin{bmatrix} 9 + (-8) & 4 + 2 \\ -1 + 6 & 3 + (-1) \\ 0 + (-5) & -4 + 5 \end{bmatrix} \text{ or } \begin{bmatrix} 1 & 6 \\ 5 & 2 \\ -5 & 1 \end{bmatrix}$$

You can multiply a matrix by a number; when you do, the number is called a **scalar.** The product of a scalar and a matrix A is defined as follows.

Scalar Product	The product of a scalar k and an $m \times n$ matrix A is an $m \times n$ matrix denoted by kA. Each element of kA equals k times the corresponding element of A.

Example 5 If $A = \begin{bmatrix} -4 & 1 & -1 \\ 3 & 7 & 0 \\ -3 & -1 & 8 \end{bmatrix}$, find $3A$.

$$3\begin{bmatrix} -4 & 1 & -1 \\ 3 & 7 & 0 \\ -3 & -1 & 8 \end{bmatrix} = \begin{bmatrix} 3(-4) & 3(1) & 3(-1) \\ 3(3) & 3(7) & 3(0) \\ 3(-3) & 3(-1) & 3(8) \end{bmatrix}$$ *Multiply each element by 3.*

$$= \begin{bmatrix} -12 & 3 & -3 \\ 9 & 21 & 0 \\ -9 & -3 & 24 \end{bmatrix}$$

You can also multiply a matrix by a matrix. For matrices A and B, you can find AB if the number of columns in A is the same as the number of rows in B.

$$\begin{bmatrix} 3 & -8 & 1 \\ 1 & 0 & 2 \end{bmatrix} \cdot \begin{bmatrix} 0 & 2 & 0 & 1 \\ -4 & 0 & -2 & 1 \\ 1 & -3 & -1 & 6 \end{bmatrix}$$

2×3 3×4

Since 3 = 3, multiplication is possible.

$$\begin{bmatrix} 5 & 3 & 1 & 0 \\ 6 & 0 & 2 & -3 \\ -5 & 3 & 2 & 2 \end{bmatrix} \cdot \begin{bmatrix} 1 & 0 & 12 & 9 \\ 0 & 0 & -4 & -8 \\ -2 & 3 & 4 & 3 \end{bmatrix}$$

3×4 3×4

Since $4 \neq 3$, multiplication is not possible.

The product of two matrices is found by multiplying columns and rows. Suppose $A = \begin{bmatrix} a_1 & b_1 \\ a_2 & b_2 \end{bmatrix}$ and $X = \begin{bmatrix} x_1 & y_1 \\ x_2 & y_2 \end{bmatrix}$. Each element of matrix AX is the product of one row of matrix A and one column of matrix X.

$$AsX = \begin{bmatrix} a_1 & b_1 \\ a_2 & b_2 \end{bmatrix} \cdot \begin{bmatrix} x_1 & y_1 \\ x_2 & y_2 \end{bmatrix} = \begin{bmatrix} a_1x_1 + b_1x_2 & a_1y_1 + b_1y_2 \\ a_2x_1 + b_2x_2 & a_2y_2 + b_2y_2 \end{bmatrix}$$

In general, the product of two matrices is defined as follows.

Product of Two Matrices

The product of an $m \times n$ matrix A and an $n \times r$ matrix B is an $m \times r$ matrix AB. The ijth element of AB is the sum of the products of the corresponding elements in the ith row of A and the jth column of B.

Example 6 Use matrices $A = \begin{bmatrix} 7 & 0 \\ 5 & 3 \end{bmatrix}$, $B = \begin{bmatrix} 3 & -3 & 6 \\ 5 & 4 & -2 \end{bmatrix}$, and $C = \begin{bmatrix} 6 & -1 & 4 \\ 2 & -2 & -1 \end{bmatrix}$ to find each product.

a. *AB*

$$AB = \begin{bmatrix} 7 & 0 \\ 5 & 3 \end{bmatrix} \cdot \begin{bmatrix} 3 & -3 & 6 \\ 5 & 4 & -2 \end{bmatrix}$$

$$AB = \begin{bmatrix} 7(3) + 0(5) & 7(-3) + 0(4) & 7(6) + 0(-2) \\ 5(3) + 3(5) & 5(-3) + 3(4) & 5(6) + 3(-2) \end{bmatrix} \text{ or } \begin{bmatrix} 21 & -21 & 42 \\ 30 & -3 & 24 \end{bmatrix}$$

b. *BC*

B is a 2×3 matrix and C is a 2×3 matrix. Since B does not have the same number of columns as C has rows, the product BC does not exist. BC is undefined.

Example 7 Real World Application

SPORTS In football, a player scores 6 points for a touchdown (TD), 3 points for kicking a field goal (FG), and 1 point for kicking the extra point after a touchdown (PAT). The chart lists the records of the top five all-time professional football scorers (as of the end of the 1997 season). Use matrix multiplication to find the number of points each player scored.

Scorer	TD	FG	PAT
George Blanda	9	335	943
Nick Lowery	0	383	562
Jan Stenerud	0	373	580
Gary Anderson	0	385	526
Morten Andersen	0	378	507

Source: *The World Almanac and Book of Facts*, 1999

Write the scorer information as a 5×3 matrix and the points per play as a 3×1 matrix. Then multiply the matrices.

$$\begin{array}{l} \\ \text{Blanda} \\ \text{Lowery} \\ \text{Stenerud} \\ \text{Anderson} \\ \text{Andersen} \end{array} \begin{array}{c} \text{TD} \quad \text{FG} \quad \text{PAT} \\ \begin{bmatrix} 9 & 335 & 943 \\ 0 & 383 & 562 \\ 0 & 373 & 580 \\ 0 & 385 & 526 \\ 0 & 378 & 507 \end{bmatrix} \end{array} \cdot \begin{array}{l} \\ \text{TD} \\ \text{FG} \\ \text{PAT} \end{array} \begin{array}{c} \text{pts} \\ \begin{bmatrix} 6 \\ 3 \\ 1 \end{bmatrix} \end{array} =$$

$$\begin{array}{l} \text{Blanda} \\ \text{Lowery} \\ \text{Stenerud} \\ \text{Anderson} \\ \text{Andersen} \end{array} \begin{bmatrix} 9(6) + 335(3) + 943(1) \\ 0(6) + 383(3) + 562(1) \\ 0(6) + 373(3) + 580(1) \\ 0(6) + 385(3) + 526(1) \\ 0(6) + 378(3) + 507(1) \end{bmatrix} = \begin{array}{l} \text{Blanda} \\ \text{Lowery} \\ \text{Stenerud} \\ \text{Anderson} \\ \text{Andersen} \end{array} \begin{bmatrix} 2002 \\ 1711 \\ 1699 \\ 1681 \\ 1641 \end{bmatrix}$$

CHECK FOR UNDERSTANDING

Communicating Mathematics

Read and study the lesson to answer each question.

1. **Write** a matrix other than the one given in Example 1 to represent the data on travel prices.
2. **Tell** the dimensions of the matrix $\begin{bmatrix} 4 & 0 & -2 & 4 \\ -1 & 3 & -1 & 5 \end{bmatrix}$.
3. **Explain** how you determine whether the sum of two matrices exists.
4. **You Decide** Sarah says that $\begin{bmatrix} 3 & 2 & 3 \\ -4 & 2 & 0 \\ 0 & -1 & 2 \\ 2 & 2 & 5 \end{bmatrix}$ is a third-order matrix. Anthony disagrees. Who is correct and why?

Guided Practice

Find the values of *x* and *y* for which each matrix equation is true.

5. $\begin{bmatrix} 2y \\ x \end{bmatrix} = \begin{bmatrix} x - 3 \\ y + 5 \end{bmatrix}$
6. $[18 \quad 24] = [4x - y \quad 12y]$
7. $[16 \quad 0 \quad 2x] = [4x \quad y \quad 8 - y]$

Use matrices X, Y, and Z to find each of the following. If the matrix does not exist, write *impossible*.

$X = \begin{bmatrix} 4 & 1 \\ -2 & 6 \end{bmatrix}$ $\quad Y = \begin{bmatrix} 0 & -3 \end{bmatrix}$ $\quad Z = \begin{bmatrix} -1 & 3 \\ 0 & -2 \end{bmatrix}$

8. $X + Z$ **9.** $Z - Y$ **10.** $Z - X$ **11.** $4X$ **12.** ZY **13.** YX

14. Advertising A newspaper surveyed companies on the annual amount of money spent on television commercials and the estimated number of people who remember seeing those commercials each week. A soft-drink manufacturer spends $40.1 million a year and estimates 78.6 million people remember the commercials. For a package-delivery service, the budget is $22.9 million for 21.9 million people. A telecommunications company reaches 88.9 million people by spending a whopping $154.9 million. Use a matrix to represent this data.

EXERCISES

Practice

Find the values of x and y for which each matrix equation is true.

15. $\begin{bmatrix} y \\ x \end{bmatrix} = \begin{bmatrix} 2x - 1 \\ y - 5 \end{bmatrix}$

16. $\begin{bmatrix} 9 & 13 \end{bmatrix} = \begin{bmatrix} x + 2y & 4x + 1 \end{bmatrix}$

17. $\begin{bmatrix} 4x \\ 5 \end{bmatrix} = \begin{bmatrix} 15 + x \\ 2y \end{bmatrix}$

18. $\begin{bmatrix} x & y \end{bmatrix} = \begin{bmatrix} 2y & 2x - 6 \end{bmatrix}$

19. $\begin{bmatrix} 27 \\ 8 \end{bmatrix} = \begin{bmatrix} 3y \\ 5x - 3y \end{bmatrix}$

20. $\begin{bmatrix} 4x - 3y \\ x + y \end{bmatrix} = \begin{bmatrix} 11 \\ 1 \end{bmatrix}$

21. $\begin{bmatrix} 2x & y & -y \end{bmatrix} = \begin{bmatrix} -10 & 3x & 15 \end{bmatrix}$

22. $\begin{bmatrix} -12 \\ 2 \\ 12y \end{bmatrix} = \begin{bmatrix} 6x \\ y + 1 \\ 10 - x \end{bmatrix}$

23. $\begin{bmatrix} x + y & 3 \\ y & 6 \end{bmatrix} = \begin{bmatrix} 0 & 2y - x \\ y^2 & 4 - 2x \end{bmatrix}$

24. $\begin{bmatrix} x^2 + 1 & 5 - y \\ x + y & y - 4 \end{bmatrix} = \begin{bmatrix} 2 & x \\ 5 & 2 \end{bmatrix}$

25. Find the values of x, y, and z for $3\begin{bmatrix} x & y - 1 \\ 4 & 3z \end{bmatrix} = \begin{bmatrix} 15 & 6 \\ 6z & 3x + y \end{bmatrix}$.

26. Solve $-2\begin{bmatrix} w + 5 & x - z \\ 3y & 8 \end{bmatrix} = \begin{bmatrix} -16 & -4 \\ 6 & 2x + 8z \end{bmatrix}$ for w, x, y, and z.

Use matrices A, B, C, D, E, and F to find each of the following. If the matrix does not exist, write *impossible*.

$A = \begin{bmatrix} 5 & 7 \\ -6 & 1 \end{bmatrix}$ $\quad B = \begin{bmatrix} 3 & 5 \\ -1 & 8 \end{bmatrix}$ $\quad C = \begin{bmatrix} 4 & -2 & 3 \\ 5 & 0 & -1 \\ 9 & 0 & 1 \end{bmatrix}$ $\quad D = \begin{bmatrix} 0 & 1 & 2 \\ -2 & 3 & 0 \\ 4 & 4 & -2 \end{bmatrix}$

$E = \begin{bmatrix} 8 & -4 & 2 \\ 3 & 1 & -5 \end{bmatrix}$ $\quad F = \begin{bmatrix} -6 & -1 & 0 \\ 1 & 4 & 0 \end{bmatrix}$

27. $A + B$ **28.** $A + C$ **29.** $D + B$ **30.** $D + C$ **31.** $B - A$

32. $C - D$ **33.** $4D$ **34.** $-2F$ **35.** $F - E$ **36.** $E - F$

37. $5A$ **38.** BA **39.** CF **40.** FC **41.** ED

42. AA **43.** $E + FD$ **44.** $-3AB$ **45.** $(BA)E$ **46.** $F - 2EC$

47. Find $3XY$ if $X = \begin{bmatrix} 2 & 4 \\ 8 & -4 \\ -2 & 6 \end{bmatrix}$ and $Y = \begin{bmatrix} 3 & -3 & 6 \\ 5 & 4 & -2 \end{bmatrix}$.

48. If $K = \begin{bmatrix} 1 & -7 \\ 3 & 2 \end{bmatrix}$ and $J = \begin{bmatrix} -4 & 5 \\ 1 & -1 \end{bmatrix}$, find $2K - 3J$.

Applications and Problem Solving

49. **Entertainment** How often do you go to the movies? The graph below shows the projected number of adults of different ages who attend movies at least once a month. Organize the information in a matrix.

Projected Movie Attendance

	Year		
Age group	1996	2000	2006
18 to 24	8485	8526	8695
25 to 34	10,102	9316	9078
35 to 44	8766	9039	8433
45 to 54	6045	6921	7900
55 to 64	2444	2741	3521
65 and older	2381	2440	2572

Source: American Demographics

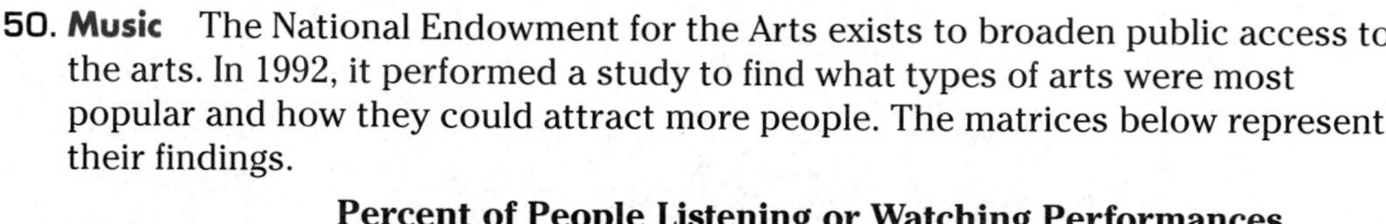

interNET CONNECTION
Data Update For the latest National Endowment for the Arts survey, visit **www.amc.glencoe.com**

50. **Music** The National Endowment for the Arts exists to broaden public access to the arts. In 1992, it performed a study to find what types of arts were most popular and how they could attract more people. The matrices below represent their findings.

Percent of People Listening or Watching Performances

1982

$$\begin{array}{r} \\ \text{Classical} \\ \text{Jazz} \\ \text{Opera} \\ \text{Musicals} \end{array} \begin{array}{c} \text{TV} \quad \text{Radio} \quad \text{Recording} \\ \begin{bmatrix} 25 & 20 & 22 \\ 18 & 18 & 20 \\ 12 & 7 & 8 \\ 21 & 4 & 8 \end{bmatrix} \end{array}$$

1992

$$\begin{array}{r} \\ \text{Classical} \\ \text{Jazz} \\ \text{Opera} \\ \text{Musicals} \end{array} \begin{array}{c} \text{TV} \quad \text{Radio} \quad \text{Recording} \\ \begin{bmatrix} 25 & 31 & 24 \\ 21 & 28 & 21 \\ 12 & 9 & 7 \\ 15 & 4 & 6 \end{bmatrix} \end{array}$$

Source: National Endowment for the Arts

a. Find the difference in arts patronage from 1982 to 1992. Express your answer as a matrix.

b. Which areas saw the greatest increase and decrease in this time?

51. **Critical Thinking** Consider the matrix equation $\begin{bmatrix} -2 & 3 \\ 4 & -5 \end{bmatrix} \cdot \begin{bmatrix} a & b \\ c & d \end{bmatrix} = \begin{bmatrix} -2 & 3 \\ 4 & -5 \end{bmatrix}$.

a. Find the values of a, b, c, and d to make the statement true.

b. If any matrix containing two columns were multiplied by the matrix containing a, b, c, and d, what would you expect the result to be? Explain.

52. Finance Investors choose different stocks to comprise a balanced portfolio. The matrix below shows the prices of one share of each of several stocks on the first business day of July, August, and September of 1998.

$$\begin{array}{l} \\ \text{Stock A} \\ \text{Stock B} \\ \text{Stock C} \\ \text{Stock D} \end{array} \begin{array}{c} \textbf{July} \quad\quad \textbf{August} \quad\quad \textbf{September} \\ \begin{bmatrix} \$33\frac{13}{16} & \$30\frac{15}{16} & \$27\frac{1}{4} \\ \$15\frac{1}{16} & \$13\frac{1}{4} & \$8\frac{3}{4} \\ \$54 & \$54 & \$46\frac{7}{16} \\ \$52\frac{1}{16} & \$44\frac{11}{16} & \$34\frac{3}{8} \end{bmatrix} \end{array}$$

a. Mrs. Numkena owns 42 shares of stock A, 59 shares of stock B, 21 shares of stock C, and 18 shares of stock D. Write a row matrix to represent Mrs. Numkena's portfolio.

b. Use matrix multiplication to find the total value of Mrs. Numkena's portfolio for each month to the nearest cent.

53. Critical Thinking Study the matrix at the right. In which row and column will 2001 occur? Explain your reasoning.

$$\begin{bmatrix} 1 & 3 & 6 & 10 & 15 & \cdots \\ 2 & 5 & 9 & 14 & 20 & \cdots \\ 4 & 8 & 13 & 19 & 26 & \cdots \\ 7 & 12 & 18 & 25 & 33 & \cdots \\ 11 & 17 & 24 & 32 & 41 & \cdots \\ 16 & 23 & 31 & 40 & 50 & \cdots \\ \vdots & \vdots & \vdots & \vdots & \vdots & \vdots \end{bmatrix}$$

54. Discrete Math Airlines and other businesses often use *finite graphs* to represent their routes. A finite graph contains points called *nodes* and segments called *edges.* In a graph for an airline, each node represents a city, and each edge represents a route between the cities. Graphs can be represented by square matrices. The elements of the matrix are the numbers of edges between each pair of nodes. Study the graph and its matrix at the right.

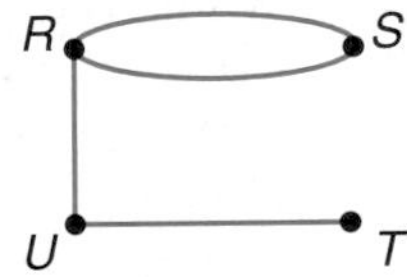

$$\begin{array}{c} \\ R \\ S \\ T \\ U \end{array} \begin{array}{c} \begin{array}{cccc} R & S & T & U \end{array} \\ \begin{bmatrix} 0 & 2 & 0 & 1 \\ 2 & 0 & 0 & 0 \\ 0 & 0 & 0 & 1 \\ 1 & 0 & 1 & 0 \end{bmatrix} \end{array}$$

a. Represent the graph with nodes *A, B, C,* and *D* at the right using a matrix.

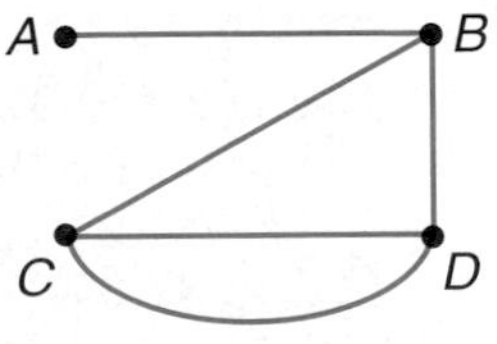

b. Equivalent graphs have the same number of nodes and edges between corresponding nodes. Can different graphs be represented by the same matrix? Explain your answer.

Mixed Review

55. Solve the system $2x + 6y + 8z = 5$, $-2x + 9y + 12z = 5$, and $4x + 6y - 4z = 3$. *(Lesson 2-2)*

56. State whether the system $4x - 2y = 7$ and $-12x + 6y = -21$ is *consistent and independent, consistent and dependent,* or *inconsistent.* *(Lesson 2-1)*

57. Graph $-6 \leq 3x - y \leq 12$. *(Lesson 1-8)*

58. Graph $f(x) = |3x| + 2$. *(Lesson 1-7)*

59. **Education** Many educators believe that taking practice tests will help students succeed on actual tests. The table below shows data gathered about students studying for an algebra test. Use the data to write a prediction equation. *(Lesson 1-6)*

Practice Test Time (minutes)	15	75	60	45	90	60	30	120	10	120
Test scores (percents)	68	87	92	73	95	83	77	98	65	94

60. Write the slope-intercept form of the equation of the line through points at (1, 4) and (5, 7). *(Lesson 1-4)*

61. Find the zero of $f(x) = 5x - 3$ *(Lesson 1-3)*

62. Find $[f \cdot g](x)$ if $f(x) = \frac{2}{5}x$ and $g(x) = 40x - 10$. *(Lesson 1-2)*

63. Given $f(x) = 4 + 6x - x^3$, find $f(14)$. *(Lesson 1-1)*

64. **SAT/ACT Practice** If $\frac{2x - 3}{x} = \frac{3 - x}{2}$, which of the following could be a value of x?

A -3 **B** -1 **C** 37 **D** 5 **E** 15

GRAPHING CALCULATOR EXPLORATION

Remember the properties of real numbers:

Properties of Addition

Commutative $a + b = b + a$

Associative $(a + b) + c = a + (b + c)$

Properties of Multiplication

Commutative $ab = ba$

Associative $(ab)c = a(bc)$

Distributive Property

$a(b + c) = ab + ac$

Do these properties hold for operations with matrices?

TRY THESE

Use matrices *A, B,* and *C* to investigate each of the properties shown above.

$A = \begin{bmatrix} -1 & 2 \\ 3 & 0 \end{bmatrix}$ $B = \begin{bmatrix} 4 & 0 \\ -2 & -3 \end{bmatrix}$ $C = \begin{bmatrix} -2 & -3 \\ 4 & 2 \end{bmatrix}$

WHAT DO YOU THINK?

1. Do these three matrices satisfy the properties of real numbers listed at the left? Explain.

2. Would these properties hold for any 2×2 matrices? Prove or disprove each statement below using variables as elements of each 2×2 matrix.
   a. Addition of matrices is commutative.
   b. Addition of matrices is associative.
   c. Multiplication of matrices is commutative.
   d. Multiplication of matrices is associative.

3. Which properties do you think hold for $n \times n$ matrices? Explain.

Extra Practice See p. A28.

GRAPHING CALCULATOR EXPLORATION

2-4A Transformation Matrices

OBJECTIVE

- Determine the effect of matrix multiplication on a vertex matrix.

An Introduction to Lesson 2-4

The coordinates of a figure can be represented by a matrix with the x-coordinates in the first row and the y-coordinates in the second. When this matrix is multiplied by certain other matrices, the result is a new matrix that contains the coordinates of the vertices of a different figure. You can use **List** and **Matrix** operations on your calculator to visualize some of these multiplications.

TRY THESE

Step 1 Enter $A = \begin{bmatrix} 0 & -1 \\ 1 & 0 \end{bmatrix}$, $B = \begin{bmatrix} -1 & 0 \\ 0 & -1 \end{bmatrix}$, and $C = \begin{bmatrix} 0 & 1 \\ -1 & 0 \end{bmatrix}$.

Step 2 Graph the triangle LMN with $L(1, -1)$, $M(2, -2)$, and $N(3, -1)$ by using **STAT PLOT**. Enter the x-coordinates in **L1** and the y-coordinates in **L2**, repeating the first point to close the figure. In **STAT PLOT**, turn **Plot 1** on and select the connected graph. After graphing the figure, press ZOOM **5** to reflect a less distorted viewing window.

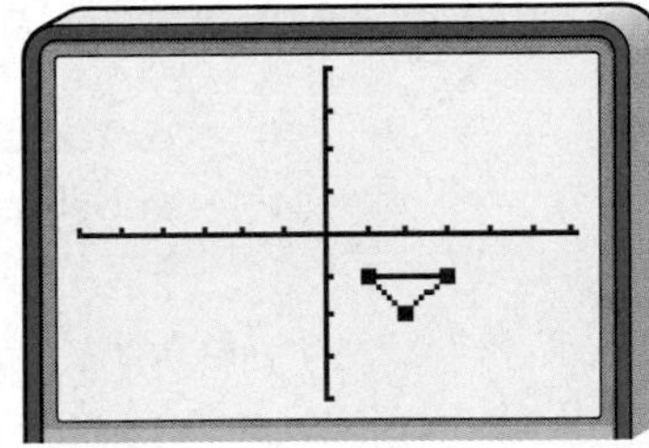

[−4.548 ..., 4.548 ...] scl: 1 by [−3, 3] scl: 1

Step 3 To transfer the coordinates from the lists to a matrix, use the **9:List▶matr** command from the **MATH** submenu in the **MATRX** menu. Store this as matrix D. Matrix D has the x-coordinates in Column 1 and y-coordinates in Column 2. But a vertex matrix has the x-coordinates in *Row* 1 and the y-coordinates in *Row 2*. This switch can be easily done by using the **2:ᵀ** (transpose) command found in the **MATH** submenu as shown in the screen.

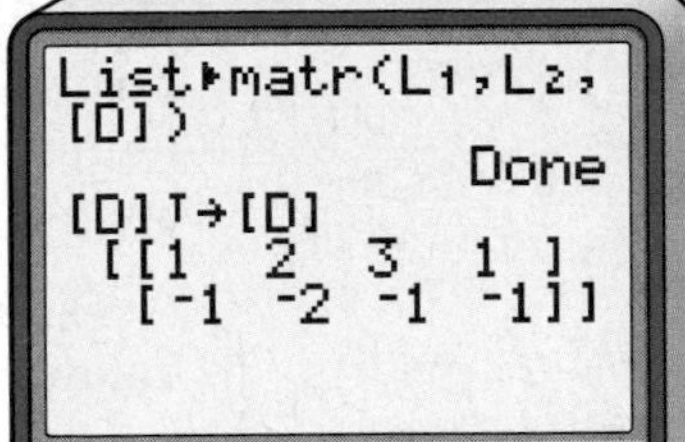

The **List▶matr** *and* **Matr▶list** *commands transfer the data column for column. That is, the data in List 1 goes to Column 1 of the matrix and vice versa.*

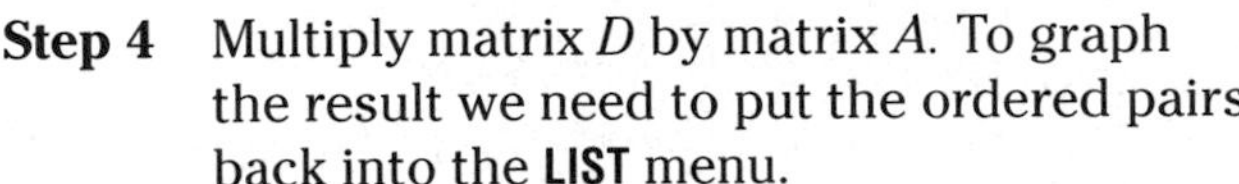

Step 4 Multiply matrix D by matrix A. To graph the result we need to put the ordered pairs back into the **LIST** menu.

- This means we need to transpose AD first. Store as new matrix E.
- Use the **8:Matr▶list** command from the math menu to store values into **L3** and **L4**.

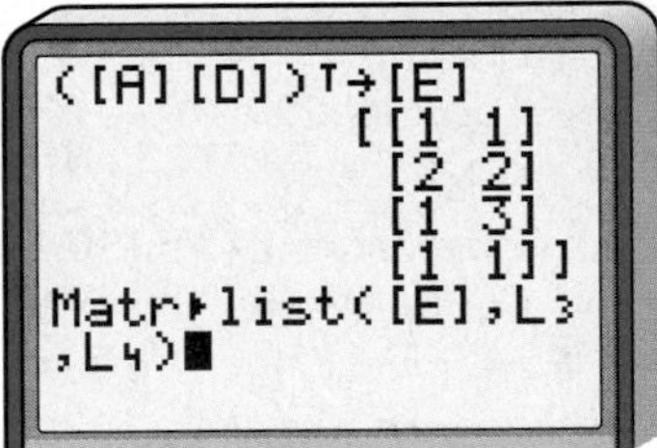

Step 5 Assign **Plot 2** as a connected graph of the **L3** and **L4** data and view the graph.

WHAT DO YOU THINK?

1. What is the relationship between the two plots?
2. Repeat Steps 4 and 5 replacing matrix A with matrix B. Compare the graphs.
3. Repeat Steps 4 and 5 replacing matrix A with matrix C. Compare the graphs.
4. Select a new figure and repeat this activity using each of the 2×2 matrices. Make a conjecture about these 2×2 matrices.

Modeling Motion with Matrices

OBJECTIVE

- Use matrices to determine the coordinates of polygons under a given transformation.

COMPUTER ANIMATION In 1995, animation took a giant step forward with the release of the first major motion picture to be created entirely on computers. Animators use computer software to create three-dimensional computer models of characters, props, and sets. These computer models describe the shape of the object as well as the motion controls that the animators use to create movement and expressions. The animation models are actually very large matrices.

Even though large matrices are used for computer animation, you can use a simple matrix to describe many of the motions called **transformations** that you learned about in geometry. Some of the transformations we will examine in this lesson are **translations** (slides), **reflections** (flips), **rotations** (turns), and **dilations** (enlargements or reductions).

An n-gon is a polygon with n sides.

A $2 \times n$ matrix can be used to express the vertices of an n-gon with the first row of elements representing the x-coordinates and the second row the y-coordinates of the vertices.

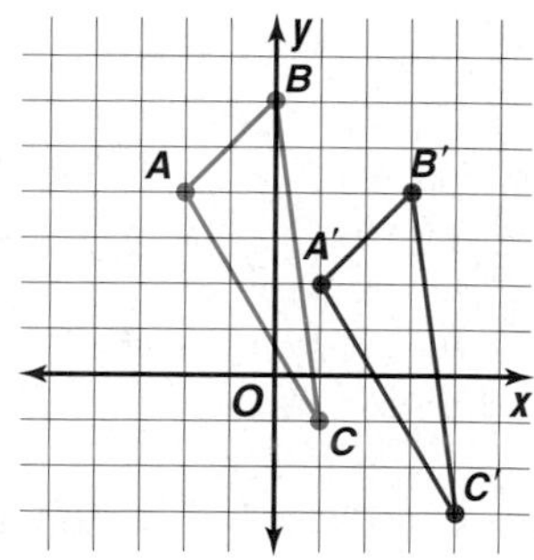

Triangle ABC can be represented by the following **vertex matrix.**

$$\begin{array}{r} \\ x\text{-coordinate} \\ y\text{-coordinate} \end{array} \begin{array}{c} \begin{array}{ccc} A & B & C \end{array} \\ \begin{bmatrix} -2 & 0 & 1 \\ 4 & 6 & -1 \end{bmatrix} \end{array}$$

Triangle $A'B'C'$ is congruent to and has the same orientation as $\triangle ABC$, but is moved 3 units right and 2 units down from $\triangle ABC$'s location. The coordinates of $\triangle A'B'C'$ can be expressed as the following vertex matrix.

$$\begin{array}{r} \\ x\text{-coordinate} \\ y\text{-coordinate} \end{array} \begin{array}{c} \begin{array}{ccc} A' & B' & C' \end{array} \\ \begin{bmatrix} 1 & 3 & 4 \\ 2 & 4 & -3 \end{bmatrix} \end{array}$$

Compare the two matrices. If you add $\begin{bmatrix} 3 & 3 & 3 \\ -2 & -2 & -2 \end{bmatrix}$ to the first matrix, you get the second matrix. Each 3 represents moving 3 units right for each x-coordinate. Likewise, each -2 represents moving 2 units down for each y-coordinate. This type of matrix is called a **translation matrix.** In this transformation, $\triangle ABC$ is the **pre-image,** and $\triangle A'B'C'$ is the **image** after the translation.

Example **1** **Suppose quadrilateral *ABCD* with vertices $A(-1, 1)$, $B(4, 0)$, $C(4, -5)$, and $D(-1, -3)$ is translated 2 units left and 4 units up.**

Note that the image under a translation is the same shape and size as the pre-image. The figures are congruent.

a. Represent the vertices of the quadrilateral as a matrix.

b. Write the translation matrix.

c. Use the translation matrix to find the vertices of $A'B'C'D'$, the translated image of the quadrilateral.

d. Graph quadrilateral *ABCD* and its image.

a. The matrix representing the coordinates of the vertices of quadrilateral *ABCD* will be a 2×4 matrix.

$$\begin{matrix} \\ x\text{-coordinate} \\ y\text{-coordinate} \end{matrix} \begin{matrix} \quad A \quad B \quad C \quad D \\ \begin{bmatrix} -1 & 4 & 4 & -1 \\ 1 & 0 & -5 & -3 \end{bmatrix} \end{matrix}$$

b. The translation matrix is $\begin{bmatrix} -2 & -2 & -2 & -2 \\ 4 & 4 & 4 & 4 \end{bmatrix}$.

c. Add the two matrices.

$$\begin{bmatrix} -1 & 4 & 4 & -1 \\ 1 & 0 & -5 & -3 \end{bmatrix} + \begin{bmatrix} -2 & -2 & -2 & -2 \\ 4 & 4 & 4 & 4 \end{bmatrix} = \begin{matrix} \quad A' \quad B' \quad C' \quad D' \\ \begin{bmatrix} -3 & 2 & 2 & -3 \\ 5 & 4 & -1 & 1 \end{bmatrix} \end{matrix}$$

d. Graph the points represented by the resulting matrix.

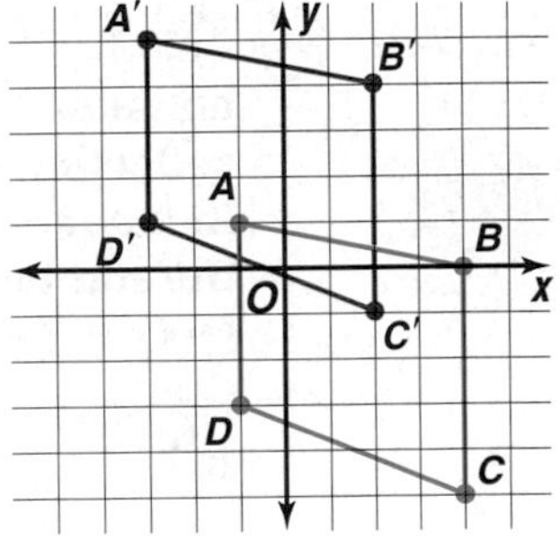

There are three lines over which figures are commonly reflected.

- the x-axis
- the y-axis, and
- the line $y = x$

The preimage and the image under a reflection are congruent.

In the figure at the right, $\triangle P'Q'R'$ is a reflection of $\triangle PQR$ over the x-axis. There is a 2×2 **reflection matrix** that, when multiplied by the vertex matrix of $\triangle PQR$, will yield the vertex matrix of $\triangle P'Q'R'$.

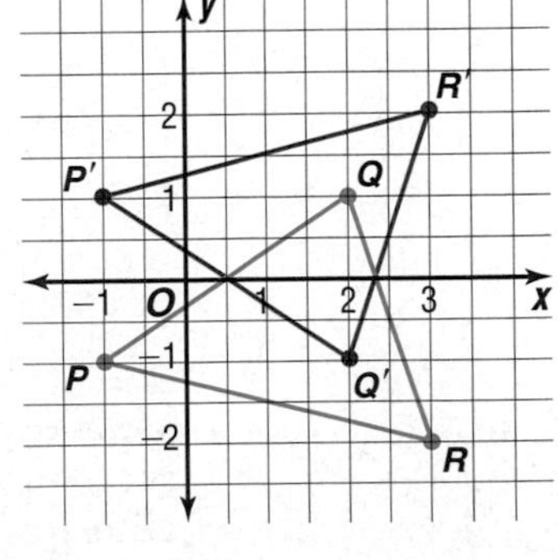

Let $\begin{bmatrix} a & b \\ c & d \end{bmatrix}$ represent the unknown square matrix.

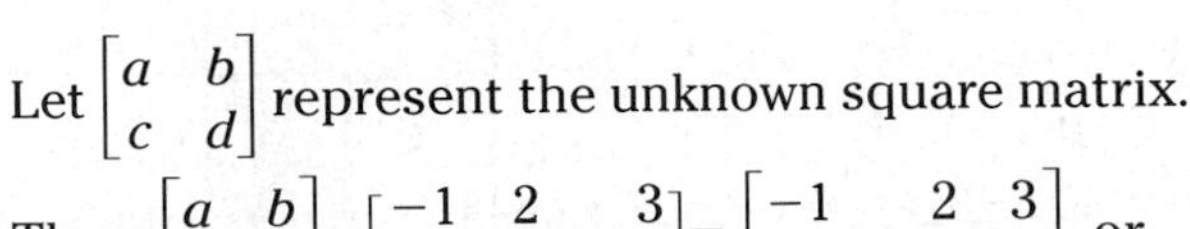

Thus, $\begin{bmatrix} a & b \\ c & d \end{bmatrix} \cdot \begin{bmatrix} -1 & 2 & 3 \\ -1 & 1 & -2 \end{bmatrix} = \begin{bmatrix} -1 & 2 & 3 \\ 1 & -1 & 2 \end{bmatrix}$, or

$$\begin{bmatrix} -a - b & 2a + b & 3a - 2b \\ -c - d & 2c + d & 3c - 2d \end{bmatrix} = \begin{bmatrix} -1 & 2 & 3 \\ 1 & -1 & 2 \end{bmatrix}.$$

Since corresponding elements of equal matrices are equal, we can write equations to find the values of the variables. These equations form two systems.

$$\begin{aligned} -a - b &= -1 \\ 2a + b &= 2 \\ 3a - 2b &= 3 \end{aligned} \qquad \begin{aligned} -c - d &= 1 \\ 2c + d &= -1 \\ 3c - 2d &= 2 \end{aligned}$$

When you solve each system of equations, you will find that $a = 1, b = 0, c = 0$, and $d = -1$. Thus, the matrix that results in a reflection over the x-axis is $\begin{bmatrix} 1 & 0 \\ 0 & -1 \end{bmatrix}$. This matrix will work for any reflection over the x-axis.

The matrices for a reflection over the y-axis or the line $y = x$ can be found in a similar manner. These are summarized below.

Reflection Matrices		
For a reflection over the:	**Symbolized by:**	**Multiply the vertex matrix by:**
x-axis	$R_{x\text{-axis}}$	$\begin{bmatrix} 1 & 0 \\ 0 & -1 \end{bmatrix}$
y-axis	$R_{y\text{-axis}}$	$\begin{bmatrix} -1 & 0 \\ 0 & 1 \end{bmatrix}$
line $y = x$	$R_{y=x}$	$\begin{bmatrix} 0 & 1 \\ 1 & 0 \end{bmatrix}$

Example 2

ANIMATION **To create an image that appears to be reflected in a mirror, an animator will use a matrix to reflect an image over the y-axis. Use a reflection matrix to find the coordinates of the vertices of a star reflected in a mirror (the y-axis) if the coordinates of the points connected to create the star are (−2, 4), (−3.5, 4), (−4, 5), (−4.5, 4), (−6, 4), (−5, 3), (−5, 1), (−4, 2), (−3, 1), and (−3, 3).**

First write the vertex matrix for the points used to define the star.

$$\begin{bmatrix} -2 & -3.5 & -4 & -4.5 & -6 & -5 & -5 & -4 & -3 & -3 \\ 4 & 4 & 5 & 4 & 4 & 3 & 1 & 2 & 1 & 3 \end{bmatrix}$$

Multiply by the y-axis reflection matrix.

$$\begin{bmatrix} -1 & 0 \\ 0 & 1 \end{bmatrix} \cdot \begin{bmatrix} -2 & -3.5 & -4 & -4.5 & -6 & -5 & -5 & -4 & -3 & -3 \\ 4 & 4 & 5 & 4 & 4 & 3 & 1 & 2 & 1 & 3 \end{bmatrix} =$$

$$\begin{bmatrix} 2 & 3.5 & 4 & 4.5 & 6 & 5 & 5 & 4 & 3 & 3 \\ 4 & 4 & 5 & 4 & 4 & 3 & 1 & 2 & 1 & 3 \end{bmatrix}$$

The vertices used to define the reflection are (2, 4), (3.5, 4), (4, 5), (4.5, 4), (6, 4), (5, 3), (5, 1), (4, 2), (3, 1), and (3, 3).

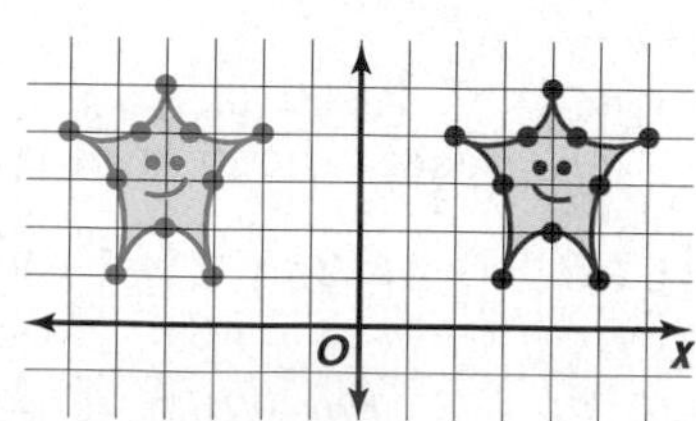

The preimage and the image under a rotation are congruent.

You may remember from geometry that a rotation of a figure on a coordinate plane can be achieved by a combination of reflections. For example, a 90° counterclockwise rotation can be found by first reflecting the image over the x-axis and then reflecting the reflected image over the line $y = x$. The **rotation matrix,** Rot_{90}, can be found by a composition of reflections. Since reflection matrices are applied using multiplication, the composition of two reflection matrices is a product. Remember that $[f \circ g](x)$ means that you find $g(x)$ first and then evaluate the result for $f(x)$. So, to define Rot_{90}, we use

$$Rot_{90} = R_{y = x} \circ R_{x\text{-axis}} \text{ or } Rot_{90} = \begin{bmatrix} 0 & 1 \\ 1 & 0 \end{bmatrix} \cdot \begin{bmatrix} 1 & 0 \\ 0 & -1 \end{bmatrix} \text{ or } \begin{bmatrix} 0 & -1 \\ 1 & 0 \end{bmatrix}.$$

Similarly, a rotation of 180° would be rotations of 90° twice or $Rot_{90} \circ Rot_{90}$. A rotation of 270° is a composite of Rot_{180} and Rot_{90}. The results of these composites are shown below.

Rotation Matrices		
For a counterclockwise rotation about the origin of:	**Symbolized by:**	**Multiply the vertex matrix by:**
90°	Rot_{90}	$\begin{bmatrix} 0 & -1 \\ 1 & 0 \end{bmatrix}$
180°	Rot_{180}	$\begin{bmatrix} -1 & 0 \\ 0 & -1 \end{bmatrix}$
270°	Rot_{270}	$\begin{bmatrix} 0 & 1 \\ -1 & 0 \end{bmatrix}$

Example 3

ANIMATION Suppose a figure is animated to spin around a certain point. Numerous rotation images would be necessary to make a smooth movement image. If the image has key points at (1, 1), (−1, 4), (−2, 4), and (−2, 3) and the rotation is about the origin, find the location of these points at the 90°, 180°, and 270° counterclockwise rotations.

First write the vertex matrix. Then multiply it by each rotation matrix.

The vertex matrix is $\begin{bmatrix} 1 & -1 & -2 & -2 \\ 1 & 4 & 4 & 3 \end{bmatrix}$.

Rot_{90} $\begin{bmatrix} 0 & -1 \\ 1 & 0 \end{bmatrix} \cdot \begin{bmatrix} 1 & -1 & -2 & -2 \\ 1 & 4 & 4 & 3 \end{bmatrix} = \begin{bmatrix} -1 & -4 & -4 & -3 \\ 1 & -1 & -2 & -2 \end{bmatrix}$

Rot_{180} $\begin{bmatrix} -1 & 0 \\ 0 & -1 \end{bmatrix} \cdot \begin{bmatrix} 1 & -1 & -2 & -2 \\ 1 & 4 & 4 & 3 \end{bmatrix} = \begin{bmatrix} -1 & 1 & 2 & 2 \\ -1 & -4 & -4 & -3 \end{bmatrix}$

Rot_{180} $\begin{bmatrix} 0 & 1 \\ -1 & 0 \end{bmatrix} \cdot \begin{bmatrix} 1 & -1 & -2 & -2 \\ 1 & 4 & 4 & 3 \end{bmatrix} = \begin{bmatrix} 1 & 4 & 4 & 3 \\ -1 & 1 & 2 & 2 \end{bmatrix}$

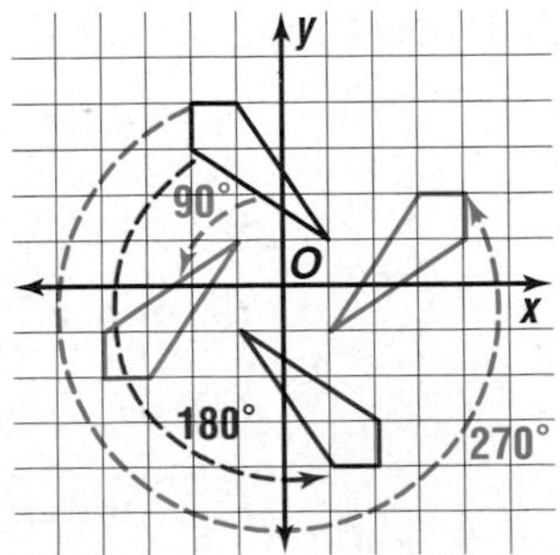

All of the transformations we have discussed have maintained the shape and size of the figure. However, a dilation changes the size of the figure. The dilated figure is similar to the original figure. Dilations using the origin as a center of projection can be achieved by multiplying the vertex matrix by the scale factor needed for the dilation. *All dilations in this lesson are with respect to the origin.*

Example 4 **A trapezoid has vertices at $L(-4, 1)$, $M(1, 4)$, $N(7, 0)$, and $P(-3, -6)$. Find the coordinates of the dilated trapezoid $L'M'N'P'$ for a scale factor of 0.5. Describe the dilation.**

First write the coordinates of the vertices as a matrix. Then do a scalar multiplication using the scale factor.

$$0.5\begin{bmatrix} -4 & 1 & 7 & -3 \\ 1 & 4 & 0 & -6 \end{bmatrix} = \begin{bmatrix} -2 & 0.5 & 3.5 & -1.5 \\ 0.5 & 2 & 0 & -3 \end{bmatrix}$$

The vertices of the image are $L'(-2, 0.5)$, $M'(0.5, 2)$, $N'(3.5, 0)$, and $P'(-1.5, -3)$.

The image has sides that are half the length of the original figure.

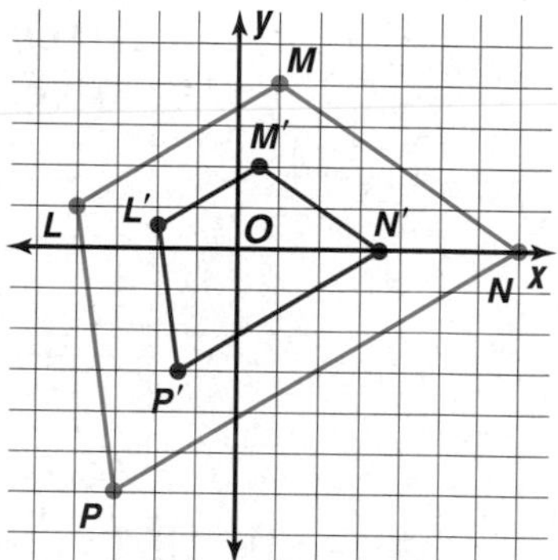

CHECK FOR UNDERSTANDING

Communicating Mathematics

Read and study the lesson to answer each question.

1. **Name** all the transformations described in this lesson. Tell how the pre-image and image are related in each type of transformation.
2. **Explain** how 90°, 180°, and 270° counterclockwise rotations correspond to clockwise rotations.
3. ***Math Journal*** **Describe** a way that you can remember the elements of the reflection matrices if you forget where the 1s, −1s, and 0s belong.
4. **Match** each matrix with the phrase that best describes its type.

a. $\begin{bmatrix} -1 & -1 \\ 1 & 1 \end{bmatrix}$

b. $\begin{bmatrix} -1 & 0 \\ 0 & 1 \end{bmatrix}$

c. $\begin{bmatrix} 0 & 1 \\ 1 & 0 \end{bmatrix}$

d. $\begin{bmatrix} 0 & -1 \\ 1 & 0 \end{bmatrix}$

(1) dilation of scale factor 2

(2) reflection over the y-axis

(3) reflection over the line $y = x$

(4) rotation of 90° counterclockwise about the origin

(5) rotation of 180° about the origin

(6) translation 1 unit left and 1 unit up

Guided Practice **Use matrices to perform each transformation. Then graph the pre-image and the image on the same coordinate grid.**

5. Triangle JKL has vertices $J(-2, 5)$, $K(1, 3)$, and $L(0, -2)$. Use scalar multiplication to find the coordinates of the triangle after a dilation of scale factor 1.5.

6. Square $ABCD$ has vertices $A(-1, 3)$, $B(3, 3)$, $C(3, -1)$, and $D(-1, -1)$. Find the coordinates of the square after a translation of 1 unit left and 2 units down.

7. Square $ABCD$ has vertices at $(-1, 2)$, $(-4, 1)$, $(-3, -2)$, and $(0, -1)$. Find the image of the square after a reflection over the y-axis.

8. Triangle PQR is represented by the matrix $\begin{bmatrix} 3 & -1 & 1 \\ 2 & 4 & -2 \end{bmatrix}$. Find the image of the triangle after a rotation of 270° counterclockwise about the origin.

9. Find the image of $\triangle LMN$ after $Rot_{180} \circ R_{y\text{-axis}}$ if the vertices are $L(-6, 4)$, $M(-3, 2)$, and $N(-1, -2)$.

10. **Physics** The wind was blowing quite strongly when Jenny was baby-sitting. She was outside with the children, and they were throwing their large plastic ball up into the air. The wind blew the ball so that it landed approximately 3 feet east and 4 feet north of where it was thrown into the air.

   a. Make a drawing to demonstrate the original location of the ball and the translation of the ball to its landing spot.

   b. If $\begin{bmatrix} x \\ y \end{bmatrix}$ represents the original location of the ball, write a matrix that represents the location of the translated ball.

EXERCISES

Practice **Use scalar multiplication to determine the coordinates of the vertices of each dilated figure. Then graph the pre-image and the image on the same coordinate grid.**

11. triangle with vertices $A(1, 1)$, $B(1, 4)$, and $C(5, 1)$; scale factor 3

12. triangle with vertices $X(0, 8)$, $Y(-5, 9)$, and $Z(-3, 2)$; scale factor $\frac{3}{4}$

13. quadrilateral $PQRS$ with vertex matrix $\begin{bmatrix} -3 & -2 & 1 & 4 \\ 0 & 2 & 3 & 2 \end{bmatrix}$; scale factor 2

14. Graph a square with vertices $A(-1, 0)$, $B(0, 1)$, $C(1, 0)$, and $D(0, -1)$ on two separate coordinate planes.

   a. On one of the coordinate planes, graph the dilation of square $ABCD$ after a dilation of scale factor 2. Label it $A'B'C'D'$. Then graph a dilation of $A'B'C'D'$ after a scale factor of 3.

   b. On the second coordinate plane, graph the dilation of square $ABCD$ after a dilation of scale factor 3. Label it $A'B'C'D'$. Then graph a dilation of $A'B'C'D'$ after a scale factor of 2.

   c. Compare the results of parts **a** and **b**. Describe what you observe.

Use matrices to determine the coordinates of the vertices of each translated figure. Then graph the pre-image and the image on the same coordinate grid.

15. triangle WXY with vertex matrix $\begin{bmatrix} -2 & 1 & 3 \\ 0 & 5 & -1 \end{bmatrix}$ translated 3 units right and 2 units down

16. quadrilateral with vertices $O(0, 0)$, $P(1, 5)$, $Q(4, 7)$, and $R(3, 2)$ translated 2 units left and 1 unit down

17. square $CDEF$ translated 3 units right and 4 units up if the vertices are $C(-3, 1)$, $D(1, 5)$, $E(5, 1)$, and $F(1, -3)$

18. Graph $\triangle FGH$ with vertices $F(4, 1)$, $G(0, 3)$, and $H(2, -1)$.
    a. Graph the image of $\triangle FGH$ after a translation of 6 units left and 2 units down. Label the image $\triangle F'G'H'$.
    b. Then translate $\triangle F'G'H'$ 1 unit right and 5 units up. Label this image $\triangle F''G''H''$.
    c. What translation would move $\triangle FGH$ to $\triangle F''G''H''$ directly?

Use matrices to determine the coordinates of the vertices of each reflected figure. Then graph the pre-image and the image on the same coordinate grid.

19. $\triangle ABC$ with vertices $A(-1, -2)$, $B(0, -4)$, and $C(2, -3)$ reflected over the x-axis

20. $R_{y\text{-axis}}$ for a rectangle with vertices $D(2, 4)$, $E(6, 2)$, $F(3, -4)$, and $G(-1, -2)$

21. a trapezoid with vertices $H(-1, -2)$, $I(-3, 1)$, $J(-1, 5)$, and $K(2, 4)$ for a reflection over the line $y = x$

Use matrices to determine the coordinates of the vertices of each rotated figure. Then graph the pre-image and the image on the same coordinate grid.

22. Rot_{90} for $\triangle LMN$ with vertices $L(1, -1)$, $M(2, -2)$, and $N(3, -1)$

23. square with vertices $O(0, 0)$, $P(4, 0)$, $Q(4, 4)$, $R(0, 4)$ rotated 180°

24. pentagon $STUVW$ with vertices $S(-1, -2)$, $T(-3, -1)$, $U(-5, -2)$, $V(-4, -4)$, and $W(-2, -4)$ rotated 270° counterclockwise

25. **Proof** Suppose $\triangle ABC$ has vertices $A(1, 3)$, $B(-2, -1)$, and $C(-1, -3)$. Use each result of the given transformation of $\triangle ABC$ to show how the matrix for that reflection or rotation is derived.
    a. $\begin{bmatrix} 1 & -2 & -1 \\ -3 & 1 & 3 \end{bmatrix}$ under $R_{x\text{-axis}}$
    b. $\begin{bmatrix} -1 & 2 & 1 \\ 3 & -1 & -3 \end{bmatrix}$ under $R_{y\text{-axis}}$
    c. $\begin{bmatrix} 3 & -1 & -3 \\ 1 & -2 & -1 \end{bmatrix}$ under $R_{y=x}$
    d. $\begin{bmatrix} -3 & 1 & 3 \\ 1 & -2 & -1 \end{bmatrix}$ under Rot_{90}
    e. $\begin{bmatrix} -1 & 2 & 1 \\ -3 & 1 & 3 \end{bmatrix}$ under Rot_{180}
    f. $\begin{bmatrix} 3 & -1 & -3 \\ -1 & 2 & 1 \end{bmatrix}$ under Rot_{270}

Given $\triangle JKL$ with vertices $J(-6, 4)$, $K(-3, 2)$, and $L(-1, -2)$. Find the coordinates of each composite transformation. Then graph the pre-image and the image on the same coordinate grid.

26. rotation of 180° followed by a translation 2 units left 5 units up

27. $R_{y\text{-axis}} \circ R_{x\text{-axis}}$

28. $Rot_{90} \circ R_{y\text{-axis}}$

Applications and Problem Solving

29. **Games** Each of the pieces on the chess board has a specific number of spaces and direction it can move. Research the game of chess and describe the possible movements for each piece as a translation matrix.

a. bishop b. knight c. king

30. **Critical Thinking** Show that a dilation with scale factor of -1 is the same result as Rot_{180}.

31. **Entertainment** The Ferris Wheel first appeared at the 1893 Chicago Exposition. Its axle was 45 feet long. Spokes radiated from it that supported 36 wooden cars, which could hold 60 people each. The diameter of the wheel itself was 250 feet. Suppose the axle was located at the origin. Find the coordinates of the car located at the loading platform. Then find the location of the car at the 90° counterclockwise, 180°, and 270° counterclockwise rotation positions.

32. **Critical Thinking** $R_{y\text{-axis}}$ gives a matrix for reflecting a figure over the y-axis. Do you think a matrix that would represent a reflection over the line $y = 4$ exists? If so, make a conjecture and verify it.

33. **Animation** Divide two sheets of grid paper into fourths by halving the length and width of the paper. Draw a simple figure on one of the pieces. On another piece, draw the figure dilated with a scale factor of 1.25. On a third piece, draw the original figure dilated with a scale factor of 1.5. On the fourth piece, draw the original figure dilated with a scale factor of 1.75. Continue dilating the original figure on each of the remaining pieces by an increase of 0.25 in scale factor each time. Put the pieces of paper in order and flip through them. What type of motion does the result of these repeated dilations animate?

34. **Critical Thinking** Write the vertex matrix for the figure graphed below.

a. Make a conjecture about the resulting figure if you multiply the vertex matrix by $\begin{bmatrix} 3 & 0 \\ 0 & 2 \end{bmatrix}$.

b. Copy the figure on grid paper and graph the resulting vertex matrix after the multiplication described.

c. How does the result compare with your conjecture? This is often called a *shear*. Why do you think it has this name?

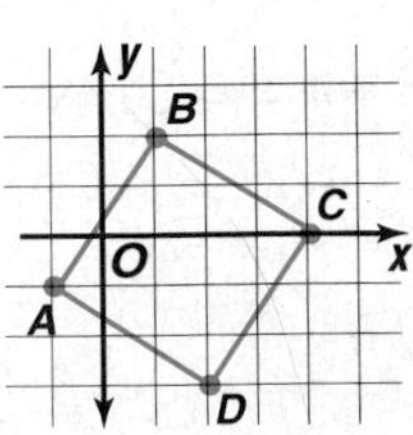

Mixed Review

35. Find $A + B$ if $A = \begin{bmatrix} 3 & 8 \\ -2 & 4 \end{bmatrix}$ and $B = \begin{bmatrix} 1 & 5 \\ -2 & 8 \end{bmatrix}$. *(Lesson 2-3)*

36. Solve the system of equations. *(Lesson 2-2)*
$x - 2y = 4.6$
$y - z = -5.6$
$x + y + z = 1.8$

37. **Sales** The Grandview Library holds an annual book sale to raise funds and dispose of excess books. One customer bought 4 hardback books and 7 paperbacks for \$5.75. The next customer paid \$4.25 for 3 hardbacks and 5 paperbacks. What are the prices for hardbacks and for paperbacks? *(Lesson 2-1)*

38. Of (0, 0), (3, 2), (−4, 2), or (−2, 4), which satisfy $x + y \geq 3$? *(Lesson 1-8)*

39. Write the standard form of the equation of the line that is parallel to the graph of $y = 4x - 8$ and passes through (−2, 1). *(Lesson 1-5)*

40. Write the slope-intercept form of the equation of the line that passes through the point at (1, 6) and has a slope of 2. *(Lesson 1-3)*

41. If $f(x) = x^3$ and $g(x) = x^2 - 3x + 7$, find $(f \cdot g)(x)$ and $\left(\frac{f}{g}\right)(x)$. *(Lesson 1-2)*

42. **SAT/ACT Practice** If $2x + y = 12$ and $x + 2y = -6$, find the value of $2x + 2y$.
A 0 **B** 4 **C** 8 **D** 12 **E** 14

MID-CHAPTER QUIZ

1. Use graphing to solve the system of equations $\frac{1}{2}x + 5y = 17$ and $3x + 2y = 18$. (Lesson 2-1)

2. Solve the system of equations $4x + y = 8$ and $6x - 2y = -9$ algebraically. (Lesson 2-1)

3. **Sales** HomeMade Toys manufactures solid pine trucks and cars and usually sells four times as many trucks as cars. The net profit from each truck is \$6 and from each car is \$5. If the company wants a total profit of \$29,000, how many trucks and cars should they sell? (Lesson 2-1)

Solve each system of equations. (Lesson 2-2)

4. $2x + y + 4z = 13$
$3x - y - 2z = -1$
$4x + 2y + z = 19$

5. $x + y = 1$
$2x - y = -2$
$4x + y + z = 8$

6. Find the values of x and y for which the matrix equation $\begin{bmatrix} y - 3 \\ y \end{bmatrix} = \begin{bmatrix} x \\ 2x \end{bmatrix}$ is true. (Lesson 2-3)

Use matrices *A* and *B* to find each of the following. If the matrix does not exist, write *impossible*. (Lesson 2-3)

$A = \begin{bmatrix} 3 & 5 & -7 \\ -1 & 0 & 4 \end{bmatrix}$ $B = \begin{bmatrix} -2 & 8 & 6 \\ 5 & -9 & 10 \end{bmatrix}$

7. $A + B$
8. BA
9. $B - 3A$

10. What is the result of reflecting a triangle with vertices at $A(a, d)$, $B(b, e)$, and $C(c, f)$ over the x-axis and then reflecting the image back over the x-axis? Use matrices to justify your answer. (Lesson 2-4)

Extra Practice See p. A28.

HISTORY of MATHEMATICS

MATRICES

Computers use matrices to solve many types of mathematical problems, but matrices have been around for a long time.

Early Evidence Around 300 B.C., as evidenced by clay tablets found by archaeologists, the Babylonians solved problems that now can be solved by using a system of linear equations. However, the exact method of solution used by the Babylonians has not been determined.

About 100 B.C.–50 B.C., in ancient China, Chapter 8 of the work *Jiuzhang suanshu (Nine Chapters of the Mathematical Art)* presented a similar problem and showed a solution on a counting board that resembles an augmented coefficient matrix.

There are three types of corn, of which three bundles of the first type, two of the second, and one of the third make 39 measures. Two of the first, three of the second, and one of the third make 34 measures. And one of the first, two of the second, and three of the third make 26 measures. How many measures of grain are contained in one bundle of each type?

Author's table

1	2	3
2	3	2
3	1	1
26	34	39

The Chinese author goes on to detail how each column can be operated on to determine the solution. This method was later credited to **Carl Friedrich Gauss.**

The Renaissance The concept of the determinant of a matrix, which you will learn about in the next lesson, appeared in Europe and Japan at almost identical times. However, **Seki** of Japan wrote about it first in 1683 with his *Method of Solving the Dissimulated Problems*. Seki's work contained matrices written in table form like those found in the ancient Chinese writings. Seki developed the pattern for determinants for 2×2, 3×3, 4×4, and 5×5 matrices and used them to solve equations, but not systems of linear equations.

In the same year in Hanover (now Germany), **Gottfried Leibniz** wrote to **Guillaume De l'Hôpital** who lived in Paris, France, about a method he had for solving a system of equations in the form $C + Ax + By = 0$. His method later became known as **Cramer's Rule.**

Margaret H. Wright

Modern Era In 1850, the word *matrix* was first used by **James Joseph Sylvester** to describe the tabular array of numbers. Sylvester actually was a lawyer who studied mathematics as a hobby. He shared his interests with **Arthur Cayley,** who is credited with the first published reference to the inverse of a matrix.

Today, computer experts like **Margaret H. Wright** use matrices to solve problems that involve thousands of variables. In her job as Distinguished Member of Technical Staff at a telecommunications company, she applies linear algebra for the solution of real-world problems.

ACTIVITIES

1. Solve the problem from the *Jiuzhang suanshu* by using a system of equations.
2. Research the types of problems solved by the Babylonians using a system of equations.
3. *interNET* CONNECTION Find out more about the personalities referenced in this article and others who contributed to the history of matrices. Visit **www.amc.glencoe.com**

Determinants and Multiplicative Inverses of Matrices

OBJECTIVES
- Evaluate determinants.
- Find inverses of matrices.
- Solve systems of equations by using inverses of matrices.

Real World Application

INVESTMENTS Marshall plans to invest $10,500 into two different bonds in order to spread out his risk. The first bond has an annual return of 10%, and the second bond has an annual return of 6%. If Marshall expects an 8.5% return from the two bonds, how much should he invest into each bond? *This problem will be solved in Example 5.*

This situation can be described by a system of equations represented by a matrix. You can solve the system by writing and solving a matrix equation.

The term *determinant is often used to mean the value of the determinant.*

Each square matrix has a **determinant.** The determinant of $\begin{bmatrix} 8 & 4 \\ 7 & 6 \end{bmatrix}$ is a number denoted by $\begin{vmatrix} 8 & 4 \\ 7 & 6 \end{vmatrix}$ or det $\begin{bmatrix} 8 & 4 \\ 7 & 6 \end{bmatrix}$. The value of a second-order determinant is defined as follows. *A matrix that has a nonzero determinant is called nonsingular.*

Second-Order Determinant The value of det $\begin{bmatrix} a_1 & b_1 \\ a_2 & b_2 \end{bmatrix}$, or $\begin{vmatrix} a_1 & b_1 \\ a_2 & b_2 \end{vmatrix}$, is $a_1b_2 - a_2b_1$.

Example 1 **Find the value of $\begin{vmatrix} 8 & 4 \\ 7 & 6 \end{vmatrix}$.**

$$\begin{vmatrix} 8 & 4 \\ 7 & 6 \end{vmatrix} = 8(6) - 7(4) \text{ or } 20$$

The **minor** of an element of any nth-order determinant is a determinant of order $(n - 1)$. This minor can be found by deleting the row and column containing the element.

$$\begin{vmatrix} a_1 & b_1 & c_1 \\ a_2 & b_2 & c_2 \\ a_3 & b_3 & c_3 \end{vmatrix} \quad \textit{The minor of } a_1 \textit{ is } \begin{vmatrix} b_2 & c_2 \\ b_3 & c_3 \end{vmatrix}.$$

One method of evaluating an nth-order determinant is expanding the determinant by minors. The first step is choosing a row, any row, in the matrix. At each position in the row, multiply the *element* times its *minor* times its *position sign,* and then add the results together for the whole row. The position signs in a matrix are alternating positives and negatives, beginning with a positive in the first row, first column.

$$\begin{bmatrix} + & - & + & \cdots \\ - & + & - & \cdots \\ + & - & + & \cdots \\ \vdots & \vdots & \vdots & \vdots \end{bmatrix}$$

Third-Order Determinant	$\begin{vmatrix} a_1 & b_1 & c_1 \\ a_2 & b_2 & c_2 \\ a_3 & b_3 & c_3 \end{vmatrix} = a_1 \begin{vmatrix} b_2 & c_2 \\ b_3 & c_3 \end{vmatrix} - b_1 \begin{vmatrix} a_2 & c_2 \\ a_3 & c_3 \end{vmatrix} + c_1 \begin{vmatrix} a_2 & b_2 \\ a_3 & b_3 \end{vmatrix}$

Example 2 **Find the value of** $\begin{vmatrix} -4 & -6 & 2 \\ 5 & -1 & 3 \\ -2 & 4 & -3 \end{vmatrix}$.

Graphing Calculator Tip

You can use the **det(** option in the **MATH** listings of the **MATRX** menu to find a determinant.

$$\begin{vmatrix} -4 & -6 & 2 \\ 5 & -1 & 3 \\ -2 & 4 & -3 \end{vmatrix} = -4 \begin{vmatrix} -1 & 3 \\ 4 & -3 \end{vmatrix} - (-6) \begin{vmatrix} 5 & 3 \\ -2 & -3 \end{vmatrix} + 2 \begin{vmatrix} 5 & -1 \\ -2 & 4 \end{vmatrix}$$

$$= -4(-9) + 6(-9) + 2(18)$$

$$= 18$$

For any $m \times m$ matrix, the identity matrix, I, must also be an $m \times m$ matrix.

The **identity matrix for multiplication** for any square matrix A is the matrix I, such that $IA = A$ and $AI = A$. A second-order matrix can be represented by $\begin{bmatrix} a_1 & b_1 \\ a_2 & b_2 \end{bmatrix}$. Since $\begin{bmatrix} a_1 & b_1 \\ a_2 & b_2 \end{bmatrix} \cdot \begin{bmatrix} 1 & 0 \\ 0 & 1 \end{bmatrix} = \begin{bmatrix} 1 & 0 \\ 0 & 1 \end{bmatrix} \cdot \begin{bmatrix} a_1 & b_1 \\ a_2 & b_2 \end{bmatrix} = \begin{bmatrix} a_1 & b_1 \\ a_2 & b_2 \end{bmatrix}$, the matrix $\begin{bmatrix} 1 & 0 \\ 0 & 1 \end{bmatrix}$ is the identity matrix for multiplication for any second-order matrix.

Identity Matrix for Multiplication	The identity matrix of nth order, I_n, is the square matrix whose elements in the main diagonal, from upper left to lower right, are 1s, while all other elements are 0s.

Multiplicative inverses exist for some matrices. Suppose A is equal to $\begin{bmatrix} a_1 & b_1 \\ a_2 & b_2 \end{bmatrix}$, a nonzero matrix of second order.

The term inverse matrix generally implies the multiplicative inverse of a matrix.

The **inverse matrix** A^{-1} can be designated as $\begin{bmatrix} x_1 & y_1 \\ x_2 & y_2 \end{bmatrix}$. The product of a matrix A and its inverse A^{-1} must equal the identity matrix, I, for multiplication.

$$\begin{bmatrix} a_1 & b_1 \\ a_2 & b_2 \end{bmatrix} \begin{bmatrix} x_1 & y_1 \\ x_2 & y_2 \end{bmatrix} = \begin{bmatrix} 1 & 0 \\ 0 & 1 \end{bmatrix}$$

$$\begin{bmatrix} a_1x_1 + b_1x_2 & a_1y_1 + b_1y_2 \\ a_2x_1 + b_2x_2 & a_2y_1 + b_2y_2 \end{bmatrix} = \begin{bmatrix} 1 & 0 \\ 0 & 1 \end{bmatrix}$$

From the previous matrix equation, two systems of linear equations can be written as follows.

$$a_1x_1 + b_1x_2 = 1 \qquad a_1y_1 + b_1y_2 = 0$$

$$a_2x_1 + b_2x_2 = 0 \qquad a_2y_1 + b_2y_2 = 1$$

By solving each system of equations, values for x_1, x_2, y_1, and y_2 can be obtained.

$$x_1 = \frac{b_2}{a_1b_2 - a_2b_1} \qquad y_1 = \frac{-b_1}{a_1b_2 - a_2b_1}$$

$$x_2 = \frac{-a_2}{a_1b_2 - a_2b_1} \qquad y_2 = \frac{a_1}{a_1b_2 - a_2b_1}$$

If a matrix A has a determinant of 0 then A^{-1} does not exist.

The denominator $a_1b_2 - a_2b_1$ is equal to the determinant of A. If the determinant of A is not equal to 0, the inverse exists and can be defined as follows.

Inverse of a Second-Order Matrix

If $A = \begin{bmatrix} a_1 & b_1 \\ a_2 & b_2 \end{bmatrix}$ and $\begin{vmatrix} a_1 & b_1 \\ a_2 & b_2 \end{vmatrix} \neq 0$, then $A^{-1} = \dfrac{1}{\begin{vmatrix} a_1 & b_1 \\ a_2 & b_2 \end{vmatrix}} \begin{bmatrix} b_2 & -b_1 \\ -a_2 & a_1 \end{bmatrix}$.

$A \cdot A^{-1} = A^{-1} \cdot A = I$, where I is the identity matrix.

Example 3 **Find the inverse of the matrix $\begin{bmatrix} 2 & -3 \\ 4 & 4 \end{bmatrix}$.**

Graphing Calculator Appendix

For keystroke instruction on how to find the inverse of a matrix, see pages A16-A17.

First, find the determinant of $\begin{bmatrix} 2 & -3 \\ 4 & 4 \end{bmatrix}$.

$\begin{vmatrix} 2 & -3 \\ 4 & 4 \end{vmatrix} = 2(4) - 4(-3)$ or 20

The inverse is $\frac{1}{20}\begin{bmatrix} 4 & 3 \\ -4 & 2 \end{bmatrix}$ or $\begin{bmatrix} \frac{1}{5} & \frac{3}{20} \\ -\frac{1}{5} & \frac{1}{10} \end{bmatrix}$. *Check to see if $A \cdot A^{-1} = A^{-1} \cdot A = 1$.*

Just as you can use the multiplicative inverse of 3 to solve $3x = -27$, you can use a matrix inverse to solve a matrix equation in the form $AX = B$. To solve this equation for X, multiply each side of the equation by the inverse of A. *When you multiply each side of a matrix equation by the same number or matrix, be sure to place the number or matrix on the left or on the right on each side of the equation to maintain equality.*

$AX = B$

$A^{-1}AX = A^{-1}B$ *Multiply each side of the equation by A^{-1}.*

$IX = A^{-1}B$ *$A^{-1} \cdot A = I$*

$X = A^{-1}B$ *$IX = X$*

Example 4 **Solve the system of equations by using matrix equations.**

$2x + 3y = -17$

$x - y = 4$

Write the system as a matrix equation.

$$\begin{bmatrix} 2 & 3 \\ 1 & -1 \end{bmatrix} \cdot \begin{bmatrix} x \\ y \end{bmatrix} = \begin{bmatrix} -17 \\ 4 \end{bmatrix}$$

To solve the matrix equation, first find the inverse of the coefficient matrix.

$$\frac{1}{\begin{vmatrix} 2 & 3 \\ 1 & -1 \end{vmatrix}} \begin{bmatrix} -1 & -3 \\ -1 & 2 \end{bmatrix} = -\frac{1}{5}\begin{bmatrix} -1 & -3 \\ -1 & 2 \end{bmatrix}$$

$\begin{vmatrix} 2 & 3 \\ 1 & -1 \end{vmatrix} = 2(-1) - (1)(3)$ or -5

Now multiply each side of the matrix equation by the inverse and solve.

$$-\frac{1}{5}\begin{bmatrix}-1 & -3\\-1 & 2\end{bmatrix}\cdot\begin{bmatrix}2 & 3\\1 & -1\end{bmatrix}\cdot\begin{bmatrix}x\\y\end{bmatrix} = -\frac{1}{5}\begin{bmatrix}-1 & -3\\-1 & 2\end{bmatrix}\cdot\begin{bmatrix}-17\\4\end{bmatrix}$$

$$\begin{bmatrix}x\\y\end{bmatrix} = \begin{bmatrix}-1\\-5\end{bmatrix}$$

The solution is $(-1, -5)$.

Example 5

INVESTMENTS Refer to the application at the beginning of the lesson. How should Marshall divide his \$10,500 investment between the bond with a 10% annual return and a bond with a 6% annual return so that he has a combined annual return on his investments of 8.5%?

First, let x represent the amount to invest in the bond with an annual return of 10%, and let y represent the amount to invest in the bond with a 6% annual return. So, $x + y = 10{,}500$ since Marshall is investing \$10,500.

Write an equation in standard form that represents the amounts invested in both bonds and the combined annual return of 8.5%. That is, the amount of interest earned from the two bonds is the same as if the total were invested in a bond that earns 8.5%.

$10\%x + 6\%y = 8.5\%(x + y)$	*Interest on 10% bond = 10%x*
$0.10x + 0.06y = 0.085(x + y)$	*Interest on 6% bond = 6%y*
$0.10x + 0.06y = 0.085x + 0.085y$	*Distributive Property*
$0.015x - 0.025y = 0$	
$3x - 5y = 0$	*Multiply by 200 to simplify the coefficients.*

Now solve the system of equations $x + y = 10{,}500$ and $3x - 5y = 0$. Write the system as a matrix equation and solve.

$$\begin{aligned}x + y &= 10{,}500\\3x - 5y &= 0\end{aligned} \longrightarrow \begin{bmatrix}1 & 1\\3 & -5\end{bmatrix}\cdot\begin{bmatrix}x\\y\end{bmatrix} = \begin{bmatrix}10{,}500\\0\end{bmatrix}$$

Multiply each side of the equation by the inverse of the coefficient matrix.

$$-\frac{1}{8}\begin{bmatrix}-5 & -1\\-3 & 1\end{bmatrix}\cdot\begin{bmatrix}1 & 1\\3 & -5\end{bmatrix}\cdot\begin{bmatrix}x\\y\end{bmatrix} = -\frac{1}{8}\begin{bmatrix}-5 & -1\\-3 & 1\end{bmatrix}\cdot\begin{bmatrix}10{,}500\\0\end{bmatrix}$$

$$\begin{bmatrix}x\\y\end{bmatrix} = \begin{bmatrix}6562.5\\3937.5\end{bmatrix}$$

The solution is $(6562.5, 3937.5)$. So, Marshall should invest \$6562.50 in the bond with a 10% annual return and \$3937.50 in the bond with a 6% annual return.

CHECK FOR UNDERSTANDING

Read and study the lesson to answer each question.

1. **Describe** the types of matrices that are considered to be nonsingular.
2. **Explain** why the matrix $\begin{bmatrix} 3 & 2 & 0 \\ 4 & -3 & 5 \end{bmatrix}$ does not have a determinant. Give another example of a matrix that does not have a determinant.
3. **Describe** the identity matrix under multiplication for a fourth-order matrix.
4. **Write** an explanation as to how you can decide whether the system of equations, $ax + cy = e$ and $bx + dy = f$, has a solution.

Guided Practice

Find the value of each determinant.

5. $\begin{vmatrix} 4 & -1 \\ -2 & 3 \end{vmatrix}$

6. $\begin{vmatrix} 12 & -26 \\ -15 & 32 \end{vmatrix}$

7. $\begin{vmatrix} 4 & 1 & 0 \\ 5 & -15 & -1 \\ -2 & 10 & 7 \end{vmatrix}$

8. $\begin{vmatrix} 6 & 4 & -1 \\ 0 & 3 & 3 \\ -9 & 0 & 0 \end{vmatrix}$

Find the inverse of each matrix, if it exists.

9. $\begin{bmatrix} -2 & 3 \\ 5 & 7 \end{bmatrix}$

10. $\begin{bmatrix} 4 & 6 \\ 6 & 9 \end{bmatrix}$

Solve each system of equations by using a matrix equation.

11. $5x + 4y = -3$
$-3x - 5y = -24$

12. $6x - 3y = 63$
$5x - 9y = 85$

13. **Metallurgy** Aluminum alloy is used in airplane construction because it is strong and lightweight. A metallurgist wants to make 20 kilograms of aluminum alloy with 70% aluminum by using two metals with 55% and 80% aluminum content. How much of each metal should she use?

EXERCISES

Practice

Find the value of each determinant.

14. $\begin{vmatrix} 3 & 4 \\ 2 & 5 \end{vmatrix}$

15. $\begin{vmatrix} -4 & -1 \\ 0 & -1 \end{vmatrix}$

16. $\begin{vmatrix} 9 & 12 \\ 12 & 16 \end{vmatrix}$

17. $\begin{vmatrix} -2 & 3 \\ -2 & 1 \end{vmatrix}$

18. $\begin{vmatrix} 13 & 7 \\ -5 & -8 \end{vmatrix}$

19. $\begin{vmatrix} -6 & 5 \\ 0 & -8 \end{vmatrix}$

20. $\begin{vmatrix} 4 & -1 & -2 \\ 0 & 2 & 1 \\ 2 & 1 & 3 \end{vmatrix}$

21. $\begin{vmatrix} 2 & -1 & 3 \\ 3 & 0 & -2 \\ 1 & -3 & 0 \end{vmatrix}$

22. $\begin{vmatrix} 8 & 9 & 3 \\ 3 & 5 & 7 \\ -1 & 2 & 4 \end{vmatrix}$

23. $\begin{vmatrix} 4 & 6 & 7 \\ 3 & -2 & -4 \\ 1 & 1 & 1 \end{vmatrix}$

24. $\begin{vmatrix} 25 & 36 & 15 \\ 31 & -12 & -2 \\ 17 & 15 & 9 \end{vmatrix}$

25. $\begin{vmatrix} 1.5 & -3.6 & 2.3 \\ 4.3 & 0.5 & 2.2 \\ -1.6 & 8.2 & 6.6 \end{vmatrix}$

26. Find det A if $A = \begin{bmatrix} 0 & 1 & -4 \\ 3 & 2 & 3 \\ 8 & -3 & 4 \end{bmatrix}$.

Find the inverse of each matrix, if it exists.

27. $\begin{bmatrix} 2 & -3 \\ -2 & -2 \end{bmatrix}$

28. $\begin{bmatrix} 2 & 0 \\ 1 & 0 \end{bmatrix}$

29. $\begin{bmatrix} 4 & 2 \\ 1 & 2 \end{bmatrix}$

30. $\begin{bmatrix} 6 & 7 \\ -6 & 7 \end{bmatrix}$

31. $\begin{bmatrix} -4 & 6 \\ 8 & -12 \end{bmatrix}$

32. $\begin{bmatrix} 9 & 13 \\ 27 & 36 \end{bmatrix}$

33. What is the inverse of $\begin{bmatrix} \frac{3}{4} & -\frac{1}{8} \\ 5 & \frac{1}{2} \end{bmatrix}$?

Solve each system by using a matrix equation.

34. $4x - y = 1$
$x + 2y = 7$

35. $9x - 6y = 12$
$4x + 6y = -12$

36. $x + 5y = 26$
$3x - 2y = -41$

37. $4x + 8y = 7$
$3x - 3y = 0$

38. $3x - 5y = -24$
$5x + 4y = -3$

39. $9x + 3y = 1$
$5x + y = 1$

Solve each matrix equation. The inverse of the coefficient matrix is given.

40. $\begin{bmatrix} 3 & -2 & 3 \\ 1 & 2 & 2 \\ -2 & 1 & -1 \end{bmatrix} \cdot \begin{bmatrix} x \\ y \\ z \end{bmatrix} = \begin{bmatrix} -4 \\ 0 \\ 1 \end{bmatrix}$, if the inverse is $\frac{1}{9}\begin{bmatrix} -4 & 1 & -10 \\ -3 & 3 & -3 \\ 5 & 1 & 8 \end{bmatrix}$.

41. $\begin{bmatrix} -6 & 5 & 3 \\ 9 & -2 & -1 \\ 3 & 1 & 1 \end{bmatrix} \cdot \begin{bmatrix} x \\ y \\ z \end{bmatrix} = \begin{bmatrix} -9 \\ 5 \\ -1 \end{bmatrix}$, if the inverse is $-\frac{1}{9}\begin{bmatrix} -1 & -2 & 1 \\ -12 & -15 & 21 \\ 15 & 21 & -33 \end{bmatrix}$.

Graphing Calculator

Use a graphing calculator to find the value of each determinant.

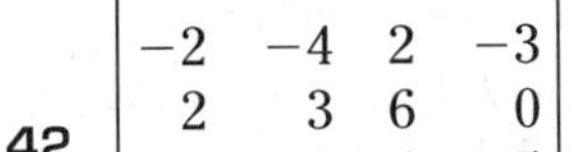

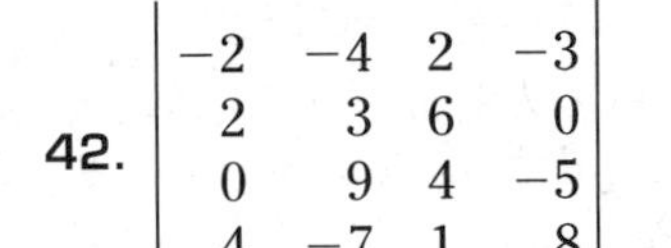

42. $\begin{vmatrix} -2 & -4 & 2 & -3 \\ 2 & 3 & 6 & 0 \\ 0 & 9 & 4 & -5 \\ 4 & -7 & 1 & 8 \end{vmatrix}$

43. $\begin{vmatrix} 2 & -9 & 1 & 8 & 4 \\ -10 & -1 & 2 & 7 & 0 \\ 0 & 4 & -6 & 1 & -8 \\ 6 & -14 & 11 & 0 & 3 \\ 5 & 1 & -3 & 2 & -1 \end{vmatrix}$

Use the algebraic methods you learned in this lesson and a graphing calculator to solve each system of equations.

44. $0.3x + 0.5y = 4.74$
$12x - 6.5y = -1.2$

45. $x - 2y + z = 7$
$6x + 2y - 2z = 4$
$4x + 6y + 4z = 14$

Applications and Problem Solving

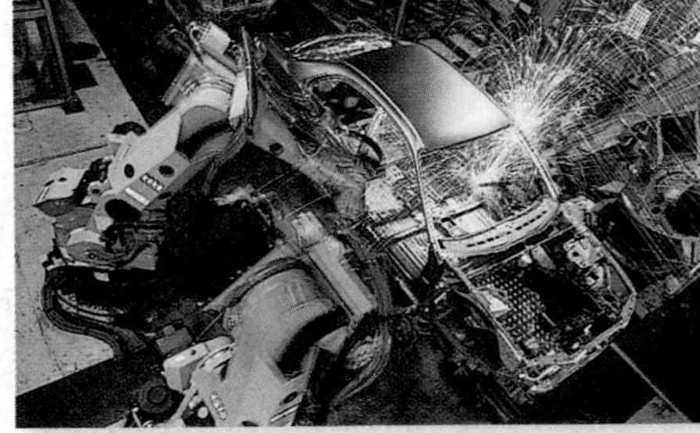

46. Industry The Flat Rock auto assembly plant in Detroit, Michigan, produces three different makes of automobiles. In 1994 and 1995, the plant constructed a total of 390,000 cars. If 90,000 more cars were made in 1994 than in 1995, how many cars were made in each year?

47. Critical Thinking Demonstrate that the expression for A^{-1} is the multiplicative inverse of A for any nonsingular second-order matrix.

48. Chemistry How many gallons of 10% alcohol solution and 25% alcohol solution should be combined to make 12 gallons of a 15% alcohol solution?

49. Critical Thinking If $A = \begin{bmatrix} a & b \\ c & d \end{bmatrix}$, does $(A^2)^{-1} = (A^{-1})^2$? Explain.

50. Geometry The area of a triangle with vertices at (a, b), (c, d), and (e, f) can be determined using the equation $A = \frac{1}{2}\begin{vmatrix} a & b & 1 \\ c & d & 1 \\ e & f & 1 \end{vmatrix}$. What is the area of a triangle with vertices at $(1, -3)$, $(0, 4)$, and $(3, 0)$? (*Hint:* You may need to use the absolute value of the determinant to avoid a negative area.)

51. Retail Suppose that on the first day of a sale, a store sold 38 complete computer systems and 53 printers. During the second day, 22 complete systems and 44 printers were sold. On day three of the sale, the store sold 21 systems and 26 printers. Total sales for these items for the three days were \$49,109, \$31,614, and \$26,353 respectively. What was the unit cost of each of these two selected items?

52. Education The following type of problem often appears on placement tests or college entrance exams.

Jessi has a total of 179 points on her last two history tests. The second test score is a 7-point improvement from the first score. What are her scores for the two tests?

Mixed Review

53. Geometry The vertices of a square are $H(8, 5)$, $I(4, 1)$, $J(0, 5)$, and $K(4, 9)$. Use matrices to determine the coordinates of the square translated 3 units left and 4 units up. *(Lesson 2-4)*

54. Multiply $\begin{bmatrix} 8 & -7 \\ -4 & 0 \end{bmatrix}$ by $\frac{3}{4}$. *(Lesson 2-3)*

55. Solve the system $x - 3y + 2z = 6$, $4x + y - z = 8$, and $-7x - 5y + 4z = -10$. *(Lesson 2-2)*

56. Graph $g(x) = -2[\![x + 5]\!]$. *(Lesson 1-7)*

57. Write the standard form of the equation of the line that is perpendicular to $y = -2x + 5$ and passes through the point at $(2, 5)$. *(Lesson 1-5)*

58. Write the point-slope form of the equation of the line that passes through the points at $(1, 5)$ and $(2, 3)$. Then write the equation in slope-intercept form. *(Lesson 1-4)*

59. Safety In 1990, the Americans with Disabilities Act (ADA) went into effect. This act made provisions for public places to be accessible to all individuals, regardless of their physical challenges. One of the provisions of the ADA is that ramps should not be steeper than a rise of 1 foot for every 12 feet of horizontal distance. *(Lesson 1-3)*

a. What is the slope of such a ramp?

b. What would be the maximum height of a ramp 18 feet long?

60. Find $[f \circ g](x)$ and $[g \circ f](x)$ if $f(x) = x^2 + 3x + 2$ and $g(x) = x - 1$. *(Lesson 1-2)*

Extra Practice See p. A29.

61. Determine if the set of points whose coordinates are (2, 3), (−3, 4), (6, 3), (2, 4), and (−3, 3) represent a function. Explain. *(Lesson 1-1)*

62. **SAT Practice Quantitative Comparison**

A if the quantity in Column A is greater.
B if the quantity in Column B is greater.
C if the two quantities are equal.
D if the relationship cannot be determined from the information given.

The radius of circle E is 3.
Point E is the midpoint of one side of equilateral triangle ACF.
Square $ABCD$ is inscribed in circle E.

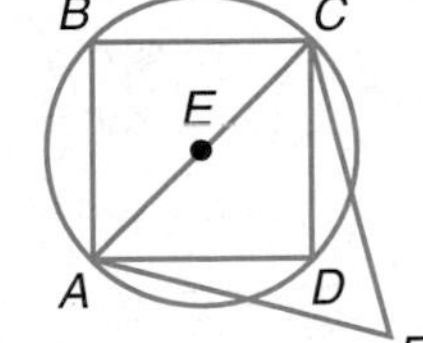

Column A	Column B
the perimeter of square $ABCD$	the perimeter of triangle ACF

CAREER CHOICES

Agricultural Manager

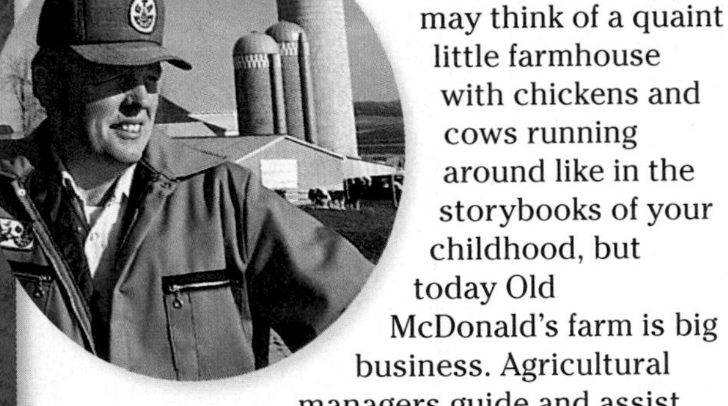

When you hear the word *agriculture,* you may think of a quaint little farmhouse with chickens and cows running around like in the storybooks of your childhood, but today Old McDonald's farm is big business. Agricultural managers guide and assist farmers and ranchers in maximizing their profits by overseeing the day-to-day activities. Their duties are as varied as there are types of farms and ranches.

An agricultural manager may oversee one aspect of the farm, as in feeding livestock on a large dairy farm, or tackle all of the activities on a smaller farm. They also may hire and supervise workers and oversee the purchase and maintenance of farm equipment essential to the farm's operation.

CAREER OVERVIEW

Degree Preferred:
Bachelor's degree in agriculture

Related Courses:
mathematics, science, finance

Outlook:
number of jobs expected to decline through 2006

Number of Farms and Average Farm Size 1975–1998
United States

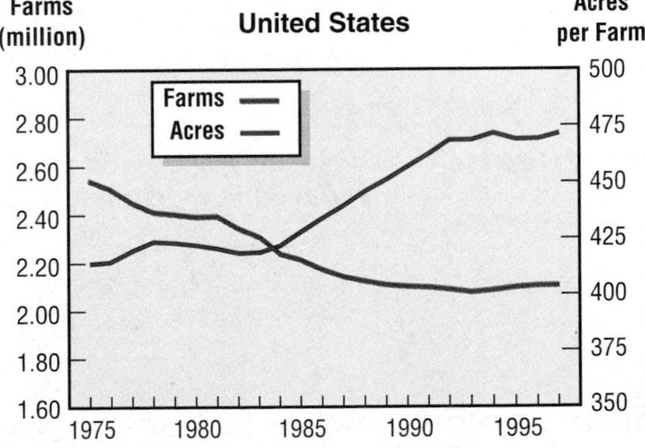

Source: NASS, Livestock & Economics Branch

interNET CONNECTION For more information on careers in agriculture, visit: **www.amc.glencoe.com** 

2-5B Augmented Matrices and Reduced Row-Echelon Form

An Extension of Lesson 2-5

OBJECTIVE

- Find reduced row-echelon form of an augmented matrix to solve systems of equations.

Another way to use matrices to solve a system of equations is to use an *augmented matrix.* An augmented matrix is composed of columns representing the coefficients of each variable and the constant term.

system of equations	*Identify the coefficients and constants.*	augmented matrix
$x - 2y + z = 7$ $3x + y - z = 2$ $2x + 3y + 2z = 7$	$\mathbf{1}x - \mathbf{2}y + \mathbf{1}z = \mathbf{7}$ $\mathbf{3}x + \mathbf{1}y - \mathbf{1}z = \mathbf{2}$ $\mathbf{2}x + \mathbf{3}y + \mathbf{2}z = \mathbf{7}$	$\begin{bmatrix} 1 & -2 & 1 & 7 \\ 3 & 1 & -1 & 2 \\ 2 & 3 & 2 & 7 \end{bmatrix}$

Each equation is always written with the constant term on the right.

A line is often drawn to separate the constants column.

Through a series of calculations that simulate the elimination methods you used in algebraically solving a system in multiple unknowns, you can find the *reduced row-echelon form* of the matrix, which is $\begin{bmatrix} 1 & 0 & 0 & c_1 \\ 0 & 1 & 0 & c_2 \\ 0 & 0 & 1 & c_3 \end{bmatrix}$, where c_1, c_2, and c_3 represent constants. The graphing calculator has a function **rref(** that will calculate this form once you have entered the augmented matrix. It is located in the **MATH** submenu when **MATRX** menu is accessed.

For example, if the augmented matrix above is stored as matrix A, you would enter the matrix name after the parenthesis and then insert a closing parenthesis before pressing ENTER. The result is shown at the right.

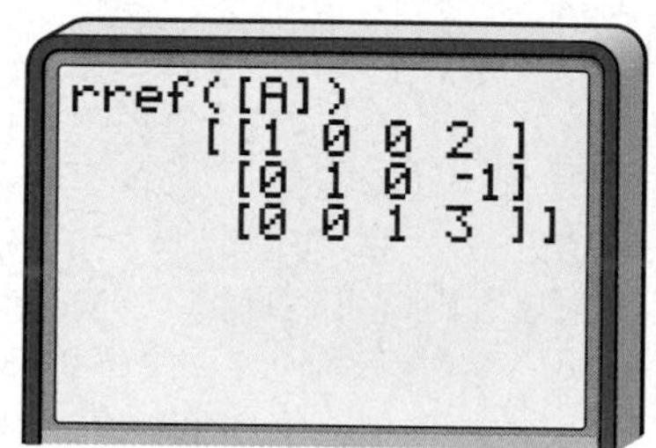

Use the following exercises to discover how this matrix is related to the solution of the system.

TRY THESE

Write an augmented matrix for each system of equations. Then find the reduced row-echelon form.

1. $2x + y - 2z = 7$
$x - 2y - 5z = -1$
$4x + y + z = -1$

2. $x + y + z - 6 = 0$
$2x - 3y + 4z - 3 = 0$
$4x - 8y + 4z - 12 = 0$

3. $w + x + y + z = 0$
$2w + x - y - z = 1$
$-w - x + y + z = 0$
$2x + y = 0$

WHAT DO YOU THINK?

4. Write the equations represented by each reduced row-echelon form of the matrix in Exercises 1-3. How do these equations related to the original system?

5. What would you expect to see on the graphing calculator screen if the constants were irrational or repeating decimals?

Solving Systems of Linear Inequalities

OBJECTIVES

- Graph systems of inequalities.
- Find the maximum or minimum value of a function defined for a polygonal convex set.

SHIPPING Package delivery services add extra charges for oversized parcels or those requiring special handling. An oversize package is one in which the sum of the length and the *girth* exceeds 84 inches. The *girth* of a package is the distance around the package. For a rectangular package, its girth is the sum of twice the width and twice the height. A package requiring special handling is one in which the length is greater than 60 inches. What size packages qualify for both oversize and special handling charges?

The situation described in the problem above can be modeled by a **system of linear inequalities.** To solve a system of linear inequalities, you must find the ordered pairs that satisfy both inequalities. One way to do this is to graph both inequalities on the same coordinate plane. The intersection of the two graphs contains points with ordered pairs in the solution set. If the graphs of the inequalities do not intersect, then the system has no solution.

Example 1

SHIPPING **What size packages qualify for both oversize and special handling charges when shipping?**

First write two inequalities that represent each type of charge. Let ℓ represent the length of a package and g represent its girth.

Oversize: $\ell + g > 84$

Special handling: $\ell > 60$

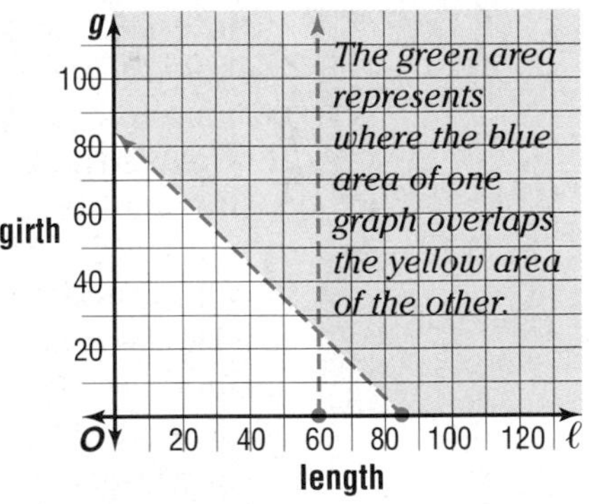

Neither of these inequalities includes the boundary line, so the lines are dashed. The graph of $\ell + g > 84$ is composed of all points above the line $\ell + g = 84$. The graph of $\ell > 60$ includes all points to the right of the line $\ell = 60$. The green area is the solution to the system of inequalities. That is, the ordered pair for any point in the green area satisfies both inequalities. For example, (90, 20) is a length greater than 90 inches and a girth of 20 inches which represents an oversize package that requires special handling.

Not every system of inequalities has a solution. For example, $y > x + 3$ and $y < x - 1$ are graphed at the right. Since the graphs have no points in common, there is no solution.

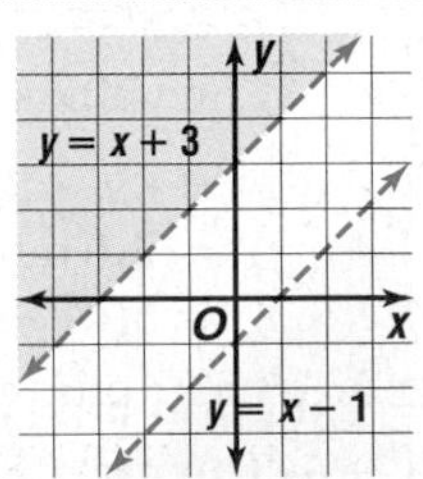

A system of more than two linear inequalities can have a solution that is a bounded set of points. A bounded set of all points on or inside a convex polygon graphed on a coordinate plane is called a **polygonal convex set.**

Example 2 **a. Solve the system of inequalities by graphing.**

$x \geq 0$
$y \geq 0$
$2x + y \leq 4$

b. Name the coordinates of the vertices of the polygonal convex set.

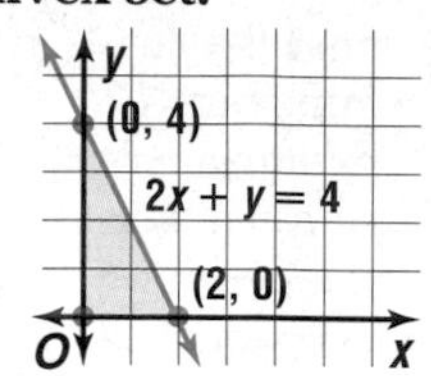

a. Since each inequality includes an equality, the boundary lines will be solid. The shaded region shows points that satisfy all three inequalities.

b. The region is a triangle whose vertices are the points at (0, 0), (0, 4) and (2, 0).

An expression whose value depends on two variables is a function of two variables. For example, the value of $6x + 7y - 9$ is a function of x and y and can be written $f(x, y) = 6x + 7y - 9$. The expression $f(3, 5)$ would then stand for the value of the function f when x is 3 and y is 5.

$$f(3, 5) = 6(3) + 7(5) - 9 \text{ or } 44.$$

Sometimes it is necessary to find the maximum or minimum value that a function has for the points in a polygonal convex set. Consider the function $f(x, y) = 5x - 3y$, with the following inequalities forming a polygonal convex set.

$y \geq 0$ $\quad -x + y \leq 2$ $\quad 0 \leq x \leq 5$ $\quad x + y \leq 6$

You may need to use algebraic methods to determine the coordinates of the vertices of the convex set.

By graphing the inequalities and finding the intersection of the graphs, you can determine a polygonal convex set of points for which the function can be evaluated. The region shown at the right is the polygonal convex set determined by the inequalities listed above. Since the polygonal convex set has infinitely many points, it would be impossible to evaluate the function for all of them. However, according to the **Vertex Theorem,** a function such as $f(x, y) = 5x - 3y$ need only be evaluated for the coordinates of the vertices of the polygonal convex boundary in order to find the maximum and minimum values.

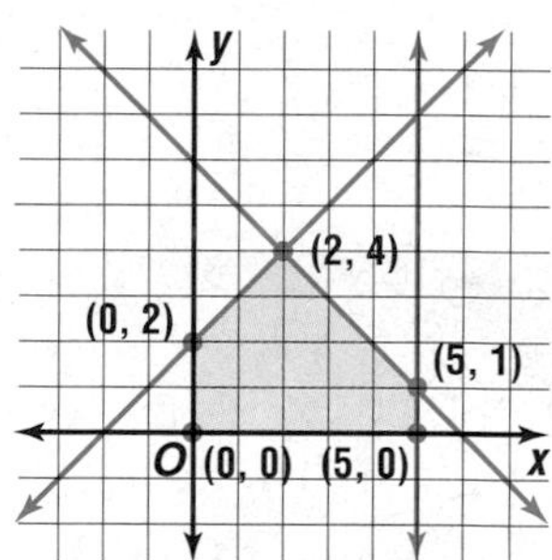

Vertex Theorem The maximum or minimum value of $f(x, y) = ax + by + c$ on a polygonal convex set occurs at a vertex of the polygonal boundary.

The value of $f(x, y) = 5x - 3y$ at each vertex can be found as follows.

$f(x, y) = 5x - 3y$
$f(0, 0) = 5(0) - 3(0) = 0$
$f(0, 2) = 5(0) - 3(2) = -6$
$f(2, 4) = 5(2) - 3(4) = -2$
$f(5, 1) = 5(5) - 3(1) = 22$
$f(5, 0) = 5(5) - 3(0) = 25$

Therefore, the maximum value of $f(x, y)$ in the polygon is 25, and the minimum is -6. The maximum occurs at (5, 0), and the minimum occurs at (0, 2).

Example 3 **Find the maximum and minimum values of $f(x, y) = x - y + 2$ for the polygonal convex set determined by the system of inequalities.**

$$x + 4y \leq 12 \qquad 3x - 2y \geq -6 \qquad x + y \geq -2 \qquad 3x - y \leq 10$$

First write each inequality in slope-intercept form for ease in graphing the boundaries.

Boundary **a**	*Boundary* **b**	*Boundary* **c**	*Boundary* **d**
$x + 4y \leq 2$	$3x - 2y \geq -6$	$x + y \geq -2$	$3x - y \leq 10$
$4y \leq -x + 12$	$-2y \geq -3x - 6$	$y \geq -x - 2$	$-y \leq -3x + 10$
$y \leq -\frac{1}{4}x + 3$	$y \leq \frac{3}{2}x + 3$		$y \geq 3x - 10$

You can use the matrix approach from Lesson 2-5 to find the coordinates of the vertices.

Graph the inequalities and find the coordinates of the vertices of the resulting polygon.

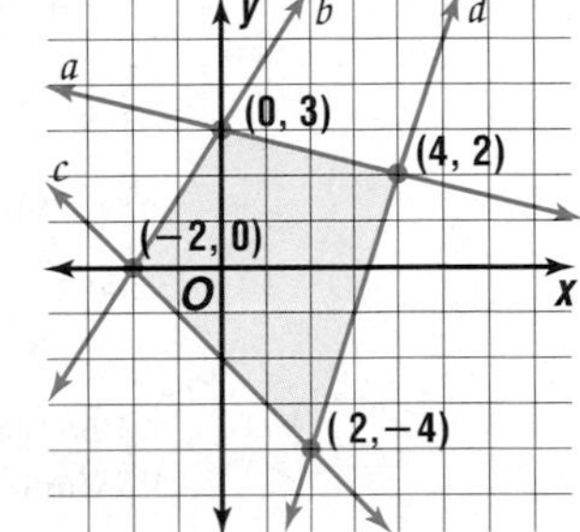

The coordinates of the vertices are $(-2, 0)$, $(2, -4)$, $(4, 2)$, $(0, 3)$.

Now evaluate the function $f(x, y) = x - y + 2$ at each vertex.

$f(-2, 0) = -2 - 0 + 2$ or 0
$f(2, -4) = 2 - (-4) + 2$ or 8
$f(4, 2) = 4 - 2 + 2$ or 4
$f(0, 3) = 0 - 3 + 2$ or -1

The maximum value of the function is 8, and the minimum value is -1.

CHECK FOR UNDERSTANDING

Communicating Mathematics

Read and study the lesson to answer each question.

1. Refer to the application at the beginning of the lesson.
   a. **Define** the *girth* of a rectangular package.
   b. **Name** some objects that might be shipped by a package delivery service and classified as oversized *and* requiring special handling.
2. **You Decide** Marcel says there is only one vertex that will yield a maximum for any given function. Tomas says that if the numbers are correct, there could be two vertices that yield the same maximum. Who is correct? Explain your answer.
3. **Determine** how many vertices of a polygonal convex set you might expect if the system defining the set contained five inequalities, no two of which are parallel.

Guided Practice

4. Solve the system of inequalities by graphing. $x + 2y \geq 4 \qquad x - y \leq 3$
5. Solve the system of inequalities by graphing. Name the coordinates of the vertices of the polygonal convex set.
   $y \geq 0 \qquad -1 \leq x \leq 7 \qquad -x + y \leq 4 \qquad x + 2y \leq 8$

Find the maximum and minimum values of each function for the polygonal convex set determined by the given system of inequalities.

6. $f(x, y) = 4x + 3y$
   $4y \leq x + 8$
   $x + y \geq 2$
   $y \geq 2x - 5$
7. $f(x, y) = 3x - 4y$
   $x - 2y \geq -7$
   $x + y \geq 8$
   $2x - y \leq 7$

8. **Business** Gina Chuez has considered starting her own custom greeting card business. With an initial start-up cost of $1500, she figures it will cost $0.45 to produce each card. In order to remain competitive with the larger greeting card companies, Gina must sell her cards for no more than $1.70 each. To make a profit, her income must exceed her costs. How many cards must she sell before making a profit?

EXERCISES

Practice

Solve each system of inequalities by graphing.

9. $y + x \geq 1$
   $y - x \leq -1$

10. $y > 1$
    $y < -3x + 3$
    $y > -3x + 1$

11. $2x + 5y < 25$
    $y < 3x - 2$
    $5x - 7y < 14$

12. Determine if $(3, -2)$ belongs to the solution set of the system of inequalities $y < \frac{1}{3}x + 5$ and $y < 2x + 1$. Verify your answer.

Solve each system of inequalities by graphing. Name the coordinates of the vertices of the polygonal convex set.

13. $y \geq -0.5x + 1$
    $y \leq -3x + 5$
    $y \leq 2x + 2$

14. $x \leq 0$
    $y + 3 \geq 0$
    $x \geq y$

15. $y \geq 0$
    $y - 5 \leq 0$
    $y + x \leq 7$
    $5x + 3y \geq 20$

16. Find the maximum and minimum values of $f(x, y) = 8x + y$ for the polygonal convex set having vertices at $(0, 0)$, $(4, 0)$, $(3, 5)$, and $(0, 5)$.

Find the maximum and minimum values of each function for the polygonal convex set determined by the given system of inequalities.

17. $f(x, y) = 3x + y$
    $x \leq 5$
    $y \geq 2$
    $2x - 5y \geq -10$

18. $f(x, y) = y - x$
    $y \leq 4 - 2x$
    $x + 2 \geq 2$
    $y \geq 0$

19. $f(x, y) = x + y$
    $y \leq 6$
    $4x - 5y \geq -10$
    $2x - 5y \leq -10$

20. $f(x, y) = 4x + 2y + 7$
    $x \geq 0$
    $y \geq 1$
    $x + y \leq 4$

21. $f(x, y) = 2x - y$
    $y \leq 4x + 6$
    $x + 4y \leq 7$
    $2x + y \leq 7$
    $x - 6y \leq 10$

22. $f(x, y) = -2x + y + 5$
    $2 \leq y \leq 8$
    $x \geq 1$
    $2x + y + 2 \leq 16$
    $y \geq 5 - x$

Applications and Problem Solving

23. **Geometry** Find the system of inequalities that will define a polygonal convex set that includes all points in the interior of a square whose vertices are $A(4, 4)$, $B(4, -4)$, $C(-4, -4)$, and $D(-4, 4)$.

24. **Critical Thinking** Write a system of more than two linear inequalities whose set of solutions is not bounded.

25. **Critical Thinking** A polygonal convex set is defined by the following system of inequalities.

    $y \leq 16 - x$
    $0 \leq 2y \leq 17$
    $3y \geq -2x + 11$
    $y \leq 3x + 1$
    $y \geq 2x - 13$
    $y \geq 7 - 2x$

    a. Determine which lines intersect and solve pairs of equations to determine the coordinates of each vertex.
    b. Find the maximum and minimum values for $f(x, y) = 5x + 6y$ in the set.

26. Business Christine's Butter Cookies sells large tins of butter cookies and small tins of butter cookies. The factory can prepare at most 200 tins of cookies a day. Each large tin of cookies requires 2 pounds of butter, and each small tin requires 1 pound of butter, with a maximum of 300 pounds of butter available each day. The profit from each day's cookie production can be estimated by the function $f(x, y) = \$6.00x + \$4.80y$, where x represents the number of large tins sold and y the number of small tins sold. Find the maximum profit that can be expected in a day.

27. Fund-raising The Band Boosters want to open a craft bazaar to raise money for new uniforms. Two sites are available. A Main Street site costs \$10 per square foot per month. The other site on High Street costs \$20 per square foot per month. Both sites require a minimum rental of 20 square feet. The Main Street site has a potential of 30 customers per square foot, while the High Street site could see 40 customers per square foot. The budget for rental space is \$1200 per month. The Band Boosters are studying their options for renting space at both sites.

a. Graph the polygonal convex region represented by the cost of renting space.

b. Determine what function would represent the possible number of customers per square foot at both locations.

c. If space is rented at both sites, how many square feet of space should the Band Boosters rent at each site to maximize the number of potential customers?

d. Suppose you were president of the Band Boosters. Would you rent space at both sites or select one of the sites? Explain your answer.

28. Culinary Arts A gourmet restaurant sells two types of salad dressing, garlic and raspberry, in their gift shop. Each batch of garlic dressing requires 2 quarts of oil and 2 quarts of vinegar. Each batch of raspberry dressing requires 3 quarts of oil and 1 quart of vinegar. The chef has 18 quarts of oil and 10 quarts of vinegar on hand for making the dressings that will be sold in the gift shop that week. If x represents the number of batches of garlic dressing sold and y represents the batches of raspberry dressing sold, the total profits from dressing sold can be expressed by the function $f(x, y) = 3x + 2y$.

a. What do you think the 3 and 2 in the function $f(x, y) = 3x + 2y$ represent?

b. How many batches of each types of dressing should the chef make to maximize the profit on sales of the dressing?

Mixed Review

29. Find the inverse of $\begin{bmatrix} 2 & 1 \\ -3 & 2 \end{bmatrix}$. *(Lesson 2-5)*

30. Graph $y < -2x + 8$. *(Lesson 1-8)*

31. **Scuba Diving** Graph the equation $d + 33 = 33p$, which relates atmospheres of pressure p to ocean depth d in feet. *(Lesson 1-3)*

32. State the domain and range of the relation $\{(16, -4), (16, 4)\}$. Is this relation a function? Explain. *(Lesson 1-1)*

33. **SAT Practice Grid-In** What is the sum of four integers whose mean is 15?

Extra Practice See p. A29.

2-7 Linear Programming

OBJECTIVES

- Use linear programming procedures to solve applications.
- Recognize situations where exactly one solution to a linear programming application may not exist.

MILITARY SCIENCE When the U.S. Army needs to determine how many soldiers or officers to put in the field, they turn to mathematics. A system called the *Manpower Long-Range Planning System* (MLRPS) enables the army to meet the personnel needs for 7- to 20-year planning periods. Analysts are able to effectively use the MLRPS to simulate gains, losses, promotions, and reclassifications. This type of planning requires solving up to 9,060 inequalities with 28,730 variables! However, with a computer, a problem like this can be solved in less than five minutes.

The Army's MLRPS uses a procedure called **linear programming.** Many practical applications can be solved by using this method. The nature of these problems is that certain **constraints** exist or are placed upon the variables, and some function of these variables must be maximized or minimized. The constraints are often written as a system of linear inequalities.

The following procedure can be used to solve linear programming applications.

Linear Programming Procedure

1. Define variables.
2. Write the constraints as a system of inequalities.
3. Graph the system and find the coordinates of the vertices of the polygon formed.
4. Write an expression whose value is to be maximized or minimized.
5. Substitute values from the coordinates of the vertices into the expression.
6. Select the greatest or least result.

In Lesson 2-6, you found the maximum and minimum values for a given function in a defined polygonal convex region. In linear programming, you must use your reasoning abilities to determine the function to be maximized or minimized and the constraints that form the region.

Example

1 **MANUFACTURING** **Suppose a lumber mill can turn out 600 units of product each week. To meet the needs of its regular customers, the mill must produce 150 units of lumber and 225 units of plywood. If the profit for each unit of lumber is \$30 and the profit for each unit of plywood is \$45, how many units of each type of wood product should the mill produce to maximize profit?**

Define variables. Let x = the units of lumber produced.
Let y = the units of plywood produced.

Write inequalities.	$x \geq 150$	*There cannot be less than 150 units of lumber produced.*
	$y \geq 225$	*There cannot be less than 225 units of plywood produced.*
	$x + y \leq 600$	*The maximum number of units produced is 600.*

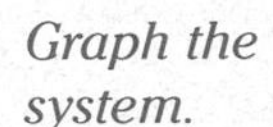

Graph the system.

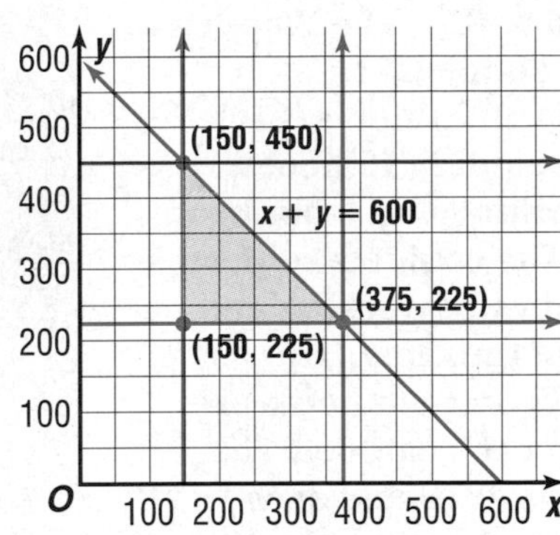

The vertices are at (150, 225), (375, 225), and (150, 450).

Write an expression. Since profit is \$30 per unit of lumber and \$45 per unit of plywood, the profit function is $P(x, y) = 30x + 45y$.

Substitute values.

$P(150, 225) = 30(150) + 45(225)$ or 14,625
$P(375, 225) = 30(375) + 45(225)$ or 21,375
$P(150, 450) = 30(150) + 45(450)$ or 24,750

Answer the problem. The maximum profit occurs when 150 units of lumber are produced and 450 units of plywood are produced.

In certain circumstances, the use of linear programming is not helpful because a polygonal convex set is not defined. Consider the graph at the right, based on the following constraints.

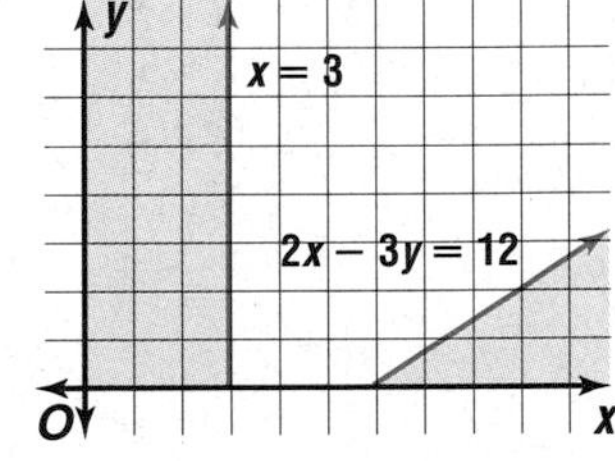

$x \geq 0$
$y \geq 0$
$x \leq 3$
$2x - 3y \geq 12$

The constraints do not define a region with any points in common in Quadrant I. When the constraints of a linear programming application cannot be satisfied simultaneously, then the problem is said to be **infeasible.**

Sometimes the region formed by the inequalities in a linear programming application is **unbounded.** In that case, an optimal solution for the problem may not exist. Consider the graph at the right. A function like $f(x, y) = x + 2y$ has a *minimum* value at (5, 3), but it is not possible to find a maximum value.

It is also possible for a linear programming application to have two or more optimal solutions. When this occurs, the problem is said to have **alternate optimal solutions.** This usually occurs when the graph of the function to be maximized or minimized is parallel to one side of the polygonal convex set.

Example 2 

SMALL BUSINESS The Woodell Carpentry Shop makes bookcases and cabinets. Each bookcase requires 15 hours of woodworking and 9 hours of finishing. The cabinets require 10 hours of woodworking and 4.5 hours of finishing. The profit is \$60 on each bookcase and \$40 on each cabinet. There are 70 hours available each week for woodworking and 36 hours available for finishing. How many of each item should be produced in order to maximize profit?

Define variables. Let b = the number of bookcases produced.
Let c = the number of cabinets produced.

Write inequalities.
$b \geq 0, c \geq 0$ — *There cannot be less than 0 bookcases or cabinets.*
$15b + 10c \leq 70$ — *No more than 70 hours of woodworking are available.*
$9b + 4.5c \leq 36$ — *No more than 36 hours of finishing are available.*

Graph the system.

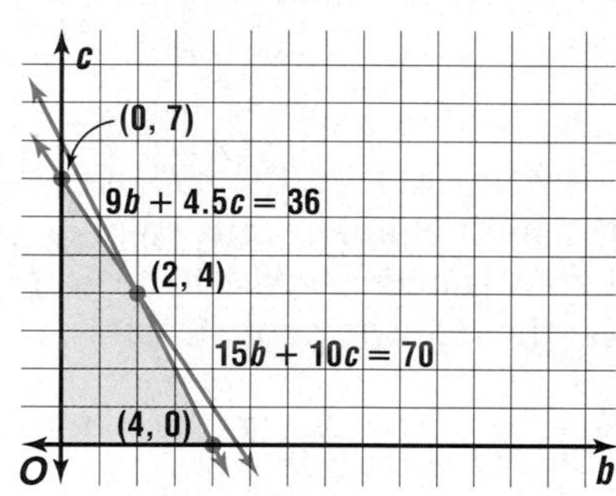

The vertices are at (0, 7), (2, 4), (4, 0), and (0, 0).

Write an expression. Since profit on each bookcase is \$60 and the profit on each cabinet is \$40, the profit function is $P(b, c) = 60b + 40c$.

Substitute values.
$P(0, 0) = 60(0) + 40(0)$ or 0
$P(0, 7) = 60(0) + 40(7)$ or 280
$P(2, 4) = 60(2) + 40(4)$ or 280
$P(4, 0) = 60(4) + 40(0)$ or 240

Answer the problem. The problem has alternate optimal solutions. The shop will make the same profit if they produce 2 bookcases and 4 cabinets as it will from producing 7 cabinets and no bookcases.

CHECK FOR UNDERSTANDING

Communicating Mathematics

Read and study the lesson to answer each question.

1. **Explain** why the inequalities $x \geq 0$ and $y \geq 0$ are usually included as constraints in linear programming applications.
2. **Discuss** the difference between the graph of the constraints when a problem is infeasible and a graph whose constraints yield an unbounded region.
3. **Write,** in your own words, the steps of the linear programming procedure.

Guided Practice

4. Graph the system of inequalities. In a problem asking you to find the maximum value of $f(x, y)$, state whether this situation is *infeasible,* has *alternate optimal solutions,* or is *unbounded.* Assume that $x \geq 0$ and $y \geq 0$.
   $0.5x + 1.5y \geq 7$
   $3x + 9y \leq 2$
   $f(x, y) = 30x + 20y$
5. **Transportation** A package delivery service has a truck that can hold 4200 pounds of cargo and has a capacity of 480 cubic feet. The service handles two types of packages: small, which weigh up to 25 pounds each and are no more than 3 cubic feet each; and large, which are 25 to 50 pounds each and are 3 to 5 cubic feet each. The delivery service charges \$5 for each small package and \$8 for each large package. Let x be the number of small packages and y be the number of large packages in the truck.
   a. Write an inequality to represent the weight of the packages in pounds the truck can carry.
   b. Write an inequality to represent the volume, in cubic feet, of packages the truck can carry.
   c. Graph the system of inequalities.
   d. Write a function that represents the amount of money the delivery service will make on each truckload.
   e. Find the number of each type of package that should be placed on a truck to maximize revenue.
   f. What is the maximum revenue per truck?
   g. In this situation, is maximizing the revenue necessarily the best thing for the company to do? Explain.

Solve each problem, if possible. If not possible, state whether the problem is *infeasible*, has *alternate optimal solutions*, or is *unbounded*.

6. **Business** The manager of a gift store is printing brochures and fliers to advertise sale items. Each brochure costs 8¢ to print, and each flier costs 4¢ to print. A brochure requires 3 pages, and a flier requires 2 pages. The manager does not want to use more than 600 pages, and she needs at least 50 brochures and 150 fliers. How many of each should she print to minimize the cost?
7. **Manufacturing** Woodland Bicycles makes two models of off-road bicycles: the Explorer, which sells for \$250, and the Grande Expedition, which sells for \$350. Both models use the same frame, but the painting and assembly time required for the Explorer is 2 hours, while the time is 3 hours for the Grande Expedition. There are 375 frames and 450 hours of labor available for production. How many of each model should be produced to maximize revenue?

8. **Business** The Grainery Bread Company makes two types of wheat bread, light whole wheat and regular whole wheat. A loaf of light whole wheat bread requires 2 cups of flour and 1 egg. A loaf of regular whole wheat uses 3 cups of flour and 2 eggs. The bakery has 90 cups of flour and 80 eggs on hand. The profit on the light bread is \$1 per loaf and on the regular bread is \$1.50 per loaf. In order to maximize profits, how many of each loaf should the bakery make?

EXERCISES

Practice

Graph each system of inequalities. In a problem asking you to find the maximum value of $f(x, y)$, state whether the situation is *infeasible*, has *alternate optimal solutions*, or is *unbounded*. In each system, assume that $x \geq 0$ and $y \geq 0$ unless stated otherwise.

9. $y \geq 6$
   $5x + 3y \leq 15$
   $f(x, y) = 12x + 3y$

10. $2x + y \geq 48$
    $x + 2y \geq 42$
    $f(x, y) = 2x + y$

11. $4x + 3y \geq 12$
    $y \leq 3$
    $x \leq 4$
    $f(x, y) = 3 + 3y$

Applications and Problem Solving

12. **Veterinary Medicine** Dr. Chen told Miranda that her new puppy needs a diet that includes at least 1.54 ounces of protein and 0.56 ounce of fat each day to grow into a healthy dog. Each cup of Good Start puppy food contains 0.84 ounce of protein and 0.21 ounce of fat. Each cup of Sirius puppy food contains 0.56 ounce of protein and 0.49 ounce of fat. If Good Start puppy food costs 36¢ per cup and Sirius costs 22¢ per cup, how much of each food should Miranda use in order to satisfy the dietary requirements at the minimum cost?
    a. Write an inequality to represent the ounces of protein required.
    b. Write an inequality to represent the ounces of fat required.
    c. Graph the system of inequalities.
    d. Write a function to represent the daily cost of puppy food.
    e. How many cups of each type of puppy food should be used in order to minimize the cost?
    f. What is the minimum cost?

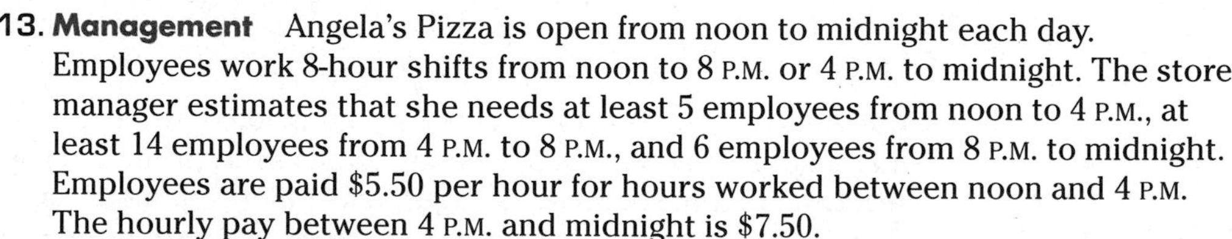

13. **Management** Angela's Pizza is open from noon to midnight each day. Employees work 8-hour shifts from noon to 8 P.M. or 4 P.M. to midnight. The store manager estimates that she needs at least 5 employees from noon to 4 P.M., at least 14 employees from 4 P.M. to 8 P.M., and 6 employees from 8 P.M. to midnight. Employees are paid \$5.50 per hour for hours worked between noon and 4 P.M. The hourly pay between 4 P.M. and midnight is \$7.50.
    a. Write inequalities to represent the number of day-shift workers, the number of night-shift workers, and the total number of workers needed.
    b. Graph the system of inequalities.
    c. Write a function to represent the daily cost of payroll.
    d. Find the number of day-shift workers and night-shift workers that should be scheduled to minimize the cost.
    e. What is the minimal cost?

Solve each problem, if possible. If not possible, state whether the problem is infeasible, has alternate optimal solutions, or is unbounded.

14. **Agriculture** The county officials in Chang Qing County, China used linear programming to aid the farmers in their choices of crops and other forms of agricultural production. This led to a 12% increase in crop profits, a 54% increase in animal husbandry profits, while improving the region's ecology. Suppose an American farmer has 180 acres on which to grow corn and soybeans. He is planting at least 40 acres of corn and 20 acres of soybeans. Based on his calculations, he can earn $150 per acre of corn and $250 per acre of soybeans.
    a. If the farmer plants at least 2 acres of corn for every acre of soybeans, how many acres of each should he plant to earn the greatest profit?
    b. What is the farmer's maximum profit?

15. **Education** Ms. Carlyle has written a final exam for her class that contains two different sections. Questions in section I are worth 10 points each, and questions in section II are worth 15 points each. Her students will have 90 minutes to complete the exam. From past experience, she knows that on average questions from section I take 6 minutes to complete and questions from section II take 15 minutes. Ms. Carlyle requires her students to answer at least 2 questions from section II. Assuming they answer correctly, how many questions from each section will her students need to answer to get the highest possible score?

16. **Manufacturing** Newline Recyclers processes used aluminum into food or drink containers. The recycling plant processes up to 1200 tons of aluminum per week. At least 300 tons must be processed for food containers, while at least 450 tons must be processed for drink containers. The profit is $17.50 per ton for processing food containers and $20 per ton for processing drink containers. What is the profit if the plant maximizes processing?

17. **Investment** Diego wants to invest up to $11,000 in certificates of deposit at First Bank and City Bank. He does not want to deposit more than $7,500 at First Bank. He will deposit at least $1,000 but not more than $7,000 at City Bank. First Bank offers 6% simple interest on deposits, while City Bank offers $6\frac{1}{2}$% simple interest. How much should Diego deposit into each account so he can earn the most interest possible in one year?

18. **Human Resources** Memorial Hospital wants to hire nurses and nurse's aides to meet patient needs at minimum cost. The average annual salary is $35,000 for a nurse and $18,000 for a nurse's aide. The hospital can hire up to 50 people, but needs to hire at least 20 to function properly. The head nurse wants at least 12 aides, but the number of nurses must be at least twice the number of aides to meet state regulations. How many nurses and nurse's aides should be hired to minimize salary costs?

19. **Manufacturing** A potato chip company makes chips to fill snack-size bags and family-size bags. In one week, production cannot exceed 2,400 units, of which at least 600 units must be for snack-size bags and at least 900 units must be for family size. The profit on a unit of snack-size bags is $12, and the profit on a unit of family-size bags is $18. How much of each type of bag must be processed to maximize profits?

20. Crafts Shelly is making soap and shampoo for gifts. She has 48 ounces of lye and 76 ounces of coconut oil and an ample supply of the other needed ingredients. She plans to make as many batches of soap and shampoo as possible. A batch of soap requires 12 ounces of lye and 20 ounces of coconut oil. Each batch of shampoo needs 6 ounces of lye and 8 ounces of coconut oil. What is the maximum number of batches of *both* soap and shampoo possible?

21. Manufacturing An electronics plant makes standard and large computer monitors on three different machines. The profit is $40 on each monitor. Use the table below to determine how many of each monitor the plant should make to maximize profits.

Manufacturing Analysis

Machine	Hours Needed to Make		Total Hours Available
	Small Monitor	Large Monitor	
A	1	2	16
B	1	1	9
C	1	4	24

22. Critical Thinking Find the area enclosed by the polygonal convex set defined by the system of inequalities $y \geq 0$, $x \leq 12$, $2x + 6y \leq 84$, $2x - 3y \leq -3$, and $8x + 3y \geq 33$.

23. Critical Thinking Mr. Perez has an auto repair shop. He offers two bargain maintenance services, an oil change and a tune-up. His profit is $12 on an oil change and $20 on a tune-up. It takes Mr. Perez 30 minutes to do an oil change and 1 hour to do a tune-up. He wants to do at least 25 oil changes per week and no more than 10 tune-ups per week. He can spend up to 30 hours each week on these two services.

a. What is the most Mr. Perez can earn performing these services?

b. After seeing the results of his linear program, Mr. Perez decides that he must make a larger profit to keep his business growing. How could the constraints be modified to produce a larger profit?

Mixed Review

24. A polygonal convex set is defined by the inequalities $y \leq -x + 5$, $y \leq x + 5$, and $y + 5x \geq -5$. Find the minimum and maximum values for $f(x, y) = \frac{1}{3}x - \frac{1}{2}y$ within the set. *(Lesson 2-6)*

25. Find the values of x and y for which $\begin{bmatrix} 4x + y \\ x \end{bmatrix} = \begin{bmatrix} 6 \\ 2y - 12 \end{bmatrix}$ is true. *(Lesson 2-3)*

26. Graph $y = 3|x - 2|$. *(Lesson 1-7)*

27. Budgeting Martina knows that her monthly telephone charge for local calls is $13.65 (excluding tax) for service allowing 30 local calls plus $0.15 for each call after 30. Write an equation to calculate Martina's monthly telephone charges and find the cost if she made 42 calls. *(Lesson 1-4)*

28. SAT/ACT Practice If $\frac{2x - 3}{x} = \frac{3 - x}{2}$, which could be a value for x?

A -3 **B** -1 **C** 37 **D** 5 **E** 15

Extra Practice See p. A29.

CHAPTER 2 STUDY GUIDE AND ASSESSMENT

VOCABULARY

additive identity matrix (p. 80)
column matrix (p. 78)
consistent (p. 67)
dependent (p. 67)
determinant (p. 98)
dilation (p. 88)
dimensions (p. 78)
element (p. 78)
elimination method (p. 68)
equal matrices (p. 79)
identity matrix for multiplication (p. 99)
image (p. 88)
inconsistent (p. 67)
independent (p. 67)
infeasible (p. 113)
inverse matrix (p. 99)
$m \times n$ matrix (p. 78)
matrix (p. 78)
minor (p. 98)
*n*th order (p. 78)
ordered triple (p. 74)
polygonal convex set (p. 108)
pre-image (p. 88)
reflection (p. 88)
reflection matrix (p. 89)
rotation (p. 88)
rotation matrix (p. 91)
row matrix (p. 78)
scalar (p. 80)
solution (p. 67)
square matrix (p. 78)
substitution method (p. 68)
system of equations (p. 67)
system of linear inequalities (p. 107)
transformation (p. 88)
translation (p. 88)
translation matrix (p. 88)
vertex matrix (p. 88)
Vertex Theorem (p. 108)
zero matrix (p. 80)

Modeling

alternate optimal solutions (p. 114)
constraints (p. 112)
linear programming (p. 112)
unbounded (p. 113)

UNDERSTANDING AND USING THE VOCABULARY

Choose the correct term from the list to complete each sentence.

1. Sliding a polygon from one location to another without changing its size, shape, or orientation is called a(n) __?__ of the figure.
2. Two matrices can be __?__ if they have the same dimensions.
3. The __?__ of $\begin{bmatrix} -1 & 4 \\ 2 & -3 \end{bmatrix}$ is -5.
4. A(n) __?__ system of equations has no solution.
5. The process of multiplying a matrix by a constant is called __?__.
6. The matrices $\begin{bmatrix} 2x \\ 0 \\ 16 \end{bmatrix}$ and $\begin{bmatrix} 8-y \\ y \\ 4x \end{bmatrix}$ are __?__ if $x = 4$ and $y = 0$.
7. A region bounded on all sides by intersecting linear inequalities is called a(n) __?__.
8. A rotation of a figure can be achieved by consecutive __?__ of the figure over given lines.
9. The symbol a_{ij} represents a(n) __?__ of a matrix.
10. Two matrices can be __?__ if the number of columns in the first matrix is the same as the number of rows in the second matrix.

added
consistent
determinant
dilation
divided
element
equal matrices
identity matrices
inconsistent
inverse
multiplied
polygonal convex set
reflections
scalar multiplication
translation

For additional review and practice for each lesson, visit: **www.amc.glencoe.com**

SKILLS AND CONCEPTS

OBJECTIVES AND EXAMPLES

Lesson 2-1 Solve systems of two linear equations.

- Solve the system of equations.

$y = -x + 2$

$3x + 4y = 2$

Substitute $-x + 2$ for y in the second equation.

$$\begin{aligned} 3x + 4y &= 2 \\ 3x + 4(-x + 2) &= 2 \\ x &= 6 \end{aligned} \qquad \begin{aligned} y &= -x + 2 \\ y &= -6 + 2 \\ y &= -4 \end{aligned}$$

The solution is $(6, -4)$.

REVIEW EXERCISES

Solve each system of equations algebraically.

11. $2y = -4x$; $y = -x - 2$

12. $y = x - 5$; $6y - x = 0$

13. $2x = 5y$; $3y + x = -1$

14. $y = 6x + 1$; $2y - 15x = -4$

15. $3x - 2y = -1$; $2x + 5y = 12$

16. $x + 5y = 20.5$; $3y - x = 13.5$

Lesson 2-2 Solve systems of equations involving three variables algebraically.

- Solve the system of equations.

$x - y + z = 5$

$-3x + 2y - z = -8$

$2x + y + z = 4$

Combine pairs of equations to eliminate z.

$$\begin{array}{r} x - y + z = 5 \\ -3x + 2y - z = -8 \\ \hline -2x + y = -3 \end{array} \qquad \begin{array}{r} -3x + 2y - z = -8 \\ 2x + y + z = 4 \\ \hline -x + 3y = -4 \end{array}$$

$-5y = 5$

$y = -1$

$-2x + (-1) = -3 \rightarrow x = 1$

$(1) - (-1) + z = 5 \rightarrow z = 3$ *Solve for z.*

The solution is $(1, -1, 3)$.

Solve each system of equations algebraically.

17. $x - 2y - 3z = 2$; $-3x + 5y + 4z = 0$; $x - 4y + 3z = 14$

18. $-x + 2y - 6z = 4$; $x + y + 2z = 3$; $2x + 3y - 4z = 5$

19. $x - 2y + z = 7$; $3x + y - z = 2$; $2x + 3y + 2z = 7$

Lesson 2-3 Add, subtract, and multiply matrices.

- Find the difference.

$$\begin{bmatrix} -4 & 7 \\ 6 & -3 \end{bmatrix} - \begin{bmatrix} 8 & -3 \\ 2 & 0 \end{bmatrix} = \begin{bmatrix} -4 - 8 & 7 - (-3) \\ 6 - 2 & -3 - 0 \end{bmatrix} = \begin{bmatrix} -12 & 10 \\ 4 & -3 \end{bmatrix}$$

Find the product.

$$-2\begin{bmatrix} 1 & -4 \\ 0 & 3 \end{bmatrix} = \begin{bmatrix} -2(1) & -2(-4) \\ -2(0) & -2(3) \end{bmatrix} = \begin{bmatrix} -2 & 8 \\ 0 & -6 \end{bmatrix}$$

Use matrices A, B, and C to find each of the following. If the matrix does not exist, write *impossible*.

$$A = \begin{bmatrix} 7 & 8 \\ 0 & -4 \end{bmatrix} \quad B = \begin{bmatrix} -3 & -5 \\ 2 & -2 \end{bmatrix} \quad C = \begin{bmatrix} 2 \\ -5 \end{bmatrix}$$

20. $A + B$

21. $B - A$

22. $3B$

23. $-4C$

24. AB

25. CB

26. $4A - 4B$

27. $AB - 2C$

OBJECTIVES AND EXAMPLES

Lesson 2-4 Use matrices to determine the coordinates of polygons under a given transformation.

Reflections

$R_{x\text{-axis}} = \begin{bmatrix} 1 & 0 \\ 0 & -1 \end{bmatrix}$

$R_{y\text{-axis}} = \begin{bmatrix} -1 & 0 \\ 0 & 1 \end{bmatrix}$

$R_{y=x} = \begin{bmatrix} 0 & 1 \\ 1 & 0 \end{bmatrix}$

Rotations (counterclockwise about the origin)

$Rot_{90} = \begin{bmatrix} 0 & -1 \\ 1 & 0 \end{bmatrix}$

$Rot_{180} = \begin{bmatrix} -1 & 0 \\ 0 & -1 \end{bmatrix}$

$Rot_{270} = \begin{bmatrix} 0 & -1 \\ -1 & 0 \end{bmatrix}$

REVIEW EXERCISES

Use matrices to perform each transformation. Then graph the pre-image and image on the same coordinate grid.

28. $A(-4, 3)$, $B(2, 1)$, $C(5, -3)$ translated 4 units down and 3 unit left
29. $W(-2, -3)$, $X(-1, 2)$, $Y(0, 4)$, $Z(1, -2)$ reflected over the x-axis
30. $D(2, 3)$, $E(2, -5)$, $F(-1, -5)$, $G(-1, 3)$ rotated 180° about the origin.
31. $P(3, -4)$, $Q(1, 2)$, $R(-1, 1)$ dilated by a scale factor of 0.5
32. triangle ABC with vertices at $A(-4, 3)$, $B(2, 1)$, $C(5, -3)$ after $Rot_{90} \circ R_{x\text{-axis}}$
33. What translation matrix would yield the same result on a triangle as a translation 6 units up and 4 units left followed by a translation 3 units down and 5 units right?

Lesson 2-5 Evaluate determinants.

Find the value of $\begin{vmatrix} -7 & 6 \\ 5 & -2 \end{vmatrix}$.

$$\begin{vmatrix} -7 & 6 \\ 5 & -2 \end{vmatrix} = (-7)(-2) - 5(6) = -16$$

Find the value of each determinant.

34. $\begin{vmatrix} -3 & 5 \\ -4 & 7 \end{vmatrix}$

35. $\begin{vmatrix} 8 & -4 \\ -6 & 3 \end{vmatrix}$

36. $\begin{vmatrix} 3 & -1 & 4 \\ 5 & -2 & 6 \\ 7 & 3 & -4 \end{vmatrix}$

37. $\begin{vmatrix} 5 & 0 & -4 \\ 7 & 3 & -1 \\ 2 & -2 & 6 \end{vmatrix}$

38. Determine whether $\begin{bmatrix} 2 & -4 & 1 \\ 3 & 8 & -2 \end{bmatrix}$ has a determinant. If so, find the value of the determinant. If not, explain.

Lesson 2-5 Find the inverse of a 2 × 2 matrix.

Find X^{-1}, if $X = \begin{bmatrix} 2 & -1 \\ 1 & -3 \end{bmatrix}$.

$$\begin{vmatrix} 2 & -1 \\ 1 & -3 \end{vmatrix} = -6 - (-1) \text{ or } -5 \text{ and } -5 \neq 0$$

$$X^{-1} = \frac{1}{ad - bc}\begin{bmatrix} d & -b \\ -c & a \end{bmatrix}$$

$$= -\frac{1}{5}\begin{bmatrix} -3 & 1 \\ -1 & 2 \end{bmatrix}$$

Find the inverse of each matrix, if it exists.

39. $\begin{bmatrix} 3 & 8 \\ -1 & 5 \end{bmatrix}$

40. $\begin{bmatrix} 5 & 2 \\ 10 & 4 \end{bmatrix}$

41. $\begin{bmatrix} -3 & 5 \\ 1 & -4 \end{bmatrix}$

42. $\begin{bmatrix} 3 & 2 \\ 5 & 7 \end{bmatrix}$

43. $\begin{bmatrix} 2 & -5 \\ 6 & 1 \end{bmatrix}$

44. $\begin{bmatrix} 2 & -4 \\ -1 & 2 \end{bmatrix}$

SKILLS AND CONCEPTS

OBJECTIVES AND EXAMPLES

Lesson 2-5 Solve systems of equations by using inverses of matrices.

- Solve the system of equations $3x - 5y = 1$ and $-2x + 2y = -2$ by using a matrix equation.

$$\begin{bmatrix} 3 & -5 \\ -2 & 2 \end{bmatrix} \cdot \begin{bmatrix} x \\ y \end{bmatrix} = \begin{bmatrix} 1 \\ -2 \end{bmatrix}$$

$$-\frac{1}{4}\begin{bmatrix} 2 & 5 \\ 2 & 3 \end{bmatrix} \cdot \begin{bmatrix} 3 & -5 \\ -2 & 2 \end{bmatrix} \cdot \begin{bmatrix} x \\ y \end{bmatrix} = -\frac{1}{4}\begin{bmatrix} 2 & 5 \\ 2 & 3 \end{bmatrix}\begin{bmatrix} 1 \\ -2 \end{bmatrix}$$

$$\begin{bmatrix} x \\ y \end{bmatrix} = \begin{bmatrix} 2 \\ 1 \end{bmatrix}$$

REVIEW EXERCISES

Solve each system by using a matrix equation.

45. $2x + 5y = 1$
$-x - 3y = 2$

46. $3x + 2y = -3$
$-6x + 4y = 6$

47. $-3x + 5y = 1$
$-2x + 4y = -2$

48. $4.6x - 2.7y = 8.4$
$2.9x + 8.8y = 74.61$

Lesson 2-6 Find the maximum or minimum value of a function defined for a polygonal convex set.

- Find the maximum and minimum values of $f(x, y) = 4y + x - 3$ for the polygonal convex set graphed at the right.

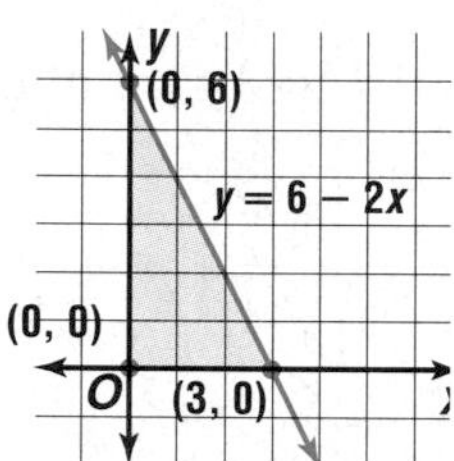

$f(0, 0) = 4(0) + 0 - 3 = -3$ *minimum*
$f(3, 0) = 4(0) + 3 - 3 = 0$
$f(0, 6) = 4(6) + 0 - 3 = 21$ *maximum*

Find the maximum and minimum values of each function for the polygonal convex set determined by the given system of inequalities.

49. $f(x, y) = 2x + 3y$
$x \geq 1$
$y \geq -2$
$y + x \leq 6$
$y \leq 10 - 2x$

50. $f(x, y) = 3x + 2y + 1$
$x \geq 0$
$y \geq 4$
$y + x \leq 11$
$2y + x \leq 18$
$x \leq 6$

Lesson 2-7 Use linear programming procedures to solve applications.

- **Linear Programming Procedure**
  1. Define variables.
  2. Write the constraints as a system of inequalities.
  3. Graph the system and find the coordinates of the vertices of the polygon formed.
  4. Write an expression to be maximized or minimized.
  5. Substitute values from the coordinates of the vertices into the expression.
  6. Select the greatest or least result.

Use linear programming to solve.

51. Transportation Justin owns a truck and a motorcycle. He can buy up to 28 gallons of gasoline for both vehicles. His truck gets 22 miles per gallon and holds up to 25 gallons of gasoline. His motorcycle gets 42 miles per gallon and holds up to 6 gallons of gasoline. How many gallons of gasoline should Justin put in each vehicle if he wants to travel the most miles possible?

APPLICATIONS AND PROBLEM SOLVING

52. **Sports** In a three-team track meet, the following numbers of first-, second-, and third-place finishes were recorded.

School	First Place	Second Place	Third Place
Broadman	2	5	5
Girard	8	2	3
Niles	6	4	1

Use matrix multiplication to find the final scores for each school if 5 points are awarded for a first place, 3 for second place, and 1 for third place. (Lesson 2-4)

53. **Geometry** The perimeter of a triangle is 83 inches. The longest side is three times the length of the shortest side and 17 inches more than one-half the sum of the other two sides. Use a system of equations to find the length of each side. (Lesson 2-5)

54. **Manufacturing** A toy manufacturer produces two types of model spaceships, the *Voyager* and the *Explorer*. Each toy requires the same three operations. Each *Voyager* requires 5 minutes for molding, 3 minutes for machining, and 5 minutes for assembly. Each *Explorer* requires 6 minutes for molding, 2 minutes for machining, and 18 minutes for assembly. The manufacturer can afford a daily schedule of not more than 4 hours for molding, 2 hours for machining, and 9 hours for assembly. (Lesson 2-7)
    a. If the profit is \$2.40 on each *Voyager* and \$5.00 on each *Explorer*, how many of each toy should be produced for maximum profit?
    b. What is the maximum daily profit?

ALTERNATIVE ASSESSMENT

OPEN-ENDED ASSESSMENT

1. Suppose that a quadrilateral $ABCD$ has been rotated 90° clockwise about the origin twice. The resulting vertices are $A'(2, 2)$, $B'(-1, 2)$, $C'(-2, -1)$, and $D'(3, 0)$.
    a. State the original coordinates of the vertices of quadrilateral $ABCD$ and how you determined them.
    b. Make a conjecture about the effect of a double rotation of 90° on any given figure.
2. If the determinant of a coefficient matrix is 0, can you use inverse matrices to solve the system of equations? Explain your answer and illustrate it with such a system of equations.

Additional Assessment See p. A57 for Chapter 2 Practice Test.

TELECOMMUNICATION

You've Got Mail!

- Research several Internet servers and make graphs that reflect the cost of using the server over a year's time period.
- Research various e-mail servers and their costs. Write and graph equations to compare the costs.
- Determine which Internet and e-mail servers best meet your needs. Write a paragraph to explain your choice. Use your graphs to support your choice.

PORTFOLIO

Devise a real-world problem that can be solved by linear programming. Define whether you are seeking a maximum or minimum. Write the inequalities that define the polygonal convex set used to determine the solution. Explain what the solution means.

Algebra Problems

TEST-TAKING TIP

Review the rules for simplifying algebraic fractions and expressions. Try to simplify expressions whenever possible.

About one third of SAT math problems and many ACT math problems involve algebra. You'll need to simplify algebraic expressions, solve equations, and solve word problems.

The word problems often deal with specific problems.

- consecutive integers
- age
- motion (*distance* = *rate* × *time*)
- investments (*principal* × *rate* = *interest income*)
- work
- coins
- mixtures

ACT EXAMPLE

1. $\frac{(xy)^3z^0}{x^3y^4} =$

   A $\frac{1}{y}$ **B** $\frac{z}{y}$ **C** z **D** xy **E** xyz

HINT Review the properties of exponents.

Solution Simplify the expression. Apply the properties of exponents.

$(xy)^3 = x^3y^3$.

Any number raised to the zero power is equal to 1. Therefore, $z^0 = 1$.

Write y^4 as y^3y^1.

Simplify.

$$\frac{(xy)^3z^0}{x^3y^4} = \frac{x^3y^3 \cdot 1}{x^3y^3y^1} = \frac{1}{y}$$

The answer is choice **A.**

SAT EXAMPLE

2. The sum of two positive consecutive integers is x. In terms of x, what is the value of the smaller of these two integers?

   A $\frac{x}{2} - 1$ **B** $\frac{x-1}{2}$ **C** $\frac{x}{2}$

   D $\frac{x+1}{2}$ **E** $\frac{x}{2} + 1$

HINT On multiple-choice questions with variables in the answer choices, you can sometimes use a strategy called "plug-in."

Solution The "plug-in" strategy uses substitution to test the choices. Suppose the two numbers were 2 and 3. Then $x = 2 + 3$ or 5. Substitute 5 for x and see which choice yields 2, the smaller of the two numbers.

Choice A: $\frac{5}{2} - 1$ This is not an integer.

Choice B: $\frac{5-1}{2} = 2$ This is the answer, but check the rest just to be sure.

Alternate Solution You can solve this problem by writing an algebraic expression for each number. Let a be the first (smallest) positive integer. Then $(a + 1)$ is the next positive integer. Write an equation for "The sum of the two numbers is x" and solve for a.

$$a + (a + 1) = x$$
$$2a + 1 = x$$
$$2a = x - 1$$
$$a = \frac{x-1}{2}$$

The answer is choice **B.**

After you work each problem, record your answer on the answer sheet provided or on a piece of paper.

Multiple Choice

1. If the product of $(1 + 2)$, $(2 + 3)$, and $(3 + 4)$ is equal to one half the sum of 20 and x, then $x =$

   A 10 **B** 85 **C** 105 **D** 190 **E** 1,210

2. $5\frac{1}{3} - 6\frac{1}{4} = ?$

   A $-\frac{11}{12}$
   B $-\frac{1}{2}$
   C $-\frac{2}{7}$
   D $\frac{1}{2}$
   E $\frac{9}{12}$

3. Mia has a pitcher containing x ounces of root beer. If she pours y ounces of root beer into each of z glasses, how much root beer will remain in the pitcher?

   A $\frac{x}{y} + z$
   B $xy - z$
   C $\frac{x}{yz}$
   D $x - yz$
   E $\frac{x}{y} - z$

4. Which of the following is equal to 0.064?

   A $\left(\frac{1}{80}\right)^2$ **B** $\left(\frac{8}{100}\right)^2$ **C** $\left(\frac{1}{8}\right)^2$
   D $\left(\frac{2}{5}\right)^3$ **E** $\left(\frac{8}{10}\right)^3$

5. A plumber charges $75 for the first thirty minutes of each house call plus $2 for each additional minute that she works. The plumber charged Mr. Adams $113 for her time. For what amount of time, in minutes, did the plumber work?

   A 38 **B** 44 **C** 49 **D** 59 **E** 64

6. If $\frac{2 + x}{5 + x} = \frac{2}{5} + \frac{2}{5}$, then $x =$

   A $\frac{2}{5}$ **B** 1 **C** 2 **D** 5 **E** 10

7. Which of the following must be true?

   I. The sum of two consecutive integers is odd.

   II. The sum of three consecutive integers is even.

   III. The sum of three consecutive integers is a multiple of 3.

   A I only
   B II only
   C I and II only
   D I and III only
   E I, II, and III

8. Jose has at least one quarter, one dime, one nickel, and one penny in his pocket. If he has twice as many pennies as nickels, twice as many nickels as dimes, and twice as many dimes as quarters, then what is the least amount of money he could have in his pocket?

   A $0.41 **B** $0.64 **C** $0.71
   D $0.73 **E** $2.51

9. **Quantitative Comparison**

   A if the quantity in Column A is greater
   B if the quantity in Column B is greater
   C if the two quantities are equal
   D if the relationship cannot be determined from the information given

Column A	Column B
$\dfrac{\frac{3}{2}}{\left(\frac{3}{2}\right)^2}$	$\frac{2}{3}$

10. **Grid-In** At a music store, the price of a CD is three times the price of a cassette tape. If 40 CDs were sold for a total of $480 and the combined sales of CDs and cassette tapes totaled $600, how many cassette tapes were sold?

*inter*NET CONNECTION **SAT/ACT Practice** For additional test practice questions, visit: **www.amc.glencoe.com**

THE NATURE OF GRAPHS

CHAPTER OBJECTIVES

- **Graph functions, relations, inverses, and inequalities.** *(Lessons 3-1, 3-3, 3-4, 3-7)*
- **Analyze families of graphs.** *(Lesson 3-2)*
- **Investigate symmetry, continuity, end behavior, and transformations of graphs.** *(Lessons 3-1, 3-2, 3-5)*
- **Find asymptotes and extrema of functions.** *(Lessons 3-6, 3-7)*
- **Solve problems involving direct, inverse, and joint variation.** *(Lesson 3-8)*

Symmetry and Coordinate Graphs

OBJECTIVES

- Use algebraic tests to determine if the graph of a relation is symmetrical.
- Classify functions as even or odd.

Real World Application

PHARMACOLOGY Designing a drug to treat a disease requires an understanding of the molecular structures of the substances involved in the disease process. The substances are isolated in crystalline form, and X rays are passed through the symmetrically-arranged atoms of the crystals. The existence of symmetry in crystals causes the X rays to be diffracted in regular patterns. These symmetrical patterns are used to determine and visualize the molecular structure of the substance. *A related problem is solved in Example 4.*

Like crystals, graphs of certain functions display special types of symmetry. For some functions with symmetrical graphs, knowledge of symmetry can often help you sketch and analyze the graphs. One type of symmetry a graph may have is **point symmetry.**

Point Symmetry Two distinct points P and P' are symmetric with respect to point M if and only if M is the midpoint of $\overline{PP'}$. Point M is symmetric with respect to itself.

When the definition of point symmetry is extended to a set of points, such as the graph of a function, then each point P in the set must have an **image point** P' that is also in the set. A figure that is symmetric with respect to a given point can be rotated 180° about that point and appear unchanged. Each of the figures below has point symmetry with respect to the labeled point.

Symmetry with respect to a given point M can be expressed as symmetry about point M.

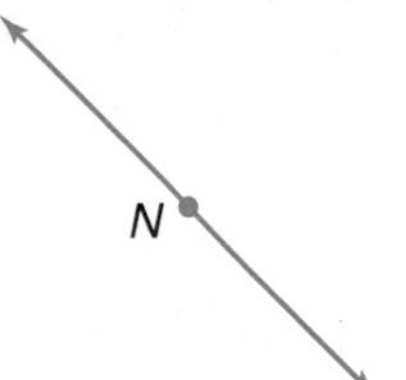

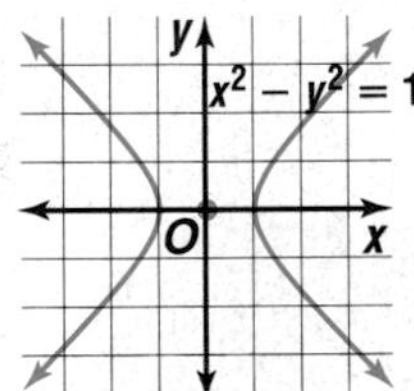

The origin is a common point of symmetry. Observe that the graphs of $f(x) = x^3$ and $g(x) = \frac{1}{x}$ exhibit symmetry with respect to the origin. Look for patterns in the table of function values beside each graph.

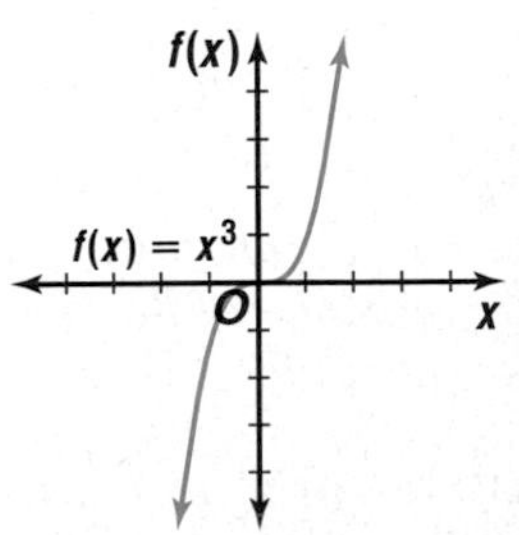

$f(x) = x^3$			
x	$f(x)$	$f(-x)$	$-f(x)$
1	1	-1	-1
2	8	-8	-8
3	27	-27	-27
4	64	-64	-64

Note that $f(-x) = -f(x)$.

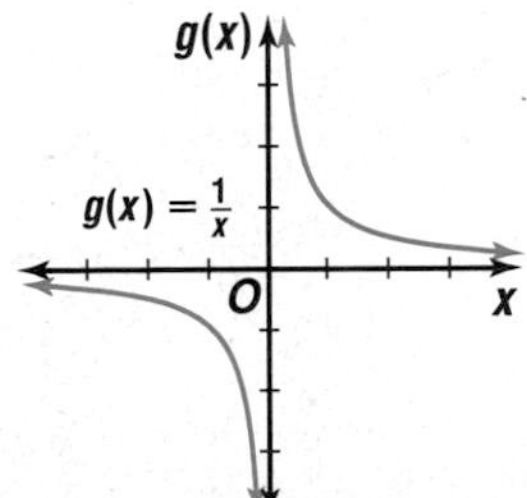

$g(x) = \frac{1}{x}$			
x	$g(x)$	$g(-x)$	$-g(x)$
1	1	-1	-1
2	$\frac{1}{2}$	$-\frac{1}{2}$	$-\frac{1}{2}$
3	$\frac{1}{3}$	$-\frac{1}{3}$	$-\frac{1}{3}$
4	$\frac{1}{4}$	$-\frac{1}{4}$	$-\frac{1}{4}$

Note that $g(-x) = -g(x)$.

The values in the tables suggest that $f(-x) = -f(x)$ whenever the graph of a function is symmetric with respect to the origin.

Symmetry with Respect to the Origin

The graph of a relation S is symmetric with respect to the origin if and only if $(a, b) \in S$ implies that $(-a, -b) \in S$. A function has a graph that is symmetric with respect to the origin if and only if $f(-x) = -f(x)$ for all x in the domain of f.

$(a, b) \in S$ means the ordered pair (a, b) belongs to the solution set S.

Example 1 demonstrates how to algebraically test for symmetry about the origin.

Example 1 **Determine whether each graph is symmetric with respect to the origin.**

a. $f(x) = x^5$

The graph of $f(x) = x^5$ appears to be symmetric with respect to the origin.

b. $g(x) = \frac{x}{1 - x}$

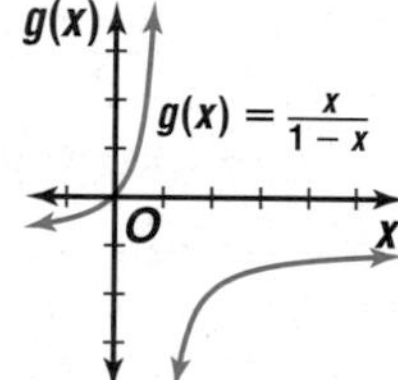

The graph of $g(x) = \frac{x}{1 - x}$ does not appear to be symmetric with respect to the origin.

We can verify these conjectures algebraically by following these two steps.

1. Find $f(-x)$ and $-f(x)$.
2. If $f(-x) = -f(x)$, the graph has point symmetry.

a. $f(x) = x^5$

Find $f(-x)$.

$f(-x) = (-x)^5$ *Replace x with −x.*

$f(-x) = -x^5$ *$(-x)^5 = (-1)^5x^5 = -1x^5$ or $-x^5$*

Find $-f(x)$.

$-f(x) = -x^5$ *Determine the opposite of the function.*

The graph of $f(x) = x^5$ is symmetric with respect to the origin because $f(-x) = -f(x)$.

b. $g(x) = \frac{x}{1 - x}$

Find $g(-x)$.

$g(-x) = \frac{-x}{1 - (-x)}$ *Replace x with −x.*

$= \frac{-x}{1 + x}$

Find $-g(x)$.

$-g(x) = -\frac{x}{1 - x}$ *Determine the opposite of the function.*

$= \frac{-x}{1 - x}$

The graph of $g(x) = \frac{x}{1 - x}$ is not symmetric with respect to the origin because $g(-x) \neq -g(x)$.

Another type of symmetry is **line symmetry.**

Line Symmetry	Two distinct points P and P' are symmetric with respect to a line ℓ if and only if ℓ is the perpendicular bisector of $\overline{PP'}$. A point P is symmetric to itself with respect to line ℓ if and only if P is on ℓ.

Each graph below has line symmetry. The equation of each line of symmetry is given. Graphs that have line symmetry can be folded along the line of symmetry so that the two halves match exactly. Some graphs, such as the graph of an ellipse, have more than one line of symmetry.

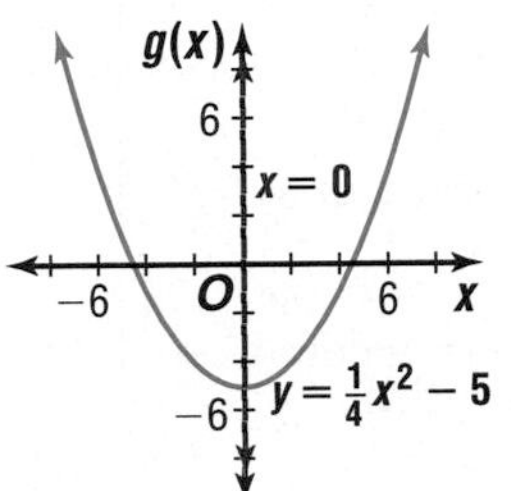

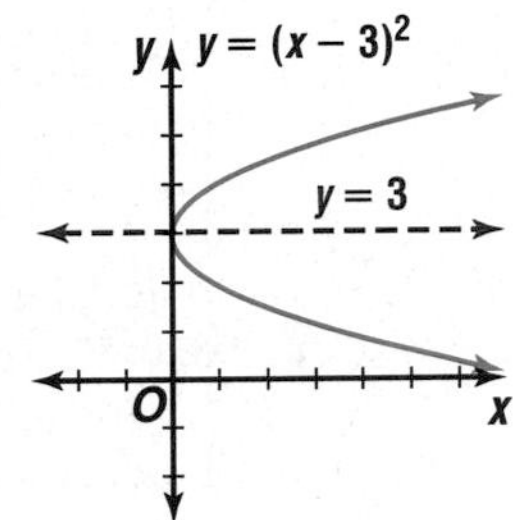

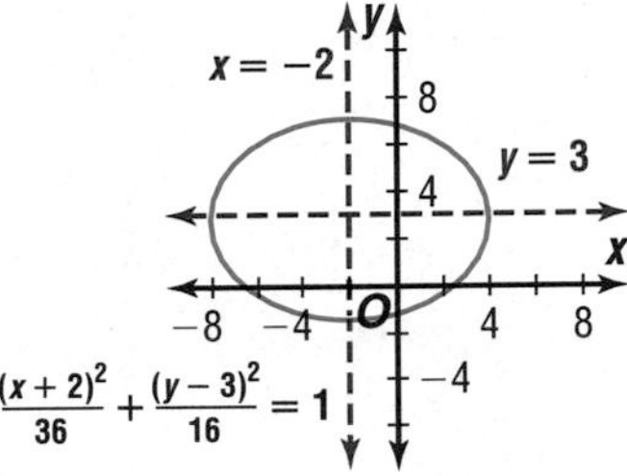

Some common lines of symmetry are the x-axis, the y-axis, the line $y = x$, and the line $y = -x$. The following table shows how the coordinates of symmetric points are related for each of these lines of symmetry. Set notation is often used to define the conditions for symmetry.

Symmetry with Respect to the:	**Definition and Test**	**Example**
x-axis	$(a, -b) \in S$ if and only if $(a, b) \in S$. *Example:* $(2, \sqrt{6})$ and $(2, -\sqrt{6})$ are on the graph. *Test:* Substituting (a, b) and $(a, -b)$ into the equation produces equivalent equations.	$x = y^2 - 4$; $(2, \sqrt{6})$; $(2, -\sqrt{6})$; $y = 0$
y-axis	$(-a, b) \in S$ if and only if $(a, b) \in S$. *Example:* $(2, 8)$ and $(-2, 8)$ are on the graph. *Test:* Substituting (a, b) and $(-a, b)$ into the equation produces equivalent equations.	$y = -x^2 + 12$; $(-2, 8)$; $(2, 8)$; $x = 0$

(continued on the next page)

Symmetry with Respect to the Line:	Definition and Test	Example
$y = x$	$(b, a) \in S$ if and only if $(a, b) \in S$. *Example:* (2, 3) and (3, 2) are on the graph. *Test:* Substituting (a, b) and (b, a) into the equation produces equivalent equations.	
$y = -x$	$(-b, -a) \in S$ if and only if $(a, b) \in S$. *Example:* (4, −1) and (1, −4) are on the graph. *Test:* Substituting (a, b) and $(-b, -a)$ into the equation produces equivalent equations.	

You can determine whether the graph of an equation has line symmetry without actually graphing the equation.

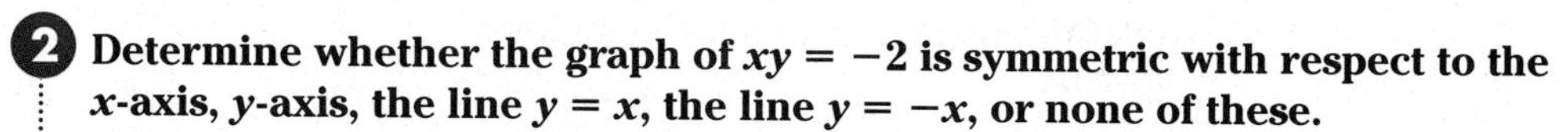

Example  **Determine whether the graph of $xy = -2$ is symmetric with respect to the x-axis, y-axis, the line $y = x$, the line $y = -x$, or none of these.**

Substituting (a, b) into the equation yields $ab = -2$. Check to see if each test produces an equation equivalent to $ab = -2$.

x-axis	$a(-b) = -2$	*Substitute $(a, -b)$ into the equation.*
	$-ab = -2$	*Simplify.*
	$ab = 2$	*Not equivalent to $ab = -2$*
y-axis	$(-a)b = -2$	*Substitute $(-a, b)$ into the equation.*
	$-ab = -2$	*Simplify.*
	$ab = 2$	*Not equivalent to $ab = -2$*
$y = x$	$(b)(a) = -2$	*Substitute (b, a) into the equation.*
	$ab = -2$	*Equivalent to $ab = -2$*
$y = -x$	$(-b)(-a) = -2$	*Substitute $(-b, -a)$ into the equation.*
	$ab = -2$	*Equivalent to $ab = -2$*

Therefore, the graph of $xy = -2$ is symmetric with respect to the line $y = x$ and the line $y = -x$. A sketch of the graph verifies the algebraic tests.

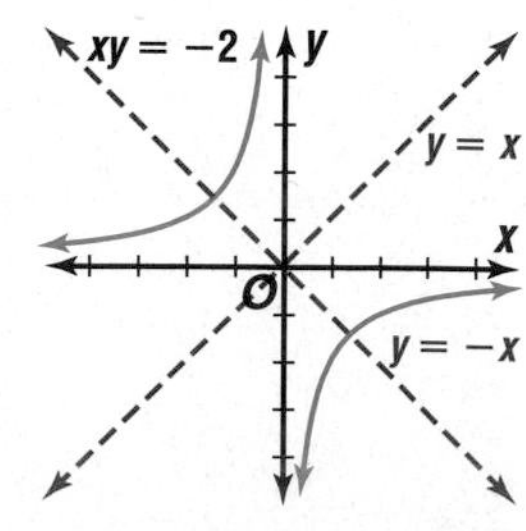

You can use information about symmetry to draw the graph of a relation.

Example 3 **Determine whether the graph of $|y| = 2 - |2x|$ is symmetric with respect to the *x*-axis, the *y*-axis, both, or neither. Use the information about the equation's symmetry to graph the relation.**

Substituting (a, b) into the equation yields $|b| = 2 - |2a|$. Check to see if each test produces an equation equivalent to $|b| = 2 - |2a|$.

x-axis $|-b| = 2 - |2a|$ *Substitute $(a, -b)$ into the equation.*
$|b| = 2 - |2a|$ *Equivalent to $|b| = 2 - |2a|$ since $|-b| = |b|$.*

y-axis $|b| = 2 - |-2a|$ *Substitute $(-a, b)$ into the equation.*
$|b| = 2 - |2a|$ *Equivalent to $|b| = 2 - |2a|$, since $|-2a| = |2a|$.*

Therefore, the graph of $|y| = 2 - |2x|$ is symmetric with respect to both the *x*-axis and the *y*-axis.

To graph the relation, let us first consider ordered pairs where $x \geq 0$ and $y \geq 0$. The relation $|y| = 2 - |2x|$ contains the same points as $y = 2 - 2x$ in the first quadrant.

Therefore, in the first quadrant, the graph of $|y| = 2 - |2x|$ is the same as the graph of $y = 2 - 2x$.

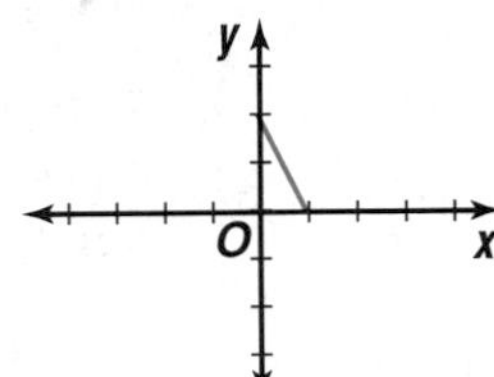

Since the graph is symmetric with respect to the *x*-axis, every point in the first quadrant has a corresponding point in the fourth quadrant.

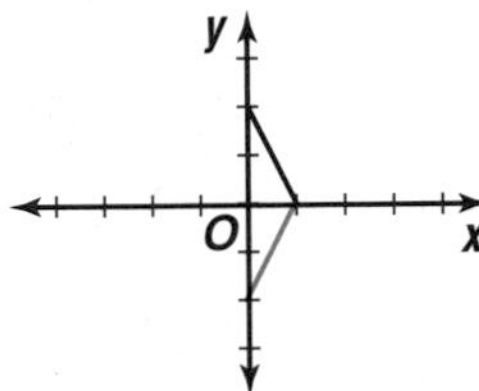

Since the graph is symmetric with respect to the *y*-axis, every point in the first and fourth quadrants has a corresponding point on the other side of the *y*-axis.

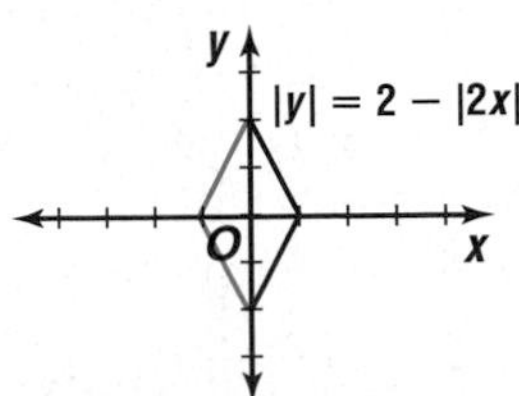

Example **4** **CRYSTALLOGRAPHY** **A crystallographer can model a cross-section of a crystal with mathematical equations. After sketching the outline on a graph, she notes that the crystal has both x-axis and y-axis symmetry. She uses the piecewise function $y = \begin{cases} 2 \text{ if } 0 \le x \le 1 \\ 3 - x \text{ if } 1 \le x \le 3 \end{cases}$ to model the first quadrant portion of the cross-section. Write piecewise equations for the remaining sides.**

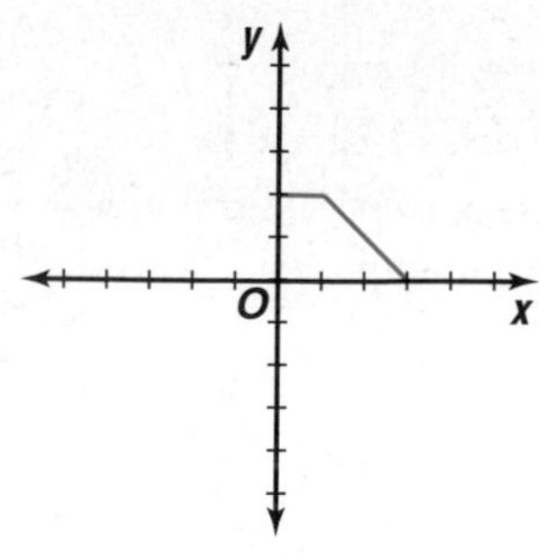

Look Back
Refer to Lesson 1-7 for more about piecewise functions.

Since the graph has x-axis symmetry, substitute $(x, -y)$ into the original equation to produce the equation for the fourth quadrant portion.

$$y = \begin{cases} 2 \text{ if } 0 \le x \le 1 \\ 3 - x \text{ if } 1 \le x \le 3 \end{cases}$$

$$-y = \begin{cases} 2 \text{ if } 0 \le x \le 1 \\ 3 - x \text{ if } 1 \le x \le 3 \end{cases}$$ *Substitute $(x, -y)$ into the equation.*

$$y = \begin{cases} -2 \text{ if } 0 \le x \le 1 \\ 3 + x \text{ if } 1 \le x \le 3 \end{cases}$$ *Solve for y.*

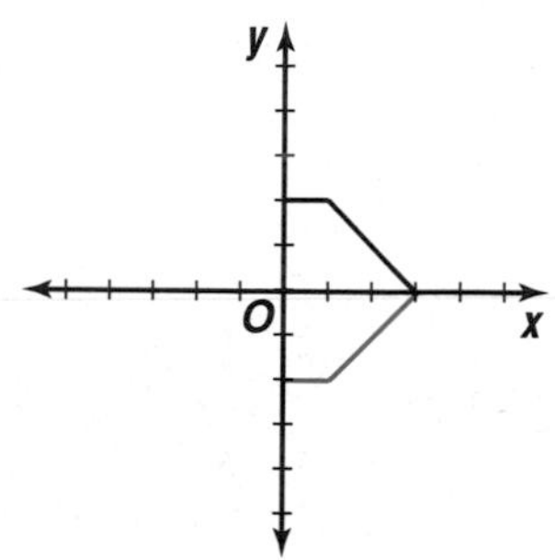

The equation $y = \begin{cases} -2 \text{ if } 0 \le x \le 1 \\ x - 3 \text{ if } 1 \le x \le 3 \end{cases}$ models the fourth quadrant portion of the cross section.

Since the graph has y-axis symmetry, substitute $(-x, y)$ into the first and fourth quadrant equations to produce the equations for the second and third quadrants.

$$y = \begin{cases} 2 \text{ if } 0 \le x \le 1 \\ 3 - x \text{ if } 1 \le x \le 3 \end{cases}$$ *Start with the first quadrant equation.*

$$y = \begin{cases} 2 \text{ if } 0 \le -x \le 1 \\ 3 - (-x) \text{ if } 1 \le -x \le 3 \end{cases}$$ *Substitute $(-x, y)$ into the equation.*

$$y = \begin{cases} 2 \text{ if } 0 \ge x \ge -1 \\ 3 + x \text{ if } -1 \ge x \ge -3 \end{cases}$$ *This is the second quadrant equation.*

$$y = \begin{cases} -2 \text{ if } 0 \le x \le 1 \\ x - 3 \text{ if } 1 \le x \le 3 \end{cases}$$ *Start with the fourth quadrant equation.*

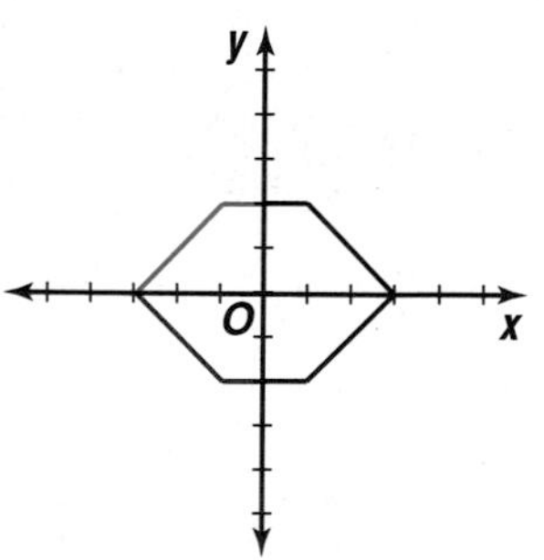

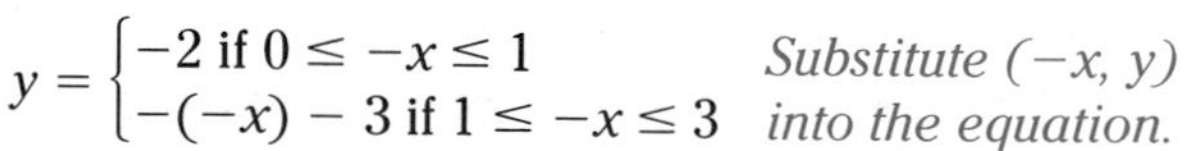

$$y = \begin{cases} -2 \text{ if } 0 \le -x \le 1 \\ -(-x) - 3 \text{ if } 1 \le -x \le 3 \end{cases}$$ *Substitute $(-x, y)$ into the equation.*

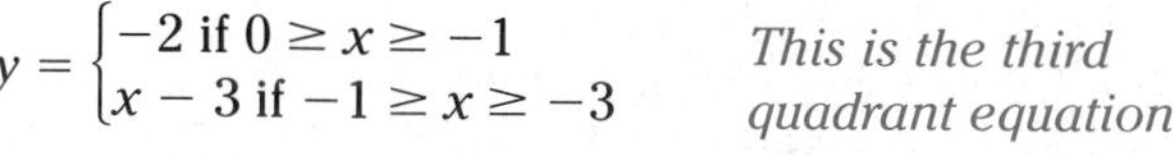

$$y = \begin{cases} -2 \text{ if } 0 \ge x \ge -1 \\ x - 3 \text{ if } -1 \ge x \ge -3 \end{cases}$$ *This is the third quadrant equation.*

The equations $y = \begin{cases} 2 \text{ if } -1 \le x \le 0 \\ 3 + x \text{ if } -3 \le x \le -1 \end{cases}$ and $y = \begin{cases} -2 \text{ if } -1 \le x \le 0 \\ x - 3 \text{ if } -3 \le x \le -1 \end{cases}$ model the second and third quadrant portions of the cross section respectively.

Graphing Calculator Programs For a graphing calculator program that determines whether a function is even, odd, or neither, visit **www.amc.glencoe.com**

Functions whose graphs are symmetric with respect to the y-axis are **even functions.** Functions whose graphs are symmetric with respect to the origin are **odd functions.** Some functions are neither even nor odd. From Example 1, $f(x) = x^5$ is an odd function, and $g(x) = \frac{x}{1-x}$ is neither even nor odd.

even functions	odd functions
$f(-x) = f(x)$	$f(-x) = -f(x)$
symmetric with respect to the y-axis	symmetric with respect to the origin

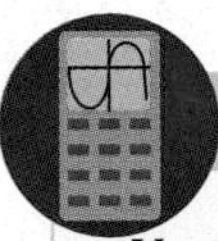

GRAPHING CALCULATOR EXPLORATION

You can use the **TRACE** function to investigate the symmetry of a function.

- Graph the function.
- Use TRACE to observe the relationship between points of the graph having opposite x-coordinates.
- Use this information to determine the relationship between $f(x)$ and $f(-x)$.

TRY THESE **Graph each function to determine how $f(x)$ and $f(-x)$ are related.**

1. $f(x) = x^8 - 3x^4 + 2x^2 + 2$
2. $f(x) = x^7 + 4x^5 - x^3$

WHAT DO YOU THINK?

3. Identify the functions in Exercises 1 and 2 as *odd, even,* or *neither* based on your observations of their graphs.

4. Verify your conjectures algebraically.

5. How could you use symmetry to help you graph an even or odd function? Give an example.

6. Explain how you could use the **ASK** option in **TBLSET** to determine the relationship between $f(x)$ and $f(-x)$ for a given function.

CHECK FOR UNDERSTANDING

Communicating Mathematics

Read and study the lesson to answer each question.

1. Refer to the tables on pages 129–130. **Identify** each graph as an even function, an odd function, or neither. Explain.
2. **Explain** how rotating a graph of an odd function 180° will affect its appearance. Draw an example.
3. Consider the graph at the right.
   a. **Determine** four lines of symmetry for the graph.
   b. How many other lines of symmetry does this graph possess?
   c. What other type of symmetry does this graph possess?

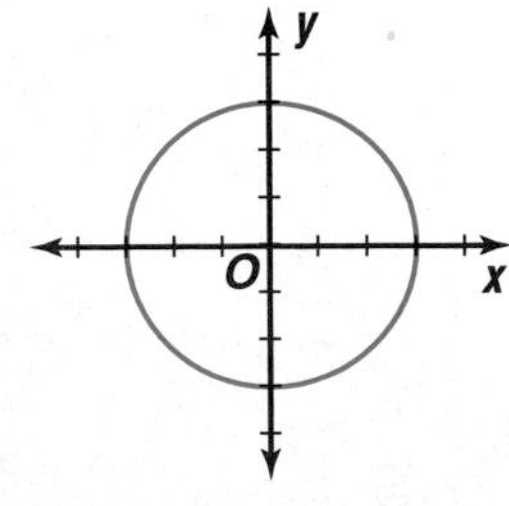

4. **Write** an explanation of how to test for symmetry with respect to the line $y = -x$.

5. **You Decide** Alicia says that any graph that is symmetric to the origin and to the y-axis must also be symmetric to the x-axis. Chet disagrees. Who is correct? Support your answer graphically and algebraically.

Guided Practice

Determine whether the graph of each function is symmetric with respect to the origin.

6. $f(x) = x^6 + 9x$

7. $f(x) = \frac{1}{5x} - x^{19}$

Determine whether the graph of each equation is symmetric with respect to the x-axis, y-axis, the line $y = x$, the line $y = -x$, or none of these.

8. $6x^2 = y - 1$

9. $x^3 + y^3 = 4$

10. Copy and complete the graph at the right so that it is the graph of an even function.

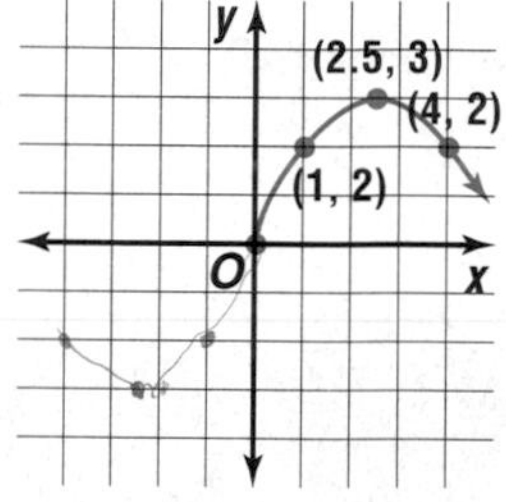

Determine whether the graph of each equation is symmetric with respect to the x-axis, the y-axis, both, or neither. Use the information about symmetry to graph the relation.

11. $y = \sqrt{2 - x^2}$

12. $|y| = x^3$

13. **Physics** Suppose the light pattern from a fog light can be modeled by the equation $\frac{x^2}{25} - \frac{y^2}{9} = 1$. One of the points on the graph of this equation is at $\left(6, \frac{3\sqrt{11}}{5}\right)$, and one of the x-intercepts is -5. Find the coordinates of three additional points on the graph and the other x-intercept.

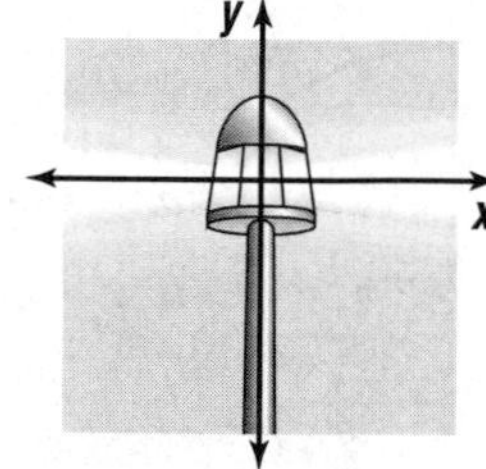

EXERCISES

Practice

Determine whether the graph of each function is symmetric with respect to the origin.

14. $f(x) = 3x$

15. $f(x) = x^3 - 1$

16. $f(x) = 5x^2 + 6x + 9$

17. $f(x) = \frac{1}{4x^7}$

18. $f(x) = -7x^5 + 8x$

19. $f(x) = \frac{1}{x} - x^{100}$

20. Is the graph of $g(x) = \frac{x^2 - 1}{x}$ symmetric with respect to the origin? Explain how you determined your answer.

Determine whether the graph of each equation is symmetric with respect to the x-axis, y-axis, the line $y = x$, the line $y = -x$, or none of these.

21. $xy = -5$

22. $x + y^2 = 1$

23. $y = -8x$

24. $y = \frac{1}{x^2}$

25. $x^2 + y^2 = 4$

26. $y^2 = \frac{4x^2}{9} - 4$

27. Which line(s) are lines of symmetry for the graph of $x^2 = \frac{1}{y^2}$?

For Exercises 28–30, refer to the graph.

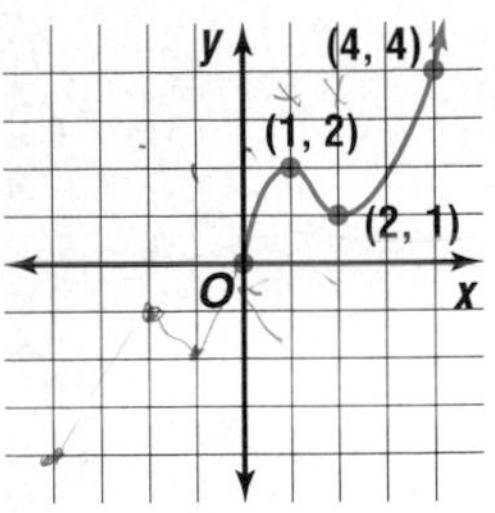

28. Complete the graph so that it is the graph of an odd function.

29. Complete the graph so that it is the graph of an even function.

30. Complete the graph so that it is the graph of a function that is neither even nor odd.

Determine whether the graph of each equation is symmetric with respect to the *x*-axis, the *y*-axis, both, or neither. Use the information about symmetry to graph the relation.

31. $y^2 = x^2$ **32.** $|x| = -3y$ **33.** $y^2 + 3x = 0$

34. $|y| = 2x^2$ **35.** $x = \pm\sqrt{12 - 8y^2}$ **36.** $|y| = xy$

37. Graph the equation $|y| = x^3 - x$ using information about the symmetry of the graph.

Applications and Problem Solving

38. Physics The path of a comet around the Sun can be modeled by a transformation of the equation $\frac{x^2}{8} + \frac{y^2}{10} = 1$.

a. Determine the symmetry in the graph of the comet's path.

b. Use symmetry to graph the equation $\frac{x^2}{8} + \frac{y^2}{10} = 1$.

c. If it is known that the comet passes through the point at $(2, \sqrt{5})$, name the coordinates of three other points through which it must pass.

39. Critical Thinking Write the equation of a graph that is symmetric with respect to the *x*-axis.

40. Geometry Draw a diagram composed of line segments that exhibits both *x*- and *y*-axis symmetry. Write equations for the boundaries.

41. Communication Radio waves emitted from two different radio towers interfere with each other's signal. The path of interference can be modeled by the equation $\frac{y^2}{12} - \frac{x^2}{16} = 1$, where the origin is the midpoint of the line segment between the two towers and the positive *y*-axis represents north. Juana lives on an east-west road 6 miles north of the *x*-axis and cannot receive the radio station at her house. At what coordinates might Juana live relative to the midpoint between the two towers?

42. Critical Thinking Must the graph of an odd function contain the origin? Explain your reasoning and illustrate your point with the graph of a specific function.

Mixed Review

43. Manufacturing A manufacturer makes a profit of $6 on a bicycle and $4 on a tricycle. Department A requires 3 hours to manufacture the parts for a bicycle and 4 hours to manufacture parts for a tricycle. Department B takes 5 hours to assemble a bicycle and 2 hours to assemble a tricycle. How many bicycles and tricycles should be produced to maximize the profit if the total time available in department A is 450 hours and in department B is 400 hours? *(Lesson 2-7)*

Extra Practice See p. A30.

44. Find AB if $A = \begin{bmatrix} 4 & 3 \\ 7 & 2 \end{bmatrix}$ and $B = \begin{bmatrix} 8 & 5 \\ 9 & 6 \end{bmatrix}$. *(Lesson 2-3)*

45. Solve the system of equations, $2x + y + z = 0$, $3x - 2y - 3z = -21$, and $4x + 5y + 3z = -2$. *(Lesson 2-2)*

46. State whether the system, $4x - 2y = 7$ and $-12x + 6y = -21$, is *consistent and independent, consistent and dependent,* or *inconsistent.* *(Lesson 2-1)*

47. Graph $0 \leq x - y \leq 2$. *(Lesson 1-8)*

48. Write an equation in slope-intercept form for the line that passes through $A(0, 2)$ and $B(-2, 16)$. *(Lesson 1-4)*

49. If $f(x) = -2x + 11$ and $g(x) = x - 6$, find $[f \circ g](x)$ and $[g \circ f](x)$. *(Lesson 1-2)*

50. **SAT/ACT Practice** What is the product of 75^3 and 75^7?

A 75^5 **B** 75^{10} **C** 150^{10} **D** 5625^{10} **E** 75^{21}

CAREER CHOICES

Biomedical Engineering

Would you like to help people live better lives? Are you interested in a career in the field of health? If you answered yes, then biomedical engineering may be the career for you. Biomedical engineers apply engineering skills and life science knowledge to design artificial parts for the human body and devices for investigating and repairing the human body. Some examples are artificial organs, pacemakers, and surgical lasers.

In biomedical engineering, there are three primary work areas: research, design, and teaching. There are also many specialty areas in this field. Some of these are bioinstrumentation, biomechanics, biomaterials, and rehabilitation engineering.

The graph shows an increase in the number of outpatient visits over the number of hospital visits. This is due in part to recent advancements in biomedical engineering.

Career Overview

Degree Preferred:
bachelor's degree in biomedical engineering

Related Courses:
biology, chemistry, mathematics

Outlook:
number of jobs expected to increase through the year 2006

Hospital Vital Signs

Total Number of Hospital Admissions and Outpatient Visits, 1965-1996 (in millions)

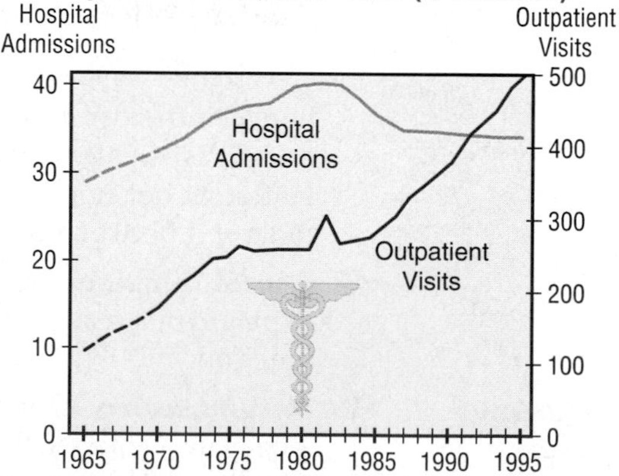

Source: *The Wall Street Journal Almanac*

***inter*NET CONNECTION** For more information on careers in biomedical engineering, visit: **www.amc.glencoe.com**

3-2

Families of Graphs

OBJECTIVES

- Identify transformations of simple graphs.
- Sketch graphs of related functions.

ENTERTAINMENT At some circuses, a human cannonball is shot out of a special cannon. In order to perform this death-defying feat safely, the maximum height and distance of the performer must be calculated accurately. Quadratic functions can be used to model the height of a projectile like a human cannonball at any time during its flight. The quadratic equation used to model height versus time is closely related to the equation of $y = x^2$. *A problem related to this is solved in Example 5.*

All parabolas are related to the graph of $y = x^2$. This makes $y = x^2$ the **parent graph** of the family of parabolas. Recall that a *family of graphs* is a group of graphs that displays one or more similar characteristics.

A parent graph is a basic graph that is transformed to create other members in a family of graphs. Some different types are shown below. Notice that with the exception of the constant function, the coefficient of x in each equation is 1.

constant function

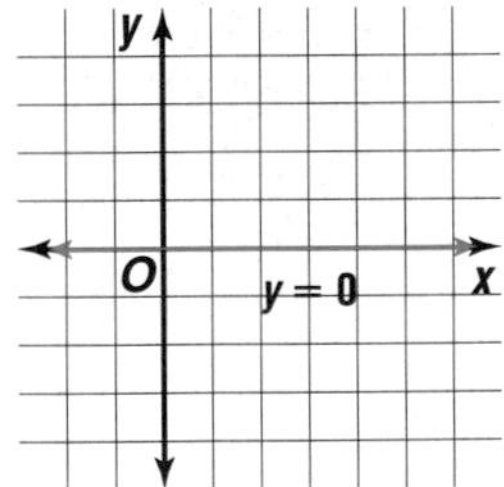

identity function

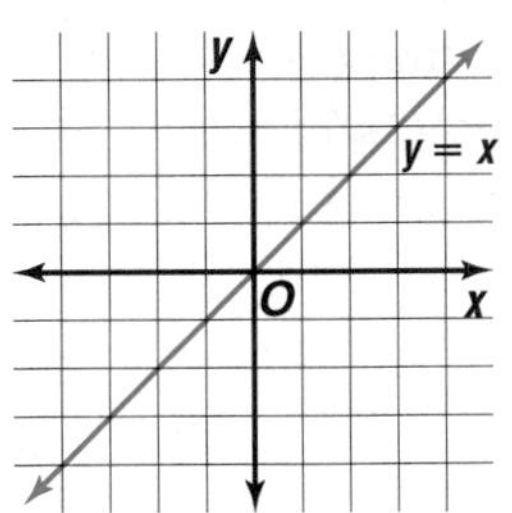

polynomial functions

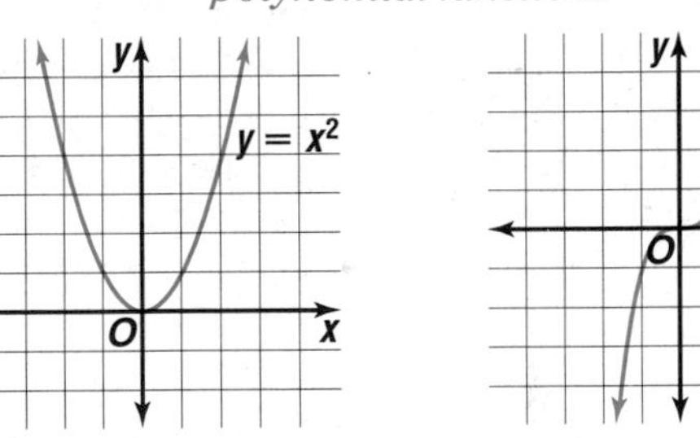

square root function

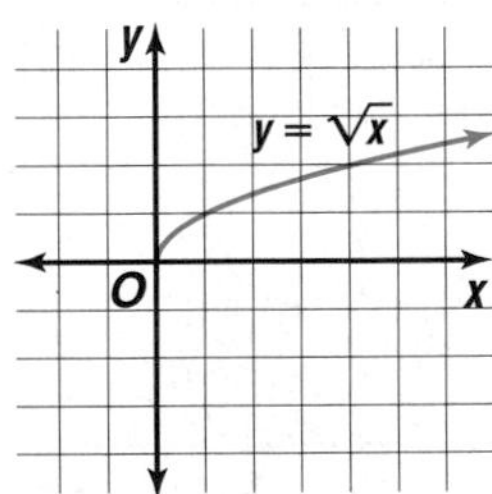

absolute value function

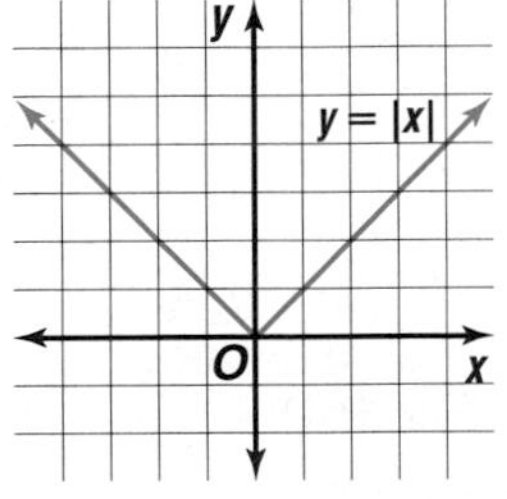

greatest integer function

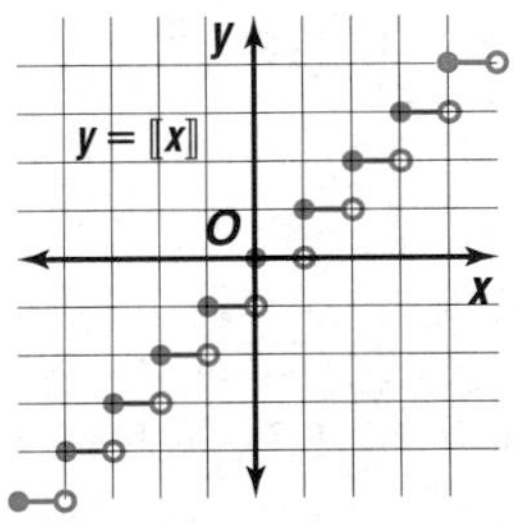

rational function

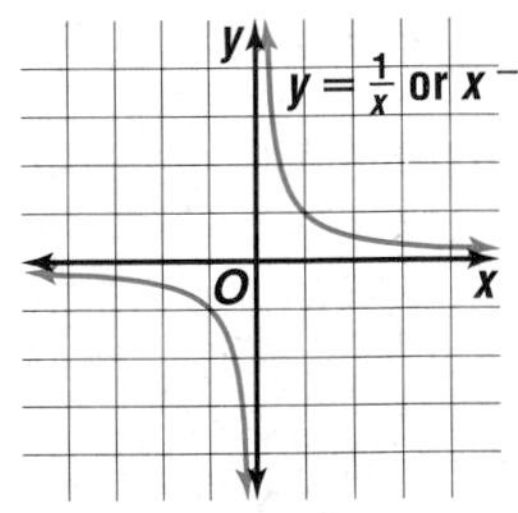

Look Back
Refer to Lesson 2-4 for more about reflections and translations.

Reflections and translations of the parent function can affect the appearance of the graph. The transformed graph may appear in a different location, but it will resemble the parent graph. A reflection flips a figure over a line called the *axis of symmetry. The axis of symmetry is also called the line of symmetry.*

Example 1 **Graph $f(x) = |x|$ and $g(x) = -|x|$. Describe how the graphs of $f(x)$ and $g(x)$ are related.**

Graphing Calculator Tip

You can use a graphing calculator to check your sketch of any function in this lesson.

x	$f(x) = \lvert x \rvert$	$g(x) = -\lvert x \rvert$
−2	2	−2
−1	1	−1
0	0	0
1	1	−1
2	2	−2

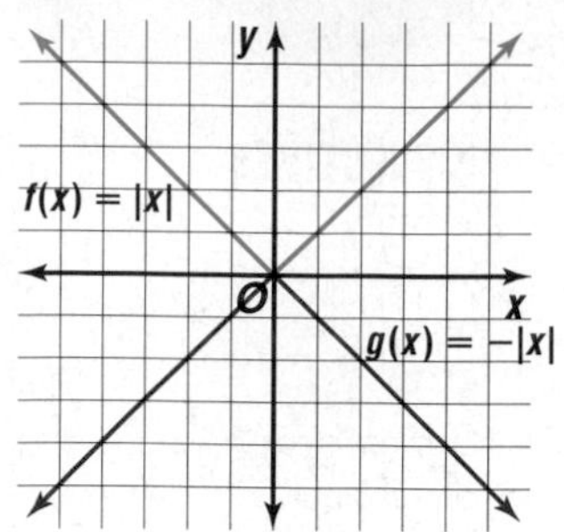

To graph both equations on the same axis, let $y = f(x)$ and $y = g(x)$.

The graph of $g(x)$ is a reflection of the graph of $f(x)$ over the x-axis. The symmetric relationship can be stated algebraically by $g(x) = -f(x)$, or $f(x) = -g(x)$. *Notice that the effect of multiplying a function by -1 is a reflection over the x-axis.*

When a constant c is added to or subtracted from a parent function, the result, $f(x) \pm c$, is a translation of the graph up or down. When a constant c is added or subtracted from x before evaluating a parent function, the result, $f(x \pm c)$, is a translation left or right.

Example 2 **Use the parent graph $y = \sqrt{x}$ to sketch the graph of each function.**

a. $y = \sqrt{x} + 2$

This function is of the form $y = f(x) + 2$. Since 2 is added to the parent function $y = \sqrt{x}$, the graph of the parent function moves up 2 units.

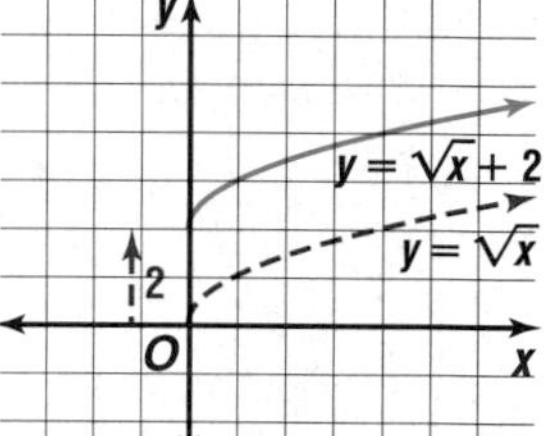

b. $y = \sqrt{x - 4}$

This function is of the form $y = f(x - 4)$. Since 4 is being subtracted from x before being evaluated by the parent function, the graph of the parent function $y = \sqrt{x}$ slides 4 units right.

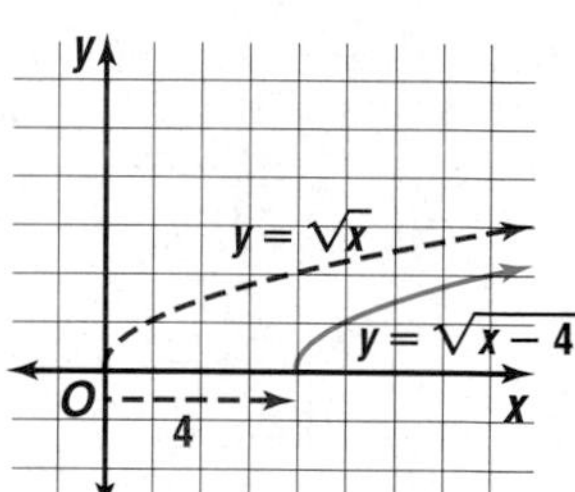

c. $y = \sqrt{x + 3} - 1$

This function is of the form $y = f(x + 3) - 1$. The addition of 3 indicates a slide of 3 units left, and the subtraction of 1 moves the parent function $y = \sqrt{x}$ down 1 unit.

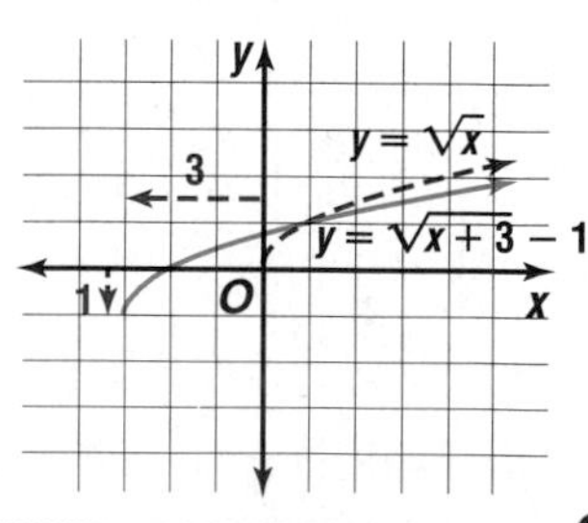

Remember that a dilation has the effect of shrinking or enlarging a figure. Likewise, when the leading coefficient of x is not 1, the function is expanded or compressed.

Example 3 **Graph each function. Then describe how it is related to its parent graph.**

a. $g(x) = 2[\![x]\!]$

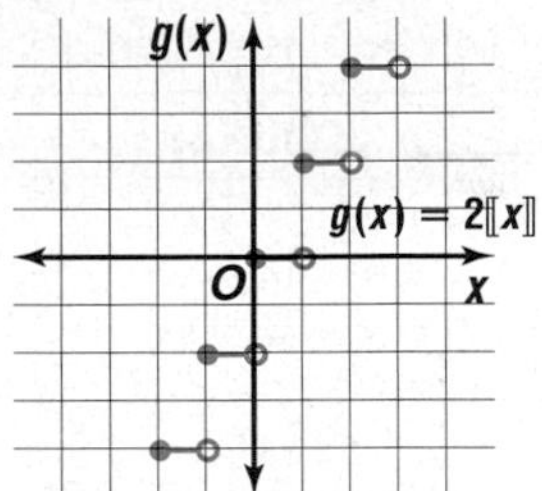

The parent graph is the greatest integer function, $f(x) = [\![x]\!]$. $g(x) = 2[\![x]\!]$ is a vertical expansion by a factor of 2. The vertical distance between the steps is 2 units.

b. $h(x) = -0.5[\![x]\!] - 4$

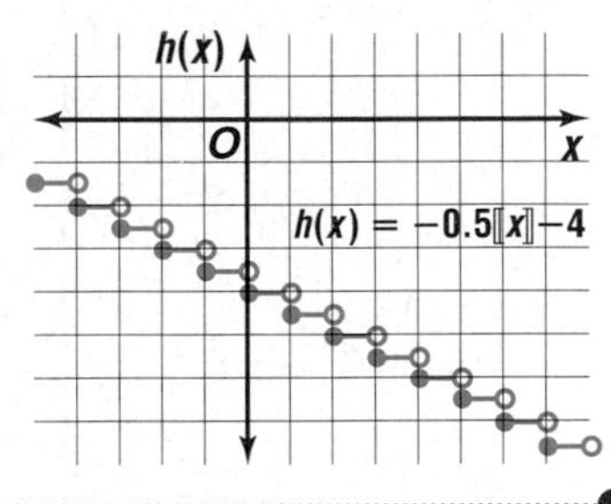

$h(x) = -0.5[\![x]\!] - 4$ reflects the parent graph over the x-axis, compresses it vertically by a factor of 0.5, and shifts the graph down 4 units. *Notice that multiplying by a positive number less than 1 compresses the graph vertically.*

The following chart summarizes the relationships in families of graphs. The parent graph may differ, but the transformations of the graphs have the same effect. Remember that more than one transformation may affect a parent graph.

Change to the Parent Function $y = f(x)$, $c > 0$	Change to Parent Graph	Examples
Reflections		
$y = -f(x)$ $y = f(-x)$	Is reflected over the x-axis. Is reflected over the y-axis.	
Translations		
$y = f(x) + c$ $y = f(x) - c$	Translates the graph c units up. Translates the graph c units down.	
$y = f(x + c)$ $y = f(x - c)$	Translates the graph c units left. Translates the graph c units right.	

(continued on the next page)

Change to the Parent Function $y = f(x)$, $c > 0$	Change to Parent Graph	Examples
Dilations		
$y = c \cdot f(x)$, $c > 1$ $y = c \cdot f(x)$, $0 < c < 1$	Expands the graph vertically. Compresses the graph vertically.	
$y = f(cx)$, $c > 1$ $y = f(cx)$, $0 < c < 1$	Compresses the graph horizontally. Expands the graph horizontally.	

Other transformations may affect the appearance of the graph. We will look at two more transformations that change the shape of the graph.

Example 4 **Observe the graph of each function. Describe how the graphs in parts b and c relate to the graph in part a.**

a. $f(x) = (x - 2)^2 - 3$

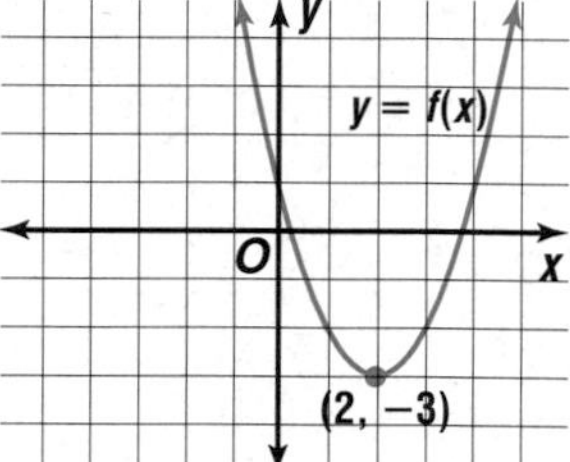

The graph of $y = f(x)$ is a translation of $y = x^2$. The parent graph has been translated right 2 units and down 3 units.

b. $y = |f(x)|$

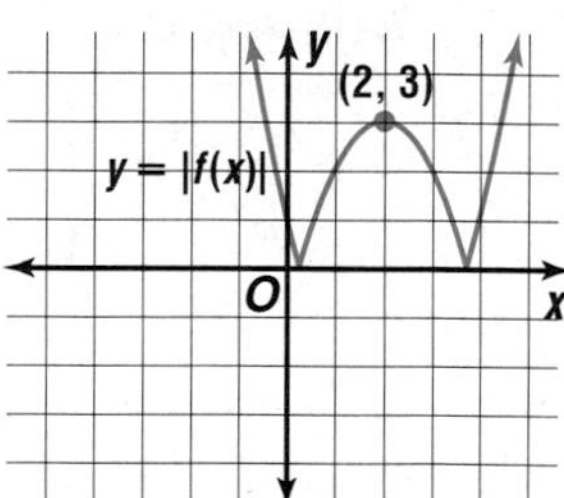

$|f(x)| = |(x - 2)^2 - 3|$

This transformation reflects any portion of the parent graph that is below the x-axis so that it is above the x-axis.

c. $y = f(|x|)$

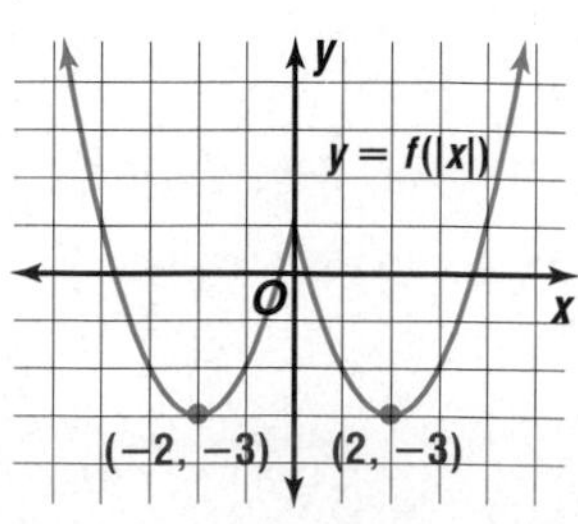

$f(|x|) = (|x| - 2)^2 - 3$

This transformation results in the portion of the parent graph on the left of the y-axis being replaced by a reflection of the portion on the right of the y-axis.

Example 5 Real World Application

ENTERTAINMENT **A traveling circus invites local schools to send math and science teams to its Science Challenge Day. One challenge is to write an equation that most accurately predicts the height of the flight of a human cannonball performer at any given time. Students collect data by witnessing a performance and examining time-lapse photographs of the flight. Using the performer's initial height of 15 feet and the photographs, one team records the data at the right. Write the equation of the related parabola that models the data.**

Time (seconds)	Height (feet)
0	15
1	39
2	47
3	39
4	15

A graph of the data reveals that a parabola is the best model for the data. The parent graph of a parabola is the graph of the equation $y = x^2$. To write the equation of the related parabola that models the data, we need to compare points located near the vertex of each graph. An analysis of the transformation these points have undergone will help us determine the equation of the transformed parabola.

From the graph, we can see that parent graph has been turned upside-down, indicating that the equation for this parabola has been multiplied by some negative constant c. Through further inspection of the graph and its data points, we can see that the vertex of the parent graph has been translated to the point (2, 47). Therefore, an equation that models the data is $y = c(x - 2)^2 + 47$.

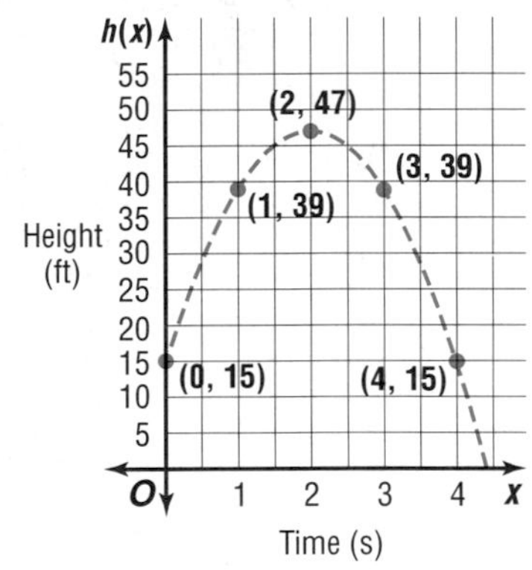

To find c, compare points near the vertex of the graphs of the parent function $f(x)$ and the graph of the data points. Look at the relationship between the differences in the y-coordinates for each set of points.

x	$f(x) = x^2$	
−2	4	
		−3
−1	1	
		−1
0	0	
		−1
1	1	
		−3
2	4	

x Time (seconds)	$h(x)$ Height (feet)	
0	15	
		24
1	39	
		8
2	47	
		8
3	39	
		24
4	15	

These differences are in a ratio of 1 to −8. This means that the graph of the parent graph has been expanded by a factor of −8. Thus, an equation that models the data is $y = -8(x - 2)^2 + 47$.

CHECK FOR UNDERSTANDING

Communicating Mathematics

Read and study the lesson to answer each question.

1. **Write** the equation of the graph obtained when the parent graph $y = x^3$ is translated 4 units left and 7 units down.
2. **Explain** the difference between the graphs of $y = (x + 3)^2$ and $y = x^2 + 3$.
3. **Name** two types of transformations for which the pre-image and the image are congruent figures.
4. **Describe** the differences between the graphs of $y = f(x)$ and $y = f(cx)$ for $c > 0$.
5. **Write** equations for the graphs of $g(x)$, $h(x)$, and $k(x)$ if the graph of $f(x) = \sqrt[3]{x}$ is the parent graph.

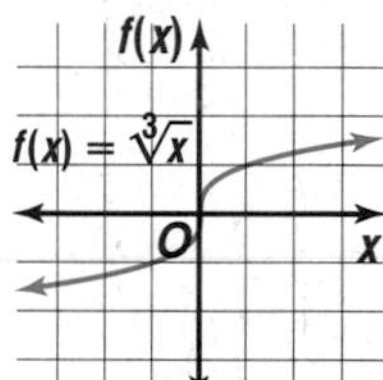

a. 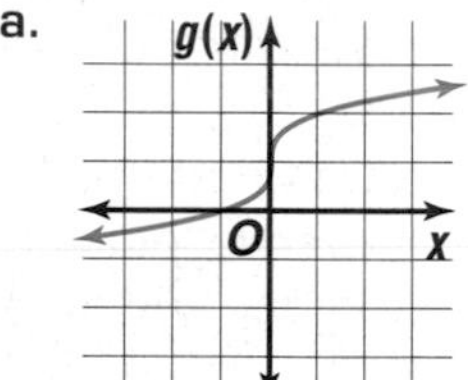

b. 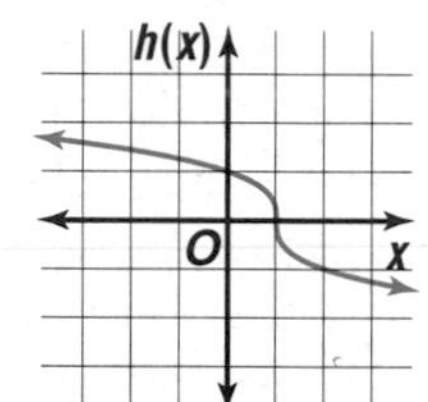

c. 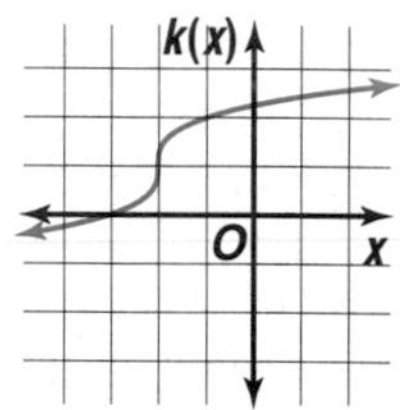

Guided Practice

Describe how the graphs of $f(x)$ and $g(x)$ are related.

6. $f(x) = |x|$ and $g(x) = |x + 4|$

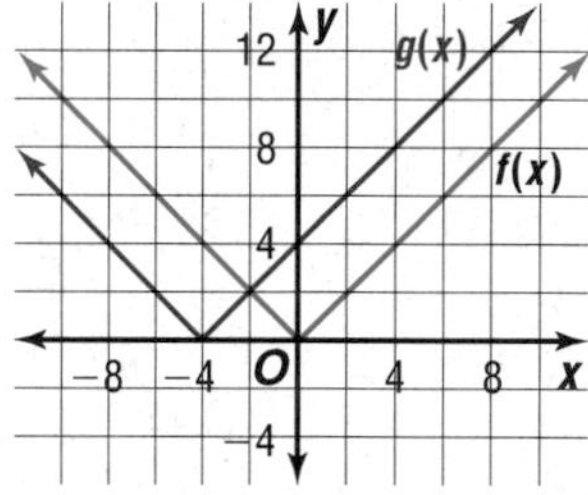

7. $f(x) = x^3$ and $g(x) = -(3x)^3$

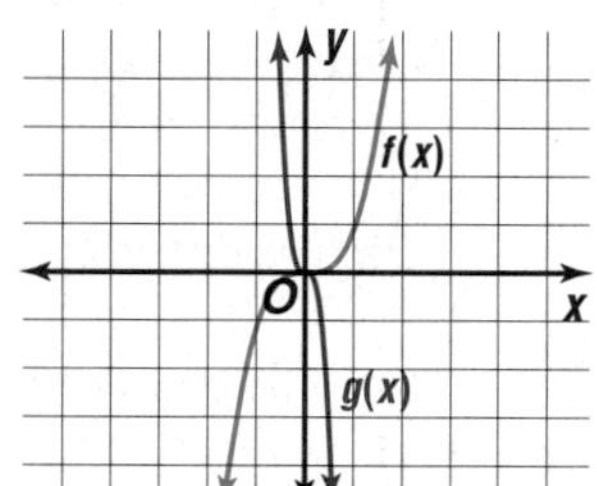

Use the graph of the given parent function to describe the graph of each related function.

8. $f(x) = x^2$
   a. $y = (0.2x)^2$
   b. $y = (x - 5)^2 - 2$
   c. $y = 3x^2 + 6$

9. $f(x) = x^3$
   a. $y = |x^3 + 3|$
   b. $y = -(2x)^3$
   c. $y = 0.75(x + 1)^3$

Sketch the graph of each function.

10. $f(x) = 2(x - 3)^3$

11. $g(x) = (0.5x)^2 - 1$

12. **Consumer Costs** The cost of labor for servicing cars at B & B Automotive is \$50 for each whole hour or for any fraction of an hour.
    a. Graph the function that describes the cost for x hours of labor.
    b. Graph the function that would show a \$25 additional charge if you decide to also get the oil changed and fluids checked.
    c. What would be the cost of servicing a car that required 3.45 hours of labor if the owner requested that the oil be changed and the fluids be checked?

EXERCISES

Practice

Describe how the graphs of $f(x)$ and $g(x)$ are related.

13. $f(x) = x$ and $g(x) = x + 6$

14. $f(x) = x^2$ and $g(x) = \frac{3}{4}x^2$

15. $f(x) = |x|$ and $g(x) = |5x|$

16. $f(x) = x^3$ and $g(x) = (x - 5)^3$

17. $f(x) = \frac{1}{x}$ and $g(x) = \frac{3}{x}$

18. $f(x) = [\![x]\!] + 1$ and $g(x) = -[\![x]\!] - 1$

19. Describe the relationship between the graphs of $f(x) = \sqrt{x}$ and $g(x) = -\sqrt{0.4x} + 3$.

Use the graph of the given parent function to describe the graph of each related function.

20. $f(x) = x^2$
    a. $y = -(1.5x)^2$
    b. $y = 4(x - 3)^2$
    c. $y = \frac{1}{2}x^2 - 5$

21. $f(x) = |x|$
    a. $y = |0.2x|$
    b. $y = 7|x| - 0.4$
    c. $y = -9|x + 1|$

22. $f(x) = x^3$
    a. $y = (x + 2)^3 - 5$
    b. $y = -(0.8x)^3$
    c. $y = \left(\frac{5}{3}x\right)^3 + 2$

23. $f(x) = \sqrt{x}$
    a. $y = \frac{1}{3}\sqrt{x + 2}$
    b. $y = \sqrt{-x} - 7$
    c. $y = 4 + 2\sqrt{x - 3}$

24. $f(x) = \frac{1}{x}$
    a. $y = \frac{1}{0.5x}$
    b. $y = \frac{1}{6x} + 8$
    c. $y = \frac{1}{|x|}$

25. $f(x) = [\![x]\!]$
    a. $y = \left[\!\left[\frac{5}{2}x\right]\!\right] - 3$
    b. $y = -0.75[\![x]\!]$
    c. $y = [\![\,|x| - 4]\!]$

26. Name the parent graph of $m(x) = |-9 + (0.75x)^2|$. Then sketch the graph of $m(x)$.

27. Write the equation of the graph obtained when the graph of $y = \frac{1}{x}$ is compressed vertically by a factor of 0.25, translated 4 units right, and then translated 3 units up.

Sketch the graph of each function.

28. $f(x) = -(x + 4) + 5$

29. $g(x) = |x^2 - 4|$

30. $h(x) = (0.5x - 1)^3$

31. $n(x) = -2.5[\![x]\!] + 3$

32. $q(x) = -4|x - 2| - 1$

33. $k(x) = -\frac{1}{2}(x - 3)^2 - 4$

34. Graph $y = f(x)$ and $y = f(|x|)$ on the same set of axes if $f(x) = (x + 3)^2 - 8$.

Graphing Calculator

Use a graphing calculator to graph each set of functions on the same screen. Name the x-intercept(s) of each function.

35. a. $y = x^2$
    b. $y = (4x - 2)^2$
    c. $y = (2x + 3)^2$

36. a. $y = x^3$
    b. $y = (3x - 2)^3$
    c. $y = (4x + 1)^3$

37. a. $y = \sqrt{x}$
    b. $y = \sqrt{2x + 5}$
    c. $y = \sqrt{5x - 3}$

Applications and Problem Solving

38. Technology Transformations can be used to create a series of graphs that appear to move when shown sequentially on a video screen. Suppose you start with the graph of $y = f(x)$. Describe the effect of graphing the following functions in succession if n has integer values from 1 to 100.

a. $y_n = f(x + 2n) - 3n$.

b. $y_n = (-1)^n f(x - n)$.

39. Critical Thinking Study the coordinates of the x-intercepts you found in the related graphs in Exercises 35–37. Make a conjecture about the x-intercept of $y = (ax + b)^n$ if $y = x^n$ is the parent function.

40. Business The standard cost of a taxi fare is $1.50 for the first unit and 25 cents for each additional unit. A unit is composed of distance (one unit equals 0.2 mile) and/or wait time (one unit equals 75 seconds). As the cab moves at more than 9.6 miles per hour, the taxi's meter clocks distance. When the cab is stopped or moving at less than 9.6 miles per hour, the meter clocks time. Thus, traveling 0.1 mile and then waiting at a stop light for 37.5 seconds generates one unit and a 25-cent charge.

a. Assuming that the cab meter rounds up to the nearest unit, write a function that would determine the cost for x units of cab fare, where $x > 0$.

b. Graph the function found in part a.

41. Geometry Suppose $f(x) = 5 - |x - 6|$.

a. Sketch the graph of $f(x)$ and calculate the area of the triangle formed by $f(x)$ and the positive x-axis.

b. Sketch the graph of $y = 2f(x)$ and calculate the area of the new triangle formed by $2f(x)$ and the positive x-axis. How do the areas of part a and part b compare? Make a conjecture about the area of the triangle formed by $y = c \cdot f(x)$ in the first quadrant if $c \geq 0$.

c. Sketch the graph of $y = f(x - 3)$ and recalculate the area of the triangle formed by $f(x - 3)$ and the positive x-axis. How do the areas of part a and part c compare? Make a conjecture about the area of the triangle formed by $y = f(x - c)$ in the first quadrant if $c \geq 0$.

42. Critical Thinking Study the parent graphs at the beginning of this lesson.

a. Select a parent graph or a modification of a parent graph that meets each of the following specifications.
(1) positive at its leftmost points and positive at its rightmost points
(2) negative at its leftmost points and positive at its rightmost points
(3) negative at its leftmost points and negative at its rightmost points
(4) positive at its leftmost points and negative at its rightmost points

b. Sketch the related graph for each parent graph that is translated 3 units right and 5 units down.

c. Write an equation for each related graph.

43. **Critical Thinking** Suppose a reflection, a translation, or a dilation were applied to an even function.

a. Which transformations would result in another even function?

b. Which transformations would result in a function that is no longer even?

Mixed Review

44. Is the graph of $f(x) = x^{17} - x^{15}$ symmetric with respect to the origin? Explain. *(Lesson 3-1)*

45. **Child Care** Elisa Dimas is the manager for the Learning Loft Day Care Center. The center offers all day service for preschool children for \$18 per day and after school only service for \$6 per day. Fire codes permit only 50 children in the building at one time. State law dictates that a child care worker can be responsible for a maximum of 3 preschool children and 5 school-age children at one time. Ms. Dimas has ten child care workers available to work at the center during the week. How many children of each age group should Ms. Dimas accept to maximize the daily income of the center? *(Lesson 2-7)*

46. **Geometry** Triangle ABC is represented by the matrix $\begin{bmatrix} 5 & 1 & -2 \\ -4 & 3 & -1 \end{bmatrix}$. Find the image of the triangle after a 90° counterclockwise rotation about the origin. *(Lesson 2-4)*

47. Find the values of x, y, and z for which $\begin{bmatrix} x^2 & 7 & 9 \\ 5 & 12 & 6 \end{bmatrix} = \begin{bmatrix} 25 & 7 & y \\ 5 & 2z & 6 \end{bmatrix}$ is true. *(Lesson 2-2)*

48. Solve the system of equations algebraically. *(Lesson 2-1)*
$6x + 5y = -14$
$5x + 2y = -3$

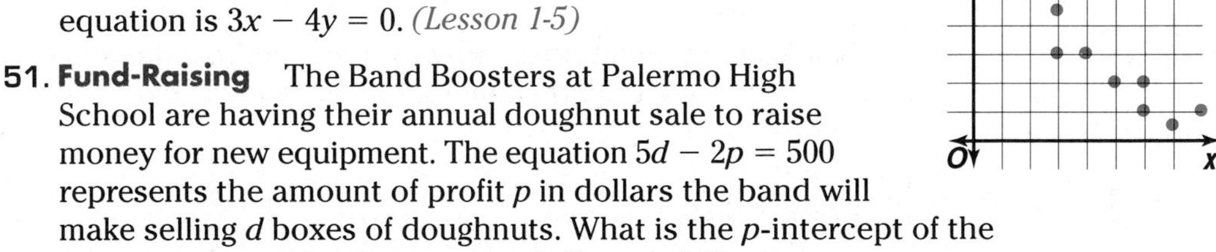

49. Describe the linear relationship implied in the scatter plot at the right. *(Lesson 1-6)*

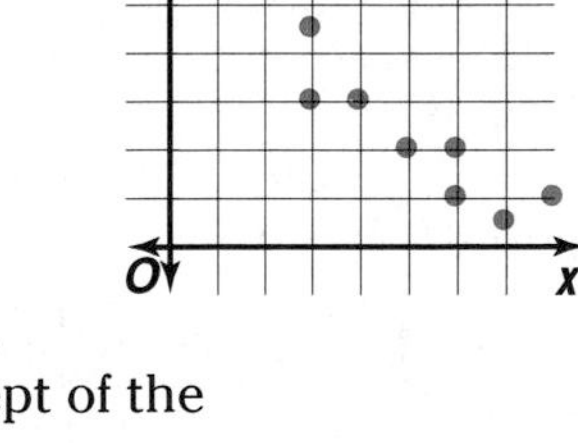

50. Find the slope of a line perpendicular to a line whose equation is $3x - 4y = 0$. *(Lesson 1-5)*

51. **Fund-Raising** The Band Boosters at Palermo High School are having their annual doughnut sale to raise money for new equipment. The equation $5d - 2p = 500$ represents the amount of profit p in dollars the band will make selling d boxes of doughnuts. What is the p-intercept of the line represented by this equation? *(Lesson 1-3)*

52. Find $[f \circ g](x)$ and $[g \circ f](x)$ if $f(x) = \frac{2}{3}x - 2$ and $g(x) = x^2 - 6x + 9$. *(Lesson 1-2)*

53. **SAT/ACT Practice** If $d = m - \frac{50}{m}$ and m is a positive number that increases in value, then d

A increases in value.
B increases, then decreases.
C remains unchanged.
D decreases in value.
E decreases, then increases.

Extra Practice See p. A30.

Graphs of Nonlinear Inequalities

OBJECTIVES
- Graph polynomial, absolute value, and radical inequalities in two variables.
- Solve absolute value inequalities.

PHARMACOLOGY Pharmacists label medication as to how much and how often it should be taken. Because oral medication requires time to take effect, the amount of medication in your body varies with time. Suppose the equation $m(x) = 0.5x^4 + 3.45x^3 - 96.65x^2 + 347.7x$ for $0 < x \leq 6$ models the number of milligrams of a certain pain reliever in the bloodstream x hours after taking 400 milligrams of it. The medicine is to be taken every 4 hours. At what times during the first 4-hour period is the level of pain reliever in the bloodstream above 300 milligrams? *This problem will be solved in Example 5.*

Problems like the one above can be solved by graphing inequalities. Graphing inequalities in two variables identifies all ordered pairs that will satisfy the inequality.

Example 1 **Determine whether (3, −4), (4, 7), (1, 1), and (−1, 6) are solutions for the inequality $y \leq (x - 2)^2 - 3$.**

Substitute the x-value and y-value from each ordered pair into the inequality.

$y \leq (x - 2)^2 - 3$
$-4 \overset{?}{\leq} (3 - 2)^2 - 3$ *(x, y) = (3, −4)*
$-4 \leq -2$ ✓ *true*

$y \leq (x - 2)^2 - 3$
$7 \overset{?}{\leq} (4 - 2)^2 - 3$ *(x, y) = (4, 7)*
$7 \leq 1$ *false*

$y \leq (x - 2)^2 - 3$
$1 \overset{?}{\leq} (1 - 2)^2 - 3$ *(x, y) = (1, 1)*
$1 \leq -2$ *false*

$y \leq (x - 2)^2 - 3$
$6 \overset{?}{\leq} (-1 - 2)^2 - 3$ *(x, y) = (−1, 6)*
$6 \leq 6$ ✓ *true*

Of these ordered pairs, (3, −4) and (−1, 6) are solutions for $y \leq (x - 2)^2 - 3$.

Look Back
Refer to Lesson 1-8 for more about graphing linear inequalities.

Similar to graphing linear inequalities, the first step in graphing nonlinear inequalities is graphing the boundary. You can use concepts from Lesson 3-2 to graph the boundary.

Example 2 **Graph $y \geq (x - 4)^3 - 2$.**

The boundary of the inequality is the graph of $y = (x - 4)^3 - 2$. To graph the boundary curve, start with the parent graph $y = x^3$. Analyze the boundary equation to determine how the boundary relates to the parent graph.

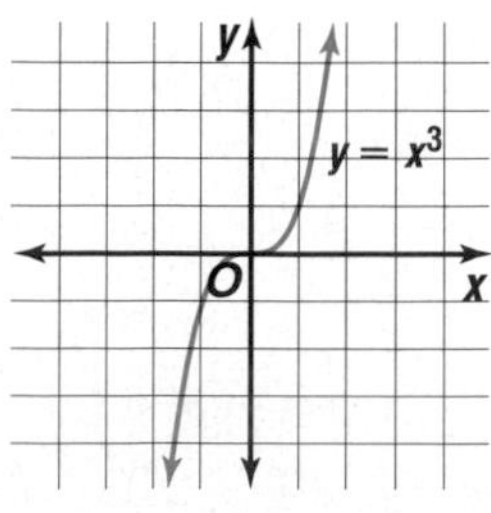

$$y = (x - 4)^3 - 2$$

move 4 units right (↑ under 4) *move 2 units down* (↑ under 2)

Graphing Calculator Appendix

For keystroke instruction on how to graph inequalities see pages A13-A15.

Since the boundary is included in the inequality, the graph is drawn as a solid curve.

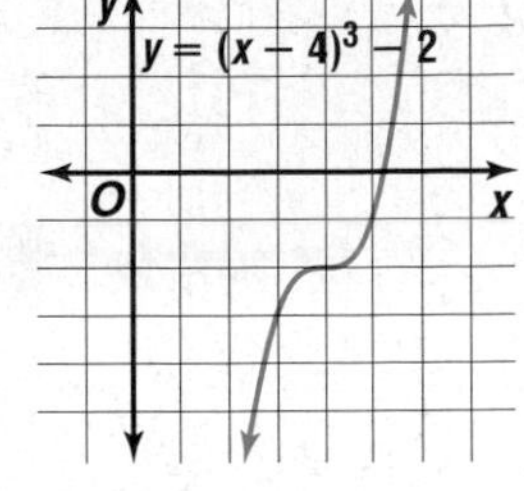

The inequality states that the y-values of the solution are greater than the y-values on the graph of $y = (x - 4)^3 - 2$. For a particular value of x, all of the points in the plane that lie above the curve have y-values greater than $y = (x - 4)^3 - 2$. So this portion of the graph should be shaded.

To verify numerically, you can test a point not on the boundary. *It is common to test (0, 0) whenever it is not on the boundary.*

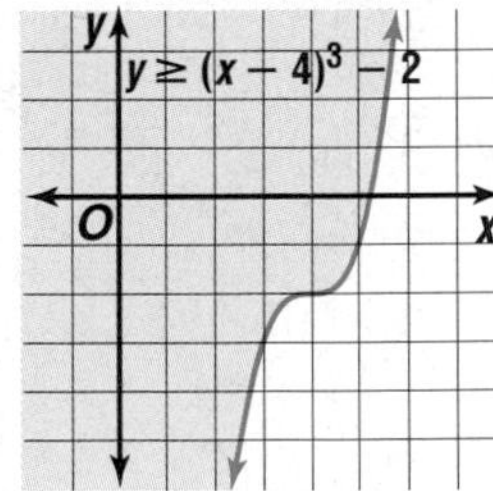

$y \geq (x - 4)^3 - 2$

$0 \overset{?}{\geq} (0 - 4)^3 - 2$ *Replace (x, y) with (0, 0).*

$0 \geq -66$ ✓ *True*

Since (0, 0) satisfies the inequality, the correct region is shaded.

The same process used in Example 2 can be used to graph inequalities involving absolute value.

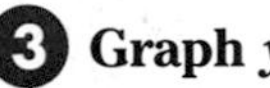

Example  **Graph $y > 3 - |x + 2|$.**

Begin with the parent graph $y = |x|$.

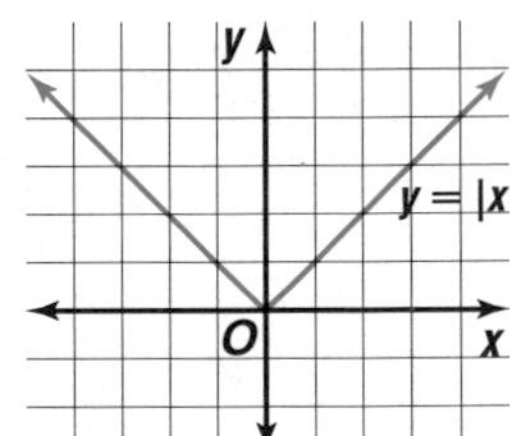

It is easier to sketch the graph of the given inequality if you rewrite it so that the absolute value expression comes first.

$y = 3 - |x + 2| \quad \rightarrow \quad y = -|x + 2| + 3$

This more familiar form tells us the parent graph is reflected over the x-axis and moved 2 units left and three units up. The boundary is not included, so draw it as a dashed line.

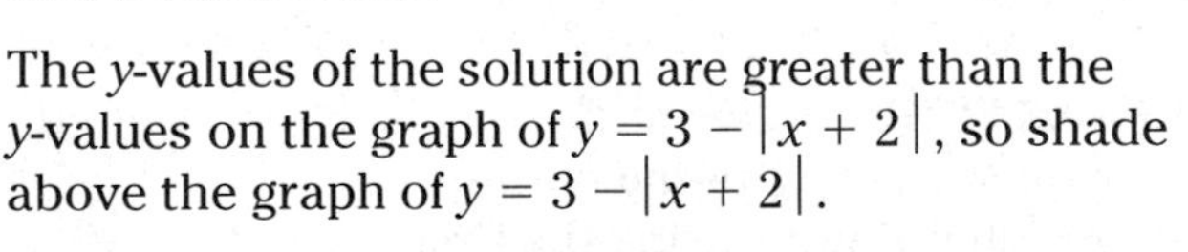

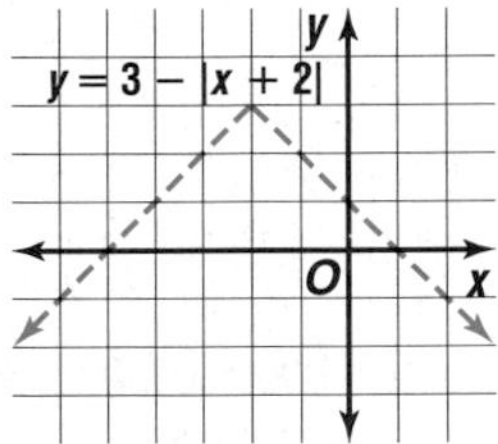

The y-values of the solution are greater than the y-values on the graph of $y = 3 - |x + 2|$, so shade above the graph of $y = 3 - |x + 2|$.

Verify by substituting (0, 0) in the inequality to obtain $0 > 1$. Since this statement is false, the part of the graph containing (0, 0) should not be shaded. Thus, the graph is correct.

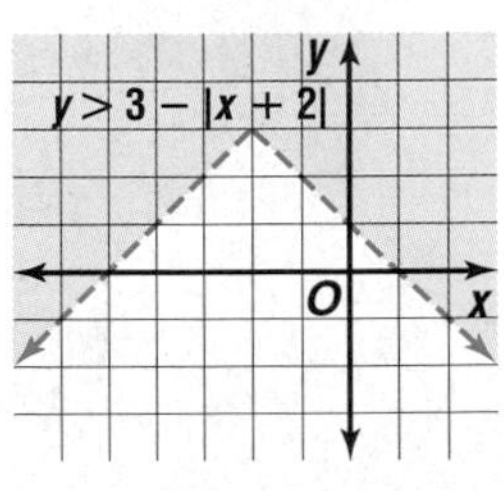

To solve absolute value inequalities algebraically, use the definition of absolute value to determine the solution set. That is, if $a < 0$, then $|a| = -a$, and if $a \geq 0$, then $|a| = a$.

Example 4 **Solve $|x - 2| - 5 < 4$.**

There are two cases that must be solved. In one case, $x - 2$ is negative, and in the other, $x - 2$ is positive.

Case 1

$$(x - 2) < 0$$
$$|x - 2| - 5 < 4$$
$$-(x - 2) - 5 < 4 \quad |x - 2| = -(x - 2)$$
$$-x + 2 - 5 < 4$$
$$-x < 7$$
$$x > -7$$

Case 2

$$(x - 2) > 0$$
$$|x - 2| - 5 < 4$$
$$x - 2 - 5 < 4 \quad |x - 2| = (x - 2)$$
$$x - 7 < 4$$
$$x < 11$$

The solution set is $\{x \mid -7 < x < 11\}$. *$\{x \mid -7 < x < 11\}$ is read "the set of all numbers x such that x is between -7 and 11."*

Verify this solution by graphing.

First, graph $y = |x - 2| - 5$. Since we are solving $|x - 2| - 5 < 4$ and $|x - 2| - 5 = y$, we are looking for a region in which $y = 4$. Therefore, graph $y = 4$ and graph it on the same set of axes.

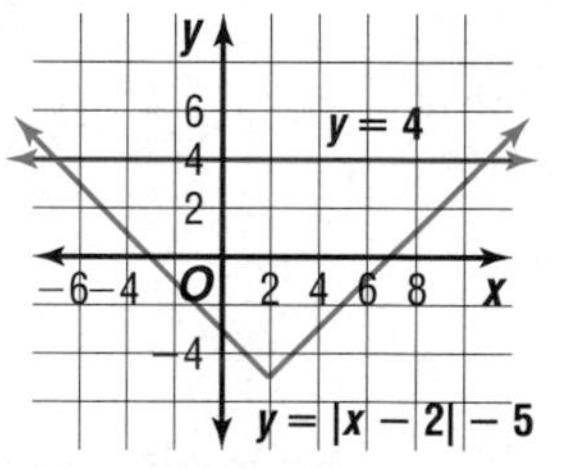

Identify the points of intersection of the two graphs. By inspecting the graph, we can see that they intersect at $(-7, 4)$ and $(11, 4)$.

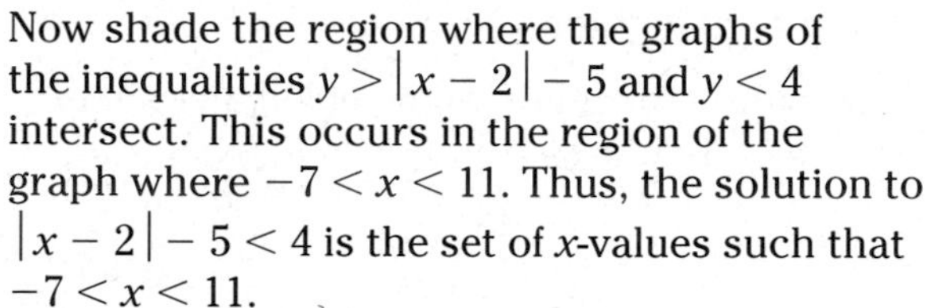

Now shade the region where the graphs of the inequalities $y > |x - 2| - 5$ and $y < 4$ intersect. This occurs in the region of the graph where $-7 < x < 11$. Thus, the solution to $|x - 2| - 5 < 4$ is the set of x-values such that $-7 < x < 11$.

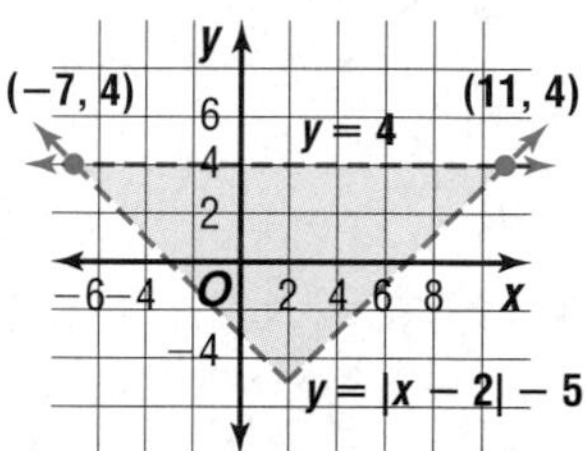

Nonlinear inequalities have applications to many real-world situations, including business, education, and medicine.

Example 5

PHARMACOLOGY **Refer to the application at the beginning of the lesson. At what times during the first 4-hour period is the amount of pain reliever in the bloodstream above 300 milligrams?**

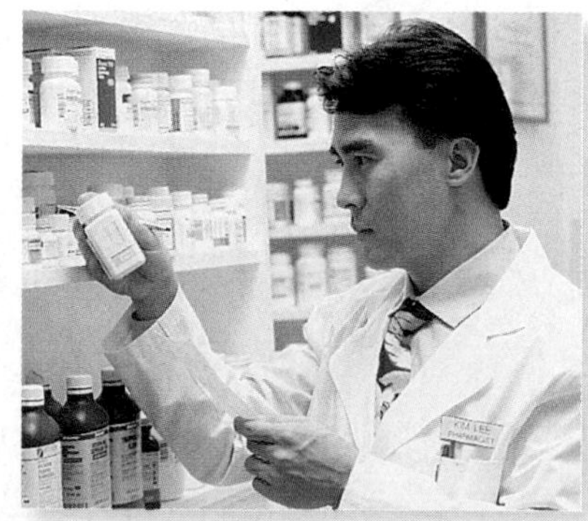

Since we need to know when the level of pain reliever in the bloodstream is above 300 milligrams, we can write an inequality.

$0.5x^4 + 3.45x^3 - 96.65x^2 + 347.7x > 300$ for $0 < x \leq 6$

Let $y = 0.5x^4 + 3.45x^3 - 96.65x^2 + 347.7x$ and $y = 300$. Graph both equations on the same set of axes using a graphing calculator.

Calculating the points of intersection, we find that the two equations intersect at about (1.3, 300) and (3.1, 300). Therefore, when $1.3 < x < 3.1$, the amount of pain reliever in the bloodstream is above 300 milligrams. That is, the amount exceeds 300 milligrams between about 1 hour 18 minutes and 3 hours 6 minutes after taking the medication.

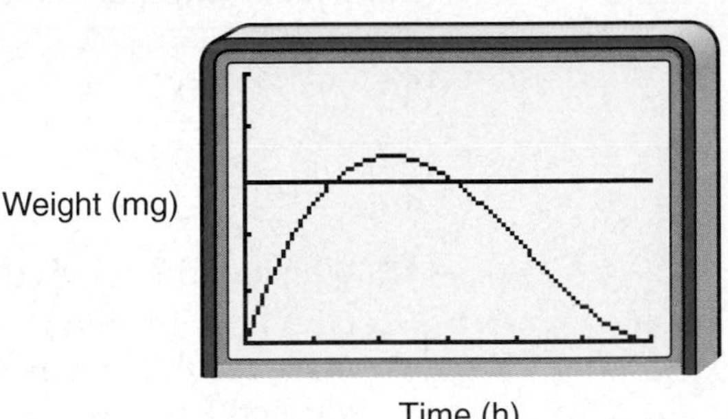

[0, 6] scl: 1 by [0, 500] scl: 100

CHECK FOR UNDERSTANDING

Communicating Mathematics

Read and study the lesson to answer each question.

1. **Describe** how knowledge of transformations can help you graph the inequality $y \leq 5 + \sqrt{x - 2}$.
2. **State** the two cases considered when solving a one-variable absolute value inequality algebraically.
3. **Write** a procedure for determining which region of the graph of an inequality should be shaded.
4. *Math Journal* **Sketch** the graphs of $y = |x - 3| + 2$ and $y = 1$ on the same set of axes. Use your sketch to solve the inequality $|x - 3| + 2 < 1$. If no solution exists, write *no solution.* Write a paragraph to explain your answer.

Guided Practice

Determine whether the ordered pair is a solution for the given inequality. Write *yes* or *no*.

5. $y \geq -5x^4 + 7x^3 + 8$, $(-1, -3)$
6. $y < |3x - 4| - 1$, $(0, 3)$

Graph each inequality.

7. $y \leq (x + 1)^3$
8. $y \leq 2(x - 3)^2$
9. $y > -|x - 4| + 2$

Solve each inequality.

10. $|x + 6| > 4$
11. $|3x - 4| \leq x$

12. **Manufacturing** The AccuData Company makes compact disks measuring 12 centimeters in diameter. The diameters of the disks can vary no more than 50 micrometers or 5×10^{-3} centimeter.

   a. Write an absolute value inequality to represent the range of diameters of compact disks.
   b. What are the largest and smallest diameters that are allowable?

EXERCISES

Practice

Determine whether the ordered pair is a solution for the given inequality. Write *yes* or *no*.

13. $y < x^3 - 4x^2 + 2$, $(1, 0)$

14. $y < |x - 2| + 7$, $(3, 8)$

15. $y > -\sqrt{x + 11} + 1$, $(-2, -1)$

16. $y < -0.2x^2 + 9x - 7$, $(10, 63)$

17. $y \leq \frac{x^2 - 6}{x}$, $(-6, -9)$

18. $y \geq 2|x|^3 - 7$, $(0, 0)$

19. Which of the ordered pairs, $(0, 0)$, $(1, 4)$, $(1, 1)$, $(-1, 0)$, and $(1, -1)$, is a solution for $y \leq \sqrt{x} + 2$? How can you use these results to determine if the graph at the right is correct?

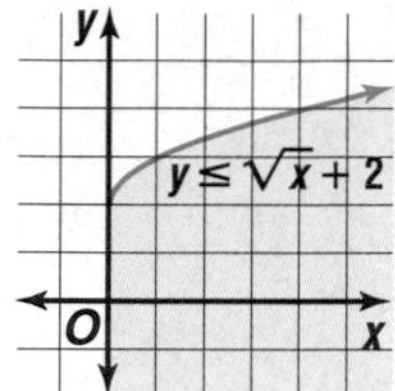

Graph each inequality.

20. $y \leq x^2 - 4$

21. $y > \sqrt{0.5x}$

22. $y < |x - 9|$

23. $y > |2x| + 3$

24. $y < (x - 5)^2$

25. $y \geq -x^3$

26. $y > -(0.4x)^2$

27. $y \leq |3(x - 4)|$

28. $y < \sqrt{x + 3} + 5$

29. $y \geq (x - 1)^2 - 3$

30. $y \geq (2x + 1)^3 + 2$

31. $y \leq -3|x - 2| + 4$

32. Sketch the graph of the inequality $y \geq x^3 - 6x^2 + 12x - 8$.

Solve each inequality.

33. $|x + 4| > 5$

34. $|3x + 12| \geq 42$

35. $|7 - 2x| - 8 < 3$

36. $|5 - x| \leq x$

37. $|5x - 8| < 0$

38. $|2x + 9| - 2x \geq 0$

39. Find all values of x that satisfy $-\frac{2}{3}|x + 5| \geq -8$.

Applications and Problem Solving

40. Chemistry Katie and Wes worked together on a chemistry lab. They determined the quantity of the unknown in their sample to be 37.5 ± 1.2 grams. If the actual quantity of unknown is x, write their results as an absolute value inequality. Solve for x to find the range of possible values of x.

41. Critical Thinking Solve $3|x - 7| < |x - 1|$.

42. Critical Thinking Find the area of the region described by $y \geq 2|x - 3| + 4$ and $x - 2y \geq -20$.

43. Education Amanda's teacher calculates grades using a weighted average. She counts homework as 10%, quizzes as 15%, projects as 20%, tests as 40%, and the final exam as 15% of the final grade. Going into the final, Amanda has scores of 90 for homework, 75 for quizzes, 76 for projects, and 80 for tests. What grade does Amanda need on the final exam if she wants to get an overall grade of at least an 80?

44. Critical Thinking Consider the equation $|(x - 3)^2 - 4| = b$. Determine the value(s) of b so that the equation has

a. no solution.

b. one solution.

c. two solutions.

d. three solutions.

e. four solutions.

45. **Business** After opening a cookie store in the mall, Paul and Carol Mason hired an consultant to provide them with information on increasing their profit. The consultant told them that their profit P depended on the number of cookies x that they sold according to the relation $P(x) \leq -0.005(x - 1200)^2 + 400$. They typically sell between 950 and 1000 cookies in a given day.

a. Sketch a graph to model this situation.

b. Explain the significance of the shaded region.

Mixed Review

46. How are the graphs of $f(x) = x^3$ and $g(x) = -2x^3$ related? *(Lesson 3-2)*

47. Determine whether the graph of $y = -\frac{1}{x^4}$ is symmetric with respect to the x-axis, y-axis, the line $y = x$, the line $y = -x$, or none of these. *(Lesson 3-1)*

48. Find the inverse of $\begin{bmatrix} 8 & -3 \\ 4 & -5 \end{bmatrix}$. *(Lesson 2-5)*

49. Multiply $\begin{bmatrix} 8 & -7 \\ -4 & 0 \end{bmatrix}$ by $\frac{3}{4}$. *(Lesson 2-3)*

50. Graph $y = 3|x| + 5$. *(Lesson 1-7).*

51. **Criminal Justice** The table shows the number of states with teen courts over a period of several years. Make a scatter plot of the data. *(Lesson 1-6)*

Year	States with Teen Courts
1976	2
1991	14
1994	17
1997	36
1999	47*

Source: American Probation and Parole Association.

*Includes District of Columbia

52. Find $[f \circ g](4)$ and $[g \circ f](4)$ for $f(x) = 5x + 9$ and $g(x) = 0.5x - 1$. *(Lesson 1-2)*

53. **SAT Practice** **Grid-In** Student A is 15 years old. Student B is one-third older. How many years ago was student B twice as old as student A?

MID-CHAPTER QUIZ

Determine whether each graph is symmetric with respect to the x-axis, the y-axis, the line $y = x$, the line $y = -x$, the origin, or none of these. (Lesson 3-1)

1. $x^2 + y^2 - 9 = 0$

2. $5x^2 + 6x - 9 = y$

3. $x = \frac{7}{y}$

4. $y = |x| + 1$

Use the graph of the given parent function to describe the graph of each related function. (Lesson 3-2)

5. $f(x) = [\![x]\!]$
   a. $y = [\![x]\!] - 2$
   b. $y = -[\![x - 3]\!]$
   c. $y = \frac{1}{4}[\![x]\!] + 1$

6. $f(x) = x^3$
   a. $y = 3x^3$
   b. $y = (0.5x)^3 - 1$
   c. $y = (x + 1)^3 + 4$

7. Sketch the graph of $g(x) = -0.5(x - 2)^2 + 3$. (Lesson 3-2)

8. Graph the inequality $y \geq \left(\frac{1}{3}x\right)^2 + 2$. (Lesson 3-3)

9. Find all values of x that satisfy $|2x - 7| < 15$. (Lesson 3-3)

10. **Technology** In September of 1999, a polling organization reported that 64% of Americans were "not very" or "not at all" concerned about the Year-2000 computer bug, with a margin of error of 3%. Write and solve an absolute value inequality to describe the range of the possible percent of Americans who were relatively unconcerned about the "Y2K bug." (Lesson 3-3)

Source: The Gallup Organization

Extra Practice See p. A30.

3-4 Inverse Functions and Relations

OBJECTIVES
- Determine inverses of relations and functions.
- Graph functions and their inverses.

METEOROLOGY The hottest temperature ever recorded in Montana was 117° F on July 5, 1937. To convert this temperature to degrees Celsius C, subtract 32° from the Fahrenheit temperature F and then multiply the result by $\frac{5}{9}$. The formula for this conversion is $C = \frac{5}{9}(F - 32)$. The coldest temperature ever recorded in Montana was −57° C on January 20, 1954. To convert this temperature to Fahrenheit, multiply the Celsius temperature by $\frac{9}{5}$ and then add 32°. The formula for this conversion is $F = \frac{9}{5}C + 32$.

The temperature conversion formulas are examples of **inverse functions.** Relations also have inverses, and these inverses are themselves relations.

Inverse Relations Two relations are inverse relations if and only if one relation contains the element (b, a) whenever the other relation contains the element (a, b).

If $f(x)$ denotes a function, then $f^{-1}(x)$ denotes the inverse of $f(x)$. However, $f^{-1}(x)$ may not necessarily be a function. To graph a function or relation and its inverse, you switch the x- and y-coordinates of the ordered pairs of the function. This results in graphs that are symmetric to each other with respect to the line $y = x$.

Example 1 **Graph $f(x) = -\frac{1}{2}|x| + 3$ and its inverse.**

To graph the function, let $y = f(x)$. To graph $f^{-1}(x)$, interchange the x- and y-coordinates of the ordered pairs of the function.

Note that the domain of one relation or function is the range of the inverse and vice versa.

$f(x) = -\frac{1}{2}\lvert x\rvert + 3$	
x	$f(x)$
−3	1.5
−2	2
−1	2.5
0	3
1	2.5
2	2
3	1.5

$f^{-1}(x)$	
x	$f^{-1}(x)$
1.5	−3
2	−2
2.5	−1
3	0
2.5	1
2	2
1.5	3

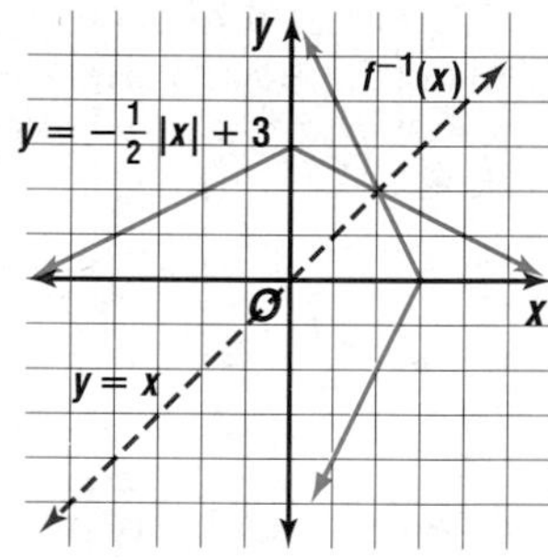

The graph of $f^{-1}(x)$ is the reflection of $f(x)$ over the line $y = x$.

Note that the inverse in Example 1 is not a function because it fails the vertical line test.

You can use the **horizontal line test** to determine if the inverse of a relation will be a function. If every horizontal line intersects the graph of the relation in at most one point, then the inverse of the relation is a function.

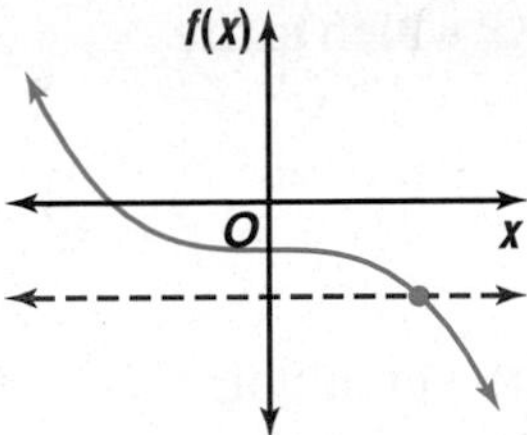

The inverse of f(x) is a function.

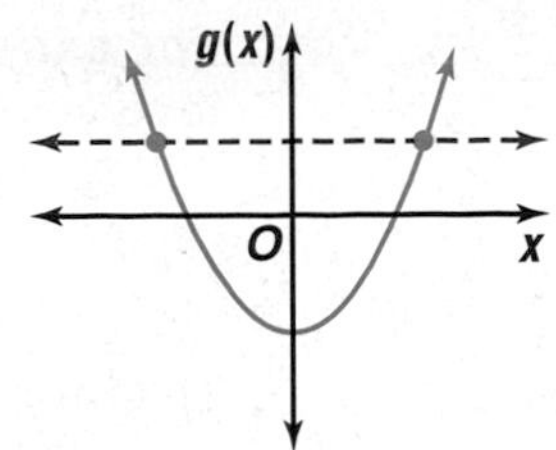

The inverse of g(x) is not a function.

You can find the inverse of a relation algebraically. First, let $y = f(x)$. Then interchange x and y. Finally, solve the resulting equation for y.

Example  **Consider $f(x) = (x + 3)^2 - 5$.**

a. Is the inverse of $f(x)$ a function?

b. Find $f^{-1}(x)$.

c. Graph $f(x)$ and $f^{-1}(x)$ using a graphing calculator.

> **Graphing Calculator Tip**
>
> You can differentiate the appearance of your graphs by highlighting the symbol in front of each equation in the **Y=** list and pressing ENTER to select line (\\), thick (\\), or dot (⋱).

a. Since the line $y = -2$ intersects the graph of $f(x)$ at more than one point, the function fails the horizontal line test. Thus, the inverse of $f(x)$ is not a function.

f(x)
O
x
y = −2
$f(x) = (x + 3)^2 - 5$

b. To find $f^{-1}(x)$, let $y = f(x)$ and interchange x and y. Then, solve for y.

$y = (x + 3)^2 - 5$	*Let $y = f(x)$.*
$x = (y + 3)^2 - 5$	*Interchange x and y.*
$x + 5 = (y + 3)^2$	*Isolate the expression containing y.*
$\pm\sqrt{x + 5} = y + 3$	*Take the square root of each side.*
$y = -3 \pm \sqrt{x + 5}$	*Solve for y.*
$f^{-1}(x) = -3 \pm \sqrt{x + 5}$	*Replace y with $f^{-1}(x)$.*

c. To graph $f(x)$ and its inverse, enter the equations
$y = (x + 3)^2 - 5$,
$y = -3 + \sqrt{x + 5}$, and
$y = -3 - \sqrt{x + 5}$ in the same viewing window.

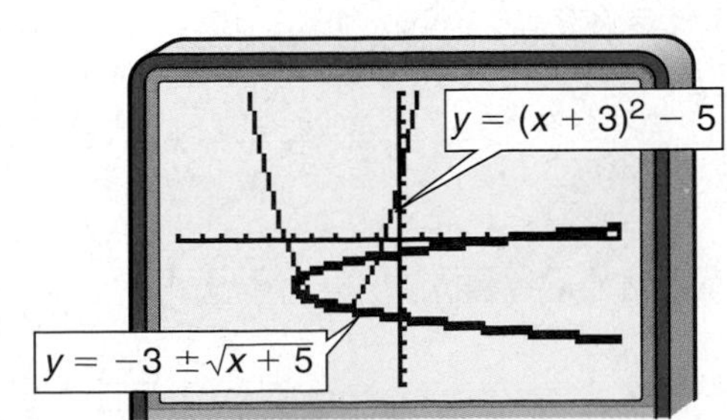

[−15.16, 15.16] scl: 2 by [−10, 10] scl: 2

You can graph a function by using the parent graph of an inverse function.

Example 3 **Graph $y = 2 + \sqrt[3]{x - 7}$.**

The parent function is $y = \sqrt[3]{x}$ which is the inverse of $y = x^3$.

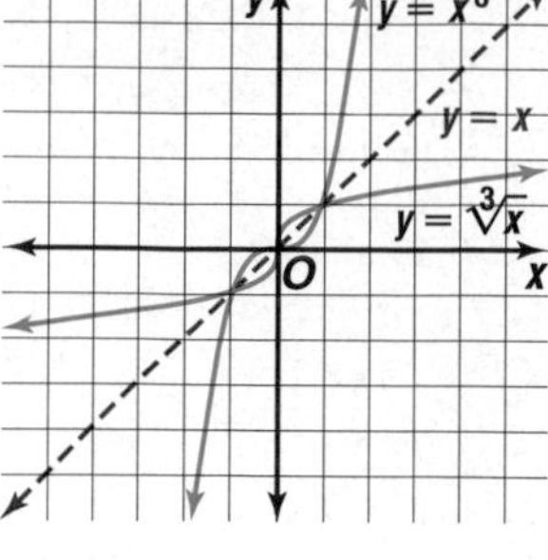

To graph $y = \sqrt[3]{x}$, start with the graph of $y = x^3$. Reflect the graph over the line $y = x$.

To graph $y = 2 + \sqrt[3]{x - 7}$, translate the reflected graph 7 units to the right and 2 units up.

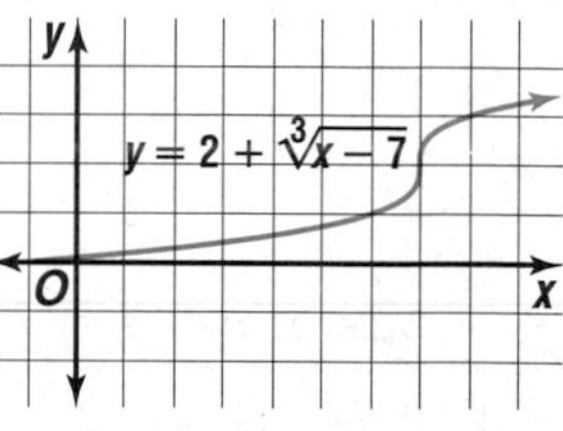

To find the inverse of a function, you use an **inverse process** to solve for y after switching variables. This inverse process can be used to solve many real-world problems.

Example 4 **FINANCE** **When the Garcias decided to begin investing, their financial advisor instructed them to set a goal. Their net pay is about 65% of their gross pay. They decided to subtract their monthly food allowance from their monthly net pay and then invest 10% of the remainder.**

a. Write an equation that gives the amount invested I as a function of their monthly gross pay G given that they allow \$450 per month for food.

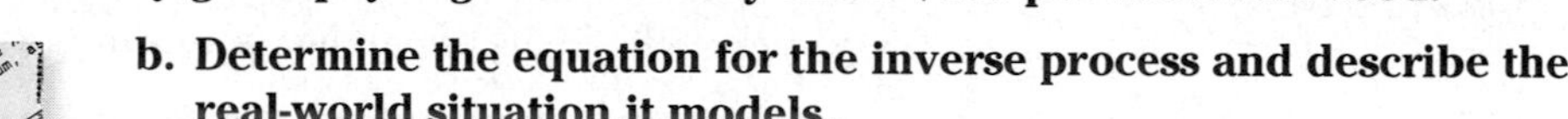

b. Determine the equation for the inverse process and describe the real-world situation it models.

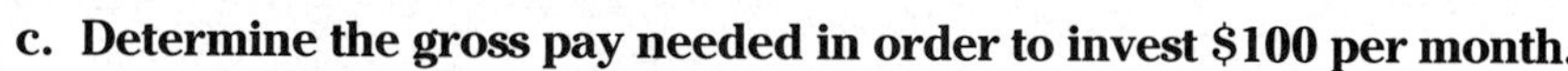

c. Determine the gross pay needed in order to invest \$100 per month.

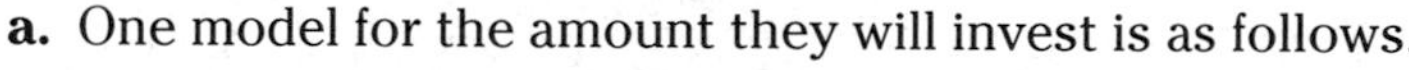

a. One model for the amount they will invest is as follows.

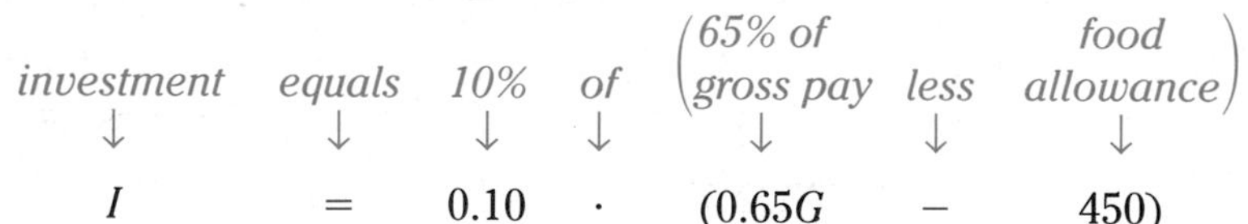

investment	*equals*	*10%*	*of*	*(65% of gross pay*	*less*	*food allowance)*
↓	↓	↓	↓	↓	↓	↓
I	$=$	0.10	$\cdot$	$(0.65G$	$-$	$450)$

b. Solve for G.

$$I = 0.10(0.65G - 450)$$

$$10I = 0.65G - 450 \quad \textit{Multiply each side by 10.}$$

$$10I + 450 = 0.65G \quad \textit{Add 450 to each side.}$$

$$\frac{10I + 450}{0.65} = G \quad \textit{Divide each side by 0.65.}$$

This equation models the gross pay needed to meet a monthly investment goal I with the given conditions.

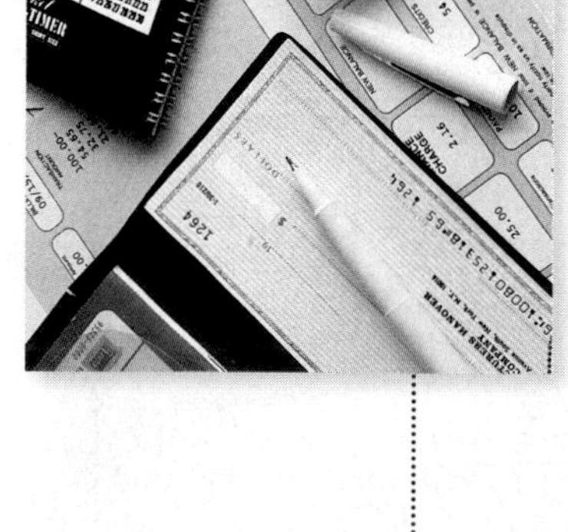

c. Substituting 100 for I gives $G = \frac{10I + 450}{0.65}$, or about \$2231. So the Garcias need to earn a monthly gross pay of about \$2231 in order to invest \$100 per month.

Look Back
Refer to Lesson 1-2 to review the composition of two functions.

If the inverse of a function is also a function, then a composition of the function and its inverse produces a unique result.

Consider $f(x) = 3x - 2$ and $f^{-1}(x) = \frac{x + 2}{3}$. You can find functions $[f \circ f^{-1}](x)$ and $[f^{-1} \circ f](x)$ as follows.

$$
\begin{aligned}
[f \circ f^{-1}](x) &= f\left(\frac{x+2}{3}\right) && f^{-1}(x) = \frac{x+2}{3} \\
&= 3\left(\frac{x+2}{3}\right) - 2 && f(x) = 3x - 2 \\
&= x
\end{aligned}
$$

$$
\begin{aligned}
[f^{-1} \circ f](x) &= f^{-1}(3x - 2) && f(x) = 3x - 2 \\
&= \frac{(3x - 2) + 2}{3} && f^{-1}(x) = \frac{x+2}{3} \\
&= x
\end{aligned}
$$

This leads to the formal definition of inverse functions.

Inverse Functions — Two functions, f and f^{-1}, are inverse functions if and only if $[f \circ f^{-1}](x) = [f^{-1} \circ f](x) = x$.

Example 5 **Given $f(x) = 4x - 9$, find $f^{-1}(x)$, and verify that f and f^{-1} are inverse functions.**

$$
\begin{aligned}
y &= 4x - 9 && f(x) = y \\
x &= 4y - 9 && \textit{Interchange } x \textit{ and } y. \\
x + 9 &= 4y && \textit{Solve for } y. \\
\frac{x+9}{4} &= y \\
\frac{x+9}{4} &= f^{-1}(x) && \textit{Replace } y \textit{ with } f^{-1}(x).
\end{aligned}
$$

Now show that $[f \circ f^{-1}](x) = [f^{-1} \circ f](x) = x$.

$$
\begin{aligned}
[f \circ f^{-1}](x) &= f\left(\frac{x+9}{4}\right) \\
&= 4\left(\frac{x+9}{4}\right) - 9 \\
&= x
\end{aligned}
$$

$$
\begin{aligned}
[f^{-1} \circ f](x) &= f^{-1}(4x - 9) \\
&= \frac{(4x - 9) + 9}{4} \\
&= x
\end{aligned}
$$

Since $[f \circ f^{-1}](x) = [f^{-1} \circ f](x) = x$, f and f^{-1} are inverse functions.

CHECK FOR UNDERSTANDING

Communicating Mathematics

Read and study the lesson to answer each question.

1. **Write** an explanation of how to determine the equation for the inverse of the relation $y = \pm\sqrt{x - 3}$.
2. **Determine** the values of n for which $f(x) = x^n$ has an inverse that is a function. Assume that n is a whole number.

3. **Find a counterexample** to this statement: The inverse of a function is also a function.

4. **Show** how you know whether the inverse of a function is also a function without graphing the inverse.

5. **You Decide** Nitayah says that the inverse of $y = 3 \pm \sqrt{x+2}$ cannot be a function because $y = 3 \pm \sqrt{x+2}$ is not a function. Is she right? Explain.

Guided Practice

Graph each function and its inverse.

6. $f(x) = |x| + 1$
7. $f(x) = x^3 + 1$
8. $f(x) = -(x-3)^2 + 1$

Find $f^{-1}(x)$. Then state whether $f^{-1}(x)$ is a function.

9. $f(x) = -3x + 2$
10. $f(x) = \frac{1}{x^3}$
11. $f(x) = (x+2)^2 + 6$

12. Graph the equation $y = 3 \pm \sqrt{x+1}$ using the parent graph $p(x) = x^2$.

13. Given $f(x) = \frac{1}{2}x - 5$, find $f^{-1}(x)$. Then verify that f and f^{-1} are inverse functions.

14. **Finance** If you deposit \$1000 in a savings account with an interest rate of r compounded annually, then the balance in the account after 3 years is given by the function $B(r) = 1000(1 + r)^3$, where r is written as a decimal.
    a. Find a formula for the interest rate, r, required to achieve a balance of B in the account after 3 years.
    b. What interest rate will yield a balance of \$1100 after 3 years?

EXERCISES

Practice

Graph each function and its inverse.

15. $f(x) = |x| + 2$
16. $f(x) = |2x|$
17. $f(x) = x^3 - 2$
18. $f(x) = x^5 - 10$
19. $f(x) = [\![x]\!]$
20. $f(x) = 3$
21. $f(x) = x^2 + 2x + 4$
22. $f(x) = -(x+2)^2 - 5$
23. $f(x) = (x+1)^2 - 4$

24. For $f(x) = x^2 + 4$, find $f^{-1}(x)$. Then graph $f(x)$ and $f^{-1}(x)$.

Find $f^{-1}(x)$. Then state whether $f^{-1}(x)$ is a function.

25. $f(x) = 2x + 7$
26. $f(x) = -x - 2$
27. $f(x) = \frac{1}{x}$
28. $f(x) = -\frac{1}{x^2}$
29. $f(x) = (x-3)^2 + 7$
30. $f(x) = x^2 - 4x + 3$
31. $f(x) = \frac{1}{x+2}$
32. $f(x) = \frac{1}{(x-1)^2}$
33. $f(x) = -\frac{2}{(x-2)^3}$

34. If $g(x) = \frac{3}{x^2 + 2x}$, find $g^{-1}(x)$.

Graph each equation using the graph of the given parent function.

35. $f(x) = \sqrt{x+5}$, $p(x) = x^2$
36. $y = 1 \pm \sqrt{x-2}$, $p(x) = x^2$
37. $f(x) = -2 - \sqrt[3]{x+3}$, $p(x) = x^3$
38. $y = 2\sqrt[5]{x-4}$, $p(x) = x^5$

Given $f(x)$, find $f^{-1}(x)$. Then verify that f and f^{-1} are inverse functions.

39. $f(x) = -\frac{2}{3}x + \frac{1}{6}$
40. $f(x) = (x-3)^3 + 4$

Applications and Problem Solving

41. Analytic Geometry The function $d(x) = |x - 4|$ gives the distance between x and 4 on the number line.

a. Graph $d^{-1}(x)$.

b. Is $d^{-1}(x)$ a function? Why or why not?

c. Describe what $d^{-1}(x)$ represents. Then explain how you could have predicted whether $d^{-1}(x)$ is a function without looking at a graph.

42. Fire Fighting The velocity v and maximum height h of water being pumped into the air are related by the equation $v = \sqrt{2gh}$ where g is the acceleration due to gravity (32 feet/second^2).

a. Determine an equation that will give the maximum height of the water as a function of its velocity.

b. The Mayfield Fire Department must purchase a pump that is powerful enough to propel water 80 feet into the air. Will a pump that is advertised to project water with a velocity of 75 feet/second meet the fire department's needs? Explain.

43. Critical Thinking

a. Give an example of a function that is its own inverse.

b. What type of symmetry must the graph of the function exhibit?

c. Would all functions with this type of symmetry be their own inverses? Justify your response.

44. Consumer Costs A certain long distance phone company charges callers 10 cents for every minute or part of a minute that they talk. Suppose that you talk for x minutes, where x is any real number greater than 0.

a. Sketch the graph of the function $C(x)$ that gives the cost of an x-minute call.

b. What are the domain and range of $C(x)$?

c. Sketch the graph of $C^{-1}(x)$.

d. What are the domain and range of $C^{-1}(x)$?

e. What real-world situation is modeled by $C^{-1}(x)$?

45. Critical Thinking Consider the parent function $y = x^2$ and its inverse $y = \pm\sqrt{x}$. If the graph of $y = x^2$ is translated 6 units right and 5 units down, what must be done to the graph of $y = \pm\sqrt{x}$ to get a graph of the inverse of the translated function? Write an equation for each of the translated graphs.

46. Physics The formula for the kinetic energy of a particle as a function of its mass m and velocity v is $KE = \frac{1}{2}mv^2$.

a. Find the equation for the velocity of a particle based on its mass and kinetic energy.

b. Use your equation from part a to find the velocity in meters per second of a particle of mass 1 kilogram and kinetic energy 15 joules.

c. Explain why the velocity of the particle is not a function of its mass and kinetic energy.

47. **Cryptography** One way to encode a message is to assign a numerical value to each letter of the alphabet and encode the message by assigning each number to a new value using a mathematical relation.

a. Does the encoding relation have to be a function? Explain.

b. Why should the graph of the encoding function pass the horizontal line test?

c. Suppose a value was assigned to each letter of the alphabet so that $1 = A, 2 = B, 3 = C, \ldots, 26 = Z$, and a message was encoded using the relation $c(x) = -2 + \sqrt{x + 3}$. What function would decode the message?

d. Try this decoding function on the following message:

1 2.899 2.123 0.449 2.796 1.464 2.243 2.123 2.690
0 2.583 0.828 1 2.899 2.123

Mixed Review

48. Solve the inequality $|2x + 4| \leq 6$. *(Lesson 3-3)*

49. State whether the figure at the right has point symmetry, line symmetry, neither, or both. *(Lesson 3-1)*

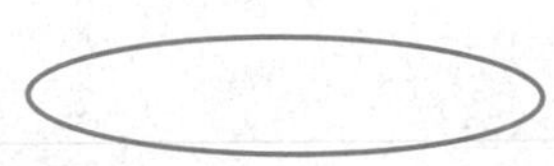

50. **Retail** Arturo Alvaré, a sales associate at a paint store, plans to mix as many gallons as possible of colors A and B. He has 32 units of blue dye and 54 units of red dye. Each gallon of color A requires 4 units of blue dye and 1 unit of red dye. Each gallon of color B requires 1 unit of blue dye and 6 units of red dye. Use linear programming to answer the following questions. *(Lesson 2-7)*

a. Let a be the number of gallons of color A and let b be the number of gallons of color B. Write the inequalities that describe this situation.

b. Find the maximum number of gallons possible.

51. Solve the system of equations $4x + 2y = 10$, $y = 6 - x$ by using a matrix equation. *(Lesson 2-5)*

52. Find the product $\frac{1}{2}\begin{bmatrix} 9 & -3 \\ -6 & 6 \end{bmatrix}$. *(Lesson 2-3)*

53. Graph $y < -2x + 8$. *(Lesson 1-8)*

54. Line ℓ_1 has a slope of $\frac{1}{4}$, and line ℓ_2 has a slope of 4. Are the lines *parallel*, *perpendicular*, or *neither*? *(Lesson 1-5)*

55. Write the slope-intercept form of the equation of the line that passes through points at (0, 7) and (5, 2). *(Lesson 1-4)*

56. **SAT/ACT Practice** In the figure at the right, if $\overline{PQ}$ is perpendicular to $\overline{QR}$, then $a + b + c + d = ?$

A 180 **B** 225 **C** 270
D 300 **E** 360

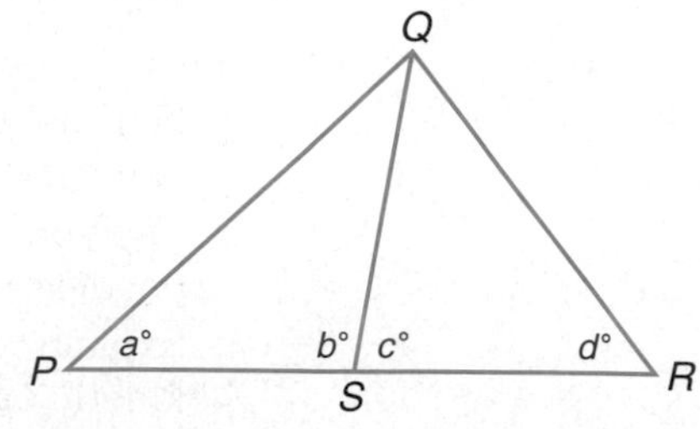

Extra Practice See p. A30.

Continuity and End Behavior

OBJECTIVES

- Determine whether a function is continuous or discontinuous.
- Identify the end behavior of functions.
- Determine whether a function is increasing or decreasing on an interval.

POSTAGE On January 10, 1999, the United States Postal Service raised the cost of a first-class stamp. After the change, mailing a letter cost $0.33 for the first ounce and $0.22 for each additional ounce or part of an ounce. The graph summarizes the cost of mailing a first-class letter. *A problem related to this is solved in Example 2.*

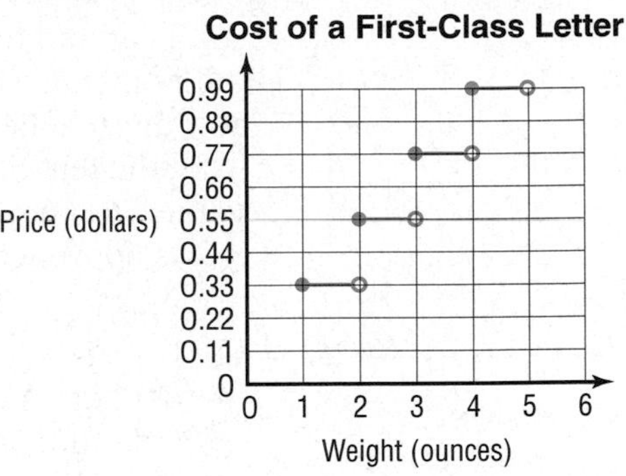

Most graphs we have studied have been smooth, continuous curves. However, a function like the one graphed above is a **discontinuous** function. That is, you cannot trace the graph of the function without lifting your pencil.

There are many types of discontinuity. Each of the functions graphed below illustrates a different type of discontinuity. That is, each function is discontinuous at some point in its domain.

The postage graph exhibits jump discontinuity.

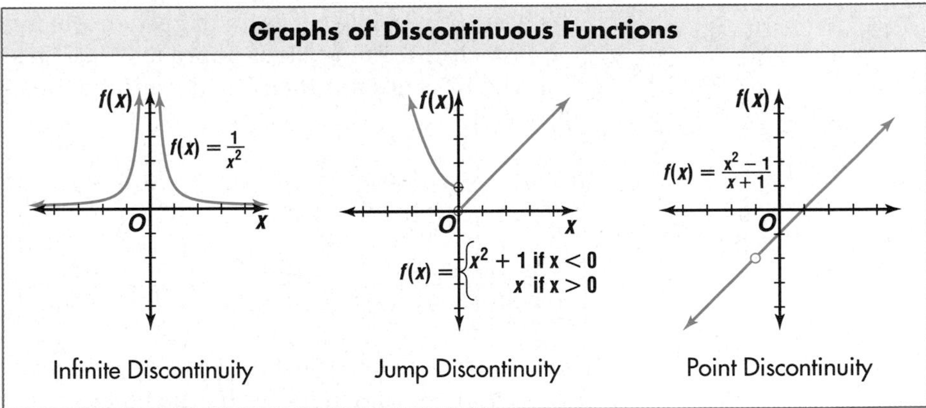

- **Infinite discontinuity** means that $|f(x)|$ becomes greater and greater as the graph approaches a given x-value.
- **Jump discontinuity** indicates that the graph stops at a given value of the domain and then begins again at a different range value for the same value of the domain.
- When there is a value in the domain for which the function is undefined, but the pieces of the graph match up, we say the function has **point discontinuity.**

There are functions that are impossible to graph in the real number system. Some of these functions are said to be **everywhere discontinuous.** An example of such a function is $f(x) = \begin{cases} 1 \text{ if } x \text{ is rational} \\ -1 \text{ if } x \text{ is irrational} \end{cases}$.

If a function is not discontinuous, it is said to be **continuous**. That is, a function is continuous at a number c if there is a point on the graph with x-coordinate c and the graph passes through that point without a break.

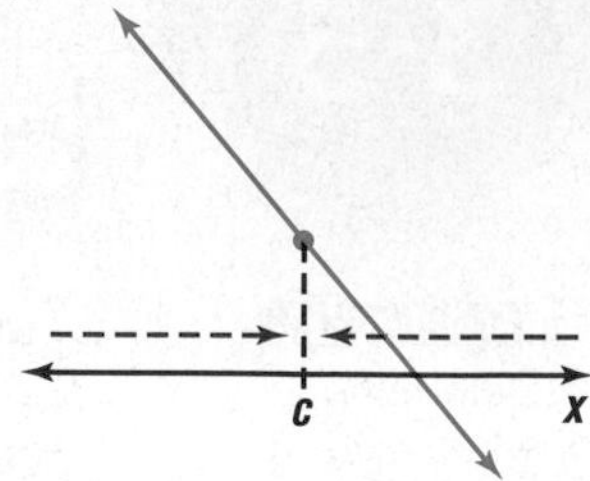

Linear and quadratic functions are continuous at all points. If we only consider x-values less than c as x approaches c, then we say x is approaching c from the left. Similarly, if we only consider x-values greater than c as x approaches c, then we say x is approaching c from the right.

Continuity Test

A function is continuous at $x = c$ if it satisfies the following conditions:
(1) the function is defined at c; in other words, $f(c)$ exists;
(2) the function approaches the same y-value on the left and right sides of $x = c$; and
(3) the y-value that the function approaches from each side is $f(c)$.

Example 1 **Determine whether each function is continuous at the given x-value.**

a. $f(x) = 3x^2 + 7;\ x = 1$

Check the three conditions in the continuity test.

(1) The function is defined at $x = 1$. In particular, $f(1) = 10$.

(2) The first table below suggests that when x is less than 1 and x approaches 1, the y-values approach 10. The second table suggests that when x is greater than 1 and x approaches 1, the y-values approach 10.

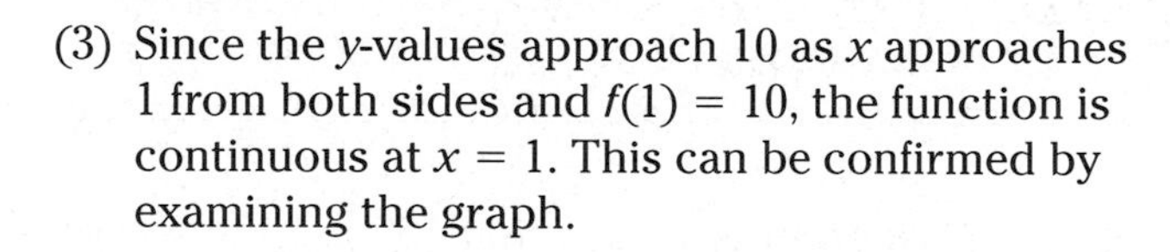

x	y = f(x)
0.9	9.43
0.99	9.9403
0.999	9.994003

x	y = f(x)
1.1	10.63
1.01	10.0603
1.001	10.006003

(3) Since the y-values approach 10 as x approaches 1 from both sides and $f(1) = 10$, the function is continuous at $x = 1$. This can be confirmed by examining the graph.

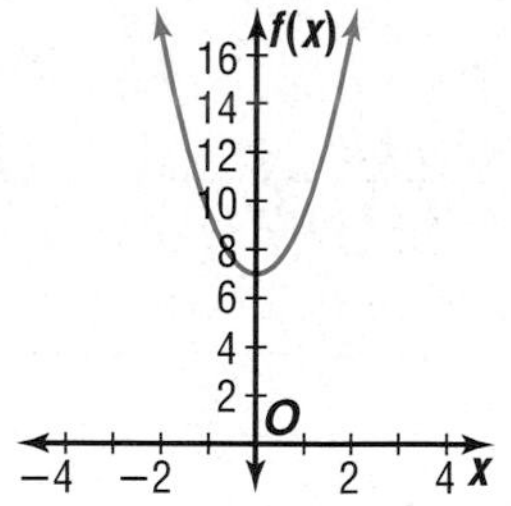

b. $f(x) = \dfrac{x - 2}{x^2 - 4};\ x = 2$

Start with the first condition in the continuity test. The function is not defined at $x = 2$ because substituting 2 for x results in a denominator of zero. So the function is discontinuous at $x = 2$. *This function has point discontinuity at $x = 2$.*

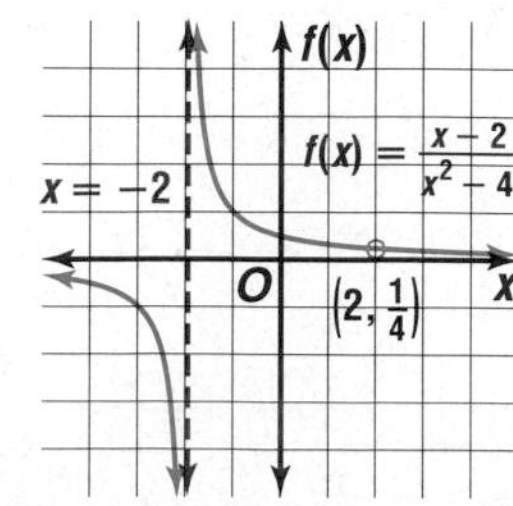

c. $f(x) = \begin{cases} \dfrac{1}{x} \text{ if } x > 1 \\ x \text{ if } x \leq 1 \end{cases}; x = 1$

The function is defined at $x = 1$.
Using the second formula we find $f(1) = 1$.

The first table suggests that $f(x)$ approaches 1 as x approaches 1 from the left. We can see from the second table that $f(x)$ seems to approach 1 as x approaches 1 from the right.

x	f(x)
0.9	0.9
0.99	0.99
0.999	0.999

x	f(x)
1.1	0.9091
1.01	0.9901
1.001	0.9990

Since the $f(x)$-values approach 1 as x approaches 1 from both sides and $f(1) = 1$, the function is continuous at $x = 1$.

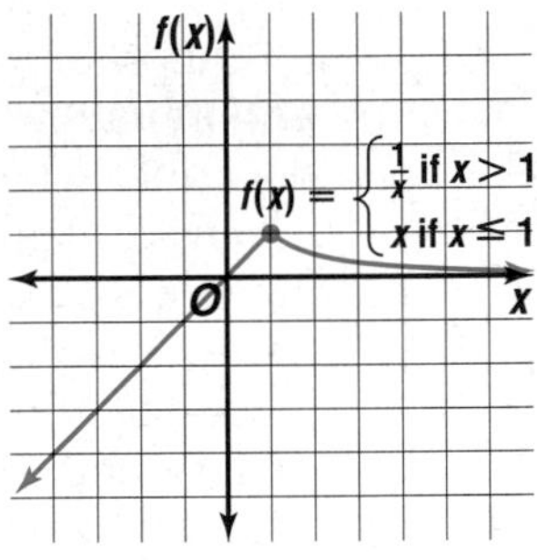

A function may have a discontinuity at one or more x-values but be continuous on an interval of other x-values. For example, the function $f(x) = \dfrac{1}{x^2}$ is continuous for $x > 0$ and $x < 0$, but discontinuous at $x = 0$.

Continuity on an Interval

A function $f(x)$ is continuous on an interval if and only if it is continuous at each number x in the interval.

In Chapter 1, you learned that a piecewise function is made from several functions over various intervals. The piecewise function

$f(x) = \begin{cases} 3x - 2 \text{ if } x > 2 \\ 2 - x \text{ if } x \leq 2 \end{cases}$ is continuous for $x > 2$

and $x < 2$ but is discontinuous at $x = 2$. The graph has a jump discontinuity. This function fails the second part of the continuity test because the values of $f(x)$ approach 0 as x approaches 2 from the left, but the $f(x)$-values approach 4 as x approaches 2 from the right.

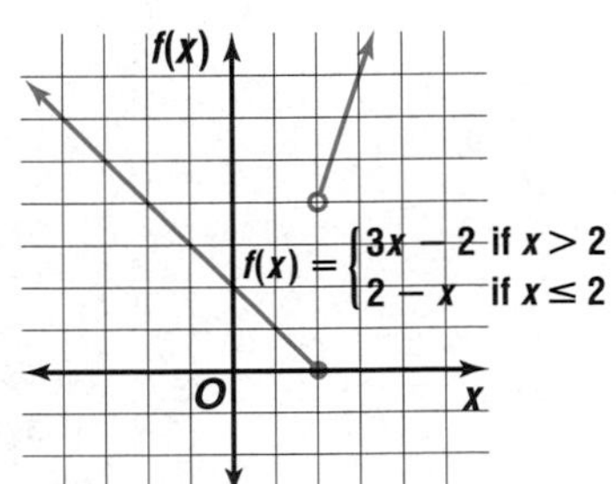

Example 2 **POSTAGE** **Refer to the application at the beginning of the lesson.**

a. Use the continuity test to show that the step function is discontinuous.

b. Explain why a continuous function would not be appropriate to model postage costs.

a. The graph of the postage function is discontinuous at each integral value of w in its domain because the function does not approach the same value from the left and the right. For example, as w approaches 1 from the left, $C(w)$ approaches 0.33 but as w approaches 1 from the right, $C(w)$ approaches 0.55.

b. A continuous function would have to achieve all real y-values (greater than or equal to 0.33.) This would be an inappropriate model for this situation since the weight of a letter is rounded to the nearest ounce and postage costs are rounded to the nearest cent.

$x \to \infty$ is read as "x approaches infinity."

Another tool for analyzing functions is **end behavior.** The end behavior of a function describes what the y-values do as $|x|$ becomes greater and greater. When x becomes greater and greater, we say that x approaches infinity, and we write $x \to \infty$. Similarly, when x becomes more and more negative, we say that x approaches negative infinity, and we write $x \to -\infty$. The same notation can also be used with y or $f(x)$ and with real numbers instead of infinity.

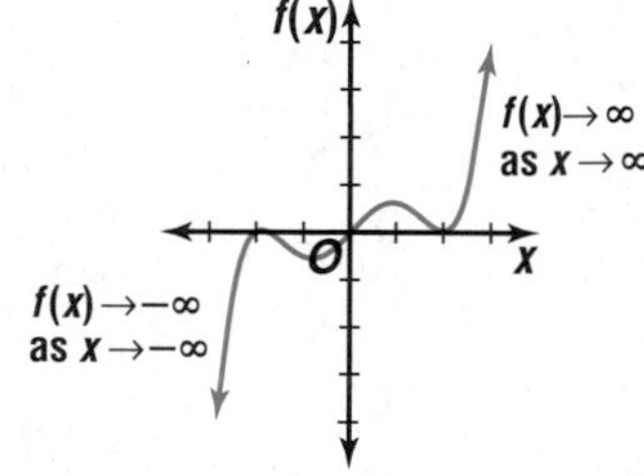

Example 3 **Describe the end behavior of $f(x) = -2x^3$ and $g(x) = -x^3 + x^2 - x + 5$.**

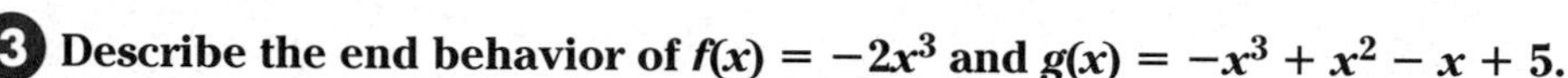

Use your calculator to create a table of function values so you can investigate the behavior of the y-values.

$f(x) = -2x^3$	
x	$f(x)$
−10,000	2×10^{12}
−1000	2×10^9
−100	2,000,000
−10	2000
0	0
10	−2000
100	−2,000,000
1000	-2×10^9
10,000	-2×10^{12}

$g(x) = x^3 + x^2 - x + 5$	
x	$g(x)$
−10,000	1.0001×10^{12}
−1000	1,001,001,005
−100	1,010,105
−10	1115
0	5
10	−905
100	−990,095
1000	−999,000,995
10,000	-9.999×10^{11}

Notice that both polynomial functions have y-values that become very large in absolute value as x gets very large in absolute value. The end behavior of $f(x)$ can be summarized by stating that as $x \to \infty$, $f(x) \to -\infty$ and as $x \to -\infty$, $f(x) \to \infty$. The end behavior of $g(x)$ is the same. *You may wish to graph these functions on a graphing calculator to verify this summary.*

In general, the end behavior of any polynomial function can be modeled by the function comprised solely of the term with the highest power of x and its coefficient. Suppose for $n \geq 0$

$$p(x) = a_n x^n + a_{n-1}x^{n-1} + a_{n-2}x^{n-2} + \cdots + a_2x^2 + a_1x + a_0.$$

Then $f(x) = a_n x^n$ has the same end behavior as $p(x)$. The following table organizes the information for such functions and provides an example of a function displaying each type of end behavior.

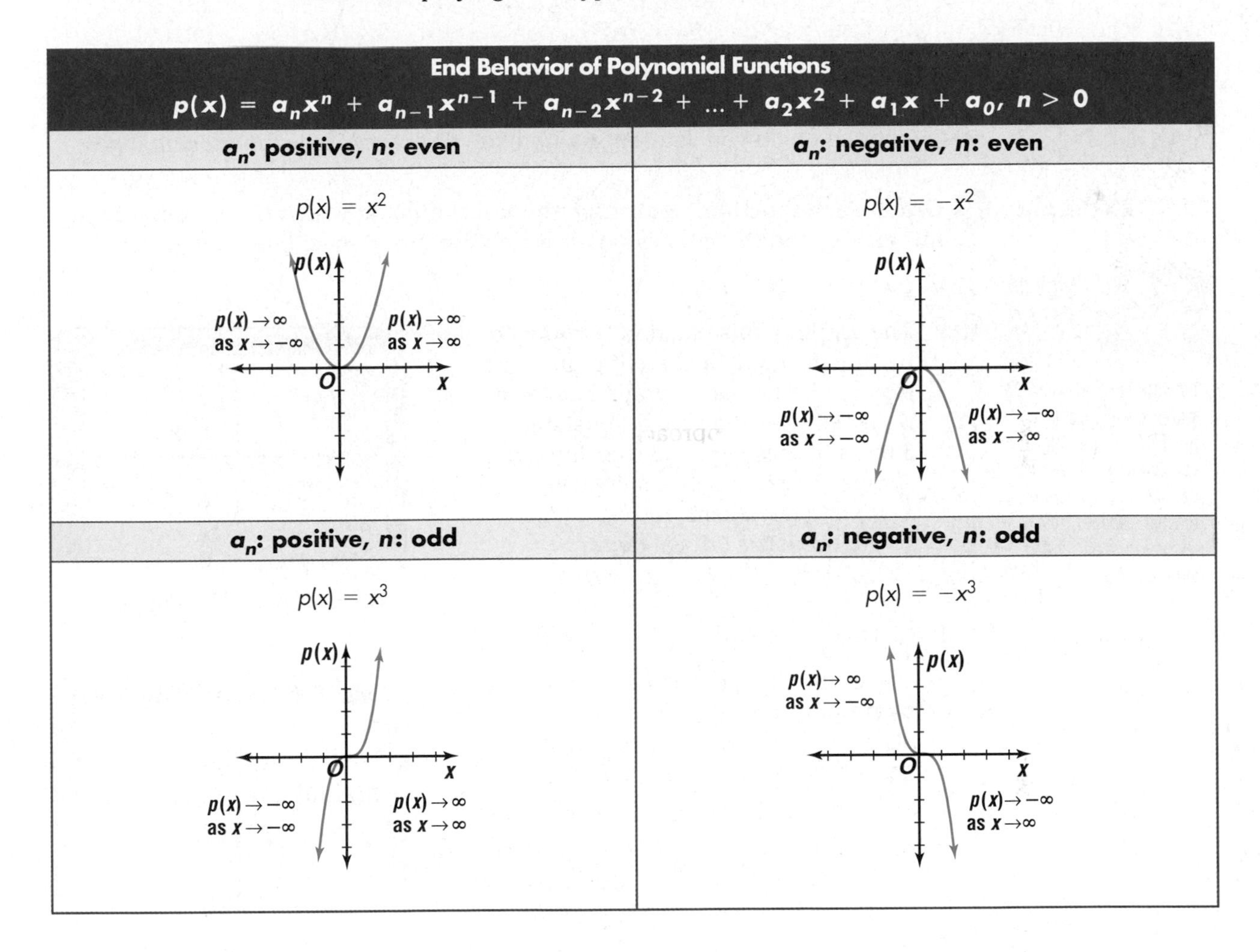

End Behavior of Polynomial Functions

$p(x) = a_n x^n + a_{n-1}x^{n-1} + a_{n-2}x^{n-2} + \ldots + a_2x^2 + a_1x + a_0,\ n > 0$

a_n: positive, n: even	a_n: negative, n: even
$p(x) = x^2$ 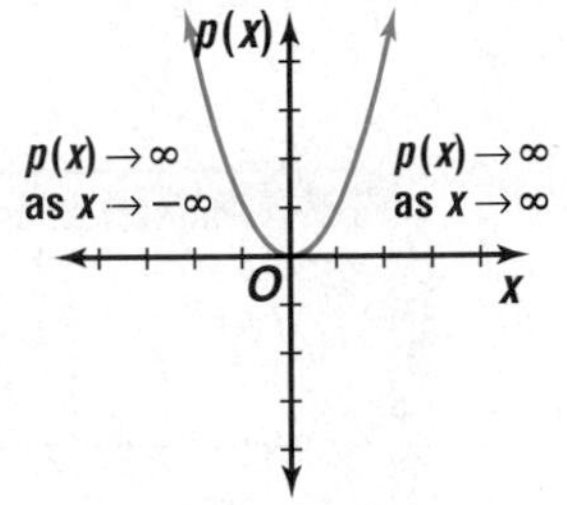 $p(x) \to \infty$ as $x \to -\infty$ $p(x) \to \infty$ as $x \to \infty$	$p(x) = -x^2$ 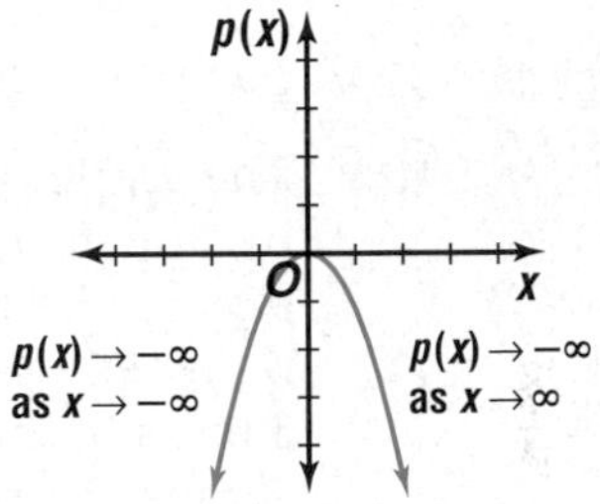 $p(x) \to -\infty$ as $x \to -\infty$ $p(x) \to -\infty$ as $x \to \infty$
a_n: positive, n: odd	**a_n: negative, n: odd**
$p(x) = x^3$ 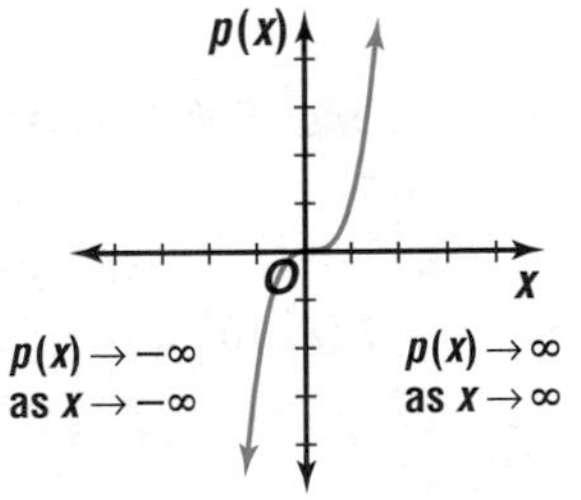 $p(x) \to -\infty$ as $x \to -\infty$ $p(x) \to \infty$ as $x \to \infty$	$p(x) = -x^3$ 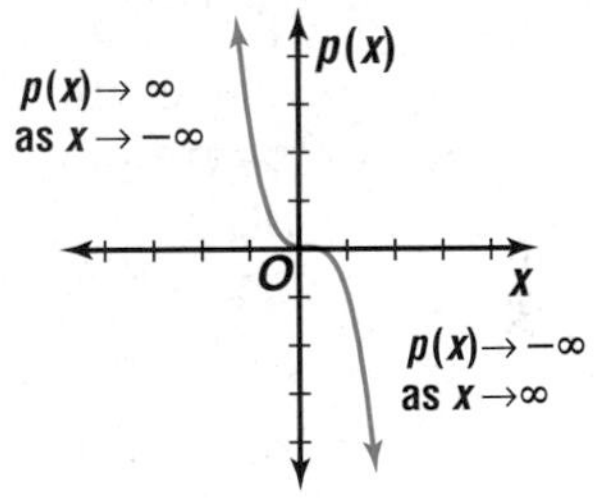 $p(x) \to \infty$ as $x \to -\infty$ $p(x) \to -\infty$ as $x \to \infty$

Another characteristic of functions that can help in their analysis is the **monotonicity** of the function. A function is said to be monotonic on an interval I if and only if the function is increasing on I or decreasing on I.

Whether a graph is increasing or decreasing is always judged by viewing a graph from left to right.

The graph of $f(x) = x^2$ shows that the function is decreasing for $x < 0$ and increasing for $x > 0$.

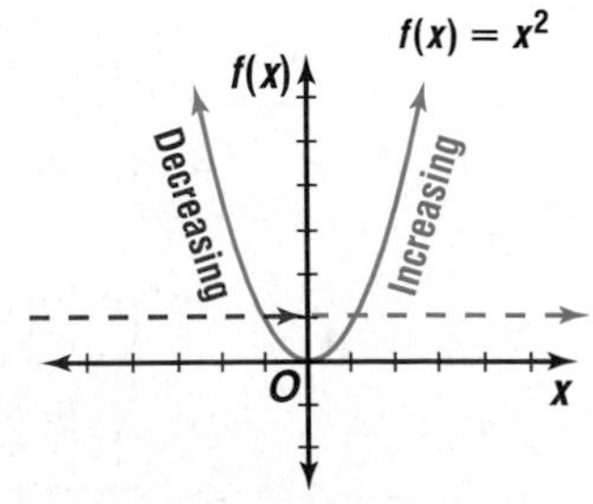

Increasing, Decreasing, and Constant Functions

A function f is increasing on an interval I if and only if for every a and b contained in I, $f(a) < f(b)$ whenever $a < b$.
A function f is decreasing on an interval I if and only if for every a and b contained in I, $f(a) > f(b)$ whenever $a < b$.
A function f remains constant on an interval I if and only if for every a and b contained in I, $f(a) = f(b)$ whenever $a < b$.

Points in the domain of a function where the function changes from increasing to decreasing or vice versa are special points called *critical points.* You will learn more about these special points in Lesson 3-6. Using a graphing calculator can help you determine where the direction of the function changes.

Graphing Calculator Tip

By watching the x- and y-values while using the **TRACE** function, you can determine approximately where a function changes from increasing to decreasing and vice versa.

Example 4 **Graph each function. Determine the interval(s) on which the function is increasing and the interval(s) on which the function is decreasing.**

a. $f(x) = 3 - (x - 5)^2$

The graph of this function is obtained by transforming the parent graph $p(x) = x^2$. The parent graph has been reflected over the x-axis, translated 5 units to the right, and translated up 3 units. The function is increasing for $x < 5$ and decreasing for $x > 5$. At $x = 5$, there is a critical point.

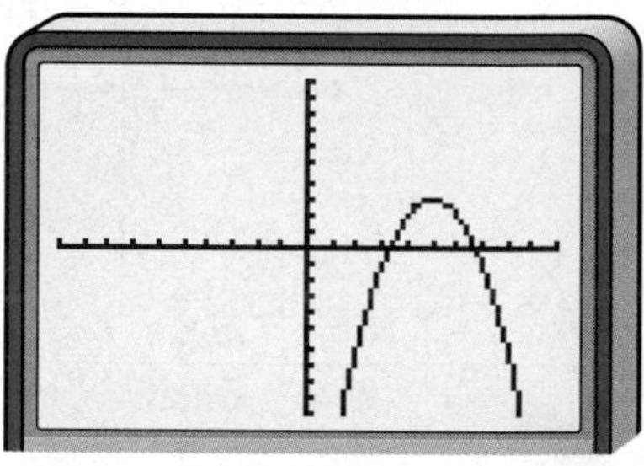

[−10, 10] scl:1 by [−10, 10] scl:1

b. $f(x) = \frac{1}{2}|x + 3| - 5$

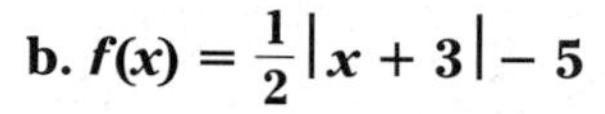

The graph of this function is obtained by transforming the parent graph $p(x) = |x|$. The parent graph has been vertically compressed by a factor of $\frac{1}{2}$, translated 3 units to the left, and translated down 5 units. This function is decreasing for $x < -3$ and increasing for $x > -3$. There is a critical point when $x = -3$.

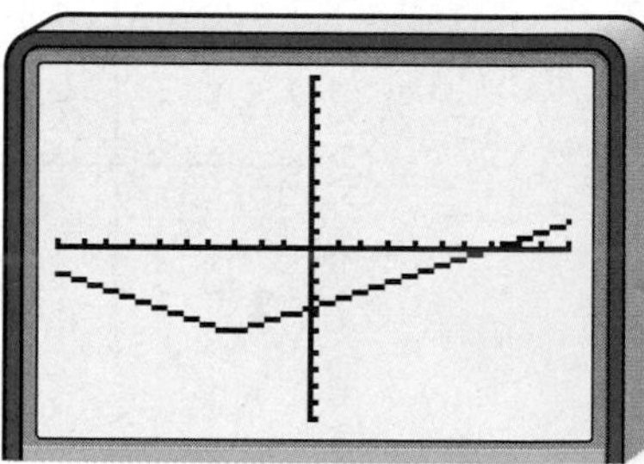

[−10, 10] scl:1 by [−10, 10] scl:1

c. $f(x) = 2x^3 + 3x^2 - 12x + 3$

This function has more than one critical point. It changes direction at $x = -2$ and $x = 1$. The function is increasing for $x < -2$. The function is also increasing for $x > 1$. When $-2 < x < 1$, the function is decreasing.

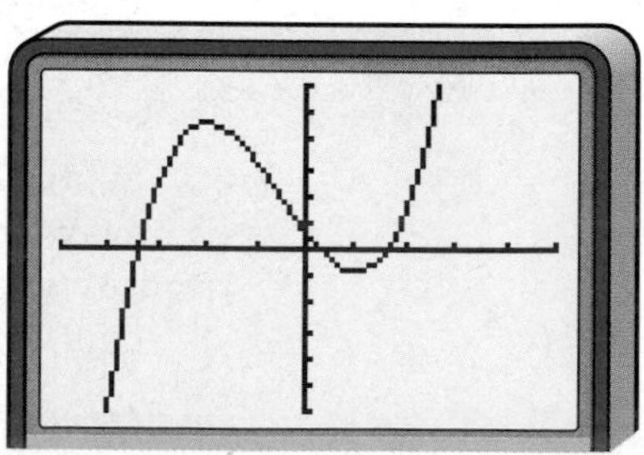

[−5, 5] scl:1 by [−30, 30] scl:5

CHECK FOR UNDERSTANDING

Communicating Mathematics

Read and study the lesson to answer each question.

1. **Explain** why the function whose graph is shown at the right is discontinuous at $x = 2$.

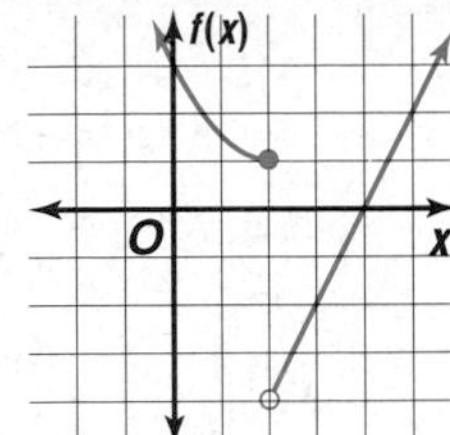

2. **Summarize** the end behavior of polynomial functions.

3. **State** whether the graph at the right has infinite discontinuity, jump discontinuity, or point discontinuity, or is continuous. Then describe the end behavior of the function.

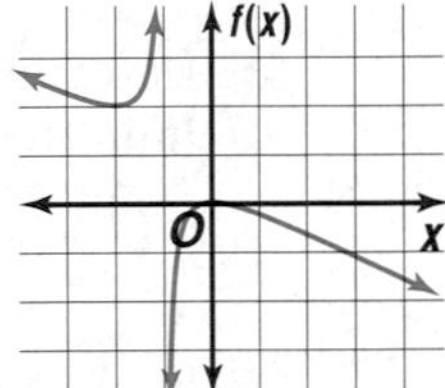

4. *Math Journal* **Write** a paragraph that compares the monotonicity of $f(x) = x^2$ with that of $g(x) = -x^2$. In your paragraph, make a conjecture about the monotonicity of the reflection over the x-axis of any function as compared to that of the original function.

Guided Practice

Determine whether each function is continuous at the given x-value. Justify your answer using the continuity test.

5. $y = \frac{x-5}{x+3}$; $x = -3$

6. $f(x) = \begin{cases} x^2 + 2 \text{ if } x < -2 \\ 3x \text{ if } x \geq -2 \end{cases}$; $x = -2$

Describe the end behavior of each function.

7. $y = 4x^5 + 2x^4 - 3x - 1$

8. $y = -x^6 + x^4 - 5x^2 + 4$

Graph each function. Determine the interval(s) for which the function is increasing and the interval(s) for which the function is decreasing.

9. $f(x) = (x + 3)^2 - 4$

10. $y = \frac{x}{x^2 + 1}$

11. **Electricity** A simple electric circuit contains only a power supply and a resistor. When the power supply is off, there is no current in the circuit. When the power supply is turned on, the current almost instantly becomes a constant value. This situation can be modeled by a graph like the one shown at the right. I represents current in amps, and t represents time in seconds.

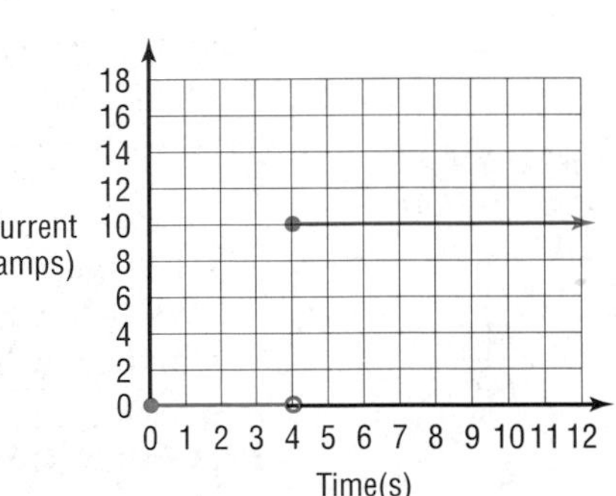

a. At what t-value is this function discontinuous?

b. When was the power supply turned on?

c. If the person who turned on the power supply left and came back hours later, what would he or she measure the current in the circuit to be?

EXERCISES

Practice

Determine whether each function is continuous at the given *x*-value. Justify your answer using the continuity test.

12. $y = x^3 - 4;\ x = 1$

13. $y = \frac{x+1}{x-2};\ x = 2$

14. $f(x) = \frac{x+3}{(x-3)^2};\ x = -3$

15. $y = \left[\!\left[\frac{1}{2}x\right]\!\right];\ x = 3$

16. $f(x) = \begin{cases} 3x + 5 \text{ if } x \le -4 \\ -x + 2 \text{ if } x > -4 \end{cases};\ x = -4$

17. $f(x) = \begin{cases} 2x + 1 \text{ if } x \ge 1 \\ 4 - x^2 \text{ if } x < 1 \end{cases};\ x = 1$

18. Determine whether the graph at the right has infinite discontinuity, jump discontinuity, or point discontinuity, or is continuous.

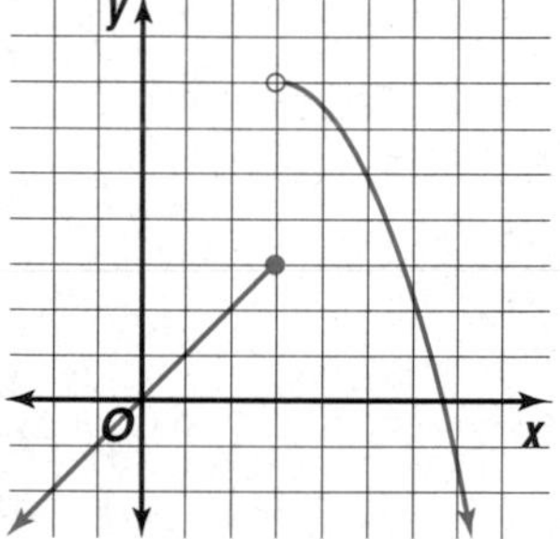

19. Find a value of x at which the function $g(x) = \frac{x-4}{x^2-3x}$ is discontinuous. Use the continuity test to justify your answer.

Describe the end behavior of each function.

20. $y = x^3 + 2x^2 + x - 1$

21. $y = 8 - x^3 - 2x^4$

22. $f(x) = x^{10} - x^9 + 5x^8$

23. $g(x) = |(x-3)^2 - 1|$

24. $y = \frac{1}{x^2}$

25. $f(x) = -\frac{1}{x^3} + 2$

Graphing Calculator

Graph each function. Determine the interval(s) for which the function is increasing and the interval(s) for which the function is decreasing.

26. $y = x^3 + 3x^2 - 9x$

27. $y = -x^3 - 2x + 1$

28. $f(x) = \frac{1}{x+1} - 4$

29. $g(x) = \frac{x^2+5}{x-2}$

30. $y = |x^2 - 4|$

31. $y = (2|x| - 3)^2 + 1$

Applications and Problem Solving

32. **Physics** The gravitational potential energy of an object is given by $U(r) = -\frac{GmM_e}{r}$, where G is Newton's gravitational constant, m is the mass of the object, M_e is the mass of Earth, and r is the distance from the object to the center of Earth. What happens to the gravitational potential energy of the object as it is moved farther and farther away from Earth?

33. **Critical Thinking** A function $f(x)$ is increasing when $0 < x < 2$ and decreasing when $x > 2$. The function has a jump discontinuity when $x = 3$ and $f(x) \to -\infty$ as $x \to \infty$.

   a. If $f(x)$ is an even function, then describe the behavior of $f(x)$ for $x < 0$. Sketch a graph of such a function.

   b. If $f(x)$ is an odd function, then describe the behavior of $f(x)$ for $x < 0$. Sketch a graph of such a function.

34. Biology One model for the population P of bacteria in a sample after t days is given by $P(t) = 1000 - 19.75t + 20t^2 - \frac{1}{3}t^3$.

a. What type of function is $P(t)$?

b. When is the bacteria population increasing?

c. When is it decreasing?

35. Employment The graph shows the minimum wage over a 43-year period in 1996 dollars adjusted for inflation.

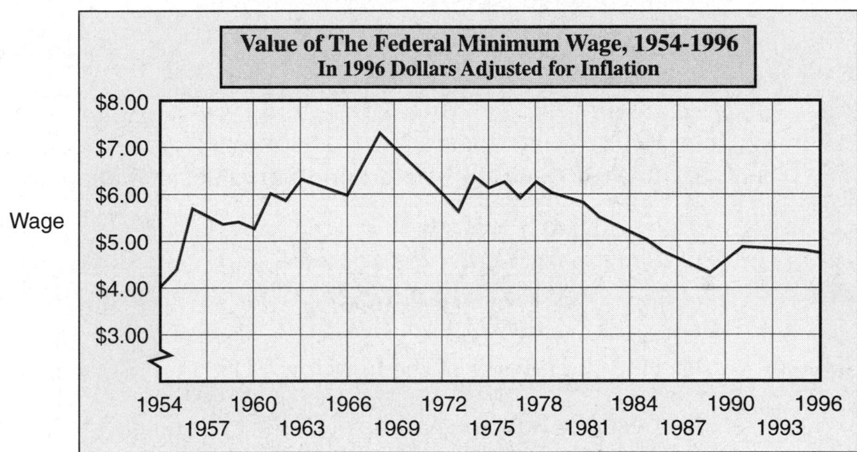

Source: Department of Labor

***inter*NET CONNECTION**

Data Update For the latest information about the minimum wage, visit **www.amc.glencoe.com.**

a. During what time intervals was the adjusted minimum wage increasing?

b. During what time intervals was the adjusted minimum wage decreasing?

36. Analytic Geometry A line is *secant* to the graph of a function if it intersects the graph in at least two distinct points. Consider the function $f(x) = -(x - 4)^2 - 3$.

a. On what interval(s) is $f(x)$ increasing?

b. Choose two points in the interval from part a. Determine the slope of the secant line that passes through those two points.

c. Make a conjecture about the slope of any secant line that passes through two points contained in an interval where a function is increasing. Explain your reasoning.

d. On what interval(s) is $f(x)$ decreasing?

e. Extend your hypothesis from part c to describe the slope of any secant line that passes through two points contained in an interval where the function is decreasing. Test your hypothesis by choosing two points in the interval from part d.

37. Critical Thinking Suppose a function is defined for all x-values and its graph passes the horizontal line test.

a. What can be said about the monotonicity of the function?

b. What can be said about the monotonicity of the inverse of the function?

38. Computers The graph at the right shows the amount of school computer usage per week for students between the ages of 12 and 18.

a. Use this set of data to make a graph of a step function. On each line segment in your graph, put the open circle at the right endpoint.

b. On what interval(s) is the function continuous?

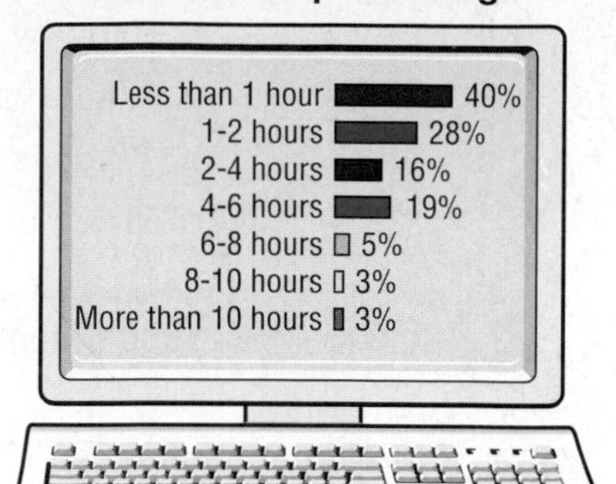

Source: Consumer Electronics Manufacturers Association

39. Critical Thinking Determine the values of a and b so that f is continuous.

$$f(x) = \begin{cases} x^2 + a \text{ if } x \geq 2 \\ bx + a \text{ if } -2 < x < 2 \\ \sqrt{-b - x} \text{ if } x \leq -2 \end{cases}$$

Mixed Review

40. Find the inverse of the function $f(x) = (x + 5)^2$. *(Lesson 3-4)*

41. Describe how the graphs of $f(x) = |x|$ and $g(x) = |x + 2| - 4$ are related. *(Lesson 3-2)*

42. Find the maximum and minimum values of $f(x, y) = x + 2y$ if it is defined for the polygonal convex set having vertices at (0, 0), (4, 0), (3, 5), and (0, 5). *(Lesson 2-6)*

43. Find the determinant of $\begin{bmatrix} 5 & -4 \\ 8 & 2 \end{bmatrix}$. *(Lesson 2-5)*

44. **Consumer Costs** Mario's Plumbing Service charges a fee of \$35 for every service call they make. In addition, they charge \$47.50 for every hour they work on each job. *(Lesson 1-4)*

a. Write an equation to represent the cost c of a service call that takes h hours to complete.

b. Find the cost of a $2\frac{1}{4}$-hour service call.

45. Find $f(-2)$ if $f(x) = 2x^2 - 2x + 8$. *(Lesson 1-1)*

46. **SAT Practice** **Quantitative Comparison**
A if quantity A is greater
B if quantity B is greater
C if the two quantities are equal
D if the relationship cannot be determined from the information given

Column A	Column B
the volume of a cube with side x, where $x > 1$	the volume of a rectangular solid with sides measuring x, $x + 1$, and $x - 1$, where $x > 1$

Extra Practice See p. A31.

GRAPHING CALCULATOR EXPLORATION

3-5B Gap Discontinuities

An Extension of Lesson 3-5

OBJECTIVE

- Construct and graph functions with gap discontinuities.

A function has a *gap discontinuity* if there is some interval of real numbers for which it is not defined. The graphs below show two types of gap discontinuities. The first function is undefined for $2 < x < 4$, and the second is undefined for $-4 \leq x \leq -2$ and for $2 \leq x \leq 3$.

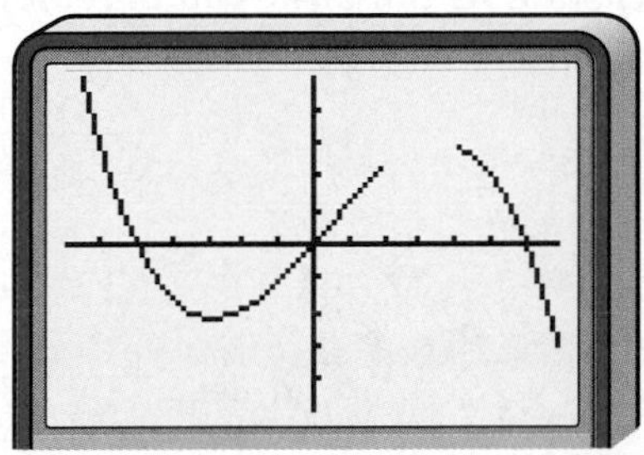

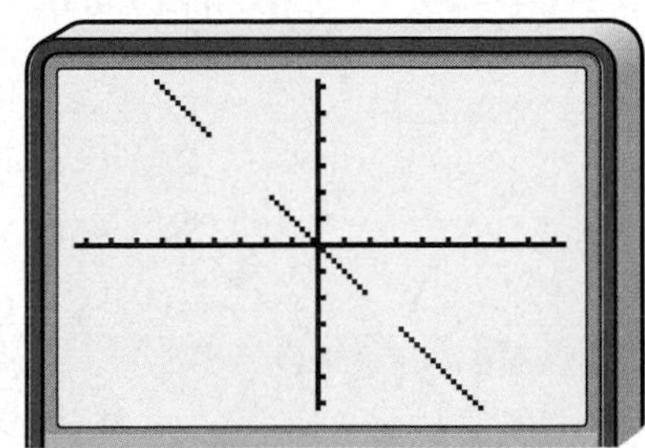

These are also called Boolean operators.

The relational and logical operations on the TEST menu are primarily used for programming. Recall that the calculator delivers a value of 1 for a true equation or inequality and a value of 0 for a false equation or inequality. Expressions that use the logical connectives ("and", "or", "not", and so on) are evaluated according to the usual truth-table rules. Enter each of the following expressions and press ENTER to confirm that the calculator displays the value shown.

Expression	Value
$3 \leq 7$	1
$-2 > -5$	1
$-4 > 6$	0

Expression	Value
$(-1 < 8)$ and $(8 < 9)$	1
$(2 > 4)$ or $(-7 \geq -3)$	0
$(4 < 3)$ and $(1 < 12)$	0

Relational and logical operations are also useful in defining functions that have point and gap discontinuities.

Example **Graph $y = x^2$ for $x \leq -1$ or $x \geq 2$.**

Enter the following expression as **Y1** on the Y= list.

$$X^2/((X \leq -1) \text{ or } (X \geq 2))$$

The function is defined as a quotient. The denominator is a Boolean statement, which has a value of 1 or 0. If the x-value for which the numerator is being evaluated is a member of the interval defined in the denominator, the denominator has a value of 1. Therefore, $\frac{f(x)}{1} = f(x)$ and that part of the function appears on the screen.

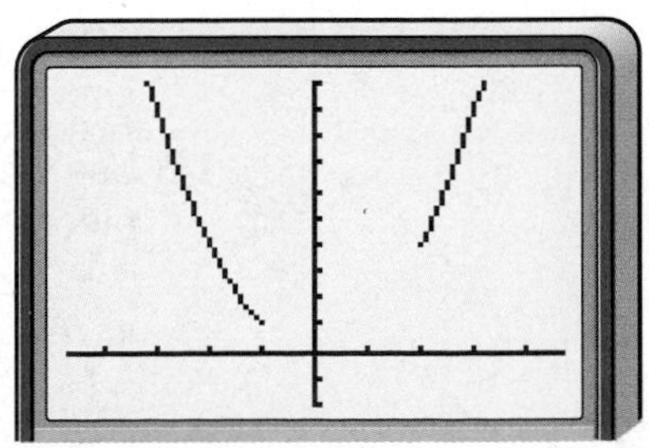

[−4.7, 4.7] scl:1 by [−2, 10] scl:1

(continued on the next page)

If the x-value is not part of the interval, the value of the denominator is 0. At these points, $\frac{f(x)}{0}$ would be undefined. Thus no graph appears on the screen for this interval.

When you use relational and logical operations to define functions, be careful how you use parentheses. Omitting parentheses can easily lead to an expression that the calculator may interpret in a way you did not intend.

TRY THESE

Graph each function and state its domain. You may need to adjust the window settings.

1. $y = \frac{x^2 - 2}{(x > 3)}$

2. $y = \frac{0.5x + 1}{((x \geq -2) \text{ and } (x \leq 4))}$

3. $y = \frac{-2x}{((x < -3) \text{ or } (x \geq 1))}$

4. $y = \frac{-0.2x^3 + 0.3x^2 - x}{((x \leq -3) \text{ or } (x > -2))}$

5. $y = \frac{|x|}{(|x| > 1)}$

6. $y = \frac{|x - 1| - |x - 3|}{(|x + 4| > 2)}$

7. $y = \frac{1.5x}{([\![x]\!] \neq 3)}$

8. $y = \frac{0.5x^2}{(([\![x]\!] \neq -2) \text{ and } ([\![x]\!] \neq 1))}$

Relational and logical operations are not the only tools available for defining and graphing functions with gap discontinuities. The square root function can easily be used for such functions. Graph each function and state its domain.

9. $y = \sqrt{(x - 1)(x - 2)(x - 3)(x - 4)}$

10. $y = \frac{x}{\frac{\sqrt{(x - 1)(x - 2)}}{\sqrt{(x - 1)(x - 2)}}}$

WHAT DO YOU THINK?

11. Suppose you want to construct a function whose graph is like that of $y = x^2$ except for "bites" removed for the values between 2 and 5 and the values between 7 and 8. What equation could you use for the function?

12. Is it possible to use the functions on the MATH NUM menu to take an infinite number of "interval bites" from the graph of a function? Justify your answer.

13. Is it possible to write an equation for a function whose graph looks just like the graph of $y = x^2$ for $x \leq -2$ and just like the graph of $y = 2x - 4$ for $x \geq 4$, with no points on the graph for values of x between -2 and 4? Justify your answer.

14. Use what you have learned about gap discontinuities to graph the following piecewise functions.

a. $f(x) = \begin{cases} -2x \text{ if } x < 0 \\ -x^4 + 2x^3 + 3x^2 + 3x \text{ if } x \geq 0 \end{cases}$

b. $g(x) = \begin{cases} (x + 4)^3 - 2 \text{ if } x < -2 \\ x^2 + 2 \text{ if } -2 \leq x \leq 2 \\ -(x - 4)^3 - 2 \text{ if } x > 2 \end{cases}$

Critical Points and Extrema

OBJECTIVE

- Find the extrema of a function.

BUSINESS America's 23 million small businesses employ more than 50% of the private workforce. Owning a business requires good management skills. Business owners should always look for ways to compete and improve their businesses. Some business owners hire an analyst to help them identify strengths and weaknesses in their operation. Analysts can collect data and develop mathematical models that help the owner increase productivity, maximize profit, and minimize waste. *A problem related to this will be solved in Example 4.*

Optimization is an application of mathematics where one searches for a maximum or a minimum quantity given a set of constraints. When maximizing or minimizing quantities, it can be helpful to have an equation or a graph of a mathematical model for the quantity to be optimized.

Recall from geometry that a line is tangent to a curve if it intersects a curve in exactly one point.

Critical points are those points on a graph at which a line drawn tangent to the curve is horizontal or vertical. A polynomial may possess three types of critical points. A critical point may be a **maximum**, a **minimum**, or a **point of inflection**. When the graph of a function is increasing to the left of $x = c$ and decreasing to the right of $x = c$, then there is a maximum at $x = c$. When the graph of a function is decreasing to the left of $x = c$ and increasing to the right of $x = c$, then there is a minimum at $x = c$. A point of inflection is a point where the graph changes its curvature as illustrated below.

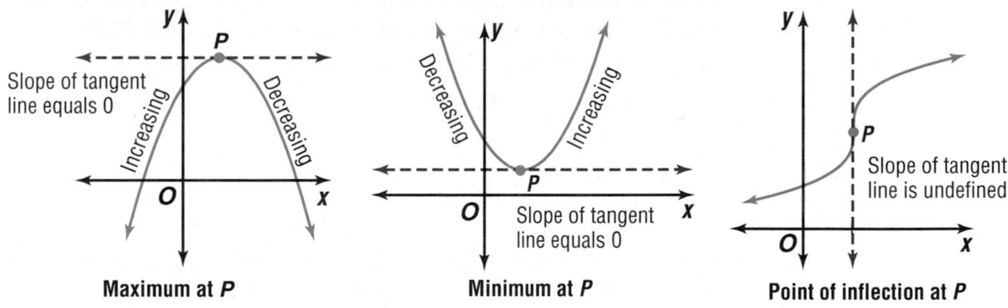

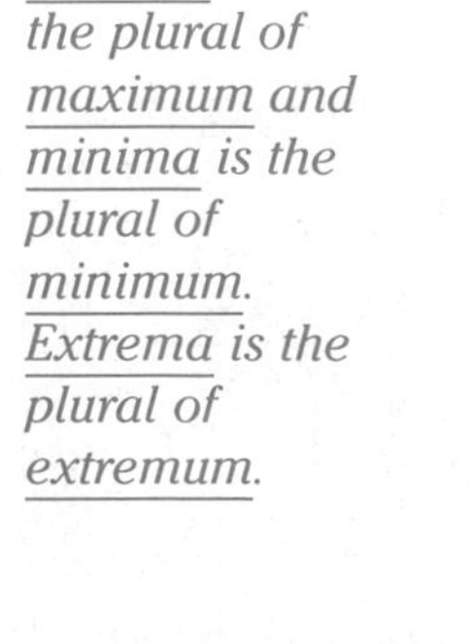

Maxima is the plural of maximum and minima is the plural of minimum. Extrema is the plural of extremum.

The graph of a function can provide a visual clue as to when a function has a maximum or a minimum value. The greatest value that a function assumes over its domain is called the **absolute maximum**. Likewise the least value of a function is the **absolute minimum**. The general term for maximum or minimum is **extremum**. The functions graphed below have absolute extrema.

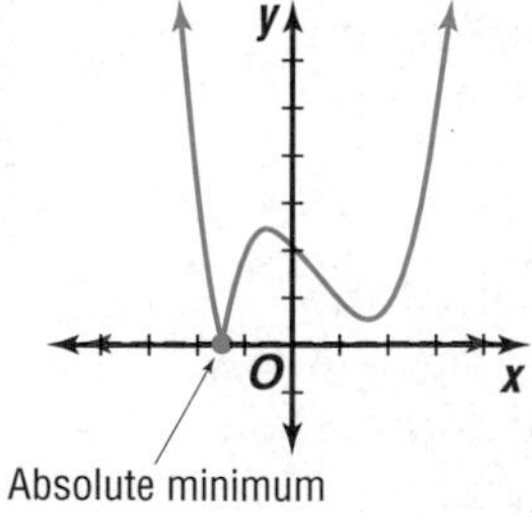

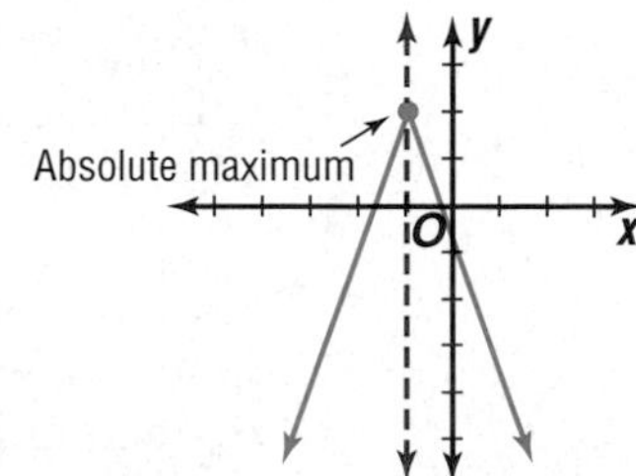

Functions can also have **relative extrema**. A **relative maximum** value of a function may not be the greatest value of f on the domain, but it is the greatest y-value on some interval of the domain. Similarly, a **relative minimum** is the least y-value on some interval of the domain. The function graphed at the right has both a relative maximum and a relative minimum.

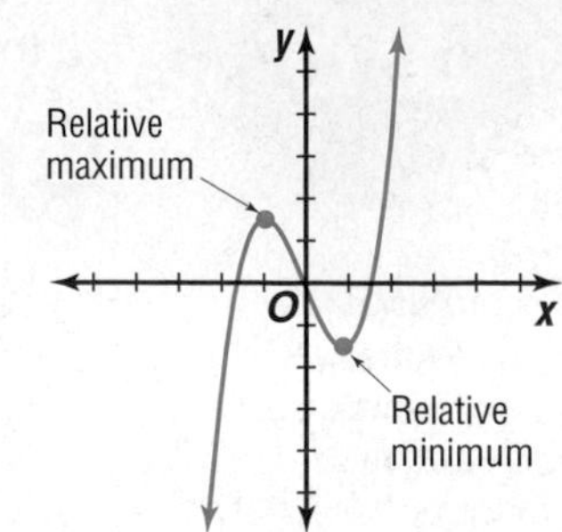

Note that extrema are *values of the function;* that is, they are the y-coordinates of each maximum and minimum point.

Example 1 **Locate the extrema for the graph of $y = f(x)$. Name and classify the extrema of the function.**

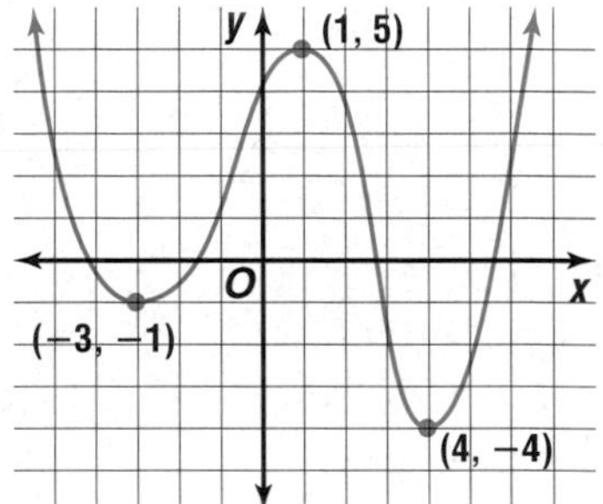

The function has a relative minimum at $(-3, -1)$.

The function has a relative maximum at $(1, 5)$.

The function has a relative minimum at $(4, -4)$.

Since the point $(4, -4)$ is the lowest point on the graph, the function appears to have an absolute minimum of -4 when $x = 4$. This function appears to have no absolute maximum values, since the graph indicates that the function increases without bound as $x \to \infty$ and as $x \to -\infty$.

A branch of mathematics called calculus can be used to locate the critical points of a function. You will learn more about this in Chapter 15. A graphing calculator can also help you locate the critical points of a polynomial function.

Example 2 **Use a graphing calculator to graph $f(x) = 5x^3 - 10x^2 - 20x + 7$ and to determine and classify its extrema.**

Use a graphing calculator to graph the function in the standard viewing window. Notice that the x-intercepts of the graph are between -2 and -1, 0 and 1, and 3 and 4. Relative maxima and minima will occur somewhere between pairs of x-intercepts.

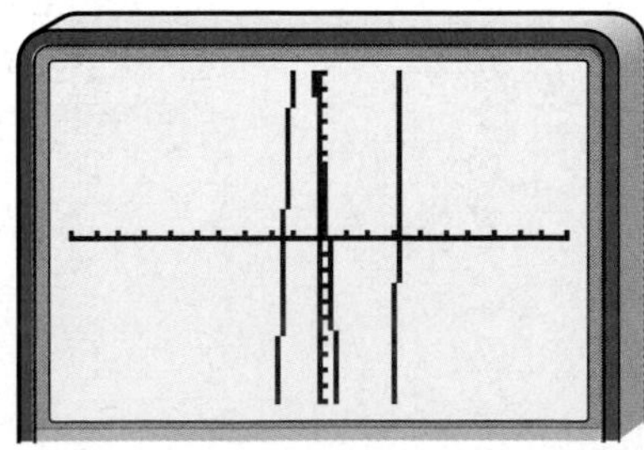

[−10, 10] scl:1 by [−10, 10] scl:1

For a better view of the graph of the function, we need to change the window to encompass the observed x-intercepts more closely.

Graphing Calculator Tip

Xmin and **Xmax** define the *x*-axis endpoints. Likewise, **Ymin** and **Ymax** define the *y*-axis endpoints.

One way to do this is to change the *x*-axis view to $-2 \leq x \leq 4$. Since the top and bottom of the graph are not visible, you will probably want to change the *y*-axis view as well. The graph at the right shows $-40 \leq y \leq 20$. From the graph, we can see there is a relative maximum in the interval $-1 < x < 0$ and a relative minimum in the interval $1 < x < 3$.

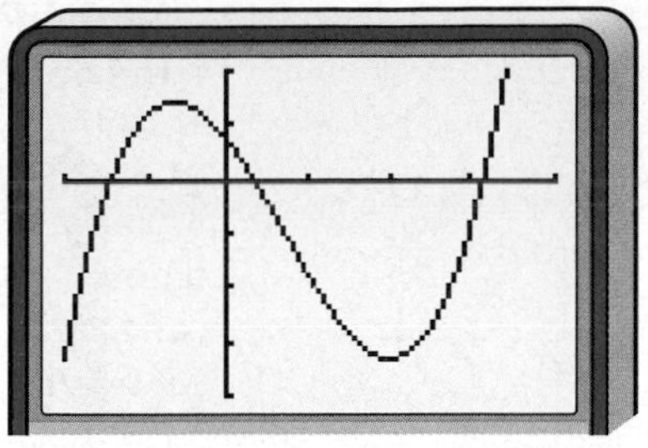

[−2,4] scl:1 by [−40, 20] scl:10

There are several methods you can use to locate these extrema more accurately.

Method 1: Use a table of values to locate the approximate greatest and least value of the function. (*Hint:* Revise **TBLSET** to begin at −2 in intervals of 0.1.)

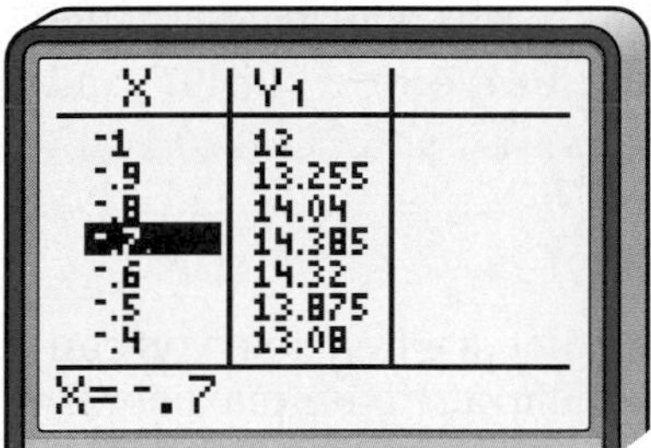

X	Y1
-1	12
-.9	13.255
-.8	14.04
-.7	14.385
-.6	14.32
-.5	13.875
-.4	13.08

X=-.7

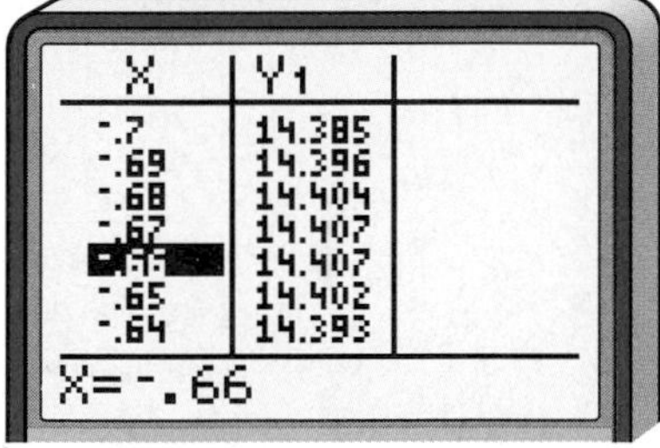

X	Y1
1.6	-30.12
1.7	-31.34
1.8	-32.24
1.9	-32.81
2	-33
2.1	-32.8
2.2	-32.16

X=2

There seems to be a relative maximum of approximately 14.385 at $x = -0.7$ and a relative minimum of -33 at $x = 2$.

You can adjust the **TBLSET** increments to hundredths to more closely estimate the *x*-value for the relative maximum. A relative maximum of about 14.407 appears to occur somewhere between the *x*-values -0.67 and -0.66. A fractional estimation of the *x*-value might be $x = -\frac{2}{3}$.

X	Y1
-.7	14.385
-.69	14.396
-.68	14.404
-.67	14.407
-.66	14.407
-.65	14.402
-.64	14.393

X=-.66

Method 2: Use the **TRACE** function to approximate the relative maximum and minimum.

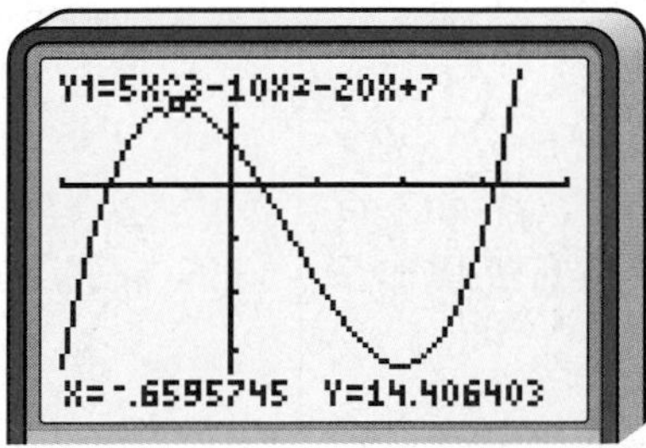

[−2, 4] scl:1 by [−40, 20] scl:10

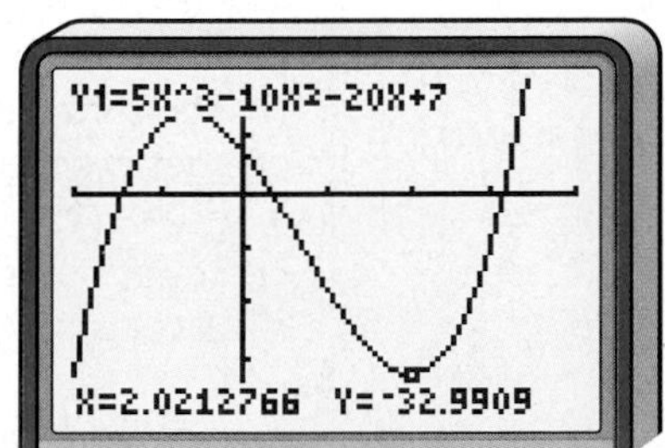

[−2, 4] scl:1 by [−40, 20] scl:10

There seems to be a maximum at $x \approx -0.66$ and a minimum at $x \approx 2.02$.

Method 3: Use **3:minimum** and **4:maximum** options on the **CALC** menu to locate the approximate relative maximum and minimum.

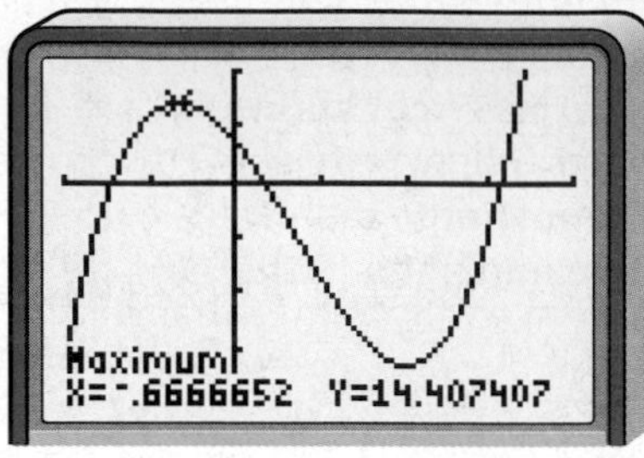

[−2,4] scl:1 by [−40, 20] scl:10

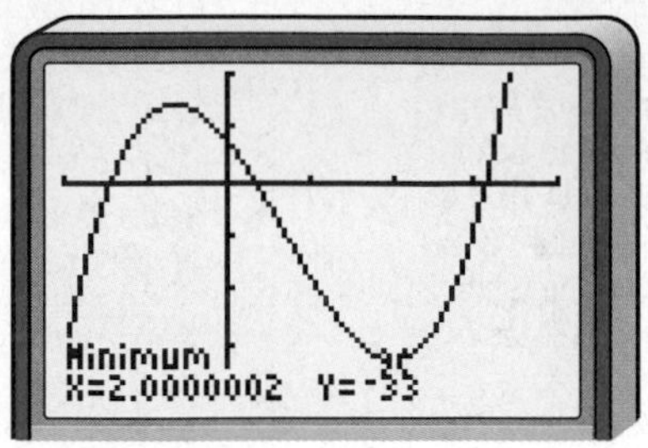

[−2,4] scl:1 by [−40, 20] scl:10

The calculator indicates a relative maximum of about 14.4 at $x \approx -0.67$ and a relative minimum of -33 at $x \approx 2.0$.

All of these approaches give approximations; some more accurate than others. From these three methods, we could estimate that a relative maximum occurs near the point at $(-0.67, 14.4)$ or $\left(-\frac{2}{3}, 14.407\right)$ and a relative minimum near the point at $(2, -33)$.

If you know a critical point of a function, you can determine if it is the location of a relative minimum, a relative maximum, or a point of inflection by testing points on both sides of the critical point. The table below shows how to identify each type of critical point.

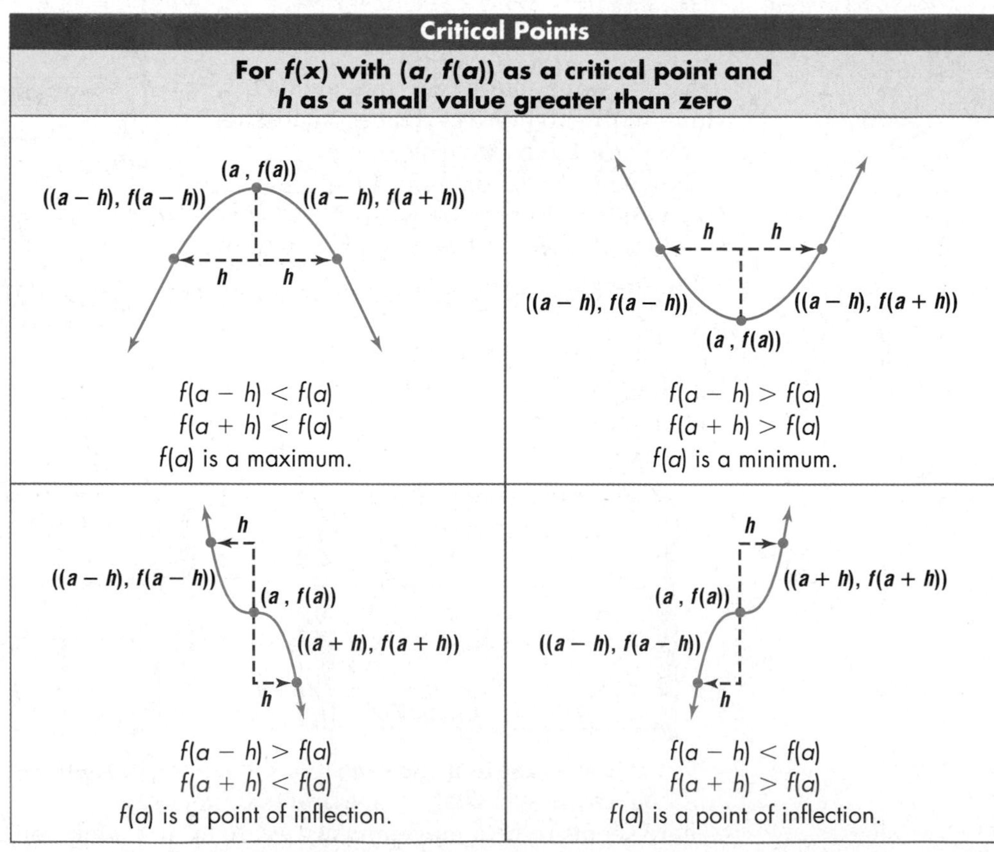

You can also determine whether a critical point is a maximum, minimum, or inflection point by examining the values of a function using a table.

Example 3 **The function $f(x) = 2x^5 - 5x^4 - 10x^3$ has critical points at $x = -1$, $x = 0$, and $x = 3$. Determine whether each of these critical points is the location of a maximum, a minimum, or a point of inflection.**

Evaluate the function at each point. Then check the values of the function around each point. Let $h = 0.1$.

x	x − 0.1	x + 0.1	f(x − 0.1)	f(x)	f(x + 0.1)	Type of Critical Point
−1	−1.1	−0.9	2.769	3	2.829	maximum
0	−0.1	0.1	0.009	0	−0.010	inflection point
3	2.9	3.1	−187.308	−189	−187.087	minimum

You can verify this solution by graphing f(x) on a graphing calculator.

You can use critical points from the graph of a function to solve real-world problems involving maximization and minimization of values.

Example 4

BUSINESS **A small business owner employing 15 people hires an analyst to help the business maximize profits. The analyst gathers data and develops the mathematical model $P(x) = \frac{1}{3}x^3 - 34x^2 + 1012x$. In this model, P is the owner's monthly profits, in dollars, and x is the number of employees. The model has critical points at $x = 22$ and $x = 46$.**

a. Determine which, if any, of these critical points is a maximum.

b. What does this critical point suggest to the owner about business operations?

c. What are the risks of following the analyst's recommendation?

a. Test values around the points. Let $h = 0.1$.

x	x − 0.1	x + 0.1	P(x − 0.1)	P(x)	P(x + 0.1)	Type of Critical Point
22	21.9	22.1	9357.21	9357.33	9357.21	maximum
46	45.9	46.1	7053.45	7053.33	7053.45	minimum

The profit will be at a maximum when the owner employs 22 people.

b. The owner should consider expanding the business by increasing the number of employees from 15 to 22.

c. It is important that the owner hire qualified employees. Hiring unqualified employees will likely cause profits to decline.

CHECK FOR UNDERSTANDING

Communicating Mathematics

Read and study the lesson to answer each question.

1. **Write** an explanation of how to determine if a critical point is a maximum, minimum, or neither.
2. **Determine** whether the point at $(1, -4)$, a critical point of the graph of $f(x) = x^3 - 3x - 2$ shown at the right, represents a relative maximum, a relative minimum, or a point of inflection. Explain your reasoning.

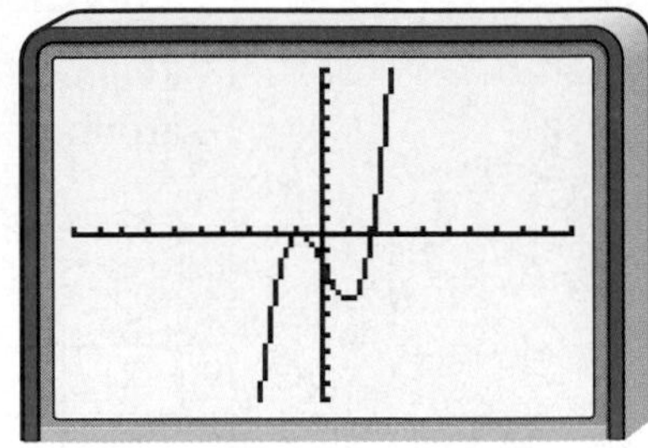

[−10, 10] scl:1 by [−10, 10] scl:1

3. **Sketch** the graph of a function that has a relative minimum at $(0, -4)$, a relative maximum at $(-3, 1)$, and an absolute maximum at $(4, 6)$.

Guided Practice

Locate the extrema for the graph of $y = f(x)$. Name and classify the extrema of the function.

4.

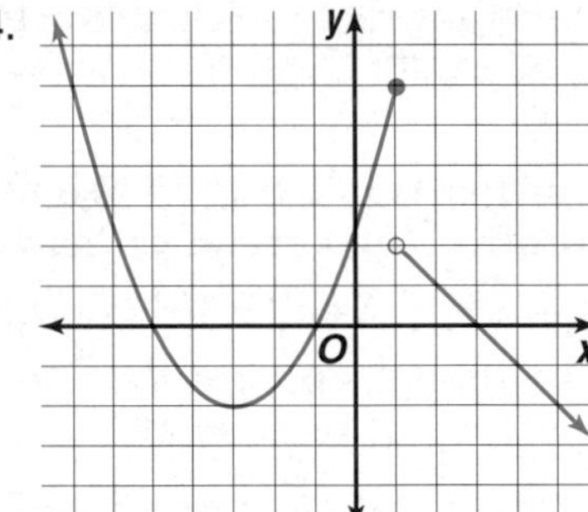

5.

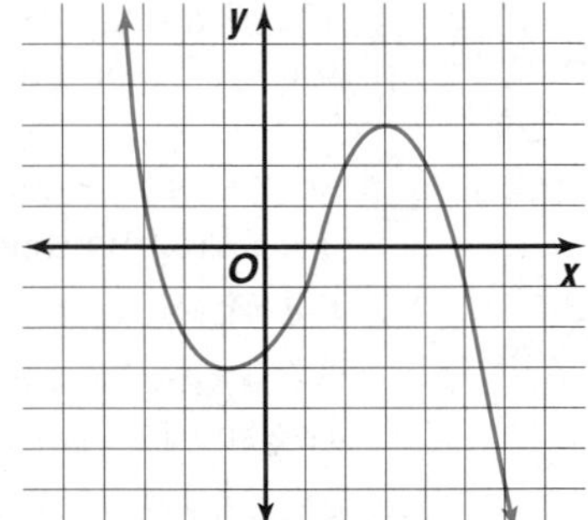

Use a graphing calculator to graph each function and to determine and classify its extrema.

6. $f(x) = 2x^5 - 5x^4$
7. $g(x) = x^4 + 3x^3 - 2$

Determine whether the given critical point is the location of a *maximum*, a *minimum*, or a *point of inflection*.

8. $y = 3x^3 - 9x - 5, x = -1$
9. $y = x^2 + 5x - 6, x = -2.5$
10. $y = 2x^3 - x^5, x = 0$
11. $y = x^6 - 3x^4 + 3x^2 - 1, x = 0$

12. **Agriculture** Malik Davis is a soybean farmer. If he harvests his crop now, the yield will average 120 bushels of soybeans per acre and will sell for \$0.48 per bushel. However, he knows that if he waits, his yield will increase by about 10 bushels per week, but the price will decrease by \$0.03 per bushel per week.
    a. If x represents the number of weeks Mr. Davis waits to harvest his crop, write and graph a function $P(x)$ to represent his profit.
    b. How many weeks should Mr. Davis wait in order to maximize his profit?
    c. What is the maximum profit?
    d. What are the risks of waiting?

EXERCISES

Practice

Locate the extrema for the graph of $y = f(x)$. Name and classify the extrema of the function.

13. 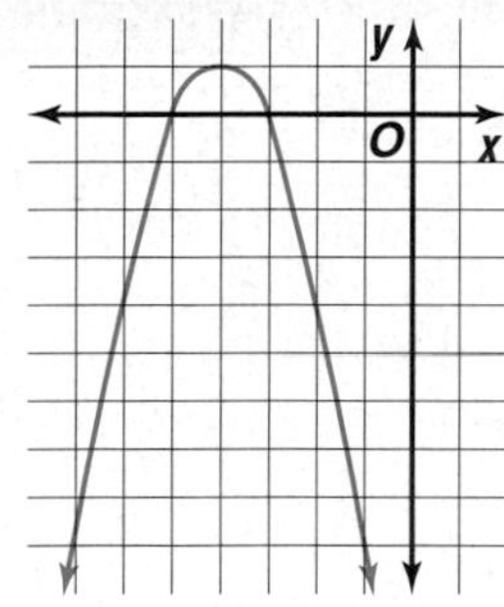

14. 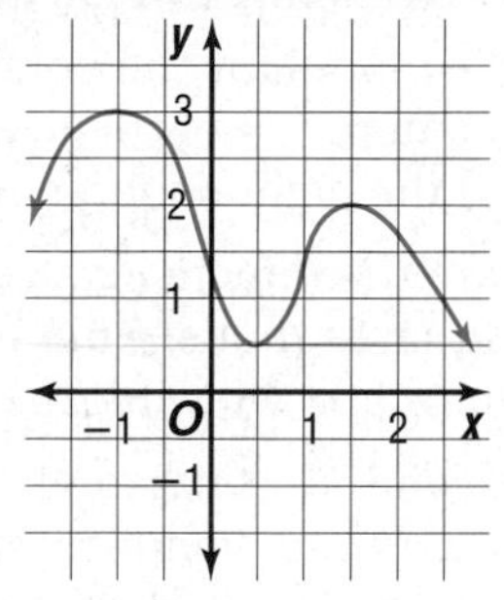

15. 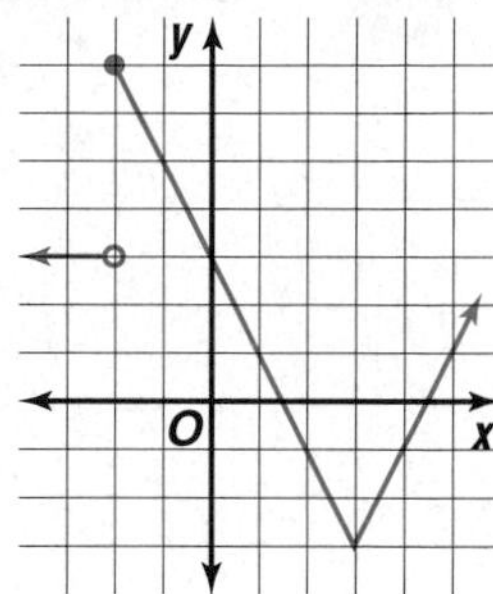

16. 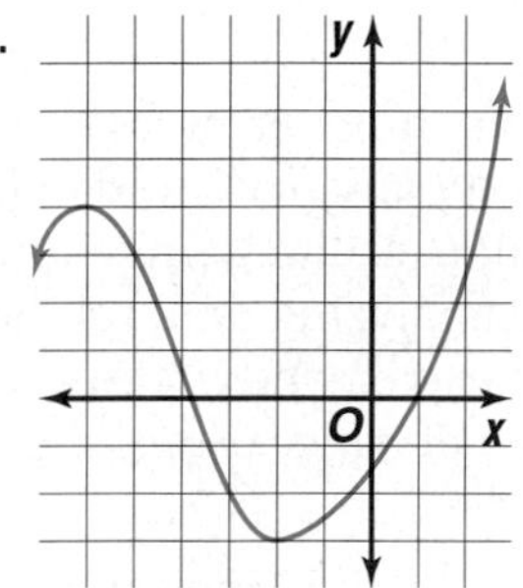

17. 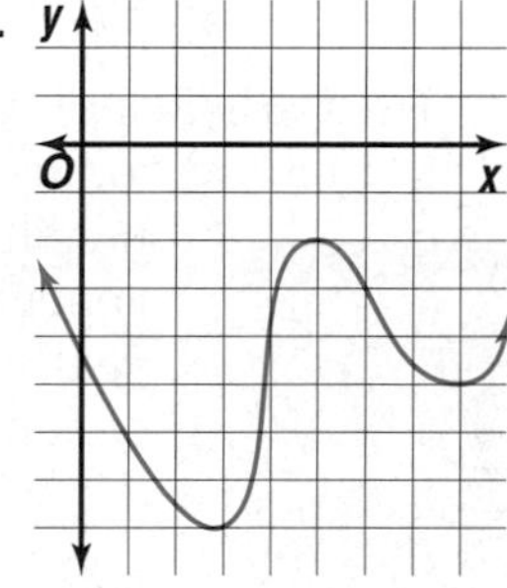

18. 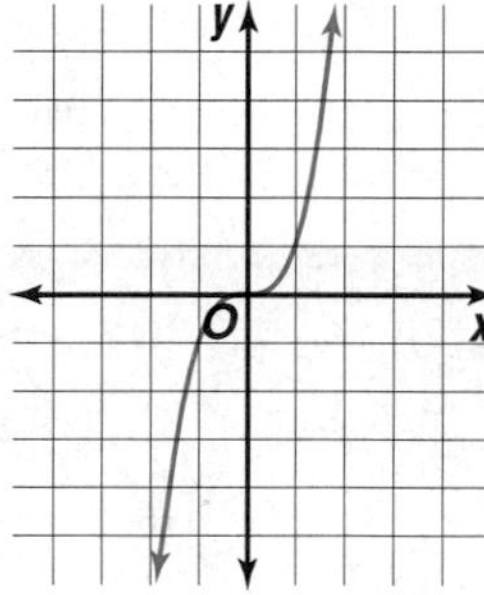

Use a graphing calculator to graph each function and to determine and classify its extrema.

19. $f(x) = -4 + 3x - x^2$

20. $V(w) = w^3 - 7w - 6$

21. $g(x) = 6x^3 + x^2 - 5x - 2$

22. $h(x) = x^4 - 4x^2 - 2$

23. $f(x) = 2x^5 + 4x^2 - 2x - 3$

24. $D(t) = t^3 + t$

25. Determine and classify the extrema of $f(x) = x^4 + 5x^3 + 3x^2 - 4x$.

Determine whether the given critical point is the location of a *maximum*, a *minimum*, or a *point of inflection*.

26. $y = x^3, x = 0$

27. $y = -x^2 + 8x - 10, x = 4$

28. $y = 2x^2 + 10x - 7, x = -2.5$

29. $y = x^4 - 2x^2 + 7, x = 0$

30. $y = \frac{1}{4}x^4 - 2x^2, x = 2$

31. $y = x^3 - 9x^2 + 27x - 27, x = 3$

32. $y = \frac{1}{3}x^3 + \frac{1}{2}x^2 - 2x + 1, x = -2$

33. $y = x^3 - x^2 + 3, x = \frac{2}{3}$

34. A function f has a relative maximum at $x = 2$ and a point of inflection at $x = -1$. Find the critical points of $y = -2f(x + 5) - 1$. Describe what happens at each new critical point.

Applications and Problem Solving

35. Manufacturing A 12.5 centimeter by 34 centimeter piece of cardboard will have eight congruent squares removed as in the diagram. The box will be folded to create a take-out hamburger box.

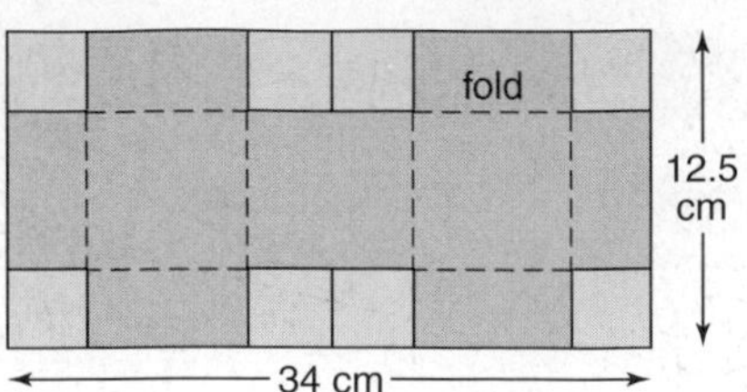

a. Find the model for the volume $V(x)$ of the box as a function of the length x of the sides of the eight squares removed.

b. What are the dimensions of each of the eight squares that should be removed to produce a box with maximum volume?

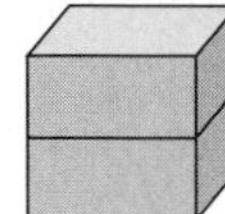

c. Construct a physical model of the box and measure its volume. Compare this result to the result from the mathematical model.

36. Business The Carlisle Innovation Company has created a new product that costs \$25 per item to produce. The company has hired a marketing analyst to help it determine a selling price for the product. After collecting and analyzing data relating selling price s to yearly consumer demand d, the analyst estimates demand for the product using the equation $d = -200s + 15{,}000$.

a. If yearly profit is the difference between total revenue and production costs, determine a selling price s, $s \geq 25$, that will maximize the company's yearly profit, P. (*Hint:* $P = sd - 25d$)

b. What are the risks of determining a selling price using this method?

37. Telecommunications A cable company wants to provide service for residents of an island. The distance from the closest point on the island's beach, point A, directly to the mainland at point B is 2 kilometers. The nearest cable station, point C, is 10 kilometers downshore from point B. It costs \$3500 per kilometer to lay the cable lines underground and \$5000 per kilometer to lay the cable lines under water. The line comes to the mainland at point M. Let x be the distance in kilometers from point B to point M.

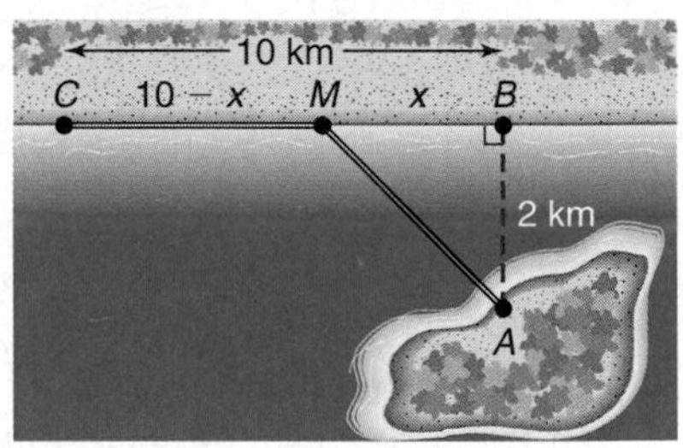

a. Write a function to calculate the cost of laying the cable.

b. At what distance x should the cable come to shore to minimize cost?

38. Critical Thinking Which families of graphs have points of inflection but no maximum or minimum points?

39. Physics When the position of a particle as a function of time t is modeled by a polynomial function, then the particle is at rest at each critical point. If a particle has a position given by $s(t) = 2t^3 - 11t^2 + 3t - 9$, find the position of the particle each time it is at rest.

40. **Critical Thinking** A cubic polynomial can have 1 or 3 critical points. Describe the possible combinations of relative maxima and minima for a cubic polynomial.

Mixed Review

41. Is $y = \frac{5x}{x^2 - 3x - 10}$ continuous at $x = 5$? Justify your answer by using the continuity test. *(Lesson 3-5)*

42. Graph the inequality $y \leq \frac{1}{5}(x - 2)^3$. *(Lesson 3-3)*

43. **Manufacturing** The Eastern Minnesota Paper Company can convert wood pulp to either newsprint or notebook paper. The mill can produce up to 200 units of paper a day, and regular customers require 10 units of notebook paper and 80 units of newsprint per day. If the profit on a unit of notebook paper is \$400 and the profit on a unit of newsprint is \$350, how much of each should the plant produce? *(Lesson 2-7)*

44. **Geometry** Find the system of inequalities that will define a polygonal convex set that includes all points in the interior of a square whose vertices are $A(-3, 4)$, $B(2, 4)$, $C(2, -1)$, and $D(-3, -1)$. *(Lesson 2-6)*

45. Find the determinant for $\begin{bmatrix} 1 & 3 \\ 2 & 5 \end{bmatrix}$. Does an inverse exist for this matrix? *(Lesson 2-5)*

46. Find $3A + 2B$ if $A = \begin{bmatrix} 4 & -2 \\ 5 & 7 \end{bmatrix}$ and $B = \begin{bmatrix} -3 & 5 \\ -4 & 3 \end{bmatrix}$. *(Lesson 2-3)*

47. **Sports** Jon played in two varsity basketball games. He scored 32 points by hitting 17 of his 1-point, 2-point, and 3-point attempts. He made 50% of his 18 2-point field goal attempts. Find the number of 1-point free throws, 2-point field goals, and 3-point field goals Jon scored in these two games. *(Lesson 2-2)*

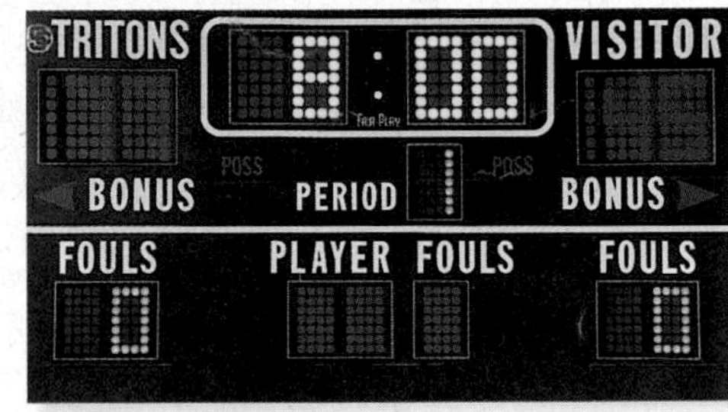

48. Graph $y + 6 \geq 4$. *(Lesson 1-8)*

49. Determine whether the graphs of $2x + 3y = 15$ and $6x = 4y + 16$ are *parallel, coinciding, perpendicular,* or *none of these.* *(Lesson 1-5)*

50. Describe the difference between a relation and a function. How do you test a graph to determine if it is the graph of a function? *(Lesson 1-1)*

51. **SAT/ACT Practice** Refer to the figure at the right. What percent of the area of rectangle *PQRS* is shaded?

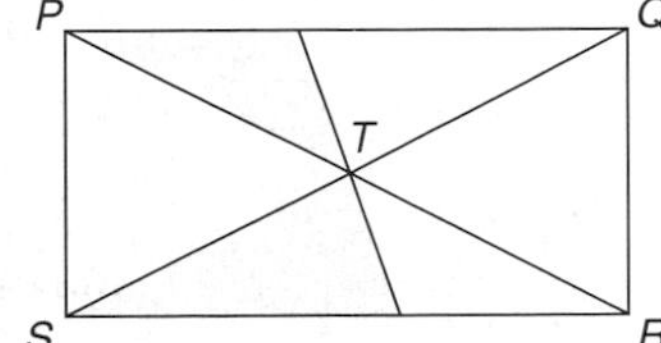

A 20%

B $33\frac{1}{3}$%

C 30%

D 25%

E 40%

Extra Practice See p. A31.

3-7 Graphs of Rational Functions

OBJECTIVES

- Graph rational functions.
- Determine vertical, horizontal, and slant asymptotes.

CHEMISTRY The creation of standard chemical solutions requires precise measurements and appropriate mixing. For example, a chemist may have 50 liters of a 10-molar sodium chloride (NaCl) solution (10 moles of NaCl per liter of water.) This solution must be diluted so it can be used in an experiment. Adding 4-molar NaCl solution (4 moles of NaCl per liter of water) to the 10-molar solution will decrease the concentration. The concentration, C, of the mixture can be modeled by $C(x) = \frac{500 + 4x}{50 + x}$, where x is the number of liters of 4-molar solution added. The graph of this function has characteristics that are helpful in understanding the situation. *A problem related to this will be solved in Example 3.*

The concentration function given above is an example of a **rational function.** A rational function is a quotient of two polynomial functions. It has the form $f(x) = \frac{g(x)}{h(x)}$, where $h(x) \neq 0$. The parent rational function is $f(x) = \frac{1}{x}$.

The graph of $f(x) = \frac{1}{x}$ consists of two branches, one in Quadrant I and the other in Quadrant III. The graph has no x- or y-intercepts. The graph of $f(x) = \frac{1}{x}$, like that of many rational functions, has branches that approach lines called **asymptotes.** In the case of $f(x) = \frac{1}{x}$, the line $x = 0$ is a **vertical asymptote.** Notice that as x approaches 0 from the right, the value of $f(x)$ increases without bound toward positive infinity (∞). As x approaches 0 from the left, the value of $f(x)$ decreases without bound toward negative infinity ($-\infty$).

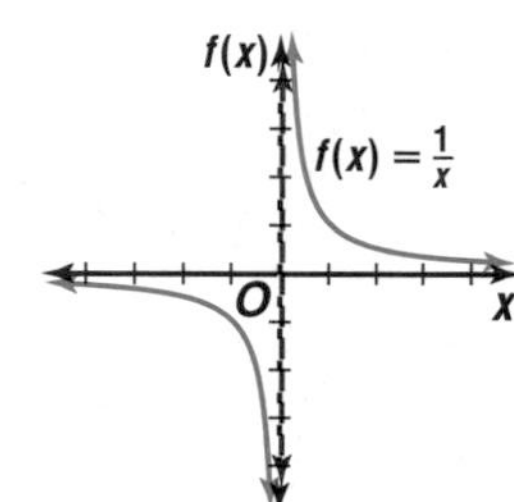

A vertical asymptote in the graph of a rational function is also called a pole of the function.

Vertical Asymptote	The line $x = a$ is a vertical asymptote for a function $f(x)$ if $f(x) \to \infty$ or $f(x) \to -\infty$ as $x \to a$ from either the left or the right.

Also notice for $f(x) = \frac{1}{x}$ that as the value of x increases and approaches positive infinity, the value of $f(x)$ approaches 0. The same type of behavior can also be observed in the third quadrant: as x decreases and approaches negative infinity, the value of $f(x)$ approaches 0. When the values of a function approach a constant value b as $x \to \infty$ or $x \to -\infty$, then the function has a **horizontal asymptote**. Thus, for $f(x) = \frac{1}{x}$, the line $f(x) = 0$ is a horizontal asymptote.

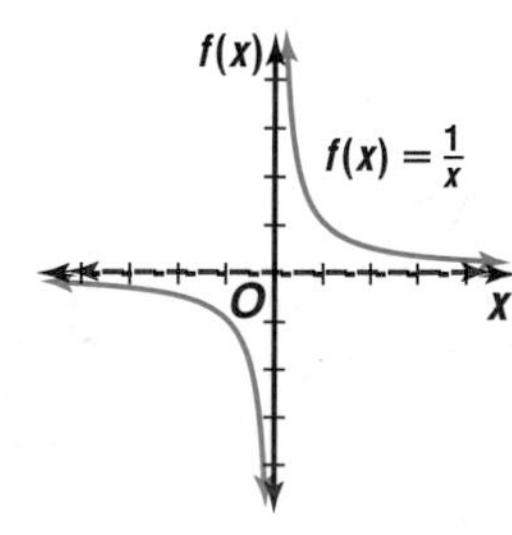

Horizontal Asymptote	The line $y = b$ is a horizontal asymptote for a function $f(x)$ if $f(x) \to b$ as $x \to \infty$ or as $x \to -\infty$.

There are several methods that may be used to determine if a rational function has a horizontal asymptote. Two such methods are used in the example below.

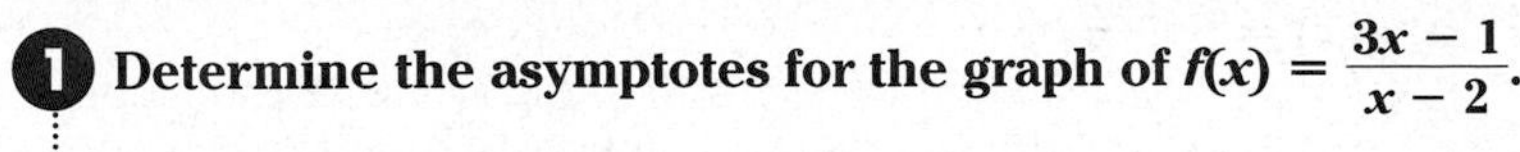

Example 1 **Determine the asymptotes for the graph of $f(x) = \frac{3x - 1}{x - 2}$.**

Graphing Calculator Tip

You can use the TRACE or TABLE function on your graphing calculator to approximate the vertical or horizontal asymptote.

Since $f(2)$ is undefined, there may be a vertical asymptote at $x = 2$. To verify that $x = 2$ is indeed a vertical asymptote, you have to make sure that $f(x) \to \infty$ or $f(x) \to -\infty$ as $x \to 2$ from either the left or the right.

x	f(x)
1.9	−47
1.99	−497
1.999	−4997
1.9999	−49997

The values in the table suggest that $f(x) \to -\infty$ as $x \to 2$ from the left, so there is a vertical asymptote at $x = 2$.

Two different methods may be used to find the horizontal asymptote.

Method 1

Let $f(x) = y$ and solve for x in terms of y. Then find where the function is undefined for values of y.

$$y = \frac{3x - 1}{x - 2}$$

$$y(x - 2) = 3x - 1 \quad \textit{Multiply each side by } (x - 2).$$

$$xy - 2y = 3x - 1 \quad \textit{Distribute.}$$

$$xy - 3x = 2y - 1$$

$$x(y - 3) = 2y - 1 \quad \textit{Factor.}$$

$$x = \frac{2y - 1}{y - 3} \quad \textit{Divide each side by } y - 3.$$

The rational expression $\frac{2y - 1}{y - 3}$ is undefined for $y = 3$. Thus, the horizontal asymptote is the line $y = 3$.

Method 2

First divide the numerator and denominator by the highest power of x.

$$y = \frac{3x - 1}{x - 2}$$

$$y = \frac{\frac{3x}{x} - \frac{1}{x}}{\frac{x}{x} - \frac{2}{x}}$$

$$y = \frac{3 - \frac{1}{x}}{1 - \frac{2}{x}}$$

As the value of x increases positively or negatively, the values of $\frac{1}{x}$ and $\frac{2}{x}$ approach zero. Therefore, the value of the entire expression approaches $\frac{3}{1}$ or 3. So, the line $y = 3$ is the horizontal asymptote.

Method 2 is preferable when the degree of the numerator is greater than that of the denominator.

The graph of this function verifies that the lines $y = 3$ and $x = 2$ are asymptotes.

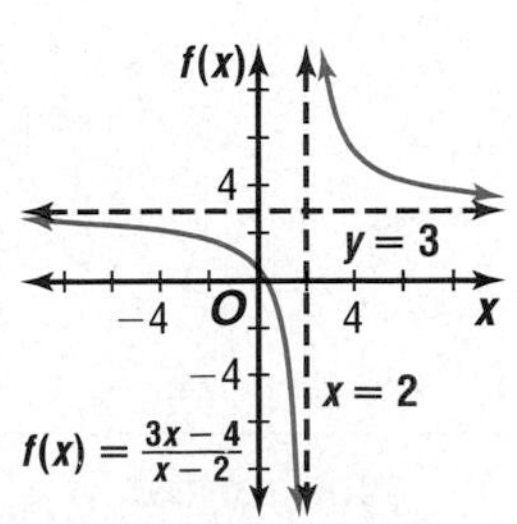

Example 2 **Use the parent graph $f(x) = \frac{1}{x}$ to graph each function. Describe the transformation(s) that take place. Identify the new location of each asymptote.**

a. $g(x) = \frac{1}{x + 5}$

To graph $g(x) = \frac{1}{x + 5}$, translate the parent graph 5 units to the left. The new vertical asymptote is $x = -5$. The horizontal asymptote, $y = 0$, remains unchanged.

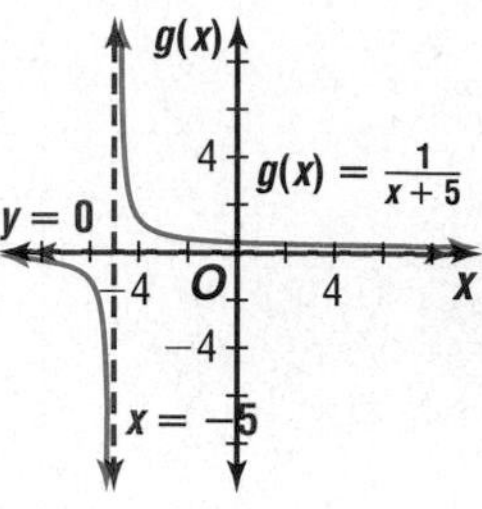

b. $h(x) = -\frac{1}{2x}$

To graph $h(x) = -\frac{1}{2x}$, reflect the parent graph over the x-axis, and compress the result horizontally by a factor of 2. This does not affect the vertical asymptote at $x = 0$. The horizontal asymptote, $y = 0$, is also unchanged.

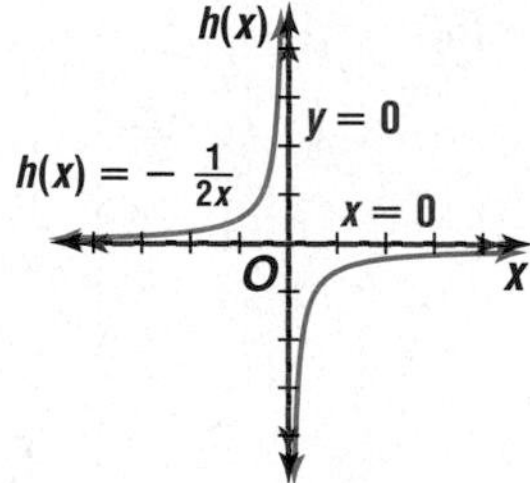

c. $k(x) = \frac{4}{x - 3}$

To graph $k(x) = \frac{4}{x - 3}$, stretch the parent graph vertically by a factor of 4 and translate the parent graph 3 units to the right. The new vertical asymptote is $x = 3$. The horizontal asymptote, $y = 0$, is unchanged.

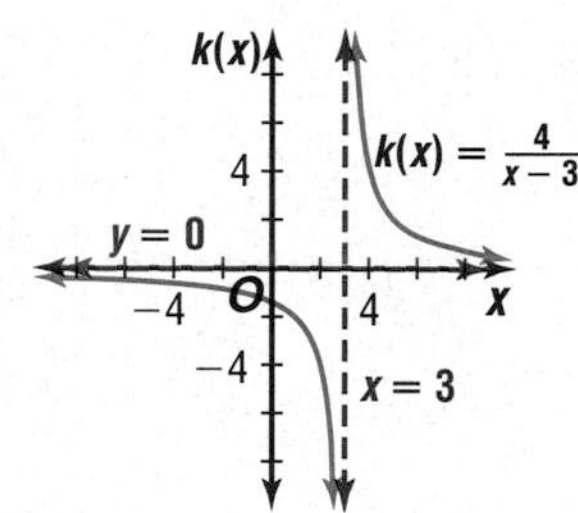

d. $m(x) = -\frac{6}{x + 2} - 4$

To graph $m(x) = -\frac{6}{x + 2} - 4$, reflect the parent graph over the x-axis, stretch the result vertically by a factor of 6, and translate the result 2 units to the left and 4 units down. The new vertical asymptote is $x = -2$. The horizontal asymptote changes from $y = 0$ to $y = -4$.

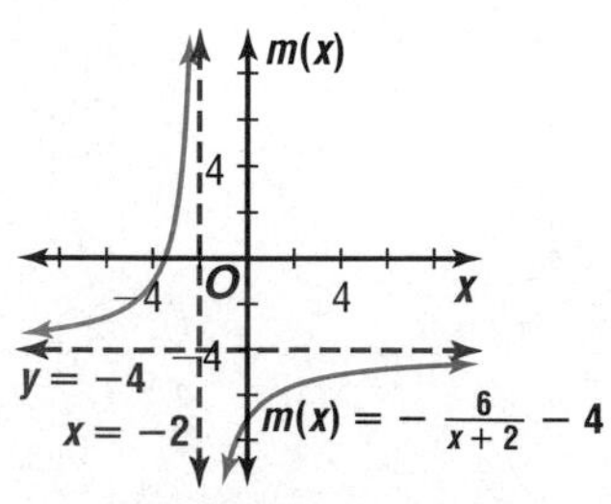

Many real-world situations can be modeled by rational functions.

Example 3 

CHEMISTRY **Refer to the application at the beginning of the lesson.**

a. Write the function as a transformation of the parent $f(x) = \frac{1}{x}$.

b. Graph the function and identify the asymptotes. Interpret their meaning in terms of the problem.

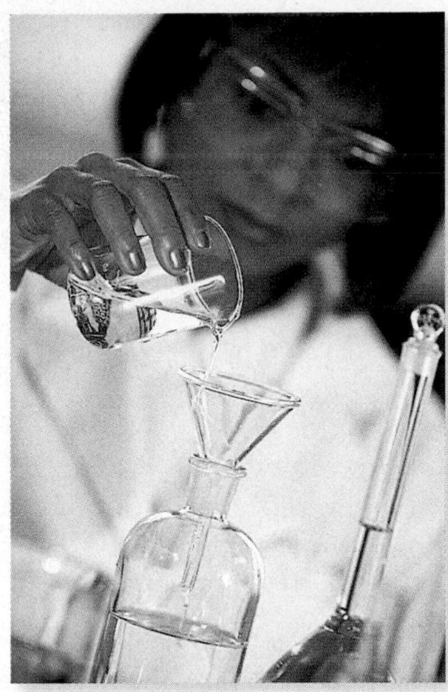

a. In order to write $C(x) = \frac{500 + 4x}{50 + x}$ as a transformation of $f(x) = \frac{1}{x}$, you must perform the indicated division of the linear functions in the numerator and denominator.

$$\begin{array}{r} 4 \\ x + 50\overline{)4x + 500} \\ \underline{4x + 200} \\ 300 \end{array} \Rightarrow 4 + \frac{300}{x + 50} \text{ or } 300\left(\frac{1}{x + 50}\right) + 4$$

Therefore, $C(x) = 300\left(\frac{1}{x + 50}\right) + 4$.

b. The graph of $C(x)$ is a vertical stretch of the parent function by a factor of 300 and a translation 50 units to the left and 4 units up. The asymptotes are $x = -50$ and $y = 4$.

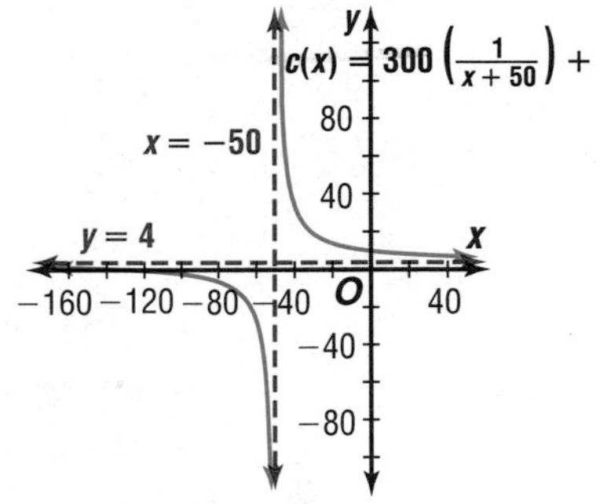

For the purposes of our model, the domain of $C(x)$ must be restricted to $x \geq 0$ since the number of liters of the 4-molar solution added cannot be negative. The vertical asymptote $x = -50$ has no meaning in this application as a result of this restriction. The horizontal asymptote indicates that as the amount of 4-molar solution that has been added grows, the overall concentration will approach a molarity of 4.

A third type of asymptote is a **slant asymptote.** Slant asymptotes occur when the degree of the numerator of a rational function is *exactly* one greater than that of the denominator. For example, in $f(x) = \frac{x^3 + 1}{x^2}$, the degree of the numerator is 3 and the degree of the denominator is 2. Therefore, this function has a slant asymptote, as shown in the graph. Note that there is also a vertical asymptote at $x = 0$. *When the degrees are the same or the denominator has the greater degree, the function has a horizontal asymptote.*

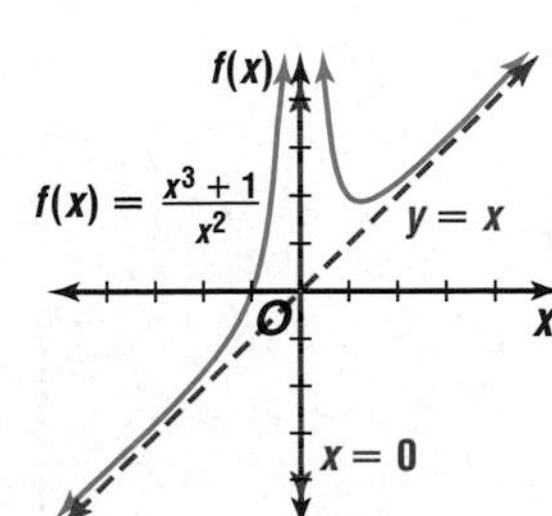

Slant Asymptote	The oblique line ℓ is a slant asymptote for a function $f(x)$ if the graph of $y = f(x)$ approaches ℓ as $x \to \infty$ or as $x \to -\infty$.

Example 4 **Determine the slant asymptote for $f(x) = \dfrac{2x^2 - 3x + 1}{x - 2}$.**

This method can also be used to determine the horizontal asymptote when the degree of the numerator is equal to or greater than that of the denominator.

First, use division to rewrite the function.

$$\begin{array}{r} 2x + 1 \\ x - 2\overline{)2x^2 - 3x + 1} \\ \underline{2x^2 - 4x} \\ x + 1 \\ \underline{x - 2} \\ 3 \end{array} \quad \Rightarrow \quad f(x) = 2x + 1 + \frac{3}{x - 2}$$

As $x \to \infty$, $\dfrac{3}{x - 2} \to 0$. So, the graph of $f(x)$ will approach that of $y = 2x + 1$. This means that the line $y = 2x + 1$ is a slant asymptote for the graph of $f(x)$. *Note that $x = 2$ is a vertical asymptote.*

The graph of this function verifies that the line $y = 2x + 1$ is a slant asymptote.

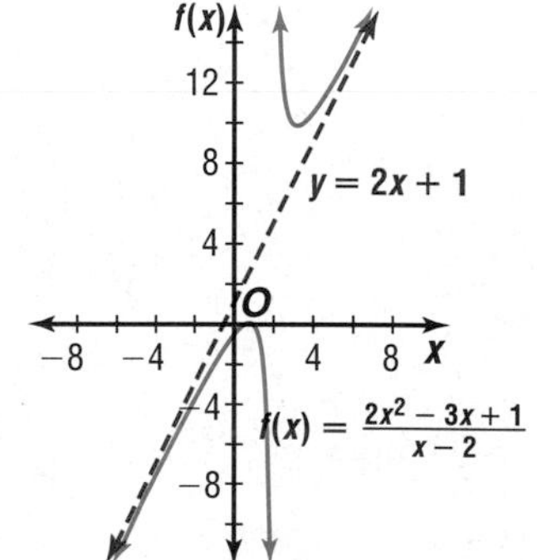

There are times when the numerator and denominator of a rational function share a common factor. Consider $f(x) = \dfrac{(x + 2)(x - 3)}{x - 3}$. Since an x-value of 3 results in a denominator of 0, you might expect there to be a vertical asymptote at $x = 3$. However, $x - 3$ is a common factor of the numerator and denominator.

Numerically, we can see that $f(x)$ has the same values as $g(x) = x + 2$ except at $x = 3$.

x	2.9	2.99	3.0	3.01	3.1
f(x)	4.9	4.99	—	5.01	5.1
g(x)	4.9	4.99	5.0	5.01	5.1

The y-values on the graph of f approach 5 from both sides but never get to 5, so the graph has point discontinuity at (3, 5).

Whenever the numerator and denominator of a rational function contain a common linear factor, a point discontinuity may appear in the graph of the function. If, after dividing the common linear factors, the same factor remains in the denominator, a vertical asymptote exists. Otherwise, the graph will have point discontinuity.

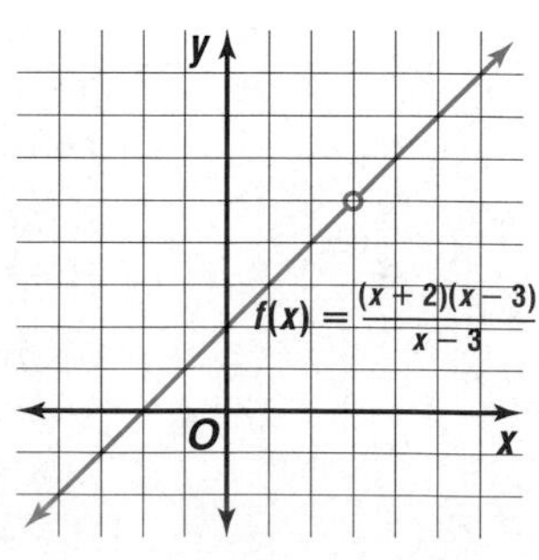

Example 5 Graph $y = \frac{(x+3)(x+1)}{x(x+3)(x-2)}$.

$$y = \frac{(x+3)(x+1)}{x(x+3)(x-2)}$$
$$= \frac{x+1}{x(x-2)},\ x \neq -3$$

Since $x + 3$ is a common factor that does not remain in the denominator after the division, there is point discontinuity at $x = -3$. Because y increases or decreases without bound close to $x = 0$ and $x = 2$, there are vertical asymptotes at $x = 0$ and $x = 2$. There is also a horizontal asymptote at $y = 0$.

The graph is the graph of $y = \frac{x+1}{x(x-2)}$, except for point discontinuity at $\left(-3, -\frac{2}{15}\right)$.

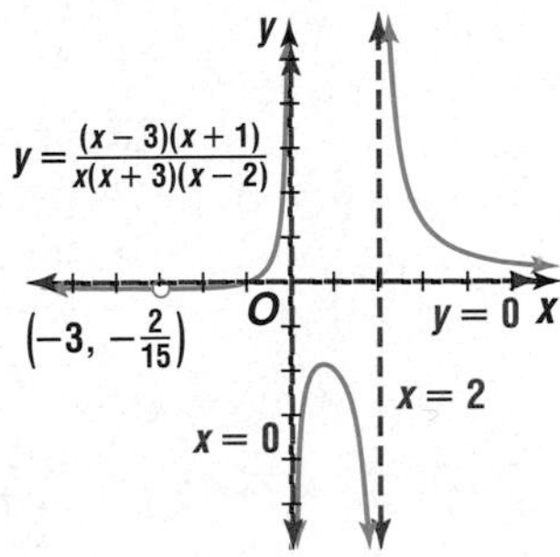

CHECK FOR UNDERSTANDING

Communicating Mathematics

Read and study the lesson to answer each question.

1. **Graph** $f(x) = \frac{1}{x}$ after it has been translated 2 units to the right and down 6 units.
   a. What are its asymptotes?
   b. Write an equation of the translated graph.
2. **Draw** a graph to illustrate each type of asymptote discussed in this lesson: vertical, horizontal, and slant.
3. **Write** an equation of a rational function that has point discontinuity.
4. **True or False:** If a value of x causes a zero in the denominator of a rational function, then there is a vertical asymptote at that x-value. Explain.

Guided Practice

Determine the equations of the vertical and horizontal asymptotes, if any, of each function.

5. $f(x) = \frac{x}{x-5}$

6. $g(x) = \frac{x^3}{(x-2)(x+1)}$

7. The graph at the right shows a transformation of $f(x) = \frac{1}{x}$. Write an equation of the function.

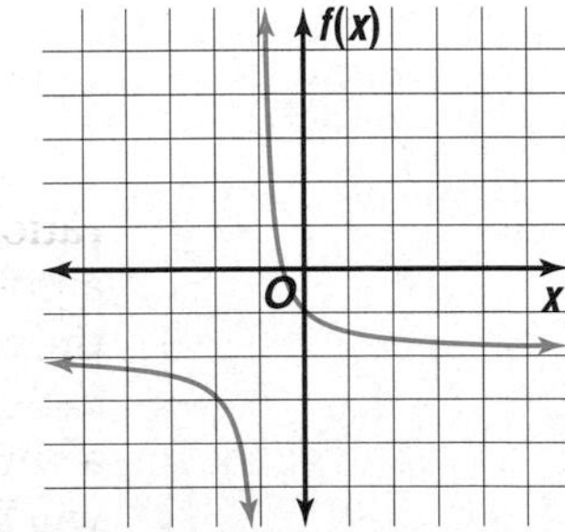

Use the parent graph $f(x) = \frac{1}{x}$ to graph each equation. Describe the transformation(s) that have taken place. Identify the new locations of the asymptotes.

8. $y = \frac{1}{x-4}$

9. $y = \frac{1}{x+2} - 1$

10. Determine the slant asymptote for the function $f(x) = \frac{3x^2 - 4x + 5}{x - 3}$.

Graph each function.

11. $y = \frac{x+2}{(x+1)(x-1)}$

12. $y = \frac{x^2 + 4x + 4}{x + 2}$

13. Chemistry The Ideal Gas Law states that the pressure P, volume V, and temperature T, of an ideal gas are related by the equation $PV = nRT$, where n is the number of moles of gas and R is a constant.

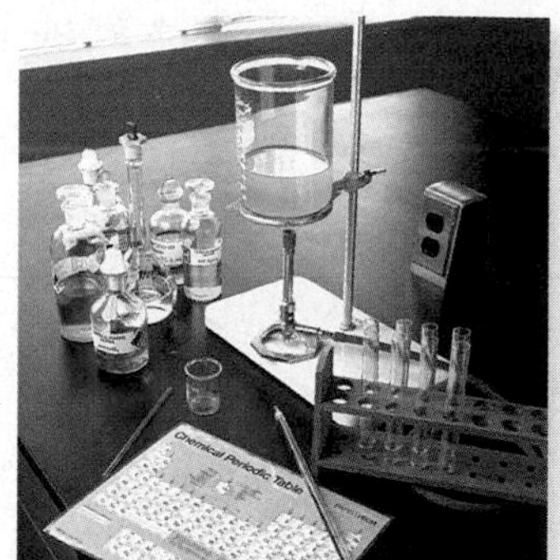

a. Sketch a graph of P versus V, assuming that T is fixed.

b. What are the asymptotes of the graph?

c. What happens to the pressure of the gas if the temperature is held fixed and the gas is allowed to occupy a larger and larger volume?

EXERCISES

Practice

Determine the equations of the vertical and horizontal asymptotes, if any, of each function.

14. $f(x) = \frac{2x}{x+4}$

15. $f(x) = \frac{x^2}{x+6}$

16. $g(x) = \frac{x-1}{(2x+1)(x-5)}$

17. $g(x) = \frac{x-2}{x^2+4x+3}$

18. $h(x) = \frac{x^2}{x^2+1}$

19. $h(x) = \frac{(x+1)^2}{x^2-1}$

20. What are the vertical and horizontal asymptotes of the function $y = \frac{x^3}{(x-2)^4}$?

Each graph below shows a transformation of $f(x) = \frac{1}{x}$. Write an equation of each function.

21.

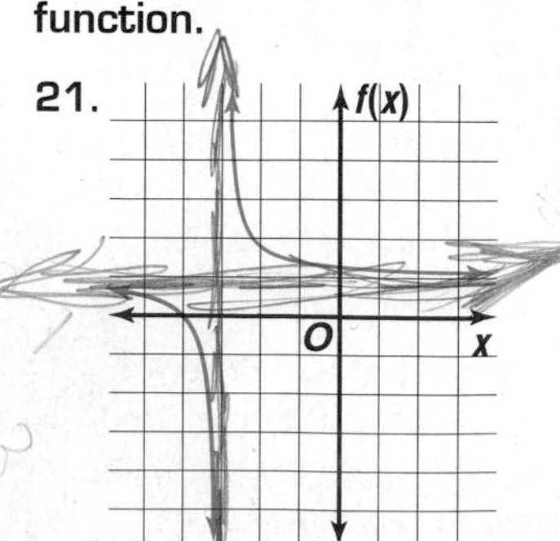

22.

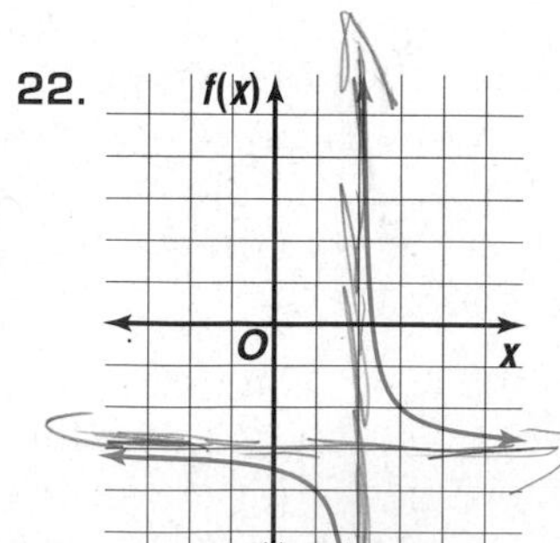

23.

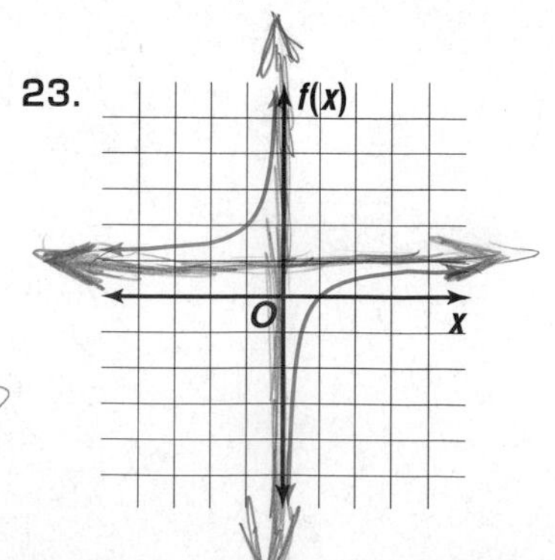

Use the parent graph $f(x) = \frac{1}{x}$ to graph each equation. Describe the transformation(s) that have taken place. Identify the new locations of the asymptotes.

24. $y = \frac{1}{x} + 3$
25. $y = \frac{2}{x - 4}$
26. $y = \frac{2}{x + 3} - 1$
27. $y = -\frac{3}{x} + 2$
28. $y = \frac{3x + 1}{x - 3}$
29. $y = \frac{-4x + 2}{x + 5}$

Determine the slant asymptote of each function.

30. $f(x) = \frac{x^2 + 3x - 3}{x + 4}$
31. $f(x) = \frac{x^2 + 3x - 4}{x}$
32. $f(x) = \frac{x^3 - 2x^2 + x - 4}{x^2 + 1}$
33. $f(x) = \frac{x^2 - 4x + 1}{2x - 3}$

34. Does the function $f(x) = \frac{x^3 - 4x^2 + 2x - 6}{x + 3}$ have a slant asymptote? If so, find an equation of the slant asymptote. If not, explain.

Graph each function.

35. $y = \frac{(x - 2)(x + 1)}{x}$
36. $y = \frac{x^2 - 4}{x - 2}$
37. $y = \frac{x + 2}{x^2 - 4}$
38. $y = \frac{(x - 2)^2(x + 1)^2}{(x - 2)(x - 1)}$
39. $y = \frac{x^2 - 6x + 9}{x^2 - x - 6}$
40. $y = \frac{x^2 - 1}{x^2 - 2x + 1}$

Applications and Problem Solving

41. **Chemistry** Suppose the chemist in the application at the beginning of the lesson had to dilute 40 liters of a 12-molar solution by adding 3-molar solution.
   a. Write the function that models the concentration of the mixture as a function of the number of liters t of 3-molar solution added.
   b. How many liters of 3-molar solution must be added to create a 10-molar solution?

42. **Electronics** Suppose the current I in an electric circuit is given by the formula $I = t + \frac{1}{10 - t}$, where t is time. What happens to the circuit as t approaches 10?

43. **Critical Thinking** Write an equation of a rational function whose graph has all of the following characteristics.
   - x-intercepts at $x = 2$ and $x = -3$;
   - a vertical asymptote at $x = 4$; and
   - point discontinuity at $(-5, 0)$.

44. **Geometry** The volume of a rectangular prism with a square base is fixed at 120 cubic feet.
   a. Write the surface area of the prism as a function $A(x)$ of the length of the side of the square x.
   b. Graph the surface area function.
   c. What happens to the surface area of the prism as the length of the side of the square approaches 0?

45. **Critical Thinking** The graph of a rational function cannot intersect a vertical asymptote, but it is possible for the graph to intersect its horizontal asymptote for small values of x. Give an example of such a rational function. (*Hint:* Let the x-axis be the horizontal asymptote of the function.)

46. **Physics** Like charges repel and unlike charges attract. Coulomb's Law states that the force F of attraction or repulsion between two charges, q_1 and q_2, is given by $F = \frac{kq_1q_2}{r^2}$, where k is a constant and r is the distance between the charges. Suppose you were to graph F as a function of r for two positive charges.

a. What asymptotes would the graph have?

b. Interpret the meaning of the asymptotes in terms of the problem.

47. **Analytic Geometry** Recall from Exercise 36 of Lesson 3-5 that a *secant* is a line that intersects the graph of a function in two or more points. Consider the function $f(x) = x^2$.

a. Find an expression for the slope of the secant through the points at (3, 9) and (a, a^2).

b. What happens to the slope of the secant as a approaches 3?

Mixed Review

48. Find and classify the extrema of the function $f(x) = -x^2 + 4x - 3$. *(Lesson 3-6)*

49. Find the inverse of $x^2 - 9 = y$. *(Lesson 3-4)*

50. Find the maximum and minimum values of $f(x, y) = y - x$ defined for the polygonal convex set having vertices at (0, 0), (4, 0), (3, 5), and (0, 5). *(Lesson 2-6)*

51. Find $-4\begin{bmatrix} -6 & 5 \\ 8 & -4 \end{bmatrix}$. *(Lesson 2-3)*

52. **Consumer Awareness** Bill and Liz are going on a vacation in Jamaica. Bill bought 8 rolls of film and 2 bottles of sunscreen for \$35.10. The next day, Liz paid \$14.30 for three rolls of film and one bottle of sunscreen. If the price of each bottle of sunscreen is the same and the price of each roll of film is the same, what is the price of a roll of film and a bottle of sunscreen? *(Lesson 2-1)*

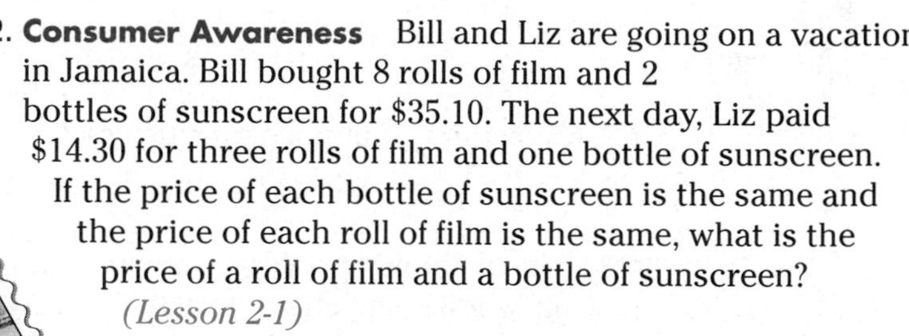

53. Of (0, 0), (3, 2), (−4, 2), or (−2, 4), which satisfy $x + y \geq 3$? *(Lesson 1-8)*

54. Write the equation $15y - x = 1$ in slope-intercept form. *(Lesson 1-4)*

55. Find $[f \circ g](x)$ and $[g \circ f](x)$ for $f(x) = 8x$ and $g(x) = 2 - x^2$. *(Lesson 1-2)*

56. **SAT/ACT Practice** Nine playing cards from the same deck are placed to form a large rectangle whose area is 180 square inches. There is no space between the cards and no overlap. What is the perimeter of this rectangle?

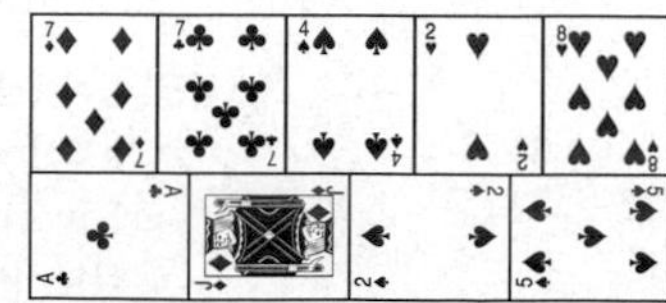

A 29 in. B 58 in. C 64 in.

D 116 in. E 210 in.

Extra Practice See p. A31.

Direct, Inverse, and Joint Variation

OBJECTIVE

- Solve problems involving direct, inverse, and joint variation.

AUTO SAFETY The faster you drive, the more time and distance you need to stop safely, and the less time you have to react. The stopping distance for a vehicle is the sum of the reaction distance d_1, how far the car travels before you hit the brakes, and the braking distance d_2, how far the car travels once the brakes are applied. According to the National Highway Traffic Safety Administration (NHTSA), under the best conditions the reaction time t of most drivers is about 1.5 seconds. Once brakes are applied, the braking distance varies directly as the square of the car's speed s. *A problem related to this will be solved in Example 2.*

The relationship between braking distance and car speed is an example of a **direct variation.** As the speed of the car increases, the braking distance also increases at a constant rate.

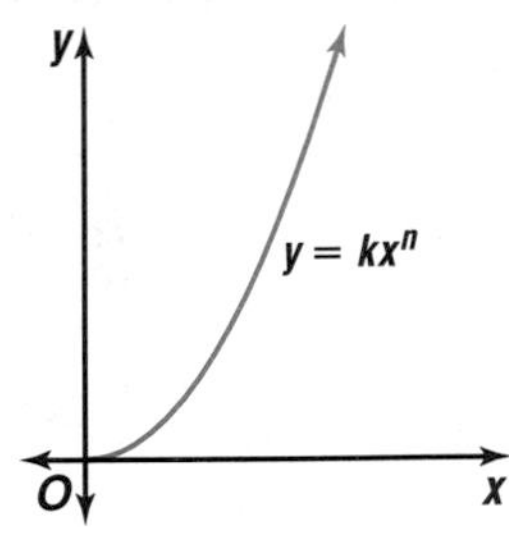

A direct variation can be described by the equation $y = kx^n$. The k in this equation is called the **constant of variation.** To express a direct variation, we say that *y varies directly as x^n.*

Direct Variation	y varies directly as x^n if there is some nonzero constant k such that $y = kx^n$, $n > 0$. k is called the *constant of variation.*

Notice when $n = 1$, the equation of direct variation simplifies to $y = kx$. This is the equation of a line through the origin written in slope-intercept form. In this special case, the constant of variation is the slope of the line.

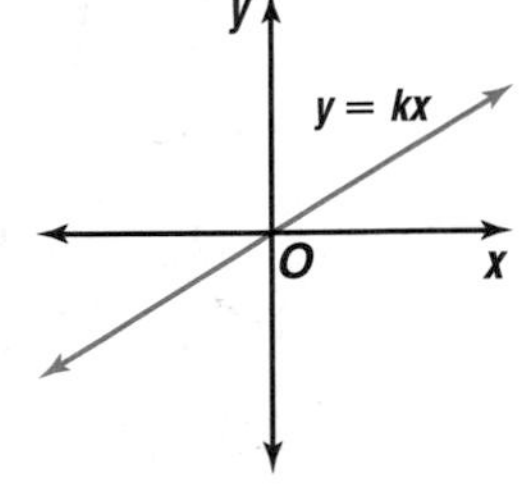

To find the constant of variation, substitute known values of x^n and y in the equation $y = kx^n$ and solve for k.

Example 1 **Suppose y varies directly as x and $y = 27$ when $x = 6$.**

a. Find the constant of variation and write an equation of the form $y = kx^n$.

b. Use the equation to find the value of y when $x = 10$.

a. In this case, the power of x is 1, so the direct variation equation is $y = kx$.

$$y = kx$$
$$27 = k(6) \quad y = 27, x = 6$$
$$\frac{27}{6} = k \quad \text{Divide each side by 6.}$$
$$4.5 = k$$

The constant of variation is 4.5. The equation relating x and y is $y = 4.5x$.

b. $y = 4.5x$

$y = 4.5(10)$ *$x = 10$*

$y = 45$

When $x = 10$, the value of y is 45.

Direct variation equations are used frequently to solve real-world problems.

Example 2 

AUTO SAFETY Refer to the application at the beginning of this lesson. The NHTSA reports an average braking distance of 227 feet for a car traveling 60 miles per hour, with a total stopping distance of 359 feet.

a. Write an equation of direct variation relating braking distance to car speed. Then graph the equation.

b. Use the equation to find the braking distance required for a car traveling 70 miles per hour.

c. Calculate the total stopping distance required for a car traveling 70 miles per hour.

a. First, translate the statement of variation into an equation of the form $y = kx^n$.

The braking distance varies directly as the square of the car's speed.

$$d_2 = k \cdot s^2$$

Then, substitute corresponding values for braking distance and speed in the equation and solve for k.

$d_2 = ks^2$

$227 = k(60)^2$ *$d_2 = 227, s = 60$*

$\frac{227}{(60)^2} = k$ *Divide each side by $(60)^2$.*

$0.063 \approx k$

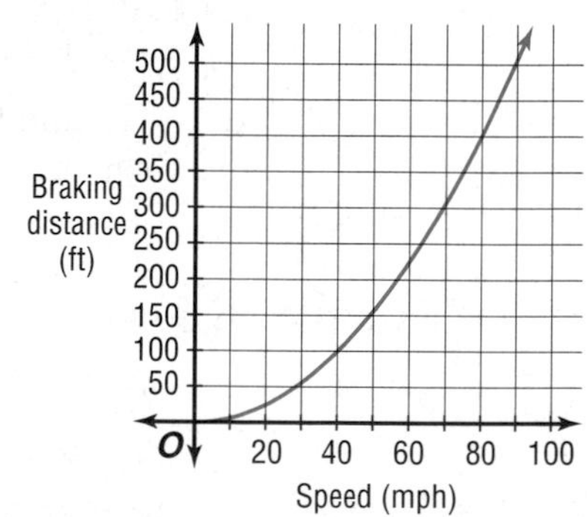

The constant of variation is approximately 0.063. Thus, the equation relating braking distance to car speed is $d_2 = 0.063s^2$. Notice the graph of the variation equation is a parabola centered at the origin and opening up.

b. Evaluate the equation of variation for $s = 70$.

$d_2 = 0.063s^2$

$d_2 = 0.063(70)^2$ *$s = 70$*

$d_2 = 308.7$ or about 309

The braking distance for a car traveling 70 mph is about 309 feet.

c. To find the total stopping distance, first calculate the reaction distance given a reaction time of 1.5 seconds and a speed of 70 miles per hour. Since time is in seconds, we need to change miles per hour to feet per second.

$$\frac{70\text{ mi}}{1\text{ h}} \cdot \frac{1\text{ h}}{60\text{ min}} \cdot \frac{1\text{ min}}{60\text{ s}} \cdot \frac{5280\text{ ft}}{1\text{ mi}} \approx 103\text{ ft/s}$$

$d_1 = st$ *reaction distance = rate of speed · time*

$d_1 = (103)(1.5)$ or about 155 feet *$s \approx 103$ ft/s; $t = 1.5$ s*

The total stopping distance is then the sum of the reaction and braking distances, $155 + 309$ or 464 feet.

If you know that y varies directly as x^n and one set of values, you can use a proportion to find the other set of corresponding values.

Suppose $y_1 = kx_1{}^n$ and $y_2 = kx_2{}^n$. Solve each equation for k.

$\frac{y_1}{x_1} = k$ and $\frac{y_2}{x_2} = k$ ➡ $\frac{y_1}{x_1} = \frac{y_2}{x_2}$ *Since both ratios equal k, the ratios are equal.*

Using the properties of proportions, you can find many other proportions that relate these same x^n- and y-values. *You will derive another of these proportions in Exercise 2.*

Example 3 **If y varies directly as the cube of x and $y = -67.5$ when $x = 3$, find x when $y = -540$.**

Use a proportion that relates the values.

$$\frac{y_1}{x_1} = \frac{y_2}{x_2}$$

$\frac{-67.5}{(3)^3} = \frac{-540}{(x_2)^3}$ *Substitute the known values.*

$-67.5(x_2)^3 = -540(3)^3$ *Cross multiply.*

$(x_2)^3 = 216$

$x_2 = \sqrt[3]{216}$ or 6 *Take the cube root of each side.*

When $y = -540$, the value of x is 6. *This is a reasonable answer for x, since as y decreased, the value of x increased.*

Many quantities are **inversely proportional** or are said to *vary inversely* with each other. This means that as one value increases the other decreases and vice versa. For example, elevation and air temperature vary inversely with each other. When you travel to a higher elevation above Earth's surface, the air temperature decreases.

Inverse Variation y varies inversely as x^n if there is some nonzero constant k such that $x^n y = k$ or $y = \frac{k}{x^n}$, $n > 0$.

Suppose y varies inversely as x such that $xy = 4$ or $y = \frac{4}{x}$. The graph of this equation is shown at the right. Notice that in this case, k is a positive value, 4, so as the value of x increases, the value of y decreases.

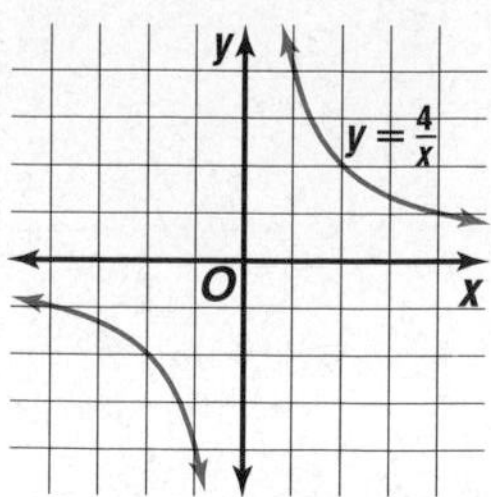

Just as with direct variation, a proportion can be used with inverse variation to solve problems where some quantities are known. The proportion shown below is only one of several that can be formed.

$$x_1^n y_1 = k \text{ and } x_2^n y_2 = k$$

$$x_1^n y_1 = x_2^n y_2 \qquad \textit{Substitution property of equality}$$

$$\frac{x_1^n}{y_2} = \frac{x_2^n}{y_1} \qquad \textit{Divide each side by } y_1y_2.$$

Other possible proportions include $\frac{y_1}{y_2} = \frac{x_2^n}{x_1^n}$, $\frac{y_2}{x_1^n} = \frac{y_1}{x_2^n}$, and $\frac{x_2^n}{x_1^n} = \frac{y_1}{y_2}$.

Example 4 **If y varies inversely as x and $y = 21$ when $x = 15$, find x when $y = 12$.**

Use a proportion that relates the values.

$$\frac{x_1}{y_2} = \frac{x_2}{y_1} \qquad n = 1$$

$$\frac{15}{12} = \frac{x_2}{21} \qquad \textit{Substitute the known values.}$$

$$12x_2 = 315 \qquad \textit{Cross multiply.}$$

$$x_2 = \frac{315}{12} \text{ or } 26.25 \qquad \textit{Divide each side by 15.}$$

When $y = 12$, the value of x is 26.25.

Another type of variation is **joint variation.** This type of variation occurs when one quantity varies directly as the product of two or more other quantities.

Joint Variation y varies jointly as x^n and z^n if there is some nonzero constant k such that $y = kx^nz^n$, where $x \neq 0$, $z \neq 0$, and $n > 0$.

Example 5 **GEOMETRY** **The volume V of a cone varies jointly as the height h and the square of the radius r of the base. Find the equation for the volume of a cone with height 6 centimeters and base diameter 10 centimeters that has a volume of 50π cubic centimeters.**

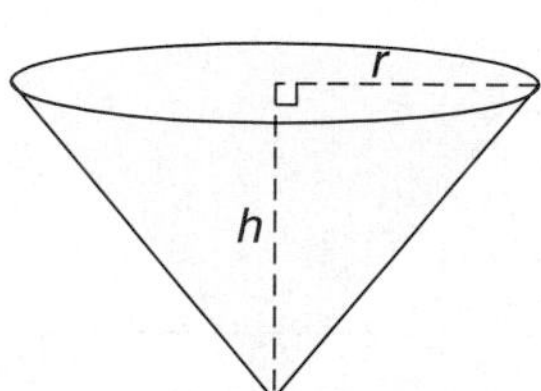

Read the problem and use the known values of V, h, and r to find the equation of joint variation.

The volume varies jointly as the height and the square of the radius.

$$V \quad = \quad k \quad \cdot \quad h \quad \cdot \quad r^2$$

$$V = khr^2$$
$$50\pi = k(6)(5)^2 \quad V = 50\pi,\ h = 6,\ r = 5$$
$$50\pi = 150k$$
$$\frac{\pi}{3} = k \quad \text{Solve for } k.$$

The equation for the volume of a cone is $V = \frac{\pi}{3}hr^2$.

Application and Problem Solving

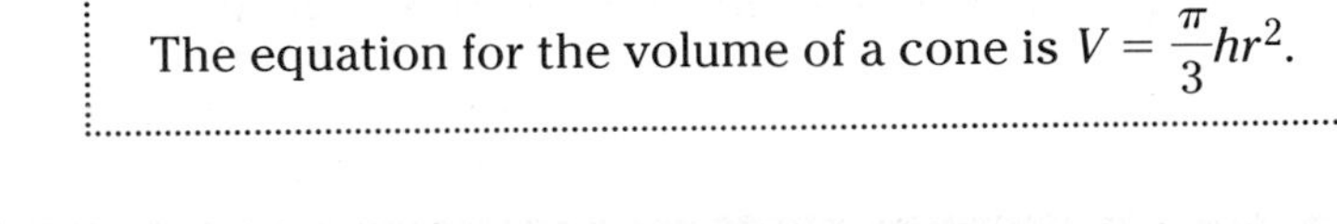

CHECK FOR UNDERSTANDING

Communicating Mathematics

Read and study the lesson to answer each question.

1. **Describe** each graph below as illustrating a direct variation, inverse variation, or neither.

a. 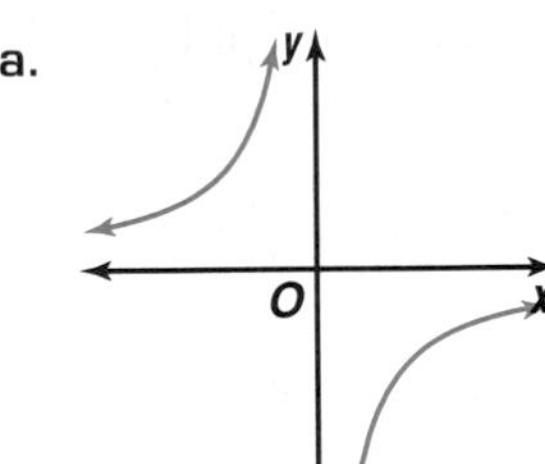

b. 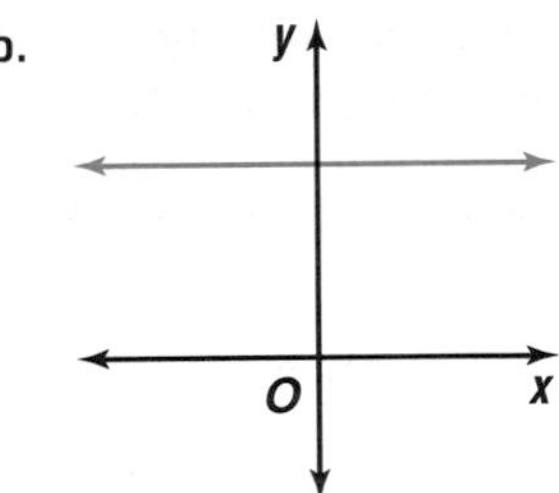

c. 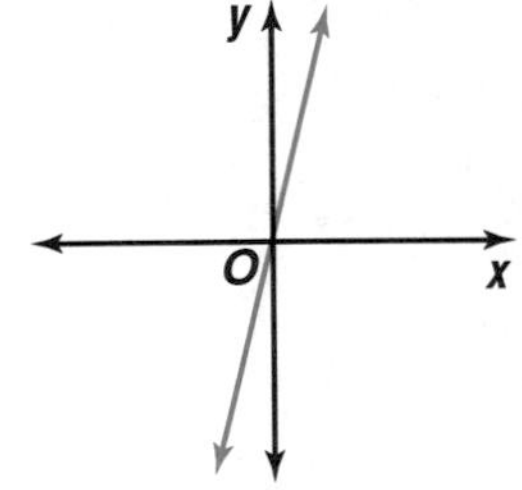

2. **Derive** a different proportion from that used in Example 3 relating x^n and y if y varies directly as x.
3. **Explain** why the graph at the right does not represent a direct variation.
4. **Write** statements relating two quantities in real life that exemplify each type of variation.

   a. direct b. inverse c. joint

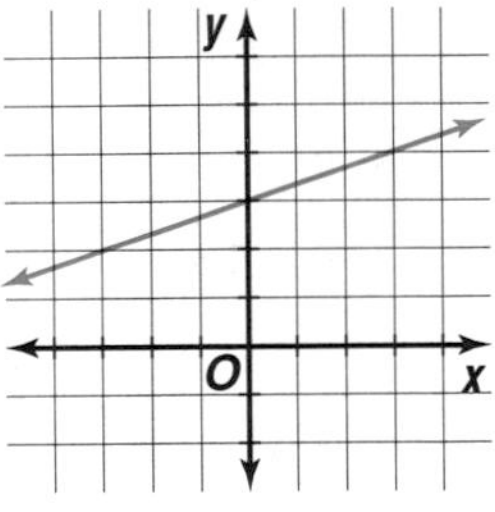

Guided Practice

Find the constant of variation for each relation and use it to write an equation for each statement. Then solve the equation.

5. If y varies inversely as x and $y = 3$ when $x = 4$, find y when $x = 15$.
6. If y varies directly as the square of x and $y = -54$ when $x = 9$, find y when $x = 6$.
7. If y varies jointly as x and the cube of z and $y = 16$ when $x = 4$ and $z = 2$, find y when $x = -8$ and $z = -3$.
8. If y varies jointly as x and z and inversely as the square of w, and $y = 3$ when $x = 3$, $z = 10$, and $w = 2$, find y when $x = 4$, $z = 20$, and $w = 4$.

Write a statement of variation relating the variables of each equation. Then name the constant of variation.

9. $\frac{x^4}{y} = 7$

10. $A = \ell w$

11. $x = \frac{-3}{y}$

SKILLS AND CONCEPTS

OBJECTIVES AND EXAMPLES

Lesson 3-1 Use algebraic tests to determine if the graph of a relation is symmetrical.

- Determine whether the graph of $f(x) = 4x - 1$ is symmetric with respect to the origin.

$f(-x) = 4(-x) - 1$
$\quad = -4x - 1$
$-f(x) = -(4x - 1)$
$\quad = -4x + 1$

The graph of $f(x) = 4x - 1$ is not symmetric with respect to the origin because $f(-x) \neq -f(x)$.

REVIEW EXERCISES

Determine whether the graph of each function is symmetric with respect to the origin.

11. $f(x) = -2x$
12. $f(x) = x^2 + 2$
13. $f(x) = x^2 - x + 3$
14. $f(x) = x^3 - 6x + 1$

Determine whether the graph of each function is symmetric with respect to the *x*-axis, *y*-axis, the line $y = x$, the line $y = -x$, or none of these.

15. $xy = 4$
16. $x + y^2 = 4$
17. $x = -2y$
18. $x^2 = \frac{1}{y}$

Lesson 3-2 Identify transformations of simple graphs.

- Describe how the graphs of $f(x) = x^2$ and $g(x) = x^2 - 1$ are related.

Since 1 is subtracted from $f(x)$, the parent function, $g(x)$ is the graph of $f(x)$ translated 1 unit down.

Describe how the graphs of $f(x)$ and $g(x)$ are related.

19. $f(x) = x^4$ and $g(x) = x^4 + 5$
20. $f(x) = |x|$ and $g(x) = |x + 2|$
21. $f(x) = x^2$ and $g(x) = 6x^2$
22. $f(x) = [\![x]\!]$ and $g(x) = \left[\!\left[\frac{3}{4}x\right]\!\right] - 4$

Lesson 3-3 Graph polynomial, absolute value, and radical inequalities in two variables.

- Graph $y < x^2 + 1$.

The boundary of the inequality is the graph of $y = x^2 + 1$. Since the boundary is not included, the parabola is dashed.

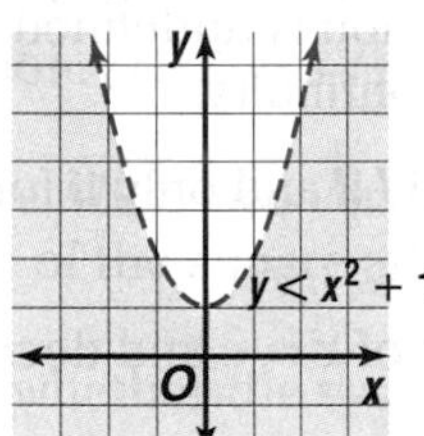

Graph each inequality.

23. $y > |x + 2|$
24. $y \leq -2x^3 + 4$
25. $y < (x + 1)^2 + 2$
26. $y \geq \sqrt{2x - 3}$

Solve each inequality.

27. $|4x + 5| > 7$
28. $|x - 3| + 2 \leq 11$

Lesson 3-4 Determine inverses of relations and functions.

- Find the inverse of $f(x) = 4(x - 3)^2$.

$y = 4(x - 3)^2$ *Let $y = f(x)$.*
$x = 4(y - 3)^2$ *Interchange x and y.*
$\frac{x}{4} = (y - 3)^2$ *Solve for y.*
$\pm\sqrt{\frac{x}{4}} = y - 3$
$3 \pm \sqrt{\frac{x}{4}} = y$ So, $f^{-1}(x) = 3 \pm \sqrt{\frac{x}{4}}$.

Graph each function and its inverse.

29. $f(x) = 3x - 1$
30. $f(x) = -\frac{1}{4}x + 5$
31. $f(x) = \frac{2}{x} + 3$
32. $f(x) = (x + 1)^2 - 4$

Find $f^{-1}(x)$. Then state whether $f^{-1}(x)$ is a function.

33. $f(x) = (x - 2)^3 - 8$
34. $f(x) = 3(x + 7)^4$

Practice

The volume varies jointly as the height and the square of the radius.

$$V \quad = \quad k \quad \cdot \quad h \quad \cdot \quad r^2$$

$$V = khr^2$$

$$50\pi = k(6)(5)^2 \quad V = 50\pi,\ h = 6,\ r = 5$$

$$50\pi = 150k$$

$$\frac{\pi}{3} = k \quad \text{Solve for } k.$$

The equation for the volume of a cone is $V = \frac{\pi}{3}hr^2$.

CHECK FOR UNDERSTANDING

Communicating Mathematics

Read and study the lesson to answer each question.

1. **Describe** each graph below as illustrating a direct variation, inverse variation, or neither.

a. 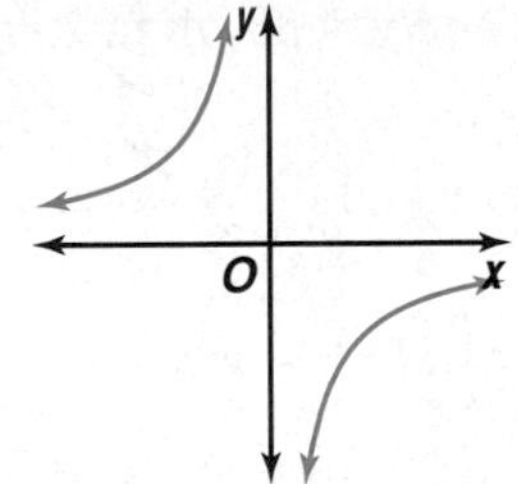

b. 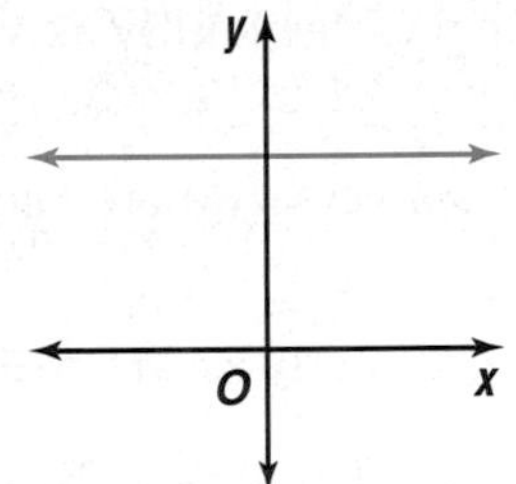

c. 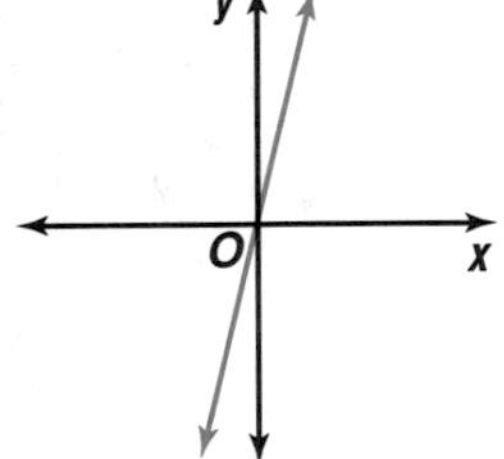

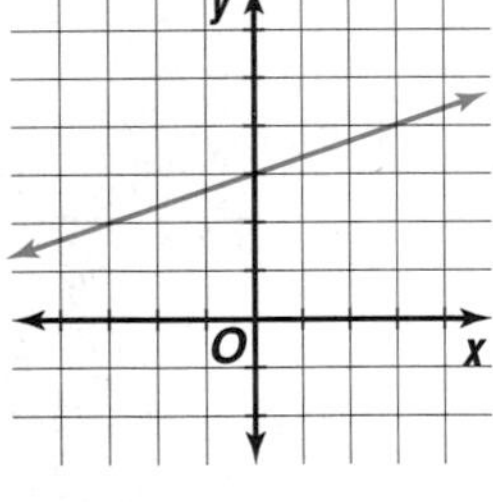

2. **Derive** a different proportion from that used in Example 3 relating x^n and y if y varies directly as x.
3. **Explain** why the graph at the right does not represent a direct variation.
4. **Write** statements relating two quantities in real life that exemplify each type of variation.

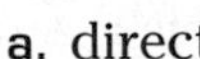

a. direct b. inverse c. joint

Guided Practice

Find the constant of variation for each relation and use it to write an equation for each statement. Then solve the equation.

5. If y varies inversely as x and $y = 3$ when $x = 4$, find y when $x = 15$.
6. If y varies directly as the square of x and $y = -54$ when $x = 9$, find y when $x = 6$.
7. If y varies jointly as x and the cube of z and $y = 16$ when $x = 4$ and $z = 2$, find y when $x = -8$ and $z = -3$.
8. If y varies jointly as x and z and inversely as the square of w, and $y = 3$ when $x = 3$, $z = 10$, and $w = 2$, find y when $x = 4$, $z = 20$, and $w = 4$.

Write a statement of variation relating the variables of each equation. Then name the constant of variation.

9. $\frac{x^4}{y} = 7$

10. $A = \ell w$

11. $x = \frac{-3}{y}$

12. **Forestry** A lumber company needs to estimate the volume of wood a load of timber will produce. The supervisor knows that the volume of wood in a tree varies jointly as the height h and the square of the tree's girth g. The supervisor observes that a tree 40 meters tall with a girth of 1.5 meters produces 288 cubic meters of wood.

a. Write an equation that represents this situation.

b. What volume of wood can the supervisor expect to obtain from 50 trees averaging 75 meters in height and 2 meters in girth?

EXERCISES

Practice

Find the constant of variation for each relation and use it to write an equation for each statement. Then solve the equation.

13. If y varies directly as x and $y = 0.3$ when $x = 1.5$, find y when $x = 6$.

14. If y varies inversely as x and $y = -2$ when $x = 25$, find x when $y = -40$.

15. Suppose y varies jointly as x and z and $y = 36$ when $x = 1.2$ and $z = 2$. Find y when $x = 0.4$ and $z = 3$.

16. If y varies inversely as the square of x and $y = 9$ when $x = 2$, find y when $x = 3$.

17. If r varies directly as the square of t and $r = 4$ when $t = \frac{1}{2}$, find r when $t = \frac{1}{4}$.

18. Suppose y varies inversely as the square root of x and $x = 1.21$ when $y = 0.44$. Find y when $x = 0.16$.

19. If y varies jointly as the cube of x and the square of z and $y = -9$ when $x = -3$ and $z = 2$, find y when $x = -4$ and $z = -3$.

20. If y varies directly as x and inversely as the square of z and $y = \frac{1}{6}$ when $x = 20$ and $z = 6$, find y when $x = 14$ and $z = 5$.

21. Suppose y varies jointly as x and z and inversely as w and $y = -3$ when $x = 2$, $z = -3$, and $w = 4$. Find y when $x = 4$, $z = -7$ and $w = -4$.

22. If y varies inversely as the cube of x and directly as the square of z and $y = -6$ when $x = 3$ and $z = 9$, find y when $x = 6$ and $z = -4$.

23. If a varies directly as the square of b and inversely as c and $a = 45$ when $b = 6$ and $c = 12$, find b when $a = 96$ and $c = 10$.

24. If y varies inversely as the square of x and $y = 2$ when $x = 4$, find x when $y = 8$.

Write a statement of variation relating the variables of each equation. Then name the constant of variation.

25. $C = \pi d$

26. $\frac{x}{y} = 4$

27. $xz^2 = \frac{3}{4}y$

28. $V = \frac{4}{3}\pi r^3$

29. $4x^2 = \frac{5}{y}$

30. $y = \frac{2}{\sqrt{x}}$

31. $A = 0.5h(b_1 + b_2)$

32. $y = \frac{x}{3z^2}$

33. $\frac{1}{7}y = \frac{x^2}{z^3}$

34. Write a statement of variation for the equation $y = \frac{kx^3 \cdot z}{w^2}$ if k is the constant of variation.

Applications and Problem Solving

35. **Physics** If you have observed people on a seesaw, you may have noticed that the heavier person must sit closer to the fulcrum for the seesaw to balance. In doing so, the heavier participant creates a rotational force, called *torque*. The torque on the end of a seesaw depends on the mass of the person and his or her distance from the seesaw's fulcrum. In order to reduce torque, one must either reduce the distance between the person and the fulcrum or replace the person with someone having a smaller mass.

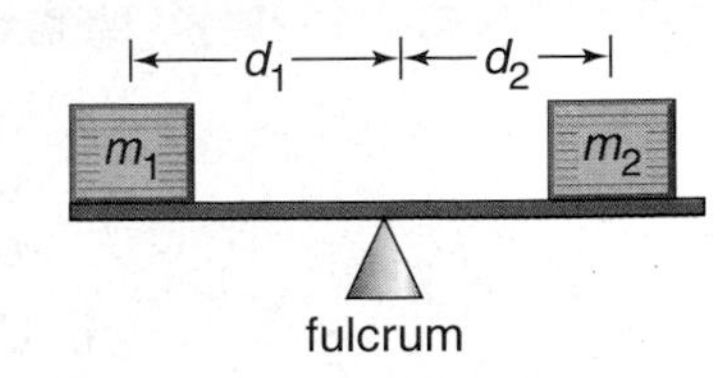

a. What type of variation describes the relationship between torque, mass, and distance? Explain.

b. Let m_1, d_1, and T_1 be the mass, distance, and torque on one side of a seesaw, and let m_2, d_2, and T_2 be the mass, distance, and torque on the other side. Derive an equation that represents this seesaw in balance.

c. A 75-pound child and a 125-pound babysitter sit at either end of a seesaw. If the child sits 3.3 meters from the fulcrum, use the equation found in part **b** to determine how far the babysitter should sit from the pivot in order to balance the seesaw.

36. **Pool Maintenance** Kai decides to empty her pool for the winter. She knows that the time t required to empty a pool varies inversely as the rate r of pumping.

a. Write an equation that represents this situation. Let k be the constant of variation.

b. In the past, Kai was able to empty her pool in 45 minutes at a rate of 800 liters per minute. She now owns a new pump that can empty the pool at a rate of 1 kiloliter per minute. How long will it take Kai to empty the pool using this new pump?

37. **Critical Thinking** Show that if y varies directly as x, then x varies directly as y.

38. **Movies** The intensity of light I, measured in *lux*, is inversely proportional to the square of the distance d between the light source and the object illuminated.

a. Write an equation that represents this situation.

b. Using a light meter, a lighting director measures the intensity of the light from a bulb hanging 6 feet overhead a circular table at 16 lux. If the table has a 5-foot diameter, what illumination reading will the director find at the edge of the table where the actors will sit? Round to the nearest tenth.

39. **Critical Thinking** If a varies directly as the square of b and inversely as the cube of c, how is the value of a changed when the values of b and c are halved? Explain.

40. Space Science Newton's Law of Universal Gravitation states that two objects attract one another with a force F that varies directly as the product of their masses, m_1 and m_2, and inversely as the square of the distance d between their centers. It is this force, called gravity, which pulls objects to Earth and keeps Earth in its orbit about the sun.

a. Write an equation that represents this situation. Let G be the constant of variation.

b. Use the chart at the right to determine the value of G if the force of attraction between Earth and the moon is 1.99×10^{20} newtons (N) and the distance between them is 3.84×10^8 meters. Be sure to include appropriate units with your answer.

	Mass
Earth	5.98×10^{24} kg
Sun	1.99×10^{30} kg
Moon	7.36×10^{22} kg

c. Using your answers to parts **a** and **b**, find the force of attraction between Earth and the sun if the distance between them is 1.50×10^{11} meters.

d. How many times greater is the force of attraction between the sun and Earth than between the moon and Earth?

41. Electricity Wires used to connect electric devices have very small resistances, allowing them to conduct currently more readily. The resistance R of a piece of wire varies directly as its length L and inversely as its cross-sectional area A. A piece of copper wire 2 meters long and 2 millimeters in diameter has a resistance of 1.07×10^{-2} ohms (Ω). Find the resistance of a second piece of copper wire 3 meters long and 6 millimeters in diameter. (*Hint*: The cross-sectional area will be πr^2.)

Mixed Review

42. Graph $y = \frac{x - 1}{x^2 - 9}$. *(Lesson 3-7)*

43. Find the inverse of $f(x) = (x - 3)^3 + 6$. Then state whether the inverse is a function. *(Lesson 3-4)*

44. Square $ABCD$ has vertices $A(1, 2)$, $B(3, -2)$, $C(-1, -4)$, and $D(-3, 0)$. Find the image of the square after a reflection over the y-axis. *(Lesson 2-4)*

45. State whether the system $4x - 2y = 7$ and $-12x + 6y = -21$ is *consistent and independent, consistent and dependent,* or *inconsistent.* *(Lesson 2-1)*

46. Graph $g(x) = \begin{cases} 2 \text{ if } x < 1 \\ -x + 5 \text{ if } x \geq 1 \end{cases}$. *(Lesson 1-7)*

47. **Education** In 1995, 23.2% of the student body at Kennedy High School were seniors. By 2000, seniors made up only 18.6% of the student body. Assuming the level of decline continues at the same rate, write a linear equation in slope-intercept form to describe the percent of seniors y in the student body in year x. *(Lesson 1-4)*

48. **SAT/ACT Practice** If $a^2b = 12^2$, and b is an odd integer, then a could be divisible by all of the following EXCEPT

A 3 B 4 C 6 D 9 E 12

Extra Practice See p. A31.

VOCABULARY

absolute maximum (p.171)
absolute minimum (p. 171)
asymptotes (p. 180)
constant function (pp. 137, 164)
constant of variation (p. 189)
continuous (p. 160)
critical point (p. 171)
decreasing function (p. 164)
direct variation (p. 189)
discontinuous (p. 159)
end behavior (p. 162)
even function (p. 133)
everywhere discontinuous (p. 159)
extremum (p. 171)
horizontal asymptote (p. 180)
horizontal line test (p. 153)
image point (p. 127)
increasing function (p. 164)
infinite discontinuity (p. 159)
inverse function (pp. 152, 155)
inverse process (p. 154)
inversely proportional (p. 191)
inverse relations (p. 152)
joint variation (p. 192)
jump discontinuity (p. 159)
line symmetry (p. 129)
maximum (p. 171)
minimum (p. 171)
monotonicity (p. 163)
odd function (p. 133)
parent graph (p. 137)
point discontinuity (p. 159)
point of inflection (p. 171)
point symmetry (p. 127)
rational function (p. 180)
relative extremum (p. 172)
relative maximum (p. 172)
relative minimum (p. 172)
slant asymptote (p. 183)
symmetry with respect to the origin (p. 128)
vertical asymptote (p. 180)

UNDERSTANDING AND USING THE VOCABULARY

Choose the correct term to best complete each sentence.

1. An (odd, even) function is symmetric with respect to the y-axis.
2. If you can trace the graph of a function without lifting your pencil, then the graph is (continuous, discontinuous).
3. When there is a value in the domain for which a function is undefined, but the pieces of the graph match up, then the function has (infinite, point) discontinuity.
4. A function f is (decreasing, increasing) on an interval I if and only if for every a and b contained in I, $f(a) > f(b)$ whenever $a < b$.
5. When the graph of a function is increasing to the left of $x = c$ and decreasing to the right of $x = c$, then there is a (maximum, minimum) at c.
6. A (greatest integer, rational) function is a quotient of two polynomial functions.
7. Two relations are (direct, inverse) relations if and only if one relation contains the element (b, a) whenever the other relation contains the element (a, b).
8. A function is said to be (monotonic, symmetric) on an interval I if and only if the function is increasing on I or decreasing on I.
9. A (horizontal, slant) asymptote occurs when the degree of the numerator of a rational expression is exactly one greater than that of the denominator.
10. (Inverse, Joint) variation occurs when one quantity varies directly as the product of two or more other quantities.

For additional review and practice for each lesson, visit: **www.amc.glencoe.com**

SKILLS AND CONCEPTS

OBJECTIVES AND EXAMPLES

Lesson 3-1 Use algebraic tests to determine if the graph of a relation is symmetrical.

- Determine whether the graph of $f(x) = 4x - 1$ is symmetric with respect to the origin.

$f(-x) = 4(-x) - 1$
$= -4x - 1$
$-f(x) = -(4x - 1)$
$= -4x + 1$

The graph of $f(x) = 4x - 1$ is not symmetric with respect to the origin because $f(-x) \neq -f(x)$.

REVIEW EXERCISES

Determine whether the graph of each function is symmetric with respect to the origin.

11. $f(x) = -2x$
12. $f(x) = x^2 + 2$
13. $f(x) = x^2 - x + 3$
14. $f(x) = x^3 - 6x + 1$

Determine whether the graph of each function is symmetric with respect to the x-axis, y-axis, the line $y = x$, the line $y = -x$, or none of these.

15. $xy = 4$
16. $x + y^2 = 4$
17. $x = -2y$
18. $x^2 = \frac{1}{y}$

Lesson 3-2 Identify transformations of simple graphs.

- Describe how the graphs of $f(x) = x^2$ and $g(x) = x^2 - 1$ are related.

Since 1 is subtracted from $f(x)$, the parent function, $g(x)$ is the graph of $f(x)$ translated 1 unit down.

Describe how the graphs of $f(x)$ and $g(x)$ are related.

19. $f(x) = x^4$ and $g(x) = x^4 + 5$
20. $f(x) = |x|$ and $g(x) = |x + 2|$
21. $f(x) = x^2$ and $g(x) = 6x^2$
22. $f(x) = [\![x]\!]$ and $g(x) = \left[\!\!\left[\frac{3}{4}x\right]\!\!\right] - 4$

Lesson 3-3 Graph polynomial, absolute value, and radical inequalities in two variables.

- Graph $y < x^2 + 1$.

The boundary of the inequality is the graph of $y = x^2 + 1$. Since the boundary is not included, the parabola is dashed.

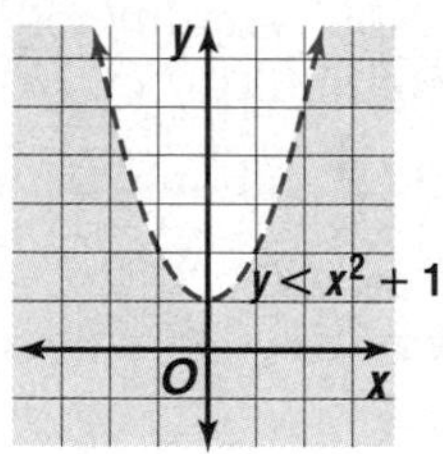

Graph each inequality.

23. $y > |x + 2|$
24. $y \leq -2x^3 + 4$
25. $y < (x + 1)^2 + 2$
26. $y \geq \sqrt{2x - 3}$

Solve each inequality.

27. $|4x + 5| > 7$
28. $|x - 3| + 2 \leq 11$

Lesson 3-4 Determine inverses of relations and functions.

- Find the inverse of $f(x) = 4(x - 3)^2$.

$y = 4(x - 3)^2$ *Let $y = f(x)$.*
$x = 4(y - 3)^2$ *Interchange x and y.*
$\frac{x}{4} = (y - 3)^2$ *Solve for y.*
$\pm\sqrt{\frac{x}{4}} = y - 3$
$3 \pm \sqrt{\frac{x}{4}} = y$ So, $f^{-1}(x) = 3 \pm \sqrt{\frac{x}{4}}$.

Graph each function and its inverse.

29. $f(x) = 3x - 1$
30. $f(x) = -\frac{1}{4}x + 5$
31. $f(x) = \frac{2}{x} + 3$
32. $f(x) = (x + 1)^2 - 4$

Find $f^{-1}(x)$. Then state whether $f^{-1}(x)$ is a function.

33. $f(x) = (x - 2)^3 - 8$
34. $f(x) = 3(x + 7)^4$

OBJECTIVES AND EXAMPLES

Lesson 3-5 Determine whether a function is continuous or discontinuous.

- Determine whether the function $y = \frac{x}{x+4}$ is continuous at $x = -4$.

  Start with the first condition of the continuity test. The function is not defined at $x = -4$ because substituting -4 for x results in a denominator of zero. So the function is discontinuous at $x = -4$.

REVIEW EXERCISES

Determine whether each function is continuous at the given *x*-value. Justify your response using the continuity test.

35. $y = x^2 + 2$; $x = 2$

36. $y = \frac{x-3}{x+1}$; $x = -1$

37. $f(x) = \begin{cases} x + 1 \text{ if } x \leq 1 \\ 2x \text{ if } x > 1 \end{cases}$; $x = 1$

OBJECTIVES AND EXAMPLES

Lesson 3-5 Identify the end behavior of functions.

- Describe the end behavior of $f(x) = 3x^4$.

  Make a chart investigating the value of f(x) for very large and very small values of x.

x	f(x)
−10,000	3×10^{16}
−1000	3×10^{12}
−100	3×10^{8}
0	0
100	3×10^{8}
1000	3×10^{12}
10,000	3×10^{16}

$y \to \infty$ as $x \to \infty$,
$y \to \infty$ as $x \to -\infty$

REVIEW EXERCISES

Describe the end behavior of each function.

38. $y = 1 - x^3$

39. $f(x) = x^9 + x^7 + 4$

40. $y = \frac{1}{x^2} + 1$

41. $y = 12x^5 + x^3 - 3x^2 + 4$

Determine the interval(s) for which the function is increasing and the interval(s) for which the function is decreasing.

42. $y = -2x^3 - 3x^2 + 12x$

43. $f(x) = |x^2 - 9| + 1$

OBJECTIVES AND EXAMPLES

Lesson 3-6 Find the extrema of a function.

- Locate the extrema for the graph of $y = f(x)$. Name and classify the extrema of the function.

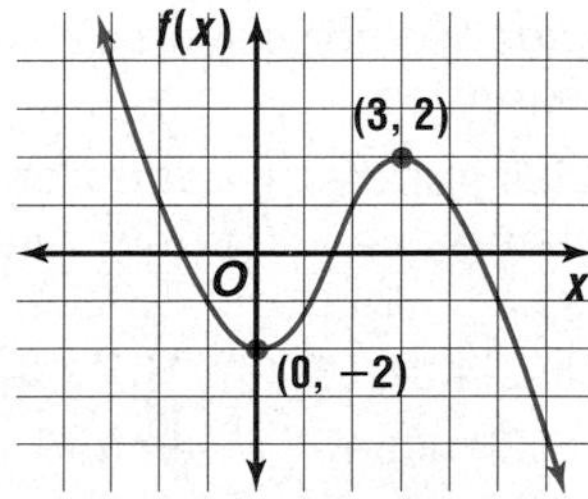

The function has a relative minimum at $(0, -2)$ and a relative maximum at $(3, 2)$.

REVIEW EXERCISES

Locate the extrema for the graph of $y = f(x)$. Name and classify the extrema of the function.

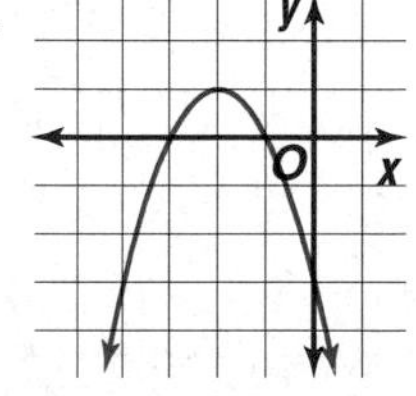

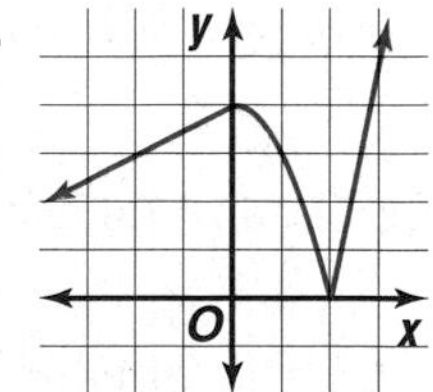

Determine whether the given critical point is the location of a *maximum*, a *minimum*, or a *point of inflection*.

46. $x^3 - 6x^2 + 9x$, $x = 3$

47. $4x^3 + 7$, $x = 0$

OBJECTIVES AND EXAMPLES

Lesson 3-7 Graph rational functions.

- The graph at the right shows a transformation of $f(x) = \frac{1}{x}$. Write an equation of the function.

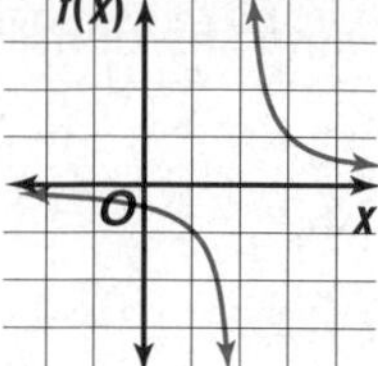

The graph of $f(x) = \frac{1}{x}$ has been translated 2 units to the right. So, the equation of the function is $f(x) = \frac{1}{x-2}$.

REVIEW EXERCISES

Each graph below shows a transformation of $f(x) = \frac{1}{x}$. Write an equation of each function.

48. 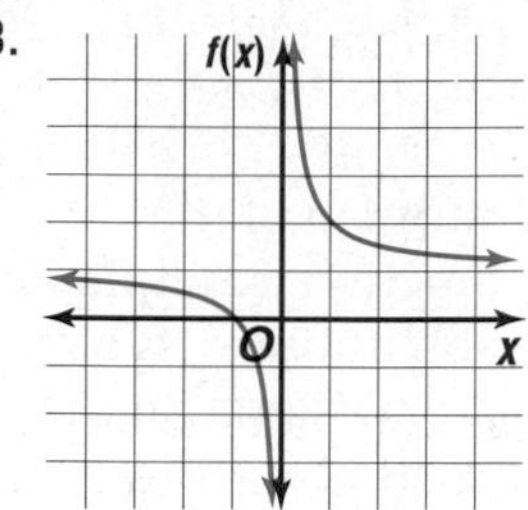

49. 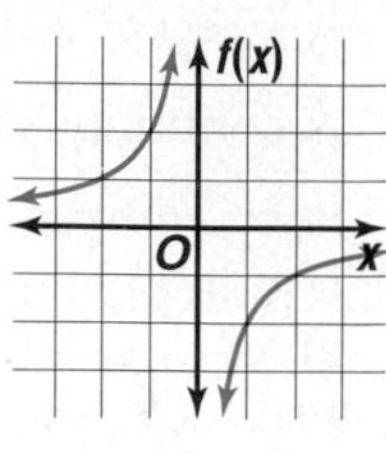

Use the parent graph $f(x) = \frac{1}{x}$ to graph each equation. Describe the transformation(s) that have taken place. Identify the new locations of the asymptotes.

50. $\frac{3}{x+2}$

51. $\frac{2x-5}{x-3}$

Lesson 3-7 Determine vertical, horizontal, and slant asymptotes.

- Determine the equation of the horizontal asymptote of $g(x) = \frac{5x-9}{x-3}$.

To find the horizontal asymptote, use division to rewrite the rational expression as a quotient.

$$x - 3\overline{)5x - 9}\quad \text{quotient } 5$$
$$\underline{5x - 15}$$
$$4$$

Therefore, $g(x) = 5 + \frac{4}{x-3}$. As $x \to \pm\infty$, the value of $\frac{4}{x-3}$ approaches 0. The value of $5 + \frac{4}{x-3}$ approaches 5. The line $y = 5$ is the horizontal asymptote.

Determine the equations of the vertical and horizontal asymptotes, if any, of each function.

52. $f(x) = \frac{x}{x-1}$

53. $g(x) = \frac{x^2+1}{x+2}$

54. $h(x) = \frac{(x-3)^2}{x^2-9}$

55. Does the function $f(x) = \frac{x^2+2x+1}{x}$ have a slant asymptote? If so, find an equation of the slant asymptote. If not, explain.

Lesson 3-8 Solve problems involving direct, inverse, and joint variation.

- If y varies inversely as the square of x and $y = 8$ when $x = 3$, find x when $y = 6$.

$$\frac{x_1^{\,n}}{y_2} = \frac{x_2^{\,n}}{y_1}$$

$$\frac{3^2}{6} = \frac{x_2}{8} \quad n = 2, x_1 = 3, y_1 = 8, y_2 = 6$$

$6x_2^{\,2} = 72$ *Cross multiply.*

$x_2^{\,2} = 12$ *Divide each side by 6.*

$x_2 = \sqrt{12}$ or $2\sqrt{3}$

Find the constant of variation and use it to write an equation for each statement. Then solve the equation.

56. If y varies jointly as x and z and $y = 5$ when $x = -4$ and $z = -2$, find y when $x = -6$ and $z = -3$.

57. If y varies inversely as the square root of x and $y = 20$ when $x = 49$, find x when $y = 10$.

58. If y varies directly as the square of x and inversely as z and $y = 7.2$ when $x = 0.3$ and $z = 4$, find y when $x = 1$ and $z = 40$.

APPLICATIONS AND PROBLEM SOLVING

59. **Manufacturing** The length of a part for a bicycle must be 6.5 ± 0.2 centimeters. If the actual length of the part is x, write an absolute value inequality to describe this situation. Then find the range of possible lengths for the part. *(Lesson 3-3)*

60. **Consumer Costs** A certain copy center charges users \$0.40 for every minute or part of a minute to use their computer scanner. Suppose that you use their scanner for x minutes, where x is any real number greater than 0. *(Lesson 3-4)*
    a. Sketch the graph of the function, $C(x)$, that gives the cost of using the scanner for x minutes.
    b. What are the domain and range of $C(x)$?
    c. Sketch the graph of $C^{-1}(x)$.
    d. What are the domain and range of $C^{-1}(x)$?
    e. What real-world situation is modeled by $C^{-1}(x)$?

61. **Sports** One of the most spectacular long jumps ever performed was by Bob Beamon of the United States at the 1968 Olympics.

His jump of 8.9027 meters surpassed the world record at that time by over half a meter! The function $h(t) = 4.6t - 4.9t^2$ describes the height of Beamon's jump (in meters) with respect to time (in seconds). *(Lesson 3-6)*
    a. Draw a graph of this function.
    b. What was the maximum height of his jump?

ALTERNATIVE ASSESSMENT

OPEN-ENDED ASSESSMENT

1. Write and then graph an equation that exhibits symmetry with respect to the
    a. x-axis.
    b. y-axis.
    c. line $y = x$.
    d. line $y = -x$.
    e. origin.

2. Write the equation of a parent function, other than the identity or constant function, after it has been translated right 4 units, reflected over the x-axis, expanded vertically by a factor of 2, and translated 1 unit up.

3. A graph has one absolute minimum, one relative minimum, and one relative maximum.
    a. Draw the graph of a function for which this is true.
    b. Name and classify the extrema of the function you graphed.

Additional Assessment See page A58 for Chapter 3 practice test.

TELECOMMUNICATION

Sorry, You are Out of Range for Your Telephone Service . . .

- Research several cellular phone services to determine their initial start-up fee, equipment fee, the monthly service charge, the charge per minute for calls, and any other charges.
- Compare the costs of the cellular phone services you researched by writing and graphing equations.

Determine which cellular phone service would best suit your needs. Write a paragraph to explain your choice. Use graphs and area maps to support your choice.

PORTFOLIO

Choose one of the functions you studied in this chapter. Describe the graph of the function and how it can be used to model a real-life situation.

THE PRINCETON REVIEW

TEST-TAKING TIP

If a problem seems to require lengthy calculations, look for a shortcut. There is probably a quicker way to solve it. Try to eliminate fractions and decimals. Try factoring.

More Algebra Problems

SAT and ACT tests include quadratic expressions and equations. You should be familiar with common factoring formulas, like the difference of two squares or perfect square trinomials.

$a^2 - b^2 = (a + b)(a - b)$
$a^2 + 2ab + b^2 = (a + b)^2$
$a^2 - 2ab + b^2 = (a - b)^2$

Some problems involve systems of equations. Simplify the equations if possible, and then add or subtract them to eliminate one of the variables.

ACT EXAMPLE

1. If $\frac{x^2 - 9}{x + 3} = 12$, then $x = ?$

A 10 **B** 15 **C** 17 **D** 19 **E** 20

HINT Look for factorable quadratics.

Solution Factor the numerator and simplify.

$$\frac{(x + 3)(x - 3)}{x + 3} = 12$$
$$x - 3 = 12 \quad \frac{x + 3}{x + 3} = 1$$
$$x = 15 \quad \textit{Add 3 to each side.}$$

The answer is choice **B.**

Alternate Solution You can also solve this type of problem, with a variable in the question and numbers in the answer choices, with a strategy called "backsolving."

Substitute each answer choice for the variable into the given expression or equation and find which one makes the statement true.

Try choice A. For $x = 10$, $\frac{10^2 - 9}{10 + 3} \neq 12$.

Try choice B. For $x = 15$, $\frac{15^2 - 9}{15 + 3} = 12$.

Therefore, choice **B** is correct.

If the number choices were large, then calculations, even using a calculator, would probably take longer than solving the problem using algebra. In this case, it is *not* a good idea to use the backsolving strategy.

SAT EXAMPLE

2. If $x = y + 1$ and $y \geq 1$, then which of the following must be equal to $x^2 - y^2$?

A $(x - y)^2$ **B** $x^2 - y - 1$ **C** $x + y$
D $x^2 - 1$ **E** $y^2 + 1$

HINT This difficult problem has variables in the answers. It can be solved by using algebra or the "Plug-in" strategy.

Solution Notice the word *must.* This means the relationship is true for all possible values of x and y.

To use the Plug-In strategy, choose a number greater than 1 for y, say 4. Then x must be 5. Since $x^2 - y^2 = 25 - 16$ or 9, check each expression choice to see if it is equal to 9 when $x = 5$ and $y = 4$.

Choice A: $(x - y)^2 = 1$
Choice B: $x^2 - y - 1 = 20$
Choice C: $x + y = 9$

Choice **C** is correct.

Alternate Solution You can also use algebraic substitution to find the answer. Recall that $x^2 - y^2 = (x + y)(x - y)$. The given equation, $x = y + 1$, includes y. Substitute $y + 1$ for x in the second term.

$$\begin{aligned} x^2 - y^2 &= (x + y)(x - y) \\ &= (x + y)[(y + 1) - y] \\ &= (x + y)(1) \\ &= x + y \end{aligned}$$

This is choice **C.**

After you work each problem, record your answer on the answer sheet provided or on a piece of paper.

Multiple Choice

1. For all $y \neq 3$, $\frac{y^2 - 9}{3y - 9} = ?$

   A y B $\frac{y+1}{8}$

   C $y + 1$ D $\frac{y}{3}$

   E $\frac{y+3}{3}$

2. If $x + y = z$ and $x = y$, then all of the following are true EXCEPT

   A $2x + 2y = 2z$

   B $x - y = 0$

   C $x - z = y - z$

   D $x = \frac{z}{2}$

   E $z - y = 2x$

3. The Kims drove 450 miles in each direction to Grandmother's house and back again. If their car gets 25 miles per gallon and their cost for gasoline was \$1.25 per gallon for the trip to Grandmother's house, but \$1.50 per gallon for the return trip, how much *more* money did they spend for gasoline returning from Grandmother's house than they spent going to Grandmother's?

   A \$2.25 B \$4.50

   C \$6.25 D \$9.00

   E \$27.00

4. If $x + 2y = 8$ and $\frac{x}{2} - y = 10$, then $x = ?$

   A -7 B 0

   C 10 D 14

   E 28

5. $\frac{900}{10} + \frac{90}{100} + \frac{9}{1000} =$

   A 90.09 B 90.099

   C 90.909 D 99.09

   E 999

6. For all x, $(10x^4 - x^2 + 2x - 8) - (3x^4 + 3x^3 + 2x + 9) = ?$

   A $7x^4 - 3x^3 - x^2 - 17$

   B $7x^4 - 4x^2 - 17$

   C $7x^4 + 3x^3 - x^2 + 4x$

   D $7x^4 + 2x^2 + 4x$

   E $13x^4 - 3x^3 + x^2 + 4x$

7. If $\frac{n}{8}$ has a remainder of 5, then which of the following has a remainder of 7?

   A $\frac{n+1}{8}$ B $\frac{n+2}{8}$

   C $\frac{n+3}{8}$ D $\frac{n+5}{8}$

   E $\frac{n+7}{8}$

8. If $x > 0$, then $\frac{\sqrt{100x^2 + 600x + 900}}{x+3} = ?$

   A 9 B 10 C 30 D 40

   E It cannot be determined from the information given.

9. **Quantitative Comparison**

   A if the quantity in Column A is greater
   B if the quantity in Column B is greater
   C if the two quantities are equal
   D if the relationship cannot be determined from the information given

   Given: $a + b = c$
   $a - c = 5$
   $b - c = 3$

Column A	Column B
c	0

10. **Grid-In** If $4x + 2y = 24$ and $\frac{7y}{2x} = 7$, then $x = ?$

interNET CONNECTION **SAT/ACT Practice** For additional test practice questions, visit: **www.amc.glencoe.com**

POLYNOMIAL AND RATIONAL FUNCTIONS

CHAPTER OBJECTIVES

- **Determine roots of polynomial equations.** *(Lessons 4-1, 4-4)*
- **Solve quadratic, rational, and radical equations and rational and radical inequalities.** *(Lessons 4-2, 4-6, 4-7)*
- **Find the factors of polynomials.** *(Lesson 4-3)*
- **Approximate real zeros of polynomial functions.** *(Lesson 4-5)*
- **Write and interpret polynomial functions that model real-world data.** *(Lesson 4-8)*

4-1 Polynomial Functions

OBJECTIVES
- Determine roots of polynomial equations.
- Apply the Fundamental Theorem of Algebra.

INVESTMENTS Many grandparents invest in the stock market for their grandchildren's college fund. Eighteen years ago, Della Brooks purchased \$1000 worth of merchandising stocks at the birth of her first grandchild Owen. Ten years ago, she purchased \$500 worth of transportation stocks, and five years ago, she purchased \$250 worth of technology stocks. The stocks will be used to help pay for Owen's college education. If the stocks appreciate at an average annual rate of 12.25%, determine the current value of the college fund. *This problem will be solved in Example 1.*

Appreciation is the increase in value of an item over a period of time. The formula for compound interest can be used to find the value of Owen's college fund after appreciation. The formula is $A = P(1 + r)^t$, where P is the original amount of money invested, r is the interest rate or rate of return (written as a decimal), and t is the time invested (in years).

Example 1

INVESTMENTS The value of Owen's college fund is the sum of the current values of his grandmother's investments.

a. Write a function in one variable that models the value of the college fund for any rate of return.

b. Use the function to determine the current value of the college fund for an average annual rate of 12.25%.

a. Let x represent $1 + r$ and $T(x)$ represent the total current value of the three stocks. The times invested, which are the exponents of x, are 18, 10, and 5, respectively.

Total	=	*merchandising*	+	*transportation*	+	*technology*
$T(x)$	=	$1000x^{18}$	+	$500x^{10}$	+	$250x^5$

b. Since $r = 0.1225$, $x = 1 + 0.1225$ or 1.1225. Now evaluate $T(x)$ for $x = 1.1225$.

$$T(x) = 1000x^{18} + 500x^{10} + 250x^5$$
$$T(1.1225) = 1000(1.1225)^{18} + 500(1.1225)^{10} + 250(1.1225)^5$$
$$T(1.1225) \approx 10{,}038.33$$

The present value of Owen's college fund is about \$10,038.33.

The expression $1000x^{18} + 500x^{10} + 250x^5$ is a **polynomial in one variable.**

Polynomial in One Variable

A polynomial in one variable, x, is an expression of the form $a_0x^n + a_1x^{n-1} + \ldots + a_{n-2}x^2 + a_{n-1}x + a_n$. The coefficients $a_0, a_1, a_2, \ldots, a_n$ represent complex numbers (real or imaginary), a_0 is not zero, and n represents a nonnegative integer.

The **degree** of a polynomial in one variable is the greatest exponent of its variable. The coefficient of the variable with the greatest exponent is called the **leading coefficient.** For the expression $1000x^{18} + 500x^{10} + 250x^5$, 18 is the degree, and 1000 is the leading coefficient.

If a function is defined by a polynomial in one variable with real coefficients, like $T(x) = 1000x^{18} + 500x^{10} + 250x^5$, then it is a **polynomial function.** If $f(x)$ is a polynomial function, the values of x for which $f(x) = 0$ are called the **zeros** of the function. *If the function is graphed, these zeros are also the x-intercepts of the graph.*

Example 2 **Consider the polynomial function $f(x) = x^3 - 6x^2 + 10x - 8$.**

a. State the degree and leading coefficient of the polynomial.

b. Determine whether 4 is a zero of $f(x)$.

Graphing Calculator Tip

To find a value of a polynomial for a given value of *x*, enter the polynomial in the **Y=** list. Then use the **1:value** option in the **CALC** menu.

a. $x^3 - 6x^2 + 10x - 8$ has a degree of 3 and a leading coefficient of 1.

b. Evaluate $f(x) = x^3 - 6x^2 + 10x - 8$ for $x = 4$. That is, find $f(4)$.

$f(4) = 4^3 - 6(4^2) + 10(4) - 8 \quad x = 4$

$f(4) = 64 - 96 + 40 - 8$

$f(4) = 0$

Since $f(4) = 0$, 4 is a zero of $f(x) = x^3 - 6x^2 + 10x - 8$.

Since 4 is a zero of $f(x) = x^3 - 6x^2 + 10x - 8$, it is also a solution for the **polynomial equation** $x^3 - 6x^2 + 10x - 8 = 0$. The solution for a polynomial equation is called a **root.** The words *zero* and *root* are often used interchangeably, but technically, you find the *zero of a function* and the *root of an equation.*

A root or zero may also be an **imaginary number** such as $3\boldsymbol{i}$. By definition, the imaginary unit $\boldsymbol{i}$ equals $\sqrt{-1}$. Since $\boldsymbol{i} = \sqrt{-1}$, $\boldsymbol{i}^2 = -1$. It also follows that $\boldsymbol{i}^3 = \boldsymbol{i}^2 \times \boldsymbol{i}$ or $-\boldsymbol{i}$ and $\boldsymbol{i}^4 = \boldsymbol{i}^2 \times \boldsymbol{i}^2$ or 1.

The imaginary numbers combined with the real numbers compose the set of **complex numbers.** A complex number is any number of the form $a + b\boldsymbol{i}$ where a and b are real numbers. If $b = 0$, then the complex number is a real number. If $a = 0$ and $b \neq 0$, then the complex number is called a **pure imaginary number.**

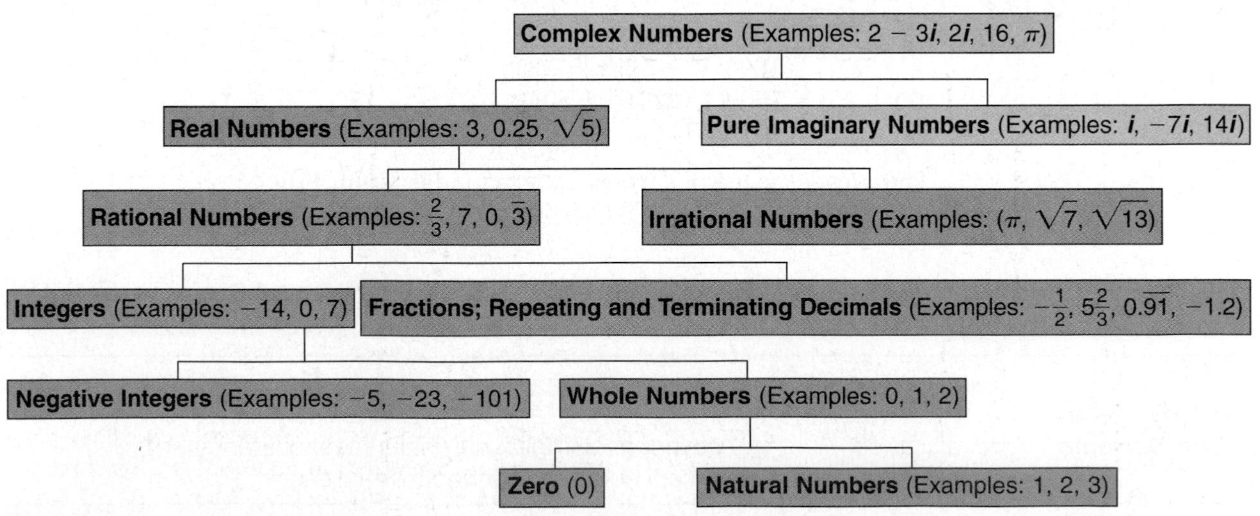

One of the most important theorems in mathematics is the **Fundamental Theorem of Algebra.**

Fundamental Theorem of Algebra	Every polynomial equation with degree greater than zero has at least one root in the set of complex numbers.

A corollary to the Fundamental Theorem of Algebra states that the degree of a polynomial indicates the number of possible roots of a polynomial equation.

Corollary to the Fundamental Theorem of Algebra	Every polynomial $P(x)$ of degree n ($n > 0$) can be written as the product of a constant k ($k \neq 0$) and n linear factors. $$P(x) = k(x - r_1)(x - r_2)(x - r_3) \ldots (x - r_n)$$ Thus, a polynomial equation of degree n has exactly n complex roots, namely $r_1, r_2, r_3, \ldots, r_n$.

The general shapes of the graphs of polynomial functions with positive leading coefficients and degree greater than 0 are shown below. These graphs also show the *maximum* number of times the graph of each type of polynomial may cross the x-axis.

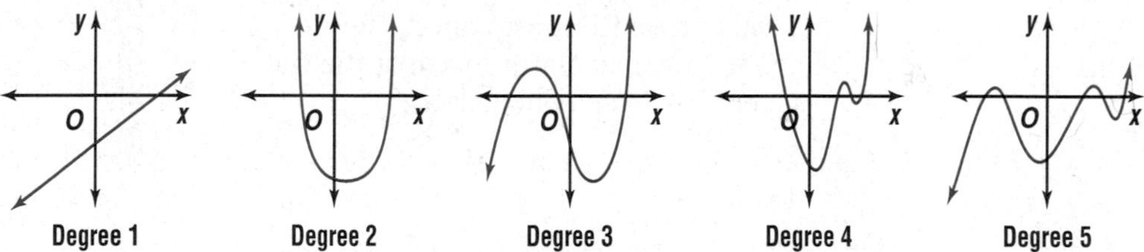

Since the x-axis only represents real numbers, imaginary roots cannot be determined by using a graph. The graphs below have the general shape of a third-degree function and a fourth-degree function. In these graphs, the third-degree function only crosses the x-axis once, and the fourth-degree function crosses the x-axis twice or not at all.

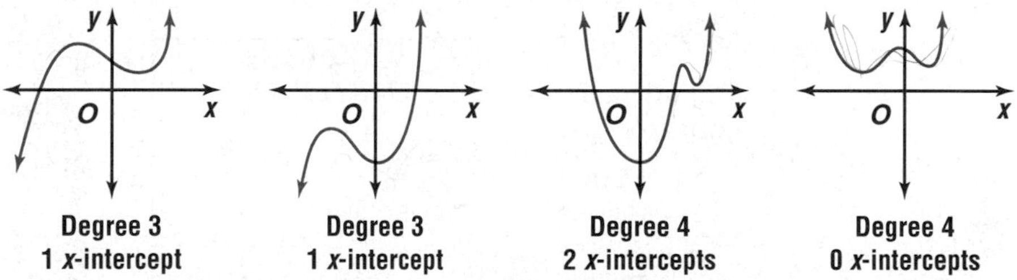

The graph of a polynomial function with odd degree must cross the x-axis at least once. The graph of a function with even degree may or may not cross the x-axis. If it does, it will cross an even number of times. Each x-intercept represents a real root of the corresponding polynomial equation.

If you know the roots of a polynomial equation, you can use the corollary to the Fundamental Theorem of Algebra to find the polynomial equation. That is, if a and b are roots of the equation, the equation must be $(x - a)(x - b) = 0$.

Examples

3 **a. Write a polynomial equation of least degree with roots 2, $4i$, and $-4i$.**

b. Does the equation have an odd or even degree? How many times does the graph of the related function cross the x-axis?

a. If $x = 2$, then $x - 2$ is a factor of the polynomial. Likewise, if $x = 4\boldsymbol{i}$ and $x = -4\boldsymbol{i}$, then $x - 4\boldsymbol{i}$ and $x - (-4\boldsymbol{i})$ are factors of the polynomial. Therefore, the linear factors for the polynomial are $x - 2$, $x - 4\boldsymbol{i}$, and $x + 4\boldsymbol{i}$. Now find the products of these factors.

$$(x - 2)(x - 4\boldsymbol{i})(x + 4\boldsymbol{i}) = 0$$

$$(x - 2)(x^2 - 16\boldsymbol{i}^2) = 0$$

$$(x - 2)(x^2 + 16) = 0 \quad -16\boldsymbol{i}^2 = -16(-1) \text{ or } 16$$

$$x^3 - 2x^2 + 16x - 32 = 0$$

A polynomial equation with roots 2, $4\boldsymbol{i}$, and $-4\boldsymbol{i}$ is $x^3 - 2x^2 + 16x - 32 = 0$.

b. The degree of this equation is 3. Thus, the equation has an odd degree since 3 is an odd number. Since two of the roots are imaginary, the graph will only cross the x-axis once. The graphing calculator image at the right verifies these conclusions.

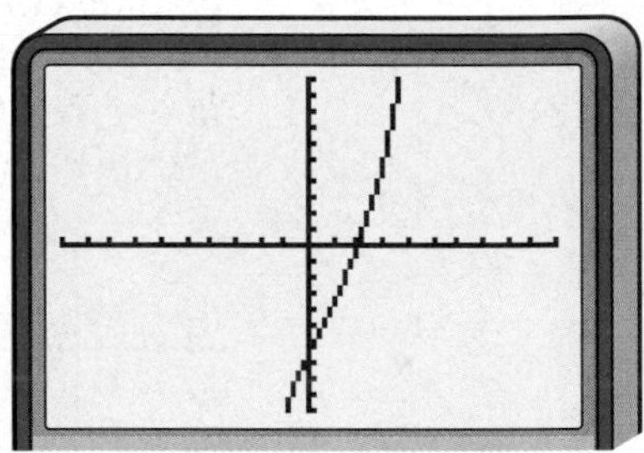

[−10, 10] scl:1 by [−50, 50] scl:5

4 **State the number of complex roots of the equation $9x^4 - 35x^2 - 4 = 0$. Then find the roots and graph the related function.**

The polynomial has a degree of 4, so there are 4 complex roots.

Factor the equation to find the roots.

$$9x^4 - 35x^2 - 4 = 0$$

$$(9x^2 + 1)(x^2 - 4) = 0$$

$$(9x^2 + 1)(x + 2)(x - 2) = 0$$

To find each root, set each factor equal to zero.

$$9x^2 + 1 = 0$$

$$x^2 = -\frac{1}{9} \quad \textit{Solve for } x^2.$$

$$x = \pm\sqrt{\frac{1}{9} \cdot (-1)} \quad \textit{Take the square root of each side.}$$

$$x = \pm\frac{1}{3}\sqrt{-1} \text{ or } \pm\frac{1}{3}\boldsymbol{i}$$

$$x + 2 = 0 \qquad x - 2 = 0$$

$$x = -2 \qquad x = 2$$

The roots are $\pm\frac{1}{3}\boldsymbol{i}$, -2, and 2.

Use a table of values or a graphing calculator to graph the function. The x-intercepts are -2 and 2.

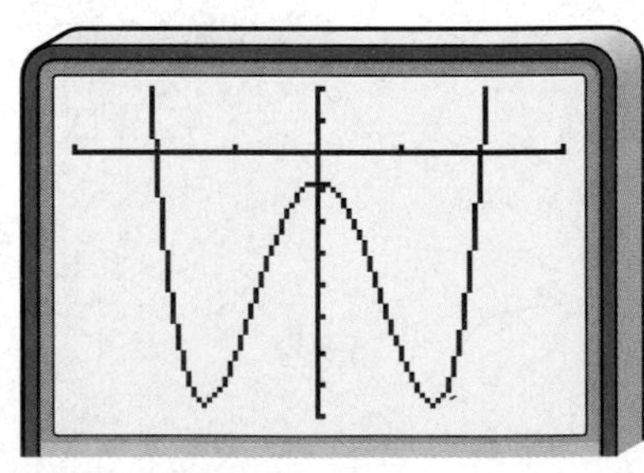

[−3, 3] scl:1 by [−40, 10] scl:5

The function has an even degree and has 2 real zeros.

Example 5 **METEOROLOGY** **A meteorologist sends a temperature probe on a small weather rocket through a cloud layer. The launch pad for the rocket is 2 feet off the ground. The height of the rocket after launching is modeled by the equation $h = -16t^2 + 232t + 2$, where h is the height of the rocket in feet and t is the elapsed time in seconds.**

a. When will the rocket be 114 feet above the ground?

b. Verify your answer using a graph.

a. $h = -16t^2 + 232t + 2$

$114 = -16t^2 + 232t + 2$ *Replace h with 114.*

$0 = -16t^2 + 232t - 112$ *Subtract 114 from each side.*

$0 = -8(2t^2 - 29t + 14)$ *Factor.*

$0 = -8(2t - 1)(t - 14)$ *Factor.*

$2t - 1 = 0$ or $t - 14 = 0$

$t = \frac{1}{2}$ $\quad$ $14 = t$

The weather rocket will be 114 feet above the ground after $\frac{1}{2}$ second and again after 14 seconds.

b. To verify the answer, graph $h(t) = -16t^2 + 232t - 112$. The graph appears to verify this solution.

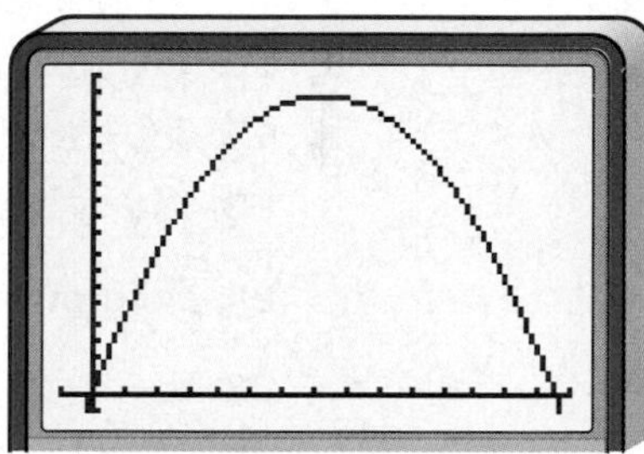

[−1, 15] scl:1 by [−50, 900] scl:50

CHECK FOR UNDERSTANDING

Communicating Mathematics

Read and study the lesson to answer each question.

1. **Write** several sentences about the relationship between zeros and roots.
2. **Explain** why zeros of a function are also the x-intercepts of its graph.
3. **Define** a complex number and tell under what conditions it will be a pure imaginary number. Write two examples and two **nonexamples** of a pure imaginary numbers.
4. **Sketch** the general graph of a sixth-degree function.

Guided Practice

State the degree and leading coefficient of each polynomial.

5. $a^3 + 6a + 14$
6. $5m^2 + 8m^5 - 2$

Determine whether each number is a root of $x^3 - 5x^2 - 3x - 18 = 0$. Explain.

7. 5
8. 6

Write a polynomial equation of least degree for each set of roots. Does the equation have an odd or even degree? How many times does the graph of the related function cross the x-axis?

9. $-5, 7$

10. $6, 2i, -2i, i, -i$

State the number of complex roots of each equation. Then find the roots and graph the related functions.

11. $x^2 - 14x + 49 = 0$

12. $a^3 + 2a^2 - 8a = 0$

13. $t^4 - 1 = 0$

14. **Geometry** A cylinder is inscribed in a sphere with a radius of 6 units as shown.

   a. Write a function that models the volume of the cylinder in terms of x. (*Hint:* The volume of a cylinder equals $\pi r^2 h$.)

   b. Write this function as a polynomial function.

   c. Find the volume of the cylinder if $x = 4$.

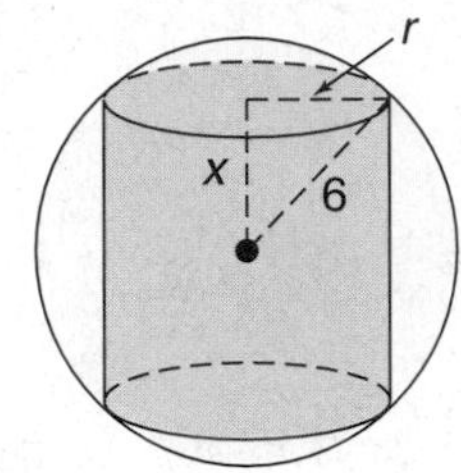

EXERCISES

Practice

State the degree and leading coefficient of each polynomial.

15. $5t^4 + t^3 - 7$

16. $3x^7 - 4x^5 + x^3$

17. $9a^2 + 5a^3 - 10$

18. $14b - 25b^5$

19. $p^5 + 7p^3 - p^6$

20. $14y + 30 + y^2$

21. Determine if $x^3 + 3x + \sqrt{5}$ is a polynomial in one variable. Explain.

22. Is $\frac{1}{a} + a^2$ a polynomial in one variable? Explain.

Determine whether each number is a root of $a^4 - 13a^2 + 12a = 0$. Explain.

23. 0

24. -1

25. 1

26. -4

27. -3

28. 3

29. Is -2 a root of $b^4 - 3b^2 - 2b + 4 = 0$?

30. Is -1 a root of $x^4 - 4x^3 - x^2 + 4x = 0$?

31. Each graph represents a polynomial function. State the number of complex zeros and the number of real zeros of each function.

a.

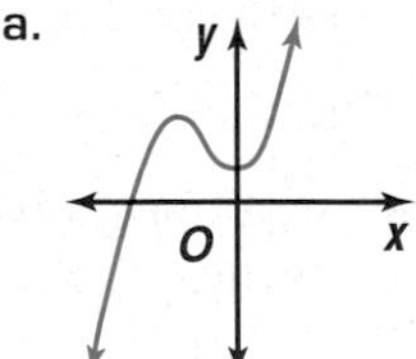

b.

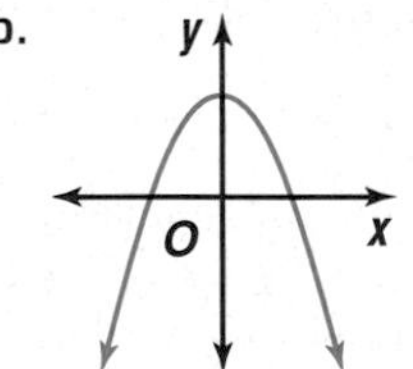

c.

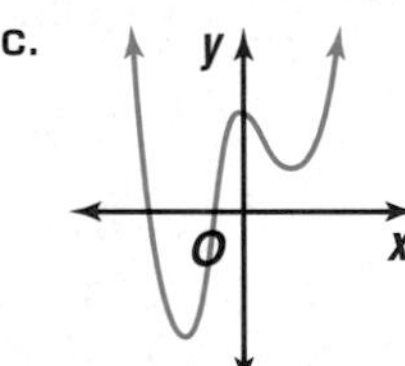

Write a polynomial equation of least degree for each set of roots. Does the equation have an odd or even degree? How many times does the graph of the related function cross the x-axis?

32. $-2, 3$

33. $-1, 1, 5$

34. $-2, -0.5, 4$

35. $-3, -2i, 2i$

36. $-5i, -i, i, 5i$

37. $-1, 1, 4, -4, 5$

38. Write a polynomial equation of least degree whose roots are -1, 1, 3, and -3.

State the number of complex roots of each equation. Then find the roots and graph the related function.

39. $x + 8 = 0$ **40.** $a^2 - 81 = 0$ **41.** $b^2 + 36 = 0$

42. $t^3 + 2t^2 - 4t - 8 = 0$ **43.** $n^3 - 9n = 0$ **44.** $6c^3 - 3c^2 - 45c = 0$

45. $a^4 + a^2 - 2 = 0$ **46.** $x^4 - 10x^2 + 9 = 0$ **47.** $4m^4 + 17m^2 + 4 = 0$

48. Solve $(u + 1)(u^2 - 1) = 0$ and graph the related polynomial function.

49. Sketch a fourth-degree equation for each situation.

a. no x-intercept b. one x-intercept c. two x-intercepts

d. three x-intercepts e. four x-intercepts f. five x-intercepts

Graphing Calculator

50. Use a graphing calculator to graph $f(x) = x^4 - 2x^2 + 1$.

a. What is the maximum number of x-intercepts possible for this function?

b. How many x-intercepts are there? Name the intercept(s).

c. Why are there fewer x-intercepts than the maximum number? (*Hint:* The factored form of the polynomial is $(x^2 - 1)^2$.)

Applications and Problem Solving

51. Classic Cars Sonia Orta invests in vintage automobiles. Three years ago, she purchased a 1953 Corvette roadster for \$99,000. Two years ago, she purchased a 1929 Pierce-Arrow Model 125 for \$55,000. A year ago she purchased a 1909 Cadillac Model Thirty for \$65,000.

a. Let x represent 1 plus the average rate of appreciation. Write a function in terms of x that models the value of the automobiles.

b. If the automobiles appreciate at an average annual rate of 15%, find the current value of the three automobiles.

52. Critical Thinking One of the zeros of a polynomial function is 1. After translating the graph of the function left 2 units, 1 is a zero of the new function. What do you know about the original function?

53. Aeronautics At liftoff, the space shuttle *Discovery* has a constant acceleration, a, of 16.4 feet per second squared. The function $d(t) = \frac{1}{2}at^2$ can be used to determine the distance from Earth for each time interval, t, after takeoff.

a. Find its distance from Earth after 30 seconds, 1 minute, and 2 minutes.

b. Study the pattern of answers to part a. If the time the space shuttle is in flight doubles, how does the distance from Earth change? Explain.

54. Construction The Santa Fe Recreation Department has a 50-foot by 70-foot area for construction of a new public swimming pool. The pool will be surrounded by a concrete sidewalk of constant width. Because of water restrictions, the pool can have a maximum area of 2400 square feet. What should be the width of the sidewalk that surrounds the pool?

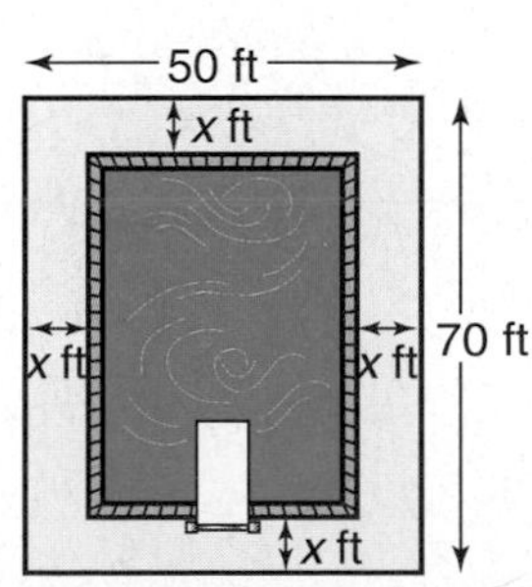

55. **Marketing** Each week, Marino's Pizzeria sells an average of 160 large supreme pizzas for \$16 each. Next week, the pizzeria plans to run a sale on these large supreme pizzas. The owner estimates that for each 40¢ decrease in the price, the store will sell approximately 16 more large pizzas. If the owner wants to sell \$4,000 worth of the large supreme pizzas next week, determine the sale price.

56. **Critical Thinking** If B and C are the real roots of $x^2 + Bx + C = 0$, where $B \neq 0$ and $C \neq 0$, find the values of B and C.

Mixed Review

57. Create a function in the form $y = f(x)$ that has a vertical asymptote at $x = -2$ and $x = 0$, and a hole at $x = 2$. *(Lesson 3-7)*

58. **Construction** Selena wishes to build a pen for her animals. He has 52 yards of fencing and wants to build a rectangular pen. *(Lesson 3-6)*
   a. Find a model for the area of the pen as a function of the length and width of the rectangle.
   b. What are the dimensions that would produce the maximum area?

59. Describe how the graphs of $y = 2x^3$ and $y = 2x^3 + 1$ are related. *(Lesson 3-2)*

60. Find the coordinates of P' if $P(4, 9)$ and P' are symmetric with respect to $M(-1, 9)$. *(Lesson 3-1)*

61. Find the determinant for $\begin{bmatrix} -15 & 5 \\ -9 & 3 \end{bmatrix}$. Tell whether an inverse exists for the matrix. *(Lesson 2-5)*

62. If $A = \begin{bmatrix} 2 & -1 \\ 3 & 4 \end{bmatrix}$ and $B = \begin{bmatrix} 3 & -9 & 2 \\ 5 & 7 & -6 \end{bmatrix}$, find AB. *(Lesson 2-3)*

63. Graph $x + 4y < 9$. *(Lesson 1-8)*

64. The slope of $\overleftrightarrow{AB}$ is 0.6. The slope of $\overleftrightarrow{CD}$ is $\frac{3}{5}$. State whether the lines are *parallel, perpendicular,* or *neither.* Explain. *(Lesson 1-5)*

65. Find $[f \circ g](x)$ and $[g \circ f](x)$ for the functions $f(x) = x^2 - 4$ and $g(x) = \frac{1}{2}x + 6$. *(Lesson 1-2)*

66. **SAT Practice Quantitative Comparison**
   A if the quantity in Column A is greater
   B if the quantity in Column B is greater
   C if the two quantities are equal
   D if the relationship cannot be determined for the information given

Year	Cars Sold by Bob's Quality Cars
1999	
2000	

In 1999, Bob's Quality Cars sold 270 more cars than in 2000.

Column A	Column B
The number of cars represents	100

Extra Practice See p. A32.

4-2 Quadratic Equations

OBJECTIVES

- Solve quadratic equations.
- Use the discriminant to describe the roots of quadratic equations.

BASEBALL On September 8, 1998, Mark McGwire of the St. Louis Cardinals broke the home-run record with his 62nd home run of the year. He went on to hit 70 home runs for the season. Besides hitting home runs, McGwire also occasionally popped out. Suppose the ball was 3.5 feet above the ground when he hit it straight up with an initial velocity of 80 feet per second. The function $d(t) = 80t - 16t^2 + 3.5$ gives the ball's height above the ground in feet as a function of time in seconds. How long did the catcher have to get into position to catch the ball after it was hit? *This problem will be solved in Example 3.*

A quadratic equation is a polynomial equation with a degree of two. Solving quadratic equations by graphing usually does not yield exact answers. Also, some quadratic expressions are not factorable over the integers. Therefore, alternative strategies for solving these equations are needed. One such alternative is solving quadratic equations by **completing the square.**

Completing the square is a useful method when the quadratic is not easily factorable. It can be used to solve any quadratic equation. Remember that, for any number b, the square of the binomial $x + b$ has the form $x^2 + 2bx + b^2$. When completing the square, you know the first term and middle term and need to supply the last term. This term equals the square of half the coefficient of the middle term. For example, to complete the square of $x^2 + 8x$, find $\frac{1}{2}(8)$ and square the result. So, the third term is 16, and the expression becomes $x^2 + 8x + 16$. Note that this technique works only if the coefficient of x^2 is 1.

Example 1 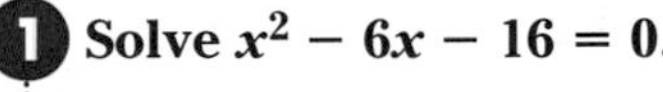

Solve $x^2 - 6x - 16 = 0$.

This equation can be solved by graphing, factoring, or completing the square.

Method 1

Solve the equation by graphing the related function $f(x) = x^2 - 6x - 16$. The zeros of the function appear to be -2 and 8.

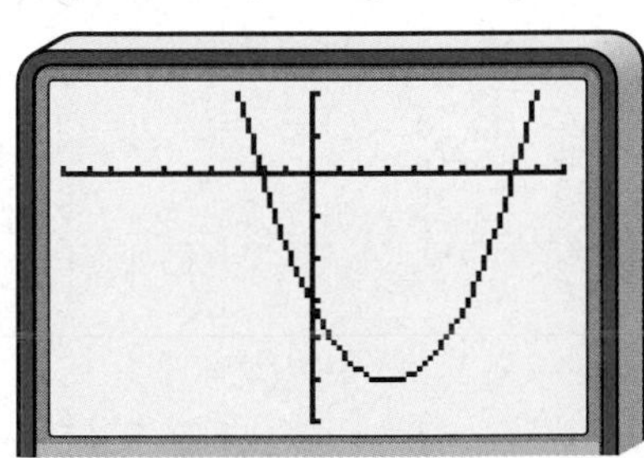

[−10, 10] scl:1 by [−30, 10] scl:5

Method 2

Solve the equation by factoring.

$x^2 - 6x - 16 = 0$

$(x + 2)(x - 8) = 0$ *Factor.*

$x + 2 = 0$ or $x - 8 = 0$

$x = -2$ $\qquad$ $x = 8$

The roots of the equation are -2 and 8.

Method 3

Solve the equation by completing the square.

$x^2 - 6x - 16 = 0$

$x^2 - 6x = 16$ *Add 16 to each side.*

$x^2 - 6x + 9 = 16 + 9$ *Complete the square by adding* $\left(\frac{-6}{2}\right)^2$ *or 9 to each side.*

$(x - 3)^2 = 25$ *Factor the perfect square trinomial.*

$x - 3 = \pm 5$ *Take the square root of each side.*

$x - 3 = 5$ or $x - 3 = -5$

$x = 8$ $\quad x = -2$

The roots of the equation are 8 and -2.

Although factoring may be an easier method to solve this particular equation, completing the square can always be used to solve any quadratic equation.

When solving a quadratic equation by completing the square, the leading coefficient must be 1. When the leading coefficient of a quadratic equation is not 1, you must first divide each side of the equation by that coefficient before completing the square.

Example 2 **Solve $3n^2 + 7n + 7 = 0$ by completing the square.**

Notice that the graph of the related function, $y = 3x^2 + 7x + 7$, does not cross the x-axis. Therefore, the roots of the equation are imaginary numbers. Completing the square can be used to find the roots of any equation, including one with no real roots.

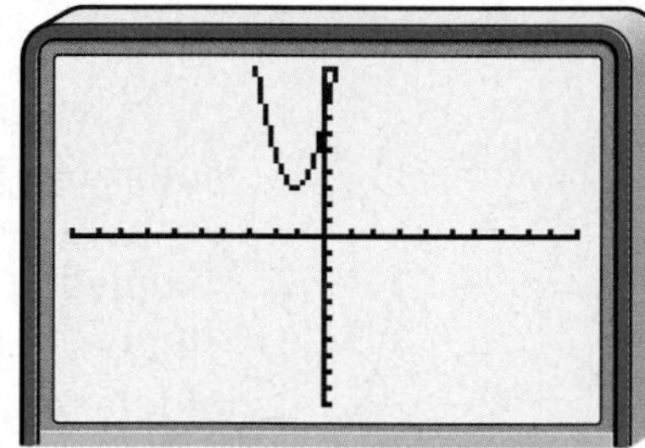

[−10, 10] scl:1 by [−10, 10] scl:1

$3n^2 + 7n + 7 = 0$

$n^2 + \frac{7}{3}n + \frac{7}{3} = 0$ *Divide each side by 3.*

$n^2 + \frac{7}{3}n = -\frac{7}{3}$ *Subtract* $\frac{7}{3}$ *from each side.*

$n^2 + \frac{7}{3}n + \frac{49}{36} = -\frac{7}{3} + \frac{49}{36}$ *Complete the square by adding* $\left(\frac{7}{6}\right)^2$ *or* $\frac{49}{36}$ *to each side.*

$\left(n + \frac{7}{6}\right)^2 = -\frac{35}{36}$ *Factor the perfect square trinomial.*

$n + \frac{7}{6} = \pm \boldsymbol{i}\frac{\sqrt{35}}{6}$ *Take the square root of each side.*

$n = -\frac{7}{6} \pm \boldsymbol{i}\frac{\sqrt{35}}{6}$ *Subtract* $\frac{7}{6}$ *from each side.*

The roots of the equation are $-\frac{7}{6} \pm \boldsymbol{i}\frac{\sqrt{35}}{6}$ or $\frac{-7 \pm \boldsymbol{i}\sqrt{35}}{6}$.

Completing the square can be used to develop a general formula for solving any quadratic equation of the form $ax^2 + bx + c = 0$. This formula is called the **Quadratic Formula.**

Quadratic Formula

The roots of a quadratic equation of the form $ax^2 + bx + c = 0$ with $a \neq 0$ are given by the following formula.

$$x = \frac{-b \pm \sqrt{b^2 - 4ac}}{2a}$$

The quadratic formula can be used to solve any quadratic equation. It is usually easier than completing the square.

Example 3 

BASEBALL Refer to the application at the beginning of the lesson. How long did the catcher have to get into position to catch the ball after if was hit?

The catcher must get into position to catch the ball before $80t - 16t^2 + 3.5 = 0$. This equation can be written as $-16t^2 + 80t + 3.5 = 0$. Use the Quadratic Formula to solve this equation.

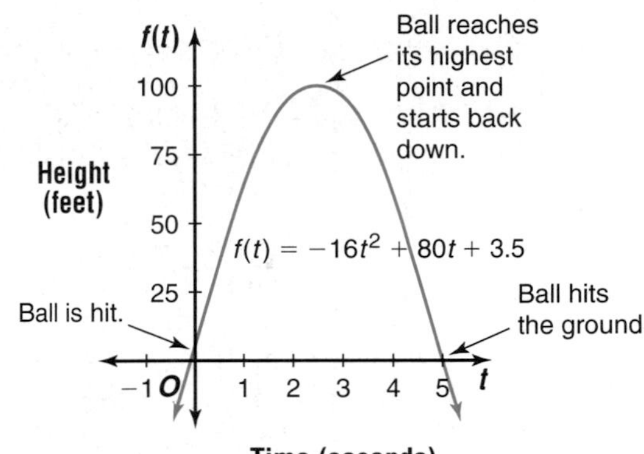

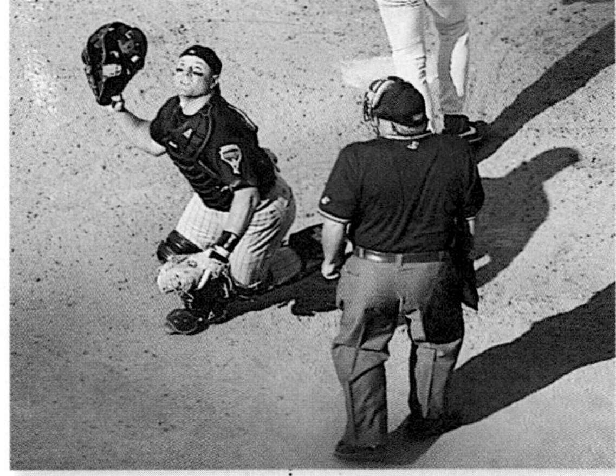

$$t = \frac{-b \pm \sqrt{b^2 - 4ac}}{2a}$$

$$t = \frac{-80 \pm \sqrt{80^2 - 4(-16)(3.5)}}{2(-16)} \quad \textit{a = −16, b = 80, and c = 3.5}$$

$$t = \frac{-80 \pm \sqrt{6624}}{-32}$$

$$t = \frac{-80 + \sqrt{6624}}{-32} \quad \text{or} \quad t = \frac{-80 - \sqrt{6624}}{-32}$$

$$t \approx -0.04 \qquad t \approx 5.04$$

The roots of the equation are about -0.04 and 5.04. Since the catcher has a positive amount of time to catch the ball, he will have about 5 seconds to get into position to catch the ball.

In the quadratic formula, the radicand $b^2 - 4ac$ is called the **discriminant** of the equation. The discriminant tells the nature of the roots of a quadratic equation or the zeros of the related quadratic function.

Discriminant	Nature of Roots/Zeros	Graph
$b^2 - 4ac > 0$	two distinct real roots/zeros	
$b^2 - 4ac = 0$	exactly one real root/zero (The one real root is actually a double root.)	
$b^2 - 4ac < 0$	no real roots/zero (two distinct imaginary roots/zeros)	

Example 4 **Find the discriminant of $x^2 - 4x + 15 = 0$ and describe the nature of the roots of the equation. Then solve the equation by using the Quadratic Formula.**

The value of the discriminant, $b^2 - 4ac$, is $(-4)^2 - 4(1)(15)$ or -44. Since the value of the discriminant is less than zero, there are no real roots.

$$x = \frac{-b \pm \sqrt{b^2 - 4ac}}{2a}$$

$$x = \frac{-(-4) \pm \sqrt{-44}}{2(1)}$$

$$x = \frac{4 \pm 2i\sqrt{11}}{2}$$

$$x = 2 \pm i\sqrt{11}$$

The roots are $2 + i\sqrt{11}$ and $2 - i\sqrt{11}$.

The graph of $y = x^2 - 4x + 15$ verifies that there are no real roots.

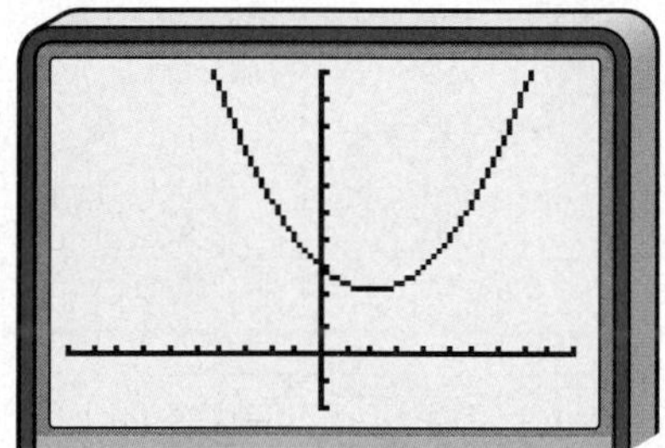

[−10, 10] scl:1 by [−10, 50] scl:5

The roots of the equation in Example 4 are the complex numbers $2 + i\sqrt{11}$ and $2 - i\sqrt{11}$. A pair of complex numbers in the form $a + bi$ and $a - bi$ are called **conjugates.** Imaginary roots of polynomial equations with real coefficients always occur in conjugate pairs. Some other examples of complex conjugates are listed below.

i and $-i$ $\qquad$ $-1 + i$ and $-1 - i$ $\qquad$ $-i\sqrt{2}$ and $i\sqrt{2}$

Complex Conjugates Theorem

Suppose a and b are real numbers with $b \neq 0$. If $a + bi$ is a root of a polynomial equation with real coefficients, then $a - bi$ is also a root of the equation. $a + bi$ and $a - bi$ are conjugate pairs.

There are four methods used to solve quadratic equations. Two methods work for any quadratic equation. One method approximates any real roots, and one method only works for equations that can be factored over the integers.

Solution Method	Situation	Examples
Graphing	Usually, only approximate solutions are shown. If roots are imaginary (discriminant is less than zero), the graph has no *x*-intercepts, and the solutions must be found by another method. Graphing is a good method to verify solutions.	$6x^2 + x - 2 = 0$ f(x) O x $f(x) = 6x^2 + x - 2$ $x = -\frac{2}{3}$ or $x = \frac{1}{2}$ $x^2 - 2x + 5 = 0$ discriminant: $(-2)^2 - 4(1)(5) = -16$ f(x) 8 4 $f(x) = x^2 - 2x + 5$ −4 −2 O 2 4 x The equation has no real roots.
Factoring	When *a*, *b*, and *c* are integers and the discriminant is a perfect square or zero, this method is useful. It cannot be used if the discriminant is less than zero.	$g^2 + 2g - 8 = 0$ discriminant: $2^2 - 4(1)(-8) = 36$ $g^2 + 2g - 8 = 0$ $(g + 4)(g - 2) = 0$ $g + 4 = 0$ or $g - 2 = 0$ $g = -4$ $g = 2$
Completing the Square	This method works for any quadratic equation. There is more room for an arithmetic error than when using the Quadratic Formula.	$r^2 + 4r - 6 = 0$ $r^2 + 4r = 6$ $r^2 + 4r + 4 = 6 + 4$ $(r + 2)^2 = 10$ $r + 2 = \pm\sqrt{10}$ $r = -2 \pm \sqrt{10}$
Quadratic Formula	This method works for any quadratic equation.	$2s^2 + 5s + 4 = 0$ $s = \frac{-5 \pm \sqrt{5^2 - 4(2)(4)}}{2(2)}$ $s = \frac{-5 \pm \sqrt{-7}}{4}$ $s = \frac{-5 \pm i\sqrt{7}}{4}$

Example 5 Solve $6x^2 + x + 2 = 0$.

Method 1: Graphing

Graph $y = 6x^2 + x + 2$.

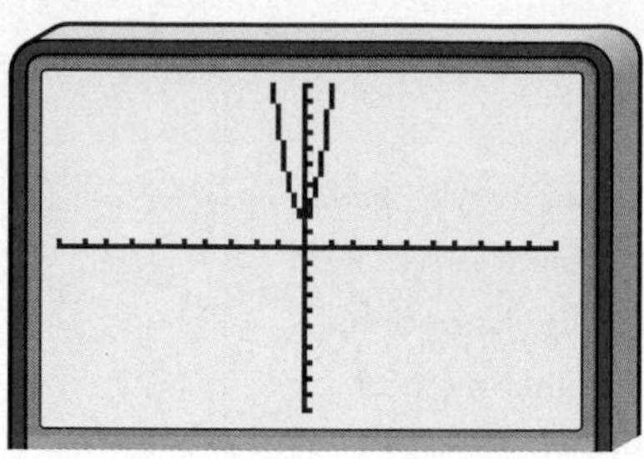

[−10, 10] scl:1 by [−50, 50] scl:1

The graph does not touch the x-axis, so there are no real roots for this equation. You cannot determine the roots from the graph.

Method 2: Factoring

Find the discriminant.

$b^2 - 4ac = 1^2 - 4(6)(2)$ or -47

The discrimimant is less than zero, so factoring cannot be used to solve the equation.

Method 3: Completing the Square

$$6x^2 + x + 2 = 0$$

$$x^2 + \frac{1}{6}x + \frac{1}{3} = 0$$

$$x^2 + \frac{1}{6}x = -\frac{1}{3}$$

$$x^2 + \frac{1}{6}x + \frac{1}{144} = -\frac{1}{3} + \frac{1}{144}$$

$$\left(x + \frac{1}{12}\right)^2 = -\frac{47}{144}$$

$$x + \frac{1}{12} = \pm i\frac{\sqrt{47}}{12}$$

$$x = -\frac{1}{12} \pm i\frac{\sqrt{47}}{12}$$

$$x = \frac{-1 \pm i\sqrt{47}}{12}$$

Completing the square works, but this method requires a lot of steps.

Method 4: Quadratic Formula

For this equation, $a = 6$, $b = 1$, $c = 2$.

$$x = \frac{-b \pm \sqrt{b^2 - 4ac}}{2a}$$

$$x = \frac{-1 \pm \sqrt{1^2 - 4(6)(2)}}{2(6)}$$

$$x = \frac{-1 \pm \sqrt{-47}}{12}$$

$$x = \frac{-1 \pm i\sqrt{47}}{12}$$

The Quadratic Formula works and requires fewer steps than completing the square.

The roots of the equation are $\frac{-1 \pm i\sqrt{47}}{12}$.

CHECK FOR UNDERSTANDING

Communicating Mathematics

Read and study the lesson to answer each question.

1. **Write** a short paragraph explaining how to solve $t^2 - 6t - 4 = 0$ by completing the square.
2. **Discuss** which method of solving $5p^2 - 13p + 7 = 0$ would be most appropriate. Explain. Then solve.

3. **Describe** the discriminant of the equation represented by each graph.

a.

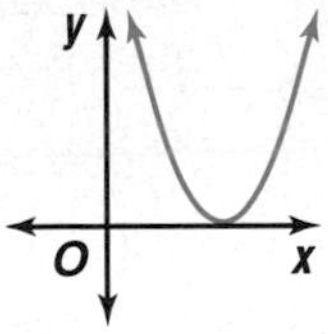

b.

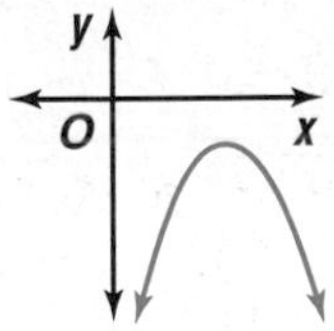

c.

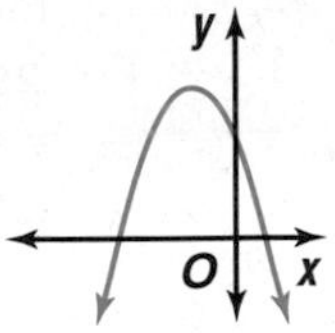

4. *Math Journal* **Solve** $x^2 + 4x - 5 = 0$ using each of the four methods discussed in this lesson. Which method do you prefer? Explain.

Guided Practice

Solve each equation by completing the square.

5. $x^2 + 8x - 20 = 0$

6. $2a^2 + 11a - 21 = 0$

Find the discriminant of each equation and describe the nature of the roots of the equation. Then solve the equation by using the Quadratic Formula.

7. $m^2 + 12m + 36 = 0$

8. $t^2 - 6t + 13 = 0$

Solve each equation.

9. $p^2 - 6p + 5 = 0$

10. $r^2 - 4r + 10 = 0$

11. **Electricity** On a cold day, a 12-volt car battery has a resistance of 0.02 ohms. The power available to start the motor is modeled by the equation $P = 12I - 0.02I^2$, where I is the current in amperes. What current is needed to produce 1600 watts of power to start the motor?

EXERCISES

Practice

Solve each equation by completing the square.

12. $z^2 - 2z - 24 = 0$

13. $p^2 - 3p - 88 = 0$

14. $x^2 - 10x + 21 = 0$

15. $d^2 - \frac{3}{4}d + \frac{1}{8} = 0$

16. $3g^2 - 12g = -4$

17. $t^2 - 3t - 7 = 0$

18. What value of c makes $x^2 - x + c$ a perfect square?

19. Describe the nature of the roots of the equation $4n^2 + 6n + 25$. Explain.

Find the discriminant of each equation and describe the nature of the roots of the equation. Then solve the equation by using the Quadratic Formula.

20. $6m^2 + 7m - 3 = 0$

21. $s^2 - 5s + 9 = 0$

22. $36d^2 - 84d + 49 = 0$

23. $4x^2 - 2x + 9 = 0$

24. $3p^2 + 4p = 8$

25. $2k^2 + 5k = 9$

26. What is the conjugate of $-7 - i\sqrt{5}$?

27. Name the conjugate of $5 - 2i$.

Solve each equation.

28. $3s^2 - 5s + 9 = 0$
29. $x^2 - 3x - 28 = 0$
30. $4w^2 + 19w - 5 = 0$
31. $4r^2 - r = 5$
32. $p^2 + 2p + 8 = 0$
33. $x^2 - 2x\sqrt{6} - 2 = 0$

Applications and Problem Solving

Real World Application

34. **Health** Normal systolic blood pressure is a function of age. For a woman, the normal systolic pressure P in millimeters of mercury (mm Hg) is modeled by $P = 0.01A^2 + 0.05A + 107$, where A is age in years.
   a. Use this model to determine the normal systolic pressure of a 25-year-old woman.
   b. Use this model to determine the age of a woman whose normal systolic pressure is 125 mm Hg.
   c. Sketch the graph of the function. Describe what happens to the normal systolic pressure as a woman gets older.

35. **Critical Thinking** Consider the equation $x^2 + 8x + c = 0$. What can you say about the value of c if the equation has two imaginary roots?

36. **Interior Design** Abey Numkena is an interior designer. She has been asked to locate an oriental rug for a new corporate office. As a rule, the rug should cover $\frac{1}{2}$ of the total floor area with a uniform width surrounding the rug.
   a. If the dimensions of the room are 12 feet by 16 feet, write an equation to model the situation.
   b. Graph the related function.
   c. What are the dimensions of the rug?

37. **Entertainment** In an action movie, a stuntwoman jumps off a building that is 50 feet tall with an upward initial velocity of 5 feet per second. The distance $d(t)$ traveled by a free falling object can be modeled by the formula $d(t) = v_0t - \frac{1}{2}gt^2$, where v_0 is the initial velocity and g represents the acceleration due to gravity. The acceleration due to gravity is 32 feet per second squared.
   a. Draw a graph that relates the woman's distance traveled with the time since the jump.
   b. Name the x-intercepts of the graph.
   c. What is the meaning of the x-intercepts of the graph?
   d. Write an equation that could be used to determine when the stuntwoman will reach the safety pad on the ground. (*Hint:* The ground is -50 feet from the starting point.)
   e. How long will it take the stuntwoman to reach the safety pad on the ground?

38. **Critical Thinking** Derive the quadratic formula by completing the square if $ax^2 + bx + c = 0$, $a \neq 0$.

Extra Practice See p. A32.

Mixed Review

39. State the number of complex roots of the equation $18a^2 + 3a - 1 = 0$. Then find the roots and graph the related function. *(Lesson 4-1)*

40. Graph $y < |x| - 2$. *(Lesson 3-5)*

41. Find the inverse of $f(x) = (x - 9)^2$. *(Lesson 3-4)*

42. Solve the system of equations, $3x + 4y = 375$ and $5x + 2y = 345$. *(Lesson 2-1)*

43. **Sales** The Computer Factory is selling a 300 MHz computer system for \$595 and a 350 MHz computer system for \$619. At this rate, what would be the cost of a 400 MHz computer system? *(Lesson 1-4)*

44. Find the slope of the line whose equation is $3y + 8x = 12$. *(Lesson 1-3)*

45. **SAT/ACT Practice** The trinomial $x^2 + x - 20$ is exactly divisible by which binomial?
A $x - 4$ **B** $x + 4$ **C** $x + 6$ **D** $x - 10$ **E** $x - 5$

CAREER CHOICES

Environmental Engineering

Would you like a career where you will constantly be learning and have the opportunity to work both outdoors and indoors? Environmental engineering has become an important profession in the past twenty-five years.

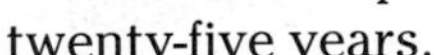

As an environmental engineer, you might design, build, or maintain systems for controlling wastes produced by cities or industry. These wastes can include solid waste, waste water, hazardous waste, or air pollutants. You could work for a private company, a consulting firm, or the Environmental Protection Agency. Opportunities for advancement in this field include becoming a supervisor or consultant. You might even have the opportunity to identify a new specialty area in the field of environmental engineering!

Career Overview

Degree Required:
Bachelor's degree in environmental engineering

Related Courses:
biology, chemistry, mathematics

Outlook:
number of jobs expected to increase though the year 2006

Emissions of Two Air Pollutants in the U.S. 1987-1996

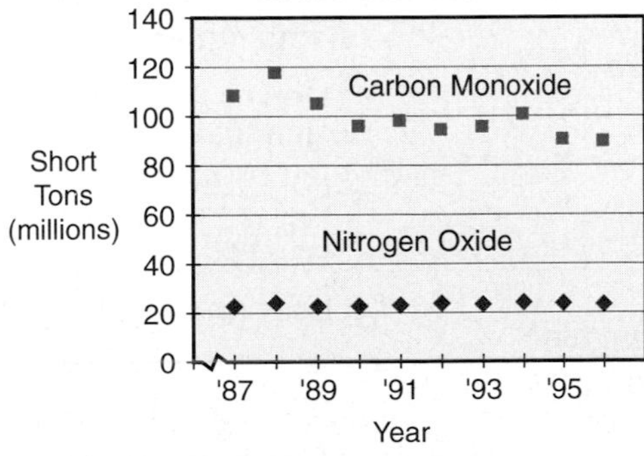

Source: *The World Almanac*

For more information about environmental engineering, visit: **www.amc.glencoe.com**

4-3 The Remainder and Factor Theorems

OBJECTIVE
- Find the factors of polynomials using the Remainder and Factor Theorems.

SKIING On December 13, 1998, Olympic champion Hermann (The Herminator) Maier won the super-G at Val d'Isere, France. His average speed was 73 meters per second. The average recreational skier skis at a speed of about 5 meters per second. Suppose you were skiing at a speed of 5 meters per second and heading downhill, accelerating at a rate of 0.8 meter per second squared. How far will you travel in 30 seconds? *This problem will be solved in Example 1.*

Hermann Maier

Consider the polynomial function $f(a) = 2a^2 + 3a - 8$. Since 2 is a factor of 8, it may be possible that $a - 2$ is a factor of $2a^2 + 3a - 8$. Suppose you use long division to divide the polynomial by $a - 2$.

$$\begin{array}{rrl} & 2a + 7 & \leftarrow \textit{quotient} \\ \textit{divisor} \rightarrow a - 2 & \overline{)2a^2 + 3a - 8} & \leftarrow \textit{dividend} \\ & \underline{2a^2 - 4a} & \\ & 7a - 8 & \\ & \underline{7a - 14} & \\ & 6 & \leftarrow \textit{remainder} \end{array}$$

From arithmetic, you may remember that the dividend equals the product of the divisor and the quotient plus the remainder. For example, $44 \div 7 = 6$ R2, so $44 = 7(6) + 2$. This relationship can be applied to polynomials.

You may want to verify that $(a - 2)(2a + 7) + 6 = 2a^2 + 3a - 8$.

$$f(a) = (a - 2)(2a + 7) + 6 \quad \Longleftrightarrow \quad f(a) = 2a^2 + 3a - 8$$

Let $a = 2$. $f(2) = (2 - 2)[2(2) + 7] + 6 = 0 + 6$ or 6 $\qquad$ $f(2) = 2(2^2) + 3(2) - 8 = 8 + 6 - 8$ or 6

Notice that the value of $f(2)$ is the same as the remainder when the polynomial is divided by $a - 2$. This example illustrates the **Remainder Theorem.**

Remainder Theorem

If a polynomial $P(x)$ is divided by $x - r$, the remainder is a constant $P(r)$, and

$$P(x) = (x - r) \cdot Q(x) + P(r),$$

where $Q(x)$ is a polynomial with degree one less than the degree of $P(x)$.

The Remainder Theorem provides another way to find the value of the polynomial function $P(x)$ for a given value of r. The value will be the remainder when $P(x)$ is divided by $x - r$.

Example 1 

SKIING Refer to the application at the beginning of the lesson. The formula for distance traveled is $d(t) = v_0t + \frac{1}{2}at^2$, where $d(t)$ is the distance traveled, v_0 is the initial velocity, t is the time, and a is the acceleration. Find the distance traveled after 30 seconds.

The distance formula becomes $d(t) = 5t + \frac{1}{2}(0.8)t^2$ or $d(t) = 0.4t^2 + 5t$. You can use one of two methods to find the distance after 30 seconds.

Method 1
Divide $0.4t^2 + 5t$ by $t - 30$.

$$\begin{array}{r}
0.4t + \ \ 17 \\
t - 30\overline{)0.4t^2 + \ \ 5t} \\
\underline{0.4t^2 - 12t} \\
17t + \ \ \ 0 \\
\underline{17t - 510} \\
510 \rightarrow D(30) = 510
\end{array}$$

Method 2
Evaluate $d(t)$ for $t = 30$.

$$\begin{aligned}
d(t) &= 0.4t^2 + 5t \\
d(30) &= 0.4(30^2) + 5(30) \\
&= 0.4(900) + 5(30) \\
&= 510
\end{aligned}$$

By either method, the result is the same. You will travel 510 meters in 30 seconds.

Long division can be very time consuming. **Synthetic division** is a shortcut for dividing a polynomial by a binomial of the form $x - r$. The steps for dividing $x^3 + 4x^2 - 3x - 5$ by $x + 3$ using synthetic division are shown below.

Step 1 Arrange the terms of the polynomial in descending powers of x. Insert zeros for any missing powers of x. Then, write the coefficients as shown.

$$\begin{array}{cccc}
x^3 & +4x^2 & -3x & -5 \\
\downarrow & \downarrow & \downarrow & \downarrow \\
1 & 4 & -3 & -5
\end{array}$$

For $x + 3$, the value of r is -3.

Step 2 Write the constant r of the divisor $x - r$. In this case, write -3.

$$\begin{array}{r|rrrr}
-3 & 1 & 4 & -3 & -5
\end{array}$$

Step 3 Bring down the first coefficient.

$$\begin{array}{r|rrrr}
-3 & 1 & 4 & -3 & -5 \\
\hline
 & 1 & & &
\end{array}$$

Step 4 Multiply the first coefficient by r. Then write the product under the next coefficient. Add.

$$\begin{array}{r|rrrr}
-3 & 1 & 4 & -3 & -5 \\
 & & -3 & & \\
\hline
\times & 1 & 1 & &
\end{array}$$

Step 5 Multiply the sum by r. Then write the product under the next coefficient. Add.

$$\begin{array}{r|rrrr}
-3 & 1 & 4 & -3 & -5 \\
 & & -3 & -3 & \\
\hline
\times & 1 & 1 & -6 &
\end{array}$$

Notice that a vertical bar separates the quotient from the remainder.

Step 6 Repeat Step 5 for all coefficients in the dividend.

$$\begin{array}{r|rrr|r}
-3 & 1 & 4 & -3 & -5 \\
 & & -3 & -3 & 18 \\
\hline
\times & 1 & 1 & -6 & 13
\end{array}$$

Step 7 The final sum represents the remainder, which in this case is 13. The other numbers are the coefficients of the quotient polynomial, which has a degree one less than the dividend. Write the quotient $x^2 + x - 6$ with remainder 13. *Check the results using long division.*

Example 2 **Divide $x^3 - x^2 + 2$ by $x + 1$ using synthetic division.**

$$\begin{array}{r|rrrr} -1 & 1 & -1 & 0 & 2 \\ & & -1 & 2 & -2 \\ \hline & 1 & -2 & 2 & \vert\ 0 \end{array}$$

Notice there is no x term. A zero is placed in this position as a placeholder.

The quotient is $x^2 - 2x + 2$ with a remainder of 0.

In Example 2, the remainder is 0. Therefore, $x + 1$ is a factor of $x^3 - x^2 + 2$. If $f(x) = x^3 - x^2 + 2$, then $f(-1) = (-1)^3 - (-1)^2 + 2$ or 0. This illustrates the **Factor Theorem,** which is a special case of the Remainder Theorem.

Factor Theorem	The binomial $x - r$ is a factor of the polynomial $P(x)$ if and only if $P(r) = 0$.

Example 3 **Use the Remainder Theorem to find the remainder when $2x^3 - 3x^2 + x$ is divided by $x - 1$. State whether the binomial is a factor of the polynomial. Explain.**

Find $f(1)$ to see if $x - 1$ is a factor.

$f(x) = 2x^3 - 3x^2 + x$

$f(1) = 2(1^3) - 3(1^2) + 1$ *Replace x with 1.*

$= 2(1) - 3(1) + 1$ or 0

Since $f(1) = 0$, the remainder is 0. So the binomial $(x - 1)$ is a factor of the polynomial by the Factor Theorem.

When a polynomial is divided by one of its binomial factors $x - r$, the quotient is called a **depressed polynomial.** A depressed polynomial has a degree less than the original polynomial. In Example 3, $x - 1$ is a factor of $2x^3 - 3x^2 + x$. Use synthetic division to find the depressed polynomial.

$$\begin{array}{r|rrrr} 1 & 2 & -3 & 1 & 0 \\ & & 2 & -1 & 0 \\ \hline & 2 & -1 & 0 & \vert\ 0 \\ & \downarrow & \downarrow & \downarrow & \\ & \multicolumn{3}{c}{2x^2 - 1x + 0} & \end{array}$$

Thus, $(2x^3 - 3x^2 + x) \div (x - 1) = 2x^2 - x$.

The depressed polynomial is $2x^2 - x$.

A depressed polynomial may also be the product of two polynomial factors, which would give you other zeros of the polynomial function. In this case, $2x^2 - x$ equals $x(2x - 1)$. So, the zeros of the polynomial function $f(x) = 2x^3 - 3x^2 + x$ are 0, $\frac{1}{2}$, and 1.

Note that the values of r where no remainder occurs are also factors of the constant term of the polynomial.

You can also find factors of a polynomial such as $x^3 + 2x^2 - 16x - 32$ by using a shortened form of synthetic division to test several values of r. In the table, the first column contains various values of r. The next three columns show the coefficients of the depressed polynomial. The fifth column shows the remainder. Any value of r that results in a remainder of zero indicates that $x - r$ is a factor of the polynomial. The factors of the original polynomial are $x + 4$, $x + 2$, and $x - 4$.

r	1	2	−16	−32
−4	1	−2	−8	0
−3	1	−1	−13	7
−2	1	0	−16	0
−1	1	1	−17	−15
0	1	2	−16	−32
1	1	3	−13	−45
2	1	4	−8	−48
3	1	5	−1	−35
4	1	6	8	0

Look at the pattern of values in the last column. Notice that when r = 1, 2, and 3, the values of f(x) decrease and then increase. This indicates that there is an x-coordinate of a relative minimum between 1 and 3.

Example 4 **Determine the binomial factors of $x^3 - 7x + 6$.**

Graphing Calculator Tip

The **TABLE** feature can help locate integral zeros. Enter the polynomial function as **Y1** in the **Y=** menu and press **TABLE**. Search the **Y1** column to find 0 and then look at the corresponding x-value.

Method 1 Use synthetic division.

r	1	0	−7	6	
−4	1	−4	9	−30	
−3	1	−3	2	0	*x + 3 is a factor.*
−2	1	−2	−3	12	
−1	1	−1	−6	12	
0	1	0	−7	6	
1	1	1	−6	0	*x − 1 is a factor.*
2	1	2	−3	0	*x − 2 is a factor.*

Method 2 Test some values using the Factor Theorem.

$f(x) = x^3 - 7x + 6$
$f(-1) = (-1)^3 - 7(-1) + 6$ or 12
$f(1) = 1^3 - 7(1) + 6$ or 0

Because $f(1) = 0$, $x - 1$ is a factor. Find the depressed polynomial.

$$\begin{array}{r|rrrr} 1 & 1 & 0 & -7 & 6 \\ & & 1 & 1 & -6 \\ \hline & 1 & 1 & -6 & 0 \end{array}$$

The depressed polynomial is $x^2 + x - 6$. Factor the depressed polynomial.

$x^2 + x - 6 = (x + 3)(x - 2)$

The factors of $x^3 + x - 6$ are $x + 3$, $x - 1$, and $x - 2$. *Verify the results.*

The Remainder Theorem can be used to determine missing coefficients.

Example 5 **Find the value of k so that the remainder of $(x^3 + 3x^2 - kx - 24) \div (x + 3)$ is 0.**

If the remainder is to be 0, $x + 3$ must be a factor of $x^3 + 3x^2 - kx - 24$. So, $f(-3)$ must equal 0.

$$\begin{aligned} f(x) &= x^3 + 3x^2 - kx - 24 \\ f(-3) &= (-3)^3 + 3(-3)^2 - k(-3) - 24 \\ 0 &= -27 + 27 + 3k - 24 \quad \text{Replace } f(-3) \text{ with } 0. \\ 0 &= 3k - 24 \\ 8 &= k \end{aligned}$$

The value of k is 8. Check using synthetic division.

$$\begin{array}{r|rrrr} -3 & 1 & 3 & -8 & -24 \\ & & -3 & 0 & 24 \\ \hline & 1 & 0 & -8 & 0 \checkmark \end{array}$$

CHECK FOR UNDERSTANDING

Communicating Mathematics

Read and study the lesson to answer each question.

1. **Explain** how the Remainder Theorem and the Factor Theorem are related.
2. **Write** the division problem illustrated by the synthetic division. What is the quotient? What is the remainder?

5	1	−4	−7	8
		5	5	−10
	1	1	−2	−2

3. **Compare** the degree of a polynomial and its depressed polynomial.
4. **You Decide** Brittany tells Isabel that if $x + 3$ is a factor of the polynomial function $f(x)$, then $f(3) = 0$. Isabel argues that if $x + 3$ is a factor of $f(x)$, then $f(-3) = 0$. Who is correct? Explain.

Guided Practice

Divide using synthetic division.

5. $(x^2 - x + 4) \div (x - 2)$
6. $(x^3 + x^2 - 17x + 15) \div (x + 5)$

Use the Remainder Theorem to find the remainder for each division. State whether the binomial is a factor of the polynomial.

7. $(x^2 + 2x - 15) \div (x - 3)$
8. $(x^4 + x^2 + 2) \div (x - 3)$

Determine the binomial factors of each polynomial.

9. $x^3 - 5x^2 - x + 5$
10. $x^3 - 6x^2 + 11x - 6$

11. Find the value of k so that the remainder of $(x^3 - 7x + k) \div (x + 1)$ is 2.
12. Let $f(x) = x^7 + x^9 + x^{12} - 2x^2$.
    a. State the degree of $f(x)$.
    b. State the number of complex zeros that $f(x)$ has.
    c. State the degree of the depressed polynomial that would result from dividing $f(x)$ by $x - a$.
    d. Find one factor of $f(x)$.
13. **Geometry** A cylinder has a height 4 inches greater than the radius of its base. Find the radius and the height to the nearest inch if the volume of the cylinder is 5π cubic inches.

*inter*NET CONNECTION

Graphing Calculator Program
For a graphing calculator program that computes the value of a function visit **www.amc.glencoe.com**

EXERCISES

Practice

Divide using synthetic division.

14. $(x^2 + 20x + 91) \div (x + 7)$
15. $(x^3 - 9x^2 + 27x - 28) \div (x - 3)$
16. $(x^4 + x^3 - 1) \div (x - 2)$
17. $(x^4 - 8x^2 + 16) \div (x + 2)$
18. $(3x^4 - 2x^3 + 5x^2 - 4x - 2) \div (x + 1)$
19. $(2x^3 - 2x - 3) \div (x - 1)$

Use the Remainder Theorem to find the remainder for each division. State whether the binomial is a factor of the polynomial.

20. $(x^2 - 2) \div (x - 1)$
21. $(x^5 + 32) \div (x + 2)$
22. $(x^4 - 6x^2 + 8) \div (x - \sqrt{2})$
23. $(x^3 - x + 6) \div (x - 2)$
24. $(4x^3 + 4x^2 + 2x + 3) \div (x - 1)$
25. $(2x^3 - 3x^2 + x) \div (x - 1)$

26. Which binomial is a factor of the polynomial $x^3 + 3x^2 - 2x - 8$?
   a. $x - 1$ b. $x + 1$ c. $x - 2$ d. $x + 2$

27. Verify that $x - \sqrt{6}$ is a factor of $x^4 - 36$.

28. Use synthetic division to find all the factors of $x^3 + 7x^2 - x - 7$ if one of the factors is $x + 1$.

Determine the binomial factors of each polynomial.

29. $x^3 + x^2 - 4x - 4$
30. $x^3 - x^2 - 49x + 49$
31. $x^3 - 5x^2 + 2x + 8$
32. $x^3 - 2x^2 - 4x + 8$
33. $x^3 + 4x^2 - x - 4$
34. $x^3 + 3x^2 + 3x + 1$

35. How many times is 2 a root of $x^6 - 9x^4 + 24x^2 - 16 = 0$?

36. Determine how many times -1 is a root of $x^3 + 2x^2 - x - 2 = 0$. Then find the other roots.

Find the value of *k* so that each remainder is zero.

37. $(2x^3 - x^2 + x + k) \div (x - 1)$
38. $(x^3 - kx^2 + 2x - 4) \div (x - 2)$
39. $(x^3 + 18x^2 + kx + 4) \div (x + 2)$
40. $(x^3 + 4x^2 - kx + 1) \div (x + 1)$

Applications and Problem Solving

41. **Bicycling** Matthew is cycling at a speed of 4 meters per second. When he starts down a hill, the bike accelerates at a rate of 0.4 meter per second squared. The vertical distance from the top of the hill to the bottom of the hill is 25 meters. Use the equation $d(t) = v_0t + \frac{1}{2}at^2$ to find how long it will take Matthew to ride down the hill.

42. **Critical Thinking** Determine a and b so that when $x^4 + x^3 - 7x^2 + ax + b$ is divided by $(x - 1)(x + 2)$, the remainder is 0.

43. **Sculpting** Esteban is preparing to start an ice sculpture. He has a block of ice that is 3 feet by 4 feet by 5 feet. Before he starts, he wants to reduce the volume of the ice by shaving off the same amount from the length, the width, and the height.
   a. Write a polynomial function to model the situation.
   b. Graph the function.
   c. He wants to reduce the volume of the ice to $\frac{3}{5}$ of the original volume. Write an equation to model the situation.
   d. How much should he take from each dimension?

44. **Manufacturing** An 18-inch by 20-inch sheet of cardboard is cut and folded to make a box for the Great Pecan Company.
   a. Write an polynomial function to model the volume of the box.
   b. Graph the function.
   c. The company wants the box to have a volume of 224 cubic inches. Write an equation to model this situation.
   d. Find a positive integer for x.

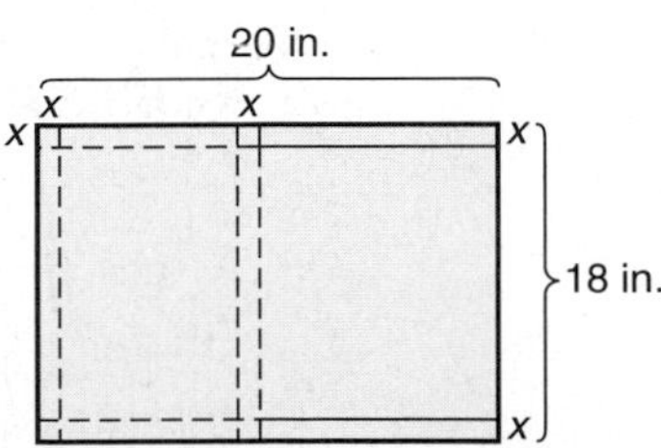

45. **Critical Thinking** Find a, b, and c for $P(x) = ax^2 + bx + c$ if $P(3 + 4\mathbf{i}) = 0$ and $P(3 - 4\mathbf{i}) = 0$.

Mixed Review

46. Solve $r^2 + 5r - 8 = 0$ by completing the square. *(Lesson 4-2)*

47. Determine whether each number is a root of $x^4 - 4x^3 - x^2 + 4x = 0$. *(Lesson 4-1)*
   a. 2 b. 0 c. −2 d. 4

48. Find the critical points of the graph of $f(x) = x^5 - 32$. Determine whether each represents a *maximum*, a *minimum*, or a *point of inflection*. *(Lesson 3-6)*

49. Describe the transformation(s) that have taken place between the parent graph of $y = x^2$ and the graph of $y = 0.5(x + 1)^2$. *(Lesson 3-2)*

50. **Business** Pristine Pipes Inc. produces plastic pipe for use in newly-built homes. Two of the basic types of pipe have different diameters, wall thickness, and strengths. The strength of a pipe is increased by mixing a special additive into the plastic before it is molded. The table below shows the resources needed to produce 100 feet of each type of pipe and the amount of the resource available each week.

Resource	Pipe A	Pipe B	Resource Availability
Extrusion Dept.	4 hours	6 hours	48 hours
Packaging Dept.	2 hours	2 hours	18 hours
Strengthening Additive	2 pounds	1 pound	16 pounds

If the profit on 100 feet of type A pipe is \$34 and of type B pipe is \$40, how much of each should be produced to maximize the profit? *(Lesson 2-7)*

51. Solve the system of equations. *(Lesson 2-2)*

$$4x + 2y + 3z = 6$$
$$2x + 7y = 3z$$
$$-3x - 9y + 13 = -2z$$

52. **Geometry** Show that the line segment connecting the midpoints of sides $\overline{TR}$ and $\overline{TI}$ is parallel to $\overline{RI}$. *(Lesson 1-5)*

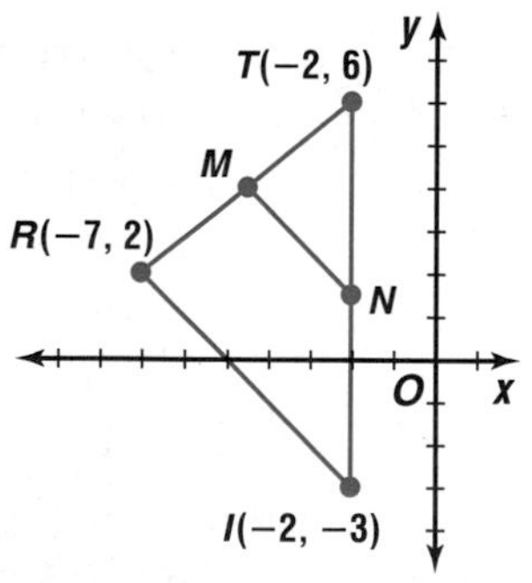

53. **SAT/ACT Practice** If $a > b$ and $c < 0$, which of the following are true?
   I. $ac < bc$
   II. $a + c > b + c$
   III. $a - c < b - c$

   A I only **B** II only **C** III only
   D I and II only **E** I, II, and III

Extra Practice See p. A32.

4-4

The Rational Root Theorem

OBJECTIVES

- Identify all possible rational roots of a polynomial equation by using the Rational Root Theorem.
- Determine the number of positive and negative real roots a polynomial function has.

CONSTRUCTION The longest and largest canal tunnel in the world was the Rove Tunnel on the Canal de Marseille au Rhone in the south of France. In 1963, the tunnel collapsed and excavation engineers were trying to duplicate the original tunnel. The height of the tunnel was 1 foot more than half its width. The length was 32 feet more than 324 times its width. The volume of the tunnel was 62,231,040 cubic feet. If the tunnel was a rectangular prism, find its original dimensions. *This problem will be solved in Example 4.*

The formula for the volume of a rectangular prism is $V = \ell wh$, where ℓ is the length, w is the width, and h is the height. From the information above, the height of the tunnel is $\frac{1}{2}w + 1$, and its length is $324w + 32$.

$$\begin{aligned} V &= \ell wh \\ 62{,}231{,}040 &= (324w + 32)w\left(\frac{1}{2}w + 1\right) && \ell = 324w + 32,\ h = \frac{1}{2}w + 1 \\ 62{,}231{,}040 &= 162w^3 + 340w^2 + 32w \\ 0 &= 162w^3 + 340w^2 + 32w - 62{,}231{,}040 \\ 0 &= 81w^3 + 170w^2 + 16w - 31{,}115{,}520 && \textit{Divide each side by 2.} \end{aligned}$$

We could use synthetic substitution to test possible zeros, but with such large numbers, this is not practical. In situations like this, the **Rational Root Theorem** can give direction in testing possible zeros.

Rational Root Theorem

Let $a_0x^n + a_1x^{n-1} + \ldots + a_{n-1}x + a_n = 0$ represent a polynomial equation of degree n with integral coefficients. If a rational number $\frac{p}{q}$, where p and q have no common factors, is a root of the equation, then p is a factor of a_n and q is a factor of a_0.

Example 1 **List the possible rational roots of $6x^3 + 11x^2 - 3x - 2 = 0$. Then determine the rational roots.**

According to the Rational Root Theorem, if $\frac{p}{q}$ is a root of the equation, then p is a factor of 2 and q is a factor of 6.

possible values of p: $\pm1, \pm2$

possible values of q: $\pm1, \pm2, \pm3, \pm6$

possible rational roots, $\frac{p}{q}$: $\pm1, \pm2, \pm\frac{1}{2}, \pm\frac{1}{3}, \pm\frac{1}{6}, \pm\frac{2}{3}$

(continued on the next page)

You can use a graphing utility to narrow down the possibilities. You know that all possible rational roots fall in the domain $-2 \le x \le 2$. So set your x-axis viewing window at $[-3, 3]$. Graph the related function $f(x) = 6x^3 + 11x^2 - 3x - 2$. A zero appears to occur at -2. Use synthetic division to check that -2 is a zero.

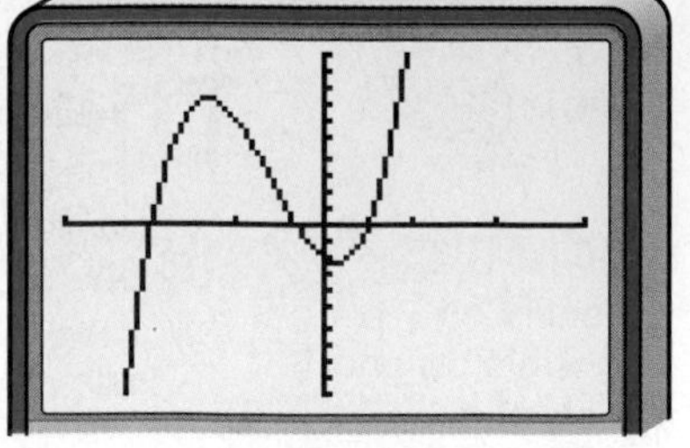

[−3, 3] scl:1 by [−10, 10] scl:1

$$\begin{array}{r|rrrr} -2 & 6 & 11 & -3 & -2 \\ & & -12 & 2 & 2 \\ \hline & 6 & -1 & -1 & 0 \end{array}$$

Thus, $6x^3 + 11x^2 - 3x - 2 = (x + 2)(6x^2 - x - 1)$. Factoring $6x^2 - x - 1$ yields $(3x + 1)(2x - 1)$. The roots of $6x^3 + 11x^2 - 3x - 2 = 0$ are -2, $-\frac{1}{3}$, and $\frac{1}{2}$.

A corollary to the Rational Root Theorem, called the **Integral Root Theorem,** states that if the leading coefficient a_0 has a value of 1, then any rational roots must be factors of a_n, $a_n \neq 0$.

Integral Root Theorem

Let $x^n + a_1x^{n-1} + \ldots + a_{n-1}x + a_n = 0$ represent a polynomial equation that has a leading coefficient of 1, integral coefficients, and $a_n \neq 0$. Any rational roots of this equation must be integral factors of a_n.

Example 2 **Find the roots of $x^3 + 8x^2 + 16x + 5 = 0$.**

There are three complex roots. According to the Integral Root Theorem, the possible rational roots of the equation are factors of 5. The possibilities are ± 5 and ± 1.

r	1	8	16	5
5	1	13	81	410
−5	1	3	1	0

← *There is a root at $x = -5$.*

The depressed polynomial is $x^2 + 3x + 1$. Use the quadratic formula to find the other two roots.

$$x = \frac{-b \pm \sqrt{b^2 - 4ac}}{2a}$$

$$x = \frac{-3 \pm \sqrt{3^2 - 4(1)(1)}}{2(1)} \qquad a = 1, b = 3, c = 1$$

$$x = \frac{-3 \pm \sqrt{5}}{2}$$

The three roots of the equation are -5, $\frac{-3 - \sqrt{5}}{2}$, and $\frac{-3 + \sqrt{5}}{2}$.

Descartes' Rule of Signs can be used to determine the possible number of positive real zeros a polynomial has. It is named after the French mathematician René Descartes, who first proved the theorem in 1637. *In Descartes' Rule of Signs, when we speak of a zero of the polynomial, we mean a zero of the corresponding polynomial function.*

Descartes' Rule of Signs

Suppose $P(x)$ is a polynomial whose terms are arranged in descending powers of the variable. Then the number of positive real zeros of $P(x)$ is the same as the number of changes in sign of the coefficients of the terms or is less than this by an even number. The number of negative real zeros of $P(x)$ is the same as the number of changes in sign of the coefficients of the terms of $P(-x)$, or less than this number by an even number.

Ignore zero coefficients when using this rule.

Example 3 **Find the number of possible positive real zeros and the number of possible negative real zeros for $f(x) = 2x^5 + 3x^4 - 6x^3 + 6x^2 - 8x + 3$. Then determine the rational zeros.**

To determine the number of possible positive real zeros, count the sign changes for the coefficients.

$$f(x) = 2x^5 + 3x^4 - 6x^3 + 6x^2 - 8x + 3$$

2 → 3 → −6 → 6 → −8 → 3

no, yes, yes, yes, yes

There are four changes. So, there are four, two, or zero positive real zeros.

To determine the number of possible negative real zeros, find $f(-x)$ and count the number of sign changes.

$$f(-x) = 2(-x)^5 + 3(-x)^4 - 6(-x)^3 + 6(-x)^2 - 8(-x) + 3$$
$$f(-x) = -2x^5 + 3x^4 + 6x^3 + 6x^2 + 8 + 3$$

−2 → 3 → 6 → 6 → 8 → 3

yes, no, no, no, no

There is one change. So, there is one negative real zero.

Determine the possible zeros.

possible values of p: $\pm 1, \pm 3$
possible values of q: $\pm 1, \pm 2$
possible rational zeros, $\frac{p}{q}$: $\pm 1, \pm 3, \pm\frac{1}{2}, \pm\frac{3}{2}$

Test the possible zeros using the synthetic division and the Remainder Theorem.

r	2	3	−6	6	−8	3	
1	2	5	−1	5	−3	0	*1 is a zero.*
−1	2	1	−7	13	−21	24	
3	2	9	21	69	199	600	
−3	2	−3	3	−3	1	0	*−3 is a zero.*

(continued on the next page)

Since there is only one negative real zero and -3 is a zero, you do not need to test any other negative possibilities.

$\frac{1}{2}$	2	4	-4	4	-6	0
$\frac{3}{2}$	2	6	3	$10\frac{1}{2}$	$7\frac{3}{4}$	$14\frac{5}{8}$

$\frac{1}{2}$ is a zero.

All possible rational roots have been considered. There are two positive zeros and one negative zero. The rational zeros for $f(x) = 2x^5 + 3x^4 - 6x^3 + 6x^2 - 8x + 3$ are -3, $\frac{1}{2}$, and 1. Use a graphing utility to check these zeros. Note that there appears to be three x-intercepts. You can use the zero function on the **CALC** menu to verify that the zeros you found are correct.

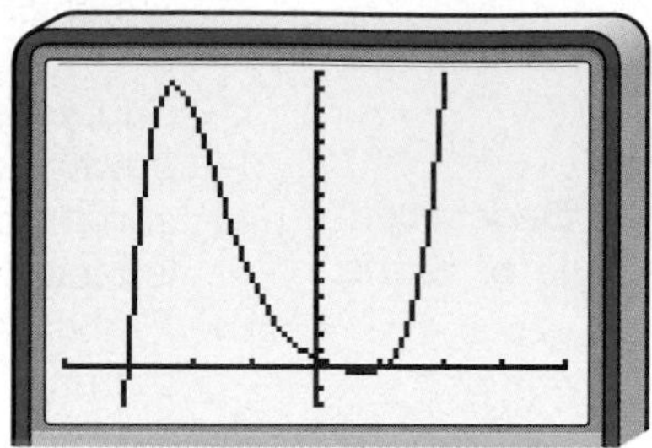

[−4, 4] scl:1 by [−10, 85] scl:5

You can use a graphing calculator to study Descartes' Rule of Signs.

GRAPHING CALCULATOR EXPLORATION

Remember that you can determine the location of the zeros of a function by analyzing its graph.

TRY THESE **Graph each function to determine how many zeros appear to exist. Use the zero function in the CALC menu to approximate each zero.**

1. $f(x) = x^4 + 4x^3 + 3x^2 - 4x - 4$
2. $f(x) = x^3 - 3x - 2$

WHAT DO YOU THINK?

3. Use Descartes' Rule of Signs to determine the possible positive and negative real zeros of each function.
4. How do your results from Exercise 3 compare with your results using the **TABLE** feature? Explain.
5. What do you think the term "double zero" means?

Because the zeros of a polynomial function are the roots of a polynomial equation, Descartes' Rule of Signs can be used to determine the types of roots of the equation.

Example  **CONSTRUCTION** **Refer to the application at the beginning of the lesson. Find the original dimensions of the Rove Tunnel.**

To determine the dimensions of the tunnel, we must solve the equation $0 = 81w^3 + 170w^2 + 16w - 31{,}115{,}520$. According to Descartes' Rule of Signs, there is one positive real root and two or zero negative real roots. Since dimensions are always positive, we are only concerned with the one positive real root.

Use a graphing utility to graph the related function $V(w) = 81w^3 + 170w^2 + 16w - 31{,}115{,}520$. Since the graph has 10 unit increments, the zero is between 70 and 80.

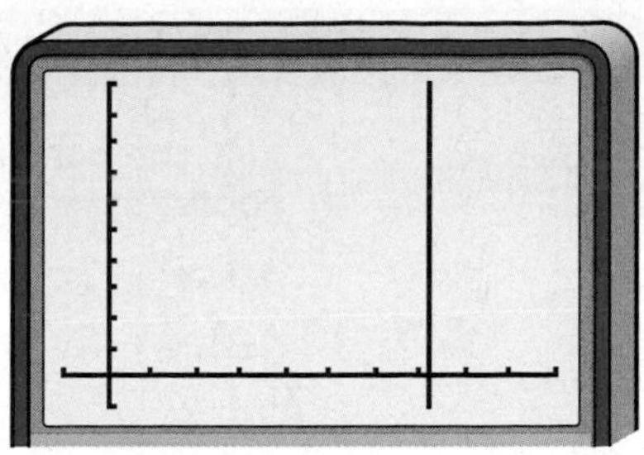

[−10, 100] scl:10 by [−10, 100] scl:10

To help determine the possible rational roots, find the prime factorization of 31,115,520.

$31{,}115{,}520 = 2^8 \times 3^2 \times 5 \times 37 \times 73$

Possible rational roots between 70 and 80 are 72 ($2^3 \times 3^2$), 73, and 74 (2×37). Use the Factor Theorem until the one zero is found.

$$V(w) = 81w^3 + 170w^2 + 16w - 31{,}115{,}520$$
$$V(72) = 81(72^3) + 170(72^2) + 16(72) - 31{,}115{,}520$$
$$V(72) = 30{,}233{,}088 + 881{,}280 + 1152 - 31{,}115{,}520$$
$$V(72) = 0$$

The original width was 72 feet, the original height was $\frac{1}{2}(72) + 1$ or 37 feet, and the original length was 324(72) + 32 or 23,360 feet, which is over 4 miles.

CHECK FOR UNDERSTANDING

Communicating Mathematics

Read and study the lesson to answer each question.

1. **Identify** the possible rational roots for the equation $x^4 - 3x^2 + 6 = 0$.
2. **Explain** why the Integral Root Theorem is a corollary to the Rational Root Theorem.
3. **Write** a polynomial function $f(x)$ whose coefficients have three sign changes. Find the number of sign changes that $f(-x)$ has. Describe the nature of the zeros.
4. *Math Journal* **Describe** several methods you could use to determine the rational zeros of a polynomial function. Which would you choose to use first? Explain.

Guided Practice

List the possible rational roots of each equation. Then determine the rational roots.

5. $x^3 - 4x^2 + x + 2 = 0$
6. $2x^3 + 3x^2 - 8x + 3 = 0$

Find the number of possible positive real zeros and the number of possible negative real zeros for each function. Then determine the rational zeros.

7. $f(x) = 8x^3 - 6x^2 - 23x + 6$
8. $f(x) = x^3 + 7x^2 + 7x - 15$

9. **Geometry** A cone is inscribed in a sphere with a radius of 15 centimeters. If the volume of the cone is 1152π cubic centimeters, find the length represented by x.

EXERCISES

Practice

List the possible rational roots of each equation. Then determine the rational roots.

10. $x^3 + 2x^2 - 5x - 6 = 0$

11. $x^3 - 2x^2 + x + 18 = 0$

12. $x^4 - 5x^3 + 9x^2 - 7x + 2 = 0$

13. $x^3 - 5x^2 - 4x + 20 = 0$

14. $2x^4 - x^3 - 6x + 3 = 0$

15. $6x^4 + 35x^3 - x^2 - 7x - 1 = 0$

16. State the number of complex roots, the number of positive real roots, and the number of negative real roots of $x^4 - 2x^3 + 7x^2 + 4x - 15 = 0$.

Find the number of possible positive real zeros and the number of possible negative real zeros for each function. Then determine the rational zeros.

17. $f(x) = x^3 - 7x - 6$

18. $f(x) = x^3 - 2x^2 - 8x$

19. $f(x) = x^3 + 3x^2 - 10x - 24$

20. $f(x) = 10x^3 - 17x^2 - 7x + 2$

21. $f(x) = x^4 + 2x^3 - 9x^2 - 2x + 8$

22. $f(x) = x^4 - 5x^2 + 4$

23. Suppose $f(x) = (x - 2)(x + 2)(x + 1)^2$.

a. Determine the zeros of the function.

b. Write $f(x)$ as a polynomial function.

c. Use Descartes' Rule of Signs to find the number of possible positive real roots and the number of possible negative real roots.

d. Compare your answers to part **a** and part **c**. Explain.

Applications and Problem Solving

24. Manufacturing The specifications for a new cardboard container require that the width for the container be 4 inches less than the length and the height be 1 inch less than twice the length.

a. Write a polynomial function that models the volume of the container in terms of its length.

b. Write an equation if the volume must be 2208 cubic inches.

c. Find the dimensions of the new container.

25. Critical Thinking Write a polynomial equation for each restriction.

a. fourth degree with no positive real roots

b. third degree with no negative real roots

c. third degree with exactly one positive root and exactly one negative real root

26. Architecture A hotel in Las Vegas, Nevada, is the largest pyramid in the United States. Prior to the construction of the building, the architects designed a scale model.

a. If the height of the scale model was 9 inches less than its length and its base is a square, write a polynomial function that describes the volume of the model in terms of its length.

b. If the volume of the model is 6300 cubic inches, write an equation describing the situation.

c. What were the dimensions of the scale model?

27. **Construction** A steel beam is supported by two pilings 200 feet apart. If a weight is placed x feet from the piling on the left, a vertical deflection d equals $0.0000008x^2(200 - x)$. How far is the weight if the vertical deflection is 0.8 feet?

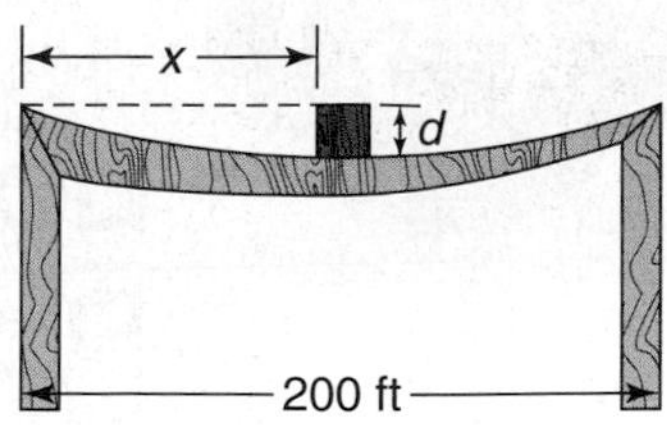

28. **Critical Thinking** Compare and contrast the graphs and zeros of $f(x) = x^3 + 3x^2 - 6x - 8$ and $g(x) = -x^3 - 3x^2 + 6x + 8$.

Mixed Review

29. Divide $x^2 - x - 56$ by $x + 7$ using synthetic division. *(Lesson 4-3)*

30. Find the discriminant of $4x^2 + 6x + 25 = 0$ and describe the nature of the roots. *(Lesson 4-2)*

31. Write the polynomial equation of least degree whose roots are 1, -1, 2, and -2. *(Lesson 4-1)*

32. **Business** The prediction equation for a set of data relating the year in which a car was rented as the independent variable to the weekly car rental fee as the dependent variable is $y = 4.3x - 8424.3$. Predict the average cost of renting a car in 2008. *(Lesson 1-6)*

33. **SAT/ACT Practice** If $\frac{2x - 3}{x} = \frac{3 - x}{2}$, which of the following could be a value for x?

A -3 B -1 C 37 D 5 E 15

MID-CHAPTER QUIZ

1. Write the polynomial equation of least degree with roots 1, -1, $2i$, and $-2i$. (Lesson 4-1)

2. State the number of complex roots of $x^3 - 11x^2 + 30x = 0$. Then find the roots. (Lesson 4-1)

3. Solve $x^2 + 5x = 150$ by completing the square. (Lesson 4-2)

4. Find the discriminant of $6b^2 - 39b + 45 = 0$ and describe the nature of the roots of the equation. Then solve the equation by using the quadratic formula. (Lesson 4-2)

5. Divide $x^3 + 3x^2 - 2x - 8$ by $x + 2$ using synthetic division. (Lesson 4-3)

6. Use the Remainder Theorem to find the remainder for $(x^3 - 4x^2 + 2x - 6) \div (x - 4)$. State whether the binomial is a factor of the polynomial. (Lesson 4-3)

7. Determine the binomial factors of $x^3 - 2x^2 - 5x + 6$. (Lesson 4-3)

8. List the possible rational roots of $x^3 + 6x^2 + 10x + 3 = 0$. Then determine the rational roots. (Lesson 4-4)

9. Find the number of possible positive zeros and the number of possible negative zeros for $F(x) = x^4 + 4x^3 + 3x^2 - 4x - 4$. Then determine the rational zeros. (Lesson 4-4)

10. **Manufacturing** The Universal Paper Product Company makes cone-shaped drinking cups. The height of each cup is 6 centimeters more than the radius. If the volume of each cup is 27π cubic centimeters, find the dimensions of the cup. (Lesson 4-4)

Extra Practice See p. A32.

4-5 Locating Zeros of a Polynomial Function

OBJECTIVE
- Approximate the real zeros of a polynomial function.

ECONOMY Layoffs at large corporations can cause the unemployment rate to increase, while low interest rates can bolster employment. From October 1997 to November 1998, the Texas economy was strong. The Texas jobless rate during that period can be modeled by the function $f(x) = -0.0003x^4 + 0.0066x^3 - 0.0257x^2 - 0.1345x + 5.35$, where x represents the number of months since October 1997 and $f(x)$ represents the unemployment rate as a percent. Use this model to predict when the unemployment will be 2.5%. *This problem will be solved in Example 4.*

The function $f(x) = -0.0003x^4 + 0.0066x^3 - 0.0257x^2 - 0.1345x + 5.35$ has four complex zeros. According to the Descartes' Rule of Signs, there are three or one positive real zeros and one negative zero. If you used a spreadsheet to evaluate the possible rational zeros, you will discover that none of the possible values is a zero of the function. This means that the zeros are not rational numbers. Another method, called the **Location Principle,** can be used to help determine the zeros of a function.

The Location Principle

Suppose $y = f(x)$ represents a polynomial function with real coefficients. If a and b are two numbers with $f(a)$ negative and $f(b)$ positive, the function has at least one real zero between a and b.

If $f(a) > 0$ and $f(b) < 0$, then the function also has at least one real zero between a and b.

This principle is illustrated by the graph. The graph of $y = f(x)$ is a continuous curve. At $x = a$, $f(a)$ is negative. At $x = b$, $f(b)$ is positive. Therefore, between the x-values of a and b, the graph must cross the x-axis. Thus, a zero exists somewhere between a and b.

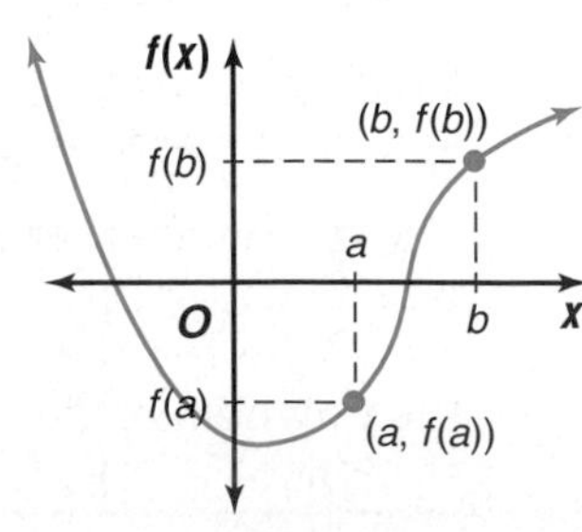

Example 1 **Determine between which consecutive integers the real zeros of $f(x) = x^3 - 4x^2 - 2x + 8$ are located.**

There are three complex zeros for this function. According to Descartes' Rule of Signs, there are two or zero positive real roots and one negative real root. You can use substitution, synthetic division, or the **TABLE** feature on a graphing calculator to evaluate the function for consecutive integral values of x.

Method 1: Synthetic Division

r	1	−4	−2	8	
−3	1	−7	19	−49	
−2	1	−6	10	−12	change in signs
−1	1	−5	3	5	
0	1	−4	−2	8	
1	1	−3	−5	3	change in signs
2	1	−2	−6	−4	
3	1	−1	−5	−7	
4	1	0	−2	0	zero
5	1	1	3	23	

Method 2: Graphing Calculator

Use the **TABLE** feature.

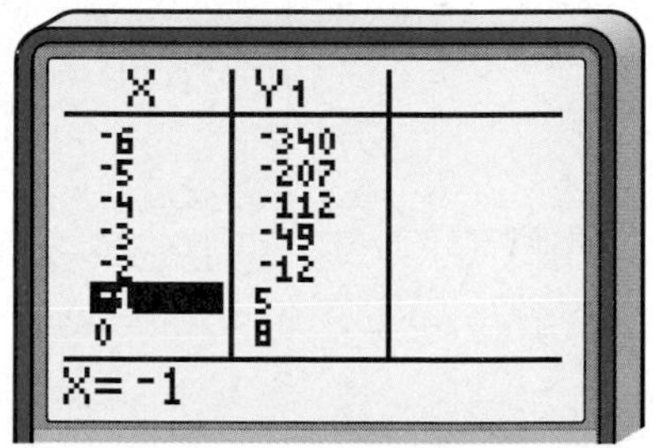

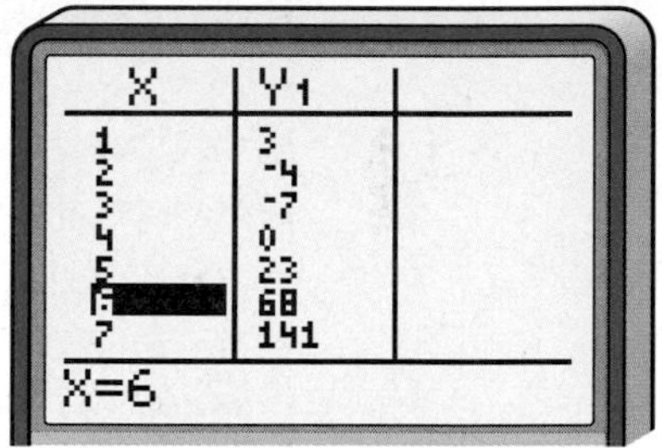

There is a zero at 4. The changes in sign indicate that there are also zeros between −2 and −1 and between 1 and 2. This result is consistent with the Descartes' Rule of Signs.

Once you know two integers between which a zero will fall, you can use substitution or a graphing calculator to approximate the zeros.

Example 2 **Approximate the real zeros of $f(x) = 12x^3 - 19x^2 - x + 6$ to the nearest tenth.**

There are three complex zeros for this function. According to Descartes' Rule of Signs, there are two or zero positive real roots and one negative real root.

Use the **TABLE** feature of a graphing calculator. There are zeros between −1 and 0, between 0 and 1, and between 1 and 2. To find the zeros to the nearest tenth, use the **TBLSET** feature changing **ΔTbl** to 0.1.

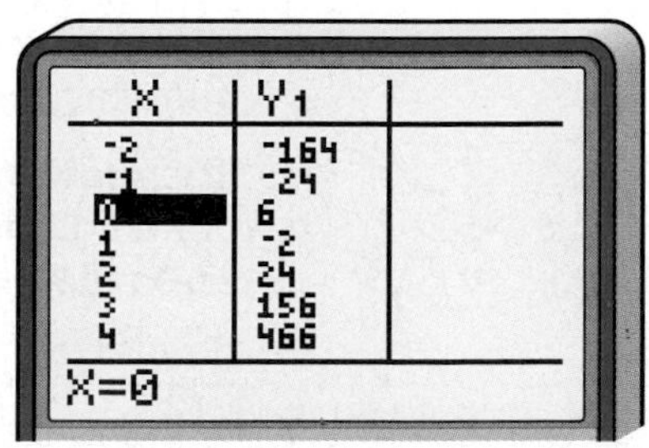

(continued on the next page)

Graphing Calculator Tip

You can check these zeros by graphing the function and selecting **zero** under the **CALC** menu.

Since 0.25 is closer to zero than −2.832, the zero is about −0.5.

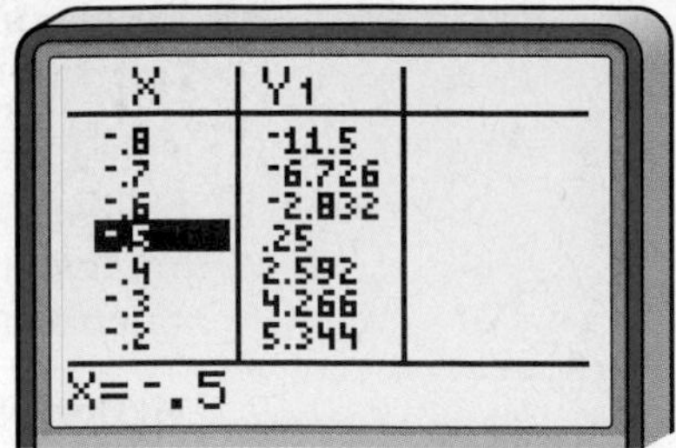

X	Y_1
-.8	-11.5
-.7	-6.726
-.6	-2.832
-.5	.25
-.4	2.592
-.3	4.266
-.2	5.344

X=-.5

Since 0.106 is closer to zero than −0.816, the zero is about 0.7.

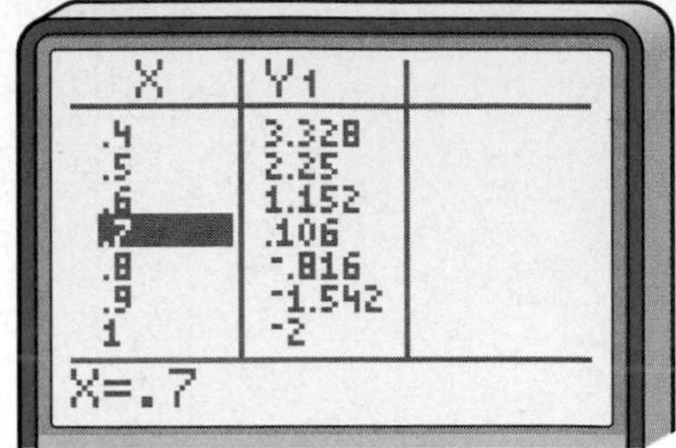

X	Y_1
.4	3.328
.5	2.25
.6	1.152
.7	.106
.8	-.816
.9	-1.542
1	-2

X=.7

Since 0.288 is closer to zero than −1.046, the zero is about 1.4.

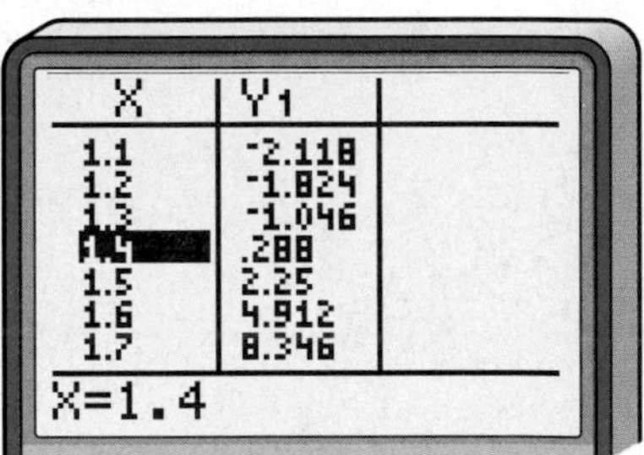

X	Y_1
1.1	-2.118
1.2	-1.824
1.3	-1.046
1.4	.288
1.5	2.25
1.6	4.912
1.7	8.346

X=1.4

The zeros are about −0.5, 0.7, and 1.4.
If you need a closer approximation, change ΔTbl to 0.01.

The **Upper Bound Theorem** will help you confirm whether you have determined all of the real zeros. An **upper bound** is an integer greater than or equal to the greatest real zero.

Upper Bound Theorem

Suppose c is a positive real number and $P(x)$ is divided by $x - c$. If the resulting quotient and remainder have no change in sign, then $P(x)$ has no real zero greater than c. Thus, c is an upper bound of the zeros of $P(x)$.

Zero coefficients are ignored when counting sign changes.

The synthetic division in Example 1 indicates that 5 is an upper bound of the zeros of $f(x) = x^3 - 4x^2 - 2x + 8$ because there are no change of signs in the quotient and remainder.

A **lower bound** is an integer less than or equal to the least real zero. A lower bound of the zeros of $P(x)$ can be found by determining an upper bound for the zeros of $P(-x)$.

Lower Bound Theorem

If c is an upper bound of the zeros of $P(-x)$, then $-c$ is a lower bound of the zeros of $P(x)$.

Examples

3 **Use the Upper Bound Theorem to find an integral upper bound and the Lower Bound Theorem to find an integral lower bound of the zeros of $f(x) = x^3 + 3x^2 - 5x - 10$.**

The Rational Root Theorem tells us that ± 1, ± 2, ± 5, and ± 10 might be roots of the polynomial equation $x^3 + 3x^2 - 5x - 10 = 0$. These possible zeros of the function are good starting places for finding an upper bound.

$f(x) = x^3 + 3x^2 - 5x - 10$				
r	1	3	−5	−10
1	1	4	−1	−11
2	1	5	5	0

$f(-x) = -x^3 + 3x^2 + 5x - 10$				
r	−1	3	5	−10
1	−1	2	7	−3
2	−1	1	7	4
3	−1	0	5	5
4	−1	−1	1	−6
5	−1	−2	−5	−35

An upper bound is 2. Since 5 is an upper bound of $f(-x)$, -5 is a lower bound of $f(x)$. This means that all real zeros of $f(x)$ can be found in the interval $-5 \leq x \leq 2$.

4 **ECONOMICS** **Refer to the application at the beginning of the lesson. Use the model to determine when the unemployment will be 2.5%.**

You need to know when $f(x)$ has a value of 2.5.

$$2.5 = -0.0003x^4 + 0.0066x^3 - 0.0257x^2 - 0.1345x + 5.35$$
$$0 = -0.0003x^4 + 0.0066x^3 - 0.0257x^2 - 0.1345x + 2.85$$

Now search for the zero of the related function,
$g(x) = -0.0003x^4 + 0.0066x^3 - 0.0257x^2 - 0.1345x + 2.85$

r	−0.0003	0.0066	−0.0257	−0.1345	2.85
17	−0.0003	0.0015	−0.0002	−0.1379	0.5057
18	−0.0003	0.0012	−0.0041	−0.2083	−0.8994

There is a zero between 17 and 18 months.

Confirm this zero using a graphing calculator. The zero is about 17.4 months. So, about 17 months after October, 1997 or March, 1999, the unemployment rate would be 2.5%. *Equations dealing with unemployment change as the economic conditions change. Therefore, long range predictions may not be accurate.*

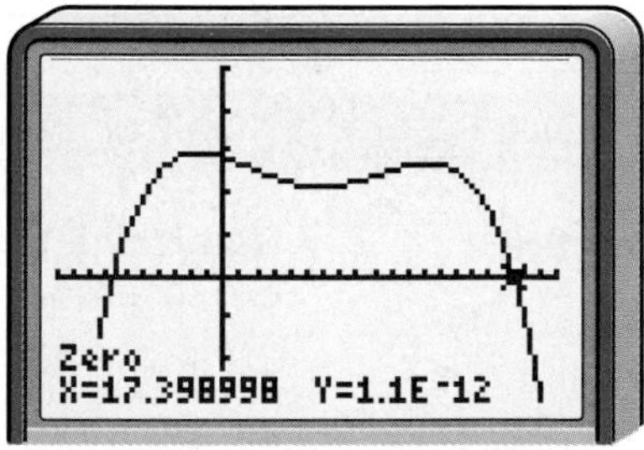

[−10, 20] scl:1 by [−3, 5] scl:1

CHECK FOR UNDERSTANDING

Communicating Mathematics

Read and study the lesson to answer each question.

1. **Write a convincing argument** that the Location Principle works. Include a labeled graph.

2. **Explain** how to use synthetic division to determine which consecutive integers the real zeros of a polynomial function are located.

3. **Describe** how you find an upper bound and a lower bound for the zeros of a polynomial function.

4. **You Decide** After looking at the table on a graphing calculator, Tiffany tells Nikki that there is a zero between -1 and 0. Nikki argues that the zero is between -2 and -1. Who is correct? Explain.

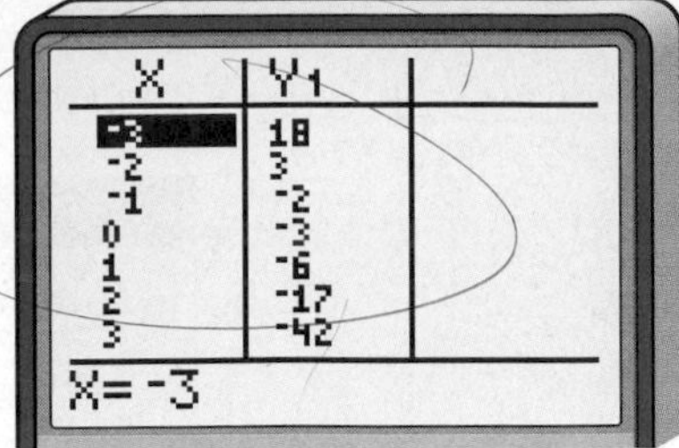

Guided Practice

Determine between which consecutive integers the real zeros of each function are located.

5. $f(x) = x^2 - 4x - 2$

6. $f(x) = x^3 - 3x^2 - 2x + 4$

Approximate the real zeros of each function to the nearest tenth.

7. $f(x) = 2x^3 - 4x^2 - 3$

8. $f(x) = x^2 + 3x + 2$

Use the Upper Bound Theorem to find an integral upper bound and the Lower Bound Theorem to find an integral lower bound of the zeros of each function.

9. $f(x) = x^4 - 8x + 2$

10. $f(x) = x^4 + x^2 - 3$

11. **Manufacturing** The It's A Snap Puzzle Company is designing new boxes for their 2000 piece 3-D puzzles. The old box measured 25 centimeters by 30 centimeters by 5 centimeters. For the new box, the designer wants to increase each dimension by a uniform amount.
    a. Write a polynomial function that models the volume of the new box.
    b. The volume of the new box must be 1.5 times the volume of the old box to hold the increase in puzzle pieces. Write an equation that models this situation.
    c. Find the dimensions of the new box.

EXERCISES

Practice

Determine between which consecutive integers the real zeros of each function are located.

12. $f(x) = x^3 - 2$

13. $f(x) = 2x^2 - 5x + 1$

14. $f(x) = x^4 - 2x^3 + x - 2$

15. $f(x) = x^4 - 8x^2 + 10$

16. $f(x) = x^3 - 3x + 1$

17. $f(x) = 2x^4 + x^2 - 3x + 3$

18. Is there a zero of $f(x) = 6x^3 + 24x^2 - 54x - 3$ between -6 and -5? Explain.

Approximate the real zeros of each function to the nearest tenth.

19. $f(x) = 3x^4 + x^2 - 1$

20. $f(x) = x^2 + 3x + 1$

21. $f(x) = x^3 - 4x + 6$

22. $f(x) = x^4 - 5x^3 + 6x^2 - x - 2$

23. $f(x) = 2x^4 - x^3 + x - 2$

24. $f(x) = x^5 - 7x^4 - 3x^3 + 2x^2 - 4x + 9$

25. Approximate the real zero of $f(x) = x^3 - 2x^2 + 5$ to the nearest hundredth.

Use the Upper Bound Theorem to find an integral upper bound and the Lower Bound Theorem to find an integral lower bound of the zeros of each function.

26. $f(x) = 3x^3 - 2x^2 + 5x - 1$

27. $f(x) = x^2 - x - 1$

28. $f(x) = x^4 - 6x^3 + 2x^2 + 6x - 13$

29. $f(x) = x^3 + 5x^2 - 3x - 20$

30. $f(x) = x^4 - 3x^3 - 2x^2 + 3x - 5$

31. $f(x) = x^5 + 5x^4 - 3x^3 + 20x^2 - 15$

32. Analyze the zeros of $f(x) = x^4 - 3x^3 - 2x^2 + 3x - 5$.

a. Determine the number of complex zeros.

b. List the possible rational zeros.

c. Determine the number of possible positive real zeros and the number of possible negative real zeros.

d. Determine the integral intervals where the zeros are located.

e. Determine an integral upper bound of the zeros and an integral lower bound of the zeros.

f. Determine the zeros to the nearest tenth.

Applications and Problem Solving

33. Population The population of Manhattan Island in New York City between 1890 to 1970 can be modeled by $P(x) = -0.78x^4 + 133x^3 - 7500x^2 + 147{,}500x + 1{,}440{,}000$, where $P(x)$ represents the population and x represents the number of years since 1890.

Population of Manhattan Island

Year	Population
1890	1,441,216
1910	2,331,542
1930	1,867,312
1950	1,960,101
1970	1,539,233

Source: U.S. Bureau of the Census

a. According to the data at the right, how valid is the model?

b. Use the model to predict the population in 1980.

c. According to this model, what happens between 1970 and 1980?

d. Do you think this model is valid for any time? Explain.

34. Critical Thinking Write a third-degree integral polynomial function with one zero at $\sqrt{2}$. State the zeros of the function. Draw a graph to support your answer.

35. Medicine A doctor tells Masa to take 60 milligrams of medication each morning for three days. The amount of medication remaining in his body on the fourth day is modeled by $M(x) = 60x^3 + 60x^2 + 60x$, where x represents the absorption rate per day. Suppose Masa has 37.44 milligrams of medication in his body on the fourth day.

a. Write an equation to model this situation.

b. Write the related function for the equation.

c. Graph this function and estimate the absorption rate.

d. Find the absorption rate of the medication.

36. Critical Thinking Write a polynomial function with an upper bound of 1 and a lower bound of -1.

37. **Ecology** In the early 1900s, the deer population of the Kaibab Plateau in Arizona experienced a rapid increase because hunters had reduced the number of natural predators. The food supply was not great enough to support the increased population, eventually causing the population to decline. The deer population for the period 1905 to 1930 can be modeled by $f(x) = -0.125x^5 + 3.125x^4 + 4000$, where x is the number of years from 1905.

a. Graph the function.

b. Use the model to determine the population in 1905.

c. Use the model to determine the population in 1920.

d. According to this model, when did the deer population become zero?

38. **Investments** Instead of investing in the stock market, many people invest in collectibles, like baseball cards. Each year, Anna uses some of the money she receives for her birthday to buy one special baseball card. For the last four birthdays, she purchased cards for \$6, \$18, \$24, and \$18. The current value of these cards is modeled by $T(x) = 6x^4 + 18x^3 + 24x^2 + 18x$, where x represents the average rate of return plus one.

a. If the cards are worth \$81.58, write an equation to model this situation.

b. Find the value of x.

c. What is the average rate of return on Anna's investments?

Mixed Review

39. Find the number of possible positive real zeros and the number of possible negative real zeros for $f(x) = 2x^3 - 5x^2 - 28x + 15$. Then determine the rational zeros. *(Lesson 4-4)*

40. **Physics** The distance $d(t)$ fallen by a free-falling body can be modeled by the formula $d(t) = v_0t - \frac{1}{2}gt^2$, where v_0 is the initial velocity and g represents the acceleration due to gravity. The acceleration due to gravity is 9.8 meters per second squared. If a rock is thrown upward with an initial velocity of +4 meters per second from the edge of the North Rim of the Grand Canyon, which is 1750 meters deep, determine how long it will take the rock to reach the bottom of the Grand Canyon. (*Hint:* The distance to the bottom of the canyon is −1750 meters from the rim.) *(Lesson 4-2)*

41. Graph $y = \frac{4x}{x - 1}$. *(Lesson 3-7)*

42. Find the value of $\begin{vmatrix} 7 & 9 \\ 3 & 6 \end{vmatrix}$. *(Lesson 2-5)*

43. Find the coordinates of the midpoint of $\overline{FG}$ given its endpoints $F(-3, -2)$ and $G(8, 4)$. *(Lesson 1-5)*

44. Name the slope and y-intercept of the graph of $x - 2y - 4 = 0$. *(Lesson 1-4)*

45. **SAT/ACT Practice** $\triangle ABC$ and $\triangle ABD$ are right triangles that share side $\overline{AB}$. $\triangle ABC$ has area x, and $\triangle ABD$ has area y. If $\overline{AD}$ is longer than $\overline{AC}$ and $\overline{BD}$ is longer than $\overline{BC}$, which of the following cannot be true?

A $y > x$ **B** $y < x$ **C** $y \geq x$ **D** $y \neq x$ **E** $\frac{y}{x} \geq 1$

Extra Practice See p. A33.

4-6 Rational Equations and Partial Fractions

OBJECTIVES

- Solve rational equations and inequalities.
- Decompose a fraction into partial fractions.

SCUBA DIVING If a scuba diver goes to depths greater than 33 feet, the function $T(d) = \frac{1700}{d - 33}$ gives the maximum time a diver can remain down and still surface at a steady rate with no decompression stops. In this function, $T(d)$ represents the dive time in minutes, and d represents the depth in feet. If a diver is planning a 45-minute dive, what is the maximum depth the diver can go without decompression stops on the way back up? *This problem will be solved in Example 1.*

To solve this problem, you need to solve the equation $45 = \frac{1700}{d - 33}$. This type of equation is called a **rational equation.** A rational equation has one or more rational expressions. One way to solve a rational equation is to multiply each side of the equation by the least common denominator (LCD).

Example 1 **SCUBA DIVING** **What is the maximum depth the diver can go without decompression stops on the way back up?**

$$T(d) = \frac{1700}{d - 33}$$

$$45 = \frac{1700}{d - 33}$$ *Replace T(d) with 45, the dive time in minutes.*

$$45(d - 33) = \frac{1700}{d - 33}(d - 33)$$ *Multiply each side by the LCD, d − 33.*

$$45d - 1485 = 1700$$

$$45d = 3185$$

$$d \approx 70.78$$

The diver can go to a depth of about 70 feet and surface without decompression stops.

Any possible solution that results in a zero in the denominator must be excluded from your list of solutions. So, in Example 1, the solution could not be 33. Always check your solutions by substituting them into the original equation.

Example 2 **Solve $a + \frac{a^2 - 5}{a^2 - 1} = \frac{a^2 + a + 2}{a + 1}$.**

$$a + \frac{a^2 - 5}{a^2 - 1} = \frac{a^2 + a + 2}{a + 1}$$

$$\left(a + \frac{a^2 - 5}{a^2 - 1}\right)(a^2 - 1) = \left(\frac{a^2 + a + 2}{\cancel{a + 1}_{\,1}}\right)\cancel{(a^2 - 1)}^{\,a - 1}$$ *Multiply each side by the LCD, $(a - 1)(a + 1)$ or $a^2 - 1$.*

$$a(a^2 - 1) + (a^2 - 5) = (a^2 + a + 2)(a - 1)$$
$$a^3 - a + a^2 - 5 = a^3 + a - 2$$
$$a^2 - 2a - 3 = 0$$
$$(a - 3)(a + 1) = 0$$ *Factor.*

$$a - 3 = 0 \qquad a + 1 = 0$$
$$a = 3 \qquad a = -1$$

When you check your solutions, you find that a cannot equal -1 because a zero denominator results. Since -1 cannot be a solution, the only solution is 3.

In order to add or subtract fractions with unlike denominators, you must first find a common denominator. Suppose you have a rational expression and you want to know what fractions were added or subtracted to obtain that expression. Finding these fractions is called decomposing the fraction into **partial fractions.** Decomposing fractions is a skill used in calculus and other advanced mathematics courses.

Example 3 **Decompose $\frac{8y + 7}{y^2 + y - 2}$ into partial fractions.**

First factor the denominator. $\qquad y^2 + y - 2 = (y - 1)(y + 2)$

Express the factored form as the sum of two fractions using A and B as numerators and the factors as denominators.

$$\frac{8y + 7}{y^2 + y - 2} = \frac{A}{y - 1} + \frac{B}{y + 2}$$

Eliminate the denominators by multiplying each side by the LCD, $(y - 1)(y - 2)$.

$$8y + 7 = A(y + 2) + B(y - 1)$$

Eliminate B by letting $y = 1$ so that $y - 1$ becomes 0.

$$8y + 7 = A(y + 2) + B(y - 1)$$
$$8(1) + 7 = A(1 + 2) + B(1 - 1)$$
$$15 = 3A$$
$$5 = A$$

Eliminate A by letting $y = -2$ so that $y + 2$ becomes 0.

$$8y + 7 = A(y + 2) + B(y - 1)$$
$$8(-2) + 7 = A(-2 + 2) + B(-2 - 1)$$
$$9 = -3B$$
$$3 = B$$

Now substitute the values for A and B to determine the partial fractions.

$$\frac{A}{y - 1} + \frac{B}{y + 2} = \frac{5}{y - 1} + \frac{3}{y + 2}$$

So, $\frac{8y + 7}{y^2 + y - 2} = \frac{5}{y - 1} + \frac{3}{y + 2}$.

Check to see if the sum of the two partial fractions equals the original fraction.

The process used to solve rational equations can be used to solve **rational inequalities.**

Example 4 **Solve $\frac{(x-2)(x-1)}{(x-3)(x-4)^2} < 0$.**

Let $f(x) = \frac{(x-2)(x-1)}{(x-3)(x-4)^2}$. On a number line, mark the zeros of $f(x)$ and the excluded values for $f(x)$ with vertical dashed lines. The zeros of $f(x)$ are the same values that make $(x-2)(x-1) = 0$. These zeros are 2 and 1. Excluded values for $f(x)$ are the values that make $(x-3)(x-4)^2 = 0$. These excluded values are 3 and 4.

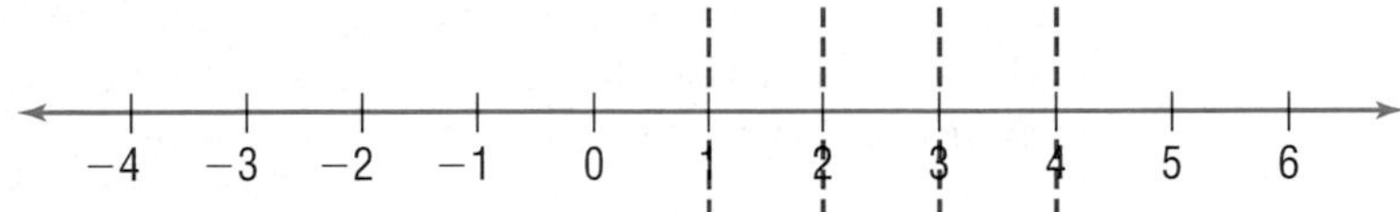

The vertical dashed lines separate the number line into intervals. Test a convenient value within each interval in the original rational inequality to see if the test value is a solution. If the value in the interval is a solution, all values are solutions. In this problem, it is not necessary to find the exact value of the expression.

For $x < 1$, test $x = 0$: $\frac{(0-2)(0-1)}{(0-3)(0-4)^2} \rightarrow \frac{(-)(-)}{(-)(-)(-)} \rightarrow -$

So in the interval $x < 1$, $f(x) < 0$. Thus, $x < 1$ is a solution.

For $1 < x < 2$, test $x = 1.5$: $\frac{(1.5-2)(1.5-1)}{(1.5-3)(1.5-4)^2} \rightarrow \frac{(-)(+)}{(-)(-)(-)} \rightarrow +$

So in the interval $1 < x < 2$, $f(x) > 0$. Thus, $1 < x < 2$ is not a solution.

For $2 < x < 3$, test $x = 2.5$: $\frac{(2.5-2)(2.5-1)}{(2.5-3)(2.5-4)^2} \rightarrow \frac{(+)(+)}{(-)(-)(-)} \rightarrow -$

So in the interval $2 < x < 3$, $f(x) < 0$. Thus, $2 < x < 3$ is a solution.

For $3 < x < 4$, test $x = 3.5$: $\frac{(3.5-2)(3.5-1)}{(3.5-3)(3.5-4)^2} \rightarrow \frac{(+)(+)}{(+)(-)(-)} \rightarrow +$

So in the interval $3 < x < 4$, $f(x) > 0$. Thus, $3 < x < 4$ is not a solution.

For $4 < x$, test $x = 5$: $\frac{(5-2)(5-1)}{(5-3)(5-4)^2} \rightarrow \frac{(+)(+)}{(+)(+)(+)} \rightarrow +$

So in the interval $x > 4$, $f(x) > 0$. Thus, $4 < x$ is not a solution.

The solution is $x < 1$ or $2 < x < 3$. This solution can be graphed on a number line.

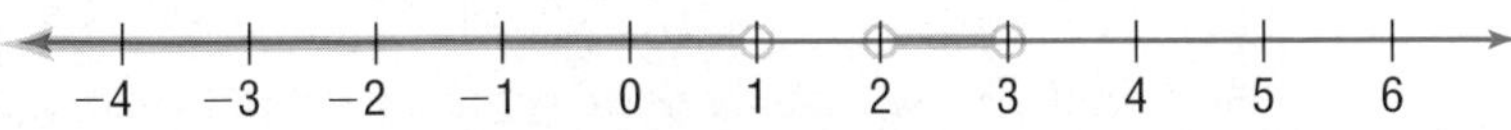

Example 5 Solve $\frac{2}{3a} + \frac{5}{6a} > \frac{3}{4}$.

The inequality can be written as $\frac{2}{3a} + \frac{5}{6a} - \frac{3}{4} > 0$. The related function is $f(a) = \frac{2}{3a} + \frac{5}{6a} - \frac{3}{4}$. Find the zeros of this function.

$$\frac{2}{3a} + \frac{5}{6a} - \frac{3}{4} = 0$$

$$\frac{2}{3a}(12a) + \frac{5}{6a}(12a) - \frac{3}{4}(12a) = 0(12a) \quad \textit{The LCD is 12a.}$$

$$8 + 10 - 9a = 0$$

$$2 = a$$

The zero is 2. The excluded value is 0. On a number line, mark these values with vertical dashed lines. The vertical dashed lines separate the number line into intervals.

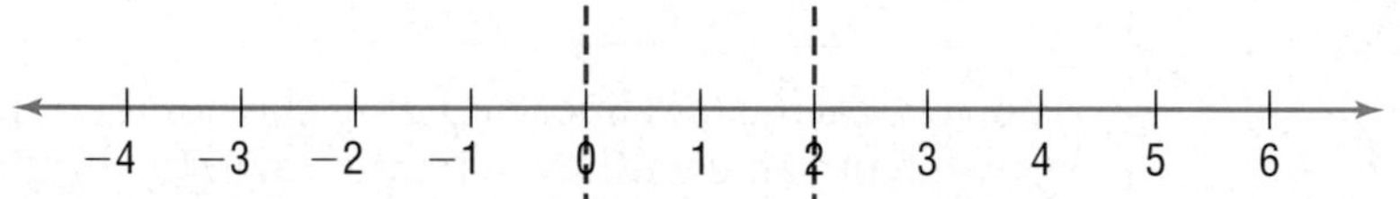

Now test a sample value in each interval to determine if the values in the interval satisfy the inequality.

For $a < 0$, test $x = -1$:

$$\frac{2}{3(-1)} + \frac{5}{6(-1)} \overset{?}{>} \frac{3}{4}$$

$$-\frac{2}{3} - \frac{5}{6} \overset{?}{>} \frac{3}{4}$$

$$-\frac{3}{2} \not> \frac{3}{4} \quad \textit{a < 0 is not a solution.}$$

For $0 < a < 2$, test $x = 1$:

$$\frac{2}{3(1)} + \frac{5}{6(1)} \overset{?}{>} \frac{3}{4}$$

$$\frac{2}{3} + \frac{5}{6} \overset{?}{>} \frac{3}{4}$$

$$\frac{3}{2} > \frac{3}{4} \quad \textit{0 < a < 2 is a solution.}$$

For $2 < a$, test $x = 3$:

$$\frac{2}{3(3)} + \frac{5}{6(3)} \overset{?}{>} \frac{3}{4}$$

$$\frac{2}{9} + \frac{5}{18} \overset{?}{>} \frac{3}{4}$$

$$\frac{1}{2} \not> \frac{3}{4} \quad \textit{2 < x is not a solution.}$$

The solution is $0 < a < 2$. This solution can be graphed on a number line.

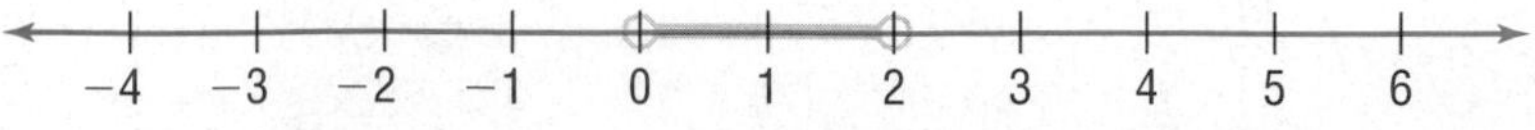

CHECK FOR UNDERSTANDING

Communicating Mathematics

Read and study the lesson to answer each question.

1. **Describe** the process used to solve $\frac{b+1}{3(b-2)} = \frac{5b}{6} + \frac{1}{b-2}$.
2. **Write** a sentence explaining why all solutions of a rational equation must be checked.
3. **Explain** what is meant by decomposing a fraction into partial fractions.
4. **Explain** why $x + \frac{2}{x-2} = 2 + \frac{2}{x-2}$ has no solution.

Guided Practice

Solve each equation.

5. $a - \frac{5}{a} = 4$
6. $\frac{9}{b+5} = \frac{3}{b-3}$
7. $\frac{t+4}{t} + \frac{3}{t-4} = \frac{-16}{t^2-4t}$
8. Decompose $\frac{3p-1}{p^2-1}$ into partial fractions.

Solve each inequality.

9. $5 + \frac{1}{x} > \frac{16}{x}$
10. $1 + \frac{5}{a-1} \leq \frac{7}{6}$
11. **Interstate Commerce** When truckers are on long-haul drives, their driving logs must reflect their average speed. Average speed is the total distance driven divided by the total time spent driving. A trucker drove 3 hours on a freeway at 60 miles per hour and then drove 20 miles in the city. The trucker's average speed was 57.14 miles per hour.
    a. Write an equation that models the situation.
    b. How long was the trucker driving in the city to the nearest hundredth of an hour?

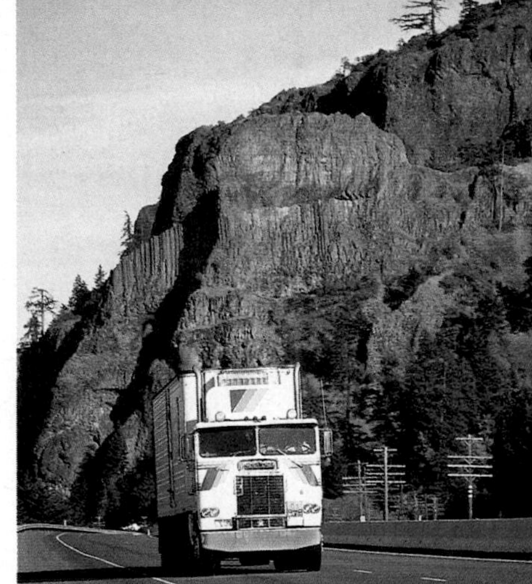

EXERCISES

Practice

Solve each equation.

12. $\frac{12}{t} + t - 8 = 0$
13. $\frac{1}{m} = \frac{m-34}{2m^2}$
14. $\frac{2}{y+2} + \frac{3}{y} = \frac{-y}{y+2}$
15. $\frac{10}{n^2-1} + \frac{2n-5}{n-1} = \frac{2n+5}{n+1}$
16. $\frac{1}{b+2} + \frac{1}{b+2} = \frac{3}{b+1}$
17. $\frac{7a}{3a+3} - \frac{5}{4a-4} = \frac{3a}{2a+2}$
18. $1 = \frac{1}{1-a} + \frac{a}{a-1}$
19. $\frac{2q}{2q+3} - \frac{2q}{2q-3} = 1$
20. $\frac{1}{3m} + \frac{6m-9}{3m} = \frac{3m-3}{4m}$
21. $\frac{-4}{x-1} = \frac{7}{2-x} + \frac{3}{x+1}$

22. Consider the equation $1 + \frac{n + 6}{n + 1} = \frac{4}{n - 2}$.

a. What is the LCD of the rational expressions?

b. What values must be excluded from the list of possible solutions.

c. What is the solution of the equation?

Decompose each expression into partial fractions.

23. $\frac{x - 6}{x^2 - 2x}$

24. $\frac{5m - 4}{m^2 - 4}$

25. $\frac{-4y}{3y^2 - 4y + 1}$

26. Find two rational expressions that have a sum of $\frac{9 - 9x}{x^2 - 9}$.

27. Consider the inequality $\frac{a - 2}{a} < \frac{a - 4}{a - 6}$.

a. What is the LCD of the rational expressions?

b. Find the zero(s) of the related function.

c. Find the excluded value(s) of the related function.

d. Solve the inequality.

Solve each inequality.

28. $\frac{2}{w} + 3 > \frac{29}{w}$

29. $\frac{(x - 3)(x - 4)}{(x - 5)(x - 6)^2} \leq 0$

30. $\frac{x^2 - 16}{x^2 - 4x - 5} \geq 0$

31. $\frac{1}{4a} + \frac{5}{8a} > \frac{1}{2}$

32. $\frac{1}{2b + 1} + \frac{1}{b + 1} > \frac{8}{15}$

33. $\frac{7}{y + 1} > 7$

34. Four times the multiplicative inverse of a number is added to the number. The result is $10\frac{2}{5}$. What is the number?

35. The ratio of $x + 2$ to $x - 5$ is greater than 30%. Solve for x.

Applications and Problem Solving

36. **Optics** The lens equation is $\frac{1}{f} = \frac{1}{d_i} + \frac{1}{d_o}$, where f is the focal length, d_i is the distance from the lens to the image, and d_o is the distance from the lens to the object. Suppose the object is 32 centimeters from the lens and the focal length is 8 centimeters.

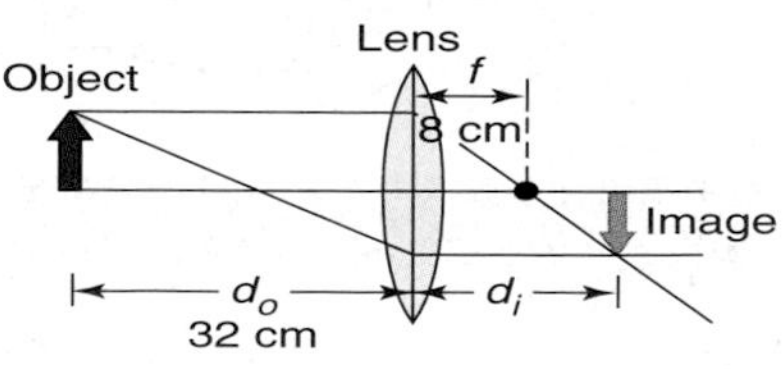

a. Write a rational equation to model the situation.

b. Find the distance from the lens to the image.

37. **Critical Thinking** Write a rational equation that cannot have 3 or −2 as a solution.

38. **Trucking** Two trucks can carry loads of coal in a ratio of 5 to 2. The smaller truck has a capacity 3 tons less than that of the larger truck. What is the capacity of the larger truck?

39. **Electricity** The diagram of an electric circuit shows three parallel resistors. If R represents the equivalent resistance of the three resistors, then $\frac{1}{R} = \frac{1}{R_1} + \frac{1}{R_2} + \frac{1}{R_3}$. In this circuit, R_1 represents twice the resistance of R_2, and R_3 equals 20 ohms. Suppose the equivalent resistance equals 10 ohms.

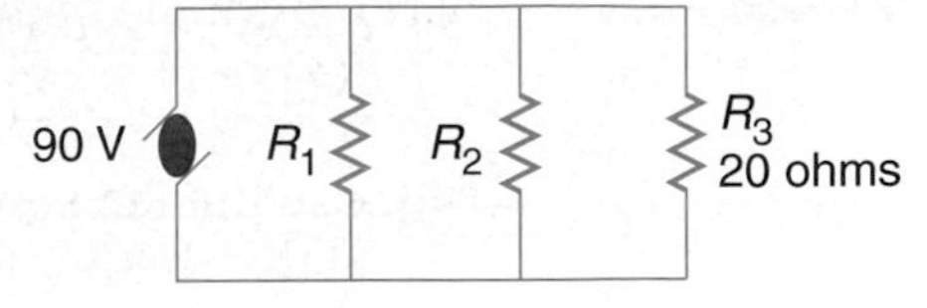

a. Write a rational equation to model the situation.

b. Find R_1 and R_2.

40. **Education** Todd has answered 11 of his last 20 daily quiz questions correctly. His baseball coach told him that he must bring his average up to at least 70% if he wants to play in the season opener. Todd vows to study diligently and answer all of the daily quiz questions correctly in the future. How many consecutive daily quiz questions must he answer correctly to bring his average up to 70%?

41. **Aviation** An aircraft flies 1062 miles with the wind at its tail. In the same amount of time, a similar aircraft flies against the wind 738 miles. If the air speed of each plane is 200 miles per hour, what is the speed of the wind? (*Hint:* Time equals the distance divided by the speed.)

42. **Critical Thinking** Solve for a if $\frac{1}{a} + \frac{1}{b} = \frac{1}{c}$.

43. **Statistics** A number x is said to be the *harmonic mean* of y and z if $\frac{1}{x}$ is the average of $\frac{1}{y}$ and $\frac{1}{z}$.

a. Write an equation whose solution is the harmonic mean of 30 and 45.

b. Find the harmonic mean of 30 and 45.

44. **Economics** Darnell drives about 15,000 miles each year. He is planning to buy a new car. The car he wants to buy averages 20 miles on one gallon of gasoline. He has decided he would buy another car if he could save at least $200 per year in gasoline expenses. Assume gasoline costs $1.20 per gallon. What is the minimum number of miles per gallon that would fulfill Darnell's criteria?

45. **Navigation** The speed of the current in Puget Sound is 5 miles per hour. A barge travels with the current 26 miles and returns in $10\frac{2}{3}$ hours. What is its speed in still water?

46. **Critical Thinking** If $\frac{3x}{5y} = 11$, find the value of $\frac{3x - 5y}{5y}$.

Puget Sound

Mixed Review

47. Determine between which consecutive integers the real zeros of $f(x) = x^3 + 2x^2 - 3x - 5$ are located. *(Lesson 4-5)*

48. Use the Remainder Theorem to find the remainder for $(x^3 - 30x) \div (x + 5)$. State whether the binomial is a factor of the polynomial. *(Lesson 4-3)*

49. State the number of complex roots of $12x^2 + 8x - 15 = 0$. Then find the roots. *(Lesson 4-1)*

50. Name all the values of x that are not in the domain of the function $f(x) = \frac{x}{|3x| - 12}$. *(Lesson 3-7)*

51. Determine if (6, 3) is a solution for $y \leq \frac{2x + 3}{x}$. *(Lesson 3-3)*

52. Determine if the graph of $y^2 = 121x^2$ is symmetric with respect to each line. *(Lesson 3-1)*

    a. the x-axis
    b. the y-axis
    c. the line $y = x$
    d. the line $y = -x$

53. **Education** The semester test in your English class consists of short answer and essay questions. Each short answer question is worth 5 points, and each essay question is worth 15 points. You may choose up to 20 questions of any type to answer. It takes 2 minutes to answer each short answer question and 12 minutes to answer each essay question. *(Lesson 2-7)*

    a. You have one hour to complete the test. Assuming that you answer all of the questions that you attempt correctly, how many of each type should you answer to earn the highest score?
    b. You have two hours to complete the test. Assuming that you answer all of the questions that you attempt correctly, how many of each type should you answer to earn the highest score?

54. Find matrix X in the equation $\begin{bmatrix} 1 & 1 \\ 1 & 1 \end{bmatrix}\begin{bmatrix} 3 & 5 \\ -3 & -5 \end{bmatrix} = X$. *(Lesson 2-3)*

55. Write the standard form of the equation of the line that is parallel to the graph of $y = 2x - 10$ and passes through the point at $(-3, 1)$. *(Lesson 1-5)*

56. **Manufacturing** It costs ABC Corporation \$3000 to produce 20 color televisions and \$5000 to produce 60 of the same color televisions. *(Lesson 1-4)*

    a. Find the cost function.
    b. Determine the fixed cost and the variable cost per unit.
    c. Sketch the graph of the cost function.

57. **SAT Practice** **Grid-In** Find the area of the shaded region in square inches.

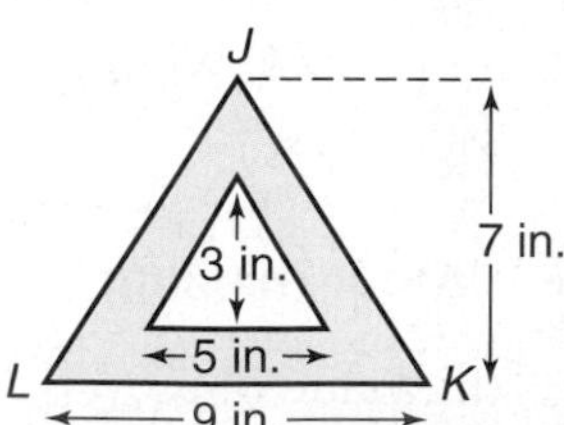

Extra Practice See p. A33.

4-7 Radical Equations and Inequalities

OBJECTIVE

- Solve radical equations and inequalities.

RECREATION A pogo stick stores energy in a spring. When a jumper compresses the spring from its natural length x_0 to a length x, the maximum height h reached by the bottom of the pogo stick is related to x_0 and x by the equation $x_0 = x + \sqrt{\frac{2mgh}{k}}$, where m is the combined mass of the jumper and the pogo stick, g is the acceleration due to gravity (9.80 meters per second squared), and k is a constant that depends on the spring. If the combined mass of the jumper and the stick is 50 kilograms, the spring compressed from its natural length of 1 meter to the length of 0.9 meter. If $k = 1.2 \times 10^4$, find the maximum height reached by the bottom of the pogo stick. *This problem will be solved in Example 1.*

Equations in which radical expressions include variables, such as the equation above, are known as **radical equations.** To solve radical equations, the first step is to isolate the radical on one side of the equation. Then raise each side of the equation to the proper power to eliminate the radical expression.

The process of raising each side of an equation to a power sometimes produces **extraneous solutions.** These are solutions that do not satisfy the original equation. Therefore, it is important to check all possible solutions in the original equation to determine if any of them should be eliminated from the solution set.

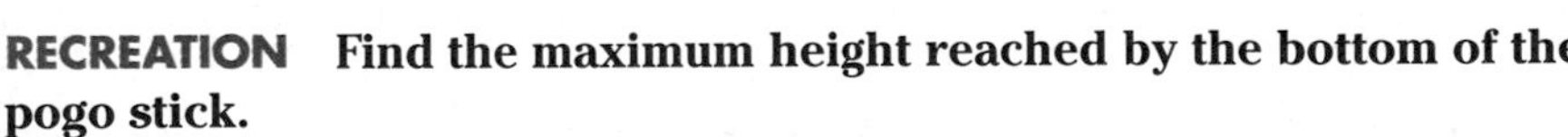

Examples **1** **RECREATION** **Find the maximum height reached by the bottom of the pogo stick.**

$$x_0 = x + \sqrt{\frac{2mgh}{k}}$$

$$1 = 0.9 + \sqrt{\frac{2 \times 50 \times 9.80 \times h}{1.2 \times 10^4}} \quad \textit{$x_0 = 1$, $x = 0.9$, $m = 50$, $g = 9.80$, $k = 1.2 \times 10^4$}$$

$$0.1 = \sqrt{\frac{980h}{1.2 \times 10^4}} \quad \textit{Isolate the radical.}$$

$$0.01 = \frac{980h}{1.2 \times 10^4} \quad \textit{Square each side.}$$

$$0.1224489796 = h \quad \textit{Solve for h.}$$

A possible solution is about 0.12. Check this solution.

(continued on the next page)

Check: $x_0 = x + \sqrt{\dfrac{2mgh}{k}}$

$1 \stackrel{?}{=} 0.9 + \sqrt{\dfrac{2 \times 50 \times 9.80 \times 0.12}{1.2 \times 10^4}}$

$1 \approx 0.9989949494$ ✔

The solution checks. The maximum height is about 0.12 meter.

2 **Solve $x = \sqrt{x + 7} + 5$.**

$x = \sqrt{x + 7} + 5$

$x - 5 = \sqrt{x + 7}$ *Isolate the radical.*

$x^2 - 10x + 25 = x + 7$ *Square each side.*

$x^2 - 11x + 18 = 0$

$(x - 9)(x - 2) = 0$ *Factor.*

$x - 9 = 0 \qquad x - 2 = 0$

$x = 9 \qquad x = 2$

Check both solutions to make sure they are not extraneous.

Check $x = 9$: $x = \sqrt{x + 7} + 5$

$9 \stackrel{?}{=} \sqrt{9 + 7} + 5$

$9 \stackrel{?}{=} \sqrt{16} + 5$

$9 \stackrel{?}{=} 4 + 5$

$9 = 9$ ✔

Check $x = 2$: $x = \sqrt{x + 7} + 5$

$2 \stackrel{?}{=} \sqrt{2 + 7} + 5$

$2 \stackrel{?}{=} \sqrt{9} + 5$

$2 \stackrel{?}{=} 3 + 5$

$2 \neq 8$

One solution checks and the other solution does not check. The solution is 9.

The same method of solution works for *n*th root equations.

Example 3 **Solve $4 = \sqrt[3]{x + 2} + 8$.**

$4 = \sqrt[3]{x + 2} + 8$

$-4 = \sqrt[3]{x + 2}$ *Isolate the cube root.*

$-64 = x + 2$ *Cube each side.* $\left(\sqrt[3]{x + 2}\right)^3 = x + 2.$

$-66 = x$

Check the solution.

Check: $4 = \sqrt[3]{x + 2} + 8$

$4 \stackrel{?}{=} \sqrt[3]{-66 + 2} + 8$

$4 \stackrel{?}{=} \sqrt[3]{-64} + 8$

$4 \stackrel{?}{=} -4 + 8$

$4 = 4$ ✔

The solution checks. The solution is -66.

If there is more than one radical in an equation, you may need to repeat the process for solving radical equations more than once until all radicals have been eliminated.

Example 4 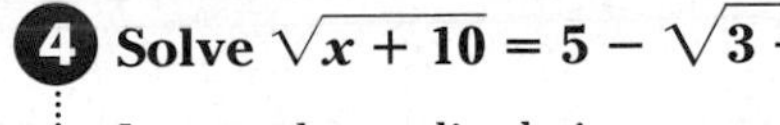 **Solve $\sqrt{x + 10} = 5 - \sqrt{3 - x}$.**

Graphing Calculator Tip

To solve a radical equation with a graphing calculator, graph each side of the equation as a separate function. The solution is any value(s) of x for points common to both graphs.

Leave the radical signs on opposite sides of the equation. If you square the sum or difference of two radical expressions, the value under the radical sign in the resulting product will be more complicated.

$$\sqrt{x + 10} = 5 - \sqrt{3 - x}$$

$$x + 10 = 25 - 10\sqrt{3 - x} + 3 - x \quad \textit{Square each side.}$$

$$2x - 18 = -10\sqrt{3 - x} \quad \textit{Isolate the radical.}$$

$$4x^2 - 72x + 324 = 100(3 - x) \quad \textit{Square each side.}$$

$$4x^2 - 72x + 324 = 300 - 100x$$

$$4x^2 + 28x + 24 = 0$$

$$x^2 + 7x + 6 = 0 \quad \textit{Divide each side by 4.}$$

$$(x + 6)(x + 1) = 0$$

$$x + 6 = 0 \qquad x + 1 = 0/$$

$$x = -6 \qquad x = -1$$

Check both solutions to make sure they are not extraneous.

Check $x = -6$:

$$\sqrt{x + 10} = 5 - \sqrt{3 - x}$$

$$\sqrt{-6 + 10} \stackrel{?}{=} 5 - \sqrt{3 - (-6)}$$

$$\sqrt{4} \stackrel{?}{=} 5 - \sqrt{9}$$

$$2 \stackrel{?}{=} 5 - 3$$

$$2 = 2 \ \checkmark$$

Check $x = -1$:

$$\sqrt{x + 10} = 5 - \sqrt{3 - x}$$

$$\sqrt{-1 + 10} \stackrel{?}{=} 5 - \sqrt{3 - (-1)}$$

$$\sqrt{9} \stackrel{?}{=} 5 - \sqrt{4}$$

$$3 \stackrel{?}{=} 5 - 2$$

$$3 = 3 \ \checkmark$$

Both solutions check. The solutions are -6 and -1.

Use the same procedures to solve **radical inequalities.**

Example 5 **Solve $\sqrt{4x + 5} \le 10$.**

$$\sqrt{4x + 5} \le 10$$

$$4x + 5 \le 100 \quad \textit{Square each side.}$$

$$4x \le 95$$

$$x \le 23.75$$

In order for $\sqrt{4x + 5}$ to be a real number, $4x + 5$ must be greater than or equal to zero.

$$4x + 5 \ge 0$$

$$4x \ge -5$$

$$x \ge -1.25$$

So the solution is $-1.25 \le x \le 23.75$. Check this solution by testing values in the intervals defined by the solution.

(continued on the next page)

On a number line, mark -1.25 and 23.75 with vertical dashed lines. The dashed lines separate the number line into intervals. Check the solution by testing values for x from each interval into the original inequality.

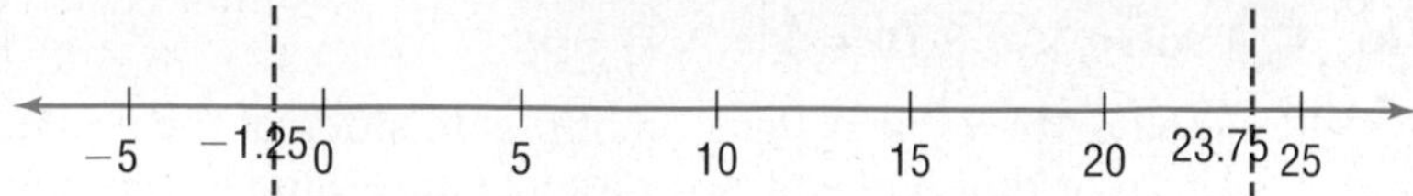

For $x \leq -1.25$, test $x = -2$: $\sqrt{4(-2) + 5} \stackrel{?}{\leq} 10$

$\sqrt{-3} \stackrel{?}{\leq} 10$ *This statement is meaningless. $x \leq -1.25$ is not a solution.*

For $-1.25 \leq x \leq 23.75$, test $x = 0$: $\sqrt{4(0) + 5} \stackrel{?}{\leq} 10$

$\sqrt{5} \stackrel{?}{\leq} 10$

$2.236 \leq 10$ *$-1.25 \leq x \leq 23.75$ is a solution.*

For $23.75 \leq x$, test $x = 25$: $\sqrt{4(25) + 5} \stackrel{?}{\leq} 10$

$\sqrt{105} \stackrel{?}{\leq} 10$

$10.247 \not\leq 10$ *$23.75 \leq x$ is not a solution.*

The solution checks and is graphed on the number line below.

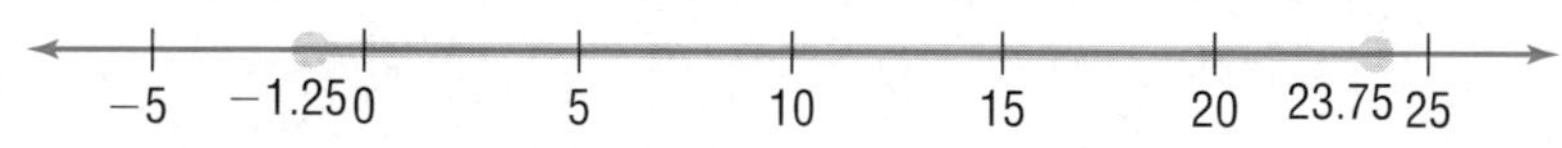

CHECK FOR UNDERSTANDING

Communicating Mathematics

Read and study the lesson to answer each question.

1. **Explain** why the first step in solving $5 + \sqrt{x + 1} = x$ should be to isolate the radical.
2. **Write** several sentences explaining why it is necessary to check for extraneous solutions in radical equations.
3. **Explain** the difference between solving an equation with one radical and solving an equation with more than one radical.

Guided Practice

Solve each equation.

4. $\sqrt{1 - 4t} = 2$
5. $\sqrt[3]{x + 4} + 12 = 3$
6. $5 + \sqrt{x - 4} = 2$
7. $\sqrt{6x - 4} = \sqrt{2x + 10}$
8. $\sqrt{a + 4} + \sqrt{a - 3} = 7$

Solve each inequality.

9. $\sqrt{5x + 4} \leq 8$
10. $3 + \sqrt{4a - 5} \leq 10$

11. **Amusement Parks** The velocity of a roller coaster as it moves down a hill is $v = \sqrt{v_0{}^2 + 64h}$, where v_0 is the initial velocity and h is the vertical drop in feet. The designer of a coaster wants the coaster to have a velocity of 90 feet per second when it reaches the bottom of the hill.
   a. If the initial velocity of the coaster at the top of the hill is 10 feet per second, write an equation that models the situation.
   b. How high should the designer make the hill?

EXERCISES

Practice

Solve each equation.

12. $\sqrt{x + 8} = 5$
13. $\sqrt{y - 7} = 4$
14. $\sqrt{8n - 5} - 1 = 2$
15. $\sqrt{x + 16} = \sqrt{x} + 4$
16. $4\sqrt{3m^2 - 15} = 4$
17. $\sqrt{9u - 4} = \sqrt{7u - 20}$
18. $\sqrt[3]{6u - 5} + 2 = -3$
19. $\sqrt{4m^2 - 3m + 2} - 2m - 5 = 0$
20. $\sqrt{k + 9} - \sqrt{k} = \sqrt{3}$
21. $\sqrt{a + 21} - 1 = \sqrt{a + 12}$
22. $\sqrt{3x + 4} - \sqrt{2x - 7} = 3$
23. $2\sqrt[3]{7b - 1} - 4 = 0$
24. $\sqrt[4]{3t} - 2 = 0$
25. $\sqrt{x + 2} - 7 = \sqrt{x + 9}$
26. $\sqrt{2x + 1} + \sqrt{2x + 6} = 5$
27. $\sqrt{3x + 10} = \sqrt{x + 11} - 1$

28. Consider the equation $\sqrt{3t - 14} + t = 6$.
    a. Name any extraneous solutions of the equation.
    b. What is the solution of the equation?

Solve each inequality.

29. $\sqrt{2x - 7} \geq 5$
30. $\sqrt{b + 4} \leq 6$
31. $\sqrt{a - 5} \leq 4$
32. $\sqrt{2x - 5} \leq 6$
33. $\sqrt[4]{5y - 9} \leq 2$
34. $\sqrt{m + 2} \leq \sqrt{3m + 4}$

35. What values of c make $\sqrt{2c - 5}$ greater than 7?

Applications and Problem Solving

36. **Physics** The time t in seconds that it takes an object at rest to fall a distance of s meters is given by the formula $t = \sqrt{\frac{2s}{g}}$. In this formula, g is the acceleration due to gravity in meters per second squared. On the moon, a rock falls 7.2 meters in 3 seconds.
    a. Write an equation that models the situation.
    b. What is the acceleration due to gravity on the moon?

37. **Critical Thinking** Solve $\sqrt{x - 5} = \sqrt[3]{x - 3}$.

38. **Driving** After an accident, police can determine how fast a car was traveling before the driver put on his breaks by using the equation $s = \sqrt{30fd}$. In this equation, s represents the speed in miles per hour, f represents the coefficient of friction, and d represents the length of the skid in feet. The coefficient of friction varies with road conditions. Suppose the coefficient of friction is 0.6.
   a. Find the speed of a car that skids 25 feet.
   b. If you were driving 35 miles per hour, how many feet would it take you to stop?
   c. If the speed is doubled, will the skid be twice as long? Explain.

39. **Physics** The period of a pendulum (the time required for one back and forth swing) can be determined by the formula $T = 2\pi\sqrt{\frac{\ell}{g}}$. In this formula, T represents the period, ℓ represents the length of the pendulum, and g represents acceleration due to gravity.
   a. Determine the period of a 1-meter pendulum on Earth if the acceleration due to gravity at Earth's surface is 9.8 meters per second squared.
   b. Suppose the acceleration due to gravity on the surface of Venus is 8.9 meters per second squared. Calculate the period of the pendulum on Venus.
   c. How must the length of the pendulum be changed to double the period?

Venus

40. **Astronomy** Johann Kepler (1571–1630) determined the relationship of the time of revolution of two planets and their average distance from the sun. This relationship can be expressed as $\frac{T_a}{T_b} = \sqrt{\left(\frac{r_a}{r_b}\right)^3}$. In this equation, T_a represents the time it takes planet a to orbit the sun, and r_a represents the average distance between planet a and the sun. Likewise, T_b represents the time it takes planet b to orbit the sun, and r_b represents the average distance between planet b and the sun. The average distance between Venus and the sun is 67,200,000 miles, and it takes Venus about 225 days to orbit the sun. If it takes Mars 687 days to orbit the sun, what is its average distance from the sun?

41. **Critical Thinking** For what values of a and b will the equation $\sqrt{2x - 9} - a = b$ have no real solution?

42. **Engineering** Engineers are often required to determine the stress on building materials. Tensile stress can be found by using the formula $T = \frac{t + c}{2} + \sqrt{\left(\frac{t - c}{2}\right)^2 + p^2}$. In this formula, T represents the tensile stress, t represents the tension, c represents the compression, and p represents the pounds of pressure per square inch. If the tensile stress is 108 pounds per square inch, the pressure is 50 pounds per square inch, and the compression is -200 pounds per square inch, what is the tension?

Mixed Review

43. Solve $\frac{a+2}{2a+1} = \frac{a}{3} + \frac{3}{4a+2}$. *(Lesson 4-6)*

44. List the possible rational roots of $x^4 + 5x^3 + 5x^2 - 5x - 6 = 0$. Then determine the rational roots. *(Lesson 4-4)*

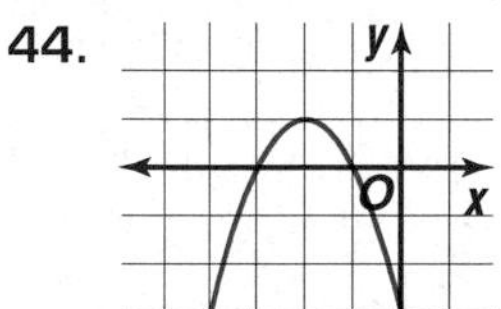

45. Determine whether each graph has *infinite discontinuity, jump discontinuity,* or *point discontinuity,* or is *continuous.* *(Lesson 3-7)*

a. 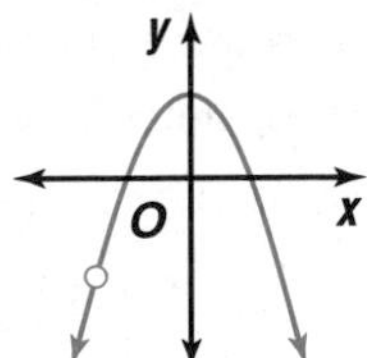

b. 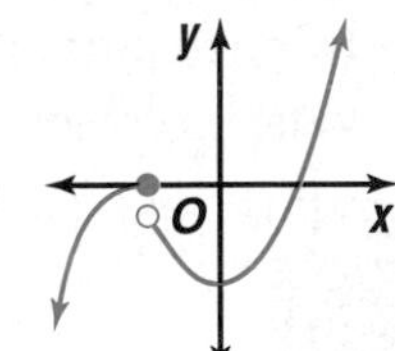

c. 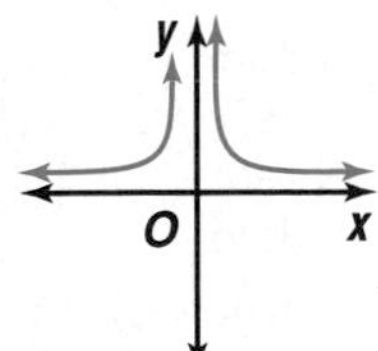

46. Music The frequency of a sound wave is called its pitch. The pitch p of a musical tone and its wavelength w are related by the equation $p = \frac{v}{w}$, where v is the velocity of sound through air. Suppose a sound wave has a velocity of 1056 feet per second. *(Lesson 3-6)*

a. Graph the equation $p = \frac{v}{w}$.

b. What lines are close to the maximum values for the pitch and the wavelength?

c. What happens to the pitch of the tone as the wavelength decreases?

d. If the wavelength is doubled, what happens to the pitch of the tone?

47. Find the product of the matrices $\begin{bmatrix} 4 & -1 & 6 \\ 4 & 0 & 2 \end{bmatrix}$ and $\begin{bmatrix} 0 & 3 \\ 2 & -2 \\ 5 & 1 \end{bmatrix}$. *(Lesson 2-3)*

48. Solve the system of equations. *(Lesson 2-2)*

$$a + b + c = 6$$
$$2a - 3b + 4c = 3$$
$$4a - 8b + 4c = 12$$

49. Education The regression equation for a set of data is $y = -3.54x + 7107.7$, where x represents the year and y represents the average number of students assigned to each advisor in a certain business school. Use the equation to predict the number of students assigned to each adviser in the year 2005. *(Lesson 1-6)*

50. Write the slope-intercept form of the equation that is perpendicular to $7y + 4x - 3 = 0$ and passes through the point with coordinates (2, 5). *(Lesson 1-5)*

51. SAT/ACT Practice In the figure at the right, four semicircles are drawn on the four sides of a rectangle. What is the total area of the shaded regions?

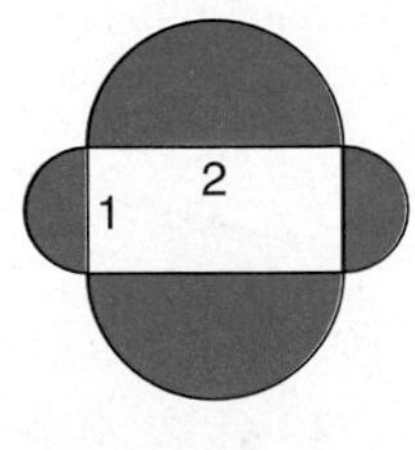

A 5π B $\frac{5\pi}{2}$ C $\frac{5\pi}{4}$

D $\frac{5\pi}{8}$ E $\frac{5\pi}{16}$

Extra Practice See p. A33.

Modeling Real-World Data with Polynomial Functions

OBJECTIVES
- Write polynomial functions to model real-world data.
- Use polynomial functions to interpret real-world data.

WASTE MANAGEMENT The average daily amount of waste generated by each person in the United States is given below. This includes all wastes such as industrial wastes, demolition wastes, and sewage.

What a Waste!

Year	1980	1985	1990	1991	1992	1993	1994	1995	1996
Pounds of Waste per Person per Day	3.7	3.8	4.5	4.4	4.5	4.5	4.5	4.4	4.3

Source: Franklin Associates, Ltd.

What polynomial function could be used to model these data? *This problem will be solved in Example 3.*

In order to model real world data using polynomial functions, you must be able to identify the general shape of the graph of each type of polynomial function.

Equation	Linear $y = ax + b$	Quadratic $y = ax^2 + bx + c$	Cubic $y = ax^3 + bx^2 + cx + d$	Quartic $y = ax^4 + bx^3 + cx^2 + dx + e$
Typical Graph				
Direction Changes	0	1	2	3

Example  **Determine the type of polynomial function that could be used to represent the data in each scatter plot.**

Look Back
You can refer to Lesson 1-6 to review scatter plots.

a.

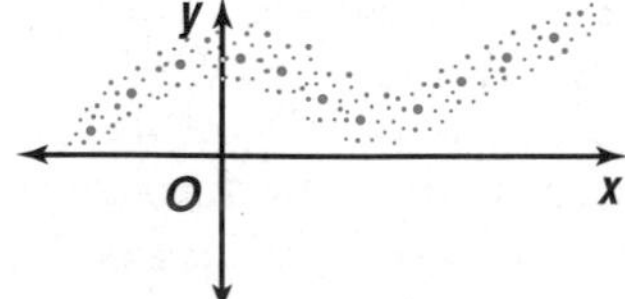

The scatter plot seems to change direction two times, so a cubic function would best fit the plot.

b.

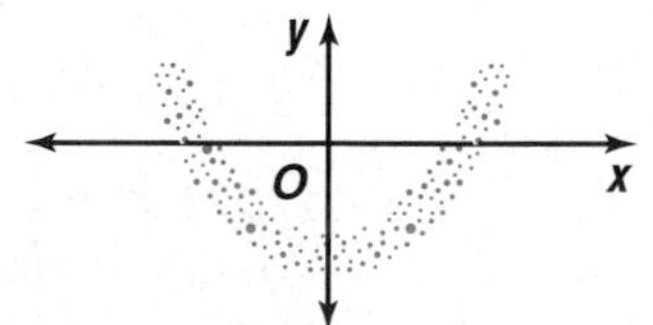

The scatter plot seems to change direction one time, so a quadratic function would best fit the plot.

You can use a graphing calculator to determine a polynomial function that models a set of data.

Example  **Use a graphing calculator to write a polynomial function to model the set of data.**

x	−1	−0.5	0	0.5	1	1.5	2	2.5	3	3.5	4
$f(x)$	−10	−6.4	−5	−5.1	−6	−6.9	−7	−5.6	−2	4.6	15

Clear the statistical memory and input the data.

L1	L2	L3 2
-1	-10	------
-.5	-6.4	
0	-5	
.5	-5.1	
1	-6	
1.5	-6.9	
2	-7	

L2(1)= -10

Using the standard viewing window, graph the data. The scatter plot seems to change direction two times, so a cubic function would best fit the scatter plot.

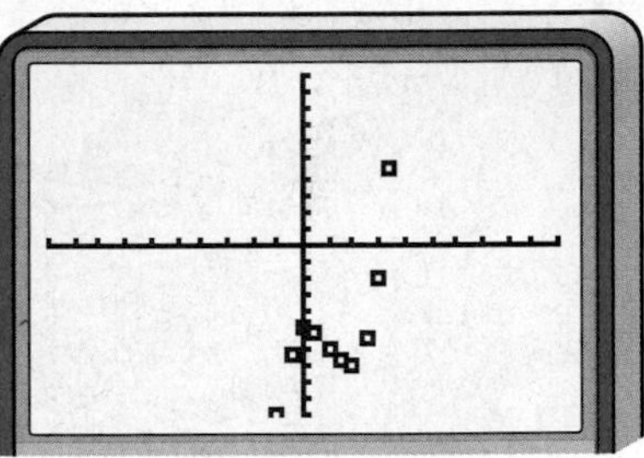

[−10, 10] scl:1 by [−10, 10] scl:1

Graphing Calculator Tip

You can change the appearance of the data points in the **STAT PLOT** menu. The data can appear as squares, dots, or + signs.

Press STAT and highlight **CALC**. Since the scatter plot seems to be a cubic function, choose **6:CubicReg**.

Press 2nd **[L1]** , 2nd **[L2]** ENTER.

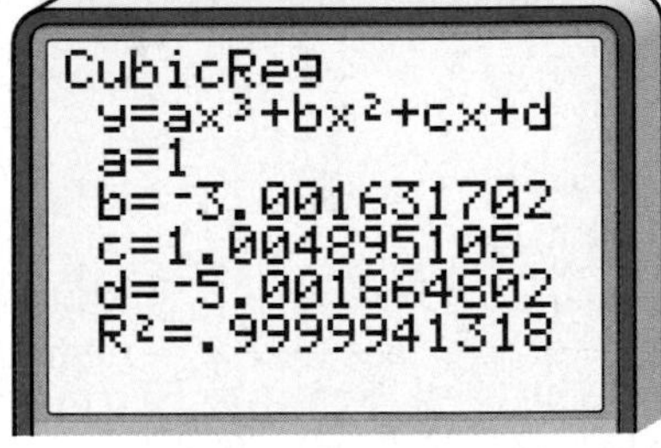

Rounding the coefficients to the nearest whole number, $f(x) = x^3 - 3x^2 + x - 5$ models the data. Since the value of the coefficient of determination r^2 is very close to 1, the function is an excellent fit. However it may not be the best model for the situation.

To check the polynomial function, enter the function in the **Y=** list. Then use the **TABLE** feature for x values from −1 with an interval of 0.5. Compare the values of y with those given in the data. The values are similar, and the polynomial function seems to be a good fit.

X	Y1	
-1	-10	
-.5	-6.375	
0	-5	
.5	-5.125	
1	-6	
1.5	-6.875	
2	-7	

X= -1

Example 3 **WASTE MANAGEMENT** **Refer to the application at the beginning of the lesson.**

a. What polynomial function could be used to model these data?

b. Use the model to predict the amount of waste produced per day in 2010.

c. Use the model to predict when the amount of waste will drop to 3 pounds per day.

a. Let **L1** be the number of years since 1980 and **L2** be the pounds of solid waste per person per day. Enter this data into the calculator.

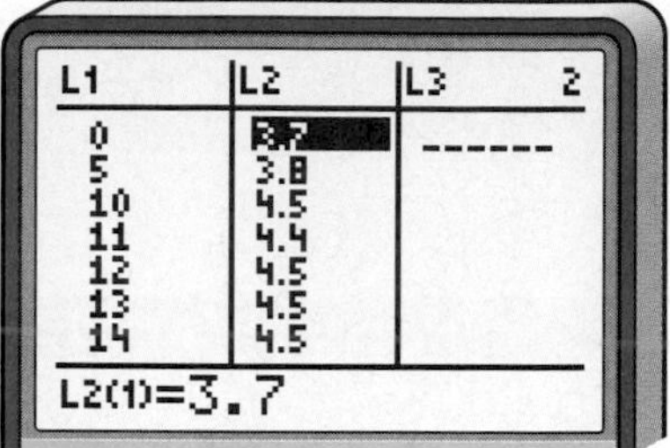

Adjust the window to an appropriate setting and graph the statistical data. The data seems to change direction one time, so a quadratic function will fit the scatter plot.

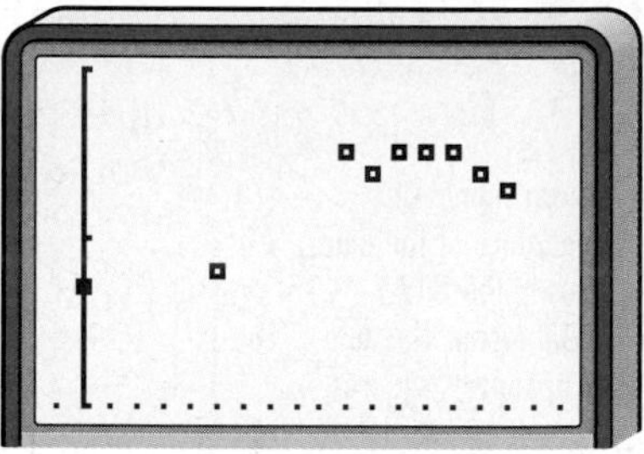

[−1, 18] scl:1 by [3, 5] scl:1

Press STAT, highlight **CALC**, and choose **5:QuadReg**.

Press 2nd [L1] , 2nd [L2] ENTER.

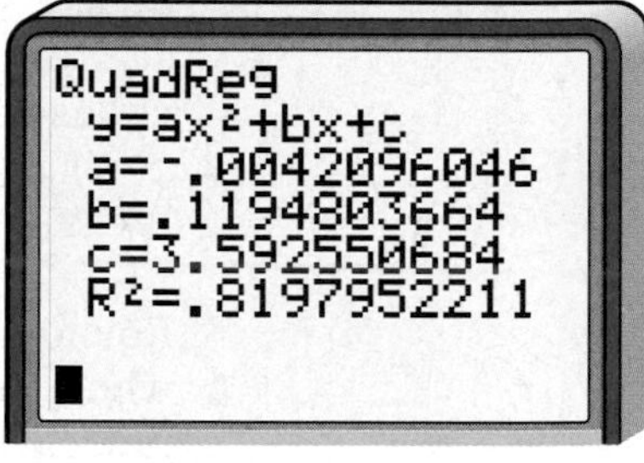

Rounding the coefficients to the nearest thousandth, $f(x) = -0.004x^2 + 0.119x + 3.593$ models the data. Since the value of the coefficient of determination r^2 is close to 1, the model is a good fit.

b. Since 2010 is 30 years after 1980, find $f(30)$.

$$f(x) = -0.004x^2 + 0.119x + 3.593$$

$$f(30) = -0.004(30^2) + 0.119(30) + 3.593$$

$$= 3.563$$

According to the model, each person will produce about 3.6 pounds of waste per day in 2010.

c. To find when the amount of waste will be reduced to 3 pounds per person per day, solve the equation $3 = -0.004x^2 + 0.119x + 3.593$.

$3 = -0.004x^2 + 0.119x + 3.593$

$0 = -0.004x^2 + 0.119x + 0.593$ *Add −3 to each side.*

Use the Quadratic Formula to solve for x.

$x = \dfrac{-0.119 \pm \sqrt{0.119^2 - 4(-0.004)(0.593)}}{2(-0.004)}$ $a = -0.004,\ b = 0.119,\ c = 0.593$

$x = \dfrac{-0.119 \pm \sqrt{0.023649}}{-0.008}$

$x \approx -4$ or 34

According to the model, 3 pounds per day per person occurs 4 years before 1980 (in 1976) or 34 years after 1980 (in 2014). Since you want to know when the amount of waste will drop to 3 pounds per day, the answer is in 2014.

To check the answer, graph the related function $f(x) = -0.004x^2 + 0.119x + 0.593$. Use the **CALC** menu to determine the zero at the right. The answer of 34 years checks.

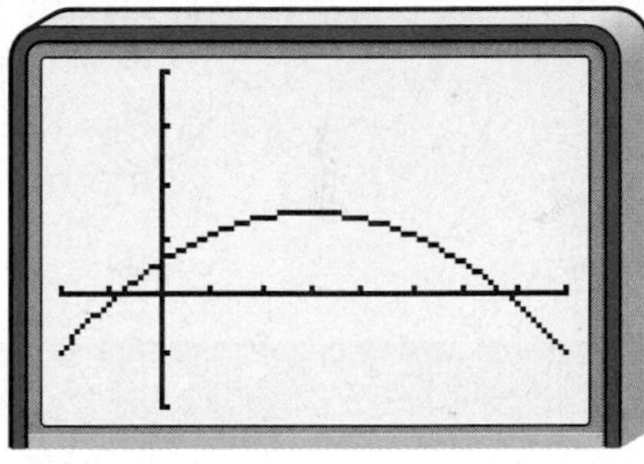

[−10, 40] scl:5 by [−2, 4] scl:1

The waste should reduce to 3 pounds per person per day about 34 years after 1980 or in 2014.

CHECK FOR UNDERSTANDING

Communicating Mathematics

Read and study the lesson to answer each question.

1. **Draw** an example of a scatter plot that can be represented by each type of function.
   a. quartic
   b. quadratic
   c. cubic
2. **Explain** why it is important to recognize the shape of the graph of each type of polynomial function.
3. **List** reasons why the amount of waste per person may vary from the model in Example 3.

Guided Practice

4. Determine the type of polynomial function that could be used to represent the data in the scatter plot.

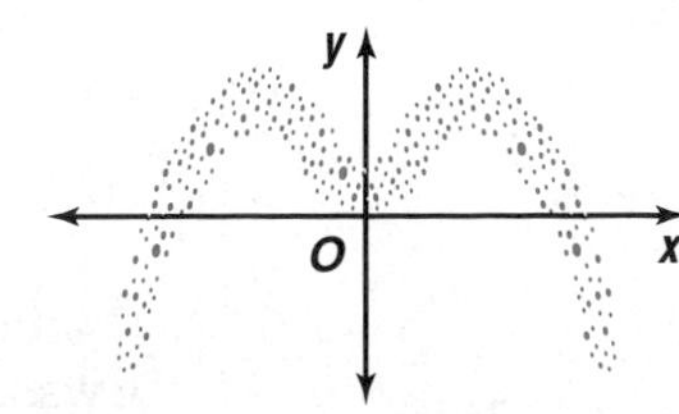

Graphing Calculator

Use a graphing calculator to write a polynomial function to model each set of data.

5.

x	−3.5	−3	−2.5	−2	−1.5	−1	−0.5	0	0.5	1	1.5	2
f(x)	103	32	−1	−11	−9	−2	3	5	4	4	12	37

6.

x	−3	−2.5	−1.5	−0.5	0	1	2	2.5	3.5
f(x)	19	11	−1	−7	−8	−5	4	11	29

7. **Population** The percent of the United States population living in metropolitan areas has increased since 1950.

Year	1950	1960	1970	1980	1990	1996
Population living in metropolitan areas	56.1%	63%	68.6%	74.8%	74.8%	79.9%

Source: *American Demographics*

a. Write a model that relates the percent as a function of the number of years since 1950.

b. Use the model to predict the percent of the population that will be living in metropolitan areas in 2010.

c. Use the model to predict what year will have 85% of the population living in metropolitan areas.

EXERCISES

Practice

Determine the type of polynomial function that could be used to represent the data in scatter plot.

8.

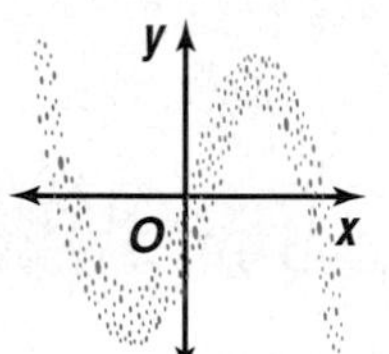

9.

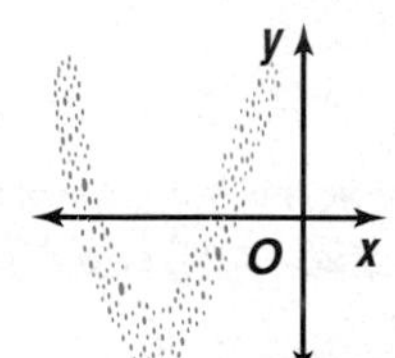

10.

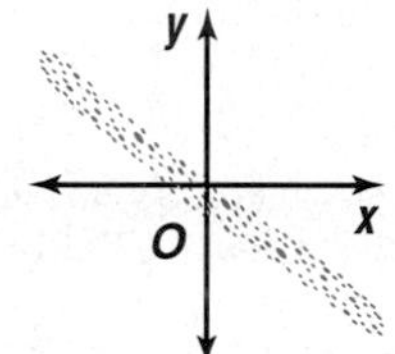

11. What type of polynomial function would be the best model for the set of data?

x	1	2	3	4	5	7	8
f(x)	15	7	2	−1	3	10	15

Graphing Calculator

Use a graphing calculator to write a polynomial function to model each set of data.

12.

x	−3	−2	−1	0	1	2	3
f(x)	8.75	7.5	6.25	5	3.75	2.5	1.25

13.

x	−2	−1	0	1	2	3
f(x)	29	2	−9	−4	17	54

14.

x	−1	−0.5	0	0.5	1	1.5	2	2.5	3	3.5
f(x)	13	3	1	2	3	3	1	−1	1	10

15.

x	5	7	8	10	11	12	15	16
f(x)	2	5	6	4	−1	−3	5	9

16.

x	30	35	40	45	50	55	60	65	70	75
f(x)	52	41	32	44	61	88	72	59	66	93

17.

x	−17	−6	−1	2	8	12	15
f(x)	51	29	−6	41	57	37	19

18. Consider the set of data.

x	−2.5	−2	−1.5	−1	−0.5	0	0.5	1	1.5
f(x)	23	11	7	6	6	5	3	2	4

a. What quadratic polynomial function models the data?
b. What cubic polynomial function models the data?
c. Which model do you think is more appropriate? Explain.

Applications and Problem Solving

Data Update
To compare current census data with predictions using your models, visit our website at **www.amc.glencoe.com**

19. **Marketing** The United States Census Bureau has projected the median age of the U.S. population to the year 2080. A fast-food chain wants to target its marketing towards customers that are about the median age.

Year	1900	1930	1960	1990	2020	2050	2080
Median age	22.9	26.5	29.5	33.0	40.2	42.7	43.9

a. Write a model that relates the median age as a function of the number of years since 1900.
b. Use the model to predict what age the fast-food chain should target in the year 2005.
c. Use the model to predict what age the fast-food chain should target in the year 2025.

20. **Critical Thinking** Write a set of data that could be best represented by a cubic polynomial function.

21. **Consumer Credit** The amount of consumer credit as a percent of disposable personal income is given below.

Year	1988	1989	1990	1991	1992	1995	1996	1997
Consumer credit	23%	24%	23%	22%	19%	21%	24%	22%

Source: *The World Almanac and Book of Facts*

a. Write a model that relates the percent of consumer credit as a function of the number of years since 1988.
b. Use the model to estimate the percent of consumer credit in 1994.

22. **Critical Thinking** What type of polynomial function should be used to model the scatter plot? Explain.

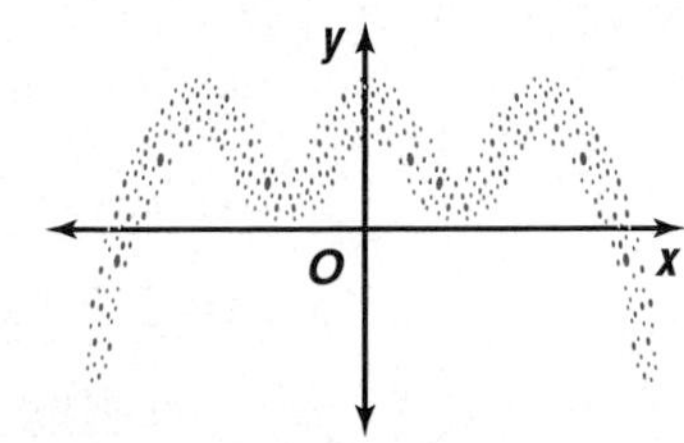

23. **Baseball** The attendance at major league baseball games for various years is listed below.

Year	1985	1987	1990	1992	1993	1994	1995	1996
Attendance (in millions)	48	53	56	57	71	50	51	62

Source: *Statistical Abstract of the United States*

a. Write a model that relates the attendance in millions as a function of the number of years since 1985.

b. Use the model to predict when the attendance will reach 71 million again.

c. In 1998, Mark McGwire and Sammy Sosa raced to break the homerun record. That year the attendance reached 70.6 million. Did the attendance for 1998 follow your model? Do you think the race to break the homerun record affected attendance? Why or why not?

24. **Communication** Cellular phones are becoming more popular. The numbers of cellular subscribers are listed in below.

Year	1983	1984	1986	1989	1990	1992	1994	1996	1997
Subscribers (in millions)	0.25	3.0	6.5	3.8	5.3	11.0	24.1	44.0	55.3

Source: *Statistical Abstract of the United States*

a. Write a model that relates the millions of subscribers as a function of the number of years since 1983.

b. According to the model, how many subscribers were there in 1995?

c. Use the model to predict when there will be 150 million subscribers.

Mixed Review

25. Solve $5 - \sqrt{b + 2} = 0$. *(Lesson 4-7)*

26. Solve $\frac{6}{p + 3} + \frac{p}{p - 3} = 1$. *(Lesson 4-6)*

27. Approximate the real zeros of $f(x) = 2x^4 - x^3 + x - 2$ to the nearest tenth. *(Lesson 4-5)*

28. **Agriculture** If Wesley Jackson harvests his apple crop now, the yield will average 120 pounds per tree. Also, he will be able to sell the apples for \$0.48 per pound. However, he knows that if he waits, his yield will increase by about 10 pounds per week, while the selling price will decrease by \$0.03 per pound per week. *(Lesson 3-6)*

a. How many weeks should Mr. Jackson wait in order to maximize the profit?

b. What is the maximum profit?

29. **SAT/ACT Practice** A college graduate goes to work for \$$x$ per week. After several months, the company falls on hard times and gives all the employees a 10% pay cut. A few months later, business picks up and the company gives all employees a 10% raise. What is the college graduate's new salary?

A \$$0.90x$ **B** \$$0.99x$ **C** \$$x$

D \$$1.01x$ **E** \$$1.11x$

Extra Practice See p. A33.

GRAPHING CALCULATOR EXPLORATION

4-8B Fitting a Polynomial Function to a Set of Points

An Extension of Lesson 4-8

OBJECTIVE

- Use matrices to find a polynomial function whose graph passes through a given set of points in the coordinate plane.

The **STAT CALC** menu of a graphing calculator has built-in programs that allow you to find a polynomial function whose graph passes through as many as five arbitrarily selected points in the coordinate plane if no two points are on the same vertical line. For example, for a set of four points, you can enter the x-coordinates in list **L1** and the y-coordinates in list **L2**. Then, select **6: CubicReg** from the **STAT CALC** menu to display the coefficients for a cubic function whose graph passes through the four points.

For any finite set of points where no two points are on the same vertical line, you can perform calculations to find a polynomial function whose graph passes exactly through those points. For a set of n points (where $n \geq 1$), the polynomial function will have a degree of $n - 1$.

Example **Find a polynomial function whose graph passes through points $A(-2, 69)$, $B(-1.5, 19.75)$, $C(-1, 7)$, $D(2, 25)$, $E(2.5, 42)$, and $F(3, 49)$.**

The equation of a polynomial function through the six points is of the form $y = ax^5 + bx^4 + cx^3 + dx^2 + ex + f$. The coordinates of each point must satisfy the equation for the function. If you substitute the x- and y-coordinates of the points into the equation, you obtain a system of six linear equations in six variables.

A: $(-2)^5a + (-2)^4b + (-2)^3c + (-2)^2d + (-2)e + 1f = 69$

B: $(-1.5)^5a + (-1.5)^4b + (-1.5)^3c + (-1.5)^2d + (-1.5)e + 1f = 19.75$

C: $(-1)^5a + (-1)^4b + (-1)^3c + (-1)^2d + (-1)e + 1f = 7$

D: $(2)^5a + (2)^4b + (2)^3c + (2)^2d + (2)e + 1f = 25$

E: $(2.5)^5a + (2.5)^4b + (2.5)^3c + (2.5)^2d + (2.5)e + 1f = 42$

F: $(3)^5a + (3)^4b + (3)^3c + (3)^2d + (3)e + 1f = 49$

Perform the following steps on a graphing calculator.

Step 1 Press [STAT], select **Edit**, and enter three lists. In list **L1**, enter the x-coordinates of the six points in the order they appear as coefficients of e. In list **L2**, enter the y-coordinates in the order they appear after the equal signs. In list **L3**, enter six 1s.

Step 2 Notice that the columns of coefficients of a, b, c, and d are powers of the column of numbers that you entered for **L1**. To solve the linear system, you need to enter the columns of coefficients as columns of a 6×6 matrix. So, press the [MATRX] key, and set the dimensions of matrix **[A]** to 6×6. Next use **9: List▶matr** from the **MATRX MATH** menu to put the columns of coefficients into matrix **[A]**. On the home screen, enter the following expression.

List▶matr (L1^5, L1^4, L1^3, L1^2, L1, L3, [A])

(continued on the next page)

Step 3 Put the column of y-coordinates (the numbers after the equals signs) into matrix **[B]**. Enter **List▶matr (L2, [B])** and press ENTER.

Step 4 Enter **[A]$^{-1}$[B]** and press ENTER. The calculator will display the solution of the linear system as a column matrix. The elements of this matrix are the coefficients for the polynomial being sought. The top element is the coefficient of x^5, the second element is the coefficient of x^4, and so on. The function is $f(x) = -x^5 + 3.5x^4 + 1.5x^3 - 4x^2 - x + 7$.

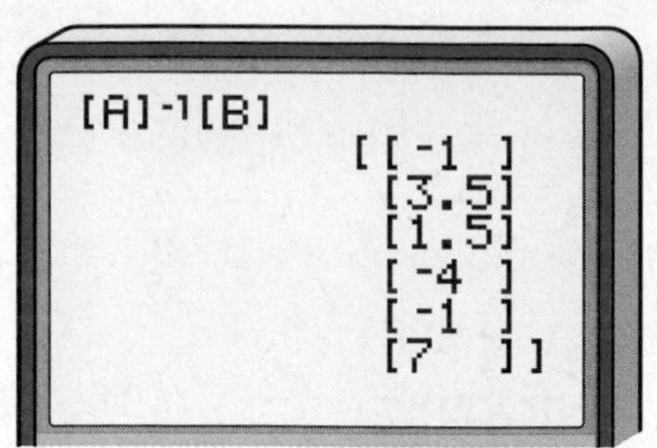

Step 5 Enter the polynomial function on the **Y=** list and display its graph. If you use **STAT PLOT** and display the data for the coordinates in **L1** and **L2** as a scatter plot, you will have visual confirmation that the graph of the polynomial function passes through the given points.

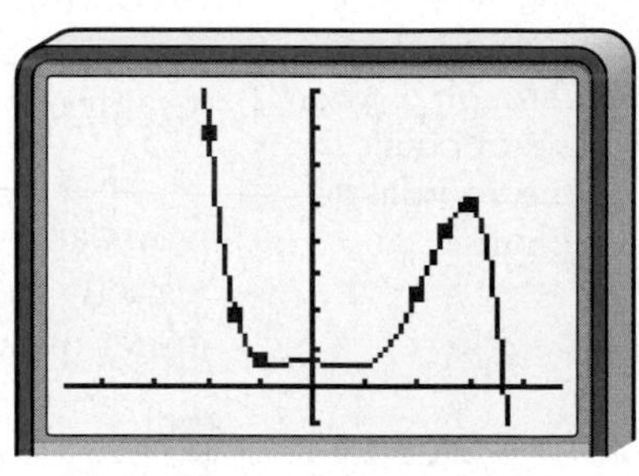

[5, 5] scl:1 by [−10, 80] scl:10

For problems of this kind, the coefficients in the solution matrix for the linear system are often decimals. You can enter the equation for the function on the **Y=** list efficiently and accurately by first changing the column matrix for the solution into a list. On the home screen, enter **Matr▶list ([A]$^{-1}$[B], 1, L4)**. Then go to the **Y=** list and enter the equation **Y1 = L4(1)X^5 + L4(2)X^4 + L4(3)X^3 + L4(4)X^2 + L4(5)X + L4(6)**.

TRY THESE

1. Use the steps given above to find a cubic polynomial function whose graph passes through points with coordinates $(-2, -23)$, $(-1, 21)$, $(0, 15)$, and $(1, 1)$.
2. Use **6: CubicReg** on the **STAT CALC** menu to fit a cubic model for the points in Exercise 1. Is the resulting cubic function the same as the one you found in Exercise 1?
3. Find a polynomial function whose graph passes through points with coordinates $(-3, 116)$, $(-2, 384)$, $(-0.4, -5.888)$, $(0, -4)$, $(1, 36)$, $(2, 256)$, and $(2.6, 34.1056)$.

WHAT DO YOU THINK?

4. How many polynomial functions have graphs that pass through a given set of points, no two of which are on the same vertical line? Explain your thinking.
5. In Step 2, why was it necessary to define the list **L3**? Would it have been possible to use **L1^0** instead of **L3** in the **List▶matr** expression used to define matrix **[A]**? Explain your thinking.

VOCABULARY

completing the square (p. 213)
complex number (p. 206)
conjugate (p. 216)
degree (p. 206)
depressed polynomial (p. 224)
Descartes' Rule of Signs (p. 231)
discriminant (p. 215)
extraneous solution (p. 251)
Factor Theorem (p. 224)
Fundamental Theorem of Algebra (p. 207)
imaginary number (p. 206)
Integral Root Theorem (p. 230)
leading coefficient (p. 206)
Location Principle (p. 236)
lower bound (p. 238)
Lower Bound Theorem (p. 238)
partial fractions (p. 244)
polynomial equation (p. 206)
polynomial function (p. 206)
polynomial in one variable (p. 205)
pure imaginary number (p. 206)
Quadratic Formula (p. 215)
radical equation (p. 251)
radical inequality (p. 253)
rational equation (p. 243)
rational inequality (p. 245)
Rational Root Theorem (p. 229)
Remainder Theorem (p. 222)
root (p. 206)
synthetic division (p. 223)
upper bound (p. 238)
Upper Bound Theorem (p. 238)
zero (p. 206)

UNDERSTANDING AND USING THE VOCABULARY

Choose the correct term from the list to complete each sentence.

1. The __?__ can be used to solve any quadratic equation.
2. The __?__ states that if the leading coefficient of a polynomial equation a_0 has a value of 1, then any rational root must be factors of a_n.
3. For a rational equation, any possible solution that results with a __?__ in the denominator must be excluded from the list of solutions.
4. The __?__ states that the binomial $x - r$ is a factor of the polynomial $P(x)$ if and only if $P(r) = 0$.
5. Descartes' Rule of Signs can be used to determine the possible number of positive real zeros a __?__ has.
6. A(n) __?__ of the zeros of $P(x)$ can be found by determining an upper bound for the zeros of $P(-x)$.
7. __?__ solutions do not satisfy the original equation.
8. Since the x-axis only represents real numbers, __?__ of a polynomial function cannot be determined by using a graph.
9. The Fundamental Theorem of Algebra states that every polynomial equation with degree greater than zero has at least one root in the set of __?__
10. A __?__ is a special polynomial equation with a degree of two.

complex numbers
complex roots
discriminant
extraneous
Factor Theorem
Integral Root Theorem
lower bound
polynomial function
quadratic equation
Quadratic Formula
radical equation
synthetic division
zero

For additional review and practice for each lesson, visit: **www.amc.glencoe.com**

SKILLS AND CONCEPTS

OBJECTIVES AND EXAMPLES

Lesson 4-1 Determine roots of polynomial equations.

- Determine whether 2 is a root of $x^4 - 3x^3 - x^2 - x = 0$. Explain.

  $f(2) = 2^4 - 3(2^3) - 2^2 - 2$

  $f(2) = 16 - 24 - 4 - 2$ or -14

  Since $f(2) \neq 0$, 2 is not a root of $x^4 - 3x^3 - x^2 - x = 0$.

REVIEW EXERCISES

Determine whether each number is a root of $a^3 - 3a^2 - 3a - 4 = 0$. Explain.

11. 0 **12.** 4 **13.** -2

14. Is -3 a root of $t^4 - 2t^2 - 3t + 1 = 0$?

15. State the number of complex roots of the equation $x^3 + 2x^2 - 3x = 0$. Then find the roots and graph the related function.

Lesson 4-2 Solve quadratic equations.

- Find the discriminant of $3x^2 - 2x - 5 = 0$ and describe the nature of the roots of the equation. Then solve the equation by using the Quadratic Formula.

  The value of the discriminant, $b^2 - 4ac$, is $(-2)^2 - 4(3)(-5)$ or 64. Since the value of the discriminant is greater than zero, there are two distinct real roots.

  $$x = \frac{-b \pm \sqrt{b^2 - 4ac}}{2a}$$

  $$x = \frac{-(-2) \pm \sqrt{64}}{2(3)}$$

  $$x = \frac{2 \pm 8}{6}$$

  $$x = -1 \text{ or } \frac{5}{3}$$

Find the discriminant of each equation and describe the nature of the roots of the equation. Then solve the equation by using the Quadratic Formula.

16. $2x^2 - 7x - 4 = 0$

17. $3m^2 - 10m + 5 = 0$

18. $x^2 - x + 6 = 0$

19. $-2y^2 + 3y + 8 = 0$

20. $a^2 + 4a + 4 = 0$

21. $5r^2 - r + 10 = 0$

Lesson 4-3 Find the factors of polynomials using the Remainder and Factor Theorems.

- Use the Remainder Theorem to find the remainder when $(x^3 + 2x^2 - 5x - 9)$ is divided by $(x + 3)$. State whether the binomial is a factor of the polynomial.

  $f(x) = x^3 + 2x^2 - 5x - 9$

  $f(-3) = (-3)^3 + 2(-3)^2 - 5(-3) - 9$

  $= -27 + 18 + 15 - 9$ or -3

  Since $f(-3) = -3$, the remainder is -3. So the binomial $x + 3$ is not a factor of the polynomial by the Remainder Theorem.

Use the Remainder Theorem to find the remainder for each division. State whether the binomial is a factor of the polynomial.

22. $(x^3 - x^2 - 10x - 8) \div (x + 2)$

23. $(2x^3 - 5x^2 + 7x + 1) \div (x - 5)$

24. $(4x^3 - 7x + 1) \div \left(x + \frac{1}{2}\right)$

25. $(x^4 - 10x^2 + 9) \div (x - 3)$

OBJECTIVES AND EXAMPLES

Lesson 4-4 Identify all possible rational roots of a polynomial equation by using the Rational Root Theorem.

- List the possible rational roots of $4x^3 - x^2 - x - 5 = 0$. Then determine the rational roots.

  If $\frac{p}{q}$ is a root of the equation, then p is a factor of 5 and q is a factor of 4.
  possible values of p: $\pm 1, \pm 5$
  possible values of q: $\pm 1, \pm 2, \pm 4$

  possible rational roots: $\pm 1, \pm 5, \pm \frac{1}{2}, \pm \frac{1}{4}, \pm \frac{5}{2}, \pm \frac{5}{4}$

  Graphing and substitution show a zero at $\frac{5}{4}$.

REVIEW EXERCISES

List the possible rational roots of each equation. Then determine the rational roots.

26. $x^3 - 2x^2 - x + 2 = 0$
27. $x^4 - x^2 + x - 1 = 0$
28. $2x^3 - 2x^2 - 2x - 4 = 0$
29. $2x^4 + 3x^3 - 6x^2 - 11x - 3 = 0$
30. $x^5 - 7x^3 + x^2 + 12x - 4 = 0$
31. $3x^3 + 7x^2 - 2x - 8 = 0$
32. $4x^3 + x^2 + 8x + 2 = 0$
33. $x^4 + 4x^2 - 5 = 0$

Lesson 4-4 Determine the number and type of real roots a polynomial function has.

- For $f(x) = 3x^4 - 9x^3 + 4x - 6$, there are three sign changes. So there are three or one positive real zeros.

  For $f(-x) = 3x^4 + 9x^3 - 4x - 6$, there is one sign change. So there is one negative real zero.

Find the number of possible positive real zeros and the number of possible negative real zeros for each function. Then determine the rational zeros.

34. $f(x) = x^3 - x^2 - 34x - 56$
35. $f(x) = 2x^3 - 11x^2 + 12x + 9$
36. $f(x) = x^4 - 13x^2 + 36$

Lesson 4-5 Approximate the real zeros of a polynomial function.

- Determine between which consecutive integers the real zeros of $f(x) = x^3 + 4x^2 + x - 2$ are located.
  Use synthetic division.

r	1	4	1	−2	
−4	1	0	1	−6	change in signs
−3	1	1	2	4	
−2	1	2	−3	4	
−1	1	3	−2	0	zero
0	1	4	1	−2	change in signs
1	1	5	6	4	

One zero is −1. Another is located between −4 and −3. The other is between 0 and 1.

Determine between which consecutive integers the real zeros of each function are located.

37. $g(x) = 3x^3 + 1$
38. $f(x) = x^2 - 4x + 2$
39. $g(x) = x^2 - 3x - 3$
40. $f(x) = x^3 - x^2 + 1$
41. $g(x) = 4x^3 + x^2 - 11x + 3$
42. $f(x) = -9x^3 + 25x^2 - 24x + 6$
43. Approximate the real zeros of $f(x) = 2x^3 + 9x^2 - 12x - 40$ to the nearest tenth.

OBJECTIVES AND EXAMPLES

Lesson 4-6 Solve rational equations and inequalities.

- Solve $\frac{1}{9} + \frac{1}{2a} = \frac{1}{a^2}$.

$$\frac{1}{9} + \frac{1}{2a} = \frac{1}{a^2}$$
$$\left(\frac{1}{9} + \frac{1}{2a}\right)(18a^2) = \left(\frac{1}{a^2}\right)(18a^2)$$
$$2a^2 + 9a = 18$$
$$2a^2 + 9a - 18 = 0$$
$$(2a - 3)(a + 6) = 0$$
$$a = \frac{3}{2} \text{ or } -6$$

REVIEW EXERCISES

Solve each equation or inequality.

44. $n - \frac{6}{n} + 5 = 0$

45. $\frac{1}{x} = \frac{x + 3}{2x^2}$

46. $\frac{5}{6} = \frac{2m}{2m + 2} - \frac{1}{3m - 3}$

47. $\frac{3}{y} - 2 < \frac{5}{y}$

48. $\frac{2}{x + 1} < 1 - \frac{1}{x - 1}$

Lesson 4-7 Solve radical equations and inequalities.

- Solve $9 + \sqrt{x - 1} = 1$.

$$9 + \sqrt{x - 1} = 1$$
$$\sqrt{x - 1} = -8$$
$$x - 1 = 64$$
$$x = 65$$

Solve each equation or inequality.

49. $5 - \sqrt{x + 2} = 0$

50. $\sqrt[3]{4a - 1} + 8 = 5$

51. $3 + \sqrt{x + 8} = \sqrt{x + 35}$

52. $\sqrt{x - 5} < 7$

53. $4 + \sqrt{2a + 7} \geq 6$

Lesson 4-8 Write polynomial functions to model real-world data.

- Determine the type of polynomial function that would best fit the data in the scatter plot.

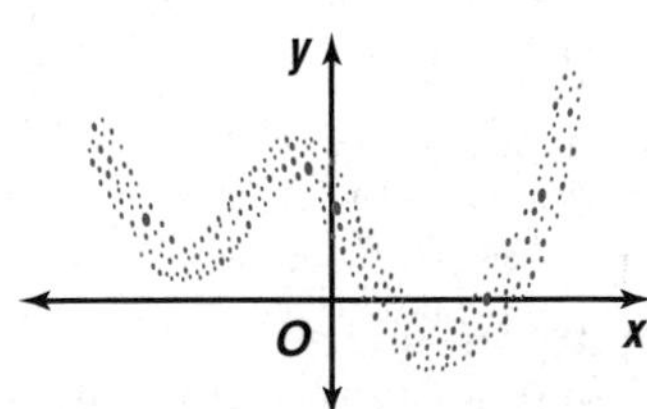

The scatter plot seems to change direction three times. So a quartic function would best fit the scatter plot.

54. Determine the type of polynomial function that would best fit the scatter plot.

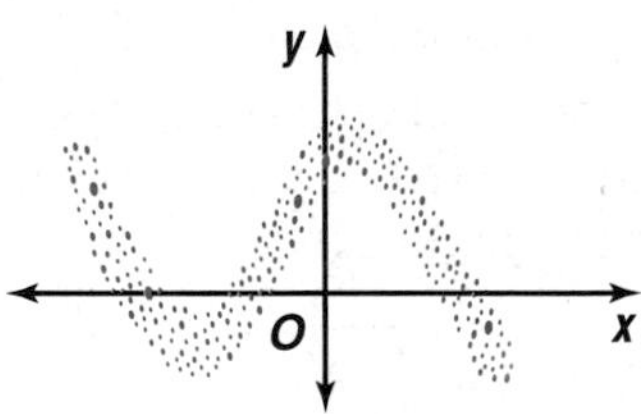

55. Write a polynomial function to model the data.

x	−3	−1	0	2	4	7
f(x)	24	6	3	9	31	94

APPLICATIONS AND PROBLEM SOLVING

56. **Entertainment** The scenery for a new children's show has a playhouse with a painted window. A special gloss paint covers the area of the window to make them look like glass. If the gloss only covers 315 square inches and the window must be 6 inches taller than it is wide, how large should the scenery painters make the window? *(Lesson 4-1)*

57. **Gardening** The length of a rectangular flower garden is 6 feet more than its width. A walkway 3 feet wide surrounds the outside of the garden. The total area of the walkway itself is 288 square feet. Find the dimensions of the garden. *(Lesson 4-2)*

58. **Medicine** Doctors can measure cardiac output in potential heart attack patients by monitoring the concentration of dye after a known amount in injected in a vein near the heart. In a normal heart, the concentration of the dye is given by $g(x) = -0.006x^4 + 0.140x^3 - 0.053x^2 + 1.79x$, where x is the time in seconds. *(Lesson 4-4)*
   a. Graph $g(x)$
   b. Find all the zeros of this function.

59. **Physics** The formula $T = 2\pi\sqrt{\frac{\ell}{g}}$ is used to find the period T of a oscillating pendulum. In this formula, ℓ is the length of the pendulum, and g is acceleration due to gravity. Acceleration due to gravity is 9.8 meters per second squared. If a pendulum has an oscillation period of 1.6 seconds, determine the length of the pendulum. *(Lesson 4-7)*

ALTERNATIVE ASSESSMENT

OPEN-ENDED ASSESSMENT

1. Write a rational equation that has at least two solutions, one which is 2. Solve your equation.
2. a. Write a radical equation that has solutions of 3 and 6, one of which is extraneous.
   b. Solve your equation. Identify the extraneous solution and explain why it is extraneous.
3. a. Write a set of data that is best represented by a cubic equation.
   b. Write a polynomial function to model the set of data.
   c. Approximate the real zeros of the polynomial function to the nearest tenth.

PORTFOLIO

Explain how you can use the leading coefficient and the degree of a polynomial equation to determine the number of possible roots of the equation.

TELECOMMUNICATION

The Pen is Mightier than the Sword!

- Gather all materials obtained from your research for the mini-projects in Chapters 1, 2, and 3. Decide what types of software would help you to prepare a presentation.
- Research websites that offer downloads of software including work processing, graphics, spreadsheet, and presentation software. Determine whether the software is a demonstration version or free shareware. Select at least two different programs for each of the four categories listed above.
- Prepare a presentation of your Unit 1 project using the software that you found. Be sure that you include graphs and maps in the presentation.

Additional Assessment See p. A59 for Chapter 4 practice test.

Coordinate Geometry Problems

The ACT test usually includes several coordinate geometry problems. You'll need to know and apply these formulas for points (x_1, y_1) and (x_2, y_2):

Midpoint	Distance	Slope
$\left(\frac{x_1 + x_2}{2}, \frac{y_1 + y_2}{2}\right)$	$\sqrt{(x_2 - x_1)^2 + (y_2 - y_1)^2}$	$\frac{y_2 - y_1}{x_2 - x_1}$

The SAT test includes problems that involve coordinate points. But they aren't easy!

THE PRINCETON REVIEW

TEST-TAKING TIP

- Draw diagrams.
- Locate points on the grid or number line.
- Eliminate any choices that are clearly incorrect.

ACT EXAMPLE

1. Point $B(4,3)$ is the midpoint of line segment AC. If point A has coordinates (0,1), then what are the coordinates of point C?

   A (−4, −1) **B** (4, 1) **C** (4, 4)

   D (8, 5) **E** (8, 9)

HINT Draw a diagram. You may be able to solve the problem without calculations.

Solution Draw a diagram showing the known quantities and the unknown point C.

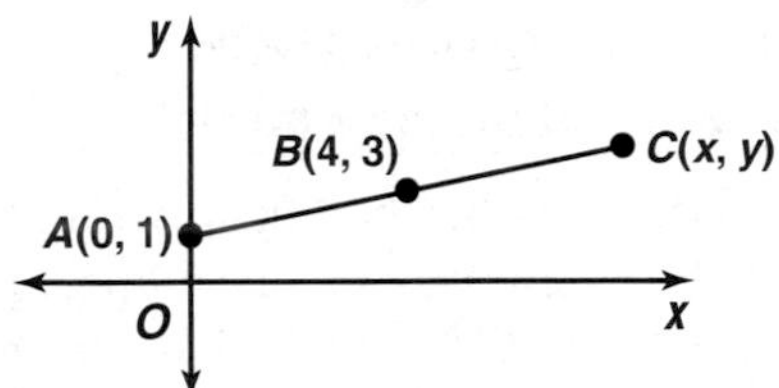

Since C lies to the right of B and the x-coordinate of A is not 4, any points with an x-coordinate of 4 or less can be eliminated. So eliminate choices A, B, and C.

Use the Midpoint Formula. Consider the x-coordinates. Write an equation in x.

$$\frac{0 + x}{2} = 4$$

$$x = 8$$

Do the same for y. $$\frac{1 + y}{2} = 3$$

$$y = 5$$

The coordinates of C are (8, 5) The answer is choice **D.**

SAT EXAMPLE

2. What is the area of square $ABCD$ in square units?

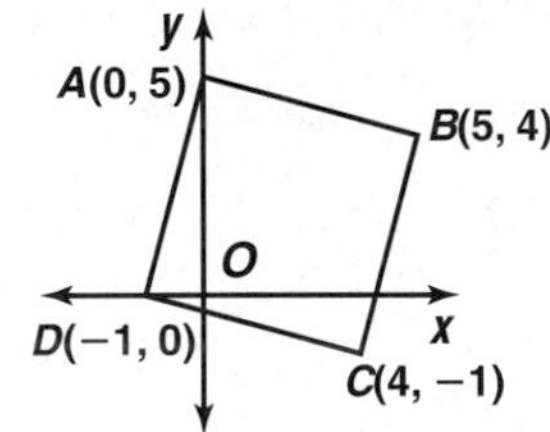

   A 25 **B** $18\sqrt{2}$ **C** 26

   D $25 + \sqrt{2}$ **E** 36

HINT Estimate the answer to eliminate impossible choices and to check your calculations.

Solution First estimate the area. Since the square's side is a little more than 5, the area is a little more than 25. Eliminate choices A and E.

To find the area, find the measure of a side and square it. Choose side AD, because points A and D have simple coordinates. Use the Distance Formula.

$$(\overline{AD})^2 = (-1 - 0)^2 + (0 - 5)^2$$

$$= (-1)^2 + (-5)^2$$

$$= 1 + 25 \text{ or } 26$$

The answer is choice **C.**

Alternate Solution You could also use the Pythagorean Theorem. Draw right triangle DOA, with the right angle at O, the origin. Then DO is 1, OA is 5, and DA is $\sqrt{26}$. So the area is 26 square units.

After you work each problem, record your answer on the answer sheet provided or on a piece of paper.

Multiple Choice

1. What is the length of the line segment whose endpoints are represented on the coordinate axis by points at $(-2, 1)$ and $(1, -3)$?

   A 3 **B** 4 **C** 5
   D 6 **E** 7

2. $\left(\frac{4}{5} \times 3\right)\left(\frac{3}{4} \times 5\right)\left(\frac{5}{3} \times 4\right) =$

   A 1 **B** 3 **C** 6 **D** 20 **E** 60

3. In the figure below, $ABCD$ is a parallelogram. What are the coordinates of point C?

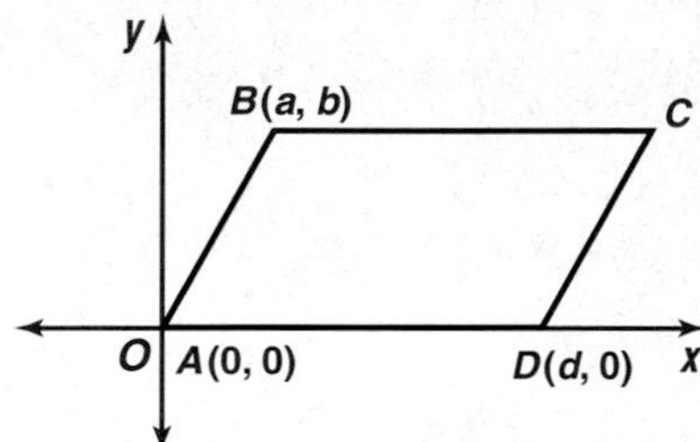

   A (x, y) **B** $(d + a, y)$
   C $(d - a, b)$ **D** $(d + x, b)$
   E $(d + a, b)$

4. A rectangular garden is surrounded by a 60-foot long fence. One side of the garden is 6 feet longer than the other. Which equation could be used to find s, the shorter side, of the garden?

   A $8s + s = 60$
   B $4s = 60 + 12$
   C $s(s + 6) = 60$
   D $2(s - 6) + 2s = 60$
   E $2(s + 6) + 2s = 60$

5. What is the slope of a line perpendicular to the line represented by the equation $3x - 6y = 12$?

   A -2 **B** $\frac{-1}{2}$ **C** $\frac{1}{3}$
   D $\frac{1}{2}$ **E** 2

6. If x is an integer, which of the following could be x^3?

   A 2.7×10^{11}
   B 2.7×10^{12}
   C 2.7×10^{13}
   D 2.7×10^{14}
   E 2.7×10^{15}

7. If $0x + 5y = 14$ and $4x - y = 2$, then what is the value of $6x + 6y$

   A 2 **B** 7 **C** 12 **D** 16 **E** 24

8. What is the midpoint of the line segment whose endpoints are represented on the coordinate grid by points at $(3, 5)$ and $(-4, 3)$?

   A $(-2, -5)$ **B** $\left(-\frac{1}{2}, 4\right)$ **C** $(1, 8)$
   D $\left(4, \frac{1}{2}\right)$ **E** $(3, 3)$

9. **Quantitative Comparison**

   A if the quantity in Column A is greater
   B if the quantity in Column B is greater
   C if the two quantities are equal
   D if the relationship cannot be determined from the information given

   $ax \neq 0$

Column A	Column B
$(x - a)^2$	$-2ax$

10. **Grid-In** Points E, F, G, and H lie on a line in that order. If $EG = \frac{5}{3}EF$ and $HF = 5FG$, then what is $\frac{EF}{HG}$?

interNET CONNECTION **SAT/ACT Practice** For additional test practice questions, visit: **www.amc.glencoe.com**

UNIT 2

Trigonometry

The simplest definition of trigonometry goes something like this:

Trigonometry is the study of angles and triangles. However, the study of trigonometry is really much more complex and comprehensive. In this unit alone, you will find trigonometric functions of angles, verify trigonometric identities, solve trigonometric equations, and graph trigonometric functions. Plus, you'll use vectors to solve parametric equations and to model motion. Trigonometry has applications in construction, geography, physics, acoustics, medicine, meteorology, and navigation, among other fields.

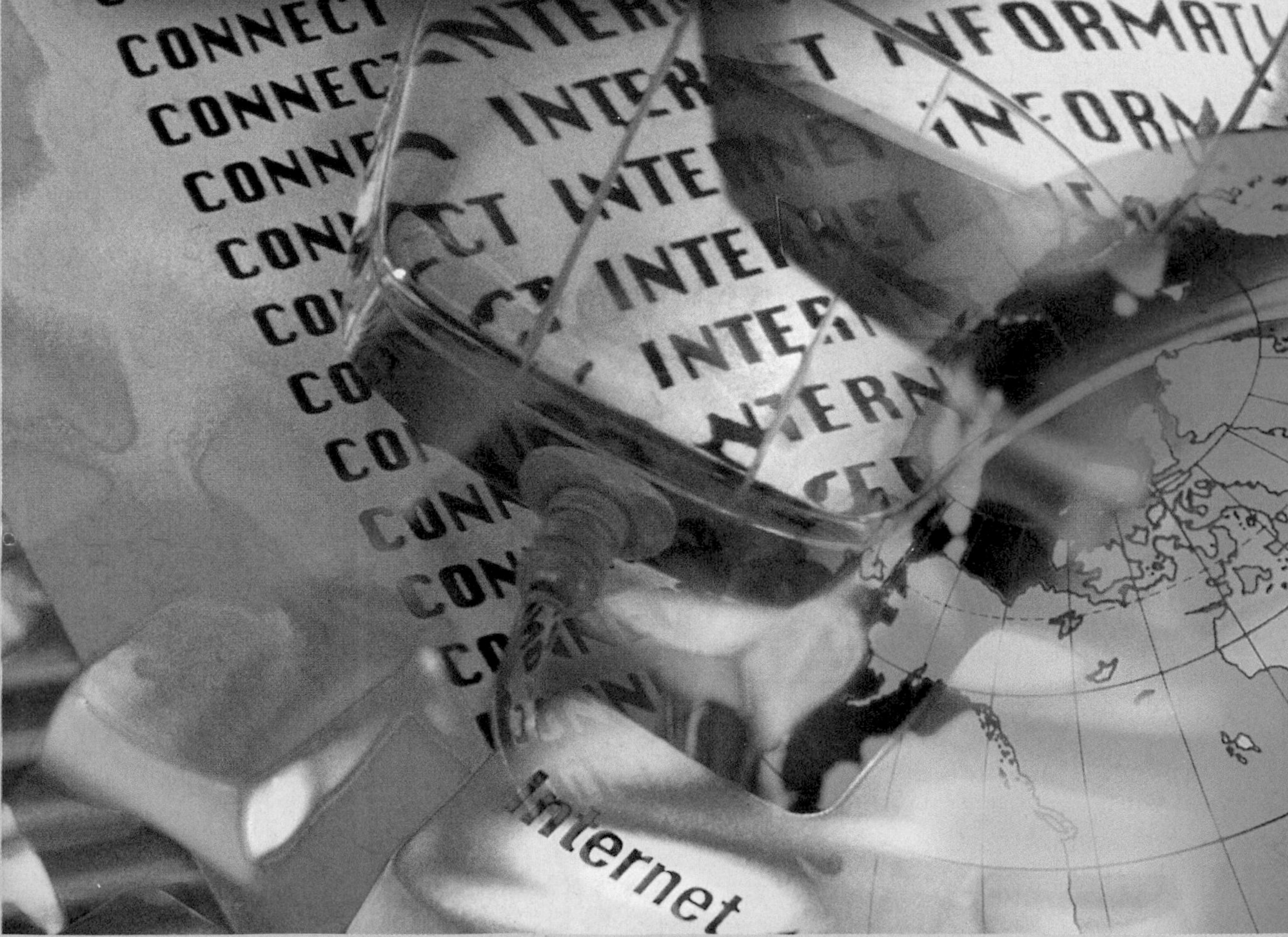

Unit 2 interNET Projects

THE CYBER CLASSROOM

From the Internet, a new form of classroom has emerged—one that exists only on the Web. Classes are now being taught exclusively over the Internet to students living across the globe. In addition, traditional classroom teachers are posting notes, lessons, and other information on web sites that can be accessed by their students. At the end of each chapter in Unit 2, you will search the Web for trigonometry learning resources.

CHAPTER 5 (page 339)

Does anybody out there know anything about trigonometry? Across the United States and the world, students are attending classes right in their own homes. What is it like to learn in this new environment? Use the Internet to find trigonometry lessons.
Math Connection: Find and compare trigonometry lessons from this textbook and the Internet. Then, select one topic from Chapter 5 and write a summary of this topic using the information you have gathered from both the textbook and the Internet.

CHAPTER 6 (page 417)

What is your sine? Mathematicians, scientists, and others share their work by means of the Internet. What are some applications of the sine or cosine function?
Math Connection: Use the Internet to find more applications of the sine or cosine function. Find data on the Internet that can be modeled by using a sine or cosine curve. Graph the data and the sine or cosine function that approximates it on the same axes.

CHAPTER 7 (page 481)

That's as clear as mud! Teachers using the Internet to deliver their courses need to provide the same clear instructions and examples to their students that they would in a traditional classroom.
Math Connection: Research the types of trigonometry sample problems given in Internet lessons. Design your own web page that includes two sample trigonometry problems.

CHAPTER 8 (page 547)

Vivid Vectors Suppose you are taking a physics class that requires you to use vectors to represent real world situations. Can you find out more about these representations of direction and magnitude?
Math Connection: Research learning sites on the Internet to find more information about vector applications. Describe three real-world situations that can be modeled by vectors. Include vector diagrams of each situation.

interNET CONNECTION For more information on the Unit Project, visit: **www.amc.glencoe.com** 

THE TRIGONOMETRIC FUNCTIONS

CHAPTER OBJECTIVES

- **Convert decimal degree measures to degrees, minutes, and seconds and vice versa.** *(Lesson 5-1)*
- **Identify angles that are coterminal with a given angle.** *(Lesson 5-1)*
- **Solve triangles.** *(Lessons 5-2, 5-4, 5-5, 5-6, 5-7, 5-8)*
- **Find the values of trigonometric functions.** *(Lessons 5-2, 5-3)*
- **Find the areas of triangles.** *(Lessons 5-6, 5-8)*

5-1 Angles and Degree Measure

OBJECTIVES

- Convert decimal degree measures to degrees, minutes, and seconds and vice versa.
- Find the number of degrees in a given number of rotations.
- Identify angles that are coterminal with a given angle.

NAVIGATION The sextant is an optical instrument invented around 1730. It is used to measure the angular elevation of stars, so that a navigator can determine the ship's current latitude. Suppose a navigator determines a ship in the Pacific Ocean to be located at north latitude 15.735°. How can this be written as degrees, minutes, and seconds? *This problem will be solved in Example 1.*

An angle may be generated by rotating one of two rays that share a fixed endpoint known as the **vertex.** One of the rays is fixed to form the **initial side** of the angle, and the second ray rotates to form the **terminal side.**

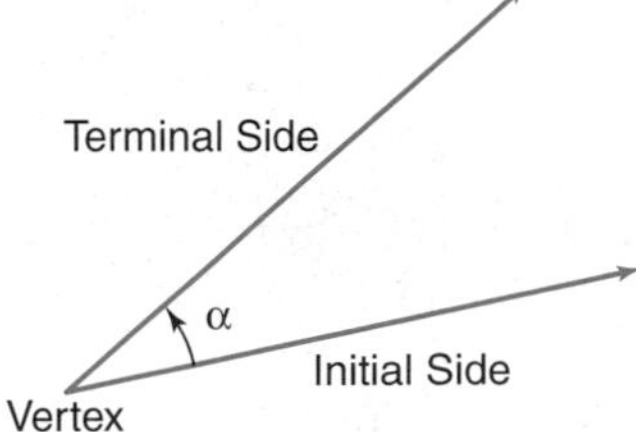

The measure of an angle provides us with information concerning the direction of the rotation and the amount of the rotation necessary to move from the initial side of the angle to the terminal side.

- If the rotation is in a counterclockwise direction, the angle formed is a *positive angle.*
- If the rotation is clockwise, it is a *negative angle.*

An angle with its vertex at the origin and its initial side along the positive x-axis is said to be in **standard position.** In the figures below, all of the angles are in standard position.

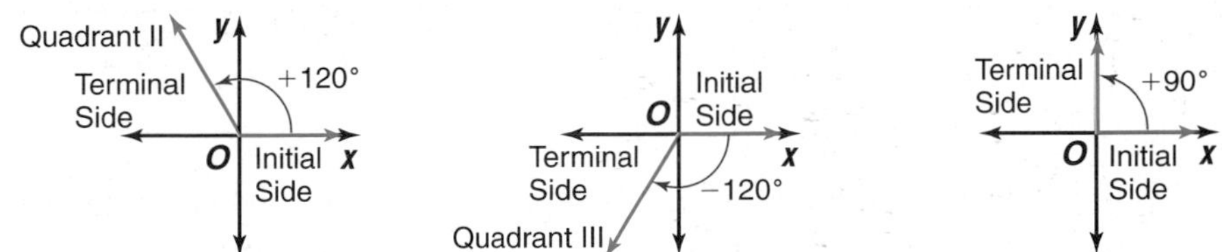

The most common unit used to measure angles is the **degree.** The concept of degree measurement is rooted in the ancient Babylonian culture. The Babylonians based their numeration system on 60 rather than 10 as we do today. In an equilateral triangle, they assigned the measure of each angle to be 60. Therefore, one sixtieth $\left(\frac{1}{60}\right)$ of the measure of the angle of an equilateral triangle was equivalent to one unit or degree (1°). The degree is subdivided into 60 equal parts known as **minutes** (1′), and the minute is subdivided into 60 equal parts known as **seconds** (1″).

Angles are used in a variety of real-world situations. For example, in order to locate every point on Earth, cartographers use a grid that contains circles through the poles, called *longitude lines,* and circles parallel to the equator, called *latitude lines.* Point P is located by traveling north from the equator through a central angle of $a°$ to a circle of latitude and then west along that circle through an angle of $b°$. Latitude and longitude can be expressed in degrees as a decimal value or in degrees, minutes, and seconds.

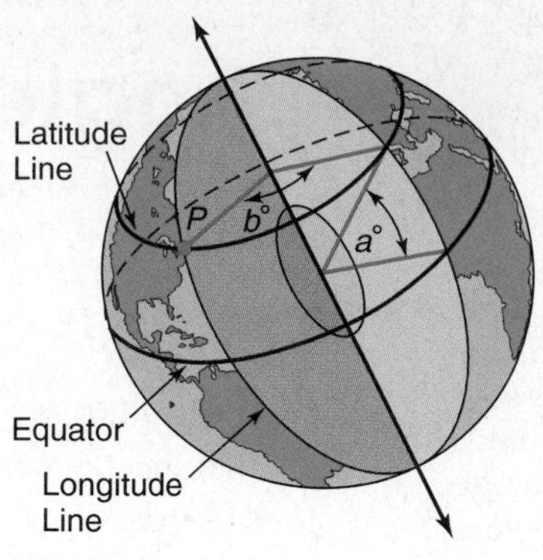

Example 1 **NAVIGATION** **Refer to the application at the beginning of the lesson.**

a. Change north latitude 15.735° to degrees, minutes, and seconds.

$15.735° = 15° + (0.735 \cdot 60)'$ *Multiply the decimal portion of the degree*
$= 15° + 44.1'$ *measure by 60 to find the number of minutes.*
$= 15° + 44' + (0.1 \cdot 60)''$ *Multiply the decimal portion of the minute*
$= 15° + 44' + 6''$ *measure by 60 to find the number of seconds.*

15.735° can be written as 15° 44′ 6″.

b. Write north latitude 39° 5′ 34″ as a decimal rounded to the nearest thousandth.

$39°\,5'\,34'' = 39° + 5'\left(\frac{1°}{60'}\right) + 34''\left(\frac{1°}{3600''}\right)$ or about 39.093°

39° 5′ 34″ can be written as 39.093°.

Graphing Calculator Tip

▶DMS on the **[ANGLE]** menu allows you to convert decimal degree values to degrees, minutes, and seconds.

If the terminal side of an angle that is in standard position coincides with one of the axes, the angle is called a **quadrantal angle.** In the figures below, all of the angles are quadrantal.

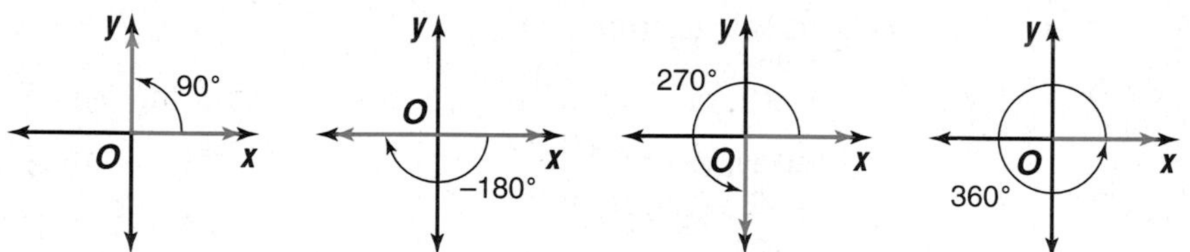

A full rotation around a circle is 360°. Measures of more than 360° represent multiple rotations.

Example 2 **Give the angle measure represented by each rotation.**

a. 5.5 rotations clockwise

$5.5 \times -360 = -1980$ *Clockwise rotations have negative measures.*

The angle measure of 5.5 clockwise rotations is −1980°.

b. 3.3 rotations counterclockwise

$3.3 \times 360 = 1188$ *Counterclockwise rotations have positive measures.*

The angle measure of 3.3 counterclockwise rotations is 1188°.

Two angles in standard position are called **coterminal angles** if they have the same terminal side. Since angles differing in degree measure by multiples of 360° are equivalent, every angle has infinitely many coterminal angles.

Coterminal Angles If α is the degree measure of an angle, then all angles measuring $\alpha + 360k°$, where k is an integer, are coterminal with α.

The symbol α is the lowercase Greek letter alpha.

Any angle coterminal with an angle of 75° can be written as $75° + 360k°$, where k is the number of rotations around the circle. The value of k is a positive integer if the rotations are counterclockwise and a negative integer if the rotations are clockwise.

Examples

3 Identify all angles that are coterminal with each angle. Then find one positive angle and one negative angle that are coterminal with the angle.

a. 45°

All angles having a measure of $45° + 360k°$, where k is an integer, are coterminal with 45°. A positive angle is $45° + 360°(1)$ or 405°. A negative angle is $45° + 360°(-2)$ or $-675°$.

b. 225°

All angles having a measure of $225° + 360k°$, where k is an integer, are coterminal with 225°. A positive angle is $225° + 360(2)°$ or 945°. A negative angle is $225° + 360(-1)°$ or $-135°$.

4 If each angle is in standard position, determine a coterminal angle that is between 0° and 360°. State the quadrant in which the terminal side lies.

a. 775°

In $\alpha + 360k°$, you need to find the value of α. First, determine the number of complete rotations (k) by dividing 775 by 360.

$$\frac{775}{360} \approx 2.152777778$$

Then, determine the number of remaining degrees (α).

Method 1

$$\alpha \approx 0.152777778 \text{ rotations} \cdot 360° \approx 55°$$

Method 2

$$\alpha + 360(2)° = 775°$$
$$\alpha + 720° = 775°$$
$$\alpha = 55°$$

The coterminal angle (α) is 55°. Its terminal side lies in the first quadrant.

b. −1297°

Use a calculator.

```
-1297/360
      -3.602777778
Ans+3
      -.6027777778
Ans*360
              -217
```

The angle is $-217°$, but the coterminal angle needs to be positive.

$360° - 217° = 143°$

The coterminal angle (α) is 143°. Its terminal side lies in the second quadrant.

If α is a nonquadrantal angle in standard position, its **reference angle** is defined as the acute angle formed by the terminal side of the given angle and the x-axis. You can use the figures and the rule below to find the reference angle for any angle α where $0° < \alpha < 360°$. If the measure of α is greater than 360° or less than 0°, it can be associated with a coterminal angle of positive measure between 0° and 360°.

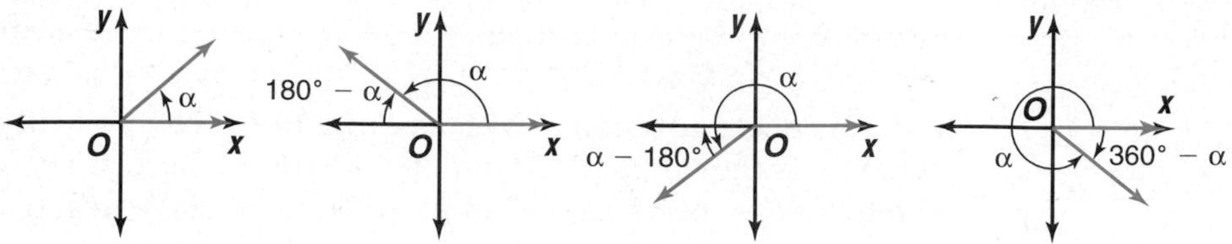

Reference Angle Rule

For any angle α, $0° < \alpha < 360°$, its reference angle α' is defined by
a. α, when the terminal side is in Quadrant I,
b. $180° - \alpha$, when the terminal side is in Quadrant II,
c. $\alpha - 180°$, when the terminal side is in Quadrant III, and
d. $360° - \alpha$, when the terminal side is in Quadrant IV.

Example 5 **Find the measure of the reference angle for each angle.**

a. 120°

Since 120° is between 90° and 180°, the terminal side of the angle is in the second quadrant.

$180° - 120° = 60°$

The reference angle is 60°.

b. −135°

A coterminal angle of −135 is 360 − 135 or 225. Since 225 is between 180° and 270°, the terminal side of the angle is in the third quadrant.

$225° - 180° = 45°$

The reference angle is 45°.

CHECK FOR UNDERSTANDING

Communicating Mathematics

Read and study the lesson to answer each question.

1. **Describe** the difference between an angle with a positive measure and an angle with a negative measure.
2. **Explain** how to write 29° 45′ 26″ as a decimal degree measure.
3. **Write** an expression for the measures of all angles that are coterminal with the angle shown.
4. **Sketch** an angle represented by 3.5 counterclockwise rotations. Give the angle measure represented by this rotation.

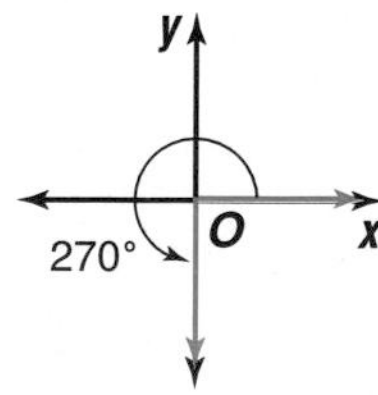

Guided Practice

Change each measure to degrees, minutes, and seconds.

5. 34.95°
6. −72.775°

Write each measure as a decimal to the nearest thousandth.

7. −128° 30′ 45″
8. 29° 6′ 6″

Give the angle measure represented by each rotation.

9. 2 rotations clockwise

10. 4.5 rotations counterclockwise

Identify all angles that are coterminal with each angle. Then find one positive angle and one negative angle that are coterminal with each angle.

11. $22°$

12. $-170°$

If each angle is in standard position, determine a coterminal angle that is between 0° and 360°. State the quadrant in which the terminal side lies.

13. $453°$

14. $-798°$

Find the measure of the reference angle for each angle.

15. $227°$

16. $-210°$

17. **Geography** Earth rotates once on its axis approximately every 24 hours. About how many degrees does a point on the equator travel through in one hour? in one minute? in one second?

EXERCISES

Practice

Change each measure to degrees, minutes, and seconds.

18. $-16.75°$

19. $168.35°$

20. $-183.47°$

21. $286.88°$

22. $27.465°$

23. $246.876°$

Write each measure as a decimal to the nearest thousandth.

24. $23° \; 14' \; 30''$

25. $-14° \; 5' \; 20''$

26. $233° \; 25' \; 15''$

27. $173° \; 24' \; 35''$

28. $-405° \; 16' \; 18''$

29. $1002° \; 30' \; 30''$

Give the angle measure represented by each rotation.

30. 3 rotations clockwise

31. 2 rotations counterclockwise

32. 1.5 rotations counterclockwise

33. 7.5 rotations clockwise

34. 2.25 rotations counterclockwise

35. 5.75 rotations clockwise

36. How many degrees are represented by 4 counterclockwise revolutions?

Identify all angles that are coterminal with each angle. Then find one positive angle and one negative angle that are coterminal with each angle.

37. $30°$

38. $-45°$

39. $113°$

40. $217°$

41. $-199°$

42. $-305°$

43. Determine the angle between $0°$ and $360°$ that is coterminal with all angles represented by $310° + 360k°$, where k is any integer.

44. Find the angle that is two counterclockwise rotations from $60°$. Then find the angle that is three clockwise rotations from $60°$.

If each angle is in standard position, determine a coterminal angle that is between 0° and 360°. State the quadrant in which the terminal side lies.

45. $400°$

46. $-280°$

47. $940°$

48. $1059°$

49. $-624°$

50. $-989°$

51. In what quadrant is the terminal side of a 1275° angle located?

Find the measure of the reference angle for each angle.

52. 327° 53. 148° 54. 563° 55. −420° 56. −197° 57. 1045°

58. Name four angles between 0° and 360° with a reference angle of 20°.

Applications and Problem Solving

59. **Technology** A computer's hard disk is spinning at 12.5 revolutions per second. Through how many degrees does it travel in a second? in a minute?

60. **Critical Thinking** Write an expression that represents all quadrantal angles.

61. **Biking** During the winter, a competitive bike rider trains on a stationary bike. Her trainer wants her to warm up for 5 to 10 minutes by pedaling slowly. Then she is to increase the pace to 95 revolutions per minute for 30 seconds. Through how many degrees will a point on the outside of the tire travel during the 30 seconds of the faster pace?

62. **Flywheels** A high-performance composite flywheel rotor can spin anywhere between 30,000 and 100,000 revolutions per minute. What is the range of degrees through which the composite flywheel can travel in a minute? Write your answer in scientific notation.

63. **Astronomy** On January 28, 1998, an x-ray satellite spotted a neutron star that spins at a rate of 62 times per second. Through how many degrees does this neutron star rotate in a second? in a minute? in an hour? in a day?

64. **Critical Thinking** Write an expression that represents any angle that is coterminal with a 25° angle, a 145° angle, and a 265° angle.

65. **Aviation** The locations of two airports are indicated on the map.
   a. Write the latitude and longitude of the Hancock County–Bar Harbor airport in Bar Harbor, Maine, as degrees, minutes, and seconds.
   b. Write the latitude and longitude of the Key West International Airport in Key West, Florida, as a decimal to the nearest thousandth.

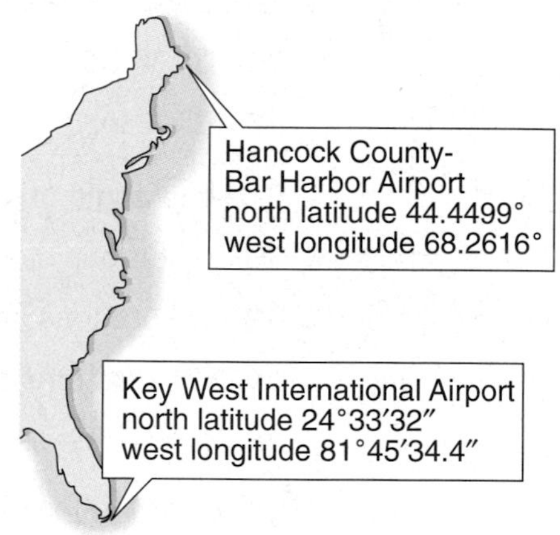

66. **Entertainment** A tower restaurant in Sydney, Australia, is 300 meters above sea level and provides a 360° panoramic view of the city as it rotates every 70 minutes. A tower restaurant in San Antonio, Texas, is 750 feet tall. It revolves at a rate of one revolution per hour.
   a. In a day, how many more revolutions does the restaurant in San Antonio make than the one in Sydney?
   b. In a week, how many more degrees does a speck of dirt on the window of the restaurant in San Antonio revolve than a speck of dirt on the window of the restaurant in Sydney?

Mixed Review

*inter*NET CONNECTION

Data Update For the latest information about motor vehicle production, visit our website at **www.amc.glencoe.com**

67. Manufacturing The percent of the motor vehicles produced in the United States since 1950 is depicted in the table at the right. *(Lesson 4-8)*

a. Write an equation to model the percent of the motor vehicles produced in the United States as a function of the number of years since 1950.

b. According to the equation, what percent of motor vehicles will be produced in the United States in the year 2010?

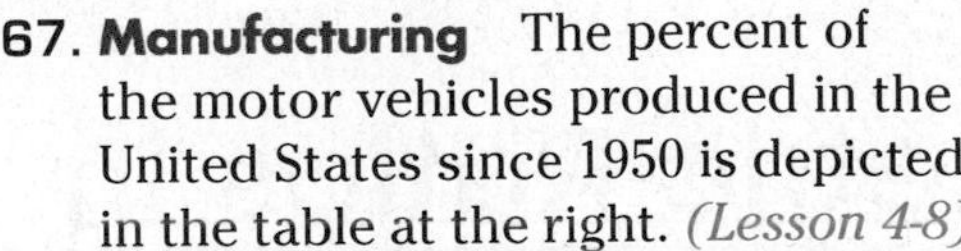

Motor Vehicle Production in the United States

Year	Percent
1950	75.7
1960	47.9
1970	28.2
1980	20.8
1990	20.1
1992	20.2
1993	23.3
1994	24.8
1997	22.7

Source: American Automobile Manufacturers Association

68. Solve $\sqrt[3]{6n + 5} - 15 = -10$. *(Lesson 4-7)*

69. Solve $\frac{x + 3}{x+2} = 2 - \frac{3}{x^2 + 5x + 6}$. *(Lesson 4-6)*

70. Use the Remainder Theorem to find the remainder if $x^3 + 8x + 1$ is divided by $x - 2$. *(Lesson 4-3)*

71. Write a polynomial equation of least degree with roots -5, -6, and 10. *(Lesson 4-1)*

72. If r varies inversely as t and $r = 18$ when $t = -3$, find r when $t = -11$. *(Lesson 3-8)*

73. Determine whether the graph of $y = \frac{x^2 - 1}{x + 1}$ has infinite discontinuity, jump discontinuity, or point discontinuity, or is continuous. Then graph the function. *(Lesson 3-7)*

74. Graph $f(x) = |(x + 1)^2 + 2|$. Determine the interval(s) for which the function is increasing and the interval(s) for which the function is decreasing. *(Lesson 3-5)*

75. Use the graph of the parent function $f(x) = \frac{1}{x}$ to describe the graph of the related function $g(x) = \frac{3}{x} - 2$. *(Lesson 3-2)*

76. Solve the system of inequalities $y \leq 5$, $3y \geq 2x + 9$, and $-3y \leq 6x - 9$ by graphing. Name the coordinates of the vertices of the convex set. *(Lesson 2-6)*

77. Find $[f \cdot g](x)$ if $f(x) = x - 0.2x$ and $g(x) = x - 0.3x$. *(Lesson 1-2)*

78. SAT/ACT Practice $\overline{AB}$ is a diameter of circle O, and $m\angle BOD = 15°$. If $m\angle EOA = 85°$, find $m\angle ECA$.

A 85° **B** 50° **C** 70°

D 35° **E** 45°

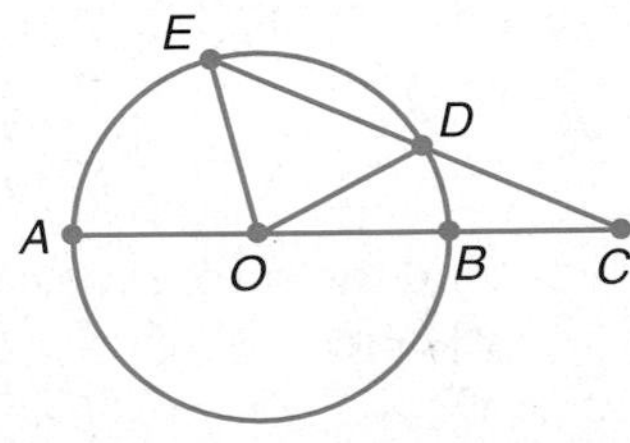

Extra Practice See p. A34.

Trigonometric Ratios in Right Triangles

OBJECTIVE

- Find the values of trigonometric ratios for acute angles of right triangles.

PHYSICS As light passes from one substance such as air to another substance such as glass, the light is bent. The relationship between the angle of incidence θ_i and the angle of refraction θ_r is given by Snell's Law, $\frac{\sin \theta_i}{\sin \theta_r} = n$, where $\sin \theta$ represents a trigonometric ratio and n is a constant called the *index of refraction.* Suppose a ray of light passes from air with an angle of incidence of 50° to glass with an angle of refraction of 32° 16′. Find the index of refraction of the glass. *This problem will be solved in Example 2.*

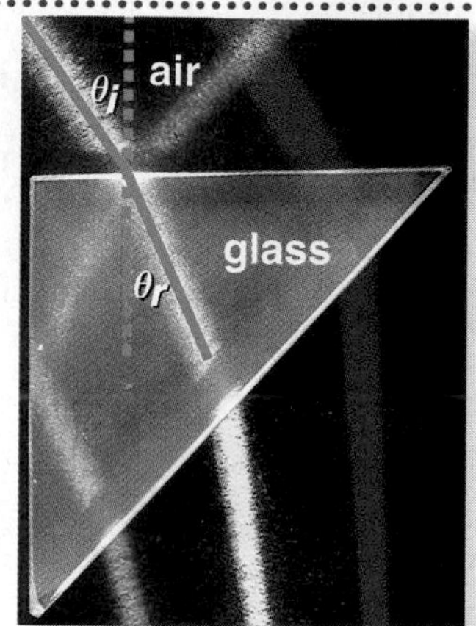

In a right triangle, one of the angles measures 90°, and the remaining two angles are *acute* and *complementary.* The longest side of a right triangle is known as the **hypotenuse** and is opposite the right angle. The other two sides are called **legs.** The leg that is a side of an acute angle is called the **side adjacent** to the angle. The other leg is the **side opposite** the angle.

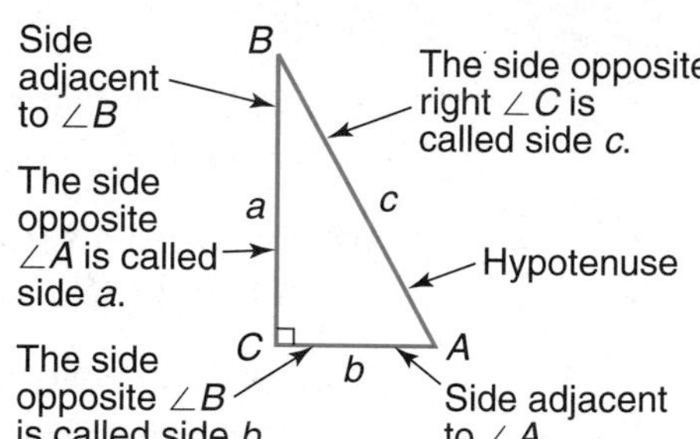

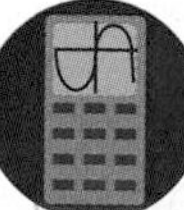

GRAPHING CALCULATOR EXPLORATION

Use a graphing calculator to find each ratio for the 22.6° angle in each triangle. Record each ratio as a decimal. Make sure your calculator is in degree mode.

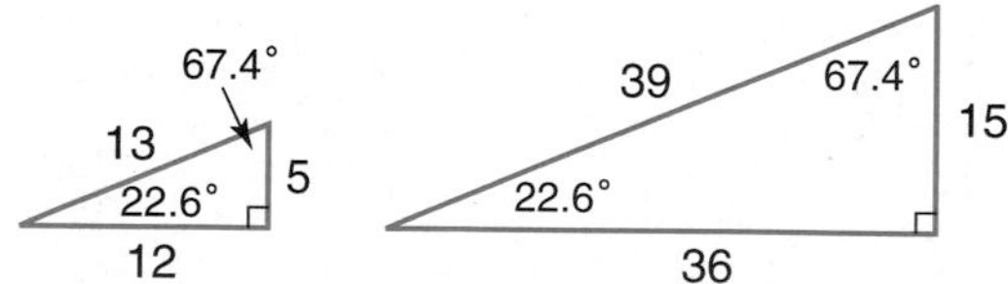

$R_1 = \frac{\text{side opposite}}{\text{hypotenuse}}$ $R_2 = \frac{\text{side adjacent}}{\text{hypotenuse}}$

$R_3 = \frac{\text{side opposite}}{\text{side adjacent}}$

Find the same ratios for the 67.4° angle in each triangle.

TRY THESE

1. Draw two other triangles that are similar to the given triangles.
2. Find each ratio for the 22.6° angle in each triangle.
3. Find each ratio for the 67.4° angle in each triangle.

WHAT DO YOU THINK?

4. Make a conjecture about R_1, R_2, and R_3 for any right triangle with a 22.6° angle.
5. Is your conjecture true for any 67.4° angle in a right triangle?
6. Do you think your conjecture is true for any acute angle of a right triangle? Why?

If two angles of a triangle are congruent to two angles of another triangle, the triangles are similar. If an acute angle of one right triangle is congruent to an acute angle of another right triangle, the triangles are similar, and the ratios of the corresponding sides are equal. Therefore, any two congruent angles of different right triangles will have equal ratios associated with them.

In right triangles, the Greek letter θ (theta) is often used to denote a particular angle.

The ratios of the sides of the right triangles can be used to define the **trigonometric ratios.** The ratio of the side opposite θ and the hypotenuse is known as the **sine.** The ratio of the side adjacent θ and the hypotenuse is known as the **cosine.** The ratio of the side opposite θ and the side adjacent θ is known as the **tangent.**

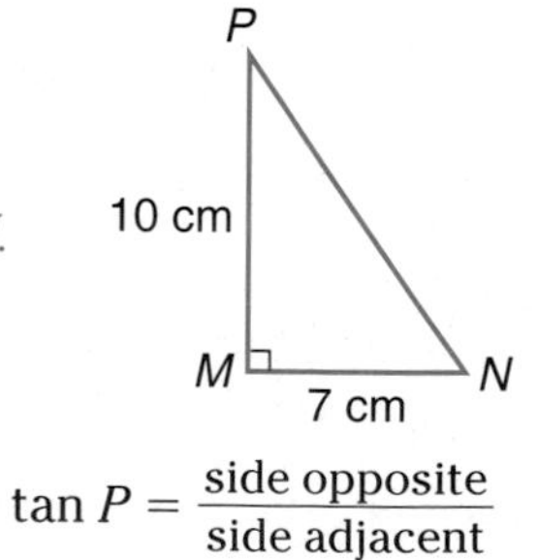

	Words	Symbol	Definition
Trigonometric Ratios	sine θ	$\sin\theta$	$\sin\theta = \frac{\text{side opposite}}{\text{hypotenuse}}$
	cosine θ	$\cos\theta$	$\cos\theta = \frac{\text{side adjacent}}{\text{hypotenuse}}$
	tangent θ	$\tan\theta$	$\tan\theta = \frac{\text{side opposite}}{\text{side adjacent}}$

Side Opposite
Hypotenuse
Side Adjacent
θ

SOH-CAH-TOA is a mnemonic device commonly used for remembering these ratios.

$$\sin\theta = \frac{\text{opposite}}{\text{hypotenuse}} \qquad \cos\theta = \frac{\text{adjacent}}{\text{hypotenuse}} \qquad \tan\theta = \frac{\text{opposite}}{\text{adjacent}}$$

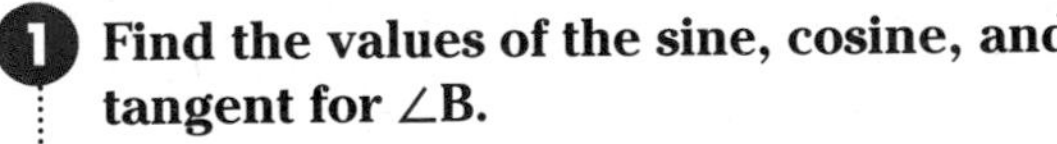

Example 1 **Find the values of the sine, cosine, and tangent for ∠B.**

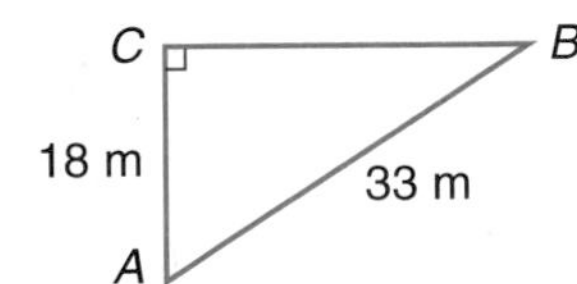

First, find the length of $\overline{BC}$

$(AC)^2 + (BC)^2 = (AB)^2$ *Pythagorean Theorem*

$18^2 + (BC)^2 = 33^2$ *Substitute 18 for AC and 33 for AB.*

$(BC)^2 = 765$

$BC = \sqrt{765}$ or $3\sqrt{85}$ *Take the square root of each side. Disregard the negative root.*

Then write each trigonometric ratio.

$$\sin B = \frac{\text{side opposite}}{\text{hypotenuse}} \qquad \cos B = \frac{\text{side adjacent}}{\text{hypotenuse}} \qquad \tan B = \frac{\text{side opposite}}{\text{side adjacent}}$$

$$\sin B = \frac{18}{33} \text{ or } \frac{6}{11} \qquad \cos B = \frac{3\sqrt{85}}{33} \text{ or } \frac{\sqrt{85}}{11} \qquad \tan B = \frac{18}{3\sqrt{85}} \text{ or } \frac{6\sqrt{85}}{85}$$

Trigonometric ratios are often simplified, but never written as mixed numbers.

In Example 1, you found the exact values of the sine, cosine, and tangent ratios. You can use a calculator to find the approximate decimal value of any of the trigonometric ratios for a given angle.

Example **2** **PHYSICS** **Refer to the application at the beginning of the lesson. Find the index of refraction of the glass.**

$$\frac{\sin \theta_i}{\sin \theta_r} = n \quad \textit{Snell's Law}$$

$$\frac{\sin 50°}{\sin 32°\ 16'} = n \quad \textit{Substitute 50° for } \theta_i \textit{ and 32° 16' for } \theta_r$$

$$\frac{0.7660444431}{0.5338605056} \approx n \quad \textit{Use a calculator to find each sine ratio.}$$

$$1.434914992 \approx n \quad \textit{Use a calculator to find the quotient.}$$

The index of refraction of the glass is about 1.4349.

Graphing Calculator Tip

If using your graphing calculator to do the calculation, make sure you are in degree mode.

In addition to the trigonometric ratios sine, cosine, and tangent, there are three other trigonometric ratios called **cosecant, secant,** and **cotangent.** These ratios are the reciprocals of sine, cosine, and tangent, respectively.

	Words	Symbol	Definition
Reciprocal Trigonometric Ratios	cosecant θ	csc θ	$\csc \theta = \frac{1}{\sin \theta}$ or $\frac{\text{hypotenuse}}{\text{side opposite}}$
	secant θ	sec θ	$\sec \theta = \frac{1}{\cos \theta}$ or $\frac{\text{hypotenuse}}{\text{side adjacent}}$
	cotangent θ	cot θ	$\cot \theta = \frac{1}{\tan \theta}$ or $\frac{\text{side adjacent}}{\text{side opposite}}$

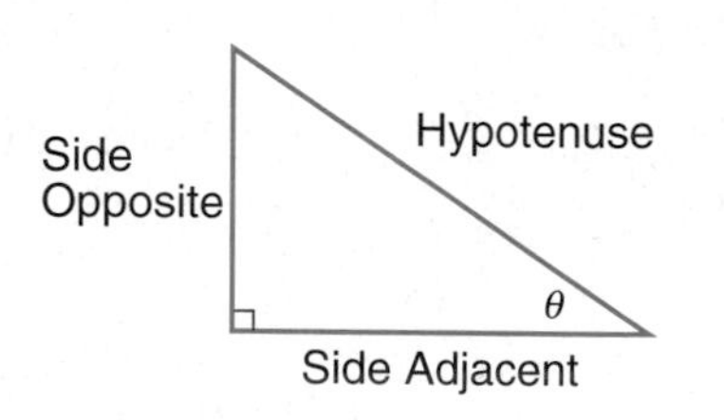

These definitions are called the 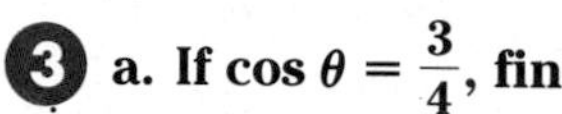 *reciprocal identities.*

Examples **3** **a. If $\cos \theta = \frac{3}{4}$, find $\sec \theta$.**

$$\sec \theta = \frac{1}{\cos \theta}$$

$$\sec \theta = \frac{1}{\frac{3}{4}} \text{ or } \frac{4}{3}$$

b. If $\csc \theta = 1.345$, find $\sin \theta$.

$$\sin \theta = \frac{1}{\csc \theta}$$

$$\sin \theta = \frac{1}{1.345} \text{ or about } 0.7435$$

4 **Find the values of the six trigonometric ratios for $\angle P$.**

First determine the length of the hypotenuse.

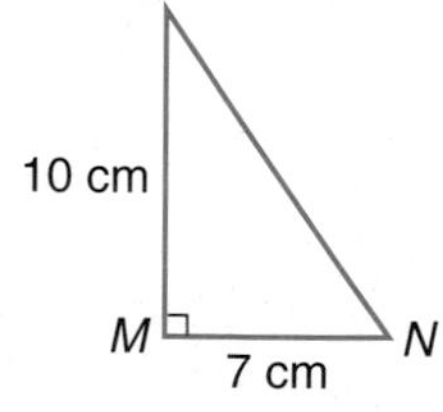

$$(MP)^2 + (MN)^2 = (NP)^2 \quad \textit{Pythagorean Theorem}$$

$$10^2 + 7^2 = (NP)^2 \quad \textit{Substitute 10 for MP and 7 for MN.}$$

$$149 = (NP)^2$$

$$\sqrt{149} = NP \quad \textit{Disregard the negative root.}$$

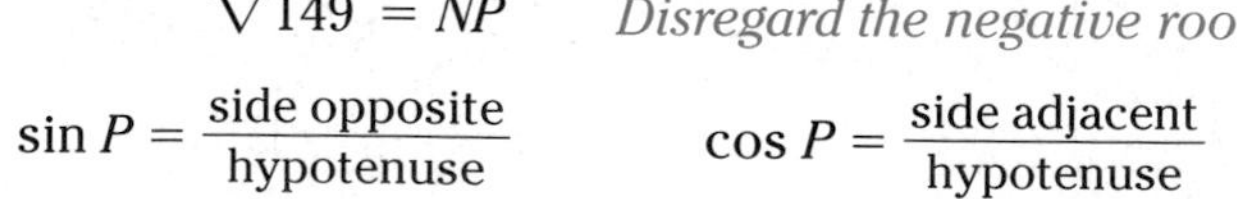

$$\sin P = \frac{\text{side opposite}}{\text{hypotenuse}} \qquad \cos P = \frac{\text{side adjacent}}{\text{hypotenuse}} \qquad \tan P = \frac{\text{side opposite}}{\text{side adjacent}}$$

$$\sin P = \frac{7}{\sqrt{149}} \text{ or } \frac{7\sqrt{149}}{149} \qquad \cos P = \frac{10}{\sqrt{149}} \text{ or } \frac{10\sqrt{149}}{149} \qquad \tan P = \frac{7}{10}$$

$$\csc P = \frac{\text{hypotenuse}}{\text{side opposite}} \qquad \sec P = \frac{\text{hypotenuse}}{\text{side adjacent}} \qquad \cot P = \frac{\text{side adjacent}}{\text{side opposite}}$$

$$\csc P = \frac{\sqrt{149}}{7} \qquad \sec P = \frac{\sqrt{149}}{10} \qquad \cot P = \frac{10}{7}$$

Consider the special relationships among the sides of $30°–60°–90°$ and $45°–45°–90°$ triangles.

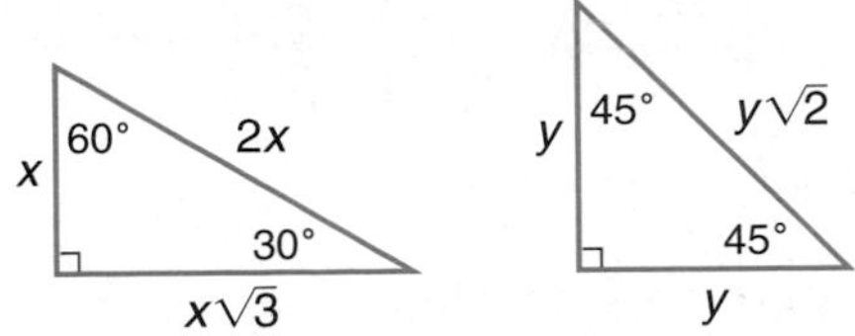

These special relationships can be used to determine the trigonometric ratios for $30°$, $45°$, and $60°$. You should memorize the sine, cosine, and tangent values for these angles.

θ	$\sin \theta$	$\cos \theta$	$\tan \theta$	$\csc \theta$	$\sec \theta$	$\cot \theta$
30°	$\frac{1}{2}$	$\frac{\sqrt{3}}{2}$	$\frac{\sqrt{3}}{3}$	2	$\frac{2\sqrt{3}}{3}$	$\sqrt{3}$
45°	$\frac{\sqrt{2}}{2}$	$\frac{\sqrt{2}}{2}$	1	$\sqrt{2}$	$\sqrt{2}$	1
60°	$\frac{\sqrt{3}}{2}$	$\frac{1}{2}$	$\sqrt{3}$	$\frac{2\sqrt{3}}{3}$	2	$\frac{\sqrt{3}}{3}$

Note that $\sin 30° = \cos 60°$ and $\cos 30° = \sin 60°$. This is an example showing that the sine and cosine are **cofunctions.** That is, if θ is an acute angle, $\sin \theta = \cos (90° - \theta)$. Similar relationships hold true for the other trigonometric ratios.

Cofunctions

$\sin \theta = \cos (90° - \theta)$ $\quad$ $\cos \theta = \sin (90° - \theta)$

$\tan \theta = \cot (90° - \theta)$ $\quad$ $\cot \theta = \tan (90° - \theta)$

$\sec \theta = \csc (90° - \theta)$ $\quad$ $\csc \theta = \sec (90° - \theta)$

CHECK FOR UNDERSTANDING

Communicating Mathematics

Read and study the lesson to answer each question.

1. **Explain** in your own words how to decide which side is opposite the given acute angle of a right triangle and which side is adjacent to the given angle.
2. **State** the reciprocal ratios of sine, cosine, and tangent.
3. **Write** each trigonometric ratio for $\angle A$ in triangle *ABC*.

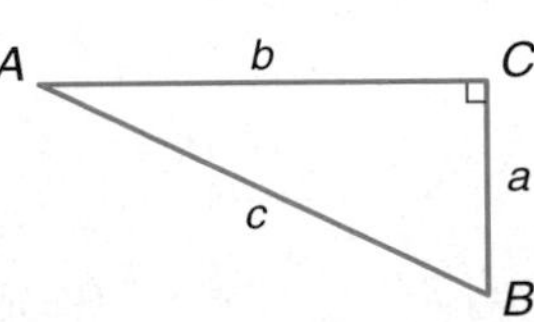

4. **Compare** $\sin A$ and $\cos B$, $\csc A$ and $\sec B$, and $\tan A$ and $\cot B$.

Guided Practice

5. Find the values of the sine, cosine, and tangent for $\angle T$.

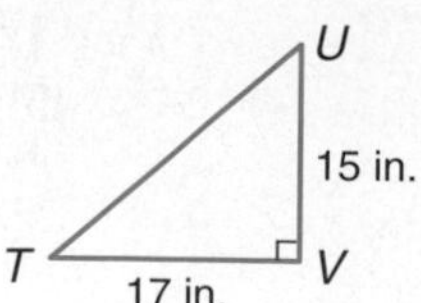

6. If $\sin \theta = \frac{2}{5}$, find $\csc \theta$.

7. If $\cot \theta = 1.5$, find $\tan \theta$.

8. Find the values of the six trigonometric ratios for $\angle P$.

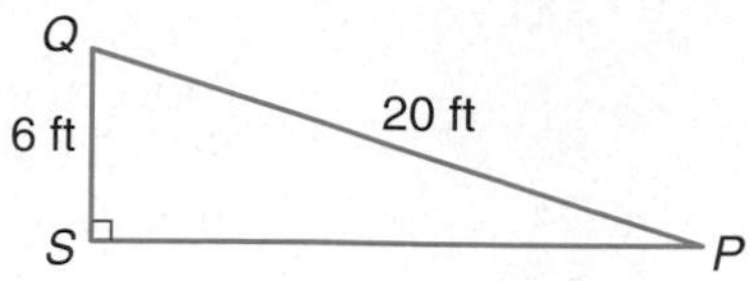

9. **Physics** You may have polarized sunglasses that eliminate glare by polarizing the light. When light is polarized, all of the waves are traveling in parallel planes. Suppose vertically polarized light with intensity I_o strikes a polarized filter with its axis at an angle of θ with the vertical. The intensity of the transmitted light I_t and θ are related by the equation $\cos \theta = \sqrt{\frac{I_t}{I_o}}$. If θ is 45°, write I_t as a function of I_o.

EXERCISES

Practice

Find the values of the sine, cosine, and tangent for each $\angle A$.

10.

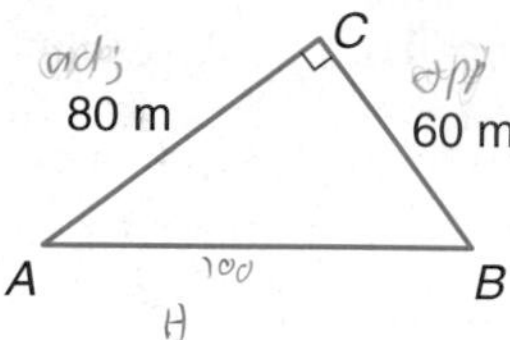

11.

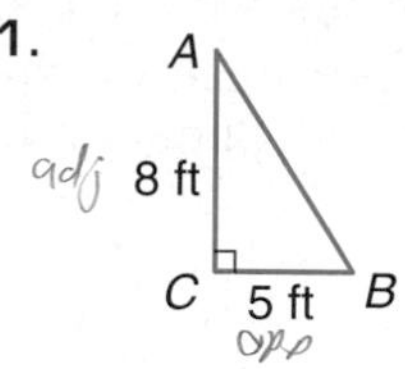

12.

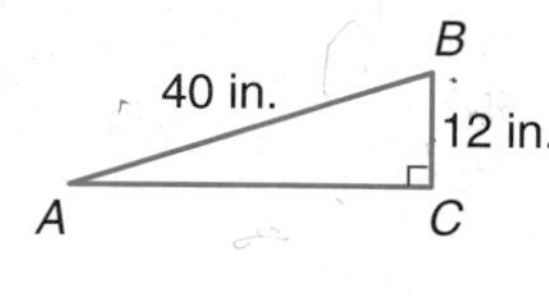

13. The slope of a line is the ratio of the change of y to the change of x. Name the trigonometric ratio of θ that equals the slope of line m.

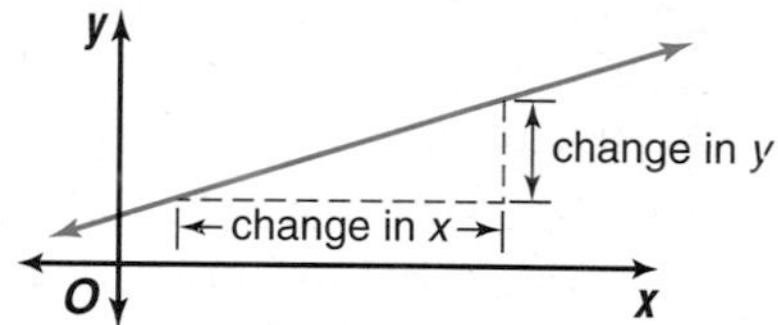

14. If $\tan \theta = \frac{1}{3}$, find $\cot \theta$.

15. If $\sin \theta = \frac{3}{7}$, find $\csc \theta$.

16. If $\sec \theta = \frac{5}{9}$, find $\cos \theta$.

17. If $\csc \theta = 2.5$, find $\sin \theta$.

18. If $\cot \theta = 0.75$, find $\tan \theta$.

19. If $\cos \theta = 0.125$, find $\sec \theta$.

Find the values of the six trigonometric ratios for each $\angle R$.

20.

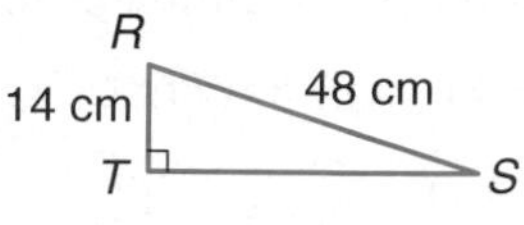

21.

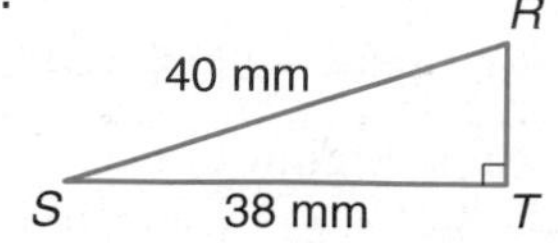

22.

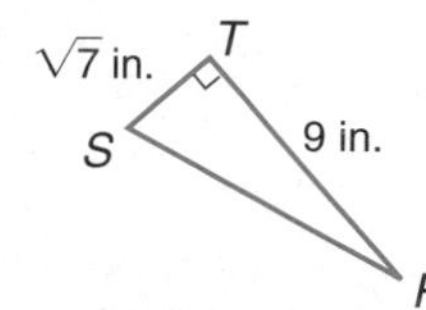

23. If $\tan \theta = 1.3$, what is the value of $\cot (90° - \theta)$?

24. Use a calculator to determine the value of each trigonometric ratio.

a. sin 52° 47′ b. cos 79° 15′ c. tan 88° 22′ 45″

d. cot 36° (*Hint*: Tangent and cotangent have a reciprocal relationship.)

Graphing Calculator

25. Use the table function on a graphing calculator to complete the table. Round values to three decimal places.

θ	72°	74°	76°	78°	80°	82°	84°	86°	88°
sin	0.951	0.961							
cos	0.309								

a. What value does sin θ approach as θ approaches 90°?

b. What value does cos θ approach as θ approaches 90°?

26. Use the table function on a graphing calculator to complete the table. Round values to three decimal places.

θ	18°	16°	14°	12°	10°	8°	6°	4°	2°
sin	0.309	0.276							
cos	0.951								
tan									

a. What value does sin θ approach as θ approaches 0°?

b. What value does cos θ approach as θ approaches 0°?

c. What value does tan θ approach as θ approaches 0°?

Applications and Problem Solving

Real World Application

27. Physics Suppose a ray of light passes from air to Lucite. The measure of the angle of incidence is 45°, and the measure of an angle of refraction is 27° 55′. Use Snell's Law, which is stated in the application at the beginning of the lesson, to find the index of refraction for Lucite.

28. Critical Thinking The sine of an acute $\angle R$ of a right triangle is $\frac{3}{7}$. Find the values of the other trigonometric ratios for this angle.

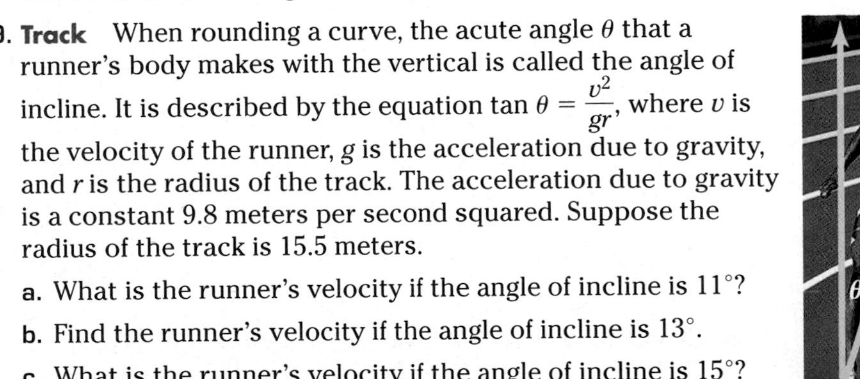

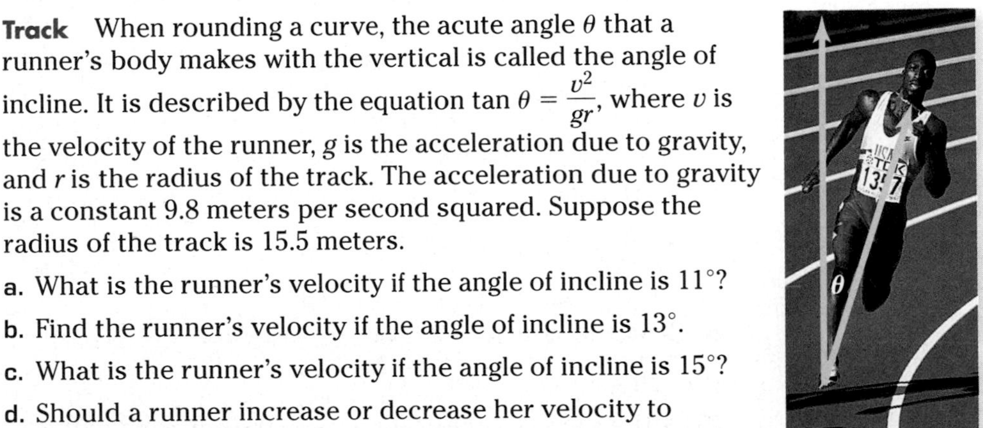

29. Track When rounding a curve, the acute angle θ that a runner's body makes with the vertical is called the angle of incline. It is described by the equation $\tan \theta = \frac{v^2}{gr}$, where v is the velocity of the runner, g is the acceleration due to gravity, and r is the radius of the track. The acceleration due to gravity is a constant 9.8 meters per second squared. Suppose the radius of the track is 15.5 meters.

a. What is the runner's velocity if the angle of incline is 11°?

b. Find the runner's velocity if the angle of incline is 13°.

c. What is the runner's velocity if the angle of incline is 15°?

d. Should a runner increase or decrease her velocity to increase his or her angle of incline?

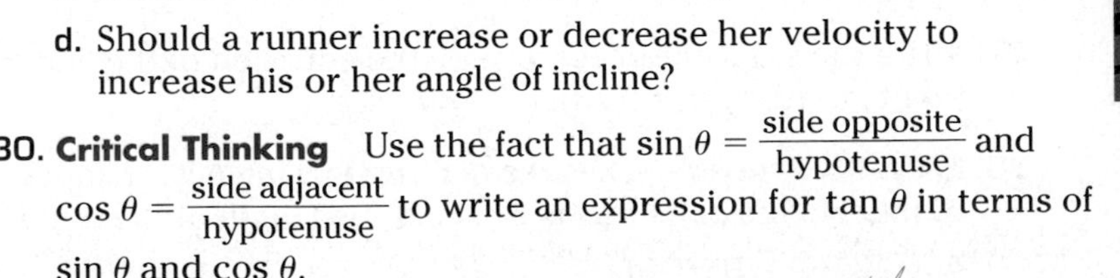

30. Critical Thinking Use the fact that $\sin \theta = \frac{\text{side opposite}}{\text{hypotenuse}}$ and $\cos \theta = \frac{\text{side adjacent}}{\text{hypotenuse}}$ to write an expression for tan θ in terms of sin θ and cos θ.

31. Architecture The angle of inclination of the sun affects the heating and cooling of buildings. The angle is greater in the summer than the winter. The sun's angle of inclination also varies according to the latitude. The sun's angle of inclination at noon equals $90° - L - 23.5° \times \cos\left[\frac{(N + 10)360}{365}\right]$. In this expression, L is the latitude of the building site, and N is the number of days elapsed in the year.

a. The latitude of Brownsville, Texas, is 26°. Find the angle of inclination for Brownsville on the first day of summer (day 172) and on the first day of winter (day 355).

b. The latitude of Nome, Alaska, is 64°. Find the angle of inclination for Nome on the first day of summer and on the first day of winter.

c. Which city has the greater change in the angle of inclination?

32. Biology An object under water is not exactly where it appears to be. The displacement x depends on the angle A at which the light strikes the surface of the water from below, the depth t of the object, and the angle B at which the light leaves the surface of the water. The measure of displacement is modeled by the equation $x = t\left(\frac{\sin (B - A)}{\cos A}\right)$. Suppose a biologist is trying to net a fish under water. Find the measure of displacement if t measures 10 centimeters, the measure of angle A is 41°, and the measure of angle B is 60°.

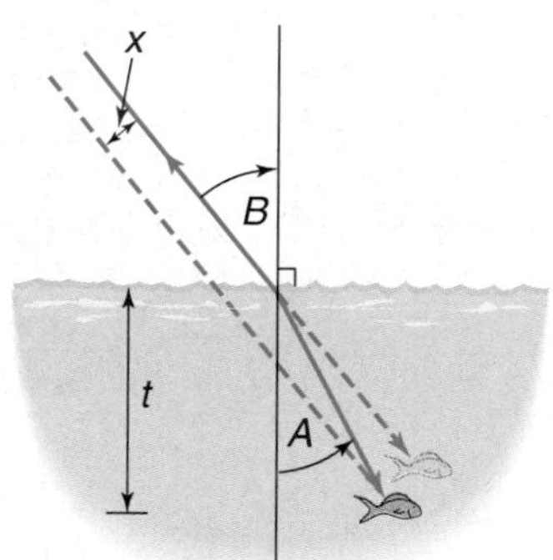

Mixed Review

33. Change 88.37° to degrees, minutes, and seconds. *(Lesson 5-1)*

34. Find the number of possible positive real zeros and the number of possible negative real zeros for $f(x) = x^4 + 2x^3 - 6x - 1$. *(Lesson 4-4)*

35. **Business** Luisa Diaz is planning to build a new factory for her business. She hires an analyst to gather data and develop a mathematical model. In the model $P(x) = 18 + 92x - 2x^2$, P is Ms. Diaz's monthly profit, and x is the number of employees needed to staff the new facility. *(Lesson 3-6)*

a. How many employees should she hire to maximize profits?

b. What is her maximum profit?

36. Find the value of $\begin{vmatrix} 7 & -3 & 5 \\ 4 & 0 & -1 \\ 8 & 2 & 0 \end{vmatrix}$. *(Lesson 2-5)*

37. Write the slope-intercept form of the equation of the line that passes through points at (2, 5) and (6, 3). *(Lesson 1-4)*

38. **SAT/ACT Practice** The area of a right triangle is 12 square inches. The ratio of the lengths of its legs is 2:3. Find the length of the hypotenuse.

A $\sqrt{13}$ in. B 26 in. C $2\sqrt{13}$ in. D 52 in. E $4\sqrt{13}$ in.

Extra Practice See p. A34.

Trigonometric Functions on the Unit Circle

OBJECTIVES

- Find the values of the six trigonometric functions using the unit circle.
- Find the values of the six trigonometric functions of an angle in standard position given a point on its terminal side.

FOOTBALL The longest punt in NFL history was 98 yards. The punt was made by Steve O'Neal of the New York Jets in 1969. When a football is punted, the angle made by the initial path of the ball and the ground affects both the height and the distance the ball will travel. If a football is punted from ground level, the maximum height it will reach is given by the formula $h = \frac{v_0^2 \sin^2 \theta}{2g}$, where v_0 is the initial velocity, θ is the measure of the angle between the ground and the initial path of the ball, and g is the acceleration due to gravity. The value of g is 9.8 meters per second squared. Suppose the initial velocity of the ball is 28 meters per second. Describe the possible maximum height of the ball if the angle is between 0° and 90°. *This problem will be solved in Example 2.*

A **unit circle** is a circle of radius 1. Consider a unit circle whose center is at the origin of a rectangular coordinate system. The unit circle is symmetric with respect to the x-axis, the y-axis, and the origin.

Consider an angle θ between 0° and 90° in standard position. Let $P(x, y)$ be the point of intersection of the angle's terminal side with the unit circle. If a perpendicular segment is drawn from point P to the x-axis, a right triangle is created. In the triangle, the side adjacent to angle θ is along the x-axis and has length x. The side opposite angle θ is the perpendicular segment and has length y. According to the Pythagorean Theorem, $x^2 + y^2 = 1$. We can find values for $\sin \theta$ and $\cos \theta$ using the definitions used in Lesson 5-2.

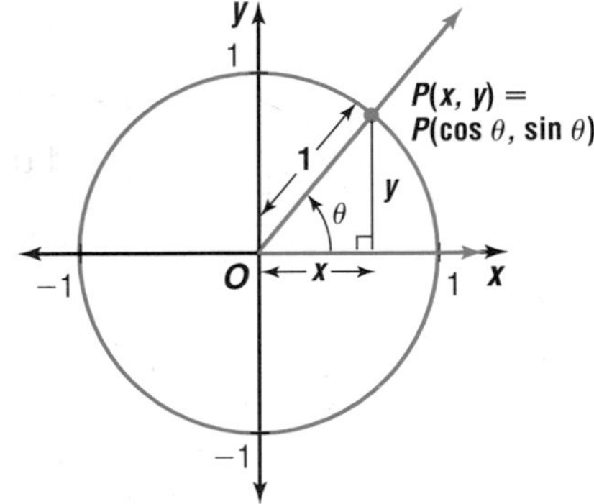

$$\sin \theta = \frac{\text{side opposite}}{\text{hypotenuse}} \qquad \cos \theta = \frac{\text{side adjacent}}{\text{hypotenuse}}$$

$$\sin \theta = \frac{y}{1} \text{ or } y \qquad \cos \theta = \frac{x}{1} \text{ or } x$$

Right triangles can also be formed for angles greater than 90°. In these cases, the reference angle is one of the acute angles. Similar results will occur. Thus, **sine** θ can be redefined as the y-coordinate and **cosine** θ can be redefined as the x-coordinate.

Sine and Cosine If the terminal side of an angle θ in standard position intersects the unit circle at $P(x, y)$, then $\cos \theta = x$ and $\sin \theta = y$.

Since there is exactly one point $P(x, y)$ for any angle θ, the relations $\cos\theta = x$ and $\sin\theta = y$ are functions of θ. Because they are both defined using the unit circle, they are often called **circular functions.**

The domain of the sine and cosine functions is the set of real numbers, since $\sin\theta$ and $\cos\theta$ are defined for any angle θ. The range of the sine and the cosine functions is the set of real numbers between -1 and 1 inclusive, since $(\cos\theta, \sin\theta)$ are the coordinates of points on the unit circle.

In addition to the sine and cosine functions, the four other **trigonometric functions** can also be defined using the unit circle.

$$\tan\theta = \frac{\text{side opposite}}{\text{side adjacent}} = \frac{y}{x} \qquad \csc\theta = \frac{\text{hypotenuse}}{\text{side opposite}} = \frac{1}{y}$$

$$\sec\theta = \frac{\text{hypotenuse}}{\text{side adjacent}} = \frac{1}{x} \qquad \cot\theta = \frac{\text{side adjacent}}{\text{side opposite}} = \frac{x}{y}$$

Since division by zero is undefined, there are several angle measures that are excluded from the domain of the tangent, cotangent, secant, and cosecant functions.

Examples **1** **Use the unit circle to find each value.**

a. cos (−180°)

The terminal side of a $-180°$ angle in standard position is the negative x-axis, which intersects the unit circle at $(-1, 0)$. The x-coordinate of this ordered pair is $\cos(-180°)$. Therefore, $\cos(-180°) = -1$.

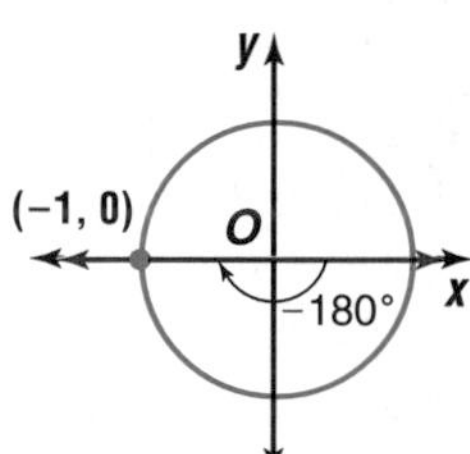

b. sec 90°

The terminal side of a 90° angle in standard position is the positive y-axis, which intersects the unit circle at (0, 1). According to the definition of secant, $\sec 90° = \frac{1}{x}$ or $\frac{1}{0}$, which is undefined. Therefore, sec 90° is undefined.

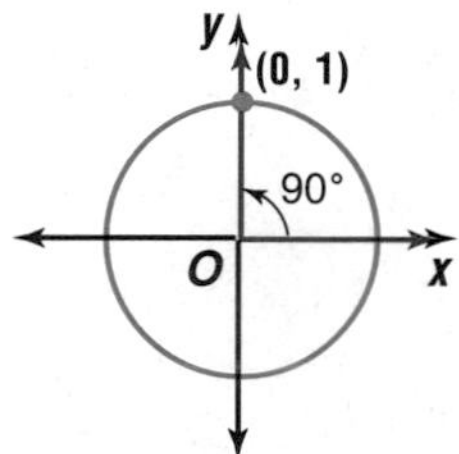

2 **FOOTBALL** **Refer to the application at the beginning of the lesson. Describe the possible maximum height of the ball if the angle is between 0° and 90°.**

The expression $\sin^2\theta$ means the square of the sine of θ or $(\sin\theta)^2$.

Find the value of h when $\theta = 0°$.

$$h = \frac{v_0^2 \sin^2\theta}{2g}$$

$$h = \frac{28^2 \sin^2 0°}{2(9.8)} \qquad v_0 = 28,\ \theta = 0°,\ g = 9.8$$

$$h = \frac{28^2(0^2)}{2(9.8)} \qquad \sin 0° = 0$$

$$h = 0$$

Find the value of h when $\theta = 90°$.

$$h = \frac{v_0^2 \sin^2 \theta}{2g}$$

$$h = \frac{28^2 \sin^2 90°}{2(9.8)} \quad v_0 = 28,\ \theta = 90°,\ g = 9.8$$

$$h = \frac{28^2(1^2)}{2(9.8)} \quad \sin 90° = 1$$

$$h = 40$$

The maximum height of the ball is between 0 meters and 40 meters.

The radius of a circle is defined as a positive value. Therefore, the signs of the six trigonometric functions are determined by the signs of the coordinates of x and y in each quadrant.

Example 3 **Use the unit circle to find the values of the six trigonometric functions for a 135° angle.**

Since 135° is between 90° and 180°, the terminal side is in the second quadrant. Therefore, the reference angle is 180° − 135° or 45°. The terminal side of a 45° angle intersects the unit circle at a point with coordinates $\left(\frac{\sqrt{2}}{2}, \frac{\sqrt{2}}{2}\right)$. Because the terminal side of a 135° angle is in the second quadrant, the x-coordinate is negative, and the y-coordinate is positive. The point of intersection has coordinates $\left(-\frac{\sqrt{2}}{2}, \frac{\sqrt{2}}{2}\right)$.

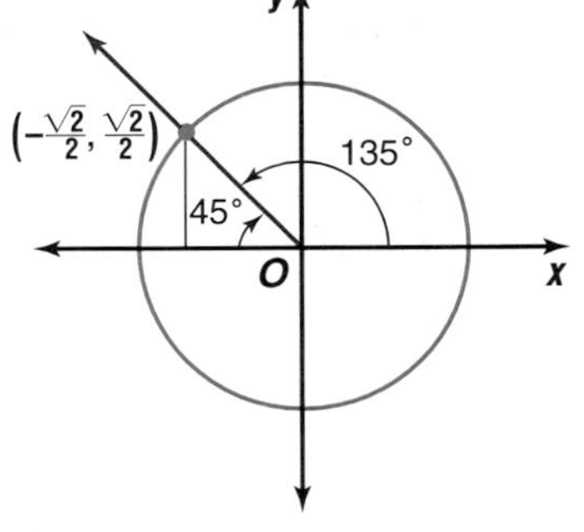

$$\sin 135° = y \qquad \cos 135° = x \qquad \tan 135° = \frac{y}{x}$$

$$\sin 135° = \frac{\sqrt{2}}{2} \qquad \cos 135° = -\frac{\sqrt{2}}{2} \qquad \tan 135° = \frac{\frac{\sqrt{2}}{2}}{-\frac{\sqrt{2}}{2}}$$

$$\tan 135° = -1$$

$$\csc 135° = \frac{1}{y} \qquad \sec 135° = \frac{1}{x} \qquad \cot 135° = \frac{x}{y}$$

$$\csc 135° = \frac{1}{\frac{\sqrt{2}}{2}} \qquad \sec 135° = \frac{1}{-\frac{\sqrt{2}}{2}} \qquad \cot 135° = \frac{\frac{-\sqrt{2}}{2}}{\frac{\sqrt{2}}{2}}$$

$$\csc 135° = \frac{2}{\sqrt{2}} \qquad \sec 135° = -\frac{2}{\sqrt{2}} \qquad \cot 135° = -1$$

$$\csc 135° = \sqrt{2} \qquad \sec 135° = -\sqrt{2}$$

$$\sin 52^\circ\ 15' = \frac{40}{r} \qquad sin = \frac{side\ opposite}{hypotenuse}$$
$$r \sin 52^\circ\ 15' = 40 \qquad \textit{Multiply each side by r.}$$
$$r = \frac{40}{\sin 52^\circ\ 15'} \qquad \textit{Divide each side by } \sin 52^\circ\ 15'.$$
$$r \approx 50.58875357 \qquad \textit{Use a calculator.}$$

The rope is about 50.6 feet long.

b. To find the distance between the bottom of the tent and the stake, you need to know the length of the side adjacent to the known angle. Use the tangent function.

$$\tan 52^\circ\ 15' = \frac{40}{d} \qquad tan = \frac{side\ opposite}{side\ adjacent}$$
$$d \tan 52^\circ\ 15' = 40 \qquad \textit{Multiply each side by d.}$$
$$d = \frac{40}{\tan 52^\circ\ 15'} \qquad \textit{Divide each side by } \tan 52^\circ\ 15'.$$
$$d \approx 30.97130911 \qquad \textit{Use a calculator.}$$

The distance between the bottom of the tent and the stake is about 31.0 feet.

You can use right triangle trigonometry to solve problems involving other geometric figures.

Example 3 **GEOMETRY** **A regular pentagon is inscribed in a circle with diameter 8.34 centimeters. The *apothem* of a regular polygon is the measure of a line segment from the center of the polygon to the midpoint of one of its sides. Find the apothem of the pentagon.**

First, draw a diagram. If the diameter of the circle is 8.34 centimeters, the radius is 8.34 ÷ 2 or 4.17 centimeters. The measure of α is 360° ÷ 10 or 36°.

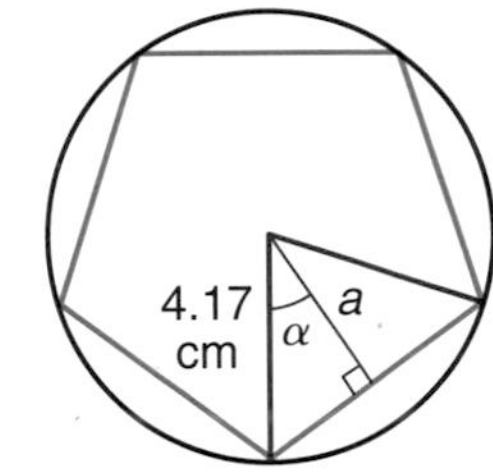

$$\cos 36^\circ = \frac{a}{4.17} \qquad cos = \frac{side\ adjacent}{hypotenuse}$$
$$4.17 \cos 36^\circ = a \qquad \textit{Multiply each side by 4.17.}$$
$$3.373600867 \approx a \qquad \textit{Use a calculator.}$$

The apothem is about 3.37 centimeters.

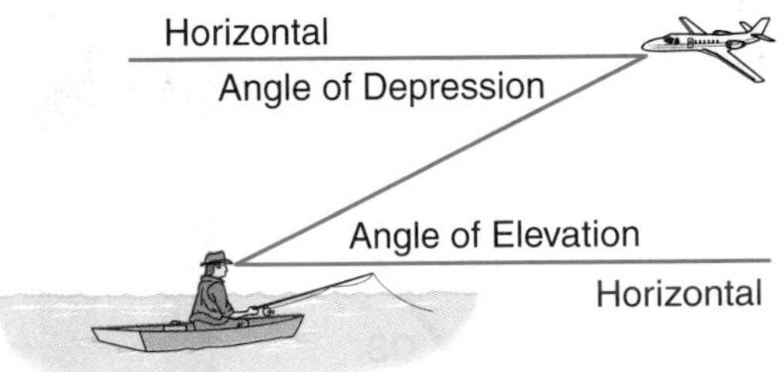

There are many other applications that require trigonometric solutions. For example, surveyors use special instruments to find the measures of **angles of elevation** and **angles of depression.** An angle of elevation is the angle between a horizontal line and the line of sight from an observer to an object at a higher level. An angle of depression is the angle between a horizontal line and the line of sight from the observer to an object at a lower level. The angle of elevation and the angle of depression are equal in measure because they are alternate interior angles.

Example 4 **SURVEYING** **On May 18, 1980, Mount Saint Helens, a volcano in Washington, erupted with such force that the top of the mountain was blown off. To determine the new height at the summit of Mount Saint Helens, a surveyor measured the angle of elevation to the top of the volcano to be 37° 46′. The surveyor then moved 1000 feet closer to the volcano and measured the angle of elevation to be 40° 30′. Determine the new height of Mount Saint Helens.**

Draw a diagram to model the situation. Let h represent the height of the volcano and x represent the distance from the surveyor's second position to the center of the base of the volcano. Write two equations involving the tangent function.

$$\tan 37°\ 46' = \frac{h}{1000 + x}$$

$$(1000 + x)\tan 37°\ 46' = h$$

$$\tan 40°\ 30' = \frac{h}{x}$$

$$x \tan 40°\ 30' = h$$

Therefore, $(1000 + x)\tan 37°\ 46' = x \tan 40°\ 30'$. Solve this equation for x.

$$(1000 + x)\tan 37°\ 46' = x \tan 40°\ 30'$$

$$1000 \tan 37°\ 46' + x \tan 37°\ 46' = x \tan 40°\ 30'$$

$$1000 \tan 37°\ 46' = x \tan 40°\ 30' - x \tan 37°\ 46'$$

$$1000 \tan 37°\ 46' = x(\tan 40°\ 30' - \tan 37°\ 46')$$

$$\frac{1000 \tan 37°\ 46'}{\tan 40°\ 30' - \tan 37°\ 46'} = x$$

$$9765.826092 \approx x$$ *Use a calculator.*

Use this value for x and the equation $x \tan 40°\ 30' = h$ to find the height of the volcano.

$$x \tan 40°\ 30' = h$$

$$9765.826092 \tan 40°\ 30' \approx h$$

$$8340.803443 \approx h$$ *Use a calculator.*

The new height of Mount Saint Helens is about 8341 feet.

CHECK FOR UNDERSTANDING

Communicating Mathematics

Read and study the lesson to answer each question.

1. **State** which trigonometric function you would use to solve each problem.
   a. If $S = 42°$ and $ST = 8$, find RS.
   b. If $T = 55°$ and $RT = 5$, find RS.
   c. If $S = 27°$ and $TR = 7$, find TS.
2. **Write** a problem that could be solved using the tangent function.

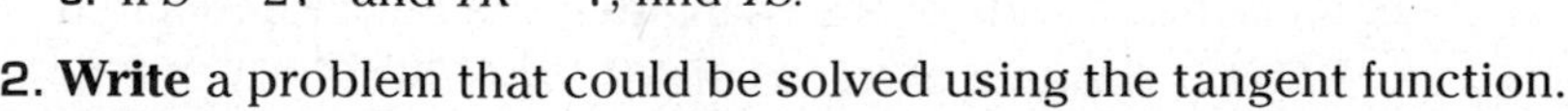

3. **Name** the angle of elevation and the angle of depression in the figure at the right. Compare the measures of these angles. Explain.

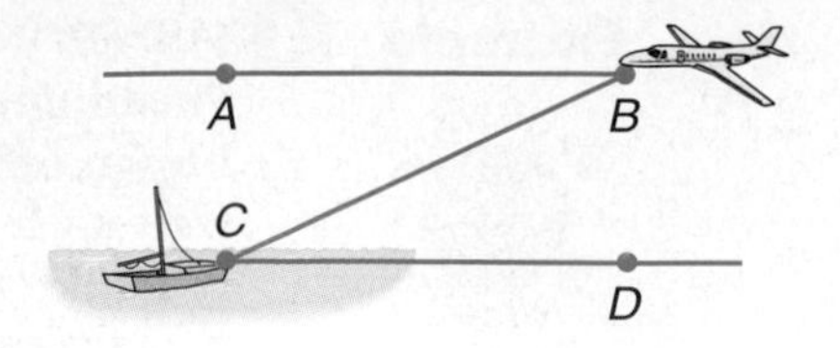

4. **Describe** a way to use trigonometry to determine the height of the building where you live.

Guided Practice **Solve each problem. Round to the nearest tenth.**

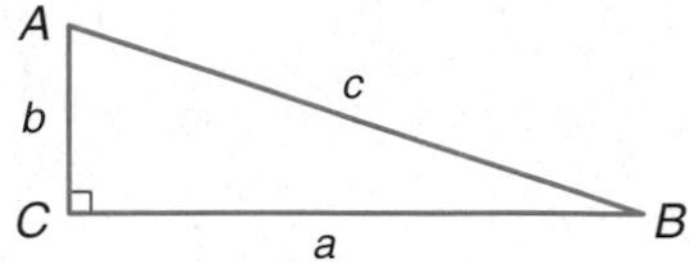

5. If $b = 13$ and $A = 76°$, find a.

6. If $B = 26°$ and $b = 18$, find c.

7. If $B = 16°\ 45'$ and $c = 13$, find a.

8. **Geometry** Each base angle of an isosceles triangle measures $55°\ 30'$. Each of the congruent sides is 10 centimeters long.
   a. Find the altitude of the triangle.
   b. What is the length of the base?
   c. Find the area of the triangle.

9. **Boating** The Ponce de Leon lighthouse in St. Augustine, Florida, is the second tallest brick tower in the United States. It was built in 1887 and rises 175 feet above sea level. How far from the shore is a motorboat if the angle of depression from the top of the lighthouse is $13°\ 15'$?

EXERCISES

Practice **Solve each problem. Round to the nearest tenth.**

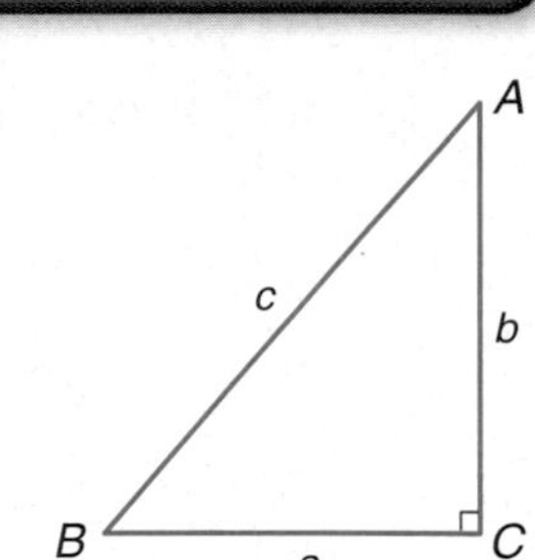

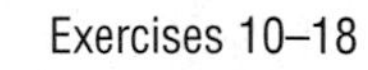
Exercises 10–18

10. If $A = 37°$ and $b = 6$, find a.

11. If $c = 16$ and $B = 67°$, find a.

12. If $B = 62°$ and $c = 24$, find b.

13. If $A = 29°$ and $a = 4.6$, find c.

14. If $a = 17.3$ and $B = 77°$, find c.

15. If $b = 33.2$ and $B = 61°$, find a.

16. If $B = 49°\ 13'$ and $b = 10$, find a.

17. If $A = 16°\ 55'$ and $c = 13.7$, find a.

18. If $a = 22.3$ and $B = 47°\ 18'$, find c.

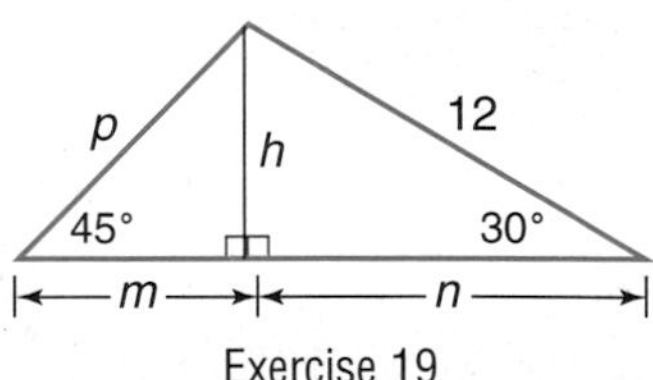

Exercise 19

19. Find h, n, m, and p. Round to the nearest tenth.

20. **Geometry** The apothem of a regular pentagon is 10.8 centimeters.
   a. Find the radius of the circumscribed circle.
   b. What is the length of a side of the pentagon?
   c. Find the perimeter of the pentagon.

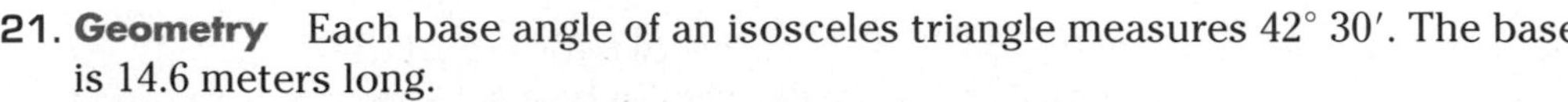

21. **Geometry** Each base angle of an isosceles triangle measures $42°\ 30'$. The base is 14.6 meters long.
   a. Find the length of a leg of the triangle.
   b. Find the altitude of the triangle.
   c. What is the area of the triangle?

22. **Geometry** A regular hexagon is inscribed in a circle with diameter 6.4 centimeters.

a. What is the apothem of the hexagon?

b. Find the length of a side of the hexagon.

c. Find the perimeter of the hexagon.

d. The area of a regular polygon equals one half times the perimeter of the polygon times the apothem. Find the area of the polygon.

Applications and Problem Solving

23. **Engineering** The escalator at St. Petersburg Metro in Russia has a vertical rise of 195.8 feet. If the angle of elevation of the escalator is 10° 21′ 36″, find the length of the escalator.

24. **Critical Thinking** Write a formula for the volume of the regular pyramid at the right in terms of α and s the length of each side of the base.

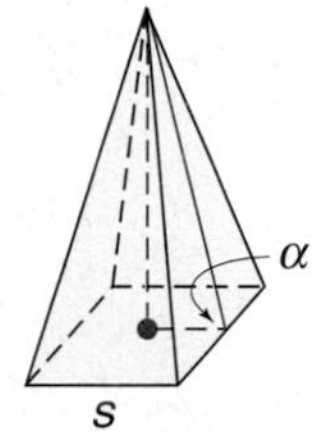

25. **Fire Fighting** The longest truck-mounted ladder used by the Dallas Fire Department is 108 feet long and consists of four hydraulic sections. Gerald Travis, aerial expert for the department, indicates that the optimum operating angle of this ladder is 60°. The fire fighters find they need to reach the roof of an 84-foot burning building. Assume the ladder is mounted 8 feet above the ground.

a. Draw a labeled diagram of the situation.

b. How far from the building should the base of the ladder be placed to achieve the optimum operating angle?

c. How far should the ladder be extended to reach the roof?

26. **Aviation** When a 757 passenger jet begins its descent to the Ronald Reagan International Airport in Washington, D.C., it is 3900 feet from the ground. Its angle of descent is 6°.

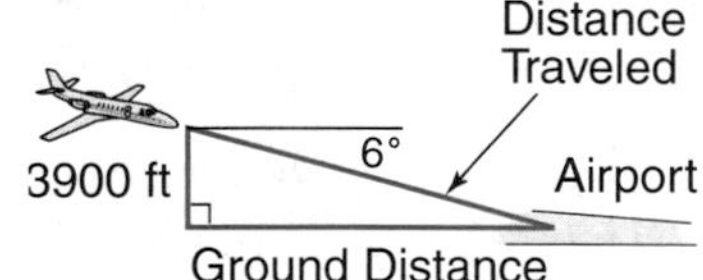

a. What is the plane's ground distance to the airport?

b. How far must the plane fly to reach the runway?

27. **Boat Safety** The Cape Hatteras lighthouse on the North Carolina coast was built in 1870 and rises 208 feet above sea level. From the top of the lighthouse, the lighthouse keeper observes a yacht and a barge along the same line of sight. The angle of depression for the yacht is 20°, and the angle of depression for the barge is 12° 30′. For safety purposes, the keeper thinks that the two sea vessels should be at least 300 feet apart. If they are less than 300 feet, she plans to sound the horn. How far apart are these vessels? Does the keeper have to sound the horn?

28. **Critical Thinking** Derive two formulas for the length of the altitude a of the triangle shown at the right, given that b, s, and θ are known. Justify each of the steps you take in your reasoning.

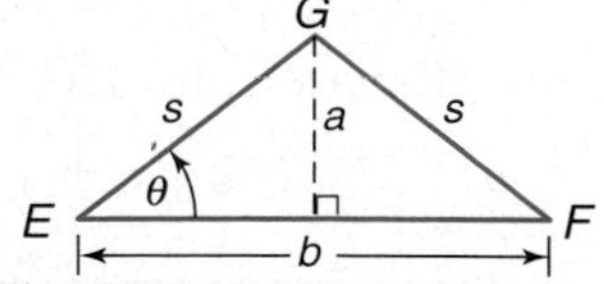

29. **Recreation** Latasha and Markisha are flying kites on a windy spring day. Latasha has released 250 feet of string, and Markisha has released 225 feet of string. The angle that Latasha's kite string makes with the horizontal is 35°. The angle that Markisha's kite string makes with the horizontal is 42°. Which kite is higher and by how much?

30. **Architecture** A flagpole 40 feet high stands on top of the Wentworth Building. From a point in front of Bailey's Drugstore, the angle of elevation for the top of the pole is 54° 54′, and the angle of elevation for the bottom of the pole is 47° 30′. How high is the building?

47°30′ 54°54′

Mixed Review

31. Find the values of the six trigonometric functions for a 120° angle using the unit circle. *(Lesson 5-3)*

32. Find the sine, cosine, and tangent ratios for $\angle P$. *(Lesson 5-2)*

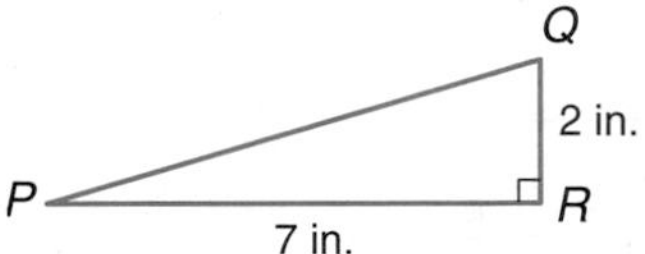

33. Write 43° 15′ 35″ as a decimal to the nearest thousandth. *(Lesson 5-1)*

34. Graph $y \leq |x + 2|$. *(Lesson 3-3)*

35. **Consumerism** Kareem and Erin went shopping for school supplies. Kareem bought 3 notebooks and 2 packages of pencils for $5.80. Erin bought 4 notebooks and 1 package of pencils for $6.20. What is the cost of one notebook? What is the cost of one package of pencils? *(Lesson 2-1)*

36. **SAT/ACT Practice** An automobile travels m miles in h hours. At this rate, how far will it travel in x hours?

A $\frac{m}{x}$ **B** $\frac{m}{xh}$ **C** $\frac{m}{h}$ **D** $\frac{mh}{x}$ **E** $\frac{mx}{h}$

MID-CHAPTER QUIZ

1. Change 34.605° to degrees, minutes, and seconds. (Lesson 5-1)

2. If a −400° angle is in standard position, determine a coterminal angle that is between 0° and 360°. State the quadrant in which the terminal side lies. (Lesson 5-1)

3. Find the six trigonometric functions for $\angle G$. (Lesson 5-2)

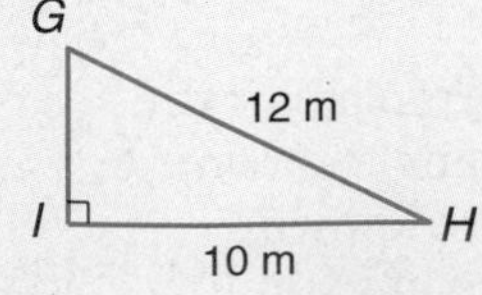

4. Find the values of the six trigonometric functions for angle θ in standard position if a point with coordinates $(2, -5)$ lies on its terminal side. (Lesson 5-3)

5. **National Landmarks** Suppose the angle of elevation of the sun is 27.8°. Find the length of the shadow made by the Washington Monument, which is 550 feet tall. (Lesson 5-4)

Extra Practice See p. A34.

5-5 Solving Right Triangles

OBJECTIVES

- Evaluate inverse trigonometric functions.
- Find missing angle measurements.
- Solve right triangles.

SECURITY A security light is being installed outside a loading dock. The light is mounted 20 feet above the ground. The light must be placed at an angle so that it will illuminate the end of the parking lot. If the end of the parking lot is 100 feet from the loading dock, what should be the angle of depression of the light? *This problem will be solved in Example 4.*

In Lesson 5-3, you learned to use the unit circle to determine the value of trigonometric functions. Some of the frequently-used values are listed below.

θ	0°	30°	45°	60°	90°	120°	135°	150°	180°
sin θ	0	$\frac{1}{2}$	$\frac{\sqrt{2}}{2}$	$\frac{\sqrt{3}}{2}$	1	$\frac{\sqrt{3}}{2}$	$\frac{\sqrt{2}}{2}$	$\frac{1}{2}$	0
cos θ	1	$\frac{\sqrt{3}}{2}$	$\frac{\sqrt{2}}{2}$	$\frac{1}{2}$	0	$-\frac{1}{2}$	$-\frac{\sqrt{2}}{2}$	$-\frac{\sqrt{3}}{2}$	-1
tan θ	0	$\frac{\sqrt{3}}{3}$	1	$\sqrt{3}$	undefined	$-\sqrt{3}$	-1	$-\frac{\sqrt{3}}{3}$	0

θ	210°	225°	240°	270°	300°	315°	330°	360°
sin θ	$-\frac{1}{2}$	$-\frac{\sqrt{2}}{2}$	$-\frac{\sqrt{3}}{2}$	-1	$-\frac{\sqrt{3}}{2}$	$-\frac{\sqrt{2}}{2}$	$-\frac{1}{2}$	0
cos θ	$-\frac{\sqrt{3}}{2}$	$-\frac{\sqrt{2}}{2}$	$-\frac{1}{2}$	0	$\frac{1}{2}$	$\frac{\sqrt{2}}{2}$	$\frac{\sqrt{3}}{2}$	1
tan θ	$\frac{\sqrt{3}}{3}$	1	$\sqrt{3}$	undefined	$-\sqrt{3}$	-1	$-\frac{\sqrt{3}}{3}$	0

Sometimes you know a trigonometric value of an angle, but not the angle. In this case, you need to use an **inverse** of the trigonometric function. The inverse of the sine function is the **arcsine relation.**

An equation such as $\sin x = \frac{\sqrt{3}}{2}$ can be written as $x = \arcsin \frac{\sqrt{3}}{2}$, which is read "$x$ is an angle whose sine is $\frac{\sqrt{3}}{2}$," or "x equals the arcsine of $\frac{\sqrt{3}}{2}$." The solution, x, consists of all angles that have $\frac{\sqrt{3}}{2}$ as the value of sine x.

Similarly, the inverse of the cosine function is the **arccosine relation,** and the inverse of the tangent function is the **arctangent relation.**

The equations in each row of the table below are equivalent. You can use these equations to rewrite trigonometric expressions.

Inverses of the Trigonometric Functions

Trigonometric Function	Inverse Trigonometric Relation
$y = \sin x$	$x = \sin^{-1} y$ or $x = \arcsin y$
$y = \cos x$	$x = \cos^{-1} y$ or $x = \arccos y$
$y = \tan x$	$x = \tan^{-1} y$ or $x = \arctan y$

Examples

1 Solve each equation.

a. $\sin x = \frac{\sqrt{3}}{2}$

If $\sin x = \frac{\sqrt{3}}{2}$, then x is an angle whose sine is $\frac{\sqrt{3}}{2}$.

$x = \arcsin \frac{\sqrt{3}}{2}$

From the table on page 305, you can determine that x equals 60°, 120°, or any angle coterminal with these angles.

b. $\cos x = -\frac{\sqrt{2}}{2}$

If $\cos x = -\frac{\sqrt{2}}{2}$, then x is an angle whose cosine is $-\frac{\sqrt{2}}{2}$.

$x = \arccos -\frac{\sqrt{2}}{2}$

From the table, you can determine that x equals 135°, 225°, or any angle coterminal with these angles.

2 Evaluate each expression. Assume that all angles are in Quadrant I.

a. $\tan\left(\tan^{-1} \frac{6}{11}\right)$

Let $A = \tan^{-1} \frac{6}{11}$. Then $\tan A = \frac{6}{11}$ by the definition of inverse. Therefore, by substitution, $\tan\left(\tan^{-1} \frac{6}{11}\right) = \frac{6}{11}$.

b. $\cos\left(\arcsin \frac{2}{3}\right)$

Let $B = \arcsin \frac{2}{3}$. Then $\sin B = \frac{2}{3}$ by the definition of inverse. Draw a diagram of the $\angle B$ in Quadrant I.

$r^2 = x^2 + y^2$ *Pythagorean Theorem*

$3^2 = x^2 + 2^2$ *Substitute 3 for r and 2 for y.*

$\sqrt{5} = x$ *Take the square root of each side. Disregard the negative root.*

Since $\cos = \frac{\text{side adjacent}}{\text{hypotenuse}}$, $\cos B = \frac{\sqrt{5}}{3}$ and $\cos\left(\arcsin \frac{2}{3}\right) = \frac{\sqrt{5}}{3}$.

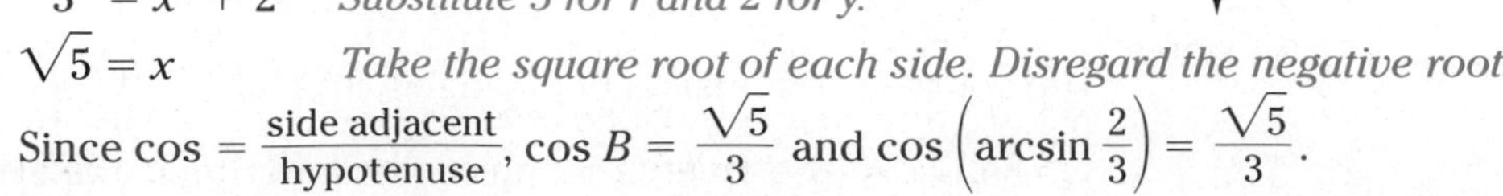

Inverse trigonometric relations can be used to find the measure of angles of right triangles. Calculators can be used to find values of the inverse trigonometric relations.

Example 3 **If $f = 17$ and $d = 32$, find E.**

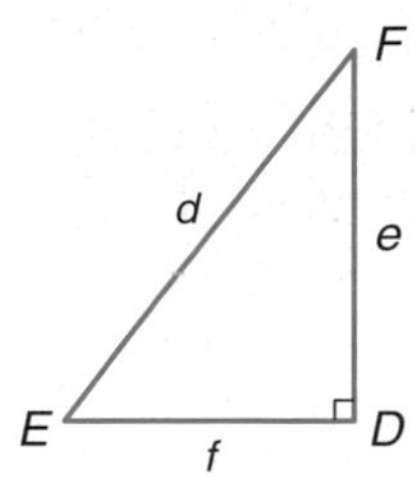

In this problem, you want to know the measure of an acute angle in a right triangle. You know the side adjacent to the angle and the hypotenuse. The cosine function relates the side adjacent to the angle and the hypotenuse.

Remember that in trigonometry the measure of an angle is symbolized by the angle vertex letter.

$\cos E = \frac{f}{d}$ $\cos = \frac{\text{side adjacent}}{\text{hypotenuse}}$

$\cos E = \frac{17}{32}$ *Substitute 17 for f and 32 for d.*

$E = \cos^{-1} \frac{17}{32}$ *Definition of inverse*

$E \approx 57.91004874$ *Use a calculator.*

Therefore, E measures about 57.9°.

Trigonometry can be used to find the angle of elevation or the angle of depression.

Example 4 **SECURITY** **Refer to the application at the beginning of the lesson. What should be the angle of depression of the light?**

The angle of depression from the light and the angle of elevation to the light are equal in measure. To find the angle of elevation, use the tangent function.

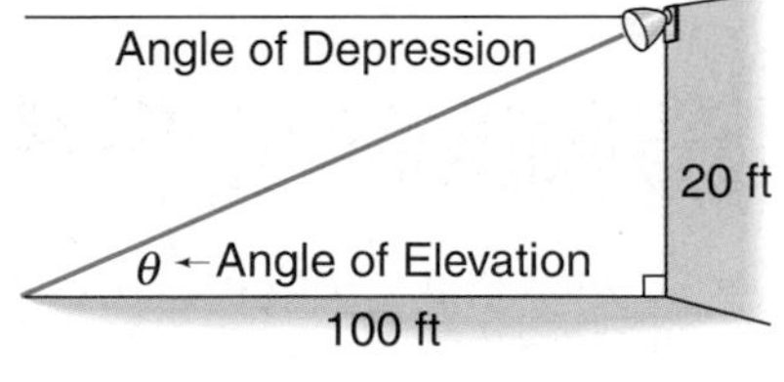

$\tan \theta = \frac{20}{100}$ $\tan = \frac{\text{side opposite}}{\text{side adjacent}}$

$\theta = \tan^{-1} \frac{20}{100}$ *Definition of inverse*

$\theta \approx 11.30993247$ *Use a calculator.*

The angle of depression should be about 11.3°.

You can use trigonometric functions and inverse relations to solve right triangles. To **solve a triangle** means to find all of the measures of its sides and angles. Usually, two measures are given. Then you can find the remaining measures.

Example **5** **Solve each triangle described, given the triangle at the right.**

a. $A = 33°$, $b = 5.8$

Find B.

$$33° + B = 90° \quad \textit{Angles A and B are complementary.}$$
$$B = 57°$$

Whenever possible, use measures given in the problem to find the unknown measures.

Find a.

$$\tan A = \frac{a}{b}$$
$$\tan 33° = \frac{a}{5.8}$$
$$5.8 \tan 33° = a$$
$$3.766564041 \approx a$$

Find c.

$$\cos A = \frac{b}{c}$$
$$\cos 33° = \frac{5.8}{c}$$
$$c \cos 33° = 5.8$$
$$c = \frac{5.8}{\cos 33°}$$
$$c \approx 6.915707098$$

Therefore, $B = 57°$, $a \approx 3.8$, and $c \approx 6.9$.

b. $a = 23$, $c = 45$

Find b.

$$a^2 + b^2 = c^2$$
$$23^2 + b^2 = 45^2$$
$$b = \sqrt{1496}$$
$$b \approx 38.67815921$$

Find A.

$$\sin A = \frac{a}{c}$$
$$\sin A = \frac{23}{45}$$
$$A = \sin^{-1}\frac{23}{45}$$
$$A \approx 30.73786867$$

Find B.

$$30.73786867 + B \approx 90$$
$$B \approx 59.26213133$$

Therefore, $b \approx 38.7$, $A \approx 30.7°$, and $B \approx 59.3°$

CHECK FOR UNDERSTANDING

Communicating Mathematics

Read and study the lesson to answer each question.

1. **Tell** whether the solution to each equation is an angle measure or a linear measurement.

   a. $\tan 34° 15' = \frac{x}{12}$ b. $\tan x = 3.284$

2. **Describe** the relationship of the two acute angles of a right triangle.

3. **Counterexample** You can usually solve a right triangle if you know two measures besides the right angle. Draw a right triangle and label two measures other than the right angle such that you cannot solve the triangle.

4. **You Decide** Marta and Rebecca want to determine the degree measure of angle ϑ if $\cos \vartheta = 0.9876$. Marta tells Rebecca to press [2nd] [COS⁻¹] .9876 on the calculator. Rebecca disagrees. She says to press [COS] .9876 [ )] [x^{-1}]. Who is correct? Explain.

Guided Practice Solve each equation if $0° \leq x \leq 360°$.

5. $\cos x = \frac{1}{2}$

6. $\tan x = \frac{-\sqrt{3}}{3}$

Evaluate each expression. Assume that all angles are in Quadrant I.

7. $\sin\left(\sin^{-1}\frac{\sqrt{3}}{2}\right)$

8. $\tan\left(\cos^{-1}\frac{3}{5}\right)$

Solve each problem. Round to the nearest tenth.

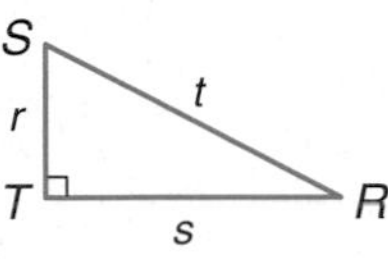

9. If $r = 7$ and $s = 10$, find R.

10. If $r = 12$ and $t = 20$, find S.

Solve each triangle described, given the triangle at the right. Round to the nearest tenth if necessary.

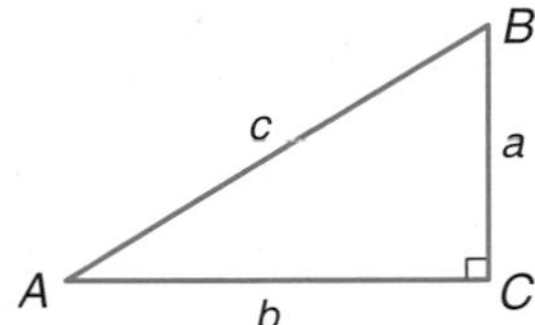

11. $B = 78°$, $a = 41$

12. $a = 11$, $b = 21$

13. $A = 32°$, $c = 13$

14. **National Monuments** In 1906, Teddy Roosevelt designated Devils Tower National Monument in northeast Wyoming as the first national monument in the United States. The tower rises 1280 feet above the valley of the Bell Fourche River.

Devils Tower National Monument

a. If the shadow of the tower is 2100 feet long at a certain time, find the angle of elevation of the sun.

b. How long is the shadow when the angle of elevation of the sun is 38°?

c. If a person at the top of Devils Tower sees a hiker at an angle of depression of 65°, how far is the hiker from the base of Devils Tower?

EXERCISES

Practice Solve each equation if $0° \leq x \leq 360°$.

15. $\sin x = 1$

16. $\tan x = -\sqrt{3}$

17. $\cos x = \frac{\sqrt{3}}{2}$

18. $\cos x = 0$

19. $\sin x = -\frac{\sqrt{2}}{2}$

20. $\tan x = -1$

21. Name four angles whose sine equals $\frac{1}{2}$.

Evaluate each expression. Assume that all angles are in Quadrant I.

22. $\cos\left(\arccos\frac{4}{5}\right)$

23. $\tan\left(\tan^{-1}\frac{2}{3}\right)$

24. $\sec\left(\cos^{-1}\frac{2}{5}\right)$

25. $\csc(\arcsin 1)$

26. $\tan\left(\cos^{-1}\frac{5}{13}\right)$

27. $\cos\left(\sin^{-1}\frac{2}{5}\right)$

Solve each problem. Round to the nearest tenth.

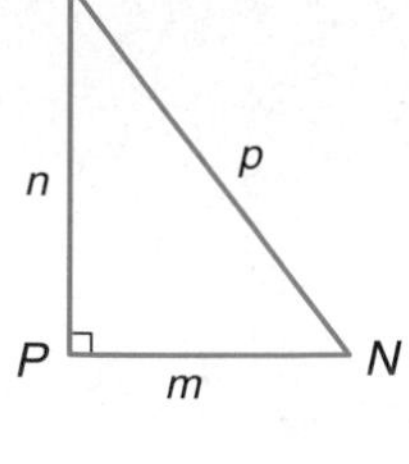

28. If $n = 15$ and $m = 9$, find N.

29. If $m = 8$ and $p = 14$, find M.

30. If $n = 22$ and $p = 30$, find M.

31. If $m = 14.3$ and $n = 18.8$, find N.

32. If $p = 17.1$ and $m = 7.2$, find N.

33. If $m = 32.5$ and $p = 54.7$, find M.

34. **Geometry** If the legs of a right triangle are 24 centimeters and 18 centimeters long, find the measures of the acute angles.

35. **Geometry** The base of an isosceles triangle is 14 inches long. Its height is 8 inches. Find the measure of each angle of the triangle.

Solve each triangle described, given the triangle at the right. Round to the nearest tenth, if necessary.

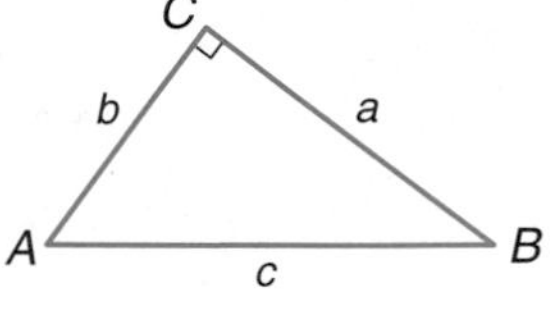

36. $a = 21, c = 30$
37. $A = 35°, b = 8$
38. $B = 47°, b = 12.5$
39. $a = 3.8, b = 4.2$
40. $c = 9.5, b = 3.7$
41. $a = 13.3, A = 51.5°$
42. $B = 33°, c = 15.2$
43. $c = 9.8, A = 14°$

Applications and Problem Solving

Real World Application

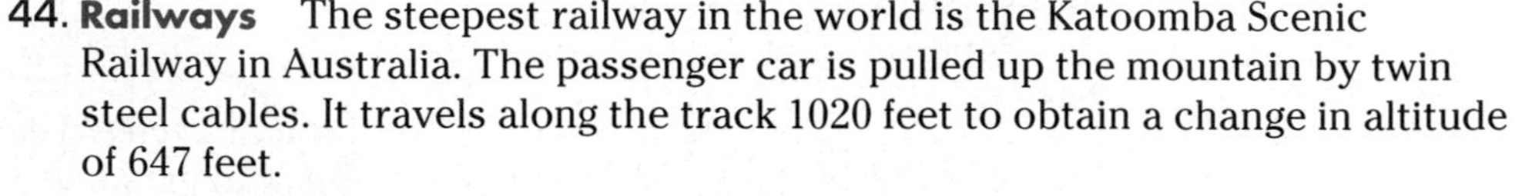

44. **Railways** The steepest railway in the world is the Katoomba Scenic Railway in Australia. The passenger car is pulled up the mountain by twin steel cables. It travels along the track 1020 feet to obtain a change in altitude of 647 feet.
    a. Find the angle of elevation of the railway.
    b. How far does the car travel in a horizontal direction?

45. **Critical Thinking** Explain why each expression is impossible.
    a. $\sin^{-1} 2.4567$
    b. $\sec^{-1} 0.5239$
    c. $\cos^{-1} (-3.4728)$

46. **Basketball** The rim of a basketball hoop is 10 feet above the ground. The free-throw line is 15 feet from the basket rim. If the eyes of a basketball player are 6 feet above the ground, what is the angle of elevation of the player's line of sight when shooting a free throw to the rim of the basket?

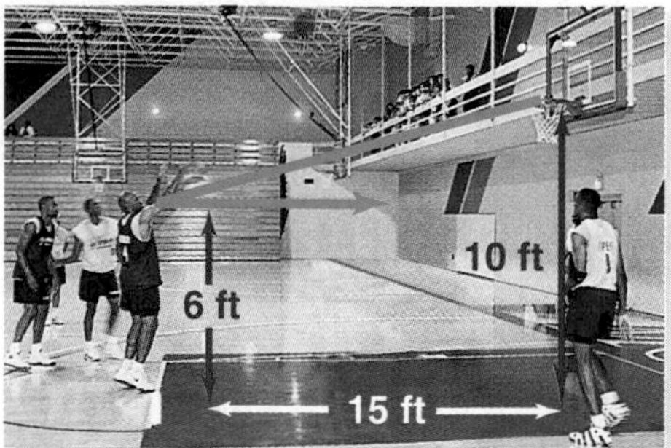

47. **Road Safety** Several years ago, a section on I-75 near Cincinnati, Ohio, had a rise of 8 meters per 100 meters of horizontal distance. However, there were numerous accidents involving large trucks on this section of highway. Civil engineers decided to reconstruct the highway so that there is only a rise of 5 meters per 100 meters of horizontal distance.
    a. Find the original angle of elevation.
    b. Find the new angle of elevation.

48. **Air Travel** At a local airport, a light that produces a powerful white-green beam is placed on the top of a 45-foot tower. If the tower is at one end of the runway, find the angle of depression needed so that the light extends to the end of the 2200-foot runway.

49. **Civil Engineering** Highway curves are usually banked or tilted inward so that cars can negotiate the curve more safely. The proper banking angle θ for a car making a turn of radius r feet at a velocity of v feet per second is given by the equation is $\tan \theta = \frac{v^2}{gr}$. In this equation, g is the acceleration due to gravity or 32 feet per second squared. An engineer is designing a curve with a radius of 1200 feet. If the speed limit on the curve will be 65 miles per hour, at what angle should the curve be banked? (*Hint*: Change 65 miles per hour to feet per second.)

50. **Physics** According to Snell's Law, $\frac{\sin \theta_i}{\sin \theta_r} = n$, where θ_i is the angle of incidence, θ_r is the angle of refraction, and n is the index of refraction. The index of refraction for a diamond is 2.42. If a beam of light strikes a diamond at an angle of incidence of 60°, find the angle of refraction.

51. **Critical Thinking** Solve the triangle. (*Hint*: Draw the altitude from Y.)

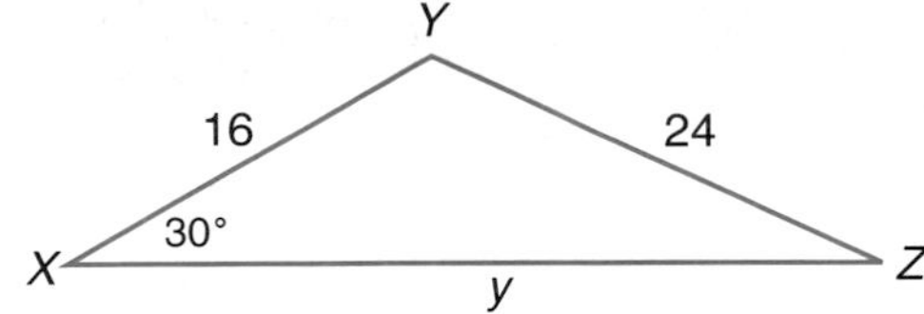

Mixed Review

52. **Aviation** A traffic helicopter is flying 1000 feet above the downtown area. To the right, the pilot sees the baseball stadium at an angle of depression of 63°. To the left, the pilot sees the football stadium at an angle of depression of 18°. Find the distance between the two stadiums. *(Lesson 5-4)*

53. Find the six trigonometric ratios for $\angle F$. *(Lesson 5-2)*

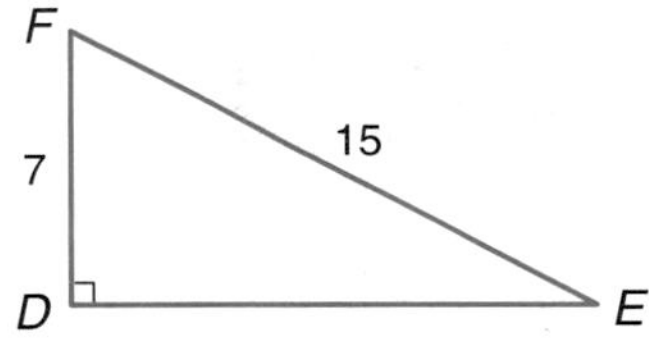

54. Approximate the real zeros of the function $f(x) = 3x^3 - 16x^2 + 12x + 6$ to the nearest tenth. *(Lesson 4-5)*

55. Determine whether the graph of $y^3 - x^2 = 2$ is symmetric with respect to the x-axis, y-axis, the graph of $y = x$, the graph of $y = -x$, or none of these. *(Lesson 3-1)*

56. Use a reflection matrix to find the coordinates of the vertices of a pentagon reflected over the y-axis if the coordinates of the vertices of the pentagon are $(-5, -3)$, $(-5, 4)$, $(-3, 6)$, $(-1, 3)$, and $(-2, -2)$. *(Lesson 2-4)*

57. Find the sum of the matrices $\begin{bmatrix} 4 & -3 & 2 \\ 8 & -2 & 0 \\ 9 & 6 & -3 \end{bmatrix}$ and $\begin{bmatrix} -2 & 2 & -2 \\ -5 & 1 & 1 \\ -7 & 2 & -2 \end{bmatrix}$. *(Lesson 2-3)*

Extra Practice See p. A35.

Pythagorean Theorem

All SAT and ACT tests contain several problems that you can solve using the Pythagorean Theorem. The **Pythagorean Theorem** states that in a right triangle, the sum of the squares of the measures of the legs equals the square of the measure of the hypotenuse.

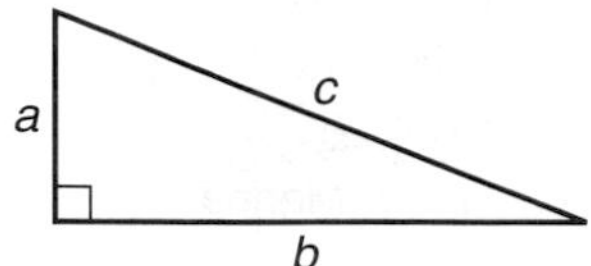

$$a^2 + b^2 = c^2$$

THE PRINCETON REVIEW

TEST-TAKING TIP

The 3-4-5 right triangle and its multiples like 6-8-10 and 9-12-15 occur frequently on the SAT and ACT. Other commonly used Pythagorean triples include 5-12-13 and 7-24-25. Memorize them.

SAT EXAMPLE

1. A 25-foot ladder is placed against a vertical wall of a building with the bottom of the ladder standing on concrete 7 feet from the base of the building. If the top of the ladder slips down 4 feet, then the bottom of the ladder will slide out how many feet?

   A 4 ft

   B 5 ft

   C 6 ft

   D 7 ft

   E 8 ft

HINT This problem does not have a diagram. So, start by drawing diagrams.

Solution The ladder placed against the wall forms a 7-24-25 right triangle. After the ladder slips down 4 feet, the new right triangle has sides that are multiples of a 3-4-5 right triangle, 15-20-25.

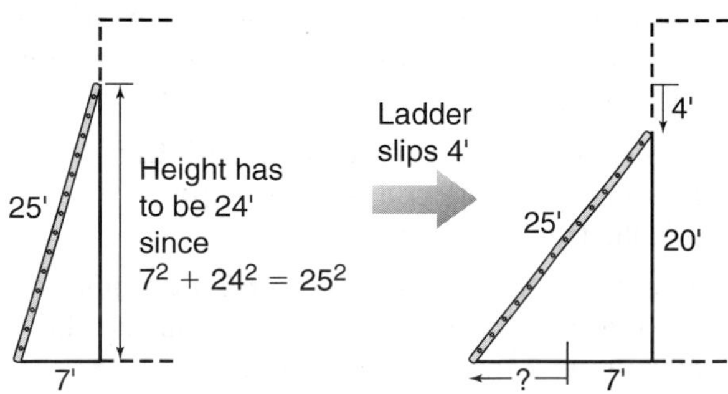

The ladder is now 15 feet from the wall. This means the ladder slipped 15 − 7 or 8 feet. The correct answer is choice **E.**

ACT EXAMPLE

2. In the figure below, right triangles *ABC* and *ACD* are drawn as shown. If $AB = 20$, $BC = 15$, and $AD = 7$, then $CD = ?$

   A 21

   B 22

   C 23

   D 24

   E 25

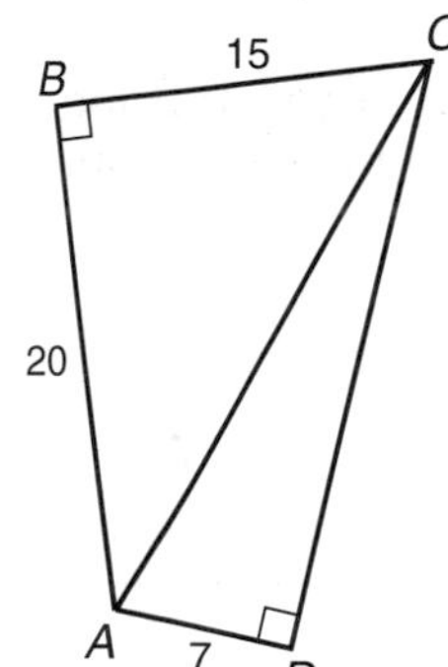

HINT Be on the lookout for problems like this one in which the application of the Pythagorean Theorem is not obvious.

Solution Notice that quadrilateral *ABCD* is separated into two right triangles, $\triangle ABC$ and $\triangle ADC$.

$\triangle ABC$ is a 15-20-25 right triangle (a multiple of the 3-4-5 right triangle). So, side $\overline{AC}$ (the hypotenuse) is 25 units long.

$\overline{AC}$ is also the hypotenuse of $\triangle ADC$. So, $\triangle ADC$ is a 7-24-25 right triangle.

Therefore, $\overline{CD}$ is 24 units long. The correct answer is choice **D.**

After working each problem, record the correct answer on the answer sheet provided or use your own paper.

Multiple Choice

1. In the figure below, $y =$

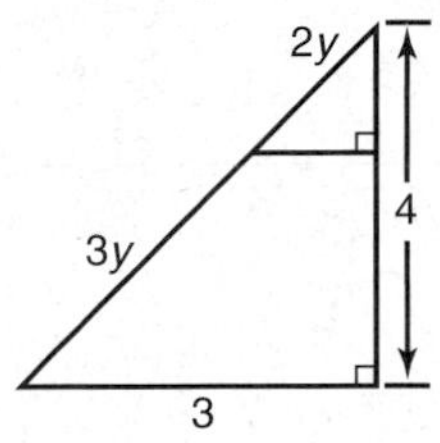

A 1 **B** 2 **C** 3 **D** 4 **E** 5

2. What graph would be created if the equation $x^2 + y^2 = 12$ were graphed in the standard (x, y) coordinate plane?

A circle **B** ellipse
C parabola **D** straight line
E 2 rays forming a "V"

3. If $999 \times 111 = 3 \times 3 \times n^2$, then which of the following could equal n?

A 9 **B** 37 **C** 111
D 222 **E** 333

4. In the figure below, $\triangle ABC$ is an equilateral triangle with $\overline{BC}$ 7 units long. If $\angle DCA$ is a right angle and $\angle D$ measures 45°, what is the length of $\overline{AD}$ in units?

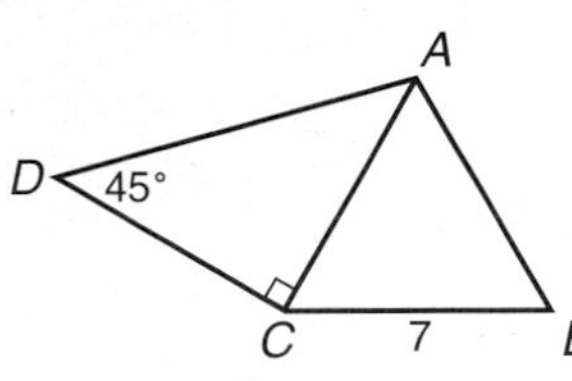

A 7 **B** $7\sqrt{2}$ **C** 14 **D** $14\sqrt{2}$
E It cannot be determined from the information given.

5. If $4 < a < 7 < b < 9$, then which of the following best defines $\frac{a}{b}$?

A $\frac{4}{9} < \frac{a}{b} < 1$ **B** $\frac{4}{9} < \frac{a}{b} < \frac{7}{9}$
C $\frac{4}{7} < \frac{a}{b} < \frac{7}{9}$ **D** $\frac{4}{7} < \frac{a}{b} < 1$
E $\frac{4}{7} < \frac{a}{b} < \frac{9}{7}$

6. A swimming pool with a capacity of 36,000 gallons originally contained 9,000 gallons of water. At 10:00 A.M. water begins to flow into the pool at a constant rate. If the pool is exactly three-fourths full at 1:00 P.M. on the same day and the water continues to flow at the same rate, what is the earliest time when the pool will be completely full?

A 1:40 P.M. **B** 2:00 P.M. **C** 2:30 P.M.
D 3:00 P.M. **E** 3:30 P.M.

7. In the figure below, what is the length of $\overline{BC}$?

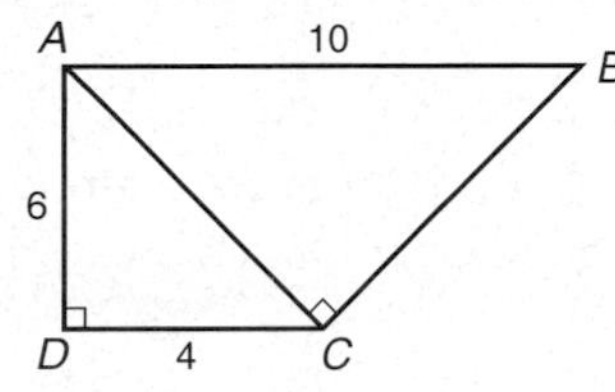

A 6 **B** $4\sqrt{3}$ **C** $2\sqrt{13}$
D 8 **E** $2\sqrt{38}$

8. If $\frac{x^2 + 7x + 12}{x + 4} = 5$, then $x =$

A 1 **B** 2 **C** 3 **D** 5 **E** 6

9. **Quantitative Comparison**
A if the quantity in Column A is greater
B if the quantity in Column B is greater
C if the two quantities are equal
D if the relationship cannot be determined for the information given

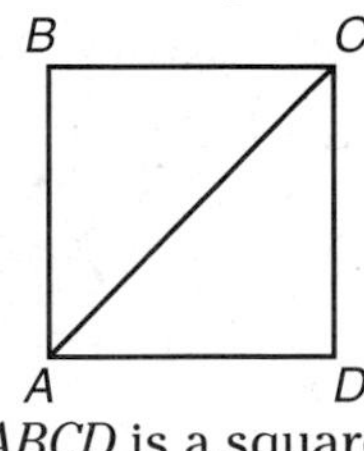

ABCD is a square.

Column A	Column B
$\frac{AC}{AD}$	2

10. **Grid-In** Segment AB is perpendicular to segment BD. Segment AB and segment CD bisect each other at point X. If $AB = 8$ and $CD = 10$, what is the length of $\overline{BD}$?

*inter***NET** CONNECTION **SAT/ACT Practice** For additional test practice questions, visit: **www.amc.glencoe.com**

Graphs of Trigonometric Functions

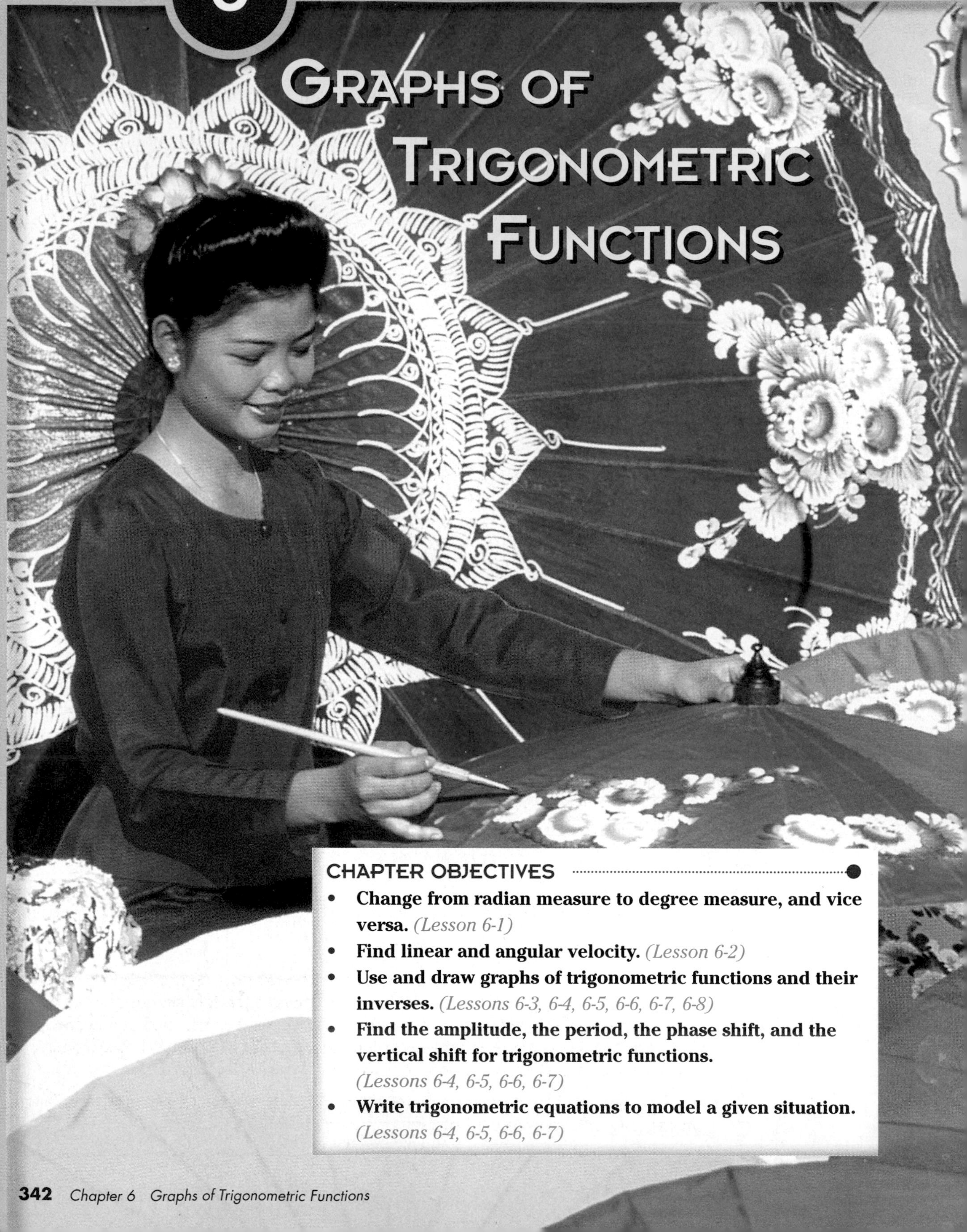

Chapter Objectives

- **Change from radian measure to degree measure, and vice versa.** *(Lesson 6-1)*
- **Find linear and angular velocity.** *(Lesson 6-2)*
- **Use and draw graphs of trigonometric functions and their inverses.** *(Lessons 6-3, 6-4, 6-5, 6-6, 6-7, 6-8)*
- **Find the amplitude, the period, the phase shift, and the vertical shift for trigonometric functions.** *(Lessons 6-4, 6-5, 6-6, 6-7)*
- **Write trigonometric equations to model a given situation.** *(Lessons 6-4, 6-5, 6-6, 6-7)*

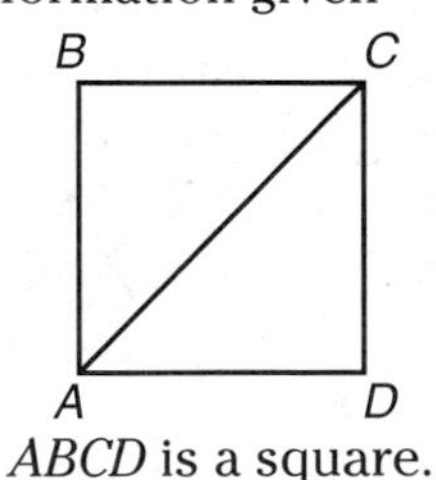

6-1

Angles and Radian Measure

OBJECTIVES

- Change from radian measure to degree measure, and vice versa.
- Find the length of an arc given the measure of the central angle.
- Find the area of a sector.

BUSINESS Junjira Putiwuthigool owns a business in Changmai, Thailand, that makes ornate umbrellas and fans. Ms. Putiwuthigool has an order for three dozen umbrellas having a diameter of 2 meters. Bamboo slats that support each circular umbrella divide the umbrella into 8 sections or sectors. Each section will be covered with a different color fabric. How much fabric of each color will Ms. Putiwuthigool need to complete the order? *This problem will be solved in Example 6.*

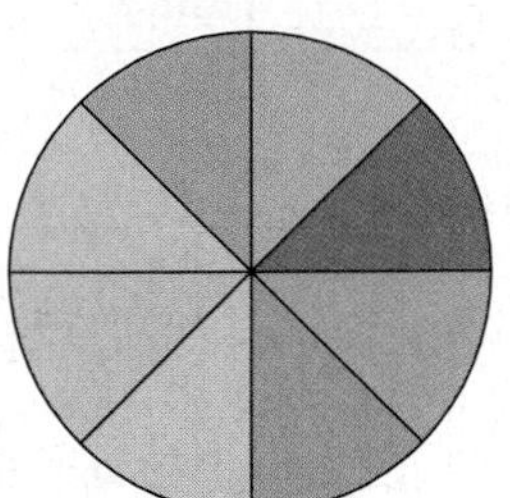

There are many real-world applications, such as the one described above, which can be solved more easily using an angle measure other than the degree. This other unit is called the **radian.**

The definition of radian is based on the concept of the unit circle. Recall that the unit circle is a circle of radius 1 whose center is at the origin of a rectangular coordinate system.

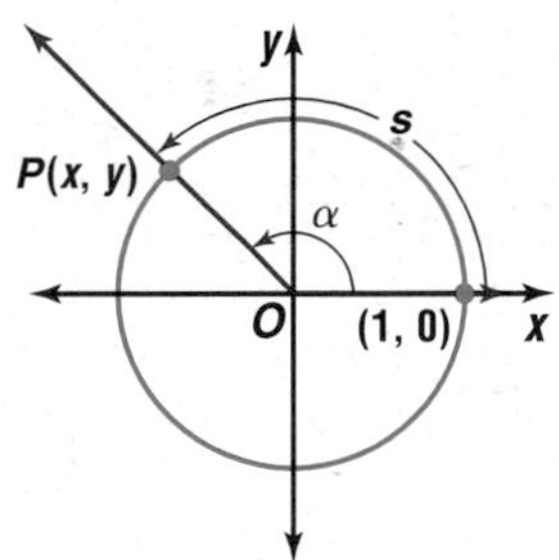

A point $P(x, y)$ is on the unit circle if and only if its distance from the origin is 1. Thus, for each point $P(x, y)$ on the unit circle, the distance from the origin is represented by the following equation.

$$\sqrt{(x-0)^2 + (y-0)^2} = 1$$

If each side of this equation is squared, the result is an equation of the unit circle.

$$x^2 + y^2 = 1$$

Consider an angle α in standard position, shown above. Let $P(x, y)$ be the point of intersection of its terminal side with the unit circle. The radian measure of an angle in standard position is defined as the length of the corresponding arc on the unit circle. Thus, the measure of angle α is s radians. Since $C = 2\pi r$, a full revolution correponds to an angle of $2\pi(1)$ or 2π radians.

There is an important relationship between radian and degree measure. Since an angle of one complete revolution can be represented either by 360° or by 2π radians, $360° = 2\pi$ radians. Thus, $180° = \pi$ radians, and $90° = \frac{\pi}{2}$ radians.

The following formulas relate degree and radian measures.

Degree/ Radian Conversion Formulas	1 radian $= \frac{180}{\pi}$ degrees or about 57.3° 1 degree $= \frac{\pi}{180}$ radians or about 0.017 radian

Angles expressed in radians are often written in terms of π. The term *radians* is also usually omitted when writing angle measures. However, the degree symbol is always used in this book to express the measure of angles in degrees.

Example 1 **a. Change 330° to radian measure in terms of π.**

$$330° = 330° \times \frac{\pi}{180°} \quad \textit{1 degree} = \frac{\pi}{180°}$$
$$= \frac{11\pi}{6}$$

b. Change $\frac{2\pi}{3}$ radians to degree measure.

$$\frac{2\pi}{3} = \frac{2\pi}{3} \times \frac{180°}{\pi} \quad \textit{1 radian} = \frac{180°}{\pi}$$
$$= 120°$$

Angles whose measures are multiples of 30° and 45° are commonly used in trigonometry. These angle measures correspond to radian measures of $\frac{\pi}{6}$ and $\frac{\pi}{4}$, respectively. The diagrams below can help you make these conversions mentally.

You may want to memorize these radian measures and their degree equivalents to simplify your work in trigonometry.

Multiples of 30° and $\frac{\pi}{6}$

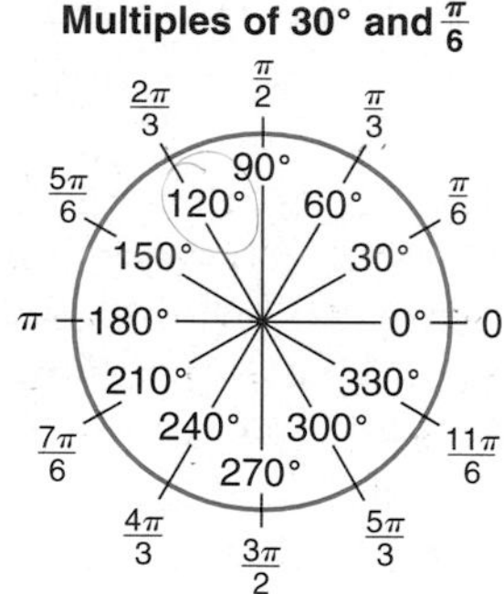

Multiples of 45° and $\frac{\pi}{4}$

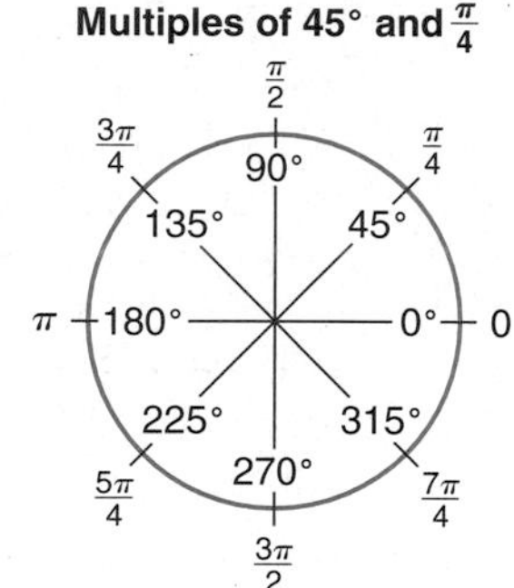

These equivalent values are summarized in the chart below.

Degrees	0	30	45	60	90	120	135	150	180	210	225	240	270	300	315	330
Radians	0	$\frac{\pi}{6}$	$\frac{\pi}{4}$	$\frac{\pi}{3}$	$\frac{\pi}{2}$	$\frac{2\pi}{3}$	$\frac{3\pi}{4}$	$\frac{5\pi}{6}$	π	$\frac{7\pi}{6}$	$\frac{5\pi}{4}$	$\frac{4\pi}{3}$	$\frac{3\pi}{2}$	$\frac{5\pi}{3}$	$\frac{7\pi}{4}$	$\frac{11\pi}{6}$

You can use reference angles and the unit circle to determine trigonometric values for angle measures expressed as radians.

Example **2** **Evaluate $\cos \frac{4\pi}{3}$.**

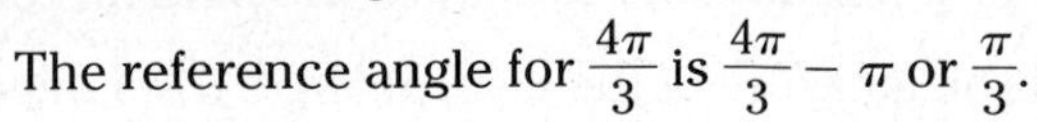

Look Back
You can refer to Lesson 5-3 to review reference angles and unit circles used to determine values of trigonometric functions.

The reference angle for $\frac{4\pi}{3}$ is $\frac{4\pi}{3} - \pi$ or $\frac{\pi}{3}$.

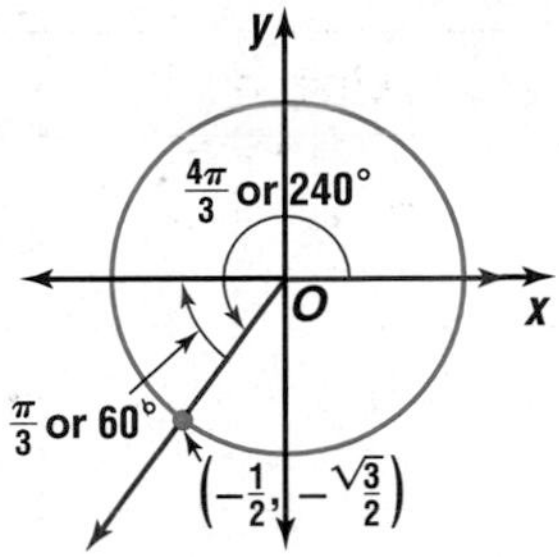

Since $\frac{\pi}{3} = 60°$, the terminal side of the angle intersects the unit circle at a point with coordinates of $\left(\frac{1}{2}, \frac{\sqrt{3}}{2}\right)$.

Because the terminal side of this angle is in the third quadrant, both coordinates are negative. The point of intersection has coordinates $\left(-\frac{1}{2}, -\frac{\sqrt{3}}{2}\right)$.

Therefore, $\cos \frac{4\pi}{3} = -\frac{1}{2}$.

Radian measure can be used to find the length of a **circular arc.** A circular arc is a part of a circle. The arc is often defined by the **central angle** that intercepts it. A central angle of a circle is an angle whose vertex lies at the center of the circle.

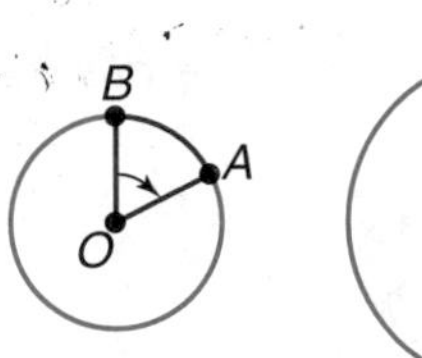

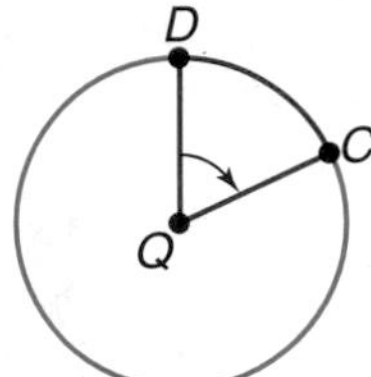

If two central angles in different circles are congruent, the ratio of the lengths of their intercepted arcs is equal to the ratio of the measures of their radii.

For example, given circles O and Q, if $\angle O \cong \angle Q$, then $\frac{m\widehat{AB}}{m\widehat{CD}} = \frac{OA}{QC}$.

Let O be the center of two concentric circles, let r be the measure of the radius of the larger circle, and let the smaller circle be a unit circle. A central angle of θ radians is drawn in the two circles that intercept $\widehat{RT}$ on the unit circle and $\widehat{SW}$ on the other circle. Suppose $\widehat{SW}$ is s units long. $\widehat{RT}$ is θ units long since it is an arc of a unit circle intercepted by a central angle of θ radians. Thus, we can write the following proportion.

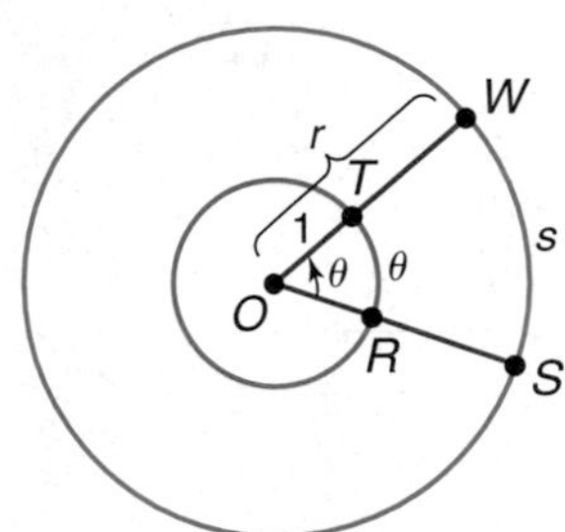

We say that an arc subtends its central angle.

$$\frac{s}{\theta} = \frac{r}{1} \text{ or } s = r\theta$$

Length of an Arc

The length of any circular arc s is equal to the product of the measure of the radius of the circle r and the radian measure of the central angle θ that it subtends.

$$s = r\theta$$

Example 3 **Given a central angle of 128°, find the length of its intercepted arc in a circle of radius 5 centimeters. Round to the nearest tenth.**

First, convert the measure of the central angle from degrees to radians.

$$128° = 128° \times \frac{\pi}{180°} \quad \textit{1 degree} = \frac{\pi}{180}$$
$$= \frac{32}{45}\pi \text{ or } \frac{32\pi}{45}$$

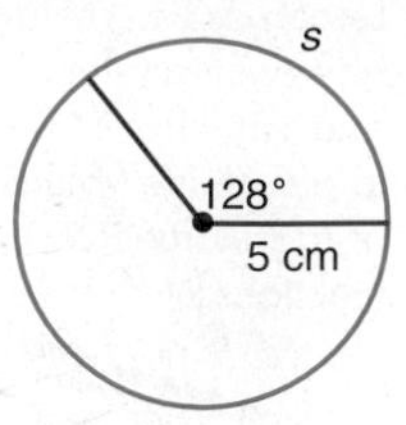

Then, find the length of the arc.

$$s = r\theta$$
$$s = 5\left(\frac{32\pi}{45}\right) \quad r = 5,\ \theta = \frac{32\pi}{45}$$
$$s \approx 11.17010721 \quad \textit{Use a calculator.}$$

The length of the arc is about 11.2 centimeters.

You can use radians to compute distances between two cities that lie on the same longitude line.

Example 4 

GEOGRAPHY **Winnipeg, Manitoba, Canada, and Dallas, Texas, lie along the 97° W longitude line. The latitude of Winnipeg is 50° N, and the latitude of Dallas is 33° N. The radius of Earth is about 3960 miles. Find the approximate distance between the two cities.**

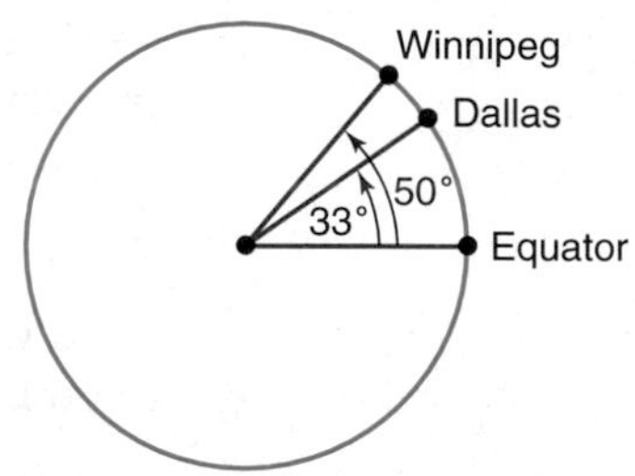

The length of the arc between Dallas and Winnipeg is the distance between the two cities. The measure of the central angle subtended by this arc is 50° − 33° or 17°.

$$17° = 17° \times \frac{\pi}{180°} \quad \textit{1 degree} = \frac{\pi}{180}$$
$$= \frac{17\pi}{180}$$

$$s = r\theta$$
$$s = 3960\left(\frac{17\pi}{180}\right) \quad r = 3960,\ \theta = \frac{17\pi}{180}$$
$$s \approx 1174.955652 \quad \textit{Use a calculator.}$$

The distance between the two cities is about 1175 miles.

A **sector** of a circle is a region bounded by a central angle and the intercepted arc. For example, the shaded portion in the figure is a sector of circle O. The ratio of the area of a sector to the area of a circle is equal to the ratio of its arc length to the circumference.

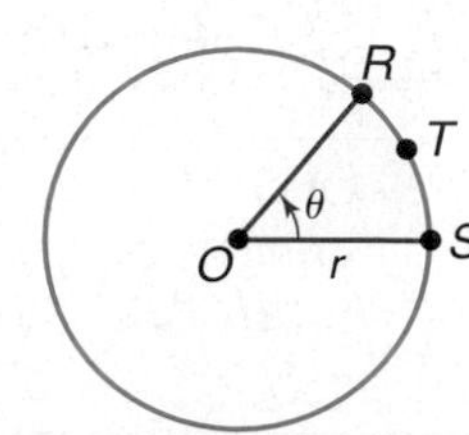

Let A represent the area of the sector.

$\frac{A}{\pi r^2} = \frac{\text{length of } \widehat{RTS}}{2\pi r}$

$\frac{A}{\pi r^2} = \frac{r\theta}{2\pi r}$ *The length of $\widehat{RTS}$ is $r\theta$.*

$A = \frac{1}{2}r^2\theta$ *Solve for A.*

Area of a Circular Sector

If θ is the measure of the central angle expressed in radians and r is the measure of the radius of the circle, then the area of the sector, A, is as follows.

$$A = \frac{1}{2}r^2\theta$$

Examples

5 **Find the area of a sector if the central angle measures $\frac{5\pi}{6}$ radians and the radius of the circle is 16 centimeters. Round to the nearest tenth.**

$A = \frac{1}{2}r^2\theta$ *Formula for the area of a circular sector*

$A = \frac{1}{2}(16^2)\left(\frac{5\pi}{6}\right)$ $r = 16,\ \theta = \frac{5\pi}{6}$

$A \approx 335.1032164$ *Use a calculator.*

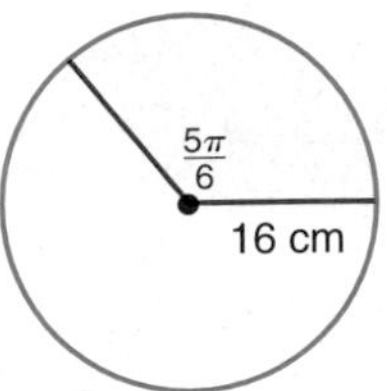

The area of the sector is about 335.1 square centimeters.

6 **BUSINESS** **Refer to the application at the beginning of the lesson. How much fabric of each color will Ms. Putiwuthigool need to complete the order?**

There are 2π radians in a complete circle and 8 equal sections or sectors in the umbrella. Therefore, the measure of each central angle is $\frac{2\pi}{8}$ or $\frac{\pi}{4}$ radians. If the diameter of the circle is 2 meters, the radius is 1 meter. Use these values to find the area of each sector.

$A = \frac{1}{2}r^2\theta$

$A = \frac{1}{2}(1^2)\left(\frac{\pi}{4}\right)$ $r = 1,\ \theta = \frac{\pi}{4}$

$A \approx 0.3926990817$ *Use a calculator.*

Since there are 3 dozen or 36 umbrellas, multiply the area of each sector by 36. Ms. Putiwuthigool needs about 14.1 square meters of each color of fabric. *This assumes that the pieces can be cut with no waste and that no extra material is needed for overlapping.*

CHECK FOR UNDERSTANDING

Communicating Mathematics

Read and study the lesson to answer each question.

1. **Draw** a unit circle and a central angle with a measure of $\frac{3\pi}{4}$ radians.
2. **Describe** the angle formed by the hands of a clock at 3:00 in terms of degrees and radians.

3. **Explain** how you could find the radian measure of a central angle subtended by an arc that is 10 inches long in a circle with a radius of 8 inches.

4. **Demonstrate** that if the radius of a circle is doubled and the measure of a central angle remains the same, the length of the arc is doubled and the area of the sector is quadrupled.

Guided Practice

Change each degree measure to radian measure in terms of π.

5. $240°$

6. $570°$

Change each radian measure to degree measure. Round to the nearest tenth, if necessary.

7. $\frac{3\pi}{2}$

8. -1.75

Evaluate each expression.

9. $\sin \frac{3\pi}{4}$

10. $\tan \frac{11\pi}{6}$

Given the measurement of a central angle, find the length of its intercepted arc in a circle of radius 15 inches. Round to the nearest tenth.

11. $\frac{5\pi}{6}$

12. $77°$

Find the area of each sector given its central angle θ and the radius of the circle. Round to the nearest tenth.

13. $\theta = \frac{2\pi}{3}, r = 1.4$

14. $\theta = 54°, r = 6$

15. **Physics** A pendulum with length of 1.4 meters swings through an angle of 30°. How far does the bob at the end of the pendulum travel as it goes from left to right?

EXERCISES

Practice

Change each degree measure to radian measure in terms of π.

16. $135°$

17. $210°$

18. $300°$

19. $-450°$

20. $-75°$

21. $1250°$

Change each radian measure to degree measure. Round to the nearest tenth, if necessary.

22. $\frac{7\pi}{12}$

23. $\frac{11\pi}{3}$

24. 17

25. -3.5

26. $-\frac{\pi}{6.2}$

27. 17.5

Evaluate each expression.

28. $\sin \frac{5\pi}{3}$

29. $\tan \frac{7\pi}{6}$

30. $\cos \frac{5\pi}{4}$

31. $\sin \frac{7\pi}{6}$

32. $\tan \frac{14\pi}{3}$

33. $\cos \left(-\frac{19\pi}{6}\right)$

Given the measurement of a central angle, find the length of its intercepted arc in a circle of radius 14 centimeters. Round to the nearest tenth.

34. $\frac{2\pi}{3}$

35. $\frac{5\pi}{12}$

36. $150°$

37. $282°$

38. $\frac{3\pi}{11}$

39. $320°$

40. The diameter of a circle is 22 inches. If a central angle measures 78°, find the length of the intercepted arc.

41. An arc is 70.7 meters long and is intercepted by a central angle of $\frac{5\pi}{4}$ radians. Find the diameter of the circle.

42. An arc is 14.2 centimeters long and is intercepted by a central angle of 60°. What is the radius of the circle?

Find the area of each sector given its central angle θ and the radius of the circle. Round to the nearest tenth.

43. $\theta = \frac{5\pi}{12}, r = 10$

44. $\theta = 90°, r = 22$

45. $\theta = \frac{\pi}{8}, r = 7$

46. $\theta = \frac{4\pi}{7}, r = 12.5$

47. $\theta = 225°, r = 6$

48. $\theta = 82°, r = 7.3$

49. A sector has arc length of 6 feet and central angle of 1.2 radians.
   a. Find the radius of the circle.
   b. Find the area of the sector.

50. A sector has a central angle of 135° and arc length of 114 millimeters.
   a. Find the radius of the circle.
   b. Find the area of the sector.

51. A sector has area of 15 square inches and central angle of 0.2 radians.
   a. Find the radius of the circle.
   b. Find the arc length of the sector.

52. A sector has area of 15.3 square meters. The radius of the circle is 3 meters.
   a. Find the radian measure of the central angle.
   b. Find the degree measure of the central angle.
   c. Find the arc length of the sector.

Applications and Problem Solving

53. **Mechanics** A wheel has a radius of 2 feet. As it turns, a cable connected to a box winds onto the wheel.
   a. How far does the box move if the wheel turns 225° in a counterclockwise direction?
   b. Find the number of degrees the wheel must be rotated to move the box 5 feet.

54. **Critical Thinking** Two gears are interconnected. The smaller gear has a radius of 2 inches, and the larger gear has a radius of 8 inches. The smaller gear rotates 330°. Through how many radians does the larger gear rotate?

55. **Physics** A pendulum is 22.9 centimeters long, and the bob at the end of the pendulum travels 10.5 centimeters. Find the degree measure of the angle through which the pendulum swings.

56. **Geography** Minneapolis, Minnesota; Arkadelphia, Arkansas; and Alexandria, Louisiana lie on the same longitude line. The latitude of Minneapolis is 45° N, the latitude of Arkadelphia is 34° N, and the latitude of Alexandria is 31° N. The radius of Earth is about 3960 miles.

a. Find the approximate distance between Minneapolis and Arkadelphia.
b. What is the approximate distance between Minneapolis and Alexandria?
c. Find the approximate distance between Arkadelphia and Alexandria.

57. **Civil Engineering** The figure below shows a stretch of roadway where the curves are arcs of circles.

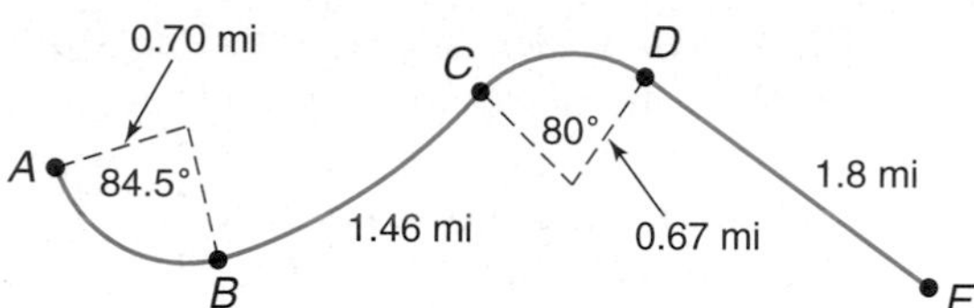

Find the length of the road from point A to point E.

58. **Mechanics** A single pulley is being used to pull up a weight. Suppose the diameter of the pulley is $2\frac{1}{2}$ feet.

a. How far will the weight rise if the pulley turns 1.5 rotations?
b. Find the number of degrees the pulley must be rotated to raise the weight $4\frac{1}{2}$ feet.

59. **Pet Care** A rectangular house is 33 feet by 47 feet. A dog is placed on a leash that is connected to a pole at the corner of the house.

a. If the leash is 15 feet long, find the area the dog has to play.
b. If the owner wants the dog to have 750 square feet to play, how long should the owner make the leash?

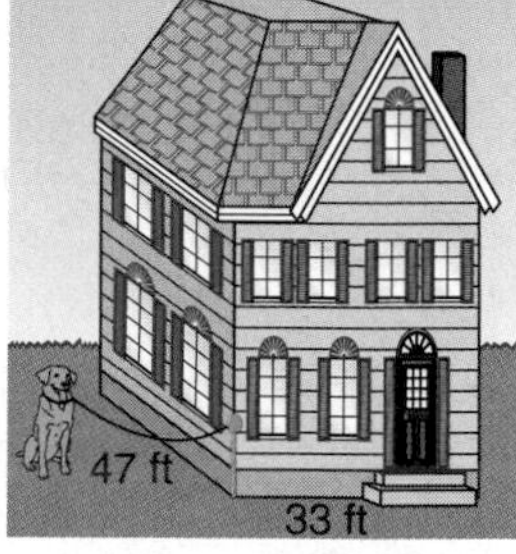

60. **Biking** Rafael rides his bike 3.5 kilometers. If the radius of the tire on his bike is 32 centimeters, determine the number of radians that a spot on the tire will travel during the trip.

61. **Critical Thinking** A *segment* of a circle is the region bounded by an arc and its chord. Consider any minor arc. If α is the radian measure of the central angle and r is the radius of the circle, write a formula for the area of the segment.

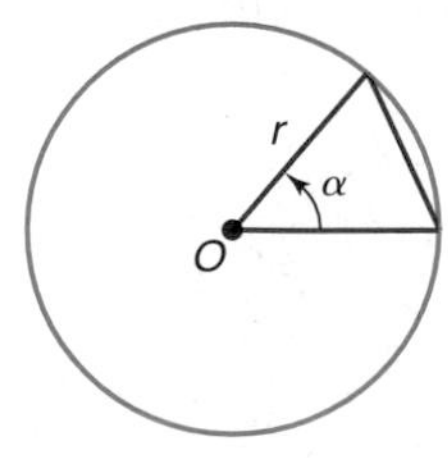

Mixed Review

62. The lengths of the sides of a triangle are 6 inches, 8 inches, and 12 inches. Find the area of the triangle. *(Lesson 5-8)*

63. Determine the number of possible solutions of $\triangle ABC$ if $A = 152°$, $b = 12$, and $a = 10.2$. If solutions exist, solve the triangle. *(Lesson 5-7)*

64. Surveying Two surveyors are determining measurements to be used to build a bridge across a canyon. The two surveyors stand 560 yards apart on one side of the canyon and sight a marker C on the other side of the canyon at angles of 27° and 38°. Find the length of the bridge if it is built through point C as shown. *(Lesson 5-6)*

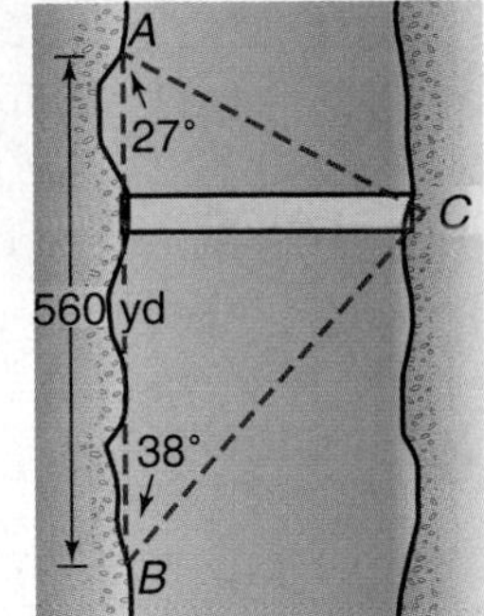

65. Suppose θ is an angle in standard position and $\tan\theta > 0$. State the quadrants in which the terminal side of θ can lie. *(Lesson 5-3)*

66. Population The population for Forsythe County, Georgia, has experienced significant growth in recent years. *(Lesson 4-8)*

Year	1970	1980	1990	1998
Population	17,000	28,000	44,000	86,000

Source: U.S. Census Bureau

a. Write a model that relates the population of Forsythe County as a function of the number of years since 1970.

b. Use the model to predict the population in the year 2020.

67. Use the Upper Bound Theorem to find an integral upper bound and the Lower Bound Theorem to find a lower bound of the zeros of $f(x) = x^4 - 3x^3 - 2x^2 + 6x + 10$. *(Lesson 4-5)*

68. Use synthetic division to determine if $x + 2$ is a factor of $x^3 + 6x^2 + 12x + 12$. Explain. *(Lesson 4-3)*

69. Determine whether the graph of $x^2 + y^2 = 16$ is symmetric with respect to the x-axis, the y-axis, the line $y = x$, or the line $y = -x$. *(Lesson 3-1)*

70. Solve the system of equations algebraically. *(Lesson 2-2)*

$4x - 2y + 3z = -6$
$3x + 3y - 2z = 2$
$5x - 4y - 3z = -75$

71. Which scatter plot shows data that has a strongly positive correlation? *(Lesson 1-6)*

a. 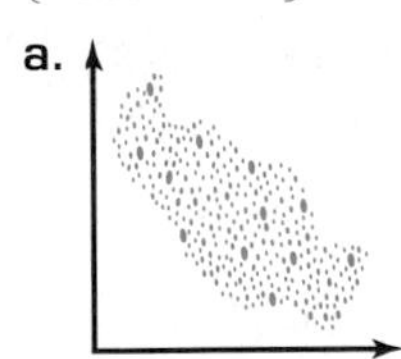

b. 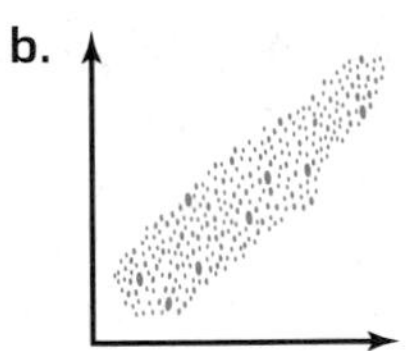

c. 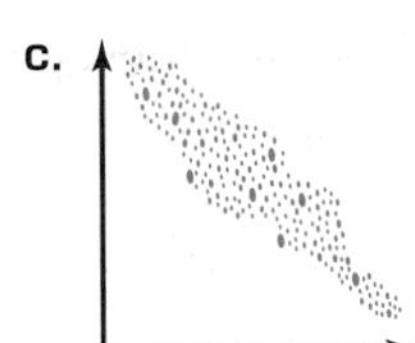

d. 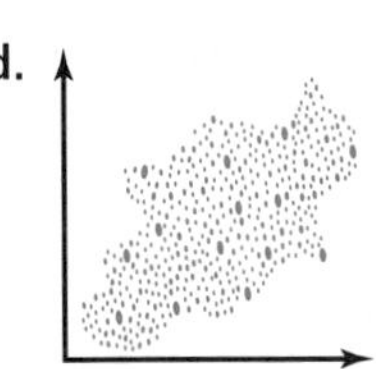

72. SAT Practice Quantitative Comparison

A if the quantity in Column A is greater
B if the quantity in Column B is greater
C if the two quantities are equal
D if the relationship cannot be determined for the information given

Given: $p > 0,\ q < 0$

Column A	Column B
$p + q$	$p - q$

Extra Practice See p. A36.

6-2 Linear and Angular Velocity

OBJECTIVE

- Find linear and angular velocity.

ENTERTAINMENT The Children's Museum in Indianapolis, Indiana, houses an antique carousel. The carousel contains three concentric circles of animals. The inner circle of animals is approximately 11 feet from the center, and the outer circle of animals is approximately 20 feet from the center. The carousel makes $2\frac{5}{8}$ rotations per minute. Determine the angular and linear velocities of someone riding an animal in the inner circle and of someone riding an animal in the same row in the outer circle. *This problem will be solved in Examples 3 and 5.*

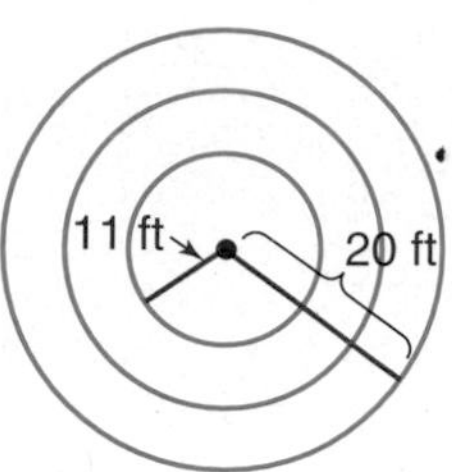

The carousel is a circular object that turns about an axis through its center. Other examples of objects that rotate about a central axis include Ferris wheels, gears, tires, and compact discs. As the carousel or any other circular object rotates counterclockwise about its center, an object at the edge moves through an angle relative to its starting position known as the **angular displacement,** or angle of rotation.

Consider a circle with its center at the origin of a rectangular coordinate system and point B on the circle rotating counterclockwise. Let the positive x-axis, or $\overrightarrow{OA}$, be the initial side of the central angle. The terminal side of the central angle is $\overrightarrow{OB}$. The angular displacement is θ. The measure of θ changes as B moves around the circle. All points on $\overrightarrow{OB}$ move through the same angle per unit of time.

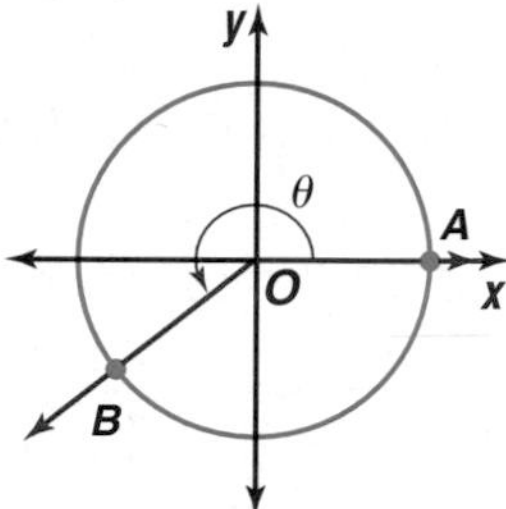

Example 1 **Determine the angular displacement in radians of 4.5 revolutions. Round to the nearest tenth.**

Each revolution equals 2π radians. For 4.5 revolutions, the number of radians is $4.5 \times 2\pi$ or 9π. 9π radians equals about 28.3 radians.

The ratio of the change in the central angle to the time required for the change is known as **angular velocity.** Angular velocity is usually represented by the lowercase Greek letter ω (omega).

Angular Velocity

If an object moves along a circle during a time of t units, then the angular velocity, ω, is given by

$$\omega = \frac{\theta}{t},$$

where θ is the angular displacement in radians.

Notice that the angular velocity of a point on a rotating object is not dependent upon the distance from the center of the rotating object.

Example 2 **Determine the angular velocity if 7.3 revolutions are completed in 5 seconds. Round to the nearest tenth.**

The angular displacement is $7.3 \times 2\pi$ or 14.6π radians.

$$\omega = \frac{\theta}{t}$$

$$\omega = \frac{14.6\pi}{5} \qquad \theta = 14.6\pi,\ t = 5$$

$$\omega \approx 9.173450548 \qquad \textit{Use a calculator.}$$

The angular velocity is about 9.2 radians per second.

To avoid mistakes when computing with units of measure, you can use a procedure called **dimensional analysis.** In dimensional analyses, unit labels are treated as mathematical factors and can be divided out.

Example 3 **ENTERTAINMENT** **Refer to the application at the beginning of the lesson. Determine the angular velocity for each rider in radians per second.**

The carousel makes $2\frac{5}{8}$ or 2.625 revolutions per minute. Convert revolutions per minute to radians per second.

$$\frac{2.625 \text{ revolutions}}{1 \text{ minute}} \times \frac{1 \text{ minute}}{60 \text{ seconds}} \times \frac{2\pi \text{ radians}}{1 \text{ revolution}} \approx 0.275 \text{ radian per second}$$

Each rider has an angular velocity of about 0.275 radian per second.

The carousel riders have the same angular velocity. However, the rider in the outer circle must travel a greater distance than the one in the inner circle. The arc length formula can be used to find the relationship between the linear and angular velocities of an object moving in a circular path. If the object moves with constant **linear velocity** (v) for a period of time (t), the distance (s) it travels is given by the formula $s = vt$. Thus, the linear velocity is $v = \frac{s}{t}$.

As the object moves along the circular path, the radius r forms a central angle of measure θ. Since the length of the arc is $s = r\theta$, the following is true.

$$s = r\theta$$

$$\frac{s}{t} = \frac{r\theta}{t} \qquad \textit{Divide each side by } t.$$

$$v = r\frac{\theta}{t} \qquad \textit{Replace } \frac{s}{t} \textit{ with } v.$$

Linear Velocity

If an object moves along a circle of radius of r units, then its linear velocity, v is given by

$$v = r\frac{\theta}{t},$$

where $\frac{\theta}{t}$ represents the angular velocity in radians per unit of time.

Since $\omega = \frac{\theta}{t}$, the formula for linear velocity can also be written as $v = r\omega$.

Examples

4 **Determine the linear velocity of a point rotating at an angular velocity of 17π radians per second at a distance of 5 centimeters from the center of the rotating object. Round to the nearest tenth.**

$v = r\omega$

$v = 5(17\pi)$ *$r = 5$, $\omega = 17\pi$*

$v \approx 267.0353756$ *Use a calculator.*

The linear velocity is about 267.0 centimeters per second.

Real World Application

5 **ENTERTAINMENT** **Refer to the application at the beginning of the lesson. Determine the linear velocity for each rider.**

From Example 3, you know that the angular velocity is about 0.275 radian per second. Use this number to find the linear velocity for each rider.

Rider on the Inner Circle

$v = r\omega$

$v \approx 11(0.275)$ *$r = 11$, $\omega = 0.275$*

$v \approx 3.025$

Rider on the Outer Circle

$v = r\omega$

$v \approx 20(0.275)$ *$r = 20$, $\omega = 0.275$*

$v \approx 5.5$

The linear velocity of the rider on the inner circle is about 3.025 feet per second, and the linear velocity of the rider on the outer circle is about 5.5 feet per second.

6 **CAR RACING** **The tires on a race car have a diameter of 30 inches. If the tires are turning at a rate of 2000 revolutions per minute, determine the race car's speed in miles per hour (mph).**

If the diameter is 30 inches, the radius is $\frac{1}{2} \times 30$ or 15 inches. This measure needs to be written in miles. The rate needs to be written in hours.

$$v = \underbrace{15 \text{ in.} \times \frac{1 \text{ ft}}{12 \text{ in.}} \times \frac{1 \text{ mi}}{5280 \text{ ft}}}_{r} \times \underbrace{\frac{2000 \text{ rev}}{1 \text{ min}} \times \frac{2\pi}{1 \text{ rev}} \times \frac{60 \text{ min}}{1 \text{ h}}}_{\omega}$$

$v \approx 178.4995826$ mph   *Use a calculator.*

The speed of the race car is about 178.5 miles per hour.

CHECK FOR UNDERSTANDING

Communicating Mathematics

Read and study the lesson to answer each question.

1. **Draw** a circle and represent an angular displacement of 3π radians.
2. **Write** an expression that could be used to change 5 revolutions per minute to radians per second.
3. **Compare and contrast** linear and angular velocity.
4. **Explain** how two people on a rotating carousel can have the same angular velocity but different linear velocity.
5. **Show** that when the radius of a circle is doubled, the angular velocity remains the same and the linear velocity of a point on the circle is doubled.

Guided Practice

Determine each angular displacement in radians. Round to the nearest tenth.

6. 5.8 revolutions
7. 710 revolutions

Determine each angular velocity. Round to the nearest tenth.

8. 3.2 revolutions in 7 seconds
9. 700 revolutions in 15 minutes

Determine the linear velocity of a point rotating at the given angular velocity at a distance *r* from the center of the rotating object. Round to the nearest tenth.

10. $\omega = 36$ radians per second, $r = 12$ inches
11. $\omega = 5\pi$ radians per minute, $r = 7$ meters
12. **Space** A geosynchronous equatorial orbiting (GEO) satellite orbits 22,300 miles above the equator of Earth. It completes one full revolution each 24 hours. Assume Earth's radius is 3960 miles.
    a. How far will the GEO satellite travel in one day?
    b. What is the satellite's linear velocity in miles per hour?

EXERCISES

Practice

Determine each angular displacement in radians. Round to the nearest tenth.

13. 3 revolutions
14. 2.7 revolutions
15. 13.2 revolutions
16. 15.4 revolutions
17. 60.7 revolutions
18. 3900 revolutions

Determine each angular velocity. Round to the nearest tenth.

19. 1.8 revolutions in 9 seconds
20. 3.5 revolutions in 3 minutes
21. 17.2 revolutions in 12 seconds
22. 28.4 revolutions in 19 seconds
23. 100 revolutions in 16 minutes
24. 122.6 revolutions in 27 minutes
25. A Ferris wheel rotates one revolution every 50 seconds. What is its angular velocity in radians per second?
26. A clothes dryer is rotating at 500 revolutions per minute. Determine its angular velocity in radians per second.

27. Change 85 radians per second to revolutions per minute (rpm).

Determine the linear velocity of a point rotating at the given angular velocity at a distance *r* from the center of the rotating object. Round to the nearest tenth.

28. $\omega = 16.6$ radians per second, $r = 8$ centimeters
29. $\omega = 27.4$ radians per second, $r = 4$ feet
30. $\omega = 6.1\pi$ radians per minute, $r = 1.8$ meters
31. $\omega = 75.3\pi$ radians per second, $r = 17$ inches
32. $\omega = 805.6$ radians per minute, $r = 39$ inches
33. $\omega = 64.5\pi$ radians per minute, $r = 88.9$ millimeters

34. A pulley is turned $120°$ per second.
   a. Find the number of revolutions per minute (rpm).
   b. If the radius of the pulley is 5 inches, find the linear velocity in inches per second.

35. Consider the tip of each hand of a clock. Find the linear velocity in millimeters per second for each hand.
   a. second hand which is 30 millimeters
   b. minute hand which is 27 millimeters long
   c. hour hand which is 18 millimeters long

Applications and Problem Solving

36. **Entertainment** The diameter of a Ferris wheel is 80 feet.
   a. If the Ferris wheel makes one revolution every 45 seconds, find the linear velocity of a person riding in the Ferris wheel.
   b. Suppose the linear velocity of a person riding in the Ferris wheel is 8 feet per second. What is the time for one revolution of the Ferris wheel?

37. **Entertainment** The Kit Carson County Carousel makes 3 revolutions per minute.
   a. Find the linear velocity in feet per second of someone riding a horse that is $22\frac{1}{2}$ feet from the center.
   b. The linear velocity of the person on the inside of the carousel is 3.1 feet per second. How far is the person from the center of the carousel?
   c. How much faster is the rider on the outside going than the rider on the inside?

38. **Critical Thinking** Two children are playing on the seesaw. The lighter child is 9 feet from the fulcrum, and the heavier child is 6 feet from the fulcrum. As the lighter child goes from the ground to the highest point, she travels through an angle of $35°$ in $\frac{1}{2}$ second.
   a. Find the angular velocity of each child.
   b. What is the linear velocity of each child?

39. **Bicycling** A bicycle wheel is 30 inches in diameter.
   a. To the nearest revolution, how many times will the wheel turn if the bicycle is ridden for 3 miles?
   b. Suppose the wheel turns at a constant rate of 2.75 revolutions per second. What is the linear speed in miles per hour of a point on the tire?

interNET CONNECTION

Research
For information about the other planets, visit **www.amc.glencoe.com**

40. **Space** The radii and times needed to complete one rotation for the four planets closest to the sun are given at the right.
 a. Find the linear velocity of a point on each planet's equator.
 b. Compare the linear velocity of a point on the equator of Mars with a point on the equator of Earth.

Planets

	Radius (kilometers)	Time for One Rotation (hours)
Mercury	2440	1407.6
Venus	6052	5832.5
Earth	6356	23.935
Mars	3375	24.623

Source: NASA

41. **Physics** A torsion pendulum is an object suspended by a wire or rod so that its plane of rotation is horizontal and it rotates back and forth around the wire without losing energy. Suppose that the pendulum is rotated θ_m radians and released. Then the angular displacement θ at time t is $\theta = \theta_m \cos \omega t$, where ω is the angular frequency in radians per second. Suppose the angular frequency of a certain torsion pendulum is π radians per second and its initial rotation is $\frac{\pi}{4}$ radians.

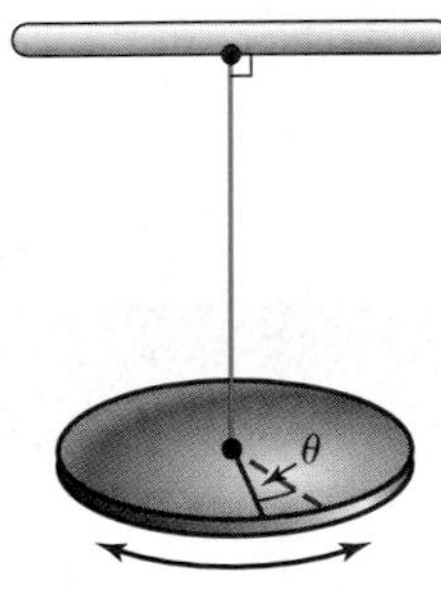

 a. Write the equation for the angular displacement of the pendulum.
 b. What are the first two values of t for which the angular displacement of the pendulum is 0?

42. **Space** Low Earth orbiting (LEO) satellites orbit between 200 and 500 miles above Earth. In order to keep the satellites at a constant distance from Earth, they must maintain a speed of 17,000 miles per hour. Assume Earth's radius is 3960 miles.
 a. Find the angular velocity needed to maintain a LEO satellite at 200 miles above Earth.
 b. How far above Earth is a LEO with an angular velocity of 4 radians per hour?
 c. Describe the angular velocity of any LEO satellite.

43. **Critical Thinking** The figure at the right is a side view of three rollers that are tangent to one another.
 a. If roller A turns counterclockwise, in which directions do rollers B and C turn?
 b. If roller A turns at 120 revolutions per minute, how many revolutions per minute do rollers B and C turn?

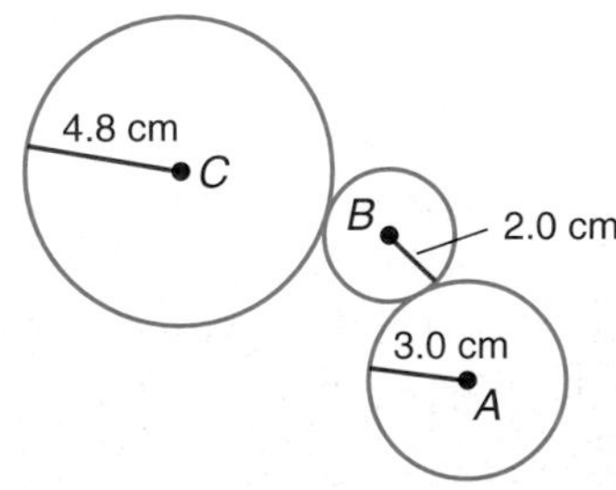

Mixed Review

44. Find the area of a sector if the central angle measures 105° and the radius of the circle is 7.2 centimeters. *(Lesson 6-1)*

45. **Geometry** Find the area of a regular pentagon that is inscribed in a circle with a diameter of 7.3 centimeters. *(Lesson 5-4)*

Extra Practice See p. A36.

46. Write 35° 20′ 55″ as a decimal to the nearest thousandth. *(Lesson 5-1)*

47. Solve $10 + \sqrt{k - 5} = 8$. *(Lesson 4-7)*

48. Write a polynomial equation of least degree with roots -4, $3\boldsymbol{i}$, and $-3\boldsymbol{i}$. *(Lesson 4-1)*

49. Graph $y > x^3 + 1$. *(Lesson 3-3)*

50. Write the slope-intercept form of the equation of the line through points at (8, 5) and (−6, 0). *(Lesson 1-4)*

51. **SAT/ACT Practice** The perimeter of rectangle $QRST$ is p, and $a = \frac{3}{4}b$. Find the value of b in terms of p.

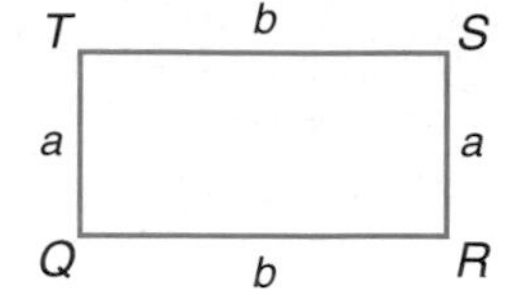

A $\frac{p}{7}$ **B** $\frac{4p}{7}$ **C** $\frac{7}{p}$ **D** $\frac{2p}{7}$ **E** $\frac{7p}{4}$

CAREER CHOICES

Audio Recording Engineer

Is music your forte? Do you enjoy being creative and solving problems? If you answered yes to these questions, you may want to consider a career as an audio recording engineer. This type of engineer is in charge of all the technical aspects of recording music, speech, sound effects, and dialogue.

Some aspects of the career include controlling the recording equipment, tackling technical problems that arise during recording, and communicating with musicians and music producers. You would need to keep up-to-date on the latest recording equipment and technology. The music producer may direct the sounds you produce through use of the equipment, or you may have the opportunity to design and perfect your own sounds for use in production.

CAREER OVERVIEW

Degree Preferred:
two- or four-year degree in audio engineering

Related Courses:
mathematics, music, computer science, electronics

Outlook:
number of jobs expected to increase at a slower pace than the average through the year 2006

Sound	Decibels
Threshold of Hearing	0
Average Whisper (4 feet)	20
Broadcast Studio (no program in progress)	30
Soft Recorded Music	36
Normal Conversation (4 feet)	60
Moderate Discotheque	90
Personal Stereo	up to 120
Percussion Instruments at a Symphony Concert	up to 130
Rock Concert	up to 140

For more information about audio recording engineering visit: **www.amc.glencoe.com**

Graphing Sine and Cosine Functions

OBJECTIVE

- Use the graphs of the sine and cosine functions.

Real World Application

METEOROLOGY The average monthly temperatures for a city demonstrate a repetitious behavior. For cities in the Northern Hemisphere, the average monthly temperatures are usually lowest in January and highest in July. The graph below shows the average monthly temperatures (°F) for Baltimore, Maryland, and Asheville, North Carolina, with January represented by 1.

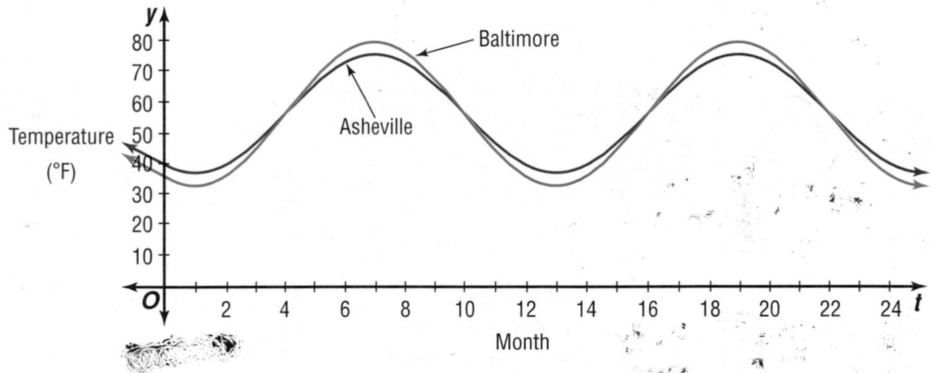

Model for Baltimore's temperature: $y = 54.4 + 22.5 \sin\left[\frac{\pi}{6}(t - 4)\right]$

Model for Asheville's temperature: $y = 54.5 + 18.5 \sin\left[\frac{\pi}{6}(t - 4)\right]$

In these equations, t denotes the month with January represented by $t = 1$.
What is the average temperature for each city for month 13?
Which city has the greater fluctuation in temperature?
These problems will be solved in Example 5.

Each year, the graph for Baltimore will be about the same. This is also true for Asheville. If the values of a function are the same for each given interval of the domain (in this case, 12 months or 1 year), the function is said to be **periodic.** The interval is the **period** of the function.

Periodic Function and Period

A function is *periodic* if, for some real number α, $f(x + \alpha) = f(x)$ for each x in the domain of f.

The least positive value of α for which $f(x) = f(x + \alpha)$ is the *period* of the function.

Example 1 **Determine if each function is periodic. If so, state the period.**

a.

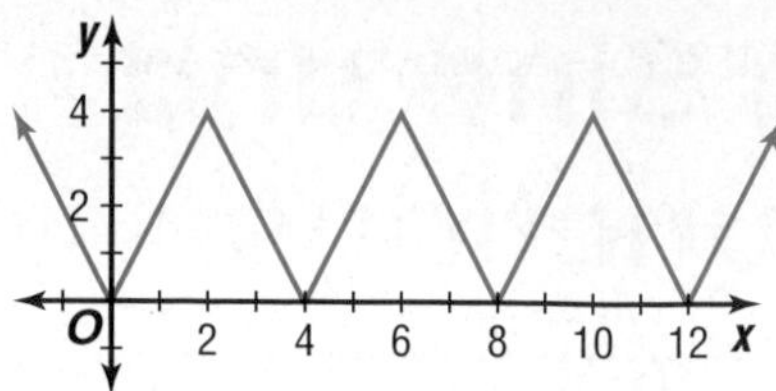

The values of the function repeat for each interval of 4 units. The function is periodic, and the period is 4.

b.

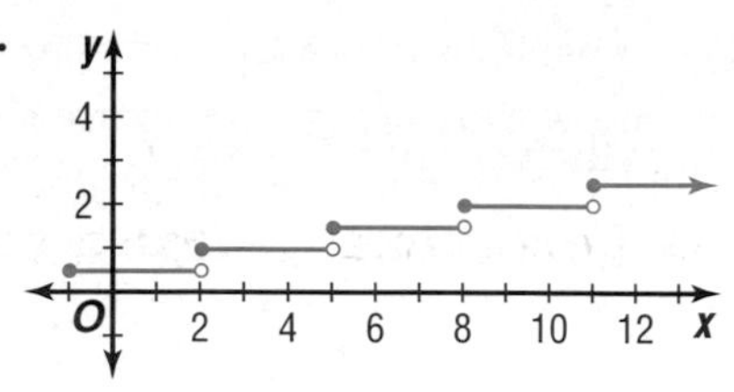

The values of the function do not repeat. The function is not periodic.

Consider the sine function. First evaluate $y = \sin x$ for domain values between -2π and 2π in multiples of $\frac{\pi}{4}$.

x	-2π	$-\frac{7\pi}{4}$	$-\frac{3\pi}{2}$	$-\frac{5\pi}{4}$	$-\pi$	$-\frac{3\pi}{4}$	$-\frac{\pi}{2}$	$-\frac{\pi}{4}$	0	$\frac{\pi}{4}$	$\frac{\pi}{2}$	$\frac{3\pi}{4}$	π	$\frac{5\pi}{4}$	$\frac{3\pi}{2}$	$\frac{7\pi}{4}$	2π
sin x	0	$\frac{\sqrt{2}}{2}$	1	$\frac{\sqrt{2}}{2}$	0	$-\frac{\sqrt{2}}{2}$	-1	$-\frac{\sqrt{2}}{2}$	0	$\frac{\sqrt{2}}{2}$	1	$\frac{\sqrt{2}}{2}$	0	$-\frac{\sqrt{2}}{2}$	-1	$-\frac{\sqrt{2}}{2}$	0

To graph $y = \sin x$, plot the coordinate pairs from the table and connect them to form a smooth curve. Notice that the range values for the domain interval $-2\pi < x < 0$ (shown in red) repeat for the domain interval between $0 < x < 2\pi$ (shown in blue). The sine function is a periodic function.

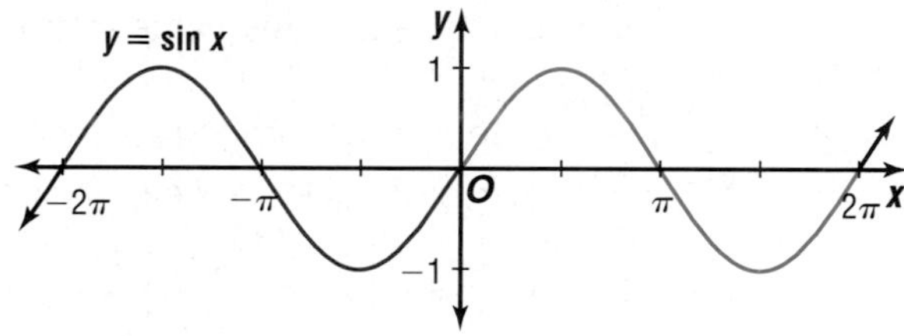

By studying the graph and its repeating pattern, you can determine the following properties of the graph of the sine function.

Properties of the Graph of $y = \sin x$

1. The period is 2π.
2. The domain is the set of real numbers.
3. The range is the set of real numbers between -1 and 1, inclusive.
4. The x-intercepts are located at πn, where n is an integer.
5. The y-intercept is 0.
6. The maximum values are $y = 1$ and occur when $x = \frac{\pi}{2} + 2\pi n$, where n is an integer.
7. The minimum values are $y = -1$ and occur when $x = \frac{3\pi}{2} + 2\pi n$, where n is an integer.

Examples

2 **Find $\sin \frac{9\pi}{2}$ by referring to the graph of the sine function.**

Because the period of the sine function is 2π and $\frac{9\pi}{2} > 2\pi$, rewrite $\frac{9\pi}{2}$ as a sum involving 2π.

$$\frac{9\pi}{2} = 4\pi + \frac{\pi}{2}$$

$$= 2\pi(2) + \frac{\pi}{2} \quad \textit{This is a form of } \frac{\pi}{2} + 2\pi n.$$

So, $\sin \frac{9\pi}{2} = \sin \frac{\pi}{2}$ or 1.

3 **Find the values of θ for which $\sin \theta = 0$ is true.**

Since $\sin \theta = 0$ indicates the x-intercepts of the function, $\sin \theta = 0$ if $\theta = n\pi$, where n is any integer.

4 **Graph $y = \sin x$ for $3\pi \leq x \leq 5\pi$.**

The graph crosses the x-axis at 3π, 4π, and 5π. It has its maximum value of 1 at $x = \frac{9\pi}{2}$, and its minimum value of -1 at $x = \frac{7\pi}{2}$. Use this information to sketch the graph.

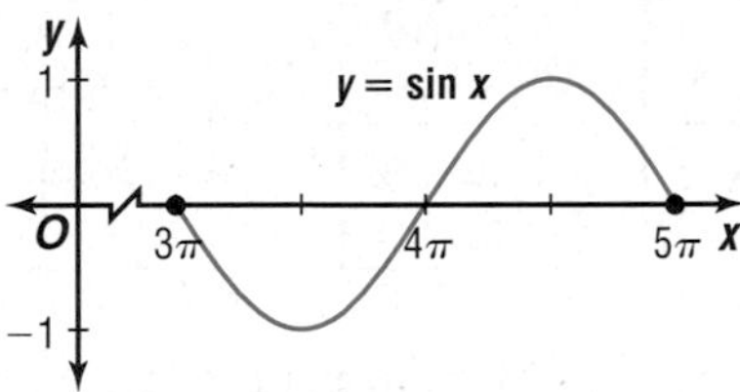

5 **METEOROLOGY** **Refer to the application at the beginning of the lesson.**

a. What is the average temperature for each city for month 13?

Month 13 is January of the second year. To find the average temperature of this month, substitute this value into each equation.

Baltimore	**Asheville**
$y = 54.4 + 22.5 \sin \left[\frac{\pi}{6}(t - 4)\right]$	$y = 54.5 + 18.5 \sin \left[\frac{\pi}{6}(t - 4)\right]$
$y = 54.4 + 22.5 \sin \left[\frac{\pi}{6}(13 - 4)\right]$	$y = 54.5 + 18.5 \sin \left[\frac{\pi}{6}(13 - 4)\right]$
$y = 54.4 + 22.5 \sin \frac{3\pi}{2}$	$y = 54.5 + 18.5 \sin \frac{3\pi}{2}$
$y = 54.4 + 22.5(-1)$	$y = 54.5 + 18.5(-1)$
$y = 31.9$	$y = 36.0$

In January, the average temperature for Baltimore is 31.9°, and the average temperature for Asheville is 36.0°.

b. Which city has the greater fluctuation in temperature? Explain.

The average temperature for January is lower in Baltimore than in Asheville. The average temperature for July is higher in Baltimore than in Asheville. Therefore, there is a greater fluctuation in temperature in Baltimore than in Asheville.

Now, consider the graph of $y = \cos x$.

x	-2π	$-\frac{7\pi}{4}$	$-\frac{3\pi}{2}$	$-\frac{5\pi}{4}$	$-\pi$	$-\frac{3\pi}{4}$	$-\frac{\pi}{2}$	$-\frac{\pi}{4}$	0	$\frac{\pi}{4}$	$\frac{\pi}{2}$	$\frac{3\pi}{4}$	π	$\frac{5\pi}{4}$	$\frac{3\pi}{2}$	$\frac{7\pi}{4}$	2π
$\cos x$	1	$\frac{\sqrt{2}}{2}$	0	$-\frac{\sqrt{2}}{2}$	-1	$-\frac{\sqrt{2}}{2}$	0	$\frac{\sqrt{2}}{2}$	1	$\frac{\sqrt{2}}{2}$	0	$-\frac{\sqrt{2}}{2}$	-1	$-\frac{\sqrt{2}}{2}$	0	$\frac{\sqrt{2}}{2}$	1

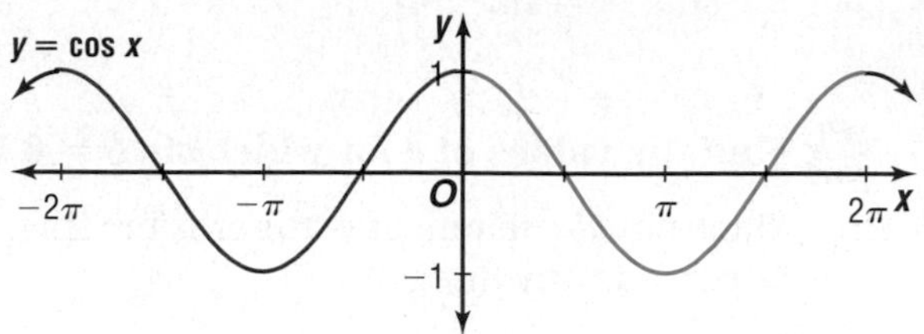

By studying the graph and its repeating pattern, you can determine the following properties of the graph of the cosine function.

Properties of the Graph of $y = \cos x$

1. The period is 2π.
2. The domain is the set of real numbers.
3. The range is the set of real numbers between -1 and 1, inclusive.
4. The x-intercepts are located at $\frac{\pi}{2} + \pi n$, where n is an integer.
5. The y-intercept is 1.
6. The maximum values are $y = 1$ and occur when $x = \pi n$, where n is an even integer.
7. The minimum values are $y = -1$ and occur when $x = \pi n$, where n is an odd integer.

Example 6 **Determine whether the graph represents $y = \sin x$, $y = \cos x$, or neither.**

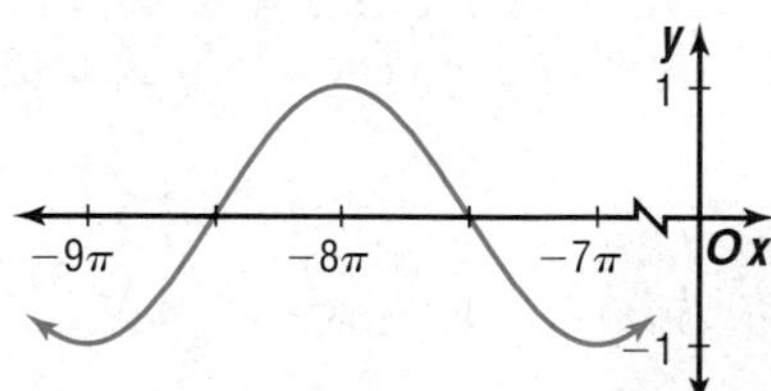

The maximum value of 1 occurs when $x = -8\pi$. *maximum of 1 when $x = \pi n \rightarrow \cos x$*

The minimum value of -1 occurs at -9π and -7π. *minimum of -1 when $x = \pi n \rightarrow \cos x$*

The x-intercepts are $-\frac{17\pi}{2}$ and $-\frac{15\pi}{2}$.

These are characteristics of the cosine function. The graph is $y = \cos x$.

Check for Understanding

Communicating Mathematics

Read and study the lesson to answer each question.

1. **Counterexample** Sketch the graph of a periodic function that is neither the sine nor cosine function. State the period of the function.
2. **Name** three values of x that would result in the maximum value for $y = \sin x$.
3. **Explain** why the cosine function is a periodic function.
4. *Math Journal* **Draw** the graphs for the sine function and the cosine function. Compare and contrast the two graphs.

Guided Practice

5. **Determine** if the function is periodic. If so, state the period.

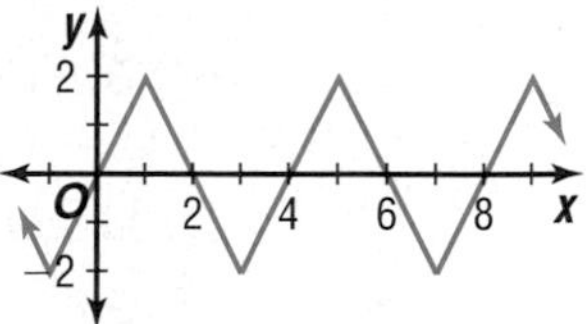

Find each value by referring to the graph of the sine or the cosine function.

6. $\cos\left(-\frac{\pi}{2}\right)$
7. $\sin \frac{5\pi}{2}$

8. Find the values of θ for which $\sin \theta = -1$ is true.

Graph each function for the given interval.

9. $y = \cos x$, $5\pi \leq x \leq 7\pi$
10. $y = \sin x$, $-4\pi \leq x \leq -2\pi$

11. Determine whether the graph represents $y = \sin x$, $y = \cos x$, or neither. Explain.

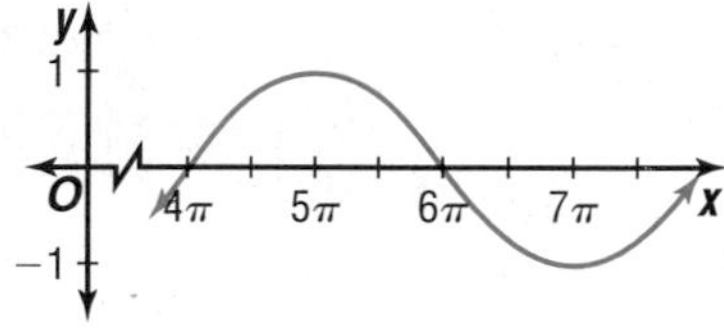

12. **Meteorology** The equation $y = 49 + 28 \sin\left[\frac{\pi}{6}(t - 4)\right]$ models the average monthly temperature for Omaha, Nebraska. In this equation, t denotes the number of months with January represented by 1. Compare the average monthly temperature for April and October.

Exercises

Practice

Determine if each function is periodic. If so state the period.

13.

14.

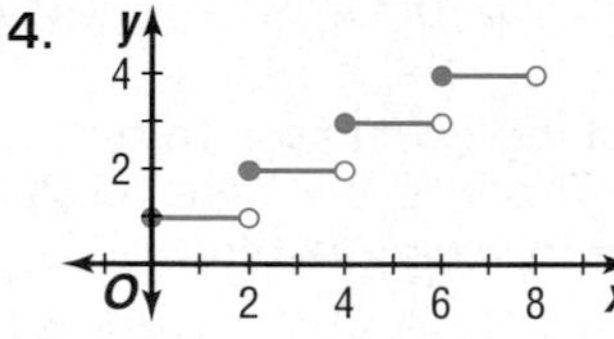

15.

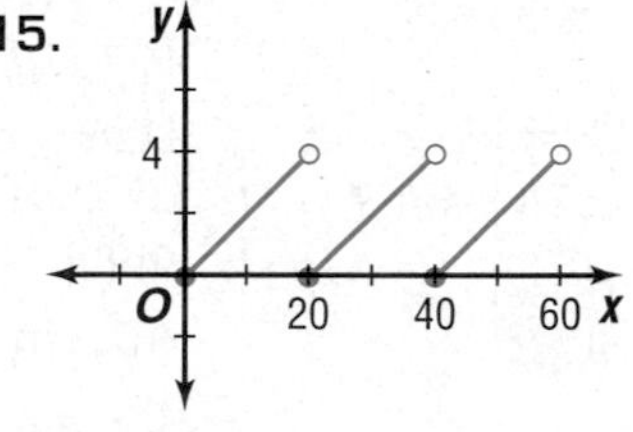

16. $y = |x + 5|$
17. $y = x^2$
18. $y = \frac{1}{x}$

Find each value by referring to the graph of the sine or the cosine function.

19. $\cos 8\pi$ **20.** $\sin 11\pi$ **21.** $\cos \frac{\pi}{2}$

22. $\sin\left(-\frac{3\pi}{2}\right)$ **23.** $\sin \frac{7\pi}{2}$ **24.** $\cos(-3\pi)$

25. What is the value of $\sin \pi + \cos \pi$?

26. Find the value of $\sin 2\pi - \cos 2\pi$.

Find the values of θ for which each equation is true.

27. $\cos \theta = -1$ **28.** $\sin \theta = 1$ **29.** $\cos \theta = 0$

30. Under what conditions does $\cos \theta = 1$?

Graph each function for the given interval.

31. $y = \sin x,\ -5\pi \leq x \leq -3\pi$ **32.** $y = \cos x,\ 8\pi \leq x \leq 10\pi$

33. $y = \cos x,\ -5\pi \leq x \leq -3\pi$ **34.** $y = \sin x,\ \frac{9\pi}{2} \leq x \leq \frac{13\pi}{2}$

35. $y = \cos x,\ -\frac{7\pi}{2} \leq x \leq -\frac{3\pi}{2}$ **36.** $y = \sin x,\ \frac{7\pi}{2} \leq x \leq \frac{11\pi}{2}$

Determine whether each graph is $y = \sin x$, $y = \cos x$, or neither. Explain.

37.

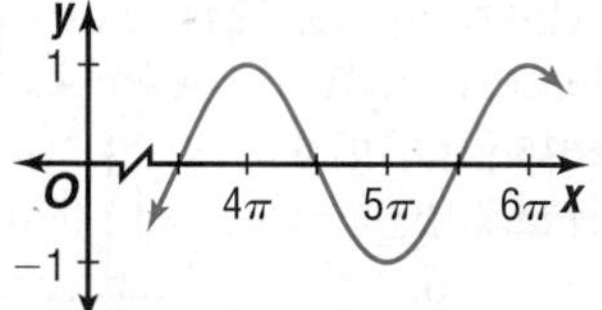

38.

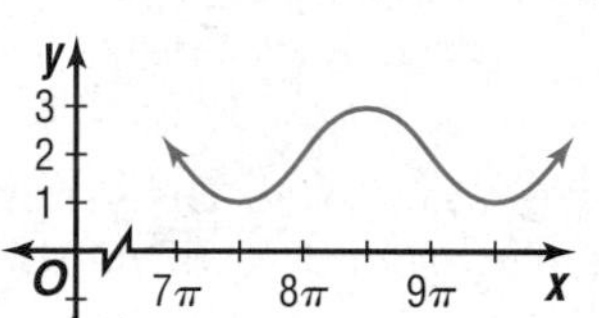

39.

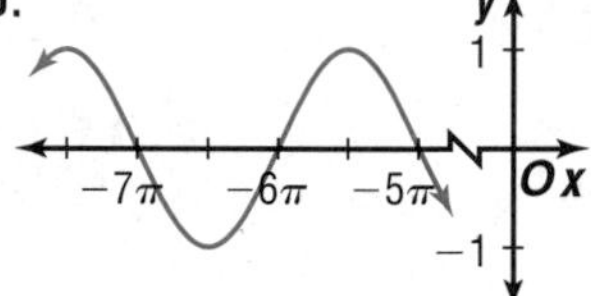

40. Describe a transformation that would change the graph of the sine function to the graph of the cosine function.

41. Name any lines of symmetry for the graph of $y = \sin x$.

42. Name any lines of symmetry for the graph of $y = \cos x$.

43. Use the graph of the sine function to find the values of θ for which each statement is true.

a. $\csc \theta = 1$ b. $\csc \theta = -1$ c. $\csc \theta$ is undefined.

44. Use the graph of the cosine function to find the values of θ for which each statement is true.

a. $\sec \theta = 1$ b. $\sec \theta = -1$ c. $\sec \theta$ is undefined.

Graphing Calculator

Use a graphing calculator to graph the sine and cosine functions on the same set of axes for $0 \leq x \leq 2\pi$. Use the graphs to find the values of x, if any, for which each of the following is true.

45. $\sin x = -\cos x$ **46.** $\sin x \leq \cos x$

47. $\sin x \cos x > 1$ **48.** $\sin x \cos x \leq 0$

49. $\sin x + \cos x = 1$ **50.** $\sin x - \cos x = 0$

Applications and Problem Solving

51. **Meteorology** The equation $y = 43 + 31 \sin\left[\frac{\pi}{6}(t - 4)\right]$ models the average monthly temperatures for Minneapolis, Minnesota. In this equation, t denotes the number of months with January represented by 1.
    a. What is the difference between the average monthly temperatures for July and January? What is the relationship between this difference and the coefficient of the sine term?
    b. What is the sum of the average monthly temperatures for July and January? What is the relationship between this sum and value of constant term?

52. **Critical Thinking** Consider the graph of $y = 2 \sin x$.
    a. What are the x-intercepts of the graph?
    b. What is the maximum value of y?
    c. What is the minimum value of y?
    d. What is the period of the function?
    e. Graph the function.
    f. How does the 2 in the equation affect the graph?

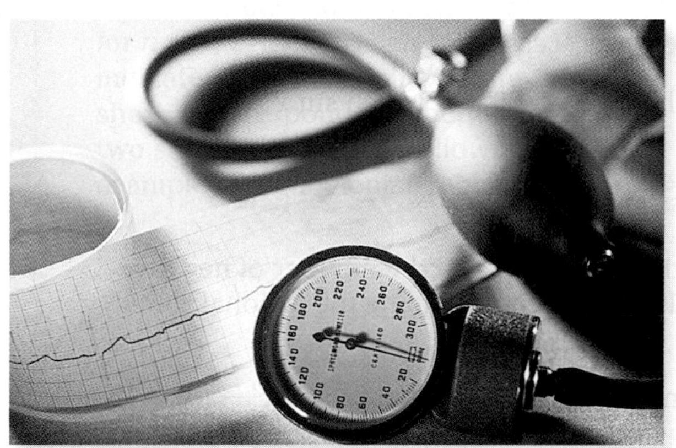

53. **Medicine** The equation $P = 100 + 20 \sin 2\pi t$ models a person's blood pressure P in millimeters of mercury. In this equation, t is time in seconds. The blood pressure oscillates 20 millimeters above and below 100 millimeters, which means that the person's blood pressure is 120 over 80. This function has a period of 1 second, which means that the person's heart beats 60 times a minute.
    a. Find the blood pressure at $t = 0$, $t = 0.25$, $t = 0.5$, $t = 0.75$, and $t = 1$.
    b. During the first second, when was the blood pressure at a maximum?
    c. During the first second, when was the blood pressure at a minimum?

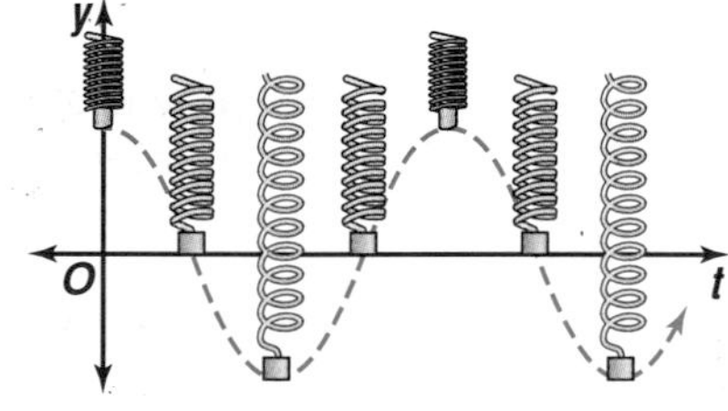

54. **Physics** The motion of a weight on a spring can be described by a modified cosine function. The weight suspended from a spring is at its equilibrium point when it is at rest. When pushed a certain distance above the equilibrium point, the weight oscillates above and below the equilibrium point. The time that it takes for the weight to oscillate from the highest point to the lowest point and back to the highest point is its period. The equation $v = 3.5 \cos\left(t\sqrt{\frac{k}{m}}\right)$ models the vertical displacement v of the weight in relationship to the equilibrium point at any time t if it is initially pushed up 3.5 centimeters. In this equation, k is the elasticity of the spring and m is the mass of the weight.
    a. Suppose $k = 19.6$ and $m = 1.99$. Find the vertical displacement after 0.9 second and after 1.7 seconds.
    b. When will the weight be at the equilibrium point for the first time?
    c. How long will it take the weight to complete one period?

Amplitude and Period of Sine and Cosine Functions

OBJECTIVES
- Find the amplitude and period for sine and cosine functions.
- Write equations of sine and cosine functions given the amplitude and period.

Real World Application

BOATING A signal buoy between the coast of Hilton Head Island, South Carolina, and Savannah, Georgia, bobs up and down in a minor squall. From the highest point to the lowest point, the buoy moves a distance of $3\frac{1}{2}$ feet. It moves from its highest point down to its lowest point and back to its highest point every 14 seconds. Find an equation of the motion for the buoy assuming that it is at its equilibrium point at $t = 0$ and the buoy is on its way down at that time. What is the height of the buoy at 8 seconds and at 17 seconds? *This problem will be solved in Example 5.*

Recall from Chapter 3 that changes to the equation of the parent graph can affect the appearance of the graph by dilating, reflecting, and/or translating the original graph. In this lesson, we will observe the vertical and horizontal expanding and compressing of the parent graphs of the sine and cosine functions.

Let's consider an equation of the form $y = A \sin \theta$. We know that the maximum absolute value of $\sin \theta$ is 1. Therefore, for every value of the product of $\sin \theta$ and A, the maximum value of $A \sin \theta$ is $|A|$. Similarly, the maximum value of $A \cos \theta$ is $|A|$. The absolute value of A is called the **amplitude** of the functions $y = A \sin \theta$ and $y = A \cos \theta$.

Amplitude of Sine and Cosine Functions

The amplitude of the functions $y = A \sin \theta$ and $y = A \cos \theta$ is the absolute value of A, or $|A|$.

The amplitude can also be described as the absolute value of one-half the difference of the maximum and minimum function values.

$$|A| = \left|\frac{A - (-A)}{2}\right|$$

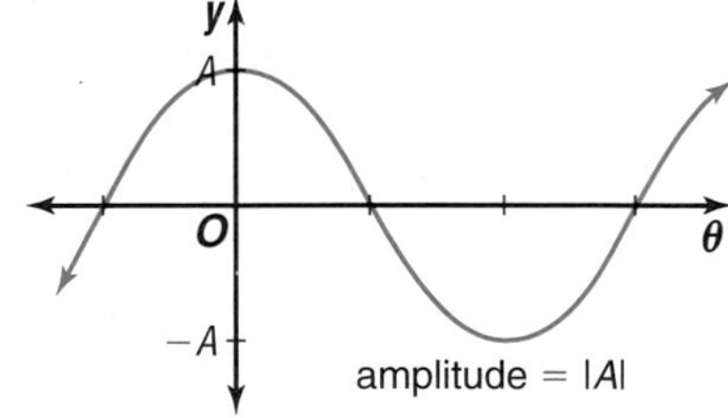

Example 1

a. State the amplitude for the function $y = 4 \cos \theta$.

b. Graph $y = 4 \cos \theta$ and $y = \cos \theta$ on the same set of axes.

c. Compare the graphs.

a. According to the definition of amplitude, the amplitude of $y = A \cos \theta$ is $|A|$. So the amplitude of $y = 4 \cos \theta$ is $|4|$ or 4.

b. Make a table of values. Then graph the points and draw a smooth curve.

θ	0	$\frac{\pi}{4}$	$\frac{\pi}{2}$	$\frac{3\pi}{4}$	π	$\frac{5\pi}{4}$	$\frac{3\pi}{2}$	$\frac{7\pi}{4}$	2π
$\cos \theta$	1	$\frac{\sqrt{2}}{2}$	0	$-\frac{\sqrt{2}}{2}$	-1	$-\frac{\sqrt{2}}{2}$	0	$\frac{\sqrt{2}}{2}$	1
$4 \cos \theta$	4	$2\sqrt{2}$	0	$-2\sqrt{2}$	-4	$-2\sqrt{2}$	0	$2\sqrt{2}$	4

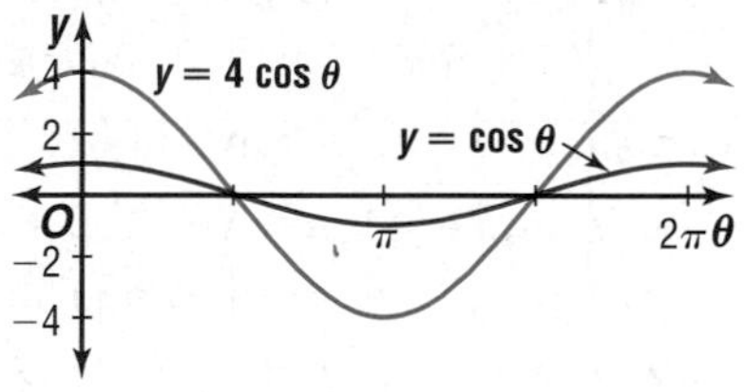

c. The graphs cross the θ-axis at $\theta = \frac{\pi}{2}$ and $\theta = \frac{3\pi}{2}$. Also, both functions reach their maximum value at $\theta = 0$ and $\theta = 2\pi$ and their minimum value at $\theta = \pi$. But the maximum and minimum values of the function $y = \cos \theta$ are 1 and -1, and the maximum and minimum values of the function $y = 4 \cos \theta$ are 4 and -4. The graph of $y = 4 \cos \theta$ is vertically expanded.

GRAPHING CALCULATOR EXPLORATION

- Select the radian mode.
- Use the domain and range values below to set the viewing window.

$-4.7 \le x \le 4.8$, **Xscl: 1** $\quad -3 \le y \le 3$, **Yscl: 1**

TRY THESE

1. Graph each function on the same screen.

a. $y = \sin x$ **b.** $y = \sin 2x$ **c.** $y = \sin 3x$

WHAT DO YOU THINK?

2. Describe the behavior of the graph of $f(x) = \sin kx$, where $k > 0$, as k increases.

3. Make a conjecture about the behavior of the graph of $f(x) = \sin kx$, if $k < 0$. Test your conjecture.

Consider an equation of the form $y = \sin k\theta$, where k is any positive integer. Since the period of the sine function is 2π, the following identity can be developed.

$$y = \sin k\theta$$
$$y = \sin (k\theta + 2\pi) \quad \textit{Definition of periodic function}$$
$$y = \sin k\left(\theta + \frac{2\pi}{k}\right) \quad k\theta + 2\pi = k\left(\theta + \frac{2\pi}{k}\right)$$

Therefore, the period of $y = \sin k\theta$ is $\frac{2\pi}{k}$. Similarly, the period of $y = \cos k\theta$ is $\frac{2\pi}{k}$.

Period of Sine and Cosine Functions

The period of the functions $y = \sin k\theta$ and $y = \cos k\theta$ is $\frac{2\pi}{k}$, where $k > 0$.

Example 2 **a. State the period for the function $y = \cos \frac{\theta}{2}$.**

b. Graph $y = \cos \frac{\theta}{2}$ and $y = \cos \theta$.

a. The definition of the period of $y = \cos k\theta$ is $\frac{2\pi}{k}$. Since $\cos \frac{\theta}{2}$ equals $\cos\left(\frac{1}{2}\theta\right)$, the period is $\frac{2\pi}{\frac{1}{2}}$ or 4π.

b.

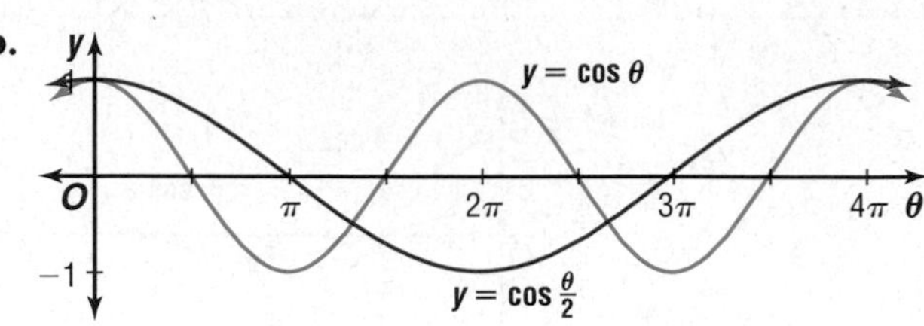

Notice that the graph of $y = \cos \frac{\theta}{2}$ is horizontally expanded.

The graphs of $y = A \sin k\theta$ and $y = A \cos k\theta$ are shown below.

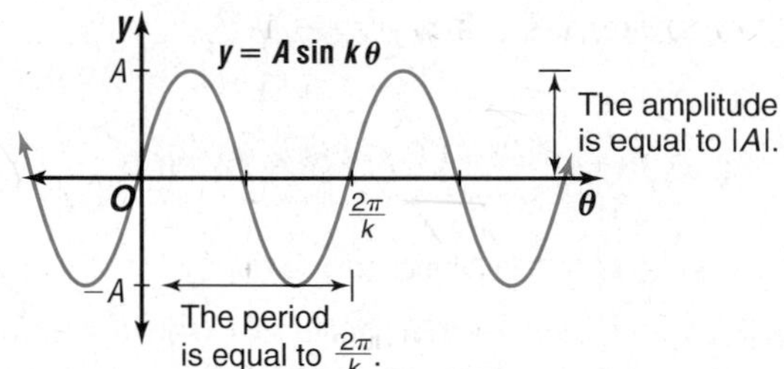

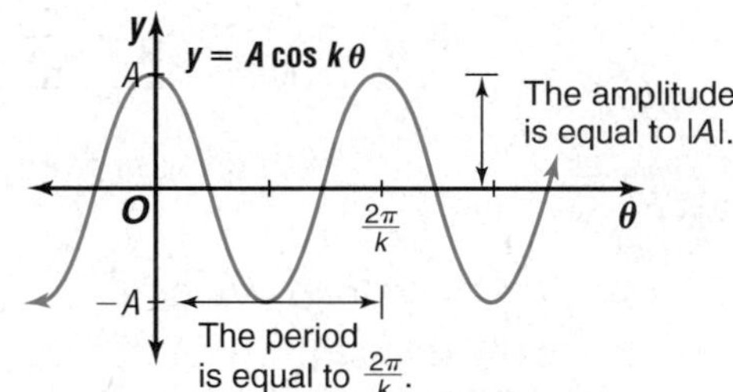

You can use the parent graph of the sine and cosine functions and the amplitude and period to sketch graphs of $y = A \sin k\theta$ and $y = A \cos k\theta$.

Example 3 **State the amplitude and period for the function $y = \frac{1}{2} \sin 4\theta$. Then graph the function.**

Since $A = \frac{1}{2}$, the amplitude is $\left|\frac{1}{2}\right|$ or $\frac{1}{2}$. Since $k = 4$, the period is $\frac{2\pi}{4}$ or $\frac{\pi}{2}$.

Use the basic shape of the sine function and the amplitude and period to graph the equation.

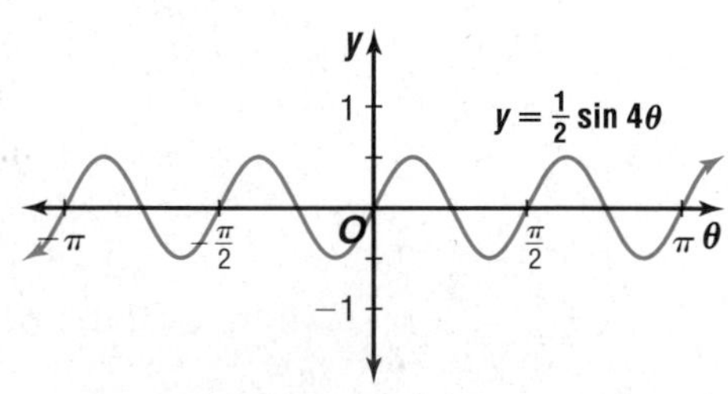

We can write equations for the sine and cosine functions if we are given the amplitude and period.

Example 4 **Write an equation of the cosine function with amplitude 9.8 and period 6π.**

The form of the equation will be $y = A \cos k\theta$. First find the possible values of A for an amplitude of 9.8.

$$|A| = 9.8$$
$$A = 9.8 \text{ or } -9.8$$

Since there are two values of A, two possible equations exist.

Now find the value of k when the period is 6π.

$$\frac{2\pi}{k} = 6\pi \qquad \textit{The period of a cosine function is } \frac{2\pi}{k}.$$
$$k = \frac{2\pi}{6\pi} \text{ or } \frac{1}{3}$$

The possible equations are $y = 9.8 \cos\left(\frac{1}{3}\theta\right)$ or $y = -9.8 \cos\left(\frac{1}{3}\theta\right)$.

Many real-world situations have periodic characteristics that can be described with the sine and cosine functions. When you are writing an equation to describe a situation, remember the characteristics of the sine and cosine graphs. If you know the function value when $x = 0$ and whether the function is increasing or decreasing, you can choose the appropriate function to write an equation for the situation.

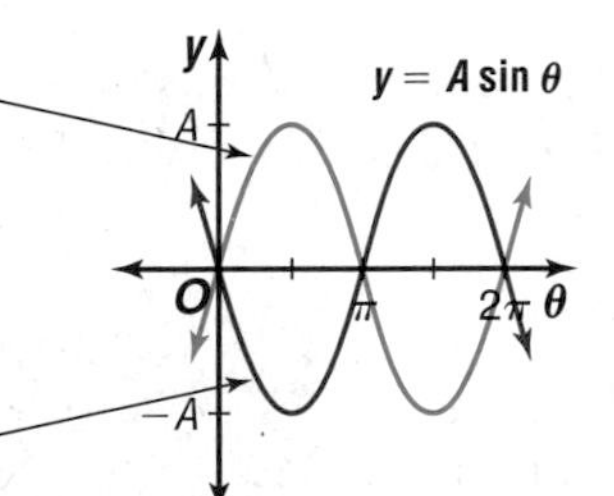

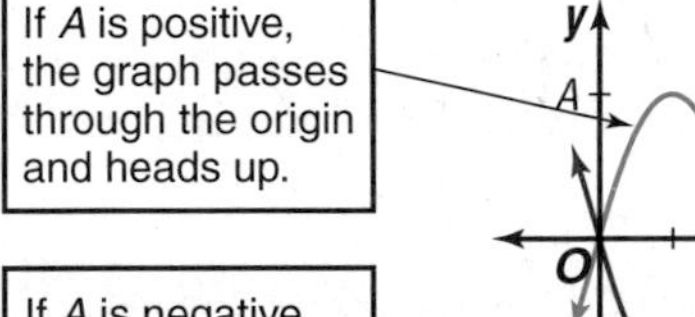
If A is positive, the graph passes through the origin and heads up.
If A is negative, the graph passes through the origin and heads down.
y = A sin θ
y
A
O
–A
π
2π
θ

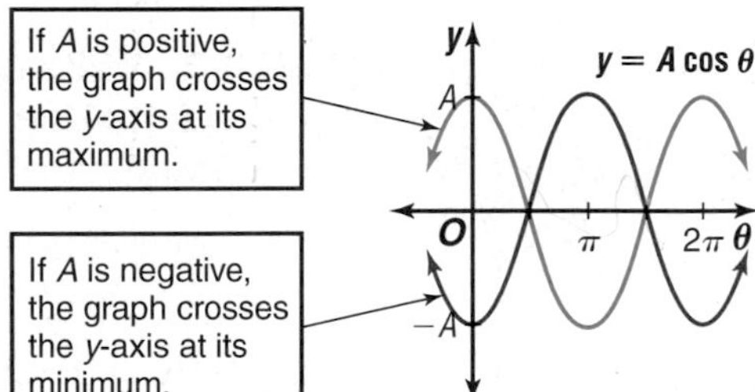
If A is positive, the graph crosses the y-axis at its maximum.
If A is negative, the graph crosses the y-axis at its minimum.
y = A cos θ
y
A
O
–A
π
2π
θ

Example 5 **BOATING** **Refer to the application at the beginning of the lesson.**

Real World Application

a. Find an equation for the motion of the buoy.

b. Determine the height of the buoy at 8 seconds and at 17 seconds.

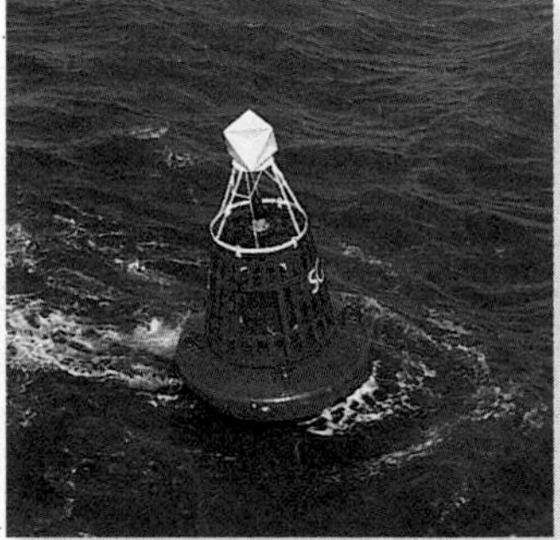

a. At $t = 0$, the buoy is at equilibrium and is on its way down. This indicates a reflection of the sine function and a negative value of A. The general form of the equation will be $y = A \sin kt$, where A is negative and t is the time in seconds.

$$A = -\left(\frac{1}{2} \times 3\frac{1}{2}\right) \qquad \frac{2\pi}{k} = 14$$
$$A = -\frac{7}{4} \text{ or } -1.75 \qquad k = \frac{2\pi}{14} \text{ or } \frac{\pi}{7}$$

An equation for the motion of the buoy is $y = -1.75 \sin \frac{\pi}{7}t$.

Graphing Calculator Tip

To find the value of y, use a calculator in radian mode.

b. Use this equation to find the location of the buoy at the given times.

At 8 seconds

$y = -1.75 \sin\left(\frac{\pi}{7} \times 8\right)$

$y \approx 0.7592965435$

At 8 seconds, the buoy is about 0.8 feet above the equilibrium point.

At 17 seconds

$y = -1.75 \sin\left(\frac{\pi}{7} \times 17\right)$

$y \approx -1.706123846$

At 17 seconds, the buoy is about 1.7 feet below the equilibrium point.

The period represents the amount of time that it takes to complete one cycle. The number of cycles per unit of time is known as the **frequency.** The period (time per cycle) and frequency (cycles per unit of time) are reciprocals of each other.

$$\text{period} = \frac{1}{\text{frequency}} \qquad \text{frequency} = \frac{1}{\text{period}}$$

The *hertz* is a unit of frequency. One hertz equals one cycle per second.

Example 6 **MUSIC** **Write an equation of the sine function that represents the initial behavior of the vibrations of the note G above middle C having amplitude 0.015 and a frequency of 392 hertz.**

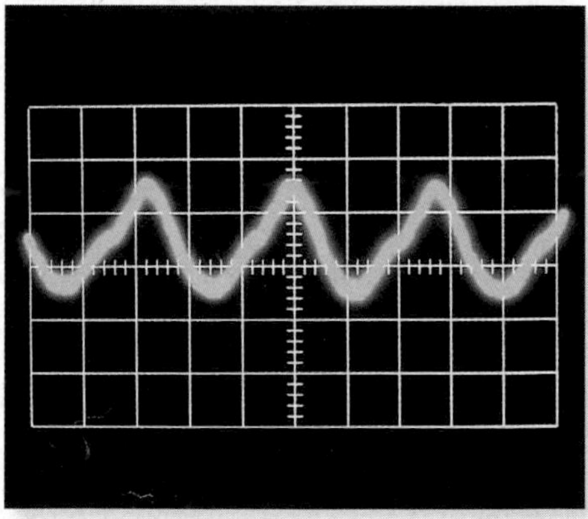

The general form of the equation will be $y = A \sin kt$, where t is the time in seconds. Since the amplitude is 0.015, $A = \pm 0.015$.

The period is the reciprocal of the frequency or $\frac{1}{392}$. Use this value to find k.

$\frac{2\pi}{k} = \frac{1}{392}$ *The period* $\frac{2\pi}{k}$ *equals* $\frac{1}{392}$.

$k = 2\pi(392)$ or 784π

One sine function that represents the vibration is $y = 0.015 \sin (784\pi \times t)$.

CHECK FOR UNDERSTANDING

Communicating Mathematics

Read and study the lesson to answer each question.

1. **Write** a sine function that has a greater maximum value than the function $y = 4 \sin 2\theta$.
2. **Describe** the relationship between the graphs of $y = 3 \sin \theta$ and $y = -3 \sin \theta$.

3. **Determine** which function has the greatest period.

A. $y = 5 \cos 2\theta$ B. $y = 3 \cos 5\theta$ C. $y = 7 \cos \frac{\theta}{2}$ D. $y = \cos \theta$

4. **Explain** the relationship between period and frequency.

5. *Math Journal* **Draw** the graphs for $y = \cos \theta$, $y = 3 \cos \theta$, and $y = \cos 3\theta$. Compare and contrast the three graphs.

Guided Practice

6. State the amplitude for $y = -2.5 \cos \theta$. Then graph the function.

7. State the period for $y = \sin 4\theta$. Then graph the function.

State the amplitude and period for each function. Then graph each function.

8. $y = 10 \sin 2\theta$

9. $y = 3 \cos 2\theta$

10. $y = 0.5 \sin \frac{\theta}{6}$

11. $y = -\frac{1}{5} \cos \frac{\theta}{4}$

Write an equation of the sine function with each amplitude and period.

12. amplitude = 0.8, period = π

13. amplitude = 7, period = $\frac{\pi}{3}$

Write an equation of the cosine function with each amplitude and period.

14. amplitude = 1.5, period = 5π

15. amplitude = $\frac{3}{4}$, period = 6

16. **Music** Write a sine equation that represents the initial behavior of the vibrations of the note D above middle C having an amplitude of 0.25 and a frequency of 294 hertz.

EXERCISES

Practice

State the amplitude for each function. Then graph each function.

17. $y = 2 \sin \theta$

18. $y = -\frac{3}{4} \cos \theta$

19. $y = 1.5 \sin \theta$

State the period for each function. Then graph each function.

20. $y = \cos 2\theta$

21. $y = \cos \frac{\theta}{4}$

22. $y = \sin 6\theta$

State the amplitude and period for each function. Then graph each function.

23. $y = 5 \cos \theta$

24. $y = -2 \cos 0.5\theta$

25. $y = -\frac{2}{5} \sin 9\theta$

26. $y = 8 \sin 0.5\theta$

27. $y = -3 \sin \frac{\pi}{2}\theta$

28. $y = \frac{2}{3} \cos \frac{3\pi}{7}\theta$

29. $y = 3 \sin 2\theta$

30. $y = 3 \cos 0.5\theta$

31. $y = -\frac{1}{3} \cos 3\theta$

32. $y = \frac{1}{3} \sin \frac{\theta}{3}$

33. $y = -4 \sin \frac{\theta}{2}$

34. $y = -2.5 \cos \frac{\theta}{5}$

35. The equation of the vibrations of the note F above middle C is represented by $y = 0.5 \sin 698\pi t$. Determine the amplitude and period for the function.

Write an equation of the sine function with each amplitude and period.

36. amplitude $= 0.4$, period $= 10\pi$

37. amplitude $= 35.7$, period $= \frac{\pi}{4}$

38. amplitude $= \frac{1}{4}$, period $= \frac{\pi}{3}$

39. amplitude $= 0.34$, period $= 0.75\pi$

40. amplitude $= 4.5$, period $= \frac{5\pi}{4}$

41. amplitude $= 16$, period $= 30$

Write an equation of the cosine function with each amplitude and period.

42. amplitude $= 5$, period $= 2\pi$

43. amplitude $= \frac{5}{8}$, period $= \frac{\pi}{7}$

44. amplitude $= 7.5$, period $= 6\pi$

45. amplitude $= 0.5$, period $= 0.3\pi$

46. amplitude $= \frac{2}{5}$, period $= \frac{3}{5}\pi$

47. amplitude $= 17.9$, period $= 16$

48. Write the possible equations of the sine and cosine functions with amplitude 1.5 and period $\frac{\pi}{2}$.

Write an equation for each graph.

49. 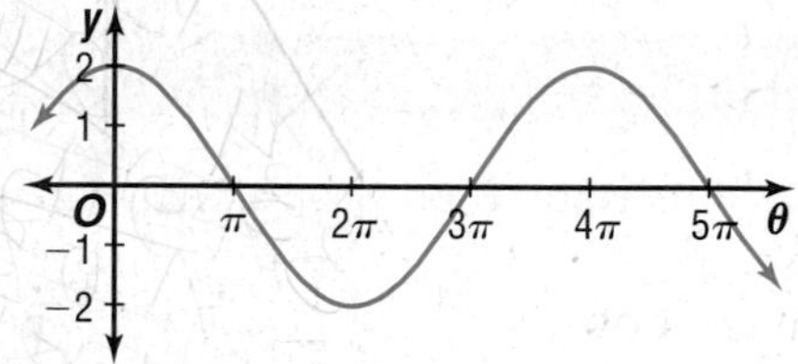

50. 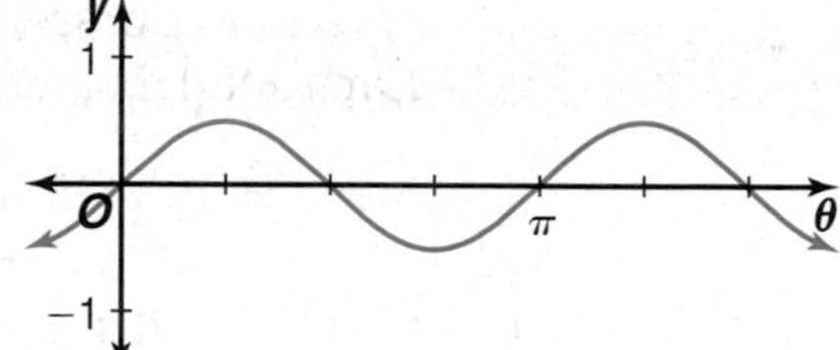

51. 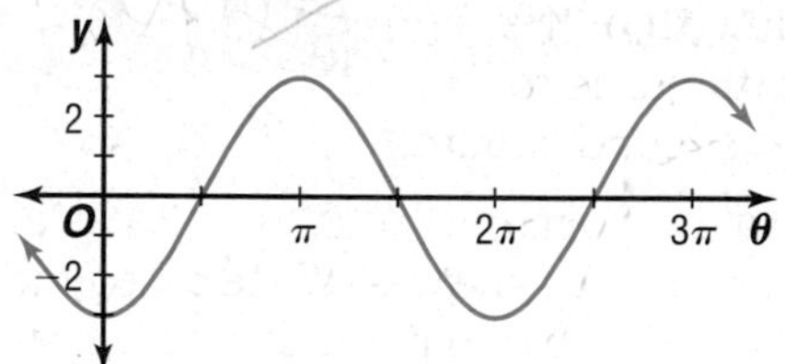

52. 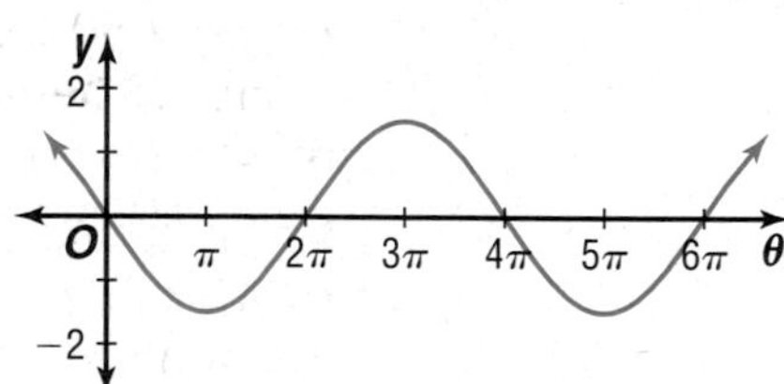

53. Write an equation for a sine function with amplitude 3.8 and frequency 120 hertz.

54. Write an equation for a cosine function with amplitude 15 and frequency 36 hertz.

55. Graph these functions on the same screen of a graphing calculator. Compare the graphs.

a. $y = \sin x$ b. $y = \sin x + 1$ c. $y = \sin x + 2$

Applications and Problem Solving

56. **Boating** A buoy in the harbor of San Juan, Puerto Rico, bobs up and down. The distance between the highest and lowest point is 3 feet. It moves from its highest point down to its lowest point and back to its highest point every 8 seconds.

 a. Find the equation of the motion for the buoy assuming that it is at its equilibrium point at $t = 0$ and the buoy is on its way down at that time.

 b. Determine the height of the buoy at 3 seconds.

 c. Determine the height of the buoy at 12 seconds.

57. **Critical Thinking** Consider the graph of $y = 2 + \sin \theta$.

 a. What is the maximum value of y?

 b. What is the minimum value of y?

 c. What is the period of the function?

 d. Sketch the graph.

58. **Music** Musical notes are classified by frequency. The note middle C has a frequency of 262 hertz. The note C above middle C has a frequency of 524 hertz. The note C below middle C has a frequency of 131 hertz.

 a. Write an equation of the sine function that represents middle C if its amplitude is 0.2.

 b. Write an equation of the sine function that represents C above middle C if its amplitude is one half that of middle C.

 c. Write an equation of the sine function that represents C below middle C if its amplitude is twice that of middle C.

59. **Physics** For a pendulum, the equation representing the horizontal displacement of the bob is $y = A \cos\left(t\sqrt{\frac{g}{\ell}}\right)$. In this equation, A is the maximum horizontal distance that the bob moves from the equilibrium point, t is the time, g is the acceleration due to gravity, and ℓ is the length of the pendulum. The acceleration due to gravity is 9.8 meters per second squared.

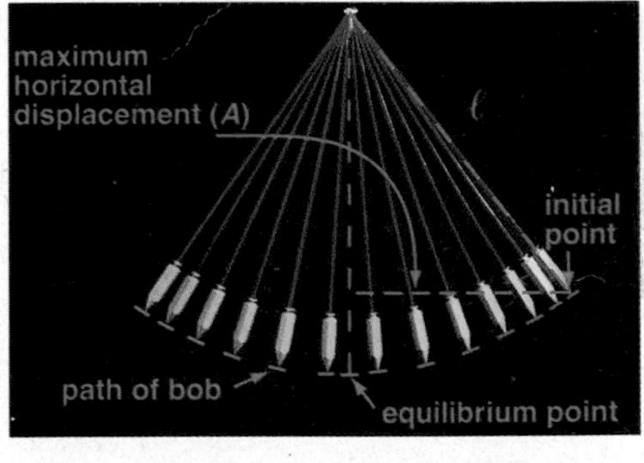

 a. A pendulum has a length of 6 meters and its bob has a maximum horizontal displacement to the right of 1.5 meters. Write an equation that models the horizontal displacement of the bob if it is at its maximum distance to the right when $t = 0$.

 b. Find the location of the bob at 4 seconds.

 c. Find the location of the bob at 7.9 seconds.

60. **Critical Thinking** Consider the graph of $y = \cos(\theta + \pi)$.

 a. Write an expression for the x-intercepts of the graph.

 b. What is the y-intercept of the graph?

 c. What is the period of the function?

 d. Sketch the graph.

61. **Physics** Three different weights are suspended from three different springs. Each spring has an elasticity coefficient of 18.5. The equation for the vertical displacement is $y = 1.5 \cos\left(t\sqrt{\frac{k}{m}}\right)$, where t is time, k is the elasticity coefficient, and m is the mass of the weight.

a. The first weight has a mass of 0.4 kilogram. Find the period and frequency of this spring.

b. The second weight has a mass of 0.6 kilogram. Find the period and frequency of this spring.

c. The third weight has a mass of 0.8 kilogram. Find the period and frequency of this spring.

d. As the mass increases, what happens to the period?

e. As the mass increases, what happens to the frequency?

Mixed Review

62. Find $\cos\left(-\frac{5\pi}{2}\right)$ by referring to the graph of the cosine function. *(Lesson 6-3)*

63. Determine the angular velocity if 84 revolutions are completed in 6 seconds. *(Lesson 6-2)*

64. Given a central angle of 73°, find the length of its intercepted arc in a circle of radius 9 inches. *(Lesson 6-1)*

65. Solve the triangle if $a = 15.1$ and $b = 19.5$. Round to the nearest tenth. *(Lesson 5-5)*

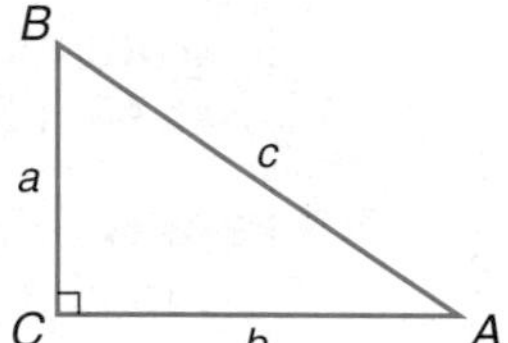

66. **Physics** The period of a pendulum can be determined by the formula $T = 2\pi\sqrt{\frac{\ell}{g}}$, where T represents the period, ℓ represents the length of the pendulum, and g represents the acceleration due to gravity. Determine the length of the pendulum if the pendulum has a period on Earth of 4.1 seconds and the acceleration due to gravity at Earth's surface is 9.8 meters per second squared. *(Lesson 4-7)*

67. Find the discriminant of $3m^2 + 5m + 10 = 0$. Describe the nature of the roots. *(Lesson 4-2)*

68. **Manufacturing** Icon, Inc. manufactures two types of computer graphics cards, Model 28 and Model 74. There are three stations, A, B, and C, on the assembly line. The assembly of a Model 28 graphics card requires 30 minutes at station A, 20 minutes at station B, and 12 minutes at station C. Model 74 requires 15 minutes at station A, 30 minutes at station B, and 10 minutes at station C. Station A can be operated for no more than 4 hours a day, station B can be operated for no more than 6 hours a day, and station C can be operated for no more than 8 hours. *(Lesson 2-7)*

a. If the profit on Model 28 is \$100 and on Model 74 is \$60, how many of each model should be assembled each day to provide maximum profit?

b. What is the maximum daily profit?

69. Use a reflection matrix to find the coordinates of the vertices of a quadrilateral reflected over the x-axis if the coordinates of the vertices of the quadrilateral are located at $(-2, -1)$, $(1, -1)$, $(3, -4)$, and $(-3, -2)$. *(Lesson 2-4)*

70. Graph $g(x) = \begin{cases} -3x \text{ if } x < -2 \\ 2 \text{ if } -2 \leq x < 3. \\ x + 1 \text{ if } x \geq 3 \end{cases}$ *(Lesson 1-7)*

71. **Fund-Raising** The regression equation of a set of data is $y = 14.7x + 140.1$, where y represents the money collected for a fund-raiser and x represents the number of members of the organization. Use the equation to predict the amount of money collected by 20 members. *(Lesson 1-6)*

72. Given that x is an integer, state the relation representing $y = x^2$ and $-4 \leq x \leq -2$ by listing a set of ordered pairs. Then state whether this relation is a function. *(Lesson 1-1)*

73. **SAT/ACT Practice** Points $RSTU$ are the centers of four congruent circles. If the area of square $RSTU$ is 100, what is the sum of the areas of the four circles?

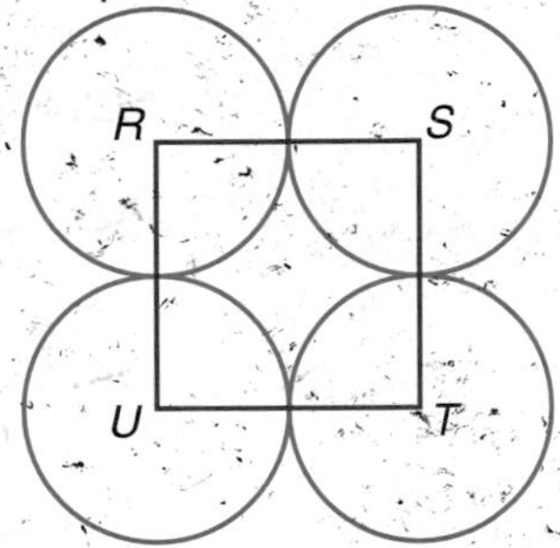

A 25π 
B 50π
C 100π
D 200π
E 400π

MID-CHAPTER QUIZ

1. Change $\frac{5\pi}{6}$ radians to degree measure. (Lesson 6-1)

2. **Mechanics** A pulley with diameter 0.5 meter is being used to lift a box. How far will the box weight rise if the pulley is rotated through an angle of $\frac{5\pi}{3}$ radians? (Lesson 6-1)

3. Find the area of a sector if the central angle measures $\frac{2\pi}{5}$ radians and the radius of the circle is 8 feet. (Lesson 6-1)

4. Determine the angular displacement in radians of 7.8 revolutions. (Lesson 6-2)

5. Determine the angular velocity if 8.6 revolutions are completed in 7 seconds. (Lesson 6-2)

6. Determine the linear velocity of a point rotating at an angular velocity of 8π radians per second at a distance of 3 meters from the center of the rotating object. (Lesson 6-2)

7. Find $\sin\left(-\frac{7\pi}{2}\right)$ by referring to the graph of the sine function. (Lesson 6-3)

8. Graph $y = \cos x$ for $7\pi \leq x \leq 9\pi$. (Lesson 6-3)

9. State the amplitude and period for the function $y = -7 \cos \frac{\theta}{3}$. Then graph the function. (Lesson 6-4)

10. Find the possible equations of the sine function with amplitude 5 and period $\frac{\pi}{3}$. (Lesson 6-4)

Extra Practice See p. A36.

6-5

Translations of Sine and Cosine Functions

OBJECTIVES

- Find the phase shift and the vertical translation for sine and cosine functions.
- Write the equations of sine and cosine functions given the amplitude, period, phase shift, and vertical translation.
- Graph compound functions.

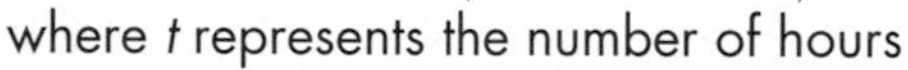

TIDES One day in March in San Diego, California, the first low tide occurred at 1:45 A.M., and the first high tide occurred at 7:44 A.M. Approximately 12 hours and 24 minutes or 12.4 hours after the first low tide occurred, the second low tide occurred. The equation that models these tides is $h = 2.9 + 2.2 \sin \left(\frac{\pi}{6.2}t - \frac{4.85\pi}{6.2}\right)$, where t represents the number of hours since midnight and h represents the height of the water. Draw a graph that models the cyclic nature of the tide. *This problem will be solved in Example 4.*

In Chapter 3, you learned that the graph of $y = (x - 2)^2$ is a horizontal translation of the parent graph of $y = x^2$. Similarly, graphs of the sine and cosine functions can be translated horizontally.

GRAPHING CALCULATOR EXPLORATION

- Select the radian mode.
- Use the domain and range values below to set the viewing window.

$-4.7 \leq x \leq 4.8$, **Xscl: 1** $\quad -3 \leq y \leq 3$, **Yscl: 1**

TRY THESE

1. Graph each function on the same screen.

a. $y = \sin x$ **b.** $y = \sin \left(x + \frac{\pi}{4}\right)$

c. $y = \sin \left(x + \frac{\pi}{2}\right)$

WHAT DO YOU THINK?

2. Describe the behavior of the graph of $f(x) = \sin (x + c)$, where $c > 0$, as c increases.

3. Make a conjecture about what happens to the graph of $f(x) = \sin (x + c)$ if $c < 0$ and continues to decrease. Test your conjecture.

A horizontal translation or shift of a trigonometric function is called a **phase shift.** Consider the equation of the form $y = A \sin (k\theta + c)$, where $A, k, c \neq 0$. To find a zero of the function, find the value of θ for which $A \sin (k\theta + c) = 0$. Since $\sin 0 = 0$, solving $k\theta + c = 0$ will yield a zero of the function.

$$k\theta + c = 0$$
$$\theta = -\frac{c}{k} \quad \textit{Solve for } \theta.$$

Therefore, $y = 0$ when $\theta = -\frac{c}{k}$. The value of $-\frac{c}{k}$ is the phase shift.

When $c > 0$: The graph of $y = A \sin (k\theta + c)$ is the graph of $y = A \sin k\theta$, shifted $\left|\frac{c}{k}\right|$ to the left.

When $c < 0$: The graph of $y = A \sin (k\theta + c)$ is the graph of $y = A \sin k\theta$, shifted $\left|\frac{c}{k}\right|$ to the right.

Phase Shift of Sine and Cosine Functions

The phase shift of the functions $y = A \sin (k\theta + c)$ and $y = A \cos (k\theta + c)$ is $-\frac{c}{k}$, where $k > 0$.
If $c > 0$, the shift is to the left.
If $c < 0$, the shift is to the right.

Example 1 **State the phase shift for each function. Then graph the function.**

a. $y = \sin (\theta + \pi)$

The phase shift of the function is $-\frac{c}{k}$ or $-\frac{\pi}{1}$, which equals $-\pi$.

To graph $y = \sin (\theta + \pi)$, consider the graph of $y = \sin \theta$. Graph this function and then shift the graph $-\pi$.

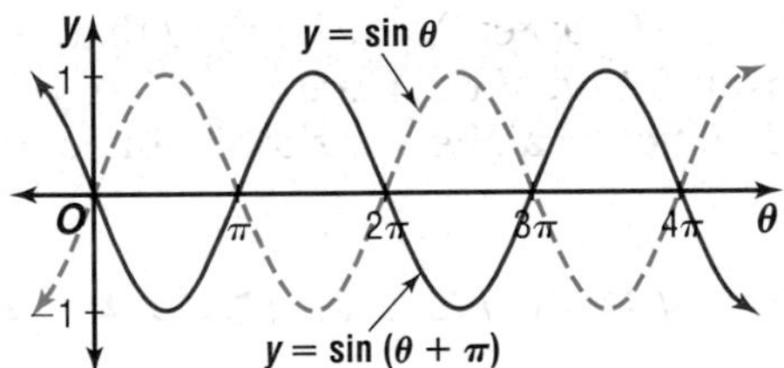

b. $y = \cos \left(2\theta - \frac{\pi}{2}\right)$

The phase shift of the function is $-\frac{c}{k}$ or $-\left(\frac{-\frac{\pi}{2}}{2}\right)$, which equals $\frac{\pi}{4}$.

To graph $y = \cos \left(2\theta - \frac{\pi}{2}\right)$, consider the graph of $y = \cos 2\theta$. The graph of $y = \cos 2\theta$ has amplitude of 1 and a period of $\frac{2\pi}{2}$ or π. Graph this function and then shift the graph $\frac{\pi}{4}$.

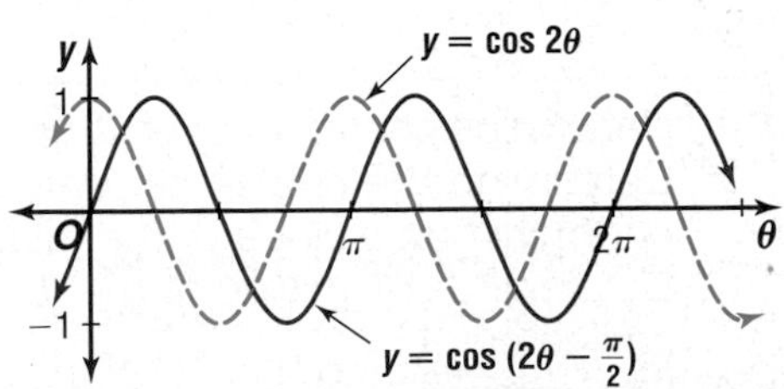

In Chapter 3, you also learned that the graph of $y = x^2 - 2$ is a vertical translation of the parent graph of $y = x^2$. Similarly, graphs of the sine and cosine functions can be translated vertically.

When a constant is added to a sine or cosine function, the graph is shifted upward or downward. If (x, y) are the coordinates of $y = \sin x$, then $(x, y + d)$ are the coordinates of $y = \sin x + d$.

A new horizontal axis known as the **midline** becomes the reference line or equilibrium point about which the graph oscillates. For the graph of $y = A \sin \theta + h$, the midline is the graph of $y = h$.

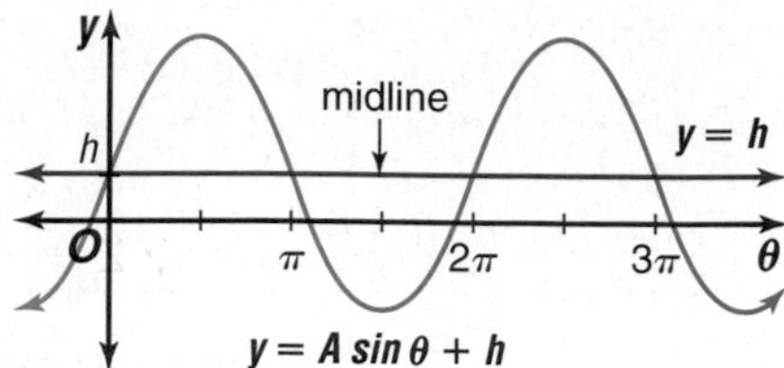

Vertical Shift of Sine and Cosine Functions

The vertical shift of the functions $y = A \sin (k\theta + c) + h$ and $y = A \cos (k\theta + c) + h$ is h. If $h > 0$, the shift is upward. If $h < 0$, the shift is downward. The midline is $y = h$.

Example 2 **State the vertical shift and the equation of the midline for the function $y = 2 \cos \theta - 5$. Then graph the function.**

The vertical shift is 5 units downward. The midline is the graph of $y = -5$.

To graph the function, draw the midline, the graph of $y = -5$. Since the amplitude of the function is $|2|$ or 2, draw dashed lines parallel to the midline which are 2 units above and below the midline. That is, $y = -3$ and $y = -7$. Then draw the cosine curve.

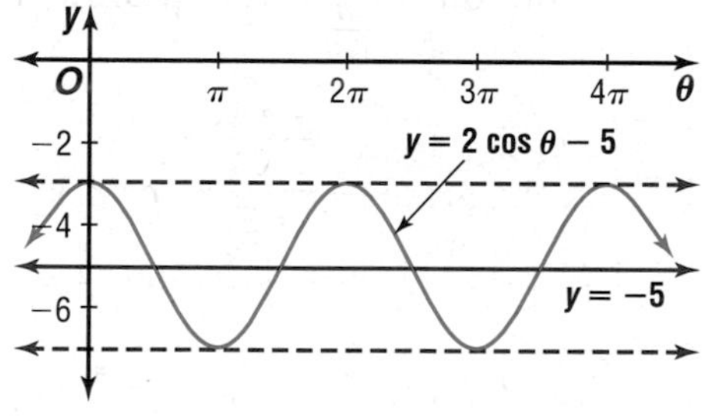

In general, use the following steps to graph any sine or cosine function.

Graphing Sine and Cosine Functions

1. Determine the vertical shift and graph the midline.
2. Determine the amplitude. Use dashed lines to indicate the maximum and minimum values of the function.
3. Determine the period of the function and graph the appropriate sine or cosine curve.
4. Determine the phase shift and translate the graph accordingly.

Example 3 **State the amplitude, period, phase shift, and vertical shift for $y = 4\cos\left(\frac{\theta}{2} + \pi\right) - 6$. Then graph the function.**

The amplitude is $|4|$ or 4. The period is $\frac{2\pi}{\frac{1}{2}}$ or 4π. The phase shift is $-\frac{\pi}{\frac{1}{2}}$ or -2π. The vertical shift is -6. Using this information, follow the steps for graphing a cosine function.

Step 1 Draw the midline which is the graph of $y = -6$.

Step 2 Draw dashed lines parallel to the midline, which are 4 units above and below the midline.

Step 3 Draw the cosine curve with period of 4π.

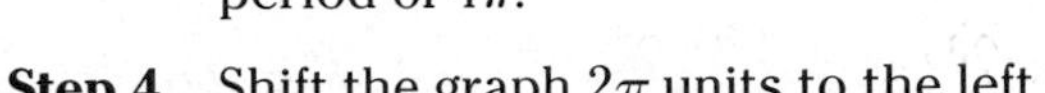

Step 4 Shift the graph 2π units to the left.

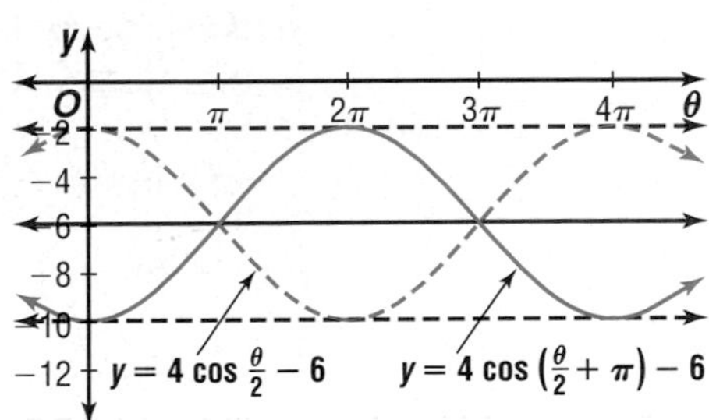

You can use information about amplitude, period, and translations of sine and cosine functions to model real-world applications.

Example 4 Real World Application

TIDES **Refer to the application at the beginning of the lesson. Draw a graph that models the San Diego tide.**

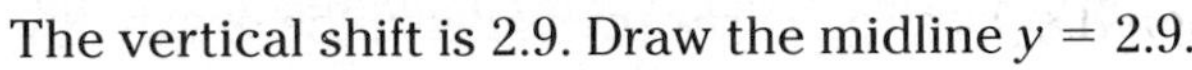

The vertical shift is 2.9. Draw the midline $y = 2.9$.

The amplitude is $|2.2|$ or 2.2. Draw dashed lines parallel to and 2.2 units above and below the midline.

The period is $\frac{2\pi}{\frac{\pi}{6.2}}$ or 12.4. Draw the sine curve with a period of 12.4.

Shift the graph $-\frac{\frac{-4.85\pi}{6.2}}{\frac{\pi}{6.2}}$ or 4.85 units.

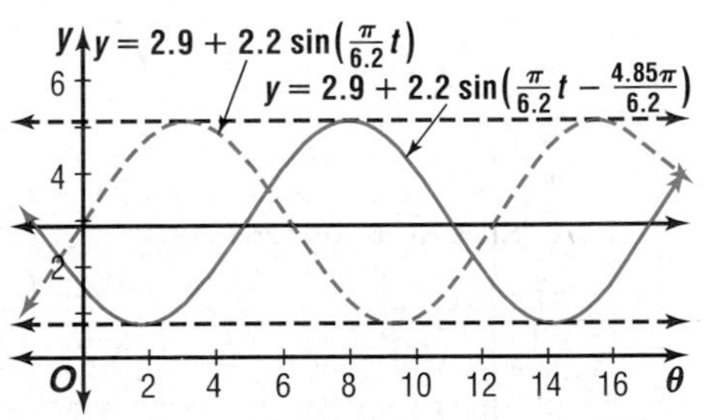

You can write an equation for a trigonometric function if you are given the amplitude, period, phase shift, and vertical shift.

Example 5 **Write an equation of a sine function with amplitude 4, period π, phase shift $-\frac{\pi}{8}$, and vertical shift 6.**

The form of the equation will be $y = A \sin (k\theta + c) + h$. Find the values of A, k, c, and h.

A: $|A| = 4$

$A = 4$ or -4

k: $\frac{2\pi}{k} = \pi$ *The period is π.*

$k = 2$

c: $-\frac{c}{k} = -\frac{\pi}{8}$ *The phase shift is $-\frac{\pi}{8}$.*

$-\frac{c}{2} = -\frac{\pi}{8}$ *$k = 2$*

$c = \frac{\pi}{4}$

h: $h = 6$

Substitute these values into the general equation. The possible equations are $y = 4 \sin \left(2\theta + \frac{\pi}{4}\right) + 6$ and $y = -4 \sin \left(2\theta + \frac{\pi}{4}\right) + 6$.

Compound functions may consist of sums or products of trigonometric functions. Compound functions may also include sums and products of trigonometric functions and other functions.

Here are some examples of compound functions.

$y = \sin x \cdot \cos x$ *Product of trigonometric functions*

$y = \cos x + x$ *Sum of a trigonometric function and a linear function*

You can graph compound functions involving addition by graphing each function separately on the same coordinate axes and then adding the ordinates. After you find a few of the critical points in this way, you can sketch the rest of the curve of the function of the compound function.

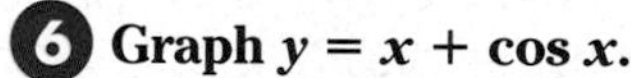

Example 6 **Graph $y = x + \cos x$.**

First graph $y = \cos x$ and $y = x$ on the same axis. Then add the corresponding ordinates of the function. Finally, sketch the graph.

x	$\cos x$	$x + \cos x$
0	1	1
$\frac{\pi}{2}$	0	$\frac{\pi}{2} + 0 \approx 1.57$
π	-1	$\pi - 1 \approx 2.14$
$\frac{3\pi}{2}$	0	$\frac{3\pi}{2} \approx 4.71$
2π	1	$2\pi + 1 \approx 7.28$
$\frac{5\pi}{2}$	0	$\frac{5\pi}{2} \approx 7.85$
3π	-1	$3\pi - 1 \approx 8.42$

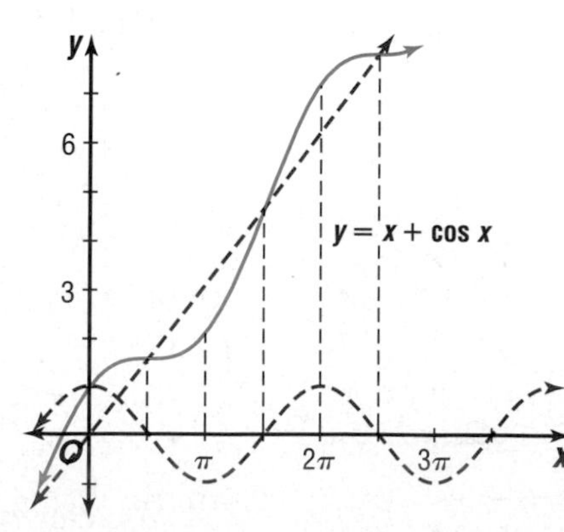

CHECK FOR UNDERSTANDING

Communicating Mathematics

Read and study the lesson to answer each question.

1. **Compare and contrast** the graphs $y = \sin x + 1$ and $y = \sin (x + 1)$.
2. **Name** the function whose graph is the same as the graph of $y = \cos x$ with a phase shift of $\frac{\pi}{2}$.
3. **Analyze** the function $y = A \sin (k\theta + c) + h$. Which variable could you increase or decrease to have each of the following effects on the graph?
   a. stretch the graph vertically
   b. translate the graph downward vertically
   c. shrink the graph horizontally
   d. translate the graph to the left.
4. **Explain** how to graph $y = \sin x + \cos x$.
5. **You Decide** Marsha and Jamal are graphing $y = \cos\left(\frac{\pi}{6}\theta - \frac{\pi}{2}\right)$. Marsha says that the phase shift of the graph is $\frac{\pi}{2}$. Jamal says that the phase shift is 3. Who is correct? Explain.

Guided Practice

6. State the phase shift for $y = 3 \cos\left(\theta - \frac{\pi}{2}\right)$. Then graph the function.
7. State the vertical shift and the equation of the midline for $y = \sin 2\theta + 3$. Then graph the function.

State the amplitude, period, phase shift, and vertical shift for each function. Then graph the function.

8. $y = 2 \sin (2\theta + \pi) - 5$
9. $y = 3 - \frac{1}{2} \cos\left(\frac{\theta}{2} - \frac{\pi}{4}\right)$
10. Write an equation of a sine function with amplitude 20, period 1, phase shift 0, and vertical shift 100.
11. Write an equation of a cosine function with amplitude 0.6, period 12.4, phase shift -2.13, and vertical shift 7.
12. Graph $y = \sin x - \cos x$.
13. **Health** If a person has a blood pressure of 130 over 70, then the person's blood pressure oscillates between the maximum of 130 and a minimum of 70.
   a. Write the equation for the midline about which this person's blood pressure oscillates.
   b. If the person's pulse rate is 60 beats a minute, write a sine equation that models his or her blood pressure using t as time in seconds.
   c. Graph the equation.

EXERCISES

Practice

State the phase shift for each function. Then graph each function.

14. $y = \sin (\theta - 2\pi)$
15. $y = \sin (2\theta + \pi)$
16. $y = 2 \cos\left(\frac{\theta}{4} + \frac{\pi}{2}\right)$

State the vertical shift and the equation of the midline for each function. Then graph each function.

17. $y = \sin\frac{\theta}{2} + \frac{1}{2}$ **18.** $y = 5\cos\theta - 4$ **19.** $y = 7 + \cos 2\theta$

20. State the horizontal and vertical shift for $y = -8\sin(2\theta - 4\pi) - 3$.

State the amplitude, period, phase shift, and vertical shift for each function. Then graph the function.

21. $y = 3\cos\left(\theta - \frac{\pi}{2}\right)$

22. $y = 6\sin\left(\theta + \frac{\pi}{3}\right) + 2$

23. $y = -2 + \sin\left(\frac{\theta}{3} - \frac{\pi}{12}\right)$

24. $y = 20 + 5\cos(3\theta + \pi)$

25. $y = \frac{1}{4}\cos\frac{\theta}{2} - 3$

26. $y = 10\sin\left(\frac{\theta}{4} - 4\pi\right) - 5$

27. State the amplitude, period, phase shift, and vertical shift of the sine curve shown at the right.

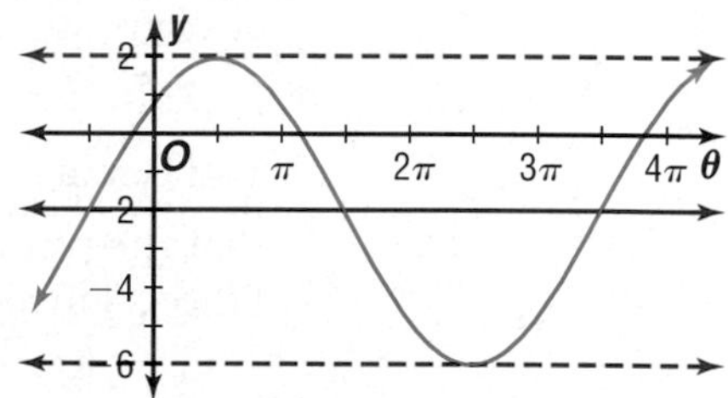

Write an equation of the sine function with each amplitude, period, phase shift, and vertical shift.

28. amplitude = 7, period = 3π, phase shift = π, vertical shift = -7

29. amplitude = 50, period = $\frac{3\pi}{4}$, phase shift = $\frac{\pi}{2}$, vertical shift = -25

30. amplitude = $\frac{3}{4}$, period = $\frac{\pi}{5}$, phase shift = π, vertical shift = $\frac{1}{4}$

Write an equation of the cosine function with each amplitude, period, phase shift, and vertical shift.

31. amplitude = 3.5, period = $\frac{\pi}{2}$, phase shift = $\frac{\pi}{4}$, vertical shift = 7

32. amplitude = $\frac{4}{5}$, period = $\frac{\pi}{6}$, phase shift = $\frac{\pi}{3}$, vertical shift = $\frac{7}{5}$

33. amplitude = 100, period = 45, phase shift = 0, vertical shift = -110

34. Write a cosine equation for the graph at the right.

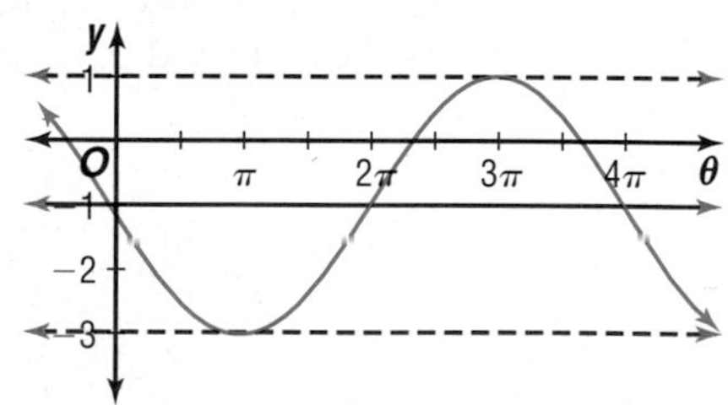

35. Write a sine equation for the graph at the right.

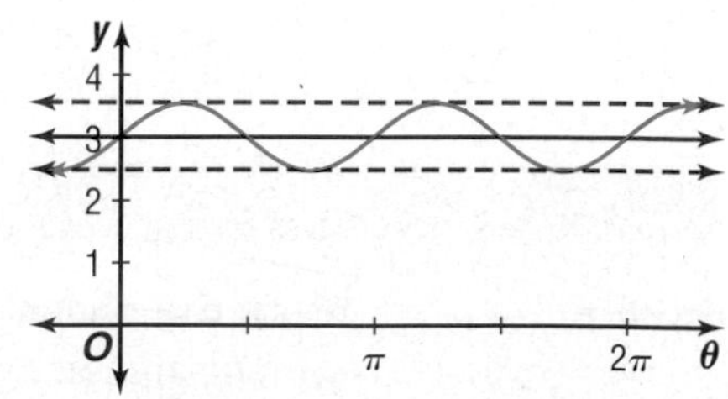

Graph each function.

36. $y = \sin x + x$ **37.** $y = \cos x - \sin x$ **38.** $y = \sin x + \sin 2x$

39. On the same coordinate plane, graph each function.

a. $y = 2 \sin x$ b. $y = 3 \cos x$ c. $y = 2 \sin x + 3 \cos x$

40. Use the graphs of $y = \cos 2x$ and $y = \cos 3x$ to graph $y = \cos 2x - \cos 3x$.

Applications and Problem Solving

41. Biology In the wild, predators such as wolves need prey such as sheep to survive. The population of the wolves and the sheep are cyclic in nature. Suppose the population of the wolves W is modeled by $W = 2000 + 1000 \sin\left(\frac{\pi t}{6}\right)$ and population of the sheep S is modeled by $S = 10{,}000 + 5000 \cos\left(\frac{\pi t}{6}\right)$ where t is the time in months.

a. What are the maximum number and the minimum number of wolves?

b. What are the maximum number and the minimum number of sheep?

c. Use a graphing calculator to graph both equations for values of t from 0 to 24.

d. During which months does the wolf population reach a maximum?

e. During which months does the sheep population reach a maximum?

f. What is the relationship of the maximum population of the wolves and the maximum population of the sheep? Explain.

42. Critical Thinking Use the graphs of $y = x$ and $y = \cos x$ to graph $y = x \cos x$.

43. Entertainment As you ride a Ferris wheel, the height that you are above the ground varies periodically. Consider the height of the center of the wheel to be the equilibrium point. Suppose the diameter of a Ferris Wheel is 42 feet and travels at a rate of 3 revolutions per minute. At the highest point, a seat on the Ferris wheel is 46 feet above the ground.

a. What is the lowest height of a seat?

b. What is the equation of the midline?

c. What is the period of the function?

d. Write a sine equation to model the height of a seat that was at the equilibrium point heading upward when the ride began.

e. According to the model, when will the seat reach the highest point for the first time?

f. According to the model, what is the height of the seat after 10 seconds?

44. Electronics In electrical circuits, the voltage and current can be described by sine or cosine functions. If the graphs of these functions have the same period, but do not pass through their zero points at the same time, they are said to have a *phase difference*. For example, if the voltage is 0 at 90° and the current is 0 at 180°, they are 90° out of phase. Suppose the voltage across an inductor of a circuit is represented by $y = 2 \cos 2x$ and the current across the component is represented by $y = \cos\left(2x - \frac{\pi}{2}\right)$. What is the phase relationship between the signals?

45. **Critical Thinking** The windows for the following calculator screens are set at [-2π, 2π] scl: 0.5π by [-2, 2] scl: 0.5. Without using a graphing calculator, use the equations below to identify the graph on each calculator screen.

$y = \cos x^2$ $\quad$ $y = \sqrt{\sin x}$ $\quad$ $y = \dfrac{\cos x}{x}$ $\quad$ $y = \sin \sqrt{x}$

a.

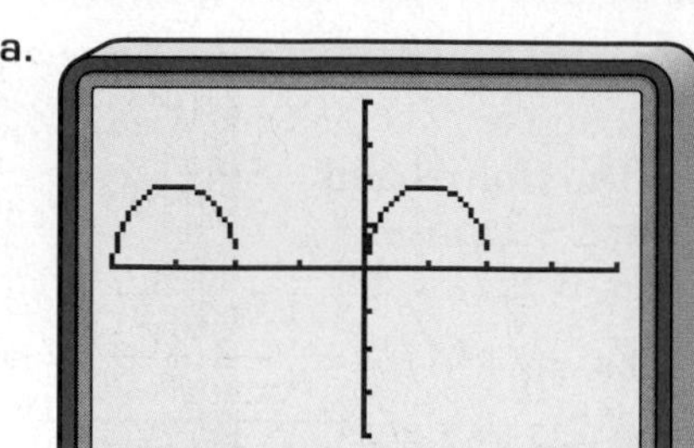

b.

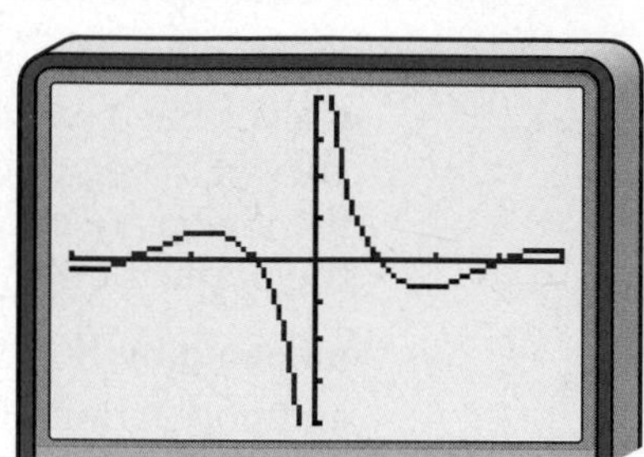

c.

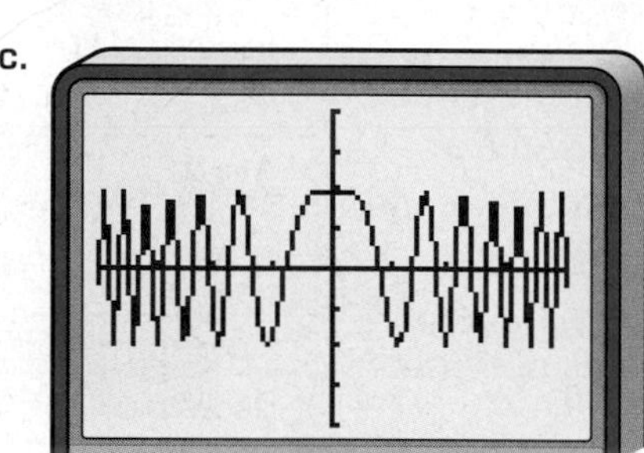

d.

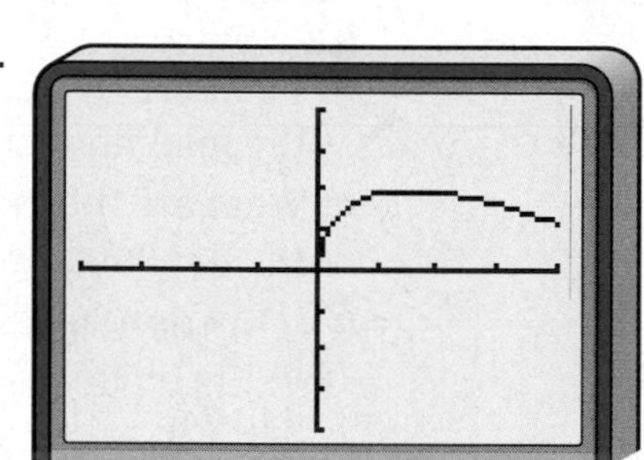

Mixed Review

46. **Music** Write an equation of the sine function that represents the initial behavior of the vibrations of the note D above middle C having amplitude 0.25 and a frequency of 294 hertz. *(Lesson 6-4)*

47. Determine the linear velocity of a point rotating at an angular velocity of 19.2 radians per second at a distance of 7 centimeters from the center of the rotating object. *(Lesson 6-2)*

48. Graph $y = \dfrac{x-3}{x-2}$. *(Lesson 3-7)*

49. Find the inverse of $f(x) = \dfrac{3}{x-1}$. *(Lesson 3-4)*

50. Find matrix X in the equation $\begin{bmatrix} 1 & 1 \\ 1 & 1 \end{bmatrix}\begin{bmatrix} 3 & 5 \\ -3 & -5 \end{bmatrix} = X$. *(Lesson 2-3)*

51. Solve the system of equations. *(Lesson 2-1)*

$3x + 5y = 4$
$14x - 35y = 21$

52. Graph $y \leq |x + 4|$. *(Lesson 1-8)*

53. Write the standard form of the equation of the line through the point at $(3, -2)$ that is parallel to the graph of $3x - y + 7 = 0$. *(Lesson 1-5)*

54. **SAT Practice** **Grid-In** A swimming pool is 75 feet long and 42 feet wide. If 7.48 gallons equals 1 cubic foot, how many gallons of water are needed to raise the level of the water 4 inches?

Extra Practice See p. A37.

6-6 Modeling Real-World Data with Sinusoidal Functions

OBJECTIVES

- Model real-world data using sine and cosine functions.
- Use sinusoidal functions to solve problems.

METEOROLOGY The table contains the times that the sun rises and sets on the fifteenth of every month in Brownsville, Texas.

Let $t = 1$ represent January 15.
Let $t = 2$ represent February 15.
Let $t = 3$ represent March 15.
⋮

Write a function that models the hours of daylight for Brownsville. Use your model to estimate the number of hours of daylight on September 30. *This problem will be solved in Example 1.*

Month	Sunrise A.M.	Sunset P.M.
January	7:19	6:00
February	7:05	6:23
March	6:40	6:39
April	6:07	6:53
May	5:44	7:09
June	5:38	7:23
July	5:48	7:24
August	6:03	7:06
September	6:16	6:34
October	6:29	6:03
November	6:48	5:41
December	7:09	5:41

Before you can determine the function for the daylight, you must first compute the amount of daylight for each day as a decimal value. Consider January 15. First, write each time in 24-hour time.

$7{:}19 \text{ A.M.} = 7{:}19$

$6{:}00 \text{ P.M.} = 6{:}00 + 12 \text{ or } 18{:}00$

Then change each time to a decimal rounded to the nearest hundredth.

$7{:}19 = 7 + \frac{19}{60} \text{ or } 7.32$

$18{:}00 = 18 + \frac{0}{60} \text{ or } 18.00$

On January 15, there will be $18.00 - 7.32$ or 10.68 hours of daylight.

Similarly, the number of daylight hours can be determined for the fifteenth of each month.

Month	Jan.	Feb.	March	April	May	June
t	1	2	3	4	5	6
Hours of Daylight	10.68	11.30	11.98	12.77	13.42	13.75

Month	July	Aug.	Sept.	Oct.	Nov.	Dec.
t	7	8	9	10	11	12
Hours of Daylight	13.60	13.05	12.30	11.57	10.88	10.53

Since there are 12 months in a year, month 13 is the same as month 1, month 14 is the same as month 2, and so on. The function is periodic. Enter the data into a graphing calculator and graph the points. The graph resembles a type of sine curve. You can write a **sinusoidal function** to represent the data. A sinusoidal function can be any function of the form $y = A \sin (k\theta + c) + h$ or $y = A \cos (k\theta + c) + h$.

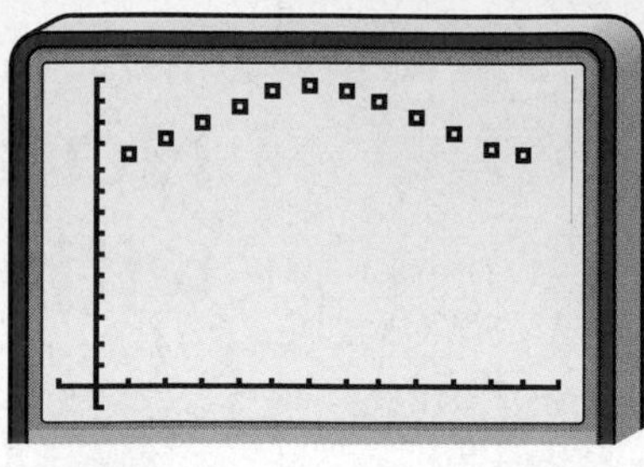

[−1, 13] scl:1 by [−1, 14] scl:1

Example 1 **METEOROLOGY** **Refer to the application at the beginning of the lesson.**

a. Write a function that models the amount of daylight for Brownsville.

b. Use your model to estimate the number of hours of daylight on September 30.

a. The data can be modeled by a function of the form $y = A \sin (kt + c) + h$, where t is the time in months. First, find A, h, and k.

interNET CONNECTION

Research
For data about amount of daylight, average temperatures, or tides, visit **www.amc.glencoe.com**

$\mathbf{A}$: $A = \dfrac{13.75 - 10.53}{2}$ or 1.61 — *A is half the difference between the most daylight (13.75 h) and the least daylight (10.53 h).*

$\mathbf{h}$: $h = \dfrac{13.75 + 10.53}{2}$ or 12.14 — *h is half the sum of the greatest value and least value.*

$\mathbf{k}$: $\dfrac{2\pi}{k} = 12$ — *The period is 12.*

$k = \dfrac{\pi}{6}$

Substitute these values into the general form of the sinusoidal function.

$y = A \sin (kt + c) + h$

$y = 1.61 \sin \left(\dfrac{\pi}{6}t + c\right) + 12.14$ — *$A = 1.61$, $k = \dfrac{\pi}{6}$, $h = 12.14$*

To compute c, substitute one of the coordinate pairs into the function.

$$y = 1.61 \sin \left(\frac{\pi}{6}t + c\right) + 12.14$$

$$10.68 = 1.61 \sin \left(\frac{\pi}{6}(1) + c\right) + 12.14 \quad \text{Use } (t, y) = (1, 10.68).$$

$$-1.46 = 1.61 \sin \left(\frac{\pi}{6} + c\right) \quad \text{Add } -12.14 \text{ to each side.}$$

$$-\frac{1.46}{1.61} = \sin \left(\frac{\pi}{6} + c\right) \quad \text{Divide each side by 1.61.}$$

$$\sin^{-1}\left(-\frac{1.46}{1.61}\right) = \frac{\pi}{6} + c \quad \text{Definition of inverse}$$

$$\sin^{-1}\left(-\frac{1.46}{1.61}\right) - \frac{\pi}{6} = c \quad \text{Add } -\frac{\pi}{6} \text{ to each side.}$$

$$-1.659305545 \approx c \quad \text{Use a calculator.}$$

The function $y = 1.61 \sin\left(\frac{\pi}{6}t - 1.66\right) + 12.14$ is one model for the daylight in Brownsville.

Graphing Calculator Tip

For keystroke instruction on how to find sine regression statistics, see page A25.

To check this answer, enter the data into a graphing calculator and calculate the **SinReg** statistics. Rounding to the nearest hundredth, $y = 1.60 \sin(0.51t - 1.60) + 12.12$. The models are similar. Either model could be used.

b. September 30 is half a month past September 15, so $t = 9.5$. Select a model and use a calculator to evaluate it for $t = 9.5$.

Model 1: Paper and Pencil

$y = 1.61 \sin\left(\frac{\pi}{6}t - 1.66\right) + 12.14$

$y = 1.61 \sin\left[\frac{\pi}{6}(9.5) - 1.66\right] + 12.14$

$y \approx 11.86349848$

Model 2: Graphing Calculator

$y = 1.60 \sin(0.51t - 1.60) + 12.12$

$y = 1.60 \sin[0.51(9.5) - 1.60] + 12.12$

$y \approx 11.95484295$

On September 30, Brownsville will have about 11.9 hours of daylight.

In general, any sinusoidal function can be written as a sine function or as a cosine function. The amplitude, the period, and the midline will remain the same. However, the phase shift will be different. To avoid a greater phase shift than necessary, you may wish to use a sine function if the function is about zero at $x = 0$ and a cosine function if the function is about the maximum or minimum at $x = 0$.

Example 2

HEALTH An average seated adult breathes in and out every 4 seconds. The average minimum amount of air in the lungs is 0.08 liter, and the average maximum amount of air in the lungs is 0.82 liter. Suppose the lungs have a minimum amount of air at $t = 0$, where t is the time in seconds.

a. Write a function that models the amount of air in the lungs.

b. Graph the function.

c. Determine the amount of air in the lungs at 5.5 seconds.

(continued on the next page)

a. Since the function has its minimum value at $t = 0$, use the cosine function. A cosine function with its minimum value at $t = 0$ has no phase shift and a negative value for A. Therefore, the general form of the model is $y = -A \cos kt + h$, where t is the time in seconds. Find A, k, and h.

A: $A = \frac{0.82 - 0.08}{2}$ or 0.37 *A is half the difference between the greatest value and the least value.*

h: $h = \frac{0.82 + 0.08}{2}$ or 0.45 *h is half the sum of the greatest value and the least value.*

k: $\frac{2\pi}{k} = 4$ *The period is 4.*

$k = \frac{\pi}{2}$

Therefore, $y = -0.37 \cos \frac{\pi}{2} t + 0.45$ models the amount of air in the lungs of an average seated adult.

b. Use a graphing calculator to graph the function.

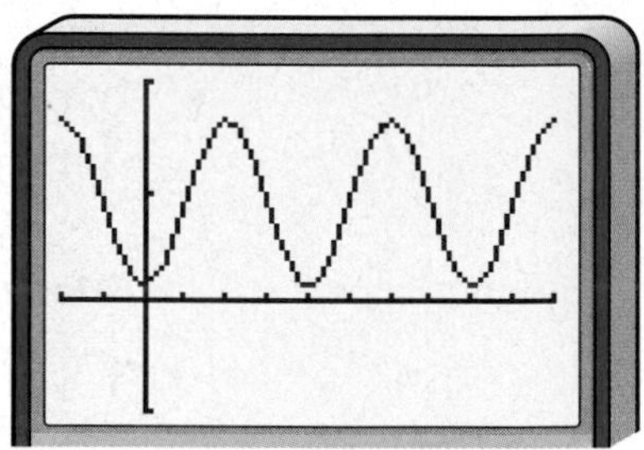

[−2, 10] scl:1 by [−0.5, 1] scl:0.5

c. Use this function to find the amount of air in the lungs at 5.5 seconds.

$y = -0.37 \cos \frac{\pi}{2} t + 0.45$

$y = -0.37 \cos \left[\frac{\pi}{2}(5.5)\right] + 0.45$

$y \approx 0.711629509$

The lungs have about 0.71 liter of air at 5.5 seconds.

CHECK FOR UNDERSTANDING

Communicating Mathematics

Read and study the lesson to answer each question.

1. **Define** sinusoidal function in your own words.
2. **Compare and contrast** real-world data that can be modeled with a polynomial function and real-world data that can be modeled with a sinusoidal function.
3. **Give** three real-world examples that can be modeled with a sinusoidal function.

Guided Practice

4. **Boating** If the equilibrium point is $y = 0$, then $y = -5 \cos\left(\frac{\pi}{6}t\right)$ models a buoy bobbing up and down in the water.
   a. Describe the location of the buoy when $t = 0$.
   b. What is the maximum height of the buoy?
   c. Find the location of the buoy at $t = 7$.

5. **Health** A certain person's blood pressure oscillates between 140 and 80. If the heart beats once every second, write a sine function that models the person's blood pressure.

6. **Meteorology** The average monthly temperatures for the city of Seattle, Washington, are given below.

Jan.	Feb.	March	April	May	June	July	Aug.	Sept.	Oct.	Nov.	Dec.
41°	44°	47°	50°	56°	61°	65°	66°	61°	54°	46°	42°

   a. Find the amplitude of a sinusoidal function that models the monthly temperatures.
   b. Find the vertical shift of a sinusoidal function that models the monthly temperatures.
   c. What is the period of a sinusoidal function that models the monthly temperatures?
   d. Write a sinusoidal function that models the monthly temperatures, using $t = 1$ to represent January.
   e. According to your model, what is the average monthly temperature in February? How does this compare to the actual average?
   f. According to your model, what is the average monthly temperature in October? How does this compare to the actual average?

EXERCISES

Applications and Problem Solving

7. **Music** The initial behavior of the vibrations of the note E above middle C can be modeled by $y = 0.5 \sin 660\pi t$.
   a. What is the amplitude of this model?
   b. What is the period of this model?
   c. Find the frequency (cycles per second) for this note.

8. **Entertainment** A rodeo performer spins a lasso in a circle perpendicular to the ground. The height of the knot from the ground is modeled by $h = -3 \cos\left(\frac{5\pi}{3}t\right) + 3.5$, where t is the time measured in seconds.
   a. What is the highest point reached by the knot?
   b. What is the lowest point reached by the knot?
   c. What is the period of the model?
   d. According to the model, find the height of the knot after 25 seconds.

9. **Biology** In a certain region with hawks as predators and rodents as prey, the rodent population R varies according to the model $R = 1200 + 300 \sin\left(\frac{\pi}{2}t\right)$, and the hawk population H varies according to the model $H = 250 + 25 \sin\left(\frac{\pi}{2}t - \frac{\pi}{4}\right)$, with t measured in years since January 1, 1970.

   a. What was the population of rodents on January 1, 1970?

   b. What was the population of hawks on January 1, 1970?

   c. What are the maximum populations of rodents and hawks? Do these maxima ever occur at the same time?

   d. On what date was the first maximum population of rodents achieved?

   e. What is the minimum population of hawks? On what date was the minimum population of hawks first achieved?

   f. According to the models, what was the population of rodents and hawks on January 1 of the present year?

10. **Waves** A leaf floats on the water bobbing up and down. The distance between its highest and lowest point is 4 centimeters. It moves from its highest point down to its lowest point and back to its highest point every 10 seconds. Write a cosine function that models the movement of the leaf in relationship to the equilibrium point.

11. **Tides** Write a sine function which models the oscillation of tides in Savannah, Georgia, if the equilibrium point is 4.24 feet, the amplitude is 3.55 feet, the phase shift is -4.68 hours, and the period is 12.40 hours.

12. **Meteorology** The mean average temperature in Buffalo, New York, is 47.5°. The temperature fluctuates 23.5° above and below the mean temperature. If $t = 1$ represents January, the phase shift of the sine function is 4.

   a. Write a model for the average monthly temperature in Buffalo.

   b. According to your model, what is the average temperature in March?

   c. According to your model, what is the average temperature in August?

13. **Meteorology** The average monthly temperatures for the city of Honolulu, Hawaii, are given below.

Jan.	Feb.	March	April	May	June	July	Aug.	Sept.	Oct.	Nov.	Dec.
73°	73°	74°	76°	78°	79°	81°	81°	81°	80°	77°	74°

a. Find the amplitude of a sinusoidal function that models the monthly temperatures.

b. Find the vertical shift of a sinusoidal function that models the monthly temperatures.

c. What is the period of a sinusoidal function that models the monthly temperatures?

d. Write a sinusoidal function that models the monthly temperatures, using $t = 1$ to represent January.

e. According to your model, what is the average temperature in August? How does this compare to the actual average?

f. According to your model, what is the average temperature in May? How does this compare to the actual average?

14. **Critical Thinking** Write a cosine function that is equivalent to $y = 3 \sin (x - \pi) + 5$.

15. **Tides** Burntcoat Head in Nova Scotia, Canada, is known for its extreme fluctuations in tides. One day in April, the first high tide rose to 13.25 feet at 4:30 A.M. The first low tide at 1.88 feet occurred at 10:51 A.M. The second high tide was recorded at 4:53 P.M.

a. Find the amplitude of a sinusoidal function that models the tides.

b. Find the vertical shift of a sinusoidal function that models the tides.

c. What is the period of a sinusoidal function that models the tides?

d. Write a sinusoidal function to model the tides, using t to represent the number of hours in decimals since midnight.

e. According to your model, determine the height of the water at 7:30 P.M.

16. **Meteorology** The table at the right contains the times that the sun rises and sets in the middle of each month in New York City, New York. Suppose the number 1 represents the middle of January, the number 2 represents the middle of February, and so on.

Month	Sunrise A.M.	Sunset P.M.
January	7:19	4:47
February	6:56	5:24
March	6:16	5:57
April	5.25	6:29
May	4:44	7:01
June	4:24	7:26
July	4:33	7:28
August	5:01	7:01
September	5:31	6:14
October	6:01	5:24
November	6:36	4:43
December	7:08	4:28

a. Find the amount of daylight hours for the middle of each month.

b. What is the amplitude of a sinusoidal function that models the daylight hours?

c. What is the vertical shift of a sinusoidal function that models the daylight hours?

d. What is the period of a sinusoidal function that models the daylight hours?

e. Write a sinusoidal function that models the daylight hours.

17. Critical Thinking The average monthly temperature for Phoenix, Arizona can be modeled by $y = 70.5 + 19.5 \sin\left(\frac{\pi}{6}t + c\right)$. If the coldest temperature occurs in January ($t = 1$), find the value of c.

18. Entertainment Several years ago, an amusement park in Sandusky, Ohio, had a ride called the Rotor in which riders stood against the walls of a spinning cylinder. As the cylinder spun, the floor of the ride dropped out, and the riders were held against the wall by the force of friction. The cylinder of the Rotor had a radius of 3.5 meters and rotated counterclockwise at a rate of 14 revolutions per minute. Suppose the center of rotation of the Rotor was at the origin of a rectangular coordinate system.

a. If the initial coordinates of the hinges on the door of the cylinder are $(0, -3.5)$, write a function that models the position of the door at t seconds.

b. Find the coordinates of the hinges on the door at 4 seconds.

19. Electricity For an alternating current, the instantaneous voltage V_R is graphed at the right. Write an equation for the instantaneous voltage.

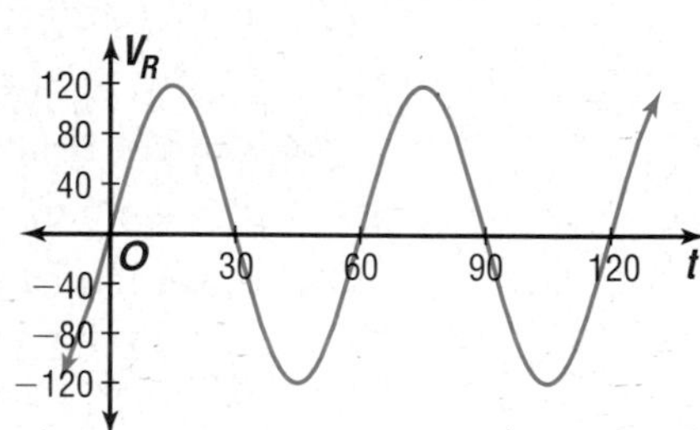

20. Meteorology Find the number of daylight hours for the middle of each month or the average monthly temperature for your community. Write a sinusoidal function to model this data.

Mixed Review

21. State the amplitude, period, phase shift, and vertical shift for $y = -3 \cos(2\theta + \pi) + 5$. Then graph the function. *(Lesson 6-5)*

22. Find the values of θ for which $\cos \theta = 1$ is true. *(Lesson 6-3)*

23. Change 800° to radians. *(Lesson 6-1)*

24. **Geometry** The sides of a parallelogram are 20 centimeters and 32 centimeters long. If the longer diagonal measures 40 centimeters, find the measures of the angles of the parallelogram. *(Lesson 5-8)*

25. Decompose $\frac{2m + 16}{m^2 - 16}$ into partial fractions. *(Lesson 4-6)*

26. Find the value of k so that the remainder of $(2x^3 + kx^2 - x - 6) \div (x + 2)$ is zero. *(Lesson 4-3)*

27. Determine the interval(s) for which the graph of $f(x) = 2|x + 1| - 5$ is increasing and the intervals for which the graph is decreasing. *(Lesson 3-5)*

28. **SAT/ACT Practice** If one half of the female students in a certain school eat in the cafeteria and one third of the male students eat there, what fractional part of the student body eats in the cafeteria?

A $\frac{5}{12}$ B $\frac{2}{5}$ C $\frac{3}{4}$ D $\frac{5}{6}$

E not enough information given

Extra Practice See p. A37.

6-7 Graphing Other Trigonometric Functions

OBJECTIVES
- Graph tangent, cotangent, secant, and cosecant functions.
- Write equations of trigonometric functions.

SECURITY A security camera scans a long, straight driveway that serves as an entrance to an historic mansion. Suppose a line is drawn down the center of the driveway. The camera is located 6 feet to the right of the midpoint of the line. Let d represent the distance along the line from its midpoint. If t is time in seconds and the camera points at the midpoint at $t = 0$, then $d = 6 \tan\left(\frac{\pi}{30}t\right)$ models the point being scanned. In this model, the distance below the midpoint is a negative. Graph the equation for $-15 \leq t \leq 15$. Find the location the camera is scanning at 5 seconds. What happens when $t = 15$? *This problem will be solved in Example 4.*

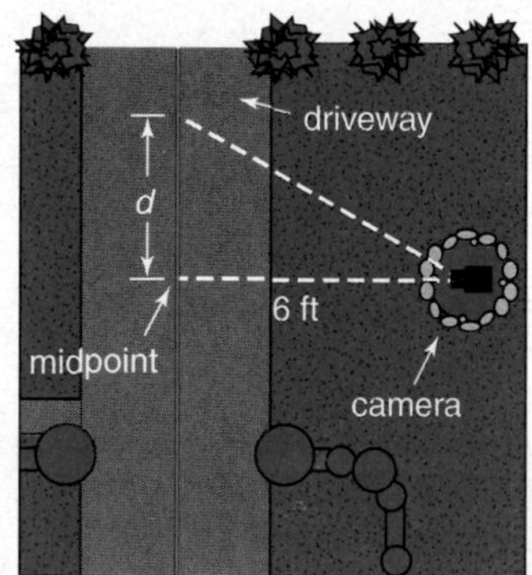

You have learned to graph variations of the sine and cosine functions. In this lesson, we will study the graphs of the tangent, cotangent, secant, and cosecant functions. Consider the tangent function. First evaluate $y = \tan x$ for multiples of $\frac{\pi}{4}$ in the interval $-\frac{3\pi}{2} \leq x \leq \frac{3\pi}{2}$.

x	$-\frac{3\pi}{2}$	$-\frac{5\pi}{4}$	$-\pi$	$-\frac{3\pi}{4}$	$-\frac{\pi}{2}$	$-\frac{\pi}{4}$	0	$\frac{\pi}{4}$	$\frac{\pi}{2}$	$\frac{3\pi}{4}$	π	$\frac{5\pi}{4}$	$\frac{3\pi}{2}$
tan x	undefined	−1	0	1	undefined	−1	0	1	undefined	−1	0	1	undefined

Look Back
You can refer to Lesson 3-7 to review asymptotes.

To graph $y = \tan x$, draw the asymptotes and plot the coordinate pairs from the table. Then draw the curves.

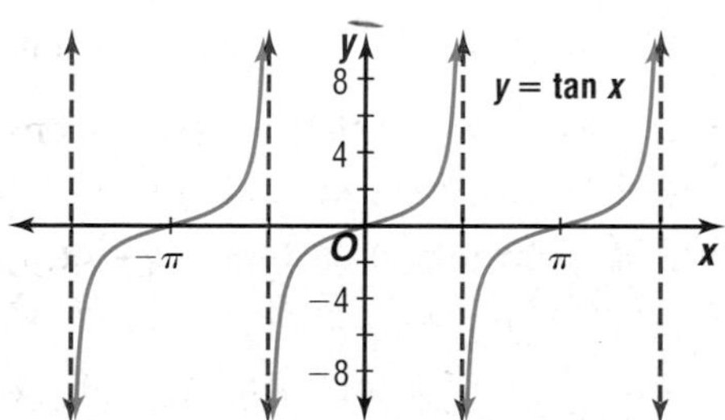

Notice that the range values for the interval $-\frac{3\pi}{2} \leq x \leq -\frac{\pi}{2}$ repeat for the intervals $-\frac{\pi}{2} \leq x \leq \frac{\pi}{2}$ and $\frac{\pi}{2} \leq x \leq \frac{3\pi}{2}$. So, the tangent function is a periodic function. Its period is π.

By studying the graph and its repeating pattern, you can determine the following properties of the graph of the tangent function.

Properties of the Graph $y = \tan x$

1. The period is π.
2. The domain is the set of real numbers except $\frac{\pi}{2}n$, where n is an odd integer.
3. The range is the set of real numbers.
4. The x-intercepts are located at πn, where n is an integer.
5. The y-intercept is 0.
6. The asymptotes are $x = \frac{\pi}{2}n$, where n is an odd integer.

Now consider the graph of $y = \cot x$ in the interval $-\pi \leq x \leq 3\pi$.

x	$-\pi$	$-\frac{3\pi}{4}$	$-\frac{\pi}{2}$	$-\frac{\pi}{4}$	0	$\frac{\pi}{4}$	$\frac{\pi}{2}$	$\frac{3\pi}{4}$	π	$\frac{5\pi}{4}$	$\frac{3\pi}{2}$	$\frac{7\pi}{4}$	2π
$\cot x$	undefined	1	0	-1	undefined	1	0	-1	undefined	1	0	-1	undefined

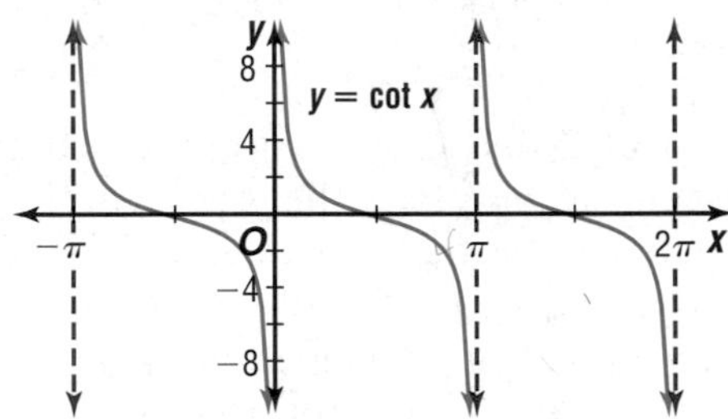

By studying the graph and its repeating pattern, you can determine the following properties of the graph of the cotangent function.

Properties of the Graph of $y = \cot x$

1. The period is π.
2. The domain is the set of real numbers except πn, where n is an integer.
3. The range is the set of real numbers.
4. The x-intercepts are located at $\frac{\pi}{2}n$, where n is an odd integer.
5. There is no y-intercept.
6. The asymptotes are $x = \pi n$, where n is an integer.

Example 1 **Find each value by referring to the graphs of the trigonometric functions.**

a. $\tan \frac{9\pi}{2}$

Since $\frac{9\pi}{2} = \frac{\pi}{2}(9)$, $\tan \frac{9\pi}{2}$ is undefined.

b. $\cot \frac{7\pi}{2}$

Since $\frac{7\pi}{2} = \frac{\pi}{2}(7)$ and 7 is an odd integer, $\cot \frac{7\pi}{2} = 0$.

The sine and cosecant functions have a reciprocal relationship. To graph the cosecant, first graph the sine function and the asymptotes of the cosecant function. By studying the graph of the cosecant and its repeating pattern, you can determine the following properties of the graph of the cosecant function.

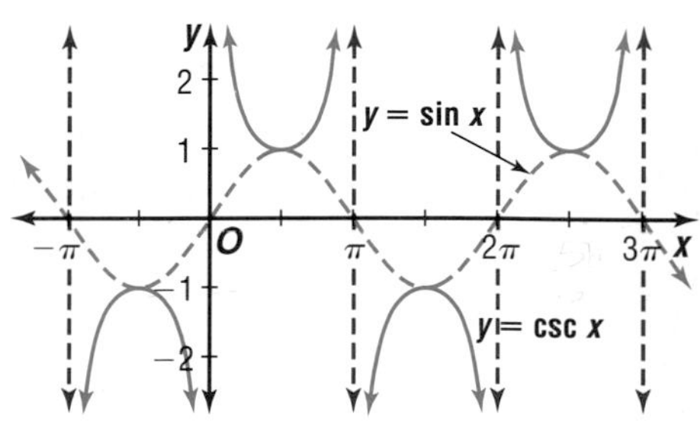

Properties of the Graph of $y = \csc x$

1. The period is 2π.
2. The domain is the set of real numbers except πn, where n is an integer.
3. The range is the set of real numbers greater than or equal to 1 or less than or equal to -1.
4. There are no x-intercepts.
5. There are no y-intercepts.
6. The asymptotes are $x = \pi n$, where n is an integer.
7. $y = 1$ when $x = \frac{\pi}{2} + 2\pi n$, where n is an integer.
8. $y = -1$ when $x = \frac{3\pi}{2} + 2\pi n$, where n is an integer.

The cosine and secant functions have a reciprocal relationship. To graph the secant, first graph the cosine function and the asymptotes of the secant function. By studying the graph and its repeating pattern, you can determine the following properties of the graph of the secant function.

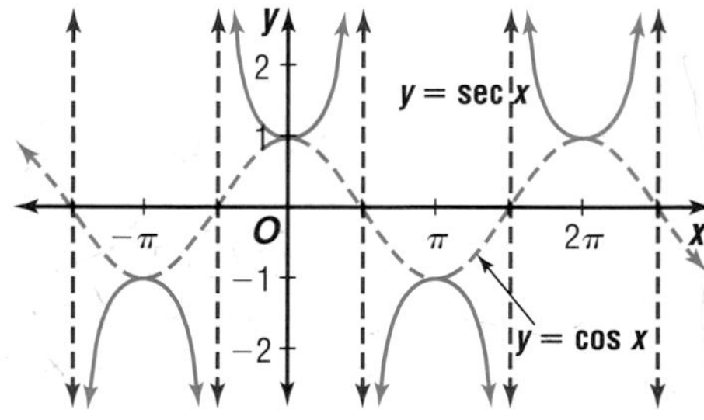

Properties of the Graph of $y = \sec x$

1. The period is 2π.
2. The domain is the set of real numbers except $\frac{\pi}{2}n$, where n is an odd integer.
3. The range is the set of real numbers greater than or equal to 1 or less than or equal to -1.
4. There are no x-intercepts.
5. The y-intercept is 1.
6. The asymptotes are $x = \frac{\pi}{2}n$, where n is an odd integer.
7. $y = 1$ when $x = \pi n$, where n is an even integer.
8. $y = -1$ when $x = \pi n$, where n is an odd integer.

Relation	Ordered Pairs	Graph	Domain	Range
$y = \cos x$	$(x, \cos x)$	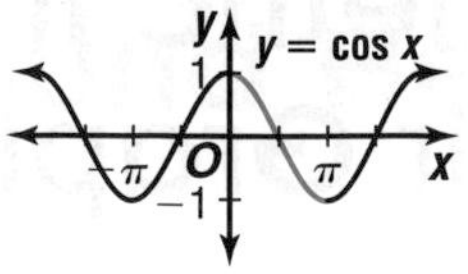 $y = \cos x$	all real numbers	$-1 \leq y \leq 1$
$y = \arccos x$	$(\cos x, x)$	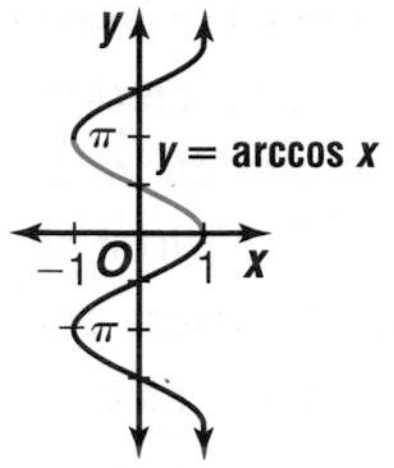 $y = \arccos x$	$-1 \leq x \leq 1$	all real numbers
$y = \tan x$	$(x, \tan x)$	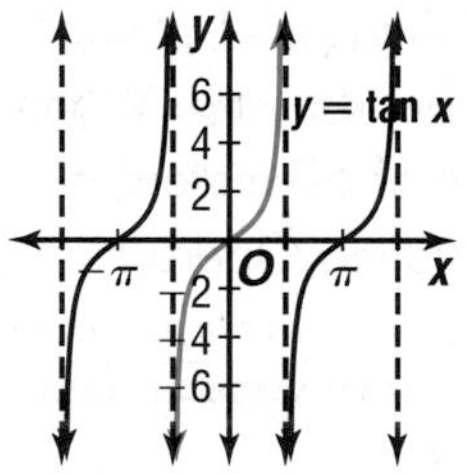 $y = \tan x$	all real numbers except $\frac{\pi}{2}n$, where n is an odd integer	all real numbers
$y = \arctan x$	$(\tan x, x)$	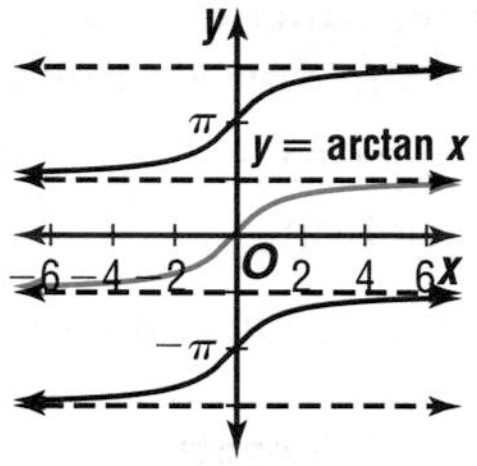 $y = \arctan x$	all real numbers	all real numbers except $\frac{\pi}{2}n$, where n is an odd integer

Notice that none of the inverses of the trigonometric functions are functions.

Consider only a part of the domain of the sine function, namely $-\frac{\pi}{2} \leq x \leq \frac{\pi}{2}$. The range then contains all of the possible values from -1 to 1. It is possible to define a new function, called Sine, whose inverse is a function.

Capital letters are used to distinguish the function with restricted domains from the usual trigonometric functions.

$$y = \text{Sin } x \text{ if and only if } y = \sin x \text{ and } -\frac{\pi}{2} \leq x \leq \frac{\pi}{2}.$$

The values in the domain of Sine are called **principal values.** Other new functions can be defined as follows.

$$y = \text{Cos } x \text{ if and only if } y = \cos x \text{ and } 0 \leq x \leq \pi.$$

$$y = \text{Tan } x \text{ if and only if } y = \tan x \text{ and } -\frac{\pi}{2} < x < \frac{\pi}{2}.$$

The graphs of $y = \text{Sin } x$, $y = \text{Cos } x$, and $y = \text{Tan } x$ are the blue portions of the graphs of $y = \sin x$, $y = \cos x$, and $y = \tan x$, respectively, shown on pages 405–406.

Note the capital "A" in the name of each inverse function.

The inverses of the Sine, Cosine, and Tangent functions are called Arcsine, Arccosine, and Arctangent, respectively. The graphs of Arcsine, Arccosine, and Arctangent are also designated in blue on pages 405–406. They are defined as follows.

Arcsine Function Given $y = \text{Sin } x$, the inverse Sine function is defined by the equation $y = \text{Sin}^{-1}\, x$ or $y = \text{Arcsin } x$.

Arccosine Function Given $y = \text{Cos } x$, the inverse Cosine function is defined by the equation $y = \text{Cos}^{-1}\, x$ or $y = \text{Arccos } x$.

Arctangent Function Given $y = \text{Tan } x$, the inverse Tangent function is defined by the equation $y = \text{Tan}^{-1}\, x$ or $y = \text{Arctan } x$.

The domain and range of these functions are summarized below.

Function	Domain	Range
$y = \text{Sin } x$	$-\frac{\pi}{2} \leq x \leq \frac{\pi}{2}$	$-1 \leq y \leq 1$
$y = \text{Arcsin } x$	$-1 \leq x \leq 1$	$-\frac{\pi}{2} \leq y \leq \frac{\pi}{2}$
$y = \text{Cos } x$	$0 \leq x \leq \pi$	$-1 \leq y \leq 1$
$y = \text{Arccos } x$	$-1 \leq x \leq 1$	$0 \leq y \leq \pi$
$y = \text{Tan } x$	$-\frac{\pi}{2} < x < \frac{\pi}{2}$	all real numbers
$y = \text{Arctan}$	all real numbers	$-\frac{\pi}{2} < y < \frac{\pi}{2}$

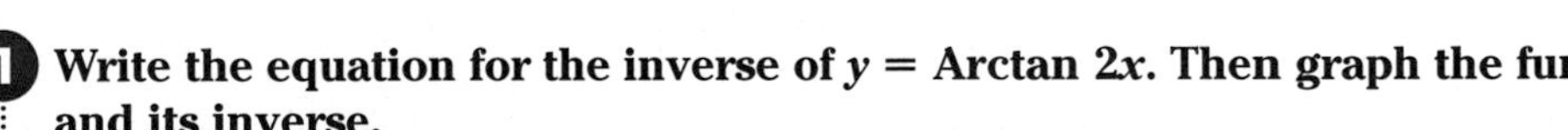

Example 1 **Write the equation for the inverse of $y = \text{Arctan } 2x$. Then graph the function and its inverse.**

$$\begin{aligned} y &= \text{Arctan } 2x \\ x &= \text{Arctan } 2y && \textit{Exchange x and y.} \\ \text{Tan } x &= 2y && \textit{Definition of Arctan function} \\ \tfrac{1}{2}\text{Tan } x &= y && \textit{Divide each side by 2.} \end{aligned}$$

Now graph the functions.

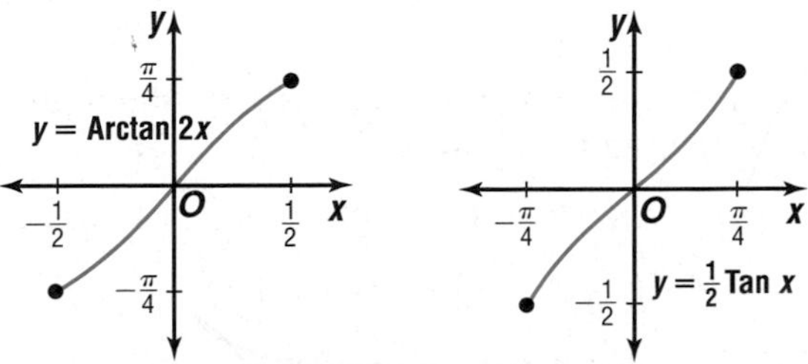

Note that the graphs are reflections of each other over the graph of $y = x$.

You can use what you know about trigonometric functions and their inverses to evaluate expressions.

Examples **2** **Find each value.**

a. $\textbf{Arcsin}\left(-\frac{\sqrt{2}}{2}\right)$

Let $\theta = \text{Arcsin}\left(-\frac{\sqrt{2}}{2}\right)$. *Think: Arcsin* $\left(-\frac{\sqrt{2}}{2}\right)$ *means that angle whose sin is* $-\frac{\sqrt{2}}{2}$.

$\text{Sin}\ \theta = -\frac{\sqrt{2}}{2}$ *Definition of Arcsin function*

$\theta = -\frac{\pi}{4}$ *Why is* θ *not* $-\frac{3\pi}{4}$*?*

b. $\textbf{Sin}^{-1}\left(\cos\frac{\pi}{2}\right)$

If $y = \cos\frac{\pi}{2}$, then $y = 0$.

$\text{Sin}^{-1}\left(\cos\frac{\pi}{2}\right) = \text{Sin}^{-1}\ 0$ *Replace* $\cos\frac{\pi}{2}$ *with 0.*

$= 0$

c. $\textbf{sin}\ (\textbf{Tan}^{-1}\ 1 - \textbf{Sin}^{-1}\ 1)$

Let $\alpha = \text{Tan}^{-1}\ 1$ and $\beta = \text{Sin}^{-1}\ 1$.

$\text{Tan}\ \alpha = 1$ $\quad$ $\text{Sin}\ \beta = 1$

$\alpha = \frac{\pi}{4}$ $\quad$ $\beta = \frac{\pi}{2}$

$$\begin{aligned}\sin\ (\text{Tan}^{-1}\ 1 - \text{Sin}^{-1}\ 1) &= \sin\ (\alpha - \beta)\\ &= \sin\left(\frac{\pi}{4} - \frac{\pi}{2}\right) && \alpha = \frac{\pi}{4}, \beta = \frac{\pi}{2}\\ &= \sin\left(-\frac{\pi}{4}\right)\\ &= -\frac{\sqrt{2}}{2}\end{aligned}$$

d. $\textbf{cos}\left[\textbf{Cos}^{-1}\left(-\frac{\sqrt{2}}{2}\right) - \frac{\pi}{2}\right]$

Let $\theta = \text{Cos}^{-1}\left(-\frac{\sqrt{2}}{2}\right)$.

$\text{Cos}\ \theta = -\frac{\sqrt{2}}{2}$ *Definition of Arccosine function*

$\theta = \frac{3\pi}{4}$

$$\begin{aligned}\cos\left[\text{Cos}^{-1}\left(-\frac{\sqrt{2}}{2}\right) - \frac{\pi}{2}\right] &= \cos\left(\theta - \frac{\pi}{2}\right)\\ &= \cos\left(\frac{3\pi}{4} - \frac{\pi}{2}\right) && \theta = \frac{3\pi}{4}\\ &= \cos\frac{\pi}{4}\\ &= \frac{\sqrt{2}}{2}\end{aligned}$$

3 **Determine if $\text{Tan}^{-1}(\tan x) = x$ is *true* or *false* for all values of x. If false, give a counterexample.**

Try several values of x to see if we can find a counterexample.

x	$\tan x$	$\text{Tan}^{-1}(\tan x)$
0	0	0
$\frac{\pi}{4}$	1	$\frac{\pi}{4}$
π	0	0

When $x = \pi$, $\text{Tan}^{-1}(\tan x) \neq x$. So $\text{Tan}^{-1}(\tan x) = x$ is not true for all values of x.

You can use a calculator to find inverse trigonometric functions. The calculator will always give the least, or principal, value of the inverse trigonometric function.

Example **4** **ENTERTAINMENT** **Refer to the application at the beginning of the lesson. When will Carla reach an altitude of 60 meters for the first time?**

First write an equation to model the height of a seat at any time t. Since the seat is at the midline point at $t = 0$, use the sine function $y = A \sin(kt + c) + h$. Find the values of A, k, c, and h.

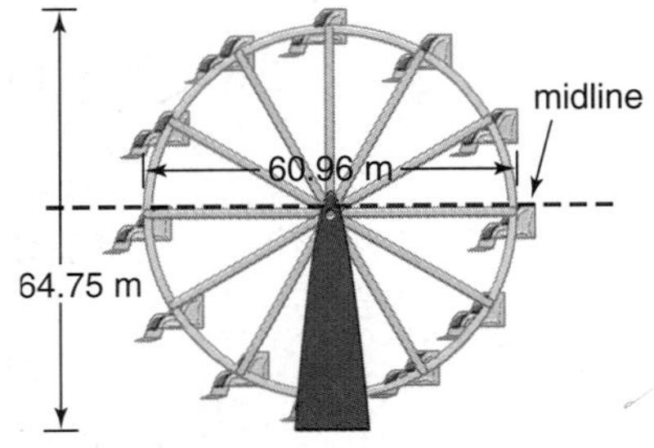

A: The value of A is the radius of the Ferris wheel.

$A = \frac{1}{2}(60.96)$ or 30.48 *The diameter is 60.96 meters.*

k: $\frac{2\pi}{k} = 4.25$ *The period is 4.25 minutes.*

$k = \frac{2\pi}{4.25}$

c: Since the seat is at the equilibrium point at $t = 0$, there is no phase shift and $c = 0$.

h: The bottom of the Ferris wheel is $64.75 - 60.96$ or 3.79 meters above the ground. So, the value of h is $30.48 + 3.79$ or 34.27.

Substitute these values into the general equation. The equation is $y = 30.48 \sin\left(\frac{2\pi}{4.25}t\right) + 34.27$. Now, solve the equation for $y = 60$.

$$60 = 30.48 \sin\left(\frac{2\pi}{4.25}t\right) + 34.27 \quad \textit{Replace y with 60.}$$

$$25.73 = 30.48 \sin\left(\frac{2\pi}{4.25}t\right) \quad \textit{Subtract 34.27 from each side.}$$

$$\frac{25.73}{30.48} = \sin\left(\frac{2\pi}{4.25}t\right) \quad \textit{Divide each side by 30.48.}$$

$$\sin^{-1}\left(\frac{25.73}{30.48}\right) = \frac{2\pi}{4.25}t \quad \textit{Definition of } \sin^{-1}$$

$$\frac{4.25}{2\pi}\sin^{-1}\left(\frac{25.73}{30.48}\right) = t \quad \textit{Multiply each side by } \frac{4.25}{2\pi}.$$

$$0.6797882017 = t \quad \textit{Use a calculator.}$$

Carla will reach an altitude of 60 meters about 0.68 minutes after 11:35 or 11:35:41.

CHECK FOR UNDERSTANDING

Communicating Mathematics

Read and study the lesson to answer each question.

1. **Compare** $y = \sin^{-1} x$, $y = (\sin x)^{-1}$, and $y = \sin (x^{-1})$.
2. **Explain** why $y = \cos^{-1} x$ is not a function.
3. **Compare and contrast** the domain and range of $y = \text{Sin } x$ and $y = \sin x$.
4. **Write** a sentence explaining how to tell if the domain of a trigonometric function is restricted.
5. **You Decide** Jake says that the period of the cosine function is 2π. Therefore, he concludes that the principal values of the domain are between 0 and 2π, inclusive. Akikta disagrees. Who is correct? Explain.

Guided Practice

Write the equation for the inverse of each function. Then graph the function and its inverse.

6. $y = \text{Arcsin } x$
7. $y = \text{Cos}\left(x + \frac{\pi}{2}\right)$

Find each value.

8. Arctan 1
9. $\cos (\text{Tan}^{-1} 1)$
10. $\cos\left[\text{Cos}^{-1}\left(\frac{\sqrt{2}}{2}\right) - \frac{\pi}{2}\right]$

Determine if each of the following is *true* or *false*. If false, give a counterexample.

11. $\sin (\text{Sin}^{-1} x) = x$ for $-1 \le x \le 1$
12. $\text{Cos}^{-1} (-x) = -\text{Cos}^{-1} x$ for $-1 \le x \le 1$

13. **Geography** Earth has been charted with vertical and horizontal lines so that points can be named with coordinates. The horizontal lines are called latitude lines. The equator is latitude line 0. Parallel lines are numbered up to $\frac{\pi}{2}$ to the north and to the south. If we assume Earth is spherical, the length of any parallel of latitude is equal to the circumference of a great circle of Earth times the cosine of the latitude angle.
    a. The radius of Earth is about 6400 kilometers. Find the circumference of a great circle.
    b. Write an equation for the circumference of any latitude circle with angle θ.
    c. Which latitude circle has a circumference of about 3593 kilometers?
    d. What is the circumference of the equator?

EXERCISES

Practice

Write the equation for the inverse of each function. Then graph the function and its inverse.

14. $y = \text{arccos } x$
15. $y = \text{Sin } x$
16. $y = \text{arctan } x$
17. $y = \text{Arccos } 2x$
18. $y = \frac{\pi}{2} + \text{Arcsin } x$
19. $y = \tan \frac{x}{2}$

20. Is $y = \text{Tan}^{-1}\left(x + \frac{\pi}{2}\right)$ the inverse of $y = \text{Tan}\left(x - \frac{\pi}{2}\right)$? Explain.

21. The principal values of the domain of the cotangent function are $0 \leq x \leq \pi$. Graph $y = \text{Cot}\ x$ and its inverse.

Find each value.

22. $\text{Sin}^{-1} 0$

23. Arccos 0

24. $\text{Tan}^{-1} \frac{\sqrt{3}}{3}$

25. $\text{Sin}^{-1}\left(\tan \frac{\pi}{4}\right)$

26. $\sin\left(2\ \text{Cos}^{-1} \frac{\sqrt{2}}{2}\right)$

27. $\cos\ (\text{Tan}^{-1} \sqrt{3})$

28. $\cos\ (\text{Tan}^{-1} 1 - \text{Sin}^{-1} 1)$

29. $\cos\left(\text{Cos}^{-1} 0 + \text{Sin}^{-1} \frac{1}{2}\right)$

30. $\sin\left(\text{Sin}^{-1} 1 - \text{Cos}^{-1} \frac{1}{2}\right)$

31. Is it possible to evaluate $\cos\left[\text{Cos}^{-1}\left(-\frac{1}{2}\right) - \text{Sin}^{-1} 2\right]$? Explain.

Determine if each of the following is *true* or *false*. If false, give a counterexample.

32. $\text{Cos}^{-1} (\cos x) = x$ for all values of x

33. $\tan\ (\text{Tan}^{-1} x) = x$ for all values of x

34. $\text{Arccos}\ x = \text{Arccos}\ (-x)$ for $-1 \leq x \leq 1$

35. $\text{Sin}^{-1} x = -\text{Sin}^{-1} (-x)$ for $-1 \leq x \leq 1$

36. $\text{Sin}^{-1} x + \text{Cos}^{-1} x = \frac{\pi}{2}$ for $-1 \leq x \leq 1$

37. $\text{Cos}^{-1} x = \frac{1}{\text{Cos}\ x}$ for all values of x

38. Sketch the graph of $y = \tan\ (\text{Tan}^{-1} x)$.

Applications and Problem Solving

39. **Meteorology** The equation $y = 54.5 + 23.5 \sin\left(\frac{\pi}{6}t - \frac{2\pi}{3}\right)$ models the average monthly temperatures of Springfield, Missouri. In this equation, t denotes the number of months with January represented by 1. During which two months is the average temperature 54.5°?

40. **Physics** The average power P of an electrical circuit with alternating current is determined by the equation $P = VI\ \text{Cos}\ \theta$, where V is the voltage, I is the current, and θ is the measure of the phase angle. A circuit has a voltage of 122 volts and a current of 0.62 amperes. If the circuit produces an average of 7.3 watts of power, find the measure of the phase angle.

41. **Critical Thinking** Consider the graphs $y = \arcsin x$ and $y = \arccos x$. Name the y coordinates of the points of intersection of the two graphs.

42. **Optics** Malus' Law describes the amount of light transmitted through two polarizing filters. If the axes of the two filters are at an angle of θ radians, the intensity I of the light transmitted through the filters is determined by the equation $I = I_0 \cos^2 \theta$, where I_0 is the intensity of the light that shines on the filters. At what angle should the axes be held so that one-eighth of the transmitted light passes through the filters?

43. Tides One day in March in Hilton Head, South Carolina, the first high tide occurred at 6:18 A.M. The high tide was 7.05 feet, and the low tide was −0.30 feet. The period for the oscillation of the tides is 12 hours and 24 minutes.

a. Determine what time the next high tide will occur.

b. Write the period of the oscillation as a decimal.

c. What is the amplitude of the sinusoidal function that models the tide?

d. If $t = 0$ represents midnight, write a sinusoidal function that models the tide.

e. At what time will the tides be at 6 feet for the first time that day?

44. Critical Thinking Sketch the graph of $y = \sin(\text{Tan}^{-1} x)$.

45. Engineering The length L of the belt around two pulleys can be determined by the equation $L = \pi D + (d - D)\,\theta + 2C \sin \theta$, where D is the diameter of the larger pulley, d is the diameter of the smaller pulley, and C is the distance between the centers of the two pulleys. In this equation, θ is measured in radians and equals $\cos^{-1} \frac{D - d}{2C}$.

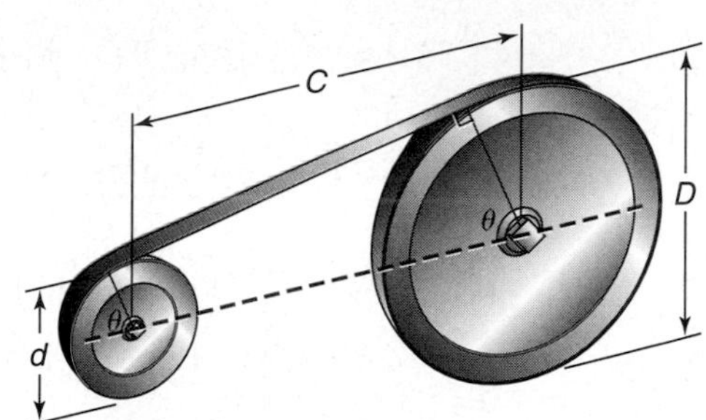

a. If $D = 6$ inches, $d = 4$ inches, and $C = 10$ inches, find θ.

b. What is the length of the belt needed to go around the two pulleys?

Mixed Review

46. What are the values of θ for which $\csc \theta$ is undefined? *(Lesson 6-7)*

47. Write an equation of a sine function with amplitude 5, period 3π, phase shift $-\pi$, and vertical shift -8. *(Lesson 6-5)*

48. Graph $y = \cos x$ for $-11\pi \le x \le -9\pi$. *(Lesson 6-3)*

49. Geometry Each side of a rhombus is 30 units long. One diagonal makes a 25° angle with a side. What is the length of each diagonal to the nearest tenth of a unit? *(Lesson 5-6)*

50. Find the measure of the reference angle for an angle of 210°. *(Lesson 5-1)*

51. List the possible rational zeros of $f(x) = 2x^3 - 9x^2 - 18x + 6$. *(Lesson 4-4)*

52. Graph $y = \frac{1}{x-2} + 3$. Determine the interval(s) for which the function is increasing and the interval(s) for which the function is decreasing. *(Lesson 3-5)*

53. Find $[f \circ g](x)$ and $[g \circ f](x)$ if $f(x) = x^3 - 1$ and $g(x) = 3x$. *(Lesson 1-2)*

54. SAT/ACT Practice Suppose every letter in the alphabet has a number value that is equal to its place in the alphabet: the letter A has a value of 1, B a value of 2, and so on. The number value of a word is obtained by adding the values of the letters in the word and then multiplying the sum by the number of letters of the word. Find the number value of the "word" *DFGH*.

A 22 **B** 44 **C** 66 **D** 100 **E** 108

Extra Practice See p. A37.

VOCABULARY

amplitude (p. 368)
angular displacement (p. 352)
angular velocity (p. 352)
central angle (p. 345)
circular arc (p. 345)
compound function (p. 382)
dimensional analysis (p. 353)
frequency (p. 372)
linear velocity (p. 353)
midline (p. 380)
period (p. 359)
periodic (p. 359)
phase shift (p. 378)
principal values (p. 406)
radian (p. 343)
sector (p. 346)
sinusoidal function (p. 388)

UNDERSTANDING AND USING THE VOCABULARY

Choose the correct term to best complete each sentence.

1. The (degree, radian) measure of an angle is defined as the length of the corresponding arc on the unit circle.
2. The ratio of the change in the central angle to the time required for the change is known as (angular, linear) velocity.
3. If the values of a function are (different, the same) for each given interval of the domain, the function is said to be periodic.
4. The (amplitude, period) of a function is one-half the difference of the maximum and minimum function values.
5. A central (angle, arc) has a vertex that lies at the center of a circle.
6. A horizontal translation of a trigonometric function is called a (phase, period) shift.
7. The length of a circular arc equals the measure of the radius of the circle times the (degree, radian) measure of the central angle.
8. The period and the (amplitude, frequency) are reciprocals of each other.
9. A function of the form $y = A \sin (k\theta + c) + h$ is a (sinusoidal, compound) function.
10. The values in the (domain, range) of Sine are called principal values.

 For additional review and practice for each lesson, visit: **www.amc.glencoe.com** 

SKILLS AND CONCEPTS

OBJECTIVES AND EXAMPLES

Lesson 6-1 Change from radian measure to degree measure, and vice versa.

- Change $-\frac{5\pi}{3}$ radians to degree measure.

$$-\frac{5\pi}{3} = \frac{5\pi}{3} \times \frac{180°}{\pi}$$
$$= -300°$$

REVIEW EXERCISES

Change each degree measure to radian measure in terms of π.

11. 60° **12.** $-75°$ **13.** 240°

Change each radian measure to degree measure. Round to the nearest tenth, if necessary.

14. $\frac{5\pi}{6}$ **15.** $-\frac{7\pi}{4}$ **16.** 2.4

Lesson 6-1 Find the length of an arc given the measure of the central angle.

- Given a central angle of $\frac{2\pi}{3}$, find the length of its intercepted arc in a circle of radius 10 inches. Round to the nearest tenth.

$s = r\theta$

$s = 10\left(\frac{2\pi}{3}\right)$

$s \approx 20.94395102$

The length of the arc is about 20.9 inches.

Given the measurement of a central angle, find the length of its intercepted arc in a circle of radius 15 centimeters. Round to the nearest tenth.

17. $\frac{3\pi}{4}$ **18.** 75°

19. 150° **20.** $\frac{\pi}{5}$

Lesson 6-2 Find linear and angular velocity.

- Determine the angular velocity if 5.2 revolutions are completed in 8 seconds. Round to the nearest tenth.

The angular displacement is $5.2 \times 2\pi$ or 10.4π radians.

$\omega = \frac{\theta}{t}$

$\omega = \frac{10.4\pi}{8}$

$\omega \approx 4.08407045$

The angular velocity is about 4.1 radians per second.

Determine each angular displacement in radians. Round to the nearest tenth.

21. 5 revolutions

22. 3.8 revolutions

23. 50.4 revolutions

24. 350 revolutions

Determine each angular velocity. Round to the nearest tenth.

25. 1.8 revolutions in 5 seconds

26. 3.6 revolutions in 2 minutes

27. 15.4 revolutions in 15 seconds

28. 50 revolutions in 12 minutes

OBJECTIVES AND EXAMPLES

Lesson 6-3 Use the graphs of the sine and cosine functions.

- Find the value of $\cos \frac{5\pi}{2}$ by referring to the graph of the cosine function.

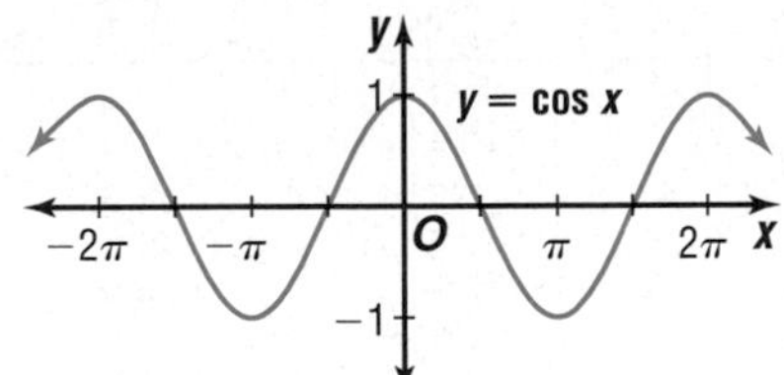

$\frac{5\pi}{2} = 2\pi + \frac{\pi}{2}$, so $\cos \frac{5\pi}{2} = \cos \frac{\pi}{2}$ or 0.

REVIEW EXERCISES

Find each value by referring to the graph of the cosine function shown at the left or sine function shown below.

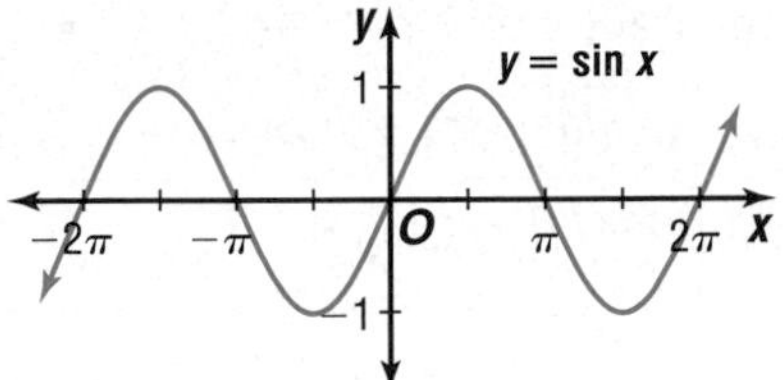

29. $\cos 5\pi$

30. $\sin 13\pi$

31. $\sin \frac{9\pi}{2}$

32. $\cos \left(-\frac{7\pi}{2}\right)$

Lesson 6-4 Find the amplitude and period for sine and cosine functions.

- State the amplitude and period for $y = -\frac{3}{4} \cos 2\theta$.

The amplitude of $y = A \cos k\theta$ is $|A|$. Since $A = -\frac{3}{4}$, the amplitude is $\left|-\frac{3}{4}\right|$ or $\frac{3}{4}$.
Since $k = 2$, the period is $\frac{2\pi}{2}$ or π.

State the amplitude and period for each function. Then graph each function.

33. $y = 4 \cos 2\theta$

34. $y = 0.5 \sin 4\theta$

35. $y = -\frac{1}{3} \cos \frac{\theta}{2}$

Lesson 6-5 Write equations of sine and cosine functions, given the amplitude, period, phase shift, and vertical translation.

- Write an equation of a cosine function with an amplitude 2, period 2π, phase shift $-\pi$, and vertical shift 2.

A: $|A| = 2$, so $A = 2$ or -2.

k: $\frac{2\pi}{k} = 2\pi$, so $k = 1$.

c: $-\frac{c}{k} = -\pi$, so $-c = -\pi$ or $c = \pi$.

h: $h = 2$

Substituting into $y = A \sin (k\theta + c) + h$, the possible equations are $y = \pm 2 \cos (\theta + \pi) + 2$.

36. Write an equation of a sine function with an amplitude 4, period $\frac{\pi}{2}$, phase shift -2π, and vertical shift -1.

37. Write an equation of a sine function with an amplitude 0.5, period π, phase shift $\frac{\pi}{3}$, and vertical shift 3.

38. Write an equation of a cosine function with an amplitude $\frac{3}{4}$, period $\frac{\pi}{4}$, phase shift 0, and vertical shift 5.

OBJECTIVES AND EXAMPLES

Lesson 6-6 Use sinusoidal functions to solve problems.

- A sinsusoidal function can be any function of the form

  $y = A \sin (k\theta + c) + h$ or

  $y = A \cos (k\theta + c) + h$.

REVIEW EXERCISES

Suppose a person's blood pressure oscillates between the two numbers given. If the heart beats once every second, write a sine function that models this person's blood pressure.

39. 120 and 80

40. 130 and 100

Lesson 6-7 Graph tangent, cotangent, secant, and cosecant functions.

- Graph $y = \tan 0.5\theta$.

  The period of this function is 2π. The phase shift is 0, and the vertical shift is 0.

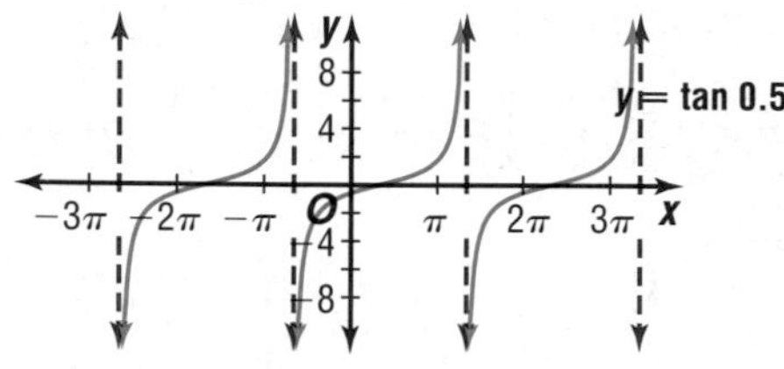

Graph each function.

41. $y = \frac{1}{3} \csc \theta$

42. $y = 2 \tan \left(3\theta + \frac{\pi}{2}\right)$

43. $y = \sec \theta + 4$

44. $y = \tan \theta - 2$

Lesson 6-8 Find the principal values of inverse trigonometric functions.

- Find $\cos (\text{Tan}^{-1} 1)$.

  Let $\alpha = \text{Tan}^{-1} 1$.

  $\text{Tan } \alpha = 1$

  $\alpha = \frac{\pi}{4}$

  $\cos \frac{\pi}{4} = \frac{\sqrt{2}}{2}$

Find each value.

45. $\text{Arctan} \left(-1\right)$

46. $\text{Sin}^{-1} 1$

47. $\text{Cos}^{-1} \left(\tan \frac{\pi}{4}\right)$

48. $\sin \left(\text{Sin}^{-1} \frac{\sqrt{3}}{2}\right)$

49. $\cos \left(\text{Arctan } \sqrt{3} + \text{Arcsin } \frac{1}{2}\right)$

APPLICATIONS AND PROBLEM SOLVING

50. **Meteorology** The mean average temperature in a certain town is 64°F. The temperature fluctuates 11.5° above and below the mean temperature. If $t = 1$ represents January, the phase shift of the sine function is 3. *(Lesson 6-6)*

   a. Write a model for the average monthly temperature in the town.

   b. According to your model, what is the average temperature in April?

   c. According to your model, what is the average temperature in July?

51. **Physics** The strength of a magnetic field is called magnetic induction. An equation for *magnetic induction* is $B = \frac{F}{IL \sin \theta}$, where F is a force on a current I which is moving through a wire of length L at an angle θ to the magnetic field. A wire within a magnetic field is 1 meter long and carries a current of 5.0 amperes. The force on the wire is 0.2 newton, and the magnetic induction is 0.04 newton per ampere-meter. What is the angle of the wire to the magnetic field? *(Lesson 6-8)*

ALTERNATIVE ASSESSMENT

OPEN-ENDED ASSESSMENT

1. The area of a circular sector is about 26.2 square inches. What are possible measures for the radius and the central angle of the sector?

2. a. You are given the graph of a cosine function. Explain how you can tell if the graph has been translated. Sketch two graphs as part of your explanation.

   b. You are given the equation of a cosine function. Explain how you can tell if the graph has been translated. Provide two equations as part of your explanation.

Additional Assessment See p. A61 for Chapter 6 practice test.

THE CYBERCLASSROOM

What Is Your Sine?

- Search the Internet to find web sites that have applications of the sine or cosine function. Find at least three different sources of information.
- Select one of the applications of the sine or cosine function. Use the Internet to find actual data that can be modeled by a graph that resembles the sine or cosine function.
- Draw a sine or cosine model of the data. Write an equation for a sinusoidal function that fits your data.

PORTFOLIO

Choose a trigonometric function you studied in this chapter. Graph your function. Write three expressions whose values can be found using your graph. Find the values of these expressions.

Trigonometry Problems

Each ACT exam contains exactly four trigonometry problems. The SAT has none! You'll need to know the trigonometric functions in a right triangle.

$$\sin\theta = \frac{\text{opposite}}{\text{hypotenuse}} \qquad \cos\theta = \frac{\text{adjacent}}{\text{hypotenuse}} \qquad \tan\theta = \frac{\text{opposite}}{\text{adjacent}}$$

Review the reciprocal functions.

$$\csc\theta = \frac{1}{\sin\theta} \qquad \sec\theta = \frac{1}{\cos\theta} \qquad \cot\theta = \frac{1}{\tan\theta}$$

Review the graphs of trigonometric functions.

THE PRINCETON REVIEW

TEST-TAKING TIP

Use the memory aid SOH-CAH-TOA. Pronounce it as *so-ca-to-a.*

SOH represents **S**ine (is) **O**pposite (over) **H**ypotenuse

CAH represents **C**osine (is) **A**djacent (over) **H**ypotenuse

TOA represents **T**angent (is) **O**pposite (over) **A**djacent

ACT EXAMPLE

1. If $\sin\theta = \frac{1}{2}$ and $90° < \theta < 180°$, then $\theta = ?$

   A 100°

   B 120°

   C 130°

   D 150°

   E 160°

HINT Memorize the sine, cosine, and tangent of special angles 0°, 30°, 45°, 60°, and 90°.

Solution Draw a diagram. Use the quadrant indicated by the size of angle θ.

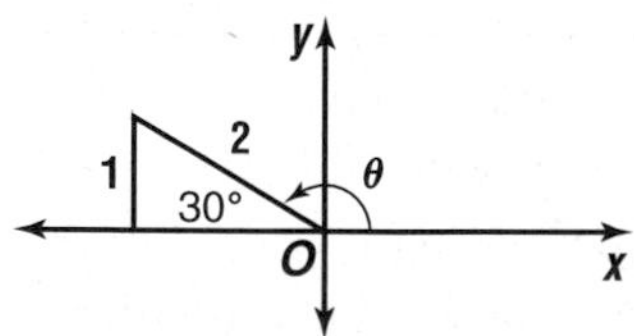

Recall that the $\sin 30° = \frac{1}{2}$. The angle inside the triangle is 30°. Then $\theta + 30° = 180°$.

If $\theta + 30° = 180°$, then $\theta = 150°$.

The answer is choice **D**.

ACT EXAMPLE

2. What is the least positive value for x where $y = \sin 4x$ reaches its maximum?

   A $\frac{\pi}{8}$

   B $\frac{\pi}{4}$

   C $\frac{\pi}{2}$

   D π

   E 2π

HINT Review the graphs of the sine and cosine functions.

Solution The least value for x where $y = \sin x$ reaches its maximum is $\frac{\pi}{2}$. If $4x = \frac{\pi}{2}$, then $x = \frac{\pi}{8}$. The answer is choice **A**.

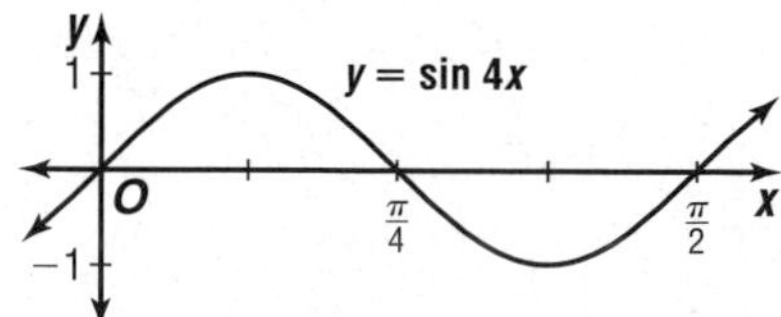

After you work each problem, record your answer on the answer sheet provided or on a piece of paper.

Multiple Choice

1. What is $\sin \theta$, if $\tan \theta = \frac{4}{3}$?

   A $\frac{3}{4}$ B $\frac{4}{5}$
   C $\frac{5}{4}$ D $\frac{5}{3}$
   E $\frac{7}{3}$

2. If the sum of two consecutive odd integers is 56, then the greater integer equals:

   A 25 B 27 C 29
   D 31 E 33

3. For all θ where $\sin \theta - \cos \theta \neq 0$, $\frac{\sin^2 \theta - \cos^2 \theta}{\sin \theta - \cos \theta}$ is equivalent to

   A $\sin \theta - \cos \theta$ B $\sin \theta + \cos \theta$
   C $\tan \theta$ D -1
   E 1

4. In the figure below, side AB of triangle ABC contains which point?

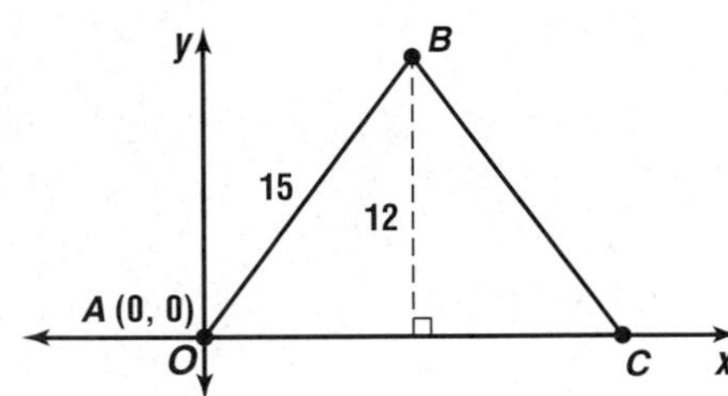

   A (3, 2) B (3, 5)
   C (4, 6) D (4, 10)
   E (6, 8)

5. Which of the following is the sum of both solutions of the equation $x^2 - 2x - 8 = 0$?

   A -6 B -4 C -2
   D 2 E 6

6. In the figure below, $\angle A$ is a right angle, AB is 3 units long, and BC is 5 units long. If $\angle C = \theta$, what is the value of $\cos \theta$?

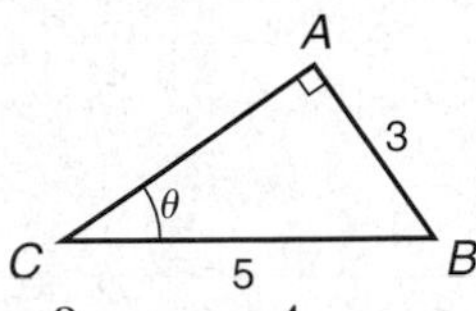

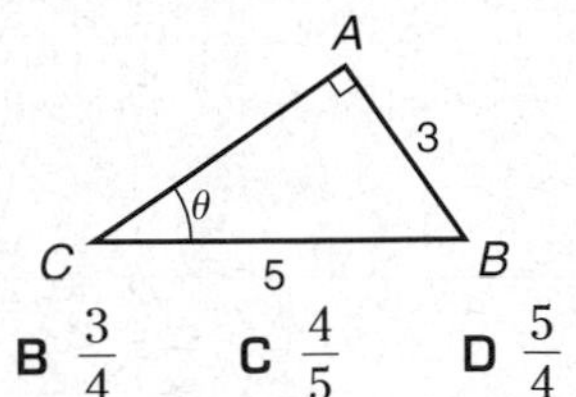

   A $\frac{3}{5}$ B $\frac{3}{4}$ C $\frac{4}{5}$ D $\frac{5}{4}$ E $\frac{5}{3}$

7. The equation $x - 7 = x^2 + y$ represents which conic?

   A parabola B circle C ellipse
   D hyperbola E line

8. If n is an integer, then which of the following must also be integers?

   I. $\frac{16n + 16}{n + 1}$

   II. $\frac{16n + 16}{16n}$

   III. $\frac{16n^2 + n}{16n}$

   A I only B II only C III only
   D I and II E II and III

9. **Quantitative Comparison**

   A if the quantity in Column A is greater
   B if the quantity in Column B is greater
   C if the two quantities are equal
   D if the relationship cannot be determined from the information given

   $x \geq 1$

Column A	Column B
$x^{(x+2)}$	$(x + 2)^x$

10. **Grid-In** In the figure, segment AD bisects $\angle BAC$, and segment DC bisects $\angle BCA$. If the measure of $\angle ADC = 100°$, then what is the measure of $\angle B$?

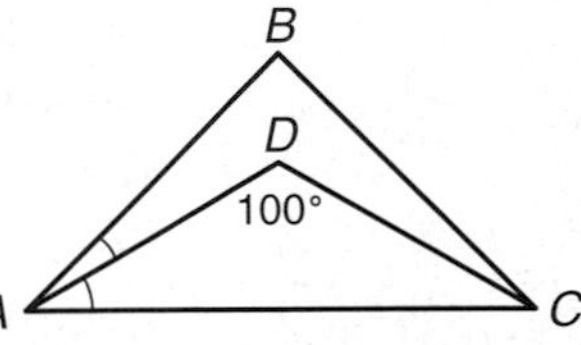

SAT/ACT Practice For additional test practice questions, visit: **www.amc.glencoe.com**

TRIGONOMETRIC IDENTITIES AND EQUATIONS

CHAPTER OBJECTIVES

- **Use reciprocal, quotient, Pythagorean, symmetry, and opposite-angle identities.** *(Lesson 7-1)*
- **Verify trigonometric identities.** *(Lessons 7-2, 7-3, 7-4)*
- **Use sum, difference, double-angle, and half-angle identities.** *(Lessons 7-3, 7-4)*
- **Solve trigonometric equations and inequalities.** *(Lesson 7-5)*
- **Write a linear equation in normal form.** *(Lesson 7-6)*
- **Find the distance from a point to a line.** *(Lesson 7-7)*

7-1

Basic Trigonometric Identities

OBJECTIVE

- Identify and use reciprocal identities, quotient identities, Pythagorean identities, symmetry identities, and opposite-angle identities.

OPTICS Many sunglasses have polarized lenses that reduce the intensity of light. When unpolarized light passes through a polarized lens, the intensity of the light is cut in half. If the light then passes through another polarized lens with its axis at an angle of θ to the first, the intensity of the light is again diminished.

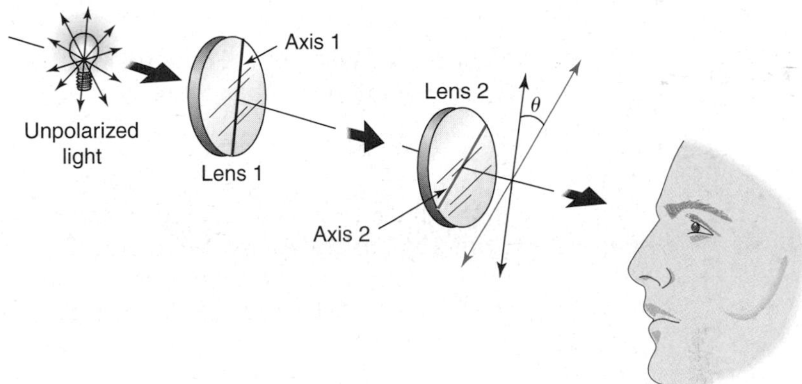

The intensity of the emerging light can be found by using the formula $I = I_0 - \frac{I_0}{\csc^2 \theta}$, where I_0 is the intensity of the light incoming to the second polarized lens, I is the intensity of the emerging light, and θ is the angle between the axes of polarization. Simplify this expression and determine the intensity of light emerging from a polarized lens with its axis at a 30° angle to the original. *This problem will be solved in Example 5.*

In algebra, variables and constants usually represent real numbers. The values of trigonometric functions are also real numbers. Therefore, the language and operations of algebra also apply to trigonometry. Algebraic expressions involve the operations of addition, subtraction, multiplication, division, and exponentiation. These operations are used to form trigonometric expressions. Each expression below is a trigonometric expression.

$$\cos x - x \qquad \sin^2 a + \cos^2 a \qquad \frac{1 - \sec A}{\tan A}$$

A statement of equality between two expressions that is true for *all* values of the variable(s) for which the expressions are defined is called an **identity.** For example, $x^2 - y^2 = (x - y)(x + y)$ is an algebraic identity. An identity involving trigonometric expressions is called a **trigonometric identity.**

If you can show that a specific value of the variable in an equation makes the equation false, then you have produced a *counterexample.* It only takes one counterexample to prove that an equation is not an identity.

Example 1 **Prove that $\sin x \cos x = \tan x$ is *not* a trigonometric identity by producing a counterexample.**

Suppose $x = \frac{\pi}{4}$.

$\sin x \cos x \stackrel{?}{=} \tan x$

$\sin \frac{\pi}{4} \cos \frac{\pi}{4} \stackrel{?}{=} \tan \frac{\pi}{4}$ *Replace x with* $\frac{\pi}{4}$.

$\left(\frac{\sqrt{2}}{2}\right)\left(\frac{\sqrt{2}}{2}\right) \stackrel{?}{=} 1$

$\frac{1}{2} \neq 1$

Since evaluating each side of the equation for the same value of x produces an inequality, the equation is not an identity.

Although producing a counterexample can show that an equation is not an identity, proving that an equation is an identity generally takes more work. Proving that an equation is an identity requires showing that the equality holds for *all* values of the variable where each expression is defined. Several fundamental trigonometric identities can be verified using geometry.

Recall from Lesson 5-3 that the trigonometric functions can be defined using the unit circle. From the unit circle, $\sin \theta = \frac{y}{1}$, or y and $\csc \theta = \frac{1}{y}$. That is, $\sin \theta = \frac{1}{\csc \theta}$. Identities derived in this manner are called **reciprocal identities.**

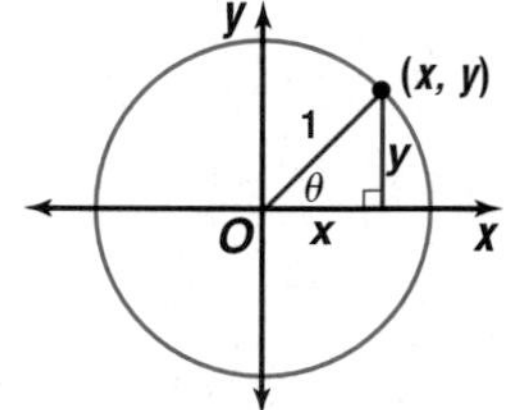

Reciprocal Identities

The following trigonometric identities hold for all values of θ where each expression is defined.

$\sin \theta = \frac{1}{\csc \theta}$ $\quad \cos \theta = \frac{1}{\sec \theta}$

$\csc \theta = \frac{1}{\sin \theta}$ $\quad \sec \theta = \frac{1}{\cos \theta}$

$\tan \theta = \frac{1}{\cot \theta}$ $\quad \cot \theta = \frac{1}{\tan \theta}$

Returning to the unit circle, we can say that $\frac{\sin \theta}{\cos \theta} = \frac{y}{x} = \tan \theta$. This is an example of a **quotient identity.**

Quotient Identities

The following trigonometric identities hold for all values of θ where each expression is defined.

$\frac{\sin \theta}{\cos \theta} = \tan \theta$ $\quad \frac{\cos \theta}{\sin \theta} = \cot \theta$

Since the triangle in the unit circle on the previous page is a right triangle, we may apply the Pythagorean Theorem: $y^2 + x^2 = 1^2$, or $\sin^2 \theta + \cos^2 \theta = 1$. Other identities can be derived from this one.

$$\sin^2 \theta + \cos^2 \theta = 1$$

$$\frac{\sin^2 \theta}{\cos^2 \theta} + \frac{\cos^2 \theta}{\cos^2 \theta} = \frac{1}{\cos^2 \theta} \quad \textit{Divide each side by } \cos^2 \theta.$$

$$\tan^2 \theta + 1 = \sec^2 \theta \quad \textit{Quotient and reciprocal identities}$$

Likewise, the identity $1 + \cot^2 \theta = \csc^2 \theta$ can be derived by dividing each side of the equation $\sin^2 \theta + \cos^2 \theta = 1$ by $\sin^2 \theta$. These are the **Pythagorean identities.**

Pythagorean Identities

The following trigonometric identities hold for all values of θ where each expression is defined.

$\sin^2 \theta + \cos^2 \theta = 1$ $\quad$ $\tan^2 \theta + 1 = \sec^2 \theta$ $\quad$ $1 + \cot^2 \theta = \csc^2 \theta$

You can use the identities to help find the values of trigonometric functions.

Example 2 **Use the given information to find the trigonometric value.**

a. If $\sec \theta = \frac{3}{2}$, find $\cos \theta$.

$$\cos \theta = \frac{1}{\sec \theta} \quad \textit{Choose an identity that involves } \cos \theta \textit{ and } \sec \theta.$$

$$= \frac{1}{\frac{3}{2}} \text{ or } \frac{2}{3} \quad \textit{Substitute } \frac{3}{2} \textit{ for } \sec \theta \textit{ and evaluate.}$$

b. If $\csc \theta = \frac{4}{3}$, find $\tan \theta$.

Since there are no identities relating $\csc \theta$ and $\tan \theta$, we must use two identities, one relating $\csc \theta$ and $\cot \theta$ and another relating $\cot \theta$ and $\tan \theta$.

$$\csc^2 \theta = 1 + \cot^2 \theta \quad \textit{Pythagorean identity}$$

$$\left(\frac{4}{3}\right)^2 = 1 + \cot^2 \theta \quad \textit{Substitute } \frac{4}{3} \textit{ for } \csc \theta.$$

$$\frac{16}{9} = 1 + \cot^2 \theta$$

$$\frac{7}{9} = \cot^2 \theta$$

$$\pm\frac{\sqrt{7}}{3} = \cot \theta \quad \textit{Take the square root of each side.}$$

Now find $\tan \theta$.

$$\tan \theta = \frac{1}{\cot \theta} \quad \textit{Reciprocal identity}$$

$$= \pm\frac{3\sqrt{7}}{7}, \text{ or about } \pm 1.134$$

To determine the sign of a function value, you need to know the quadrant in which the angle terminates. The signs of function values in different quadrants are related according to the symmetries of the unit circle. Since we can determine the values of tan A, cot A, sec A, and csc A in terms of sin A and/or cos A with the reciprocal and quotient identities, we only need to investigate sin A and cos A.

Case	Relationship between angles A and B	Diagram	Conclusion
1	The angles differ by a multiple of 360°. $B - A = 360k°$ or $B = A + 360\,k°$	$A + 360k°$; y; (a, b); A; O; x	Since A and $A + 360k°$ are coterminal, they share the same value of sine and cosine.
2	The angles differ by an odd multiple of 180°. $B - A = 180°(2k - 1)$ or $B = A + 180°(2k - 1)$	$A + 180°\,(2k - 1)$; y; (a, b); A; O; x; $(-a, -b)$	Since A and $A + 180°(2k - 1)$ have terminal sides in diagonally opposite quadrants, the values of both sine and cosine change sign.
3	The sum of the angles is a multiple of 360°. $A + B = 360k°$ or $B = 360k° - A$	$360k° - A$; y; (a, b); A; O; x; $(-a, -b)$	Since A and $360k° - A$ lie in vertically adjacent quadrants, the sine values are opposite but the cosine values are the same.
4	The sum of the angles is an odd multiple of 180°. $A + B = 180°\,(2k - 1)$ or $B = 180°(2k - 1) - A$	$(-a, b)$; y; (a, b); $180°\,(2k - 1) - A$; A; O; x	Since A and $180°(2k - 1) - A$ lie in horizontally adjacent quadrants, the sine values are the same but the cosine values are opposite.

These general rules for sine and cosine are called **symmetry identities.**

Symmetry Identities The following trigonometric identities hold for any integer k and all values of A.

Case 1:	$\sin(A + 360k°) = \sin A$	$\cos(A + 360k°) = \cos A$
Case 2:	$\sin(A + 180°(2k - 1)) = -\sin A$	$\cos(A + 180°(2k - 1)) = -\cos A$
Case 3:	$\sin(360k° - A) = -\sin A$	$\cos(360k° - A) = \cos A$
Case 4:	$\sin(180°(2k - 1) - A) = \sin A$	$\cos(180°(2k - 1) - A) = -\cos A$

To use the symmetry identities with radian measure, replace 180° with π and 360° with 2π.

Example 3 **Express each value as a trigonometric function of an angle in Quadrant I.**

a. sin 600°

Relate 600° to an angle in Quadrant I.

$600° = 60° + 3(180°)$ *600° and 60° differ by an odd multiple of 180°.*

$\sin 600° = \sin(60° + 3(180°))$ *Case 2, with A = 60° and k = 2*

$= -\sin 60°$

b. $\sin \frac{19\pi}{4}$

The sum of $\frac{19\pi}{4}$ and $\frac{\pi}{4}$, which is $\frac{20\pi}{4}$ or 5π, is an odd multiple of π.

$\frac{19\pi}{4} = 5\pi - \frac{\pi}{4}$

$\sin \frac{19\pi}{4} = \sin\left(5\pi - \frac{\pi}{4}\right)$ *Case 4, with $A = \frac{\pi}{4}$ and k = 3*

$= \sin \frac{\pi}{4}$

c. cos (−410°)

The sum of −410° and 50° is a multiple of 360°.

$-410° = -360° - 50°$

$\cos(-410°) = \cos(-360° - 50°)$ *Case 3, with A = 50° and k = −1*

$= \cos 50°$

d. $\tan \frac{37\pi}{6}$

$\frac{37\pi}{6}$ and $\frac{\pi}{6}$ differ by a multiple of 2π.

$\frac{37\pi}{6} = 3(2\pi) + \frac{\pi}{6}$ *Case 1, with $A = \frac{\pi}{6}$ and k = 3*

$\tan \frac{37\pi}{6} = \frac{\sin \frac{37\pi}{6}}{\cos \frac{37\pi}{6}}$ *Rewrite using a quotient identity since the symmetry identities are in terms of sine and cosine.*

$= \frac{\sin\left(3(2\pi) + \frac{\pi}{6}\right)}{\cos\left(3(2\pi) + \frac{\pi}{6}\right)}$

$= \frac{\sin \frac{\pi}{6}}{\cos \frac{\pi}{6}}$ or $\tan \frac{\pi}{6}$ *Quotient identity*

Case 3 of the Symmetry Identities can be written as the **opposite-angle identities** when $k = 0$.

Opposite-Angle Identities

The following trigonometric identities hold for all values of A.

$$\sin(-A) = -\sin A$$

$$\cos(-A) = \cos A$$

The basic trigonometric identities can be used to simplify trigonometric expressions. Simplifying a trigonometric expression means that the expression is written using the fewest trigonometric functions possible and as algebraically simplified as possible. This may mean writing the expression as a numerical value.

Examples

4 **Simplify $\sin x + \sin x \cot^2 x$.**

$\sin x + \sin x \cot^2 x = \sin x\,(1 + \cot^2 x)$ *Factor.*

$= \sin x \csc^2 x$ *Pythagorean identity: $1 + \cot^2 x = \csc^2 x$*

$= \sin x \cdot \dfrac{1}{\sin^2 x}$ *Reciprocal identity*

$= \dfrac{1}{\sin x}$

$= \csc x$ *Reciprocal identity*

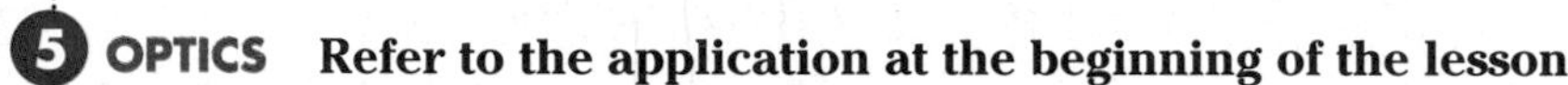

5 **OPTICS** **Refer to the application at the beginning of the lesson.**

a. Simplify the formula $I = I_0 - \dfrac{I_0}{\csc^2 \theta}$.

b. Use the simplified formula to determine the intensity of light that passes through a second polarizing lens with axis at 30° to the original.

a. $I = I_0 - \dfrac{I_0}{\csc^2 \theta}$

$I = I_0 - I_0 \sin^2 \theta$ *Reciprocal identity*

$I = I_0(1 - \sin^2 \theta)$ *Factor.*

$I = I_0 \cos^2 \theta$ *$1 - \sin^2 \theta = \cos^2 \theta$*

b. $I = I_0 \cos^2 30°$

$I = I_0\left(\dfrac{\sqrt{3}}{2}\right)^2$

$I = \dfrac{3}{4} I_0$

The light has three-fourths the intensity it had before passing through the second polarizing lens.

CHECK FOR UNDERSTANDING

Communicating Mathematics

Read and study the lesson to answer each question.

1. **Find a counterexample** to show that the equation $1 - \sin x = \cos x$ is not an identity.
2. **Explain** why the Pythagorean and opposite-angle identities are so named.
3. **Write** two reciprocal identities, one quotient identity, and one Pythagorean identity, each of which involves $\cot \theta$.
4. **Prove** that $\tan(-A) = -\tan A$ using the quotient and opposite-angle identities.
5. **You Decide** Claude and Rosalinda are debating whether an equation from their homework assignment is an identity. Claude says that since he has tried ten specific values for the variable and all of them worked, it must be an identity. Rosalinda explained that specific values could only be used as counterexamples to prove that an equation is not an identity. Who is correct? Explain your answer.

Guided Practice

Prove that each equation is *not* a trigonometric identity by producing a counterexample.

6. $\sin \theta + \cos \theta = \tan \theta$
7. $\sec^2 x + \csc^2 x = 1$

Use the given information to determine the exact trigonometric value.

8. $\cos \theta = \frac{2}{3}, 0° < \theta < 90°$; $\sec \theta$
9. $\cot \theta = -\frac{\sqrt{5}}{2}, \frac{\pi}{2} < \theta < \pi$; $\tan \theta$
10. $\sin \theta = -\frac{1}{5}, \pi < \theta < \frac{3\pi}{2}$; $\cos \theta$
11. $\tan \theta = -\frac{4}{7}, 270° < \theta < 360°$; $\sec \theta$

Express each value as a trigonometric function of an angle in Quadrant I.

12. $\cos \frac{7\pi}{3}$
13. $\csc(-330°)$

Simplify each expression.

14. $\frac{\csc \theta}{\cot \theta}$
15. $\cos x \csc x \tan x$
16. $\cos x \cot x + \sin x$

17. **Physics** When there is a current in a wire in a magnetic field, a force acts on the wire. The strength of the magnetic field can be determined using the formula $B = \frac{F \csc \theta}{I\ell}$, where F is the force on the wire, I is the current in the wire, ℓ is the length of the wire, and θ is the angle the wire makes with the magnetic field. Physics texts often write the formula as $F = I\ell B \sin \theta$. Show that the two formulas are equivalent.

EXERCISES

Practice

Prove that each equation is not a trigonometric identity by producing a counterexample.

18. $\sin \theta \cos \theta = \cot \theta$
19. $\frac{\sec \theta}{\tan \theta} = \sin \theta$
20. $\sec^2 x - 1 = \frac{\cos x}{\csc x}$
21. $\sin x + \cos x = 1$
22. $\sin y \tan y = \cos y$
23. $\tan^2 A + \cot^2 A = 1$

24. Find a value of θ for which $\cos\left(\theta + \frac{\pi}{2}\right) \neq \cos \theta + \cos \frac{\pi}{2}$.

Use the given information to determine the exact trigonometric value.

25. $\sin \theta = \frac{2}{5}$, $0° < \theta < 90°$; $\csc \theta$

26. $\tan \theta = \frac{\sqrt{3}}{4}$, $0 < \theta < \frac{\pi}{2}$; $\cot \theta$

27. $\sin \theta = \frac{1}{4}$, $0 < \theta < \frac{\pi}{2}$; $\cos \theta$

28. $\cos \theta = -\frac{2}{3}$, $90° < \theta < 180°$; $\sin \theta$

29. $\csc \theta = \frac{\sqrt{11}}{3}$, $\frac{\pi}{2} < \theta < \pi$; $\cot \theta$

30. $\sec \theta = -\frac{5}{4}$, $90° < \theta < 180°$; $\tan \theta$

31. $\sin \theta = -\frac{1}{3}$, $180° < \theta < 270°$; $\tan \theta$

32. $\tan \theta = \frac{2}{3}$, $\pi < \theta < \frac{3\pi}{2}$; $\cos \theta$

33. $\sec \theta = -\frac{7}{5}$, $180° < \theta < 270°$; $\sin \theta$

34. $\cos \theta = \frac{1}{8}$, $\frac{3\pi}{2} < \theta < 2\pi$; $\tan \theta$

35. $\cot \theta = -\frac{4}{3}$, $270° < \theta < 360°$; $\sin \theta$

36. $\cot \theta = -8$, $\frac{3\pi}{2} < \theta < 2\pi$; $\csc \theta$

37. If A is a second quadrant angle, and $\cos A = -\frac{\sqrt{3}}{4}$, find $\frac{\sec^2 A - \tan^2 A}{2 \sin^2 A + 2 \cos^2 A}$.

Express each value as a trigonometric function of an angle in Quadrant I.

38. $\sin 390°$

39. $\cos \frac{27\pi}{8}$

40. $\tan \frac{19\pi}{5}$

41. $\csc \frac{10\pi}{3}$

42. $\sec (-1290°)$

43. $\cot (-660°)$

Simplify each expression.

44. $\frac{\sec x}{\tan x}$

45. $\frac{\cot \theta}{\cos \theta}$

46. $\frac{\sin (\theta + \pi)}{\cos (\theta - \pi)}$

47. $(\sin x + \cos x)^2 + (\sin x - \cos x)^2$

48. $\sin x \cos x \sec x \cot x$

49. $\cos x \tan x + \sin x \cot x$

50. $(1 + \cos \theta)(\csc \theta - \cot \theta)$

51. $1 + \cot^2 \theta - \cos^2 \theta - \cos^2 \theta \cot^2 \theta$

52. $\frac{\sin x}{1 + \cos x} + \frac{\sin x}{1 - \cos x}$

53. $\cos^4 \alpha + 2 \cos^2 \alpha \sin^2 \alpha + \sin^4 \alpha$

Applications and Problem Solving

54. **Optics** Refer to the equation derived in Example 5. What angle should the axes of two polarizing lenses make in order to block all light from passing through?

55. **Critical Thinking** Use the unit circle definitions of sine and cosine to provide a geometric interpretation of the opposite-angle identities.

56. **Dermatology** It has been shown that skin cancer is related to sun exposure. The rate W at which a person's skin absorbs energy from the sun depends on the energy S, in watts per square meter, provided by the sun, the surface area A exposed to the sun, the ability of the body to absorb energy, and the angle θ between the sun's rays and a line perpendicular to the body. The ability of an object to absorb energy is related to a factor called the *emissivity, e*, of the object. The emissivity can be calculated using the formula $e = \frac{W \sec \theta}{AS}$.

a. Solve this equation for W. Write your answer using only $\sin \theta$ or $\cos \theta$.

b. Find W if $e = 0.80$, $\theta = 40°$, $A = 0.75\text{ m}^2$, and $S = 1000\text{ W/m}^2$.

57. **Physics** A skier of mass m descends a θ-degree hill at a constant speed. When Newton's Laws are applied to the situation, the following system of equations is produced.

$$F_N - mg \cos \theta = 0$$

$$mg \sin \theta - \mu_k F_N = 0$$

where g is the acceleration due to gravity, F_N is the normal force exerted on the skier, and μ_k is the coefficient of friction. Use the system to define μ_k as a function of θ.

58. **Geometry** Show that the area of a regular polygon of n sides, each of length a, is given by $A = \frac{1}{4} na^2 \cot \left(\frac{180°}{n}\right)$.

59. **Critical Thinking** The circle at the right is a unit circle with its center at the origin. $\overleftrightarrow{AB}$ and $\overleftrightarrow{CD}$ are tangent to the circle. State the segments whose measures represent the ratios $\sin \theta$, $\cos \theta$, $\tan \theta$, $\sec \theta$, $\cot \theta$, and $\csc \theta$. Justify your answers.

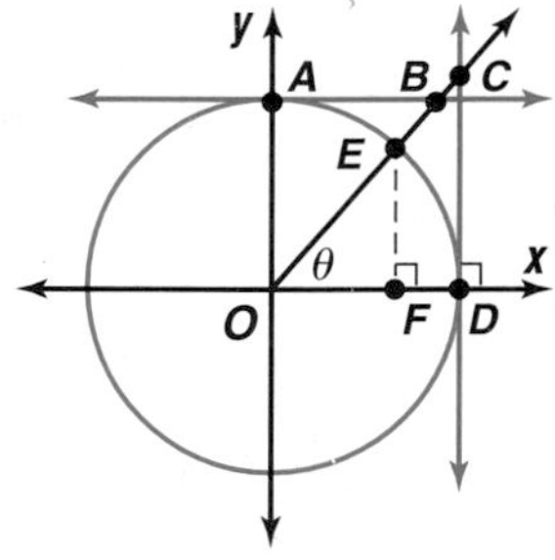

Mixed Review

60. Find $\text{Cos}^{-1}\left(-\frac{\sqrt{2}}{2}\right)$. *(Lesson 6-8)*

61. Graph $y = \cos\left(x - \frac{\pi}{6}\right)$. *(Lesson 6-5)*

62. **Physics** A pendulum 20 centimeters long swings 3°30′ on each side of its vertical position. Find the length of the arc formed by the tip of the pendulum as it swings. *(Lesson 6-1)*

63. Angle C of $\triangle ABC$ is a right angle. Solve the triangle if $A = 20°$ and $c = 35$. *(Lesson 5-4)*

64. Find all the rational roots of the equation $2x^3 + x^2 - 8x - 4 = 0$. *(Lesson 4-4)*

65. Solve $2x^2 + 7x - 4 = 0$ by completing the square. *(Lesson 4-2)*

66. Determine whether $f(x) = 3x^3 + 2x - 5$ is continuous or discontinuous at $x = 5$. *(Lesson 3-5)*

Extra Practice See p. A38.

67. Solve the system of equations algebraically. *(Lesson 2-2)*
$x + y - 2z = 3$
$-4x - y - z = 0$
$-x - 5y + 4z = 11$

68. Write the slope-intercept form of the equation of the line that passes through points at (5, 2) and $(-4, 4)$. *(Lesson 1-4)*

69. **SAT/ACT Practice** Triangle ABC is inscribed in circle O, and $\overleftrightarrow{CD}$ is tangent to circle O at point C. If $m\angle BCD = 40°$, find $m\angle A$.

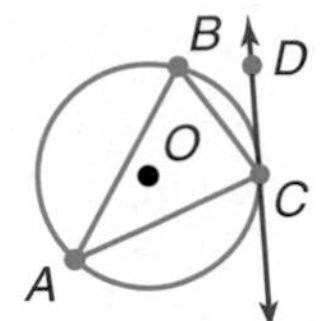

A 60° **B** 50° **C** 40° **D** 30° **E** 20°

CAREER CHOICES

Cartographer

Do maps fascinate you? Do you like drawing, working with computers, and geography? You may want to consider a career in cartography. As a cartographer, you would make maps, charts, and drawings. Cartography has changed a great deal with modern technology. Computers and satellites have become powerful new tools in making maps. As a cartographer, you may work with manual drafting tools as well as computer software designed for making maps.

The image at the right shows how a cartographer uses a three-dimensional landscape to create a two-dimensional topographic map.

There are several areas of specialization in the field of cartography. Some of these include making maps from political boundaries and natural features, making maps from aerial photographs, and correcting original maps.

Career Overview

Degree Preferred:
bachelor's degree in engineering or a physical science

Related Courses:
mathematics, geography, computer science, mechanical drawing

Outlook:
slower than average through 2006

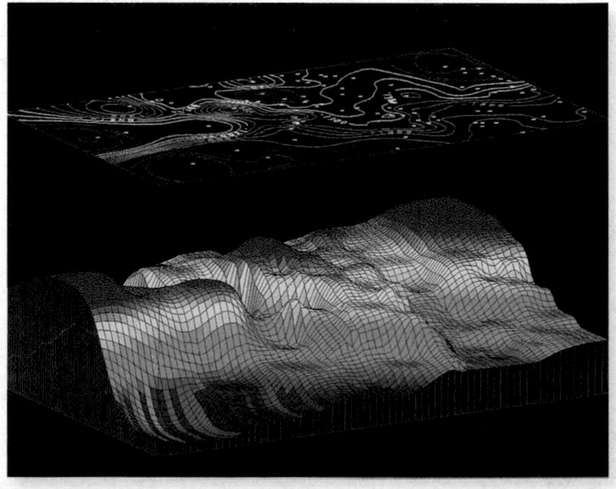

interNET CONNECTION For more information on careers in cartography, visit: **www.amc.glencoe.com**

Verifying Trigonometric Identities

OBJECTIVES

- Use the basic trigonometric identities to verify other identities.
- Find numerical values of trigonometric functions.

PROBLEM SOLVING While working on a mathematics assignment, a group of students derived an expression for the length of a ladder that, when held horizontally, would turn from a 5-foot wide corridor into a 7-foot wide corridor. They determined that the maximum length ℓ of a ladder that would fit was given by $\ell(\theta) = \dfrac{7 \sin \theta + 5 \cos \theta}{\sin \theta \cos \theta}$, where θ is the angle that the ladder makes with the outer wall of the 5-foot wide corridor. When their teacher worked the problem, she concluded that $\ell(\theta) = 7 \sec \theta + 5 \csc \theta$. Are the two expressions for $\ell(\theta)$ equivalent? *This problem will be solved in Example 2.*

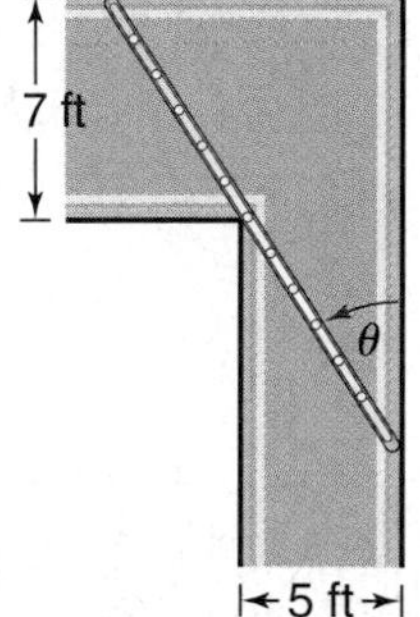

Verifying trigonometric identities algebraically involves transforming one side of the equation into the same form as the other side by using the basic trigonometric identities and the properties of algebra. Either side may be transformed into the other side, or both sides may be transformed separately into forms that are the same.

Suggestions for Verifying Trigonometric Identities

- Transform the more complicated side of the equation into the simpler side.
- Substitute one or more basic trigonometric identities to simplify expressions.
- Factor or multiply to simplify expressions.
- Multiply expressions by an expression equal to 1.
- Express all trigonometric functions in terms of sine and cosine.

You cannot add or subtract quantities from each side of an unverified identity, nor can you perform any other operation on each side, as you often do with equations. An unverified identity is not an equation, so the properties of equality do not apply.

Example 1 **Verify that $\sec^2 x - \tan x \cot x = \tan^2 x$ is an identity.**

Since the left side is more complicated, transform it into the expression on the right.

$$\sec^2 x - \tan x \cot x \stackrel{?}{=} \tan^2 x$$

$$\sec^2 x - \tan x \cdot \frac{1}{\tan x} \stackrel{?}{=} \tan^2 x \qquad \textit{cot } x = \frac{1}{\tan x}$$

$$\sec^2 x - 1 \stackrel{?}{=} \tan^2 x \qquad \textit{Multiply.}$$

$$\tan^2 x + 1 - 1 \stackrel{?}{=} \tan^2 x \qquad \sec^2 x = \tan^2 x + 1$$

$$\tan^2 x = \tan^2 x \qquad \textit{Simplify.}$$

We have transformed the left side into the right side. The identity is verified.

Examples **2** **PROBLEM SOLVING** **Verify that the two expressions for $\ell(\theta)$ in the application at the beginning of the lesson are equivalent. That is, verify that $\dfrac{7 \sin \theta + 5 \cos \theta}{\sin \theta \cos \theta} = 7 \sec \theta + 5 \csc \theta$ is an identity.**

Begin by writing the right side in terms of sine and cosine.

$$\frac{7 \sin \theta + 5 \cos \theta}{\sin \theta \cos \theta} \stackrel{?}{=} 7 \sec \theta + 5 \csc \theta$$

$$\frac{7 \sin \theta + 5 \cos \theta}{\sin \theta \cos \theta} \stackrel{?}{=} \frac{7}{\cos \theta} + \frac{5}{\sin \theta} \qquad \textit{$\sec \theta = \frac{1}{\cos \theta}$, $\csc \theta = \frac{1}{\sin \theta}$}$$

$$\frac{7 \sin \theta + 5 \cos \theta}{\sin \theta \cos \theta} \stackrel{?}{=} \frac{7 \sin \theta}{\sin \theta \cos \theta} + \frac{5 \cos \theta}{\sin \theta \cos \theta} \qquad \textit{Find a common denominator.}$$

$$\frac{7 \sin \theta + 5 \cos \theta}{\sin \theta \cos \theta} = \frac{7 \sin \theta + 5 \cos \theta}{\sin \theta \cos \theta} \qquad \textit{Simplify.}$$

The students and the teacher derived equivalent expressions for $\ell(\theta)$, the length of the ladder.

3 **Verify that $\dfrac{\sin A}{\csc A} + \dfrac{\cos A}{\sec A} = \csc^2 A - \cot^2 A$ is an identity.**

Since the two sides are equally complicated, we will transform each side independently into the same form.

$$\frac{\sin A}{\csc A} + \frac{\cos A}{\sec A} \stackrel{?}{=} \csc^2 A - \cot^2 A$$

$$\frac{\sin A}{\frac{1}{\sin A}} + \frac{\cos A}{\frac{1}{\cos A}} \stackrel{?}{=} (1 + \cot^2 A) - \cot^2 A \qquad \textit{Quotient identities; Pythagorean identity}$$

$$\sin^2 A + \cos^2 A \stackrel{?}{=} 1 \qquad \textit{Simplify.}$$

$$1 = 1 \qquad \textit{$\sin^2 A + \cos^2 A = 1$}$$

The techniques that you use to verify trigonometric identities can also be used to simplify trigonometric equations. Sometimes you can change an equation into an equivalent equation involving a single trigonometric function.

Example **4** **Find a numerical value of one trigonometric function of x if $\dfrac{\cot x}{\cos x} = 2$.**

You can simplify the trigonometric expression on the left side by writing it in terms of sine and cosine.

$$\frac{\cot x}{\cos x} = 2$$

$$\frac{\frac{\cos x}{\sin x}}{\cos x} = 2 \qquad \textit{$\cot x = \frac{\cos x}{\sin x}$}$$

$$\frac{\cos x}{\sin x} \cdot \frac{1}{\cos x} = 2 \qquad \textit{Definition of division}$$

$$\frac{1}{\sin x} = 2 \quad \textit{Simplify.}$$

$$\csc x = 2 \quad \frac{1}{\sin x} = \csc x$$

Therefore, if $\frac{\cot x}{\cos x} = 2$, then $\csc x = 2$.

You can use a graphing calculator to investigate whether an equation may be an identity.

GRAPHING CALCULATOR EXPLORATION

- Graph both sides of the equation as two separate functions. For example, to test $\sin^2 x = (1 - \cos x)(1 + \cos x)$, graph $y_1 = \sin^2 x$ and $y_2 = (1 - \cos x)(1 + \cos x)$ on the same screen.
- If the two graphs do not match, then the equation is not an identity.
- If the two sides appear to match in every window you try, then the equation may be an identity.

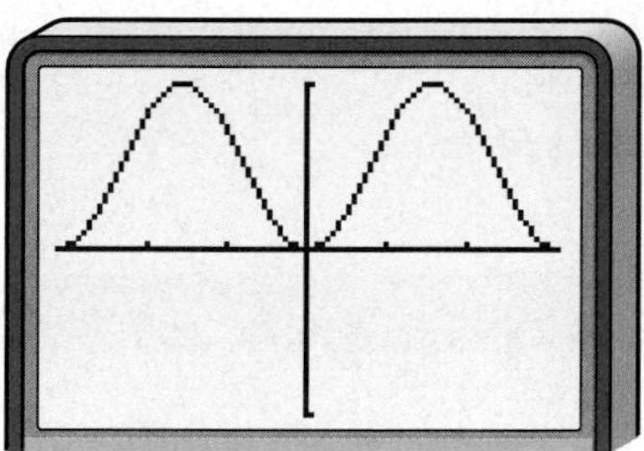

$[-\pi, \pi]$ scl:1 by $[-1, 1]$ scl:1

TRY THESE **Determine whether each equation could be an identity. Write *yes* or *no*.**

1. $\sin x \csc x - \sin^2 x = \cos^2 x$
2. $\sec x + \csc x = 1$
3. $\sin x - \cos x = \dfrac{1}{\csc x - \sec x}$

WHAT DO YOU THINK?

4. If the two sides appear to match in every window you try, does that prove that the equation is an identity? Justify your answer.
5. Graph the function $f(x) = \dfrac{\sec x - \cos x}{\tan x}$. What simpler function could you set equal to $f(x)$ in order to obtain an identity?

CHECK FOR UNDERSTANDING

Communicating Mathematics

Read and study the lesson to answer each question.

1. **Write** a trigonometric equation that is not an identity. Explain how you know it is not an identity.
2. **Explain** why you cannot square each side of the equation when verifying a trigonometric identity.
3. **Discuss** why both sides of a trigonometric identity are often rewritten in terms of sine and cosine.

4. *Math Journal* **Create** your own trigonometric identity that contains at least three different trigonometric functions. Explain how you created it. Give it to one of your classmates to verify. Compare and contrast your classmate's approach with your approach.

Guided Practice **Verify that each equation is an identity.**

5. $\cos x = \dfrac{\cot x}{\csc x}$

6. $\dfrac{1}{\tan x + \sec x} = \dfrac{\cos x}{\sin x + 1}$

7. $\csc \theta - \cot \theta = \dfrac{1}{\csc \theta + \cot \theta}$

8. $\sin \theta \tan \theta = \sec \theta - \cos \theta$

9. $(\sin A - \cos A)^2 = 1 - 2 \sin^2 A \cot A$

Find a numerical value of one trigonometric function of *x*.

10. $\tan x = \dfrac{1}{4} \sec x$

11. $\cot x + \sin x = -\cos x \cot x$

12. **Optics** The amount of light that a source provides to a surface is called the *illuminance*. The illuminance E in foot candles on a surface that is R feet from a source of light with intensity I candelas is $E = \dfrac{I \cos \theta}{R^2}$, where θ is the measure of the angle between the direction of the light and a line perpendicular to the surface being illuminated. Verify that $E = \dfrac{I \cot \theta}{R^2 \csc \theta}$ is an equivalent formula.

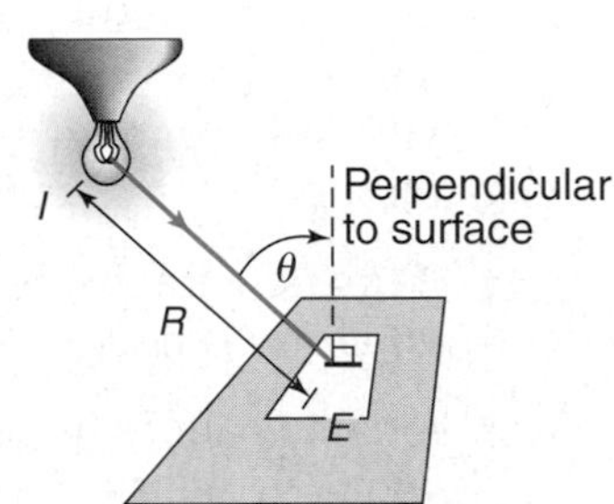

EXERCISES

Practice **Verify that each equation is an identity.**

13. $\tan A = \dfrac{\sec A}{\csc A}$

14. $\cos \theta = \sin \theta \cot \theta$

15. $\sec x - \tan x = \dfrac{1 - \sin x}{\cos x}$

16. $\dfrac{1 + \tan x}{\sin x + \cos x} = \sec x$

17. $\sec x \csc x = \tan x + \cot x$

18. $\sin \theta + \cos \theta = \dfrac{2 \sin^2 \theta - 1}{\sin \theta - \cos \theta}$

19. $(\sin A + \cos A)^2 = \dfrac{2 + \sec A \csc A}{\sec A \csc A}$

20. $(\sin \theta - 1)(\tan \theta + \sec \theta) = -\cos \theta$

21. $\dfrac{\cos y}{1 - \sin y} = \dfrac{1 + \sin y}{\cos y}$

22. $\cos \theta \cos (-\theta) - \sin \theta \sin (-\theta) = 1$

23. $\csc x - 1 = \dfrac{\cot^2 x}{\csc x + 1}$

24. $\cos B \cot B = \csc B - \sin B$

25. $\sin \theta \cos \theta \tan \theta + \cos^2 \theta = 1$

26. $(\csc x - \cot x)^2 = \dfrac{1 - \cos x}{1 + \cos x}$

27. $\sin x + \cos x = \dfrac{\cos x}{1 - \tan x} + \dfrac{\sin x}{1 - \cot x}$

28. Show that $\sin \theta + \cos \theta + \tan \theta \sin \theta = \sec \theta + \cos \theta \tan \theta$.

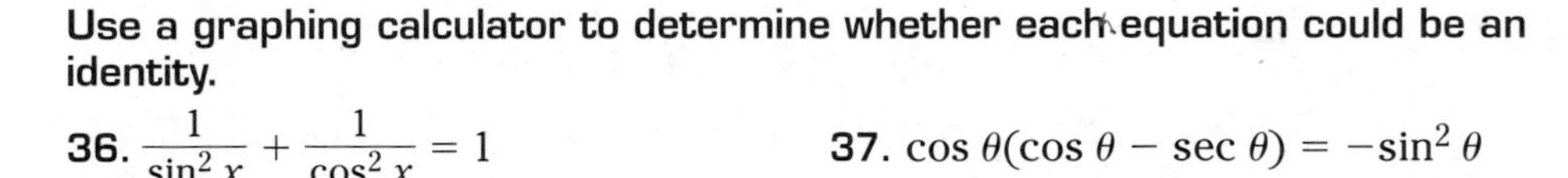

Find a numerical value of one trigonometric function of *x*.

29. $\dfrac{\csc x}{\cot x} = \sqrt{2}$

30. $\dfrac{1 + \tan x}{1 + \cot x} = 2$

31. $\dfrac{1}{\cot x} - \dfrac{\sec x}{\csc x} = \cos x$

32. $\dfrac{1 + \cos x}{\sin x} + \dfrac{\sin x}{1 + \cos x} = 4$

33. $\cos^2 x + 2 \sin x - 2 = 0$

34. $\csc x = \sin x \tan x + \cos x$

35. If $\dfrac{\tan^3 \theta - 1}{\tan \theta - 1} - \sec^2 \theta - 1 = 0$, find $\cot \theta$.

Graphing Calculator

Use a graphing calculator to determine whether each equation could be an identity.

36. $\dfrac{1}{\sin^2 x} + \dfrac{1}{\cos^2 x} = 1$

37. $\cos \theta(\cos \theta - \sec \theta) = -\sin^2 \theta$

38. $2 \sin A + (1 - \sin A)^2 = 2 - \cos^2 A$

39. $\dfrac{\sin^3 x - \cos^3 x}{\sin x - \cos x} = \sin^2 x + \cos^2 x$

Applications and Problem Solving

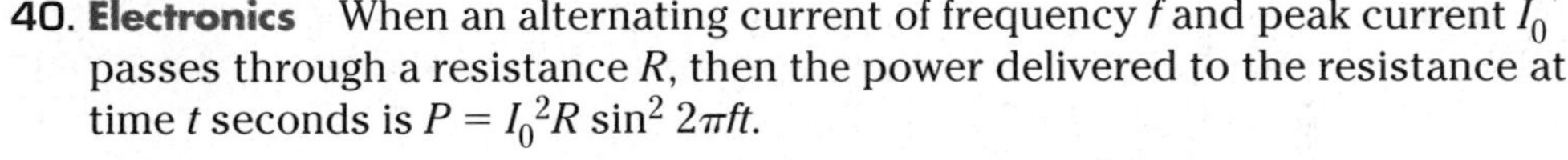

40. **Electronics** When an alternating current of frequency f and peak current I_0 passes through a resistance R, then the power delivered to the resistance at time t seconds is $P = I_0^2 R \sin^2 2\pi ft$.

    a. Write an expression for the power in terms of $\cos^2 2\pi ft$.

    b. Write an expression for the power in terms of $\csc^2 2\pi ft$.

41. **Critical Thinking** Let $x = \frac{1}{2} \tan \theta$ where $-\frac{\pi}{2} < \theta < \frac{\pi}{2}$. Write $f(x) = \dfrac{x}{\sqrt{1 + 4x^2}}$ in terms of a single trigonometric function of θ.

42. **Spherical Geometry** Spherical geometry is the geometry that takes place on the surface of a sphere. A line segment on the surface of the sphere is measured by the angle it subtends at the center of the sphere. Let a, b, and c be the sides of a right triangle on the surface of the sphere. Let the angles opposite those sides be α, β, and $\gamma = 90°$, respectively. The following equations are true:

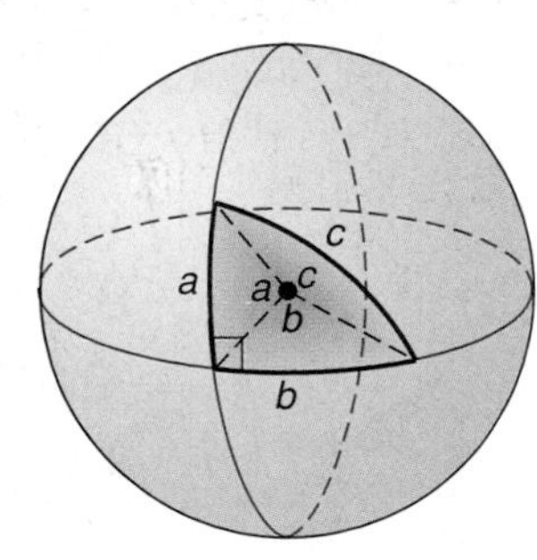

β is the Greek letter beta and γ is the Greek letter gamma.

$$\sin a = \sin \alpha \sin c$$

$$\cos b = \frac{\cos \beta}{\sin \alpha}$$

$$\cos c = \cos a \cos b.$$

Show that $\cos \beta = \tan a \cot c$.

43. **Physics** When a projectile is fired from the ground, its height y and horizontal displacement x are related by the equation $y = \dfrac{-gx^2}{2{v_0}^2\cos^2 \theta} + \dfrac{x \sin \theta}{\cos \theta}$, where v_0 is the initial velocity of the projectile, θ is the angle at which it was fired, and g is the acceleration due to gravity. Rewrite this equation so that $\tan \theta$ is the only trigonometric function that appears in the equation.

44. **Critical Thinking** Consider a circle O with radius 1. $\overline{PA}$ and $\overline{TB}$ are each perpendicular to $\overline{OB}$. Determine the area of $ABTP$ as a product of trigonometric functions of θ.

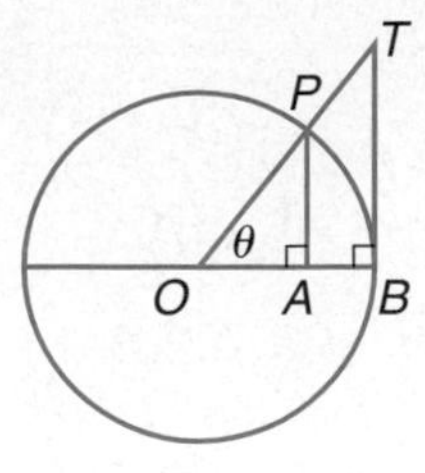

45. **Geometry** Let a, b, and c be the sides of a triangle. Let α, β, and γ be the respective opposite angles. Show that the area A of the triangle is given by $A = \dfrac{a^2 \sin \beta \sin \gamma}{2 \sin (\beta + \gamma)}$.

Mixed Review

46. Simplify $\dfrac{\tan x + \cos x + \sin x \tan x}{\sec x + \tan x}$. *(Lesson 7-1)*

47. Write an equation of a sine function with amplitude 2, period 180°, and phase shift 45°. *(Lesson 6-5)*

48. Change $\dfrac{15\pi}{16}$ radians to degree measure to the nearest minute. *(Lesson 6-1)*

49. Solve $\sqrt[3]{3y - 1} - 2 = 0$. *(Lesson 4-7)*

50. Determine the equations of the vertical and horizontal asymptotes, if any, of $f(x) = \dfrac{3x}{x + 1}$. *(Lesson 3-7)*

51. **Manufacturing** The Simply Sweats Corporation makes high quality sweatpants and sweatshirts. Each garment passes through the cutting and sewing departments of the factory. The cutting and sewing departments have 100 and 180 worker-hours available each week, respectively. The fabric supplier can provide 195 yards of fabric each week. The hours of work and yards of fabric required for each garment are shown in the table below. If the profit from a sweatshirt is \$5.00 and the profit from a pair of sweatpants is \$4.50, how many of each should the company make for maximum profit? *(Lesson 2-7)*

Simply Sweats Corporation

"Quality Sweatpants and Sweatshirts"

Clothing	Cutting	Sewing	Fabric
Shirt	1 h	2.5 h	1.5 yd
Pants	1.5 h	2 h	3 yd

52. State the domain and range of the relation $\{(16, -4), (16, 4)\}$. Is this relation a function? Explain. *(Lesson 1-1)*

53. **SAT/ACT Practice** Divide $\dfrac{a - b}{a + b}$ by $\dfrac{b - a}{b + a}$.

A 1 **B** $\dfrac{(a - b)^2}{(a + b)^2}$ **C** $\dfrac{1}{a^2 - b^2}$

D -1 **E** 0

Extra Practice See p. A38.

Sum and Difference Identities

OBJECTIVE

- Use the sum and difference identities for the sine, cosine, and tangent functions.

BROADCASTING Have you ever had trouble tuning in your favorite radio station? Does the picture on your TV sometimes appear blurry? Sometimes these problems are caused by *interference*. Interference can result when two waves pass through the same space at the same time. The two kinds of interference are:

- *constructive interference,* which occurs if the amplitude of the sum of the waves is greater than the amplitudes of the two component waves, and
- *destructive interference,* which occurs if the amplitude of the sum is less than the amplitudes of the component waves.

What type of interference results when a signal modeled by the equation $y = 20 \sin(3t + 45°)$ is combined with a signal modeled by the equation $y = 20 \sin(3t + 225°)$? *This problem will be solved in Example 4.*

Consider two angles α and β in standard position. Let the terminal side of α intersect the unit circle at point $A(\cos \alpha, \sin \alpha)$. Let the terminal side of β intersect the unit circle at $B(\cos \beta, \sin \beta)$. We will calculate $(AB)^2$ in two different ways.

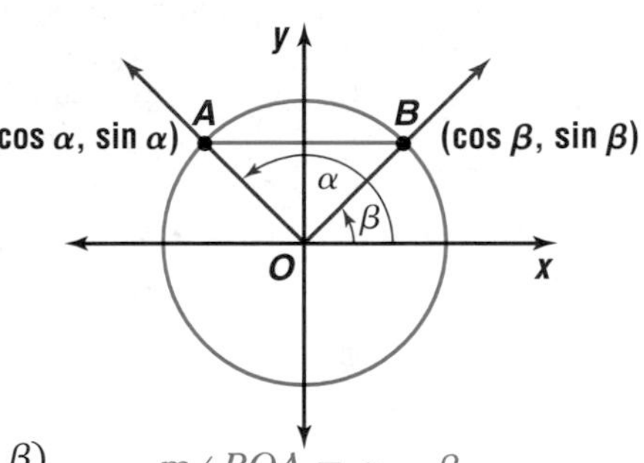

Look Back
You can refer to Lesson 5-8 to review the Law of Cosines.

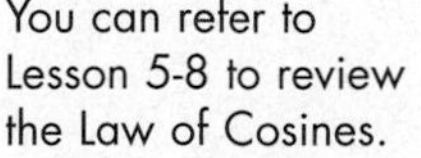

First use the Law of Cosines.

$(AB)^2 = (OA)^2 + (OB)^2 - 2(OA)(OB) \cos (\alpha - \beta)$

$(AB)^2 = 1^2 + 1^2 - 2(1)(1) \cos (\alpha - \beta)$ *OA = OB = 1*

$(AB)^2 = 2 - 2 \cos (\alpha - \beta)$ *Simplify.*

Now use the distance formula.

$(AB)^2 = (\cos \alpha - \cos \beta)^2 + (\sin \alpha - \sin \beta)^2$

$(AB)^2 = \cos^2 \alpha - 2 \cos \alpha \cos \beta + \cos^2 \beta + \sin^2 \alpha - 2 \sin \alpha \sin \beta + \sin^2 \beta$

$(AB)^2 = (\cos^2 \alpha + \sin^2 \alpha) + (\cos^2 \beta + \sin^2 \beta) - 2(\cos \alpha \cos \beta + \sin \alpha \sin \beta)$

$(AB)^2 = 1 + 1 - 2(\cos \alpha \cos \beta + \sin \alpha \sin \beta)$ *$\cos^2 a + \sin^2 a = 1$*

$(AB)^2 = 2 - 2(\cos \alpha \cos \beta + \sin \alpha \sin \beta)$ *Simplify.*

Set the two expressions for $(AB)^2$ equal to each other.

$2 - 2 \cos (\alpha - \beta) = 2 - 2(\cos \alpha \cos \beta + \sin \alpha \sin \beta)$

$-2 \cos (\alpha - \beta) = -2(\cos \alpha \cos \beta + \sin \alpha \sin \beta)$ *Subtract 2 from each side.*

$\cos (\alpha - \beta) = \cos \alpha \cos \beta + \sin \alpha \sin \beta$ *Divide each side by −2.*

This equation is known as the **difference identity for cosine.**

The **sum identity for cosine** can be derived by substituting $-\beta$ for β in the difference identity.

$$\begin{aligned}\cos(\alpha + \beta) &= \cos(\alpha - (-\beta))\\ &= \cos\alpha\cos(-\beta) + \sin\alpha\sin(-\beta)\\ &= \cos\alpha\cos\beta + \sin\alpha(-\sin\beta) && \textit{\cos(-\beta) = \cos\beta;\ \sin(-\beta) = -\sin\beta}\\ &= \cos\alpha\cos\beta - \sin\alpha\sin\beta\end{aligned}$$

Sum and Difference Identities for the Cosine Function

If α and β represent the measures of two angles, then the following identities hold for all values of α and β.

$$\cos(\alpha \pm \beta) = \cos\alpha\cos\beta \mp \sin\alpha\sin\beta$$

Notice how the addition and subtraction symbols are related in the sum and difference identities.

You can use the sum and difference identities and the values of the trigonometric functions of common angles to find the values of trigonometric functions of other angles. Note that α and β can be expressed in either degrees or radians.

Example 1 **a. Show by producing a counterexample that $\cos(x + y) \neq \cos x + \cos y$.**

b. Show that the sum identity for cosine is true for the values used in part a.

a. Let $x = \frac{\pi}{3}$ and $y = \frac{\pi}{6}$. First find $\cos(x + y)$ for $x = \frac{\pi}{3}$ and $y = \frac{\pi}{6}$.

$$\begin{aligned}\cos(x + y) &= \cos\left(\frac{\pi}{3} + \frac{\pi}{6}\right) && \textit{Replace x with } \tfrac{\pi}{3} \textit{ and y with } \tfrac{\pi}{6}.\\ &= \cos\frac{\pi}{2} && \tfrac{\pi}{3} + \tfrac{\pi}{6} = \tfrac{\pi}{2}\\ &= 0\end{aligned}$$

Now find $\cos x + \cos y$.

$$\begin{aligned}\cos x + \cos y &= \cos\frac{\pi}{3} + \cos\frac{\pi}{6} && \textit{Replace x with } \tfrac{\pi}{3} \textit{ and y with } \tfrac{\pi}{6}.\\ &= \frac{1}{2} + \frac{\sqrt{3}}{2} \text{ or } \frac{1 + \sqrt{3}}{2}\end{aligned}$$

So, $\cos(x + y) \neq \cos x + \cos y$.

b. Show that $\cos(x + y) = \cos x\cos y - \sin x\sin y$ for $x = \frac{\pi}{3}$ and $y = \frac{\pi}{6}$.

First find $\cos(x + y)$. From part a, we know that $\cos\left(\frac{\pi}{3} + \frac{\pi}{6}\right) = 0$.

Now find $\cos x\cos y - \sin x\sin y$.

$$\begin{aligned}\cos x\cos y - \sin x\sin y &= \cos\frac{\pi}{3}\cos\frac{\pi}{6} - \sin\frac{\pi}{3}\sin\frac{\pi}{6} && \textit{Substitute for x and y.}\\ &= \left(\frac{1}{2}\right)\left(\frac{\sqrt{3}}{2}\right) - \left(\frac{\sqrt{3}}{2}\right)\left(\frac{1}{2}\right)\\ &= 0\end{aligned}$$

Thus, the sum identity for cosine is true for $x = \frac{\pi}{3}$ and $y = \frac{\pi}{6}$.

Example 2 **Use the sum or difference identity for cosine to find the exact value of $\cos 735°$.**

$735° = 2(360°) + 15°$ *Symmetry identity, Case 1*

$\cos 735° = \cos 15°$

$\cos 15° = \cos (45° - 30°)$ *45° and 30° are two common angles that differ by 15°.*

$= \cos 45° \cos 30° + \sin 45° \sin 30°$ *Difference identity for cosine*

$$= \frac{\sqrt{2}}{2} \cdot \frac{\sqrt{3}}{2} + \frac{\sqrt{2}}{2} \cdot \frac{1}{2}$$

$$= \frac{\sqrt{6} + \sqrt{2}}{4}$$

Therefore, $\cos 735° = \frac{\sqrt{6} + \sqrt{2}}{4}$.

We can derive sum and difference identities for the sine function from those for the cosine function. Replace α with $\frac{\pi}{2}$ and β with s in the identities for $\cos (\alpha \pm \beta)$. The following equations result.

These equations are examples of cofunction identities.

$$\cos\left(\frac{\pi}{2} + s\right) = -\sin s \qquad \cos\left(\frac{\pi}{2} - s\right) = \sin s$$

Replace s with $\frac{\pi}{2} + s$ in the equation for $\cos\left(\frac{\pi}{2} + s\right)$ and with $\frac{\pi}{2} - s$ in the equation for $\cos\left(\frac{\pi}{2} - s\right)$ to obtain the following equations.

These equations are other cofunction identities.

$$\cos s = \sin\left(\frac{\pi}{2} + s\right) \qquad \cos s = \sin\left(\frac{\pi}{2} - s\right)$$

Replace s with $(\alpha + \beta)$ in the equation for $\cos\left(\frac{\pi}{2} - s\right)$ to derive an identity for the sine of the sum of two real numbers.

$$\cos\left[\frac{\pi}{2} - (\alpha + \beta)\right] = \sin (\alpha + \beta)$$

$$\cos\left[\left(\frac{\pi}{2} - \alpha\right) - \beta\right] = \sin (\alpha + \beta)$$

$$\cos\left(\frac{\pi}{2} - \alpha\right)\cos \beta + \sin\left(\frac{\pi}{2} - \alpha\right)\sin \beta = \sin\left(\alpha + \beta\right) \quad \textit{Identity for } \cos (\alpha - \beta)$$

$$\sin \alpha \cos \beta + \cos \alpha \sin \beta = \sin (\alpha + \beta) \quad \textit{Substitute.}$$

This equation is known as the **sum identity for sine.**

The **difference identity for sine** can be derived by replacing β with $(-\beta)$ in the sum identity for sine.

$$\sin [\alpha + (-\beta)] = \sin \alpha \cos (-\beta) + \cos \alpha \sin (-\beta)$$

$$\sin (\alpha - \beta) = \sin \alpha \cos \beta - \cos \alpha \sin \beta$$

Sum and Difference Identities for the Sine Function

If α and β represent the measures of two angles, then the following identities hold for all values of α and β.

$$\sin (\alpha \pm \beta) = \sin \alpha \cos \beta \pm \cos \alpha \sin \beta$$

Examples **3** **Find the value of sin $(x - y)$ if $0 < x < \frac{\pi}{2}$, $0 < y < \frac{\pi}{2}$, $\sin x = \frac{9}{41}$, and $\sin y = \frac{7}{25}$.**

In order to use the difference identity for sine, we need to know cos x and cos y. We can use a Pythagorean identity to determine the necessary values.

$\sin^2 \alpha + \cos^2 \alpha = 1 \Rightarrow \cos^2 \alpha = 1 - \sin^2 \alpha$ *Pythagorean identity*

Since we are given that the angles are in Quadrant I, the values of sine and cosine are positive. Therefore, $\cos \alpha = \sqrt{1 - \sin^2 \alpha}$.

$$\cos x = \sqrt{1 - \left(\frac{9}{41}\right)^2}$$
$$= \sqrt{\frac{1600}{1681}} \text{ or } \frac{40}{41}$$
$$\cos y = \sqrt{1 - \left(\frac{7}{25}\right)^2}$$
$$= \sqrt{\frac{576}{625}} \text{ or } \frac{24}{25}$$

Now substitute these values into the difference identity for sine.

$$\sin (x - y) = \sin x \cos y - \cos x \sin y$$
$$= \left(\frac{9}{41}\right)\left(\frac{24}{25}\right) - \left(\frac{40}{41}\right)\left(\frac{7}{25}\right)$$
$$= -\frac{64}{1025} \text{ or about } -0.0624$$

4 **BROADCASTING** **Refer to the application at the beginning of the lesson. What type of interference results when signals modeled by the equations $y = 20 \sin (3t + 45°)$ and $y = 20 \sin (3t + 225°)$ are combined?**

Add the two sine functions together and simplify.

$$20 \sin (3t + 45°) + 20 \sin (3t + 225°)$$
$$= 20(\sin 3t \cos 45° + \cos 3t \sin 45°) + 20(\sin 3t \cos 225° + \cos 3t \sin 225°)$$
$$= 20\left[(\sin 3t)\left(\frac{\sqrt{2}}{2}\right) + (\cos 3t)\left(\frac{\sqrt{2}}{2}\right)\right] + 20\left[(\sin 3t)\left(-\frac{\sqrt{2}}{2}\right) + (\cos 3t)\left(-\frac{\sqrt{2}}{2}\right)\right]$$
$$= 10\sqrt{2} \sin 3t + 10\sqrt{2} \cos 3t - 10\sqrt{2} \sin 3t - 10\sqrt{2} \cos 3t$$
$$= 0$$

The interference is destructive. The signals cancel each other completely.

You can use the sum and difference identities for the cosine and sine functions to find sum and difference identities for the tangent function.

Sum and Difference Identities for the Tangent Function

If α and β represent the measures of two angles, then the following identities hold for all values of α and β.

$$\tan(\alpha \pm \beta) = \frac{\tan\alpha \pm \tan\beta}{1 \mp \tan\alpha\tan\beta}$$

You will be asked to derive these identities in Exercise 47.

Example 5 **Use the sum or difference identity for tangent to find the exact value of tan 285°.**

$\tan 285° = \tan(240° + 45°)$ — *240° and 45° are common angles whose sum is 285°.*

$= \dfrac{\tan 240° + \tan 45°}{1 - \tan 240° \tan 45°}$ — *Sum identity for tangent*

$= \dfrac{\sqrt{3} + 1}{1 - (\sqrt{3})(1)}$ — *Multiply by $\dfrac{1+\sqrt{3}}{1+\sqrt{3}}$ to simplify.*

$= -2 - \sqrt{3}$

You can use sum and difference identities to verify other identities.

Example 6 **Verify that $\csc\left(\dfrac{3\pi}{2} + A\right) = -\sec A$ is an identity.**

Transform the left side since it is more complicated.

$\csc\left(\dfrac{3\pi}{2} + A\right) \stackrel{?}{=} -\sec A$

$\dfrac{1}{\sin\left(\dfrac{3\pi}{2} + A\right)} \stackrel{?}{=} -\sec A$ — *Reciprocal identity: $\csc x = \dfrac{1}{\sin x}$*

$\dfrac{1}{\sin\dfrac{3\pi}{2}\cos A + \cos\dfrac{3\pi}{2}\sin A} \stackrel{?}{=} -\sec A$ — *Sum identity for sine*

$\dfrac{1}{(-1)\cos A + (0)\sin A} \stackrel{?}{=} -\sec A$ — *$\sin\dfrac{3\pi}{2} = -1$; $\cos\dfrac{3\pi}{2} = 0$*

$-\dfrac{1}{\cos A} \stackrel{?}{=} -\sec A$ — *Simplify.*

$-\sec A = -\sec A$ — *Reciprocal identity*

CHECK FOR UNDERSTANDING

Communicating Mathematics

Read and study the lesson to answer each question.

1. **Describe** how you would convince a friend that $\sin(x + y) \neq \sin x + \sin y$.
2. **Explain** how to use the sum and difference identities to find values for the secant, cosecant, and cotangent functions of a sum or difference.

3. **Write** an interpretation of the identity $\sin(90° - A) = \cos A$ in terms of a right triangle.

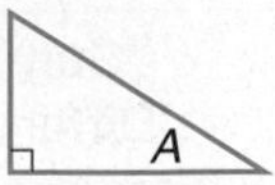

4. **Derive** a formula for $\cot(\alpha + \beta)$ in terms of $\cot \alpha$ and $\cot \beta$.

Guided Practice

Use sum or difference identities to find the exact value of each trigonometric function.

5. $\cos 165°$

6. $\tan \frac{\pi}{12}$

7. $\sec 795°$

Find each exact value if $0 < x < \frac{\pi}{2}$ and $0 < y < \frac{\pi}{2}$.

8. $\sin(x - y)$ if $\sin x = \frac{4}{9}$ and $\sin y = \frac{1}{4}$

9. $\tan(x + y)$ if $\csc x = \frac{5}{3}$ and $\cos y = \frac{5}{13}$

Verify that each equation is an identity.

10. $\sin(90° + A) = \cos A$

11. $\tan\left(\theta + \frac{\pi}{2}\right) = -\cot \theta$

12. $\sin(x - y) = \frac{1 - \cot x \tan y}{\csc x \sec y}$

ω is the Greek letter omega.

13. **Electrical Engineering** Analysis of the voltage in certain types of circuits involves terms of the form $\sin(n\omega_0 t - 90°)$, where n is a positive integer, ω_0 is the frequency of the voltage, and t is time. Use an identity to simplify this expression.

EXERCISES

Practice

Use sum or difference identities to find the exact value of each trigonometric function.

14. $\cos 105°$

15. $\sin 165°$

16. $\cos \frac{7\pi}{12}$

17. $\sin \frac{\pi}{12}$

18. $\tan 195°$

19. $\cos\left(-\frac{\pi}{12}\right)$

20. $\tan 165°$

21. $\tan \frac{23\pi}{12}$

22. $\sin 735°$

23. $\sec 1275°$

24. $\csc \frac{5\pi}{12}$

25. $\cot \frac{113\pi}{12}$

Find each exact value if $0 < x < \frac{\pi}{2}$ and $0 < y < \frac{\pi}{2}$.

26. $\sin(x + y)$ if $\cos x = \frac{8}{17}$ and $\sin y = \frac{12}{37}$

27. $\cos(x - y)$ if $\cos x = \frac{3}{5}$ and $\cos y = \frac{4}{5}$

28. $\tan(x - y)$ if $\sin x = \frac{8}{17}$ and $\cos y = \frac{3}{5}$

29. $\cos(x + y)$ if $\tan x = \frac{5}{3}$ and $\sin y = \frac{1}{3}$

30. $\tan(x + y)$ if $\cot x = \frac{6}{5}$ and $\sec y = \frac{3}{2}$

31. $\sec(x - y)$ if $\csc x = \frac{5}{3}$ and $\tan y = \frac{12}{5}$

32. If α and β are two angles in Quadrant I such that $\sin \alpha = \frac{1}{5}$ and $\cos \beta = \frac{2}{7}$, find $\sin(\alpha - \beta)$.

33. If x and y are acute angles such that $\cos x = \frac{1}{3}$ and $\cos y = \frac{3}{4}$, what is the value of $\cos(x + y)$?

Verify that each equation is an identity.

34. $\cos\left(\frac{\pi}{2} + x\right) = -\sin x$

35. $\cos(60° + A) = \sin(30° - A)$

36. $\sin(A + \pi) = -\sin A$

37. $\cos(180° + x) = -\cos x$

38. $\tan(x + 45°) = \frac{1 + \tan x}{1 - \tan x}$

39. $\sin(A + B) = \frac{\tan A + \tan B}{\sec A \sec B}$

40. $\cos(A + B) = \frac{1 - \tan A \tan B}{\sec A \sec B}$

41. $\sec(A - B) = \frac{\sec A \sec B}{1 + \tan A \tan B}$

42. $\sin(x + y)\sin(x - y) = \sin^2 x - \sin^2 y$

Applications and Problem Solving

43. **Electronics** In an electric circuit containing a capacitor, inductor, and resistor the voltage drop across the inductor is given by $V_L = I_0 \omega L \cos\left(\omega t + \frac{\pi}{2}\right)$, where I_0 is the peak current, ω is the frequency, L is the inductance, and t is time. Use the sum identity for cosine to express V_L as a function of $\sin \omega t$.

44. **Optics** The index of refraction for a medium through which light is passing is the ratio of the velocity of light in free space to the velocity of light in the medium. For light passing symmetrically through a glass prism, the index of refraction n is given by the equation $n = \frac{\sin\left[\frac{1}{2}(\alpha + \beta)\right]}{\sin \frac{\beta}{2}}$, where α is the deviation angle and β is the angle of the apex of the prism as shown in the diagram. If $\beta = 60°$, show that $n = \sqrt{3}\sin\frac{\alpha}{2} + \cos\frac{\alpha}{2}$.

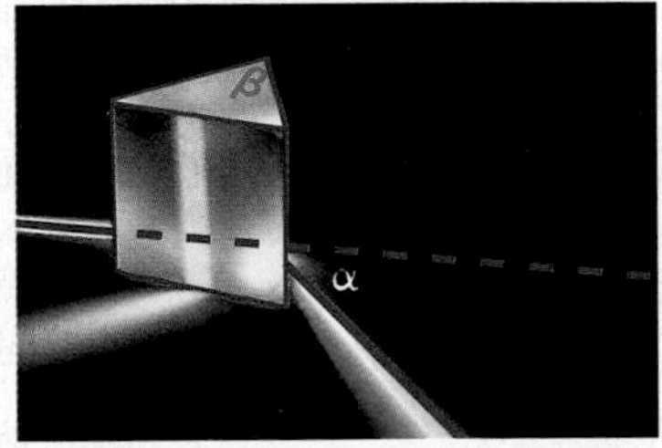

45. Critical Thinking Simplify the following expression without expanding any of the sums or differences.

$$\sin\left(\frac{\pi}{3} - A\right)\cos\left(\frac{\pi}{3} + A\right) - \cos\left(\frac{\pi}{3} - A\right)\sin\left(\frac{\pi}{3} + A\right)$$

46. Calculus In calculus, you will explore the *difference quotient* $\frac{f(x + h) - f(x)}{h}$.

a. Let $f(x) = \sin x$. Write and expand an expression for the difference quotient.

b. Set your answer from part a equal to y. Let $h = 0.1$ and graph.

c. What function has a graph similar to the graph in part b?

47. Critical Thinking Derive the sum and difference identities for the tangent function.

48. Critical Thinking Consider the following theorem.
If A, B, and C are the angles of a nonright triangle, then
$\tan A + \tan B + \tan C = \tan A \tan B \tan C$.

a. Choose values for A, B, and C. Verify that the conclusion is true for your specific values.

b. Prove the theorem.

Mixed Review

49. Verify the identity $\sec^2 x = \frac{1 - \cos^2 x}{1 - \sin^2 x} + \csc^2 x - \cot^2 x$. *(Lesson 7-2)*

50. If $\sin \theta = -\frac{1}{8}$ and $\pi < \theta < \frac{3\pi}{2}$, find $\tan \theta$. *(Lesson 7-1)*

51. Find $\sin\left(\text{Arctan } \sqrt{3}\right)$. *(Lesson 6-8)*

52. Find the values of θ for which $\csc \theta$ is undefined. *(Lesson 6-7)*

53. **Weather** The average seasonal high temperatures for Greensboro, North Carolina, are given in the table. Write a sinusoidal function that models the temperatures, using $t = 1$ to represent winter. *(Lesson 6-6)*

Winter	Spring	Summer	Fall
50°	70°	86°	71°

Source: Rand McNally & Company

54. State the amplitude, period, and phase shift for the function $y = 8 \cos (\theta - 30°)$. *(Lesson 6-5)*

55. Find the value of $\sin (-540°)$. *(Lesson 6-3)*

56. **Geometry** A sector has arc length of 18 feet and a central angle measuring 2.9 radians. Find the radius and the area of the sector. *(Lesson 6-1)*

57. **Navigation** A ship at sea is 70 miles from one radio transmitter and 130 miles from another. The angle formed by the rays from the ship to the transmitters measures 130°. How far apart are the transmitters? *(Lesson 5-8)*

58. Determine the number of possible solutions for a triangle if $A = 120°$, $b = 12$, and $a = 4$. *(Lesson 5-7)*

59. Photography A photographer observes a 35-foot totem pole that stands vertically on a uniformly-sloped hillside and the shadow cast by it at different times of day. At a time when the angle of elevation of the sun is 37°12′, the shadow of the pole extends directly down the slope. This is the effect that the photographer is seeking. If the hillside has an angle of inclination of 6°40′, find the length of the shadow. *(Lesson 5-6)*

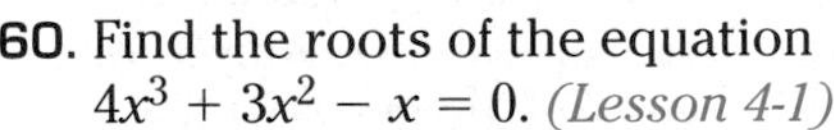

60. Find the roots of the equation $4x^3 + 3x^2 - x = 0$. *(Lesson 4-1)*

61. Solve $|x + 1| > 4$. *(Lesson 3-3)*

62. Find the value of the determinant $\begin{vmatrix} -1 & -2 \\ 3 & -6 \end{vmatrix}$. *(Lesson 2-5)*

63. If $f(x) = 3x^2 - 4$ and $g(x) = 5x + 1$, find $f \circ g(4)$ and $g \circ f(4)$. *(Lesson 1-2)*

64. SAT Practice **Quantitative Comparison**

A if the quantity in column A is greater
B if the quantity in column B is greater
C if the two quantities are equal
D if the relationship cannot be determined from the information given

Column A	Column B
$(-8)^{62}$	$(-8)^{75}$

MID-CHAPTER QUIZ

Use the given information to determine the exact trigonometric value. (Lesson 7-1)

1. $\sin \theta = \frac{2}{7}, 0 < \theta < \frac{\pi}{2}$; $\cot \theta$

2. $\tan \theta = -\frac{4}{3}, 90° < \theta < 180°$; $\cos \theta$

3. Express $\cos \frac{19\pi}{4}$ as a trigonometric function of an angle in Quadrant I. (Lesson 7-1)

Verify that each equation is an identity. (Lesson 7-2)

4. $\frac{1}{1 + \tan^2 x} + \frac{1}{1 + \cot^2 x} = 1$

5. $\frac{\csc^2 \theta + \sec^2 \theta}{\sec^2 \theta} = \csc^2 \theta$

Verify that each equation is an identity. (Lessons 7-2 and 7-3)

6. $\cot x \sec x \sin x = 2 - \tan x \cos x \csc x$

7. $\tan (\alpha - \beta) = \frac{1 - \cot \alpha \tan \beta}{\cot \alpha + \tan \beta}$

8. Use a sum or difference identity to find the exact value of cos 75°. (Lesson 7-3)

Find each exact value if $0 < x < \frac{\pi}{2}$ and $0 < y < \frac{\pi}{2}$. (Lesson 7-3)

9. $\cos (x + y)$ if $\sin x = \frac{2}{3}$ and $\sin y = \frac{3}{4}$

10. $\tan (x - y)$ if $\tan x = \frac{5}{4}$ and $\sec y = 2$

Extra Practice See p. A38.

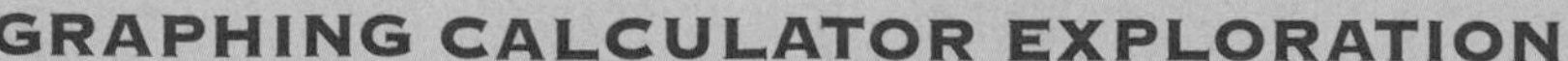

GRAPHING CALCULATOR EXPLORATION

7-3B Reduction Identities

An Extension of Lesson 7-3

OBJECTIVE

- Identify equivalent values for trigonometric functions involving quadrantal angles.

In Chapter 5, you learned that any trigonometric function of an acute angle is equal to the *cofunction* of the complement of the angle. For example, $\sin \alpha = \cos (90° - \alpha)$. This is a part of a large family of identities called the **reduction identities.** These identities involve adding and subtracting the quandrantal angles, 90°, 180°, and 270°, from the angle measure to find equivalent values of the trigonometric function. You can use your knowledge of phase shifts and reflections to find the components of these identities.

Example

Find the values of the sine and cosine functions for $\alpha - 90°$, $\alpha - 180°$, and $\alpha - 270°$ that are equivalent to $\sin \alpha$.

$\alpha - 90°$

Graph $y = \sin \alpha$, $y = \sin (\alpha - 90°)$, and $y = \cos (\alpha - 90°)$, letting X in degree mode represent α. *You can select different display formats to help you distinguish the three graphs.*

You may recall from Chapter 6 that a phase shift of 90° right for the cosine function results in the sine function.

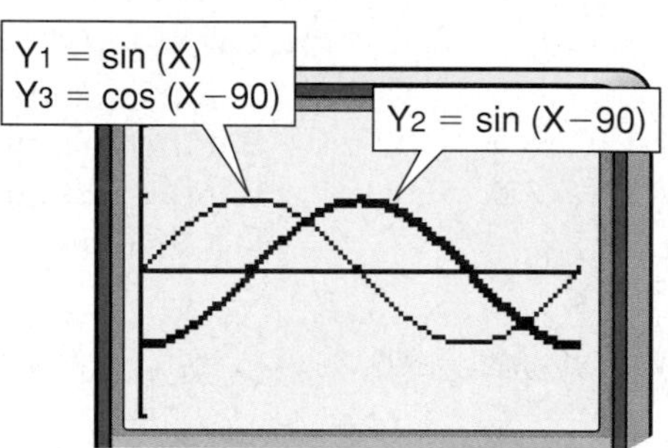

[0, 360] scl:90 by [−2, 2] scl:1

Note that the graph of $y = \cos (X - 90)$ is the same as the graph of $y = \sin X$. This suggests that $\sin \alpha = \cos (\alpha - 90°)$. *Remember that an identity must be proved algebraically. A graph does not prove an identity.*

$\alpha - 180°$

Graph $y = \sin \alpha$, $y = \sin (\alpha - 180°)$, and $y = \cos (\alpha - 180°)$ using X to represent α.

Discount $y = \cos (\alpha - 180°)$ as a possible equivalence because it would involve a phase shift, which would change the actual value of the angle being considered.

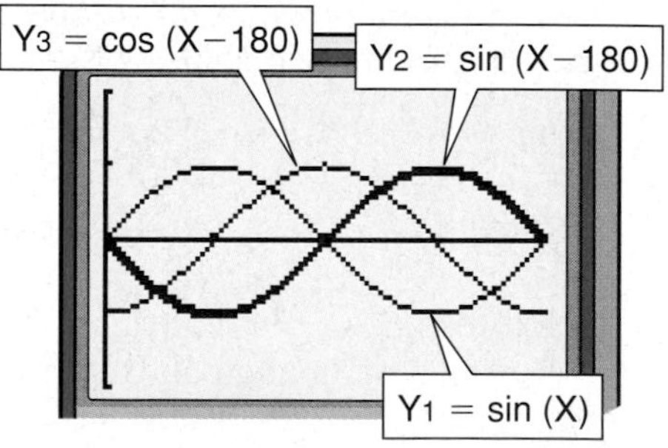

[0, 360] scl:90 by [−2, 2] scl:1

Note that the graph of $\sin (\alpha - 180°)$ is a mirror reflection of $\sin \alpha$. Remember that a reflection over the x-axis results in the mapping $(x, y) \rightarrow (x, -y)$. So to obtain a graph that is identical to $y = \sin \alpha$, we need the reflection of $y = \sin (\alpha - 180°)$ over the x-axis, or $y = -\sin (\alpha - 180°)$. Thus, $\sin \alpha = -\sin (\alpha - 180°)$. Graph the two equations to investigate this equality.

$\alpha - 270°$

In this case, $\sin(\alpha - 270°)$ is a phase shift, so ignore it. The graph of $\cos(\alpha - 270°)$ is a reflection of $\sin \alpha$ over the x-axis. So, $\sin \alpha = -\cos(\alpha - 270°)$.

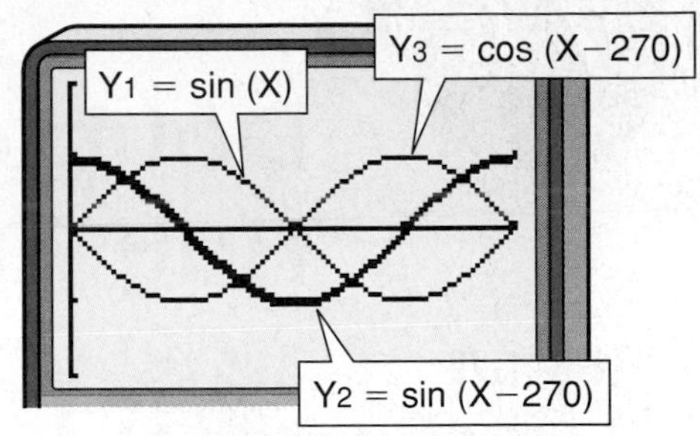

[0, 360] scl:90 by [−2, 2] scl:1

The family of reduction identities also contains the relationships among the other cofunctions of tangent and cotangent and secant and cosecant. In addition to $\alpha - 90°$, $\alpha - 180°$, and $\alpha - 270°$ angle measures, the reduction identities address other measures such as $90° \pm \alpha$, $180° \pm \alpha$, $270° \pm \alpha$, and $360° \pm \alpha$.

TRY THESE

Copy and complete each statement with the proper trigonometric functions.

1. $\cos \alpha = \underline{\ ?\ }(\alpha - 90°) = \underline{\ ?\ }(\alpha - 180°) = \underline{\ ?\ }(\alpha - 270°)$

2. $\tan \alpha = \underline{\ ?\ }(\alpha - 90°) = \underline{\ ?\ }(\alpha - 180°) = \underline{\ ?\ }(\alpha - 270°)$

3. $\cot \alpha = \underline{\ ?\ }(\alpha - 90°) = \underline{\ ?\ }(\alpha - 180°) = \underline{\ ?\ }(\alpha - 270°)$

4. $\sec \alpha = \underline{\ ?\ }(\alpha - 90°) = \underline{\ ?\ }(\alpha - 180°) = \underline{\ ?\ }(\alpha - 270°)$

5. $\csc \alpha = \underline{\ ?\ }(\alpha - 90°) = \underline{\ ?\ }(\alpha - 180°) = \underline{\ ?\ }(\alpha - 270°)$

WHAT DO YOU THINK?

6. Suppose the expressions involving subtraction in Exercises 1-5 were changed to sums.

a. Copy and complete each statement with the proper trigonometric functions.

(1) $\sin \alpha = \underline{\ ?\ }(\alpha + 90°) = \underline{\ ?\ }(\alpha + 180°) = \underline{\ ?\ }(\alpha + 270°)$

(2) $\cos \alpha = \underline{\ ?\ }(\alpha + 90°) = \underline{\ ?\ }(\alpha + 180°) = \underline{\ ?\ }(\alpha + 270°)$

(3) $\tan \alpha = \underline{\ ?\ }(\alpha + 90°) = \underline{\ ?\ }(\alpha + 180°) = \underline{\ ?\ }(\alpha + 270°)$

(4) $\cot \alpha = \underline{\ ?\ }(\alpha + 90°) = \underline{\ ?\ }(\alpha + 180°) = \underline{\ ?\ }(\alpha + 270°)$

(5) $\sec \alpha = \underline{\ ?\ }(\alpha + 90°) = \underline{\ ?\ }(\alpha + 180°) = \underline{\ ?\ }(\alpha + 270°)$

(6) $\csc \alpha = \underline{\ ?\ }(\alpha + 90°) = \underline{\ ?\ }(\alpha + 180°) = \underline{\ ?\ }(\alpha + 270°)$

b. How do the identities in part a compare to those in Exercises 1-5?

7. a. Copy and complete each statement with the proper trigonometric functions.

(1) $\sin \alpha = \underline{\ ?\ }(90° - \alpha) = \underline{\ ?\ }(180° - \alpha) = \underline{\ ?\ }(270° - \alpha)$

(2) $\cos \alpha = \underline{\ ?\ }(90° - \alpha) = \underline{\ ?\ }(180° - \alpha) = \underline{\ ?\ }(270° - \alpha)$

(3) $\tan \alpha = \underline{\ ?\ }(90° - \alpha) = \underline{\ ?\ }(180° - \alpha) = \underline{\ ?\ }(270° - \alpha)$

(4) $\cot \alpha = \underline{\ ?\ }(90° - \alpha) = \underline{\ ?\ }(180° - \alpha) = \underline{\ ?\ }(270° - \alpha)$

(5) $\sec \alpha = \underline{\ ?\ }(90° - \alpha) = \underline{\ ?\ }(180° - \alpha) = \underline{\ ?\ }(270° - \alpha)$

(6) $\csc \alpha = \underline{\ ?\ }(90° - \alpha) = \underline{\ ?\ }(180° - \alpha) = \underline{\ ?\ }(270° - \alpha)$

b. How do the identities in part a compare to those in Exercise 6a?

8. a. How did reduction identities get their name?

b. If you needed one of these identities, but could not remember it, what other type(s) of identities could you use to derive it?

Double-Angle and Half-Angle Identities

OBJECTIVE

- Use the double- and half-angle identities for the sine, cosine, and tangent functions.

ARCHITECTURE Mike MacDonald is an architect who designs water fountains. One part of his job is determining the placement of the water jets that shoot the water into the air to create arcs. These arcs are modeled by parabolic functions. When a stream of water is shot into the air with velocity v at an angle of θ with the horizontal, the model predicts that the water will travel a horizontal distance of $D = \frac{v^2}{g} \sin 2\theta$ and reach a maximum height of $H = \frac{v^2}{2g} \sin^2 \theta$, where g is the acceleration due to gravity. The ratio of H to D helps determine the total height and width of the fountain. Express $\frac{H}{D}$ as a function of θ. *This problem will be solved in Example 3.*

It is sometimes useful to have identities to find the value of a function of twice an angle or half an angle. We can substitute θ for both α and β in $\sin(\alpha + \beta)$ to find an identity for $\sin 2\theta$.

$$\begin{aligned}\sin 2\theta &= \sin(\theta + \theta) \\ &= \sin\theta\cos\theta + \cos\theta\sin\theta \quad \textit{Sum identity for sine} \\ &= 2\sin\theta\cos\theta\end{aligned}$$

The same method can be used to find an identity for $\cos 2\theta$.

$$\begin{aligned}\cos 2\theta &= \cos(\theta + \theta) \\ &= \cos\theta\cos\theta - \sin\theta\sin\theta \quad \textit{Sum identity for cosine} \\ &= \cos^2\theta - \sin^2\theta\end{aligned}$$

If we substitute $1 - \cos^2\theta$ for $\sin^2\theta$ or $1 - \sin^2\theta$ for $\cos^2\theta$, we will have two alternate identities for $\cos 2\theta$.

$$\cos 2\theta = 2\cos^2\theta - 1$$

$$\cos 2\theta = 1 - 2\sin^2\theta$$

These identities may be used if θ is measured in degrees or radians. So, θ may represent either a degree measure or a real number.

The tangent of a double angle can be found by substituting θ for both α and β in $\tan(\alpha + \beta)$.

$$\tan 2\theta = \tan(\theta + \theta)$$

$$= \frac{\tan\theta + \tan\theta}{1 - \tan\theta\tan\theta} \quad \textit{Sum identity for tangent}$$

$$= \frac{2\tan\theta}{1 - \tan^2\theta}$$

Double-Angle Identities

If θ represents the measure of an angle, then the following identities hold for all values of θ.

$$\sin 2\theta = 2\sin\theta\cos\theta$$

$$\cos 2\theta = \cos^2\theta - \sin^2\theta$$

$$\cos 2\theta = 2\cos^2\theta - 1$$

$$\cos 2\theta = 1 - 2\sin^2\theta$$

$$\tan 2\theta = \frac{2\tan\theta}{1 - \tan^2\theta}$$

Example 1 **If $\sin\theta = \frac{2}{3}$ and θ has its terminal side in the first quadrant, find the exact value of each function.**

a. $\sin 2\theta$

To use the double-angle identity for $\sin 2\theta$, we must first find $\cos\theta$.

$$\sin^2\theta + \cos^2\theta = 1$$

$$\left(\frac{2}{3}\right)^2 + \cos^2\theta = 1 \quad \sin\theta = \frac{2}{3}$$

$$\cos^2\theta = \frac{5}{9}$$

$$\cos\theta = \frac{\sqrt{5}}{3}$$

Then find $\sin 2\theta$.

$$\sin 2\theta = 2\sin\theta\cos\theta$$

$$= 2\left(\frac{2}{3}\right)\left(\frac{\sqrt{5}}{3}\right) \quad \sin\theta = \frac{2}{3};\ \cos\theta = \frac{\sqrt{5}}{3}$$

$$= \frac{4\sqrt{5}}{9}$$

b. $\cos 2\theta$

Since we know the values of $\cos\theta$ and $\sin\theta$, we can use any of the double-angle identities for cosine.

$$\cos 2\theta = \cos^2\theta - \sin^2\theta$$

$$= \left(\frac{\sqrt{5}}{3}\right)^2 - \left(\frac{2}{3}\right)^2 \quad \cos\theta = \frac{\sqrt{5}}{3};\ \sin\theta = \frac{2}{3}$$

$$= \frac{1}{9}$$

c. tan 2θ

We must find tan θ to use the double-angle identity for tan 2θ.

$$\tan\theta = \frac{\sin\theta}{\cos\theta}$$

$$= \frac{\frac{2}{3}}{\frac{\sqrt{5}}{3}} \quad \textit{\sin\theta = \frac{2}{3}, \cos\theta = \frac{\sqrt{5}}{3}}$$

$$= \frac{2}{\sqrt{5}} \text{ or } \frac{2\sqrt{5}}{5}$$

Then find tan 2θ.

$$\tan 2\theta = \frac{2\tan\theta}{1-\tan^2\theta}$$

$$= \frac{2\left(\frac{2\sqrt{5}}{5}\right)}{1-\left(\frac{2\sqrt{5}}{5}\right)^2} \quad \tan\theta = \frac{2\sqrt{5}}{5}$$

$$= \frac{\frac{4\sqrt{5}}{5}}{\frac{1}{5}} \text{ or } 4\sqrt{5}$$

d. cos 4θ

Since $4\theta = 2(2\theta)$, use a double-angle identity for cosine again.

$$\cos 4\theta = \cos 2(2\theta)$$

$$= \cos^2(2\theta) - \sin^2(2\theta) \quad \textit{Double-angle identity}$$

$$= \left(\frac{1}{9}\right)^2 - \left(\frac{4\sqrt{5}}{9}\right)^2 \quad \cos 2\theta = \frac{1}{9}, \sin 2\theta = \frac{4\sqrt{5}}{9} \text{ (parts a and b)}$$

$$= -\frac{79}{81}$$

We can solve two of the forms of the identity for cos 2θ for cos θ and sin θ, respectively, and the following equations result.

$\cos 2\theta = 2\cos^2\theta - 1$ **Solve for cos θ.** $\cos\theta = \pm\sqrt{\frac{1+\cos 2\theta}{2}}$

$\cos 2\theta = 1 - 2\sin^2\theta$ **Solve for sin θ.** $\sin\theta = \pm\sqrt{\frac{1-\cos 2\theta}{2}}$

We can replace 2θ with α and θ with $\frac{\alpha}{2}$ to derive the half-angle identities.

$$\tan \frac{\alpha}{2} = \frac{\sin \frac{\alpha}{2}}{\cos \frac{\alpha}{2}}$$

$$= \frac{\pm\sqrt{\frac{1 - \cos \alpha}{2}}}{\pm\sqrt{\frac{1 + \cos \alpha}{2}}} \text{ or } \pm\sqrt{\frac{1 - \cos \alpha}{1 + \cos \alpha}}$$

Half-Angle Identities

If α represents the measure of an angle, then the following identities hold for all values of α.

$$\sin \frac{\alpha}{2} = \pm\sqrt{\frac{1 - \cos \alpha}{2}}$$

$$\cos \frac{\alpha}{2} = \pm\sqrt{\frac{1 + \cos \alpha}{2}}$$

$$\tan \frac{\alpha}{2} = \pm\sqrt{\frac{1 - \cos \alpha}{1 + \cos \alpha}},\ \cos \alpha \neq -1$$

Unlike with the double-angles identities, you must determine the sign.

Example 2 **Use a half-angle identity to find the exact value of each function.**

a. $\sin \frac{7\pi}{12}$

$$\sin \frac{7\pi}{12} = \sin \frac{\frac{7\pi}{6}}{2}$$

$$= \sqrt{\frac{1 - \cos \frac{7\pi}{6}}{2}}$$

Use $\sin \frac{\alpha}{2} = \pm\sqrt{\frac{1 - \cos \alpha}{2}}$. *Since* $\frac{7\pi}{12}$ *is in Quadrant II, choose the positive sine value.*

$$= \sqrt{\frac{1 - \left(-\frac{\sqrt{3}}{2}\right)}{2}}$$

$$= \frac{\sqrt{2 + \sqrt{3}}}{2}$$

b. $\cos 67.5°$

$$\cos 67.5° = \cos \frac{135°}{2}$$

$$= \sqrt{\frac{1 + \cos 135°}{2}}$$

Use $\cos \frac{\alpha}{2} = \pm\sqrt{\frac{1 + \cos \alpha}{2}}$. *Since 67.5° is in Quadrant I, choose the positive cosine value.*

$$= \sqrt{\frac{1 - \frac{\sqrt{2}}{2}}{2}}$$

$$= \frac{\sqrt{2 - \sqrt{2}}}{2}$$

Double- and half-angle identities can be used to simplify trigonometric expressions.

Example 3 **ARCHITECTURE** **Refer to the application at the beginning of the lesson.**

a. Find and simplify $\frac{H}{D}$.

b. What is the ratio of the maximum height of the water to the horizontal distance it travels for an angle of 27°?

a. $$\frac{H}{D} = \frac{\frac{v^2}{2g}\sin^2\theta}{\frac{v^2}{g}\sin 2\theta}$$

$$= \frac{\sin^2\theta}{2\sin 2\theta} \quad \textit{Simplify.}$$

$$= \frac{\sin^2\theta}{4\sin\theta\cos\theta} \quad \sin 2\theta = 2\sin\theta\cos\theta$$

$$= \frac{1}{4}\cdot\frac{\sin\theta}{\cos\theta} \quad \textit{Simplify.}$$

$$= \frac{1}{4}\tan\theta \quad \textit{Quotient identity: } \frac{\sin\theta}{\cos\theta} = \tan\theta$$

Therefore, the ratio of the maximum height of the water to the horizontal distance it travels is $\frac{1}{4}\tan\theta$.

b. When $\theta = 27°$, $\frac{H}{D} = \frac{1}{4}\tan 27°$, or about 0.13.

For an angle of 27°, the ratio of the maximum height of the water to the horizontal distance it travels is about 0.13.

The double- and half-angle identities can also be used to verify other identities.

Example 4 **Verify that $\frac{\cos 2\theta}{1 + \sin 2\theta} = \frac{\cot\theta - 1}{\cot\theta + 1}$ is an identity.**

$$\frac{\cos 2\theta}{1 + \sin 2\theta} \stackrel{?}{=} \frac{\cot\theta - 1}{\cot\theta + 1}$$

$$\frac{\cos 2\theta}{1 + \sin 2\theta} \stackrel{?}{=} \frac{\frac{\cos\theta}{\sin\theta} - 1}{\frac{\cos\theta}{\sin\theta} + 1} \quad \textit{Reciprocal identity: } \cot\theta = \frac{\cos\theta}{\sin\theta}$$

$$\frac{\cos 2\theta}{1 + \sin 2\theta} \stackrel{?}{=} \frac{\cos\theta - \sin\theta}{\cos\theta + \sin\theta} \quad \textit{Multiply numerator and denominator by } \sin\theta.$$

$$\frac{\cos 2\theta}{1 + \sin 2\theta} \stackrel{?}{=} \frac{\cos\theta - \sin\theta}{\cos\theta + \sin\theta}\cdot\frac{\cos\theta + \sin\theta}{\cos\theta + \sin\theta} \quad \textit{Multiply each side by 1.}$$

$$\frac{\cos 2\theta}{1 + \sin 2\theta} \stackrel{?}{=} \frac{\cos^2\theta - \sin^2\theta}{\cos^2\theta + 2\cos\theta\sin\theta + \sin^2\theta} \quad \textit{Multiply.}$$

$$\frac{\cos 2\theta}{1 + \sin 2\theta} \stackrel{?}{=} \frac{\cos^2 \theta - \sin^2 \theta}{1 + 2\cos\theta \sin\theta}$$ *Simplify.*

$$\frac{\cos 2\theta}{1 + \sin 2\theta} = \frac{\cos 2\theta}{1 + \sin 2\theta}$$ *Double-angle identities: $\cos^2 \theta - \sin^2 \theta = \cos 2\theta$, $2 \cos \theta \sin \theta = \sin 2\theta$*

CHECK FOR UNDERSTANDING

Communicating Mathematics

Read and study the lesson to answer each question.

1. **Write** a paragraph about the conditions under which you would use each of the three identities for $\cos 2\theta$.
2. **Derive** the identity $\sin \frac{\alpha}{2} = \pm\sqrt{\frac{1 - \cos \alpha}{2}}$ from $\cos 2\theta = 1 - 2 \sin^2 \theta$.
3. **Name** the quadrant in which the terminal side lies.
   a. x is a second quadrant angle. In which quadrant does $2x$ lie?
   b. $\frac{x}{2}$ is a first quadrant angle. In which quadrant does x lie?
   c. $2x$ is a second quadrant angle. In which quadrant does $\frac{x}{2}$ lie?
4. **Provide a counterexample** to show that $\sin 2\theta = 2 \sin \theta$ is not an identity.
5. **You Decide** Tamika calculated the exact value of $\sin 15°$ in two different ways. Using the difference identity for sine, $\sin 15°$ was $\frac{\sqrt{6} - \sqrt{2}}{4}$. When she used the half-angle identity, $\sin 15°$ equaled $\frac{\sqrt{2 - \sqrt{3}}}{2}$. Which answer is correct? Explain.

Guided Practice

Use a half-angle identity to find the exact value of each function.

6. $\sin \frac{\pi}{8}$
7. $\tan 165°$

Use the given information to find $\sin 2\theta$, $\cos 2\theta$, and $\tan 2\theta$.

8. $\sin \theta = \frac{2}{5}, 0° < \theta < 90°$
9. $\tan \theta = \frac{4}{3}, \pi < \theta < \frac{3\pi}{2}$

Verify that each equation is an identity.

10. $\tan 2\theta = \frac{2}{\cot \theta - \tan \theta}$
11. $1 + \frac{1}{2} \sin 2A = \frac{\sec A + \sin A}{\sec A}$
12. $\sin \frac{x}{2} \cos \frac{x}{2} = \frac{\sin x}{2}$

13. **Electronics** Consider an AC circuit consisting of a power supply and a resistor. If the current in the circuit at time t is $I_0 \sin \omega t$, then the power delivered to the resistor is $P = I_0^2 R \sin^2 \omega t$, where R is the resistance. Express the power in terms of $\cos 2\omega t$.

EXERCISES

Practice

Use a half-angle identity to find the exact value of each function.

14. $\cos 15°$
15. $\sin 75°$
16. $\tan \frac{5\pi}{12}$
17. $\sin \frac{3\pi}{8}$
18. $\cos \frac{7\pi}{12}$
19. $\tan 22.5°$
20. If θ is an angle in the first quadrant and $\cos \theta = \frac{1}{4}$, find $\tan \frac{\theta}{2}$.

Use the given information to find sin 2θ, cos 2θ, and tan 2θ.

21. $\cos \theta = \frac{4}{5}$, $0° < \theta < 90°$
22. $\sin \theta = \frac{1}{3}$, $0 < \theta < \frac{\pi}{2}$
23. $\tan \theta = -2$, $\frac{\pi}{2} < \theta < \pi$
24. $\sec \theta = -\frac{4}{3}$, $90° < \theta < 180°$
25. $\cot \theta = \frac{3}{2}$, $180° < \theta < 270°$
26. $\csc \theta = -\frac{5}{2}$, $\frac{3\pi}{2} < \theta < 2\pi$
27. If α is an angle in the second quadrant and $\cos \alpha = -\frac{\sqrt{2}}{3}$, find $\tan 2\alpha$.

Verify that each equation is an identity.

28. $\csc 2\theta = \frac{1}{2} \sec \theta \csc \theta$
29. $\cos A - \sin A = \frac{\cos 2A}{\cos A + \sin A}$
30. $(\sin \theta + \cos \theta)^2 - 1 = \sin 2\theta$
31. $\cos x - 1 = \frac{\cos 2x - 1}{2(\cos x + 1)}$
32. $\sec 2\theta = \frac{\cos^2 \theta + \sin^2 \theta}{\cos^2 \theta - \sin^2 \theta}$
33. $\tan \frac{A}{2} = \frac{\sin A}{1 + \cos A}$
34. $\sin 3x = 3 \sin x - 4 \sin^3 x$
35. $\cos 3x = 4 \cos^3 x - 3 \cos x$

Applications and Problem Solving

36. **Architecture** Refer to the application at the beginning of the lesson. If the angle of the water is doubled, what is the ratio of the new maximum height to the original maximum height?

37. **Critical Thinking** Circle O is a unit circle. Use the figure to prove that $\tan \frac{1}{2}\theta = \frac{\sin \theta}{1 + \cos \theta}$.

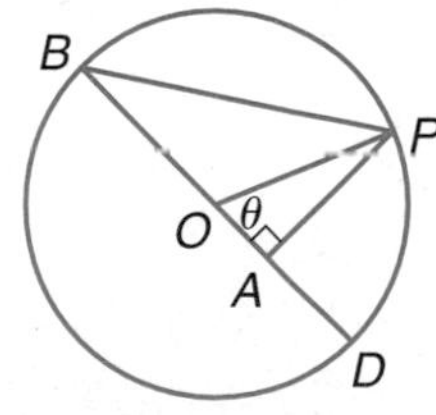

38. **Physics** Suppose a projectile is launched with velocity v at an angle θ to the horizontal from the base of a hill that makes an angle α with the horizontal ($\theta > \alpha$). Then the range of the projectile, measured along the slope of the hill, is given by $R = \frac{2v^2 \cos \theta \sin (\theta - \alpha)}{g \cos^2 \alpha}$. Show that if $\alpha = 45°$, then $R = \frac{v^2\sqrt{2}}{g}(\sin 2\theta - \cos 2\theta - 1)$.

interNET CONNECTION

Research
For the latitude and longitude of world cities, and the distance between them, visit: **www.amc.glencoe.com**

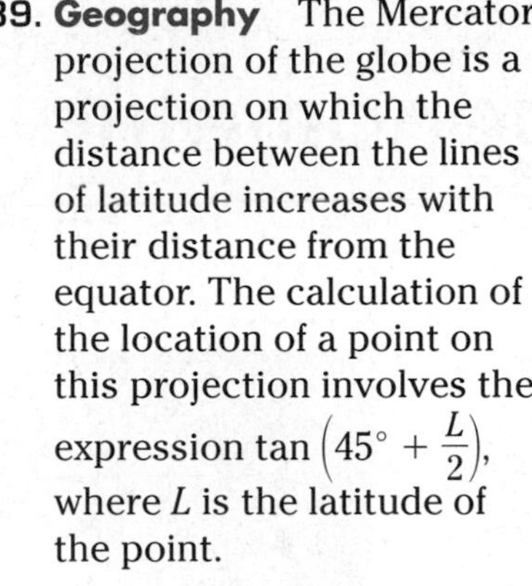

39. **Geography** The Mercator projection of the globe is a projection on which the distance between the lines of latitude increases with their distance from the equator. The calculation of the location of a point on this projection involves the expression $\tan\left(45^\circ + \frac{L}{2}\right)$, where L is the latitude of the point.

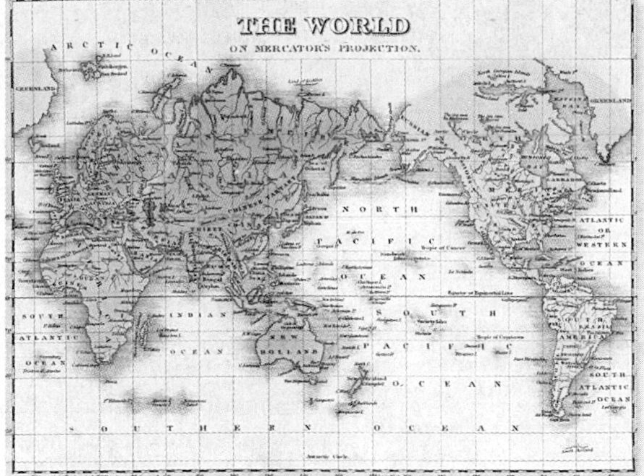

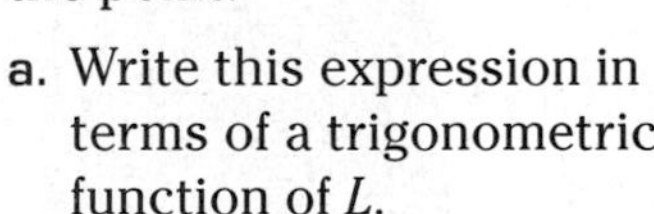

a. Write this expression in terms of a trigonometric function of L.

b. Find the value of this expression if $L = 60^\circ$.

40. **Critical Thinking** Determine the tangent of angle α in the figure.

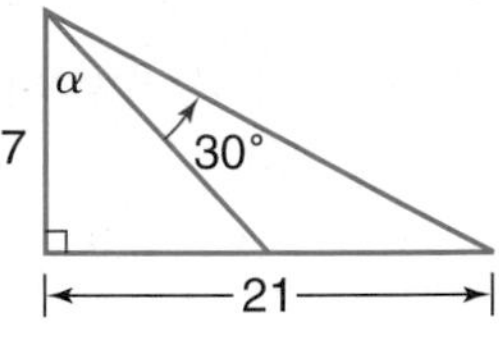

Mixed Review

41. Find the exact value of $\sec \frac{\pi}{12}$. *(Lesson 7-3)*

42. Show that $\sin x^2 + \cos x^2 = 1$ is not an identity. *(Lesson 7-1)*

43. Find the degree measure to the nearest tenth of the central angle of a circle of radius 10 centimeters if the measure of the subtended arc is 17 centimeters. *(Lesson 6-1)*

44. **Surveying** To find the height of a mountain peak, points A and B were located on a plain in line with the peak, and the angle of elevation was measured from each point. The angle at A was $36°40'$, and the angle at B was $21°10'$. The distance from A to B was 570 feet. How high is the peak above the level of the plain? *(Lesson 5-4)*

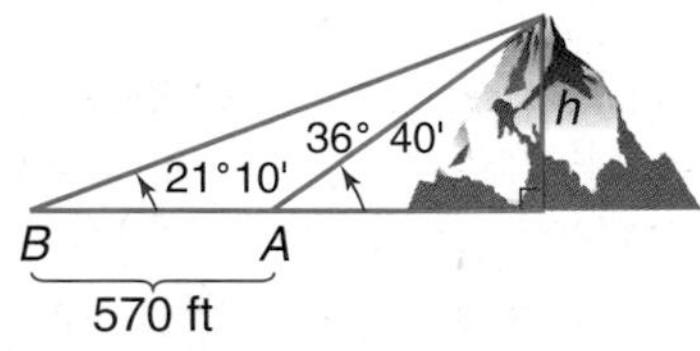

45. Write a polynomial equation of least degree with roots -3, 0.5, 6, and 2. *(Lesson 4-1)*

46. Graph $y = 2x + 5$ and its inverse. *(Lesson 3-4)*

47. Solve the system of equations. *(Lesson 2-1)*
$x + 2y = 11$
$3x - 5y = 11$

48. **SAT Practice** **Grid-In** If $(a - b)^2 = 64$, and $ab = 3$, find $a^2 + b^2$.

Extra Practice See p. A39.

7-5 Solving Trigonometric Equations

OBJECTIVE
- Solve trigonometric equations and inequalities.

ENTERTAINMENT When you ride a Ferris wheel that has a diameter of 40 meters and turns at a rate of 1.5 revolutions per minute, the height above the ground, in meters, of your seat after t minutes can be modeled by the equation $h = 21 - 20 \cos 3\pi t$. How long after the ride starts will your seat first be 31 meters above the ground? *This problem will be solved in Example 4.*

So far in this chapter, we have studied a special type of trigonometric equation called an identity. Trigonometric identities are equations that are true for all values of the variable for which both sides are defined. In this lesson, we will examine another type of trigonometric equation. These equations are true for only certain values of the variable. Solving these equations resembles solving algebraic equations.

Most trigonometric equations have more than one solution. If the variable is not restricted, the periodic nature of trigonometric functions will result in an infinite number of solutions. Also, many trigonometric expressions will take on a given value twice every period.

If the variable is restricted to two adjacent quadrants, a trigonometric equation will have fewer solutions. These solutions are called **principal values.** For $\sin x$ and $\tan x$, the principal values are in Quadrants I and IV. So x is in the interval $-90° \leq x \leq 90°$. For $\cos x$, the principal values are in Quadrants I and II, so x is in the interval $0° \leq x \leq 180°$.

Example 1 **Solve $\sin x \cos x - \frac{1}{2} \cos x = 0$ for principal values of x. Express solutions in degrees.**

$$\sin x \cos x - \frac{1}{2} \cos x = 0$$

$$\cos x\left(\sin x - \frac{1}{2}\right) = 0 \quad \textit{Factor.}$$

$\cos x = 0$ or $\sin x - \frac{1}{2} = 0$ *Set each factor equal to 0.*

$x = 90°$

$\sin x = \frac{1}{2}$

$x = 30°$

The principal values are 30° and 90°.

If an equation cannot be solved easily by factoring, try writing the expressions in terms of only one trigonometric function. Remember to use your knowledge of identities.

Example 2 **Solve $\cos^2 x - \cos x + 1 = \sin^2 x$ for $0 \leq x < 2\pi$.**

This equation can be written in terms of $\cos x$ only.

$\cos^2 x - \cos x + 1 = \sin^2 x$

$\cos^2 x - \cos x + 1 = 1 - \cos^2 x$ *Pythagorean identity: $\sin^2 x = 1 - \cos^2 x$*

$2\cos^2 x - \cos x = 0$

$\cos x(2\cos x - 1) = 0$ *Factor.*

$\cos x = 0$ or $2\cos x - 1 = 0$

$x = \frac{\pi}{2}$ or $x = \frac{3\pi}{2}$

$\cos x = \frac{1}{2}$

$x = \frac{\pi}{3}$ or $x = \frac{5\pi}{3}$

The solutions are $\frac{\pi}{3}, \frac{\pi}{2}, \frac{3\pi}{2}$, and $\frac{5\pi}{3}$.

As indicated earlier, most trigonometric equations have infinitely many solutions. When all of the values of x are required, the solution should be represented as $x + 360k°$ or $x + 2\pi k$ for $\sin x$ and $\cos x$ and $x + 180k°$ or $x + \pi k$ for $\tan x$, where k is any integer.

Example 3 **Solve $2\sec^2 x - \tan^4 x = -1$ for all real values of x.**

A Pythagorean identity can be used to write this equation in terms of $\tan x$ only.

$2\sec^2 x - \tan^4 x = -1$

$2(1 + \tan^2 x) - \tan^4 x = -1$ *Pythagorean identity: $\sec^2 x = 1 + \tan^2 x$*

$2 + 2\tan^2 x - \tan^4 x = -1$ *Simplify.*

$\tan^4 x - 2\tan^2 x - 3 = 0$

$(\tan^2 x - 3)(\tan^2 x + 1) = 0$ *Factor.*

$\tan^2 x - 3 = 0$ or $\tan^2 x + 1 = 0$

$\tan^2 x = 3$

$\tan^2 x = -1$ *This part gives no solutions since $\tan^2 x \geq 0$.*

$\tan x = \pm\sqrt{3}$

$x = \frac{\pi}{3} + \pi k$ or $x = -\frac{\pi}{3} + \pi k$, where k is any integer.

When a problem asks for real values of x, use radians.

The solutions are $\frac{\pi}{3} + \pi k$ and $-\frac{\pi}{3} + \pi k$.

There are times when a general expression for all of the solutions is helpful for determining a specific solution.

Example **4** **ENTERTAINMENT** **Refer to the application at the beginning of the lesson. How long after the Ferris wheel starts will your seat first be 31 meters above the ground?**

$h = 21 - 20 \cos 3\pi t$

$31 = 21 - 20 \cos 3\pi t$ *Replace h with 31.*

$-\frac{1}{2} = \cos 3\pi t$

$\frac{2\pi}{3} + 2\pi k = 3\pi t$ or $\frac{4\pi}{3} + 2\pi k = 3\pi t$ *where k is any integer*

$\frac{2}{9} + \frac{2}{3}k = t$ or $\frac{4}{9} + \frac{2}{3}k = t$

The least positive value for t is obtained by letting $k = 0$ in the first expression. Therefore, $t = \frac{2}{9}$ of a minute or about 13 seconds.

You can solve some trigonometric inequalities using the same techniques as for algebraic inequalities. The unit circle can be useful when deciding which angles to include in the answer.

Example **5** **Solve $2 \sin \theta + 1 > 0$ for $0 \leq \theta < 2\pi$.**

$2 \sin \theta + 1 > 0$

$\sin \theta > -\frac{1}{2}$ *Solve for sin θ.*

In terms of the unit circle, we need to find points with y-coordinates greater than $-\frac{1}{2}$.

The values of θ for which $\sin \theta = -\frac{1}{2}$ are $\frac{7\pi}{6}$ and $\frac{11\pi}{6}$. The figure shows that the solution of the inequality is $0 \leq \theta < \frac{7\pi}{6}$ or $\frac{11\pi}{6} < \theta < 2\pi$.

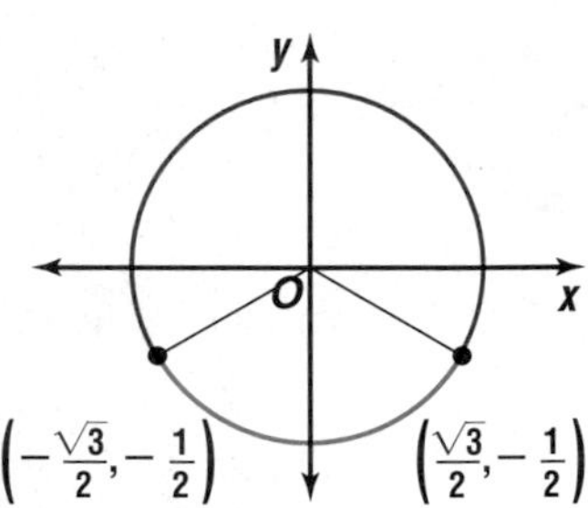

GRAPHING CALCULATOR EXPLORATION

Some trigonometric equations and inequalities are difficult or impossible to solve with only algebraic methods. A graphing calculator is helpful in such cases.

TRY THESE **Graph each side of the equation as a separate function.**

1. $\sin x = 2 \cos x$ for $0 \leq x \leq 2\pi$
2. $\tan 0.5x = \cos x$ for $-2\pi \leq x \leq 2\pi$
3. Use the **CALC** menu to find the intersection point(s) of the graphs in Exercises 1 and 2.

WHAT DO YOU THINK?

4. What do the values in Exercise 3 represent? How could you verify this conjecture?
5. Graph $y = 2 \cos x - \sin x$ for $0 \leq x \leq 2\pi$.
   a. How could you use the graph to solve the equation $\sin x = 2 \cos x$? How does this solution compare with those found in Exercise 3?
   b. What equation would you use to apply this method to $\tan 0.5x = \cos x$?

CHECK FOR UNDERSTANDING

Communicating Mathematics

Read and study the lesson to answer each question.

1. **Explain** the difference between a trigonometric identity and a trigonometric equation that is not an identity.
2. **Explain** why many trigonometric equations have infinitely many solutions.
3. **Write** all the solutions to a trigonometric equation in terms of $\sin x$, given that the solutions between 0° and 360° are 45° and 135°.
4. *Math Journal* **Compare and contrast** solving trigonometric equations with solving linear and quadratic equations. What techniques are the same? What techniques are different? How many solutions do you expect? Do you use a graphing calculator in a similar manner?

Guided Practice

Solve each equation for principal values of x. Express solutions in degrees.

5. $2 \sin x + 1 = 0$
6. $2 \cos x - \sqrt{3} = 0$

Solve each equation for $0° \leq x < 360°$.

7. $\sin x \cot x = \frac{\sqrt{3}}{2}$
8. $\cos 2x = \sin^2 x - 2$

Solve each equation for $0 \leq x < 2\pi$.

9. $3 \tan^2 x - 1 = 0$
10. $2 \sin^2 x = 5 \sin x + 3$

Solve each equation for all real values of x.

11. $\sin^2 2x + \cos^2 x = 0$
12. $\tan^2 x + 2 \tan x + 1 = 0$
13. $\cos^2 x + 3 \cos x = -2$
14. $\sin 2x - \cos x = 0$
15. Solve $2 \cos \theta + 1 < 0$ for $0 \leq \theta < 2\pi$.

16. **Physics** The work done in moving an object through a displacement of d meters is given by $W = Fd \cos \theta$, where θ is the angle between the displacement and the force F exerted. If Lisa does 1500 joules of work while exerting a 100-newton force over 20 meters, at what angle was she exerting the force?

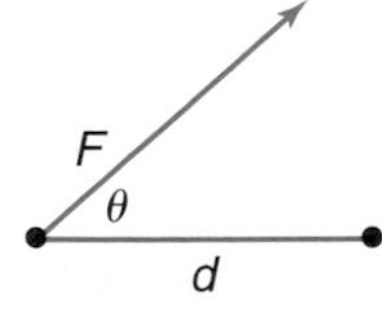

EXERCISES

Practice

Solve each equation for principal values of x. Express solutions in degrees.

17. $\sqrt{2} \sin x - 1 = 0$
18. $2 \cos x + 1 = 0$
19. $\sin 2x - 1 = 0$
20. $\tan 2x - \sqrt{3} = 0$
21. $\cos^2 x = \cos x$
22. $\sin x = 1 + \cos^2 x$

Solve each equation for $0° \leq x < 360°$.

23. $\sqrt{2} \cos x + 1 = 0$
24. $\cos x \tan x = \frac{1}{2}$
25. $\sin x \tan x - \sin x = 0$
26. $2 \cos^2 x + 3 \cos x - 2 = 0$
27. $\sin 2x = -\sin x$
28. $\cos (x + 45°) + \cos (x - 45°) = \sqrt{2}$

29. Find all solutions to $2 \sin \theta \cos \theta + \sqrt{3} \sin \theta = 0$ in the interval $0° \leq \theta < 360°$.

Solve each equation for $0 \leq x < 2\pi$.

30. $(2 \sin x - 1)(2 \cos^2 x - 1) = 0$

31. $4 \sin^2 x + 1 = -4 \sin x$

32. $\sqrt{2} \tan x = 2 \sin x$

33. $\sin x = \cos 2x - 1$

34. $\cot^2 x - \csc x = 1$

35. $\sin x + \cos x = 0$

36. Find all values of θ between 0 and 2π that satisfy $-1 - 3 \sin \theta = \cos 2\theta$.

Solve each equation for all real values of *x*.

37. $\sin x = -\frac{1}{2}$

38. $\cos x \tan x - 2 \cos^2 x = -1$

39. $3 \tan^2 x = \sqrt{3} \tan x$

40. $2 \cos^2 x = 3 \sin x$

41. $\frac{1}{\cos x - \sin x} = \cos x + \sin x$

42. $2 \tan^2 x - 3 \sec x = 0$

43. $\sin x \cos x = \frac{1}{2}$

44. $\cos^2 x - \sin^2 x = \frac{\sqrt{3}}{2}$

45. $\sin^4 x - 1 = 0$

46. $\sec^2 x + 2 \sec x = 0$

47. $\sin x + \cos x = 1$

48. $2 \sin x + \csc x = 3$

Solve each inequality for $0 \leq \theta < 2\pi$.

49. $\cos \theta \leq -\frac{\sqrt{3}}{2}$

50. $\cos \theta - \frac{1}{2} > 0$

51. $\sqrt{2} \sin \theta - 1 < 0$

Graphing Calculator

Solve each equation graphically on the interval $0 \leq x < 2\pi$.

52. $\tan x = 0.5$

53. $\sin x - \frac{x}{2} = 0$

54. $\cos x = 3 \sin x$

Applications and Problem Solving

55. Optics When light passes through a small slit, it is diffracted. The angle θ subtended by the first diffraction minimum is related to the wavelength λ of the light and the width D of the slit by the equation $\sin \theta = \frac{\lambda}{D}$. Consider light of wavelength 550 nanometers (5.5×10^{-7} m). What is the angle subtended by the first diffraction minimum when the light passes through a slit of width 3 millimeters?

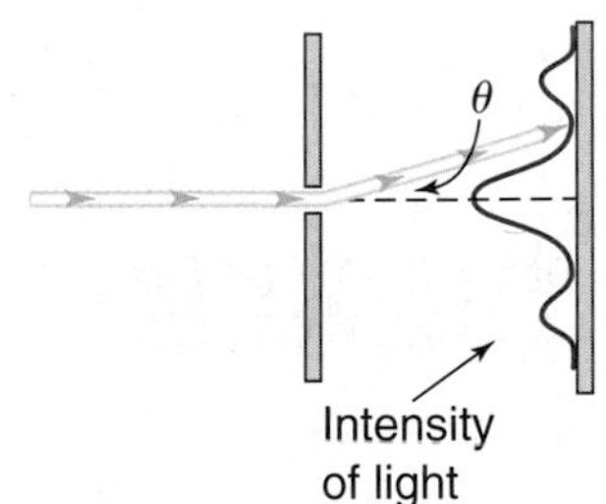

56. Critical Thinking Solve the inequality $\sin 2x < \sin x$ for $0 \leq x < 2\pi$ without a calculator.

57. Physics The range of a projectile that is launched with an initial velocity v at an angle of θ with the horizontal is given by $R = \frac{v^2}{g} \sin 2\theta$, where g is the acceleration due to gravity or 9.8 meters per second squared. If a projectile is launched with an initial velocity of 15 meters per second, what angle is required to achieve a range of 20 meters?

58. **Gemology** The sparkle of a diamond is created by *refracted* light. Light travels at different speeds in different mediums. When light rays pass from one medium to another in which they travel at a different velocity, the light is bent, or refracted. According to Snell's Law, $n_1 \sin i = n_2 \sin r$, where n_1 is the index of refraction of the medium the light is exiting, n_2 is the index of refraction of the medium the light is entering, i is the angle of incidence, and r is the angle of refraction.
    a. The index of refraction of a diamond is 2.42, and the index of refraction of air is 1.00. If a beam of light strikes a diamond at an angle of 35°, what is the angle of refraction?
    b. Explain how a gemologist might use Snell's Law to determine if a diamond is genuine.

59. **Music** A wave traveling in a guitar string can be modeled by the equation $D = 0.5 \sin(6.5x) \sin(2500t)$, where D is the displacement in millimeters at the position x meters from the left end of the string at time t seconds. Find the first positive time when the point 0.5 meter from the left end has a displacement of 0.01 millimeter.

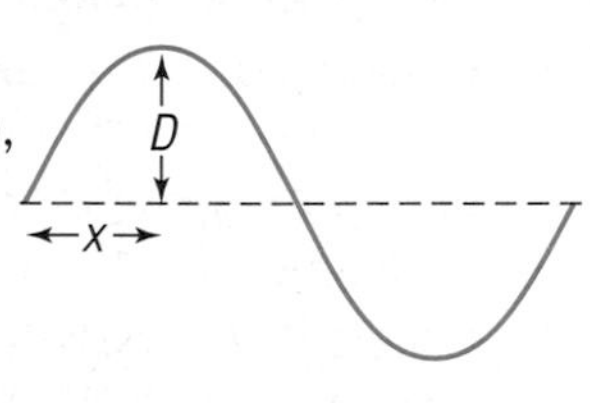

60. **Critical Thinking** How many solutions in the interval $0° \leq x < 360°$ should you expect for the equation $a \sin(bx + c) + d = d + \frac{a}{2}$, if $a \neq 0$ and b is a positive integer?

61. **Geometry** The point $P(x, y)$ can be rotated θ degrees counterclockwise about the origin by multiplying the matrix $\begin{bmatrix} x \\ y \end{bmatrix}$ on the left by the rotation matrix $R_\theta = \begin{bmatrix} \cos\theta & -\sin\theta \\ \sin\theta & \cos\theta \end{bmatrix}$. Determine the angle required to rotate the point $P(3, 4)$ to the point $P'(\sqrt{17}, 2\sqrt{2})$.

Mixed Review

62. Find the exact value of cot 67.5°. *(Lesson 7-4)*

63. Find a numerical value of one trigonometric function of x if $\frac{\tan x}{\sec x} = \frac{\sqrt{2}}{5}$. *(Lesson 7-2)*

64. Graph $y = \frac{2}{3} \cos \theta$. *(Lesson 6-4)*

65. **Transportation** A boat trailer has wheels with a diameter of 14 inches. If the trailer is being pulled by a car going 45 miles per hour, what is the angular velocity of the wheels in revolutions per second? *(Lesson 6-2)*

66. Use the unit circle to find the value of csc 180°. *(Lesson 5-3)*

67. Determine the binomial factors of $x^3 - 3x - 2$. *(Lesson 4-3)*

68. Graph $y = x^3 - 3x + 5$. Find and classify its extrema. *(Lesson 3-6)*

69. Find the values of x and y for which $\begin{bmatrix} 3x + 4 \\ 6 \end{bmatrix} = \begin{bmatrix} 16 \\ 2y \end{bmatrix}$ is true. *(Lesson 2-3)*

70. Solve the system $x - y + z = 1$, $2x + y + 3z = 5$, and $x + y - z = 11$. *(Lesson 2-2)*

71. Graph $g(x) = |x + 3|$. *(Lesson 1-7)*

72. **SAT/ACT Practice** If $AC = 6$, what is the area of triangle ABC?
    A 1 B $\sqrt{6}$ C 3
    D 6 E 12

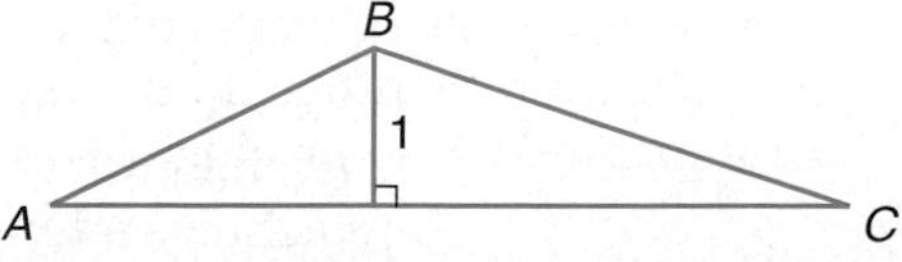

Extra Practice See p. A39.

HISTORY of MATHEMATICS

TRIGONOMETRY

Trigonometry was developed in response to the needs of astronomers. In fact, it was not until the thirteenth century that trigonometry and astronomy were treated as separate disciplines.

Early Evidence The earliest use of trigonometry was to calculate the heights of objects using the lengths of shadows. Egyptian mathematicians produced tables relating the lengths of shadows at particular times of day as early as the thirteenth century B.C.

The Greek mathematician **Hipparchus** (190–120 B.C.), is generally credited with laying the foundations of trigonometry. Hipparchus is believed to have produced a twelve-book treatise on the construction of a table of chords. This table related the lengths of chords of a circle to the angles subtended by those chords. In the diagram, $\angle AOB$ would be compared to the length of chord $\overline{AB}$.

In about 100 A.D., the Greek mathematician **Menelaus**, credited with the first work on spherical trigonometry, also produced a treatise on chords in a circle. **Ptolemy**, a Babylonian mathematician, produced yet another book of chords, believed to have been adapted from Hipparchus' treatise. He used an identity similar to $\sin^2 x + \cos^2 x = 1$, except that it was relative to chords. He also used the formulas $\sin (x + y) = \sin x \cos y + \cos x \sin y$ and $\frac{a}{\sin A} = \frac{b}{\sin B} = \frac{c}{\sin C}$ as they related to chords.

In about 500 A.D., **Aryabhata**, a Hindu mathematician, was the first person to use the sine function as we know it today. He produced a table of sines and called the sine *jya.* In 628 A.D., another Hindu mathematician, **Brahmagupta**, also produced a table of sines.

The Renaissance Many mathematicians developed theories and applications of trigonometry during this time period. **Nicolas Copernicus** (1473–1543) published a book highlighting all the trigonometry necessary for astronomy at that time. During this period, the sine and versed sine were the most important trigonometric functions. Today, the versed sine, which is defined as $\text{versin } x = 1 - \cos x$, is rarely used.

Copernicus

Modern Era Mathematicians of the 1700s, 1800s, and 1900s worked on more sophisticated trigonometric ideas such as those relating to complex variables and hyperbolic functions. Renowned mathematicians who made contributions to trigonometry during this era were **Bernoulli**, **Cotes**, **DeMoivre**, **Euler**, and **Lambert**.

Today architects, such as Dennis Holloway of Santa Fe, New Mexico, use trigonometry in their daily work. Mr. Holloway is particularly interested in Native American designs. He uses trigonometry to determine the best angles for the walls of his buildings and for finding the correct slopes for landscaping.

ACTIVITIES

1. Draw a circle of radius 5 centimeters. Make a table of chords for angles of measure 10° through 90°(use 10° intervals). The table headings should be "angle measure" and "length of chord." (In the diagram of circle O, you are using $\angle AOB$ and chord $\overline{AB}$.)
2. Find out more about personalities referenced in this article and others who contributed to the history of trigonometry. Visit **www.amc.glencoe.com**

7-6 Normal Form of a Linear Equation

OBJECTIVES
- Write the standard form of a linear equation given the length of the normal and the angle it makes with the x-axis.
- Write linear equations in normal form.

TRACK AND FIELD When a discus thrower releases the discus, it travels in a path that is tangent to the circle traced by the discus while the thrower was spinning around. Suppose the center of motion is the origin in the coordinate system. If the thrower spins so that the discus traces the unit circle and the discus is released at (0.96, 0.28), find an equation of the line followed by the discus. *This problem will be solved in Example 2.*

You are familiar with several forms for the equation of a line. These include slope-intercept form, point-slope form, and standard form. The usefulness of each form in a particular situation depends on how much relevant information each form provides upon inspection. In this lesson, you will learn about the **normal form** of a linear equation. The normal form uses trigonometry to provide information about the line.

ϕ is the Greek letter phi.

In general, a **normal line** is a line that is perpendicular to another line, curve, or surface. Given a line in the xy-plane, there is a normal line that intersects the given line and passes through the origin. The angle between this normal line and the x-axis is denoted by ϕ. The normal form of the equation of the given line is written in terms of ϕ and the length p of the segment of the normal line from the given line to the origin.

Suppose ℓ is a line that does not pass through the origin and p is the length of the normal from the origin. Let C be the point of intersection of ℓ with the normal, and let ϕ be the positive angle formed by the x-axis and $\overline{OC}$. Draw $\overline{MC}$ perpendicular to the x-axis. Since ϕ is in standard position, $\cos \phi = \frac{OM}{p}$ or $OM = p \cos \phi$ and $\sin \phi = \frac{MC}{p}$ or $MC = p \sin \phi$. So $\frac{p \sin \phi}{p \cos \phi}$ or $\frac{\sin \phi}{\cos \phi}$ is the slope of $\overline{OC}$. Since ℓ is perpendicular to $\overline{OC}$, the slope of ℓ is the negative reciprocal of the slope of $\overline{OC}$, or $-\frac{\cos \phi}{\sin \phi}$.

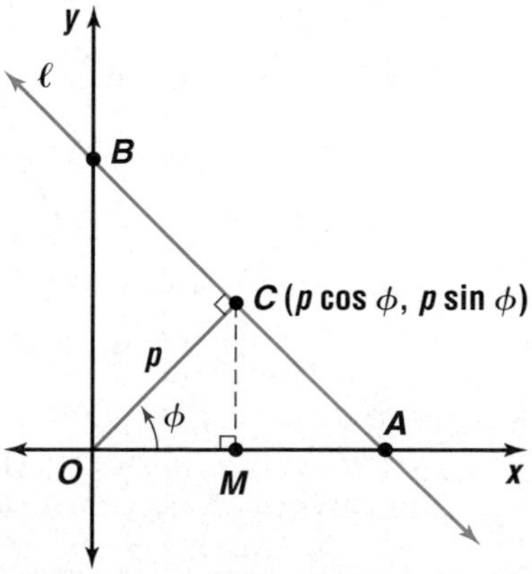

Look Back
You can refer to Lesson 1-4 to review the point-slope form.

Since ℓ contains C, we can use the point-slope form to write an equation of line ℓ.

$$y - y_1 = m(x - x_1)$$

$$y - p \sin \phi = -\frac{\cos \phi}{\sin \phi}(x - p \cos \phi) \quad \textit{Substitute for } m, x_1, \textit{ and } y_1.$$

$$y \sin \phi - p \sin^2 \phi = -x \cos \phi + p \cos^2 \phi \quad \textit{Multiply each side by } \sin \phi.$$

$$x \cos \phi + y \sin \phi = p(\sin^2 \phi + \cos^2 \phi)$$

$$x \cos \phi + y \sin \phi - p = 0 \quad \sin^2 \phi + \cos^2 \phi = 1$$

Normal Form

The normal form of a linear equation is

$$x \cos \phi + y \sin \phi - p = 0,$$

where p is the length of the normal from the line to the origin and ϕ is the positive angle formed by the positive x-axis and the normal.

You can write the standard form of a linear equation if you are given the values of ϕ and p.

Examples

1 **Write the standard form of the equation of a line for which the length of the normal segment to the origin is 6 and the normal makes an angle of 150° with the positive x-axis.**

$$x \cos \phi + y \sin \phi - p = 0 \quad \textit{Normal form}$$

$$x \cos 150° + y \sin 150° - 6 = 0 \quad \phi = 150° \textit{ and } p = 6$$

$$-\frac{\sqrt{3}}{2}x + \frac{1}{2}y - 6 = 0$$

$$\sqrt{3}x - y + 12 = 0 \quad \textit{Multiply each side by } -2.$$

The equation is $\sqrt{3}x - y + 12 = 0$.

2 **TRACK AND FIELD** **Refer to the application at the beginning of the lesson.**

a. Determine an equation of the path of the discus if it is released at (0.96, 0.28).

b. Will the discus strike an official at (-30, 40)? Explain your answer.

a. From the figure, we see that $p = 1$, $\sin \phi = 0.28$ or $\frac{7}{25}$, and $\cos \phi = 0.96$ or $\frac{24}{25}$.

An equation of the line is $\frac{24}{25}x + \frac{7}{25}y - 1 = 0$, or $24x + 7y - 25 = 0$.

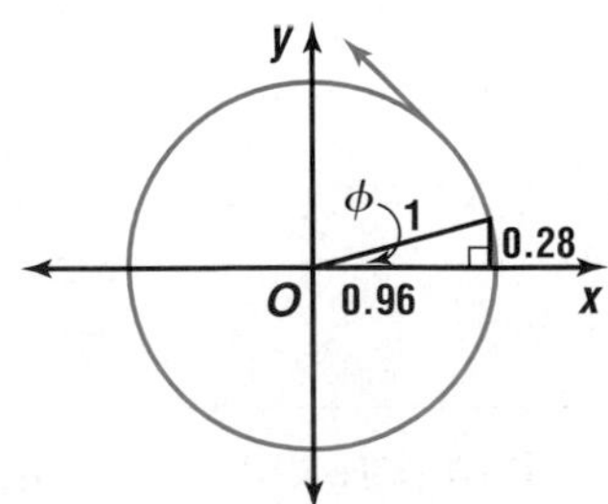

b. The y-coordinate of the point on the line with an x-coordinate of -30 is $\frac{745}{7}$ or about 106. So the discus will not strike the official at $(-30, 40)$.

We can transform the standard form of a linear equation, $Ax + By + C = 0$, into normal form if the relationship between the coefficients in the two forms is known. The equations will represent the same line if and only if their corresponding coefficients are proportional. If $\frac{A}{\cos \phi} = \frac{B}{\sin \phi} = \frac{C}{-p}$, then you can solve to find expressions for $\cos \phi$ and $\sin \phi$ in terms of p and the coefficients.

$$\frac{A}{\cos \phi} = \frac{C}{-p} \Rightarrow -\frac{Ap}{C} = \cos \phi \quad \text{or} \quad \cos \phi = -\frac{Ap}{C}$$
$$\frac{B}{\sin \phi} = \frac{C}{-p} \Rightarrow -\frac{Bp}{C} = \sin \phi \quad \text{or} \quad \sin \phi = -\frac{Bp}{C}$$

We can divide $\sin \phi = -\frac{Bp}{C}$ by $\cos \phi = -\frac{Ap}{C}$, where $\cos \phi \neq 0$.

$$\frac{\sin \phi}{\cos \phi} = \frac{-\frac{Bp}{C}}{-\frac{Ap}{C}}$$

$$\tan \phi = \frac{B}{A} \qquad \textit{\frac{\sin \phi}{\cos \phi} = \tan \phi}$$

Refer to the diagram at the right. Consider an angle ϕ in standard position such that $\tan \phi = \frac{B}{A}$.

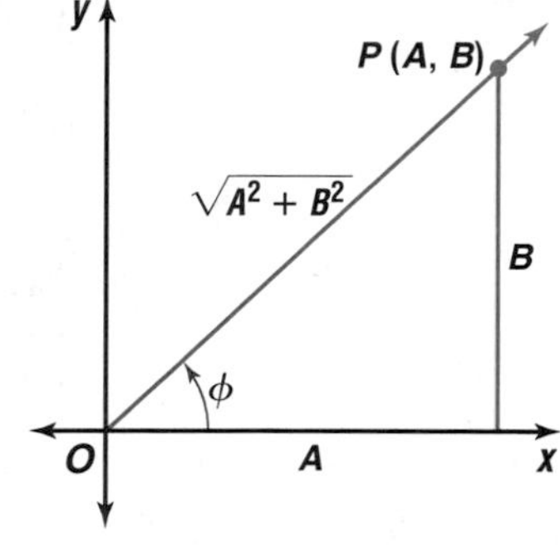

The length of $\overline{OP}$ is $\sqrt{A^2 + B^2}$.

So, $\sin \phi = \frac{B}{\pm\sqrt{A^2 + B^2}}$ and $\cos \phi = \frac{A}{\pm\sqrt{A^2 + B^2}}$.

Since we know that $\frac{B}{\sin \phi} = \frac{C}{-p}$, we can substitute

to get the result $\frac{B}{\frac{B}{\pm\sqrt{A^2 + B^2}}} = \frac{C}{-p}$.

Therefore, $p = \frac{C}{\pm\sqrt{A^2 + B^2}}$.

If $C = 0$, the sign is chosen so that $\sin \phi$ is positive; that is, the same sign as that of B.

The $\pm$ sign is used since p is a measure and must be positive in the equation $x \cos \phi + y \sin \phi - p = 0$. Therefore, the sign must be chosen as opposite of the sign of C. That is, if C is positive, use $-\sqrt{A^2 + B^2}$, and, if C is negative, use $\sqrt{A^2 + B^2}$.

Substitute the values for $\sin \phi$, $\cos \phi$, and p into the normal form.

$$\frac{Ax}{\pm\sqrt{A^2 + B^2}} + \frac{By}{\pm\sqrt{A^2 + B^2}} - \frac{C}{\pm\sqrt{A^2 + B^2}} = 0$$

Notice that the standard form is closely related to the normal form.

Changing the Standard Form to Normal Form

The standard form of a linear equation, $Ax + By + C = 0$, can be changed to normal form by dividing each term by $\pm\sqrt{A^2 + B^2}$. The sign is chosen opposite the sign of C.

If the equation of a line is in normal form, you can find the length of the normal, p units, directly from the equation. You can find the angle ϕ by using the relation $\tan \phi = \frac{\sin \phi}{\cos \phi}$. However, you must find the quadrant in which the normal lies to find the correct angle for ϕ. When the equation of a line is in normal form, the coefficient of x is equal to $\cos \phi$, and the coefficient of y is equal to $\sin \phi$. Thus, the correct quadrant can be determined by studying the signs of $\cos \phi$ and $\sin \phi$. For example, if $\sin \phi$ is negative and $\cos \phi$ is positive, the normal lies in the fourth quadrant.

Example 3 **Write each equation in normal form. Then find the length of the normal and the angle it makes with the positive *x*-axis.**

a. $6x + 8y + 3 = 0$

Since C is positive, use $-\sqrt{A^2 + B^2}$ to determine the normal form.

$-\sqrt{A^2 + B^2} = -\sqrt{6^2 + 8^2}$ or -10

The normal form is $\frac{6}{-10}x + \frac{8}{-10}y + \frac{3}{-10} = 0$, or $-\frac{3}{5}x - \frac{4}{5}y - \frac{3}{10} = 0$.

Therefore, $\sin \phi = -\frac{4}{5}$, $\cos \phi = -\frac{3}{5}$, and $p = \frac{3}{10}$. Since $\sin \phi$ and $\cos \phi$ are both negative, ϕ must lie in the third quadrant.

$$\tan \phi = \frac{-\frac{4}{5}}{-\frac{3}{5}} \text{ or } \frac{4}{3} \qquad \textit{$\tan \phi = \frac{\sin \phi}{\cos \phi}$}$$

$$\phi \approx 233° \qquad \textit{Add 180° to the arctangent to get the angle in Quadrant III.}$$

The normal segment to the origin has length 0.3 unit and makes an angle of 233° with the positive x-axis.

b. $-x + 4y - 6 = 0$

Since C is negative, use $\sqrt{A^2 + B^2}$ to determine the normal form.

$\sqrt{A^2 + B^2} = \sqrt{(-1)^2 + 4^2}$ or $\sqrt{17}$

The normal form is $-\frac{1}{\sqrt{17}}x + \frac{4}{\sqrt{17}}y - \frac{6}{\sqrt{17}} = 0$ or $-\frac{\sqrt{17}}{17}x + \frac{4\sqrt{17}}{17}y - \frac{6\sqrt{17}}{17} = 0$. Therefore, $\sin \phi = \frac{4\sqrt{17}}{17}$, $\cos \phi = -\frac{\sqrt{17}}{17}$, and $p = \frac{6\sqrt{17}}{17}$. Since $\sin \phi > 0$ and $\cos \phi < 0$, ϕ must lie in the second quadrant.

$$\tan \phi = \frac{\frac{4\sqrt{17}}{17}}{-\frac{\sqrt{17}}{17}} \text{ or } -4 \qquad \textit{$\tan \phi = \frac{\sin \phi}{\cos \phi}$}$$

$$\phi \approx 104° \qquad \textit{Add 180° to the arctangent to get the angle in Quadrant II.}$$

The normal segment to the origin has length $\frac{6\sqrt{17}}{17} \approx 1.46$ units and makes an angle of 104° with the positive x-axis.

CHECK FOR UNDERSTANDING

Communicating Mathematics

Read and study the lesson to answer each question.

1. **Define** the geometric meaning of the word *normal.*
2. **Describe** how to write the normal form of the equation of a line when $p = 10$ and $\phi = 30°$.
3. **Refute or defend** the following statement. *Determining the normal form of the equation of a line is like finding the equation of a tangent line to a circle of radius p.*
4. **Write** each form of the equation of a line that you have learned. Compare and contrast the information that each provides upon inspection. Create a sample problem that would require you to use each form.

Guided Practice

Write the standard form of the equation of each line given *p*, the length of the normal segment, and ϕ, the angle the normal segment makes with the positive *x*-axis.

5. $p = 10, \phi = 30°$
6. $p = \sqrt{3}, \phi = 150°$
7. $p = 5\sqrt{2}, \phi = \frac{7\pi}{4}$

Write each equation in normal form. Then find the length of the normal and the angle it makes with the positive *x*-axis.

8. $4x + 3y = -10$
9. $y = -3x + 2$
10. $\sqrt{2}x - \sqrt{2}y = 6$

11. **Transportation** An airport control tower is located at the origin of a coordinate system where the coordinates are measured in miles. An airplane radios in to report its direction and location. The controller determines that the equation of the plane's path is $3x - 4y = 8$.

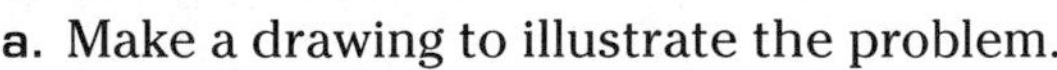

   a. Make a drawing to illustrate the problem.
   b. What is the closest the plane will come to the tower?

CHECK FOR UNDERSTANDING

Practice

Write the standard form of the equation of each line given *p*, the length of the normal segment, and ϕ, the angle the normal segment makes with the positive *x*-axis.

12. $p = 15, \phi = 60°$
13. $p = 12, \phi = \frac{\pi}{4}$
14. $p = 3\sqrt{2}, \phi = 135°$
15. $p = 2\sqrt{3}, \phi = \frac{5\pi}{6}$
16. $p = 2, \phi = \frac{\pi}{2}$
17. $p = 5, \phi = 210°$
18. $p = 5, \phi = \frac{4\pi}{3}$
19. $p = \frac{3}{2}, \phi = 300°$
20. $p = 4\sqrt{3}, \phi = \frac{11\pi}{6}$

Write each equation in normal form. Then find the length of the normal and the angle it makes with the positive *x*-axis.

21. $5x + 12y + 65 = 0$
22. $x + y = 1$
23. $3x - 4y = 15$
24. $y = 2x - 4$
25. $x = 3$
26. $-\sqrt{3}x - y = 2$
27. $y - 2 = \frac{1}{4}(x + 20)$
28. $\frac{x}{3} = y - 4$
29. $\frac{x}{20} + \frac{y}{24} = 1$

30. Write the standard form of the equation of a line if the point on the line nearest to the origin is at (6, 8).

31. The point nearest to the origin on a line is at (4, −4). Find the standard form of the equation of the line.

Applications and Problem Solving

32. Geometry The three sides of a triangle are tangent to a unique circle called the *incircle*. On the coordinate plane, the incircle of $\triangle ABC$ has its center at the origin. The lines whose equations are $x + 4y = 6\sqrt{17}$, $2x + \sqrt{5}y = -18$, and $2\sqrt{2}x = y + 18$ contain the sides of $\triangle ABC$. What is the length of the radius of the incircle?

33. History Ancient slingshots were made from straps of leather that cradled a rock until it was released. One would spin the slingshot in a circle, and the initial path of the released rock would be a straight line tangent to the circle at the point of release.

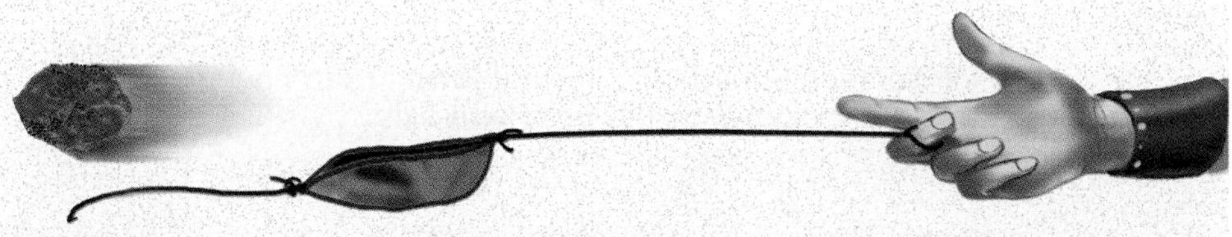

The rock will travel the greatest distance if it is released when the angle between the normal to the path and the horizontal is −45°. The center of the circular path is the origin and the radius of the circle measures 1.25 feet.

a. Draw a labeled diagram of the situation.

b. Write an equation of the initial path of the rock in standard form.

34. Critical Thinking Consider a line ℓ with positive x- and y-intercepts. Suppose ℓ makes an angle of θ with the positive x-axis.

a. What is the relationship between θ and ϕ, the angle the normal line makes with the positive x-axis?

b. What is the slope of ℓ in terms of θ?

c. What is the slope of the normal line in terms of θ?

d. What is the slope of ℓ in terms of ϕ?

35. Analytic Geometry Armando was trying to determine how to rotate the graph of a line $\delta°$ about the origin. He hypothesized that the following would be an equation of the new line.

$$x \cos(\phi + \delta) + y \sin(\phi + \delta) - p = 0$$

a. Write the normal form of the line $5x + 12y - 39 = 0$.

b. Determine ϕ. Replace ϕ by $\phi + 90°$ and graph the result.

c. Choose another value for δ, not divisible by 90°, and test Armando's conjecture.

d. Write an argument as to whether Armando is correct. Include a graph in your argument.

36. Critical Thinking Suppose two lines intersect to form an acute angle α. Suppose that each line has a positive y-intercept and that the x-intercepts of the lines are on opposite sides of the origin.

a. How are the angles ϕ_1 and ϕ_2 that the respective normals make with the positive x-axis related?

b. Write an equation involving $\tan \alpha$, $\tan \phi_1$, and $\tan \phi_2$.

37. Engineering The village of Plentywood must build a new water tower to meet the needs of its residents. This means that each of the water mains must be connected to the new tower. On a map of the village, with units in hundreds of feet, the water tower was placed at the origin. Each of the existing water mains can be described by an equation. These equations are $5x - y = 15$, $3x + 4y = 36$, and $5x - 2y = -20$. The cost of laying new underground pipe is \$500 per 100 feet. Find the lowest possible cost of connecting the existing water mains to the new tower.

Mixed Review

38. Solve $2 \cos^2 x + 7 \cos x - 4 = 0$ for $0 \leq x < 2\pi$. *(Lesson 7-5)*

39. If x and y are acute angles such that $\cos x = \frac{1}{6}$ and $\cos y = \frac{2}{3}$, find $\sin(x + y)$. *(Lesson 7-3)*

40. Graph $y = \sin 4\theta$. *(Lesson 6-4)*

41. **Engineering** A metallic ring used in a sprinkler system has a diameter of 13.4 centimeters. Find the length of the metallic cross brace if it subtends a central angle of 26°20′. *(Lesson 5-8)*

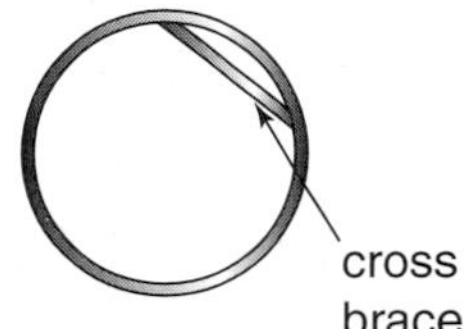

42. Solve $\frac{x}{x-5} + \frac{17}{25 - x^2} = \frac{1}{x+5}$. *(Lesson 4-6)*

43. **Manufacturing** A porcelain company produces collectible thimble sets that contain 8 thimbles in a box that is 4 inches by 6 inches by 2 inches. To celebrate the company's 100th anniversary, they wish to market a deluxe set of 8 large thimbles. They would like to increase each of the dimensions of the box by the same amount to create a new box with a volume that is 1.5 times the volume of the original box. What should be the dimensions of the new box for the large thimbles? *(Lesson 4-5)*

44. Find the maximum and minimum values of the function $f(x, y) = 3x - y + 4$ for the polygonal convex set determined by the system of inequalities. *(Lesson 2-6)*

$x + y \leq 8$
$y \geq 3$
$x \geq 2$

45. Solve $\begin{bmatrix} -1 & 2 \\ -4 & 3 \end{bmatrix} \cdot \begin{bmatrix} x \\ y \end{bmatrix} = \begin{bmatrix} 0 \\ 15 \end{bmatrix}$. *(Lesson 2-5)*

46. **SAT/ACT Practice** If $\frac{a}{b} = \frac{4}{5}$, what is the value of $2a + b$?

A 3 B 13 C 14 D 26

E cannot be determined from the given information

Extra Practice See p. A39.

Distance From a Point to a Line

OBJECTIVES

- Find the distance from a point to a line.
- Find the distance between two parallel lines.
- Write equations of lines that bisect angles formed by intersecting lines.

OPTICS When light waves strike a surface, they are reflected in such a way that the angle of incidence equals the angle of reflection. Consider a flat mirror situated in a coordinate system such that light emanating from the point at (5, 4) strikes the mirror at (3, 0) and then passes through the point at (7, −6). Determine an equation of the line on which the mirror lies. Use the equation to determine the angle the mirror makes with the x-axis. *This problem will be solved in Example 4.*

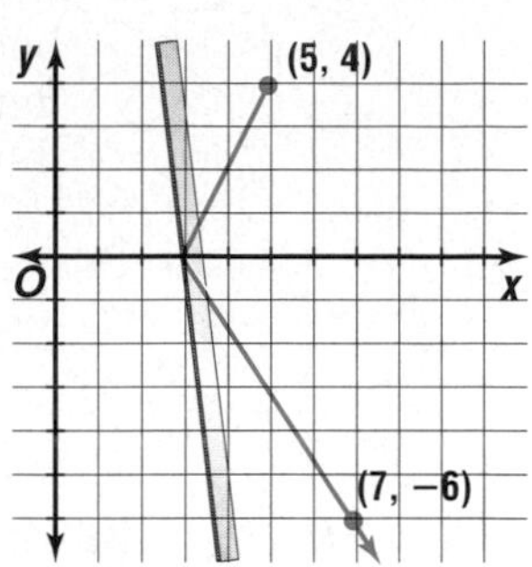

The normal form of a linear equation can be used to find the distance from a point to a line. Let $\overleftrightarrow{RS}$ be a line in the coordinate plane, and let $P(x_1, y_1)$ be a point not on $\overleftrightarrow{RS}$. P may lie on the same side of $\overleftrightarrow{RS}$ as the origin does or it may lie on the opposite side. If a line segment joining P to the origin does not intersect $\overleftrightarrow{RS}$, point P is on the same side of the line as the origin. Construct $\overleftrightarrow{TV}$ parallel to $\overleftrightarrow{RS}$ and passing through P. The distance d between the parallel lines is the distance from P to $\overleftrightarrow{RS}$. So that the derivation is valid in both cases, we will use a negative value for d if point P and the origin are on the same side of $\overleftrightarrow{RS}$.

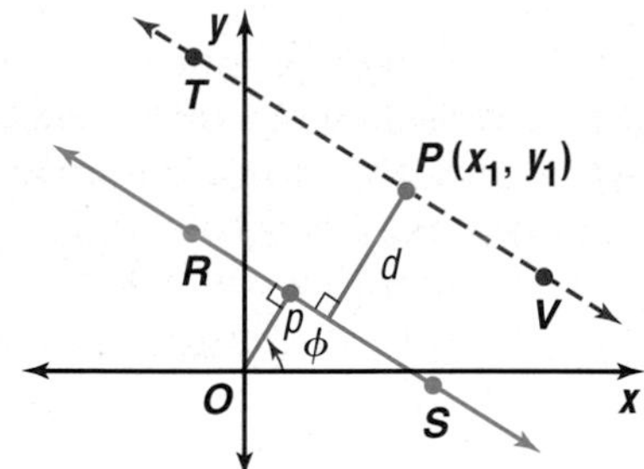

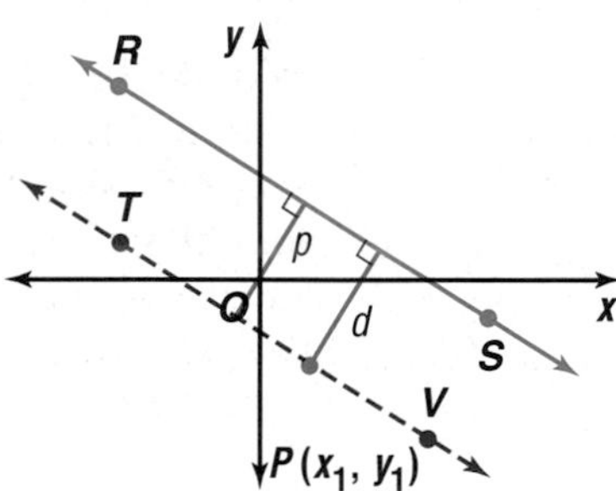

interNET CONNECTION

Graphing Calculator Programs
To download a graphing calculator program that computes the distance from a point to a line, visit: **www.amc.glencoe.com**

Let $x \cos \phi + y \sin \phi - p = 0$ be the equation of $\overleftrightarrow{RS}$ in normal form. Since $\overleftrightarrow{TV}$ is parallel to $\overleftrightarrow{RS}$, they have the same slope. The equation for $\overleftrightarrow{TV}$ can be written as $x \cos \phi + y \sin \phi - (p + d) = 0$. Solve this equation for d.

$$d = x \cos \phi + y \sin \phi - p$$

Since $P(x_1, y_1)$ is on $\overleftrightarrow{TV}$, its coordinates satisfy this equation.

$$d = x_1 \cos \phi + y_1 \sin \phi - p$$

We can use an equivalent form of this expression to find d when the equation of a line is in standard form.

Distance from a Point to a Line

The following formula can be used to find the distance from a point at (x_1, y_1) to a line with equation $Ax + By + C = 0$.

$$d = \frac{Ax_1 + By_1 + C}{\pm\sqrt{A^2 + B^2}}$$

The sign of the radical is chosen opposite the sign of C.

The distance will be positive if the point and the origin are on opposite sides of the line. The distance will be treated as negative if the origin is on the same side of the line as the point. If you are solving an application problem, the absolute value of d will probably be required.

Example 1 **Find the distance between $P(4, 5)$ and the line with equation $8x + 5y = 20$.**

First rewrite the equation of the line in standard form.

$8x + 5y = 20 \Rightarrow 8x + 5y - 20 = 0$

Then use the formula for the distance from a point to a line.

$$d = \frac{Ax_1 + By_1 + C}{\pm\sqrt{A^2 + B^2}}$$

$$d = \frac{8(4) + 5(5) - 20}{\pm\sqrt{8^2 + 5^2}} \qquad A = 8, B = 5, C = -20, x_1 = 4, y_1 = 5$$

$$d = \frac{37}{\sqrt{89}} \text{ or about } 3.92 \qquad \textit{Since C is negative, use } +\sqrt{A^2 + B^2}.$$

Therefore, P is approximately 3.92 units from the line $8x + 5y = 20$. Since d is positive, P is on the opposite side of the line from the origin.

You can use the formula for the distance from a point to a line to find the distance between two parallel lines. To do this, choose any point on one of the lines and use the formula to find the distance from that point to the other line.

Example 2 **Find the distance between the lines with equations $6x - 2y = 7$ and $y = 3x + 4$.**

Since $y = 3x + 4$ is in slope-intercept form, we know that it passes through the point at $(0, 4)$. Use this point to find the distance to the other line.

The standard form of the other equation is $6x - 2y - 7 = 0$.

$$d = \frac{Ax_1 + By_1 + C}{\pm\sqrt{A^2 + B^2}}$$

$$d = \frac{6(0) - 2(4) - 7}{\pm\sqrt{6^2 + (-2)^2}} \qquad A = 6, B = -2, C = -7, x_1 = 0, y_1 = 4$$

$$d = -\frac{15}{\sqrt{40}} \text{ or about } -2.37 \qquad \textit{Since C is negative, use } +\sqrt{A^2 + B^2}.$$

The distance between the lines is about 2.37 units.

VECTORS AND PARAMETRIC EQUATIONS

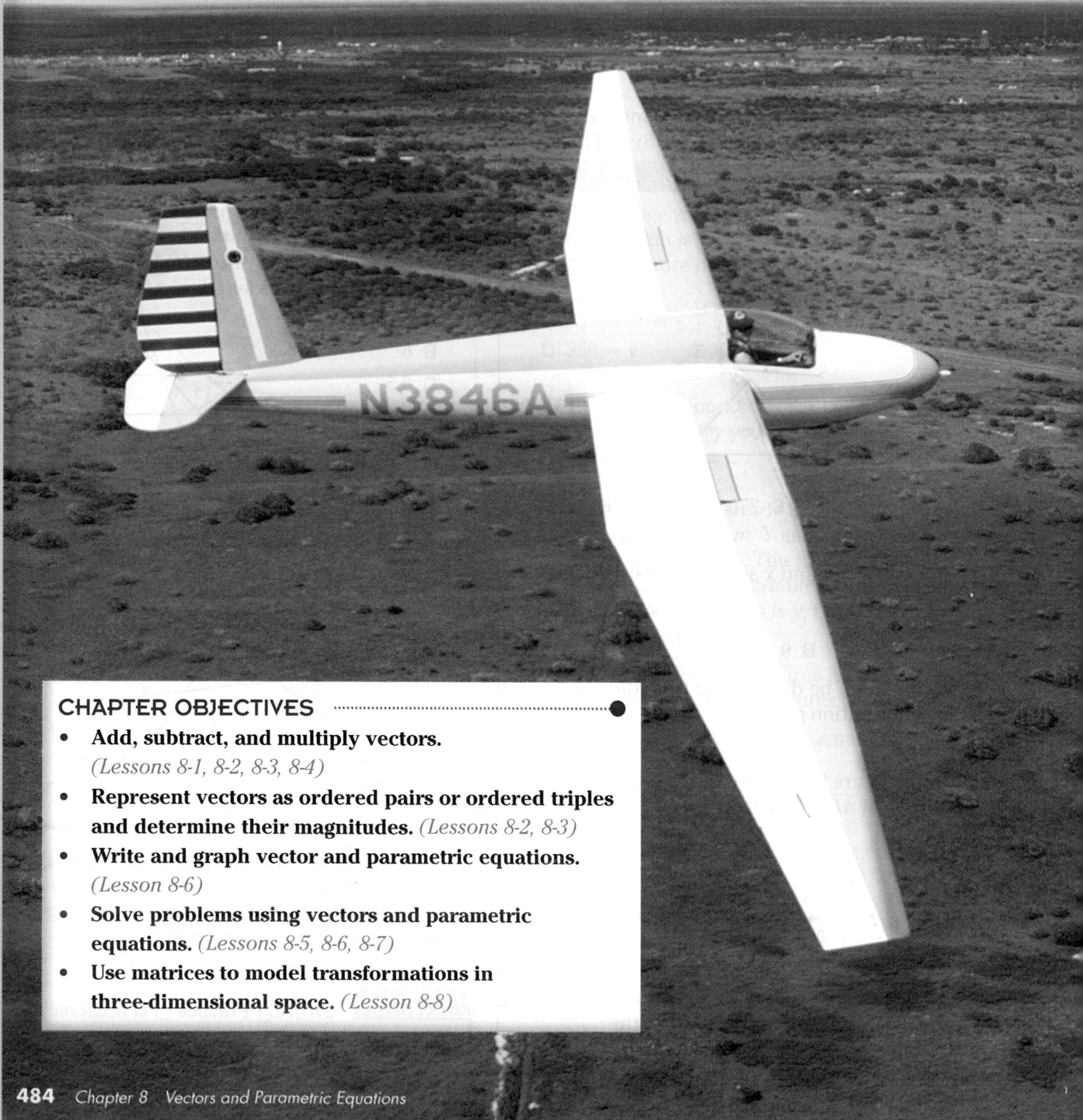

CHAPTER OBJECTIVES

- **Add, subtract, and multiply vectors.** *(Lessons 8-1, 8-2, 8-3, 8-4)*
- **Represent vectors as ordered pairs or ordered triples and determine their magnitudes.** *(Lessons 8-2, 8-3)*
- **Write and graph vector and parametric equations.** *(Lesson 8-6)*
- **Solve problems using vectors and parametric equations.** *(Lessons 8-5, 8-6, 8-7)*
- **Use matrices to model transformations in three-dimensional space.** *(Lesson 8-8)*

8-1 Geometric Vectors

OBJECTIVES

- Find equal, opposite, and parallel vectors.
- Add and subtract vectors geometrically.

AERONAUTICS An advanced glider known as a sailplane, has high maneuverability and glide capabilities. The Ventus 2B sailplane placed first at the World Soaring Contest in New Zealand. At one competition, a sailplane traveled forward at a rate of 8 m/s, and it descended at a rate of 4 m/s. *A problem involving this situation will be solved in Example 3.*

The velocity of a sailplane can be represented mathematically by a **vector.** A vector is a quantity that has both *magnitude* and *direction.* A vector is represented geometrically by a directed line segment.

A directed line segment with an initial point at P and a terminal point at Q is shown at the right. The length of the line segment represents the **magnitude** of the vector. The direction of the arrowhead indicates the direction of the vector. The vectors can be denoted by $\vec{\mathbf{a}}$ or $\overrightarrow{PQ}$. The magnitude of $\vec{\mathbf{a}}$ is denoted by $|\vec{\mathbf{a}}|$.

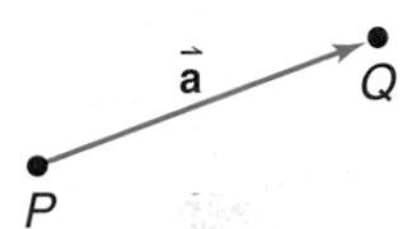

If a vector has its initial point at the origin, it is in **standard position.** The **direction** of the vector is the directed angle between the positive x-axis and the vector. The direction of $\vec{\mathbf{b}}$ is 45°.

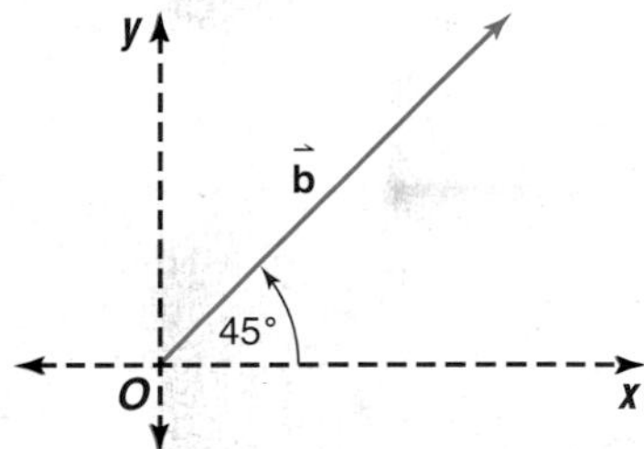

If both the initial point and the terminal point are at the origin, the vector is the **zero vector** and is denoted by $\vec{\mathbf{0}}$. The magnitude of the zero vector is 0, and it can be in any direction.

Example 1 **Use a ruler and protractor to determine the magnitude (in centimeters) and the direction of $\vec{\mathbf{n}}$.**

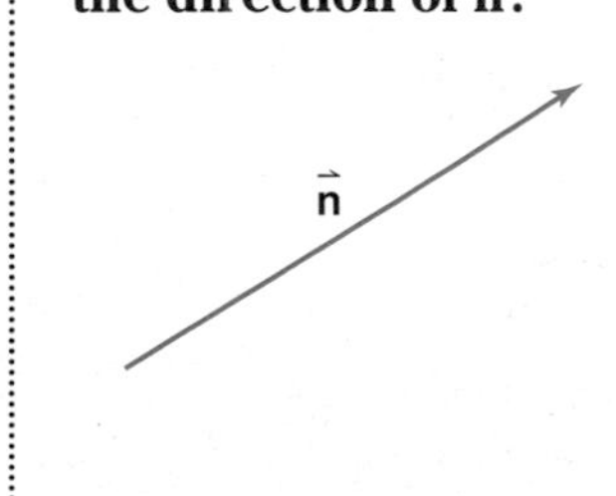

Sketch the vector in standard position and measure the magnitude and direction. The magnitude is 3.6 centimeters, and the direction is 30°.

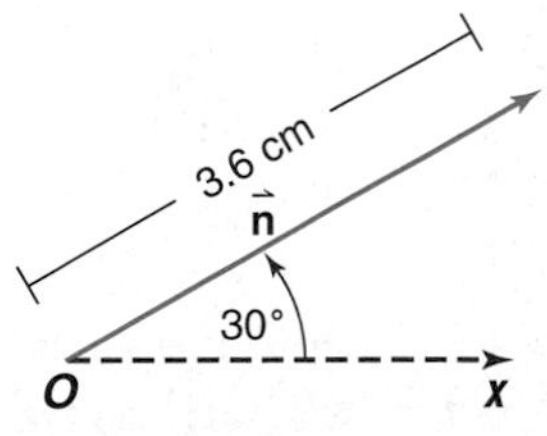

Two vectors are **equal** if and only if they have the same direction and the same magnitude.

Six vectors are shown at the right.

- $\vec{z}$ and $\vec{y}$ are equal since they have the same direction and $|\vec{z}| = |\vec{y}|$.
- $\vec{v}$ and $\vec{u}$ are equal.
- $\vec{x}$ and $\vec{w}$ have the same direction but $|\vec{x}| \neq |\vec{w}|$, so $\vec{x} \neq \vec{w}$.
- $|\vec{v}| = |\vec{y}|$, but they have different directions, so they are not equal.

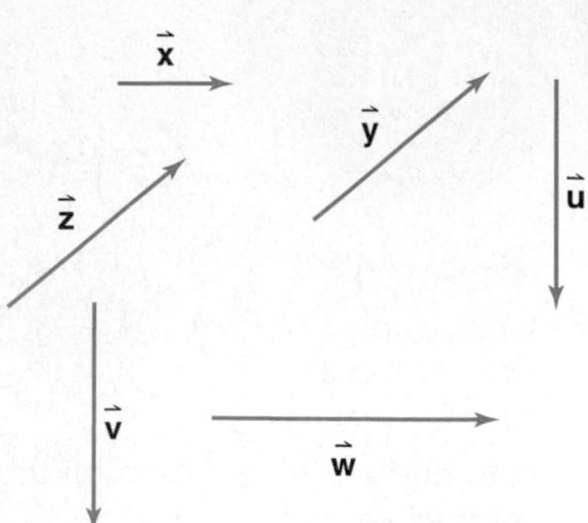

The sum of two or more vectors is called the **resultant** of the vectors. The resultant can be found using either the *parallelogram method* or the *triangle method.*

<table>
<tr><th>Parallelogram Method</th><th>Triangle Method</th></tr>
<tr><td>Draw the vectors so that their initial points coincide. Then draw lines to form a complete parallelogram. The diagonal from the initial point to the opposite vertex of the parallelogram is the resultant.</td><td>Draw the vectors one after another, placing the initial point of each successive vector at the terminal point of the previous vector. Then draw the resultant from the initial point of the first vector to the terminal point of the last vector.</td></tr>
<tr><td>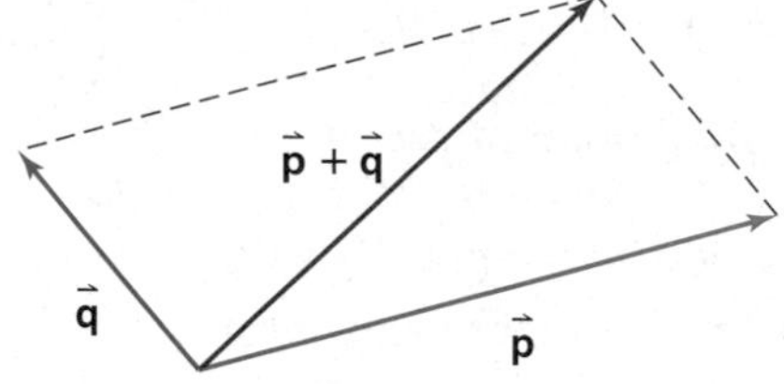
</td><td>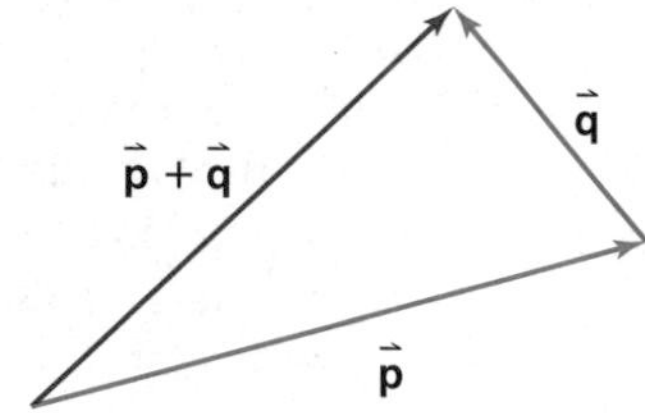
</td></tr>
<tr><td>The parallelogram method cannot be used to find the sum of a vector and itself.</td><td>This method is also called the tip-to-tail method.</td></tr>
</table>

Example 2 Find the sum of $\vec{v}$ and $\vec{w}$ using:

a. the parallelogram method.

b. the triangle method.

c. Compare the resultants found in both methods.

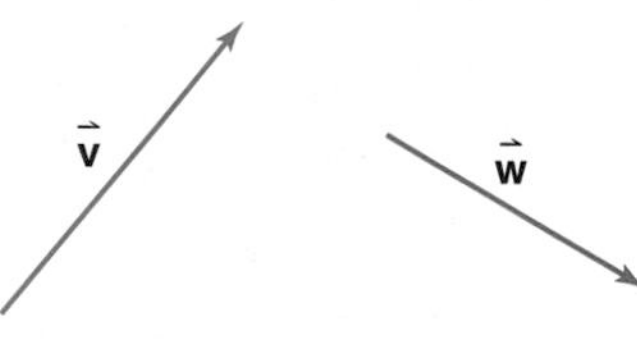

a. Copy $\vec{v}$ then copy $\vec{w}$ placing the initial points together.

Form a parallelogram that has $\vec{v}$ and $\vec{w}$ as two of its sides. Draw dashed lines to represent the other two sides.

The resultant is the vector from the vertex of $\vec{v}$ and $\vec{w}$ to the opposite vertex of the parallelogram.

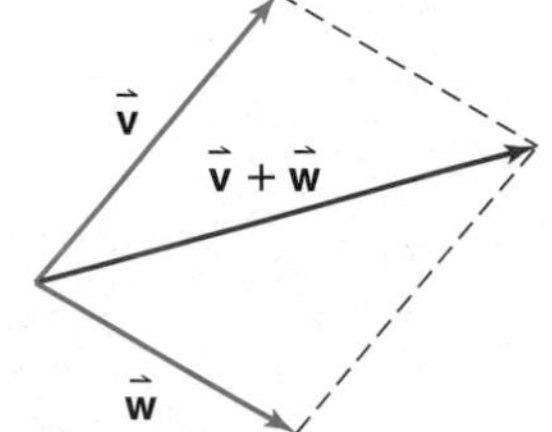

b. Copy $\vec{v}$, then copy $\vec{w}$ so that the initial point of $\vec{w}$ is on the terminal point of $\vec{v}$. (The tail of $\vec{v}$ connects to the tip of $\vec{w}$.)

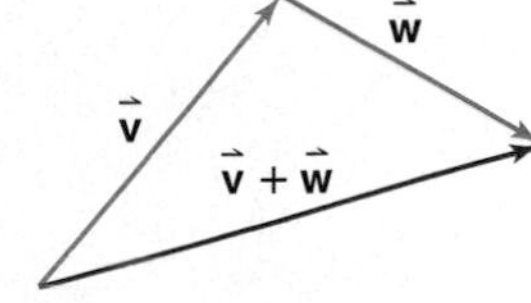

The resultant is the vector from the initial point of $\vec{v}$ to the terminal point of $\vec{w}$.

c. Use a ruler and protractor to measure the magnitude and direction of each resultant. The resultants of both methods have magnitudes of 3.5 centimeters and directions of 20°. So, the resultants found in both methods are equal.

Vectors can be used to solve real-world applications.

Example 3 **AERONAUTICS** **Refer to the application at the beginning of the lesson. At the European Championships, a sailplane traveled forward at 8 m/s and descended at 4 m/s. Determine the magnitude of the resultant velocity of the sailplane.**

Let 1 centimeter represent 2 m/s. Draw two vectors, $\vec{f}$ and $\vec{d}$, to represent the forward velocity and the descending velocity of the sailplane, respectively.

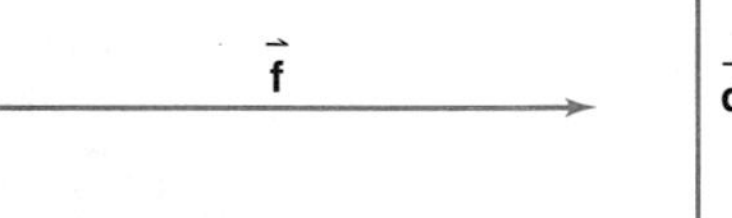

Use the parallelogram method. Copy $\vec{f}$. Then copy $\vec{d}$, placing the initial point at the initial point of $\vec{f}$. Draw dashed lines to represent the other two sides of the parallelogram.

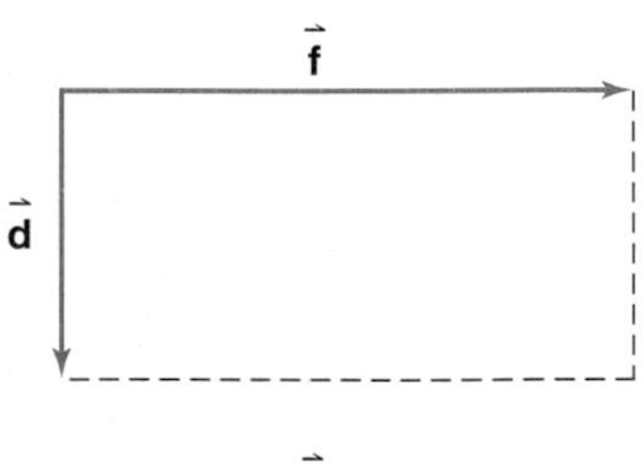

The resultant velocity of the sailplane is the vector $\vec{p}$ from the vertex of $\vec{f}$ and $\vec{d}$ to the opposite vertex of the parallelogram.

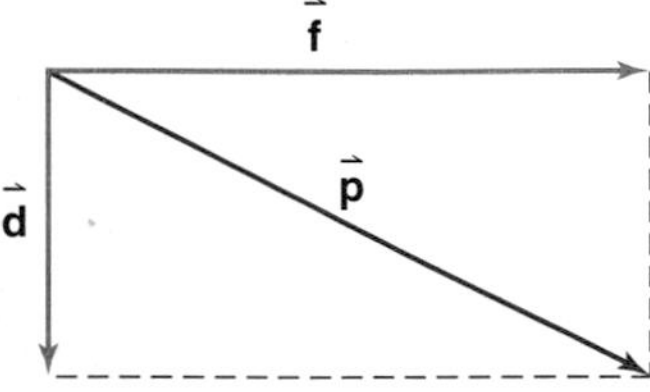

Measure the resultant, 4.5 centimeters. Multiply the resultant's magnitude by 2 to determine the magnitude of the resultant velocity of the sailplane. *Why?*

4.5 cm × 2 m/s = 9 m/s

The sailplane is moving at 9 m/s.

Two vectors are **opposites** if they have the same magnitude and opposite directions. In the diagram, $\vec{g}$ and $\vec{h}$ are opposites, as are $\vec{i}$ and $\vec{j}$. The opposite of $\vec{g}$ is denoted by $-\vec{g}$. You can use opposite vectors to subtract vectors. To find $\vec{g} - \vec{h}$, find $\vec{g} + (-\vec{h})$.

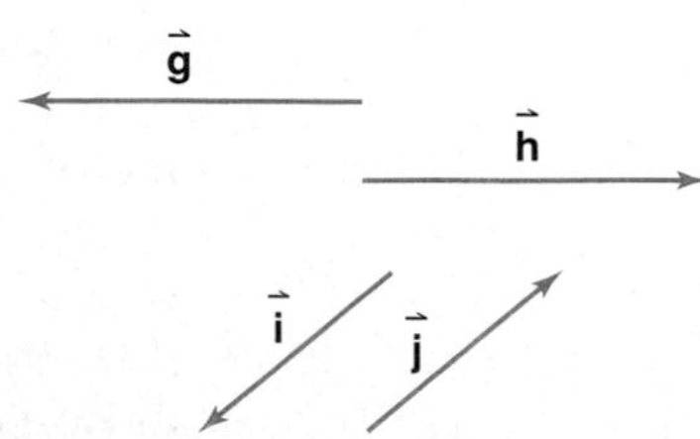

A quantity with only magnitude is called a **scalar quantity.** Examples of scalars include mass, length, time, and temperature. The numbers used to measure scalar quantities are called **scalars.**

The product of a scalar k and a vector $\vec{\mathbf{a}}$ is a vector with the same direction as $\vec{\mathbf{a}}$ and a magnitude of $k|\vec{\mathbf{a}}|$, if $k > 0$. If $k < 0$, the vector has the opposite direction of $\vec{\mathbf{a}}$ and a magnitude of $k|\vec{\mathbf{a}}|$. In the figure at the right, $\vec{\mathbf{b}} = 3\vec{\mathbf{a}}$ and $\vec{\mathbf{c}} = -2\vec{\mathbf{a}}$.

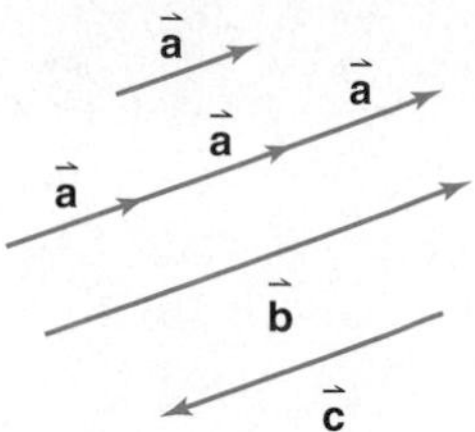

Example 4 **Use the triangle method to find $2\vec{\mathbf{v}} - \frac{1}{2}\vec{\mathbf{w}}$.**

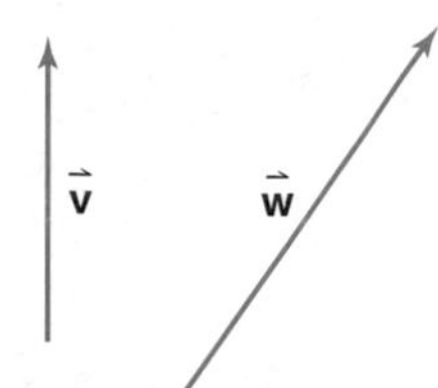

Rewrite the expression as a sum.

$$2\vec{\mathbf{v}} - \frac{1}{2}\vec{\mathbf{w}} = 2\vec{\mathbf{v}} + \left(-\frac{1}{2}\vec{\mathbf{w}}\right)$$

Draw a vector twice the magnitude of $\vec{\mathbf{v}}$ to represent $2\vec{\mathbf{v}}$. Draw a vector with the opposite direction to $\vec{\mathbf{w}}$ and half its magnitude to represent $-\frac{1}{2}\vec{\mathbf{w}}$.

Place the initial point of $-\frac{1}{2}\vec{\mathbf{w}}$ on the terminal point of $2\vec{\mathbf{v}}$. *(Tip-to-tail method)*

Then $2\vec{\mathbf{v}} - \frac{1}{2}\vec{\mathbf{w}}$ has the initial point of $2\vec{\mathbf{v}}$ and the terminal point of $-\frac{1}{2}\vec{\mathbf{w}}$.

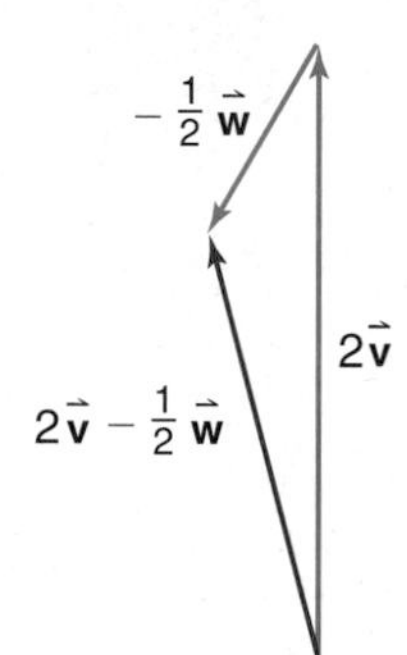

Two or more vectors are **parallel** if and only if they have the same or opposite directions. Vectors $\vec{\mathbf{i}}$ and $\vec{\mathbf{j}}$ have opposite direction and they are parallel. Vector $\vec{\boldsymbol{\ell}}$ and $\vec{\mathbf{k}}$ are parallel and have the same direction.

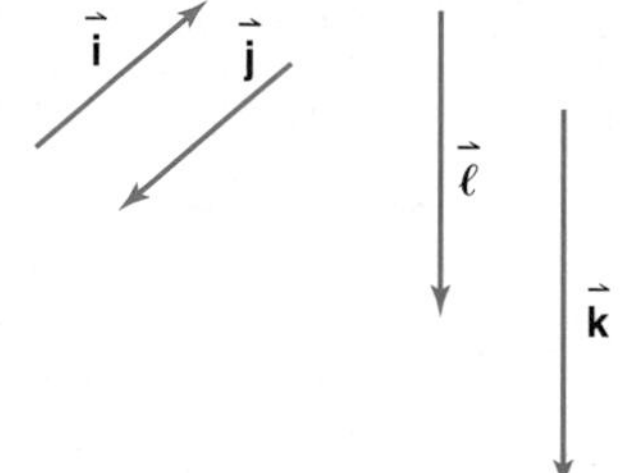

Two or more vectors whose sum is a given vector are called **components** of the given vector. Components can have any direction. Often it is useful to express a vector as the sum of two perpendicular components. In the figure at the right, $\vec{\mathbf{p}}$ is the vertical component of $\vec{\mathbf{x}}$, and $\vec{\mathbf{q}}$ is the horizontal component of $\vec{\mathbf{x}}$.

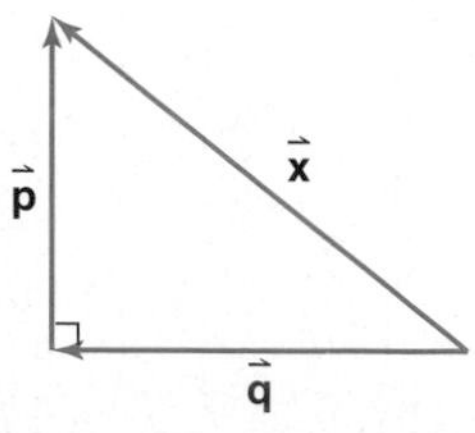

Example 5 

NAVIGATION **A ship leaving port sails for 75 miles in a direction 35° north of due east. Find the magnitude of the vertical and horizontal components.**

Draw $\vec{s}$. Then draw a horizontal vector through the initial point of $\vec{s}$. Label the resulting angle 35°. Draw a vertical vector through the terminal point of $\vec{s}$. The vectors will form a right triangle, so you can use the sine and cosine ratios to find the magnitude of the components.

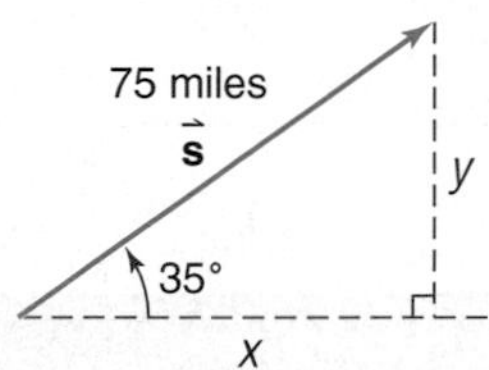

$$\sin 35° = \frac{y}{75} \qquad \cos 35° = \frac{x}{75}$$

$$y = 75 \sin 35° \qquad x = 75 \cos 35°$$

$$y \approx 43 \qquad x \approx 61$$

The magnitude of the vertical component is approximately 43 miles, and the magnitude of the horizontal component is approximately 61 miles.

Vectors are used in physics to represent velocity, acceleration, and forces acting upon objects.

Example 6 

CONSTRUCTION **A piling for a high-rise building is pushed by two bulldozers at exactly the same time. One bulldozer exerts a force of 1550 pounds in a westerly direction. The other bulldozer pushes the piling with a force of 3050 pounds in a northerly direction.**

a. What is the magnitude of the resultant force upon the piling, to the nearest ten pounds?

b. What is the direction of the resulting force upon the piling, to the nearest ten pounds?

a. Let $\vec{x}$ represent the force for bulldozer 1.
Let $\vec{y}$ represent the force for bulldozer 2.

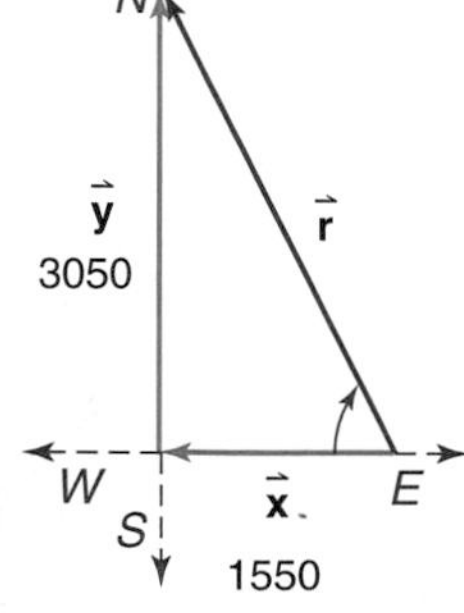

Draw the resultant $\vec{r}$. This represents the total force acting upon the piling. Use the Pythagorean Theorem to find the magnitude of the resultant.

$$c^2 = a^2 + b^2$$

$$\vec{r}^{\,2} = 1550^2 + 3050^2$$

$$|\vec{r}| = \sqrt{(1550)^2 + (3050)^2}$$

$$|\vec{r}| \approx 3421$$

The magnitude of the resultant force upon the piling is about 3420 pounds.

(continued on the next page)

b. Let a be the measure of the angle $\vec{\mathbf{r}}$ makes with $\vec{\mathbf{x}}$.

The direction of the resultant can be found by using the tangent ratio.

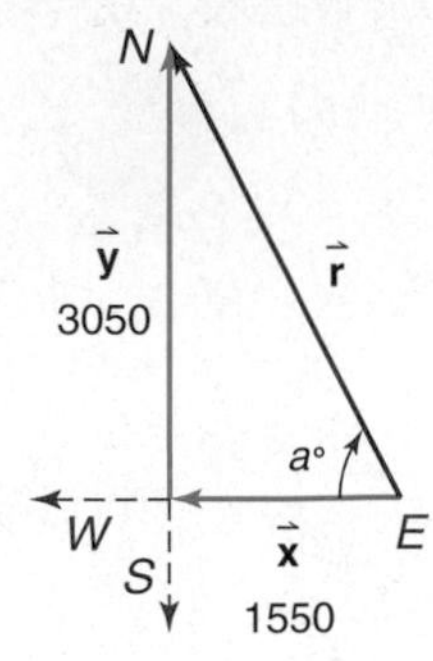

$$\tan a = \frac{|\vec{\mathbf{y}}|}{|\vec{\mathbf{x}}|}$$

$$\tan a = \frac{3050}{1550}$$

$$a = 63°$$ *Take $\tan^{-1}$ of each side.*

The resultant makes an angle of 63° with $\vec{\mathbf{x}}$. The direction of the resultant force upon the piling is $90 - 63$ or 27° west of north. *Direction of vectors is often expressed in terms of position relative to due north, due south, due east, or due west, or as a navigational angle measures clockwise from due north.*

CHECK FOR UNDERSTANDING

Communicating Mathematics

Read and study the lesson to answer each question.

1. **Draw** a diagram showing the resultant of two vectors, and describe how you obtained it.
2. **Compare** a line segment and a vector.
3. **Describe** a real-world situation involving vectors that could be represented by the diagram at the right.

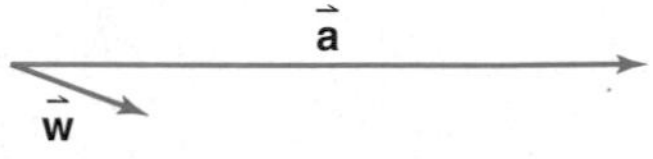

4. **Tell** whether $\overrightarrow{RS}$ is the same as $\overrightarrow{SR}$. Explain.

Guided Practice

Use a ruler and a protractor to determine the magnitude (in centimeters) and direction of each vector.

5. 

6. 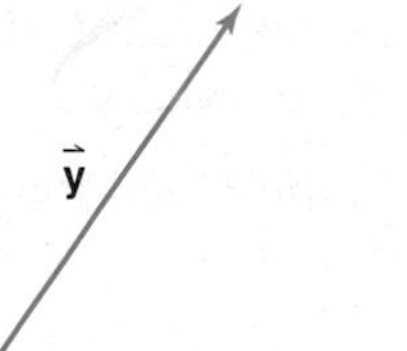

7. 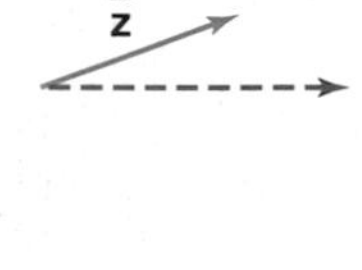

Use $\vec{\mathbf{x}}$, $\vec{\mathbf{y}}$, and $\vec{\mathbf{z}}$ above to find the magnitude and direction of each resultant.

8. $\vec{\mathbf{x}} + \vec{\mathbf{y}}$
9. $\vec{\mathbf{x}} - \vec{\mathbf{y}}$
10. $4\vec{\mathbf{y}} + \vec{\mathbf{z}}$
11. the difference of a vector twice as long as $\vec{\mathbf{z}}$ and a vector one third the magnitude of $\vec{\mathbf{x}}$
12. Find the magnitude of the horizontal and vertical components of $\vec{\mathbf{y}}$.
13. **Aviation** An airplane is flying due west at a velocity of 100 m/s. The wind is blowing out of the south at 5 m/s.
    a. Draw a labeled diagram of the situation.
    b. What is the magnitude of airplane's resultant velocity?

EXERCISES

Practice

Use a ruler and a protractor to determine the magnitude (in centimeters) and direction of each vector.

14. $\vec{r}$

15. 

16. 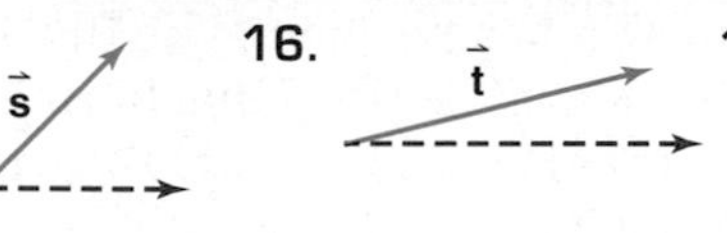

17. 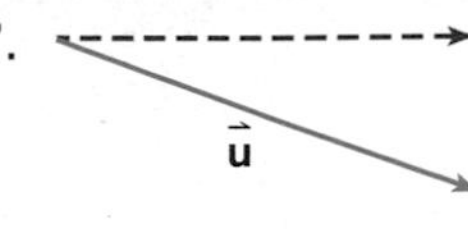

Use $\vec{r}$, $\vec{s}$, $\vec{t}$, and $\vec{u}$ above to find the magnitude and direction of each resultant.

18. $\vec{r} + \vec{s}$
19. $\vec{s} + \vec{t}$
20. $\vec{s} + \vec{u}$
21. $\vec{u} - \vec{r}$
22. $\vec{r} - \vec{t}$
23. $2\vec{r}$
24. $3\vec{s}$
25. $3\vec{u} - 2\vec{s}$
26. $\vec{r} + \vec{t} + \vec{u}$
27. $\vec{r} + \vec{s} - \vec{u}$
28. $2\vec{s} + \vec{u} - \frac{1}{2}\vec{r}$
29. $\vec{r} - 2\vec{t} - \vec{s} + 3\vec{u}$
30. three times $\vec{t}$ and twice $\vec{u}$

Find the magnitude of the horizontal and vertical components of each vector shown for Exercises 14–17.

31. $\vec{r}$
32. $\vec{s}$
33. $\vec{t}$
34. $\vec{u}$

35. The magnitude of $\vec{m}$ is 29.2 meters, and the magnitude of $\vec{n}$ is 35.2 meters. If $\vec{m}$ and $\vec{n}$ are perpendicular, what is the magnitude of their sum?

36. In the parallelogram drawn for the parallelogram method, what does the diagonal between the two terminal points of the vectors represent? Explain your answer.

37. Is addition of vectors commutative? Justify your answer. (*Hint:* Find the sum of two vectors, $\vec{r} + \vec{s}$ and $\vec{s} + \vec{r}$, using the triangle method.)

Applications and Problem Solving

38. **Physics** Three workers are pulling on ropes attached to a tree stump as shown in the diagram. Find the magnitude and direction of the resultant force on the tree. *A newton (N) is a unit of force used in physics. A force of one newton will accelerate a one-kilogram mass at a rate of one meter per second squared.*

60 newtons

35 newtons

60°

40 newtons

39. **Critical Thinking** Does $|\vec{a} + \vec{b}|$ always, sometimes, or never equal $|\vec{a}| + |\vec{b}|$? Draw a diagram to justify your answer.

40. **Toys** Belkis is pulling a toy by exerting a force of 1.5 newtons on a string attached to the toy.
    a. The string makes an angle of 52° with the floor. Find the vertical and horizontal components of the force.
    b. If Belkis raises the string so that it makes a 78° angle with the floor, what are the magnitudes of the horizontal and vertical components of the force?

41. **Police Investigation** Police officer Patricia Malloy is investigating an automobile accident. A car slid off an icy road at a 40° angle with the center line of the road with an initial velocity of 47 miles per hour. Use the drawing to determine the initial horizontal and vertical components of the car's velocity.

42. **Critical Thinking** If $\vec{\mathbf{a}}$ is a vector and k is a scalar, is it possible that $\vec{\mathbf{a}} = k\vec{\mathbf{a}}$? Under what conditions is this true?

43. **Ranching** Matsuko has attached two wires to a corner fence post to complete a horse paddock. The wires make an angle of 90° with each other. If each wire pulls on the post with a force of 50 pounds, what is the resultant force acting on the pole?

44. **Physics** Mrs. Keaton is the director of a local art gallery. She needs to hang a painting weighing 24 pounds with a wire whose parts form a 120° angle with each other. Find the pull on each part of the wire.

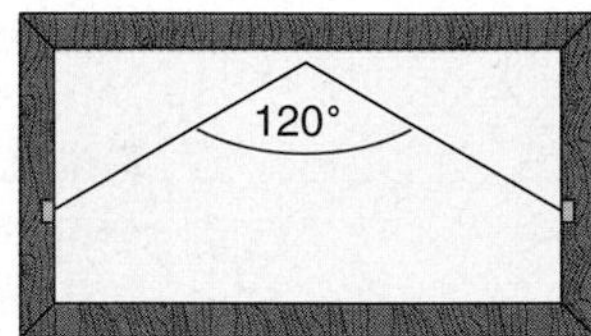

Mixed Review

45. Find the equation of the line that bisects the acute angle formed by the lines $x - y + 2 = 0$ and $y - 5 = 0$. *(Lesson 7-7)*

46. Verify that $\csc \theta \cos \theta \tan \theta = 1$ is an identity. *(Lesson 7-2)*

47. Find the values of θ for which $\tan \theta = 1$ is true. *(Lesson 6-7)*

48. Use the graphs of the sine and cosine functions to find the values of x for which $\sin x + \cos x = -1$. *(Lesson 6-5)*

49. **Geometry** The base angles of an isosceles triangle measure 18°29′, and the altitude to the base is 5 centimeters long. Find the length of the base and the lengths of the congruent sides. *(Lesson 5-2)*

50. **Manufacturing** A manufacturer produces packaging boxes for other companies. The largest box they currently produce has a height two times its width and a length one more that its width. They wish to produce a larger packaging box where the height will be increased by 2 feet and the width and length will be increased by 1 foot. The volume of the new box is 160 cubic feet. Find the dimensions of the original box. *(Lesson 4-4)*

51. Determine the equations of the vertical and horizontal asymptotes, if any, of $g(x) = \frac{x + 2}{(x - 1)(x + 3)}$. *(Lesson 3-7)*

52. **SAT/ACT Practice** **Grid-In** Three times the least of three consecutive odd integers is three greater than two times the greatest. Find the greatest of the three integers.

Extra Practice See p. A40.

8-2

Algebraic Vectors

OBJECTIVES

- Find ordered pairs that represent vectors.
- Add, subtract, multiply, and find the magnitude of vectors algebraically.

EMERGENCY MEDICINE Paramedics Paquita Gonzalez and Trevor Howard are moving a person on a stretcher. Ms. Gonzalez is pushing the stretcher with a force of 135 newtons at 58° with the horizontal, while Mr. Howard is pulling the stretcher with a force of 214 newtons at 43° with the horizontal. What is the magnitude of the force exerted on the stretcher? *This problem will be solved in Example 3.*

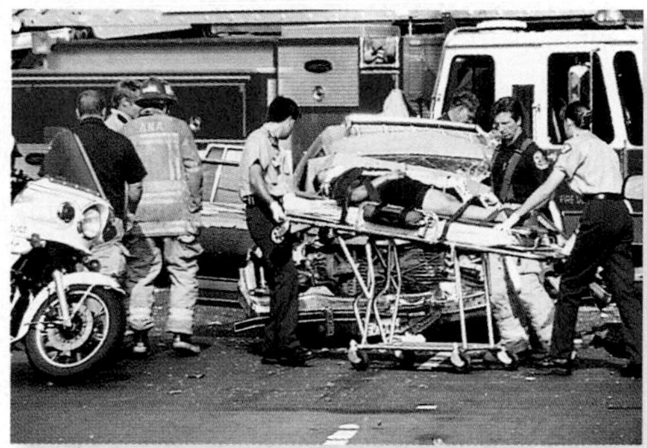

We can find the magnitude and direction of the resultant by drawing vectors to represent the forces on the stretcher. However, drawings can be inaccurate and quite confusing for more complex systems of vectors. In cases where more accuracy is necessary or where the system is more complicated, we can perform operations on vectors algebraically.

Vectors can be represented algebraically using ordered pairs of real numbers. For example, the ordered pair $\langle 3, 5 \rangle$ can represent a vector in standard position. That is, its initial point is at the origin and its terminal point is at (3, 5). You can think of this vector as the resultant of a horizontal vector with a magnitude of 3 units and a vertical vector with magnitude of 5 units.

Since vectors with the same magnitude and direction are equal, many vectors can be represented by the same ordered pair. Each vector on the graph can be represented by the ordered pair $\langle 3, 5 \rangle$. The initial point of a vector can be any point in the plane. In other words, a vector does not have to be in standard position to be expressed algebraically.

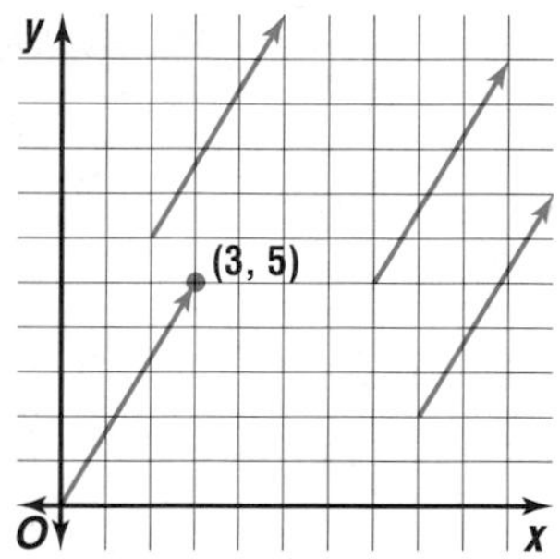

Assume that P_1 and P_2 are any two distinct points in the coordinate plane. Drawing the horizontal and vertical components of $\overrightarrow{P_1P_2}$ yields a right triangle. So, the magnitude of $\overrightarrow{P_1P_2}$ can be found by using the Pythagorean Theorem.

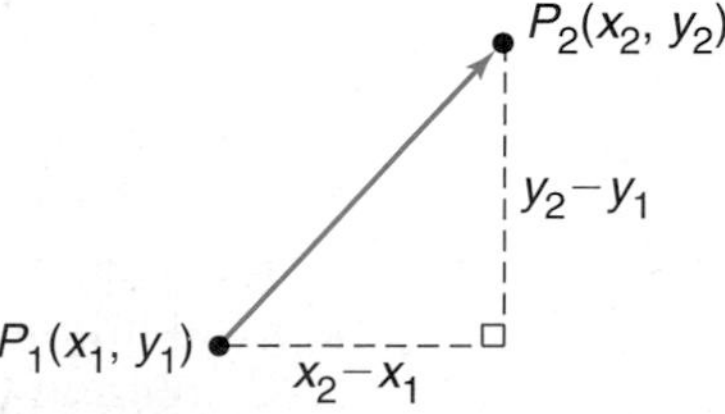

Representation of a Vector as an Ordered Pair	Let $P_1(x_1, y_1)$ be the initial point of a vector and $P_2(x_2, y_2)$ be the terminal point. The ordered pair that represents $\overrightarrow{P_1P_2}$ is $\langle x_2 - x_1, y_2 - y_1 \rangle$. Its magnitude is given by $|\overrightarrow{P_1P_2}| = \sqrt{(x_2 - x_1)^2 + (y_2 - y_1)^2}$.

Example 1 **Write the ordered pair that represents the vector from $X(-3, 5)$ to $Y(4, -2)$. Then find the magnitude of $\overrightarrow{XY}$.**

First, represent $\overrightarrow{XY}$ as an ordered pair.

$\overrightarrow{XY} = \langle 4 - (-3), -2 - 5 \rangle$ or $\langle 7, -7 \rangle$

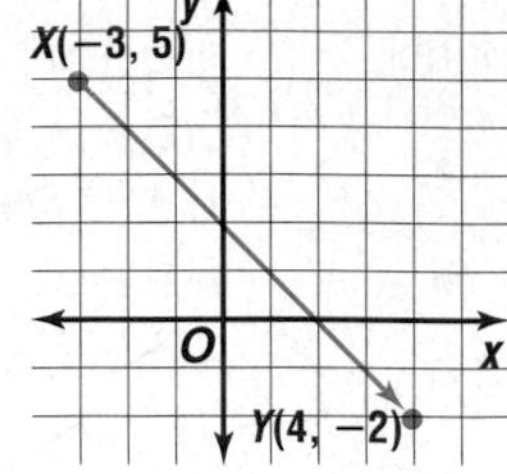

Then, determine the magnitude of $\overrightarrow{XY}$.

$$|\overrightarrow{XY}| = \sqrt{[4 - (-3)]^2 + (-2 - 5)^2}$$
$$= \sqrt{7^2 + 7^2}$$
$$= \sqrt{98} \text{ or } 7\sqrt{2}$$

$\overrightarrow{XY}$ is represented by the ordered pair $\langle 7, -7 \rangle$ and has a magnitude of $7\sqrt{2}$ units.

When vectors are represented by ordered pairs, they can be easily added, subtracted, or multiplied by a scalar. The rules for these operations on vectors are similar to those for matrices. In fact, vectors in a plane can be represented as row matrices of dimension 1×2.

Vector Operations	The following operations are defined for $\vec{a}\ \langle a_1, a_2 \rangle$, $\vec{b}\ \langle b_1, b_2 \rangle$, and any real number k. Addition: $\vec{a} + \vec{b} = \langle a_1, a_2 \rangle + \langle b_1, b_2 \rangle = \langle a_1 + b_1, a_2 + b_2 \rangle$ Subtraction: $\vec{a} - \vec{b} = \langle a_1, a_2 \rangle - \langle b_1, b_2 \rangle = \langle a_1 - b_1, a_2 - b_2 \rangle$ Scalar multiplication: $k\vec{a} = k\langle a_1, a_2 \rangle = \langle ka_1, ka_2 \rangle$

Example 2 **Let $\vec{m} = \langle 5, -7 \rangle$, $\vec{n} = \langle 0, 4 \rangle$, and $\vec{p} = \langle -1, 3 \rangle$. Find each of the following.**

a. $\vec{m} + \vec{p}$

$$\vec{m} + \vec{p} = \langle 5, -7 \rangle + \langle -1, 3 \rangle$$
$$= \langle 5 + (-1), -7 + 3 \rangle$$
$$= \langle 4, -4 \rangle$$

b. $\vec{m} - \vec{n}$

$$\vec{m} - \vec{n} = \langle 5, -7 \rangle - \langle 0, 4 \rangle$$
$$= \langle 5 - 0, -7 - 4 \rangle$$
$$= \langle 5, -11 \rangle$$

c. $7\vec{p}$

$$7\vec{p} = 7\langle -1, 3 \rangle$$
$$= \langle 7 \cdot (-1), 7 \cdot 3 \rangle$$
$$= \langle -7, 21 \rangle$$

d. $2\vec{m} + 3\vec{n} - \vec{p}$

$$2\vec{m} + 3\vec{n} - \vec{p} = 2\langle 5, -7 \rangle + 3\langle 0, 4 \rangle - \langle -1, 3 \rangle$$
$$= \langle 10, -14 \rangle + \langle 0, 12 \rangle - \langle -1, 3 \rangle$$
$$= \langle 11, -5 \rangle$$

If we write the vectors described in the application at the beginning of the lesson as ordered pairs, we can find the resultant vector easily using vector addition. Using ordered pairs to represent vectors allows for a more accurate solution than using geometric representations.

Example 3 **EMERGENCY MEDICINE** **Refer to the application at the beginning of the lesson. What is the magnitude of the force exerted on the stretcher?**

Draw a diagram of the situation. Let $\vec{G}_1$ represent the force Ms. Gonzalez exerts, and let $\vec{G}_2$ represent the force Mr. Howard exerts.

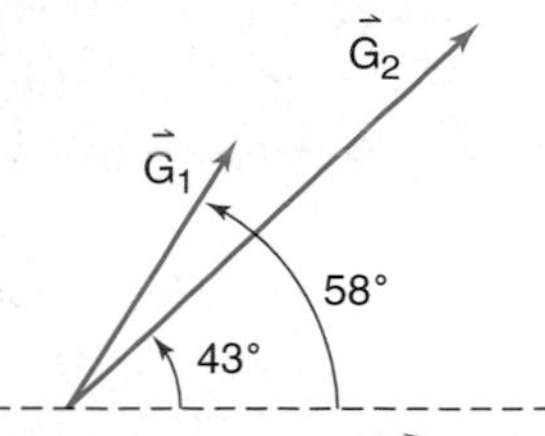

Write each vector as an ordered pair by finding its horizontal and vertical components. Let $\vec{G}_{1x}$ and $\vec{G}_{1y}$ represent the x- and y-components of $\vec{G}_1$. Let $\vec{G}_{2x}$ and $\vec{G}_{2y}$ represent the x- and y-components of $\vec{G}_2$.

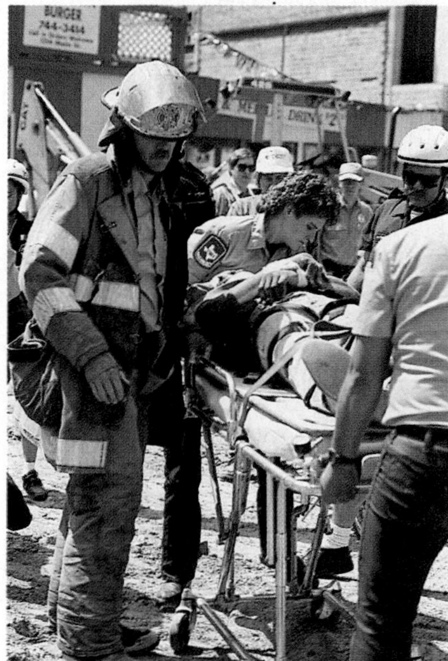

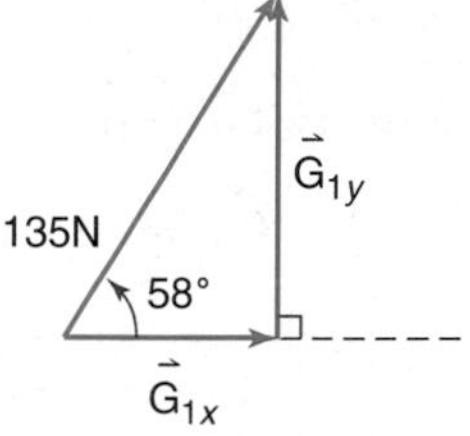

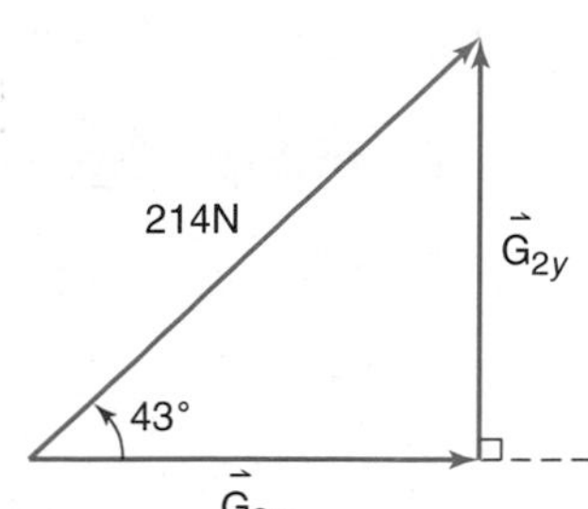

$$\cos 58° = \frac{|\vec{G}_{1x}|}{135}$$
$$|\vec{G}_{1x}| = 135 \cos 58°$$
$$|\vec{G}_{1x}| \approx 71.5$$
$$\sin 58° = \frac{|\vec{G}_{1y}|}{135}$$
$$|\vec{G}_{1y}| = 135 \sin 58°$$
$$|\vec{G}_{1y}| \approx 114.5$$
$$\vec{G}_1 \approx \langle 71.5, 114.5 \rangle$$

$$\cos 43° = \frac{|\vec{G}_{2x}|}{214}$$
$$|\vec{G}_{2x}| = 214 \cos 43°$$
$$|\vec{G}_{2x}| \approx 156.5$$
$$\sin 43° = \frac{|\vec{G}_{2y}|}{214}$$
$$|\vec{G}_{2y}| = 214 \sin 43°$$
$$|\vec{G}_{2y}| \approx 145.9$$
$$\vec{G}_2 \approx \langle 156.5, 145.9 \rangle$$

Find the sum of the vectors.

$$\vec{G}_1 + \vec{G}_2 \approx \langle 71.5, 114.5 \rangle + \langle 156.5, 145.9 \rangle$$
$$\approx \langle 228.0, 260.4 \rangle$$

The net force on the stretcher is the magnitude of the sum.

$$|\vec{G}_1 + \vec{G}_2| \approx \sqrt{(228.0)^2 + (260.4)^2} \text{ or about } 346$$

The net force on the stretcher is about 346 newtons.

To convert newtons to pounds, divide the number of newtons by 4.45. So, the net force on the stretcher can also be expressed as 346 ÷ 4.45 or 77.8 pounds.

A vector that has a magnitude of one unit is called a **unit vector.** A unit vector in the direction of the positive x-axis is represented by $\vec{\mathbf{i}}$, and a unit vector in the direction of the positive y-axis is represented by $\vec{\mathbf{j}}$. So, $\vec{\mathbf{i}} = \langle 1, 0 \rangle$ and $\vec{\mathbf{j}} = \langle 0, 1 \rangle$.

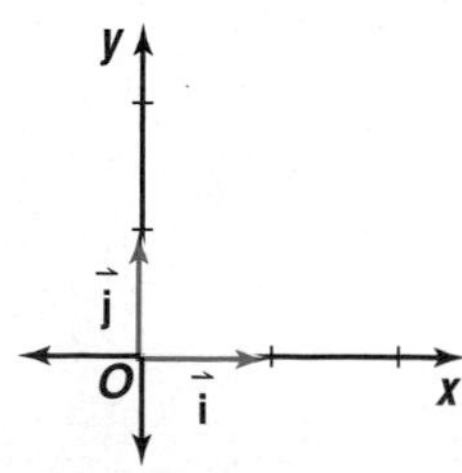

Any vector $\vec{\mathbf{a}} = \langle a_1, a_2 \rangle$ can be expressed as $a_1\vec{\mathbf{i}} + a_2\vec{\mathbf{j}}$.

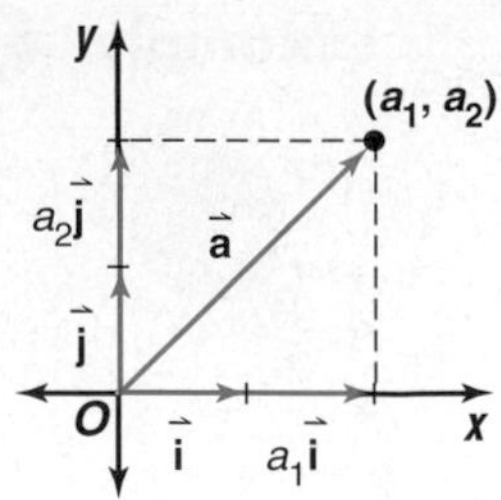

$$
\begin{aligned}
a_1\vec{\mathbf{i}} + a_2\vec{\mathbf{j}} &= a_1\langle 1, 0\rangle + a_2\langle 0, 1\rangle && \vec{i} = \langle 1, 0\rangle \text{ and } \vec{j} = \langle 0, 1\rangle \\
&= \langle a_1, 0\rangle + \langle 0, a_2\rangle && \text{Scalar product} \\
&= \langle a_1 + 0, 0 + a_2\rangle && \text{Addition of vectors} \\
&= \langle a_1, a_2\rangle
\end{aligned}
$$

Since $\langle a_1, a_2\rangle = \vec{\mathbf{a}}$, $a_1\vec{\mathbf{i}} + a_2\vec{\mathbf{j}} = \vec{\mathbf{a}}$. Therefore, any vector that is represented by an ordered pair can also be written as the sum of unit vectors. The zero vector can be represented as $\vec{\mathbf{0}} = \langle 0, 0\rangle = 0\vec{\mathbf{i}} + 0\vec{\mathbf{j}}$.

Example 4 **Write $\overrightarrow{AB}$ as the sum of unit vectors for $A(4, -1)$ and $B(6, 2)$.**

First, write $\overrightarrow{AB}$ as an ordered pair.

$$
\begin{aligned}
\overrightarrow{AB} &= \langle 6 - 4, 2 - (-1)\rangle \\
&= \langle 2, 3\rangle
\end{aligned}
$$

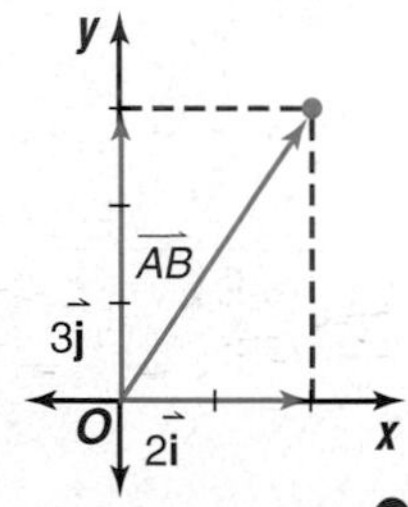

Then, write $\overrightarrow{AB}$ as the sum of unit vectors.

$$\overrightarrow{AB} = 2\vec{\mathbf{i}} + 3\vec{\mathbf{j}}$$

CHECK FOR UNDERSTANDING

Communicating Mathematics

Read and study the lesson to answer each question.

1. **Find a counterexample** If $|\vec{\mathbf{a}}| = 10$ and $|\vec{\mathbf{b}}| = 10$, then $\vec{\mathbf{a}}$ and $\vec{\mathbf{b}}$ are equal vectors.
2. **Describe** how to find $|\overrightarrow{XY}|$ using the graph at the right.

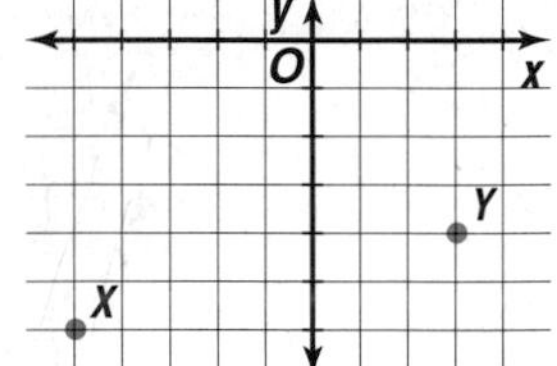

3. **You Decide** Lina showed Jacqui the representation of $\langle 2, -5\rangle$ as a unit vector as follows.

$$
\begin{aligned}
\langle 2, -5\rangle &= \langle 2, 0\rangle + \langle 0, -5\rangle \\
&= (5)\langle 1, 0\rangle + (-2)\langle 0, 1\rangle \\
&= 5\vec{\mathbf{i}} + (-2)\vec{\mathbf{j}} \\
&= 5\vec{\mathbf{i}} - 2\vec{\mathbf{j}}
\end{aligned}
$$

Jacqui said that her work was incorrect. Who is right? Explain.

Guided Practice

Write the ordered pair that represents $\overrightarrow{MP}$. Then find the magnitude of $\overrightarrow{MP}$.

4. $M(2, -1)$, $P(-3, 4)$
5. $M(5, 6)$, $P(0, 5)$
6. $M(-19, 4)$, $P(4, 0)$

Find an ordered pair to represent $\vec{\mathbf{t}}$ in each equation if $\vec{\mathbf{u}} = \langle -1, 4\rangle$ and $\vec{\mathbf{v}} = \langle 3, -2\rangle$.

7. $\vec{\mathbf{t}} = \vec{\mathbf{u}} + \vec{\mathbf{v}}$
8. $\vec{\mathbf{t}} = \frac{1}{2}\vec{\mathbf{u}} - \vec{\mathbf{v}}$
9. $\vec{\mathbf{t}} = 4\vec{\mathbf{u}} + 6\vec{\mathbf{v}}$
10. $\vec{\mathbf{t}} = -8\vec{\mathbf{u}}$

Find the magnitude of each vector. Then write each vector as the sum of unit vectors.

11. $\langle 8, -6 \rangle$

12. $\langle -7, -5 \rangle$

13. Construction The Walker family is building a cabin for vacationing. Mr. Walker and his son Terrell have erected a scaffold to stand on while they build the walls of the cabin. As they stand on the scaffold Terrell pulls on a rope attached to a support beam with a force of 400 newtons (N) at an angle of 65° with the horizontal. Mr. Walker pulls with a force of 600 newtons at an angle of 110° with the horizontal. What is the magnitude of the combined force they exert on the log?

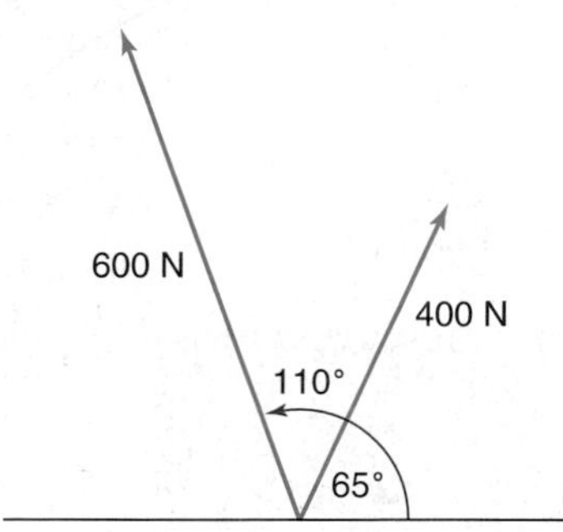

EXERCISES

Practice

Write the ordered pair that represents $\overrightarrow{YZ}$. Then find the magnitude of $\overrightarrow{YZ}$.

14. $Y(4, 2)$, $Z(2, 8)$

15. $Y(-5, 7)$, $Z(-1, 2)$

16. $Y(-2, 5)$, $Z(1, 3)$

17. $Y(5, 4)$, $Z(0, -3)$

18. $Y(3, 1)$, $Z(0, 4)$

19. $Y(-4, 12)$, $Z(1, 19)$

20. $Y(5, 0)$, $Z(7, 6)$

21. $Y(14, -23)$, $Z(23, -14)$

22. Find an ordered pair that represents the vector from $A(31, -33)$ to $B(36, -45)$. Then find the magnitude of $\overrightarrow{AB}$.

Find an ordered pair to represent $\vec{a}$ in each equation if $\vec{b} = \langle 6, 3 \rangle$ and $\vec{c} = \langle -4, 8 \rangle$.

23. $\vec{\mathbf{a}} = \vec{\mathbf{b}} + \vec{\mathbf{c}}$

24. $\vec{\mathbf{a}} = 2\vec{\mathbf{b}} + \vec{\mathbf{c}}$

25. $\vec{\mathbf{a}} = \vec{\mathbf{b}} + 2\vec{\mathbf{c}}$

26. $\vec{\mathbf{a}} = 2\vec{\mathbf{b}} + 3\vec{\mathbf{c}}$

27. $\vec{\mathbf{a}} = -\vec{\mathbf{b}} + 4\vec{\mathbf{c}}$

28. $\vec{\mathbf{a}} = \vec{\mathbf{b}} - 2\vec{\mathbf{c}}$

29. $\vec{\mathbf{a}} = 3\vec{\mathbf{b}}$

30. $\vec{\mathbf{a}} = -\frac{1}{2}\vec{\mathbf{c}}$

31. $\vec{\mathbf{a}} = 6\vec{\mathbf{b}} + 4\vec{\mathbf{c}}$

32. $\vec{\mathbf{a}} = 0.4\vec{\mathbf{b}} - 1.2\vec{\mathbf{c}}$

33. $\vec{\mathbf{a}} = \frac{1}{3}\left(2\vec{\mathbf{b}} - 5\vec{\mathbf{c}}\right)$

34. $\vec{\mathbf{a}} = \left(3\vec{\mathbf{b}} + \vec{\mathbf{c}}\right) + 5\vec{\mathbf{b}}$

35. For $\vec{\mathbf{m}} = \langle -5, -6 \rangle$ and $\vec{\mathbf{n}} = \langle 6, -9 \rangle$ find the sum of the vector three times the magnitude of $\vec{\mathbf{m}}$ and the vector two and one half times the magnitude of the opposite of $\vec{\mathbf{n}}$.

Find the magnitude of each vector. Then write each vector as the sum of unit vectors.

36. $\langle 3, 4 \rangle$

37. $\langle 2, -3 \rangle$

38. $\langle -6, -11 \rangle$

39. $\langle 3.5, 12 \rangle$

40. $\langle -4, 1 \rangle$

41. $\langle -16, -34 \rangle$

42. Write $\overrightarrow{ST}$ as the sum of unit vectors for points $S(-9, 2)$ and $T(-4, -3)$.

43. Prove that addition of vectors is associative.

Applications and Problem Solving

44. **Recreation** In the 12th Bristol International Kite Festival in September 1997 in England, Peter Lynn set a record for flying the world's biggest kite, which had a lifting surface area of 630 square meters. Suppose the wind is blowing against the kite with a force of 100 newtons at an angle 20° above the horizontal.
    a. Draw a diagram representing the situation.
    b. How much force is lifting the kite?

45. **Surfing** During a weekend surfboard competition, Kiyoshi moves at a 30° angle toward the shore. The velocity component toward the shore is 15 mph.
    a. Make a labeled diagram to show Kiyoshi's velocity and the velocity components.
    b. What is Kiyoshi's velocity?

46. **Critical Thinking** Suppose the points Q, R, S, and T are noncollinear, and $\overrightarrow{QR} + \overrightarrow{ST} = \vec{\mathbf{0}}$.
    a. What is the relationship between $\overrightarrow{QR}$ and $\overrightarrow{ST}$?
    b. What is true of the quadrilateral with vertices Q, R, S, and T?

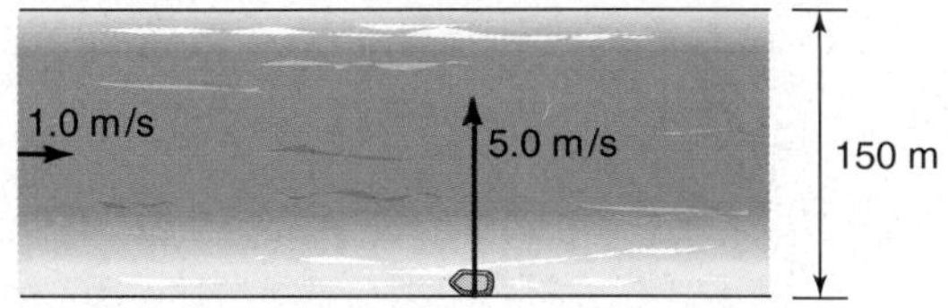

47. **River Rafting** The Soto family is rafting on the Colorado River. Suppose that they are on a stretch of the river that is 150 meters wide, flowing south at a rate of 1.0 m/s. In still water their raft travels 5.0 m/s.
    a. How long does it take them to travel from one bank to the other if they head for a point directly across the river?
    b. How far downriver will the raft land?
    c. What is the velocity of the raft relative to the shore?

48. **Critical Thinking** Show that any vector $\vec{\mathbf{v}}$ can be written as $\langle |\vec{\mathbf{v}}| \cos \theta, |\vec{\mathbf{v}}| \sin \theta \rangle$.

Mixed Review

49. State whether $\overrightarrow{PQ}$ and $\overrightarrow{RS}$ are *equal*, *opposite*, *parallel*, or *none of these* for points $P(8, -7)$, $Q(-2, 5)$, $R(8, -7)$, and $S(7, 0)$. *(Lesson 8-1)*

50. Find the distance from the graph of $3x - 7y - 1 = 0$ to the point at $(-1, 4)$. *(Lesson 7-7)*

51. Use the sum or difference identities to find the exact value of sin 255°. *(Lesson 7-3)*

52. Write an equation of the sine function that has an amplitude of 17, a period of 45°, and a phase shift of −60°. *(Lesson 6-5)*

53. **Geometry** Two sides of a triangle are 400 feet and 600 feet long, and the included angle measures 46°20′. Find the perimeter and area of the triangle. *(Lesson 5-8)*

54. Use the Upper Bound Theorem to find an integral upper bound and the Lower Bound Theorem to find the lower bound of the zeros of the function $f(x) = 3x^2 - 2x + 1$. *(Lesson 4-5)*

Extra Practice See p. A40.

55. Using a graphing calculator to graph $y = x^3 - x^2 + 3$. Determine and classify its extrema. *(Lesson 3-6)*

56. Describe the end behavior of $f(x) = x^2 + 3x + 1$. *(Lesson 3-5)*

57. **SAT Practice Quantitative Comparison**

A if the quantity in Column A is greater
B if the quantity in Column B is greater
C if the two quantities are equal
D if the relationship cannot be determined from the information given

Column A	Column B
$7x + 1$	$7x - 1$

CAREER CHOICES

Aerospace Engineering

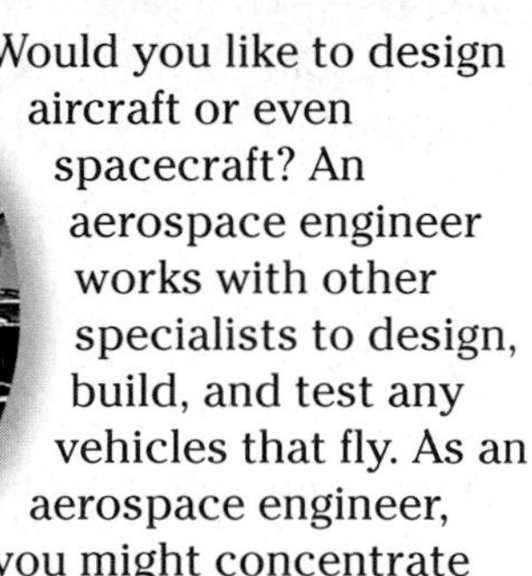

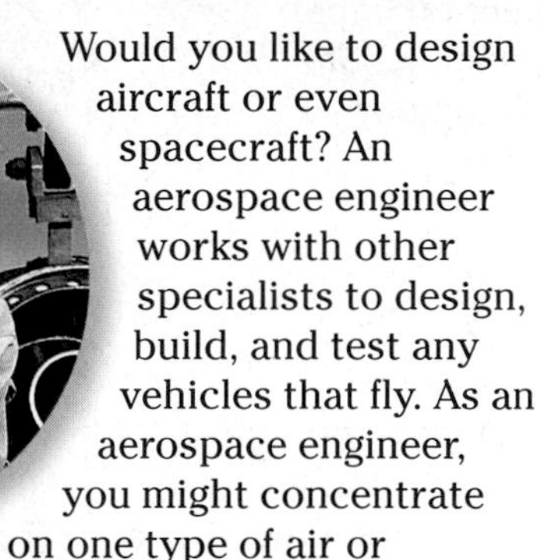

Would you like to design aircraft or even spacecraft? An aerospace engineer works with other specialists to design, build, and test any vehicles that fly. As an aerospace engineer, you might concentrate on one type of air or spacecraft, or you might work on specific components of these crafts.

There are also a number of specialties within the field of aerospace engineering including analytical engineering, stress engineering, materials aerospace engineering, and marketing and sales aerospace engineering. One of the aspects of aerospace engineering is the analysis of airline accidents to determine if structural defects in design were the cause of the accident.

CAREER OVERVIEW

Degree Preferred:
bachelor's degree in aeronautical or aerospace engineering

Related Courses:
mathematics, science, computer science, mechanical drawing

Outlook:
slower than average through the year 2006

Airline Departures and Fatal Accidents

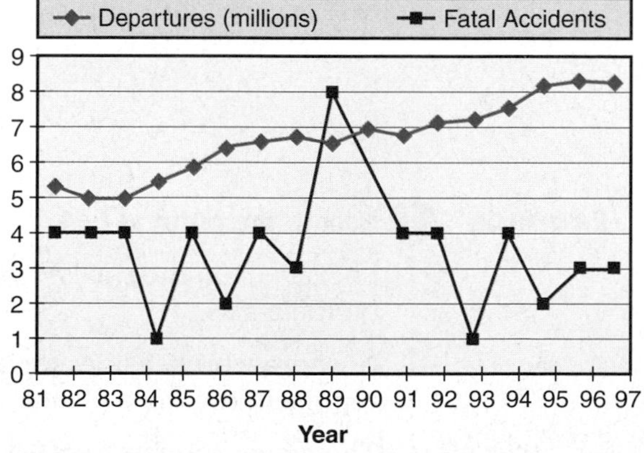

For more information on careers in aerospace engineering, visit: **www.amc.glencoe.com**

8-3 Vectors in Three-Dimensional Space

OBJECTIVES
- Add and subtract vectors in three-dimensional space.
- Find the magnitude of vectors in three-dimensional space.

Real World Application

ENTOMOLOGY Entomology (en tuh MAHL uh jee) is the study of insects. Entomologists often use time-lapse photography to study the flying behavior of bees. They have discovered that worker honeybees let other workers know about new sources of food by rapidly vibrating their wings and performing a dance. Researchers have been able to create three-dimensional models using vectors to represent the flying behavior of bees. *You will use vectors in Example 3 to model a bee's path.*

Vectors in three-dimensional space can be described by coordinates in a way similar to the way we describe vectors in a plane. Imagine three real number lines intersecting at the zero point of each, so that each line is perpendicular to the plane determined by the other two. To show this arrangement on paper, a figure with the x-axis appearing to come out of the page is used to convey the feeling of depth. The axes are named the x-axis, y-axis, and z-axis.

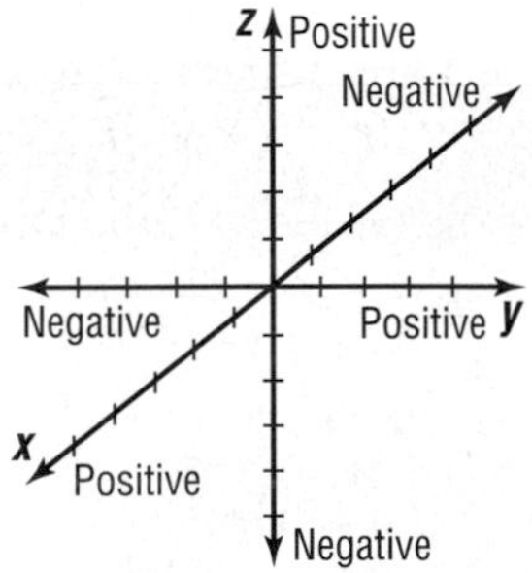

Each point in space corresponds to an ordered triple of real numbers. To locate a point P with coordinates (x_1, y_1, z_1), first find x_1 on the x-axis, y_1 on the y-axis, and z_1 on the z-axis. Then imagine a plane perpendicular to the x-axis at x_1 and planes perpendicular to the y- and z-axes at y_1 and z_1, respectively. The three planes will intersect at point P.

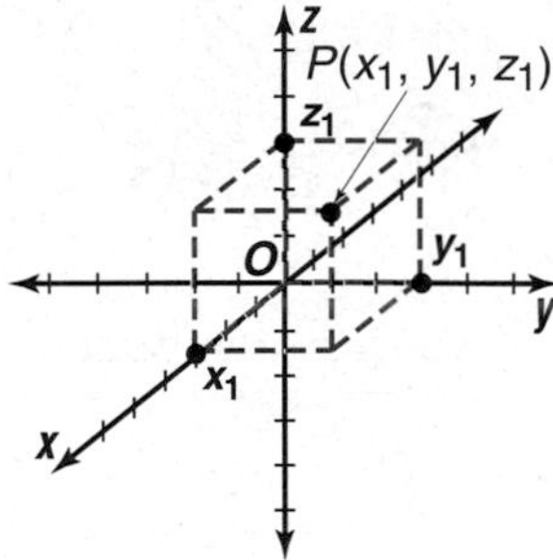

Example 1 **Locate the point at (−4, 3, 2).**

Locate −4 on the x-axis, 3 on the y-axis, and 2 on the z-axis.

Now draw broken lines for parallelograms to represent the three planes.

The planes intersect at (−4, 3, 2).

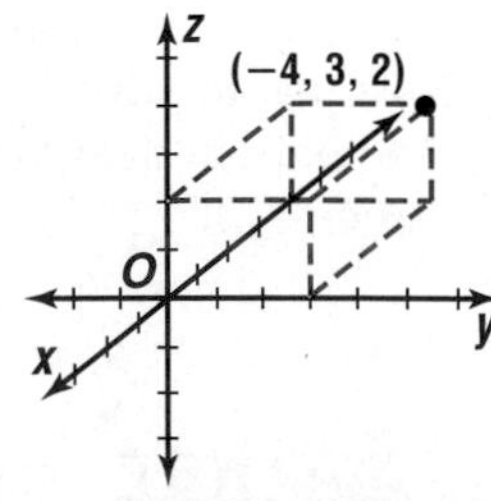

Ordered triples, like ordered pairs, can be used to represent vectors. The geometric interpretation is the same for a vector in space as it is for a vector in a plane. A directed line segment from the origin O to $P(x, y, z)$ is called vector $\overrightarrow{OP}$, corresponding to vector $\langle x, y, z \rangle$.

An extension of the formula for the distance between two points in a plane allows us to find the distance between two points in space. The distance from the origin to a point (x, y, z) is $\sqrt{x^2 + y^2 + z^2}$. So the magnitude of vector $\langle x, y, z \rangle$ is $\sqrt{x^2 + y^2 + z^2}$. This can be adapted to represent any vector in space.

Representation of a Vector as an Ordered Triple

Suppose $P_1(x_1, y_1, z_1)$ is the initial point of a vector in space and $P_2(x_2, y_2, z_2)$ is the terminal point. The ordered triple that represents $\overrightarrow{P_1P_2}$ is $\langle x_2 - x_1, y_2 - y_1, z_2 - z_1 \rangle$. Its magnitude is given by $|\overrightarrow{P_1P_2}| = \sqrt{(x_2 - x_1)^2 + (y_2 - y_1)^2 + (z_2 - z_1)^2}$.

Examples

2 **Write the ordered triple that represents the vector from $X(5, -3, 2)$ to $Y(4, -5, 6)$.**

$$\begin{aligned}\overrightarrow{XY} &= (4, -5, 6) - (5, -3, 2)\\ &= \langle 4 - 5, -5 - (-3), 6 - 2 \rangle\\ &= \langle -1, -2, 4 \rangle\end{aligned}$$

3 **ENTOMOLOGY** **Refer to the application at the beginning of the lesson. Suppose the flight of a honeybee passed through points at (0, 3, 3) and (5, 0, 4), in which each unit represents a meter. What is the magnitude of the displacement the bee experienced in traveling between these two points?**

$$\begin{aligned}\text{magnitude} &= \sqrt{(x_2 - x_1)^2 + (y_2 - y_1)^2 + (z_2 - z_1)^2}\\ &= \sqrt{(5 - 0)^2 + (0 - 3)^2 + (4 - 3)^2} \qquad \langle x_1, y_1, z_1 \rangle = \langle 0, 3, 3 \rangle,\ \langle x_2, y_2, z_2 \rangle = \langle 5, 0, 4 \rangle\\ &= \sqrt{25 + 9 + 1}\\ &= 5.9\end{aligned}$$

The magnitude of the displacement is about 5.9 meters.

Operations on vectors represented by ordered triples are similar to those on vectors represented by ordered pairs.

Example

4 **Find an ordered triple that represents $3\vec{p} - 2\vec{q}$ if $\vec{p} = \langle 3, 0, 4 \rangle$ and $\vec{q} = \langle 2, 1, -1 \rangle$.**

$$\begin{aligned}3\vec{p} - 2\vec{q} &= 3\langle 3, 0, 4 \rangle - 2\langle 2, 1, -1 \rangle \qquad \vec{p} = \langle 3, 0, 4 \rangle, \vec{q} = \langle 2, 1, -1 \rangle\\ &= \langle 9, 0, 12 \rangle - \langle 4, 2, -2 \rangle\\ &= \langle 5, -2, 14 \rangle\end{aligned}$$

Three unit vectors are used as components of vectors in space. The unit vectors on the x-, y-, and z-axes are $\vec{\mathbf{i}}$, $\vec{\mathbf{j}}$, and $\vec{\mathbf{k}}$ respectively, where $\vec{\mathbf{i}} = \langle 1, 0, 0 \rangle$, $\vec{\mathbf{j}} = \langle 0, 1, 0 \rangle$, and $\vec{\mathbf{k}} = \langle 0, 0, 1 \rangle$. The vector $\vec{\mathbf{a}} = \langle a_1, a_2, a_3 \rangle$ is shown on the graph at the right. The component vectors of $\vec{\mathbf{a}}$ along the three axes are $a_1\vec{\mathbf{i}}$, $a_2\vec{\mathbf{j}}$, and $a_3\vec{\mathbf{k}}$. Vector $\vec{\mathbf{a}}$ can be written as the sum of unit vectors; that is, $\vec{\mathbf{a}} = a_1\vec{\mathbf{i}} + a_2\vec{\mathbf{j}} + a_3\vec{\mathbf{k}}$.

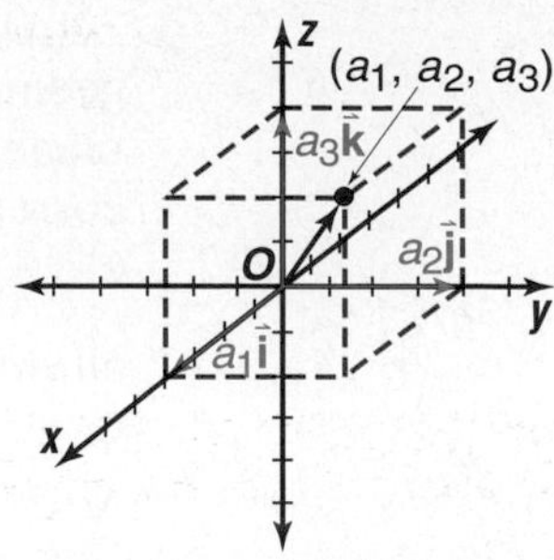

Example **5** **Write $\overrightarrow{AB}$ as the sum of unit vectors for $A(5, 10, -3)$ and $B(-1, 4, -2)$.**

First, express $\overrightarrow{AB}$ as an ordered triple. Then, write the sum of the unit vectors $\vec{\mathbf{i}}$, $\vec{\mathbf{j}}$, and $\vec{\mathbf{k}}$.

$$\begin{aligned}\overrightarrow{AB} &= (-1, 4, -2) - (5, 10, -3)\\ &= \langle -1 - 5, 4 - 10, -2 - (-3) \rangle\\ &= \langle -6, -6, 1 \rangle\\ &= -6\vec{\mathbf{i}} - 6\vec{\mathbf{j}} + \vec{\mathbf{k}}\end{aligned}$$

CHECK FOR UNDERSTANDING

Communicating Mathematics

Read and study the lesson to answer each question.

1. **Explain** the process you would use to locate $\overrightarrow{AB} = 2\vec{\mathbf{i}} + 3\vec{\mathbf{j}} + 4\vec{\mathbf{k}}$ in space. Then sketch the vector on a coordinate system.
2. **Describe** the information you need to find the components of a three-dimensional vector from its given magnitude.
3. **You Decide** Marshall wrote the vector $\vec{\mathbf{v}} = \langle 1, -4, 0 \rangle$ as the sum of unit vectors $\vec{\mathbf{i}} + 4\vec{\mathbf{j}} + \vec{\mathbf{k}}$. Denise said that $\vec{\mathbf{v}}$ should be written as $\vec{\mathbf{i}} - 4\vec{\mathbf{j}} + \vec{\mathbf{k}}$. Who is correct? Explain.

Guided Practice

4. Locate point $G(4, -1, 7)$ in space. Then find the magnitude of a vector from the origin to G.

Write the ordered triple that represents $\overrightarrow{RS}$. Then find the magnitude of $\overrightarrow{RS}$.

5. $R(-2, 5, 8)$, $S(3, 9, -3)$
6. $R(3, 7, -1)$, $S(10, -4, 0)$

Find an ordered triple to represent $\vec{\mathbf{a}}$ in each equation if $\vec{\mathbf{f}} = \langle 1, -3, -8 \rangle$ and $\vec{\mathbf{g}} = \langle 3, 9, -1 \rangle$.

7. $\vec{\mathbf{a}} = 3\vec{\mathbf{f}} + \vec{\mathbf{g}}$
8. $\vec{\mathbf{a}} = 2\vec{\mathbf{g}} - 5\vec{\mathbf{f}}$

Write $\overrightarrow{EF}$ as the sum of unit vectors.

9. $E(-5, -2, 4)$, $F(6, -6, 6)$
10. $E(-12, 15, -9)$, $F(-12, 17, -22)$

11. **Physics** Suppose that during a storm the force of the wind blowing against a skyscraper can be expressed by the vector $\langle 132, 3454, 0 \rangle$, where each measure in the ordered triple represents the force in newtons. What is the magnitude of this force?

EXERCISES

Practice

Locate point *B* in space. Then find the magnitude of a vector from the origin to *B*.

12. $B(4, 1, -3)$ 13. $B(7, 2, 4)$ 14. $B(10, -3, 15)$

Write the ordered triple that represents $\overrightarrow{TM}$. Then find the magnitude of $\overrightarrow{TM}$.

15. $T(2, 5, 4), M(3, 1, -4)$
16. $T(-2, 4, 7), M(-3, 5, 2)$
17. $T(2, 5, 4), M(3, 1, 0)$
18. $T(3, -5, 6), M(-1, 1, 2)$
19. $T(-5, 8, 3), M(-2, -1, -6)$
20. $T(0, 6, 3), M(1, 4, -3)$

21. Write the ordered triple to represent $\overrightarrow{CJ}$. Then find the magnitude of $\overrightarrow{CJ}$.

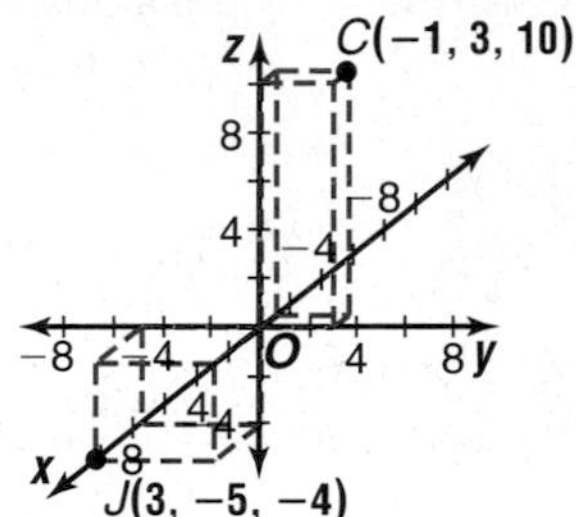

Find an ordered triple to represent $\vec{\mathbf{u}}$ in each equation if $\vec{\mathbf{v}} = \langle 4, -3, 5 \rangle$, $\vec{\mathbf{w}} = \langle 2, 6, -1 \rangle$, and $\vec{\mathbf{z}} = \langle 3, 0, 4 \rangle$.

22. $\vec{\mathbf{u}} = 6\vec{\mathbf{w}} + 2\vec{\mathbf{z}}$
23. $\vec{\mathbf{u}} = \frac{1}{2}\vec{\mathbf{v}} - \vec{\mathbf{w}} + 2\vec{\mathbf{z}}$
24. $\vec{\mathbf{u}} = \frac{3}{4}\vec{\mathbf{v}} - \vec{\mathbf{w}}$
25. $\vec{\mathbf{u}} = 3\vec{\mathbf{v}} - \frac{2}{3}\vec{\mathbf{w}} + 2\vec{\mathbf{z}}$
26. $\vec{\mathbf{u}} = 0.75\vec{\mathbf{v}} + 0.25\vec{\mathbf{w}}$
27. $\vec{\mathbf{u}} = -4\vec{\mathbf{w}} + \vec{\mathbf{z}}$

28. Find an ordered triple to represent the sum $\frac{2}{3}\vec{\mathbf{f}} + 3\vec{\mathbf{g}} - \frac{2}{5}\vec{\mathbf{h}}$, if $\vec{\mathbf{f}} = \langle -3, 4.5, -1 \rangle$, $\vec{\mathbf{g}} = \langle -2, 1, 6 \rangle$, and $\vec{\mathbf{h}} = \langle 6, -3, -3 \rangle$.

Write $\overrightarrow{LB}$ as the sum of unit vectors.

29. $L(2, 2, 7), B(5, -6, 2)$
30. $L(-6, 1, 0), B(-4, 5, -1)$
31. $L(9, 7, -11), B(7, 3, -2)$
32. $L(12, 2, 6), B(-8, 7, -5)$
33. $L(-1, 2, -4), B(-8, 5, -10)$
34. $L(-9, 12, -5), B(6, 5, -5)$

35. Show that $|\overrightarrow{G_1G_2}| = |\overrightarrow{G_2G_1}|$.

36. If $\vec{\mathbf{m}} = \langle m_1, m_2, m_3 \rangle$, then $-\vec{\mathbf{m}}$ is defined as $\langle -m_1, -m_2, -m_3 \rangle$. Show that $|-\vec{\mathbf{m}}| = |\vec{\mathbf{m}}|$

Applications and Problem Solving

37. **Physics** An object is in equilibrium if the magnitude of the resultant force on it is zero. Two forces on an object are represented by $\langle 3, -2, 4\rangle$ and $\langle 6, 2, 5\rangle$. Find a third vector that will place the object in equilibrium.

38. **Critical Thinking** Find the midpoint of $\vec{v}$ that extends from $V(2, 3, 6)$ to $W(4, 5, 2)$.

39. **Computer Games** Nate Rollins is designing a computer game. In the game, a knight is standing at point (1, 4, 0) watching a wizard sitting at the top of a tree. In the computer screen, the tree is one unit high, and its base is at (2, 4, 0). Find the displacement vector for each situation.

 a. from the origin to the knight
 b. from the bottom of the tree to the knight

40. **Critical Thinking** Find the vector $\vec{c}$ that must be added to $\vec{a} = \langle 1, 3, 1\rangle$ to obtain $\vec{b} = \langle 3, 1, 5\rangle$.

Dr. Chiaki Mukai

41. **Aeronautics** Dr. Chiaki Mukai is Japan's first female astronaut. Suppose she is working inside a compartment shaped like a cube with sides 15 feet long. She realizes that the tool she needs is diagonally in the opposite corner of the compartment.

 a. Draw a diagram of the situation described above.
 b. What is the minimum distance she has to glide to secure the tool?
 c. At what angle to the floor must she launch herself?

42. **Chemistry** Dr. Alicia Sanchez is a researcher for a pharmaceutical firm. She has graphed the structure of a molecule with atoms having positions $A = (2, 0, 0)$, $B = \left(1, \sqrt{3}, 0\right)$, and $C = \left(1, \frac{1}{3}, \frac{2\sqrt{2}}{3}\right)$. She needs to have every atom in this molecule equidistant from each other. Has she achieved this goal? Explain why or why not.

Mixed Review

43. Find the sum of the vectors $\langle 3, 5\rangle$ and $\langle -1, 2\rangle$ algebraically. *(Lesson 8-2)*

44. Find the coordinates of point D such that $\overline{AB}$ and $\overline{CD}$ are equal vectors for points $A(5, 2)$, $B(-3, 3)$, and $C(0, 0)$. *(Lesson 8-1)*

45. Verify that $\cot X = (\sin 2X) \div (1 - \cos 2X)$ is an identity. *(Lesson 7-4)*

46. If $\cos \theta = \frac{2}{3}$ and $0° \leq \theta \leq 90°$, find $\sin \theta$. *(Lesson 7-1)*

47. State the amplitude and period for the function $y = 6 \sin \frac{\theta}{2}$. *(Lesson 6-4)*

48. **Physics** If a pulley is rotating at 16 revolutions per minute, what is its rate in radians per second? *(Lesson 6-2)*

49. Determine if $(7, -2)$ is a solution for $y < 4x^2 - 3x + 5$. Explain. *(Lesson 3-3)*

50. **SAT/ACT Practice** You have added the same positive quantity to the numerator and denominator of a fraction. The result is

 A greater than the original fraction.
 B less than the original fraction.
 C equal to the original fraction.
 D one-half the original fraction.
 E cannot be determined from the information given.

Extra Practice See p. A40.

8-4 Perpendicular Vectors

OBJECTIVES
- Find the inner and cross products of two vectors.
- Determine whether two vectors are perpendicular.

AUTO RACING In addition to being an actor, Emilio Estevez is an avid stock car driver. Drivers in stock car races must constantly shift gears to maneuver their cars into competitive positions. A gearshift rotates about the connection to the transmission. The rate of change of the rotation of the gearshift depends on the magnitude of the force exerted by the driver and on the perpendicular distance of its line of action from the center of rotation. *You will use vectors in Example 4 to determine the force on a gearshift.*

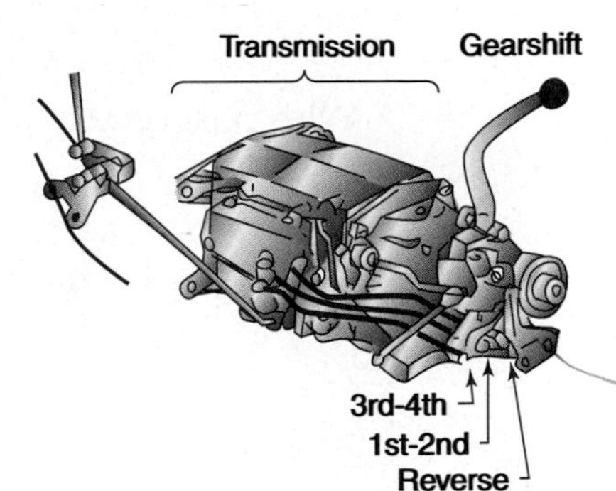

In physics, *torque* is the measure of the effectiveness of a force in turning an object about a pivot point. We can use perpendicular vectors to find the torque of a force.

Let $\vec{\mathbf{a}}$ and $\vec{\mathbf{b}}$ be perpendicular vectors, and let $\overrightarrow{BA}$ be a vector between their terminal points as shown. Then the magnitudes of $\vec{\mathbf{a}}$, $\vec{\mathbf{b}}$, and $\overrightarrow{BA}$ must satisfy the Pythagorean Theorem.

$$|\overrightarrow{BA}|^2 = |\vec{\mathbf{a}}|^2 + |\vec{\mathbf{b}}|^2$$

Now use the definition of magnitude of a vector to evaluate $|\overrightarrow{BA}|^2$.

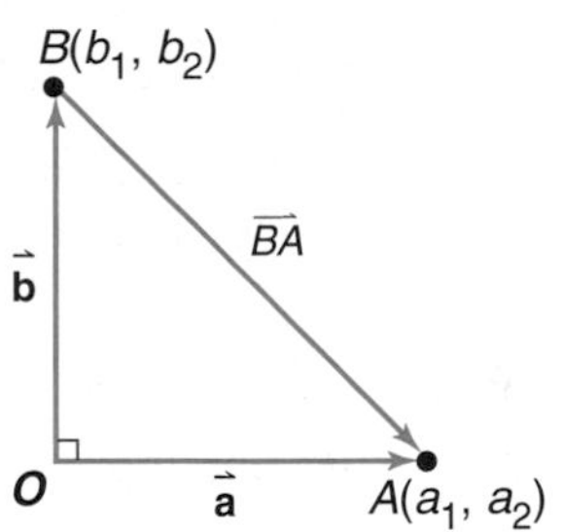

$|\overrightarrow{BA}| = \sqrt{(a_1 - b_1)^2 + (a_2 - b_2)^2}$ *Definition of magnitude*

$|\overrightarrow{BA}|^2 = (a_1 - b_1)^2 + (a_2 - b_2)^2$ *Square each side.*

$|\overrightarrow{BA}|^2 = a_1{}^2 - 2a_1b_1 + b_1{}^2 + a_2{}^2 - 2a_2b_2 + b_2{}^2$ *Simplify.*

$|\overrightarrow{BA}|^2 = (a_1{}^2 + a_2{}^2) + (b_1{}^2 + b_2{}^2) - 2(a_1b_1 + a_2b_2)$ *Group the squared terms.*

$|\overrightarrow{BA}|^2 = |\vec{\mathbf{a}}|^2 + |\vec{\mathbf{b}}|^2 - 2(a_1b_1 + a_2b_2)$ *$|\vec{a}|^2 = a_1{}^2 + a_2{}^2$, $|\vec{b}|^2 = b_1{}^2 + b_2{}^2$*

Compare the resulting equation with the original one. $|\overrightarrow{BA}|^2 = |\vec{\mathbf{a}}|^2 + |\vec{\mathbf{b}}|^2$ if and only if $a_1b_1 + a_2b_2 = 0$.

The expression $a_1b_1 + a_2b_2$ is frequently used in the study of vectors. It is called the **inner product** of $\vec{\mathbf{a}}$ and $\vec{\mathbf{b}}$.

Inner Product of Vectors in a Plane

If $\vec{a}$ and $\vec{b}$ are two vectors, $\langle a_1, a_2\rangle$ and $\langle b_1, b_2\rangle$, the inner product of $\vec{a}$ and $\vec{b}$ is defined as $\vec{a} \cdot \vec{b} = a_1b_1 + a_2b_2$.

$\vec{a} \cdot \vec{b}$ is read "a dot b" and is often called the dot product.

Two vectors are perpendicular if and only if their inner product is zero. That is, vectors $\vec{a} = \langle a_1, a_2\rangle$ and $\vec{b} = \langle b_1, b_2\rangle$ are perpendicular if $\vec{a} \cdot \vec{b} = 0$.

$$\vec{a} \cdot \vec{b} = a_1b_1 + a_2b_2$$

Let $a_1b_1 + a_2b_2 = 0$ if

$$a_1b_1 = -a_2b_2$$

$$\frac{a_1}{a_2} = -\frac{b_2}{b_1} \qquad a_2 \neq 0,\ b_1 \neq 0$$

Since the ratio of the components can also be thought of as the slopes of the line on which the vectors lie, the slopes are opposite reciprocals. Thus, the lines and the vectors are perpendicular.

Example 1 **Find each inner product if $\vec{p} = \langle 7, 14\rangle$, $\vec{q} = \langle 2, -1\rangle$ and $\vec{m} = \langle 3, 5\rangle$. Are any pair of vectors perpendicular?**

a. $\vec{p} \cdot \vec{q}$

$$\begin{aligned}\vec{p} \cdot \vec{q} &= 7(2) + 14(-1)\\ &= 14 - 14\\ &= 0\end{aligned}$$

$\vec{p}$ and $\vec{q}$ are perpendicular.

b. $\vec{p} \cdot \vec{m}$

$$\begin{aligned}\vec{p} \cdot \vec{m} &= 7(3) + 14(5)\\ &= 21 + 70\\ &= 91\end{aligned}$$

$\vec{p}$ and $\vec{m}$ are not perpendicular.

c. $\vec{q} \cdot \vec{m}$

$$\begin{aligned}\vec{q} \cdot \vec{m} &= 2(3) + (-1)(5)\\ &= 6 - 5\\ &= 1\end{aligned}$$

$\vec{q}$ and $\vec{m}$ are not perpendicular.

The inner product of vectors in space is similar to that of vectors in a plane.

Inner Product of Vectors in Space

If $\vec{a} = \langle a_1, a_2, a_3\rangle$ and $\vec{b} = \langle b_1, b_2, b_3\rangle$, then $\vec{a} \cdot \vec{b} = a_1b_1 + a_2b_2 + a_3b_3$.

Just as in a plane, two vectors in space are perpendicular if and only if their inner product is zero.

Example 2 **Find the inner product of $\vec{a}$ and $\vec{b}$ if $\vec{a} = \langle -3, 1, 1\rangle$ and $\vec{b} = \langle 2, 8, -2\rangle$. Are $\vec{a}$ and $\vec{b}$ perpendicular?**

$$\begin{aligned}\vec{a} \cdot \vec{b} &= (-3)(2) + (1)(8) + (1)(-2)\\ &= -6 + 8 + (-2)\\ &= 0\end{aligned}$$

$\vec{a}$ and $\vec{b}$ are perpendicular since their inner product is zero.

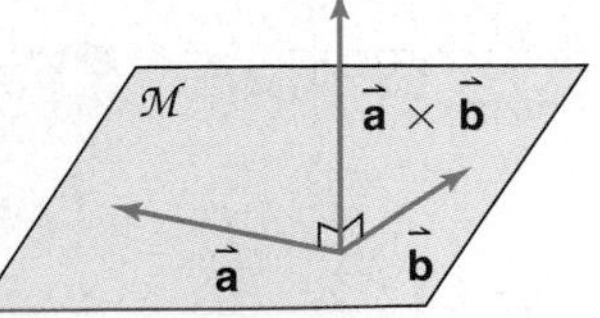

Another important product involving vectors in space is the **cross product.** Unlike the inner product, the cross product of two vectors is a vector. This vector does not lie in the plane of the given vectors, but is perpendicular to the plane containing the two vectors. In other words, the cross product of two vectors is perpendicular to both vectors. The cross product of $\vec{\mathbf{a}}$ and $\vec{\mathbf{b}}$ is written $\vec{\mathbf{a}} \times \vec{\mathbf{b}}$.

$\vec{\mathbf{a}} \times \vec{\mathbf{b}}$ is perpendicular to $\vec{\mathbf{a}}$.
$\vec{\mathbf{a}} \times \vec{\mathbf{b}}$ is perpendicular to $\vec{\mathbf{b}}$.

Cross Product of Vectors in Space

If $\vec{\mathbf{a}} = \langle a_1, a_2, a_3 \rangle$ and $\vec{\mathbf{b}} = \langle b_1, b_2, b_3 \rangle$, then the cross product of $\vec{\mathbf{a}}$ and $\vec{\mathbf{b}}$ is defined as follows.

$$\vec{\mathbf{a}} \times \vec{\mathbf{b}} = \begin{vmatrix} a_2 & a_3 \\ b_2 & b_3 \end{vmatrix} \vec{\mathbf{i}} - \begin{vmatrix} a_1 & a_3 \\ b_1 & b_3 \end{vmatrix} \vec{\mathbf{j}} + \begin{vmatrix} a_1 & a_2 \\ b_1 & b_2 \end{vmatrix} \vec{\mathbf{k}}$$

Look Back
Refer to Lesson 2-5 to review determinants and expansion by minors.

An easy way to remember the coefficients of $\vec{\mathbf{i}}, \vec{\mathbf{j}}, \vec{\mathbf{k}}$ is to write a determinant as shown and expand by minors using the first row.

$$\begin{vmatrix} \vec{\mathbf{i}} & \vec{\mathbf{j}} & \vec{\mathbf{k}} \\ a_1 & a_2 & a_3 \\ b_1 & b_2 & b_3 \end{vmatrix}$$

Example 3 **Find the cross product of $\vec{\mathbf{v}}$ and $\vec{\mathbf{w}}$ if $\vec{\mathbf{v}} = \langle 0, 3, 1 \rangle$ and $\vec{\mathbf{w}} = \langle 0, 1, 2 \rangle$. Verify that the resulting vector is perpendicular to $\vec{\mathbf{v}}$ and $\vec{\mathbf{w}}$.**

$$\vec{\mathbf{v}} \times \vec{\mathbf{w}} = \begin{vmatrix} \vec{\mathbf{i}} & \vec{\mathbf{j}} & \vec{\mathbf{k}} \\ 0 & 3 & 1 \\ 0 & 1 & 2 \end{vmatrix}$$

$$= \begin{vmatrix} 3 & 1 \\ 1 & 2 \end{vmatrix} \vec{\mathbf{i}} - \begin{vmatrix} 0 & 1 \\ 0 & 2 \end{vmatrix} \vec{\mathbf{j}} + \begin{vmatrix} 0 & 3 \\ 0 & 1 \end{vmatrix} \vec{\mathbf{k}} \quad \textit{Expand by minors.}$$

$$= 5\vec{\mathbf{i}} - 0\vec{\mathbf{j}} + 0\vec{\mathbf{k}}$$
$$= 5\vec{\mathbf{i}} \text{ or } \langle 5, 0, 0 \rangle$$

Find the inner products $\langle 5, 0, 0 \rangle \cdot \langle 0, 3, 1 \rangle$ and $\langle 5, 0, 0 \rangle \cdot \langle 0, 1, 2 \rangle$.

$5(0) + 0(3) + 0(1) = 0 \qquad 5(0) + 0(1) + 0(2) = 0$

Since the inner products are zero, the cross product $\vec{\mathbf{v}} \times \vec{\mathbf{w}}$ is perpendicular to both $\vec{\mathbf{v}}$ and $\vec{\mathbf{w}}$.

In physics, the torque $\vec{\mathbf{T}}$ about a point A created by a force $\vec{\mathbf{F}}$ at a point B is given by $\vec{\mathbf{T}} = \overrightarrow{AB} \times \vec{\mathbf{F}}$. The magnitude of $\vec{\mathbf{T}}$ represents the torque in foot-pounds.

Example **4** **AUTO RACING** **Refer to the application at the beginning of the lesson. Suppose Emilio Estevez is applying a force of 25 pounds along the positive z-axis to the gearshift of his car. If the center of the connection of the gearshift is at the origin, the force is applied at the point (0.75, 0, 0.27). Find the torque.**

Emilio Estevez

We need to find $|\vec{\mathbf{T}}|$, the torque of the force at (0.75, 0, 0.27) where each value is the distance from the origin in feet and $\vec{\mathbf{F}}$ represents the force in pounds.

To find the magnitude of $\vec{\mathbf{T}}$, we must first find $\overrightarrow{AB}$ and $\vec{\mathbf{F}}$.

$$\overrightarrow{AB} = (0.75, 0, 0.27) - (0, 0, 0)$$
$$= \langle 0.75 - 0, 0 - 0, 0.27 - 0\rangle \text{ or } \langle 0.75, 0, 0.27\rangle$$

Any upward force is measured along the z-axis, so $\vec{\mathbf{F}} = 25\vec{\mathbf{k}}$ or $\langle 0, 0, 25\rangle$.

Now, find $\vec{\mathbf{T}}$.

$$\vec{\mathbf{T}} = \overrightarrow{AB} \times \vec{\mathbf{F}}$$

$$= \begin{vmatrix} \vec{\mathbf{i}} & \vec{\mathbf{j}} & \vec{\mathbf{k}} \\ 0.75 & 0 & 0.27 \\ 0 & 0 & 25 \end{vmatrix}$$

$$= \begin{vmatrix} 0 & 0.27 \\ 0 & 25 \end{vmatrix}\vec{\mathbf{i}} - \begin{vmatrix} 0.75 & 0.27 \\ 0 & 25 \end{vmatrix}\vec{\mathbf{j}} + \begin{vmatrix} 0.75 & 0 \\ 0 & 0 \end{vmatrix}\vec{\mathbf{k}}$$

$$= 0\vec{\mathbf{i}} - 18.75\vec{\mathbf{j}} + 0\vec{\mathbf{k}} \text{ or } \langle 0, -18.75, 0\rangle$$

Find the magnitude of $\vec{\mathbf{T}}$.

$$|\vec{\mathbf{T}}| = \sqrt{0^2 + (-18.75)^2 + 0^2}$$
$$= \sqrt{(-18.75)^2} \text{ or } 18.75$$

The torque is 18.75 foot-pounds.

CHECK FOR UNDERSTANDING

Communicating Mathematics

Read and study the lesson to answer each question.

1. **Compare** $\vec{\mathbf{v}} \times \vec{\mathbf{w}}$ and $\vec{\mathbf{w}} \times \vec{\mathbf{v}}$ for $\vec{\mathbf{v}} = \langle -1, 0, 3\rangle$ and $\vec{\mathbf{w}} = \langle 1, 2, 4\rangle$.
2. **Show** that the cross product of a three-dimensional vector with itself is the zero vector.
3. *Math Journal* Could the inner product of a nonzero vector and itself ever be zero? Explain why or why not.

Guided Practice

Find each inner product and state whether the vectors are perpendicular. Write *yes* or *no*.

4. $\langle 5, 2\rangle \cdot \langle -3, 7\rangle$
5. $\langle -8, 2\rangle \cdot \langle 4.5, 18\rangle$
6. $\langle -4, 9, 8\rangle \cdot \langle 3, 2, -2\rangle$

Find each cross product. Then verify that the resulting vector is perpendicular to the given vectors.

7. $\langle 1, -3, 2\rangle \times \langle -2, 1, -5\rangle$

8. $\langle 6, 2, 10\rangle \times \langle 4, 1, 9\rangle$

9. Find a vector perpendicular to the plane containing the points (0, 1, 2), (−2, 2, 4), and (−1, −1, −1).

10. Mechanics Tikiro places a wrench on a nut and applies a downward force of 32 pounds to tighten the nut. If the center of the nut is at the origin, the force is applied at the point (0.65, 0, 0.3). Find the torque.

EXERCISES

Practice

Find each inner product and state whether the vectors are perpendicular. Write *yes* or *no*.

11. $\langle 4, 8\rangle \cdot \langle 6, -3\rangle$

12. $\langle 3, 5\rangle \cdot \langle 4, -2\rangle$

13. $\langle 5, -1\rangle \cdot \langle -3, 6\rangle$

14. $\langle 7, 2\rangle \cdot \langle 0, -2\rangle$

15. $\langle 8, 4\rangle \cdot \langle 2, 4\rangle$

16. $\langle 4, 9, -3\rangle \cdot \langle -6, 7, 5\rangle$

17. $\langle 3, 1, 4\rangle \cdot \langle 2, 8, -2\rangle$

18. $\langle -2, 4, 8\rangle \cdot \langle 16, 4, 2\rangle$

19. $\langle 7, -2, 4\rangle \cdot \langle 3, 8, 1\rangle$

20. Find the inner product of $\vec{a}$ and $\vec{b}$, $\vec{b}$ and $\vec{c}$, and $\vec{a}$ and $\vec{c}$ if $\vec{a} = \langle 3, 12\rangle$, $\vec{b} = \langle 8, -2\rangle$, and $\vec{c} = \langle 3, -2\rangle$. Are any of the pairs perpendicular? If so, which one(s)?

Find each cross product. Then verify that the resulting vector is perpendicular to the given vectors.

21. $\langle 0, 1, 2\rangle \times \langle 1, 1, 4\rangle$

22. $\langle 5, 2, 3\rangle \times \langle -2, 5, 0\rangle$

23. $\langle 3, 2, 0\rangle \times \langle 1, 4, 0\rangle$

24. $\langle 1, -3, 2\rangle \times \langle 5, 1, -2\rangle$

25. $\langle -3, -1, 2\rangle \times \langle 4, -4, 0\rangle$

26. $\langle 4, 0, -2\rangle \times \langle -7, 1, 0\rangle$

27. Prove that for any vector $\vec{a}$, $\vec{a} \times (-\vec{a}) = 0$.

28. Use the definition of cross products to prove that for any vectors $\vec{a}$, $\vec{b}$, and $\vec{c}$, $\vec{a} \times (\vec{b} + \vec{c}) = (\vec{a} \times \vec{b}) + (\vec{a} \times \vec{c})$.

Find a vector perpendicular to the plane containing the given points.

29. (0, −2, 2), (1, 2, −3), and (4, 0, −1)

30. (−2, 1, 0), (−3, 0, 0), and (5, 2, 0)

31. (0, 0, 1), (1, 0, 1), and (−1, −1, −1)

32. Explain whether the equation $\vec{m} \times \vec{n} = \vec{n} \times \vec{m}$ is true. Discuss your reasoning.

Applications and Problem Solving

33. Physiology Whenever we lift an object, a torque is applied by the biceps muscle on the lower arm. The elbow acts as the axis of rotation through the joint. Suppose the muscle is attached 4 centimeters from the joint and you exert a force of 600 N lifting an object 30° to the horizontal.

a. Make a labeled diagram showing this situation.

b. What is the torque about the elbow?

34. **Critical Thinking** Let $\vec{\mathbf{x}} = \langle 2, 3, 0\rangle$ and $\vec{\mathbf{y}} = \langle -1, 1, 4\rangle$. Find the area of the triangle whose vertices are the origin and the endpoints of $\vec{\mathbf{x}}$ and $\vec{\mathbf{y}}$.

35. **Business Management** Mr. Toshiro manages a company that supplies a variety of domestic and imported nuts to supermarkets. He received an order for 120 bags of cashews, 310 bags of walnuts, and 60 bags of Brazil nuts. The prices per bag for each type are \$29, \$18, and \$21, respectively.

    a. Represent the number of bags ordered and the cost as vectors.

    b. Using what you have learned in the lesson about vectors, compute the value of the order.

36. **Physics** The work done by a force $\vec{\mathbf{F}}$, that displaces an object through a distance d is defined as $\vec{\mathbf{F}} \cdot \vec{\mathbf{d}}$. This can also be expressed as $|\vec{\mathbf{F}}|\,|\vec{\mathbf{d}}| \cos\theta$. Alexa is pushing a construction barrel up a ramp 4 feet long into the back of a truck. She is using a force of 120 pounds in a horizontal direction and the ramp is 45° from the horizontal.

    a. Sketch a drawing of the situation.

    b. How much work is Alexa doing?

37. **Architecture** Steve Herr is an architect in Minneapolis, Minnesota. His latest project is designing a park. On the blueprint, the park is determined by a plane which contains the points at (1, 0, 3), (2, 5, 0), and (3, 1, 4). One of the features of the park is a monument that must be perpendicular to the ground.

    a. Find a nonzero vector, representing the monument, perpendicular to the plane defined by the given points.

    b. Explain how you know that the vector is perpendicular to the plane defining the park.

38. **Geometry** A parallelepiped is a prism whose opposite faces are all parallelograms.

    a. Determine the volume of the parallelepiped using the expression $\vec{\mathbf{p}} \cdot (\vec{\mathbf{q}} \times \vec{\mathbf{r}})$.

    b. Write a 3×3 matrix using the three vectors and calculate its determinant. How does this value compare to your answer in part a?

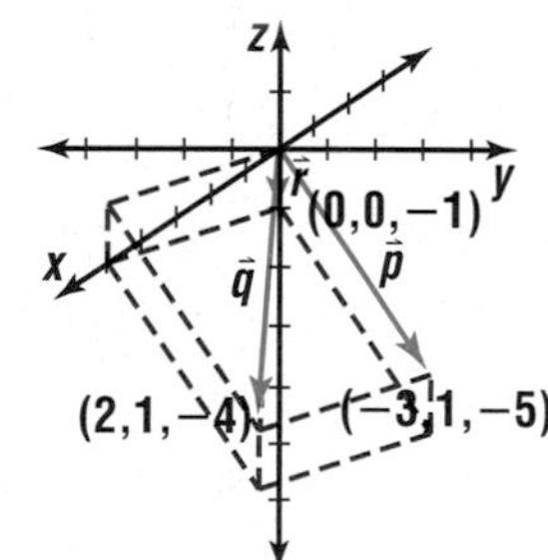

39. **Critical Thinking** If $\vec{\mathbf{v}} = \langle 1, 2\rangle$, $\vec{\mathbf{w}} = \langle -1, 2\rangle$, and $\vec{\mathbf{u}} = \langle 5, 12\rangle$, for what scalar k will the vector $k\vec{\mathbf{v}} + \vec{\mathbf{w}}$ be perpendicular to $\vec{\mathbf{u}}$?

40. **Proof** Let $\vec{\mathbf{a}} = \langle a_1, a_2\rangle$ and $\vec{\mathbf{b}} = \langle b_1, b_2\rangle$. Use the Law of Cosines to show that if the measure of the angle between $\vec{\mathbf{a}}$ and $\vec{\mathbf{b}}$, θ, is any value, then the inner product $\vec{\mathbf{a}} \cdot \vec{\mathbf{b}} = |\vec{\mathbf{a}}|\,|\vec{\mathbf{b}}| \cos\theta$. (*Hint:* Refer to the proof of inner product using the Pythagorean Theorem on page 505.)

Mixed Review

41. Given points $A(3, 3, -1)$ and $B(5, 3, 2)$, find an ordered triple that represents $\overrightarrow{AB}$. *(Lesson 8-3)*

42. Write the ordered pair that represents $\overrightarrow{DE}$ for the points on the graph. Then find the magnitude of $\overrightarrow{DE}$. *(Lesson 8-2)*

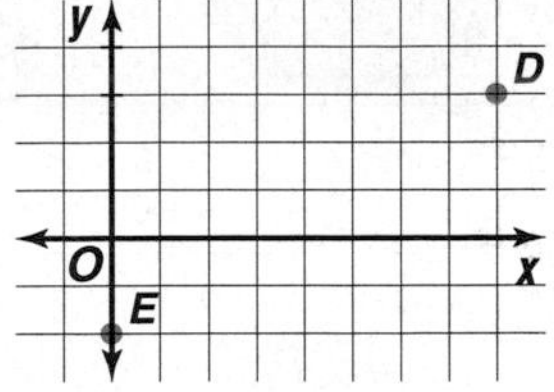

43. Write the equation $4x + y = 6$ in normal form. Then find p, the measure of the normal, and ϕ, the angle it makes with the positive x-axis. *(Lesson 7-6)*

44. Solve $\triangle ABC$ if $A = 36°$, $b = 13$, and $c = 6$. Round angle measures to the nearest minute and side measures to the nearest tenth. *(Lesson 5-8)*

45. **Utilities** A utility pole is braced by a cable attached to it at the top and anchored in a concrete block at ground level, a distance of 4 meters from the base of the pole. If the angle between the cable and the ground is 73°, find the height of the pole and the length of the cable to the nearest tenth of a meter. *(Lesson 5-4)*

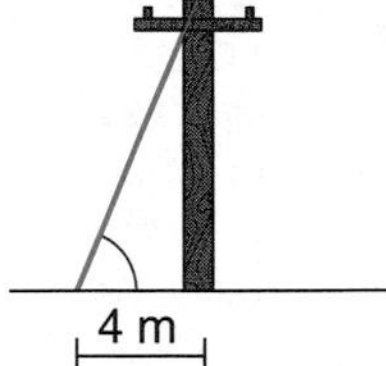

46. Solve $3+\sqrt{3x-4} \geq 10$. *(Lesson 4-7)*

47. **SAT/ACT Practice** Let x be an integer greater than 1. What is the least value of x for which $a^2 = b^3 = x$ for some integers a and b?
A 81 **B** 64 **C** 4 **D** 2 **E** 9

MID-CHAPTER QUIZ

Use a ruler and a protractor to draw a vector with the given magnitude and direction. Then find the magnitude of the horizontal and vertical components of the vector. (Lesson 8-1)

1. 2.3 centimeters, 46°
2. 27 millimeters, 245°

Write the ordered pair or ordered triple that represents $\overrightarrow{CD}$. Then find the magnitude of $\overrightarrow{CD}$. (Lessons 8-2 and 8-3)

3. $C(-9, 2)$, $D(-4, -3)$
4. $C(3, 7, -1)$, $D(5, 7, 2)$

Find an ordered pair or ordered triple to represent $\vec{r}$ in each equation if $\vec{s} = \langle 4, -3 \rangle$, $\vec{t} = \langle -6, 2 \rangle$, $\vec{u} = \langle 1, -3, -8 \rangle$, and $\vec{v} = \langle 3, 9, -1 \rangle$. (Lessons 8-2 and 8-3)

5. $\vec{r} = \vec{t} - 2\vec{s}$
6. $\vec{r} = 3\vec{u} + \vec{v}$

Find each inner product and state whether the vectors are perpendicular. Write *yes* or *no*. (Lesson 8-4)

7. $\langle 3, 6 \rangle \cdot \langle -4, 2 \rangle$
8. $\langle 3, -2, 4 \rangle \cdot \langle 1, -4, 0 \rangle$

9. Find the cross product $\langle 1, 3, 2 \rangle \times \langle 2, -1, -1 \rangle$. Then verify that the resulting vector is perpendicular to the given vectors. (Lesson 8-4)

10. **Entomology** Suppose the flight of a housefly passed through points at $(2, 0, 4)$ and $(7, 4, 6)$, in which each unit represents a meter. What is the magnitude of the displacement the housefly experienced in traveling between these points? (Lesson 8-3)

Extra Practice See p. A41.

GRAPHING CALCULATOR EXPLORATION

8-4B Finding Cross Products

An Extension of Lesson 8-4

OBJECTIVE

- Use a graphing calculator program to obtain the components of the cross product of two vectors in space.

The following graphing calculator program allows you to input the components of two vectors $\vec{\mathbf{p}} = \langle a, b, c \rangle$ and $\vec{\mathbf{q}} = \langle x, y, z \rangle$ in space and obtain the components of their cross product.

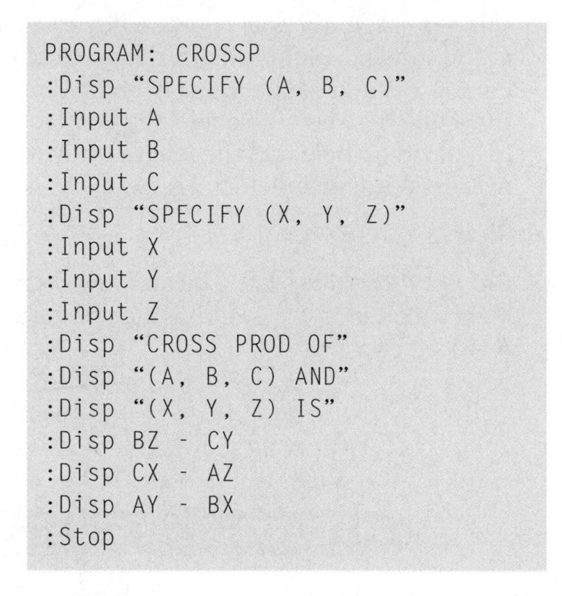

```
PROGRAM: CROSSP
:Disp "SPECIFY (A, B, C)"
:Input A
:Input B
:Input C
:Disp "SPECIFY (X, Y, Z)"
:Input X
:Input Y
:Input Z
:Disp "CROSS PROD OF"
:Disp "(A, B, C) AND"
:Disp "(X, Y, Z) IS"
:Disp BZ - CY
:Disp CX - AZ
:Disp AY - BX
:Stop
```

***inter*NET CONNECTION**

Graphing Calculator Programs
To download this program, visit **www.amc.glencoe.com**

Enter this program on your calculator and use it as needed to complete the following exercises.

TRY THESE

Use a graphing calculator to determine each cross product.

1. $\langle 7, 9, -1 \rangle \times \langle 3, -4, -5 \rangle$

2. $\langle 8, 14, 0 \rangle \times \langle -2, 6, 12 \rangle$

3. $\langle 5, 5, 5 \rangle \times \langle 4, 4, 4 \rangle$

4. $\langle -3, 2 -1 \rangle \times \langle 6, -3, 7 \rangle$

5. $\langle 1, 6, 0 \rangle \times \langle 2, 5, 0 \rangle$

6. $\langle 4, 0, -2 \rangle \times \langle 6, 0, -13 \rangle$

It can be proved that if $\vec{\mathbf{u}}$ and $\vec{\mathbf{v}}$ are adjacent sides of a parallelogram, then the area of the parallelogram is $|\vec{\mathbf{u}} \times \vec{\mathbf{v}}|$. Find the area of the parallelogram having the given vectors as adjacent sides.

7. $\vec{\mathbf{u}} = \langle -2, 4, 1 \rangle, \vec{\mathbf{v}} = \langle 0, 6, 3 \rangle$

8. $\vec{\mathbf{u}} = \langle 1, 3, 2 \rangle, \vec{\mathbf{v}} = \langle 9, 7, 5 \rangle$

WHAT DO YOU THINK?

9. What instructions could you insert at the end of the program to have the program display the magnitude of the vectors resulting from the cross products of the vectors above?

Applications with Vectors

OBJECTIVE

- Solve problems using vectors and right triangle trigonometry.

SPORTS Ty Murray was a recent World Championship Bull Rider in the National Finals Rodeo in Las Vegas, Nevada. One of the most dangerous tasks, besides riding the bulls, is the job of getting the bulls back into their pens after each event. Experienced handlers, dressed as clowns, expertly rope the animals without harming them.

Ty Murray has completed his competition ride and the two rodeo clowns are restraining the bull to return it to the paddocks. Suppose one clown is exerting a force of 270 newtons due north and the other is pulling with a force of 360 newtons due east. What is the resultant force on the bull? *This problem will be solved in Example 1.*

Vectors can represent the forces exerted by the handlers on the bull. Physicists resolve this force into component vectors. Vectors can be used to represent any quantity, such as a force, that has magnitude and direction. Velocity, weight, and gravity are some of the quantities that physicists represent with vectors.

Example

1 **SPORTS** **Use the information above to describe the resultant force on the bull.**

a. Draw a labeled diagram that represents the forces.

Let $\vec{\mathbf{F}}_1$ and $\vec{\mathbf{F}}_2$ represent the forces exerted by the clowns. Then $\vec{\mathbf{F}}$ represents the resultant. Let θ represent the angle $\vec{\mathbf{F}}$ makes with the east-west or x-axis.

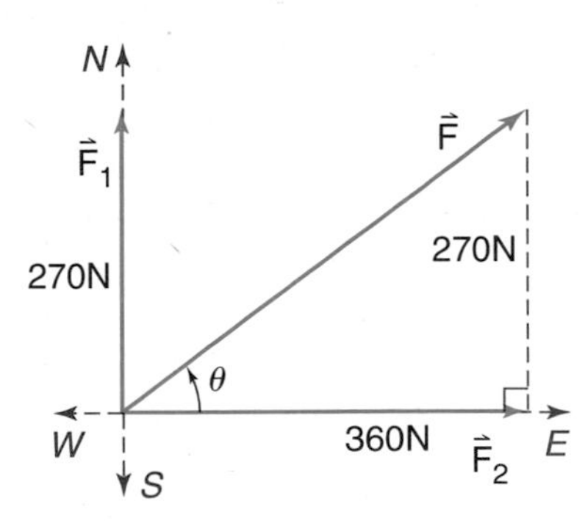

b. Determine the resultant force exerted on the bull by the two clowns.

$$|\vec{\mathbf{F}}|^2 = |\vec{\mathbf{F}}_1|^2 + |\vec{\mathbf{F}}_2|^2$$

$$|\vec{\mathbf{F}}|^2 = (270)^2 + (360)^2$$

$$|\vec{\mathbf{F}}|^2 = 202{,}500$$

$$\sqrt{|\vec{\mathbf{F}}|^2} = \sqrt{202{,}500} \text{ or } 450$$

The resultant force on the bull is 450 newtons.

1 pound ≈ 4.45 newtons so 450 N ≈ 101.12 lb

c. Find the angle the resultant force makes with the east-west axis.

Use the tangent ratio.

$$\tan \theta = \frac{270}{360}$$

$$\theta = \tan^{-1}\frac{270}{360}$$

$$\theta \approx 36.9^\circ \text{ north of due east}$$

The resultant force is applied at an angle of 36.9° north of due east.

In physics, if a constant force $\vec{\mathbf{F}}$ displaces an object an amount represented by $\vec{\mathbf{d}}$, it does *work* on the object. The amount of work is given by $W = \vec{\mathbf{F}} \cdot \vec{\mathbf{d}}$.

Example 2 **Alvaro works for a package delivery service. Suppose he is pushing a cart full of packages weighing 100 pounds up a ramp 8 feet long at an incline of 25°. Find the work done by gravity as the cart moves the length of the ramp. Assume that friction is not a factor.**

First draw a labeled diagram representing the forces involved. Let $\overrightarrow{OQ}$ represent the force of gravity, or weight. The weight has a magnitude of 100 pounds and its direction is down. The corresponding unit vector is $0\vec{\mathbf{i}} - 100\vec{\mathbf{j}}$. So, $\vec{\mathbf{F}} = 0\vec{\mathbf{i}} - 100\vec{\mathbf{j}}$. The application of the force is $\overrightarrow{OP}$, and it has a magnitude of 8 feet.

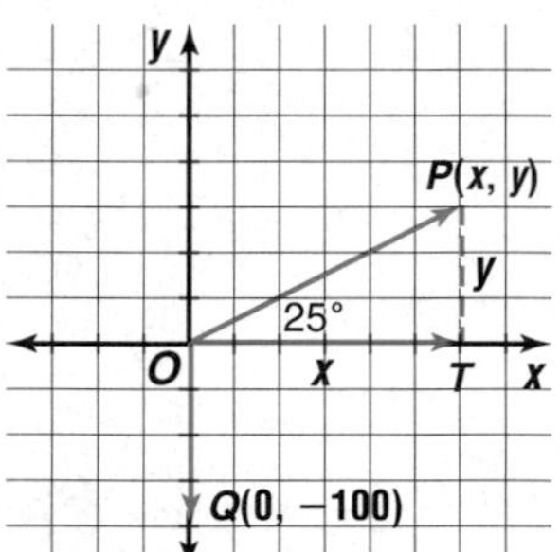

Write $\overrightarrow{OP}$ as $\vec{\mathbf{d}} = x\vec{\mathbf{i}} + y\vec{\mathbf{j}}$ and use trigonometry to find x and y.

$$\cos 25^\circ = \frac{x}{8} \qquad \sin 25^\circ = \frac{y}{8}$$

$$x = 8 \cos 25^\circ \qquad y = 8 \sin 25^\circ$$

$$x \approx 7.25 \qquad y \approx 3.38$$

Then, $\vec{\mathbf{d}} = 7.25\vec{\mathbf{i}} + 3.38\vec{\mathbf{j}}$

Apply the formula for determining the work done by gravity.

$$W = \vec{\mathbf{F}} \cdot \vec{\mathbf{d}}$$

$$W = \langle 0\vec{\mathbf{i}} - 100\vec{\mathbf{j}} \rangle \cdot \langle 7.25\vec{\mathbf{i}} + 3.38\vec{\mathbf{j}} \rangle$$

$$W = 0 - 338 \text{ or } -338$$

Work done by gravity is negative when an object is lifted or raised. As the cart moves the length of the ramp, the work done by gravity is −338 ft-lb.

Sometimes there is no motion when several forces are at work on an object. This situation, when the forces balance one another, is called *equilibrium*. Recall that an object is in equilibrium if the resultant force on it is zero.

Example 3 **Ms. Davis is hanging a sign for her restaurant. The sign is supported by two lightweight support bars as shown in the diagram. If the bars make a 30° angle with each other, and the sign weighs 200 pounds, what are the magnitudes of the forces exerted by the sign on each support bar?**

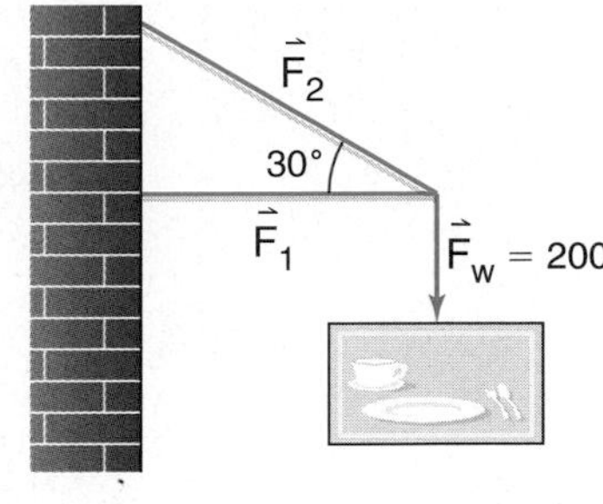

$\vec{\mathbf{F}}_1$ represents the force exerted on bar 1 by the sign, $\vec{\mathbf{F}}_2$ represents the force exerted on bar 2 by the sign, and $\vec{\mathbf{F}}_w$ represents the weight of the sign.

Remember that equal vectors have the same magnitude and direction. So by drawing another vector from the initial point of $\vec{\mathbf{F}}_1$ to the terminal point of $\vec{\mathbf{F}}_w$, we can use the sine and cosine ratios to determine $|\vec{\mathbf{F}}_1|$ and $|\vec{\mathbf{F}}_2|$.

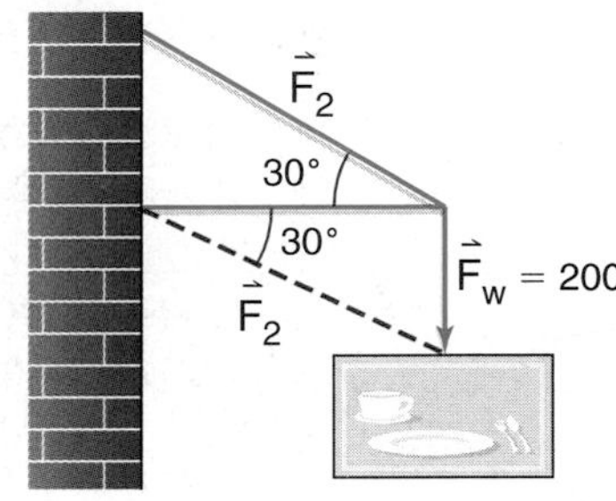

$$\sin 30° = \frac{200}{|\vec{\mathbf{F}}_2|}$$

$$|\vec{\mathbf{F}}_2| = \frac{200}{\sin 30°}$$ *Divide each side by sin 30° and multiply each side by $|\vec{F}_2|$.*

$$|\vec{\mathbf{F}}_2| = 400$$

Likewise, $|\vec{\mathbf{F}}_1|$ can be determined by using cos 30°.

$$\cos 30° = \frac{|\vec{\mathbf{F}}_1|}{400}$$

$$|\vec{\mathbf{F}}_1| = 400 \cos 30°$$

$$|\vec{\mathbf{F}}_1| \approx 346$$

The sign exerts a force of about 346 pounds on bar 1 and a force of 400 pounds on bar 2.

If the vectors representing forces at equilibrium are drawn tip-to-tail, they will form a polygon.

Example 4 **A lighting system for a theater is supported equally by two cables suspended from the ceiling of the theater. The cables form a 140° angle with each other. If the lighting system weighs 950 pounds, what is the force exerted by each of the cables on the lighting system?**

Draw a diagram of the situation. Then draw the vectors tip-to-tail.

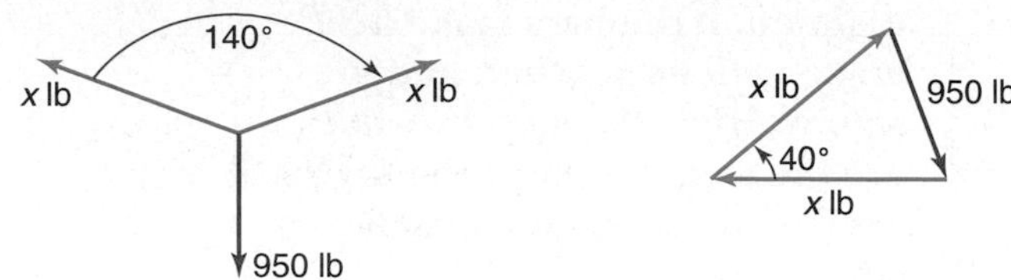

Look Back
Refer to Lessons 5-6 through 5-8 to review Law of Sines and Law of Cosines.

Since the triangle is isosceles, the base angles are congruent. Thus, each base angle measures $\frac{180° - 40°}{2}$ or 70°. We can use the Law of Sines to find the force exerted by the cables.

$$\frac{950}{\sin 40°} = \frac{x}{\sin 70°} \quad \textit{Law of Sines}$$

$$x = \frac{950 \sin 70°}{\sin 40°}$$

$$x \approx 1388.81$$

The force exerted by each cable is about 1389 pounds.

CHECK FOR UNDERSTANDING

Communicating Mathematics

Read and study the lesson to answer each question.

1. **Determine** which would require more force—pushing an object up to the top of an incline, or lifting it straight up. Assume there is no friction.
2. **Describe** how the force exerted by the cables in Example 4 is affected if the angle between the cables is increased.
3. **Explain** what it means for forces to be in equilibrium.

Guided Practice

4. Make a sketch to show the forces acting on a ship traveling at 23 knots at an angle of 17° with the current.
5. Find the magnitude and direction of the resultant vector for the diagram.

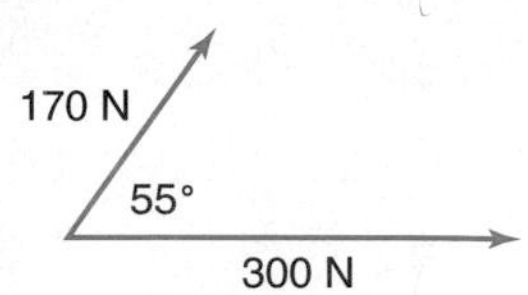

6. A 100-newton force and a 50-newton force act on the same object. The angle between the forces measures 90°. Find the magnitude of the resultant force and the angle between the resultant force and the 50-pound force.

7. Denzel pulls a wagon along level ground with a force of 18 newtons on the handle. If the handle makes an angle of 40° with the horizontal, find the horizontal and vertical components of the force.

8. A 33-newton force at 90° and a 44-newton force at 60° are exerted on an object. What is the magnitude and direction of a third force that produces equilibrium on the object?

9. **Transportation** Two ferry landings are directly across a river from each other. A ferry that can travel at a speed of 12 miles per hour in still water is attempting to cross directly from one landing to the other. The current of the river is 4 miles per hour.

   a. Make a sketch of the situation.

   b. If a heading of 0° represents the line between the two landings, at what angle should the ferry's captain head?

EXERCISES

Practice

Make a sketch to show the given vectors.

10. a force of 42 newtons acting on an object at an angle of 53° with the ground

11. an airplane traveling at 256 miles per hour at an angle of 27° from the wind

12. a force of 342 pounds acting on an object while a force of 454 pounds acts on the same object an angle of 94° with the first force

Find the magnitude and direction of the resultant vector for each diagram.

13.

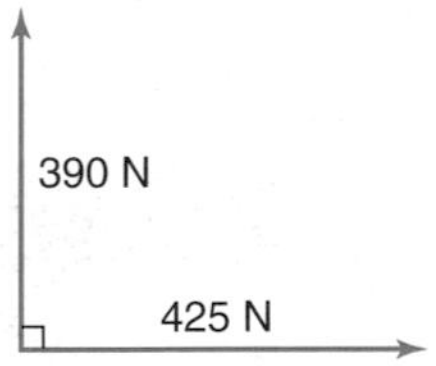

14.

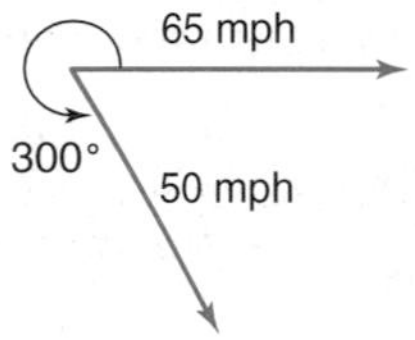

15.

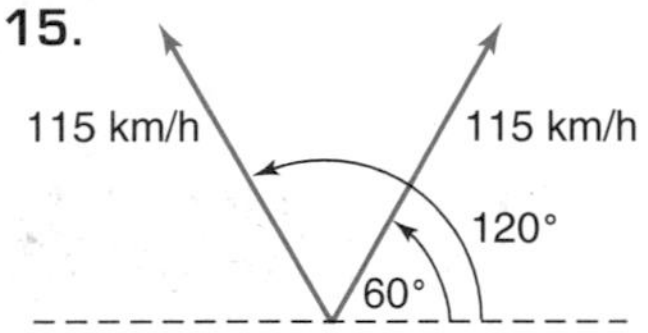

16. What would be the force required to push a 100-pound object along a ramp that is inclined 10° with the horizontal?

17. What is the magnitude and direction of the resultant of a 105-newton force along the x-axis and a 110-newton force at an angle of 50° to one another?

18. To keep a 75-pound block from sliding down an incline, a 52.1-pound force is exerted on the block along the incline. Find the angle that the incline makes with the horizontal.

19. Find the magnitude and direction of the resultant of two forces of 250 pounds and 45 pounds at angles of 25° and 250° with the x-axis, respectively.

20. Three forces with magnitudes of 70 pounds, 40 pounds, and 60 pounds act on an object at angles 330°, 45°, and 135°, respectively, with the positive x-axis. Find the direction and magnitude of the resultant of these forces.

21. A 23-newton force acting at 60° above the horizontal and a second 23-newton force acting at 120° above the horizontal act concurrently on a point. What is the magnitude and direction of a third force that produces equilibrium?

22. An object is placed on a ramp and slides to the ground. If the ramp makes an angle of 40° with the ground and the object weighs 25 pounds, find the acceleration of the object. Assume that there is no friction.

23. A force of 36 newtons pulls an object at an angle of 20° north of due east. A second force pulls on the object with a force of 48 newtons at an angle of 42° south of due west. Find the magnitude and direction of the resultant force.

24. Three forces in a plane act on an object. The forces are 70 pounds, 115 pounds and 135 pounds. The 70 pound force is exerted along the positive x-axis. The 115 pound force is applied below the x-axis at a 120° angle with the 70 pound force. The angle between the 115-pound and 135-pound forces is 75°, and between the 135-pound and 70-pound forces is 165°.

a. Make a diagram showing the forces.

b. Are the vectors in equilibrium? If not, find the magnitude and the direction of the resultant force.

Applications and Problem Solving

25. **Physics** While pulling a stalled car, a tow truck's cable makes an angle of 50° above the road. If the tension on the tow truck's cable is 1600 newtons, how much work is done by the truck on the car pulling it 1.5 kilometers down the road?

26. **Critical Thinking** The handle of a lawnmower you are pushing makes an angle of 60° with the ground.

a. How could you increase the horizontal forward force you are applying without increasing the total force?

b. What are some of the disadvantages of doing this?

27. **Boating** A sailboat is headed east at 18 mph relative to the water. A current is moving the water south at 3 mph.

a. What is the angle of the path of the sailboat?

b. What is the sailboat's speed with respect to the ocean floor?

28. **Physics** Suzanne is pulling a wagon loaded with gardening bricks totaling 100 kilograms. She is applying a force of 100 newtons on the handle at 25° with the ground. What is the horizontal force on the wagon?

29. **Entertainment** A unicyclist is performing on a tightrope at a circus. The total weight of the performer and her unicycle is 155 pounds. How much tension is being exerted on each part of the cable?

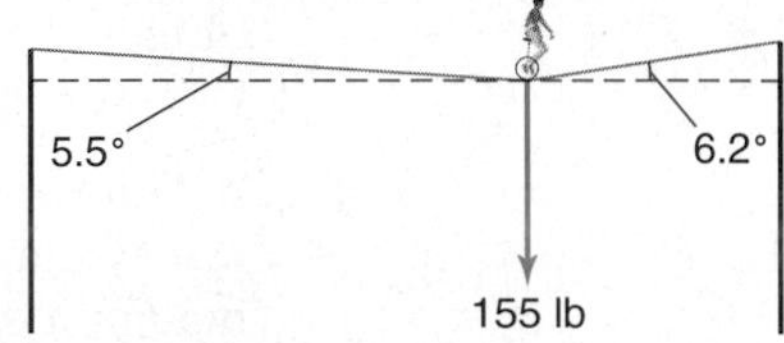

30. Critical Thinking Chaz is using a rope tied to his tractor to remove an old tree stump from a field. Which method given below—a or b—will result in the greatest force applied to the stump? Assume that the tractor will exert the same amount of force using either method. Explain your answer.

a. Tie the rope to the stump and pull.

b. Tie one end to the stump and the other end to a nearby pole. Then pull on the rope perpendicular to it at a point about halfway between the two.

31. Travel A cruise ship is arriving at the port of Miami from the Bahamas. Two tugboats are towing the ship to the dock. They exert a force of 6000 tons along the axis of the ship. Find the tension in the towlines if each tugboat makes a 20° angle with the axis of the ship.

32. Physics A painting weighing 25 pounds is supported at the top corners by a taut wire that passes around a nail embedded in the wall. The wire forms a 120° angle at the nail. Assuming that the wire rests on the nail at its midpoint, find the pull on the wires.

Mixed Review

33. Find the inner product of $\vec{\mathbf{u}}$ and $\vec{\mathbf{v}}$ if $\vec{\mathbf{u}} = \langle 9, 5, 3 \rangle$ and $\vec{\mathbf{v}} = \langle -3, 2, 5 \rangle$. Are the vectors perpendicular? *(Lesson 8-4)*

34. If $A = (12, -5, 18)$ and $B = (0, -11, 21)$, write the ordered triple that represents $\overrightarrow{AB}$. *(Lesson 8-3)*

35. Sports Sybrina Floyd hit a golf ball on an approach shot with an initial velocity of 100 feet per second. The distance a golf ball travels is found by the formula $d = \frac{2v_0^2}{g} \sin \theta \cos \theta$, where v_0 is the initial velocity, g is the acceleration due to gravity, and θ is the measure of the angle that the initial path of the ball makes with the ground. Find the distance Ms. Floyd's ball traveled if the measure of the angle between the initial path of the ball and the ground is 65° and the acceleration due to gravity is 32 ft/s^2. *(Lesson 7-4)*

36. Write a polynomial function to model the set of data. *(Lesson 4-8)*

x	−1.5	−1	−0.5	0	0.5	1	1.5	2	2.5
f(x)	9	8	6.5	4.3	2	2.7	4.1	6.1	7.8

37. Food Processing A meat packer makes a kind of sausage using beef, pork, cereal, fat, water, and spices. The minimum cereal content is 12%, the minimum fat content is 15%, the minimum water content is 6.5%, and the spices are 0.5%. The remaining ingredients are beef and pork. There must be at least 30% beef content and at least 20% pork content for texture. The beef content must equal or exceed the pork content. The cost of all of the ingredients except beef and pork is $32 per 100 pounds. Beef can be purchased for $140 per 100 pounds and pork for $90 per 100 pounds. Find the combination of beef and pork for the minimum cost. What is the minimum cost per 100 pounds? *(Lesson 2-6)*

38. SAT/ACT Practice Let $*x$ be defined as $*x = x^3 - x$. What is the value of $*4 - *(-3)$?

A 84 **B** 55 **C** −10

D 22 **E** 4

Vectors and Parametric Equations

OBJECTIVES
- Write vector and parametric equations of lines.
- Graph parametric equations.

AIR RACING Air racing's origins date back to the early 1900s when aviation began. One of the most important races is the National Championship Air Race held every year in Reno, Nevada. In the competition, suppose pilot Bob Hannah passes the starting gate at a speed of 410.7 mph. Fifteen seconds later, Dennis Sanders flies by at 412.9 mph. They are racing for the next 500 miles. *This information will be used in Example 4.*

The relative positions of the airplanes are dependent upon their velocities at a given moment in time and their starting positions. **Vector equations** and equations known as **parametric equations** allow us to model that movement.

In Lesson 1-4, you learned how to write the equation of a line in the coordinate plane. For objects that are moving in a straight line, there is a more useful way of writing equations for the line describing the object's path using vectors.

If a line passes through the points P_1 and P_2 and is parallel to the vector $\vec{\mathbf{a}} = \langle a_1, a_2 \rangle$, the vector $\overrightarrow{P_1P_2}$ is also parallel to $\vec{\mathbf{a}}$. Thus, $\overrightarrow{P_1P_2}$ must be a scalar multiple of $\vec{\mathbf{a}}$. Using the scalar t, we can write the equation $\overrightarrow{P_1P_2} = t\vec{\mathbf{a}}$. Notice that both sides of the equation are vectors. This is called the vector equation of the line. Since $\vec{\mathbf{a}}$ is parallel to the line, it is called a **direction vector.** The scalar t is called a **parameter.**

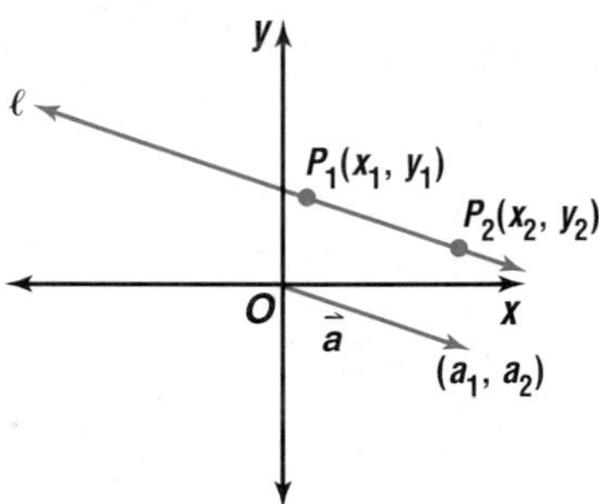

Vector Equation of a Line

A line through $P_1(x_1, y_1)$ parallel to the vector $\vec{\mathbf{a}} = \langle a_1, a_2 \rangle$ is defined by the set of points $P_1(x_1, y_1)$ and $P_2(x_2, y_2)$ such that $\overrightarrow{P_1P_2} = t\vec{\mathbf{a}}$ for some real number t. Therefore, $\langle x_2 - x_1, y_2 - y_1 \rangle = t\langle a_1, a_2 \rangle$.

In Lesson 8-2, you learned that the vector from (x_1, y_1) to (x, y) is $\langle x - x_1, y - y_1 \rangle$.

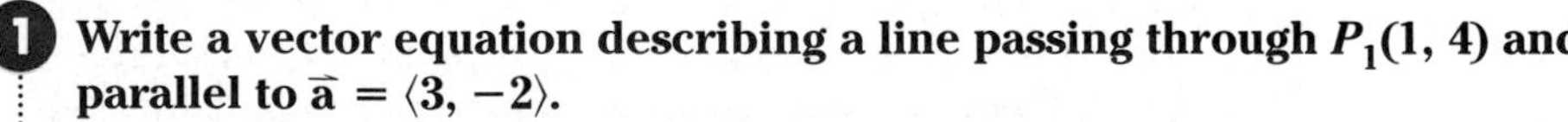

Example 1 **Write a vector equation describing a line passing through $P_1(1, 4)$ and parallel to $\vec{\mathbf{a}} = \langle 3, -2 \rangle$.**

Let the line ℓ through $P_1(1, 4)$ be parallel to $\vec{\mathbf{a}}$. For any point $P(x, y)$ on ℓ, $\overrightarrow{P_1P} = \langle x - 1, y - 4 \rangle$. Since P_1P is on ℓ and is parallel to $\vec{\mathbf{a}}$, $\overrightarrow{P_1P} = t\vec{\mathbf{a}}$, for some value t. By substitution, we have $\langle x - 1, y - 4 \rangle = t\langle 3, -2 \rangle$.

Therefore, the equation $\langle x - 1, y - 4 \rangle = t\langle 3, -2 \rangle$ is a vector equation describing all of the points (x, y) on ℓ parallel to $\vec{\mathbf{a}}$ through $P_1(1, 4)$.

A vector equation like the one in Example 1 can be used to describe the coordinates for a point on the line for any value of the parameter t. For example, when $t = 4$ we can write the equation $\langle x - 1, y - 4 \rangle = 4\langle 3, -2 \rangle$ or $(12, -8)$. Then write the equation $x - 1 = 12$ and $y - 4 = -8$ to find the ordered pair $(13, -4)$. Likewise, when $t = 0$, the ordered pair $(1, 4)$ results. The parameter t often represents time. (In fact, that is the reason for the choice of the letter t.) An object moving along the line described by the vector equation $\langle x - 1, y - 4 \rangle = t\langle 3, 2 \rangle$ will be at the point $(1, 4)$ at time $t = 0$ and will be at the point $(13, -4)$ at time $t = 4$.

As you have seen, the vector equation $\langle x - x_1, y - y_1 \rangle = t\langle a_1, a_2 \rangle$ can be written as two equations relating the horizontal and vertical components of these two vectors separately.

$$x - x_1 = ta_1 \qquad\qquad y - y_1 = ta_2$$

$$x = x_1 + ta_1 \qquad\qquad y = y_1 + ta_2$$

The resulting equations, $x = x_1 + ta_1$ and $y = y_1 + ta_2$, are known as parametric equations of the line through $P_1(x_1, y_1)$ parallel to $\vec{\mathbf{a}} = \langle a_1, a_2 \rangle$.

Parametric Equations of a Line

A line through $P_1(x_1, y_1)$ that is parallel to the vector $\vec{\mathbf{a}} = \langle a_1, a_2 \rangle$ has the following parametric equations, where t is any real number.

$$x = x_1 + ta_1$$
$$y = y_1 + ta_2$$

If we know the coordinates of a point on a line and its direction vector, we can write its parametric equations.

Example 2 **Find the parametric equations for a line parallel to $\vec{\mathbf{q}} = \langle 6, -3 \rangle$ and passing through the point at $(-2, -4)$. Then make a table of values and graph the line.**

Use the general form of the parametric equations of a line with $\langle a_1, a_2 \rangle = \langle 6, -3 \rangle$ and $\langle x_1, y_1 \rangle = \langle -2, -4 \rangle$.

$x = x_1 + ta_1$
$x = -2 + t(6)$
$x = -2 + 6t$

$y = y_1 + ta_2$
$y = -4 + t(-3)$
$y = -4 - 3t$

Now make a table of values for t. Evaluate each expression to find values for x and y. Then graph the line.

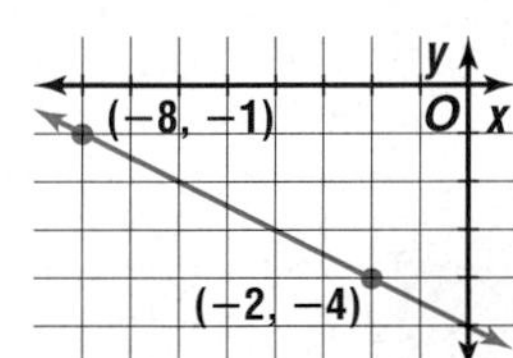

t	x	y
−1	−8	−1
0	−2	−4
1	4	−7
2	10	−10

Notice in Example 2 that each value of t establishes an ordered pair (x, y) whose graph is a point. As you have seen, these points can be considered the position of an object at various times t. Evaluating the parametric equations for a value of t gives us the coordinates of the position of the object after t units of time.

If the slope-intercept form of the equation of a line is given, we can write parametric equations for that line.

Example 3 **Write parametric equations of $y = -4x + 7$.**

In the equation $y = -4x + 7$, x is the independent variable, and y is the dependent variable. In parametric equations, t is the independent variable, and x and y are dependent variables. If we set the independent variables x and t equal, we can write two parametric equations in terms of t.

$x = t$
$y = -4t + 7$

Parametric equations for the line are $x = t$ and $y = -4t + 7$.

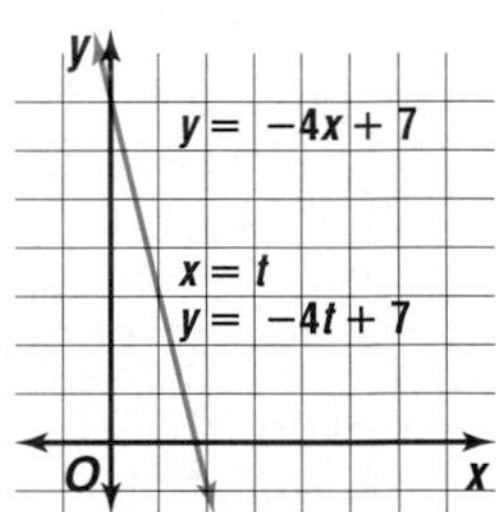

By making a table of values for t and evaluating each expression to find values for x and y and graphing the line, the parametric equations $x = t$ and $y = -4t + 7$ describe the same line as $y = -4x + 7$.

We can use vector equations and parametric equations to model physical situations, such as the National Championship Air Race, where t represents time in hours.

Example 4 **AIR RACING** **Refer to the application at the beginning of the lesson. Use parametric equations to model the situation. Assume that both planes maintain a constant speed.**

a. How long is it until the second plane overtakes the first?

b. How far have both planes traveled when the second plane overtakes the first?

a. First, write a set of parametric equations to represent each airplane's position at t hours.

Airplane 1: $x = 410.7t$ *$x = vt$*
Airplane 2: $x = 412.9(t - 0.0042)$ *15 s ≈ 0.0042 h*

Since the time at which the second plane overtakes the first is when they have traveled the same distance, set the two expressions for x equal to each other.

$$410.7t \approx 412.9(t - 0.0042)$$
$$410.7t \approx 412.9t - 1.73418$$
$$1.73418 \approx 2.2t$$
$$0.788 \approx t$$

In about 0.788 hour or 47 minutes, the second plane overtakes the first.

b. Use the time to find the distance traveled when the planes pass.

$x = 410.7t$ — *Since the speeds are given in tenths of mph, we round the final answer to the nearest tenth.*

$x = 410.7(0.788)$

$x = 323.6316$ or about 323.6 miles

The planes have traveled about 323.6 miles when the second plane overtakes the first plane.

We can also write the equation of a line in slope-intercept or standard form if we are given the parametric equations of the line.

Example 5 **Write an equation in slope-intercept form of the line whose parametric equations are $x = -5 + 4t$ and $y = 2 - 3t$.**

Solve each parametric equation for t.

$$x = -5 + 4t \qquad y = 2 - 3t$$

$$x + 5 = 4t \qquad y - 2 = -3t$$

$$\frac{x + 5}{4} = t \qquad \frac{y - 2}{3} = t$$

Use substitution to write an equation for the line without the variable t.

$\frac{x + 5}{4} = \frac{(y - 2)}{-3}$ *Substitution*

$(x + 5)(-3) = 4(y - 2)$ *Cross multiply.*

$-3x - 15 = 4y - 8$ *Simplify.*

$y = -\frac{3}{4}x - \frac{7}{4}$ *Solve for y.*

CHECK FOR UNDERSTANDING

Communicating Mathematics

Read and study the lesson to answer each question.

1. **Describe** the graph of the parametric equations $x = 3 + 4t$ and $y = -1 + 2t$.
2. **Explain** how to find the parametric equations for the line through the point at (3, 6), parallel to the vector $\vec{\mathbf{v}} = \vec{\mathbf{i}} + 2\vec{\mathbf{j}}$.
3. **Describe** the line having parametric equations $x = 1 + t$ and $y = -t$. Include the slope of the line and the y-intercept in your description.

Guided Practice

Write a vector equation of the line that passes through point P and is parallel to $\vec{\mathbf{a}}$. Then write parametric equations of the line.

4. $P(-4, 11)$, $\vec{\mathbf{a}} = \langle -3, 8 \rangle$
5. $P(1, 5)$, $\vec{\mathbf{a}} = \langle -7, 2 \rangle$

Write parametric equations of each line with the given equation.

6. $3x + 2y = 5$
7. $4x - 6y = -12$

Write an equation in slope-intercept form of the line with the given parametric equations.

8. $x = -4t + 3$
$y = 5t - 3$

9. $x = 9t$
$y = 4t + 2$

10. Set up a table of values and then graph the line whose parametric equations are $x = 2 + 4t$ and $y = -1 + t$.

11. Sports A wide receiver catches a ball and begins to run for the endzone following a path defined by $\langle x - 5, y - 50\rangle = t\langle 0, -10\rangle$. A defensive player chases the receiver as soon as he starts running following a path defined by $\langle x - 10, y - 54\rangle = t\langle -0.9, -10.72\rangle$.

a. Write parametric equations for the path of each player.

b. If the receiver catches the ball on the 50-yard line (y = 50), will he reach the goal line (y = 0) before the defensive player catches him?

EXERCISES

Practice

Write a vector equation of the line that passes through point P and is parallel to $\vec{\mathbf{a}}$. Then write parametric equations of the line.

12. $P(5, 7), \vec{\mathbf{a}} = \langle 2, 0\rangle$

13. $P(-1, 4), \vec{\mathbf{a}} = \langle 6, -10\rangle$

14. $P(-6, 10), \vec{\mathbf{a}} = \langle 3, 2\rangle$

15. $P(1, 5), \vec{\mathbf{a}} = \langle -7, 2\rangle$

16. $P(1, 0), \vec{\mathbf{a}} = \langle -2, -4\rangle$

17. $P(3, -5), \vec{\mathbf{a}} = \langle -2, 5\rangle$

Write parametric equations of each line with the given equation.

18. $y = 4x - 5$

19. $-3x + 4y = 7$

20. $2x - y = 3$

21. $9x + y = -1$

22. $2x + 3y = 11$

23. $-4x + y = -2$

24. Write parametric equations for the line passing through the point at $(-2, 5)$ and parallel to the line $3x - 6y = -8$.

Write an equation in slope-intercept form of the line with the given parametric equations.

25. $x = 2t$
$y = 1 - t$

26. $x = -7 + \frac{1}{2}t$
$y = 3t$

27. $x = 4t - 11$
$y = t + 3$

28. $x = 4t - 8$
$y = 3 + t$

29. $x = 3 + 2t$
$y = -1 + 5t$

30. $x = 8$
$y = 2t + 1$

31. A line passes through the point at $(11, -4)$ and is parallel to $\vec{\mathbf{a}} = \langle 3, 7\rangle$.

a. Write a vector equation of the line.

b. Write parametric equations for the line.

c. Use the parametric equations to write the equation of the line in slope-intercept form.

Graphing Calculator

Use a graphing calculator to set up a table of values and then graph each line from its parametric form.

32. $x = 5t$
$y = -4 - t$

33. $x = 3t + 5$
$y = 1 + t$

34. $x = 1 + t$
$y = 1 - t$

Applications and Problem Solving

35. **Geometry** A line in a plane is defined by the parametric equations $x = 2 + 3t$ and $y = 4 + 7t$.
    a. What part of this line is obtained by assigning only non-negative values to t?
    b. How should t be defined to give the part of the line to the left of the y-axis?

36. **Critical Thinking** Graph the parametric equations $x = \cos^2 t$ and $y = \sin^2 t$.

37. **Navigation** During a military training exercise, an unmanned target drone has been detected on a radar screen following a path represented by the vector equation $\langle x, y\rangle = (3, 4) + t\langle -1, 0\rangle$. A surface to air missile is launched to intercept it and destroy it. This missile is following a trajectory represented by $\langle x, y\rangle = (2, 2) + t\langle 1, 2\rangle$.
    a. Write the parametric equations for the path of the target drone and the missile.
    b. Will the missile intercept the target drone?

38. **Astronomy** Astronomers have traced the path of two asteroids traveling through space. At a particular time t, the position of asteroid Ceres can be represented by $(-1 + t, 4 - t, -1 + 2t)$. Asteroid Pallas' path at any time t, can be expressed by $(-7 + 2t, -6 + 2t, -1 + t)$.
    a. Write the parametric equations for the path of each asteroid.
    b. Do the paths of the asteroids cross? If so, where?
    c. Following these paths, will the asteroids collide? If so, where?

39. **Critical Thinking** Find the parametric equations for the line passing through points at $\left(-\frac{1}{3}, 1, 1\right)$ and $(0, 5, -8)$. (*Hint:* Equations are needed for x, y, and z)

Mixed Review

40. An airplane flies at 150 km/h and heads 30° south of east. A 50-km/h wind blows in the direction 25° west of south. Find the ground speed and direction of the plane. *(Lesson 8-5)*

41. Find the inner product $\langle 1, 3\rangle \cdot \langle 3, -2\rangle$ and state whether the vectors are perpendicular. Write *yes* or *no*. *(Lesson 8-4)*

42. Solve $\triangle ABC$ if $A = 40°$, $b = 16$, and $a = 9$. *(Lesson 5-7)*

43. Approximate the real zero(s) of $f(x) = x^5 + 3x^3 - 4$ to the nearest tenth. *(Lesson 4-5)*

44. Find the inverse of the function $y = \frac{3}{2}x - 2$. *(Lesson 3-4)*

45. Write the standard form of the equation of the line that is parallel to the graph of $y = x - 8$ and passes through the point at $(-3, 1)$. *(Lesson 1-5)*

46. **SAT/ACT Practice** A pulley having a 9-inch diameter is belted to a pulley having a 6-inch diameter, as shown in the figure. If the larger pulley runs at 120 rpm (revolutions per minute), how fast does the smaller pulley run?

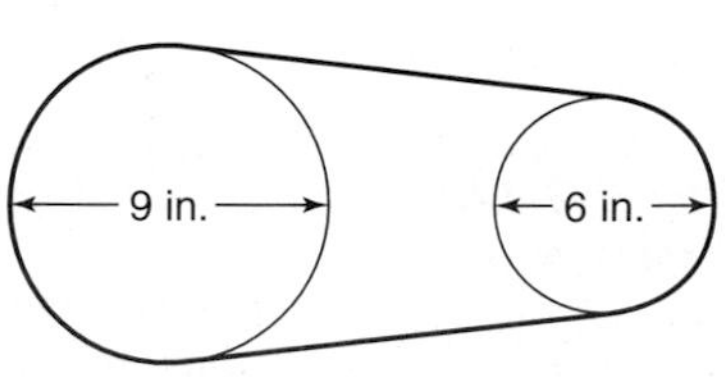

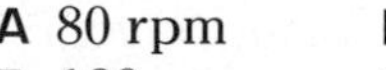

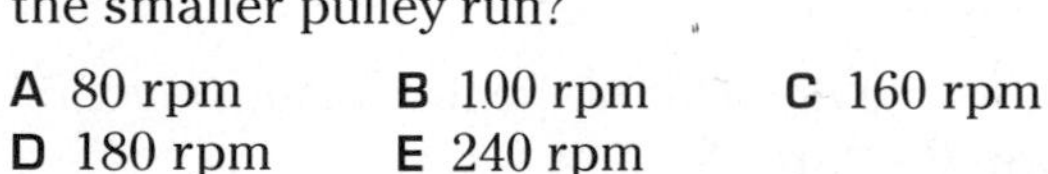

A 80 rpm **B** 100 rpm **C** 160 rpm
D 180 rpm **E** 240 rpm

Extra Practice See p. A41.

GRAPHING CALCULATOR EXPLORATION

8-6B Modeling With Parametric Equations

An Extension of Lesson 8-6

OBJECTIVE
- Investigate the use of parametric equations.

You can use a graphing calculator and parametric equations to investigate real-world situations, such as simulating a 500-mile, two airplane race. Suppose the first plane averages a speed of 408.7 miles per hour for the entire race. The second plane starts 30 seconds after the first and averages 418.3 miles per hour.

TRY THIS

Write a set of parametric equations to represent each plane's position at t hours. This will simulate the race on two parallel race courses so that it is easy to make a visual comparison. Use the formula $x = vt$.

Plane 1: $x = 408.7t$ — *$x = vt$*
$y = 1$ — *y represents the position of plane 1 on course 1.*

Plane 2: $x = 418.3(t - 0.0083)$ — *Remember that plane 2 started 30 seconds or 0.0083 hour later.*
$y = 2$ — *Plane 2 is on course 2.*

Set the graphing calculator to parametric and simultaneous modes. Next, set the viewing window to the following values: **Tmin** = 0, **Tmax** = 5, **Tstep** = .005, **Xmin** = 0, **Xmax** = 500, **Xscl** = 10, **Ymin** = 0, **Ymax** = 5, and **Yscl** = 1. Then enter the parametric equations on the **Y=** menu. Then press GRAPH to "see the race."

Which plane finished first? Notice that the line on top reached the edge of the screen first. That line represented the position of the second plane, so plane 2 finished first. You can confirm the conclusion using the TRACE function. The plane with the smaller t-value when x is about 500 is the winner. Note that when x_1 is about 500, $t = 1.225$ and when x_2 is about 500, $t = 1.205$.

WHAT DO YOU THINK?

1. How long does it take the second plane to overtake the first?
2. How far have the two planes traveled when the second plane overtakes the first?
3. How much would the pilot of the first plane have to increase the plane's speed to win the race?

8-7

Modeling Motion Using Parametric Equations

OBJECTIVES

- Model the motion of a projectile using parametric equations.
- Solve problems related to the motion of a projectile, its trajectory, and range.

SPORTS Suppose a professional football player kicks a football with an initial velocity of 29 yards per second at an angle of 68° to the horizontal. Suppose a kick returner catches the ball 5 seconds later. How far has the ball traveled horizontally and what is its vertical height at that time? *This problem will be solved in Example 2.*

Objects that are launched, like a football, are called *projectiles*. The path of a projectile is called its *trajectory*. The horizontal distance that a projectile travels is its *range*. Physicists describe the motion of a projectile in terms of its position, velocity, and acceleration. All these quantities can be represented by vectors.

The figure at the right describes the initial trajectory of a punted football as it leaves the kicker's foot. The magnitude of the initial velocity $|\vec{\mathbf{v}}|$ and the direction θ of the ball can be described by a vector that can be expressed as the sum of its horizontal and vertical components $\vec{\mathbf{v}}_x$ and $\vec{\mathbf{v}}_y$.

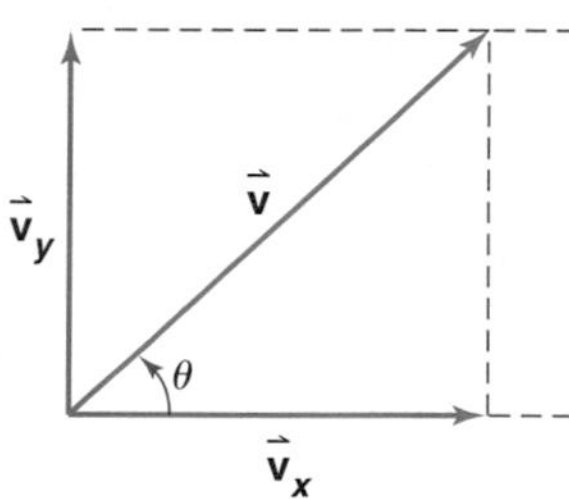

As the ball moves, gravity will act on it in the vertical direction. The horizontal component is unaffected by gravity. So, discounting air resistance, the horizontal speed is constant throughout the flight of the ball.

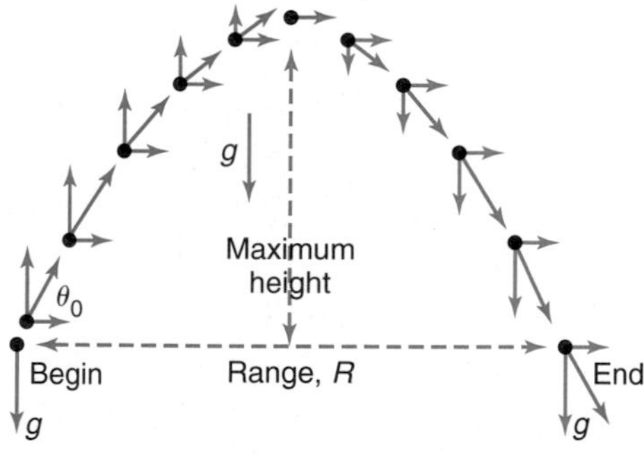

The vertical component of the velocity of the ball is large and positive at the beginning, decreasing to zero at the top of its trajectory, then increasing in the negative direction as it falls. When the ball returns to the ground, its vertical speed is the same as when it left the kicker's foot, but in the opposite direction.

Parametric equations can represent the position of the ball relative to the starting point in terms of the parameter of time.

In order to find parametric equations that represent the path of a projectile like a football, we must write the horizontal $\vec{\mathbf{v}}_x$ and vertical $\vec{\mathbf{v}}_y$ components of the initial velocity.

$$\cos\theta = \frac{|\vec{\mathbf{v}}_x|}{|\vec{\mathbf{v}}|} \qquad \sin\theta = \frac{|\vec{\mathbf{v}}_y|}{|\vec{\mathbf{v}}|}$$

$$|\vec{\mathbf{v}}_x| = |\vec{\mathbf{v}}|\cos\theta \qquad |\vec{\mathbf{v}}_y| = |\vec{\mathbf{v}}|\sin\theta$$

Example 1 **Find the initial horizontal velocity and vertical velocity of a stone kicked with an initial velocity of 16 feet per second at an angle of 38° with the ground.**

$$|\vec{\mathbf{v}}_x| = |\vec{\mathbf{v}}|\cos\theta \qquad |\vec{\mathbf{v}}_y| = |\vec{\mathbf{v}}|\sin\theta$$

$$|\vec{\mathbf{v}}_x| = 16\cos 38^\circ \qquad |\vec{\mathbf{v}}_y| = 16\sin 38^\circ$$

$$|\vec{\mathbf{v}}_x| \approx 13 \qquad |\vec{\mathbf{v}}_y| \approx 10$$

The initial horizontal velocity is about 13 feet per second and the initial vertical velocity is about 10 feet per second.

Because horizontal velocity is unaffected by gravity, it is the magnitude of the horizontal component of the initial velocity. Therefore, the horizontal position of a projectile, after t seconds is given by the following equation.

$$\underbrace{\textit{horizontal distance}} = \underbrace{\textit{horizontal velocity}} \cdot \underbrace{\textit{time}}$$

$$x \quad = \quad |\vec{\mathbf{v}}|\cos\theta \quad \cdot \quad t$$

$$x = t|\vec{\mathbf{v}}|\cos\theta$$

Since vertical velocity is affected by gravity, we must adjust the vertical component of initial velocity. By subtracting the vertical displacement due to gravity from the vertical displacement caused by the initial velocity, we can determine the height of the projectile after T seconds. The height, in feet or meters, of a free-falling object affected by gravity is given by the equation $h = \frac{1}{2}gt^2$, where $g \approx 9.8$ m/s^2 or 32 ft/s^2 and t is the time in seconds.

$$\underbrace{\textit{vertical displacement}} = \underbrace{\textit{displacement due to initial velocity}} - \underbrace{\textit{displacement due to gravity}}$$

$$y \quad = \quad (|\vec{\mathbf{v}}|\sin\theta)t \quad - \quad \frac{1}{2}gt^2$$

$$y = t|\vec{\mathbf{v}}|\sin\theta - \frac{1}{2}gt^2$$

Therefore, the path of a projectile can be expressed in terms of parametric equations.

Parametric Equations for the Path of a Projectile

If a projectile is launched at an angle of θ with the horizontal with an initial velocity of magnitude $|\vec{\mathbf{v}}|$, the path of the projectile may be described by these equations, where t is time and g is acceleration due to gravity.

$$x = t|\vec{\mathbf{v}}| \cos \theta$$
$$y = t|\vec{\mathbf{v}}| \sin \theta - \frac{1}{2}gt^2$$

Example 2

SPORTS Refer to the application at the beginning of the lesson.

a. How far has the ball traveled horizontally and what is its vertical height at that time?

b. Suppose the kick returner lets the ball hit the ground instead of catching it. What is the hang time, the elapsed time between the moment the ball is kicked and the time it hits the ground?

a. Write the position of the ball as a pair of parametric equations defining the path of the ball for any time t in seconds.

$x = t|\vec{\mathbf{v}}| \cos \theta$ $\qquad$ $y = t|\vec{\mathbf{v}}| \sin \theta - \frac{1}{2}gt^2$

The initial velocity of 29 yards per second must be expressed as 87 feet per second as gravity is expressed in terms of feet per second squared.

$x = t(87) \cos 68°$ $\qquad$ $y = t(87) \sin 68° - \frac{1}{2}(32)t^2$

$x = 87t \cos 68°$ $\qquad$ $y = 87t \sin 68° - 16t^2$

Find x and y when $t = 5$.

$x = 87(5) \cos 68°$ $\qquad$ $y = 87(5) \sin 68° - 16(5)^2$

$x \approx 163$ $\qquad$ $y \approx 3$

After 5 seconds, the football has traveled about 163 feet or $54\frac{1}{3}$ yards horizontally and is about 3 feet or 1 yard above the ground.

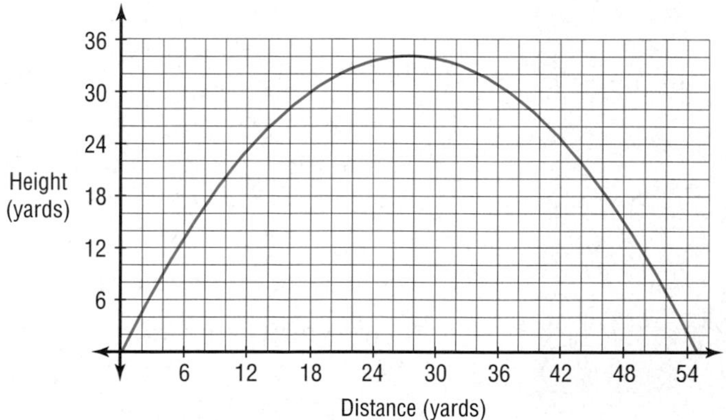

b. Determine when the vertical height is 0 using the equation for y.

$y = 87t \sin 68° - 16t^2$

$y = 80.66t - 16t^2$

$y = t(80.66 - 16t)$ $\qquad$ *Factor.*

(continued on the next page)

Now let $y = 0$ and solve for t.

$$0 = 80.66 - 16t$$
$$16t = 80.66$$
$$t = 5.04$$

The hang time is about 5 seconds.

The parametric equations describe the path of an object that is launched from ground level. Some objects are launched from above ground level. For example, a baseball may be hit at a height of 3 feet. So, you must add the initial vertical height to the expression for y. This accounts for the fact that at time 0, the object will be above the ground.

Example 3

SOFTBALL Kaci Clark led the Women's Pro Softball League in strikeouts in 1998. Suppose she throws the ball at an angle of 5.2° with the horizontal at a speed of 67 mph. The distance from the pitcher's mound to home plate is 43 feet. If Kaci releases the ball 2.7 feet above the ground, how far above the ground is the ball when it crosses home plate?

First, write parametric equations that model the path of the softball. Remember to convert 67 mph to about 98.3 feet per second.

$$x = t|\vec{\mathbf{v}}| \cos \theta \qquad y = t|\vec{\mathbf{v}}| \sin \theta - \frac{1}{2}gt^2 + h$$
$$x = t(98.3)\cos 5.2° \qquad y = t(98.3)\sin 5.2° - \frac{1}{2}(32)t^2 + 2.7$$
$$x = 98.3t \cos 5.2° \qquad y = 98.3t \sin 5.2° - 16t^2 + 2.7$$

Kaci Clark

Then, find the amount of time that it will take the baseball to travel 43 feet horizontally. This will be the moment when it crosses home plate.

$$43 = 98.3t \cos 5.2°$$
$$t = \frac{43}{98.3 \cos 5.2°}$$
$$t \approx 0.439$$

The softball will cross home plate in about 0.44 second.

To find the vertical position of the ball at that time, find y when $t = 0.44$.

$$y = 98.3t \sin 5.2° - 16t^2 + 2.7$$
$$y = 98.3(0.44) \sin 5.2° - 16(0.44)^2 + 2.7$$
$$y \approx 3.522$$

The softball will be about 3.5 feet above home plate.

CHECK FOR UNDERSTANDING

Communicating Mathematics

Read and study the lesson to answer each question.

1. **Describe** situations in which a projectile travels vertically. At what angle with the horizontal must the projectile be launched?
2. **Describe** the vertical velocities of a projectile at its launch and its landing.
3. **Explain** the effect the angle of a golf club's head has on the angle of initial velocity of a golf ball.

Guided Practice

4. Find the initial vertical velocity of a stone thrown with an initial velocity of 50 feet per second at an angle of 40° with the horizontal.
5. Find the initial horizontal velocity of a discus thrown with an initial velocity of 20 meters per second at an angle of 50° with the horizontal.

Find the initial horizontal velocity and vertical velocity for each situation.

6. A soccer ball is kicked with an initial velocity of 45 feet per second at an angle of 32° with the horizontal.
7. A stream of water is shot from a sprinkler head with an initial velocity of 7.5 meters per second at an angle of 20° with the ground.
8. **Meteorology** An airplane flying at an altitude of 3500 feet is dropping research probes into the eye of a hurricane. The path of the plane is parallel to the ground at the time the probes are released with an initial velocity of 300 mph.
   a. Write the parametric equations that represent the path of the probes.
   b. Sketch the graph describing the path of the probes.
   c. How long will it take the probes to reach the ground?
   d. How far will the probes travel horizontally before they hit the ground?

EXERCISES

Practice

Find the initial horizontal velocity and vertical velocity for each situation.

9. a javelin thrown at 65 feet per second at an angle of 60° with the horizontal
10. an arrow released at 47 meters per second at an angle of 10.7° with the horizontal
11. a cannon shell fired at 1200 feet per second at a 42° angle with the ground
12. a can is kicked with an initial velocity of 17 feet per second at an angle of 28° with the horizontal
13. a golf ball hit with an initial velocity of 69 yards per second at 37° with the horizontal
14. a kangaroo leaves the ground at an angle of 19° at 46 kilometers per hour

Applications and Problem Solving

Real World Application

15. **Golf** Professional golfer Nancy Lopez hits a golf ball with a force to produce an initial velocity of 175 feet per second at an angle of 35° above the horizontal. She estimates the distance to the hole to be 225 yards.
    a. Write the position of the ball as a pair of parametric equations.
    b. Find the range of the ball.

16. **Projectile Motion** What is the relationship between the angle at which a projectile is launched, the time it stays in the air, and the distance it covers?

17. **Physics** Alfredo and Kenishia are performing a physics experiment in which they will launch a model rocket. The rocket is supposed to release a parachute 300 feet in the air, 7 seconds after liftoff. They are firing the rocket at a 78° angle from the horizontal.
    a. Find the initial velocity of the rocket.
    b. To protect other students from the falling rockets, the teacher needs to place warning signs 50 yards from where the parachute is released. How far should the signs be from the point where the rockets are launched?

18. **Critical Thinking** Is it possible for a projectile to travel in a circular arc rather than a parabolic arc? Explain your answer.

19. **Critical Thinking** Janelle fired a projectile and measured its range. She hypothesizes that if the magnitude of the initial velocity is doubled and the angle of the velocity remains the same, the projectile will travel twice as far as it did before. Do you agree with her hypothesis? Explain.

20. **Aviation** Commander Patrick Driscoll flies with the U.S. Navy's Blue Angels. Suppose he places his aircraft in a 45° dive at an initial speed of 800 km/h.
    a. Write parametric equations to represent the descent path of the aircraft.
    b. How far has the aircraft descended after 2.5 seconds?
    c. What is the average rate at which the plane is losing altitude during the first 2.5 seconds?

21. **Entertainment** The "Human Cannonball" is shot out of a cannon with an initial velocity of 70 mph 10 feet above the ground at an angle of 35°.
    a. What is the maximum range of the cannon?
    b. How far from the launch point should a safety net be placed if the "Human Cannonball" is to land on it at a point 8 feet above the ground?
    c. How long is the flight of the "Human Cannonball" from the time he is launched to the time he lands in the safety net?

22. **Critical Thinking** If a circle of radius 1 unit is rolled along the x-axis at a rate of 1 unit per second, then the path of a point P on the circle is called a *cycloid*.
    a. Sketch what you think a cycloid must look like. (*Hint:* Use a coin or some other circular object to simulate the situation.)

    b. The parametric equations of a cycloid are $x = t - \sin t$ and $y = 1 - \cos t$ where t is measured in radians. Use a graphing calculator to graph the cycloid. Set your calculator in radian mode. An appropriate range is **Tmin** = 0, **Tmax** = 18.8, **Tstep** = 0.2, **Xmin** = −6.5, **Xmax** = 25.5, **Xscl** = 2, **Ymin** = −8.4, **Ymax** = 12.9, and **Yscl** = 1. Compare the results with your sketch.

23. Entertainment The Independence Day fireworks at Memorial Park are fired at an angle of 82° with the horizontal. The technician firing the shells expects them to explode about 300 feet in the air 4.8 seconds after they are fired.

a. Find the initial velocity of a shell fired from ground level.

b. Safety barriers will be placed around the launch area to protect spectators. If the barriers are placed 100 yards from the point directly below the explosion of the shells, how far should the barriers be from the point where the fireworks are launched?

24. Baseball Derek Jeter, shortstop for the New York Yankees, comes to bat with runners on first and third bases. Greg Maddux, pitcher for the Atlanta Braves, throws a slider across the plate about waist high, 3 feet above the ground. Derek Jeter hits the ball with an initial velocity of 155 feet per second at an angle of 22° above the horizontal. The ball travels straight at the 420 foot mark on the center field wall which is 15 feet high.

a. Write parametric equations that describe the path of the ball.

b. Find the height of the ball after it has traveled 420 feet horizontally. Will the ball clear the fence for a home run, or will the center fielder be able to catch it?

c. If there were no outfield seats, how far would the ball travel before it hits the ground?

Mixed Review

25. Write an equation of the line with parametric equations $x = 11 - t$ and $y = 8 - 6t$ in slope-intercept form. *(Lesson 8-6)*

26. Food Industry Fishmongers will often place ice over freshly-caught fish that are to be shipped to preserve the freshness. Suppose a block of ice with a mass of 300 kilograms is held on an ice slide by a rope parallel to the slide. The slide is inclined at an angle of 22°. *(Lesson 8-5)*

a. What is the pull on the rope?

b. What is the force on the slide?

27. If $b = 17.4$ and $c = 21.9$, find the measure of angle A to the nearest degree. *(Lesson 5-5)*

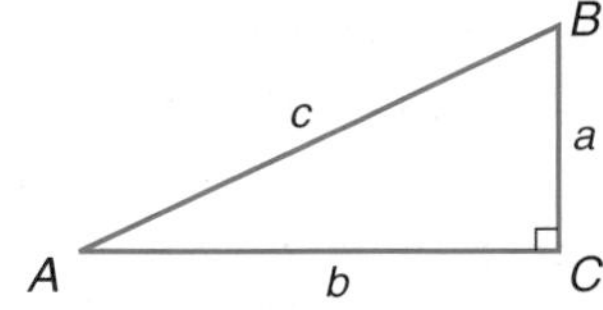

28. Solve the system of equations. *(Lesson 2-2)*

$2x - y + z = 2$

$x + 3y - 2z = -3.25$

$-4x - 5y + z = 2.5$

29. SAT/ACT Practice What is the area of the shaded region?

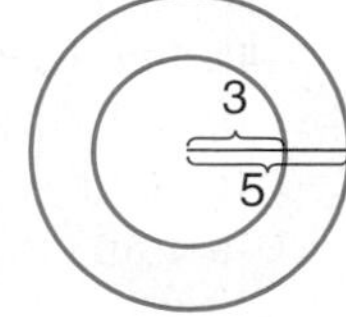

A 19π **B** 16π

C 9π **D** 4π

E 2π

Extra Practice See p. A41.

HISTORY of MATHEMATICS

ANGLE MEASURE

You know that the direction of a vector is the directed angle, usually measured in degrees, between the positive x-axis and the vector. How did that system originate? How did other angle measurement systems develop?

Early Evidence The division of a circle into 360° is based upon a unit of distance used by the **Babylonians,** which was equal to about 7 miles. They related time and miles by observing the time that it took for a person to travel one of their units of distance. An entire day was 12 "time-miles," so one complete revolution of the sky by the sun was divided into 12 units. They also divided each time-mile into 30 units. So 12×30 or 360 units represented a complete revolution of the sun. Other historians feel that the Babylonians used 360° in a circle because they had a base 60 number system, and thought that there were 360 days in a year.

Ptolemy

The Greeks adopted the 360° circle. **Ptolemy** (85-165) used the degree system in his work in astronomy, introducing symbols for degrees, minutes and seconds in his mathematical work, *Almagest,* though they differed from our modern symbols. When his work was translated into Arabic, sixtieths were called "first small parts," sixtieths of sixtieths were called "second small parts," and so on. The later Latin translations were "partes minutae primae," now our "minutes," and "partes minutae secundae," now our "seconds."

The Renaissance The degree symbol (°) became more widely used after the publication of a book by Dutch mathematician, **Gemma Frisius** in 1569.

Modern Era Radian angle measure was introduced by **James Thomson** in 1873. The radian was developed to simplify formulas used in trigonometry and calculus. Radian measure can be given as an exact quantity not just an approximation.

Other measurement units for angles have been developed to meet a specific need. These include mils, decimally- or centesimally-divided degrees, and millicycles.

Today mechanical engineers, like **Mark Korich,** use angle measure in designing electric and hybrid vehicles. Mr. Korich, a senior staff engineer for an auto manufacturer, calculates the draft angle that allows molded parts to pull away from the mold more easily. He also uses vectors to calculate the magnitude and direction of forces acting upon the vehicle under normal conditions.

ACTIVITIES

1. Greek mathematician, **Eratosthenes,** calculated the circumference of Earth using proportion and angle measure. By using the angle of a shadow cast by a rod in Alexandria, he determined that $\angle AOB$ was equal to $7°12'$. He knew that the distance from Alexandria to Syene was 5000 stadia (singular stadium). One stadium equals 500 feet. Use the diagram to find the circumference of Earth. Compare your result to the actual circumference of Earth.

2. Research one of the famous construction problems of the Greeks—trisecting an angle using a straightedge and compass. Try to locate a solution to this problem.

3. *inter*NET CONNECTION Find out more about those who contributed to the history of angle measure. Visit **www.amc.glencoe.com**

Transformation Matrices in Three-Dimensional Space

OBJECTIVE

- Transform three-dimensional figures using matrix operations to describe the transformation.

COMPUTER ANIMATION Chris Wedge of Blue Sky Studios, Inc. used software to create the film that won the 1999 Academy Award for Animated Short Film. The computer software allows Mr. Wedge to draw three-dimensional objects and manipulate or transform them to create motion, color, and light direction. The mathematical processes used by the computer are very complex. *A problem related to animation will be solved in Example 2.*

Look Back
Refer to Lesson 2-3 to review ordered pairs and matrices.

Basic movements in three-dimensional space can be described using vectors and transformation matrices. Recall that a point at (x, y) in a two-dimensional coordinate system can be represented by the matrix $\begin{bmatrix} x \\ y \end{bmatrix}$. This idea can be extended to three-dimensional space. A point at (x, y, z) can be expressed as the matrix $\begin{bmatrix} x \\ y \\ z \end{bmatrix}$.

Example 1 **Find the coordinates of the vertices of the rectangular prism and represent them as a vertex matrix.**

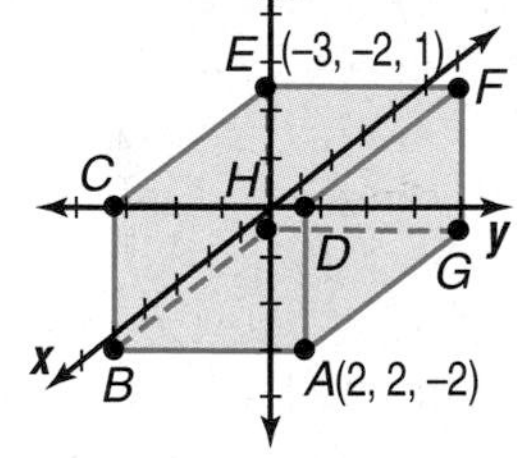

$\overrightarrow{AE} = \langle -3 - 2, -2 - 2, 1 - (-2) \rangle$ or $\langle -5, -4, 3 \rangle$

You can use the coordinates of $\overrightarrow{AE}$ to find the coordinates of the other vertices.

You could also find the coordinates of B, G, and H first, then add 3 to the z-coordinates of A, B, and G to find the coordinates of D, C, and F.

$B(2, 2 + (-4), -2) = B(2, -2, -2)$
$C(2, 2 + (-4), -2 + 3) = C(2, -2, 1)$
$D(2, 2, -2 + 3) = D(2, 2, 1)$
$F(2 + (-5), 2, -2 + 3) = F(-3, 2, 1)$
$G(2 + (-5), 2, -2) = G(-3, 2, -2)$
$H(2 + (-5), 2 + (-4), -2) = H(-3, -2, -2)$

The vertex matrix for the prism is
$$\begin{matrix} & A & B & C & D & E & F & G & H \\ \begin{matrix} x \\ y \\ z \end{matrix} & \multicolumn{8}{c}{} \end{matrix}$$
$$\begin{matrix} x \\ y \\ z \end{matrix}\begin{bmatrix} 2 & 2 & 2 & 2 & -3 & -3 & -3 & -3 \\ 2 & -2 & -2 & 2 & -2 & 2 & 2 & -2 \\ -2 & -2 & 1 & 1 & 1 & 1 & -2 & -2 \end{bmatrix}$$
(columns A, B, C, D, E, F, G, H).

In Lesson 2-4, you learned that certain matrices could transform a polygon on a coordinate plane. Likewise, transformations of three-dimensional figures, called **polyhedra,** can also be represented by certain matrices. A polyhedron (singular of *polyhedra*) is a closed three-dimensional figure made up of flat polygonal regions.

Example 2 **COMPUTER ANIMATION** 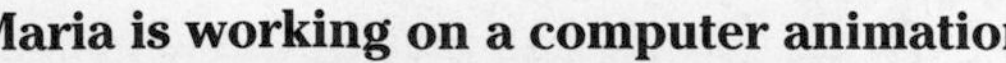 **Maria is working on a computer animation project.**

She needs to translate a prism using the vector $\vec{a} = \langle 3, 3, 0 \rangle$. The vertices of the prism have the following coordinates.

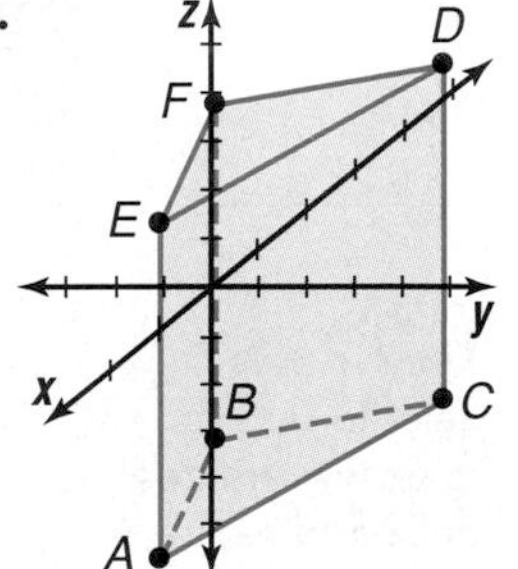

$A(2, 1, -4)$	$B(-1, -1, -4)$	$C(-2, 3, -4)$
$D(-2, 3, 3)$	$E(2, 1, 3)$	$F(-1, -1, 3)$

a. Write a matrix that will have such an effect on the figure.

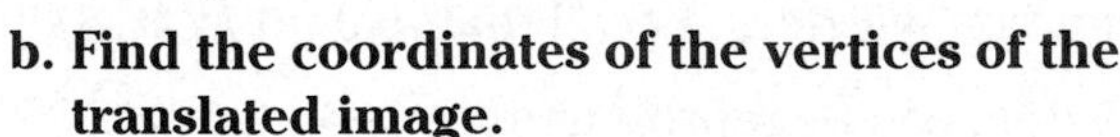

b. Find the coordinates of the vertices of the translated image.

c. Graph the translated image.

a. To translate the prism by the vector $\vec{a} = \langle 3, 3, 0 \rangle$, we must first add 3 to each of the x- and y-coordinates. The z-coordinates remain the same. The translation matrix can be written as

$$\begin{bmatrix} 3 & 3 & 3 & 3 & 3 & 3 \\ 3 & 3 & 3 & 3 & 3 & 3 \\ 0 & 0 & 0 & 0 & 0 & 0 \end{bmatrix}.$$

b. Write the vertices of the prism in a 6×3 matrix. Then add it to the translation matrix to find the vertices of the translated image.

$$\begin{bmatrix} 3 & 3 & 3 & 3 & 3 & 3 \\ 3 & 3 & 3 & 3 & 3 & 3 \\ 0 & 0 & 0 & 0 & 0 & 0 \end{bmatrix} + \begin{array}{c} \begin{matrix} A & \;\; B & \;\; C & \;\; D & \;\; E & \;\; F \end{matrix} \\ \begin{bmatrix} 2 & -1 & -2 & -2 & 2 & -1 \\ 1 & -1 & 3 & 3 & 1 & -1 \\ -4 & -4 & -4 & 3 & 3 & 3 \end{bmatrix} \end{array}$$

$$= \begin{array}{c} \begin{matrix} A' & B' & C' & D' & E' & F' \end{matrix} \\ \begin{bmatrix} 5 & 2 & 1 & 1 & 5 & 2 \\ 4 & 2 & 6 & 6 & 4 & 2 \\ -4 & -4 & -4 & 3 & 3 & 3 \end{bmatrix} \end{array}$$

c. Draw the graph of the image.

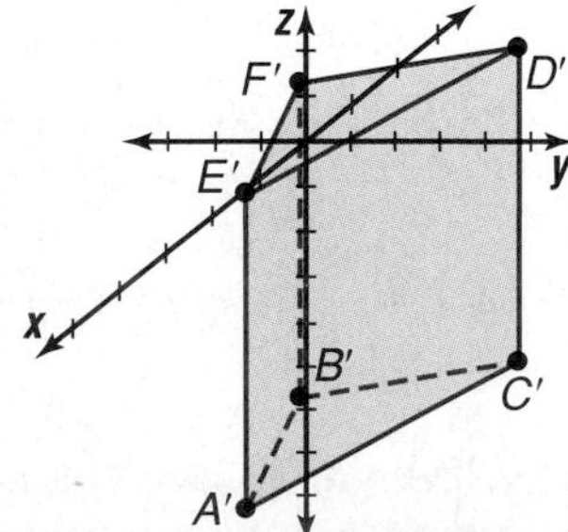

Recall that certain 2×2 matrices could be used to reflect a plane figure across an axis. Likewise, certain 3×3 matrices can be used to reflect three-dimensional figures in space.

Example 3 **Let M represent the vertex matrix of the rectangular prism in Example 1.**

a. Find TM if $T = \begin{bmatrix} 1 & 0 & 0 \\ 0 & 1 & 0 \\ 0 & 0 & -1 \end{bmatrix}$.

b. Graph the resulting image.

c. Describe the transformation represented by matrix T.

a. First find TM.

$$TM = \begin{bmatrix} 1 & 0 & 0 \\ 0 & 1 & 0 \\ 0 & 0 & -1 \end{bmatrix} \cdot \begin{bmatrix} 2 & 2 & 2 & 2 & -3 & -3 & -3 & -3 \\ 2 & -2 & -2 & 2 & -2 & 2 & 2 & -2 \\ -2 & -2 & 1 & 1 & 1 & 1 & -2 & -2 \end{bmatrix}$$

$$TM = \begin{matrix} A' & B' & C' & D' & E' & F' & G' & H' \end{matrix}$$
$$TM = \begin{bmatrix} 2 & 2 & 2 & 2 & -3 & -3 & -3 & -3 \\ 2 & -2 & -2 & 2 & -2 & 2 & 2 & -2 \\ 2 & 2 & -1 & -1 & -1 & -1 & 2 & 2 \end{bmatrix}$$

b. Then graph the points represented by the resulting matrix.

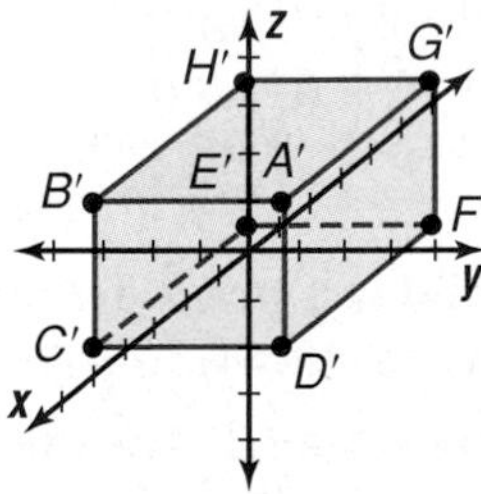

c. The transformation matrix $\begin{bmatrix} 1 & 0 & 0 \\ 0 & 1 & 0 \\ 0 & 0 & -1 \end{bmatrix}$ reflects the image of each vertex over the xy-plane. This results in a reflection of the prism when the new vertices are connected by segments.

The transformation matrix in Example 3 resulted in a reflection over the xy-plane. Similar transformations will result in reflections over the xz- and yz-planes. These transformations are summarized in the chart on the next page.

Reflection Matrices		
For a reflection over the:	**Multiply the vertex matrix by:**	**Resulting image**
yz-plane	$R_{yz\text{-}plane} = \begin{bmatrix} -1 & 0 & 0 \\ 0 & 1 & 0 \\ 0 & 0 & 1 \end{bmatrix}$	
xz-plane	$R_{xz\text{-}plane} = \begin{bmatrix} 1 & 0 & 0 \\ 0 & -1 & 0 \\ 0 & 0 & 1 \end{bmatrix}$	
xy-plane	$R_{xy\text{-}plane} = \begin{bmatrix} 1 & 0 & 0 \\ 0 & 1 & 0 \\ 0 & 0 & -1 \end{bmatrix}$	

One other transformation of two-dimensional figures that you have studied is the *dilation.* A dilation with scale factor k, for $k \neq 0$, can be represented by the matrix $D = \begin{bmatrix} k & 0 & 0 \\ 0 & k & 0 \\ 0 & 0 & k \end{bmatrix}$.

Example 4 **A parallelepiped is a prism whose faces are all parallelograms as shown in the graph.**

a. Find the vertex matrix for the transformation D where $k = 2$.

b. Draw a graph of the resulting figure.

c. What effect does transformation D have on the original figure?

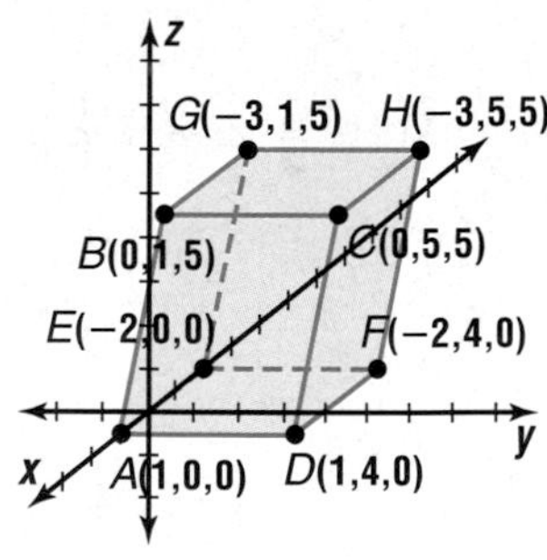

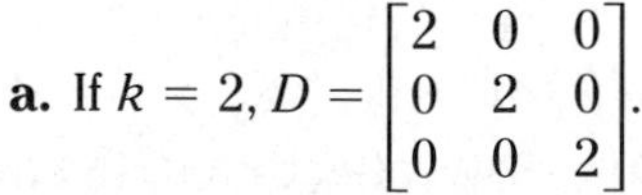

a. If $k = 2$, $D = \begin{bmatrix} 2 & 0 & 0 \\ 0 & 2 & 0 \\ 0 & 0 & 2 \end{bmatrix}$.

Find the coordinates of the vertices of the parallelepiped. Write them as vertex matrix P.

$$P = \begin{array}{c} \\ \\ \end{array}\begin{matrix} A & B & C & D & E & F & G & H \end{matrix}$$
$$P = \begin{bmatrix} 1 & 0 & 0 & 1 & -2 & -2 & -3 & -3 \\ 0 & 1 & 5 & 4 & 0 & 4 & 1 & 5 \\ 0 & 5 & 5 & 0 & 0 & 0 & 5 & 5 \end{bmatrix}$$

Then, find the product of D and P.

$$DP = \begin{bmatrix} 2 & 0 & 0 \\ 0 & 2 & 0 \\ 0 & 0 & 2 \end{bmatrix} \cdot \begin{bmatrix} 1 & 0 & 0 & 1 & -2 & -2 & -3 & -3 \\ 0 & 1 & 5 & 4 & 0 & 4 & 1 & 5 \\ 0 & 5 & 5 & 0 & 0 & 0 & 5 & 5 \end{bmatrix}$$

$$\begin{matrix} & A' & B' & C' & D' & E' & F' & G' & H' \end{matrix}$$

$$= \begin{bmatrix} 2 & 0 & 0 & 2 & -4 & -4 & -6 & -6 \\ 0 & 2 & 10 & 8 & 0 & 8 & 2 & 10 \\ 0 & 10 & 10 & 0 & 0 & 0 & 10 & 10 \end{bmatrix}$$

b. Graph the transformation.

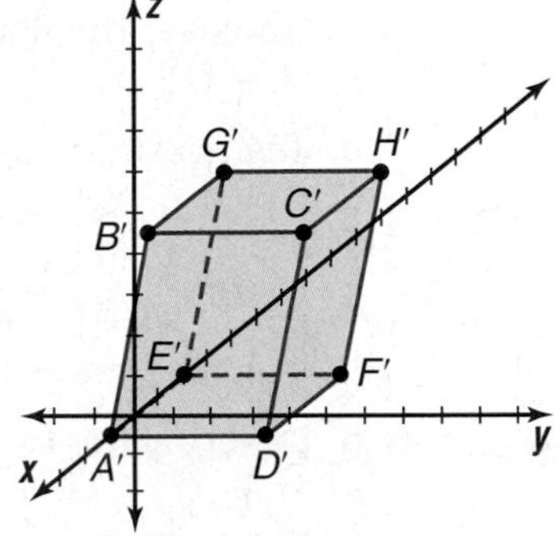

c. The transformation matrix D is a dilation. The dimensions of the prism have increased by a factor of 2.

CHECK FOR UNDERSTANDING

Communicating Mathematics

Read and study the lesson to answer each question.

1. **Describe** the transformation that matrix $T = \begin{bmatrix} -2 & 0 & 0 \\ 0 & 2 & 0 \\ 0 & 0 & 2 \end{bmatrix}$ produces on a three-dimensional figure.

2. **Write** a transformation matrix that represents the translation shown at the right.

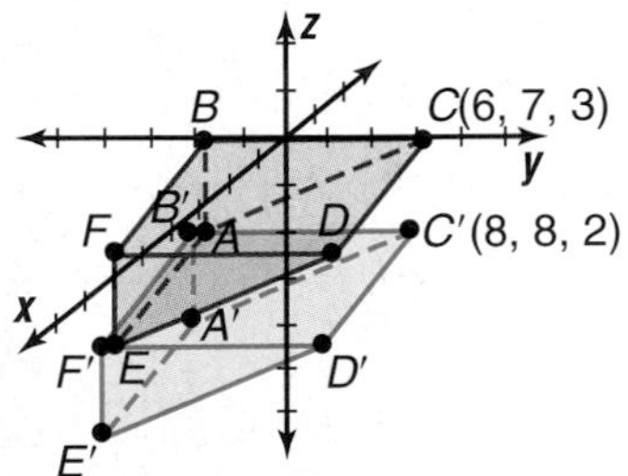

3. **Determine** if multiplying a vertex matrix by the transformation matrix

$$T = \begin{bmatrix} -1 & 0 & 0 \\ 0 & 1 & 0 \\ 0 & 0 & -1 \end{bmatrix}$$ produces the same result as multiplying by

$$U = \begin{bmatrix} 1 & 0 & 0 \\ 0 & 1 & 0 \\ 0 & 0 & -1 \end{bmatrix}$$ and then by $$V = \begin{bmatrix} -1 & 0 & 0 \\ 0 & 1 & 0 \\ 0 & 0 & 1 \end{bmatrix}.$$

4. *Math Journal* Using a chart, describe the effects reflection, translation, and dilation have on

a. the figure's orientation on the coordinate system.

b. the figure's size.

c. the figure's shape.

Guided Practice

5. Refer to the rectangular prism at the right.

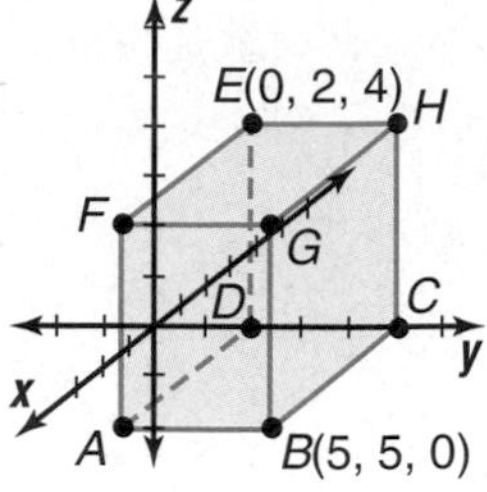

a. Write the matrix for the figure.

b. Write the resulting matrix if you translate the figure using the vector $\langle 4, -1, 2 \rangle$.

c. Transform the figure using the matrix

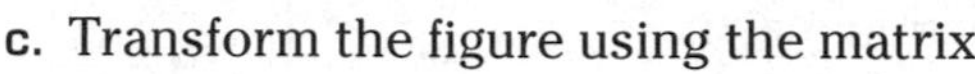

$\begin{bmatrix} 1 & 0 & 0 \\ 0 & -1 & 0 \\ 0 & 0 & 1 \end{bmatrix}$. Graph the image and describe the result.

d. Describe the transformation on the figure resulting from its product with the matrix $\begin{bmatrix} 0.5 & 0 & 0 \\ 0 & 0.5 & 0 \\ 0 & 0 & 0.5 \end{bmatrix}$.

6. **Architecture** Trevor is revising a design for a playground that contains a piece of equipment shaped like a rectangular prism. He needs to enlarge the prism 4 times its size and move it along the x-axis 2 units.

a. What are the transformation matrices?

b. Graph the rectangular prism and its image after the transformation.

EXERCISES

Practice

Write the matrix for each prism.

7.

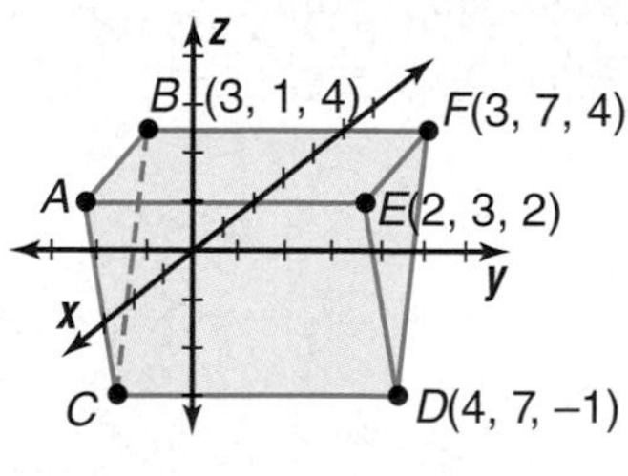

8.

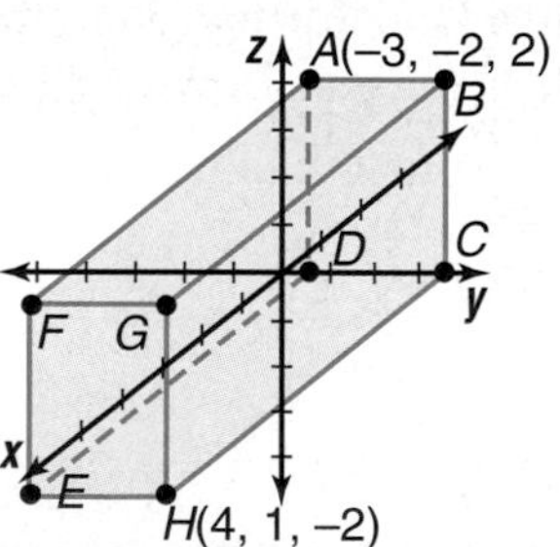

9.

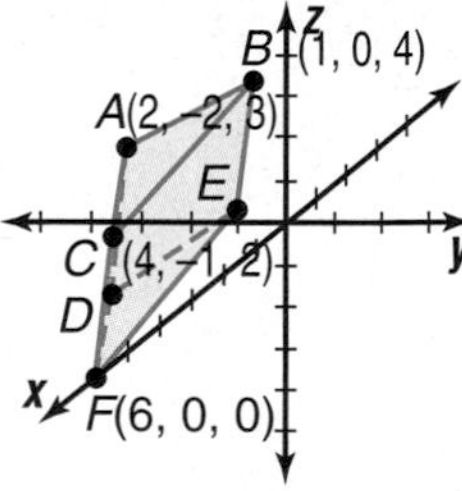

Use the prism at the right for Exercises 10-15. The vertices are $A(0, 0, 1)$, $B(0, 3, 2)$, $C(0, 5, 5)$, $D(0, 0, 4)$, $E(2, 0, 1)$, $F(2, 0, 4)$, $G(2, 3, 5)$, and $H(2, 3, 2)$. Translate the prism using the given vectors. Graph each image and describe the result.

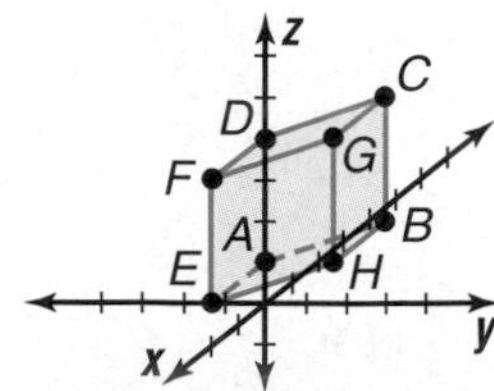

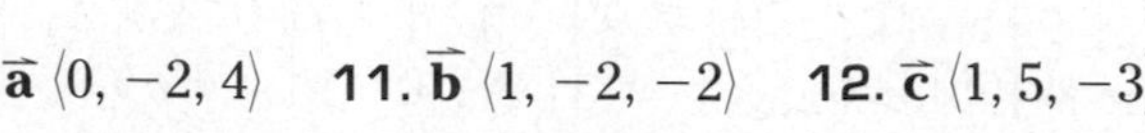

10. $\vec{a}\ \langle 0, -2, 4 \rangle$ **11.** $\vec{b}\ \langle 1, -2, -2 \rangle$ **12.** $\vec{c}\ \langle 1, 5, -3 \rangle$

Refer to the figure for Exercises 10-12. Transform the figure using each matrix. Graph each image and describe the result.

13. $\begin{bmatrix} 1 & 0 & 0 \\ 0 & 1 & 0 \\ 0 & 0 & 1 \end{bmatrix}$

14. $\begin{bmatrix} 1 & 0 & 0 \\ 0 & -1 & 0 \\ 0 & 0 & -1 \end{bmatrix}$

15. $\begin{bmatrix} -1 & 0 & 0 \\ 0 & -1 & 0 \\ 0 & 0 & -1 \end{bmatrix}$

Describe the result of the product of a vertex matrix and each matrix.

16. $\begin{bmatrix} 2 & 0 & 0 \\ 0 & 2 & 0 \\ 0 & 0 & 2 \end{bmatrix}$

17. $\begin{bmatrix} 3 & 0 & 0 \\ 0 & 3 & 0 \\ 0 & 0 & -3 \end{bmatrix}$

18. $\begin{bmatrix} -0.75 & 0 & 0 \\ 0 & -0.75 & 0 \\ 0 & 0 & -0.75 \end{bmatrix}$

19. A transformation in three-dimensional space can also be represented by $T(x, y, z) \rightarrow (2x, 2y, 5z)$.

    a. Determine the transformation matrix.

    b. Describe the transformation that such a matrix will have on a figure.

Applications and Problem Solving

20. **Marine Biology** A researcher studying a group of dolphins uses the matrix

$$\begin{bmatrix} 20 & 136 & 247 & 302 & 351 \\ -58 & -71 & -74 & -83 & -62 \\ 27 & 53 & 59 & 37 & 52 \end{bmatrix},$$

with the ship at the origin, to track the dolphins' movement. Later, the researcher will translate the matrix to a fixed reference point using the vector $\langle 23.6, 72, 0 \rangle$.

    a. Write the translation matrix used for the transformation.

    b. What is the resulting matrix?

    c. Describe the result of the translation.

21. **Critical Thinking** Write a transformational matrix that would first reflect a rectangular prism over the *yz*-plane and then reduce its dimensions by half.

22. **Meteorology** At the National Weather Service Center in Miami, Florida, meteorologists test models to forecast weather phenomena such as hurricanes and tornadoes. One weather disturbance, wind shear, can be modeled using a transformation on a cube. (*Hint:* You can use any size cube for the model.)

    a. Write a matrix to transform a cube into a slanted parallelepiped.

    b. Graph a cube and its image after the transformation.

23. **Geometry** Suppose a cube is transformed by two matrices,

first by $U = \begin{bmatrix} -1 & 0 & 0 \\ 0 & -1 & 0 \\ 0 & 0 & -1 \end{bmatrix}$, and then by $T = \begin{bmatrix} 1 & 0 & 2 \\ 0 & 2 & 0 \\ 0 & 0 & 2 \end{bmatrix}$.

Describe the resulting image.

24. Critical Thinking Matrix T maps a point $P(x, y, z)$ to the point $P'(3x, 2y, x - 4z)$. Write a 3×3 matrix for T.

25. Seismology Seismologists classify movements in the earth's crust by determining the direction and amount of movement which has taken place on a fault. Some of the classifications are shown using *block diagrams.*

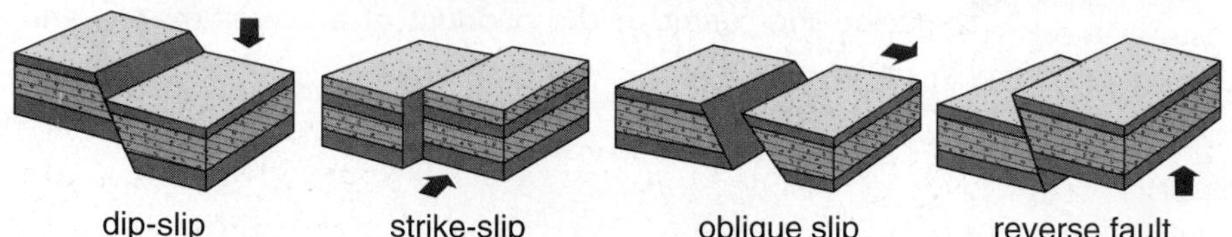

a. A seismologist used the matrix below to describe a particular feature along a fault.

$$\begin{bmatrix} 123.9 & -41.3 & 201.7 & 73.8 & -129.4 & 36.4 \\ 86.4 & 144.2 & -29.9 & -84.2 & 95.5 & -125.5 \\ 206.5 & 247.8 & 262.7 & 213.2 & -165.2 & -84.1 \end{bmatrix}$$

After a series of earthquakes he determined the matrix describing the feature had changed to the following matrix.

$$\begin{bmatrix} 123.9 & -41.3 & 201.7 & 73.8 & -129.4 & 36.4 \\ 88.0 & 145.8 & -28.3 & -82.6 & 97.1 & -123.9 \\ 205.3 & 246.6 & 261.5 & 212.0 & -166.4 & -85.3 \end{bmatrix}$$

What classification of movement bests describes the transformation that has occurred?

b. Determine the matrix that describes the movement that occurred.

Mixed Review

26. Physics Jaimie and LaShawna are standing at the edge of a cliff that is 150 feet high. At the same time, LaShawna drops a stone and Jaimie throws a stone horizontally at a velocity of 35 ft/s. *(Lesson 8-7)*

a. About how far apart will the stones be when they land?

b. Will the stones land at the same time? Explain.

27. Write an equation in slope-intercept form of the line with parametric equations $x = -5t - 1$ and $y = 2t + 10$. *(Lesson 8-6)*

28. Evaluate $\sec\left(\cos^{-1}\frac{2}{5}\right)$ if the angle is in Quadrant I. *(Lesson 6-8)*

29. Medicine Maria was told to take 80 milligrams of medication each morning for three days. The amount of medicine in her body on the fourth day is modeled by $M(x) = 80x^3 + 80x^2 + 80x$, where x represents the absorption rate per day. Suppose Maria has 24.2 milligrams of the medication present in her body on the fourth day. Find the absorption rate of the medication. *(Lesson 4-5)*

30. SAT/ACT Practice Which of the following equations are equivalent?

I. $2x + 4y = 8$ II. $3x + 6y = 12$

III. $4x + 8y = 8$ IV. $6x + 12y = 16$

A I and II only **B** I and IV only **C** II and IV only

D III and IV only **E** I, II, and III

Extra Practice See p. A41.

VOCABULARY

component (p. 488)
cross product (p. 507)
direction (p. 485)
direction vector (p. 520)
dot product (p. 506)
equal vectors (p. 485)
inner product (p. 505)
magnitude (p. 485)
opposite vectors (p. 487)
parallel vectors (p. 488)
parameter (p. 520)
parametric equation (p. 520)
polyhedron (p. 535)
resultant (p. 486)
scalar (p. 488)
scalar quantity (p. 488)
standard position (p. 485)
unit vector (p. 495)
vector (p. 485)
vector equation (p. 520)
zero vector (p. 485)

UNDERSTANDING AND USING THE VOCABULARY

Choose the correct term from the list to complete each sentence.

1. The __?__ of two or more vectors is the sum of the vectors.
2. A(n) __?__ vector has a magnitude of one unit.
3. Two vectors are equal if and only if they have the same direction and __?__.
4. The __?__ product of two vectors is a vector.
5. Two vectors in space are perpendicular if and only if their __?__ product is zero.
6. Velocity can be represented mathematically by a(n) __?__.
7. Vectors with the same direction and different magnitudes are __?__.
8. A vector with its initial point at the origin is in __?__ position.
9. A __?__ vector is used to describe the slope of a line.
10. Two or more vectors whose sum is a given vector are called __?__ of the given vector.

components
cross
direction
equal
inner
magnitude
parallel
parameter
resultant
scalar
standard
unit
vector

interNET CONNECTION For additional review and practice for each lesson, visit: **www.amc.glencoe.com**

SKILLS AND CONCEPTS

OBJECTIVES AND EXAMPLES

Lesson 8-1 Add and subtract vectors geometrically.

- Find the sum of $\vec{a}$ and $\vec{b}$ using the parallelogram method.

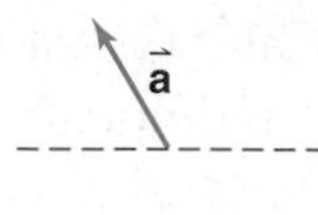

b

Place the initial points of $\vec{a}$ and $\vec{b}$ together. Draw dashed lines to form a complete parallelogram. The diagonal from the initial points to the opposite vertex of the parallelogram is the resultant.

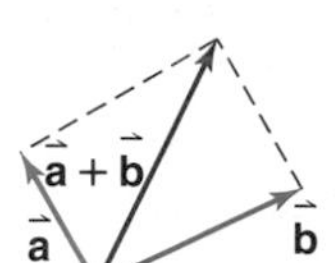

REVIEW EXERCISES

Use a ruler and a protractor to determine the magnitude (in centimeters) and direction of each vector.

11.

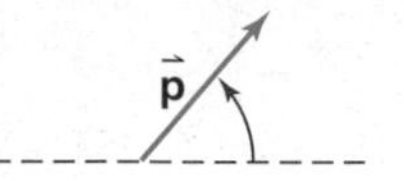

12.

q

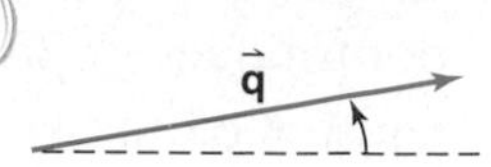

Use $\vec{p}$ and $\vec{q}$ above to find the magnitude and direction of each resultant.

13. $\vec{p} + \vec{q}$
14. $2\vec{p} + \vec{q}$
15. $3\vec{p} - \vec{q}$
16. $4\vec{p} - \vec{q}$

Find the magnitude of the horizontal and vertical components of each vector shown for Exercises 11 and 12.

17. $\vec{p}$
18. $\vec{q}$

Lesson 8-2 Find ordered pairs that represent vectors, and sums and products of vectors.

- Write the ordered pair that represents the vector from $M(3, 1)$ to $N(-7, 4)$. Then find the magnitude of $\overrightarrow{MN}$.

$\overrightarrow{MN} = \langle -7 - 3, 4 - 1\rangle$ or $\langle -10, 3\rangle$

$|\overrightarrow{MN}| = \sqrt{(-7-3)^2 + (4-1)^2}$ or $\sqrt{109}$

Find $\vec{a} + \vec{b}$ if $\vec{a} = \langle 1, -5\rangle$ and $\vec{b} = \langle -2, 4\rangle$.

$\vec{a} + \vec{b} = \langle 1, -5\rangle + \langle -2, 4\rangle$

$= \langle 1 + (-2), -5 + 4\rangle$ or $\langle -1, -1\rangle$

Write the ordered pair that represents $\overrightarrow{CD}$. Then find the magnitude of $\overrightarrow{CD}$.

19. $C(2, 3), D(7, 15)$
20. $C(-2, 8), D(4, 12)$
21. $C(2, -3), D(0, 9)$
22. $C(-6, 4), D(-5, -4)$

Find an ordered pair to represent $\vec{u}$ in each equation if $\vec{v} = \langle 2, -5\rangle$ and $\vec{w} = \langle 3, -1\rangle$.

23. $\vec{u} = \vec{v} + \vec{w}$
24. $\vec{u} = \vec{v} - \vec{w}$
25. $\vec{u} = 3\vec{v} + 2\vec{w}$
26. $\vec{u} = 3\vec{v} - 2\vec{w}$

Lesson 8-3 Find the magnitude of vectors in three-dimensional space.

- Write the ordered triple that represents the vector from $R(-2, 0, 8)$ to $S(5, -4, -1)$. Then find the magnitude of $\overrightarrow{RS}$.

$\overrightarrow{RS} = \langle 5 - (-2), -4 - 0, -1 - 8\rangle$

$= \langle 7, -4, -9\rangle$

$|\overrightarrow{RS}| = \sqrt{(5-(-2))^2 + (-4-0)^2 + (-1-8)^2}$

$= \sqrt{49 + 16 + 81}$ or about 12.1

Write the ordered triple that represents $\overrightarrow{EF}$. Then find the magnitude of $\overrightarrow{EF}$.

27. $E(2, -1, 4), F(6, -2, 1)$
28. $E(9, 8, 5), F(-1, 5, 11)$
29. $E(-4, -3, 0), F(2, -1, 7)$
30. $E(3, 7, -8), F(-4, 0, 5)$

Find an ordered triple to represent $\vec{u}$ in each equation if $\vec{v} = \langle -1, 7, -4\rangle$ and $\vec{w} = \langle 4, -1, 5\rangle$.

31. $\vec{u} = 2\vec{w} - 5\vec{v}$
32. $\vec{u} = 0.25\vec{v} + 0.4\vec{w}$

OBJECTIVES AND EXAMPLES

Lesson 8-4 Find the inner and cross products of vectors.

- Find the inner product of $\vec{a}$ and $\vec{b}$ if $\vec{a} = \langle 3, -1, 7\rangle$ and $\vec{b} = \langle 0, -2, -4\rangle$. Are $\vec{a}$ and $\vec{b}$ perpendicular?

$\vec{a} \cdot \vec{b} = 3(0) + (-1)(-2) + 7(-4)$ or -26

$\vec{a}$ and $\vec{b}$ are not perpendicular since their inner product is not zero.

Find $\vec{c} \times \vec{d}$ if $\vec{c} = \langle -2, 1, 1\rangle$ and $\vec{d} = \langle 1, -3, 0\rangle$.

$$\vec{c} \times \vec{d} = \begin{vmatrix} \vec{i} & \vec{j} & \vec{k} \\ -2 & 1 & 1 \\ 1 & -3 & 0 \end{vmatrix}$$

$$= \begin{vmatrix} 1 & 1 \\ -3 & 0 \end{vmatrix}\vec{i} - \begin{vmatrix} -2 & 1 \\ 1 & 0 \end{vmatrix}\vec{j} + \begin{vmatrix} -2 & 1 \\ 1 & -3 \end{vmatrix}\vec{k}$$

$$= 3\vec{i} + 1\vec{j} + 5\vec{k} \text{ or } \langle 3, 1, 5\rangle$$

Determine if $\vec{a}$ is perpendicular to $\vec{b}$ and $\vec{c}$ if $\vec{a} = \langle 3, 1, 5\rangle$, $\vec{b} = \langle -2, 1, 1\rangle$, and $\vec{c} = \langle 1, -3, 0\rangle$.

$\vec{a} \cdot \vec{b} = 3(-2) + 1(1) + 5(1)$ or 0

$\vec{a} \cdot \vec{c} = 3(1) + 1(-3) + 5(0)$ or 0

Since the inner products are zero, $\vec{a}$ is perpendicular to both $\vec{b}$ and $\vec{c}$.

REVIEW EXERCISES

Find each inner product and state whether the vectors are perpendicular. Write *yes* or *no*.

33. $\langle 5, -1\rangle \cdot \langle -2, 6\rangle$
34. $\langle 2, 6\rangle \cdot \langle 3, -4\rangle$
35. $\langle 4, 1, -2\rangle \cdot \langle 3, -4, 4\rangle$
36. $\langle 2, -1, 4\rangle \cdot \langle 6, -2, 1\rangle$
37. $\langle 5, 2, -10\rangle \cdot \langle 2, -4, -4\rangle$

Find each cross product. Then verify if the resulting vector is perpendicular to the given vectors.

38. $\langle 5, -2, 5\rangle \times \langle -1, 0, -3\rangle$
39. $\langle -2, -3, 1\rangle \times \langle 2, 3, -4\rangle$
40. $\langle -1, 0, 4\rangle \times \langle 5, 2, -1\rangle$
41. $\langle 7, 2, 1\rangle \times \langle 2, 5, 3\rangle$

42. Find a vector perpendicular to the plane containing the points (1, 2, 3), (−4, 2, −1), and (5, −3, 0).

OBJECTIVES AND EXAMPLES

Lesson 8-5 Solve problems using vectors and right triangle trigonometry.

- Find the magnitude and direction of the resultant vector for the diagram.

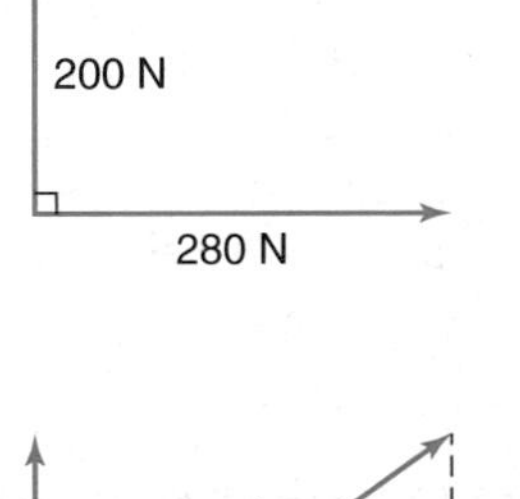

$|\vec{r}|^2 = 200^2 + 280^2$

$|\vec{r}|^2 = 118{,}400$

$|\vec{r}| = \sqrt{118{,}400}$

≈ 344 N

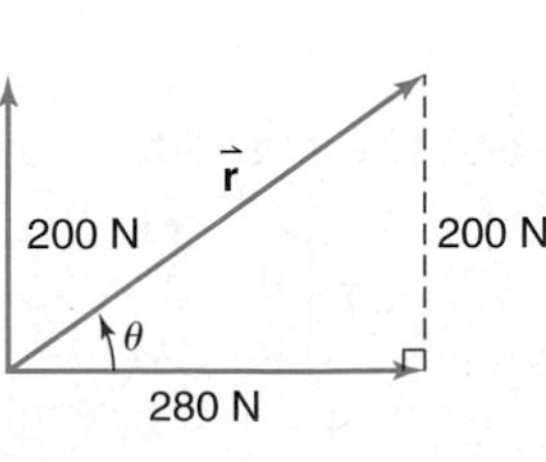

$\tan\theta = \frac{200}{280}$

$\theta = \tan^{-1}\frac{200}{280}$

$\approx 35.6°$

REVIEW EXERCISES

Find the magnitude and direction of the resultant vector for each diagram.

43.

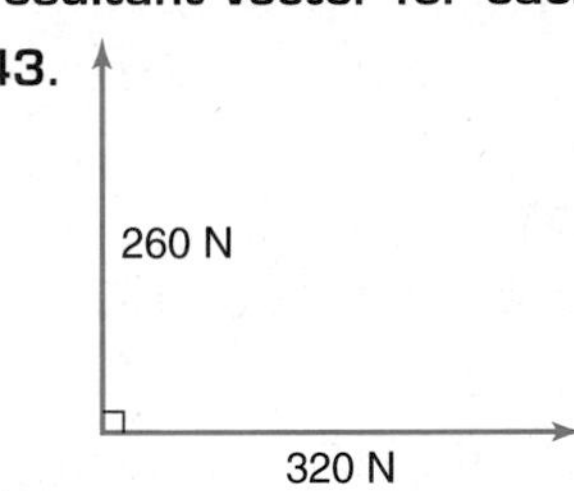

44.

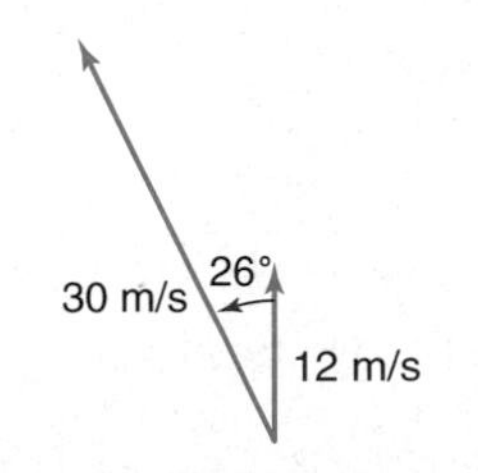

OBJECTIVES AND EXAMPLES

Lesson 8-6 Write vector and parametric equations of lines.

- Write a vector equation of the line that passes through $P(-6, 3)$ and is parallel to $\vec{\mathbf{v}} = \langle 1, 4 \rangle$. Then write parametric equations of the line.

  The vector equation is $\langle x + 6, y - 3 \rangle = t\langle 1, 4 \rangle$.

  Write the parametric equations of a line with $(a_1, a_2) = (-6, 3)$ and $(x_1, y_2) = (1, 4)$.

  $x = x_1 + ta_1$ $\quad$ $y = y_1 + ta_2$
  $x = -6 + t(1)$ $\quad$ $y = 3 + t(4)$
  $x = -6 + t$ $\quad$ $y = 3 + 4t$

REVIEW EXERCISES

Write a vector equation of the line that passes through point P and is parallel to $\vec{\mathbf{v}}$. Then write parametric equations of the line.

45. $P(3, -5)$, $\vec{\mathbf{v}} = \langle 4, 2 \rangle$
46. $P(-1, 9)$, $\vec{\mathbf{v}} = \langle -7, -5 \rangle$
47. $P(4, 0)$, $\vec{\mathbf{v}} = \langle 3, -6 \rangle$

Write parametric equations of each line with the given equation.

48. $y = -8x - 7$
49. $y = -\frac{1}{2}x + \frac{5}{2}$

Lesson 8-7 Model the motion of a projectile using parametric equations.

- Find the initial horizontal and vertical velocity for an arrow released at 52 m/s at an angle of 12° with the horizontal.

  $|\vec{\mathbf{v}}_x| = |\vec{\mathbf{v}}| \cos \theta$ $\quad$ $|\vec{\mathbf{v}}_y| = |\vec{\mathbf{v}}| \sin \theta$
  $|\vec{\mathbf{v}}_x| = 52 \cos 12°$ $\quad$ $|\vec{\mathbf{v}}_y| = 52 \sin 12°$
  $|\vec{\mathbf{v}}_x| \approx 50.86$ $\quad$ $|\vec{\mathbf{v}}_y| \approx 10.81$

  The initial horizontal velocity is about 50.86 m/s and the initial vertical velocity is about 10.81 m/s.

Find the initial horizontal and vertical velocity for each situation.

50. a stone thrown with an initial velocity of 15 feet per second at an angle of 55° with the horizontal
51. a baseball thrown with an initial velocity of 13.2 feet per second at an angle of 66° with the horizontal
52. a soccer ball kicked with an initial velocity of 18 meters per second at an angle of 28° with the horizontal

Lesson 8-8 Transform three-dimensional figures using matrix operations to describe the transformation.

- A prism needs to be translated using the vector $\langle 0, 1, 2 \rangle$. Write a matrix that will have such an effect on a figure.

$$T = \begin{bmatrix} 0 & 0 & 0 & 0 & 0 & 0 & 0 & 0 \\ 1 & 1 & 1 & 1 & 1 & 1 & 1 & 1 \\ 2 & 2 & 2 & 2 & 2 & 2 & 2 & 2 \end{bmatrix}$$

Use the prism for Exercises 53 and 54.

53. Translate the figure using the vector $\vec{\mathbf{n}} = \langle 2, 0, 3 \rangle$. Graph the image and describe the result.

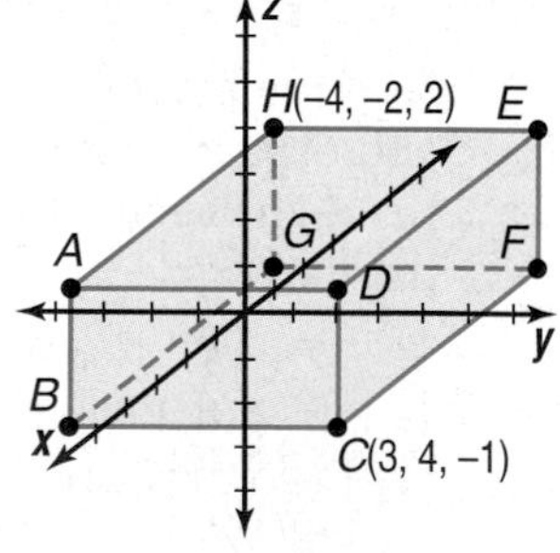

54. Transform the figure using the matrix $M = \begin{bmatrix} 1 & 0 & 0 \\ 0 & -1 & 0 \\ 0 & 0 & 1 \end{bmatrix}$. Graph the image and describe the result.

APPLICATIONS AND PROBLEM SOLVING

55. **Physics** Karen uses a large wrench to change the tire on her car. She applies a downward force of 50 pounds at an angle of 60° one foot from the center of the lug nut. Find the torque. *(Lesson 8-4)*

56. **Sports** Bryan punts a football with an initial velocity of 38 feet per second at an angle of 40° from the horizontal. If the ball is 2 feet above the ground when it is kicked, how high is it after 0.5 second? *(Lesson 8-7)*

57. **Navigation** A boat that travels at 16 km/h in calm water is sailing across a current of 3 km/h on a river 250 meters wide. The boat makes an angle of 35° with the current heading into the current. *(Lesson 8-5)*
    a. Find the resultant velocity of the boat.
    b. How far upstream is the boat when it reaches the other shore?

58. **Physics** Mario and Maria are moving a stove. They are applying forces of 70 N and 90 N at an angle of 30° to each other. If the 90 N force is applied along the x-axis find the magnitude and direction of the resultant of these forces. *(Lesson 8-5)*

ALTERNATIVE ASSESSMENT

OPEN-ENDED ASSESSMENT

1. The ordered pair $\langle -3, 2\rangle$ represents $\overrightarrow{XY}$.
    a. Give possible coordinates for X and Y. Show that your coordinates are correct.
    b. Find the magnitude of $\overrightarrow{XY}$. Did you need to know the coordinates for X and Y to find this magnitude? Explain.
2. a. $\overrightarrow{PQ}$ and $\overrightarrow{RS}$ are parallel. Give ordered pairs for P, Q, R, and S for which this is true. Explain how you know $\overrightarrow{PQ}$ and $\overrightarrow{RS}$ are parallel.
    b. $\vec{\mathbf{a}}$ and $\vec{\mathbf{b}}$ are perpendicular. Give ordered pairs to represent $\vec{\mathbf{a}}$ and $\vec{\mathbf{b}}$. Explain how you know that $\vec{\mathbf{a}}$ and $\vec{\mathbf{b}}$ are perpendicular.

PORTFOLIO

Devise a real-world problem that can be solved using vectors. Explain why vectors are needed to solve your problem. Solve your problem. Be sure to show and explain all of your work.

THE CYBERCLASSROOM

Vivid Vectors!

- Search the Internet to find websites that have lessons about vectors and their applications. Find at least three different sources of information.
- Make diagrams of examples that use vectors to solve problems by combining the ideas you found in Internet lessons and your textbook.
- Prepare a presentation to summarize the Internet Project for Unit 2. Design the presentation as a webpage. Use appropriate software to create the webpage.

Additional Assessment See p. A63 for Chapter 8 practice test.

Geometry Problems—Perimeter, Area, and Volume

Many SAT and ACT problems use perimeter, circumference, and area. A few problems use volume. Even though the formulas are often given on the test, it is more efficient if you memorize them.

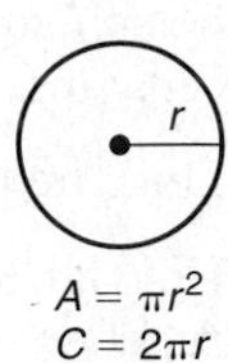

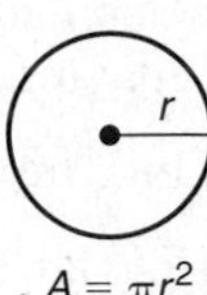

$A = \pi r^2$
$C = 2\pi r$

ℓ w

$A = \ell w$

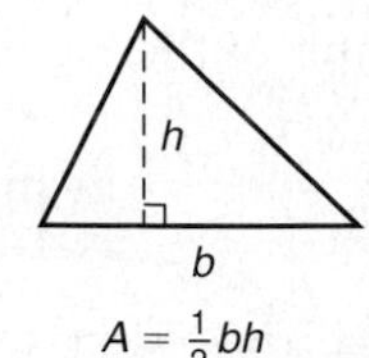

$A = \frac{1}{2}bh$

THE PRINCETON REVIEW

TEST-TAKING TIP

Recall that the value of π is about 3.14. When you need to estimate a quantity that involves π, use 3 as a rough approximation.

SAT EXAMPLE

1. In the figure below, the radius of circle A is twice the radius of the circle B and four times the radius of C. If the sum of the areas of the circles is 84π, what is the measure of $\overline{AC}$?

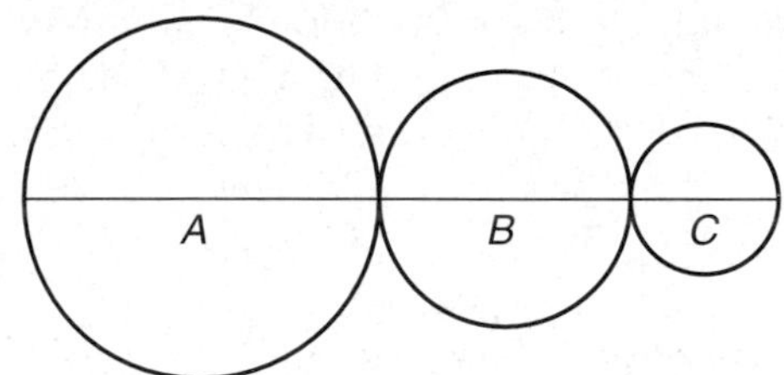

HINT Use a variable when necessary. Choose it carefully.

Solution This is a grid-in problem. You must find the length of $\overline{AC}$. This segment contains the radii of the circles. Let r be the radius of circle C. Then the radius of circle B is $2r$, and the radius of circle A is $4r$.

You know the total area. Write an equation for the sum of the three areas.

$$\text{Area } A + \text{Area } B + \text{Area } C = 84\pi$$
$$\pi(4r)^2 + \pi(2r)^2 + \pi(r)^2 = 84\pi$$
$$16r^2\pi + 4r^2\pi + r^2\pi = 84\pi$$
$$21r^2\pi = 84\pi$$
$$r^2 = 4$$
$$r = 2$$

The radius of C is 2; the radius of B is 4; the radius of A is 8. Recall that the problem asked for the length of $\overline{AC}$. Use the diagram to see which lengths to add—one radius of A, two radii of B, and one radius of C.

$$AC = 8 + 4 + 4 + 2 \text{ or } 18$$

The answer is 18.

ACT EXAMPLE

2. In the figure below, if $AB = 27$, $CD = 20$, and the area of $\triangle ADC = 240$, what is the area of polygon $ABCD$ in square units?

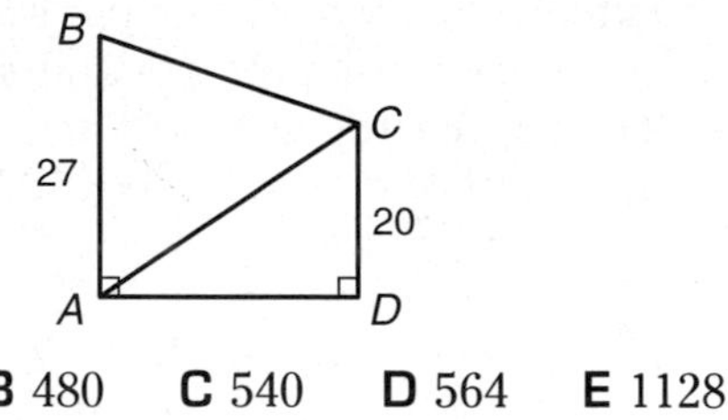

A 420 **B** 480 **C** 540 **D** 564 **E** 1128

HINT You may write in the test booklets. Mark all information you discover on the figure.

Solution The polygon is made up of two triangles, so find the area of each triangle. You know that the area of $\triangle ADC$ is 240 and its height is 20.

$$A = \frac{1}{2}bh$$
$$240 = \frac{1}{2}(b)(20)$$
$$240 = 10b$$
$$24 = b \qquad \text{So, } AD = 24.$$

Now find the area of $\triangle ABC$. The height of this triangle can be measured along $\overline{AD}$. So, $h = 24$, and the height is the measure of $\overline{AB}$, or 27. The area of $\triangle ABC = \frac{1}{2}(27)(24)$ or 324. You need to add the areas of the two triangles.

$$240 + 324 = 564$$

The answer is choice **D.**

After you work each problem, record your answer on the answer sheet provided or on a piece of paper.

Multiple Choice

1. In parallelogram $ABCD$ below, $BD = 3$ and $CD = 5$. What is the area of $ABCD$?

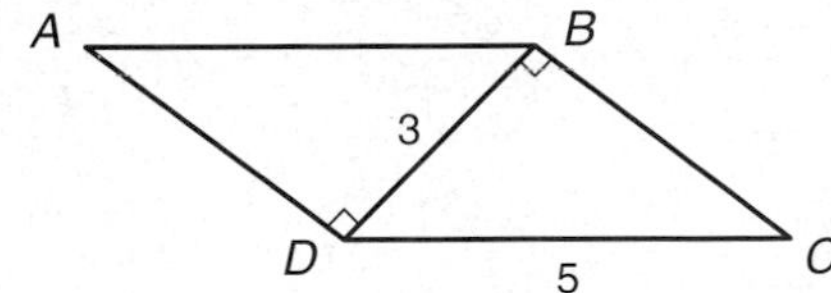

A 12 B 15 C 18 D 20
E It cannot be determined from the information given.

2. In square $ABCD$ below, what is the equation of circle Q that is circumscribed around the square?

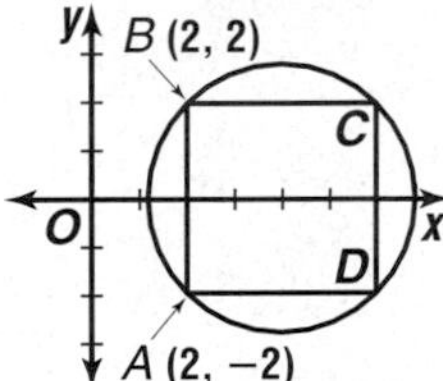

A $(x-4)^2 + y^2 = 4$ B $(x-4)^2 + y^2 = 8$
C $(x+4)^2 + y^2 = 8$ D $(x-4)^2 + y^2 = 32$
E $(x+4)^2 + y^2 = 32$

3. If the perimeter of the rectangle $ABCD$ is equal to p, and $x = \frac{2}{3}y$, what is the value of y in terms of p?

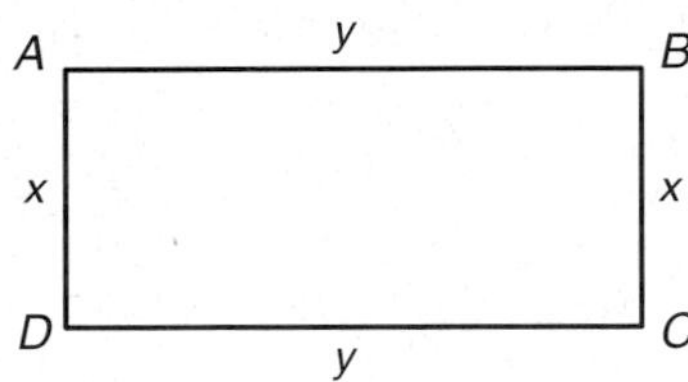

A $\frac{p}{10}$ B $\frac{3p}{10}$ C $\frac{p}{3}$ D $\frac{2p}{5}$ E $\frac{3p}{5}$

4. If two sides of a triangle have lengths of 40 and 80, which of the following cannot be the length of the third side?

A 40 B 41 C 50 D 80 E 81

5. $\sqrt[3]{x^2} \cdot \sqrt[9]{x^3} =$

A $x^{\frac{2}{9}}$ B $x^{\frac{1}{3}}$ C $x^{\frac{1}{2}}$ D $x^{\frac{2}{3}}$ E x

6. The figure below is made of three concentric semi-circles. What is the area of the shaded region in square units?

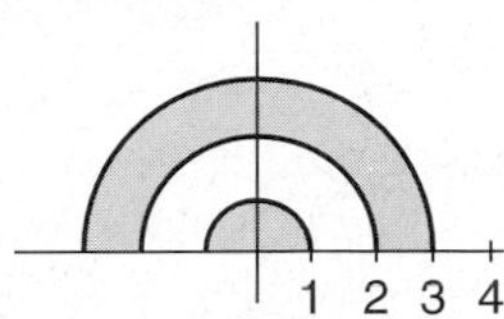

A 3π B $\frac{9}{2}\pi$ C 6π D 7π E 9π

7. $\frac{1}{5} + \frac{2}{25} + \frac{3}{50} =$

A 0.170 B 0.240 C 0.320
D 0.340 E 0.463

8. A 30-inch by 40-inch rectangular surface is to be completely covered with 1-inch square tiles, which cannot overlap one another and cannot overhang. If white tiles are to cover the interior and red tiles are to form a 1-inch wide border along the edge of the surface, how many red tiles will be needed?

A 70 B 136 C 140 D 142 E 144

9. **Quantitative Comparison**

A if the quantity in Column A is greater
B if the quantity in Column B is greater
C if the two quantities are equal
D if the relationship cannot be determined from the information given

Column A	Column B
The area of a triangle with base $3x$ and height $2x$.	The area of a circle with radius x.

10. **Grid-In**
In the figure, if $AE = 1$, what is the sum of the area of $\triangle ABC$ and the area of $\triangle CDE$?

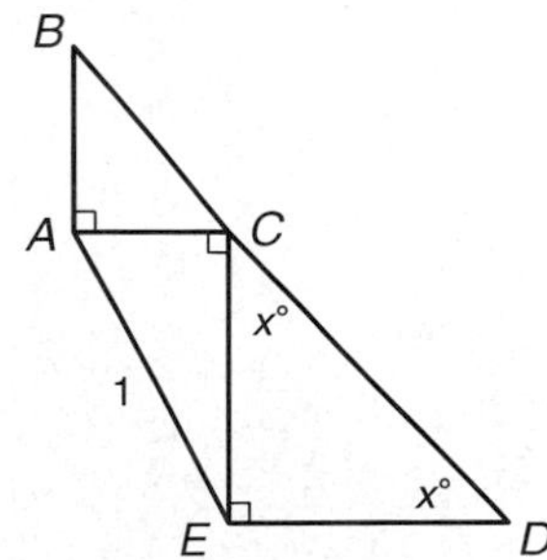

interNET CONNECTION **SAT/ACT Practice** For additional test practice questions, visit: **www.amc.glencoe.com**

UNIT 3 Advanced Functions and Graphing

You are now ready to apply what you have learned in earlier units to more complex functions. The three chapters in this unit contain very different topics; however, there are some similarities among them. The most striking similarity is that all three chapters require that you have had some experience with graphing. As you work through this unit, try to make connections between what you have already studied and what you are currently studying. This will help you use the skills you have already mastered more effectively.

Unit 3 interNET Projects

SPACE—THE FINAL FRONTIER

People have been fascinated by space since the beginning of time. Until 1961, however, human beings have been bound to Earth, unable to feel and experience life in space. Current space programs undertaken by NASA are exploring our solar system and beyond using sophisticated unmanned satellites such as the Hubble Space Telescope in orbit around Earth and the Mars Global Surveyor in orbit around Mars. In these projects, you will look at some interesting mathematics related to space—the final frontier. At the end of each chapter in Unit 3, you will be given specific tasks to explore space using the Internet.

CHAPTER 9
(page 611)

From Point A to Point B In Chapter 9, you will learn about the polar coordinate system, which is quite different from the rectangular coordinate system. Do scientists use the rectangular coordinate system or the polar coordinate system as they record the position of objects in space? Or, do they use some other system?

Math Connection: Research coordinate systems by using the Internet. Write a summary that describes each coordinate system that you find. Include diagrams and any information about converting between systems.

CHAPTER 10
(page 691)

Out in Orbit! What types of orbits do planets, artificial satellites, or space exploration vehicles have? Can orbits be modeled by the conic sections?

Math Connection: Use the Internet to find data about the orbit of a space vehicle, satellite, or planet. Make a scale drawing of the object's orbit labeling important features and dimensions. Then, write a summary describing the orbit of the object, being sure to discuss which conic section best models the orbit.

CHAPTER 11
(page 753)

Kepler is Still King! Johannes Kepler (1571–1630) was an important mathematician and scientist of his time. He observed the planets and stars and developed laws for the motion of those bodies. His laws are still used today. It is truly amazing how accurate his laws are considering the primitive observation tools that he used.

Math Connection: Research Kepler's Laws by using the Internet. Kepler's Third Law relates the distance of planets from the sun and the period of each planet. Use the Internet to find the distance each planet is from the sun and to find each planet's period. Verify Kepler's Third Law.

interNET CONNECTION For more information on the Unit Project, visit: **www.amc.glencoe.com**

Polar Coordinates and Complex Numbers

CHAPTER OBJECTIVES

- **Graph polar equations.** *(Lessons 9-1, 9-2, 9-4)*
- **Convert between polar and rectangular coordinates.** *(Lessons 9-3, 9-4)*
- **Add, subtract, multiply, and divide complex numbers in rectangular and polar forms.** *(Lessons 9-5, 9-7)*
- **Convert between rectangular and polar forms of complex numbers.** *(Lesson 9-6)*
- **Find powers and roots of complex numbers.** *(Lesson 9-8)*

9-1 Polar Coordinates

OBJECTIVES

- Graph points in polar coordinates.
- Graph simple polar equations.
- Determine the distance between two points with polar coordinates.

SURVEYING Before large road construction projects, or even the construction of a new home, take place, a surveyor maps out characteristics of the land. A surveyor uses a device called a *theodolite* to measure angles. The precise locations of various land features are determined using distances and the angles measured with the theodolite. While mapping out a level site, a surveyor identifies a landmark 450 feet away and 30° to the left and another landmark 600 feet away and 50° to the right. What is the distance between the two landmarks? *This problem will be solved in Example 5.*

Recording the position of an object using the distance from a fixed point and an angle made with a fixed ray from that point uses a **polar coordinate system.** When surveyors record the locations of objects using distances and angles, they are using **polar coordinates.**

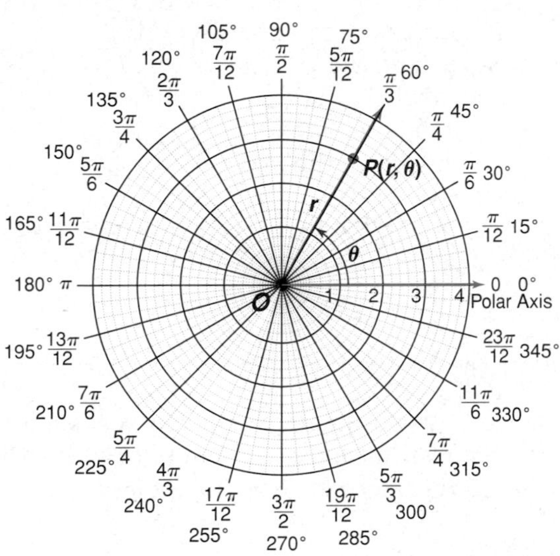

In a polar coordinate system, a fixed point O is called the **pole** or origin. The **polar axis** is usually a horizontal ray directed toward the right from the pole. The location of a point P in the polar coordinate system can be identified by polar coordinates in the form (r, θ). If a ray is drawn from the pole through point P, the distance from the pole to point P is $|r|$. The measure of the angle formed by $\overrightarrow{OP}$ and the polar axis is θ. The angle can be measured in degrees or radians. *This grid is sometimes called the polar plane.*

Consider positive and negative values of r.

Suppose $r > 0$. Then θ is the measure of any angle in standard position that has $\overrightarrow{OP}$ as its terminal side.

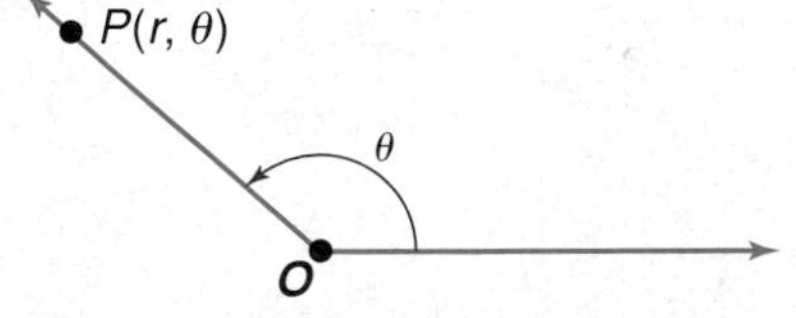

Suppose $r < 0$. Then θ is the measure of any angle that has the ray opposite $\overrightarrow{OP}$ as its terminal side.

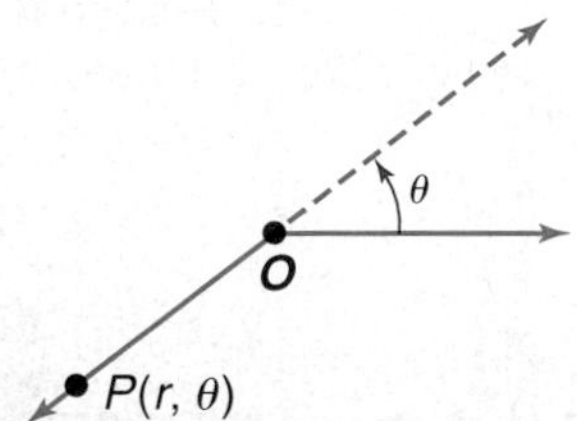

Example 1 **Graph each point.**

a. $P(3, 60°)$

On a polar plane, sketch the terminal side of a $60°$ angle in standard position.

Since r is positive ($r = 3$), find the point on the terminal side of the angle that is 3 units from the pole.

Notice that point P is on the third circle from the pole.

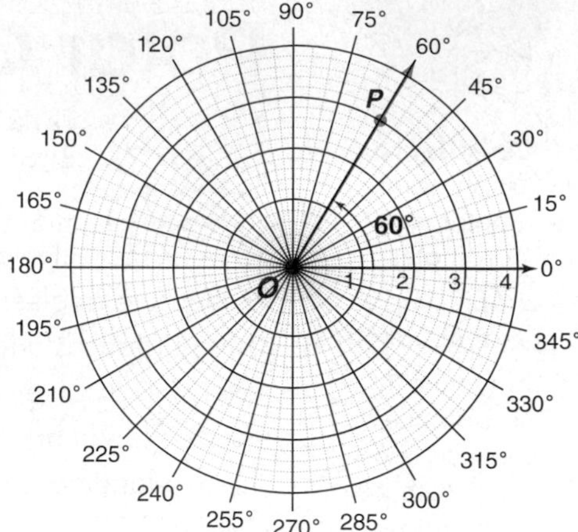

b. $Q\left(-1.5, \frac{7\pi}{6}\right)$

Sketch the terminal side of an angle measuring $\frac{7\pi}{6}$ radians in standard position.

Since r is negative, extend the terminal side of the angle in the opposite direction. Find the point Q that is 1.5 units from the pole along this extended ray.

Notice that point Q is halfway between the first and second circles from the pole.

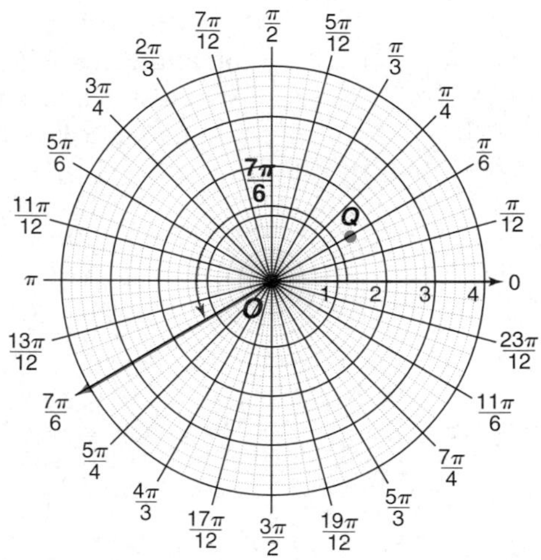

As you have seen, the r-coordinate can be any real value. The angle θ can also be negative. If $\theta > 0$, then θ is measured counterclockwise from the polar axis. If $\theta < 0$, then θ is measured clockwise from the polar axis.

Example 2 **Graph $R(-2, -135°)$.**

Negative angles are measured clockwise. Sketch the terminal side of an angle of $-135°$ in standard position.

Since r is negative, the point $R(-2, -135°)$ is 2 units from the pole along the ray opposite the terminal side that is already drawn.

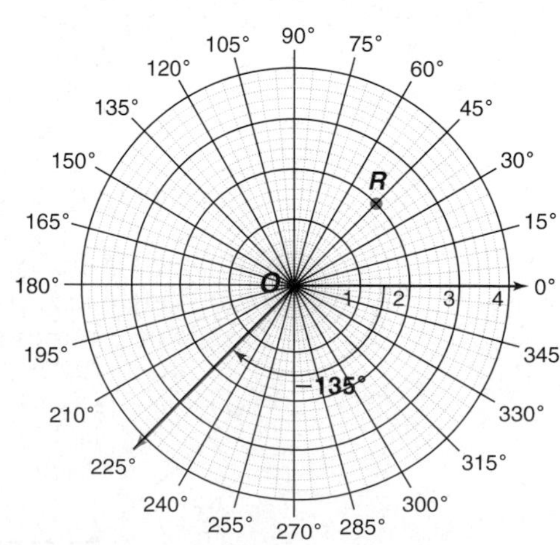

In Example 2, the point $R(-2, -135°)$ lies in the polar plane 2 units from the pole on the terminal side of a 45° angle in standard position. This means that point R could also be represented by the coordinates (2, 45°). In general, the polar coordinates of a point are not unique. Every point can be represented by infinitely many pairs of polar coordinates. This happens because any angle in standard position is coterminal with infinitely many other angles.

If a point has polar coordinates (r, θ), then it also has polar coordinates $(r, \theta + 2\pi)$ in radians, or $(r, \theta + 360°)$ in degrees. In fact, you can add any integer multiple of 2π to θ and find another pair of polar coordinates for the same point. If you use the opposite r-value, the angle will change by π, giving $(-r, \theta + \pi)$ as another ordered pair for the same point. You can then find even more polar coordinates for the same point by adding multiples of 2π to $\theta + \pi$. The following graphs illustrate six of the different ways to name the polar coordinates of the same point.

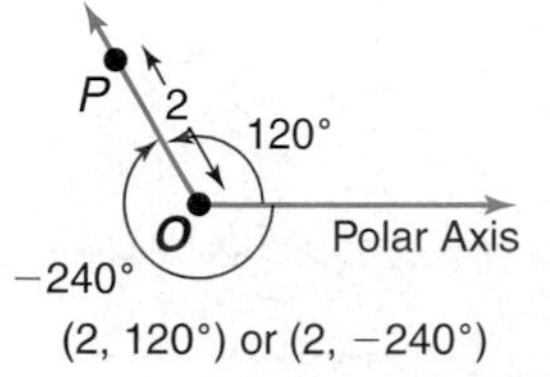

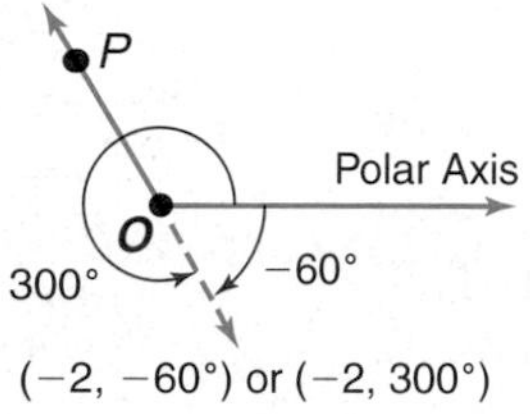

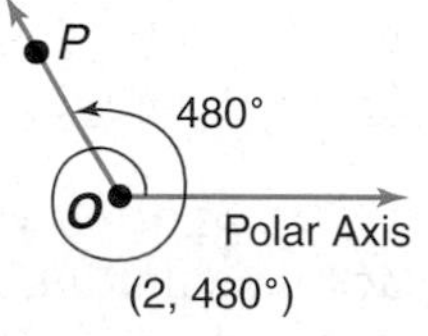

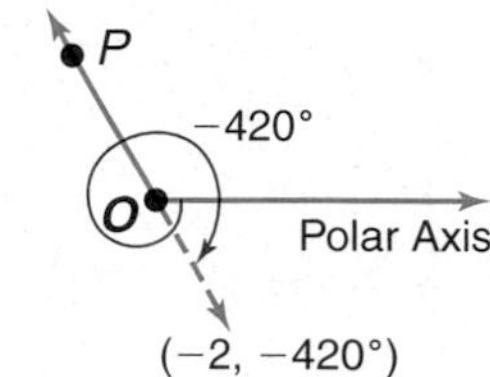

Here is a summary of all the ways to represent a point in polar coordinates.

Multiple Representations of (r, θ)

If a point P has polar coordinates (r, θ), then P can also be represented by polar coordinates $(r, \theta + 2\pi k)$ or $(-r, \theta + (2k + 1)\pi)$, where k is any integer.

In degrees, the representations are $(r, \theta + 360k°)$ and $(-r, \theta + (2k + 1)180°)$. For every angle, there are infinitely many representations.

Example 3 **Name four different pairs of polar coordinates that represent point S on the graph with the restriction that $-360° \leq \theta \leq 360°$.**

One pair of polar coordinates for point S is (3, 150°).

To find another representation, use $(r, \theta + 360k°)$ with $k = -1$.
$(3, 150° + 360(-1)°) = (3, -210°)$

To find additional polar coordinates, use $(-r, \theta + (2k + 1)180°)$.
$(-3, 150° + 180°) = (-3, 330°)$ $k = 0$

To find a fourth pair, use $(-r, \theta + (2k + 1)180°)$ with $k = -1$.
$(-3, 150° + (-1)180°) = (-3, -30°)$

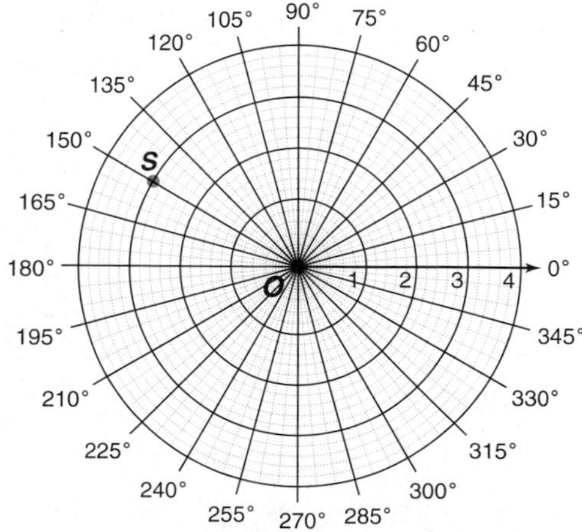

Therefore, (3, 150°), (3, −210°), (−3, 330°), and (−3, −30°) all represent the same point in the polar plane.

An equation expressed in terms of polar coordinates is called a **polar equation.** For example, $r = 2 \sin \theta$ is a polar equation. A **polar graph** is the set of all points whose coordinates (r, θ) satisfy a given polar equation.

You already know how to graph equations in the *Cartesian,* or *rectangular, coordinate system.* Graphs of equations involving constants like $x = 2$ and $y = -3$ are considered basic in the Cartesian coordinate system. Similarly, the polar coordinate system has some basic graphs. Graphs of the polar equations $r = k$ and $\theta = k$, where k is a constant, are considered basic.

Example 4 **Graph each polar equation.**

a. $r = 3$

The solutions to $r = 3$ are the ordered pairs (r, θ) where $r = 3$ and θ is any real number. Some examples are $(3, 0)$, $\left(3, \frac{\pi}{4}\right)$, and $(3, \pi)$. In other words, the graph of this equation is the set of all ordered pairs of the form $(3, \theta)$. Any point that is 3 units from the pole is included in this graph. The graph is the circle centered at the origin with radius 3.

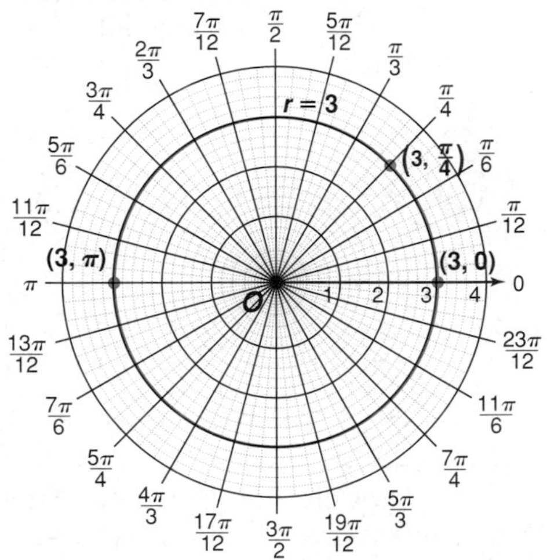

b. $\theta = \frac{3\pi}{4}$

The solutions to this equation are the ordered pairs (r, θ) where $\theta = \frac{3\pi}{4}$ and r is any real number. Some examples are $\left(1, \frac{3\pi}{4}\right)$, $\left(-2, \frac{3\pi}{4}\right)$, and $\left(3, \frac{3\pi}{4}\right)$. In other words, the graph of this equation is the set of all ordered pairs of the form $\left(r, \frac{3\pi}{4}\right)$. The graph is the line that includes the terminal side of the angle $\frac{3\pi}{4}$ in standard position.

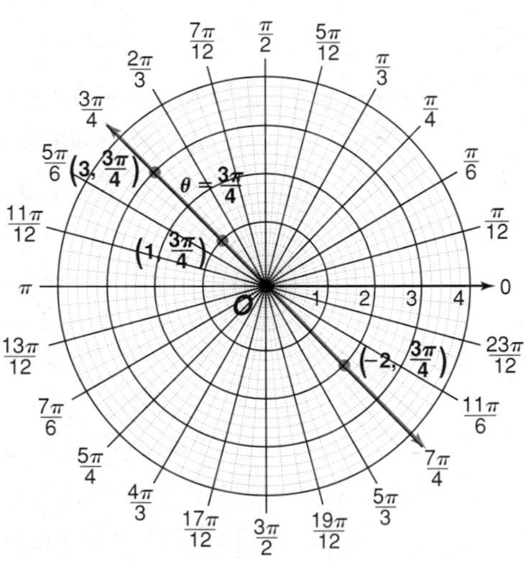

Just as you can find the distance between two points in a rectangular coordinate plane, you can derive a formula for the distance between two points in a polar plane.

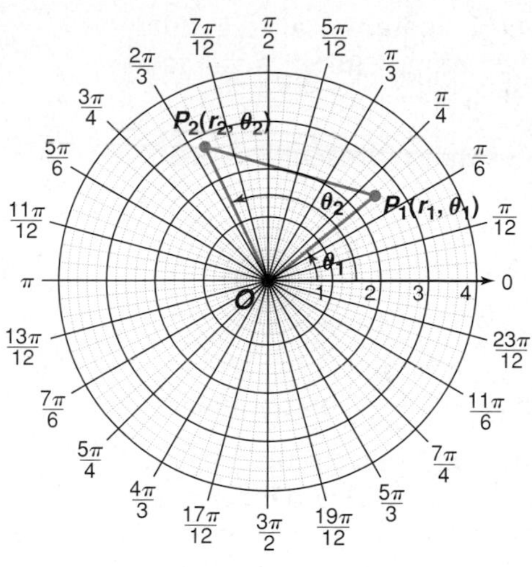

Given two points $P_1(r_1, \theta_1)$ and $P_2(r_2, \theta_2)$ in the polar plane, draw $\triangle P_1OP_2$. $\angle P_1OP_2$ has measure $\theta_2 - \theta_1$. Apply the Law of Cosines to $\triangle P_1OP_2$.

$$(P_1P_2)^2 = (OP_1)^2 + (OP_2)^2 - 2(OP_1)(OP_2)\cos(\theta_2 - \theta_1)$$

Now substitute r_1 for OP_1 and r_2 for OP_2.

$$(P_1P_2)^2 = r_1^{\,2} + r_2^{\,2} - 2r_1r_2\cos(\theta_2 - \theta_1)$$

$$P_1P_2 = \sqrt{r_1^{\,2} + r_2^{\,2} - 2r_1r_2\cos(\theta_2 - \theta_1)}$$

Distance Formula in Polar Plane

If $P_1(r_1, \theta_1)$ and $P_2(r_2, \theta_2)$ are two points in the polar plane, then

$$P_1P_2 = \sqrt{r_1^{\,2} + r_2^{\,2} - 2r_1r_2\cos(\theta_2 - \theta_1)}.$$

Example 5

SURVEYING Refer to the application at the beginning of the lesson. What is the distance between the two landmarks?

Set up a coordinate system so that turning to the left is a positive angle and turning to the right is a negative angle. With this convention, the landmarks are at $L_1(450, 30°)$ and $L_2(600, -50°)$.

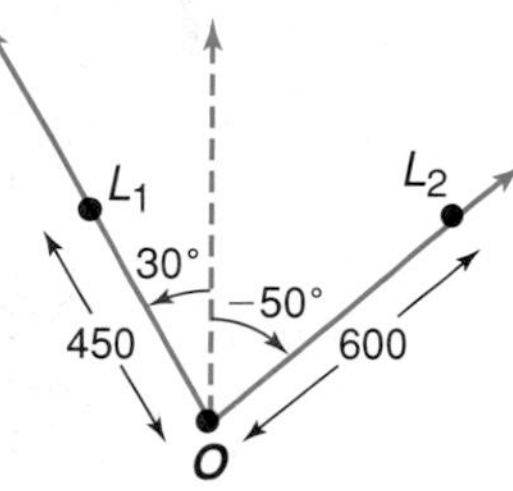

Now use the Polar Distance Formula.

$$P_1P_2 = \sqrt{r_1^{\,2} + r_2^{\,2} - 2r_1r_2\cos(\theta_2 - \theta_1)}$$

$$L_1L_2 = \sqrt{450^2 + 600^2 - 2(450)(600)\cos(-50° - 30°)} \qquad (r_1, \theta_1) = (450, 30°),\ (r_2, \theta_2) = (600, -50°)$$

$$\approx 684.6 \text{ feet}$$

The landmarks are about 685 feet apart.

CHECK FOR UNDERSTANDING

Communicating Mathematics

Read and study the lesson to answer each question.

1. **Explain** why a point in the polar plane cannot be named by a unique ordered pair (r, θ).
2. **Explain** how to graph (r, θ) if $r < 0$ and $\theta > 0$.
3. **Name** two values of θ such that $(-4, \theta)$ represents the same point as $(4, 120°)$.
4. **Explain** why the graph in Example 4a is also the graph of $r = -3$.
5. **Describe** the polar coordinates of the pole.

Guided Practice Graph each point.

6. $A(1, 135°)$ 7. $B\left(2.5, -\frac{\pi}{6}\right)$ 8. $C(-3, -120°)$ 9. $D\left(-2, \frac{13\pi}{6}\right)$

10. Name four different pairs of polar coordinates that represent the point at $\left(-2, \frac{\pi}{6}\right)$.

Graph each polar equation.

11. $r = 1$ 12. $\theta = -\frac{\pi}{3}$ 13. $r = 3.5$

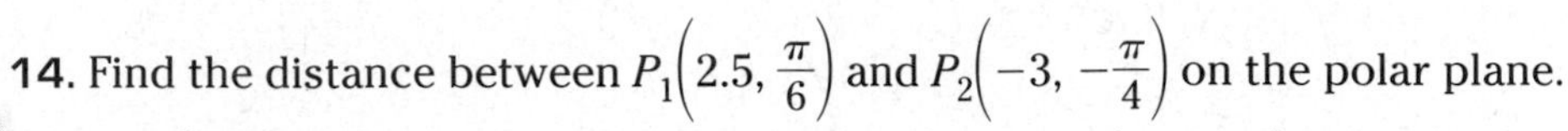

14. Find the distance between $P_1\left(2.5, \frac{\pi}{6}\right)$ and $P_2\left(-3, -\frac{\pi}{4}\right)$ on the polar plane.

15. **Gardening** A lawn sprinkler can cover the part of a circular region determined by the polar inequalities $-30° \leq \theta \leq 210°$ and $0 \leq r \leq 20$, where r is measured in feet.
   a. Sketch a graph of the region that the sprinkler can cover.
   b. Find the area of the region.

EXERCISES

Practice Graph each point.

16. $E(2, 30°)$ 17. $F\left(1, \frac{\pi}{2}\right)$ 18. $G(5, 240°)$ 19. $H\left(\frac{1}{2}, \frac{3\pi}{4}\right)$

20. $J\left(1.5, -\frac{\pi}{4}\right)$ 21. $K\left(\frac{5}{2}, -210°\right)$ 22. $L\left(3, -\frac{\pi}{6}\right)$ 23. $M(2, -90°)$

24. $N(-1, 120°)$ 25. $P\left(-0.5, -\frac{11\pi}{6}\right)$ 26. $Q\left(-2, \frac{25\pi}{3}\right)$ 27. $R\left(-\frac{7}{2}, 1050°\right)$

28. List four pairs of polar coordinates that represent the point S in the graph. Use both radians and degrees.

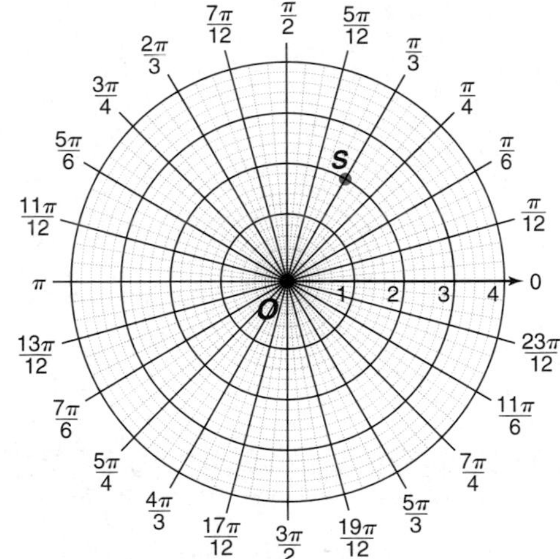

Name four other pairs of polar coordinates for each point.

29. $T(1.5, 180°)$ 30. $U\left(-1, \frac{\pi}{3}\right)$ 31. $V(4, 315°)$

Graph each polar equation.

32. $r = 1.5$ 33. $\theta = \frac{5\pi}{4}$ 34. $r = 2$

35. $\theta = 30°$ 36. $\theta = -150°$ 37. $\theta = -\frac{\pi}{4}$

38. $\theta = 840°$ 39. $r = 0$ 40. $r = -1$

41. Write a polar equation for the circle centered at the origin with radius $\sqrt{2}$.

Find the distance between the points with the given polar coordinates.

42. $P_1(4, 170°)$ and $P_2(6, 105°)$

43. $P_1\left(1, \frac{\pi}{6}\right)$ and $P_2\left(5, \frac{3\pi}{4}\right)$

44. $P_1\left(-2.5, \frac{\pi}{8}\right)$ and $P_2\left(-1.75, -\frac{2\pi}{5}\right)$

45. $P_1(1.3, -47°)$ and $P_2(-3.6, -62°)$

46. Find an ordered pair of polar coordinates to represent the point whose rectangular coordinates are $(-3, 4)$.

Applications and Problem Solving

47. **Web Page Design** When designing websites with circular graphics, it is often convenient to use polar coordinates, sometimes called "pizza coordinates" in this context. If the origin is at the center of the screen, what are the polar equations of the lines that cut the region into the six congruent slices shown?

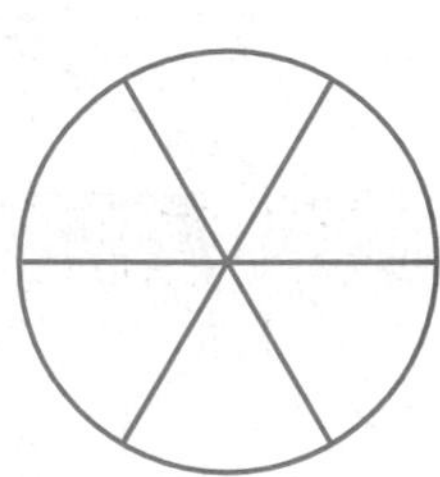

48. **Critical Thinking** Prove that if $P_1(r_1, \theta)$ and $P_2(r_2, \theta)$ are two points in the polar plane, then the distance formula reduces to $P_1P_2 = |r_1 - r_2|$.

49. **Sailing** A graph of the maximum speed of a sailboat versus the angle of the wind is called a "performance curve" or just a "polar." The graph at the right is the polar for a particular boat when the wind speed is 20 knots. θ represents the angle of the wind in degrees, and r is the maximum speed of the boat in knots.

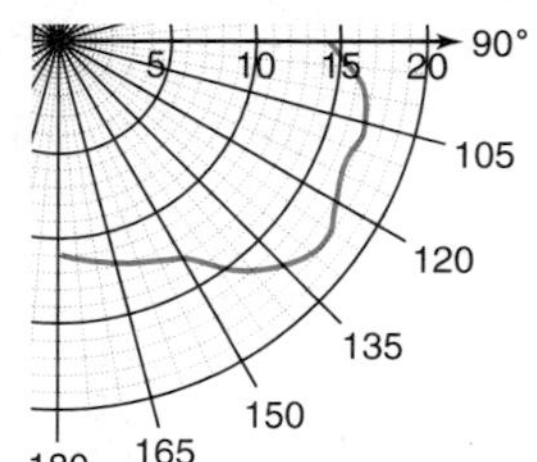

a. What is the maximum speed when $\theta = 120°$?

b. What is the maximum speed when $\theta = 150°$?

50. **Acoustics** Polar coordinates can be used to model a concert amphitheater. Suppose the performer is placed at the pole and faces the direction of the polar axis. The seats have been built to occupy the region with $-\frac{\pi}{3} \leq \theta \leq \frac{\pi}{3}$ and $0.25 \leq r \leq 3$, where r is measured in hundreds of feet.

a. Sketch this region in the polar plane.

b. How many seats are there if each person has 6 square feet of space?

51. **Critical Thinking** Explain why the order of the points used in the distance formula is irrelevant. That is, why can you choose either point to be P_1 and the other to be P_2?

U.S. Navy Blue Angels

52. **Aviation** Two jets at the same altitude are being tracked on an air traffic controller's radar screen. The coordinates of the planes are $(5, 310°)$ and $(6, 345°)$, with r measured in miles.
    a. Sketch a graph of this situation.
    b. If regulations prohibit jets from passing within three miles of each other, are these planes in violation? Explain.

Mixed Review

53. **Transportation** Two docks are directly across a river from each other. A boat that can travel at a speed of 8 miles per hour in still water is attempting to cross directly from one dock to the other. The current of the river is 3 miles per hour. At what angle should the captain head? *(Lesson 8-5)*

54. Find the inner product $\langle 3, -2, 4\rangle \cdot \langle 1, -4, 0\rangle$. Then state whether the vectors are perpendicular. Write *yes* or *no*. *(Lesson 8-4)*

55. Find the distance from the line with equation $y = 9x - 3$ to the point at $(-3, 2)$. *(Lesson 7-7)*

56. Simplify the expression $\dfrac{1 - \sin^2 \alpha}{\sin^2 \alpha}$. *(Lesson 7-1)*

57. Find Arccos $\dfrac{\sqrt{3}}{2}$. *(Lesson 6-8)*

58. State the amplitude and period for the function $y = 5 \cos 4\theta$. *(Lesson 6-4)*

59. Determine the number of possible solutions and, if a solution exists, solve $\triangle ABC$ if $A = 30°$, $b = 18.6$, and $a = 9.3$. Round to the nearest tenth. *(Lesson 5-7)*

60. Find the number of possible positive real zeros and the number of possible negative real zeros of the function $f(x) = x^3 - 4x^2 + 4x - 1$. Then determine the rational zeros. *(Lesson 4-4)*

61. Determine the slant asymptote of $f(x) = \dfrac{x^2 + 2x - 3}{x + 5}$. *(Lesson 3-7)*

62. Determine whether the graph of $f(x) = x^4 + 3x^2 + 2$ is symmetric with respect to the x-axis, the y-axis, both, or neither. *(Lesson 3-1)*

63. Find the value of the determinant $\begin{vmatrix} -2 & 4 & -1 \\ 1 & -1 & 0 \\ -3 & 4 & 5 \end{vmatrix}$. *(Lesson 2-5)*

64. Given that x is an integer, state the relation representing $y = 11 - x$ and $-3 \le x \le 0$ by listing a set of ordered pairs. Then state whether this relation is a function. *(Lesson 1-1)*

65. **SAT/ACT Practice** The circumference of the circle is 50π. What is the length of the diagonal $\overline{AB}$ of the square inscribed in the circle?

A $12\frac{1}{2}$ 
B $10\sqrt{2}$
C 25
D $25\sqrt{2}$
E 50

A

B

Extra Practice See p. A42.

Graphs of Polar Equations

OBJECTIVE
- Graph polar equations.

AUDIO TECHNOLOGY One way to describe the ability of a microphone to pick up sounds from different directions is to examine its polar pattern. A polar coordinate system is set up with the microphone at the origin. θ is used to locate a source of sound that moves in a horizontal circle around the microphone, and r measures the amplitude of the signal that the microphone detects. The polar graph of r as a function of θ is called the *polar pattern* of the microphone. A cardioid microphone is a microphone whose polar pattern is shaped like the graph of the equation $r = 2.5 + 2.5 \cos \theta$. A graph of this polar pattern provides information about the microphone. *A problem related to this will be solved in Example 3.*

In Lesson 9-1, you learned to graph polar equations of the form $r = k$ and $\theta = k$, where k is a constant. In this lesson, you will learn how to graph more complicated types of polar equations. When you graph any type of polar equation for the first time, you can create a table of values for r and θ. Then you can use the table to plot points in the polar plane.

Example 1 **Graph $r = \sin \theta$.**

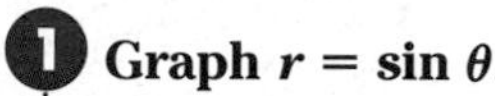

Make a table of values. Round the values of r to the nearest tenth. Graph the ordered pairs and connect them with a smooth curve.

θ	$\sin \theta$	(r, θ)
0°	0	(0, 0°)
30°	0.5	(0.5, 30°)
45°	0.7	(0.7, 45°)
60°	0.9	(0.9, 60°)
90°	1	(1, 90°)
120°	0.9	(0.9, 120°)
135°	0.7	(0.7, 135°)
150°	0.5	(0.5, 150°)
180°	0	(0, 180°)
210°	−0.5	(−0.5, 210°)
225°	−0.7	(−0.7, 225°)
240°	−0.9	(−0.9, 240°)
270°	−1	(−1, 270°)
300°	−0.9	(−0.9, 300°)
315°	−0.7	(−0.7, 315°)
330°	−0.5	(−0.5, 330°)
360°	0	(0, 360°)

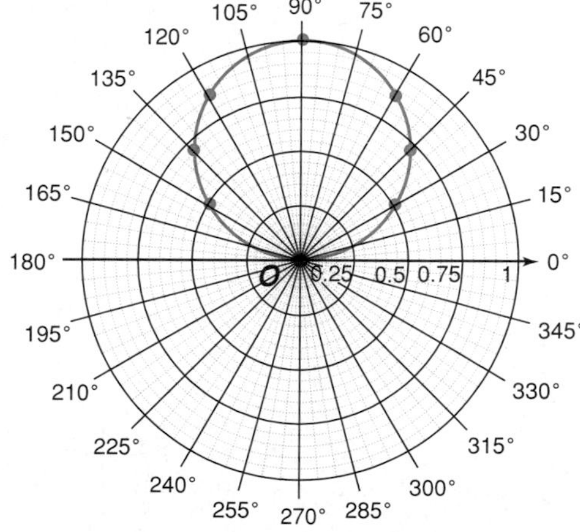

Notice that the ordered pairs obtained when $180° \leq \theta \leq 360°$ represent the same points as the ordered pairs obtained when $0° \leq \theta \leq 180°$.

The graph of $r = \cos\theta$ is shown at the right. Notice that this graph and the graph in Example 1 are circles with a diameter of 1 unit. Both pass through the origin. As with the families of graphs you studied in Chapter 3, you can alter the position and shape of a polar graph by multiplying the function by a number or by adding to it. You can also multiply θ by a number or add a number to it in order to alter the graph. However, the changes in the graphs of polar equations can be quite different from those you studied in Chapter 3.

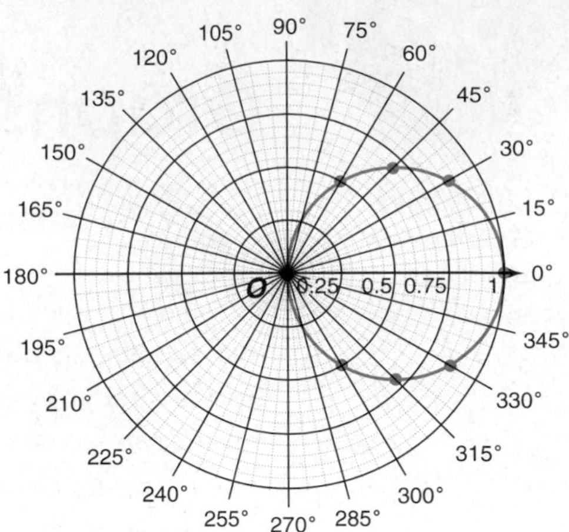

Examples 2 **Graph each polar equation.**

a. $r = 3 - 5\cos\theta$

θ	$3 - 5\cos\theta$	(r, θ)
0	−2	(−2, 0)
$\frac{\pi}{6}$	−1.3	$\left(-1.3, \frac{\pi}{6}\right)$
$\frac{\pi}{3}$	0.5	$\left(0.5, \frac{\pi}{3}\right)$
$\frac{\pi}{2}$	3	$\left(3, \frac{\pi}{2}\right)$
$\frac{2\pi}{3}$	5.5	$\left(5.5, \frac{2\pi}{3}\right)$
$\frac{5\pi}{6}$	7.3	$\left(7.3, \frac{5\pi}{6}\right)$
π	8	$(8, \pi)$
$\frac{7\pi}{6}$	7.3	$\left(7.3, \frac{7\pi}{6}\right)$
$\frac{4\pi}{3}$	5.5	$\left(5.5, \frac{4\pi}{3}\right)$
$\frac{3\pi}{2}$	3	$\left(3, \frac{3\pi}{2}\right)$
$\frac{5\pi}{3}$	0.5	$\left(0.5, \frac{5\pi}{3}\right)$
$\frac{11\pi}{6}$	−1.3	$\left(-1.3, \frac{11\pi}{6}\right)$
2π	−2	$(-2, 2\pi)$

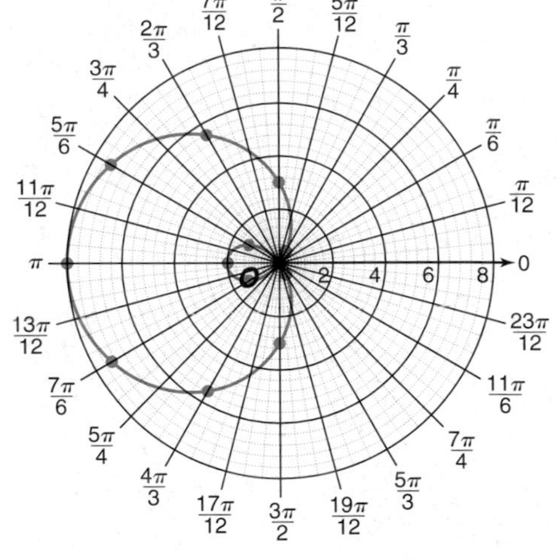

This type of curve is called a **limaçon.** Some limaçons have inner loops like this one. Other limaçons come to a point, have a dimple, or just curve outward.

b. $r = 3 \sin 2\theta$

θ	$3 \sin 2\theta$	(r, θ)
0	0	(0, 0)
$\frac{\pi}{6}$	2.6	$\left(2.6, \frac{\pi}{6}\right)$
$\frac{\pi}{4}$	3	$\left(3, \frac{\pi}{4}\right)$
$\frac{\pi}{3}$	2.6	$\left(2.6, \frac{\pi}{3}\right)$
$\frac{\pi}{2}$	0	$\left(0, \frac{\pi}{2}\right)$
$\frac{2\pi}{3}$	−2.6	$\left(-2.6, \frac{2\pi}{3}\right)$
$\frac{3\pi}{4}$	−3	$\left(-3, \frac{3\pi}{4}\right)$
$\frac{5\pi}{6}$	−2.6	$\left(-2.6, \frac{5\pi}{6}\right)$
π	0	$(0. \pi)$
$\frac{7\pi}{6}$	2.6	$\left(2.6, \frac{7\pi}{6}\right)$
$\frac{5\pi}{4}$	3	$\left(3, \frac{5\pi}{4}\right)$
$\frac{4\pi}{3}$	2.6	$\left(2.6, \frac{4\pi}{3}\right)$

θ	$3 \sin 2\theta$	(r, θ)
$\frac{3\pi}{2}$	0	$\left(0, \frac{3\pi}{2}\right)$
$\frac{5\pi}{3}$	−2.6	$\left(-2.6, \frac{5\pi}{3}\right)$
$\frac{7\pi}{4}$	−3	$\left(-3, \frac{7\pi}{4}\right)$
$\frac{11\pi}{6}$	−2.6	$\left(-2.6, \frac{11\pi}{6}\right)$

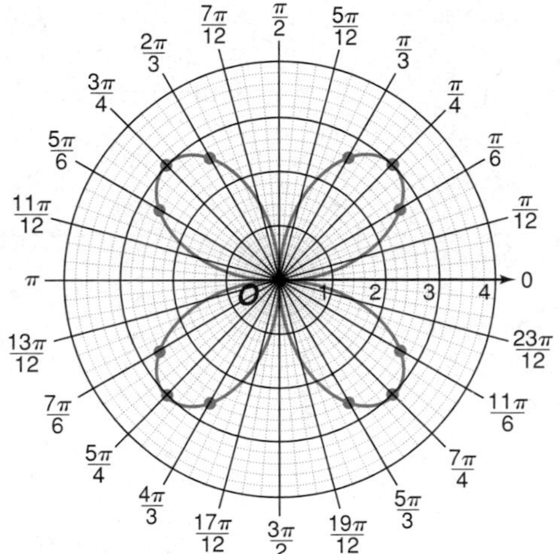

This type of curve is called a **rose.**

3 AUDIO TECHNOLOGY **Refer to the application at the beginning of the lesson. The polar pattern of a microphone can be modeled by the polar equation $r = 2.5 + 2.5 \cos \theta$.**

a. Sketch a graph of the polar pattern.

b. Describe what the polar pattern tells you about the microphone.

a.

θ	$2.5 + 2.5 \cos \theta$	(r, θ)
0°	5	(5, 0°)
30°	4.7	(4.7, 30°)
60°	3.8	(3.8, 60°)
90°	2.5	(2.5, 90°)
120°	1.3	(1.3, 120°)
150°	0.3	(0.3, 150°)
180°	0	(0, 180°)
210°	0.3	(0.3, 210°)
240°	1.3	(1.3, 240°)
270°	2.5	(2.5, 270°)
300°	3.8	(3.8, 300°)
330°	4.7	(4.7, 330°)

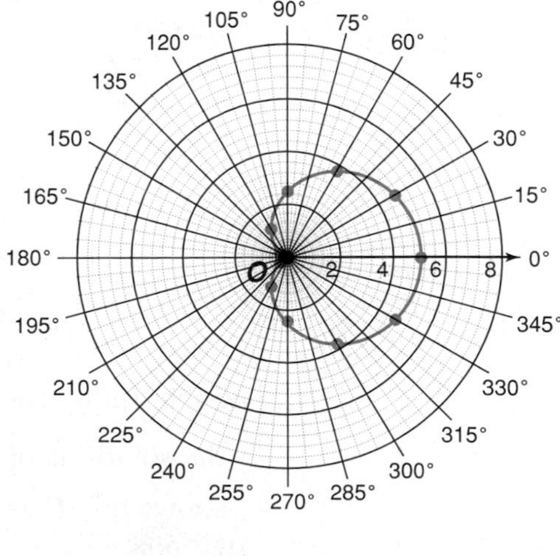

(continued on the next page)

This type of curve is called a **cardioid.** A cardioid is a special type of limaçon.

b. The polar pattern indicates that the microphone will pick up only very loud sounds from behind it. It will detect much softer sounds from in front.

Limaçons, roses, and cardioids are examples of **classical curves.** The classical curves are summarized in the chart below.

Classical Curves					
Curve	rose	lemniscate *(pronounced lehm NIHS kuht)*	limaçon *(pronounced lee muh SOHN)*	cardioid *(pronounced KARD ee oyd)*	spiral of Archimedes *(pronounced ar kih MEED eez)*
Polar Equation	$r = a \cos n\theta$ $r = a \sin n\theta$ *n is a positive integer.*	$r^2 = a^2 \cos 2\theta$ $r^2 = a^2 \sin 2\theta$	$r = a + b \cos \theta$ $r = a + b \sin \theta$	$r = a + a \cos \theta$ $r = a + a \sin \theta$	$r = a\theta$ (θ in radians)
General Graph					

It is possible to graph more than one polar equation at a time on a polar plane. However, the points where the graphs intersect do not always represent common solutions to the equations, since every point can be represented by infinitely many polar coordinates.

Example 4 **Graph the system of polar equations. Solve the system using algebra and trigonometry and compare the solutions to those on your graph.**

$$\boldsymbol{r = 3 - 3 \sin \theta}$$
$$\boldsymbol{r = 4 - \sin \theta}$$

Graphing Calculator Tip

A graphing calculator can graph polar equations in its polar mode. You can use the simultaneous graphing mode to help determine if the graphs pass through the same point at the same time.

To solve the system of equations, substitute $3 - 3 \sin \theta$ for r in the second equation.

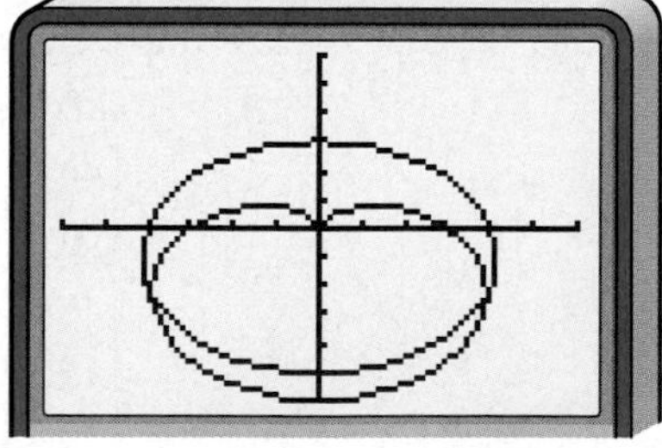

[−6, 6] scl:1 by [−6, 6] scl:1

$$3 - 3 \sin \theta = 4 - \sin \theta$$
$$-2 \sin \theta = 1$$
$$\sin \theta = -\frac{1}{2}$$
$$\theta = \frac{7\pi}{6} \text{ or } \theta = \frac{11\pi}{6}$$

Substituting each angle into either of the original equations gives $r = 4.5$, so the solutions of the system are $\left(4.5, \frac{7\pi}{6}\right)$ and $\left(4.5, \frac{11\pi}{6}\right)$. Tracing the curves shows that these solutions correspond with the intersection points of the graphs.

CHECK FOR UNDERSTANDING

Communicating Mathematics

Read and study the lesson to answer each question.

1. **Write** a polar equation whose graph is a rose.
2. **Determine** the maximum value of r in the equation $r = 3 + 5 \sin \theta$. What is the minimum value of r?
3. **State** the reason that algebra and trigonometry do not always find all the points of intersection of the graphs of polar equations.
4. **You Decide** Linh and Barbara were working on their homework together. Linh said that she thought that when graphing polar equations, you only need to generate points for which $0 \leq \theta \leq \pi$ because other values of θ would just generate the same points. Barbara said she thought she remembered an example from class where values of θ ranging from 0 to 4π had to be considered. Who is correct? Explain.

Guided Practice

Graph each polar equation. Identify the type of curve each represents.

5. $r = 1 + \sin \theta$
6. $r = 2 - 3 \sin \theta$
7. $r = \cos 2\theta$
8. $r = 1.5\theta$
9. Graph the system of polar equations $r = 2 \sin \theta$ and $r = 2 \cos 2\theta$. Solve the system using algebra and trigonometry. Assume $0 \leq \theta < 2\pi$.
10. **Biology** The chambered nautilus is a mollusk whose shell can be modeled by the polar equation $r = 2\theta$.
    a. Graph this equation for $0 \leq \theta \leq 2\pi$.
    b. Determine an approximate interval for θ that would result in a graph that models the chambered nautilus shown in the photo.

EXERCISES

Practice

Graph each polar equation. Identify the type of curve each represents.

11. $r = -3 \sin \theta$
12. $r = 3 + 3 \cos \theta$
13. $r = 3\theta$
14. $r^2 = 4 \cos 2\theta$
15. $r = 2 \sin 3\theta$
16. $r = -2 \sin 3\theta$
17. $r = \frac{5}{2}\theta$
18. $r = -5 + 3 \cos \theta$
19. $r = -2 - 2 \sin \theta$
20. $r^2 = 9 \sin 2\theta$
21. $r = \sin 4\theta$
22. $r = 2 + 2 \cos \theta$
23. Write an equation for a rose with 3 petals.
24. What is an equation for a spiral of Archimedes that passes through $A\left(\frac{\pi}{4}, \frac{\pi}{2}\right)$?

Graph each system of polar equations. Solve the system using algebra and trigonometry. Assume $0 \leq \theta < 2\pi$.

25. $r = 3$
    $r = 2 + \cos \theta$
26. $r = 1 + \cos \theta$
    $r = 1 - \cos \theta$
27. $r = 2 \sin \theta$
    $r = 2 \sin 2\theta$

Graphing Calculator

Graph each system using a graphing calculator. Find the points of intersection. Round coordinates to the nearest tenth. Assume $0 \leq \theta < 2\pi$.

28. $r = 1$
$r = 2 \cos 2\theta$

29. $r = 3 + 3 \sin \theta$
$r = 2$

30. $r = 2 + 2 \cos \theta$
$r = 3 + \sin \theta$

Applications and Problem Solving

31. Textiles Patterns in fabric can often be created by modifying a mathematical graph. The pattern at the right can be modeled by a lemniscate.

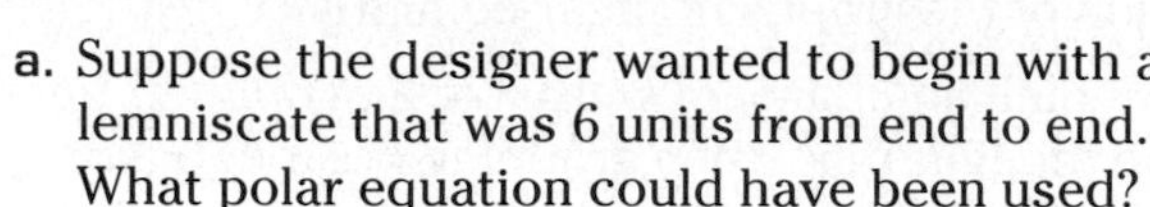

a. Suppose the designer wanted to begin with a lemniscate that was 6 units from end to end. What polar equation could have been used?

b. What polar equation could have been used to generate a lemniscate that was 8 units from end to end?

32. Audio Technology Refer to the application at the beginning of the lesson and Example 3. Another microphone has a polar pattern that can be modeled by the polar equation $r = 3 + 2 \cos \theta$. Graph this polar pattern and compare it to the pattern of the cardioid microphone.

33. Music The curled part at the end of a violin is called the *scroll*. The scroll in the picture appears to curl around twice. For what interval of θ-values will the graph of $r = \theta$ model this violin scroll?

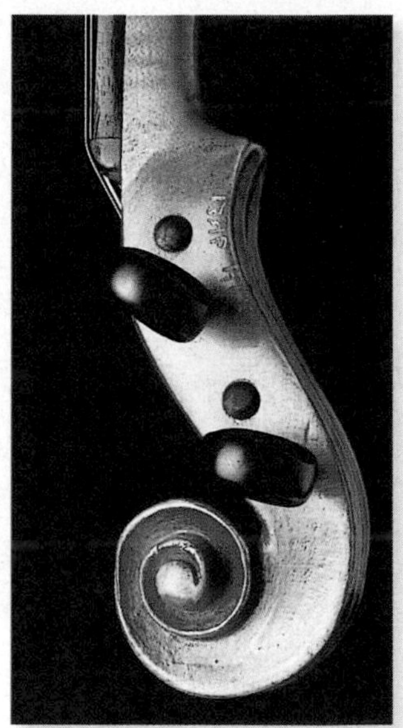

34. Critical Thinking

a. Graph $r = \cos \frac{\theta}{n}$ for $n = 2, 4, 6,$ and 8. Predict the shape of the graph of $r = \cos \frac{\theta}{10}$.

b. Graph $r = \cos \frac{\theta}{n}$ for $n = 3, 5, 7,$ and 9. Predict the shape of the graph of $r = \cos \frac{\theta}{11}$.

35. Communication Suppose you want to use your computer to make a heart-shaped card for a friend. Write a polar equation that you could use to generate the shape of a heart. Make sure the graph of your equation points in the right direction.

36. Critical Thinking The general form for a limaçon is $r = a + b \cos \theta$ or $r = a + b \sin \theta$.

a. When will a limaçon have an inner loop?

b. When will it have a dimple?

c. When will it have neither an inner loop nor a dimple? (This is called a *convex limaçon.*)

37. Critical Thinking Describe the transformation necessary to obtain the graph of each equation from the graph of the polar function $r = f(\theta)$.

a. $r = f(\theta - \alpha)$ (α is a constant.)
b. $r = f(-\theta)$
c. $r = -f(\theta)$
d. $r = cf(\theta)$ (c is a constant, $c > 0$.)

Mixed Review

38. Find four other pairs of polar coordinates that represent the same point as (4, 45°). *(Lesson 9-1)*

39. Find the cross product of $\vec{v}\langle 2, 3, 0\rangle$ and $\vec{w}\langle -1, 2, 4\rangle$. Verify that the resulting vector is perpendicular to $\vec{v}$ and $\vec{w}$. *(Lesson 8-4)*

40. Use a ruler and protractor to determine the magnitude (in centimeters) and direction of the vector shown at the right. *(Lesson 8-1)*

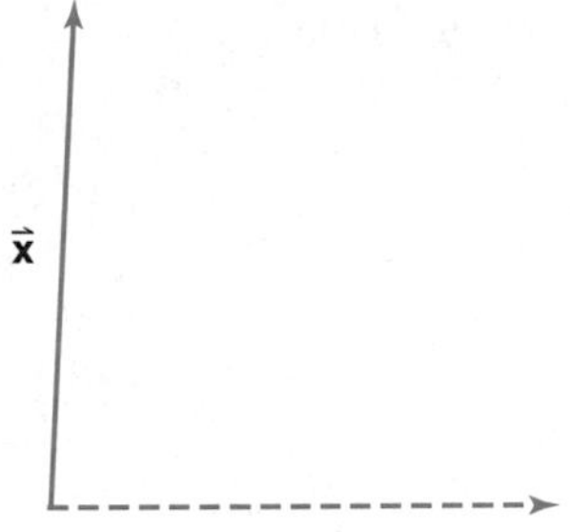

41. Verify that $\dfrac{\sin^2 x}{\cos^4 x + \cos^2 x \sin^2 x} = \tan^2 x$ is an identity. *(Lesson 7-2)*

42. Solve $\triangle ABC$ if $A = 21°15'$, $B = 49°40'$, and $c = 28.9$. Round angle measures to the nearest minute and side measures to the nearest tenth. *(Lesson 5-6)*

43. **Travel** Adita is trying to decide where to go on vacation. He prefers not to fly, so he wants to take a bus or a train. The table below shows the round-trip fares for trips from his home in Kansas City, Missouri to various cities. Represent this data with a matrix. *(Lesson 2-3)*

Round–Trip Fares to Kansas City

	New York	Los Angeles	Miami
Bus	\$240	\$199	\$260
Train	\$254	\$322	\$426

44. **SAT Practice Quantitative Comparison**
A if the quantity in Column A is greater
B if the quantity in Column B is greater
C if the two quantities are equal
D if the relationship cannot be determined from the information given

Column A	Column B
$\dfrac{\frac{1}{8} + \frac{6}{4}}{\frac{3}{16}}$	$\dfrac{\frac{2}{12} + \frac{1}{3}}{\frac{12}{16}}$

Polar and Rectangular Coordinates

OBJECTIVE

- Convert between polar and rectangular coordinates.

PHYSIOLOGY A laboratory has designed a voice articulation program that synthesizes speech by controlling the speech articulators: the jaw, tongue, lips, and so on. To create a mathematical model that is manageable, each of these speech articulators is identified by a point.

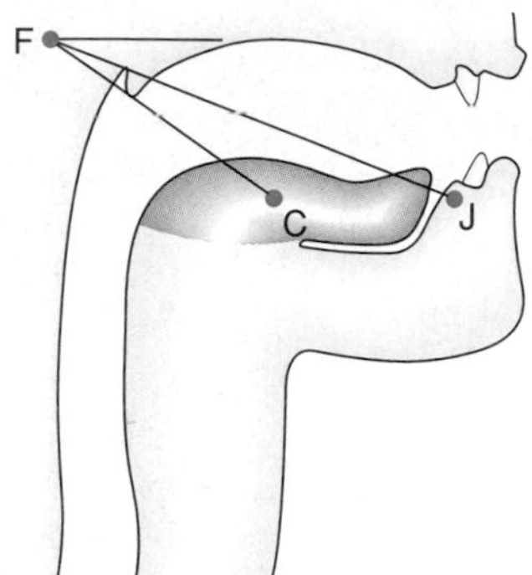

J = the edge of the jaw
C = the center of the tongue
F = a fixed point about which the jaw rotates

The positions of these points are given in polar coordinates with F as the pole and the horizontal as the polar axis. Changing these positions alters the sounds that are synthesized. *A problem related to this will be solved in Example 2.*

For some real-world phenomena, it is useful to be able to convert between polar coordinates and rectangular coordinates.

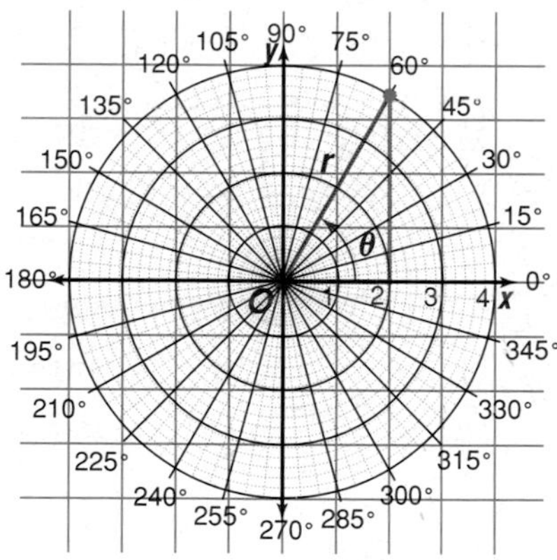

Suppose a rectangular coordinate system is superimposed on a polar coordinate system so that the origins coincide and the x-axis aligns with the polar axis. Let P be any point in the plane.

Polar coordinates: $P(r, \theta)$
Rectangular coordinates: $P(x, y)$

Trigonometric functions can be used to convert polar coordinates to rectangular coordinates.

Converting Polar Coordinates to Rectangular Coordinates

The rectangular coordinates (x, y) of a point named by the polar coordinates (r, θ) can be found by using the following formulas.

$$x = r \cos \theta$$
$$y = r \sin \theta$$

You will derive these formulas in Exercise 42.

Examples **1** **Find the rectangular coordinates of each point.**

a. $P\left(5, \frac{\pi}{3}\right)$

For $P\left(5, \frac{\pi}{3}\right)$, $r = 5$ and $\theta = \frac{\pi}{3}$.

$$x = r\cos\theta \qquad\qquad y = r\sin\theta$$
$$= 5\cos\frac{\pi}{3} \qquad\qquad = 5\sin\frac{\pi}{3}$$
$$= 5\left(\frac{1}{2}\right) \text{ or } \frac{5}{2} \qquad\qquad = 5\left(\frac{\sqrt{3}}{2}\right) \text{ or } \frac{5\sqrt{3}}{2}$$

The rectangular coordinates of P are $\left(\frac{5}{2}, \frac{5\sqrt{3}}{2}\right)$ or (2.5, 4.33) to the nearest hundredth.

b. $Q(-13, -70°)$

For $Q(-13, -70°)$, $r = -13$ and $\theta = -70°$.

$$x = r\cos\theta \qquad\qquad y = r\sin\theta$$
$$= -13\cos(-70°) \qquad\qquad = -13\sin(-70°)$$
$$\approx -13(0.34202) \qquad\qquad \approx -13(-0.93969)$$
$$\approx -4.45 \qquad\qquad \approx 12.22$$

The rectangular coordinates of Q are approximately $(-4.45, 12.22)$.

2 **PHYSIOLOGY** **Refer to the application at the beginning of the lesson. Suppose the computer model assigns polar coordinates (7.5, 330°) to point J and (4.5, 310°) to point C in order to create a particular sound. Each unit represents a centimeter. Is the center of the tongue above or below the edge of the jaw? by how far?**

Real World Application

Find the rectangular coordinates of each point.

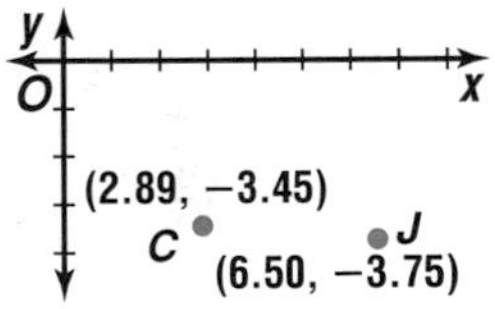

For $J(7.5, 330°)$, $r = 7.5$ and $\theta = 330°$.

$$x = r\cos\theta \qquad\qquad y = r\sin\theta$$
$$= 7.5\cos 330° \qquad\qquad = 7.5\sin 330°$$
$$\approx 6.50 \qquad\qquad = -3.75$$

$J(7.5, 330°) \rightarrow J(6.50, -3.75)$

For $C(4.5, 310°)$, $r = 4.5$ and $\theta = 310°$.

$$x = r\cos\theta \qquad\qquad y = r\sin\theta$$
$$= 4.5\cos 310° \qquad\qquad = 4.5\sin 310°$$
$$\approx 2.89 \qquad\qquad \approx -3.45$$

$C(4.5, 310°) \rightarrow C(2.89, -3.45)$

Since $-3.45 > -3.75$, the center of the tongue is above the edge of the jaw when this sound is made. Subtracting the y-coordinates, we see that the center of the tongue is about 0.3 centimeter, or 3 millimeters, higher than the edge of the jaw.

If a point is named by the rectangular coordinates (x, y), you can find the corresponding polar coordinates by using the Pythagorean Theorem and the Arctangent function. Since the Arctangent function only determines angles in the first and fourth quadrants, you must add π radians to the value of θ for points with coordinates (x, y) that lie in the second or third quadrants.

When x is zero, $\theta = \pm\frac{\pi}{2}$. Why?

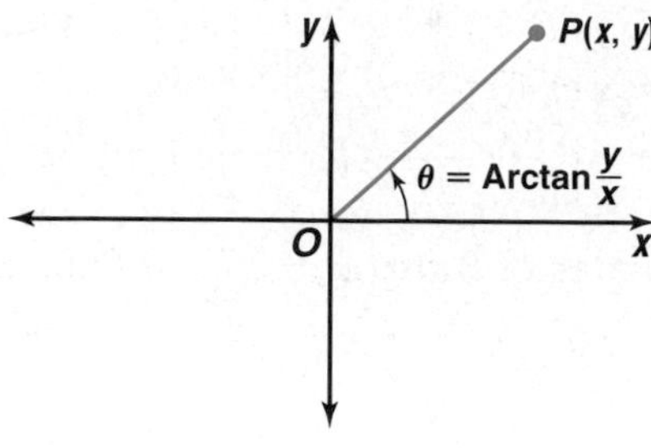

When $x > 0$, $\theta = \text{Arctan}\ \frac{y}{x}$.

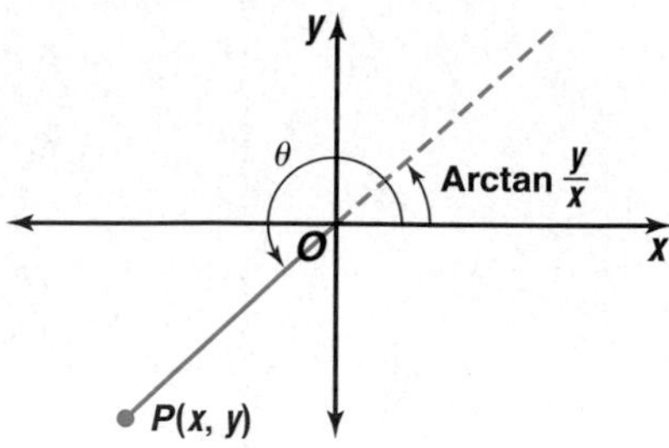

When $x < 0$, $\theta = \text{Arctan}\ \frac{y}{x} + \pi$.

Converting Rectangular Coordinates to Polar Coordinates

The polar coordinates (r, θ) of a point named by the rectangular coordinates (x, y) can be found by the following formulas.

$$r = \sqrt{x^2 + y^2}$$

$$\theta = \text{Arctan}\ \frac{y}{x}, \text{ when } x > 0$$

$$\theta = \text{Arctan}\ \frac{y}{x} + \pi, \text{ when } x < 0$$

Example 3 **Find the polar coordinates of $R(-8, -12)$.**

For $R(-8, -12)$, $x = -8$ and $y = -12$.

$$\begin{aligned} r &= \sqrt{x^2 + y^2} \\ &= \sqrt{(-8)^2 + (-12)^2} \\ &= \sqrt{208} \\ &\approx 14.42 \end{aligned}$$

$$\begin{aligned} \theta &= \text{Arctan}\ \frac{y}{x} + \pi \quad\quad x < 0 \\ &= \text{Arctan}\ \frac{-12}{-8} + \pi \\ &= \text{Arctan}\ \frac{3}{2} + \pi \\ &\approx 4.12 \end{aligned}$$

The polar coordinates of R are approximately (14.42, 4.12). *Other polar coordinates can also represent this point.*

The conversion equations can also be used to convert equations from one coordinate system to the other.

Examples 4 **Write the polar equation $r = 6 \cos \theta$ in rectangular form.**

$$\begin{aligned} r &= 6 \cos \theta \\ r^2 &= 6r \cos \theta && \textit{Multiply each side by } r. \\ x^2 + y^2 &= 6x && r^2 = x^2 + y^2,\ r\cos\theta = x \end{aligned}$$

5 **Write the rectangular equation $(x - 3)^2 + y^2 = 9$ in polar form.**

$$
\begin{aligned}
(x - 3)^2 + y^2 &= 9 \\
(r\cos\theta - 3)^2 + (r\sin\theta)^2 &= 9 && x = r\cos\theta,\ y = r\sin\theta \\
r^2\cos^2\theta - 6r\cos\theta + 9 + r^2\sin^2\theta &= 9 && \textit{Multiply.} \\
r^2\cos^2\theta - 6r\cos\theta + r^2\sin^2\theta &= 0 && \textit{Subtract 9 from each side.} \\
r^2\cos^2\theta + r^2\sin^2\theta &= 6r\cos\theta && \textit{Isolate squared terms.} \\
r^2(\cos^2\theta + \sin^2\theta) &= 6r\cos\theta && \textit{Factor.} \\
r^2(1) &= 6r\cos\theta && \textit{Pythagorean Identity} \\
r^2 &= 6r\cos\theta \\
r &= 6\cos\theta && \textit{Simplify.}
\end{aligned}
$$

CHECK FOR UNDERSTANDING

Communicating Mathematics

Read and study the lesson to answer each question.

1. **Write** the polar coordinates of the point in the graph at the right.

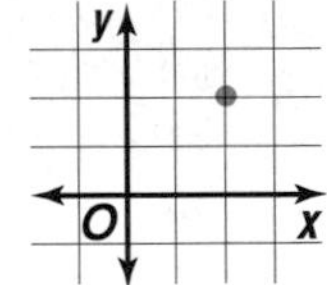

2. **Explain** why you have to consider what quadrant a point lies in when converting from rectangular coordinates to polar coordinates.
3. **Determine** the polar equation for $x = 2$.
4. *Math Journal* **Write** a paragraph explaining how to convert from polar coordinates to rectangular coordinates and vice versa. Include a diagram with your explanation.

Guided Practice

Find the polar coordinates of each point with the given rectangular coordinates. Use $0 \le \theta < 2\pi$ and $r \ge 0$.

5. $\left(-\sqrt{2}, \sqrt{2}\right)$
6. $(-2, -5)$

Find the rectangular coordinates of each point with the given polar coordinates.

7. $\left(-2, \frac{4\pi}{3}\right)$
8. $(2.5, 250°)$

Write each rectangular equation in polar form.

9. $y = 2$
10. $x^2 + y^2 = 16$

Write each polar equation in rectangular form.

11. $r = 6$
12. $r = -\sec\theta$

13. **Acoustics** The polar pattern of a cardioid lavalier microphone is given by $r = 2 + 2\cos\theta$.
    a. Graph the polar pattern.
    b. Will the microphone detect a sound that originates from the point with rectangular coordinates $(-2, 0)$? Explain.

EXERCISES

Practice

Find the polar coordinates of each point with the given rectangular coordinates. Use $0 \leq \theta < 2\pi$ and $r \geq 0$.

14. $(2, -2)$ **15.** $(0, 1)$ **16.** $(1, \sqrt{3})$

17. $\left(-\frac{1}{4}, -\frac{\sqrt{3}}{4}\right)$ **18.** $(3, 8)$ **19.** $(4, -7)$

Find the rectangular coordinates of each point with the given polar coordinates.

20. $\left(3, \frac{\pi}{2}\right)$ **21.** $\left(\frac{1}{2}, \frac{3\pi}{4}\right)$ **22.** $\left(-1, -\frac{\pi}{6}\right)$

23. $(-2, 270°)$ **24.** $(4, 210°)$ **25.** $(14, 130°)$

Write each rectangular equation in polar form.

26. $x = -7$ **27.** $y = 5$

28. $x^2 + y^2 = 25$ **29.** $x^2 + y^2 = 2y$

30. $x^2 - y^2 = 1$ **31.** $x^2 + (y - 2)^2 = 4$

Write each polar equation in rectangular form.

32. $r = 2$ **33.** $r = -3$

34. $\theta = \frac{\pi}{3}$ **35.** $r = 2 \csc \theta$

36. $r = 3 \cos \theta$ **37.** $r^2 \sin 2\theta = 8$

38. Write the equation $y = x$ in polar form.

39. What is the rectangular form of $r = \sin \theta$?

Applications and Problem Solving

40. Surveying A surveyor identifies a landmark at the point with polar coordinates $(325, 70°)$. What are the rectangular coordinates of this point?

41. Machinery An arc of a spiral of Archimedes is used to create a disc that drives the spindle on a sewing machine. Note from the figure that the spinning disc drives the rod that moves the piston. The outline of the disc can be modeled by the graph of $r = \frac{\theta}{6}$ for $\frac{\pi}{4} \leq \theta \leq \frac{5\pi}{4}$ and its reflection in the line $\theta = \frac{\pi}{4}$. How far does the piston move from right to left?

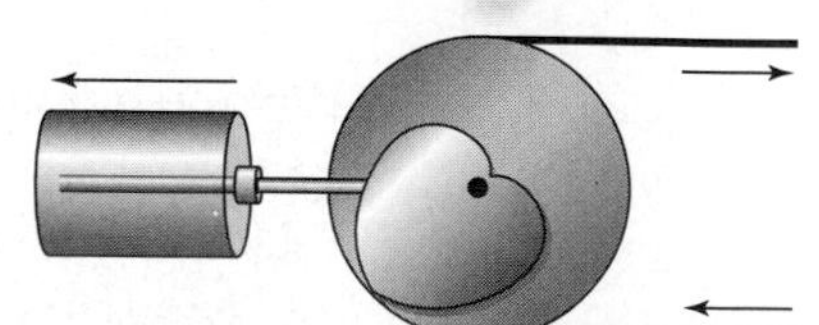

42. Critical Thinking Write a convincing argument to prove that, when converting polar coordinates to rectangular coordinates, the formulas $x = r \cos \theta$ and $y = r \sin \theta$ are true. Include a labeled drawing in your answer.

43. **Irrigation** A sod farm can use a combined Cartesian and polar coordinate system to identify points in the field. Points have coordinates of the form (x, y, r, θ). The sprinkler heads are spaced 25 meters apart. The Cartesian coordinates x and y, which are positive integers, indicate how many sprinkler heads to go across and up in the grid. The polar coordinates give the location of the point relative to the sprinkler head. If a point has coordinates (4, 3, 2, 120°), then how far to the east and north of the origin is that point?

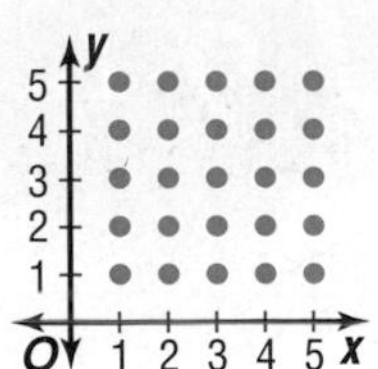

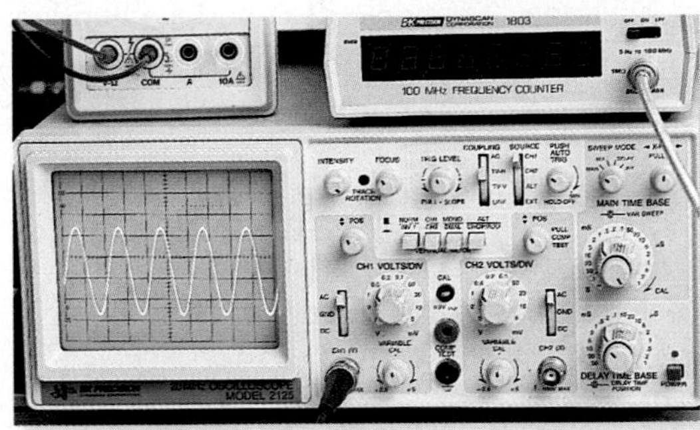

44. **Electrical Engineering** When adding sinusoidal expressions like $4 \sin (3.14t + 20°)$ and $5 \sin (3.14t + 70°)$, electrical engineers sometimes use *phasors*, which are vectors in polar form.
   a. The phasors for these two expressions are written as $4\angle 20°$ and $5\angle 70°$. Find the rectangular forms of these vectors.
   b. Add the two vectors you found in part a.
   c. Write the sum of the two vectors in phasor notation.
   d. Write the sinusoidal expression corresponding to the phasor in part c.

45. **Critical Thinking** Identify the type of graph generated by the polar equation $r = 2a \sin \theta + 2a \cos \theta$. Write the equivalent rectangular equation.

Mixed Review

46. Graph the polar equation $r = 2 + \sin \theta$. *(Lesson 9-2)*

47. Name four different pairs of polar coordinates that represent the point at $(-2, 45°)$. *(Lesson 9-1)*

48. **Aviation** An airplane flies at an air speed of 425 mph on a heading due south. It flies against a headwind of 50 mph from a direction 30° east of south. Find the airplane's ground speed and direction. *(Lesson 8-5)*

49. Solve $\sin^2 A = \cos A - 1$ for principal values of A. *(Lesson 7-5)*

50. Graph $y = 2 \cos \theta$. *(Lesson 6-4)*

51. Use the unit circle to find the exact value of cos 210°. *(Lesson 5-3)*

52. Write a polynomial function to model the set of data. *(Lesson 4-8)*

x	−10	−7	−4	−1	2	5	8	11	14
f(x)	−15	−9.2	−6.9	−3	−0.1	2	1.1	−2.3	−4.5

53. Use synthetic division to divide $x^5 - 3x^2 - 20$ by $x - 2$. *(Lesson 4-3)*

54. Write the equation of best fit for a set of data using the ordered pairs (17, 145) and (25, 625). *(Lesson 1-6)*

55. **SAT/ACT Practice** x, y, and z are different positive integers. $\frac{x}{y}$ and $\frac{y}{z}$ are also positive integers. Which of the following cannot be a positive integer?

A $\frac{x}{z}$ B $(x)(y)$ C $\frac{z}{x}$ D $(x + y)z$ E $(x - z)y$

Extra Practice See p. A42.

9-4

Polar Form of a Linear Equation

OBJECTIVES

- Write the polar form of a linear equation.
- Graph the polar form of a linear equation.

BIOLOGY Scientists hope that by studying the behavior of flies, they will gain insight into the genetics and brain functions of flies. Karl Götz of the Max Planck Institute in Tübingen, Germany, designed the Buridan Paradigm, which is a device that tracks the path of a fruit fly walking between two visual landmarks. The position of the fly is recorded in both rectangular and polar coordinates. If the fly walks in a straight line from the point with polar coordinates (6, 2) to the point with polar coordinates (2, −0.5), what is the equation of the path of the fly? How close did the fly come to the origin? *This problem will be solved in Example 4.*

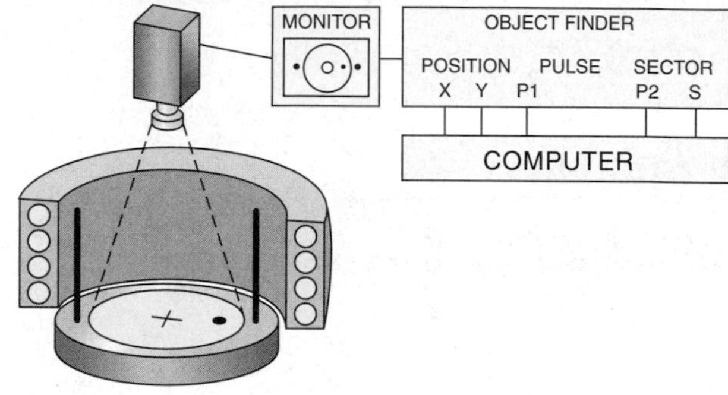

Look Back
You can refer to Lesson 7-6 to review normal form.

The polar form of the equation for a line ℓ is closely related to the normal form, which is $x \cos \phi + y \sin \phi - p = 0$. We have learned that $x = r \cos \theta$ and $y = r \sin \theta$. The polar form of the equation of line ℓ can be obtained by substituting these values into the normal form.

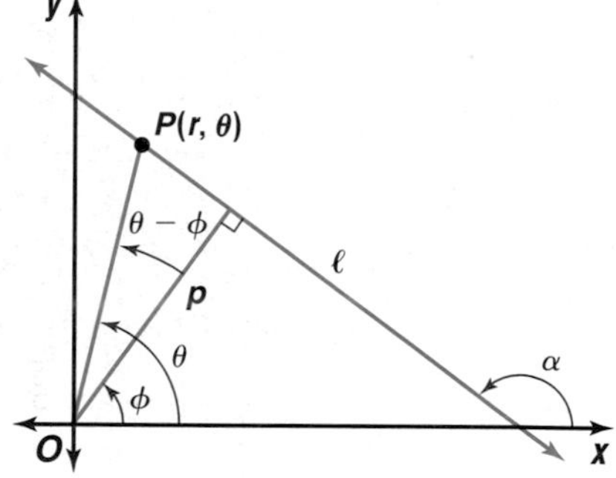

$$x \cos \phi + y \sin \phi - p = 0$$
$$(r \cos \theta) \cos \phi + (r \sin \theta) \sin \phi - p = 0$$
$$r(\cos \theta \cos \phi + \sin \theta \sin \phi) = p$$
$$r \cos (\theta - \phi) = p \quad \textit{cos θ cos ϕ + sin θ sin ϕ = cos (θ − ϕ)}$$

Polar Form of a Linear Equation

The polar form of a linear equation, where p is the length of the normal and ϕ is the positive angle between the positive x-axis and the normal, is

$$p = r \cos (\theta - \phi).$$

In the polar form of a linear equation, θ and r are variables, and p and ϕ are constants. Values for p and ϕ can be obtained from the normal form of the standard equation of a line. Remember to choose the value for ϕ according to the quadrant in which the normal lies.

Example 1 **Write each equation in polar form.**

a. $\mathbf{5x + 12y = 26}$

The standard form of the equation is $5x + 12y - 26 = 0$. First, write the equation in normal form to find the values of p and ϕ. To convert to normal form, we need to find the value of $\pm\sqrt{A^2 + B^2}$.

$\pm\sqrt{A^2 + B^2} = \pm\sqrt{5^2 + 12^2}$, or ± 13

Since C is negative, use $+13$. The normal form of the equation is

$\frac{5}{13}x + \frac{12}{13}y - 2 = 0.$ *The normal form is $x \cos \phi + y \sin \phi - p = 0$.*

We can see from the normal form that $p = 2$, $\cos \phi = \frac{5}{13}$, and $\sin \phi = \frac{12}{13}$.

Since $\cos \phi$ and $\sin \phi$ are both positive, the normal lies in the first quadrant.

$\tan \phi = \frac{\sin \phi}{\cos \phi}$

$\tan \phi = \frac{\frac{12}{13}}{\frac{5}{13}}$

$\tan \phi = \frac{12}{5}$

$\phi \approx 1.18$ *Use the Arctangent function.*

Substitute the values for p and ϕ into the polar form.

$p = r \cos (\theta - \phi) \rightarrow 2 = r \cos (\theta - 1.18)$

The polar form of $5x + 12y = 26$ is $2 = r \cos (\theta - 1.18)$.

b. $\mathbf{2x - 7y = -5}$

The standard form of this equation is $2x - 7y + 5 = 0$.

So, $\pm\sqrt{A^2 + B^2} = \sqrt{\pm 2^2 + (-7)^2}$, or $\pm\sqrt{53}$.

Since C is positive, use $-\sqrt{53}$. Then the normal form of the equation is

$-\frac{2}{\sqrt{53}}x + \frac{7}{\sqrt{53}}y - \frac{5}{\sqrt{53}} = 0.$

We see that $p = \frac{5}{\sqrt{53}}$ or $\frac{5\sqrt{53}}{53}$, $\cos \phi = -\frac{2}{\sqrt{53}}$, and $\sin \phi = \frac{7}{\sqrt{53}}$.

Since $\cos \phi < 0$ but $\sin \phi > 0$, the normal lies in the second quadrant.

$\tan \phi = \frac{\sin \phi}{\cos \phi}$

$\tan \phi = -\frac{7}{2}$ $\quad \frac{7}{\sqrt{53}} \div \left(-\frac{2}{\sqrt{53}}\right) = \frac{7}{\sqrt{53}} \cdot \left(-\frac{\sqrt{53}}{2}\right)$

$\phi \approx 106°$ *Add 180° to the Arctangent value.*

Substitute the values for p and ϕ into the polar form.

$p = r \cos (\theta - \phi) \rightarrow \frac{5\sqrt{53}}{53} = r \cos (\theta - 106°)$

The polar form of the equation is $\frac{5\sqrt{53}}{53} = r \cos (\theta - 106°)$.

The polar form of a linear equation can be converted to rectangular form by using the angle sum and difference identities for cosine and the polar coordinate conversion equations.

Example 2 **Write $2 = r\cos(\theta - 60°)$ in rectangular form.**

$2 = r\cos(\theta - 60°)$

$2 = r(\cos\theta\cos 60° + \sin\theta\sin 60°)$ *Difference identity for cosine*

$2 = r\left(\frac{1}{2}\cos\theta + \frac{\sqrt{3}}{2}\sin\theta\right)$ $\cos 60° = \frac{1}{2}, \sin 60° = \frac{\sqrt{3}}{2}$

$2 = \frac{1}{2}r\cos\theta + \frac{\sqrt{3}}{2}r\sin\theta$ *Distributive Property*

$2 = \frac{1}{2}x + \frac{\sqrt{3}}{2}y$ *Polar to rectangular conversion equations*

$4 = x + \sqrt{3}y$ *Multiply each side by 2.*

$0 = x + \sqrt{3}y - 4$ *Subtract 4 from each side.*

The rectangular form of $2 = r\cos(\theta - 60°)$ is $x + \sqrt{3}y - 4 = 0$.

The polar form of a linear equation can be graphed by preparing a table of coordinates and then graphing the ordered pairs in the polar system.

Examples 3 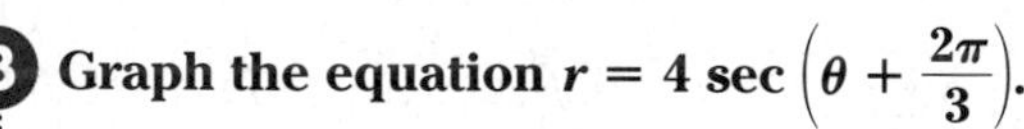

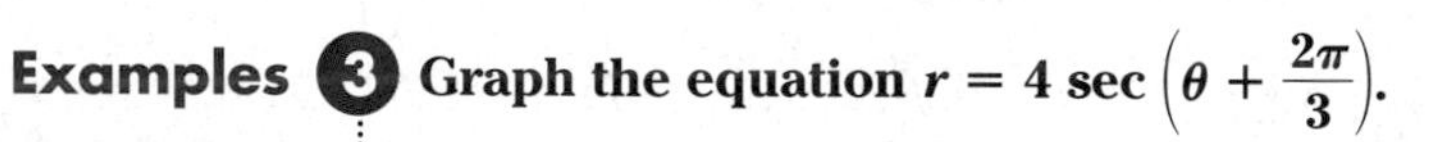

Graph the equation $r = 4\sec\left(\theta + \frac{2\pi}{3}\right)$.

Graphing Calculator Tip

You can use the **TABLE** feature with Δ**Tbl** $= \frac{\pi}{6}$ to generate these values.

Write the equation in the form $r = \frac{4}{\cos\left(\theta + \frac{2\pi}{3}\right)}$.

Use your calculator to make a table of values.

θ	0	$\frac{\pi}{6}$	$\frac{\pi}{3}$	$\frac{\pi}{2}$	$\frac{2\pi}{3}$	$\frac{5\pi}{6}$	π
r	-8	-4.6	-4	-4.6	-8	undefined	8

Graph the ordered pairs on a polar plane.

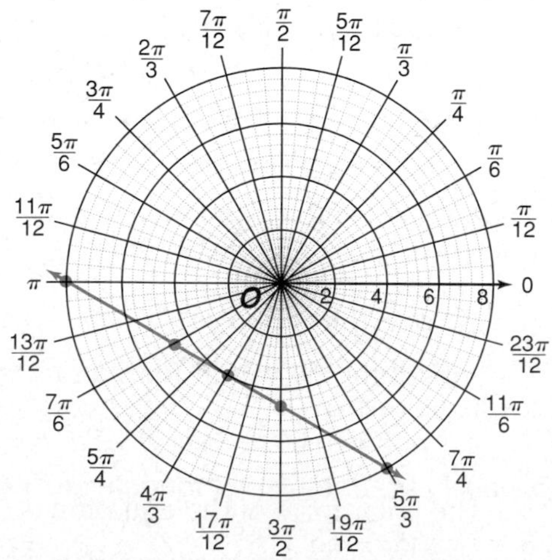

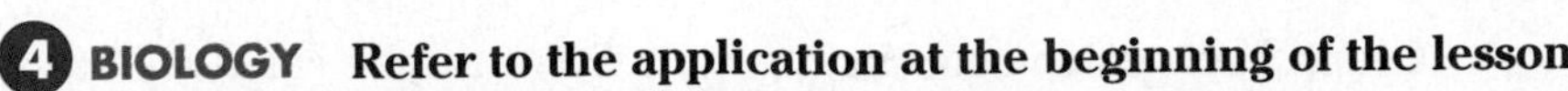

4 BIOLOGY Refer to the application at the beginning of the lesson.

a. If the fly walks in a straight line from the point with polar coordinates (6, 2) to the point with polar coordinates (2, −0.5), what is the equation of the path of the fly?

b. If r is measured in centimeters, how close did the fly come to the origin?

a. The two points must satisfy the equation $p = r\cos(\theta - \phi)$. Substitute both ordered pairs into this form to create a system of equations.

$p = 6\cos(2 - \phi)$ $\quad (r, \theta) = (6, 2)$
$p = 2\cos(-0.5 - \phi)$ $\quad (r, \theta) = (2, -0.5)$

Substituting either expression for p into the other equation results in the equation $6\cos(2 - \phi) = 2\cos(-0.5 - \phi)$.

A graphing calculator shows that there are two solutions to this equation between 0 and 2π: $\phi \approx 0.59$ and $\phi \approx 3.73$.

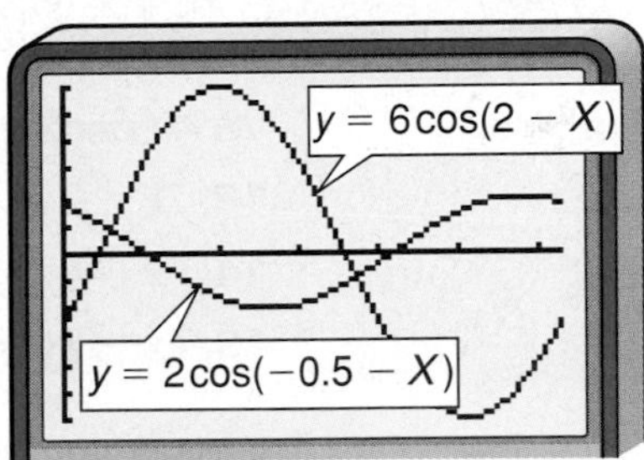

[0, 2π] scl:1 by [−6, 6] scl:1

Substituting these values into $p = 6\cos(2 - \phi)$ yields $p \approx 0.93$ and $p \approx -0.93$, respectively. Since p, the length of the normal, must be positive, we use $\phi \approx 0.59$ and $p \approx 0.93$.

Therefore, the polar form of the equation of the fly's path is $0.93 = r\cos(\theta - 0.59)$.

b. The closest that the fly came to the origin is the value of p, 0.93 centimeter or 9.3 millimeters.

CHECK FOR UNDERSTANDING

Communicating Mathematics

Read and study the lesson to answer each question.

1. **State** the general form for the polar equation of a line. Explain the significance of each part of the equation.
2. **Determine** what value of θ will result in $r = p$ in the polar equation of a line.
3. **Explain** how to find ϕ and the length of the normal for a rectangular equation of the form $x = k$. Write the polar form of the equation.
4. **Explain** why it is a good idea to find several ordered pairs when graphing the polar equation of a line, even though only two points are needed to determine a line.

Guided Practice

Write each equation in polar form. Round ϕ to the nearest degree.

5. $3x - 4y - 10 = 0$
6. $-2x + 4y = 9$

Write each equation in rectangular form.

7. $3 = r\cos(\theta - 60°)$
8. $r = 2\sec\left(\theta + \frac{\pi}{4}\right)$

Graph each polar equation.

9. $3 = r\cos\left(\theta - \frac{\pi}{3}\right)$

10. $r = 2\sec(\theta + 45°)$

11. Aviation An air traffic controller is looking at a radar screen with the control tower at the origin.

a. If the path of an airplane can be modeled by the equation $5 = r\cos\left(\theta - \frac{5\pi}{6}\right)$, then what are the polar coordinates of the plane when it comes the closest to the tower?

b. Sketch a graph of the path of the plane.

EXERCISES

Practice

Write each equation in polar form. Round ϕ to the nearest degree.

12. $7x - 24y + 100 = 0$

13. $21x + 20y = 87$

14. $6x - 8y = 21$

15. $3x + 2y - 5 = 0$

16. $4x - 5y = 10$

17. $-x + 3y = 7$

Write each equation in rectangular form.

18. $6 = r\cos(\theta - 120°)$

19. $4 = r\cos\left(\theta + \frac{\pi}{4}\right)$

20. $2 = r\cos(\theta + \pi)$

21. $1 = r\cos(\theta - 330°)$

22. $r = 11\sec\left(\theta + \frac{7\pi}{6}\right)$

23. $r = 5\sec(\theta - 60°)$

Graph each polar equation.

24. $6 = r\cos(\theta - 45°)$

25. $1 = r\cos\left(\theta - \frac{\pi}{6}\right)$

26. $2 = r\cos(\theta + 60°)$

27. $3 = r\cos(\theta + 90°)$

28. $r = 3\sec\left(\theta + \frac{\pi}{3}\right)$

29. $r = 4\sec\left(\theta - \frac{\pi}{4}\right)$

30. Write the polar form of the equation of the line that passes through points with rectangular coordinates $(4, -1)$ and $(-2, 3)$.

31. What is the polar form of the equation of the line that passes through points with polar coordinates $\left(3, \frac{\pi}{4}\right)$ and $\left(2, \frac{7\pi}{6}\right)$?

Applications and Problem Solving

32. Biology Refer to the application at the beginning of the lesson. Suppose the Buridan Paradigm tracks a fruit fly whose path is modeled by the polar equation $r = 6\sec(\theta - 15°)$, where r is measured in centimeters.

a. How close did the fly come to the origin?

b. What were the polar coordinates of the fly when it was closest to the origin?

33. Critical Thinking Write the polar forms of the equations of two lines, neither of which is vertical, such that the lines intersect at a 90° angle and have normal segments of length 2.

34. Robotics The diagram at the right shows a robot with a telescoping arm. The "hand" at the end of such an arm is called the manipulator. What polar equation should the arm be programmed to follow in order to move the manipulator in a straight line from the point with rectangular coordinates (5, 4) to the point with rectangular coordinates (15, 4)?

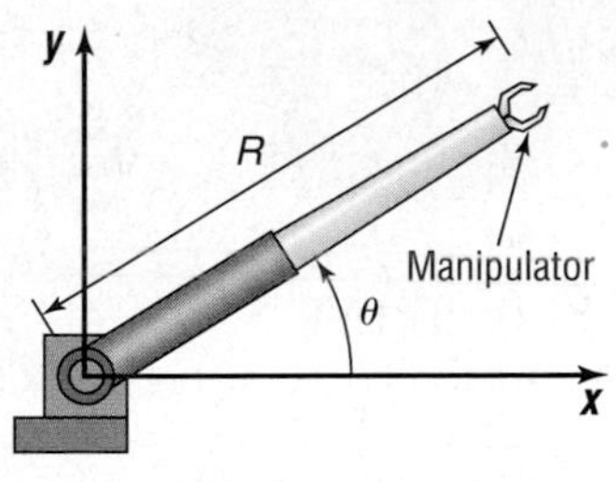

35. Surveying A surveyor records the locations of points in a plot of land by means of polar coordinates. In a circular plot of radius 500 feet, stakes are placed at (125, 130°) and (300, 70°), where r is measured in feet. The stakes are at the same elevation.

a. Draw a diagram of the plot of land. Include the locations of the two stakes.

b. Find the polar equation of the line determined by the stakes.

36. Critical Thinking Show that $k = r \sin(\theta - \alpha)$ is also the equation of a line in polar coordinates. Identify the significance of k and α in the graph.

37. Design A carnival ride designer wants to create a Ferris wheel with a 40-foot radius. The designer wants the interior of the circle to have lines of lights that form a regular pentagon. Find the polar equation of the line that contains (40, 0°) and (40, 72°).

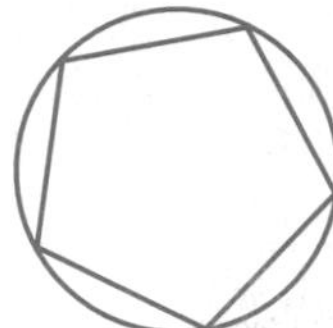

Mixed Review

38. Write the polar equation $r = 6$ in rectangular form. *(Lesson 9-3)*

39. Identify the type of curve represented by the equation $r = \sin 6\theta$. *(Lesson 9-2)*

40. Write parametric equations of the line with equation $x - 3y = 6$. *(Lesson 8-6)*

41. **Gardening** A sprinkler is set to rotate 65° and spray a distance of 6 feet. What is the area of the gound being watered? *(Lesson 6-1)*

42. Use the unit circle to find the value of sin 360° *(Lesson 5-3)*

43. Solve $2x^3 + 5x^2 - 12x = 0$. *(Lesson 4-1)*

44. **SAT Practice** **Grid-In** If $c + d = 12$ and $c^2 - d^2 = 48$, then $c - d = ?$

MID-CHAPTER QUIZ

Graph each polar equation. (Lesson 9-1)

1. $r = 4$
2. $\theta = \dfrac{2\pi}{3}$

Graph each polar equation. (Lesson 9-2)

3. $r = 3 + 3 \sin \theta$
4. $r = \cos 2\theta$

Find the polar coordinates of each point with the given rectangular coordinates. (Lesson 9-3)

5. $P(-\sqrt{2}, -\sqrt{2})$
6. $Q(0, -4)$

7. Write the polar form of the equation $x^2 + y^2 = 36$. (Lesson 9-3)

8. Find the rectangular form of $r = 2 \csc \theta$. (Lesson 9-3)

Write each equation in polar form. Round ϕ to the nearest degree. (Lesson 9-4)

9. $5x - 12y = -3$
10. $-2x - 6y = 2$

Extra Practice See p. A42.

9-5

Simplifying Complex Numbers

OBJECTIVE

- Add, subtract, multiply, and divide complex numbers in rectangular form.

DYNAMICAL SYSTEMS

Dynamical systems is a branch of mathematics that studies constantly changing systems like the stock market, the weather, and population. In many cases, one can catch a glimpse of the system at some point in time, but the forces that act on the system cause it to change quickly. By analyzing how a dynamical system changes over time, it may be possible to predict the behavior of the system in the future. One of the basic mathematical models of a dynamical system is iteration of a complex function. *A problem related to this will be solved in Example 4.*

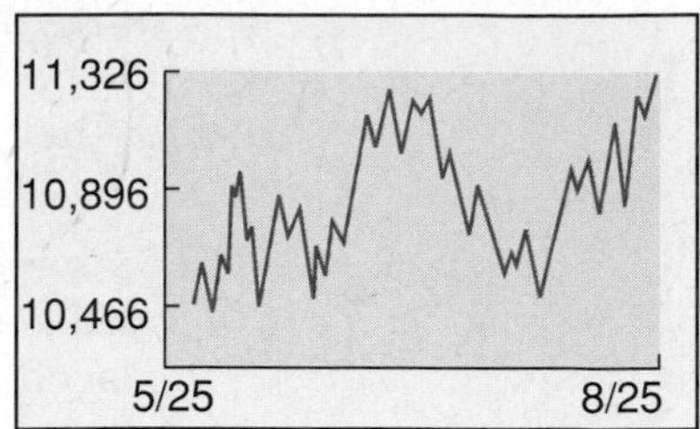

A Typical Graph of the Stock Market

Recall that complex numbers are numbers of the form $a + bi$, where a and b are real numbers and $\boldsymbol{i}$, the imaginary unit, is defined by $\boldsymbol{i}^2 = -1$. The first few powers of $\boldsymbol{i}$ are shown below.

$i^1 = i$	$i^2 = -1$	$i^3 = i^2 \cdot i = -i$	$i^4 = (i^2)^2 = 1$
$i^5 = i^4 \cdot i = i$	$i^6 = i^4 \cdot i^2 = -1$	$i^7 = i^4 \cdot i^3 = -i$	$i^8 = (i^2)^4 = 1$

Notice the repeating pattern of the powers of $\boldsymbol{i}$.

$$i, -1, -i, 1, i, -1, -i, 1$$

In general, the value of $\boldsymbol{i}^n$, where n is a whole number, can be found by dividing n by 4 and examining the remainder as summarized in the table at the right.

You can also simplify any integral power of $\boldsymbol{i}$ by rewriting the exponent as a multiple of 4 plus a positive remainder.

To find the value of i^n, let R be the remainder when n is divided by 4.	
if R = 0	$i^n = 1$
if R = 1	$i^n = i$
if R = 2	$i^n = -1$
if R = 3	$i^n = -i$

Example 1 **Simplify each power of $\boldsymbol{i}$.**

a. i^{53}

Method 1

$53 \div 4 = 13 \text{ R1}$

If R = 1, $i^n = i$.

$i^{53} = i$

Method 2

$i^{53} = (i^4)^{13} \cdot i$

$= (1)^{13} \cdot i$

$= i$

b. i^{-13}

Method 1

$-13 \div 4 = -4 \text{ R3}$

If R = 3, $i^n = -i$.

$i^{-13} = -i$

Method 2

$i^{-13} = (i^4)^{-4} \cdot i^3$

$= (1)^{-4} \cdot i^3$

$= -i$

The complex number $a + bi$, where a and b are real numbers, is said to be in **rectangular form.** a is called the **real part** and b is called the **imaginary part.** If $b = 0$, the complex number is a real number. If $b \neq 0$, the complex number is an **imaginary number.** If $a = 0$ and $b \neq 0$, as in $4i$, then the complex number is a **pure imaginary number.** Complex numbers can be added and subtracted by performing the chosen operation on both the real and imaginary parts.

Example 2 **Simplify each expression.**

Graphing Calculator Tip

Some calculators have a complex number mode. In this mode, they can perform complex number arithmetic.

a. $(5 - 3i) + (-2 + 4i)$

$$(5 - 3i) + (-2 + 4i) = [5 + (-2)] + [-3i + 4i]$$
$$= 3 + i$$

b. $(10 - 2i) - (14 - 6i)$

$$(10 - 2i) - (14 - 6i) = 10 - 2i - 14 + 6i$$
$$= -4 + 4i$$

The product of two or more complex numbers can be found using the same procedures you use when multiplying binomials.

Example 3 **Simplify $(2 - 3i)(7 - 4i)$.**

$(2 - 3i)(7 - 4i) = 7(2 - 3i) - 4i(2 - 3i)$ *Distributive property*
$= 14 - 21i - 8i + 12i^2$ *Distributive property*
$= 14 - 21i - 8i + 12(-1)$ $i^2 = -1$
$= 2 - 29i$

Iteration is the process of repeatedly applying a function to the output produced by the previous input. When using complex numbers with functions, it is traditional to use z for the independent variable.

Example 4 **DYNAMICAL SYSTEMS** **If $f(z) = (0.5 + 0.5i)z$, find the first five iterates of f for the initial value $z_0 = 1 + i$. Describe any pattern that you see.**

$f(z) = (0.5 + 0.5i)z$

$f(1 + i) = (0.5 + 0.5i)(1 + i)$ *Replace z with 1 + i.*
$= 0.5 + 0.5i + 0.5i + 0.5i^2$
$= i$ $z_1 = i$

$f(i) = (0.5 + 0.5i)i$
$= 0.5i + 0.5i^2$
$= -0.5 + 0.5i$ $z_2 = -0.5 + 0.5i$

$f(-0.5 + 0.5i) = (0.5 + 0.5i)(-0.5 + 0.5i)$
$= -0.25 + 0.25i - 0.25i + 0.25i^2$
$= -0.5$ $z_3 = -0.5$

(continued on the next page)

Graphing Calculator Programs To download a graphing calculator program that performs complex iteration, visit: **www.amc.glencoe.com**

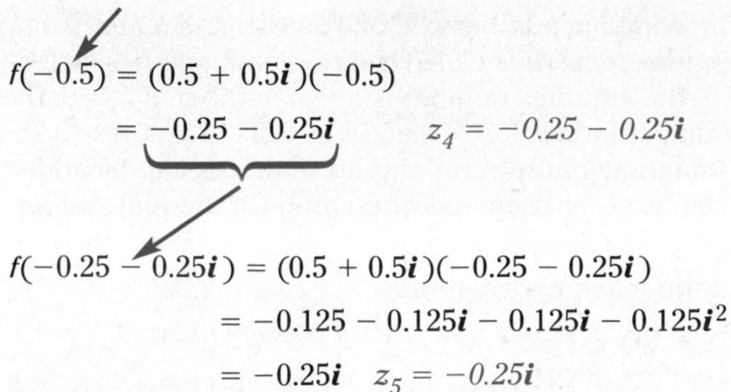

$$f(-0.5) = (0.5 + 0.5\boldsymbol{i})(-0.5)$$
$$= -0.25 - 0.25\boldsymbol{i} \qquad z_4 = -0.25 - 0.25\boldsymbol{i}$$

$$f(-0.25 - 0.25\boldsymbol{i}) = (0.5 + 0.5\boldsymbol{i})(-0.25 - 0.25\boldsymbol{i})$$
$$= -0.125 - 0.125\boldsymbol{i} - 0.125\boldsymbol{i} - 0.125\boldsymbol{i}^2$$
$$= -0.25\boldsymbol{i} \qquad z_5 = -0.25\boldsymbol{i}$$

The first five iterates of $1 + \boldsymbol{i}$ are $\boldsymbol{i}$, $-0.5 + 0.5\boldsymbol{i}$, -0.5, $-0.25 - 0.25\boldsymbol{i}$, and $-0.25\boldsymbol{i}$. The absolute values of the nonzero real and imaginary parts (1, 0.5, 0.25) stay the same for two steps and then are halved.

Two complex numbers of the form $a + b\boldsymbol{i}$ and $a - b\boldsymbol{i}$ are called **complex conjugates.** Recall that if a quadratic equation with real coefficients has complex solutions, then those solutions are complex conjugates. Complex conjugates also play a useful role in the division of complex numbers. To simplify the quotient of two complex numbers, multiply the numerator and denominator by the conjugate of the denominator. The process is similar to rationalizing the denominator in an expression like $\frac{1}{3 + \sqrt{2}}$.

Example 5 **Simplify $(5 - 3\boldsymbol{i}) \div (1 - 2\boldsymbol{i})$.**

$$(5 - 3\boldsymbol{i}) \div (1 - 2\boldsymbol{i}) = \frac{5 - 3\boldsymbol{i}}{1 - 2\boldsymbol{i}}$$

$$= \frac{5 - 3\boldsymbol{i}}{1 - 2\boldsymbol{i}} \cdot \frac{1 + 2\boldsymbol{i}}{1 + 2\boldsymbol{i}} \qquad \textit{Multiply by 1; } 1 + 2\boldsymbol{i} \textit{ is the conjugate of } 1 - 2\boldsymbol{i}.$$

$$= \frac{5 + 10\boldsymbol{i} - 3\boldsymbol{i} - 6\boldsymbol{i}^2}{1 - 4\boldsymbol{i}^2}$$

$$= \frac{5 + 7\boldsymbol{i} - 6(-1)}{1 - (-4)} \qquad \boldsymbol{i}^2 = -1$$

$$= \frac{11 + 7\boldsymbol{i}}{5}$$

$$= \frac{11}{5} + \frac{7}{5}\boldsymbol{i} \qquad \textit{Write the answer in the form } a + b\boldsymbol{i}.$$

The list below summarizes the operations with complex numbers presented in this lesson.

Operations with Complex Numbers

For any complex numbers $a + bi$ and $c + di$, the following are true.

$$(a + bi) + (c + di) = (a + c) + (b + d)i$$
$$(a + bi) - (c + di) = (a - c) + (b - d)i$$
$$(a + bi)(c + di) = (ac - bd) + (ad + bc)i$$
$$\frac{a + bi}{c + di} = \frac{ac + bd}{c^2 + d^2} + \frac{bc - ad}{c^2 + d^2}i$$

CHECK FOR UNDERSTANDING

Communicating Mathematics

Read and study the lesson to answer each question.

1. **Describe** how to simplify any integral power of $\boldsymbol{i}$.
2. **Draw** a Venn diagram to show the relationship between real, pure imaginary, and complex numbers.
3. **Explain** why it is useful to multiply by the conjugate of the denominator over itself when simplifying a fraction containing complex numbers.
4. **Write** a quadratic equation that has two complex conjugate solutions.

Guided Practice

Simplify.

5. i^{-6}
6. $i^{10} + i^2$
7. $(2 + 3i) + (-6 + i)$
8. $(2.3 + 4.1i) - (-1.2 - 6.3i)$
9. $(2 + 4i) + (-1 + 5i)$
10. $(-2 - i)^2$
11. $\dfrac{i}{1 + 2i}$

Look Back
You can refer to Lessons 8-1 and 8-2 to review vectors, components, and resultants.

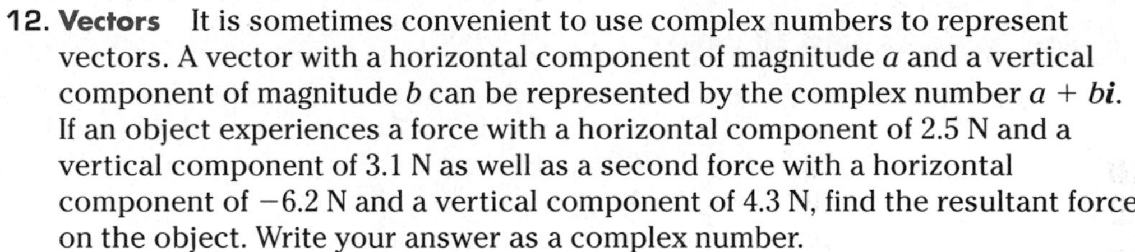

12. **Vectors** It is sometimes convenient to use complex numbers to represent vectors. A vector with a horizontal component of magnitude a and a vertical component of magnitude b can be represented by the complex number $a + bi$. If an object experiences a force with a horizontal component of 2.5 N and a vertical component of 3.1 N as well as a second force with a horizontal component of -6.2 N and a vertical component of 4.3 N, find the resultant force on the object. Write your answer as a complex number.

EXERCISES

Practice

Simplify.

13. i^6
14. i^{19}
15. i^{1776}
16. $i^9 + i^{-5}$
17. $(3 + 2i) + (-4 + 6i)$
18. $(7 - 4i) + (2 - 3i)$
19. $\left(\frac{1}{2} + i\right) - (2 - i)$
20. $(-3 - i) - (4 - 5i)$
21. $(2 + i)(4 + 3i)$
22. $(1 + 4i)^2$
23. $\left(1 + \sqrt{7}i\right)\left(-2 - \sqrt{5}i\right)$
24. $\left(2 + \sqrt{-3}\right)\left(-1 + \sqrt{-12}\right)$
25. $\dfrac{2 + i}{1 + 2i}$
26. $\dfrac{3 - 2i}{-4 - i}$
27. $\dfrac{5 - i}{5 + i}$
28. Write a quadratic equation with solutions i and $-i$.
29. Write a quadratic equation with solutions $2 + i$ and $2 - i$.

Simplify.

30. $(2 - i)(3 + 2i)(1 - 4i)$
31. $(-1 - 3i)(2 + 2i)(1 - 2i)$
32. $\dfrac{\frac{1}{2} + \sqrt{3}i}{1 - \sqrt{2}i}$
33. $\dfrac{2 - \sqrt{2}i}{3 + \sqrt{6}i}$
34. $\dfrac{3 + i}{(2 + i)^2}$
35. $\dfrac{(1 + i)^2}{(-3 + 2i)^2}$

Applications and Problem Solving

36. **Electricity** *Impedance* is a measure of how much hindrance there is to the flow of charge in a circuit with alternating current. The impedance Z depends on the resistance R, the reactance due to capacitance X_C, and the reactance due to inductance X_L in the circuit. The impedance is written as the complex number $Z = R + (X_L - X_C)\boldsymbol{j}$. (Electrical engineers use $\boldsymbol{j}$ to denote the imaginary unit.) In the first part of a particular series circuit, the resistance is 10 ohms, the reactance due to capacitance is 2 ohms, and the reactance due to inductance is 1 ohm. In the second part of the circuit, the respective values are 3 ohms, 1 ohm, and 1 ohm.

a. Write complex numbers that represent the impedances in the two parts of the circuit.

b. Add your answers from part a to find the total impedance in the circuit.

c. The *admittance* of an AC circuit is a measure of how well the circuit allows current to flow. Admittance is the reciprocal of impedance. That is, $S = \frac{1}{Z}$. The units for admittance are siemens. Find the admittance in a circuit with an impedance of $6 + 3\boldsymbol{j}$ ohms.

37. **Critical Thinking**

a. Solve the equation $x^2 + 8\boldsymbol{i}x - 25 = 0$.

b. Are the solutions complex conjugates?

c. How does your result in part b compare with what you already know about complex solutions to quadratic equations?

d. Check your solutions.

38. **Critical Thinking** Sometimes it is useful to separate a complex function into its real and imaginary parts. Substitute $z = x + y\boldsymbol{i}$ into the function $f(z) = z^2$ to write the equation of the function in terms of x and y only. Simplify your answer.

39. **Dynamical Systems** Find the first five iterates for the given function and initial value.

a. $f(z) = \boldsymbol{i}z$, $z_0 = 2 - \boldsymbol{i}$

b. $f(z) = (0.5 - 0.866\boldsymbol{i})z$, $z_0 = 1 + 0\boldsymbol{i}$

40. **Critical Thinking** Simplify $(1 + 2\boldsymbol{i})^{-3}$.

41. **Physics** One way to derive the equation of motion in a spring-mass system is to solve a *differential equation.* The solutions of such a differential equation typically involve expressions of the form $\cos \beta t + \boldsymbol{i} \sin \beta t$. You generally expect solutions that are real numbers in such a situation, so you must use algebra to eliminate the imaginary numbers. Find a relationship between the constants c_1 and c_2 such that $c_1(\cos 2t + \boldsymbol{i} \sin 2t) + c_2(\cos 2t - \boldsymbol{i} \sin 2t)$ is a real number for all values of t.

Mixed Review

42. Write the equation $6x - 2y = -3$ in polar form. *(Lesson 9-4)*

43. Graph the polar equation $r = 4\theta$. *(Lesson 9-2)*

44. Write a vector equation of the line that passes through $P(-3, 6)$ and is parallel to $\vec{\mathbf{v}}\langle 1, -4\rangle$. *(Lesson 8-6)*

45. Find an ordered triple to represent $\vec{u}$ if $\vec{u} = \frac{1}{4}\vec{v} - 2\vec{w}$, $\vec{v} = \langle -8, 6, 4 \rangle$, and $\vec{w} = \langle 2, -6, 3 \rangle$. *(Lesson 8-3)*

46. If α and β are measures of two first quadrant angles, find $\cos(\alpha + \beta)$ if $\tan \alpha = \frac{4}{3}$ and $\cot \beta = \frac{5}{12}$. *(Lesson 7-3)*

47. A twig floats on the water, bobbing up and down. The distance between its highest and lowest points is 7 centimeters. It moves from its highest point down to its lowest point and back up to its highest point every 12 seconds. Write a cosine function that models the movement of the twig in relationship to the equilibrium point. *(Lesson 6-6)*

48. **Surveying** A surveyor finds that the angle of elevation from a certain point to the top of a cliff is 60°. From a point 45 feet farther away, the angle of elevation to the top of the cliff is 52°. How high is the cliff to the nearest foot? *(Lesson 5-4)*

49. What type of polynomial function would be the best model for the set of data? *(Lesson 4-8)*

x	−3	−1	1	3	5	7	9	11
f(x)	−4	−2	3	8	6	1	−3	−8

50. **Construction** A community wants to build a second pool at their community park. Their original pool has a width 5 times its depth and a length 10 times its depth. They wish to make the second pool larger by increasing the width of the original pool by 4 feet, increasing the length by 6 feet, and increasing the depth by 2 feet. The volume of the new pool will be 3420 cubic feet. Find the dimensions of the original pool. *(Lesson 4-4)*

51. If y varies jointly as x and z and $y = 80$ when $x = 5$ and $z = 8$, find y when $x = 16$ and $z = 2$. *(Lesson 3-8)*

52. If $f(x) = 7 - x^2$, find $f^{-1}(x)$. *(Lesson 3-4)*

53. Find the maximum and minimum values of the function $f(x, y) = -2x + y$ for the polygonal convex set determined by the system of inequalities. *(Lesson 2-6)*

$x \leq 6$
$y \geq 1$
$y - x \leq 2$

54. Solve the system of equations. *(Lesson 2-2)*

$x + 2y - 7z = 14$
$-x - 3y + 5z = -21$
$5x - y + 2z = -7$

55. **SAT/ACT Practice** If $BC = BD$ in the figure, what is the value of $x + 40$?

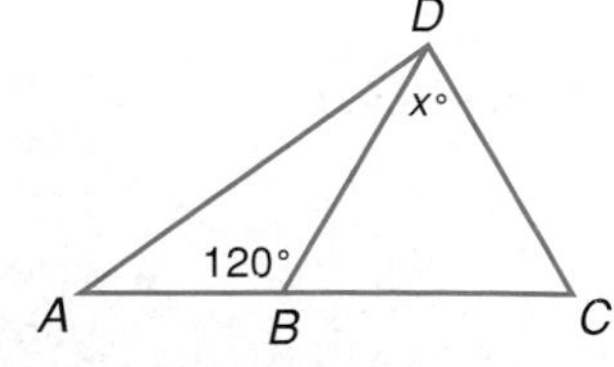

A 100 B 80 C 60 D 40

E cannot be determined from the information given

Extra Practice See p. A43.

9-6

The Complex Plane and Polar Form of Complex Numbers

OBJECTIVES
- Graph complex numbers in the complex plane.
- Convert complex numbers from rectangular to polar form and vice versa.

FRACTALS One of the standard ways to generate a fractal involves iteration of a quadratic function. If the function $f(z) = z^2$ is iterated using a complex number as the initial input, there are three possible outcomes. The terms of the sequence of outputs, called the *orbit,* may

- increase in absolute value,
- decrease toward 0 in absolute value, or
- always have an absolute value of 1.

One way to analyze the behavior of the orbit is to graph the numbers in the complex plane. Plot the first five members of the orbit of $z_0 = 0.9 + 0.3\boldsymbol{i}$ under iteration by $f(z) = z^2$. *This problem will be solved in Example 3.*

Recall that $a + b\boldsymbol{i}$ is referred to as the rectangular form of a complex number. The rectangular form is sometimes written as an ordered pair, (a, b). Two complex numbers in rectangular form are equal if and only if their real parts are equal and their imaginary parts are equal.

Example 1 **Solve the equation $2x + y + 3i = 9 + xi - yi$ for x and y, where x and y are real numbers.**

$2x + y + 3\boldsymbol{i} = 9 + x\boldsymbol{i} - y\boldsymbol{i}$

$(2x + y) + 3\boldsymbol{i} = 9 + (x - y)\boldsymbol{i}$ *On each side of the equation, group the real parts and the imaginary parts.*

$2x + y = 9$ and $x - y = 3$ *Set the corresponding parts equal to each other.*

$x = 4$ and $y = 1$ *Solve the system of equations.*

Complex numbers can be graphed in the **complex plane.** The complex plane has a real axis and an imaginary axis. The real axis is horizontal, and the imaginary axis is vertical. The complex number $a + b\boldsymbol{i}$ is graphed as the ordered pair (a, b) in the complex plane. The complex plane is sometimes called the **Argand plane.**

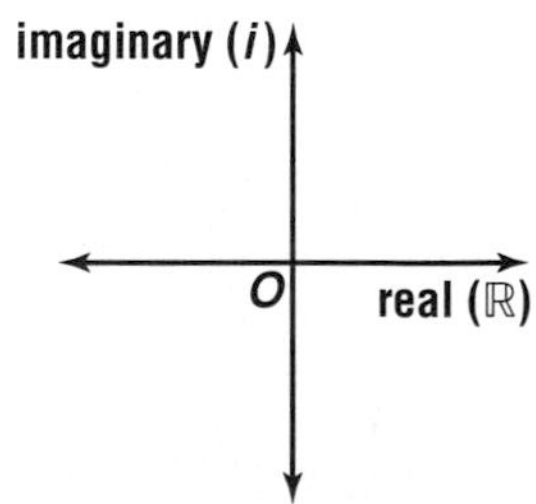

Recall that the absolute value of a real number is its distance from zero on the number line. Similarly, the **absolute value** of a complex number is its distance from zero in the complex plane. When $a + b\boldsymbol{i}$ is graphed in the complex plane, the distance from zero can be calculated using the Pythagorean Theorem.

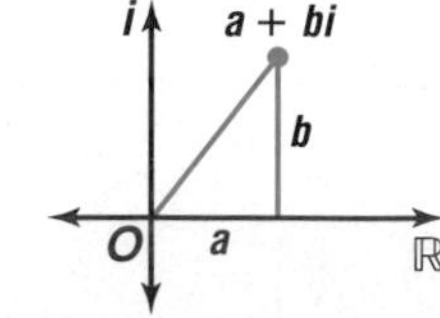

Absolute Value of a Complex Number

If $z = a + bi$, then $|z| = \sqrt{a^2 + b^2}$.

Examples **2** **Graph each number in the complex plane and find its absolute value.**

a. $z = 3 + 2i$

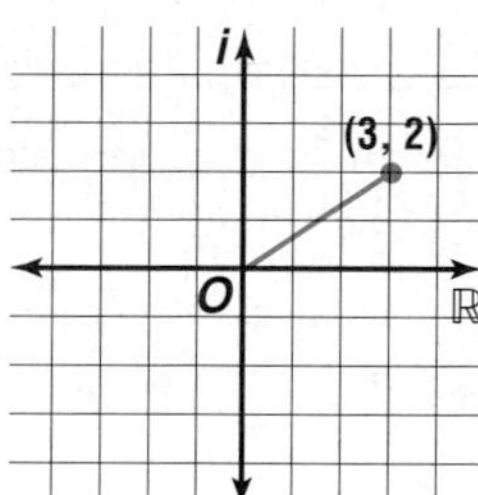

$$z = 3 + 2i$$
$$|z| = \sqrt{3^2 + 2^2}$$
$$= \sqrt{13}$$

b. $z = 4i$

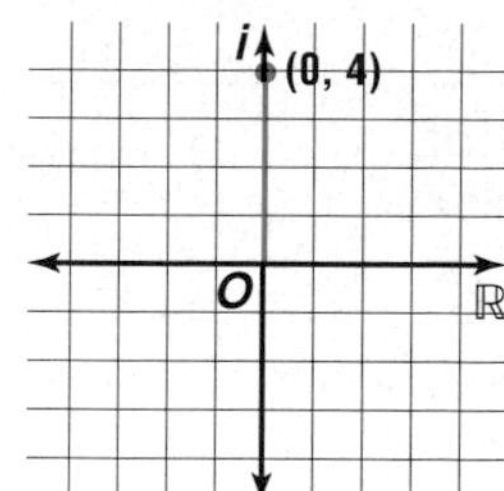

$$z = 0 + 4i$$
$$|z| = \sqrt{0^2 + 4^2}$$
$$= 4$$

3 **FRACTALS** **Refer to the application at the beginning of the lesson. Plot the first five members of the orbit of $z_0 = 0.9 + 0.3i$ under iteration by $f(z) = z^2$.**

Real World Application

First, calculate the first five members of the orbit. Round the real and imaginary parts to the nearest hundredth.

$z_1 = 0.72 + 0.54i$ $\quad z_1 = f(z_0)$

$z_2 = 0.23 + 0.78i$ $\quad z_2 = f(z_1)$

$z_3 = -0.55 + 0.35i$ $\quad z_3 = f(z_2)$

$z_4 = 0.18 - 0.39i$ $\quad z_4 = f(z_3)$

$z_5 = -0.12 - 0.14i$ $\quad z_5 = f(z_4)$

Then graph the numbers in the complex plane. The iterates approach the origin, so their absolute values decrease toward 0.

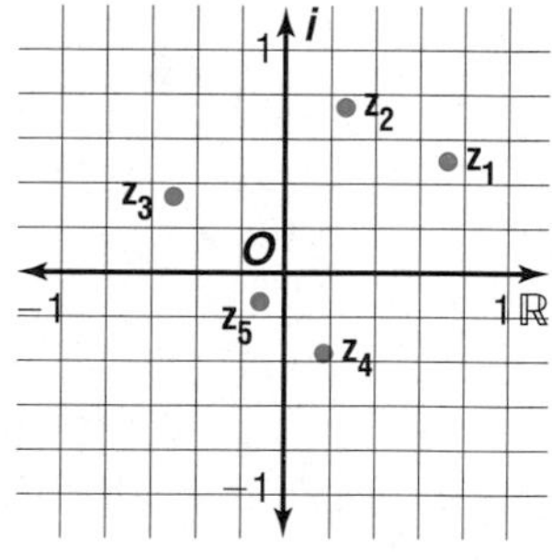

So far we have associated the complex number $a + bi$ with the rectangular coordinates (a, b). You know from Lesson 9-1 that there are also polar coordinates (r, θ) associated with the same point. In the case of a complex number, r represents the absolute value, or **modulus,** of the complex number. The angle θ is called the **amplitude** or **argument** of the complex number. Since θ is not unique, it may be replaced by $\theta + 2\pi k$, where k is any integer.

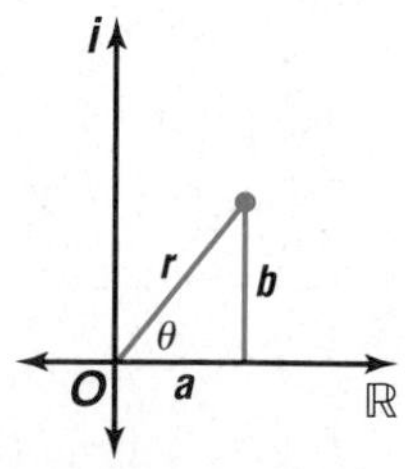

As with other rectangular coordinates, complex coordinates can be written in polar form by substituting $a = r\cos\theta$ and $b = r\sin\theta$.

$$\begin{aligned} z &= a + b\boldsymbol{i} \\ &= r\cos\theta + (r\sin\theta)\boldsymbol{i} \\ &= r(\cos\theta + \boldsymbol{i}\sin\theta) \end{aligned}$$

This form of a complex number is often called the **polar** or **trigonometric form.**

Polar Form of a Complex Number

The polar form or trigonometric form of the complex number $a + bi$ is

$$r\,(\cos\theta + \boldsymbol{i}\sin\theta).$$

$r(\cos\theta + \boldsymbol{i}\sin\theta)$ is often abbreviated as $r \text{ cis } \theta$.

Values for r and θ can be found by using the same process you used when changing rectangular coordinates to polar coordinates. For $a + b\boldsymbol{i}$, $r = \sqrt{a^2 + b^2}$ and $\theta = \text{Arctan}\,\frac{b}{a}$ if $a > 0$ or $\theta = \text{Arctan}\,\frac{b}{a} + \pi$ if $a < 0$. The amplitude θ is usually expressed in radian measure, and the angle is in standard position along the polar axis.

Example 4 **Express each complex number in polar form.**

a. $\mathbf{-3 + 4i}$

First, plot the number in the complex plane.

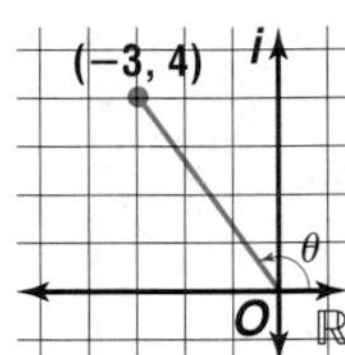

Then find the modulus.

$r = \sqrt{(-3)^2 + 4^2}$ or 5

Now find the amplitude. Notice that θ is in Quadrant II.

$$\begin{aligned} \theta &= \text{Arctan}\,\frac{4}{-3} + \pi \\ &\approx 2.21 \end{aligned}$$

Therefore, $-3 + 4\boldsymbol{i} \approx 5(\cos 2.21 + \boldsymbol{i}\sin 2.21)$ or $5 \text{ cis } 2.21$.

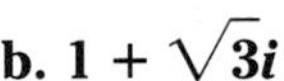

b. $\mathbf{1 + \sqrt{3}i}$

First, plot the number in the complex plane.

Then find the modulus.

$r = \sqrt{1^2 + \left(\sqrt{3}\right)^2}$ or 2

Now find the amplitude. Notice that θ is in Quadrant I.

$\theta = \text{Arctan}\,\frac{\sqrt{3}}{1}$ or $\frac{\pi}{3}$

Therefore, $1 + \sqrt{3}\boldsymbol{i} = 2\left(\cos\frac{\pi}{3} + \boldsymbol{i}\sin\frac{\pi}{3}\right)$ or $2 \text{ cis } \frac{\pi}{3}$.

You can also graph complex numbers in polar form.

Example **5** **Graph $4\left(\cos \frac{11\pi}{6} + i \sin \frac{11\pi}{6}\right)$. Then express it in rectangular form.**

In the polar form of this complex number, the value of r is 4, and the value of θ is $\frac{11\pi}{6}$. Plot the point with polar coordinates $\left(4, \frac{11\pi}{6}\right)$.

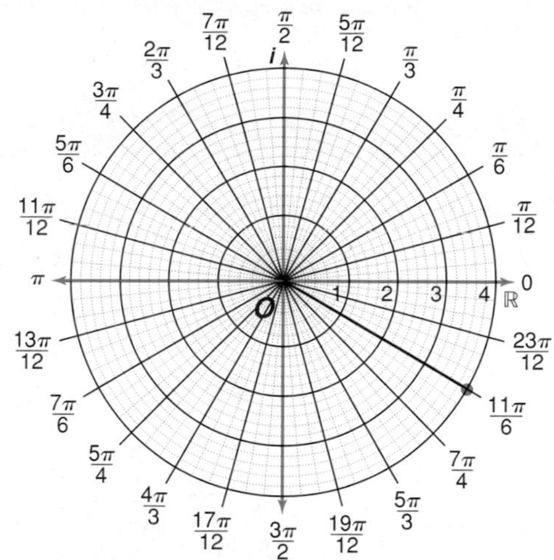

To express the number in rectangular form, simplify the trigonometric values:

$4\left(\cos \frac{11\pi}{6} + i \sin \frac{11\pi}{6}\right)$

$= 4\left(\frac{\sqrt{3}}{2} + i\left(-\frac{1}{2}\right)\right)$

$= 2\sqrt{3} - 2i$

CHECK FOR UNDERSTANDING

Communicating Mathematics

Read and study the lesson to answer each question.

1. **Explain** how to find the absolute value of a complex number.
2. **Write** the polar form of i.
3. **Find a counterexample** to the statement $|z_1 + z_2| = |z_1| + |z_2|$ *for all complex numbers* z_1 *and* z_2.
4. *Math Journal* Your friend is studying complex numbers at another school at the same time that you are. She learned that the absolute value of a complex number is the square root of the product of the number and its conjugate. You know that this is not how you learned it. Write a letter to your friend explaining why this method gives the same answer as the method you know. Use algebra, but also include some numerical examples of both techniques.

Guided Practice

5. Solve the equation $2x + y + xi + yi = 5 + 4i$ for x and y, where x and y are real numbers.

Graph each number in the complex plane and find its absolute value.

6. $-2 - i$
7. $1 + \sqrt{2}i$

Express each complex number in polar form.

8. $2 - 2i$
9. $4 + 5i$
10. -2

Graph each complex number. Then express it in rectangular form.

11. $4\left(\cos \frac{\pi}{3} + i \sin \frac{\pi}{3}\right)$
12. $2(\cos 3 + i \sin 3)$
13. $\frac{3}{2}(\cos 2\pi + i \sin 2\pi)$

14. Graph the first five members of the orbit of $z_0 = -0.25 + 0.75i$ under iteration by $f(z) = z^2 + 0.5$.

15. **Vectors** The force on an object is represented by the complex number $10 + 15i$, where the components are measured in newtons.

a. What is the magnitude of the force?

b. What is the direction of the force?

EXERCISES

Practice

Solve each equation for x and y, where x and y are real numbers.

16. $2x - 5yi = 12 + 15i$

17. $1 + (x + y)i = y + 3xi$

18. $4x + yi - 5i = 2x - y + xi + 7i$

Graph each number in the complex plane and find its absolute value.

19. $2 + 3i$

20. $3 - 4i$

21. $-1 - 5i$

22. $-3i$

23. $-1 + \sqrt{5}i$

24. $4 + \sqrt{2}i$

25. Find the modulus of $z = -4 + 6i$.

Express each complex number in polar form.

26. $3 + 3i$

27. $-1 - \sqrt{3}i$

28. $6 - 8i$

29. $-4 + i$

30. $20 - 21i$

31. $-2 + 4i$

32. 3

33. $-4\sqrt{2}$

34. $-2i$

Graph each complex number. Then express it in rectangular form.

35. $3\left(\cos \frac{\pi}{4} + i \sin \frac{\pi}{4}\right)$

36. $\cos\left(-\frac{\pi}{6}\right) + i \sin\left(-\frac{\pi}{6}\right)$

37. $2\left(\cos \frac{4\pi}{3} + i \sin \frac{4\pi}{3}\right)$

38. $10(\cos 6 + i \sin 6)$

39. $2\left(\cos \frac{5\pi}{4} + i \sin \frac{5\pi}{4}\right)$

40. $2.5(\cos 1 + i \sin 1)$

41. $5(\cos 0 + i \sin 0)$

42. $3(\cos \pi + i \sin \pi)$

Graph the first five members of the orbit of each initial value under iteration by the given function.

43. $z_0 = -0.5 + i, f(z) = z^2 + 0.5$

44. $z_0 = \frac{\sqrt{2}}{2} + \frac{\sqrt{2}}{2}i, f(z) = z^2$

45. Graph the first five iterates of $z_0 = 0.5 - 0.5i$ under $f(z) = z^2 - 0.5$.

Applications and Problem Solving

46. **Electrical Engineering** Refer to Exercise 44 in Lesson 9-3. Consider a circuit with alternating current that contains two voltage sources in series. Suppose these two voltages are given by $v_1(t) = 40 \sin (250t + 30°)$ and $v_2(t) = 60 \sin (250t + 60°)$, where t represents time, in seconds.

a. The phasors for these two voltage sources are written as $40\angle 30°$ and $60\angle 60°$, respectively. Convert these phasors to complex numbers in rectangular form. (Use j as the imaginary unit, as electrical engineers do.)

b. Add these two complex numbers to find the total voltage in the circuit.

c. Write a sinusoidal function that gives the total voltage in the circuit.

47. **Critical Thinking** How are the polar forms of complex conjugates alike? How are they different?

48. **Electricity** A series circuit contains two sources of impedance, one of $10(\cos 0.7 + \boldsymbol{j} \sin 0.7)$ ohms and the other of $16(\cos 0.5 + \boldsymbol{j} \sin 0.5)$ ohms.
    a. Convert these complex numbers to rectangular form.
    b. Add your answers from part a to find the total impedance in the circuit.
    c. Convert the total impedance back to polar form.

49. **Transformations** Certain operations with complex numbers correspond to geometric transformations in the complex plane. Describe the transformation applied to point z to obtain point w in the complex plane for each of the following operations.
    a. $w = z + (2 - 3\boldsymbol{i})$
    b. $w = \boldsymbol{i} \cdot z$
    c. $w = 3z$
    d. w is the conjugate of z

50. **Critical Thinking** Choose any two complex numbers, z_1 and z_2, in rectangular form.
    a. Find the product z_1z_2.
    b. Write z_1, z_2, and z_1z_2 in polar form.
    c. Repeat this procedure with a different pair of complex numbers.
    d. Make a conjecture about the product of two complex numbers in polar form.

Mixed Review

51. Simplify $(6 - 2\boldsymbol{i})(-2 + 3\boldsymbol{i})$. *(Lesson 9-5)*

52. Find the rectangular coordinates of the point with polar coordinates $(-3, -135°)$. *(Lesson 9-3)*

53. Find the magnitude of the vector $\langle -3, 7 \rangle$, and write the vector as a sum of unit vectors. *(Lesson 8-2)*

54. Use a sum or difference identity to find tan 105°. *(Lesson 7-3)*

55. **Mechanics** A pulley of radius 18 centimeters turns at 12 revolutions per second. What is the linear velocity of the belt driving the pulley in meters per second? *(Lesson 6-2)*

56. If $a = 12$ and $c = 18$ in $\triangle ABC$, find the measure of angle A to the nearest tenth of a degree. *(Lesson 5-5)*

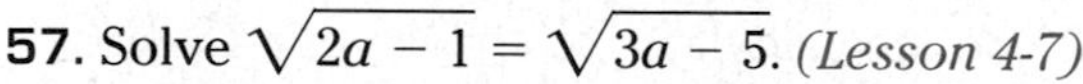

57. Solve $\sqrt{2a - 1} = \sqrt{3a - 5}$. *(Lesson 4-7)*

58. Without graphing, describe the end behavior of the graph of $y = 2x^2 + 2$. *(Lesson 3-5)*

59. **SAT/ACT Practice** A person is hired for a job that pays \$500 per month and receives a 10% raise in each following month. In the fourth month, how much will that person earn?
    A \$550 **B** \$600.50 **C** \$650.50 **D** \$665.50 **E** \$700

Extra Practice See p. A43.

GRAPHING CALCULATOR EXPLORATION

9-6B Geometry in the Complex Plane

OBJECTIVE

- Explore geometric relationships in the complex plane.

An Extension of Lesson 9-6

Many geometric figures and relationships can be described by using complex numbers. To show points on figures, you can store the real and imaginary parts of the complex numbers that correspond to the points in lists **L1** and **L2** and use **STAT PLOT** to graph the points.

TRY THESE

1. Store $-1 + 2i$ as M and $1 + 5i$ as N. Now consider complex numbers of the form $(1 - T)M + TN$, where T is a real number. You can generate several numbers of this form and store their real and imaginary parts in **L1** and **L2**, respectively, by entering the following instructions on the home screen.

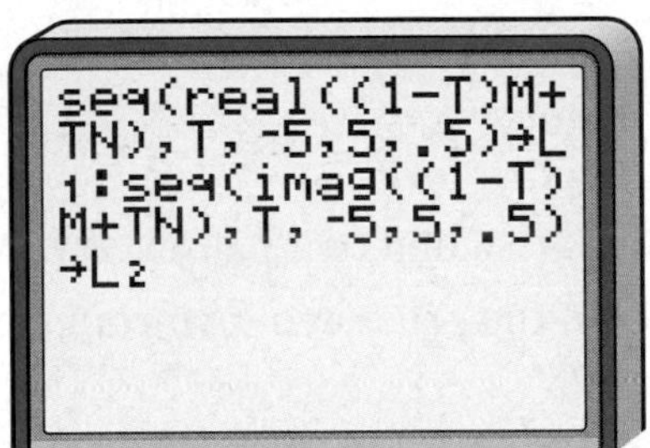

seq(*is in the* ***LIST OPS*** *menu.*
real(*and* ***imag(*** *are in the* ***MATH CPX*** *menu.*

Use a graphing window of $[-10, 10]$ scl:1 by $[-25, 25]$ scl:5. Turn on **Plot 1** and use a scatter plot to display the points defined in **L1** and **L2**. What do you notice about the points in the scatter plot?

2. Are the original numbers M and N shown in the scatter plot? Explain.

3. Repeat Exercise 1 storing $-1 + 1.5i$ as M and $-2 - i$ as N. Describe your results.

4. Repeat Exercises 1 and 2 for several complex numbers M and N of your choice. (You may need to change the window settings.) Then make a conjecture about where points of the form $(1 - T)M + TN$ are located in relation to M and N.

5. Suppose K, M, and N are three noncollinear points in the complex plane. Where will you find all the points that can be expressed in the form $aK + bM + cN$, where a, b, and c are nonnegative real numbers such that $a + b + c = 1$? Use the calculator to check your answer.

WHAT DO YOU THINK?

6. In Exercises 1-4, where is $(1 - T)M + TN$ in relation to M and N if the value of T is between 0 and 1?

7. Where in the complex plane will you find the complex numbers z that satisfy the equation $|z - (1 - i)| = 5$?

8. What equation models the points in the complex plane that lie on the circle of radius 2 that is centered at the point $2 + 3i$?

9-7 Products and Quotients of Complex Numbers in Polar Form

OBJECTIVE

- Find the product and quotient of complex numbers in polar form.

ELECTRICITY Complex numbers can be used in the study of electricity, specifically alternating current (AC). There are three basic quantities to consider:

- the *current I*, measured in amperes,
- the *impedance Z* to the current, measured in ohms, and
- the *electromotive force E* or *voltage*, measured in volts.

These three quantities are related by the equation $E = I \cdot Z$. Current, impedance, and voltage can be expressed as complex numbers. Electrical engineers use j as the imaginary unit, so they write complex numbers in the form $a + bj$. For the total impedance $a + bj$, the real part a represents the opposition to current flow due to resistors, and the imaginary part b is related to the opposition due to inductors and capacitors. If a circuit has a total impedance of $2 - 6j$ ohms and a voltage of 120 volts, find the current in the circuit. *This problem will be solved in Example 3.*

Multiplication and division of complex numbers in polar form are closely tied to geometric transformations in the complex plane. Let $r_1(\cos \theta_1 + \boldsymbol{i} \sin \theta_1)$ and $r_2(\cos \theta_2 + \boldsymbol{i} \sin \theta_2)$ be two complex numbers in polar form. A formula for the product of the two numbers can be derived by multiplying the two numbers directly and simplifying the result.

$$
\begin{aligned}
& r_1(\cos \theta_1 + \boldsymbol{i} \sin \theta_1) \cdot r_2(\cos \theta_2 + \boldsymbol{i} \sin \theta_2) \\
&= r_1r_2(\cos \theta_1 \cos \theta_2 + \boldsymbol{i} \cos \theta_1 \sin \theta_2 + \boldsymbol{i} \sin \theta_1 \cos \theta_2 + \boldsymbol{i}^2 \sin \theta_1 \sin \theta_2) \\
&= r_1r_2[(\cos \theta_1 \cos \theta_2 - \sin \theta_1 \sin \theta_2) + \boldsymbol{i}(\sin \theta_1 \cos \theta_2 + \cos \theta_1 \sin \theta_2)] \quad \textit{i}^2 = -1 \\
&= r_1r_2[\cos (\theta_1 + \theta_2) + \boldsymbol{i} \sin (\theta_1 + \theta_2)] \quad \textit{Sum identities for cosine and sine}
\end{aligned}
$$

Product of Complex Numbers in Polar Form

$$r_1(\cos \theta_1 + \boldsymbol{i} \sin \theta_1) \cdot r_2(\cos \theta_2 + \boldsymbol{i} \sin \theta_2) = r_1r_2[\cos (\theta_1 + \theta_2) + \boldsymbol{i} \sin (\theta_1 + \theta_2)]$$

Notice that the modulus (r_1r_2) of the product of the two complex numbers is the product of their moduli. The amplitude $(\theta_1 + \theta_2)$ of the product is the sum of the amplitudes.

Example 1 **Find the product $3\left(\cos \frac{7\pi}{6} + i \sin \frac{7\pi}{6}\right) \cdot 2\left(\cos \frac{2\pi}{3} + i \sin \frac{2\pi}{3}\right)$. Then express the product in rectangular form.**

Find the modulus and amplitude of the product.

$$\begin{aligned} r &= r_1r_2 & \theta &= \theta_1 + \theta_2 \\ &= 3(2) & &= \frac{7\pi}{6} + \frac{2\pi}{3} \\ &= 6 & &= \frac{11\pi}{6} \end{aligned}$$

The product is $6\left(\cos \frac{11\pi}{6} + i \sin \frac{11\pi}{6}\right)$.

Now find the rectangular form of the product.

$$6\left(\cos \frac{11\pi}{6} + i \sin \frac{11\pi}{6}\right) = 6\left(\frac{\sqrt{3}}{2} - \frac{1}{2}i\right) \quad \textit{$\cos \frac{11\pi}{6} = \frac{\sqrt{3}}{2}$, $\sin \frac{11\pi}{6} = -\frac{1}{2}$}$$

$$= 3\sqrt{3} - 3i$$

The rectangular form of the product is $3\sqrt{3} - 3i$.

Suppose the quotient of two complex numbers is expressed as a fraction. A formula for this quotient can be derived by rationalizing the denominator. To rationalize the denominator, multiply both the numerator and denominator by the same value so that the resulting new denominator does not contain imaginary numbers.

$$\frac{r_1(\cos \theta_1 + i \sin \theta_1)}{r_2(\cos \theta_2 + i \sin \theta_2)}$$

$$= \frac{r_1(\cos \theta_1 + i \sin \theta_1)}{r_2(\cos \theta_2 + i \sin \theta_2)} \cdot \frac{(\cos \theta_2 - i \sin \theta_2)}{(\cos \theta_2 - i \sin \theta_2)} \quad \textit{$\cos \theta_2 - i \sin \theta_2$ is the conjugate of $\cos \theta_2 + i \sin \theta_2$.}$$

$$= \frac{r_1}{r_2} \cdot \frac{(\cos \theta_1 \cos \theta_2 + \sin \theta_1 \sin \theta_2) + i(\sin \theta_1 \cos \theta_2 - \cos \theta_1 \sin \theta_2)}{\cos^2 \theta_2 + \sin^2 \theta_2}$$

$$= \frac{r_1}{r_2}[\cos (\theta_1 - \theta_2) + i \sin (\theta_1 - \theta_2)] \quad \textit{Trigonometric identities}$$

Quotient of Complex Numbers in Polar Form

$$\frac{r_1(\cos \theta_1 + i \sin \theta_1)}{r_2(\cos \theta_2 + i \sin \theta_2)} = \frac{r_1}{r_2}[\cos (\theta_1 - \theta_2) + i \sin (\theta_1 - \theta_2)]$$

Notice that the modulus $\left(\frac{r_1}{r_2}\right)$ of the quotient of two complex numbers is the quotient of their moduli. The amplitude $(\theta_1 - \theta_2)$ of the quotient is the difference of the amplitudes.

Example 2 **Find the quotient $12\left(\cos \frac{\pi}{4} + i \sin \frac{\pi}{4}\right) \div 4\left(\cos \frac{3\pi}{2} + i \sin \frac{3\pi}{2}\right)$. Then express the quotient in rectangular form.**

Find the modulus and amplitude of the quotient.

$$r = \frac{r_1}{r_2} \qquad \theta = \theta_1 - \theta_2$$
$$= \frac{12}{4} \qquad = \frac{\pi}{4} - \frac{3\pi}{2}$$
$$= 3 \qquad = -\frac{5\pi}{4}$$

The quotient is $3\left[\cos\left(-\frac{5\pi}{4}\right) + i \sin\left(-\frac{5\pi}{4}\right)\right]$.

Now find the rectangular form of the quotient.

$$3\left[\cos\left(-\frac{5\pi}{4}\right) + i \sin\left(-\frac{5\pi}{4}\right)\right] = 3\left(-\frac{\sqrt{2}}{2} + \frac{\sqrt{2}}{2}i\right) \quad \textit{$\cos\left(-\frac{5\pi}{4}\right) = -\frac{\sqrt{2}}{2}$, $\sin\left(-\frac{5\pi}{4}\right) = \frac{\sqrt{2}}{2}$}$$
$$= -\frac{3\sqrt{2}}{2} + \frac{3\sqrt{2}}{2}i$$

The rectangular form of the quotient is $-\frac{3\sqrt{2}}{2} + \frac{3\sqrt{2}}{2}i$.

You can use products and quotients of complex numbers in polar form to solve the problem presented at the beginning of the lesson.

Example 3 **ELECTRICITY If a circuit has an impedance of $2 - 6j$ ohms and a voltage of 120 volts, find the current in the circuit.**

Express each complex number in polar form.

$120 = 120(\cos 0 + j \sin 0)$

$2 - 6j \approx \sqrt{40}[\cos(-1.25) + j \sin(-1.25)]$ *$r = \sqrt{2^2 + (-6)^2} = \sqrt{40}$ or $2\sqrt{10}$,*

$\approx 2\sqrt{10}[\cos(-1.25) + j \sin(-1.25)]$ *$\theta = \text{Arctan}\frac{-6}{2}$ or -1.25*

Substitute the voltage and impedance into the equation $E = I \cdot Z$.

$$E = I \cdot Z$$
$$120(\cos 0 + j \sin 0) = I \cdot 2\sqrt{10}[\cos(-1.25) + j \sin(-1.25)]$$
$$\frac{120(\cos 0 + j \sin 0)}{2\sqrt{10}[\cos(-1.25) + j \sin(-1.25)]} = I$$
$$6\sqrt{10}(\cos 1.25 + j \sin 1.25) = I$$

Now express the current in rectangular form.

$I = 6\sqrt{10}(\cos 1.25 + j \sin 1.25)$

$\approx 5.98 + 18.01j$ *Use a calculator. $6\sqrt{10} \cos 1.25 \approx 5.98$, $6\sqrt{10} \sin 1.25 \approx 18.01$*

The current is about $6 + 18j$ amps.

CHECK FOR UNDERSTANDING

Communicating Mathematics

Read and study the lesson to answer each question.

1. **Explain** how to find the quotient of two complex numbers in polar form.
2. **Describe** how to square a complex number in polar form.
3. **List** which operations with complex numbers you think are easier in rectangular form and which you think are easier in polar form. Defend your choices with examples.

Guided Practice

Find each product or quotient. Express the result in rectangular form.

4. $2\left(\cos\frac{\pi}{2} + \boldsymbol{i}\sin\frac{\pi}{2}\right) \cdot 2\left(\cos\frac{3\pi}{2} + \boldsymbol{i}\sin\frac{3\pi}{2}\right)$
5. $3\left(\cos\frac{\pi}{6} + \boldsymbol{i}\sin\frac{\pi}{6}\right) \div 4\left(\cos\frac{2\pi}{3} + \boldsymbol{i}\sin\frac{2\pi}{3}\right)$
6. $4\left(\cos\frac{9\pi}{4} + \boldsymbol{i}\sin\frac{9\pi}{4}\right) \div 2\left[\cos\left(-\frac{\pi}{2}\right) + \boldsymbol{i}\sin\left(-\frac{\pi}{2}\right)\right]$
7. $\frac{1}{2}\left(\cos\frac{\pi}{3} + \boldsymbol{i}\sin\frac{\pi}{3}\right) \cdot 6\left(\cos\frac{5\pi}{6} + \boldsymbol{i}\sin\frac{5\pi}{6}\right)$
8. Use polar form to find the product $\left(2 + 2\sqrt{3}\boldsymbol{i}\right) \cdot \left(-3 + \sqrt{3}\boldsymbol{i}\right)$. Express the result in rectangular form.
9. **Electricity** Determine the voltage in a circuit when there is a current of $2\left(\cos\frac{11\pi}{6} + \boldsymbol{j}\sin\frac{11\pi}{6}\right)$ amps and an impedance of $3\left(\cos\frac{\pi}{3} + \boldsymbol{j}\sin\frac{\pi}{3}\right)$ ohms.

EXERCISES

Practice

Find each product or quotient. Express the result in rectangular form.

10. $4\left(\cos\frac{\pi}{3} + \boldsymbol{i}\sin\frac{\pi}{3}\right) \cdot 7\left(\cos\frac{2\pi}{3} + \boldsymbol{i}\sin\frac{2\pi}{3}\right)$
11. $6\left(\cos\frac{3\pi}{4} + \boldsymbol{i}\sin\frac{3\pi}{4}\right) \div 2\left(\cos\frac{\pi}{4} + \boldsymbol{i}\sin\frac{\pi}{4}\right)$
12. $\frac{1}{2}\left(\cos\frac{\pi}{3} + \boldsymbol{i}\sin\frac{\pi}{3}\right) \div 3\left(\cos\frac{\pi}{6} + \boldsymbol{i}\sin\frac{\pi}{6}\right)$
13. $5(\cos\pi + \boldsymbol{i}\sin\pi) \cdot 2\left(\cos\frac{3\pi}{4} + \boldsymbol{i}\sin\frac{3\pi}{4}\right)$
14. $6\left[\cos\left(-\frac{\pi}{3}\right) + \boldsymbol{i}\sin\left(-\frac{\pi}{3}\right)\right] \cdot 3\left(\cos\frac{5\pi}{6} + \boldsymbol{i}\sin\frac{5\pi}{6}\right)$
15. $3\left(\cos\frac{7\pi}{3} + \boldsymbol{i}\sin\frac{7\pi}{3}\right) \div \left(\cos\frac{\pi}{2} + \boldsymbol{i}\sin\frac{\pi}{2}\right)$
16. $2(\cos 240° + \boldsymbol{i}\sin 240°) \cdot 3(\cos 60° + \boldsymbol{i}\sin 60°)$
17. $\sqrt{2}\left(\cos\frac{7\pi}{4} + \boldsymbol{i}\sin\frac{7\pi}{4}\right) \div \frac{\sqrt{2}}{2}\left(\cos\frac{3\pi}{4} + \boldsymbol{i}\sin\frac{3\pi}{4}\right)$
18. $3(\cos 4 + \boldsymbol{i}\sin 4) \cdot 0.5(\cos 2.5 + \boldsymbol{i}\sin 2.5)$

19. $4[\cos(-2) + \boldsymbol{i}\sin(-2)] \div (\cos 3.6 + \boldsymbol{i}\sin 3.6)$

20. $20\left(\cos\frac{7\pi}{6} + \boldsymbol{i}\sin\frac{7\pi}{6}\right) \div 15\left(\cos\frac{11\pi}{3} + \boldsymbol{i}\sin\frac{11\pi}{3}\right)$

21. $2\left(\cos\frac{3\pi}{4} + \boldsymbol{i}\sin\frac{3\pi}{4}\right) \cdot \sqrt{2}\left(\cos\frac{\pi}{2} + \boldsymbol{i}\sin\frac{\pi}{2}\right)$

22. Find the product of $2\left(\cos\frac{\pi}{3} + \boldsymbol{i}\sin\frac{\pi}{3}\right)$ and $6\left[\cos\left(-\frac{\pi}{6}\right) + \boldsymbol{i}\sin\left(-\frac{\pi}{6}\right)\right]$. Write the answer in rectangular form.

23. If $z_1 = 4\left(\cos\frac{5\pi}{3} + \boldsymbol{i}\sin\frac{5\pi}{3}\right)$ and $z_2 = \frac{1}{2}\left(\cos\frac{\pi}{3} + \boldsymbol{i}\sin\frac{\pi}{3}\right)$, find $\frac{z_1}{z_2}$ and express the result in rectangular form.

Use polar form to find each product or quotient. Express the result in rectangular form.

24. $(2 - 2\boldsymbol{i}) \cdot (-3 + 3\boldsymbol{i})$

25. $\left(\sqrt{2} - \sqrt{2}\boldsymbol{i}\right) \cdot \left(-3\sqrt{2} - 3\sqrt{2}\boldsymbol{i}\right)$

26. $\left(\sqrt{3} - \boldsymbol{i}\right) \div \left(2 - 2\sqrt{3}\boldsymbol{i}\right)$

27. $\left(-4\sqrt{2} + 4\sqrt{2}\boldsymbol{i}\right) \div (6 + 6\boldsymbol{i})$

Applications and Problem Solving

28. **Electricity** Find the current in a circuit with a voltage of 13 volts and an impedance of $3 - 2\boldsymbol{j}$ ohms.

29. **Electricity** Find the impedance in a circuit with a voltage of 100 volts and a current of $4 - 3\boldsymbol{j}$ amps.

30. **Critical Thinking** Given z_1 and z_2 graphed at the right, graph $z_1 z_2$ and $\frac{z_1}{z_2}$ without actually calculating them.

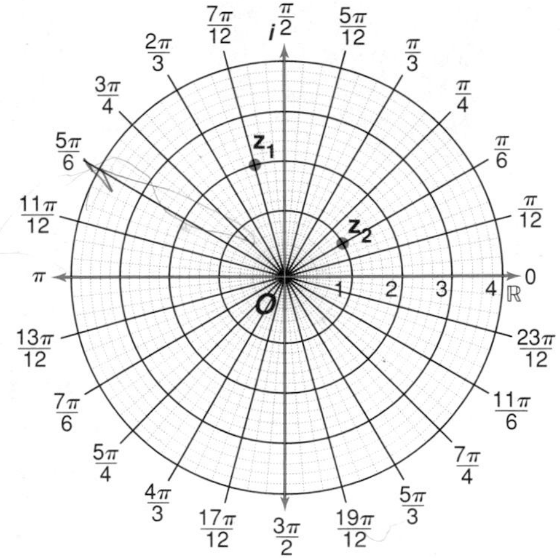

31. **Transformations**
   a. Describe the transformation applied to the graph of the complex number z if z is multiplied by $\cos\theta + \boldsymbol{i}\sin\theta$.
   b. Describe the transformation applied to the graph of the complex number z if z is multiplied by $\frac{1}{2} + \frac{\sqrt{3}}{2}\boldsymbol{i}$.

32. **Critical Thinking** Find the quadratic equation $az^2 + bz + c = 0$ such that $a = 1$ and the solutions are $3\left(\cos\frac{\pi}{3} + \boldsymbol{i}\sin\frac{\pi}{3}\right)$ and $2\left(\cos\frac{5\pi}{6} + \boldsymbol{i}\sin\frac{5\pi}{6}\right)$.

Mixed Review

33. Express $5 - 12\boldsymbol{i}$ in polar form. *(Lesson 9-6)*

34. Write the equation $r = 5\sec\left(\theta - \frac{5\pi}{6}\right)$ in rectangular form. *(Lesson 9-4)*

35. **Physics** A prop for a play is supported equally by two wires suspended from the ceiling. The wires form a 130° angle with each other. If the prop weighs 23 pounds, what is the tension in each of the wires? *(Lesson 8-5)*

Extra Practice See p. A43.

36. Solve $\cos 2x + \sin x = 1$ for principal values of x. *(Lesson 7-5)*

37. Write the equation for the inverse of $y = \cos x$. *(Lesson 6-8)*

38. **SAT/ACT Practice** In the figure, the perimeter of square *BCDE* is how much smaller than the perimeter of rectangle *ACDF*?

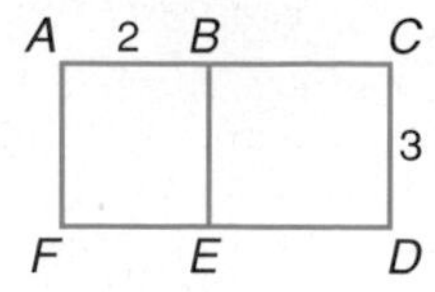

A 2 **B** 3 **C** 4

D 6 **E** 16

CAREER CHOICES

Astronomer

Have you ever gazed into the sky at night hoping to spot a constellation? Do you dream of having your own telescope? If you enjoy studying about the universe, then a career in astronomy may be just for you. Astronomers collect and analyze data about the universe including stars, planets, comets, asteroids, and even artificial satellites. As an astronomer, you may collect information by using a telescope or spectrometer here on earth, or you may use information collected by spacecraft and satellites.

Most astronomers specialize in one branch of astronomy such as astrophysics or celestial mechanics. Astronomers often teach in addition to conducting research. Astronomers located throughout the world are prime sources of information for NASA and other countries' space programs.

CAREER OVERVIEW

Degree Preferred:
at least a bachelor's degree in astronomy or physics

Related Courses:
mathematics, physics, chemistry, computer science

Outlook:
average through the year 2006

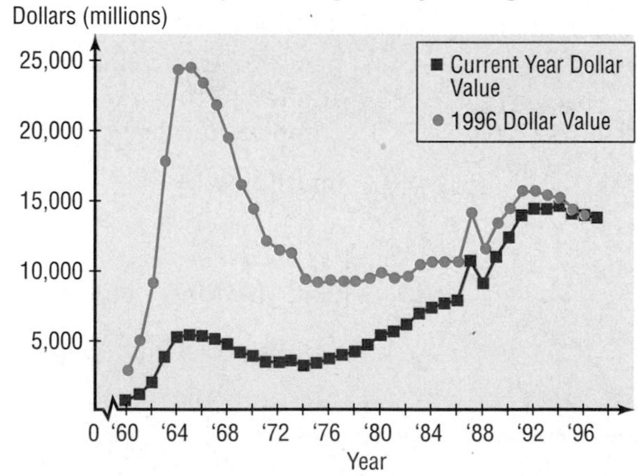

Source: National Aeronautics and Space Administration

interNET CONNECTION For more information on careers in astronomy, visit: **www.amc.glencoe.com** 

Powers and Roots of Complex Numbers

OBJECTIVE

- Find powers and roots of complex numbers in polar form using De Moivre's Theorem.

COMPUTER GRAPHICS Many of the computer graphics that are referred to as fractals are graphs of Julia sets, which are named after the mathematician Gaston Julia. When a function like $f(z) = z^2 + c$, where c is a complex constant, is iterated, points in the complex plane can be classified according to their behavior under iteration.

- Points that escape to infinity under iteration belong to the **escape set** of the function.
- Points that do not escape belong to the **prisoner set.**

The **Julia set** is the boundary between the escape set and the prisoner set. Is the number $w = 0.6 - 0.5\boldsymbol{i}$ in the escape set or the prisoner set of the function $f(z) = z^2$? *This problem will be solved in Example 6.*

You can use the formula for the product of complex numbers to find the square of a complex number.

$$\begin{aligned}[r(\cos\theta + \boldsymbol{i}\sin\theta)]^2 &= [r(\cos\theta + \boldsymbol{i}\sin\theta)] \cdot [r(\cos\theta + \boldsymbol{i}\sin\theta)]\\ &= r^2[\cos(\theta + \theta) + \boldsymbol{i}\sin(\theta + \theta)]\\ &= r^2(\cos 2\theta + \boldsymbol{i}\sin 2\theta)\end{aligned}$$

Other powers of complex numbers can be found using De Moivre's Theorem.

De Moivre's Theorem

$$[r(\cos\theta + \boldsymbol{i}\sin\theta)]^n = r^n(\cos n\theta + \boldsymbol{i}\sin n\theta)$$

You will be asked to prove De Moivre's Theorem in Chapter 12.

Example 1 **Find $\left(2 + 2\sqrt{3}\boldsymbol{i}\right)^6$.**

First, write $2 + 2\sqrt{3}\boldsymbol{i}$ in polar form. Note that its graph is in the first quadrant of the complex plane.

$$\begin{aligned} r &= \sqrt{2^2 + \left(2\sqrt{3}\right)^2} & \theta &= \text{Arctan}\ \frac{2\sqrt{3}}{2}\\ &= \sqrt{4 + 12} & &= \text{Arctan}\ \sqrt{3}\\ &= 4 & &= \frac{\pi}{3}\end{aligned}$$

(continued on the next page)

The polar form of $2 + 2\sqrt{3}\boldsymbol{i}$ is $4\left(\cos\frac{\pi}{3} + \boldsymbol{i}\sin\frac{\pi}{3}\right)$.

Now use De Moivre's Theorem to find the sixth power.

$$\begin{aligned}
(2 + 2\sqrt{3}\boldsymbol{i})^6 &= \left[4\left(\cos\frac{\pi}{3} + \boldsymbol{i}\sin\frac{\pi}{3}\right)\right]^6 \\
&= 4^6\left[\cos 6\left(\frac{\pi}{3}\right) + \boldsymbol{i}\sin 6\left(\frac{\pi}{3}\right)\right] \\
&= 4096(\cos 2\pi + \boldsymbol{i}\sin 2\pi) \\
&= 4096(1 + 0\boldsymbol{i}) \quad \textit{Write the result in rectangular form.} \\
&= 4096
\end{aligned}$$

Therefore, $\left(2 + 2\sqrt{3}\boldsymbol{i}\right)^6 = 4096$.

De Moivre's Theorem is valid for all rational values of n. Therefore, it is also useful for finding negative powers of complex numbers and roots of complex numbers.

Example 2 **Find** $\left(\frac{\sqrt{3}}{2} - \frac{1}{2}\boldsymbol{i}\right)^{-5}$.

First, write $\frac{\sqrt{3}}{2} - \frac{1}{2}\boldsymbol{i}$ in polar form. Note that its graph is in the fourth quadrant of the complex plane.

$$r = \sqrt{\left(\frac{\sqrt{3}}{2}\right)^2 + \left(-\frac{1}{2}\right)^2} \qquad \theta = \text{Arctan}\,\frac{-\frac{1}{2}}{\frac{\sqrt{3}}{2}}$$

$$= \sqrt{\frac{3}{4} + \frac{1}{4}} \text{ or } 1 \qquad = \text{Arctan}\left(-\frac{\sqrt{3}}{3}\right) \text{ or } -\frac{\pi}{6}$$

The polar form of $\frac{\sqrt{3}}{2} - \frac{1}{2}\boldsymbol{i}$ is $1\left[\cos\left(-\frac{\pi}{6}\right) + \boldsymbol{i}\sin\left(-\frac{\pi}{6}\right)\right]$.

Use De Moivre's Theorem to find the negative 5th power.

$$\begin{aligned}
\left(\frac{\sqrt{3}}{2} - \frac{1}{2}\boldsymbol{i}\right)^{-5} &= \left[1\left(\cos\left(-\frac{\pi}{6}\right) + \boldsymbol{i}\sin\left(-\frac{\pi}{6}\right)\right)\right]^{-5} \\
&= 1^{-5}\left[\cos(-5)\left(-\frac{\pi}{6}\right) + \boldsymbol{i}\sin(-5)\left(-\frac{\pi}{6}\right)\right] \quad \textit{De Moivre's Theorem} \\
&= 1\left(\cos\frac{5\pi}{6} + \boldsymbol{i}\sin\frac{5\pi}{6}\right) \quad \textit{Simplify.} \\
&= -\frac{\sqrt{3}}{2} + \frac{1}{2}\boldsymbol{i} \quad \textit{Write the answer in rectangular form.}
\end{aligned}$$

Recall that positive real numbers have two square roots and that the positive one is called the principal square root. In general, all nonzero complex numbers have p distinct pth roots. That is, they each have two square roots, three cube roots, four fourth roots, and so on. The principal pth root of a complex number is given by:

$$(a + b\boldsymbol{i})^{\frac{1}{p}} = [r(\cos\theta + \boldsymbol{i}\sin\theta)]^{\frac{1}{p}}$$
$$= r^{\frac{1}{p}}\left(\cos\frac{\theta}{p} + \boldsymbol{i}\sin\frac{\theta}{p}\right).$$

When finding a principal root, the interval $-\pi < \theta \leq \pi$ is used.

Example 3 **Find $\sqrt[3]{8\boldsymbol{i}}$.**

$\sqrt[3]{8\boldsymbol{i}} = (0 + 8\boldsymbol{i})^{\frac{1}{3}}$ *$a = 0$, $b = 8$*

$= \left[8\left(\cos\frac{\pi}{2} + \boldsymbol{i}\sin\frac{\pi}{2}\right)\right]^{\frac{1}{3}}$ *Polar form; $r = \sqrt{0^2 + 8^2}$ or 8, $\theta = \frac{\pi}{2}$ since $a = 0$.*

$= 8^{\frac{1}{3}}\left[\cos\left(\frac{1}{3}\right)\left(\frac{\pi}{2}\right) + \boldsymbol{i}\sin\left(\frac{1}{3}\right)\left(\frac{\pi}{2}\right)\right]$ *De Moivre's Theorem*

$= 2\left(\cos\frac{\pi}{6} + \boldsymbol{i}\sin\frac{\pi}{6}\right)$

$= 2\left(\frac{\sqrt{3}}{2} + \frac{1}{2}\boldsymbol{i}\right)$ or $\sqrt{3} + \boldsymbol{i}$ *This is the principal cube root.*

The following formula generates all of the pth roots of a complex number. It is based on the identities $\cos\theta = \cos(\theta + 2n\pi)$ and $\sin\theta = \sin(\theta + 2n\pi)$, where n is any integer.

The p Distinct pth Roots of a Complex Number

The p distinct pth roots of $a + bi$ can be found by replacing n with 0, 1, 2, ..., $p - 1$, successively, in the following equation.

$$(a + bi)^{\frac{1}{p}} = (r[\cos(\theta + 2n\pi) + i\sin(\theta + 2n\pi)])^{\frac{1}{p}}$$
$$= r^{\frac{1}{p}}\left(\cos\frac{\theta + 2n\pi}{p} + i\sin\frac{\theta + 2n\pi}{p}\right)$$

Example 4 **Find the three cube roots of $-2 - 2i$.**

First, write $-2 - 2\boldsymbol{i}$ in polar form.

$r = \sqrt{(-2)^2 + (-2)^2}$ or $2\sqrt{2}$ $\qquad \theta = \text{Arctan}\,\frac{-2}{-2} + \pi$ or $\frac{5\pi}{4}$

$-2 - 2\boldsymbol{i} = 2\sqrt{2}\left[\cos\left(\frac{5\pi}{4} + 2n\pi\right) + \boldsymbol{i}\sin\left(\frac{5\pi}{4} + 2n\pi\right)\right]$ *n is any integer.*

Now write an expression for the cube roots.

$(-2 - 2\boldsymbol{i})^{\frac{1}{3}} = \left(2\sqrt{2}\left[\cos\left(\frac{5\pi}{4} + 2n\pi\right) + \boldsymbol{i}\sin\left(\frac{5\pi}{4} + 2n\pi\right)\right]\right)^{\frac{1}{3}}$

$= \sqrt{2}\left[\cos\left(\frac{\frac{5\pi}{4} + 2n\pi}{3}\right) + \boldsymbol{i}\sin\left(\frac{\frac{5\pi}{4} + 2n\pi}{3}\right)\right]$ *$\left(2\sqrt{2}\right)^{\frac{1}{3}} = \left(2^{\frac{3}{2}}\right)^{\frac{1}{3}} = 2^{\frac{1}{2}}$ or $\sqrt{2}$*

(continued on the next page)

Let $n = 0$, 1, and 2 successively to find the cube roots.

Let n = 0. $\sqrt{2}\left[\cos\left(\dfrac{\frac{5\pi}{4} + 2(0)\pi}{3}\right) + \boldsymbol{i}\sin\left(\dfrac{\frac{5\pi}{4} + 2(0)\pi}{3}\right)\right]$

$= \sqrt{2}\left(\cos\frac{5\pi}{12} + \boldsymbol{i}\sin\frac{5\pi}{12}\right)$

$\approx 0.37 + 1.37\boldsymbol{i}$

Let n = 1. $\sqrt{2}\left[\cos\left(\dfrac{\frac{5\pi}{4} + 2(1)\pi}{3}\right) + \boldsymbol{i}\sin\left(\dfrac{\frac{5\pi}{4} + 2(1)\pi}{3}\right)\right]$

$= \sqrt{2}\left(\cos\frac{13\pi}{12} + \boldsymbol{i}\sin\frac{13\pi}{12}\right)$

$\approx -1.37 - 0.37i$

Let n = 2. $\sqrt{2}\left[\cos\left(\dfrac{\frac{5\pi}{4} + 2(2)\pi}{3}\right) + \boldsymbol{i}\sin\left(\dfrac{\frac{5\pi}{4} + 2(2)\pi}{3}\right)\right]$

$= \sqrt{2}\left(\cos\frac{21\pi}{12} + \boldsymbol{i}\sin\frac{21\pi}{12}\right)$

$= 1 - \boldsymbol{i}$

The cube roots of $-2 - 2\boldsymbol{i}$ are approximately $0.37 + 1.37\boldsymbol{i}$, $-1.37 - 0.37\boldsymbol{i}$, and $1 - \boldsymbol{i}$. *These roots can be checked by multiplication.*

GRAPHING CALCULATOR EXPLORATION

The p distinct pth roots of a complex number can be approximated using the parametric mode on a graphing calculator. For a particular complex number $r(\cos\theta + \boldsymbol{i}\sin\theta)$ and a particular value of p:

- Select the **Radian** and **Par** modes.
- Select the viewing window.

  Tmin $= \frac{\theta}{p}$, Tmax $= \frac{\theta}{p} + 2\pi$, Tstep $= \frac{2\pi}{p}$,

  Xmin $= -r^{\frac{1}{p}}$, Xmax $= r^{\frac{1}{p}}$, Xscl $= 1$,

  Ymin $= -r^{\frac{1}{p}}$, Ymax $= r^{\frac{1}{p}}$, and Yscl $= 1$.
- Enter the parametric equations $X_{1T} = r^{\frac{1}{p}}\cos T$ and $Y_{1T} = r^{\frac{1}{p}}\sin T$.
- Graph the equations.
- Use TRACE to locate the roots.

TRY THESE

1. Approximate the cube roots of 1.
2. Approximate the fourth roots of $\boldsymbol{i}$.
3. Approximate the fifth roots of $1 + \boldsymbol{i}$.

WHAT DO YOU THINK?

4. What geometric figure is formed when you graph the three cube roots of a complex number?
5. What geometric figure is formed when you graph the fifth roots of a complex number?
6. Under what conditions will the complex number $a + b\boldsymbol{i}$ have a root that lies on the positive real axis?

You can also use De Moivre's Theorem to solve some polynomial equations.

Examples **5** **Solve $x^5 - 32 = 0$. Then graph the roots in the complex plane.**

The solutions to this equation are the same as those of the equation $x^5 = 32$. That means we have to find the fifth roots of 32.

$32 = 32 + 0\boldsymbol{i}$ *$a = 32$, $b = 0$*

$= 32(\cos 0 + \boldsymbol{i} \sin 0)$ *Polar form; $r = \sqrt{32^2 + 0^2}$ or 32, $\theta = \text{Arctan } \frac{0}{32}$ or 0*

Now write an expression for the fifth roots.

$$32^{\frac{1}{5}} = [32(\cos(0 + 2n\pi) + \boldsymbol{i} \sin(0 + 2n\pi))]^{\frac{1}{5}}$$

$$= 2\left(\cos \frac{2n\pi}{5} + \boldsymbol{i} \sin \frac{2n\pi}{5}\right)$$

Let $n = 0, 1, 2, 3$, and 4 successively to find the fifth roots, x_1, x_2, x_3, x_4, and x_5.

Let $n = 0$. $\quad x_1 = 2(\cos 0 + \boldsymbol{i} \sin 0) = 2$

Let $n = 1$. $\quad x_2 = 2\left(\cos \frac{2\pi}{5} + \boldsymbol{i} \sin \frac{2\pi}{5}\right) \approx 0.62 + 1.90\boldsymbol{i}$

Let $n = 2$. $\quad x_3 = 2\left(\cos \frac{4\pi}{5} + \boldsymbol{i} \sin \frac{4\pi}{5}\right) \approx -1.62 + 1.18\boldsymbol{i}$

Let $n = 3$. $\quad x_4 = 2\left(\cos \frac{6\pi}{5} + \boldsymbol{i} \sin \frac{6\pi}{5}\right) \approx -1.62 - 1.18\boldsymbol{i}$

Let $n = 4$. $\quad x_5 = 2\left(\cos \frac{8\pi}{5} + \boldsymbol{i} \sin \frac{8\pi}{5}\right) \approx 0.62 - 1.90\boldsymbol{i}$

The solutions of $x^5 - 32 = 0$ are 2, $0.62 \pm 1.90\boldsymbol{i}$, and $-1.62 \pm 1.18\boldsymbol{i}$.

The solutions are graphed at the right. Notice that the points are the vertices of a regular pentagon. The roots of a complex number are cyclical in nature. That means, when the roots are graphed on the complex plane, the roots are equally spaced around a circle.

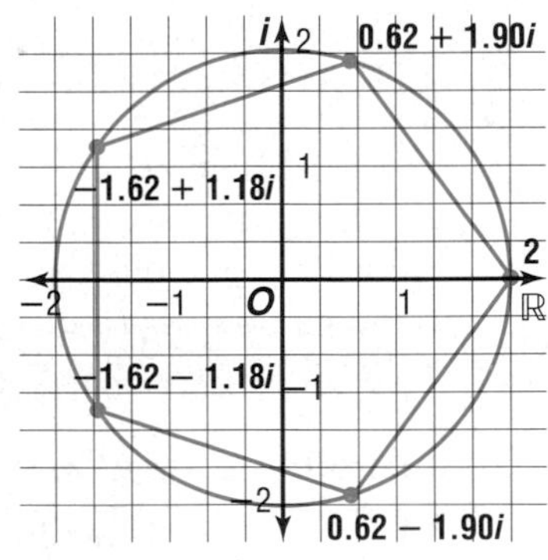

6 **COMPUTER GRAPHICS** **Refer to the application at the beginning of the lesson. Is the number $w = 0.6 - 0.5i$ in the escape set or the prisoner set of the function $f(z) = z^2$?**

Real World Application

Iterating this function requires you to square complex numbers, so you can use De Moivre's Theorem.

Write w in polar form.   *$r = \sqrt{0.6^2 + (-0.5)^2}$ or about 0.78*

$w = 0.78[\cos(-0.69) + \boldsymbol{i} \sin(-0.69)]$ *$\theta = \text{Arctan } \frac{-0.5}{0.6}$ or about -0.69*

(continued on the next page)

Graphing Calculator Programs For a program that draws Julia sets, visit: **www.amc.glencoe.com**

Now iterate the function.

$$\begin{aligned} w_1 &= f(w) \\ &= w^2 \\ &= (0.78[\cos(-0.69) + \boldsymbol{i}\sin(-0.69)])^2 \\ &= 0.78^2[\cos 2(-0.69) + \boldsymbol{i}\sin 2(-0.69)] && \textit{De Moivre's Theorem} \\ &= 0.12 - 0.60\boldsymbol{i} && \textit{Use a calculator to approximate the rectangular form.} \end{aligned}$$

$$\begin{aligned} w_2 &= f(w_1) \\ &= w_1^{\,2} \\ &= (0.78^2[\cos 2(-0.69) + \boldsymbol{i}\sin 2(-0.69)])^2 && \textit{Use the polar form of } w_1. \\ &= 0.78^4[\cos 4(-0.69) + \boldsymbol{i}\sin 4(-0.69)] \\ &= -0.34 - 0.14\boldsymbol{i} \end{aligned}$$

$$\begin{aligned} w_3 &= f(w_2) \\ &= w_2^{\,2} \\ &= (0.78^4[\cos 4(-0.69) + \boldsymbol{i}\sin 4(-0.69)])^2 && \textit{Use the polar form of } w_2. \\ &= 0.78^8[\cos 8(-0.69) + \boldsymbol{i}\sin 8(-0.69)] \\ &= 0.10 + 0.09\boldsymbol{i} \end{aligned}$$

The moduli of these iterates are 0.78^2, 0.78^4, 0.78^8, and so on. These moduli will approach 0 as the number of iterations increases. This means the graphs of the iterates approach the origin in the complex plane, so $w = 0.6 - 0.5\boldsymbol{i}$ is in the prisoner set of the function.

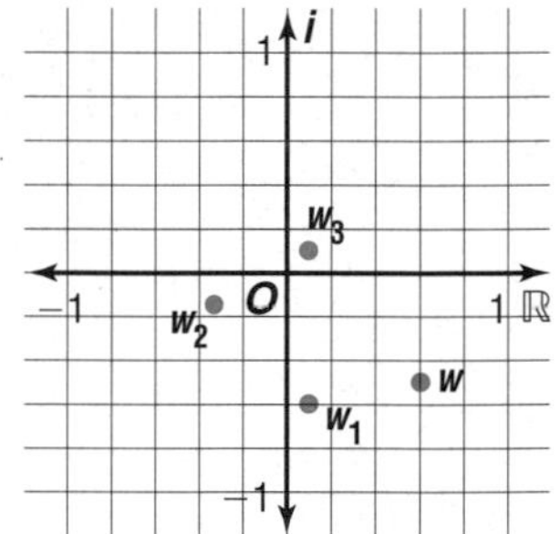

CHECK FOR UNDERSTANDING

Communicating Mathematics

Read and study the lesson to answer each question.

1. **Evaluate** the product $(1 + \boldsymbol{i})(1 + \boldsymbol{i})(1 + \boldsymbol{i})(1 + \boldsymbol{i})(1 + \boldsymbol{i})$ by traditional multiplication. Compare the results with the results using De Moivre's Theorem on $(1 + \boldsymbol{i})^5$. Which method do you prefer?

2. **Explain** how to use De Moivre's Theorem to find the reciprocal of a complex number in polar form.

3. **Graph** all the fourth roots of a complex number if $a + a\boldsymbol{i}$ is one of the fourth roots. Assume a is positive.

4. **You Decide** Shembala says that if $a \neq 0$, then $(a + a\boldsymbol{i})^2$ must be a pure imaginary number. Arturo disagrees. Who is correct? Use polar form to explain.

Guided Practice

Find each power. Express the result in rectangular form.

5. $\left(\sqrt{3} - i\right)^3$
6. $(3 - 5i)^4$

Find each principal root. Express the result in the form $a + bi$ with a and b rounded to the nearest hundredth.

7. $i^{\frac{1}{6}}$
8. $(-2 - i)^{\frac{1}{3}}$

Solve each equation. Then graph the roots in the complex plane.

9. $x^4 + i = 0$
10. $2x^3 + 4 + 2i = 0$

11. **Fractals** Refer to the application at the beginning of the lesson. Is $w = 0.8 - 0.7i$ in the prisoner set or the escape set for the function $f(z) = z^2$? Explain.

EXERCISES

Practice

Find each power. Express the result in rectangular form.

12. $\left[3\left(\cos\frac{\pi}{6} + i\sin\frac{\pi}{6}\right)\right]^3$
13. $\left[2\left(\cos\frac{\pi}{4} + i\sin\frac{\pi}{4}\right)\right]^5$
14. $(-2 + 2i)^3$
15. $\left(1 + \sqrt{3}i\right)^4$
16. $(3 - 6i)^4$
17. $(2 + 3i)^{-2}$
18. Raise $2 + 4i$ to the fourth power.

Find each principal root. Express the result in the form $a + bi$ with a and b rounded to the nearest hundredth.

19. $\left[32\left(\cos\frac{2\pi}{3} + i\sin\frac{2\pi}{3}\right)\right]^{\frac{1}{5}}$
20. $(-1)^{\frac{1}{4}}$
21. $(-2 + i)^{\frac{1}{4}}$
22. $(4 - i)^{\frac{1}{3}}$
23. $(2 + 2i)^{\frac{1}{3}}$
24. $(-1 - i)^{\frac{1}{4}}$
25. Find the principal square root of i.

Solve each equation. Then graph the roots in the complex plane.

26. $x^3 - 1 = 0$
27. $x^5 + 1 = 0$
28. $2x^4 - 128 = 0$
29. $3x^4 + 48 = 0$
30. $x^4 - (1 + i) = 0$
31. $2x^4 + 2 + 2\sqrt{3}i = 0$

Graphing Calculator

Use a graphing calculator to find all of the indicated roots.

32. fifth roots of $10 - 9i$
33. sixth roots of $2 + 4i$
34. eighth roots of $36 + 20i$

Applications and Problem Solving

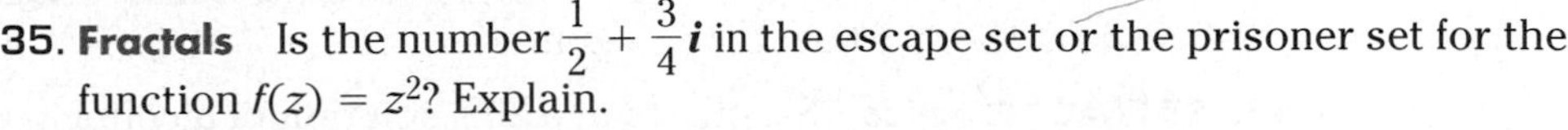

35. **Fractals** Is the number $\frac{1}{2} + \frac{3}{4}i$ in the escape set or the prisoner set for the function $f(z) = z^2$? Explain.

36. **Critical Thinking** Suppose $w = a + bi$ is one of the 31st roots of 1.
    a. What is the maximum value of a?
    b. What is the maximum value of b?

37. **Design** Gloribel works for an advertising agency. She wants to incorporate a hexagon design into the artwork for one of her proposals. She knows that she can locate the vertices of a regular hexagon by graphing the solutions to the equation $x^6 - 1 = 0$ in the complex plane. What are the solutions to this equation?

38. **Computer Graphics** Computer programmers can use complex numbers and the complex plane to implement geometric transformations. If a programmer starts with a square with vertices at (2, 2), (−2, 2), (−2, −2), and (2, −2), each of the vertices can be stored as a complex number in polar form. Complex number multiplication can be used to rotate the square 45° counterclockwise and dilate it so that the new vertices lie at the midpoints of the sides of the original square.
    a. What complex number should the programmer multiply by to produce this transformation?
    b. What happens if the original vertices are multiplied by the square of your answer to part a?

39. **Critical Thinking** Explain why the sum of the imaginary parts of the p distinct pth roots of any positive real number must be zero.

Mixed Review

40. Find the product $2\left(\cos \frac{\pi}{6} + \boldsymbol{i} \sin \frac{\pi}{6}\right) \cdot 3\left(\cos \frac{5\pi}{3} + \boldsymbol{i} \sin \frac{5\pi}{3}\right)$. Express the result in rectangular form. *(Lesson 9-7)*

41. Simplify $(2 - 5\boldsymbol{i}) + (-3 + 6\boldsymbol{i}) - (-6 + 2\boldsymbol{i})$. *(Lesson 9-5)*

42. Write parametric equations of the line with equation $y = -2x + 7$. *(Lesson 8-6)*

43. Use a half-angle identity to find the exact value of $\cos 22.5°$. *(Lesson 7-4)*

44. Solve triangle ABC if $A = 81°15'$ and $b = 28$. Round angle measures to the nearest minute and side measures to the nearest tenth. *(Lesson 5-4)*

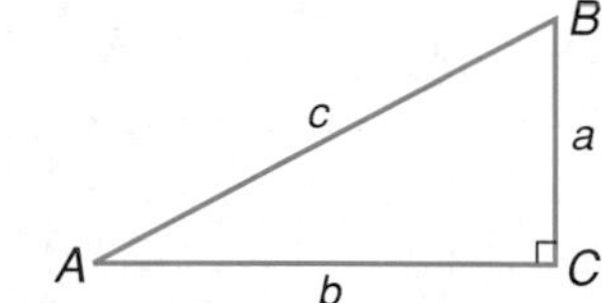

45. **Manufacturing** The Precious Animal Company must produce at least 300 large stuffed bears and 400 small stuffed bears per day. At most, the company can produce a total of 1200 bears per day. The profit for each large bear is \$9.00, and the profit for each small bear is \$5.00. How many of each type of bear should be produced each day to maximize profit? *(Lesson 2-7)*

46. **SAT/ACT Practice** Six quarts of a 20% solution of alcohol in water are mixed with 4 quarts of a 60% solution of alcohol in water. The alcoholic strength of the mixture is

A 36% **B** 40% **C** 48% **D** 60% **E** 80%

Extra Practice See p. A43.

VOCABULARY

- absolute value of a complex number (p. 586)
- amplitude of a complex number (p. 587)
- Argand plane (p. 586)
- argument of a complex number (p. 587)
- cardioid (p. 563)
- complex conjugates (p. 582)
- complex number (p. 580)
- complex plane (p. 586)
- escape set (p. 599)
- imaginary number (p. 581)
- imaginary part (p. 581)
- iteration (p. 581)
- Julia set (p. 599)
- lemniscate (p. 564)
- limaçon (p. 562)
- modulus (p. 587)
- polar axis (p. 553)
- polar coordinates (p. 553)
- polar equation (p. 556)
- polar form of a complex number (p. 588)
- polar graph (p. 556)
- polar plane (p. 553)
- pole (p. 553)
- prisoner set (p. 599)
- pure imaginary number (p. 581)
- real part (p. 581)
- rectangular form of a complex number (p. 581)
- rose (p. 563)
- spiral of Archimedes (p. 564)
- trigonometric form of a complex number (p. 588)

UNDERSTANDING AND USING THE VOCABULARY

Choose the correct term to best complete each sentence.

1. The (absolute value, conjugate) of a complex number is its distance from zero in the complex plane.
2. (Complex, Polar) coordinates give the position of an object using distances and angles.
3. Points that do not escape under iteration belong to the (escape, prisoner) set.
4. The process of repeatedly applying a function to the output produced by the previous input is called (independent, iteration).
5. A complex number in the form bi where $b \neq 0$ is a (pure imaginary, real) number.
6. A (cardioid, rose) is a special type of limaçon.
7. The complex number $a + bi$, where a and b are real numbers, is in (polar, rectangular) form.
8. The (spiral of Archimedes, lemniscate) has a polar equation of the form $r = a\theta$.
9. The complex plane is sometimes called the (Argand, polar) plane.
10. The polar form of a complex number is given by $r(\cos\theta + \boldsymbol{i}\sin\theta)$, where r represents the (modulus, amplitude) of the complex number.

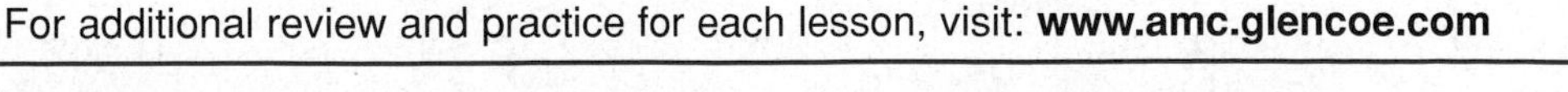

For additional review and practice for each lesson, visit: **www.amc.glencoe.com**

SKILLS AND CONCEPTS

OBJECTIVES AND EXAMPLES

Lesson 9-1 Graph points in polar coordinates.

- Graph $P\left(-2, \frac{5\pi}{6}\right)$.

  Extend the terminal side of the angle measuring $\frac{5\pi}{6}$. Find the point P that is 2 units from the pole along the ray opposite the terminal side.

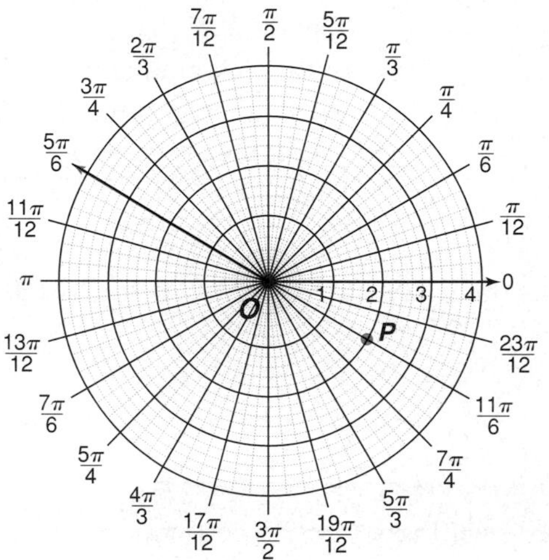

REVIEW EXERCISES

Graph each point.

11. $A(-3, 50°)$
12. $B(1.5, -110°)$
13. $C\left(2, \frac{\pi}{4}\right)$
14. $D\left(-3, \frac{\pi}{2}\right)$
15. Name four other pairs of polar coordinates for the point at $(4, 225°)$.

Graph each polar equation.

16. $r = \sqrt{7}$
17. $r = -2$
18. $\theta = -80°$
19. $\theta = \frac{3\pi}{4}$

Lesson 9-2 Graph polar equations.

- Graph $r = 3 + 3 \sin \theta$.

  The graph below can be made from a table of values. This graph is a cardioid.

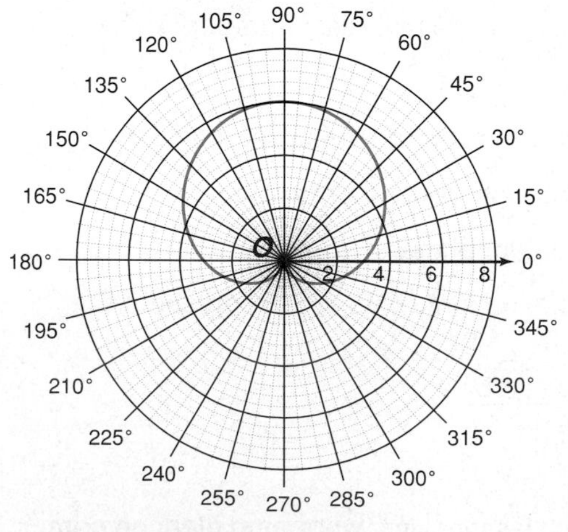

Graph each polar equation. Identify the type of curve each represents.

20. $r = 7 \cos \theta$
21. $r = 5\theta$
22. $r = 2 + 4 \cos \theta$
23. $r = 6 \sin 2\theta$

OBJECTIVES AND EXAMPLES

Lesson 9-3 Convert between polar and rectangular coordinates.

- Find the rectangular coordinates of $C\left(-3, \frac{\pi}{6}\right)$.

For $C\left(-3, \frac{\pi}{6}\right)$, $r = -3$ and $\theta = \frac{\pi}{6}$.

$$x = r\cos\theta = -3\cos\frac{\pi}{6} = -\frac{3\sqrt{3}}{2} \qquad y = r\sin\theta = -3\sin\frac{\pi}{6} = -\frac{3}{2}$$

The rectangular coordinates of C are $\left(-\frac{3\sqrt{3}}{2}, -\frac{3}{2}\right)$, or about $(-2.6, -1.5)$.

REVIEW EXERCISES

Find the rectangular coordinates of each point with the given polar coordinates.

24. $(6, 45°)$
25. $(2, 330°)$
26. $\left(-2, \frac{3\pi}{4}\right)$
27. $\left(1, \frac{\pi}{2}\right)$

Find the polar coordinates of each point with the given rectangular coordinates. Use $0 \le \theta < 2\pi$ and $r \ge 0$.

28. $\left(-\sqrt{3}, -3\right)$
29. $(5, 5)$
30. $(-3, 1)$
31. $(4, 2)$

Lesson 9-4 Write the polar form of a linear equation.

- Write $\sqrt{2}x - \sqrt{2}y = 6$ in polar form.

The standard form of the equation is $\sqrt{2}x - \sqrt{2}y - 6 = 0$.
Since $\sqrt{A^2 + B^2} = \sqrt{2 + 2}$ or 2, the normal form is $\frac{\sqrt{2}}{2}x - \frac{\sqrt{2}}{2}y - 3 = 0$.

$$\tan\phi = \frac{\sin\phi}{\cos\phi}$$
$$\tan\phi = -1$$
$$\phi = -45° \text{ or } 315°$$

The polar form is $3 = r\cos(\theta - 315°)$.

Write each equation in polar form. Round ϕ to the nearest degree.

32. $2x + y = -3$
33. $y = -3x - 4$

Write each equation in rectangular form.

34. $3 = r\cos\left(\theta - \frac{\pi}{3}\right)$
35. $4 = r\cos\left(\theta + \frac{\pi}{2}\right)$

Lesson 9-5 Add, subtract, multiply, and divide complex numbers in rectangular form.

- Simplify $(2 - 4\boldsymbol{i})(5 + 2\boldsymbol{i})$.

$$\begin{aligned}(2 - 4\boldsymbol{i})(5 + 2\boldsymbol{i}) &= 2(5 + 2\boldsymbol{i}) - 4\boldsymbol{i}(5 + 2\boldsymbol{i}) \\ &= 10 + 4\boldsymbol{i} - 20\boldsymbol{i} - 8\boldsymbol{i}^2 \\ &= 10 - 16\boldsymbol{i} - 8(-1) \\ &= 18 - 16\boldsymbol{i}\end{aligned}$$

Simplify.

36. $\boldsymbol{i}^{10} + \boldsymbol{i}^{25}$
37. $(2 + 3\boldsymbol{i}) - (4 - 4\boldsymbol{i})$
38. $(2 + 7\boldsymbol{i}) + (-3 - \boldsymbol{i})$
39. $\boldsymbol{i}^3(4 - 3\boldsymbol{i})$
40. $(\boldsymbol{i} - 7)(-\boldsymbol{i} + 7)$
41. $\frac{4 + 2\boldsymbol{i}}{5 - 2\boldsymbol{i}}$
42. $\frac{5 + \boldsymbol{i}}{1 - \sqrt{2}\boldsymbol{i}}$

OBJECTIVES AND EXAMPLES

Lesson 9-6 Convert complex numbers from rectangular to polar form and vice versa.

- Express $5 - 2\boldsymbol{i}$ in polar form.

  Find the modulus.

  $r = \sqrt{5^2 + (-2)^2}$ or $\sqrt{29}$

  Find the amplitude. Since $5 - 2\boldsymbol{i}$ is in Quadrant IV in the complex plane, θ is in Quadrant IV.

  $\theta = \text{Arctan}\,\frac{-2}{5}$ or about -0.38

  $5 - 2\boldsymbol{i} = \sqrt{29}[\cos(-0.38) + \boldsymbol{i}\sin(-0.38)]$

REVIEW EXERCISES

Express each complex number in polar form.

43. $2 + 2\boldsymbol{i}$
44. $1 - 3\boldsymbol{i}$
45. $-1 + \sqrt{3}\boldsymbol{i}$
46. $-6 - 4\boldsymbol{i}$
47. $-4 - \boldsymbol{i}$
48. 4
49. $-2\sqrt{2}$
50. $3\boldsymbol{i}$

Graph each complex number. Then express it in rectangular form.

51. $2\left(\cos\frac{\pi}{6} + \boldsymbol{i}\sin\frac{\pi}{6}\right)$
52. $3\left(\cos\frac{5\pi}{3} + \boldsymbol{i}\sin\frac{5\pi}{3}\right)$

Lesson 9-7 Find the product and quotient of complex numbers in polar form.

- Find the product of $4\left(\cos\frac{\pi}{2} + \boldsymbol{i}\sin\frac{\pi}{2}\right)$ and $3\left(\cos\frac{3\pi}{4} + \boldsymbol{i}\sin\frac{3\pi}{4}\right)$. Then express the product in rectangular form.

  $r = r_1r_2 = 4(3) = 12$ $\qquad$ $\theta = \theta_1 + \theta_2 = \frac{\pi}{2} + \frac{3\pi}{4} = \frac{5\pi}{4}$

  The product is $12\left(\cos\frac{5\pi}{4} + \boldsymbol{i}\sin\frac{5\pi}{4}\right)$.

  Now find the rectangular form of the product.

  $$12\left(\cos\frac{5\pi}{4} + \boldsymbol{i}\sin\frac{5\pi}{4}\right) = 12\left(-\frac{\sqrt{2}}{2} - \frac{\sqrt{2}}{2}\boldsymbol{i}\right) = -6\sqrt{2} - 6\sqrt{2}\boldsymbol{i}$$

Find each product or quotient. Express the result in rectangular form.

53. $4\left(\cos\frac{\pi}{3} + \boldsymbol{i}\sin\frac{\pi}{3}\right) \cdot 3\left(\cos\frac{\pi}{3} + \boldsymbol{i}\sin\frac{\pi}{3}\right)$
54. $8\left(\cos\frac{\pi}{4} + \boldsymbol{i}\sin\frac{\pi}{4}\right) \cdot 4\left(\cos\frac{\pi}{2} + \boldsymbol{i}\sin\frac{\pi}{2}\right)$
55. $2(\cos 2 + \boldsymbol{i}\sin 2) \cdot 5(\cos 0.5 + \boldsymbol{i}\sin 0.5)$
56. $8\left(\cos\frac{7\pi}{6} + \boldsymbol{i}\sin\frac{7\pi}{6}\right) \div 2\left(\cos\frac{5\pi}{3} + \boldsymbol{i}\sin\frac{5\pi}{3}\right)$
57. $6\left(\cos\frac{\pi}{2} + \boldsymbol{i}\sin\frac{\pi}{2}\right) \div 4\left(\cos\frac{\pi}{6} + \boldsymbol{i}\sin\frac{\pi}{6}\right)$
58. $2.2(\cos 1.5 + \boldsymbol{i}\sin 1.5) \div 4.4(\cos 0.6 + \boldsymbol{i}\sin 0.6)$

Lesson 9-8 Find powers and roots of complex numbers in polar form using De Moivre's Theorem.

- Find $(-3 + 3\boldsymbol{i})^4$.

  $$(-3 + 3\boldsymbol{i})^4 = \left[3\sqrt{2}\left(\cos\frac{3\pi}{4} + \boldsymbol{i}\sin\frac{3\pi}{4}\right)\right]^4 = \left(3\sqrt{2}\right)^4(\cos 3\pi + \boldsymbol{i}\sin 3\pi) = 324(-1 + 0) \text{ or } -324$$

Find each power. Express the result in rectangular form.

59. $(2 + 2\boldsymbol{i})^8$
60. $\left(\sqrt{3} - \boldsymbol{i}\right)^7$
61. $(-1 + \boldsymbol{i})^4$
62. $(-2 - 2\boldsymbol{i})^3$

Find each principal root. Express the result in the form $a + bi$ with a and b rounded to the nearest hundredth.

63. $\boldsymbol{i}^{\frac{1}{4}}$
64. $\left(\sqrt{3} + \boldsymbol{i}\right)^{\frac{1}{3}}$

APPLICATIONS AND PROBLEM SOLVING

65. **Chemistry** An electron moves about the nucleus of an atom at such a high speed that if it were visible to the eye, it would appear as a cloud. Identify the classical curve represented by the electron cloud below. *(Lesson 9-2)*

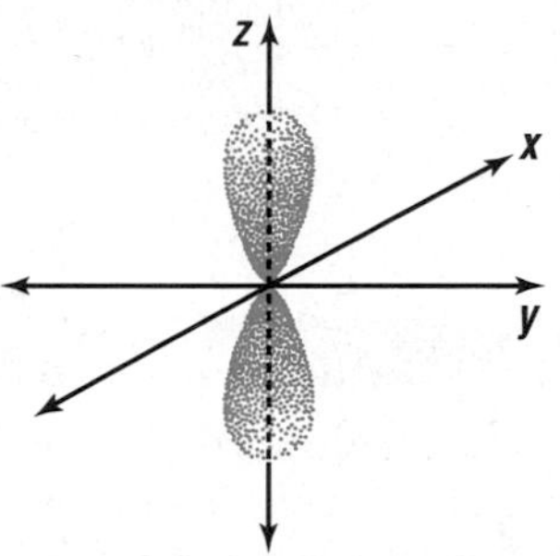

66. **Surveying** A surveyor identifies a landmark at the point with rectangular coordinates (75, 125). What are the polar coordinates of this point? *(Lesson 9-3)*

67. **Navigation** A submarine sonar is tracking a ship. The path of the ship is being coded as the equation $r \cos\left(\theta - \frac{\pi}{2}\right) + 5 = 0$. Find the rectangular equation of the path of the ship. *(Lesson 9-4)*

68. **Electricity** Find the current in a circuit with a voltage of $50 + 180\boldsymbol{j}$ volts and an impedance of $4 + 5\boldsymbol{j}$ ohms. *(Lesson 9-7)*

ALTERNATIVE ASSESSMENT

OPEN-ENDED ASSESSMENT

1. The simplest form of the sum of two complex numbers is $7 - 4\boldsymbol{i}$.
   a. Give examples of two complex numbers in which $a \neq 0$ and $b \neq 0$ with this sum.
   b. Are the two complex numbers you chose in part a the only two with this sum? Explain.

2. The absolute value of a complex number is $\sqrt{17}$.
   a. Give a complex number in which $a \neq 0$ and $b \neq 0$ with this absolute value.
   b. Is the complex number you chose in part a the only one with this absolute value? Explain.

Additional Assessment See p. A64 for Chapter 9 practice test.

WORLD WIDE WEB
Unit 3 interNET Project
SPACE—THE FINAL FRONTIER

From Point *A* to Point *B*

- Search the Internet to find types of coordinate systems that are used to locate objects in space. Find at least two different types.
- Write a summary that describes each coordinate system that you found. Compare them to rectangular and polar coordinates. Include diagrams illustrating how to use each coordinate system and any information you found about converting between systems.

PORTFOLIO

Choose one of the classical curves you studied in this chapter. Give the possible general forms of the polar equation for your curve and sketch the general graph. Then write and graph a specific polar equation.

Geometry Problems—Lines, Angles, and Arcs

THE PRINCETON REVIEW

TEST-TAKING TIP
Use the square corner of a sheet of paper to estimate angle measure. Fold the corner in half to form a 45° angle.

SAT and ACT geometry problems often combine triangles and quadrilaterals with angles and parallel lines. Problems that deal with circles often include arcs and angles.

Review these concepts.

- Angles: vertical, supplementary, complementary
- Parallel Lines: transversal, alternate interior angles
- Circles: inscribed angle, central angle, arc length, tangent line

SAT EXAMPLE

1. If four lines intersect as shown in the figure, $x + y =$

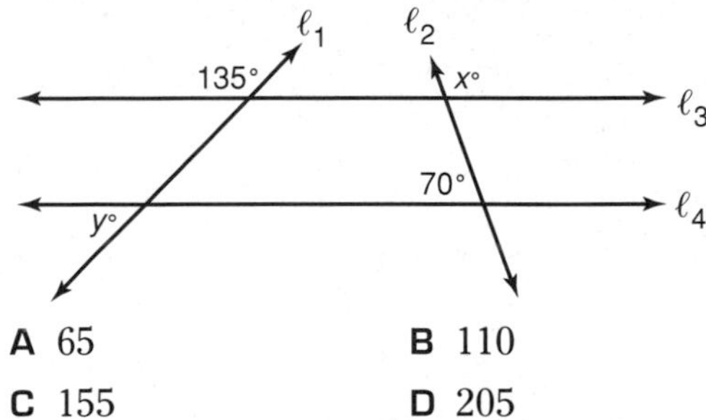

A 65 **B** 110

C 155 **D** 205

E It cannot be determined from the information given.

HINT Do not assume anything from a figure. In this figure, ℓ_3 and ℓ_4 look parallel, but that information is not given.

Solution Notice that the lines intersect to form a quadrilateral. The sum of the measures of the interior angles of a quadrilateral is 360°. Use vertical angles to determine the angles in the quadrilateral.

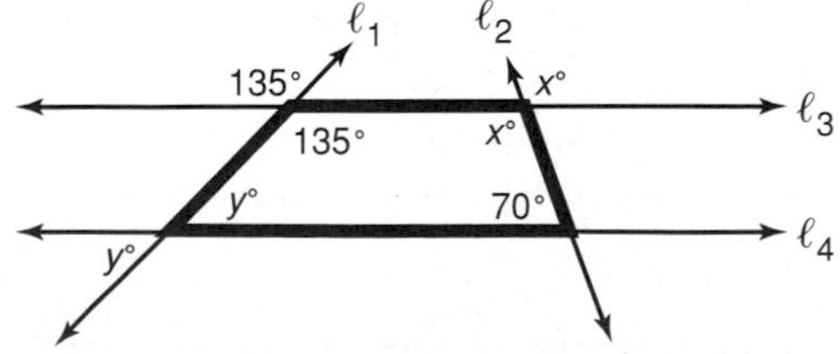

Write an equation for the sum of the angles.

$$135 + x + 70 + y = 360$$
$$205 + x + y = 360$$
$$x + y = 155$$

The answer is choice **C**.

ACT EXAMPLE

2. In the figure below, *ABCD* is a square inscribed in the circle centered at *O*. If $\overline{OB}$ is 6 units long, how many units long is minor arc $\overset{\frown}{BC}$?

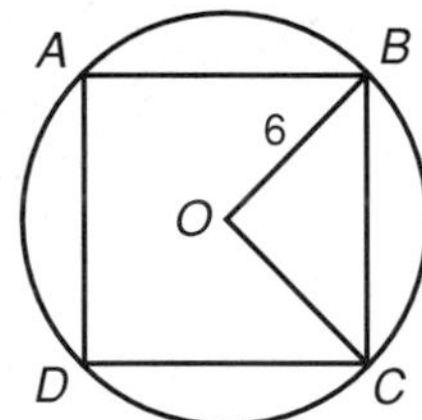

A $\frac{3\pi}{2}$ **B** 3π **C** 6π **D** 12π **E** 36π

HINT Consider *all* of the information given and implied. For example, a square's diagonals are perpendicular.

Solution Figure *ABCD* is a square, so its angles are each 90°, and its diagonals are perpendicular. So $\angle BOC = 90°$. This means that minor arc $\overset{\frown}{BC}$ is one-fourth of the circle.

Look at the answer choices. They all include π. Use the formula for the circumference of a circle.

$$C = 2\pi r$$
$$= 2\pi(6) \text{ or } 12\pi$$

The length of minor arc $\overset{\frown}{BC}$ is one-fourth of the circumference of the circle.

$\frac{1}{4}C = \frac{1}{4}(12\pi)$ or 3π

The answer is choice **B**.

After you work each problem, record your answer on the answer sheet provided or on a piece of paper.

Multiple Choice

1. In the figure below, line *L* is parallel to line *M*. Line *N* intersects both *L* and *M*, with angles *a*, *b*, *c*, *d*, *e*, *f*, *g*, and *h* as shown. Which of the following lists includes all of the angles that are supplementary to $\angle a$?

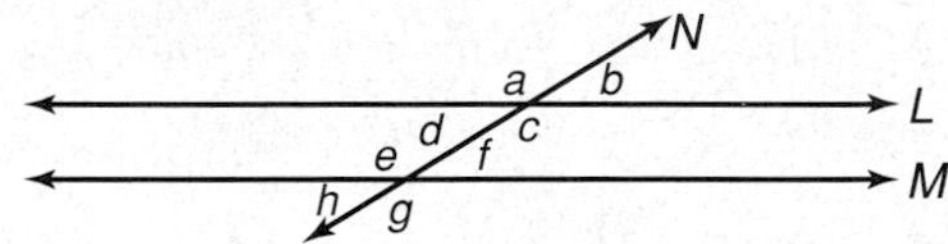

A *b*, *d*, *f*, *h* **B** *c*, *e*, *g* **C** *b*, *d*, *c*

D *e*, *f*, *g*, *h* **E** *d*, *c*, *h*, *g*

2. In the figure below, what is the area of $\triangle ABC$ in terms of *x*?

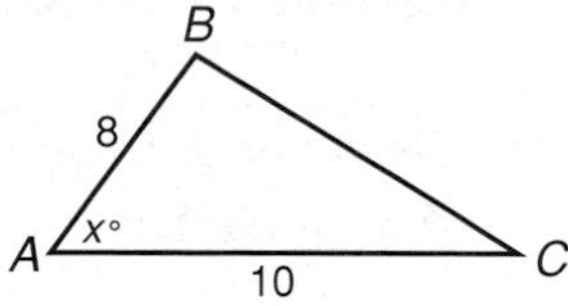

A $10 \sin x$ **B** $40 \sin x$ **C** $80 \sin x$

D $40 \cos x$ **E** $80 \cos x$

3. If *PQRS* is a parallelogram and $\overline{MN}$ is a line segment, then *x* must equal

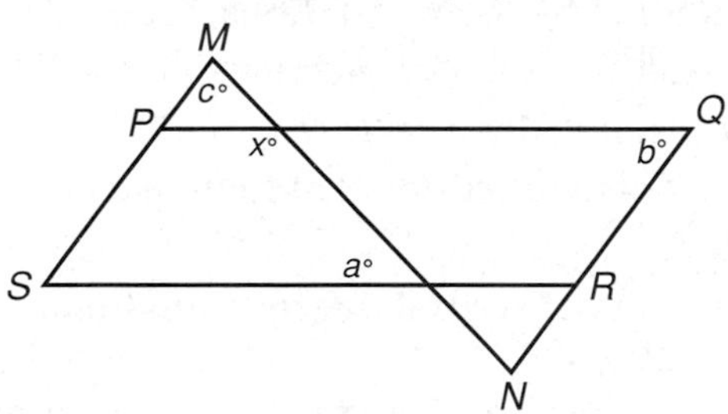

A $180 - b$ **B** $180 - c$ **C** $a + b$

D $a + c$ **E** $b + c$

4. If a rectangular swimming pool has a volume of 16,500 cubic feet, a depth of 10 feet, and a length of 75 feet, what is the width of the pool, in feet?

A 22 **B** 26 **C** 32 **D** 110 **E** 1650

5. $\frac{1}{10^{100}} - \frac{1}{10^{99}} =$

A $\frac{-9}{10^{100}}$ **B** $\frac{-1}{10^{100}}$ **C** $\frac{1}{10^{100}}$ **D** $\frac{1}{10}$ **E** $\frac{9}{10}$

6. In the figure, what is the sum of the degree measures of the marked angles?

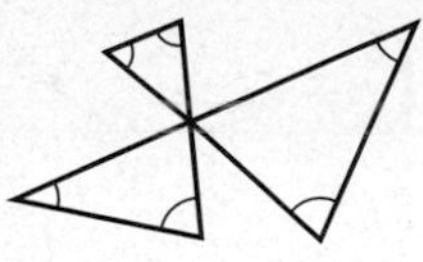

A 180 **B** 270 **C** 360 **D** 540

E It cannot be determined from the information given.

7. If $5x^2 + 6x = 70$ and $5x^2 - 6y = 10$, then what is the value of $10x + 10y$?

A 10 **B** 20 **C** 60 **D** 80 **E** 100

8. In the figure below, if $\overline{AB} \parallel \overline{CD}$, then what is the value of *y*? *Figure not drawn to scale.*

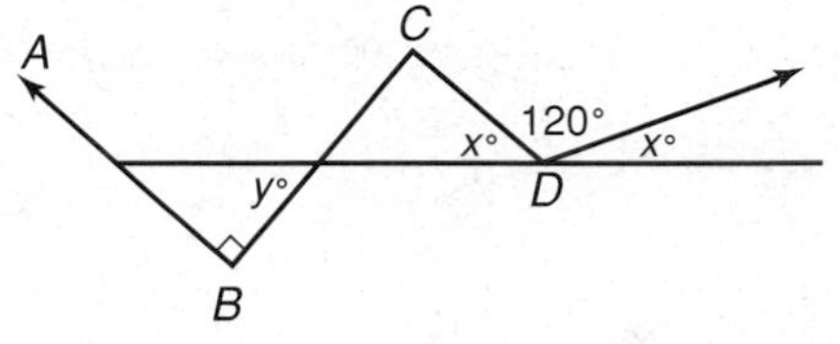

A 30 **B** 60 **C** 90 **D** 120 **E** 150

9. **Quantitative Comparison**

A if the quantity in Column A is greater
B if the quantity in Column B is greater
C if the two quantities are equal
D if the relationship cannot be determined from the information given

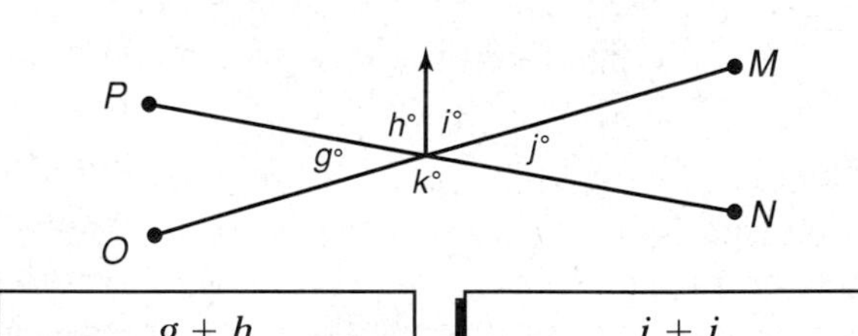

Column A	Column B
$g + h$	$i + j$

10. **Grid-In** If ℓ_1 is parallel to ℓ_2 in the figure below, what is the value of *y*?

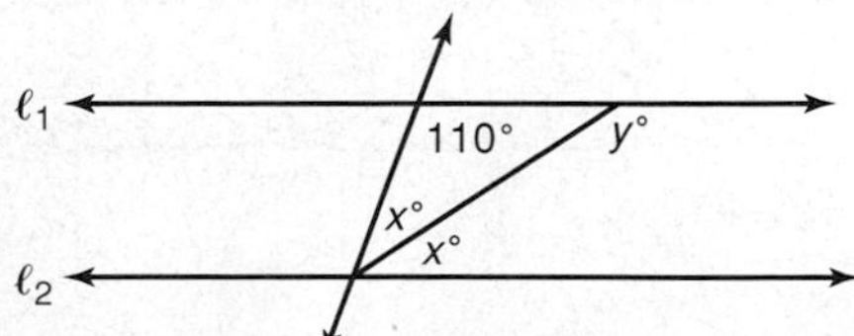

interNET CONNECTION **SAT/ACT Practice** For additional test practice questions, visit: **www.amc.glencoe.com**

CONICS

CHAPTER OBJECTIVES

- **Use analytic methods to prove geometric relationships.** *(Lesson 10-1)*
- **Use the standard and general forms of the equations of circles, parabolas, ellipses, and hyperbolas.** *(Lessons 10-2, 10-3, 10-4, and 10-5)*
- **Graph circles, parabolas, ellipses, and hyperbolas.** *(Lessons 10-2, 10-3, 10-4, and 10-5)*
- **Find the eccentricity of conic sections.** *(Lessons 10-2, 10-3, 10-4, and 10-5)*
- **Recognize conic sections by their equations.** *(Lesson 10-6)*
- **Find parametric equations for conic sections defined by rectangular equations and vice versa.** *(Lesson 10-6)*
- **Find the equations of conic sections that have been translated or rotated.** *(Lesson 10-7)*
- **Graph and solve systems of second-degree equations and inequalities.** *(Lesson 10-8)*

Introduction to Analytic Geometry

OBJECTIVES

- Find the distance and midpoint between two points on a coordinate plane.
- Prove geometric relationships among points and lines using analytical methods.

Real World Application

SEARCH AND RESCUE The *Absaroka Search Dogs* has provided search teams, each of which consists of a canine and its handler, to assist in lost person searches throughout the Montana and Wyoming area since 1986. While the dogs use their highly sensitive noses to detect the lost individual, the handlers use land navigation skills to insure that they do not become lost themselves. A handler needs to be able to read a map and calculate distances. *A problem related to this will be solved in Example 2.*

The distance between two points on a number line can be found by using absolute value. Let A and B be two points with coordinates a and b, respectively.

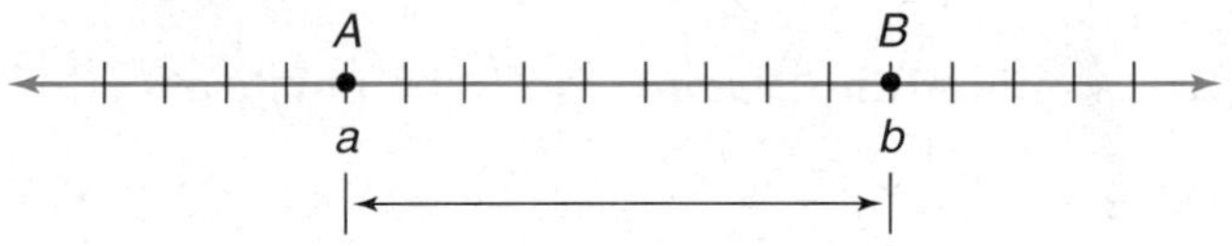

Distance between A and B $\Rightarrow$ $|a - b|$ or $|b - a|$

The distance between two points in the coordinate plane can also be found. Consider points $J(-4, 1)$ and $K(5, 9)$. To find JK, first choose a point L such that $\overline{JL}$ is parallel to the x-axis and $\overline{KL}$ is parallel to the y-axis. In this case, L has coordinates (5, 1). Since K and L lie along the line $x = 5$, KL is equal to the absolute value of the difference in the y-coordinates of K and L, $|9 - 1|$. Similarly, JL is equal to the absolute value of the difference in the x-coordinates of J and L, $|5 - (-4)|$.

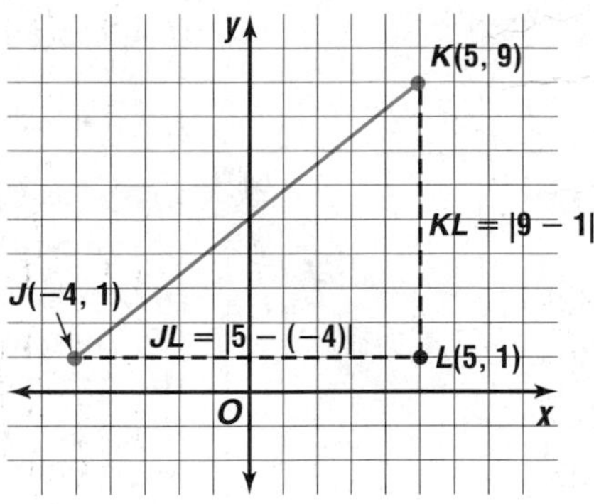

Since $\triangle JKL$ is a right triangle, JK can be found using the Pythagorean Theorem.

$(JK)^2 = (KL)^2 + (JL)^2$	*Pythagorean Theorem*
$JK = \sqrt{(KL)^2 + (JL)^2}$	*Take the positive square root of each side.*
$JK = \sqrt{\|9 - 1\|^2 + \|5 - (-4)\|^2}$	$KL = \|9 - 1\|$, $JL = \|15 - (-4)\|$
$JK = \sqrt{8^2 + 9^2}$	$\|9 - 1\|^2 = 8^2$, $\|15 - (-4)\|^2 = 9^2$
$JK = \sqrt{145}$ or about 12	

$\overline{JK}$ is about 12 units long.

From this specific case, we can derive a formula for the distance between any two points. In the figure, assume (x_1, y_1) and (x_2, y_2) represent the coordinates of any two points in the plane.

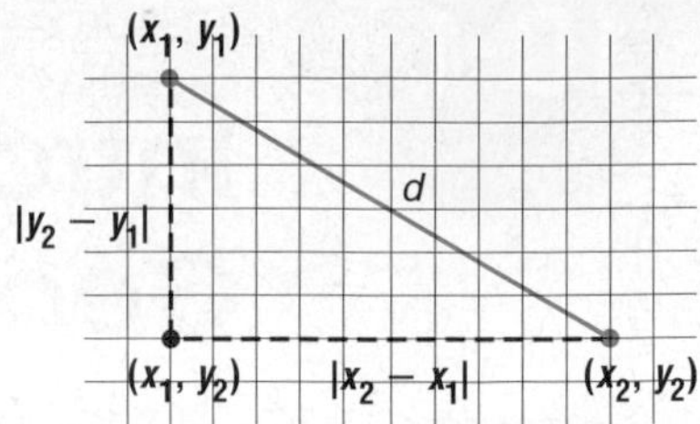

$d = \sqrt{|x_2 - x_1|^2 + |y_2 - y_1|^2}$ *Pythagorean Theorem*

$d = \sqrt{(x_2 - x_1)^2 + (y_2 - y_1)^2}$ *Why does $|x_2 - x_1|^2 + |y_2 - y_1|^2 = (x_2 - x_1)^2 + (y_2 - y_1)^2$?*

Distance Formula for Two Points

The distance, d units, between two points with coordinates (x_1, y_1) and (x_2, y_2) is given by $d = \sqrt{(x_2 - x_1)^2 + (y_2 - y_1)^2}$.

Examples

1 **Find the distance between points at (−3, 7) and (2, −5).**

$d = \sqrt{(x_2 - x_1)^2 + (y_2 - y_1)^2}$ *Distance Formula*

$d = \sqrt{(2 - (-3))^2 + (-5 - 7)^2}$ *Let $(x_1, y_1) = (-3, 7)$ and $(x_2, y_2) = (2, -5)$.*

$d = \sqrt{5^2 + (-12)^2}$

$d = \sqrt{169}$ or 13

The distance is 13 units.

2 **SEARCH AND RESCUE** **Refer to the application at the beginning of the lesson. Suppose a backpacker lies injured in the region shown on the map at the right. Each side of a square on the grid represents 15 meters. An Absaroka team searching for the missing individual is located at (−1.5, 4.0) on the map grid while the injured person is located at (2.0, 2.8). How far is the search team from the missing person?**

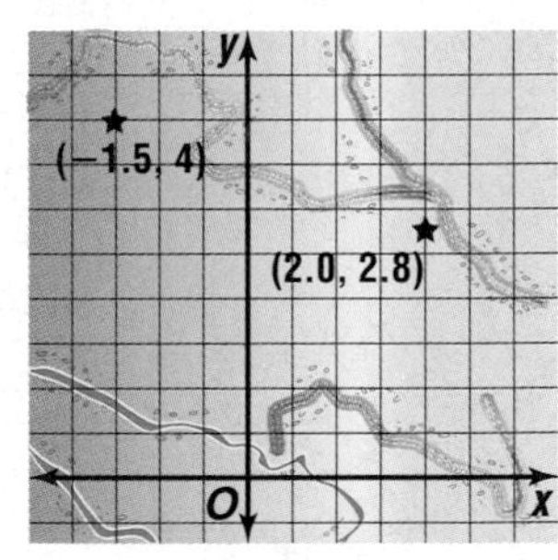

Use the distance formula to find the distance between (2.0, 2.8) and (−1.5, 4.0).

$d = \sqrt{(x_2 - x_1)^2 + (y_2 - y_1)^2}$

$d = \sqrt{(-1.5 - 2.0)^2 + (4.0 - 2.8)^2}$ *Let $(x_1, y_1) = (2.0, 2.8)$ and $(x_2, y_2) = (-1.5, 4.0)$.*

$d = 13.69$ or 3.7

The map distance is 3.7 units. Each unit equals 15 kilometers. So, the actual distance is about 3.7(15) or 55.5 kilometers.

You can use the Distance Formula and what you know about slope to investigate geometric figures on the coordinate plane.

Example **3** **Determine whether quadrilateral *ABCD* with vertices *A*(3, 2), *B*(2, −4), *C*(−2, −3), and *D*(−1, 3) is a parallelogram.**

Recall that a quadrilateral is a parallelogram if one pair of opposite sides are parallel and congruent.

First, graph the figure. $\overline{DA}$ and $\overline{CB}$ are one pair of opposite sides.

To determine if $\overline{DA} \parallel \overline{CB}$, find the slopes of $\overline{DA}$ and $\overline{CB}$.

Look Back
Refer to Lesson 1-3 to review the slope formula.

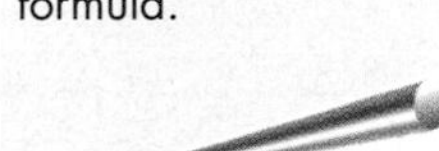

slope of $\overline{DA}$

$$m = \frac{y_2 - y_1}{x_2 - x_1} \quad \textit{Slope formula}$$

$$= \frac{2 - 3}{3 - (-1)} \quad \textit{Let } (x_1, y_1) = (-1, -3) \textit{ and } (x_2, y_2) = (3, 2).$$

$$= -\frac{1}{4}$$

slope of $\overline{CB}$

$$m = \frac{y_2 - y_1}{x_2 - x_1} \quad \textit{Slope formula}$$

$$= \frac{-4 - (-3)}{2 - (-2)} \quad \textit{Let } (x_1, y_1) = (-2, -3) \textit{ and } (x_2, y_2) = (2, -4).$$

$$= -\frac{1}{4}$$

Their slopes are equal. Therefore, $\overline{DA} \parallel \overline{CB}$.

To determine if $\overline{DA} \cong \overline{CB}$, use the distance formula to find DA and CB.

$$DA = \sqrt{(x_2 - x_1)^2 + (y_2 - y_1)^2} = \sqrt{[3 - (-1)]^2 + (2 - 3)^2} = \sqrt{17}$$

$$CB = \sqrt{(x_2 - x_1)^2 + (y_2 - y_1)^2} = \sqrt{[2 - (-2)]^2 + [-4 - (-3)]^2} = \sqrt{17}$$

The measures of $\overline{DA}$ and $\overline{CB}$ are equal. Therefore, $\overline{DA} \cong \overline{CB}$.

Since $\overline{DA} \parallel \overline{CB}$ and $\overline{DA} \cong \overline{CB}$, quadrilateral $ABCD$ is a parallelogram. *You can also check your work by showing $\overline{DC} \parallel \overline{AB}$ and $\overline{DC} \cong \overline{AB}$.*

In addition to finding the distance between two points, you can use the coordinates of two points to find the midpoint of the segment between the points. In the figure at the right, the midpoint of $\overline{P_1P_2}$ is P_m. Notice that the x-coordinate of P_m is the average of the x-coordinates of P_1 and P_2. The y-coordinate of P_m is the average of the y-coordinates of P_1 and P_2.

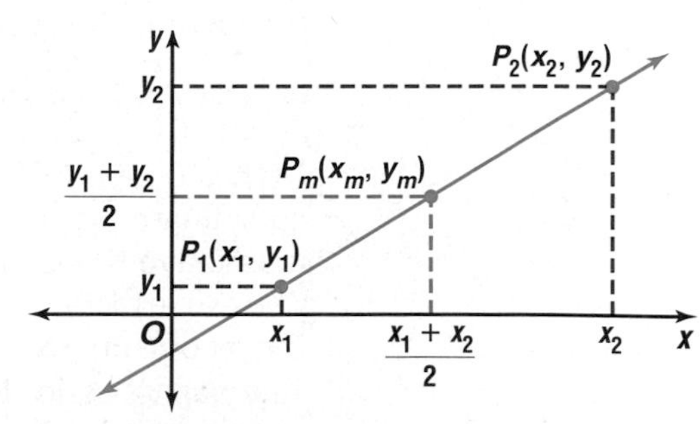

Midpoint of a Line Segment

If the coordinates of P_1 and P_2 are (x_1, y_1) and (x_2, y_2), respectively, then the midpoint of $\overline{P_1P_2}$ has coordinates $\left(\frac{x_1 + x_2}{2}, \frac{y_1 + y_2}{2}\right)$.

Example 4 **Find the coordinates of the midpoint of the segment that has endpoints at (−2, 4) and (6, −5).**

Let (−2, 4) be (x_1, y_1) and (6, −5) be (x_2, y_2). Use the Midpoint Formula.

$$\left(\frac{x_1 + x_2}{2}, \frac{y_1 + y_2}{2}\right) = \left(\frac{-2 + 6}{2}, \frac{4 + (-5)}{2}\right)$$
$$= \left(2, -\frac{1}{2}\right)$$

The midpoint of the segment is at $\left(2, -\frac{1}{2}\right)$.

Many theorems from plane geometry can be more easily proven by analytic methods. That is, they can be proven by placing the figure in a coordinate plane and using algebra to express and draw conclusions about the geometric relationships. The study of coordinate geometry from an algebraic perspective is called **analytic geometry.**

When using analytic methods to prove theorems from geometry, the position of the figure in the coordinate plane can be arbitrarily selected as long as size and shape are preserved. This means that the figure may be translated, rotated, or reflected from its original position. For polygons, one vertex is usually located at the origin, and one side coincides with the x-axis, as shown below.

In a right triangle, the legs are on the axes.

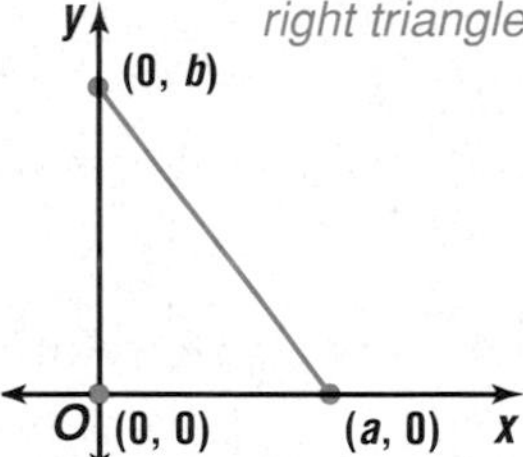

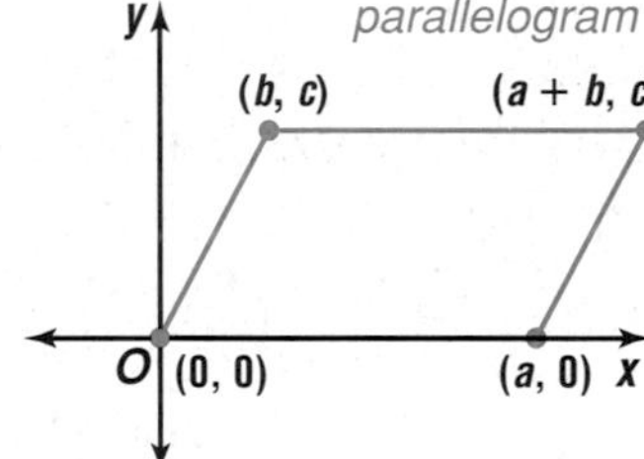

Example 5 **Prove that the measure of the median of a trapezoid is equal to one half of the sum of the measures of the two bases.**

In trapezoid $ABCD$, choose two vertices as $A(0, 0)$ and $B(a, 0)$. Since $\overline{AB} \parallel \overline{DC}$, $\overline{DC}$ lies on a horizontal grid line of the coordinate plane. Therefore, C and D must have the same y-coordinate. Choose arbitrary letters to represent the y-coordinates, and the two x-coordinates; in this case, $D(b, c)$ and $C(d, c)$. Let E be the midpoint of $\overline{AD}$, and let F be the midpoint of $\overline{BC}$.

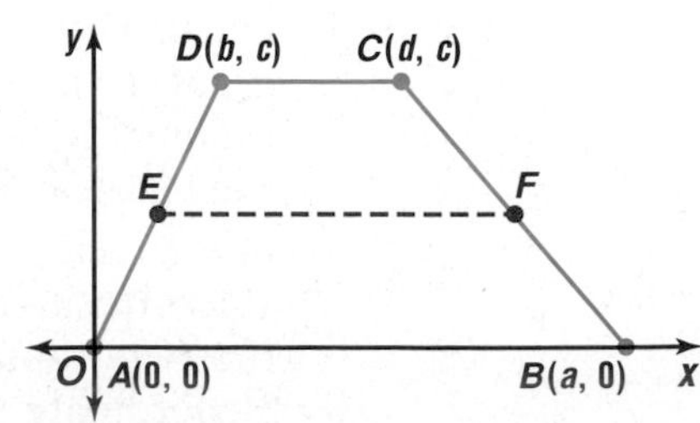

Now, find the coordinates of E and F by using the Midpoint Formula.

The coordinates of E are $\left(\frac{b+0}{2}, \frac{c+0}{2}\right)$ or $\left(\frac{b}{2}, \frac{c}{2}\right)$.

The coordinates of F are $\left(\frac{d+a}{2}, \frac{c+0}{2}\right)$ or $\left(\frac{d+a}{2}, \frac{c}{2}\right)$.

Then find the measures of each base and the median by using the Distance Formula.

$DC = \sqrt{(d-b)^2 + (c-c)^2}$ or $d - b$

$AB = \sqrt{(a-0)^2 + (0-0)^2}$ or a

$EF = \sqrt{\left(\frac{d+a}{2} - \frac{b}{2}\right)^2 + \left(\frac{c}{2} - \frac{c}{2}\right)^2}$ or $\frac{1}{2}(d - b + a)$

Calculate one half of the sum of the measures of the bases.

$\frac{1}{2}(DC + AB) = \frac{1}{2}(d - b + a)$ $DC = d - b, AB = a$

Since both $\frac{1}{2}(DC + AB)$ and EF equal $\frac{1}{2}(d - b + a)$, it follows that $\frac{1}{2}(DC + AB) = EF$. Therefore, the measure of the median of a trapezoid is one half of the sum of the measures of its bases.

CHECK FOR UNDERSTANDING

Communicating Mathematics

Read and study the lesson to answer each question.

1. **Explain** why only the positive square root is considered when applying the distance formula.
2. **Describe** how can you show that a midpoint of a segment is equidistant from its endpoints given the coordinates of each point.
3. **Determine** whether each diagram represents an isosceles triangle. Explain your reasoning.

a.
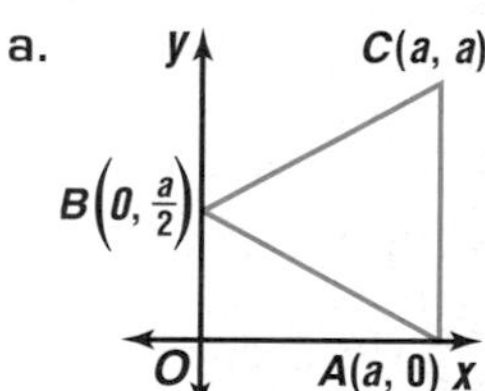

b.
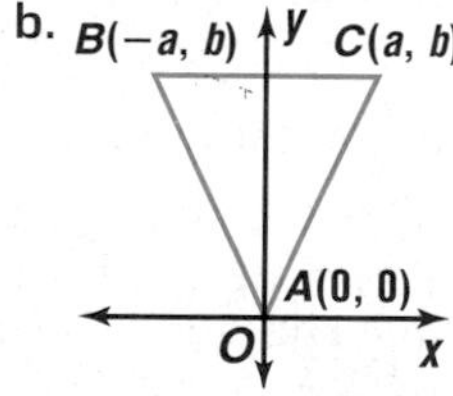

c.
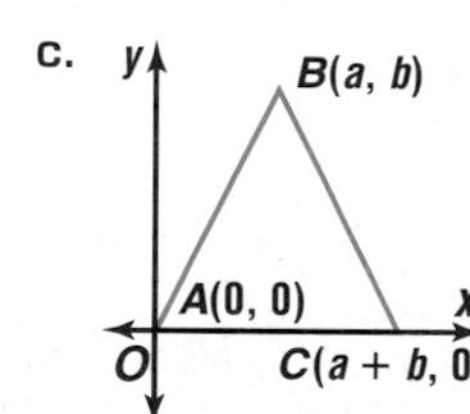

4. **Describe** four different ways of proving that a quadrilateral is a parallelogram if you are given the coordinates of its vertices.

Guided Practice

Find the distance between each pair of points with the given coordinates. Then, find the coordinates of the midpoint of the segment that has endpoints at the given coordinates.

5. (5, 1), (5, 11)
6. (0, 0), (−4, −3)
7. (−2, 2), (0, 4)

8. Determine whether the quadrilateral $ABCD$ with vertices $A(3, 4)$, $B(6, 2)$, $C(8, 7)$, and $D(5, 9)$ is a parallelogram. Justify your answer.

9. Determine whether the triangle XYZ with vertices $X(-3, 2)$, $Y(-1, -6)$, and $Z(5, 0)$ is isosceles. Justify your answer.

10. Consider rectangle $ABCD$.
    a. Draw and label rectangle $ABCD$ on the coordinate plane.
    b. Prove that $\overline{AC} \cong \overline{BD}$.
    c. Suppose the diagonals intersect at point E. Prove that $\overline{AE} \cong \overline{EC}$ and $\overline{BE} \cong \overline{ED}$.
    d. What can you conclude about the diagonals of a rectangle? Explain.

Crew Stadium, Columbus, Ohio

11. **Sports** The dimensions of a soccer field are120 yards by 80 yards. A player kicks the ball from a corner to his teammate at the center of the playing field. Suppose the kicker is located at the origin.
    a. Find the ordered pair that represents the location of the kicker's teammate.
    b. Find the distance the ball travels.

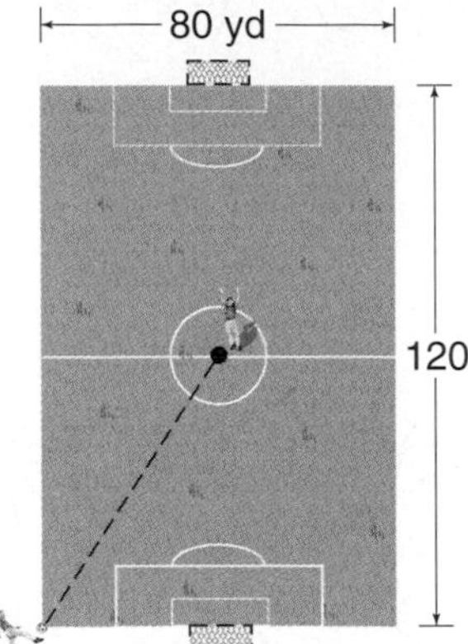

EXERCISES

Practice

Find the distance between each pair of points with the given coordinates. Then, find the coordinates of the midpoint of the segment that has endpoints at the given coordinates.

12. $(-1, 1), (4, 13)$
13. $(1, 3), (-1, -3)$
14. $(8, 0), (0, 8)$
15. $(-1, -6), (5, -3)$
16. $\left(3\sqrt{2}, -5\right), \left(7\sqrt{2}, -1\right)$
17. $(a, 7), (a, -9)$
18. $(6 + r, s), (r - 2, s)$
19. $(c, d), (c + 2, d - 1)$
20. $(w - 2, w), (w, 4w)$

interNET CONNECTION
Graphing Calculator Programs
To download a graphing calculator program that determines the distance and midpoint between two points, visit **www.amc.glencoe.com**

21. Find all values of a so that the distance between points at $(a, -9)$ and $(-2a, 7)$ is 20 units.

22. If $M\left(-3, \frac{5}{2}\right)$ is the midpoint of $\overline{CD}$ and C has coordinates $(4, -1)$, find the coordinates of D.

Determine whether the quadrilateral having vertices with the given coordinates is a parallelogram.

23. $(-2, 3), (-3, -2), (2, -3), (3, 2)$
24. $(4, 11), (8, 14), (4, 19), (0, 15)$

25. Collinear points lie on the same line. Find the value of k for which the points $(15, 1)$, $(-3, -8)$, and $(3, k)$ are collinear.

26. Determine whether the points $A(-3, 0)$, $B\left(-1, 2\sqrt{3}\right)$, and $C(1, 0)$ are the vertices of an equilateral triangle. Justify your answer.

27. Show that points $E(2, 5)$, $F(4, 4)$, $G(2, 0)$, and $H(0, 1)$ are the vertices of a rectangle.

Prove using analytic methods. Be sure to include a coordinate diagram.

28. The measure of the line segment joining the midpoints of two sides of a triangle is equal to one-half the measure of the third side.

29. The diagonals of an isosceles trapezoid are congruent.

30. The medians to the congruent sides of an isosceles triangle are congruent. (*Hint*: A median of a triangle is a segment connecting a vertex to the midpoint of the side opposite the vertex.)

31. The diagonals of a parallelogram bisect each other.

32. The line segments joining the midpoints of consecutive sides of any quadrilateral form a parallelogram.

Applications and Problem Solving

33. **Geometry** The vertices of a rectangle are at $(-3, 1)$, $(-1, 3)$, $(3, -1)$, and $(1, -3)$. Find the area of the rectangle.

34. **Web Page Design** Many Internet Web pages are designed so that when the cursor is positioned over a specified area of an image, lines of text are displayed. The programming string "$x \mid y \mid$ width $\mid$ height" defines the location of the bottom left corner of the designated region using x and y coordinates and then defines the width and height of the region in pixels.
   a. If the center of a Web page is located at the origin, graph the two regions defined by $-22 \mid 12 \mid 10 \mid 8$ and $31 \mid -10 \mid 8 \mid 10$.
   b. Suppose the two regions are to be no less than 40 pixels apart. Calculate the distance between the regions at their closest points to determine if this criteria is met.

35. **Critical Thinking** Prove analytically that the segments joining midpoints of consecutive sides of an isosceles trapezoid form a rhombus. Include a coordinate diagram with your proof.

36. **Landscaping** The diagram shows the plans made by a landscape artist for a homeowner's 16-meter by 16-meter backyard. The homeowner has requested that the rosebushes, fountain, and garden bench be placed so that they are no more than 14 meters apart.

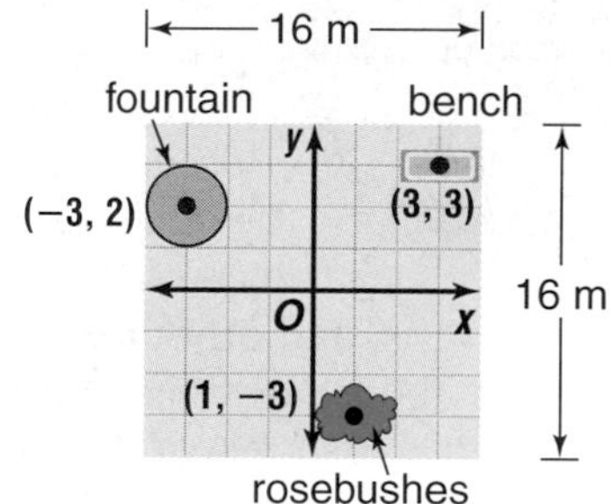

   a. Has the landscape artist met the homeowner's requirements? Explain.
   b. The homeowner has purchased a sundial to be placed midway between the fountain and the rosebushes. Determine the coordinates indicating where the sundial should be placed.

37. **Critical Thinking** Consider point $M(t, 3t - 12)$.
   a. Prove that for all values of t, M is equidistant from $A(0, 3)$ and $B(9, 0)$.
   b. Describe the figure formed by the points M for all values of t. What is the relationship between the figure and points A and B?

Mixed Review

38. Find $(-5 + 12\boldsymbol{i})^2$. *(Lesson 9-8)*

Extra Practice See p. A44.

39. **Physics** Suppose that during a storm, the force of the wind blowing against a skyscraper can be expressed by the vector ⟨115, 2018, 0⟩, where each measure in the ordered triple represents the force in Newtons. What is the magnitude of this force? *(Lesson 8-3)*

40. Verify that $2 \sec^2 x = \frac{1}{1 + \sin x} + \frac{1}{1 - \sin x}$ is an identity. *(Lesson 7-2)*

41. A circle has a radius of 12 inches. Find the degree measure of the central angle subtended by an arc 11.5 inches long. *(Lesson 6-1)*

42. Find sin 390°. *(Lesson 5-3)*

43. Solve $z^2 - 8z = -14$ by completing the square. *(Lesson 4-2)*

44. **SAT Practice** **Grid-In** If $x^2 = 16$ and $y^2 = 4$, what is the greatest possible value of $(x - y)^2$?

CAREER CHOICES

Meteorologist

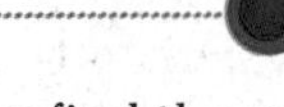

If you find the weather intriguing, then you may want to investigate a career in meteorology. Meteorologists spend their time studying weather and forecasting changes in the weather. They analyze charts, weather maps, and other data to make predictions about future weather patterns. Meteorologists also research different types of weather, such as tornadoes and hurricanes, and may even teach at universities.

As a meteorologist, you may choose to specialize in one of several areas such as climatololgy, operational meteorology, or industrial meteorology. As a meteorologist, you might even be seen on television forecasting the weather for your area!

CAREER OVERVIEW

Degree Preferred:
Bachelor's degree in meteorology

Related Courses:
mathematics, geography, physics, computer science

Outlook:
slower than average job growth through the year 2006

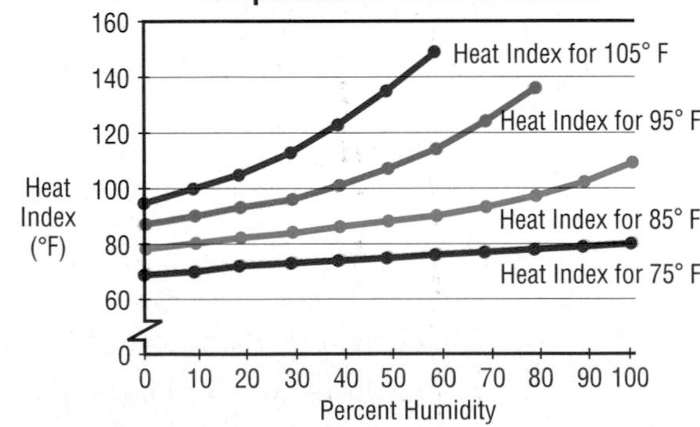

Source: *The World Almanac 1999*

interNET CONNECTION For more information on careers in meteorology, visit: **www.amc.glencoe.com**

Circles

OBJECTIVES
- Use and determine the standard and general forms of the equation of a circle.
- Graph circles.

SEISMOLOGY Portable autonomous digital seismographs (PADSs) are used to investigate the strong ground motions produced by the aftershocks of large earthquakes. Suppose a PADS is deployed 2 miles west and 3.5 miles south of downtown Olympia, Washington, to record the aftershocks of a recent earthquake. While there, the PADS detects and records the seismic activity of another quake located 24 miles away. What are all the possible locations of this earthquake's epicenter? *This problem will be solved in Example 2.*

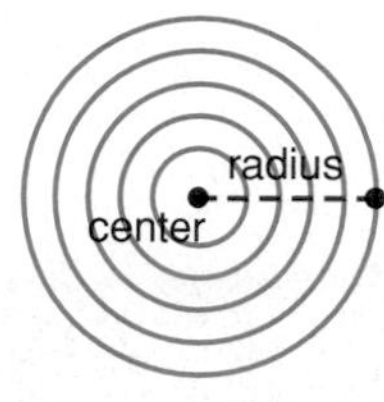

The pattern of the shock waves from an earthquake form **concentric** circles. A **circle** is the set of all points in the plane that are equidistant from a given point in the plane, called the **center.** The distance from the center to any point on the circle is called the **radius** of the circle. Concentric circles have the same center but not necessarily the same radius.

A circle is one type of **conic section.** Conic sections, which include circles, parabolas, ellipses and hyperbolas, were first studied in ancient Greece sometime between 600 and 300 B.C. The Greeks were largely concerned with the properties, not the applications, of conics. In the seventeenth century, applications of conics became prominent in the development of calculus.

Conic sections are used to describe all of the possible ways a plane and a double right cone can intersect. In forming the four basic conics, the plane does not pass through the vertex of the cone.

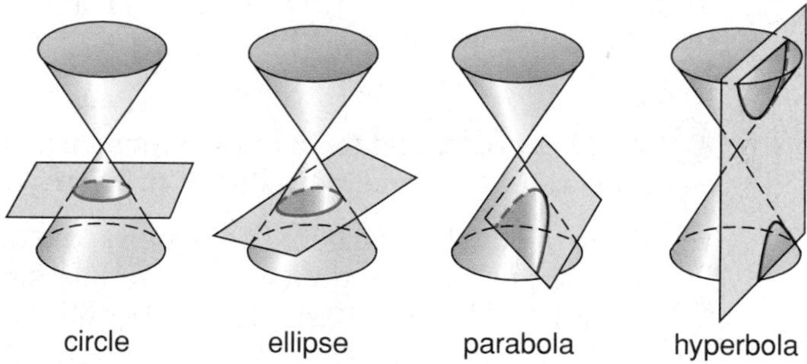

When the plane does pass through the vertex of a conical surface, as illustrated below, the resulting figure is called a **degenerate conic.** A degenerate conic may be a point, line, or two intersecting lines.

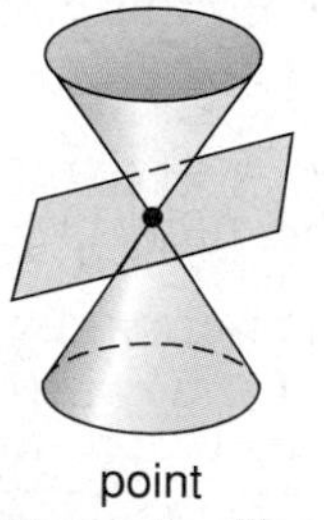
point
(degenerate ellipse)

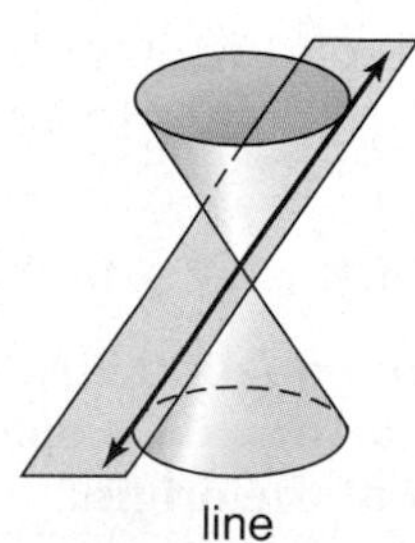
line
(degenerate parabola)

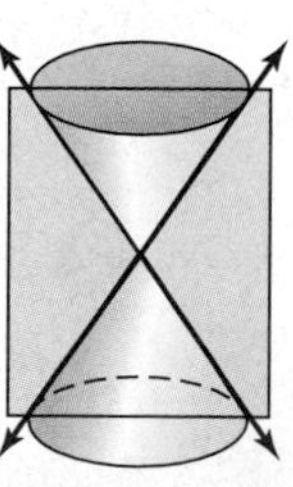
intersecting lines
(degenerate hyperbola)

Radius can also refer to the line segment from the center to any point on the circle.

In the figure at the right, the center of the circle is at the origin. By drawing a perpendicular from any point $P(x, y)$ on the circle but not on an axis to the x-axis, you form a right triangle. The Pythagorean Theorem can be used to write an equation that describes every point on a circle whose center is located at the origin.

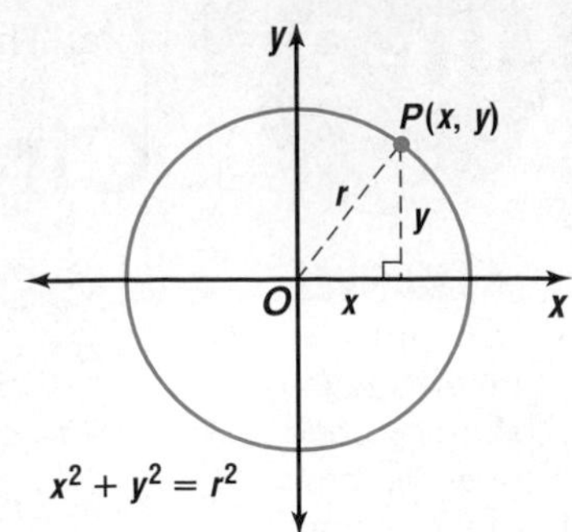

$$x^2 + y^2 = r^2 \quad \textit{Pythagorean Theorem}$$

This is the equation for the parent graph of all circles.

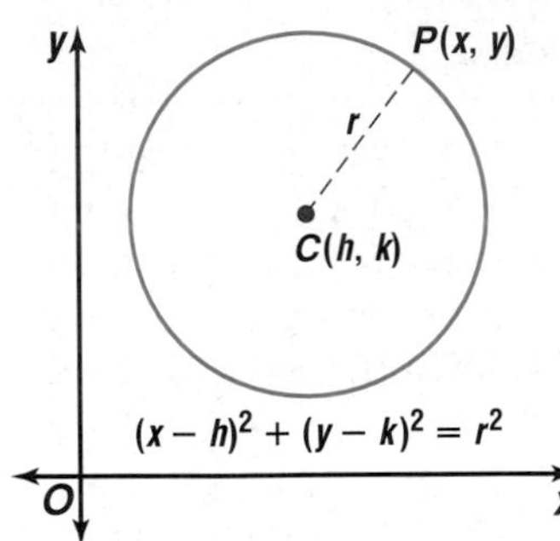

Suppose the center of this circle is translated from the origin to $C(h, k)$. You can use the distance formula to write the equation for this translated circle.

$$d = \sqrt{(x_2 - x_1)^2 + (y_2 - y_1)^2} \quad \textit{Distance formula}$$

$$r = \sqrt{(x - h)^2 + (y - k)^2} \quad d = r,\ (x_2, y_2) = (x, y), \text{ and } (x_1, y_1) = (h, k)$$

$$r^2 = (x - h)^2 + (y - k)^2 \quad \textit{Square each side.}$$

This equation is the standard form of the equation of a circle.

Standard Form of the Equation of a Circle

The standard form of the equation of a circle with radius r and center at (h, k) is

$$(x - h)^2 + (y - k)^2 = r^2.$$

Examples

Write the standard form of the equation of the circle that is tangent to the x-axis and has its center at $(3, -2)$. Then graph the equation.

Since the circle is tangent to the x-axis, the distance from the center to the x-axis is the radius. The center is 2 units below the x-axis. Therefore, the radius is 2.

$$(x - h)^2 + (y - k)^2 = r^2 \quad \textit{Standard form}$$

$$(x - 3)^2 + [y - (-2)]^2 = 2^2 \quad h = 3,\ k = -2,\ r = 2$$

$$(x - 3)^2 + (y + 2)^2 = 4$$

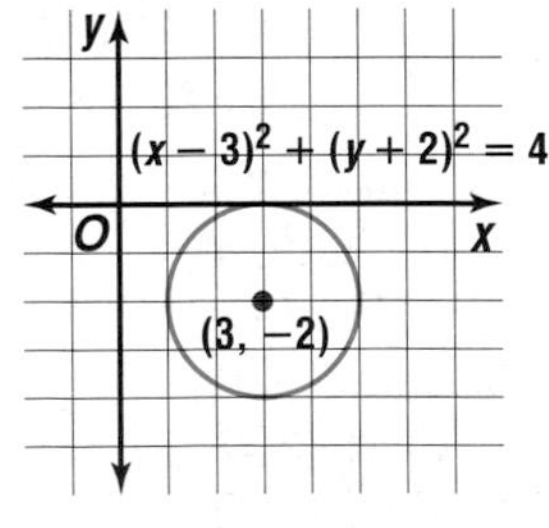

The standard form of the equation for this circle is $(x - 3)^2 + (y + 2)^2 = 4$.

SEISMOLOGY **Refer to the application at the beginning of the lesson.**

a. Write an equation for the set of points representing all possible locations of the earthquake's epicenter. Let downtown Olympia, Washington, be located at the origin.

b. Graph the equation found in part a.

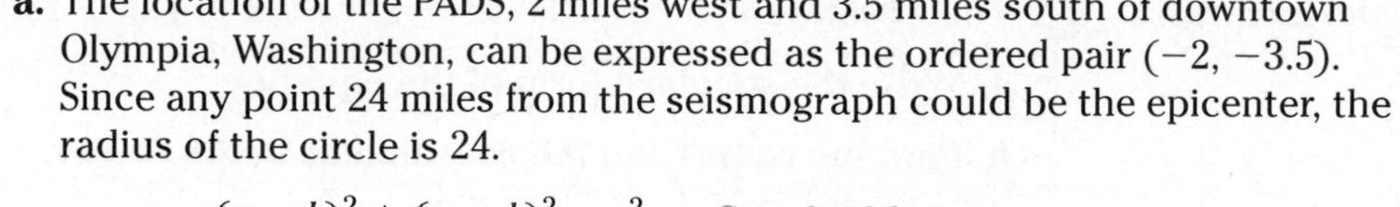

a. The location of the PADS, 2 miles west and 3.5 miles south of downtown Olympia, Washington, can be expressed as the ordered pair (−2, −3.5). Since any point 24 miles from the seismograph could be the epicenter, the radius of the circle is 24.

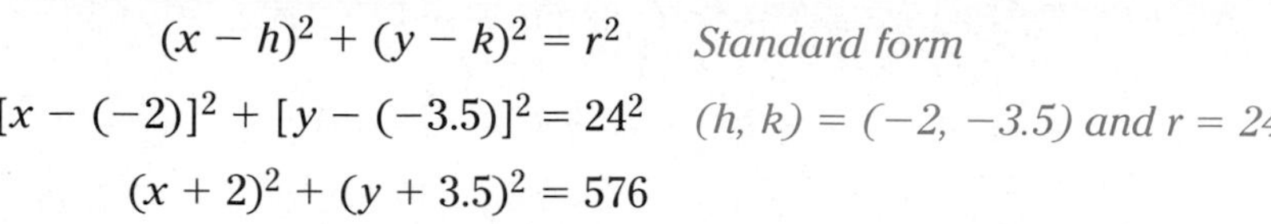

$$(x - h)^2 + (y - k)^2 = r^2 \quad \textit{Standard form}$$

$$[x - (-2)]^2 + [y - (-3.5)]^2 = 24^2 \quad (h, k) = (-2, -3.5) \text{ and } r = 24$$

$$(x + 2)^2 + (y + 3.5)^2 = 576$$

b. The location of the epicenter lies on the circle with equation $(x + 2)^2 + (y + 3.5)^2 = 576$.

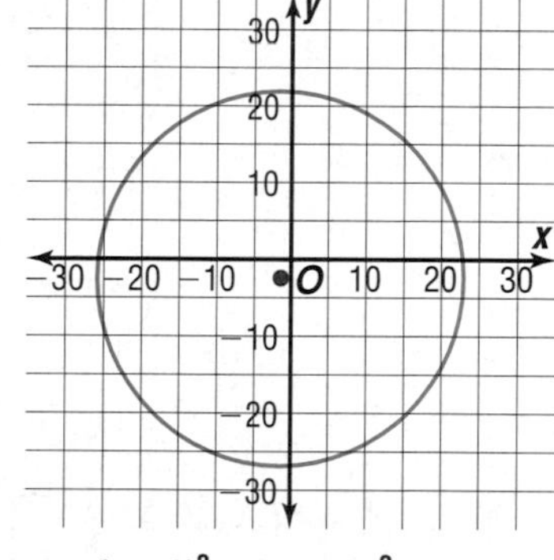

$(x + 2)^2 + (y + 3.5)^2 = 576$

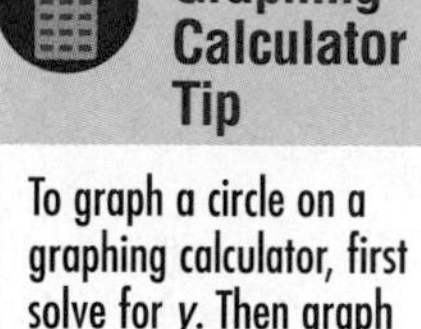

Graphing Calculator Tip

To graph a circle on a graphing calculator, first solve for *y*. Then graph the two resulting equations on the same screen by inserting {1, −1} in front of the equation for the symbol ±. Use **5:ZSquare** in the **ZOOM** menu to make the graph look like a circle

The standard form of the equation of a circle can be expanded to obtain a general form of the equation.

$$(x - h)^2 + (y - k)^2 = r^2 \quad \textit{Standard form}$$

$$(x^2 - 2hx + h^2) + (y^2 - 2ky + k^2) = r^2 \quad \text{Expand } (x - h)^2 \text{ and } (y - k)^2.$$

$$x^2 + y^2 + (-2h)x + (-2k)y + (h^2 + k^2) - r^2 = 0$$

Since h, k, and r are constants, let D, E, and F equal $-2h$, $-2k$ and $(h^2 + k^2) - r^2$, respectively.

$$x^2 + y^2 + Dx + Ey + F = 0$$

This equation is called the general form of the equation of a circle.

General Form of the Equation of a Circle

The general form of the equation of a circle is

$$x^2 + y^2 + Dx + Ey + F = 0,$$

where D, E, and F are constants.

Notice that the coefficients of x^2 and y^2 in the general form must be 1. If those coefficients are not 1, division can be used to transform the equation so that they are 1. Also notice that there is no term containing the product of the variables, xy.

When the equation of a circle is given in general form, it can be rewritten in standard form by completing the square for the terms in x and the terms in y.

Example 3 **The equation of a circle is $2x^2 + 2y^2 - 4x + 12y - 18 = 0$.**

a. Write the standard form of the equation.

b. Find the radius and the coordinates of the center.

c. Graph the equation.

a.

$2x^2 + 2y^2 - 4x + 12y - 18 = 0$	
$x^2 + y^2 - 2x + 6y - 9 = 0$	*Divide each side by 2.*
$(x^2 - 2x + ?) + (y^2 + 6y + ?) = 9$	*Group to form perfect square trinomials.*
$(x^2 - 2x + 1) + (y^2 + 6y + 9) = 9 + 1 + 9$	*Complete the square.*
$(x - 1)^2 + (y + 3)^2 = 19$	*Factor the trinomials.*
$(x - 1)^2 + (y + 3)^2 = (\sqrt{19})^2$	*Express 19 as $(\sqrt{19})^2$ to show that $r = \sqrt{19}$.*

b. The center of the circle is located at $(1, -3)$, and the radius is $\sqrt{19}$.

c. Plot the center at $(-1, 3)$. The radius of $\sqrt{19}$ is approximately equal to 4.4.

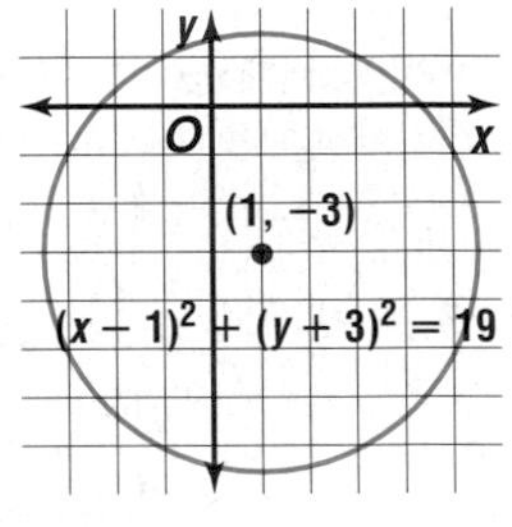

From geometry, you know that any two points in the coordinate plane determine a unique line. It is also true that any three noncollinear points in the coordinate plane determine a unique circle. The equation of this circle can be found by substituting the coordinates of the three points into the general form of the equation of a circle and solving the resulting system of three equations.

Example 4 **Write the standard form of the equation of the circle that passes through the points at (5, 3), (−2, 2), and (−1, −5). Then identify the center and radius of the circle.**

Substitute each ordered pair for (x, y) in $x^2 + y^2 + Dx + Ey + F = 0$, to create a system of equations.

$$(5)^2 + (3)^2 + D(5) + E(3) + F = 0 \quad (x, y) = (5, 3)$$
$$(-2)^2 + (2)^2 + D(-2) + E(2) + F = 0 \quad (x, y) = (-2, 2)$$
$$(-1)^2 + (-5)^2 + D(-1) + E(-5) + F = 0 \quad (x, y) = (-1, -5)$$

Simplify the system of equations.
$$5D + 3E + F + 34 = 0$$
$$-2D + 2E + F + 8 = 0$$
$$-D - 5E + F + 26 = 0$$

Look Back
Refer to Lesson 2-2 to review solving systems of three equations.

The solution to the system is $D = -4$, $E = 2$, and $F = -20$.

The general form of the equation of the circle is $x^2 + y^2 - 4x + 2y - 20 = 0$. After completing the square, the standard form is $(x - 2)^2 + (y + 1)^2 = 25$. The center of the circle is at $(2, -1)$, and its radius is 5.

Check for Understanding

Communicating Mathematics

Read and study the lesson to answer each question.

1. **Explain** how to convert the general form of the equation of a circle to the standard form of the equation of a circle.
2. **Write** the equations of five concentric circles with different radii whose centers are at $(-4, 9)$.
3. **Describe** how you might determine the equation of a circle if you are given the endpoints of the circle's diameter.
4. **Find a counterexample** to this statement: The graph of any equation of the form $x^2 + y^2 + Dx + Ey + F = 0$ is a circle.
5. **You Decide** Kiyo says that you can take the square root of each side of an equation. Therefore, he decides that $(x - 3)^2 + (y - 1)^2 = 49$ and $(x - 3) + (y - 1) = 7$ are equivalent equations. Ramon says that the equations are not equivalent. Who is correct? Explain.

Guided Practice

Write the standard form of the equation of each circle described. Then graph the equation.

6. center at $(0, 0)$, radius 9
7. center at $(-1, 4)$ and tangent to $x = 3$

Write the standard form of each equation. Then graph the equation.

8. $x^2 + y^2 - 4x + 14y - 47 = 0$
9. $2x^2 + 2y^2 - 20x + 8y + 34 = 0$

Write the standard form of the equation of the circle that passes through points with the given coordinates. Then identify the center and radius.

10. $(0, 0)$, $(4, 0)$, $(0, 4)$
11. $(1, 3)$, $(5, 5)$, $(5, 3)$

Write the equation of the circle that satisfies each set of conditions.

12. The circle passes through the point at $(1, 5)$ and has its center at $(-2, 1)$.
13. The endpoints of a diameter are at $(-2, 6)$ and at $(10, -10)$.

14. **Space Science** Apollo 8 was the first manned spacecraft to orbit the moon at an average altitude of 185 kilometers above the moon's surface. Determine an equation to model the orbit of the Apollo 8 command module if the radius of the moon is 1740 kilometers. Let the center of the moon be at the origin.

← Apollo 8 crew: (from left) James A. Lovell, Jr., William A. Anders, Frank Borman

Exercises

Practice

Write the standard form of the equation of each circle described. Then graph the equation.

15. center at $(0, 0)$, radius 5
16. center at $(-4, 7)$, radius $\sqrt{3}$
17. center at $(-1, -3)$, radius $\frac{\sqrt{2}}{2}$
18. center at $(-5, 0)$, radius $\frac{9}{2}$
19. center at $(6, 1)$, tangent to the y-axis
20. center at $(3, -2)$, tangent to $y = 2$

Write the standard form of each equation. Then graph the equation.

21. $36 - x^2 = y^2$

22. $x^2 + y^2 + y = \frac{3}{4}$

23. $x^2 + y^2 - 4x + 12y + 30 = 0$

24. $2x^2 + 2y^2 + 2x - 4y = -1$

25. $6x^2 - 12x + 6y^2 + 36y = 36$

26. $16x^2 + 16y^2 - 8x - 32y = 127$

27. Write $x^2 + y^2 + 14x + 24y + 157 = 0$ in standard form. Then graph the equation.

interNET CONNECTION

Graphing Calculator Programs
For a graphing calculator program that determines the radius and the coordinates of the center of a circle from an equation written in general form, visit **www.amc.glencoe.com**

Write the standard form of the equation of the circle that passes through the points with the given coordinates. Then identify the center and radius.

28. $(0, -1), (-3, -2), (-6, -1)$

29. $(7, -1), (11, -5), (3, -5)$

30. $(-2, 7), (-9, 0), (-10, -5)$

31. $(-2, 3), (6, -5), (0, 7)$

32. $(4, 5), (-2, 3), (-4, -3)$

33. $(1, 4), (2, -1), (-3, 0)$

34. Write the standard form of the equation of the circle that passes through the origin and points at (2.8, 0) and (5, 2).

Write the equation of the circle that satisfies each set of conditions.

35. The circle passes through the origin and has its center at $(-4, 3)$.

36. The circle passes through the point (5, 6) and has its center at (2, 3).

37. The endpoints of a diameter are at (2, 3) and at $(-6, -5)$.

38. The points at $(-3, 4)$ and (2, 1) are the endpoints of a diameter.

39. The circle is tangent to the line with equation $x + 3y = -2$ and has its center at (5, 1).

40. The center of the circle is on the x-axis, its radius is 1, and it passes through the point at $\left(\frac{\sqrt{2}}{2}, \frac{\sqrt{2}}{2}\right)$.

Graphing Calculator

41. A rectangle is inscribed in a circle centered at the origin with diameter 12.

a. Write the equation of the circle that meets these conditions.

b. Write the dimensions of the rectangle in terms of x.

c. Write a function $A(x)$ that represents the area of the rectangle.

d. Use a graphing calculator to graph the function $y = A(x)$.

e. Find the value of x, to the nearest tenth, that maximizes the area of the rectangle. What is the maximum area of the rectangle?

42. Select the standard viewing window and then select **ZSquare** from the ZOOM menu.

a. Select **9:Circle(** from the **DRAW** menu and then enter 2 , 3 , 4) . Then press ENTER .

b. Describe what appears on the viewing screen.

c. Write the equation for this graph.

d. Use what you have learned to write the command to graph the equation $(x + 4)^2 + (y - 2) = 36$. Then graph the equation.

Applications and Problem Solving

43. Sports Cindy is taking an archery class and decides to practice her skills at home. She attaches the target shown at right to a bale of hay. The circles on the target are concentric and equally spaced apart.

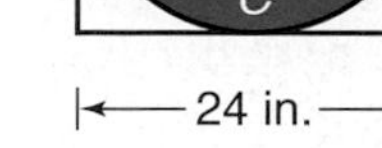

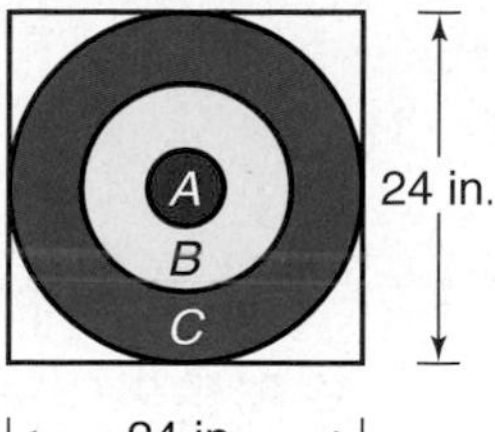

a. If the common center of the circles is located at the origin, write an equation that models the largest circle.

b. If the smallest circle is modeled by the equation $x^2 + y^2 = 6.25$, find the area of the region marked B.

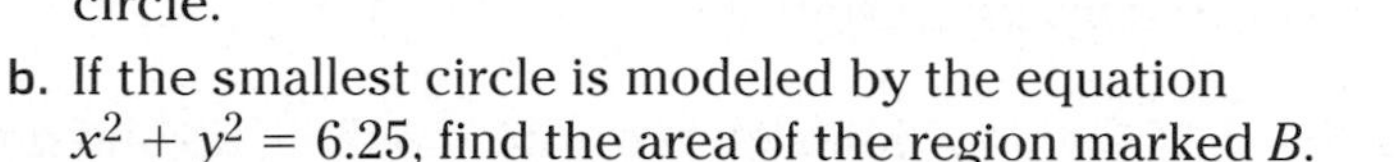

44. Geometry Write the equation of a circle that circumscribes the triangle whose sides are the graphs of $4x - 7y = 27$, $x - 5y + 3 = 0$, and $2x + 3y - 7 = 0$.

45. Critical Thinking Consider a family of circles in which $h = k$ and the radius is 2. Let k be any real number.

a. Write the equation of the family of circles.

b. Graph three members of this family on the same set of axes.

c. Write a description of all members of this family of circles.

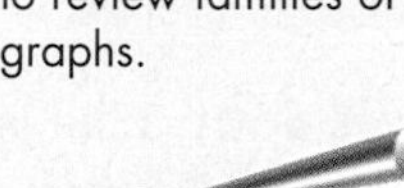

Look Back
Refer to Lesson 3-2 to review families of graphs.

46. Transportation A moving truck 7 feet wide and 13 feet high is approaching a semi-circular brick archway at an apartment complex. The base of the archway is 28 feet wide. The road under the archway is divided, allowing for two-way traffic.

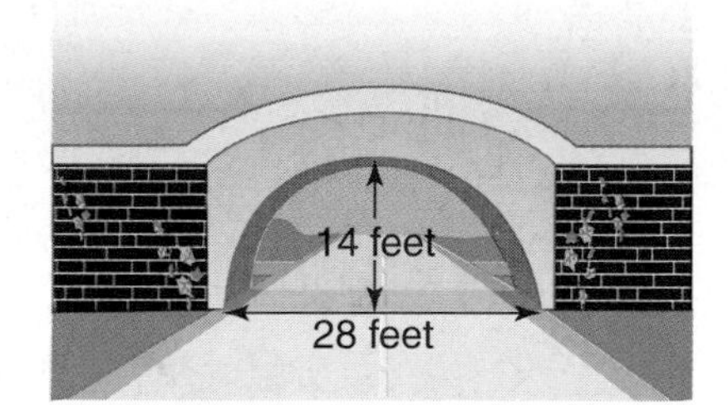

a. Write an equation, centered at the origin, of the archway that models its shape.

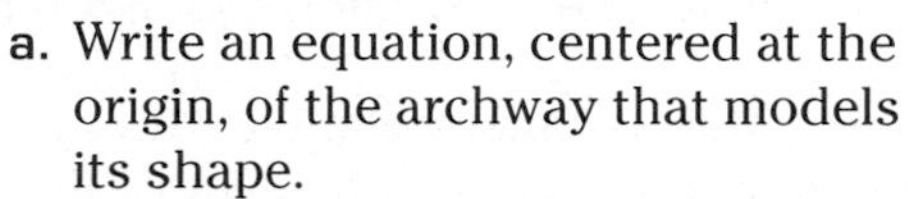

b. If the truck remains just to the right of the median, will it be able to pass under the archway without damage? Explain.

47. Critical Thinking Find the radius and the coordinates of the center of a circle defined by the equation $x^2 + y^2 - 8x + 6y + 25 = 0$. Describe the graph of this circle.

48. Agriculture One method of irrigating crops is called the center pivot system. This system rotates a sprinkler pipe from the center of the field to be irrigated. Suppose a farmer places one of these units at the center of a square plot of land 2500 feet on each side. With the center of this plot at the origin, the irrigator sends out water far enough to reach a point located at (475, 1140).

a. Find an equation representing the farthest points the water can reach.

b. Find the area of the land that receives water directly.

c. About what percent of the farmer's plot does not receive water directly?

49. Critical Thinking Consider points $A(3, 4)$, $B(-3, -4)$, and $P(x, y)$.

a. Write an equation for all x and y for which $\overline{PA} \perp \overline{PB}$.

b. What is the relationship between points A, B, and P if $\overline{PA} \perp \overline{PB}$?

Mixed Review

50. Find the distance between points at $(4, -3)$ and $(-2, 6)$. *(Lesson 10-1)*

51. Simplify $(2 + \mathbf{i})(3 - 4\mathbf{i})(1 + 2\mathbf{i})$. *(Lesson 9-5)*

52. **Sports** Patrick kicked a football with an initial velocity of 60 ft/s at an angle of 60° to the horizontal. After 0.5 seconds, how far has the ball traveled horizontally and vertically. *(Lesson 8-7)*

53. **Toys** A toy boat floats on the water bobbing up and down. The distance between its highest and lowest point is 5 centimeters. It moves from its highest point down to its lowest point and back up to its highest point every 20 seconds. Write a cosine function that models the movement of the boat in relationship to the equilibrium point. *(Lesson 6-6)*

54. Find the area to the nearest square unit of $\triangle ABC$ if $a = 15$, $b = 25$, and $c = 35$. *(Lesson 5-8)*

55. **Amusement Parks** The velocity of a roller coaster as it moves down a hill is $\sqrt{v_0^2 + 64h}$. The designer of a coaster wants the coaster to have a velocity of 95 feet per second when it reaches the bottom of the hill. If the initial velocity of the coaster at the top of the hill is 15 feet per second, how high should the designer make the hill? *(Lesson 4-7)*

56. Determine whether the graph of $y = 6x^4 - 3x^2 + 1$ is symmetric with respect to the x-axis, the y-axis, the line $y = x$, the line $y = -x$, or the origin. *(Lesson 3-1)*

57. **Business** A pharmaceutical company manufactures two drugs. Each case of drug A requires 3 hours of processing time and 1 hour of curing time per week. Each case of drug B requires 5 hours of processing time and 5 hours of curing time per week. The schedule allows 55 hours of processing time and 45 hours of curing time weekly. The company must produce no more than 10 cases of drug A and no more than 9 cases of drug B. *(Lesson 2-7)*
   a. If the company makes a profit of \$320 on each case of drug A and \$500 on each case of drug B, how many cases of each drug should be produced in order to maximize profit?
   b. Find the maximum profit.

58. **SAT/ACT Practice** An owner plans to divide and enclose a rectangular property with dimensions x and y into rectangular regions as shown at the right. In terms of x and y, what is the total length of fence needed if every line segment represents a section of fence and there is no overlap between sections of fence.

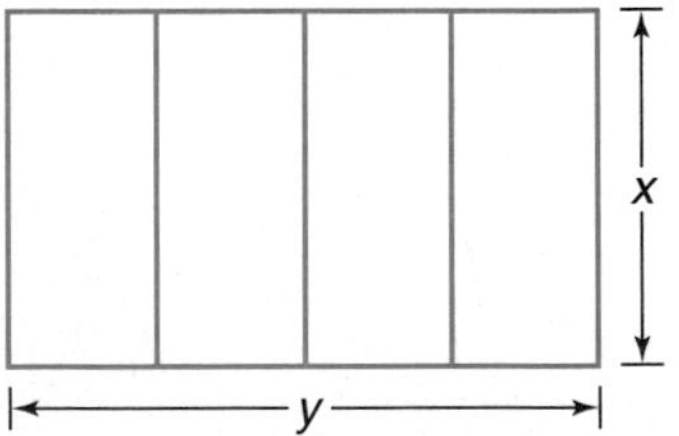

A $5x + 2y$
B $5x + 8y$
C $4x + 2y$
D $4xy$
E xy

Extra Practice See p. A44.

10-3

Ellipses

OBJECTIVES

- Use and determine the standard and general forms of the equation of an ellipse.
- Graph ellipses.

MEDICAL TECHNOLOGY To eliminate kidney stones, doctors sometimes use a medical tool called a *lithotripter,* (*LITH-uh-trip-tor*) which means "stone crusher." A lithotripter is a device that uses ultra-high-frequency shock waves moving through water to break up the stone. After x-raying a patient's kidney to precisely locate and measure the stone, the lithotripter is positioned so that the shock waves reflect off the inner surface of the elliptically-shaped tub and break up the stone. *A problem related to this will be solved in Example 1.*

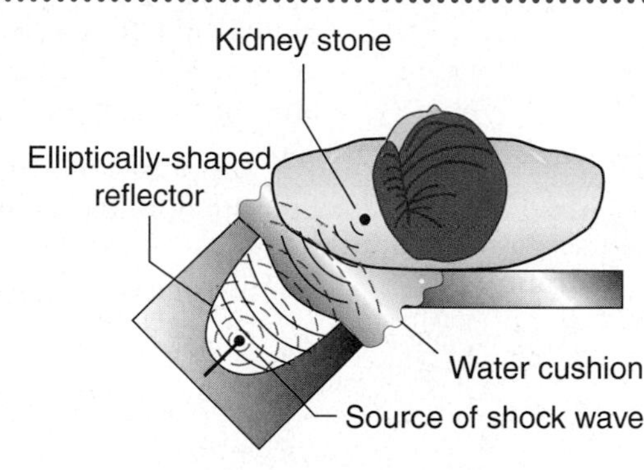

An **ellipse** is the set of all points in the plane, the sum of whose distances from two fixed points, called **foci,** is constant. In the figure at the right, F_1 and F_2 are the foci, and the midpoint C of the line segment joining the foci is called the **center** of the ellipse. P and Q are any two points on the ellipse. By definition, $PF_1 + PF_2 = QF_1 + QF_2$.

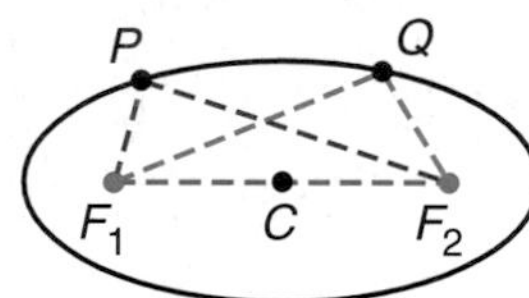

Foci (FOH sigh) is the plural of focus.

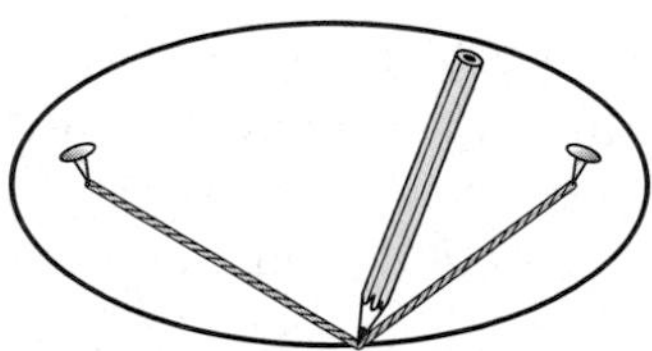

To help visualize this definition, imagine tacking two ends of a string at the foci and using a pencil to trace a curve as it is held tight against the string. The curve which results will be an ellipse since the sum of the distances to the foci, the total length of the string, remains constant.

The parent graph of an ellipse, shown at the right, is centered at the origin. An ellipse has two axes of symmetry, in this case, the x-axis and the y-axis. Notice that the ellipse intersects each axis of symmetry two times. The longer line segment, $\overline{AD}$, which contains the foci, is called the **major axis.** The shorter segment, $\overline{BE}$, is called the **minor axis.** The endpoints of each axis are the **vertices** of the ellipse.

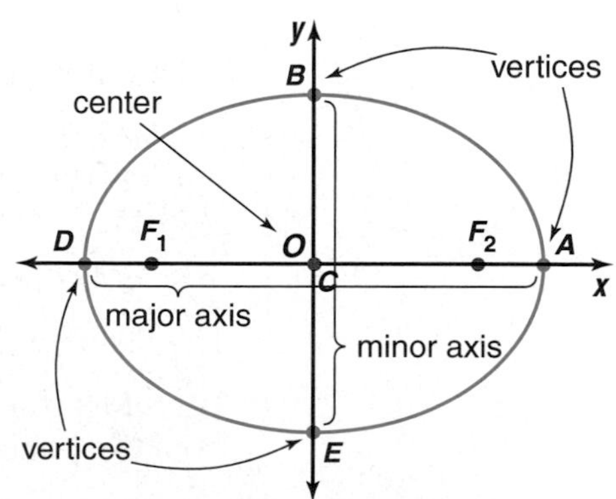

The center separates each axis into two congruent segments. Suppose we let b represent the length of the **semi-minor axis** $\overline{BC}$ and a represent the length of the **semi-major axis** $\overline{CA}$. The foci are located along the major axis, c units from the center. There is a special relationship among the values a, b, and c.

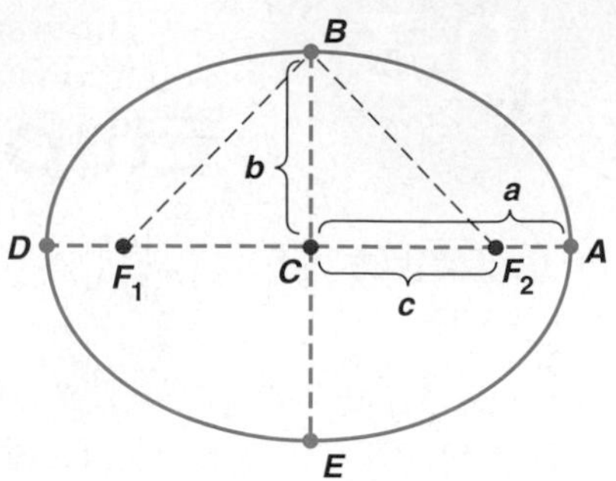

Suppose you draw $\overline{BF_1}$ and $\overline{BF_2}$. The lengths of these segments are equal because $\triangle BF_1C \cong \triangle BF_2C$. Since B and A are two points on the ellipse, you can use the definition of an ellipse to find BF_2.

$BF_1 + BF_2 = AF_1 + AF_2$	*Definition of ellipse*
$BF_1 + BF_2 = AF_1 + DF_1$	$AF_2 = DF_1$
$BF_1 + BF_2 = AD$	*Segment addition:* $AF_1 + DF_1 = AD$
$BF_1 + BF_2 = 2a$	*Substitution:* $AD = 2a$
$2(BF_2) = 2a$	$BF_1 = BF_2$
$BF_2 = a$	

Since $BF_2 = a$ and $\triangle BCF_2$ is a right triangle, $b^2 + c^2 = a^2$ by the Pythagorean Theorem.

Example **1** **MEDICAL TECHNOLOGY** **Refer to the application at the beginning of this lesson. Suppose the reflector of a mobile lithotripter is 24 centimeters wide and 24 centimeters deep. How far, to the nearest hundredth of a centimeter, should the shock wave emitter be placed from a patient's kidney stone?**

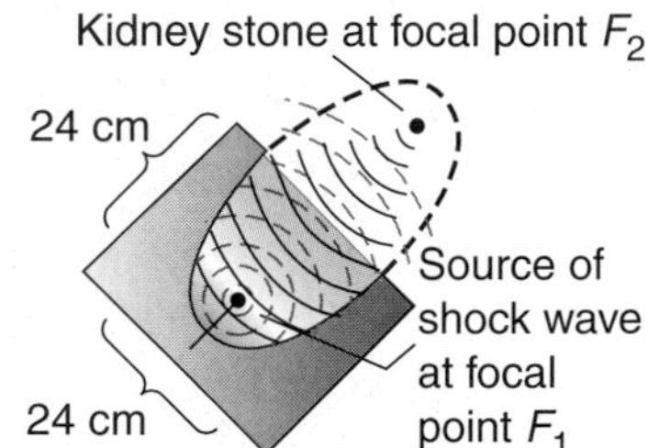

For the lithotripter to break up the stone, the shock wave emitter must be positioned at one focal point of the ellipse and the kidney stone at the other. To determine the distance between the emitter and the kidney stone, we must first find the lengths of the semi-major and semi-minor axes.

The semi-major axis in this ellipse is equal to the depth of the reflector, 24 centimeters. So, $a = 24$.

The semi-minor axis in this ellipse is half the width of the reflector, 12 centimeters. So, $b = 12$.

To find the focal length of the ellipse, use the formula $c^2 = a^2 - b^2$.

$c^2 = a^2 - b^2$	
$c^2 = (24)^2 - (12)^2$	$a = 24$ *and* $b = 12$
$c^2 = 432$	*Simplify.*
$c = \sqrt{432}$ or approximately 20.8	*Take the square root of each side.*

The distance between the two foci of the ellipse is $2c$ or 41.6.

Therefore, the emitter should be placed approximately 41.6 centimeters away from the patient's kidney along the ellipse's major axis.

The standard form of the equation of an ellipse can be derived from the definition and the distance formula. Consider the special case when the center is at the origin.

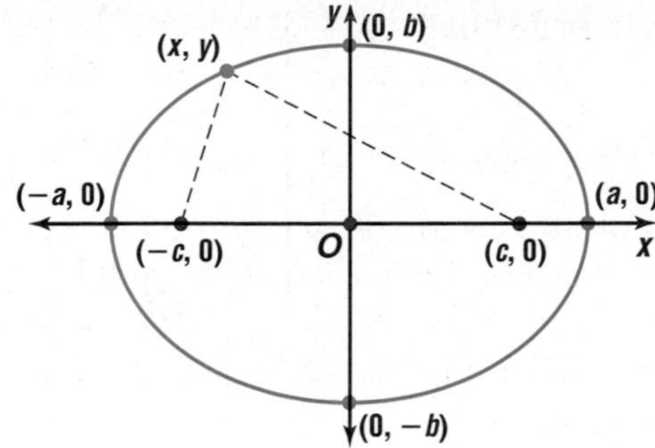

Suppose the foci are at $F_1(-c, 0)$ and $F_2(c, 0)$, and (x, y) is any point on the ellipse. By definition, the sum of the distances from a point at (x, y) to the foci is constant. To find a value for this constant, let (x, y) be one of the vertices on the x-axis, for example, $(a, 0)$. Let d_1 and d_2 be the distances from the point at $(a, 0)$ to F_1 and F_2, respectively. Use the Distance Formula.

$$d_1 = \sqrt{[a - (-c)]^2 + (0 - 0)^2} \qquad d_2 = \sqrt{(a - c)^2 + (0 - 0)^2}$$
$$= \sqrt{(a + c)^2} \qquad = \sqrt{(a - c)^2}$$
$$= a + c \qquad = a - c$$

$d_1 + d_2 = a + c + a - c$ or $2a$ Therefore, one value for the constant is $2a$.

Using the Distance Formula and this value for the constant, a general equation for any point at (x, y) is $2a = \sqrt{(x + c)^2 + y^2} + \sqrt{(x - c)^2 + y^2}$.

$$2a = \sqrt{(x + c)^2 + y^2} + \sqrt{(x - c)^2 + y^2}$$

$$\sqrt{(x + c)^2 + y^2} = 2a - \sqrt{(x - c)^2 + y^2} \quad \textit{Isolate a radical.}$$

$$(x + c)^2 + y^2 = 4a^2 - 4a\sqrt{(x - c)^2 + y^2} + (x - c)^2 + y^2 \quad \textit{Square each side.}$$

$$a^2 - xc = a\sqrt{(x - c)^2 + y^2} \quad \textit{Simplify.}$$

$$a^4 - 2a^2xc + x^2c^2 = a^2[(x - c)^2 + y^2] \quad \textit{Square each side again.}$$

$$x^2(a^2 - c^2) + a^2y^2 = a^2(a^2 - c^2) \quad \textit{Simplify.}$$

$$\frac{x^2}{a^2} + \frac{y^2}{a^2 - c^2} = 1 \quad \textit{Divide each side by } a^2(a^2 - c^2).$$

$$\frac{x^2}{a^2} + \frac{y^2}{b^2} = 1 \quad b^2 = a^2 - c^2$$

The resulting equation, $\frac{x^2}{a^2} + \frac{y^2}{b^2} = 1$, is the equation of an ellipse whose center is the origin and whose foci are on the x-axis. When the foci are on the y-axis, the equation is of the form $\frac{y^2}{a^2} + \frac{x^2}{b^2} = 1$.

The standard form of the equation of an ellipse with a center other than the origin is a translation of the parent graph to a center at (h, k). The table on the next page gives the standard form, graph, and general description of the equation of each type of ellipse.

Standard Form of the Equation of an Ellipse	Orientation	Description
$\frac{(x-h)^2}{a^2} + \frac{(y-k)^2}{b^2} = 1$, where $c^2 = a^2 - b^2$	y, O, x, (h, k), y = k, x = h	Center: (h, k) Foci: $(h \pm c, k)$ Major axis: $y = k$ Major axis vertices: $(h \pm a, k)$ Minor axis: $x = h$ Minor axis vertices: $(h, k \pm b)$
$\frac{(y-k)^2}{a^2} + \frac{(x-h)^2}{b^2} = 1$, where $c^2 = a^2 - b^2$	y, O, x, (h, k), y = k, x = h	Center: (h, k) Foci: $(h, k \pm c)$ Major axis: $x = h$ Major axis vertices: $(h, k \pm a)$ Minor axis: $y = k$ Minor axis vertices: $(h \pm b, k)$

Example 2 **Consider the ellipse graphed at the right.**

a. Write the equation of the ellipse in standard form.

b. Find the coordinates of the foci.

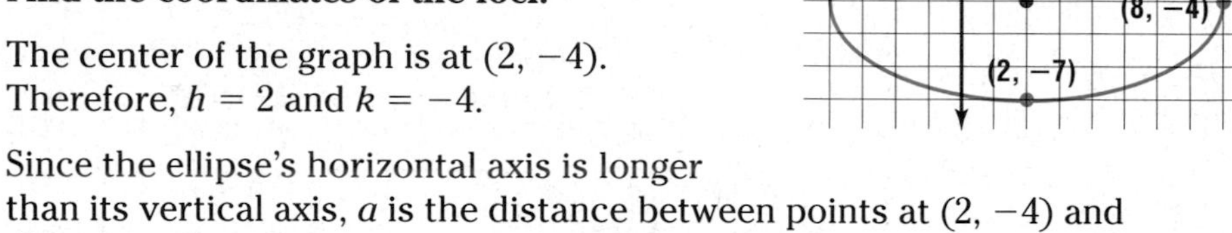

a. The center of the graph is at $(2, -4)$. Therefore, $h = 2$ and $k = -4$.

Since the ellipse's horizontal axis is longer than its vertical axis, a is the distance between points at $(2, -4)$ and $(8, -4)$ or 6. The value of b is the distance between points at $(2, -4)$ and $(2, -7)$ or 3.

Therefore, the standard form of the equation of this ellipse is

$$\frac{(x-2)^2}{6^2} + \frac{(y+4)^2}{3^2} = 1 \text{ or } \frac{(x-2)^2}{36} + \frac{(y+4)^2}{9} = 1.$$

b. Using the equation $c = \sqrt{a^2 - b^2}$, we find that $c = 5$. The foci are located on the horizontal axis, 5 units from the center of the ellipse. Therefore, the foci have coordinates $(7, -4)$ and $(-3, -4)$.

In all ellipses, $a^2 \geq b^2$. You can use this information to determine the orientation of the major axis from the equation. If a^2 is the denominator of the x^2 term, the major axis is parallel to the x-axis. If a^2 is the denominator of the y^2 term, the major axis is parallel to the y-axis.

Example 3 **For the equation $\frac{(y-3)^2}{25} + \frac{(x+4)^2}{9} = 1$, find the coordinates of the center, foci, and vertices of the ellipse. Then graph the equation.**

Determine the values of a, b, c, h, and k.

Since $a^2 > b^2$, $a^2 = 25$ and $b^2 = 9$.

$a^2 = 25$	$b^2 = 9$	$c = \sqrt{a^2 - b^2}$
$a = \sqrt{25}$ or 5	$b = \sqrt{9}$ or 3	$c = \sqrt{25 - 9}$ or 4
$x - h = x + 4$	$y - k = y - 3$	
$h = -4$	$k = 3$	

Since a^2 is the denominator of the y term, the major axis is parallel to the y-axis.

Graphing Calculator Tip

You can graph an ellipse by first solving for y. Then graph the two resulting equations on the same screen.

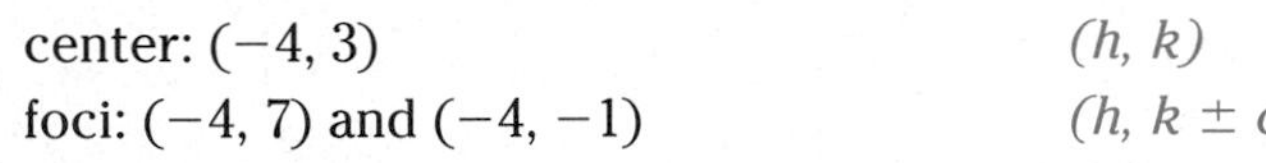

center: $(-4, 3)$ — *(h, k)*

foci: $(-4, 7)$ and $(-4, -1)$ — *$(h, k \pm c)$*

major axis vertices: $(-4, 8)$ and $(-4, -2)$ — *$(h, k \pm a)$*

minor axis vertices: $(-1, 3)$ and $(-7, 3)$ — *$(h \pm b, k)$*

Graph these ordered pairs. Other points on the ellipse can be found by substituting values for x and y. Complete the ellipse.

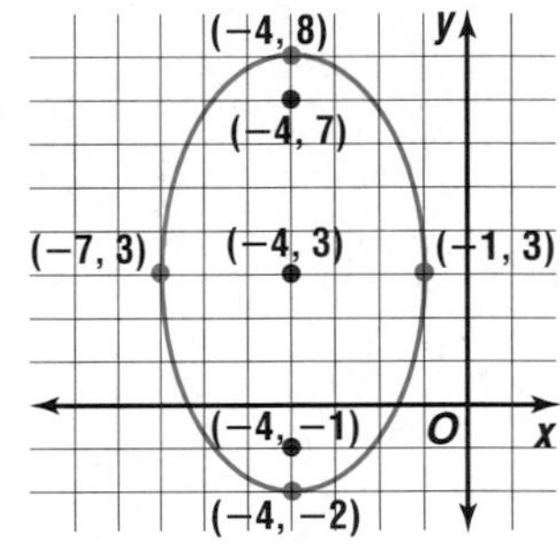

As with circles, the standard form of the equation of an ellipse can be expanded to obtain the general form. The result is a second-degree equation of the form $Ax^2 + Cy^2 + Dx + Ey + F = 0$, where $A \neq 0$ and $C \neq 0$, and A and C have the same sign. An equation in general form can be rewritten in standard form to determine the center at (h, k), the measure of the semi-major axis, a, and the measure of the semi-minor axis, b.

Example 4 **Find the coordinates of the center, the foci, and the vertices of the ellipse with the equation $4x^2 + 9y^2 - 40x + 36y + 100 = 0$. Then graph the equation.**

First write the equation in standard form.

$$4x^2 + 9y^2 - 40x + 36y + 100 = 0$$

$$4(x^2 - 10x + ?) + 9(y^2 + 4y + ?) = -100 + ? + ?$$ *The GCF of the x terms is 4. The GCF of the y terms is 9.*

$$4(x^2 - 10x + 25) + 9(y^2 + 4y + 4) = -100 + 4(25) + 9(4)$$ *Complete the square.*

$$4(x - 5)^2 + 9(y + 2)^2 = 36$$ *Factor.*

$$\frac{(x-5)^2}{9} + \frac{(y+2)^2}{4} = 1$$ *Divide each side by 36.*

(continued on the next page)

Since $a^2 > b^2$, $a^2 = 9$ and $b^2 = 4$. Thus, $a = 3$ and $b = 2$.

Since $c^2 = a^2 - b^2$, $c = \sqrt{5}$.

Since a^2 is the denominator of the x term, the major axis is parallel to the x-axis.

center: $(5, -2)$ — *(h, k)*
foci: $\left(5 \pm \sqrt{5}, -2\right)$ — *$(h \pm c, k)$*
major axis vertices: $(8, -2)$ and $(2, -2)$ — *$(h \pm a, k)$*
minor axis vertices: $(5, 0)$ and $(5, -4)$ — *$(h, k \pm b)$*

Sketch the ellipse.

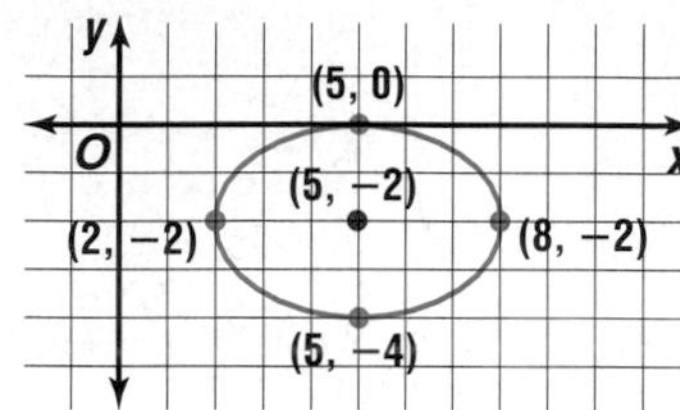

The **eccentricity** of an ellipse, denoted by e, is a measure that describes the shape of an ellipse. It is defined as $e = \frac{c}{a}$. Since $0 < c < a$, you can divide by a to show that $0 < e < 1$.

$0 < c < a$

$0 < \frac{c}{a} < 1$ *Divide by a.*

$0 < e < 1$ *Replace $\frac{c}{a}$ with e.*

The table shows the relationship between the value of e, the location of the foci, and the shape of the ellipse.

Value of *e*	Location of Foci	Graph
close to 0	near center of ellipse	$e = \frac{1}{5}$
close to 1	far from center of ellipse	$e = \frac{5}{6}$

Sometimes, you may need to find the value of b when you know the values of a and e. In any ellipse, $b^2 = a^2 - c^2$ and $\frac{c}{a} = e$. By using the two equations, it can be shown that $b^2 = a^2(1 - e^2)$. *You will derive this formula in Exercise 4.*

All of the planets in our solar system have elliptical orbits with the sun as one focus. These orbits are often described by their eccentricity.

Example 5 

ASTRONOMY **Of the nine planetary orbits in our solar system, Pluto's has the greatest eccentricity, 0.248. Astronomers have determined that the orbit is about 29.646 AU (astronomical units) from the sun at its closest point to the sun (perihelion). The length of the semi-major axis is about 39.482 AU.**

1 AU = the average distance between the sun and Earth, about 9.3×10^7 miles

a. Sketch the orbit of Pluto showing the sun in its position.

b. Find the length of the semi-minor axis of the orbit.

c. Find the distance of Pluto from the sun at its farthest point (aphelion).

a. The sketch at the right shows the sun as a focus for the elliptical orbit.

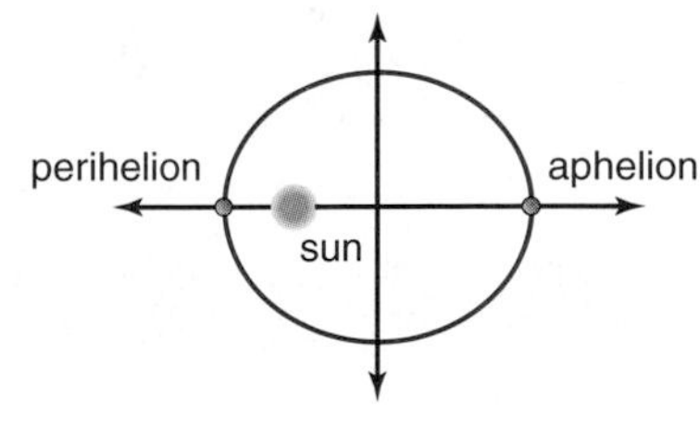

b. b is the length of the semi-minor axis.

$$b^2 = a^2(1 - e^2)$$
$$b = \sqrt{a^2(1 - e^2)}$$
$$b = \sqrt{(39.482)^2(1 - 0.248^2)} \quad a = 39.482,\ e = 0.248$$
$$b \approx 38.249 \text{ AU}$$

c. *aphelion = length of major axis − distance from sun to perihelion*

$$d = 2(39.482) - 29.646$$
$$= 49.318$$

Pluto is about 49 AU from the sun at its aphelion.

CHECK FOR UNDERSTANDING

Communicating Mathematics

Read and study the lesson to answer each question.

1. **Write** the equation of an ellipse centered at the origin, with $a = 8$, $b = 5$, and the major axis on the y-axis.
2. **Explain** how to determine whether the foci of an ellipse lie on the horizontal or vertical axis of an ellipse.
3. **Describe** the result when the foci and center of an ellipse coincide and give the eccentricity of such an ellipse.
4. **Derive** the equation $b^2 = a^2(1 - e^2)$.
5. **You Decide** Manuel says that the graph of $3y - 36 = 2x^2 - 18x$ is an ellipse with its major axis parallel to the y-axis. Shanice disagrees. Who is correct? Explain your answer.

Guided Practice

6. Write the equation of the ellipse graphed at the right in standard form. Then find the coordinates of its foci.

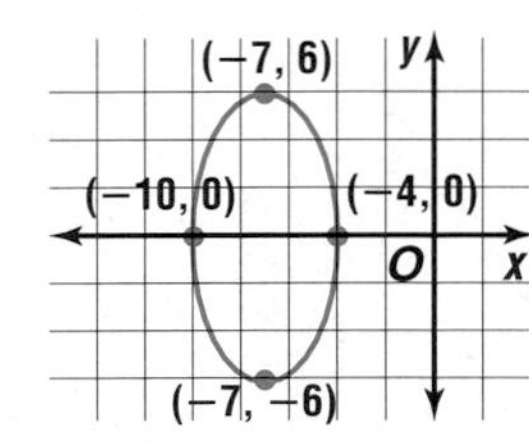

For the equation of each ellipse, find the coordinates of the center, foci, and vertices. Then graph the equation.

7. $\dfrac{x^2}{36} + \dfrac{y^2}{4} = 1$
8. $\dfrac{x^2}{81} + \dfrac{(y - 4)^2}{49} = 1$
9. $25x^2 + 9y^2 + 100x - 18y = 116$
10. $9x^2 + 4y^2 - 18x + 16y = 11$

Write the equation of the ellipse that meets each set of conditions.

11. The center is at $(-2, -3)$, the length of the vertical major axis is 8 units, and the length of the minor axis is 2 units.
12. The foci are located at $(-1, 0)$ and $(1, 0)$ and $a = 4$.
13. The center is at $(1, 2)$, the major axis is parallel to the x-axis, and the ellipse passes through points at $(1, 4)$ and $(5, 2)$.
14. The center is at $(3, 1)$, the vertical semi-major axis is 6 units long, and $e = \frac{1}{3}$.
15. **Astronomy** The elliptical orbit of Mars has its foci at $(0.141732, 0)$ and $(-0.141732, 0)$, where 1 unit equals 1 AU. The length of the major axis is 3.048 AU. Determine the equation that models Mars' elliptical orbit.

EXERCISES

Practice

Write the equation of each ellipse in standard form. Then find the coordinates of its foci.

16. 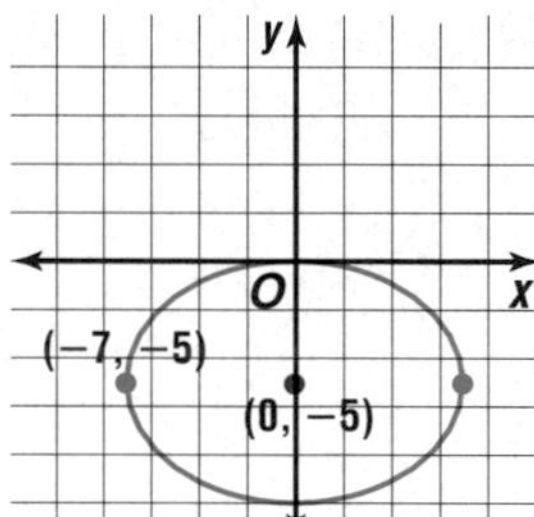

17. 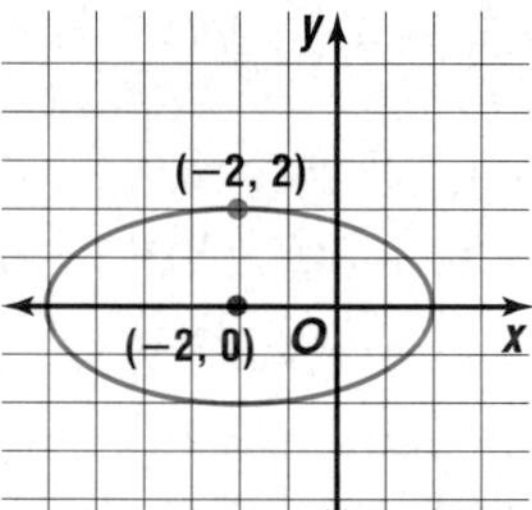

18. 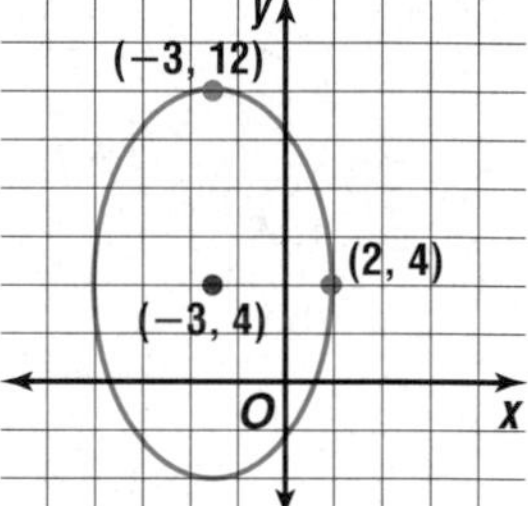

For the equation of each ellipse, find the coordinates of the center, foci, and vertices. Then graph the equation.

19. $\frac{(x+2)^2}{1} + \frac{(y-1)^2}{4} = 1$
20. $\frac{(x-6)^2}{100} + \frac{(y-7)^2}{121} = 1$
21. $\frac{(x-4)^2}{16} + \frac{(y+6)^2}{9} = 1$
22. $\frac{x^2}{4} + \frac{y^2}{9} = 1$
23. $4x^2 + y^2 - 8x + 6y + 9 = 0$
24. $16x^2 + 25y^2 - 96x - 200y = -144$
25. $3x^2 + y^2 + 18x - 2y + 4 = 0$
26. $6x^2 - 12x + 6y^2 + 36y = 36$
27. $18y^2 + 12x^2 - 144y - 48x = -120$
28. $4y^2 - 8y + 9x^2 - 54x + 49 = 0$
29. $49x^2 + 16y^2 + 160y - 384 = 0$
30. $9y^2 + 108y + 4x^2 - 56x = -484$

Write the equation of the ellipse that meets each set of conditions.

31. The center is at $(-3, -1)$, the length of the horizontal semi-major axis is 7 units, and the length of the semi-minor axis is 5 units.
32. The foci are at $(-2, 0)$ and $(2, 0)$, and $a = 7$.
33. The length of the semi-minor axis is $\frac{3}{4}$ the length of the horizontal semi-major axis, the center is at the origin, and $b = 6$.
34. The semi-major axis has length $2\sqrt{13}$ units, and the foci are at $(-1, 1)$ and $(-1, -5)$.
35. The endpoints of the major axis are at $(-11, 5)$ and $(7, 5)$. The endpoints of the minor axis are at $(-2, 9)$ and $(-2, 1)$.

36. The foci are at $(1, -1)$ and $(1, 5)$, and the ellipse passes through the point at $(4, 2)$.

37. The center is the origin, $\frac{1}{2} = \frac{c}{a}$, and the length of the horizontal semi-major axis is 10 units.

38. The ellipse is tangent to the x- and y-axes and has its center at $(4, -7)$.

39. The ellipse has its center at the origin, $a = 2$, and $e = \frac{3}{4}$.

40. The foci are at $(3, 5)$ and $(1, 5)$, and the ellipse has eccentricity 0.25.

41. The ellipse has a vertical major axis of 20 units, its center is at $(3, 0)$, and $e = \frac{7}{10}$.

42. The center is at $(1, -1)$, one focus is located at $\left(1, -1 + \sqrt{5}\right)$, and the ellipse has eccentricity $\frac{\sqrt{5}}{3}$.

Graphing Calculator

Graph each equation. Then use the TRACE function to approximate the coordinates of the vertices to the nearest integer.

43. $x^2 + 4y^2 - 6x + 24y = -41$

44. $4x^2 + y^2 - 8x - 2y = -1$

45. $4x^2 + 9y^2 - 16x + 18y = 11$

46. $25y^2 + 16x^2 - 150y + 32x = 159$

Applications and Problem Solving

47. **Entertainment** *Elliptipool* is a billiards game that use an elliptically-shaped pool table with only one pocket in the surface. A cue ball and a target ball are used in play. The object of the game is to strike the target ball with the cue ball so that the target ball rolls into the pocket after one bounce off the side. Suppose the cue ball and target ball can be placed anywhere on the half of the table opposite the pocket. The pool table shown at the right is 4 feet wide and 6 feet long. The pocket is located $\sqrt{5}$ feet from the center of the table along the ellipse's major axis. Assuming no spin is placed on either ball and the target ball is struck squarely, where should each be placed to have the best chance of hitting the target ball into the pocket? Explain your reasoning.

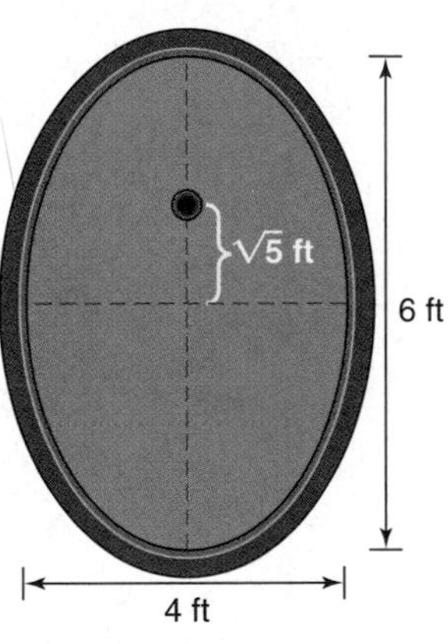

48. **Critical Thinking** As the foci of an ellipse move farther apart with the major axis fixed, what figure does the ellipse approach? Justify your answer.

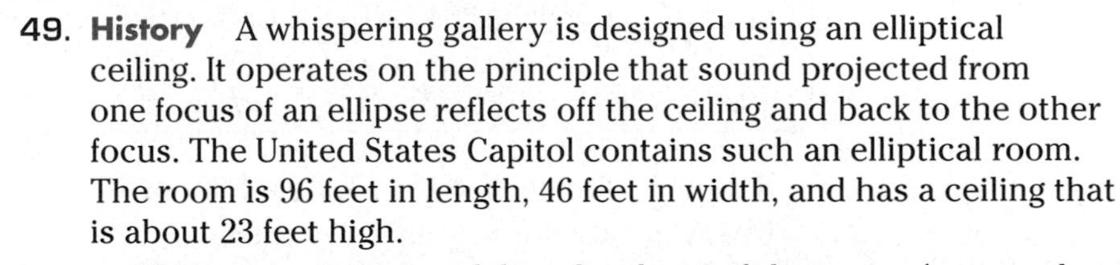

49. **History** A whispering gallery is designed using an elliptical ceiling. It operates on the principle that sound projected from one focus of an ellipse reflects off the ceiling and back to the other focus. The United States Capitol contains such an elliptical room. The room is 96 feet in length, 46 feet in width, and has a ceiling that is about 23 feet high.
   a. Write an equation modeling the shape of the room. Assume that is centered at the origin and the major axis is horizontal.
   b. John Quincy Adams is known to have overheard conversations being held at the opposing party leader's desk by standing in a certain spot in this room. Describe two possible places where Adams might have stood to overhear.
   c. About how far did Adams stand from the desk?

50. Geometry The square has an area of 1 square unit. If the square is stretched horizontally by a factor of a and compressed vertically by a factor of b, the area of the rectangle formed is ab square units.

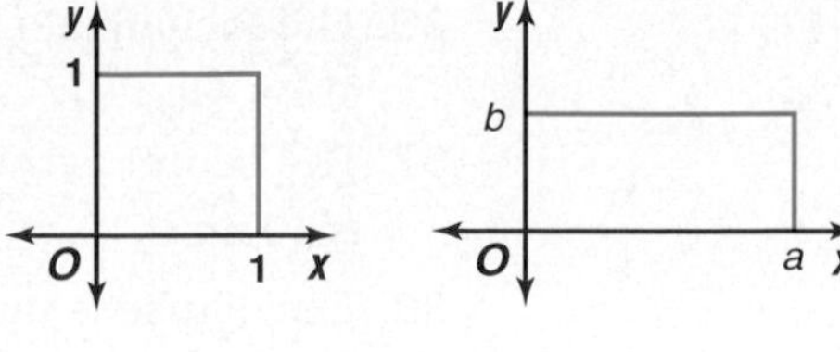

a. The area of a circle with equation $x^2 + y^2 = r^2$ is πr^2. Develop a formula for the area of an ellipse with equation $\frac{x^2}{a^2} + \frac{y^2}{b^2} = 1$.

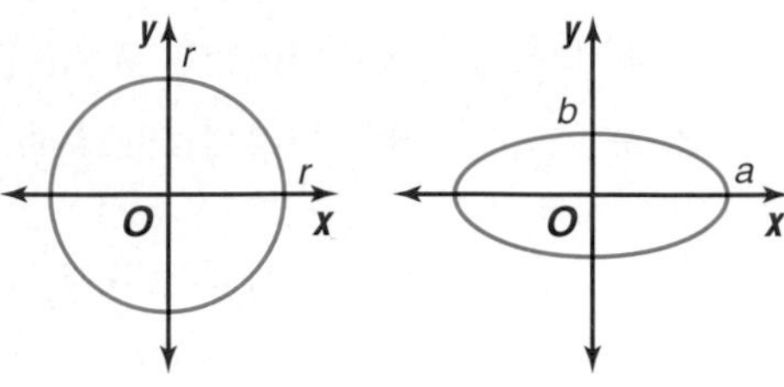

b. Use the formula found in part **a** to find the area of the ellipse $\frac{x^2}{9} + \frac{y^2}{4} = 1$.

51. Critical Thinking Show that the ellipse $\frac{x^2}{a^2} + \frac{y^2}{b^2} = 1$ is symmetric with respect to the origin.

52. Construction The arch of a fireplace is to be constructed in the shape of a semi-ellipse. The opening is to have a height of 3 feet at the center and a width of 8 feet along the base. To sketch the outline of the fireplace, the contractor uses an 8-foot string tied to two thumbtacks.

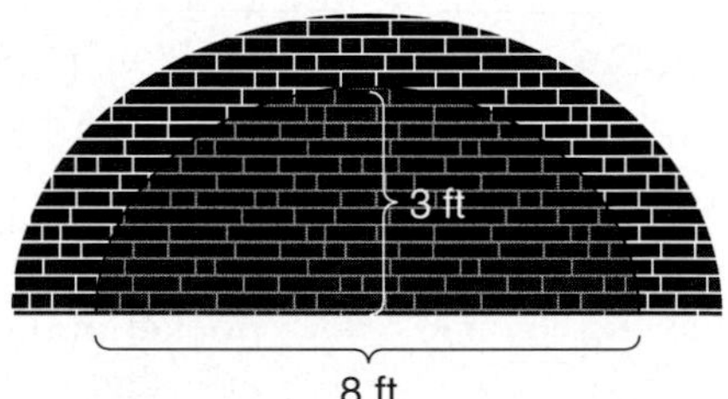

a. Where should the thumbtacks be placed?

b. Explain why this technique works.

53. Astronomy The satellites orbiting Earth follow elliptical paths with the center of Earth as one focus. The table below lists data on five satellites that have orbited or currently orbit Earth.

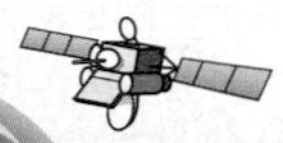

Satellites Orbiting Earth

Name	Launch Date	Time or expected time aloft	Semi-major axis a (km)	Eccentricity
Sputnik I	Oct. 1957	57.6 days	6955	0.052
Vanguard I	Mar. 1958	300 years	8872	0.208
Skylab 4	Nov. 1973	84.06 days	6808	0.001
GOES 4	Sept. 1980	10^6 years	42,166	0.0003
Intelsat 5	Dec. 1980	10^6 years	41,803	0.007

a. Which satellite has the most circular orbit? Explain your reasoning.

b. Soviet satellite Sputnik I was the first artificial satellite to orbit Earth. If the radius of Earth is approximately 6357 kilometers, find the greatest distance Sputnik I orbited from the surface of Earth to the nearest kilometer.

Mixed Review

54. Write the standard form of the equation of the circle that passes through points at $(0, -9)$, $(7, -2)$, and $(-5, -10)$. Identify the circle's center and radius. Then graph the equation. *(Lesson 10-2)*

55. Determine whether the quadrilateral with vertices at points $(-1, -2)$, $(5, -4)$, $(4, 1)$, and $(-5, 4)$ is a parallelogram. *(Lesson 10-1)*

56. If $\sin \theta = \frac{7}{8}$ and the terminal side of θ is in the first quadrant, find $\cos 2\theta$. *(Lesson 7-4)*

57. Write an equation of the cosine function with an amplitude of 4, a period of 180°, and a phase shift of 20°. *(Lesson 6-5)*

58. Solve $\triangle ABC$ if $C = 121°32'$, $B = 42°5'$, and $a = 4.1$. Round angle measures to the nearest minute and side measures to the nearest tenth. *(Lesson 5-6)*

59. Use the Remainder Theorem to find the remainder for the quotient $(x^4 - 4x^3 - 2x^2 - 1) \div (x - 5)$. Then, state whether the binomial is a factor of the polynomial. *(Lesson 4-3)*

60. Determine whether the point at $(-2, -16)$ is the location of a *minimum,* a *maximum,* or a *point of inflection* for the function $x^2 + 4x - 12$. *(Lesson 3-6)*

61. Sketch the graph of $g(x) = |x - 2|$. *(Lesson 3-2)*

62. **Motion** A ceiling fan has four evenly spaced blades that are each 2 feet long. Suppose the center of the fan is located at the origin and the blades of the fan lie in the x- and y-axis. Imagine a fly lands on the tip of the blade along the positive x-axis. Find the location of the fly where it landed and its location after a 90°, 180°, and 270° counterclockwise rotation of the fan. *(Lesson 2-4)*

63. **SAT Practice** In parallelogram $QRST$, $a = b$ and $c = d$. What is the value of p?

A 45 **B** 60 **C** 90 **D** 120

E Cannot be determined from the information given

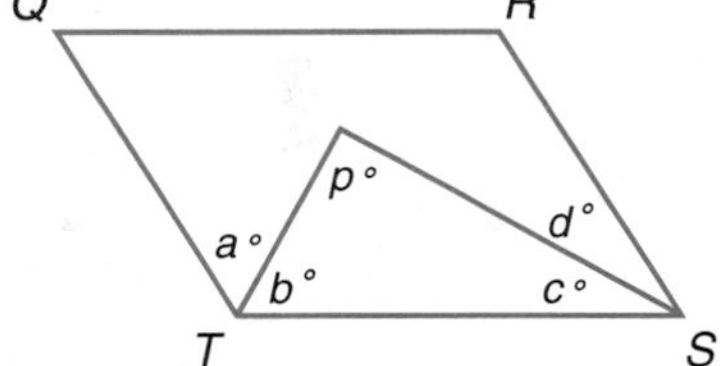

GRAPHING CALCULATOR EXPLORATION

To graph an ellipse, such as the equation $\frac{(x-3)^2}{18} + \frac{(y+2)^2}{32} = 1$, on a graphing calculator, you must first solve for y.

$$\frac{(x-3)^2}{18} + \frac{(y+2)^2}{32} = 1$$

$$32(x-3)^2 + 18(y-2)^2 = 576$$

$$18(y-2)^2 = 576 - 32(x-3)^2$$

$$(y-2)^2 = \frac{576 - 32(x-3)^2}{18}$$

So, $y = \pm\sqrt{\frac{576 - 32(x-3)^2}{18}} + 2.$

The result is two equations. To graph both equations as Y_1, replace $\pm$ with $\{1, -1\}$.

Like other graphs, there are families of ellipses. Changing certain values in the equation of an ellipse creates a new member of that family.

TRY THESE For each situation, make a conjecture about the behavior of the graph. Then verify by graphing the original equation and the modified equation on the same screen using a square window.

1. x is replaced by $(x - 4)$.
2. x is replaced by $(x + 4)$.
3. y is replaced by $(y - 4)$.
4. y is replaced by $(y + 4)$.
5. 32 is switched with 18.

WHAT DO YOU THINK?

6. Describe the effects of replacing x in the equation of an ellipse with $(x \pm c)$ for $c > 0$.
7. Describe the effects of replacing y in the equation of an ellipse with $(y \pm c)$ for $c > 0$.
8. In the equation $\frac{x^2}{a^2} + \frac{y^2}{b^2} = 1$, describe the effect of interchanging a and b.

Extra Practice See p. A44.

10-4

Hyperbolas

OBJECTIVES

- Use and determine the standard and general forms of the equation of a hyperbola.
- Graph hyperbolas.

NAVIGATION Since World War II, ships have used the LORAN (*LO*ng *RA*nge *N*avigation) system as a means of navigation independent of visibility conditions. Two stations, located a great distance apart, simultaneously transmit radio pulses to ships at sea. Since a ship is usually closer to one station than the other, the ship receives these pulses at slightly different times. By measuring the time differential and by knowing the speed of the radio waves, a ship can be located on a conic whose foci are the positions of the two stations. *A problem related to this will be solved in Example 4.*

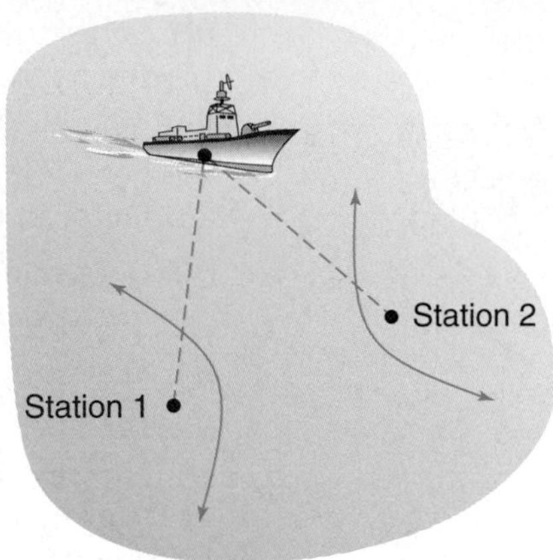

A **hyperbola** is the set of all points in the plane in which the difference of the distances from two distinct fixed points, called **foci,** is constant. That is, if F_1 and F_2 are the foci of a hyperbola and P and Q are any two points on the hyperbola, $|PF_1 - PF_2| = |QF_1 - QF_2|$.

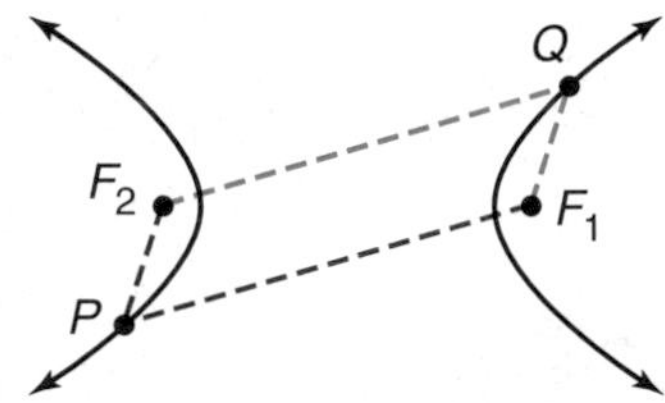

The **center** of a hyperbola is the midpoint of the line segment whose endpoints are the foci. The point on each branch of the hyperbola that is nearest the center is called a **vertex.**

Look Back
Refer to Lesson 3-7 to review asymptotes.

The **asymptotes** of a hyperbola are lines that the curve approaches as it recedes from the center. As you move farther out along the branches, the distance between points on the hyperbola and the asymptotes approaches zero.

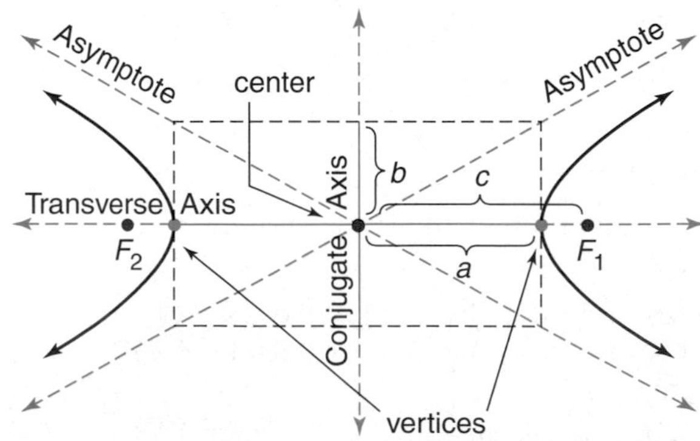

Note that $c > a$ for the hyperbola.

A hyperbola has two axes of symmetry. The line segment connecting the vertices is called the **transverse axis** and has a length of $2a$ units. The segment perpendicular to the transverse axis through the center is called the **conjugate axis** and has length $2b$ units.

For a hyperbola, the relationship among a, b, and c is represented by $a^2 + b^2 = c^2$. The asymptotes contain the diagonals of the rectangle guide, which is $2a$ units by $2b$ units. The point at which the diagonals meet coincides with the center of the hyperbola.

The standard form of the equation of a hyperbola with its origin as its center can be derived from the definition and the Distance Formula. Suppose the foci are on the x-axis at $(c, 0)$ and $(-c, 0)$ and the coordinates of any point on the hyperbola are (x, y).

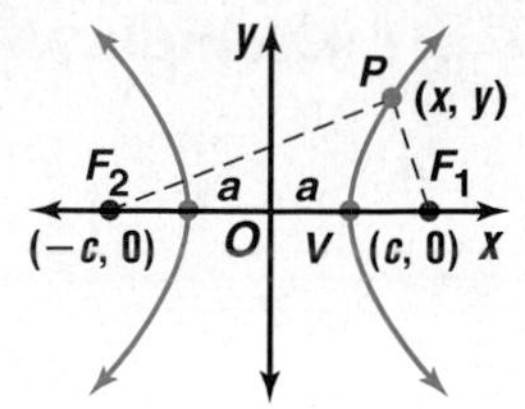

$\lvert PF_2 - PF_1 \rvert = \lvert VF_2 - VF_1 \rvert$	*Definition of hyperbola*
$\left\lvert \sqrt{(x + c)^2 + y^2} - \sqrt{(x + c)^2 + y^2} \right\rvert = \lvert c + a - (c - a) \rvert$	*Distance Formula*
$\sqrt{(x + c)^2 + y^2} - \sqrt{(x + c)^2 + y^2} = 2a$	*Simplify.*
$\sqrt{(x - c)^2 + y^2} = 2a + \sqrt{(x + c)^2 + y^2}$	*Isolate a radical.*
$(x - c)^2 + y^2 = 4a^2 + 4a\sqrt{(x + c)^2 + y^2} + (x + c)^2 + y^2$	*Square each side.*
$-4xc - 4a^2 = 4a\sqrt{(x + c)^2 + y^2}$	*Simplify.*
$xc + a^2 = -a\sqrt{(x + c)^2 + y^2}$	*Divide each side by −4.*
$x^2c^2 + 2a^2xc + a^4 = a^2x^2 + 2a^2xc + a^2c^2 + a^2y^2$	*Square each side.*
$(c^2 - a^2)x^2 - a^2y^2 = a^2(c^2 - a^2)$	*Simplify.*
$\frac{x^2}{a^2} - \frac{y^2}{c^2 - a^2} = 1$	*Divide by $a^2(c^2 - a^2)$.*
$\frac{x^2}{a^2} - \frac{y^2}{b^2} = 1$	*By the Pythagorean Theorem, $c^2 - a^2 = b^2$.*

If the foci are on the y-axis, the equation is $\frac{y^2}{a^2} - \frac{x^2}{b^2} = 1$.

As with the other graphs we have studied in this chapter, the standard form of the equation of a hyperbola with center other than the origin is a translation of the parent graph to a center at (h, k).

Standard Form of the Equation of a Hyperbola	**Orientation**	**Description**
$\frac{(x - h)^2}{a^2} - \frac{(y - k)^2}{b^2} = 1$, where $b^2 = c^2 - a^2$	$\frac{(x - h)^2}{a^2} - \frac{(y - k)^2}{b^2} = 1$; $y = k$; (h, k); O; $x = h$; x; y	center: (h, k) foci: $(h \pm c, k)$ vertices: $(h \pm a, k)$ equation of transverse axis: $y = k$ *(parallel to x-axis)*
$\frac{(y - k)^2}{a^2} - \frac{(x - h)^2}{b^2} = 1$, where $b^2 = c^2 - a^2$	$\frac{(y - k)^2}{a^2} - \frac{(x - h)^2}{b^2} = 1$; $y = k$; (h, k); $x = h$; O; x; y	center: (h, k) foci: $(h, k \pm c)$ vertices: $(h, k \pm a)$ equation of transverse axis: $x = h$ *(parallel to y-axis)*

Example **1** **Find the equation of the hyperbola with foci at (7, 1) and (−3, 1) whose transverse axis is 8 units long.**

A sketch of the graph is helpful. Let F_1 and F_2 be the foci, let V_1 and V_2 be the vertices, and let C be the center.

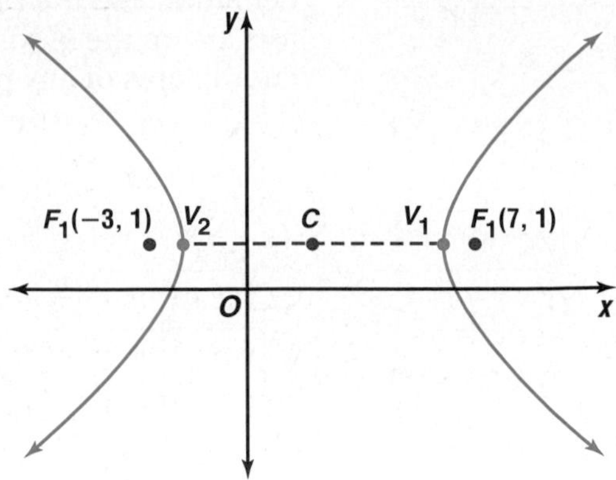

To locate the center, find the midpoint of $\overline{F_1F_2}$.

$$\left(\frac{7 + (-3)}{2}, \frac{1 + 1}{2}\right) \text{ or } (2, 1)$$

Thus, $h = 2$ and $k = 1$ since (2, 1) is the center.

The transverse axis is 8 units long. Thus, $2a = 8$ or $a = 4$. So $a^2 = 16$.

Use the equation $b^2 = c^2 - a^2$ to find b^2.

$b^2 = c^2 - a^2$

$b^2 = 25 - 16$ *$c = 5, a = 4$*

$b^2 = 9$

Recall that c is the distance from the center to a focus. Here $c = CF_1$ or 5.

Use the standard form when the transverse axis is parallel to the x-axis.

$$\frac{(x - h)^2}{a^2} - \frac{(y - k)^2}{b^2} = 1 \Rightarrow \frac{(x - 2)^2}{16} - \frac{(y - 1)^2}{9} = 1$$

Before graphing a hyperbola, it is often helpful to graph the asymptotes. As noted in the beginning of the lesson, the asymptotes contain the diagonals of the rectangle guide defined by the transverse and conjugate axes. While not part of the graph, a sketch of this $2a$ by $2b$ rectangle provides an easy way to graph the asymptotes of the hyperbola. Suppose the center of a hyperbola is the origin and the transverse axis lies along the x-axis.

From the figure at the right, we can see that the asymptotes have slopes equal to $\pm\frac{b}{a}$. Since both lines have a y-intercept of 0, the equations for the asymptotes are $y = \pm\frac{b}{a}x$.

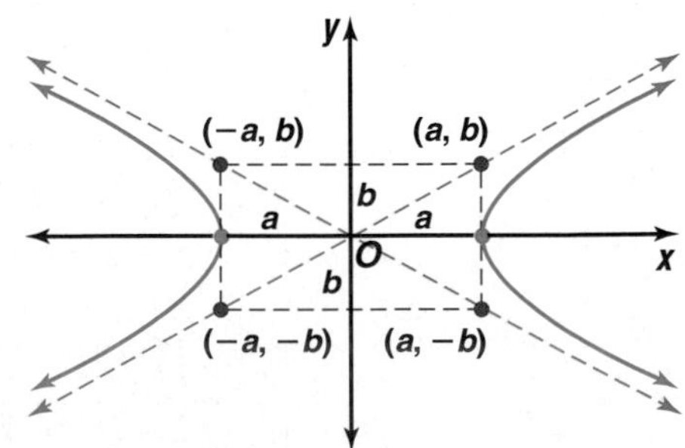

If the hyperbola were oriented so that the traverse axis was parallel to the y-axis, the slopes of the asymptotes would be $\pm\frac{a}{b}$. Thus, the equations of the asymptotes would be $y = \pm\frac{a}{b}x$.

The equations of the asymptotes of any hyperbola can be determined by a translation of the graph to a center at (h, k).

Equations of the Asymptotes of a Hyperbola	$y - k = \pm\frac{b}{a}(x - h)$, for a hyperbola with standard form $\frac{(x-h)^2}{a^2} - \frac{(y-k)^2}{b^2} = 1$	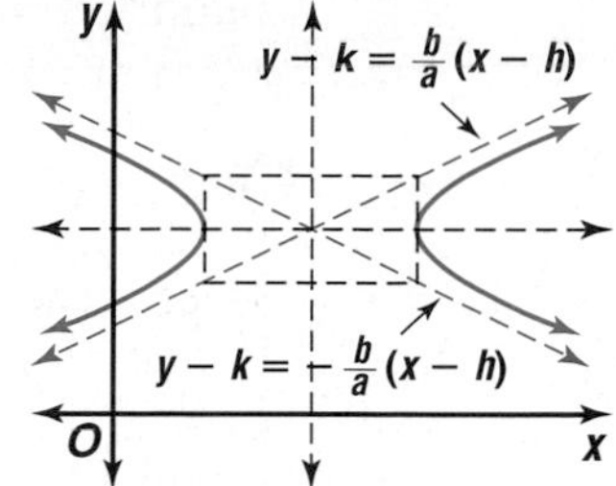
	$y - k = \pm\frac{a}{b}(x - h)$, for a hyperbola with standard form $\frac{(y-k)^2}{a^2} - \frac{(x-h)^2}{b^2} = 1$	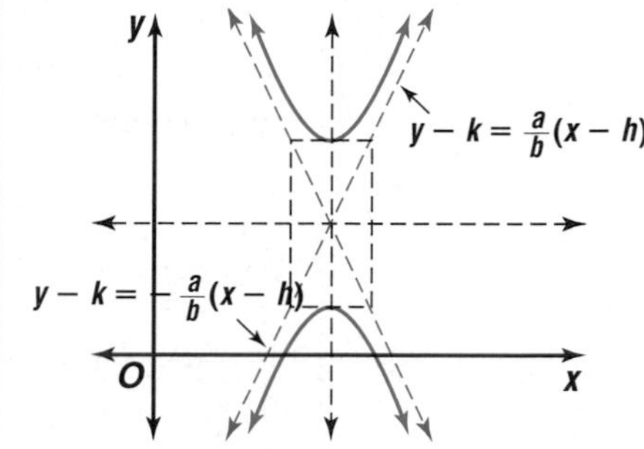

Example **2** **Find the coordinates of the center, foci, and vertices, and the equations of the asymptotes of the graph of $\frac{(y+4)^2}{36} - \frac{(x-2)^2}{25} = 1$. Then graph the equation.**

Since the y terms are in the first expression, the hyperbola has a vertical transverse axis.

From the equation, $h = 2$, $k = -4$, $a = 6$, and $b = 5$. The center is at $(2, -4)$.

The equations of the asymptotes are $y + 4 = \pm\frac{6}{5}(x - 2)$.

The vertices are at $(h, k \pm a)$ or $(2, 2)$ and $(2, -10)$.

Since $c^2 = a^2 + b^2$, $c = \sqrt{61}$. Thus, the foci are at $\left(2, -4 + \sqrt{61}\right)$ and $\left(2, -4 - \sqrt{61}\right)$.

Graph the center, vertices, and the rectangle guide. Next graph the asymptotes. Then sketch the hyperbola.

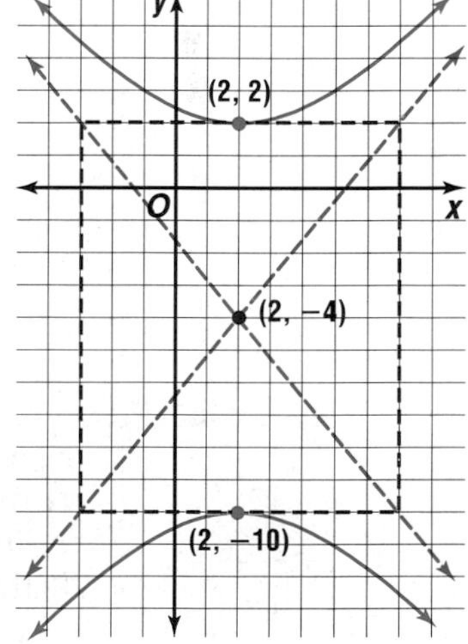

Graphing Calculator Tip

You can graph a hyperbola on a graphing calculator by first solving for y and then graphing the two resulting equations on the same screen.

By expanding the standard form for a hyperbola, you can determine the general form $Ax^2 + Cy^2 + Dx + Ey + F = 0$ where $A \neq 0$, $C \neq 0$, and A and C have different signs.

As with other general forms we have studied, the general form of a hyperbola can be rewritten in standard form. While it is important to be able to recognize the equation of a hyperbola in general form, the standard form provides important information about the hyperbola that makes it easier to graph.

Example 3 **Find the coordinates of the center, foci, and vertices, and the equations of the asymptotes of the graph of $9x^2 - 4y^2 - 54x - 40y - 55 = 0$. Then graph the equation.**

Write the equation in standard form. Use the same process you used with ellipses.

$$9x^2 - 4y^2 - 54x - 40y - 55 = 0$$

$$9x^2 - 54x - 4y^2 - 40y = 55 \quad \textit{Rearrange terms.}$$

$$9(x^2 - 6x + ?) - 4(y^2 + 10y + ?) = 55 + ? + ? \quad \textit{Factor GCF for each variable.}$$

$$9(x^2 - 6x + 9) - 4(y^2 + 10y + 25) = 55 + 9(9) - 4(25) \quad \textit{Complete the square.}$$

$$9(x - 3)^2 - 4(y + 5)^2 = 36 \quad \textit{Factor.}$$

$$\frac{(x-3)^2}{4} - \frac{(y+5)^2}{9} = 1 \quad \textit{Divide each side by 36.}$$

The center is at $(3, -5)$. Since the x terms are in the first expression, the transverse axis is horizontal.

$a = 2, b = 3, c = \sqrt{13}$

The foci are at $\left(3 - \sqrt{13}, -5\right)$ and $\left(3 + \sqrt{13}, -5\right)$.

The vertices are at $(1, -5)$ and $(5, -5)$.

The asymptotes have equations $y + 5 = \pm\frac{3}{2}(x - 3)$.

Graph the vertices and the rectangle guide. Next graph the asymptotes. Then sketch the hyperbola.

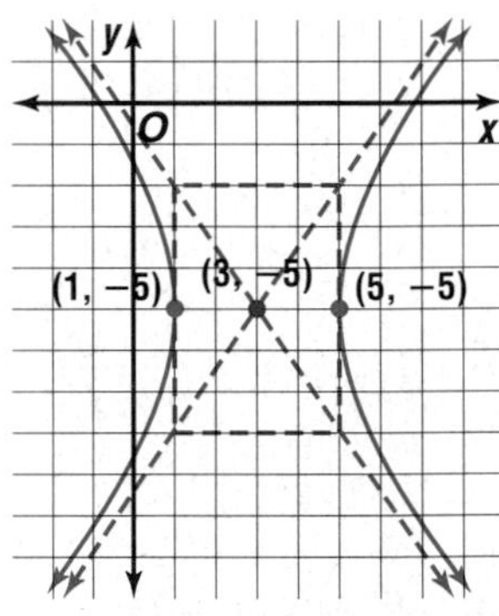

One interesting application of hyperbolas is in navigation.

Example 4

NAVIGATION **Refer to the application at the beginning of the lesson. Suppose LORAN stations *A* and *B* are located 400 miles apart along a straight shore, with *A* due west of *B*. A ship approaching the shore receives radio pulses from the stations and is able to determine that it is 100 miles farther from station *A* than it is from station *B*.**

a. Find the equation of the hyperbola on which the ship is located.

b. Find the exact coordinates of the ship if it is 60 miles from shore.

a. First set up a rectangular coordinate system with the origin located midway between station A and station B.

The stations are located at the foci of the hyperbola, so $c = 200$.

The difference of the distances from the ship to each station is 100 miles. By definition of a hyperbola, this difference equals $2a$, so $a = 50$. The vertices of the hyperbola are located on the same axis as the foci, so the vertices of the hyperbola the ship is on are at $(-50, 0)$ and $(50, 0)$.

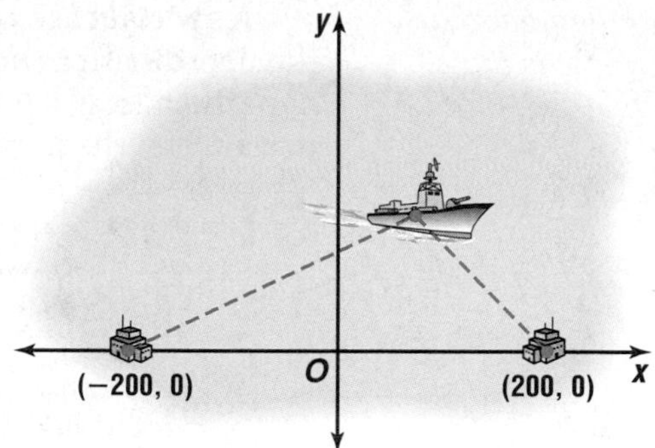

Since the hyperbola's transverse axis is the x-axis, the form of the equation of this hyperbola is $\frac{x^2}{a^2} - \frac{y^2}{b^2} = 1$. Using the equation $b^2 = c^2 - a^2$, we can find the value for b^2.

$b^2 = c^2 - a^2$

$b^2 = 200^2 - 50^2$ *$a = 50, c = 200$*

$b^2 = 37{,}500$

Thus, the equation of the hyperbola is $\frac{x^2}{2500} - \frac{y^2}{37{,}500} = 1$

b. If the ship is 60 miles from shore, let $y = 60$ in the equation of the hyperbola and solve for x.

$$\frac{x^2}{2500} - \frac{60^2}{37{,}500} = 1$$

$$\frac{x^2}{2500} = 1 + \frac{60^2}{37{,}500}$$

$$\frac{x^2}{2500} = 1.096$$

$$x^2 = 2500(1.096)$$

$$x = \pm\sqrt{2740} \approx \pm 52.3$$

Since the ship is closer to station B than station A, we use the positive value of x to locate the ship at coordinates $(52.3, 60)$.

In the standard form of the equation of a hyperbola, if $a = b$, the graph is an **equilateral hyperbola.** Replacing a with b in the equations of the asymptotes of a hyperbola with a horizontal transverse axis reveals a property of equilateral hyperbolas.

$y - k = \frac{b}{a}(x - h)$ $\qquad$ $y - k = -\frac{b}{a}(x - h)$

$y - k = \frac{b}{b}(x - h)$ $\qquad$ *Let $a = b$* $\qquad$ $y - k = -\frac{b}{b}(x - h)$

$y - k = (x - h)$ $\qquad$ $y - k = -(x - h)$

The slopes of the equations of the two asymptotes are negative reciprocals, 1 and -1. Thus, the asymptotes of an equilateral hyperbola are perpendicular.

Remember that $xy = c$ is the general equation that models inverse variation.

A special case of the equilateral hyperbola is a **rectangular hyperbola,** where the coordinate axes are the asymptotes. The general equation of a rectangular hyperbola is $xy = c$, where c is a nonzero constant. The sign of the constant c determines the location of the branches of the hyperbola.

Rectangular Hyperbola: $xy = c$	
Value of c	**Location of branches of hyperbola**
Positive	Quadrants I and III
Negative	Quadrants II and IV

Example 5 Graph $xy = 16$.

Since c is positive, the hyperbola lies in the first and third quadrants.

The transverse axis is along the graph of $y = x$.

The coordinates of the vertices must satisfy the equation of the hyperbola and also their graph must be points on the traverse axis. Thus, the vertices are at $(4, 4)$ and $(-4, -4)$.

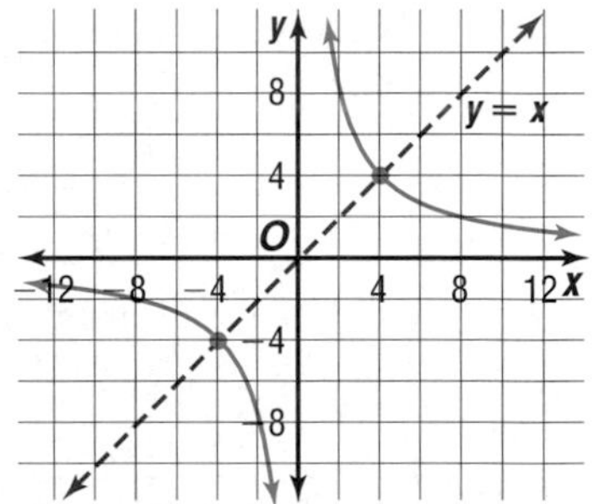

Like an ellipse, the shape of a hyperbola is determined by its eccentricity, which is again defined as $e = \frac{c}{a}$. However, in a hyperbola, $0 < a < c$. So, $0 < 1 < e$ or $e > 1$. The table below shows the relationship between the value of the e and the shape of the hyperbola.

Value of e	Graph
close to 1	$e = \frac{9}{8}$ (y, O, x)
not close to 1	$e = \frac{13}{5}$ (y, O, x)

Since $c^2 = a^2 + b^2$ in hyperbolas, it can be shown that $b^2 = a^2(e^2 - 1)$. *You will derive this formula in Exercise 3.*

Example 6 **Write the equation of the hyperbola with center at (3, −1), a focus at (3, −4), and eccentricity $\frac{3}{2}$.**

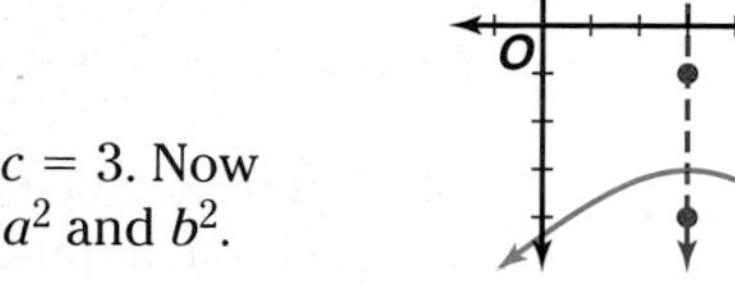

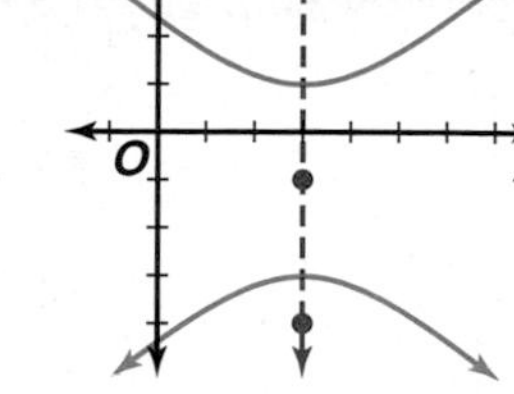

Sketch the graph using the points given. Since the center and focus have the same x-coordinate, the transverse axis is vertical. Use the form $\frac{(y-k)^2}{a^2} - \frac{(x-h)^2}{b^2} = 1$.

The focus is 3 units below the center, so $c = 3$. Now use the eccentricity to find the values of a^2 and b^2.

$e = \frac{c}{a}$

$\frac{3}{2} = \frac{3}{a}$ *c = 3 and e = $\frac{3}{2}$*

$2 = a$

$4 = a^2$

$b^2 = a^2(e^2 - 1)$

$b^2 = 4\left(\frac{9}{4} - \frac{4}{4}\right)$ *$a^2 = 4$ and e = $\frac{3}{2}$*

$b^2 = 5$

The equation is $\frac{(y+1)^2}{4} - \frac{(x-3)^2}{5} = 1$.

CHECK FOR UNDERSTANDING

Communicating Mathematics

Read and study the lesson to answer each question.

1. **Compare and contrast** the standard forms of the equations of hyperbolas and ellipses.
2. **Determine** which of the following equations matches the graph of the hyperbola at right.

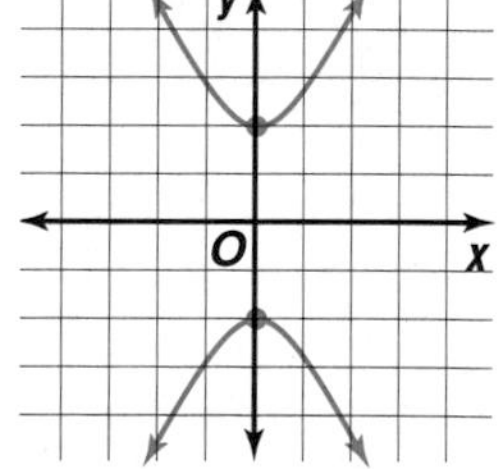

   a. $\frac{x^2}{4} - y^2 = 1$ b. $\frac{y^2}{4} - x^2 = 1$ c. $x^2 - \frac{y^2}{4} = 1$
3. **Derive** the equation $b^2 = a^2(e^2 - 1)$ for a hyperbola.
4. *Math Journal* **Write** an explanation of how to determine whether the transverse axis of a hyperbola is horizontal or vertical.

Guided Practice

For the equation of each hyperbola, find the coordinates of the center, the foci, and the vertices and the equations of the asymptotes of its graph. Then graph the equation.

5. $\frac{x^2}{25} - \frac{y^2}{4} = 1$
6. $\frac{(y-3)^2}{16} - \frac{(x-2)^2}{4} = 1$
7. $y^2 - 5x^2 + 20x = 50$

8. Write the equation of the hyperbola graphed at the right.

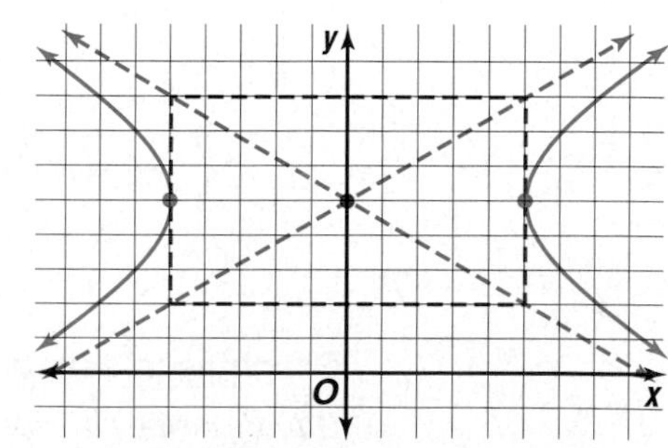

9. Graph $xy = -9$.

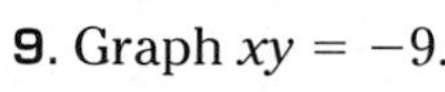

Write an equation of the hyperbola that meets each set of conditions.

10. The center is at $(1, -4)$, $a = 5$, $b = 2$, and it has a horizontal transverse axis.
11. The length of the conjugate axis is 6 units, and the vertices are at $(3, 4)$ and $(3, 0)$.
12. The hyperbola is equilateral and has foci at $(0, 6)$ and $(0, -6)$.
13. The eccentricity of the hyperbola is $\frac{5}{3}$, and the foci are at $(10, 0)$ and $(-10, 0)$.
14. **Aviation** Airplanes are equipped with signal devices to alert rescuers to their position. Suppose a downed plane sends out radio pulses that are detected by two receiving stations, *A* and *B*. The stations are located 130 miles apart along a stretch of I-40, with *A* due west of *B*. The two stations are able to determine that the plane is 50 miles farther from station *B* than from station *A*.
    a. Determine the equation of the hyperbola centered at the origin on which the plane is located.
    b. Graph the equation, indicating on which branch of the hyperbola the plane is located.
    c. If the pilot estimates that the plane is 6 miles from I-40, find the exact coordinates of its position.

EXERCISES

Practice

For the equation of each hyperbola, find the coordinates of the center, the foci, and the vertices and the equations of the asymptotes of its graph. Then graph the equation.

15. $\frac{x^2}{100} - \frac{y^2}{16} = 1$
16. $\frac{x^2}{9} - \frac{(y-5)^2}{81} = 1$
17. $\frac{x^2}{4} - \frac{y^2}{49} = 1$
18. $\frac{(y-7)^2}{64} - \frac{(x+1)^2}{4} = 1$
19. $x^2 - 4y^2 + 6x - 8y = 11$
20. $-4x^2 + 9y^2 - 24x - 90y + 153 = 0$
21. $16y^2 - 25x^2 - 96y + 100x - 356 = 0$
22. $36x^2 - 49y^2 - 72x - 294y = 2169$

23. Graph the equation $25y^2 - 9x^2 - 100y - 72x - 269 = 0$. Label the center, foci and the equations of the asymptotes.

Write the equation of each hyperbola.

24.

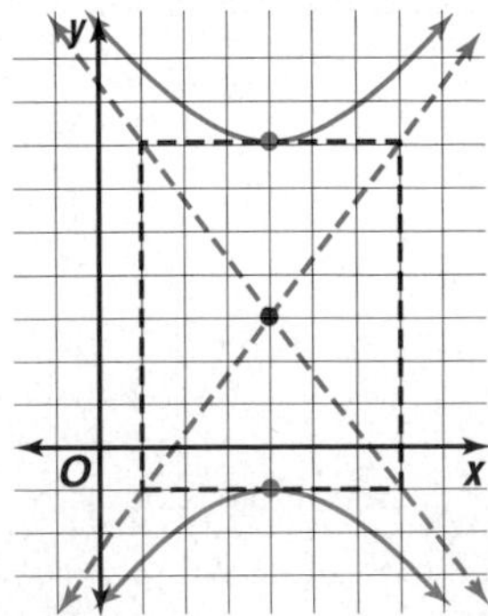

25.

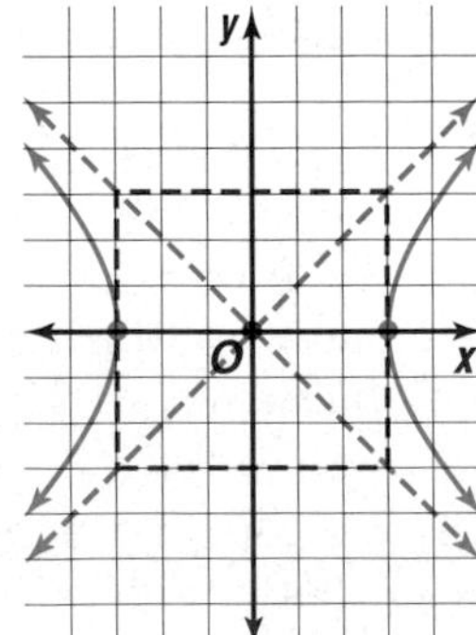

26.

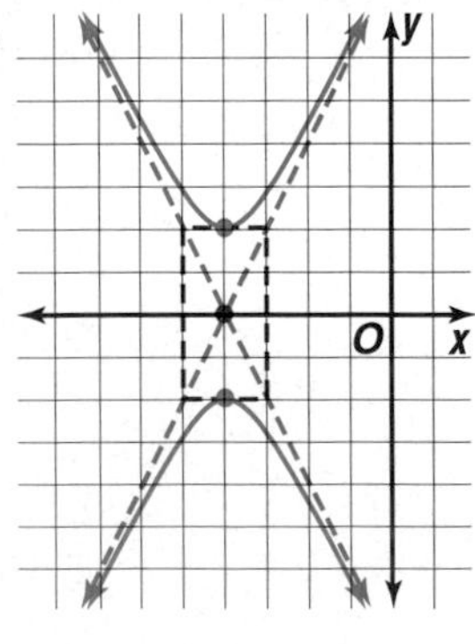

Graph each equation.

27. $xy = 49$
28. $xy = -36$
29. $4xy = -25$
30. $9xy = 16$

Write an equation of the hyperbola that meets each set of conditions.

31. The center is at $(4, -2)$, $a = 2$, $b = 3$, and it has a vertical transverse axis.
32. The vertices are at $(0, 3)$ and $(0, -3)$, and a focus is at $(0, -9)$.
33. The length of the transverse axis is 6 units, and the foci are at $(5, 2)$ and $(-5, 2)$.
34. The length of the conjugate axis is 8 units, and the vertices are at $(-3, 9)$ and $(-3, -5)$.
35. The hyperbola is equilateral and has foci at $(8, 0)$ and $(-8, 0)$.
36. The center is at $(-3, 1)$, one focus is at $(2, 1)$, and the eccentricity is $\frac{5}{4}$.
37. A vertex is at $(4, 5)$, the center is at $(4, 2)$, and an equation of one asymptote is $4y + 4 = 3x$.
38. The equation of one asymptote is $3x - 11 = 2y$. The hyperbola has its center at $(3, -1)$ and a vertex at $(5, -1)$.
39. The hyperbola has foci at $(0, 8)$ and $(0, -8)$ and eccentricity $\frac{4}{3}$.
40. The hyperbola has eccentricity $\frac{6}{5}$ and foci at $(10, -3)$ and $(-2, -3)$.
41. The hyperbola is equilateral and has foci at $(9, 0)$ and $(-9, 0)$.
42. The slopes of the asymptotes are ± 2, and the foci are at $(1, 5)$ and $(1, -3)$.

Applications and Problem Solving

43. **Chemistry** According to Boyle's Law, the pressure P (in kilopascals) exerted by a gas varies inversely as the volume V (in cubic decimeters) of a gas if the temperature remains constant. That is, $PV = c$. Suppose the constant for oxygen at 25°C is 505.
    a. Graph the function $PV = c$ for $c = 505$.
    b. Determine the volume of oxygen if the pressure is 101 kilopascals.
    c. Determine the volume of oxygen if the pressure is 50.5 kilopascals.
    d. Study your results for parts **b** and **c**. If the pressure is halved, make a conjecture about the effect on the volume of gas.

44. **Critical Thinking** Prove that the eccentricity of all equilateral hyperbolas is $\sqrt{2}$.

45. **Nuclear Power** A nuclear cooling tower is a *hyperboloid,* that is, a hyperbola rotated around its conjugate axis. Suppose the hyperbola used to generate the hyperboloid modeling the shape of the cooling tower has an eccentricity of $\frac{5}{3}$.
    a. If the cooling tower is 150 feet wide at its narrowest point, determine an equation of the hyperbola used to generate the hyperboloid.
    b. If the tower is 450 feet tall, the top is 100 feet above the center of the hyperbola, and the base is 350 feet below the center, what is the radius of the top and the base of the tower?

46. Forestry Two ranger stations located 4 miles apart observe a lightning strike. A ranger at station A reports hearing the sound of thunder 2 seconds prior to a ranger at station B. If sound travels at 1100 feet per second, determine the equation of the hyperbola on which the lightning strike was located. Place the two ranger stations on the x-axis with the midpoint between the two stations at the origin. The transverse axis is horizontal.

47. Critical Thinking A hyperbola has foci $F_1(-6, 0)$ and $F_1(6, 0)$. For any point $P(x, y)$ on the hyperbola, $|PF_1 - PF_2| = 10$. Write the equation of the hyperbola in standard form.

48. Analytic Geometry Two hyperbolas in which the transverse axis of one is the conjugate axis of the other are called *conjugate hyperbolas.* In equations of conjugate hyperbolas, the x^2 and y^2 terms are reversed. For example, $\frac{x^2}{16} - \frac{y^2}{9} = 1$ and $\frac{y^2}{9} - \frac{x^2}{16} = 1$ are equations of conjugate hyperbolas.

a. Graph $\frac{x^2}{16} - \frac{y^2}{9} = 1$ and $\frac{y^2}{9} - \frac{x^2}{16} = 1$ on the same coordinate plane.

b. What is true of the asymptotes of conjugate hyperbolas?

c. Write the equation of the conjugate hyperbola for $\frac{(x-3)^2}{16} - \frac{(y-2)^2}{25} = 1$.

d. Graph the conjugate hyperbolas in part **c**.

Mixed Review

49. Write the equation of the ellipse that has a semi-major axis length of 4 units and foci at (2, 3) and (2, −3). *(Lesson 10-3)*

50. Write $x^2 + y^2 - 4x + 14y - 28 = 0$ in standard form. Then graph the equation. *(Lesson 10-2)*

51. Show that the points with coordinates (−1, 3), (3, 6), (6, 2), and (2, −1) are the vertices of a square. *(Lesson 10-1)*

52. Name three different pairs of polar coordinates that represent point R. Assume $-360° \leq \theta \leq 360°$. *(Lesson 9-1)*

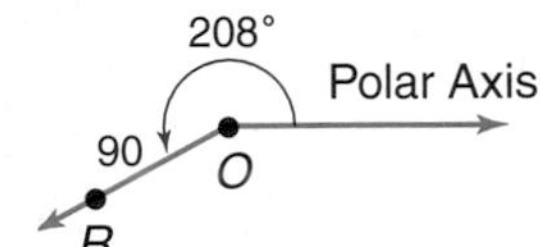

53. Find the inner product of vectors (4, −1, 8) and (−5, 2, 2). Are the vectors perpendicular? Explain. *(Lesson 8-4)*

54. Write the standard from of the equation of the line that has a normal 3 units long and makes an angle of 60° with the positive x-axis. *(Lesson 7-6)*

55. **Aviation** An airplane flying at an altitude of 9000 meters passes directly overhead. Fifteen seconds later, the angle of elevation to the plane is 60°. How fast is the airplane flying? *(Lesson 5-4)*

56. Approximate the real zeros of the function $f(x) = 4x^4 + 5x^3 - x^2 + 1$ to the nearest tenth. *(Lesson 4-5)*

57. **SAT/ACT Practice** If r and s are integers and $r + s = 0$, which of the following must be true?

I. $r^3 > s^3$ II. $r^3 = s^3$ III. $r^4 = s^4$

A I only
B II only
C III only
D I and II only
E I and III only

Extra Practice See p. A45.

10-5 Parabolas

OBJECTIVES

- Use and determine the standard and general forms of the equation of a parabola.
- Graph parabolas.

ENERGY The Odeillo Solar Furnace, located in southern France, uses a series of 63 flat mirrors, arranged on terraces on a hillside, to reflect the sun's rays on to a large parabolic mirror. These computer-controlled mirrors tilt to track the sun and ensure that its rays are always reflected to the central parabolic mirror.

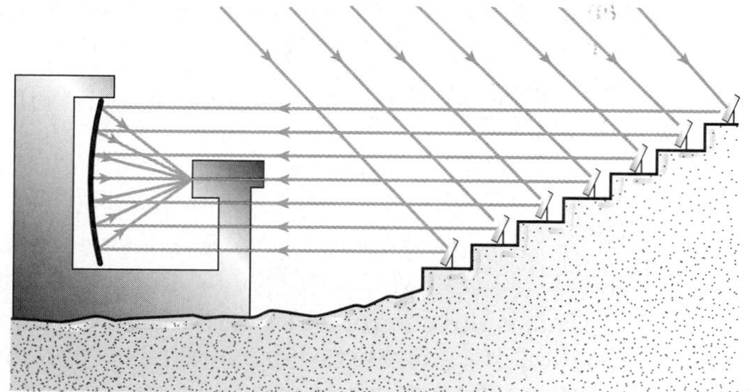

This mirror in turn reflects the sun's rays to the focal point where a furnace is mounted on a tower. The concentrated energy generates temperatures of up to 6870°F. If the width of the Odeillo parabolic mirror is 138 feet and the furnace is located 58 feet from the center of the mirror, how deep is the mirror? *This problem will be solved in Example 2.*

In Chapter 4, you learned that the graphs of quadratic equations like $x = y^2$ or $y = x^2$ are called *parabolas.* A parabola is defined as the set of all points in a plane that are the same distance from a given point, called the **focus,** and a given line, called the **directrix.** *Remember that the distance from a point to a line is the length of the segment from the point perpendicular to the line.*

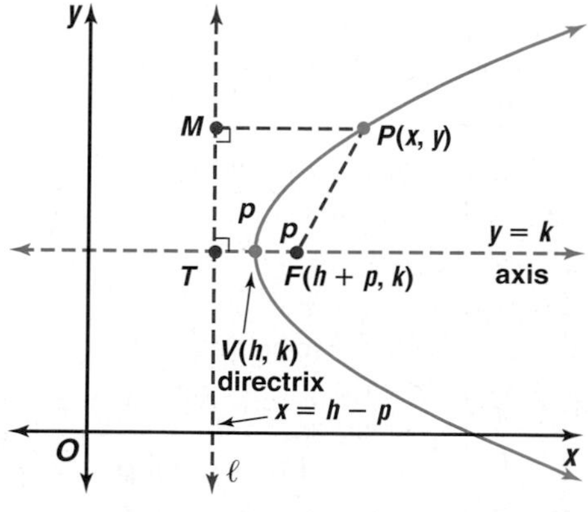

In the figure at the left, F is the focus of the parabola and ℓ is the directrix. This parabola is symmetric with respect to the line $y = k$, which passes through the focus. This line is called the **axis of symmetry,** or, more simply, the *axis* of the parabola. The point at which the axis intersects the parabola is called the **vertex.**

Suppose the vertex V has coordinates (h, k). Let p be the distance from the focus to the vertex, FV. By the definition of a parabola, the distance from any point on the parabola to the focus must equal the distance from that point to the directrix. So, if $FV = p$, then $VT = p$. The coordinates of F are $(h + p, k)$, and the equation of the directrix is $x = h - p$.

Now suppose that $P(x, y)$ is any point on the parabola other than the vertex. From the definition of a parabola, you know that $PF = PM$. Since M lies on the directrix, the coordinates of M are $(h - p, y)$.

For PF, let $F(h + p, k)$ be (x_1, y_1) and $P(x, y)$ be (x_2, y_2). Then for PM, let M be (x_1, y_1). You can use the Distance Formula to determine the equation for the parabola.

$$PF = PM$$

$$\sqrt{[x - (h + p)]^2 + (y - k)^2} = \sqrt{[x - (h - p)]^2 + (y - y)^2}$$

$$[x - (h + p)]^2 + (y - k)^2 = [x - (h - p)]^2 \quad \textit{Square each side.}$$

$$x^2 - 2x(h + p) + (h + p)^2 + (y - k)^2 = x^2 - 2x(h - p) + (h - p)^2$$

This equation can be simplified to obtain the equation

$$(y - k)^2 = 4p(x - h).$$

When p is positive, the parabola opens to the right.
When p is negative, the parabola opens to the left.

Unlike the equations of other conic sections, the equation of a parabola has only one squared term.

This is the equation of a parabola whose directrix is parallel to the y-axis. The equation of a parabola whose directrix is parallel to the x-axis can be obtained by switching the terms in the parentheses of the previous equation.

$$(x - h)^2 = 4p(y - k)$$

When p is positive, the parabola opens upward.
When p is negative, the parabola opens downward.

Standard Form of the Equation of a Parabola	Orientation when $p > 0$	Description
$(y - k)^2 = 4p(x - h)$		vertex: (h, k) focus: $(h + p, k)$ axis of symmetry: $y = k$ directrix: $x = h - p$ opening: right if $p > 0$ left if $p < 0$
$(x - h)^2 = 4p(y - k)$		vertex: (h, k) focus: $(h, k + p)$ axis of symmetry: $x = h$ directrix: $y = k - p$ opening: upward if $p > 0$ downward if $p < 0$

Example 1 **Consider the equation $y^2 = 8x + 48$.**

a. Find the coordinates of the focus and the vertex and the equations of the directrix and the axis of symmetry.

b. Graph the equation of the parabola.

a. First, write the equation in the form $(y - k)^2 = 4p(x - h)$.

$$y^2 = 8x + 48$$
$$y^2 = 8(x + 6) \quad \textit{Factor.}$$
$$(y - 0)^2 = 4(2)(x + 6) \quad 4p = 8, \text{ so } p = 2$$

In this form, we can see that $h = -6$, $k = 0$, and $p = 2$. We can use this to find the desired information.

Graphing Calculator Tip

You can graph a parabola that opens to the right or to left by first solving for y and then graphing the two resulting equations on the same screen.

Vertex: $(-6, 0)$ — (h, k)

Directrix: $x = -8$ — $x = h - p$

Focus: $(-4, 0)$ — $(h + p, k)$

Axis of Symmetry: $y = 0$ — $y = k$

The axis of symmetry is the x-axis. Since p is positive, the parabola opens to the right.

b. Graph the directrix, the vertex, and the focus. To determine the shape of the parabola, graph several other ordered pairs that satisfy the equation and connect them with a smooth curve.

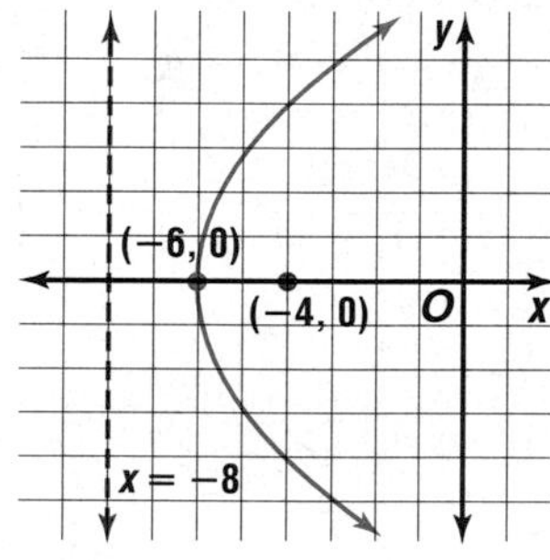

One useful property of parabolic mirrors is that all light rays traveling parallel to the mirror's axis of symmetry will be reflected by the parabola to the focus.

Example 2 **ENERGY** **Refer to the application at the beginning of the lesson.**

a. Find and graph the equation of a parabola that models the shape of the Odeillo mirror.

b. Find the depth of the parabolic mirror.

a. The shape of the mirror can be modeled by a parabola with vertex at the origin and opening to the right. The general equation of such a parabola is $y^2 = 4px$, where p is the focal length. Given a focal length of 58 feet, we can derive the model equation.

$$y^2 = 4px$$
$$y^2 = 4(58)x \quad p = 58$$
$$y^2 = 232x$$

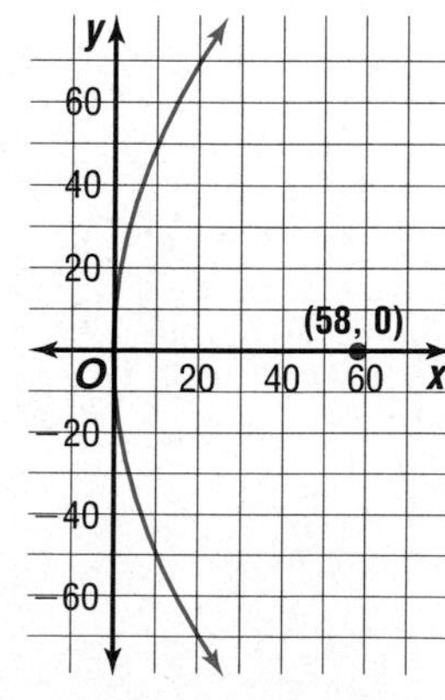

b. With the mirror's vertex at the origin, the distance from the vertex to one edge of the mirror is half the overall width of the mirror, $\frac{1}{2}$(138 feet) or 69 feet.

Use the model equation to find the depth x of the mirror when the distance from the center is 69 feet.

$$y^2 = 232x$$
$$(69)^2 = 232x \quad \textit{y = 69}$$
$$4761 = 232x$$
$$x = \frac{4761}{232} \text{ or about } 20.5$$

The mirror is about 20.5 feet deep.

You can use the same process you used with circles to rewrite the standard form of the equation of a parabola in general form. *You will derive the general form in Exercise 37.*

General Form for the Equation of a Parabola

The general form of the equation of a parabola is
$y^2 + Dx + Ey + F = 0$, when the directrix is parallel to the y-axis, or
$x^2 + Dx + Ey + F = 0$, when the directrix is parallel to the x-axis.

It is necessary to convert an equation in general form to standard form to determine the coordinates of the vertex (h, k) and the distance from the vertex to the focus p.

Example 3 **Consider the equation $2x^2 - 8x + y + 6 = 0$.**

a. Write the equation in standard form.

b. Find the coordinates of the vertex and focus and the equations for the directrix and the axis of symmetry.

c. Graph the equation of the parabola.

a. Since x is squared, the directrix of this parabola is parallel to the x-axis.

$$2x^2 - 8x + y + 6 = 0$$
$$2x^2 - 8x = -y - 6 \quad \textit{Isolate the x terms and the y terms.}$$
$$2(x^2 - 4x + ?) = -y - 6 + ? \quad \textit{The GCF of the x terms is 2.}$$
$$2(x^2 - 4x + 4) = -y - 6 + 2(4) \quad \textit{Complete the square.}$$
$$2(x - 2)^2 = -(y - 2) \quad \textit{Simplify and factor.}$$
$$(x - 2)^2 = -\frac{1}{2}(y - 2) \quad \textit{Divide each side by 2.}$$

The standard form of the equation is $(x - 2)^2 = -\frac{1}{2}(y - 2)$.

b. Since $4p = -\frac{1}{2}$, $p = -\frac{1}{8}$.

vertex: (2, 2) *(h, k)* focus: $\left(2, \frac{15}{8}\right)$ *(h, k + p)*

directrix: $y = \frac{17}{8}$ *y = k − p* axis of symmetry: $x = 2$ *x = h*

c. Now sketch the graph of the parabola using the information found in part b.

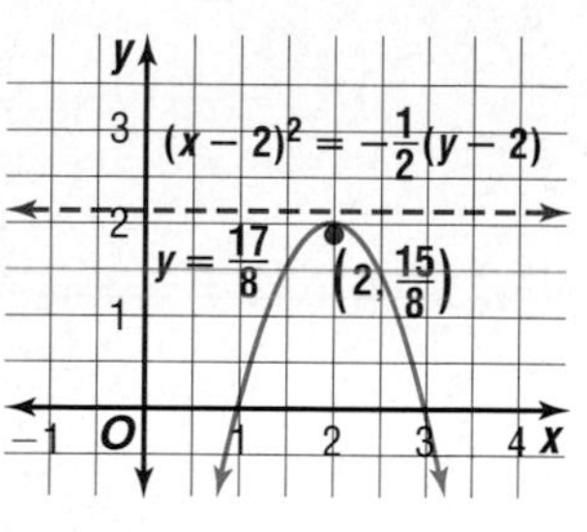

Parabolas are often used to demonstrate maximum or minimum points in real-world situations.

Example 4 

AERONAUTICS NASA's KC-135A aircraft flies in parabolic arcs to simulate the weightlessness experienced by astronauts in space. The aircraft starts its ascent at 24,000 feet. During the ascent, all on board experience 2g's or twice the pull of Earth's gravity. As the aircraft approaches its maximum height, the engines are stopped, and the aircraft is allowed to free fall at a precisely determined angle. Zero gravity is achieved for 25 seconds as the plane reaches the top of the parabola and begins its descent. After this 25-second period, the engines are throttled to bring the aircraft out of the dive. If the height of the aircraft in feet (*y*) versus time in seconds (*x*) is modeled by the equation $x^2 - 65x + 0.11y - 2683.75 = 0$, what is the maximum height achieved by the aircraft during its parabolic flight?

First, write the equation in standard form.

$$x^2 - 65x + 0.11y - 2683.75 = 0$$

$$x^2 - 65x = -0.11y + 2683.75 \quad \textit{Isolate the x terms and y terms.}$$

$$x^2 - 65x + 1056.25 = -0.11y + 2683.75 + 1056.25 \quad \textit{Complete the square.}$$

$$(x - 32.5)^2 = -0.11y + 3740$$

$$(x - 32.5)^2 = -0.11(y - 34{,}000)$$

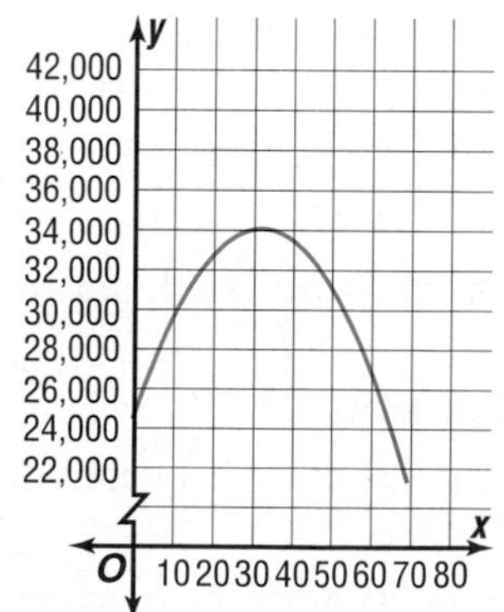

The vertex of the parabola is at (32.5, 34,000).

Remember that the vertex is the maximum or minimum point of a parabola. Since the parabola opens downward, the vertex is the maximum.

The *x*-coordinate of the vertex, 32.5, represents 32.5 seconds after the aircraft began the parabolic maneuver. The *y*-coordinate, 34,000, represents a maximum height of 34,000 feet.

All conics can be defined using the focus-directrix definition presented in this lesson. A conic section is defined to be the **locus** of points such that, for any point P in the locus, the ratio of the distance between that point and a fixed point F to the distance between that point and a fixed line ℓ, is constant. As we have seen, the point F is called the focus, and the line ℓ is the directrix. That ratio is the eccentricity of the curve, and its value can be used to determine the conic's classification. In the case of a parabola, $e = 1$. As shown previously, if $0 < e < 1$, the conic is an ellipse. If $e = 0$, the conic is a circle, and if $e > 1$, the conic is a hyperbola.

parabola	ellipse	hyperbola
$e = 1$	$e < 1, e \neq 0$	$e > 1$
M, P, F, ℓ; $e = \frac{PF}{PM}$, $e = 1$	P, M, F′, F, ℓ', ℓ; $e = \frac{PF}{PM}$, $e < 1$	F′, M, P, F, ℓ', ℓ; $e = \frac{PF}{PM}$, $e > 1$

For those conics having more than one focus and directrix, F′ and ℓ′ represent alternates that define the same conic.

CHECK FOR UNDERSTANDING

Communicating Mathematics

Read and study the lesson to answer each question.

1. **Explain** a way in which you might distinguish the equation of a parabola from the equation of a hyperbola.
2. **Write** the equation of the graph shown at the right.
3. **Describe** the relationships among the vertex, focus, directrix and axis of symmetry of a parabola.
4. **Write** the equation in standard form of a parabola with vertex at $(-4, 5)$, opening to the left, and with a focus 5 units from its vertex.
5. **Identify** each of the following conic sections given their eccentricities.

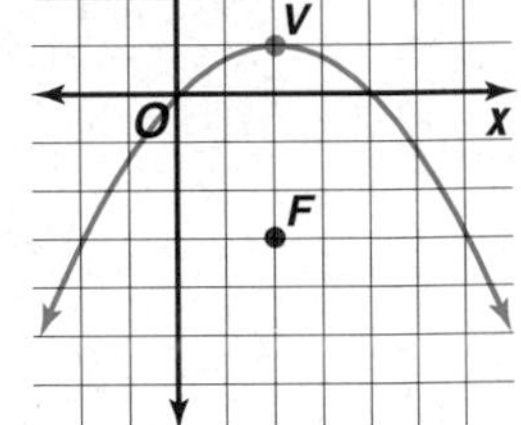

a. $e = \frac{1}{2}$ b. $e = 1$ c. $e = 1.25$ d. $e = 0$

Guided Practice

For the equation of each parabola, find the coordinates of the vertex and focus, and the equations of the directrix and axis of symmetry. Then graph the equation.

6. $x^2 = 12(y - 1)$
7. $y^2 - 4x + 2y + 5 = 0$
8. $x^2 + 8x + 4y + 8 = 0$

Write the equation of the parabola that meets each set of conditions. Then graph the equation.

9. The vertex is at the origin, and the focus is at $(0, -4)$.
10. The parabola passes through the point at $(2, -1)$, has its vertex at $(-7, 5)$, and opens to the right.
11. The parabola passes through the point at $(5, 2)$, has a vertical axis, and has a minimum at $(4, -3)$.

12. **Sports** In 1998, Sammy Sosa of the Chicago Cubs was in a homerun race with Mark McGwire of the St. Louis Cardinals. One day, Mr. Sosa popped a baseball straight up at an initial velocity v_0 of 56 feet per second. Its distance s above the ground after t seconds is described by $s = v_0t - 16t^2 + 3$.
   a. Graph the function $s = v_0t - 16t^2 + 3$ for the given initial velocity.
   b. Find the maximum height achieved by the ball.
   c. If the ball is allowed to fall to the ground, how many seconds, to the nearest tenth, is it in the air?

EXERCISES

Practice

For the equation of each parabola, find the coordinates of the vertex and focus, and the equations of the directrix and axis of symmetry. Then graph the equation.

13. $y^2 = 8x$
14. $x^2 = -4(y - 3)$
15. $(y - 6)^2 = 4x$
16. $y^2 + 12x = 2y - 13$
17. $y - 2 = x^2 - 4x$
18. $x^2 + 10x + 25 = -8y + 24$
19. $y^2 - 2x + 14y = -41$
20. $y^2 - 2y - 12x + 13 = 0$
21. $2x^2 - 12y - 16x + 20 = 0$
22. $3x^2 - 30y - 18x + 87 = 0$
23. Consider the equation $2y^2 + 16y + 16x + 64 = 0$. Identify the coordinates of the vertex and focus and the equations of the directrix and axis of symmetry. Then graph the equation.

Write the equation of the parabola that meets each set of conditions. Then graph the equation.

24. The vertex is at $(-5, 1)$, and the focus is at $(2, 1)$.
25. The equation of the axis is $y = 6$, the focus is at $(0, 6)$, and $p = -3$.
26. The focus is at $(4, -1)$, and the equation of the directrix is $y = -5$.
27. The parabola passes through the point at $(5, 2)$, has a vertical axis, and has a maximum at $(4, 3)$.
28. The parabola passes through the point at $(-3, 1)$, has its vertex at $(-2, -3)$, and opens to the left.
29. The focus is at $(-1, 7)$, the length from the focus to the vertex is 2 units, and the function has a minimum.
30. The parabola has a vertical axis and passes through points at $(1, -7)$, $(5, -3)$, and $(7, -4)$.
31. The parabola has a horizontal axis and passes through the origin and points at $(-1, 2)$ and $(3, -2)$.
32. The parabola's directrix is parallel to the x-axis, and the parabola passes through points at $(1, 1)$, $(0, 9)$, and $(2, 1)$.

Applications and Problem Solving

33. Automotive Automobile headlights contain parabolic reflectors, which work on the principle that light placed at the focus of a parabola will reflect off the mirror-like surface in lines parallel to the axis of symmetry. Suppose a bulb is placed at the focus of a headlight's reflector, which is 2 inches from the vertex.

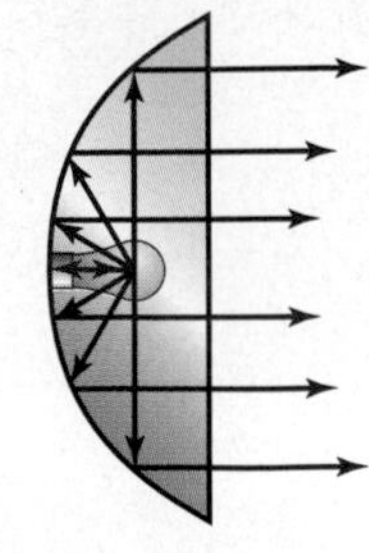

a. If the depth of the headlight is to be 4 inches, what should the diameter of the headlight be at its opening?

b. Find the diameter of the headlight at its opening if the depth is increased by 25%.

34. Business An airline has been charging $140 per seat for a one-way flight. This flight has been averaging 110 passengers but can transport up to 180 passengers. The airline is considering a decrease in the price for a one-way ticket during the winter months. The airline estimates that for each $10 decrease in the ticket price, they will gain approximately 20 passengers per flight.

a. Based on these estimates, what ticket price should the airline charge to achieve the greatest income on an average flight?

b. New estimates reveal that the increase in passengers per flight is closer to 10 for each $10 decrease in the original ticket price. To maximize income, what should the new ticket price be?

35. Critical Thinking Consider the standard form of the equation of a parabola in which the vertex is known but the value of p is not known.

a. As $|p|$ becomes greater, what happens to the shape of the parabola?

b. As $|p|$ becomes smaller, what happens to the shape of the parabola?

36. Construction The Golden Gate Bridge in San Francisco, California, is a catenary suspension bridge, which is very similar in appearance to a parabola. The main span cables are suspended between two towers that are 4200 feet apart and 500 feet above the roadway. The cable extends 10 feet above the roadway midway between the two towers.

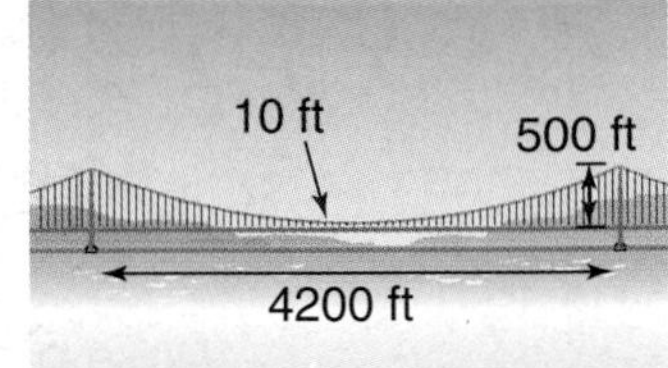

a. Find an equation that models the shape of the cable.

b. How far from the roadway is the cable 720 feet from the bridge's center?

37. Critical Thinking Using the standard form of the equation of a parabola, derive the general form of the equation of a parabola.

38. Critical Thinking The *latus rectum* of a parabola is the line segment through the focus that is perpendicular to the axis and has endpoints on the parabola. The length of the latus rectum is $|4p|$ units, where p is the distance from the vertex to the focus.

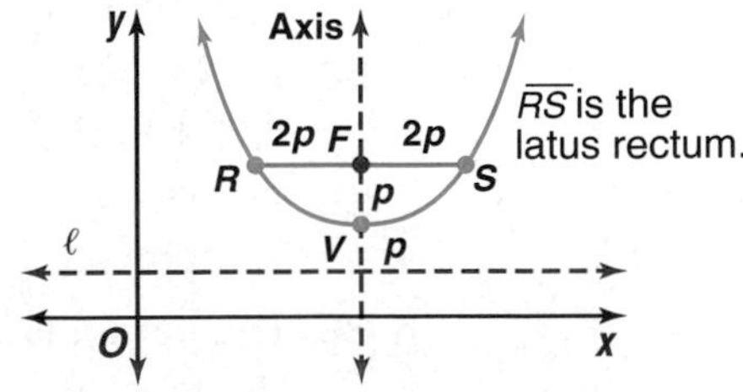

a. Write the equation of a parabola with vertex at $(-3, 2)$, axis $y = 2$, and latus rectum 8 units long.

b. The latus rectum of the parabola with equation $(x - 1)^2 = -16(y - 4)$ coincides with the diameter of a circle. Write the equation of the circle.

Mixed Review

39. Find the coordinates of the center, foci, and vertices, and the equations of the asymptotes of the graph of $\frac{(y-3)^2}{25} - \frac{(x-2)^2}{16} = 1$. Then graph the equation. *(Lesson 10-4)*

40. Find the coordinates of the center, foci, and vertices of the ellipse whose equation is $4x^2 + 25y^2 + 250y + 525 = 0$. Then graph the ellipse. *(Lesson 10-3)*

41. Graph $r = 12 \cos 2\theta$. *(Lesson 9-2)*

42. Find the values of θ for which $\cos \theta = 1$ is true. *(Lesson 6-3)*

43. **Geometry** A regular hexagon is inscribed in a circle with a radius 6.4 centimeters long. Find the apothem; that is, the distance from the center of the circle to the midpoint of a side. *(Lesson 5-4)*

44. Describe the end behavior of $g(x) = \frac{4}{x^2 + 1}$. *(Lesson 3-5)*

45. **SAT/ACT Practice** Triangle QRS has sides of lengths 14, 19, and t, where t is the length of the longest side. If t is the cube of an integer, what is the perimeter of the triangle?

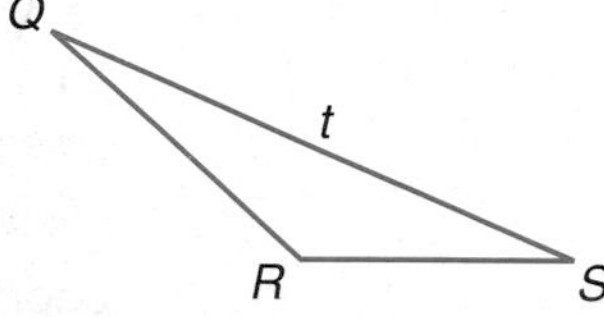

A 41 **B** 58 **C** 60 **D** 69 **E** 76

MID-CHAPTER QUIZ

1. Given: $A(3, 3)$, $B(6, 9)$, and $C(9, 3)$ (Lesson 10-1)
   a. Show that these points form an isosceles triangle.
   b. Determine the perimeter of the triangle to the nearest hundredth.

2. Determine the midpoint of the diagonals of the rectangle with vertices $A(-4, 9)$, $B(5, 9)$, $C(5, 5)$, and $D(-4, 5)$. (Lesson 10-1)

3. Find the coordinates of the center and radius of the circle with equation $x^2 + y^2 - 6y - 8x = -16$. Then graph the circle. (Lesson 10-2)

4. Write the equation of the circle with center at $(-5, 2)$ and radius $\sqrt{7}$. (Lesson 10-2)

5. **Astronomy** A satellite orbiting Earth follows an elliptical path with the center of Earth as one focus. The eccentricity of the orbit is 0.16, and the major axis is 10,440 miles long. (Lesson 10-3)
   a. If the mean diameter of Earth is 7920 miles, find the greatest and least distance of the satellite from the surface of Earth.
   b. Assuming that the center of the ellipse is the origin and the foci lie on the x-axis, write the equation of the orbit of the satellite.

6. Identify the center, vertices, and foci of the ellipse with equation $9x^2 + 25y^2 - 72x + 250y + 544 = 0$. Then graph the equation. (Lesson 10-3)

7. Identify the center, vertices, foci, and equations of the asymptotes of the graph of the hyperbola with equation $3y^2 + 24y - x^2 - 2x + 41 = 0$. Then graph the equation. (Lesson 10-4)

8. Write the equation of a hyperbola that passes through the point at $(4, 2)$ and has asymptotes with equations $y = 2x$ and $y = -2x + 4$. (Lesson 10-4)

9. Identify the vertex, focus, and equations of the axis of symmetry and directrix for the parabola with equation $y^2 - 4x + 2y + 5 = 0$. Then graph the equation. (Lesson 10-5)

10. Write the equation of the parabola that passes through the point at $(9, -2)$, has its vertex at $(5, -1)$, and opens downward. (Lesson 10-5)

Extra Practice See p. A45.

10-6

Rectangular and Parametric Forms of Conic Sections

OBJECTIVES

- Recognize conic sections in their rectangular form by their equations.
- Find a rectangular equation for a curve defined parametrically and vice versa.

TRANSPORTATION The first self-propelled boats on the western rivers of the United States were the paddlewheels. This boat used a steam engine to turn one or more circular wheels that had a paddle attached to the end of each spoke. In 1811, Robert Fulton and Nicholas Roosevelt built the first paddlewheel large enough for commercial use on the Ohio and Mississippi Rivers. By the end of the 19th century, paddlewheel boats had fought wars and carried people and cargo on nearly every river in the United States. Despite advancements in technology, paddlewheels are still in use today, though mainly for sentimental reasons. *You will solve a problem related to this in Exercise 40.*

We have determined a general equation for each conic section we have studied. All of these equations are forms of the general equation for conic sections.

General Equation for Conic Sections

The equation of a conic section can be written in the form

$$Ax^2 + Bxy + Cy^2 + Dx + Ey + F = 0,$$

where A, B, and C are not all zero.

The graph of a second-degree equation in two variables always represents a conic or degenerate case, unless the equation has no graph at all in the real number plane. Most of the conic sections that we have studied have axes that are parallel to the coordinate axes. The general equations of these conics have no xy term; thus, $B = 0$. The one conic section we have discussed whose axes are not parallel to the coordinate axes is the hyperbola whose equation is $xy = k$. In its equation, $B \neq 0$.

To identify the conic section represented by a given equation, it is helpful to write the equation in standard form. However, when $B = 0$, you can also identify the conic section by how the equation compares to the general equation. The table on the next page summarizes the standard forms and differences among the general forms.

General Form: $Ax^2 + Bxy + Cy^2 + Dx + Ey + F = 0$		
Conic Section	**Standard Form of Equation**	**Variation of General Form of Conic Equations**
circle	$(x-h)^2 + (y-k)^2 = r^2$	$A = C$
parabola	$(y-k)^2 = 4p(x-h)$ or $(x-h)^2 = 4p(y-k)$	Either A or C is zero.
ellipse	$\frac{(x-h)^2}{a^2} + \frac{(y-k)^2}{b^2} = 1$ or $\frac{(y-k)^2}{a^2} + \frac{(x-h)^2}{b^2} = 1$	A and C have the same sign and $A \neq C$.
hyperbola	$\frac{(x-h)^2}{a^2} - \frac{(y-k)^2}{b^2} = 1$ or $\frac{(y-k)^2}{a^2} - \frac{(x-h)^2}{b^2} = 1$	A and C have opposite signs.
	$xy = k$	$A = C = D = E = 0$

The circle is actually a special form of the ellipse, where $a^2 = b^2 = r^2$.

Remember that graphs can also be degenerate cases.

Example 1 **Identify the conic section represented by each equation.**

a. $6y^2 + 3x - 4y - 12 = 0$

$A = 0$ and $C = 6$. Since $A = 0$, the conic is a parabola.

b. $3y^2 - 2x^2 + 5y - x - 15 = 0$

$A = -2$ and $C = 3$. Since A and C have different signs, the conic is a hyperbola.

c. $9x^2 + 27y^2 - 6x - 108y + 82 = 0$

$A = 9$ and $C = 27$. Since A and C have the same signs and are not equal, the conic is an ellipse.

d. $4x^2 + 4y^2 + 5x + 2y - 150 = 0$

$A = 4$ and $C = 4$. Since $A = C$, the conic is a circle.

So far we have discussed equations of conic sections in their rectangular form. Some conic sections can also be described parametrically.

Look Back
You can refer to Lesson 8-6 to review writing and graphing parametric equations.

The general form for a set of parametric equations is

$$x = f(t) \text{ and } y = g(t), \text{ where } t \text{ is in some interval } I.$$

As t varies over I in some order, a curve containing points (x, y) is traced out in a certain direction.

A parametric equation can be transformed into its more familiar rectangular form by eliminating the parameter t from the parametric equations.

Example **2** **Graph the curve defined by the parametric equations $x = 4t^2$ and $y = 3t$, where $-2 \le t \le 2$. Then identify the curve by finding the corresponding rectangular equation.**

Make a table of values assigning values for t and evaluating each expression to find values for x and y.

t	x	y	(x, y)
−2	16	−6	(16, −6)
−1	4	−3	(4, −3)
0	0	0	(0, 0)
1	4	3	(4, 3)
2	16	6	(16, 6)

Then graph the curve.

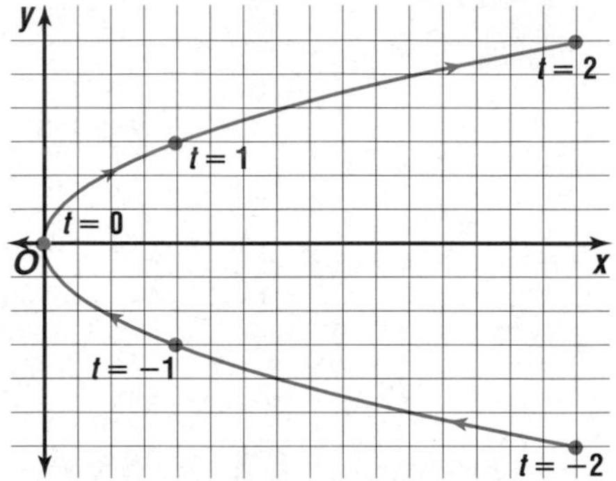

Notice the arrows indicating the direction in which the curve is traced for increasing values of t.

The graph appears to be part of a parabola. To identify the curve accurately, find the corresponding rectangular equation by eliminating t from the given parametric equations.

First, solve the equation $y = 3t$ for t.

$$y = 3t$$
$$\frac{y}{3} = t \qquad \textit{Solve for } t.$$

Then substitute $\frac{y}{3}$ for t in the equation $x = 4t^2$.

$$x = 4t^2$$
$$x = 4\left(\frac{y}{3}\right)^2 \qquad t = \frac{y}{3}$$
$$x = \frac{4y^2}{9}$$

The equation $x = \frac{4y^2}{9}$ is the equation of a parabola with vertex at (0, 0) and its axis of symmetry along the x-axis. Notice that the domain of the rectangular equation is $x \ge 0$, which is greater than that of its parametric representation. By restricting the domain to $0 \le x \le 16$, our rectangular representation matches our parametric representation for the graph.

Some parametric equations require the use of trigonometric identities to eliminate the parameter t.

Example 3 **Find the rectangular equation of the curve whose parametric equations are $x = 2 \cos t$ and $y = 2 \sin t$, where $0 \leq t \leq 2\pi$. Then graph the equation using arrows to indicate how the graph is traced.**

Solve the first equation for cos *t* and the second equation for sin *t*.

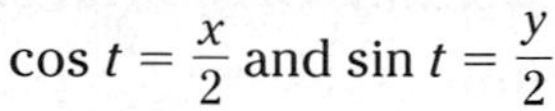

$$\cos t = \frac{x}{2} \text{ and } \sin t = \frac{y}{2}$$

Graphing Calculator Appendix

For keystroke instructions on how to graph parametric equations, see page A21.

Use the trigonometric identity $\cos^2 t + \sin^2 t = 1$ to rewrite the equation to eliminate *t*.

$$\cos^2 t + \sin^2 t = 1$$

$$\left(\frac{x}{2}\right)^2 + \left(\frac{y}{2}\right)^2 = 1 \quad \textit{Substitution}$$

$$\frac{x^2}{4} + \frac{y^2}{4} = 1$$

$$x^2 + y^2 = 4 \quad \textit{Multiply each side by 4.}$$

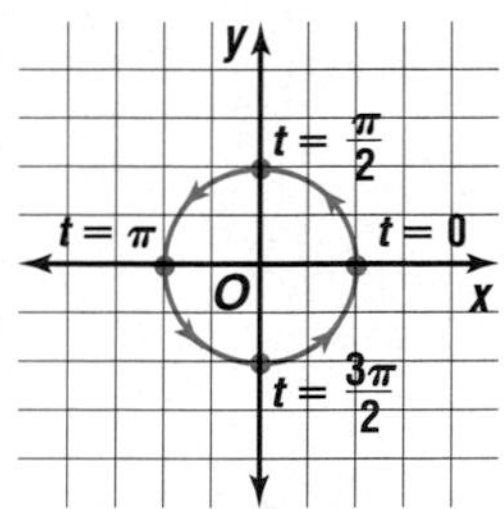

This is the equation of a circle with center at (0, 0) and radius 2. As *t* increases from $t = 0$ to $t = 2\pi$, we see that the curve is traced in a counterclockwise motion.

You can also use substitution to find the parametric equations for a given conic section. If the conic section is defined as a function, $y = f(x)$, one way of finding the parametric equations is by letting $x = t$ and $y = f(t)$, where *t* is in the domain of *f*.

Example 4 **Find parametric equations for the equation $y = x^2 + 3$.**

Let $x = t$. Then $y = t^2 + 3$. Since the domain of the function $f(t)$ is all real numbers, the parametric equations are $x = t$ and $y = t^2 + 3$, where $-\infty < t < \infty$.

GRAPHING CALCULATOR EXPLORATION

The graph of the parametric equations $x = \cos t$ and $y = \sin t$, where $0 \leq t \leq 2\pi$ is the unit circle. Interchanging the trigonometric functions or changing the coefficients can alter the graph's size and shape as well as its starting point and the direction in which it is traced. Watch while the graph is being drawn to see the effects.

TRY THESE

1. Graph the parametric equations $x = -\cos t$ and $y = \sin t$, where $0 \leq t \leq 2\pi$.
   a. Where does the graph start?
   b. In which direction is the graph traced?
2. Graph the parametric equations $x = \sin t$ and $y = \cos t$, where $0 \leq t \leq 2\pi$.
   a. Where does the graph start?
   b. In which direction is the graph traced?
3. Graph $x = 2 \cos t$, and $y = 3 \sin t$, where $0 \leq t \leq 2\pi$. What is the shape of the graph?

WHAT DO YOU THINK?

4. What is the significance of the number *a* in the equations $x = a \cos t$ and $y = a \sin t$, where $0 \leq t \leq 2\pi$?
5. What is the result of changing the interval to $0 \leq t \leq 4\pi$ in Exercises 1-3?

Parametric equations are particularly useful in describing the motion of an object along a curved path.

Example 5 **ASTRONOMY** **The orbit of Saturn around the sun is modeled by the equation $\frac{x^2}{(9.50)^2} + \frac{y^2}{(9.48)^2} = 1$. It takes Saturn approximately 30 Earth years to complete one revolution of its orbit.**

a. Find parametric equations that model the motion of Saturn beginning at (9.50, 0) and moving counterclockwise around the sun.

b. Use the parametric equations to determine Saturn's position after 18 years.

a. From the given equation, you can determine that the orbital path of Saturn is an ellipse with a major axis of 9.50 AU and a minor axis of 9.48 AU.

Like a circle, the parametric representation for an ellipse involves the use of sines and cosines. The parametric representation for the given equation is an ellipse with $x = 9.5$ and $y = 0$ when $t = 0$, so the following equations are true.

$$\frac{x}{9.50} = \cos \omega t \qquad \text{and} \qquad \frac{y}{9.48} = \sin \omega t$$

You can verify that by using the equations $x = 9.50$ and $y = 0$ when $t = 0$ and $\cos^2 \omega t + \sin^2 \omega t = 1$.

To move counterclockwise, the motion will have to begin with the value of x decreasing and y increasing, so $\omega > 0$. Since Saturn completes an orbit in 30 Earth years, the sine and cosine functions have a period $\frac{2\pi}{\omega} = 30$, so $\omega = \frac{\pi}{15}$.

Thus, the parametric equations corresponding to the rectangular equation $\frac{x^2}{(9.50)^2} + \frac{y^2}{(9.48)^2} = 1$ are $x = 9.50 \cos \frac{\pi}{15}t$ and $y = 9.48 \sin \frac{\pi}{15}t$, where $0 \leq t \leq 30$.

You can verify the equations above using a graphing calculator to trace the ellipse.

b. The position of Saturn after 18 years is found by letting $t = 18$ in both parametric equations.

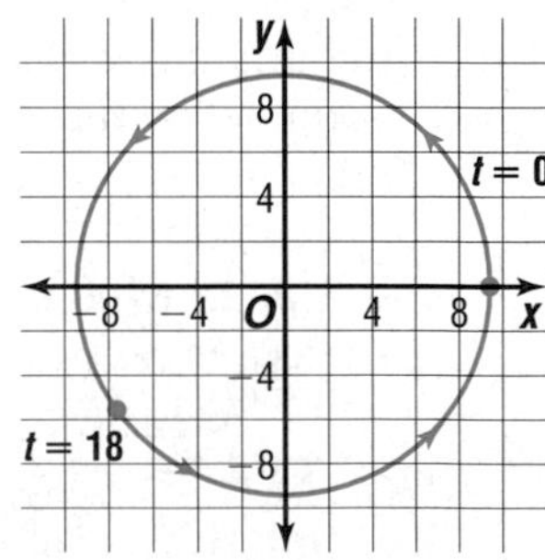

$x = 9.50 \cos\left(\frac{\pi}{15}t\right)$

$x = 9.50 \cos\left[\frac{\pi}{15}(18)\right]$ *t = 18*

$x \approx -7.69$

$y = 9.48 \sin\left(\frac{\pi}{15}t\right)$

$y = 9.48 \sin\left[\frac{\pi}{15}(18)\right]$ *t = 18*

$y \approx -5.57$

Eighteen years later, Saturn is located at $(-7.69, -5.57)$.

CHECK FOR UNDERSTANDING

Communicating Mathematics

Read and study the lesson to answer each question.

1. **Compare and contrast** the general form of the equations of the four conic sections we have studied.
2. **Give** new restrictions on the parameter t in Example 2 so that the domains of the rectangular and parametric equations are the same.
3. **Write** the rectangular equation of a parabola with vertex at the origin and opening to the left. Then write the parametric equations that correspond to that parabola.

Guided Practice

Identify the conic section represented by each equation. Then write the equation in standard form and graph the equation.

4. $x^2 + 9y^2 + 2x - 18y + 1 = 0$
5. $y^2 - 8x = -8$
6. $x^2 - 4x - y^2 - 5 - 4y = 0$
7. $x^2 - 6x + y^2 - 12y + 41 = 0$

Find the rectangular equation of the curve whose parametric equations are given. Then graph the equation using arrows to indicate orientation.

8. $x = t, y = -t^2 - 6t + 2; -\infty < t < \infty$
9. $x = 2 \cos t, y = 3 \sin t; 0 \leq t \leq 2\pi$

Find parametric equations for each rectangular equation.

10. $y = 2x^2 - 5x$
11. $x^2 + y^2 = 36$

12. **Astronomy** Some comets traveling at great speeds follow parabolic paths with the sun as their focus. Suppose the motion of a certain comet is modeled by the parametric equations $x = \frac{t^2}{80}$, and $y = t$ for $-\infty < t < \infty$. Find the rectangular equation that models the comet's path.

EXERCISES

Practice

Identify the conic section represented by each equation. Then write the equation in standard form and graph the equation.

13. $x^2 - 4y - 6x + 9 = 0$
14. $x^2 - 8x + y^2 + 6y + 24 = 0$
15. $x^2 - 3y^2 + 2x - 24y - 41 = 0$
16. $9x^2 + 25y^2 - 54x - 50y - 119 = 0$
17. $x^2 = y + 8x - 16$
18. $2xy = 3$
19. $5x^2 + 2y^2 - 40x - 20y + 110 = 0$
20. $x^2 - 8x + 11 = -y^2$
21. $8y^2 - 9x^2 - 16y + 36x - 100 = 0$
22. $4y^2 + 4y + 8x = 15$

23. Identify the conic section represented by $-4y^2 + 10x = 16y - x^2 - 5$. Write the equation in standard form and then graph the equation.
24. In the general equation of a conic, $A = C = 2$, $B = 0$, $D = -8$, $E = 12$, and $F = 6$. Write the equation in standard form. Then graph the equation.

Find the rectangular equation of the curve whose parametric equations are given. Then graph the equation using arrows to indicate orientation.

25. $x = t, y = 2t^2 - 4t + 1; -\infty < t < \infty$
26. $x = \cos 2t, y = \sin 2t; 0 \leq t \leq 2\pi$
27. $x = -\cos t, y = \sin t; 0 \leq t \leq 2\pi$
28. $x = 3 \sin t, y = 2 \cos t; 0 \leq t \leq 2\pi$
29. $x = -\sin 2t, y = 2 \cos 2t; 0 \leq t \leq \pi$
30. $x = 2t - 1, y = \sqrt{t}; 0 \leq t \leq 4$

31. Find a rectangular equation for the curve whose parametric equations are $x = -3 \cos 2t$ and $y = 3 \sin 2t$, $0 \le t \le 2\pi$.

Find parametric equations for each rectangular equation.

32. $x^2 + y^2 = 25$
33. $x^2 + y^2 - 16 = 0$
34. $\frac{x^2}{4} + \frac{y^2}{25} = 1$
35. $\frac{y^2}{16} + x^2 = 1$
36. $y = x^2 - 4x + 7$
37. $x = y^2 + 2y - 1$

38. Find parametric equations for the rectangular equation $(y + 3)^2 = 4(x - 2)$.

Graphing Calculator

39. Consider the rectangular equation $x = \sqrt{y}$.
   a. By using different choices for t, find two different parametric representations of this equation.
   b. Graph the rectangular equation by hand. Then use a graphing calculator to sketch the graphs of each set of parametric equations.
   c. Are your graphs from part **b** the same?
   d. What does this suggest about parametric representations of rectangular equations?

Applications and Problem Solving

40. **Transportation** A riverboat's paddlewheel has a diameter of 12 feet and at full speed, makes one clockwise revolution in 2 seconds.
   a. Write a rectangular equation to model the shape of the paddlewheel.
   b. Write parametric equations describing the position of a point A on the paddlewheel for any given time t. Assume that at $t = 0$, A is at the very top of the wheel.
   c. How far will point A, which is a fixed point on the wheel, move in 1 minute?

41. **Critical Thinking** Identify the graph of each equation using the method described in this lesson. Then identify the graph of each equation after first rewriting the equation in standard form and solving for y. Explain the discrepancies, if any, in your answers.
   a. $2x^2 + 5y^2 = 0$
   b. $x^2 + y^2 - 4x - 6y + 13 = 0$
   c. $y^2 - 9x^2 = 0$

42. **Critical Thinking** Explain why a substitution of $x = t^2$ is not appropriate when trying to find a parametric representation of $y = x^2 - 5$?

43. **Timing** The path traced by the tip of the second-hand of a clock can be modeled by the equation of a circle in parametric form.

   a. If the radius of the clock is 6 inches, find an equation in rectangular form that models the shape of the clock.
   b. Find parametric equations that describe the motion of the tip as it moves from 12 o'clock noon to 12 o'clock noon of the next day.
   c. Simulate the motion described by graphing the equations on a graphing calculator.

44. **Framing** Portraits are often framed so that the opening through which the picture is seen is an ellipse. These oval mats must be custom cut using an oval cutter whose design relies upon the parametric equations of an ellipse. The elliptical compass at the right consists of a stick with a pencil attached to one end and two pivot holes at the other. Through these holes, the stick is anchored to two small blocks, one of which can slide horizontally and the other vertically in its groove. Use the diagram of the elliptical compass at right to verify that $x = a \cos t$ and $y = b \sin t$. (*Hint*: Draw an extra vertical and an extra horizontal line to create right triangles and then use trigonometry.)

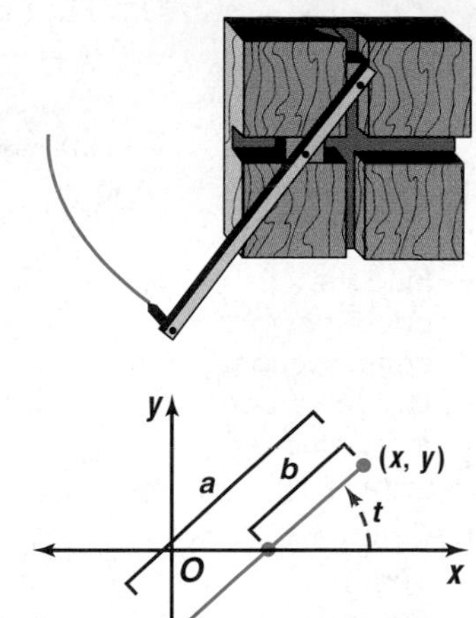

Mixed Review

45. Find the coordinates of the vertex, focus, and the equations of the axis of symmetry and directrix of the parabola with equation $x^2 - 12y + 10x = -25$. Then graph the equation. *(Lesson 10-5).*

46. Graph $xy = -25$. *(Lesson 10-4)*

47. Write $3x^2 + 3y^2 - 18x + 12y = 9$ in standard form. Then graph the equation. *(Lesson 10-2)*

48. A 30-pound force is applied to an object at an angle of 60° with the horizontal. Find the magnitude of the horizontal and vertical components of the force. *(Lesson 8-1)*

49. **Statistics** The prediction equation $y = -0.13x + 37.8$ gives the fuel economy y for a car with horsepower x. Is the equation a better predictor for Car 1, which has a horsepower of 135 and average 19 miles per gallon, or for Car 2, which has a horsepower of 245 and averages 16 miles per gallon? Explain. *(Lesson 7-7)*

50. Find the value of $\sin\left(2 \text{ Sin}^{-1} \frac{1}{2}\right)$. *(Lesson 6-8)*

51. Find the area to the nearest square unit of $\triangle ABC$ if $a = 48$, $b = 32$, and $c = 44$. *(Lesson 5-8)*

52. Solve $\sqrt{2y - 3} - \sqrt{2y + 3} = -1$. *(Lesson 4-7)*

53. If y varies jointly as x and z and $y = 16$ when $x = 5$ and $z = 2$, find y when $x = 8$ and $z = 3$. *(Lesson 3-8)*

54. Find the determinant of $\begin{bmatrix} 5 & 9 \\ 7 & -3 \end{bmatrix}$. Then state whether an inverse exists for the matrix. *(Lesson 2-5)*

55. Write the point-slope form of the equation of the line through the points $(-6, 4)$ and $(3, 7)$. Then write the equation in slope-intercept form. *(Lesson 1-4)*

56. **SAT/ACT Practice** For all values where $x \neq y$, let $x \# y$ represent the lesser of the numbers x and y, and let $x @ y$ represent the greater of the number x and y. What is the value of $(1 \# 4) @ (2 \# 3)$?

A 1 **B** 2 **C** 3 **D** 4 **E** 5

Extra Practice See p. A45.

10-7 Transformations of Conics

OBJECTIVES

- Find the equations of conic sections that have been translated or rotated.
- Graph rotations and/or translations of conic equations.
- Identify the equations of conic sections using the discriminant.
- Find the angle of rotation for a given equation.

SPORTS At the 1996 Olympics in Atlanta, Georgia, U.S. high school student Kim Rhode won the gold medal in the women's double trap shooting event, being staged at the Olympics for the first time. The double trap event consists of firing double barrel shotguns at flying clay targets that are launched two at a time out of a house located 14.6 to 24.7 meters in front of the contestants. Targets are thrown out of the house in a random arc, but always at the same height. *A problem related to this will be solved in Example 1.*

Thus far, we have used a transformation called a translation to show how the parent graph of each of the conic sections is translated to a center other than the origin. For example, the equation of the circle $x^2 + y^2 = r^2$ becomes $(x - h)^2 + (y - k)^2 = r^2$ for a center of (h, k). A translation of a set of points with respect to (h, k) is often written as follows.

$$T_{(h, k)} \quad \Longleftrightarrow \quad \text{translation with respect to } (h, k)$$

Example 1

SPORTS Refer to the application above. A video game simulating the sport of double trap allows a player to shoot at two elliptically-shaped targets released from a house at the bottom of the screen. With the house located at the origin, a target at its initial location is modeled by the equation $\frac{x^2}{16} + \frac{y^2}{4} = 1$. Suppose a player misses one of the two targets released and the center of the target leaves the screen at the point (24, 30). Find an equation that models the shape and position of the target with its center translated to this point.

To write the equation of $\frac{x^2}{16} + \frac{y^2}{4} = 1$ for $T_{(24, 30)}$, let $h = 24$ and $k = 30$. Then replace x with $x - h$ and y with $y - k$.

$x^2 \Rightarrow (x - 24)^2$

$y^2 \Rightarrow (y - 30)^2$

Thus, the translated equation is $\frac{(x - 24)^2}{16} + \frac{(y - 30)^2}{4} = 1$

The graph shows the parent ellipse and its translation.

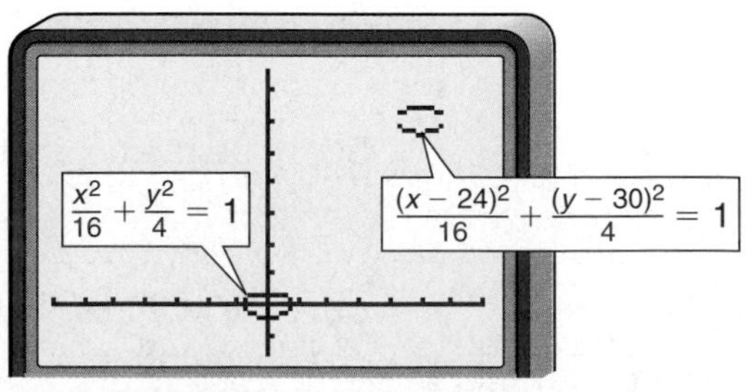

[−35, 35] scl:5 by [−8.09, 38.09] scl:5

Look Back
You can refer to Lesson 2-4 to review rotation.

Another type of transformation you have studied is a rotation. Except for hyperbolas whose equations are of the form $xy = k$, all of the conic sections we have studied thus far have been oriented with their axes parallel to the coordinate axes. In the general form of these conics $Ax^2 + Bxy + Cy^2 + Dx + Ey + F = 0$, $B = 0$. Whenever $B \neq 0$, then the axes of the conic section are not parallel to the coordinate axes. That is, the graph is rotated.

The figures below show an ellipse whose center is the origin and its rotation. Notice that the angle of rotation has the same measure as the angles formed by the positive x-axis and the major axis and the positive y-axis and the minor axis.

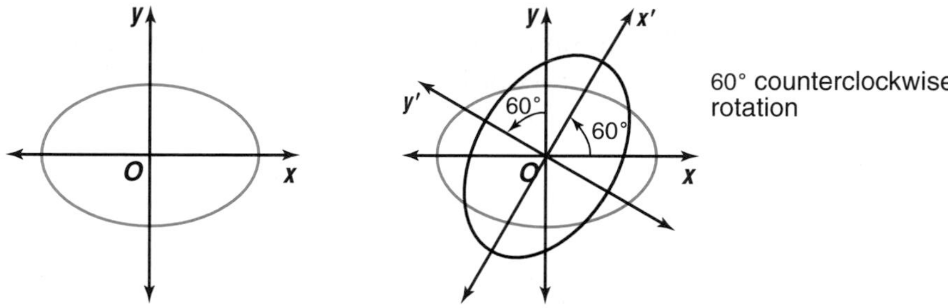

The coordinates of the points of a rotated figure can be found by using a rotation matrix.

A positive value of θ indicates a counterclockwise rotation. A negative value of θ indicates a clockwise rotation.

A rotation of θ about the origin can be described by the matrix

$$\begin{bmatrix} \cos\theta & -\sin\theta \\ \sin\theta & \cos\theta \end{bmatrix}.$$

Let $P(x, y)$ be a point on the graph of a conic section. Then let $P'(x', y')$ be the image of P after a counterclockwise rotation of θ. The values of x' and y' can be found by matrix multiplication.

$$\begin{bmatrix} x' \\ y' \end{bmatrix} = \begin{bmatrix} \cos\theta & -\sin\theta \\ \sin\theta & \cos\theta \end{bmatrix} \cdot \begin{bmatrix} x \\ y \end{bmatrix}$$

The inverse of the rotation matrix represents a rotation of $-\theta$. Multiply each side of the equation by the inverse rotation matrix to solve for x and y.

$$\begin{bmatrix} \cos\theta & \sin\theta \\ -\sin\theta & \cos\theta \end{bmatrix} \cdot \begin{bmatrix} x' \\ y' \end{bmatrix} = \begin{bmatrix} \cos\theta & \sin\theta \\ -\sin\theta & \cos\theta \end{bmatrix} \cdot \begin{bmatrix} \cos\theta & -\sin\theta \\ \sin\theta & \cos\theta \end{bmatrix} \cdot \begin{bmatrix} x \\ y \end{bmatrix}$$

$$\begin{bmatrix} x'\cos\theta + y'\sin\theta \\ -x'\sin\theta + y'\cos\theta \end{bmatrix} = \begin{bmatrix} 1 & 0 \\ 0 & 1 \end{bmatrix} \cdot \begin{bmatrix} x \\ y \end{bmatrix}$$

$$\begin{bmatrix} x'\cos\theta + y'\sin\theta \\ -x'\sin\theta + y'\cos\theta \end{bmatrix} = \begin{bmatrix} x \\ y \end{bmatrix}$$

The result is two equations that can be used to determine the equation of a conic with respect to a rotation of θ.

Rotation Equations

To find the equation of a conic section with respect to a rotation of θ, replace

x with $x'\cos\theta + y'\sin\theta$
and y with $-x'\sin\theta + y'\cos\theta$.

Systems of second-degree equations are useful in solving real-world problems involving more than one parameter.

Example 2 **SALES** **During the month of January, Photo World collected \$2700 from the sale of a certain camera. After lowering the price by \$15, the store sold 30 more cameras and took in \$3375 from the sale of this camera the next month.**

a. Write a system of second-degree equations to model this situation.

b. Find the price of the camera during each month.

c. Use a graphing calculator to check your solution.

a. From the information in the problem, we can write two equations, each of which is the equation of a conic section.

Let x = number of cameras sold
Let y = price per camera in January.

Sales in January: $xy = 2700$
Sales in February: $(x + 30)(y - 15) = 3375$

These are equations of hyperbolas.

b. To solve the system algebraically, use substitution. You can rewrite the equation of the first hyperbola as $y = \frac{2700}{x}$. Before substituting, expand the left-hand side of the second equation and simplify the equation.

$$
\begin{aligned}
(x + 30)(y - 15) - 3375 &= 0 \\
xy - 15x + 30y - 450 - 3375 &= 0 && \textit{Expand } (x - 30)(y + 15). \\
xy - 15x + 30y - 3825 &= 0 && \textit{Simplify.} \\
2700 - 15x + 30\left(\frac{2700}{x}\right) - 3825 &= 0 && xy = 2700,\ y = \frac{2700}{x} \\
-15x - 1125 + \frac{81{,}000}{x} &= 0 && \textit{Simplify.} \\
-15x^2 - 1125x + 81{,}000 &= 0 && \textit{Multiply each side by } x. \\
x^2 + 75x - 5400 &= 0 && \textit{Divide each side by } -15. \\
(x - 45)(x + 120) &= 0 && \textit{Factor.}
\end{aligned}
$$

$$x - 45 = 0 \quad \text{or} \quad x + 120 = 0$$
$$x = 45 \qquad\qquad x = -120$$

Since the number of cameras sold cannot be negative, the store sold 45 cameras during January.

The price of each camera sold during January was $\$\frac{2700}{45}$ or \$60, and the price per camera in February was \$60 − 15 or \$45.

c. Solve each equation for y. Then, graph the equations on the same screen.

$$y = \frac{2700}{x}$$

$$y = \frac{3375}{x + 30} + 15$$

[−200, 200] scl:50 by [−200, 200] scl:50

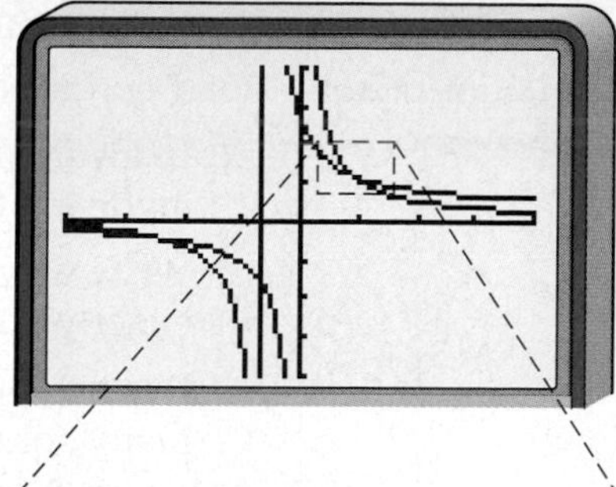

Use ZOOM to enlarge the section of the graph containing the intersection in the first quadrant. Use the **CALC: intersect** function to find the coordinates of the solution, (45, 60).

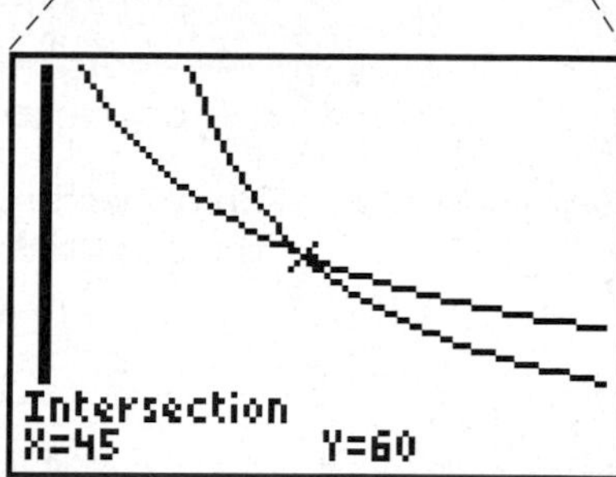

Look Back
You can refer to Lesson 2-6 to review solving systems of linear inequalities.

Previously you learned how to graph different types of inequalities by graphing the corresponding equation and then testing points in the regions of the graph to find solutions for the inequality. The same process is used when graphing systems of inequalities involving second-degree equations.

Example 3 **Graph the solutions for the system of inequalities.**
$x^2 + 4y^2 \leq 4$
$x^2 > y^2 + 1$

First graph $x^2 + 4y^2 \leq 4$. The ellipse should be a solid curve. Test a point either inside or outside the ellipse to see if its coordinates satisfy the inequality.

Test (0, 0):

$$x^2 + y^2 \leq 4$$
$$0^2 + 4(0)^2 \stackrel{?}{\leq} 4 \quad (x, y) = (0, 0)$$
$$0 \leq 4 \ \checkmark$$

Since (0, 0) satisfies the inequality, shade the interior of the ellipse. Then graph $x^2 \geq y^2 + 1$. The hyperbola should be dashed. Test a point inside the branches of the hyperbola or outside its branches. *Since a hyperbola is symmetric, you need not test points within both branches.*

Test (2, 0):

$$x^2 > y^2 + 1$$
$$2^2 \stackrel{?}{>} 0^2 + 1 \quad (x, y) = (2, 0)$$
$$4 > 1 \ \checkmark$$

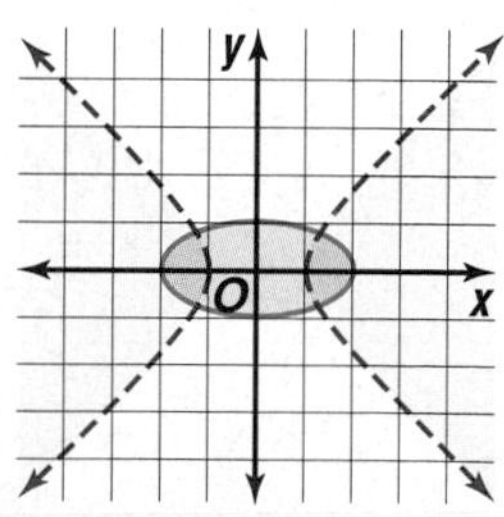

Since (2, 0) satisfies the inequality, the regions inside the branches should be shaded. The intersection of the two graphs, which is shown in green, represents the solution of the system.

Check for Understanding

Communicating Mathematics

Read and study the lesson to answer each question.

1. **Draw** figures illustrating each of the possible numbers of solutions to a system involving the equations of a parabola and a hyperbola.
2. **Write** a system of equations involving two different conic sections that has exactly one solution, the origin.
3. **Describe** the graph of a system of second-degree equations having infinitely many solutions.
4. *Math Journal* **Write** a paragraph explaining how to solve a system of second-degree inequalities.

Guided Practice

Solve each system of equations algebraically. Round to the nearest tenth. Check the solutions by graphing each system.

5. $\frac{(x-1)^2}{20} + \frac{(y-1)^2}{5} = 1$
   $x - y = 0$
6. $x^2 + y^2 = 16$
   $x + 2y = 10$
7. $9x^2 - 4y^2 = 36$
   $x^2 + y^2 = 4$
8. $x^2 = y$
   $xy = 1$

Graph each system of inequalities.

9. $x^2 + y^2 \geq 16$
   $x + y \leq 2$
10. $(x-5)^2 + 2y < 10$
    $y - 9 \geq -2x$
11. $x^2 + y^2 \leq 100$
    $x^2 + y^2 \geq 25$

12. **Gardening** A garden contains two square flowerbeds. The total area of the flowerbeds is 680 square feet, and the second bed has 288 more square feet than the first.
    a. Write a system of second-degree equations that models this situation.
    b. Graph the system found in part **a** and estimate the solution.
    c. Solve the system algebraically to find the length of each flowerbed within the garden.

Exercises

Practice

Solve each system of equations algebraically. Round to the nearest tenth. Check the solutions by graphing each system.

13. $x - 1 = 0$
    $y^2 = 49 - x^2$
14. $xy = 2$
    $x^2 = 3 + y^2$
15. $4x^2 + y^2 = 25$
    $-1 = 2x + y$
16. $x - y = 2$
    $x^2 = 100 - y^2$
17. $x - y = 0$
    $\frac{(x-1)^2}{9} - y^2 = 1$
18. $3x^2 = 9 - y^2$
    $x^2 + 2y^2 = 10$
19. $(y-1)^2 = 4 + x$
    $x + y = -1$
20. $x^2 + y^2 = 13$
    $xy + 6 = 0$

21. $x^2 + 4y^2 = 36$
$x^2 + y - 3 = 0$

22. $x^2 = 16 - y^2$
$2y - x + 3 = 0$

23. Find the coordinates of the point(s) of intersection for the graphs of $x^2 = 25 - 9y^2$ and $xy = -4$.

Graph each system of inequalities.

24. $x + y^2 \leq 9$
$y + x^2 \leq 0$

25. $x^2 + 4y^2 < 16$
$x^2 \leq y^2 + 4$

26. $x^2 + y^2 \leq 36$
$x + y^2 > 0$

27. $y^2 < 81 - 9x^2$
$16 \leq x^2 + y^2$

28. $y + 4 < (x - 3)^2$
$y^2 + x \geq 5$

29. $16x^2 + 49y^2 \leq 784$
$49x^2 + 16y^2 \geq 784$

30. $x - (y - 1)^2 \leq 0$
$4y^2 \geq x^2 - 16$

31. $y - x^2 < 2$
$4x^2 + 9y^2 > 36$

32. $x > \frac{2}{y}$
$16x^2 - 25y^2 \geq 400$

33. Graph the solution to the system $(x + 3)^2 + (y + 2)^2 \geq 36$ and $x + 3 = 0$.

Write the system of equations or inequalities represented by each graph.

34.
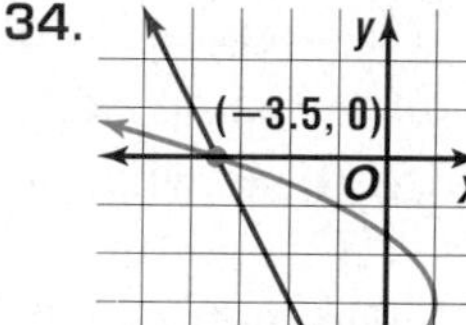

35.
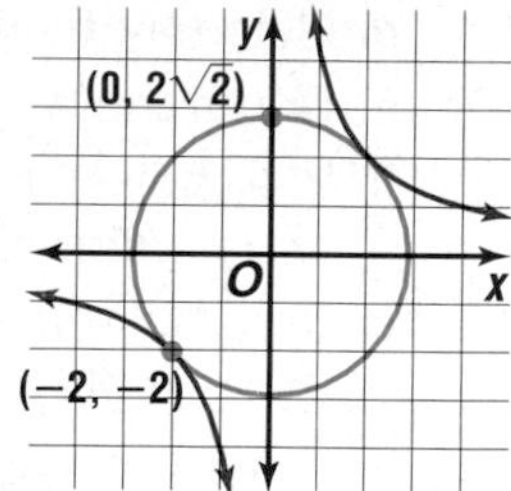

36.
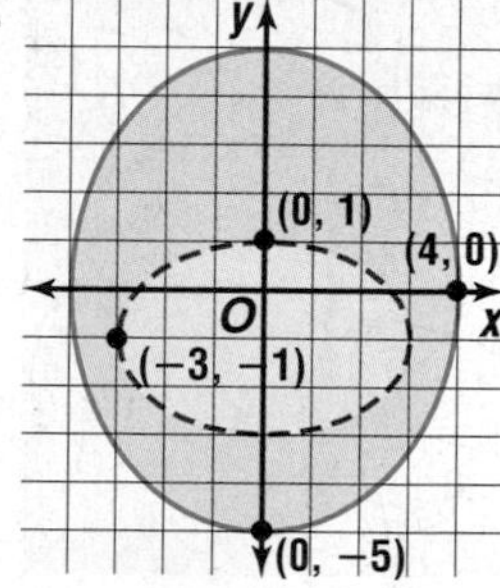

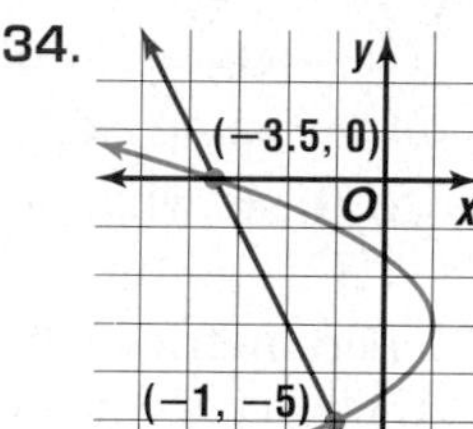

Applications and Problem Solving

37. **Construction** Carrie has 150 meters of fencing material to make a pen for her bird dog. She wants to form a rectangular pen with an area of 800 square meters. What will be the dimensions of her pen?
   a. Let x be the width of the field and y be its length. Write a system of equations that models this situation.
   b. How many solutions are possible for this type of system?
   c. Graph the system to estimate the dimensions of the pen.
   d. Solve this system algebraically, rounding the dimensions to the nearest tenth of a meter.

38. **Engineering** The Transport and Road Research Laboratory in Great Britain proposes the use of parabolic speed bumps 4 inches in height and 1 foot in width.
   a. Write a system of second-degree inequalities that models a cross-section of this speed bump. Locate the vertex of the speed bump at (0, 4).
   b. Graph the system found in part **a**.
   c. If the height of the speed bump is decreased to 3 inches, write a system of equations to model this new cross-section.

39. **Critical Thinking** Solve the system $x = -y + 1$, $xy = -12$, and $y^2 = 25 - x^2$ algebraically. Then graph the system to verify your solution(s).

40. **Seismology** Each of three stations in a seismograph network has detected an earthquake in their region. Seismograph readings indicate that the epicenter of the earthquake is 50 kilometers from the first station, 40 kilometers from the second station, and 13 kilometers from the third station. On a map in which each grid represents one square kilometer, the first station is located at the origin, the second station at (0, 30), and the third station at (35, 18).

a. Write a system of second-degree equations that models this situation.

b. Graph the system and use the graph to approximate the location of the epicenter.

c. Solve the system of equations algebraically to determine the location of the epicenter.

41. **Critical Thinking** Find the value of k so that the graphs of $x = 2y^2$ and $x + 3y = k$ are tangent to each other.

42. **Entertaiment** In a science fiction movie, astronomers track a large incoming asteroid and predict that it will strike Earth with disastrous results. Suppose a certain latitude of Earth's surface is modeled by $x^2 + y^2 = 40$ and the path of the asteroid is modeled by $x = 0.25y^2 + 5$.

a. Graph the two equations on the same axes.

b. Will the asteroid strike Earth? If so, what are the coordinates of the point of impact?

c. Describe this situation with parametric equations. Assume both the asteroid and Earth's surface are moving counterclockwise.

d. Graph the equations found in part **c** using a graphing calculator. Use a window that shows complete graphs of both Earth's surface and the asteroid's path.

Mixed Review

43. Find the equation of $\frac{x^2}{9} + y^2 = 1$ after a 30° rotation about the origin. *(Lesson 10-7)*

44. Write an equation in standard form of the line with the parametric equations $x = 4t + 1$ and $y = 5t - 7$. *(Lesson 8-6)*

45. Simplify $4 \csc \theta \cos \theta \tan \theta$. *(Lesson 7-1)*

46. **Mechanics** A pulley of radius 10 centimeters turns at 5 revolutions per second. Find the linear velocity of the belt driving the pulley in meters per second. *(Lesson 6-2)*

47. Determine between which consecutive integers the real zeros of the function $f(x) = x^3 - 4$ are located. *(Lesson 4-5)*

48. Graph $y = (x + 2)^2 - 3$ and its inverse. *(Lesson 3-4)*

49. Is the relation {(4, 0), (3, 0), (5, −2), (4, −3), (0, −13)} a function? Explain. *(Lesson 1-1)*

50. **SAT/ACT Practice** In the figure at the right, two circles are tangent to each other and each is tangent to three sides of the rectangle. If the radius of each circle is 2, what is the area of the shaded region?

A $32 - 12\pi$ **B** $16 - 8\pi$ **C** $16 - 6\pi$

D $8 - 6\pi$ **E** $32 - 8\pi$

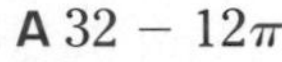

Extra Practice See p. A45.

GRAPHING CALCULATOR EXPLORATION

10-8B Shading Areas on a Graph

An Extension of Lesson 10-8

OBJECTIVE

- Graph a system of second-degree inequalities using the **Shade(** command.

The **Shade(** command can be used to shade areas between the graphs of two equations. To shade an area on a graph, select **7:Shade(** from the **DRAW** menu. The instruction is pasted to the home screen. The argument, or restrictive information, for this command is as follows.

Shade(*lowerfunc,upperfunc,Xleft,Xright,pattern,patres*)

This command draws the lower function, *lowerfunc,* and the upper function, *upperfunc,* in terms of *X* on the current graph and shades the area that is specifically above *lowerfunc* and below *upperfunc.* This means that only the areas between the two functions defined are shaded.

Xleft and *Xright,* if included, specify left and right boundaries for the shading. *Xleft* and *Xright* must be numbers between **Xmin** and **Xmax**, which are the defaults.

The parameter *pattern* specifies one of four shading patterns.

pattern = **1** vertical lines (default)
pattern = **2** horizontal lines
pattern = **3** 45° lines with positive slope
pattern = **4** 45° lines with negative slope

The parameter *patres* specifies one of eight shading resolutions.

patres = **1** shades every pixel (default)
patres = **2** shades every second pixel
patres = **3** shades every third pixel
patres = **4** shades every fourth pixel
patres = **5** shades every fifth pixel
patres = **6** shades every sixth pixel
patres = **7** shades every seventh pixel
patres = **8** shades every eighth pixel

In a system of second-degree inequalities, this technique can be used to shade the interior region of a conic that is not a function.

Example **1** **Graph the solutions for the system of inequalities below.**

$y \geq -x^2 + 2$

$x^2 + 9y^2 \leq 36$

The boundary equation of the first inequality, $y = -x^2 + 2$, is a function. This inequality is graphed by first entering the equation $y = -x^2 + 2$ into the **Y=** list. Since the test point (0, 3) satisfies the inequality, set the graph style to ◥ (shade above).

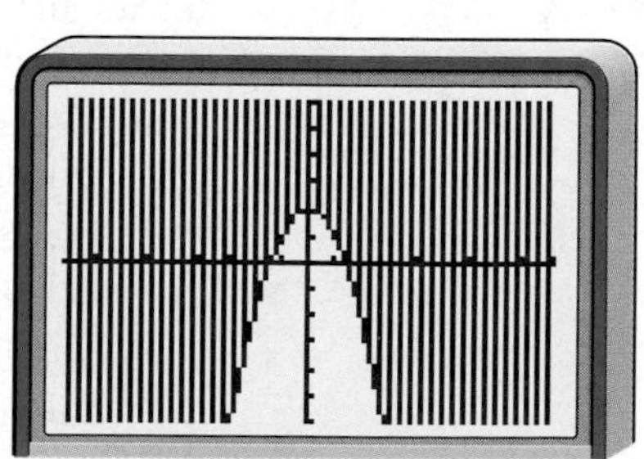

[−9.1, 9.1] scl:1 by [−6, 6] scl:1

(continued on the next page)

The boundary equation for the second inequality is $x^2 + 9y^2 = 36$, which is defined using two functions, $y = -\sqrt{\frac{36 - x^2}{9}}$ and $y = \sqrt{\frac{36 - x^2}{9}}$. The lower function is $y = -\sqrt{\frac{36 - x^2}{9}}$, and the upper function is $y = \sqrt{\frac{36 - x^2}{9}}$. The two halves of the ellipse intersect at $x = -6$ and $x = 6$.

The expression to shade the area between the two halves of the ellipse is shown at the right.

```
Shade(-√((36-X²)
/9),√((36-X²)/9)
,-6,6,3,4)
```

Pressing ENTER will compute the graph as shown below.

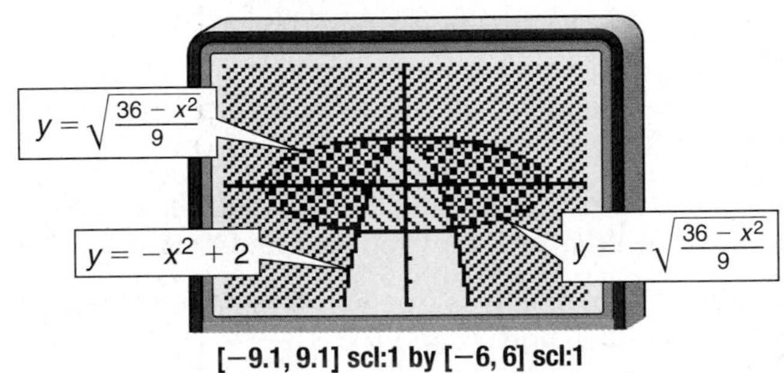

[−9.1, 9.1] scl:1 by [−6, 6] scl:1

The solution set for this system of inequalities is the darker region in which the shadings for the two inequalities over lap.

To clear the any **SHADE(** commands from the viewing window, select **1:CLRDRAW** from the **DRAW** menu and then press ENTER. Remember to also clear any functions defined in the **Y=** list.

TRY THESE

Use the shade feature to graph each system of second-degree inequalities.

1. $y \leq x^2 - 5$
   $9y^2 - x^2 \leq 36$
2. $x^2 + y^2 \leq 16$
   $x \geq y^2 - 4$
3. $16x^2 + 25y^2 \leq 400$
   $25x^2 - 16y^2 \geq 400$
4. $8x^2 + 32y^2 \leq 256$
   $32x^2 + 8y^2 \leq 256$

WHAT DO YOU THINK?

5. Recall that the **Shade(** command can only shade the area between two functions.
   - **a.** To shade just the solution set for the system of inequalities in the example problem, how many regions would need a separate **Shade(** command?
   - **b.** How could you determine the approximate domain intervals for each region?
   - **c.** List and then execute three **Shade(** commands to shade the region representing the solution set for the example problem.
6. Use the **Shade(** command to create a “real-world” picture. Make a list of each command needed to create the picture, as well as a sketch of what the finished picture should look like.

VOCABULARY

analytic geometry (p. 618)
asymptotes (p. 642)
axis of symmetry (p. 653)
center (p. 623, 642)
circle (p. 623)
concentric (p. 623)
conic section (p. 623)
conjugate axis (p. 642)
degenerate conic (p. 623)
directrix (p. 653)
eccentricity (p. 636)
ellipse (p. 631)
equilateral hyperbola (p. 647)
focus (p. 631, 642, 653)
hyperbola (p. 642)
locus (p. 658)
major axis (p. 631)
minor axis (p. 631)
radius (p. 623)
rectangular hyperbola (p. 648)
semi-major axis (p. 632)
semi-minor axis (p. 632)
transverse axis (p. 642)
vertex (p. 631, 642, 653)

UNDERSTANDING AND USING THE VOCABULARY

State whether each statement is *true* or *false*. If false, replace the underlined word(s) to make a true statement.

1. Circles, ellipses, parabolas, and hyperbolas are all examples of <u>conic sections</u>.
2. Circles that have the same <u>radius</u> are concentric circles.
3. The line segment connecting the vertices of a hyperbola is called the <u>conjugate</u> axis.
4. The foci of an ellipse are located along the <u>major</u> axis of the ellipse.
5. In the general form of a <u>circle</u>, A and C have opposite signs.
6. A parabola is symmetric with respect to its <u>vertex</u>.
7. The shape of an ellipse is described by a measure called <u>eccentricity</u>.
8. A <u>hyperbola</u> is the set of all points in a plane that are the same distance from a given point and a given line.
9. The general equation of a <u>rectangular</u> hyperbola, where the coordinate axes are the asymptotes, is $xy = c$.
10. A point is the degenerate form of a <u>parabola</u> conic.

For additional review and practice for each lesson, visit: **www.amc.glencoe.com**

SKILLS AND CONCEPTS

OBJECTIVES AND EXAMPLES

Lesson 10-1 Find the distance and midpoint between two points on a coordinate plane.

- Find the distance between points at (3, 8) and (−5, 10).

$$d = \sqrt{(x_2 - x_1)^2 + (y_2 - y_1)^2}$$
$$= \sqrt{(-5 - 3)^2 + (10 - 8)^2}$$
$$= \sqrt{68} \text{ or } 2\sqrt{17}$$

REVIEW EXERCISES

Find the distance between each pair of points with the given coordinates. Then, find the midpoint of the segment that has endpoints at the given coordinates.

11. (1, −6), (−3, −4)
12. (a, b), $(a + 3, b + 4)$
13. Determine whether the points $A(-5, -2)$, $B(3, 4)$, $C(10, 3)$, and $D(2, -3)$ are the vertices of a parallelogram. Justify your answer.

Lesson 10-2 Determine the standard form of the equation of a circle and graph it.

- Write $x^2 + y^2 - 4x + 2y - 4 = 0$ in standard form. Then graph the equation.

$$x^2 + y^2 - 4x + 2y - 4 = 0$$
$$x^2 - 4x + ? + y^2 + 2y + ? = 4$$
$$x^2 - 4x + 4 + y^2 + 2y + 1 = 4 + 4 + 1$$
$$(x - 2)^2 + (y + 1)^2 = 9$$

The center of the circle is located at (2, −1), and the radius is 3.

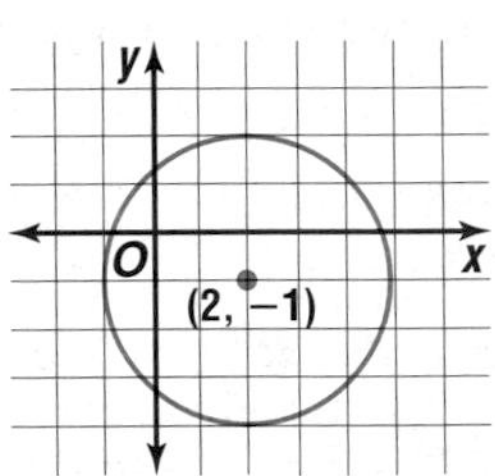

Write the standard form of the equation of each circle described. Then graph the equation.

14. center at (0, 0), radius $3\sqrt{3}$
15. center at (2, 1), tangent to the y-axis

Write the standard form of each equation. Then graph the equation.

16. $x^2 + y^2 = 6y$
17. $x^2 + 14x + y^2 + 6y = 23$
18. $3x^2 + 3y^2 + 6x + 12y - 60 = 0$
19. Write the standard form of the equation of the circle that passes through points at (1, 1), (−2, 2), and (−5, 1). Then identify the center and radius.

Lesson 10-3 Determine the standard form of the equation of an ellipse and graph it.

- For the equation, $\frac{(x + 1)^2}{9} + \frac{(y - 3)^2}{16} = 1$, find the coordinates of the center, foci, and vertices of the ellipse. Then graph the equation.

center: (−1, 3)
foci: $\left(-1, 3 + \sqrt{7}\right)$, $\left(-1, 3 - \sqrt{7}\right)$
vertices: (2, 3), (−4, 3), (−1, −1), (−1, 7)

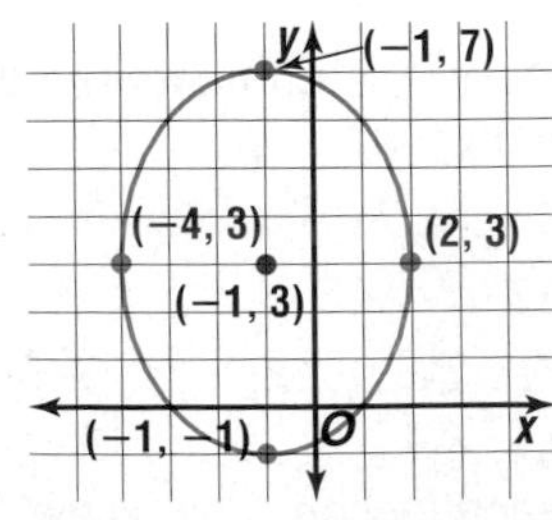

For the equation of each ellipse, find the coordinates of the center, foci, and vertices. Then graph the equation.

20. $\frac{(x - 5)^2}{16} + \frac{(y - 2)^2}{36} = 1$
21. $4x^2 + 25y^2 - 24x + 50y = 39$
22. $6x^2 + 4y^2 + 24x - 32y + 64 = 0$
23. $x^2 + 4y^2 + 124 = 8x + 48y$
24. Write the equation of an ellipse centered at (−4, 1) with a vertical semi-major axis 9 units long and a semi-minor axis 6 units long.

OBJECTIVES AND EXAMPLES

Lesson 10-4 Determine the standard and general forms of the equation of a hyperbola and graph it.

- Find the coordinates of the center, foci, and vertices, and the equations of the asymptotes of the graph of $\frac{(y-2)^2}{4} - (x-5)^2 = 1$. Then graph the equation.

center: $(5, 2)$

foci: $\left(5, 2 + \sqrt{5}\right)$, $\left(5, 2 - \sqrt{5}\right)$

vertices: $(5, 4)$, $(5, 0)$

asymptotes: $y - 2 = \pm 2(x - 5)$

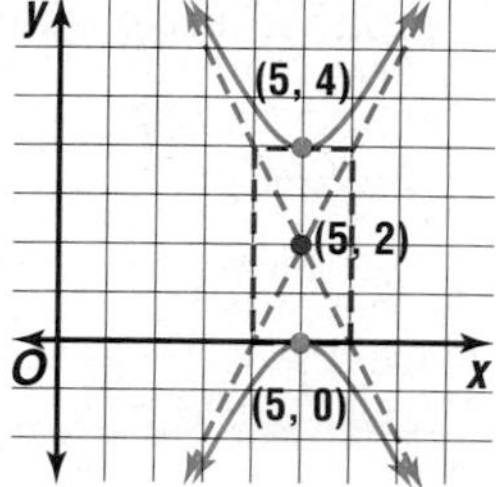

REVIEW EXERCISES

For the equation of each hyperbola, find the coordinates of the center, the foci, and the vertices and the equations of the asymptotes of its graph. Then graph the equation.

25. $\frac{x^2}{25} - \frac{y^2}{16} = 1$

26. $\frac{(y+5)^2}{36} - \frac{(x-1)^2}{9} = 1$

27. $x^2 - 4y^2 - 16y = 20$

28. $9x^2 - 16y^2 - 36x - 96y + 36 = 0$

29. Graph $xy = 9$.

Write an equation of the hyperbola that meets each set of conditions.

30. The length of the conjugate axis is 10 units, and the vertices are at $(1, -1)$ and $(1, 5)$.

31. The vertices are at $(-2, -3)$ and $(6, -3)$, and a focus is at $(-4, -3)$.

Lesson 10-5 Determine the standard and general forms of the equation of a parabola and graph it.

- For the equation $(x + 1)^2 = 2(y - 3)$, find the coordinates of the vertex and focus, and the equations of the directrix and axis of symmetry. Then graph the equation.

vertex: $(-1, 3)$

focus: $\left(-1, \frac{7}{2}\right)$

directrix: $y = \frac{5}{2}$

axis of symmetry: $x = -1$

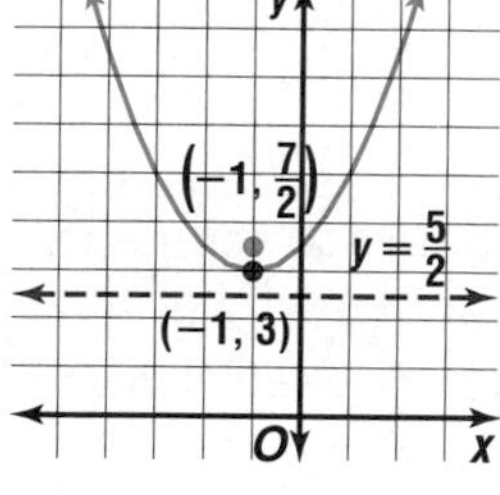

For the equation of each parabola, find the coordinates of the vertex and focus, and the equations of the directrix and axis of symmetry. Then graph the equation.

32. $(x - 5)^2 = 8(y - 3)$

33. $(y + 2)^2 = -16(x - 1)$

34. $y^2 + 6y - 4x = -25$

35. $x^2 + 4x = y - 8$

Write an equation of the parabola that meets each set of conditions.

36. The parabola passes through the point at $(-3, 7)$, has its vertex at $(-1, 3)$, and opens to the left.

37. The focus is at $(5, 2)$, and the equation of the directrix is $y = -4$.

Lesson 10-6 Recognize conic sections in their rectangular form by their equations.

- Identify the conic section represented by the equation $2x^2 - 3x - y + 4 = 0$.

$A = 2$ and $C = 0$

Since $C = 0$, the conic is a parabola.

Identify the conic section represented by each equation.

38. $5x^2 - 7x + 2y^2 = 10$

39. $xy = 5$

40. $2x^2 + 4x + 2y^2 - 6y + 16 = 0$

41. $4y^2 + 6x - 5y = 20$

OBJECTIVES AND EXAMPLES

Lesson 10-6 Find a rectangular equation for a curve defined parametrically and vice versa.

- Find the rectangular equation of the curve whose parametric equations are $x = 3 \sin t$ and $y = \cos t$, where $0 \leq t \leq 2\pi$. Then graph the equation using arrows to indicate orientation.

$x = 3\sin t \Rightarrow \sin t = \frac{x}{3}$

$y = \cos t \Rightarrow \cos t = y$

$\sin^2 t + \cos^2 t = 1$

$\left(\frac{x}{3}\right)^2 + y^2 = 1$

$\frac{x^2}{9} + y^2 = 1$

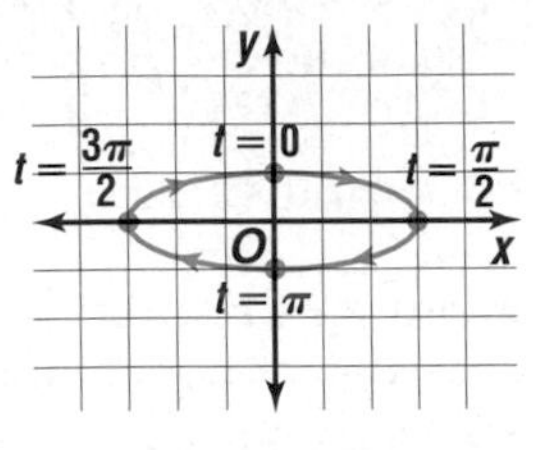

REVIEW EXERCISES

Find the rectangular equation of the curve whose parametric equations are given. Then graph the equation using arrows to indicate orientation.

42. $x = t, y = -t^2 + 3, -\infty \leq t \leq \infty$

43. $x = \cos 4t, y = \sin 4t, 0 \leq t \leq \frac{\pi}{2}$

44. $x = 2 \sin t, y = 3 \cos t, 0 \leq t \leq 2\pi$

45. $x = \sqrt{t}, y = \frac{t}{2} - 1, 0 \leq t \leq 9$

Find parametric equations for each rectangular equation.

46. $y = 2x^2 + 4$

47. $x^2 + y^2 = 49$

48. $\frac{x^2}{36} + \frac{y^2}{81} = 1$

49. $x = -y^2$

Lesson 10-7 Find the equations of conic sections that have been translated or rotated and find the angle of rotation for a given equation.

- To find the equation of a conic section with respect to a rotation of θ, replace

x with $x' \cos \theta + y' \sin \theta$
and y with $-x' \sin \theta + y' \cos \theta$.

Identify the graph of each equation. Write an equation of the translated or rotated graph in general form.

50. $4x^2 + 9y^2 = 36, \theta = \frac{\pi}{6}$

51. $y^2 - 4x = 0, \theta = 45°$

52. $4x^2 - 16(y - 1)^2 = 64$ for $T_{(1, -2)}$

Identify the graph of each equation. Then find θ to the nearest degree.

53. $6x^2 + 2\sqrt{3}xy + 8y^2 = 45$

54. $x^2 - 6xy + 9y^2 = 7$

Lesson 10-8 Graph and solve systems of second degree equations and inequalities.

- Solve the system of equations $x^2 - y = -1$ and $x^2 - 3y^2 = -11$ algebraically.

$$\begin{array}{r} x^2 \quad\quad\ - y + \ 1 = 0 \\ -x^2 + 3y^2 \quad\quad - 11 = 0 \\ \hline 3y^2 - y - 10 = 0 \end{array}$$

$y = -\frac{5}{3}$ or $y = 2$

Substituting we find the solutions to be (1, 2) and (−1, 2). The graph shows these solutions to be true.

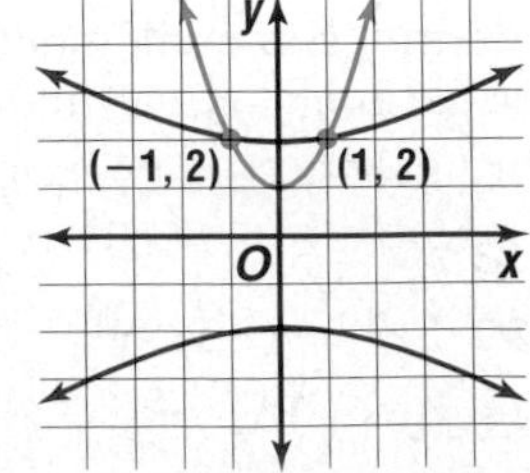

Solve each system of equations algebraically. Round to the nearest tenth. Check the solutions by graphing each system.

55. $(x - 1)^2 + 4(y - 1)^2 = 20$
$x = y$

56. $2x - y = 0$
$y^2 = 49 + x^2$

57. $x^2 - 4x - 4y = 4$
$(x - 2)^2 + 4y = 0$

58. $x^2 + y^2 = 12$
$xy = -4$

Graph each system of inequalities.

59. $x^2 + y \leq 4$
$y^2 - x \leq 0$

60. $xy \geq 9$
$x^2 + y^2 < 36$

61. $x^2 \leq 16 - y^2$
$36y^2 > 324 - 9x^2$

62. $x^2 - 4y \geq 8$
$4y^2 - 25x^2 \geq 100$

APPLICATIONS AND PROBLEM SOLVING

63. **Gardening** Migina bought a new sprinkler that covers part or all of a circular area. With the center of the sprinkler as the origin, the sprinkler sends out water far enough to reach a point located at (12, 16). *(Lesson 10-2)*

    a. Find an equation representing the farthest points the sprinkler can reach.

    b. Migina's backyard is 40 feet wide and 50 feet long. If Migina waters her backyard without moving the sprinkler, what percent of her backyard will not be watered directly?

64. **Astronomy** A satellite orbiting Earth follows an elliptical path with Earth at its center. The eccentricity of the orbit is 0.2, and the major axis is 12,000 miles long. Assuming that the center of the ellipse is the origin and the foci lie on the x-axis, write the equation of the orbit of the satellite. *(Lesson 10-6)*

65. **Carpentry** For a remodeling project, a carpenter is building a picture window that is topped with an arch in the shape of a semi-ellipse. The width of the window is to be 7 feet, and the height of the arch is to be 3 feet. To sketch the arch above the window, the carpenter uses a 7-foot string attached to two thumbtacks. Approximately where should the thumbtacks be placed? *(Lesson 10-3)*

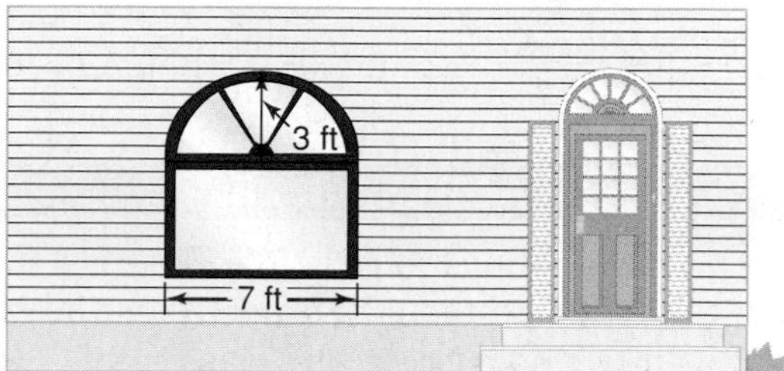

ALTERNATIVE ASSESSMENT

OPEN-ENDED ASSESSMENT

1. An ellipse has its center at the origin and an eccentricity of $\frac{1}{9}$. What is a possible equation for the ellipse?
2. A parabola has an axis of symmetry of $x = 2$ and a focus of (2, 5). What is a possible equation for the parabola in standard form?

PORTFOLIO

Choose one of the conic sections you studied in this chapter. Explain why it is a conic section and describe how you graph it.

Additional Assessment See page A65 for Chapter 10 practice test.

SPACE—THE FINAL FRONTIER

Out in Orbit!

- Search the internet for a satellite, space vehicle, or planet that travels in an orbit around a planet or star.
- Find data on the orbit of the satellite, space vehicle, or planet. This information should include the closest and farthest distance of that object from the planet it is orbiting.
- Make a scale drawing of the orbit of the satellite, space vehicle, or planet. Label important features and dimensions.
- Write a summary describing the orbit of the satellite, space vehicle, or planet. Be sure to discuss which conic section best models the orbit.

Ratio and Proportion Problems

Several problems on the SAT and ACT involve ratios or proportions. The ratio of x to y can be expressed in several ways.

$$\frac{x}{y} \qquad x{:}y \qquad x \text{ to } y$$

Think of a ratio as comparing parts of a whole. If the ratio of boys to girls in a class is 2:1, then one part is 2, one part is 1, and the whole is 3. The fraction of boys in the class is $\frac{2}{3}$.

Memorize the property of proportions.

$$\text{If } \frac{a}{b} = \frac{c}{d}, \text{ then } ad = bc.$$

THE PRINCETON REVIEW

A ratio compares a part to a part. A fraction compares a part to a whole.

Ratios only tell you the *relative* sizes of quantities, not the actual quantities.

When setting up a proportion, label quantities to prevent careless errors.

ACT EXAMPLE

1. The ratio of boys to girls in a class is 4 to 5. If there are a total of 27 students in the class, how many boys are in the class?

A 4 **B** 9 **C** 12

D 14 **E** 17

HINT Notice what question is asked. Is it a ratio, a fraction, or a number?

Solution The ratio is 4 to 5, so the whole must be 9. The fraction of boys is $\frac{4}{9}$. The total number of students is 27, so the number of boys is $\frac{4}{9}$ of 27 or 12. The answer is choice **C**.

Alternate Solution Another method is to use a 'ratio box' to record the numbers and guide your calculations.

Boys	Girls	Whole
4	5	9
		27

In the Whole column, 9 must be multiplied by 3 to get the total of 27. So multiply the 4 by 3 to get the number of boys.

Boys	Girls	Whole
4	5	9
12		27

This is answer choice **C**.

SAT EXAMPLE

2. If 2 packages contain a total of 12 doughnuts, how many doughnuts are there in 5 packages?

A 12 **B** 24 **C** 30

D 36 **E** 60

HINT In a proportion, one ratio equals another ratio.

Solution Write a proportion.

$$\begin{matrix} \text{packages} \longrightarrow \\ \text{doughnuts} \longrightarrow \end{matrix} \frac{2}{12} = \frac{5}{x} \begin{matrix} \longleftarrow \text{packages} \\ \longleftarrow \text{doughnuts} \end{matrix}$$

Cross multiply.

$$\frac{2}{12} = \frac{5}{x}$$
$$2(x) = 5(12)$$
$$x = 30$$

The answer is choice **C**.

Alternate Solution You can also solve this problem without using a proportion. Since 2 packages contain 12 doughnuts, each package must contain $12 \div 2$ or 6 doughnuts. Then five packages will contain 5×6 or 30 doughnuts.

This is answer choice **C**.

After you work each problem, record your answer on the answer sheet provided or on a piece of paper.

Multiple Choice

1. In a jar of red and green jelly beans, the ratio of green jelly beans to red jelly beans is 5:3. If the jar contains a total of 160 jelly beans, how many of them are red?

   A 30 **B** 53 **C** 60
   D 100 **E** 160

2. If $a^2b = 12^2$ and b is an odd integer, then a could be divisible by all of the following EXCEPT

   A 3 **B** 4 **C** 6 **D** 9 **E** 12

3. In the figure below, $\angle A$ and $\angle ADC$ are right angles, the length of $\overline{AD}$ is 7 units, the length of $\overline{AB}$ is 10 units, and the length of $\overline{DC}$ is 6 units. What is the area, in square units, of $\triangle DCB$?

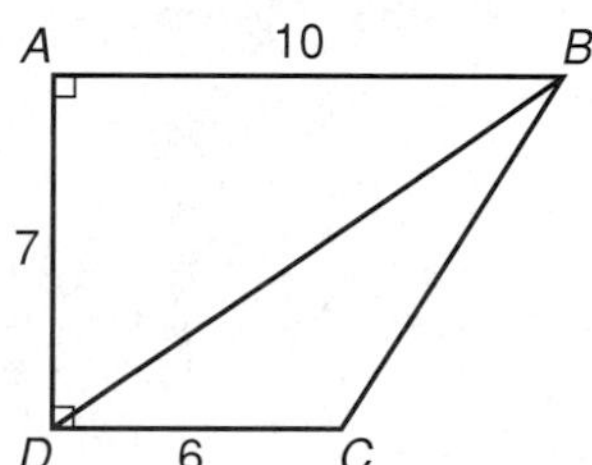

   A 21
   B 24
   C $3\sqrt{149}$
   D 142
   E 210

4. A science class has a ratio of girls to boys of 4 to 3. If the class has a total of 35 students, how many more girls are there than boys?

   A 20 **B** 15 **C** 7 **D** 5 **E** 1

5. In the figure below, $\overline{AC} \parallel \overline{ED}$. If the length of $BD = 3$, what is the length of BE?

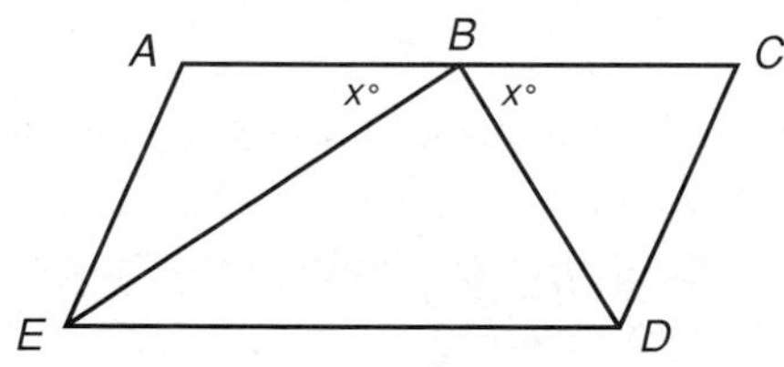

Note: Figure not drawn to scale.

   A 3 **B** 4 **C** 5 **D** $3\sqrt{3}$
   E It cannot be determined from the information given.

6. What is the slope of the line that contains points at (6, 4) and (13, 5)?

   A $\frac{1}{8}$ **B** $-\frac{1}{9}$ **C** $\frac{1}{7}$ **D** 1 **E** 7

7. In $\triangle ABC$ below, if AC is equal to 8, then BC is equal to

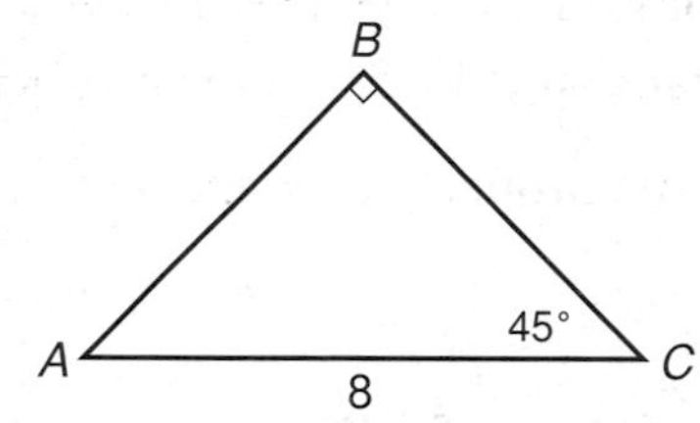

   A $8\sqrt{2}$ **B** 8 **C** 6
   D $4\sqrt{2}$ **E** $3\sqrt{2}$

8. The ratio of $\frac{1}{7}$ to $\frac{1}{5}$ is equal to the ratio of 100 to

   A $\frac{20}{7}$ **B** 20 **C** 35 **D** 100 **E** 140

9. **Quantitative Comparison**

   A if the quantity in Column A is greater
   B if the quantity in Column B is greater
   C if the two quantities are equal
   D if the relationship cannot be determined from the information given

Column A	Column B
The ratio of nickels to dimes in Jar A, where there are 4 more nickels than dimes.	The ratio of nickels to dimes in Jar B, where there are 4 more dimes than nickels.

10. **Grid-In** Twenty bottles contain a total of 8 liters of apple juice. If each bottle contains the same amount of apple juice, how much juice (in liters) is in each bottle?

interNET CONNECTION **SAT/ACT Practice** For additional test practice questions, visit: **www.amc.glencoe.com**

Exponential and Logarithmic Functions

CHAPTER OBJECTIVES

- **Simplify and evaluate expressions containing rational and irrational exponents.** *(Lessons 11-1, 11-2)*
- **Use and graph exponential functions and inequalities.** *(Lessons 11-2, 11-3)*
- **Evaluate expressions and graph and solve equations involving logarithms.** *(Lesson 11-4)*
- **Model real-world situations and solve problems using common and natural logarithms.** *(Lessons 11-5, 11-6, 11-7)*

11-1 Real Exponents

OBJECTIVES
- Use the properties of exponents.
- Evaluate and simplify expressions containing rational exponents.
- Solve equations containing rational exponents.

AEROSPACE On July 4, 1997, the Mars Pathfinder Lander touched down on Mars. It had traveled 4.013×10^8 kilometers from Earth. Two days later, the Pathfinder's Sojourner rover was released and transmitted data from Mars until September 27, 1997.

The distance Pathfinder traveled is written in **scientific notation.** A number is in scientific notation when it is in the form $a \times 10^n$ where $1 \leq a < 10$ and n is an integer. Working with numbers in scientific notation requires an understanding of the definitions and properties of integral exponents. For any real number b and a positive integer n, the following definitions hold.

Definition	Example
If $n = 1$, $b^n = b$.	$17^1 = 17$
If $n > 1$, $b^n = b \cdot b \cdot b \cdot \ldots \cdot b$. *n factors*	$15^4 = 15 \cdot 15 \cdot 15 \cdot 15$ or 50,625
If $b = 0$, $b^0 = 1$.	$400{,}785^0 = 1$
If $b \neq 0$, $b^{-n} = \frac{1}{b^n}$.	$7^{-3} = \frac{1}{7^3}$ or $\frac{1}{343}$

Example 1

AEROSPACE At their closest points, Mars and Earth are approximately 7.5×10^7 kilometers apart.

a. Write this distance in standard form.

$7.5 \times 10^7 = 7.5\,(10 \times 10 \times 10 \times 10 \times 10 \times 10 \times 10)$ or 75,000,000

b. How many times farther is the distance Mars Pathfinder traveled than the minimum distance between Earth and Mars?

Let n represent the number of times farther the distance is.

$(7.5 \times 10^7)n = 4.013 \times 10^8$

$n = \frac{4.013 \times 10^8}{7.5 \times 10^7}$ or 5.4

GRAPHING CALCULATOR EXPLORATION

Recall that if the graphs of two equations coincide, the equations are equivalent.

TRY THESE Graph each set of equations on the same screen. Use the graphs and tables to determine whether Y1 is equivalent to Y2 or Y3.

1. Y1 $= x^2 \cdot x^3$, Y2 $= x^5$, Y3 $= x^6$
2. Y1 $= (x^2)^3$, Y2 $= x^5$, Y3 $= x^6$

WHAT DO YOU THINK?

3. Make a conjecture about the value of $a^m \cdot a^n$.
4. Make a conjecture about the value of $(a^m)^n$.
5. Use the graphing calculator to investigate the value of an expression like $\left(\frac{a}{b}\right)^m$. What do you observe?

The Graphing Calculator Exploration leads us to the following properties of exponents. You can use the definitions of exponents to verify the properties.

Properties of Exponents		
Suppose *m* and *n* are positive integers and *a* and *b* are real numbers. Then the following properties hold.		
Property	**Definition**	**Example**
Product	$a^m a^n = a^{m+n}$	$16^3 \cdot 16^7 = 16^{3+7}$ or 16^{10}
Power of a Power	$(a^m)^n = a^{mn}$	$(9^3)^2 = 9^{3 \cdot 2}$ or 9^6
Power of a Quotient	$\left(\frac{a}{b}\right)^m = \frac{a^m}{b^m}$, where $b \neq 0$	$\left(\frac{3}{4}\right)^5 = \frac{3^5}{4^5}$ or $\frac{243}{1024}$
Power of a Product	$(ab)^m = a^m b^m$	$(5x)^3 = 5^3 \cdot x^3$ or $125x^3$
Quotient	$\frac{a^m}{a^n} = a^{m-n}$, where $a \neq 0$	$\frac{15^6}{15^2} = 15^{6-2}$ or 15^4

Examples

2 Evaluate each expression.

a. $\frac{2^4 \cdot 2^8}{2^5}$

$$\frac{2^4 \cdot 2^8}{2^5} = 2^{(4+8)-5} \quad \textit{Product and Quotient Properties}$$
$$= 2^7$$
$$= 128$$

b. $\left(\frac{2}{5}\right)^{-1}$

$$\left(\frac{2}{5}\right)^{-1} = \frac{1}{\frac{2}{5}} \quad b^{-n} = \frac{1}{b^n}$$
$$= \frac{5}{2}$$

3 Simplify each expression.

a. $(s^2t^3)^5$

$$(s^2t^3)^5 = (s^2)^5(t^3)^5 \quad \textit{Power of a Product}$$
$$= s^{(2 \cdot 5)}t^{(3 \cdot 5)} \quad \textit{Power of a Power}$$
$$= s^{10}t^{15}$$

b. $\frac{x^3y}{(x^4)^3}$

$$\frac{x^3y}{(x^4)^3} = \frac{x^3y}{x^{12}} \quad \textit{Power of a Power}$$
$$= x^{(-3-12)}y \quad \textit{Quotient Property}$$
$$= x^{-9}y \text{ or } \frac{y}{x^9}$$

Expressions with rational exponents can be defined so that the properties of integral exponents are still valid. Consider the expressions $3^{\frac{1}{2}}$ and $1^{\frac{1}{3}}$. Extending the properties of integral exponents gives us the following equations.

$$3^{\frac{1}{2}} \cdot 3^{\frac{1}{2}} = 3^{\frac{1}{2}+\frac{1}{2}}$$
$$= 3^1 \text{ or } 3$$

By definition, $\sqrt{3} \cdot \sqrt{3} = 3$.
Therefore, $3^{\frac{1}{2}}$ and $\sqrt{3}$ are equivalent.

$$12^{\frac{1}{3}} \cdot 12^{\frac{1}{3}} \cdot 12^{\frac{1}{3}} = 12^{\frac{1}{3}+\frac{1}{3}+\frac{1}{3}}$$
$$= 12^1 \text{ or } 12$$

We know that $\sqrt[3]{12} \cdot \sqrt[3]{12} \cdot \sqrt[3]{12} = 12$.
Therefore, $12^{\frac{1}{3}}$ and $\sqrt[3]{12}$ are equivalent.

Exploring other expressions can reveal the following properties.

- If n is an odd number, then $b^{\frac{1}{n}}$ is the nth root of b.
- If n is an even number and $b \geq 0$, then $b^{\frac{1}{n}}$ is the non-negative nth root of b.
- If n is an even number and $b < 0$, then $b^{\frac{1}{n}}$ does not represent a real number, but a complex number.

In general, let $y = b^{\frac{1}{n}}$ for a real number b and a positive integer n. Then, $y^n = \left(b^{\frac{1}{n}}\right)^n = b^{\frac{n}{n}}$ or b. But $y^n = b$ if and only if $y = \sqrt[n]{b}$. Therefore, we can define $b^{\frac{1}{n}}$ as follows.

Definition of $b^{\frac{1}{n}}$	For any real number $b \geq 0$ and any integer $n > 1$, $b^{\frac{1}{n}} = \sqrt[n]{b}$. This also holds when $b < 0$ and n is odd.

In this chapter, b will be a real number greater than or equal to 0 so that we can avoid complex numbers that occur by taking an even root of a negative number.

The properties of integral exponents given on page 696 can be extended to rational exponents.

Examples

4 Evaluate each expression.

a. $\mathbf{125^{\frac{1}{3}}}$

$125^{\frac{1}{3}} = (5^3)^{\frac{1}{3}}$ *Rewrite 125 as 5^3.*

$= 5^{\frac{3}{3}}$ *Power of a Power*

$= 5$

b. $\mathbf{\sqrt{14} \cdot \sqrt{7}}$

$\sqrt{14} \cdot \sqrt{7} = 14^{\frac{1}{2}} \cdot 7^{\frac{1}{2}}$ *$\sqrt[n]{b} = b^{\frac{1}{n}}$*

$= (2 \cdot 7)^{\frac{1}{2}} \cdot 7^{\frac{1}{2}}$ *$14 = 2 \cdot 7$*

$= 2^{\frac{1}{2}} \cdot 7^{\frac{1}{2}} \cdot 7^{\frac{1}{2}}$ *Power of a Product*

$= 2^{\frac{1}{2}} \cdot 7$ *Product Property*

$= 7\sqrt{2}$

5 Simplify each expression.

a. $\mathbf{(81c^4)^{\frac{1}{4}}}$

$(81c^4)^{\frac{1}{4}} = (3^4c^4)^{\frac{1}{4}}$ *$81 = 3^4$*

$= (3^4)^{\frac{1}{4}} \cdot (c^4)^{\frac{1}{4}}$ *Power of a Product*

$= 3^{\frac{4}{4}} \cdot c^{\frac{4}{4}}$ *Power of a Power*

$= 3|c|$

b. $\mathbf{\sqrt[6]{9x^3}}$

$\sqrt[6]{9x^3} = \sqrt[6]{9} \cdot \sqrt[6]{x^3}$ *Power of a Product*

$= 9^{\frac{1}{6}} \cdot (x^3)^{\frac{1}{6}}$ *$b^{\frac{1}{n}} = \sqrt[n]{b}$*

$= (3^2)^{\frac{1}{6}} \cdot (x^3)^{\frac{1}{6}}$ *$9 = 3^2$*

$= 3^{\frac{1}{3}} \cdot x^{\frac{1}{2}}$ *Power of a Power*

$= \sqrt[3]{3} \cdot \sqrt{x}$ *$b^{\frac{1}{n}} = \sqrt[n]{b}$*

Rational exponents with numerators other than 1 can be evaluated by using the same properties. Study the two methods of evaluating $64^{\frac{5}{6}}$ below.

Method 1

$$46^{\frac{5}{6}} = \left(46^{\frac{1}{6}}\right)^5$$
$$= \left(\sqrt[6]{46}\right)^5$$

Method 2

$$46^{\frac{5}{6}} = (46^5)^{\frac{1}{6}}$$
$$= \sqrt[6]{46^5}$$

Therefore, $\left(\sqrt[6]{46}\right)^5$ and $\sqrt[6]{46^5}$ both equal $46^{\frac{5}{6}}$.

In general, we define $b^{\frac{m}{n}}$ as $\left(b^{\frac{1}{n}}\right)^m$ or $(b^m)^{\frac{1}{n}}$. Now apply the definition of $b^{\frac{1}{n}}$ to $\left(b^{\frac{1}{n}}\right)^m$ and $(b^m)^{\frac{1}{n}}$.

$$\left(b^{\frac{1}{n}}\right)^m = \left(\sqrt[n]{b}\right)^m \qquad\qquad (b^m)^{\frac{1}{n}} = \sqrt[n]{b^m}$$

Rational Exponents

For any nonzero number b, and any integers m and n with $n > 1$, and m and n have no common factors

$$b^{\frac{m}{n}} = \sqrt[n]{b^m} = \left(\sqrt[n]{b}\right)^m$$

except where $\sqrt[n]{b}$ is not a real number.

Examples

6 **Evaluate each expression.**

a. $625^{\frac{3}{4}}$

$625^{\frac{3}{4}} = (5^4)^{\frac{3}{4}}$ *Write 625 as 5^4.*
$= 5^3$ *Power of a Product*
$= 125$

b. $\dfrac{16^{\frac{3}{4}}}{16^{\frac{1}{4}}}$

$\dfrac{16^{\frac{3}{4}}}{16^{\frac{1}{4}}} = 16^{\frac{3}{4} - \frac{1}{4}}$ *Quotient Property*
$= 16^{\frac{1}{2}}$ or 4

7 **a. Express $\sqrt[3]{64s^9t^{15}}$ using rational exponents.**

$\sqrt[3]{64s^9t^{15}} = (64s^9t^{15})^{\frac{1}{3}}$ $\quad b^{\frac{1}{n}} = \sqrt[n]{b}$
$= 64^{\frac{1}{3}}s^{\frac{9}{3}}t^{\frac{15}{3}}$ *Power of a Product*
$= 4s^3t^5$

b. Express $12x^{\frac{2}{3}}y^{\frac{1}{2}}$ using a radical.

$12x^{\frac{2}{3}}y^{\frac{1}{2}} = 12(x^4y^3)^{\frac{1}{6}}$ *Power of a Product*
$= 12\sqrt[6]{x^4y^3}$

When you simplify a radical, use the product property to factor out the nth roots and use the smallest index possible for the radical. Remember that you should use caution when evaluating even roots to avoid negative values that would result in an imaginary number.

Example 8 **Simplify $\sqrt{r^7s^{25}t^3}$.**

For $r^7s^{25}t^3$ to be nonnegative, none or exactly two of the variables must be negative. Check the final answer to determine if an absolute value is needed.

$$\sqrt{r^7s^{25}t^3} = (r^7s^{25}t^3)^{\frac{1}{2}} \qquad b^{\frac{1}{n}} = \sqrt[n]{b}$$
$$= r^{\frac{7}{2}}s^{\frac{25}{2}}t^{\frac{3}{2}} \qquad \textit{Power of a Product}$$
$$= r^{\frac{6}{2}}r^{\frac{1}{2}}s^{\frac{24}{2}}s^{\frac{1}{2}}t^{\frac{2}{2}}t^{\frac{1}{2}} \qquad \textit{Product Property}$$
$$= |r|^3s^{12}|t|\sqrt{rst}$$

Use $|r|$ and $|t|$ since rst must be nonnegative and there is no indication of which variables are negative.

You can also use the properties of exponents to solve equations containing rational exponents.

Example 9 **Solve $734 = x^{\frac{3}{4}} + 5$.**

$$734 = x^{\frac{3}{4}} + 5$$
$$729 = x^{\frac{3}{4}} \qquad \textit{Subtract 5 from each side.}$$
$$729^{\frac{4}{3}} = \left(x^{\frac{3}{4}}\right)^{\frac{4}{3}} \qquad \textit{Raise each side to the } \tfrac{4}{3} \textit{ power.}$$
$$6561 = x \qquad \textit{Use a calculator.}$$

Graphing Calculator Tip

Use the [^] calculator key to enter rational exponents. For example, to evaluate $729^{\frac{4}{3}}$, press

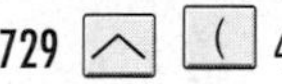

729 [^] [(] 4 [÷] 3 [)].

An expression can also have an irrational exponent, but what does it represent? Consider the expression $2^{\sqrt{3}}$. Since $1.7 < \sqrt{3} < 1.8$, it follows that $2^{1.7} < 2^{\sqrt{3}} < 2^{1.8}$. Closer and closer approximations for $\sqrt{3}$ allow us to find closer and closer approximations for $2^{\sqrt{3}}$. Therefore, we can define a value for a^x when x is an irrational number.

Irrational Exponents

If x is an irrational number and $b > 0$, then b^x is the real number between b^{x_1} and b^{x_2} for all possible choices of rational numbers x_1 and x_2, such that $x_1 < x < x_2$.

A calculator can be used to approximate the value of an expression with irrational exponents. For example, $2^{\sqrt{3}} \approx 3.321997085$. While operations with irrational exponents are rarely used, evaluating such expressions is useful in graphing exponential equations.

CHECK FOR UNDERSTANDING

Communicating Mathematics

Read and study the lesson to answer each question.

1. **State** whether -4^{-2} and $(-4)^{-2}$ represent the same quantity. Explain.
2. **Explain** why rational exponents are not defined when the denominator of the exponent in lowest terms is even and the base is negative.
3. **You Decide** Laura says that a number written in scientific notation with an exponent of -10 is between 0 and 1. Lina says that a number written in scientific notation with an exponent of -10 is a negative number with a very large absolute value. Who is correct and why?

Guided Practice

Evaluate each expression.

4. 5^{-4}
5. $\left(\frac{9}{16}\right)^{-2}$
6. $216^{\frac{1}{3}}$
7. $\sqrt{27} \cdot \sqrt{3}$
8. $32^{\frac{3}{5}}$

Simplify each expression.

9. $(3a^{-2})^3 \cdot 3a^5$
10. $\sqrt{m^3n^2} \cdot \sqrt{m^4n^5}$
11. $\sqrt{\frac{8^n \cdot 2^7}{4^{-n}}}$
12. $(2x^4y^8)^{\frac{1}{2}}$

Express using rational exponents.

13. $\sqrt{169x^5}$
14. $\sqrt[4]{a^2b^3c^4d^5}$

Express using a radical.

15. $6^{\frac{1}{4}}b^{\frac{3}{4}}c^{\frac{1}{4}}$
16. $15x^{\frac{1}{3}}y^{\frac{1}{5}}$
17. Simplify $\sqrt[3]{p^4q^6r^5}$
18. Solve $y^{\frac{4}{5}} = 34$.

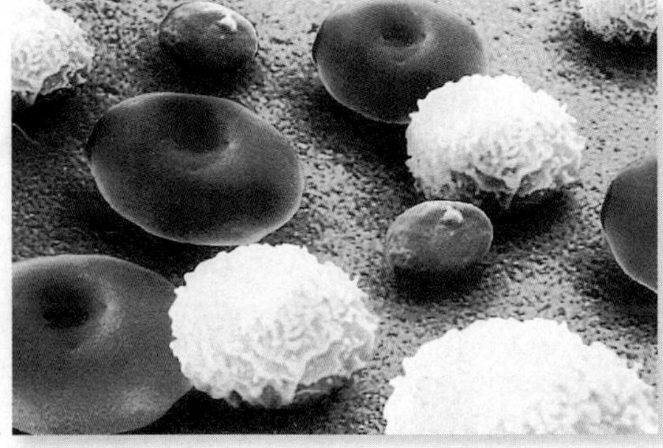

19. **Biology** Red blood cells are circular-shaped cells that carry oxygen through the bloodstream. The radius of a red blood cell is about 3.875×10^{-7} meters. Find the area of a red blood cell.

EXERCISES

Practice

Evaluate each expression.

20. $(-6)^{-4}$
21. -6^{-4}
22. $(5 \cdot 3)^2$
23. $\frac{2^4}{2^{-1}}$
24. $\left(\frac{7}{8}\right)^{-3}$
25. $(3^{-1} + 3^{-2})^{-1}$
26. $81^{\frac{1}{2}}$
27. $729^{\frac{1}{3}}$
28. $\frac{27}{27^{\frac{2}{3}}}$
29. $2^{\frac{1}{2}} \cdot 12^{\frac{1}{2}}$
30. $64^{\frac{1}{12}}$
31. $16^{-\frac{1}{4}}$
32. $\frac{(3^7)(9^4)}{\sqrt{27^6}}$
33. $(\sqrt[3]{216})^2$
34. $81^{\frac{1}{2}} - 81^{-\frac{1}{2}}$
35. $\frac{1}{\sqrt[7]{(-128)^4}}$

Simplify each expression.

36. $(3n^2)^3$
37. $(y^2)^{-4} \cdot y^8$
38. $(4y^4)^{\frac{3}{2}}$
39. $(27p^3q^6r^{-1})^{\frac{1}{3}}$
40. $[(2x)^4]^{-2}$
41. $(36x^6)^{\frac{1}{2}}$
42. $\left(\dfrac{b^{2n}}{b^{-2n}}\right)^{\frac{1}{2}}$
43. $\dfrac{2n}{4n^{\frac{1}{2}}}$
44. $\left(3m^{\frac{1}{2}} \cdot 27n^{\frac{1}{4}}\right)^4$
45. $\left(\dfrac{f^{-16}}{256g^4h^{-4}}\right)^{-\frac{1}{4}}$
46. $\sqrt[6]{x^2\left(x^{\frac{3}{4}} + x^{-\frac{3}{4}}\right)}$
47. $\left(2x^{\frac{1}{4}}y^{\frac{1}{3}}\right)\left(3x^{\frac{1}{4}}y^{\frac{2}{3}}\right)$
48. Show that $\sqrt[m]{\sqrt[n]{a}} = \sqrt[mn]{a}$.

Express using rational exponents.

49. $\sqrt{m^6n}$
50. $\sqrt{xy^3}$
51. $\sqrt[3]{8x^3y^6}$
52. $17\sqrt[7]{x^{14}y^7z^{12}}$
53. $\sqrt[5]{a^{10}b^2} \cdot \sqrt[4]{c^2}$
54. $60\sqrt[8]{r^{80}s^{56}t^{27}}$

Express using a radical.

55. $16^{\frac{1}{5}}$
56. $(7a)^{\frac{5}{8}}b^{\frac{3}{8}}$
57. $p^{\frac{2}{3}}q^{\frac{1}{2}}r^{\frac{1}{3}}$
58. $\dfrac{2^{\frac{2}{3}}}{2^{\frac{1}{3}}}$
59. $13a^{\frac{1}{7}}b^{\frac{1}{3}}$
60. $(n^3m^9)^{\frac{1}{2}}$
61. What is the value of x in the equation $x = \sqrt[3]{(-245)^{-\frac{1}{5}}}$ to the nearest hundredth?

Simplify each expression.

62. $\sqrt{d^3e^2f^2}$
63. $\sqrt[3]{a^5b^7c}$
64. $\sqrt{20x^3y^6}$

Solve each equation.

65. $14.2 = x^{-\frac{3}{2}}$
66. $724 = 15a^{\frac{5}{2}} + 12$
67. $\frac{1}{8}\sqrt{x^5} = 3.5$

Applications and Problem Solving

68. **Aerospace** Mars has an aproximate diameter of 6.794×10^3 kilometers. Assume that Mars is a perfect sphere. Use the formula for the volume of a sphere,
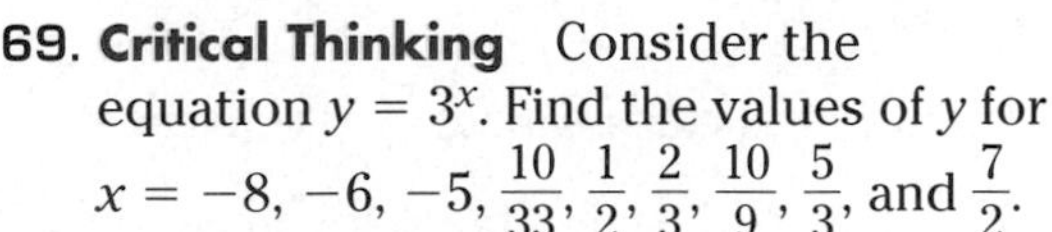
$V = \frac{4}{3}\pi r^3$ to determine the volume of Mars.

Mars

69. **Critical Thinking** Consider the equation $y = 3^x$. Find the values of y for $x = -8, -6, -5, \frac{10}{33}, \frac{1}{2}, \frac{2}{3}, \frac{10}{9}, \frac{5}{3}$, and $\frac{7}{2}$.
    a. What is the value of y if $x < 0$?
    b. What is the value of y if $0 < x < 1$?
    c. What is the value of y if $x > 1$?
    d. Write a conjecture about the relationship between the value of the base and the value of the power if the exponent is greater than or less than 1. Justify your answer.

70. **Chemistry** The nucleus of an atom is the center portion of the atom that contains most of its mass. A formula for the radius r of a nucleus of an atom is $r = (1.2 \times 10^{-15})A^{\frac{1}{3}}$ meters, where A is the mass number of the nucleus. Suppose the radius of a nucleus is approximately 2.75×10^{-15} meters. Is the atom boron with a mass number of 11, carbon with a mass number of 12, or nitrogen with a mass number of 14?

71. Critical Thinking Find the solutions for $32^{(x^2 + 4x)} = 16^{(x^2 + 4x + 3)}$.

72. Meteorology Have you ever heard the meteorologist on the news tell the day's windchill factor? A model that approximates the windchill temperature for an air temperature 5°F and a wind speed of s miles per hour is $C = 69.2(0.94^s) - 50$.

a. Copy and complete the table.

b. How does the effect of a 5-mile per hour increase in the wind speed when the wind is light compare to the effect of a 5-mile per hour increase in the wind speed when the wind is heavy?

Wind Speed	Windchill
5	
10	
15	
20	
25	
30	

73. Communication Geosynchronous satellites stay over a single point on Earth. To do this, they rotate with the same period of time as Earth. The distance r of an object from the center of Earth, that has a period of t seconds, is given by $r^3 = \frac{GM_e t^2}{4\pi^2}$, where G is 6.67×10^{-11} N m^2/kg^2, M_e is the mass of Earth, and t is time in seconds. The mass of Earth is 5.98×10^{24} kilograms.

a. How many meters is a geosynchronous communications satellite from the center of Earth?

b. If the radius of Earth is approximately 6380 kilometers, how many kilometers is the satellite above the surface of Earth?

74. Critical Thinking Use the definition of an exponent to verify each property.

a. $a^m a^n = a^{m+n}$ b. $(a^m)^n = a^{mn}$ c. $(ab)^m = a^m b^m$

d. $\left(\frac{a}{b}\right)^m = \frac{a^m}{b^m}$, where $b \neq 0$ e. $\frac{a^m}{a^n} = a^{m-n}$, where $a \neq 0$

Mixed Review

75. Graph the system of inequalities $x^2 + y^2 \leq 9$ and $x^2 + y^2 \geq 4$. *(Lesson 10-8)*

76. Find the coordinates of the focus and the equation of the directrix of the parabola with equation $y^2 = 12x$. Then graph the equation. *(Lesson 10-5)*

77. Find $(2\sqrt{3} + 2\boldsymbol{i})^{\frac{1}{5}}$. Express the answer in the form $a + b\boldsymbol{i}$ with a and b to the nearest hundredth. *(Lesson 9-8)*

78. Graph $r^2 = 9 \cos 2\theta$. Identify the classical curve it represents. *(Lesson 9-2)*

79. **Baseball** Lashon hits a baseball 3 feet above the ground with a force that produces an initial velocity of 105 feet per second at an angle of 42° above the horizontal. What is the elapsed time between the moment the ball is hit and the time it hits the ground? *(Lesson 8-7)*

80. Find an ordered triple that represents $\overrightarrow{TC}$ for $T(3, -4, 6)$ and $C(2, 6, -5)$. Then find the magnitude of $\overrightarrow{TC}$. *(Lesson 8-3)*

81. Find the numerical value of one trigonometric function of S if $\tan S \cos S = \frac{1}{2}$. *(Lesson 7-2)*

82. Find the values of θ for which $\cot \theta = 0$ is true. *(Lesson 6-7)*

83. **Agriculture** A center-pivot irrigation system with a 75-meter radial arm completes one revolution every 6 hours. Find the linear velocity of a nozzle at the end of the arm. *(Lesson 6-2)*

Extra Practice See p. A46.

84. Find the values of x in the interval $0° \leq x \leq 360°$ for which $x = \arccos 0$. *(Lesson 5-5)*

85. State the number of complex roots of the equation $x^3 - 25x = 0$. Then find the roots and graph. *(Lesson 4-1)*

86. Use a graphing calculator to graph $f(x) = 2x^2 - 3x + 3$ and to determine and classify its extrema. *(Lesson 3-6)*

87. **SAT/ACT Practice** If all painters work at the same rate and 8 painters can paint a building in 48 hours, how many hours will it take 16 painters to paint the building?

A 96 **B** 72 **C** 54 **D** 36 **E** 24

CAREER CHOICES

Food Technologist

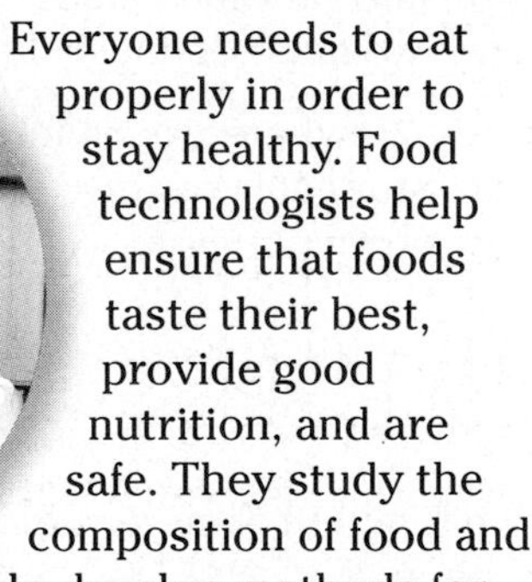

Everyone needs to eat properly in order to stay healthy. Food technologists help ensure that foods taste their best, provide good nutrition, and are safe. They study the composition of food and help develop methods for processing, preserving, and packaging foods. Food technologists usually specialize in some aspect of food technology such as improving taste or increasing nutritional value.

Food safety is one very important area in this field. Foods need to be handled, processed, and packaged under strict conditions to insure that they are safe for human consumption. Food technologists generally work either in private industry or for the government.

Career Overview

Degree Preferred:
bachelor's degree in Food Science, Food Management, or Human Nutrition

Related Courses:
mathematics, biology, chemistry, physics

Outlook:
better than average through the year 2006

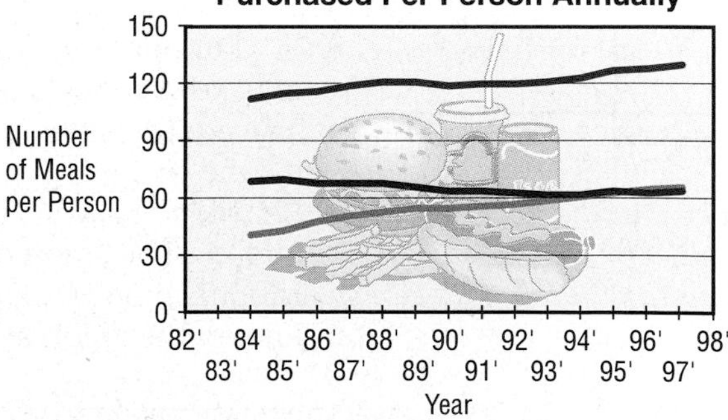

interNET CONNECTION For more information on careers in food technology, visit: **www.amc.glencoe.com** 

11-2

Exponential Functions

OBJECTIVES

- Graph exponential functions and inequalities.
- Solve problems involving exponential growth and decay.

ENTOMOLOGY In recent years, beekeepers have experienced a serious decline in the honeybee population in the United States. One of the causes for the decline is the arrival of varroa mites. Experts estimate that as much as 90% of the wild bee colonies have been wiped out.

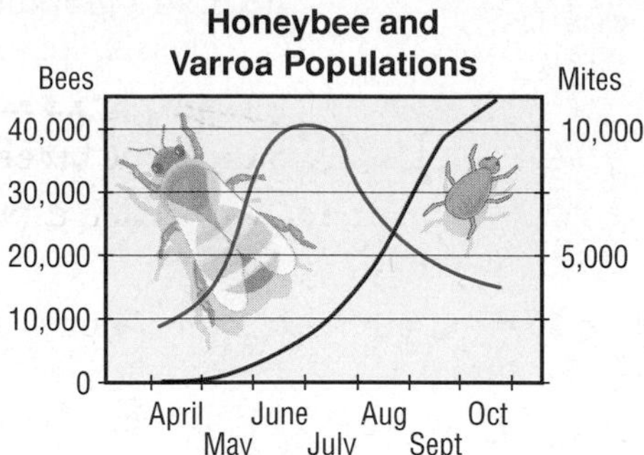

The graph shows typical honeybee and varroa populations over several months. A graph of the varroa population growth from April to September resembles an *exponential curve. A problem related to this will be solved in Example 2.*

You have evaluated functions in which the variable is the base and the exponent is any real number. For example, $y = x^5$ has x as the base and 5 as the power. Such a function is known as a **power function.** Functions of the form $y = b^x$, in which the base b is a positive real number and the exponent is a variable are known as **exponential functions.** In the previous lesson, the expression b^x was defined for integral and rational values of x. In order for us to graph $y = b^x$ with a continuous curve, we must define b^x for irrational values of x.

Consider the graph of $y = 2^x$, where x is an integer. This is a function since there is a unique y-value for each x-value.

x	−4	−3	−2	−1	0	1	2	3	4	5
$\mathbf{2^x}$	$\frac{1}{16}$	$\frac{1}{8}$	$\frac{1}{4}$	$\frac{1}{2}$	1	2	4	8	16	32

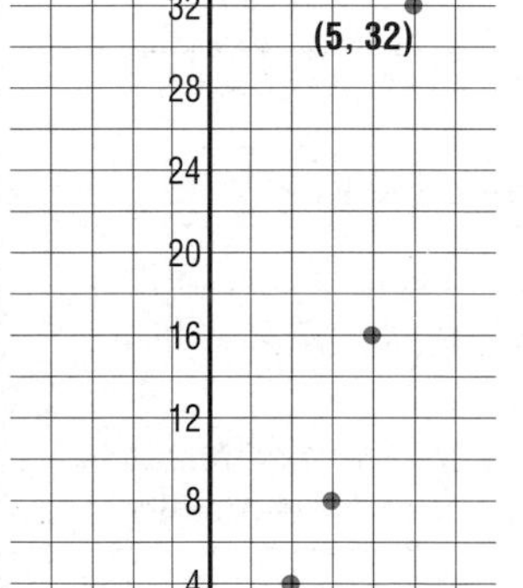

Notice that the vertical scale is condensed.

The graph suggests that the function is increasing. That is, for any values x_1 and x_2, if $x_1 < x_2$, then $2^{x_1} < 2^{x_2}$.

Suppose the domain of $y = 2^x$ is expanded to include all rational numbers. The additional points graphed seem to "fill in" the graph of $y = 2^x$. That is, if k is between x_1 and x_2, then 2^k is between 2^{x_1} and 2^{x_2}. The graph of $y = 2^x$, when x is a rational number, is indicated by the broken line on the graph at the right.

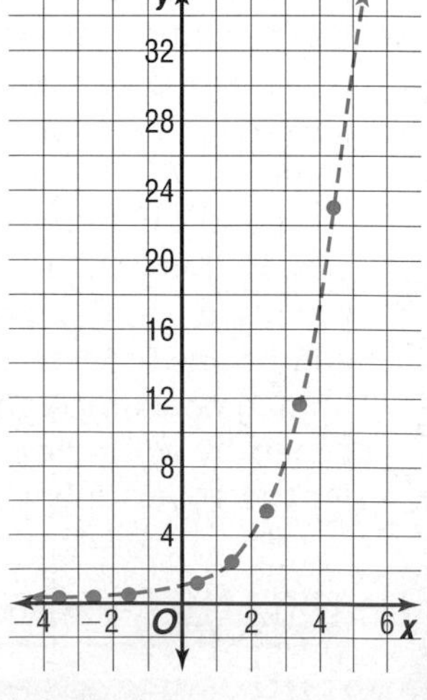

x	−3.5	−2.5	−1.5	−0.5	0.5	1.5	2.5	3.5	4.5
$\mathbf{2^x}$	0.09	0.18	0.35	0.71	1.41	2.83	5.66	11.31	22.63

Values in the table are approximate.

In Lesson 11-1, you learned that exponents can also be irrational. By including irrational values of x, we can explore the graph of $y = b^x$ for the domain of all real numbers.

GRAPHING CALCULATOR EXPLORATION

TRY THESE **Graph $y = b^x$ for $b = 0.5$, 0.75, 2, and 5 on the same screen.**

1. What is the range of each exponential function?
2. What point is on the graph of each function?
3. What is the end behavior of each graph?
4. Do the graphs have any asymptotes?

WHAT DO YOU THINK?

5. Is the range of every exponential function the same? Explain.
6. Why is the point at (0, 1) on the graph of every exponential function?
7. For what values of a is the graph of $y = a^x$ increasing and for what values is the graph decreasing? Explain.
8. Explain the existence or absence of the asymptotes in the graph of an exponential function.

The equations graphed in the Graphing Calculator Exploration demonstrate many properties of exponential graphs.

Characteristics of graphs of $y = b^x$

	$b > 1$	$0 < b < 1$
Domain	all real numbers	all real numbers
Range	all real numbers > 0	all real numbers > 0
y-intercept	(0, 1)	(0, 1)
behavior	continuous, one-to-one, and increasing	continuous, one-to-one, and decreasing
Horizontal asymptote	negative x-axis	positive x-axis
Vertical asymptote	none	none

When $b = 1$ the graph of $y = b^x$ is the horizontal line $y = 1$.

As with other types of graphs, $y = b^x$ represents a parent graph. The same techniques used to transform the graphs of other functions you have studied can be applied to graphs of exponential functions.

Examples

To graph an exponential function using paper and pencil, you can use a calculator to find each range value for each domain value.

1 **a. Graph the exponential functions $y = 4^x$, $y = 4^x + 2$, and $y = 4^x - 3$ on the same set of axes. Compare and contrast the graphs.**

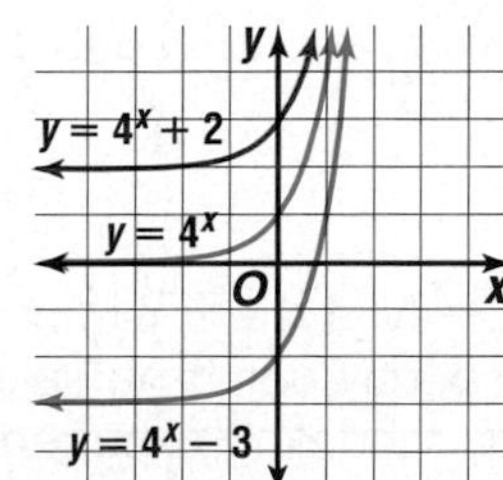

All of the graphs are continuous, increasing, and one-to-one. They have the same domain, and no vertical asymptote.

The y-intercept and the horizontal asymptotes for each graph are different from the parent graph $y = 4^x$. While $y = 4^x$ and $y = 4^x + 2$ have no x-intercept, $y = 4^x - 3$ has an x-intercept.

b. Graph the exponential functions $y = \left(\frac{1}{5}\right)^x$, $y = 6\left(\frac{1}{5}\right)^x$, and $y = -2\left(\frac{1}{5}\right)^x$ on the same set of axes. Compare and contrast the graphs.

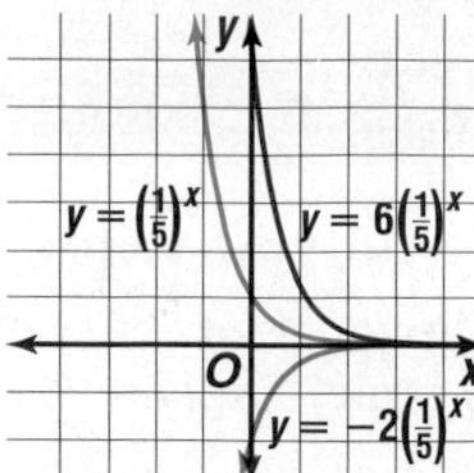

All of the graphs are decreasing, continuous, and one-to-one. They have the same domain and horizontal asymptote. They have no vertical asymptote or x-intercept.

The y-intercepts for each graph are different from the parent graph $y = \left(\frac{1}{5}\right)^x$.

2 PHYSICS **According to Newton's Law of Cooling, the difference between the temperature of an object and its surroundings decreases in time exponentially. Suppose a certain cup of coffee is 95°C and it is in a room that is 20°C. The cooling for this particular cup can be modeled by the equation $y = 75(0.875)^t$ where y is the temperature difference and t is time in minutes.**

a. Find the temperature of the coffee after 15 minutes.

b. Graph the cooling function.

a. $y = 75(0.875)^t$

$y = 75(0.875)^{15}$ *$t = 15$*

$y \approx 10.12003603$

The difference is about 10.1° C. So the coffee is 95° C − 10.1° C or 84.9° C.

b.

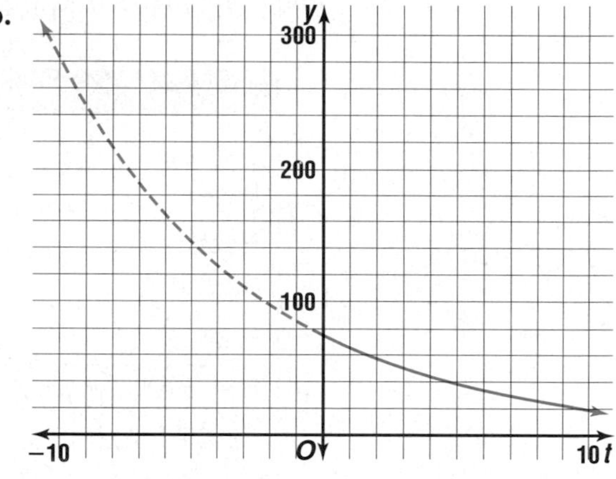

In a situation like cooling when a quantity loses value exponentially over time, the value exhibits what is called **exponential decay.** Many real-world situations involve quantities that increase exponentially over time. For example, the balance in a savings or money market account and the population of people or animals in a region often demonstrate what is called **exponential growth.** As you saw in Example 1, you can use the general form of an exponential function to describe exponential growth or decay. When you know the rate at which the growth or decay is occurring, the following equation may be used.

Exponential Growth or Decay	The equation $N = N_0(1 + r)^t$, where N is the final amount, N_0 is the initial amount, r is the rate of growth or decay per time period, and t is the number of time periods, is used for modeling exponential growth or decay.

Example **3** **ENTOMOLOGY** **Refer to the application at the beginning of the lesson. Suppose that a researcher estimates that the initial population of varroa in a colony is 500. They are increasing at a rate of 14% per week. What is the expected population in 22 weeks?**

$N = N_0(1 + r)^t$

$N = 500(1 + 0.14)^{22}$ *$N_0 = 500$, $r = 0.14$, $t = 22$*

$N \approx 8930.519719$ *Use a calculator.*

There will be about 8931 varroa in the colony in 22 weeks.

The general equation for exponential growth is modified for finding the balance in an account that earns compound interest.

Compound Interest

The compound interest equation is $A = P\left(1 + \frac{r}{n}\right)^{nt}$, where P is the principal or initial investment, A is the final amount of the investment, r is the annual interest rate, n is the number of times interest is paid, or compounded each year, and t is the number of years.

Example **4** **FINANCE** **Determine the amount of money in a money market account providing an annual rate of 5% compounded daily if Marcus invested $2000 and left it in the account for 7 years.**

$A = P\left(1 + \frac{r}{n}\right)^{nt}$

$A = 2000\left(1 + \frac{0.05}{365}\right)^{365 \cdot 7}$ *$P = 2000$, $r = 0.05$, $n = 365$, $t = 7$*

$A \approx 2838.067067$ *Use a calculator.*

After 7 years, the $2000 investment will have a value of $2838.06.

The value 2838.067067 is rounded to 2838.06 as banks generally round down when they are paying interest.

Graphing exponential inequalities is similar to graphing other inequalities.

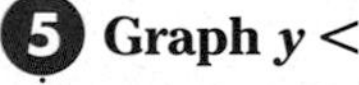

Example **5** **Graph $y < 2^x + 1$**

First, graph $y = 2^x + 1$. Since the points on this curve are not in the solution of the inequality, the graph of $y = 2^x + 1$ is shown as a dashed curve.

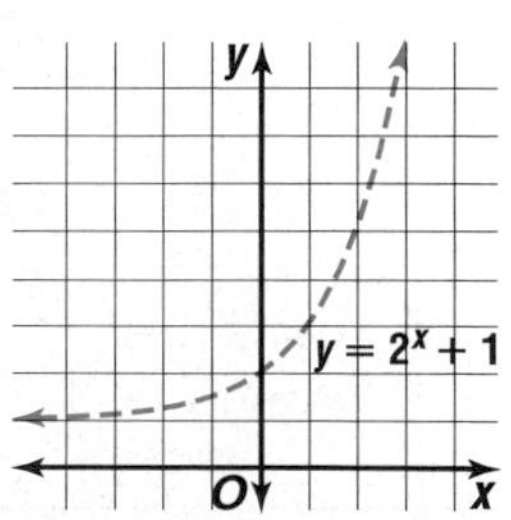

(continued on the next page)

Then, use (0, 0) as a test point to determine which area to shade.

$y < 2^x + 1 \Rightarrow 0 < 2^0 + 1$
$0 < 1 + 1$
$0 < 2$

Since (0, 0) satisfies the inequality, the region that contains (0, 0) should be shaded.

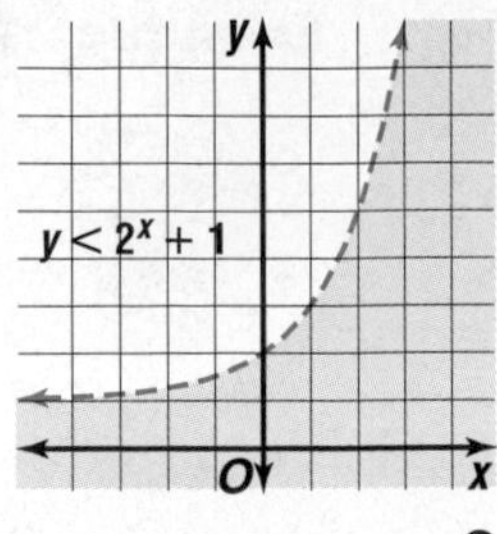

CHECK FOR UNDERSTANDING

Communicating Mathematics

Read and study the lesson to answer each question.

1. **Tell** whether $y = x^4$ is an exponential or a power function. Justify your answer.
2. **Compare and contrast** the graphs of $y = b^x$ when $b > 1$ and when $0 < b < 1$.
3. **Write** about how you can tell whether an exponential function represents exponential growth or exponential decay.
4. **Describe** the differences between the graphs of $y = 4^x$ and $y = 4^x - 3$.

Guided Practice

Graph each exponential function or inequality.

5. $y = 3^x$
6. $y = 3^{-x}$
7. $y > 2^x - 4$

8. **Business** Business owners keep track of the value of their assets for tax purposes. Suppose the value of a computer depreciates at a rate of 25% a year. Determine the value of a laptop computer two years after it has been purchased for $3750.

9. **Demographics** In the 1990 U.S. Census, the population of Los Angeles County was 8,863,052. By 1997, the population had increased to 9,145,219.
   a. Find the yearly growth rate by dividing the average change in population by the 1990 population.
   b. Assuming the growth rate continues at a similar rate, predict the number of people who will be living in Los Angeles County in 2010.

EXERCISES

Practice

Graph each exponential function or inequality.

10. $y = 2^x$
11. $y = -2^x$
12. $y = 2^{-x}$
13. $y = 2^{x + 3}$
14. $y = -2^{x + 3}$
15. $y > -4^x + 2$
16. $y = \left(\frac{1}{5}\right)^x$
17. $y \leq \left(\frac{1}{2}\right)^x$
18. $y < 2^{x - 4}$

Match each equation to its graph.

19. $y = 0.01^x$
20. $y = 5^{-x}$
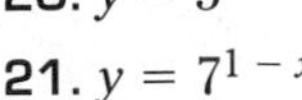
21. $y = 7^{1 - x}$
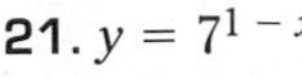

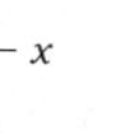

A
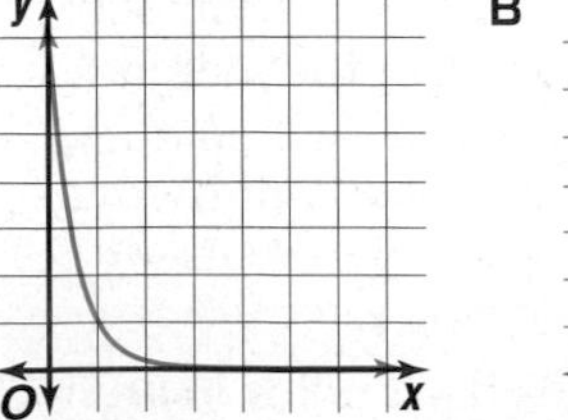

B
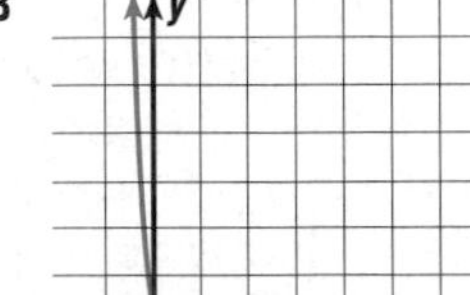

C
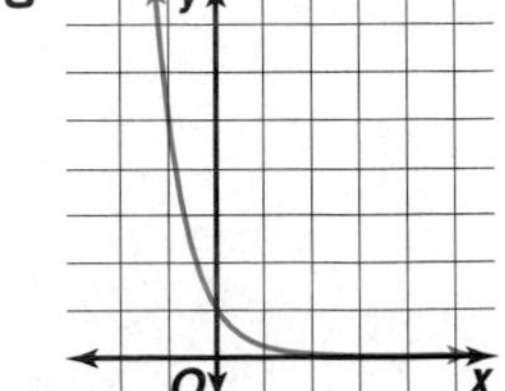

Graphing Calculator

22. Graph the functions $y = 5^x$, $y = -5^x$, $y = 5^{-x}$, $y = 5^x + 2$, $y = 5^x - 2$, and $y = 10^x$ on the same screen.
    a. Compare $y = -5^x$, $y = 5^{-x}$ to the parent graph $y = 5^x$. Describe the transformations of the functions.
    b. What transformation occurs with the graphs of $y = 5^x + 2$ and $y = 5^x - 2$?
    c. The function $y = 10^x$ can be expressed as $y = (2 \cdot 5)^x$. Compare this function to $y = 5^x$. Is the graph of $y = 5^{2x}$ the same as $y = 10^x$? Explain.

23. Without graphing, describe how each pair of graphs is related. Then use a graphing calculator to check your descriptions.
    a. $y = 6^x$ and $y = 6^x + 4$
    b. $y = -3^x$ and $y = 3^x$
    c. $y = 7^x$ and $y = 7^{-x}$
    d. $y = 2^x$ and $y = \left(\frac{1}{2}\right)^x$

Applications and Problem Solving

24. **Employment** Average national teachers' salaries can be modeled using the equation $y = 9.25(1.06)^n$, where y is the salary in thousands of dollars and n is the number of years since 1970.
    a. Graph the function.
    b. Using this model, what can a teacher expect to have as a salary in the year 2020?

25. **Aviation** When kerosene is purified to make jet fuel, pollutants are removed by passing the kerosene through a special clay filter. Suppose a filter is fitted in a pipe so that 15% of the impurities are removed for every foot that the kerosene travels.
    a. Write an exponential function to model the percent of impurity left after the kerosene travels x feet.
    b. Graph the function.
    c. About what percent of the impurity remains after the kerosene travels 12 feet?
    d. Will the impurities ever be completely removed? Explain.

26. **Demographics** Find the projected population of each location in 2015.
    a. In Honolulu, Hawaii, the population was 836,231 in 1990. The average yearly rate of growth is 0.7%.
    b. The population in Kings County, New York has demonstrated an average decrease of 0.45% over several years. The population in 1997 was 2,240,384.
    c. Janesville, Wisconsin had a population of 139,420 in 1980 and 139,510 in 1990.
    d. The population in Cedar Rapids, Iowa was 169,775 in the 1980 U.S. Census and 168,767 in the 1990 U.S. Census.

27. **Biology** Scientists who study Atlantic salmon have found that the oxygen consumption of a yearling salmon O is given by the function $O = 100\left(3^{\frac{3s}{5}}\right)$, where s is the speed that the fish is traveling in feet per second.

    a. What is the oxygen consumption of a fish that is traveling at 5 feet per second?
    b. If a fish has traveled 4.2 miles in an hour, what is its oxygen consumption?

Graphing Calculator Programs
For a graphing calculator program that calculates loan payments, visit **www.amc.glencoe.com**

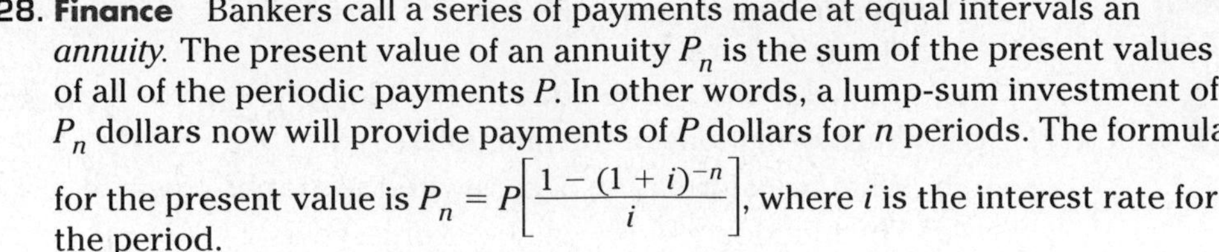

28. **Finance** Bankers call a series of payments made at equal intervals an *annuity*. The present value of an annuity P_n is the sum of the present values of all of the periodic payments P. In other words, a lump-sum investment of P_n dollars now will provide payments of P dollars for n periods. The formula for the present value is $P_n = P\left[\frac{1 - (1 + i)^{-n}}{i}\right]$, where i is the interest rate for the period.
    a. Use the present value formula to find the monthly payment you would pay on a home mortgage if the present value is \$121,000, the annual interest rate is 7.5%, and payments will be made for 30 years. (*Hint*: First find the interest rate for the period.)
    b. How much is the monthly mortgage payment if the borrowers choose a loan with a 20-year term and an interest rate of 7.25%?
    c. How much will be paid in interest over the life of each mortgage?
    d. Explain why a borrower might choose each mortgage.

29. **Finance** The future value of an annuity F_n is the sum of all of the periodic payments P and all of the accumulated interest. The formula for the future value is $F_n = P\left[\frac{(1 + i)^n - 1}{i}\right]$, where i is the interest rate for the period.
    a. When Connie Hockman began her first job at the age of 22, she started saving for her retirement. Each year she places \$4000 in an account that will earn an average 4.75% annual interest until she retires at 65. How much will be in the account when she retires?
    b. If Ms. Hockman had invested in an account that earns an average of 5.25% annual interest, how much more would her account be worth?

30. **Critical Thinking** Explain when the exponential function $y = a^x$ is undefined for $a < 0$.

31. **Investments** The number of times that interest is compounded has a dramatic effect on the total interest earned on an investment.
    a. How much interest would you earn in one year on an \$1000 investment earning 5% interest if the interest is compounded once, twice, four times, twelve times, or 365 times in the year?
    b. If you are making an investment that you will leave in an account for one year, which account should you choose to get the highest return?

Account	Rate	Compounded
Statement Savings	5.1%	Yearly
Money Market Savings	5.05%	Monthly
Super Saver	5%	Daily

    c. Suppose you are a bank manager determining rates on savings accounts. If the account with interest compounded annually offers 5% interest, what interest rate should be offered on an account with interest compounded daily in order for the interest earned on equal investments to be the same?

Mixed Review

32. Simplify $4x^2(4x)^{-2}$. *(Lesson 11-1)*

33. Write the rectangular equation $y = 15$ in polar form. *(Lesson 9-4)*

34. Find the inner product of vectors $\langle -3, 9\rangle$ and $\langle 2, 1\rangle$. Are the vectors perpendicular? Explain. *(Lesson 8-4)*

35. Find $\frac{1}{3}\left(\cos \frac{7\pi}{8} + \boldsymbol{i} \sin \frac{7\pi}{8}\right) \cdot 3\sqrt{3}\left[\cos\left(-\frac{\pi}{4}\right) + \boldsymbol{i} \sin\left(-\frac{\pi}{4}\right)\right]$. Then express the product in rectangular form with a and b to the nearest hundredth. *(Lesson 9-7)*

36. **Sports** Suppose a baseball player popped a baseball straight up at an initial velocity v_0 of 72 feet per second. Its distance s above the ground after t seconds is described by $s = v_0t - 16t^2 + 4$. Find the maximum height of the ball. *(Lesson 10-5)*

37. **Art** Leigh Ann is drawing a sketch of a village. She wants the town square to be placed midway between the library and the fire station. Suppose the ordered pair for the library is (−7, 6) and the ordered pair for the fire station is (12, 8). *(Lesson 10-1)*
    a. Draw a graph that represents this situation.
    b. Determine the coordinates indicating where the town square should be placed.

38. Verify that $\sin^4 A + \cos^2 A = \cos^4 A + \sin^2 A$ is an identity. *(Lesson 7-2)*

39. **Mechanics** A circular saw 18.4 centimeters in diameter rotates at 2400 revolutions per second. What is the linear velocity at which a saw tooth strikes the cutting surface in centimeters per second? *(Lesson 6-2)*

40. **Travel** Martina went to Acapulco, Mexico, on a vacation with her parents. One of the sights they visited was a cliff-diving exhibition into the waters of the Gulf of Mexico. Martina stood at a lookout site on top of a 200-foot cliff. A team of medical experts was in a boat below in case of an accident. The angle of depression to the boat from the top of the cliff was 21°. How far is the boat from the base of the cliff? *(Lesson 5-4)*

41. **Salary** Diane has had a part time job as a Home Chef demonstrator for 9 years. Her yearly income is listed in the table. Write a model that relates the income as a function of the number of years since 1990. *(Lesson 4-8)*

Year	1990	1991	1992	1993	1994	1995	1996	1997	1998
Income ($)	4012	6250	7391	8102	8993	9714	10,536	11,362	12,429

42. Use the parent graph $f(x) = \frac{1}{x}$ to graph $f(x) = \frac{1}{x + 3}$. Describe the transformation(s) that have taken place. Identify the new locations of the asymptotes. *(Lesson 3-7)*

43. **SAT/ACT Practice** In the figure, $\overline{AB}$ is the diameter of the smaller circle, and $\overline{AC}$ is the diameter of the larger circle. If the distance from B to C is 16 inches, then the circumference of the larger circle is approximately how many inches greater than the circumference of the smaller circle?

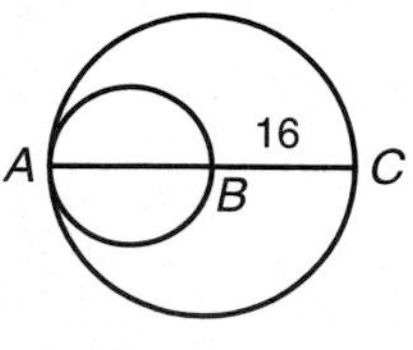

A 8 **B** 16 **C** 25 **D** 35 **E** 50

Extra Practice See p. A46.

The Number *e*

OBJECTIVES

- Use the exponential function $y = e^x$.

MEDICINE Swiss entomologist Dr. Paul Mueller was awarded the Nobel Prize in medicine in 1948 for his work with the pesticide DDT. Dr. Mueller discovered that DDT is effective against insects that destroy agricultural crops, mosquitoes that transmit malaria and yellow fever, as well as lice that carry typhus.

It was later discovered that DDT presented a risk to humans. Effective January 1, 1973, the United States Environmental Protection Agency banned all uses of DDT. More than 1.0×10^{10} kilograms of DDT had been used in the U.S. before the ban. How much will remain in the environment in 2005? *This problem will be solved in Example 1.*

DDT degrades into harmless materials over time. To find the amount of a substance that decays exponentially remaining after a certain amount of time, you can use the following formula for exponential growth or decay, which involves the number ***e***.

Exponential Growth or Decay (in terms of *e*)

$N = N_0e^{kt}$, where N is the final amount, N_0 is the initial amount, k is a constant and t is time.

The number e in the formula is not a variable. It is a special irrational number. This number is the sum of the infinite series shown below.

$$e = 1 + \frac{1}{1} + \frac{1}{1 \cdot 2} + \frac{1}{1 \cdot 2 \cdot 3} + \frac{1}{1 \cdot 2 \cdot 3 \cdot 4} + \cdots + \frac{1}{1 \cdot 2 \cdot 3 \cdot \cdots \cdot n} + \cdots$$

The following computation for e is correct to three decimal places.

$$\begin{aligned} e &= 1 + \frac{1}{1} + \frac{1}{1 \cdot 2} + \frac{1}{1 \cdot 2 \cdot 3} + \frac{1}{1 \cdot 2 \cdot 3 \cdot 4} + \frac{1}{1 \cdot 2 \cdot 3 \cdot 4 \cdot 5} + \\ &\quad \frac{1}{1 \cdot 2 \cdot 3 \cdot 4 \cdot 5 \cdot 6} + \frac{1}{1 \cdot 2 \cdot 3 \cdot 4 \cdot 5 \cdot 6 \cdot 7} \\ &= 1 + 1 + \frac{1}{2} + \frac{1}{6} + \frac{1}{24} + \frac{1}{120} + \frac{1}{720} + \frac{1}{5040} \\ &= 1 + 1 + 0.5 + 0.16667 + 0.04167 + 0.00833 + \\ &\quad 0.00139 + 0.000198 \\ &= 2.718 \end{aligned}$$

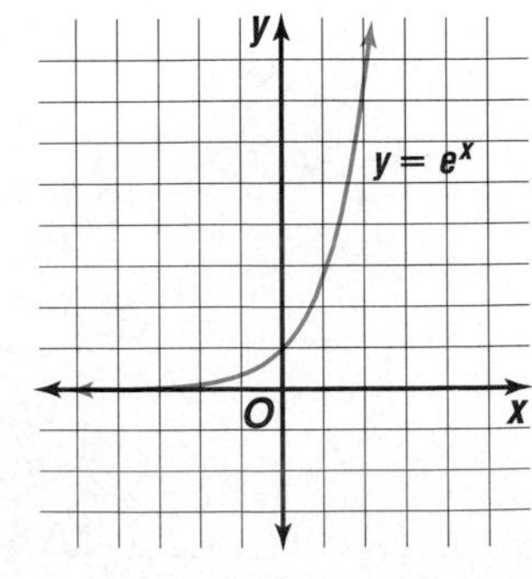

The function $y = e^x$ is one of the most important exponential functions. The graph of $y = e^x$ is shown at the right.

Example 1 **MEDICINE** **Refer to the application at the beginning of the lesson. Assume that there were 1.0×10^9 kilograms of DDT in the environment in 1973 and that for DDT, $k = -0.0211$.**

a. Write a function to model the amount of DDT remaining in the environment.

b. Find the amount of DDT that will be in the environment in 2005.

c. Graph the function and use the graph to verify your answer in part b.

Graphing Calculator Tip

To graph an equation or evaluate an expression involving e raised to a power, use the second function of ln on a calculator.

Example 1 is an example of chemical decay.

a. $y = ne^{kt}$

$y = (1 \times 10^9)e^{-0.0211t}$

$y = (10^9)e^{-0.0211t}$

b. In 2005, it will have been 2005 − 1973 or 32 years since DDT was banned. Thus, $t = 32$.

$y = (10^9)e^{-0.0211t}$

$y = (10^9)e^{-0.0211(32)}$ *$t = 32$*

$y \approx (10^9)0.5090545995$

c. Use a graphing calculator to graph the function.

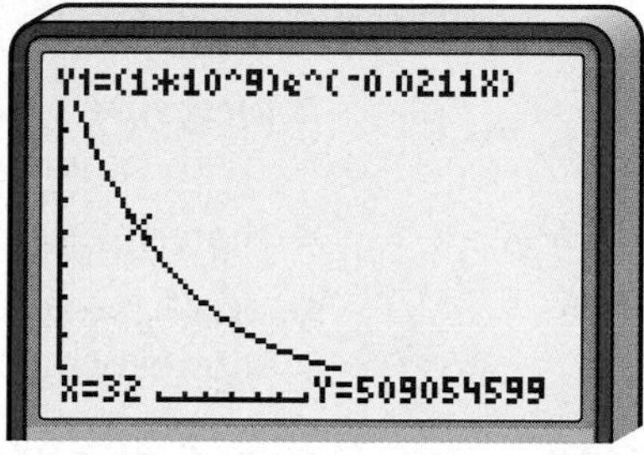

[−0, 200] scl:10, [0, $(1 \times 10^9]$ scl:1×10^8

If there were 1×10^9 kilograms of DDT in the environment in 1973, there will be about $0.51(1 \times 10^9)$ or 5.1×10^8 kilograms remaining in 2005.

Some banks offer accounts that compound the interest continuously. The formula for finding continuously compounded interest is different from the one used for interest that is compounded a specific number of times each year.

Continuously Compounded Interest

The equation $A = Pe^{rt}$, where P is the initial amount, A is the final amount, r is the annual interest rate, and t is time in years, is used for calculating interest that is compounded continuously.

Example 2 **FINANCE** **Compare the balance after 25 years of a \$10,000 investment earning 6.75% interest compounded continuously to the same investment compounded semiannually.**

In both cases, $P = 10{,}000$, $r = 0.0675$, $t = 25$. When the interest is compounded semiannually, $n = 2$. Use a calculator to evaluate each expression.

Continuously	Semiannually
$A = Pe^{rt}$	$A = P\left(1 + \frac{r}{n}\right)^{nt}$
$A = 10{,}000(e)^{(0.0675 \cdot 25)}$	$A = 10{,}000\left(1 + \frac{0.0675}{2}\right)^{2 \cdot 25}$
$A = 54{,}059.49$	$A = 52{,}574.62$

The same principal invested over the same amount of time yields \$54,059.49 if compounded continuously and \$52,574.62 when compounded twice a year. You would earn \$54059.49 − \$52574.62 = \$1484.87 more by choosing the account that compounds continuously.

CHECK FOR UNDERSTANDING

Communicating Mathematics

Read and study the lesson to answer each question.

1. **Tell** which equation is represented by the graph.

   a. $y = e^x$ b. $y = e^{-x}$

   c. $y = -e^x$ d. $y = -e^{-x}$

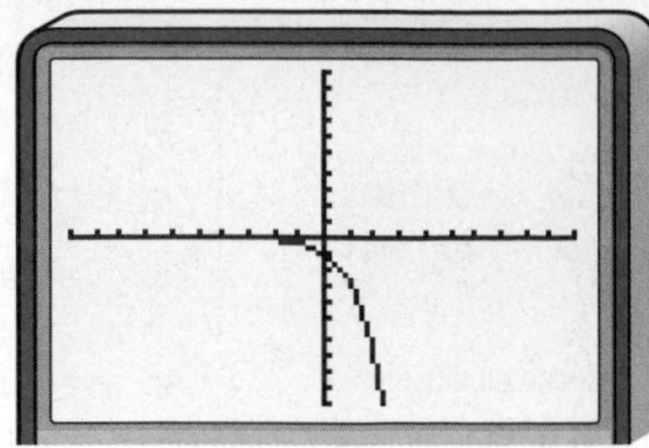

[−10, 10] scl:1 by [−10, 10] scl:1

2. **Describe** the value of k when the equation $N = N_0e^{kt}$ represents exponential growth and when it represents exponential decay.
3. **Describe** a situation that could be modeled by the equation $A = 3000e^{0.055t}$.
4. **State** the domain and range of the function $f(x) = e^x$.
5. *Math Journal* **Write** a sentence or two to explain the difference between interest compounded continuously and interest compounded monthly.

Guided Practice

6. **Demographics** Bakersfield, California was founded in 1859 when Colonel Thomas Baker planted ten acres of alfalfa for travelers going from Visalia to Los Angeles to feed their animals. The city's population can be modeled by the equation $y = 33{,}430e^{0.0397t}$, where t is the number of years since 1950.
   a. Has Bakersfield experienced growth or decline in population?
   b. What was Bakersfield's population in 1950?
   c. Find the projected population of Bakersfield in 2010.
7. **Financial Planning** The Kwans are saving for their daughter's college education. If they deposit $12,000 in an account bearing 6.4% interest compounded continuously, how much will be in the account when Ann goes to college in 12 years?

EXERCISES

Applications and Problem Solving

8. **Psychology** Without further study, as time passes you forget things you have learned. The Ebbinghaus model of human memory gives the percent p of acquired knowledge that a person retains after t weeks. The formula is $p = (100 - a)e^{-bt} + a$, where a and b vary from one person to another. If $a = 18$ and $b = 0.6$ for a certain student, how much information will the student retain two weeks after learning a new topic?
9. **Physics** Newton's Law of Cooling expresses the relationship between the temperature of a cooling object y and the time t elapsed since cooling began. This relationship is given by $y = ae^{-kt} + c$, where c is the temperature of the medium surrounding the cooling object, a is the difference between the initial temperature of the object and the surrounding temperature, and k is a constant related to the cooling object.
   a. The initial temperature of a liquid is 160°F. When it is removed from the heat, the temperature in the room is 76°F. For this object, $k = 0.23$. Find the temperature of the liquid after 15 minutes.
   b. Alex likes his coffee at a temperature of 135°. If he pours a cup of 170°F coffee in a 72°F room and waits 5 minutes before drinking, will his coffee be too hot or too cold? Explain. For Alex's cup, $k = 0.34$.

10. Civil Engineering Suspension bridges can span distances far longer than any other kind of bridge. The roadway of a suspension bridge is suspended from huge cables. When a flexible cable is suspended between two points, it forms a *catenary curve.*

a. Use a graphing calculator to graph the catenary $y = \frac{e^x + e^{-x}}{2}$.

b. What kind of symmetry is displayed by the graph?

11. Banking If your bank account earns interest that is compounded more than one time per year, the effective annual yield E is the interest rate that would give the same amount of interest earnings if the interest were compounded once per year. To find the effective annual yield, divide the interest earned by the principal.

a. Copy and complete the table to find the effective annual yield for each account if the principal is \$1000, the annual interest rate is 8%, and the term is one year.

Interest Compounded	Interest	Effective Annual Yield
Annually		
Semi-annually		
Quarterly		
Monthly		
Daily		
Continuously		

b. Which type of compounding provides the greatest effective annual yield?

c. If P represents the principal and A is the total value of the investment, the value of an investment is $A = P(1 + E)$. Find a formula for the effective annual yield for an account with interest compounded n times per year.

d. Write a formula for the effective annual yield of an account with interest compounded continuously.

12. Sociology Sociologists have found that information spreads among a population at an exponential rate. Suppose that the function $y = 525(1 - e^{-0.038t})$ models the number of people in a town of 525 people who have heard news within t hours of its distribution.

a. How many people will have heard about the opening of a new grocery store within 24 hours of the announcement?

b. Graph the function on a graphing calculator. When will 90% of the people have heard about the grocery store opening?

13. Customer Service The service-time distribution describes the probability P that the service time of the customer will be no more than t hours. If m is the mean number of customers serviced in an hour, then $P = 1 - e^{-mt}$.

a. Suppose a computer technical support representative can answer calls from 6 customers in an hour. What is the probability that a customer will be on hold less than 30 minutes?

b. A credit card customer service department averages 34 calls per hour. Use a graphing calculator to determine the amount of time after which it is 50% likely that a customer has been served?

14. **Critical Thinking** In 1997, inventor and amateur mathematician Harlan Brothers discovered some simple algebraic expressions that approximate e.

a. Use Brothers' expression $\left(\frac{2x+1}{2x-1}\right)^x$ for $x = 10$, $x = 100$, and $x = 1000$.

b. Compare each approximation to the value of e stored in a calculator. If each digit of the calculator value is correct, how accurate is each approximation?

c. Is the approximation always greater than e, always less than e, or sometimes greater than and sometimes less than e?

15. **Marketing** The probability P that a person has responded to an advertisement can be modeled by the exponential equation $P = 1 - e^{-0.047t}$ where t is the number of days since the advertisement began to appear in the media.

a. What are the probabilities that a person has responded after 5 days, 20 days, and 90 days?

b. Use a graphing calculator to graph the function. Use the graph to find when the probability that an individual has responded to the advertisement is 75%.

c. If you were planning a marketing campaign, how would you use this model to plan the introduction of new advertisements?

16. **Critical Thinking** Consider $f(x) = \frac{e^x}{e^x + c}$ if c is a constant greater than 0.

a. What is the domain of the function?

b. What is the range of the function?

c. How does the value of c affect the graph?

Mixed Review

17. **Finance** A *sinking fund* is a fund into which regular payments are made in order to pay off a debt when it is due. The Gallagher Construction Company foresees the need to buy a new cement truck in four years. At that time, the truck will probably cost \$120,000. The firm sets up a sinking fund in order to accumulate the money. They will pay semiannual payments into a fund with an APR of 7%. Use the formula for the future value of an annuity $F_n = P\left[\frac{(1+i)^n - 1}{i}\right]$, where F_n is the future value of the annuity, P is the payment amount, i is the interest rate for the period, and n is the number of payments, to find the payment amount. *(Lesson 11-2)*

18. Express $x^{\frac{8}{5}} y^{\frac{3}{5}} z^{\frac{1}{5}}$ using radicals. *(Lesson 11-1)*

19. **Communications** A satellite dish tracks a satellite directly overhead. Suppose the equation $y = 6x^2$ models the shape of the dish when it is oriented in this position. Later in the day, the dish is observed to have rotated approximately 45°. Find an equation that models the new orientation of the dish. *(Lesson 10-7)*

20. Express $-5 - \boldsymbol{i}$ in polar form. *(Lesson 9-6)*

21. **Physics** Tony is pushing a cart weighing 150 pounds up a ramp 10 feet long at an incline of 28°. Find the work done to push the cart the length of the ramp. Assume that friction is not a factor. *(Lesson 8-5)*

22. **Safety** A model relating the average number of crimes reported at a shopping mall as a function of the number of years since 1991 is $y = -1.5x^2 + 13.3x + 19.4$. According to this model, what is the number of crimes for the year 2001? *(Lesson 4-8)*

Extra Practice See p. A46.

23. Solve $\sqrt{2x + 3} = 4$. *(Lesson 4-7)*

24. Solve $|3x + 2| \leq 6$. *(Lesson 3-3)*

25. Quadrilateral *JKLM* has vertices at $J(-3, -2)$, $K(-2, 6)$, $L(2, 5)$ and $M(3, -1)$. Find the coordinates of the dilated quadrilateral $J'K'L'M'$ for a scale factor of 3. Describe the dilation. *(Lesson 2-4)*

26. Find the values of x and y for which $\begin{bmatrix} 4x + y \\ x \end{bmatrix} = \begin{bmatrix} 6 \\ 2y - 12 \end{bmatrix}$ is true. *(Lesson 2-3)*

27. State the domain and range of the relation $\{(2, 7), (-4, 5), (5, 7)\}$. Is this relation a function? *(Lesson 1-1)*

28. **SAT/ACT Practice** An artist wants to shade exactly 18 of the 36 smallest triangles in the pattern, including the one shown. If no two shaded triangles can have a side in common, which of the triangles indicated must *not* be shaded?

A 1 **B** 2 **C** 3 **D** 4 **E** 5

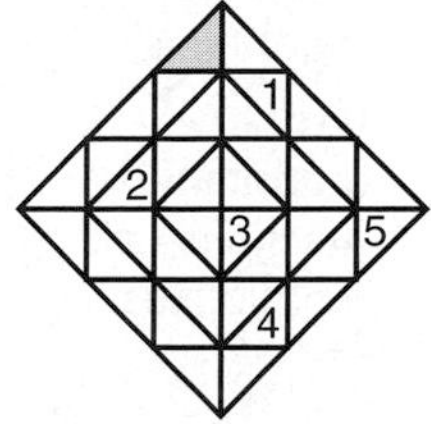

MID-CHAPTER QUIZ

Evaluate each expression. (Lesson 11-1)

1. $64^{\frac{1}{2}}$

2. $\left(\sqrt[3]{343}\right)^{-2}$

3. Simplify $\left(\dfrac{8x^3y^{-6}}{27w^6z^{-9}}\right)^{\frac{1}{3}}$

4. Express $\sqrt{a^6b^3}$ using rational exponents. (Lesson 11-1)

5. Express $(125a^2b^3)^{\frac{1}{3}}$ using radicals. (Lesson 11-1)

6. **Architecture** A soap bubble will enclose the maximum space with a minimum amount of surface material. Architects have used this principle to create buildings that enclose a great amount of space with a small amount of building material. If a soap bubble has a surface area of A, then its volume V is given by the equation $V = 0.094\sqrt{A^3}$. Find the surface area of a bubble with a volume of 1.75×10^2 cubic millimeters. (Lesson 11-1)

7. **Demographics** In 1990, the population of Houston, Texas was 1,637,859. In 1998, the population was 1,786,691. Predict the population of Houston in 2014. (Lesson 11-2)

8. **Finance** Determine the amount of money in a money market account at an annual rate of 5.2% compounded quarterly if Mary invested \$3500 and left it in the account for $3\frac{1}{2}$ years. (Lesson 11-2)

9. **Forestry** The yield in millions of cubic feet y of trees per acre is given by $y = 6.7e^{\frac{-48.1}{t}}$ for a forest that is t years old.
   a. Find the yield after 15 years.
   b. Find the yield after 50 years.
   (Lesson 11-3)

10. **Biology** It has been observed that the rate of growth of a population of organisms will increase until the population is half its maximum and then the rate will decrease. If M is the maximum population and b and c are constants determined by the type of organism, the population n after t years is given by $n = \dfrac{M}{1 + be^{-ct}}$. A certain organism yields the values of $M = 200$, $b = 20$, and $c = 0.35$. What is the population after 2, 15, and 60 years? (Lesson 11-3)

11-4

Logarithmic Functions

OBJECTIVES

- Evaluate expressions involving logarithms.
- Solve equations and inequalities involving logarithms.
- Graph logarithmic functions and inequalities.

CHEMISTRY Radioactive materials decay, or change to a non-radioactive material, in a predictable way. Archeologists use the decaying property of Carbon-14 in dating artifacts like dinosaur bones. Geologists use Thorium-230 in determining the age of rock formations, and medical researchers conduct tests using Arsenic-74.

A radioactive material's half-life is the time it takes for half of a given amount of the material to decay and can range from less than a second to billions of years. The half-life for various elements is shown in the table at the right.

Element	Half-Life
Arsenic-74	17.5 days
Carbon-14	5730 years
Polonium-194	0.5 second
Thorium-230	80,000 years
Thorium-232	14 billion years
Thorium-234	25 days

How long would it take for 256,000 grams of Thorium-234 to decay to 1000 grams? Radioactive decay can be modeled by the equation $N = N_0\left(\frac{1}{2}\right)^t$ where N is the final amount of a substance, N_0 is the initial amount, and t represents the number of half-lifes. If you want to find the number of 25-day half-lifes that will pass, you would have to use an inverse function. *This problem will be solved in Example 4.*

Since the graphs of exponential functions pass the horizontal line test, their inverses are also functions. As with inverses of other functions, you can find the inverse of an exponential function by interchanging the x- and y- values in the ordered pairs of the function.

$f(x)$: $y = 5^x$		$f^{-1}(x)$: $x = 5^y$	
x	y	x	y
−3	0.008	0.008	−3
−2	0.004	0.004	−2
−1	0.2	0.2	−1
0	1	1	0
1	0	0	1
2	25	25	2
3	125	125	3

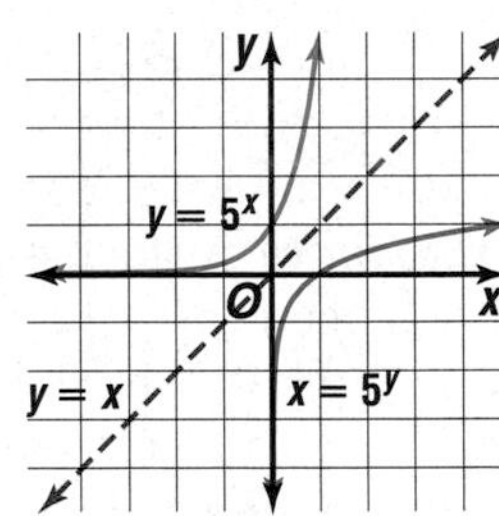

The function $f^{-1}(x)$ can be defined as $x = 5^y$. The ordered pairs can be used to sketch the graphs $y = 5^x$ and $x = 5^y$ on the same axes. *Note that they are reflections of each other over the line $y = x$.*

This example can be applied to a general statement that the inverse of $y = a^x$ is $x = a^y$. In the function $x = a^y$, y is called the **logarithm** of x. It is usually written as $y = \log_a x$ and is read "y equals the log, base a, of x." The function $y = \log_a x$ is called a **logarithmic function.**

Logarithmic Function	The logarithmic function $y = \log_a x$, where $a > 0$ and $a \neq 1$, is the inverse of the exponential function $y = a^x$. So, $y = \log_a x$ if and only if $x = a^y$.

Each logarithmic equation corresponds to an equivalent exponential equation. That is, the logarithmic equation $y = \log_a x$ is equivalent to $a^y = x$.

Since $2^5 = 32$, $y = 5$. Thus, $\log_2 32 = 5$.

Example 1 **Write each equation in exponential form.**

a. $\log_{125} 25 = \frac{2}{3}$

The base is 125, and the exponent is $\frac{2}{3}$.

$25 = 125^{\frac{2}{3}}$

b. $\log_8 2 = \frac{1}{3}$

The base is 8, and the exponent is $\frac{1}{3}$.

$2 = 8^{\frac{1}{3}}$

You can also write an exponential function as a logarithmic function.

Example 2 **Write each equation in logarithmic form.**

a. $4^3 = 64$

The base is 4, and the exponent or logarithm is 3.

$\log_4 64 = 3$

b. $3^{-3} = \frac{1}{27}$

The base is 3, and the exponent or logarithm is -3.

$\log_3 \frac{1}{27} = -3$

Using the fact that if $a^u = a^v$ then $u = v$, you can evaluate a logarithmic expression to determine its logarithm.

Example 3 **Evaluate the expression $\log_7 \frac{1}{49}$.**

Let $x = \log_7 \frac{1}{49}$.

$$x = \log_7 \frac{1}{49}$$

$7^x = \frac{1}{49}$ *Definition of logarithm.*

$7^x = (49)^{-1}$ $a^{-m} = \frac{1}{a^m}$

$7^x = (7^2)^{-1}$ $7^2 = 49$

$7^x = 7^{-2}$ $(a^m)^n = a^{mn}$

$x = -2$ *If* $a^u = a^v$ *then* $u = v$.

Example 4 

CHEMISTRY **Refer to the application at the beginning of the lesson. How long would it take for 256,000 grams of Thorium-234, with a half-life of 25 days, to decay to 1000 grams?**

$$N = N_0\left(\frac{1}{2}\right)^t \qquad N = N_0(1 + r)^t \text{ for } r = \frac{1}{2}$$

$$1000 = 256{,}000\left(\frac{1}{2}\right)^t \qquad N = 1000,\ N_0 = 256{,}000$$

$$\frac{1}{256} = \left(\frac{1}{2}\right)^t \qquad \textit{Divide each side by 256,000.}$$

$$\log_{\frac{1}{2}} \frac{1}{256} = t \qquad \textit{Write the equation in logarithmic form.}$$

$$\log_{\frac{1}{2}} \left(\frac{1}{2}\right)^8 = t \qquad 256 = 2^8$$

$$\log_{\frac{1}{2}} \left(\frac{1}{2}\right)^8 = t \qquad \left(\frac{1}{b^n}\right) = \left(\frac{1}{b}\right)^n$$

$$\left(\frac{1}{2}\right)^8 = \left(\frac{1}{2}\right)^t \qquad \textit{Definition of logarithm}$$

$$8 = t \qquad \text{It will take 8 half-lifes or 200 days.}$$

Since the logarithmic function and the exponential function are inverses of each other, both of their compositions yield the identity function. Let $f(x) = \log_a x$ and $g(x) = a^x$. For $f(x)$ and $g(x)$ to be inverses, it must be true that $f(g(x)) = x$ and $g(f(x)) = x$.

Look Back
Refer to Lesson 1-2 to review composition of functions.

$$f(g(x)) \stackrel{?}{=} x \qquad g(f(x)) \stackrel{?}{=} x.$$

$$f(a^x) \stackrel{?}{=} x \qquad g(\log_a x) \stackrel{?}{=} x$$

$$\log_a a^x \stackrel{?}{=} x \qquad a^{\log_a x} \stackrel{?}{=} x$$

$$x = x \qquad x = x$$

The properties of logarithms can be derived from the properties of exponents.

Properties of Logarithms

Suppose *m* and *n* are positive numbers, *b* is a positive number other than 1, and *p* is any real number. Then the following properties hold.

Property	Definition	Example
Product	$\log_b mn = \log_b m + \log_b n$	$\log_3 9x = \log_3 9 + \log_3 x$
Quotient	$\log_b \frac{m}{n} = \log_b m - \log_b n$	$\log_{\frac{1}{4}} \frac{4}{5} = \log_{\frac{1}{4}} 4 - \log_{\frac{1}{4}} 5$
Power	$\log_b m^p = p \cdot \log_b m$	$\log_2 8^x = x \cdot \log_2 8$
Equality	If $\log_b m = \log_b n$, then $m = n$.	$\log_8 (3x - 4) = \log_8 (5x + 2)$ so, $3x - 4 = 5x + 2$

Each of these properties can be verified using the properties of exponents. For example, suppose we want to prove the Product Property. Let $x = \log_b m$ and $y = \log_b n$. Then by definition $b^x = m$ and $b^y = n$.

You will be asked to prove other properties in Exercises 2 and 61.

$$\log_b mn = \log_b (b^x \cdot b^y) \qquad b^x = m,\ b^y = n$$

$$= \log_b (b^{x+y}) \qquad \textit{Product Property of Exponents}$$

$$= x + y \qquad \textit{Definition of logarithm}$$

$$= \log_b m + \log_b n \qquad \textit{Substitution}$$

Equations can be written involving logarithms. Use the properties of logarithms and the definition of logarithms to solve these equations.

Example 5 **Solve each equation.**

a. $\log_p 64^{\frac{1}{3}} = \frac{1}{2}$

$$\log_p 64^{\frac{1}{3}} = \frac{1}{2}$$

$$p^{\frac{1}{2}} = 64^{\frac{1}{3}} \quad \textit{Definition of logarithm.}$$

$$\sqrt{p} = \sqrt[3]{64} \quad b^{\frac{m}{n}} = \sqrt[n]{b^m}$$

$$\sqrt{p} = 4$$

$$(\sqrt{p})^2 = (4)^2 \quad \textit{Square each side.}$$

$$p = 16$$

b. $\log_4 (2x + 11) = \log_4 (5x - 4)$

$$\log_4 (2x + 11) = \log_4 (5x - 4)$$

$$2x + 11 = 5x - 4 \quad \textit{Power of Equality Property}$$

$$-3x = -15$$

$$x = 5$$

c. $\log_{11} x + \log_{11} (x + 1) = \log_{11} 6$

$$\log_{11} x + \log_{11} (x + 1) = \log_{11} 6$$

$$\log_{11} [x(x + 1)] = \log_{11} 6 \quad \textit{Product Property}$$

$$(x^2 + x) = 6 \quad \textit{Power of Equality Property}$$

$$x^2 + x - 6 = 0$$

$$(x - 2)(x + 3) = 0 \quad \textit{Factor.}$$

$$x - 2 = 0 \quad \text{or} \quad x + 3 = 0$$

$$x = 2 \qquad\qquad x = -3$$

By substituting $x = 2$ and $x = -3$ into the equation, we find that $x = -3$ is undefined for the equation $\log_{11} x + \log_{11} (x + 1) = \log_{11} 6$. When $x = -3$ we get an extraneous solution. So, $x = 2$ is the correct solution.

You can graph a logarithmic function by rewriting the logarithmic function as an exponential function and constructing a table of values.

Example 6 **Graph $y = \log_3 (x + 1)$.**

The graph of $y = \log_3 (x + 1)$ is a horizontal translation of the graph of $y = \log_3 x$.

The equation $y = \log_3 (x + 1)$ can be written as $3^y = x + 1$. Choose values for y and then find the corresponding values of x.

y	x + 1	x	(x, y)
−3	0.037	−0.963	(−0.963, −3)
−2	0.11	−0.89	(−0.89, −2)
−1	0.33	−0.67	(−0.67, −1)
0	1	0	(0, 0)
1	3	2	(2, 1)
2	9	8	(8, 2)
3	27	26	(26, 3)

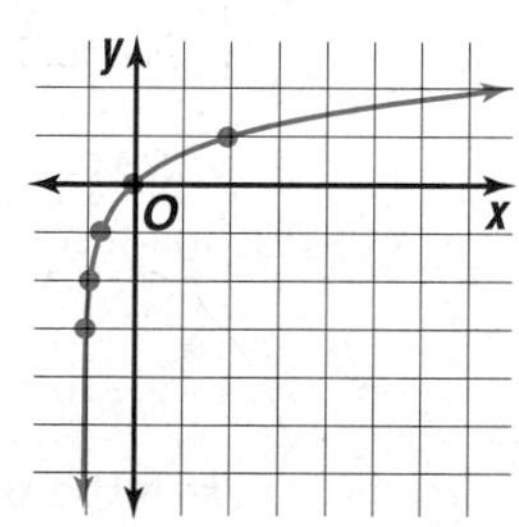

You can graph logarithmic inequalities using the same techniques as shown in Example 6. Choose a test point to determine which region to shade.

Example 7 **Graph $y \leq \log_5 x - 2$.**

The boundary for the inequality $y \leq \log_5 x - 2$ can be written as $y = \log_5 x - 2$. Rewrite this equation in exponential form.

$$y = \log_5 x - 2$$
$$y + 2 = \log_5 x$$
$$5^{y+2} = x$$

The graph of $y = \log_5 x - 2$ is a vertical translation of the graph of $y = \log_5 x$.

Use a table of values to graph the boundary.

y	y + 2	x	(x, y)
−5	−3	0.008	(0.008, −5)
−4	−2	0.25	(0.25, −4)
−3	−1	0.2	(0.2, −3)
−2	0	1	(1, −2)
−1	1	5	(5, −1)
0	2	25	(25, 0)
1	3	125	(125, 1)

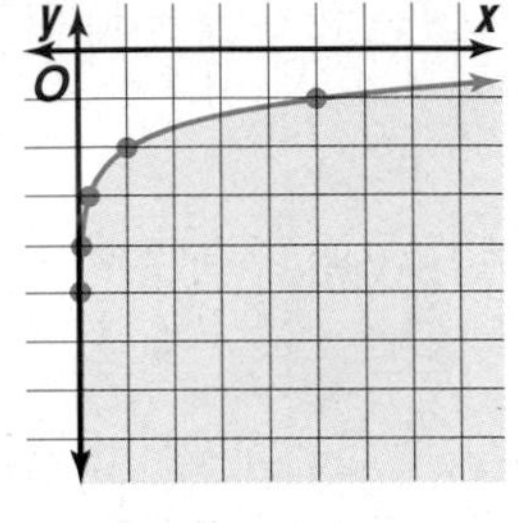

Test a point, for example (0, 0), to determine which region to shade.

$5^{y+2} \leq 0 \rightarrow 5^{0+2} \leq 0$ *False*

Shade the region that does not contain the point at (0, 0).

Remember that you must exclude values for which the log is undefined. For example, for $y \geq \log_5 x - 2$, the negative values for x must be omitted from the domain and no shading would occur in that area.

CHECK FOR UNDERSTANDING

Communicating Mathematics

Read and study the lesson to answer each question.

1. **Compare and contrast** the graphs of $y = 3^x$ and $y = \log_3 x$.
2. **Show** that the Power Property of logarithms is valid.
3. **State** the difference between the graph of $y = \log_5 x$ and $y = \log_{\frac{1}{5}} x$.
4. **You Decide** Courtney and Sean are discussing the expansion of $\log_b mn$. Courtney states that $\log_b mn$ can be written as $\log_b m \times \log_b n$. Sean states that $\log_b mn$ can be written as $\log_b m + \log_b n$. Who is correct? Explain your answer.
5. **Explain** the relationship between the decay formula $N = N_0(1 + r)^t$ and the formula for determining half-life, $N = N_0\left(\frac{1}{2}\right)^t$.

Guided Practice

Write each equation in exponential form.

6. $\log_9 27 = \frac{3}{2}$

7. $\log_{\frac{1}{25}} 5 = -\frac{1}{2}$

Write each equation in logarithmic form.

8. $7^{-6} = y$

9. $8^{-\frac{2}{3}} = \frac{1}{4}$

Evaluate each expression.

10. $\log_2 \frac{1}{16}$

11. $\log_{10} 0.01$

12. $\log_7 \frac{1}{343}$

Solve each equation.

13. $\log_2 x = 5$

14. $\log_7 n = \frac{2}{3} \log_7 8$

15. $\log_6 (4x + 4) = \log_6 64$

16. $2 \log_6 4 - \frac{1}{4} \log_6 16 = \log_6 x$

Graph each equation or inequality.

17. $y = \log_{\frac{1}{2}} x$

18. $y \geq \log_6 x$

19. **Biology** The generation time for bacteria is the time that it takes for the population to double. The generation time G can be found using experimental data and the formula $G = \frac{t}{3.3 \log_b f}$, where t is the time period, b is the number of bacteria at the beginning of the experiment, and f is the number of bacteria at the end of the experiment. The generation time for mycobacterium tuberculosis is 16 hours. How long will it take four of these bacteria to multiply into 1024 bacteria?

EXERCISES

Practice

Write each equation in exponential form.

20. $\log_{27} 3 = \frac{1}{3}$

21. $\log_{16} 4 = \frac{1}{2}$

22. $\log_7 \frac{1}{2401} = -4$

23. $\log_4 32 = \frac{5}{2}$

24. $\log_e 65.98 = x$

25. $\log_{\sqrt{6}} 36 = 4$

Write each equation in logarithmic form.

26. $81^{\frac{1}{2}} = 9$

27. $36^{\frac{3}{2}} = 216$

28. $\left(\frac{1}{8}\right)^{-3} = 512$

29. $6^{-2} = \frac{1}{36}$

30. $16^0 = 1$

31. $x^{1.238} = 14.36$

Evaluate each expression.

32. $\log_8 64$

33. $\log_{125} 5$

34. $\log_2 32$

35. $\log_4 128$

36. $\log_9 9^6$

37. $\log_{49} 343$

38. $\log_8 16$

39. $\log_{\sqrt{8}} 4096$

40. $10^{4 \log_{10} 2}$

Solve each equation.

41. $\log_x 49 = 2$

42. $\log_3 3x = \log_3 36$

43. $\log_6 x + \log_6 9 = \log_6 54$

44. $\log_8 48 - \log_8 w = \log_8 6$

45. $\log_6 216 = x$

46. $\log_5 0.04 = x$

47. $\log_{10} \sqrt[3]{10} = x$

48. $\log_{12} x = \frac{1}{2} \log_{12} 9 + \frac{1}{3} \log_{12} 27$

49. $\log_5(x + 4) + \log_5 8 = \log_5 64$

50. $\log_4(x - 3) + \log_4(x + 3) = 2$

51. $\frac{1}{2}(\log_7 x + \log_7 8) = \log_7 16$

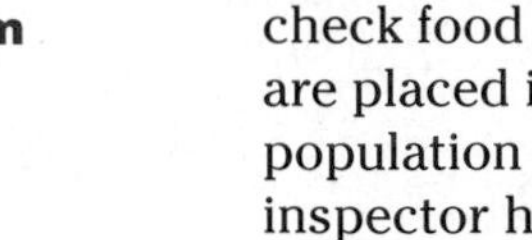

52. $2 \log_5(x - 2) = \log_5 36$

Graph each equation or inequality.

53. $y = \log_4 x$

54. $y = 3 \log_2 x$

55. $y = \log_5(x - 1)$

56. $y \leq \log_2 x$

57. $y \geq 2 \log_2 x$

58. $y > \log_{10}(x + 1)$

Applications and Problem Solving

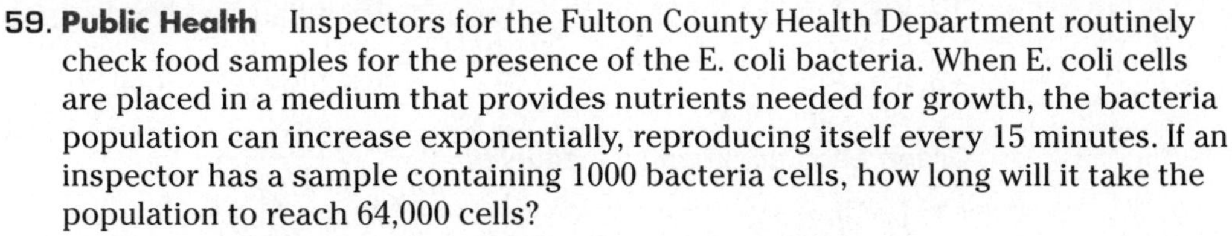

59. **Public Health** Inspectors for the Fulton County Health Department routinely check food samples for the presence of the E. coli bacteria. When E. coli cells are placed in a medium that provides nutrients needed for growth, the bacteria population can increase exponentially, reproducing itself every 15 minutes. If an inspector has a sample containing 1000 bacteria cells, how long will it take the population to reach 64,000 cells?

60. **Critical Thinking** Using the definition of a logarithm where $y = \log_a x$, explain why the base a cannot equal 1.

61. **Proof** Prove the quotient property of logarithms, $\log_b \frac{m}{n} = \log_b m - \log_b n$, using the definition of a logarithm.

62. **Finance** Latasha plans to invest \$2500 in an account that compounds quarterly, hoping to at least double her money in 10 years.
   a. Write an inequality that can be used to find the interest rate at which Latasha should invest her money.
   b. What is the lowest interest rate that will allow her to meet her goal?
   c. Suppose she only wants to invest her money for 7 years. What interest rate would allow her to double her money?

63. **Photography** The aperture setting of a camera, or *f*-stop, controls the amount of light exposure on film. Each step up in *f*-stop setting allows twice as much light exposure as the previous setting. The formula $n = \log_2 \frac{1}{p}$ where p is the fraction of sunlight, represents the change in the *f*-stop setting n to use in less light.

   a. A nature photographer sets her camera's *f*-stop at *f*/6.7 while taking outdoor pictures in direct sunlight. If the amount of sunlight on a cloudy day is $\frac{1}{4}$ as bright as direct sunlight, how many *f*-stop settings should she move to accommodate less light?
   b. If she moves down 3 *f*-stop settings from her original setting, is she allowing more or less light into the camera? What fraction of daylight is she accommodating?

64. **Critical Thinking** Show that $\log_a x = \frac{\log_b x}{\log_b a}$.

65. **Meteorology** Atmospheric pressure decreases as altitude above sea level increases. Atmospheric pressure P, measured in pounds per square inch, at altitude h miles can be represented by the logarithmic function $\log_{2.72} \frac{P}{14.7} = -0.02h$.
   a. Graph the function for atmospheric pressure.
   b. The elevation of Denver, Colorado, is about 1 mile. What is the atmospheric pressure in Denver?
   c. The lowest elevation on Earth is in the Mariana Trench in the Atlantic Ocean about 6.8 miles below sea level. If there were no water at this point, what would be the atmospheric pressure at this elevation?

66. **Radiation Safety** Radon is a naturally ocurring radioactive gas which can collect in poorly ventilated structures. Radon gas is formed by the decomposition of radium-226 which has a half-life of 1622 years. The half-life of radon is 3.82 days. Suppose a house basement contained 38 grams of radon gas when a family moved in. If the source of radium producing the radon gas is removed so that the radon gas eventually decays, how long will it take until there is only 6.8 grams of radon gas present?

Mixed Review

67. Find the value of $e^{4.243}$ to the nearest ten thousandth. *(Lesson 11-3)*

68. **Finance** What is the monthly principal and interest payment on a home mortgage of \$90,000 for 30 years at 11.5%? *(Lesson 11-2)*

69. Identify the conic section represented by $9x^2 - 18x + 4y^2 - 16y - 11 = 0$. Then write the equation in standard form and graph the equation. *(Lesson 10-6)*

70. **Engineering** A wheel in a motor is turning counterclockwise at 2 radians per second. There is a small hole in the wheel 3 centimeters from its center. Suppose a model of the wheel is drawn on a rectangular coordinate system with the wheel centered at the origin. If the hole has initial coordinates (3, 0), what are its coordinates after t seconds? *(Lesson 10-6)*

71. Find the lengths of the sides of a triangle whose vertices are $A(-1, 3)$, $B(-1, -3)$, and $C(3, 0)$. *(Lesson 10-1)*

72. Find the product $5\left(\cos \frac{3\pi}{4} + \boldsymbol{i} \sin \frac{3\pi}{4}\right) \cdot 2\left(\cos \frac{2\pi}{3} + \boldsymbol{i} \sin \frac{2\pi}{3}\right)$. Then express it in rectangular form. *(Lesson 9-7)*

73. **Electricity** A circuit has a current of $(3 - 4\boldsymbol{j})$ amps and an impedance of $(12 + 7\boldsymbol{j})$ ohms. Find the voltage of this circuit. *(Lesson 9-5)*

74. State whether $\overrightarrow{AB}$ and $\overrightarrow{CD}$ are *opposite, parallel,* or *neither of these* for $A(-2, 5)$, $B(3, -1)$, $C(2, 6)$, and $D(7, 0)$. *(Lesson 8-1)*

75. Find $\cos (A + B)$ if $\cos A = \frac{5}{13}$ and $\cos B = \frac{35}{37}$ and A and B are first quadrant angles. *(Lesson 7-3)*

76. **Weather** The maximum normal daily temperatures in each season for New Orleans, Louisiana, are given below.

Winter	Spring	Summer	Fall
64°	78°	90°	79°

Source: Rand McNally & Company

Write a sinusoidal function that models the temperatures, using $t = 1$ to represent winter. *(Lesson 6-6)*

77. Solve $\triangle ABC$ if $C = 105°18'$, $a = 6.11$, and $b = 5.84$. *(Lesson 5-8)*

78. **SAT/ACT Practice** If $PQRS$ is a parallelogram and MN is a segment, then x must equal

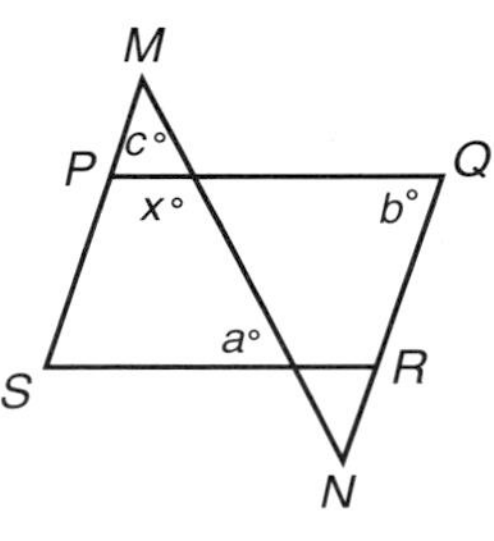

A $180 - b$

B $180 - c$

C $a + b$

D $a + c$

E $b + c$

Extra Practice See p. A46.

UNIT 4 Discrete Mathematics

Discrete mathematics is a branch of mathematics that deals with finite or discontinuous quantities. Usually discrete mathematics is defined in terms of its key topics. These include graphs, certain functions, logic, combinatorics, sequences, iteration, algorithms, recursion, matrices, and induction. Some of these topics have already been introduced in this book. For example, linear functions are continuous, while step functions are discrete. As you work through Unit 4, you will construct models, discover and use algorithms, and examine exciting new concepts as you solve real-world problems.

Unit 4 *inter*NET Projects

THE UNITED STATES CENSUS BUREAU

Did you know that the very first United States Census was conducted in 1790? The United States Census Bureau completed this task in less than nine months. This is remarkable considering they had to do so without benefit of cars, telephones, or computers! As the U.S. population grew, it took more and more time to complete the census. The 1880 census took seven years to finish. Luckily, advancements in technology have helped the people employed by the Census Bureau analyze the data and prepare reports in a reasonable amount of time. At the end of each chapter in Unit 4, you will be given tasks to explore the data collected by the Census Bureau.

CHAPTER 12 (page 833)

That's a lot of people! In addition to determining how many people are in the United States every ten years, the Census Bureau also attempts to estimate the population at any particular time and in the future. On their web site, you can see estimates of the U. S. and world population for the current day. How does the Census Bureau estimate population?
Math Connection: Use the Internet to find population data. Model the population of the U.S. by using an arithmetic and then a geometric sequence. Predict the population for a future date using your models.

CHAPTER 13 (page 885)

Radically random! During a census year, the Census Bureau collects many type of data about people. This information includes age, ethnic background, and income, to name just a few. What types of data about the people of the United States can be found using the Internet?
Math Connection: Use data from the Internet to find the probability that a randomly-selected person in the U.S. belongs to a particular age group.

CHAPTER 14 (page 937)

More and more models! Did you know that in 1999 a person was born in the United States about every eight seconds? In that same year, about one person died every 15 seconds, and one person migrated every 20 seconds. Birth, deaths, and other factors directly affect the population. In Chapter 12, you modeled the U.S. population using sequences. What other types of population models could you use to predict population growth?
Math Connection: Use data from the Internet to write and graph several functions representing the U.S. population growth. Predict the population for a future data using your models.

*inter*NET CONNECTION For more information on the Unit Project, visit: **www.amc.glencoe.com**

Chapter 12

SEQUENCES AND SERIES

CHAPTER OBJECTIVES

- **Identify and find *n*th terms of arithmetic, geometric, and infinite sequences.** *(Lessons 12-1, 12-2)*
- **Find sums of arithmetic, geometric, and infinite series.** *(Lessons 12-1, 12-2, 12-3)*
- **Determine whether a series is convergent or divergent.** *(Lesson 12-4)*
- **Use sigma notation.** *(Lesson 12-5)*
- **Use the Binomial Theorem to expand binomials.** *(Lesson 12-6)*
- **Evaluate expressions using exponential, trigonometric, and iterative series.** *(Lessons 12-7, 12-8)*
- **Use mathematical induction to prove the validity of mathematical statements.** *(Lesson 12-9)*

Arithmetic Sequences and Series

OBJECTIVES
- Find the *n*th term and arithmetic means of an arithmetic sequence.
- Find the sum of *n* terms of an arithmetic series.

REAL ESTATE Ofelia Gonzales sells houses in a new development. She makes a commission of \$3750 on the sale of her first house. To encourage aggressive selling, Ms. Gonzales' employer promises a \$500 increase in commission for each additional house sold. Thus, on the sale of her next house, she will earn \$4250 commission. How many houses will Ms. Gonzales have to sell for her total commission in one year to be at least \$65,000? *This problem will be solved in Example 6.*

The set of numbers representing the amount of money earned for each house sold is an example of a **sequence.** A sequence is a function whose domain is the set of natural numbers. The **terms** of a sequence are the range elements of the function. The first term of a sequence is denoted a_1, the second term is a_2, and so on up to the nth term a_n.

Look Back
Refer to Lesson 4-1 for more about natural numbers.

Symbol	a_1	a_2	a_3	a_4	a_5	a_6	a_7	a_8	a_9	a_{10}
Term	3	$2\frac{1}{2}$	2	$1\frac{1}{2}$	1	$\frac{1}{2}$	0	$-\frac{1}{2}$	-1	$-1\frac{1}{2}$

The sequence given in the table above is an example of an **arithmetic sequence.** The difference between successive terms of an arithmetic sequence is a constant called the **common difference,** denoted d. In the example above, $d = \frac{1}{2}$.

Arithmetic Sequence

An arithmetic sequence is a sequence in which each term after the first, a_1, is equal to the sum of the preceding term and the common difference, d. The terms of the sequence can be represented as follows.

$$a_1,\ a_1 + d,\ a_1 + 2d,\ \ldots$$

To find the next term in an arithmetic sequence, first find the common difference by subtracting any term from its succeeding term. Then add the common difference to the last term to find the next term in the sequence.

Example 1 **Find the next four terms in the arithmetic sequence $-5, -2, 1, \ldots$**

First, find the common difference.

$a_2 - a_1 = -2 - (-5)$ or 3
$a_3 - a_2 = 1 - (-2)$ or 3
Find the difference between pairs of consecutive terms to verify the common difference.

The common difference is 3.

Add 3 to the third term to get the fourth term, and so on.

$a_4 = 1 + 3$ or 4 $\quad a_5 = 4 + 3$ or 7 $\quad a_6 = 7 + 3$ or 10 $\quad a_7 = 10 + 3$ or 13

The next four terms are 4, 7, 10, and 13.

By definition, the nth term is also equal to $a_{n-1} + d$, where a_{n-1} is the $(n - 1)$th term. That is, $a_n = a_{n-1} + d$. This type of formula is called a **recursive formula.** This means that each succeeding term is formulated from one or more previous terms.

The nth term of an arithmetic sequence can also be found when only the first term and the common difference are known. Consider an arithmetic sequence in which $a = -3.7$ and $d = 2.9$. Notice the pattern in the way the terms are formed.

first term	a_1	a	-3.7
second term	a_2	$a + d$	$-3.7 + 1(2.9) = -0.8$
third term	a_3	$a + 2d$	$-3.7 + 2(2.9) = 2.1$
fourth term	a_4	$a + 3d$	$-3.7 + 3(2.9) = 5.0$
fifth term	a_5	$a + 4d$	$-3.7 + 4(2.9) = 7.9$
⋮	⋮	⋮	⋮
nth term	a_n	$a + (n - 1)d$	$-3.7 + (n - 1)2.9$

The nth Term of an Arithmetic Sequence

The nth term of an arithmetic sequence with first term a_1 and common difference d is given by $a_n = a_1 + (n - 1)d$.

Notice that the preceding formula has four variables: a_n, a_1, n, and d. If any three of these are known, the fourth can be found.

Examples

2 Find the 47th term in the arithmetic sequence −4, −1, 2, 5, ...

First, find the common difference.

$a_2 - a_1 = -1 - (-4)$ or 3 $\quad a_3 - a_2 = 2 - (-1)$ or 3 $\quad a_4 - a_3 = 5 - 2$ or 3

The common difference is 3.

Then use the formula for the nth term of an arithmetic sequence.

$$a_n = a_1 + (n - 1)d$$
$$a_{47} = -4 + (47 - 1)3 \quad n = 47,\ a_1 = -4, \text{ and } d = 3$$
$$a_{47} = 134$$

3 Find the first term in the arithmetic sequence for which $a_{19} = 42$ and $d = -\frac{2}{3}$.

$$a_n = a_1 + (n - 1)d$$
$$a_{19} = a_1 + (19 - 1)\left(-\frac{2}{3}\right) \quad n = 19 \text{ and } d = -\frac{2}{3}$$
$$42 = a_1 + (-12) \quad a_{19} = 42$$
$$a_1 = 54$$

Sometimes you may know two terms of an arithmetic sequence that are not in consecutive order. The terms between any two nonconsecutive terms of an arithmetic sequence are called **arithmetic means.** In the sequence below, 38 and 49 are the arithmetic means between 27 and 60.

5, 16, 27, 38, 49, 60

Example 4 **Write an arithmetic sequence that has five arithmetic means between 4.9 and 2.5.**

The sequence will have the form 4.9, ___?___, ___?___, ___?___, ___?___, ___?___, 2.5. Note that 2.5 is the 7th term of the sequence or a_7.

First, find the common difference, using $n = 7$, $a_7 = 2.5$, and $a_1 = 4.9$

$$a_n = a_1 + (n - 1)d$$
$$2.5 = 4.9 + (7 - 1)d$$
$$2.5 = 4.9 + 6d$$
$$d = -0.4$$

Then determine the arithmetic means.

$$a_2 = 4.9 + (-0.4) \text{ or } 4.5$$
$$a_3 = 4.5 + (-0.4) \text{ or } 4.1$$
$$a_4 = 4.1 + (-0.4) \text{ or } 3.7$$
$$a_5 = 3.7 + (-0.4) \text{ or } 3.3$$
$$a_6 = 3.3 + (-0.4) \text{ or } 2.9$$

The sequence is 4.9, 4.5, 4.1, 3.7, 3.3, 2.9, 2.5.

An indicated sum is 1 + 2 + 3 + 4. The sum 1 + 2 + 3 + 4 is 10.

An **arithmetic series** is the indicated sum of the terms of an arithmetic sequence. The lists below show some examples of arithmetic sequences and their corresponding arithmetic series.

Arithmetic Sequence	Arithmetic Series
$-9, -3, 3, 9, 15$	$-9 + (-3) + 3 + 9 + 15$
$3, \frac{5}{2}, 2, \frac{3}{2}, 1, \frac{1}{2}$	$3 + \frac{5}{2} + 2 + \frac{3}{2} + 1 + \frac{1}{2}$
$a_1, a_2, a_3, a_4, \ldots, a_n$	$a_1 + a_2 + a_3 + a_4 + \cdots + a_n$

The symbol S_n, called the ***n*th partial sum**, is used to represent the sum of the first n terms of a series. To develop a formula for S_n for a finite arithmetic series, a series can be written in two ways and added term by term, as shown below. The second equation for S_n given below is obtained by reversing the order of the terms in the series.

$$S_n = a_1 + (a_1 + d) + (a_1 + 2d) + \cdots + (a_n - 2d) + (a_n - d) + a_n$$
$$+ S_n = a_n + (a_n - d) + (a_n - 2d) + \cdots + (a_1 + 2d) + (a_1 + d) + a_1$$
$$2S_n = (a_1 + a_n) + (a_1 + a_n) + (a_1 + a_n) + \cdots + (a_1 + a_n) + (a_1 + a_n) + (a_1 + a_n)$$
$$2S_n = n(a_1 + a_n)$$ *There are n terms in the series, all of which are $(a_1 + a_n)$.*

Therefore, $S_n = \frac{n}{2}(a_1 + a_n)$.

Sum of a Finite Arithmetic Series

The sum of the first n terms of an arithmetic series is given by $S_n = \frac{n}{2}(a_1 + a_n)$.

Example 5 **Find the sum of the first 60 terms in the arithmetic series 9 + 14 + 19 + ⋯ + 304.**

$$S_n = \frac{n}{2}(a_1 + a_n)$$
$$S_{60} = \frac{60}{2}(9 + 304)$$ *$n = 60$, $a_1 = 9$, $a_{60} = 304$*
$$= 9390$$

When the value of the last term, a_n, is not known, you can still determine the sum of the series. Using the formula for the nth term of an arithmetic sequence, you can derive another formula for the sum of a finite arithmetic series.

$$S_n = \frac{n}{2}(a_1 + a_n)$$

$$S_n = \frac{n}{2}[a_1 + (a_1 + (n - 1)d)] \quad a_n = a_1 + (n - 1)d$$

$$S_n = \frac{n}{2}[2a_1 + (n - 1)d]$$

Example 6 **REAL ESTATE** **Refer to the application at the beginning of the lesson. How many houses will Ms. Gonzales have to sell for her total commission in one year to be at least $65,000?**

Let S_n = the amount of her desired commission, $65,000.
Let a_1 = the first commission, $3750.
In this example, $d = 500$.

We want to find n, the number of houses that Ms. Gonzales has to sell to have a total commission greater than or equal to $65,000.

$$S_n = \frac{n}{2}[2a_1 + (n - 1)d]$$

$$65{,}000 = \frac{n}{2}[2(3750) + (n - 1)(500)] \quad S_n = 65{,}000,\ a_1 = 3750$$

$$130{,}000 = n(7500 + 500n - 500) \quad \text{Multiply each side by 2.}$$

$$130{,}000 = 7000n + 500n^2 \quad \text{Simplify.}$$

$$0 = 500n^2 + 7000n - 130{,}000$$

$$0 = 5n^2 + 70n - 1300 \quad \text{Divide each side by 100.}$$

$$n = \frac{-70 \pm \sqrt{70^2 - 4(5)(-1300)}}{2(5)} \quad \text{Use the Quadratic Formula.}$$

$$n = \frac{-70 \pm \sqrt{30{,}900}}{10}$$

$$n \approx 10.58 \text{ and } -24.58 \quad -24.58 \text{ is not a possible answer.}$$

Ms. Gonzales must sell 11 or more houses for her total commission to be at least $65,000.

CHECK FOR UNDERSTANDING

Communicating Mathematics

Read and study the lesson to answer each question.

1. **Write** the first five terms of the sequence defined by $a_n = 6 - 4n$. Is this an arithmetic sequence? Explain.
2. Consider the arithmetic sequence defined by $a_n = \frac{5 - 2n}{2}$.
   a. **Graph** the first five terms of the sequence. Let n be the x-coordinate and a_n be the y-coordinate, and connect the points.
   b. **Describe** the graph found in part **a**.
   c. **Find** the common difference of the sequence and determine its relationship to the graph found in part **a**.

3. Refer to Example 6.

   a. **Explain** why −24.58 is *not* a possible answer.

   b. **Determine** how much money Ms. Gonzales will make if she sells 10 houses.

4. **Describe** the common difference for an arithmetic sequence in which the terms are decreasing.

5. **You Decide** Ms. Brooks defined two sequences, $a_n = (-1)^n$ and $b_n = (-2)^n$, for her class. She asked the class to determine if they were arithmetic sequences. Latonya said the second was an arithmetic sequence and that the first was not. Diana thought the reverse was true. Who is correct? Explain.

Guided Practice

Find the next four terms in each arithmetic sequence.

6. 6, 11, 16, ...
7. −15, −7, 1, ...
8. $a - 6, a - 2, a + 2, \ldots$

For Exercises 9-15, assume that each sequence or series is arithmetic.

9. Find the 17th term in the sequence for which $a_1 = 10$ and $d = -3$.
10. Find n for the sequence for which $a_n = 37$, $a_1 = -13$, and $d = 5$.
11. What is the first term in the sequence for which $d = -2$ and $a_7 = 3$?
12. Find d for the sequence for which $a_1 = 100$ and $a_{12} = 34$.
13. Write a sequence that has two arithmetic means between 9 and 24.
14. What is the sum of the first 35 terms in the series $7 + 9 + 11 + \cdots$?
15. Find n for a series for which $a_1 = 30$, $d = -4$, and $S_n = -210$.

16. **Theater Design** The right side of the orchestra section of the Nederlander Theater in New York City has 19 rows, and the last row has 27 seats. The numbers of seats in each row increase by 1 as you move toward the back of the section. How many seats are in this section of the theater?

EXERCISES

Practice

Find the next four terms in each arithmetic sequence.

17. 5, −1, −7, ...
18. −18, −7, 4, ...
19. 3, 4.5, 6, ...
20. 5.6, 3.8, 2, ...
21. $b, b + 4, b + 8, \ldots$
22. $-x, 0, x, \ldots$
23. $5n, -n, -7n, \ldots$
24. $5 + k, 5, 5 - k, \ldots$
25. $2a - 5, 2a + 2, 2a + 9, \ldots$

26. Determine the common difference and find the next three terms of the arithmetic sequence $3 + \sqrt{7}, 5, 7 - \sqrt{7}, \ldots$.

For Exercises 27-34, assume that each sequence or series is arithmetic.

27. Find the 25th term in the sequence for which $a_1 = 8$ and $d = 3$.
28. Find the 18th term in the sequence for which $a_1 = 1.4$ and $d = 0.5$.
29. Find n for the sequence for which $a_n = -41$, $a_1 = 19$, and $d = -5$.
30. Find n for the sequence for which $a_n = 138$, $a_1 = -2$, and $d = 7$.
31. What is the first term in the sequence for which $d = -3$, and $a_{15} = 38$?
32. What is the first term in the sequence for which $d = \frac{1}{3}$ and $a_7 = 10\frac{2}{3}$?
33. Find d for the sequence in which $a_1 = 6$ and $a_{14} = 58$.
34. Find d for the sequence in which $a_1 = 8$ and $a_{11} = 26$.

For Exercises 35-49, assume that each sequence or series is arithmetic.

35. What is the eighth term in the sequence $-4 + \sqrt{5}, -1 + \sqrt{5}, 2 + \sqrt{5}, \ldots$?

36. What is the twelfth term in the sequence $5 - \boldsymbol{i}, 6, 7 + \boldsymbol{i}, \ldots$?

37. Find the 33rd term in the sequence 12.2, 10.5, 8.8, ...

38. Find the 79th term in the sequence $-7, -4, -1, \ldots$.

39. Write a sequence that has one arithmetic mean between 12 and 21.

40. Write a sequence that has two arithmetic means between -5 and 4.

41. Write a sequence that has two arithmetic means between $\sqrt{3}$ and 12.

42. Write a sequence that has three arithmetic means between 2 and 5.

43. Find the sum of the first 11 terms in the series $\frac{3}{2} + 1 + \frac{1}{2} + \ldots$.

44. Find the sum of the first 100 terms in the series $-5 - 4.8 - 4.6 - \ldots$.

45. Find the sum of the first 26 terms in the series $-19 - 13 - 7 - \ldots$.

46. Find n for a series for which $a_1 = -7$, $d = 1.5$, and $S_n = -14$.

47. Find n for a series for which $a_1 = -3$, $d = 2.5$, and $S_n = 31.5$.

48. Write an expression for the nth term of the sequence 5, 7, 9, ...

49. Write an expression for the nth term of the sequence $6, -2, -10, \ldots$.

Applications and Problem Solving

Real World Application

50. **Keyboarding** Antonio has found that he can input statistical data into his computer at the rate of 2 data items faster each half hour he works. One Monday, he starts work at 9:00 A.M., inputting at a rate of 3 data items per minute. At what rate will Antonio be inputting data into the computer by lunchtime (noon)?

51. **Critical Thinking** Show that if x, y, z, and w are the first four terms of an arithmetic sequence, then $x + w - y = z$.

52. **Construction** The Arroyos are planning to build a brick patio that approximates the shape of a trapezoid. The shorter base of the trapezoid needs to start with a row of 5 bricks, and each row must increase by 2 bricks on each side until there are 25 rows. How many bricks do the Arroyos need to buy?

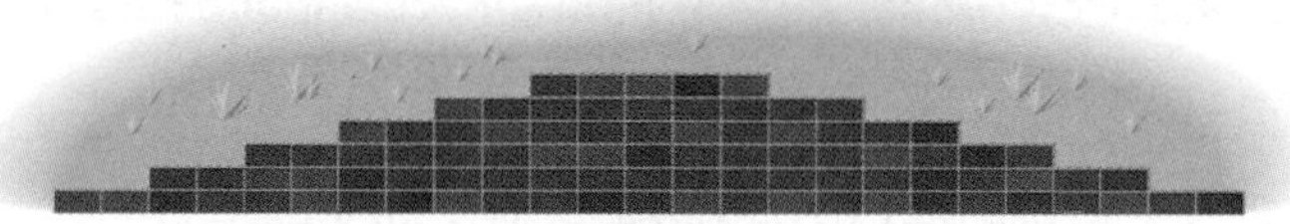

53. **Critical Thinking** The measures of the angles of a convex polygon form an arithmetic sequence. The least measurement in the sequence is 85°. The greatest measurement is 215°. Find the number of sides in this polygon.

54. **Geometry** The sum of the interior angles of a triangle is 180°.
    a. What are the sums of the interior angles of polygons with 4, 5, 6, and 7 sides?
    b. Show that these sums (beginning with the triangle) form an arithmetic sequence.
    c. Find the sum of the interior angles of a 35-sided polygon.

55. Critical Thinking Consider the sequence of odd natural numbers.

a. What is S_5?

b. What is S_{10}?

c. Make a conjecture as to the pattern that emerges concerning the sum. Write an algebraic proof verifying your conjecture.

56. Sports At the 1998 Winter X-Games held in Crested Butte, Colorado, Jennie Waara, from Sweden, won the women's snowboarding slope-style competition. Suppose that in one of the qualifying races, Ms. Waara traveled 5 feet in the first second, and the distance she traveled increased by 7 feet each subsequent second. If Ms. Waara reached the finish line in 15 seconds, how far did she travel?

57. Entertainment A radio station advertises a contest with ten cash prizes totaling \$5510. There is to be a \$100 difference between each successive prize. Find the amounts of the least and greatest prizes the radio station will award.

58. Critical Thinking Some sequences involve a pattern but are not arithmetic. Find the sum of the first 72 terms in the sequence 6, 8, 2, …, where $a_n = a_{n-1} - a_{n-2}$.

Mixed Review

59. Personal Finance If Parker Hamilton invests \$100 at 7% compounded continuously, how much will he have at the end of 15 years? *(Lesson 11-3)*

60. Find the coordinates of the center, foci, and vertices of the ellipse whose equation is $4x^2 + 25y^2 + 250y + 525 = 0$. Then graph the equation. *(Lesson 10-3)*

61. Find $6\left(\cos \frac{5\pi}{8} + \boldsymbol{i} \sin \frac{5\pi}{8}\right) \div 12\left(\cos \frac{\pi}{2} + \boldsymbol{i} \sin \frac{\pi}{2}\right)$. Then express the quotient in rectangular form. *(Lesson 9-7)*

62. Find the inner product of $\vec{\mathbf{u}}$ and $\vec{\mathbf{v}}$ if $\vec{\mathbf{u}} = \langle 2, -1, 3\rangle$ and $\vec{\mathbf{v}} = \langle 5, 3, 0\rangle$. *(Lesson 8-4)*

63. Write the standard form of the equation of a line for which the length of the normal is 5 units and the normal makes an angle of 30° with the positive x-axis. *(Lesson 7-6)*

64. Graph $y = \sec 2\theta - 3$. *(Lesson 6-7)*

65. Solve triangle ABC if $B = 19°32'$ and $c = 4.5$. Round angle measures to the nearest minute and side measures to the nearest tenth. *(Lesson 5-5)*

B
c
a
A
b
C

66. Find the discriminant of $4p^2 - 3p + 2 = 0$. Describe the nature of its roots. *(Lesson 4-2)*

67. Determine the slant asymptote of $f(x) = \frac{x^2 - 4x + 2}{x - 3}$. *(Lesson 3-7)*

68. Triangle ABC is represented by the matrix $\begin{bmatrix} -2 & 0 & 1 \\ 1 & 3 & -4 \end{bmatrix}$. Find the image of the triangle after a rotation of 270° counterclockwise about the origin. *(Lesson 2-4)*

69. SAT/ACT Practice If $a - 4b = 15$ and $4a - b = 15$, then $a - b = ?$

A 3 **B** 4 **C** 6 **D** 15 **E** 30

Extra Practice See p. A48.

Geometric Sequences and Series

OBJECTIVES

- Find the nth term and geometric means of a geometric sequence.
- Find the sum of n terms of a geometric series.

ACCOUNTING Bertha Blackwell is an accountant for a small company. On January 1, 1996, the company purchased \$50,000 worth of office copiers. Since this equipment is a company asset, Ms. Blackwell needs to determine how much the copiers are presently worth. She estimates that copiers depreciate at a rate of 45% per year. What value should Ms. Blackwell assign the copiers on her 2001 year-end accounting report? *This problem will be solved in Example 3.*

The following sequence is an example of a **geometric sequence.**

10, 2, 0.4, 0.08, 0.016, ... *Can you find the next term?*

The ratio of successive terms in a geometric sequence is a constant called the **common ratio,** denoted r.

Geometric Sequence

A geometric sequence is a sequence in which each term after the first, a_1, is the product of the preceding term and the common ratio, r. The terms of the sequence can be represented as follows, where a_1 is nonzero and r is not equal to 1 or 0.

$$a_1, a_1r, a_1r^2, \ldots.$$

You can find the next term in a geometric sequence as follows.

- First divide any term by the preceding term to find the common ratio.
- Then multiply the last term by the common ratio to find the next term in the sequence.

Example 1 **Determine the common ratio and find the next three terms in each sequence.**

a. $\mathbf{1, -\frac{1}{2}, \frac{1}{4}, \ldots}$

First, find the common ratio.

$a_2 \div a_1 = -\frac{1}{2} \div 1$ or $-\frac{1}{2}$ $\qquad$ $a_3 \div a_2 = \frac{1}{4} \div \left(-\frac{1}{2}\right)$ or $-\frac{1}{2}$

The common ratio is $-\frac{1}{2}$.

Multiply the third term by $-\frac{1}{2}$ to get the fourth term, and so on.

$a_4 = \frac{1}{4} \cdot \left(-\frac{1}{2}\right)$ or $-\frac{1}{8}$ $\qquad$ $a_5 = -\frac{1}{8} \cdot \left(-\frac{1}{2}\right)$ or $\frac{1}{16}$ $\qquad$ $a_6 = \frac{1}{16} \cdot \left(-\frac{1}{2}\right)$ or $-\frac{1}{32}$

The next three terms are $-\frac{1}{8}, \frac{1}{16}$, and $-\frac{1}{32}$.

b. $r - 1, -3r + 3, 9r - 9, \ldots$

First, find the common ratio.

$a_2 \div a_1 = \frac{-3r + 3}{r - 1}$

$a_2 \div a_1 = \frac{-3(r - 1)}{r - 1}$ *Factor.*

$a_2 \div a_1 = -3$ *Simplify.*

$a_3 \div a_2 = \frac{9r - 9}{-3r + 3}$

$a_3 \div a_2 = \frac{9(r - 1)}{-3(r - 1)}$ *Factor.*

$a_3 \div a_2 = -3$ *Simplify.*

The common ratio is -3.

Multiply the third term by -3 to get the fourth term, and so on.

$a_4 = -3(9r - 9)$ or $-27r + 27$ $a_5 = -3(-27r + 27)$ or $81r - 81$

$a_6 = -3(81r - 81)$ or $-243r + 243$

The next three terms are $-27r + 27$, $81r - 81$, and $-243r + 243$.

As with arithmetic sequences, geometric sequences can also be defined recursively. By definition, the nth term is also equal to $a_{n-1}r$, where a_{n-1} is the $(n - 1)$th term. That is, $a_n = a_{n-1}r$.

Since successive terms of a geometric sequence can be expressed as the product of the common ratio and the previous term, it follows that each term can be expressed as the product of a_1 and a power of r. The terms of a geometric sequence for which $a_1 = -5$ and $r = 7$ can be represented as follows.

first term	a_1	a_1	-5
second term	a_2	a_1r	$-5 \cdot 7^1 = -35$
third term	a_3	a_1r^2	$-5 \cdot 7^2 = -245$
fourth term	a_4	a_1r^3	$-5 \cdot 7^3 = -1715$
fifth term	a_5	a_1r^4	$-5 \cdot 7^4 = -12{,}005$
⋮	⋮	⋮	⋮
nth term	a_n	ar^{n-1}	$-5 \cdot 7^{n-1}$

The nth Term of a Geometric Sequence

The nth term of a geometric sequence with first term a_1 and common ratio r is given by $a_n = a_1r^{n-1}$.

Example 2 **Find an approximation for the 23rd term in the sequence 256, −179.2, 125.44, ...**

First, find the common ratio.

$a_2 \div a_1 = -179.2 \div 256$ or -0.7 $a_3 \div a_2 = 125.44 \div (-179.2)$ or -0.7

The common ratio is -0.7.

Then, use the formula for the nth term of a geometric sequence.

$a_n = a_1r^{n-1}$

$a_{23} = 256(-0.7)^{23-1}$ *$n = 23$, $a_1 = 256$, $r = -0.7$*

$a_{23} \approx 0.1000914188$ *Use a calculator.*

The 23rd term is about 0.1.

Geometric sequences can represent growth or decay.

- For a common ratio greater than 1, a sequence may model growth. Applications include compound interest, appreciation of property, and population growth.
- For a positive common ratio less than 1, a sequence may model decay. Applications include some radioactive behavior and depreciation.

Example 3 **ACCOUNTING** **Refer to the application at the beginning of the lesson. Compute the value of the copiers at the end of the year 2001.**

Since the copiers were purchased at the beginning of the first year, the original purchase price of the copiers represents a_1. If the copiers depreciate at a rate of 45% per year, then they retain 100 − 45 or 55% of their value each year.

Use the formula for the nth term of a geometric sequence to find the value of the copiers six years later or a_7.

$a_n = a_1 r^{n-1}$

$a_7 = 50{,}000 \cdot (0.55)^{7-1}$ *$a_1 = 50{,}000$, $r = 0.55$, $n = 7$*

$a_7 \approx 1384.032031$ *Use a calculator.*

Ms. Blackwell should list the value of the copiers on her report as $1384.03.

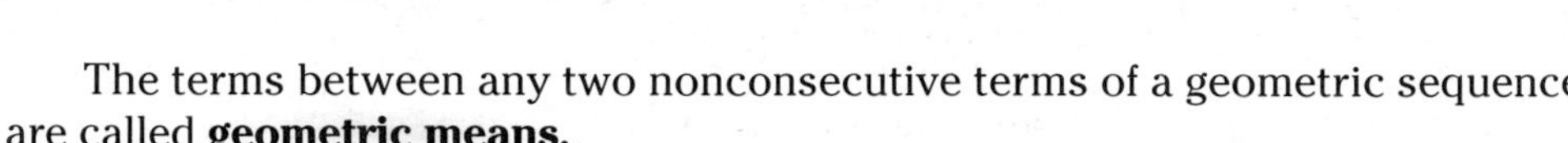

The terms between any two nonconsecutive terms of a geometric sequence are called **geometric means.**

Example 4 **Write a sequence that has two geometric means between 48 and −750.**

This sequence will have the form 48, __?__, __?__, −750.

First, find the common ratio.

$a_n = a_1 r^{n-1}$

$a_4 = a_1 r^3$ *Since there will be four terms in the sequence, $n = 4$.*

$-750 = 48r^3$ *$a_4 = -750$ and $a_1 = 48$*

$\frac{-125}{8} = r^3$ *Divide each side by 48 and simplify.*

$\sqrt[3]{-\frac{125}{8}} = r$ *Take the cube root of each side.*

$-2.5 = r$

Then, determine the geometric sequence.

$a_2 = 48(-2.5)$ or -120 $a_3 = -120(-2.5)$ or 300

The sequence is 48, −120, 300, −750.

A **geometric series** is the indicated sum of the terms of a geometric sequence. The lists below show some examples of geometric sequences and their corresponding series.

Geometric Sequence	Geometric Series
3, 9, 27, 81, 243	3 + 9 + 27 + 81 + 243
$16, 4, 1, \frac{1}{4}, \frac{1}{16}$	$16 + 4 + 1 + \frac{1}{4} + \frac{1}{16}$
$a_1, a_2, a_3, a_4, \ldots, a_n$	$a_1 + a_2 + a_3 + a_4 + \cdots + a_n$

To develop a formula for the sum of a finite geometric sequence, S_n, write an expression for S_n and for rS_n, as shown below. Then subtract rS_n from S_n and solve for S_n.

$$\begin{aligned} S_n &= a_1 + a_1r + a_1r^2 + \cdots + a_1r^{n-2} + a_1r^{n-1} \\ rS_n &= \qquad a_1r + a_1r^2 + \cdots + a_1r^{n-2} + a_1r^{n-1} + a_1r^n \\ \hline S_n - rS_n &= a_1 - a_1r^n \quad \textit{Subtract.} \\ S_n(1 - r) &= a_1 - a_1r^n \quad \textit{Factor.} \\ S_n &= \frac{a_1 - a_1r^n}{1 - r} \quad \textit{Divide each side by } 1 - r,\ r \neq 1. \end{aligned}$$

Sum of a Finite Geometric Series

The sum of the first n terms of a finite geometric series is given by $S_n = \frac{a_1 - a_1r^n}{1 - r}$.

Example 5 **Find the sum of the first ten terms of the geometric series $16 - 48 + 144 - 432 + \cdots$.**

The formula for the sum of a geometric series can also be written as $S_n = a_1 \frac{1 - r^n}{1 - r}$.

First, find the common ratio.

$a_2 \div a_1 = -48 \div 16$ or -3 $\qquad$ $a_4 \div a_3 = -432 \div 144$ or -3

The common ratio is -3.

$$S_n = \frac{a_1 - a_1r^n}{1 - r}$$

$$S_{10} = \frac{16 - 16(-3)^{10}}{1 - (-3)} \quad n = 10,\ a_1 = 16,\ r = -3.$$

$$S_{10} = -236{,}192 \quad \textit{Use a calculator.}$$

The sum of the first ten terms of the series is $-236{,}192$.

Banks and other financial institutions use compound interest to determine earnings in accounts or how much to charge for loans. The formula for compound interest is $A = P\left(1 + \frac{r}{n}\right)^{tn}$, where

A = the account balance,
P = the initial deposit or amount of money borrowed,
r = the annual percentage rate (APR),
n = the number of compounding periods per year, and
t = the time in years.

Suppose at the beginning of each quarter you deposit \$25 in a savings account that pays an APR of 2% compounded quarterly. Most banks post the interest for each quarter on the last day of the quarter. The chart below lists the additions to the account balance as a result of each successive deposit through the rest of the year. Note that $1 + \frac{r}{n} = 1 + \frac{0.02}{4}$ or 1.005.

Date of Deposit	$A = P\left(1 + \frac{r}{n}\right)^{tn}$	1st Year Additions (to the nearest penny)
January 1	\$25 $(1.005)^4$	\$25.50
April 1	\$25 $(1.005)^3$	\$25.38
July 1	\$25 $(1.005)^2$	\$25.25
October 1	\$25 $(1.005)^1$	\$25.13
Account balance at the end of one year		\$101.26

The chart shows that the first deposit will gain interest through all four compounding periods while the second will earn interest through only three compounding periods. The third and last deposits will earn interest through two and one compounding periods, respectively. The sum of these amounts, \$101.26, is the balance of the account at the end of one year. This sum also represents a finite geometric series where $a_1 = 25.13$, $r = 1.005$, and $n = 4$.

$$25.13 + 25.13(1.005) + 25.13(1.005)^2 + 25.13(1.005)^3$$

Example 6

INVESTMENTS Hiroshi wants to begin saving money for college. He decides to deposit \$500 at the beginning of each quarter (January 1, April 1, July 1, and October 1) in a savings account that pays an APR of 6% compounded quarterly. The interest for each quarter is posted on the last day of the quarter. Determine Hiroshi's account balance at the end of one year.

The interest is compounded each quarter. So $n = 4$ and the interest rate per period is 6% ÷ 4 or 1.5%. The common ratio r for the geometric series is then $1 + 0.015$, or 1.015.

The first term a_1 in this series is the account balance at the end of the first quarter. Thus, $a_1 = 500(1.015)$ or 507.5.

Apply the formula for the sum of a geometric series.

$$S_n = \frac{a_1 - a_1 r^n}{1 - r}$$

$$S_4 = \frac{507.5 - 507.5(1.015)^4}{1 - 1.015} \quad n = 4,\ r = 1.015$$

$$S_4 \approx 2076.13$$

Hiroshi's account balance at the end of one year is \$2076.13.

CHECK FOR UNDERSTANDING

Communicating Mathematics

Read and study the lesson to answer each question.

1. **Compare and contrast** arithmetic and geometric sequences.
2. **Show** that the sequence defined by $a_n = (-3)^{n+1}$ is a geometric sequence.
3. **Explain** why the first term in a geometric sequence must be nonzero.
4. **Find a counterexample** for the statement "The sum of a geometric series cannot be less than its first term."
5. **Determine** whether the given terms form a finite geometric sequence. Write *yes* or *no* and then explain your reasoning.
   a. 3, 6, 18 b. $\sqrt{3}, 3, \sqrt{27}$ c. $x^{-2}, x^{-1}, 1 ; x \neq 1$
6. Refer to Example 3.
   a. **Make a table** to represent the situation. In the first row, put the number of years, and in the second row, put the value of the computers.
   b. **Graph** the numbers in the table. Let years be the x-coordinate, let value be the y-coordinate, and connect the points.
   c. **Describe** the graph found in part **b**.

Guided Practice

Determine the common ratio and find the next three terms of each geometric sequence.

7. $\frac{2}{3}$, 4, 24, ...
8. 2, 3, $\frac{9}{2}$, ...
9. 1.8, −7.2, 28.8, ...

For Exercises 10–14, assume that each sequence or series is geometric.

10. Find the seventh term of the sequence 7, 2.1, 0.63, ...
11. If $r = 2$ and $a_5 = 24$, find the first term of the sequence.
12. Find the first three terms of the sequence for which $a_4 = 2.5$ and $r = 2$.
13. Write a sequence that has two geometric means between 1 and 27.
14. Find the sum of the first nine terms of the series $0.5 - 1 + 2 - \cdots$.

15. **Investment** Mika Rockwell invests in classic cars. He recently bought a 1978 convertible valued at $20,000. The value of the car is predicted to appreciate at a rate of 3.5% per year. Find the value of the car after 10, 20, and 40 years, assuming that the rate of appreciation remains constant.

EXERCISES

Practice

Determine the common ratio and find the next three terms of each geometric sequence.

16. 10, 2, 0.4, ...
17. 8, −20, 50, ...
18. $\frac{2}{9}, \frac{2}{3}$, 2, ...
19. $\frac{3}{4}, \frac{3}{10}, \frac{3}{25}$, ...
20. −7, 3.5, −1.75, ...
21. $3\sqrt{2}, 6, 6\sqrt{2}$, ...
22. $9, 3\sqrt{3}, 3$, ...
23. $i, -1, -i$, ...
24. t^8, t^5, t^2, ...

25. The first term of a geometric sequence is $\frac{a}{b^2}$, and the common ratio is $\frac{b}{a^2}$. Find the next five terms of the geometric sequence.

For Exercises 26–40, assume that each sequence or series is geometric.

26. Find the fifth term of a sequence whose first term is 8 and common ratio is $\frac{3}{2}$.

27. Find the sixth term of the sequence $\frac{1}{2}, -\frac{3}{8}, \frac{9}{32}, \ldots$.

28. Find the seventh term of the sequence 40, 0.4, 0.004, ….

29. Find the ninth term of the sequence $\sqrt{5}, \sqrt{10}, 2\sqrt{5}, \ldots$.

30. If $r = 4$ and $a_6 = 192$, what is the first term of the sequence?

31. If $r = -\sqrt{2}$ and $a_5 = 32\sqrt{2}$, what is the first term of the sequence?

32. Find the first three terms of the sequence for which $a_5 = -6$ and $r = -\frac{1}{3}$.

33. Find the first three terms of the sequence for which $a_5 = 0.32$ and $r = 0.2$.

34. Write a sequence that has three geometric means between 256 and 81.

35. Write a sequence that has two geometric means between −2 and 54.

36. Write a sequence that has one geometric mean between $\frac{4}{7}$ and 7.

37. What is the sum of the first five terms of the series $\frac{5}{3} + 5 + 15 + \cdots$?

38. What is the sum of the first six terms of the series $65 + 13 + 2.6 + \cdots$?

39. Find the sum of the first ten terms of the series $1 - \frac{3}{2} + \frac{9}{4} - \cdots$.

40. Find the sum of the first eight terms of the series $2 + 2\sqrt{3} + 6 + \cdots$.

Applications and Problem Solving

41. **Biology** A cholera bacterium divides every half-hour to produce two complete cholera bacteria.
    a. If an initial colony contains a population of b_0 bacteria, write an equation that will determine the number of bacteria present after t hours.
    b. Suppose a petri dish contains 30 cholera bacteria. Use the equation from part **a** to determine the number of bacteria present 5 hours later.
    c. What assumptions are made in using the formula found in part **a**?

42. **Critical Thinking** Consider the geometric sequence with $a_4 = 4$ and $a_7 = 12$.
    a. Find the common ratio and the first term of the sequence.
    b. Find the 28th term of the sequence.

43. **Consumerism** High Tech Electronics advertises a weekly installment plan for the purchase of a popular brand of big screen TV. The buyer pays $5 at the end of the first week, $5.50 at the end of the second week, $6.05 at the end of the third week, and so on for one year.
    a. What will the payments be at the end of the 10th, 20th, and 40th weeks?
    b. Find the total cost of the TV.
    c. Why is the cost found in part **b** not entirely accurate?

44. **Statistics** A number x is said to be the *harmonic mean* of y and z if $\frac{1}{x}$ is the average of $\frac{1}{y}$ and $\frac{1}{z}$.
    a. Find the harmonic mean of 5 and 8.
    b. 8 is the harmonic mean of 20 and another number. What is the number?

45. Critical Thinking In a geometric sequence, $a_1 = -2$ and every subsequent term is defined by $a_n = -3a_{n-1}$, where $n > 1$. Find the nth term in the sequence in terms of n.

46. Genealogy Wei-Ling discovers through a research of her Chinese ancestry that one of her fifteenth-generation ancestors was a famous military leader. How many descendants does this ancestor have in the fifteenth-generation, assuming each descendent had an average of 2.5 children?

47. Personal Finance Tonisha is about to begin her junior year in high school and is looking ahead to her college career. She estimates that by the time she is ready to enter a university she will need at least \$750 to purchase a computer system that will meet her needs. To avoid purchasing the computer on credit, she opens a savings account earning an APR of 2.4%, compounded monthly, and deposits \$25 at the beginning of each month.

a. Find the balance of the account at the end of the first month.

b. If Tonisha continues this deposit schedule for the next two years, will she have enough money in her account to buy the computer system? Explain.

c. Find the least amount of money Tonisha can deposit each month and still have enough money to purchase the computer.

48. Critical Thinking Use algebraic methods to determine which term 6561 is of the geometric sequence $\frac{1}{81}, \frac{1}{27}, \frac{1}{9}, \ldots$.

Mixed Review

49. Banking Gloria Castaneda has \$650 in her checking account. She is closing out the account by writing one check for cash against it each week. The first check is for \$20, the second is for \$25, and so on. Each check exceeds the previous one by \$5. In how many weeks will the balance in Ms. Castaneda account be \$0 if there is no service charge? *(Lesson 12-1)*

50. Find the value of $\log_{11} 265$ using the change of base formula. *(Lesson 11-5)*

51. Graph the system $xy \geq 2$ and $x - 3y = 2$. *(Lesson 10-8)*

52. Write $3x - 5y + 5 = 0$ in polar form. *(Lesson 9-4)*

53. Write parametric equations of the line $3x + 4y = 5$. *(Lesson 8-6)*

54. If $\csc \theta = 3$ and $0° \leq \theta \leq 90°$, find $\sin \theta$. *(Lesson 7-1)*

55. Weather The maximum normal daily temperatures in each season for Lincoln, Nebraska, are given below. Write a sinusoidal function that models the temperatures, using $t = 1$ to represent winter. *(Lesson 6-6)*

Normal Daily Temperatures for Lincoln, Nebraska

Winter	Spring	Summer	Fall
36°	61°	86°	65°

Source: Rand McNally & Company

56. Given $A = 43°$, $b = 20$, and $a = 11$, do these measurements determine one triangle, two triangles, or no triangle? *(Lesson 5-7)*

57. SAT Practice **Grid-In** If n and m are integers, and $-(n^2) \leq -\sqrt{49}$ and $m = n + 1$, what is the least possible value of mn?

Extra Practice See p. A48.

12-3

Infinite Sequences and Series

OBJECTIVES

- Find the limit of the terms of an infinite sequence.
- Find the sum of an infinite geometric series.

ECONOMICS On January 28, 1999, Florida governor Jeb Bush proposed a tax cut that would allow the average family to keep an additional \$96. The *marginal propensity to consume (MPC)* is defined as the percentage of a dollar by which consumption increases when income rises by a dollar. Suppose the *MPC* for households and businesses in 1999 was 75%. What would be the total amount of money spent in the economy as a result of just one family's tax savings? *This problem will be solved in Example 5.*

Governor Jeb Bush

Transaction	Expenditure	Terms of Sequence
1	$96(0.75)^1$	72
2	$96(0.75)^2$	54
3	$96(0.75)^3$	40.50
4	$96(0.75)^4$	30.76
5	$96(0.75)^5$	22.78
⋮	⋮	⋮
10	$96(0.75)^{10}$	5.41
⋮	⋮	⋮
100	$96(0.75)^{100}$	3.08×10^{-11}
⋮	⋮	⋮
500	$96(0.75)^{500}$	3.26×10^{-61}
⋮	⋮	⋮
n	$96(0.75)^n$	ar^n

Study the table at the left. Transaction 1 represents the initial expenditure of \$96(0.75) or \$72 by a family. The businesses receiving this money, Transaction 2, would in turn spend 75%, and so on. We can write a geometric sequence to model this situation with $a_1 = 72$ and $r = 0.75$. Thus, the geometric sequence representing this situation is

$$72, 54, 40.50, 30.76, 22.78, \ldots.$$

In theory, the sequence above can have infinitely many terms. Thus, it is called an **infinite sequence.** As n increases, the terms of the sequence decrease and get closer and closer to zero. The terms of the modeling sequence will never actually become zero; however, the terms approach zero as n increases without bound.

Consider the infinite sequence $1, \frac{1}{2}, \frac{1}{3}, \frac{1}{4}, \frac{1}{5}, \ldots$, whose nth term, a_n, is $\frac{1}{n}$. Several terms of this sequence are graphed at the right.

Notice that the terms approach a value of 0 as n increases. Zero is called the **limit** of the terms in this sequence.

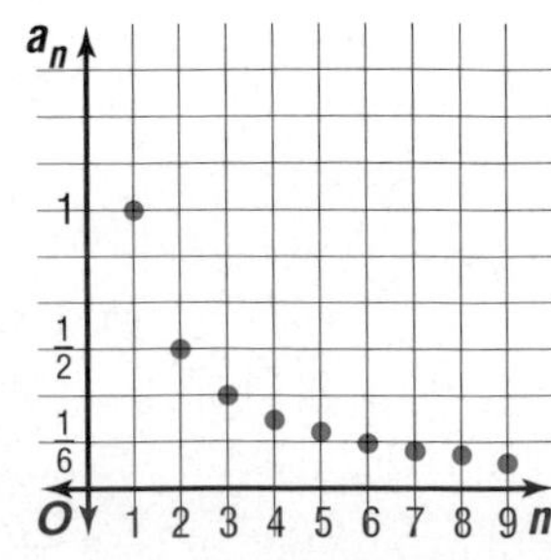

This limit can be expressed as follows.

$$\lim_{n\to\infty} \frac{1}{n} = 0 \quad \infty \textit{ is the symbol for infinity.}$$

This is read "the limit of 1 over n, as n approaches infinity, equals zero."

In fact, when any positive power of n appears only in the denominator of a fraction and n approaches infinity, the limit equals zero.

$$\lim_{n\to\infty} \frac{1}{n^r} = 0, \text{ for } r > 0$$

If a general expression for the nth term of a sequence is known, the limit can usually be found by substituting large values for n. Consider the following infinite geometric sequence.

$$7, \frac{7}{4}, \frac{7}{16}, \frac{7}{64}, \frac{7}{256}, \ldots$$

This sequence can be defined by the general expression $a_n = 7\left(\frac{1}{4}\right)^{n-1}$.

$$a_{10} = 7\left(\frac{1}{4}\right)^{10-1} \approx 2.67 \times 10^{-5}$$

$$a_{50} = 7\left(\frac{1}{4}\right)^{50-1} \approx 2.21 \times 10^{-25}$$

$$a_{100} = 7\left(\frac{1}{4}\right)^{100-1} \approx 4.36 \times 10^{-60}$$

Notice that as the value of n increases, the value for a_n appears to approach 0, suggesting $\lim_{n\to\infty} 7\left(\frac{1}{4}\right)^{n-1} = 0$.

Example 1 **Estimate the limit of $\frac{9}{5}, \frac{16}{4}, \frac{65}{27}, \ldots, \frac{7n^2+2}{2n^2+3n}, \ldots$.**

The 50th term is $\frac{7(50)^2+2}{2(50)^2+3(50)}$, or about 3.398447.

The 100th term is $\frac{7(100)^2+2}{2(100)^2+3(100)}$, or about 3.448374.

The 500th term is $7\frac{(500)^2+2}{2(500)^2+3(500)}$, or about 3.489535.

The 1000th term is $\frac{7(1000)^2+2}{2(1000)^2+3(1000)}$, or 3.494759.

Notice that as $n \to \infty$, the values appear to approach 3.5, suggesting $\lim_{n\to\infty} \frac{7n^2+2}{2n^2+3n} = 3.5$.

For sequences with more complicated general forms, applications of the following limit theorems, which we will present without proof, can make the limit easier to find.

Theorems for Limits

If the $\lim_{n\to\infty} a_n$ exists, $\lim_{n\to\infty} b_n$ exists, and c is a constant, then the following theorems are true.

Limit of a Sum $\quad \lim_{n\to\infty} (a_n + b_n) = \lim_{n\to\infty} a_n + \lim_{n\to\infty} b_n$

Limit of a Difference $\quad \lim_{n\to\infty} (a_n - b_n) = \lim_{n\to\infty} a_n - \lim_{n\to\infty} b_n$

Limit of a Product $\quad \lim_{n\to\infty} a_n \cdot b_n = \lim_{n\to\infty} a_n \cdot \lim_{n\to\infty} b_n$

Limit of a Quotient $\quad \lim_{n\to\infty} \frac{a_n}{b_n} = \frac{\lim_{n\to\infty} a_n}{\lim_{n\to\infty} b_n}$, where $\lim_{n\to\infty} b_n \neq 0$

Limit of a Constant $\quad \lim_{n\to\infty} c_n = c$, where $c_n = c$ for each n

---

The form of the expression for the nth term of a sequence can often be altered to make the limit easier to find.

Example 2 **Find each limit.**

a. $\lim_{n\to\infty} \frac{(1 + 3n^2)}{n^2}$

$$\lim_{n\to\infty} \frac{(1 + 3n^2)}{n^2} = \lim_{n\to\infty}\left(\frac{1}{n^2} + 3\right)$$ *Rewrite as the sum of two fractions and simplify.*

$$= \lim_{n\to\infty} \frac{1}{n^2} + \lim_{n\to\infty} 3$$ *Limit of a Sum*

$$= 0 + 3 \text{ or } 3$$ *$\lim_{n\to\infty} \frac{1}{n^2} = 0$ and $\lim_{n\to\infty} 3 = 3$*

Note that the Limit of a Sum theorem only applies here because $\lim_{n\to\infty} \frac{1}{n^2}$ and $\lim_{n\to\infty} 3$ each exist.

Thus, the limit is 3.

b. $\lim_{n\to\infty} \frac{5n^2 + n - 4}{n^2 + 1}$

The highest power of n in the expression is n^2. Divide each term in the numerator and the denominator by n^2. *Why does doing this produce an equivalent expression?*

$$\lim_{n\to\infty} \frac{5n^2 + n - 4}{n^2 + 1} = \lim_{n\to\infty} \frac{\frac{5n^2}{n^2} + \frac{n}{n^2} - \frac{4}{n^2}}{\frac{n^2}{n^2} + \frac{1}{n^2}}$$

$$= \lim_{n\to\infty} \frac{5 + \frac{1}{n} - \frac{4}{n^2}}{1 + \frac{1}{n^2}}$$ *Simplify.*

$$= \frac{\lim_{n\to\infty} 5 + \lim_{n\to\infty} \frac{1}{n} - \lim_{n\to\infty} 4 \cdot \lim_{n\to\infty} \frac{1}{n^2}}{\lim_{n\to\infty} 1 + \lim_{n\to\infty} \frac{1}{n^2}}$$ *Apply limit theorems.*

$$= \frac{5 + 0 - 4 \cdot 0}{1 + 0} \text{ or } 5$$ *$\lim_{n\to\infty} 5 = 5$, $\lim_{n\to\infty} \frac{1}{n} = 0$, $\lim_{n\to\infty} 4 = 4$, $\lim_{n\to\infty} 1 = 1$, and $\lim_{n\to\infty} \frac{1}{n^2} = 0a$*

Thus, the limit is 5.

Limits do not exist for all infinite sequences. If the absolute value of the terms of a sequence becomes arbitrarily great or if the terms do not approach a value, the sequence has no limit. Example 3 illustrates both of these cases.

Example 3 **Find each limit.**

a. $\lim_{n\to\infty} \frac{2 + 5n + 4n^2}{2n}$

$$\lim_{n\to\infty} \frac{2 + 5n + 4n^2}{2n} = \lim_{n\to\infty} \left(\frac{1}{n} + \frac{5}{2} + 2n\right)$$ *Simplify.*

Note that $\lim_{n\to\infty} \frac{1}{n} = 0$ and $\lim_{n\to\infty} \frac{5}{2} = \frac{5}{2}$, but $2n$ becomes increasingly large as n approaches infinity. Therefore, the sequence has no limit.

b. $\lim_{n\to\infty} \frac{(-1)^n n}{8n + 1}$

Begin by rewriting $\frac{(-1)^n n}{8n + 1}$ as $(-1)^n \cdot \frac{n}{8n + 1}$. Now find $\lim_{n\to\infty} \frac{n}{8n + 1}$.

$$\lim_{n\to\infty} \frac{n}{8n + 1} = \lim_{n\to\infty} \frac{\frac{n}{n}}{\frac{8n}{n} + \frac{1}{n}}$$ *Divide the numerator and denominator by n.*

$$= \lim_{n\to\infty} \frac{1}{8 + \frac{1}{n}}$$ *Simplify.*

$$= \frac{\lim_{n\to\infty} 1}{\lim_{n\to\infty} 8 + \lim_{n\to\infty} \frac{1}{n}}$$ *Apply limit theorems.*

$$= \frac{1}{8}$$ *$\lim_{n\to\infty} 1 = 1$, $\lim_{n\to\infty} 8 = 8$, and $\lim_{n\to\infty} \frac{1}{n} = 0$*

When n is even, $(-1)^n = 1$. When n is odd, $(-1)^n = -1$. Thus, the odd-numbered terms of the sequence described by $\frac{(-1)^n n}{8n + 1}$ approach $-\frac{1}{8}$, and the even-numbered terms approach $+\frac{1}{8}$. Therefore, the sequence has no limit.

An **infinite series** is the indicated sum of the terms of an infinite sequence. Consider the series $\frac{1}{5} + \frac{1}{25} + \frac{1}{125} + \cdots$. Since this is a geometric series, you can find the sum of the first 100 terms by using the formula $S_n = \frac{a_1 - a_1 r^n}{1 - r}$, where $r = \frac{1}{5}$.

$$S_{100} = \frac{\frac{1}{5} - \frac{1}{5}\left(\frac{1}{5}\right)^{100}}{1 - \frac{1}{5}}$$

$$= \frac{\frac{1}{5} - \frac{1}{5}\left(\frac{1}{5}\right)^{100}}{\frac{4}{5}}$$

$$= \frac{5}{4}\left[\frac{1}{5} - \frac{1}{5}\left(\frac{1}{5}\right)^{100}\right] \text{ or } \frac{1}{4} - \frac{1}{4}\left(\frac{1}{5}\right)^{100}$$

Since $\left(\frac{1}{5}\right)^{100}$ is very close to 0, S_{100} is nearly equal to $\frac{1}{4}$. No matter how many terms are added, the sum of the infinite series will never exceed $\frac{1}{4}$, and the difference from $\frac{1}{4}$ gets smaller as $n \to \infty$. Thus, $\frac{1}{4}$ is the sum of the infinite series.

Sum of an Infinite Series

If S_n is the sum of the first n terms of a series, and S is a number such that $S - S_n$ approaches zero as n increases without bound, then the sum of the infinite series is S.

$$\lim_{n\to\infty} S_n = S$$

If the sequence of partial sums S_n has a limit, then the corresponding infinite series has a sum, and the nth term a_n of the series approaches 0 as $n \to \infty$. If $\lim\limits_{n\to\infty} a_n \neq 0$, the series has no sum. If $\lim\limits_{n\to\infty} a_n = 0$, the series may or may not have a sum.

The formula for the sum of the first n terms of a geometric series can be written as follows.

$$S_n = a_1 \frac{(1 - r^n)}{1 - r}, r \neq 1$$

Recall that $|r| < 1$ means $-1 < r < 1$.

Suppose $n \to \infty$; that is, the number of terms increases without limit. If $|r| > 1$, r^n increases without limit as $n \to \infty$. However, when $|r| < 1$, r^n approaches 0 as $n \to \infty$. Under this condition, the above formula for S_n approaches a value of $\frac{a_1}{1 - r}$.

Sum of an Infinite Geometric Series

The sum S of an infinite geometric series for which $|r| < 1$ is given by

$$S = \frac{a_1}{1 - r}.$$

Example **4** **Find the sum of the series $21 - 3 + \frac{3}{7} - \cdots$.**

In the series, $a_1 = 21$ and $r = -\frac{1}{7}$. Since $|r| < 1$, $S = \frac{a_1}{1 - r}$.

$$S = \frac{a_1}{1 - r}$$

$$= \frac{21}{1 - \left(-\frac{1}{7}\right)} \qquad a_1 = 21,\ r = -\frac{1}{7}$$

$$= \frac{147}{8} \text{ or } 18\frac{3}{8}$$

The sum of the series is $18\frac{3}{8}$.

In economics, finding the sum of an infinite series is useful in determining the overall effect of economic trends.

Example **5** **ECONOMICS** **Refer to the application at the beginning of the lesson. What would be the total amount of money spent in the economy as a result of just one family's tax savings?**

For the geometric series modeling this situation, $a_1 = 72$ and $r = 0.75$.

Since $|r| < 1$, the sum of the series is equal to $\frac{a_1}{1 - r}$.

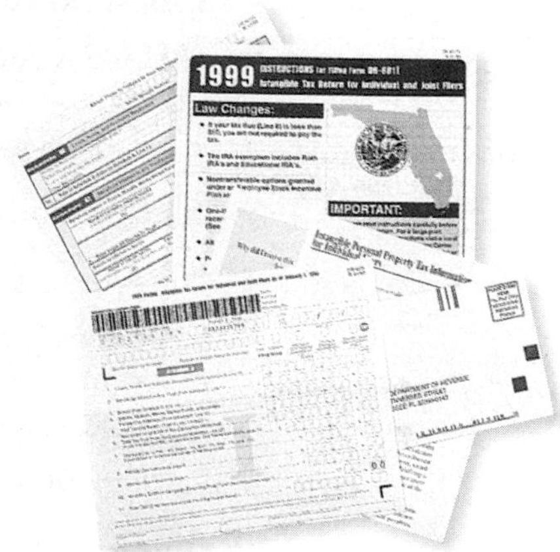

$$S = \frac{a_1}{1 - r}$$

$$= \frac{72}{1 - 0.75} \text{ or } 288$$

Therefore, the total amount of money spent is \$288.

You can use what you know about infinite series to write repeating decimals as fractions. The first step is to write the repeating decimal as an infinite geometric series.

Example  **Write $0.\overline{762}$ as a fraction.**

$$0.\overline{762} = \frac{762}{1000} + \frac{762}{1{,}000{,}000} + \frac{762}{1{,}000{,}000{,}000} + \cdots$$

In this series, $a_1 = \frac{762}{1000}$ and $r = \frac{1}{1000}$.

(continued on the next page)

$$S = \frac{a_1}{1 - r}$$

$$= \frac{\frac{762}{1000}}{1 - \frac{1}{1000}}$$

$$= \frac{762}{999} \text{ or } \frac{254}{333}$$

Thus, $0.762762 \cdots = \frac{254}{333}$. *Check this with a calculator.*

CHECK FOR UNDERSTANDING

Communicating Mathematics

Read and study the lesson to answer each question.

1. Consider the sequence given by the general expression $a_n = \frac{n - 1}{n}$.
   a. **Graph** the first ten terms of the sequence with the term number on the x-axis and the value of the term on the y-axis.
   b. **Describe** what happens to the value of a_n as n increases.
   c. **Make a conjecture** based on your observation in part **a** as to the limit of the sequence as n approaches infinity.
   d. **Apply** the techniques presented in the lesson to evaluate $\lim_{n \to \infty} \frac{n - 1}{n}$. How does your answer compare to your conjecture made in part **c**?

2. Consider the infinite geometric sequence given by the general expression r^n.
   a. **Determine** the limit of the sequence for $r = \frac{1}{2}$, $r = \frac{1}{4}$, $r = 1$, $r = 2$, and $r = 5$.
   b. **Write** a general rule for the limit of the sequence, placing restrictions on the value of r.

3. **Give an example** of an infinite geometric series having no sum.

4. **You Decide** Tyree and Zonta disagree on whether the infinite sequence described by the general expression $2n - 3$ has a limit. Tyree says that after dividing by the highest-powered term, the expression simplifies to $2 - \frac{3}{n}$, which has a limit of 2 as n approaches infinity. Zonta says that the sequence has no limit. Who is correct? Explain.

Guided Practice

Find each limit, or state that the limit does not exist and explain your reasoning.

5. $\lim_{n \to \infty} \frac{1}{5^n}$

6. $\lim_{n \to \infty} \frac{5 - n^2}{2n}$

7. $\lim_{n \to \infty} \frac{3n - 6}{7n}$

Write each repeating decimal as a fraction.

8. $0.\overline{7}$

9. $5.\overline{126}$

Find the sum of each infinite series, or state that the sum does not exist and explain your reasoning.

10. $-6 + 3 - \frac{3}{2} + \cdots$ **11.** $\frac{3}{4} + \frac{1}{4} + \frac{1}{12} + \cdots$ **12.** $\sqrt{3} + 3 + \sqrt{27} + \cdots$

13. Entertainment Pete's Pirate Ride operates like the bob of a pendulum. On its longest swing, the ship travels through an arc 75 meters long. Each successive swing is two-fifths the length of the preceding swing. If the ride is allowed to continue without intervention, what is the total distance the ship will travel before coming to rest?

EXERCISES

Practice

Find each limit, or state that the limit does not exist and explain your reasoning.

14. $\lim_{n\to\infty} \frac{7-2n}{5n}$ **15.** $\lim_{n\to\infty} \frac{n^3-2}{n^2}$ **16.** $\lim_{n\to\infty} \frac{6n^2+5}{3n^2}$

17. $\lim_{n\to\infty} \frac{9n^3+5n-2}{2n^3}$ **18.** $\lim_{n\to\infty} \frac{(3n+4)(1-n)}{n^2}$ **19.** $\lim_{n\to\infty} \frac{8n^2+5n+2}{3+2n}$

20. $\lim_{n\to\infty} \frac{4-3n+n^2}{2n^3-3n^2+5}$ **21.** $\lim_{n\to\infty} \frac{n}{3^n}$ **22.** $\lim_{n\to\infty} \frac{(-2)^n n}{4+n}$

23. Find the limit of the sequence described by the general expression $\frac{5n+(-1)^n}{n^2}$, or state that the limit does not exist. Explain your reasoning.

Write each repeating decimal as a fraction.

24. $0.\overline{4}$ **25.** $0.\overline{51}$ **26.** $0.\overline{370}$

27. $6.\overline{259}$ **28.** $0.1\overline{5}$ **29.** $0.2\overline{63}$

30. Explain why the sum of the series $0.2 + 0.02 + 0.002 + \cdots$ exists. Then find the sum.

Find the sum of each series, or state that the sum does not exist and explain your reasoning.

31. $16 + 12 + 9 + \cdots$ **32.** $5 + 7.5 + 11.25 + \cdots$ **33.** $10 + 5 + 2.5 + \cdots$

34. $6 + 5 + 4 + \cdots$ **35.** $\frac{1}{8} + \frac{1}{4} + \frac{1}{2} + \cdots$ **36.** $-\frac{2}{3} + \frac{1}{9} - \frac{1}{54} + \cdots$

37. $\frac{6}{5} + \frac{4}{5} + \frac{8}{15} + \cdots$ **38.** $\sqrt{5} + 1 + \frac{\sqrt{5}}{5} + \cdots$ **39.** $8 - 4\sqrt{3} + 6 - \cdots$

Applications and Problem Solving

40. Physics A basketball is dropped from a height of 35 meters and bounces $\frac{2}{5}$ of the distance after each fall.

a. Find the first five terms of the infinite series representing the vertical distance traveled by the ball.

b. What is the total vertical distance the ball travels before coming to rest? (*Hint:* Rewrite the series found in part **a** as the sum of two infinite geometric series.)

41. Critical Thinking Consider the sequence whose nth term is described by $\frac{n^2}{2n+1} - \frac{n^2}{2n-1}$.

a. Explain why $\lim_{n\to\infty}\left(\frac{n^2}{2n+1} - \frac{n^2}{2n-1}\right) \neq \lim_{n\to\infty}\frac{n^2}{2n+1} - \lim_{n\to\infty}\frac{n^2}{2n-1}$.

b. Find $\lim_{n\to\infty}\left(\frac{n^2}{2n+1} - \frac{n^2}{2n-1}\right)$.

42. Engineering Francisco designs a toy with a rotary flywheel that rotates at a maximum speed of 170 revolutions per minute. Suppose the flywheel is operating at its maximum speed for one minute and then the power supply to the toy is turned off. Each subsequent minute thereafter, the flywheel rotates two-fifths as many times as in the preceding minute. How many complete revolutions will the flywheel make before coming to a stop?

43. Critical Thinking Does $\lim_{n\to\infty} \cos\frac{n\pi}{2}$ exist? Explain.

44. Medicine A certain drug developed to fight cancer has a half-life of about 2 hours in the bloodstream. The drug is formulated to be administered in doses of D milligrams every 6 hours. The amount of each dose has yet to be determined.

a. What fraction of the first dose will be left in the bloodstream before the second dose is administered?

b. Write a general expression for the geometric series that models the number of milligrams of drug left in the bloodstream after the nth dose.

c. About what amount of medicine is present in the bloodstream for large values of n?

d. A level of more than 350 milligrams of this drug in the bloodstream is considered toxic. Find the largest possible dose that can be given repeatedly over a long period of time without harming the patient.

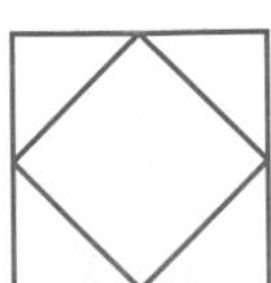

45. Geometry If the midpoints of a square are joined by straight lines, the new figure will also be a square.

a. If the original square has a perimeter of 20 feet, find the perimeter of the new square. (*Hint:* Use the Pythagorean Theorem.)

b. If this process is continued to form a sequence of "nested" squares, what will be the sum of the perimeters of all the squares?

46. Technology Since the mid-1980s, the number of computers in schools has steadily increased. The graph below shows the corresponding decline in the student-computer ratio.

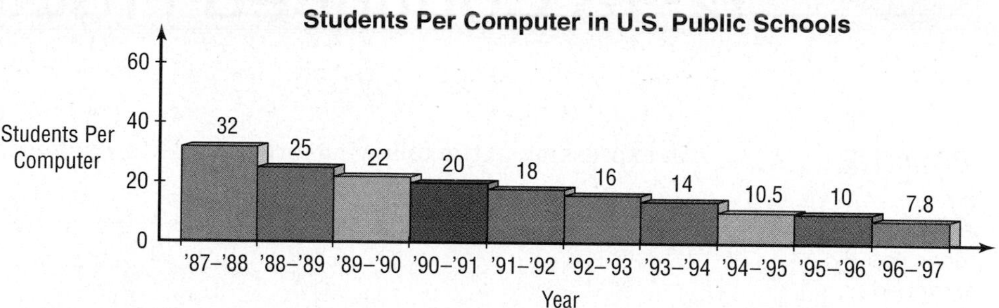

Source: *QED's Technology in Public Schools*, 16th Edition

Another publication states that the average number of students per computer in U.S. public schools can be estimated by the sequence model $a_n = 35.812791(0.864605)^n$, *for* $n = 1, 2, 3, \ldots$, with the 1987-1988 school year corresponding to $n = 1$.

a. Find the first ten terms of the model. Round your answers to the nearest tenth.

b. Use the model to estimate the average number of students having to share a computer during the 1995-1996 school year. How does this estimate compare to the actual data given in the graph?

c. Make a prediction as to the average number of students per computer for the 2000-2001 school year.

d. Does this sequence approach a limit? If so, what is the limit?

e. Realistically, will the student computer ratio ever reach this limit? Explain.

Mixed Review

47. The first term of a geometric sequence is -3, and the common ratio is $\frac{2}{3}$. Find the next four terms of the sequence. *(Lesson 12-2)*

48. Find the 16th term of the arithmetic sequence for which $a_1 = 1.5$ and $d = 0.5$. *(Lesson 12-1)*

49. Name the coordinates of the center, foci, and vertices, and the equation of the asymptotes of the hyperbola that has the equation $x^2 - 4y^2 - 12x - 16y = -16$. *(Lesson 10-4)*

50. Graph $r = 6 \cos 3\theta$. *(Lesson 9-2)*

51. **Navigation** A ship leaving port sails for 125 miles in a direction 20° north of due east. Find the magnitude of the vertical and horizontal components. *(Lesson 8-1)*

52. Use a half-angle identity to find the exact value of cos 112.5°. *(Lesson 7-4)*

53. Graph $y = \cos x$ on the interval $-180° \leq x \leq 360°$. *(Lesson 6-3)*

54. List all possible rational zeros of the function $f(x) = 8x^3 + 3x - 2$. *(Lesson 4-4)*

55. **SAT/ACT Practice** If $a = 4b + 26$, and b is a positive integer, then a could be divisible by all of the following EXCEPT

A 2 **B** 4 **C** 5 **D** 6 **E** 7

Extra Practice See p. A49.

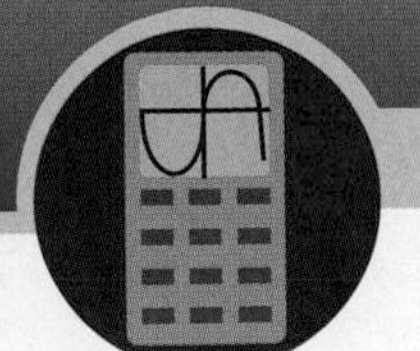

GRAPHING CALCULATOR EXPLORATION

12-3B Continued Fractions

An Extension of Lesson 12-3

OBJECTIVE

- Explore sequences generated by continued fractions.

An expression of the following form is called a *continued fraction.*

$$a_1 + \cfrac{b_1}{a_2 + \cfrac{b_2}{a_3 + \cfrac{b_3}{a_4 + \cdots}}}$$

By using only a finite number of "decks" and values of a_n and b_n that follow regular patterns, you can often obtain a sequence of terms that approaches a limit, which can be represented by a simple expression. For example, if all of the numbers a_n and b_n are equal to 1, then the continued fraction gives rise to the following sequence.

$$1, 1 + \frac{1}{1}, 1 + \cfrac{1}{1 + \cfrac{1}{1}}, 1 + \cfrac{1}{1 + \cfrac{1}{1 + \cfrac{1}{1}}}, \ldots$$

It can be shown that the terms of this sequence approach the limit $\frac{1 + \sqrt{5}}{2}$. This number is often called the *golden ratio.*

The golden ratio is closely related to the Fibonnaci sequence, which you will learn about in Lesson 12-7.

Now consider the following more general sequence.

$$A, A + \frac{1}{A}, A + \cfrac{1}{A + \cfrac{1}{A}}, A + \cfrac{1}{A + \cfrac{1}{A + \cfrac{1}{A}}}, \ldots$$

To help you visualize what this sequence represents, suppose $A = 5$. The sequence becomes $5, 5 + \frac{1}{5}, 5 + \cfrac{1}{5 + \cfrac{1}{5}}, 5 + \cfrac{1}{5 + \cfrac{1}{5 + \cfrac{1}{5}}}, \ldots$ or $5, \frac{26}{5}, \frac{135}{26}, \frac{701}{135}, \ldots$.

A calculator approximation of this sequence is 5, 5.2, 5.192307692, 5,192592593, … .

Each term of the sequence is the sum of A and the reciprocal of the previous term. The program at the right calculates the value of the nth term of the above sequence for $n \geq 3$ and a specified value of A.

When you run the program it will ask you to input values for A and N.

```
PROGRAM: CFRAC
: Prompt A
: Disp "INPUT TERM"
: Disp "NUMBER N, N ≥ 3"
: Prompt N
: 1 → K
: A + 1/A → C
: Lbl 1
: A + 1/C → C
: K + 1 → K
: If K< N − 1
: Then: Goto 1
: Else: Disp C
```

interNET CONNECTION

Graphing Calculator Programs
To download this graphing calculator program, visit our website at **www.amc.glencoe.com**

TRY THESE

Enter the program into your calculator and use it for the exercises that follow.

1. What output is given when the program is executed for $A = 1$ and $N = 10$?

2. With $A = 1$, determine the least value of N necessary to obtain an output that agrees with the calculator's nine decimal approximation of $\frac{1 + \sqrt{5}}{2}$.

3. Use algebra to show that the continued fraction $1 + \cfrac{1}{1 + \cfrac{1}{1 + \cdots}}$ has a value of $\frac{1 + \sqrt{5}}{2}$. (*Hint:* If $x = 1 + \cfrac{1}{1 + \cfrac{1}{1 + \cdots}}$, then $x = 1 + \frac{1}{x}$. Solve this last equation for x.)

4. Find the exact value of $3 + \cfrac{1}{3 + \cfrac{1}{3 + \cdots}}$.

5. Execute the program with $A = 3$ and $N = 40$. How does this output compare to the decimal approximation of the expression found in Exercise 4?

6. Find a radical expression for $A + \cfrac{1}{A + \cfrac{1}{A + \cdots}}$.

7. Write a modified version of the program that calculates the nth term of the following sequence for $n \geq 3$.

$$A,\ A + \frac{B}{2A},\ A + \cfrac{B}{2A + \cfrac{B}{2A}},\ A + \cfrac{B}{2A + \cfrac{B}{2A + \cfrac{B}{2A}}},\ \ldots$$

8. Choose several positive integer values for A and B and compare the program output with the decimal approximation of $\sqrt{A^2 + B}$ for several values of n, for $n \geq 3$. Describe your observations.

9. Use algebra to show that for $A > 0$ and $B > 0$, $A + \cfrac{B}{2A + \cfrac{B}{2A + \cdots}}$ has a value of $\sqrt{A^2 + B}$.

$$\left(\textit{Hint: } \text{If } x = A + \cfrac{B}{2A + \cfrac{B}{2A + \cdots}},\ \text{then } x + A = 2A + \cfrac{B}{2A + \cfrac{B}{2A + \cdots}}.\right)$$

WHAT DO YOU THINK?

10. If you execute the original program for $A = 1$ and $N = 20$ and then execute it for $A = -1$ and $N = 20$, how will the two outputs compare?

11. What values can you use for A and B in the program for Exercise 7 in order to approximate $\sqrt{15}$?

Convergent and Divergent Series

OBJECTIVE

- Determine whether a series is convergent or divergent.

HISTORY The Greek philosopher Zeno of Elea (c. 490–430 B.C.) proposed several perplexing riddles, or paradoxes. One of Zeno's paradoxes involves a race on a 100-meter track between the mythological Achilles and a tortoise. Zeno claims that even though Achilles can run twice as fast as the tortoise, if the tortoise is given a 10-meter head start, Achilles will never catch him. Suppose Achilles runs 10 meters per second and the tortoise a remarkable 5 meters per second. By the time Achilles has reached the 10-meter mark, the tortoise will be at 15 meters. By the time Achilles reaches the 15-meter mark, the tortoise will be at 17.5 meters, and so on. Thus, Achilles is always behind the tortoise and never catches up.

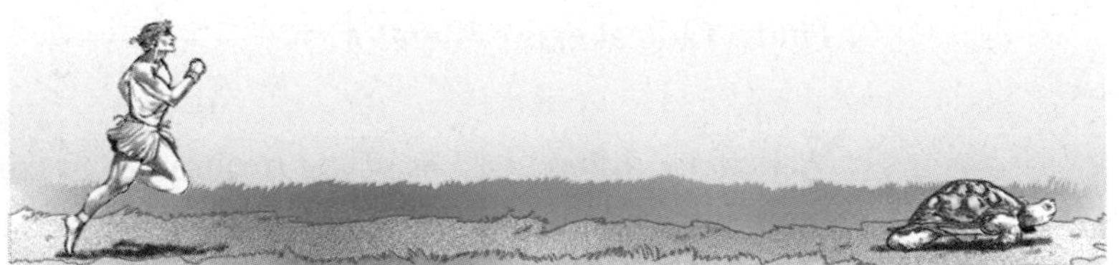

Is Zeno correct? Let us look at the distance between Achilles and the tortoise after specified amounts of time have passed. Notice that the distance between the two contestants will be zero as n approaches infinity since $\lim_{n \to \infty} \frac{10}{2^n} = 0$.

Time (seconds)	Distance Apart (meters)
0	10
1	$\frac{10}{2} = 5$
$1 + \frac{1}{2} = \frac{3}{2}$	$\frac{10}{4} = 2.5$
$1 + \frac{1}{2} + \frac{1}{4} = \frac{7}{4}$	$\frac{10}{8} = 1.25$
$1 + \frac{1}{2} + \frac{1}{4} + \frac{1}{8} = \frac{15}{8}$ ⋮	$\frac{10}{16} = 0.625$ ⋮
$1 + \frac{1}{2} + \frac{1}{4} + \frac{1}{8} + \cdots$	$\frac{10}{2^n}$

To disprove Zeno's conclusion that Achilles will never catch up to the tortoise, we must show that there is a time value for which this 0 difference can be achieved. In other words, we need to show that the infinite series $1 + \frac{1}{2} + \frac{1}{4} + \frac{1}{8} + \cdots$ has a sum, or limit. *This problem will be solved in Example 5.*

Starting with a time of 1 second, the partial sums of the time series form the sequence $1, \frac{3}{2}, \frac{7}{4}, \frac{15}{18}, \ldots$. As the number of terms used for the partial sums increases, the value of the partial sums also increases. If this sequence of partial sums approaches a limit, the related infinite series is said to **converge.** If this sequence of partial sums does not have a limit, then the related infinite series is said to **diverge.**

Convergent and Divergent Series If an infinite series has a sum, or limit, the series is convergent. If a series is not convergent, it is divergent.

Example 1 **Determine whether each arithmetic or geometric series is convergent or divergent.**

There are many series that begin with the first few terms shown in this example. In this chapter, always assume that the expression for the general term is the simplest one possible.

a. $-\frac{1}{2} + \frac{1}{4} - \frac{1}{8} + \frac{1}{16} - \cdots$

This is a geometric series with $r = -\frac{1}{2}$. Since $|r| < 1$, the series has a limit. Therefore, the series is convergent.

b. $2 + 4 + 8 + 16 + \cdots$

This is a geometric series with $r = 2$. Since $|r| > 1$, the series has no limit. Therefore, the series is divergent.

c. $10 + 8.5 + 7 + 5.5 + \cdots$

This is an arithmetic series with $d = -1.5$. Arithmetic series do not have limits. Therefore, the series is divergent.

When a series is neither arithmetic nor geometric, it is more difficult to determine whether the series is convergent or divergent. Several different techniques can be used. One test for convergence is the **ratio test.** This test can only be used when all terms of a series are positive. The test depends upon the ratio of consecutive terms of a series, which must be expressed in general form.

Ratio Test Let a_n and a_{n+1} represent two consecutive terms of a series of positive terms. Suppose $\lim_{n\to\infty} \frac{a_{n+1}}{a_n}$ exists and that $r = \lim_{n\to\infty} \frac{a_{n+1}}{a_n}$. The series is convergent if $r < 1$ and divergent if $r > 1$. If $r = 1$, the test provides no information.

The ratio test is especially useful when the general form for the terms of a series contains powers.

Example 2 **Use the ratio test to determine whether each series is convergent or divergent.**

a. $\frac{1}{2} + \frac{2}{4} + \frac{3}{8} + \frac{4}{16} + \cdots$

First, find a_n and a_{n+1}. $a_n = \frac{n}{2^n}$ and $a_{n+1} = \frac{n+1}{2^{n+1}}$

Then use the ratio test. $r = \lim_{n\to\infty} \frac{\frac{n+1}{2^{n+1}}}{\frac{n}{2^n}}$

(continued on the next page)

$$r = \lim_{n\to\infty} \frac{n+1}{2^{n+1}} \cdot \frac{2^n}{n}$$ *Multiply by the reciprocal of the divisor.*

$$r = \lim_{n\to\infty} \frac{1}{2} \cdot \frac{n+1}{n}$$ $\frac{2^n}{2^{n+1}} = \frac{1}{2}$

$$r = \lim_{n\to\infty} \frac{1}{2} \cdot \lim_{n\to\infty} \frac{n+1}{n}$$ *Limit of a Product*

$$r = \frac{1}{2} \lim_{n\to\infty} \frac{1 + \frac{1}{n}}{1}$$ *Divide by the highest power of n and then apply limit theorems.*

$$r = \frac{1}{2} \cdot \frac{1+0}{1} \text{ or } \frac{1}{2}$$ Since $r < 1$, the series is convergent.

b. $\frac{1}{2} + \frac{2}{3} + \frac{3}{4} + \frac{4}{5} + \cdots$

$a_n = \frac{n}{n+1}$ and $a_{n+1} = \frac{n+1}{(n+1)+1}$ or $\frac{n+1}{n+2}$

$$r = \lim_{n\to\infty} \frac{\frac{n+1}{n+2}}{\frac{n}{n+1}}$$

$$r = \lim_{n\to\infty} \frac{n^2+2n+1}{n^2+2n}$$ $\frac{n+1}{n+2} \cdot \frac{n+1}{n} = \frac{n^2+2n+1}{n^2+2n}$

$$r = \lim_{n\to\infty} \frac{1 + \frac{2}{n} + \frac{1}{n^2}}{1 + \frac{2}{n}}$$ *Divide by the highest power of n and apply limit theorems.*

$$r = \frac{1+0+0}{1+0} \text{ or } 1$$ Since $r = 1$, the test provides no information.

The ratio test is also useful when the general form of the terms of a series contains products of consecutive integers.

Example 3 **Use the ratio test to determine whether the series**

$1 + \frac{1}{1 \cdot 2} + \frac{1}{1 \cdot 2 \cdot 3} + \frac{1}{1 \cdot 2 \cdot 3 \cdot 4} + \cdots$ **is convergent or divergent.**

First find the nth term and $(n + 1)$th term. Then, use the ratio test.

$a_n = \frac{1}{1 \cdot 2 \cdot \cdots \cdot n}$ and $a_{n+1} = \frac{1}{1 \cdot 2 \cdot \cdots \cdot (n+1)}$

$$r = \lim_{n\to\infty} \frac{\frac{1}{1 \cdot 2 \cdot \cdots \cdot (n+1)}}{\frac{1}{1 \cdot 2 \cdot \cdots \cdot n}}$$

$$r = \lim_{n\to\infty} \frac{1 \cdot 2 \cdot \cdots \cdot n}{1 \cdot 2 \cdot \cdots \cdot (n+1)}$$ *Note that $1 \cdot 2 \cdot \cdots \cdot (n+1) = 1 \cdot 2 \cdot \cdots \cdot n \cdot (n+1)$.*

$$r = \lim_{n\to\infty} \frac{1}{n+1} \text{ or } 0$$ *Simplify and apply limit theorems.*

Since $r < 1$, the series is convergent.

When the ratio test does not determine if a series is convergent or divergent, other methods must be used.

Example 4 **Determine whether the series $1 + \frac{1}{2} + \frac{1}{3} + \frac{1}{4} + \frac{1}{5} + \cdots$ is convergent or divergent.**

Suppose the terms are grouped as follows. Beginning after the second term, the number of terms in each successive group is doubled.

$$(1) + \left(\frac{1}{2}\right) + \left(\frac{1}{3} + \frac{1}{4}\right) + \left(\frac{1}{5} + \frac{1}{6} + \frac{1}{7} + \frac{1}{8}\right) + \left(\frac{1}{9} + \cdots + \frac{1}{16}\right) + \cdots$$

Notice that the first enclosed expression is greater than $\frac{1}{2}$, and the second is equal to $\frac{1}{2}$. Beginning with the third expression, each sum of enclosed terms is greater than $\frac{1}{2}$. Since there are an unlimited number of such expressions, the sum of the series is unlimited. Thus, the series is divergent.

A series can be compared to other series that are known to be convergent or divergent. The following list of series can be used for reference.

Summary of Series for Reference

1. **Convergent:** $a_1 + a_1r + a_1r^2 + \cdots + a_1r^{n-1} + \cdots, |r| < 1$
2. **Divergent:** $a_1 + a_1r + a_1r^2 + \cdots + a_1r^{n-1} + \cdots, |r| > 1$
3. **Divergent:** $a_1 + (a_1 + d) + (a_1 + 2d) + (a_1 + 3d) + \cdots$
4. **Divergent:** $1 + \frac{1}{2} + \frac{1}{3} + \frac{1}{4} + \frac{1}{5} + \cdots + \frac{1}{n} + \cdots$ *This series is known as the harmonic series.*
5. **Convergent:** $1 + \frac{1}{2^p} + \frac{1}{3^p} + \cdots + \frac{1}{n^p} + \cdots, p > 1$

If a series has all positive terms, the **comparison test** can be used to determine whether the series is convergent or divergent.

Comparison Test

- A series of positive terms is convergent if, for $n > 1$, each term of the series is equal to or less than the value of the corresponding term of some convergent series of positive terms.
- A series of positive terms is divergent if, for $n > 1$, each term of the series is equal to or greater than the value of the corresponding term of some divergent series of positive terms.

Example 5 **Use the comparison test to determine whether the following series are convergent or divergent.**

a. $\frac{4}{5} + \frac{4}{7} + \frac{4}{9} + \frac{4}{11} + \cdots$

The general term of this series is $\frac{4}{2n + 3}$. The general term of the divergent series $1 + \frac{1}{2} + \frac{1}{3} + \frac{1}{4} + \frac{1}{5} + \cdots$ is $\frac{1}{n}$. Since for all $n > 1$, $\frac{4}{2n + 3} > \frac{1}{n}$, the series $\frac{4}{5} + \frac{4}{7} + \frac{4}{9} + \frac{4}{11} + \cdots$ is also divergent.

b. $\frac{1}{1^2} + \frac{1}{3^2} + \frac{1}{5^2} + \frac{1}{7^2} + \cdots$

The general term of the series is $\frac{1}{(2n-1)^2}$. The general term of the convergent series $1 + \frac{1}{2^2} + \frac{1}{3^2} + \frac{1}{4^2} + \cdots$ is $\frac{1}{n^2}$. Since $\frac{1}{(2n-1)^2} \leq \frac{1}{n^2}$ for all n, the series $\frac{1}{1^2} + \frac{1}{3^2} + \frac{1}{5^2} + \frac{1}{7^2} + \cdots$ is also convergent.

With a better understanding of convergent and divergent infinite series, we are now ready to tackle Zeno's paradox.

Example 6 **HISTORY** **Refer to the application at the beginning of the lesson. To disprove Zeno's conclusion that Achilles will never catch up to the tortoise, we must show that the infinite time series $1 + 0.5 + 0.25 + \cdots$ has a limit.**

To show that the series $1 + 0.5 + 0.25 + \cdots$ has a limit, we need to show that the series is convergent.

The general term of this series is $\frac{1}{2^n}$. Try using the ratio test for convergence of a series.

$a_n = \frac{1}{2^n}$ and $a_{n+1} = \frac{1}{2^{n+1}}$

$$r = \frac{\frac{1}{2^{n+1}}}{\frac{1}{2^n}}$$

$$= \frac{1}{2} \qquad \frac{1}{2^{n+1}} \cdot \frac{2^n}{1} = \frac{1}{2}$$

Since $r < 1$, the series converges and therefore has a sum. Thus, there is a time value for which the distance between Achilles and the tortoise will be zero. *You will determine how long it takes him to do so in Exercise 34.*

CHECK FOR UNDERSTANDING

Communicating Mathematics

Read and study the lesson to answer each question.

1. a. **Write** an example, of an infinite geometric series in which $|r| > 1$.
   b. **Determine** the 25th, 50th, and 100th terms of your series.
   c. **Identify** the sum of the first 25, 50, and 100 terms of your series.
   d. **Explain** why this type of infinite geometric series does not converge.
2. **Estimate** the sum S_n of the series whose partial sums are graphed at the right.

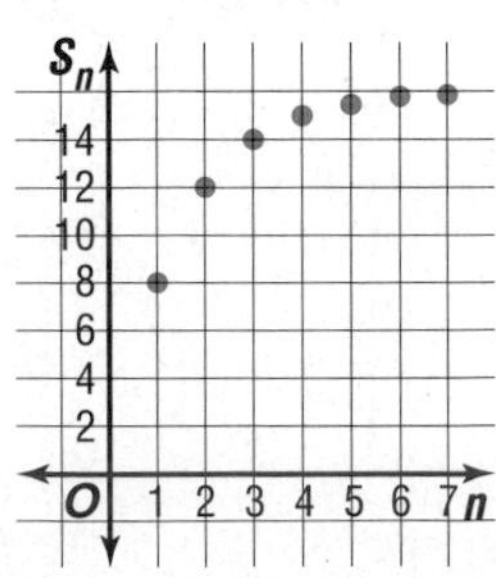

3. Consider the infinite series $\frac{1}{3} + \frac{2^2}{3^2} + \frac{3^2}{3^3} + \frac{4^2}{3^4} + \cdots$.

a. **Sketch** a graph of the first eight partial sums of this series.

b. **Make** a conjecture based on the graph found in part **a** as to whether the series is convergent or divergent.

c. **Determine** a general term for this series.

d. **Write a convincing argument** using the general term found in part **c** to support the conjecture you made in part **b**.

4. *Math Journal* **Make a list** of the methods presented in this lesson and in the previous lesson for determining convergence or divergence of an infinite series. Be sure to indicate any restrictions on a method's use. Then number your list as to the order in which these methods should be applied.

Guided Practice

Use the ratio test to determine whether each series is *convergent* or *divergent*.

5. $\frac{1}{2} + \frac{2}{2^2} + \frac{3}{2^3} + \cdots$

6. $\frac{3}{4} + \frac{7}{8} + \frac{11}{12} + \frac{15}{16} + \cdots$

7. Use the comparison test to determine whether the series $\frac{2}{1} + \frac{3}{2} + \frac{4}{3} + \cdots$ is *convergent* or *divergent.*

Determine whether each series is *convergent* or *divergent*.

8. $\frac{1}{4} + \frac{5}{16} + \frac{3}{8} + \frac{7}{16} + \cdots$

9. $\frac{1}{2 + 1^2} + \frac{1}{2 + 2^2} + \frac{1}{2 + 3^2} + \cdots$

10. $\frac{1}{1 \cdot 2} + \frac{1}{2 \cdot 2^2} + \frac{1}{3 \cdot 2^3} + \cdots$

11. $4 + 3 + \frac{9}{4} + \cdots$

12. **Ecology** An underground storage container is leaking a toxic chemical. One year after the leak began, the chemical had spread 1500 meters from its source. After two years, the chemical had spread 900 meters more, and by the end of the third year, it had reached an additional 540 meters.

a. If this pattern continues, how far will the spill have spread from its source after 10 years?

b. Will the spill ever reach the grounds of a school located 4000 meters away from the source? Explain.

EXERCISES

Practice

Use the ratio test to determine whether each series is *convergent* or *divergent*.

13. $\frac{4}{3} + \frac{4}{9} + \frac{4}{27} + \frac{4}{81} + \cdots$

14. $\frac{2}{5} + \frac{4}{10} + \frac{8}{15} + \cdots$

15. $2 + \frac{4}{2^2} + \frac{8}{3^2} + \frac{16}{4^2} + \cdots$

16. $\frac{2}{2 \cdot 3} + \frac{2}{3 \cdot 4} + \frac{2}{4 \cdot 5} + \cdots$

17. $1 + \frac{3}{1 \cdot 2 \cdot 3} + \frac{5}{1 \cdot 2 \cdot 3 \cdot 4 \cdot 5} + \cdots$

18. $5 + \frac{5^2}{1 \cdot 2} + \frac{5^3}{1 \cdot 2 \cdot 3} + \cdots$

19. Use the ratio test to determine whether the series $\frac{2 \cdot 4}{2} + \frac{4 \cdot 6}{4} + \frac{6 \cdot 8}{8} + \frac{8 \cdot 10}{16} + \cdots$ is convergent or divergent.

Use the comparison test to determine whether each series is *convergent* or *divergent*.

20. $\frac{1}{2^2} + \frac{1}{4^2} + \frac{1}{6^2} + \cdots$

21. $\frac{1}{2} + \frac{1}{9} + \frac{1}{28} + \frac{1}{65} + \cdots$

22. $\frac{1}{2} + \frac{2}{3} + \frac{3}{4} + \cdots$

23. $\frac{5}{3} + \frac{5}{4} + 1 + \frac{5}{6} + \cdots$

24. Use the comparison test to determine whether the series $\frac{1}{3} + \frac{1}{5} + \frac{1}{9} + \frac{1}{17} \cdots$ is *convergent* or *divergent.*

Determine whether each series is *convergent* or *divergent*.

25. $\frac{1}{2} - \frac{3}{8} + \frac{9}{32} - \cdots$

26. $3 + \frac{5}{3} + \frac{7}{5} + \cdots$

27. $\frac{1}{5 + 1^2} + \frac{1}{5 + 2^2} + \frac{1}{5 + 3^2} + \cdots$

28. $1 + \frac{1}{\sqrt{2}} + \frac{1}{\sqrt{3}} + \frac{1}{\sqrt{4}} + \cdots$

29. $\frac{4\pi}{3} + \frac{5\pi}{6} + \frac{\pi}{3} + \cdots$

30. $\frac{1}{4} + \frac{3}{8} + \frac{5}{16} + \frac{7}{32} + \cdots$

Applications and Problem Solving

31. **Economics** The MagicSoft software company has a proposal to the city council of Alva, Florida, to relocate there. The proposal claims that the company will generate \$3.3 million for the local economy by the \$1 million in salaries that will be paid. The city council estimates that 70% of the salaries will be spent in the local community, and 70% of that money will again be spent in the community, and so on.
    a. According to the city council's estimates, is the claim made by MagicSoft accurate? Explain.
    b. What is the correct estimate of the amount generated to the local economy?

32. **Critical Thinking** Give an example of a series $a_1 + a_2 + a_3 + \cdots + a_n + \cdots$ that diverges, but when its terms are squared, the resulting series $a_1^2 + a_2^2 + a_3^2 + \cdots + a_n^2 + \cdots$ converges.

33. **Cellular Growth** Leticia Cox is a biochemist. She is testing two different types of drugs that induce cell growth. She has selected two cultures of 1000 cells each. To culture A, she administers a drug that raises the number of cells by 200 each day and every day thereafter. Culture B gets a drug that increases cell growth by 8% each day and everyday thereafter.
    a. Assuming no cells die, how many cells will have grown in each culture by the end of the seventh day?
    b. At the end of one month's time, which drug will prove to be more effective in promoting cell growth? Explain.

34. **Critical Thinking** Refer to Example 6 of this lesson. The sequence of partial sums, $S_1, S_2, S_3, \ldots, S_n, \ldots$, for the time series is $1, \frac{3}{2}, \frac{7}{4}, \frac{15}{8}, \ldots$.
    a. Find a general expression for the nth term of this sequence.
    b. To determine how long it takes for Achilles to catch-up to the tortoise, find the sum of the infinite time series. (*Hint:* Recall from the definition of the sum S of an infinite series that $\lim_{n \to \infty} S_n = S$.)

35. Clocks The hour and minute hands of a clock travel around its face at different speeds, but at certain times of the day, the two hands coincide. In addition to noon and midnight, the hands also coincide at times occurring between the hours. According the figure at the right, it is 4:00.

a. When the minute hand points to 4, what fraction of the distance between 4 and 5 will the hour hand have traveled?

b. When the minute hand reaches the hour hand's new position, what additional fraction will the hour hand have traveled?

c. List the next two terms of this series representing the distance traveled by the hour hand as the minute hand "chases" its position.

d. At what time between the hours of 4 and 5 o'clock will the two hands coincide?

Mixed Review

36. Evaluate $\lim_{n \to \infty} \frac{4n^2 + 5}{3n^2 - 2n}$. *(Lesson 12-3)*

37. Find the ninth term of the geometric sequence $\sqrt{2}, 2, 2\sqrt{2}, \ldots$. *(Lesson 12-2)*

38. Form an arithmetic sequence that has five arithmetic means between −11 and 19. *(Lesson 12-1)*

39. Solve $45.9 = e^{0.075t}$ *(Lesson 11-6)*

40. **Navigation** A submarine sonar is tracking a ship. The path of the ship is recorded as $6 = 12r \cos(\theta - 30°)$. Find the linear equation of the path of the ship. *(Lesson 9-4)*

41. Find an ordered pair that represents $\overrightarrow{AB}$ for $A(8, -3)$ and $B(5, -1)$. *(Lesson 8-2)*

42. **SAT/ACT Practice** How many numbers from 1 to 200 inclusive are equal to the cube of an integer?

A one **B** two **C** three **D** four **E** five

MID-CHAPTER QUIZ

1. Find the 19th term in the sequence for which $a_1 = 11$ and $d = -2$. (Lesson 12-1)

2. Find S_{20} for the arithmetic series for which $a_1 = -14$ and $d = 6$. (Lesson 12-1)

3. Form a sequence that has two geometric means between 56 and 189. (Lesson 12-2)

4. Find the sum of the first eight terms of the series $3 - 6 + 12 - \cdots$. (Lesson 12-2)

5. Find $\lim_{n \to \infty} \frac{n^2 + 2n - 5}{n^2 - 1}$ or explain why the limit does not exist. (Lesson 12-3)

6. **Recreation** A bungee jumper rebounds 55% of the height jumped. If a bungee jump is made using a cord that stretches 250 feet, find the total distance traveled by the jumper before coming to rest. (Lesson 12-3)

7. Find the sum of the following series. $\frac{1}{25} + \frac{1}{250} + \frac{1}{2500} + \cdots$. (Lesson 12-3)

Determine whether each series is *convergent* or *divergent*. (Lesson 12-4)

8. $\frac{1}{10} + \frac{2}{100} + \frac{6}{1000} + \frac{24}{10{,}000} + \cdots$

9. $\frac{6}{5} + \frac{2}{5} + \frac{2}{15} + \cdots$

10. **Finance** Ms. Fuentes invests $500 quarterly (January 1, April 1, July 1, and October 1) in a retirement account that pays an APR of 12% compounded quarterly. Interest for each quarter is posted on the last day of the quarter. Determine the value of her investment at the end of the year. (Lesson 12-2)

Extra Practice See p. A49.

Sigma Notation and the *n*th Term

OBJECTIVE

- Use sigma notation.

MANUFACTURING Manufacturers are required by the Environmental Protection Agency to meet certain emission standards. If these standards are not met by a preassigned date, the manufacturer is fined. To encourage swift compliance, the fine increases a specified amount each day until the manufacturer is able to pass inspection. Suppose a manufacturing plant is charged $2000 for not meeting its January 1st deadline. The next day it is charged $2500, the next day $3000, and so on, until it passes inspection on January 21st. What is the total amount of fines owed by the manufacturing plant? *This problem will be solved in Example 2.*

In mathematics, the uppercase Greek letter sigma, Σ, is often used to indicate a sum or series. A series like the one indicated above may be written using **sigma notation.**

maximum value of n → $\sum_{n=1}^{k} a_n$ ← *expression for general term*

starting value of n → $n=1$

Other variables besides n may be used for the index of summation.

The variable n used with the sigma notation is called the **index of summation.**

Sigma Notation of a Series

For any sequence a_1, a_2, a_3, ..., the sum of the first k terms may be written $\sum_{n=1}^{k} a_n$, which is read "the summation from $n = 1$ to k of a_n." Thus,

$$\sum_{n=1}^{k} a_n = a_1 + a_2 + a_3 + \cdots + a_k, \text{ where } k \text{ is an integer value.}$$

Example  **Write each expression in expanded form and then find the sum.**

a. $\sum_{n=1}^{4} (n^2 - 3)$

First, write the expression in expanded form.

$$\sum_{n=1}^{4} (n^2 - 3) = \underset{n=1}{(1^2 - 3)} + \underset{n=2}{(2^2 - 3)} + \underset{n=3}{(3^2 - 3)} + \underset{n=4}{(4^2 - 3)}$$

Now find the sum.

Method 1 Simplify the expanded form.
$(1^2 - 3) + (2^2 - 3) + (3^2 - 3) + (4^2 - 3) = -2 + 1 + 6 + 13$ or 18

Method 2 Use a graphing calculator.

You can combine **sum(** with **seq(** to obtain the sum of a finite sequence. In the **LIST** menu, select the **MATH** option to locate the **sum(** command. The **seq(** command can be found in the **OPS** option of the **LIST** menu.

expression *starting value* *step size*

sum(seq(N² – 3 , N , 1 , 4 , 1))

index of summation *maximum value*

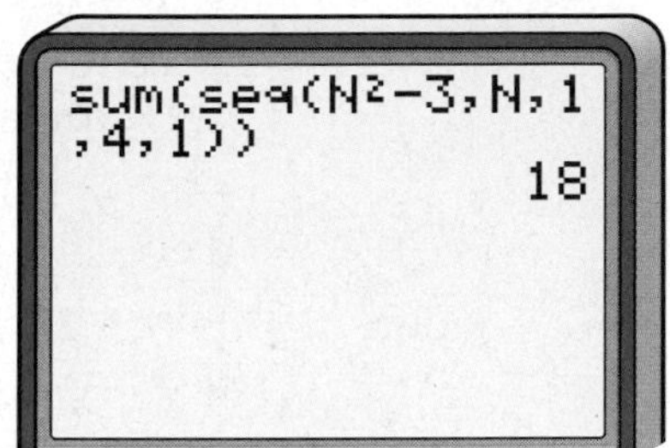

b. $\sum_{n=1}^{\infty} 5\left(-\frac{2}{7}\right)^{n-1}$

$$\sum_{n=1}^{\infty} 5\left(-\frac{2}{7}\right)^{n-1} = \underset{n=1}{5\left(-\frac{2}{7}\right)^{1-1}} + \underset{n=2}{5\left(-\frac{2}{7}\right)^{2-1}} + \underset{n=3}{5\left(-\frac{2}{7}\right)^{3-1}} + \underset{n=4}{5\left(-\frac{2}{7}\right)^{4-1}} + \cdots$$
$$= 5 + \frac{10}{7} + \frac{20}{49} + \frac{40}{343} + \cdots$$

This is an infinite geometric series. Use the formula $S = \frac{a_1}{1 - r}$.

$S = \frac{5}{1 - \left(-\frac{2}{7}\right)}$ *$a_1 = 5$, $r = -\frac{2}{7}$*

$S = \frac{35}{9}$

Therefore, $\sum_{n=1}^{\infty} 5\left(-\frac{2}{7}\right)^{n-1} = \frac{35}{9}$.

A series in expanded form can be written using sigma notation if a general formula can be written for the *n*th term of the series.

Example 2 **MANUFACTURING** **Refer to the application at the beginning of this lesson.**

a. How much is the company fined on the 20th day?

b. What is the total amount in fines owed by the manufacturing plant?

c. Represent this sum using sigma notation.

a. Since this sequence is arithmetic, we can use the formula for the *n*th term of an arithmetic sequence to find the amount of the fine charged on the 20th day.

$a_n = a_1 + (n - 1)d$
$a_{20} = 2000 + (20 - 1)500$ *$a_1 = 2000$, $n = 20$, and $d = 500$*
$= 11{,}500$

The fine on the 20th day will be \$11,500.

b. To determine the total amount owed in fines, we can use the formula for the sum of an arithmetic series. The plant will not be charged for the day it passes inspection, so it is assessed 20 days in fines.

$$S_n = \frac{n}{2}(a_1 + a_n)$$

$$= \frac{20}{2}(2000 + 11{,}500) \quad n = 20,\ a_1 = 2000, \text{ and } a_{20} = 11{,}500$$

$$= 135{,}000$$

The plant must pay a total of \$135,000 in fines.

c. To determine the fine for the nth day, we can again use the formula for the nth term of an arithmetic sequence.

$$a_n = a_1 + (n - 1)d$$

$$= 2000 + (n - 1)500 \quad a_1 = 2000 \text{ and } d = 500$$

$$= 2000 + 500n - 500$$

$$= 500n + 1500$$

The fine on the nth day is \$500n + \$1500.

Since \$2000 is the fine on the first day and \$12,000 is the fine on the 20th day, the index of summation goes from $n = 1$ to $n = 20$.

Therefore, $2000 + 2500 + 3000 + \cdots + 11{,}500 = \sum_{n=1}^{20} (500n + 1500)$ or $\sum_{n=1}^{20} 500(n + 3)$.

When using sigma notation, it is not always necessary that the sum start with the index equal to 1.

Example 3 **Express the series 15 + 24 + 35 + 48 + ⋯ + 143 using sigma notation.**

Notice that each term is 1 less than a perfect square. Thus, the nth term of the series is $n^2 - 1$. Since $4^2 - 1 = 15$ and $12^2 - 1 = 143$, the index of summation goes from $n = 4$ to $n = 12$.

Therefore, $15 + 24 + 35 + 48 + \cdots + 143 = \sum_{n=4}^{12} (n^2 - 1)$.

As you have seen, not all sequences are arithmetic or geometric. Some important sequences are generated by products of consecutive integers. The product $n(n - 1)(n - 2) \cdots 3 \cdot 2 \cdot 1$ is called ***n* factorial** and symbolized $n!$.

***n* Factorial**

The expression $n!$ (n factorial) is defined as follows for n, an integer greater than zero.

$$n! = n(n - 1)(n - 2) \cdots 1$$

By definition 0! = 1.

Factorial notation can be used to express the general form of a series.

Example 4 **Express the series $\frac{2}{2} - \frac{4}{6} + \frac{6}{24} - \frac{8}{120} + \frac{10}{720}$ using sigma notation.**

The sequence representing the numerators is 2, 4, 6, 8, 10. This is an arithmetic sequence with a common difference of 2. Thus the nth term can be represented by $2n$.

Because the series has alternating signs, one factor for the general term of the series is $(-1)^{n+1}$. Thus, when n is odd, the terms are positive, and when n is even, the terms are negative.

The sequence representing the denominators is 2, 6, 24, 120, 720. This sequence is generated by factorials.

$2! = 2$
$3! = 6$
$4! = 24$
$5! = 120$
$6! = 720$

Therefore, $\frac{2}{2} - \frac{4}{6} + \frac{6}{24} - \frac{8}{120} + \frac{10}{720} = \sum_{n=1}^{5} \frac{(-1)^{n+1}2n}{(n+1)!}$.

You can check this answer by substituting values of n into the general term.

CHECK FOR UNDERSTANDING

Communicating Mathematics

Read and study the lesson to answer each question.

1. **Find a counterexample** to the following statement: "The summation notion used to represent a series is unique."
2. In Example 4 of this lesson, the alternating signs of the series were represented by a factor of $(-1)^{n+1}$.
   a. **Write** a different factor that could have been used in the general form of the nth term of the series.
   b. **Determine** a factor that could be used if the alternating signs of the series began with a negative first term.
3. Consider the series $\sum_{j=2}^{10} (-2j + 1)$.
   a. **Identify** the number of terms in this series.
   b. **Write** a formula that determines the number of terms t in a finite series if the index of summation has a minimum value of a and a maximum value of b.
   c. **Use** the formula in part **b** to identify the number of terms in the series $\sum_{k=-2}^{3} \frac{1}{k+3}$.
   d. **Verify** your answer in part **c** by writing the series $\sum_{k=-2}^{3} \frac{1}{k+3}$ in expanded form.

Guided Practice

Write each expression in expanded form and then find the sum.

4. $\sum_{n=1}^{6} (n - 3)$
5. $\sum_{k=2}^{5} 4k$
6. $\sum_{a=0}^{4} \frac{1}{2^a}$
7. $\sum_{p=0}^{\infty} 5\left(\frac{3}{4}\right)^p$

Express each series using sigma notation.

8. $5 + 10 + 15 + 20 + 25$

9. $2 + 4 + 10 + 28$

10. $2 - 4 - 10 - 16$

11. $\frac{3}{4} + \frac{3}{8} + \frac{3}{16} + \frac{3}{32} + \cdots$

12. $-3 + 9 - 27 + \cdots$

13. Aviation Each October Albuquerque, New Mexico, hosts the Balloon Fiesta. In 1998, 873 hot air balloons participated in the opening day festivities. One of these balloons rose 389 feet after 1 minute. Because the air in the balloon was not reheated, each succeeding minute the balloon rose 63% as far as it did the previous minute.

a. Use sigma notation to represent the height of the balloon above the ground after one hour. Then calculate the total height of the balloon after one hour to the nearest foot.

b. What was the maximum height achieved by this balloon?

EXERCISES

Practice

Write each expression in expanded form and then find the sum.

14. $\sum_{n=1}^{4} (2n - 7)$

15. $\sum_{a=2}^{5} 5a$

16. $\sum_{b=3}^{8} (6 - 4b)$

17. $\sum_{k=2}^{6} (k + k^2)$

18. $\sum_{n=5}^{8} \frac{n}{n-4}$

19. $\sum_{j=4}^{8} 2^j$

20. $\sum_{m=0}^{3} 3^{m-1}$

21. $\sum_{r=1}^{3} \left(\frac{1}{2} + 4^r\right)$

22. $\sum_{i=3}^{5} (0.5)^{-i}$

23. $\sum_{k=3}^{7} k!$

24. $\sum_{p=0}^{\infty} 4(0.75)^p$

25. $\sum_{n=1}^{\infty} 4\left(\frac{2}{5}\right)^n$

26. Write $\sum_{n=2}^{5} n + \boldsymbol{i}^n$ in expanded form. Then find the sum.

Express each series using sigma notation.

27. $6 + 9 + 12 + 15$

28. $1 + 4 + 16 + \cdots + 256$

29. $8 + 10 + 12 + \cdots + 24$

30. $-8 + 4 - 2 + 1$

31. $10 + 50 + 250 + 1250$

32. $13 + 9 + 5 + 1$

33. $\frac{1}{9} + \frac{1}{14} + \frac{1}{19} + \cdots + \frac{1}{49}$

34. $\frac{2}{3} + \frac{4}{5} + \frac{8}{7} + \frac{16}{9} + \cdots$

35. $4 - 9 + 16 - 25 + \cdots$

36. $5 + 5 + \frac{5}{2} + \frac{5}{6} + \frac{5}{24} + \cdots$

37. $-32 + 16 - 8 + 4 - \cdots$

38. $2 + \frac{6}{2} + \frac{24}{3} + \frac{120}{4} + \cdots$

39. $\frac{1}{5} + \frac{2}{7} + \frac{3}{11} + \frac{4}{19} + \frac{5}{35} + \cdots$

40. $\frac{3}{9 \cdot 2} + \frac{8}{27 \cdot 6} + \frac{15}{81 \cdot 24} + \cdots$

41. Express the series $\frac{\sqrt{2}}{3} + \frac{2}{6} + \frac{\sqrt{8}}{18} + \frac{4}{72} + \frac{\sqrt{32}}{360} + \cdots$ using sigma notation.

Simplify. Assume that n and m are positive integers, $a > b$, and $a > 2$.

42. $\frac{(a-2)!}{a!}$

43. $\frac{(a+1)!}{(a-2)!}$

44. $\frac{(a+b)!}{(a+b-1)!}$

Graphing Calculator

45. Use a graphing calculator to find the sum of $\sum_{n=1}^{100} \frac{8n^3 - 2n^2 + 5}{n^4}$. Round to the nearest hundredth.

Applications and Problem Solving

46. Advertising A popular shoe manufacturer is planning to market a new style of tennis shoe in a city of 500,000 people. Using a prominent professional athlete as their spokesperson, the company's ad agency hopes to induce 35% of the people to buy the product. The ad agency estimates that these satisfied customers will convince 35% of 35% of 500,000 to buy a pair of shoes, and those will persuade 35% of 35% of 35% of 500,000, and so on.

a. Model this situation using sigma notation.

b. Find the total number of people that will buy the product as a result of the advertising campaign.

c. What percentage of the population is this?

d. What important assumption does the advertising agency make in proposing the figure found in part **b** to the shoe manufacturer?

47. Critical Thinking Solve each equation for x.

a. $\sum_{n=1}^{6} (x - 3n) = -3$ b. $\sum_{n=0}^{5} n(n - x) = 25$

48. Critical Thinking Determine whether each equation is *true* or *false*. Explain your answer.

a. $\sum_{k=3}^{7} 3^k + \sum_{b=7}^{9} 3^b = \sum_{a=3}^{9} 3^a$ b. $\sum_{n=2}^{8} (2n - 3) = \sum_{m=3}^{9} (2m - 5)$

c. $2\sum_{n=3}^{7} n^2 = \sum_{n=3}^{7} 2n^2$ d. $\sum_{k=1}^{10} (5 + k) = \sum_{p=0}^{9} (4 + p)$

49. Word Play An *anagram* is a word or phrase that is made by rearranging the letters of another word or phrase. Consider the word "SILENT."

a. How many different arrangements of the letters in this word are possible? Write this number as a factorial. (*Hint:* First solve a simpler problem to see a pattern, such as how many different arrangements are there of just 2 letters? 3 letters?)

b. If a friend gives you a hint and tells you that an anagram of this word starts with "L," how many different arrangements still remain?

c. Your friend gives you one more hint. The last letter in the anagram is "N." Determine how many possible arrangements remain and then determine the anagram your friend is suggesting.

50. Chess A standard chess board contains 64 small black or white squares. These squares make up many other larger squares of various sizes.

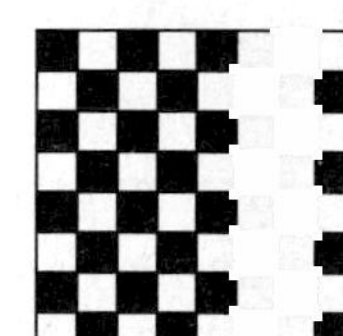

a. How many 8×8 squares are there on a standard 8×8 chessboard? How many 7×7 squares?

b. Continue this list until you have accounted for all 8 sizes of squares.

c. Use sigma notation to represent the total number of squares found on an 8×8 chessboard. Then calculate this sum.

Mixed Review

51. Use the comparison test to determine whether the series $\frac{3}{3} + \frac{3}{4} + \frac{3}{5} + \frac{3}{6} + \cdots$ is convergent or divergent. *(Lesson 12-4)*

52. Chemistry A vacuum pump removes 20% of the air in a sealed jar on each stroke of its piston. The jar contains 21 liters of air before the pump starts. After how many strokes will only 42% of the air remain? *(Lesson 12-3)*

Extra Practice See p. A49.

53. Find the first four terms of the geometric sequence for which $a_5 = 32\sqrt{2}$ and $r = -\sqrt{2}$. *(Lesson 12-2)*

54. Evaluate $\log_{10} 0.001$. *(Lesson 11-4)*

55. Write the standard form of the equation of the circle that passes through points at $(0, 9)$, $(-7, 2)$, and $(0, -5)$. *(Lesson 10-2)*

56. Simplify $(\sqrt{2} + \boldsymbol{i})(4\sqrt{2} + \boldsymbol{i})$. *(Lesson 9-5)*

57. **Sports** Find the initial vertical and horizontal velocities of a javelin thrown with an initial velocity of 59 feet per second at an angle of 63° with the horizontal. *(Lesson 8-7)*

58. Find the equation of the line that bisects the obtuse angle formed by the graphs of $2x - 3y + 9 = 0$ and $x + 4y + 4 = 0$. *(Lesson 7-7)*

59. **SAT/ACT Practice** If $\frac{5 + m}{9 + m} = \frac{2}{3}$, then $m = ?$

A 8 **B** 6 **C** 5 **D** 3 **E** 2

CAREER CHOICES

Operations Research Analyst

In the changing economy of today, it is difficult to start and maintain a successful business. A business operator needs to be sure that the income from the business exceeds the expenses. Sometimes, businesses need the services of an operations research analyst. If you enjoy mathematics and solving tough problems, then you may want to consider a career as an operations research analyst.

In this occupation, you would gather many types of data about a business and analyze that data using mathematics and statistics. Examples of ways you might assist a business are: help a retail store determine the best store layout, help a bank in processing deposits more efficiently, or help a business set prices. Most operations research analysts work for private industry, private consulting firms, or the government.

Career Overview

Degree Preferred:
bachelor's degree in applied mathematics

Related Courses:
mathematics, statistics, computer science, English

Outlook:
faster than average through the year 2006

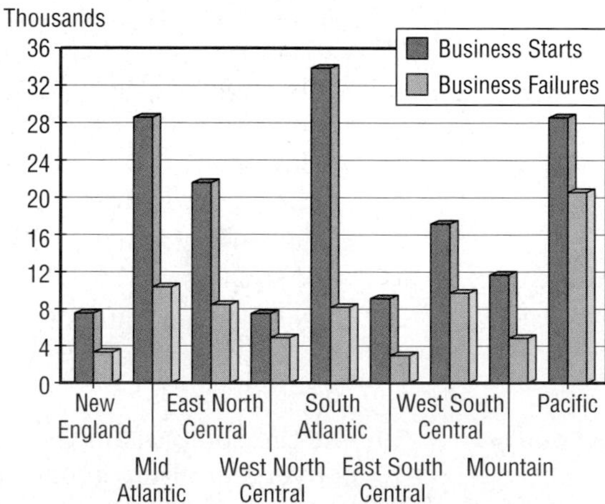

interNET CONNECTION For more information on careers in operations research, visit: **www.amc.glencoe.com**

12-6 The Binomial Theorem

OBJECTIVE

- Use the Binomial Theorem to expand binomials.

FAMILY On November 19, 1997, Bobbie McCaughey, an Iowa seamstress, gave birth by Caesarian section to seven babies. The birth of the septuplets was the first of its kind in the United States since 1985. The babies, born after just 30 weeks of pregnancy, weighed from 2 pounds 5 ounces to 3 pounds 4 ounces. *A problem related to this will be solved in Example 2.*

Recall that a binomial, such as $x + y$, is an algebraic expression involving the sum of two unlike terms, in this case x and y. Just as there are patterns in sequences and series, there are numerical patterns in the expansion of powers of binomials. Let's examine the expansion of $(x + y)^n$ for $n = 0$ to $n = 5$. You already know a few of these expansions and the others can be obtained using algebraic properties.

$$(x + y)^0 = 1x^0y^0$$
$$(x + y)^1 = 1x^1y^0 + 1x^0y^1$$
$$(x + y)^2 = 1x^2y^0 + 2x^1y^1 + 1x^0y^2$$
$$(x + y)^3 = 1x^3y^0 + 3x^2y^1 + 3x^1y^2 + 1x^0y^3$$
$$(x + y)^4 = 1x^4y^0 + 4x^3y^1 + 6x^2y^2 + 4x^1y^3 + 1x^0y^4$$
$$(x + y)^5 = 1x^5y^0 + 5x^4y^1 + 10x^3y^2 + 10x^2y^3 + 5x^1y^4 + 1x^0y^5$$

Patterns in the Binomial Expansion of $(x + y)^n$

The following patterns can be observed in the expansion of $(x + y)^n$.

1. The expansion of $(x + y)^n$ has $n + 1$ terms.
2. The first term is x^n, and the last term is y^n.
3. In successive terms, the exponent of x decreases by 1, and the exponent of y increases by 1.
4. The degree of each term (the sum of the exponents of the variables) is n.
5. In any term, if the coefficient is multiplied by the exponent of x and the product is divided by the number of that term, the result is the coefficient of the following term.
6. The coefficients are symmetric. That is, the first term and the last term have the same coefficient. The second term and the second from the last term have the same coefficient, and so on.

If just the coefficients of these expansions are extracted and arranged in a triangular array, they form a pattern called **Pascal's triangle.** As you examine the triangle shown below, note that if two consecutive numbers in any row are added, the sum is a number in the following row. These three numbers also form a triangle.

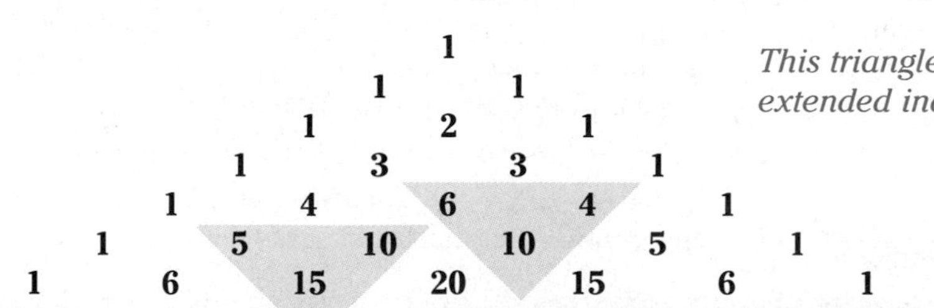

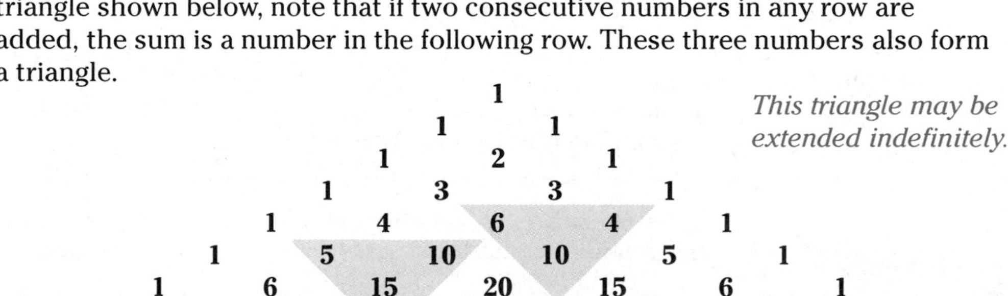

This triangle may be extended indefinitely.

Example **1** **Use Pascal's triangle to expand each binomial.**

a. $(x + y)^6$

First, write the series without the coefficients. Recall that the expression should have 6 + 1 or 7 terms, with the first term being x^6 and the last term being y^6. Also note that the exponents of x should decrease from 6 to 0 while the exponents of y should increase from 0 to 6, while the degree of each term is 6.

$$x^6 + x^5y + x^4y^2 + x^3y^3 + x^2y^4 + xy^5 + y^6 \quad y^0 = 1,\ x^0 = 1$$

Then, use the numbers in the seventh row of Pascal's triangle as the coefficients of the terms. *Why is the seventh row used instead of the sixth row?*

$$\begin{array}{ccccccc} 1 & 6 & 15 & 20 & 15 & 6 & 1 \\ \downarrow & \downarrow & \downarrow & \downarrow & \downarrow & \downarrow & \downarrow \end{array}$$

$$(x + y)^6 = x^6 + 6x^5y + 15x^4y^2 + 20x^3y^3 + 15x^2y^4 + 6xy^5 + y^6$$

b. $(3x + 2y)^7$

Extend Pascal's triangle to the eighth row.

$$\begin{array}{ccccccccccccccc} & 1 & & 6 & & 15 & & 20 & & 15 & & 6 & & 1 & \\ 1 & & 7 & & 21 & & 35 & & 35 & & 21 & & 7 & & 1 \end{array}$$

Then, write the expression and simplify each term. Replace each x with $3x$ and y with $2y$.

$$\begin{aligned}(3x + 2y)^7 &= (3x)^7 + 7(3x)^6(2y) + 21(3x)^5(2y)^2 + 35(3x)^4(2y)^3 + 35(3x)^3(2y)^4 + \\ &\quad 21(3x)^2(2y)^5 + 7(3x)(2y)^6 + (2y)^7 \\ &= 2187x^7 + 10{,}206x^6y + 20{,}412x^5y^2 + 22{,}680x^4y^3 + 15{,}120x^3y^4 + \\ &\quad 6048x^2y^5 + 1344xy^6 + 128y^7\end{aligned}$$

You can use Pascal's triangle to solve real-world problems in which there are only two outcomes for each event. For example, you can determine the distribution of answers on true-false tests, the combinations of heads and tails when tossing a coin, or the possible sequences of boys and girls in a family.

Example **2** **FAMILY Refer to the application at the beginning of the lesson. Of the seven children born to the McCaughey's, at least three were boys. How many of the possible groups of boys and girls have at least three boys?**

Let g represent girls and b represent boys. To find the number of possible groups, expand $(g + b)^7$. Use the eighth row of Pascal's triangle for the ' expansion.

$$g^7 + 7g^6b + 21g^5b^2 + 35g^4b^3 + 35g^3b^4 + 21g^2b^5 + 7gb^6 + b^7$$

To have at least three boys means that there could be 3, 4, 5, 6, or 7 boys. The total number of ways to have at least three boys is the same as the sum of the coefficients of g^4b^3, g^3b^4, g^2b^5, gb^6, and b^7. This sum is 35 + 35 + 21 + 7 + 1 or 99.

Thus, there are 99 possible groups of boys and girls in which there are at least three boys.

The general expansion of $(x + y)^n$ can also be determined by the **Binomial Theorem.**

Binomial Theorem If n is a positive integer, then the following is true.

$$(x + y)^n = x^n + nx^{n-1}y + \frac{n(n-1)}{1 \cdot 2}x^{n-2}y^2 + \frac{n(n-1)(n-2)}{1 \cdot 2 \cdot 3}x^{n-3}y^3 + \cdots + y^n$$

Example 3 **Use the Binomial Theorem to expand $(2x - y)^6$.**

The expansion will have seven terms. Find the first four terms using the sequence 1, $\frac{6}{1}$ or 6, $\frac{6 \cdot 5}{1 \cdot 2}$ or 15, $\frac{6 \cdot 5 \cdot 4}{1 \cdot 2 \cdot 3}$ or 20. Then use symmetry to find the remaining terms, 15, 6, and 1.

$$\begin{aligned}(2x - y)^6 &= (2x)^6 + 6(2x)^5(-y) + 15(2x)^4(-y)^2 + 20(2x)^3(-y)^3 + 15(2x)^2(-y)^4 + \\ &\quad 6(2x)(-y)^5 + (-y)^6 \\ &= 64x^6 - 192x^5y + 240x^4y^2 - 160x^3y^3 + 60x^2y^4 - 12xy^5 + y^6\end{aligned}$$

An equivalent form of the Binomial Theorem uses both sigma and factorial notation. It is written as follows, where n is a positive integer and r is a positive integer or zero.

$$(x + y)^n = \sum_{r=0}^{n} \frac{n!}{r!(n-r)!}x^{n-r}y^r$$

You can use this form of the Binomial Theorem to find individual terms of an expansion.

Example 4 **Find the fifth term of $(4a + 3b)^7$.**

$$(4a + 3b)^7 = \sum_{r=0}^{7} \frac{7!}{r!(7-r)!}(4a)^{7-r}(3b)^r$$

To find the fifth term, evaluate the general term for $r = 4$. *Since r increases from 0 to n, r is one less than the number of the term.*

$$\begin{aligned}\frac{7!}{r!(7-r)!}(4a)^{7-r}(3b)^r &= \frac{7!}{4!(7-4)!}(4a)^{7-4}(3b)^4 \\ &= \frac{7 \cdot 6 \cdot 5 \cdot 4!}{4!\,3!}(4a)^3(3b)^4 \\ &= 181{,}440a^3b^4\end{aligned}$$

The fifth term of $(4a + 3b)^7$ is $181{,}440a^3b^4$.

CHECK FOR UNDERSTANDING

Communicating Mathematics

Read and study the lesson to answer each question.

1. a. **Calculate** the sum of the numbers in each row of Pascal's triangle for $n = 0$ to 5.
   b. **Write** an expression to represent the sum of the numbers in the nth row of Pascal's triangle.

2. Examine the expansion of $(x - y)^n$ for $n = 3, 4,$ and 5.
   a. **Identify** the sign of the second term for each expansion.
   b. **Identify** the sign of the third term for each expansion.
   c. **Explain** how to determine the sign of a term without writing out the entire expansion.
3. **Restate** what is meant by the observation that each term in a binomial expansion has degree n.
4. If ax^7y^b is a term from the expansion of $(x + y)^{12}$, **describe** how to determine its coefficient a and missing exponent b without writing the entire expansion.

Guided Practice

5. Use Pascal's triangle to expand $(c + d)^5$.

Use the Binomial Theorem to expand each binomial.

6. $(a + 3)^6$
7. $(5 - y)^3$
8. $(3p - 2q)^4$

Find the designated term of each binomial expansion.

9. 6th term of $(a - b)^7$
10. 4th term of $\left(x + \sqrt{3}\right)^9$
11. **Coins** A coin is flipped five times. Find the number of possible sets of heads and tails that have each of the following.
    a. 0 heads
    b. 2 heads
    c. at least 4 heads
    d. at most 3 heads

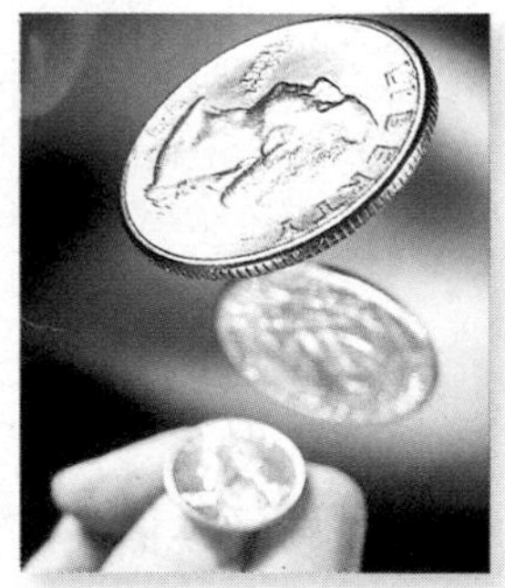

EXERCISES

Practice

Use Pascal's triangle to expand each binomial.

12. $(a + b)^8$
13. $(n - 4)^6$
14. $(3c - d)^4$
15. Expand $(2 + a)^9$ using Pascal's triangle.

Use the Binomial Theorem to expand each binomial.

16. $(d + 2)^7$
17. $(3 - x)^5$
18. $(4a + b)^4$
19. $(2x - 3y)^3$
20. $\left(3m + \sqrt{2}\right)^4$
21. $\left(\sqrt{c} - 1\right)^6$
22. $\left(\frac{1}{2}n + 2\right)^5$
23. $\left(3a + \frac{2}{3}b\right)^4$
24. $(p^2 + q)^8$
25. Expand $(xy - 2z^3)^6$ using the Binomial Theorem.

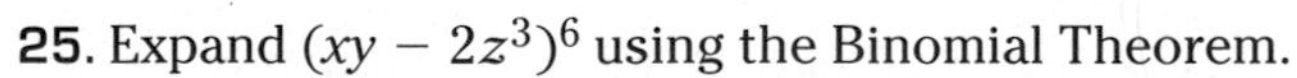

Find the designated term of each binomial expansion.

26. 5th term of $(x + y)^9$
27. 4th term of $\left(a - \sqrt{2}\right)^8$
28. 4th term of $(2a - b)^7$
29. 7th term of $(3c + 2d)^9$
30. 8th term of $\left(\frac{1}{2}x - y\right)^{10}$
31. 6th term of $(2p - 3q)^{11}$
32. Find the middle term of the expansion of $\left(\sqrt{x} + \sqrt{y}\right)^8$.

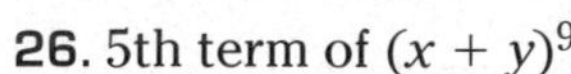

Applications and Problem Solving

33. **Business** A company decides to form a recycling committee to find a more efficient means of recycling waste paper. The committee is to be composed of eight employees. Of these eight employees, at least four women are to be on the committee. How many of the possible groups of men and women have at least four women?

34. **Critical Thinking** Describe a strategy that uses the Binomial Theorem to expand $(a + b + c)^{12}$.

35. **Education** Rafael is taking a test that contains a section of 12 true-false questions.
    a. How many of the possible groups of answers to these questions have exactly 8 correct answers of false?
    b. How many of the possible groups of answers to these questions have at least 6 correct answers of true?

36. **Critical Thinking** Find a term in the expansion of $\left(3x^2 - \frac{1}{4x}\right)^6$ that does not contain the variable x.

37. **Numerical Analysis** Before the invention of modern calculators and computers, mathematicians searched for ways to shorten lengthy calculations such as $(1.01)^4$.
    a. Express 1.01 as a binomial.
    b. Use the binomial representation of 1.01 found in part **a** and the Binomial Theorem to calculate the value of $(1.01)^4$ to eight decimal places.
    c. Use a calculator to estimate $(1.01)^4$ to eight decimal places. Compare this value to the value found in part **b**.

Mixed Review

38. Write $\sum_{k=2}^{7} 5 - 2k$ in expanded form and then find the sum. *(Lesson 12-5)*

39. Use the ratio test to determine whether the series $2 + \frac{2^2}{2!} + \frac{2^3}{3!} + \frac{2^4}{4!} + \cdots$ is *convergent* or *divergent*. *(Lesson 12-4)*

40. Find the sum of $\frac{2}{3} + \frac{1}{3} + \frac{1}{6} + \frac{1}{12} + \cdots$ or explain why one does not exist. *(Lesson 12-3)*

41. **Finance** A bank offers a home mortgage for an annual interest rate of 8%. If a family decides to mortgage a $150,000 home over 30 years with this bank, what will the monthly payment for the principal and interest on their mortgage be? *(Lesson 11-2)*

42. Write $\overrightarrow{MK}$ as the sum of unit vectors if $M(-2, 6, 3)$ and $K(4, 8, -2)$. *(Lesson 8-3)*

43. **Construction** A highway curve, in the shape of an arc of a circle, is 0.25 mile. The direction of the highway changes 45° from one end of the curve to the other. Find the radius of the circle in feet that the curve follows. *(Lesson 6-1)*

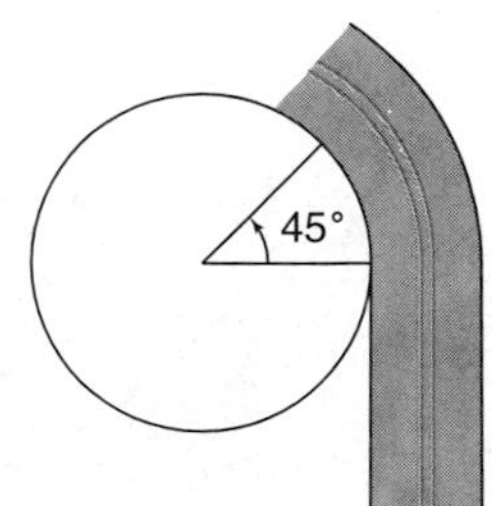

44. **SAT/ACT Practice** If b is a prime integer such that $3b > 10 > \frac{5}{6}b$, which of the following is a possible value of b?

A 2 B 3 C 4 D 11 E 13

Extra Practice See p. A50.

Special Sequences and Series

OBJECTIVES

- Approximate e^x, trigonometric values, and logarithms of negative numbers by using series.
- Use Euler's Formula to write the exponential form of a complex number.

NATURE An important sequence found in nature can be seen in the spiral of a Nautilus shell. To see this sequence follow procedure below.

- Begin by placing two small squares with side length 1 next to each other.
- Below both of these, draw a square with side length 2.
- Now draw a square that touches both a unit square and the 2-square. This square will have sides 3 units long.
- Add another square that touches both the 2-square and the 3-square. This square will have sides of 5 units.
- Continue this pattern around the picture so that each new square has sides that are as long as the sum of the latest two square's sides.

The side lengths of these squares form what is known as the **Fibonacci sequence:** 1, 1, 2, 3, 5, 8, 13, ... A spiral that closely models the Nautilus spiral can be drawn by first rearranging the squares so that the unit squares are in the interior and then connecting quarter circles, each of whose radius is a side of a new square. This spiral is known as the *Fibonacci spiral. A problem related to the Fibonacci sequence will be solved in Example 1.*

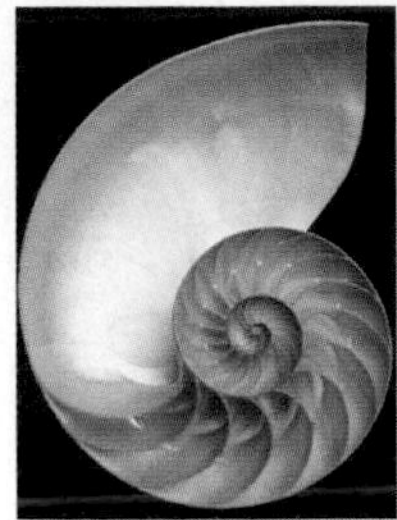

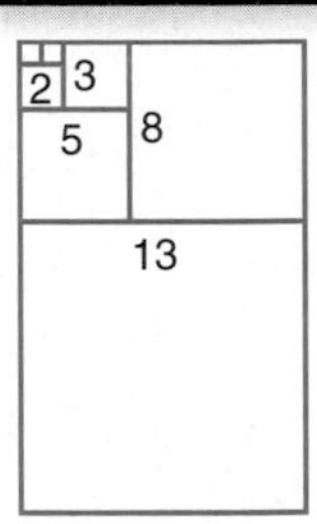

rearrange

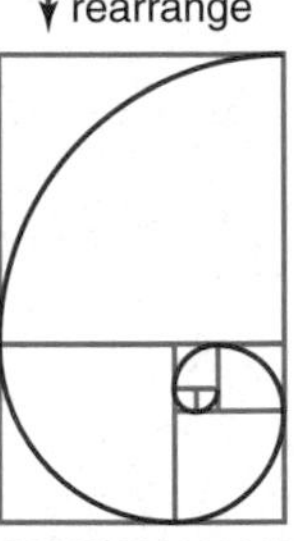

The Fibonacci sequence describes many patterns of numbers found in nature. This sequence was presented by an Italian mathematician named Leonardo de Pisa, also known as Fibonacci (pronounced *fih buh NACH ee*), in 1201. The first two numbers in the sequence are 1, that is, $a_1 = 1$ and $a_2 = 1$. As you have seen in the example above, adding the two previous terms generates each additional term in the sequence.

The original problem that Fibonacci investigated in 1202 involved the reproductive habits of rabbits under ideal conditions.

Example 1 **NATURE Suppose a newly born pair of rabbits are allowed to breed in a controlled environment. How many rabbit pairs will there be after one year if the following assumptions are made?**

- **A male and female rabbit can mate at the age of one month.**
- **At the end of its second month, a female rabbit can produce another pair of rabbits (one male, one female).**
- **The rabbits never die.**
- **The female always produces one new pair every month from the second month on.**

Based on these assumptions, at the end of the first month, there will be one pair of rabbits. And at the end of the second month, the female produces a new pair, so now there are two pairs of rabbits.

Month	Number of Pairs
1	1
2	1
3	2
4	3
5	5

The following table shows the pattern.

Month	1	2	3	4	5	6	7	8	9	10	11	12
Number of Pairs	1	1	2	3	5	8	13	21	34	55	89	144

There will be 144 pairs of rabbits during the twelfth month.

Each term in the sequence is the sum of the two previous terms.

Another important series is the series that is used to define the irrational number e. The Swiss mathematician Leonhard Euler (pronounced OY ler), published a work in which he developed this irrational number. It has been suggested that in his honor the number is called e, the Euler number. The number can be expressed as the sum of the following infinite series.

$$e = 1 + \frac{1}{1!} + \frac{1}{2!} + \frac{1}{3!} + \frac{1}{4!} + \frac{1}{5!} + \cdots + \frac{1}{n!} + \cdots$$

The Binomial Theorem can be used to derive the series for e as follows. Let k be any positive integer and apply the Binomial Theorem to $\left(1 + \frac{1}{k}\right)^k$.

$$\left(1 + \frac{1}{k}\right)^k = 1 + k\left(\frac{1}{k}\right) + \frac{k(k-1)}{2!}\left(\frac{1}{k}\right)^2 + \frac{k(k-1)(k-2)}{3!}\left(\frac{1}{k}\right)^3 + \cdots$$

$$+ \frac{k(k-1)(k-2)\cdots 1}{k!}\left(\frac{1}{k}\right)^k$$

$$= 1 + 1 + \frac{1\left(1 - \frac{1}{k}\right)}{2!} + \frac{1\left(1 - \frac{1}{k}\right)\left(1 - \frac{2}{k}\right)}{3!} + \cdots + \frac{1\left(1 - \frac{1}{k}\right)\left(1 - \frac{2}{k}\right)\cdots\frac{1}{k}}{k!}$$

Then, find the limit of $\left(1 + \frac{1}{k}\right)^k$ as k increases without bound.

Recall that $\lim_{k \to \infty} \frac{1}{k} = 0.$

$$\lim_{k \to \infty}\left(1 + \frac{1}{k}\right)^k = 1 + 1 + \frac{1}{2!} + \frac{1}{3!} + \frac{1}{4!} + \frac{1}{5!} + \cdots$$

As $k \to \infty$*, the number of terms in the sum becomes infinite.*

Thus, e can be defined as follows.

$$e = \lim_{k \to \infty}\left(1 + \frac{1}{k}\right)^k \quad \text{or} \quad e = 1 + 1 + \frac{1}{2!} + \frac{1}{3!} + \frac{1}{4!} + \frac{1}{5!} + \cdots$$

The value of e^x can be approximated by using the following series. This series is often called the **exponential series.**

Exponential Series

$$e^x = \sum_{n=0}^{\infty} \frac{x^n}{n!} = 1 + x + \frac{x^2}{2!} + \frac{x^3}{3!} + \frac{x^4}{4!} + \frac{x^5}{5!} + \cdots$$

Example 2 **Use the first five terms of the exponential series and a calculator to approximate the value of $e^{2.03}$ to the nearest hundredth.**

$$e^x \approx 1 + x + \frac{x^2}{2!} + \frac{x^3}{3!} + \frac{x^4}{4!}$$

$$e^{2.03} \approx 1 + 2.03 + \frac{(2.03)^2}{2!} + \frac{(2.03)^3}{3!} + \frac{(2.03)^4}{4!}$$

$$\approx 1 + 2.03 + 2.06045 + 1.394237833 + 0.7075757004$$

$$\approx 7.19$$

Euler's name is associated with a number of important mathematical relationships. Among them is the relationship between the exponential series and a series called the **trigonometric series.** The trigonometric series for $\cos x$ and $\sin x$ are given below.

Trigonometric Series

$$\cos x = \sum_{n=0}^{\infty} \frac{(-1)^n x^{2n}}{(2n)!} = 1 - \frac{x^2}{2!} + \frac{x^4}{4!} - \frac{x^6}{6!} + \frac{x^8}{8!} - \cdots$$

$$\sin x = \sum_{n=0}^{\infty} \frac{(-1)^n x^{2n+1}}{(2n+1)!} = x - \frac{x^3}{3!} + \frac{x^5}{5!} - \frac{x^7}{7!} + \frac{x^9}{9!} - \cdots$$

These two series are convergent for all values of x. By replacing x with any angle measure expressed in radians and carrying out the computations, approximate values of the trigonometric functions can be found to any desired degree of accuracy.

Example  **Use the first five terms of the trigonometric series to approximate the value of $\cos \frac{\pi}{3}$ to four decimal places.**

$$\cos x \approx 1 - \frac{x^2}{2!} + \frac{x^4}{4!} - \frac{x^6}{6!} + \frac{x^8}{8!}$$

$$\cos \frac{\pi}{3} \approx 1 - \frac{(1.0472)^2}{2!} + \frac{(1.0472)^4}{4!} - \frac{(1.0472)^6}{6!} + \frac{(1.0472)^8}{8!} \quad x = \frac{\pi}{3} \text{ or about } 1.0472$$

$$\cos \frac{\pi}{3} \approx 1 - 0.54831 + 0.05011 - 0.00183 + 0.00004$$

$$\cos \frac{\pi}{3} \approx 0.5004 \quad \textit{Compare this result to the actual value.}$$

GRAPHING CALCULATOR EXPLORATION

In this Exploration you will examine polynomial functions that can be used to approximate $\sin x$. The graph below shows the graphs of $f(x) = \sin x$ and $g(x) = x - \frac{x^3}{3!} + \frac{x^5}{5!}$ on the same screen.

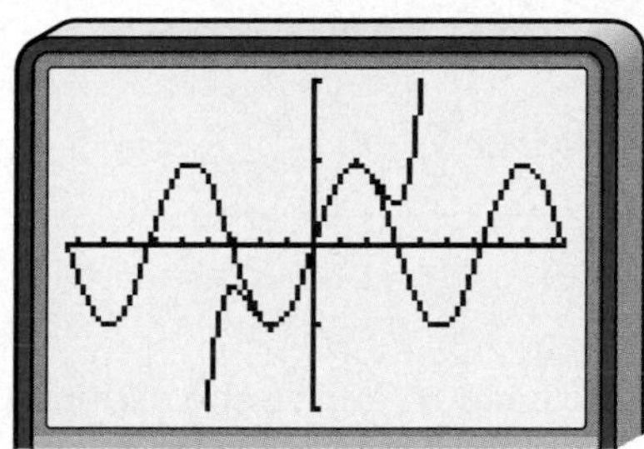

$[-3\pi, 3\pi]$ scl:1 by $[-2, 2]$ scl:1

TRY THESE

1. Use TRACE to help you write an inequality describing the x-values for which the graphs seem very close together.
2. In absolute value, what are the greatest and least differences between the values of $f(x)$ and $g(x)$ for the values of x described by the inequality you wrote in Exercise 1?
3. Repeat Exercises 1 and 2 using $h(x) = x - \frac{x^3}{3!} + \frac{x^5}{5!} - \frac{x^7}{7!}$ instead of $g(x)$.
4. Repeat Exercises 1 and 2 using $k(x) = x - \frac{x^3}{3!} + \frac{x^5}{5!} - \frac{x^7}{7!} + \frac{x^9}{9!}$.

WHAT DO YOU THINK?

5. Are the intervals for which you get good approximations for $\sin x$ larger or smaller for polynomials that have more terms?
6. What term should be added to $k(x)$ to obtain a polynomial with six terms that gives good approximations to $\sin x$?

Another very important formula is derived by replacing x by $i\alpha$ in the exponential series, where i is the imaginary unit and α is the measure of an angle in radians.

$$e^{i\alpha} = 1 + i\alpha + \frac{(i\alpha)^2}{2!} + \frac{(i\alpha)^3}{3!} + \frac{(i\alpha)^4}{4!} + \cdots$$

$$e^{i\alpha} = 1 + i\alpha - \frac{\alpha^2}{2!} - i\frac{\alpha^3}{3!} + \frac{\alpha^4}{4!} + \cdots \quad i^2 = -1,\ i^3 = -i,\ i^4 = 1$$

Group the terms according to whether they contain the factor i.

$$e^{i\alpha} = \left(1 - \frac{\alpha^2}{2!} + \frac{\alpha^4}{4!} - \frac{\alpha^6}{6!} + \cdots\right) + i\left(\alpha - \frac{\alpha^3}{3!} + \frac{\alpha^5}{5!} - \frac{\alpha^7}{7!} + \cdots\right)$$

Notice that the real part is exactly $\cos \alpha$ and the imaginary part is exactly $\sin \alpha$. This relationship is called **Euler's Formula.**

Euler's Formula $e^{i\alpha} = \cos \alpha + i \sin \alpha$

If $-i\alpha$ had been substituted for x rather than $i\alpha$, the result would have been $e^{-i\alpha} = \cos \alpha - i \sin \alpha$.

Euler's Formula can be used to write a complex number, $a + bi$, in its exponential form, $re^{i\theta}$.

$$\begin{aligned} a + bi &= r(\cos \theta + i \sin \theta) \\ &= re^{i\theta} \end{aligned}$$

Example 4 **Write $1 + \sqrt{3}i$ in exponential form.**

Look Back
Refer to Lesson 9-6 to review the polar form of complex numbers.

Write the polar form of $1 + \sqrt{3}i$. Recall that $a + bi = r(\cos \theta + i \sin \theta)$, where $r = \sqrt{a^2 + b^2}$ and $\theta = \text{Arctan}\,\frac{b}{a}$ when $a > 0$.

$r = \sqrt{(1)^2 + (\sqrt{3})^2}$ or 2, and $\theta = \text{Arctan}\,\frac{\sqrt{3}}{1}$ or $\frac{\pi}{3}$ *$a = 1$ and $b = \sqrt{3}$*

$$\begin{aligned} 1 + \sqrt{3}i &= 2\left(\cos \frac{\pi}{3} + i \sin \frac{\pi}{3}\right) \\ &= 2e^{i\frac{\pi}{3}} \end{aligned}$$

Thus, the exponential form of $1 + \sqrt{3}i$ *is* $2e^{i\frac{\pi}{3}}$.

The equations for $e^{i\alpha}$ and $e^{-i\alpha}$ can be used to derive the exponential values of $\sin \alpha$ and $\cos \alpha$.

$$\begin{aligned} e^{i\alpha} - e^{-i\alpha} &= (\cos \alpha + i \sin \alpha) - (\cos \alpha - i \sin \alpha) \\ e^{i\alpha} - e^{-i\alpha} &= 2i \sin \alpha \\ \sin \alpha &= \frac{e^{i\alpha} - e^{-i\alpha}}{2i} \end{aligned}$$

$$\begin{aligned} e^{i\alpha} + e^{-i\alpha} &= (\cos \alpha + i \sin \alpha) + (\cos \alpha - i \sin \alpha) \\ e^{i\alpha} + e^{-i\alpha} &= 2 \cos \alpha \\ \cos \alpha &= \frac{e^{i\alpha} + e^{-i\alpha}}{2} \end{aligned}$$

From your study of logarithms, you know that there is no real number that is the logarithm of a negative number. However, you can use a special case of Euler's Formula to find a complex number that is the natural logarithm of a negative number.

$$\begin{aligned} e^{i\alpha} &= \cos \alpha + i \sin \alpha \\ e^{i\pi} &= \cos \pi + i \sin \pi && \textit{Let } \alpha = \pi. \\ e^{i\pi} &= -1 + i(0) \\ e^{i\pi} &= -1 && \textit{So } e^{i\pi} + 1 = 0. \end{aligned}$$

If you take the natural logarithm of both sides of $e^{i\pi} = -1$, you can obtain a value for ln (-1).

$$\ln e^{i\pi} = \ln(-1)$$
$$i\pi = \ln(-1)$$

Thus, the natural logarithm of a negative number $-k$, for $k > 0$, can be defined using $\ln(-k) = \ln(-1)k$ or $\ln(-1) + \ln k$, a complex number.

Example 5 **Evaluate ln (-270).**

$\ln(-270) = \ln(-1) + \ln(270)$ *Use a calculator to compute ln (270).*

$\approx i\pi + 5.5984$

Thus, $\ln(-270) \approx i\pi + 5.5984$. *The logarithm is a complex number.*

CHECK FOR UNDERSTANDING

Communicating Mathematics

Read and study the lesson to answer each question.

1. **Explain** why in Example 2 of this lesson the exponential series gives an approximation of 7.19 for $e^{2.03}$, while a calculator gives an approximation of 7.61.
2. **Estimate** for what values of x the series $1 + x + \frac{x^2}{2} + \frac{x^3}{6}$ gives a good approximation of the function e^x using the graph at the right.

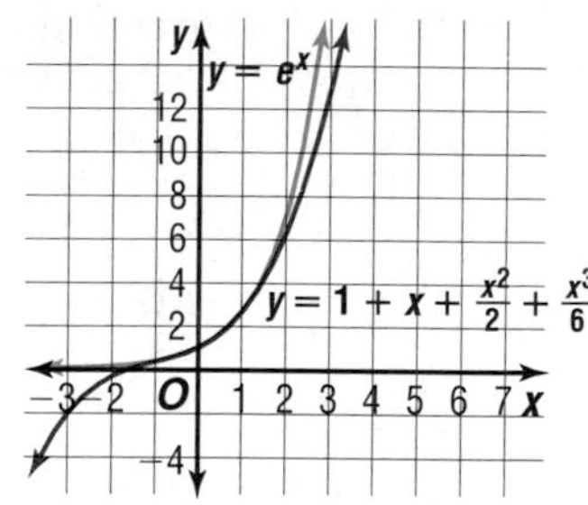

3. **List** at least two reasons why Fibonacci's rabbit problem in Example 1 is not realistic.
4. **Write** a recursive formula for the terms of the Fibonacci sequence.

Guided Practice

Find each value to four decimal places.

5. $\ln(-7)$
6. $\ln(-0.379)$

Use the first five terms of the exponential series and a calculator to approximate each value to the nearest hundredth.

7. $e^{0.8}$
8. $e^{1.36}$
9. Use the first five terms of the trigonometric series to approximate the value of sin π to four decimal places. Then, compare the approximation to the actual value.

Write each complex number in exponential form.

10. $\sqrt{2}\left(\cos \frac{3\pi}{4} + i \sin \frac{3\pi}{4}\right)$
11. $-1 + \sqrt{3}i$

12. **Investment** The Cyberbank advertises an Advantage Plus savings account with a 6% interest rate compounded continuously. Morgan is considering opening up an Advantage Plus account, because she needs to double the amount she will deposit over 5 years.
   a. If Morgan deposits P dollars into the account, approximate the return on her investment after 5 years using three terms of the exponential series. (*Hint:* The formula for continuous compounding interest is $A = Pe^{rt}$.)
   b. Will Morgan double her money in the desired amount of time? Explain.
   c. Check the approximation found in part **a** using a calculator. How do the two answers compare?

EXERCISES

Practice

Find each value to four decimal places.

13. $\ln(-4)$
14. $\ln(-3.1)$
15. $\ln(-0.25)$
16. $\ln(-0.033)$
17. $\ln(-238)$
18. $\ln(-1207)$

Use the first five terms of the exponential series and a calculator to approximate each value to the nearest hundredth.

19. $e^{1.1}$
20. $e^{-0.2}$
21. $e^{4.2}$
22. $e^{0.55}$
23. $e^{3.5}$
24. $e^{2.73}$

Use the first five terms of the trigonometric series to approximate the value of each function to four decimal places. Then, compare the approximation to the actual value.

25. $\cos \pi$
26. $\sin \frac{\pi}{4}$
27. $\cos \frac{\pi}{6}$

28. Approximate the value of $\sin \frac{\pi}{2}$ to four decimal places using the first five terms of the trigonometric series.

Write each complex number in exponential form.

29. $5\left(\cos \frac{5\pi}{3} + \boldsymbol{i} \sin \frac{5\pi}{3}\right)$
30. $\boldsymbol{i}$
31. $1 + \boldsymbol{i}$
32. $\sqrt{3} + \boldsymbol{i}$
33. $-\sqrt{2} + \sqrt{2}\boldsymbol{i}$
34. $-4\sqrt{3} - 4\boldsymbol{i}$

35. Write the expression $3 + 3\boldsymbol{i}$ in exponential form.

Applications and Problem Solving

36. **Research** Explore the meaning of the term *transcendental number.*

37. **Critical Thinking** Show that for all real numbers x, $\sin x = \frac{e^{ix} - e^{-ix}}{2\boldsymbol{i}}$ and $\cos x = \frac{e^{ix} + e^{-ix}}{2}$.

38. **Research** Investigate and write a one-page paper on the occurrence of Fibonacci numbers in plants.

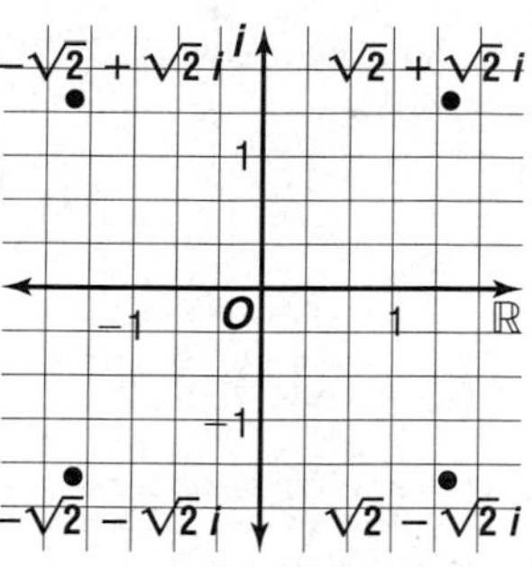

39. **Critical Thinking** Examine Pascal's triangle. What relationship can you find between Pascal's triangle and the Fibonacci sequence?

Research For more information about the Fibonacci sequence, visit: **www.amc.glencoe.com**

40. **Number Theory** Consider the Fibonacci sequence 1, 1, 2, 3, ..., $F_{n-2} + F_{n-1}$.

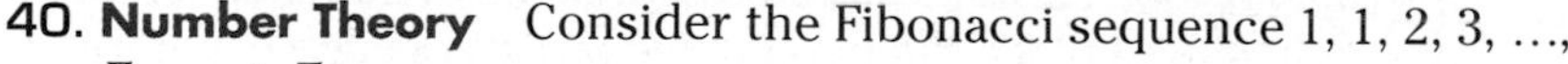

a. Find $\frac{F_n}{F_{n-1}}$ for the second through eleventh terms of the Fibonacci sequence.

b. Is this sequence of ratios arithmetic, geometric, or neither?

c. Sketch a graph of the terms found in part **a**. Let n be the x-coordinate and $\frac{F_n}{F_{n-1}}$ be the y-coordinate, and connect the points.

d. Based on the graph found in part c, does this sequence appear to approach a limit? If so, use the last term found to approximate this limit to three decimal places.

e. The *golden ratio* has a value of approximately 1.61804. How does the limit of the sequence $\frac{F_n}{F_{n-1}}$ compare to the golden ratio?

f. Research the term *golden ratio.* Write several paragraphs on the history of the golden ratio and describe its application to art and architecture.

41. **Investment** When Cleavon turned 5 years old, his grandmother decided it was time to start saving for his college education. She deposited \$5000 in a special account earning 5% interest compounded continuously. By the time Cleavon begins college at the age of 18, his grandmother estimates that her grandson will need \$40,000 for college tuition.

a. Approximate Cleavon's savings account balance on his 18th birthday using five terms of the exponential series.

b. Will the account have sufficient funds for Cleavon's college tuition by the time he is ready to start college? Explain.

c. Use a calculator to compute how long it will take the account to accumulate \$40,000. How old would Cleavon be by this time?

d. To have at least \$40,000 when Cleavon is 18 years old, how much should his grandmother have invested when he was 5 years old to the nearest dollar?

42. **Critical Thinking** Find a pattern, if one exists, for the following types of Fibonacci numbers.

a. even numbers

b. multiples of 3

Mixed Review

43. Use the Binomial Theorem to expand $(2x + y)^6$. *(Lesson 12-6)*

44. Express the series $2 + 4 + 8 + \cdots + 64$ using sigma notation. *(Lesson 12-5)*

45. **Number Theory** If you tear a piece of paper that is 0.005 centimeter thick in half, and place the two pieces on top of each other, the height of the pile of paper is 0.01 centimeter. Let's call this the second pile. If you tear these two pieces of paper in half, the third pile will have four pieces of paper in it. *(Lesson 12-2)*
   a. How high is the third pile? the fourth pile?
   b. Write a formula to determine how high the nth pile is.
   c. Use the formula to determine in theory how high the 10th pile would be and how high the 100th pile would be.

46. Evaluate $\frac{8^{\frac{2}{3}}}{8^{\frac{1}{3}}}$. *(Lesson 11-1)*

47. Write an equation of the parabola in general form that has a horizontal axis and passes through points at (0, 0), (2, −1), and (4, −4). *(Lesson 10-5)*

48. Find the quotient of $16\left(\cos \frac{\pi}{8} + \boldsymbol{i} \sin \frac{\pi}{8}\right)$ divided by $4\left(\cos \frac{\pi}{4} + \boldsymbol{i} \sin \frac{\pi}{4}\right)$. *(Lesson 9-7)*

49. **Physics** Two forces, one of 30 N and the other of 50 N, act on an object. If the angle between the forces is 40°, find the magnitude and the direction of the resultant force. *(Lesson 8-5)*

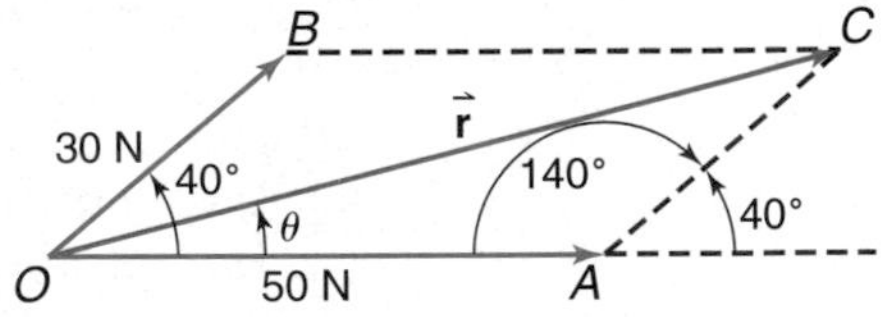

50. **Carpentry** Carpenters use circular sanders to smooth rough surfaces, such as wood or plaster. The disk of a sander has a radius of 6 inches and is rotating at a speed of 5 revolutions per second. *(Lesson 6-2)*
   a. Find the angular velocity of the sander disk in radians per second.
   b. What is the linear velocity of a point on the edge of the sander disk in feet per second?

51. **Education** You may answer up to 30 questions on your final exam in history class. It consists of multiple-choice and essay questions. Two 48-minute class periods have been set aside for taking the test. It will take you 1 minute to answer each multiple-choice question and 12 minutes for each essay question. Correct answers of multiple-choice questions earn 5 points and correct essay answers earn 20 points. If you are confident that you will answer all of the questions you attempt correctly, how many of each type of questions should you answer to receive the highest score? *(Lesson 2-7)*

52. **SAT/ACT Practice** In triangle ABC, if $BD = 5$, what is the length of $\overline{BC}$?
   A 3
   B 5
   C $\sqrt{39}$
   D $\sqrt{70}$
   E $\sqrt{126}$

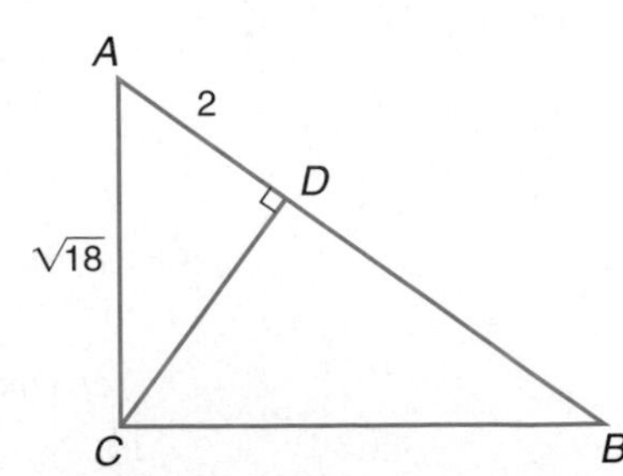

Extra Practice See p. A50.

12-8

Sequences and Iteration

OBJECTIVES

- Iterate functions using real and complex numbers.

ECOLOGY The population of grizzly bears on the high Rocky Mountain Front near Choteau, Montana, has a growth factor of 1.75. The maximum population of bears that can be sustained in the area is 500 bears, and the current population is 240. Write an equation to model the population. Use the equation to find the population of bears at the end of fifteen years. *This problem will be solved in Example 1.*

The population of a species in a defined area changes over time. Changes in the availability of food, good or bad weather conditions, the amount of hunting allowed, disease, and the presence or absence of predators can all affect the population of a species. We can use a mathematical equation to model the changes in a population.

One such model is the *Verhulst population model.* The model uses the recursive formula $p_{n+1} = p_n + rp_n(1 - p_n)$, where n represents the number of time periods that have passed, p_n represents the percent of the maximum sustainable population that exists at time n, and r is a growth factor.

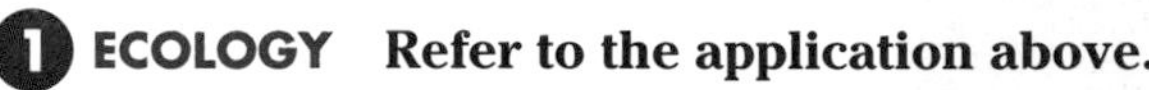

Example 1 **ECOLOGY** **Refer to the application above.**

a. Write an equation to model the population.

For this problem, the Verhulst population model can be used to represent the changes in the population of a species.

$p_{n+1} = p_n + rp_n(1 - p_n)$

$p_{n+1} = p_n + 1.75p_n(1 - p_n)$ *$r = 1.75$*

b. Find the population of bears at the end of fifteen years.

The initial percent of the maximum sustainable population can be represented by the ratio $p_0 = \frac{240}{500}$ or 0.48. This means that the current population is 48% of the maximum sustainable population.

Now, we can find the first few iterates as follows.

$p_1 = 0.48 + 1.75(0.48)(1 - 0.48)$ or 0.9168 *$n = 1$, $p_0 = 0.48$, $r = 1.75$*

$(0.9168)(500) \approx 458$ bears

$p_2 = 0.9168 + 1.75(0.9168)(1 - 0.9168)$ or 1.0503 *$n = 2$, $p_1 = 0.9168$, $r = 1.75$*

$(1.0503)(500) \approx 525$ bears

$p_3 = 1.0503 + 1.75(1.0503)(1 - 1.0503)$ or 0.9578 *$n = 3$, $p_2 = 1.0503$, $r = 1.75$*

$(0.9578)(500) \approx 479$ bears

(continued on the next page)

The table shows th

n	4	5
number of bears	514	489

At the end of the fir

Notice that to determ
for a year, you must use
learned that this process
iteration. Each output is
function value $f(x_0)$ of th
the function performed c

Example 2 BANKING Selina An
of 6.3%. Find the bala
her initial balance is

The balance of a savin
end of a period of time
where p_n is the princip
period of time.

$p_1 = p_0 + rp_0$
$p_1 = 4210 + (0.063)(42$
$p_1 = \$4475.23$

$p_2 = 4475.23 + (0.063)$
$p_2 \approx \$4757.17$

$p_3 = 4757.17 + (0.063)$
$p_3 \approx \$5056.87$

The balance in Ms. An

Look Back
You can review graphing numbers in the complex plane in Lesson 9-6

Remember that every
a real part and an imagin
number $a + bi$ has been
plane at the right. The ho
complex plane represent
the number, and the verti
imaginary part.

Functions can be iter

EXERCISES

Practice

Find the first four iterates of each function using the given initial value. If necessary, round answers to the nearest hundredth.

11. $f(x) = 3x - 7; x_0 = 4$ **12.** $f(x) = x^2; x_0 = -2$ **13.** $f(x) = (x - 5)^2; x_0 = 4$

14. $f(x) = x^2 - 1; x_0 = -1$ **15.** $f(x) = 2x^2 - x; x_0 = 0.1$

16. Find the first ten iterates of $f(x) = \frac{2}{x}$ for each initial value.
a. $x_0 = 1$ **b.** $x_0 = 4$ **c.** $x_0 = 7$
d. What do you observe about the iterates of this function?

Find the first three iterates of the function $f(z) = 2z + (3 - 2i)$ for each initial value.

17. $z_0 = 5i$ **18.** $z_0 = 4$ **19.** $z_0 = 1 + 2i$

20. $z_0 = -1 - 2i$ **21.** $z_0 = 6 + 2i$ **22.** $z_0 = 0.3 - i$

23. Find the first three iterates of the function $f(z) = 3z - 2i$ for $z_0 = \frac{1}{3} + \frac{2}{3}i$.

Find the first three iterates of the function $f(z) = z^2 + c$ for each given value of c and each initial value.

24. $c = -1; z_0 = 0 - i$ **25.** $c = 1 - 3i; z_0 = i$

26. $c = 3 + 2i; z_0 = 1$ **27.** $c = -4i; z_0 = 1 + i$

28. $c = 0; z_0 = \frac{\sqrt{2}}{2} - \frac{\sqrt{2}}{2}i$ **29.** $c = 2 + 3i; z_0 = 1 - i$

Applications and Problem Solving

30. Banking Amelia has a savings account that has an annual yield of 5.2%. Find the balance of the account after each of the first five years if her initial balance is \$2000.

31. Ecology The population of elk on the Bridger Range in the Rocky Mountains of western Montana has a growth factor of 2.5. The population in 1984 was 10% of the maximum population sustainable. What percent of the maximum sustainable population should be present in 2002?

32. Critical Thinking If $f(z) = z^2 + c$ is iterated with an initial value of $2 + 3i$ and $z_1 = -1 + 15i$, find c.

33. Critical Thinking In Exercise 16, find an initial value that produces iterates that all have the same value.

34. Research Investigate the applications of fractal geometry to agriculture. How is fractal geometry being used in this field? Write several paragraphs about your findings.

35. Critical Thinking
a. Use a calculator to find $\sqrt{2}$, $\sqrt{\sqrt{2}}$, $\sqrt{\sqrt{\sqrt{2}}}$, and $\sqrt{\sqrt{\sqrt{\sqrt{2}}}}$.
b. Define an iterative function $f(z)$ that models the situation described in part a.
c. Determine the limit of $f(z)$ as the number of iterations approaches infinity.
d. Determine the limit of $f(z)$ as the number of iterations approaches infinity for integral values of $z_0 > 0$.

Mixed Review

36. Write $2\left(\cos \frac{\pi}{3} + \boldsymbol{i} \sin \frac{\pi}{3}\right)$ in exponential form. *(Lesson 12-7)*

37. Find the fifth term in the binomial expansion of $(2a - 3b)^8$. *(Lesson 12-6)*

38. State whether the series $\frac{1}{2} + \frac{1}{8} + \frac{1}{32} + \cdots$ is *convergent* or *divergent.* Explain your reasoning. *(Lesson 12-4)*

39. Nuclear Power A nuclear cooling tower has an eccentricity of $\frac{7}{5}$. At its narrowest point the cooling tower is 130 feet wide. Determine the equation of the hyperbola used to generate the hyperboloid of the cooling tower. *(Lesson 10-4)*

40. Optics A beam of light strikes a diamond at an angle of 42°. What is the angle of refraction if the index of refraction of a diamond is 2.42 and the index of refraction of air is 1.00? (Use Snell's Law, $n_1 \sin I = n_2 \sin r$, where n_1 is the index of refraction of the medium the light is exiting, n_2 is the index of refraction of the medium it is entering, I is the angle of incidence, and r is the angle of refraction.) *(Lesson 7-5)*

41. Air Traffic Safety The traffic pattern for airplanes into San Diego's airport is over the heart of downtown. Therefore, there are restrictions on the heights of new construction. The owner of an office building wishes to erect a microwave tower on top of the building. According to the architect's design, the angle of elevation from a point on the ground to the top of the 40-foot tower is 56°. The angle of elevation from the ground point to the top of the building is 42°. *(Lesson 5-4)*

a. Draw a sketch of this situation.

b. The maximum allowed height of any structure is 100 feet. Will the city allow the building of this tower? Explain.

42. Determine whether the graph has infinite discontinuity, jump discontinuity, or point discontinuity, or it is continuous. *(Lesson 3-5)*

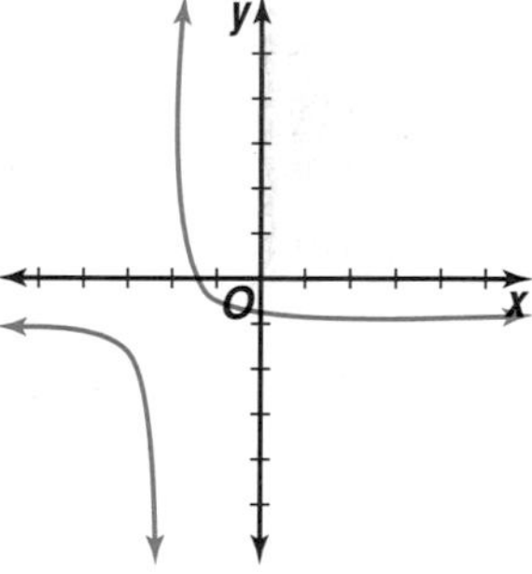

43. Find the maximum and minimum values of the function $f(x, y) = 2x + 8y + 10$ for the polygonal convex set determined by the following system of inequalities. *(Lesson 2-6)*

$x \geq 3$
$x \leq 8$
$5 \leq y \leq 9$
$x + y \leq 14$

44. SAT/ACT Practice Two students took a science test and received different scores between 10 and 100. If H equals the higher score and L equals the lower score, and the difference between the two scores equals the average of the two scores, what is the value of $\frac{H}{L}$?

A $\frac{3}{2}$ B 2 C $\frac{5}{2}$ D 3

E It cannot be determined from the information given.

Extra Practice See p. A50.

12-9

Mathematical Induction

OBJECTIVE

- Use mathematical induction to prove the validity of mathematical statements.

BUSINESS Felipe and Emily work a booth on the midway at the state fair. Before the fair opens, they discover that their cash supply contains only \$5 and \$10 bills. Rene volunteers to go and get a supply of \$1s, but before leaving, Rene remarks that she could have given change for any amount greater than \$4 had their cash supply contained only \$2 and \$5 bills. Felipe replies, "Even if we had \$2 bills, you would still need \$1 bills." Rene disagrees and says that she can prove that she is correct. *This problem will be solved in Example 4.*

A method of proof called **mathematical induction** can be used to prove certain conjectures and formulas. Mathematical induction depends on a recursive process that works much like an unending line of dominoes arranged so that if any one domino falls the next one will also fall.

Suppose the first domino is knocked over.	Condition 1
The first will knock down the second.	Condition 2
The second will knock down the third.	Condition 3
The third will knock down the fourth.	Condition 4
⋮	⋮

Thus, the whole line of dominos will eventually fall.

Mathematical induction operates in a similar manner. If a statement S_k implies the truth of S_{k+1} and the statement S_1 is true, then the chain reaction follows like an infinite set of tumbling dominoes.

S_1 is true.	Condition 1
S_1 implies that S_2 is true.	Condition 2
S_2 implies that S_3 is true.	Condition 3
S_3 implies that S_4 is true.	Condition 4
⋮	⋮

In general, the following steps are used to prove a conjecture by mathematical induction.

Proof by Mathematical Induction

1. First, verify that the conjecture S_n is valid for the first possible case, usually $n = 1$. *This is called the anchor step.*
2. Then, assume that S_n is valid for $n = k$, and use this assumption to prove that it is also valid for $n = k + 1$. *This is called the induction step.*

Thus, since S_n is valid for $n = 1$ (or any other first case), it is valid for $n = 2$. Since it is valid for $n = 2$, it is valid for $n = 3$, and so on, indefinitely.

Mathematical induction can be used to prove summation formulas.

Example 1 **Prove that the sum of the first n positive integers is $\frac{n(n+1)}{2}$.**

Here S_n is defined as $1 + 2 + 3 + \cdots + n = \frac{n(n+1)}{2}$.

1. First, verify that S_n is valid for $n = 1$.

Since the first positive integer is 1 and $\frac{1(1+1)}{2} = 1$, the formula is valid for $n = 1$.

2. Then assume that S_n is valid for $n = k$.

$S_k \Rightarrow 1 + 2 + 3 + \cdots + k = \frac{k(k+1)}{2}$ *Replace n with k.*

Next, prove that it is also valid for $n = k + 1$. *Add (k + 1) to both sides of S_k.*

$S_{k+1} \Rightarrow 1 + 2 + 3 + \cdots + k + (k + 1) = \frac{k(k+1)}{2} + (k + 1)$

We can simplify the right side by adding $\frac{k(k+1)}{2} + (k + 1)$.

(k + 1) is a common factor

$$S_{k+1} \Rightarrow 1 + 2 + 3 + \cdots + k + (k + 1) = \frac{k(k+1) + 2(k+1)}{2}$$
$$= \frac{(k+1)(k+2)}{2}$$

If $k + 1$ is substituted into the original formula $\left(\frac{n(n+1)}{2}\right)$, the same result is obtained.

$$\frac{(k+1)(k+1+1)}{2} = \frac{(k+1)(k+2)}{2}$$

Thus, if the formula is valid for $n = k$, it is also valid for $n = k + 1$. Since S_n is valid for $n = 1$, it is also valid for $n = 2$, $n = 3$, and so on. That is, the formula for the sum of the first n positive integers holds.

Mathematical induction is also used to prove properties of divisibility. Recall that an integer p is *divisible* by an integer q if $p = qr$ for some integer r.

Example 2 **Prove that $4^n - 1$ is divisible by 3 for all positive integers n.**

Using the definition of divisibility, we can state the conjecture as follows:

Notice that here S_n does not represent a sum but a conjecture.

$S_n \Rightarrow 4^n - 1 = 3r$ for some integer r

1. First verify that S_n is valid for $n = 1$.

$S_1 \Rightarrow 4^1 - 1 = 3$. Since 3 is divisible by 3, S_n is valid for $n = 1$.

2. Then assume that S_n is valid for $n = k$ and use this assumption to prove that it is also valid for $n = k + 1$.

$S_k \Rightarrow 4^k - 1 = 3r$ for some integer r *Assume S_k is true.*
$S_{k+1} \Rightarrow 4^{k+1} - 1 = 3t$ for some integer t *Show that S_{k+1} must follow.*

(continued on the next page)

For this proof, rewrite the left-hand side of S_k so that it matches the left-hand side of S_{k+1}.

$4^k - 1 = 3r$	S_k
$4(4^k - 1) = 4(3r)$	*Multiply each side by 4.*
$4^{k+1} - 4 = 12r$	*Simplify.*
$4^{k+1} - 1 = 12r + 3$	*Add 3 to each side.*
$4^{k+1} - 1 = 3(4r + 3)$	*Factor.*

Let $t = 4r + 3$, an integer. Then $4^{k+1} - 1 = 3t$

We have shown that if S_k is valid, then S_{k+1} is also valid. Since S_n is valid for $n = 1$, it is also valid for $n = 2$, $n = 3$, and so on. Hence, $4^n - 1$ is divisible by 3 for all positive integers n.

There is no "fixed" way of completing Step 2 for a proof by mathematical induction. Often, each problem has its own special characteristics that require a different technique to complete the proof. You may have to multiply the numerator and denominator of an expression by the same quantity, factor or expand an expression in a special way, or see an important relationship with the distributive property.

Example 3 **Prove that $6^n - 2^n$ is divisible by 4 for all positive integers n.**

Begin by restating the conjecture using the definition of divisibility.

$S_n \Rightarrow 6^n - 2^n = 4r$ for some integer r

1. Verify that S_n is valid for $n = 1$

$S_1 \Rightarrow 6^1 - 2^1 = 4$. Since 4 is divisible by 4, S_n is valid for $n = 1$.

2. Assume that S_n is valid for $n = k$ and use this to prove that it is also valid for $n = k + 1$.

$S_k \Rightarrow 6^k - 2^k = 4r$ for some integer r	*Assume S_k is true*
$S_{k+1} \Rightarrow 6^{k+1} - 2^{k+1} = 4t$ for some integer t	*Show that S_{k+1} must follow.*

Begin by rewriting S_k as $6^k = 2^k + 4r$. Now rewrite the left-hand side of this expression so that it matches the left-hand side of S_{k+1}.

$6^k = 2^k + 4r$	S_k
$6^k \cdot 6 = (2^k + 4r)(2 + 4)$	*Multiply each side by a quantity equal to 6.*
$6^{k+1} = 2^{k+1} + 4(2^k) + 8r + 16r$	*Distributive Property*
$6^{k+1} - 2^{k+1} = 2^{k+1} + 4(2^k) + 8r + 16r - 2^{k+1}$	*Subtract 2^{k+1} from each side.*
$6^{k+1} - 2^{k+1} = 4 \cdot 2^k + 24r$	*Simplify.*
$6^{k+1} - 2^{k+1} = 4(2^k + 6r)$	*Factor.*

Let $t = 2^k + 6r$, an integer. Then $6^{k+1} - 2^{k+1} = 4t$.

We have shown that if S_k is valid, then S_{k+1} is also valid. Since S_n is valid for $n = 1$, it is also valid for $n = 2$, $n = 3$, and so on. Hence, $6^n - 2^n$ is divisible by 4 for all positive integers n.

Let's look at another proof that requires some creative thinking to complete.

Example 4 **BUSINESS** **Refer to the application at the beginning of the lesson. Use mathematical induction to prove that Emily can give change in \$2s and \$5s for amounts of \$4, \$5, \$6, ..., \$n.**

Let n = the dollar amount for which a customer might need change.
Let a = the number of \$2 bills used to make the exchange.
Let b = the number of \$5 bills used to make the exchange.

$S \Rightarrow n = 2a + 5b$

Notice that the first possible case for n is not always 1.

1. First, verify that S_n is valid for $n = 4$. In other words, are there values for a and b such that $4 = 2a + 5b$?

   Yes. If $a = 2$ and $b = 0$, then $4 = 2(2) + 5(0)$, so S_n is valid for the first case.

2. Then, assume that S_n is valid for $n = k$ and prove that it is also valid for $n = k + 1$.

   $S_k \Rightarrow k = 2a + 5b$

   $S_{k+1} \Rightarrow k + 1 = 2a + 5b + 1$ *Add 1 to each side.*

   $= 2a + 5b + 6 - 5$ *$1 = 6 - 5$*

   $= 2(a + 3) + 5(b - 1)$ *$2a + 6 = 2(a + 3)$; $5b - 5 = 5(b - 1)$*

   Notice that this expression is true for all $a \geq 0$ and $b \geq 1$, since the expression $b - 1$ must be nonnegative. Therefore, we must also consider the case where $b = 0$.

   If $b = 0$, then $n = 2a + 5(0) = 2a$. Assuming S_n to be true for $n = k$, we must show that it is also true for $k + 1$.

   $S_k \Rightarrow k = 2a$

   $S_{k+1} \Rightarrow k + 1 = 2a + 1$

   $= 2(a - 2) + 4 + 1$ *$2a = 2(a - 2) + 4$*

   $= 2(a - 2) + 1 \cdot 5$ *$4 + 1 = 1 \cdot 5$*

   Notice that this expression is only true for all $a \geq 2$, since the expression $a - 2$ must be nonnegative. However, we have already considered the case where $a = 0$.

   Thus, it can be concluded that since the conjecture is true for $k = 4$, it also valid for $n = k + 1$. Therefore, Emily can make change for amounts of \$4, \$5, \$6, ... \$n using only \$2 and \$5 bills.

CHECK FOR UNDERSTANDING

Communicating Mathematics

Read and study the lesson to answer each question.

1. **Explain** why it is necessary to prove the $n + 1$ case in the process of mathematical induction.
2. **Describe** a method you might use to show that a conjecture is false.
3. Consider the series $3 + 5 + 7 + 9 + \cdots + 2n + 1$.
   a. **Write a conjecture** for the general formula S_n for the sum of the first n terms.
   b. Verify S_n for $n = 1, 2,$ and 3.
   c. Write S_k and S_{k+1}.
4. **Restate** the conjecture that $8^n - 1$ is divisible by 7 for all positive integers n using the definition of divisibility.
5. *Math Journal* The "domino effect" presented in this lesson is just one way to illustrate the principle behind mathematical induction. **Write** a paragraph describing another situation in real life that illustrates this principle.

Guided Practice

Use mathematical induction to prove that each proposition is valid for all positive integral values of *n*.

6. $3 + 5 + 7 + \cdots + (2n + 1) = n(n + 2)$
7. $2 + 2^2 + 2^3 + \cdots + 2^n = 2(2^n - 1)$
8. $\frac{1}{2} + \frac{1}{2^2} + \frac{1}{2^3} + \cdots + \frac{1}{2^n} = 1 - \frac{1}{2^n}$
9. $3^n - 1$ is divisible by 2.

10. **Number Theory** The ancient Greeks were very interested in number patterns. Triangular numbers are numbers that can be represented by a triangular array of dots, with n dots on each side. The first three triangular numbers are 1, 3, and 6.

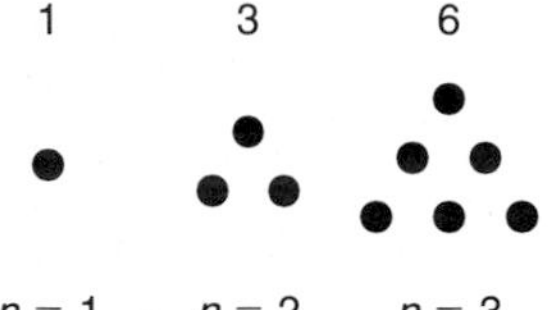

   a. Find the next five triangular numbers.
   b. Write a general formula for the nth term of this sequence.
   c. Prove that the sum of first n triangular numbers can be found using the formula $\frac{n(n + 1)(n + 2)}{6}$.

EXERCISES

Practice

For Exercises 11-19, use mathematical induction to prove that each proposition is valid for all positive integral values of *n*.

11. $1 + 5 + 9 + \cdots + (4n - 3) = n(2n - 1)$
12. $1 + 4 + 7 + \cdots + (3n - 2) = \frac{n(3n - 1)}{2}$
13. $-\frac{1}{2} - \frac{1}{4} - \frac{1}{8} - \cdots - \frac{1}{2^n} = \frac{1}{2^n} - 1$
14. $1 + 8 + 27 + \cdots + n^3 = \frac{n^2(n + 1)^2}{4}$
15. $1^2 + 3^2 + 5^2 + \cdots + (2n - 1)^2 = \frac{n(2n - 1)(2n + 1)}{3}$

16. $1 + 2 + 4 + \cdots + 2^{n-1} = 2^n - 1$
17. $7^n + 5$ is divisible by 6
18. $8^n - 1$ is divisible by 7
19. $5^n - 2^n$ is divisible by 3

20. Prove $S_n \Rightarrow a + (a + d) + (a + 2d) + \cdots + [a + (n - 1)d] = \frac{n}{2}[2a + (n - 1)d]$.
21. Prove $S_n \Rightarrow \frac{1}{1 \cdot 2} + \frac{1}{2 \cdot 3} + \frac{1}{3 \cdot 4} + \cdots + \frac{1}{n(n + 1)} = \frac{n}{n + 1}$.
22. Prove $S_n \Rightarrow 2^{2n + 1} + 3^{2n + 1}$ is divisible by 5.

Applications and Problem Solving

23. **Historical Proof** In 1730, Abraham De Moivre proposed the following theorem for finding the power of a complex number written in polar form.

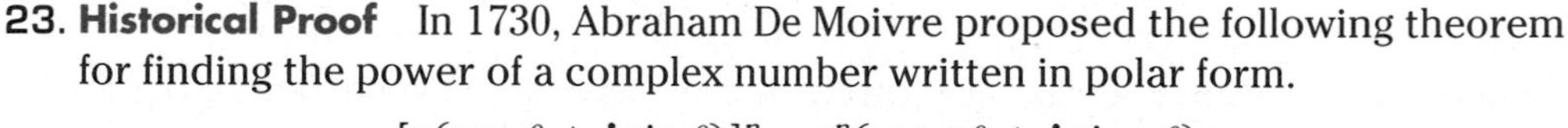

$$[r(\cos \theta + \boldsymbol{i} \sin \theta)]^n = r^n(\cos n\theta + \boldsymbol{i} \sin n\theta)$$

Prove that De Moivre's Theorem is valid for any positive integer n.

24. **Number Patterns** Consider the pattern of dots shown below.

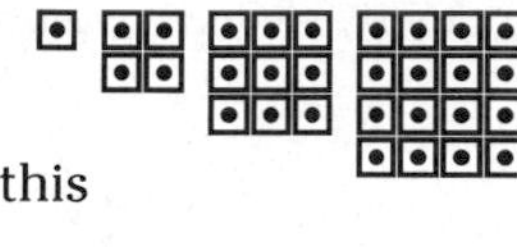

a. Find the next figure in this pattern.
b. Write a sequence to represent the number of dots added to the previous figure to create the next figure in this pattern. The first term in your sequence should be 1.
c. Find the general form for the nth term of the sequence found in part **b**.
d. Determine a formula that will calculate the total number of dots for the nth figure in this dot pattern.
e. Prove that the formula found in part d is correct using mathematical induction.

25. **Critical Thinking** Prove that $n^2 + 5n$ is divisible by 2 for all positive integral values of n.

26. **Club Activites** Melissa is the activities director for her school's science club and needs to coordinate a "get to know you" activity for the group's first meeting. The activity must last at least 40 minutes but no more than 60 minutes. Melissa has chosen an activity that will require each participant to interact with every other participant in attendance only once and estimates that if her directions are followed, each interaction should take approximately 30 seconds.

a. If n people participate in Melissa's activity, develop a formula to calculate the total number of such interactions that should take place.
b. Prove that the formula found in part a is valid for all positive integral values of n.
c. If 15 people participate, will Melissa's activity meet the guidelines provided to her? Explain.

27. Critical Thinking Use mathematical induction to prove that the Binomial Theorem is valid for all positive integral values of n.

28. Number Theory Consider the following statement: $0.99999 \cdots = 1$.

a. Write $0.\overline{9}$ as an infinite series.

b. Write an expression for the nth term of this series.

c. Write a formula for the sum of the first n terms of this series.

d. Prove that the formula you found in part **c** is valid for all positive integral values of n using mathematical induction.

e. Use the formula found in part **c** to prove that $0.99999 \cdots = 1$. (*Hint:* Use what you know about the limits of infinite sequences.)

Mixed Review

29. Find the first three iterates of the function $f(z) = 2z + \boldsymbol{i}$ for $z_0 = 4 - \boldsymbol{i}$. *(Lesson 12-8)*

30. Find d for the arithmetic sequence for which $a_1 = -6$ and $a_{29} = 64$. *(Lesson 12-1)*

31. Write the standard form of the equation $25x^2 + 4y^2 - 100x - 40y + 100 = 0$. Then identify the conic section this equation represents. *(Lesson 10-6)*

32. Geology A *drumlin* is an elliptical streamlined hill whose shape can be expressed by the equation $r = \ell \cos k\theta$ for $-\frac{\pi}{2k} \leq \theta \leq \frac{\pi}{2k}$, where ℓ is the length of the drumlin and $k > 1$ is a parameter that is the ratio of the length to the width. Find the area in square centimeters, $A = \frac{\ell^2\pi}{4k}$, of a drumlin modeled by the equation $r = 250 \cos 4\theta$. *(Lesson 9-3)*

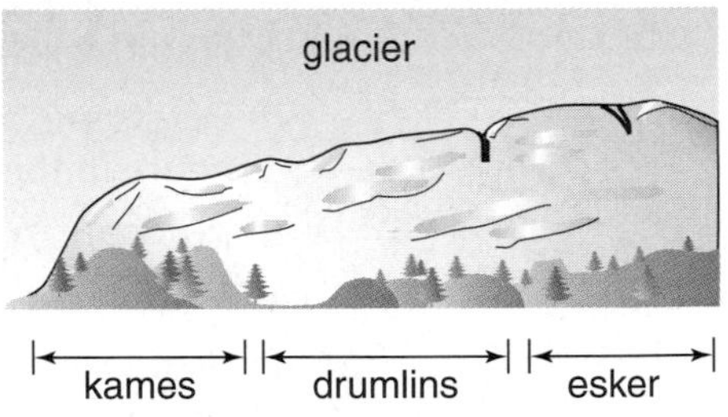

33. Write an equation of a sine function with amplitude $\frac{3}{4}$ and period π. *(Lesson 6-4)*

34. Find the values of x in the interval $0° \leq x \leq 360°$ for which $x = \cos^{-1}\left(-\frac{\sqrt{3}}{2}\right)$. *(Lesson 5-5)*

35. SAT Practice **Quantitative Comparison**

A if the quantity in Column A is greater
B if the quantity in Column B is greater
C if the two quantities are equal
D if the relationship cannot be determined from the information given

The area of the square is 25. Points A, B, C, and D are on the square. $ABCD$ is not a square.

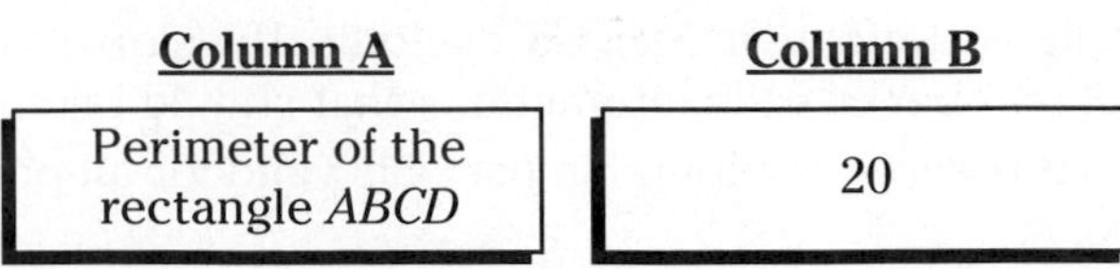

Column A	Column B
Perimeter of the rectangle $ABCD$	20

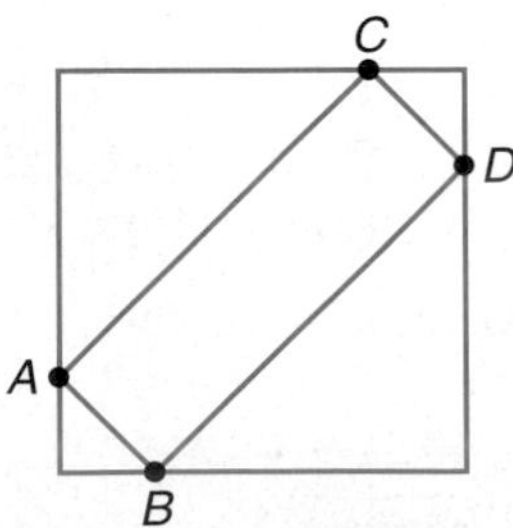

Extra Practice See p. A50.

VOCABULARY

arithmetic mean (p. 760)
arithmetic sequence (p. 759)
arithmetic series (p. 761)
Binomial Theorem (p. 803)
common difference (p. 759)
common ratio (p. 766)
comparison test (p. 789)
convergent series (p. 786)
divergent series (p. 786)
escaping point (p. 818)
Euler's Formula (p. 809)
exponential series (p. 808)
Fibonacci sequence (p. 806)
fractal geometry (p. 817)
geometric mean (p. 768)
geometric sequence (p. 766)
geometric series (p. 769)
index of summation (p. 794)
infinite sequence (p. 774)
infinite series (p. 778)
limit (p. 774)
mathematical induction (p. 822)
n factorial (p. 796)
*n*th partial sum (p. 761)
orbit (p. 817)
Pascal's Triangle (p. 801)
prisoner point (p. 818)
ratio test (p. 787)
recursive formula (p. 760)
sequence (p. 759)
sigma notation (p. 794)
term (p. 759)
trigonometric series (p. 808)

UNDERSTANDING AND USING THE VOCABULARY

Choose the letter of the term that best matches each statement or phrase.

1. each succeeding term is formulated from one or more previous terms
2. ratio of successive terms in a geometric sequence
3. used to determine convergence
4. can have infinitely many terms
5. $e^{ix} = \cos \alpha + \boldsymbol{i} \sin \alpha$
6. the terms between any two nonconsecutive terms of a geometric sequence
7. indicated sum of the terms of an arithmetic sequence
8. $n! = n(n - 1)(n - 2) \cdot \cdots \cdot 1$
9. used to demonstrate the validity of a conjecture based on the truth of a first case, the assumption of truth of a *k*th case, and the demonstration of truth for the $(k + 1)$th case
10. an infinite series with a sum or limit

a. term
b. mathematical induction
c. arithmetic series
d. recursive formula
e. *n* factorial
f. geometric mean
g. sigma notation
h. convergent
i. common ratio
j. infinite sequence
k. Euler's Formula
l. prisoner point
m. ratio test
n. limit

For additional review and practice for each lesson, visit: **www.amc.glencoe.com**

SKILLS AND CONCEPTS

OBJECTIVES AND EXAMPLES

Lesson 12-1 Find the nth term and arithmetic means of an arithmetic sequence.

- Find the 35th term in the arithmetic sequence $-5, -1, 3, \ldots$.
  Begin by finding the common difference d.
  $d = -1 - (-5)$ or 4
  Use the formula for the nth term.
  $a_n = a_1 + (n - 1)d$
  $a_{35} = -5 + (35 - 1)(4)$ or 131

REVIEW EXERCISES

11. Find the next four terms of the arithmetic sequence 3, 4.3, 5.6, ...

12. Find the 20th term of the arithmetic sequence for which $a_1 = 5$ and $d = -3$.

13. Form an arithemetic sequence that has three arithmetic means between 6 and -4.

Lesson 12-1 Find the sum of n terms of an arithmetic series.

- The sum S_n of the first n terms of an arithmetic series is given by
  $S_n = \frac{n}{2}(a_1 + a_n)$.

14. What is the sum of the first 14 terms in the arithmetic series $-30 - 23 - 16 - \cdots$?

15. Find n for the arithmetic series for which $a_1 = 2$, $d = 1.4$, and $S_n = 250.2$.

Lesson 12-2 Find the nth term and geometric means of a geometric sequence.

- Find an approximation for the 12th term of the sequence $-8, 4, -2, 1, \ldots$.
  First, find the common ratio.
  $a_2 \div a_1 = 4 \div (-8)$ or -0.5
  Use the formula for the nth term.
  $a_{12} = -8(-0.5)^{12-1}$ *$a_n = a_1r^{n-1}$*
  $= -8(-0.5)^{11}$ or about 0.004

16. Find the next three terms of the geometric sequence 49, 7, 1, ...

17. Find the 15th term of the geometric sequence for which $a_1 = 2.2$ and $r = 2$.

18. If $r = 0.2$ and $a_7 = 8$, what is the first term of the geometric sequence?

19. Write a geometric sequence that has three geometric means between 0.2 and 125.

Lesson 12-2 Find the sum of n terms of a geometric series.

- Find the sum of the first 12 terms of the geometric series $4 + 10 + 25 + 62.5 + \cdots$.
  First find the common ratio.
  $a_2 \div a_1 = 10 \div 4$ or 2.5
  Now use the formula for the sum of a finite geometric series.
  $S_n = \dfrac{a_1 - a_1r^n}{1 - r}$
  $S_{12} = \dfrac{4 - 4(2.5)^{12}}{1 - 2.5}$ *$n = 12, a_1 = 4, r = 2.5$*
  $S_{12} \approx 158{,}943.05$ *Use a calculator.*

20. What is the sum of the first nine terms of the geometric series $1.2 - 2.4 + 4.8 - \cdots$?

21. Find the sum of the first eight terms of the geometric series $4 + 4\sqrt{2} + 8 + \cdots$.

OBJECTIVES AND EXAMPLES

Lesson 12-3 Find the limit of the terms and the sum of an infinite geometric series.

- Find $\lim_{n \to \infty} \frac{2n^2 + 5}{3n^2}$.

$$\lim_{n \to \infty} \frac{2n^2 + 5}{3n^2} = \lim_{n \to \infty}\left(\frac{2}{3} + \frac{5}{3n^2}\right)$$
$$= \lim_{n \to \infty} \frac{2}{3} + \lim_{n \to \infty} \frac{5}{3} \cdot \lim_{n \to \infty} \frac{1}{n^2}$$
$$= \frac{2}{3} + \frac{5}{3} \cdot 0$$

Thus, the limit is $\frac{2}{3}$.

REVIEW EXERCISES

Find each limit, or state that the limit does not exist and explain your reasoning.

22. $\lim_{n \to \infty} \frac{3n}{4n + 1}$ **23.** $\lim_{n \to \infty} \frac{6n - 3}{n}$

24. $\lim_{n \to \infty} \frac{2^n n^3}{3n^3}$ **25.** $\lim_{n \to \infty} \frac{4n^3 - 3n}{n^4 - 4n^3}$

26. Write $5.\overline{123}$ as a fraction.

27. Find the sum of the infinite series $1260 + 504 + 201.6 + 80.64 + \cdots$, or state that the sum does not exist and explain your reasoning.

Lesson 12-4 Determine whether a series is convergent or divergent.

- Use the ratio test to determine whether the series $3 + \frac{3^2}{2!} + \frac{3^3}{3!} + \frac{3^4}{4!}$ is convergent or divergent.

The nth term a_n of this series has a general form of $\frac{3^n}{n!}$ and $a_{n+1} = \frac{3^{n+1}}{(n+1)!}$. Find $\lim_{n \to \infty} \frac{a_{n+1}}{a_n}$.

$$r = \lim_{n \to \infty} \frac{\frac{3^{n+1}}{(n+1)!}}{\frac{3^n}{n!}}$$

$$r = \lim_{n \to \infty} \left[\frac{3^{n+1}}{(n+1)!} \cdot \frac{n!}{3^n}\right]$$

$$r = \lim_{n \to \infty} \frac{3}{n + 1} \text{ or } 0$$

Since $r < 0$, the series is convergent.

28. Use the ratio test to determine whether the series $\frac{1}{5} + \frac{2^2}{5^2} + \frac{3^2}{5^3} + \frac{4^2}{5^4} + \cdots$ is *convergent* or *divergent.*

29. Use the comparison test determine whether the series $\frac{6}{1} + \frac{7}{2} + \frac{8}{3} + \frac{9}{4} + \cdots$ is *convergent* or *divergent.*

30. Determine whether the series $2 + 1 + \frac{2}{3} + \frac{1}{2} + \frac{2}{5} + \frac{1}{3} + \frac{2}{7} + \cdots$ is *convergent* or *divergent.*

Lesson 12-5 Use sigma notation.

- Write $\sum_{n=1}^{3} (n^2 - 1)$ in expanded form and then find the sum.

$$\sum_{n=1}^{3} (n^2 - 1) = (1^2 - 1) + (2^2 - 1) + (3^2 - 1)$$
$$= 0 + 3 + 8 \text{ or } 11$$

Write each expression in expanded form and then find the sum.

31. $\sum_{a=5}^{9} (3a - 3)$ **32.** $\sum_{k=1}^{\infty} (0.4)^k$

Express each series using sigma notation

33. $-1 + 1 + 3 + 5 + \cdots$

34. $2 + 5 + 10 + 17 + \cdots + 82$

OBJECTIVES AND EXAMPLES

Lesson 12-6 Use the Binomial Theorem to expand binomials.

- Find the fourth term of $(2x - y)^6$.

$$(2x - y)^6 = \sum_{r=0}^{6} \frac{6!}{r!(6 - r)!}(2x)^{6-r}(-y)^r$$

To find the fourth term, evaluate the general term for $r = 3$.

$$\frac{6!}{3!(6-3)!}(2x)^{6-3}(-y)^3$$

$$= \frac{6 \cdot 5 \cdot 4 \cdot 3!}{3! \cdot 3!}(2x)^3(-y^3) \text{ or } -160x^3y^3$$

REVIEW EXERCISES

Use the Binomial Theorem to expand each binomial.

35. $(a - 4)^6$
36. $(2r + 3s)^4$

Find the designated term of each binomial expansion.

37. 5th term of $(x - 2)^{10}$
38. 3rd term of $(4m + 1)^8$
39. 8th term of $(x + 3y)^{10}$
40. 6th term of $(2c - d)^{12}$

Lesson 12-7 Use Euler's Formula to write the exponential form of a complex number.

- Write $\sqrt{3} - \boldsymbol{i}$ in exponential form.

Write the polar form of $\sqrt{3} - \boldsymbol{i}$.

$r = \sqrt{(\sqrt{3})^2 + (-1)^2}$ or 2, and

$\theta = \text{Arctan } \frac{-1}{\sqrt{3}}$ or $\frac{5\pi}{6}$

$$\sqrt{3} - \boldsymbol{i} = 2\left(\cos \frac{5\pi}{6} + \boldsymbol{i} \sin \frac{5\pi}{6}\right) = 2e^{\boldsymbol{i}\frac{5\pi}{6}}$$

Write each expression or complex number in exponential form.

41. $2\left(\cos \frac{3\pi}{4} + \boldsymbol{i} \sin \frac{3\pi}{4}\right)$
42. $4\boldsymbol{i}$
43. $2 - 2\boldsymbol{i}$
44. $3\sqrt{3} + 3\boldsymbol{i}$

Lesson 12-8 Iterate functions using real and complex numbers.

- Find the first three iterates of the function $f(z) = 2z$ if the initial value is $3 - \boldsymbol{i}$.

$z_0 = 3 - \boldsymbol{i}$
$z_1 = 2(3 - \boldsymbol{i})$ or $6 - 2\boldsymbol{i}$
$z_2 = 2(6 - 2\boldsymbol{i})$ or $12 - 4\boldsymbol{i}$
$z_3 = 2(12 - 4\boldsymbol{i})$ or $24 - 8\boldsymbol{i}$

Find the first four iterates of each function using the given initial value. If necessary, round your answers to the nearest hundredth.

45. $f(x) = 6 - 3x, x_0 = 2$
46. $f(x) = x^2 + 4, x_0 = -3$

Find the first three iterates of the function $f(z) = 0.5z + (4 - 2i)$ for each initial value.

47. $z_0 = 4\boldsymbol{i}$
48. $z_0 = -8$
49. $z_0 = -4 + 6\boldsymbol{i}$
50. $z_0 = 12 - 8\boldsymbol{i}$

Lesson 12-9 Use mathematical induction to prove the validity of mathematical statements.

- Proof by mathematical induction:
  1. First, verify that the conjecture S_n is valid for the first possible case, usually $n = 1$.
  2. Then, assume that S_n is valid for $n = k$ and use this assumption to prove that it is also valid for $n = k + 1$.

Use mathematical induction to prove that each proposition is valid for all positive integral values of *n*.

51. $1 + 2 + 3 + \cdots + n = \frac{n(n+1)}{2}$
52. $3 + 8 + 15 + \cdots + n(n - 2) = \frac{n(n+1)(2n+7)}{6}$
53. $9^n - 4^n$ is divisible by 5.

APPLICATIONS AND PROBLEM SOLVING

54. **Physics** If an object starting at rest falls in a vacuum near the surface of Earth, it will fall 16 feet during the first second, 48 feet during the next second, 80 feet during the third second, and so on. *(Lesson 12-1)*
    a. How far will the object fall during the twelfth second?
    b. How far will the object have fallen after twelve seconds?

55. **Budgets** A major corporation plans to cut the budget on one of its projects by 3 percent each year. If the current budget for the project to be cut is $160 million, what will the budget for that project be in 10 years? *(Lesson 12-2)*

56. **Geometry** If the midpoints of the sides of an equilateral triangle are joined by straight lines, the new figure will also be an equilateral triangle. *(Lesson 12-3)*
    a. If the original triangle has a perimeter of 6 units, find the perimeter of the new triangle.

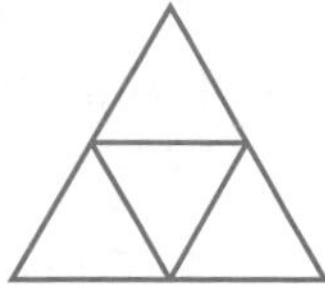

    b. If this process is continued to form a sequence of "nested" triangles, what will be the sum of the perimeters of all the triangles?

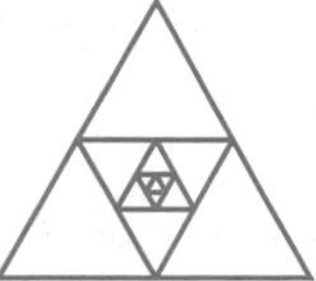

ALTERNATIVE ASSESSMENT

OPEN-ENDED ASSESSMENT

1. A sequence has a common difference of 3.
    a. Is this sequence arithmetic or geometric? Explain.
    b. Form a sequence that has this common difference. Write a recursive formula for your sequence.
2. Write a general expression for an infinite sequence that has no limit. Explain your reasoning.

PORTFOLIO

Explain the difference between a convergent and a divergent series. Give an example of each type of series and show why it is that type of series.

Additional Assessment See p. A67 for Chapter 12 practice test.

THE UNITED STATES CENSUS BUREAU

That's a lot of people!

- Use the Internet to find the population of the United States from at least 1900 through 2000. Write a sequence using the population for each ten-year interval, for example, 1900, 1910, and so on.
- Write a formula for an arithmetic sequence that provides a reasonable model for the population sequence.
- Write a formula for a geometric sequence that provides a reasonable model for the population sequence.
- Use your models to predict the U.S. population for the year 2050. Write a one-page paper comparing the arithmetic and geometric sequences you used to model the population data. Discuss which formula you think best models the data.

TEST-TAKING TIP
With common percents, like 10%, 25%, or 50%, it is faster to use the fraction equivalents and mental math than a calculator.

Percent Problems

A few common words and phrases used in percent problems, along with their translations into mathematical expressions, are listed below.

what ➡ x (variable)	What percent of A is B?	➡ $\frac{x}{100} \cdot A = B$
of ➡ $\times$ (multiply)	What is A percent more than B?	➡ $x = B + \frac{A}{100} \cdot B$
is ➡ $=$ (equals)	What is A percent less than B?	➡ $x = B - \frac{A}{100} \cdot B$

percent of change (increase or decrease) ➡ $\frac{\text{amount of change}}{\text{original amount}} \times 100$

ACT EXAMPLE

1. If c is positive, what percent of $3c$ is 9?

A $\frac{c}{100}\%$ **B** $\frac{300}{c}\%$ **C** $\frac{9}{c}\%$ **D** 3% **E** $\frac{c}{3}\%$

HINT Use variables just as you would use numbers.

Solution Start by translating the question into an equation.

What percent of $3c$ is 9?

$$\frac{x}{100} \cdot 3c = 9$$

Now solve the equation for x, not c.

$\frac{x}{100} = \frac{9}{3c}$ *Divide each side by 3c.*

$\frac{x}{100} = \frac{3}{c}$ *Simplify.*

$x = \frac{300}{c}$ *Solve for x.*

The answer is choice **B**.

Alternate Solution "Plug in" 3 for c. The question becomes "what percent of 9 is 9?" The answer is 100%, so check each expression choice to see if it is equal to 100% when $c = 3$.

Choice A: $\frac{c}{100}\% = \frac{3}{100}\%$

Choice B: $\frac{300}{c}\% = 100\%$

Choice **B** is correct.

SAT EXAMPLE

2. An electronics store offers a 25% discount on all televisions during a sale week. How much must a customer pay for a television marked at $240?

A $60 **B** $300 **C** $230.40

D $180 **E** $215

HINT A discount is a decrease in the price of an item. So, the question asked is "What is 25% less than 240?"

Solution Start by translating this question into an equation.

What is 25% less than 240? ➡ $x = 240 - \frac{25}{100} \cdot 240$

Now simplify the right-hand side of the equation.

$$x = 240 - \frac{1}{4} \cdot 240 \text{ or } 180$$

Choice **D** is correct.

Alternate Solution If there is a 25% discount, a customer will pay (100 − 25)% or 75% of the marked price.

What is 75% of the marked price? ➡ $x = 0.75 \cdot 240$ or 180

The answer is choice **D**.

After you work each problem, record your answer on the answer sheet provided or on a piece of paper.

Multiple Choice

1. Shanika has a collection of 80 tapes. If 40% of her records are jazz tapes and the rest are blues tapes, how many blues tapes does she have?

   A 32 B 40 C 42 D 48 E 50

2. If ℓ_1 is parallel to ℓ_2 in the figure below, what is the value of x?

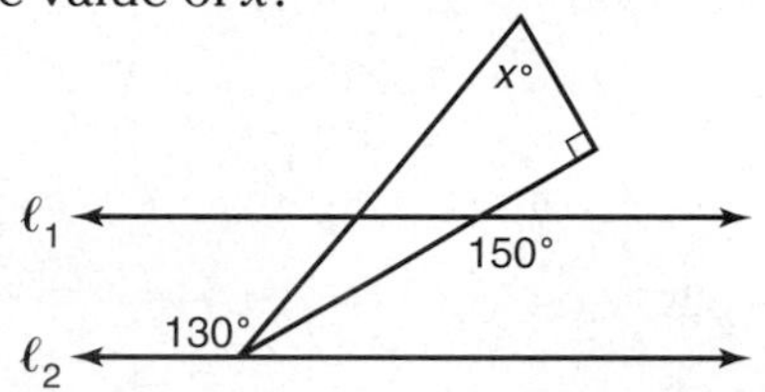

   A 20 B 50 C 70 D 80 E 90

3. There are k gallons of gasoline available to fill a tank. After d gallons have been pumped, then, in terms of k and d, what percent of the gasoline has been pumped?

   A $\frac{100d}{k}\%$ B $\frac{k}{100d}\%$ C $\frac{100k}{d}\%$

   D $\frac{k}{100(k-d)}\%$ E $\frac{100(k-d)}{k}\%$

4. In 1985, Andrei had a collection of 48 baseball caps. Since then he has given away 13 caps, purchased 17 new caps, and traded 6 of his caps to Pierre for 8 of Pierre's caps. Since 1985, what has been the net percent increase in Andrei's collection?

   A 6% B $12\frac{1}{2}\%$ C $16\frac{2}{3}\%$

   D 25% E $28\frac{1}{2}\%$

5. In the figure below, $AB = AC$ and AD is a line segment. What is the value of $x - y$?

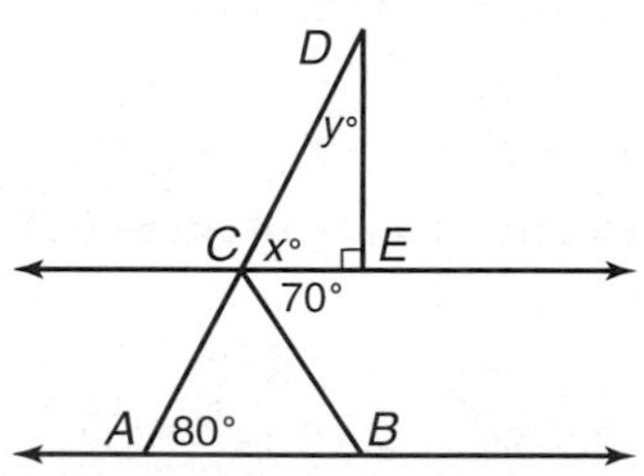

   Note: Figure is NOT drawn to scale.

   A 10 B 20 C 30 D 70 E 90

6. At the beginning of 2000, the population of Rockville was 204,000, and the population of Springfield was 216,000. If the population of each city increased by exactly 20% in 2000, how many more people lived in Springfield than in Rockville at the end of 2000?

   A 9,600 B 10,000 C 12,000

   D 14,400 E 20,000

7. In the figure, the slope of $\overline{AC}$ is $-\frac{1}{6}$, and $m\angle C = 30°$. What is the length of $\overline{BC}$?

   y
   B (6, 10)
   A
   (5, 4)
   C
   O
   x

   A $\sqrt{37}$

   B $\sqrt{111}$

   C 2

   D $2\sqrt{37}$

   E It cannot be determined from the information given.

8. If $x + 6 > 0$ and $1 - 2x > -1$, then x could equal each of the following EXCEPT ?

   A -6 B -4 C -2 D 0 E $\frac{1}{2}$

9. **Quantitative Comparison**
   A if the quantity in Column A is greater;
   B if the quantity in Column B is greater;
   C if the two quantities are equal;
   D if the relationship cannot be determined from the information given

Column A	Column B
The percent increase from 99 to 100.	The percent decrease from 100 to 99.

10. **Grid-In** One fifth of the cars in a parking lot are blue, and $\frac{1}{2}$ of the blue cars are convertibles. If $\frac{1}{4}$ of the convertibles in the parking lot are blue, then what percent of the cars in the lot are neither blue nor convertibles?

interNET CONNECTION **SAT/ACT Practice** For additional test practice questions, visit: **www.amc.glencoe.com**

Combinatorics and Probability

CHAPTER OBJECTIVES

- **Solve problems involving combinations and permutations.** *(Lessons 13-1, 13-2)*
- **Distinguish between independent and dependent events and between mutually exclusive and mutually inclusive events.** *(Lessons 13-1, 13-4, 13-5)*
- **Find probabilities.** *(Lessons 13-3, 13-4, 13-5, 13-6)*
- **Find odds for the success and failure of an event.** *(Lesson 13-3)*

13-1

Permutations and Combinations

OBJECTIVES

- Solve problems related to the Basic Counting Principle.
- Distinguish between dependent and independent events.
- Solve problems involving permutations or combinations.

EDUCATION Ivette is a freshman at the University of Miami. She is planning her fall schedule for next year. She has a choice of three mathematics courses, two science courses, and two humanities courses. She can only select one course from each area. How many course schedules are possible?

Let M_1, M_2, and M_3 represent the three math courses, S_1 and S_2 the science courses, and H_1 and H_2 the humanities courses. Once Ivette makes a selection from the three mathematics courses she has two choices for her science course. Then, she has two choices for humanities. A **tree diagram** is often used to show all the choices.

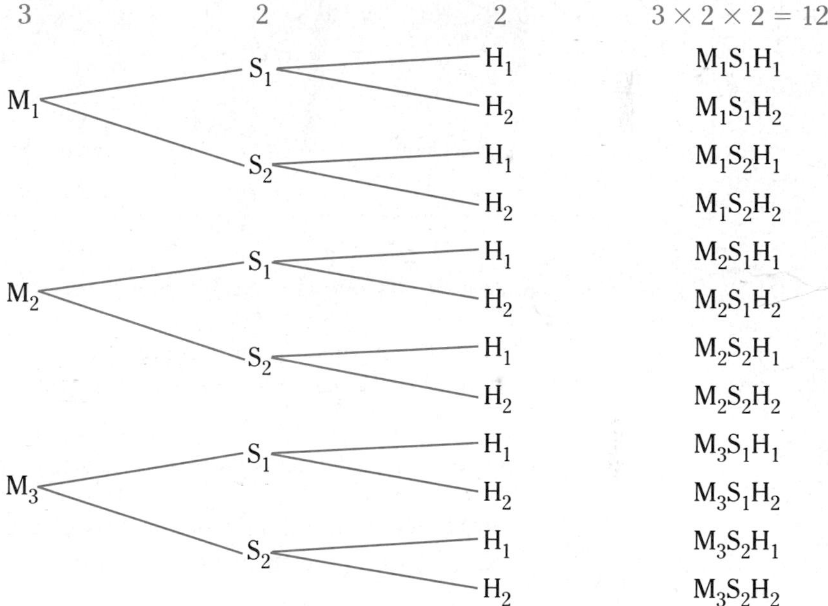

Ivette has 12 possible schedules from which to choose.

The choice of selecting a mathematics course does *not* affect the choice of ways to select a science or humanities course. Thus, these three choices are called **independent events.** Events that do affect each other are called **dependent events.** An example of dependent events would be the order in which runners finish a race. The first place runner affects the possibilities for second place, the second place runner affects the possibilities for third place, and so on.

The branch of mathematics that studies different possibilities for the arrangement of objects is called **combinatorics.** The example of choosing possible course schedules illustrates a rule of combinatorics known as the **Basic Counting Principle.**

Basic Counting Principle	Suppose one event can be chosen in p different ways, and another independent event can be chosen in q different ways. Then the two events can be chosen successively in $p \cdot q$ ways.

This principle can be extended to any number of independent events. For example, in the previous application, the events are chosen in $p \times q \times r$ or $3 \times 2 \times 2$ different ways.

Example 1 **Vickie works for a bookstore. Her manager asked her to arrange a set of five best-sellers for a display. The display is to be set up as shown below. The display set is made up of one book from each of 5 categories. There are 4 nonfiction, 4 science fiction, 3 history, 3 romance, and 3 mystery books from which to choose.**

a. Are the choices for each book independent or dependent events?

Since the choice of one type of book does not affect the choice of another type of book, the events are independent.

b. How many different ways can Vickie choose the books for the display?

Vickie has four choices for the first spot in the display, four choices for the second spot, and three choices for each of the next three spots.

1st spot		2nd spot		3rd spot		4th spot		5th spot
4	·	4	·	3	·	3	·	3

This can be represented as $4 \cdot 4 \cdot 3 \cdot 3 \cdot 3$ or 432 different arrangements.

There are 432 possible ways for Vickie to choose books for the display.

The arrangement of objects in a certain order is called a **permutation.** In a permutation, the order of the objects is very important. The symbol $P(n, n)$ denotes the number of permutations of n objects taken all at once. The symbol $P(n, r)$ denotes the number of permutations of n objects taken r at a time.

Permutations	The number of permutations of n objects, taken n at a time is defined as $P(n, n) = n!$.
$P(n, n)$ and $P(n, r)$	The number of permutations of n objects, taken r at a time is defined as $P(n, r) = \frac{n!}{(n - r)!}$.

Recall that $n!$ is read "n factorial" and, $n! = n(n - 1)(n - 2)\ldots(1)$.

Example 2 **During a judging of a horse show at the Fairfield County Fair, there are three favorite horses: Rye Man, Oiler, and Sea of Gus.**

a. Are the selection of first, second and third place from the three horses independent or dependent events?

b. Assuming there are no ties and the three favorites finish in the top three places, how many ways can the horses win first, second and third places?

a. The choice of a horse for first place does affect the choice for second and third places. For example, if Rye Man is first, then is impossible for it to finish second or third. Therefore, the events are dependent.

b. Since order is important, this situation is a permutation.

Method 1: Tree diagram

There are three possibilities for first place, two for second, and one for third as shown in the tree diagram below. If Rye Man finishes first, then either Oiler or Sea of Gus will finish second. If Oiler finishes second, then Sea of Gus must finish third. Likewise, if Sea of Gus finishes second, then Oiler finishes third.

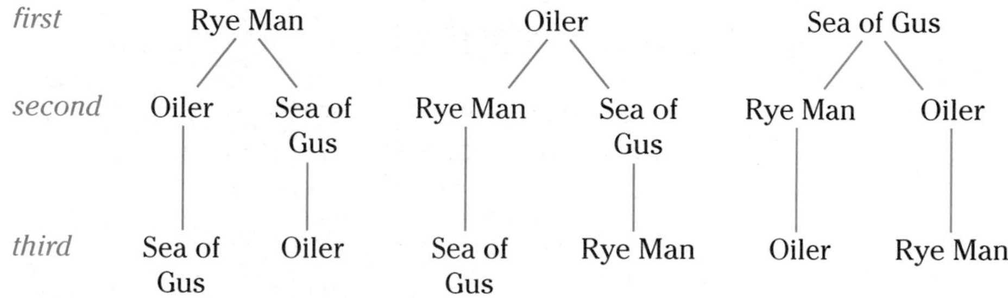

There are 6 possible ways the horses can win.

Method 2: Permutation formula

This situation depicts three objects taken three at a time and can be represented as $P(3, 3)$.

$$\begin{aligned} P(3, 3) &= 3! \\ &= 3 \cdot 2 \cdot 1 \text{ or } 6 \end{aligned}$$

Thus, there are 6 ways the horses can win first, second, and third place.

Example 3 **The board of directors of B.E.L.A. Technology Consultants is composed of 10 members.**

a. How many different ways can all the members sit at the conference table as shown?

b. In how many ways can they elect a chairperson, vice-chairperson, treasurer, and secretary, assuming that one person cannot hold more than one office?

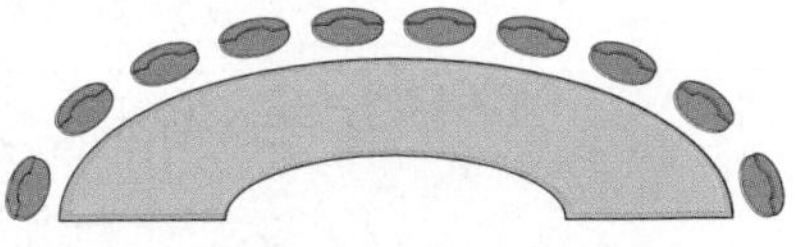

a. Since order is important, this situation is a permutation. Also, the 10 members are being taken all at once so the situation can be represented as $P(10, 10)$.

$$P(10, 10) = 10!$$
$$= 10 \cdot 9 \cdot 8 \cdot 7 \cdot 6 \cdot 5 \cdot 4 \cdot 3 \cdot 2 \cdot 1 \text{ or } 3{,}628{,}800$$

There are 3,628,800 ways that the 10 board members can sit at the table.

b. This is a permutation of 10 people being chosen 4 at a time.

$$P(10, 4) = \frac{10!}{(10 - 4)!}$$
$$= \frac{10 \cdot 9 \cdot 8 \cdot 7 \cdot 6!}{6!}$$
$$= 5040$$

There are 5040 ways in which the offices can be filled.

Suppose that in the situation presented in Example 1, Vickie needs to select three types of books from the five types available. There are $P(5, 3)$ or 60 possible arrangements. She can arrange them as shown in the table below.

Arrangement	Type		
1	nonfiction	science fiction	history
2	nonfiction	history	science fiction
3	nonfiction	romance	mystery
4	nonfiction	mystery	romance
5	science fiction	nonfiction	history
6	science fiction	history	nonfiction
7	science fiction	romance	mystery
8	science fiction	mystery	romance
9	history	nonfiction	science fiction
10	history	science fiction	nonfiction
⋮	⋮	⋮	⋮
60	romance	mystery	nonfiction

Note that arrangements 1, 2, 5, 6, 9 and 10 contain the same three types of books. In each group of three books, there are 3! or 6 ways they can be arranged. Thus, if order is disregarded, there are $\frac{60}{3!}$ or 10 different groups of three types of books that can be selected from the five types. In this situation, called a **combination,** the order in which the books are selected is *not* a consideration.

A combination of n objects taken r at a time is calculated by dividing the number of permutations by the number of arrangements containing the same elements and is denoted by $C(n, r)$.

Combination

C(n, r)

The number of combinations of n objects taken r at a time is defined as

$$C(n, r) = \frac{n!}{(n - r)!\, r!}.$$

The main difference between a permutation and a combination is whether order is considered (as in permutation) or not (as in combination). For example, for objects E, F, G, and H taken two at a time, the permutations and combinations are listed below.

Permutations				Combinations	
EF	FE	GE	HE	EF	FG
EG	FG	GF	HF	EG	FH
EH	FH	GH	HG	EH	GH

Note that in permutations, EF is different from FE. But in combinations, EF is the same as FE.

Example 4 **ART** **In 1999, The National Art Gallery in Washington, D.C., opened an exhibition of the works of John Singer Sargent (1856–1925). The gallery's curator wanted to select four paintings out of twenty on display to showcase the work of the artist. How many groups of four paintings can be chosen?**

Since order is not important, the selection is a combination of 20 objects taken 4 at a time. It can be represented as $C(20, 4)$.

$$\begin{aligned} C(20, 4) &= \frac{20!}{(20 - 4)!\, 4!} \\ &= \frac{20!}{16!\, 4!} \\ &= \frac{20 \cdot 19 \cdot 18 \cdot 17 \cdot 16!}{16!\, 4!} \\ &= \frac{20 \cdot 19 \cdot 18 \cdot 17}{4 \cdot 3 \cdot 2 \cdot 1} \\ &= 4845 \end{aligned}$$

Express 20! as $20 \cdot 19 \cdot 18 \cdot 17 \cdot 16!$ since $\frac{16!}{16!} = 1$.

There are 4845 possible groups of paintings.

"Oyster Gatherers of Cancale," 1878

Example 5 **At Grant Senior High School, there are 15 names on the ballot for junior class officers. Five will be selected to form a class committee.**

a. How many different committees of 5 can be formed?

b. In how many ways can a committee of 5 be formed if each student has a different responsibility?

c. If there are 8 girls and 7 boys on the ballot, how many committees of 2 boys and 3 girls can be formed?

a. Order is not important in this situation, so the selection is a combination of 15 people chosen 5 at a time.

$$C(15, 5) = \frac{15!}{(15-5)!\,5!}$$

$$= \frac{15!}{10!\,5!}$$

$$= \frac{15 \cdot 14 \cdot 13 \cdot 12 \cdot 11 \cdot 10!}{10!\,5!} \text{ or } 3003$$

There are 3003 different ways to form the committees of 5.

b. Order has to be considered in this situation because each committee member has a different responsibility.

$$P(15, 5) = \frac{15!}{(15-5)!}$$

$$= \frac{15!}{10!} \text{ or } 360{,}360$$

There are 360,360 possible committees.

c. Order is not important. There are three questions to consider.
How many ways can 2 boys be chosen from 7?
How many ways can 3 girls be chosen from 8?
Then, how many ways can 2 boys and 3 girls be chosen together?

Since the events are independent, the answer is the product of the combinations $C(7, 2)$ and $C(8, 3)$.

$$C(7, 2) \cdot C(8, 3) = \frac{7!}{(7-2)!\,2!} \cdot \frac{8!}{(8-3)!\,3!}$$

$$= \frac{7!}{5!\,2!} \cdot \frac{8!}{5!\,3!}$$

$$= 21 \cdot 56 \text{ or } 1176$$

There are 1176 possible committees.

CHECK FOR UNDERSTANDING

Communicating Mathematics

Read and study the lesson to answer each question.

1. **Compare and contrast** permutations and combinations.
2. **Write** an expression for the number of ways, out of a standard 52-card deck, that 5-card hands can have 2 jacks and 3 queens.

3. **You Decide** Ms. Sloan asked her students how many ways 5 patients in a hospital could be assigned to 7 identical private rooms. Anita said that the problem dealt with computing $C(7, 5)$. Sam disagreed, saying that $P(7, 5)$ was the correct way to answer the question. Who is correct? Why?

4. **Draw** a tree diagram to illustrate all of the possible T-shirts available that come in sizes small, medium, large, and extra large and in the colors blue, green and gray.

Guided Practice

5. A restaurant offers the choice of an entrée, a vegetable, a dessert, and a drink for a lunch special. If there are 4 entrees, 3 vegetables, 5 desserts and 5 drinks available to choose from, how many different lunches are available?

6. Are choosing a movie to see and choosing a snack to buy dependent or independent events?

Find each value.

7. $P(6, 6)$
8. $P(5, 3)$
9. $\frac{P(12, 8)}{P(6, 4)}$
10. $C(7, 4)$
11. $C(20, 15)$
12. $C(4, 3) \cdot C(5, 2)$

13. If a group of 10 students sits in the same row in an auditorium, how many possible ways can they be arranged?

14. How many baseball lineups of 9 players can be formed from a team that has 15 members if all players can play any position?

15. **Postal Service** The U.S. Postal Service uses 5-digit ZIP codes to route letters and packages to their destinations.
    a. How many ZIP codes are possible if the numbers 0 through 9 are used for each of the 5 digits?
    b. Suppose that when the first digit is 0, the second, third, and fourth digits cannot be 0. How many 5-digit ZIP codes are possible if the first digit is 0?
    c. In 1983, the U.S. Postal Service introduced the ZIP +4, which added 4 more digits to the existing 5-digit ZIP codes. Using the numbers 0 through 9, how many additional ZIP codes were possible?

EXERCISES

Practice

16. If you toss a coin, then roll a die, and then spin a 4-colored spinner with equal sections, how many outcomes are possible?

17. How many ways can 7 classes be scheduled, if each class is offered in each of 7 periods?

18. Find the number of different 7-digit telephone numbers where:
    a. the first digit cannot be zero.
    b. only even digits are used.
    c. the complete telephone numbers are multiples of 10.
    d. the first three digits are 593 in that order.

State whether the events are *independent* or *dependent*.

19. selecting members for a team

20. tossing a penny, rolling a die, then tossing a dime

21. deciding the order in which to complete your homework assignments

Find each value.

22. $P(8, 8)$
23. $P(6, 4)$
24. $P(5, 3)$
25. $P(7, 4)$
26. $P(9, 5)$
27. $P(10, 7)$
28. $\dfrac{P(6, 3)}{P(4, 2)}$
29. $\dfrac{P(6, 4)}{P(5, 3)}$
30. $\dfrac{P(6, 3) \cdot P(7, 5)}{P(9, 6)}$
31. $C(5, 3)$
32. $C(10, 5)$
33. $C(4, 2)$
34. $C(12, 4)$
35. $C(9, 9)$
36. $C(14, 7)$
37. $C(3, 2) \cdot C(8, 3)$
38. $C(7, 3) \cdot C(8, 5)$
39. $C(5, 1) \cdot C(4, 2) \cdot C(8, 2)$

40. A pizza shop has 14 different toppings from which to choose. How many different 4-topping pizzas can be made?

41. If you make a fruit salad using 5 different fruits and you have 14 different varieties from which to choose, how many different fruit salads can you make?

42. How many different 12-member juries can be formed from a group of 18 people?

43. A bag contains 3 red, 5 yellow, and 8 blue marbles. How many ways can 2 red, 1 yellow, and 2 blue marbles be chosen?

44. How many different ways can 11 paintings be displayed on a wall?

45. From a standard 52-card deck, find how many 5-card hands are possible that have:
   a. 3 hearts and 2 clubs.
   b. 1 ace, 2 jacks, and 2 kings.
   c. all face cards.

Applications and Problem Solving

46. **Home Security** A home security company offers a security system that uses the numbers 0 through 9, inclusive, for a 5-digit security code.
   a. How many different security codes are possible?
   b. If no digits can be repeated, how many security codes are available?
   c. Suppose the homeowner does not want to use 0 as one of the digits and wants only two of the digits to be odd. How many codes can be formed if the digits can be repeated? If no repetitions are allowed, how many codes are available?

47. **Baseball** How many different 9-player teams can be fielded if there are 3 players who can only play catcher, 4 players who can only play first base, 6 players who can only pitch, and 14 players who can play in any of the remaining 6 positions?

48. **Transportation** In a train yard, there are 12 flatcars, 10 tanker cars, 15 boxcars, and 5 livestock cars.
   a. If the cars must be connected according to their final destinations, how many different ways can they be arranged?
   b. How many ways can the train be made up if it is to have 30 cars?
   c. How many trains can be formed with 3 livestock cars, 6 flatcars, 6 tanker cars, and 5 boxcars?

49. **Critical Thinking** Prove $P(n, n - 1) = P(n, n)$.

50. **Entertainment** Three couples have reserved seats for a Broadway musical. Find how many different ways they can sit if:
   a. there are no seating restrictions.
   b. two members of each couple wish to sit together.

51. **Botany** A researcher with the U.S. Department of Agriculture is conducting an experiment to determine how well certain crops can survive adverse weather conditions. She has gathered 6 corn plants, 3 wheat plants, and 2 bean plants. She needs to select four plants at random for the experiment.

a. In how many ways can this be done?

b. If exactly 2 corn plants must be included, in how many ways can the plants be selected?

52. **Geometry** How many lines are determined by 10 points, no 3 of which are collinear?

53. **Critical Thinking** There are 6 permutations of the digits 1, 6, and 7.

167 176 617 671 716 761

The average of these six numbers is $\frac{3108}{6} = 518$ which is equal to $37(1 + 6 + 7)$. If the digits are 0, 4, and 7, then the average of the six permutations is $\frac{2442}{6} = 407$ or $37(0 + 4 + 7)$.

a. Use this pattern to find the average of the six permutations of 2, 5, and 9.

b. Will this pattern hold for all sets of three digits? If so, prove it.

Mixed Review

54. **Banking** Cynthia has a savings account that has an annual yield of 5.8%. Find the balance of the account after each of the first three years if her initial balance is $2140. *(Lesson 12-8)*

55. Find the sum of the first ten terms of the series $1^3 + 2^3 + 3^3 + \cdots$. *(Lesson 12-5)*

56. Solve $7.1^x = 83.1$ using logarithms. Round to the nearest hundredth. *(Lesson 11-6)*

57. Find the value of x to the nearest tenth such that $x = e^{0.346}$. *(Lesson 11-3)*

58. **Communications** A satellite dish tracks a satellite directly overhead. Suppose the graph of the equation $y = 4x^2$ models the shape of the dish when it is oriented in this position. Later in the day, the dish is observed to have rotated approximately 45°. Find an equation that models the new orientation of the dish. *(Lesson 10-7)*

59. Graph the system of polar equations $r = 2$, and $r = 2 \cos 2\theta$. Then solve the system and compare the common solutions. *(Lesson 9-2)*

60. Find the initial vertical and horizontal velocities of a rock thrown with an initial velocity of 28 feet per second at an angle of 45° with the horizontal. *(Lesson 8-7)*

61. Solve $\sin 2x + 2 \sin x = 0$ for $0° \le x \le 360°$. *(Lesson 7-5)*

62. State the amplitude, period, and phase shift for the function $y = 8 \cos(\theta - 30°)$. *(Lesson 6-5)*

63. Given the triangle at the right, solve the triangle if $A = 27°$ and $b = 15.2$. Round angle measures to the nearest degree and side measures to the nearest tenth. *(Lesson 5-5)*

B
c
a
A
C
b

64. **SAT/ACT Practice** What is the number of degrees through which the hour hand of a clock moves in 2 hours 12 minutes?

A 66° B 72° C 126° D 732° E 792°

Extra Practice See p. A51.

Permutations with Repetitions and Circular Permutations

OBJECTIVES
- Solve problems involving permutations with repetitions.
- Solve problems involving circular permutations.

MARKETING Marketing professionals sometimes investigate the number of permutations and arrangements of letters to create company or product names. For example, the company JATACO was derived from the first initials of the owners Alan, Anthony, John, and Thomas. Suppose five high school students have developed a web site to help younger students better understand first year algebra. They decided to use the initials of their first names to create the title of their web site. The initials are: E, L, O, B, O. How many different five-letter words can be created with these letters?

Each "O" is marked a different color to differentiate it from the other. Some of the possible arrangements are listed below.

EOLBO	EOOLB	EOOBL	EOBOL
LEOOB	OBELO	OBLEO	BLOOE

The five letters can be arranged in $P(5, 5)$ or 5! ways. However, several of these 120 arrangements are the same unless the Os are colored. So without coloring the Os, there are repetitions in the 5! possible arrangements.

Permutations with Repetitions

The number of permutations of n objects of which p are alike and q are alike is

$$\frac{n!}{p!\,q!}.$$

Using this formula, we find there are only $\frac{5!}{2!}$ or 60 permutations of the five letters of which 2 are Os.

Example 1 **How many eight-letter patterns can be formed from the letters of the word *parabola*?**

The eight letters can be arranged in $P(8, 8)$ or 8! ways. However, several of these 40,320 arrangements have the same appearance since a appears 3 times.

The number of permutations of 8 letters of which 3 are the same is $\frac{8!}{3!}$ or 6720.

There are 6720 different eight-letter patterns that can be formed from the letters of the word *parabola*.

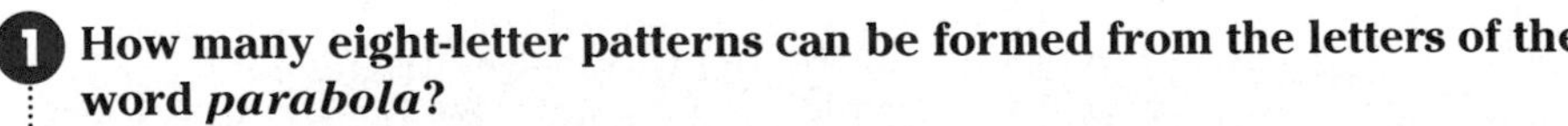

Example 2 **How many eleven-letter patterns can be formed from the letters of the word *Mississippi*?**

$$\frac{11!}{4!\,4!\,2!} = \frac{11 \cdot 10 \cdot 9 \cdot 8 \cdot 7 \cdot 6 \cdot 5 \cdot 4 \cdot 3 \cdot 2 \cdot 1}{4 \cdot 3 \cdot 2 \cdot 1 \cdot 4 \cdot 3 \cdot 2 \cdot 1 \cdot 2 \cdot 1}$$ *There are 11 letters in Mississippi.* *4 i's* *4 s's* *2 p's*

$$= 34{,}650$$

There are 34,650 eleven-letter patterns.

So far, you have been studying objects that are arranged in a line. Consider the problem of making distinct arrangements of six children riding on a merry-go-round at a playground. How many riding arrangements are possible?

Let the numbers 1 through 6 represent the children. Four possible arrangements are shown below.

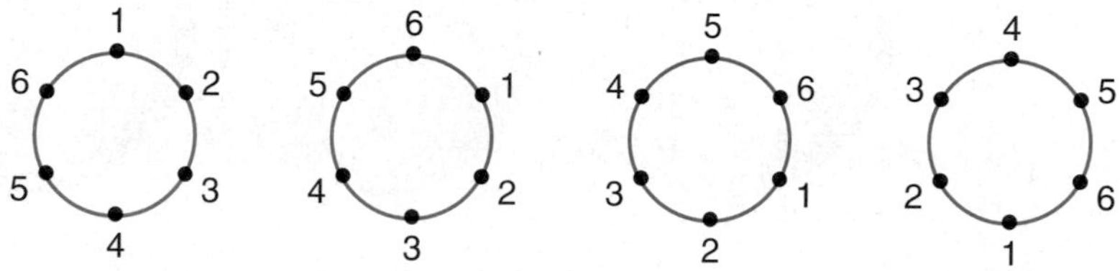

When objects are arranged in a circle, some of the arrangements are alike. In the situation above, these similar arrangements fall into groups of six, each of which can be found by rotating the circle $\frac{1}{6}$ of a revolution. Thus, the number of distinct arrangements around the circular merry-go-round is $\frac{1}{6}$ of the total number of arrangements if the children stood in a line.

$$\frac{1}{6} \cdot 6! = \frac{6 \cdot 5 \cdot 4 \cdot 3 \cdot 2 \cdot 1}{6}$$
$$= 5 \cdot 4 \cdot 3 \cdot 2 \cdot 1$$
$$= 5! \text{ or } (6 - 1)!$$

Thus, there are $(6 - 1)!$ or 120 distinct arrangements of the 6 children around the merry-go-round.

Circular Permutations If n objects are arranged in a circle, then there are $\frac{n!}{n}$ or $(n - 1)!$ permutations of the n objects around the circle.

If the circular object looks the same when it is turned over, such as a plain key ring, then the number of permutations must be divided by 2.

Example 3 **At the Family Friendly Restaurant, nine bowls of food are placed on a circular, revolving tray in the center of the table. You can serve yourself from each of the bowls.**

a. Is the arrangement of the bowls on the tray a linear or circular permutation? Explain.

The arrangement is a circular permutation since the bowls form a circle on the tray and there is no reference point.

b. How many ways can the bowls be arranged?

There are nine bowls so the number of arrangements can be described by $(9 - 1)!$ or $8!$

$8! = 8 \cdot 7 \cdot 6 \cdot 5 \cdot 4 \cdot 3 \cdot 2 \cdot 1$ or 40,320

There are 40,320 ways in which the bowls can be arranged on the tray.

Suppose a CD changer holds 4 CDs on a circular platter. Let each circle below represent the platter and the labeled points represent each CD. The arrow indicates which CD will be played.

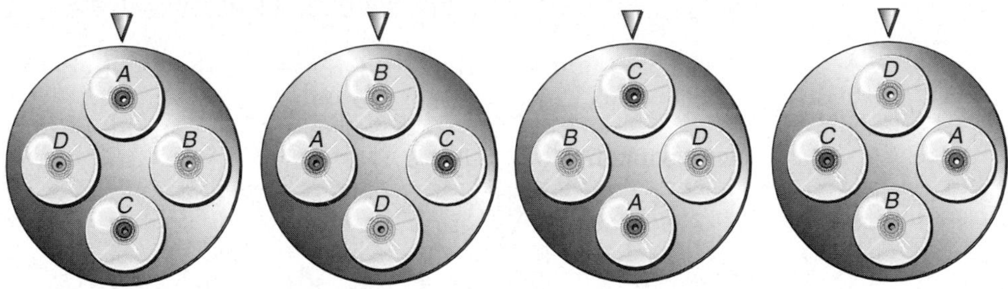

These arrangements are different. In each one, a different CD is being played. Thus, there are $P(4, 4)$ or 24 arrangements relative to the playing position.

If n objects are arranged relative to a fixed point, then there are $n!$ permutations. Circular arrangements with fixed points of reference are treated as linear permutations.

Example 4 **Seven people are to be seated at a round table where one person is seated next to a window.**

a. Is the arrangement of the people around the table a linear or circular permutation? Explain.

b. How many possible arrangements of people relative to the window are there?

a. Since the people are seated around a table with a fixed reference point, the arrangement is a linear permutation.

b. There are seven people with a fixed reference point. So there are 7! or 5040 ways in which the people can be seated around the table.

CHECK FOR UNDERSTANDING

Communicating Mathematics

Read and study the lesson to answer each question.

1. **Write** an explanation as to why a circular permutation is not computed the same as a linear permutation.
2. **Describe** two real-world situations involving permutations with repetitions.
3. Provide a **counterexample** for the following statement.
   The number of permutations for n objects in a circular arrangement is $(n - 1)!$.

Guided Practice

How many different ways can the letters of each word be arranged?

4. *kangaroo*
5. *classical*
6. In how many ways can 2 red lights, 4 yellow lights, 5 blue lights, 1 green light, and 2 pink lights be arranged on a string of lights?

Determine whether each arrangement of objects is a linear or circular permutation. Then determine the number of arrangements for each situation.

7. 11 football players in a huddle
8. 8 jewels on a necklace
9. 12 decorative symbols around the face of a clock
10. 5 beads strung on a string arranged in a square pattern relative to a knot in the string

11. **Communication** Morse code is a system of dots, dashes, and spaces that telegraphers in the United States and Canada once used to send messages by wire. How many different arrangements are there of 5 dots, 2 dashes, and 2 spaces?

EXERCISES

Practice

How many different ways can the letters of each word be arranged?

12. *pizzeria*
13. *California*
14. *calendar*
15. *centimeter*
16. *trigonometry*
17. *Tennessee*

18. How many different 7-digit phone numbers can have the digits 7, 3, 5, 2, 7, 3, and 2?
19. Five country CDs and four rap CDs are to be placed in a display window. How many ways can they be arranged if they are placed by category?
20. The table below shows the initial letter for each of the 50 states. How many different ways can you arrange the initial letters of the states?

Initial Letter	A	C	D	F	G	H	I	K	L	M	N	O	P	R	S	T	U	V	W
Number of States	4	3	1	1	1	1	4	2	1	8	8	3	1	1	2	2	1	1	3

Determine whether each arrangement of objects is a linear or circular permutation. Then determine the number of arrangements for each situation.

21. 12 gondolas on a Ferris wheel
22. a stack of 6 pennies, 3 nickels, 7 dimes, and 10 quarters
23. the placement of 9 specialty departments along the outside perimeter of a supermarket
24. a family of 5 seated around a rectangular table
25. 8 tools on a utility belt

Determine whether each arrangement of objects is a linear or circular permutation. Then determine the number of arrangements for each situation.

26. 6 houses on a cul-de-sac relative to the incoming street
27. 10 different beads on a string
28. a waiter placing 9 drinks along the edge of a circular tray
29. 14 keys on a key ring
30. 20 wooden dowels used as spokes for a wagon wheel
31. 32 horses on the outside edge of a carousel
32. 25 sections of a circular stadium relative to the main entrance

Applications and Problem Solving

33. **Biology** A biologist needs to determine the number of possible arrangements of 4 kinds of molecules in a chain. If the chain contains 8 molecules with 2 of each kind, how many different arrangements are possible?

34. **Geometry** Suppose 7 points on the circle at the right are selected at random.
    a. Using the letters A through G, how many ways can the points be named on the circle?
    b. Relative to the point which lies on the x-axis, how many arrangements are possible?

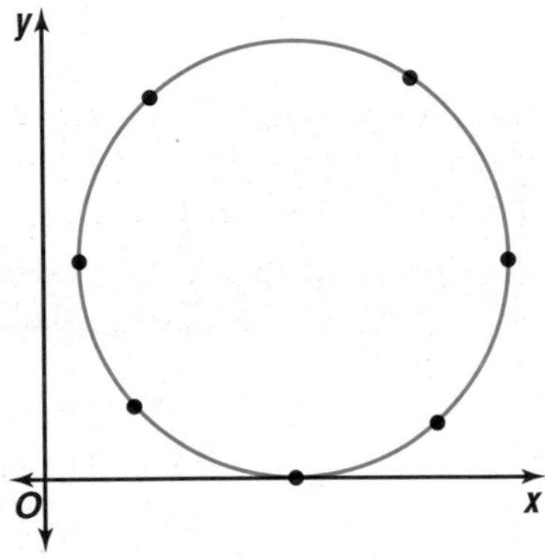

35. **Money** Trish has a penny, 3 nickels, 4 dimes, and 3 quarters in her pocket. How many different arrangements are possible if she removes one coin at a time from her pocket?

36. **Critical Thinking** An anagram is a word or phrase made from another by rearranging its letters. For example, *now* can be changed to *won*. Consider the phrase "*calculating rules.*"
    a. How many different ways can the letters in *calculating* be arranged?
    b. Rearrange the letters and the space in the phrase to form the name of a branch of mathematics.

37. **Auto Racing** Most stock car races are held on oval-shaped tracks with cars from various manufacturers. Let F, C, and P represent three auto manufacturers.
    a. Suppose for one race 20 F cars, 14 C cars, and 9 P cars qualified to be in a race. How many different starting line-ups based on manufacturer are possible?
    b. If there are 43 cars racing, how many different ways could the cars be arranged on the track?
    c. Relative to the leader of the race, how many different ways could the cars be arranged on the track?

38. **Critical Thinking** To break a code, Zach needs to find how many symbols there are in a particular sequence. He is told that there are 3 x's and some dashes. He is also told that there are 35 linear permutations of the symbols. What is the total number of symbols in the code sequence?

Mixed Review

39. **Food** Classic Pizza offers pepperoni, mushrooms, sausage, onions, peppers, and olives as toppings for their 7-inch pizza. How many different 3-topping pizzas can be made? *(Lesson 13-1)*

Extra Practice See p. A51.

40. Use the Binomial Theorem to expand $(5x - 1)^3$. *(Lesson 12-6)*

41. Solve $x < \log_2 413$ using logarithms. Round to the nearest hundredth. *(Lesson 11-5)*

42. Write the equation of the parabola with vertex at $(6, -1)$ and focus at $(3, -1)$. *(Lesson 10-5)*

43. Simplify $2(4 - 3\boldsymbol{i})(7 - 2\boldsymbol{i})$. *(Lesson 9-5)*

44. Find the cross product of $\vec{\mathbf{v}}\ \langle 2, 0, 3\rangle$ and $\vec{\mathbf{w}}\ \langle 2, 5, 0\rangle$. Verify that the resulting vector is perpendicular to $\vec{\mathbf{v}}$ and $\vec{\mathbf{w}}$. *(Lesson 8-4)*

45. **Manufacturing** A knife is held at a 45° angle to the vertical on a 16-inch diameter sharpening wheel. How far above the wheel must a lamp be placed so it will not be showered with sparks? *(Lesson 7-6)*

46. **SAT/ACT Practice** If $x^2 = 36$, then 2^{x-1} could equal

A 4 **B** 6 **C** 8 **D** 16 **E** 32

CAREER CHOICES

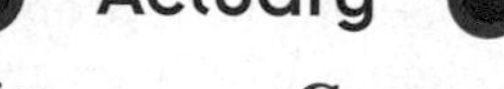

Actuary

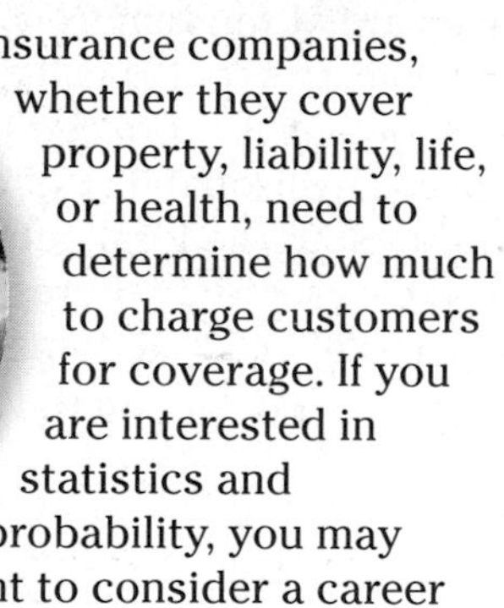

Insurance companies, whether they cover property, liability, life, or health, need to determine how much to charge customers for coverage. If you are interested in statistics and probability, you may want to consider a career as an actuary.

Actuaries use statistical methods to determine the probability of such events as death, injury, unemployment, and property damage or loss. An actuary must estimate the number and amount of claims that may be made in order for the insurance company to set its insurance coverage rates for its customers. As an actuary, you can specialize in property and liability or life and health statistics. Most actuaries work for insurance companies, consulting firms, or the government.

Career Overview

Degree Preferred:
bachelor's degree in actuarial science or mathematics

Related Courses:
mathematics, statistics, computer science, business courses

Outlook:
faster than average through the year 2006

Population of Various Age Groups 1960-1990

Population (millions): 0, 10, 20, 30, 40, 50, 60, 70, 80, 90
Year: 1960, 1970, 1980, 1990
Legend: Under 18 years; 18-34 years; 35-64 years; Over 65 years

interNET CONNECTION For more information on careers in actuarial science, visit: **www.amc.glencoe.com**

13-3 Probability and Odds

OBJECTIVES
- Find the probability of an event.
- Find the odds for the success and failure of an event.

MARKET RESEARCH To determine television ratings, Nielsen Media Research estimates how many people are watching any given television program. This is done by selecting a sample audience, having them record their viewing habits in a journal, and then counting the number of viewers for each program. There are about 100 million households in the U.S., and only 5000 are selected for the sample group. What is the probability of any one household being selected to participate? *This problem will be solved in Example 1.*

When we are uncertain about the occurrence of an event, we can measure the chances of its happening with **probability.** For example, there are 52 possible outcomes when selecting a card at random from a standard deck of playing cards. The set of all outcomes of an event is called the **sample space.** A desired outcome, drawing the king of hearts for example, is called a **success.** Any other outcome is called a **failure.** The probability of an event is the ratio of the number of ways an event can happen to the total number of outcomes in the sample space, which is the sum of successes and failures. There is one way to draw a king of hearts, and there are a total of 52 outcomes when selecting a card from a standard deck. So, the probability of selecting the king of hearts is $\frac{1}{52}$.

Probability of Success and of Failure

If an event can succeed in s ways and fail in f ways, then the probability of success $P(s)$ and the probability of failure $P(f)$ are as follows.

$$P(s) = \frac{s}{s+f} \qquad P(f) = \frac{f}{s+f}$$

Example 1

MARKET RESEARCH What is the probability of any one household being chosen to participate for the Nielsen Media Research group?

Use the probability formula. Since 5000 households are selected to participate $s = 5000$. The denominator, $s + f$, represents the total number of households, those selected, s, and those not selected, f. So, $s + f = 100{,}000{,}000$.

$$P(5000) = \frac{5000}{100{,}000{,}000} \text{ or } \frac{1}{20{,}000} \qquad P(s) = \frac{s}{s+f}$$

The probability of any one household being selected is $\frac{1}{20{,}000}$ or 0.005%.

An event that cannot fail has a probability of 1. An event that cannot succeed has a probability of 0. Thus, the probability of success $P(s)$ is always between 0 and 1 inclusive. That is, $0 \leq P(s) \leq 1$.

Example 2 **A bag contains 5 yellow, 6 blue, and 4 white marbles.**

a. What is the probability that a marble selected at random will be yellow?

b. What is the probability that a marble selected at random will *not* be white?

a. The probability of selecting a yellow marble is written $P(\text{yellow})$. There are 5 ways to select a yellow marble from the bag, and 6 + 4 or 10 ways not to select a yellow marble. So, $s = 5$ and $f = 10$.

$$P(\text{yellow}) = \frac{5}{5 + 10} \text{ or } \frac{1}{3} \quad P(s) = \frac{s}{s + f}$$

The probability of selecting a yellow marble is $\frac{1}{3}$

b. There are 4 ways to select a white marble. So there are 11 ways not to select a white marble.

$$P(\text{not white}) = \frac{11}{4 + 11} \text{ or } \frac{11}{15}$$

The probability of *not* selecting a white marble is $\frac{11}{15}$.

The counting methods you used for permutations and combinations are often used in determining probability.

Example 3 **A circuit board with 20 computer chips contains 4 chips that are defective. If 3 chips are selected at random, what is the probability that all 3 are defective?**

There are $C(4, 3)$ ways to select 3 out of 4 defective chips, and $C(20, 3)$ ways to select 3 out of 20 chips.

$$P(\text{3 defective chips}) = \frac{C(4, 3)}{C(20, 3)} \begin{array}{l} \leftarrow \textit{ways of selecting 3 defective chips} \\ \leftarrow \textit{ways of selecting 3 chips} \end{array}$$

$$= \frac{\frac{4!}{1!\,3!}}{\frac{20!}{17!\,3!}} \text{ or } \frac{1}{285}$$

The probability of selecting three defective computer chips is $\frac{1}{285}$.

The sum of the probability of success and the probability of failure for any event is always equal to 1.

$$P(s) + P(f) = \frac{s}{s + f} + \frac{f}{s + f}$$

$$= \frac{s + f}{s + f} \text{ or } 1$$

This property is often used in finding the probability of events. For example, the probability of drawing a king of hearts is $P(s) = \frac{1}{52}$, so the probability of not drawing the king of hearts is $P(f) = 1 - \frac{1}{52}$ or $\frac{51}{52}$. Because their sum is 1, $P(s)$ and $P(f)$ are called **complements.**

Example 4 **The CyberToy Company has determined that out of a production run of 50 toys, 17 are defective. If 5 toys are chosen at random, what is the probability that at least 1 is defective?**

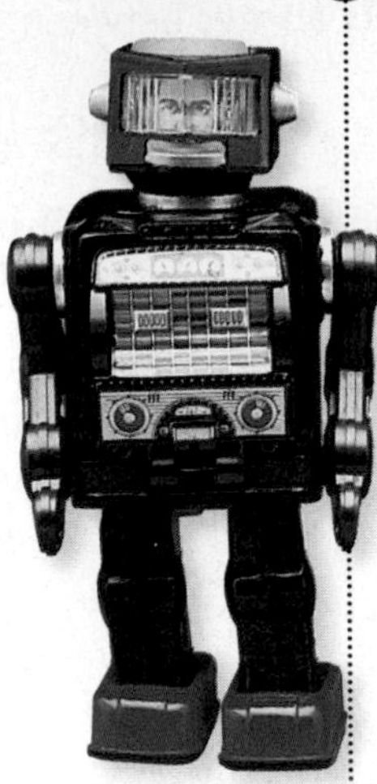

The complement of selecting at least 1 defective toy is selecting no defective toys. That is, $P(\text{at least 1 defective toy}) = 1 - P(\text{no defective toys})$.

$$P(\text{at least 1 defective toy}) = 1 - P(\text{no defective toys}).$$

$$= 1 - \frac{C(33, 5)}{C(50, 5)} \quad \begin{matrix} \leftarrow \textit{ways of selecting 5 defective toys} \\ \leftarrow \textit{ways of selecting 5 toys} \end{matrix}$$

$$= 1 - \frac{237{,}336}{2{,}118{,}760}$$

$$\approx 0.8879835375 \quad \textit{Use a calculator.}$$

The probability of selecting at least 1 defective toy is about 89%.

Another way to measure the chance of an event occurring is with **odds.** The probability of success of an event and its complement are used when computing the odds of an event.

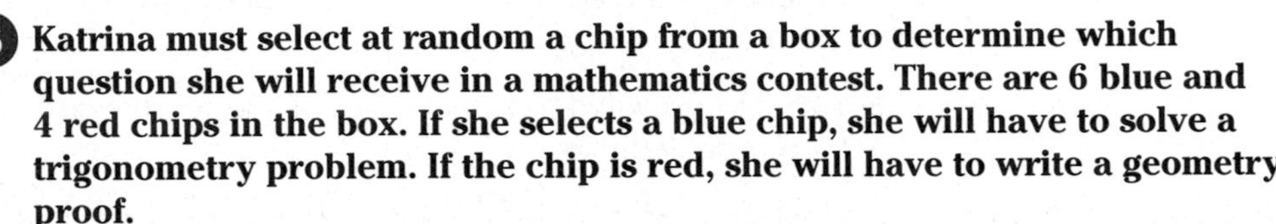

Odds

The odds of the successful outcome of an event is the ratio of the probability of its success to the probability of its failure.

$$\text{Odds} = \frac{P(s)}{P(f)}$$

Example 5 **Katrina must select at random a chip from a box to determine which question she will receive in a mathematics contest. There are 6 blue and 4 red chips in the box. If she selects a blue chip, she will have to solve a trigonometry problem. If the chip is red, she will have to write a geometry proof.**

a. What is the probability that Katrina will draw a red chip?

b. What are the odds that Katrina will have to write a geometry proof?

a. The probability that Katrina will select a red chip is $\frac{4}{10}$ or $\frac{2}{5}$.

b. To find the odds that Katrina will have to write a geometry proof, you need to know the probability of a successful outcome and of a failing outcome.

Let s represent selecting a red chip and f represent not selecting a red chip.

$$P(s) = \frac{2}{5} \qquad P(f) = 1 - \frac{2}{5} \text{ or } \frac{3}{5}$$

Now find the odds.

$$\frac{P(s)}{P(f)} = \frac{\frac{2}{5}}{\frac{3}{5}} \text{ or } \frac{2}{3}$$

The odds that Katrina will choose a red chip and thus have to write a geometry proof is $\frac{2}{3}$. *The ratio $\frac{2}{3}$ is read "2 to 3."*

Sometimes when computing odds, you must find the sample space first. This can involve finding permutations and combinations.

Example 6 **Twelve male and 16 female students have been selected as equal qualifiers for 6 college scholarships. If the awarded recipients are to be chosen at random, what are the odds that 3 will be male and 3 will be female?**

First, determine the total number of possible groups.

$C(12, 3)$ *number of groups of 3 males*
$C(16, 3)$ *number of groups of 3 females*

Using the Basic Counting Principle we can find the number of possible groups of 3 males and 3 females.

$C(12, 3) \cdot C(16, 3) = \frac{12!}{9!\,3!} \cdot \frac{16!}{13!\,3!}$ or 123,200 possible groups

The total number of groups of 6 recipients out of the 28 who qualified is $C(28, 6)$ or 376,740. So, the number of groups that do not have 3 males and 3 females is $376{,}740 - 123{,}200$ or 253,540.

Finally, determine the odds.

$P(s) = \frac{123{,}200}{376{,}740}$ $\qquad P(f) = \frac{253{,}540}{376{,}740}$

$$\text{odds} = \frac{\frac{123{,}200}{376{,}740}}{\frac{253{,}540}{376{,}740}} \text{ or } \frac{880}{1811}$$

Thus, the odds of selecting a group of 3 males and 3 females are $\frac{880}{1811}$ or close to $\frac{1}{2}$.

CHECK FOR UNDERSTANDING

Communicating Mathematics

Read and study the lesson to answer each question.

1. **Explain** how you would interpret $P(E) = \frac{1}{2}$.
2. Find two examples of the use of probability in newspapers or magazines. **Describe** how probability concepts are applied.
3. **Write** about the difference between the probability of the successful outcome of an event and the odds of the successful outcome of an event.
4. **You Decide** Mika has figured that his odds of winning the student council election are 3 to 2. Geraldo tells him that, based on those odds, the probability of his winning is 60%. Mika disagreed. Who is correct? Explain your answer.

Guided Practice

A box contains 3 tennis balls, 7 softballs, and 11 baseballs. One ball is chosen at random. Find each probability.

5. P(softball)
6. P(not a baseball)
7. P(golf ball)
8. In an office, there are 7 women and 4 men. If one person is randomly called on the phone, find the probability the person is a woman.

Of 7 kittens in a litter, 4 have stripes. Three kittens are picked at random. Find the odds of each event.

9. All three have stripes.
10. Only 1 has stripes.
11. One is not striped.

12. **Meteorology** A local weather forecast states that the probability of rain on Saturday is 80%. What are the odds that it will not rain Saturday? (*Hint*: Rewrite the percent as a fraction.)

EXERCISES

Practice

Using a standard deck of 52 cards, find each probability. *The face cards include kings, queens, and jacks.*

13. P(face card)
14. P(a card of 6 or less)
15. P(a black, non-face card)
16. P(not a face card)

One flower is randomly taken from a vase containing 5 red flowers, 2 white flowers, and 3 pink flowers. Find each probability.

17. P(red)
18. P(white)
19. P(not pink)
20. P(red or pink)

Jacob has 10 rap, 18 rock, 8 country, and 4 pop CDs in his music collection. Two are selected at random. Find each probability.

21. P(2 pop)
22. P(2 country)
23. P(1 rap and 1 rock)
24. P(not rock)

25. A number cube is thrown two times. What is the probability of rolling 2 fives?

A box contains 1 green, 2 yellow, and 3 red marbles. Two marbles are drawn at random without replacement. What are the odds of each event occurring?

26. drawing 2 red marbles
27. not drawing yellow marbles
28. drawing 1 green and 1 red
29. drawing two different colors

Of 27 students in a class, 11 have blue eyes, 13 have brown eyes, and 3 have green eyes. If 3 students are chosen at random what are the odds of each event occurring?

30. all three have blue eyes
31. 2 have brown and 1 has blue eyes
32. no one has brown eyes
33. only 1 has green eyes

34. The odds of winning a prize in a raffle with one raffle ticket are $\frac{1}{249}$. What is the probability of winning with one ticket?

35. The probability of being accepted to attend a state university is $\frac{4}{5}$. What are the odds of being accepted to this university?

36. From a deck of 52 cards, 5 cards are drawn. What are the odds of having three cards of one suit and the other two cards be another suit?

Applications and Problem Solving

37. **Weather** During a particular hurricane, hurricane trackers determine that the odds of it hitting the South Carolina coast are 1 to 4. What is the probability of this happening?

38. **Baseball** At one point in the 1999 season, Ken Griffey, Jr. had a batting average of 0.325. What are the odds that he would hit the ball the next time he came to bat?

39. **Security** Kim uses a combination lock on her locker that has 3 wheels, each labeled with 10 digits from 0 to 9. The combination is a particular sequence with no digits repeating.
   a. What is the probability of someone guessing the correct combination?
   b. If the digits can be repeated, what are the odds against someone guessing the combination?

40. **Critical Thinking** Spencer is carrying out a survey of the bear population at Yellowstone National Park. He spots two bears—one has a light colored coat and the other has a dark coat.
   a. Assume that there are equal numbers of male and female bears in the park. What is the probability that both bears are male?
   b. If the lighter colored bear is male, what are the odds that both are male?

41. **Testing** Ms. Robinson gives her precalculus class 20 study problems. She will select 10 to answer on an upcoming test. Carl can solve 15 of the problems.
   a. Find the probability that Carl can solve all 10 problems on the test.
   b. Find the odds that Carl will know how to solve 8 of the problems.

42. **Mortality Rate** During 1990, smoking was linked to 418,890 deaths in the United States. The graph shows the diseases that caused these smoking-related deaths.
   a. Find the probability that a smoking-related death was the result of either cardiovascular disease or cancer.
   b. Determine the odds against a smoking-related death being caused by cancer.

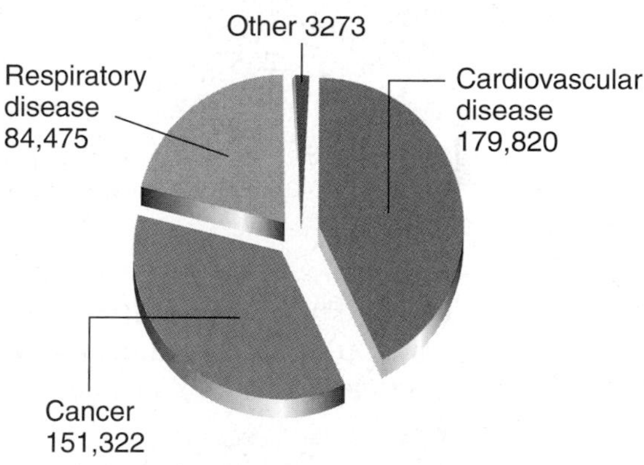

43. **Critical Thinking** A plumber cuts a pipe in two pieces at a point selected at random. What is the probability that the length of the longer piece of pipe is at least 8 times the length of the shorter piece of pipe?

Mixed Review

44. A food vending machine has 6 different items on a revolving tray. How many different ways can the items be arranged on the tray? *(Lesson 13-2)*

45. The Foxtrail Condominium Association is electing board members. How many groups of 4 can be chosen from the 10 candidates who are running? *(Lesson 13-1)*

46. Find S_{14} for the arithmetic series for which $a_1 = 3.2$ and $d = 1.5$. *(Lesson 12-1)*

47. Simplify $7^{\log_7 2x}$. *(Lesson 11-4)*

48. **Landscaping** Carolina bought a new sprinkler to water her lawn. The sprinkler rotates 360° while spraying a stream of water. Carolina places the sprinkler in her yard so the ordered pair that represents its location is (7, 2), and the sprinkler sends out water that just barely reaches the point at (10, −8). Find an equation representing the farthestmost points the water can reach. *(Lesson 10-2)*

49. Find the product $3(\cos \pi + \mathbf{i} \sin \pi) \cdot 2\left(\cos \frac{\pi}{4} + \mathbf{i} \sin \frac{\pi}{4}\right)$. Then express it in rectangular form. *(Lesson 9-7)*

50. Find an ordered pair to represent $\vec{\mathbf{u}}$ if $\vec{\mathbf{u}} = \vec{\mathbf{v}} + \vec{\mathbf{w}}$, if $\vec{\mathbf{v}} = \langle 3, -5 \rangle$ and $\vec{\mathbf{w}} = \langle -4, 2 \rangle$. *(Lesson 8-2)*

51. **SAT Practice** **Quantitative Comparison**

A if quantity A is greater
B if quantity B is greater
C if the quantities are equal
D if there is not enough information to determine the relationship

Column A	Column B
the area of a square with sides *s* units long	the area of an equilateral triangle with sides 2*s* units long

MID-CHAPTER QUIZ

Find each value. (Lesson 13-1)

1. $P(15, 5)$.
2. $C(20, 9)$.

3. Regular license plates in Ohio have three letters followed by four digits. How many different license plate arrangements are possible? (Lesson 13-1)

4. Suppose there are 12 runners competing in the finals of a track event. Awards are given to the top five finishers. How many top-five arrangements are possible? (Lesson 13-1)

5. An ice cream shop has 18 different flavors of ice cream, which can be ordered in a cup, sugar cone, or waffle cone. There is also a choice of six toppings. How many two-scoop servings with a topping are possible? (Lesson 13-1)

6. How many nine-letter patterns can be formed from the letters in the word *quadratic*? (Lesson 13-2)

7. How many different arrangements can be made with ten pieces of silverware laid in a row if three are identical spoons, four are identical forks, and three are identical knives? (Lesson 13-2)

8. Eight children are riding a merry-go-round. How many ways can they be seated? (Lesson 13-2)

9. Two cards are drawn at random from a standard deck of 52 cards. What is the probability that both are hearts? (Lesson 13-3)

10. A bowl contains four apples, three bananas, three oranges, and two pears. If two pieces of fruit are selected at random, what are the odds of selecting an orange and a banana? (Lesson 13-3)

Extra Practice See p. A51.

Probabilities of Compound Events

OBJECTIVES

- Find the probability of independent and dependent events.
- Identify mutually exclusive events.
- Find the probability of mutually exclusive and inclusive events.

TRANSPORTATION According to U.S. Department of Transportation statistics, the top ten airlines in the United States arrive on time 80% of the time. During their vacation, the Hiroshi family has direct flights to Washington, D.C., Chicago, Seattle, and San Francisco on different days. What is the probability that all their flights arrived on time?

Since the flights occur on different days, the four flights represent independent events. Let A represent an on-time arrival of an airplane.

$$P(\text{all flights on time}) = \underbrace{P(A)}_{\text{Flight 1}} \cdot \underbrace{P(A)}_{\text{Flight 2}} \cdot \underbrace{P(A)}_{\text{Flight 3}} \cdot \underbrace{P(A)}_{\text{Flight 4}}$$

$$= (0.80)^4 \quad A = 0.80$$

$$\approx 0.4096 \text{ or about } 41\%$$

Thus, the probability of all four flights arriving on time is about 41%.

This problem demonstrates that the probability of more than one independent event is the product of the probabilities of the events.

Probability of Two Independent Events

If two events, A and B, are independent, then the probability of both events occurring is the product of each individual probability.

$$P(A \text{ and } B) = P(A) \cdot P(B)$$

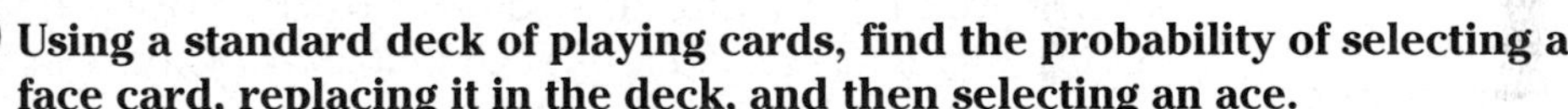

Example 1 **Using a standard deck of playing cards, find the probability of selecting a face card, replacing it in the deck, and then selecting an ace.**

Let A represent a face card for the first card drawn from the deck, and let B represent the ace in the second selection.

$P(A) = \frac{12}{52}$ or $\frac{3}{13}$ $\quad$ *12 face cards / 52 cards in a standard deck*

$P(B) = \frac{4}{52}$ or $\frac{1}{13}$ $\quad$ *4 aces / 52 cards in a standard deck*

The two draws are independent because when the card is returned to the deck, the outcome of the second draw is not affected by the first one.

$$P(A \text{ and } B) = P(A) \cdot P(B)$$

$$= \frac{3}{13} \cdot \frac{1}{13} \text{ or } \frac{3}{169}$$

The probability of selecting a face card first, replacing it, and then selecting an ace is $\frac{3}{169}$.

Example 2 

OCCUPATIONAL HEALTH Statistics collected in a particular coal-mining region show that the probability that a miner will develop black lung disease is $\frac{5}{11}$. Also, the probability that a miner will develop arthritis is $\frac{1}{5}$. If one health problem does not affect the other, what is the probability that a randomly-selected miner will not develop black lung disease but will develop arthritis?

The events are independent since having black lung disease does not affect the existence of arthritis.

$$P(\text{not black lung disease and arthritis}) = [1 - P(\text{black lung disease})] \cdot P(\text{arthritis})$$
$$= \left(1 - \frac{5}{11}\right) \cdot \frac{1}{5} \text{ or } \frac{6}{55}$$

The probability that a randomly-selected miner will not develop black lung disease but will develop arthritis is $\frac{6}{55}$.

What do you think the probability of selecting two face cards would be if the first card drawn were not placed back in the deck? Unlike the situation in Example 1, these events are dependent because the outcome of the first event affects the second event. This probability is also calculated using the product of the probabilities.

first card	*second card*
$P(\text{face card}) = \frac{12}{52}$	$P(\text{face card}) = \frac{11}{51}$

$$P(\text{two face cards}) = \frac{12}{52} \cdot \frac{11}{51} \text{ or } \frac{11}{221}$$

Notice that when a face card is removed from the deck, not only is there one less face card, but also one less card in the deck.

Thus, the probability of selecting two face cards from a deck without replacing the cards is $\frac{11}{221}$ or about $\frac{1}{20}$.

Probability of Two Dependent Events

If two events, A and B, are dependent, then the probability of both events occurring is the product of each individual probability.

$$P(A \text{ and } B) = P(A) \cdot P(B \text{ following } A)$$

Example 3 **Tasha has 3 rock, 4 country, and 2 jazz CDs in her car. One day, before she starts driving, she pulls 2 CDs from her CD carrier without looking.**

a. Determine if the events are independent or dependent.

b. What is the probability that both CDs are rock?

a. The events are dependent. This event is equivalent to selecting one CD, not replacing it, then selecting another CD.

b. Determine the probability.

$P(\text{rock, rock}) = P(\text{rock}) \cdot P(\text{rock following first rock selection})$

$P(\text{rock, rock}) = \frac{3}{9} \cdot \frac{2}{8} \text{ or } \frac{1}{12}$

The probability that Tasha will select two rock CDs is $\frac{1}{12}$.

There are times when two events cannot happen at the same time. For example, when tossing a number cube, what is the probability of tossing a 2 *or* a 5? In this situation, both events cannot happen at the same time. That is, the events are **mutually exclusive.** The probability of tossing a 2 or a 5 is $P(2) + P(5)$, which is $\frac{1}{6} + \frac{1}{6}$ or $\frac{2}{6}$.

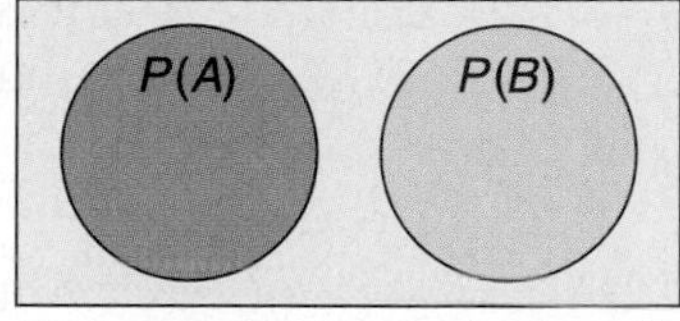

Events A and B are mutually exclusive.

Note that the two events do not overlap, as shown in the Venn diagram. So, the probability of two mutually exclusive events occurring can be represented by the sum of the areas of the circles.

Probability of Mutually Exclusive Events

If two events, A and B, are mutually exclusive, then the probability that either A or B occurs is the sum of their probabilities.

$$P(A \text{ or } B) = P(A) + P(B)$$

Example 4 **Lenard is a contestant in a game where if he selects a blue ball or a red ball he gets an all-expenses paid Caribbean cruise. Lenard must select the ball at random from a box containing 2 blue, 3 red, 9 yellow, and 10 green balls. What is the probability that he will win the cruise?**

These are mutually exclusive events since Lenard cannot select a blue and a red ball at the same time. Find the sum of the individual probabilities.

$P(\text{blue or red}) = P(\text{blue}) + P(\text{red})$

$= \frac{2}{24} + \frac{3}{24}$ or $\frac{5}{24}$ *$P(\text{blue}) = \frac{2}{24}$, $P(\text{red}) = \frac{3}{24}$*

The probability that Lenard will win the cruise is $\frac{5}{24}$.

What is the probability of rolling two number cubes, in which the first number cube shows a 2 or the sum of the number cubes is 6 or 7? Since each number cube can land six different ways, and two number cubes are rolled, the sample space can be represented by making a chart. A **reduced sample space** is the subset of a sample space that contains only those outcomes that satisfy a given condition.

		Second Number Cube					
		1	2	3	4	5	6
First Number Cube	1	(1, 1)	(1, 2)	(1, 3)	(1, 4)	(1, 5)	(1, 6)
	2	(2, 1)	(2, 2)	(2, 3)	(2, 4)	(2, 5)	(2, 6)
	3	(3, 1)	(3, 2)	(3, 3)	(3, 4)	(3, 5)	(3, 6)
	4	(4, 1)	(4, 2)	(4, 3)	(4, 4)	(4, 5)	(4, 6)
	5	(5, 1)	(5, 2)	(5, 3)	(5, 4)	(5, 5)	(5, 6)
	6	(6, 1)	(6, 2)	(6, 3)	(6, 4)	(6, 5)	(6, 6)

It is possible to have the first number cube show a 2 *and* have the sum of the two number cubes be 6 or 7. Therefore, these events are not mutually exclusive. They are called **inclusive events.** In this case, you must adjust the formula for mutually exclusive events.

Note that the circles in the Venn diagram overlap. This area represents the probability of both events occurring at the same time. When the areas of the two circles are added, this overlapping area is counted twice. Therefore, it must be subtracted to find the correct probability of the two events.

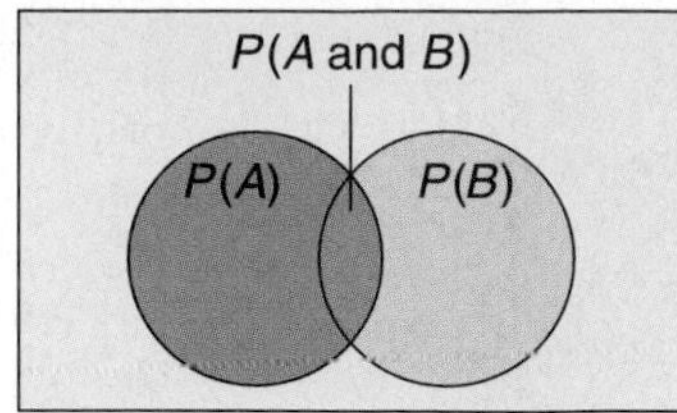

Events A and B are inclusive events.

Let A represent the event "the first number cube shows a 2".
Let B represent the event "the sum of the two number cubes is 6 or 7".

$P(A) = \frac{6}{36}$ $\qquad$ $P(B) = \frac{11}{36}$

Note that (2, 4) and (2, 5) are counted twice, both as the first cube showing a 2 and as a sum of 6 or 7. To find the correct probability, you must subtract P(2 and sum of 6 or 7).

$$P(\text{2 or sum of 6 or 7}) = \underbrace{\frac{6}{36}}_{P(2)} + \underbrace{\frac{11}{36}}_{P(\text{sum of 6 or 7})} - \underbrace{\frac{2}{36}}_{P(\text{2 and sum of 6 or 7})} \text{ or } \frac{15}{36}$$

The probability of the first number cube showing a 2 or the sum of the number cubes being 6 or 7 is $\frac{15}{36}$ or $\frac{5}{12}$.

Probability of Inclusive Events

If two events, A and B, are inclusive, then the probability that either A or B occurs is the sum of their probabilities decreased by the probability of both occurring.

$$P(A \text{ or } B) = P(A) + P(B) - P(A \text{ and } B)$$

Examples

5 Kerry has read that the probability for a driver's license applicant to pass the road test the first time is $\frac{5}{6}$. He has also read that the probability of passing the written examination on the first attempt is $\frac{9}{10}$. The probability of passing both the road and written examinations on the first attempt is $\frac{4}{5}$.

a. Determine if the events are mutually exclusive or mutually inclusive.

Since it is possible to pass both the road examination and the written examination, these events are mutually inclusive.

b. What is the probability that Kerry can pass either examination on his first attempt?

$P(\text{passing road exam}) = \frac{5}{6}$ $\qquad$ $P(\text{passing written exam}) = \frac{9}{10}$

$P(\text{passing both exams}) = \frac{4}{5}$

$P(\text{passing either examination}) = \frac{5}{6} + \frac{9}{10} - \frac{4}{5} = \frac{56}{60} \text{ or } \frac{14}{15}$

The probability that Kerry will pass either test on his first attempt is $\frac{14}{15}$.

6 **There are 5 students and 4 teachers on the school publications committee. A group of 5 members is being selected at random to attend a workshop on school newspapers. What is the probability that the group attending the workshop will have at least 3 students?**

At least 3 students means the groups may have 3, 4, or 5 students. It is not possible to select a group of 3 students, a group of 4 students, and a group of 5 students in the same 5-member group. Thus, the events are mutually exclusive.

$$P(\text{at least 3 students}) = P(\text{3 students}) + P(\text{4 students}) + P(\text{5 students})$$
$$= \frac{C(5, 3) \cdot C(4, 2)}{C(9, 5)} + \frac{C(5, 4) \cdot C(4, 1)}{C(9, 5)} + \frac{C(5, 5) \cdot C(4, 0)}{C(9, 5)}$$
$$= \frac{60}{126} + \frac{20}{126} + \frac{1}{126} \text{ or } \frac{9}{14}$$

The probability of at least 3 students going to the workshop is $\frac{9}{14}$.

CHECK FOR UNDERSTANDING

Communicating Mathematics

Read and study the lesson to answer each question.

1. **Describe** the difference between independent and dependent events.
2. a. **Draw** a Venn diagram to illustrate the event of selecting an ace or a diamond from a deck of cards.
   b. Are the events mutually exclusive? Explain why or why not.
   c. **Write** the formula you would use to determine the probability of these events.
3. *Math Journal* **Write** an example of two mutually exclusive events and two mutually inclusive events in your own life. **Explain** why the events are mutually exclusive or inclusive.

Guided Practice

Determine if each event is *independent* or *dependent*. Then determine the probability.

4. the probability of rolling a sum of 7 on the first toss of two number cubes and a sum of 4 on the second toss
5. the probability of randomly selecting two navy socks from a drawer that contains 6 black and 4 navy socks
6. There are 2 bottles of fruit juice and 4 bottles of sports drink in a cooler. Without looking, Desiree chose a bottle for herself and then one for a friend. What is the probability of choosing 2 bottles of the sports drink?

Determine if each event is *mutually exclusive* or *mutually inclusive*. Then determine each probability.

7. the probability of choosing a penny or a dime from 4 pennies, 3 nickels, and 6 dimes
8. the probability of selecting a boy or a blonde-haired person from 12 girls, 5 of whom have blonde hair, and 15 boys, 6 of whom have blonde hair
9. the probability of drawing a king or queen from a standard deck of cards

In a bingo game, balls numbered 1 to 75 are placed in a bin. Balls are randomly drawn and not replaced. Find each probability for the first 5 balls drawn.

10. P(selecting 5 even numbers)

11. P(selecting 5 two digit numbers)

12. P(5 odd numbers or 5 multiples of 4)

13. P(5 even numbers or 5 numbers less than 30)

14. **Business** A furniture importer has ordered 100 grandfather clocks from an overseas manufacturer. Four clocks are damaged in shipment, but the packaging shows no signs of damage. If a dealer buys 6 of the clocks without examining them first, what is the probability that none of the 6 clocks is damaged?

15. **Sports** A baseball team's pitching staff has 5 left-handed and 8 right-handed pitchers. If 2 pitchers are randomly chosen to warm up, what is the probability that at least one of them is right-handed? (*Hint:* Consider the order when selecting one right-handed and one left-handed pitcher.)

EXERCISES

Practice

Determine if each event is *independent* or *dependent.* Then determine the probability.

16. the probability of selecting a blue marble, not replacing it, then a yellow marble from a box of 5 blue marbles and 4 yellow marbles

17. the probability of randomly selecting two oranges from a bowl of 5 oranges and 4 tangerines, if the first selection is replaced

18. A green number cube and a red number cube are tossed. What is the probability that a 4 is shown on the green number cube and a 5 is shown on the red number cube?

19. the probability of randomly taking 2 blue notebooks from a shelf which has 4 blue and 3 black notebooks

20. A bank contains 4 nickels, 4 dimes, and 7 quarters. Three coins are removed in sequence, without replacement. What is the probability of selecting a nickel, a dime, and a quarter in that order?

21. the probability of removing 13 cards from a standard deck of cards and have all of them be red

22. the probability of randomly selecting a knife, a fork, and a spoon in that order from a kitchen drawer containing 8 spoons, 8 forks, and 12 table knives

23. the probability of selecting three different-colored crayons from a box containing 5 red, 4 black, and 7 blue crayons, if each crayon is replaced

24. the probability that a football team will win its next four games if the odds of winning each game are 4 to 3

For Exercises 25-33, determine if each event is *mutually exclusive* or *mutually inclusive.* Then determine each probability.

25. the probability of tossing two number cubes and either one shows a 4

26. the probability of selecting an ace or a red card from a standard deck of cards

27. the probability that if a card is drawn from a standard deck it is red or a face card

28. the probability of randomly picking 5 puppies of which at least 3 are male puppies, from a group of 5 male puppies and 4 female puppies.

29. the probability of two number cubes being tossed and showing a sum of 6 or a sum of 9.

30. the probability that a group of 6 people selected at random from 7 men and 7 women will have at least 3 women

31. the probability of at least 4 tails facing up when 6 coins are dropped on the floor

32. the probability that two cards drawn from a standard deck will both be aces or both will be black

33. from a collection of 6 rock and 5 rap CDs, the probability that at least 2 are rock from 3 randomly selected

Find the probability of each event using a standard deck of cards.

34. P(all red cards) if 5 cards are drawn without replacement

35. P(both kings or both aces) if 2 cards are drawn without replacement

36. P(all diamonds) if 10 cards are selected with replacement

37. P(both red or both queens) if 2 cards are drawn without replacement

There are 5 pennies, 7 nickels, and 9 dimes in an antique coin collection. If two coins are selected at random and the coins are not replaced, find each probability.

38. P(2 pennies)

39. P(2 nickels or 2 silver-colored coins)

40. P(at least 1 nickel)

41. P(2 dimes or 1 penny and 1 nickel)

There are 5 male and 5 female students in the executive council of the Douglas High School honor society. A committee of 4 members is to be selected at random to attend a conference. Find the probability of each group being selected.

42. P(all female)

43. P(all female or all male)

44. P(at least 3 females)

45. P(at least 2 females and at least 1 male)

Applications and Problem Solving

46. **Computers** A survey of the members of the Piper High School Computer Club shows that $\frac{2}{5}$ of the students who have home computers use them for word processing, $\frac{1}{3}$ use them for playing games, and $\frac{1}{4}$ use them for both word processing and playing games. What is the probability that a student with a home computer uses it for word processing or playing games?

47. **Weather** A weather forecaster states that the probability of rain is $\frac{3}{5}$, the probability of lightning is $\frac{2}{5}$, and the probability of both is $\frac{1}{5}$. What is the probability that a baseball game will be cancelled due to rain or lightning?

48. **Critical Thinking** Felicia and Martin are playing a game where the number cards from a single suit are selected. From this group, three cards are then chosen at random. What is the probability that the sum of the value of the cards will be an even number?

49. **City Planning** There are six women and seven men on a committee for city services improvement. A subcommittee of five members is being selected at random to study the feasibility of modernizing the water treatment facility. What is the probability that the committee will have at least three women?

50. **Medicine** A study of two doctors finds that the probability of one doctor correctly diagnosing a medical condition is $\frac{93}{100}$ and the probability the second doctor will correctly diagnose a medical condition is $\frac{97}{100}$. What is the probability that at least one of the doctors will make a correct diagnosis?

51. **Disaster Relief** During the 1999 hurricane season, Hurricanes Dennis, Floyd, and Irene caused extensive flooding and damage in North Carolina. After a relief effort, 2500 people in one supporting community were surveyed to determine if they donated supplies or money. Of the sample, 812 people said they donated supplies and 625 said they donated money. Of these people, 375 people said they donated both. If a member of this community were selected at random, what is the probability that this person donated supplies or money?

52. **Critical Thinking** If events A and B are inclusive, then $P(A \text{ or } B) = P(A) + P(B) - P(A \text{ and } B)$.
    a. Draw a Venn diagram to represent $P(A \text{ or } B \text{ or } C)$.
    b. Write a formula to find $P(A \text{ or } B \text{ or } C)$.

53. **Product Distribution** Ms. Kameko is the shipping manager of an Internet-based audio and video store. Over the past few months, she has determined the following probabilities for items customers might order.

Item	Probability
Action video	$\frac{4}{7}$
Pop/rock CD	$\frac{1}{2}$
Romance DVD	$\frac{5}{11}$
Action video and pop/rock CD	$\frac{2}{9}$
Pop/rock CD and romance DVD	$\frac{1}{7}$
Action video and romance DVD	$\frac{1}{4}$
Action video, pop/rock CD, and romance DVD	$\frac{1}{44}$

What is the probability, rounded to the nearest hundredth, that a customer will order an action video, pop/rock CD, or a romance DVD?

54. **Critical Thinking** There are 18 students in a classroom. The students are surveyed to determine their birthday (month and day only). Assume that 366 birthdays are possible.
    a. What is the probability of any two students in the classroom having the same birthday?
    b. Write an inequality that can be used to determine the probability of any two students having the same birthday to be greater than $\frac{1}{2}$.
    c. Are there enough students in the classroom to have the probability in part **a** be greater than $\frac{1}{2}$? If not, at least how many more students would there need to be?

55. **Automotive Repairs** An auto club's emergency service has determined that when club members call to report that their cars will not start, the probability that the engine is flooded is $\frac{1}{2}$, the probability that the battery is dead is $\frac{2}{5}$, and the probability that both the engine is flooded and the battery is dead is $\frac{1}{10}$.

a. Are the events mutually exclusive or mutually inclusive?

b. Draw a Venn Diagram to represent the events.

c. What is the probability that the next member to report that a car will not start has a flooded engine or a dead battery?

Mixed Review

56. Two number cubes are tossed and their sum is 6. Find the probability that each cube shows a 3. *(Lesson 13-3)*

57. How many ways can 7 people be seated around a table? *(Lesson 13-2)*

58. **Sports** Ryan plays basketball every weekend. He averages 12 baskets per game out of 20 attempts. He has decided to try to make 15 baskets out of 20 attempts in today's game. How many ways can Ryan make 15 out of 20 baskets? *(Lesson 12-6)*

59. **Ecology** An underground storage container is leaking a toxic chemical. One year after the leak began, the chemical has spread 1200 meters from its source. After two years, the chemical has spread 480 meters more, and by the end of the third year it has reached an additional 192 meters. If this pattern continues, will the spill reach a well dug 2300 meters away? *(Lesson 12-4)*

60. Solve $12^{x+2} = 3^{x-4}$. *(Lesson 11-5)*

61. **Entertainment** A theater has been staging children's plays during the summer. The average attendance at each performance is 400 people and the cost of a ticket is $3. Next summer, they would like to increase the cost of the tickets, while maximizing their profits. The director estimates that for every $1 increase in ticket price, the attendance at each performance will decrease by 20. What price should the director propose to maximize their income, and what maximum income might be expected? *(Lesson 10-5)*

62. **Geology** A drumlin is an elliptical streamlined hill whose shape can be expressed by the equation $r = \ell \cos k\theta$ for $-\frac{\pi}{2k} \le \theta \le \frac{\pi}{2k}$, where ℓ is the length of the drumlin and $k > 1$ is a parameter that is the ratio of the length to the width. Suppose the area of a drumlin is 8270 square yards and the formula for area is $A = \frac{\ell^2\pi}{4k}$. Find the length of a drumlin modeled by $r = \ell \cos 7\theta$. *(Lesson 9-3)*

63. Write a vector equation describing a line passing through $P(1, -5)$ and parallel to $\vec{\mathbf{v}} = \langle -2, -4 \rangle$. *(Lesson 8-6)*

64. Solve $2 \tan x - 4 = 0$ for principal values of x. *(Lesson 7-5)*

65. **SAT/ACT Practice** If $a = 45$, which of the following statements must be true?

I. $\overline{AD} \parallel \overline{BC}$

II. ℓ_3 bisects $\angle ABC$.

III. $b = 45$

A None **B** I only

C I and II only **D** I and III only

E I, II, and III

Extra Practice See p. A52.

13-5 Conditional Probability

OBJECTIVE

- Find the probability of an event given the occurrence of another event.

MEDICINE Danielle Jones works in a medical research laboratory where a drug that promotes hair growth in balding men is being tested. The results of the preliminary tests are shown in the table.

	Number of Subjects	
	Using Drug	**Using Placebo**
Hair growth	1600	1200
No hair growth	800	400

Ms. Jones needs to find the probability that a subject's hair growth was a result of using the experimental drug. *This problem will be solved in Example 1.*

The probability of an event under the condition that some preceding event has occurred is called **conditional probability.** The conditional probability that event A occurs given that event B occurs can be represented by $P(A|B)$. *$P(A|B)$ is read "the probability of A given B."*

Conditional Probability

The conditional probability of event A, given event B, is defined as

$$P(A|B) = \frac{P(A \text{ and } B)}{P(B)} \text{ where } P(B) \neq 0.$$

Example 1

MEDICINE **Refer to the application above. What is the probability that a test subject's hair grew, given that he used the experimental drug?**

Let H represent hair growth and D represent experimental drug usage. We need to find $P(H|D)$.

$$P(H|D) = \frac{P(\text{used experimental drug and had hair growth})}{P(\text{used experimental drug})}$$

$$P(H|D) = \frac{\frac{1600}{4000}}{\frac{2400}{4000}} \quad \begin{array}{l} \leftarrow P(\text{used experimental drug and had hair growth}) = \frac{1600}{4000} \\ \leftarrow P(\text{used experimental drug}) = \frac{1600 + 800}{4000} \end{array}$$

$$P(H|D) = \frac{1600}{2400} \text{ or } \frac{2}{3}$$

The probability that a subject's hair grew, given that they used the experimental drug is $\frac{2}{3}$.

Example 2 **Denette tosses two coins. What is the probability that she has tossed 2 heads, given that she has tossed at least 1 head?**

Let event A be that the two coins come up heads.

Let event B be that there is at least one head.

$P(B) = \frac{3}{4}$ *Three of the four outcomes have at least one head.*

$P(A \text{ and } B) = \frac{1}{4}$ *One of the four outcomes has two heads.*

$$P(A \mid B) = \frac{P(A \text{ and } B)}{P(B)}$$

$$= \frac{\frac{1}{4}}{\frac{3}{4}}$$

$$= \frac{1}{4} \cdot \frac{4}{3} \text{ or } \frac{1}{3}$$

The probability of tossing two heads, given that at least one toss was a head is $\frac{1}{3}$.

Sample spaces and reduced sample spaces can be used to help determine the outcomes that satisfy a given condition.

Example 3 **Alfonso is conducting a survey of families with 3 children. If a family is selected at random, what is the probability that the family will have exactly 2 boys if the second child is a boy?**

The sample space is $S = \{BBB, BBG, BGB, BGG, GBB, GBG, GGB, GGG\}$ and includes all of the possible outcomes for a family with three children.

Determine the reduced sample spaces that satisfy the given conditions that there are exactly 2 boys and that the second child is a boy.

The condition that there are exactly 2 boys reduces the sample space to exclude the outcomes where there are 1, 3, or no boys.

Let X represent the event that there are two boys.

$$X = \{BBG, BGB, GBB\}$$
$$P(X) = \frac{3}{8}$$

The condition that the second child is a boy reduces the sample space to exclude the outcomes where the second child is a girl.

Let Y represent the event that the second child is a boy.

$$Y = \{BBB, BBG, GBB, GBG\}$$
$$P(Y) = \frac{4}{8} \text{ or } \frac{1}{2}$$

$(X \text{ and } Y)$ is the intersection of X and Y. $(X \text{ and } Y) = \{BBG, GBB\}$.
So, $P(X \text{ and } Y) = \frac{2}{8}$ or $\frac{1}{4}$.

(continued on the next page)

$$P(X|Y) = \frac{P(X \text{ and } Y)}{P(Y)}$$
$$= \frac{\frac{1}{4}}{\frac{1}{2}}$$
$$= \frac{1}{4} \cdot \frac{2}{1} \text{ or } \frac{1}{2}$$

The probability that a family with 3 children selected at random will have exactly 2 boys, given that the second child is a boy, is $\frac{1}{2}$.

In some situations, event A is a subset of event B. When this occurs, the probability that both event A and event B, $P(A \text{ and } B)$, occur is the same as the probability of event A occurring. Thus, in these situations $P(A|B) = \frac{P(A)}{P(B)}$.

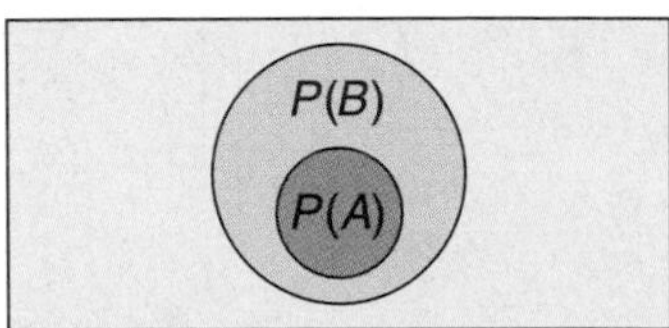

Event A is a subset of event B.

Example 4 **A 12-sided dodecahedron has the numerals 1 through 12 on its faces. The die is rolled once, and the number on the top face is recorded. What is the probability that the number is a multiple of 4 if it is known that it is even?**

Let A represent the event that the number is a multiple of 4. Thus, $A = \{4, 8, 12\}$.

$$P(A) = \frac{3}{12} \text{ or } \frac{1}{4}$$

Let B represent the event that the number is even. So, $B = \{2, 4, 6, 8, 10, 12\}$.

$$P(B) = \frac{6}{12} \text{ or } \frac{1}{2}$$

In this situation, A is a subset of B.

$P(A \text{ and } B) = P(A) = \frac{1}{4}$ $\qquad$ $P(B) = \frac{1}{2}$

$$P(A|B) = \frac{P(A)}{P(B)}$$
$$= \frac{\frac{1}{4}}{\frac{1}{2}} \text{ or } \frac{1}{2}$$

The probability that a multiple of 4 is rolled, given that the number is even, is $\frac{1}{2}$.

CHECK FOR UNDERSTANDING

Communicating Mathematics

Read and study the lesson to answer each question.

1. **Explain** the relationship between conditional probability and the probability of two independent events.

2. **Describe** the sample space for P(face card) if the card drawn is black.

3. *Math Journal* Find two real-world examples that use conditional probability. Explain how you know conditional probability is used.

Guided Practice **Find each probability.**

4. Two number cubes are tossed. Find the probability that the numbers showing on the cubes match given that their sum is greater than five.

5. One card is drawn from a standard deck of cards. What is the probability that it is a queen if it is known to be a face card?

Three coins are tossed. Find the probability that they all land heads up for each known condition.

6. the first coin shows a head

7. at least one coin shows a head

8. at least two coins show heads

A pair of number cubes is thrown. Find each probability given that their sum is greater than or equal to 9.

9. P(numbers match)

10. P(sum is even)

11. P(numbers match or sum is even)

12. **Medicine** To test the effectiveness of a new vaccine, researchers gave 100 volunteers the conventional treatment and gave 100 other volunteers the new vaccine. The results are shown in the table below.

a. What is the probability that the disease is prevented in a volunteer chosen at random?

b. What is the probability that the disease is prevented in a volunteer who was given the new vaccine?

c. What is the probability that the disease is prevented in a volunteer who was not given the new vaccine?

Treatment	Disease Prevented	Disease Not Prevented
New Vaccine	68	32
Conventional Treatment	62	38

13. **Currency** A dollar-bill changer in a snack machine was tested with 100 \$1-bills. Twenty-five of the bills were counterfeit. The results of the test are shown in the chart at the right.

Bill	Accepted	Rejected
Legal	69	6
Counterfeit	1	24

a. What is the probability that a bill accepted by the changer is legal?

b. What is the probability that a bill is rejected given that it is legal?

c. What is the probability that a counterfeit bill is not rejected?

EXERCISES

Practice

Find each probability.

14. Two coins are tossed. What is the probability that one coin shows heads if it is known that at least one coin is tails?
15. A city council consists of six Democrats, two of whom are women, and six Republicans, four of whom are men. A member is chosen at random. If the member chosen is a man, what is the probability that he is a Democrat?
16. A bag contains 4 red chips and 4 blue chips. Another bag contains 2 red chips and 6 blue chips. A chip is randomly selected from one of the bags, and found to be blue. What is the probability that the chip is from the first bag?
17. Two boys and two girls are lined up at random. What is the probability that the girls are separated if a girl is at an end?
18. A five-digit number is formed from the digits 1, 2, 3, 4, and 5. What is the probability that the number ends in the digits 52, given that it is even?
19. Two game tiles, numbered 1 through 9, are selected at random from a box without replacement. If their sum is even, what is the probability that both numbers are odd?

A card is chosen at random from a standard deck of cards. Find each probability given that the card is black.

20. *P*(ace)
21. *P*(4)
22. *P*(face card)
23. *P*(queen of hearts)
24. *P*(6 of clubs)
25. *P*(jack or ten)

A container holds 3 green marbles and 5 yellow marbles. One marble is randomly drawn and discarded. Then a second marble is drawn. Find each probability.

26. the second marble is green, given that the first marble was green
27. the second marble is yellow, given that the first marble was green
28. the second marble is yellow, given that the first marble was yellow

Three fish are randomly removed from an aquarium that contains a trout, a bass, a perch, a catfish, a walleye, and a salmon. Find each probability.

29. *P*(salmon, given bass)
30. *P*(not walleye, given trout and perch)
31. *P*(bass and perch, given not catfish)
32. *P*(perch and trout, given neither bass nor walleye)

In Mr. Hewson's homeroom, 60% of the students have brown hair, 30% have brown eyes, and 10% have both brown hair and eyes. A student is excused early to go to a doctor's appointment.

33. If the student has brown hair, what is the probability that the student also has brown eyes?
34. If the student has brown eyes, what is the probability that the student does not have brown hair?
35. If the student does not have brown hair, what is the probability that the student does not have brown eyes?

In a game played with a standard deck of cards, each face card has a value of 10 points, each ace has a value of 1 point, and each number card has a value equal to its number. Two cards are drawn at random.

36. At least one card is an ace. What is the probability that the sum of the cards is 7 or less?

37. One card is the queen of diamonds. What is the probability that the sum of the cards is greater than 18?

Applications and Problem Solving

38. Health Care At Park Medical Center, in a sample group, there are 40 patients diagnosed with lung cancer, and 30 patients who are chronic smokers. Of these, there are 25 patients who have lung cancer and smoke.

a. Draw a Venn diagram to represent the situation.

b. If the medical center currently has 200 patients, and one of them is randomly selected for a medical study, what is the probability that the patient has lung cancer, given that the patient smokes?

39. Business The manager of a computer software store wants to know whether people who come in and ask questions are more likely to make a purchase than the average person. A survey of 500 people exiting the store found that 250 people bought something, 120 asked questions and bought something, and 30 people asked questions but did not buy anything. Based on the survey, determine whether a person who asks questions is more likely to buy something than the average person.

40. Critical Thinking In a game using two number cubes, a sum of 10 has not turned up in the past few rolls. A player believes that a roll of 10 is "due" to come up. Analyze the player's thinking.

41. Testing Winona's chances of passing a precalculus exam are $\frac{4}{5}$ if she studies, and only $\frac{1}{5}$ if she decides to take it easy. She knows that $\frac{2}{3}$ of her class studied for and passed the exam. What is the probability that Winona studied for it?

42. Manufacturing Three computer chip companies manufacture a product that enhances the 3-D graphic capacities of computer displays. The table below shows the number of functioning and defective chips produced by each company during one day's manufacturing cycle.

Company	Number of functioning chips	Number of defective chips
CyberChip Corp.	475	25
3-D Images, Inc.	279	21
MegaView Designs	180	20

a. What is the probability that a randomly selected chip is defective?

b. What is the probability that a defective chip came from 3-D Images, Inc.?

c. What is the probability that a randomly selected chip is functioning?

d. If you were a computer manufacturer, which company would you select to produce the most reliable graphic chip? Why?

43. **Critical Thinking** The probability of an event A is equal to the probability of the same event, given that event B has already occurred. Prove that A and B are independent events.

Mixed Review

44. **City Planning** There are 6 women and 7 men on the committee for city park enhancement. A subcommittee of five members is being selected at random to study the feasibility of redoing the landscaping in one of the parks. What is the probability that the committee will have at least three women? *(Lesson 13-4)*

45. Suppose there are 9 points on a circle. How many 4-sided closed figures can be formed by joining any 4 of these points? *(Lesson 13-1)*

46. Write $\sum_{b=1}^{\infty} 3(0.5)^b$ in expanded form. Then find the sum. *(Lesson 12-5)*

47. Compare and contrast the graphs of $y = 3^x$ and $y = -3^x$ *(Lesson 11-2)*

48. Graph the system of inequalities. *(Lesson 10-8)*
$x^2 + y^2 \leq 81$
$x^2 + y^2 \geq 64$

49. **Navigation** A submarine sonar is tracking a ship. The path of the ship is recorded as $r \cos\left(\theta - \frac{\pi}{2}\right) + 5 = 0$. Find the linear equation of the path of the ship. *(Lesson 9-4)*

50. Graph the line whose parametric equations are $x = 4t$, and $y = 3 + 2t$. *(Lesson 8-6)*

51. Find the area of the sector of a circle of radius 8 feet, given its central angle is 98°. Round your answer to the nearest tenth. *(Lesson 6-1)*

52. When the angle of elevation of the sun is 27°, the shadow of a tree is 25 meters long. How tall is the tree? Round your answer to the nearest tenth. *(Lesson 5-4)*

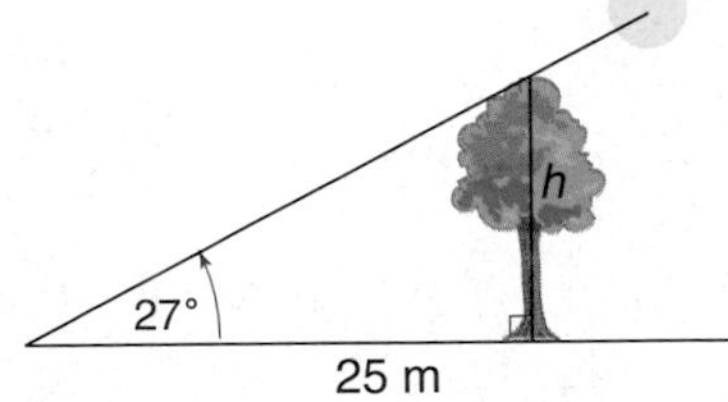

53. **Photography** A photographer has a frame that is 3 feet by 4 feet. She wants to mat a group photo such that there is a uniform width of mat surrounding the photo. If the area of the photo is 6 square feet, find the width of the mat. *(Lesson 4-2)*

54. Find the value of x at which $f(x) = \frac{5}{x^2 - 4}$ is discontinuous. Use the continuity test to justify your answer. *(Lesson 3-5)*

55. **SAT/ACT Practice** In parallelogram $ABCD$, the ratio of the shaded area to the unshaded area is

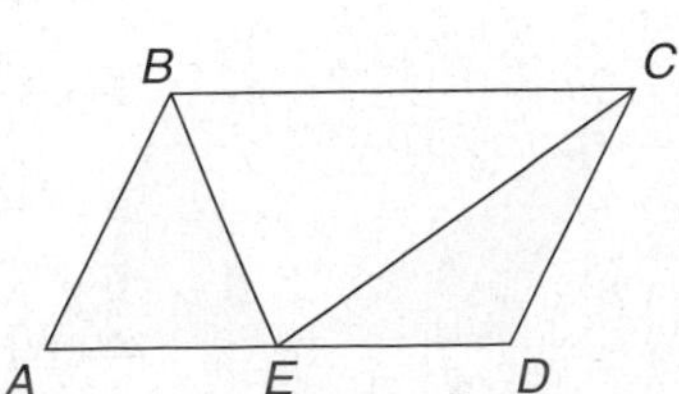

A 1:2 **B** 1:1
C 4:3 **D** 2:1
E It cannot be determined from the information given.

Extra Practice See p. A52.

The Binomial Theorem and Probability

OBJECTIVE
- Find the probability of an event by using the Binomial Theorem.

LANDSCAPING Managers at the Eco-Landscaping Company know that a mahogany tree they plant has a survival rate of about 90% if cared for properly. If 10 trees are planted in the last phase of a landscaping project, what is the probability that 7 of the trees will survive? *This problem will be solved in Example 3.*

We can examine a simpler form of this problem. Suppose that there are only 5 trees to be planted. What is the probability that 4 will survive? The number of ways that this can happen is $C(5, 4)$ or 5.

Let S represent the probability of a tree surviving.
Let D represent the probability of a tree dying.

Look Back
Refer to Lesson 12-6 to review binomial expansions and the Binomial Theorem.

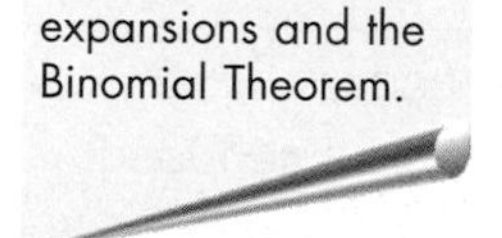

Since this situation has two outcomes, we can represent it using the binomial expansion of $(S + D)^5$. The terms of the expansion can be used to find the probabilities of each combination of the survival and death of the trees.

$$(S + D)^5 = 1S^5 + 5S^4D + 10S^3D^2 + 10S^2D^3 + 5SD^4 + 1D^5$$

coefficient	term	meaning
$C(5, 5) = 1$	$1S^5$	1 way to have all 5 trees survive
$C(5, 4) = 5$	$5S^4D$	5 ways to have 4 trees survive and 1 die
$C(5, 3) = 10$	$10S^3D^2$	10 ways to have 3 trees survive and 2 die
$C(5, 2) = 10$	$10S^2D^3$	10 ways to have 2 trees survive and 3 die
$C(5, 1) = 5$	$5SD^4$	5 ways to have 1 tree survive and 4 die
$C(5, 0) = 1$	$1D^5$	1 way to have all 5 trees die

The probability of a tree surviving is 0.9. So, the probability of a tree not surviving is $1 - 0.9$ or 0.1. The probability of having 4 trees survive out of 5 can be determined as follows.

Use $5S^4D$ since this term represents 4 trees surviving and 1 tree dying.

$5S^4D = 5(0.9)^4(0.1)$ *Substitute 0.9 for S and 0.1 for D*

$5S^4D = 5(0.6561)0.1$

$5S^4D = 0.3281$ or about $\frac{1}{3}$

Thus, the probability of having 4 trees survive is about $\frac{1}{3}$.

Other probabilities can be determined from the expansion of $(S + D)^5$. For example, what is the probability that at least 2 trees out of the 5 trees planted will die?

Example 1 **LANDSCAPING** **Refer to the application at the beginning of the lesson. Five mahogany trees are planted. What is the probability that at least 2 trees die?**

The third, forth, fifth, and sixth terms represent the conditions that two or more trees die. So, the probability of this happening is the sum of the probabilities of those terms.

$$
\begin{aligned}
P(\text{at least 2 trees die}) \\
&= 10S^3D^2 + 10S^2D^3 + 5SD^4 + 1D^5 \\
&= 10(0.9)^3(0.1)^2 + 10(0.9)^2(0.1)^3 + 5(0.9)(0.1)^4 + (0.1)^5 \\
&= 10(0.729)(0.01) + 10(0.81)(0.001) + 5(0.9)(0.0001) + (0.00001) \\
&= 0.0729 + 0.0081 + 0.00045 + 0.00001 \\
&= 0.0815
\end{aligned}
$$

The probability that at least 2 trees die is about 8%.

Problems that can be solved using the binomial expansion are called **binomial experiments.**

Conditions of a Binomial Experiment

A binomial experiment exists if and only if these conditions occur.

- Each trial has exactly two outcomes, or outcomes that can be reduced to two outcomes.
- There must be a fixed number of trials.
- The outcomes of each trial must be independent.
- The probabilities in each trial are the same.

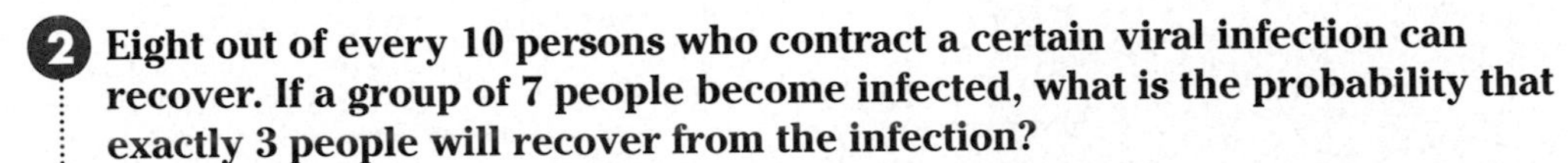

Example 2 **Eight out of every 10 persons who contract a certain viral infection can recover. If a group of 7 people become infected, what is the probability that exactly 3 people will recover from the infection?**

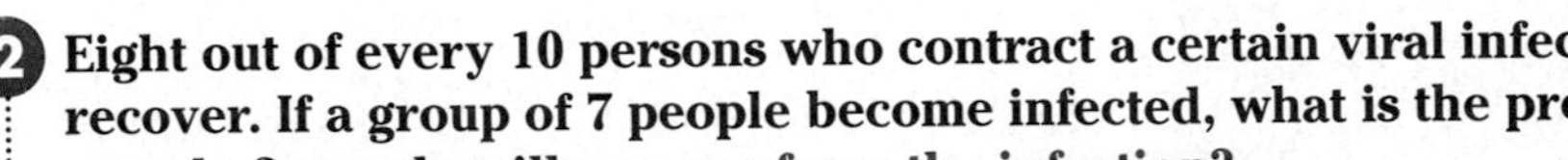

There are 7 people involved, and there are only 2 possible outcomes, recovery R or not recovery N. These events are independent, so this is a binomial experiment.

When $(R + N)^7$ is expanded, the term R^3N^4 represents 3 people recovering and 4 people not recovering from the infection. The coefficient of R^3N^4 is $C(7, 3)$ or 35.

$$
\begin{aligned}
P(\text{exactly 3 people recovering}) &= 35(0.8)^3(0.2)^4 \qquad R = 0.8,\ N = 1 - 0.8 \text{ or } 0.2 \\
&= 35(0.512)(0.0016) \\
&= 0.028672
\end{aligned}
$$

The probability that exactly 3 of the 7 people will recover from the infection is 2.9%.

The Binomial Theorem can be used to find the probability when the number of trials makes working with the binomial expansion unrealistic.

Example 3 

LANDSCAPING Refer to the application at the beginning of the lesson. What is the probability that 7 of the 10 trees planted will survive?

Let S be the probability that a tree will survive.
Let D be the probability that a tree will die.

Since there are 10 trees, we can use the Binomial Theorem to find any term in the expression $(S + D)^{10}$.

$$(S + D)^{10} = \sum_{r=0}^{10} \frac{10!}{r!(10 - r)!} S^{10-r}D^r$$

Look Back
Refer to Lesson 12-5 to review sigma notation.

Having 7 trees survive means that 3 will die. So the probability can be found using the term where $r = 3$, the fourth term.

$$\begin{aligned}\frac{10!}{3!\,(10 - 3)!} S^7D^3 &= 120S^7D^3\\ &= 120(0.9)^7(0.1)^3\\ &= 120(0.4782969)(0.001) \text{ or } 0.057395628\end{aligned}$$

The probability of exactly 7 trees surviving is about 5.7%.

So far, the probabilities we have found have been **theoretical probabilities.** These are determined using mathematical methods and provide an idea of what to expect in a given situation. **Experimental probability** is determined by performing experiments and observing and interpreting the outcomes. One method for finding experimental probability is a **simulation.** In a simulation, a device such as a graphing calculator is used to model the event.

GRAPHING CALCULATOR EXPLORATION

You can use a graphing calculator to simulate a binomial experiment. Consider the following situation.

Robby wins 2 out of every 3 chess matches he plays with Marlene. What is the probability that he wins exactly 5 of the next 6 matches?

TRY THIS

To simulate this situation, enter **int(3*rand)** and press ENTER. Note: **(int(** and **rand** can be found in the menus accessed by pressing MATH. This will randomly generate the numbers 0, 1, or 2. Robby wins if the outcome is 0 or 1. Robby loses if 2 comes up.

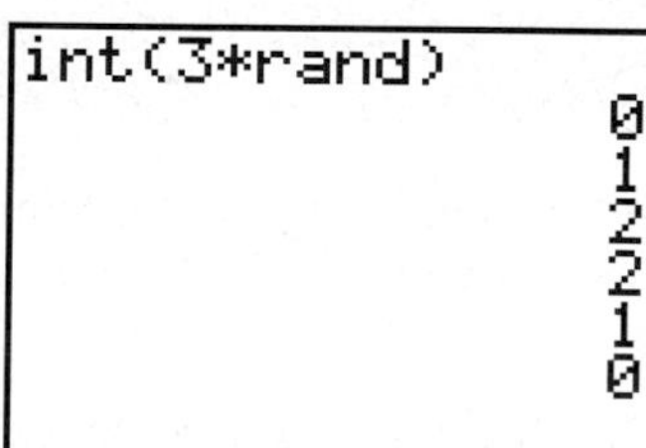

In the simulation, one repetition of the complete binomial experiment consists of six trials or six presses of the ENTER key. Try 40 repetitions.

WHAT DO YOU THINK?

1. What is the sample space?
2. What is P(Robby wins)?
3. In the simulation, with what probability did Robby win exactly 5 times?
4. Using the formula for computing binomial probabilities, what is the probability of Robby winning exactly five games?
5. Why do you think there is a difference between the simulation (experimental probability) and the probability computed using the formula (theoretical probability)?
6. What would you do to have the experimental probability approximate the theoretical probability?

CHECK FOR UNDERSTANDING

Communicating Mathematics

Read and study the lesson to answer each question.

1. **Explain** whether or not each situation represents a binomial experiment.
   a. the probability of winning in a game where a number cube is tossed, and if 1, 2, or 3 comes up you win.
   b. the probability of drawing two red marbles from a jar containing 10 red, 30 blue, and 5 yellow marbles.
   c. the probability of drawing a jack from a standard deck of cards, knowing that the card is red.
2. **Write** an explanation of experimental probability. Give a real-world example that uses experimental probability.
3. **Describe** how to find the probability of getting exactly 2 correct answers on a true/false quiz that has 5 questions.

Guided Practice

Find each probability if a number cube is tossed five times.

4. P(only one 4)
5. P(no more than two 4s)
6. P(at least three 4s)
7. P(exactly five 4s)

Jasmine Myers, a weather reporter for Channel 6, is forecasting a 30% chance of rain for today and the next four days. Find each probability.

8. P(not having rain on any day)
9. P(having rain on exactly one day)
10. P(having rain no more than three days)

11. **Cooking** In cooking class, 1 out of 5 soufflés that Sabrina makes will collapse. She is preparing 6 soufflés to serve at a party for her parents. What is the probability that exactly 4 of them do not collapse?
12. **Finance** A stock broker is researching 13 independent stocks. An investment in each will either make or lose money. The probability that each stock will make money is $\frac{5}{8}$. What is the probability that exactly 10 of the stocks will make money?

EXERCISES

Practice

Isabelle carries lipstick tubes in a bag in her purse. The probability of pulling out the color she wants is $\frac{2}{3}$. Suppose she uses her lipstick 4 times in a day. Find each probability.

13. P(never the correct color)
14. P(correct at least 3 times)
15. P(no more than 3 times correct)
16. P(correct exactly 2 times)

Maura guesses at all 10 questions on a true/false test. Find each probability.

17. P(7 correct)
18. P(at least 6 correct)
19. P(all correct)
20. P(at least half correct)

The probability of tossing a head on a bent coin is $\frac{1}{3}$. Find each probability if the coin is tossed 4 times.

21. P(4 heads)
22. P(3 heads)
23. P(at least 2 heads)

Kyle guesses at all of the 10 questions on his multiple choice test. Find each probability if each question has 4 choices.

24. P(6 correct answers)
25. P(half answers correct)
26. P(from 3 to 5 correct answers)

If a thumbtack is dropped, the probability of its landing point up is $\frac{2}{5}$. Mrs. Davenport drops 10 tacks while putting up the weekly assignment sheet on the bulletin board. Find each probability.

27. P(all point up)
28. P(exactly 3 point up)
29. P(exactly 5 point up)
30. P(at least 6 point up)

Find each probability if three coins are tossed.

31. P(3 heads or 3 tails)
32. P(at least 2 heads)
33. P(exactly 2 tails)

Graphing Calculator

34. Enter the expression **6 nCr X** into the **Y=** menu. The **nCr** command is found in the probability section of the **MATH** menu. Use the **TABLE** feature to observe the results.
    a. How do these results compare with the expansion of $(a + b)^6$?
    b. How would you change the expression to find the expansion of $(a + b)^8$?

35. **Sports** A football team is scheduled to play 16 games in its next season. If there is a 70% probability the team will win each game, what is the probability that the team will win at least 12 of its games? (*Hint*: Use the information from Exercise 34.)

Applications and Problem Solving

36. **Military Science** During the Gulf War in 1990–1991, SCUD missiles hit 20% of their targets. In one incident, six missiles were fired at a fuel storage installation.
    a. Describe what success means in this case, and state the number of trials and the probability of success on each trial.
    b. Find the probability that between 2 and 6 missiles hit the target.

37. **Critical Thinking** Door prizes are given at a party through a drawing. Four out of 10 tickets are given to men who will attend, and 6 out of 10 tickets are distributed to women. Each person will receive only one ticket. Ten tickets will be drawn at random with replacement. What is the probability that all winners will be the same sex?

38. **Medicine** Ten percent of African-Americans are carriers of the genetic disease sickle-cell anemia. Find each probability for a random sample of 20 African-Americans.
    a. P(all carry the disease)
    b. P(exactly half have the disease)

39. **Airlines** A commuter airline has found that 4% of the people making reservations for a flight will not show up. As a result, the airline decides to sell 75 seats on a plane that has 73 seats (overbooking). What is the probability that for every person who shows up for the flight there will be a seat available?

40. **Sales** Luis is an insurance agent. On average, he sells 1 policy for every 2 prospective clients he meets. On a particular day, he calls on 4 clients. He knows that he will not receive a bonus if the sales are less than or equal to three policies. What is the probability that he will not get a bonus?

41. **Critical Thinking** Trina is waiting for her friend who is late. To pass the time, she takes a walk using the following rules. She tosses a fair coin. If it falls heads, she walks 10 meters north. If it falls tails, she walks 10 meters south. She repeats this process every 10 meters and thus executes what is called a random walk. What is the probability that after 100 meters of walking she will be at one of the following points?
   a. P(back at her starting point)
   b. P(within 10 meters of the starting point)
   c. P(exactly 20 meters from the starting point)

Mixed Review

42. A pair of number cubes is thrown. Find the probability that their sum is less than 9 if both cubes show the same number. *(Lesson 13-5)*

43. A letter is picked at random from the alphabet. Find the probability that the letter is contained in the word *house* or in the word *phone*. *(Lesson 13-4)*

44. **Physical Science** Dry air expands as it moves upward into the atmosphere. For each 1000 feet that it moves upward, the air cools 5° F. Suppose the temperature at ground level is 80° F. *(Lesson 12-1)*
   a. Write a sequence representing the temperature decrease per 1000 feet.
   b. If n is the height of the air in thousands of feet, write a formula for the temperature T in terms of n.
   c. What is the ground level temperature if the air at 40,000 feet is −125°?

45. Solve $3^{x-1} = 6^{-x}$ using logarithms. Round to the nearest hundredth. *(Lesson 11-6)*

46. Name the coordinates of the center, foci, and vertices of the ellipse with the equation $\frac{x^2}{49} + \frac{(y+3)^2}{25} = 1$. *(Lesson 10-3)*

47. Express $\sqrt{2}\left(\cos -\frac{\pi}{2} + \boldsymbol{i} \sin -\frac{\pi}{2}\right)$ in rectangular form. *(Lesson 9-6)*

48. Find the ordered pair that represents $\overrightarrow{WX}$ if $W(8, -3)$ and $X(6, 5)$. Then find the magnitude of $\overrightarrow{WX}$. *(Lesson 8-2)*

49. **Geometry** The sides of a parallelogram are 55 cm and 71 cm long. Find the length of each diagonal if the larger angle measures 106°. *(Lesson 5-8)*

50. Use the Remainder Theorem to find the remainder when $x^4 + 12x^3 + 21x^2 - 62x - 72$ is divided by $x + 4$. State whether the binomial is a factor of the polynomial. *(Lesson 4-3)*

51. **SAT Practice Grid-In** A word processor uses a sheet of paper that is 9 inches wide by 12 inches long. It leaves a 1-inch margin on each side and a 1.5-inch margin on the top and bottom. What fraction of the page is used for text?

Extra Practice See p. A52.

VOCABULARY

Basic Counting Principle (p. 837)
binomial experiments (p. 876)
circular permutation (p. 847)
combination (p. 841)
combinatorics (p. 837)
complements (p. 853)
conditional probability (p. 868)
dependent event (p. 837)
experimental probability (p. 877)
failure (p. 852)
inclusive event (p. 863)
independent event (p. 837)
mutually exclusive (p. 862)
odds (p. 854)
permutation (p. 838)
permutation with repetition (p. 846)
probability (p. 852)
reduced sample space (p. 862)
sample space (p. 852)
simulation (p. 877)
success (p. 852)
theoretical probability (p. 877)
tree diagram (p. 837)

UNDERSTANDING AND USING THE VOCABULARY

Choose the correct term to best complete each sentence.

1. Events that do not affect each other are called (dependent, independent) events.
2. In probability, any outcome other than the desired outcome is called a (failure, success).
3. The sum of the probability of an event and the probability of the complement of the event is always (0, 1).
4. The (odds, probability) of an event occurring is the ratio of the number of ways the event can succeed to the sum of the number of ways the event can succeed and the number of ways the event can fail.
5. The arrangement of objects in a certain order is called a (combination, permutation).
6. A (permutation with repetitions, circular permutation) specifically deals with situations in which some objects that are alike.
7. Two (inclusive, mutually exclusive) events cannot happen at the same time.
8. A (sample space, Venn diagram) is the set of all possible outcomes of an event.
9. The probability of an event A given that event B has occurred is called a (conditional, inclusive) probability.
10. The branch of mathematics that studies different possibilities for the arrangement of objects is called (statistics, combinatorics).

For additional review and practice for each lesson, visit: **www.amc.glencoe.com**

SKILLS AND CONCEPTS

OBJECTIVES AND EXAMPLES

Lesson 13-1 Solve problems related to the Basic Counting Principle.

- How many possible ways can a group of eight students line up to buy tickets to a play?

  There are eight choices for the first spot in line, seven choices for the second spot, six for the third spot, and so on.

  $8 \cdot 7 \cdot 6 \cdot 5 \cdot 4 \cdot 3 \cdot 2 \cdot 1 = 40{,}320$

  There are 40,320 ways for the students to line up.

REVIEW EXERCISES

11. How many different ways can three books be arranged in a row on a shelf?
12. How many different ways can the digits 1, 2, 3, 4, and 5 be arranged to create a password?
13. How many ways can six teachers be assigned to teach six different classes, if each teacher can teach any of the classes?

Lesson 13-1 Solve problems involving permutations and combinations.

- From a choice of 3 meat toppings and 4 vegetable toppings, how many 5-topping pizzas are possible?

  Since order is not important, the selection is a combination of 7 objects taken 5 at a time, or $C(7, 5)$.

  $C(7, 5) = \frac{7!}{(7-5)!\,5!} = 21$

  There are 21 possible 5-topping pizzas.

Find each value.

14. $P(6, 3)$
15. $P(8, 6)$
16. $C(5, 3)$
17. $C(11, 8)$
18. $\frac{P(6, 3)}{P(5, 3)}$
19. $C(5, 5) \cdot C(3, 2)$
20. How many ways can 6 different books be placed on a shelf if the only dictionary must be on an end?
21. From a group of 3 men and 7 women, how many committees of 2 men and 2 women can be formed?

Lesson 13-2 Solve problems involving permutations with repetitions.

- How many ways can the letters of *Tallahassee* be arranged?

  There are 3 *a*'s, 2 *l*'s, 2 *s*'s, and 2 *e*'s. So the number of possible arrangements is $\frac{11!}{3!2!2!2!}$ or 831,600 ways.

How many different ways can the letters of each word be arranged?

22. *level*
23. *Cincinnati*
24. *graduate*
25. *banana*
26. How many different 9-digit Social Security numbers can have the digits 2, 9, 5, 5, 0, 7, 0, 5, and 8.

OBJECTIVES AND EXAMPLES

Lesson 13-3 Find the probability of an event.

- Find the probability of randomly selecting 3 red pencils from a box containing 5 red, 3 blue, and 4 green pencils.

  There are $C(5, 3)$ ways to select 3 out of 5 red pencils and $C(12, 3)$ ways to select 3 out of 12 pencils.

  $$P(\text{3 red pencils}) = \frac{C(5, 3)}{C(12, 3)}$$
  $$= \frac{\frac{5!}{2!\,3!}}{\frac{12!}{9!\,3!}}$$
  $$= \frac{12}{220} \text{ or } \frac{1}{22}$$

REVIEW EXERCISES

A bag containing 7 pennies, 4 nickels, and 5 dimes. Three coins are drawn at random. Find each probability.

27. P(3 pennies)

28. P(2 pennies and 1 nickel)

29. P(3 nickels)

30. P(1 nickel and 2 dimes)

Lesson 13-3 Find the odds for the success and failure of an event.

- Find the odds of randomly selecting 3 red pencils from a box containing 5 red, 3 blue, and 4 green pencils.

  $$P(\text{3 red pencils}) = P(s) = \frac{1}{22}$$
  $$P(\text{not 3 red pencils}) = P(f) = 1 - \frac{1}{22} \text{ or } \frac{21}{22}$$
  $$\text{Odds} = \frac{P(s)}{P(f)} = \frac{\frac{1}{22}}{\frac{21}{22}}$$
  $$= \frac{1}{21} \text{ or } 1{:}21$$

Refer to the bag of coins used for Exercises 27-30. Find the odds of each event occurring.

31. 3 pennies

32. 2 pennies and 1 nickel

33. 3 nickels

34. 1 nickel and 2 dimes

Lesson 13-4 Find the probability of independent and dependent events.

- Three yellow and 5 black marbles are placed in a bag. What is the probability of drawing a black marble, replacing it, and then drawing a yellow marble?

  $$P(\text{black}) = \frac{5}{8} \qquad P(\text{yellow}) = \frac{3}{8}$$
  $$P(\text{black and yellow}) = P(\text{black}) \cdot P(\text{yellow})$$
  $$= \frac{5}{8} \cdot \frac{3}{8} = \frac{15}{64}$$

Determine if each event is *independent* or *dependent*. Then determine the probability.

35. the probability of rolling a sum of 2 on the first toss of two number cubes and a sum of 6 on the second toss

36. the probability of randomly selecting two yellow markers from a box that contains 4 yellow and 6 pink markers

SKILLS AND CONCEPTS

OBJECTIVES AND EXAMPLES

Lesson 13-4 Find the probability of mutually exclusive and inclusive events.

- On a school board, 2 of the 4 female members are over 40 years of age, and 5 of the 6 male members are over 40. If one person did not attend the meeting, what is the probability that the person was a male or a member over 40?

$$P(\text{male or over } 40) = P(\text{male}) + P(\text{over } 40) - P(\text{male \& over } 40)$$
$$= \frac{6}{10} + \frac{7}{10} - \frac{5}{10} \text{ or } \frac{4}{5}$$

REVIEW EXERCISES

A box contains slips of paper numbered from 1 to 14. One slip of paper is drawn at random. Find each probability.

37. P(selecting a prime number or a multiple of 4)
38. P(selecting a multiple of 2 or a multiple of 3)
39. P(selecting a 3 or a 4)
40. P(selecting an 8 or a number less than 8)

Lesson 13-5 Find the probability of an event given the occurrence of another event.

- A coin is tossed 3 times. What is the probability that at the most 2 heads are tossed given that at least 1 head has been tossed?

Let event A be that at most 2 heads are tossed.
Let event B be that there is at least 1 head.

$$P(A \mid B) = \frac{P(A \text{ and } B)}{P(B)}$$
$$= \frac{\frac{6}{8}}{\frac{7}{8}} \text{ or } \frac{6}{7}$$

Two number cubes are tossed.

41. What is the probability that the sum of the numbers shown on the cubes is less than 5 if exactly one cube shows a 1?
42. What is the probability that the numbers shown on the cubes are different given that their sum is 8?
43. What is the probability that the numbers shown on the cubes match given that their sum is greater than or equal to 5?

Lesson 13-6 Find the probability of an event by using the Binomial Theorem.

- If you guess the answers on all 8 questions of a true/false quiz, what is the probability that exactly 5 of your answers will be correct?

$$(p + q)^8 = \sum_{r=0}^{8} \frac{8!}{r!(8-r)!} p^{8-r}q^r$$
$$= \frac{8!}{5!(8-5)!}\left(\frac{1}{2}\right)^5\left(\frac{1}{2}\right)^3$$
$$= \frac{56}{256} \text{ or } \frac{7}{32}$$

Find each probability if a coin is tossed 4 times.

44. P(exactly 1 head)
45. P(no heads)
46. P(2 heads and 2 tails)
47. P(at least 3 tails)

APPLICATIONS AND PROBLEM SOLVING

48. **Travel** Five people, including the driver, can be seated in Nate's car. Nate and 6 of his friends want to go to a movie. How many different groups of friends can ride in Nate's car on the first trip if the car is full? *(Lesson 13-1)*

49. Sommer has 7 different keys. How many ways can she place these keys on the key ring shown below? *(Lesson 13-2)*

50. **Quality Control** A collection of 15 memory chips contains 3 chips that are defective. If 2 memory chips are selected at random, what is the probability that at least one of them is good? *(Lesson 13-3)*

51. **Gift Exchange** The Burnette family is drawing names from a bag for a gift exchange. There are 7 males and 8 females in the family. If someone draws their own name, then they must draw again before replacing their name. *(Lesson 13-4)*
    a. Reba draws the first name. What is the probability that Reba will draw a female's name that is not her own?
    b. What is the probability that Reba will draw her own name, not replace it, and then draw a male's name?

ALTERNATIVE ASSESSMENT

OPEN-ENDED ASSESSMENT

1. The probability of two independent events occurring is $\frac{1}{12}$. If the probability of one of the events occurring is $\frac{1}{2}$, is it possible to find the probability of the other event? If so, find the probability and give an example of a situation for which this probability could apply. If not, explain why not.

2. Perry says "A permutation is the same as a combination." How would you explain to Perry that his statement is incorrect?

PORTFOLIO

Choose one of the types of probability you studied in this chapter. Describe a situation in which this type of probability would be used. Explain why no other type of probability should be used in this situation.

THE UNITED STATES CENSUS BUREAU

Radically Random!

- Use the Internet to find the population of the United States by age groups or ethnic background for the most recent census.
- Make a table or spreadsheet of the data.
- Suppose that a person was selected at random from all the people in the United States to answer some survey questions. Find the probability that the person was from each one of the age or ethnic groups you used for your table or spreadsheet.
- Write a summary describing how you calculated the probabilities. Include a graph with your summary. Discuss why someone might be interested in your findings.

Additional Assessment See p. A68 for Chapter 13 practice test.

TEST-TAKING TIP
For problems involving combinations, either use the formula or make a list.
Example:
$C(5, 3) = \frac{5 \cdot 4 \cdot 3}{3!} = 10$

Probability and Combination Problems

Both the ACT and SAT contain probability problems. You'll need to know these concepts:

Combinations   Permutations   Tree Diagram

Outcomes   Probability

Memorize the definition of the probability of an event:

$$P(\text{event}) = \frac{\text{number of favorable outcomes}}{\text{total number of possible outcomes}}$$

SAT EXAMPLE

1. A bag contains 4 red balls, 10 green balls, and 6 yellow balls. If three balls are removed at random and no ball is returned to the bag after removal, what is the probability that all three balls will be green?

A $\frac{1}{2}$ **B** $\frac{1}{8}$ **C** $\frac{3}{20}$ **D** $\frac{2}{19}$ **E** $\frac{3}{8}$

HINT Calculate the probability of two independent events by multiplying the probability of the first event by the probability of the second event.

Solution Use the definition of the probability of an event. Calculate the probability of getting a green ball each time a ball is removed. The first time a ball is removed there are a total of 20 balls and 10 of them are green. So the probability of removing a green ball as the first ball is $\frac{10}{20}$ or $\frac{1}{2}$. Now there are just 19 balls and 9 of them are green. The probability of removing a green ball is $\frac{9}{19}$. When the third ball is removed, there are 18 balls and 8 of them are green, so the probability of removing a green ball is $\frac{8}{18}$ or $\frac{4}{9}$. To find the probability of removing green balls as the first *and* the second *and* the third balls chosen, multiply the three probabilities.

$$\frac{1}{2} \times \frac{9}{19} \times \frac{4}{19} = \frac{2}{19}$$

The answer is choice **D**.

ACT EXAMPLE

2. If you toss 3 fair coins, what is the probability of getting exactly 2 heads?

A $\frac{1}{3}$ **B** $\frac{3}{8}$

C $\frac{1}{2}$ **D** $\frac{2}{3}$

E $\frac{7}{8}$

HINT Start by listing all the possible outcomes. You can do this since the numbers are small.

Solution Make a list and then count the possible outcomes.

HHH, HHT, HTH, HTT,

TTH, THT, THH, TTT

There are 8 possible outcomes for 3 coins. Since the coins are fair, these are equally likely outcomes.

The favorable outcomes are those that include exactly 2 heads: HHT, HTH, THH. There are 3 favorable outcomes. Give the answer.

$$P(A) = \frac{\text{number of successful outcomes}}{\text{total number of outcomes}}$$

$$P(\text{exactly 2 heads}) = \frac{3}{8}$$

The answer is choice **B**.

After you work each problem, record your answer on the answer sheet provided or on a piece of paper.

Multiple Choice

1. A coin was flipped 20 times and came up heads 10 times and tails 10 times. If the first and the last flips were both heads, what is the greatest number of consecutive heads that could have occurred?

   A 1 **B** 2 **C** 8
   D 9 **E** 10

2. In the figure below, *ABCD* is a parallelogram. What must be the coordinates of Point *C*?

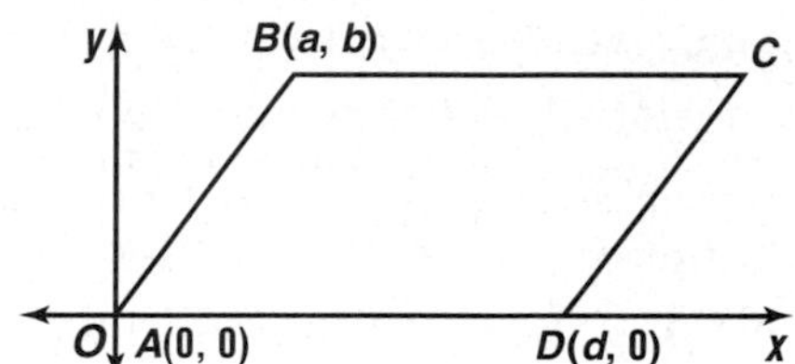

   A (x, y) **B** $(d + a, y)$ **C** $(d - a, b)$
   D $(d + x, b)$ **E** $(d + a, b)$

3. In a plastic jar there are 5 red marbles, 7 blue marbles, and 3 green marbles. How many green marbles need to be added to the jar in order to double the probability of selecting a green marble?

   A 2 **B** 3 **C** 5 **D** 6 **E** 7

4. The average of 5 numbers is 20. If one of the numbers is 18, then what is the sum of the other four numbers?

   A 2 **B** 20.5 **C** 82
   D 90 **E** 100

5. If the sum of x and y is an even number, and the sum of x and z is an even number, and z is an odd number, then which of the following must be true?
   I. y is an even number
   II. $y + z$ is an even number
   III. y is an odd number

   A I only **B** II only **C** III only
   D I and II **E** II and III

6. A bag contains only white and blue marbles. The probability of selecting a blue marble is $\frac{1}{5}$. The bag contains 200 marbles. If 100 white marbles are added to the bag, what is the probability of selecting a white marble?

   A $\frac{2}{15}$ **B** $\frac{7}{15}$ **C** $\frac{8}{15}$ **D** $\frac{4}{5}$ **E** $\frac{13}{15}$

7. In the figure below, $\ell_1 \parallel \ell_2$. Which of the labeled angles must be equal to each other?

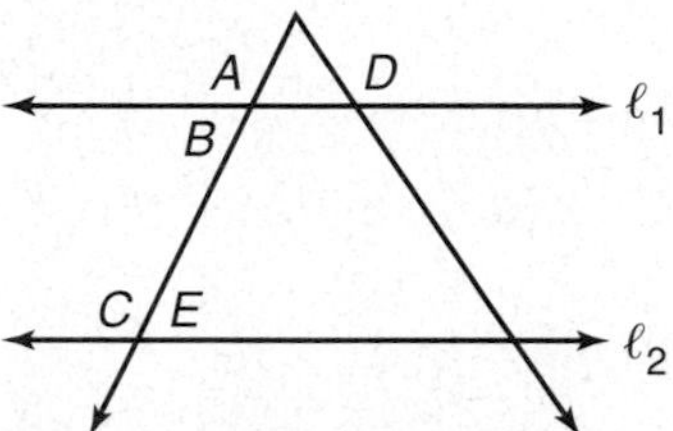

   A *A* and *C* **B** *D* and *E* **C** *A* and *B*
   D *D* and *B* **E** *C* and *B*

8. What is the probability of drawing a diamond from a well-shuffled standard deck of playing cards?

   A $\frac{1}{52}$ **B** $\frac{1}{13}$ **C** $\frac{1}{4}$
   D $\frac{4}{13}$ **E** 1

9. **Quantitative Comparison**

   A if the quantity in Column A is greater
   B if the quantity in Column B is greater
   C if the two quantities are equal
   D if the relationship cannot be determined from the information given

Column A	Column B
The number of distinct 3-player teams that can be drawn from a pool of 5 players.	12

10. **Grid-In** Six cards are numbered 0 through 5. Two are selected without replacement. What is the probability that the sum of the cards is 4?

*inter*NET CONNECTION **SAT/ACT Practice** For additional test practice questions, visit: **www.amc.glencoe.com**

Statistics and Data Analysis

CHAPTER OBJECTIVES

- **Make and use bar graphs, histograms, frequency distribution tables, stem-and-leaf plots, and box-and-whisker plots.** *(Lessons 14-1, 14-2, 14-3)*
- **Find the measures of central tendency and the measures of variability.** *(Lessons 14-2, 14-3)*
- **Use the normal distribution curve.** *(Lesson 14-4)*
- **Find the standard error of the mean to predict the true mean of a population with a certain level of confidence.** *(Lesson 14-5)*

14-1

The Frequency Distribution

OBJECTIVES

- Draw, analyze, and use bar graphs and histograms.
- Organize data into a frequency distribution table.

FOOTBALL The AFL-NFL World Championship Game, as it was originally called, became the Super Bowl in 1969. The graph below shows the first 34 Super Bowl winners. What team has won the most Super Bowls?

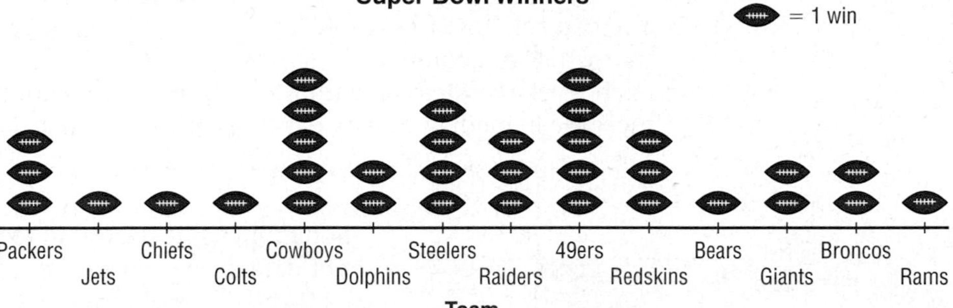

By looking at the graph, you can quickly determine that the Dallas Cowboys and the San Francisco 49ers have both won five Super Bowls.

A graph is often used to provide a picture of statistical data. One advantage of using a graph to show data is that a person can easily see any relationships or patterns that may exist. The number of Super Bowl victories for various teams is depicted as a **line plot.** A line plot uses symbols to show frequency. A **bar graph** can show the same information by using bars to indicate the frequency.

A **back-to-back bar graph** is a special bar graph that shows the comparisons of two sets of related data. A back-to-back graph is plotted on a coordinate system with the horizontal scale repeated in each direction from the central axis.

Example **1** **ECONOMICS** **The following data relates the amount of education with the median weekly earnings of a full-time worker 25 years old or older for the years 1980 and 1997.**

Median Weekly Earnings

	Less than 4 Years of High School	High School Diploma	1 to 3 Years of College	College Degree
1980	$222	$266	$304	$376
1997	$321	$461	$518	$779

Source: *The Wall Street Journal 1999 Almanac*

a. Make a back-to-back bar graph that represents the data.

b. Describe any trends indicated by the graph.

(continued on the next page)

a. Let the level of education be the central axis. Draw a horizontal axis that is scaled \$0 to \$800 in each direction. Let the left side of the graph represent the earnings from 1980 and the right side of the graph be those from 1997. Draw the bars to the appropriate length for the data.

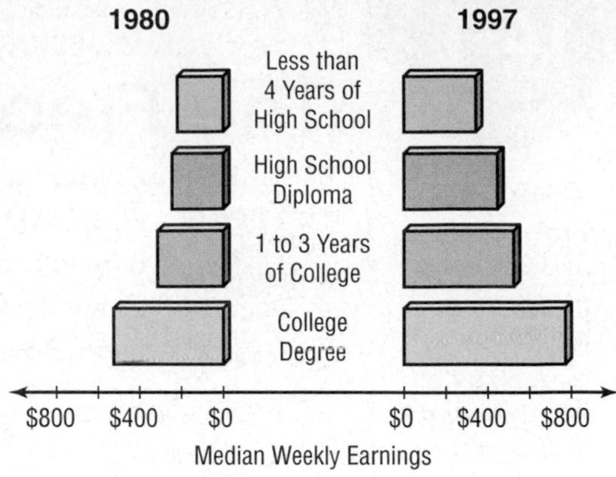

b. You can see from the graph that when you compare each level of education with the next, more education resulted in a greater increase in median weekly earnings in 1997 than in 1980.

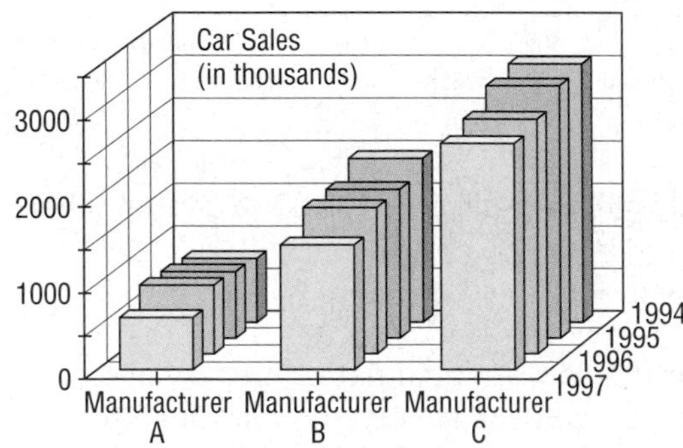

Sometimes it is desirable to show three aspects of a set of data at the same time. To present data in this way, a **three-dimensional bar graph** is often used. The graph at the left represents the retail sales in thousands of passenger cars in the United States for three major domestic car manufacturers during the years 1994 to 1997. The grid defines the car and year. The height of each bar represents the number of cars sold each year.

Sometimes the amount of data you wish to represent in a bar graph is too great for each item of data to be considered individually. In this case, a **frequency distribution** is a convenient system for organizing the data. A number of classes are determined, and all values in a class are tallied and grouped together. To determine the number of classes, first find the **range.** The range of a set of data is the difference between the greatest and the least values in the set.

Retail Management Testing Scores

Scores	Frequency
60–70	9
70–80	10
80–90	12
90–100	3

The intervals are often named by a range of values. In the table, the interval described by 60-70 means all the test scores s such that $60 \leq s < 70$. The **class interval** is the range of each class. The class intervals in a frequency distribution should all be equal. In the table, the range for each class interval is 10.

Class intervals are often multiples of 5.

The **class limits** of a set of data organized in a frequency distribution are the upper and lower values in each interval. The class limits in the testing data above are 60, 70, 80, 90, and 100. The **class marks** are the midpoints of the classes; that is the average of the upper and lower limit for each interval. The class mark for the interval 60–70 is $\frac{60 + 70}{2}$ or 65.

The difference in consecutive class marks is the same as the class interval.

The most common way of displaying frequency distributions is by using a **histogram.** A histogram is a type of bar graph in which the width of each bar represents a class interval and the height of the bar represents the frequency in that interval. Histograms usually have fewer than ten intervals.

Example 2 

FOOTBALL The winning scores for the first 34 Super Bowls are 35, 33, 16, 23, 16, 24, 14, 24, 16, 21, 32, 27, 35, 31, 27, 26, 27, 38, 38, 46, 39, 42, 20, 55, 20, 37, 52, 30, 49, 27, 35, 31, 34, and 23.

a. Find the range of the data.

b. Determine an appropriate class interval.

c. Find the class marks.

d. Construct a frequency distribution of the data.

e. Draw a histogram of the data.

f. What conclusions can you determine from the graph?

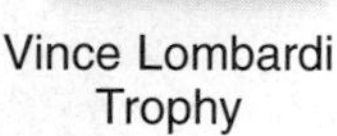

Vince Lombardi Trophy

a. The range of the data is 55 − 14 or 41.

b. An appropriate class interval is 10 points, beginning with 10 points and ending with 60 points. There will be five classes.

c. The class marks are the averages of the class limits of each interval. The class marks are 15, 25, 35, 45, and 55.

d. Make a table listing class limits. Use tallies to determine the number of scores in each interval.

Winning Score	Tallies	Frequency
10–20	||||	4
20–30	𝍸 𝍸 ||	12
30–40	𝍸 𝍸 |||	13
40–50	|||	3
50–60	||	2

e. Label the horizontal axis with the class limits. The vertical axis should be labeled from 0 to a value that will allow for the greatest frequency. Draw the bars side by side so that the height of each bar corresponds to its interval's frequency.

Winning Scores at the Super Bowl

Frequency (0–14); Winning Score (0, 10, 20, 30, 40, 50, 60)

You can also use a graphing calculator to create the histogram. In statistics mode, enter the class marks in the **L1** list and the frequency in the **L2** list. Set the window using the class interval for **Xscl**, and select the histogram as the type of graph.

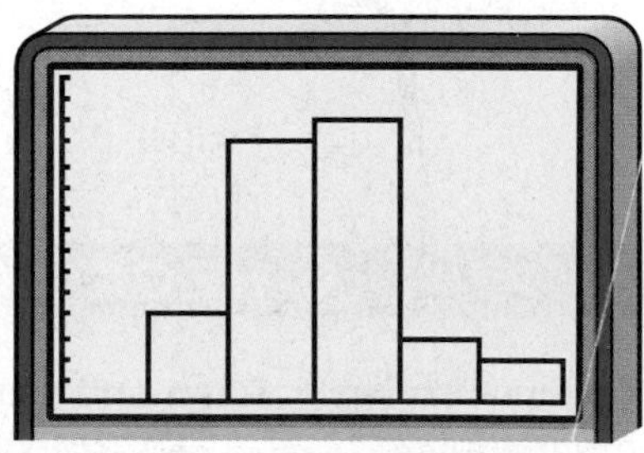

[0, 60] scl:10 by [0, 15] scl:1

f. The winning score at the Super Bowl tends to be between 20 and 40 points.

Graphing Calculator Appendix

For keystroke instruction on how to create a histogram, see pages A23-A24.

Another type of graph can be created from a histogram. A broken line graph, often called a **frequency polygon,** can be drawn by connecting the class marks on the graph. The class marks are graphed as the midpoints of the top edge of each bar. The frequency polygon for the histogram in Example 2 is shown at the right.

Winning Scores at the Super Bowl

Frequency: 0, 2, 4, 6, 8, 10, 12, 14

Winning Score: 0, 10, 20, 30, 40, 50, 60

Example 3 HEALTH **A graduate student researching the effect of smoking on blood pressure collected the following readings of systolic blood pressure from 30 people within a control group.**

125, 145, 110, 126, 128, 180, 177, 176, 156, 144, 182, 205, 191, 140, 138, 126, 154, 163, 172, 159, 174, 151, 142, 160, 147, 143, 158, 129, 132, 137

a. Find an appropriate class interval. Then name the class limits and the class marks.

b. Construct a frequency distribution.

c. Use a graphing calculator to draw a frequency polygon.

a. The range of the data is 205 − 110 or 95. An appropriate class interval is 15 units. The class limits are 105, 120, 135, 150, 165, 180, 195, and 210. The class marks are 112.5, 127.5, 142.5, 157.5, 172.5, 187.5, and 202.5.

b.

Systolic Blood Pressure	Tallies	Frequency
105–120	|	1
120–135	~~||||~~ |	6
135–150	~~||||~~ |||	8
150–165	~~||||~~ ||	7
165–180	||||	4
180–195	|||	3
195–210	|	1

c. In statistics mode, enter the class marks in the **L1** list and the frequency in the **L2** list. Set the window using the class interval for **Xscl**, and select the line graph as the type of graph.

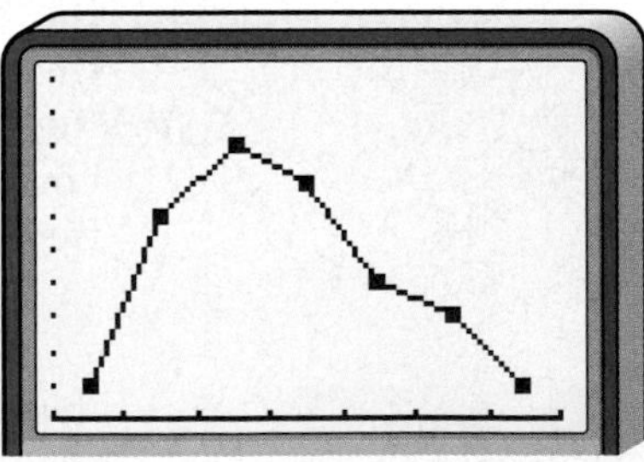

[105, 210] scl:15 by [0, 10] scl:1

For keystroke instruction on how to create a frequency polygon, see page A24.

CHECK FOR UNDERSTANDING

Communicating Mathematics

Read and study the lesson to answer each question.

1. **Compare and contrast** line plot, bar graph, histogram, and frequency polygon.
2. **Explain** how to construct a frequency distribution.

3. **Determine** which class intervals would be appropriate for the data. Explain.

55, 72, 51, 47, 73, 81, 74, 88, 83, 47, 58, 66, 64, 71, 73, 84, 61, 89, 73, 82

a. 1 b. 5 c. 10 d. 20 e. 30

4. *Math Journal* **Select** three graphs from newspapers or magazines. For each graph, write what conclusions might be drawn from the graph.

Guided Practice

5. **Population** The table gives the percent of the U.S. population by age group.

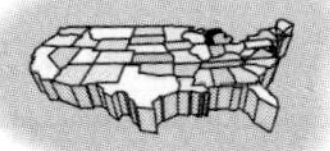

Percent of U.S. Population

Age	0-9	10-19	20-29	30-39	40-49	50-59	60-69	70+
1900	14.8%	14.1%	16.3%	16.8%	12.6%	8.8%	8.6%	8.0%
1999	14.2%	14.4%	13.3%	15.6%	15.2%	10.7%	7.3%	9.3%

Source: U.S. Bureau of the Census

a. Make a back-to-back bar graph of the data.
b. Describe any trends indicated by the graph.

6. **History** The ages of the first 42 presidents when they first took office are listed.

57, 61, 57, 57, 58, 57, 61, 54, 68, 51, 49, 64, 50, 48, 65, 52, 56, 46, 54, 49, 50, 47, 55, 55, 54, 42, 51, 56, 55, 51, 54, 51, 60, 62, 43, 55, 56, 61, 52, 69, 64, 46

a. Find the range of the data.
b. Determine an appropriate class interval.
c. What are the class limits?
d. Find the class marks.
e. Construct a frequency distribution of the data.
f. Draw a histogram of the data.
g. Name the interval or intervals that describe the age of most presidents.

EXERCISES

Applications and Problem Solving

7. **Sales** As customers come to the cash register at an electronics store, the sales associate asks them to give their ZIP code. During one hour, a sales associate gets the following responses.

43221, 43212, 43026, 43220, 43214, 43026, 43229, 43229, 43220, 45414, 43220, 43221, 43212, 43220, 43212, 43220, 43221, 43221, 43214, 43026

a. Make a line plot showing how many times each ZIP code was recorded.
b. Which ZIP code was recorded most frequently?
c. Why would a store want this type of information?

8. **Transportation** The average number of minutes men and women drivers spend behind the wheel daily is given below.

Age	16–19	20–34	35–49	50–64	65+
Men	58	81	86	88	73
Women	56	65	67	61	55

Source: Federal Highway Administration and the American Automobile Manufacturers Association

a. Make a back-to-back bar graph of the data.
b. What conclusions can you draw from the graph?

EXERCISES

Practice

Find the mean, median, and mode of each set of data.

10. {140, 150, 160, 170}

11. {3, 3, 6, 12, 3}

12. {21, 19, 17, 19}

13. {5, 8, 18, 5, 3, 18, 14, 15}

14. {64, 87, 62, 87, 63, 98, 76, 54, 87, 58, 70, 76}

15. {6, 9, 11, 11, 12, 7, 6, 11, 5, 8, 10, 6}

16. Crates of books are being stored for later use. The weights of the crates in pounds are 142, 160, 151, 139, 145, 117, 172, 155, and 124.

a. What is the mean of their weights?

b. Find the median of their weights.

c. If 5 pounds is added to each crate, how will the mean and median be affected?

Find the mean, median, and mode of the data represented by each stem-and-leaf plot.

17.

stem	leaf
3	5 8 8 9
4	4 5 5 5 8
5	7 7 9

3|5 = 35

18.

stem	leaf
5	2 4 6
6	0 1 7 8 9
7	1 6
8	0 2 6
9	1

5|2 = 5.2

19.

stem	leaf
9	0 1 7 8 9
10	5 6 9
11	3 8 8 8
12	0 5 5

9|0 = 900

20. Make a stem-and-leaf plot of the following ages of people attending a family picnic.

15, 55, 35, 46, 28, 35, 25, 17, 30, 30, 27, 35,
15, 25, 25, 20, 20, 15, 20, 17, 15, 25, 10

21. The store manager of a discount department store is studying the weekly wages of the part-time employees. The table profiles the employees.

Weekly Wages	Frequency
\$130–\$140	11
\$140–\$150	24
\$150–\$160	30
\$160–\$170	10
\$170–\$180	13
\$180–\$190	8
\$190–\$200	4

a. Find the sum of the wages in each class.

b. What is the sum of all of the wages in the frequency distribution?

c. Find the number of employees in the frequency distribution.

d. What is the mean weekly wage in the frequency distribution?

e. Find the median class of the frequency distribution.

f. Estimate the median weekly wage in the frequency distribution.

g. Explain why both the mean and median are good measures of central tendency in this situation.

22. Find the value of x so that the mean of $\{2, 4, 5, 8, x\}$ is 7.5.

23. What is the value of x so that the mean of $\{x, 2x - 1, 2x, 3x + 1\}$ is 6?

24. Find the value of x so that the median of $\{11, 2, 3, 3.2, 13, 14, 8, x\}$ is 8.

25. The frequency distribution of the verbal scores on the SAT test for students at Kennedy High School is shown below.

Scores	Number of Students	Scores	Number of Students
200–250	9	500–550	18
250–300	14	550–600	12
300–350	23	600–650	7
350–400	30	650–700	3
400–450	33	700–750	1
450–500	28	750–800	1

a. What is the mean of the verbal scores at Kennedy High School?

b. What is the median class of the frequency distribution?

c. Estimate the median of the verbal scores at Kennedy High School.

Applications and Problem Solving

26. **Weather** The growing season in Tennessee is the period from May to September. The table at the right shows the normal rainfall for those months.

a. Find the mean, median, and mode of this data.

b. Suppose Tennessee received heavy rain in May totaling 8.2 inches. If this figure were used for May, how would the measures of central tendency be affected?

c. If September were eliminated from the period, how would this affect the measures of central tendency?

Normal Rainfall for Tennessee (inches)	
May	4.8
June	3.6
July	3.9
August	3.6
September	3.7

27. **Critical Thinking** Find a set of numbers that satisfies each list of conditions.

a. The mean, median, and mode are all the same number.

b. The mean is greater than the median.

c. The mode is 10 and the median is greater than the mean.

d. The mean is 6, the median is $5\frac{1}{2}$, and the mode is 9.

28. **Government** As of 1999, the number of members in the House of Representatives for each state is given below.

AL	7	HI	2	MA	10	NM	3	SD	1
AK	1	ID	2	MI	16	NY	31	TN	9
AZ	6	IL	20	MN	8	NC	12	TX	30
AR	4	IN	10	MS	5	ND	1	UT	3
CA	52	IA	5	MO	9	OH	19	VT	1
CO	6	KS	4	MT	1	OK	6	VA	11
CT	6	KY	6	NE	3	OR	5	WA	9
DE	1	LA	7	NV	2	PA	21	WV	3
FL	23	ME	2	NH	2	RI	2	WI	9
GA	11	MD	8	NJ	13	SC	6	WY	1

a. Make a stem-and-leaf plot of the number of representatives.

b. Find the mean of the data.

c. What is the median of the data?

d. Find the mode of the data.

e. What is a representative average for the number of members in the House of Representatives per state? Explain.

Data Update For the latest information about the number of goals scored in hockey, visit **www.amc. glencoe.com**

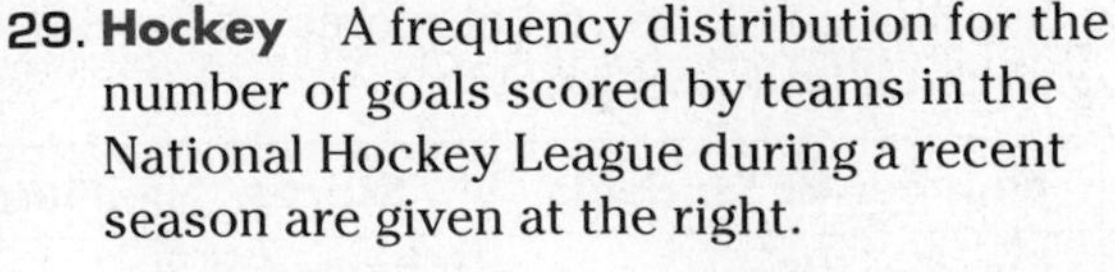

29. **Hockey** A frequency distribution for the number of goals scored by teams in the National Hockey League during a recent season are given at the right.

    a. Use the frequency chart to estimate the mean of the number of goals scored by a team.

    b. What is the median class of the frequency distribution?

    c. Use the frequency chart to estimate the median of the number of goals scored by a team.

National Hockey League Goals

Goals	Number of Teams
160–180	1
180–200	6
200–220	10
220–240	6
240–260	3
260–280	1

Source: National Hockey League

    d. The actual numbers of goals scored are listed below. Find the mean and median of the data.

    268, 248, 245, 242, 239, 239, 237, 236, 231, 230, 217, 215, 214, 211, 210, 210, 207, 205, 202, 200, 196, 194, 192, 190, 189, 184, 179

    e. How do the measures of central tendency found by using the frequency chart compare with the measures of central tendency found by using the actual data?

30. **Critical Thinking** A one-meter rod is suspended at its middle so that it balances. Suppose one-gram weights are hung on the rod at the following distances from one end.

    5 cm   20 cm   37 cm   44 cm   52 cm   68 cm   71 cm   85 cm

    The rod does not balance at the 50-centimeter mark.

    a. Where must a one-gram weight be hung so that the rod will balance at the 50-centimeter mark?

    b. Where must a two-gram weight be hung so that the rod will balance at the 50-centimeter mark?

31. **Salaries** The salaries of the ten employees at the XYZ Corporation are listed below.

    $54,000, $75,000, $55,000, $62,000, $226,000, $65,000, $59,000, $61,000, $162,000, $59,000

    a. What is the mean of the salaries?

    b. Find the median of the salaries.

    c. Find the mode of the salaries.

    d. What measure of central tendency might an employee use when asking for a raise?

    e. What measure of central tendency might management use to argue against a raise for an employee?

    f. What measure of central tendency do you think is most representative of the data? Why?

    g. Suppose you are an employee of the company making $75,000. Write a convincing argument that you deserve a raise.

32. **Education** The grade point averages for a graduating class are listed in the frequency table below.

Grade Point Averages	1.75–2.25	2.25–2.75	2.75–3.25	3.25–3.75	3.75–4.25
Frequency	12	15	31	37	5

a. What is the estimated mean of the data?

b. Estimate the median of the data.

33. **Basketball** Jackson High School just announced the members of its varsity basketball team for the year. Kwan, who is 5′ 9″ tall, is the only sophomore to make the team. The other basketball team members are 5′ 11″, 6′ 0″, 5′ 7″, 6′ 3″, 6′ 1″, 6′ 6″, 5′ 8″, 5′ 9″ and 6′ 2″. How does Kwan compare with the other team members?

Mixed Review

34. **Highway Safety** The maximum speed limits in miles per hour for interstate highways for the fifty states are given below. Construct a frequency polygon of the data. *(Lesson 14-1)*
70, 65, 75, 70, 70, 75, 65, 65, 70, 70, 55, 75, 65, 65, 65, 70, 65, 70, 65, 65, 65, 70, 70, 70, 70, 65, 75, 75, 65, 65, 75, 65, 70, 70, 65, 75, 65, 65, 65, 65, 75, 65, 70, 75, 65, 65, 70, 70, 65, 75

Source: National Motorists Association

35. Determine if the following event is *independent* or *dependent*. Then determine the probability. *(Lesson 13-4)*

the probability of randomly selecting two fitness magazines at one time from a basket containing 6 news magazines, 3 fitness magazines, and 2 sports magazines

36. Use the ratio test to determine if the series $\frac{1}{3} + \frac{2}{3^2} + \frac{3}{3^3} + \cdots + \frac{n}{3^n} + \cdots$ is *convergent* or *divergent*. *(Lesson 12-4)*

37. **Investments** An annuity pays 6%. What is the future value of the annuity if \$1500 is deposited into the account every 6 months for 10 years? *(Lesson 11-2)*

38. Graph the system of inequalities. *(Lesson 10-8)*

$3x + y^2 \leq 18$

$x^2 + y^2 \geq 9$

39. **SAT Practice Quantitative Comparison**

A if the quantity in Column A is greater

B if the quantity in Column B is greater

C if the quantities are equal

D if the relationship cannot be determined for the information given

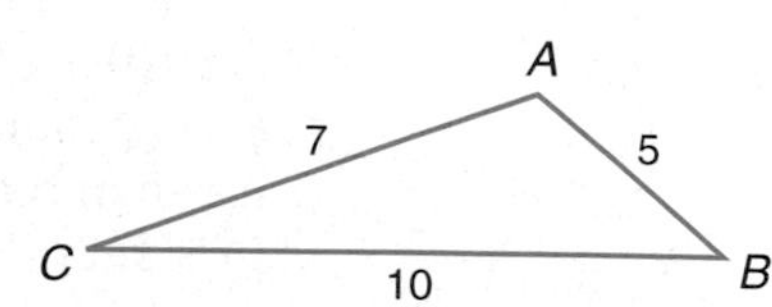

Column A	Column B
Perimeter of Triangle *ABC*	Area of Triangle *ABC*

Extra Practice See p. A53.

Measures of Variability

OBJECTIVES

- Find the interquartile range, the semi-interquartile range, mean deviation, and standard deviation of a set of data.
- Organize and compare data using box-and-whisker plots.

EDUCATION Are you planning to attend college? If so, do you know which school you are going to attend? There are several factors influencing students' decisions concerning which college to attend. Two of those factors may be the cost of tuition and the size of the school. The table lists some of the largest colleges with their total enrollment and cost for in-state tuition and fees.

College Enrollment and Tuition

College	Enrollment, 1997-1998	Tuition and Fees ($), 1997-1998
University of Texas	47,476	2866
The Ohio State University	45,462	3687
Penn State University	37,718	5832
University of Georgia	29,693	2838
Florida State University	28,285	1988
University of Southern California	27,874	20,480
Virginia Tech	24,481	4147
North Carolina State University	24,141	2232
Texas Tech University	24,075	2414
University of South Carolina	22,836	3534
University of Nebraska	22,393	2769
Colorado State University	21,970	2933
University of Illinois	21,645	4364
Auburn University (AL)	21,498	2610
University of Kentucky	20,925	2736
Kansas State University	20,325	2467
University of Oklahoma	19,886	2311
Cornell University (NY)	18,001	21,914
University of Alaska	17,090	2294

Source: College Entrance Examination Board

Data Update For the latest information about college enrollment and tuition, visit **www.amc.glencoe.com**

You will solve problems related to this in Examples 1-4.

Measures of central tendency, such as the mean, median, and mode, are statistics that describe certain important characteristics of data. However, they do not indicate anything about the variability of the data. For example, 50 is the mean of both {0, 50, 100} and {40, 50, 60}. The variability is much greater in the first set of data than in the second, since $100 - 0$ is much greater than $60 - 40$.

One **measure of variability** is the *range*. Use the information in the table above to find the range of enrollment.

$$47{,}476 - 17{,}090 = 30{,}386.$$

University of Texas (47,476) — University of Alaska (17,090)

The range of enrollment is 30,386 students.

If the median is a member of the set of data, that item of data is excluded when calculating the first and third quartile points.

If the data have been arranged in order and the median is found, the set of data is divided into two groups. Then if the median of each group is found, the data is divided into four groups. Each of these groups is called a **quartile.** There are three quartile points, Q_1, Q_2, and Q_3, that denote the breaks in the data for each quartile. The median is the second quartile point Q_2. The medians of the two groups defined by the median are the first quartile point Q_1 and the third quartile point Q_3.

One fourth of the data is less than the first quartile point Q_1, and three fourths of the data is less than the third quartile point Q_3. The difference between the first quartile point and third quartile point is called the **interquartile range.** When the interquartile range is divided by 2, the quotient is called the **semi-interquartile range.**

Semi-Interquartile Range

If a set of data has first quartile point Q_1 and third quartile point Q_3, the semi-interquartile range Q_R can be found as follows.

$$Q_R = \frac{Q_3 - Q_1}{2}$$

Example 1 **EDUCATION** **Refer to the application at the beginning of the lesson.**

a. Find the interquartile range of the college enrollments and state what it represents.

b. Find the semi-interquartile range of the college enrollments.

a. First, order the data from least to greatest, and identify Q_1, Q_2, and Q_3.

17,090 18,001 19,886 20,325 $\underset{}{\overset{Q_1}{\underline{20{,}925}}}$ 21,498 21,645 21,970 22,393 $\overset{Q_2}{\underline{22{,}836}}$

24,075 24,141 24,481 27,874 $\overset{Q_3}{\underline{28{,}285}}$ 29,693 37,718 45,462 47,476

The interquartile range is 28,285 − 20,925 or 7360. This means that the middle half of the student enrollments are between 28,285 and 20,925 and are within 7360 of each other.

b. The semi-interquartile range is $\frac{7360}{2}$ or 3680. The halfway point between Q_1 and Q_3 can be found by adding the semi-interquartile range to Q_1. That is, 3680 + 20,925 or 24,605. Since $24{,}605 > Q_2$, this indicates the data is more clustered between Q_1 and Q_2 than between Q_2 and Q_3.

Graphing Calculator Tip

Enter the data into **L1** and use the **SortA(** command to reorder the list from least to greatest.

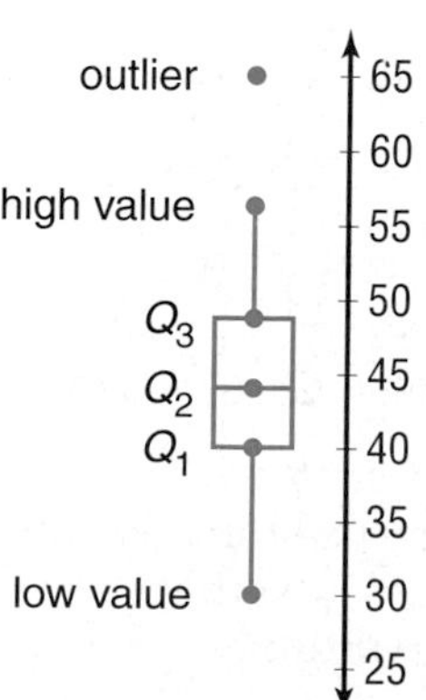

Box-and-whisker plots are used to summarize data and to illustrate the variability of the data. These plots graphically display the median, quartiles, interquartile range, and extreme values in a set of data. They can be drawn vertically, as shown at the right, or horizontally. A box-and-whisker plot consists of a rectangular box with the ends, or **hinges,** located at the first and third quartiles. The segments extending from the ends of the box are called **whiskers.** The whiskers stop at the extreme values of the set, unless the set contains **outliers.** Outliers are extreme values that are more than 1.5 times the interquartile range beyond the upper or lower quartiles. Outliers are represented by single points. If an outlier exists, each whisker is extended to the last value of the data that is not an outlier.

The dimensions of the box-and-whisker plot can help you characterize the data. Each whisker and each small box contains 25% of the data. If the whisker or box is short, the data are concentrated over a narrower range of values. The longer the whisker or box, the larger the range of the data in that quartile. Thus, the box-and-whisker is a pictorial representation of the variability of the data.

Example 2 **EDUCATION** **Refer to the application at the beginning of the lesson. Draw a box-and-whisker plot for the enrollments.**

In Example 1, you found that Q_1 is 20,925, Q_2 is 22,836, and Q_3 is 28,285. The extreme values are the least value 17,090 and the greatest value 47,476.

Draw a number line and plot the quartiles, the median, and the extreme values. Draw a box to show the interquartile range. Draw a segment through the median to divide the box into two smaller boxes.

15,000 20,000 25,000 30,000 35,000 40,000 45,000 50,000

Before drawing the whiskers, determine if there are any outliers. From Example 1, we know that the interquartile range is 7360. An outlier is any value that lies more than 1.5(7360) or 11,040 units below Q_1 or above Q_3.

$Q_1 - 1.5(7360) = 20{,}925 - 11{,}040 = 9885$

$Q_3 + 1.5(7360) = 28{,}285 + 11{,}040 = 39{,}325$

The lower extreme 17,090 is within the limits. However, 47,476 and 45,462 are not within the limits. They are outliers. Graph these points on the plot. Then draw the left whisker from 17,090 to 20,925 and the right whisker from 28,285 to the greatest value that is not an outlier, 37,718.

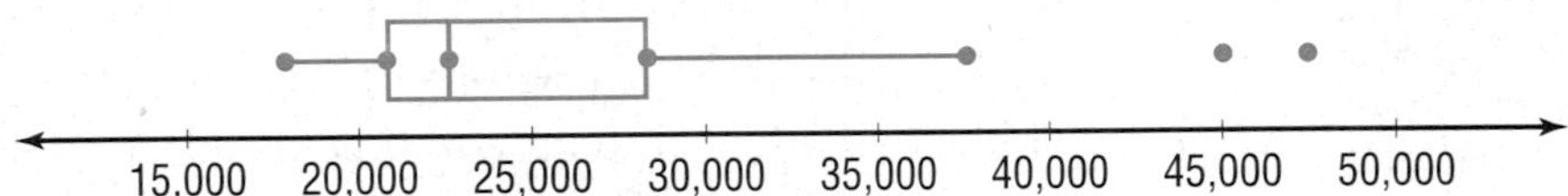

The box-and-whisker plot shows that the two lower quartiles of data are fairly concentrated. However, the upper quartile of data is more diverse.

Another measure of variability can be found by examining deviation from the mean, symbolized by $X_i - \overline{X}$. The sum of the deviations from the mean is zero. That is, $\sum_{i=1}^{n} (X_i - \overline{X}) = 0$. For example, the mean of the data set {14, 16, 17, 20, 33} is 20. The sum of the deviations from the mean is shown in the table.

X_i	$\overline{X}$	$X_i - \overline{X}$
14	20	−6
16	20	−4
17	20	−3
20	20	0
33	20	13
$\sum_{i=1}^{5} (X_i - \overline{X})$		0

To indicate how far individual items vary from the mean, we use the absolute values of the deviation. The arithmetic mean of the absolute values of the deviations from the mean of a set of data is called the **mean deviation,** symbolized by *MD*.

Mean Deviation

If a set of data has n values given by X_i, such that $1 \leq i \leq n$, with arithmetic mean $\overline{X}$, then the mean deviation MD can be found as follows.

$$MD = \frac{1}{n}\sum_{i=1}^{n} |X_i - \overline{X}|$$

In sigma notation for statistical data, i is always an integer and not the imaginary unit.

Example 3 

EDUCATION **Refer to the application at the beginning of the lesson. Find the mean deviation of the enrollments.**

There are 19 college enrollments listed, and the mean is $\frac{1}{19}\sum_{i=1}^{19} X_i$ or about 26,093.37.

Method 1: Sigma notation

$$MD \approx \frac{1}{19}\sum_{i=1}^{19} |X_i - 26{,}093.37|$$

$$MD \approx \frac{1}{19}(|47{,}476 - 26{,}093.37| + |45{,}462 - 26{,}093.37| + \cdots + |17{,}090 - 26{,}093.37|)$$

$$MD \approx \frac{1}{19}(|21{,}382.63| + |19{,}368.63| + \cdots + |-9003.37|)$$

$$MD \approx 6310.29$$

The mean deviation of the enrollments is about 6310.29. This means that the enrollments are an average of about 6310.29 above or below the mean enrollment of 26,093.37.

Method 2: Graphing Calculator

Enter the data for the enrollments into **L1**. At the home screen, enter the following formula.

sum(abs(L1 − 26093.37))/19

The calculator determines the difference between the scores and the mean, takes the absolute value, adds the absolute values of the differences, and divides by 19. This verifies the calculation in Method 1.

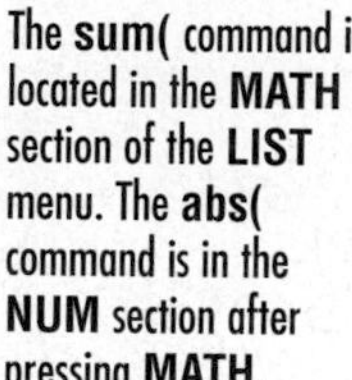

Graphing Calculator Tip

The **sum(** command is located in the **MATH** section of the **LIST** menu. The **abs(** command is in the **NUM** section after pressing **MATH.**

A measure of variability that is often associated with the arithmetic mean is the **standard deviation.** Like the mean deviation, the standard deviation is a measure of the average amount by which individual items of data deviate from the arithmetic mean of all the data. Each individual deviation can be found by subtracting the arithmetic mean from each individual value, $X_i - \overline{X}$. Some of these differences will be negative, but if they are squared, the results are positive. The standard deviation is the square root of the mean of the squares of the deviation from the arithmetic mean.

Standard Deviation

If a set of data has n values, given by X_i such that $1 \leq i \leq n$, with arithmetic mean $\overline{X}$, the standard deviation σ can be found as follows.

$$\sigma = \sqrt{\frac{1}{n}\sum_{i=1}^{n}(X_i - \overline{X})^2}$$

σ is the lowercase Greek letter sigma.

The standard deviation is the most important and widely used measure of variability. Another statistic used to describe the spread of data about the mean is **variance.** The variance, denoted σ^2, is the mean of the squares of the deviations from $\overline{X}$. The standard deviation is the positive square root of the variance.

Example 4 **EDUCATION** **Refer to the application at the beginning of the lesson. Find the standard deviation of the enrollments.**

Method 1: Standard Deviation Formula

There are 19 college enrollments listed, and the mean is about 26,093.37.

$$\sigma \approx \sqrt{\frac{1}{19}\sum_{i=1}^{19}(X_i - 26{,}093.37)^2}$$

$$\sigma \approx \sqrt{\frac{1}{19}(47{,}476 - 26{,}093.37)^2 + (45{,}462 - 26{,}093.37)^2 + \cdots + (17{,}090 - 26{,}093.37)^2}$$

$$\sigma \approx \sqrt{\frac{1}{19}(21{,}382.63)^2 + (19{,}368.63)^2 + \cdots + (-9003.37)^2}$$

$$\sigma \approx 8354.59$$

The standard deviation is about 8354.59. Since the mean of the enrollments is about 26,093.37 and the standard deviation is about 8354.59, the data have a great amount of variability.

Method 2: Graphing Calculator

Enter the data in **L1**. Use the **CALC** menu after pressing STAT to find the 1-variable statistics.

The standard deviation, indicated by **σx**, is the fifth statistic listed.

The mean ($\bar{x}$) is 26,093.36842 and the standard deviation is 8354.5913383, which agree with the calculations using the formulas.

When studying the standard deviation of a set of data, it is important to consider the mean. For example, compare a standard deviation of 5 with a mean of 10 to a standard deviation of 5 with a mean of 1000. The latter indicates very little variation, while the former indicates a great deal of variation since 5 is 50% of 10 while 5 is only 0.5% of 1000.

The standard deviation of a frequency distribution is the square root of the mean of the squares of the deviations of the class marks from the mean of the frequency data, weighted by the frequency of each interval.

Standard Deviation of the Data in a Frequency Distribution

If $X_1, X_2, \ldots, X_k$ are the class marks in a frequency distribution with k classes, and $f_1, f_2, \ldots, f_k$ are the corresponding frequencies, then the standard deviation σ of the data in the frequency distribution is found as follows.

$$\sqrt{\frac{\sum_{i=1}^{k}(X_i - \overline{X})^2 \cdot f_i}{\sum_{i=1}^{k} f_i}}$$

The standard deviation of a frequency distribution is an approximate number.

Example **5** **ECONOMICS** **Use the frequency distribution data below to find the arithmetic mean and the standard deviation of the price-earnings ratios of 100 manufacturing stocks.**

Method 1: Using Formulas

Class Limits	Class Marks (X)	f	$f \cdot X$	$(X - \overline{X})$	$(X - \overline{X})^2$	$(X - \overline{X})^2 \cdot f$
−0.5–4.5	2.0	5	10	−8	64	320
4.5–9.5	7.0	54	378	−3	9	486
9.5–14.5	12.0	25	300	2	4	100
14.5–19.5	17.0	13	221	7	49	637
19.5–24.5	22.0	0	0	12	144	0
24.5–29.5	27.0	1	27	17	289	289
29.5–34.5	32.0	2	64	22	484	968
		100	1000			2800

The mean $\overline{X}$ is $\frac{1000}{100}$ or 10.

The standard deviation σ is $\sqrt{\frac{2800}{100}}$ or approximately 5.29.

Since the mean number of price-earnings ratios is 10 and the standard deviation is 5.29, this indicates a great amount of variability in the data.

Method 2: Graphing Calculator

Enter the class marks in the **L1** list and the frequency in the **L2** list.

Use the **CALC** menu after pressing STAT to find the 1-variable statistics. Then type **L1**, **L2** and press ENTER.

The calculator confirms the standard deviation is about 5.29.

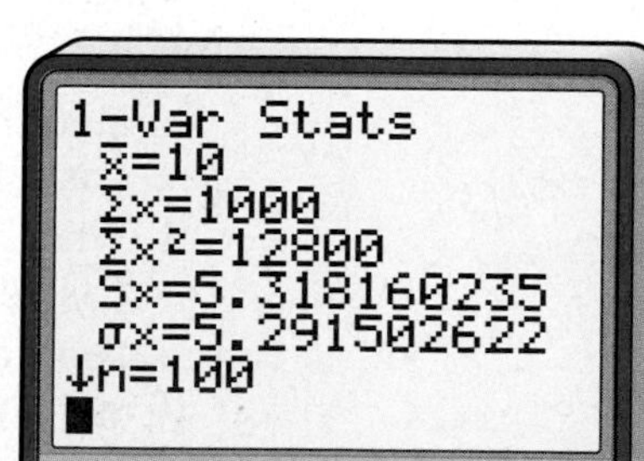

CHECK FOR UNDERSTANDING

Communicating Mathematics

Read and study the lesson to answer each question.

1. **Describe** the data shown in the box-and-whisker plot below. Include the quartile points, interquartile range, semi-interquartile range, and any outliers.

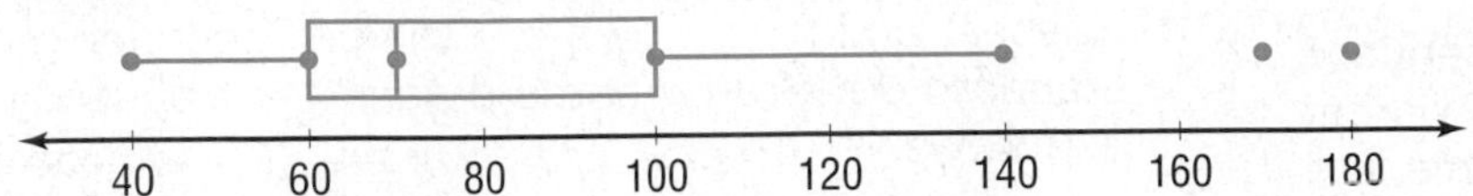

2. **Explain** how to find the variance of a set of data if you know the standard deviation.
3. **Compare and contrast** mean deviation and standard deviation.
4. *Math Journal* **Draw** a box-and-whisker plot for data you found in a newspaper or magazine. What conclusions can you derive from the plot?

Guided Practice

5. Find the interquartile range and the semi-interquartile range of {17, 28, 44, 37, 28, 42, 21, 41, 35, 25}. Then draw a box-and-whisker plot.
6. Find the mean deviation and the standard deviation of {\$4.45, \$5.50, \$5.50, \$6.30, \$7.80, \$11.00, \$12.20, \$17.20}
7. Find the arithmetic mean and the standard deviation of the frequency distribution at the right.

Class Limits	Frequency
0–10,000	15
10,000–20,000	30
20,000–30,000	50
30,000–40,000	60
40,000–50,000	30
50,000–60,000	15

8. **Meteorology** The following table gives the normal maximum daily temperature for Los Angeles and Las Vegas.

Normal Maximum Daily Temperatures

	January	February	March	April	May	June
Los Angeles	65.7	65.9	65.5	67.4	69.0	71.9
Las Vegas	57.3	63.3	68.8	77.5	87.8	100.3

	July	August	September	October	November	December
Los Angeles	75.3	76.6	76.6	74.4	70.3	65.9
Las Vegas	105.9	103.2	94.7	82.1	67.4	57.5

Source: National Oceanic and Atmosphere Administration

a. Find the mean, median, and standard deviation for the temperatures in Los Angeles.
b. What are the mean, median, and standard deviation for the temperatures in Las Vegas?
c. Draw a box-and-whisker plot for the temperatures for each city.
d. Which city has a smaller variability in temperature?
e. What might cause one city to have a greater variability in temperature than another?

EXERCISES

Practice

Find the interquartile range and the semi-interquartile range of each set of data. Then draw a box-and-whisker plot.

9. {30, 28, 24, 24, 22, 22, 21, 17, 16, 15}

10. {7, 14, 18, 72, 13, 15, 19, 8, 17, 28, 11, 15, 24}

11. {15.1, 9.0, 8.5, 5.8, 6.2, 8.5, 10.5, 11.5, 8.8, 7.6}

12. Use a graphing calculator to draw a box-and-whisker plot for {7, 1, 11, 5, 4, 8, 12, 15, 9, 6, 5, 9}?

Find the mean deviation and the standard deviation of each set of data.

13. {200, 476, 721, 579, 152, 158}

14. {5.7, 5.7, 5.6, 5.5, 5.3, 4.9, 4.4, 4.0, 4.0, 3.8}

15. {369, 398, 381, 392, 406, 413, 376, 454, 420, 385, 402, 446}

16. Find the variance of {34, 55, 91, 13, 22}.

Find the arithmetic mean and the standard deviation of each frequency distribution.

17.

Class Limits	Frequency
1–5	2
5–9	8
9–13	15
13–17	6
17–21	38
21–25	31
25–29	13
29–33	7

18.

Class Limits	Frequency
53–61	3
61–69	7
69–77	11
77–85	38
85–93	19
93–101	12

19.

Class Limits	Frequency
70–90	2
90–110	11
110–130	39
130–150	17
150–170	9
170–190	7

Applications and Problem Solving

20. Geography There are seven navigable rivers that feed into the Ohio River. The lengths of these rivers are given at the right.

a. Find the median of the lengths.

b. Name the first quartile point and the third quartile point.

c. Find the interquartile range.

d. What is the semi-interquartile range?

e. Are there any outliers? If so, name them.

f. Make a box-and-whisker plot of the lengths of the rivers.

g. Use the box-and-whisker plot to discuss the variability of the data.

Length of Rivers Feeding into the Ohio River

River	Length
Monongahela	129 miles
Allegheny	325 miles
Kanawha	97 miles
Kentucky	259 miles
Green	360 miles
Cumberland	694 miles
Tennessee	169 miles

Source: *The Universal Almanac*

21. Critical Thinking Write a set of numerical data that could be represented by the box-and-whisker plot at the right.

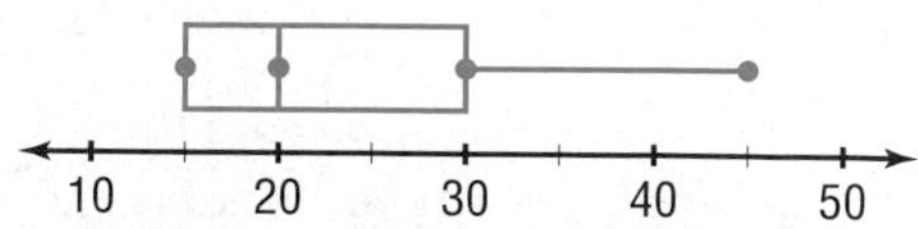

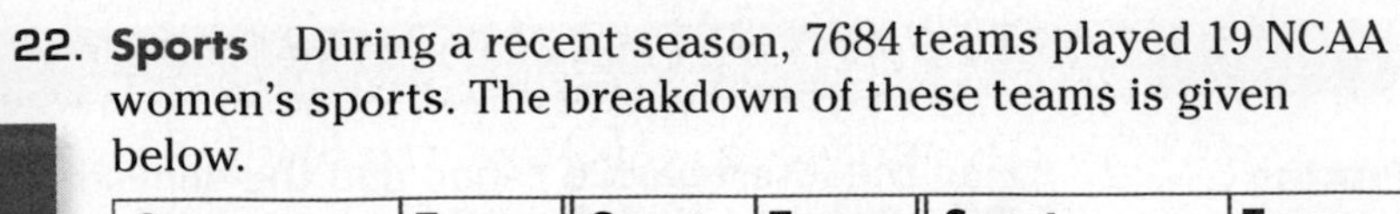

22. **Sports** During a recent season, 7684 teams played 19 NCAA women's sports. The breakdown of these teams is given below.

Sport	Teams	Sport	Teams	Sport	Teams
Basketball	966	Lacrosse	182	Swimming	432
Cross Country	838	Rowing	97	Tennis	859
Fencing	42	Skiing	40	Track, Indoor	528
Field Hockey	228	Soccer	691	Track, Outdoor	644
Golf	282	Softball	770	Volleyball	923
Gymnastics	91	Squash	26	Water Polo	23
Ice Hockey	22				

Source: The National Collegiate Athletic Association

a. What is the median of the number of women's teams playing a sport?
b. Find the first quartile point and the third quartile point.
c. What is the interquartile range and semi-interquartile range?
d. Are there any outliers? If so, name them.
e. Make a box-and-whisker of the number of women's teams playing a sport.
f. What is the mean of the number of women's teams playing a sport?
g. Find the mean deviation of the data.
h. Find the variance of the data.
i. What is the standard deviation of the data?
j. Discuss the variability of the data.

23. **Education** Refer to the data on the college tuition and fees in the application at the beginning of the lesson.

a. What are the quartile points of the data?
b. Find the interquartile range.
c. Name any outliers.
d. Make a box-and-whisker plot of the data.
e. What is the mean deviation of the data?
f. Find the standard deviation of the data.
g. Discuss the variability of the data.

24. **Government** The number of times the first 42 presidents vetoed bills are listed below.

2, 0, 0, 7, 1, 0, 12, 1, 0, 10, 3, 0, 0, 9, 7, 6, 29, 93, 13, 0, 12, 414, 44, 170, 42, 82, 39, 44, 6, 50, 37, 635, 250, 181, 21, 30, 43, 66, 31, 78, 44, 25

a. Make a box-and-whisker plot of the number of vetoes.
b. Find the mean deviation of the data.
c. What is the variance of the data?
d. What is the standard deviation of the data?
e. Describe the variability of the data.

25. **Entertainment** The frequency distribution shows the average audience rating for the top fifty network television shows for one season.

Audience Rating	8–10	10–12	12–14	14–16	16–18	18–20	20–22
Frequency	26	12	6	2	2	0	2

Source: Nielsen Media Research

a. Find the arithmetic mean of the audience ratings.
b. What is the standard deviation of the audience ratings?

26. Critical Thinking Is it possible for the variance to be less than the standard deviation for a set of data? If so, explain when this will occur. When would the variance be equal to the standard deviation for a set of data?

27. Research Find the number of students attending each school in your county. Make a box-and-whisker plot of the data. Determine various measures of variability and discuss the variability of the data.

Mixed Review

28. Consider the data represented by the stem-and-leaf plot at the right. *(Lesson 14-2)*

a. What is the mean of the data?

b. Find the median of the data.

c. What is the mode of the data?

stem	leaf
4	4 4 9
5	4 5
6	2 2 4 5 9
7	1 4 5 6 7 8 9
8	0 2 4 5 6 7 8 9 9 9
9	0 2 3 3 5 6 8 9

5|4 = 5.4

29. Fund-Raising Twelve students are selling programs at the Grove City High School to raise money for the athletic department. The numbers of programs sold by each student are listed below. *(Lesson 14-1)*

51, 27, 55, 54, 68, 60, 39, 46, 46, 53, 57, 23

a. Find the range of the number of programs sold.

b. Determine an appropriate class interval.

c. What are the class limits?

d. Construct a frequency distribution of the data.

e. Draw a histogram of the data.

30. Food Service Suppose nine salad toppings are placed on a circular, revolving tray. How many ways can the salad items be arranged? *(Lesson 13-2)*

31. Find the first three iterates of the function $f(x) = 0.5x - 1$ using $x_0 = 8$. *(Lesson 12-8)*

32. SAT/ACT Practice A carpenter divides a board that is 7 feet 9 inches long into three equal parts. What is the length of each part?

A 2 ft $6\frac{1}{3}$ in. **B** 2 ft $8\frac{1}{3}$ in. **C** 2 ft 7 in.

D 2 ft 8 in. **E** 2 ft 9 in.

MID-CHAPTER QUIZ

The scores for an exam given in physics class are given below.

82, 77, 84, 98, 93, 71, 76, 64, 89, 95, 78, 89, 65, 88, 54, 96, 87, 92, 80, 85, 93, 89, 55, 62, 79, 90, 86, 75, 99, 62

1. What is an appropriate class interval for the test scores? (Lesson 14-1)
2. Construct a frequency distribution of the test scores. (Lesson 14-1)
3. Draw a histogram of the test scores. (Lesson 14-1)
4. Make a stem-and-leaf plot of the test scores. (Lesson 14-2)
5. What is the mean of the test scores? (Lesson 14-2)
6. Find the median of the test scores. (Lesson 14-2)
7. Find the mode of the test scores. (Lesson 14-2)
8. Make a box-and-whisker plot of the test scores. (Lesson 14-3)
9. What is the mean deviation of the test scores? (Lesson 14-3)
10. Discuss the variability of the data. (Lesson 14-3)

Extra Practice See p. A54.

The Normal Distribution

OBJECTIVES
- Use the normal distribution curve.

TESTING The class of 1996 was the first class to take the adjusted Scholastic Assessment Test. The test was adjusted so that the median of the scores for the verbal section and the math section would be 500. For each section, the lowest score is 200 and the highest is 800. Suppose the verbal and math scores follow the *normal distribution.* What percent of the students taking the test would have a math score between 375 and 625? *This problem will be solved in Example 4.*

A frequency polygon displays a limited number of data and may not represent an entire population. To display the frequency of an entire population, a smooth curve is used rather than a polygon.

If the curve is symmetric, then information about the measures of central tendency can be gathered from the graph. Study the graphs below.

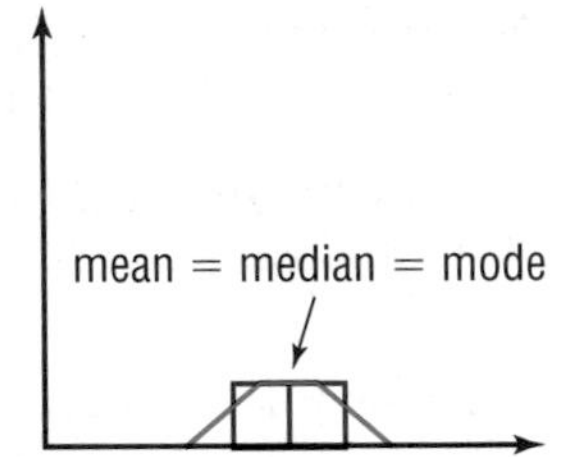

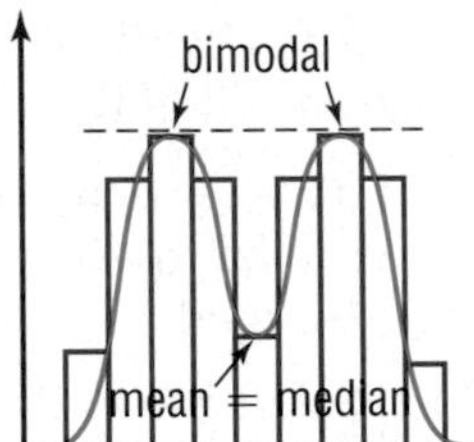

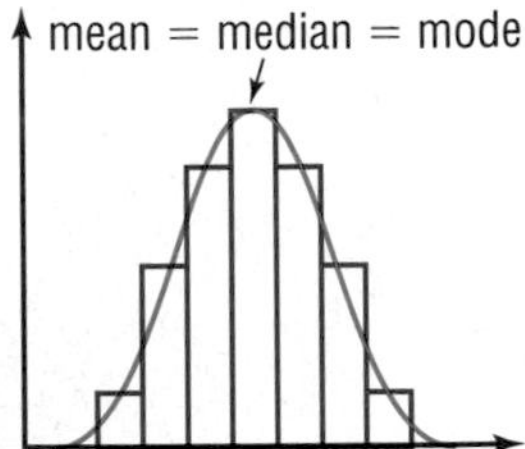

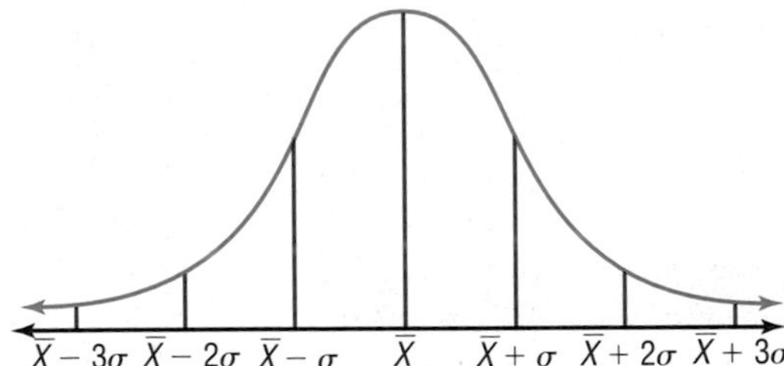

A **normal distribution** is a frequency distribution that often occurs when there is a large number of values in a set of data. The graph of this distribution is a symmetric, bell-shaped curve, shown at the left. This is known as a **normal curve.** The shape of the curve indicates that the frequencies in a normal distribution are concentrated around the center portion of the distribution. A small portion of the population occurs at the extreme values.

In a normal distribution, small deviations are much more frequent than large ones. Negative deviations and positive deviations occur with the same frequency. The points on the horizontal axis represent values that are a certain number of standard deviations from the mean $\overline{X}$. In the curve shown above, each interval represents one standard deviation. So, the section from $\overline{X}$ to $\overline{X} + \sigma$ represents those values between the mean and one standard deviation greater than the mean, the section from $\overline{X} + \sigma$ to $\overline{X} + 2\sigma$ represents the interval one standard deviation greater than the mean to two standard deviations greater than the mean, and so on. The total area under the normal curve and above the horizontal axis represents the total probability of the distribution, which is 1.

Example 1 **MEDICINE** **The average healing time of a certain type of incision is 240 hours with a standard deviation of 20 hours. Sketch a normal curve that represents the frequency of healing times.**

First, find the values defined by the standard deviation in a normal distribution.

$\overline{X} - 1\sigma = 240 - 1(20)$ or 220
$\overline{X} - 2\sigma = 240 - 2(20)$ or 200
$\overline{X} - 3\sigma = 240 - 3(20)$ or 180

$\overline{X} + 1\sigma = 240 + 1(20)$ or 260
$\overline{X} + 2\sigma = 240 + 2(20)$ or 280
$\overline{X} + 3\sigma = 240 + 3(20)$ or 300

Sketch the general shape of a normal curve. Then, replace the horizontal scale with the values you have calculated.

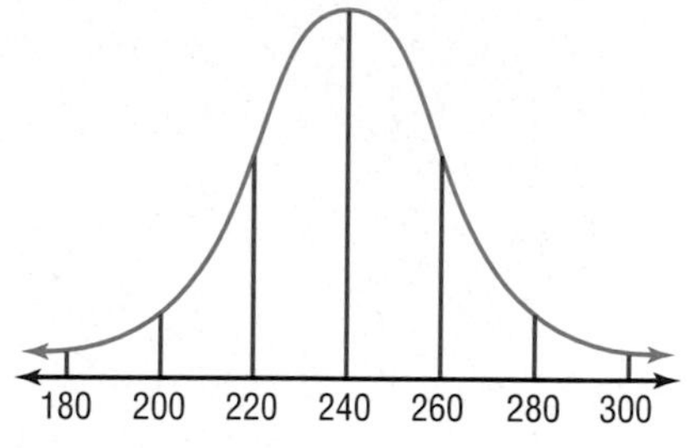

The tables below give the fractional parts of a normally distributed set of data for selected areas about the mean. The letter t represents the number of standard deviations from the mean (that is, $\overline{X} \pm t\sigma$). When $t = 1$, t represents 1 standard deviation above and below the mean.

P represents the fractional part of the data that lies in the interval $\overline{X} \pm t\sigma$. The percent of the data within these limits is $100P$.

t	P
0.0	0.000
0.1	0.080
0.2	0.159
0.3	0.236
0.4	0.311
0.5	0.383
0.6	0.451
0.7	0.516
0.8	0.576

t	P
0.9	0.632
1.0	0.683
1.1	0.729
1.2	0.770
1.3	0.807
1.4	0.838
1.5	0.866
1.6	0.891
1.65	0.900

t	P
1.7	0.911
1.8	0.929
1.9	0.943
1.96	0.950
2.0	0.955
2.1	0.964
2.2	0.972
2.3	0.979
2.4	0.984

t	P
2.5	0.988
2.58	0.990
2.6	0.991
2.7	0.993
2.8	0.995
2.9	0.996
3.0	0.997
3.5	0.9995
4.0	0.9999

The P value also corresponds to the probability that a randomly selected member of the sample lies within t standard deviation units of the mean. For example, suppose the mean of a set of data is 85 and the standard deviation is 5.

Boundaries:

$$\overline{X} - t\sigma \text{ to } \overline{X} + t\sigma$$
$$85 - t(5) \text{ to } 85 + t(5)$$
$$85 - 1(5) \text{ to } 85 + 1(5)$$
$$80 \text{ to } 90$$

68.3% of the values in this set of data lie within one standard deviation of 85; that is, between 80 and 90.

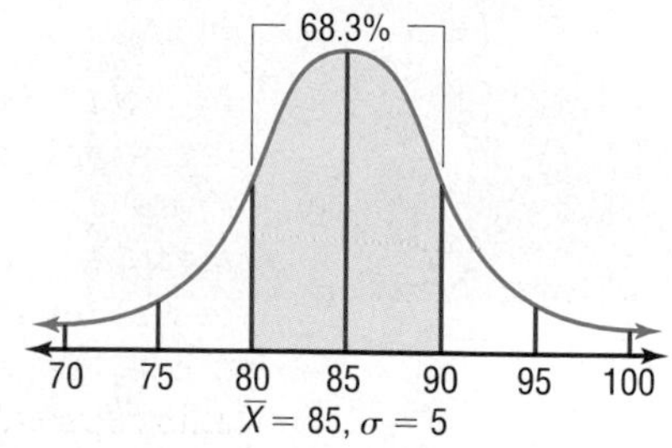

If you randomly select one item from the sample, the probability that the one you pick will be between 80 and 90 is 0.683. If you repeat the process 1000 times, approximately 68.3% (about 683) of those selected will be between 80 and 90.

Thus, normal distributions have the following properties.

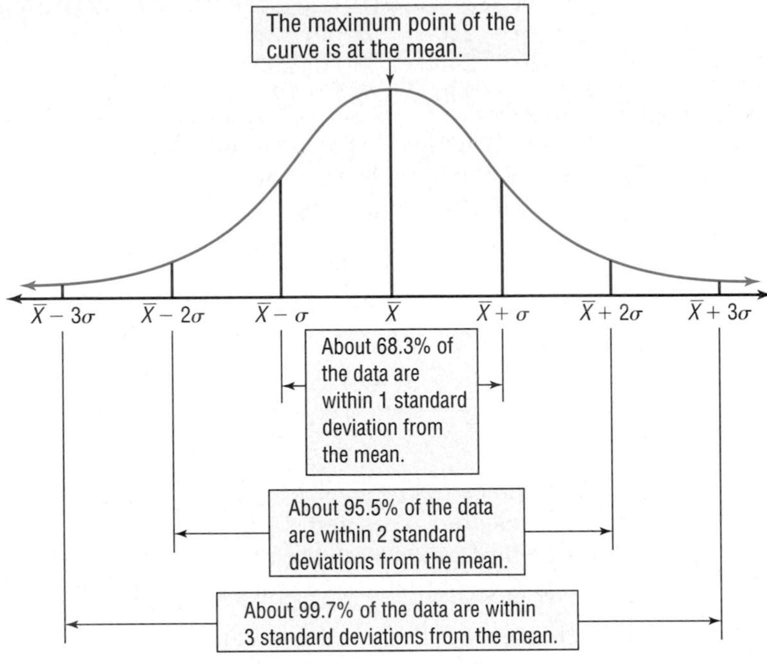

Example 2 **MEDICINE** **Refer to Example 1. Suppose a hospital has treated 2000 patients in the past five years having this type of incision. Estimate how many patients healed in each of the following intervals.**

a. 220–260 hours

The interval 220–260 hours represents $\overline{X} \pm 1\sigma$, which represents a probability of 68.3%.

$$68.3\%(2000) = 1366$$

Approximately 1366 patients took between 220 and 260 hours to heal.

b. 200–280 hours

The interval 200–280 hours represents $\overline{X} \pm 2\sigma$, which represents a probability of 95.5%.

$$95.5\%(2000) = 1910$$

Approximately 1910 patients took between 200 and 280 hours to heal.

c. 180–300 hours

The interval 180–300 hours represents $\overline{X} \pm 3\sigma$, which represents a probability of 99.7%.

$$99.7\%(2000) = 1994$$

Approximately 1994 patients took between 180 and 300 hours to heal.

If you know the mean and the standard deviation, you can find a range of values for a given probability.

Example 3 **Find the upper and lower limits of an interval about the mean within which 45% of the values of a set of normally distributed data can be found if $\overline{X} = 110$ and $\sigma = 15$.**

Use the table on page 919 to find the value of t that most closely approximates $P = 0.45$. For $t = 0.6$, $P = 0.451$. Choose $t = 0.6$. Now find the limits.

$$\overline{X} \pm t\sigma = 110 \pm 0.6(15) \quad \textit{\overline{X} = 110, t = 0.6, \sigma = 15}$$

$$= 101 \text{ and } 119$$

The interval in which 45% of the data lies is 101–119.

If you know the mean and standard deviation, you can also find the percent of the data that lies within a given range of values.

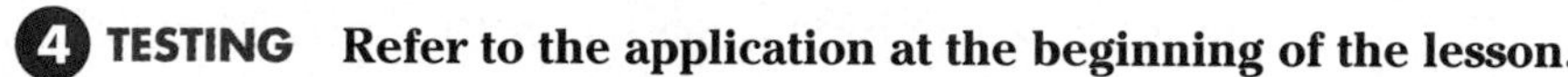

Example 4 **TESTING** **Refer to the application at the beginning of the lesson.**

a. Determine the standard deviation.

b. What percent of the students taking the test would have a math score between 375 and 625?

c. What is the probability that a senior chosen at random has a math score between 550 and 650?

a. For a normal distribution, both the median and the mean are 500. All of the scores are between 200 and 800. Therefore, all of the scores must be within 300 points from the mean. In a normal distribution, 0.9999 of the data is within 4 standard deviations of the mean. If the scores are to be a normal distribution, the standard deviation should be $\frac{300}{4}$ or 75.

b. Write each of the limits in terms of the mean.

$$375 = 500 - 125 \text{ and } 625 = 500 + 125$$

Therefore, $\overline{X} \pm t\sigma = 500 \pm 125$ and $t\sigma = 125$. Solve for t.

$$t\sigma = 125$$
$$t(75) = 125 \quad \sigma = 75$$
$$t \approx 1.7$$

If $t = 1.7$, then $P = 0.911$. *Use the table on page 919.*

About 91.1% of the students taking the test would have a math score between 375 and 625.

c. The graph shows that 550–650 does not define an interval that can be represented by $\overline{X} \pm t\sigma$. However, the interval can be defined as the difference between the intervals 500–650 and 500–550.

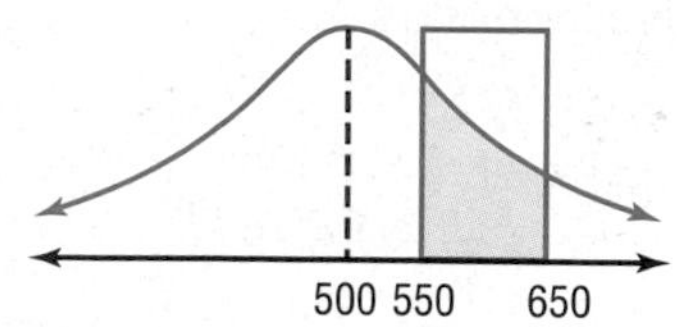

(continued on the next page)

First, find the probability that the score is between the mean 500 and the upper limit 650.

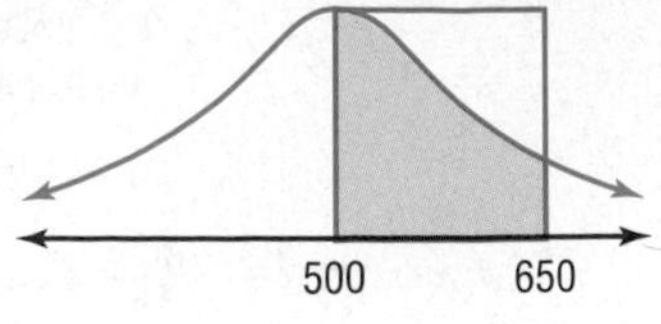

$$\bar{X} + t\sigma = 650$$
$$500 + t(75) = 650$$
$$t = 2$$

The value of P that corresponds to $t = 2$ is 0.955.

$P = 0.955$ describes the probability that a student's score falls $\pm 2(75)$ points about the mean, or between 350 and 650, but we are only considering half that interval. So, the probability that a student's score is between 500 and 650 is $\frac{1}{2}(0.955)$ or about 0.478.

Next, find the probability that a score is between the mean and the lower limit 550.

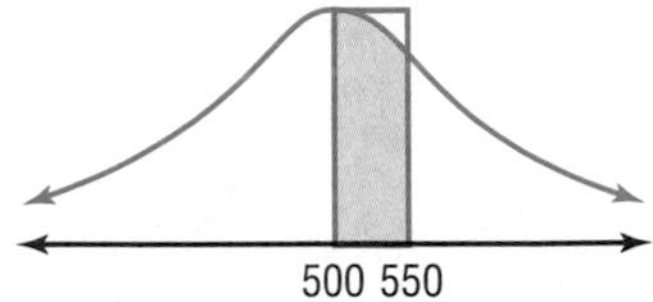

$$\bar{X} + t\sigma = 550$$
$$500 + t(75) = 550$$
$$t \approx 0.7$$

For $t = 0.7$, $P = 0.516$.

Likewise, we will only consider half of this probability or 0.258.

Now find the probability that a student's score falls in the interval 550–650.

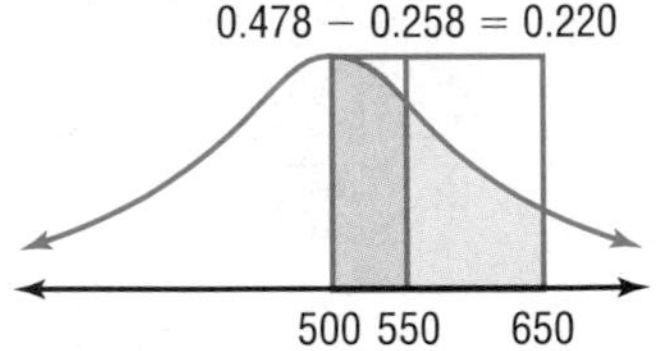

$$P(550\text{–}650) = P(500\text{–}650) - P(500\text{–}550)$$
$$P \approx 0.478 - 0.258$$
$$P \approx 0.220 \text{ or } 22\%$$

The probability that a student's score is between 550 and 650 is about 22%.

Students who take the SAT or ACT tests will receive a score as well as a **percentile.** The percentile indicates how the student's score compares with other students taking the test.

Percentile

The nth percentile of a set of data is the value in the set such that n percent of the data is less than or equal to that value.

Therefore if a student scores in the 65th percentile, this means that 65% of the students taking the test scored the same or less than that student.

CHECK FOR UNDERSTANDING

Communicating Mathematics

Read and study the lesson to answer each question.

1. **Compare** the median, mean, and mode of a set of normally distributed data.
2. **Write** an expression for the interval that is within 1.5 standard deviations from the mean.
3. **Sketch** a normal curve with a mean of 75 and a standard deviation of 10 and a normal curve with a mean of 75 and a standard deviation of 5. Which curve displays less variability?

4. **Counterexample** Draw a curve that represents data which is not normally distributed.

5. **Name** the percentile that describes the median.

Guided Practice

6. The mean of a set of normally distributed data is 550 and the standard deviation is 35.
   a. Sketch a curve that represents the frequency distribution.
   b. What percent of the data is between 515 and 585?
   c. Name the interval about the mean in which about 99.7% of the data are located.
   d. If there are 200 values in the set of data, how many would be between 480 and 620?

7. A set of 500 values is normally distributed with a mean of 24 and a standard deviation of 2.
   a. What percent of the data is in the interval 22-26?
   b. What percent of the data is in the interval 20.5-27.5?
   c. Find the interval about the mean that includes 50% of the data.
   d. Find the interval about the mean that includes 95% of the data.

8. **Education** In her first semester of college, Salali earned a grade of 82 in chemistry and a grade of 90 in speech.
   a. The mean of the chemistry grades was 73, and the standard deviation was 3. Draw a normal distribution for the chemistry grades.
   b. The mean of the speech grades was 80, and the standard deviation was 5. Draw a normal distribution for the speech grades.
   c. Which of Salali's grades is relatively better based on standard deviation from the mean? Explain.

EXERCISES

Practice

9. The mean of a set of normally distributed data is 12 and the standard deviation is 1.5.
   a. Sketch a curve that represents the frequency distribution.
   b. Name the interval about the mean in which about 68.3% of the data are located.
   c. What percent of the data is between 7.5 and 16.5?
   d. What percent of the data is between 9 and 15?

10. Suppose 200 values in a set of data are normally distributed.
    a. How many values are within one standard deviation of the mean?
    b. How many values are within two standard deviations of the mean?
    c. How many values fall in the interval between the mean and one standard deviation above the mean?

11. A set of data is normally distributed with a mean of 82 and a standard deviation of 4.
    a. Find the interval about the mean that includes 45% of the data.
    b. Find the interval about the mean that includes 80% of the data.
    c. What percent of the data is between 76 and 88?
    d. What percent of the data is between 80.5 and 83.5?

12. The mean of a set of normally distributed data is 402, and the standard deviation is 36.
   a. Find the interval about the mean that includes 25% of the data.
   b. What percent of the data is between 387 and 417?
   c. What percent of the data is between 362 and 442?
   d. Find the interval about the mean that includes 45% of the data.
13. A set of data is normally distributed with a mean of 140 and a standard deviation of 20.
   a. What percent of the data is between 100 and 150?
   b. What percent of the data is between 150 and 180?
   c. Find the value that defines the 75th percentile.
14. The mean of a set of normally distributed data is 6, and the standard deviation is 0.35.
   a. What percent of the data is between 6.5 and 7?
   b. What percent of the data is between 5.5 and 6.2?
   c. Find the limit above which 90% of the data lies.

Applications and Problem Solving

15. **Probability** Tossing six coins is a binomial experiment.
   a. Find each probability.
      P(no tails) P(one tail) P(two tails) P(three tails)
      P(four tails) P(five tails) P(six tails)
   b. Assume that the experiment was repeated 64 times. Make a bar graph showing how many times you would expect each outcome to occur.
   c. Use the bar graph to determine the mean number of tails.
   d. Find the standard deviation of the number of tails.
   e. Compare the bar graph to a normal distribution.
16. **Critical Thinking** Consider the percentile scores on a standardized test.
   a. Describe the 92nd percentile in terms of standard deviation.
   b. Name the percentile of a student whose score is 0.8 standard deviation above the mean.
17. **Health** The lengths of babies born in City Hospital in the last year are normally distributed. The mean length is 20.4 inches, and the standard deviation is 0.8 inch. Trey was 22.3 inches long at birth.
   a. What percent of the babies born at City Hospital were longer than Trey?
   b. What percent of the babies born at City Hospital were shorter than Trey?
18. **Business** The length of time a brand of CD players can be used before needing service is normally distributed. The mean length of time is 61 months, and the standard deviation is 5 months. The manufacturer plans to issue a guarantee that it will replace any CD player that breaks within a certain length of time. If the manufacturer does not want to replace any more than 2% of the CD players, how many months should they limit the guarantee?
19. **Education** A college professor plans to grade a test on a curve. The mean score on the test is 65, and the standard deviation is 7. The professor wants 15% A's, 20% B's, 30% C's, 20% D's and 15% F's. Assume the grades are normally distributed.
   a. What is the lowest score for an A?
   b. Find the lowest passing score.
   c. What is the interval for the B's?

20. Critical Thinking Describe the frequency distribution represented by each graph.

a.　　b.　　c.　　d.

21. Industry A machine is used to fill cans of cola. The amount of cola dispensed into each can varies slightly. Suppose the amount of cola dispensed into the cans is normally distributed.

a. If at least 95% of the cans must have between 350 and 360 milliliters of cola, find the greatest standard deviation that can be allowed.

b. What percent of the cans will have between 353 and 357 milliliters of cola?

Mixed Review

22. Nutrition The numbers of Calories in one serving of twenty different cereals are listed below. *(Lesson 14-3)*

110, 110, 330, 200, 88, 110, 88, 110, 165, 390,
150, 440, 536, 200, 110, 165, 88, 147, 110, 165

a. What is the median of the data?

b. Find the first quartile point and the third quartile point.

c. Find the interquartile range of the data.

d. What is the semi-quartile range of the data?

e. Draw a box-and-whisker plot of the data.

23. Find the mean, median, and mode of {33, 42, 71, 19, 42, 45, 79, 48, 55}. *(Lesson 14-2)*

24. Write an equation for a secant function with a period of $\frac{\pi}{2}$, a phase shift of $-\pi$, and a vertical shift of 3. *(Lesson 6-7)*

25. Education The numbers of students attending Wilder High School during various years are listed below. *(Lesson 4-8)*

Year	1965	1970	1975	1980	1985	1990	1995	2000
Enrollment	365	458	512	468	426	401	556	629

a. Write an equation that models the enrollment as a function of the number of years since 1965.

b. Use the model to predict the enrollment in 2015.

26. SAT/ACT Practice What is the value of x in the figure at the right?

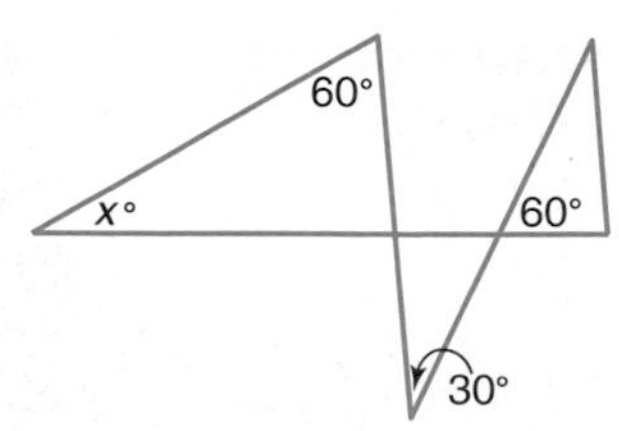

A 50　　**B** 45

C 40　　**D** 35

E 30

Extra Practice See p. A54.

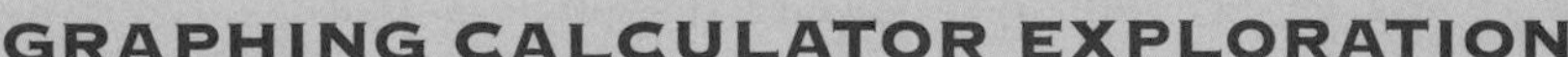

GRAPHING CALCULATOR EXPLORATION

14-4B The Standard Normal Curve

An Extension of Lesson 14-4

WHAT YOU'LL LEARN

- Use the standard normal curve to study properties of normal distributions.

The graph shown at the right is known as the *standard normal curve.* The standard normal curve is the graph of $f(x) = \frac{1}{\sqrt{2\pi}}e^{-\frac{x^2}{2}}$. You can use a graphing calculator to investigate properties of this function and its graph. Enter the function for the normal curve in the **Y=** list of a graphing calculator.

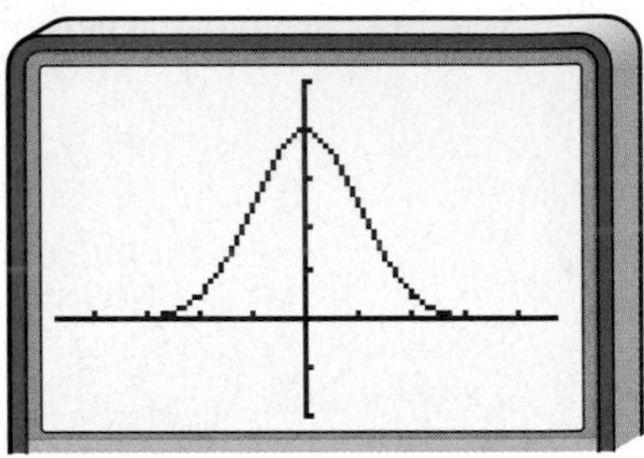

[−4.7, 4.7] scl:1 by [−0.2, 0.5] scl:0.1

TRY THESE

1. What can you say about the function and its graph? Be sure to include information about the domain, the range, symmetry, and end behavior.
2. The standard normal curve models a probability distribution. As a result, probabilities for intervals of x-values are equal to areas of regions bounded by the curve, the x-axis, and the vertical lines through the endpoints of the intervals. The calculator can approximate the areas of such regions. To find the area of the region bounded by the curve, the x-axis, and the vertical lines $x = -1$ and $x = 1$, go to the **CALC** menu and select **7:∫f(x) dx**. Move the cursor to the point where $x = -1$. Press ENTER. Then move the cursor to the point where $x = 1$ and press ENTER. The calculator will shade the region described above and display its approximate area. What number does the calculator display for the area of the shaded region?
3. Refer to the diagram on page 920. For normal distributions, about what percent of the data are within one standard deviation from the mean? How is this number related to the area you found in Exercise 2?
4. Enter 2nd **[DRAW]** 1. This causes the calculator to clear the shading and redisplay the graph. Find the area of the region bounded by the curve, the x-axis, and the vertical lines $x = -2$ and $x = 2$.
5. Find the area of the region bounded by the curve, the x-axis, and the vertical lines $x = -3$ and $x = 3$.
6. How do your answers for Exercises 4 and 5 compare to the percents in the diagram on page 920?

WHAT DO YOU THINK?

7. Without using a calculator, estimate the area of the region bounded by the curve, the x-axis, and the vertical lines $x = -4$ and $x = 4$ to four decimal places.
8. Change the graphing window to **Xmin** $= -47$ and **Xmax** $= 47$. Find the area of the region bounded by the curve, the x-axis, and the vertical lines $x = -20$ and $x = 20$. Do you think that your answer is the exact area for the region? Explain.

Sample Sets of Data

OBJECTIVE

- Find the standard error of the mean to predict the true mean of a population with a certain level of confidence.

EDUCATION Rosalinda Perez is doing some research for her doctoral thesis. She wants to determine the mean amount of money spent by a school district to educate one student for a year. Since there are 14,883 school districts in the United States, she cannot contact every school district. She randomly contacts 100 of these districts to find out how much money they spend per pupil. Using these 100 values, she computes the mean expenditure to be $6130 with a standard deviation of $1410. What is the standard error of the mean? *This problem will be solved in Example 1.*

In statistics, the word **population** does not always refer to a group of people. A population is the entire set of items or individuals in the group being considered. In the application above, the population is the 14,883 school districts. Rarely will a researcher find 100% of a population accessible as a source of data. Therefore, a **random sample** of the population is selected so that it is representative of the entire population. A random sample will give each member of the population the same chance of being selected as a member of the sample.

If the sample is random, the characteristics of the population pertinent to the study should be found in the sample in about the same ratio as they exist in the total population. If Ms. Perez randomly selected the school districts in the sample, the sample will probably include some large school districts, some small school districts, some urban school districts, and some rural school districts. It would probably also include school districts from all parts of the United States.

Based upon a random sample of the population, certain inferences can be made about the population in general. The major purpose of **inferential statistics** is to use the information gathered in a sample to make predictions about a population.

Ms. Perez is not sure that the mean school district expenditure per student is a truly representative mean of all school districts. She uses another sample of 100 districts. This time she finds the mean of the per pupil expenditures to be $6080 with a standard deviation of $1390.

The discrepancies that Ms. Perez found in her two samples are common when taking random samples. Large companies and statistical organizations often take many samples to find the "average" they are seeking. For this reason, a sample mean is assumed to be near its true population mean, symbolized by μ. The standard deviation of the distribution of the sample means is known as the **standard error of the mean.**

μ is the lowercase Greek letter mu.

Standard Error of the Mean

If a sample of data has N values and σ is the standard deviation, the standard error of the mean $\sigma_{\overline{X}}$ is

$$\sigma_{\overline{X}} = \frac{\sigma}{\sqrt{N}}.$$

The symbol $\sigma_{\overline{X}}$ is read "sigma sub x bar."

The standard error of the mean is a measurement of how well the mean of a sample selected at random estimates the true mean.

- If the standard error of the mean is small, then the sample means are closer to (or approximately the same value as) the true mean.
- If the standard error of the mean is large, many of the sample means would be far from the true mean.

The greater the number of items or subjects in the sample, the closer the sample mean reflects the true mean and the smaller the standard error of the mean.

Example 1 Real World Application

EDUCATION **Refer to the application at the beginning of the lesson. What is the standard error of the mean?**

For the sample, $N = 100$. Find $\sigma_{\overline{X}}$.

$$\sigma_{\overline{X}} = \frac{1410}{\sqrt{100}} \text{ or } 141$$

The standard error of the mean of the per pupil expenditures is 141. Although 141 is a fairly large number, it is not extremely large compared to the mean 6130. The mean is a good approximation of the true mean.

The sampling error is the difference between the population mean μ and the sample mean $\overline{X}$.

The sample mean is only an estimate of the true mean of the population. Sample means of various random samples of the same population are normally distributed about the true mean with the standard error of the mean as a measure of their variability. Thus, the standard error of the mean behaves like the standard deviation. Using the standard error of the mean and the sample mean, we can state a range about the sample mean in which we think the true mean lies. Probabilities of the occurrence of sample means and true means may be determined by referring to the tables on page 919.

Example 2

EDUCATION **Refer to the application at the beginning of the lesson. Using Ms. Perez's first sample, determine the interval of per pupil expenditures such that the probability is 95% that the mean expenditure of the entire population lies within the interval.**

When $P = 95\%$ or 0.95, $t = 1.96$. *Refer to the tables on page 919.*

To find the range, use a technique similar to finding the interval for a normal distribution.

$$\begin{aligned}\overline{X} \pm t\sigma_{\overline{X}} &= 6130 \pm 1.96(141)\\ &= 5853.64 \text{ to } 6406.36\end{aligned}$$

The probability is 95% that the true mean μ is within the interval of \$5853.64 and \$6406.36.

The probability of the true mean being within a certain range of a sample mean may be expressed as a **level of confidence.** The most commonly used levels of confidence are 1% and 5%. A 1% level of confidence means that there is less than a 1% chance that the true mean differs from the sample mean by a certain amount. That is, you are 99% confident that the true mean is within a certain range of the sample mean. A 5% level of confidence means that the probability of the true mean being within a certain range of the sample mean is 95%.

If a higher level of confidence is desired for the same number of values, accuracy must be sacrificed by providing a larger interval. However, if the number of values in the sample is larger, the interval for a given level of confidence is smaller.

Example 3 **AUTOMOTIVE ENGINEERING** **The number of miles a certain sport utility vehicle can travel on open highway on one gallon of gasoline is normally distributed. You are to take a sample of vehicles, test them, and record the miles per gallon. You wish to have a 1% level of confidence that the interval containing the mean miles per gallon of the sample also contains the true mean.**

a. Twenty-five sports utility vehicles are selected and tested. From this sample, the average miles per gallon is 22 with a standard deviation of 4 miles per gallon. Determine the interval about the sample mean that has a 1% level of confidence.

b. Four hundred sports utility vehicles are randomly selected and their miles per gallon are recorded. From this sample, the average miles per gallon is 22 with a standard deviation of 4 miles per gallon. Determine the interval about the sample mean that has a 1% level of confidence.

c. What happens when the number of items in the sample is increased?

a. A 1% level of confidence is given when $P = 99\%$. When $P = 0.99$, $t = 2.58$.

Find $\sigma_{\overline{X}}$ $\qquad \sigma_{\overline{X}} = \frac{4}{\sqrt{25}}$ or 0.8

Find the range. $\qquad \overline{X} \pm t\sigma_{\overline{X}} = 22 \pm 2.58(0.8)$

$\overline{X} \pm t\sigma_{\overline{X}} = 19.936$ and 24.064

Thus, the interval about the sample mean is 19.936 miles per gallon to 24.064 miles per gallon.

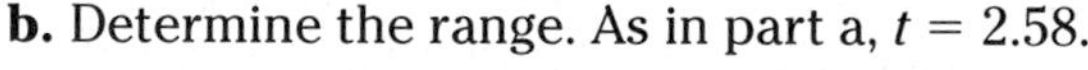

b. Determine the range. As in part a, $t = 2.58$.

Find $\sigma_{\overline{X}}$. $\qquad \sigma_{\overline{X}} = \frac{4}{\sqrt{400}}$ or 0.2

Find the range. $\qquad \overline{X} \pm t\sigma_{\overline{X}} = 22 \pm 2.58(0.2)$

$\overline{X} \pm t\sigma_{\overline{X}} = 21.484$ and 22.516

Thus, the interval about the sample mean is 21.484 miles per gallon to 22.516 miles per gallon.

c. By increasing the number of items in the sample, but with the same 1% confidence level, the range decreased substantially, from about $24.064 - 19.036$ or 4.128 miles per gallon to about $22.516 - 21.484$ or 1.032 miles per gallon.

Check for Understanding

Communicating Mathematics

Read and study the lesson to answer each question.

1. **Compare and contrast** a population and a sample of the population.
2. **Describe** how to determine the standard error of the mean.
3. **Explain** how you can find a smaller interval for the true mean and still have the same level of confidence.
4. **You Decide** As part of a class project, Karen, Tyler, and Mark need to determine the average number of hours that the students in their high school study each school night. Karen suggests they ask the students in their senior English class. Tyler suggests they ask every twentieth student as they enter the school in the morning. Mark suggests they ask the members of the track team. Whose sample should they use? Explain.

Guided Practice

Find the standard error of the mean for each sample.

5. $\sigma = 73, N = 100$
6. $\sigma = 3.4, N = 250$

7. If $\sigma = 5$, $N = 36$, and $\overline{X} = 45$, find the interval about the sample mean that has a 1% level of confidence.

8. If $\sigma = 5.6$, $N = 300$, and $\overline{X} = 55$, find the interval about the sample mean that has a 5% level of confidence.

9. **Employment** An employment agency requires all clients to take an aptitude test. They randomly selected 150 clients and recorded the amount of time each client took to complete the test. The average time needed to complete the test was 27.5 minutes with a standard deviation of 3.5 minutes.
   a. What is the standard error of the mean?
   b. Find the interval about the sample mean that reflects a 50% chance that the true mean lies within that interval.
   c. Find the interval about the sample mean that has a 1% level of confidence.

Exercises

Practice

Find the standard error of the mean for each sample.

10. $\sigma = 1.8, N = 81$
11. $\sigma = 5.8, N = 250$
12. $\sigma = 7.8, N = 140$
13. $\sigma = 14, N = 700$
14. $\sigma = 2.7, N = 130$
15. $\sigma = 13.5, N = 375$

16. If the standard deviation of a sample set of data is 5.6 and the standard error of the mean is 0.056, how many values are in the sample set?

For each sample, find the interval about the sample mean that has a 1% level of confidence.

17. $\sigma = 5.3, N = 50, \overline{X} = 335$
18. $\sigma = 40, N = 64, \overline{X} = 200$
19. $\sigma = 12, N = 200, \overline{X} = 80$
20. $\sigma = 11.12, N = 1000, \overline{X} = 110$

21. The mean height of a sample of 100 high school seniors is 68 inches with a standard deviation of 4 inches. Determine the interval of heights such that the probability is 90% that the mean height of the entire population lies within that interval.

For each sample, find the interval about the sample mean that has a 5% level of confidence.

22. $\sigma = 2.4, N = 100, \overline{X} = 24$

23. $\sigma = 17.1, N = 350, \overline{X} = 4526$

24. $\sigma = 28, N = 370, \overline{X} = 678$

25. $\sigma = 0.67, N = 80, \overline{X} = 5.38$

26. The following is a frequency distribution of the time in minutes required for a shopper to get through the checkout line at a certain discount store on a weekend. The distribution is a random sample of the weekend shoppers at the store.

Number of Minutes	3–5	5–7	7–9	9–11	11–13	13–15	15–17	17–19	19–21
Frequency	1	3	5	12	17	13	7	4	2

a. What is the mean of the data in the frequency distribution?

b. Find the standard deviation of the data.

c. What is the standard error of the mean?

d. Find the interval about the sample mean such that the probability is 0.95 that the true mean lies within the interval.

e. Find the probability that the mean of the population will be less than one minute from the mean of the sample.

27. The standard deviation of the blood pressure of 45 women ages 40 to 50 years old is 12. What is the probability that the mean blood pressure of the random sample will differ by more than 3 points from the mean blood pressure reading for all women in that age bracket?

Applications and Problem Solving

28. **Botany** A botanist is studying the effects of a drought on the size of acorns produced by the oak trees. A random sample of 50 acorns reveals a mean diameter of 16.2 millimeters and a standard deviation of 1.4 millimeters.

a. Find the standard error of the mean.

b. What is the interval about the sample mean that has a 5% level of confidence?

c. Find the interval about the sample mean that gives a 99% chance that the true mean lies within the interval.

d. Find the interval about the sample mean such that the probability is 0.80 that the true mean lies within the interval.

29. **Advertising** The Brite Light Company wishes to include the average lifetime of its light bulbs in its advertising. One hundred light bulbs are randomly selected and illuminated. The time for each bulb to burn out is recorded. From this sample, the average life is 350 hours with a standard deviation of 45 hours.

a. Find the standard error of the mean.

b. Determine the interval about the sample mean that has a 1% level of confidence.

c. If you want to avoid false advertising charges, what number would you use as the average lifetime of the bulbs? Explain.

30. Critical Thinking There is a probability of 0.99 that the average life of a disposable hand-warming package is between 9.7936 and 10.2064 hours. The standard deviation of the sample is 0.8 hour. What is the size of the sample used to determine these values?

31. Entertainment In a certain town, a random sample of 10 families was interviewed about their television viewing habits. The mean number of hours that these families had their televisions on per day was 4.1 hours. The standard deviation was 1.8 hours.

a. What is the standard error of the mean?

b. Make a 5% level of confidence statement about the mean number of hours that families of this town have their televisions turned on per day.

c. What inferences about the television habits of the entire city can be drawn from this data? Explain.

32. Quality Control The Simply Crackers Company selects a random sample of 50 snack packages of their cheese crackers. The mean number of crackers in a package is 42.7 with a standard deviation of 3.2.

a. Find the standard error of the mean.

b. Determine the interval of number of crackers such that the probability is 50% that the mean number of crackers lies within the interval.

c. If the true population mean should be 43 crackers per package, should the company be concerned about this sample? Explain.

33. Critical Thinking The lifetimes of 1600 batteries used in radios are tested. With a 5% level of confidence, the true average life of the batteries is from 746.864 to 753.136 hours.

a. What is the mean life of a battery in the sample?

b. Find the standard deviation of the life of the batteries in the sample.

Mixed Review

34. Tires The lifetimes of a certain type of car tire are normally distributed. The mean lifetime is 40,000 miles with a standard deviation of 5000 miles. Consider a sample of 10,000 tires. *(Lesson 14-4)*

a. How many tires would you expect to last between 35,000 and 45,000 miles?

b. How many tires would you expect to last between 30,000 and 40,000 miles?

c. How many tires would you expect to last less than 40,000 miles?

d. How many tires would you expect to last more than 50,000 miles?

e. How many tires would you expect to last less than 25,000 miles?

35. Find the mean deviation and the standard deviation of $\{44, 72, 58, 61, 71, 49, 55, 68\}$. *(Lesson 14-3)*

36. Find the sum of the first ten terms of the series $\frac{1}{16} + \frac{1}{4} + 1 + \cdots$. *(Lesson 12-2)*

37. Write $x = y$ in polar form. *(Lesson 9-3)*

38. Solve $\tan x + \cot x = 2$ for principal values of x. *(Lesson 7-5)*

39. SAT/ACT Practice $*x$ is defined such that $*x = x^2 - 2x$. What is the value of $*2 - *1$?

A -1 B 0 C 1 D 2 E 4

Extra Practice See p. A54.

VOCABULARY

arithmetic mean (p. 897)
back-to-back bar graph (p. 889)
bar graph (p. 889)
bimodal (p. 899)
box-and-whisker plot (p. 909)
class interval (p. 890)
class limits (p. 890)
class mark (p. 890)
cumulative frequency distribution (p. 902)
frequency distribution (p. 890)
frequency polygon (p. 892)
hinge (p. 909)
histogram (p. 890)
inferential statistics (p. 927)
interquartile range (p. 909)
leaf (p. 899)
level of confidence (p. 929)
line plot (p. 889)
mean (p. 897)
mean deviation (p. 910)
measure of central tendency (p. 897)
measure of variability (p. 908)
median (p. 897)
median class (p. 902)
mode (p. 897)
normal curve (p. 918)
normal distribution (p. 918)
outlier (p. 909)
percentile (p. 922)
population (p. 927)
quartile (p. 909)
random sample (p. 927)
range (p. 890)
semi-interquartile range (p. 909)
standard deviation (p. 911)
standard error of the mean (p. 927)
stem (p. 899)
stem-and-leaf plot (p. 899)
three-dimensional bar graph (p. 890)
variance (p. 912)
whisker (p. 909)

UNDERSTANDING AND USING THE VOCABULARY

Choose the term from the list above that best completes each statement.

1. A ___?___ is a display that visually shows the quartile points and the extreme values of a set of data.
2. The ___?___ of a set of data is the middle value if there are an odd number of values.
3. The standard deviation of the distribution of the sample means is known as the ___?___.
4. The ___?___ of a set of data is the difference between the greatest and the least values in the set.
5. A statistic that describes the center of a set of data is called an average or ___?___.
6. A ___?___ is the entire set of items or individuals in the group being considered.
7. Data with two modes are ___?___.
8. The major purpose of ___?___ is to use the information gathered in a sample to make predictions about a population.
9. A ___?___ is the most common way of displaying a frequency distribution.
10. A measure of variability often associated with the arithmetic mean is the ___?___.

For additional review and practice for each lesson, visit: **www.amc.glencoe.com** 

SKILLS AND CONCEPTS

OBJECTIVES AND EXAMPLES

Lesson 14-1 Draw, analyze, and use bar graphs and histograms.

- Draw a histogram of the data below.

Scores	Frequency
60–70	2
70–80	8
80–90	11
90–100	6

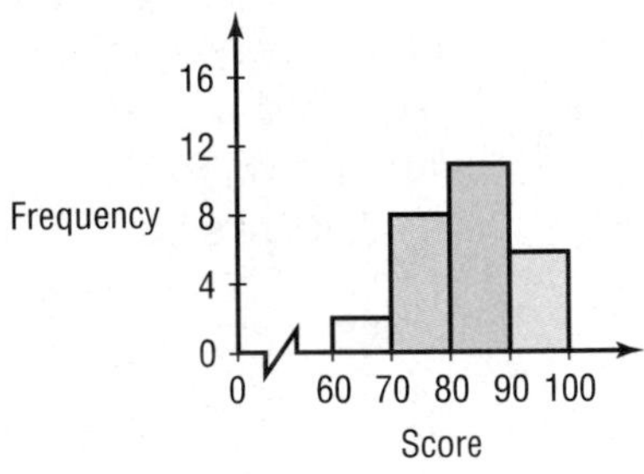

REVIEW EXERCISES

The table below gives the weight in ounces of the popular women's tennis shoes.

Weight (ounces)	Number of Shoes
9.0–10.0	2
10.0–11.0	18
11.0–12.0	5
12.0–13.0	2
13.0–14.0	3

11. What is the range of the data?

12. What are the class marks?

13. Draw a histogram of the data.

Lesson 14-2 Find the mean, median, and mode of a set of data.

- Find the mean, median, and mode of the set {46, 47, 59, 49, 50, 48, 58, 56, 58, 54, 53}.

$$\frac{1}{11}\sum_{i=1}^{11} X_i = \frac{46 + 47 + \cdots + 54 + 53}{11} \text{ or } 53$$

The mean is 53.

To find the median, order the data.

46, 47, 48, 49, 50, 53, 54, 56, 58, 58, 64

Since there are an odd number of data, the median is the middle value. The median is 53.

The most frequent value in this set of data is 58. So, the mode is 58.

Find the mean, median, and mode of each set of data.

14. {4, 8, 2, 4, 5, 5, 6, 7, 4}

15. {250, 200, 160, 240, 200}

16. {19, 11, 13, 15, 16}

17. {6.6, 6.3, 6.8, 6.6, 6.7, 5.9, 6.4, 6.3}

18.

stem	leaf
12	2 8
13	0 1 3 5
14	1 6

12|2 = 122

OBJECTIVES AND EXAMPLES

Lesson 14-3 Find the interquartile range, the semi-interquartile range, mean deviation, and standard deviation of data.

- interquartile range: $Q_3 - Q_1$

  semi-interquartile range: $Q_R = \frac{Q_3 - Q_1}{2}$

  mean deviation: $MD = \frac{1}{n}\sum_{i=1}^{n} |X_i - \overline{X}|$

  standard deviation: $\sigma = \sqrt{\frac{1}{n}\sum_{i=1}^{n} (X_i - \overline{X})^2}$

REVIEW EXERCISES

A number cube is tossed 10 times with the following results.

5 1 5 4 2 3 6 2 5 1

19. Find the interquartile range.
20. Find the semi-interquartile range.
21. Find the mean deviation.
22. Find the standard deviation.

Lesson 14-4 Use the normal distribution curve.

- **A Normal Distribution**

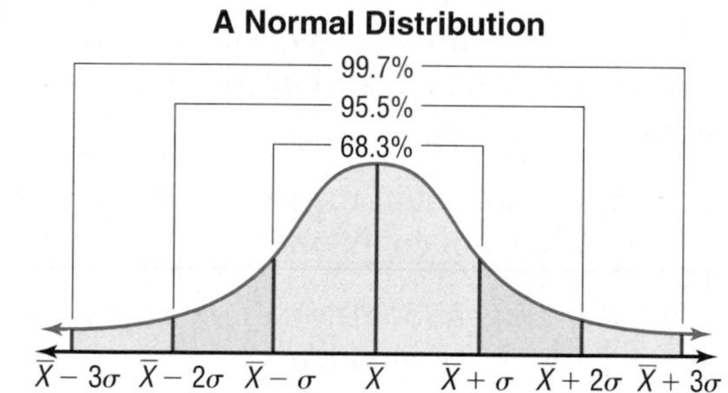

A set of data is normally distributed with a mean of 75 and a standard deviation of 6. What percent of the data is between 69 and 81?

The values within one standard deviation of the mean are between 75 − 6, or 69, and 75 + 6, or 81. So, 68.3% of the data is between 69 and 81.

The mean of a set of normally distributed data is 88 and the standard deviation is 5.

23. What percent of the data is in the interval 78–98?
24. Find the probability that a value selected at random from the data lies in the interval 86–90.
25. Find the interval about the mean that includes 90% of the data.

Suppose 150 values in a data set are normally distributed.

26. How many values are within one standard deviation of the mean?
27. How many values are within two standard deviations of the mean?
28. How many values fall in the interval between the mean and one standard deviation above the mean?

OBJECTIVES AND EXAMPLES

Lesson 14-5 Find the standard error of the mean to predict the true mean of a population with a certain level of confidence.

- Find the standard error of the mean for $\sigma = 12$ and $N = 100$. If $\overline{X} = 75$, find the range for a 1% level of confidence.

$$\sigma_{\overline{X}} = \frac{\sigma}{\sqrt{N}}$$

$$\sigma_{\overline{X}} = \frac{12}{\sqrt{100}} \text{ or } 1.2$$

The standard error of the mean is 1.2.

A 1% level of confidence is given when $P = 99\%$.

When $P = 0.99$, $t = 2.58$.

Use $\sigma_{\overline{X}} = 1.2$ to find the range.

$$\overline{X} \pm t\sigma_{\overline{X}} = 75 \pm (2.58)(1.2)$$
$$= 71.90 \text{ to } 78.10$$

Thus, the interval about the mean is 71.90 to 78.10.

REVIEW EXERCISES

Find the standard error of the mean for each sample.

29. $\sigma = 1.5$, $N = 90$

30. $\sigma = 4.9$, $N = 120$

31. $\sigma = 25$, $N = 400$

32. $\sigma = 18$, $N = 25$

For each sample, find the interval about the sample mean that has a 1% level of confidence.

33. $\sigma = 15$, $N = 50$, $\overline{X} = 100$

34. $\sigma = 30$, $N = 15$, $\overline{X} = 90$

35. $\sigma = 24$, $N = 200$, $\overline{X} = 40$

In a random sample of 200 adults, it was found that the average number of hours per week spent cleaning their home was 1.8, with a standard deviation of 0.5.

36. Find the standard error of the mean.

37. Find the range about the mean such that the probability is 0.90 that the true mean lies within the range.

38. Find the range about the sample mean that has a 5% level of confidence.

39. Find the range about the sample mean that has a 1% level of confidence.

40. **Entertainment** In a random sample of 100 families, the children watched television an average of 4.6 hours a day. The standard deviation is 1.4 hours. Find the range about the sample mean so that a probability of 0.90 exists that the true mean will lie within the range.

APPLICATIONS AND PROBLEM SOLVING

41. **Safety** The numbers of job-related injuries at a construction site for each month of 1999 are listed below. *(Lesson 14-2)*

10	13	15	39	21	24
19	16	39	17	23	25

   a. Make a stem-and-leaf plot of the numbers of injuries.

   b. What is the mean number of the data?

   c. Find the median of the data.

   d. Find the mode of the data.

42. The height of members of the boys basketball team are normally distributed. The mean height is 75 inches, and the standard deviation is 2 inches. Randall is 80 inches tall. What percent of the boys on the basketball team are taller than Randall? *(Lesson 14-4)*

ALTERNATIVE ASSESSMENT

OPEN-ENDED ASSESSMENT

1. The mean of a set of five pieces of data is 15, and the median is 10. When one certain value is added to the set, the mean stays the same but the median changes.

   a. Find a set of data for which this is true.

   b. What value can be added to your set so that the mean stays the same but the median changes?

2. Find some data in a newspaper or magazine. Use what you have learned in this chapter to analyze the data.

PORTFOLIO

Choose one of the types of data displays you studied in this chapter. Describe a situation in which this type of display would be used. Explain why the type of display you chose is the best one to use in this situation.

Additional Assessment See p. A69 for a practice Chapter 14 test.

THE UNITED STATES CENSUS BUREAU

More and more models!

- Use the data you collected for the project in Chapter 12. Display the data in a table or use software to prepare a spreadsheet of the data.
- Use computer software or a graphing calculator to find at least three models for the population data. Draw a graph of each function.
- Compare your function models for the population data. Use your models to predict the U.S. population for the year 2050. Determine which one you think best fits the data.
- Write a one-page paper comparing the arithmetic and geometric sequences you wrote for Chapter 12 with the function models. Discuss which one model you think best fits the population data and give your estimate for the population in 2050.

THE PRINCETON REVIEW

TEST-TAKING TIP

If a problem includes a graph, look carefully at the graph including its labels and units. Then read the question.

Two or three questions may refer to the same graph.

Statistics and Data Analysis Problems

On the SAT and ACT exams, you will calculate the mean (average), median, and mode of a data set.

The SAT and ACT exams may include a one or two questions on interpreting graphs. The most common graphs are bar graphs, circle graphs, line graphs, stem-and-leaf plots, histograms, and frequency tables.

ACT EXAMPLES

Questions 1 and 2 refer to the following graph.

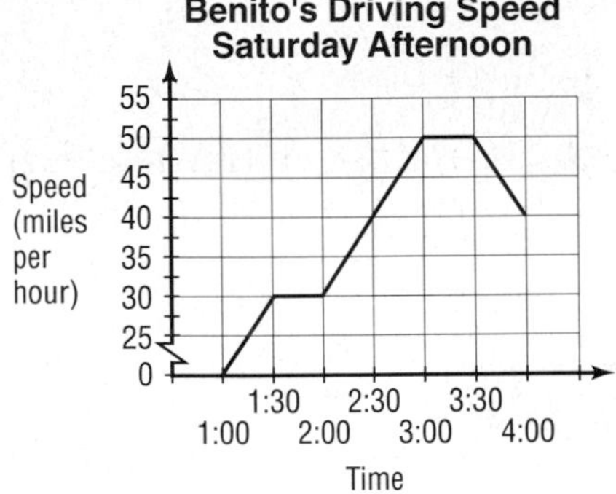

1. For what percent of the time was Benito driving 40 miles per hour or faster?

A 20 **B** 25 **C** $33\frac{1}{3}$ **D** 40 **E** 50

HINT Watch for different units of measure.

Solution Benito drove a total of 3 hours. He drove 40 miles per hour or faster for 1 hour and 30 minutes or $1\frac{1}{2}$ hours. The fraction of the time he drove 40 mph or more is $\frac{\frac{3}{2}}{3}$ or $\frac{1}{2}$, which equals 50%. The answer is choice **E**.

2. How far, in miles, did Benito drive between 1:30 and 2:00?

A 0 **B** 15 **C** 20 **D** 30

E It cannot be determined from the information given.

Solution rate × time = distance

$$\left(30 \text{ mph}\right)\left(\frac{1}{2} \text{ hour}\right) = 15 \text{ miles}$$

The answer is choice **B**.

SAT EXAMPLE

3. For $x = 0$, $x = 1$, and $x = 2$, Set $A = \{x, x + 3, 3x, x^2\}$. What is the mode of Set A?

A 0 **B** 1 **C** 2 **D** 2.5 **E** 3

HINT Look carefully at the given information and at the form of the answer choices (numbers, variables, and so on.)

Solution Notice that the answer choices are numbers. But Set A is defined using variable expressions. First determine the actual data of set A. Consider each value of x, one at a time. Substitute the value for x into each element of Set A.

For $x = 0$: $x = 0$, $x + 3 = 3$, $3x = 0$, and $x^2 = 0$.

When $x = 0$, $A = \{0, 3, 0, 0\}$.

When $x = 1$, $A = \{1, 4, 3, 1\}$.

When $x = 2$, $A = \{2, 5, 6, 4\}$.

Thus, $A = \{0, 0, 0, 1, 1, 2, 3, 3, 4, 4, 5, 6\}$. The element 0 occurs three times and no other element occurs as many times. So the mode of Set A is 0. The answer is choice **A**.

You might notice that choice D, 2.5, is the value of the median set A.

After you work each problem, record your answer on the answer sheet provided or on a piece of paper.

Multiple Choice

1. Based on the graph below, which worker had the greatest percent increase in income from week 1 to week 2?

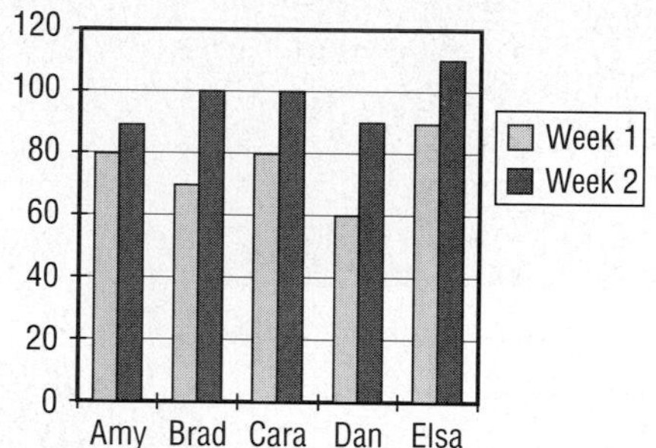

A Amy **B** Brad **C** Cara
D Dan **E** Elsa

2. If $a = b + bc$, then in terms of c, what does $\frac{b}{a}$ equal?

A $\frac{1}{c+2}$ **B** $\frac{1}{c+1}$ **C** $\frac{1}{c}$
D c **E** $c + 1$

3. If 0.1% of m is equal to 10% of n, then m is what percent of $10n$?

A 1/1000% **B** 10% **C** 100%
D 1000% **E** 10,000%

4. S is the set of all positive numbers n such that $n < 100$ and $\sqrt{n}$ is an integer. What is the median value of the members of set S?

A 5 **B** 5.5 **C** 25
D 50 **E** 99

5. In the figure, D, B, and E are collinear. What is the measure of $\angle ABC$?

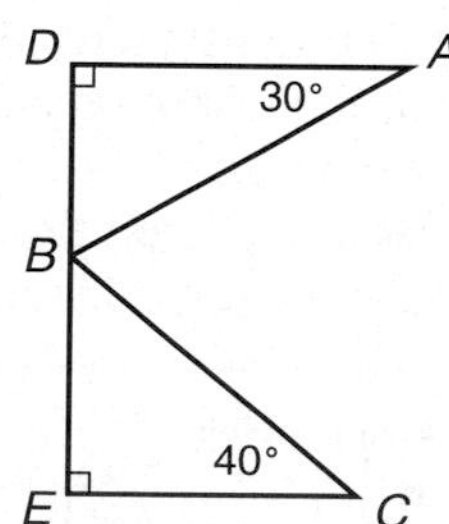

A 20° **B** 35° **C** 50° **D** 60° **E** 70°

6. How many of the scores 10, 20, 30, 35, 35, and 50 are greater than the arithmetic mean of the scores?

A 0 **B** 1 **C** 2
D 3 **E** 4

7. In $\triangle ABC$, what is the ratio $\frac{\tan A}{\text{area } \triangle ABC}$?

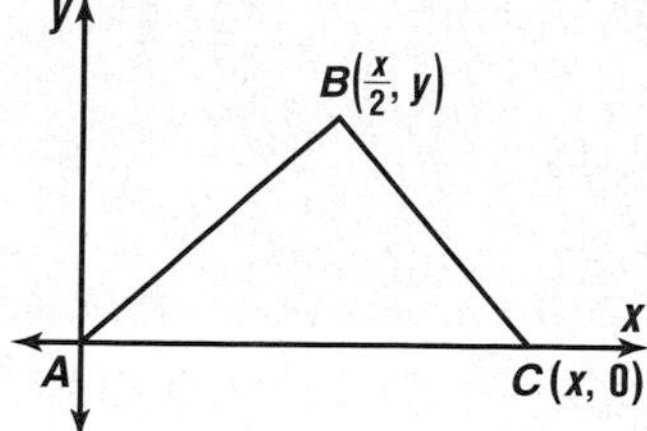

A $\frac{1}{2y^2}$ **B** $\frac{1}{y^2}$ **C** $\frac{2}{x^2}$ **D** $\frac{4}{x^2}$ **E** $\frac{x^2}{4}$

8. Based on the data in the table below, how many employees can this company expect to have by 2003?

Year	1997	1998	1999	2000	2001
Number of Employees	1900	2200	2500	2800	3100

A 3100 **B** 3400 **C** 3550
D 3700 **E** 4000

9. **Quantitative Comparison**

A if the quantity in Column A is greater
B if the quantity in Column B is greater
C if the two quantities are equal
D if the relationship cannot be determined from the information given

Set A: {2, −1, 7, −4, 11, 3}
Set B: {10, 5, −3, 4, 7, −8}

Column A	Column B
The median of Set A.	The mean of Set B.

10. **Grid-In** What is the arithmetic mean of the ten numbers below?

−820, −65, −32, 0, 1, 2, 3, 32, 65, 820

*inter*NET CONNECTION **SAT/ACT Practice** For additional test practice questions, visit: **www.amc.glencoe.com**

UNIT 5

Calculus

Calculus is one of the most important areas of mathematics. There are two branches of calculus, *differential* calculus and *integral* calculus. Differential calculus deals mainly with variable, or changing, quantities. Integral calculus deals mainly with finding sums of infinitesimally small quantities. This generally involves finding a limit.

Chapter 15, the only chapter in Unit 5, provides an overview of some aspects and applications of calculus.

Chapter 15 Introduction to Calculus

CHAPTER OBJECTIVES

- **Evaluate limits of functions.** *(Lesson 15-1)*
- **Find derivatives and antiderivatives of polynomial functions.** *(Lessons 15-2, 15-4)*
- **Evaluate definite integrals using limits and the Fundamental Theorem of Calculus.** *(Lessons 15-3, 15-4)*

Unit 5 *inter*NET Project

DISEASES

Did you know that many communicable diseases have been virtually eliminated as the result of vaccinations? In 1954, Jonas Salk invented a vaccine for polio. Polio was a dreaded disease from about 1942 to 1954. In 1952, there were 60,000 cases reported. As a result of Salk's miraculous discovery, there were only 5 cases of polio reported in the United States in 1996. In this project, you will look at data about diseases in the United States.

CHAPTER 15 (page 981)

Miracles of Science! Even though many diseases that once disabled or even killed many people have been controlled, the treatment or cure for many other diseases still eludes researches. Use the Internet to find data on a particular disease.

Math Connection: Model the data with at least two functions. Predict the course of the disease in the future using your model.

*inter*NET CONNECTION For more information on the Unit Project, visit: **www.amc.glencoe.com**

15-1 Limits

OBJECTIVES

- Calculate limits of polynomial and rational functions algebraically.
- Evaluate limits of functions using a calculator.

SPORTS In football, if the length of a penalty exceeds half the distance to the offending team's goal line, then the ball is moved only half the distance to the goal line. Suppose one team has the ball at the other team's 10-yard line. The other team, in an effort to prevent a touchdown, repeatedly commits penalties. After the first penalty, the ball would be moved to the 5-yard line.

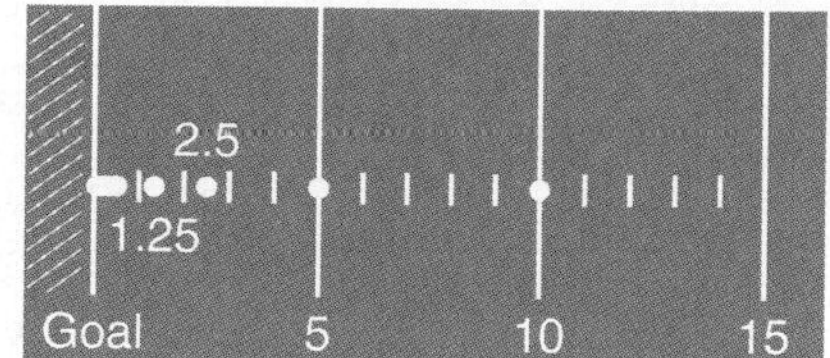

The results of the subsequent penalties are shown in the table. Assuming the penalties could continue indefinitely, would the ball ever actually cross the goal line?

Penalty	1st	2nd	3rd	...
Yard Line	5	2.5	1.25	...

The ball will never reach the goal line, but it will get closer and closer after each penalty. As you saw in Chapter 12, a number that the terms of a sequence approach, without necessarily reaching it, is called a **limit.** In the application above, the limit is the goal line or 0-yard line. The idea of a limit also exists for functions.

Limit of a Function

If there is a number L such that the value of $f(x)$ gets closer and closer to L as x gets closer to a number a, then L is called the limit of $f(x)$ as x approaches a.

In symbols, $L = \lim_{x \to a} f(x)$.

Example 1 **Consider the graph of the function $y = f(x)$ shown at the right. Find each pair of values.**

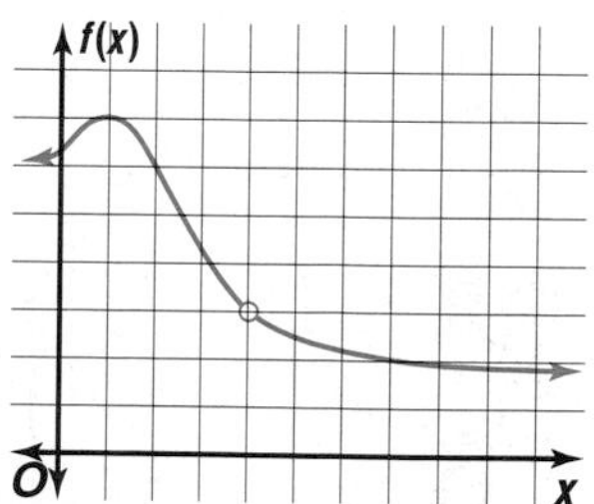

a. $f(2)$ **and** $\lim_{x \to 2} f(x)$

At the point on the graph where the x-coordinate is 2, the y-coordinate is 6. So, $f(2) = 6$.

Look at points on the graph whose x-coordinates are close to, but not equal to, 2. Notice that the closer x is to 2, the closer y is to 6. So, $\lim_{x \to 2} f(x) = 6$.

b. $f(4)$ **and** $\lim_{x \to 4} f(x)$

The hole in the graph indicates that the function does not have a value when $x = 4$. That is, $f(4)$ is undefined.

Look at points on the graph whose x-coordinates are close to, but not equal to, 4. The closer x is to 4, the closer y is to 3. So, $\lim_{x \to 4} f(x) = 3$.

You can see from Example 1 that sometimes $f(a)$ and $\lim_{x \to a} f(x)$ are the same, but at other times they are different. In Lesson 3-5, you learned about continuous functions and how to determine whether a function is continuous or discontinuous for a given value. We can use the definition of continuity to make a statement about limits.

Limit of a Continuous Function

$f(x)$ is continuous at a if and only if

$$\lim_{x \to a} f(x) = f(a).$$

Examples of continuous functions include polynomials as well as the functions $\sin x$, $\cos x$, and a^x. Also, $\log_a x$ is continuous if $x > 0$.

Example 2 **Evaluate each limit.**

a. $\lim_{x \to 3} (x^3 - 5x^2 + 7x - 10)$

Since $f(x) = x^3 - 5x^2 + 7x - 10$ is a polynomial function, it is continuous at every number. So the limit as x approaches 3 is the same as the value of $f(x)$ at $x = 3$.

$$\begin{aligned} \lim_{x \to 3} (x^3 - 5x^2 + 7x - 10) &= 3^3 - 5 \cdot 3^2 + 7 \cdot 3 - 10 \quad \textit{Replace } x \textit{ with 3.} \\ &= 27 - 45 + 21 - 10 \\ &= -7 \end{aligned}$$

The limit of $x^3 - 5x^2 + 7x - 10$ as x approaches 3 is -7.

b. $\lim_{x \to \pi} \frac{\cos x}{x}$

Since the denominator of $\frac{\cos x}{x}$ is not 0 at $x = \pi$, the function is continuous at $x = \pi$.

$$\begin{aligned} \lim_{x \to \pi} \frac{\cos x}{x} &= \frac{\cos \pi}{\pi} \quad \textit{Replace } x \textit{ with } \pi. \\ &= \frac{-1}{\pi} \quad \cos \pi = -1 \end{aligned}$$

The limit of $\frac{\cos x}{x}$ as x approaches π is $-\frac{1}{\pi}$.

Limits can also be used to model real-world situations in which values approach a given value.

Example 3 **PHYSICS**

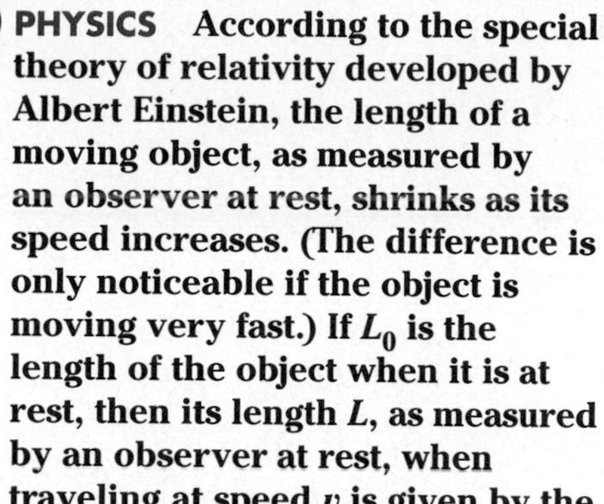

According to the special theory of relativity developed by Albert Einstein, the length of a moving object, as measured by an observer at rest, shrinks as its speed increases. (The difference is only noticeable if the object is moving very fast.) If L_0 is the length of the object when it is at rest, then its length L, as measured by an observer at rest, when traveling at speed v is given by the formula $L = L_0\sqrt{1 - \frac{v^2}{c^2}}$, where c is the speed of light. If the space shuttle were able to approach the speed of light, what would happen to its length?

interNET CONNECTION

Research For more information about relativity, visit: **www.amc.glencoe.com**

We need to find $\lim\limits_{v \to c} L_0\sqrt{1 - \frac{v^2}{c^2}}$.

$$\begin{aligned}\lim_{v \to c} L_0\sqrt{1 - \frac{v^2}{c^2}} &= L_0\sqrt{1 - \frac{c^2}{c^2}} && \textit{Replace } v \textit{ with } c\textit{, the speed of light.}\\ &= L_0\sqrt{0}\\ &= 0\end{aligned}$$

The closer the speed of the shuttle is to the speed of light, the closer the length of the shuttle, as seen by an observer at rest, gets to 0.

When a function is not continuous at the x-value in question, it is more difficult to evaluate the limit. Consider the function $f(x) = \frac{x^2 - 9}{x - 3}$. This function is not continuous at $x = 3$, because the denominator is 0 when $x = 3$. To compute $\lim\limits_{x \to 3} f(x)$, apply algebraic methods to decompose the function into a simpler one.

$$\begin{aligned}\frac{x^2 - 9}{x - 3} &= \frac{(x + 3)(x - 3)}{x - 3} && \textit{Factor.}\\ &= x + 3, x \neq 3 && \textit{Simplify.}\end{aligned}$$

When computing the limit, we are only interested in x-values close to 3. What happens when $x = 3$ is irrelevant, so we can replace $f(x)$ with the simpler expression $x + 3$.

$$\begin{aligned}\lim_{x \to 3} \frac{x^2 - 9}{x - 3} &= \lim_{x \to 3} (x + 3)\\ &= 3 + 3 \text{ or } 6\end{aligned}$$

The graph of $f(x)$ indicates that this answer is correct. As x gets closer to 3, the y-coordinates get closer and closer to, but never equal, 6. The limit is 6.

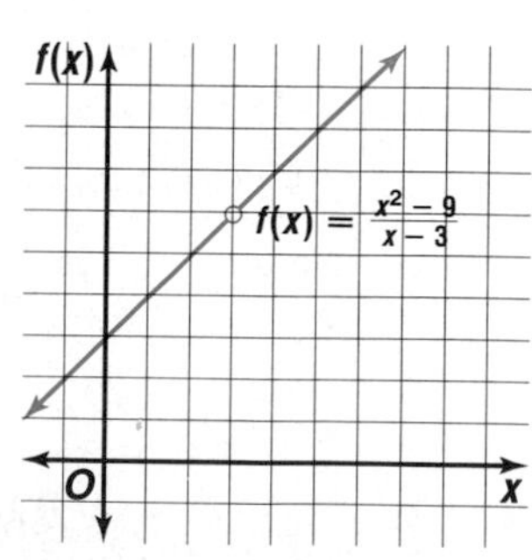

Example 4 **Evaluate each limit.**

a. $\lim_{x \to 4} \frac{x^2 - 2x - 8}{x^2 - 4x}$

$$\lim_{x \to 4} \frac{x^2 - 2x - 8}{x^2 - 4x} = \lim_{x \to 4} \frac{(x + 2)(x - 4)}{x(x - 4)}$$

$$= \lim_{x \to 4} \frac{x + 2}{x}$$

$$= \frac{4 + 2}{4} \text{ or } \frac{3}{2} \quad \textit{Replace } x \textit{ with } 4.$$

b. $\lim_{h \to 0} \frac{h^3 - 4h^2 - 6h}{h}$

$$\lim_{h \to 0} \frac{h^3 - 4h^2 - 6h}{h} = \lim_{h \to 0} \frac{h(h^2 - 4h - 6)}{h}$$

$$= \lim_{h \to 0} (h^2 - 4h - 6)$$

$$= 0^2 - 4 \cdot 0 - 6 \text{ or } -6 \quad \textit{Replace } h \textit{ with } 0.$$

Sometimes algebra is not sufficient to find a limit. A calculator may be useful. Consider the problem of finding $\lim_{x \to 0} \frac{\sin x}{x}$, where x is in radians. The function is not continuous at $x = 0$, so the limit cannot be found by replacing x with 0. On the other hand, the function cannot be simplified to help make the limit easier to find. You can use a calculator to compute values of the function $\frac{\sin x}{x}$ for x-values that get closer and closer to 0 from either side (that is, both less than 0 and greater than 0).

Graphing Calculator Tip

Enter the function in the **Y=** menu and set **Indpnt** to **Ask** in the **TBLSET** menu to help generate these values.

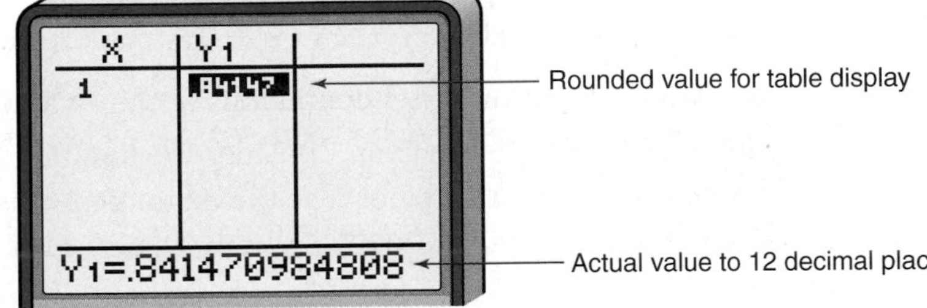

The tables below show the expression evaluated for values of x that approach 0.

x	$\frac{\sin x}{x}$
1	0.841470984808
0.1	0.998334166468
0.01	0.999983333417
0.001	0.999999833333
0.0001	0.999999998333

x	$\frac{\sin x}{x}$
−1	0.841470984808
−0.1	0.998334166468
−0.01	0.999983333417
−0.001	0.999999833333
−0.0001	0.999999998333

As x gets closer and closer to 0, from either side, the value of $\frac{\sin x}{x}$ gets closer and closer to 1. That is, $\lim_{x \to 0} \frac{\sin x}{x} = 1$.

Example 5 **Evaluate each limit.**

a. $\lim_{x \to 0} \frac{1 - \cos x}{x^2}$ **(x is in radians.)**

A graphing calculator or spreadsheet can generate more decimal places for the expression than shown here.

x	$\frac{1 - \cos x}{x^2}$
1	0.45970
0.1	0.49958
0.01	0.499996
0.001	0.49999996

x	$\frac{1 - \cos x}{x^2}$
−1	0.45970
−0.1	0.49958
−0.01	0.499996
−0.001	0.49999996

As x approaches 0, the value of $\frac{1 - \cos x}{x^2}$ gets closer to 0.5, so $\lim_{x \to 0} \frac{1 - \cos x}{x^2} = 0.5$.

b. $\lim_{x \to 1} \frac{\ln x}{x - 1}$

x	$\frac{\ln x}{x - 1}$
0.9	1.0536
0.99	1.0050
0.999	1.0005

x	$\frac{\ln x}{x - 1}$
1.1	0.95310
1.01	0.99503
1.001	0.99950

The closer x is to 1, the closer $\frac{\ln x}{x - 1}$ is to 1, so $\lim_{x \to 1} \frac{\ln x}{x - 1} = 1$.

Using a calculator is not a foolproof way of evaluating $\lim_{x \to a} f(x)$. You may only analyze the values of $f(x)$ for a few values of x near a. However, the function may do something unexpected as x gets even closer to a. You should use algebraic methods whenever possible to find limits.

GRAPHING CALCULATOR EXPLORATION

You can use a graphing calculator to find a limit, with less work than an ordinary scientific calculator. To find $\lim_{x \to a} f(x)$, first graph the equation $y = f(x)$. Then use ZOOM and TRACE to locate a point on the graph whose x-coordinate is as close to a as you like. The y-coordinate should be close to the value of the limit.

TRY THESE **Evaluate each limit.**

1. $\lim_{x \to 0} \frac{e^x - 1}{x}$

2. $\lim_{x \to 2} \frac{x^2 - 4}{x^2 - 3x + 2}$

WHAT DO YOU THINK?

3. If you graph $y = \frac{\ln x}{x - 1}$ and use TRACE, why doesn't the calculator tell you what y is when $x = 1$?

4. Solve Exercise 2 algebraically. Do you get the same answer as you got from the graphing calculator?

5. Will the graphing calculator give you the exact answer for every limit problem? Explain.

CHECK FOR UNDERSTANDING

Communicating Mathematics

Read and study the lesson to answer each question.

1. **Define** the expression *limit of f(x) as x approaches a* in your own words.
2. **Describe** the difference between $f(1)$ and $\lim_{x \to 1} f(x)$ and explain when they would be the same number.
3. *Math Journal* **Write** a description of the three methods in this lesson for computing $\lim_{x \to a} f(x)$. Explain when each method would be used and include examples.

Guided Practice

4. Use the graph of $y = f(x)$ to find $\lim_{x \to 0} f(x)$ and $f(0)$.

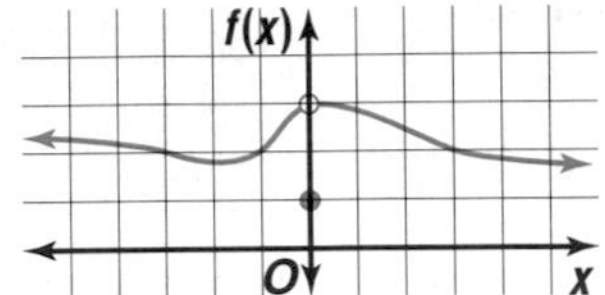

Evaluate each limit.

5. $\lim_{x \to 2} (-4x^2 + 2x - 5)$
6. $\lim_{x \to 0} (1 + x + 2^x - \cos x)$
7. $\lim_{x \to -2} \dfrac{x + 2}{x^2 - 4}$
8. $\lim_{x \to 0} \dfrac{x^2 - 3x}{x^3 + 4x}$
9. $\lim_{x \to 3} \dfrac{x^2 + 3x - 10}{x^2 + 5x + 6}$
10. $\lim_{x \to -2} \dfrac{2x^2 + 5x + 2}{x^2 + x - 2}$

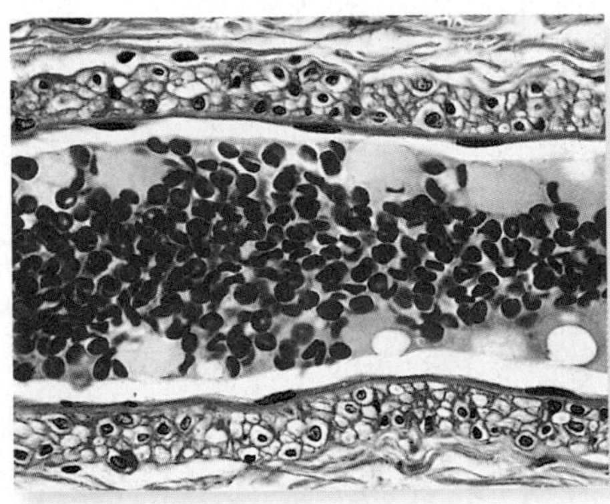

11. **Hydraulics** The velocity of a molecule of liquid flowing through a pipe depends on the distance of the molecule from the center of the pipe. The velocity, in inches per second, of a molecule is given by the function $v(r) = k(R^2 - r^2)$, where r is the distance of the molecule from the center of the pipe in inches, R is the radius of the pipe in inches, and k is a constant. Suppose for a particular liquid and a particular pipe that $k = 0.65$ and $R = 0.5$.
    a. Graph $v(r)$.
    b. Determine the limiting velocity of molecules closer and closer to the wall of the pipe.

EXERCISES

Practice

Use the graph of $y = f(x)$ to find each value.

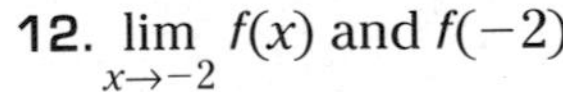

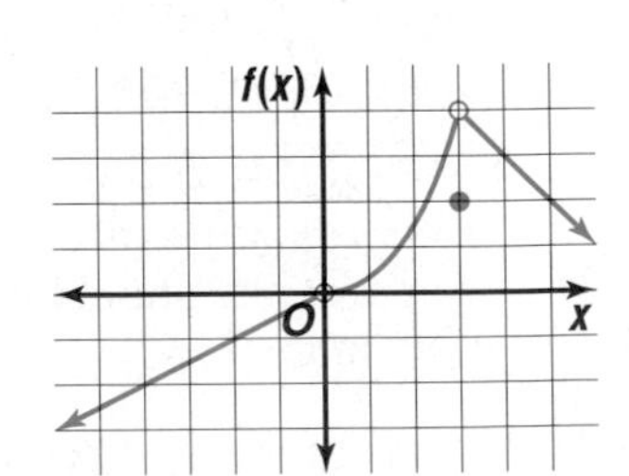

12. $\lim_{x \to -2} f(x)$ and $f(-2)$
13. $\lim_{x \to 0} f(x)$ and $f(0)$
14. $\lim_{x \to 3} f(x)$ and $f(3)$

f(x)
O
x

Evaluate each limit.

15. $\lim_{x \to 2} (-4x^2 - 3x + 6)$
16. $\lim_{x \to -1} (-x^3 + 3x^2 - 4)$

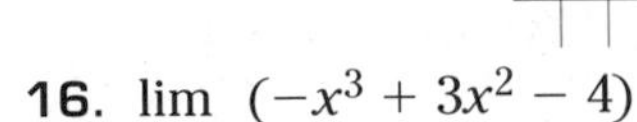

17. $\lim_{x \to \pi} \dfrac{\sin x}{x}$
18. $\lim_{x \to 0} (x + \cos x)$

19. $\lim_{x \to 5} \frac{x^2 - 25}{x - 5}$

20. $\lim_{n \to 0} \frac{2n^2}{n}$

21. $\lim_{x \to 3} \frac{x^2 - 3x}{x^2 + 2x - 15}$

22. $\lim_{x \to 1} \frac{x^3 + 3x^2 - 4x + 8}{x + 6}$

23. $\lim_{h \to -2} \frac{h^2 + 4h + 4}{h + 2}$

24. $\lim_{x \to 3} \frac{2x^2 - 3x}{x^3 - 2x^2 + x + 6}$

25. $\lim_{x \to 0} \frac{x^3 - x^2 + 2x}{x^3 + 4x^2 - 2x}$

26. $\lim_{x \to 0} \frac{x \cos x}{x^2 + x}$

27. $\lim_{x \to 0} \frac{(x + 2)^2 - 4}{x}$

28. $\lim_{x \to -2} \frac{(x + 1)^2 - 1}{x + 2}$

29. $\lim_{x \to -2} \frac{x^3 + 8}{x^2 - 4}$

30. $\lim_{x \to 4} \frac{2x - 8}{x^3 - 64}$

31. $\lim_{x \to 1} \frac{\frac{1}{x} - 1}{x - 1}$

32. $\lim_{x \to 4} \frac{x - 4}{\sqrt{x} - 2}$

33. Find the limit as h approaches 0 of $\frac{2h^3 - h^2 + 5h}{h}$.

34. What value does the function $g(x) = \frac{x + \pi}{\cos(x + \pi)}$ approach as x approaches 0?

Graphing Calculator

Use a graphing calculator to find the value of each limit. (Use radians with trigonometric functions.)

35. $\lim_{x \to 0} \frac{\tan 2x}{x}$

36. $\lim_{x \to 1} \frac{\ln x}{\ln(2x - 1)}$

37. $\lim_{x \to 1} \frac{1 - \sqrt{x}}{x - 1}$

38. $\lim_{x \to 0} \frac{3x - \sin 3x}{x^2 \sin x}$

Applications and Problem Solving

Real World Application

39. **Geometry** The area of an ellipse with semi-major axis a is $\pi a\sqrt{a^2 - c^2}$, where c is the distance from the foci to the center. Find the limit of the area of the ellipse as c approaches 0. Explain why the answer makes sense.

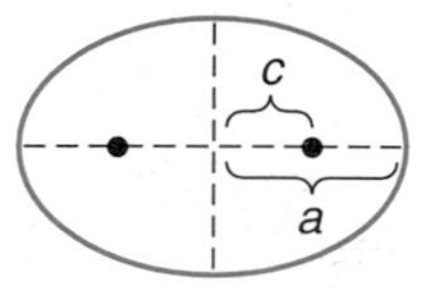

40. **Biology** If a population of bacteria doubles every 10 hours, then its initial hourly growth rate is $\lim_{t \to 0} \frac{2^{\frac{t}{10}} - 1}{t}$, where t is the time in hours. Use a calculator to approximate the value of this limit to the nearest hundredth. Write your answer as a percent.

41. **Critical Thinking** Does $\lim_{x \to 0} \sin\left(\frac{1}{x}\right)$ exist? That is, can you say $\lim_{x \to 0} \sin\left(\frac{1}{x}\right) = L$ for some real number L? Explain why or why not.

42. **Critical Thinking** You saw in Example 5 that $\lim_{x \to 0} \frac{1 - \cos x}{x^2} = 0.5$. That is, for values of x close to 0, $\frac{1 - \cos x}{x^2} \approx 0.5$. Solving for $\cos x$, we get $\cos x \approx 1 - \frac{x^2}{2}$.

a. Copy and complete the table by using a calculator. Round to six decimal places, if necessary.

x	1	0.5	0.1	0.01	0.001
$\cos x$					
$1 - \frac{x^2}{2}$					

b. Is it correct to say that for values of x close to 0, the expression $1 - \frac{x^2}{2}$ is a good approximation for $\cos x$? Explain.

43. Physics When an object, such as a bowling ball, is dropped near Earth's surface, the distance $d(t)$ (in feet) that the object falls in t seconds is given by $d(t) = 16t^2$. Its velocity (in feet per second) after 2 seconds is given by $\lim_{t \to 2} \frac{d(t) - d(2)}{t - 2}$. Evaluate this limit algebraically to find the velocity of the bowling ball after 2 seconds. *You will learn more about the relationship between distance and velocity in Lesson 15-2.*

44. Critical Thinking Yoshi decided that $\lim_{x \to 0} (1 + x)^{\frac{1}{x}}$ is 0, because as x approaches 0, the base of the exponential expression approaches 1, and 1 to any power is 1.

a. Use a calculator to help deduce the exact value of $\lim_{x \to 0} (1 + x)^{\frac{1}{x}}$.

b. Explain where Yoshi's reasoning was wrong.

Mixed Review

45. Botany A random sample of fifty acorns from an oak tree in the park reveals a mean diameter of 16.2 millimeters and a standard deviation of 1.4 millimeters. Find the range about the sample mean that gives a 99% chance that the true mean lies within it. *(Lesson 14-5)*

46. Tess is running a carnival game that involves spinning a wheel. The wheel has the numbers 1 to 10 on it. What is the probability of 7 never coming up in five spins of the wheel? *(Lesson 13-6)*

47. Find the third term of $(x - 3y)^5$. *(Lesson 12-6)*

48. Simplify $(16y^8)^{\frac{3}{4}}$. *(Lesson 11-1)*

49. Write the equation of the ellipse if the endpoints of the major axis are at $(1, -2)$ and $(9, -2)$ and the endpoints of the minor axis are at $(5, 1)$ and $(5, -5)$. *(Lesson 10-3)*

50. Graph the polar equation $r = -3$. *(Lesson 9-1)*

51. Write the ordered pair that represents $\overrightarrow{WX}$ for $W(4, 0)$ and $X(-3, -6)$. Then find the magnitude of $\overrightarrow{WX}$. *(Lesson 8-2)*

52. Transportation A car is being driven at 65 miles per hour. The car's tires have a diameter of 25 inches. What is the angular velocity of the wheels in revolutions per second? *(Lesson 6-2)*

53. Use the unit circle to find the value of csc 270°. *(Lesson 5-3)*

54. Determine the rational roots of the equation $12x^4 - 11x^3 - 54x^2 - 18x + 8 = 0$. *(Lesson 4-4)*

55. Without graphing, describe the end behavior of the function $y = 4x^5 - 2x^2 + 4$. *(Lesson 3-5)*

56. Find the value of the determinant $\begin{vmatrix} -1 & -2 \\ 3 & -6 \end{vmatrix}$. *(Lesson 2-5)*

57. Geometry Determine whether the figure with vertices at $(0, 3)$, $(8, 4)$, $(2, -5)$, and $(10, -4)$ is a parallelogram. Explain. *(Lesson 1-5)*

58. SAT Practice Grid-In If $2^n = 8$, what is the value of 3^{n+2}?

Extra Practice See p. A55.

GRAPHING CALCULATOR EXPLORATION

15-2A The Slope of a Curve

A Preview of Lesson 15-2

OBJECTIVE

- Approximate the slope of a curve.

Recall from Chapter 1 that the slope of a line is a measure of its steepness. The slope of a line is given by the formula $m = \frac{y_2 - y_1}{x_2 - x_1}$, where (x_1, y_1) and (x_2, y_2) are the coordinates of two distinct points on the line.

What about the slope of a *curve*? A general curve does not have the same steepness at every point, but if you look at one particular point on the graph, there will be a certain steepness at that point. How would you calculate this "slope" at a particular point?

The answer lies in an important fact about curves: the graphs of most functions are "locally linear." This means that if you look at them up close, they appear to be lines. You are familiar with this phenomenon in everyday life—the surface of Earth looks flat, even though we know it is a giant sphere.

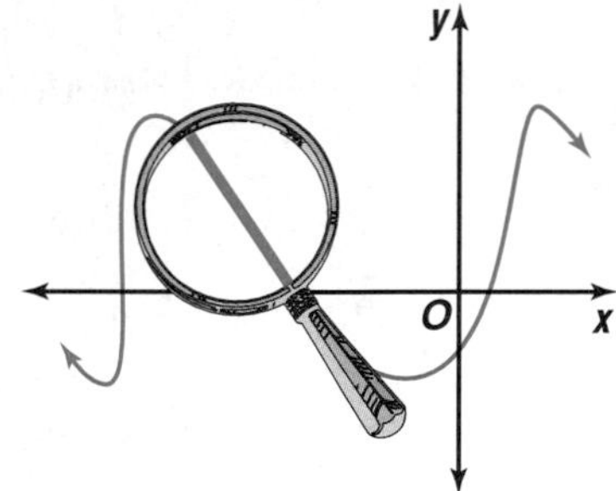

You can use ZOOM on a graphing calculator to look very closely at the graph of a function.

Example 1 **Find the slope of the graph of $y = x^2$ at (1, 1).**

Graph the equation $y = x^2$. Use the window [0, 2] by [0, 2] so that (1, 1) is at the center. Zoom in on the graph four times, using (1, 1) as the center each time. The graph should then look like the screen below. This graph is so straight that it has no visible curvature.

To approximate the slope of the graph, you can use TRACE to identify the approximate coordinates of two points on the curve. Then use the formula for slope. For example, use the coordinates (1, 1) and (1.0000831, 1.0001662).

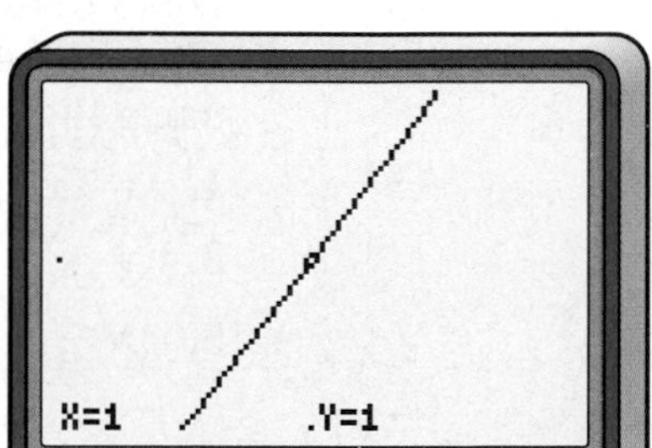

$$m \approx \frac{1.0001662 - 1}{1.0000831 - 1}$$

$$\approx \frac{0.0001662}{0.0000831}$$

$$\approx 2$$

The slope at (1, 1) is approximately 2.

You can also have the calculator find its own approximation for the slope.

Example 2 **Find the slope of the graph of $y = \frac{x^2 + 1}{x}$ at (0.5, 2.5).**

Method 1: Slope Formula

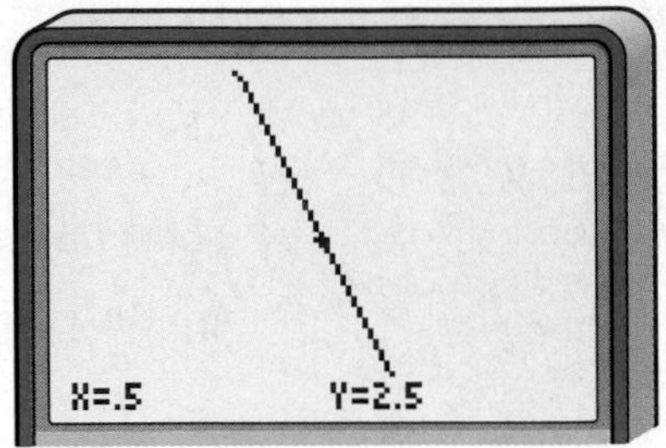

Graph the equation $y = \frac{x^2 + 1}{x}$. Use the window [0, 1] by [2, 3] so that (0.5, 2.5) is the center. Zooming in four times results in the screen shown at the right.

The **TRACE** feature shows that the point at (0.50004156, 2.4998753) is on the graph. Use these coordinates and (0.5, 2.5) to compute an approximate slope.

$$m \approx \frac{2.4998753 - 2.5}{0.50004156 - 0.5}$$
$$\approx -3.00048123$$

Our approximation to the slope is -3.00048123, which is quite close to -3.

Method 2: Calculator Computation

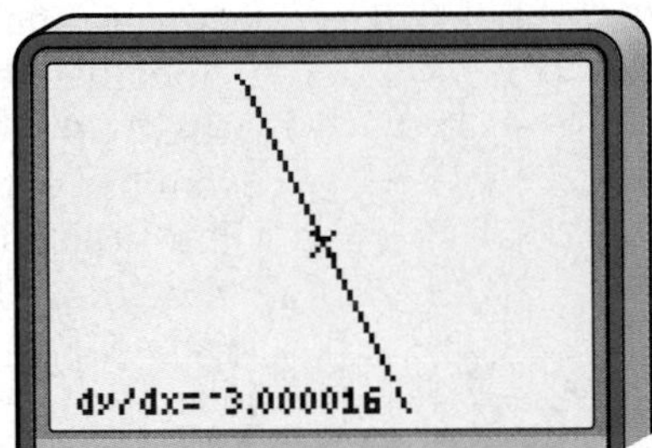

To have the calculator find an approximation, apply the ***dy/dx*** feature from the **CALC** menu at (0.5, 2.5). The calculator display is shown at the right. This also suggests that the exact value of the slope might be -3.

When you zoom in to measure the slope, you will not always obtain the exact answer. No matter how far you zoom in on the graph of a nonlinear function, the graph is never truly straight, whether it appears to be or not. Your calculation of an approximate slope may not exactly match the calculator's value for dy/dx. Sometimes your algebraic approximation may be more accurate. Other times the calculator's approximation may be more accurate.

TRY THESE

Zoom in to find the slope of the graph of each function at the given point. (Zoom in at least four times before calculating the slope.) Check your answer using the calculator's *dy/dx* feature.

1. $y = 2x^2$; (1, 2)

2. $y = \sin x$; (0, 0)

3. $y = \sqrt{x}$; (1, 1)

4. $y = 4x^4 - x^2$; (0.5, 0)

5. $y = \frac{1}{x - 3}$; (4, 1)

6. $y = \frac{x + 1}{x - 2}$; (1, −2)

WHAT DO YOU THINK?

7. For what type of function are the methods described in this lesson guaranteed to always give the exact slope?

8. What is the slope of a polynomial curve at a maximum or minimum point?

9. Graph $y = e^x$. Use the ***dy/dx*** feature to approximate the slope of the curve at several different points. What do you notice about the values of y and dy/dx?

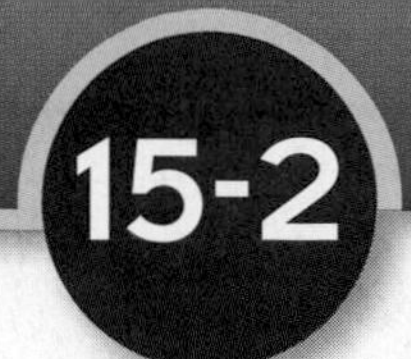

Derivatives and Antiderivatives

OBJECTIVES

- Find derivatives and antiderivatives of polynomial functions.
- Use derivatives and antiderivatives in applications.

ROCKETRY Scott and Jabbar are testing a homemade rocket in Jabbar's back yard. The boys want to keep a record of the rocket's performance so they will know if it improves when they change the design. In physics class they learned that after the rocket uses up its fuel, the rocket's height above the ground is given by the equation $H(t) = H_0 + v_0t - 16t^2$, where H_0 is the height of the rocket (in feet) when the fuel is used up, v_0 is the rocket's velocity (in feet per second) at that time, and t is the elapsed time (in seconds) since the fuel was used up. Determine the velocity of the rocket when the fuel ran out and the maximum height the rocket reached. *This problem will be solved in Example 3.*

To solve this type of problem, we need to find the **derivative** of the function H. The derivative is related to the idea of a **tangent line** from geometry. A line tangent to a curve at a point on the curve is the line that passes through that point and has a slope equal to the slope of the curve at that point. The derivative of a function $f(x)$ is another function, $f'(x)$, that gives the slope of the tangent line to $y = f(x)$ at any point.

$f'(x)$ is read "f-prime of x."

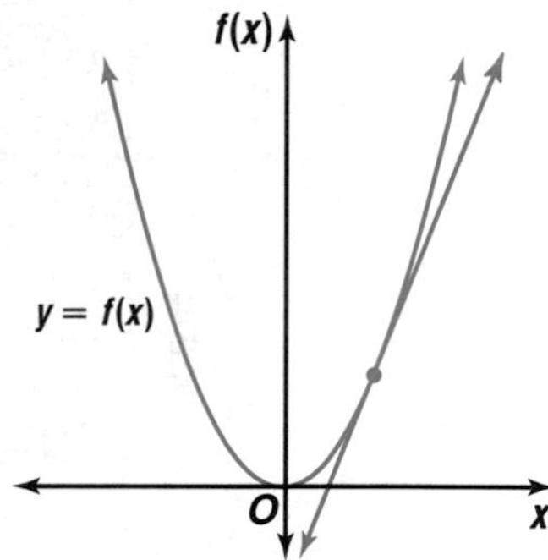

Consider the graph of $y = f(x)$ and a point $(x, f(x))$ on the graph. If the number h is close to 0, the point on the graph with x-coordinate $x + h$ will be close to $(x, f(x))$. The y-coordinate of this second point is $f(x + h)$.

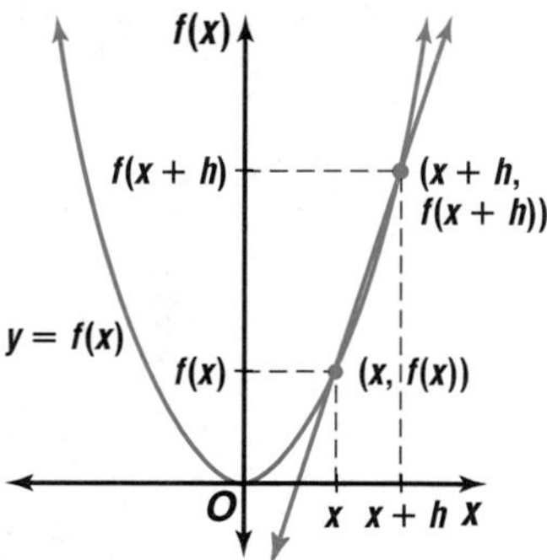

Now consider the line through the points $(x, f(x))$ and $(x + h, f(x + h))$. A line that intersects a graph in two points like this is called a **secant line.** The slope of this secant line is

$$m = \frac{f(x + h) - f(x)}{(x + h) - x} \text{ or } \frac{f(x + h) - f(x)}{h}.$$

If we make h closer and closer to 0, the point $(x + h, f(x + h))$ will get closer and closer to the original point $(x, f(x))$, so the secant line will look more and more like a tangent line. This means we can compute the slope of the tangent line by finding $\lim_{h \to 0} \frac{f(x + h) - f(x)}{h}$. This limit is the derivative of the function $f(x)$.

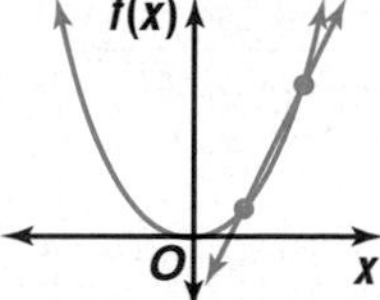

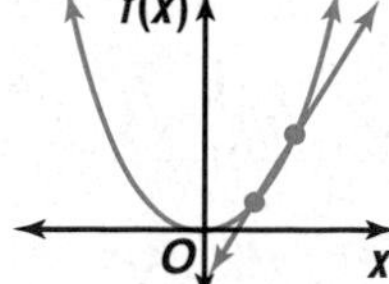

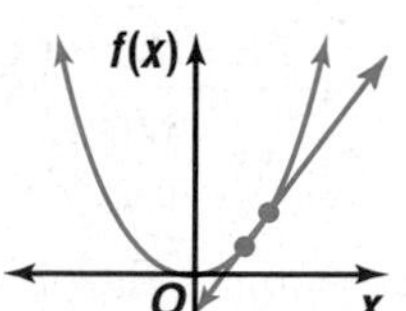

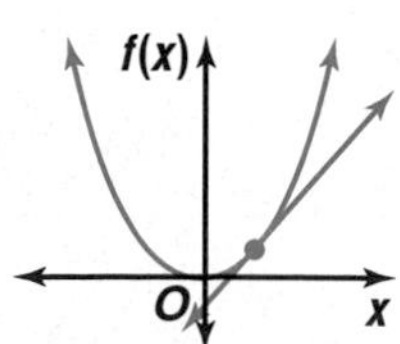

h approaches 0.

Derivative of a Function	The derivative of the function $f(x)$ is the function $f'(x)$ given by $$f'(x) = \lim_{h \to 0} \frac{f(x + h) - f(x)}{h}.$$

$\frac{dy}{dx}$ is read "dy, dx." This notation emphasizes that the derivative is a limit of slope, which is a change in y divided by a change in x.

The process of finding the derivative is called **differentiation.** Another common notation for $f'(x)$ is $\frac{dy}{dx}$. The following chart summarizes the information about tangent lines and secant lines.

Type of Line	Points of Intersection with Graph	Example	Slope
Tangent	1	$f(x)$; $(x, f(x))$; O; x; $y = f(x)$	$\frac{dy}{dx} = f'(x) = \lim_{h \to 0} \frac{f(x + h) - f(x)}{h}$
Secant	2	$f(x)$; $(x + h, f(x + h))$; $(x, f(x))$; O; x; $y = f(x)$	$m = \frac{f(x + h) - f(x)}{h}$

Example 1 **a. Find an expression for the slope of the tangent line to the graph of $y = x^2 - 4x + 2$ at any point. That is, compute $\frac{dy}{dx}$.**

b. Find the slopes of the tangent lines when $x = 0$ and $x = 3$.

a. Find and simplify $\frac{f(x + h) - f(x)}{h}$, where $f(x) = x^2 - 4x + 2$.

First, find $f(x + h)$.

$$\begin{aligned} f(x + h) &= (x + h)^2 - 4(x + h) + 2 && \textit{Replace } x \textit{ with } x + h \textit{ in } f(x). \\ &= x^2 + 2xh + h^2 - 4x - 4h + 2 \end{aligned}$$

Now find $\frac{f(x + h) - f(x)}{h}$.

$$\begin{aligned} \frac{f(x + h) - f(x)}{h} &= \frac{x^2 + 2xh + h^2 - 4x - 4h + 2 - (x^2 - 4x + 2)}{h} \\ &= \frac{2xh + h^2 - 4h}{h} && \textit{Simplify.} \\ &= \frac{h(2x + h - 4)}{h} && \textit{Factor.} \\ &= 2x + h - 4 && \textit{Divide by } h. \end{aligned}$$

Now find the limit of $2x + h - 4$ as h approaches 0 to compute $\frac{dy}{dx}$.

$$\frac{dy}{dx} = f'(x)$$
$$= \lim_{h \to 0} \frac{f(x+h) - f(x)}{h}$$
$$= \lim_{h \to 0} (2x + h - 4)$$
$$= 2x + 0 - 4$$
$$= 2x - 4$$

In the limit, only h approaches 0. x is fixed.

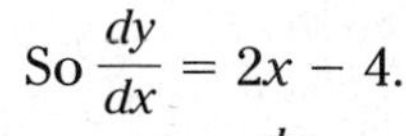

So $\frac{dy}{dx} = 2x - 4$.

b. At $x = 0$, $\frac{dy}{dx} = 2(0) - 4$ or -4. The slope of the tangent line at $x = 0$ is -4.

At $x = 3$, $\frac{dy}{dx} = 2(3) - 4$ or 2. The slope of the tangent line at $x = 3$ is 2.

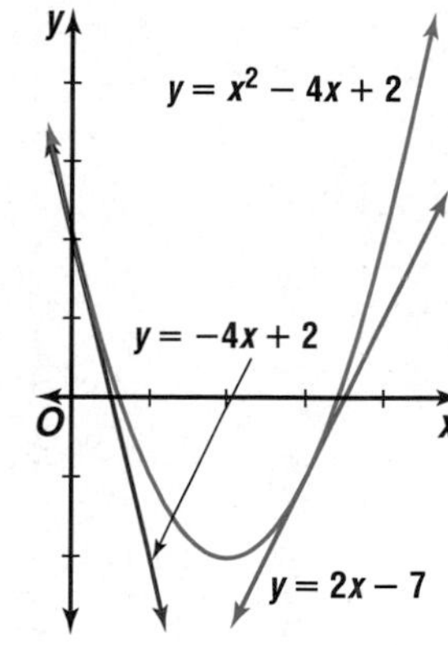

To find the derivatives of polynomials, you can use the following rules.

Derivative Rules

Constant Rule:	The derivative of a constant function is zero. If $f(x) = c$, then $f'(x) = 0$.
Power Rule:	If $f(x) = x^n$, where n is a rational number, then $f'(x) = nx^{n-1}$.
Constant Multiple of a Power Rule:	If $f(x) = cx^n$, where c is a constant and n is a rational number, then $f'(x) = cnx^{n-1}$.
Sum and Difference Rule:	If $f(x) = g(x) \pm h(x)$, then $f'(x) = g'(x) \pm h'(x)$.

Example 2 **Find the derivative of each function.**

a. $f(x) = x^6$

$f'(x) = 6x^{6-1}$ *Power Rule*
$= 6x^5$

b. $f(x) = x^2 - 4x + 2$

$f(x) = x^2 - 4x + 2$
$= x^2 - 4x^1 + 2$ *Rewrite x as a power.*

$f'(x) = 2x^{2-1} - 4 \cdot 1x^{1-1} + 0$ *Use all four rules.*
$= 2x^1 - 4x^0$
$= 2x - 4$ $\quad$ *$x^0 = 1$*

c. $f(x) = 2x^4 - 7x^3 + 12x^2 - 8x - 10$

$f'(x) = 2 \cdot 4x^3 - 7 \cdot 3x^2 + 12 \cdot 2x - 8 \cdot 1 - 0$
$= 8x^3 - 21x^2 + 24x - 8$

d. $\mathbf{f(x) = x^3 (x^2 + 5)}$

$$f(x) = x^3 (x^2 + 5)$$
$$= x^5 + 5x^3 \quad \textit{Multiply to write the function as a polynomial.}$$

$$f'(x) = 5x^4 + 5 \cdot 3x^2$$
$$= 5x^4 + 15x^2$$

e. $\mathbf{f(x) = (x^2 + 4)^2}$

$$f(x) = (x^2 + 4)^2$$
$$= x^4 + 8x^2 + 16 \quad \textit{Square to write the function as a polynomial.}$$

$$f'(x) = 4x^3 + 8 \cdot 2x + 0$$
$$= 4x^3 + 16x$$

Suppose $s(t)$ is the displacement of a moving object at time t. For example, $s(t)$ might be the object's altitude or its distance from its starting point. Then the derivative, denoted $s'(t)$ or $\frac{ds}{dt}$, is the velocity of the object at time t. Velocity is usually denoted by $v(t)$.

Example 3 **ROCKETRY** **Refer to the application at the beginning of the lesson. Suppose Scott's stopwatch shows that the rocket reached its highest point 5.3 seconds after its fuel was exhausted. Jabbar's stopwatch says that the rocket hit the ground 12.7 seconds after the fuel ran out.**

a. How fast was the rocket moving at the instant its fuel ran out?

b. What was the maximum height of the rocket?

a. We have to find the value of v_0. This value cannot be found directly from the height function $H(t)$ because H_0 is still unknown. Instead we use the velocity function $v(t)$ and what we can deduce about the velocity of the rocket at its highest point.

$$H(t) = H_0 + v_0 t - 16t^2$$
$$v(t) = H'(t) \quad \textit{The velocity of the rocket is the derivative of its height.}$$
$$= 0 + v_0 \cdot 1 - 16 \cdot 2t \quad \textit{H_0 and v_0 are constants; t is the variable.}$$
$$= v_0 - 32t$$

When the rocket was at its highest point, it was neither rising nor falling, so its velocity was 0. Substituting $v(t) = 0$ and $t = 5.3$ into the equation $v(t) = v_0 - 32t$ yields $0 = v_0 - 32(5.3)$, or $v_0 = 169.6$.

The velocity of the rocket was 169.6 ft/s when the fuel ran out.

b. We can now write the equation for the height of the rocket as $H(t) = H_0 + 169.6t - 16t^2$. When the rocket hit the ground, its height $H(t)$ was 0, so we substitute $H(t) = 0$ and $t = 12.7$ into the height equation.

$$H(t) = H_0 + 169.6t - 16t^2$$
$$0 = H_0 + 169.6(12.7) - 16(12.7)^2 \quad \textit{$H(t) = 0$, $t = 12.7$}$$
$$16(12.7)^2 - 169.6(12.7) = H_0 \quad \textit{Solve for H_0.}$$
$$H_0 = 426.72$$

The height of the rocket can now be written as $H(t) = 426.72 + 169.6t - 16t^2$. To find the maximum height of the rocket, which occurred at $t = 5.3$, compute $H(5.3)$.

$$\begin{aligned} H(t) &= 426.72 + 169.6t - 16t^2 \\ H(5.3) &= 426.72 + 169.6(5.3) - 16(5.3)^2 \quad \textit{Replace } t \textit{ with } 5.3. \\ &= 876.16 \end{aligned}$$

The maximum height of the rocket was about 876 feet.

Finding the **antiderivative** of a function is the inverse of finding the derivative. That is, instead of finding the derivative of $f(x)$, you are trying to find a function whose derivative is $f(x)$. For a function $f(x)$, the antiderivative is often denoted by $F(x)$. The relationship between the two functions is $F'(x) = f(x)$.

Example 4 Find the antiderivative of the function $f(x) = 2x$.

We are looking for a function whose derivative is $2x$. You may recall from previous examples that the function x^2 fits that description. The derivative of x^2 is $2x^{2-1}$, or $2x$.

However, x^2 is not the only function that works. The function $G(x) = x^2 + 1$ is another, since its derivative is $G'(x) = 2x + 0$ or $2x$. Another answer is $H(x) = x^2 + 17$, and still another is $J(x) = x^2 - 6$. In fact, adding any constant, positive or negative, to x^2 does not change the fact that the derivative is $2x$.

So there is an endless list of answers, all of which can be summarized by the expression $x^2 + C$, where C is any constant. So for the function $f(x) = 2x$, we say the antiderivative is $F(x) = x^2 + C$.

As with derivatives, there are rules for finding antiderivatives.

Antiderivative Rules

Power Rule:	If $f(x) = x^n$, where n is a rational number other than -1, the antiderivative is $F(x) = \frac{1}{n+1}x^{n+1} + C$.
Constant Multiple of a Power Rule:	If $f(x) = kx^n$, where n is a rational number other than -1 and k is a constant, the antiderivative is $F(x) = k \cdot \frac{1}{n+1}x^{n+1} + C$.
Sum and Difference Rule:	If the antiderivatives of $f(x)$ and $g(x)$ are $F(x)$ and $G(x)$, respectively, then the antiderivative of $f(x) \pm g(x)$ is $F(x) \pm G(x)$.

Example 5 **Find the antiderivative of each function.**

a. $\boldsymbol{f(x) = 3x^7}$

$F(x) = 3 \cdot \frac{1}{7+1} x^{7+1} + C$ *Constant Multiple of a Power Rule*

$= \frac{3}{8} x^8 + C$

b. $\boldsymbol{f(x) = 4x^2 - 7x + 5}$

$f(x) = 4x^2 - 7x + 5$

$= 4x^2 - 7x^1 + 5x^0$ *Rewrite the function so that each term has a power of x.*

$F(x) = 4 \cdot \frac{1}{3} x^3 + C_1 - \left(7 \cdot \frac{1}{2} x^2 + C_2\right) + 5 \cdot \frac{1}{1} x^1 + C_3$ *Constant Multiple of a Power and Sum and Difference Rules*

$= \frac{4}{3} x^3 - \frac{7}{2} x^2 + 5x + C$ *Let $C = C_1 - C_2 + C_3$.*

c. $\boldsymbol{f(x) = x(x^2 + 2)}$

$f(x) = x(x^2 + 2)$

$= x^3 + 2x$ *Multiply to write the function as a polynomial.*

$F(x) = \frac{1}{4} x^4 + C_1 + 2 \cdot \frac{1}{2} x^2 + C_2$ *Use all three antiderivative rules.*

$= \frac{1}{4} x^4 + x^2 + C$ *Let $C = C_1 + C_2$.*

In real-world situations, the derivative of a function is often called the **rate of change** of the function because it measures how fast the function changes. If you are given the derivative or rate of change of a function, you can find the antiderivative to recover the original function. If given additional information, you may also be able to find a value for the constant *C*.

Example 6

CENSUS **Data on the growth of world population provided by the U. S. Census Bureau can be used to create a model of Earth's population growth. According to this model, the rate of change of the world's population since 1950 is given by** $\boldsymbol{p(t) = -0.012t^2 + 48t - 47{,}925}$**, where *t* is the calendar year and *p(t)* is in millions of people per year.**

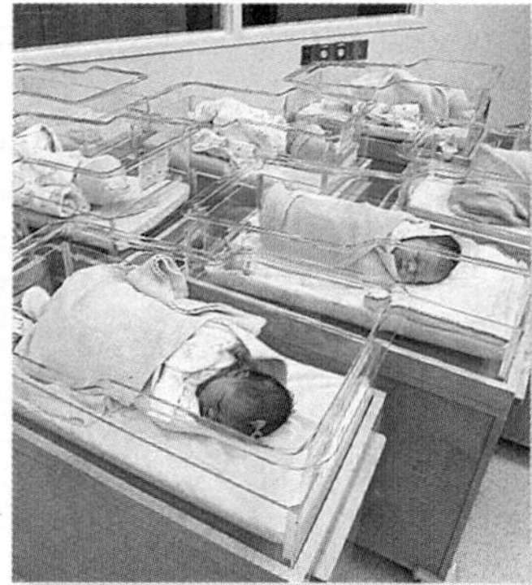

a. Given that the population in 2000 was about 6000 million people, find an equation for *P(t)*, the total population as a function of the calendar year.

b. Use the equation from part a to predict the world population in 2050.

interNET CONNECTION

Data Update
For the latest information about the population of the U.S. and the world, visit: **www.amc.glencoe.com**

a. $P(t)$ is the antiderivative of $p(t)$.

$p(t) = -0.012t^2 + 48t - 47{,}925$

$P(t) = -0.012 \cdot \frac{1}{3} t^3 + 48 \cdot \frac{1}{2} t^2 - 47{,}925t + C$ *Antiderivative rules*

$= -0.004t^3 + 24t^2 - 47{,}925t + C$

To find C, substitute 2000 for t and 6000 for $P(t)$.

$$6000 = -0.004(2000)^3 + 24(2000)^2 - 47{,}925(2000) + C$$
$$6000 = -32{,}000{,}000 + 96{,}000{,}000 - 95{,}850{,}000 + C$$
$$C = 31{,}856{,}000 \quad \textit{Solve for C.}$$

Substituting this value of C into our formula for $P(t)$ gives $P(t) = -0.004t^3 + 24t^2 - 47{,}925t + 31{,}856{,}000$. Of all the antiderivatives of $p(t)$, this is the only one that gives the proper population for the year 2000.

b. Substitute 2050 for t.

$$P(t) = -0.004t^3 + 24t^2 - 47{,}925t + 31{,}856{,}000$$
$$P(2050) = -0.004(2050)^3 + 24(2050)^2 - 47{,}925(2050) + 31{,}856{,}000$$
$$= 9250$$

According to the model, the world population in 2050 should be about 9250 million, or 9.25 billion.

CHECK FOR UNDERSTANDING

Communicating Mathematics

Read and study the lesson to answer each question.

1. **Write** two different sentences that describe the relationship between the functions $4x^3$ and x^4, one using the word *derivative,* the other using the word *antiderivative.*
2. **Explain** why the Power Rule for antiderivatives is not valid when $n = -1$.
3. *Math Journal* **Write** a paragraph explaining the difference between $f(x + h)$ and $f(x) + h$. What answer would you always get if you mistakenly used $f(x) + h$ when finding a derivative using the definition?

Guided Practice

Use the definition of derivative to find the derivative of each function.

4. $f(x) = 3x + 2$
5. $f(x) = x^2 + x$

Use the derivative rules to find the derivative of each function.

6. $f(x) = 2x^2 - 3x + 5$
7. $f(x) = -x^3 - 2x^2 + 3x + 6$
8. $f(x) = 3x^4 + 2x^3 - 3x - 2$
9. Find the slope of the tangent line to the graph of $y = x^2 + 2x + 3$ at the point where $x = 1$.

Find the antiderivative of each function.

10. $f(x) = x^2$
11. $f(x) = x^3 + 4x^2 - x - 3$
12. $f(x) = 5x^5 + 2x^3 - x^2 + 4$

13. **Business** The Better Book Company finds that the cost, in dollars, to print x copies of a book is given by the function $C(x) = 1000 + 10x - 0.001x^2$. The derivative $C'(x)$ is called the *marginal cost function.* The marginal cost is the approximate cost of printing one more book after x copies have been printed. What is the marginal cost when 1000 books have been printed?

EXERCISES

Practice

Use the definition of derivative to find the derivative of each function.

14. $f(x) = 2x$
15. $f(x) = 7x + 4$
16. $f(x) = -3x$
17. $f(x) = -4x - 9$
18. $f(x) = 2x^2 + 5x$
19. $f(x) = x^3 + 5x^2 + 6$

Use the derivative rules to find the derivative of each function.

20. $f(x) = 8x$
21. $f(x) = 2x + 6$
22. $f(x) = \frac{1}{3}x + \frac{4}{5}$
23. $f(x) = -3x^2 + 2x + 9$
24. $f(x) = \frac{1}{2}x^2 - x - 2$
25. $f(x) = x^3 - 2x^2 + 5x - 6$
26. $f(x) = 3x^4 + 7x^3 - 2x^2 + 7x - 12$
27. $f(x) = (x^2 + 3)(2x - 7)$
28. $f(x) = (2x + 4)^2$
29. $f(x) = (3x - 4)^3$

30. Find $f'(x)$ for the function $f(x) = \frac{2}{3}x^3 + \frac{1}{3}x^2 - x - 9$.

Find the slope of the tangent line to the graph of each equation at $x = 1$.

31. $y = x^3$
32. $y = x^3 - 7x^2 + 4x + 9$
33. $y = (x + 1)(x - 2)$
34. $y = (5x^2 + 7)^2$

Find the antiderivative of each function.

35. $f(x) = x^6$
36. $f(x) = 3x + 4$
37. $f(x) = 4x^2 - 6x + 7$
38. $f(x) = 12x^2 - 6x + 1$
39. $f(x) = 8x^3 + 5x^2 - 9x + 3$
40. $f(x) = \frac{1}{4}x^4 - \frac{2}{3}x^2 + 4$
41. $f(x) = (2x + 3)(3x - 7)$
42. $f(x) = x^4(x + 2)^2$
43. $f(x) = \dfrac{x^3 + 4x^2 + x}{x}$
44. $f(x) = \dfrac{2x^2 - 5x - 3}{x - 3}$

45. Find a function whose derivative is $f(x) = (x^3 - 1)(x^2 + 1)$.

Applications and Problem Solving

46. **Motion** Acceleration is the rate at which the velocity of a moving object changes. That is, acceleration is the derivative of velocity. If time is measured in seconds and velocity in feet per second, then acceleration is measured in feet per second squared, or ft/s^2. Suppose a car is moving with velocity $v(t) = 15 + 4t + \frac{1}{8}t^2$. *Feet per second squared is feet per second per second.*
    a. Find the car's velocity at $t = 12$.
    b. Find the car's acceleration at $t = 12$.
    c. Interpret your answer to part b in words.
    d. Suppose $s(t)$ is the car's distance, in feet, from its starting point. Find an equation for $s(t)$.
    e. Find the distance the car travels in the first 12 seconds.

47. **Critical Thinking** Use the definition of derivative to find the derivative of $f(x) = \frac{1}{x}$.

48. **Economics** The graph shows the annual spending on health care in the U.S. for the years 1992 to 2006 (using projections for the years after 1998.) Let $T(y)$ be the total annual spending on health care in year y.
    a. Estimate $T(2003)$ and describe what it measures.
    b. Estimate $T'(2003)$ and describe what it measures.

Health Care Spending (Trillions of Dollars)

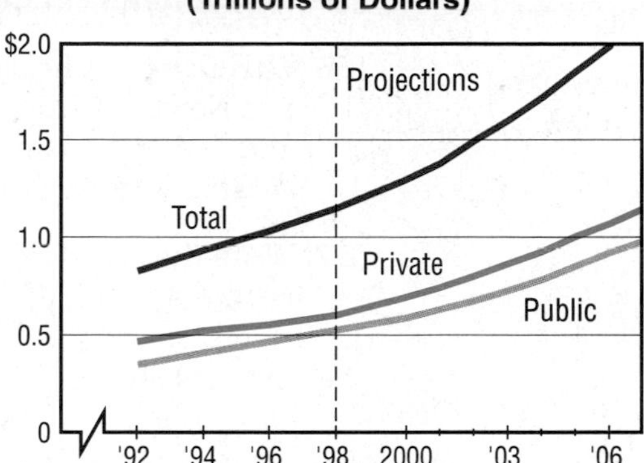

Source: Health Care Financing Administration

49. **Sports** Suppose a punter kicks a football so that the upward component of its velocity is 80 feet per second. If the ball is 3 feet off the ground when it is kicked, then the height of the ball, in feet, t seconds after it is kicked is given by $h(t) = 3 + 80t - 16t^2$.
    a. Find the upward velocity $v(t)$ of the football.
    b. How fast is the ball travelling upward 1 second after it is kicked?
    c. Find the time when the ball reaches its maximum height.
    d. What is the maximum height of the ball?

50. **Critical Thinking** The derivative of the function $f(x) = e^x$ is *not* xe^{x-1}. (e^x is an exponential function, so the Power Rule for derivatives does not apply.) Use the definition of derivative to find the correct derivative. (*Hint*: You will need a calculator to evaluate a limit that arises in the computation.)

51. **Business** Joaquin and Marva are selling lemonade. The higher the price they charge for a cup of lemonade, the fewer cups they sell. They have found that when they charge p cents for a cup of lemonade, they sell $100 - 2p$ cups in a day.
    a. Find a formula for the function $r(p)$ that gives their total daily revenue.
    b. Find the price that Joaquin and Marva should charge to generate the highest possible revenue.

Mixed Review

52. Evaluate $\lim_{x \to 3} \frac{x^2 - 2x - 3}{x - 3}$. *(Lesson 15-1)*

53. **Nutrition** The amounts of sodium, in milligrams, present in the top brands of peanut butter are given below. *(Lesson 14-3)*

195	210	180	225	225	225	195
225	203	225	195	195	188	191
210	233	225	248	225	210	240
180	225	240	180	225	240	240
195	189	178	255	225	225	225
194	210	225	195	188	205	

a. Make a box-and-whisker plot of the data.
b. Write a paragraph describing the variability of the data.

54. A pair of dice is tossed. Find the probability that their sum is greater than 7 given that the numbers match. *(Lesson 13-5)*

55. The first term of a geometric sequence is 9, and the common ratio is $-\frac{1}{3}$. Find the sixth term of the sequence. *(Lesson 12-2)*

56. **Chemistry** A beaker of water has been heated to 210°F in a room that is 74°F. Use Newton's Law of Cooling, $y = ae^{-kt} + c$, with $a = 136$°F, $k = 0.06 \text{ min}^{-1}$, and $c = 74$°F to find the temperature of the water after half an hour. *(Lesson 11-3)*

57. Write the standard form of the equation of the circle that passes through points at $(2, -1)$, $(-3, 0)$, and $(1, 4)$. *(Lesson 10-2)*

58. Express $5\left(\cos \frac{5\pi}{6} + \boldsymbol{i} \sin \frac{5\pi}{6}\right)$ in rectangular form. *(Lesson 9-6)*

59. Write parametric equations of the line passing through $P(-3, -2)$ and parallel to $\vec{\mathbf{v}} = \langle 8, 3 \rangle$. *(Lesson 8-6)*

60. Graph $y = -3 \sin(\theta - 45°)$. *(Lesson 6-5)*

61. **Surveying** A surveying crew is studying a housing project for possible relocation for the airport expansion. They are located on the ground, level with the houses. If the distance to one of the houses is 253 meters and the distance to the other is 319 meters, what is the distance between the houses if the angle subtended by them at the point of observation is 42°12′? *(Lesson 5-8)*

62. List the possible rational roots of $2x^3 + 3x^2 - 8x + 3 = 0$. Then determine the rational roots. *(Lesson 4-4)*

63. **SAT/ACT Practice**
In the figure, $x + y + z = ?$
A 0 **B** 90 **C** 180
D 270 **E** 360

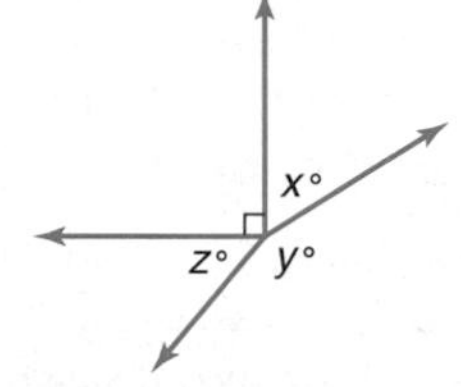

MID-CHAPTER QUIZ

Evaluate each limit. (Lesson 15-1)

1. $\lim_{x \to -3} (2x^2 - 4x + 6)$
2. $\lim_{x \to 2} \frac{x^2 - 9x + 14}{2x^2 - 7x + 6}$
3. $\lim_{x \to 0} \frac{\sin 2x}{x}$
4. Use the definition of derivative to find the derivative of $f(x) = x^2 - 3$. (Lesson 15-2)

Use the derivative rules to find the derivative of each function. (Lesson 15-2)

5. $f(x) = \pi$
6. $f(x) = 3x^2 - 5x + 2$

7. **Medicine** If $R(M)$ measures the reaction of the body to an amount M of medicine, then $R'(M)$ measures the sensitivity of the body to the medicine. Find $R'(M)$ if $R(M) = M^2\left(\frac{C}{2} - \frac{M}{3}\right)$ where C is a constant.

Find the antiderivative of each function. (Lesson 15-2)

8. $f(x) = -x^2 + 7x - 6$
9. $f(x) = 2x^3 + x^2 + 8$
10. $f(x) = -2x^4 + 6x^3 - 2x - 5$

Extra Practice See p. A55.

Area Under a Curve

OBJECTIVES

- Find values of integrals of polynomial functions.
- Find areas under graphs of polynomial functions.

BUSINESS The derivative of a cost function is called a *marginal cost function.* A shoe company determines that the marginal cost function for a particular type of shoe is $f(x) = 20 - 0.004x$, where x is the number of pairs of shoes manufactured and $f(x)$ is in dollars. If the company is already producing 2000 pairs of this type of shoe per day, how much more would it cost them to increase production to 3000 pairs per day? *This problem will be solved in Example 3.*

Problems like the one above can be solved using **integrals.** To understand integrals, we must first examine the area between the graph of a polynomial function and the x-axis for an interval from $x = a$ to $x = b$.

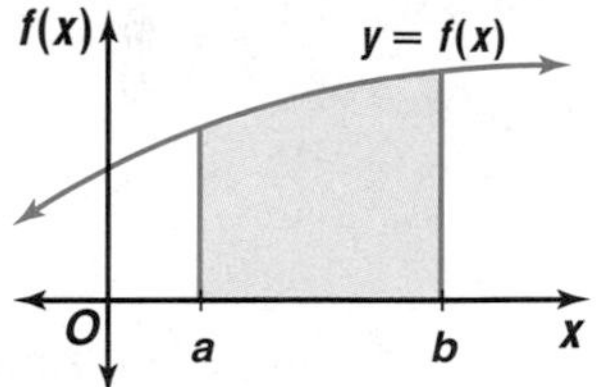

One way to estimate this area is by filling the region with rectangles, whose areas we know how to compute. If the boundary of the region is curved, the rectangles will not fit the region exactly, but you can use them for approximation. You can use rectangles of any width.

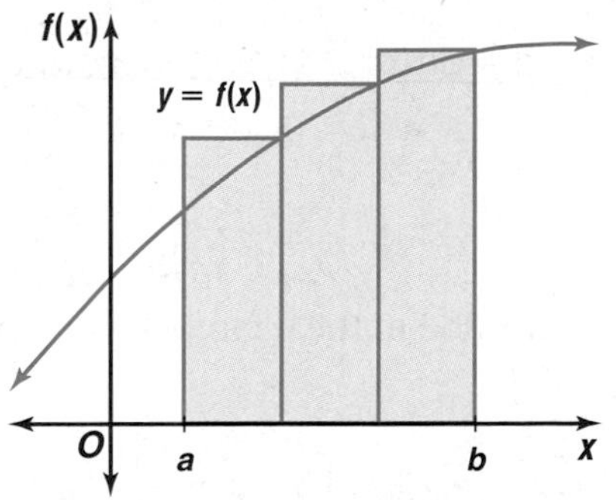

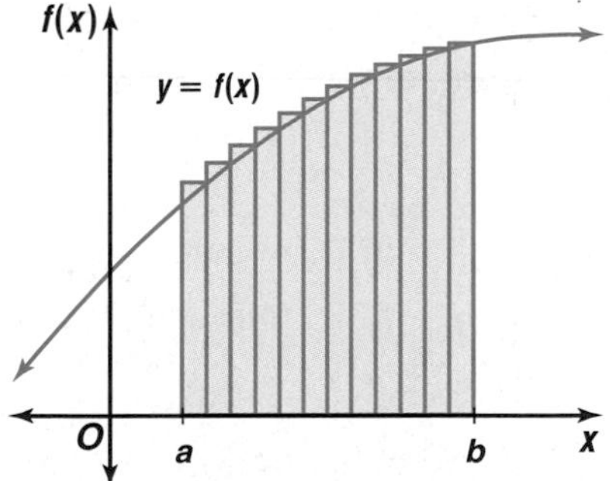

interNET CONNECTION

Graphing Calculator Programs
To download a program that uses rectangles to approximate the area under a curve, visit: **www.amc.glencoe.com**

Notice from the figures above that the thinner the rectangles are, the better they fit the region, and the better their total area approximates the area of the region. If you were to continue making the rectangles thinner and thinner, their total area would approach the exact area of the region. That is, the area of a region under the graph of a function is the limit of the total area of the rectangles as the widths of the rectangles approach 0.

In the figure below, the interval from a to b has been subdivided into n equal subintervals. A rectangle has been drawn on each subinterval. Each rectangle touches the graph at its upper right corner; the first touches at the x-coordinate x_1, the second touches at the x-coordinate x_2, and so on, with the last rectangle touching at the x-coordinate b, which is also denoted by x_n for consistency.

The height of the first rectangle is $f(x_1)$, the height of the second is $f(x_2)$, and so on, with the height of the last rectangle being $f(x_n)$. The length of the entire interval from a to b is $b - a$, so the width of each of the n rectangles must be $\frac{b-a}{n}$. This common width is traditionally denoted Δx. *Δx is read "delta x."*

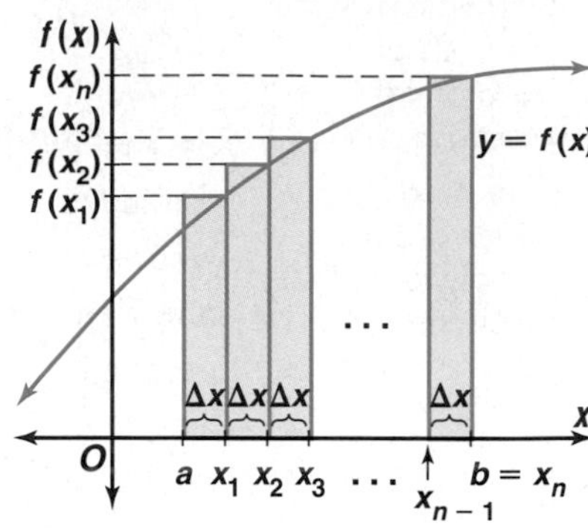

Look Back
You can refer to Lesson 12-5 to review sigma notation.

The area of the first rectangle is $f(x_1)\Delta x$, the area of the second rectangle is $f(x_2)\Delta x$, and so on. The total area A_n of the n rectangles is given by the sum of the areas.

$$A_n = f(x_1)\Delta x + f(x_2)\Delta x + \cdots + f(x_n)\Delta x$$
$$= \sum_{i=1}^{n} f(x_i)\Delta x \quad \textit{i is the index of summation, not the imaginary unit.}$$

$\int_a^b f(x)\,dx$ is read "the integral of $f(x)$ from a to b."

To make the width of the rectangles approach 0, we let the number of rectangles approach infinity. Therefore, the exact area of the region under the graph of the function is $\lim_{n\to\infty} A_n$, or $\lim_{n\to\infty} \sum_{i=1}^{n} f(x_i)\Delta x$. This limit is called a **definite integral** and is denoted $\int_a^b f(x)\,dx$.

Definite Integral

$$\int_a^b f(x)\,dx = \lim_{n\to\infty} \sum_{i=1}^{n} f(x_i)\Delta x \text{ where } \Delta x = \frac{b-a}{n}.$$

The process of finding the area under a curve is called **integration.** The following formulas will be needed in the examples and exercises.

$$1 + 2 + 3 + \cdots + n = \frac{n(n+1)}{2}$$

$$1^2 + 2^2 + 3^2 + \cdots + n^2 = \frac{n(n+1)(2n+1)}{6}$$

$$1^3 + 2^3 + 3^3 + \cdots + n^3 = \frac{n^2(n+1)^2}{4}$$

$$1^4 + 2^4 + 3^4 + \cdots + n^4 = \frac{6n^5 + 15n^4 + 10n^3 - n}{30}$$

$$1^5 + 2^5 + 3^5 + \cdots + n^5 = \frac{2n^6 + 6n^5 + 5n^4 - n^2}{12}$$

Before beginning the examples, we will derive a formula for x_i. The width Δx of each rectangle is the distance between successive x_i-values. Study the labels below the x-axis.

$\Delta x \quad x_1 \quad \Delta x \quad x_2 \quad \Delta x \quad x_3 \quad \Delta x \quad x_4 \quad \cdots \quad x_n$

$a \quad a + \Delta x \quad a + 2\Delta x \quad a + 3\Delta x \quad a + 4\Delta x \quad \cdots \quad a + n\Delta x$

We see that $x_i = a + i\Delta x$. This formula will work when finding the area under the graph of any function.

Example 1 **Use limits to find the area of the region between the graph of $y = x^2$ and the x-axis from $x = 0$ to $x = 1$. That is, find $\int_0^1 x^2\,dx$.**

First find Δx.

$$\Delta x = \frac{b - a}{n}$$

$$= \frac{1 - 0}{n} \text{ or } \frac{1}{n}$$

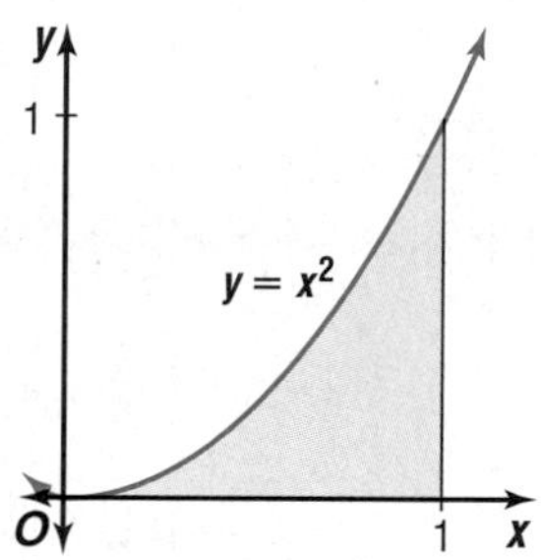

Then find x_i.

$$x_i = a + i\Delta x$$

$$= 0 + i \cdot \frac{1}{n} \text{ or } \frac{i}{n}$$

Now we can calculate the integral that gives the area.

$$\int_0^1 x^2\,dx = \lim_{n\to\infty} \sum_{i=1}^{n} (x_i)^2\,\Delta x \qquad f(x_i) = x_i^2$$

$$= \lim_{n\to\infty} \sum_{i=1}^{n} \left(\frac{i}{n}\right)^2 \left(\frac{1}{n}\right) \qquad x_i = \frac{i}{n},\ \Delta x = \frac{1}{n}$$

$$= \lim_{n\to\infty} \sum_{i=1}^{n} \frac{i^2}{n^3} \qquad \textit{Multiply.}$$

$$= \lim_{n\to\infty} \left(\frac{1^2}{n^3} + \frac{2^2}{n^3} + \cdots + \frac{n^2}{n^3}\right)$$

$$= \lim_{n\to\infty} \frac{1}{n^3}(1^2 + 2^2 + \cdots + n^2) \qquad \textit{Factor.}$$

$$= \lim_{n\to\infty} \frac{1}{n^3} \cdot \frac{n(n + 1)(2n + 1)}{6} \qquad 1^2 + 2^2 + \cdots + n^2 = \frac{n(n + 1)(2n + 1)}{6}$$

$$= \lim_{n\to\infty} \frac{2n^2 + 3n + 1}{6n^2} \qquad \textit{Multiply.}$$

$$= \lim_{n\to\infty} \frac{1}{6}\left(2 + \frac{3}{n} + \frac{1}{n^2}\right) \qquad \textit{Factor and divide by } n^2.$$

$$= \left(\lim_{n\to\infty} \frac{1}{6}\right)\left[\lim_{n\to\infty} 2 + \left(\lim_{n\to\infty} 3\right)\left(\lim_{n\to\infty} \frac{1}{n}\right) + \lim_{n\to\infty} \frac{1}{n^2}\right] \qquad \textit{Limit theorems from Chapter 12}$$

$$= \frac{1}{6}[2 + (3)(0) + 0] \text{ or } \frac{1}{3} \qquad \lim_{n\to\infty} \frac{1}{n} = 0,\ \lim_{n\to\infty} \frac{1}{n^2} = 0$$

The area of the region is $\frac{1}{3}$ square unit.

Example 2 **Use limits to find the area of the region between the graph of $y = x^3$ and the x-axis from $x = 2$ to $x = 4$.**

First, find the area under the graph from $x = 0$ to $x = 4$. Then subtract from it the area under the graph from $x = 0$ to $x = 2$. In other words,

$$\int_2^4 x^3\,dx = \int_0^4 x^3\,dx - \int_0^2 x^3\,dx.$$

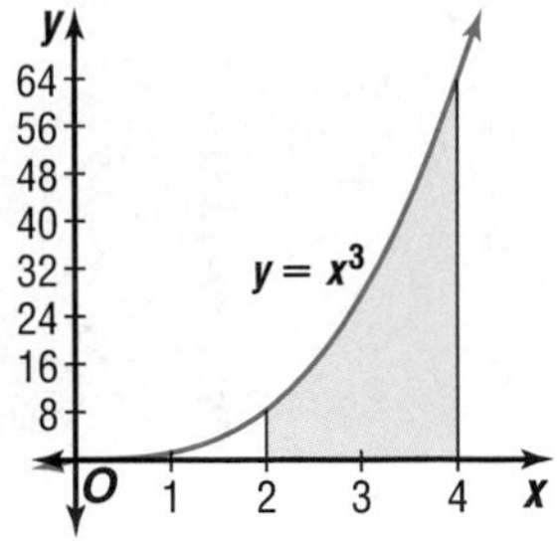

For $\int_0^4 x^3\,dx$, $a = 0$ and $b = 4$, so $\Delta x = \frac{4}{n}$ and $x_i = \frac{4i}{n}$.

$$\int_0^4 x^3\,dx = \lim_{n\to\infty} \sum_{i=1}^{n} (x_i)^3\,\Delta x \qquad f(x_i) = x_i^3$$

$$= \lim_{n\to\infty} \sum_{i=1}^{n} \left(\frac{4i}{n}\right)^3 \cdot \frac{4}{n} \qquad x_i = \frac{4i}{n},\ \Delta x = \frac{4}{n}$$

$$= \lim_{n\to\infty} \sum_{i=1}^{n} \frac{256i^3}{n^4}$$

$$= \lim_{n\to\infty}\left(\frac{256\cdot 1^3}{n^4} + \frac{256\cdot 2^3}{n^4} + \cdots + \frac{256\cdot n^3}{n^4}\right)$$

$$= \lim_{n\to\infty} \frac{256}{n^4}\cdot(1^3 + 2^3 + \cdots + n^3)$$

$$= \lim_{n\to\infty} \frac{256}{n^4}\cdot\frac{n^2(n+1)^2}{4} \qquad 1^3 + 2^3 + \cdots + n^3 = \frac{n^2(n+1)^2}{4}$$

$$= \lim_{n\to\infty} \frac{64n^2 + 128n + 64}{n^2}$$

$$= \lim_{n\to\infty}\left(64 + \frac{128}{n} + \frac{64}{n^2}\right) \qquad \text{Divide by } n^2.$$

$$= 64 + 0 + 0 \text{ or } 64$$

For $\int_0^2 x^3\,dx$, $a = 0$ and $b = 2$, so $\Delta x = \frac{2}{n}$ and $x_i = \frac{2i}{n}$.

$$\int_0^2 x^3\,dx = \lim_{n\to\infty} \sum_{i=1}^{n} (x_i)^3\,\Delta x$$

$$= \lim_{n\to\infty} \sum_{i=1}^{n} \left(\frac{2i}{n}\right)^3 \cdot \frac{2}{n} \qquad x_i = \frac{2i}{n},\ \Delta x = \frac{2}{n}$$

$$= \lim_{n\to\infty} \sum_{i=1}^{n} \frac{16i^3}{n^4}$$

$$= \lim_{n\to\infty}\left(\frac{16\cdot 1^3}{n^4} + \frac{16\cdot 2^3}{n^4} + \cdots + \frac{16\cdot n^3}{n^4}\right)$$

$$= \lim_{n\to\infty} \frac{16}{n^4}\cdot(1^3 + 2^3 + \cdots + n^3)$$

$$= \lim_{n\to\infty} \frac{16}{n^4}\cdot\frac{n^2(n+1)^2}{4} \qquad 1^3 + 2^3 + \cdots + n^3 = \frac{n^2(n+1)^2}{4}$$

$$= \lim_{n\to\infty} \frac{4n^2 + 8n + 4}{n^2}$$

$$= \lim_{n\to\infty}\left(4 + \frac{8}{n} + \frac{4}{n^2}\right) \quad \textit{Divide by } n^2.$$

$$= 4 + 0 + 0 \text{ or } 4$$

The area of the region between the graph of $y = x^3$ and the x-axis from $x = 2$ to $x = 4$ is $64 - 4$, or 60 square units.

In physics, when the velocity of an object is graphed with respect to time, the area under the curve represents the displacement of the object. In business, the area under the graph of a marginal cost function from $x = a$ to $x = b$ represents the amount it would cost to increase production from a units to b units.

Example 3 **BUSINESS** **Refer to the application at the beginning of the lesson. How much would it cost the shoe company to increase production from 2000 pairs per day to 3000 pairs per day?**

Since f(x) is a linear function, we can calculate the value directly, without subtracting integrals as in Example 2.

The cost is given by $\int_{2000}^{3000} f(x)\,dx$ where $f(x) = 20 - 0.004x$ is the marginal cost function.

$a = 2000$ and $b = 3000$, so $\Delta x = \frac{1000}{n}$ and $x_i = 2000 + \frac{1000i}{n}$.

$$\int_{2000}^{3000} f(x)\,dx = \lim_{n\to\infty}\sum_{i=1}^{n} f(x_i)\Delta x$$

$$= \lim_{n\to\infty}\sum_{i=1}^{n}(20 - 0.004x_i)\Delta x \quad f(x_i) = 20 - 0.004x_i$$

$$= \lim_{n\to\infty}\sum_{i=1}^{n}\left[20 - 0.004\left(2000 + \frac{1000i}{n}\right)\right]\cdot\frac{1000}{n}$$

$$= \lim_{n\to\infty}\sum_{i=1}^{n}\left(12 - \frac{4i}{n}\right)\cdot\frac{1000}{n} \quad \textit{Simplify.}$$

$$= \lim_{n\to\infty}\frac{1000}{n}\left[\left(12 - \frac{4\cdot 1}{n}\right) + \left(12 - \frac{4\cdot 2}{n}\right) + \cdots + \left(12 - \frac{4\cdot n}{n}\right)\right]$$

$$= \lim_{n\to\infty}\frac{1000}{n}\cdot\left[12n - \frac{4}{n}(1 + 2 + \cdots + n)\right] \quad \textit{Combine and factor.}$$

$$= \lim_{n\to\infty}\frac{1000}{n}\cdot\left[12n - \frac{4}{n}\cdot\frac{n(n+1)}{2}\right] \quad 1 + 2 + \cdots + n = \frac{n(n+1)}{2}$$

$$= \lim_{n\to\infty}\frac{1000}{n}\cdot(10n - 2) \quad \textit{Simplify.}$$

$$= \lim_{n\to\infty}\frac{10{,}000n - 2000}{n} \quad \textit{Multiply.}$$

$$= \lim_{n\to\infty}\left(10{,}000 - \frac{2000}{n}\right) \quad \textit{Divide by } n.$$

$$= 10{,}000 - 0, \text{ or } 10{,}000$$

The increase in production would cost the company \$10,000.

CHECK FOR UNDERSTANDING

Communicating Mathematics

Read and study the lesson to answer each question.

1. **Write** an equation of a function for which you would need the formula for $1^4 + 2^4 + 3^4 + \cdots + n^4$ to find the area under the graph.
2. **Describe** the steps involved in finding the area under the graph of $y = f(x)$ between $x = a$ and $x = b$.
3. **You Decide** Rita says that when you use rectangles that touch the graph of a function at their upper right corners, the total area of the rectangles will always be greater than the area under the curve because the rectangles stick out above the curve. Lorena disagrees. Who is correct? Explain.

Guided Practice

4. Use a limit to find the area of the shaded region in the graph at the right.

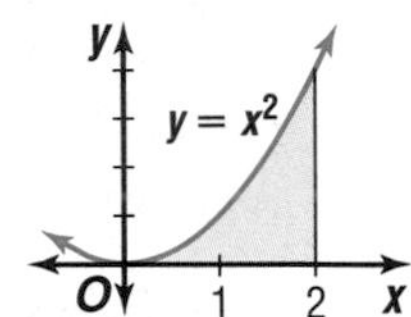

Use limits to find the area between each curve and the x-axis for the given interval.

5. $y = x^2$ from $x = 1$ to $x = 3$
6. $y = x^3$ from $x = 0$ to $x = 1$

Use limits to evaluate each integral.

7. $\int_0^6 x^2\,dx$
8. $\int_0^3 x^3\,dx$

9. **Physics** Neglecting air resistance, an object in free fall accelerates at 32 feet per second squared. So the velocity of the object t seconds after being dropped is $32t$ feet per second. Suppose a ball is dropped from the top of the Sears Tower.
   a. Use integration to find how far the ball would fall in the first six seconds.
   b. Refer to the graph at the right. Would the ball hit the ground within ten seconds of being dropped? Explain your reasoning.

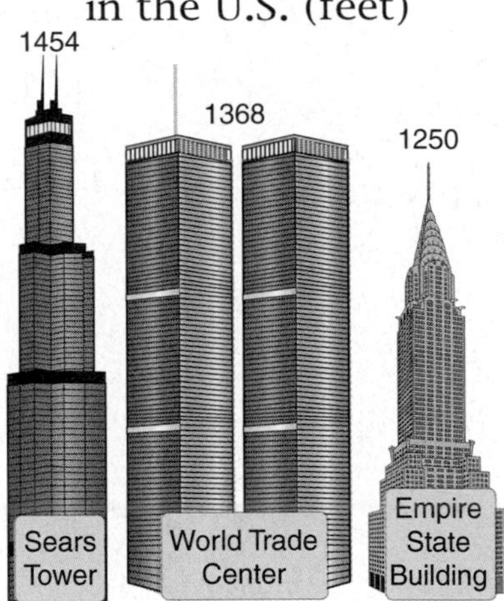

EXERCISES

Practice

Use limits to find the area of the shaded region in each graph.

10.

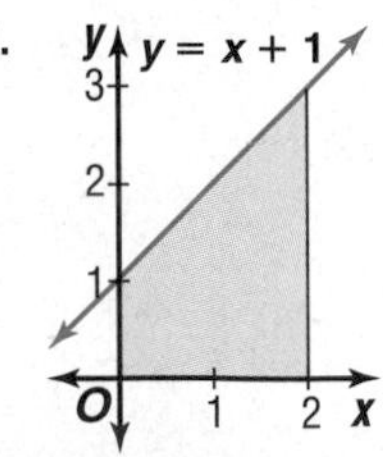

11.

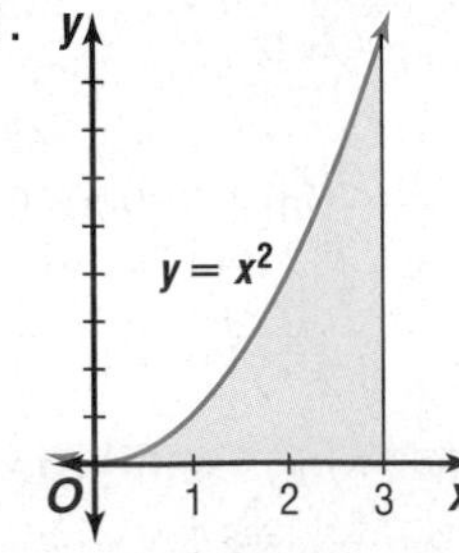

12.

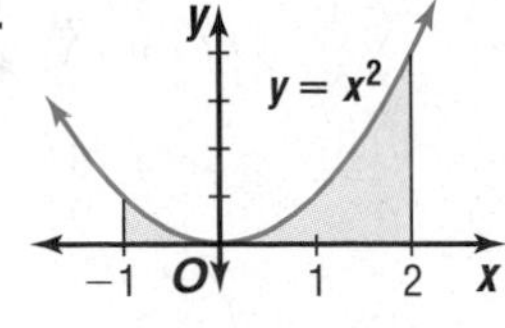

Use limits to find the area between each curve and the *x*-axis for the given interval.

13. $y = x$ from $x = 1$ to $x = 3$

14. $y = x^2$ from $x = 0$ to $x = 5$

15. $y = 2x^3$ from $x = 1$ to $x = 5$

16. $y = x^4$ from $x = 0$ to $x = 5$

17. $y = x^2 + 6x$ from $x = 0$ to $x = 4$

18. $y = x^2 - x + 1$ from $x = 0$ to $x = 3$

19. Write a limit that gives the area under the graph of $y = \sin x$ from $x = 0$ to $x = \pi$. (Do *not* evaluate the limit.)

Use limits to evaluate each integral.

20. $\int_0^2 8x\,dx$

21. $\int_1^4 (x + 2)\,dx$

22. $\int_0^4 x^2\,dx$

23. $\int_3^5 8x^3\,dx$

24. $\int_1^4 (x^2 + 4x - 2)\,dx$

25. $\int_0^2 (x^5 + x^2)\,dx$

26. Find the integral of x^3 from 0 to 5.

Applications and Problem Solving

27. **Sewing** A patch in the shape of the region shown at the right is to be sewn onto a flag. If each unit in the coordinate system represents one foot, how much material is required for the patch?

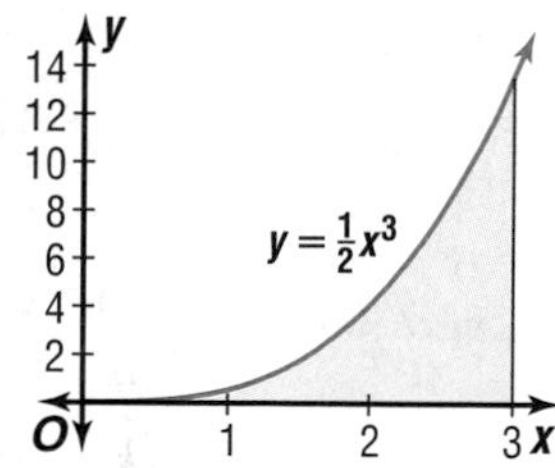

28. **Business** Suppose the Auburn Widget Corporation finds that the marginal cost function associated with producing x widgets is $f(x) = 80 - 2x$ dollars.
    a. Refer to Exercise 13 of Lesson 15-2. Use the marginal cost function to approximate the cost for the company to produce one more widget when the production level is 20 widgets.
    b. How much would it cost the company to double its production from 20 widgets to 40 widgets?

29. **Mining** In order to distribute stress, mine tunnels are sometimes rounded. Suppose that the vertical cross sections of a tunnel can be modeled by the parabola $y = 6 - 0.06x^2$. If x and y are measured in feet, how much rock would have to be moved to make such a tunnel that is 100 feet long?

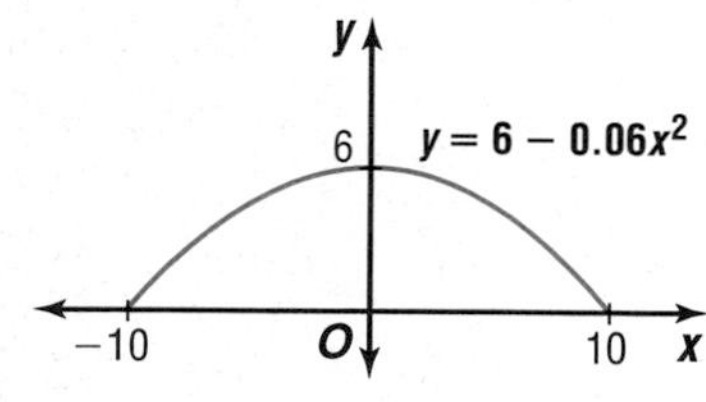

30. **Critical Thinking** Find the area of the region enclosed by the line $y = x$ and the parabola $y = x^2$.

31. Budgets If the function $r(t)$ gives the rate at which a family spends money, then the total money spent between times $t = a$ and $t = b$ is $\int_a^b r(t)\,dt$. A local electric company in Alabama, where electric bills are generally low in winter and very high in summer, offers customers the option of paying a flat monthly fee for electricity throughout the year so that customers can avoid enormous summertime bills. The company has found that in past years the Johnson family's rate of electricity spending can be modeled by $r(t) = 50 + 36t - 3t^2$ dollars per month, where t is the number of months since the beginning of the year.

a. Sketch a graph of the function $r(t)$ for $0 \leq t \leq 12$.

b. Find the total amount of money the Johnsons would spend on electricity during a full year.

c. If the Johnsons choose the option of paying a flat monthly fee, how much should the electric company charge them each month?

32. Sports A sprinter is trying to decide between two strategies for running a race. She can put a lot of energy into an initial burst of speed, which gives her a velocity of $v(t) = 3.5t - 0.25t^2$ meters per second after t seconds, or she can save her energy for more acceleration at the end so that her velocity is given by $v(t) = 1.2t + 0.03t^2$.

a. Graph the two velocity functions on the same set of axes for $0 \leq t \leq 10$.

b. Use integration to determine which velocity results in a greater distance covered in a 10-second race.

33. Critical Thinking Find the value of $\int_{-r}^{r} \sqrt{r^2 - x^2}\,dx$, where r is a constant.

Mixed Review

34. Find the derivative of $f(x) = -3x^3 + x^2 - 7x$. *(Lesson 15-2)*

35. Evaluate $\lim_{x \to 2} \frac{x - 2}{x + 2}$. *(Lesson 15-1)*

36. Solve the equation $\log_{\frac{1}{3}} x = -3$. *(Lesson 11-4)*

37. Find an ordered triple to represent $\vec{u}$ if $\vec{u} = \vec{v} + \vec{w}$, $\vec{v} = \langle 2, -5, -3 \rangle$, and $\vec{w} = \langle -3, 4, -7 \rangle$. Then write $\vec{u}$ as the sum of unit vectors. *(Lesson 8-3)*

38. If $\sin r = \frac{3}{5}$ and r is in the first quadrant, find $\cos 2r$. *(Lesson 7-4)*

39. State the amplitude and period for the function $y = \frac{1}{2} \sin 10\theta$. *(Lesson 6-4)*

40. **Manufacturing** A cereal manufacturer wants to make a cardboard cereal box of maximum volume. The function representing the volume of the box is $v(x) = -0.7x^3 + 5x^2 + 7x$, where x is the width of the box in centimeters. Find the width of the box that will maximize the volume. *(Lesson 3-6)*

41. **SAT/ACT Practice** Triangle ABC has sides that are 6, 8, and 10 inches long. A rectangle that has an area equal to that of the triangle has a width of 3 inches. Find the perimeter of the rectangle in inches.

A 30 **B** 24 **C** 22 **D** 16 **E** 11

Extra Practice See p. A55.

HISTORY of MATHEMATICS

CALCULUS

Calculus is fundamental to solving problems in the sciences and engineering. Two basic tools of calculus are differentiation and integration. Some of the basic ideas of calculus began to develop over 2000 years ago, but a usable form was not developed until the seventeenth century.

Early Evidence Several ideas basic to the development of calculus are the concepts of limit, infinite processes, and approximation. The Egyptians and Babylonians solved problems, such as finding the areas of circles and the volumes of pyramids, by methods resembling calculus. In about 450 B.C., **Zeno** of Elea posed problems, often called Zeno's Paradoxes, dealing with infinity. In trying to deal with these paradoxes, **Eudoxus** (about 370 B.C.), a Greek, proposed his "method of exhaustion," which is based on the idea of infinite processes. An example of this method is to show that the difference in area between a circle and an inscribed polygon can be made smaller and smaller by increasing the number of sides of the polygon.

The Renaissance Mathematicians and scientists, such as **Johann Kepler** (1571–1630), **Pierre Fermat** (1601–1665), **Gilles Roberval** (1602–1675), and **Bonaventura Cavalieri** (1598–1647), used the concept of summing an infinite number of strips to find the area under a curve. Cavalieri called this the "method of indivisibles." The use of coordinates and the development of analytic geometry by Fermat and **Renè Descartes** (1596–1650) aided in the further development of calculus.

Modern Era Most historians name **Gottfried Leibniz** (1646–1716) and **Isaac Newton** (1642–1727) as coinventors of calculus. They worked independently at approximately the same time on ideas which evolved into what is known as calculus today. In the argument over which mathematician developed calculus first, it seems that Newton had the ideas first, but did not publish them until after Leibniz made his ideas public. However, the notation used by Leibniz was more understandable than that of Newton, and much of it is still in use.

Tahani R. Amer

Today aerospace engineers like **Tahani R. Amer** use calculus in many aspects of their jobs. In her job at the NASA Langley Research Center, she uses calculus for characterizing pressure measurements taken during wind tunnel tests of experimental aircraft and for working with optical measurements.

ACTIVITIES

1. Demonstrate the method of exhaustion. Draw three circles of equal radii. In the first circle, inscribe a triangle, in the second a square, and in the third a pentagon. Find the difference between the area of each circle and its inscribed polygon.

2. Fermat discovered a simple method for finding the maximum and minimum points of polynomial curves. Consider the curve $y = 2x^3 - 5x^2 + 4x - 7$. If another point has abscissa $x + E$, then the ordinate is $2(x + E)^3 - 5(x + E)^2 + 4(x + E) - 7$. He set this expression equal to the original function and arrived at the equation $(6x^2 - 10x + 4)E + (6x - 5)E^2 + 2E^3 = 0$. Finish Fermat's method. Divide each term by E. Then let E be 0. What is the relationship between the roots of the resulting equation and the derivative of $2x^3 - 5x^2 + 4x - 7$?

3. *inter*NET CONNECTION Find out more about persons referenced in this article and others who contributed to the history of calculus. Visit **www.amc.glencoe.com**

The Fundamental Theorem of Calculus

OBJECTIVES

- Use the Fundamental Theorem of Calculus to evaluate definite integrals of polynomial functions.
- Find indefinite integrals of polynomial functions.

CONSTRUCTION Two construction contractors have been hired to clean the Gateway Arch in St. Louis. The Arch is very close to a parabola in shape, 630 feet high and 630 feet across at the bottom. Using the point on the ground directly below the apex of the Arch as the origin, the equation of the Arch is approximately $y = 630 - \frac{x^2}{157.5}$. One contractor's first idea for approaching the project is to build scaffolding in the entire space under the Arch, so that the cleaning crew can easily climb up and down to any point on the Arch. The other contractor thinks there is too much space under the Arch to make the scaffolding practical. To settle the matter, the contractors want to find out how much area there is under the Arch. *This problem will be solved in Example 4.*

interNET CONNECTION

Research For more information about the dimensions and shape of the Gateway Arch, visit: **www.amc.glencoe.com**

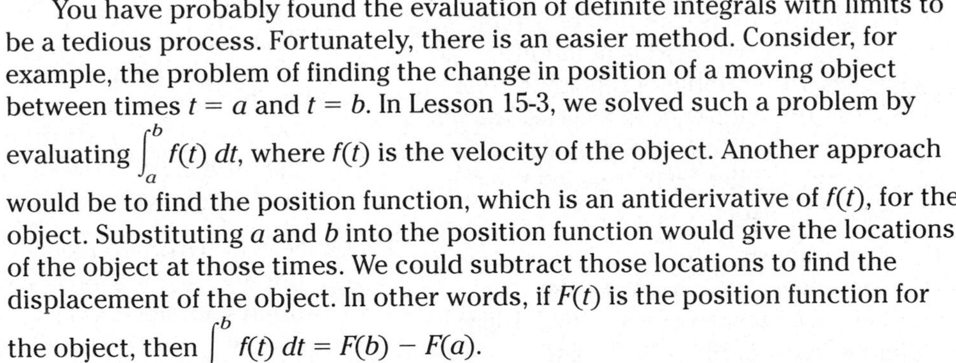

You have probably found the evaluation of definite integrals with limits to be a tedious process. Fortunately, there is an easier method. Consider, for example, the problem of finding the change in position of a moving object between times $t = a$ and $t = b$. In Lesson 15-3, we solved such a problem by evaluating $\int_a^b f(t)\,dt$, where $f(t)$ is the velocity of the object. Another approach would be to find the position function, which is an antiderivative of $f(t)$, for the object. Substituting a and b into the position function would give the locations of the object at those times. We could subtract those locations to find the displacement of the object. In other words, if $F(t)$ is the position function for the object, then $\int_a^b f(t)\,dt = F(b) - F(a)$.

The above relationship is actually true for *any* continuous function $f(x)$. This connection between definite integrals and antiderivatives is so important that it is called the **Fundamental Theorem of Calculus.**

Fundamental Theorem of Calculus

If $F(x)$ is the antiderivative of the continuous function $f(x)$, then

$$\int_a^b f(x)\,dx = F(b) - F(a).$$

The Fundamental Theorem of Calculus provides a way to evaluate the definite integral $\int_a^b f(x)\,dx$ if an antiderivative $F(x)$ can be found. A vertical line on the right side is used to abbreviate $F(b) - F(a)$. Thus, the principal statement of the theorem may be written as follows.

$$\int_a^b f(x)\,dx = F(x)\,\Big|_a^b = F(b) - F(a)$$

Example 1 **Evaluate $\int_2^4 x^3\,dx$.**

The antiderivative of $f(x) = x^3$ is $F(x) = \frac{1}{4}x^4 + C$.

$$\int_2^4 x^3\,dx = \frac{1}{4}x^4 + C\Big|_2^4 \quad \textit{Fundamental Theorem of Calculus}$$

$$= \left(\frac{1}{4}\cdot 4^4 + C\right) - \left(\frac{1}{4}\cdot 2^4 + C\right) \quad \textit{Let } x = 4 \textit{ and 2 and subtract.}$$

$$= 64 - 4 \text{ or } 60$$

Notice how much easier this example was than Example 2 of Lesson 15-3. Also notice that C was eliminated during the calculation. This always happens when you use the Fundamental Theorem to evaluate a definite integral. So in this situation you can neglect the constant term when writing the antiderivative.

Due to the connection between definite integrals and antiderivatives, the antiderivative of $f(x)$ is often denoted by $\int f(x)\,dx$. $\int f(x)\,dx$ is called the **indefinite integral** of $f(x)$.

It is helpful to rewrite the antiderivative rules in terms of indefinite integrals.

Antiderivative Rules		
	Power Rule:	$\int x^n\,dx = \frac{1}{n+1}x^{n+1} + C$, where n is a rational number and $n \neq -1$.
	Constant Multiple of a Power Rule:	$\int kx^n\,dx = k\cdot\frac{1}{n+1}x^{n+1} + C$, where k is a constant, n is a rational number, and $n \neq -1$.
	Sum and Difference Rule:	$\int [f(x) \pm g(x)]\,dx = \int f(x)\,dx \pm \int g(x)\,dx$

Example 2 **Evaluate each indefinite integral.**

a. $\int 5x^2\,dx$

$$\int 5x^2\,dx = 5\cdot\frac{1}{3}x^3 + C \quad \textit{Constant Multiple of a Power Rule}$$

$$= \frac{5}{3}x^3 + C \quad \textit{Simplify.}$$

b. $\int (4x^5 + 7x^2 - 4x)\,dx$

$$\int (4x^5 + 7x^2 - 4x)\,dx = 4\cdot\frac{1}{6}x^6 + 7\cdot\frac{1}{3}x^3 - 4\cdot\frac{1}{2}x^2 + C \quad \textit{Remember } x = x^1.$$

$$= \frac{2}{3}x^6 + \frac{7}{3}x^3 - 2x^2 + C \quad \textit{Simplify.}$$

Examples **3** **Find the area of the shaded region.**

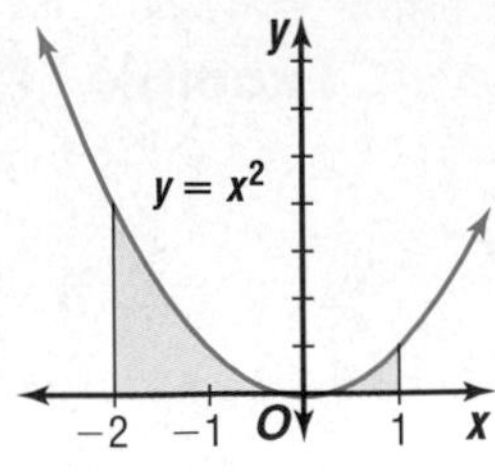

The area is given by $\int_{-2}^{1} x^2\,dx$.

The antiderivative of $f(x) = x^2$ is $F(x) = \frac{1}{3}x^3 + C$.

$$\int_{-2}^{1} x^2\,dx = \frac{1}{3}x^3 \Big|_{-2}^{1}$$ *+ C is not needed with a definite integral.*

$$= \frac{1}{3}(1)^3 - \frac{1}{3}(-2)^3$$ *Let x = 1 and −2 and subtract.*

$$= 3$$

The area of the region is 3 square units.

4 **CONSTRUCTION** **Refer to the application at the beginning of the lesson. What is the area under the Gateway Arch?**

The area is given by $\int_{-315}^{315} \left(630 - \frac{x^2}{157.5}\right) dx$.

315 and −315 are the x-intercepts of the parabola that models the Arch.

$$\int_{-315}^{315} \left(630 - \frac{x^2}{157.5}\right) dx = \int_{-315}^{315} \left(630 - \frac{1}{157.5}x^2\right) dx$$ *Rewrite the function.*

$$= 630x - \frac{1}{157.5} \cdot \frac{1}{3}x^3 \Big|_{-315}^{315}$$ *Antiderivative; + C not needed.*

$$= \left(630 \cdot 315 - \frac{1}{472.5}(315)^3\right)$$ *Let x = 315 and −315 and subtract.*

$$- \left(630 \cdot (-315) - \frac{1}{472.5}(-315)^3\right)$$

$$= 132{,}300 - (-132{,}300) \text{ or } 264{,}600$$

The area under the Arch is 264,600 square feet.

CHECK FOR UNDERSTANDING

Communicating Mathematics

Read and study the lesson to answer each question.

1. **Explain** the difference between $\int f(x)\,dx$ and $\int_a^b f(x)\,dx$.
2. **Find a counterexample** to the statement $\int_a^b f(x)g(x)\,dx = \int_a^b f(x)\,dx \cdot \int_a^b g(x)\,dx$ *for all a and b and all functions f(x) and g(x).*
3. **Explain** why the "+ C" is not needed in the antiderivative when evaluating a definite integral.
4. **You Decide** Cole says that when evaluating a definite integral, the order in which you substitute a and b into the antiderivative and subtract does not matter. Rose says it does matter. Who is correct? Explain.

Guided Practice Evaluate each indefinite integral.

5. $\int (2x^2 - 4x + 3)\,dx$

6. $\int (x^3 + 3x + 1)\,dx$

7. Find the area of the shaded region in the graph at the right.

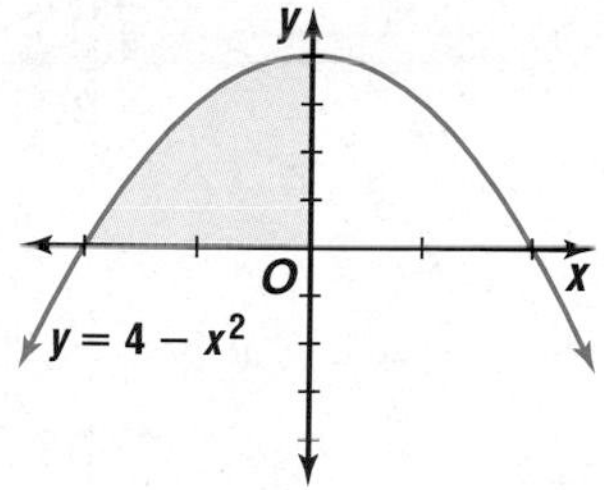

Find the area between each curve and the x-axis for the given interval.

8. $y = x^4$ from $x = 0$ to $x = 2$

9. $y = x^2 + 4x + 4$ from $x = -1$ to $x = 1$

Evaluate each definite integral.

10. $\int_1^3 2x^3\,dx$

11. $\int_1^4 (x^2 - x + 6)\,dx$

12. $\int_0^2 (-2x^2 + 3x + 2)\,dx$

13. $\int_2^4 (x^3 + x - 6)\,dx$

14. **Physics** The work, in joules (J), required to stretch a certain spring a distance of ℓ meters beyond its natural length is given by $W = \int_0^{\ell} 500x\,dx$. How much work is required to stretch the spring 10 centimeters beyond its natural length?

EXERCISES

Practice Evaluate each indefinite integral.

15. $\int x^5\,dx$

16. $\int 6x^7\,dx$

17. $\int (x^2 - 2x + 4)\,dx$

18. $\int (-3x^2 - x + 6)\,dx$

19. $\int (x^4 + 2x^2 - 3)\,dx$

20. $\int (4x^5 - 6x^3 + 7x^2 - 8)\,dx$

21. Find the antiderivative of $x^2 - 6x + 3$.

Find the area of the shaded region in each graph.

22. 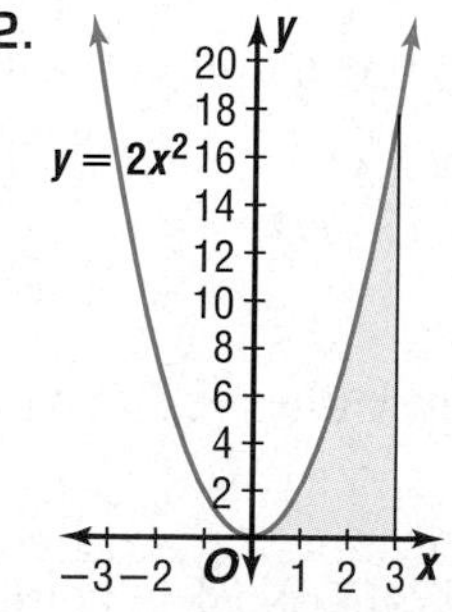

23. 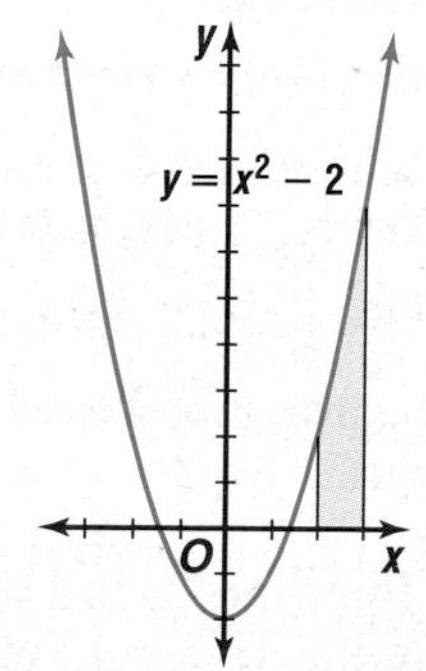

24. 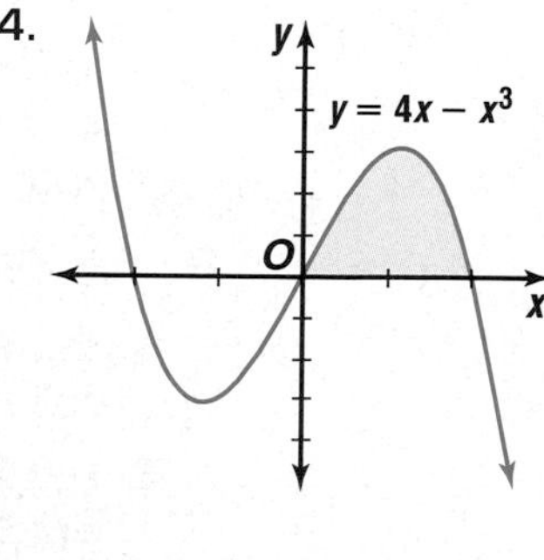

Find the area between each curve and the *x*-axis for the given interval.

25. $y = x^3$ from $x = 0$ to $x = 4$

26. $y = 3x^6$ from $x = -1$ to $x = 1$

27. $y = x^2 - 2x$ from $x = -2$ to $x = 0$

28. $y = -x^2 + 2x + 3$ from $x = 1$ to $x = 3$

29. $y = x^3 + x$ from $x = 0$ to $x = 1$

30. $y = x^3 + 8x + 10$ from $x = -1$ to $x = 3$

Evaluate each definite integral.

31. $\int_0^7 6x^2\,dx$

32. $\int_2^4 3x^4\,dx$

33. $\int_{-1}^3 (x + 4)\,dx$

34. $\int_1^5 (3x^2 - 2x + 1)\,dx$

35. $\int_1^3 (x^3 - x^2)\,dx$

36. $\int_0^1 (x^4 + 2x^2 + 1)\,dx$

37. $\int_{-1}^0 (x^4 - x^3)\,dx$

38. $\int_0^2 (x^3 + x + 1)\,dx$

39. $\int_{-2}^5 (x^2 - 3x + 8)\,dx$

40. $\int_1^3 (x + 3)(x - 1)\,dx$

41. $\int_2^3 (x - 1)^3\,dx$

42. $\int_0^1 \frac{x^2 - x - 2}{x - 2}\,dx$

43. Find the integral of $x(4x^2 + 1)$ from 0 to 2.

44. What is the integral of $(x + 1)(3x + 2)$ from -1 to 1?

45. The integral $\int_0^{n+0.5} x^k\,dx$ gives a fairly close, quick estimate of the sum of the series $\sum_{i=1}^{n} i^k$. Use the integral to estimate each sum and then find the actual sum.

a. $\sum_{i=1}^{20} i^3$

b. $\sum_{i=1}^{100} i^2$

Applications and Problem Solving

46. **Physics** The work (in joules) required to pump all of the water out of a 10 meter by 5 meter by 2 meter swimming pool is given by $\int_0^2 490{,}000x\,dx$. Evaluate this integral to find the required work.

47. **Critical Thinking**

a. Suppose $f(x)$ is a function whose graph is *below* the *x*-axis for $a \leq x \leq b$. What can you say about the values of $f(x)$, $\sum_{i=1}^{n} f(x_i)\Delta x$, and $\int_a^b f(x)\,dx$?

b. Evaluate $\int_0^2 (x^2 - 5)\,dx$.

c. What is the area between the graph of $y = x^2 - 5$ and the *x*-axis from $x = 0$ to $x = 2$?

48. **Critical Thinking** Find the value of $\int_2^5 (3x - 6)\,dx$ *without* using limits or the Fundamental Theorem of Calculus.

49. **Stock Market** The *average value* of a function $f(x)$ over the interval $a \le x \le b$ is defined to be $\frac{1}{b-a}\int_a^b f(x)\,dx$. A stock market analyst has determined that the price of the stock of the Acme Corporation over the year 2001 can be modeled by the function $f(x) = 75 + 8x - \frac{1}{2}x^2$, where x is the time, in months, since the beginning of 2001, and $f(x)$ is in dollars.
   a. Sketch a graph of $f(x)$ from $x = 0$ to $x = 12$.
   b. Find the average value of the Acme Corporation stock over the first half of 2001.
   c. Find the average value of the stock over the second half of 2001.

50. **Geometry** The volume of a sphere of radius R can be found by slicing the sphere vertically and then integrating the areas of the resulting circular cross sections. (The cross section in the figure is a circle of radius $\sqrt{R^2 - x^2}$.) This process results in the integral $\int_{-R}^{R} (\pi R^2 - \pi x^2)\,dx$. Evaluate this integral to obtain the expression for the volume of a sphere of radius R.

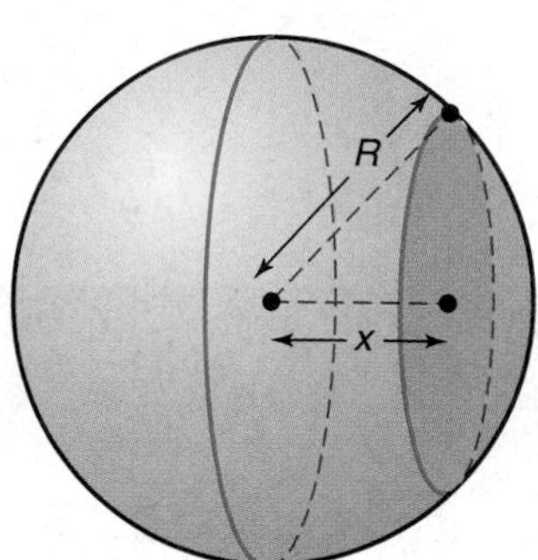

51. **Space Exploration** The weight of an object that is at a distance x from the center of Earth can be written as kx^{-2}, where k is a constant that depends on the mass of the object. The energy required to move the object from $x = a$ to $x = b$ is the integral of its weight, that is, $\int_a^b kx^{-2}\,dx$. Suppose a Lunar Surveying Module (LSM), designed to analyze the surface of the moon, weighs 1000 newtons on the surface of Earth.
   a. Find k for the LSM. Use 6.4×10^6 meters for the radius of Earth.
   b. Find the energy required to lift the LSM from Earth's surface to the moon, 3.8×10^8 meters from the center of Earth.

Mixed Review

52. Use a limit to evaluate $\int_0^2 \frac{1}{2}x^2\,dx$. *(Lesson 15-3)*

53. Find the derivative of $f(x) = 2x^6 - 3x^2 + 2$. *(Lesson 15-2)*

54. **Education** The scores of a national achievement test are normally distributed with a mean of 500 and a standard deviation of 100. What percent of those who took the test had a score more than 100 points above or below the mean? *(Lesson 14-4)*

55. Fifty tickets, numbered consecutively from 1 to 50 are placed in a box. Four tickets are drawn without replacement. What is the probability that four odd numbers are drawn? *(Lesson 13-4)*

56. **Banking** Find the amount accumulated if \$600 is invested at 6% for 15 years and interest is compounded continuously. *(Lesson 11-3)*

57. Write an equation of the parabola with vertex at $(6, -1)$ and focus at $(3, -1)$. *(Lesson 10-5)*

58. Find $2\sqrt{2}\left(\cos\frac{2\pi}{3} + \boldsymbol{i}\sin\frac{2\pi}{3}\right) \div \sqrt{2}\left(\cos\frac{\pi}{3} + \boldsymbol{i}\sin\frac{\pi}{3}\right)$. Then express the result in rectangular form. *(Lesson 9-7)*

Extra Practice See p. A55.

59. Find the initial vertical velocity of a stone thrown with an initial velocity of 45 feet per second at an angle of 52° with the horizontal. *(Lesson 8-7)*

60. **SAT Practice** **Quantitative Comparison** In the circle with center X, $\widehat{AE}$ is the shortest of the five unequal arcs.

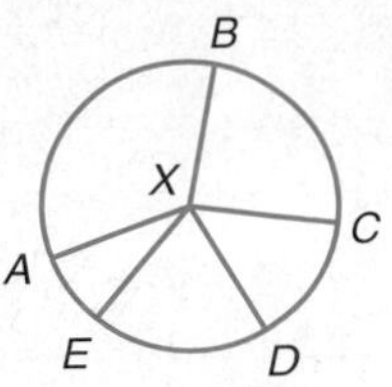

A if the quantity in column A is greater
B if the quantity in column B is greater
C if the two quantities are equal
D if the relationship cannot be determined from the given information

Column A	Column B
72	The measure of angle AXE

CAREER CHOICES

  Mathematician 

Algebra, geometry, trigonometry, statistics, calculus—if you enjoy studying these subjects, then a career in mathematics may be for you. As a mathematician, you would have several options for employment.

First, a theoretical mathematician develops new principles and discovers new relationships, which may be purely abstract in nature. Applied mathematicians use new ideas generated by theoretical mathematicians to solve problems in many fields, including science, engineering and business. Mathematicians may work in related fields such as computer science, engineering, and business. As a mathematician, you can become an elementary or secondary teacher if you obtain a teaching certificate. An advanced degree is required to teach at the college level.

Career Overview

Degree Preferred:
bachelor's degree in mathematics

Related Courses:
mathematics, science, computer science

Outlook:
increased demand for teachers and math-related occupations through the year 2006

The Average Teacher Salary Compared to the Average Experience Level of Teachers

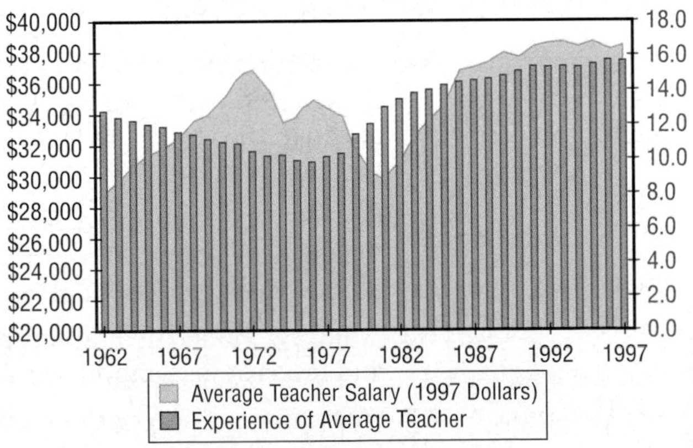

Source: American Federation of Teachers

For more information on careers in mathematics, visit: **www.amc.glencoe.com** 

VOCABULARY

antiderivative (p. 955)
definite integral (p. 962)
derivative (p. 951)
differentiation (p. 952)
Fundamental Theorem of Calculus (p. 970)
indefinite integral (p. 971)
integral (p. 961)
integration (p. 962)
limit (p. 941)
rate of change (p. 956)
secant line (p. 951)
slope of a curve (p. 949)
tangent line (p. 951)

UNDERSTANDING AND USING THE VOCABULARY

State whether each sentence is *true* or *false.* If false, replace the underlined word(s) to make a true statement.

1. $f(a)$ and $\lim_{x \to a} f(x)$ are always the same.
2. The process of finding the area under a curve is called integration.
3. The inverse of finding the derivative of a function is finding the definite integral.
4. The Fundamental Theorem of Calculus can be used to evaluate a definite integral.
5. A line that intersects a graph in two points is called a tangent line.
6. A line that passes through a point on a curve and has a slope equal to the slope of the curve at that point is called a secant line.
7. The conjugate of a function $f(x)$ is another function $f'(x)$ that gives the slope of the tangent line to $y = f(x)$ at any point.
8. If you look at one particular point on the graph of a curve, there is a certain steepness, called the slope, at that point.
9. The derivative of a function can also be called the domain of the function because it measures how fast the function changes.
10. The process of finding a limit is called differentiation.

For additional review and practice for each lesson, visit: **www.amc.glencoe.com**

SKILLS AND CONCEPTS

OBJECTIVES AND EXAMPLES

Lesson 15-1 Calculate limits of polynomial and rational functions algebraically.

- Consider the graph of the function $y = f(x)$ shown below. Find $f(-3)$ and $\lim_{x \to -3} f(x)$.

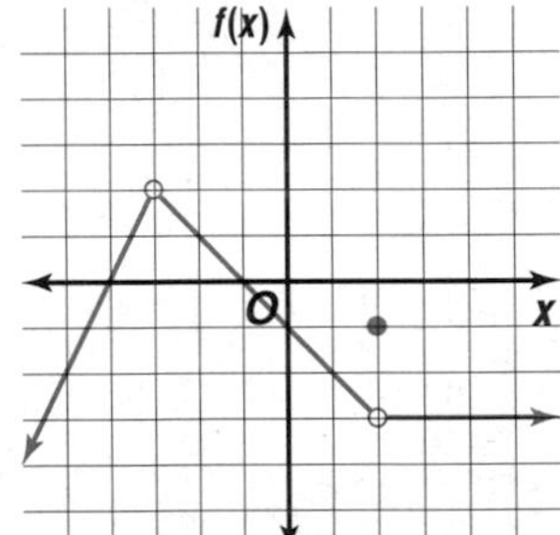

There is no point on the graph with an x-coordinate of -3, so $f(-3)$ is undefined.

Look at points on the graph whose x-coordinates are close to, but not equal to, -3. The closer x is to -3, the closer y is to 2. So, $\lim_{x \to -3} f(x) = 2$.

Evaluate each limit.

a. $\lim_{x \to 2} \frac{x^2 + x - 6}{x^2 + 3x}$

$$\begin{aligned}\lim_{x \to 2} \frac{x^2 + x - 6}{x^2 + 3x} &= \lim_{x \to 2} \frac{(x + 3)(x - 2)}{x(x + 3)} \\ &= \lim_{x \to 2} \frac{x - 2}{x} \\ &= \frac{2 - 2}{2} \\ &= 0\end{aligned}$$

b. $\lim_{x \to 0} \frac{x \cos x}{x}$

$$\begin{aligned}\lim_{x \to 0} \frac{x \cos x}{x} &= \lim_{x \to 0} \cos x \\ &= \cos 0 \\ &= 1\end{aligned}$$

REVIEW EXERCISES

11. Refer to the graph of $y = f(x)$ at the left. Find $f(2)$ and $\lim_{x \to 2} f(x)$.

Evaluate each limit.

12. $\lim_{x \to -2} (x^3 - x^2 - 5x + 6)$

13. $\lim_{x \to 0} (2x - \cos x)$

14. $\lim_{x \to 1} \frac{x^2 - 36}{x + 6}$

15. $\lim_{x \to 0} \frac{5x^2}{2x}$

16. $\lim_{x \to 4} \frac{x^2 + 2x}{x^2 - 3x - 10}$

17. $\lim_{x \to 0} (x + \sin x)$

18. $\lim_{x \to 0} \frac{x^2 + x \cos x}{2x}$

19. $\lim_{x \to 2} \frac{x^3 + 2x^2 - 4x - 8}{x^2 - 4}$

20. $\lim_{x \to 0} \frac{(x - 3)^2 - 9}{2x}$

21. $\lim_{x \to 5} \frac{x^2 - 9x + 20}{x^2 - 5x}$

OBJECTIVES AND EXAMPLES

Lesson 15-2 Find derivatives of polynomial functions.

- Find the derivative of each function.

a. $f(x) = 3x^4 + 2x^3 - 7x - 5$

$$f'(x) = 3 \cdot 4x^3 + 2 \cdot 3x^2 - 7 \cdot 1 - 0$$
$$= 12x^3 + 6x^2 - 7$$

b. $f(x) = 2x^3(x^2 + 1)$

First, multiply to write the function as a polynomial.

$$f(x) = 2x^3(x^2 + 1)$$
$$= 2x^5 + 2x^3$$

Then find the derivative.

$$f'(x) = 10x^4 + 6x^2$$

REVIEW EXERCISES

Use the definition of derivative to find the derivative of each function.

22. $f(x) = 2x + 1$
23. $f(x) = 4x^2 + 3x - 5$
24. $f(x) = x^3 - 3x$

Use the derivative rules to find the derivative of each function.

25. $f(x) = 2x^6$
26. $f(x) = -3x + 7$
27. $f(x) = 3x^2 - 5x$
28. $f(x) = \frac{1}{4}x^2 - x + 4$
29. $f(x) = \frac{1}{2}x^4 - 2x^3 + \frac{1}{3}x - 4$
30. $f(x) = (x + 3)(x + 4)$
31. $f(x) = 5x^3(x^4 - 3x^2)$
32. $f(x) = (x - 2)^3$

Lesson 15-2 Find antiderivatives of polynomial functions.

- Find the antiderivative of each function.

a. $f(x) = 5x^2$

$$F(x) = 5 \cdot \frac{1}{2+1}x^{2+1} + C$$
$$= \frac{5}{3}x^3 + C$$

b. $f(x) = -2x^3 + 6x^2 - 5x + 4$

$$F(x) = -2 \cdot \frac{1}{4}x^4 + 6 \cdot \frac{1}{3}x^3 -$$
$$5 \cdot \frac{1}{2}x^2 + 4 \cdot x + C$$
$$= -\frac{1}{2}x^4 + 2x^3 - \frac{5}{2}x^2 + 4x + C$$

Find the antiderivative of each function.

33. $f(x) = 8x$
34. $f(x) = 3x^2 + 2$
35. $f(x) = -\frac{1}{2}x^3 + 2x^2 - 3x - 2$
36. $f(x) = x^4 - 5x^3 + 2x - 6$
37. $f(x) = (x - 4)(x + 2)$
38. $f(x) = \frac{x^2 - x}{x}$

OBJECTIVES AND EXAMPLES

Lesson 15-3 Find areas under graphs of polynomial functions.

- Use limits to find the area of the region between the graph of $y = 3x^2$ and the x-axis from $x = 0$ to $x = 1$.

$$\int_0^1 3x^2\,dx = \lim_{n\to\infty} \sum_{i=1}^{n} 3(x_i)^2 \Delta x$$
$$= \lim_{n\to\infty} \sum_{i=1}^{n} 3\left(\frac{i}{n}\right)^2\left(\frac{1}{n}\right)$$
$$= \lim_{n\to\infty} \sum_{i=1}^{n} \frac{3i^2}{n^3}$$
$$= \lim_{n\to\infty} \frac{3}{n^3}(1^2 + 2^2 + \cdots + n^2)$$
$$= \lim_{n\to\infty} \frac{3}{n^3} \cdot \frac{n(n+1)(2n+1)}{6}$$
$$= \lim_{n\to\infty} \frac{1}{2}\left(2 + \frac{3}{n} + \frac{1}{n^2}\right)$$
$$= 1 + 0 + 0 \text{ or } 1 \text{ unit}^2$$

REVIEW EXERCISES

Use limits to find the area between each curve and the x-axis for the given interval.

39. $y = 2x$ from $x = 0$ to $x = 2$

40. $y = x^3$ from $x = 0$ to $x = 1$

41. $y = x^2$ from $x = 3$ to $x = 4$

42. $y = 6x^2$ from $x = 1$ to $x = 2$

Lesson 15-4 Use the Fundamental Theorem of Calculus to evaluate definite integrals of polynomial functions.

- Evaluate $\int_4^7 (x^2 - 3)\,dx$.

$$\int_4^7 (x^2 - 3)\,dx$$
$$= \frac{1}{3}x^3 - 3x \Big|_4^7$$
$$= \left(\frac{1}{3}\cdot 7^3 - 3\cdot 7\right) - \left(\frac{1}{3}\cdot 4^3 - 3\cdot 4\right)$$
$$= 84$$

Evaluate each definite integral.

43. $\int_2^4 6x\,dx$

44. $\int_{-3}^2 3x^2\,dx$

45. $\int_{-2}^2 (3x^2 - x + 3)\,dx$

46. $\int_0^4 (x + 2)(2x + 3)\,dx$

Lesson 15-4 Find indefinite integrals of polynomial functions.

- Evaluate $\int (6x^2 - 4x)\,dx$.

$$\int (6x^2 - 4x)\,dx = 6\cdot\frac{1}{3}x^3 - 4\cdot\frac{1}{2}x^2 + C$$
$$= 2x^3 - 2x^2 + C$$

Evaluate each indefinite integral.

47. $\int 6x^4\,dx$

48. $\int (-3x^2 + 2x)\,dx$

49. $\int (x^2 + 5x - 2)\,dx$

50. $\int (3x^5 + 4x^4 - 7x)\,dx$

APPLICATIONS AND PROBLEM SOLVING

51. **Physics** The kinetic energy of an object with mass m is given by the formula $k(t) = \frac{1}{2} m \cdot v(t)^2$, where $v(t)$ is the velocity of the object at time t. Suppose $v(t) = \frac{50}{1 + t^2}$ for all $t \geq 0$. What does the kinetic energy of the object approach as time approaches 100? *(Lesson 15-1)*

52. **Business** The controller for an electronics company has used the production figures for the last few months to determine that the function $c(x) = -9x^5 + 135x^3 + 10{,}000$ approximates the cost of producing x thousands of one of their products. Find the marginal cost if they are now producing 2600 units. *(Lesson 15-2)*

53. **Motion** An advertisement for a sports car claims that the car can accelerate from 0 to 60 miles per hour in 5 seconds. *(Lesson 15-2)*
    a. Find the acceleration of the sports car in feet per second squared, assuming that it is constant.
    b. Write an equation for the velocity of the sports car at t seconds.
    c. Write an equation for the distance traveled in t seconds.

ALTERNATIVE ASSESSMENT

OPEN-ENDED ASSESSMENT

1. The limit of a continuous function $f(x)$ as x approaches 1 is 5. Give an example of a function for which this is true. Show why the limit of your function as x approaches 1 is 5.
2. The area of the region between the graph of the function $g(x)$ and the x-axis from $x = 0$ to $x = 1$ is 4. Give an example of a function for which this is true. Show that the area of the region between the graph of your function and the x-axis from $x = 0$ to 1 is in fact 4.

PORTFOLIO

Explain the difference between a definite integral and an indefinite integral. Give an example of each.

Now that you have completed your work in this book, review your portfolio entries for each chapter. Make any necessary changes or corrections. Add a table of contents to your portfolio at this time.

DISEASES

- Use the Internet to find the number of cases reported or the number of deaths for one particular disease for a period of at least 10 years. Some possible diseases you might choose to research are measles, tuberculosis, or AIDS. Make a table or spreadsheet of the data.
- Use computer software or a graphing calculator to find at least two polynomial functions that model the data. Find the derivative for each of your function models. What does the derivative represent?
- Use each model to predict the cases or deaths from the disease in the year 2010. Write a one-page paper comparing the models. Discuss which model you think best fits the data. Include any limitations of the model.

Additional Assessment See p. A70 for Chapter 15 Practice Test.

CHAPTER 15 SAT & ACT Preparation

Special Function and Counting Problems

> THE PRINCETON REVIEW
>
> Special symbols can appear in the question or in the answer choices or both.
>
> Read the explanation thoroughly and work carefully.

The SAT includes function problems that use special symbols like ⊕ or # or ⊗. (The ACT does not contain this type of problem.)

Here's a simple example: If $x \# y = x + 2y$, then what is $2 \# 5$?
To find $2 \# 5$, replace x with 2 and y with 5. Thus, $2 \# 5 = 2 + 2(5) = 12$.
The SAT may also include problems that involve counting regions, surfaces, or intersections. The questions usually ask for the maximum or minimum number.

SAT EXAMPLE

1. Let $\boxed{x}$ be defined for all positive integers x as the product of the distinct prime factors of x. What is the value of

$$\frac{\boxed{6}}{\boxed{81}}?$$

HINT The SAT often combines two mathematical concepts in one problem. For example, this problem combines a special function and prime factors.

Solution Carefully read the definition of $\boxed{x}$. Recall the meaning of "distinct prime factors." Write the prime factorization of each number, identify which prime factors are distinct, and then find the product.

Start with the first number, 6. $6 = 2 \times 3$. Both 2 and 3 are distinct prime factors. The product of the distinct prime factors is 6.

Do the same with 81. $81 = 3 \times 3 \times 3 \times 3$. There is just one distinct prime factor, 3. So the product of the distinct prime factors is also 3.

Finally, substitute the values for $\boxed{6}$ and $\boxed{81}$ into the fraction.

$\frac{\boxed{6}}{\boxed{81}} = \frac{6}{3} = 2$. The answer is 2.

Grid-in this answer on your answer sheet.

SAT EXAMPLE

2. The figure below is a square separated into two non-overlapping regions. What is the greatest number of non-overlapping regions that can be made by drawing any two additional straight lines?

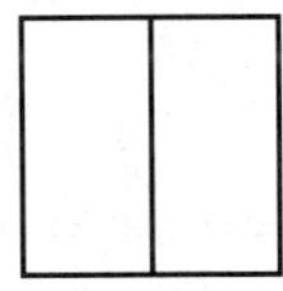

A 4 **B** 5 **C** 6 **D** 7 **E** 8

HINT Watch out for "obvious" answers on difficult problems (those numbered 18 or higher). They are usually wrong answers.

Solution Draw right on your test booklet.

The most obvious ways to draw two more lines are shown at the right.

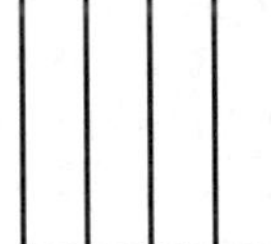

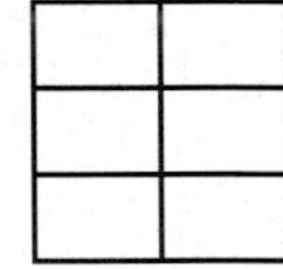

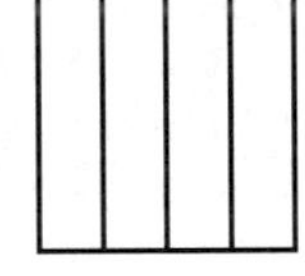

The first figure has 4 regions; the second figure has 6 regions. So you can immediately eliminate answer choices A and B.

For the maximum number of regions, it is likely that the lines will *not* be parallel, as they are in the figures above.

Draw the two lines with the fewest possible criteria: not horizontal, not vertical, not parallel, and not perpendicular.

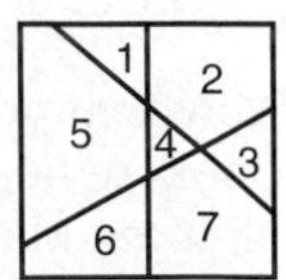

There are 7 regions. The answer is choice **D**.

After you work each problem, record your answer on the answer sheet provided or on a piece of paper.

Multiple Choice

1. If $x \otimes y = \frac{1}{x - y}$, what is the value of $\frac{1}{2} \otimes \frac{1}{3}$?

   A 6 **B** $\frac{6}{5}$ **C** $\frac{1}{6}$

   D -1 **E** -6

2. If one side of a triangle is twice as long as a second side of length x, then the perimeter of the triangle can be:

   A $2x$ **B** $3x$ **C** $4x$

   D $5x$ **E** $6x$

3. If 3 parallel lines are cut by 3 nonparallel lines, what is the maximum number of intersections possible?

   A 9 **B** 10 **C** 11

   D 12 **E** 13

4. In the figure below, if segment $\overline{WZ}$ and segment $\overline{XY}$ are diameters with lengths of 12, what is the area of the shaded region?

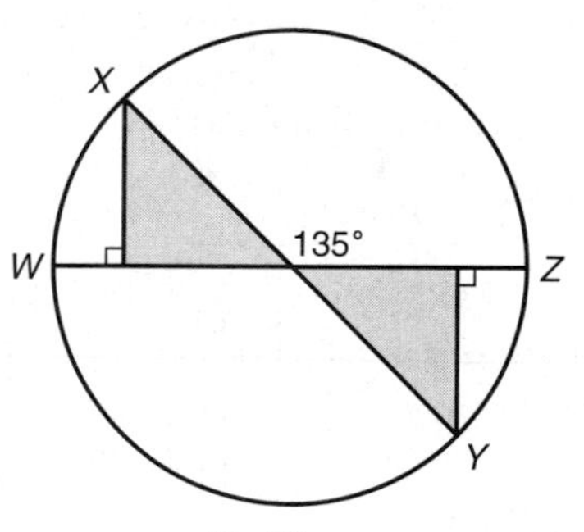

   A 9 **B** 18 **C** 36

   D 54 **E** 108

5. Which of the following represents the values of x that are solutions of the inequality $x^2 < x + 6$?

   A $x > -2$

   B $x < 3$

   C $-2 < x < 3$

   D $-3 < x < 2$

   E $x < -2 \cup x > 3$

6. $\boxed{x} = \frac{1}{2}x$ if x is composite.

   $\boxed{x} = 3x$ if x is prime.

   What is the value of $\boxed{5} + \boxed{16}$?

   A 21 **B** 23 **C** 31

   D 46 **E** 69

7. What is the average of all the integers from 1 to 20 inclusive?

   A 9.5 **B** 10 **C** 10.5

   D 20 **E** 21

8. All faces of a cube with a 4-meter edge are painted blue. If the cube is then cut into cubes with 1-meter edges, how many of the 1-meter cubes have blue paint on exactly one face?

   A 24 **B** 36 **C** 48

   D 60 **E** 72

9. **Quantitative Comparison**

   A if the quantity in Column A is greater

   B if the quantity in Column B is greater

   C if the two quantities are equal

   D if the relationship cannot be determined from the information given

   For all numbers n, let $\{n\}$ be defined as $n^3 + 2n^2 - n$.

Column A	Column B
$\{x\}$	$\{x + 1\}$

10. **Grid-In** Let $\boxed{x}$ be defined for all positive integers x as the product of the distinct prime factors of x. What is the value of

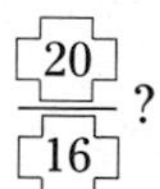

SAT/ACT Practice For additional test practice questions, visit: **www.amc.glencoe.com**

STUDENT HANDBOOK

Skills

Graphing Calculator Appendix

Reference

1 Introduction to the Graphing Calculator

This section introduces you to some commonly-used keys and menus of the calculator.

Setting Preferences

MODE The MODE key allows you to select your preferences in many aspects of calculation and graphing. Many of these settings are rarely changed in common usage. This screen shows the default mode settings.

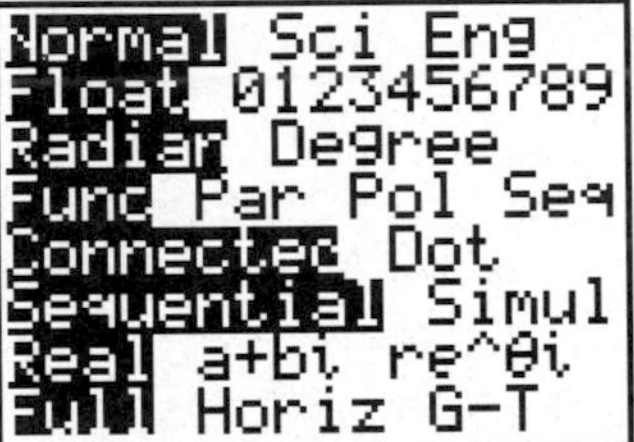

← type of numeric notation
← number of decimal places in results
← unit of angle measure used
← type of graph (function, parametric, polar, sequence)
← whether to connect graphed points
← real, rectangular complex, or polar complex number system
← graph occupies full screen, top of screen with HOME screen below, or left side of screen with TABLE on right

To change the preferences, use the arrow keys to highlight your choice and press ENTER.

FORMAT The **FORMAT** menu is the second function of ZOOM and sets preferences for the appearance of your graphing screen. The default screen is shown below.

← rectangular or polar coordinate system
← whether to display the cursor coordinates on screen
← whether to show a grid pattern on screen
← whether to show the axes
← whether to label the axes
← whether to show the equation being graphed

You can change your preferences in the **FORMAT** menu in the same way you change MODE settings.

Using Menus

Many keys on the calculator access menus from which you can select a function, command, or setting. Some keys access multiple menus. You can use the right and left arrow keys to scroll through the different menu names located at the top of the screen. As each menu name is highlighted, the choices on the screen change. The screens on the next page show various menus accessed by using MATH.

Math menu

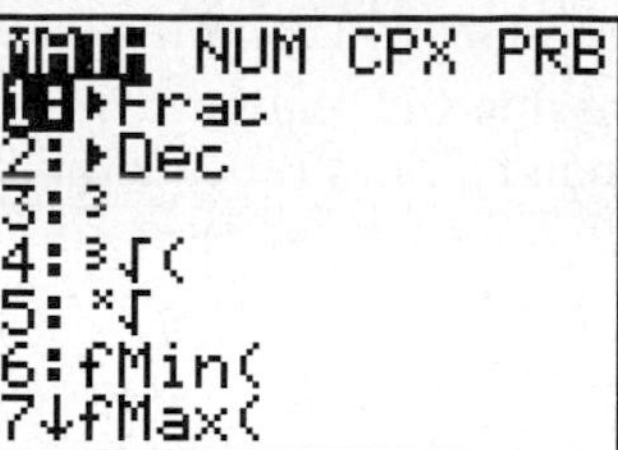

Number menu

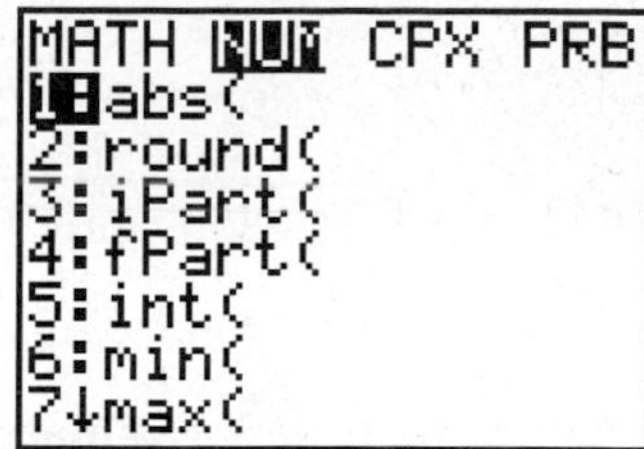

Complex Number menu

Probability menu

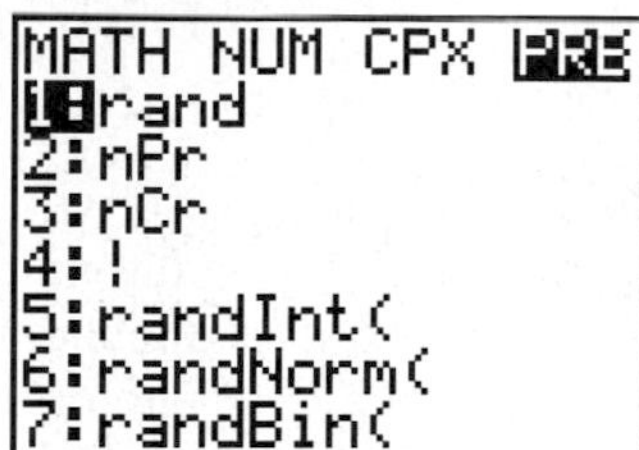

To select a choice in a menu, either use the arrow keys to highlight your choice and press ENTER or simply press the number or letter of your selection. Notice that entry 7 in the first screen has a down arrow instead of a colon after the 7. This signifies there are more entries in the menu.

Alternate Function Keys

Whenever an alternate function is indicated in the keystrokes of this appendix, we will use brackets to show that the function is listed above a key.

Above most keys are one or two additional labels representing commands, menus, letters, lists, or operational symbols. These are accessed by using 2nd or ALPHA.

- 2nd accesses the commands on the left above the key. Note that these commands and 2nd are the same color.
- ALPHA accesses the commands on the right above each key. These commands and ALPHA are also the same color.
- Pressing 2nd ALPHA engages the **[A-LOCK]** or Alpha Lock command. This enables you to select consecutive ALPHA commands without pressing ALPHA before each command. This is especially useful when entering programs.

Each letter accessed by using ALPHA can be used to enter words or labels on the screen, but can also be used as a variable. A value can be stored to each variable.

Computation

A graphing calculator is also a scientific calculator. That is, it follows the order of operations when evaluating entries. Unlike some scientific calculators, the graphing calculator displays every entry in the expression.

Before pressing ENTER to evaluate the expression, you can use the arrow keys to scroll through the expression to make corrections. Corrections can be made in three ways.

- Use DEL to delete any unwanted entries.
- Use 2nd **[INS]** to insert omitted entries.
- "Type" over an incorrect entry. This overprints any entries and does not shift the entries to the right as a word processor does.

If you have an expression that you wish to evaluate repeatedly with a change in one part of the expression, you can press 2nd **[ENTRY]** after you have pressed ENTER and the expression will reappear. You can edit it for your next computation. The **ENTRY** command always repeats the last entered expression. You cannot scroll back through previous expressions you have evaluated.

Example 1 **Evaluate** $\dfrac{\sqrt{3^2 - 4(6) + [5 - (-12)]^3}}{6}$.

Press: 2nd [√‾] 3 x^2 — 4 (6) + (5 — (–) 12) ^ 3) ÷ 6 ENTER

The minus key and the negative key are different keys.

Note that the square root function automatically includes a left parenthesis. You must enter the right parenthesis to indicate the end of the expression under the radical sign. If you have the decimal in the **Float** mode, as many as 10 digits may appear in the answer.

```
√(3²-4(6)+(5--12
)^3)/6
          11.66428547
```

2 **Evaluate each expression if $a = 4$, $b = -5$, $c = 2$, $d = \frac{2}{3}$, and $e = -1.5$.**

a. $abc - 3de^4$ **b.** $\dfrac{e + 4a}{c^2 + 8b}$

For a series of expressions that use the same values for the variables, it is often helpful to store the value for each variable into the calculator. You can combine several commands in one line by using the colon after each command. The following commands save the values for variables *a*, *b*, *c*, *d* and *e*.

Press: 4 STO▸ ALPHA [A] ALPHA [:] (–) 5 STO▸ ALPHA [B] ALPHA [:] 2 STO▸ ALPHA [C] ALPHA [:] 2 ÷ 3 STO▸ ALPHA [D] ALPHA [:] (–) 1.5 STO▸ ALPHA [E] ENTER

a. Method 1: Using stored values

ALPHA [A] ALPHA [B] ALPHA [C] — 3 ALPHA [D] ALPHA [E] ^ 4 ENTER

Method 2: Entering computations

4 × (–) 5 × 2 — 3 × 2 ÷ 3 × ((–) 1.5) ^ 4 ENTER

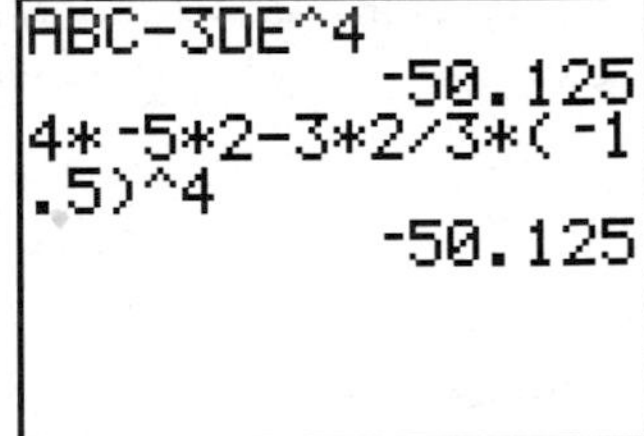

b. Method 1: Using stored values

(ALPHA [E] + 4 ALPHA [A]) ÷ (ALPHA [C] x^2 + 8 ALPHA [B]) ENTER

Method 2: Entering computations

((–) 1.5 + 4 × 4) ÷ (2 x^2 + 8 × (–) 5) ENTER

```
(E+4A)/(C²+8B)
          -.4027777778
(-1.5+4*4)/(2²+8
*-5)
          -.4027777778
```

2 Graphing Functions

Most functions can be graphed by using the [Y=] key. The viewing window most often used for non-trigonometric functions is the standard viewing window [−10, 10] scl:1 by [−10, 10] scl:1, which can be accessed by selecting **6:ZStandard** on the **ZOOM** menu. Then the window can be adjusted so that a **complete graph** can be viewed. A complete graph is one that shows the basic characteristics of the parent graph.

Example 1 **Linear Functions** A complete linear graph shows the x- and y-intercepts.

a. Graph $y = 3x - 4$ in the standard viewing window.

Press: [Y=] 3 [X,T,θ,n] [−] 4 [ZOOM] 6

If your calculator is already set for the standard viewing window, press [GRAPH] *instead of* [ZOOM] *6.*

Both the x- and y-intercepts of the linear graph are viewable in this window, so the graph is complete.

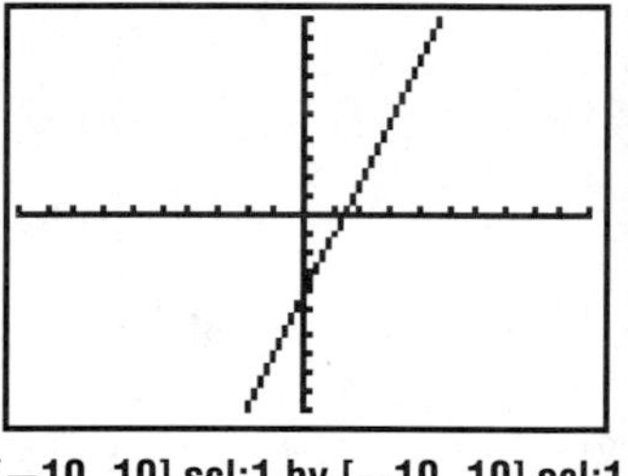

[−10, 10] scl:1 by [−10, 10] scl:1

b. Graph $y = -2(x + 5) - 2$.

Press: [Y=] [(−)] 2 [(] [X,T,θ,n] [+] 5 [)] [−] 2 [GRAPH]

When this equation is graphed in the standard viewing window (Figure 1), a complete graph is not visible. The graph indicates that the y-intercept is less than −10. You can experiment with the **Ymin** setting or you can rewrite the equation in $y = mx + b$ form, which would be $y = -2x - 12$. The y-intercept is −12, so **Ymin** should be less than −12. Remember that **Xmax** and **Ymax** can be less than 10 so that your screen is less compressed. Use the **WINDOW** menu to change the parameters, or settings, and press [GRAPH] to view the result. There are many windows that will enable you to view the complete graph (Figure 2).

Figure 1

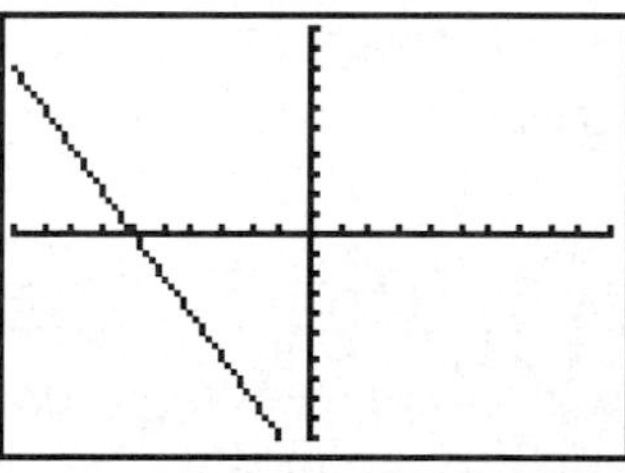

[−10, 10] scl:1 by [−10, 10] scl:1

Figure 2

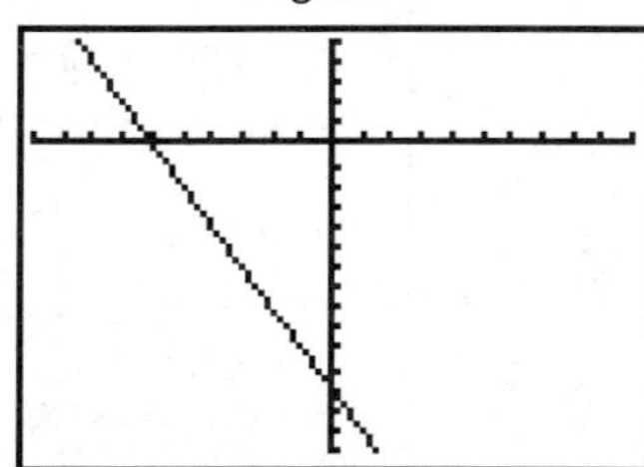

[−10, 10] scl:1 by [−15, 5] scl:1

2 **Quadratic Functions** When graphing quadratic functions, a complete graph includes the vertex of the parabola and enough of the graph to determine if it opens upward or downward.

(continued on the next page)

Graph $y = 4(x - 3)^2 + 4$.

Press: [Y=] 4 [(] [X,T,θ,n] [—] 3 [)] [x^2] [+] 4 [GRAPH]

While the standard viewing window shows a complete graph, you may want to change the viewing window to see more of the graph.

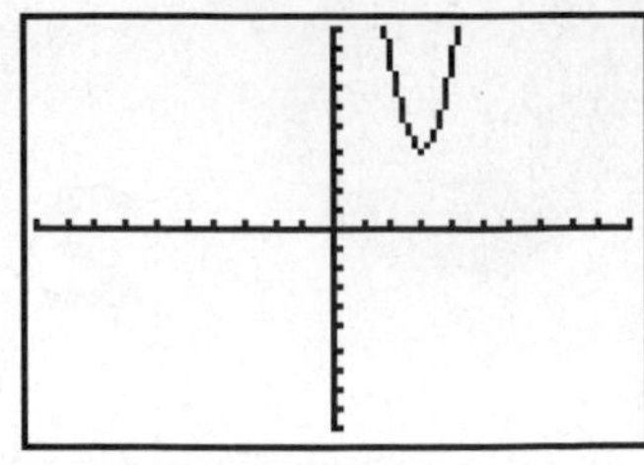

[−10, 10] scl:1 by [−10, 10] scl:1

Example 3 **Polynomial Functions** The graphs of other polynomial functions are complete when their maxima, minima, and x-intercepts are visible.

a. Graph $y = 5x^3 + 4x^2 - 2x + 4$.

Press: [Y=] 5 [X,T,θ,n] [^] 3 [+] 4 [X,T,θ,n] [x^2] [—] 2 [X,T,θ,n] [+] 4 [GRAPH]

A complete graph is shown in the standard viewing window. You may want to redefine your window to observe the intercepts, maximum, and minimum points more closely.

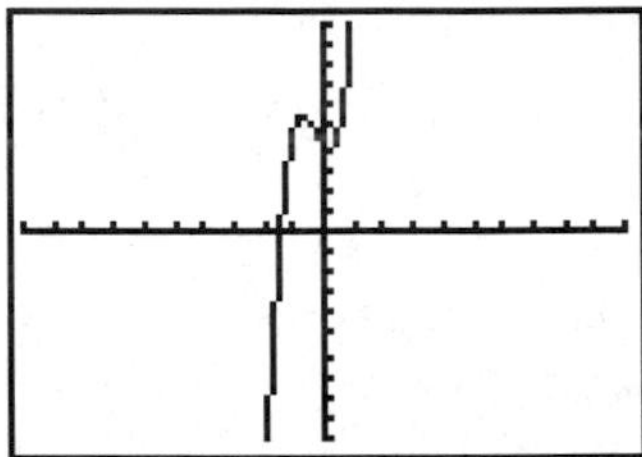

[−10, 10] scl:1 by [−10, 10] scl:1

b. Graph $y = x^4 - 13x^2 + 36$.

Press: [Y=] [X,T,θ,n] [^] 4 [—] 13 [X,T,θ,n] [x^2] [+] 36 [GRAPH]

The standard viewing window (Figure 1) does not show a complete graph. It seems that only the y parameters need to be adjusted. Experiment to find a window that is suitable. Figure 2 shows a sample.

Figure 1

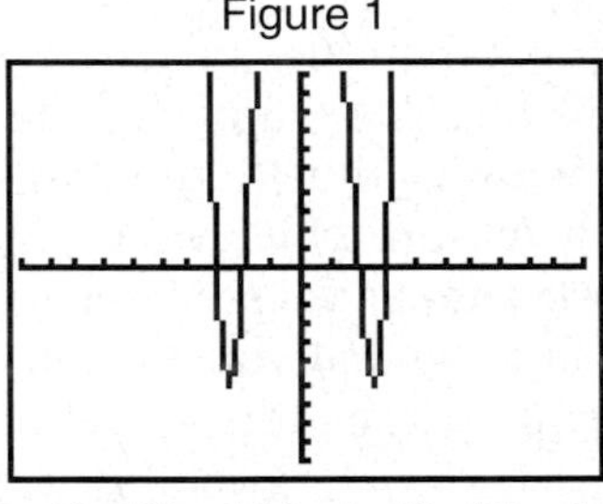

[−10, 10] scl:1 by [−10, 10] scl:1

Figure 2

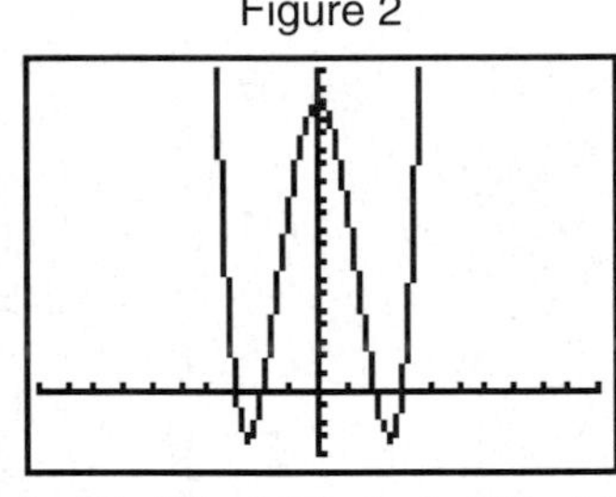

[−10, 10] scl:1 by [−8, 40] scl:2

4 **Exponential Functions** A complete graph of an exponential function shows the curvature of the graph and the y-value that it approaches.

Graph $y = 9^{2+x}$.

Press: [Y=] 9 [^] [(] 2 [+] [X,T,θ,n] [)] [GRAPH]

Note that you must use parentheses to group the terms that make up the exponent.

A complete graph seems to appear in the second quadrant of the standard viewing window. Vary the **WINDOW** settings to view the graph more closely.

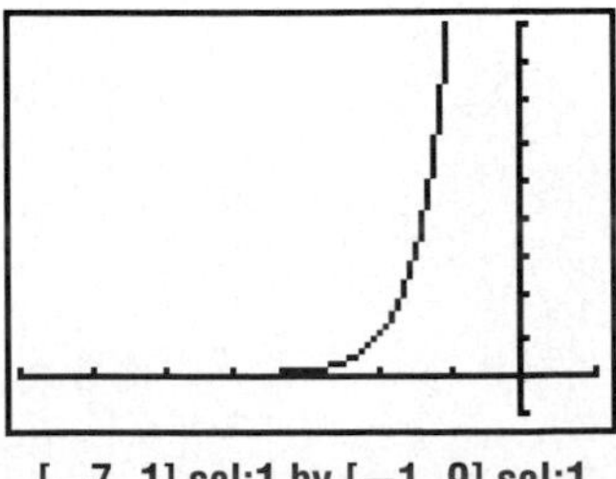

[−7, 1] scl:1 by [−1, 9] scl:1

Example 5 **Logarithmic Functions** A complete graph of a logarithmic function shows the curvature of the graph and the values, or locations, of the asymptotes that the curve approaches.

a. Graph $y = \log(x + 6)$.

Press: [Y=] [LOG] [X,T,θ,n] [+] 6 [)] [GRAPH]

An entire graph appears in the standard viewing window, but is very small. Redefine the **WINDOW** parameters for y, so that the graph is more visible.

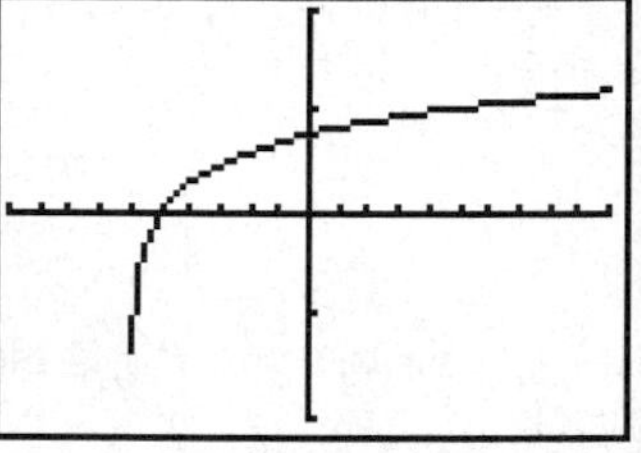

[−10, 10] scl:1 by [−2, 2] scl:1

b. Graph $y = \log_4 x$.

To graph a logarithmic function with a base other than 10, you must first change the function by using the change of base formula, $\log_a x = \frac{\log x}{\log a}$.

Press: [Y=] [LOG] [X,T,θ,n] [)] [÷] [LOG] 4 [)] [GRAPH]

An entire graph appears in the standard viewing window, but is very small. Redefine the **WINDOW** parameters, so that the graph is more visible.

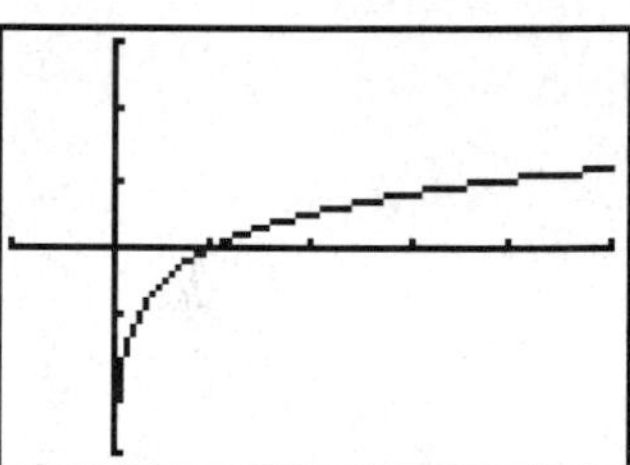

[−1, 5] scl:1 by [−3, 3] scl:1

You can graph multiple functions on a single screen. Each function is denoted by **Y1=**, **Y2=**, **Y3=**, and so on, in the **Y=** menu. To graph more than one function, press [ENTER] at the end of each function you are entering and the cursor will move to the next function to be entered.

Example 6 **Systems of Equations**

Graph $y = 0.5x + 4$ and $y = 2x^2 - 5x + 1$.

Press: [Y=] 0.5 [X,T,θ,n] [+] 4 [ENTER] 2 [X,T,θ,n] [x^2] [−] 5 [X,T,θ,n] [+] 1 [GRAPH]

The standard viewing window shows that the line and parabola intersect in two points.

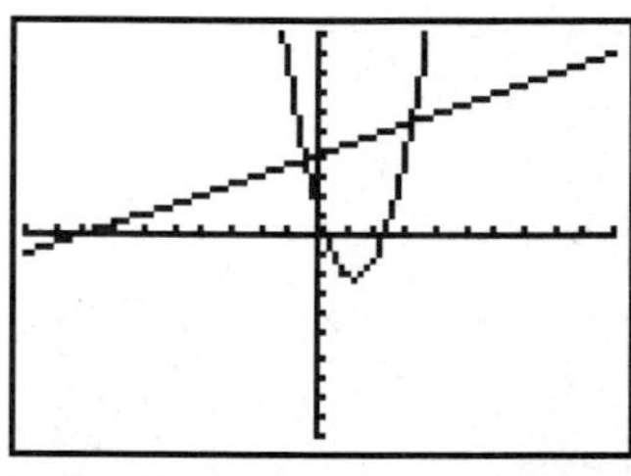

[−10, 10] scl:1 by [−10, 10] scl:1

GRAPHING CALCULATOR APPENDIX

3 Analyzing Functions

In addition to graphing a function, you can use other tools on a graphing calculator to analyze functions. One of those tools is a function table.

Example 1 **How to Use a Table** You may complete a table manually or automatically. To create a table for one or more functions, you must first enter each function into the **Y=** list. Then set up and create the table.

The function values are the dependent variable values.

a. Use a table to evaluate the function $y = 4x^2 - 2x + 7$ for $\{-9, -4, 0, 1, 5\}$.

In this case you only need to evaluate the function for selected values, so use the **TBLSET** menu to have the calculator ask for the values of the independent variable (domain) and find the function value (range) automatically.

Press: Y= 4 X,T,θ,n x^2 — 2 X,T,θ,n +
7 2nd **[TBLSET]** ▼ ▼ ▶ ENTER
2nd **[TABLE]** (–) 9 ENTER (–)
4 ENTER 0 ENTER 1 ENTER 5 ENTER

X	Y1	
-9	349	
-4	79	
0	7	
1	9	
5	97	

X=

b. Use a table to evaluate the functions $y = 5x^2 - x + 1$ and $y = 6 - x^3$ for the integers from -3 to 3, inclusive.

When you want to evaluate a function for a range of values, have the calculator find both the values of the independent variable and the function values automatically. In **Table Setup**, enter the initial number of the domain as the **TblStart** value and the increment between the values of the independent variable as **ΔTbl**. Entering more than one function in the **Y=** list allows you to evaluate all of the functions in one table.

Press: Y= 5 X,T,θ,n x^2 — X,T,θ,n +
1 ENTER 6 — X,T,θ,n ^ 3 2nd
[TBLSET] (–) 3 ENTER 1 ENTER ENTER
2nd **[TABLE]**

X	Y1	Y2
-3	49	33
-2	23	14
-1	7	7
0	1	6
1	5	5
2	19	-2
3	43	-21

X= -3

Once you create a table, you can scroll through the values using the arrow keys.

ZOOM allows you to quickly adjust the viewing window of a graph in different ways. The effect of each choice on the **ZOOM** menu is shown on the next page.

1: ZBox	Allows you to draw a box to define the viewing window
2: Zoom In	Magnifies the graph around the cursor
3: Zoom Out	Views more of a graph around the cursor
4: ZDecimal	Sets **ΔX** and **ΔY** to 0.1
5: ZSquare	Sets equal-sized pixels on the x-and y-axes
6: ZStandard	Sets the standard viewing window, $[-10, 10]$ scl:1 by $[-10, 10]$ scl:1
7: ZTrig	Sets the built-in trig window, $\left[-\frac{47}{24}\pi, \frac{47}{24}\pi\right]$ scl: $\frac{\pi}{2}$ by $[-4, 4]$ scl:1 for radians or $[-352.5. 352.5]$ scl:90 by $[-4, 4]$ scl: 1 for degrees
8: ZInteger	Sets integer values on both the x-and y-axes
9: ZoomStat	Sets values for displaying all of the data in the current stat lists
0: ZoomFit	Fits Ymin and Ymax to show all function values for Xmin to Xmax

Example 2 Using ZOOM to Graph in the Standard and Square Windows

Graph the circle with equation $x^2 + y^2 = 16$ in the standard viewing window. Then use ZSquare to view the graph in a square screen.

First solve the equation for y in order to enter it into the **Y=** list.

$x^2 + y^2 = 16 \rightarrow y = \pm\sqrt{16 - x^2}$

The two pieces of the graph can be entered at one time using $\{-1, 1\}$. This expression tells the calculator to graph -1 and 1 times the function.

Press: Y= 2nd [{] (-) 1 , 1 2nd [}] 2nd [√] 16 − X,T,θ,n x²) ZOOM 6

The circle is distorted when viewed in the standard viewing window.

Press: ZOOM 5

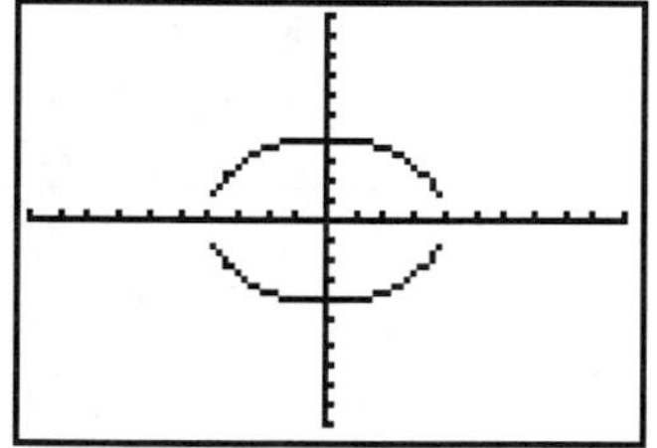

[−10, 10] scl:1 by [−10, 10] scl:1

Using **ZSquare** makes the circle appear as a circle.

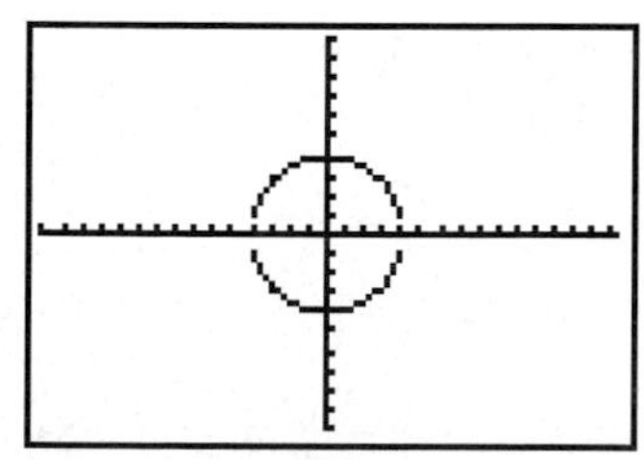

[−15.16, 15.16] scl:1 by [−10, 10] scl:1

Example 3 **Using [ZOOM] to Zoom In and Out** **Graph $y = 0.5x^3 - 3x^2 - 12$ in the standard viewing. Zoom out to view a complete graph. Then zoom in to approximate the *y*-intercept of the graph to the nearest whole number.**

Press: [Y=] 0.5 [X,T,θ,n] [^] 3 [−] 3 [X,T,θ,n] [x^2] [−] 12 [ZOOM] 6

The complete graph is not shown in the standard viewing window. (Figure 1) When you zoom out or in, the calculator allows you to choose the point around which it will zoom. Zooming out around the origin once allows a complete graph to be shown. (Figure 2)

Press: [ZOOM] 3 [ENTER]

Figure 1

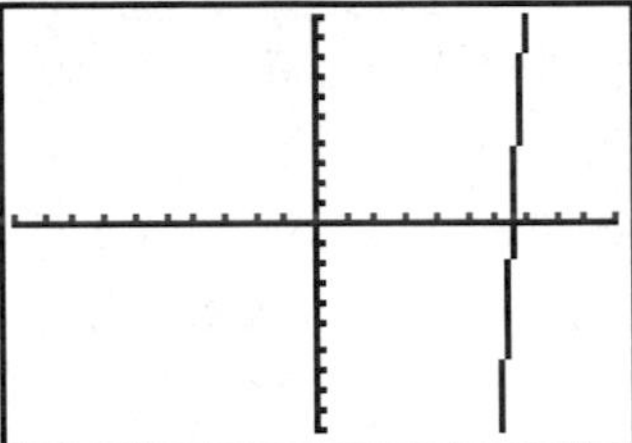

[−10, 10] scl:1 by [−10, 10] scl:1

Figure 2

[−40, 40] scl:1 by [−40, 40] scl:1

Now zoom in to approximate the *y*-intercept. Choose a point close to the intercept by using the arrow keys.

Press: [ZOOM] 2 [ENTER]

The *y*-intercept appears to be about −12. Zooming in again may allow you to make a closer approximation.

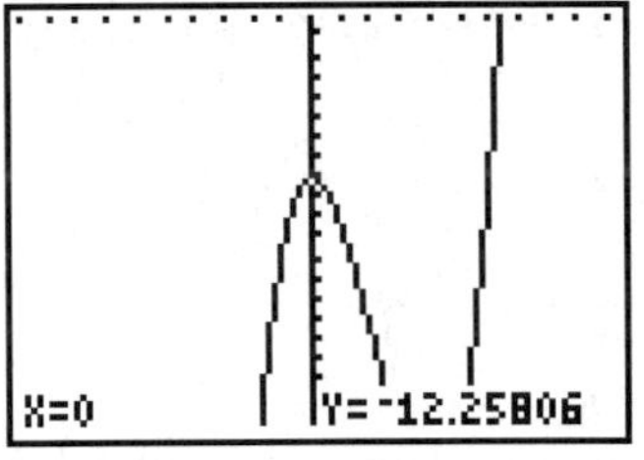

[−10, 10] scl:1 by [−24.19, −4.19] scl:1

The [TRACE] feature allows you to move the cursor along a graph and display the coordinates of the points on the graph.

Example 4 **Using [TRACE]** **Graph $y = 4x + 2$ and $y = -3x^2 - x + 5$. Use the TRACE feature to approximate the coordinates of the intersection of the graphs in the first quadrant. Then evaluate $y = -3x^2 - x + 5$ for $x = 1.7$.**

Make sure that **CoordOn** *is highlighted in the* **FORMAT** *menu to display the cursor coordinates as you trace.*

Press: [Y=] 4 [X,T,θ,n] [+] 2 [ENTER] [(−)] 3 [X,T,θ,n] [x^2] [−] [X,T,θ,n] [+] 5 [TRACE]

Move the cursor along the graphs using [◀] and [▶].

Pressing [2nd] [◀] or [2nd] [▶] moves the cursor more quickly. If your cursor moves off of the screen, the calculator will automatically update the viewing window so that the cursor is visible. Use [▼] and [▲] to move from one function to the other. The intersection is at about (0.4, 4).

To evaluate a function for a value and move to that point, place the cursor on the function graph. Then enter the value and press [ENTER]. When $x = 1.7$, $y = -5.37$ for $y = -3x^2 - x + 5$.

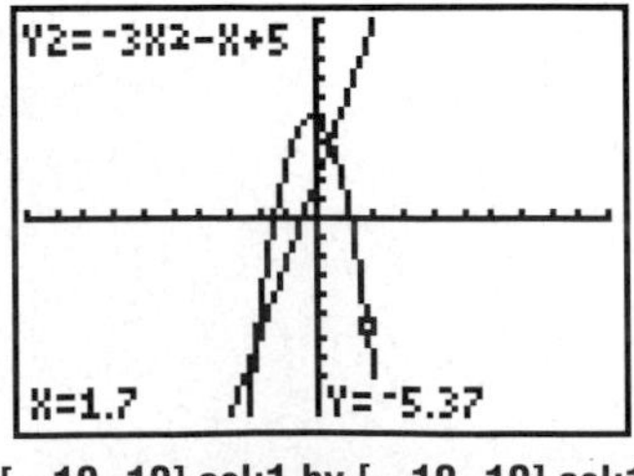

[−10, 10] scl:1 by [−10, 10] scl:1

Using TRACE to locate the intersection points of the graphs of two functions gives you an approximation of the coordinates. For more accurate coordinates, you can use the **intersect** option on the **CALC** menu.

Example 5 **Finding Intersection Points** **Use 5:intersection on the CALC menu to find the coordinates of the intersection of the graphs of $y = 4x + 2$ and $y = -3x^2 - x + 5$.**

If you do not have the functions graphed, enter the functions into the **Y=** list and press GRAPH . Then find the coordinates of the intersection.

Press: 2nd **[CALC]** 5

The intersection of the graphs must appear on the screen to find the coordinates when using **intersect**.

Place the cursor on one graph and press ENTER . Then move the cursor to the other graph and press ENTER . To guess at the intersection, move the cursor to a point close to the intersection or enter an x-value and press ENTER . If there is more than one intersection point, the calculator will find the one closest to your guess. The cursor will move to the intersection point and the coordinates will be displayed.

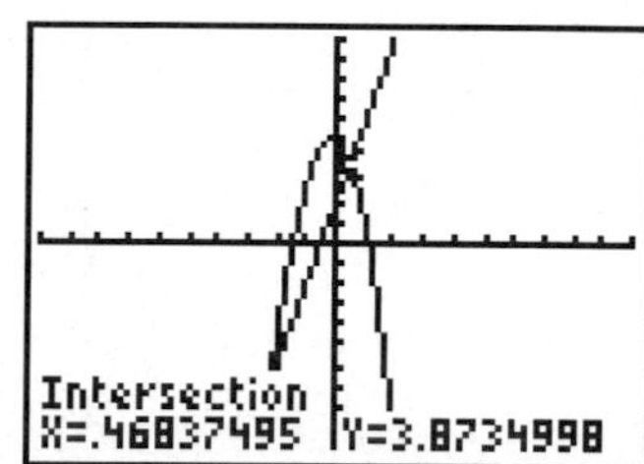

[−10, 10] scl:1 by [−10, 10] scl:1

The **CALC** menu also allows you to find the zeros of a function.

Example 6 **Finding Zeros** **Find the zeros of $f(x) = -2x^4 + 3x^2 + 2x + 5$.**

Press: Y= (−) 2 X,T,θ,n ^ 4 + 3 X,T,θ,n x^2 + 2 X,T,θ,n + 5 2nd **[CALC]** 2

The calculator can find one zero at a time. Use the arrow keys or enter a value to choose the left bound for the interval in which the calculator will search for the zero and press ENTER . Choose the right bound and press ENTER . Select a point near the zero using the arrow keys or by entering a value and press ENTER . Repeat with another interval to find the other zero. The zeros of this function are about −1.42 and 1.71.

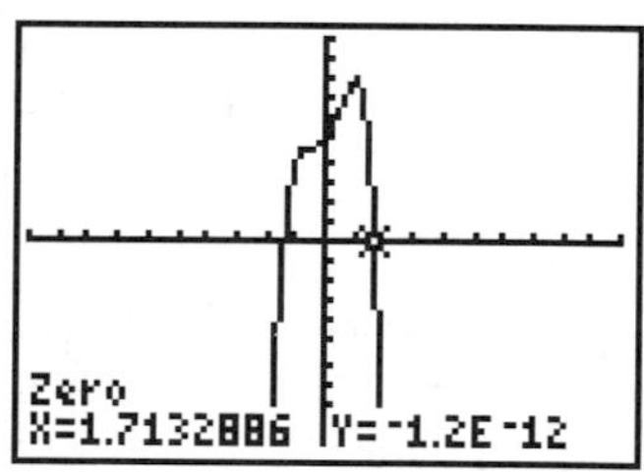

[−10, 10] scl:1 by [−10, 10] scl:1

Real-world application problems often require you to find the relative minimum or maximum of a function. You can use **3:minimum** or **4:maximum** features on the **CALC** menu of a graphing calculator to solve these problems.

GRAPHING CALCULATOR APPENDIX

Example 7 **Finding Maxima and Minima** **Determine the relative minimum and the relative maximum for the graph of $f(x) = 4x^3 - 6x + 5$.**

First graph the function.

Press: [Y=] 4 [X,T,θ,n] [^] 3 [–] 6 [X,T,θ,n] [+] 5 [ZOOM] 6

To find the relative minimum press [2nd] [CALC] 3.

Similar to finding a zero, choose the left and right bound of the interval and guess the minimum or maximum. The point at about (0.71, 2.17) is a relative minimum.

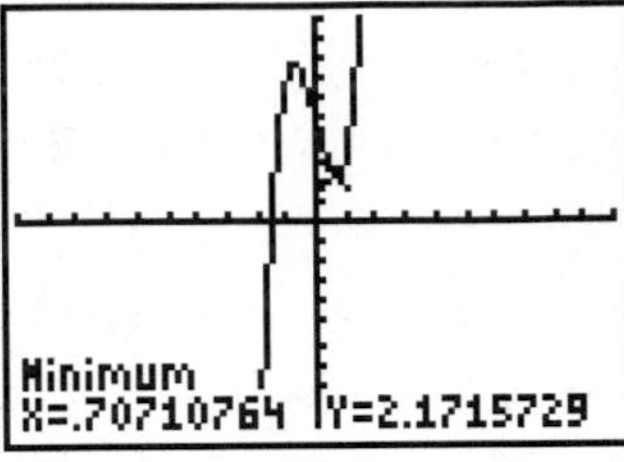

[−10, 10] scl:1 by [−10, 10] scl:1

Use a similar method to find the relative maximum, by pressing [2nd] [CALC] 4. The point at about (−0.71, 7.83) is a relative maximum.

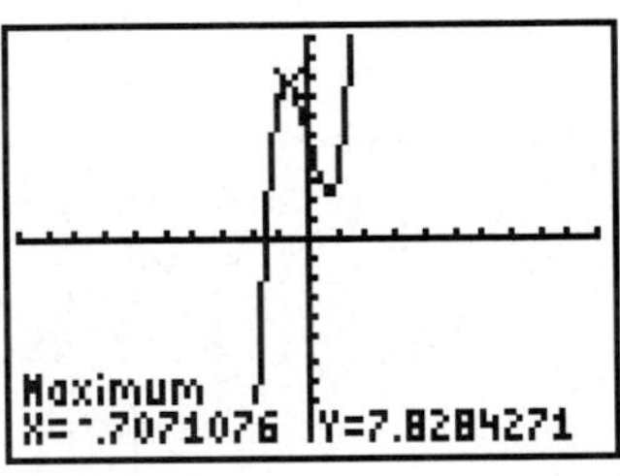

[−10, 10] scl:1 by [−10, 10] scl:1

4 Graphing Inequalities

Most linear and nonlinear inequalities can be graphed using the Y= key and selecting the appropriate graph style in the **Y=** editor. To select the appropriate graph style, select the graph style icon in the first column of the **Y=** editor and press ENTER repeatedly to rotate through the graph styles.

- To shade the area above a graph, select the Above style icon, ◥.
- To shade the area below a graph select the Below style icon, ◣.

Before graphing an inequality, clear any functions in the **Y=** list by pressing Y= and then using the arrow keys and the CLEAR key to select and clear all functions. If you do not wish to clear a function, you can turn that particular graph off by using the arrow keys to position the cursor over that function's = sign and then pressing ENTER to change the selection status.

Example 1 Linear Inequalities

a. Graph $y \leq 2x - 3$ in the standard viewing window.

First enter the boundary equation $y = 2x - 3$ into the **Y=** list.

Press: Y= 2 X,T,θ,n — 3

Next, press the ◀ key until the icon before = flashes. Press ENTER until the icon changes to the Below style icon, ◣, for "$y \leq$".

Finally, if your calculator is not already set for the standard viewing window, press ZOOM 6. Otherwise, press GRAPH.

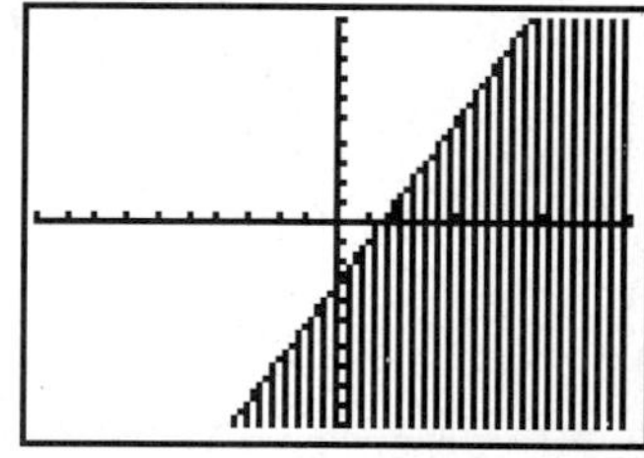

[−10, 10] scl:1 by [−10, 10] scl:1

b. Graph $y \geq -4x + 5$ in the standard viewing window.

Press: Y= (−) 4 X,T,θ,n + 5

Next, press the ◀ key until the icon before = flashes. Then press ENTER until the icon changes to the Above style icon, ◥, since the inequality asks for "$y \geq$".

Finally, press GRAPH.

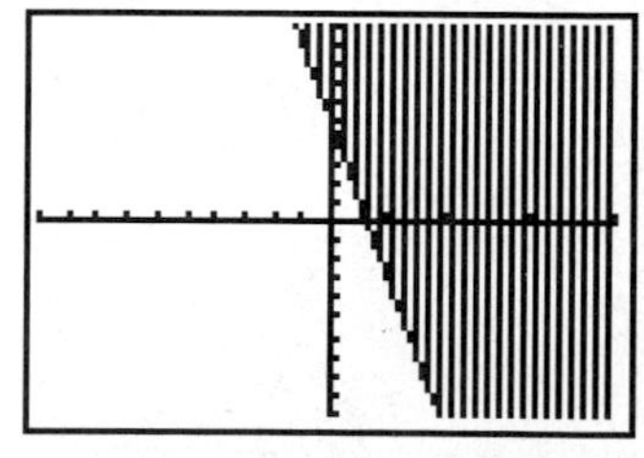

[−10, 10] scl:1 by [−10, 10] scl:1

Example 2 **Nonlinear Inequalities** The procedure for graphing nonlinear inequalities is the same as that of graphing linear inequalities.

a. Graph $y \leq 0.25x^2 - 4$ in the standard viewing window.

Press: Y= 0.25 X,T,θ,n x^2 — 4

Next, select the Below style icon, ◣, since the inequality asks for "$y \leq$", and then press GRAPH.

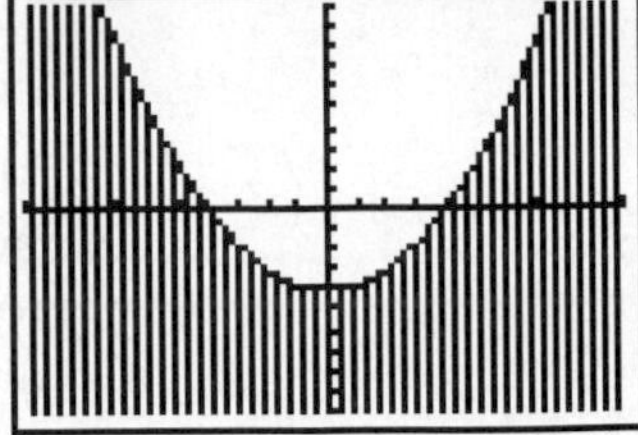

[−10, 10] scl:1 by [−10, 10] scl:1

b. Graph $y \geq 0.2x^4 - 3x^2 + 4$.

Press: Y= 0.2 X,T,θ,n ^ 4 — X,T,θ,n ^ 2 + 4

Next, select the Above style icon, ◥, since the inequality asks for "$y \geq$". Then press GRAPH.

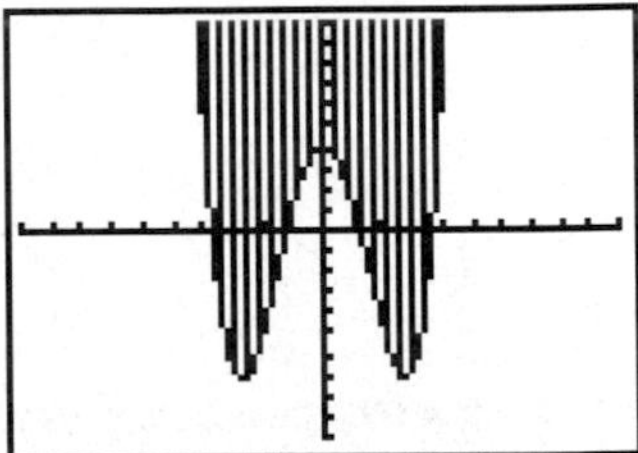

[−10, 10] scl:1 by [−10, 10] scl:1

c. Graph $y \leq \sqrt{x + 2} + 4$.

Press: Y= 2nd [√] X,T,θ,n + 2) + 4

Next, select the Below style icon, ◣, since the inequality asks for "$y \leq$". Then press GRAPH.

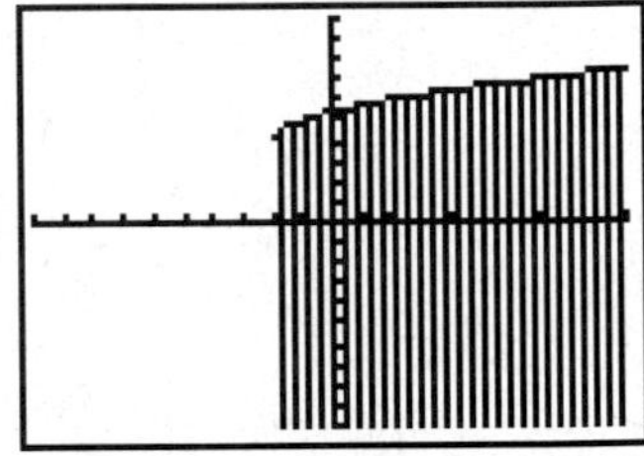

[−10, 10] scl:1 by [−10, 10] scl:1

d. Graph $y \geq 3^x - 5$.

Press: Y= 3 ^ X,T,θ,n — 5

Next, select the Above style icon, ◥, since the inequality asks for "$y \geq$". Then press GRAPH.

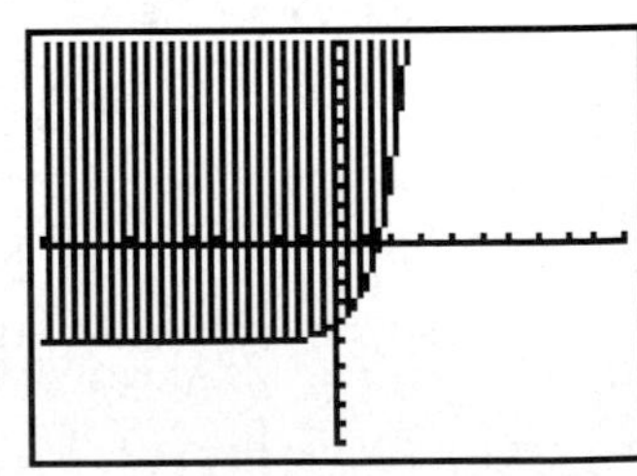

[−10, 10] scl:1 by [−10, 10] scl:1

Graphing systems of inequalities on a graphing calculator is similar to graphing systems of equations.

Example 3 **Graph the system of inequalities.**

$y \geq 2x - 5$

$y \leq x^2 - 4x + 1$

Method 1: Shading Options in **Y=**

Press: [Y=] 2 [X,T,θ,n] [—] 5 [ENTER] [X,T,θ,n] [x^2] [—] 4 [X,T,θ,n] [+] 1

Select the Above style icon, ◥, for $y \geq 2x - 5$ and the Below style icon, ◣, for $y \leq x^2 - 4x + 1$. Then press [GRAPH].

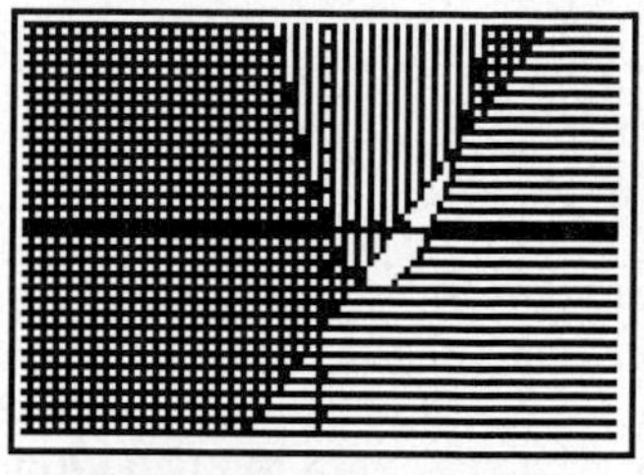

[−10, 10] scl:1 by [−10, 10] scl:1

Notice that the first inequality is indicated using vertical lines and the second inequality uses horizontal lines. The solution to the system is shown by the intersection of the shaded areas.

Method 2: Using the Shade Command

The **Shade(** *command can only be used with two inequalities which can be written with "$y \leq$" in one inequality and "$y \geq$" in the other.*

Some systems of inequalities can be graphed by using the **Shade(** command and entering a function for a lower boundary and a function for the upper boundary of the inequality. The calculator first graphs both functions and then shades above the first function entered and below the second function entered.

Before graphing an inequality using the **Shade(** command, clear any graphics from the viewing window by pressing [2nd] **[DRAW]** 1 [ENTER]. Also clear any equations in the **Y=** list. If not already there, return to the home screen by pressing [2nd] **[QUIT]**.

Press: [2nd] **[DRAW]** 7 2 [X,T,θ,n] [—] 5 [,] [X,T,θ,n] [x^2] [—] 4 [X,T,θ,n] [+] 1 [)] [ENTER]

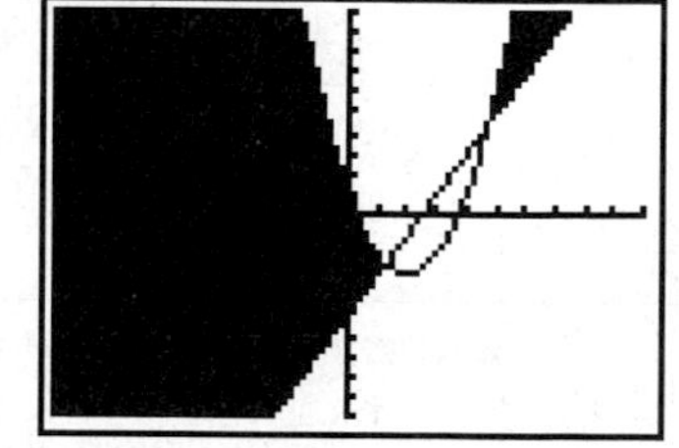

[−10, 10] scl:1 by [−10, 10] scl:1

For more shading pattern options, see page 685.

5 Matrices

A graphing calculator can perform operations with matrices. It can also find determinants and inverses of matrices. The **MATRX** menus are accessed using 2nd **[MATRX]**.

There are three menus in the **MATRX** menu.

- The **NAMES** menu lists the matrix locations available. There are ten matrix variables, [A] through [J].
- The **MATH** menu lists the matrix functions available.
- The **EDIT** menu allows you to define matrices.

A matrix with dimension 2×3 indicates a matrix with 2 rows and 3 columns. Depending on available memory, a matrix may have up to 99 rows or columns.

Example 1 **Entering a Matrix** **Enter matrix** $A = \begin{bmatrix} 1 & 3 \\ 2 & -2 \end{bmatrix}$.

To enter a matrix into your calculator, choose the **EDIT** menu and select the matrix name. Then enter the dimensions and elements of the matrix.

Press: 2nd [MATRX] ◀ ENTER 2 ENTER 2 ENTER 1 ENTER 3 ENTER 2 ENTER (-) 2

Press 2nd **[QUIT]** to return to the **HOME** screen. Then press 2nd **[MATRX]** ENTER ENTER to display the matrix.

```
[A]
          [[1  3 ]
           [2 -2]]
```

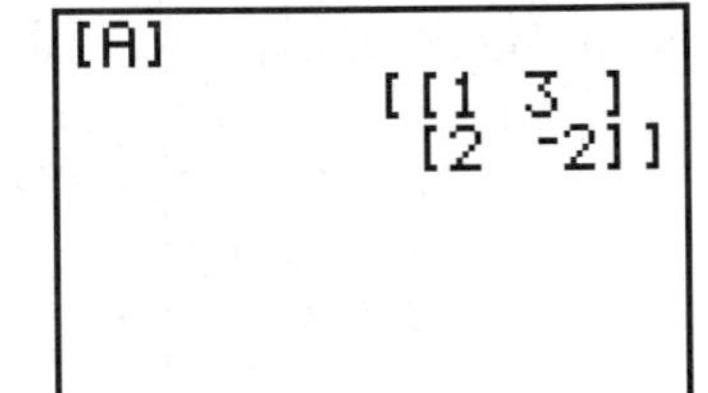

You can find the determinant and inverse of a matrix very quickly with a graphing calculator.

Example 2 **Determinant and Inverse of a Matrix**

a. Find the determinant of matrix *A*.

Press: 2nd **[MATRX]** ▶ 1 2nd **[MATRX]** 1 ENTER

The determinant of matrix A is -8.

```
det([A]
                    -8
```

b. Find the inverse of matrix *A*.

Press: 2nd **[MATRX]** 1 x^{-1} ENTER

$A^{-1} = \begin{bmatrix} 0.25 & 0.375 \\ 0.25 & -0.125 \end{bmatrix}$

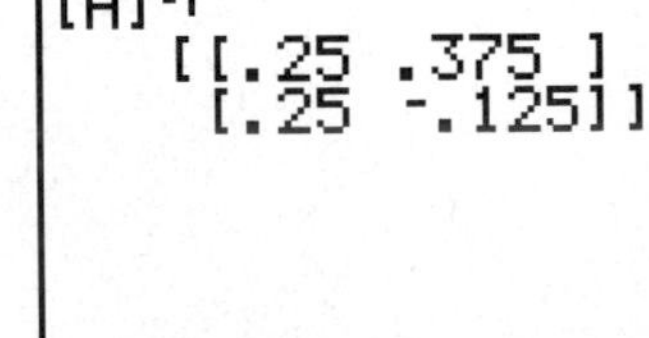

Example 3 Operations with Matrices

Enter matrix $B = \begin{bmatrix} 2 & 3 & 6 \\ 4 & -8 & 5 \end{bmatrix}$. **Then perform each operation.**

a. $\frac{1}{2}B$

First, enter matrix B.

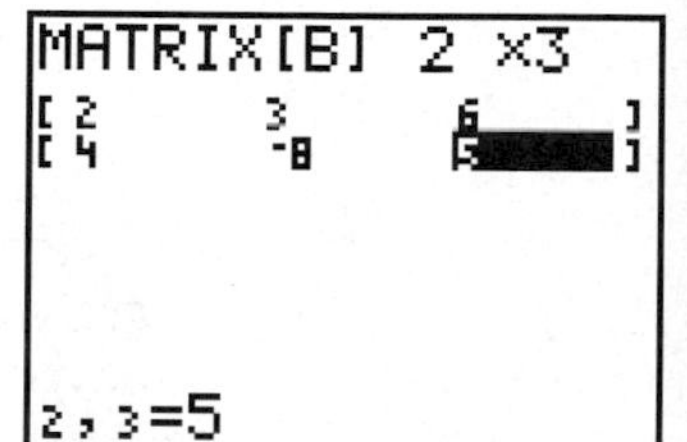

Press: [2nd] [MATRX] [◀] 2 2 [ENTER]
3 [ENTER] 2 [ENTER] 3 [ENTER] 6 [ENTER]
4 [ENTER] [(–)] 8 [ENTER] 5 [2nd] [QUIT]

```
MATRIX[B] 2 ×3
[ 2    3    6   ]
[ 4   -8    5   ]

2,3=5
```

Then find $\frac{1}{2}B$.

Press: .5 [2nd] [MATRX] 2 [ENTER]

$$\frac{1}{2}B = \begin{bmatrix} 1 & 1.5 & 3 \\ 2 & -4 & 2.5 \end{bmatrix}$$

```
.5[B]
   [[1  1.5  3  ]
    [2  -4   2.5]]
```

b. AB

Press: [2nd] [MATRX] 1 [2nd] [MATRX] 2 [ENTER]

$$AB = \begin{bmatrix} 14 & -21 & 21 \\ -4 & 22 & 2 \end{bmatrix}$$

```
[A][B]
   [[14  -21  21]
    [-4  22   2 ]]
```

c. A^2

Press: [2nd] [MATRX] 1 [x^2] [ENTER]

$$A^2 = \begin{bmatrix} 7 & -3 \\ -2 & 10 \end{bmatrix}$$

```
[A]²
          [[7   -3]
           [-2  10]]
```

d. $AB + B$

Press: [2nd] [MATRX] 1 [2nd] [MATRX] 2 [+]
[2nd] [MATRX] 2 [ENTER]

$$AB + B = \begin{bmatrix} 16 & -18 & 27 \\ 0 & 14 & 7 \end{bmatrix}$$

```
[A][B]+[B]
   [[16  -18  27]
    [0   14   7 ]]
```

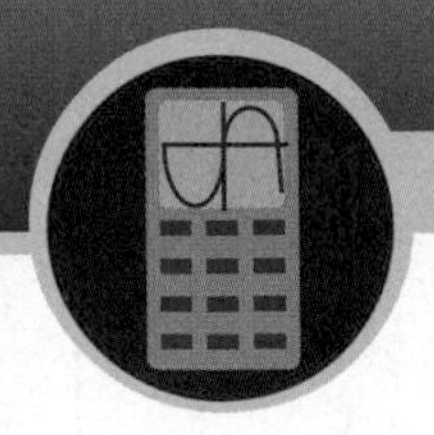

6 Graphing Trigonometric Functions

Trigonometric functions and the inverses of trigonometric functions can be graphed using [Y=]. The functions and their inverses can be graphed in degrees or radians. You must set the calculator in **Radian** or **Degree** mode. The standard viewing window for trigonometric functions can be set by pressing [ZOOM] **7:Trig**, which automatically adjusts the x- and y-axes scales for degrees or radians.

Example 1 a. **Using Degrees Graph $y = \cos x$.**

First, set the calculator in degree mode by pressing [MODE] [▼] [▼] [▶] [ENTER].

Now enter and graph the function. Press [Y=] [COS] [X,T,θ,n] [)] [ZOOM] 7.

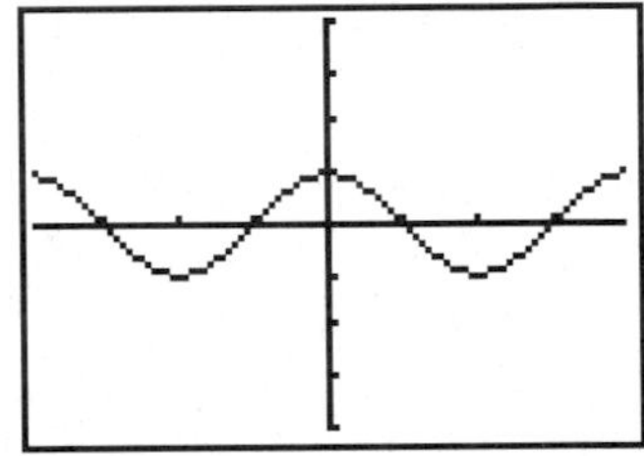

[−352.5, 352.5] scl:90 by [−4, 4] scl:1

b. **Using Radians Graph $y = \sin x$.**

Change to radian mode by pressing [MODE] [▼] [▼] [ENTER]. Press [Y=] [CLEAR] to delete the function entered in part **a**. Then press [SIN] [X,T,θ,n] [)] to enter the new function. Next, press [ZOOM] 7 to set the viewing window to accommodate radian mode and graph the function.

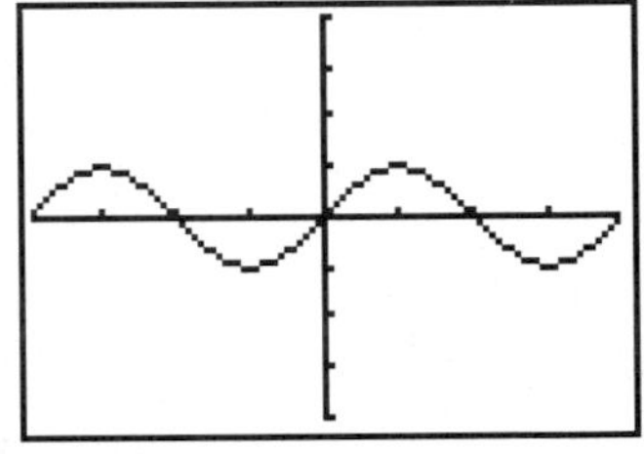

$[-2\pi, 2\pi]$ scl:$\frac{\pi}{2}$ by [−4, 4] scl:1

c. **Amplitude, Period, and Phase Shift Graph $y = 2 \sin\left(\frac{1}{2}x - 60°\right)$ using the viewing window [−540, 540] scl:90 by [−3, 3] scl:1. Then state the amplitude, period, and phase shift.**

Make sure the calculator is in degree mode.

Press: [Y=] 2 [SIN] [X,T,θ,n] [÷] 2 [−] 60 [)] [GRAPH]

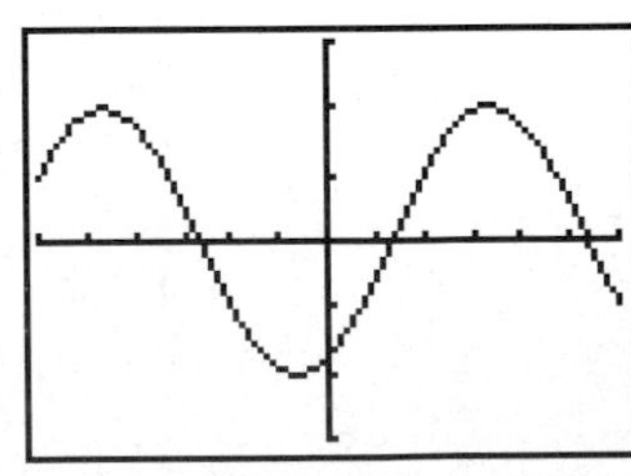

[−540, 540] scl:90 by [−3, 3] scl:1

The amplitude of the function is $\frac{|-2 - 2|}{2}$ or 2. The period is $\frac{360}{\frac{1}{2}}$ or 720°, and the phase shift is $\frac{-60}{\frac{1}{2}}$ or −120°.

d. Graph $y = 3x + \sin x$.

Use radian mode and the viewing window $[-2\pi, 2\pi]$ scl: $\frac{\pi}{2}$ by $[-6\pi, 6\pi]$ scl: $\frac{\pi}{2}$. You can enter the viewin\g window using 2nd $[\pi]$. The calculator will determine a decimal approximation.

Press: Y= 3 X,T,θ,n + SIN X,T,θ,n GRAPH

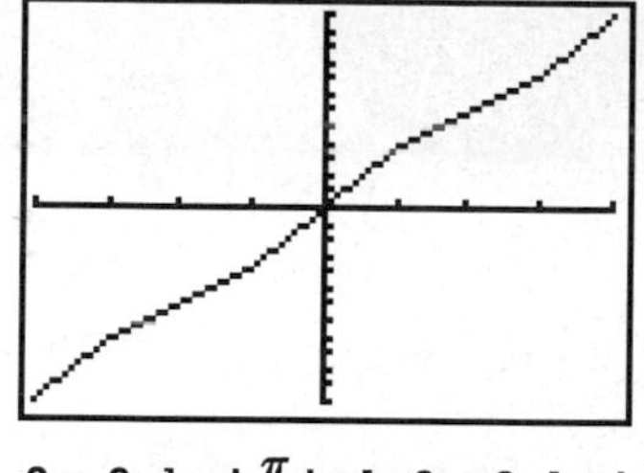

$[-2\pi, 2\pi]$ scl:$\frac{\pi}{2}$ by $[-6\pi, 6\pi]$ scl:$\frac{\pi}{2}$

Because a graphing calculator only graphs functions, the graph of each inverse is limited to the domain for which the inverse of the function is defined.

Example 2 Inverses of Trigonometric Functions

a. Graph $y = \text{Arcsin } x$.

Choose an appropriate viewing window for degree or radian mode.

Press: Y= 2nd $[\sin^{-1}]$ X,T,θ,n) GRAPH

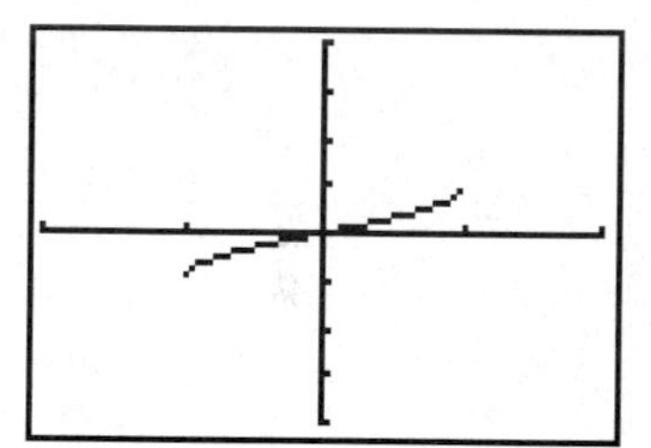

$[-2, 2]$ scl:1 by $[-360, 360]$ scl:90 or $[-2, 2]$ scl:1 by $[-2\pi, 2\pi]$ scl:$\frac{\pi}{2}$

b. Graph $y = \text{Cos}^{-1}\left(\frac{\sqrt{2}}{2} + x\right)$.

Press: Y= 2nd $[\cos^{-1}]$ 2nd $[\sqrt{\ }]$ 2) ÷ 2 + X,T,θ,n) GRAPH

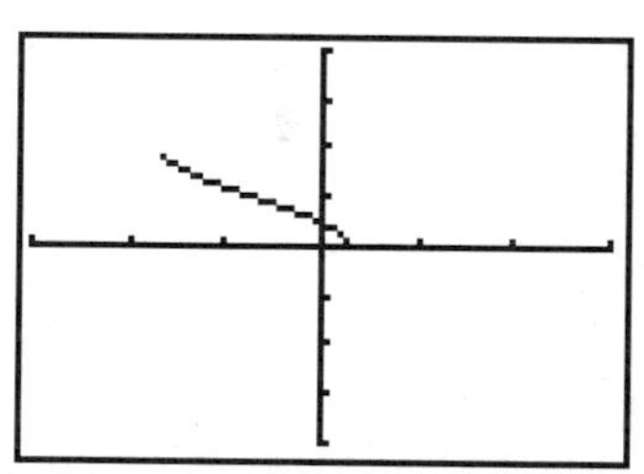

$[-3, 3]$ scl:1 by $[-360, 360]$ scl:90 or $[-3, 3]$ scl:1 by $[-2\pi, 2\pi]$ scl:$\frac{\pi}{2}$

c. Graph $y = \cos(2\ \text{Sin}^{-1} x)$.

Use the viewing window $[-3, 3]$ scl:1 by $[-2, 2]$ scl:1.

Press Y= COS 2 2nd $[\sin^{-1}]$ X,T,θ,n)) GRAPH

The result is the same whether using degree or radian mode.

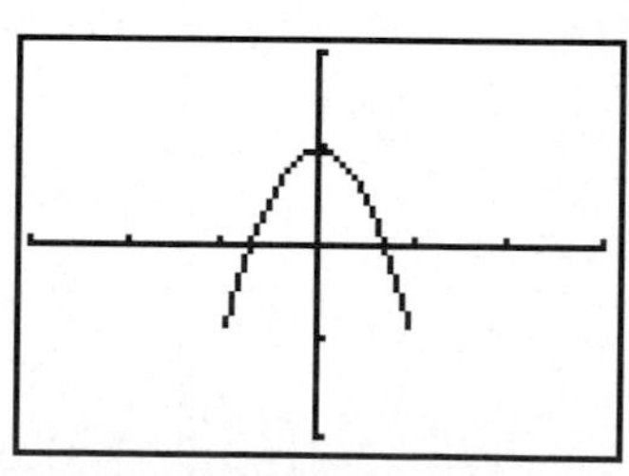

$[-3, 3]$ scl:1 by $[-2, 2]$ scl:1

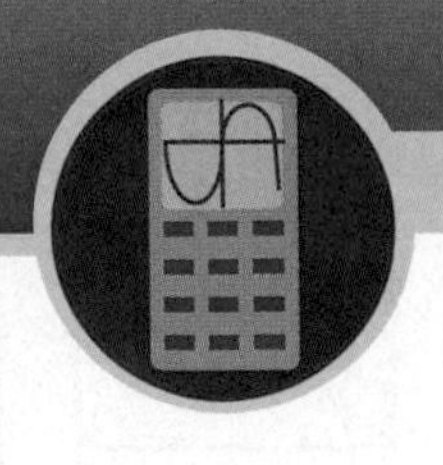

7 Graphing Special Functions

Most special functions can be graphed using the Y= key. The absolute value function **abs(** and the greatest integer function **int(** can be found in the **MATH NUM** menu.

Example 1 Absolute Value **Graph $y = 2|x - 4|$.**

Press Y= 2 MATH ▶ 1 X,T,θ,n − 4) ZOOM 6 to graph the function in the standard viewing window.

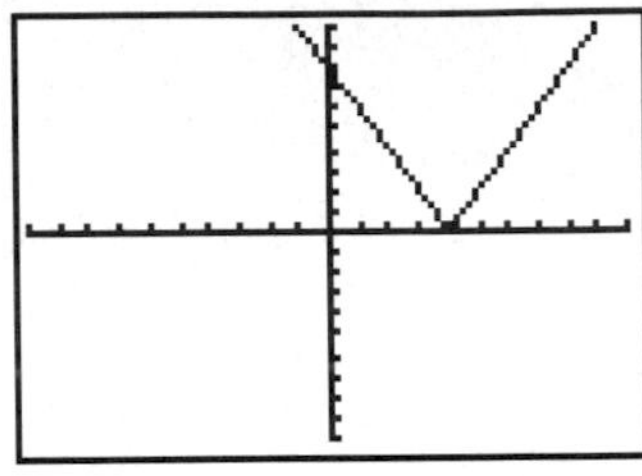

[−10, 10] scl:1 by [−10, 10] scl:1

2 Greatest Integer Function **Graph $y = [\![x - 1.5]\!]$.**

First, make sure the calculator is set for dot plotting rather than the connected plotting used in most other functions. Press MODE, highlight **Dot**, and press ENTER.

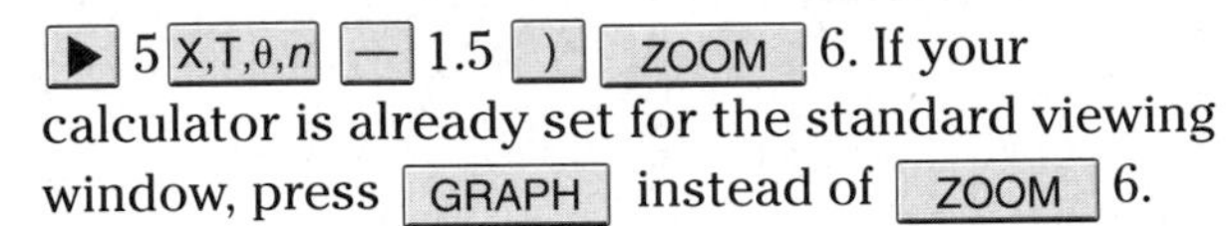

Then, enter the function. Press Y= MATH ▶ 5 X,T,θ,n − 1.5) ZOOM 6. If your calculator is already set for the standard viewing window, press GRAPH instead of ZOOM 6.

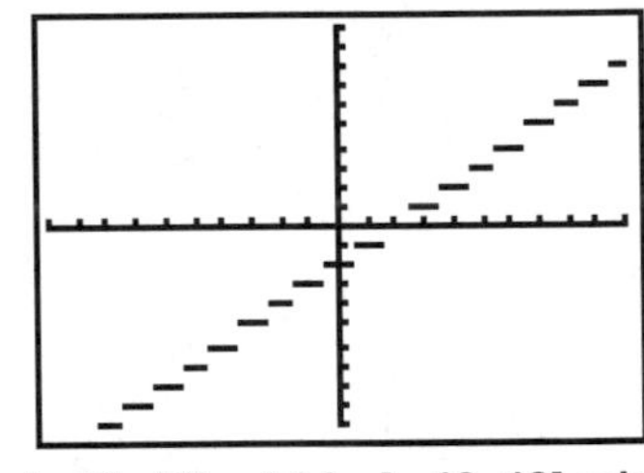

[−10, 10] scl:1 by [−10, 10] scl:1

The **TEST** menu allows you to graph other piecewise functions. Enter the pieces of the function as a sum of the products of each piece of the function and its domain. For example, $y = \begin{cases} 5 \text{ if } x < 2 \\ 4x \text{ if } x > 2 \end{cases}$ is entered as **(5)(X<2) + (4X)(X>2)** in the **Y=** menu.

Example 3 Piecewise Function **Graph $y = \begin{cases} 3 \text{ if } x \le -3 \\ 1 + x \text{ if } -3 < x \le 2. \\ 9 - 2x \text{ if } x > 2 \end{cases}$**

Place the calculator in **Dot** mode. Then enter the function in the **Y=** list using the **TEST** menu options.

Press: Y= (3) (X,T,θ,n 2nd [TEST] 6 (−) 3) + (1 + X,T,θ,n) ((−) 3 2nd [TEST] 5 X,T,θ,n) (X,T,θ,n 2nd [TEST] 6 2) + (9 − 2 X,T,θ,n) (X,T,θ,n 2nd [TEST] 3 2) ZOOM 6

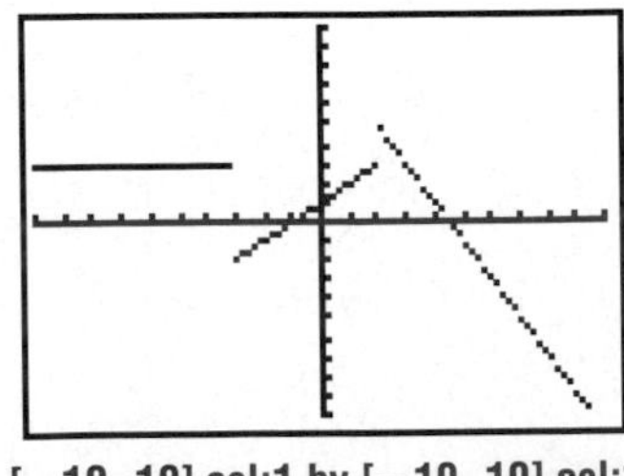

[−10, 10] scl:1 by [−10, 10] scl:1

8 Parametric and Polar Equations

Parametric and polar equations can be graphed using the Y= key by selecting the **Par** or **Pol** setting in the **MODE** menu.

Example 1 **Parametric Equations** **Graph the parametric equations $x = -3 + 2t$ and $y = 5 - 4t$.**

First, set the calculator to parametric mode by pressing MODE ▼ ▼ ▼ ▶ ENTER. Then press WINDOW 0 ENTER 10 ENTER 0.5 ENTER (–) 10 ENTER 10 ENTER 1 ENTER (–) 10 ENTER 10 ENTER 1 ENTER to set the viewing window.

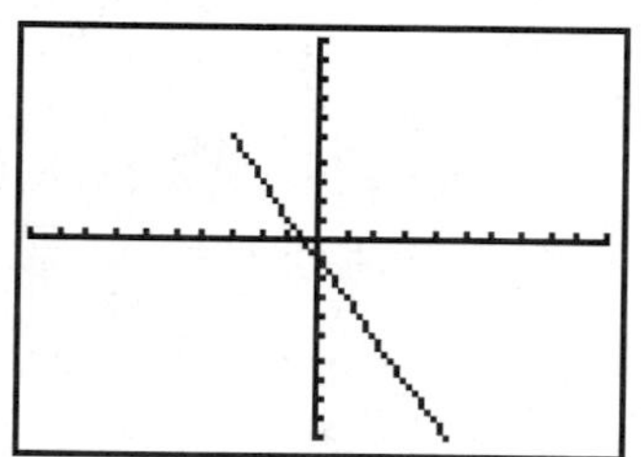

[−10, 10] tstep:0.5, [−10, 10] scl:1 by [−10, 10] scl:1

Enter the parametric equations.

Press: Y= (–) 3 + 2 X,T,θ,n ENTER 5 – 4 X,T,θ,n GRAPH

2 **Polar Equations** **Graph $r = 2 \cos \theta$.**

Set the calculator in polar mode. Press MODE ▼ ▼ ▼ ▶ ▶ ENTER, and make sure the calculator is in radian mode. Then set the viewing window by pressing WINDOW 0 ENTER 2nd [π] ENTER 0.05 ENTER (–) 5 ENTER 5 ENTER 1 ENTER (–) 2 ENTER 2 ENTER 1 ENTER.

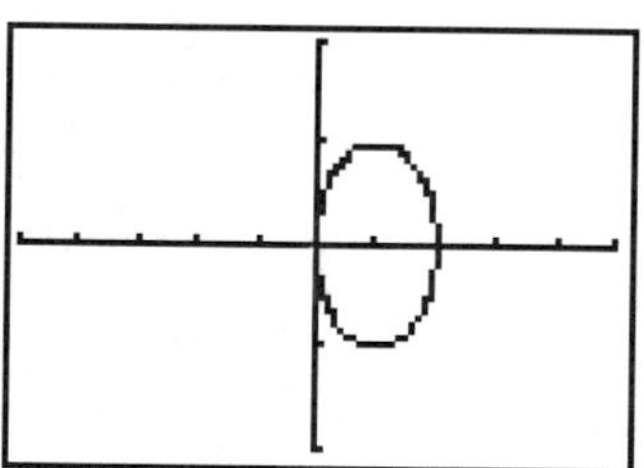

[0, 2π] θstep:0.05, [−5, 5] scl:1 by [−2, 2] scl:1

Enter the equation. Press Y= 2 COS X,T,θ,n) GRAPH.

The graph appears to be an ellipse, but pressing ZOOM 5 to set a square viewing window shows that it is actually a circle.

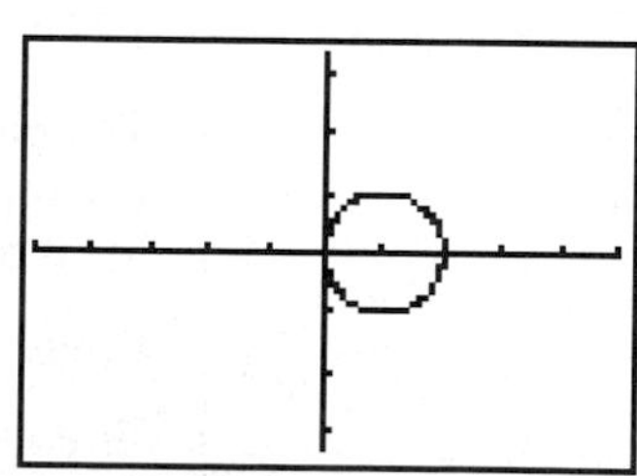

[0, π] θstep:0.05, [−5, 5] scl:1 by [−3.29787234, 3.29787234] scl:1

9 Statistics and Statistical Graphs

A graphing calculator allows you to enter a set of data and generate statistics and statistical graphs. Before you enter data values, make sure you clear the **Y=** list, **L1** and **L2**, and the graphics screen. Clear the **Y=** list by pressing [Y=] [CLEAR]. Use the [▼] key to select additional equations and clear them also. To clear **L1** and **L2**, press [STAT] 4 [2nd] **[L1]** [,] **[L2]** [ENTER]. If you need to clear the graphics screen, press [2nd] **[DRAW]** 1 [ENTER].

Example 1 **Enter Data into Lists** **Enter the following data into a graphing calculator.**

49 53 54 54 56 55 57 61 51 58 41 59 54 50 60 44

Press: [STAT] 1 53 [ENTER] 54 [ENTER] 54 [ENTER] 56 [ENTER] 55 [ENTER] 57 [ENTER] 61 [ENTER] 51 [ENTER] 58 [ENTER] 41 [ENTER] 59 [ENTER] 54 [ENTER] 50 [ENTER] 60 [ENTER] 44 [ENTER]

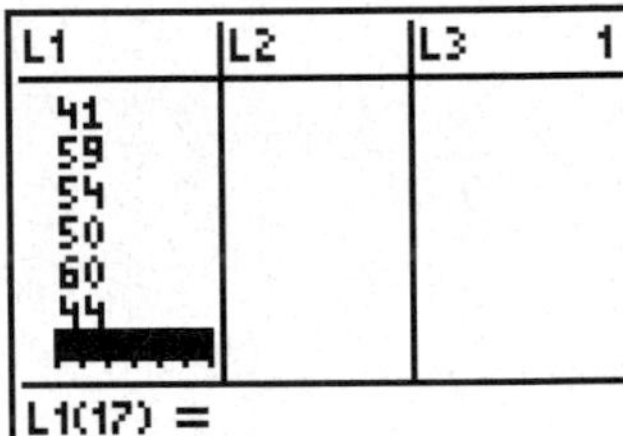

You can use the up and down arrow keys to scroll through the list.

Example 2 **Find Mean, Median, and Mode** **Find the mean, median, and mode of the data in Example 1.**

Press [STAT] [▶] 1 [ENTER]. This function displays many statistics about the data. **X̄** denotes the mean. Scroll down to find the median.

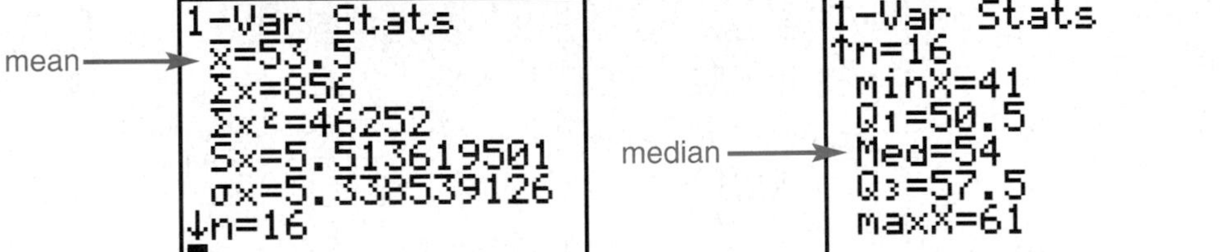

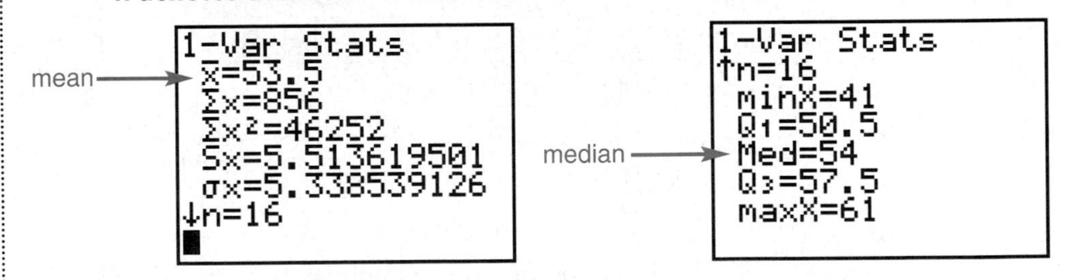

The calculator does not have a function to determine the mode. You can find the mode by examining the data. First sort the data to write them in order from least to greatest.

Press: [STAT] 2 [2nd] **[L1]** [)] [ENTER]

Then scroll through the data by pressing [STAT] 1 and using the [▲] and [▼] keys. You will find that the mode is 54.

Examples

3 Box-and-Whisker Plots

a. Draw a box-and-whisker plot for the data.

30.2	29.0	26.2	25.8	23.8	43.0	19.8	19.4
26.0	46.6	26.8	22.8	35.4	25.2	12.2	31.4

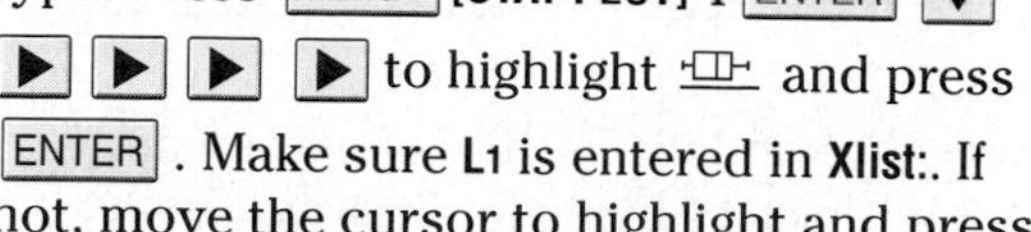

Set the viewing window. Next, set the plot type. Press 2nd [STAT PLOT] 1 ENTER ▼ ▶ ▶ ▶ ▶ to highlight and press ENTER. Make sure L1 is entered in **Xlist:**. If not, move the cursor to highlight and press 2nd [L1] ENTER. Then, enter the data into L1. Press STAT 1 30.2 ENTER 29 ENTER … 31.4 ENTER GRAPH.

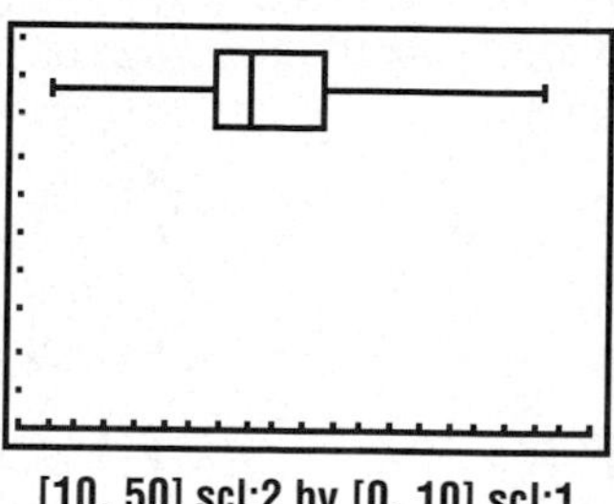

[10, 50] scl:2 by [0, 10] scl:1

b. Draw a box-and-whisker plot with outliers using the data.

Without clearing the lists or graphic screen, press 2nd [STAT PLOT] 2 ENTER ▼ ▶ ▶ ▶, to highlight, and press ENTER. Make sure L1 is entered in **Xlist:**. Press GRAPH.

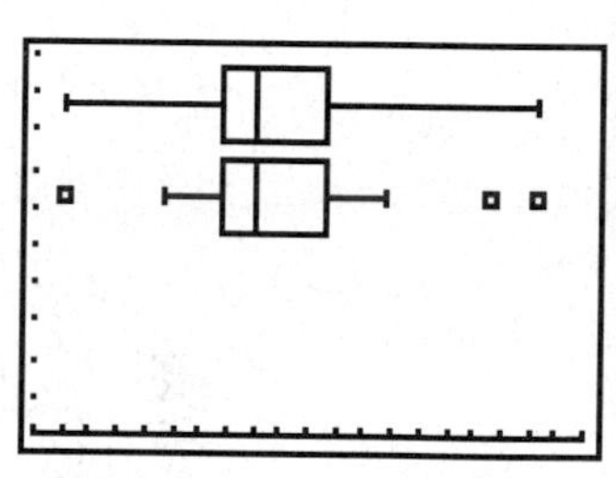

[10, 50] scl:2 by [0, 10] scl:1

c. Find the upper and lower quartiles, the median, and the outliers.

Press TRACE and use ◀ and ▶ to move the cursor along the graph. The values will be displayed. For this data, the upper quartile is 30.8, the lower quartile is 23.3, the median is 26.1, and the outliers are 43 and 46.6.

4 Histograms

a. Use the data on the number of public libraries in each state and Washington, D.C., to make a histogram.

273	102	159	196	1030	235	244	30	27	428	366	49	141
772	427	554	372	188	322	273	187	491	659	361	243	346
110	283	78	238	455	92	1067	352	86	684	192	201	640
74	180	134	284	753	96	204	308	309	174	451	74	

Enter the data in L1. Press STAT 1 273 ENTER 102 ENTER … 74 ENTER.

Set the viewing window. Choose **Xmin**, **Xmax**, and **Xscl** to determine the number of bars in the histogram. For this data, the least value is 27 and the greatest is 1067. If **Xmin** = 0, **Xmax** = 1100, and **Xscl** = 100, the histogram will have 11 bars each representing an interval of 100.

Choose the type of graph. Press 2nd [STAT PLOT] 1 ENTER ▼ ▶ ▶ ENTER ▼ 2nd [L1] ENTER ▼ 1 ENTER. Then press GRAPH to draw the histogram.

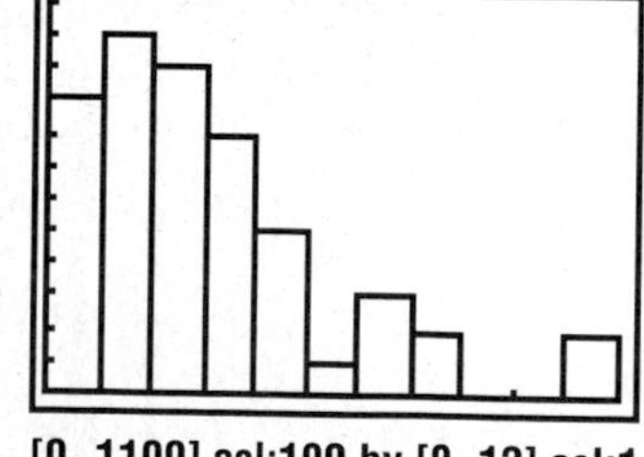

[0, 1100] scl:100 by [0, 12] scl:1

Make sure you have cleared the **Y=** *list,* **L1** *and* **L2**, *and the graphic screen.*

b. Use the data in the table to draw a histogram and its frequency polygram.

Class Limits	Frequency
155–170	7
170–185	15
185–200	34
200–215	38
215–230	42
230–245	35
245–260	33
260–275	21
275–290	18
290–305	6
305–320	3

Enter the class marks as **L1**. Press STAT 1 162.5 ENTER 177.5 ENTER ... 312.5 ENTER.

Move the cursor to **L2**. Enter the frequencies. Press 7 ENTER 15 ENTER ... 3 ENTER 2nd **[QUIT]**.

Set the viewing window. Use the minimum and maximum of the class limits for **Xmin** and **Xmax**. Use the size of the intervals for **Xscl**. Choose the *y*-axis values to show the complete histogram.

Set the plot type. Press 2nd **[STAT PLOT]** 1 ENTER ▼ ▶ ▶ ENTER ▼ 2nd **[L1]** ENTER ▼ 2nd **[L2]** ENTER. Then press GRAPH .(Figure 1)

Without clearing the lists or graphic screen, press 2nd **[STAT PLOT]** 1 ENTER ▼ ▶ to highlight ⊵, and press ENTER. Make sure **L1** is entered in **Xlist:** and **L2** is the **Ylist**. Choose ▫ as the mark. Press GRAPH.

Figure 1

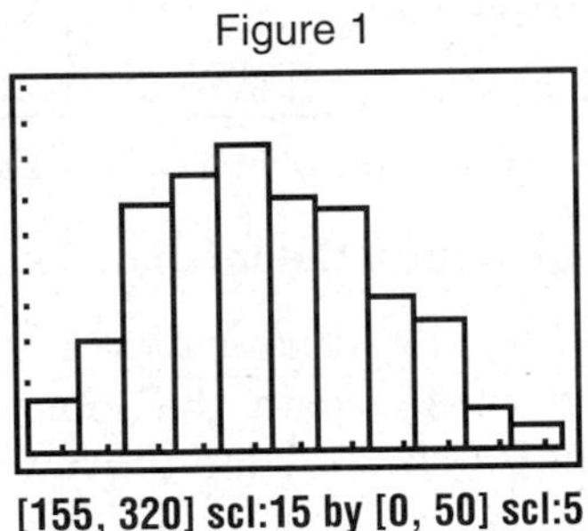

[155, 320] scl:15 by [0, 50] scl:5

Figure 2

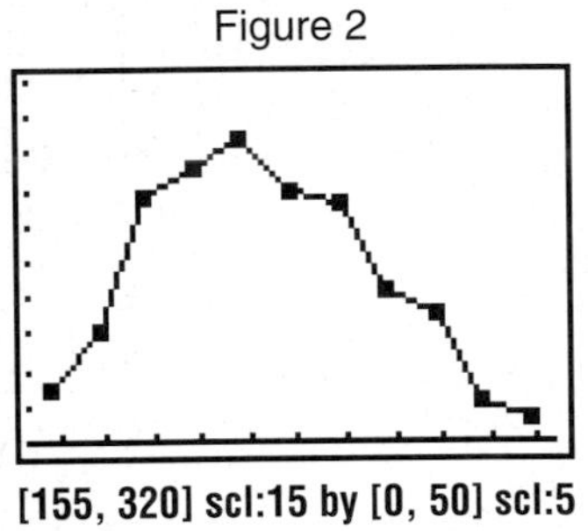

[155, 320] scl:15 by [0, 50] scl:5

Example 5 **Scatter Plot, Connected Line Scatter Plot, and Regression Line**

a. Use these data to draw a scatter plot: (20.0, 5.2), (10.2, 1.9), (7.3, 1.6), (6.8, 2.6), (5.9, 1.0), (2.6, 0.7), (2.8, 0.35), (2.7, 0.15).

Clear previous data and graphs and set the viewing window. Enter the *x*-values into **L1** and the *y*-values into **L2**. Then draw the scatter plot by pressing 2nd **[STAT PLOT]** 1 ENTER ▼ to highlight ⋰ and press ENTER. Make sure **L1** is the **Xlist:** and **L2** is the **Ylist:**. Then press GRAPH.

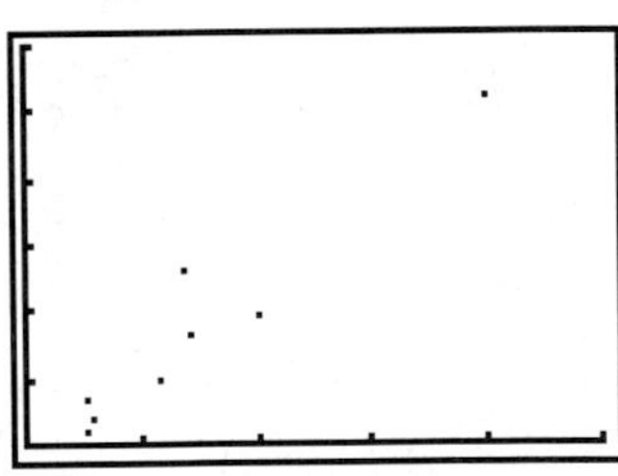

[0, 25] scl:5 by [0, 6] scl:1

b. Use the data to draw a line graph.

Press 2nd **[STAT PLOT]** 1 ▼ ▶ to highlight ⊵ and press ENTER GRAPH.

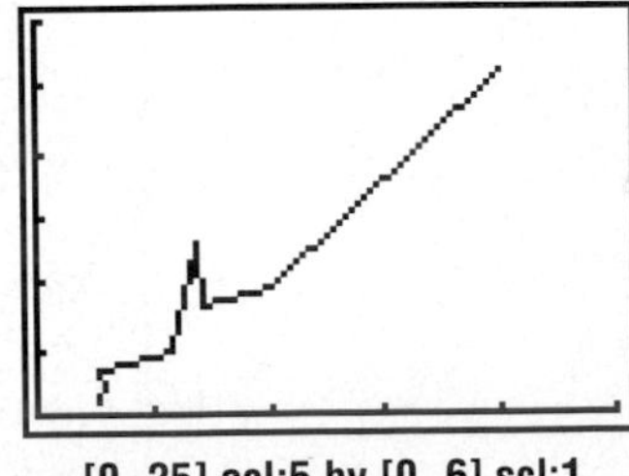

[0, 25] scl:5 by [0, 6] scl:1

c. Draw a regression line for the data in the table.

Set the plot to display a scatter plot by pressing [2nd] **[STAT PLOT]** 1 [▼] [ENTER] [2nd] **[QUIT]**.

To calculate the coefficients of regression press [STAT] [▶] 4 [ENTER].

```
LinReg
 y=ax+b
 a=.2695895968
 b=-.2771341868
 r²=.9089138336
 r=.9533697255
```

Then, write the equation of the regression line. You can automatically enter the regression equation in the **Y=** list.

Press [Y=] [VARS] 5 [▶] [▶] 1. Finally, graph the regression line by pressing [GRAPH].

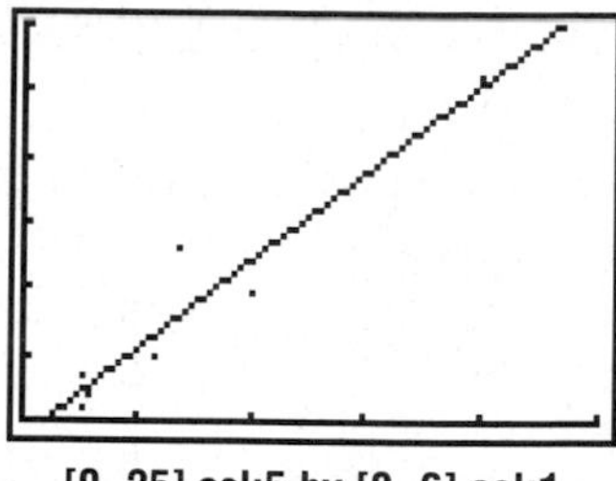

[0, 25] scl:5 by [0, 6] scl:1

There are also regression models for analyzing data that are not linear built into the calculator.

Example 6 **Nonlinear Regression** **Find a sine regression equation to model the data in the table. Graph the data and the regression equation.**

x	1	2	3	4	5	6	7	8	9	10	11	12
y	39	42	45	48	54	59	63	64	59	52	44	40

Enter the data into lists **L1** and **L2**. Press [STAT] 1 [ENTER] 2 [ENTER] ... 12 [ENTER] [▶] 39 [ENTER] 42 [ENTER] ... 40 [ENTER].

Find the regression statistics.

Press: [STAT] [▶] [ALPHA] **[C]** [ENTER]

Enter the regression equation into the **Y=** list.

Press: [Y=] [VARS] 5 [▶] [▶] 1

```
SinReg
 y=a*sin(bx+c)+d
 a=11.83736682
 b=.5543464048
 c=-2.449519342
 d=51.35457494
```

Then format the scatter plot to graph the data by pressing [2nd] **[STAT PLOT]** 1 [▼] [ENTER]. Make sure that **L1** is chosen as the **Xlist** and **L2** is chosen as the **Ylist**. Set the viewing window. Press [GRAPH] to see the scatter plot and the graph of the regression equation.

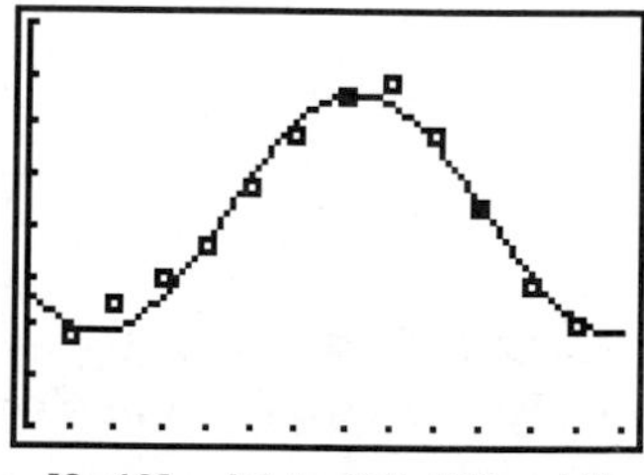

[0, 13] scl:1 by [30, 70] scl:5

Also see pages 389, 739, 741–744 for other examples of nonlinear regression.

EXTRA PRACTICE

Lesson 1-1 *(Pages 5–11)*

State the domain and range of each relation. Then state whether the relation is a function. Write *yes* or *no*.

1. $\{(1, 4), (-2, 2), (2, 2), (1, -4)\}$
2. $\{(0.5, 3), (-0.5, 3), (3, 0.5), (-3, 0.5)\}$
3. $\{(2, 2), (5, 7), (-1, 1), (0, 3), (7, 5)\}$
4. $\{(3.2, 4), (2.3, -4), (2, 3), (3.2, -1)\}$

Evaluate each function for the given value.

5. $f(4)$ if $f(x) = 4x - 2$
6. $g(-3)$ if $g(x) = 2x^2 - x + 5$
7. $h(1.5)$ if $h(x) = \frac{3}{2x}$
8. $k(5m)$ if $k(x) = |3x^2 - 3|$

Lesson 1-2 *(Pages 13–19)*

1. Given $f(x) = 2x - 1$ and $g(x) = x^2 + 3x - 1$, find $f(x) + g(x)$, $f(x) - g(x)$, $f(x) \cdot g(x)$, and $\left(\frac{f}{g}\right)(x)$.

Find $[f \circ g](x)$ and $[g \circ f](x)$ for each $f(x)$ and $g(x)$.

2. $f(x) = 3 - x$
   $g(x) = 4x^2$
3. $f(x) = \frac{1}{3}x - 1$
   $g(x) = x + 9$
4. $f(x) = -2x$
   $g(x) = 2x^3 - x^2 + x - 1$

Lesson 1-3 *(Pages 20–25)*

Find the zero of each function. If no zero exists, write *none*. Then graph the function.

1. $f(x) = x - 2$
2. $f(x) = -3x + 4$
3. $f(x) = -1$
4. $f(x) = 4x$
5. $f(x) = 2x - 1$
6. $f(x) = -x - 5$

Lesson 1-4 *(Pages 26–31)*

Write an equation in slope-intercept form for each line described.

1. slope = 2, y-intercept = 1
2. slope = -1, passes through $(1, 2)$
3. slope = $-\frac{1}{4}$, y-intercept = -3
4. slope = 0, passes through $(-2, -4)$
5. passes through $A(2, 1)$ and $B(-2, 3)$
6. x-intercept = -1, y-intercept = 6
7. the x-axis
8. slope = 1.5, x-intercept = 10

Lesson 1-5 *(Pages 32–37)*

1. Are the graphs of $y = 4x - 2$ and $y = -4x + 3$ *parallel, coinciding, perpendicular,* or *none of these*? Explain.

Write the standard form of the equation of the line that is parallel to the graph of the given equation and passes through the point with the given coordinates.

2. $y = x + 1$; $(0, -2)$
3. $y = 2x - 2$; $(1, 3)$
4. $y = -1$; $(-4, 12)$

Write the standard form of the equation of the line that is perpendicular to the graph of the given equation and passes through the point with the given coordinates.

5. $3x - y = 5$; $(-2, 6)$
6. $x = 10$; $(12, -15)$
7. $5x - 2y = 1$; $(3, -7)$

Lesson 1-6 *(Pages 38–44)*

The table shows the number of students enrolled in U.S. public elementary and secondary schools for several school years.

Students Enrolled in U.S. Public Elementary and Secondary Schools								
School Year	1989–1990	1990–1991	1991–1992	1992–1993	1993–1994	1994–1995	1995–1996	2010–2011
Enrollment (thousands)	40.5	41.2	42.0	42.8	43.5	44.1	44.8	?

Source: *The World Almanac*, 1999

1. Graph the data on a scatter plot. Use the last year of the school year for your graph.
2. Use two ordered pairs to write the equation of a best-fit line.
3. Use a graphing calculator to find an equation of the regression line for the data. What is the correlation value?
4. If the equation of the regression line shows a moderate or strong relationship, predict the missing value. Explain whether you think the prediction is reliable.

Lesson 1-7 *(Pages 45–51)*

Graph each function.

1. $f(x) = \begin{cases} x + 2 \text{ if } x \leq -1 \\ 2x \text{ if } x > -1 \end{cases}$
2. $g(x) = |x + 2|$
3. $h(x) = [\![x]\!] - 1$
4. $f(x) = \begin{cases} -2 \text{ if } -2 \leq x \leq -1 \\ x \text{ if } -1 < x \leq 2 \\ 3x \text{ if } x > 2 \end{cases}$
5. $g(x) = |5 - 2x|$
6. $k(x) = 2[\![x]\!]$

Lesson 1-8 *(Pages 52–56)*

Graph each inequality.

1. $y > 2$
2. $x + y \leq 3$
3. $-y > x + 1$
4. $4x + 2y \leq 6$
5. $-1 \leq x + y \leq 4$
6. $y \leq |x|$

Lesson 2-1 *(Pages 67–72)*

Solve each system of equations by graphing. Then state whether the system is *consistent and independent, consistent and dependent,* or *inconsistent.*

1. $2x - y - 1 = 0$
   $3y = 6x - 3$
2. $y - 3x = 8$
   $x + y = 4$
3. $3x - y = -1$
   $2y - 6x = -4$

Solve each system of equations algebraically.

4. $5x + 2y = 1$
   $x + 2y = 5$
5. $2x + 4y = 8$
   $2x + 3y = 8$
6. $8x + 2y = 2$
   $3x - 4y = -23$

Lesson 2-2 *(Pages 73–77)*

Solve each system of equations.

1. $x + y = 6$
   $x + z = -2$
   $y + z = 2$
2. $x + 2y - z = -7$
   $2x - 2y - z = 6$
   $x + y - 2z = -6$
3. $2x - 3y + z = 1$
   $x + y - z = -4$
   $3x - 2y + 2z = 3$

Lesson 2-3 *(Pages 78–86)*

Find the values of x and y for which each of the following equations is true.

1. $\begin{bmatrix} 2x + y \\ x + 2y \end{bmatrix} = \begin{bmatrix} -1 \\ 1 \end{bmatrix}$
2. $\begin{bmatrix} x + 2y \\ 2x - 2y \end{bmatrix} = \begin{bmatrix} 5 \\ -2 \end{bmatrix}$
3. $\begin{bmatrix} 3x \\ 4y \end{bmatrix} = \begin{bmatrix} y - 7 \\ 5x \end{bmatrix}$

Use matrices A, B, C, D, and E to find each of the following. If the matrix does not exist, write *impossible.*

$A = \begin{bmatrix} 4 & -1 \\ 1 & 5 \\ 2 & 6 \end{bmatrix}$ $B = \begin{bmatrix} 2 & 0 & -3 \\ 4 & -3 & 2 \end{bmatrix}$ $C = \begin{bmatrix} 7 & -5 \\ 0 & 1 \\ 8 & 4 \end{bmatrix}$ $D = \begin{bmatrix} -4 & 1 \\ 2 & 3 \end{bmatrix}$ $E = \begin{bmatrix} 1 & -5 \\ -3 & 2 \end{bmatrix}$

4. $A + C$
5. $D - E$
6. $4B$
7. $D + B$
8. $2C + 3A$
9. AC
10. ED
11. $BC - D$

Lesson 2-4 *(Pages 88–96)*

Use matrices to perform each transformation. Then graph the pre-image and image on the same coordinate grid.

1. Triangle ABC has vertices $A(5, -3)$, $B(2, 4)$, and $C(-2, -6)$. Use scalar multiplication to find the coordinates of the triangle after a dilation of scale factor 0.5.
2. Quadrilateral $JKLM$ has vertices $J(6, -2)$, $K(-3, -5)$, $L(-4, 7)$, and $M(1, 5)$. Find the coordinates of the quadrilateral after a translation of 2 units left and 3 units down.
3. Triangle NPQ is represented by the matrix $\begin{bmatrix} -5 & 3 & 0 \\ 2 & 8 & -4 \end{bmatrix}$. Find the coordinates of the image of the triangle after a reflection over the y-axis.
4. Square $WXYZ$ has vertices $W(-3, 3)$, $X(-6, 2)$, $Y(-5, -1)$, and $Z(-2, 0)$. Find the coordinates of the image of the square after a rotation of 180° counterclockwise about the origin.
5. Triangle FGH has vertices $F(-4, 2)$, $G(2, -3)$, and $H(-6, -7)$. Find the coordinates of the image of $\triangle FGH$ after $Rot_{90} \circ R_{y = x}$.

Lesson 2-5 *(Pages 98–105)*

Find the value of each determinant.

1. $\begin{vmatrix} 3 & 7 \\ -11 & 2 \end{vmatrix}$
2. $\begin{vmatrix} -3 & -5 \\ -7 & -2 \end{vmatrix}$
3. $\begin{vmatrix} -5 & 0 \\ -\frac{1}{2} & -6 \end{vmatrix}$
4. $\begin{vmatrix} -1 & 0 & 2 \\ -3 & 1 & -2 \\ 5 & -1 & -3 \end{vmatrix}$
5. $\begin{vmatrix} -1 & 3 & 2 \\ 4 & -2 & 1 \\ 3 & -3 & -4 \end{vmatrix}$
6. $\begin{vmatrix} 4 & 0 & -1 \\ 5 & 3 & 6 \\ -2 & -5 & 2 \end{vmatrix}$

Find the multiplicative inverse of each matrix, if it exists.

7. $\begin{bmatrix} 2 & 1 \\ 1 & 3 \end{bmatrix}$
8. $\begin{bmatrix} 10 & 0 \\ 5 & 4 \end{bmatrix}$
9. $\begin{bmatrix} 5 & -6 \\ -3 & 4 \end{bmatrix}$
10. $\begin{bmatrix} 3 & -5 \\ 6 & 1 \end{bmatrix}$

Solve each system by using a matrix equation.

11. $3x + 2y = 22$
    $x - 2y = -6$
12. $4x - 2y = -6$
    $3x + y = -7$
13. $2x + y = 4$
    $4x - 3y = 13$

Lesson 2-6 *(Pages 107–111)*

Solve each system of inequalities by graphing. Name the vertices of each polygonal convex set. Then, find the maximum and minimum values of each function for the set.

1. $x \leq 3$
   $y \geq 1$
   $y \leq 2x + 1$
   $f(x, y) = 4x + 3y$
2. $y \geq 2$
   $0 \leq x \leq 4$
   $y \leq x + 3$
   $f(x, y) = 2x - y$
3. $y \leq -x$
   $y \geq 2x - 4$
   $|x| \leq 1$
   $f(x, y) = -x - y$

Lesson 2-7 *(Pages 112–118)*

Solve each problem, if possible. If not possible, state whether the problem is *infeasible*, has *alternate optimal solutions*, or is *unbounded*.

1. **Business** The area of a parking lot is 600 square meters. A car requires 6 square meters of space and a bus requires 30 square meters. The attendant can handle only 60 vehicles. If a car is charged \$3.00 and a bus is charged \$8.00, find how many of each type of vehicle should be accepted to maximize income.
2. **Manufacturing** Based on survey results, Yummy Ice Cream concluded that it should make at least twice as many gallons of black walnut flavor as chocolate mint flavor. One distributor wants to order at least 20,000 gallons of the chocolate mint flavor. The company has all of the ingredients to produce both flavors, but it has only 45,000 gallon size containers available. If each gallon of ice cream sells for \$2.95, how many gallons of each type flavor should the company produce?
3. **Business** Cathy places greeting cards from two different companies on a display rack that can hold 90 cards. She has agreed to display at least 40 of company A's cards on the rack and at least 25 of company B's cards. Cathy makes a profit of \$0.30 on each card she sells from company A and \$0.32 on each card she sells from company B. How many cards should Cathy display from each company to maximize her profit?
4. **Personal Finance** Kristin makes \$5 an hour at a video store and \$7 an hour at a landscaping company. She must work at least 4 hours per week at the video store, and the total number of hours she works at both jobs in a week cannot be greater than 15. What is the maximum amount Kristin could earn (before deductions) in a week?

Lesson 3-1 *(Pages 127–136)*

Determine whether the graph of each function is symmetric with respect to the origin.

1. $f(x) = -4x$
2. $f(x) = x^2 + 3$
3. $f(x) = \frac{1}{3x^3}$

Determine whether the graph of each equation is symmetric with respect to the *x*-axis, *y*-axis, the line $y = x$, the line $y = -x$, or none of these.

4. $xy = 2$
5. $y + x^2 = 3$
6. $y^2 = \frac{2x^2}{7} + 1$
7. $|x| = 4y$
8. $y = 3x$
9. $y = \pm\sqrt{x^2 - 1}$

Lesson 3-2 *(Pages 137–145)*

Describe how the graphs of $f(x)$ and $g(x)$ are related.

1. $f(x) = \frac{1}{x}$ and $g(x) = \frac{1}{x} + 2$
2. $f(x) = x^2$ and $g(x) = 3x^2$
3. $f(x) = |x|$ and $g(x) = |x + 4| - 3$
4. $f(x) = x^3$ and $g(x) = \frac{1}{2}(x - 1)^3$
5. Write the equation of the graph obtained when the graph of $y = [\![x]\!]$ is expanded horizontally by a factor of 2 reflected over the *x*-axis, and then translated 5 units down.

Lesson 3-3 *(Pages 146–151)*

Graph each inequality.

1. $y \leq x^2 + 1$
2. $y > |x + 4|$
3. $y \leq -\sqrt{x - 2}$
4. $y < 2|x - 1|$
5. $y \leq |x| + 2$
6. $y > (x + 2)^2 - 1$

Solve each inequality.

7. $|x - 2| \leq 3$
8. $|4x - 2| \geq 18$
9. $|5 - 2x| < 9$
10. $|x + 1| - 3 > 1$
11. $|2x + 3| < 27$
12. $|3x + 4| - 3x \geq 0$

Lesson 3-4 *(Pages 152–158)*

Graph each function and its inverse.

1. $f(x) = |x| - 2$
2. $f(x) = x^2 + 1$
3. $f(x) = -1$

Find $f^{-1}(x)$. Then state whether $f^{-1}(x)$ is a function.

4. $f(x) = 4x - 5$
5. $f(x) = -2x + 2$
6. $f(x) = x^2 + 6$
7. $f(x) = (x - 2)^2$
8. $f(x) = -\frac{x}{2}$
9. $f(x) = \frac{1}{x - 4}$
10. $f(x) = x^2 + 8x - 2$
11. $f(x) = x^3 + 4$
12. $f(x) = -\frac{3}{(x + 1)^2}$

Lesson 3-5 *(Pages 159–168)*

Determine whether each function is continuous at the given *x*-value. Justify your response using the continuity test.

1. $y = x^2 - 3;\ x = 2$
2. $y = \frac{x}{x+3};\ x = -3$
3. $f(x) = \frac{x-2}{(x-1)^2};\ x = 1$
4. $f(x) = \begin{cases} x - 2 \text{ if } x \geq 3 \\ 2x - 5 \text{ if } x < 3 \end{cases};\ x = 3$
5. Determine whether the graph at the right has infinite discontinuity, jump discontinuity, or point discontinuity, or is continuous.

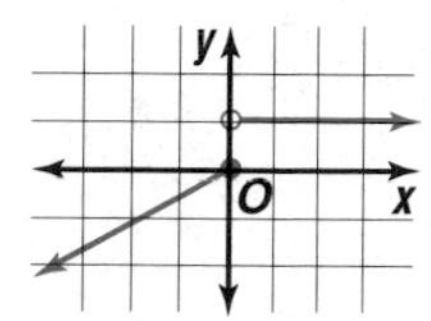

Describe the end behavior of each function.

6. $y = x^3 + 3x^2 + x - 2$
7. $y = 5 - x^4$
8. $f(x) = \frac{3}{x^2}$

Lesson 3-6 *(Pages 171–179)*

Locate the extrema for the graph of $y = f(x)$. Name and classify the extrema of the function.

1.

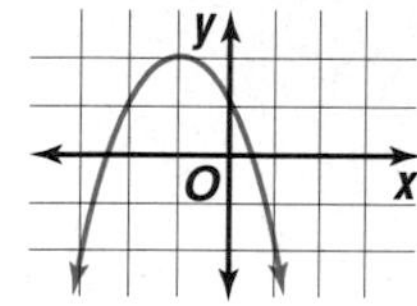

2.

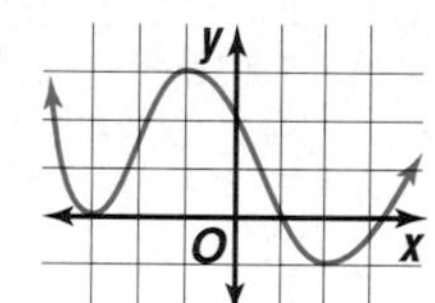

Lesson 3-7 *(Pages 180–188)*

Determine the equations of the vertical and horizontal asymptotes, if any, of each function.

1. $f(x) = \frac{3x}{x-2}$
2. $g(x) = \frac{2x^2}{x+3}$
3. $h(x) = \frac{x-5}{x^2 + 6x + 5}$
4. Does the function $f(x) = \frac{x^2 + 2x + 1}{x - 3}$ have a slant asymptote? If so, find an equation of the slant asymptote. If not, explain.

Lesson 3-8 *(Pages 189–196)*

Find the constant of variation for each relation and use it to write an equation for each statement. Then solve the equation.

1. If y varies directly as x, and $y = 8$ when $x = 2$, find y when $x = 9$.
2. If g varies directly as w, and $g = 10$ when $w = -3$, find w when $g = 4$.
3. If t varies inversely as r, and $r = 14$ when $t = -6$, find r when $t = -7$.
4. If y varies jointly as x and z, and $y = 60$ when $x = 5$, and $z = 4$, find y when $x = 5$ and $z = 10$.
5. Suppose y varies inversely as the square of x and $x = 3$ when $y = 27$. Find y when $x = 5$.
6. Suppose a varies jointly as b and the cube of c and $a = -36$ when $b = 3$ and $c = 2$. Find a when $b = 5$ and $c = 3$.

Lesson 4-1 *(Pages 205–212)*

Determine whether each number is a root of $x^3 - 7x^2 + 2x + 40 = 0$. Explain.

1. -2
2. 1
3. 2
4. 5

Write the polynomial equation of least degree for each set of roots. Does the equation have an odd or even degree? How many times does the graph of the related function cross the x-axis?

5. $3, 4$
6. $-2, -1, 2$
7. $-1.5, -1, 1$
8. $-2, -i, i$
9. $-3i, -i, i, 3i$
10. $-1, 1, 2, 3$

Lesson 4-2 *(Pages 213–221)*

Solve each equation by completing the square.

1. $x^2 - 4x - 5 = 0$
2. $x^2 + 6x + 8 = 0$
3. $m^2 + 3m - 2 = 0$
4. $2a^2 - 8a - 6 = 0$
5. $h^2 - 12h = 4$
6. $x^2 - 9x + 1 = 0$

Find the discriminant of each equation and describe the nature of the roots of the equation. Then solve the equation by using the Quadratic Formula.

7. $4x^2 - 3x - 7 = 0$
8. $w^2 + 2w - 10 = 0$
9. $12t^2 - 5t + 6 = 0$
10. $x^2 + 6x - 13 = 0$
11. $4n^2 - 4n + 1 = 0$
12. $4x^2 + 6x = 15$

Lesson 4-3 *(Pages 222–228)*

Divide using synthetic division.

1. $(x^2 + 10x + 8) \div (x + 2)$
2. $(x^3 - 3x^2 + 4x - 1) \div (x - 1)$
3. $(x^3 - 3x - 5) \div (x + 1)$
4. $(x^4 - 2x^3 - 7x^2 - 3x - 4) \div (x - 4)$

Use the Remainder Theorem to find the remainder for each division. State whether the binomial is a factor of the polynomial.

5. $(x^2 + 2x - 8) \div (x + 4)$
6. $(x^3 + 12) \div (x - 1)$
7. $(4x^3 + 2x^2 + 6x + 1) \div (x + 1)$
8. $(x^4 - 4x^2 + 16) \div (x - 4)$

Lesson 4-4 *(Pages 229–235)*

List the possible rational roots of each equation. Then determine the rational roots.

1. $x^3 + 2x^2 - 5x - 6 = 0$
2. $2x^4 - x^3 + 2x^2 - 3x + 1 = 0$
3. $x^3 + x^2 - 2 = 0$
4. $6x^4 + x^3 + 22x^2 + 4x - 8 = 0$

Find the number of possible positive real zeros and the number of possible negative real zeros for each function. Then determine the rational zeros.

5. $f(x) = x^3 - 4x^2 - x + 4$
6. $f(x) = x^4 + x^3 + 3x^2 - 5x + 10$
7. $f(x) = 4x^3 - 7x + 3$
8. $f(x) = x^4 - x^3 + 4x - 4$

Lesson 4-5 *(Pages 236–242)*

Determine between which consecutive integers the real zeros of each function are located.

1. $f(x) = 2x^2 - 4x - 5$
2. $f(x) = x^3 - 5$
3. $f(x) = x^4 - x^2 + 4x + 2$

Approximate the real zeros of each function to the nearest tenth.

4. $f(x) = 4x^4 - 6x^2 - 2x + 1$
5. $f(x) = 2x^5 + 3x^4 - 12x + 4$
6. $f(x) = -2x^4 + 5$

Lesson 4-6 *(Pages 243–250)*

Solve each equation or inequality.

1. $\frac{6}{x} + x = 5$
2. $\frac{7}{y-1} - \frac{4}{y} = \frac{y}{y-1}$
3. $\frac{5}{r+1} - \frac{4}{r-1} = \frac{1}{r^2-1}$
4. $2 = \frac{1}{2-t} + \frac{4}{t-2}$
5. $\frac{1}{3w} + \frac{4}{5w} \leq \frac{1}{15}$
6. $\frac{x-2}{x} < \frac{x-4}{x-6}$

Lesson 4-7 *(Pages 251–257)*

Solve each equation or inequality.

1. $\sqrt{2 + 3t} = 4$
2. $4 - \sqrt{x - 2} = 1$
3. $\sqrt[3]{y - 7} + 10 = 2$
4. $\sqrt{a + 1} - 5 = \sqrt{a - 6}$
5. $\sqrt{2x + 3} \leq 2$
6. $\sqrt[4]{6a - 2} > 4$

Lesson 4-8 *(Pages 258–264)*

Use a graphing calculator to write a polynomial function to model each set of data.

1.

x	−4	−2	0	2	4	6	8
f(x)	−5	−3.5	−2	−0.5	1	2.5	4

2.

x	−2	−1.5	−1	−0.5	0	0.5	1
f(x)	0	2.375	3	2.625	2	1.875	3

3.

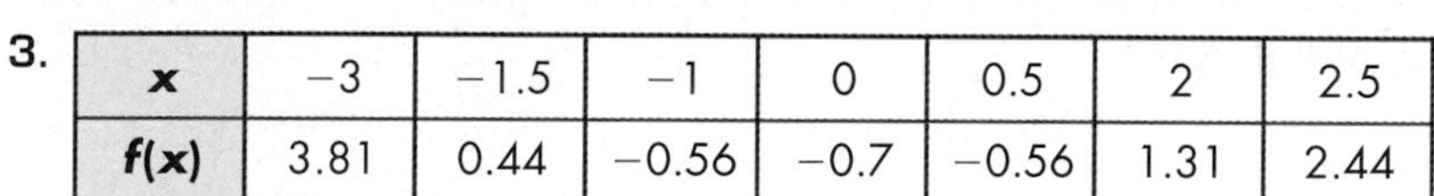

x	−3	−1.5	−1	0	0.5	2	2.5
f(x)	3.81	0.44	−0.56	−0.7	−0.56	1.31	2.44

4. **Families** The average numbers of households in the United States from 1991 to 1997 are listed below.

Year	1991	1992	1993	1994	1995	1996	1997
Number of Households (in thousands)	94.3	95.7	96.4	97.1	98.9	99.6	100.0

Source: *The 1999 World Almanac and Book of Facts*

a. Write a model that relates the number of households as a function of the number of years since 1991.

b. Use the model to predict the number of U.S. households in the year 2006.

Lesson 5-1 *(Pages 277–283)*

Change each measure to degrees, minutes, and seconds.

1. 13.75°
2. 75.72°
3. −29.44°
4. 87.81°

Write each measure as a decimal to the nearest thousandth.

5. 144° 12′ 30″
6. −38° 15′ 10″
7. −107° 12′ 45″
8. 51° 14′ 32″

If each angle is in standard position, determine a coterminal angle that is between 0° and 360°. State the quadrant in which the terminal side lies.

9. 850°
10. −65°
11. 1012°
12. 578°

Find the measure of the reference angle for each angle.

13. 126°
14. −480°
15. 642°
16. 1154°

Lesson 5-2 *(Pages 284–290)*

1. If $\tan\theta = \frac{5}{6}$, find $\cot\theta$.
2. If $\csc\theta = 2.5$, find $\sin\theta$.

Find the values of the six trigonometric functions for each $\angle A$.

3. 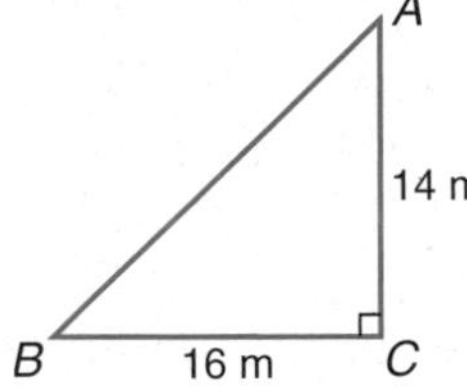

4. 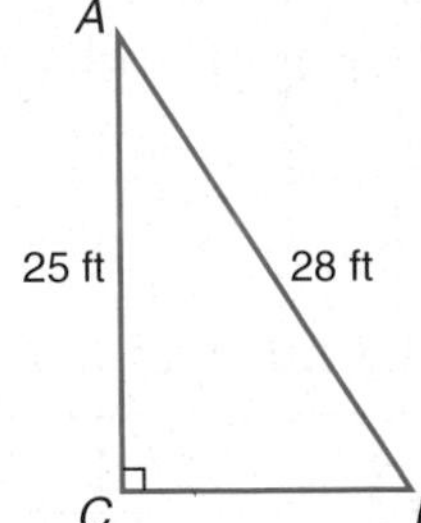

5. 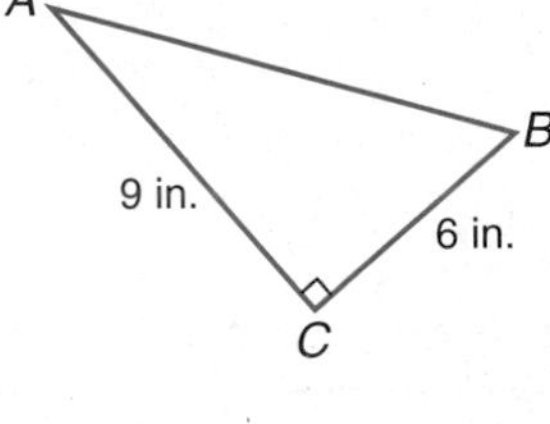

Lesson 5-3 *(Pages 291–298)*

1. If $\tan\theta = 0$, what is $\cot\theta$?
2. Find two values of θ for which $\cos\theta = 0$.

Find the values of the six trigonometric functions for angle θ in standard position if a point with the given coordinates lies on its terminal side.

3. (−1, −2)
4. (−2, 2)
5. (5, 2)
6. (−4, 3)

Lesson 5-4 *(Pages 299–304)*

Solve each problem. Round to the nearest tenth.

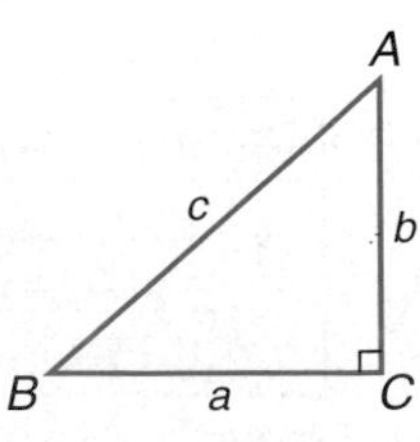

1. If $A = 38°$ and $b = 15$, find a.
2. If $c = 19$ and $B = 87°$, find a.
3. If $a = 16.5$ and $B = 65.4°$, find c.
4. If $B = 42° 30'$ and $b = 12$, find a.
5. If $B = 75°$ and $c = 5.8$, find b.
6. A statue 20 feet high stands on top of a base. From a point in front of the statue, the angle of elevation to the top of the statue is 48°, and the angle of elevation to the bottom of the statue is 42°. How tall is the base?

Lesson 5-5 *(Pages 305–312)*

Evaluate each expression. Assume that all angles are in Quadrant I.

1. $\sin\left(\arcsin \frac{3}{4}\right)$
2. $\sec\left(\cos^{-1} \frac{1}{2}\right)$
3. $\cot(\tan^{-1} 1)$

Solve each problem. Round to the nearest tenth.

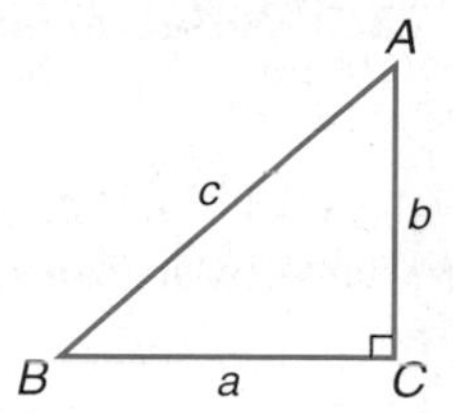

4. If $a = 38$ and $b = 25$, find A.
5. If $c = 19$ and $b = 17$, find B.
6. If $a = 24$ and $c = 30$, find B.
7. If $c = 12.6$ and $a = 9.2$, find B.
8. If $b = 36.5$ and $a = 28.4$, find A.

Lesson 5-6 *(Pages 313–318)*

Solve each triangle. Round to the nearest tenth.

1. $A = 75°, B = 50°, a = 7$
2. $A = 97°, C = 42°, c = 12$
3. $B = 49°, C = 32°, a = 10$
4. $A = 22°, C = 41°, b = 25$

Find the area of each triangle. Round to the nearest tenth.

5. $A = 34°, b = 12, c = 6$
6. $B = 56.8°, A = 87°, c = 6.8$
7. $a = 8, B = 60°, C = 75°$
8. $A = 43°, b = 16, c = 12$

Lesson 5-7 *(Pages 320–326)*

Find all solutions for each triangle. If no solution exists, write *none*. Round to the nearest tenth.

1. $a = 5, b = 10, A = 145°$
2. $A = 25°, a = 6, b = 10$
3. $B = 56°, b = 34, c = 50$
4. $A = 45°, B = 85°, c = 15$

Lesson 5-8 *(Pages 327–332)*

Solve each triangle. Round to the nearest tenth.

1. $b = 6, c = 8, A = 62°$
2. $a = 9, b = 7, c = 12$
3. $B = 48°, c = 18, a = 14$
4. $a = 14.2, b = 24.5, C = 85.3°$

Find the area of each triangle. Round to the nearest tenth.

5. $a = 4, b = 7, c = 10$
6. $a = 4, b = 6, c = 5$
7. $a = 12.4, b = 8.6, c = 14.2$
8. $a = 150, b = 124, c = 190$

9. **Geometry** The lengths of two sides of a parallelogram are 35 inches and 28 inches. One angle measures 110°.
   a. Find the length of the longest diagonal.
   b. Find the area of the parallelogram.

Lesson 6-1 *(Pages 343–351)*

Change each degree measure to radian measure in terms of π.

1. $120°$
2. $280°$
3. $-440°$
4. $-150°$

Change each radian measure to degree measure. Round to the nearest tenth.

5. $\frac{8\pi}{3}$
6. $\frac{5\pi}{12}$
7. -2
8. 10.5

Evaluate each expression.

9. $\sin \frac{5\pi}{6}$
10. $\sin \frac{4\pi}{3}$
11. $\cos \frac{9\pi}{4}$
12. $\cos \left(-\frac{3\pi}{2}\right)$

13. The diameter of a circle is 10 inches. If a central angle measures $80°$, find the length of the intercepted arc.

Lesson 6-2 *(Pages 352–358)*

Determine each angular displacement in radians. Round to the nearest tenth.

1. 5 revolutions
2. 3.8 revolutions
3. 14.2 revolutions

Determine each angular velocity. Round to the nearest tenth.

4. 2.1 revolutions in 5 seconds
5. 1.5 revolutions in 2 minutes
6. 15.8 revolutions in 18 seconds
7. 140 revolutions in 20 minutes

8. A children's Ferris wheel rotates one revolution every 30 seconds. What is its angular velocity in radians per second?

Lesson 6-3 *(Pages 359–366)*

Find each value by referring to the graph of the sine or cosine function.

1. $\cos 4\pi$
2. $\sin 8\pi$
3. $\sin \frac{3\pi}{2}$

Graph each function for the given interval.

4. $y = \sin x,\ -4\pi \le x \le -2\pi$
5. $y = \cos x,\ -\frac{9\pi}{2} \le x \le -\frac{5\pi}{2}$

Lesson 6-4 *(Pages 368–377)*

State the amplitude and period for each function. Then graph each function.

1. $y = 2 \cos \theta$
2. $y = -3 \sin 0.5\theta$
3. $y = \frac{1}{2} \cos \frac{\theta}{4}$

Write an equation of the sine function with each amplitude and period.

4. amplitude $= 0.5$, period $= 6\pi$
5. amplitude $= 2$, period $= \frac{\pi}{3}$

Write an equation of the cosine function with each amplitude and period.

6. amplitude $= \frac{3}{5}$, period $= 4\pi$
7. amplitude $= 0.25$, period $= 8$

Lesson 6-5 *(Pages 378–386)*

State the phase shift for each function. Then graph each function.

1. $y = \sin(2\theta - \pi)$
2. $y = 2\cos(\theta + 2\pi)$
3. $y = \sin\left(\frac{\theta}{2} + \frac{\pi}{2}\right)$

Write an equation of the sine function with each amplitude, period, phase shift, and vertical shift.

4. amplitude $= 2$, period $= 2\pi$, phase shift $= \pi$, vertical shift $= -1$
5. amplitude $= 0.5$, period $= \frac{\pi}{4}$, phase shift $= 0$, vertical shift $= 3$

Write an equation of the cosine function with each amplitude, period, phase shift, and vertical shift.

6. amplitude $= 20$, period $= \frac{\pi}{2}$, phase shift $= 2\pi$, vertical shift $= 4$
7. amplitude $= \frac{3}{4}$, period $= 10$, phase shift $= 0$, vertical shift $= \frac{1}{2}$

Lesson 6-6 *(Pages 387–394)*

1. **Meteorology** The equation $d = 2.7\sin(0.5m - 1.4) + 12.1$ models the amount of daylight in Cincinnati, Ohio, for any given day. In this equation $m = 1$ represents the middle of January, $m = 2$ represents the middle of February, etc.
   a. What is the least amount of daylight in Cincinnati?
   b. What is the greatest amount of daylight in Cincinnati?
   c. Find the number of hours of daylight in the middle of October.
2. **Waves** A buoy floats on the water bobbing up and down. The distance between its highest and lowest point is 6 centimeters. It moves from its highest point down to its lowest point and back to its highest point every 14 seconds. Write a cosine function that models the movement to the equilibrium point.

Lesson 6-7 *(Pages 395–403)*

Graph each function.

1. $y = \cot\left(\theta - \frac{\pi}{4}\right)$
2. $y = \sec\theta + 2$
3. $y = \csc(2\theta + 2\pi)$

Lesson 6-8 *(Pages 405–412)*

Find each value.

1. $\text{Cos}^{-1}\, 0$
2. Arcsin 0
3. $\cos(\text{Tan}^{-1}\, 1)$
4. $\text{Cos}^{-1}\left(\tan\frac{3\pi}{4}\right)$
5. $\sin\left(\text{Cos}^{-1}\frac{1}{2} + \text{Sin}^{-1}\, 0\right)$
6. $\cos\left(2\,\text{Sin}^{-1}\frac{\sqrt{3}}{2}\right)$

Lesson 7-1 *(Pages 421–430)*

Use the given information to determine the exact trigonometric value.

1. $\cos\theta = \frac{1}{4}$, $0° < \theta < 90°$; $\csc\theta$
2. $\cot\theta = -\frac{\sqrt{6}}{3}$, $\frac{\pi}{2} < \theta < \pi$; $\tan\theta$

Express each value as a trigonometric function of an angle in Quadrant I.

3. $\cos\frac{13\pi}{6}$
4. $\tan(-315°)$
5. $\csc(-930°)$

Simplify each expression.

6. $\frac{\tan\theta}{\sin\theta}$
7. $\cot\theta\tan\theta + \sin\theta\sec\theta$
8. $(1 + \sin x)(1 - \sin x)$
9. $\frac{\cot x\sin x}{\csc x\cos x}$

Lesson 7-2 *(Pages 431–436)*

Verify that each equation is an identity.

1. $\csc^2\theta = \cot^2\theta + \sin\theta\csc\theta$
2. $\frac{\sec\theta - \csc\theta}{\csc\theta\sec\theta} = \sin\theta - \cos\theta$
3. $\sin^2 x + \cos^2 x = \sec^2 x - \tan^2 x$
4. $\sec A - \cos A = \tan A\sin A$

Find a numerical value of one trigonometric function of *x*.

5. $\frac{\cot x}{\csc x} = 1$
6. $2\tan x\sin x + 2\cos x = \csc x$

Lesson 7-3 *(Pages 437–445)*

Use sum or difference identities to find the exact value of each trigonometric function.

1. $\cos 75°$
2. $\sin 105°$
3. $\tan\frac{\pi}{12}$
4. $\tan\frac{7\pi}{12}$
5. $\sec\frac{29\pi}{12}$
6. $\cot 375°$

Find each exact value if $0 < x < \frac{\pi}{2}$ and $0 < y < \frac{\pi}{2}$.

7. $\sin(x + y)$ if $\cos x = \frac{2}{5}$ and $\sin y = \frac{3}{4}$
8. $\cos(x - y)$ if $\cos x = \frac{5}{12}$ and $\cos y = \frac{11}{12}$
9. $\tan(x + y)$ if $\cot x = \frac{4}{3}$ and $\sec y = \frac{5}{4}$
10. $\sec(x - y)$ if $\tan x = \frac{7}{6}$ and $\csc y = \frac{8}{5}$

Lesson 7-4 *(Pages 448–455)*

Use a half-angle identity to find the exact value of each function.

1. $\sin 15°$
2. $\cos 75°$
3. $\tan \frac{\pi}{12}$
4. $\cos 22.5°$
5. $\sin \frac{5\pi}{12}$
6. $\tan 112.5°$

Use the given information to find $\sin 2\theta$, $\cos 2\theta$, and $\tan 2\theta$.

7. $\cos \theta = \frac{2}{7}$, $0° < \theta < 90°$
8. $\sin \theta = \frac{2}{3}$, $0 < \theta < \frac{\pi}{2}$
9. $\tan \theta = -3$, $90° < \theta < 180°$
10. $\csc \theta = -\frac{3}{2}$, $\frac{3\pi}{2} < \theta < 2\pi$

Lesson 7-5 *(Pages 456–461)*

Solve each equation for $0° \leq x < 360°$.

1. $4 \cos^2 x - 2 = 0$
2. $\sin^2 x \csc x - 1 = 0$
3. $\sqrt{3} \cot x = 2 \cos x$
4. $3 \cos^2 x = 6 \cos x - 3$

Lesson 7-6 *(Pages 463–469)*

Write the standard form of the equation of each line given p, the length of the normal segment, and ϕ, the angle the normal segment makes with the positive x-axis.

1. $p = 12$, $\phi = 30°$
2. $p = 2$, $\phi = \frac{\pi}{3}$
3. $p = \frac{1}{2}$, $\phi = 150°$

Write each equation in normal form. Then find the length of the normal and the angle it makes with the positive x-axis.

4. $4x + 10y + 10 = 0$
5. $x - y = 2$
6. $2x + 3y = 12$

Lesson 7-7 *(Pages 470–476)*

Find the distance between the point with the given coordinates and the line with the given equation.

1. $(2, 1)$, $3x - 2y = 2$
2. $(3, 0)$, $2x + 4y - 2 = 0$
3. $(-1, -4)$, $-3x + y = 1$
4. $(4, -2)$, $y = \frac{2}{3}x - 2$

Find the distance between the parallel lines with the given equations.

5. $x + 2y = -2$
   $x + 2y = 4$
6. $y = \frac{2}{3}x + 3$
   $3y - 2x = 7$

7. **Geometry** Find the height to the nearest tenth of a unit of a trapezoid with parallel bases that lie on lines with equations $2x + 5y = -1$ and $2x + 5y = 4$.

Lesson 8-1 *(Pages 485–492)*

Use a ruler and a protractor to determine the magnitude (in centimeters) and direction of each vector.

1. 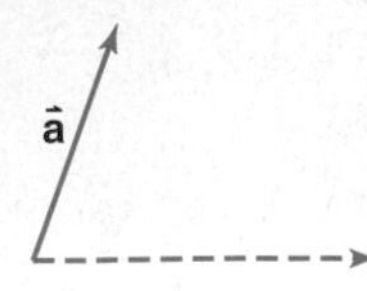

2. 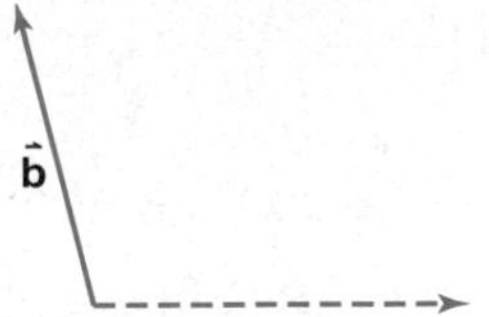

3. 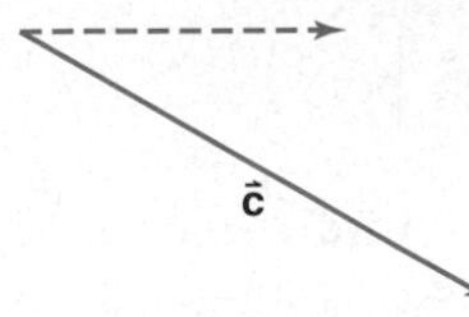

Use $\vec{a}$, $\vec{b}$, and $\vec{c}$ above to find the magnitude and direction of each resultant.

4. $\vec{a} + \vec{b}$
5. $\vec{b} + \vec{c}$
6. $\vec{a} + \vec{c}$
7. $\vec{a} - \vec{b}$
8. $2\vec{c}$
9. $2\vec{c} - \vec{b}$

Find the magnitude of the horizontal and vertical components of each vector shown for Exercises 1–3.

10. $\vec{a}$
11. $\vec{b}$
12. $\vec{c}$

Lesson 8-2 *(Pages 493–499)*

Find the ordered pair that represents $\overrightarrow{AB}$. Then find the magnitude of $\overrightarrow{AB}$.

1. $A(3, 6), B(4, 1)$
2. $A(-1, 3), B(-2, 2)$
3. $A(0, -4), B(-1, -8)$
4. $A(1, 10), B(3, -9)$
5. $A(-6, 0), B(-3, -6)$
6. $A(4, -5), B(0, 7)$

Find the magnitude of each vector and write each vector as the sum of unit vectors.

7. $\langle 5, 6 \rangle$
8. $\langle -2, 4 \rangle$
9. $\langle -10, -5 \rangle$
10. $\langle 2.5, 6 \rangle$
11. $\langle 2, -6 \rangle$
12. $\langle -15, -12 \rangle$

Lesson 8-3 *(Pages 500–504)*

Find an ordered triple to represent $\vec{p}$ in each equation if $\vec{q} = \langle 1, 2, -1 \rangle$, $\vec{r} = \langle -2, 2, 4 \rangle$, and $\vec{s} = \langle -4, -3, 0 \rangle$.

1. $\vec{p} = 2\vec{q} + 3\vec{s}$
2. $\vec{p} = \vec{q} - \frac{1}{2}\vec{r} + \vec{s}$
3. $\vec{p} = -2\vec{r} + \vec{s}$
4. $\vec{p} = \frac{3}{4}\vec{s} + 2\vec{q}$
5. **Physics** If vectors working on an object are in equilibrium, then their resultant is zero. Two forces on an object are represented by $\langle 2, -4, 1 \rangle$ and $\langle 5, 4, 3 \rangle$. Find a third vector that will place the object in equilibrium.

Lesson 8-4 *(Pages 505–511)*

Find each inner product and state whether the vectors are perpendicular. Write *yes* or *no*.

1. $\langle 3, 4\rangle \cdot \langle 2, 5\rangle$
2. $\langle -3, 2\rangle \cdot \langle 4, 6\rangle$
3. $\langle -5, 3\rangle \cdot \langle 2, -3\rangle$
4. $\langle 8, 6\rangle \cdot \langle -2, -3\rangle$
5. $\langle 3, 4, 0\rangle \cdot \langle 4, -3, 6\rangle$
6. $\langle 4, 5, 1\rangle \cdot \langle -1, -2, 3\rangle$

Find each cross product. Then verify that the resulting vector is perpendicular to the given vectors.

7. $\langle 1, 0, 3\rangle \times \langle 1, 1, 2\rangle$
8. $\langle 3, 0, 4\rangle \times \langle -1, 5, 2\rangle$
9. $\langle -1, 1, 0\rangle \times \langle 2, 1, 3\rangle$
10. $\langle -1, -3, 2\rangle \times \langle 6, -1, -2\rangle$

Lesson 8-5 *(Pages 513–519)*

Find the magnitude and direction of the resultant vector for each diagram.

1.
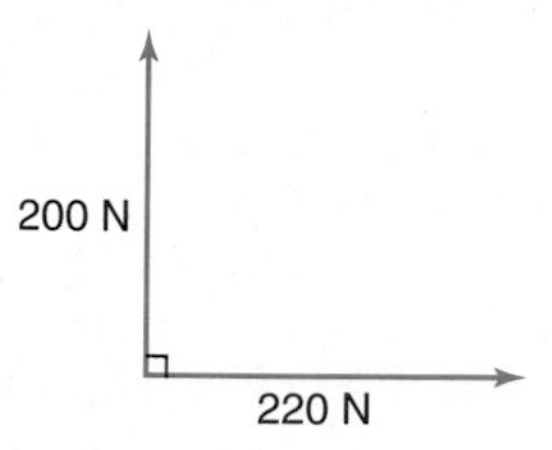

2.
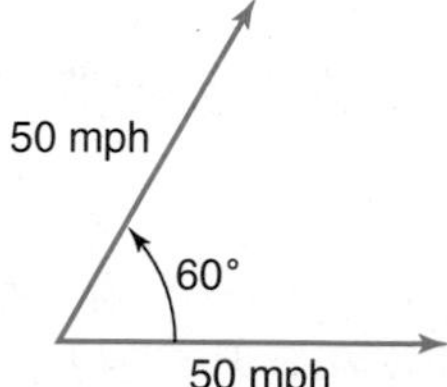

3.
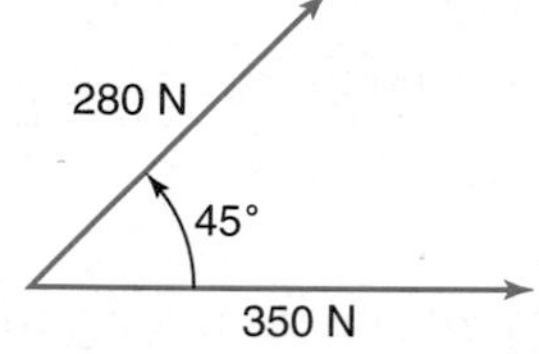

4. A 90 Newton force and a 110 Newton force act on the same object. The angle between the forces measures 90°. Find the magnitude of the resulting force.

Lesson 8-6 *(Pages 520–525)*

Write a vector equation of the line that passes through point *P* and is parallel to $\vec{\mathbf{a}}$. Then write parametric equations of the line.

1. $P(2, 3)$, $\vec{\mathbf{a}} = \langle 1, 0\rangle$
2. $P(-1, -4)$, $\vec{\mathbf{a}} = \langle 5, 2\rangle$
3. $P(-3, 6)$, $\vec{\mathbf{a}} = \langle -2, 4\rangle$
4. $P(3, 0)$, $\vec{\mathbf{a}} = \langle 0, -1\rangle$

Write an equation in slope-intercept form of the line with the given parametric equations.

5. $x = 3t$
   $y = 2 + t$
6. $x = -1 + 2t$
   $y = 4t$
7. $x = 3t - 10$
   $y = t - 1$

Lesson 8-7 *(Pages 527–533)*

1. **Sports** A golf ball is hit with an initial velocity of 70 yards per second at 34° with the horizontal. Find the initial vertical and horizontal velocity for the ball.
2. **Sports** An outfielder catches a fly ball and then throws it to third base to tag the runner. The outfielder releases the ball at an initial velocity of 75 feet per second at an angle of 25° with the horizontal. Assume the ball is released 5 feet above the ground.
   a. Write two parametric equations that represent the path of the ball.
   b. How far will the ball travel horizontally before hitting the ground?
   c. What is the maximum height of the trajectory?

Lesson 8-8 *(Pages 535–542)*

Describe the result of the product of a vertex matrix and each matrix.

1. $\begin{bmatrix} 4 & 0 & 0 \\ 0 & 4 & 0 \\ 0 & 0 & -4 \end{bmatrix}$
2. $\begin{bmatrix} 0.5 & 0 & 0 \\ 0 & 0.5 & 0 \\ 0 & 0 & 0.5 \end{bmatrix}$
3. $\begin{bmatrix} -1.5 & 0 & 0 \\ 0 & 1.5 & 0 \\ 0 & 0 & 1.5 \end{bmatrix}$

Lesson 9-1 *(Pages 553–560)*

Graph each point.

1. $K(4, 45°)$
2. $M\left(2, \frac{\pi}{6}\right)$
3. $N\left(\frac{3}{2}, -240°\right)$
4. $P\left(-1.5, \frac{5\pi}{6}\right)$

Graph each polar equation.

5. $r = 2$
6. $\theta = 60°$
7. $r = -2.5$
8. $\theta = \frac{7\pi}{6}$
9. Write a polar equation for the circle centered at the origin with radius $\sqrt{5}$.

Lesson 9-2 *(Pages 561–567)*

Graph each polar equation. Identify the type of curve each represents.

1. $r = -2 \sin \theta$
2. $r = 4\theta$
3. $r = 2 + 2 \cos \theta$
4. Write an equation for a rose with 5 petals.

Lesson 9-3 *(Pages 568–573)*

Find the polar coordinates of each point with the given rectangular coordinates. Use $0 \leq \theta < 2\pi$ and $r \geq 0$.

1. $(1, -1)$
2. $(3, 0)$
3. $(2, \sqrt{2})$

Find the rectangular coordinates of each point with the given polar coordinates.

4. $\left(2, \frac{\pi}{4}\right)$
5. $\left(\frac{1}{4}, \frac{\pi}{2}\right)$
6. $(5, 240°)$

Write each rectangular equation in polar form.

7. $x = -2$
8. $y = 6$
9. $x^2 + y^2 = 36$
10. $x^2 + y^2 = 3y$

Write each polar equation in rectangular form.

11. $r = 4$
12. $r = 4 \cos \theta$

Lesson 9-4 *(Pages 574–579)*

Write each equation in polar form. Round ϕ to the nearest degree.

1. $6x - 5y + 6 = 0$
2. $3x + 9y = 90$

Write each equation in rectangular form.

3. $8 = r \cos (\theta - 30°)$
4. $1 = r \cos (\theta + \pi)$
5. Graph the polar equation $3 = r \cos (\theta - 30°)$.

Lesson 9-5 *(Pages 580–585)*

Simplify.

1. i^{-10}
2. i^{17}
3. i^{1000}
4. $i^{12} + i^{-4}$
5. $(4 - i) + (-3 + 5i)$
6. $(6 + 6i) - (2 + 4i)$
7. $(3 + i)(5 - 3i)$
8. $(2 + 5i)^2$
9. $(1 - \sqrt{2}i)(-3 - \sqrt{8}i)$
10. $\dfrac{4 + i}{1 - i}$
11. $\dfrac{6 + 2i}{-2 + i}$
12. $\dfrac{(i - 2)^2}{4 + 2i}$

Lesson 9-6 *(Pages 586–591)*

1. Solve $4x - 6yi = 14 + 12i$ for x and y, where x and y are real numbers.

Graph each number in the complex plane and find its absolute value.

2. $4 + i$
3. $-5i$
4. $2 - \sqrt{3}i$

Express each complex number in polar form.

5. $4 + 4i$
6. $-2 + i$
7. $4 - \sqrt{2}i$

8. **Electricity** The impedance in one part of a series circuit is $5(\cos 0.9 + j \sin 0.9)$ ohms and in the second part of the circuit it is $8(\cos 0.4 + j \sin 0.4)$ ohms.
   a. Convert these complex numbers to rectangular form.
   b. Add your answers from part **a** to find the total impedance in the circuit.
   c. Convert the total impedance back to polar form.

Lesson 9-7 *(Pages 593–598)*

Find each product or quotient. Express the result in rectangular form.

1. $6\left(\cos \frac{\pi}{2} + i \sin \frac{\pi}{2}\right) \cdot 4\left(\cos \frac{\pi}{4} + i \sin \frac{\pi}{4}\right)$
2. $3\left(\cos \frac{3\pi}{4} + i \sin \frac{3\pi}{4}\right) \div \frac{1}{2}(\cos \pi + i \sin \pi)$
3. $5(\cos 135° + i \sin 135°) \cdot 2(\cos 45° + i \sin 45°)$

Lesson 9-8 *(Pages 599–606)*

Find each power. Express the result in rectangular form.

1. $\left[4\left(\cos \frac{\pi}{2} + i \sin \frac{\pi}{2}\right)\right]^4$
2. $(12i - 5)^3$

Find each principal root. Express the result in the form $a + bi$ with a and b rounded to the nearest hundredth.

3. $(1 + i)^{\frac{1}{3}}$
4. $(-1)^{\frac{1}{5}}$

EXTRA PRACTICE

EXTRA PRACTICE

Lesson 10-1 *(Pages 615–622)*

Find the distance between each pair of points with the given coordinates. Then, find the coordinates of the midpoint of the segment that has endpoints at the given coordinates.

1. $(-2, 2), (4, 5)$
2. $(-3, 6), (8, -1)$
3. $(r, 6), (r, -2)$
4. If $M(-5, 8)$ is the midpoint of $\overline{AB}$ and B has coordinates $(6, 2)$, find the coordinates of A.
5. Determine whether the quadrilateral having vertices at $(5, 10)$, $(5, 2)$, $(2, 8)$, and $(2, 5)$ is a parallelogram.
6. Andrea's garden is 50 feet long and 40 feet wide. Andrea makes a path from one corner of the garden to a water fountain at the center. Suppose the corner of the garden is the origin and the garden is in the first quadrant.
   a. Find the ordered pair that represents the location of the water fountain.
   b. Find the length of the path.

Lesson 10-2 *(Pages 623–630)*

Write the standard form of the equation of each circle described. Then graph the equation.

1. center at $(-2, 2)$, radius $\sqrt{2}$
2. center at $(0, -4)$, tangent to the x-axis

Write the standard form of each equation. Then graph the equation.

3. $x^2 = 49 - y^2$
4. $x^2 + y^2 + 6x - 8y + 18 = 0$

Write the standard form of the equation of the circle that passes through the points with the given coordinates. Then identify the center and the radius.

5. $(2, -2), (0, -4), (-2, -2)$
6. $(-1, 3), (-4, 6), (-7, 3)$
7. Write the equation of the circle that passes through the point $(2, -5)$ and has its center at $(4, 0)$.

Lesson 10-3 *(Pages 631–641)*

Write the equation of each ellipse in standard form. Then find the coordinates of its foci.

1. 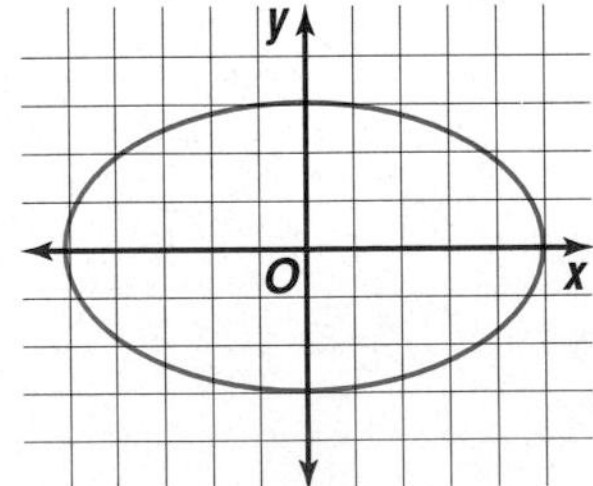

2. 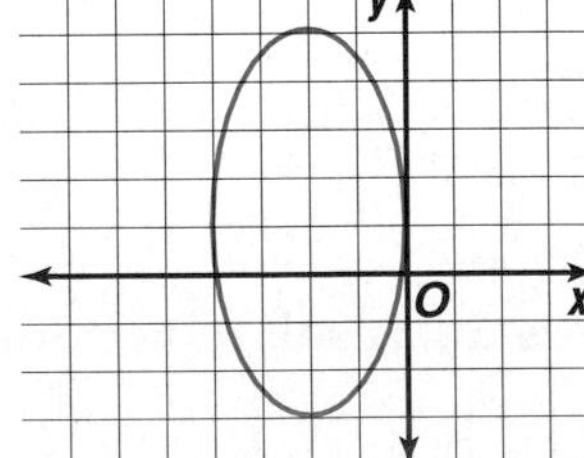

3. For the equation of the ellipse $\frac{(x-4)^2}{64} + \frac{(y-1)^2}{16} = 1$, find the coordinates of the center, foci, and vertices. Then graph the equation.

Lesson 10-4 *(Pages 642–652)*

1. Graph the equation $\frac{(y-2)^2}{49} - \frac{(x-1)^2}{9} = 1$. Label the center, foci, and the equations of the asymptotes.
2. Graph the equation $xy = 16$.

Write an equation of the hyperbola that meets each set of conditions.

3. The center is at $(-4, 3)$, $a = 3$, $b = 2$, and it has a horizontal transverse axis.
4. The foci are at $(2, -3)$, and $(2, 7)$ and the vertices are at $(2, -1)$ and $(2, 5)$.

Lesson 10-5 *(Pages 653–661)*

For the equation of each parabola, find the coordinates of the vertex and focus, and the equations of the directrix and axis of symmetry. Then graph the equation.

1. $y^2 = 4x$
2. $x^2 - 4x + 4 = 12y - 12$

Write the equation of the parabola that meets each set of conditions.

3. The vertex is at $(-2, 3)$ and the focus is at $(0, 3)$.
4. The focus is at $(0, -2)$ and the equation of the directrix is $y = -3$.

Lesson 10-6 *(Pages 662–669)*

Identify the conic section represented by each equation. Then write the equation in standard form.

1. $x^2 + y^2 - 8x + 2y + 13 = 0$
2. $x^2 - 4y^2 + 10x - 16y = -5$
3. $y^2 - 5x - 6y + 9 = 0$
4. $x^2 + 2y^2 + 2x + 8y = 15$

Lesson 10-7 *(Pages 670–677)*

Identify the graph of each equation. Write an equation of the translated or rotated graph in general form.

1. $x^2 + y^2 = 9$ for $T_{(1, -1)}$
2. $4x^2 + y^2 = 16$ for $T_{(-3, -2)}$
3. $49x^2 - 16y^2 = 784$; $\theta = \frac{\pi}{4}$
4. $4x^2 - 25y^2 = 64$; $\theta = 90°$

Identify the graph of each equation. Then find θ to the nearest degree.

5. $2y^2 + 3y - 2\sqrt{2}xy + x^2 - 1 = 0$
6. $15x^2 + 5xy + 5y^2 + 9 = 0$

Lesson 10-8 *(Pages 678–684)*

Solve each system of equations algebraically. Round to the nearest tenth. Check the solutions by graphing each system.

1. $xy = 3$
   $x^2 - y^2 = 8$
2. $x - y = 4$
   $x^2 = 10y^2 + 10$

EXTRA PRACTICE

Lesson 11-1 *(Pages 695–703)*

Evaluate each expression.

1. $(-12)^{-2}$
2. -12^{-2}
3. $(4 \cdot 6)^3$
4. $\left(\frac{2}{3}\right)^4$
5. $\frac{16}{16^{\frac{1}{2}}}$
6. $27^{\frac{1}{2}} \cdot 20^{\frac{1}{2}}$
7. $(\sqrt[4]{625})^2$
8. $\frac{1}{\sqrt[3]{(15)^6}}$

Simplify each expression.

9. $(2a^4)^2$
10. $(x^4)^3 \cdot x^5$
11. $((3f)^{-2})^3$
12. $\left(\frac{c^{-3a}}{c^{4a}}\right)^2$
13. $(2n^{\frac{1}{3}} \cdot 3n^{\frac{1}{3}})^6$
14. $\left(\frac{h^6}{216h^{-3}}\right)^{-\frac{1}{3}}$
15. $\sqrt[3]{z^4(z^4)^{\frac{1}{2}}}$
16. $(4r^2t^5)(16r^4t^8)^{\frac{1}{4}}$

Express using rational exponents.

17. $\sqrt{a^3b^5}$
18. $\sqrt[3]{64m^9n^6}$
19. $15\sqrt[3]{r^{12}t^2}$
20. $\sqrt[8]{256x^2y^{16}}$

Lesson 11-2 *(Pages 704–711)*

Graph each exponential function.

1. $y = 3^x$
2. $y = 3^{-x}$
3. $y = -3^{x+1}$

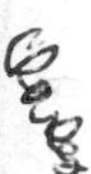

Lesson 11-3 *(Pages 712–717)*

1. **Psychology** The Ebbinghaus Model for human memory gives the percent p of acquired knowledge that a person retains after an amount of time. The formula is $p = (100 - a)e^{-bt} + a$, where t is the time in weeks, and a and b vary from one person to another. If $a = 18$ and $b = 0.6$ for a certain student, how much information will the student retain two weeks after learning a new topic?
2. **Physics** Newton's Law of Cooling expresses the relationship between the temperature in degrees Fahrenheit of a cooling object y and the time elapsed since cooling began t in minutes. This relationship is given by $y = ae^{-kt} + c$, where c is the temperature surrounding the medium. Suppose vegetable soup is heated to 210°F in the microwave. If the room temperature is 70°F, what will the temperature of the soup be after 10 minutes? Assume that $a = 140$ and $k = 0.01$.
3. **Forestry** The yield y in millions of cubic feet of trees per acre for a forest stand that is t years old is given by $y = 6.7\, e^{\frac{-48.1}{t}}$.
   a. Find the yield after 15 years.
   b. Find the yield after 50 years.
4. **Banking** Compare the balance after 20 years of a \$5000 investment earning 5.8% compounded continuously to the same investment compounded semiannually.

Lesson 11-4 *(Pages 718–725)*

Write each equation in exponential form.

1. $\log_{16} 2 = \frac{1}{4}$
2. $\log_{\frac{1}{2}} 8 = -3$
3. $\log_4 \frac{1}{4} = -1$

Write each equation in logarithmic form.

4. $8^{-2} = x$
5. $x^5 = 32$
6. $\left(\frac{1}{4}\right)^{-2} = 16$

Evaluate each expression.

7. $\log_5 \frac{1}{5}$
8. $\log_3 27$
9. $\log_{36} 6$

Solve each equation.

10. $\log_3 y = 4$
11. $\log_5 r = \log_5 8$
12. $\log_5 35 - \log_5 d = \log_5 5$
13. $\log_4 \sqrt{4} = x$
14. $\log_4 (2x + 3) = \log_4 15$
15. $4 \log_8 2 + \frac{1}{3}\log_8 27 = \log_8 a$

Lesson 11-5 *(Pages 726–732)*

Given that log 5 = 0.6990, log 8 = 0.9031, and log 14 = 1.1461, evaluate each logarithm.

1. $\log 5000$
2. $\log 0.0008$
3. $\log 0.14$

Find the value of each logarithm using the change of base formula.

4. $\log_3 81$
5. $\log_6 12$
6. $\log_5 29$

Solve each equation.

7. $3^x = 45$
8. $6^x = 2^{x-1}$
9. $5 \log y = \log 32$

Lesson 11-6 *(Pages 733–737)*

Evaluate each expression.

1. $\ln 35$
2. $\ln 0.562$
3. $\text{antiln } 1.2354$

Convert each logarithm to a natural logarithm and evaluate.

4. $\log_{15} 10$
5. $\log_3 14$
6. $\log_8 350$

Use natural logarithms to solve each equation or inequality.

7. $5^x = 90$
8. $7^{x+2} = 5.25$
9. $4^x = 4\sqrt{3}$
10. $6e^x = 48$
11. $50.2 < e^{0.2x}$
12. $16 = 10(1 + e^x)$

Lesson 11-7 *(Pages 740–748)*

Find the amount of time required for an amount to double at the given rate if the interest is compounded continuously.

1. 4.5%
2. 6%
3. 8.125%
4. **Biology** The data below give the number of bacteria found in a certain culture.

Time (hours)	0	1	2	3	4
Bacteria	6	7	12	20	32

a. Find an exponential model for the data.

b. Write the equation from part **a** in terms of base e.

c. Use the model to estimate the doubling time for the culture.

EXTRA PRACTICE

EXTRA PRACTICE

Lesson 12-1 *(Pages 759–765)*

Find the next four terms in each arithmetic sequence.

1. 7, 3, −1, ...
2. 0.5, −1, −2.5, ...
3. −14, −8, −2, ...
4. 3, 2.8, 2.6, ...
5. $4x, -x, -6x, \ldots$
6. $2y - 4, 2y - 2, 2y, \ldots$

For Exercises 7–13, assume that each sequence or series is arithmetic.

7. Find the 16th term in the sequence for which $a_1 = 2$ and $d = 5$.
8. Find n for the sequence for which $a_n = -20$, $a_1 = 6$ and $d = -2$.
9. What is the first term in the sequence for which $d = 4$ and $a_{12} = 42$?
10. Find d for the sequence in which $a_1 = 7$ and $a_{13} = 30$.
11. What is the 24th term in the sequence 10.5, 10, 9.5, ...?
12. Find the sum of the first 12 terms in the series $2 + 2.8 + 3.6 + \cdots$.
13. Find n for the series for which $a_1 = -4$, $d = 4$, and $S_n = 80$.

Lesson 12-2 *(Pages 766–773)*

Determine the common ratio and find the next three terms of each geometric sequence.

1. 14, 7, 3.5, ...
2. −2, 4, −8, ...
3. $\frac{2}{3}, \frac{1}{2}, \frac{3}{8}, \ldots$
4. 10, −5, 2.5, ...
5. $8, 8\sqrt{2}, 16, \ldots$
6. $a^{10}, a^8, a^6, \ldots$

For Exercises 7–11, assume that each sequence or series is geometric.

7. Find the sixth term of a sequence whose first term is 9 and common ratio is 2.
8. If $r = 4$ and $a_8 = 100$, what is the first term of the sequence?
9. Find the first three terms of the sequence for which $a_5 = 10$ and $r = -\frac{1}{2}$.
10. Write a sequence that has two geometric means between 4 and 256.
11. What is the sum of the first six terms of the series $3 + 9 + 27 + \cdots$?

12. **Biology** A certain bacteria divides every 15 minutes to produce two complete bacteria.
    a. If an initial colony contains a population of b_0 bacteria, write an equation that will determine the number of bacteria b_t present after t hours.
    b. Suppose a petri dish contains 12 bacteria. Use the equation found in part **a** to determine the number of bacteria present 4 hours later.

Lesson 12-3 *(Pages 774–783)*

Find each limit, or state that the limit does not exist and explain your reasoning.

1. $\lim_{n\to\infty} \frac{4+2n}{3n}$
2. $\lim_{n\to\infty} \frac{n^4-3n}{n^3}$
3. $\lim_{n\to\infty} \frac{8n^2+6n-2}{4n^2}$
4. $\lim_{n\to\infty} \frac{4n^2-2n+1}{n^2+2}$
5. $\lim_{n\to\infty} \frac{n^3-n^2+4}{5+2n^3}$
6. $\lim_{n\to\infty} \frac{2^n n}{2+n}$

Write each repeating decimal as a fraction.

7. $0.\overline{09}$
8. $0.1\overline{3}$
9. $7.\overline{407}$

Find the sum of the series, or state that the sum does not exist and explain your reasoning.

10. $\frac{1}{20}+\frac{1}{40}+\frac{1}{80}+\cdots$
11. $\frac{2}{7}+\frac{4}{7}+\frac{8}{7}+\cdots$

Lesson 12-4 *(Pages 786–793)*

Use the ratio test to determine whether each series is *convergent* or *divergent*.

1. $1^2+2^2+4^2+8^2+\cdots$
2. $\frac{1}{3}+\frac{2}{3}+1+\frac{4}{3}\cdots$
3. $1+\frac{1}{3\cdot 8}+\frac{1}{9\cdot 27}+\frac{1}{27\cdot 64}+\cdots$
4. $4+2+1+\frac{1}{2}+\cdots$

Use the comparison test to determine whether each series is *convergent* or *divergent*.

5. $\frac{7}{7}+\frac{7}{13}+\frac{7}{19}+\frac{7}{25}+\cdots$
6. $\frac{1}{2^2}+\frac{1}{4^2}+\frac{1}{6^2}+\frac{1}{8^2}+\cdots$

Determine whether each series is *convergent* or *divergent*.

7. $1+\frac{1}{2^0+1}+\frac{1}{2^1+1}+\frac{1}{2^2+1}+\frac{1}{2^3+1}+\cdots$
8. $\frac{2}{3}+\frac{4}{4}+\frac{8}{5}+\frac{16}{6}+\cdots$

Lesson 12-5 *(Pages 794–800)*

Write each expression in expanded form and then find the sum.

1. $\sum_{n=1}^{5}(3n-1)$
2. $\sum_{a=3}^{6}4a$
3. $\sum_{k=3}^{7}(k^2-2)$
4. $\sum_{j=4}^{8}\frac{j}{j+3}$
5. $\sum_{p=0}^{4}3^p$
6. $\sum_{n=1}^{\infty}2\cdot\left(\frac{3}{4}\right)^n$

Express each series using sigma notation.

7. $5+8+11+14$
8. $-8-12-16-\cdots-40$
9. $\frac{1}{4}+0+4+\cdots+65{,}536$
10. $1+2+6+24+\cdots$

EXTRA PRACTICE

Lesson 12-6 *(Pages 801–805)*

Use Pascal's triangle to expand each binomial.

1. $(2 + x)^4$
2. $(n + m)^5$
3. $(4a - b)^3$

Use the Binomial Theorem to expand each binomial.

4. $(m - 3)^6$
5. $(2r + s)^4$
6. $(5x - 4y)^3$

Find the designated term of each binomial expression.

7. 6th term of $(x + y)^8$
8. 5th term of $(b + \sqrt{3})^7$
9. 3rd term of $(4z - w)^{10}$
10. 8th term of $(2h - k)^{12}$

Lesson 12-7 *(Pages 806–814)*

Find each value to four decimal places.

1. ln (-3)
2. ln (-4.6)
3. ln (-0.75)

Use the first five terms of the exponential series and a calculator to approximate each value to the nearest hundredth.

4. $e^{1.2}$
5. $e^{-0.7}$
6. $e^{3.65}$

Use the first five terms of the trigonometric series to approximate the value of each function to four decimal places. Then, compare the approximation to the actual value.

7. $\cos \frac{\pi}{4}$
8. $\sin \frac{\pi}{6}$
9. $\cos \frac{\pi}{3}$

Lesson 12-8 *(Pages 815–821)*

Find the first four iterates of each function using the given initial value. If necessary, round your answers to the nearest hundredth.

1. $f(x) = 2x; x_0 = -2$
2. $f(x) = x^2; x_0 = 4$

Find the first three iterates of the function $f(z) = 0.5z + i$ for each initial value.

3. $z_0 = 2i$
4. $z_0 = 4 + 4i$

5. **Banking** Mendella has a savings account that has an annual yield of 5.4%. Find the balance of the account after each of the first five years if her initial balance is \$4000.

Lesson 12-9 *(Pages 822–828)*

Use mathematical induction to prove that each proposition is valid for all positive integral values of *n*.

1. $2 + 4 + 6 + \cdots + 2n = n(n + 1)$
2. $1 + 3 + 6 + \cdots + \frac{n(n + 1)}{2} = \frac{n(n + 1)(n + 2)}{6}$

3. Prove that $5^n - 1$ is even for all positive integers n.

Lesson 13-1 *(Pages 837–845)*

1. If you roll two 6-sided number cubes and then spin a 6-colored spinner with equal sections, how many outcomes are possible?
2. How many ways can 8 books be arranged on a shelf?

State whether the events are *independent* or *dependent*.

3. tossing three coins, then rolling a die
4. selecting members for a committee
5. deciding the order in which to answer your e-mail messages

Find each value.

6. $P(5, 5)$
7. $P(8, 3)$
8. $P(4, 1)$
9. $P(10, 9)$
10. $P(9, 6)$
11. $P(7, 3)$
12. $\frac{P(5, 2)}{P(2, 1)}$
13. $\frac{P(8, 6)}{P(7, 4)}$
14. $\frac{P(5, 2) \cdot P(8, 4)}{P(10, 1)}$
15. $C(4, 2)$
16. $C(10, 7)$
17. $C(6, 5)$
18. $C(4, 3) \cdot C(7, 3)$
19. $C(3, 1) \cdot C(8, 7)$
20. $C(9, 5) \cdot C(4, 3)$

Lesson 13-2 *(Pages 846–851)*

How many different ways can the letters of each word be arranged?

1. *mailbox*
2. *textbook*
3. *almanac*
4. *dictionary*
5. How many different 4-digit access codes can have the digits 5, 7, 2, and 7?

Determine whether each arrangement of objects is a *linear* or *circular* permutation. Then determine the number of arrangements for each situation.

6. 4 friends seated around a square table
7. 9 charms on a charm bracelet with no clasp
8. a stack of 5 books on a table

Lesson 13-3 *(Pages 852–859)*

Using a standard deck of 52 cards, find each probability.

1. P(ace)
2. P(a card of 5 or less)
3. P(a red face card)
4. P(not a queen)

One pencil is randomly taken from a box containing 8 red pencils, 4 green pencils, and 2 blue pencils. Find each probability.

5. P(blue)
6. P(green)
7. P(not red)
8. P(red or green)

A bag contains 2 white, 4 yellow, and 10 red markers. Two markers are drawn at random without replacement. What are the odds of each event occurring?

9. drawing 2 yellow markers
10. not drawing red markers
11. drawing 1 white and 1 yellow
12. drawing two different colors

Lesson 13-4 *(Pages 860–867)*

Determine if each event is *independent* or *dependent*. Then determine the probability.

1. the probability of selecting a red marble, not replacing it, then a green marble from a box of 6 red marbles and 2 green marbles
2. the probability of randomly selecting two dimes from a bag containing 10 dimes and 8 pennies, if the first selection is replaced
3. There are two traffic lights along the route that Laura drives from home to work. One traffic light is red 50% of the time. The next traffic light is red 60% of the time. The lights operate on separate timers. Find the probability that these lights will both be red on Laura's way from home to work.

Determine if each event is *mutually exclusive* or *mutually inclusive*. Then determine each probability.

4. the probability of tossing two number cubes and either one shows a 5
5. the probability of selecting a card from a standard deck of cards and the card is a 10 or an ace

Lesson 13-5 *(Pages 868–874)*

A jar contains 4 blue paper clips and 8 red paper clips. One paper clip is randomly drawn and discarded. Then a second paper clip is drawn. Find each probability.

1. the second paper clip is blue, given that the first paper clip was red
2. the second paper clip is blue, given that the first paper clip was blue
3. A pair of number cubes is thrown. Find the probability that the numbers of the dice match given that their sum is greater than 7.
4. A pair of number cubes is thrown. Find the probability that their sum is greater than 7 given that the numbers match.
5. One box contains 3 red balls and 4 white balls. A second box contains 5 red balls and 3 white balls. A box is selected at random and one ball is withdrawn. If the ball is white, what is the probability that it was taken from the second box?

Lesson 13-6 *(Pages 875–882)*

Find each probability if a coin is tossed three times.

1. *P*(all heads)
2. *P*(exactly 2 tails)
3. *P*(at least 2 heads)

Jojo MacMahon plays for the Worthington Wolves softball team. She is now batting 0.200 (meaning 200 hits in 1000 times at bat). Find the probability for the next five times at bat.

4. *P*(exactly 1 hit)
5. *P*(exactly 3 hits)
6. *P*(at least 4 hits)

Lesson 14-1 *(Pages 889–896)*

The daily grams of fat consumed by 30 adults who participated in a random survey are listed below.

45, 22, 36, 30, 59, 29, 28, 45, 55, 38, 36, 40, 35, 62, 69,
28, 45, 38, 39, 45, 40, 42, 62, 51, 42, 60, 29, 26, 60, 70

1. What is the range of the data?
2. Determine an appropriate class interval.
3. Name the class limits.
4. What are the class marks?
5. Construct a frequency distribution of the data.
6. Draw a histogram of the data.
7. Name the interval or intervals that describe the grams of fat consumed by most adults who participated in the survey.

Lesson 14-2 *(Pages 897–907)*

Find the mean, median, and mode of each set of data.

1. {130, 190, 180, 150}
2. {18, 19, 18, 16, 17, 15}
3. {25, 38, 36, 42, 30, 28}
4. {2, 5, 9, 10, 3, 4, 6, 9, 5, 1}
5. {2.5, 5.6, 6, 7, 2.3, 6.4, 6.5, 7, 8, 10, 4, 5.6}

Find the mean, median, and mode of the data represented by each stem-and-leaf plot.

6.

stem	leaf
1	4 5 5 6
2	0 1 4 7 8
3	6 9 9

1|4 = 14

7.

stem	leaf
3	0 4 6
5	2 4 4 6 6 7
6	2 3 6 8
7	0 1 6 7
8	2

3|0 = 3.0

8.

stem	leaf
8	0 2 3 9
9	6 7 8
10	4 5 5 8
11	1 7 8

8|0 = 800

9. Make a stem-and-leaf plot of the following number of hours worked by employees at a restaurant.

36, 17, 24, 39, 44, 37, 28, 29, 40, 55,
35, 34, 42, 29, 26, 24, 12, 19, 34, 23

10. Find the value of x so that the mean of {4, 5, 6, 9, 10, x} is 8.
11. Find the value of x so that the median of {4, 3, 19, 16, 4, 7, 12, x} is 7.5.

Lesson 14-3 *(Pages 908–917)*

Find the interquartile range and semi-interquartile range of each set of data. Then draw a box-and-whisker plot.

1. {45, 39, 44, 39, 51, 38, 59, 35, 58, 79, 40}
2. {2, 6, 4.5, 4, 3, 8, 3, 8, 10, 4, 2.5, 7.3, 4, 8, 1, 2.2}

Find the mean deviation and the standard deviation of each set of data.

3. {150, 220, 180, 200, 175, 180, 250, 212, 195}
4. {3.5, 4.2, 3.7, 5.5, 2.9, 1.4, 2.4, 2, 3, 5.3, 4.6}

5. **Sports** The numbers of hours per week members of the North High School basketball team spent practicing, either as a team or individually, are listed below.

   15, 18, 16, 20, 22, 18, 19, 20, 24, 18, 16, 18

   a. Find the median number of practice hours.
   b. Name the first quartile point and the third quartile point.
   c. Find the interquartile range.
   d. What is the semi-interquartile range?
   e. Are there any outliers? If so, name them.
   f. Make a box-and-whisker plot of the data.

Lesson 14-4 *(Pages 918–925)*

1. The mean of a set of normally distributed data is 10 and the standard deviation is 2.
   a. Find the interval about the mean that includes 25% of the data.
   b. What percent of the data is between 8 and 14?
   c. What percent of the data is between 7 and 10?
   d. Find the interval about the mean that includes 80% of the data.
2. Suppose 400 values in a set of data are normally distributed.
   a. How many values are within one standard deviation of the mean?
   b. How many values are within two standard deviations of the mean?
   c. How many values fall in the interval between the mean and one standard deviation above the mean?

Lesson 14-5 *(Pages 927–932)*

Find the standard error of the mean for each sample.

1. $\sigma = 1.2, N = 90$
2. $\sigma = 3.4, N = 100$
3. $\sigma = 12.4, N = 240$

For each sample, find the interval about the sample mean that has a 1% level of confidence.

4. $\sigma = 4.2, N = 40, \overline{X} = 150$
5. $\sigma = 10, N = 78, \overline{X} = 320$

Lesson 15-1 *(Pages 941–948)*

Evaluate each limit.

1. $\lim_{x \to 4} (x^2 + 2x - 2)$
2. $\lim_{x \to -1} (-x^4 + x^3 - 2x + 1)$
3. $\lim_{x \to 0} (x + \sin x)$
4. $\lim_{x \to -4} \frac{x^2 - 16}{x + 4}$
5. $\lim_{x \to -2} \frac{x^2 + 5x + 6}{x^2 + x - 2}$
6. $\lim_{x \to 2} \frac{3x + 9}{x^2 - 5x - 24}$

Lesson 15-2 *(Pages 951–960)*

Use the definition of derivative to find the derivative of each function.

1. $f(x) = 5x$
2. $f(x) = 9x - 2$

Use the derivative rules to find the derivative of each function.

3. $f(x) = \frac{1}{2}x + \frac{2}{3}$
4. $f(x) = x^2 + 4x + 8$

Find the antiderivative of each function.

5. $f(x) = x^5$
6. $f(x) = 2x^2 - 8x + 2$
7. $f(x) = \frac{1}{5}x^3 - \frac{3}{4}x - 1$
8. $f(x) = \frac{x^3 - 2x^2 + x}{x}$

Lesson 15-3 *(Pages 961–968)*

Use limits to evaluate each integral.

1. $\int_0^3 5x\,dx$
2. $\int_1^5 (x + 1)\,dx$
3. $\int_0^2 (x^2 + 4x + 4)\,dx$

4. **Business** A T-shirt company determines that the marginal cost function for one of their T-shirts is $f(x) = 6 - 0.002x$, where x is the number of T-shirts manufactured and $f(x)$ is in dollars. If the company is already producing 1500 of this type of T-shirt per day, how much more would it cost them to increase production to 2000 T-shirts per day?

Lesson 15-4 *(Pages 970–976)*

Evaluate each indefinite integral.

1. $\int x^6\,dx$
2. $\int 5x^4\,dx$
3. $\int (x^2 - x + 5)\,dx$
4. $\int (-4x^4 + x^2 - 6)\,dx$

Evaluate each definite integral.

5. $\int_{-2}^2 14x^6\,dx$
6. $\int_0^6 (x + 2)\,dx$
7. $\int_2^4 (x^2 - 4)\,dx$
8. $\int_4^5 (x - 4)(x + 2)\,dx$

CHAPTER 1 TEST

State the domain and range of each relation. Then state whether the relation is a function. Write *yes* or *no*.

1. $\{(-1, 2), (0, 5), (3, 4), (2, 4)\}$
2. $\{(6, 7), (7, 2), (4, -2), (6, -3), (-5, 0)\}$

Find each value if $f(x) = x - 3x^2$.

3. $f(4)$
4. $f(-7)$
5. $f(a + 2)$
6. **Physics** The equation of the illuminance, E, of a small light source is $E = \frac{P}{4\pi d^2}$, where P is the luminous flux (in lumens, lm) of the source and d is its distance from the surface. Suppose the luminous flux of a desk lamp is 1200 lm.
   a. What is the illuminance (in lm/m^2) on a desktop if the lamp is 0.9 m above it?
   b. Name the values of d, if any, that are not in the domain of the given function.
7. If $f(x) = x^2 - 7$ and $g(x) = x + 3$, find $f(x) + g(x)$, $f(x) - g(x)$, $f(x) \cdot g(x)$, and $\left(\frac{f}{g}\right)(x)$.

Find $[f \circ g](x)$ and $[g \circ f](x)$ for each $f(x)$ and $g(x)$.

8. $f(x) = x + 1$
   $g(x) = 4x - 5$
9. $f(x) = 5x$
   $g(x) = 2x^2 + 6$

Graph each equation or inequality.

10. $y = 3x - 6$
11. $2x + y = 1$
12. $y + 4x \leq 12$
13. $y > |x| + 2$

Write an equation in slope-intercept form for each line described.

14. slope = $\frac{5}{3}$, passes through $(-1, 3)$
15. passes through $(0, 4)$ and $(8, -2)$

Write the standard forms of the equations of the lines that are parallel and perpendicular to the graph of the given equation and pass through the point with the given coordinates.

16. $y = 4x - 1$; $(0, 2)$
17. $x - 5y = 3$; $(-1, 2)$
18. Graph $g(x) = [\![x + 1]\!]$
19. Graph $f(x) = \begin{cases} 2 \text{ if } x \leq -3 \\ x + 1 \text{ if } -3 < x \leq 1. \\ \frac{1}{2}x \text{ if } x > 1 \end{cases}$
20. **Grades** The table below shows the statistics grades and the economics grades for a group of college students at the end of a semester.

Statistics Grade	95	51	79	47	82	52	67	78	66
Economics Grade	88	70	70	67	71	80	68	85	90

   a. Graph the data on a scatter plot.
   b. Use two ordered pairs to write the equation of a best-fit line.

CHAPTER 2 TEST

1. Solve the system of equations by graphing. Then state whether the system is *consistent and independent, consistent and dependent,* or *inconsistent.*
$2x - y = -1$
$x + y = 4$

Solve each system of equations algebraically.

2. $3x + y = 7$
$5x + 2y = 12$

3. $4x - 5y = -2$
$3x + 2y = -13$

4. $4x + 6y - 3z = 20$
$x - 5y + z = -15$
$-7x + y + 2z = 1$

5. Find the values of x and y for which $\begin{bmatrix} y \\ 3x - 4 \end{bmatrix} = \begin{bmatrix} 12 - 2x \\ 2y \end{bmatrix}$ is true.

Use matrices *A*, *B*, and *C* to find each of the following. If the matrix does not exist, write *impossible*.

$A = \begin{bmatrix} -5 & 1 \\ 4 & 2 \end{bmatrix}$ $\quad B = \begin{bmatrix} -2 & 3 \\ -3 & 0 \end{bmatrix}$ $\quad C = \begin{bmatrix} -1 & 0 & 3 \\ 3 & 2 & -4 \end{bmatrix}$

6. $A + B$

7. $-3C$

8. $2B - A$

9. BC

Use matrices to perform each transformation. Then graph the pre-image and image on the same coordinate grid.

10. Triangle ABC has vertices $A(1, -3)$, $B(-4, 3)$, and $C(6, 2)$. Use scalar multiplication to find the coordinates of the triangle after a dilation of scale factor 1.5.

11. Quadrilateral $FGHJ$ has vertices $F(3, 2)$, $G(1, -5)$, $H(-6, 1)$, and $J(-3, 5)$. Find the coordinates of the quadrilateral after a reflection over the x-axis.

Find the value of each determinant.

12. $\begin{vmatrix} 5 & 7 \\ 3 & 6 \end{vmatrix}$

13. $\begin{vmatrix} 2 & 4 \\ -10 & 9 \end{vmatrix}$

14. $\begin{vmatrix} 1 & 2 & -1 \\ 3 & 0 & 1 \\ 4 & 1 & -2 \end{vmatrix}$

Find the inverse of each matrix, if it exists.

15. $\begin{bmatrix} 2 & 1 \\ 1 & 3 \end{bmatrix}$

16. $\begin{bmatrix} 3 & -4 \\ 4 & -2 \end{bmatrix}$

17. $\begin{bmatrix} -5 & 4 \\ -15 & 12 \end{bmatrix}$

18. Solve the system by using a matrix equation. $2x - y = 7$
$3x + y = 3$

19. Find the maximum and minimum values of $f(x, y) = 2x - y$ for the polygonal convex set determined by the system of inequalities.
$x \leq 2,\ y \geq 0,\ y \leq \frac{1}{2}x + 3,\ y \geq 3x - 4$

20. **Advertising** Ruff, a dog food manufacturer, wants to advertise both in a magazine and on radio. The magazine charges \$100 per ad and requires the purchase of at least three ads. The radio station charges \$200 per commercial minute and requires the purchase of at least four minutes. Each magazine ad reaches 12,000 people while each commercial minute reaches 16,000 people. Ruff can spend at most \$1300 on advertising. How many ads and commercial minutes should the manufacturing company purchase to reach the most people?

CHAPTER 3 TEST

Determine whether the graph of each equation is symmetric with respect to the *x*-axis, *y*-axis, the line $y = x$, the line $y = -x$, or none of these.

1. $y = 2x + 1$
2. $y = -\frac{2}{x^2}$
3. $x = y^2 + 3$
4. $xy = 5$

Describe how the graphs of $f(x)$ and $g(x)$ are related.

5. $f(x) = x^2$ and $g(x) = -(x + 3)^2$
6. $f(x) = |x|$ and $g(x) = 4|x| - 2$

Graph each inequality.

7. $y > |x - 4|$
8. $y \leq 2x^2 + 3$

Find $f^{-1}(x)$ and state whether $f^{-1}(x)$ is a function. Then graph the function and its inverse.

9. $f(x) = 5x - 4$
10. $f(x) = \frac{3}{x - 2}$

Determine whether each function is continuous at the given *x*-value. Justify your answer using the continuity test.

11. $\frac{x^2}{x - 2}; x = 2$
12. $f(x) = \begin{cases} x + 1 \text{ if } x > 0 \\ 1 \text{ if } x \leq 0 \end{cases}; x = 0$

13. Describe the end behavior of $y = x^4 - 2x^2 - 1$.

Determine whether the given critical point is the location of a *maximum*, a *minimum*, or a *point of inflection*.

14. $y = x^2 - 8x + 4, x = 4$
15. $y = x^3 - 3x^2 + x - 1, x = 1$

Determine the equations of the vertical and horizontal asymptotes, if any, of each function.

16. $f(x) = \frac{4x}{x - 1}$
17. $g(x) = \frac{x}{x^2 - 4}$

18. **Chemistry** A chemist pours 4 molar acid solution into 250 mL of *a* 1 molar acid solution. After *x* mL of the 4 molar solution have been added, the concentration of the mixture is given by $C(x) = \frac{0.25 + 0.004x}{0.25 + 0.001x}$.
    a. Find the horizontal asymptote of the graph of $C(x)$.
    b. What is the chemical interpretation of the horizontal asymptote?

Find the constant of variation for each relation and use it to write an equation for each statement. Then solve the equation.

19. If y varies directly as x and $y = 0.5$ when $x = 2$, find y when $x = 10$.
20. If y varies inversely as the square of x and $y = 8$ when $x = 3$, find x when $y = 18$.

CHAPTER 4 TEST

1. Write a polynomial equation of least degree with roots 4, i, and $-i$.

Find the discriminant of each equation and describe the nature of the roots of the equation. Then solve the equation by using the Quadratic Formula.

2. $n^2 - 5n + 4 = 0$
3. $z^2 - 7z - 3 = 0$
4. $2a^2 - 5a + 4 = 0$

Divide using synthetic division.

5. $(2x^3 - 3x^2 + 3x - 4) \div (x - 2)$
6. $(x^4 - 5x^3 - 13x^2 + 53x + 60) \div (x + 1)$

Use the Remainder Theorem to find the remainder for each division. State whether the binomial is a factor of the polynomial.

7. $(x^3 + 8x^2 + 2x - 11) \div (x + 2)$
8. $(4x^4 - 2x^2 + x - 3) \div (x - 1)$

Find the number of possible positive real zeros and the number of possible negative real zeros for each function. Then determine the rational zeros.

9. $f(x) = 6x^3 + 11x^2 - 3x - 2$
10. $f(x) = x^4 + x^3 - 9x^2 - 17x - 8$

Approximate the real zeros of each function to the nearest tenth.

11. $f(x) = x^2 - 3x - 3$
12. $f(x) = x^3 - x + 1$

Use the Upper Bound Theorem to find an integral upper bound and the Lower Bound Theorem to find an integral lower bound of the zeros of each function.

13. $f(x) = x^3 + 3x^2 - 5x - 9$
14. $f(x) = 2x^4 + 3x^3 - x^2 + x + 1$

Solve each equation or inequality.

15. $\frac{1}{80} + \frac{1}{a} = \frac{1}{10}$
16. $\frac{4}{x - 2} = \frac{3}{x^2 - 4} - \frac{1}{4}$
17. $\frac{5}{x + 2} > \frac{5}{x} + \frac{2}{3x}$
18. $\sqrt{y - 2} - 3 = 0$
19. $\sqrt{2x + 2} = \sqrt{3x - 5}$
20. $\sqrt{11 - 10m} > 9$

Decompose each expression into partial fractions.

21. $\frac{5z - 11}{2z^2 + z - 6}$
22. $\frac{7x^2 + 18x - 1}{(x^2 - 1)(x + 2)}$

23. What type of polynomial function would be the best model for the set of data?

x	−3	−2	−1	0	1	2	3
f(x)	26	16	6	4	3	8	14

24. **Manufacturing** The volume of a fudge tin must be 120 cubic centimeters. The tin is 7 centimeters longer than it is wide and six times longer than it is tall. Find the dimensions of the tin.

25. **Travel** A car travels 500 km in the same time that a freight train travels 350 km. The speed of the car is 30 km/h more than the speed of the train. Find the speed of the freight train.

CHAPTER 5 TEST

If each angle is in standard position, determine a coterminal angle that is between 0° and 360°. State the quadrant in which the terminal side lies.

1. 995°
2. −234°
3. 410°
4. −1245°

5. Find the values of the six trigonometric ratios for $\angle R$.

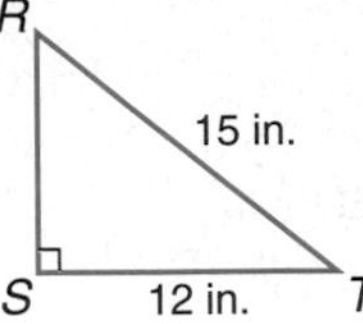

Use the unit circle to find each value.

6. tan 60°
7. sec 270°
8. sin (−405°)

Find the values of the six trigonometric functions for angle θ in standard position if a point with the given coordinates lies on its terminal side.

9. (3, 5)
10. (−4, 2)
11. (0, −3)

Solve each problem. Round to the nearest tenth.

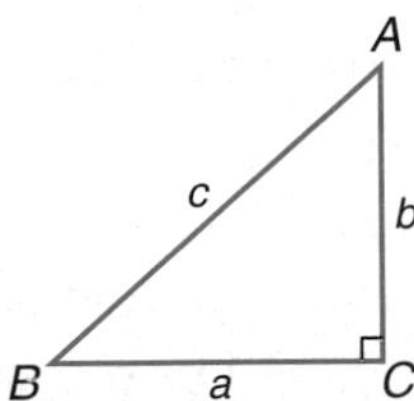

12. If $b = 42$ and $A = 77°$, find c.
13. If $a = 13$ and $B = 27°$, find b.
14. If $c = 14$ and $A = 32° \ 17'$, find a.
15. If $a = 23$ and $c = 37$, find B.
16. If $a = 3$ and $b = 11$, find A.

17. **Recreation** A kite is fastened to the ground by a string that is 65 meters long. If the angle of elevation of the kite is 70°, how far is the kite above the ground?

Find the area of each triangle. Round to the nearest tenth.

18. $A = 36°$, $a = 24$, $C = 87°$
19. $b = 56.4$, $c = 92.5$, $A = 58.4°$

20. **Geometry** An isosceles triangle has a base of 22 centimeters and a vertex angle measuring 36°. Find its perimeter.

Find all solutions for each triangle. If no solutions exist, write *none*. Round to the nearest tenth.

21. $a = 64$, $c = 90$, $C = 98°$
22. $a = 9$, $b = 20$, $A = 31°$

Solve each triangle. Round to the nearest tenth.

23. $a = 13$, $b = 7$, $c = 15$
24. $a = 20$, $c = 24$, $B = 47°$

25. **Navigation** A ship at sea is 70 miles from one radio transmitter and 130 miles from another. The measurement of the angle between signals is 130°. How far apart are the transmitters?

CHAPTER 6 TEST

Change each degree measure to radian measure in terms of π.

1. $225°$

2. $480°$

Evaluate each expression.

3. $\sin \frac{5\pi}{6}$

4. $\tan \frac{5\pi}{4}$

5. Determine the linear velocity of a point rotating at an angular velocity of 7.1 radians per second at a distance of 12 centimeters from the center of the rotating object.

Graph each function for the given interval.

6. $y = \sin x, -\pi \leq x \leq \pi$

7. $y = \cos x, -\pi \leq x \leq \pi$

State the amplitude and period for each function.

8. $y = 3 \cos 4\theta$

9. $y = -2 \sin \frac{3\pi}{2}\theta$

10. Write an equation of the sine function with amplitude 4, period 4π, phase shift 0, and vertical shift -2.

11. Write an equation of the cosine function with amplitude 0.5, period $\frac{\pi}{2}$, phase shift $\frac{\pi}{4}$, and vertical shift 1.

Graph each function.

12. $y = 3 \cos \frac{\theta}{2}$

13. $y = \tan \left(2\theta - \frac{\pi}{4}\right)$

Write the equation for the inverse of each function. Then graph the function and its inverse.

14. $y = \text{Arccsc } x$

15. $y = \tan x$

Find each value.

16. $\sin \left(\text{Arccos } \frac{1}{2}\right)$

17. $\tan \left(\pi + \text{Sin}^{-1} \frac{1}{2}\right)$

18. **Astronomy** The linear velocity of Earth's moon is about 2300 mph. If the average distance from the center of Earth to the center of the Moon is 240,000 miles, how long does it take the Moon to make one revolution about Earth? Assume the orbit is circular.

19. **Meteorology** The average monthly temperatures for the city of Chicago, Illinois, are given below.

Jan.	Feb.	March	April	May	June	July	Aug.	Sept.	Oct.	Nov.	Dec.
21°	25°	37°	49°	59°	69°	73°	72°	64°	53°	40°	27°

Write a sinusoidal function that models the monthly temperatures, using $t = 1$ to represent January.

20. **Civil Engineering** Jake is designing a curve for a new highway. He uses the equation $\tan \theta = \frac{v^2}{rg}$, where the radius r will be 1200 feet. The curve will be designed for a maximum velocity v of 65 mph, and the acceleration g due to gravity is 32 ft/s^2. At what angle θ should the curve be banked? (Express your answer in radians.)

CHAPTER 7 TEST

Use the given information to determine the exact trigonometric value.

1. $\sin \theta = \frac{1}{3}$, $0° < \theta < 90°$; $\cos \theta$
2. $\sec \theta = -2$, $\frac{\pi}{2} < \theta < \pi$; $\tan \theta$
3. $\sin \theta = -\frac{4}{5}$, $\pi < \theta < \frac{3\pi}{2}$; $\sec \theta$
4. $\csc \theta = -\frac{5}{3}$, $270° < \theta < 360°$; $\cos \theta$
5. Express $\tan(-420°)$ as a trigonometric function of an angle in Quadrant I.

Verify that each equation is an identity.

6. $\tan \theta (\cot \theta + \tan \theta) = \sec^2 \theta$
7. $\sin^2 A \cot^2 A = (1 - \sin A)(1 + \sin A)$
8. $\frac{\sec x}{\sin x} - \frac{\sin x}{\cos x} = \cot x$
9. $\frac{\cos x}{1 + \sin x} + \frac{\cos x}{1 - \sin x} = 2 \sec x$
10. $\csc (A - B) = \frac{\sec B}{\sin A - \cos A \tan B}$
11. $\cot 2\theta = \frac{1}{2} \cot \theta - \frac{1}{2} \tan \theta$

Use sum or difference identities to find the exact value of each trigonometric function.

12. $\sin 255°$
13. $\tan \frac{5\pi}{12}$
14. If $\cos \theta = \frac{3}{4}$ and $270° < \theta < 360°$, find $\sin 2\theta$, $\cos 2\theta$, and $\tan 2\theta$.
15. Use a half-angle identity to find the exact value of $\cos 22.5°$.

Solve each equation for principal values of *x*. Express solutions in degrees.

16. $\tan^2 x = \sqrt{3} \tan x$
17. $\cos 2x - \cos x = 0$

Solve each equation for $0° \leq x < 360°$.

18. $\sin x - \cos x = 0$
19. $2 \cos^2 x + 3 \sin x = 3$

Write each equation in normal form. Then find the length of the normal and the angle it makes with the positive *x*-axis.

20. $y = x + 3$
21. $5 + 5y = 10x$

Find the distance between the point with the given coordinates and the line with the given equation.

22. $(-5, 8)$, $2x + y = 6$
23. $(-6, 8)$, $3x + 4y + 2 = 0$
24. Find an equation of the line that bisects the acute angles formed by the graphs of $5x + 2y = 7$ and $y = -\frac{3}{4}x + 1$.
25. **Physics** The range of a projected object is the distance that it travels from the point where it is released. In the absence of air resistance, a projectile released at an angle of inclination θ with an initial velocity of v_0 has a range of $R = \frac{{v_0}^2}{g} \sin 2\theta$, where g is the acceleration due to gravity. Find the range of a projectile with an initial velocity of 88 feet per second if $\sin \theta = \frac{3}{5}$ and $\cos \theta = \frac{4}{5}$. The acceleration due to gravity is 32 feet per second squared.

CHAPTER 8 TEST

Use a ruler and a protractor to determine the magnitude (in centimeters) and direction of each vector.

1. 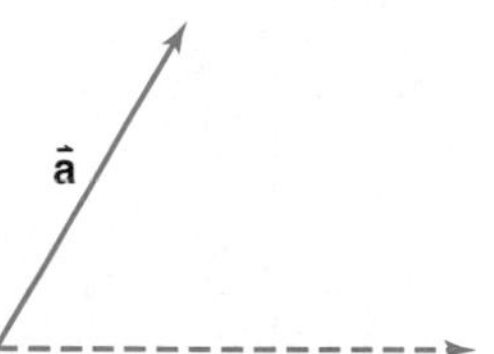

2. 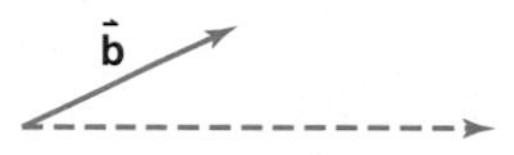

Use $\vec{a}$ and $\vec{b}$ above to find the magnitude and direction of each resultant.

3. $\vec{a} + \vec{b}$
4. $2\vec{b} - \vec{a}$

Write the order pair or ordered triple that represents $\overrightarrow{AB}$. Then find the magnitude of $\overrightarrow{AB}$.

5. $A(3, 6), B(-1, 9)$
6. $A(-2, 7), B(3, 10)$
7. $A(2, -4, 5), B(9, -3, 7)$
8. $A(-4, -8, -2), B(-8, -10, 2)$

Let $\vec{r} = \langle -1, 3, 4\rangle$ and $\vec{s} = \langle 4, 3, -6\rangle$.

9. Find $\vec{r} - \vec{s}$.
10. Find $3\vec{s} - 2\vec{r}$.
11. Find $\vec{r} + 3\vec{s}$.
12. Find $|\vec{r}|$.
13. Find $|\vec{s}|$.
14. Write $\vec{r}$ as the sum of unit vectors.
15. Write $\vec{s}$ as the sum of unit vectors.
16. Find $\vec{r} \cdot \vec{s}$.
17. Find $\vec{r} \times \vec{s}$.
18. Is $\vec{r}$ perpendicular to $\vec{s}$?
19. Find the magnitude and direction of the resultant of two forces of 125 pounds and 60 pounds at angles of 40° and 165° with the x-axis, respectively.
20. Write parametric equations for the line that passes through the point $P(3, 11)$ and is parallel to $\vec{a} = \langle 2, -5\rangle$.
21. Write an equation in slope-intercept form of the line with the parametric equations.
$$x = 2t + 3$$
$$y = t + 1$$
22. Find the initial horizontal velocity and vertical velocity of a hockey puck shot at 100 mph at an angle of 2° with the ice.
23. Describe the result of the product of a vertex matrix and the matrix $\begin{bmatrix} -4 & 0 & 0 \\ 0 & 4 & 0 \\ 0 & 0 & 4 \end{bmatrix}$.
24. **Physics** A downward force of 110 pounds is applied to the end of a 1.5-foot lever. What is the resulting torque about the axis of rotation if the angle of the lever to the horizontal is 60°?
25. **Gardening** Tei uses a sprinkler to water his garden. The sprinkler discharges water with a velocity of 28 feet per second. If the angle of the water with the ground is 35°, how far will the water travel in the horizontal direction? The acceleration due to gravity is 32 feet per second squared.

CHAPTER 9 TEST

Graph each point.

1. $A(2.5, 140°)$
2. $B\left(-2, \frac{5\pi}{4}\right)$
3. $C\left(3, -\frac{\pi}{6}\right)$

Graph each polar equation.

4. $\theta = \frac{3\pi}{2}$
5. $r = -4$
6. $r = 6 \cos 3\theta$
7. $r = 2 + 2 \sin \theta$

Find the polar coordinates of each point with the given rectangular coordinates. Use $0 \le \theta < 2\pi$ and $r \ge 0$.

8. $(2, 2)$
9. $(-6, 0)$
10. $(-2, -3)$

Find the rectangular coordinates of each point with the given polar coordinates.

11. $\left(3, -\frac{5\pi}{4}\right)$
12. $\left(2, \frac{7\pi}{6}\right)$
13. $(-4, 1.4)$

Write each rectangular equation in polar form.

14. $y = -3$
15. $x^2 + y^2 = 3x$

Write each polar equation in rectangular form.

16. $r = 7$
17. $5 = r \cos (\theta - 45°)$

Write each equation in polar form. Round ϕ to the nearest degree.

18. $5x + 3y = -3$
19. $2x - 4y = 1$

Simplify.

20. $\boldsymbol{i}^{93}$
21. $(2 - 5\boldsymbol{i}) + (-2 + 4\boldsymbol{i})$
22. $-6\boldsymbol{i} - (-3 + 2\boldsymbol{i})$
23. $(3 + 5\boldsymbol{i})(3 - 2\boldsymbol{i})$
24. $(1 - 3\boldsymbol{i})(2 - \boldsymbol{i})(1 + 2\boldsymbol{i})$
25. $\frac{6 - 2\boldsymbol{i}}{2 + \boldsymbol{i}}$

Express each complex number in polar form.

26. $-4 + 4\boldsymbol{i}$
27. -5

Find each product or quotient. Express the result in rectangular form.

28. $4\left(\cos \frac{3\pi}{2} + \boldsymbol{i} \sin \frac{3\pi}{2}\right) \cdot 3\left(\cos \frac{\pi}{4} + \boldsymbol{i} \sin \frac{\pi}{4}\right)$
29. $2\sqrt{3}\left(\cos \frac{2\pi}{3} + \boldsymbol{i} \sin \frac{2\pi}{3}\right) \div \sqrt{3}\left(\cos \frac{\pi}{6} + \boldsymbol{i} \sin \frac{\pi}{6}\right)$
30. Find $(1 - \boldsymbol{i})^8$.
31. Find $\sqrt[3]{-27\boldsymbol{i}}$.
32. Solve the equation $x^3 - \boldsymbol{i} = 0$. Then graph the roots in the complex plane.
33. **Electricity** The current in a circuit is $8(\cos 307° + \boldsymbol{j} \sin 307°)$ amps and the impedance is $20(\cos 115° + \boldsymbol{j} \sin 115°)$ ohms. Find the polar form of the voltage in the circuit. (*Hint:* $E = I \cdot Z$)

CHAPTER 10 TEST

Find the distance between each pair of points with the given coordinates. Then, find the midpoint of the segment that has endpoints at the given coordinates.

1. $(-1, 2), (3, 1)$
2. $(3k, k + 1), (2k, k - 1)$
3. Write the standard form of the equation of the circle that passes through $(-6, -4)$ and has its center at $(-8, 3)$.
4. Find the coordinates of the center, foci, and vertices of the ellipse with equation $\frac{x^2}{6} + \frac{y^2}{10} = 1$.
5. Write an equation of the ellipse centered at the origin that has a horizontal major axis, $e = \frac{1}{2}$, and $c = \frac{1}{2}$.
6. Find the coordinates of the center, foci, and the vertices and the equations of the asymptotes of the graph of the hyperbola with equation $\frac{(y-2)^2}{16} - \frac{(x+4)^2}{7} = 1$.
7. Write an equation of the hyperbola that has eccentricity $\frac{3}{2}$ and foci at $(-5, -2)$ and $(-5, 4)$.
8. Find the coordinates of the vertex and focus, and the equations of the directrix and axis of symmetry of the parabola with equation $(y + 3)^2 = 8x$.
9. Write an equation of the parabola that has a focus at $(3, -5)$ and a directrix with equation $y = -2$.

Identify the conic section represented by each equation. Then write the equation in standard form and graph the equation.

10. $x^2 + 6x - 8y - 7 = 0$
11. $4x^2 - y^2 = 1$
12. $x^2 + y^2 + 4x - 12y + 36 = 0$
13. $3x^2 - 16y^2 - 18x + 128y - 37 = 0$
14. $9x^2 - y^2 - 90x + 8y + 200 = 0$
15. $2x^2 - 13y^2 + 5 = 0$
16. $y^2 - 2x + 10y + 27 = 0$
17. $x^2 + 2y^2 + 2x - 12y + 11 = 0$

Find the rectangular equation of the curve whose parametric equations are given.

18. $x = t, y = 2t^2 + t, -\infty < t < \infty$
19. $x = 2 \cos t, y = 2 \sin t, 0 \le t \le 2\pi$

Identify the graph of each equation. Write an equation of the translated or rotated graph in general form.

20. $4(x + 1)^2 + (y - 3)^2 = 36$ for $T_{(3, -5)}$
21. $2x^2 - y^2 = 8, \theta = 60°$
22. Solve the system of equations algebraically. Round to the nearest tenth.
    $x^2 + 4y^2 = 4$
    $(x - 1)^2 + y^2 = 1$
23. Graph the system of inequalities.
    $x^2 + y^2 - 2x - 4y + 1 \ge 0$
    $x^2 - 4y - 2x + 5 \ge 0$
24. **Geometry** Write the standard form of the equation of the circle that passes through $(0, 1)$, $(-2, 3)$, and $(4, 5)$. Then identify the center and radius.
25. **Technology** Mirna bought a motion detector light and installed it in the center of her backyard. It can detect motion within a circle defined by the equation $x^2 + y^2 = 90$. If a person walks northeast through the backyard along a line defined by the equation $y = 2x - 3$, at what point will the person set off the motion detector?

CHAPTER 11 TEST

Evaluate each expression.

1. $343^{\frac{2}{3}}$
2. $64^{-\frac{1}{3}}$

Simplify each expression.

3. $((2a)^3)^{-2}$
4. $(x^{\frac{3}{2}}y^2z^{\frac{5}{4}})^4$
5. Express $\sqrt[3]{27a^6b^{12}}$ using rational exponents.
6. Express $m^{\frac{1}{2}}n^{\frac{2}{3}}$ using a radical.

Graph each exponential function.

7. $y = \left(\frac{1}{3}\right)^{x-2}$
8. $y = 5^{x+1}$

Write each equation in exponential form.

9. $\log_4 2 = \frac{1}{2}$
10. $\log_{\frac{1}{6}} 216 = -3$

Write each equation in logarithmic form.

11. $5^4 = 625$
12. $8^5 = m$

Solve each equation.

13. $\log_x 32 = -5$
14. $\log_5 (2x) = \log_5 (3x - 4)$
15. $3.6^x = 72.4$
16. $6^{x-1} = 8^{2-x}$

Find the value of each logarithm using the change of base formula.

17. $\log_4 15$
18. $\log_3 0.9375$

Evaluate each expression.

19. $\log_{81} 3$
20. $\log 542$
21. $\ln 0.248$
22. antiln (-1.9101)

23. Find the amount of time required for an amount to double at a rate of 5.4% if the interest is compounded continuously.
24. **Biology** A certain bacteria will triple in 6 hours. If the final count is 8 times the original count, how much time has passed?
25. **Electricity** The charge on a discharging capacitor is given by $q(t) = Qe^{-\frac{t}{RC}}$, where Q is the initial charge on the capacitor with capacitance C, R is the resistance in the circuit, and t is time, in seconds. If $Q = 5 \times 10^{-6}$ coulomb, $R = 3$ ohms, and $C = 2 \times 10^{-6}$ farad, find the time required for the charge on the capacitor to decrease to 1×10^{-6} coulomb.

CHAPTER 12 TEST

For Exercises 1–4, assume that each sequence or series is arithmetic.

1. Find the next four terms of the sequence 2, 4.5, 7, … .
2. Find the 24th term of the sequence $-6, -1, 4, \ldots$.
3. Write a sequence that has three arithmetic means between -4 and 8.
4. Find n for the series in which $a_1 = 12$, $d = 5$, and $S_n = 345$.

For Exercises 5–7, assume that each sequence or series is geometric.

5. Determine the common ratio and find the next three terms of the sequence $\frac{1}{4}, \frac{1}{10}, \frac{1}{25}, \ldots$.
6. Write a sequence that has three geometric means between 16 and 1.
7. Find the sum of the first 10 terms of the series $\frac{5}{2} + 5 + 10 + \cdots$.

Find each limit, or state that the limit does not exist and explain your reasoning.

8. $\lim\limits_{n\to\infty} \frac{n^3 + 3}{3n^2 + 1}$

9. $\lim\limits_{n\to\infty} \frac{n^3 + 4}{2n^3 + 3n}$

Determine whether each series is *convergent* or *divergent*.

10. $\frac{1}{3 \cdot 1^2} + \frac{1}{3 \cdot 2^2} + \frac{1}{3 \cdot 3^2} + \cdots$

11. $\frac{1}{6} + \frac{1}{3} + \frac{1}{2} + \frac{2}{3} + \cdots$

Express each series using sigma notation.

12. $5 + 10 + 15 + \cdots + 95$

13. $6 + 9 + \frac{27}{2} + \frac{81}{4} + \cdots$

14. Use the Binomial Theorem to expand $(2a - 3b)^5$.

Find the designated term of each binomial expansion.

15. sixth term of $(a + 2)^{10}$

16. fifth term of $(3x - y)^8$

17. Write $-2 + 2\boldsymbol{i}$ in exponential form.

18. Find the first three iterates of the function $f(z) = 3z + (2 - \boldsymbol{i})$ for the initial value $z_0 = 2\boldsymbol{i}$.

19. Use mathematical induction to prove that the following proposition is valid for all positive integral values of n.

$$2 \cdot 3 + 4 \cdot 5 + 6 \cdot 7 + \cdots + 2n(2n + 1) = \frac{n(n + 1)(4n + 5)}{3}$$

20. **Finance** Melissa deposits $200 into an investment every 3 months. The investment pays an APR of 8% and interest is compounded quarterly. Melissa makes each payment at the beginning of the quarter and the interest is posted at the end of the quarter. What will be the total value of the investment at the end of 10 years?

CHAPTER 13 TEST

Find each value.

1. $P(6, 2)$
2. $P(7, 5)$
3. $C(8, 3)$
4. $C(5, 4)$

5. The letters r, s, t, u, and v are to be used to form five-letter patterns. How many patterns can be formed if repetitions are not allowed?
6. **Employment** Five people have applied for three different positions in a store. If each person is qualified for each position, in how many ways can the positions be filled?
7. How many ways can 7 people be seated at a round table relative to each other?
8. **Sports** How many baseball teams can be formed from 15 players if 3 only pitch and the others play any of the remaining 8 positions?

A bag contains 4 red and 6 white marbles.

9. How many ways can 5 marbles be selected if exactly 2 must be red?
10. If two marbles are chosen at random, find $P(2 \text{ white})$.

11. **Sports** The probability that the Pirates will win a game against the Hornets is $\frac{1}{4}$. What are the odds that the Pirates will beat the Hornets?
12. Five cards are dealt from a standard deck of cards. What is the probability that they are all from the same suit?
13. Find the probability of getting a sum of 8 on the first throw of two number cubes and a sum of 4 on the second throw.
14. **Communication** Callie is having a new phone line installed. What is the probability that the final 3 digits in the telephone number will all be odd?
15. A bag contains 3 red, 4 white, and 5 blue marbles. If 3 marbles are selected at random, what is the probability that all are red or all are blue?
16. A card is drawn from a standard deck of cards. What is the probability of selecting an ace or a black card?
17. A four-digit number is formed from the digits 7, 3, 3, 2, and 2. If the number formed is odd, what is the probability that the 3s are together?
18. Two different numbers are selected at random from the numbers 1 through 9. If their product is even, what is the probability that both numbers are even?
19. Five bent coins are tossed. The probability of heads is $\frac{2}{3}$ for each of them. What is the probability that no more than 2 coins will show heads?
20. **Archery** While shooting arrows, Akira can hit the center of the target 4 out of 5 times. What is the probability that he will hit it exactly 4 out of the next 7 times?

CHAPTER 14 TEST

The number of absences for a random sample of 80 high school students at Dover High School last school year are given below.

6	16	12	7	7	9	13	12	7	7
19	4	9	6	4	11	13	10	16	20
10	17	11	12	6	9	10	14	3	8
8	12	13	8	8	11	12	1	11	5
11	16	13	5	10	1	10	8	15	10
13	9	20	5	9	15	11	18	12	14
10	16	8	10	2	11	19	10	12	17
14	6	9	12	10	14	8	9	7	9

1. What is the range of the data?
2. What is an appropriate class interval?
3. What are the class limits?
4. What are the class marks?
5. Construct a frequency distribution of the data.
6. Draw a histogram of the data.
7. Find the mean of the data.
8. Find the median of the data.

A small metal object is weighed on a laboratory balance by each of 15 students in a physics class. The weight of the object in grams is reported as 2.341, 2.347, 2.338, 2.350, 2.344, 2.342, 2.345, 2.348, 2.340, 2.345, 2.343, 2.344, 2.347, 2.341, and 2.344.

9. Find the median of the data.
10. Find the first quartile point and the third quartile point.
11. What is the semi-interquartile range?
12. Make a box-and-whisker plot of the data.
13. What is the mean deviation of the data?
14. What is the standard deviation of the data?

The mean of a set of normally distributed data is 24 and the standard deviation is 2.8.

15. Find the interval about the mean that includes 68.3% of the data.
16. Find the interval about the mean that includes 90% of the data.
17. What percent of the data is between 18.4 and 32.4?
18. What percent of the data is between 24 and 29.6?

Suppose a random sample of data has $\sigma = 3.6$, $N = 400$, and $\overline{X} = 57$.

19. Find the standard error of the mean.
20. Find the interval about the sample mean that has a 5% level of confidence.

CHAPTER 15 TEST

Use the graph of $y = f(x)$ to find each value.

1. $\lim_{x \to -1} f(x)$ and $f(-1)$
2. $\lim_{x \to 2} f(x)$ and $f(2)$

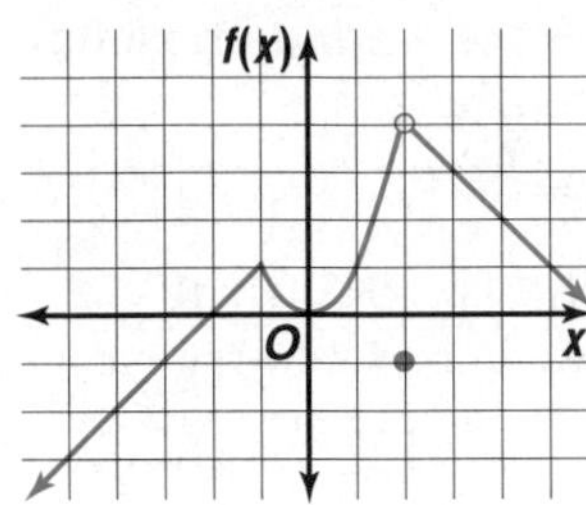

Evaluate each limit.

3. $\lim_{x \to 0} \dfrac{x^2 + 3x}{x}$
4. $\lim_{x \to 1} \dfrac{x^2 - 3x + 2}{x - 1}$
5. $\lim_{x \to 3} \dfrac{x^2 - 9}{x^3 - 27}$
6. $\lim_{x \to 1} \dfrac{x^2 - 2x + 3}{3x^2 - 5}$
7. Use the definition of derivative to find the derivative of $f(x) = x^2 - 2x$.

Use the derivative rules to find the derivative of each function.

8. $f(x) = \frac{1}{2}x^2 - 7x + 1$
9. $f(x) = 4x^3 - 4$
10. $f(x) = 6x^4 - 2x^2 - 30$
11. $f(x) = 2x^5 - 4x^3 + \frac{2}{5}x^2 - 6$
12. $f(x) = 2x^4(x^3 + 3x^2)$
13. $f(x) = (x + 3)^2$

Find the antiderivative of each function.

14. $f(x) = -2x + 6$
15. $f(x) = -x^3 + 4x^2 - x + 4$
16. $f(x) = \frac{1}{2}x^3 - \frac{2}{7}x + 5$
17. $f(x) = \dfrac{x^3 - 4x^2 + x}{x}$
18. Use a limit to find the area between the graph of $y = x^3$ and the x-axis from $x = 0$ to $x = 2$.
19. Find the area of the shaded region in the graph at the right.

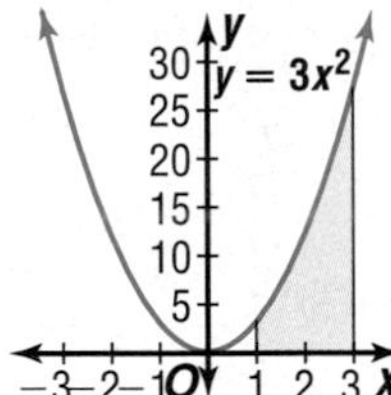

Evaluate each definite integral.

20. $\int_0^1 (2x + 3)\,dx$
21. $\int_1^3 (-x^2 + x + 6)\,dx$

Evaluate each indefinite integral.

22. $\int (1 - 2x)\,dx$
23. $\int (3x^2 + 4x + 7)\,dx$
24. **Baseball** A baseball is hit so that its height, in feet, above the ground after t seconds is given by $h(t) = 3 + 95t - 16t^2$. Find the vertical velocity of the ball after 2 seconds.
25. **Geometry** The volume of a cone of radius r and height h can be calculated using the formula $V = 2\pi \int_0^r \left(-\frac{h}{r}x^2 + hx\right) dx$. Use this formula to find the volume of a cone with radius 3 units and height 2 units.

GLOSSARY

For commonly-used formulas, see inside back cover.

A

abscissa (5) the first element of an ordered pair

absolute maximum (171) a point that represents the maximum value a function assumes over its domain

absolute minimum (171) a point that represents the minimum value a function assumes over its domain

absolute value function (47) a piecewise function, written as $f(x) = |x|$, where $f(x) \geq 0$ for all values of x

absolute value of a complex number (586) The absolute value of the complex number $a + bi$ is $|a + bi| = \sqrt{a^2 + b^2}$.

additive identity matrix (80) If an $m \times n$ matrix I is a zero matrix, then I is an additive identity matrix, such that for any $m \times n$ matrix A, $A + I = A$.

alternate optimal solutions (114) When there are two or more optimal solutions for a linear programming problem, the problem is said to have alternate optimal solutions.

ambiguous case (320) Given two sides and a nonincluded angle, there exist two possible triangles with the given measures.

amplitude (368) $|A|$ for functions in the form $y = A \sin k\alpha$ or $y = A \cos k\alpha$

amplitude of a complex number (587) the angle θ when a complex number is written in the form $r(\cos\theta + \boldsymbol{i}\sin\theta)$

analytic geometry (618) the study of coordinate geometry from an algebraic perspective

angle of depression (300) the angle between a horizontal line and the line of sight from the observer to an object at a lower level

angle of elevation (300) the angle between a horizontal line and the line of sight from the observer to an object at a higher level

angular displacement (352) As any circular object rotates counterclockwise about its center, an object at the edge moves through an angle relative to its starting position known as the angular displacement, or angle of rotation.

angular velocity (352) the change in the central angle with respect to time as an object moves along a circular path

antiderivative (955) $F(x)$ is an antiderivative of $f(x)$ if and only if $F'(x) = f(x)$.

antilogarithm (729) If $\log x = a$, then x is called the antilogarithm of a, abbreviated antilog a.

apothem (300) a segment from the center of a regular polygon perpendicular to a side of the polygon

arccosine (305) the inverse of $y = \cos x$, written as $x = \arccos y$

arcsine (305) the inverse of $y = \sin x$, written as $x = \arcsin y$

arctangent (305) the inverse of $y = \tan x$ written as $x = \arctan y$

argument of a complex number (587) The argument of the complex number $r(\cos\theta + \boldsymbol{i}\sin\theta)$ is the angle θ.

arithmetic mean (758; 897) 1. the terms between any two nonconsecutive terms of an arithmetic sequence 2. a measure of central tendency found by dividing the sum of all values by the number of values

arithmetic sequence (757) a sequence in which the difference between successive terms is a constant

arithmetic series (759) the indicated sum of the terms of an arithmetic sequence

asymptote (180) a line that a graph approaches but never intersects

axis of symmetry (653) a line about which a figure is symmetric

B

back-to-back bar graph (889) a graph plotted on a two-quadrant coordinate system with the horizontal scale repeated in each direction from the central axis used to show comparisons

bar graph (889) a graphic form using bars to make comparisons of statistics

Basic Counting Principle (837) If event M can occur in m ways and is followed by an event N that can occur in n ways, then the event M followed by the event N can occur in $m \cdot n$ ways.

best-fit line (38) the graph of a prediction equation

bimodal data (899) data with two modes

binomial experiment (878) a problem that can be solved using binomial expansion

boundary (52) a line or curve that separates the coordinate plane into two regions

box-and-whisker plot (909) a diagram that graphically displays the median, quartiles, extreme values, and outliers in a set of data

cardioid (563) the graph of a polar equation of the form $r = a \pm a \sin \theta$ or $r = a \pm a \cos \theta$

center of a circle See circle.

center of a hyperbola (642) the midpoint of the segment whose endpoints are the foci

central angle (345) an angle whose vertex lies at the center of a circle

characteristic (727) the part of the logarithm of a number which is the exponent of 10 used to write the number in scientific notation

circle (623) the locus of all points in a plane at a given distance, called the radius, from a fixed point on the plane, called the center

circular arc (345) a part of a circle that is intercepted by a central angle of the circle

circular functions (292) functions defined using a unit circle

circular permutation (847) a circular arrangement of objects in a certain order

classic curves (564) special curves formed by graphing polar equations

class interval (890) the range of each class in a frequency distribution

class limits (890) the upper and lower values in each class in a frequency distribution

class marks (890) the means of the class limits in a frequency distribution

coinciding lines (32) The graphs of two equations that represent the same line are coinciding lines.

column matrix (78) a matrix that has only one column

combination (841) an arrangement of objects where the order is not a consideration

combinatorics (837) the investigation of the different possibilities for the arrangement of objects

common difference (757) the difference between the successive terms of an arithmetic sequence

common logarithms (726) logarithms that use 10 as the base

common ratio (764) the ratio of successive terms of a geometric sequence

comparison test (787) a method to test convergence in an infinite series

complements (853) Two events are complements if and only if the sum of their probabilities is 1.

completing the square (213) a process used to create a perfect square trinomial

complex conjugates (216, 582) The conjugate of the complex number $a + bi$ is $a - bi$.

complex number (206, 580) any number that can be written in the form $a + bi$, where a and b are real numbers and i is the imaginary unit

complex plane (586) The complex number $a + bi$ is graphed as the ordered pair (a, b) in the complex plane. The real axis is horizontal, and the imaginary axis is vertical.

components of a vector (488) two or more vectors whose sum is the given vector

composite (15) Given functions f and g, the composite function $f \circ g$ can be described by $[f \circ g](x) = f(g(x))$.

composition of functions See composite.

compound function (382) a function consisting of the sums or products of trigonometric functions

concentric circles (623) circles with the same center

conditional probability (868) the probability of an event under the condition that some preceding event has occurred

conic section (623) a curve determined by the intersection of a plane with a double right cone

conjugate See complex conjugates.

conjugate axis (642) the segment perpendicular to the transverse axis of a hyperbola through its center

consistent (67) A system of equations that has at least one solution is called consistent.

constant function (22, 137, 164) a function of the form $f(x) = b$

constant of variation (189) the constant k used with direct or inverse variation

constraints (112) conditions given to variables, often expressed as linear inequalities

continuous (160) A function is said to be continuous at point (x_1, y_1) if it is defined at that point and passes through that point without a break.

converge (784) If a sequence has a limit, then the related infinite series is said to converge.

convergent series (784) an infinite series that has a sum or limit

correlation coefficient (40) a value that describes the nature of a set of data. The more closely the data fit a line, the closer the correlation coefficient, r, approaches 1 or –1.

cosecant (286, 292) For any angle, with measure α, a point $P(x, y)$ on its terminal side, $r = \sqrt{x^2 + y^2}$, $\csc \alpha = \frac{r}{y}$.

cosine (285, 291) For any angle, with measure α, a point $P(x, y)$ on its terminal side, $r = \sqrt{x^2 + y^2}$, $\cos \alpha = \frac{x}{r}$.

cotangent (286, 292) For any angle, with measure α, a point $P(x, y)$ on its terminal side, $r = \sqrt{x^2 + y^2}$, $\cot \alpha = \frac{x}{y}$.

coterminal angles (279) two angles in standard position that have the same terminal side

counterexample (421) an example used to show that a given statement is not always true

critical points (171) points at which the nature of a graph changes

cross product (507) The cross product of $\vec{\mathbf{a}}$ and $\vec{\mathbf{b}}$ if $\vec{\mathbf{a}} = (a_1, a_2, a_3)$ and $\vec{\mathbf{b}} = (b_1, b_2, b_3)$ is defined as follows.

$$\vec{\mathbf{a}} \times \vec{\mathbf{b}} = \begin{vmatrix} a_2 & a_3 \\ b_2 & b_3 \end{vmatrix} \vec{\mathbf{i}} - \begin{vmatrix} a_1 & a_3 \\ b_1 & b_3 \end{vmatrix} \vec{\mathbf{j}} + \begin{vmatrix} a_1 & a_2 \\ b_1 & b_2 \end{vmatrix} \vec{\mathbf{k}}$$

cumulative frequency distribution (902) the sum of the frequency of a class and the frequencies of previous classes

D

decreasing (164) A function *f* is decreasing on an interval *I* if and only if for every *a* and *b* contained in *I*, $f(a) > f(b)$ whenever $a < b$.

definite integral (962) an integral that has lower and upper bounds

degenerate conic (623) the intersection of a plane with a double right cone resulting in a point, a line, or two intersecting lines

degree (277) the measure of an angle that is $\frac{1}{360}$ of a complete rotation in the positive direction

degree of a polynomial in one variable (206) the greatest exponent of the variable of the polynomial

dependent events (837) events that affect each other

dependent system (67) A system of equations that has infinitely many solutions is called dependent.

depressed polynomial (224) the quotient when a polynomial is divided by one of its factors

derivative of *f*(*x*) (953) the function $f'(x)$, which is defined as $f'(x) = \lim_{h \to 0} \frac{f(x+h) - f(x)}{h}$

determinant (98) a square array of numbers having a numerical value; the numerical value of the square array of numbers

differentiation (952) the process of finding derivatives

dilation (88) a transformation in which a figure is enlarged or reduced

dimensional analysis (353) a procedure in which unit labels are treated as mathematical factors and can be divided out

dimensions of a matrix (78) the number of rows, *m*, and the number of columns, *n*, of the matrix written as $m \times n$

direction of a vector (485) the directed angle between the positive *x*-axis and the vector

direction vector (520) a vector used to describe the slope of a line

directrix See parabola.

direct variation (189) *y* varies directly as x^n if there is some nonzero constant *k* such that $y = kx^n$, $n > 0$. The variable *k* is called the constant of variation.

discontinuous (159) A function is said to be discontinuous at point (x_1, y_1) if there is a break in the graph of the function at that point.

discriminant (215) in the quadratic formula, the expression under the radical sign, $b^2 - 4ac$

diverge (784) If a sequence does not have a limit, then the related infinite series is said to diverge.

divergent series (784) an infinite series that does not have a sum or limit

domain (5) the set of all abscissas of the ordered pairs of a relation

dot product See inner product.

doubling time (740) the amount of time it takes a quantity to reach twice its initial amount

E

eccentricity (636) the ratio of the distance between any point of a conic section and a fixed point to the distance between the same point of the conic section to a fixed line

element of a matrix (78) any value in the array of values

elimination method (68) a technique used to solve a system of equations

ellipse (631) the locus of all points in a plane such that the sum of the distances from two given points in the plane, called foci, is constant

end behavior (162) the behavior of $f(x)$ as $|x|$ becomes very large

equilateral hyperbola (647) a hyperbola with perpendicular asymptotes

equal matrices (79) Two matrices are equal if and only if they have the same dimensions and their corresponding elements are identical.

equal vectors (485) Two vectors are equal if and only if they have the same direction and the same magnitude.

escape set (599) the set of initial values for which the iterates of a function approach infinity

escaping point (816) If the iterates of a function approach infinity for some initial value, the initial point is called an escaping point.

even function (133) a function hose graph is symmetric with respect to the *y*-axis

everywhere discontinuous (159) A function that is impossible to graph in the real number system is said to be everywhere discontinuous.

linear velocity (353) distance traveled per unit of time

line plot (889) a display of statistical data on a number line so that patterns and variability in data can be determined

line symmetry (129) Two distinct points P and P' are symmetric with respect to a line ℓ if and only if ℓ is the perpendicular bisector of $\overline{PP'}$. A point P is symmetric to itself with respect to a line ℓ if and only if P is on ℓ.

locus (658) a set of points and only those points that satisfy a given set of conditions

logarithm (718) In the function $x = a^y$, y is called the logarithm, base a, of x.

logarithmic function (718) $y = \log_a x$, $a > 0$ and $a \neq 1$, which is the inverse of the exponential function $y = a^x$.

lower bound (238) the integer less than or equal to the least real zero of the polynomial $P(x)$

M

***m* × *n* matrix** (78) a matrix with m rows and n columns

magnitude of a vector (485) the length of the directed line segment

major axis (631) the axis of symmetry of an ellipse which contains the foci

mantissa (727) the common logarithm of a number between 1 and 10

mathematical induction (820) a method of proof that depends on a recursive process

matrix (78) any rectangular array of terms called elements

maximum (121) a critical point of a graph where the curve changes from an increasing curve to a decreasing curve

mean (897) a measure of central tendency found by dividing the sum of all values by the number of values

mean deviation (910) the arithmetic mean of the absolute value of the deviations from the mean of a set of data

measure of central tendency (897) a number that represents the center or middle of a set of data

measure of variability (908) a number that represents the variability or diversity of a set of data

median (897) the middle value of a set of data that has been arranged into an ordered sequence

median class (902) in a frequency distribution, the class in which the median of the data is located

midline (380) a horizontal axis used as the reference line for vertical shifts of the graphs of sine and cosine functions

minimum (171) a critical point of a graph where the curve changes from a decreasing curve to an increasing curve

minor (98) The minor of an element of an nth-order matrix is the determinant of $(n - 1)$th order found by deleting the row and column containing the element.

minor axis (631) the axis of symmetry of an ellipse which does not contain the foci

minute (277) a unit of angle measure that is $\frac{1}{60}$ of a degree

mode (897) the item of data that appears more frequently than any other in the set

model (27) an equation used to approximate a real-world set of data

modulus (587) the number r when a complex number is written in the form $r(\cos\theta + \boldsymbol{i}\sin\theta)$

monotonicity (163) A function is said to be monotonic on an interval I if and only if the function is increasing on I or decreasing on I.

mutually exclusive events (862) two events whose outcomes can never be the same

N

natural logarithm (733) logarithms that use e as the base, written $\ln x$

***n* factorial** (794) written $n!$, for n, an integer greater than zero, the product $n(n - 1)(n - 2) \cdots 1$

nonlinear regression (741) the process of fitting an equation to nonlinear data

nonsingular matrix (98) a matrix with a nonzero determinant

normal curve (918) a symmetric bell-shaped graph of a normal distribution

normal distribution (918) A frequency distribution that often occurs when there is a large number of values in a set of data: about 68% of the values are within one standard deviation of the mean, 95% of the values are within two standard deviations from the mean, and 99% of the values are within three standard deviations.

normal form (463) the equation of a line that is written in terms of the length of the normal from the line to the origin

normal line (463) a line that is perpendicular to another line, curve, or surface

***n*th order matrix** (78) a square matrix with n rows and n columns

***n*th partial sum** (759) the sum of the first n terms of a series

odd function (133) a function whose graph is symmetric with respect to the origin

odds (854) the ratio of the probability of the success of an event to the probability of its compliment

opposite vectors (487) two vectors that have the same magnitude and opposite directions

orbit (815) the graph of the sequence of successive iterates

ordered triple (74; 500) 1. the solution of a system of equations in three variables 2. coordinates of the location of a point in space

ordinate (5) the second element of an ordered pair

outlier (909) a value of a set of data that is more than 1.5 interquartile ranges beyond the upper or lower quartiles

parabola (653) the locus of all points in a given plane that are the same distance from a given point, called the focus, and a given line, called the directrix

parallel lines (32) nonvertical coplanar lines that have equal slopes; any two coplanar vertical lines

parallel vectors (488) two vectors that have the same or opposite directions

parameter (520) the independent variable t in the vector equation of a line

parametric equation of a line (520) the vector equation $\langle x - x_1, y - y_1\rangle = t\langle a_1, a_2\rangle$ written as the two equations $x = x_1 + ta_1$ and $y = y_1 + ta_2$

parent graph (137) an anchor graph from which other graphs in the family are derived

partial fraction (244) one of the fractions that was added or subtracted to result in a given rational expression

Pascal's triangle (799) a triangular array of numbers such that the $(n + 1)^{th}$ row is the coefficient of the terms of the expansion $(x + y)^n$ for $n = 0, 1, 2 \ldots$

percentile (922) The nth percentile of a set of data is the value of the data that is equal to or greater than n percent of the data.

periodic function (359) A function is periodic if, for some real number α, $f(x + \alpha) = f(x)$ for each x in the domain of f. The least positive value of α for which $f(x) = f(x + \alpha)$ is the *period* of the function.

permutation (838) the arrangement of objects in a certain order

permutation with repetition (846) the arrangement of objects in a certain order in which some of the objects are alike

perpendicular lines (34) any two nonvertical lines, the product of whose slopes is -1; any vertical line and any horizontal line

phase shift (378) the least value of $|k\theta + c|$, for which the trigonometric function $f(k\theta + c) = 0$

piecewise function (45) a function in which different equations are used for different intervals of the domain

point discontinuity (159) When there is a value in the domain for which $f(x)$ is undefined, but the pieces of the graph match up, then $f(x)$ has point discontinuity.

point of inflection (171) a critical point of a graph where the graph changes its curvature from concave down to concave up or vise versa

point-slope form (28) the equation of the line that contains the point with coordinates (x_1, y_1) and having slope m written in the form $y - y_1 = m(x - x_1)$

point symmetry (127) Two distinct points P and P' are symmetric with respect to point M, if and only if M is the midpoint of $\overline{PP'}$. Point M is symmetric with respect to itself.

polar axis (553) a ray whose initial point is the pole

polar coordinate system (553) a grid of concentric circles and their center, which is called the pole, whose radii are integral multiples of 1

polar equation (556) an equation that uses polar coordinates

polar form (588) the complex number $x + \boldsymbol{y}\boldsymbol{i}$ written as $r(\cos\theta + \boldsymbol{i}\sin\theta)$ where $r = \sqrt{x^2 + y^2}$ and $\theta = \text{Arctan}\,\frac{y}{x}$ where $x > 0$ and $\theta = \text{Arctan}\,\frac{y}{x} + \pi$ where $x < 0$

polar graph (556) the representation of the solution set which is the set of points whose coordinates (r, θ) satisfy a given polar equation

polar plane See polar coordinate system.

pole (180) 1. See polar coordinate system. 2. vertical asymptote of a rational function

polygonal convex set (108) the solution of a system of linear inequalities

polyhedron (535) a closed three-dimensional figure made up of flat polygonal regions

polynomial equation (206) a polynomial that is set equal to zero

polynomial function (206) a function $y = P(x)$ where $P(x)$ is a polynomial in one variable

polynomial in one variable (205) an expression of the form $a_0x^n + a_1x^{n-1} + \cdots + a_{n-1}x + a_n$ where the coefficients $a_0, a_1, \ldots, a_n$ represent complex numbers, a_0 is not zero, and n represents a nonnegative integer

population (927) the entire set of items or individuals in the group being considered

power function (704) a function in the form $y = x^b$, where b is a real number

prediction equation (38) an equation suggested by the points of a scatter plot used to predict other points

pre-image (88) the graph of an object before a transformation

principal values (406, 456) the unique solutions of a trigonometric equation if the values of the function are restricted to two adjacent quadrants

prisoner point (816) If the iterates do not approach infinity for some initial value, that point is called a prisoner point.

prisoner set (599) the set of initial values for which the iterates of a function do not approach infinity

probability (852) the measure of the chance of a desired outcome happening

pure imaginary number (206, 581) the complex number $a + bi$ when $a = 0$ and $b \neq 0$

Q

quadrantal angle (278) an angle in standard position whose terminal side coincides with one of the axes

quartile (909) one of four groupings of a set of data determined by the median of the set and the medians of the sets determined by the median

R

radian (343) the measure of a central angle whose sides intercept an arc that is the same length as the radius of the circle

radical equation (251) an equation that contains a radical expression with the variable in the radicand

radical inequality (253) an inequality that contains a radical expression with the variable in the radicand

radius See circle.

random sample (927) a sample in which every member of the population has an equal chance to be selected

range (890) the difference of the greatest and least values in a set of data

range of a relation (5) the set of all ordinates of the ordered pairs of a relation

rate of change (956) the derivative of a function when applied to real-world applications

rational equation (243) an equation that consists of one or more rational expressions

rational function (180) the quotient of two polynomials in the form $f(x) = \frac{g(x)}{h(x)}$, where $h(x) \neq 0$

rational inequality (245) an inequality that consists of one or more rational expressions

ratio test (785) a method to test convergence in an infinite series

real part of a complex number (581) a in the complex number $a + bi$

rectangular form (581) a complex number written as $x + yi$, where x is the real part and yi is the imaginary part

rectangular hyperbola (648) A special case of the equilateral hyperbola, where the coordinate axes are the asymptotes. The general equation of a rectangular hyperbola is $xy = c$, where c is a nonzero constant.

recursive formula (758) a formula used for determining the next term of a sequence using one or more of the previous terms

reduced sample space (862) the subset of a sample space that contains only those outcomes that satisfy a given condition

reduction identity (446) identity that involves adding and subtracting the quandrantal angles, 90°, 180°, and 270°, from the angle measure to find equivalent values of a trigonometric function

reference angle (280) the acute angle formed by the terminal side of an angle in standard position and the x-axis

reflection (88) a linear transformation that flips a figure over a line called the line of symmetry

reflection matrix (89) a matrix used to reflect an object over a line or plane

regression line (40) a best-fit line

relation (5) a set of ordered pairs

relative extremum (172) a point that represents the maximum or minimum for a certain interval

relative maximum (172) a point that represents the maximum for a certain interval

relative minimum (172) a point that represents the minimum for a certain interval

resultant of vectors (486) the sum of two or more vectors

root (206) a solution of the equation $P(x) = 0$

rose (563) The graph of a polar equation of the form $r = a \cos n\theta$ or $r = a \sin n\theta$.

rotation (88) a transformation in which an object is moved around a center point

rotation matrix (91) a matrix used to rotate an object

row matrix (78) a matrix that has only one row

S

sample space (852) the set of all possible outcomes of an event

scalar (80, 488) a real number

scalar quantity (488) A quantity with only magnitude is called a scalar quantity.

scatter plot (38) a visual representation of data

scientific notation (695) the expression of a number in the form $a \times 10n$, where $1 \leq a < 10$ and n is an integer

secant (286, 292) For any angle, with measure α, a point $P(x, y)$ on its terminal side, $r = \sqrt{x^2 + y^2}$, $\sec \alpha = \frac{r}{x}$.

secant line (953) a line that intersects a curve at two or more points

second (277) a unit of angle measure that is $\frac{1}{60}$ of a minute

sector of a circle (346) a region bounded by a central angle and the intercepted arc

semi-interquartile range (909) one-half the interquartile range of a set of data

semi-major axis (632) one of the two segments into which the center of an ellipse divides the major axis

semi-minor axis (632) one of the two segments into which the center of an ellipse divides the minor axis

sequence (757) a set of numbers in a specific order

side adjacent (284) the side of a right triangle that is a side of an acute angle in the triangle

side opposite (284) the side of a right triangle that is opposite an acute angle in the triangle

sigma notation (792) For any sequence $a_1, a_2, a_3, \ldots$, the sum of the first k terms may be written $\sum_{n=1}^{k} a_n$, which is read "the summation from $n = 1$ to k of a_n." Thus, $\sum_{n=1}^{k} a_n = a_1 + a_2 + a_3 + \cdots + a_k$, where k is an integer value.

simulation (877) a technique used to model probability experiments for real-world applications

sine (285, 291) For any angle, with measure α, a point $P(x, y)$ on its terminal side, $r = \sqrt{x^2 + y^2}$, $\cos \alpha = \frac{y}{r}$.

sinusoidal function (388) a function of the form $y = A \sin (k\theta + c) + h$ or $y = A \cos (k\theta + c) + h$

slant asymptote (183) The oblique line ℓ is a slant asymptote for a function $f(x)$ if the graph of $f(x)$ approaches ℓ as $x \to \infty$ or as $x \to -\infty$.

slope-intercept form (21) the equation of a line with slope, m, and y-intercept, b, written in the form $y = mx + b$

slope of a curve (949) the slope of a line tangent to a particular point on the graph of a curve

slope of a line (20-21) the value $m = \frac{y_2 - y_1}{x_2 - x_1}$ where (x_1, y_1) and (x_2, y_2), $x_2 \neq x_1$, are two points of the line

solution (67) an ordered pair representing the solution common to both equations in a system of equations

solve a triangle (307) to find all of the measures of a triangle's sides and angles

spiral of Archimedes (564) a function of the form $r = n\theta$

square matrix (78) a matrix with the same number of rows as columns

standard deviation (911) a measure of the average amount by which individual items of data deviate from the arithmetic mean of all the data

standard error of the mean (927) the standard deviation of the distribution of a sample mean

standard form (21) a linear equation written in the form $Ax + By + C = 0$, where A, B, and C are real numbers and A and B are not both zero

standard position (277) an angle with its vertex at the origin and its initial side along the positive x-axis

standard position of a vector (485) If a vector has its initial point at the origin, it is in standard position.

stem-and-leaf plot (899) a display of numerical data for which each value is separated into two numbers, the stem which consists of the digits in the greatest common place value, and the leaf which contains the other digits of each item of data

step function (46) a function whose graph is a series of disjoint lines or steps

substitution method (68) a method for solving a system of equations

success (850) the desired outcome of an event

symmetry identity (424) trigonometric identities related to the symmetries of the unit circle

symmetry with respect to the origin (128) The graph of a relation S is symmetric with respect to the origin if and only if $(a, b) \in S$ implies that $(-a, -b) \in S$. A function has a graph that is symmetric with respect to the origin if and only if $f(-x) = -f(x)$ for all x in the domain of f.

synthetic division (223) a method used to divide a polynomial by a binomial

system of equations (67) a set of equations with the same variables

system of linear inequalities (107) a set of inequalities with the same variables

T

tangent (953; 285, 292) 1. a line that intersects a curve at exactly one point 2. For any angle, with measure α, a point $P(x, y)$ on its terminal side, $r = \sqrt{x^2 + y^2}$, $\tan \alpha = \frac{y}{x}$.

term (757) a number in a sequence or series

terminal side of an angle (277) a ray of an angle that rotates about the vertex

theoretical probability (877) probability determined using mathematical methods to model outcomes of a given situation

three-dimensional bar graph (890) a graphic form using bars to make comparisons among multiple aspects of statistical data

transformation (88) functions that map points of a graph onto its image

translation (88) a linear transformation that slides the graph vertically and/or horizontally on the coordinate plane, but does not change its shape

translation matrix (88) the matrix used to represent the translation of a set of points with respect to (h, k) which is equal to $\begin{bmatrix} h & h & h & h \\ k & k & k & k \end{bmatrix}$

transverse axis (642) the line segment that has as its endpoints the vertices of a hyperbola

tree diagram (837) a diagram used to show the total number of possible outcomes of an event

trigonometric form See polar form.

trigonometric functions (292) For any angle, with measure α, a point $P(x, y)$ on its terminal side, $r = \sqrt{x^2 + y^2}$, the trigonometric functions of α are as follows.

$\sin \alpha = \frac{y}{r}$ $\cos \alpha = \frac{x}{r}$ $\tan \alpha = \frac{y}{x}$

$\csc \alpha = \frac{r}{y}$ $\sec \alpha = \frac{x}{r}$ $\cot \alpha = \frac{x}{y}$

trigonometric identity (421) an equation involving a trigonometric function that is true for all values of the variable. *See inside back cover for complete list of identities.*

trigonometric ratio (285) a ratio of the sides of a right triangle used to define the sine, cosine, and tangent ratios of the triangle

trigonometric series (806) infinite series that define the trigonometric functions sine and cosine

U

unbounded (113) The solution of a linear programming problem is unbounded if the region defined by the constraints is infinitely large.

unit circle (291) a circle of radius 1 unit whose center is at the origin of a rectangular coordinate system

unit vector (496) a vector of length 1 that is parallel to the x-, y-, or z-axis

upper bound (238) the integer greater than or equal to the greatest zero of the polynomial function

variance (912) the mean of the squares of the deviations from the arithmetic mean

vector (485) a quantity, or directed distance, that has both magnitude and direction

vector equation (520) A line through $P(x, y)$ parallel to the vector $\vec{\mathbf{a}} = \langle a_1, a_2 \rangle$ is defined by the set of points $P_1(x_1, y_1)$ and $P(x, y)$ such that $\overrightarrow{P_1P} = t\vec{\mathbf{a}}$ for some real number t. Therefore, $\langle x - x_1, y - y_1 \rangle = t\langle a_1, a_2 \rangle$.

vertex (277; 535) 1. the common endpoint of two rays forming an angle 2. the common endpoints of the sides of a polygon or where three or more edges of a polyhedron intersect

vertex matrix (88) a matrix used to represent the coordinates of the vertices of an object

vertex of a conic section (631, 642, 653) a point at which a conic section intersects its axis of symmetry

vertical asymptote (180) The line $x = a$ is a vertical asymptote for a function $f(x)$ if $f(x) \to \infty$ or $f(x) \to -\infty$ as $x \to a$ from either the left or the right.

vertical line test (7) a test used to determine if a relation is a function

whisker (909) the segments extending from the ends of the box in a box-and-whisker plot

***x*-intercept** (20) the x-coordinate of the point at which the graph of an equation crosses the x-axis

***y*-intercept** (20) the y-coordinate of the point at which the graph of an equation crosses the y-axis

zero (22, 206) a value of x for which $f(x) = 0$

zero matrix (80) a matrix whose elements are all zero

zero vector (485) a vector with initial and terminal points at the origin

SELECTED ANSWERS

CHAPTER 1 LINEAR RELATIONS AND FUNCTIONS

Pages 9–12 Lesson 1-1

5. $y = x - 4$

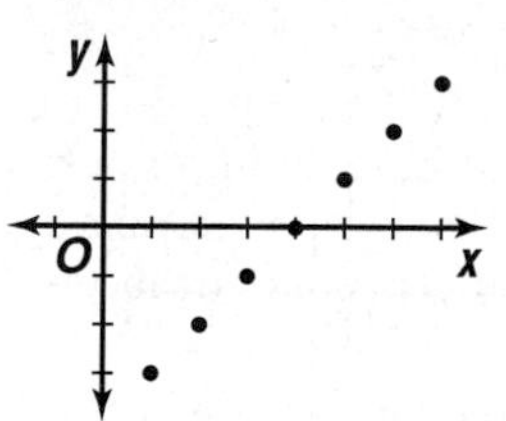

x	y
1	−3
2	−2
3	−1
4	0
5	1
6	2
7	3

7. {(−6, 1), (−4, 0), (−2, −4), (1, 3), (4, 3)}; D = {−6, −4, −2, 1, 4}; R = {−4, 0, 1, 3}

9.

x	y
1	−5
2	−5
3	−5
4	−5
5	−5
6	−5
7	−5
8	−5

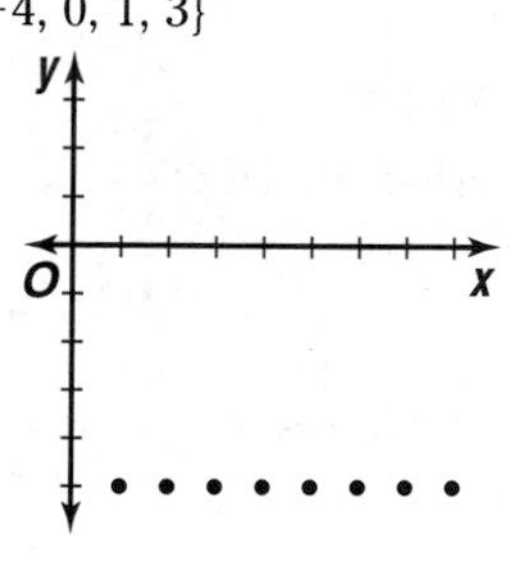

11. {−3, 3, 6};{−6, −2, 0, 4}; no; 6 is matched with two members of the range. **13.** −84

15. $x \geq -1$

17. $y = 3x$

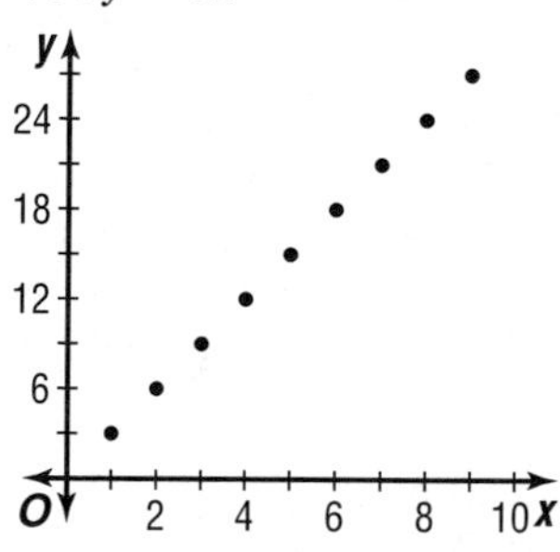

x	y
1	3
2	6
3	9
4	12
5	15
6	18
7	21
8	24
9	27

19. $y = 8 + x$

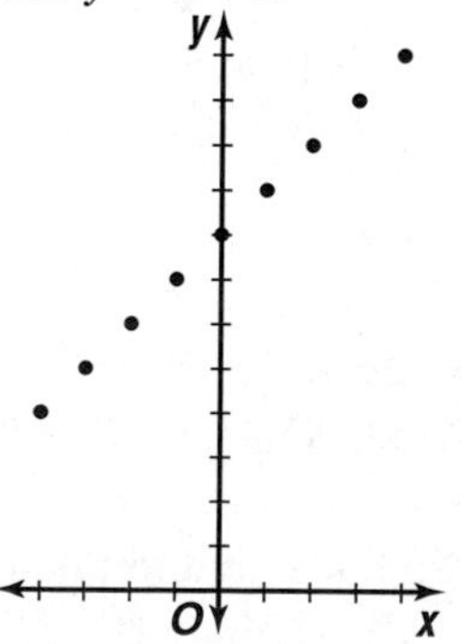

x	y
−4	4
−3	5
−2	6
−1	7
0	8
1	9
2	10
3	11
4	12

21. {(−10, 0), (−5, 0), (0, 0), (5, 0)}; D = {−10, −5, 0, 5}; R = {0} **23.** {(−3, −2), (−1, 1), (0, 0), (1, 1)}; D = {−3, −1, 0, 1}; R = {−2, 0, 1}

25. {(3, −4), (3, −2), (3, 0), (3, 1), (3, 3)}; D = {3}; R = {−4, −2, 0, 1, 3}

27.

x	y
1	−1
2	−2
3	−3
4	−4
5	−5
6	−6

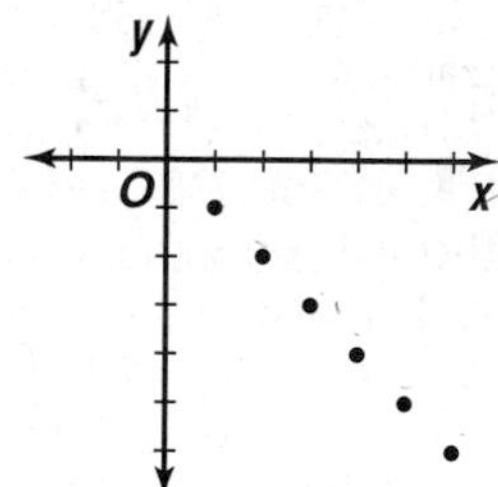

29.

x	y
1	0
2	3
3	6
4	9
5	12

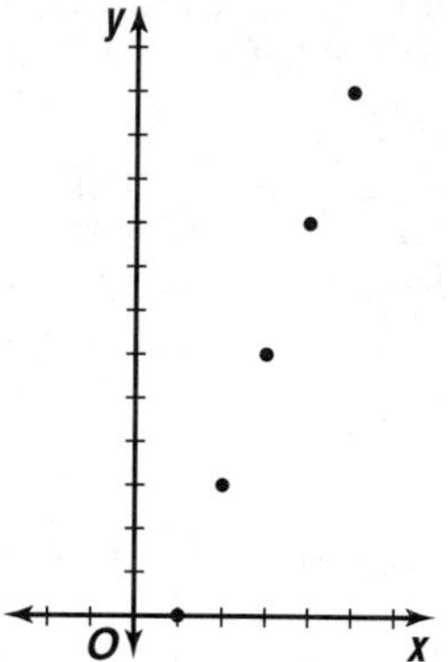

31.

x	y
4	2
4	−2

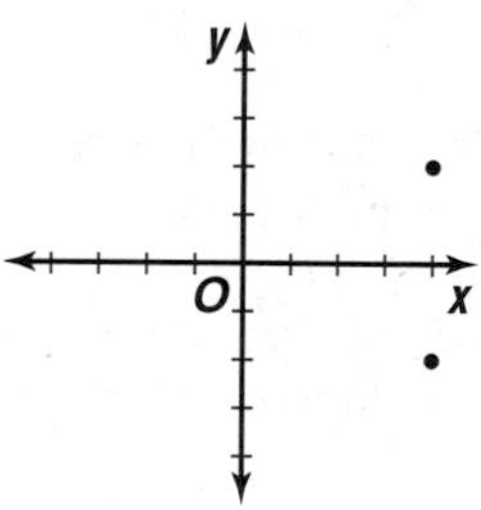

33. {1}; {−6, −2, 0, 4}; no; The x-value 1 is paired with more than one y-value. **35.** {0, 2, 5}; {−8, −2, 0, 2, 8}; no; The x-values 2 and 5 are paired with more than one y-value. **37.** {−9, 2, 8, 9}; {−3, 0, 8}; yes; Each x-value is paired with exactly one y-value. **39.** domain: {−3, −2, −1, 1, 2, 3}; range: {−1, 1, 2, 3}; A function because each x-value is paired with exactly one y-value. **41.** 9 **43.** 2 **45.** $2n^2 - 5n + 12$ **47.** $|25m^2 - 13|$ **49.** $x \leq -3$ or $x \geq 3$ **51a.** $x \neq 1$ **51b.** $x \neq -5$ **51c.** $x \neq -2, 2$ **53.** $3x^3 + 4x - 7$ **55a.** 14,989,622.9 m; 59,958,491.6 m; 419,709,441.2 m; 1,768,775,502 m **55b.** 23,983,396.64 m **57.** B

Pages 17–19 Lesson 1-2

5. $3x^2 + 6x + 4$; $3x^2 + 2x - 14$; $6x^3 + 35x^2 + 26x - 45$; $\frac{3x^2 + 4x - 5}{2x + 9}, x \neq -\frac{9}{2}$ **7.** $2x^2 - 4x - 3$; $4x^2 - 16x + 15$ **9.** 5, 11, 23 **11.** $x^2 - x + 9$; $x^2 - 3x - 9$; $x^3 + 7x^2 - 18x$; $\frac{x^2 - 2x}{x + 9}, x \neq -9$

13. $\frac{x^3 - 2x^2 - 35x + 3}{x - 7}, x \neq 7 - \frac{x^3 - 2x^2 - 35x - 3}{x - 7}$, $x \neq 7$; $\frac{3x^2 + 15x}{x - 7}, x \neq 7$; $\frac{3}{x^3 - 2x^2 - 35x}, x \neq -5, 0,$ or 7

15. $x^2 + 8x + 7$; $x^2 - 5$ **17.** $3x^2 - 4$; $3x^2 - 24x + 48$ **19.** $2x^3 + 2x^2 + 2$; $8x^3 + 4x^2 + 1$

21. $\frac{x}{x - 1}, x \neq 1$; $\frac{1}{x}, x \neq 0$ **23.** $x \neq 7$ **25.** 7, 2, 7

27. 2, 2, 2 **29.** Yes; If $f(x)$ and $g(x)$ are both lines, they can be represented as $f(x) = m_1x + b_1$ and $g(x) = m_2x + b_2$. Then $[f \circ g](x) = m_1(m_2x + b_2) + b_1 = m_1m_2x + m_1b_2 + b_1$. Since m_1 and m_2 are constants, m_1m_2 is a constant. Similarly, m_1, b_2, and b_1 are constants, so $m_1b_2 + b_1$ is a constant. Thus, $[f \circ g](x)$ is a linear function if $f(x)$ and $g(x)$ are both linear. **31a.** $h[f(x)]$, because you must subtract before figuring the bonus.

31b. \$3750 **33a.** $v(p) = \frac{7p}{47}$ **33b.** $r(v) = 0.84v$

33c. $r(p) = \frac{147p}{1175}$ **33d.** \$52.94, \$28.23, \$99.72

35. $\{(-1, 8), (0, 4), (2, -6), (5, -9)\}$; $D = \{-1, 0, 2, 5\}$; $R = \{-9, -6, 4, 8\}$ **37.** $3\frac{11}{16}$ **39.** C

13.

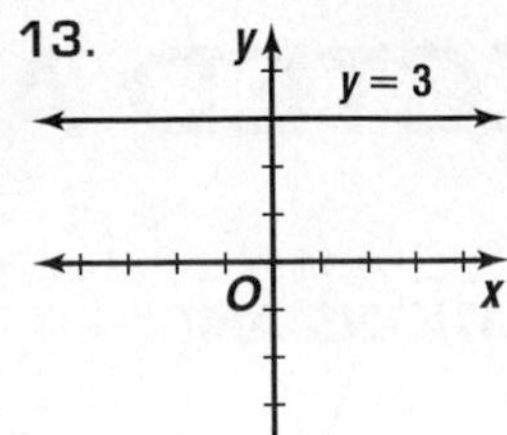

15.

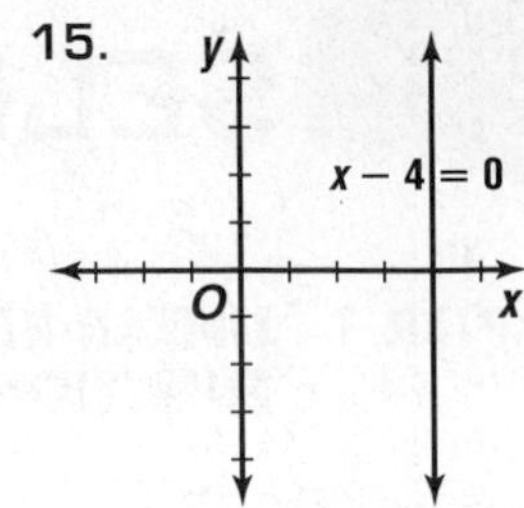

17.

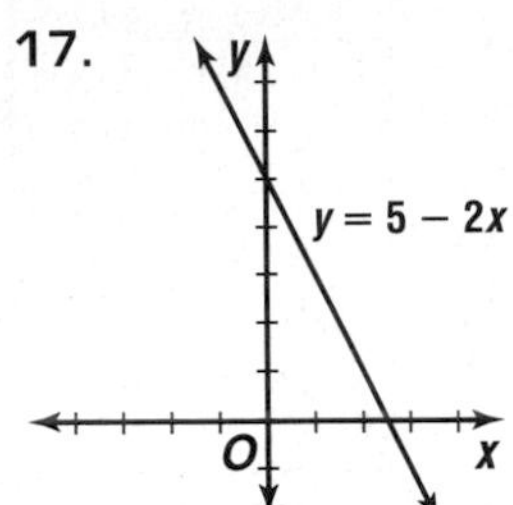

19.

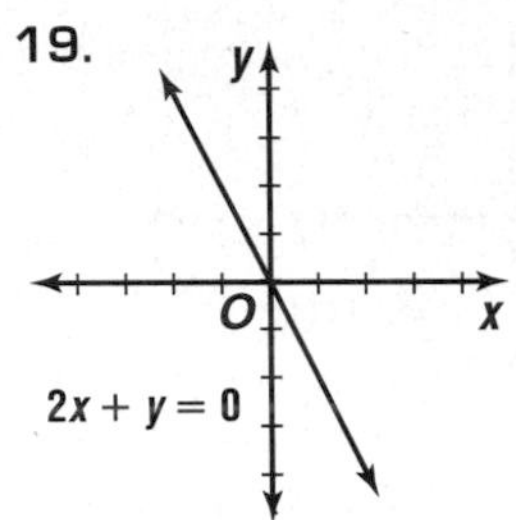

21.

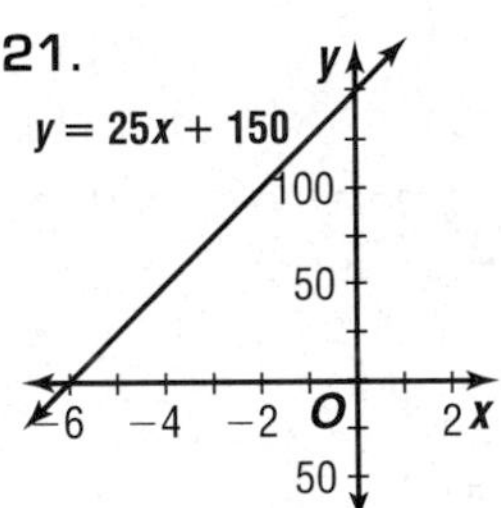

23.

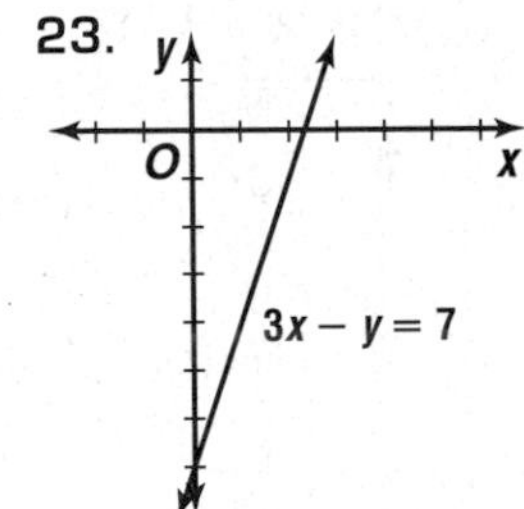

Pages 23–25 Lesson 1-3

5.

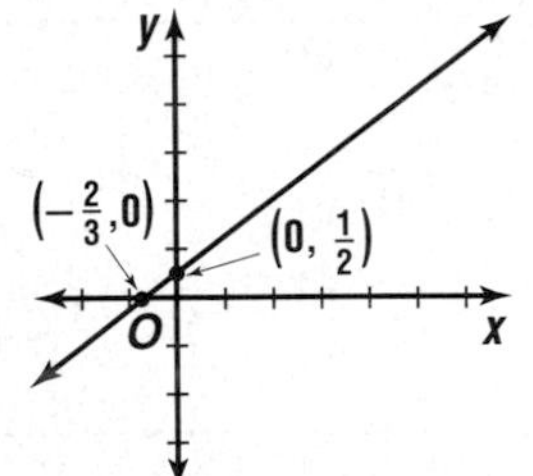

7.

9. −12

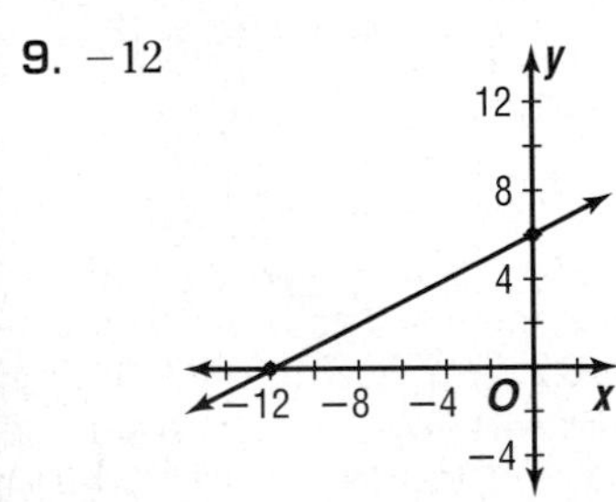

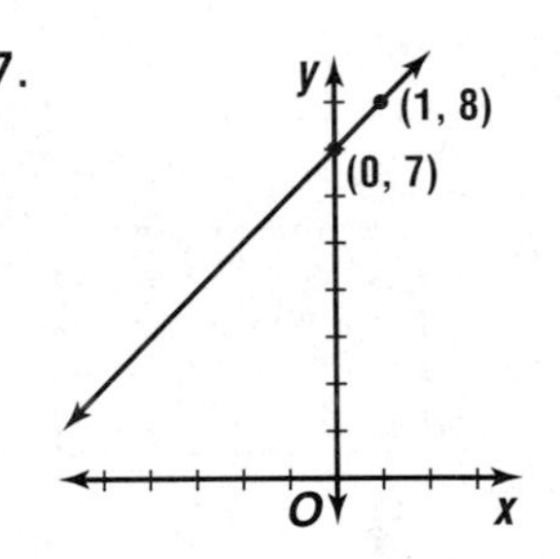

11a. (38.500, 173), (44.125, 188) **11b.** 2.667

11c. For each 1-centimeter increase in the length of a man's tibia, there is a 2.667-centimeter increase in the man's height.

25. 3

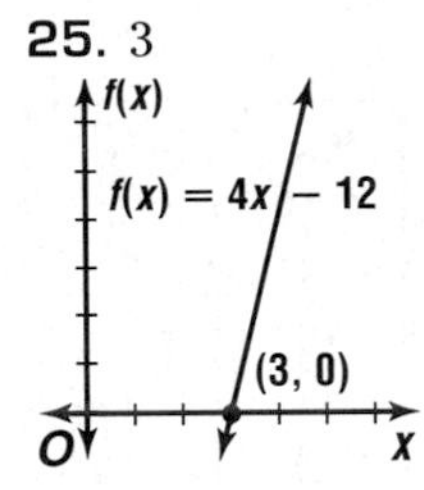

27. 0

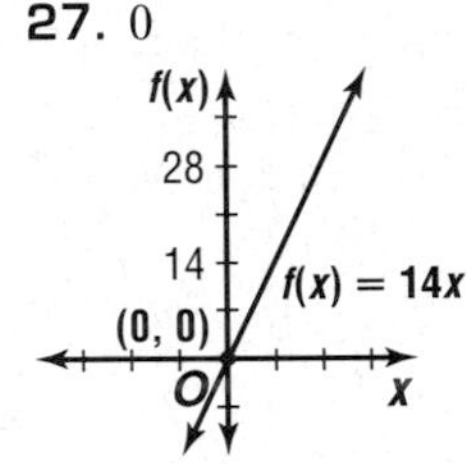

29. $\frac{8}{5}$

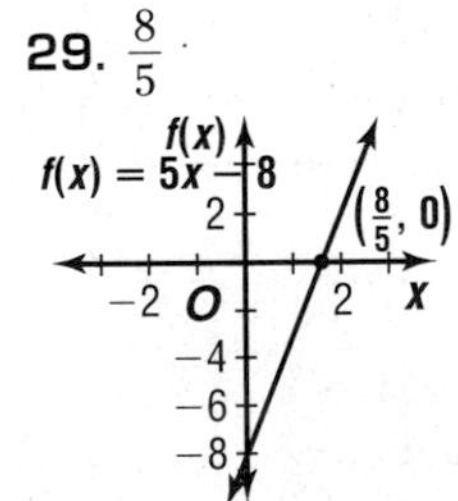

31. 2

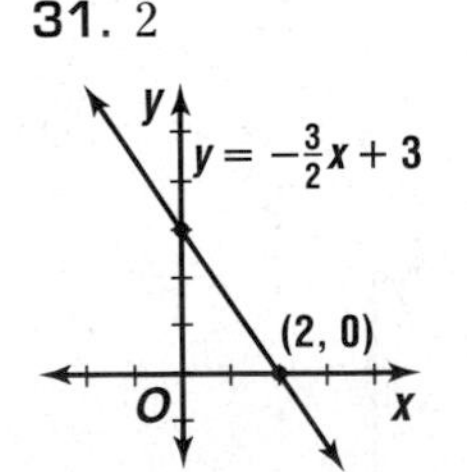

33a. 0.4 ohm **33b.** 2.4 volts **35a.** $\frac{1}{4}$ **35b.** For each 1-degree increase in the temperature, there is a $\frac{1}{4}$-pascal increase in the pressure.

35c.

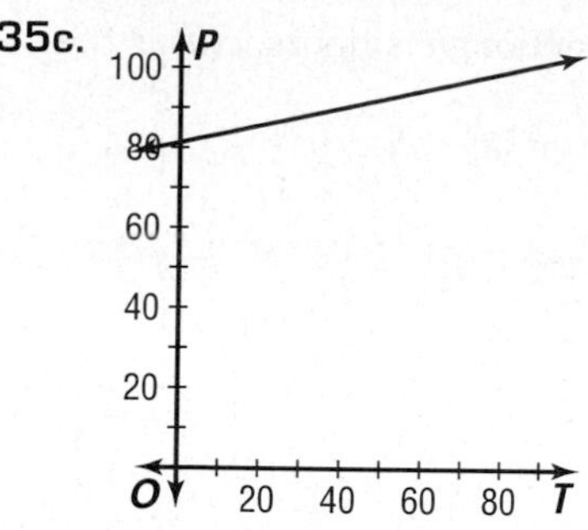

37a. 36; The software has no monetary value after 36 months. **37b.** −290; For every 1-month change in the number of months, there is a \$290 decrease in the value of the software.

37c.

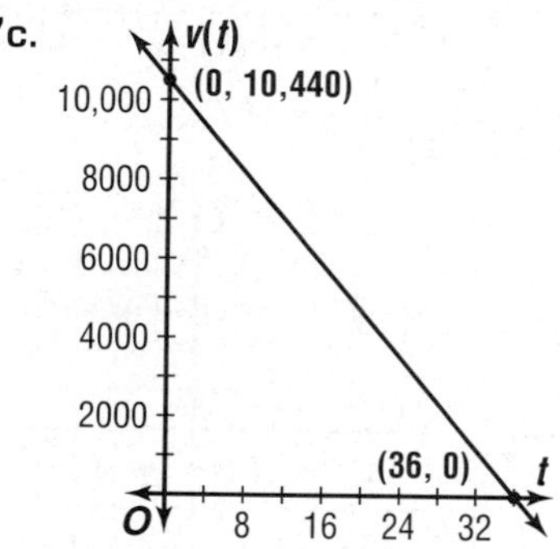

39a. 0.86 **39b.** \$1552.30 **39c.** 0.14 **39d.** \$252.70 **41a.** $d(p) = 0.88p$ **41b.** $r(d) = d - 100$ **41c.** $r(p) = 0.88p - 100$ **41d.** \$603.99, \$779.99, \$1219.99 **43.** −671 **45.** {(−3, 14), (−2, 13), (−1, 12), (0, 11)}, yes

Pages 29–31 Lesson 1-4

7. $y = 4x - 10$ **9.** $y = 2$ **11.** $y = 5x - 2$ **13.** $y = -\frac{3}{4}x$ **15.** $y = 6x - 19$ **17.** $y = -\frac{4}{9}x + \frac{49}{9}$ **19.** $y = 1$ **21.** $x = 0$ **23.** $x + 2y + 10 = 0$ **25a.** $t = 2 + \frac{x - 7000}{2000}$ **25b.** about 5.7 weeks **27a.** Sample answer: Using (20, 28) and (27, 37), $y = \frac{9}{7}x + \frac{16}{7}$ **27b.** Using sample answer from part a, 26.7 mpg **27c.** Sample answer: The estimate is close but not exact since only two points were used to write the equation. **29.** Yes; the slope of the line through (5, 9) and (−3, 3) is $\frac{3 - 9}{-3 - 5}$ or $\frac{3}{4}$. The slope of the line through (−3, 3), and (1, 6) is $\frac{6 - 3}{1 - (-3)}$ or $\frac{3}{4}$. Since these two lines would have the same slope and would share a point, their equations would be the same. Thus, they are the same line and all three points are collinear. **31a.** \$6111 billion **31b.** The rate is the slope. **33.** $x^5 - 3x^4 + 7x^3$, $\frac{x^3}{x^2 - 3x + 7}$ **35.** A

Pages 35–37 Lesson 1-5

5. none of these **7.** parallel **9.** $5x - y - 16 = 0$ **11.** parallelogram **13.** parallel **15.** perpendicular **17.** perpendicular **19.** coinciding **21.** None of these; the slopes are neither the same nor opposite reciprocals. **23.** $4x - 9y - 183 = 0$ **25.** $x + 5y + 15 = 0$ **27.** $y + 13 = 0$ **29a.** 4 **29b.** $-\frac{49}{4}$ **31.** $x - 5y - 29 = 0$; $x = 7$; $x + 5y + 15 = 0$ **33a.** No; the lines that represent the situation do not coincide. **33b.** Yes; the lines that represent the situation coincide. **35.** $y = -2x + 7$

37.

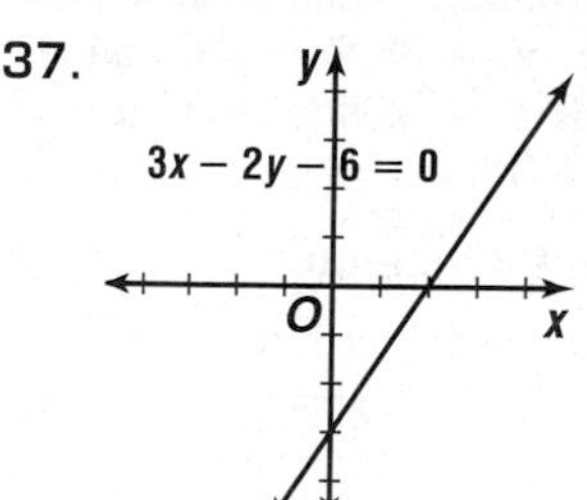

39. Sample answer: {(2, 4), (2, −4), (1, 2), (1, −2), (0, 0)}; because the x-values 1 and 2 are paired with more than one y-value

Pages 41–44 Lesson 1-6

5a.

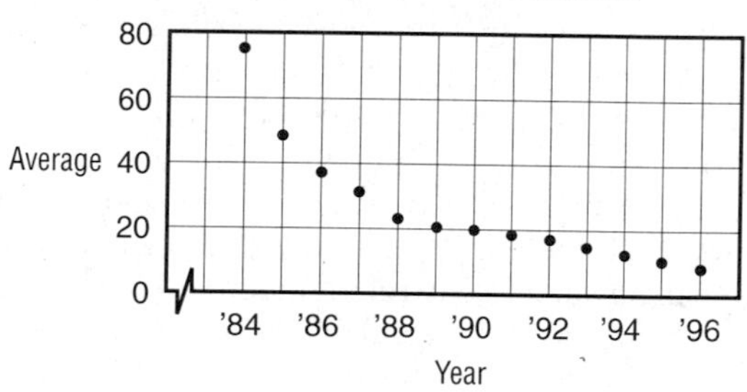

5b. Sample answer: Using (1987, 32) and (1996, 7.8), $y = -2.69x + 5377.03$ **5c.** $y = -6.28x + 12{,}530.14$; $r \approx -0.82$ **5d.** 1995; No; In 1995 there were 10 students per computer.

7a.

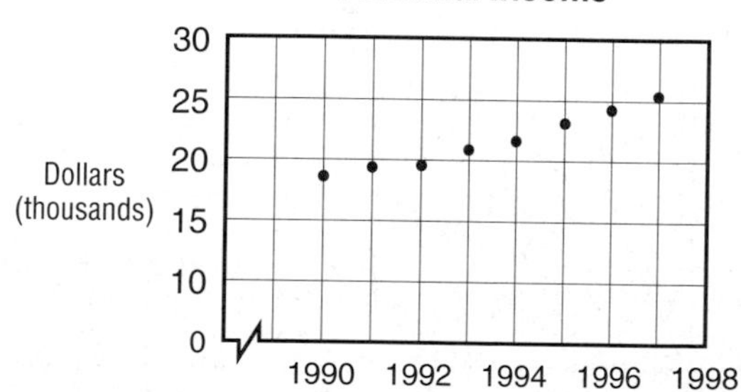

7b. Sample answer: Using (1991, 19,100) and (1995, 23,233), $y = 1058.25x - 2{,}087{,}875.75$ **7c.** $y = 1052.32x - 2{,}076{,}129.64$; $r \approx 0.99$ **7d.** \$33,771.96; Yes, r shows a strong relationship.

9a.

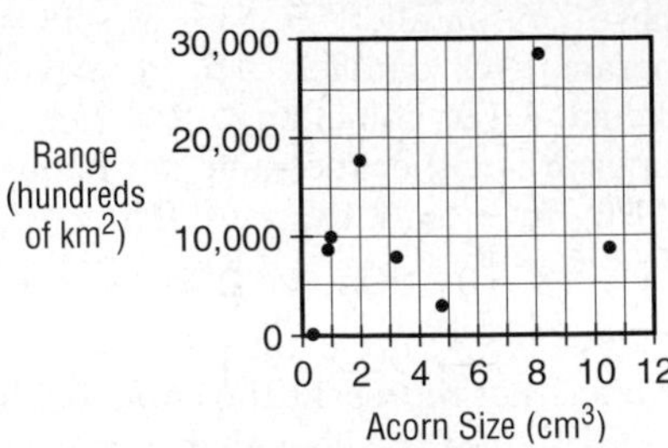

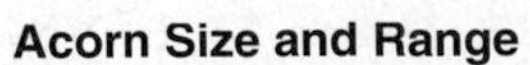

9b. Sample answer: Using (0.3, 233) and (3.4, 7900), $y = 2473.23x - 508.97$ **9c.** $y = 885.82 + 6973.14$; $r \approx 0.38$ **9d.** The correlation value does not show a strong or moderate relationship.

11a.

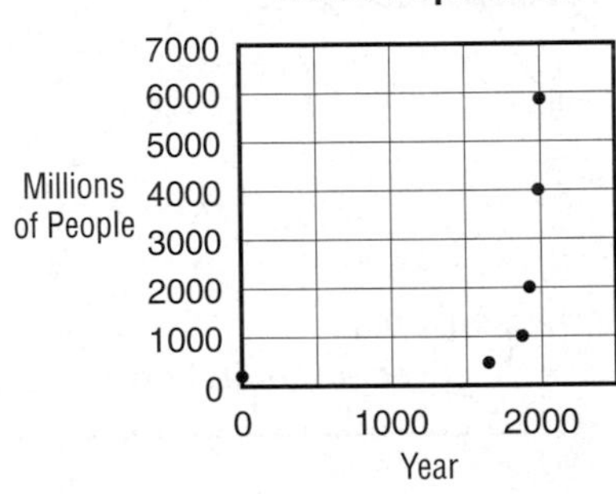

11b. Using (1, 200) and (1998, 5900), $y = 2.85x + 197.14$ **11c.** $y = 1.62x - 277.53$; $r \approx 0.56$ **11d.** 2979 million; No, the correlation value is not showing a very strong relationship. **13.** The rate of growth, which is the slope of the graphs of the regression equations, for the women is less than that of the men's rate of growth. If that trend continues, the men's median salary will always be more than the women's. **15.** $6x + y + 22 = 0$ **17.** $x^3 + 3x^2 + 3x + 1$; $x^3 + 1$ **19.** C

Pages 48–51 Lesson 1-7

5.

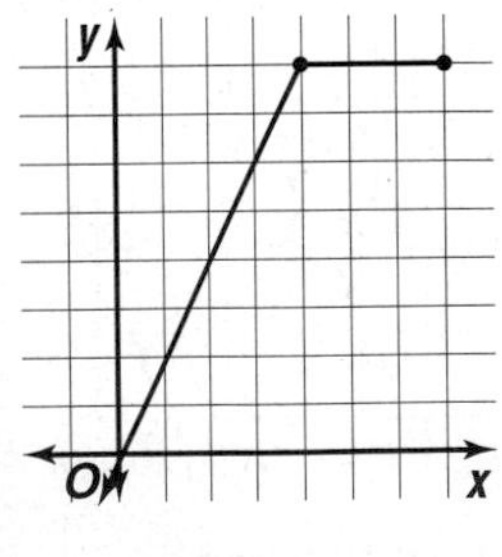

7.

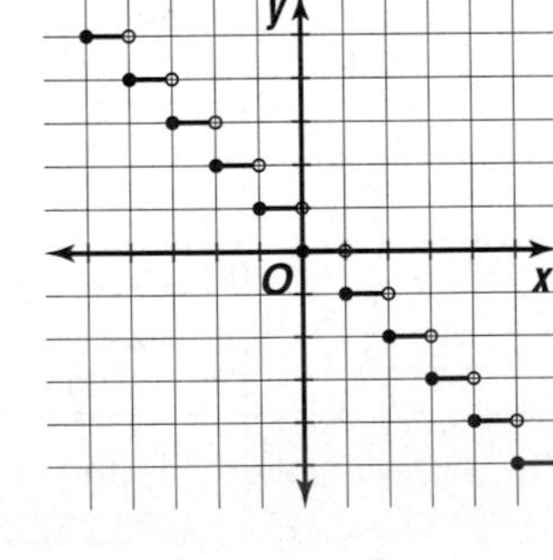

9. greatest integer function; h is hours, $c(h)$ is the cost, $c(h) = \begin{cases} 50h \text{ if } [\![h]\!] = h \\ 50[\![h + 1]\!] \text{ if } [\![h]\!] < h \end{cases}$

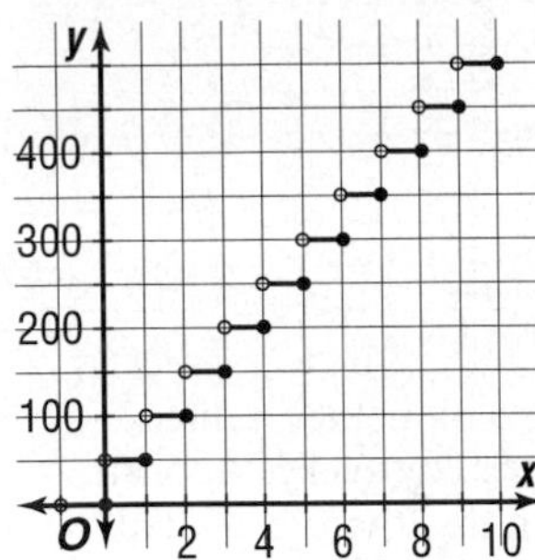

11.

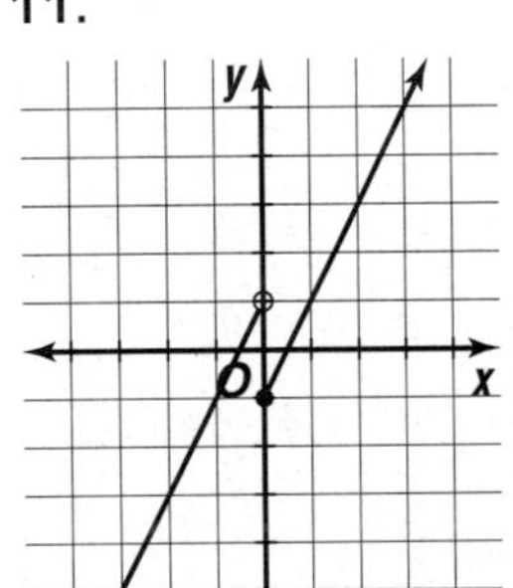

13.

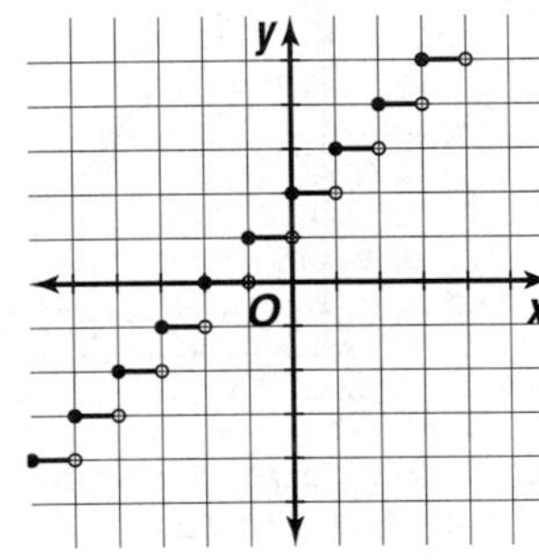

15.

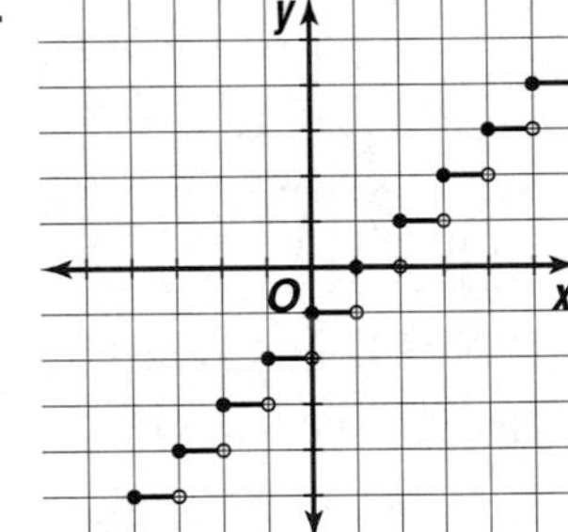

17.

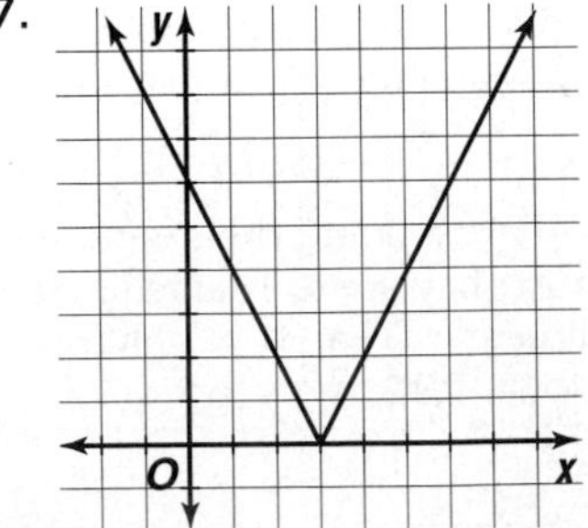

19.

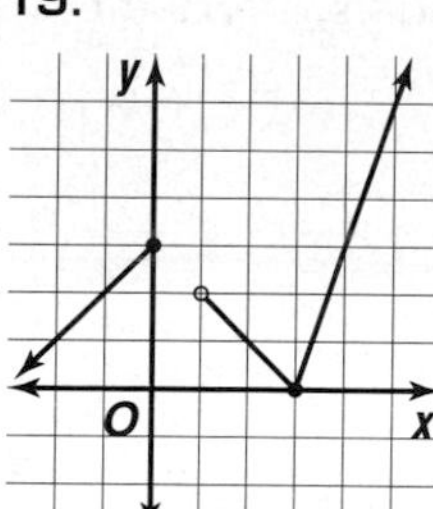

21.

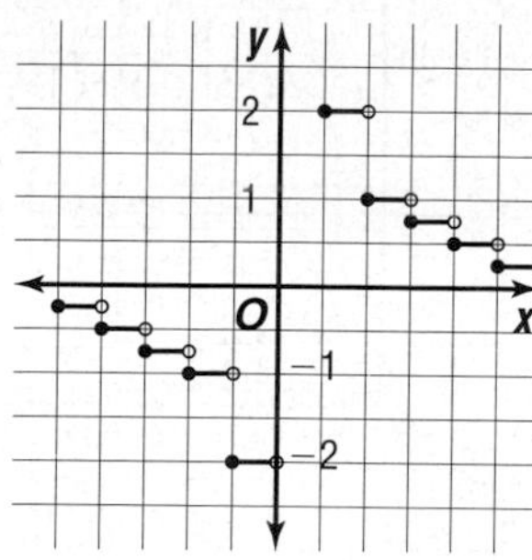

23. step; t is the time in hours, $c(t)$ is the cost in dollars, $c(t) = \begin{cases} 6 \text{ if } t \leq \frac{1}{2} \\ 10 \text{ if } \frac{1}{2} < t \leq 1 \\ 16 \text{ if } 1 < t \leq 2 \\ 24 \text{ if } 2 < t \leq 24 \end{cases}$

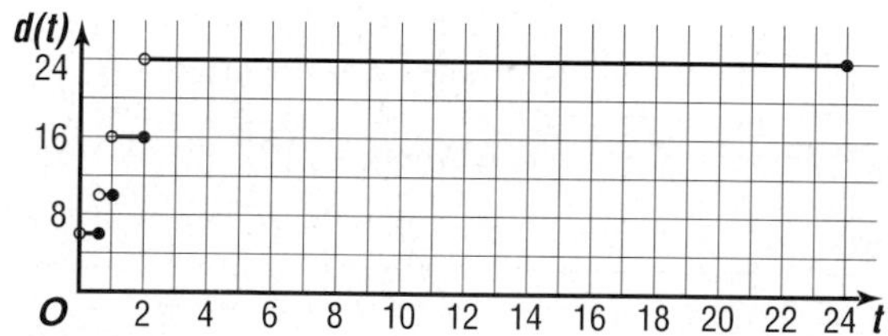

25. w is the weight in pounds, $d(w)$ is the discrepancy, $d(w) = |1 - w|$

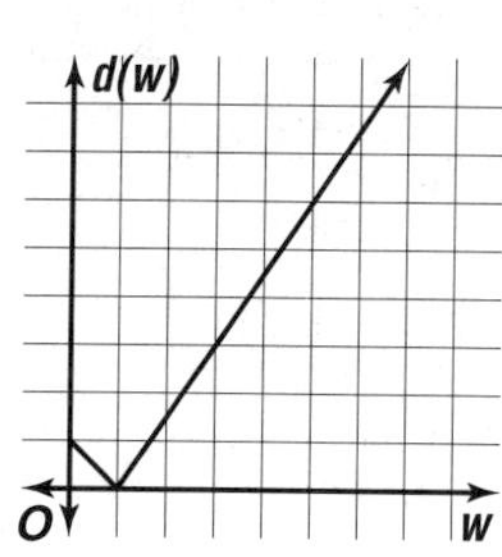

27. If n is any integer, then all ordered pairs (x, y) where x and y are both in the interval $[n, n + 1)$ are solutions. **29a.** step

29b. $t(x) = \begin{cases} 6\% \text{ if } x \leq \$10{,}000 \\ 8\% \text{ if } \$10{,}000 < x \leq \$20{,}000 \\ 9.5\% \text{ if } x > \$20{,}000 \end{cases}$

29c.

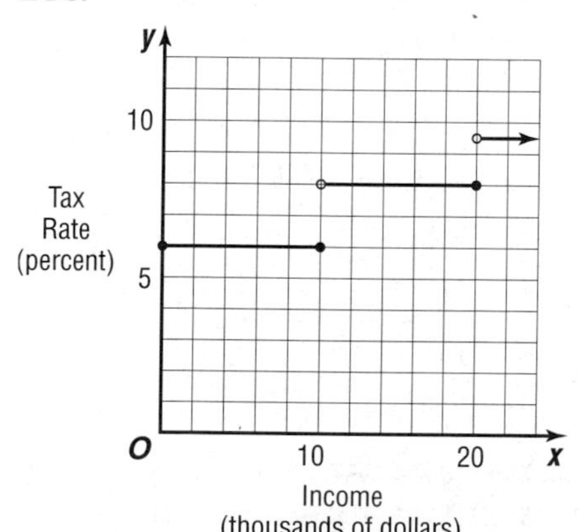

29d. 9.5%

31a.

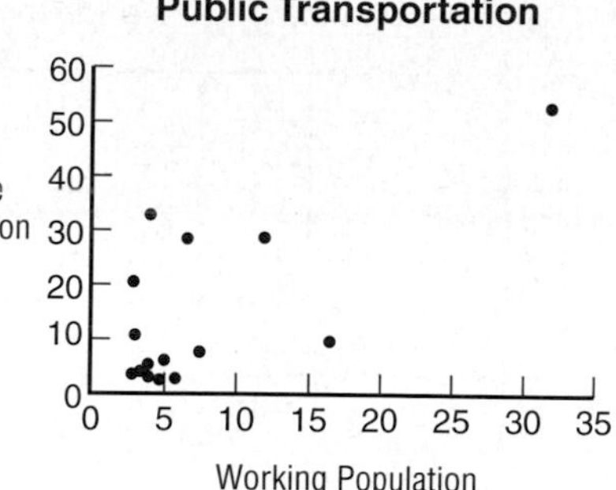

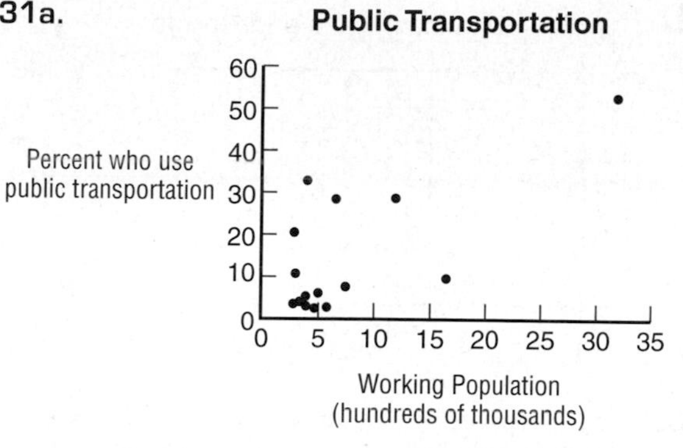

31b. Sample answer: Using (3,183,088, 53.4) and (362,777, 3.3), $y = 0.0000178x - 3.26$ **31c.** $y = 0.0000136x + 4.55$, $r \approx 0.68$ **31d.** 8.73%; No, the actual value is 22%. **33a.** (39, 29), (32, 15) **33b.** 2 **33c.** The average number of points scored each minute. **35.** $47.92 **37.** A

Pages 55–56 Lesson 1-8

5.

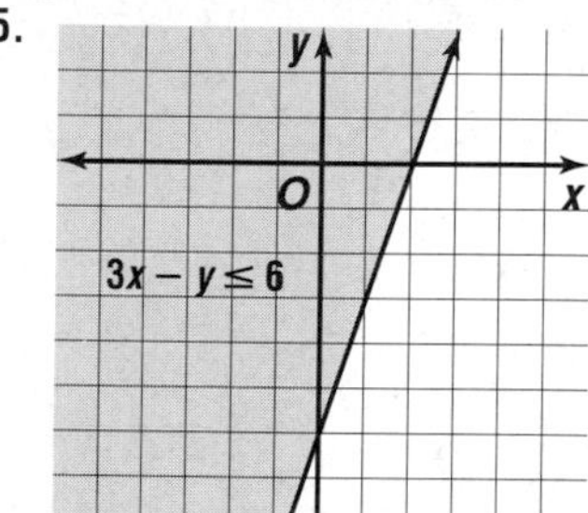

7.

9.

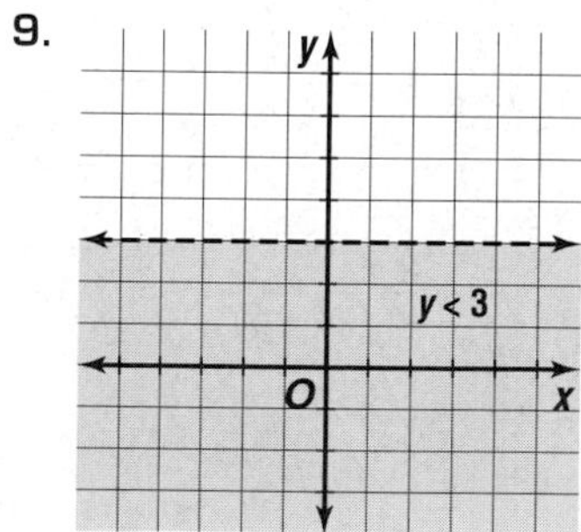

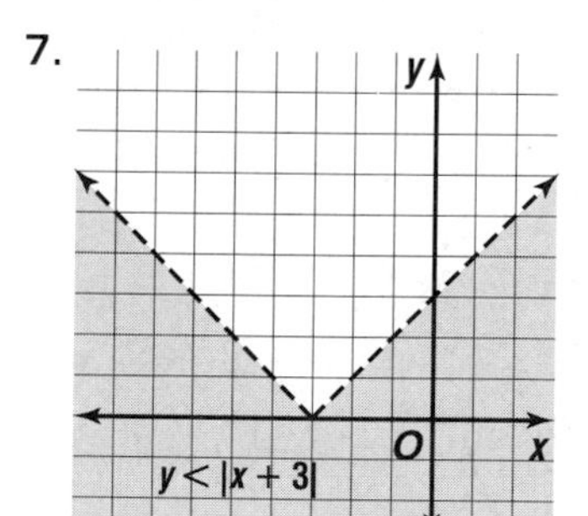

SELECTED ANSWERS

11.

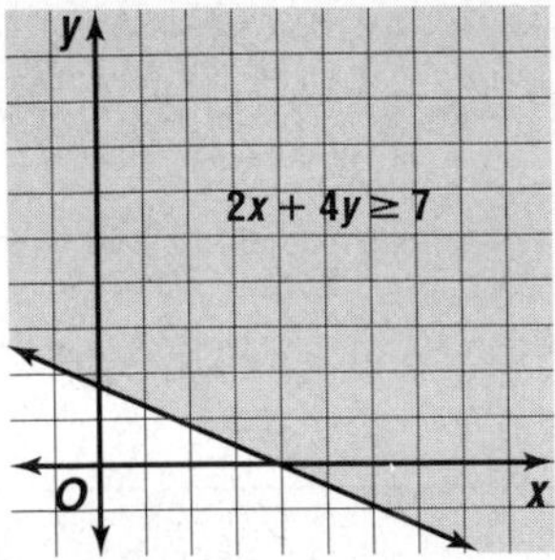

13.

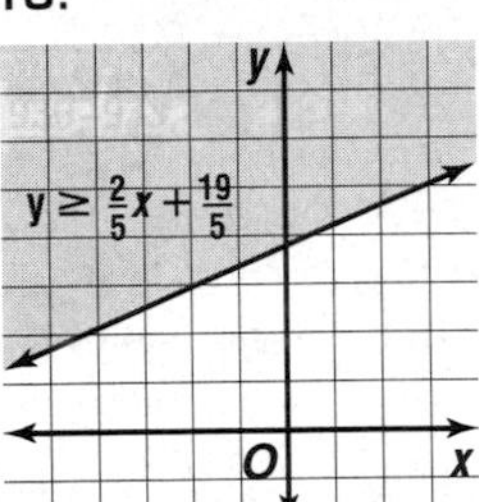

15.

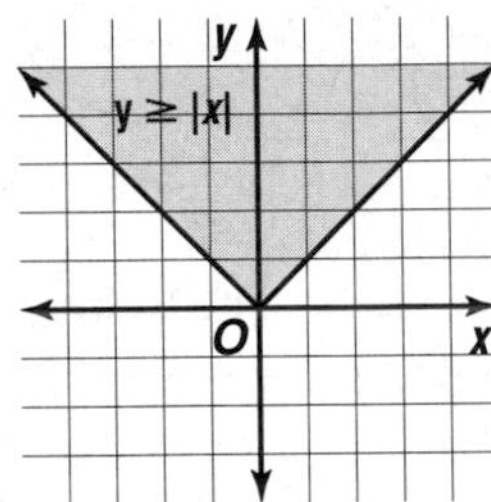

17.

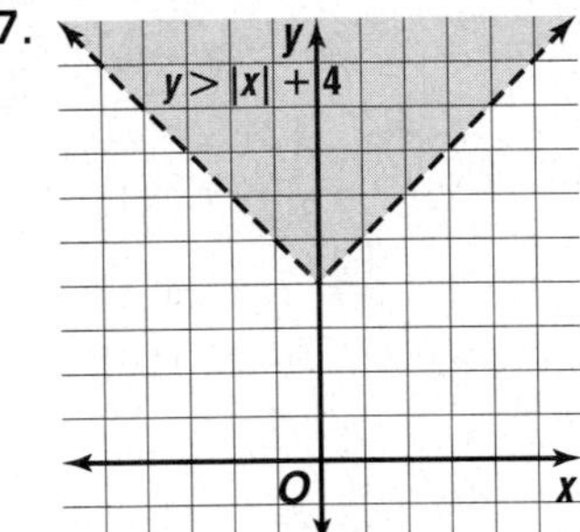

19.

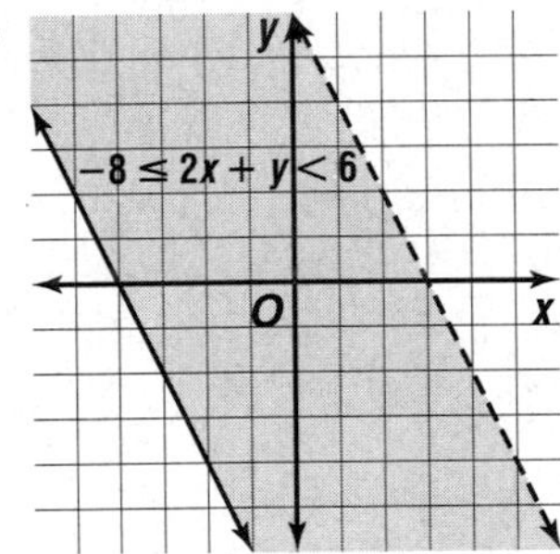

21.

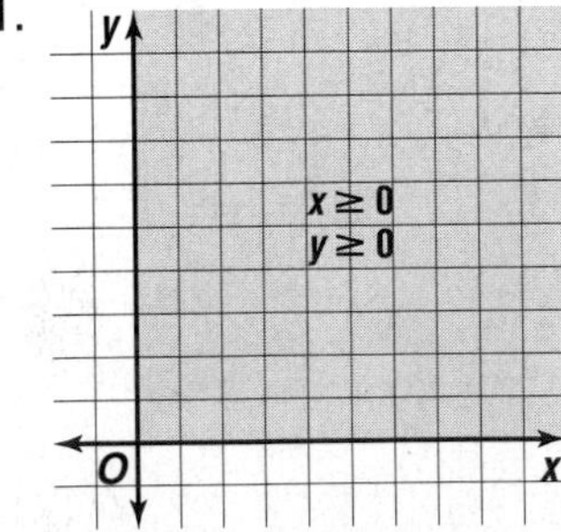

23a. $8x + 10y \le 480$

23b.

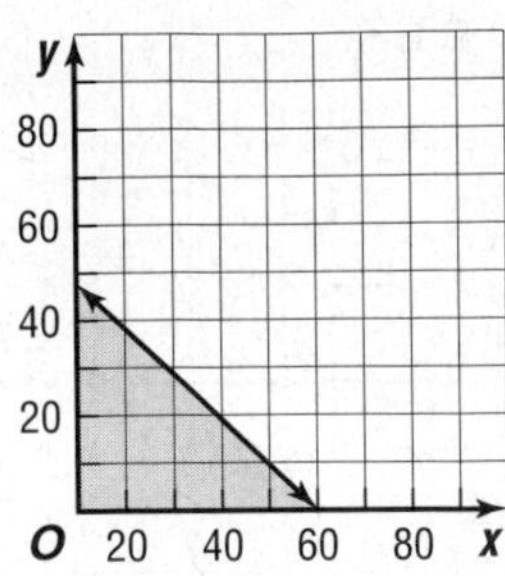

23c. Sample answer: (0, 48), (60, 0), (45, 6) **23d.** Sample answer: Using complex computer programs and systems of inequalities. **25a.** points in the first and third quadrants **25b.** If x and y satisfy the inequality, then either $x \ge 0$ and $y \ge 0$ or $x \le 0$ and $y \le 0$. If $x \ge 0$ and $y \ge 0$, then $|x| = x$ and $|y| = y$. Thus, $|x| + |y| = x + y$. Since $x + y$ is positive, $|x + y| = x + y$. If $x \le 0$ and $y \le 0$, then $|x| = -x$ and $|y| = -y$. Then $|x| + |y| = -x + (-y)$ or $-(x + y)$. Since both x and y are negative, $(x + y)$ is negative, and $|x + y| = -(x + y)$.
27a. $0.6(220 - a) \le r \le 0.9(220 - a)$

27b.

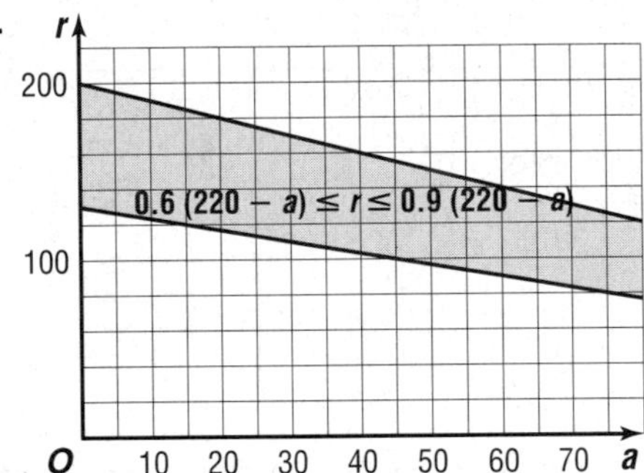

29a. $3x - y - 2 = 0$ **29b.** $x + 3y + 6 = 0$
31a. (0, 23), (16, 48); 1.5625 **31b.** the average change in the temperature per hour

Pages 57–61 Chapter 1 Study Guide and Assessment

1. c **3.** d **5.** i **7.** h **9.** e **11.** 10 **13.** 57
15. $\frac{6}{5}$ **17.** $|m^2 + 5m + 4|$ **19.** $x^2 + 5x - 2$; $x^2 + 3x + 2$; $x^3 + 2x^2 - 8x$; $\frac{x^2 + 4x}{x - 2}, x \ne 2$
21. $x^2 + 8x + 16$; $x^2 + 6x + 8$; $x^3 + 11x^2 + 40x + 48$; $x + 3, x \ne -4$ **23.** $\frac{x^3 - 8x^2 + 16x + 4}{x - 4}$, $x \ne 4$; $\frac{x^3 - 8x^2 + 16x - 4}{x - 4}$, $x \ne 4$; $4x, x \ne 4$; $\frac{x^3 - 8x^2 + 16x}{4}, x \ne 4$ **25.** $1.5x^2 + 5$; $0.75x^2 + 15x + 75$ **27.** $x^2 - x + 7$; $x^2 + 11x + 31$
29. $-2x^2 - 7$; $2x^2 - 12x + 28$

31.

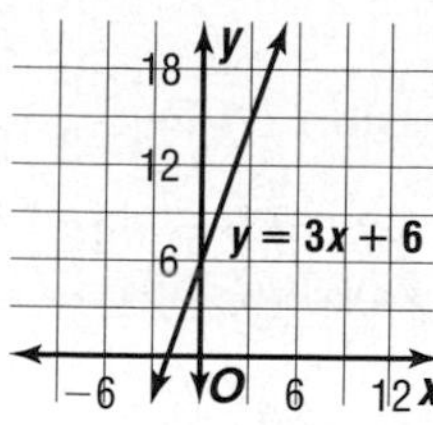

33.

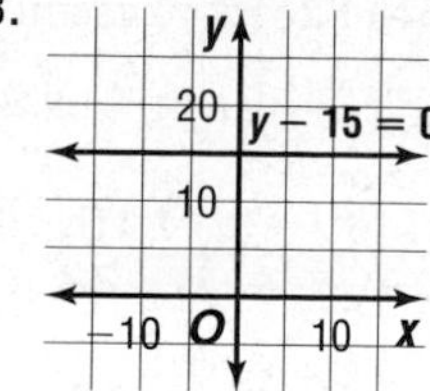

35.

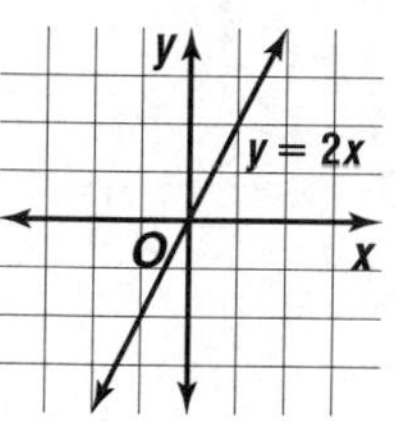

37.

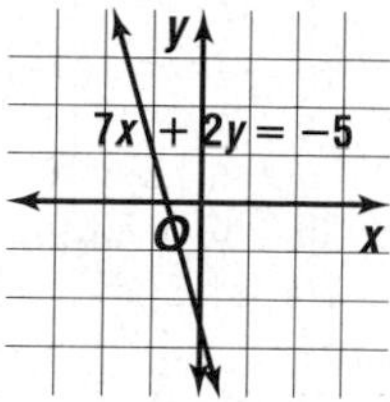

39. $y = 2x - 3$ **41.** $y = \frac{1}{2}x + \frac{9}{2}$ **43.** $y = 4x - 4$ **45.** $y = 0$ **47.** $x - y = 0$ **49.** $2x + y + 4 = 0$ **51.** $x + 2y - 9 = 0$

53a.

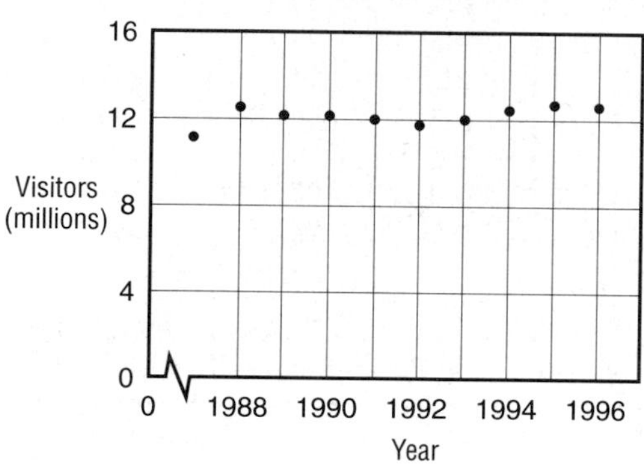

53b. Sample answer: Using (1987, 10,434) and (1996, 12,909), $y = 275x - 535{,}991$ **53c.** $y = 147.8x - 282{,}157.4$; $r \approx 0.61$ **53d.** 14,181,600 visitors; Sample answer: This is not a good prediction, because the r-value does not indicate a strong relationship.

55.

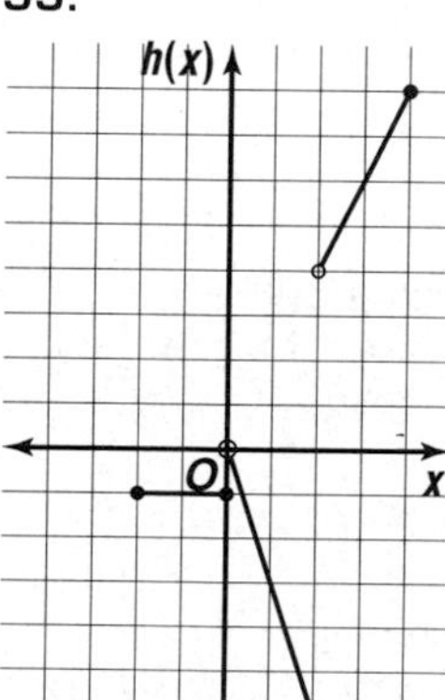

57.

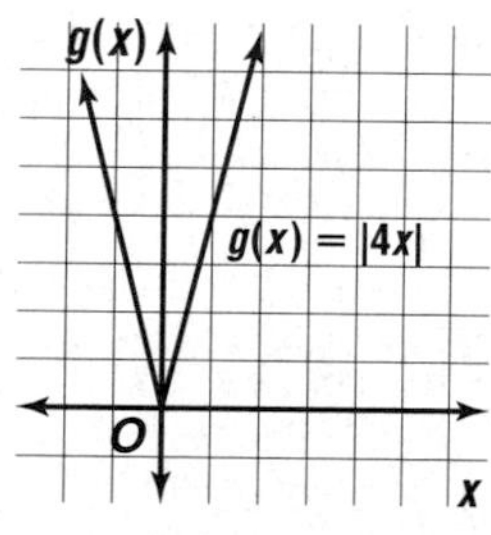

59.

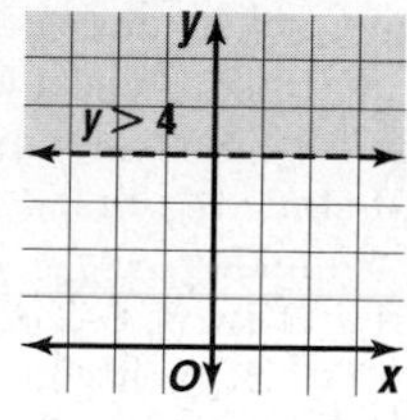

61.

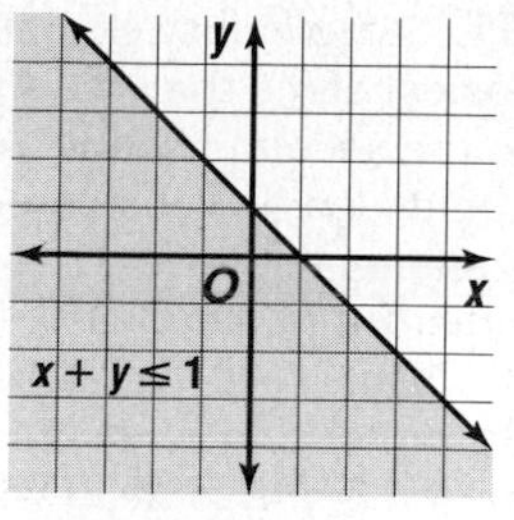

63.

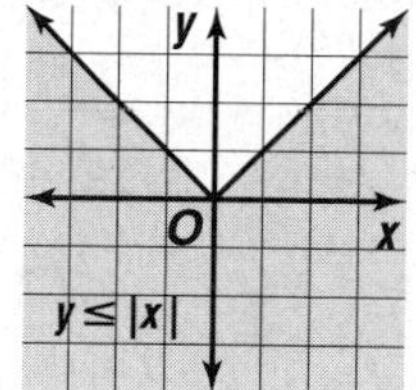

65.

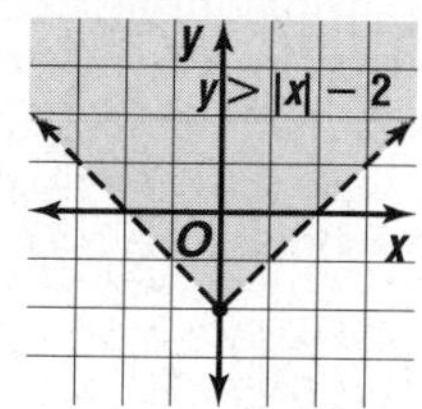

67a. 10 m, 40 m, 90 m, 160 m, 250 m **67b.** Yes, each element of the domain is paired with exactly one element of the range. **69.** $y = -0.284x + 12.964$; The correlation is moderately negative, so the regression line is somewhat representative of the data.

Page 65 Chapter 1 SAT and ACT Practice
1. D **3.** B **5.** A **7.** A **9.** C

Chapter 2 Systems of Equations and Inequalities

Pages 70–72 Lesson 2-1
5. (1, 3)

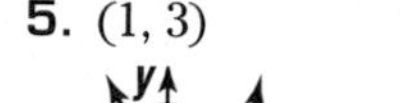
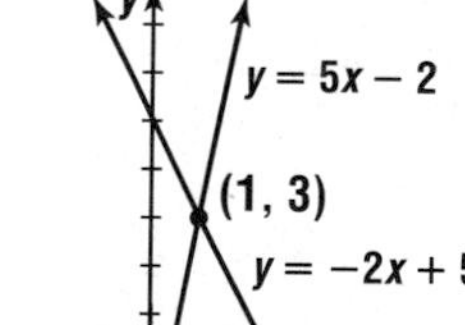
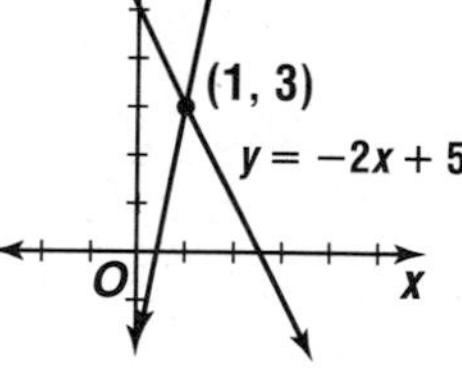

7. (2, −5) **9.** (6, 4) **11.** consistent and independent **13.** consistent and dependent **15.** (4, −3)

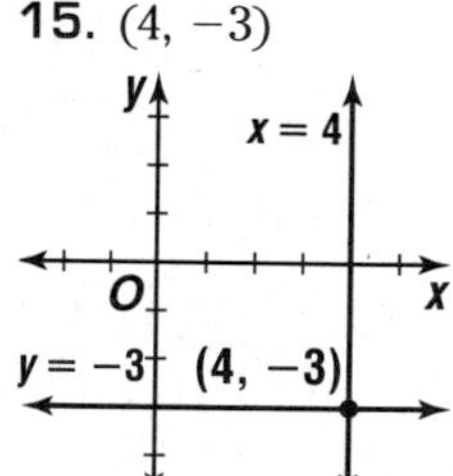

17. no solution

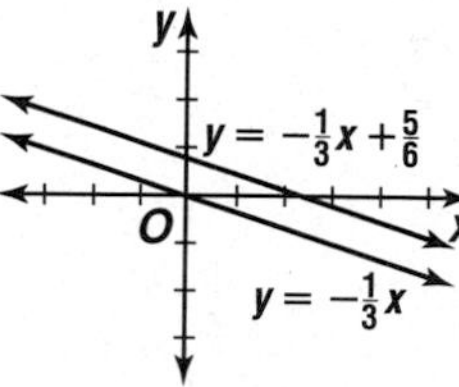

19. (0, 3)

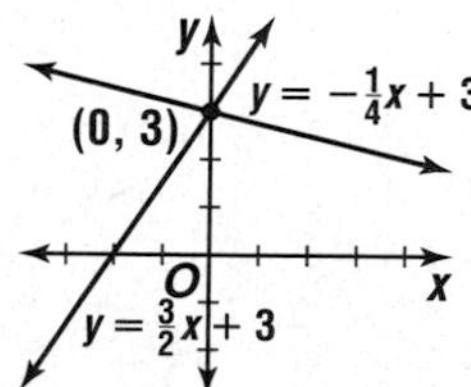

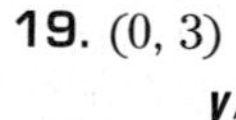

21. (3, 21) **23.** (5.25, 0.75) **25.** (5, 2) **27.** $\left(\frac{1}{3}, \frac{2}{3}\right)$ **29.** $\left(-\frac{6}{43}, -\frac{64}{43}\right)$

31. Sample answer: Elimination could be considered easiest since the first equation multiplied by 2 added to the second equation eliminates b; Substitution could also be considered easiest since the first equation can be written as $a = b$, making substitution very easy; $(-3, -3)$. **33a.** 6, 6, 8; 6, 6, 8 **33b.** isosceles **35a.** (7, 5.95) **35b.** If you drink 7 servings of soft drink, the price for each option is the same. If you drink fewer than 7 servings of soft drink during that week, the disposable cup price is better. If you drink more than 7 servings of soft drink, the refillable mug price is better. **35c.** Over a year's time, the refillable mug would be more economical. **37.** $1500

39.

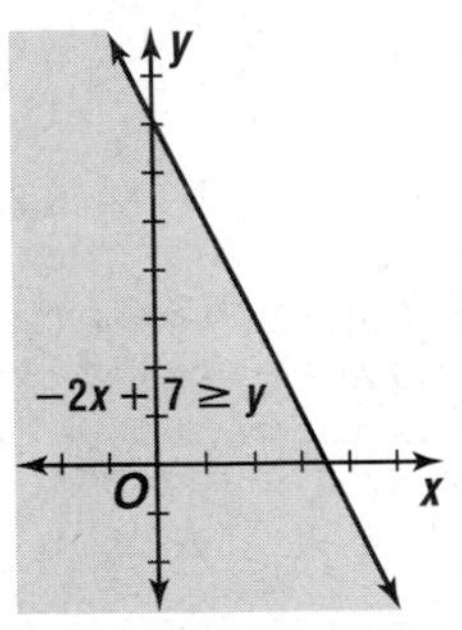

41. $y = 2x + 6$
43. $3x + 1$
45. A

Pages 76–77 Lesson 2-2

5. $(7, -1, 1)$ **7.** acceleration: -32 ft/s^2, initial velocity: 56 ft/s, initial height: 35 ft **9.** $(-2, 2, 4)$ **11.** $(-6, -4, 7)$ **13.** no solution **15.** $(11, -17, 14)$ **17.** $(-4, 10, 7)$ **19.** International Fund = \$1200; Fixed Assets Fund = \$200; company stock = \$600 **21.** $(-32, 138, 2)$ **23.** $(1, 1, 1), (-2, -2, -2)$

25.

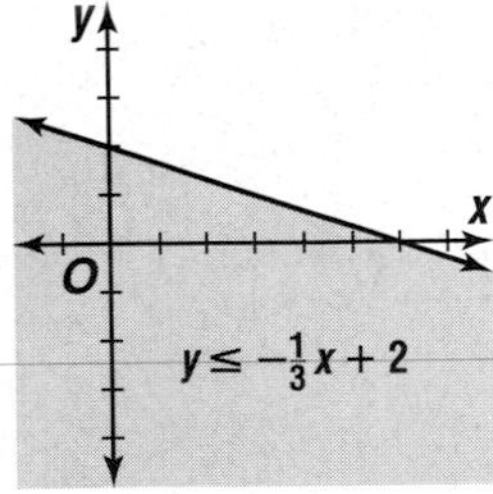

27a. $C(x) = 50x + 2000$
27b. \$2000, \$50

27c.

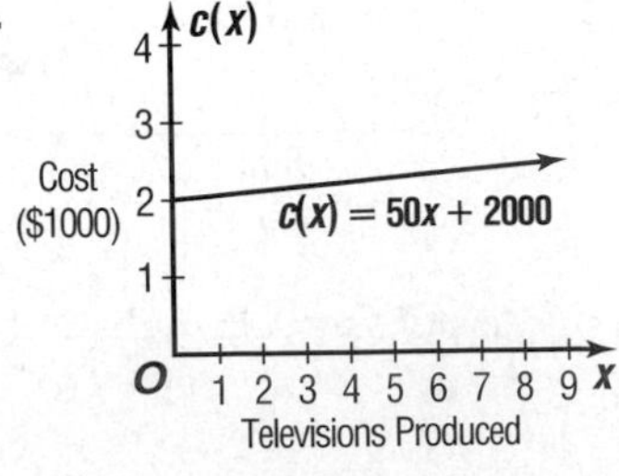

Pages 82–86 Lesson 2-3

5. (7, 2) **7.** (4, 0) **9.** impossible **11.** $\begin{bmatrix} 16 & 4 \\ -8 & 24 \end{bmatrix}$

13. $\begin{bmatrix} 6 & -18 \end{bmatrix}$ **15.** (6, 11) **17.** (5, 2.5) **19.** (7, 9) **21.** $(-5, -15)$ **23.** $(-1, 1)$ **25.** (5, 3, 2)

27. $\begin{bmatrix} 8 & 12 \\ -7 & 9 \end{bmatrix}$ **29.** impossible **31.** $\begin{bmatrix} -2 & -2 \\ 5 & 7 \end{bmatrix}$

33. $\begin{bmatrix} 0 & 4 & 8 \\ -8 & 12 & 0 \\ 16 & 16 & -8 \end{bmatrix}$ **35.** $\begin{bmatrix} -14 & 3 & -2 \\ -2 & 3 & 5 \end{bmatrix}$

37. $\begin{bmatrix} 25 & 35 \\ -30 & 5 \end{bmatrix}$ **39.** impossible

41. $\begin{bmatrix} 16 & 4 & 12 \\ -22 & -14 & 16 \end{bmatrix}$ **43.** $\begin{bmatrix} 10 & -13 & -10 \\ -5 & 14 & -3 \end{bmatrix}$

45. $\begin{bmatrix} -42 & 86 & -160 \\ -421 & 213 & -111 \end{bmatrix}$ **47.** $\begin{bmatrix} 78 & 30 & 12 \\ 12 & -120 & 168 \\ 72 & 90 & -72 \end{bmatrix}$

49. Sample answer:

	1996	2000	2006
18 to 24	8485	8526	8695
25 to 34	10,102	9316	9078
35 to 44	8766	9039	8433
45 to 54	6045	6921	7900
55 to 64	2444	2741	3521
65 and older	2381	2440	2572

51a. $a = 1, b = 0, c = 0, d = 1$ **51b.** a matrix equal to the original one **53.** The numbers in the first row are the triangular numbers. If you look at the diagonals in the matrix, the triangular numbers are the end numbers. To find the diagonal that contains 2001, find the smallest triangular number that is greater than or equal to 2001. The formula for the nth triangular number is $\frac{n(n+1)}{2}$. Solve $\frac{n(n+1)}{2} \geq 2001$. The solution is 63. So the 63rd entry in the first row is $\frac{63(63+1)}{2} = 2016$. Since $2016 - 2001 = 15$, we must count 15 places backward along the diagonal to locate 2001 in the matrix. This movement takes us from the position (row, column) = (1, 63) to $(1 + 15, 63 - 15) = (16, 48)$. **55.** $\left(\frac{1}{2}, \frac{1}{3}, \frac{1}{4}\right)$

57.

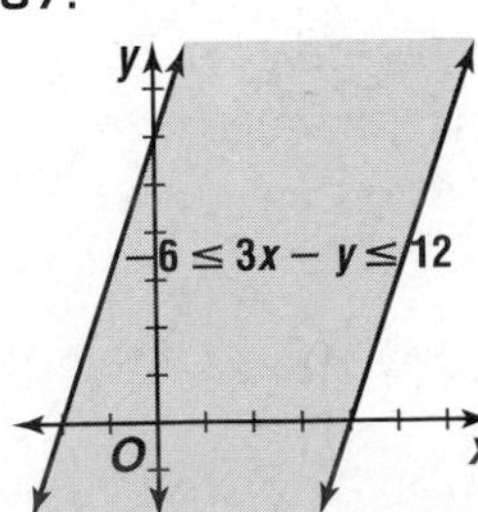

59. Sample answer: $y = 0.36x + 61.4$ **61.** $\frac{3}{5}$ **63.** -2656

Pages 93–96 Lesson 2-4

5. $J'(-3, 7.5)$, $K'(1.5, 4.5)$, $L'(0, -3)$

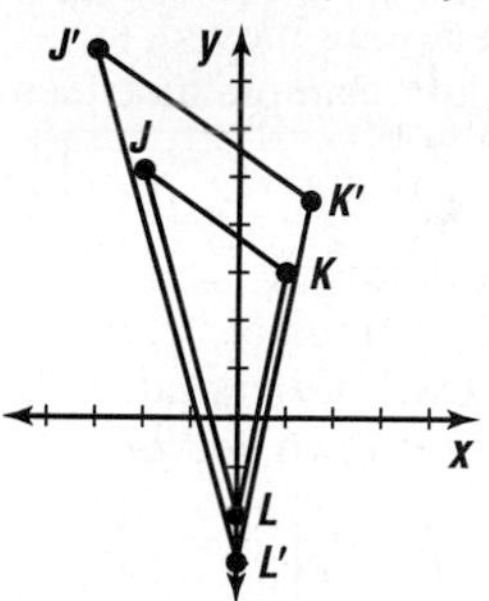

7. $A'(1, 2)$, $B'(4, 1)$, $C'(3, -2)$, $D'(0, -1)$

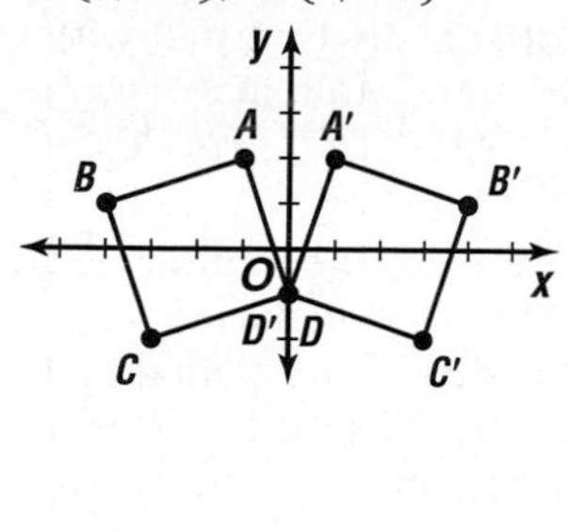

9. $L'(-6, -4)$, $M'(-3, -2)$, $N'(-1, 2)$

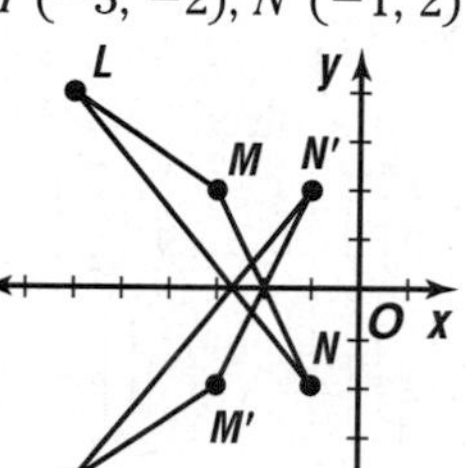

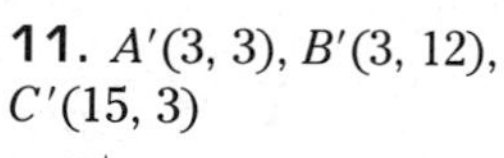

11. $A'(3, 3)$, $B'(3, 12)$, $C'(15, 3)$

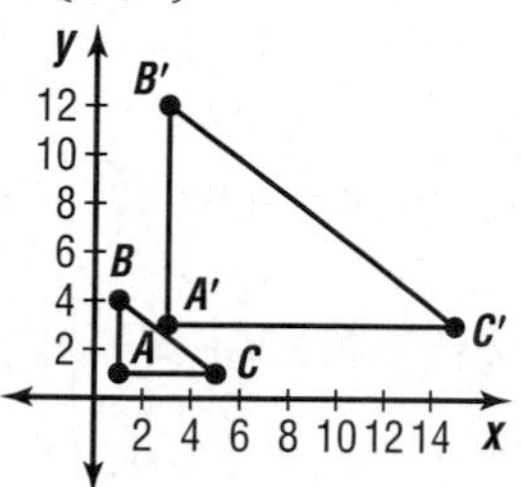

13. $P'(-6, 0)$, $Q'(-4, 4)$, $R'(2, 6)$, $S'(8, 4)$

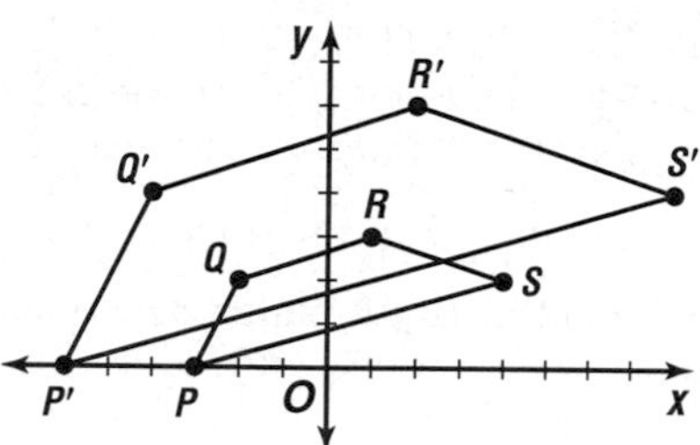

15. $W'(1, -2)$, $X'(4, 3)$, $Y'(6, -3)$

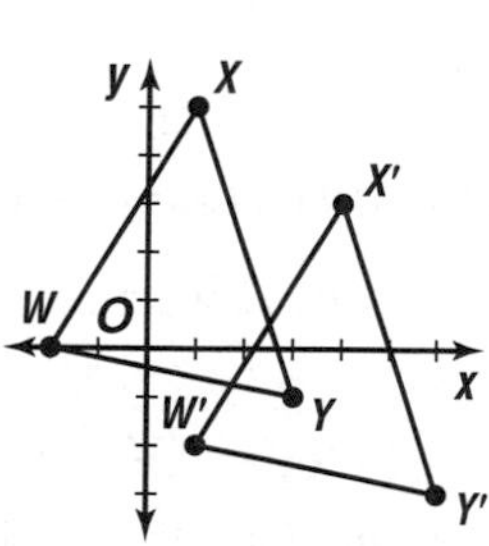

17. $C'(0, 5)$, $D'(4, 9)$, $E'(8, 5)$, $F'(4, 1)$

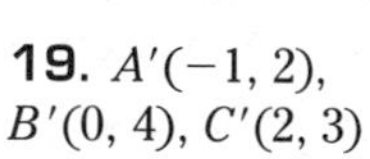

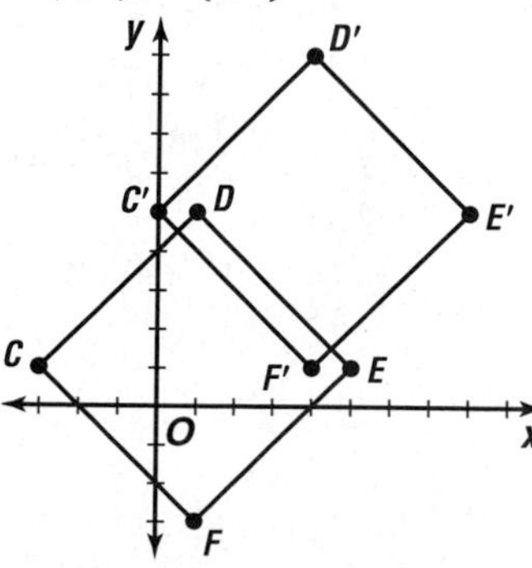

19. $A'(-1, 2)$, $B'(0, 4)$, $C'(2, 3)$

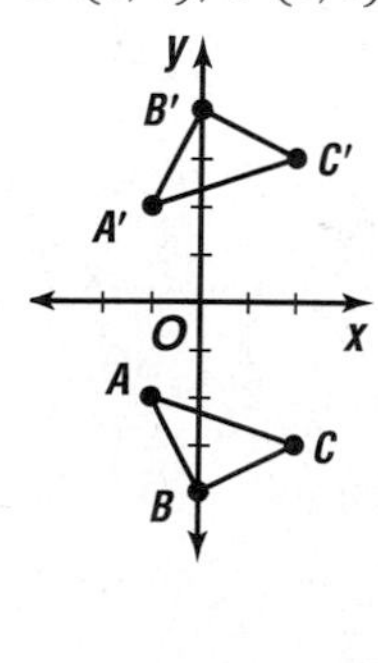

21. $H(-2, -1)$, $I'(1, -3)$, $J'(5, -1)$, $K'(4, 2)$

23. $O'(0, 0)$, $P'(-4, 0)$, $Q'(-4, -4)$, $R'(0, -4)$

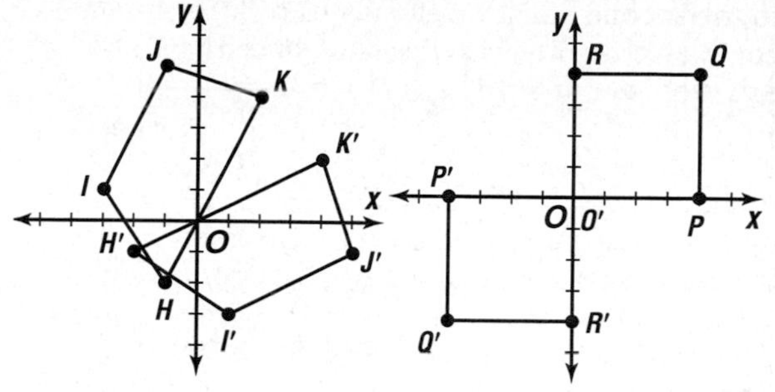

25a. Let $\begin{bmatrix} a & b \\ c & d \end{bmatrix} = R_{x\text{-axis}}$.

$$\begin{bmatrix} a & b \\ c & d \end{bmatrix} \cdot \begin{bmatrix} 1 & -2 & -1 \\ 3 & -1 & -3 \end{bmatrix} = \begin{bmatrix} 1 & -2 & -1 \\ -3 & 1 & 3 \end{bmatrix}$$

$$\begin{bmatrix} a + 3b & -2a - b & -a - 3b \\ c + 3d & -2c - d & -c - 3d \end{bmatrix} = \begin{bmatrix} 1 & -2 & -1 \\ -3 & 1 & 3 \end{bmatrix}$$

$a + 3b = 1$ $\quad -2a - b = -2$ $\quad -a - 3b = -1$

$c + 3d = -3$ $\quad -2c - d = 1$ $\quad -c - 3d = 3$

Thus, $a = 1$, $b = 0$, $c = 0$, and $d = -1$. By substitution, $R_{x\text{-axis}} = \begin{bmatrix} 1 & 0 \\ 0 & -1 \end{bmatrix}$.

25b. Let $\begin{bmatrix} a & b \\ c & d \end{bmatrix} = R_{y\text{-axis}}$.

$$\begin{bmatrix} a & b \\ c & d \end{bmatrix} \cdot \begin{bmatrix} 1 & -2 & -1 \\ 3 & -1 & -3 \end{bmatrix} = \begin{bmatrix} -1 & -2 & 1 & 1 \\ 3 & & & -3 \end{bmatrix}$$

$$\begin{bmatrix} a + 3b & -2a - b & -a - 3b \\ c + 3d & -2c - d & -c - 3d \end{bmatrix} = \begin{bmatrix} -1 & 2 & 1 \\ 3 & -1 & -3 \end{bmatrix}$$

$a + 3b = -1$ $\quad -2a - b = 2$ $\quad -a - 3b = 1$

$c + 3d = 3$ $\quad -2c - d = -1$ $\quad -c - 3d = -3$

Thus, $a = -1$, $b = 0$, $c = 0$, and $d = 1$. By substitution, $R_{y\text{-axis}} = \begin{bmatrix} -1 & 0 \\ 0 & 1 \end{bmatrix}$.

25c. Let $\begin{bmatrix} a & b \\ c & d \end{bmatrix} = R_{y = x}$.

$$\begin{bmatrix} a & b \\ c & d \end{bmatrix} \cdot \begin{bmatrix} 1 & -2 & -1 \\ 3 & -1 & -3 \end{bmatrix} = \begin{bmatrix} 3 & -1 & -3 \\ 1 & -2 & -1 \end{bmatrix}$$

$$\begin{bmatrix} a + 3b & -2a - b & -a - 3b \\ c + 3d & -2c - d & -c - 3d \end{bmatrix} = \begin{bmatrix} 3 & -1 & -3 \\ 1 & -2 & -1 \end{bmatrix}$$

$a + 3b = 3$ $\quad -2a - b = -1$ $\quad -a - 3b = -3$

$c + 3d = 1$ $\quad -2c - d = -2$ $\quad -c - 3d = -1$

Thus, $a = 0$, $b = 1$, $c = 1$, and $d = 0$. By substitution, $R_{y = x} = \begin{bmatrix} 0 & 1 \\ 1 & 0 \end{bmatrix}$.

25d. Let $\begin{bmatrix} a & b \\ c & d \end{bmatrix} = Rot_{90}$.

$$\begin{bmatrix} a & b \\ c & d \end{bmatrix} \cdot \begin{bmatrix} 1 & -2 & -1 \\ 3 & -1 & -3 \end{bmatrix} = \begin{bmatrix} -3 & 1 & 3 \\ 1 & -2 & -1 \end{bmatrix}$$

$$\begin{bmatrix} a + 3b & -2a - b & -a - 3b \\ c + 3d & -2c - d & -c - 3d \end{bmatrix} = \begin{bmatrix} -3 & 1 & 3 \\ 1 & -2 & -1 \end{bmatrix}$$

$a + 3b = -3 \quad -2a - b = -1 \quad -a - 3b = 3$
$c + 3d = 1 \quad -2c - d = -2 \quad -c - 3d = -1$

Thus, $a = 0$, $b = -1$, $c = 1$, and $d = 0$. By substitution, $Rot_{90} = \begin{bmatrix} 0 & -1 \\ 1 & 0 \end{bmatrix}$.

25e. Let $\begin{bmatrix} a & b \\ c & d \end{bmatrix} = Rot_{180}$.

$$\begin{bmatrix} a & b \\ c & d \end{bmatrix} \cdot \begin{bmatrix} 1 & -2 & -1 \\ 3 & -1 & -3 \end{bmatrix} = \begin{bmatrix} -1 & 2 & 1 \\ -3 & 1 & 3 \end{bmatrix}$$

$$\begin{bmatrix} a + 3b & -2a - b & -a - 3b \\ c + 3d & -2c - d & -c - 3d \end{bmatrix} = \begin{bmatrix} -1 & 2 & 1 \\ -3 & 1 & 3 \end{bmatrix}$$

$a + 3b = -1 \quad -2a - b = 2 \quad -a - 3b = 1$
$c + 3d = -3 \quad -2c - d = 1 \quad -c - 3d = 3$

Thus, $a = -1$, $b = 0$, $c = 0$, and $d = -1$. By substitution, $Rot_{180} = \begin{bmatrix} -1 & 0 \\ 0 & -1 \end{bmatrix}$.

25f. Let $\begin{bmatrix} a & b \\ c & d \end{bmatrix} = Rot_{270}$.

$$\begin{bmatrix} a & b \\ c & d \end{bmatrix} \cdot \begin{bmatrix} 1 & -2 & -1 \\ 3 & -1 & -3 \end{bmatrix} = \begin{bmatrix} 3 & -1 & -3 \\ -1 & 2 & 1 \end{bmatrix}$$

$$\begin{bmatrix} a + 3b & -2a - b & -a - 3b \\ c + 3d & -2c - d & -c - 3d \end{bmatrix} = \begin{bmatrix} 3 & -1 & -3 \\ -1 & 2 & 1 \end{bmatrix}$$

$a + 3b = 3 \quad -2a - b = -1 \quad -a - 3b = -3$
$c + 3d = -1 \quad -2c - d = 2 \quad -c - 3d = 1$

Thus, $a = 0$, $b = 1$, $c = -1$, and $d = 0$. By substitution, $Rot_{270} = \begin{bmatrix} 0 & 1 \\ -1 & 0 \end{bmatrix}$.

27. $J''(6, -4)$, $K''(3, -2)$, $L''(1, 2)$

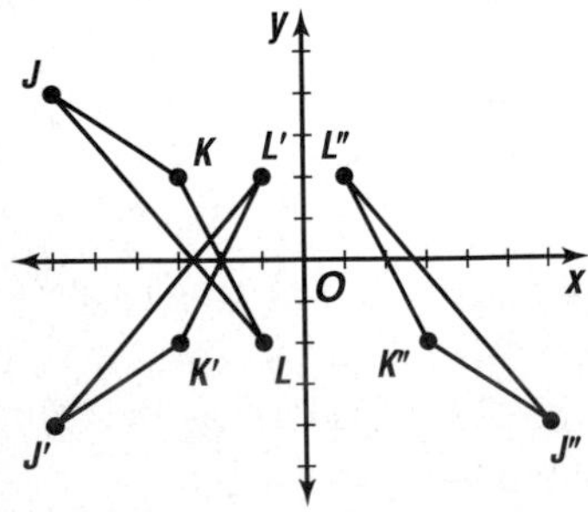

29a. The bishop moves along a diagonal until it encounters the edge of the board or another piece. The line along which it moves changes vertically and horizontally by 1 unit with each square moved, so the translation matrices are scalars. Sample matrices are $c\begin{bmatrix} 1 & 1 \\ 1 & 1 \end{bmatrix}$, $c\begin{bmatrix} 1 & 1 \\ -1 & -1 \end{bmatrix}$, $c\begin{bmatrix} -1 & -1 \\ 1 & 1 \end{bmatrix}$, and $c\begin{bmatrix} -1 & -1 \\ -1 & -1 \end{bmatrix}$, where c is the number of squares moved.

29b. The knight moves in combinations of 2 vertical-1 horizontal or 1 vertical-2 horizontal squares. These can be either up or down, left or right. Sample matrices are $\begin{bmatrix} 1 & 1 \\ 2 & 2 \end{bmatrix}$, $\begin{bmatrix} 1 & 1 \\ -2 & -2 \end{bmatrix}$, $\begin{bmatrix} -1 & -1 \\ 2 & 2 \end{bmatrix}$, $\begin{bmatrix} -1 & -1 \\ -2 & -2 \end{bmatrix}$, $\begin{bmatrix} 2 & 2 \\ 1 & 1 \end{bmatrix}$, $\begin{bmatrix} 2 & 2 \\ -1 & -1 \end{bmatrix}$, $\begin{bmatrix} -2 & -2 \\ 1 & 1 \end{bmatrix}$, and $\begin{bmatrix} -2 & -2 \\ -1 & -1 \end{bmatrix}$.

29c. The king can move 1 unit in any direction. The matrices describing this are $\begin{bmatrix} 1 & 1 \\ 0 & 0 \end{bmatrix}$, $\begin{bmatrix} -1 & -1 \\ 0 & 0 \end{bmatrix}$, $\begin{bmatrix} 0 & 0 \\ 1 & 1 \end{bmatrix}$, $\begin{bmatrix} 0 & 0 \\ -1 & -1 \end{bmatrix}$, $\begin{bmatrix} 1 & 1 \\ 1 & 1 \end{bmatrix}$, $\begin{bmatrix} 1 & 1 \\ -1 & -1 \end{bmatrix}$, $\begin{bmatrix} -1 & -1 \\ 1 & 1 \end{bmatrix}$, and $\begin{bmatrix} -1 & -1 \\ -1 & -1 \end{bmatrix}$.

31. $(0, -125)$; $(125, 0)$, $(0, 125)$, $(-125, 0)$ **33.** The repeated dilations animate the growth of something from small to larger similar to a lens zooming into the origin. **35.** $\begin{bmatrix} 4 & 13 \\ -4 & 12 \end{bmatrix}$

37. hardbacks \$1, paperbacks \$0.25

39. $4x - y + 9 = 0$ **41.** $x^5 - 3x^4 + 7x^3$, $\dfrac{x^3}{x^2 - 3x + 7}$

Pages 102–105 Lesson 2-5

5. 10 **7.** -413 **9.** $-\frac{1}{29}\begin{bmatrix} 7 & -3 \\ -5 & -2 \end{bmatrix}$

11. $\left(-\frac{111}{13}, \frac{129}{13}\right)$ **13.** 8 kg of the metal with 55% aluminum and 12 kg of the metal with 80% aluminum

15. 4 **17.** 4 **19.** 48 **21.** -37 **23.** 1

25. 175.668 **27.** $-\frac{1}{10}\begin{bmatrix} -2 & 3 \\ 2 & 2 \end{bmatrix}$ **29.** $\frac{1}{6}\begin{bmatrix} 2 & -2 \\ -1 & 4 \end{bmatrix}$

31. does not exist **33.** $\begin{bmatrix} \frac{1}{2} & \frac{1}{8} \\ -5 & \frac{3}{4} \end{bmatrix}$ **35.** $(0, -2)$

37. $\left(\frac{7}{12}, \frac{7}{12}\right)$ **39.** $\left(\frac{1}{3}, -\frac{2}{3}\right)$ **41.** $\left(\frac{2}{9}, -\frac{4}{3}, -\frac{1}{3}\right)$

43. 30,143 **45.** $(2, -1, 3)$

47. Let $A = \begin{bmatrix} a_1 & b_1 \\ a_2 & b_2 \end{bmatrix}$ and $I = \begin{bmatrix} 1 & 0 \\ 0 & 1 \end{bmatrix}$.

$$A^{-1} = \begin{bmatrix} \frac{b_2}{a_1b_2 - a_2b_1} & \frac{-b_1}{a_1b_2 - a_2b_1} \\ \frac{-a_2}{a_1b_2 - a_2b_1} & \frac{a_1}{a_1b_2 - a_2b_1} \end{bmatrix}$$

$$AA^{-1} = \begin{bmatrix} \frac{a_1b_2 - a_2b_1}{a_1b_2 - a_2b_1} & \frac{-a_1b_1 + b_1a_1}{a_1b_2 - a_2b_1} \\ \frac{a_2b_2 - a_2b_2}{a_1b_2 - a_2b_1} & \frac{a_1b_2 - a_2b_1}{a_1b_2 - a_2b_1} \end{bmatrix}$$

$$= \begin{bmatrix} 1 & 0 \\ 0 & 1 \end{bmatrix} = I$$

Thus, $AA^{-1} = I$.

49. Yes

$A = \begin{bmatrix} a & b \\ c & d \end{bmatrix}$. Does $(A^2)^{-1} = (A^{-1})^2$?

$$A^2 = \begin{bmatrix} a^2 + bc & ab + bd \\ ac + cd & bc + d^2 \end{bmatrix}$$

$$(A^2)^{-1} = \frac{1}{a^2d^2 - 2abcd + b^2d^2}\begin{bmatrix} bc + d^2 & -ab - bd \\ -ac - cd & a^2 + bc \end{bmatrix}$$

$$A^{-1} = \frac{1}{ad - bc}\begin{bmatrix} d & -b \\ -c & a \end{bmatrix} = \begin{bmatrix} \frac{d}{ad - bc} & \frac{-b}{ad - bc} \\ \frac{-c}{ad - bc} & \frac{a}{ad - bc} \end{bmatrix}$$

$$(A^{-1})^2 = \frac{1}{a^2d^2 - 2abcd + b^2d^2}\begin{bmatrix} bc + d^2 & -ab - bd \\ -ac - cd & a^2 + bc \end{bmatrix}$$

Thus, $(A^2)^{-1} = (A^{-1})^2$.

51. computer system: \$959, printer: \$239

53. $H'(5, 9)$, $I'(1, 5)$, $J'(-3, 9)$, $K'(1, 13)$

55. infinitely many solutions 57. $x - 2y + 8 = 0$

59a. $\frac{1}{12}$ or approximately 0.0833 59b. 1.5 ft

61. No, more than one member of the range is paired with the same member of the domain.

Pages 109–111 Lesson 2-6

5. $(-1, 0)$, $(-1, 3)$, $(0, 4)$, $(7, 0.5)$, $(7, 0)$

7. 3, −11

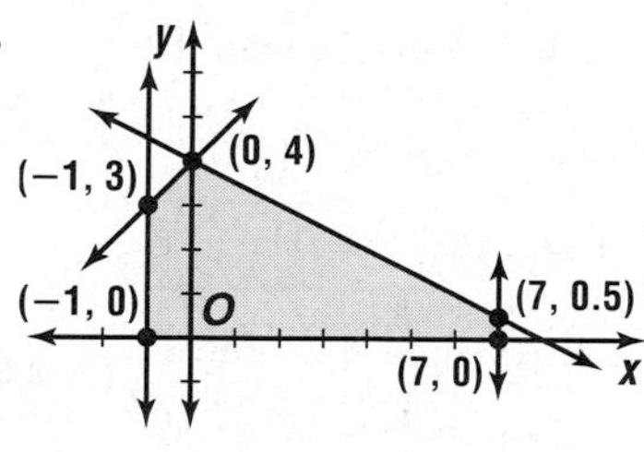

9.

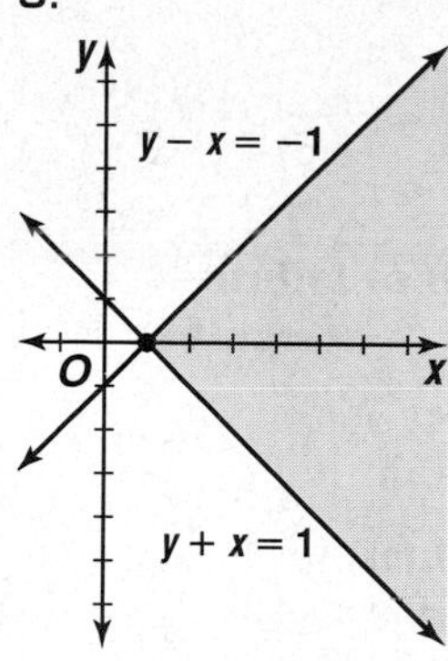

11.

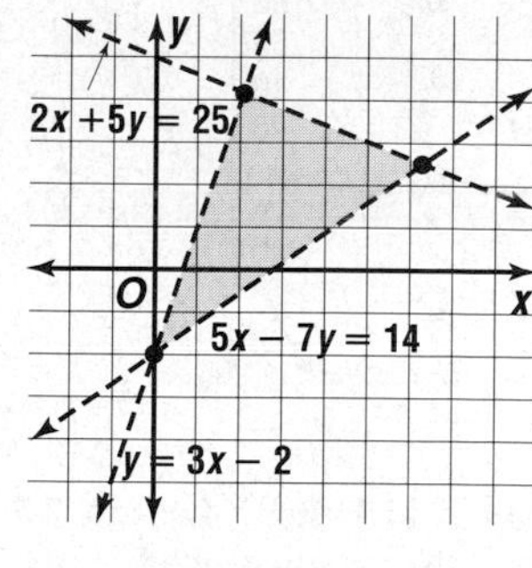

13. $\left(\frac{3}{5}, 3\frac{1}{5}\right)$, $\left(-\frac{2}{5}, 1\frac{1}{5}\right)$, $\left(1\frac{3}{5}, \frac{1}{5}\right)$

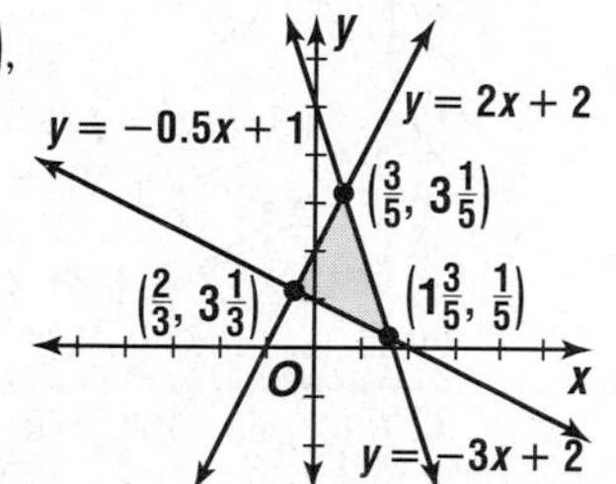

15. (2, 5), (7, 0), (4, 0), (1, 5)

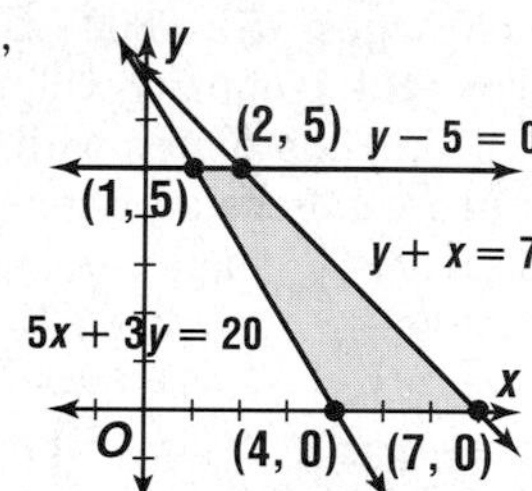

17. 19, 2 19. 16, 2 21. 9, −4 23. $x \le 4$, $x \ge -4$, $y \le 4$, $y \ge -4$ 25a. vertices: $\left(5\frac{1}{2}, 0\right)$, $\left(6\frac{1}{2}, 0\right)$, $\left(9\frac{2}{3}, 6\frac{1}{3}\right)$, $\left(7\frac{1}{2}, 8\frac{1}{2}\right)$, $\left(2\frac{1}{2}, 8\frac{1}{2}\right)$, $\left(1\frac{1}{5}, 4\frac{3}{5}\right)$, $\left(2\frac{1}{2}, 2\right)$ 25b. max at $\left(7\frac{1}{2}, 8\frac{1}{2}\right) = 88\frac{1}{2}$; min at $\left(2\frac{1}{2}, 2\right) = 24\frac{1}{2}$

27a.

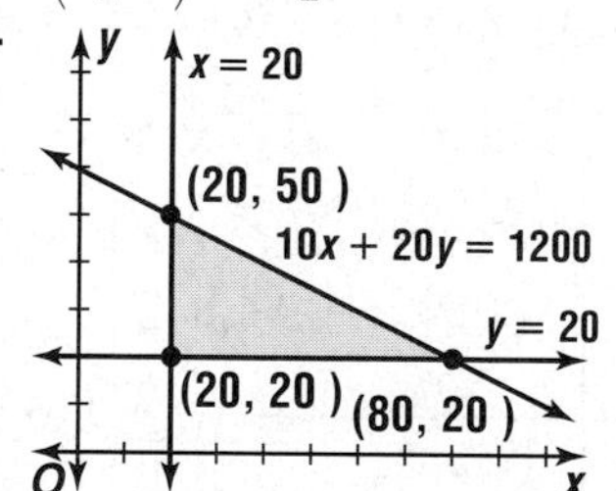

27b. $f(x, y) = 30x + 40y$ 27c. 80 ft^2 at the Main St. site and 20 ft^2 at the High St. site 27d. The maximum number of customers can be reached by renting 120 ft^2 at Main St. 29. $\frac{1}{7}\begin{bmatrix} 2 & -1 \\ 3 & 2 \end{bmatrix}$

31.

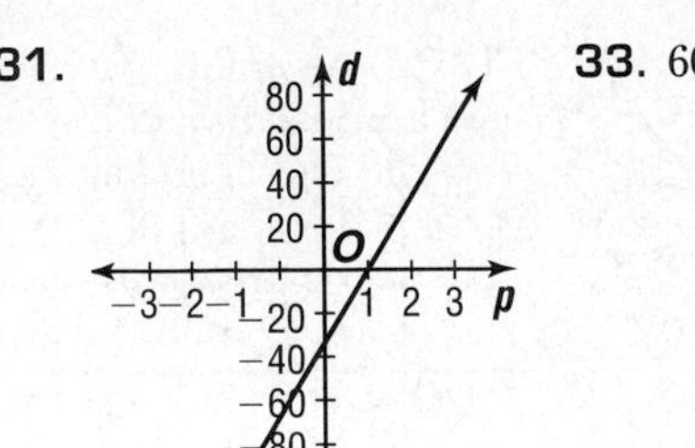

33. 60

Pages 115–118 Lesson 2-7

5a. $25x + 50y \leq 4200$ **5b.** $3x + 5y \leq 480$

5c. 

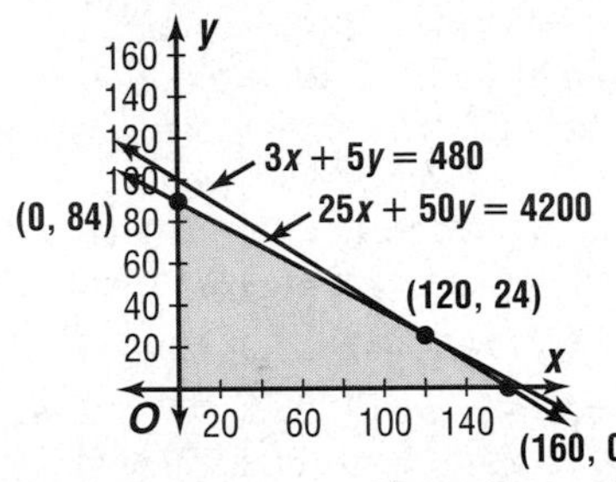

5d. $P(x, y) = 5x + 8y$ **5e.** 160 small packages, 0 large packages **5f.** \$800 **5g.** No. If revenue is maximized, the company will not deliver any large packages, and customers with large packages to ship will probably choose another carrier for all of their business. **7.** 225 Explorers, 0 Grande Expeditions

9. infeasible **11.** alternate optimal solutions

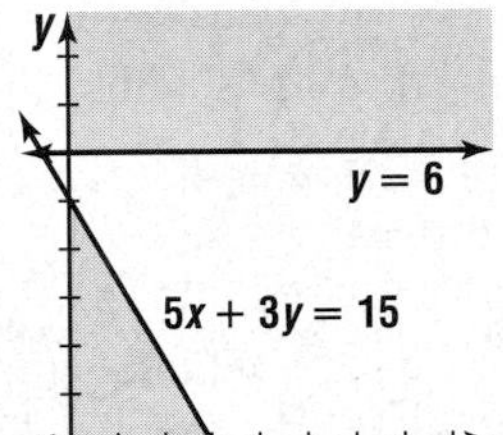

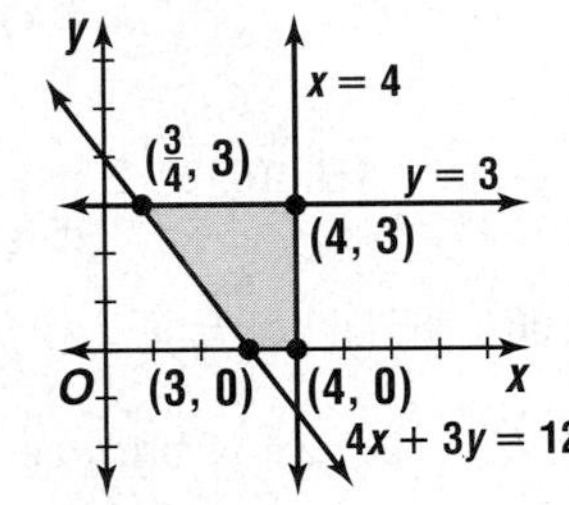

13a. Let d = the number of day-shift workers and n = the number of night-shift workers. $d \geq 5$; $n \geq 6$; $d + n \geq 14$

13b.

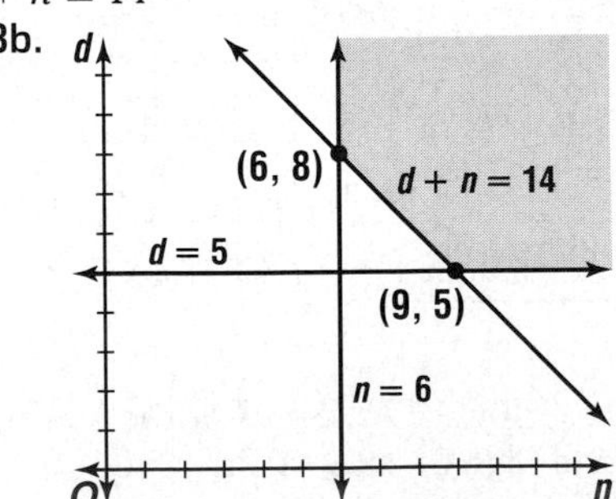

13c. $C(n, d) = 52d + 60n$ **13d.** 8 day-shift and 6 night-shift workers **13e.** \$776 **15.** 10 section-I questions, 2 section-II questions **17.** \$4000 in First Bank, \$7000 in City Bank **19.** 600 units of snack size, 1800 units of family-size

21. alternate optimal solutions **23a.** \$720 **23b.** Sample answer: Spend more than 30 hours per week on these services. **25.** (0, 6) **27.** Sample answer: $C = \$13.65 + \$0.15(n - 30)$; \$15.45

Pages 119–123 Chapter 2 Study Guide and Assessment

1. translation **3.** determinant **5.** scalar multiplication **7.** polygonal convex set

9. element **11.** (2, −4) **13.** $\left(-\frac{5}{11}, -\frac{2}{11}\right)$

15. (1, 2) **17.** (−10, −6, 0) **19.** (2, −1, 3)

21. $\begin{bmatrix} -10 & -13 \\ 2 & 2 \end{bmatrix}$ **23.** $\begin{bmatrix} -8 \\ 20 \end{bmatrix}$ **25.** impossible

27. impossible

29. $W'(-2, 3)$, $X'(-1, -2)$, $Y'(0, -4)$, $Z'(1, 2)$

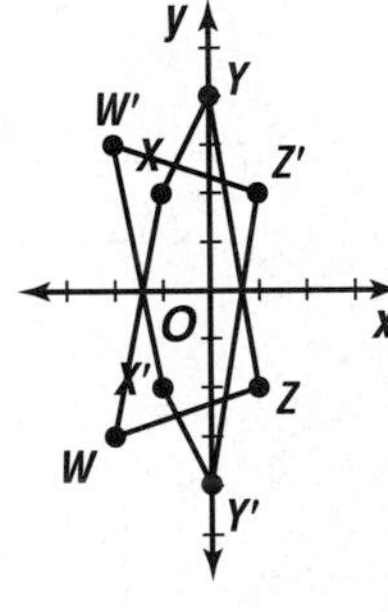

31. $P'(6, -8)$, $Q'(2, 4)$, $R'(-2, 2)$

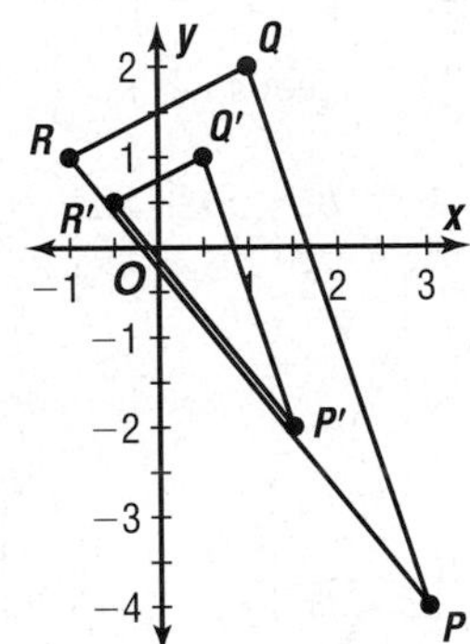

33. $\begin{bmatrix} 1 & 1 & 1 \\ 3 & 3 & 3 \end{bmatrix}$ **35.** 0 **37.** 160 **39.** $\frac{1}{23}\begin{bmatrix} 5 & -8 \\ 1 & 3 \end{bmatrix}$

41. $\frac{1}{7}\begin{bmatrix} -4 & -5 \\ -1 & -3 \end{bmatrix}$ **43.** $\frac{1}{32}\begin{bmatrix} 1 & 5 \\ -6 & 2 \end{bmatrix}$ **45.** (13, −5)

47. (−7, −4) **49.** 17, −4 **51.** 22 gallons in the truck and 6 gallons in the motorcycle **53.** 39 in., 31 in., 13 in.

Page 125 Chapter 2 SAT and ACT Practice

1. D **3.** D **5.** C **7.** D **9.** C

Chapter 3 The Nature of Graphs

Pages 134–136 Lesson 3-1

7. yes **9.** $y = x$ **11.** y-axis

13. x-intercept: (5, 0); other points: $\left(-6, \frac{3\sqrt{11}}{5}\right)$, $\left(6, -\frac{3\sqrt{11}}{5}\right)$, $\left(-6, -\frac{3\sqrt{11}}{5}\right)$

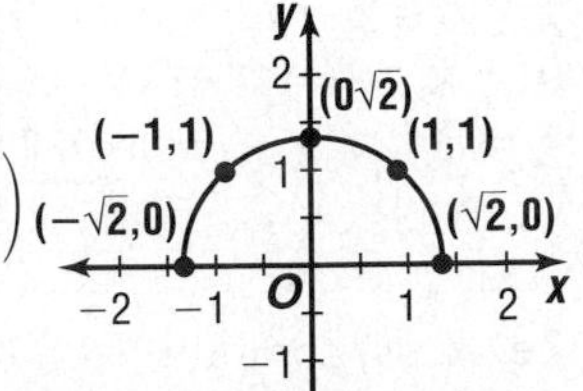

15. no

17. yes **19.** no

21. $y = x$ and $y = -x$

23. none of these **25.** all **27.** x-axis and y-axis, $y = x$, and $y = -x$

29.

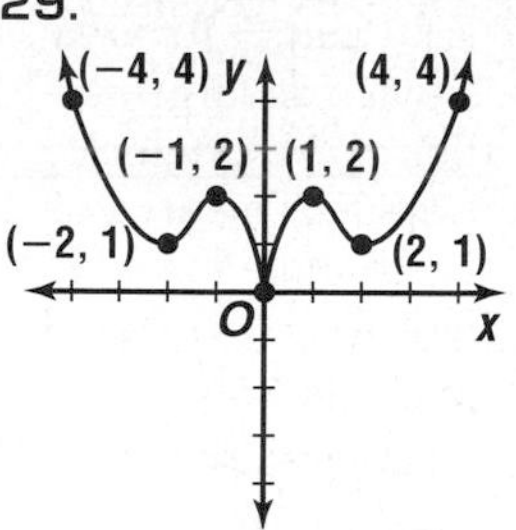

31. both

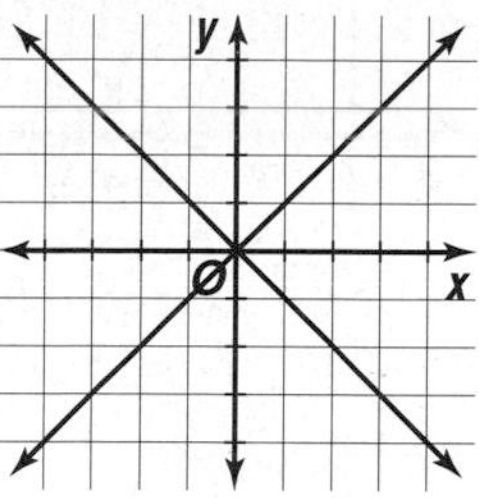

33. x-axis

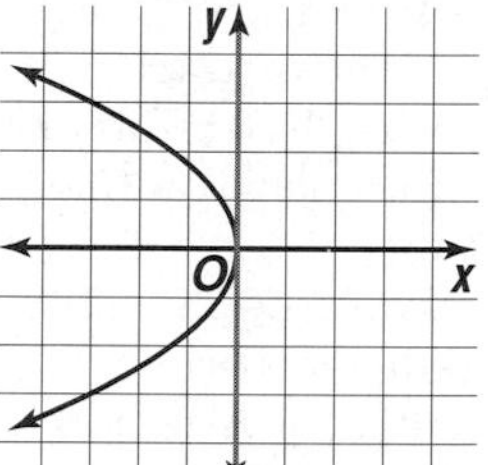

35. both

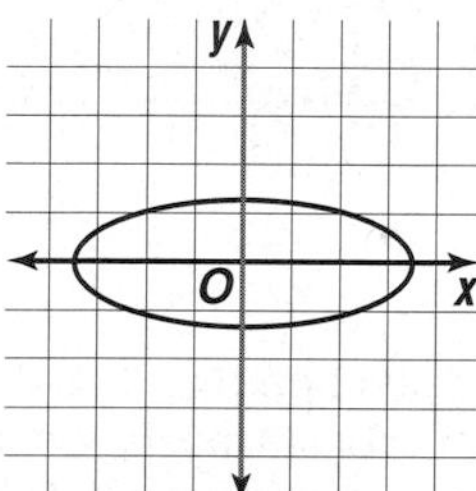

37. The equation $|y| = x^3 - x$ is symmetric about the x-axis.

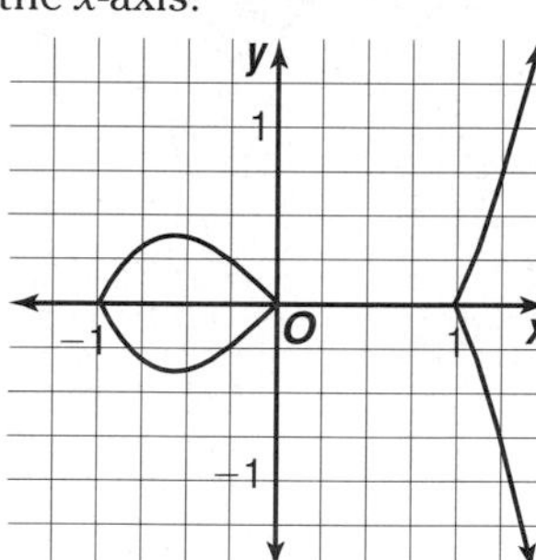

39. Sample answer: $y = 0$
41. $(4\sqrt{2}, 6)$ or $(-4\sqrt{2}, 6)$
43. 50 bicycles, 75 tricycles
45. $(-2, -3, 7)$

47.

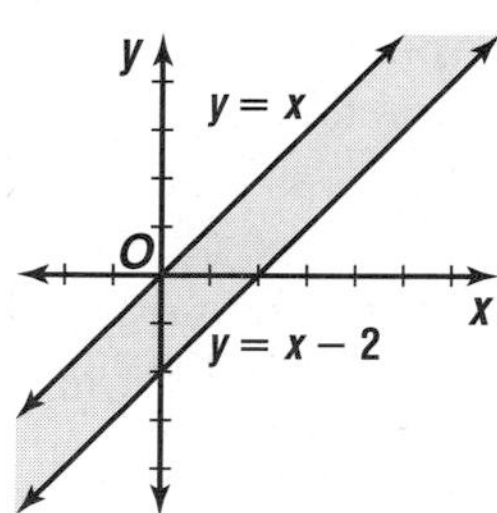

49. $-2x + 23, -2x + 5$

Pages 142–145 Lesson 3-2

7. $g(x)$ is the graph of $f(x)$ compressed horizontally by a factor of $\frac{1}{3}$, reflected over the x-axis.
9a. translated up 3 units, portion of graph below x-axis reflected over the x-axis **9b.** reflected over the x-axis, compressed horizontally by a factor of $\frac{1}{2}$
9c. translated left 1 unit, compressed vertically by a factor of 0.75

11.

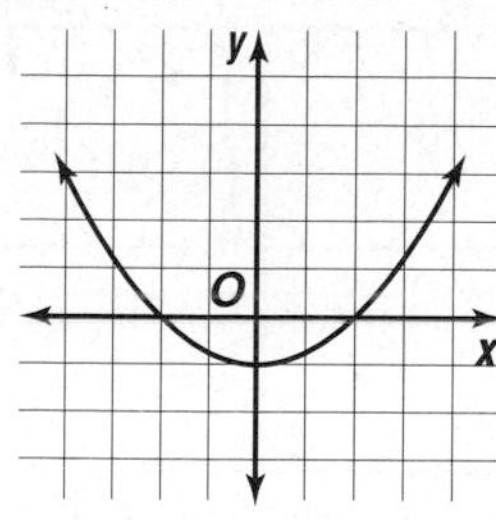

13. The graph of $g(x)$ is a translation of the graph of $f(x)$ up 6 units. **15.** The graph of $g(x)$ is the graph of $f(x)$ compressed horizontally by a factor of $\frac{1}{5}$. **17.** The graph of $g(x)$ is the graph of $f(x)$ expanded vertically by a factor of 3. **19.** $g(x)$ is the graph of $f(x)$ reflected over the x-axis, expanded horizontally by a factor of 2.5, translated up 3 units. **21a.** expanded horizontally by a factor of 5 **21b.** expanded vertically by a factor of 7, translated down 0.4 units **21c.** reflected across the x-axis, expanded vertically by a factor of 9, translated left 1 unit **23a.** compressed vertically by a factor of $\frac{1}{3}$, translated left 2 units **23b.** reflected over the y-axis, translated down 7 units **23c.** translated right 3 units and up 4 units, expanded vertically by a factor of 2 **25a.** compressed horizontally by a factor of $\frac{2}{5}$, translated down 3 units
25b. reflected over the y-axis, compressed vertically by a factor of 0.75 **25c.** The portion of the parent graph on left of the y-axis is replaced by a reflection of the portion on the right of the y-axis. The new image is then translated 4 units right.

27. $y = \frac{0.25}{x - 4} + 3$ **29.**

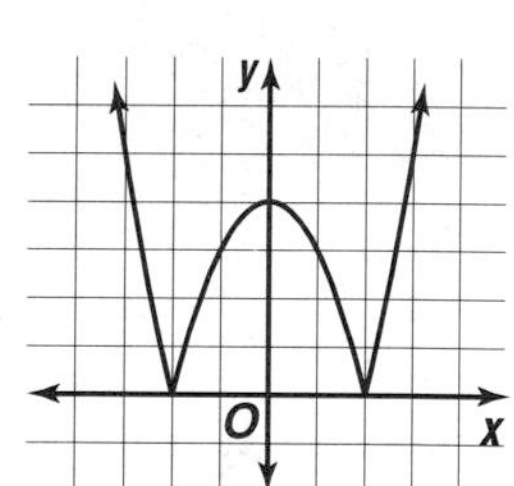

31.

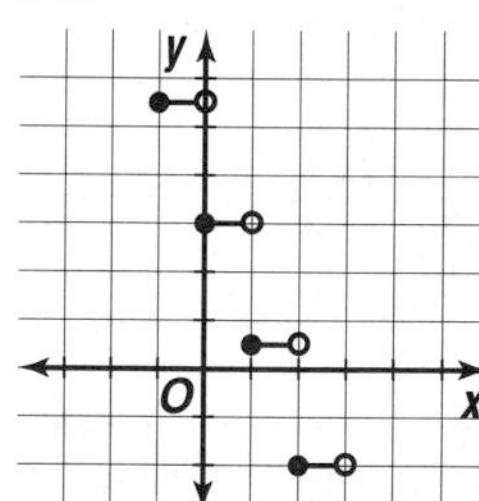

33.

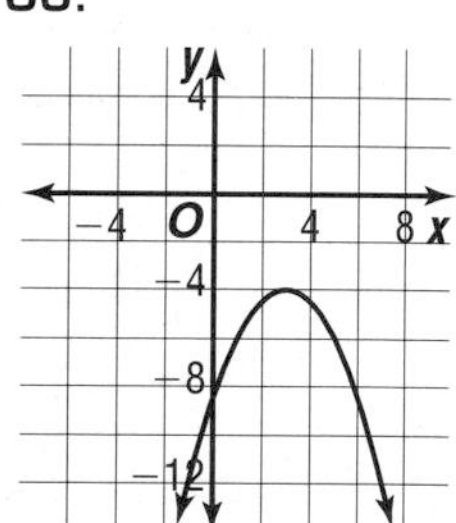

35a. 0

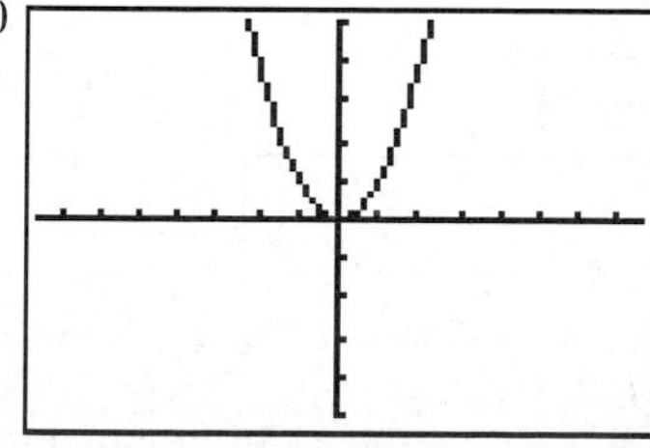

[−7.6, 7.6] scl:1 by [−5, 5] scl:1

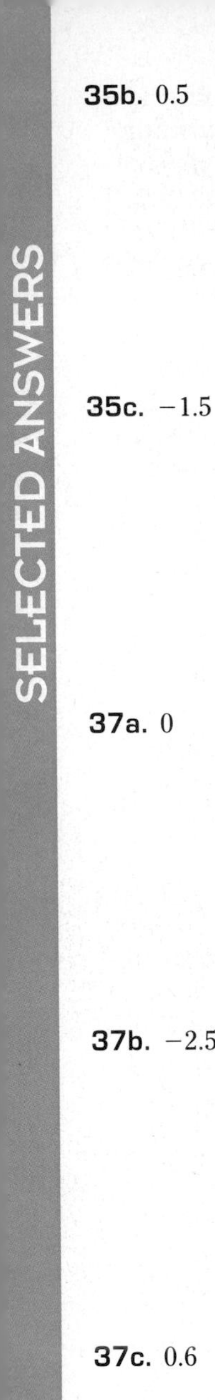

35b. 0.5

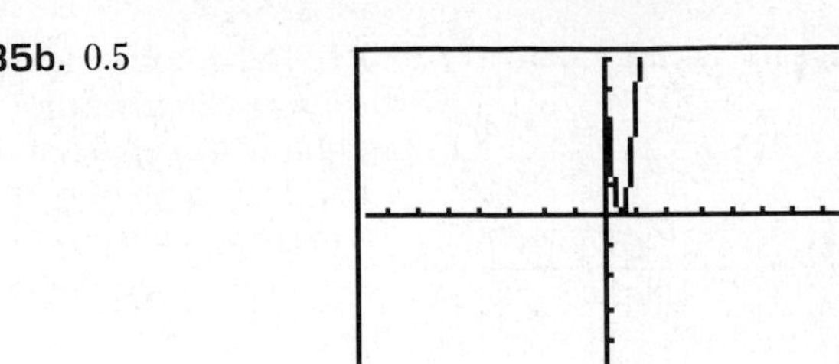

[−7.6, 7.6] scl:1 by [−5, 5] scl:1

35c. −1.5

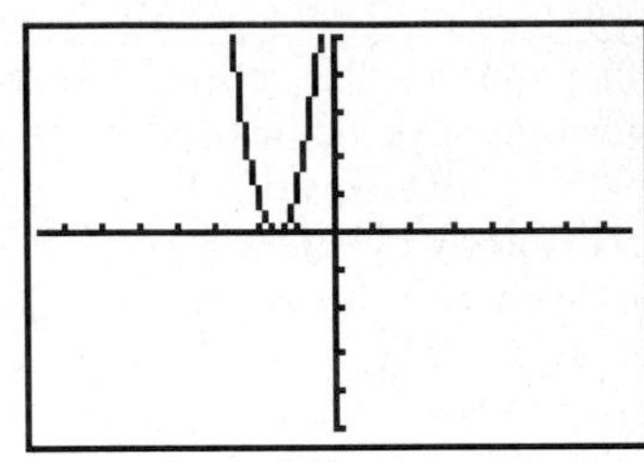

[−7.6, 7.6] scl:1 by [−5, 5] scl:1

37a. 0

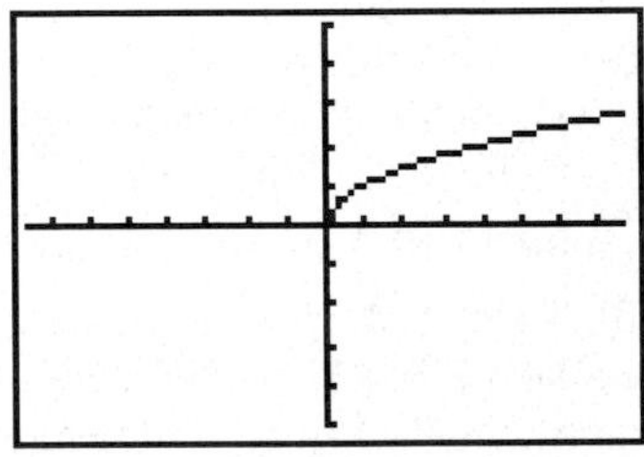

[−7.6, 7.6] scl:1 by [−5, 5] scl:1

37b. −2.5

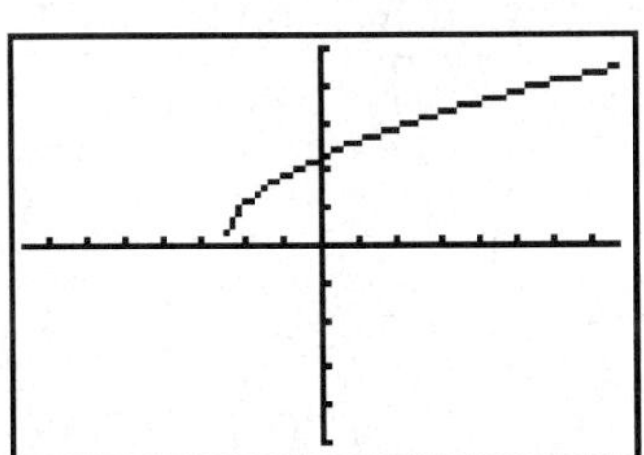

[−7.6, 7.6] scl:1 by [−5, 5] scl:1

37c. 0.6

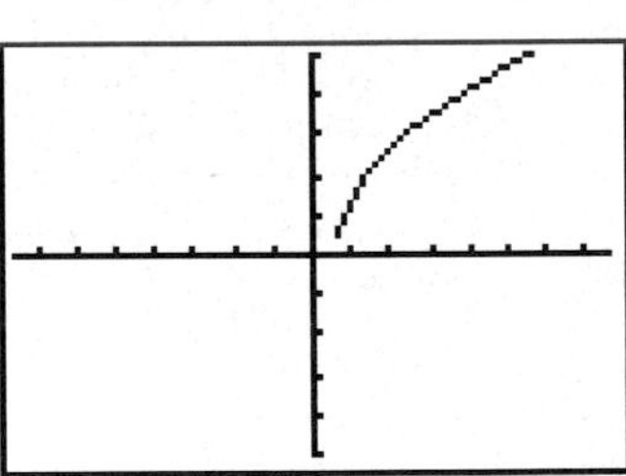

[−7.6, 7.6] scl:1 by [−5, 5] scl:1

39. The x-intercept will be $-\frac{b}{a}$.

41a. 25 units^2

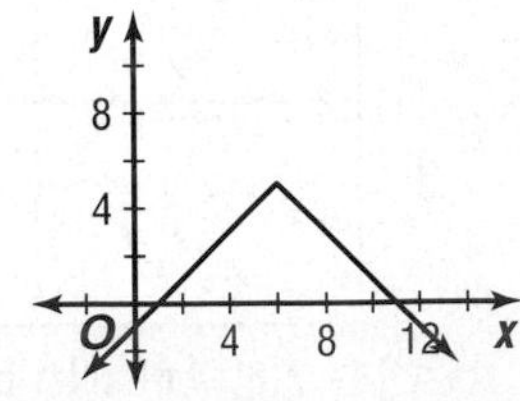

41b.

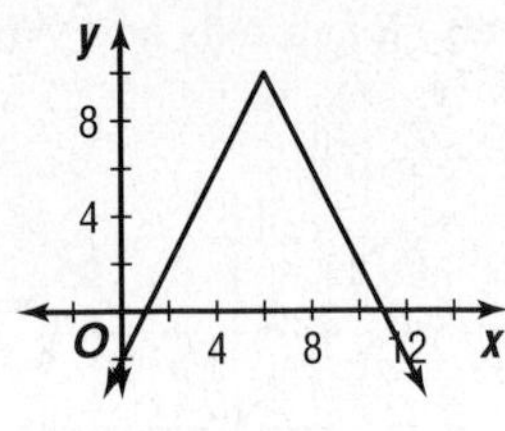

The area of the triangle is $\frac{1}{2}(10)(10)$ or 50 units^2. Its area is twice as large as that of the original triangle. The area of the triangle formed by $y = c \cdot f(x)$ would be $25c$ units^2.

41c.

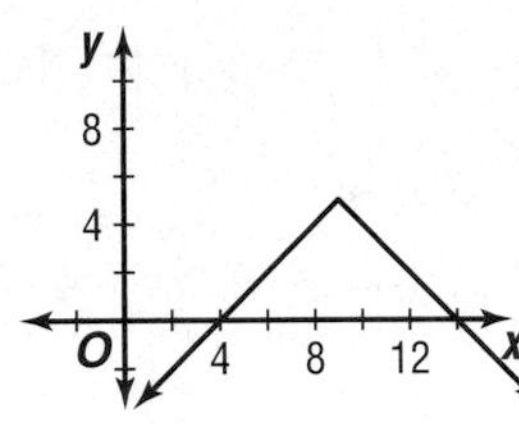

The area of the triangle is $\frac{1}{2}(10)(5)$ or 25 units^2. Its area is the same as that of the original triangle. The area of the triangle formed by $y = f(x + c)$ would be 25 units^2.

43a. reflection over the x-axis, reflection over the y-axis, vertical translation, horizontal compression or expansion, and vertical expansion or compression **43b.** horizontal translation **45.** 30 preschoolers and 20 school-age **47.** $x = \pm 5, y = 9, z = 6$ **49.** The graph implies a negative linear relationship. **51.** −250 **53.** A

Pages 149–151 Lesson 3-3

5. yes **7.**

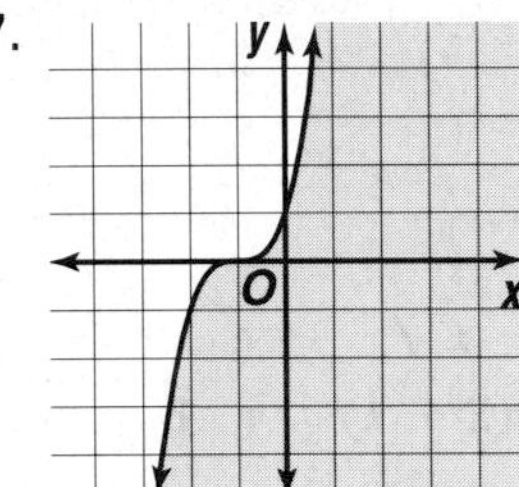

9.

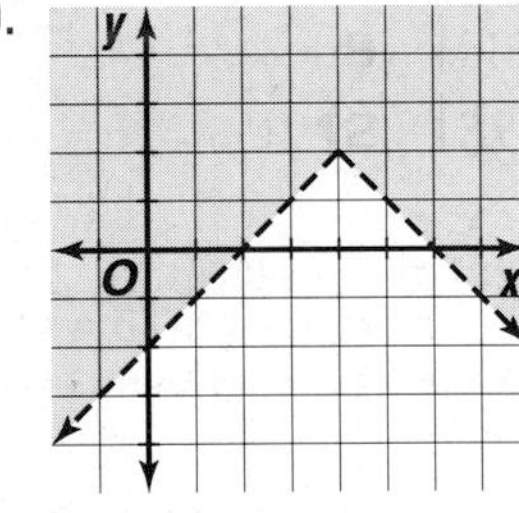

11. $\{x \mid 1 \leq x \leq 2\}$ **13.** no **15.** yes **17.** yes **19.** (0, 0), (1, 1), and (1, −1); if these points are in the shaded region and the other points are not, then the graph is correct.

21.

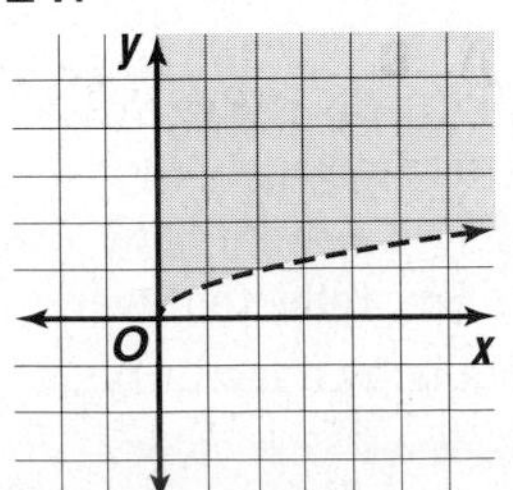

23.

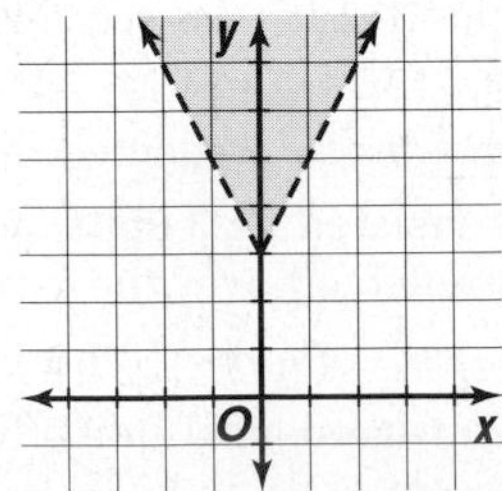

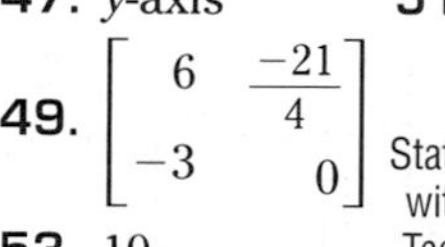

25.

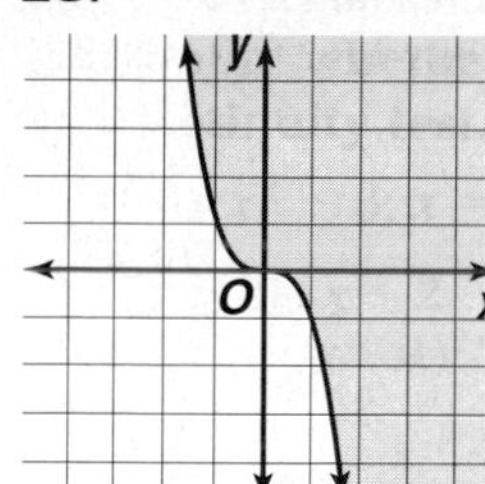

27.

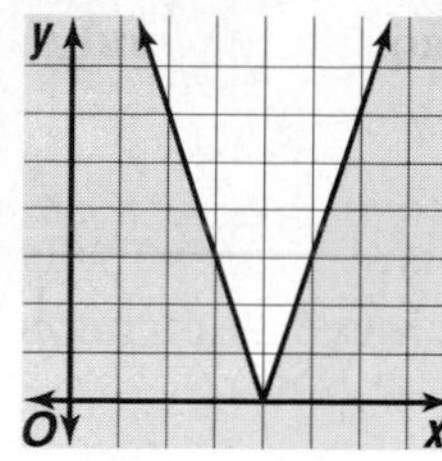

29.

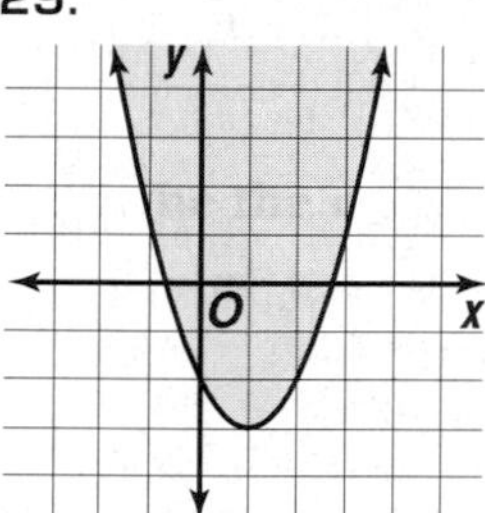

31.

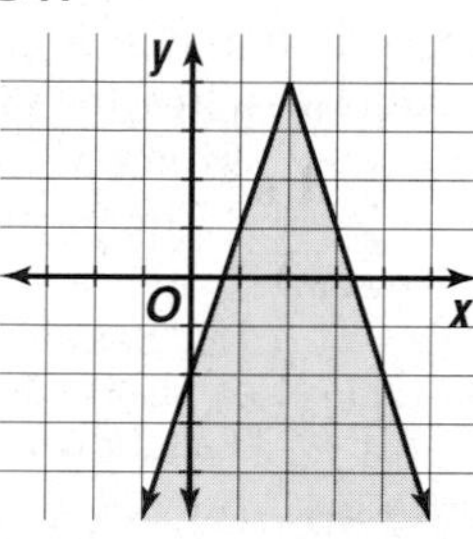

33. $\{x \mid x < -9 \text{ or } x > 1\}$ **35.** $\{x \mid -2 < x < 9\}$
37. no solution **39.** $\{x \mid -17 \leq x \leq 7\}$
41. $\{x \mid 5.5 < x < 10\}$ **43.** $x \geq 83\frac{2}{3}$

45a.

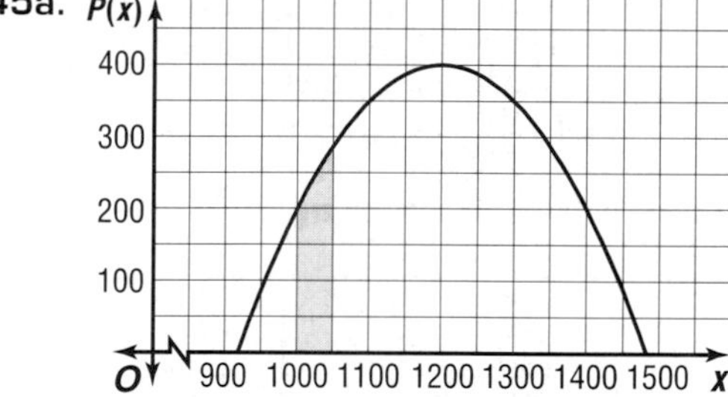

45b. The shaded region shows all points (x, y) where x represents the number of cookies sold and y represents the possible profit made for a given week.

47. y-axis

49. $\begin{bmatrix} 6 & \frac{-21}{4} \\ -3 & 0 \end{bmatrix}$

51.

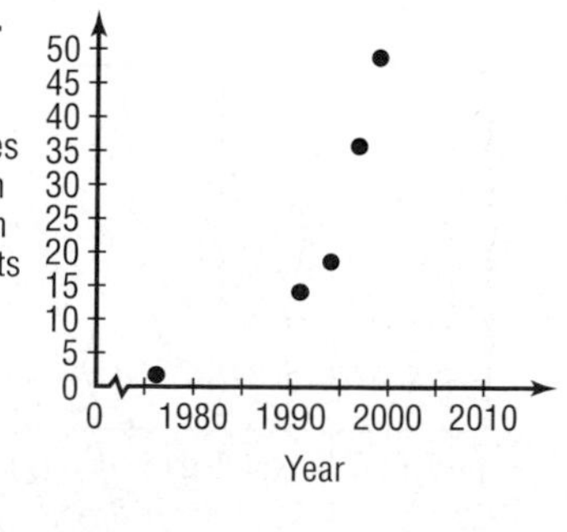

53. 10

Pages 156–158 Lesson 3-4

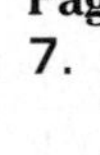

7.

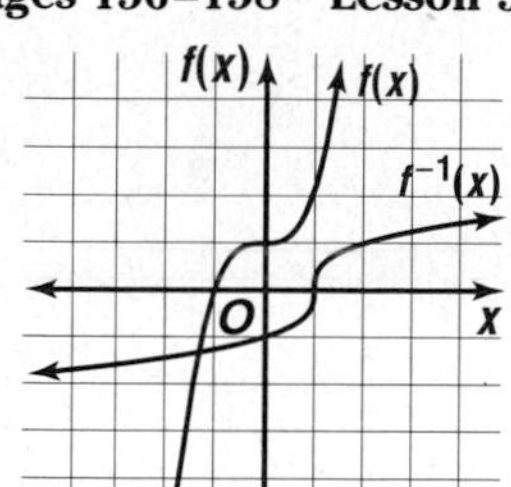

9. $f^{-1}(x) = -\frac{1}{3}x + \frac{2}{3}$; $f^{-1}(x)$ is a function.
11. $f^{-1}(x) = -2 \pm \sqrt{x - 6}$; $f^{-1}(x)$ is not a function.

13. $f^{-1}(x) = 2x + 10$; $[f \circ f^{-1}](x) = f(2x + 10) = \frac{1}{2}(2x + 10) - 5 = x$

$[f^{-1} \circ f](x) = f^{-1}\left(\frac{1}{2}x - 5\right) = 2\left(\frac{1}{2}x - 5\right) + 10 = x$

Since $[f \circ f^{-1}](x) = [f^{-1} \circ f](x) = x$, f and f^{-1} are inverse functions.

15.

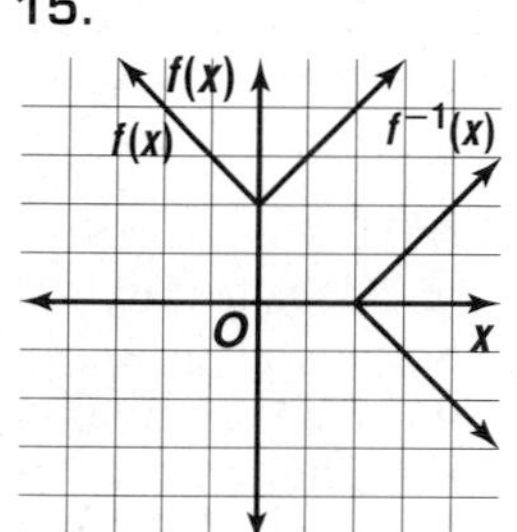

17.

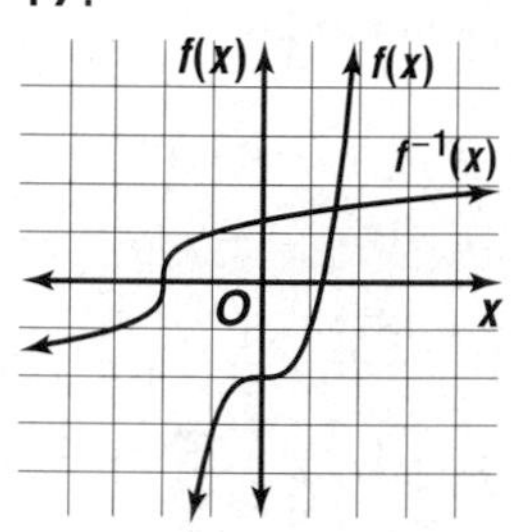

19.

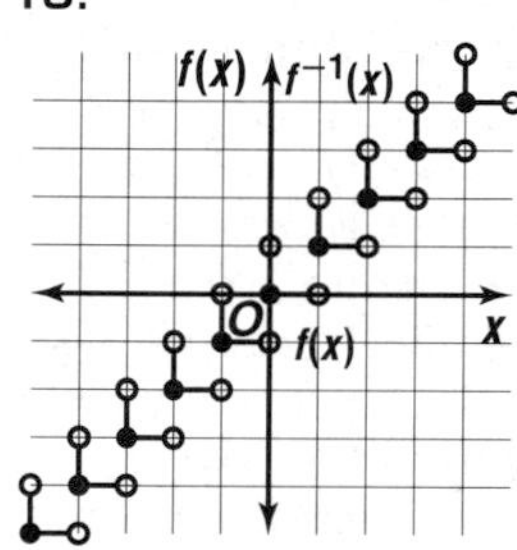

21.

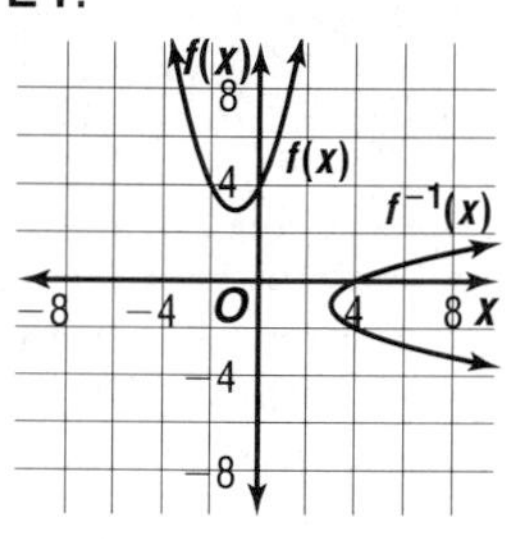

23.

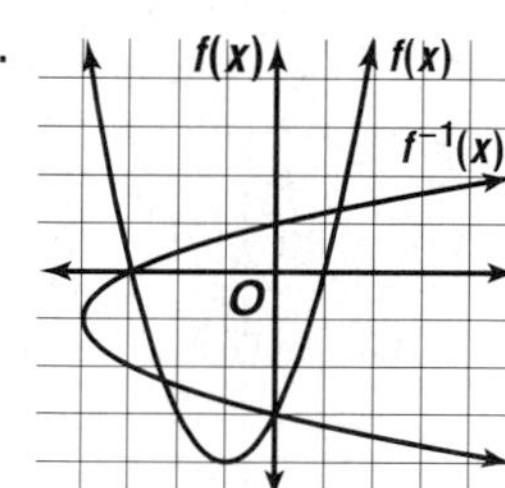

25. $f^{-1}(x) = \frac{x - 7}{2}$; $f^{-1}(x)$ is a function.
27. $f^{-1}(x) = \frac{1}{x}$; $f^{-1}(x)$ is a function.
29. $f^{-1}(x) = 3 \pm \sqrt{x - 7}$; $f^{-1}(x)$ is not a function.

31. $f^{-1}(x) = \frac{1}{x} - 2$; $f^{-1}(x)$ is a function.
33. $f^{-1}(x) = 2 - \sqrt[3]{\frac{2}{x}}$; $f^{-1}(x)$ is a function.

35.

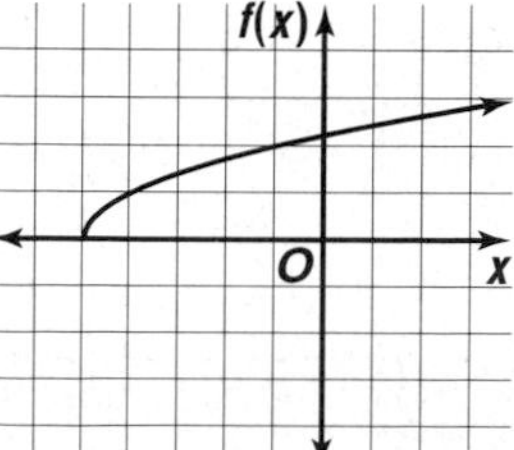

37.

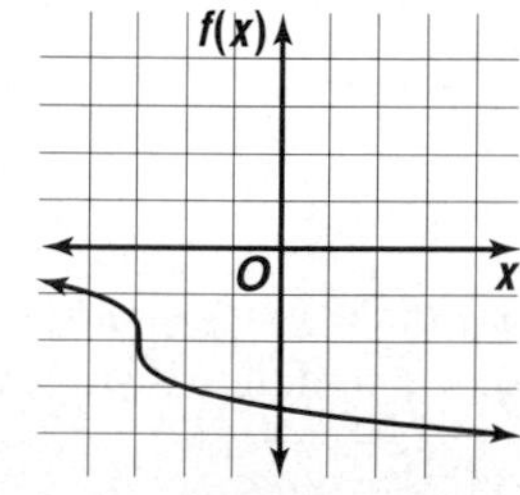

39. $f^{-1}(x) = -\frac{3}{2}x + \frac{1}{4}$

$$[f \circ f^{-1}](x) = f\left(-\frac{3}{2}x + \frac{1}{4}\right)$$
$$= -\frac{2}{3}\left(-\frac{3}{2}x + \frac{1}{4}\right) + \frac{1}{6}$$
$$= x - \frac{1}{6} + \frac{1}{6}$$
$$= x$$

$$[f^{-1} \circ f](x) = f^{-1}\left(-\frac{2}{3}x + \frac{1}{6}\right)$$
$$= -\frac{3}{2}\left(-\frac{2}{3}x + \frac{1}{6}\right) + \frac{1}{4}$$
$$= x - \frac{1}{4} + \frac{1}{4}$$
$$= x$$

Since $[f \circ f^{-1}](x) = [f^{-1} \circ f](x) = x$, f and f^{-1} are inverse functions.

41a.

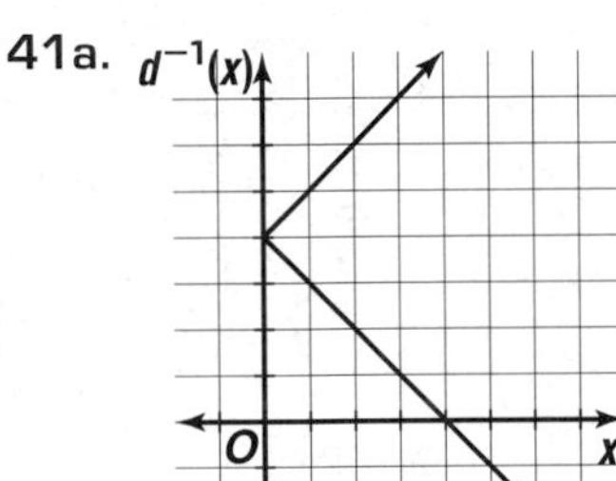

41b. No; the graph of $d(x)$ fails the horizontal line test. **41c.** $d^{-1}(x)$ gives the numbers that are 4 units from x on the number line. There are always two such numbers, so d^{-1} associates two values with each x-value. Hence, $d^{-1}(x)$ is not a function. **43a.** Sample answer: $y = -x$. **43b.** The graph of the function must be symmetric about the line $y = x$. **43c.** Yes, because the line $y = x$ is the axis of symmetry *and* the reflection line. **45.** It must be translated up 6 units and 5 units to the left; $y = (x - 6)^2 - 5$, $y = 6 \pm \sqrt{x + 5}$. **47a.** Yes. If the encoded message is not unique, it may not be decoded properly. **47b.** The inverse of the encoding function must be a function so that the encoded message may be decoded.
47c. $(x + 2)^2 - 3$
47d. FUNCTIONS ARE FUN **49.** both
51. $(-1, 7)$
53.

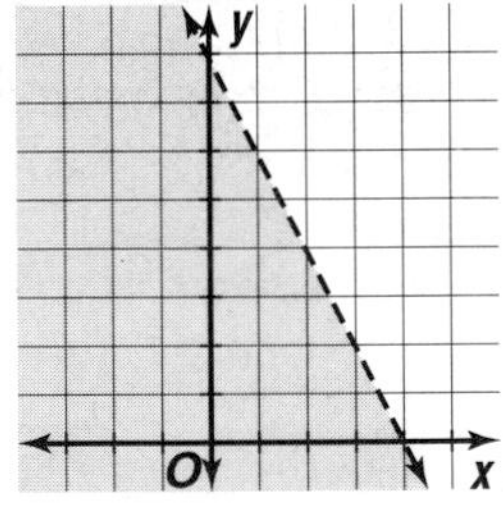

55. $y = -x + 7$

Pages 165–168 Lesson 3-5
5. No. y is undefined when $x = -3$.
7. $y \to \infty$ as $x \to \infty$, $y \to -\infty$ as $x \to -\infty$

9.

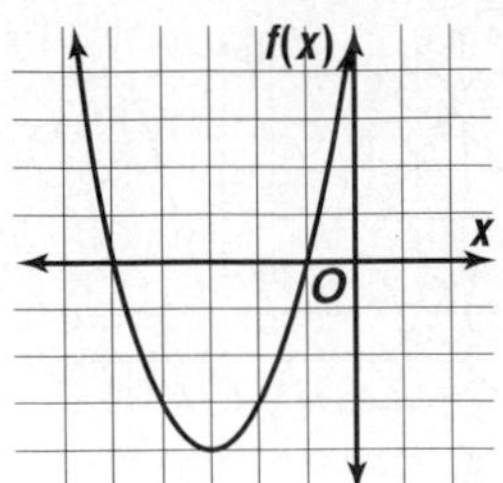

Decreasing for $x < -3$.
Increasing for $x > -3$.

11a. $t = 4$ **11b.** when $t = 4$ **11c.** 10 amps
13. No. The function is undefined when $x = 2$.
15. Yes. The function is defined when $x = 3$; the function approaches 1 (in fact is equal to 1) as x approaches 3 from both sides; and $y = 1$ when $x = 3$. **17.** Yes. The function is defined when $x = 1$; $f(x)$ approaches 3 as x approaches 1 from both sides; and $f(1) = 3$. **19.** Sample answer: $x = 0$. $g(x)$ is undefined when $x = 0$. **21.** $y \to -\infty$ as $x \to \infty$, $y \to -\infty$ as $x \to -\infty$ **23.** $y \to \infty$ as $x \to \infty$, $y \to \infty$ as $x \to -\infty$ **25.** $f(x) \to 2$ as $x \to \infty$, $f(x) \to 2$ as $x \to -\infty$

27.

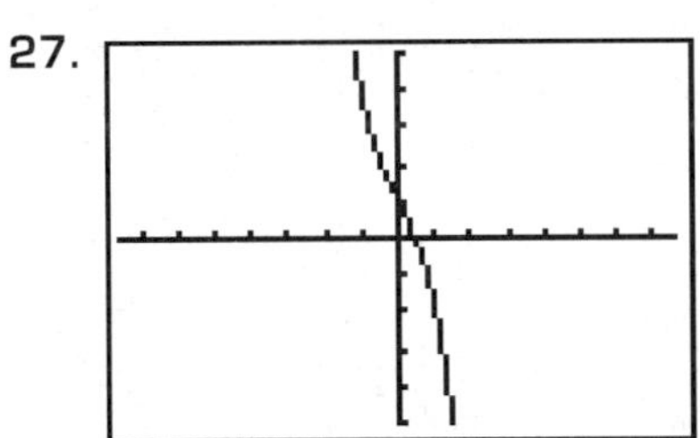

[−7.6, 7.6] scl:1 by [−5, 5] scl:1

decreasing for all x

29.

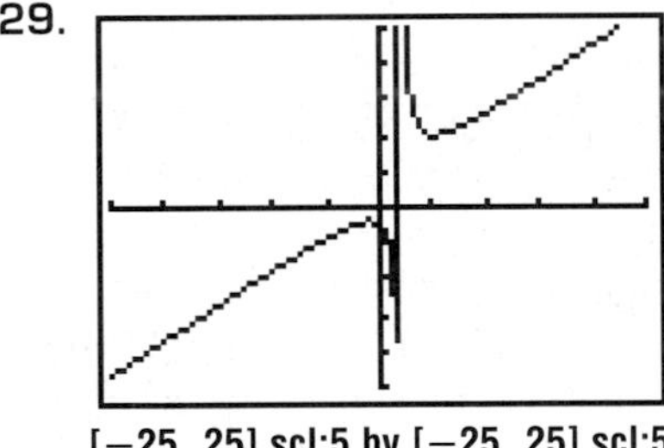

[−25, 25] scl:5 by [−25, 25] scl:5

increasing for $x < -1$ and $x > 5$; decreasing for $-1 < x < 2$ and $2 < x < 5$

31.

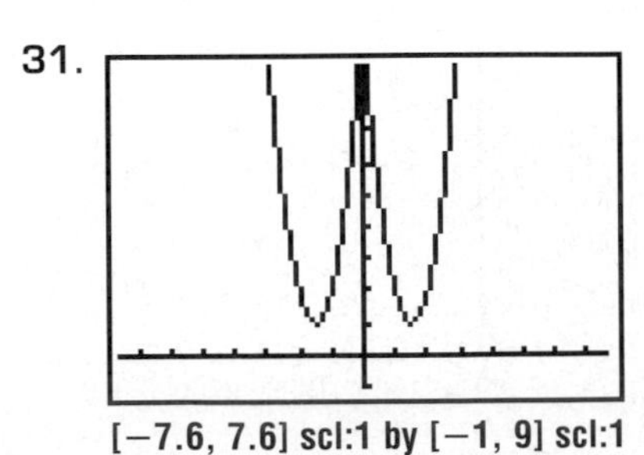

[−7.6, 7.6] scl:1 by [−1, 9] scl:1

decreasing for $x < -\frac{3}{2}$ and $0 < x < \frac{3}{2}$; increasing for $-\frac{3}{2} < x < 0$ and $x > \frac{3}{2}$

33a. f is decreasing for $-2 < x < 0$ and increasing for $x < -2$. f has jump discontinuity when $x = -3$ and $f(x) \to -\infty$ as $x \to -\infty$.

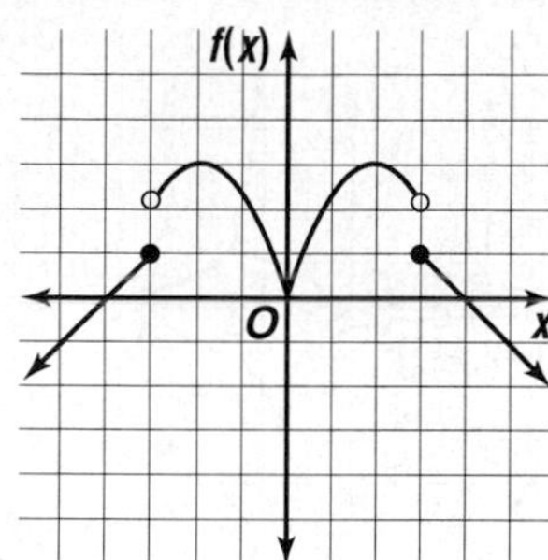

33b. f is increasing for $-2 < x < 0$ and decreasing for $x < -2$. f has a jump discontinuity when $x = -3$ and $f(x) \to \infty$ as $x \to -\infty$.

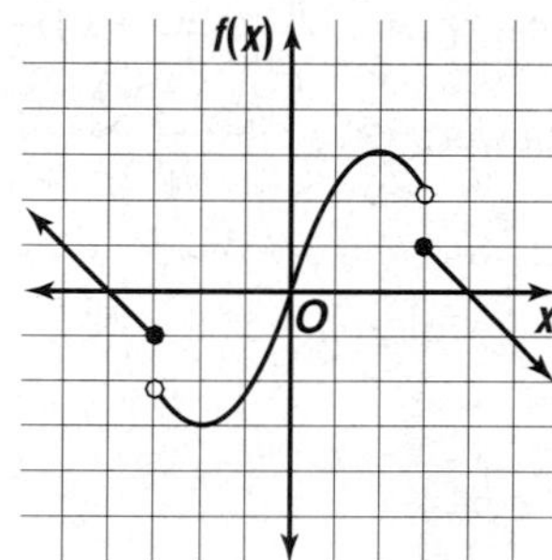

35a. 1954–1956, 1958–1959, 1960–1961, 1962–1963, 1966–1968, 1973–1974, 1975–1976, 1977–1978, 1989–1991 **35b.** 1956–1958, 1959–1960, 1961–1962, 1963–1966, 1968–1973, 1974–1975, 1976–1977, 1978–1989, 1991–1996 **37a.** The function must be monotonic. **37b.** The inverse must be monotonic. **39.** $a = 4$, $b = 2$ **41.** The graph of $g(x)$ is the graph of $f(x)$ translated left 2 units and down 4 units. **43.** 42 **45.** 20

Pages 176–179 Lesson 3-6
5. rel. min.: $(-1, -3)$; rel. max.: $(3, 3)$
7. rel. min.: $(-2.25, -10.54)$

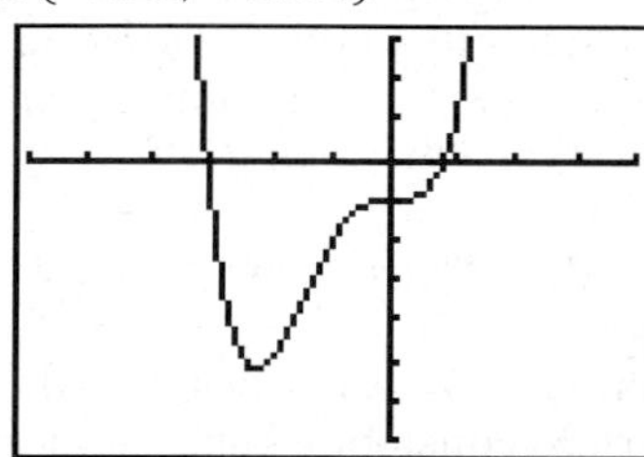

[−6, 4] scl:1 by [−14, 6] scl:2

9. min. **11.** min. **13.** abs. max.: $(-4, 1)$
15. rel. max.: $(-2, 7)$; abs. min.: $(3, -3)$
17. abs. min.: $(3, -8)$; rel. max.: $(5, -2)$; rel. min.: $(8, -5)$

19. abs. max.: $(1.5, -1.75)$

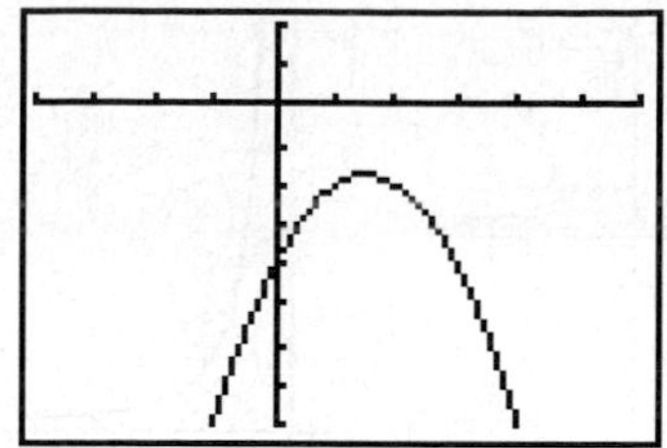

[−5, 5] scl:1 by [−8, 2] scl:1

21. rel. max.: $(-0.59, 0.07)$, rel. min.: $(0.47, -3.51)$

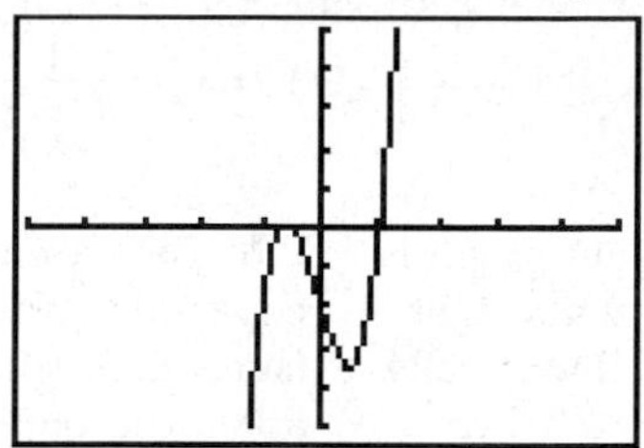

[−5, 5] scl:1 by [−5, 5] scl:1

23. rel. max.: $(-1, 1)$; rel. min.: $(0.25, -3.25)$

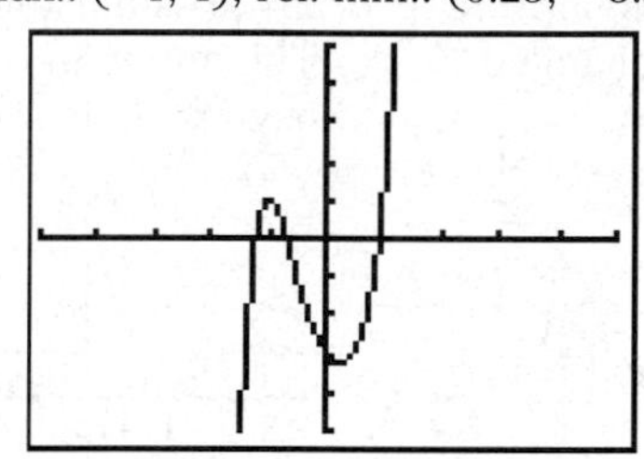

[−5, 5] scl:1 by [−5, 5] scl:1

25. abs. min.: $(-3.18, -15.47)$; rel. min.: $(0.34, -0.80)$; rel. max.: $(-0.91, 3.04)$ **27.** max. **29.** max. **31.** pt. of inflection **33.** min. **35a.** $V(x) = 2x(12.5 - 2x)(17 - 2x)$ **35b.** 2.37 cm by 2.37 cm **37a.** $f(x) = 5000\sqrt{x^2 + 4} + 3500(10 - x)$ **37b.** 1.96 km from point B **39.** The particle is at rest when $t \approx 0.14$ and when $t \approx 3.52$. Its position at these times are $s(0.14) \approx -8.79$ and $s(3.52) \approx -47.51$. **41.** No; the function is undefined when $x = 5$. **43.** 120 units of notebook and 80 units of newsprint **45.** -1, yes **47.** 5 free throws, 9 2-point field goals, 3 3-point field goals **49.** perpendicular **51.** D

Pages 185–188 Lesson 3-7

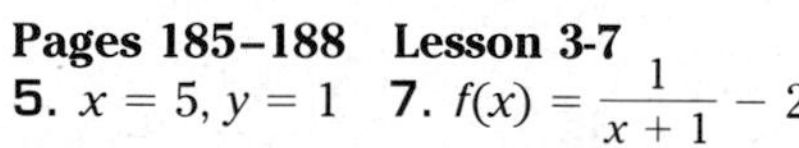

5. $x = 5$, $y = 1$ **7.** $f(x) = \frac{1}{x + 1} - 2$
9. The parent graph is translated 2 units to the left and down 1 unit. The vertical asymptote is now at $x = -2$ and the horizontal asymptote is now $y = -1$.

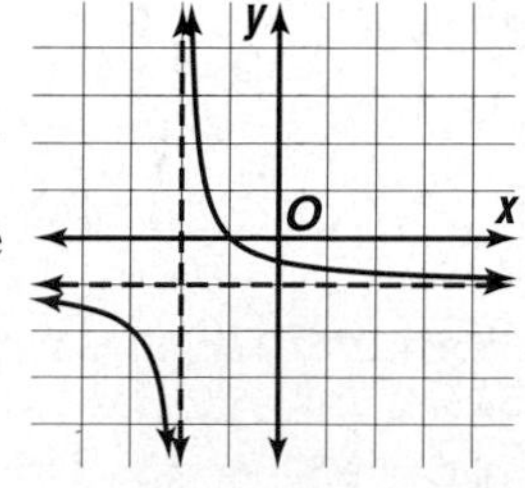

11.

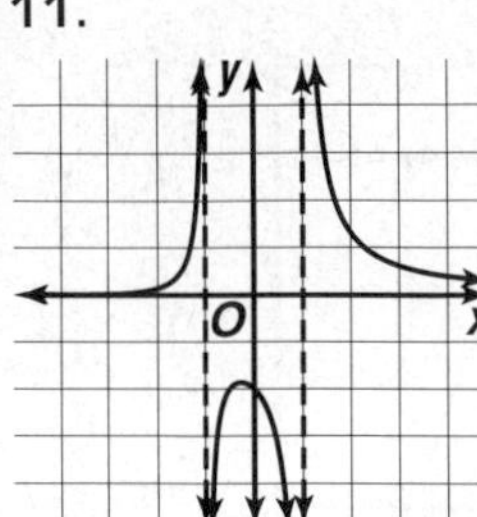

13a.

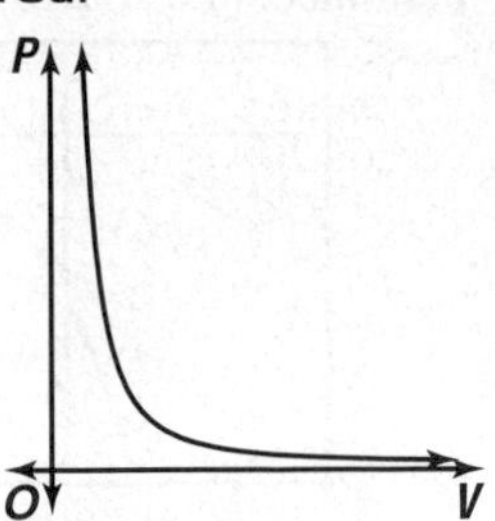

13b. $P = 0, V = 0$ **13c.** The pressure approaches 0. **15.** $x = -6$ **17.** $x = -1, x = -3, y = 0$ **19.** $x = 1, y = 1$ **21.** $f(x) = \frac{1}{x+3} + 1$ **23.** $f(x) = -\frac{1}{x} + 1$

25. The parent graph is translated 4 units right and expanded vertically by a factor of 2. The vertical asymptote is now $x = 4$. The horizontal asymptote, $y = 0$, is unchanged.

27. The parent graph is expanded vertically by a factor of 3, reflected about the x-axis, and translated 2 units up. The vertical asymptote, $x = 0$, is unchanged. The horizontal asymptote is now $y = 2$.

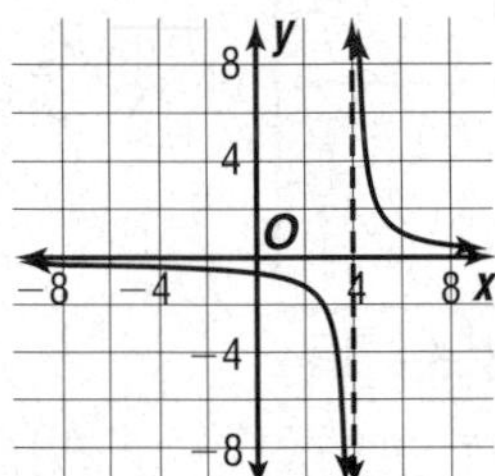

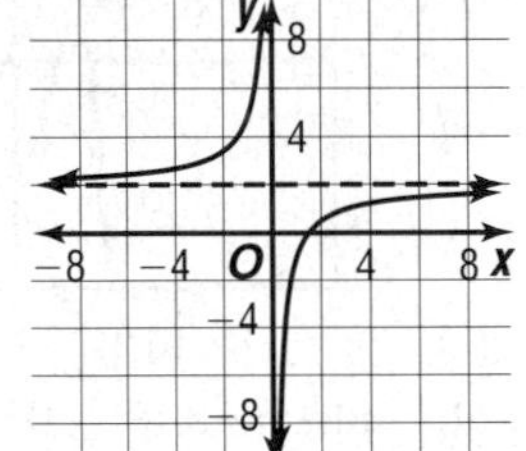

29. The parent graph is translated 5 units left. The translated graph is expanded vertically by a factor of 22 and then translated 4 units down. The vertical asymptote is $x = -5$ and the horizontal asymptote is $y = -4$.

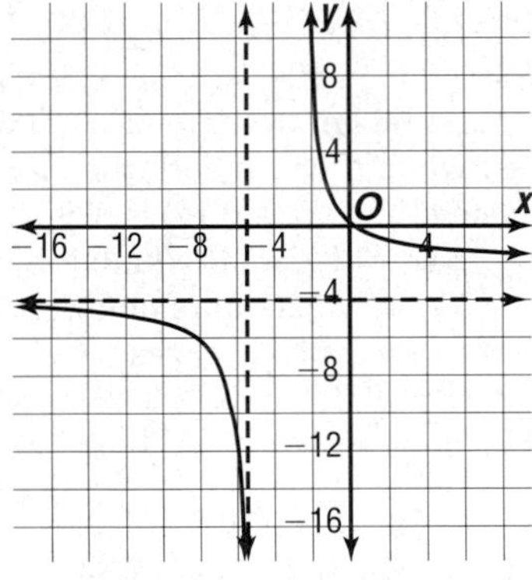

31. $y = x + 3$
33. $y = \frac{1}{2}x - \frac{5}{4}$

35.

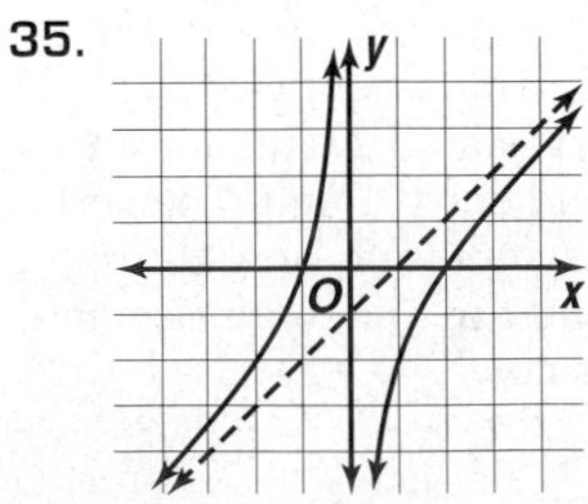

37.

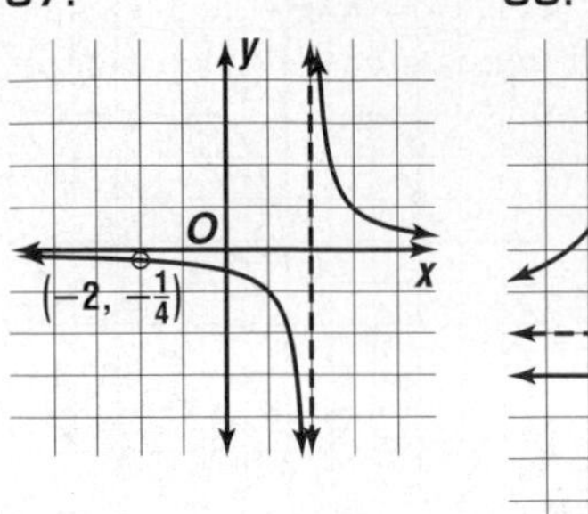

39.

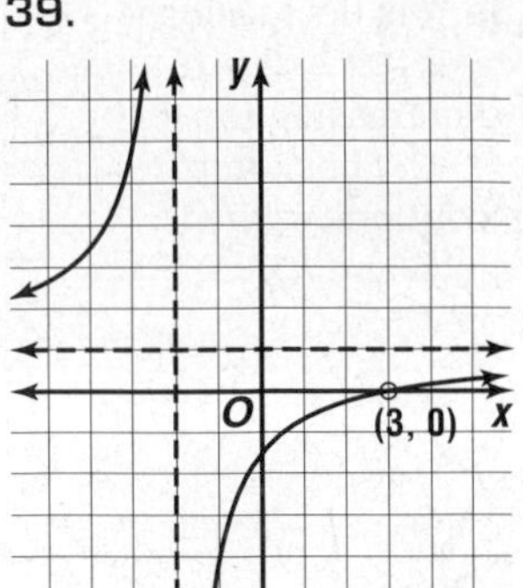

41a. $C(t) = \frac{480 + 3t}{40 + t}$ **41b.** 11.43 L **43.** Sample answer: $f(x) = \frac{(x-2)(x+3)(x+5)^2}{(x-4)(x+5)}$ **45.** Sample answer: $f(x) = \frac{x}{x^2+1}$ **47a.** $\frac{a^2-9}{a-3}$ **47b.** The slope approaches 6. **49.** $y = \pm\sqrt{x+9}$ **51.** $\begin{bmatrix} 24 & -20 \\ -32 & 16 \end{bmatrix}$ **53.** (3, 2) **55.** $16 - 8x^2$, $2 - 64x^2$

Pages 193–196 Lesson 3-8

5. 12, $xy = 12$; $\frac{4}{5}$ **7.** 0.5; $y = 0.5xz^3$; 108 **9.** y varies directly as x^4, $\frac{1}{7}$ **11.** y varies inversely as x; -3 **13.** 0.2; $y = 0.2x$; 1.2 **15.** 15, $y = 15xz$; 18 **17.** 16; $r = 16t^2$; 1 **19.** $\frac{1}{12}$; $y = \frac{1}{12}x^3z^2$; -48 **21.** 2; $y = \frac{2xz}{w}$; 14 **23.** 15; $a = \frac{15b^2}{c}$; ± 8 **25.** C varies directly as d; π **27.** y varies jointly as x and the square of z; $\frac{4}{3}$ **29.** y varies inversely as the square of x, $\frac{5}{4}$ **31.** A varies jointly as h and the quantity $b_1 + b_2$; 0.5 **33.** y varies directly as x^2 and inversely as the cube of z; 7 **35a.** Joint variation; to reduce torque one must either reduce the distance or reduce the mass on the end of the fulcrum. Thus, torque varies directly as the mass and the distance from the fulcrum. Since there is more than one quantity in direct variation with the torque on the seesaw, the variation is joint.
35b. $T_1 = km_1d_1$ and $T_2 = km_2d_2$

$$T_1 = T_2$$
$$km_1d_1 = km_2d_2 \quad \textit{Substitution property of equality}$$
$$m_1d_1 = m_2d_2$$

35c. 1.98 meters **37.** If y varies directly as x then there is a nonzero constant k such that $y = kx$. Solving for x, we find $x = \frac{1}{k}y$. $\frac{1}{k}$ is a nonzero constant, so x varies directly as y. **39.** a is doubled

$$a = \frac{kb^2}{c^3}; \; a = \frac{k\left(\frac{1}{2}b\right)^2}{\left(\frac{1}{2}c\right)^3}; \; \frac{a = \frac{1}{4}kb^2}{\frac{1}{8}c^3}; \; a = 2\,\frac{kb^2}{c^3}$$

41. $1.78 \times 10^{-3}\ \Omega$ **43.** $f^{-1}(x) = \sqrt[3]{x - 6} + 3$; $f^{-1}(x)$ is a function. **45.** consistent and dependent **47.** $y = -0.92x + 1858.60$

Pages 197–201 Chapter 3 Study Guide and Assessment

1. even **3.** point **5.** maximum **7.** inverse **9.** slant **11.** yes **13.** no **15.** $y = x$ and $y = -x$ **17.** none **19.** $g(x)$ is a translation of the graph of $f(x)$ up 5 units. **21.** $g(x)$ is the graph of $f(x)$ expanded vertically by a factor of 6.

23.

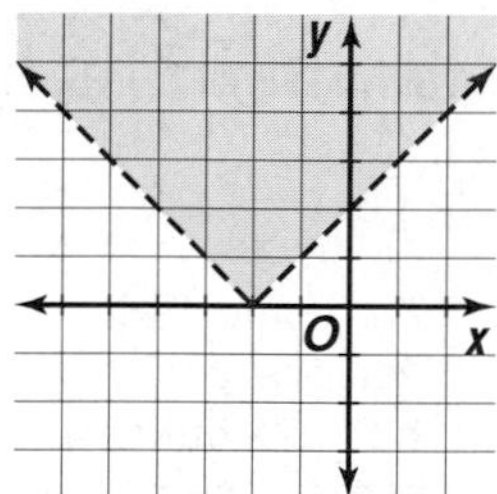

25.

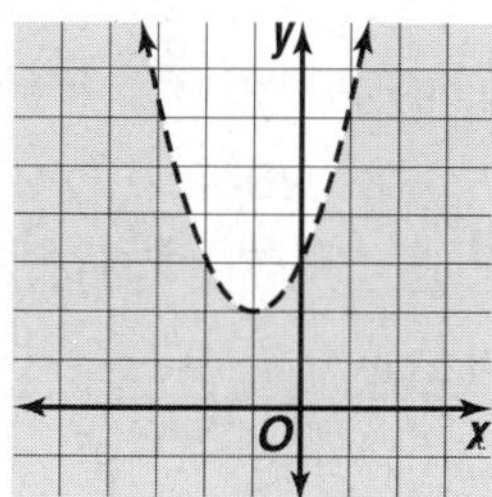

27. $\{x \mid x < -3 \text{ or } x > 0.5\}$

29.

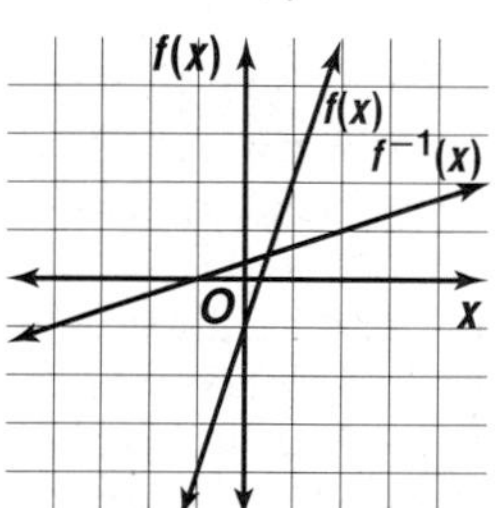

31.

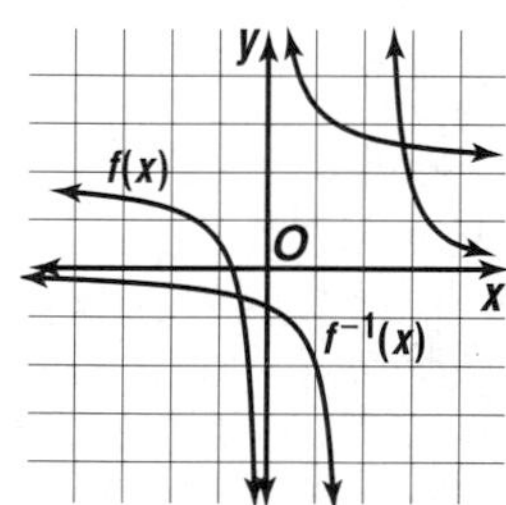

33. $f^{-1}(x) = \sqrt[3]{x + 8} + 2$; yes **35.** Yes. The function is defined when $x = 2$; the function approaches 6 as x approaches 2 from both sides; and $y = 6$ when $x = 2$. **37.** Yes. The function is defined when $x = 1$; the function approaches 2 as x approaches 1 from both sides; and $y = 2$ when $x = 1$. **39.** $y \to \infty$ as $x \to \infty$, $y \to -\infty$ as $x \to -\infty$ **41.** $y \to \infty$ as $x \to \infty$, $y \to -\infty$ as $x \to -\infty$ **43.** decreasing for $x < -3$ and $0 < x < 3$; increasing for $-3 < x < 0$ and $x > 3$ **45.** rel. max.: (0, 4), rel. min.: (2, 0) **47.** pt. of inflection **49.** $f(x) = -\dfrac{2}{x}$

51. The parent graph is translated 3 units right and then translated 2 units up. The vertical asymptote is now $x = 3$ and the horizontal asymptote is $y = 2$.

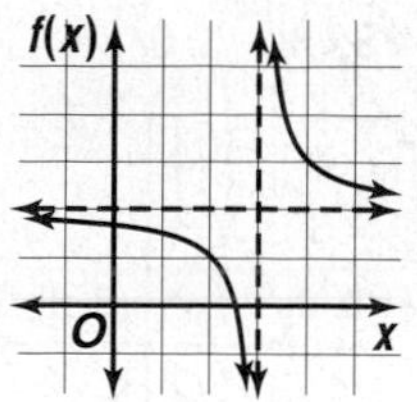

53. $x = -2$
55. yes; $y = x + 2$
57. 140; $y = \dfrac{140}{\sqrt{x}}$; 196
59. $|x - 6.5| \le 0.2$; $6.3 \le x \le 6.7$
61a.

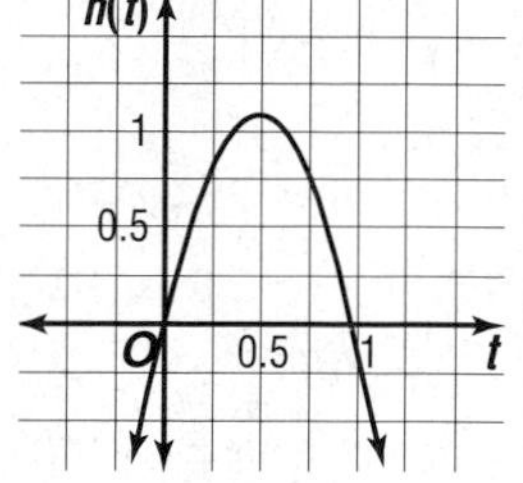

61b. 1.08 m

Page 203 Chapter 3 SAT and ACT Practice

1. E **3.** B **5.** C **7.** B **9.** B

Chapter 4 Polynomial and Rational Functions

Pages 209–212 Lesson 4-1

5. 3; 1 **7.** no; $f(5) = -33$ **9.** $x^2 - 2x - 35 = 0$; even; 2

11. 2; 7, 7

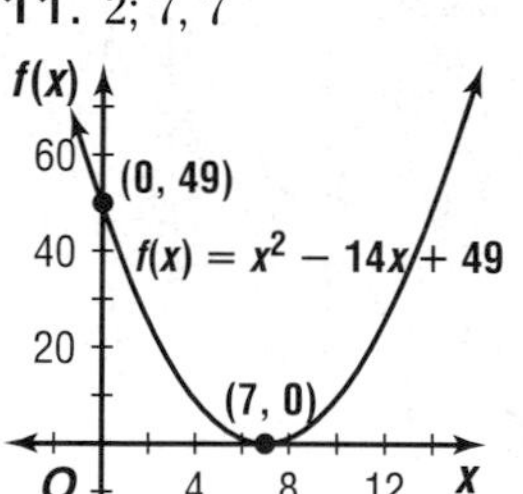

13. 4; −1, 1, $\boldsymbol{i}$, $-\boldsymbol{i}$

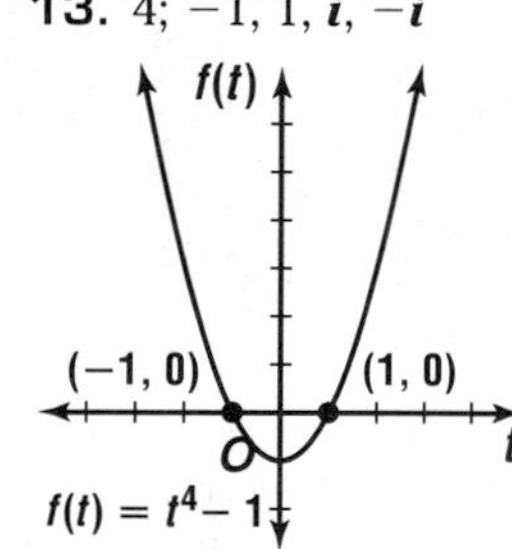

15. 4; 5 **17.** 3; 5 **19.** 6; −1 **21.** Yes; the coefficients are complex numbers and the exponents of the variable are nonnegative integers. **23.** yes; $f(0) = 0$ **25.** yes; $f(1) = 0$ **27.** no; $f(-3) = -72$ **29.** no **31a.** 3; 1 **31b.** 2; 2 **31c.** 4; 2 **33.** $x^3 - 5x^2 - x + 5 = 0$; odd; 3 **35.** $x^3 + 3x^2 + 4x + 12 = 0$; odd; 1 **37.** $x^5 - 5x^4 - 17x^3 + 85x^2 + 16x - 80 = 0$; odd; 5

39. 1; −8

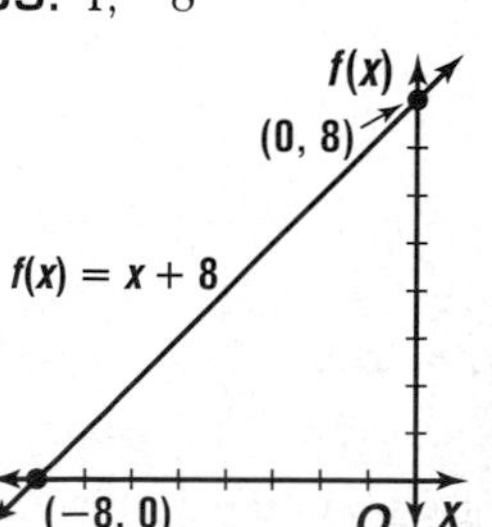

41. 2; $\pm 6\boldsymbol{i}$

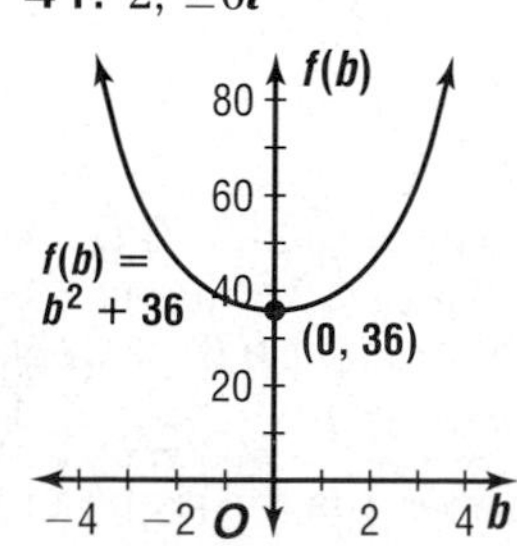

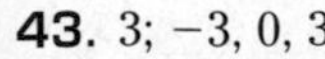

43. 3; −3, 0, 3

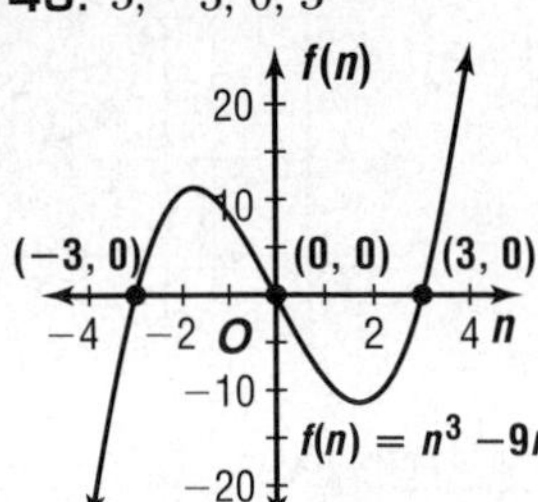

45. 4; −1, 1, $-\sqrt{2}\boldsymbol{i}$, $\sqrt{2}\boldsymbol{i}$

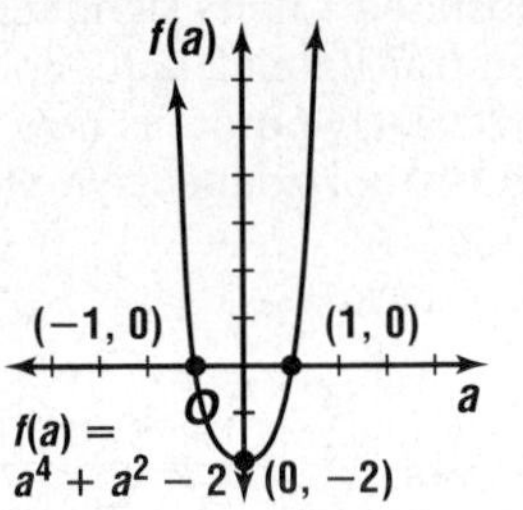

47. 4; $-0.5\boldsymbol{i}$, $0.5\boldsymbol{i}$, $-2\boldsymbol{i}$, $2\boldsymbol{i}$

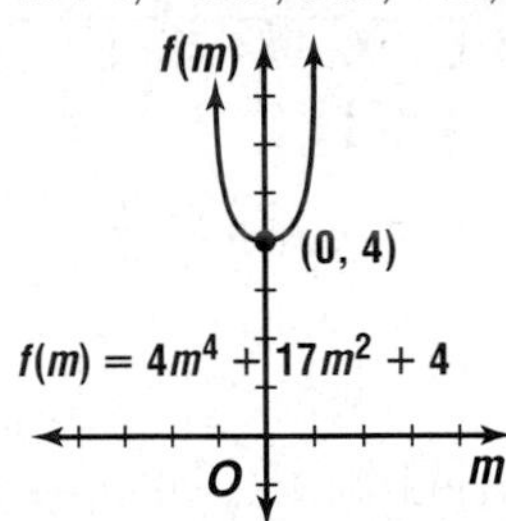

49a.

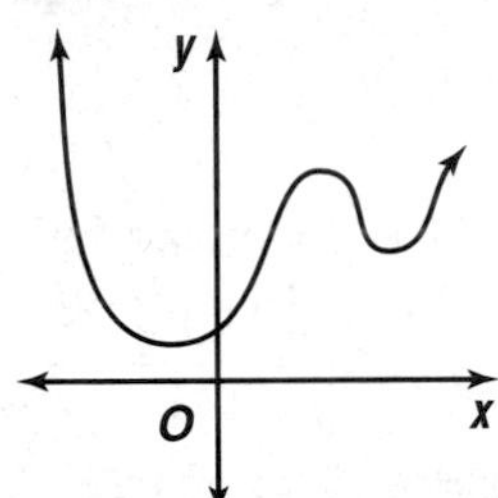

49b.

49c.

49d.

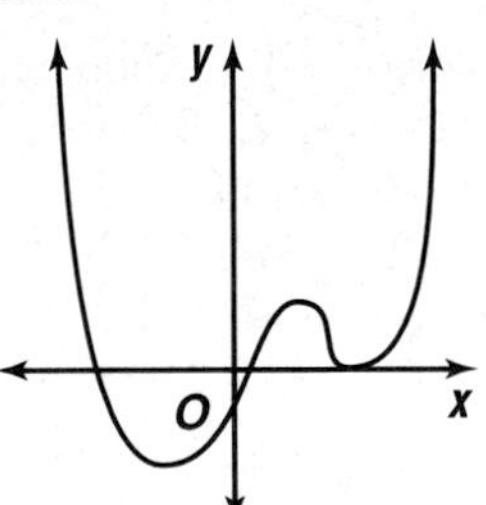

49e.

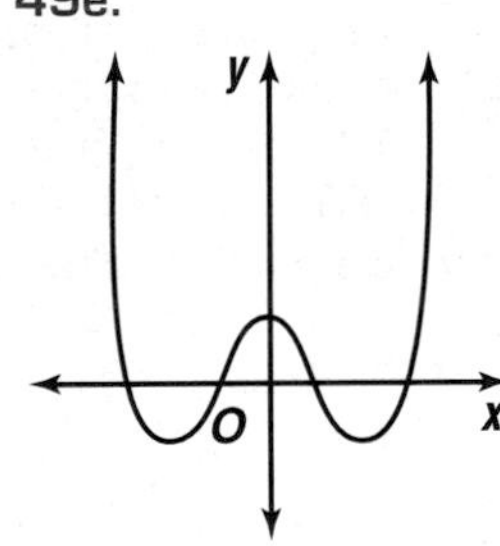

49f. not possible **51a.** $V(x) = 99{,}000x^3 + 55{,}000x^2 + 65{,}000x$ **51b.** about $298,054.13 **53a.** 7380 ft; 29,520 ft; 118,080 ft **53b.** It quadruples; $(2t)^2 = 4t^2$. **55.** $10 **57.** $y = \frac{x-2}{x(x+2)(x-2)}$ **59.** The graph of $y = 2x^3 + 1$ is the graph of $y = 2x^3$ shifted 1 unit up. **61.** 0; no

63.

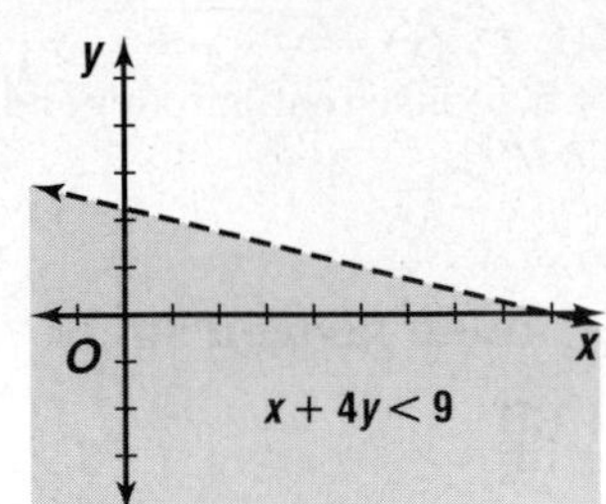

65. $\frac{1}{4}x^2 + 6x + 32$; $\frac{1}{2}x^2 + 4$

Pages 219–221 Lesson 4-2

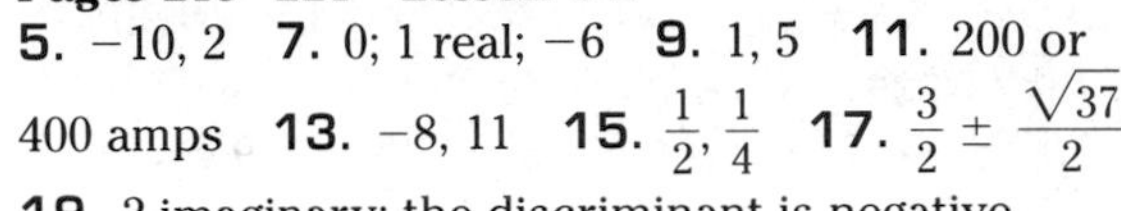

5. −10, 2 **7.** 0; 1 real; −6 **9.** 1, 5 **11.** 200 or 400 amps **13.** −8, 11 **15.** $\frac{1}{2}, \frac{1}{4}$ **17.** $\frac{3}{2} \pm \frac{\sqrt{37}}{2}$
19. 2 imaginary; the discriminant is negative.
21. −11; 2 imaginary; $\frac{5 \pm \boldsymbol{i}\sqrt{11}}{2}$ **23.** −140; 2 imaginary; $\frac{1 \pm \boldsymbol{i}\sqrt{35}}{4}$ **25.** 97; 2 real; $\frac{-5 \pm \sqrt{97}}{4}$
27. $5 + 2\boldsymbol{i}$ **29.** −4, 7 **31.** $-1, \frac{5}{4}$
33. $\sqrt{6} \pm 2\sqrt{2}$ **35.** $c > 16$
37a.

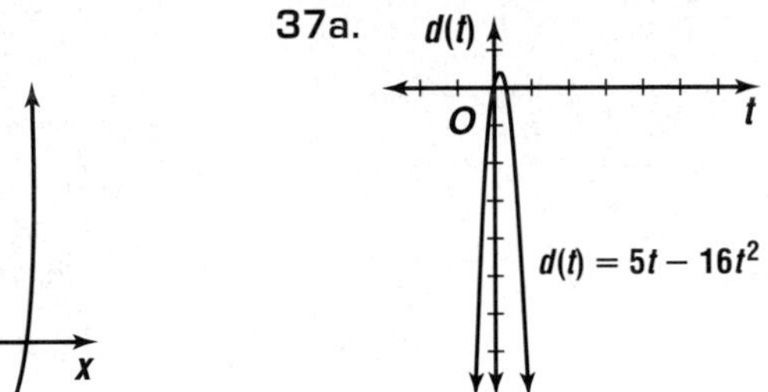

37b. 0 and about 0.3 **37c.** The x-intercepts indicate when the woman is at the same height as the beginning of the jump. **37d.** $-50 = 5t - 16t^2$
37e. about 1.93 s

39. 2; $-\frac{1}{3}, \frac{1}{6}$

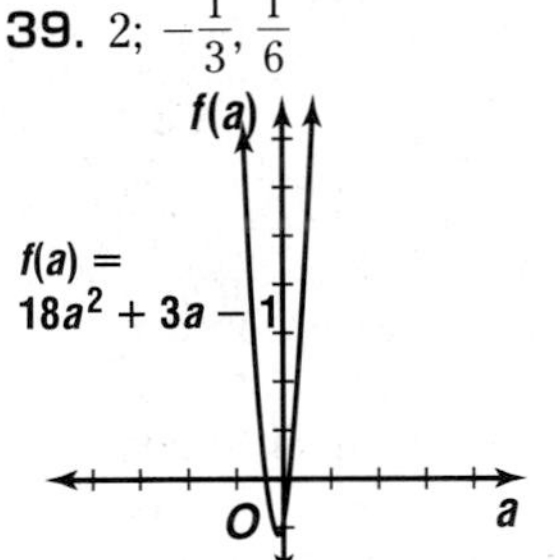

41. $f^{-1}(x) = \pm\sqrt{x + 9}$
43. $643
45. A

Pages 226–228 Lesson 4-3

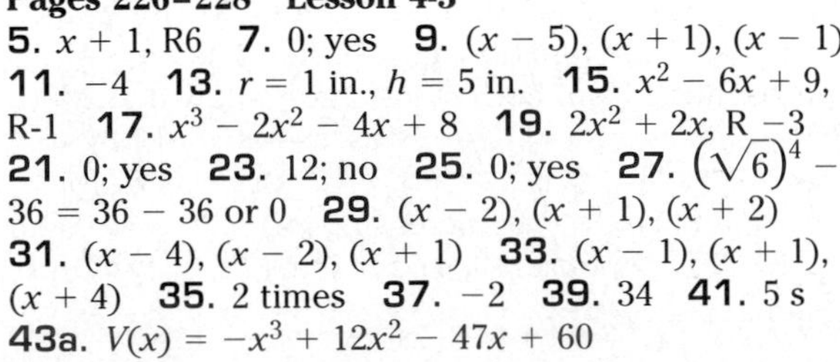

5. $x + 1$, R6 **7.** 0; yes **9.** $(x - 5)$, $(x + 1)$, $(x - 1)$ **11.** −4 **13.** $r = 1$ in., $h = 5$ in. **15.** $x^2 - 6x + 9$, R-1 **17.** $x^3 - 2x^2 - 4x + 8$ **19.** $2x^2 + 2x$, R −3 **21.** 0; yes **23.** 12; no **25.** 0; yes **27.** $(\sqrt{6})^4 - 36 = 36 - 36$ or 0 **29.** $(x - 2)$, $(x + 1)$, $(x + 2)$ **31.** $(x - 4)$, $(x - 2)$, $(x + 1)$ **33.** $(x - 1)$, $(x + 1)$, $(x + 4)$ **35.** 2 times **37.** −2 **39.** 34 **41.** 5 s **43a.** $V(x) = -x^3 + 12x^2 - 47x + 60$

43b.

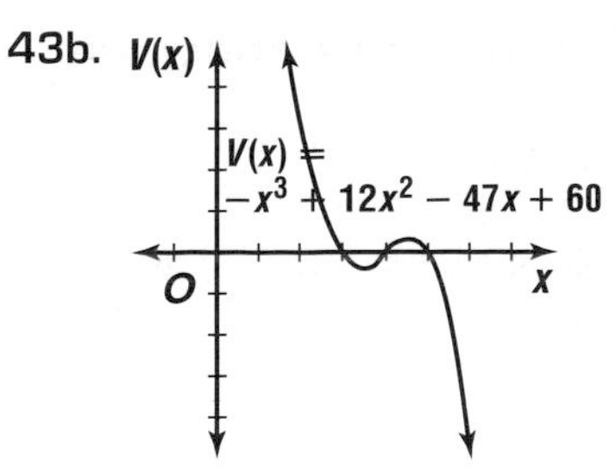

43c. $36 = -x^3 + 12x^2 - 47x + 60$ **43d.** about 0.60 ft

45. $a = 1, b = -6, c = 25$ **47a.** no **47b.** yes **47c.** no **47d.** yes **49.** wider than parent graph and moved 1 unit left **51.** $\left(-\frac{91}{11}, \frac{68}{11}, \frac{98}{11}\right)$ **53.** D

Page 233–235 Lesson 4-4

5. $\pm1, \pm2$; 1 **7.** 2 or 0; 1; $-1\frac{1}{2}, \frac{1}{4}, 2$ **9.** 9 cm **11.** $\pm1, \pm2, \pm3, \pm6, \pm9, \pm18$; -2 **13.** $\pm1, \pm2, \pm4, \pm5, \pm10, \pm20$; $-2, 2, 5$ **15.** $\pm1, \pm\frac{1}{2}, \pm\frac{1}{3}, \pm\frac{1}{6}$; $-\frac{1}{3}, \frac{1}{2}$ **17.** 1; 2 or 0; $-2, -1, 3$ **19.** 1; 2 or 0; $-4, -2, 3$ **21.** 2 or 0; 2 or 0; $-4, -1, 1, 2$ **23a.** $2, -2, -1$ **23b.** $f(x) = x^4 + 2x^3 - 3x^2 - 8x - 4$ **23c.** 1; 3 or 1 **23d.** There are 2 negative zeros, but according to Descartes' rule of signs, there should be 3 or 1. This is because, -1 is actually a zero twice. **25a.** Sample answer: $x^4 + x^3 + x^2 + x + 3 = 0$ **25b.** Sample answer: $x^3 - x^2 - 2 = 0$ **25c.** Sample answer: $x^3 - x = 0$ **27.** 100 ft **29.** $x - 8$ **31.** $x^4 - 5x^2 + 4 = 0$ **33.** A

Pages 240–242 Lesson 4-5

5. 4 and 5, -1 and 0 **7.** 2.3 **9.** Sample answers: 2; 0 **11a.** $V(x) = x^3 + 60x^2 + 1025x + 3750$ **11b.** $5625 = x^3 + 60x^2 + 1025x + 3750$ **11c.** about 26.7 cm by 31.7 cm by 6.7 cm **13.** 0 and 1, 2 and 3 **15.** -3 and -2, -2 and -1, 1 and 2, 2 and 3 **17.** no real zeros **19.** $-0.7, 0.7$ **21.** -2.5 **23.** $-1, 1$ **25.** -1.24 **27.** Sample answers: 2; -1 **29.** Sample answers: 2; -6 **31.** Sample answers: 1; -7 **33a.** The model is fairly close, although it is less accurate for 1950 and 1970. **33b.** $-253{,}800$ **33c.** The population becomes 0. **33d.** No; there are still many people living in Manhattan. **35a.** $37.44 = 60x^3 + 60x^2 + 60x$ **35b.** $f(x) = 60x^3 + 60x^2 + 60x - 37.44$

35c.

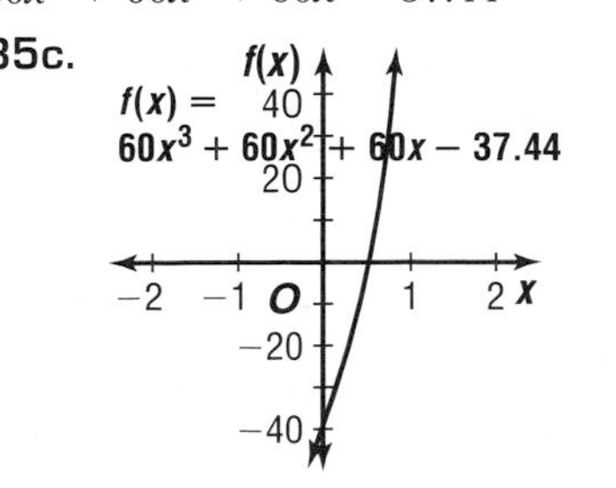

about $\frac{1}{2}$ **35d.** 0.4

37a.

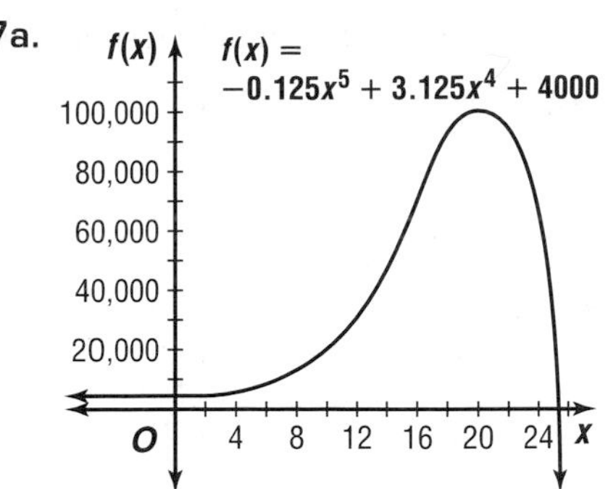

37b. 4000 deer **37c.** about 67,281 deer **37d.** in 1930 **39.** 2 or 0; 1; $-3, 0.5, 5$

41.

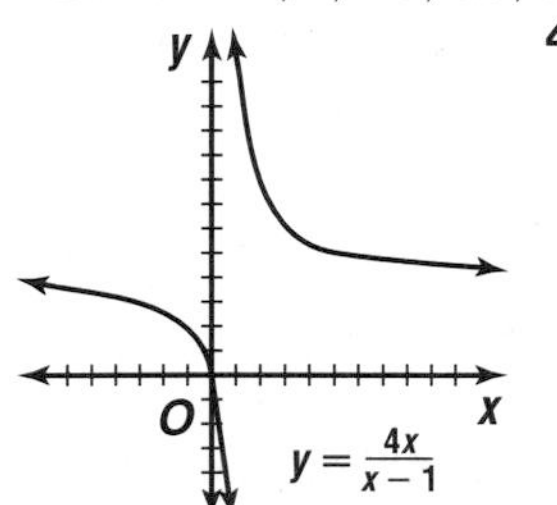

43. (2.5, 1) **45.** B

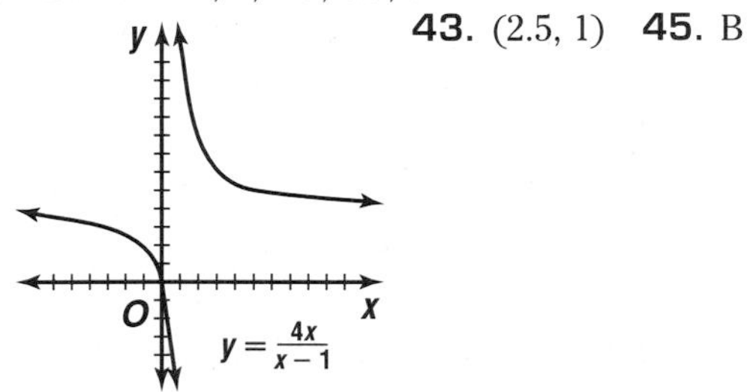

Pages 247–250 Lesson 4-6

5. $-1, 5$ **7.** -3 **9.** $x < 0, x > 3$ **11a.** $\frac{3 \times 60 + 20}{3 + x} = 57.14$ **11b.** 0.50 h **13.** -34 **15.** $\frac{5}{3}$ **17.** $-\frac{1}{2}, 3$ **19.** $\frac{-3 \pm 3\sqrt{2}}{2}$ **21.** $\frac{5}{13}$ **23.** $\frac{3}{x} + \frac{-2}{x - 2}$ **25.** $\frac{2}{3y - 1} + \frac{-2}{y - 1}$ **27a.** $a(a - 6)$ **27b.** 3 **27c.** 0, 6 **27d.** $0 < a < 3, 6 < a$ **29.** $x \le 3, 4 \le x < 5$ **31.** $0 < a < \frac{7}{4}$ **33.** $-1 < y < 0$ **35.** $x < -5$, or $x > 5$ **37.** Sample answer: $\frac{x}{x - 3} = \frac{1}{x + 2}$ **39a.** $\frac{1}{10} = \frac{1}{2r} + \frac{1}{r} + \frac{1}{20}$ **39b.** 60 ohms, 30 ohms **41.** 36 mph **43a.** $\frac{1}{x} = \frac{1}{2}\left(\frac{1}{30} + \frac{1}{45}\right)$ **43b.** 36 **45.** 8 mph **47.** -3 and -2, -2 and -1, 1 and 2 **49.** 2; $\frac{5}{6}, -\frac{3}{2}$ **51.** no **53a.** 18 short answer and 2 essay for a score of 120 points **53b.** 12 short answer and 8 essay for a score of 180 points **55.** $2x - y + 7 = 0$ **57.** 24

Pages 254–257 Lesson 4-7

5. -733 **7.** 3.5 **9.** $-0.8 \le x \le 12$ **11a.** $90 = \sqrt{100 + 64h}$ **11b.** 125 ft **13.** 71 **15.** 0 **17.** no real solution **19.** -1 **21.** 4 **23.** $\frac{9}{7}$ **25.** no real solution **27.** -2 **29.** $x \ge 16$ **31.** $5 \le a \le 21$ **33.** $1.8 \le y \le 5$ **35.** $c > 27$ **37.** about 7.88 **39a.** about 2.01 s **39b.** about 2.11 s **39c.** It must be multiplied by 4. **41.** $a + b < 0$ **43.** $\frac{3}{2}$ **45a.** point discontinuity

47. $2\pi k, \frac{\pi}{2} + 2\pi k$ **49.** $\frac{5\pi}{6} \le \theta \le \frac{7\pi}{6}$
51. $0 \le \theta < \frac{\pi}{4}$ or $\frac{3\pi}{4} < \theta < 2\pi$ **53.** 0, 1.8955
55. 0.01° **57.** 30.29° or 59.71° **59.** 0.0013 s
61. 341.32° **63.** Sample answer: $\sin x = \frac{\sqrt{2}}{5}$
65. about 18 rps **67.** $(x - 2)(x + 1)(x + 1)$
69. (4, 3) **71.**

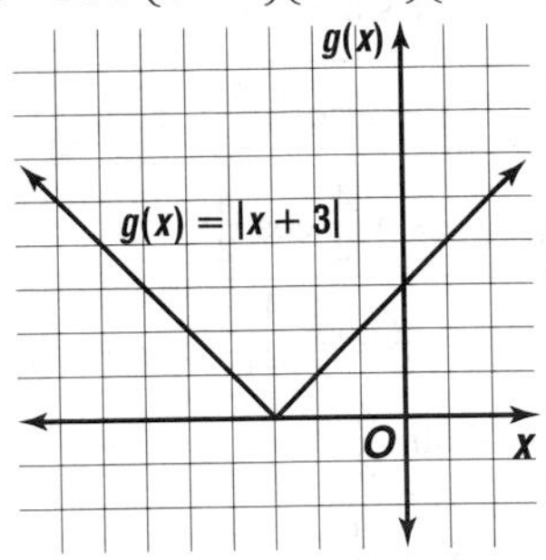

Pages 467–469 Lesson 7-6

5. $\sqrt{3}x + y - 20 = 0$ **7.** $x - y - 10 = 0$
9. $\frac{3\sqrt{10}}{10}x + \frac{\sqrt{10}}{10}y - \frac{\sqrt{10}}{5} = 0; \frac{\sqrt{10}}{5}$; 18°
11a.

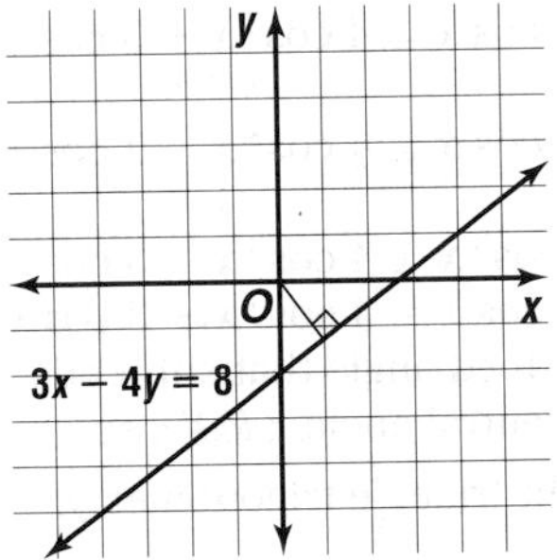

11b. 1.6 miles **13.** $\sqrt{2}x + \sqrt{2}y - 24 = 0$
15. $\sqrt{3}x - y + 4\sqrt{3} = 0$ **17.** $\sqrt{3}x + y + 10 = 0$
19. $x - \sqrt{3}y - 3 = 0$ **21.** $-\frac{5}{13}x - \frac{12}{13}y - 5 = 0$; 5; 247° **23.** $\frac{3}{5}x - \frac{4}{5}y - 3 = 0$; 3; 307°
25. $x - 3 = 0$; 3; 0° **27.** $-\frac{\sqrt{17}}{17}x + \frac{4\sqrt{17}}{17}y - \frac{28\sqrt{17}}{17} = 0; \frac{28\sqrt{17}}{17}$; 104° **29.** $\frac{6\sqrt{61}}{61}x + \frac{5\sqrt{61}}{61}y - \frac{120\sqrt{61}}{61} = 0; \frac{120\sqrt{61}}{61}$; 40°
31. $x - y - 8 = 0$
33a.

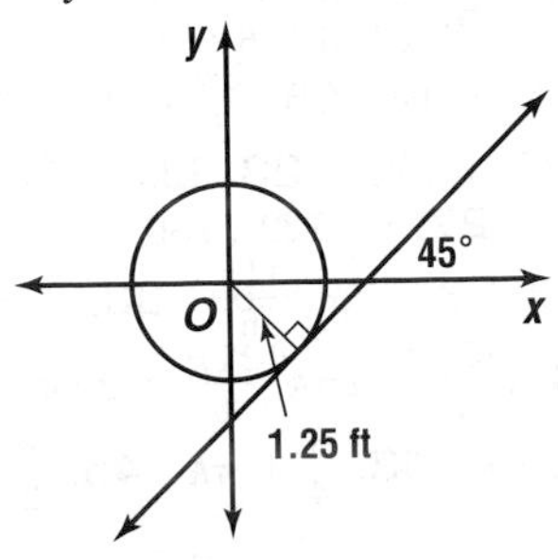

33b. $\sqrt{2}x - \sqrt{2}y - 2.5 = 0$ **35a.** $\frac{5}{13}x + \frac{12}{13}y - 3 = 0$ **35b.** $\theta = 67°$

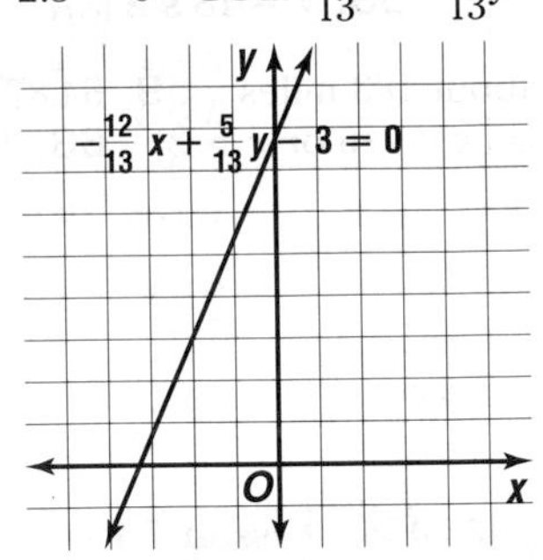

35d. The line with normal form $x \cos \phi + y \sin \phi - p = 0$ makes an angle of ϕ with the positive x-axis and has a normal of length p. The graph of Armando's equation is a line whose normal makes an angle of $\phi + \delta$ with the x-axis and also has length p. Therefore, the graph of Armando's equation is the graph of the original line rotated $\delta°$ counterclockwise about the origin. Armando is correct. **37.** \$6927.82 **39.** $\frac{2\sqrt{35} + \sqrt{5}}{18}$
41. 3.05 cm **43.** 4.5 in. by 6.5 in. by 2.5 in.
45. (−6, −3)

Pages 474–476 Lesson 7-7

5. $\frac{2\sqrt{13}}{13}$ **7.** $\frac{2\sqrt{34}}{17}$ **9.** $(20 - 6\sqrt{13})x - (30 + 8\sqrt{13})y - 40 - 5\sqrt{13} = 0; (20 + 6\sqrt{13})x + (8\sqrt{13} - 30)y - 40 + 5\sqrt{13} = 0$ **11.** $\frac{21}{5}$
13. $\frac{3\sqrt{5}}{5}$ **15.** 0 **17.** $\frac{\sqrt{10}}{10}$ **19.** $\frac{6\sqrt{41}}{41}$
21. $\frac{\sqrt{10}}{5}$ **23.** $\frac{8\sqrt{13}}{13}$ **25.** $x + 8y = 0; 16x - 2y - 65 = 0$ **27.** $(2\sqrt{10} + 3\sqrt{13})x + (\sqrt{13} - 3\sqrt{10})y + 3\sqrt{10} + 2\sqrt{13} = 0; (-2\sqrt{10} + 3\sqrt{13})x + (\sqrt{13} + 3\sqrt{10})y - 3\sqrt{10} + 2\sqrt{13} = 0$ **29.** 1.09 m **31.** $\frac{34}{5}, \frac{34\sqrt{53}}{53}, \frac{17\sqrt{26}}{26}$
33. $-\frac{2\sqrt{53}}{53}x + \frac{7\sqrt{53}}{53}y - \frac{5\sqrt{53}}{53} = 0$
35.

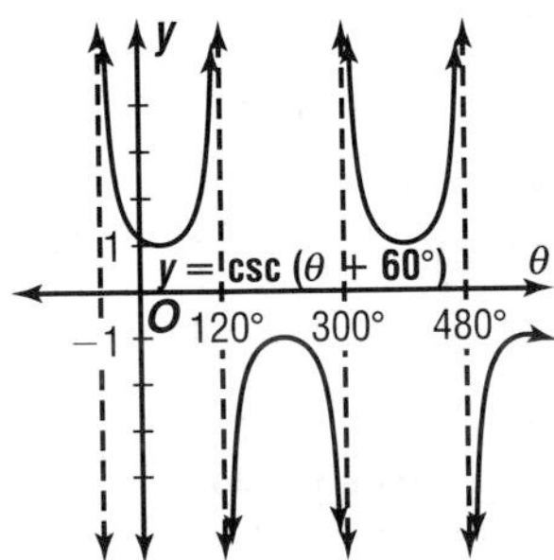

37. about 2.8 s **39.** (−6, −2, −5)

Pages 477–481 Chapter 7 Study Guide and Assessment

1. b **3.** d **5.** i **7.** h **9.** e **11.** 2 **13.** $\frac{4}{5}$
15. $\sin x$

17.
$$\frac{1-\cos\theta}{1+\cos\theta} \stackrel{?}{=} (\csc\theta - \cot\theta)^2$$
$$\frac{1-\cos\theta}{1+\cos\theta} \stackrel{?}{=} \left(\frac{1}{\sin\theta} - \frac{\cos\theta}{\sin\theta}\right)^2$$
$$\frac{1-\cos\theta}{1+\cos\theta} \stackrel{?}{=} \frac{(1-\cos\theta)^2}{\sin^2\theta}$$
$$\frac{1-\cos\theta}{1+\cos\theta} \stackrel{?}{=} \frac{(1-\cos\theta)^2}{1-\cos^2\theta}$$
$$\frac{1-\cos\theta}{1+\cos\theta} \stackrel{?}{=} \frac{(1-\cos\theta)^2}{(1-\cos\theta)(1+\cos\theta)}$$
$$\frac{1-\cos\theta}{1+\cos\theta} = \frac{1-\cos\theta}{1+\cos\theta}$$

19.
$$\frac{\sin^4 x - \cos^4 x}{\sin^2 x} \stackrel{?}{=} 1 - \cot^2 x$$
$$\frac{(\sin^2 x - \cos^2 x)(\sin^2 x + \cos^2 x)}{\sin^2 x} \stackrel{?}{=} 1 - \cot^2 x$$
$$\frac{\sin^2 x - \cos^2 x}{\sin^2 x} \stackrel{?}{=} 1 - \cot^2 x$$
$$1 - \frac{\cos^2 x}{\sin^2 x} \stackrel{?}{=} 1 - \cot^2 x$$
$$1 - \cot^2 x = 1 - \cot^2 x$$

21. $\frac{\sqrt{6}+\sqrt{2}}{4}$ **23.** $-2+\sqrt{3}$ **25.** $-\frac{180+82\sqrt{5}}{61}$

27. $\frac{\sqrt{2-\sqrt{2}}}{2}$ **29.** $2-\sqrt{3}$ **31.** $-\frac{7}{25}$

33. $-\frac{336}{625}$ **35.** 0°, 90°, 270° **37.** $\pi k, \frac{\pi}{4} + 2\pi k, \frac{3\pi}{4} + 2\pi k$ **39.** $2\pi k$ **41.** $y - 5 = 0$ **43.** $x + y + 8 = 0$ **45.** $-\frac{3\sqrt{13}}{13}x + \frac{2\sqrt{13}}{13}y - \frac{5\sqrt{13}}{26} = 0$; $\frac{5\sqrt{13}}{26}$; 146° **47.** $-\frac{\sqrt{2}}{10}x + \frac{7\sqrt{2}}{10}y - \frac{\sqrt{2}}{2} = 0$; $\frac{\sqrt{2}}{2}$; 98° **49.** $\frac{23\sqrt{13}}{13}$ **51.** $\frac{21\sqrt{10}}{10}$ **53.** $\frac{14}{5}$

55. $\frac{9\sqrt{13}}{13}$ **57.** $(-\sqrt{34} - 3\sqrt{10})x + (3\sqrt{34} + 5\sqrt{10})y - 2\sqrt{34} - 15\sqrt{10} = 0$; $(-\sqrt{34} + 3\sqrt{10})x + (3\sqrt{34} - 5\sqrt{10})y - 2\sqrt{34} + 15\sqrt{10} = 0$ **59.** 1431 ft

Page 483 Chapter 7 SAT and ACT Practice

1. B **3.** D **5.** B **7.** A **9.** C

Chapter 8 Vectors and Parametric Equations

Pages 490–492 Lesson 8-1

5. 1.2 cm, 120° **7.** 1.4 cm, 20° **9.** 2.6 cm, 210°
11. 2.9 cm, 12°
13a.
5 100

13b. ≈100.12 m/s **15.** 1.4 cm, 45° **17.** 3.0 cm, 340° **19.** 3.4 cm, 25° **21.** 5.5 cm, 324° **23.** 5.2 cm, 128° **25.** 8.2 cm, 322° **27.** 5.4 cm, 133° **29.** 3.4 cm, 301° **31.** −1.60 cm, 2.05 cm **33.** 2.04 cm, 0.51 cm **35.** 45.73 m
37. Yes; Sample answer:

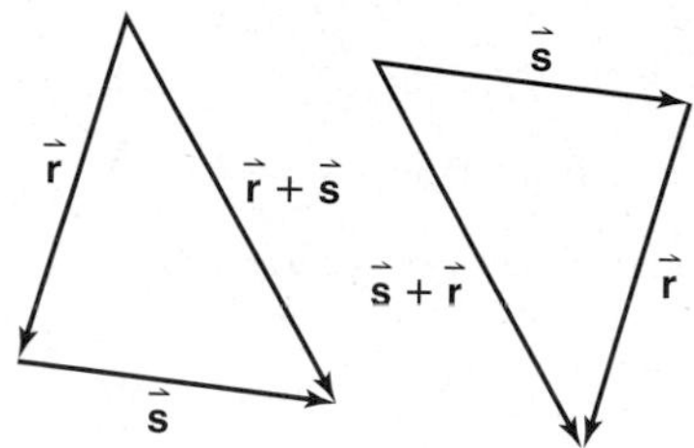

39. Sometimes;

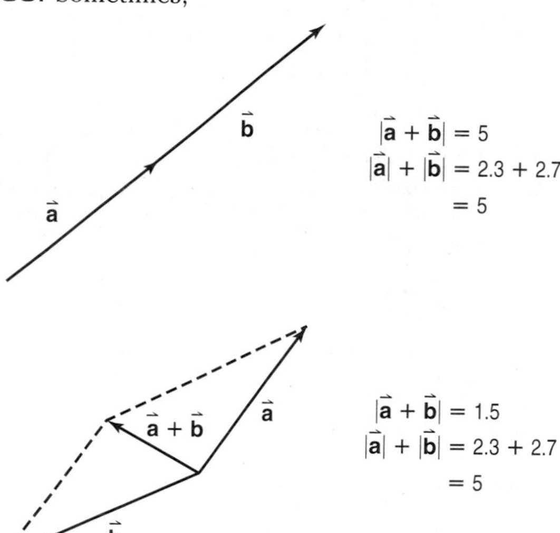

41. 36 mph, 30 mph **43.** 71 lb
45. $x - (1+\sqrt{2})y + 2 + 5\sqrt{2} = 0$ **47.** $\frac{\pi}{4} + \pi n$ where n is an integer **49.** 15.8 cm; 29.9 cm
51. $x = -3, x = 1, y = 0$

Pages 496–499 Lesson 8-2

5. $\langle -5, -1\rangle, \sqrt{26}$ **7.** $\langle 2, 2\rangle$ **9.** $\langle 14, 4\rangle$ **11.** 10, $8\vec{i} - 6\vec{j}$ **13.** ≈927 N **15.** $\langle 4, -5\rangle, \sqrt{41}$
17. $\langle -5, -7\rangle, \sqrt{74}$ **19.** $\langle 5, 7\rangle, \sqrt{74}$ **21.** $\langle 9, 9\rangle, 9\sqrt{2}$ **23.** $\langle 2, 11\rangle$ **25.** $\langle -2, 19\rangle$ **27.** $\langle -22, 29\rangle$
29. $\langle 18, 9\rangle$ **31.** $\langle 20, 50\rangle$ **33.** $\left\langle \frac{32}{3}, -\frac{34}{3}\right\rangle$
35. $\langle -30, 4.5\rangle$ **37.** $\sqrt{13}, 2\vec{i} - 3\vec{j}$
39. 12.5, $3.5\vec{i} + 12\vec{j}$ **41.** $2\sqrt{353}, -16\vec{i} - 34\vec{j}$
43.
$$\begin{aligned}(\vec{v}_1 + \vec{v}_2) + \vec{v}_3 &= [\langle a, b\rangle + \langle c, d\rangle] + \langle e, f\rangle \\ &= \langle a + c, b + d\rangle + \langle e, f\rangle \\ &= \langle (a + c) + e, (b + d) + f\rangle \\ &= \langle a + (c + e), b + (d + f)\rangle \\ &= \langle a, b\rangle + \langle c + e, d + f\rangle \\ &= \langle a, b\rangle + [\langle c, d\rangle + \langle e, f\rangle] \\ &= \vec{v}_1 + (\vec{v}_2 + \vec{v}_3)\end{aligned}$$

45a.

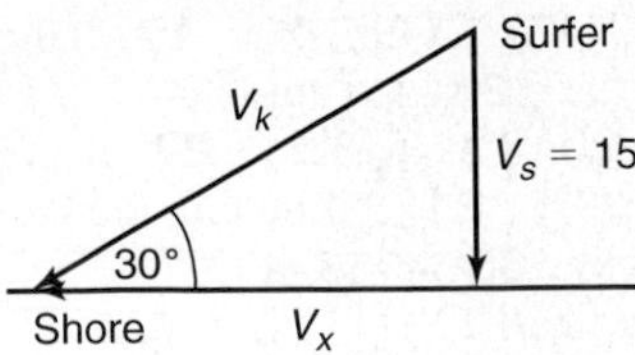

45b. 30 mph **47a.** 30 s **47b.** 30 m **47c.** 5.1 m/s **49.** None **51.** $\frac{-\sqrt{6}-\sqrt{2}}{4}$ **53.** ≈1434 ft; ≈86,751 sq ft **55.** max: (0, 3), min: (0.67, 2.85) **57.** A

Pages 502–504 Lesson 8-3

5. $\langle 5, 4, -11\rangle, 9\sqrt{2}$ **7.** $\langle 6, 0, -25\rangle$ **9.** $11\vec{\mathbf{i}} - 4\vec{\mathbf{j}} + 2\vec{\mathbf{k}}$ **11.** ≈3457 N

13. $\sqrt{69}$

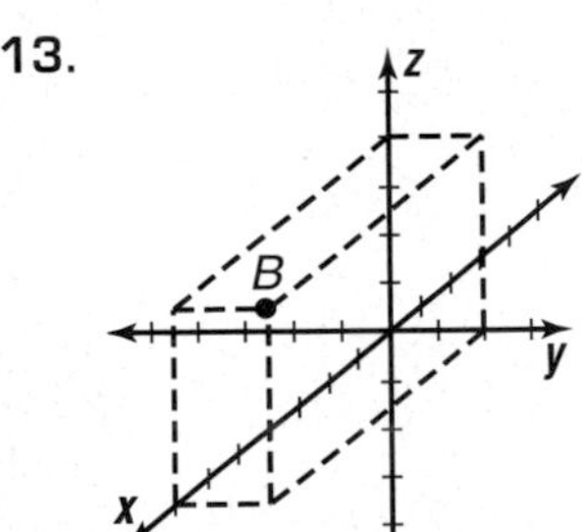

15. $\langle 1, -4, -8\rangle, 9$ **17.** $\langle 1, -4, -4\rangle, \sqrt{33}$ **19.** $\langle 3, -9, -9\rangle, 3\sqrt{19}$ **21.** $\langle 4, -8, -14\rangle, 2\sqrt{69}$ **23.** $\left\langle 6, -7\frac{1}{2}, 11\frac{1}{2}\right\rangle$ **25.** $\left\langle 16\frac{2}{3}, -13, 23\frac{2}{3}\right\rangle$ **27.** $\langle -5, -24, 8\rangle$ **29.** $3\vec{\mathbf{i}} - 8\vec{\mathbf{j}} - 5\vec{\mathbf{k}}$ **31.** $-2\vec{\mathbf{i}} - 4\vec{\mathbf{j}} + 9\vec{\mathbf{k}}$ **33.** $-7\vec{\mathbf{i}} + 3\vec{\mathbf{j}} - 6\vec{\mathbf{k}}$ **35.** $|\overrightarrow{G_1G_2}| = \sqrt{(x_2-x_1)^2 + (y_2-y_1)^2 + (z_2-z_1)^2} = \sqrt{(x_1-x_2)^2 + (y_1-y_2)^2 + (z_1-z_2)^2} = |\overrightarrow{G_2G_1}|$ because $(x-y)^2 = (y-x)^2$ for all real numbers x and y. **37.** $\langle -9, 0, -9\rangle$ **39a.** $\vec{\mathbf{i}} + 4\vec{\mathbf{j}}$ **39b.** $-\vec{\mathbf{i}}$

41a.

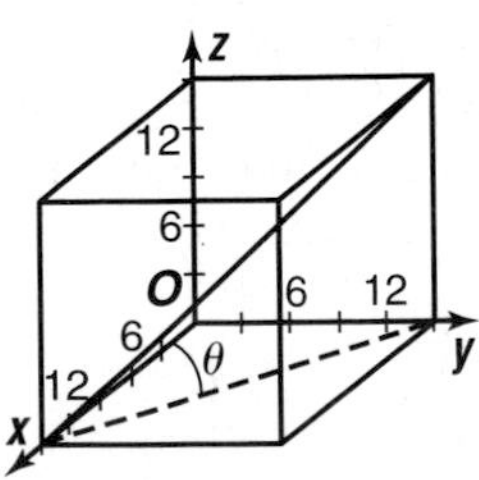

41b. About 26 ft **41c.** $\theta = 35.25°$ **43.** $\langle 2, 7\rangle$

45.
$$\frac{\sin 2X}{1-\cos 2X} = \cot X$$
$$\frac{2\sin X\cos X}{1-\cos^2 X+\sin^2 X} = \cot X$$
$$\frac{2\sin X\cos X}{2\sin^2 X} = \cot X$$
$$\frac{\cos X}{\sin X} = \cot X$$
$$\cot X = \cot X$$

47. $6, 4\pi$ **49.** yes, because substituting 7 for x and -2 for y results in the inequality $-2 < 180$ which is true.

Pages 508–511 Lesson 8-4

5. 0, yes **7.** $\langle 13, 1, -5\rangle$, yes; $\langle 13, 1, -5\rangle \cdot \langle 1, -3, 2\rangle = 13(1) + 1(-3) + (-5)(2) = 13 - 3 - 10 = 0$; $\langle 13, 1, -5\rangle \cdot \langle -2, 1, -5\rangle = 13(-2) + 1(1) + (-5)(-5) = -26 + 1 + 25 = 0$ **9.** Sample answer: $\langle 1, -8, 5\rangle$ **11.** 0, yes **13.** −21, no **15.** 32, no **17.** 6, no **19.** 9, no **21.** $\langle 2, 2, -1\rangle$, yes; $\langle 2, 2, -1\rangle \cdot \langle 0, 1, 2\rangle = 2(0) + 2(1) + (-1)(2) = 2 + 2 - 2 = 0$; $\langle 2, 2, -1\rangle \cdot \langle 1, 1, 4\rangle = 2(1) + 2(1) + (-1)(4) = 2 + 2 - 4 = 0$ **23.** $\langle 0, 0, 10\rangle$, yes; $\langle 0, 0, 10\rangle \cdot \langle 3, 2, 0\rangle = 0(3) + 0(2) + 10(0) = 0 + 0 + 0 = 0$; $\langle 0, 0, 10\rangle \cdot \langle 1, 4, 0\rangle = 0(1) + 0(4) + 10(0) = 0 + 0 + 0 = 0$ **25.** $\langle 8, 8, 16\rangle$, yes; $\langle 8, 8, 16\rangle \cdot \langle -3, -1, 2\rangle = 8(-3) + 8(-1) + 16(2) = -24 - 8 + 32 = 0$; $\langle 8, 8, 16\rangle \cdot \langle 4, -4, 0\rangle = 8(4) + 8(-4) + 16(0) = 32 - 32 + 0 = 0$ **27.** Sample answer: Let $\vec{\mathbf{v}} = \langle v_1, v_2, v_3\rangle$ and $-\vec{\mathbf{v}} = \langle -v_1, -v_2, -v_3\rangle$

$$\vec{\mathbf{v}} \times (-\vec{\mathbf{v}}) = \begin{bmatrix} \vec{\mathbf{i}} & \vec{\mathbf{j}} & \vec{\mathbf{k}} \\ v_1 & v_2 & v_3 \\ -v_1 & -v_2 & -v_3 \end{bmatrix}$$

$$= \begin{bmatrix} v_2 & v_3 \\ -v_2 & -v_3 \end{bmatrix}\vec{\mathbf{i}} - \begin{bmatrix} v_1 & v_3 \\ -v_1 & -v_3 \end{bmatrix}\vec{\mathbf{j}} + \begin{vmatrix} v_1 & v_2 \\ -v_1 & -v_2 \end{vmatrix}\vec{\mathbf{k}}$$

$$= 0\vec{\mathbf{i}} - 0\vec{\mathbf{j}} + 0\vec{\mathbf{k}} = \vec{0}$$

29. Sample answer: $\langle -2, -17, -14\rangle$ **31.** Sample answer: $\langle 0, 2, -1\rangle$

33a.

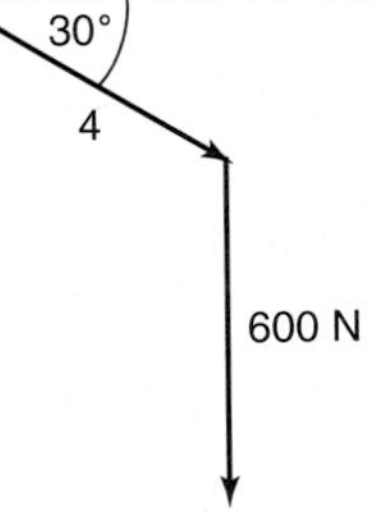

33b. 21 N · m

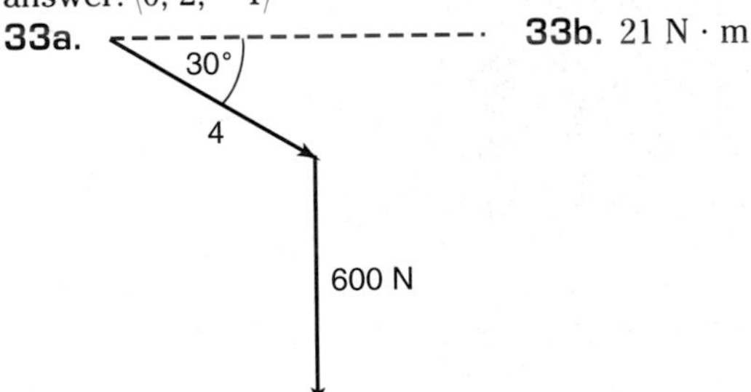

35a. $\vec{\mathbf{o}} = \langle 120, 310, 60\rangle$, $\vec{\mathbf{c}} = \langle 29, 18, 21\rangle$ **35b.** \$10,320 **37a.** Sample answer: $\langle 8, -7, -9\rangle$ **37b.** The cross product of two vectors is always a vector perpendicular to the two vectors and the plane in which they lie. **39.** $-\frac{19}{29}$ **41.** $\langle 2, 0, 3\rangle$

43. $\frac{4}{\sqrt{17}}x + \frac{1}{\sqrt{17}}y - \frac{6}{\sqrt{17}} = 0$; $\frac{6}{\sqrt{17}} \approx 1.46$ units; 76° **45.** 13.1 meters; 13.7 meters **47.** B

Pages 516–519 Lesson 8-5

5. 421.19 N, 19.3° **7.** 13.79 N, 11.57 N

9a.

4 mph

12 mph

9b. ≈ 19.5°

11.

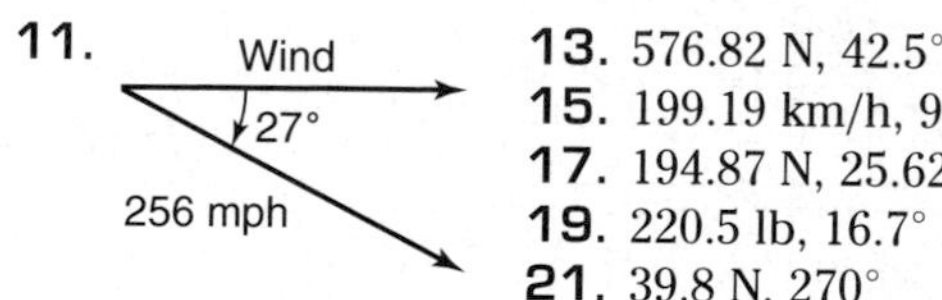

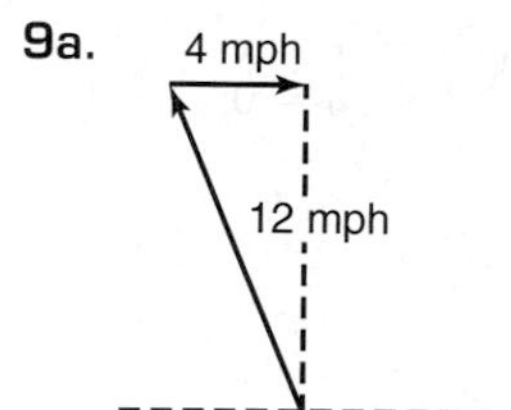

13. 576.82 N, 42.5°
15. 199.19 km/h, 90°
17. 194.87 N, 25.62°
19. 220.5 lb, 16.7°
21. 39.8 N, 270°
23. 19.9 N, 5.3° west of south **25.** 1,542,690 N · m
27a. 9.5° south of east **27b.** 18.2 mph
29. Left side: 760 lb, Right side: 761 lb
31. 3192.5 tons **33.** −2, no **35.** 239.4 ft
37. 30% beef, 20% pork; $76

Pages 523–525 Lesson 8-6

5. $\langle x - 1, y - 5\rangle = t\langle -7, 2\rangle$; $x = 1 - 7t$, $y = 5 + 2t$
7. $x = t$, $y = \frac{2}{3}t + 2$ **9.** $y = \frac{4}{9}x + 2$ **11a.** Receiver: $x = 5$, $y = 50 - 10t$; Defensive player: $x = 10 - 0.9t$, $y = 54 - 10.72t$ **11b.** yes
13. $\langle x + 1, y - 4\rangle = t\langle 6, -10\rangle$; $x = -1 + 6t$, $y = 4 - 10t$ **15.** $\langle x - 1, y - 5\rangle = t\langle -7, 2\rangle$; $x = 1 - 7t$, $y = 5 + 2t$ **17.** $\langle x - 3, y + 5\rangle = t\langle -2, 5\rangle$; $x = 3 - 2t$, $y = -5 + 5t$ **19.** $x = t$, $y = \frac{3}{4}t + \frac{7}{4}$ **21.** $x = t$, $y = -9t - 1$ **23.** $x = t$, $y = 4t - 2$
25. $y = -\frac{1}{2}x + 1$ **27.** $y = \frac{1}{4}x + \frac{23}{4}$ **29.** $y = \frac{5}{2}x - \frac{17}{2}$ **31a.** $\langle x - 11, y + 4\rangle = t\langle 3, 7\rangle$ **31b.** $x = 3t + 11$, $y = 7t - 4$ **31c.** $y = \frac{7}{3}x - \frac{89}{3}$

33.

T	X1T	Y1T
1.0000	8.0000	2.0000
2.0000	11.000	3.0000
3.0000	14.000	4.0000
4.0000	17.000	5.0000
5.0000	20.000	6.0000
6.0000	23.000	7.0000
14.000	47.000	15.000

Y1T=2

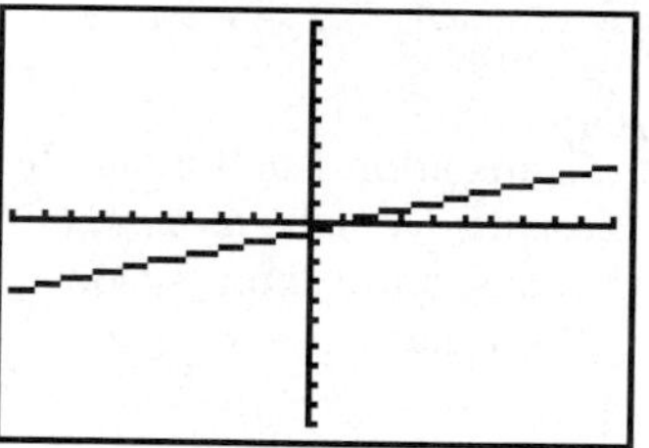

[−10, 10] tstep:1 [−20, 20] Xscl:2 [−20, 20] Yscl:2

35a. Right of point (2, 4) **35b.** $t < -\frac{2}{3}$
37a. Target drone: $x = 3 - t$, $y = 4$; Missile: $x = 2 + t$, $y = 2 + 2t$ **37b.** No **39.** $x = -\frac{1}{3} + \frac{1}{3}t$, $y = 1 + 4t$, $z = 1 - 9t$ **41.** −3, no **43.** 1 **45.** $x - y + 4 = 0$

Pages 531–533 Lesson 8-7

5. 12.86 m/s **7.** 7.05 m/s, 2.57 m/s **9.** 32.5 ft/s, 56.29 ft/s **11.** 891.77 ft/s, 802.96 ft/s
13. 55.11 yd/s, 41.53 yd/s **15a.** $x = 175t \cos 35°$, $y = 175t \sin 35° - 16t^2$ **15b.** 899.32 ft or 299.77 yd
17a. 158.32 ft/s **17b.** 127 yd. **19.** Sample answer: No, the projectile will travel four times as far. **21a.** 323.2 ft **21b.** 312.4 ft **21c.** 3.71 s
23a. 140.7 ft/s **23b.** 131.3 yd **25.** $y = 6x - 58$
27. 37° **29.** B

Pages 540–542 Lesson 8-8

5a. $\begin{bmatrix} 5 & 5 & 0 & 0 & 0 & 5 & 5 & 0 \\ 2 & 5 & 5 & 2 & 2 & 2 & 5 & 5 \\ 0 & 0 & 0 & 0 & 4 & 4 & 4 & 4 \end{bmatrix}$

5b. $\begin{bmatrix} 9 & 9 & 4 & 4 & 4 & 9 & 9 & 4 \\ 1 & 4 & 4 & 1 & 1 & 1 & 4 & 4 \\ 2 & 2 & 2 & 2 & 6 & 6 & 6 & 6 \end{bmatrix}$

5c.

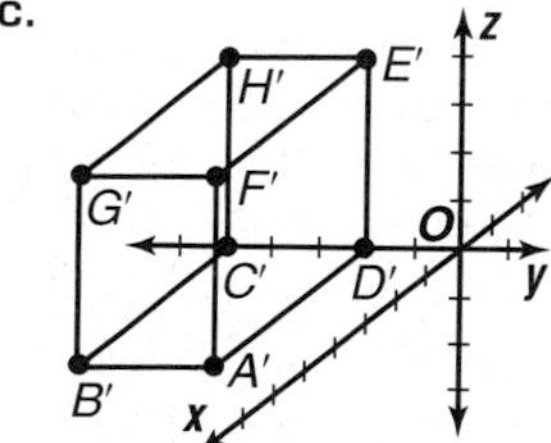

Reflection over the xz-plane.

5d. The dimensions of the resulting figure are half the original.

7. $\begin{bmatrix} 2 & 3 & 4 & 4 & 2 & 3 \\ -3 & 1 & 1 & 7 & 3 & 7 \\ 2 & 4 & -1 & -1 & 2 & 4 \end{bmatrix}$

9. $\begin{bmatrix} 2 & 1 & 4 & 4 & 3 & 6 \\ -2 & 0 & -1 & -1 & 1 & 0 \\ 3 & 4 & 2 & 1 & 2 & 0 \end{bmatrix}$

11.

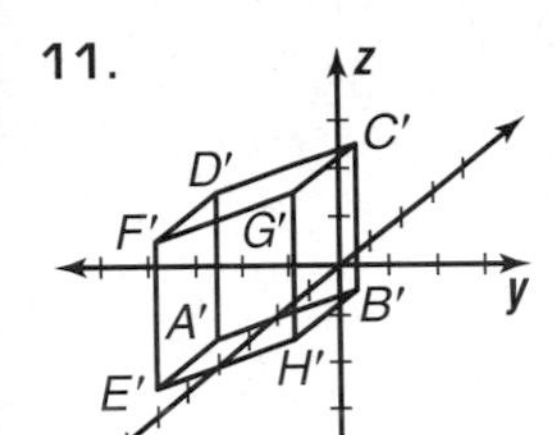

Translation 1 unit along the x-axis, -2 units along the y-axis, and -2 units along the z-axis.

13.

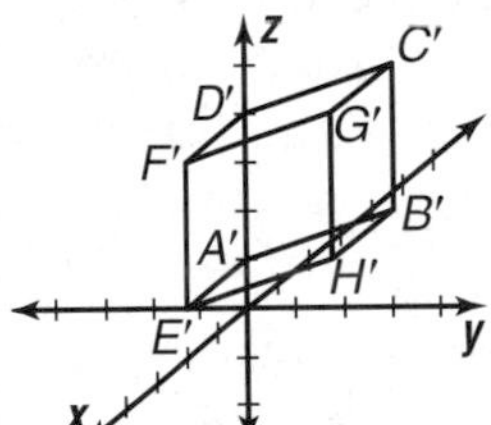

No change

15.

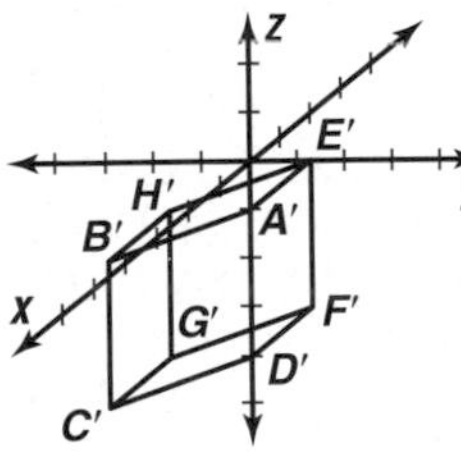

Reflection across all three coordinate planes

17. The figure is three times the original size and reflected over the xy-plane.

19a. $\begin{bmatrix} 2 & 0 & 0 \\ 0 & 2 & 0 \\ 0 & 0 & 5 \end{bmatrix}$ **19b.** The transformation will magnify the x- and y-dimensions two-fold, and the z-dimension 5-fold.

21. $\begin{bmatrix} -0.5 & 0 & 0 \\ 0 & 0.5 & 0 \\ 0 & 0 & 0.5 \end{bmatrix}$ **23.** The first transformation reflects the figure over all three coordinate planes. The second transformation stretches the dimensions along y- and z-axes and skews it along the xy-plane.
25a. dip-slip

25b. $\begin{bmatrix} 0 & 0 & 0 & 0 & 0 & 0 \\ 1.6 & 1.6 & 1.6 & 1.6 & 1.6 & 1.6 \\ -1.2 & -1.2 & -1.2 & -1.2 & -1.2 & -1.2 \end{bmatrix}$

27. $y = -\frac{2}{5}x + \frac{48}{5}$ **29.** ≈ 0.2

Pages 543–547 Chapter 8 Study Guide and Assessment

1. resultant **3.** magnitude **5.** inner **7.** parallel **9.** direction **11.** 1.5 cm; 50° **13.** 4.1 cm; 23° **15.** 2.5 cm; 98° **17.** 0.8 cm; 1 cm **19.** $\langle 5, 12 \rangle$; 13 **21.** $\langle -2, 12 \rangle$; $2\sqrt{37}$ **23.** $\langle 5, -6 \rangle$ **25.** $\langle 12, -17 \rangle$ **27.** $\langle 4, -1, -3 \rangle$; $\sqrt{26}$ **29.** $\langle 6, 2, 7 \rangle$; $\sqrt{89}$ **31.** $\langle 13, -37, 30 \rangle$ **33.** -16; no **35.** 0; yes **37.** 42; no **39.** $\langle 9, -6, 0 \rangle$; yes; $\langle 9, -6, 0 \rangle \cdot \langle -2, -3, 1 \rangle = 9(-2) + (-6)(-3) + 0(1) = -18 + 18 + 0 = 0$ $\langle 9, -6, 0 \rangle \cdot \langle 2, 3, -4 \rangle = 9(2) + (-6)(3) + 0(-4) = 18 - 18 + 0 = 0$ **41.** $\langle 1, -19, 31 \rangle$; yes; $\langle 1, -19, 31 \rangle \cdot \langle 7, 2, 1 \rangle = \langle 1(7) + (-19)(2) + 31(1) \rangle = 7 - 38 + 31 = 0$; $\langle 1, -19, 31 \rangle \cdot \langle 2, 5, 3 \rangle = 1(2) + (-19)(5) + 31(3) = 2 - 95 + 95 = 0$ **43.** 412.31 N; 39.09° **45.** $\langle x - 3, y + 5 \rangle = t\langle 4, 2 \rangle$; $x = 3 + 4t$, $y = -5 + 2t$ **47.** $\langle x - 4, y \rangle = t\langle 3, -6 \rangle$; $x = 4 + 3t$, $y = -6t$ **49.** $x = t$, $y = -\frac{1}{2}t + \frac{5}{2}$ **51.** 5.37 ft/s, 12.06 ft/s

53.

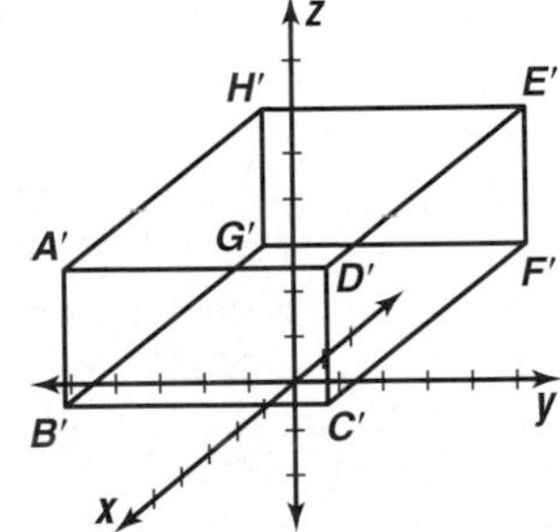

moves 2 units along x-axis and 3 units along the z-axis. **55.** 25 lb-ft **57a.** 13.7 km/h **57b.** 275.3 m

Page 549 Chapter 8 SAT and ACT Practice

1. A **3.** B **5.** E **7.** D **9.** B

Chapter 9 Polar Coordinates and Complex Numbers

Pages 558–560 Lesson 9-1

7.

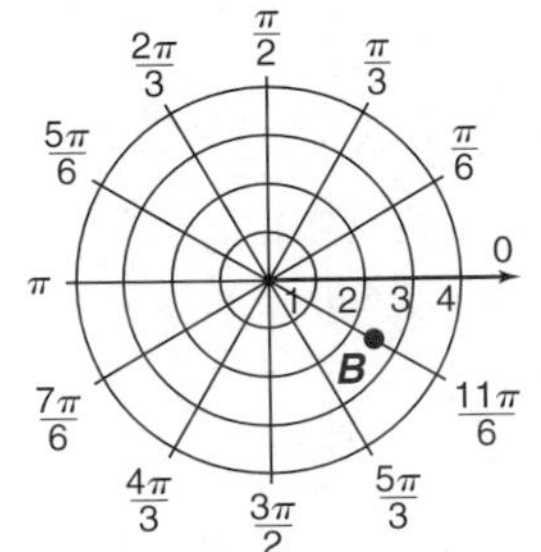

9.

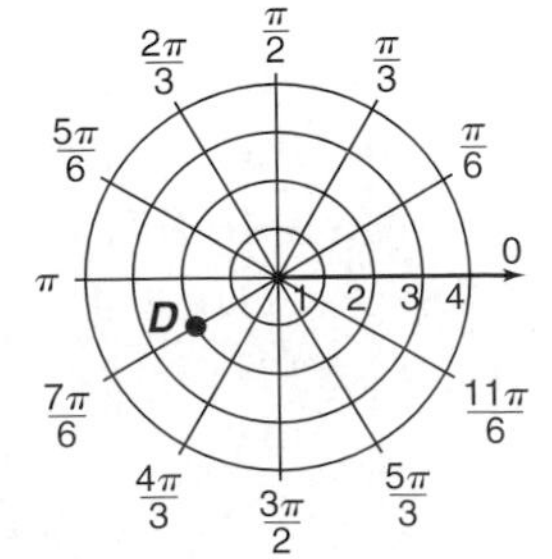

11. **13.**

15a.

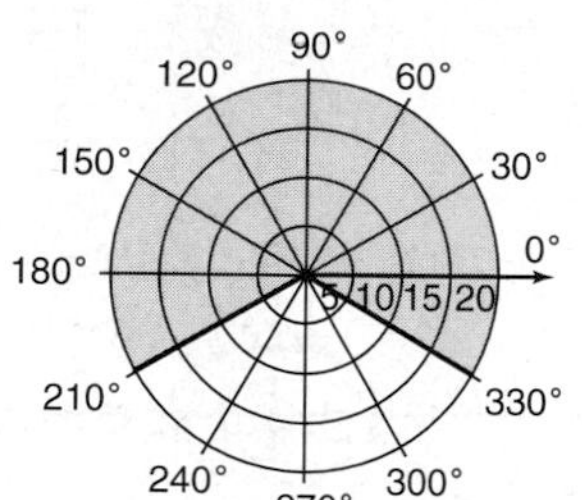

15b. about 838 ft^2

17.

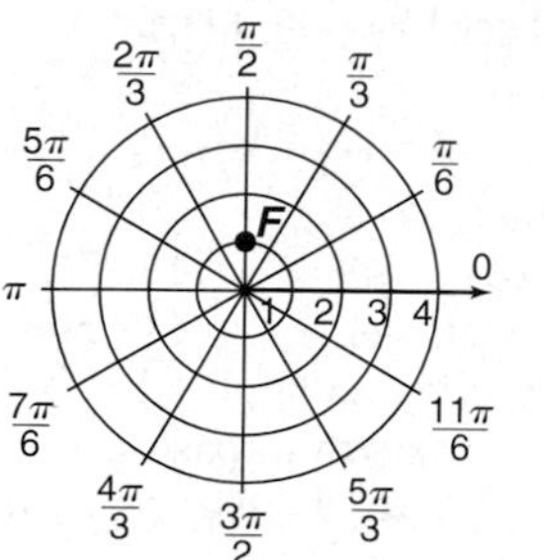

19.

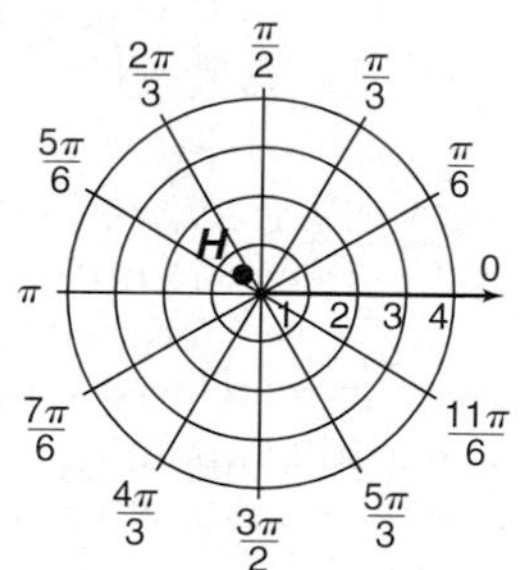

21.

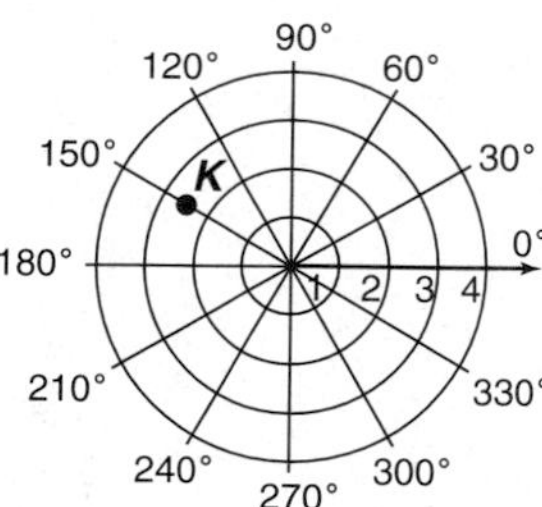

23.

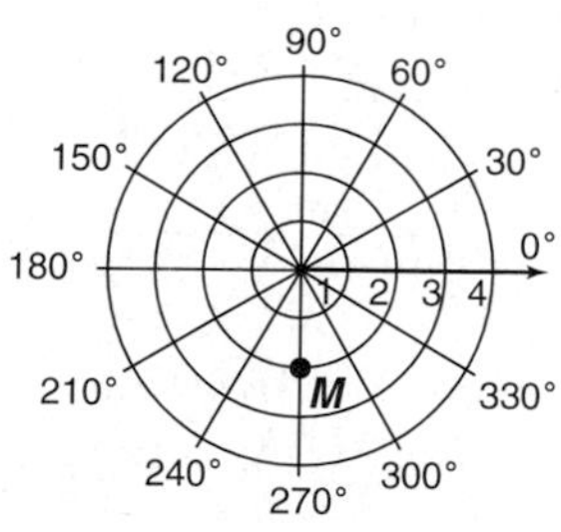

25.

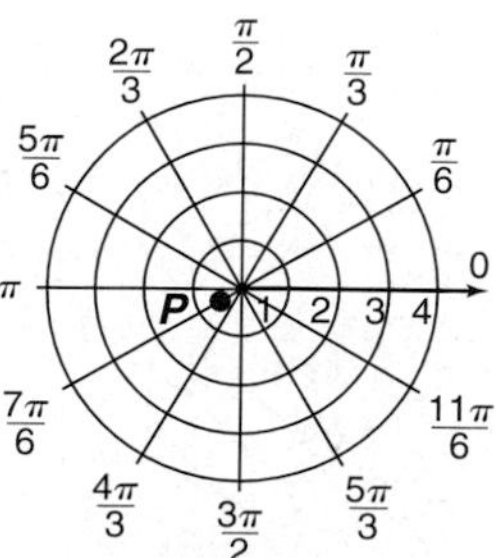

27.

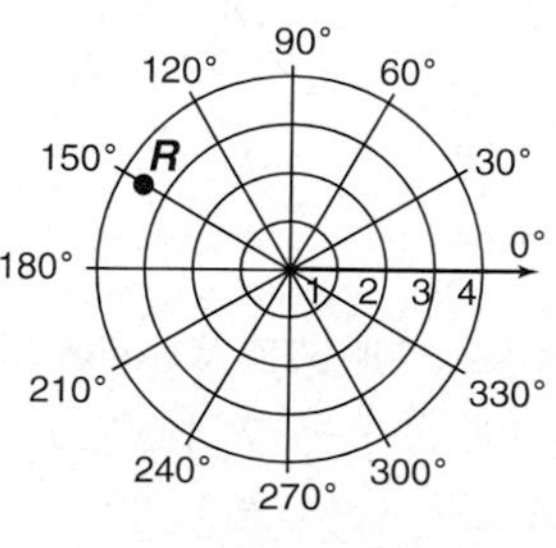

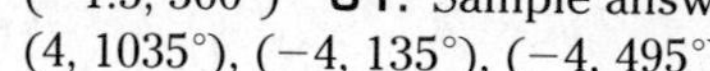

29. Sample answer: (1.5, 540°), (1.5, 900°), (−1.5, 0°), (−1.5, 360°) **31.** Sample answer: (4, 675°), (4, 1035°), (−4, 135°), (−4, 495°)

33.

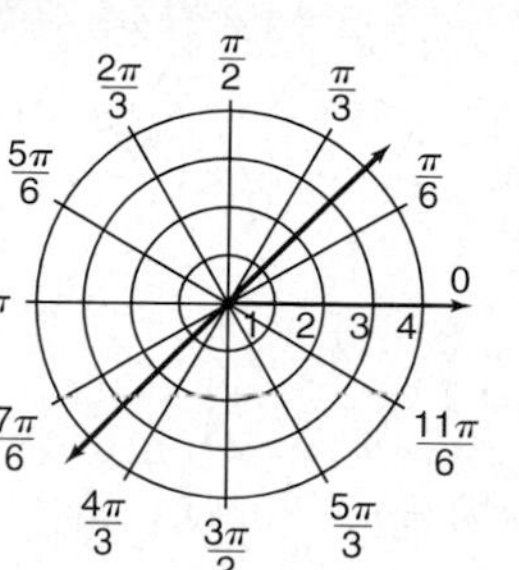

35.

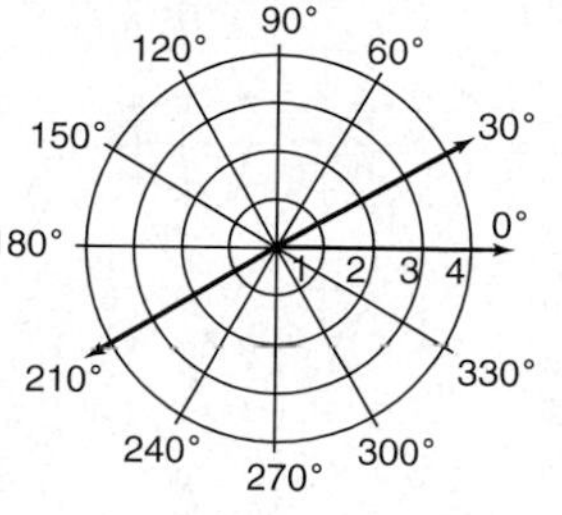

37.

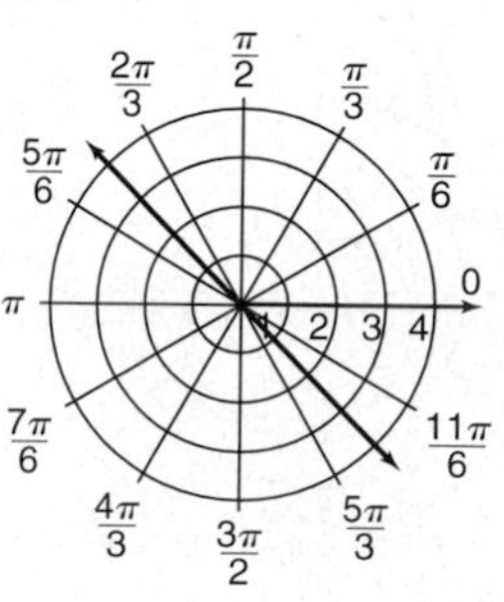

39.

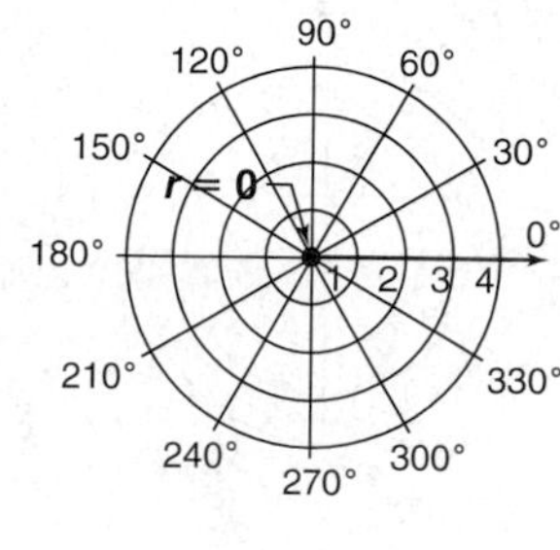

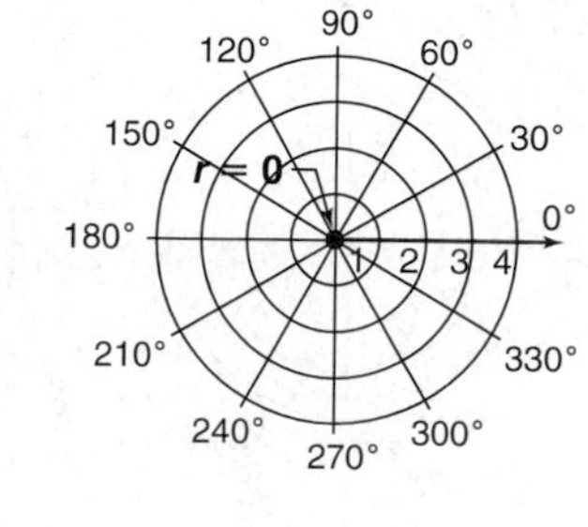

41. $r = \sqrt{2}$ or $r = -\sqrt{2}$ **43.** 5.35 **45.** 4.87 **47.** $\theta = 0°, \theta = 60°, \theta = 120°$ **49a.** 17 knots **49b.** 13 knots **51.** The distance formula is symmetric with respect to (r_1, θ_1) and (r_2, θ_2). That is,

$\sqrt{r_2^{\,2} + r_1^{\,2} - 2r_2r_1 \cos(\theta_1 - \theta_2)} =$
$\sqrt{r_1^{\,2} + r_2^{\,2} - 2r_1r_2 \cos[-(\theta_2 - \theta_1)]} =$
$\sqrt{r_1^{\,2} + r_2^{\,2} - 2r_1r_2 \cos(\theta_2 - \theta_1)}$

53. about 22.0° **55.** $\frac{16\sqrt{82}}{41}$ **57.** 30° **59.** one; $B = 90°, C = 60°, c = 16.1$ **61.** $y = x - 3$ **63.** −11 **65.** E

Pages 565–567 Lesson 9-2

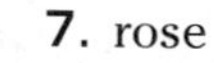

5. cardioid **7.** rose

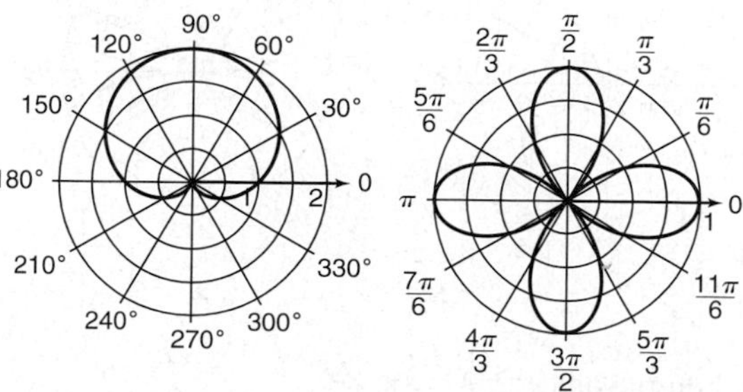

9. 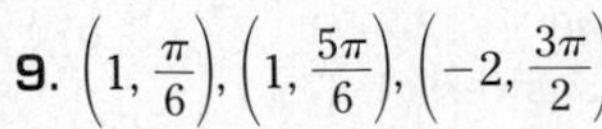
$\left(1, \frac{\pi}{6}\right), \left(1, \frac{5\pi}{6}\right), \left(-2, \frac{3\pi}{2}\right)$

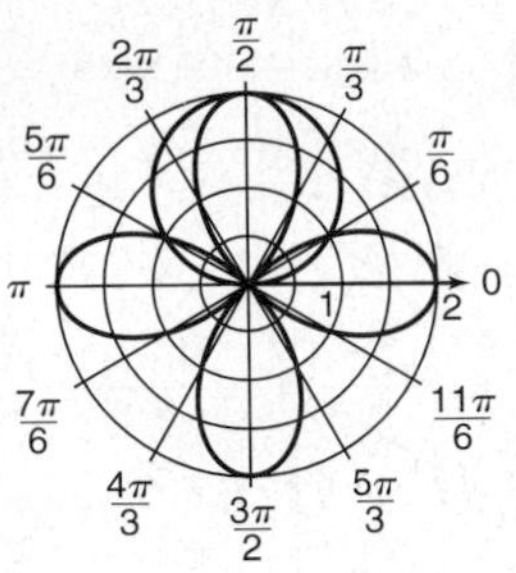

11. circle

13. spiral of Archimedes

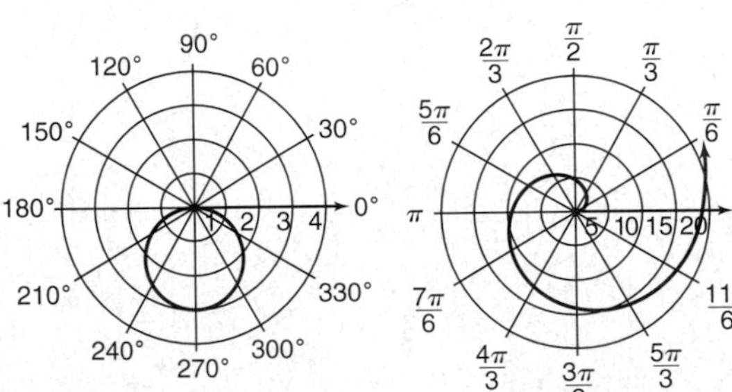

15. rose

17. spiral of Archimedes

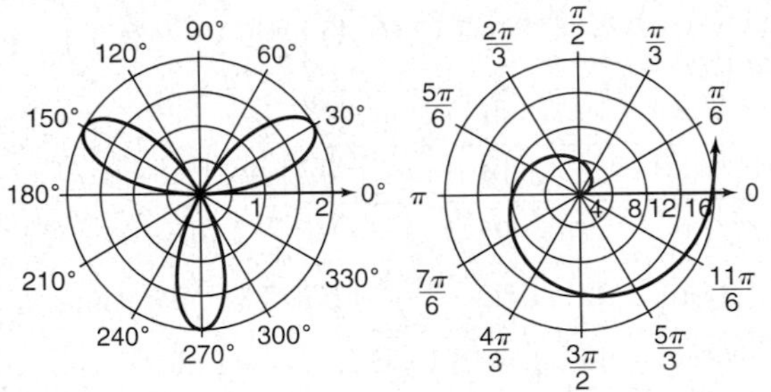

19. cardioid

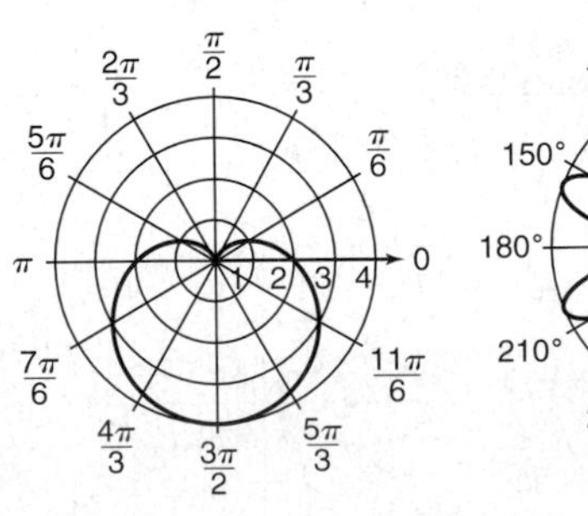

21. rose

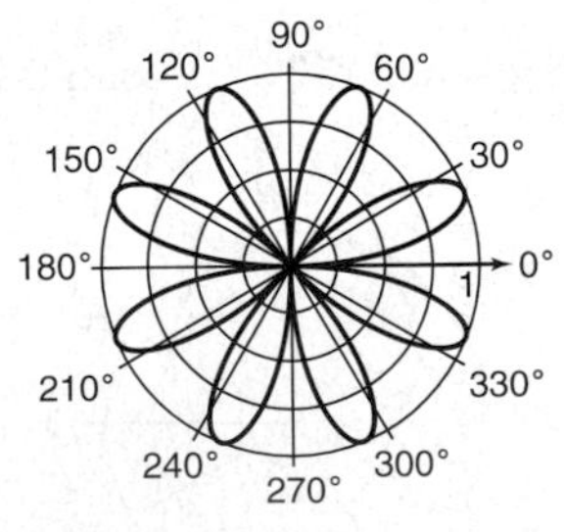

23. Sample answer: $r = \sin 3\theta$

25. $(3, 0)$

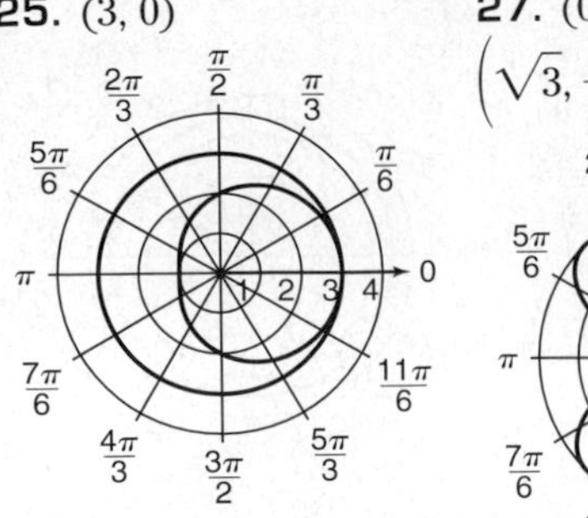

27. $(0, 0), (0, \pi),$
$\left(\sqrt{3}, \frac{\pi}{3}\right), \left(-\sqrt{3}, \frac{5\pi}{3}\right)$

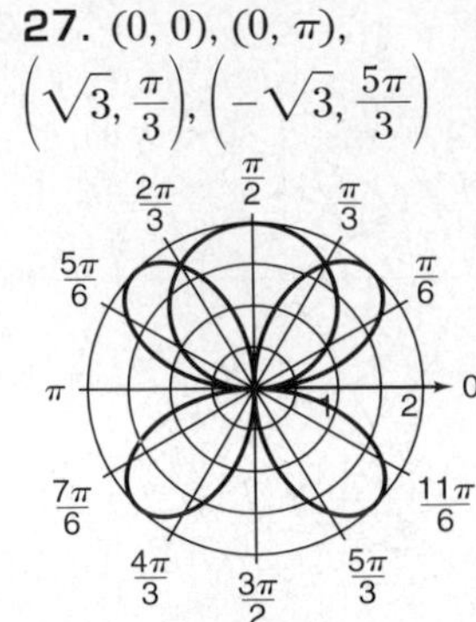

29. $(2, 3.5), (2, 5.9)$

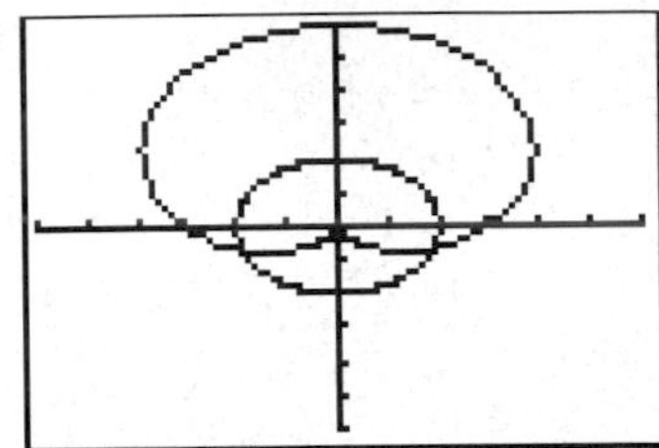

[−6, 6] scl:1 by [−6, 6] scl:1

31a. $r^2 = 9 \cos 2\theta$ or $r^2 = 9 \sin 2\theta$
31b. $r^2 = 16 \cos 2\theta$ or $r^2 = 16 \sin 2\theta$
33. $0 \leq \theta \leq 4\pi$ 35. Sample answer: $r = -1 - \sin \theta$
37a. counterclockwise rotation by an angle of α
37b. reflection about the polar axis or x-axis
37c. reflection about the origin 37d. dilation by a factor of c
39. $\langle 12, -8, 7\rangle$; $\langle 2, 3, 0\rangle \cdot \langle 12, -8, 7\rangle = 0$, $\langle -1, 2, 4\rangle \cdot \langle 12, -8, 7\rangle = 0$

41.
$$\frac{\sin^2 x}{\cos^4 x + \cos^2 x \sin^2 x} \stackrel{?}{=} \tan^2 x$$
$$\frac{\sin^2 x}{\cos^2 x(\cos^2 x + \sin^2 x)} \stackrel{?}{=} \tan^2 x$$
$$\frac{\sin^2 x}{(\cos^2 x)(1)} \stackrel{?}{=} \tan^2 x$$
$$\frac{\sin^2 x}{\cos^2 x} \stackrel{?}{=} \tan^2 x$$
$$\tan^2 x = \tan^2 x$$

43.
$$\begin{array}{l} \\ \text{Bus} \\ \text{Train} \end{array} \begin{array}{c} \begin{array}{ccc} \text{NY} & \text{LA} & \text{Miami} \end{array} \\ \begin{bmatrix} \$240 & \$199 & \$260 \\ \$254 & \$322 & \$426 \end{bmatrix} \end{array}$$

Pages 571–573 Lesson 9-3

5. $\left(2, \frac{3\pi}{4}\right)$ 7. $(1, \sqrt{3})$ 9. $r = 2 \csc \theta$
11. $x^2 + y^2 = 36$

13a.

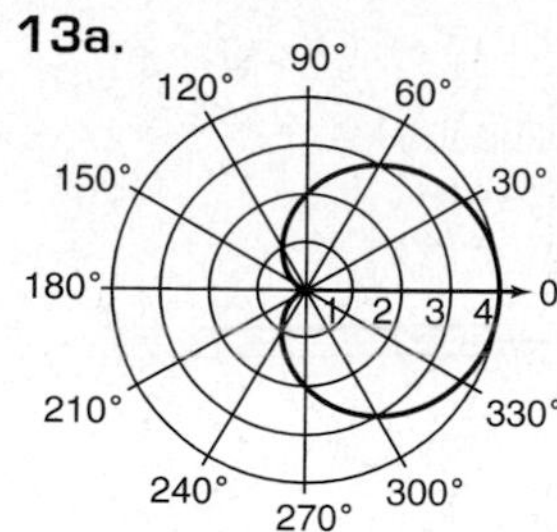

13b. No. The given point is on the negative x-axis, directly behind the microphone. The polar pattern indicates that the microphone does not pick up any sound from this direction.

15. $\left(1, \frac{\pi}{2}\right)$ **17.** $\left(\frac{1}{2}, \frac{4\pi}{3}\right)$ **19.** (8.06, 5.23)
21. $\left(-\frac{\sqrt{2}}{4}, \frac{\sqrt{2}}{4}\right)$ **23.** (0, 2) **25.** (−9.00, 10.72)
27. $r = 5 \csc \theta$ **29.** $r = 2 \sin \theta$ **31.** $r = 4 \sin \theta$
33. $x^2 + y^2 = 9$ **35.** $y = 2$ **37.** $xy = 4$
39. $x^2 + y^2 = y$ **41.** 0.52 unit **43.** 75 m east; 118.30 m north **45.** circle centered at (a, a) with radius $\sqrt{2}|a|$; $(x - a)^2 + (y - a)^2 = 2a^2$.
47. Sample answer: (−2, 405°), (−2, 765°), (2, 225°), (2, 585°) **49.** 0° **51.** $-\frac{\sqrt{3}}{2}$ **53.** $x^4 + 2x^3 + 4x^2 + 5x + 10$ **55.** C

Pages 577–579 Lesson 9-4

5. $2 = r \cos(\theta - 307°)$ **7.** $x + \sqrt{3}y - 6 = 0$
9.

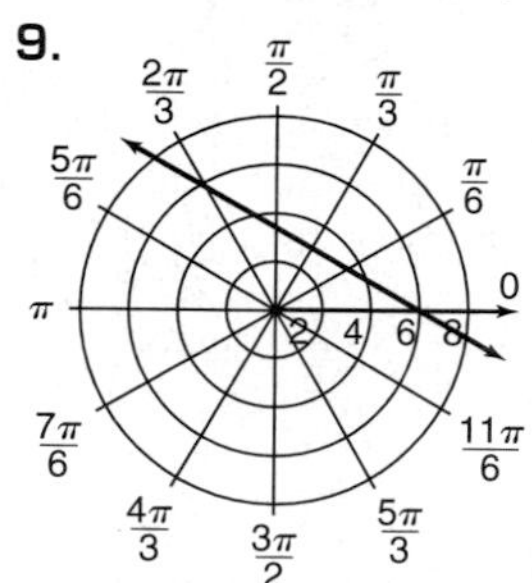

11a. $\left(5, \frac{5\pi}{6}\right)$
11b.

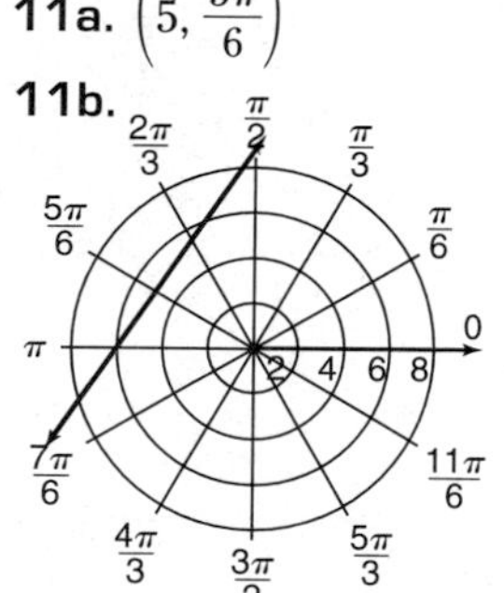

13. $3 = r \cos(\theta - 44°)$ **15.** $\frac{5\sqrt{13}}{13} = r \cos(\theta - 34°)$
17. $\frac{7\sqrt{10}}{10} = r \cos(\theta - 108°)$ **19.** $\sqrt{2}x - \sqrt{2}y - 8 = 0$ **21.** $\sqrt{3}x - y - 2 = 0$
23. $x + \sqrt{3}y - 10 = 0$
25.

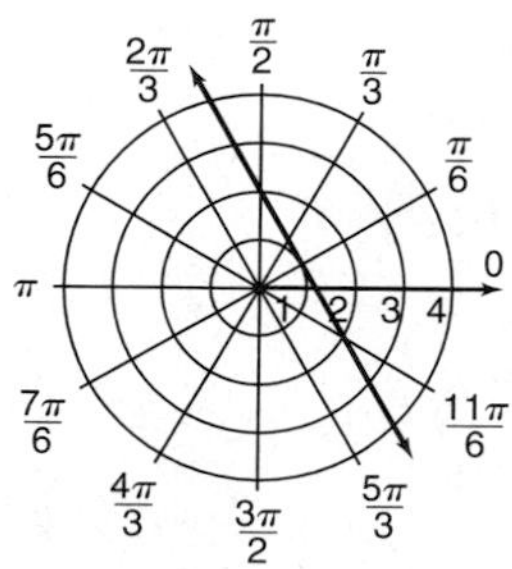

27.

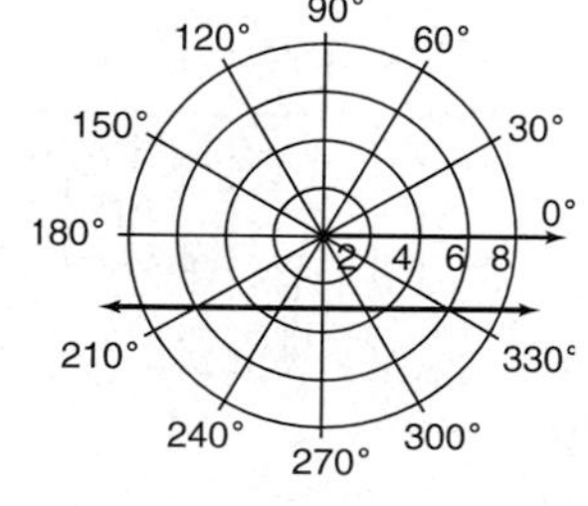

29.

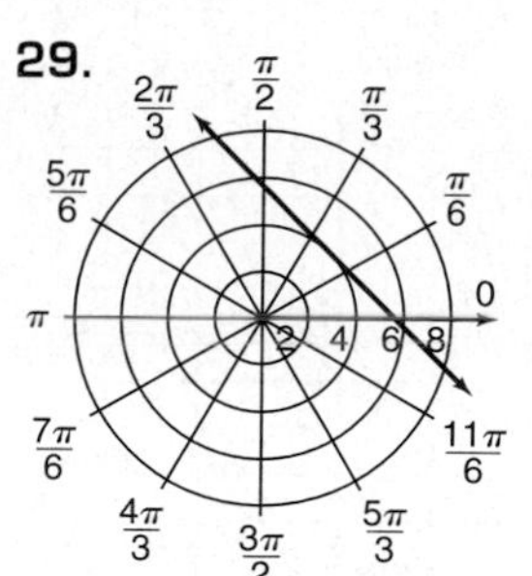

31. $0.31 = r \cos(\theta - 2.25)$
33. Sample answer: $2 = r \cos(\theta - 45°)$ and $2 = r \cos(\theta - 135°)$

35a.

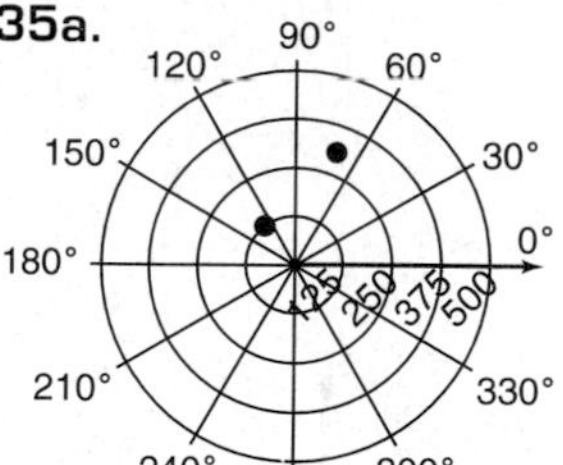

35b. $124.43 = r \cos(\theta - 135°)$
37. $32.36 = r \cos(\theta - 36°)$
39. rose
41. about 20.42 ft²
43. $0, \frac{3}{2}, -4$

Pages 583–585 Lesson 9-5

5. −1 **7.** $-4 + 4i$ **9.** $1 + 9i$ **11.** $\frac{2}{5} + \frac{1}{5}i$
13. −1 **15.** 1 **17.** $-1 + 8i$ **19.** $-\frac{3}{2} + 2i$
21. $5 + 10i$ **23.** $(-2 + \sqrt{35}) + (-2\sqrt{7} - \sqrt{5})i$
25. $\frac{4}{5} - \frac{3}{5}i$ **27.** $\frac{12}{13} - \frac{5}{13}i$ **29.** $x^2 - 4x + 5 = 0$
31. $-12 - 16i$ **33.** $\left(\frac{2}{5} - \frac{2\sqrt{3}}{15}\right) + \left(-\frac{\sqrt{2}}{5} - \frac{2\sqrt{6}}{15}\right)i$ **35.** $-\frac{24}{169} + \frac{10}{169}i$
37a. $\pm 3 - 4i$ **37b.** No. **37c.** The solutions need not be complex conjugates because the coefficients in the equation are not all real.
37d.
$$(3 - 4i)^2 + 8i(3 - 4i) - 25 \stackrel{?}{=} 0$$
$$-7 - 24i + 24i + 32 - 25 \stackrel{?}{=} 0$$
$$0 = 0$$

$$(-3 - 4i)^2 + 8i(-3 - 4i) - 25 \stackrel{?}{=} 0$$
$$-7 + 24i - 24i + 32 - 25 \stackrel{?}{=} 0$$
$$0 = 0$$

39a. $1 + 2i, -2 + i, -1 - 2i, 2 - i, 1 + 2i$
39b. $0.5 - 0.866i, -0.500 - 0.866i, -1.000 - 0.000i, -0.500 + 0.866i, 0.500 + 0.866i$ **41.** $c_1 = c_2$
43.

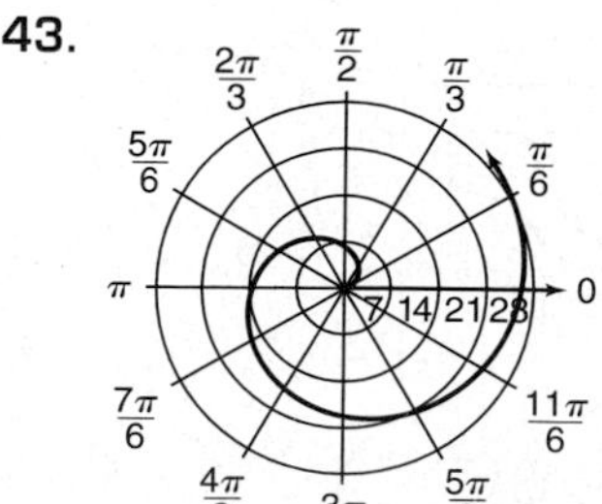

45. $\left\langle -6, \frac{27}{2}, -5 \right\rangle$
47. $y = 3.5 \cos\left(\frac{\pi}{6}t\right)$
49. quadratic
51. 64 **53.** 3, −11
55. A

SELECTED ANSWERS

Pages 589–591 Lesson 9-6

5. $x = 1, y = 3$ **7.** $\sqrt{3}$

9. $\sqrt{41}(\cos 0.90 + \boldsymbol{i} \sin 0.90)$

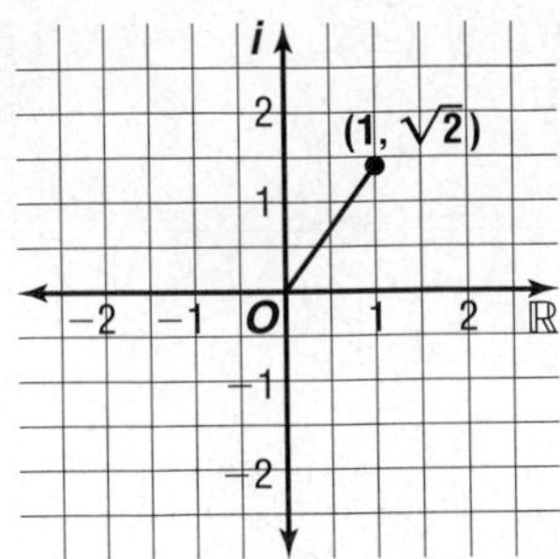

11. $2 + 2\sqrt{3}\boldsymbol{i}$ **13.** 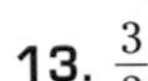 $\frac{3}{2}$

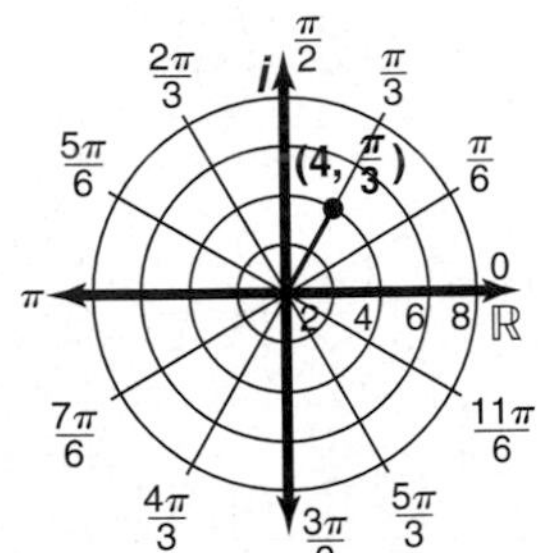

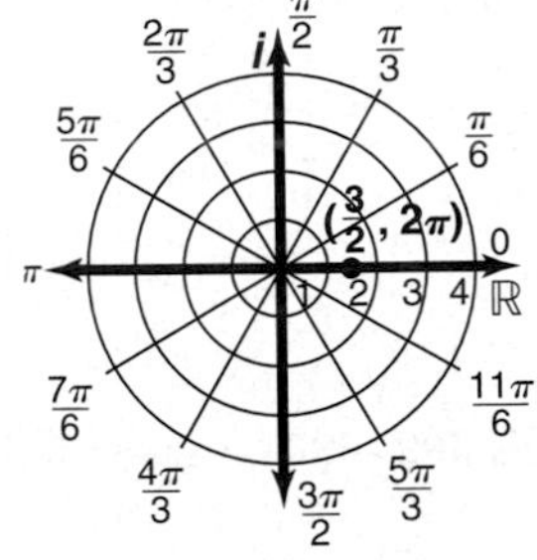

15a. about 18.03 N **15b.** about 56.31°

17. $x = \frac{1}{2}, y = 1$

19. $\sqrt{13}$

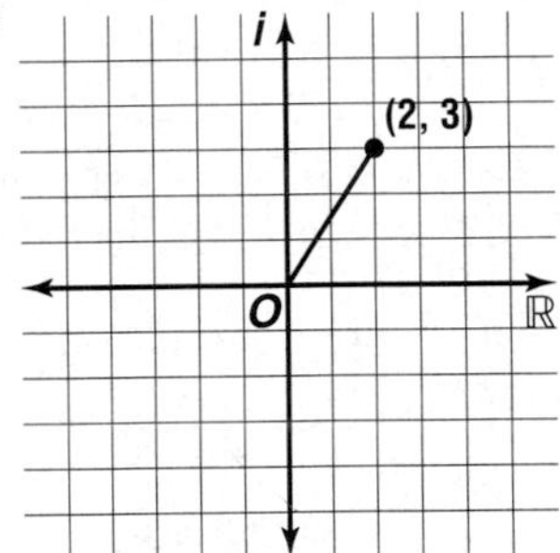

21. $\sqrt{26}$

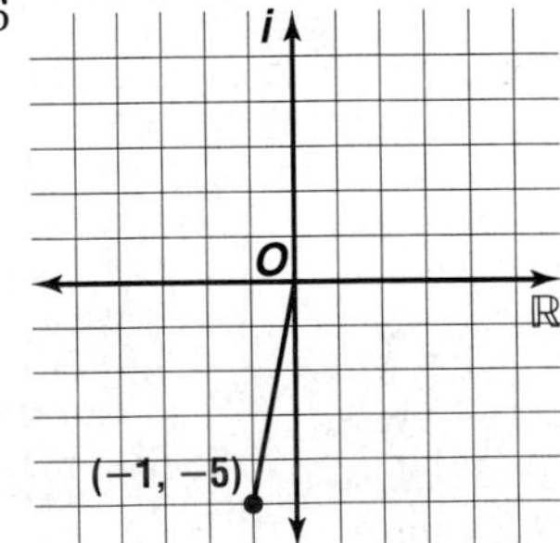

23. $\sqrt{6}$

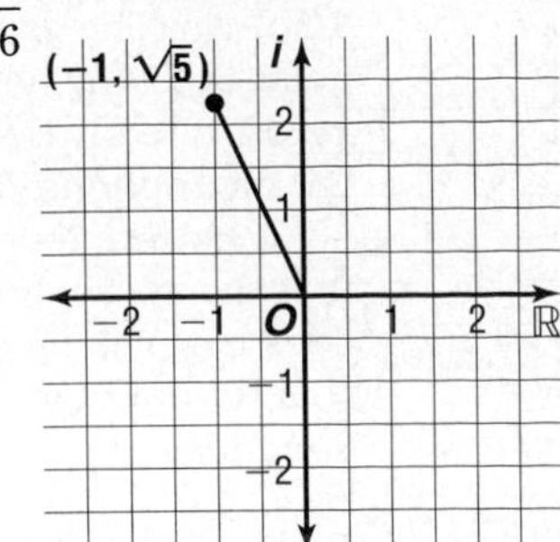

25. $2\sqrt{13}$ **27.** $2\left(\cos \frac{4\pi}{3} + \boldsymbol{i} \sin \frac{4\pi}{3}\right)$

29. $\sqrt{17}(\cos 2.90 + \boldsymbol{i} \sin 2.90)$

31. $2\sqrt{5}(\cos 2.03 + \boldsymbol{i} \sin 2.03)$

33. $4\sqrt{2}(\cos \pi + \boldsymbol{i} \sin \pi)$

35. $\frac{3\sqrt{2}}{2} + \frac{3\sqrt{2}}{2}\boldsymbol{i}$ **37.** $-1 - \sqrt{3}\boldsymbol{i}$

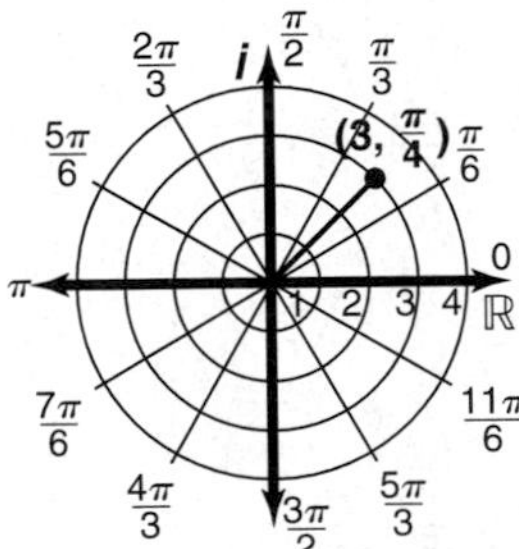

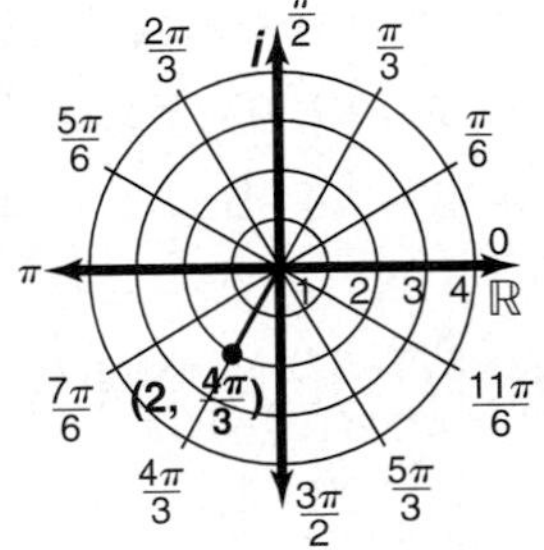

39. $-\sqrt{2} - \sqrt{2}\boldsymbol{i}$ **41.** 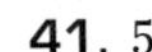 5

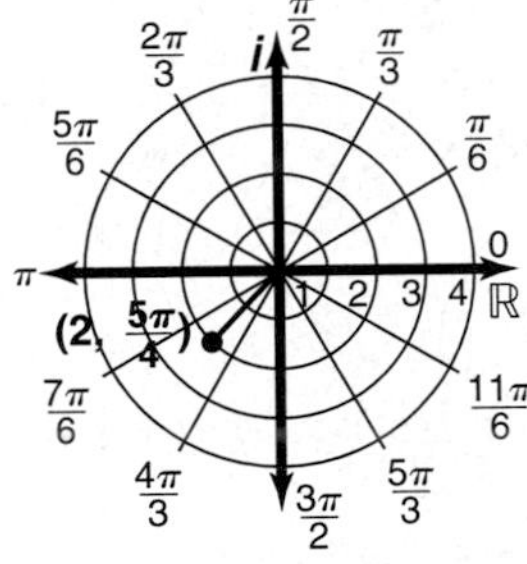

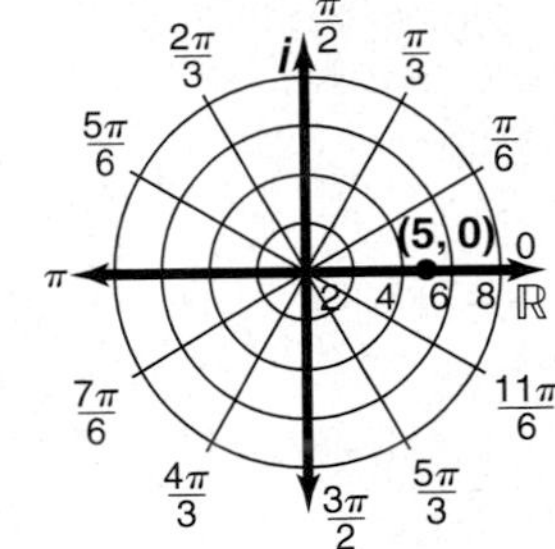

43.

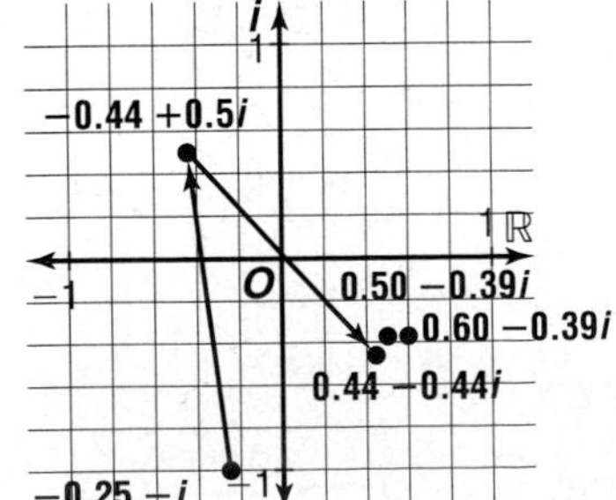

45.

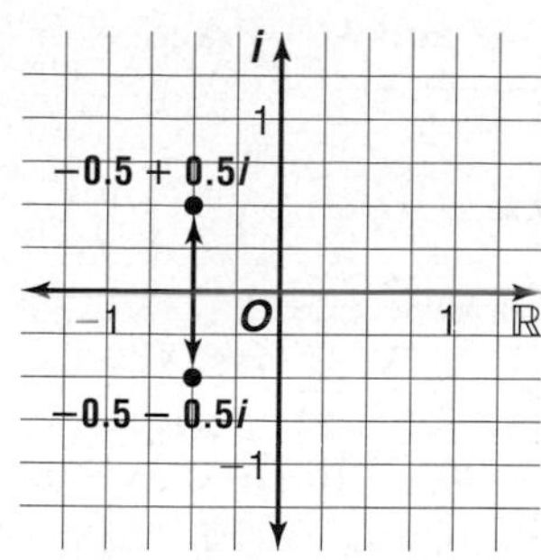

47. The moduli are the same, but the amplitudes are opposites. **49a.** Translate 2 units to the right and down 3 units. **49b.** Rotate 90° counterclockwise about the origin. **49c.** Dilate by a factor of 3. **49d.** Reflect about the real axis. **51.** $-6 + 22i$ **53.** $\sqrt{58}, -3\vec{i} + 7\vec{j}$ **55.** about 13.57 m/s **57.** 4 **59.** D

Pages 596–598 Lesson 9-7

5. $-\frac{3}{4}i$ **7.** $-\frac{3\sqrt{3}}{2} - \frac{3}{2}i$

9. $6\left(\cos\frac{\pi}{6} + j\sin\frac{\pi}{6}\right)$ volts **11.** $3i$

13. $5\sqrt{2} - 5\sqrt{2}i$ **15.** $\frac{3\sqrt{3}}{2} - \frac{3}{2}i$

17. -2 **19.** $3.10 + 2.53i$ **21.** $-2 - 2i$

23. $-4 - 4\sqrt{3}i$ **25.** -12 **27.** $\frac{2\sqrt{2}}{3}i$

29. $16 + 12j$ ohms **31a.** The point is rotated counterclockwise about the origin by an angle of θ. **31b.** The point is rotated 60° counterclockwise about the origin. **33.** $13(\cos 5.11 + i\sin 5.11)$ **35.** about 27.21 lb **37.** $y = \arccos x$

Pages 605–606 Lesson 9-8

5. $-8i$ **7.** $0.97 + 0.26i$

9. $0.38 + 0.92i$, $-0.92 + 0.38i$, $-0.38 - 0.92i$, $0.92 - 0.38i$

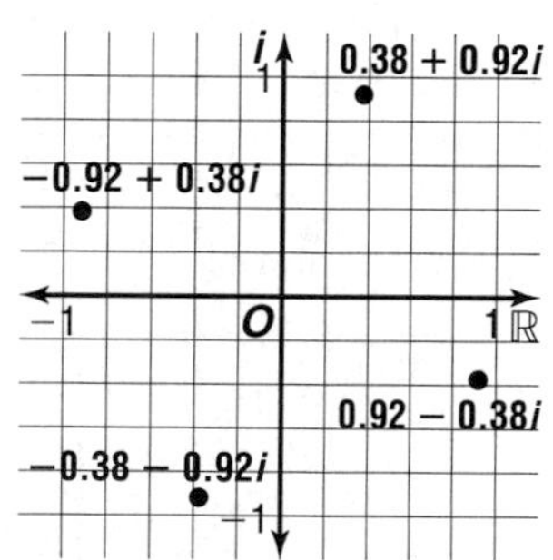

11. Escape set; the iterates escape to infinity. **13.** $-16\sqrt{2} - 16\sqrt{2}i$ **15.** $-8 - 8\sqrt{3}i$ **17.** $-0.03 - 0.07i$ **19.** $1.83 + 0.81i$ **21.** $0.96 + 0.76i$ **23.** $1.37 + 0.37i$ **25.** $0.71 + 0.71i$

27. $0.81 + 0.59i$, $-0.31 + 0.95i$, -1, $-0.31 - 0.95i$, $0.81 - 0.59i$

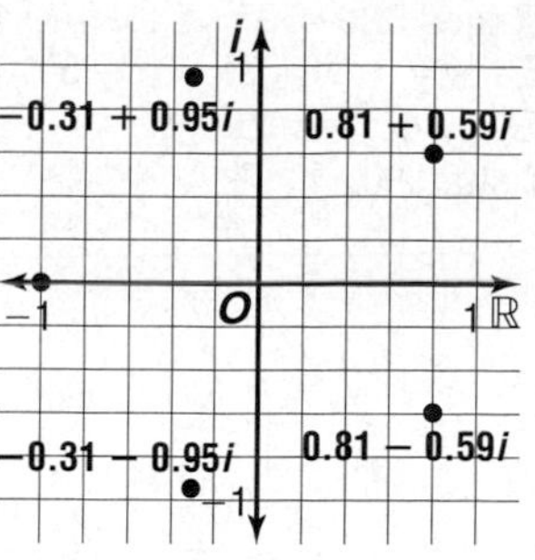

29. $\sqrt{2} + \sqrt{2}i$, $-\sqrt{2} + \sqrt{2}i$, $-\sqrt{2} - \sqrt{2}i$, $\sqrt{2} - \sqrt{2}i$

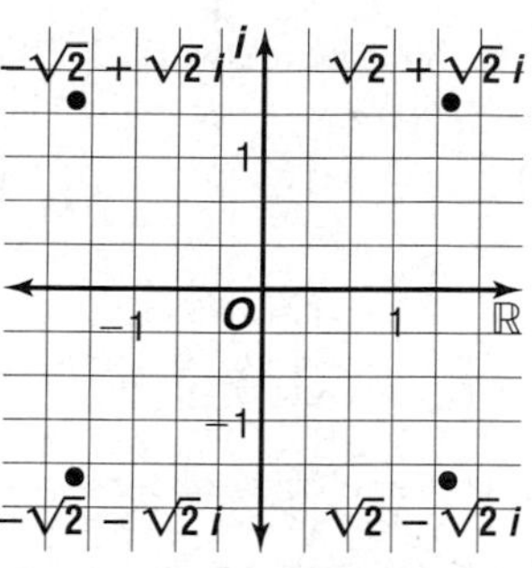

31. $0.59 + 1.03i$, $-1.03 + 0.59i$, $-0.59 - 1.03i$, $1.03 - 0.59i$

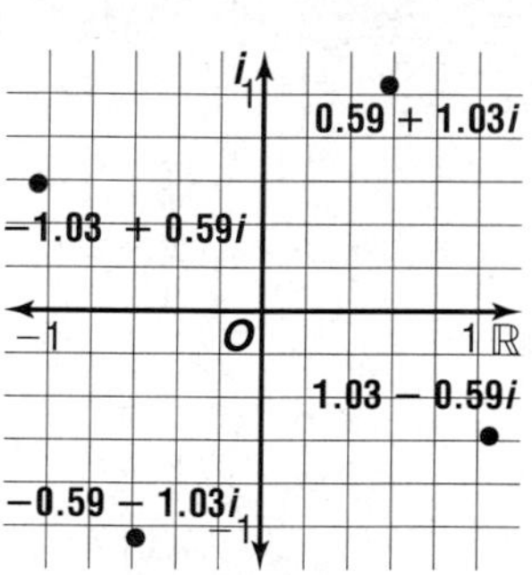

33. $1.26 + 0.24i, 0.43 + 1.21i, -0.83 + 0.97i, -1.26 - 0.24i, -0.43 - 1.21i, 0.83 - 0.97i$ **35.** Prisoner set; the iterates approach 0. **37.** $1, \frac{1}{2} + \frac{\sqrt{3}}{2}i, -\frac{1}{2} + \frac{\sqrt{3}}{2}i, -1, -\frac{1}{2} - \frac{\sqrt{3}}{2}i, \frac{1}{2} - \frac{\sqrt{3}}{2}i$ **39.** The roots are the vertices of a regular polygon. Since one of the roots must be a positive real number, a vertex of the polygon lies on the positive real axis and the polygon is symmetric about the real axis. This means the non-real complex roots occur in conjugate pairs. Since the imaginary part of the sum of two complex conjugates is 0, the imaginary part of the sum of all the roots must be 0. **41.** $5 - i$

43. $\frac{\sqrt{2 + \sqrt{2}}}{2}$ **45.** 800 large bears, 400 small bears

Pages 607–611 Chapter 9 Study Guide and Assessment

1. absolute value **3.** prisoner **5.** pure imaginary **7.** rectangular **9.** Argand

11. **13.**

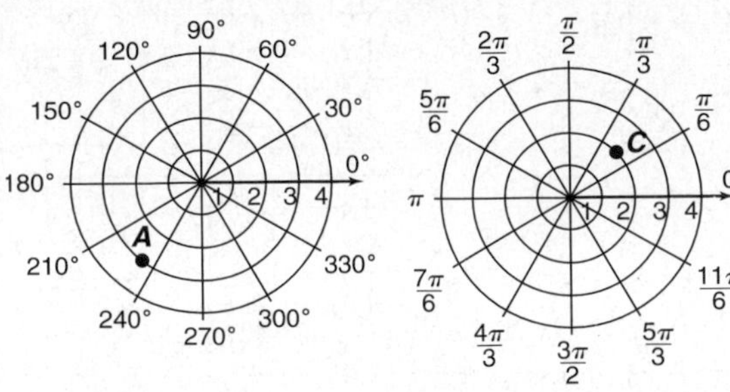

15. Sample answer: (4, 585°), (4, 945°), (−4, 45°), (−4, 405°)

17.

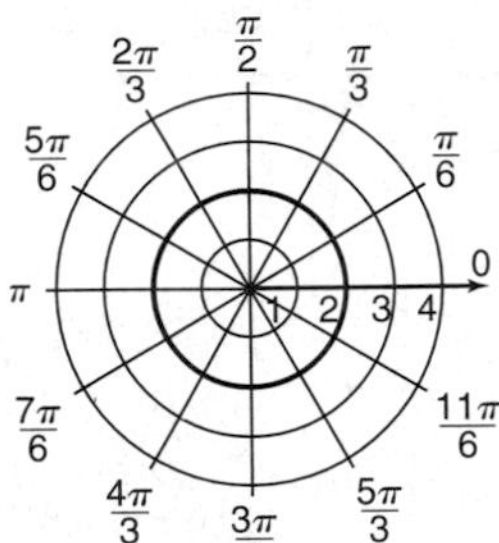

19.

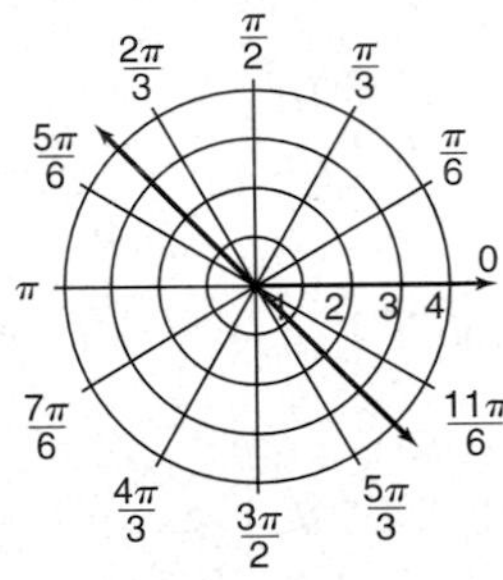

21. spiral of Archimedes

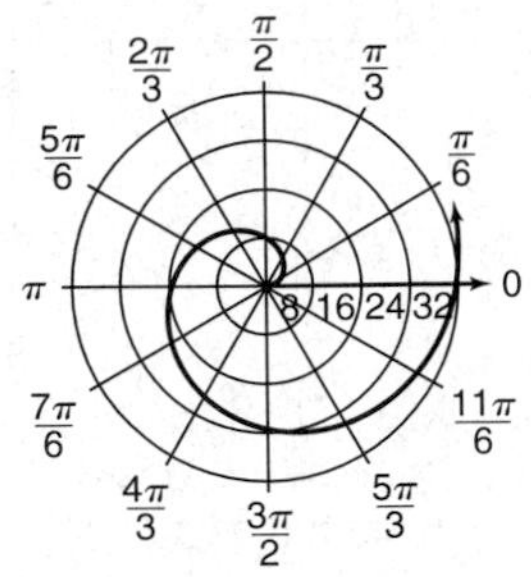

23. rose

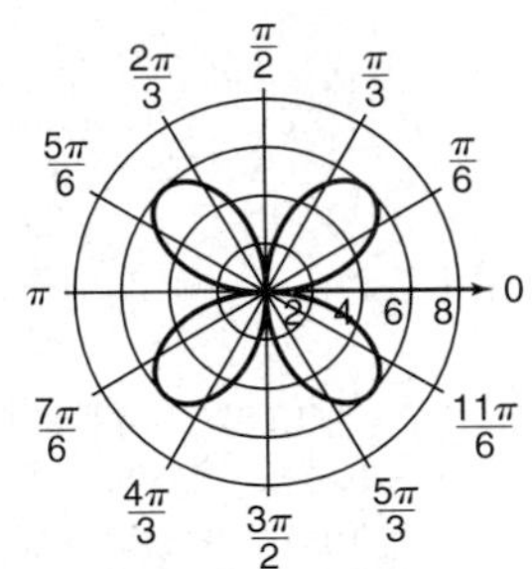

25. $(\sqrt{3}, -1)$ **27.** (0, 1) **29.** $\left(5\sqrt{2}, \frac{\pi}{4}\right)$

31. (4.47, 0.46) **33.** $\frac{2\sqrt{10}}{5} = r\cos(\theta - 198°)$

35. $y + 4 = 0$ **37.** $-2 + 7\boldsymbol{i}$ **39.** $-3 - 4\boldsymbol{i}$

41. $\frac{16}{29} + \frac{18}{29}\boldsymbol{i}$ **43.** $2\sqrt{2}\left(\cos\frac{\pi}{4} + \boldsymbol{i}\sin\frac{\pi}{4}\right)$

45. $2\left(\cos\frac{2\pi}{3} + \boldsymbol{i}\sin\frac{2\pi}{3}\right)$ **47.** $\sqrt{17}(\cos 3.39 + \boldsymbol{i}\sin 3.39)$ **49.** $2\sqrt{2}(\cos\pi + \boldsymbol{i}\sin\pi)$

51. $\sqrt{3} + \boldsymbol{i}$

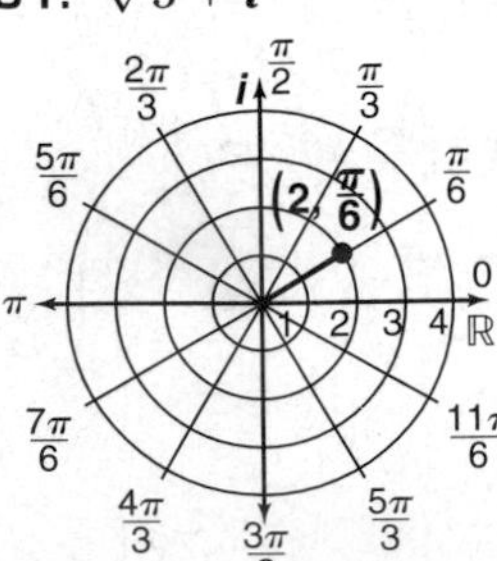

53. $-6 + 6\sqrt{3}\boldsymbol{i}$
55. $-8.01 + 5.98\boldsymbol{i}$
57. $\frac{3}{4} + \frac{3\sqrt{3}}{4}\boldsymbol{i}$
59. 4096 **61.** −4
63. $0.92 + 0.38\boldsymbol{i}$
65. lemniscate
67. $y = -5$

Page 613 Chapter 9 SAT and ACT Practice
1. A **3.** E **5.** A **7.** E **9.** D

Chapter 10 Conics

Pages 620–622 Lesson 10-1
5. 10, (5, 6) **7.** $2\sqrt{2}$, (−1, 3) **9.** yes; $\overline{XY} \cong \overline{XZ}$, since $XY = 2\sqrt{17}$ and $DC = 2\sqrt{17}$, therefore $\triangle XYZ$ is isosceles. **11a.** (40, 60) **11b.** $20\sqrt{13}$ yd or about 72 yards **13.** $2\sqrt{10}$; (0, 0) **15.** $3\sqrt{5}$; (2, −4.5) **17.** 16; (*a*, −1) **19.** $\sqrt{5}$; $\left(c + 1, d - \frac{1}{2}\right)$ **21.** $a = \pm 4$ **23.** yes **25.** −5 **27.** $\overline{EF} \cong \overline{HG}$ since $EF = \sqrt{5}$ and $HG = \sqrt{5}$. $\overline{EF} \parallel \overline{HG}$ since the slope of $\overline{EF}$ is $-\frac{1}{2}$ and the slope of $\overline{HG}$ is $-\frac{1}{2}$. Thus the points form a parallelogram. $\overline{EF} \perp \overline{FG}$ since the product of the slopes of $\overline{EF}$ and $\overline{FG}$, $-\frac{1}{2} \cdot \frac{2}{1}$, is −1. Therefore, the points form a rectangle. **29.** In trapezoid *ABCD*, let *A* and *B* have coordinates (0, 0) and (*b*, 0), respectively. To make the trapezoid isosceles, let *C* have coordinates (*b* − *a*, *c*) and let *D* have coordinates (*a*, *c*).

y
D(b − a, c) C(a, c)
O A(0, 0) B(b, 0) x

$AC = \sqrt{(a - 0)^2 + (c - 0)^2} = \sqrt{a^2 + c^2}$
$BD = \sqrt{(b - a - b)^2 + (c - 0)^2} = \sqrt{a^2 + c^2}$
$AC = \sqrt{a^2 + c^2} = \sqrt{a^2 + c^2} = BD$, so the diagonals of an isosceles trapezoid are congruent.

31. Let *A* and *B* have coordinates (0, 0) and (*b*, 0) respectively. To make a parallelogram, let *C* have coordinates (*a* + *b*, *c*) and let *D* have coordinates (*a*, *c*). The midpoint of $\overline{BD}$ is $\left(\frac{a + b}{2}, \frac{0 + c}{2}\right)$ or $\left(\frac{a + b}{2}, \frac{c}{2}\right)$.

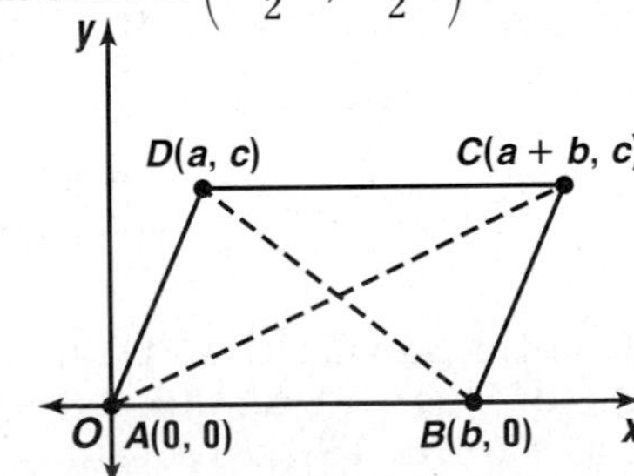

The midpoint of $\overline{AC}$ is $\left(\frac{a + b + 0}{2}, \frac{c + 0}{2}\right)$ or $\left(\frac{a + b}{2}, \frac{c}{2}\right)$. Since the diagonals have the same midpoint, the diagonals bisect each other.

33. 16 square units **35.** Let the vertices of the isosceles trapezoid have the coordinates $A(0, 0)$, $B(2a, 0)$, $C(2a - 2c, 2b)$, $D(2c, 2b)$. The coordinates of the midpoints are: $P(a, 0)$, $Q(2a - c, b)$, $R(a, 2b)$, and $S(c, b)$.

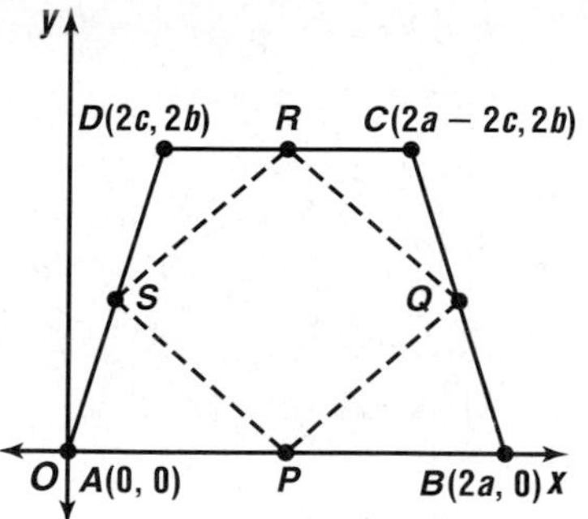

$$PQ = \sqrt{(2a - c - a)^2 + (b - 0)^2}$$
$$= \sqrt{(a - c)^2 + b^2}$$
$$QR = \sqrt{(2a - c - a)^2 + (b - 2b)^2}$$
$$= \sqrt{(a - c)^2 + b^2}$$
$$RS = \sqrt{(a - c)^2 + (2b - b)^2}$$
$$= \sqrt{(a - c)^2 + b^2}$$
$$PS = \sqrt{(a - c)^2 + (0 - b)^2}$$
$$= \sqrt{(a - c)^2 + b^2}$$

So, all of the sides are congruent and quadrilateral $PQRS$ is a rhombus.

37a.

$$MA = \sqrt{t^2 + (3t - 15)^2}$$
$$= \sqrt{t^2 + 9t^2 - 90t + 225}$$
$$= \sqrt{10t^2 - 90t + 225}$$
$$MB = \sqrt{(t - 9)^2 + (3t - 12)^2}$$
$$= \sqrt{t^2 - 18t + 81 + 9t^2 - 72t + 144}$$
$$= \sqrt{10t^2 - 90t + 225}$$
$$MA = MB$$
$$\sqrt{10t^2 - 90t + 225} = \sqrt{10t^2 - 90t + 225}$$

Since the above equation is a true statement, t can take on any real values. **37b.** A line; this line is the perpendicular bisector of $\overline{AB}$. **39.** about 2021 N **41.** 54.9° **43.** $4 \pm \sqrt{2}$

Pages 627–630 Lesson 10-2

7. $(x + 1)^2 + (y - 4)^2 = 16$

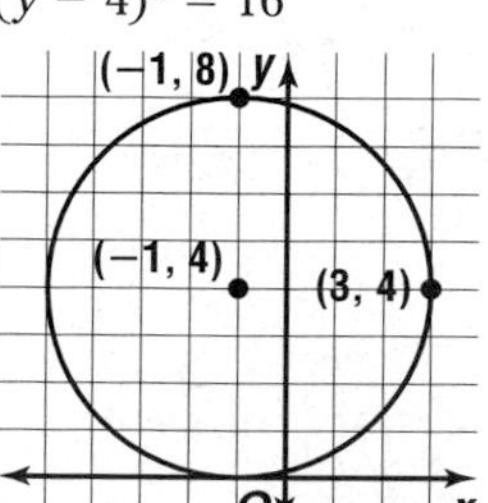

9. $(x - 5)^2 + (y + 2)^2 = 12$

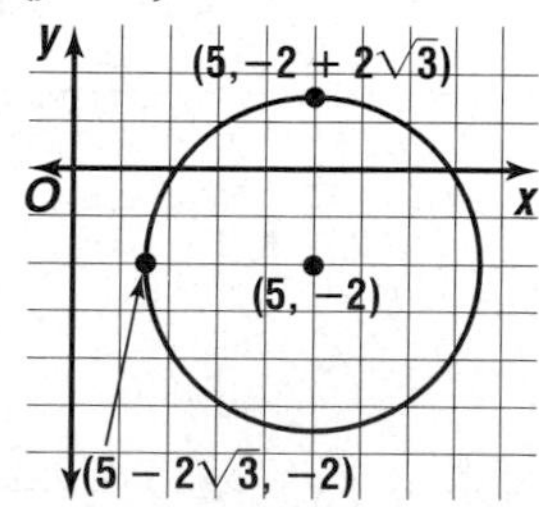

11. $(x - 3)^2 + (y - 4)^2 = 5$; $(3, 4)$; $\sqrt{5}$
13. $(x - 4)^2 + (y + 2)^2 = 100$
15. $x^2 + y^2 = 25$

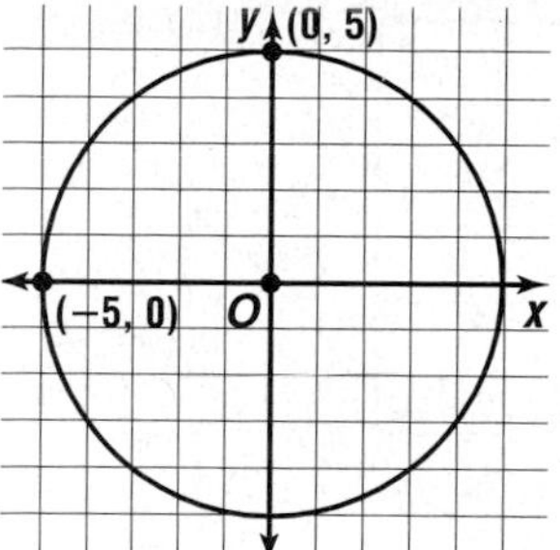

17. $(x + 1)^2 + (y + 3)^2 = \frac{1}{2}$

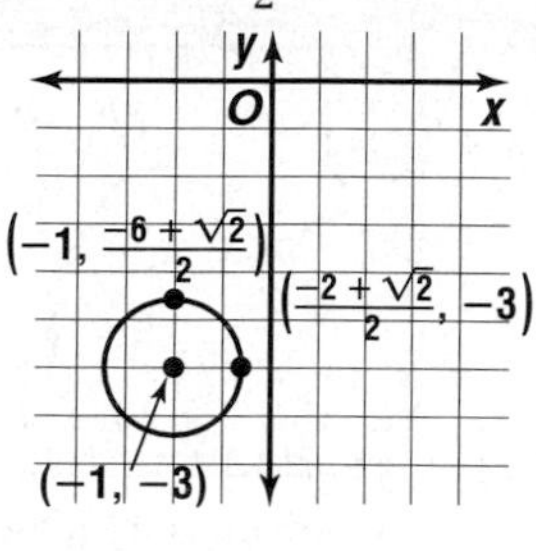

19. $(x - 6)^2 + (y - 1)^2 = 36$

21. $x^2 + y^2 = 36$

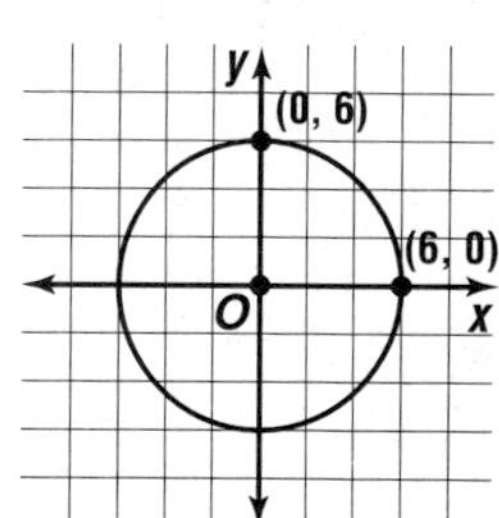

23. $(x - 2)^2 + (y + 6)^2 = 10$

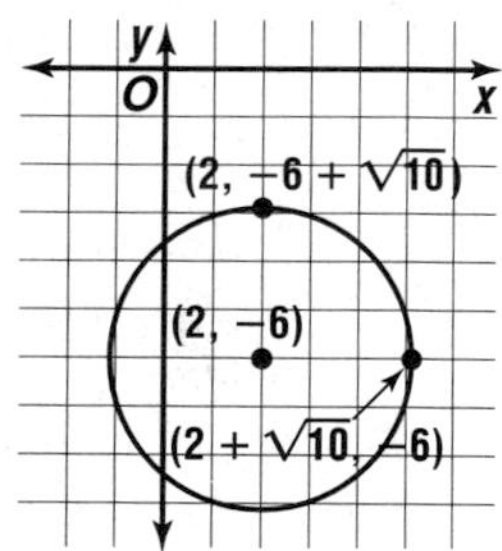

25. $(x - 1)^2 + (y + 3)^2 = 16$

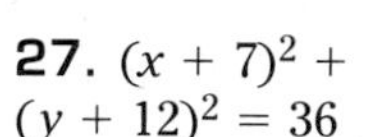

27. $(x + 7)^2 + (y + 12)^2 = 36$

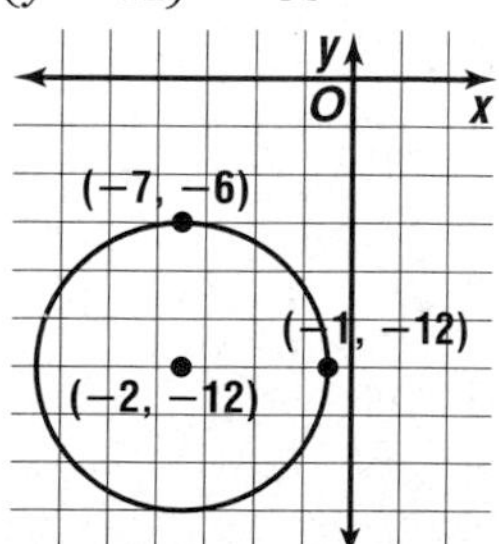

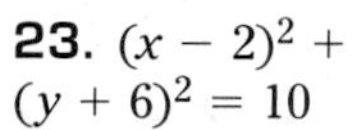

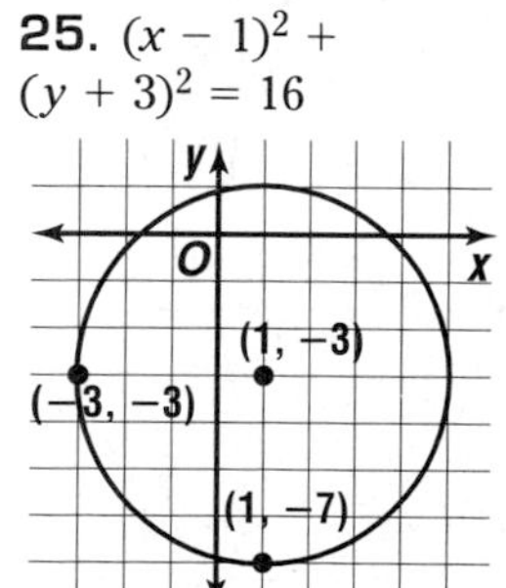

29. $(x - 7)^2 + (y + 5)^2 = 16$; $(7, -5)$; 4
31. $(x - 5)^2 + (y - 2)^2 = 50$; $(5, 2)$, $5\sqrt{2}$
33. $\left(x + \frac{1}{6}\right)^2 + \left(y - \frac{7}{6}\right)^2 = \frac{169}{18}$; $\left(-\frac{1}{6}, \frac{7}{6}\right)$; $\frac{13\sqrt{2}}{6}$

35. $(x + 4)^2 + (y - 3)^2 = 25$ **37.** $(x + 2)^2 + (y + 1)^2 = 32$ **39.** $(x - 5)^2 + (y - 1)^2 = 10$
41a. $x^2 + y^2 = 36$ **41b.** $2x$ by $2\sqrt{36 - x^2}$
41c. $A(x) = 4x\sqrt{36 - x^2}$

41d.

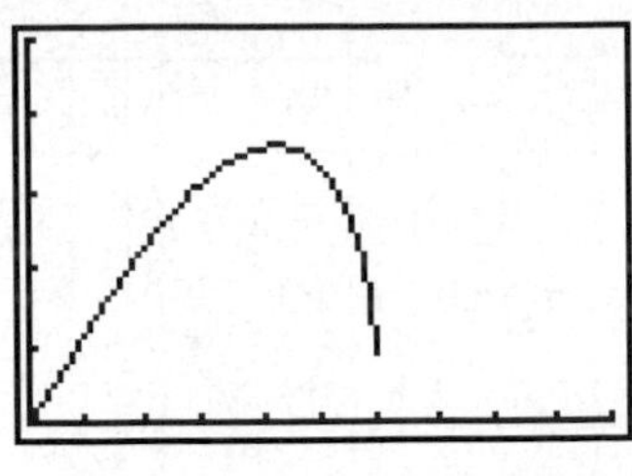

[0, 10] scl:1 by [0, 100] scl:20

41e. 4.2; 72 units^2 **43a.** $x^2 + y^2 = 144$
43b. about 145.50 in.^{2} **45a.** $(x - k)^2 + (y - k)^2 = 4$
45b. Sample graph:

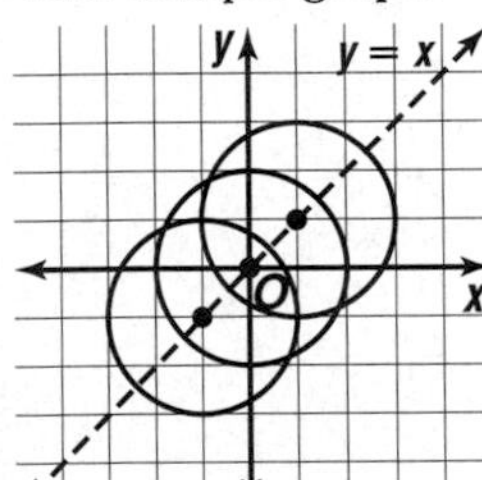

45c. All the circles in this family have a radius of 2 and centers located on the line $y = x$. **47.** radius: 0; center: (4, −3); graph is a point located at (4, −3)
49a. $x^2 + y^2 = 25$

49b. If $\overline{PA} \perp \overline{PB}$, then A, P, and B are on the circle $x^2 + y^2 = 25$. **51.** $20 + 15\boldsymbol{i}$ **53.** $y = 2.5 \cos\left(\frac{\pi}{10}t\right)$
55. 137.5 ft **57a.** 10 cases of drug A, 5 cases of drug B **57b.** $5700

Pages 637–641 Lesson 10-3

7. center: (0, 0); foci: $(\pm 4\sqrt{2}, 0)$; vertices: $(\pm 6, 0)$, $(0, \pm 2)$

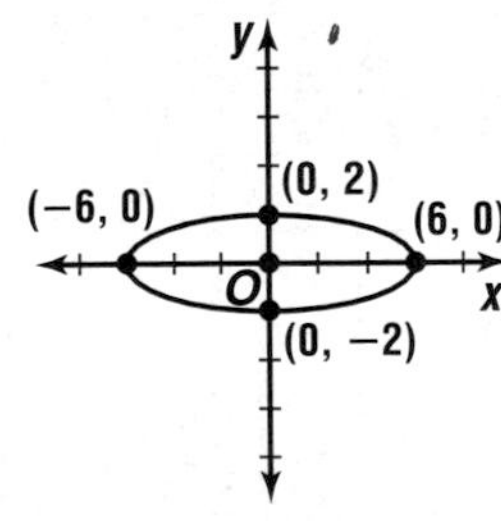

9. center: (−2, 1); foci: (−2, 5), (−2, −3); vertices: (−2, 6), (−2, −4), (1, 1), (−5, 1)

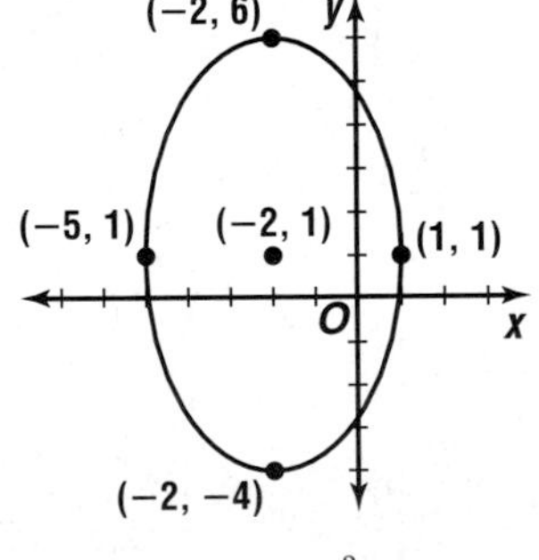

11. $\frac{(y + 3)^2}{16} + \frac{(x + 2)^2}{1} = 1$ **13.** $\frac{(x - 1)^2}{16} + \frac{(y - 2)^2}{4} = 1$ **15.** $\frac{x^2}{1.524^2} + \frac{y^2}{1.517^2} = 1$
17. $\frac{(x + 2)^2}{16} + \frac{y^2}{4} = 1$; foci: $\left(-2 \pm 2\sqrt{3}, 0\right)$

19. center: (−2, 1); foci: $\left(-2, 1 \pm \sqrt{3}\right)$; vertices: (−2, 3), (−2, −1), (−1, 1), (−3, 1)

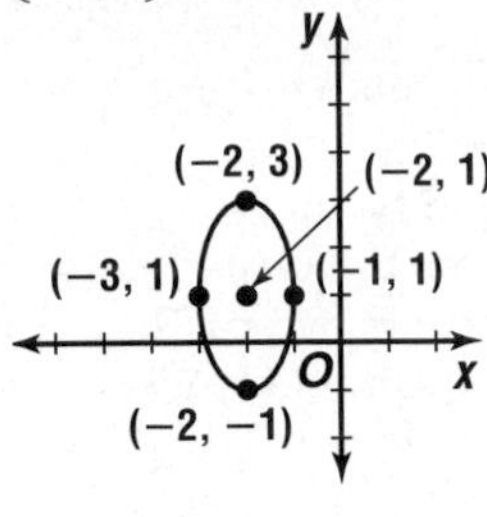

21. center: (4, −6), foci: $(4 \pm \sqrt{7}, -6)$; vertices: (0, −6), (8, −6), (4, −3), (4, −9)

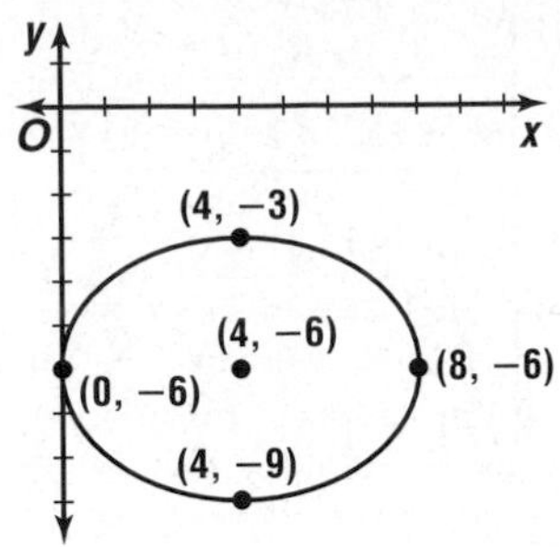

23. center: (1, −3); foci: $\left(1, -3 \pm \sqrt{3}\right)$; vertices: (2, −3), (0, −3), (1, −1), (1, −5)

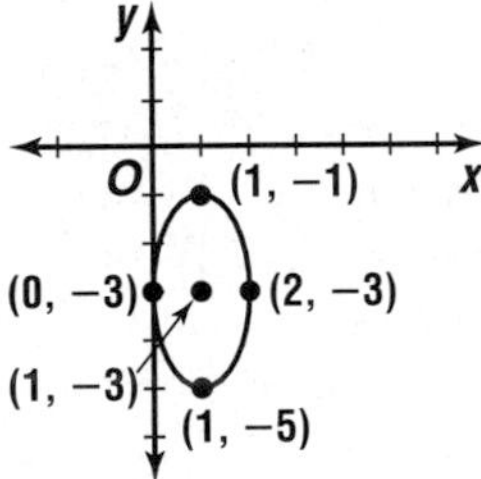

25. center: (−3, 1); foci: (−3, 5), (−3, −3); vertices: $\left(-3 \pm 2\sqrt{2}, 1\right)$, $\left(-3, 1 \pm 2\sqrt{6}\right)$

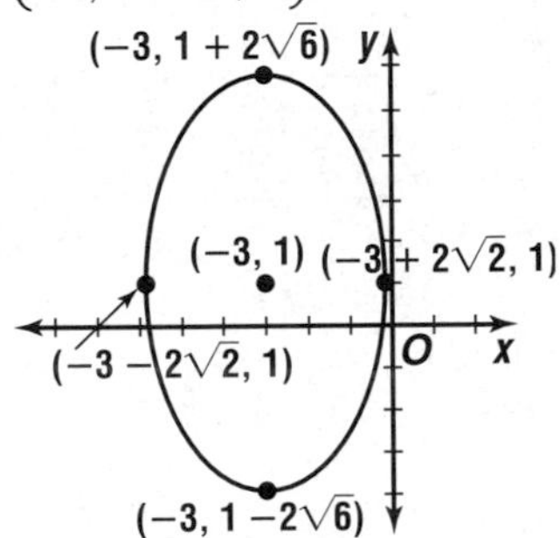

27. center: (2, 4); foci: $\left(2 \pm \sqrt{6}, 4\right)$; vertices: $\left(2 \pm 3\sqrt{2}, 4\right)$, $\left(2, 4 \pm 2\sqrt{3}\right)$

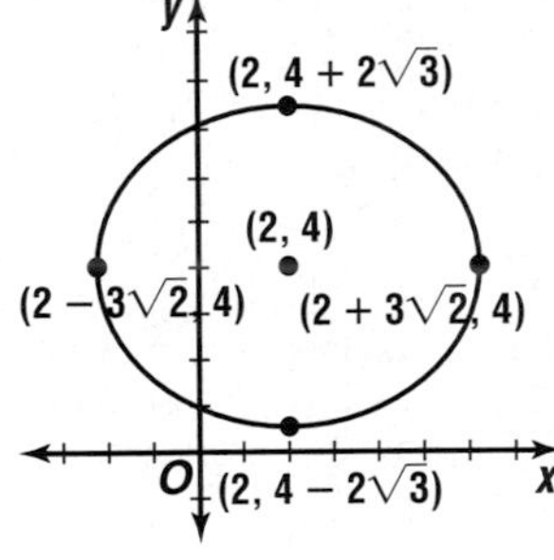

29. center: (0, −5); foci: $\left(0, -5 \pm \sqrt{33}\right)$; vertices: (±4, −5), (0, −12), (0, 2)

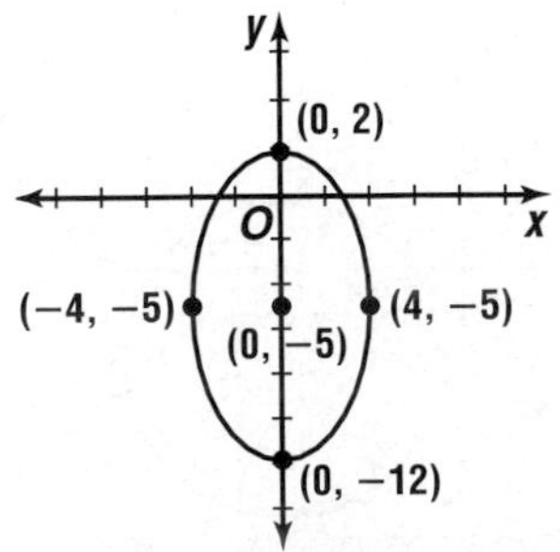

31. $\frac{(x+3)^2}{49} + \frac{(y+1)^2}{25} = 1$ **33.** $\frac{x^2}{64} + \frac{y^2}{36} = 1$
35. $\frac{(x+2)^2}{81} + \frac{(y-5)^2}{16} = 1$ **37.** $\frac{x^2}{100} + \frac{y^2}{75} = 1$
39. $\frac{y^2}{4} + \frac{x^2}{1.75} = 1$ or $\frac{y^2}{1.75} + \frac{x^2}{4} = 1$
41. $\frac{y^2}{100} + \frac{(x-3)^2}{51} = 1$ **43.** (5, −3), (1, −3), (3, −2), (3, −4) **45.** (−1, −1), (5, −1), (2, −3), (2, 1) **47.** The target ball should be placed opposite the pocket, $\sqrt{5}$ feet from the center along the major axis of the ellipse. The cue ball can be placed anywhere on the side opposite the pocket. The ellipse has semi-major axis of length 3 ft and a semi-minor axis of length 2 ft. Using the equation $c^2 = a^2 - b^2$, the focus of the ellipse is found to be $\sqrt{5}$ ft from the center of the ellipse. Thus the hole is located at one focus of the ellipse. The reflective properties of an ellipse should insure that a ball placed $\sqrt{5}$ ft from the center of the ellipse and hit so that it rebounds once off the wall, should fall into the pocket at the other focus of the ellipse.
49a. $\frac{x^2}{2304} + \frac{y^2}{529} = 1$ **49b.** about 42 ft on either side of the center along the major axis **49c.** about 84 ft **51.** If (x, y) is a point on the ellipse, then show that $(-x, -y)$ is also on the ellipse.

$$\frac{x^2}{a^2} + \frac{y^2}{b^2} = 1$$

$$\frac{(-x)^2}{a^2} + \frac{(-y)^2}{b^2} = 1 \quad \textit{Replace } x \textit{ with } -x \textit{ and } y \textit{ with } -y.$$

$$\frac{x^2}{a^2} + \frac{y^2}{b^2} = 1 \quad (-x)^2 = x^2 \textit{ and } (-y)^2 = y^2$$

Thus $(-x, -y)$ is also a point on the ellipse and the ellipse is therefore symmetric with respect to the origin.
53a. GOES 4; its eccentricity is closest to 0 **53b.** 960 km
55. no
57. $y = \pm 4 \cos (2x - 40°)$
59. 74, no
61.

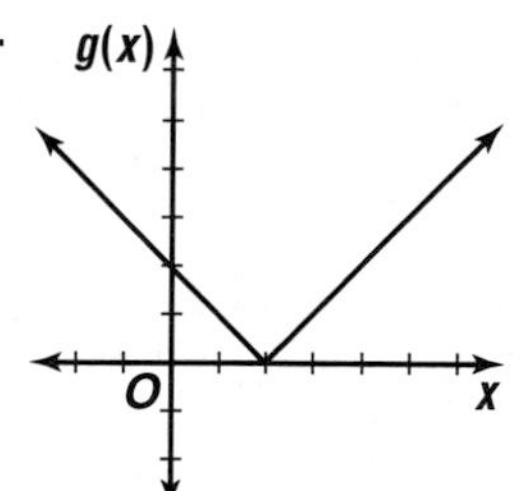

63. C

Pages 649–652 Lesson 10-4
5. center: (0, 0);
foci: $(\pm\sqrt{29}, 0)$;
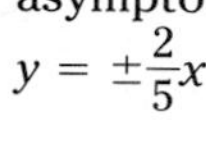
vertices: (±5, 0);
asymptotes:
$y = \pm\frac{2}{5}x$

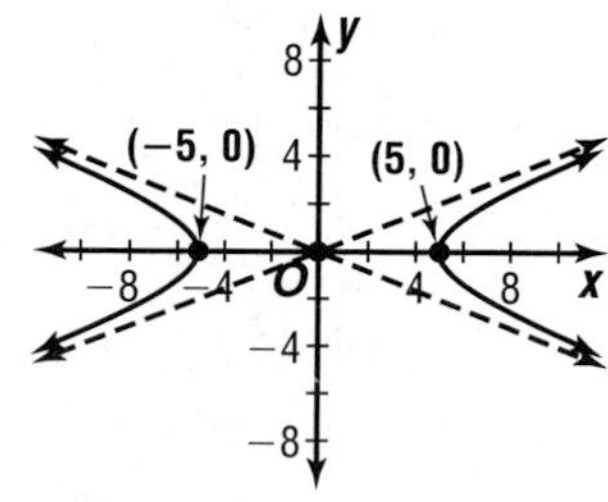

7. center: (2, 0); foci: (2, ±6); vertices: $(2, \pm\sqrt{30})$; asymptotes: $y = \pm\sqrt{5}(x - 2)$

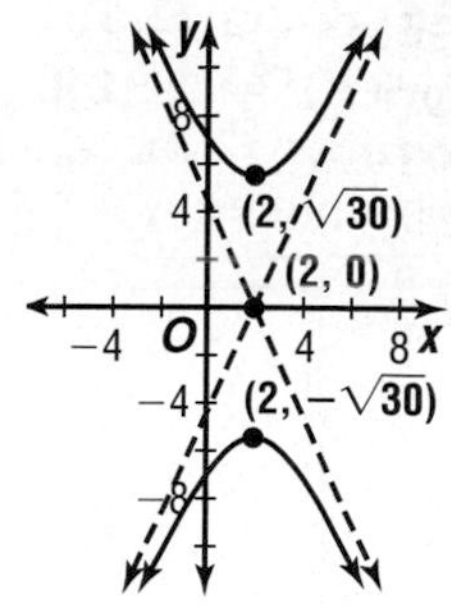

9.

11. $\frac{(y-2)^2}{4} - \frac{(x-3)^2}{9} = 1$
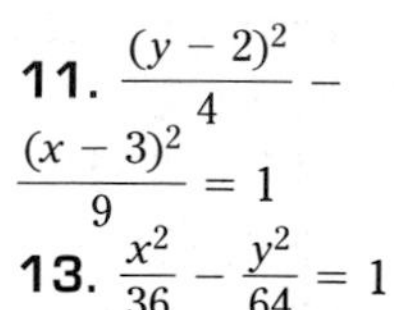
13. $\frac{x^2}{36} - \frac{y^2}{64} = 1$

15. center: (0, 0); foci: $(\pm 2\sqrt{29}, 0)$; vertices: (±10, 0); asymptotes: $y = \pm\frac{2}{5}x$

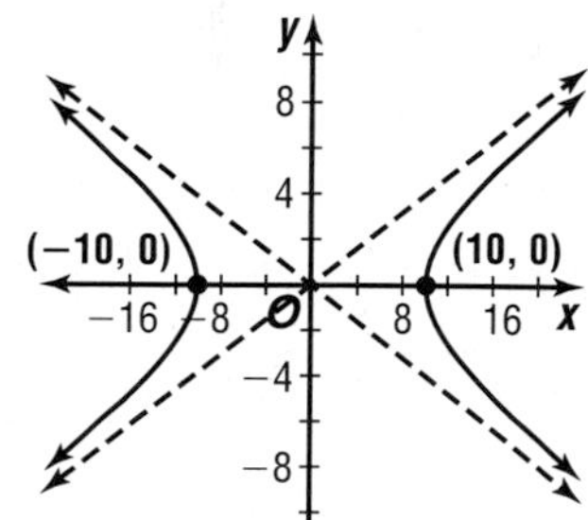

17. center: (0, 0), foci: $(\pm\sqrt{53}, 0)$; vertices: (±2, 0); asymptotes: $y = \pm\frac{7}{2}x$
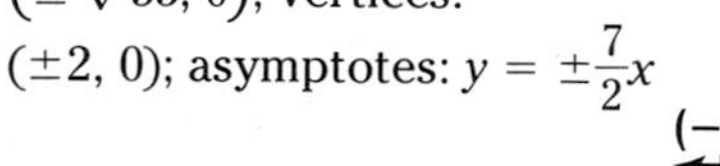

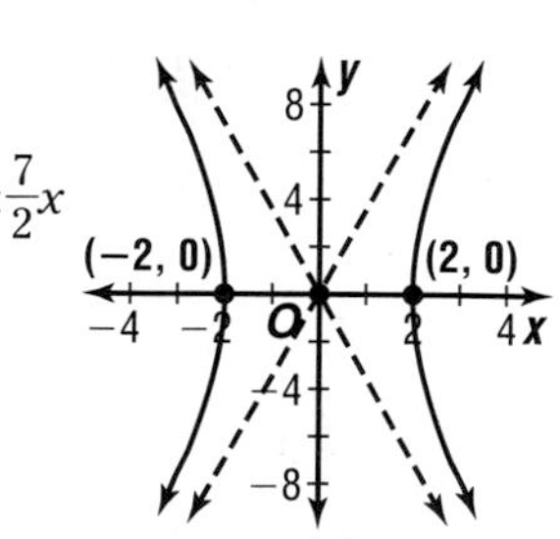

19. center: (−3, −1); foci: $(-3 \pm 2\sqrt{5}, -1)$, vertices: (1, −1), (−7, −1); asymptotes: $y + 1 = \pm\frac{1}{2}(x + 3)$

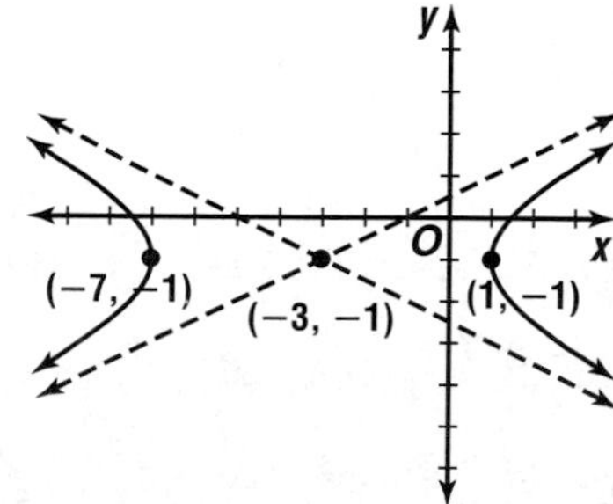

21. center: $(2, 3)$; foci: $\left(2, 3 \pm \sqrt{41}\right)$, vertices: $(2, 8)$, $(2, -2)$; asymptotes: $y - 3 = \pm\frac{5}{4}(x - 2)$

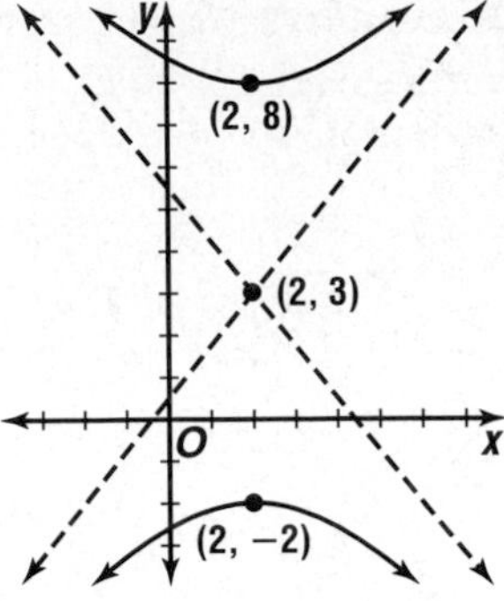

23. center: $(-4, 2)$; foci: $(-4, 2 \pm \sqrt{34})$; vertices: $(-4, 5)$, $(-4, -1)$; asymptotes: $y - 2 = \pm\frac{3}{5}(x - 4)$

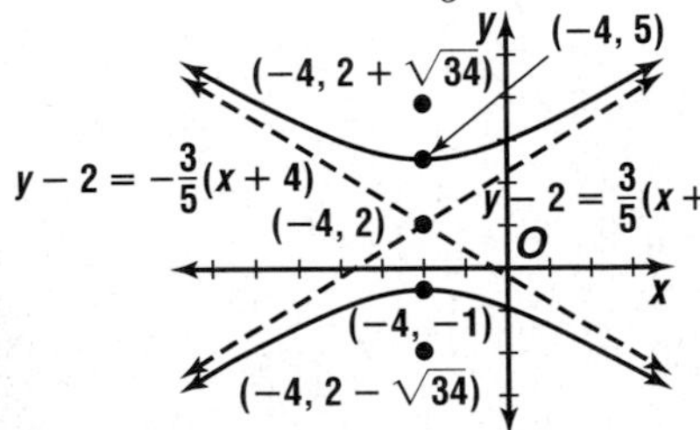

25. $\frac{x^2}{9} - \frac{y^2}{9} = 1$

27.

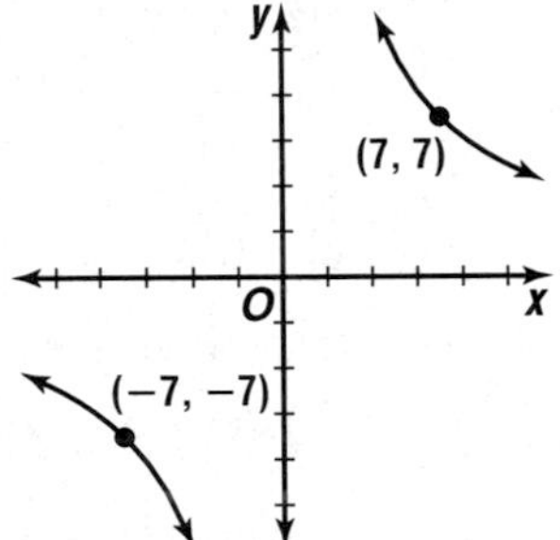

29.

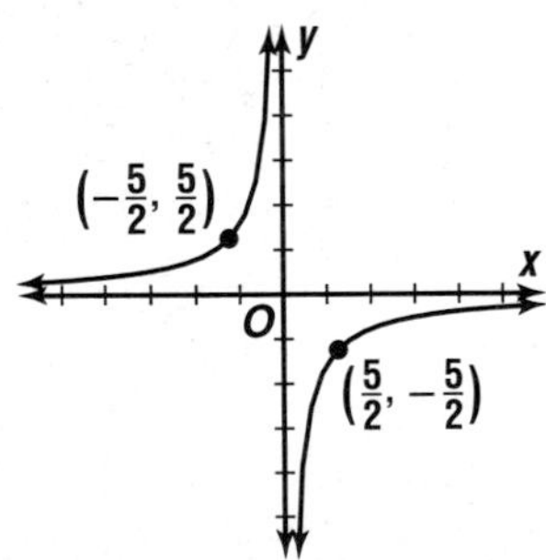

31. $\frac{(y + 2)^2}{4} - \frac{(x - 4)^2}{9} = 1$ **33.** $\frac{x^2}{9} - \frac{(y - 2)^2}{16} = 1$

35. $\frac{x^2}{32} - \frac{y^2}{32} = 1$ **37.** $\frac{(y - 2)^2}{9} - \frac{(x - 4)^2}{16} = 1$

39. $\frac{y^2}{36} - \frac{x^2}{28} = 1$ **41.** $\frac{2x^2}{81} - \frac{2y^2}{81} = 1$

43a.

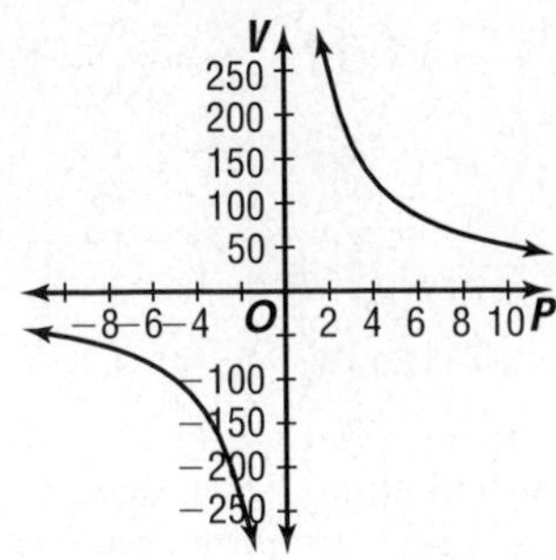

43b. 5.0 dm^3 **43c.** 10.0 dm^3 **43d.** $V = 2$(original V) **45a.** $\frac{x^2}{75^2} - \frac{y^2}{100^2} = 1$ **45b.** top: 106.07 ft; base: 273.00 ft **47.** $\frac{x^2}{25} - \frac{y^2}{11} = 1$

49. $\frac{y^2}{16} + \frac{(x - 2)^2}{7} = 1$

51.

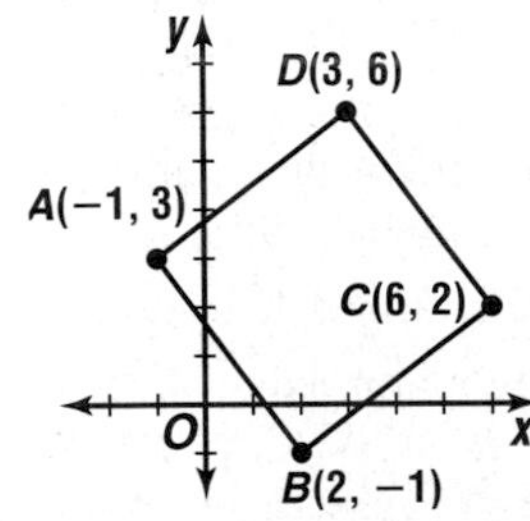

$AB = \sqrt{(2 + 1)^2 + (-1 - 3)^2} = 5$
$BC = \sqrt{(2 - 6)^2 + (-1 - 2)^2} = 5$
$CD = \sqrt{(6 - 3)^2 + (2 - 6)^2} = 5$
$AD = \sqrt{(3 + 1)^2 + (6 - 3)^2} = 5$
Thus, $ABCD$ is a rhombus. The slope of $\overline{AD} = \frac{6 - 3}{3 + 1}$ or $\frac{3}{4}$ and the slope of $\overline{AB} = \frac{3 + 1}{-1 - 2}$ or $-\frac{4}{3}$. Thus, $\overline{AD}$ is perpendicular to $\overline{AB}$ and $ABCD$ is a square. **53.** -6; No, the inner product of the two vectors is not zero. **55.** about 346 m/s **57.** C

Pages 658–661 Lesson 10-5

7. vertex: $(1, -1)$; focus: $(2, -1)$; directrix: $x = 0$; axis of symmetry: $y = -1$

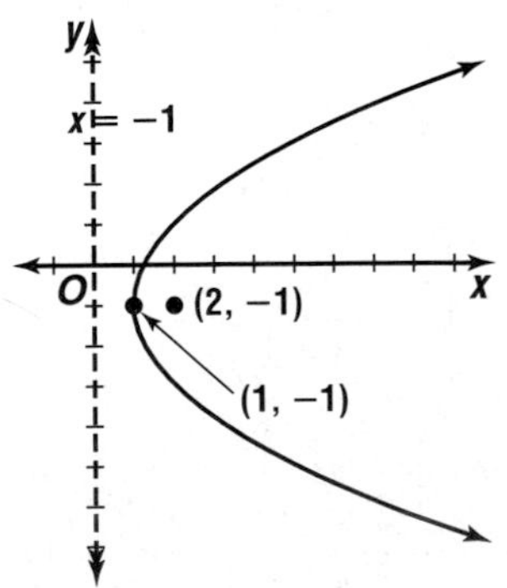

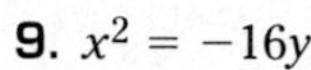
9. $x^2 = -16y$

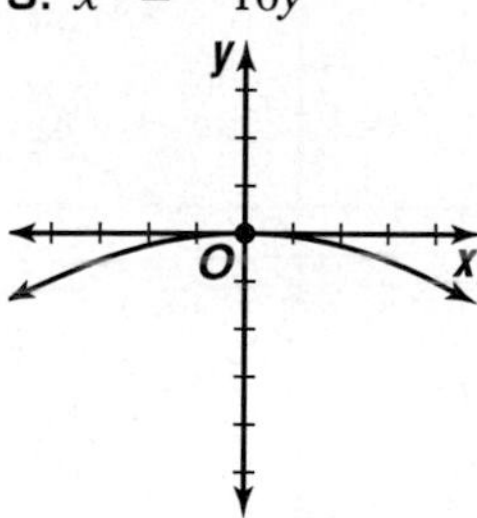

11. $(x - 4)^2 = \frac{1}{5}(y + 3)$

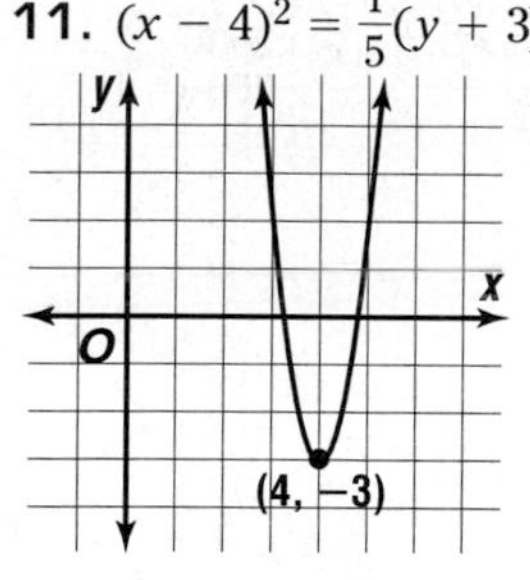

13. vertex: $(0, 0)$; focus: $(2, 0)$; directrix: $x = -2$; axis of symmetry: $y = 0$

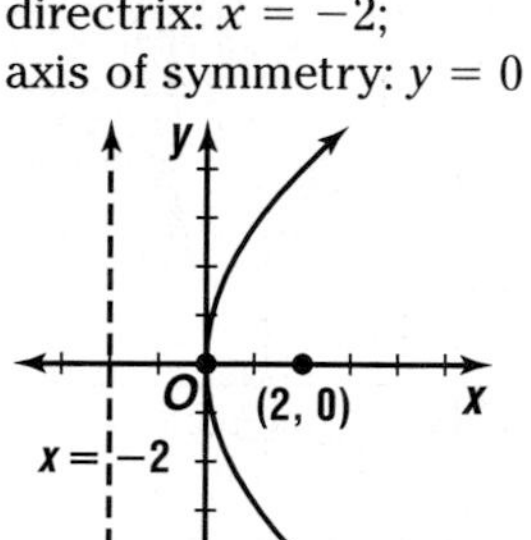

15. vertex: $(0, 6)$; focus: $(1, 6)$; directrix: $x = -1$; axis of symmetry: $y = 6$

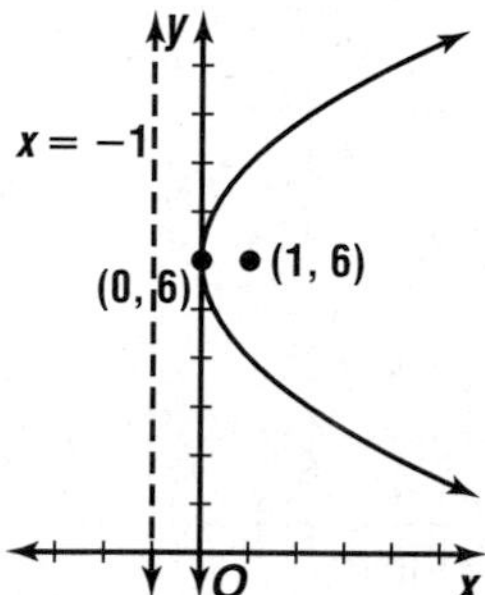

17. vertex: $(2, -2)$; focus: $\left(2, -\frac{7}{4}\right)$; directrix: $y = -\frac{9}{4}$; axis of symmetry: $x = 2$

19. vertex: $(-4, -7)$; focus: $\left(-\frac{7}{2}, -7\right)$; directrix: $x = -\frac{9}{2}$; axis of symmetry: $y = -7$

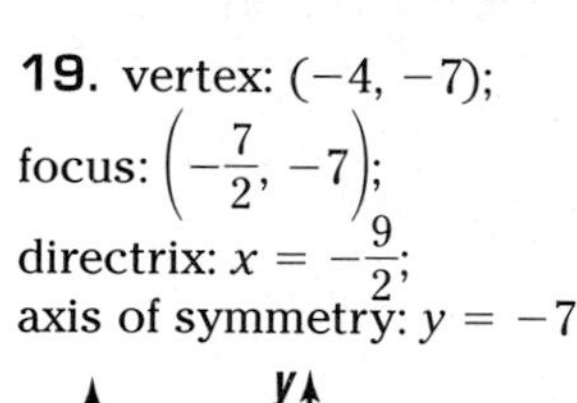

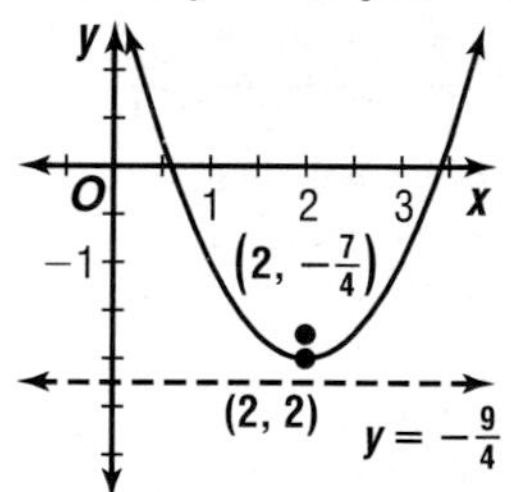

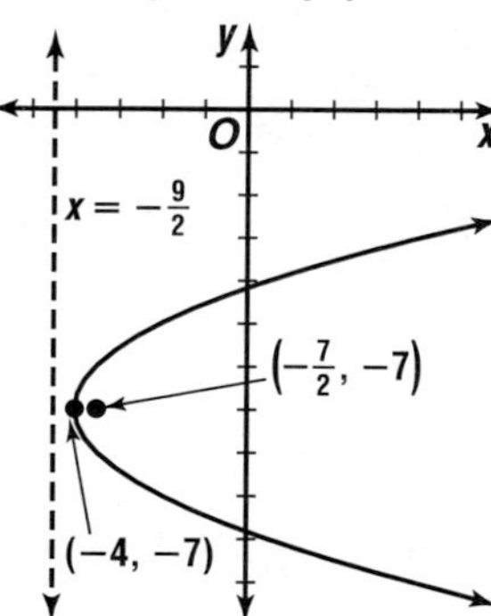

21. vertex: $(4, -1)$; focus: $\left(4, \frac{1}{2}\right)$, directrix: $y = -\frac{5}{2}$; axis of symmetry:

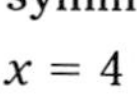
$x = 4$

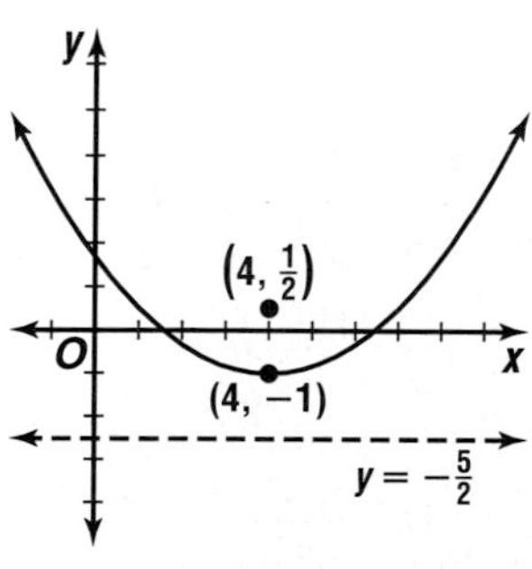

23. vertex: $(-2, -4)$; focus: $(-4, -4)$; directrix: $x = 0$; axis of symmetry: $y = -4$

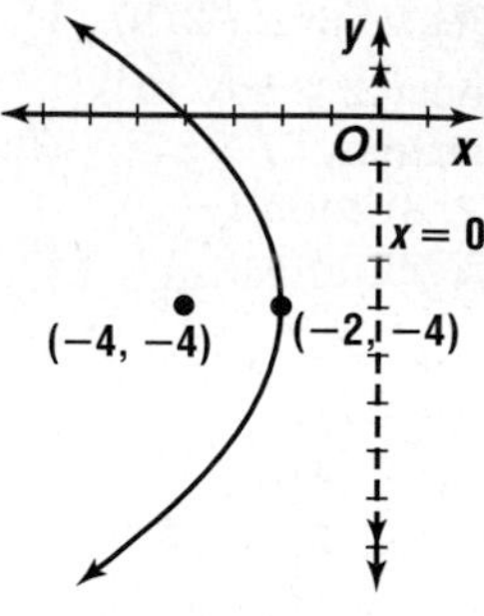

25. $(y - 6)^2 = -12(x - 3)$

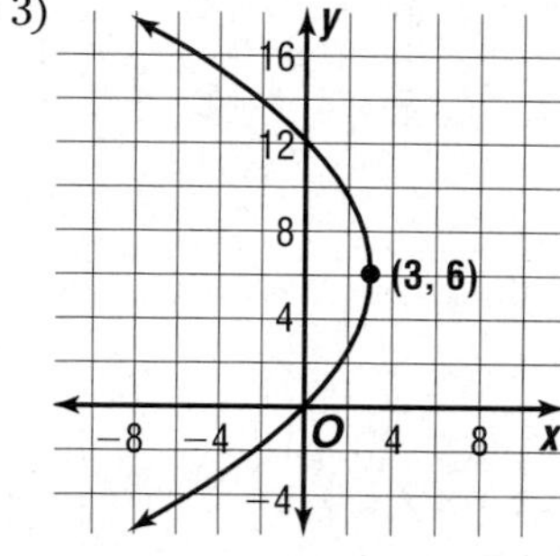

27. $(x - 4)^2 = -(y - 3)$

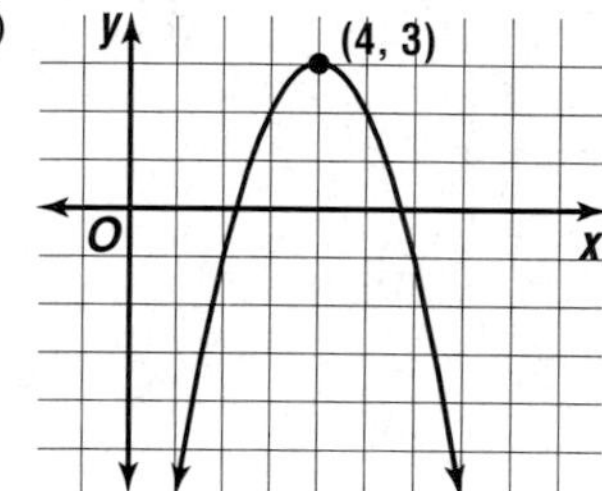

29. $(x + 1)^2 = 8(y - 5)$ **31.** $(y - 2)^2 = 4(x + 1)$

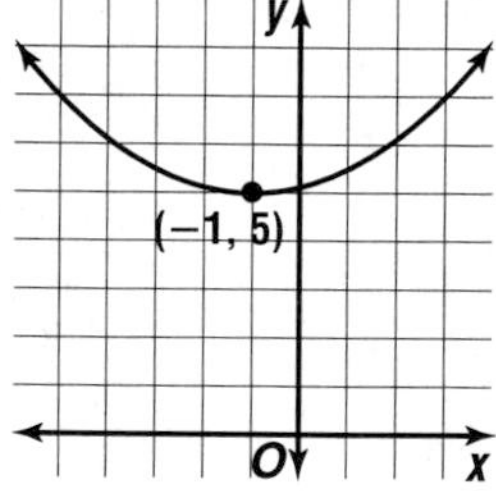

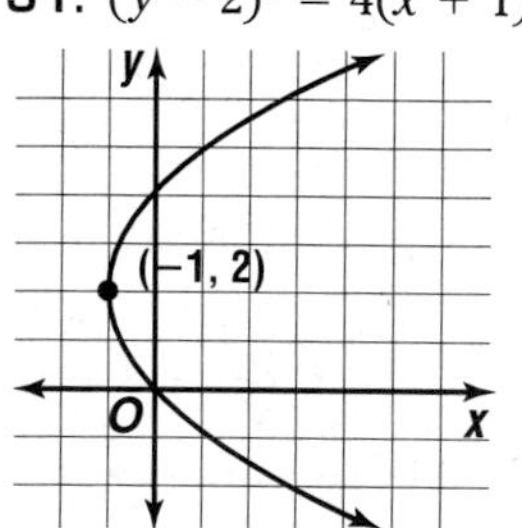

33a. $8\sqrt{2}$ in. **33b.** $4\sqrt{10}$ in. **35a.** The opening becomes narrower. **35b.** The opening becomes wider.

39. center: (2, 3), foci: $\left(2, 3 \pm \sqrt{41}\right)$; vertices: (2, 8) and (2, −2); asymptotes: $y - 3 = \pm\frac{5}{4}(x - 2)$

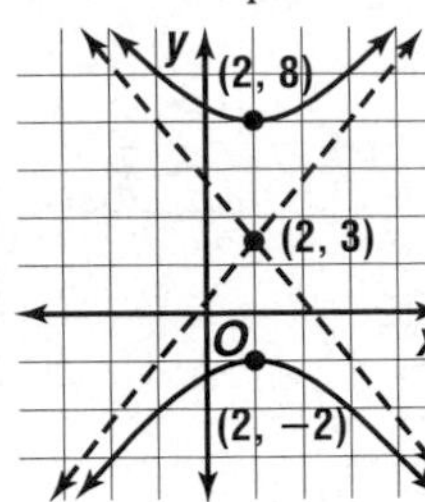

41.

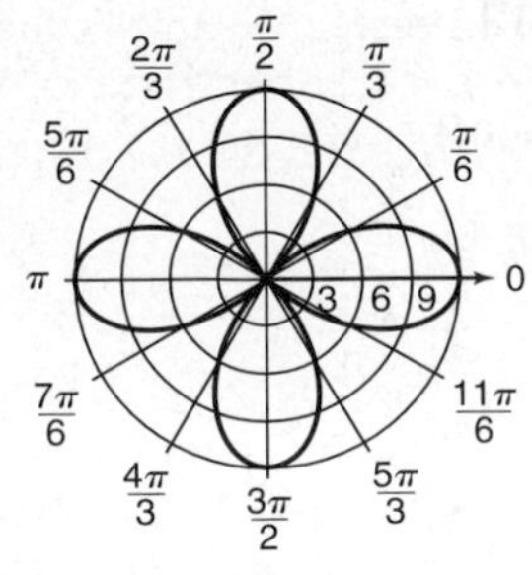

43. 5.5 cm

45. C

15. hyperbola;

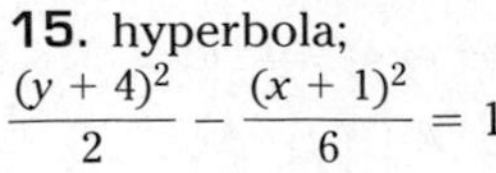

$$\frac{(y + 4)^2}{2} - \frac{(x + 1)^2}{6} = 1$$

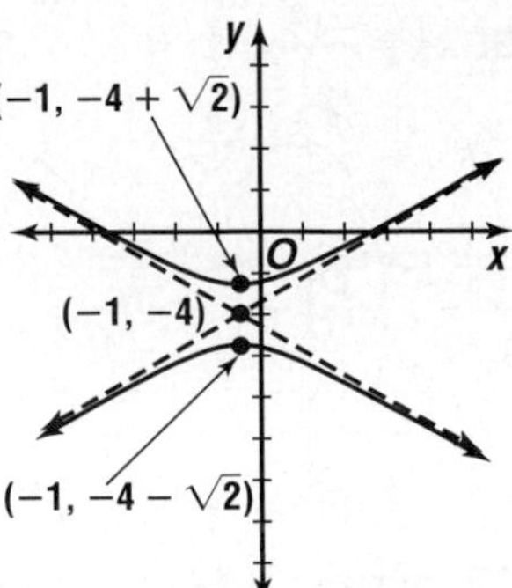

17. parabola; $(x - 4)^2 = y$

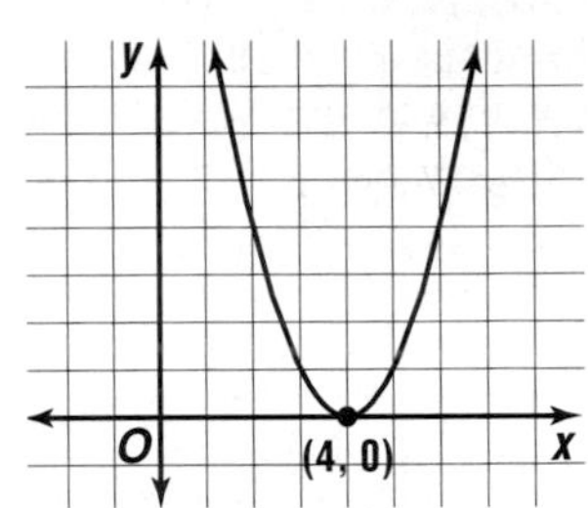

Pages 667–669 Lesson 10-6

5. parabola; $y^2 = 8(x - 1)$

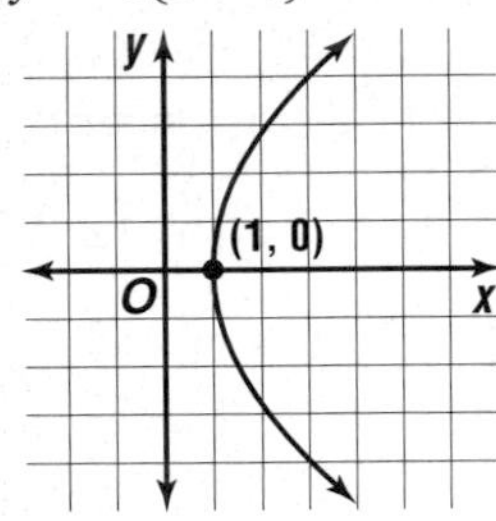

7. circle; $(x - 3)^2 + (y - 6)^2 = 4$

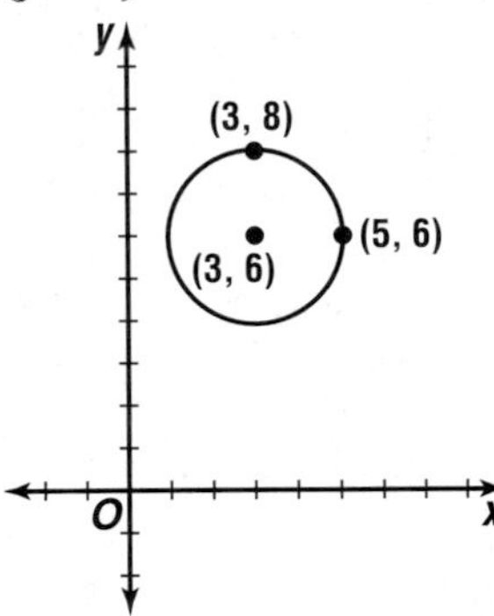

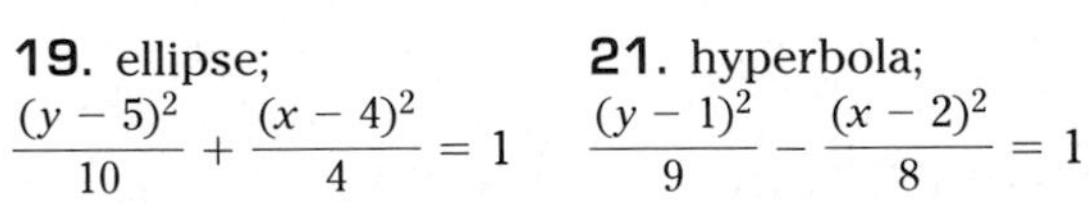

19. ellipse; $\frac{(y - 5)^2}{10} + \frac{(x - 4)^2}{4} = 1$

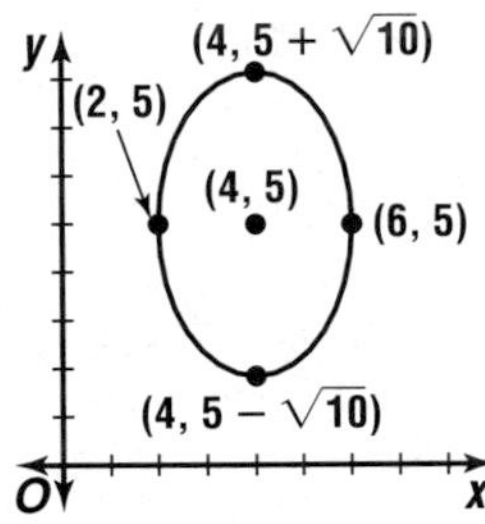

21. hyperbola;

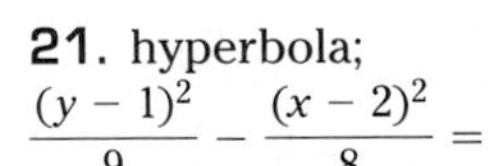

$$\frac{(y - 1)^2}{9} - \frac{(x - 2)^2}{8} = 1$$

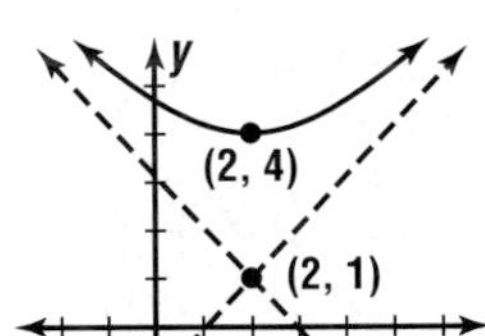

23. hyperbola; $\frac{(x + 5)^2}{4} - \frac{(y + 2)^2}{1} = 1$

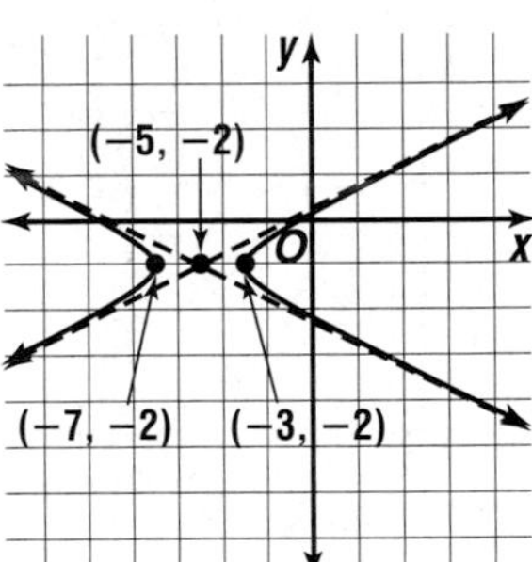

9. $\frac{x^2}{4} + \frac{y^2}{9} = 1$

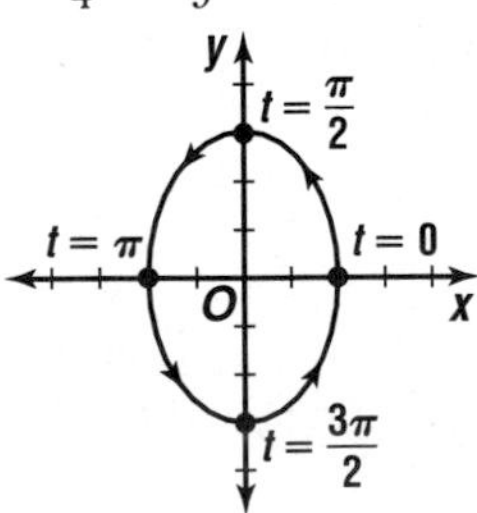

11. Sample answer: $x = 6 \cos t$, $y = 6 \sin t$, $0 \leq t \leq 2\pi$

13. parabola; $(x - 3)^2 = 4y$

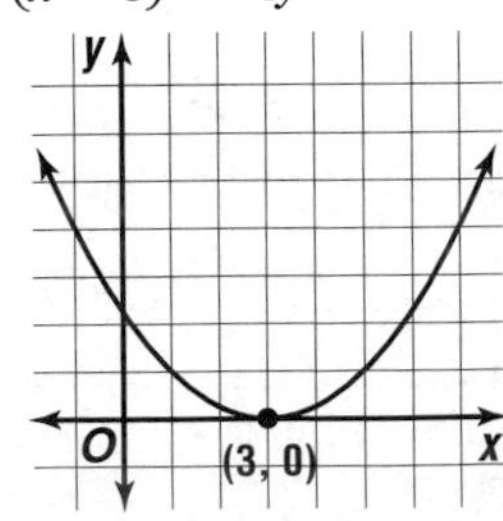

25. $y = 2x^2 - 4x + 1$

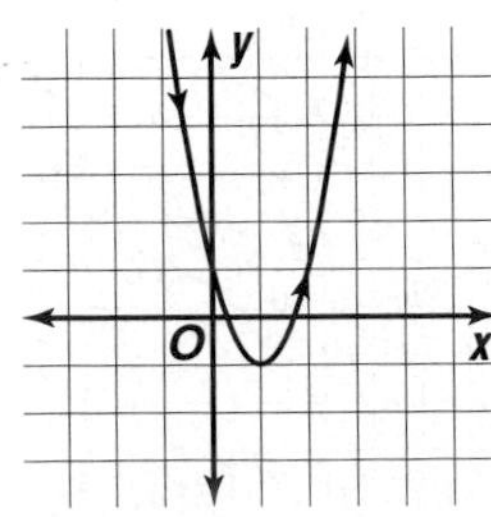

27. $x^2 + y^2 = 1$

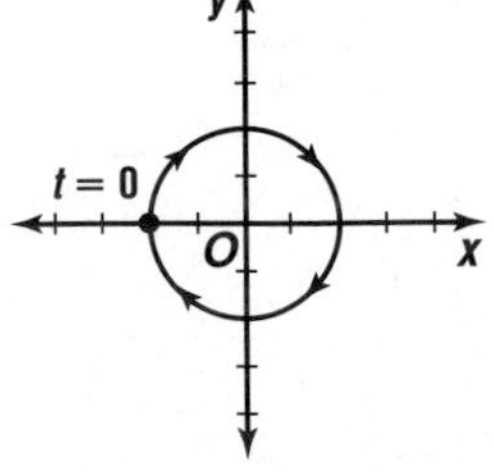

29. $x^2 + \frac{y^2}{4} = 1$

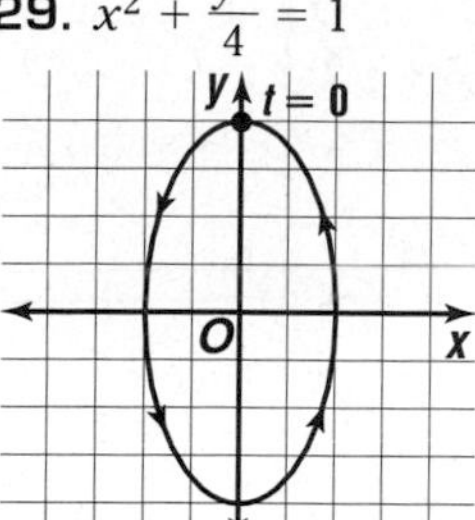

31. $x^2 + y^2 = 9$

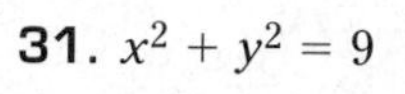

33. Sample answer: $x = 4 \cos t$, $y = 4 \sin t$, $0 \leq t \leq 2\pi$ **35.** Sample answer: $x = \cos t$, $y = 4 \sin t$, $0 \leq t \leq 2\pi$ **37.** Sample answer: $x = t^2 + 2t - 1$, $y = t$, $-\infty < t < \infty$ **39a.** Answers will vary. Sample answers: $x = t$, $y = t^2$, $t \geq 0$; $x = \sqrt{t}$, $y = t$, $t \geq 0$.

39b.

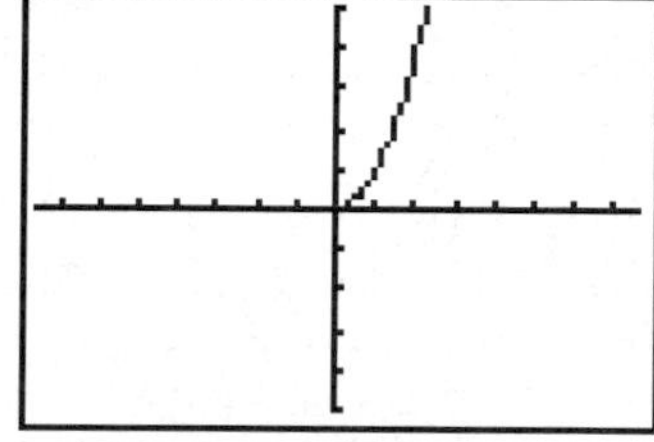

Tmin: [0, 5] step: 0.1
[−7.58, 7.58] scl:1 by [−5, 5] scl:1

39c. yes **39d.** There is usually more than one parametric representation for the graph of a rectangular equation. **41a.** Ellipse; point at (0, 0); the equation is that of a degenerate ellipse. **41b.** Circle; point at (2, 3); the equation is that of a degenerate circle. **41c.** Hyperbola; two intersecting lines $y = \pm 3x$; the equation is that of a degenerate hyperbola. **43a.** $x^2 + y^2 = 36$ **43b.** $x = 6 \sin t$, $y = 6 \cos t$, $0 \leq t \leq 4\pi$

43c.

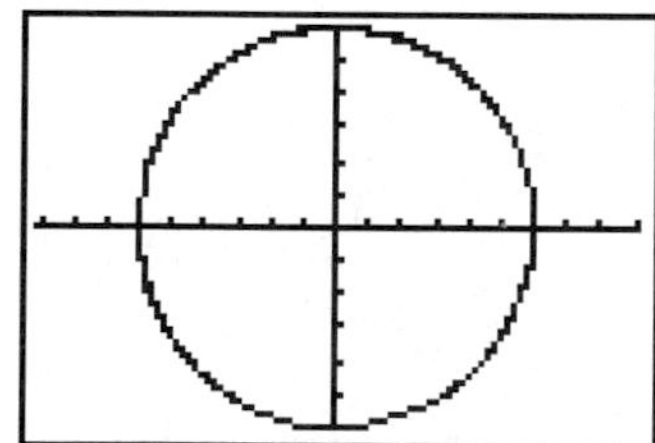

Tmin: [0, 4π] step: 0.1
[29.10, 9.10] scl:1 by [26, 6] scl:1

45. vertex: $(-5, 0)$; focus: $(-5, 3)$; axis of symmetry: $x = -5$, directrix: $y = -3$

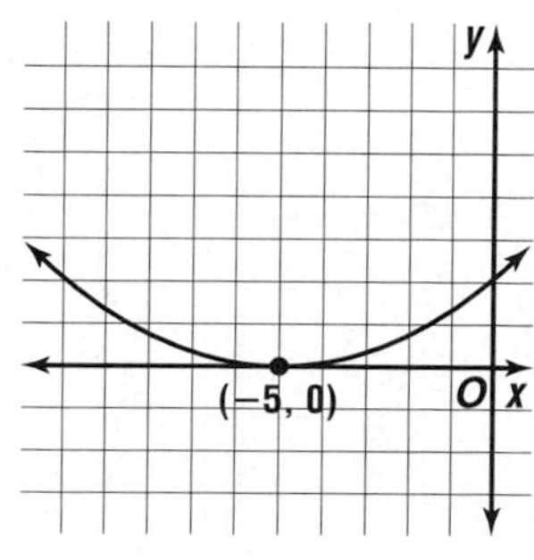

47.

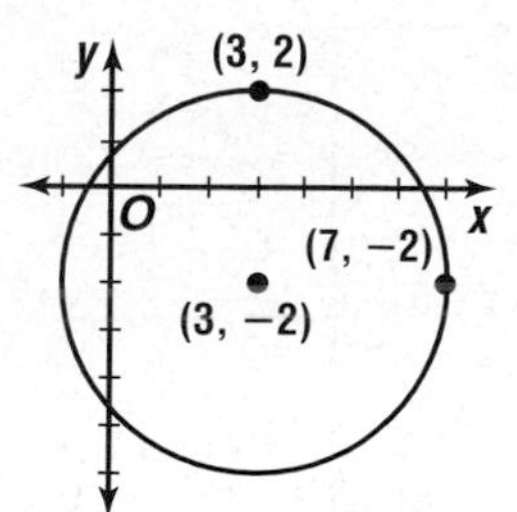

49. Car 1; the point (135, 19) is about 9 units closer to the line $y = -0.13x + 37.8$ than the point (245, 16). **51.** 685 units^2 **53.** 38.4 **55.** $y - 4 = \frac{1}{3}(x + 6)$ or $y - 7 = \frac{1}{3}(x - 3)$, $y = \frac{1}{3}x + 6$

Pages 675–677 Lesson 10-7

5. circle; $x^2 + y^2 - 6x - 4y + 6 = 0$ **7.** hyperbola; $(x')^2 - 2\sqrt{3}x'y' - (y')^2 + 18 = 0$ **9.** ellipse; 19° **11.** point

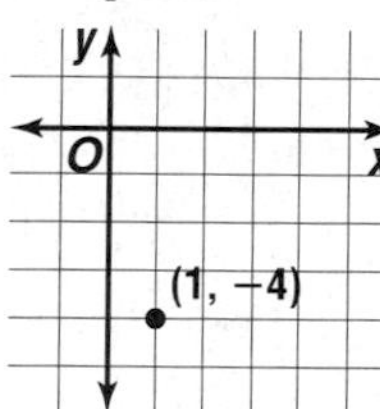

13. parabola; $3x^2 - 14x - y + 18 = 0$ **15.** ellipse; $3x^2 + y^2 + 6x - 6y + 3 = 0$ **17.** hyperbola; $9x^2 - 25y^2 + 250y - 850 = 0$ **19.** parabola; $(y')^2 + 8x' = 0$

21. parabola; $(x')^2 - 2\sqrt{3}x'y' + 3(y')^2 + 16\sqrt{3}x' + 16y' = 0$ **23.** circle; $2(x')^2 + 2(y')^2 - 5x' - 5\sqrt{3}y' - 6 = 0$ **25.** $23(x')^2 + 2\sqrt{3}x'y' + 21(y')^2 - 120 = 0$ **27.** hyperbola; −6° **29.** ellipse; −18° **31.** parabola; −30°

33. line

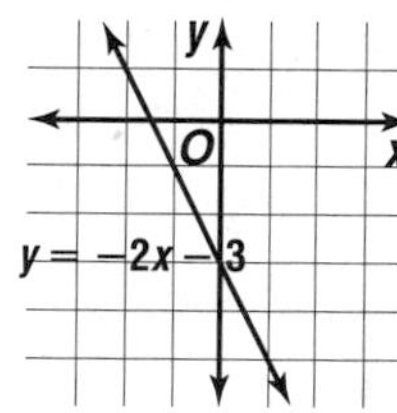

35. intersecting lines

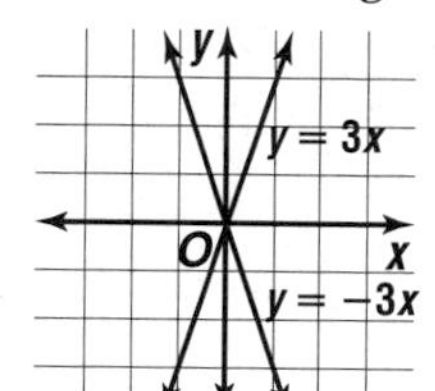

37.

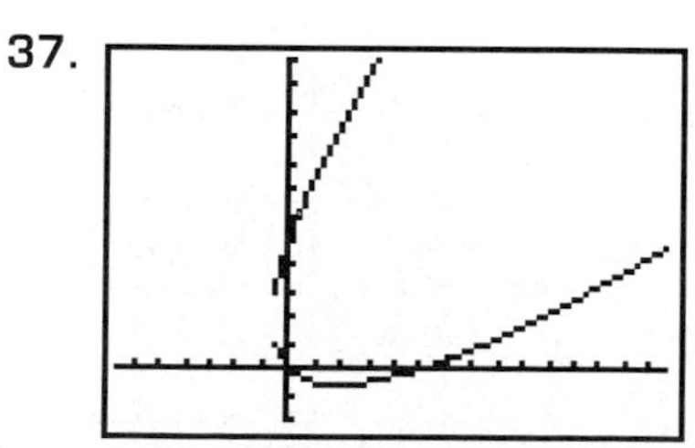

[−6.61, 14.6] scl:1 by [−2, 12] scl:1

39.

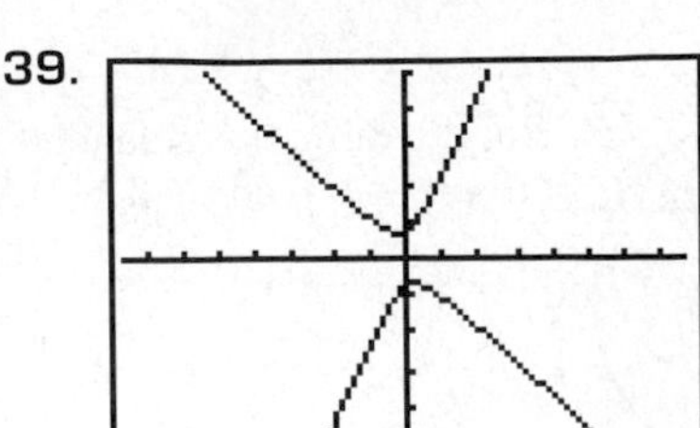

[−7.58, 7.58] scl:1 by [−5, 5] scl:1

41.

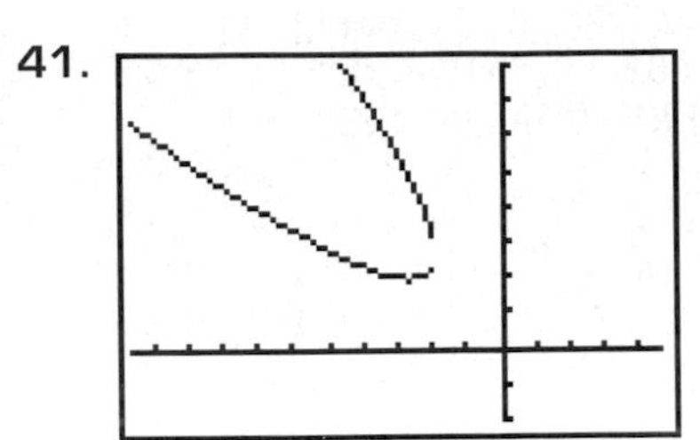

[−10.58, 4.58] scl:1 by [−2, 8] scl:1

43a. $T_{(1320,\ 1320)}$ **43b.** $(x - 1320)^2 + (y - 1320)^2 = 1{,}742{,}400$ **45.** Let $x = x'\cos\theta + y'\sin\theta$ and $y = -x'\sin\theta + y'\cos\theta$.

$$x^2 + y^2 = r^2$$
$$(x'\cos\theta + y'\sin\theta)^2 + (-x'\sin\theta + y'\cos\theta)^2 = r^2$$
$$(x')^2\cos^2\theta + x'y'\cos\theta\sin\theta + (y')^2\sin^2\theta + (x')^2\sin^2\theta - x'y'\cos\theta\sin\theta + (y')^2\cos\theta = r^2$$
$$[(x')^2 + (y')^2]\cos^2\theta + [(x')^2 + (y')^2]\sin^2\theta = r^2$$
$$[(x')^2 + (y')^2](\cos^2\theta + \sin^2\theta) = r^2$$
$$[(x')^2 + (y')^2](1) = r^2$$
$$(x')^2 + (y')^2 = r^2$$

47a. $-30°$ **47b.** $\frac{(x')^2}{3} + \frac{(y')^2}{2} = 1$

49. hyperbola

51.

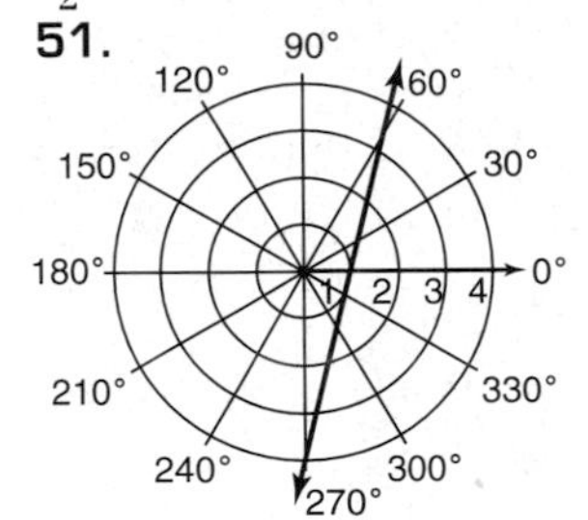

53. cos 70°

55. $\frac{-1}{y + 2} + \frac{3}{y + 1}$

57. $\left(\frac{3}{4}, -\frac{2}{3}, \frac{1}{2}\right)$

59. B

Pages 682–684 Lesson 10-8

5. (3, 3), (−1, −1)

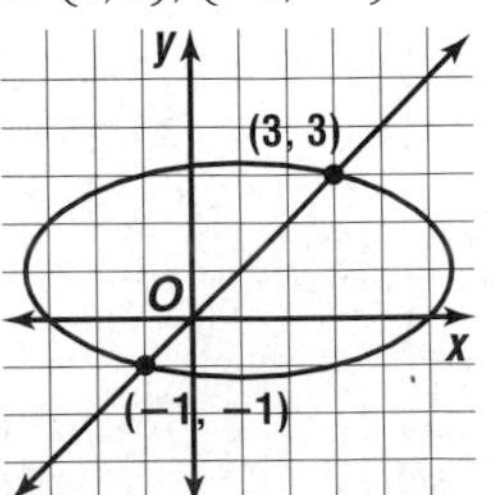

7. (±2, 0)

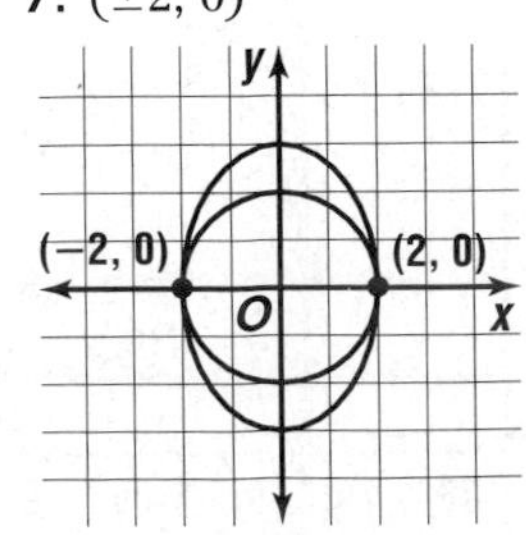

9.

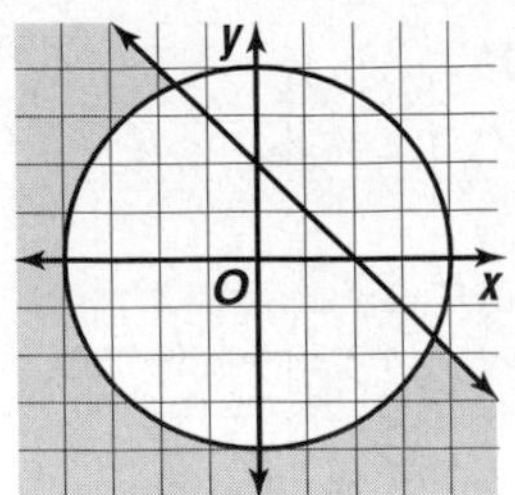

11.

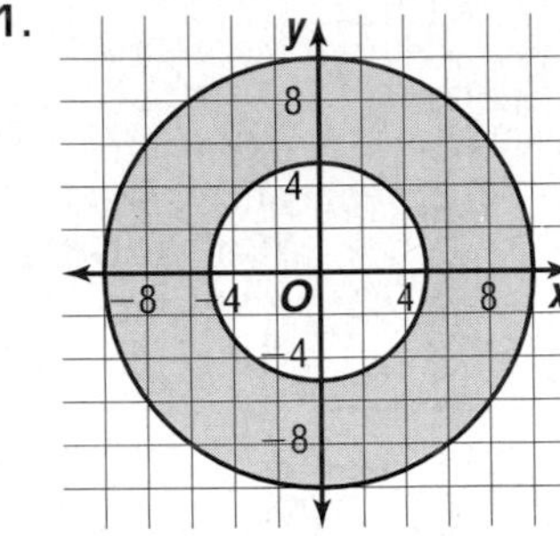

13. (1, ±6.9)

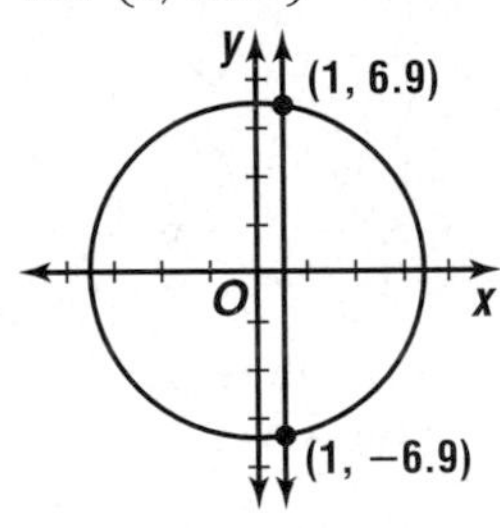

15. (1.5, −4), (−2, 3)

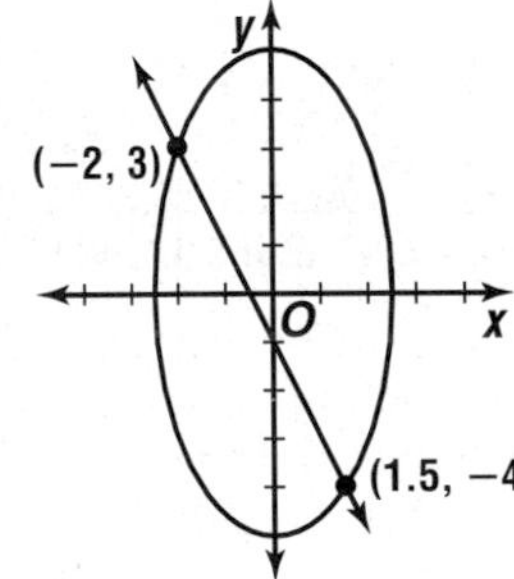

17. no solution

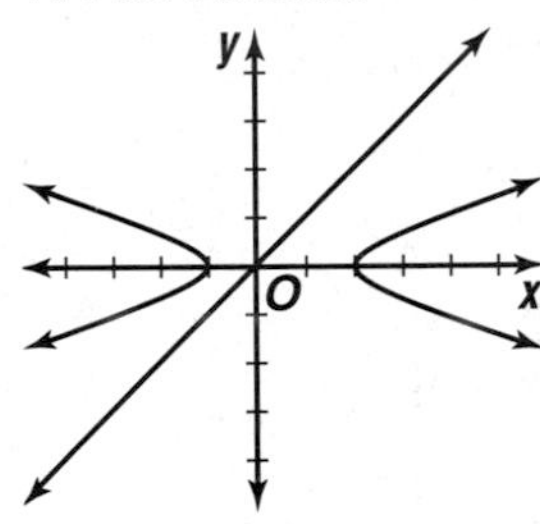

19. (0, −1), (−3, 2)

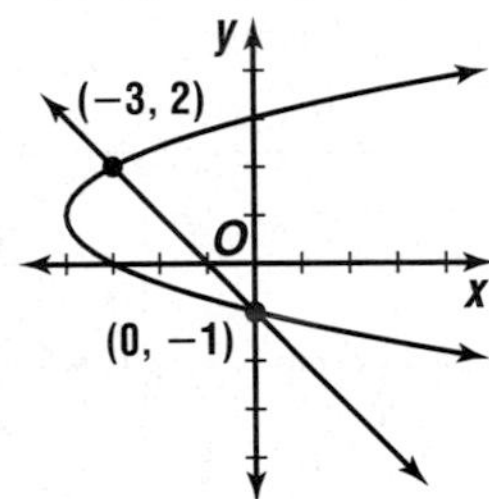

21. (0, 3), (±2.4, −2.8)

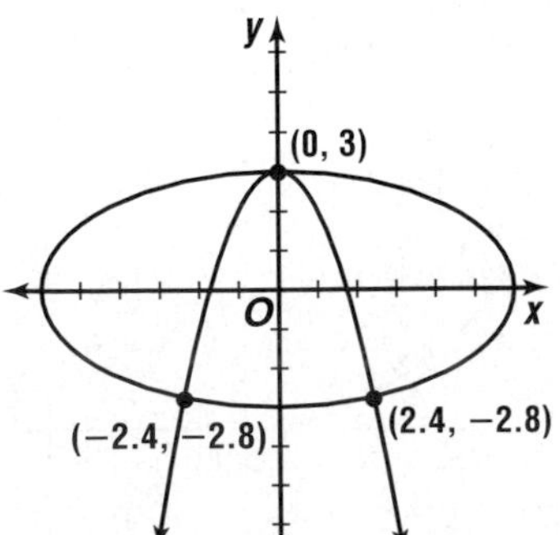

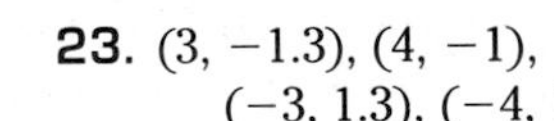

23. (3, −1.3), (4, −1), (−3, 1.3), (−4, 1)

25.

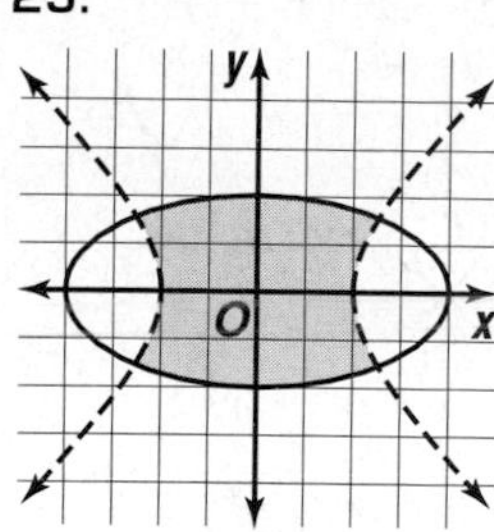

27.

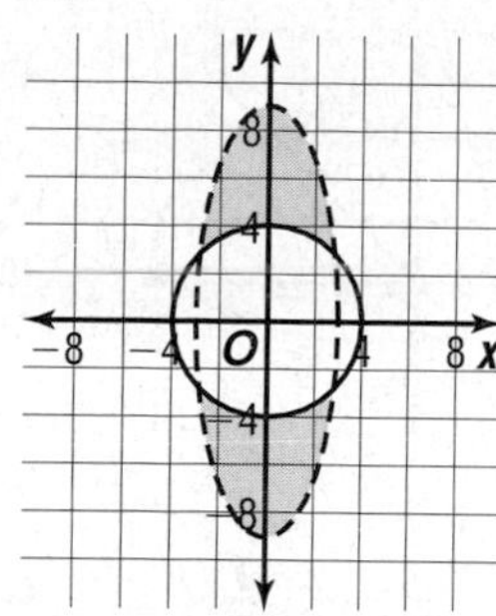

29.

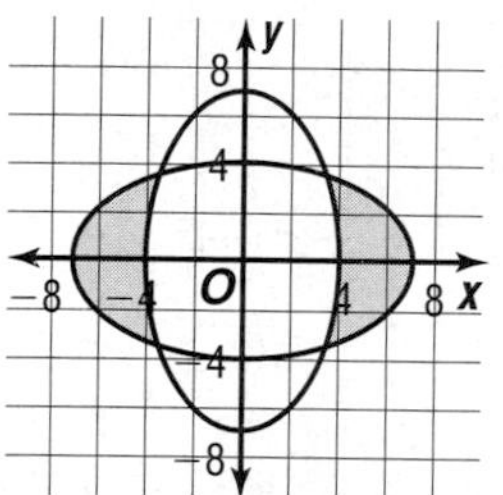

31.

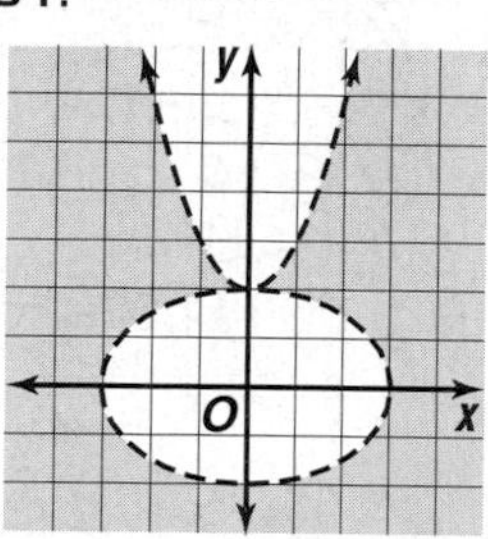

33.

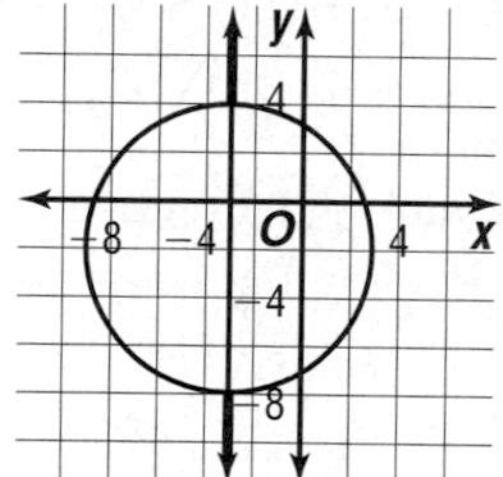

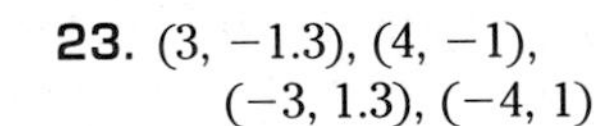

35. $x^2 + y^2 = 8$, $xy = 4$ **37a.** $2x + 2y = 150$; $xy = 800$ **37b.** 0, 1, 2

37c.

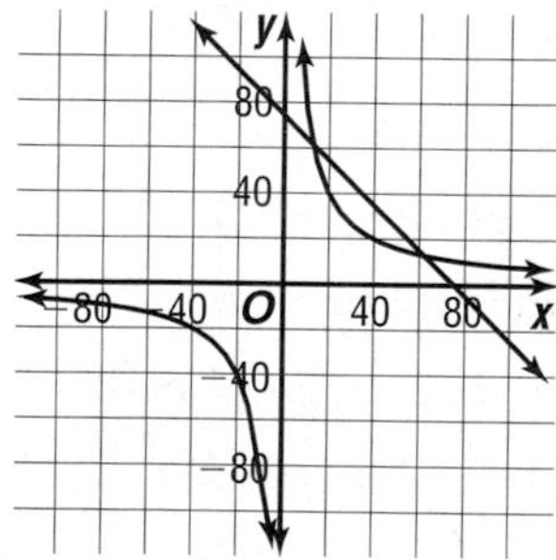

37d. 12.9 m by 62.1 m or 62.1 m by 12.9 m

39. (4, −3), (−3, 4)

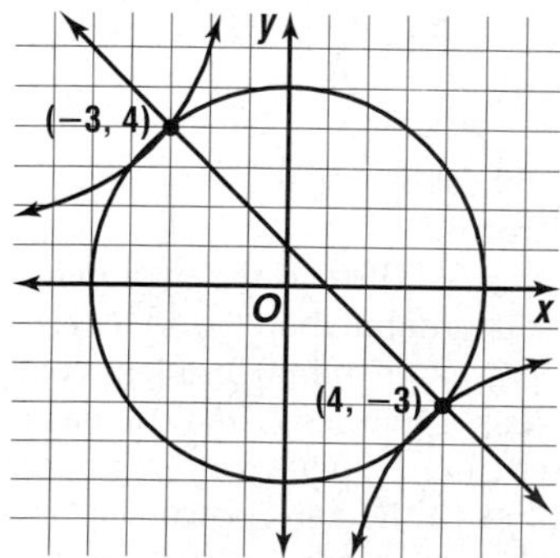

41. $-\frac{9}{8}$ **43.** $3(x')^2 - 4\sqrt{3}x'y' + 7(y')^2 - 9 = 0$ **45.** 4 **47.** 1 and 2 **49.** No; the domain value 4 is mapped to two elements in the range, 0 and −3.

Pages 687–691 Chapter 10 Study Guide and Assessment

1. true **3.** false; transverse **5.** false, hyperbola **7.** true **9.** true **11.** $2\sqrt{5}$; (−1, −5) **13.** yes; $AB = DC = 10$ and $BC = AD = 5\sqrt{2}$. Since opposite sides of quadrilateral *ABCD* are congruent, *ABCD* is a parallelogram.

15. $(x - 2)^2 + (y - 1)^2 = 4$

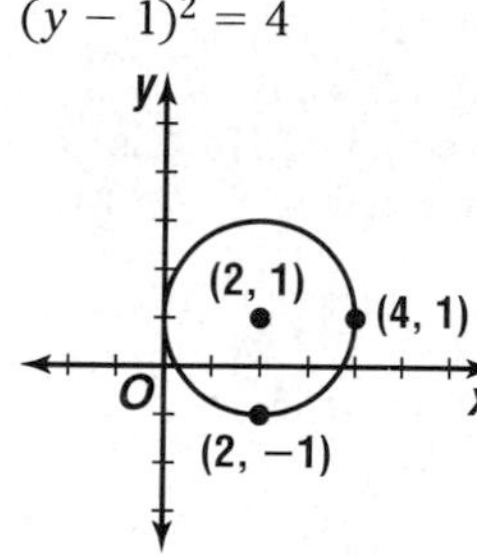

17. $(x + 7)^2 + (y + 3)^2 = 81$

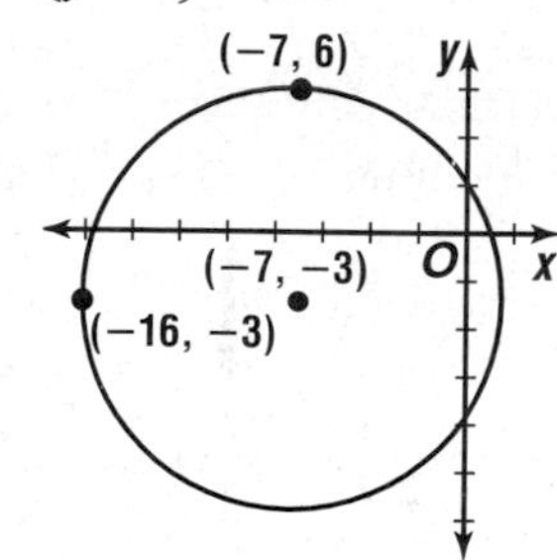

19. $(x + 2)^2 + (y + 3)^2 = 25$; (−2, −3); 5

21. center: (3, −1), foci: $(3 \pm \sqrt{21}, -1)$, vertices: (3, 1), (8, −1), (3, −3), (−2, −1)

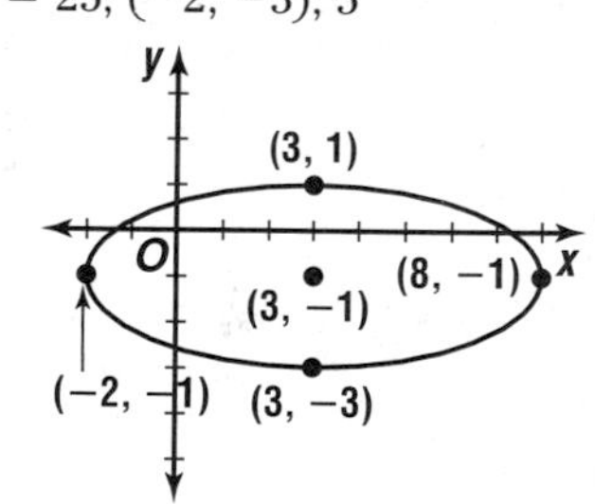

23. center: (4, 6); foci: $\left(4 \pm 3\sqrt{3}\right)$; vertices: (−2, 6), (10, 6), (4, 3), (4, 9)

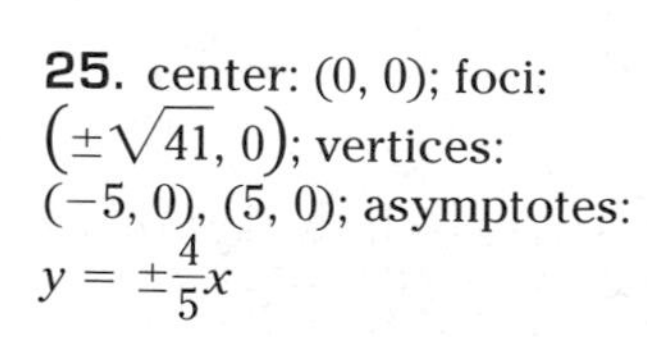

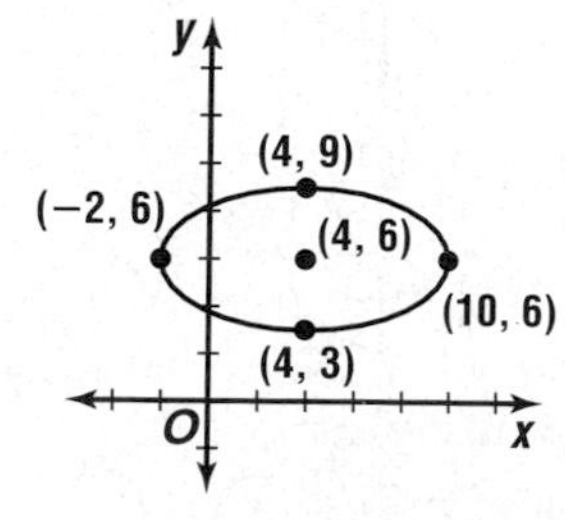

25. center: (0, 0); foci: $\left(\pm\sqrt{41}, 0\right)$; vertices: (−5, 0), (5, 0); asymptotes: $y = \pm\frac{4}{5}x$

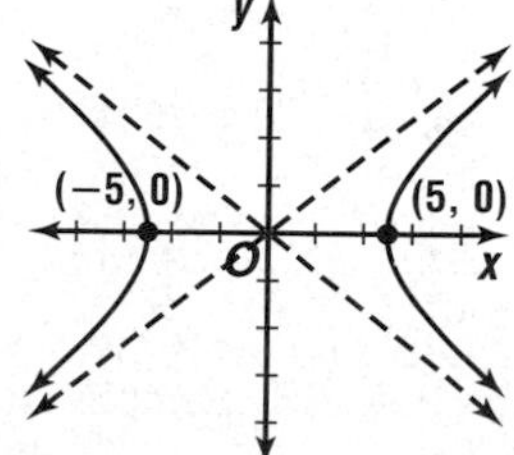

SELECTED ANSWERS

27. center: $(0, -2)$; foci: $\left(\pm\sqrt{5}, -2\right)$; vertices: $(-2, -2)$, $(2, -2)$ asymptotes: $y + 2 = \pm\frac{1}{2}x$

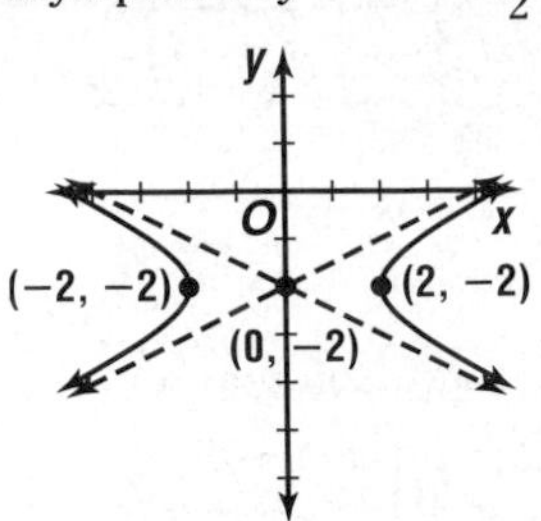

29.

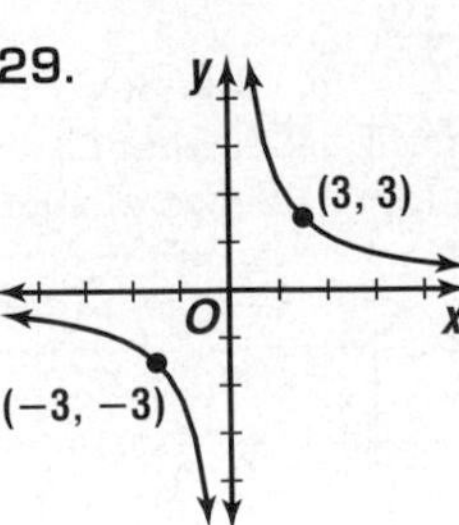

31. $\frac{(x-2)^2}{16} - \frac{(y+3)^2}{20} = 1$

33. vertex: $(-3, -2)$, focus: $(-3, -2)$, directrix: $x = 5$; axis of symmetry: $y = -2$

35. vertex: $(-2, 4)$, focus: $(-2, 4.25)$, directrix: $y = 3.75$; axis of symmetry: $x = -2$

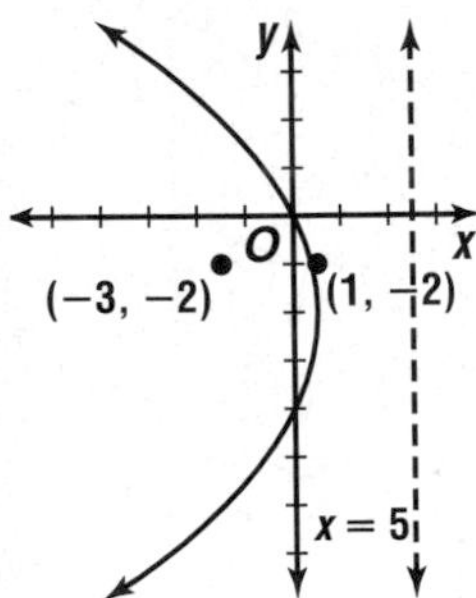

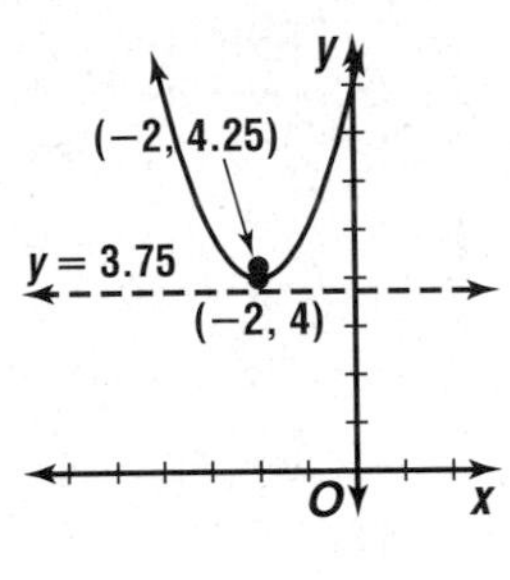

37. $(x - 5)^2 = 12(y + 1)$ **39.** equilateral hyperbola **41.** parabola

43. $x^2 + y^2 = 1$

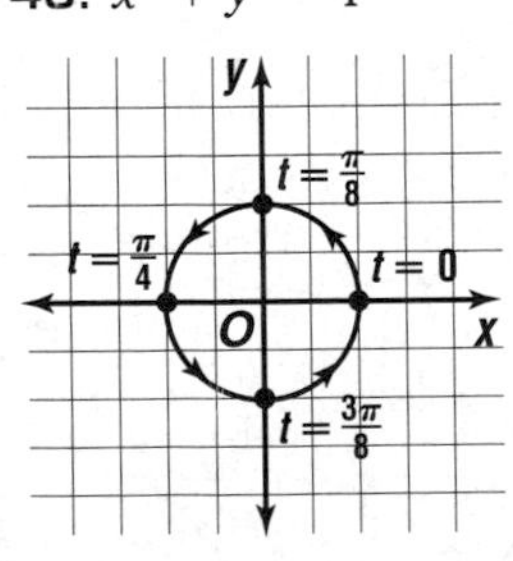

45. $y = \frac{x^2}{2} - 1$

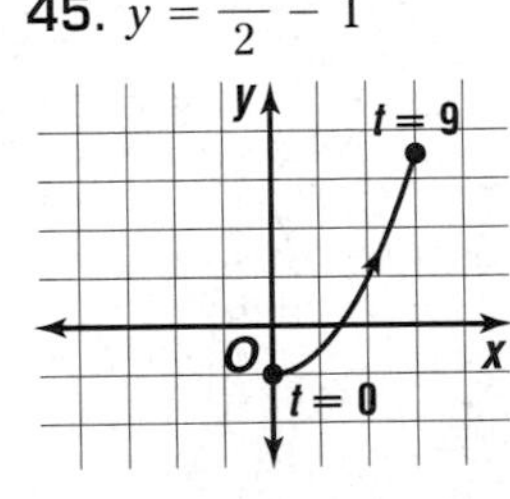

47. Sample answer: $x = 7 \sin t, y = 7 \cos t$, $0 \leq t \leq 2\pi$ **49.** Sample answer: $x = -t^2, y = t$, $-\infty < t < \infty$ **51.** parabola; $(x')^2 - 2x'y' + (y')^2 - 4\sqrt{2}x' - 4\sqrt{2}y' = 0$ **53.** ellipse; $-30°$

55. $(3, 3)$, $(-1, -1)$

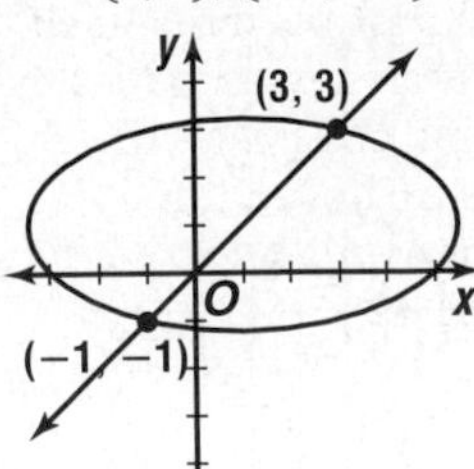

57. $(0, -1)$ $(4, -1)$

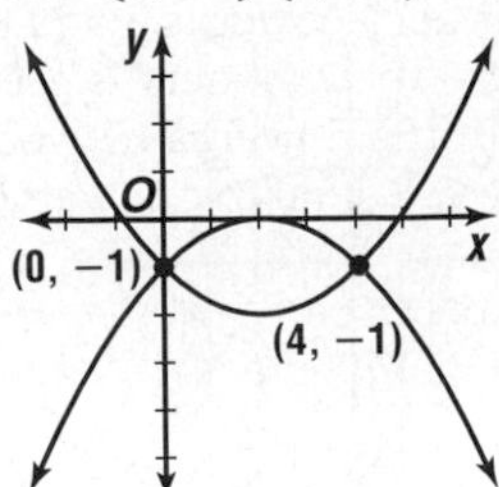

59.

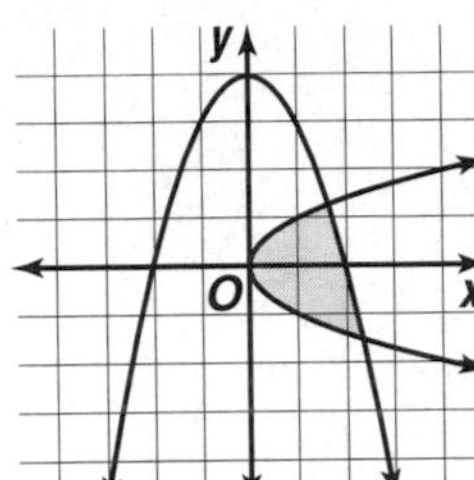

61.

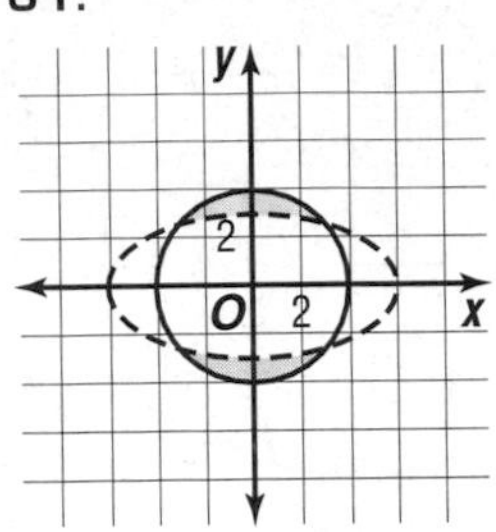

63a. $x^2 + y^2 = 400$ **63b.** about 37% **65.** about 1.8 feet from the center

Page 693 Chapter 10 SAT and ACT Practice
1. C **3.** A **5.** A **7.** D **9.** A

Chapter 11 Exponential and Logarithmic Functions

Pages 700–703 Lesson 11-1
5. $\frac{256}{81}$ **7.** 9 **9.** $81a^{-1}$ or $\frac{81}{a}$ **11.** $2^{2n+3}\sqrt{2^{n+1}}$ **13.** $13x^{\frac{5}{2}}$ **15.** $\sqrt[4]{6b^3c}$ **17.** $pq^2r\sqrt[3]{pr^2}$ **19.** 4.717×10^{-13} m^2 **21.** $-\frac{1}{1296}$ **23.** 32 **25.** $\frac{9}{4}$ **27.** 9 **29.** $2\sqrt{6}$ **31.** $\frac{1}{2}$ **33.** 36 **35.** $\frac{1}{16}$ **37.** 1 **39.** $3pq^2r^{-\frac{1}{3}}$ **41.** $6|x|^3$ **43.** $\frac{\sqrt{n}}{2}$ **45.** $4f^4|g||h|^{-1}$ or $\frac{4f^4|g|}{|h|}$ **47.** $6x^{\frac{1}{2}}y$ **49.** $|m|^3n^{\frac{1}{2}}$ **51.** $2xy^2$ **53.** $a^2b^{\frac{2}{5}}|c|^{\frac{1}{2}}$ **55.** $\sqrt[5]{16}$ **57.** $\sqrt[6]{p^4q^3r^2}$ **59.** $13\sqrt[21]{a^3b^7}$ **61.** -0.69 **63.** $ab^2\sqrt[3]{a^2bc}$ **65.** 0.17 **67.** 3.79 **69a.** $0 < y < 1$ **69b.** $1 < y < 3$ **69c.** $y > 3$ **69d.** If the exponent is less than 0, the power is greater than 0 and less than 1. If the exponent is greater than 0 and less than 1, the power is greater than 1 and less than the base. If the exponent is greater than 1, the power is greater than the base. Any number to the zero power is 1. Thus, if the exponent is less than zero, the power is less than 1. A power of a positive number is never

negative, so the power is greater than 0. Any number to the zero power is 1 and to the first power is itself. Thus, if the exponent is greater than zero and less than 1, the power is between 1 and the base. Any number to the first power is itself. Thus, if the exponent is greater than 1, the power is greater than the base. **71.** 2, −6 **73a.** 42,250,474.31 m **73b.** 35,870 km

75.

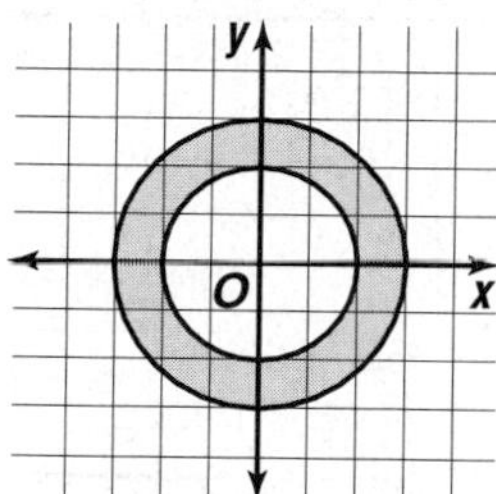

77. $1.31 + 0.14i$ **79.** about 4.43 s **81.** Sample answer: $\sin S = \frac{1}{2}$ **83.** 25π m/h **85.** 3; −5, 0, 5 **87.** E

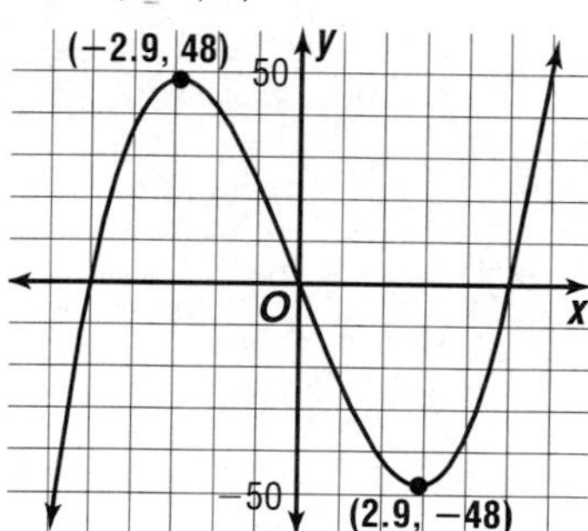

Pages 708–711 Lesson 11-2

5.

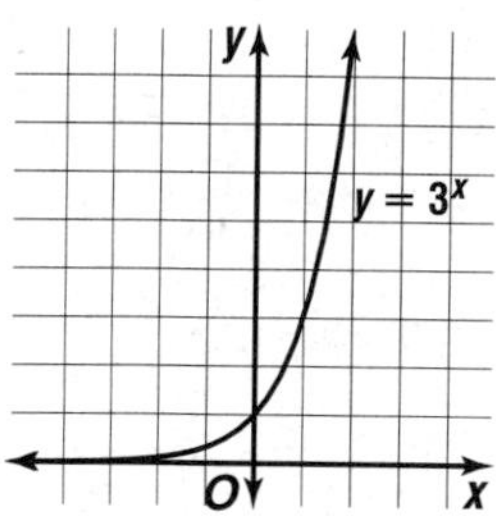

7.

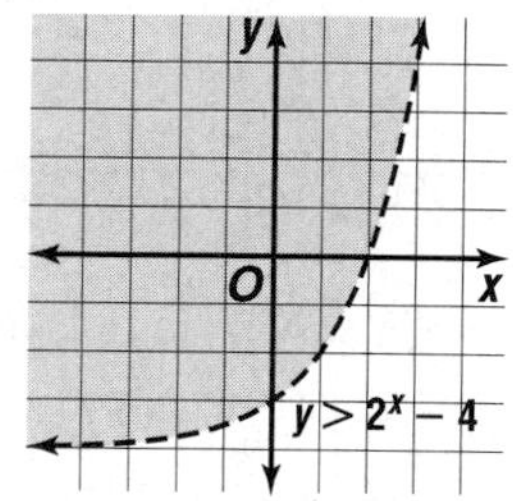

9a. 0.45% **9b.** 9,695,766

11.

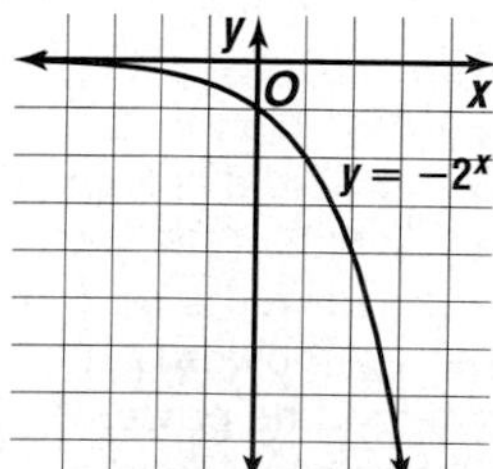

13.

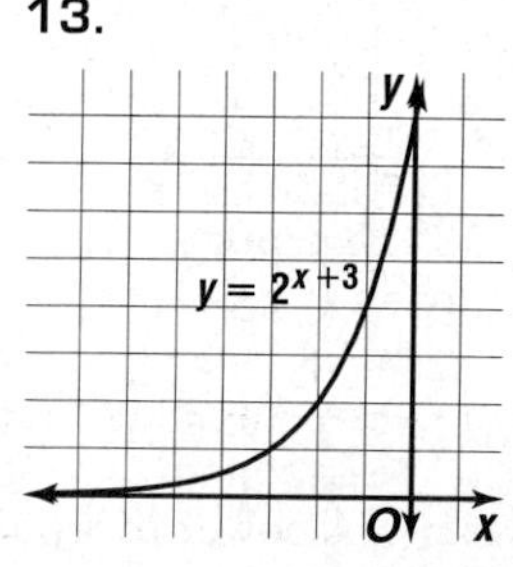

15.

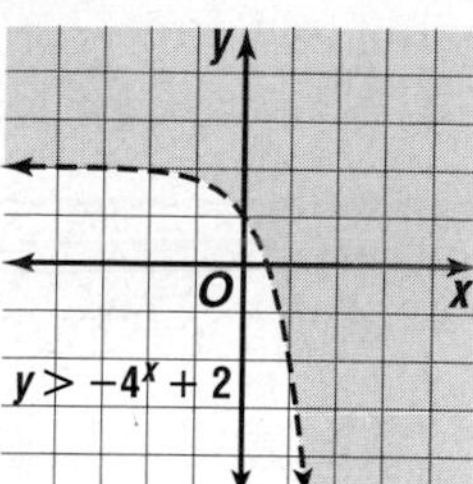

17.

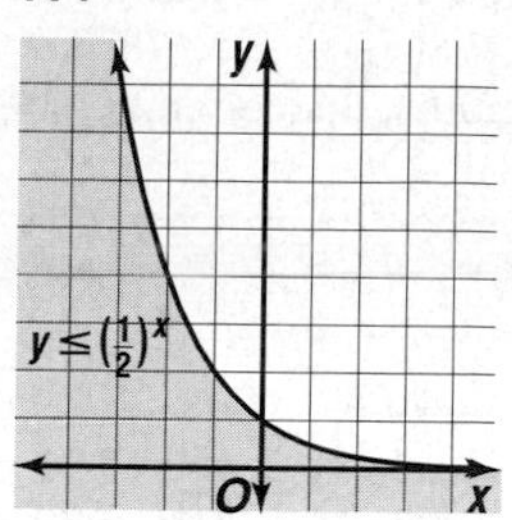

19. B **21.** A **23a.** The graph of $y = 6^x + 4$ is shifted up four units from the graph of $y = 6^x$. **23b.** The graph of $y = -3^x$ is a reflection of the graph of $y = 3^x$ across the x-axis. **23c.** The graph of $y = 7^{-x}$ is a reflection of the graph of $y = 7^x$ across the y-axis. **23d.** The graph of $y = \left(\frac{1}{2}\right)^x$ is a reflection of the graph of $y = 2^x$ across the y-axis. **25a.** $y = (0.85)^x$

25b.

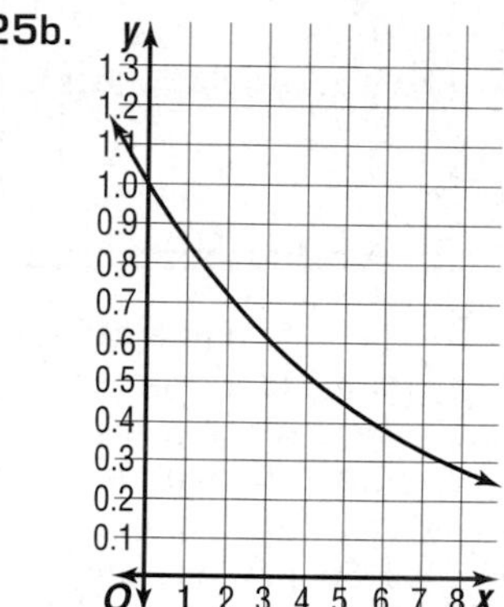

25c. 14% **25d.** No; the graph has an asymptote at $y = 0$, so the percent of impurities y will never reach 0. **27a.** 2700 units **27b.** 5800 units **29a.** $535,215.92 **29b.** $76,376.20 **31a.** $50; $50.63; $50.94; $51.16; $51.26 **31b.** Money Market Savings **31c.** 4.88% **33.** $15 = r \sin \theta$

35. $\sqrt{3}\left(\cos \frac{5\pi}{8} + i \sin \frac{5\pi}{8}\right)$; $-0.66 + 1.60i$

37a.

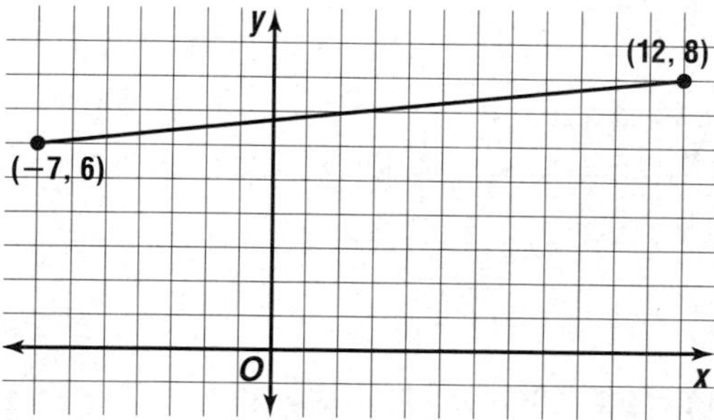

37b. (2.5, 7) **39.** 139,000 cm/s **41.** Sample answer: $y = 948.4x + 4960.6$ **43.** E

Pages 714–717 Lesson 11-3

7. \$25,865.41 **9a.** 78.7°F **9b.** Too cold; after 5 minutes, his coffee will be about 90°F.

11a.

Interest Compounded	Interest	Effective Annual Yield
Annually	\$80.00	8%
Semi-annually	\$81.60	8.16%
Quarterly	\$82.43	8.243%
Monthly	\$83.00	8.3%
Daily	\$83.28	8.328%
Continually	\$83.29	8.329%

11b. continuously **11c.** $E = \left(1 + \frac{r}{n}\right)^n - 1$

11d. $E = e^r - 1$ **13a.** 95% **13b.** about 1.2 min

15a. 20.9%; 60.9%; 98.5%

15b. about 29 days

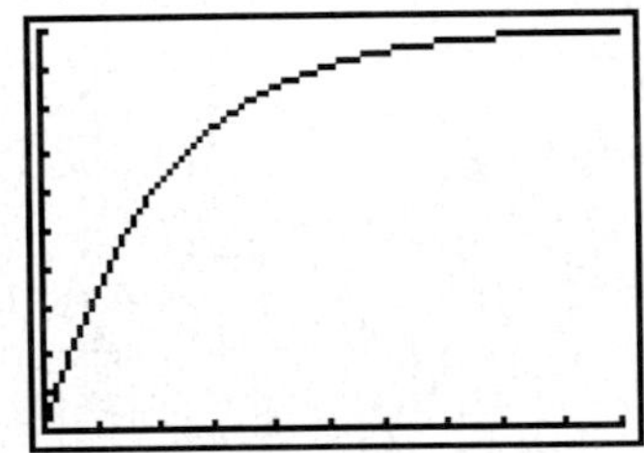

15c. Sample answer: The probability that a person who is going to respond has responded approaches 100% as t approaches infinity. New ads may be introduced after a high percentage of those who will respond have responded. The graph appears to level off after about 50 days. So, new ads can be introduced after an ad has run about 50 days.

17. \$13,257.20 **19.** $6x^2 + 12xy + 6y^2 + \sqrt{2}x - \sqrt{2}y = 0$ **21.** 704.2 ft · lb **23.** $\frac{13}{2}$ **25.** $J'(-9, -6)$, $K'(-6, 18)$, $L'(6, 15)$, $M'(9, -3)$; the dilated image has sides that are 3 times the length of the original figure. **27.** $\{-4, 2, 5\}$; $\{5, 7\}$; yes

Pages 722–724 Lesson 11-4

7. $\left(\frac{1}{25}\right)^{-\frac{1}{2}} = 5$ **9.** $\log_8 \frac{1}{4} = -\frac{2}{3}$ **11.** −2 **13.** 32

15. 15

17.

$y = \log_{\frac{1}{2}} x$

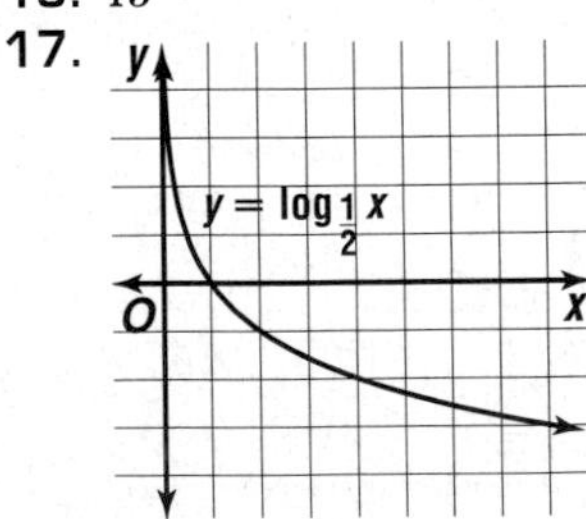

19. 264 h **21.** $16^{\frac{1}{2}} = 4$ **23.** $4^{\frac{5}{2}} = 32$

25. $(\sqrt{6})^4 = 36$ **27.** $\log_{36} 216 = \frac{3}{2}$

29. $\log_6 \frac{1}{36} = -2$ **31.** $\log_x 14.36 = 1.238$ **33.** $\frac{1}{3}$

35. 3.5 **37.** 1.5 **39.** 8 **41.** 7 **43.** 6 **45.** 3

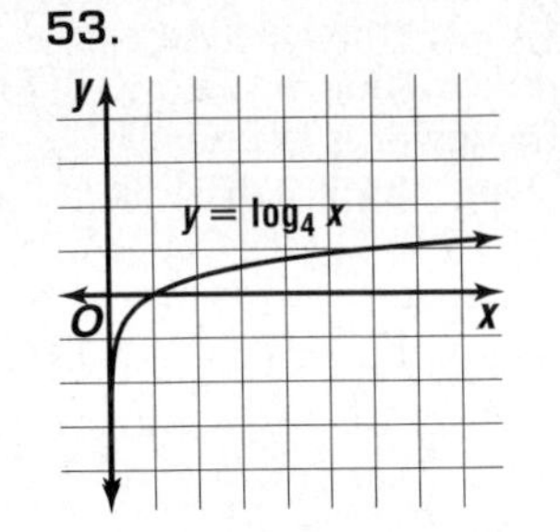

47. $\frac{1}{3}$ **49.** 4 **51.** 32

53.

$y = \log_4 x$

55.

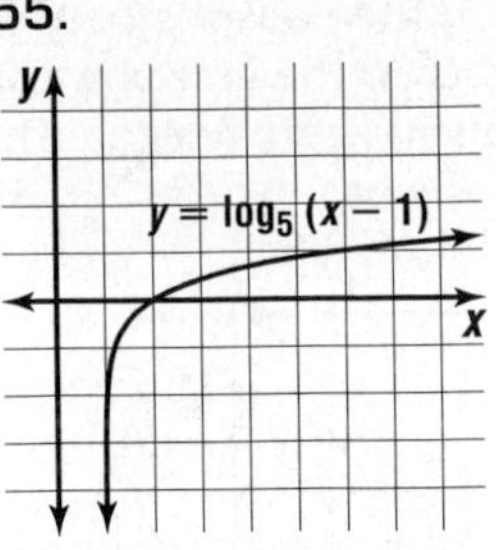

57.

$y \geq 2 \log_2 x$

59. 90 min

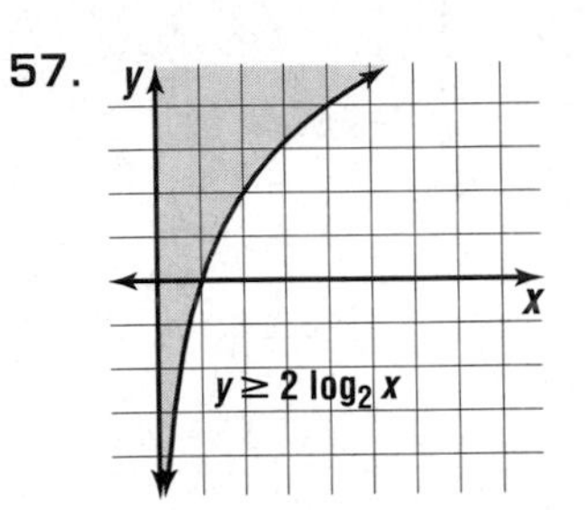

61. Let $\log_b m = x$ and $\log_b n = y$.
So, $b^x = m$ and $b^y = n$.

$\frac{m}{n} = \frac{b^x}{b^y} = b^{x-y}$

$\frac{m}{n} = b^{x-y}$

$\log_b \frac{m}{n} = x - y$

$\log_b \frac{m}{n} = \log_b m - \log_b n$

63a. 2 **63b.** less light; $\frac{1}{8}$

65a.

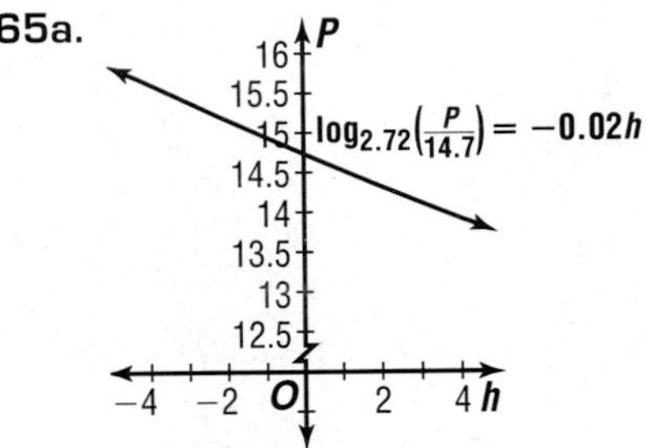

65b. 14.4 psi **65c.** 16.84 psi **67.** 69.6164

69. ellipse, $\frac{(x-1)^2}{4} + \frac{(y-2)^2}{9} = 1$

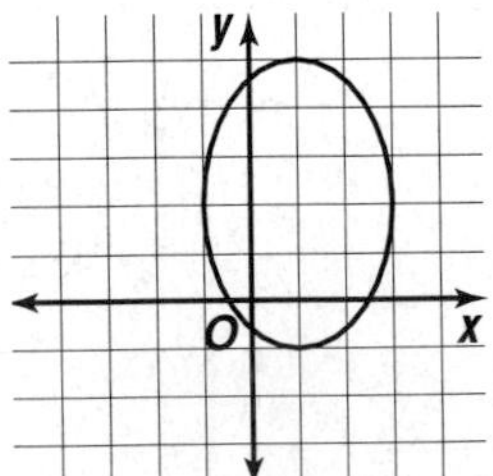

71. $AB = 6, BC = 5, AC = 5$ **73.** $64 - 27j$ volts

75. $\frac{31}{481}$ **77.** $c = 9.5, A = 38°20', B = 36°22'$

Pages 730–732 Lesson 11-5

5. 4.9031 **7.** −2.0915 **9.** 74,816.95 **11.** 1.1632 **13.** 7.83 **15.** $x < 2.97$ **17.** 5.5850 **19.** 5.6021 **21.** 0.0792 **23.** 1.5563 **25.** −2.3188 **27.** 3.2553 **29.** 2.9515 **31.** 2.001 **33.** 2.1745 **35.** 4 **37.** 0.7124 **39.** −3.9069 **41.** 18.6377 **43.** 0.3434 **45.** 0.2076 **47.** $1 < x < 6$ **49.** $x \geq 3.8725$ **51.** $x < 3.6087$

53.

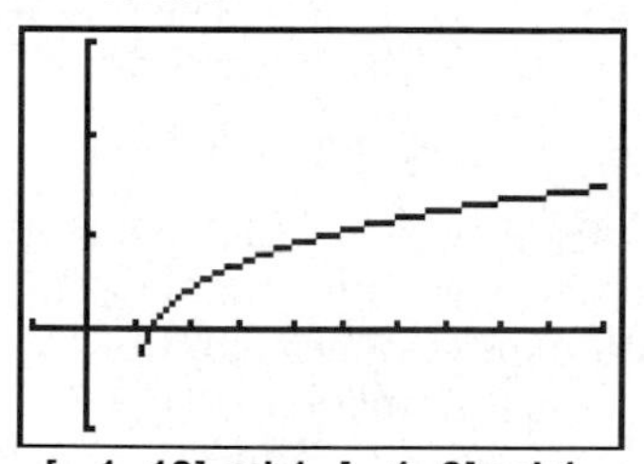

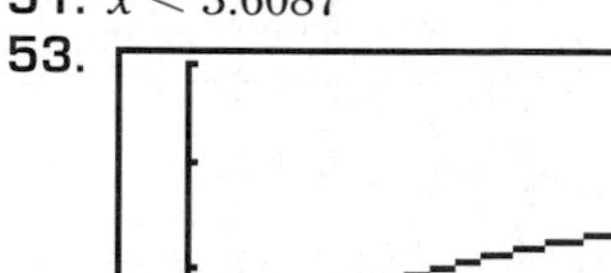

[−1, 10] scl:1, [−1, 3] scl:1

55. 0.3210 **57.** 2 **59a.** 1.58 **59b.** 0.0219 **61.** Sample answer: x is between 2 and 3 because 372 is between 100 and 1000, and log 100 = 2 and log 1000 = 3. **63.** 3819 yr **65.** 3 **67.** $a\sqrt[3]{ab^2c^2}$ **69.** $(2\sqrt{5}, -11)$

71.

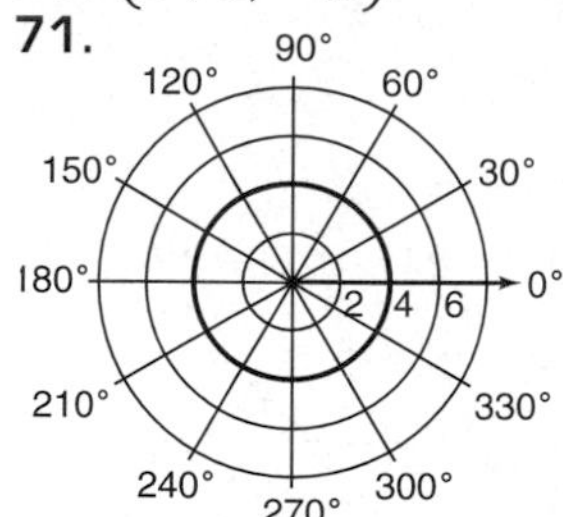

73. 31.68 cm^2 **75.** Neither; the graph of the function is not symmetric with respect to either the origin or the y-axis.

Pages 735–737 Lesson 11-6

5. −4.7217 **7.** −1.5606 **9.** 3.0339 **11.** 0.9635 **13.** $x < 1.3863$ **15.** 13.57

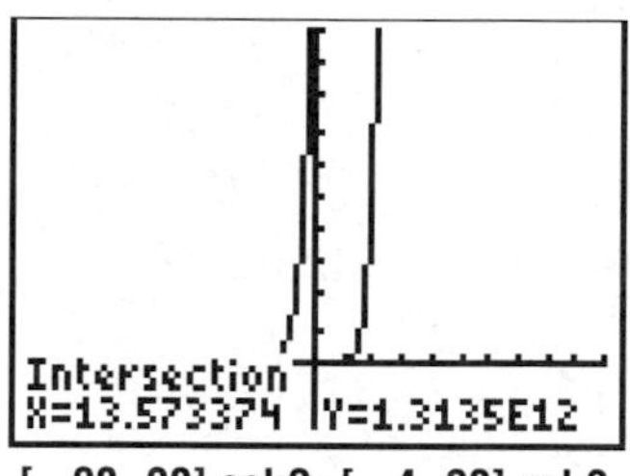

[−20, 20] scl:2, [−4, 20] scl:2

17a. 503.1 torrs **17b.** 4.2 km **19.** −0.2705 **21.** 0.9657 **23.** 2.2322 **25.** 1.2134 **27.** 0.9966 **29.** 0.2417 **31.** 2.2266 **33.** −0.3219 **35.** 1.7593 **37.** 4.7549 **39.** 1.3155 **41.** $40.9933 < t$ **43.** −0.3466 **45.** $x \leq 1.7657$ **47.** $x \geq 144.9985$ **49.** −7.64

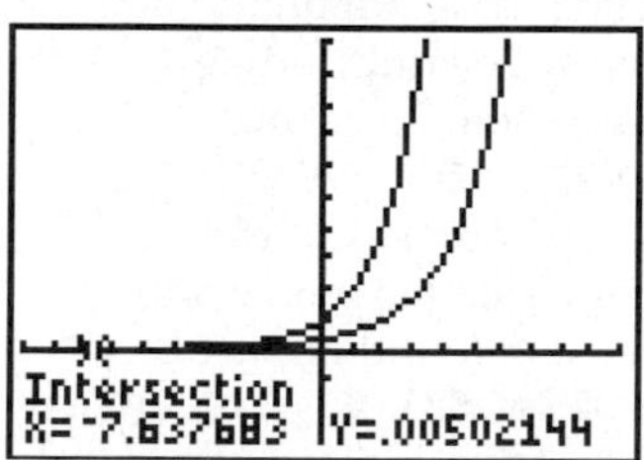

[−10, 10] scl:1, [−3, 10] scl:1

51. 2.14

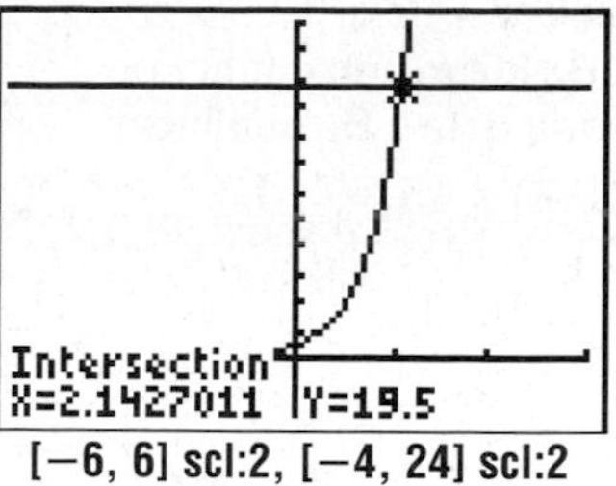

[−6, 6] scl:2, [−4, 24] scl:2

53. $x \geq 0.37$

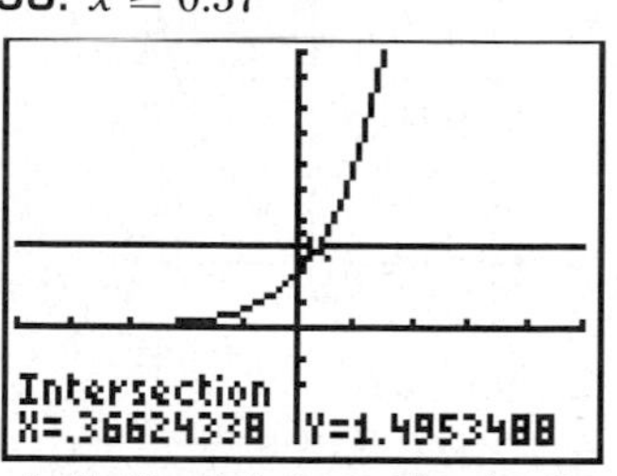

[−5, 5] scl:1, [−2, 5] scl:0.5

55. 324 hr **57.** 0 or −1.0986 **59.** ≈ 70%

61. y is a logarithmic function of x. The pattern in the table can be determined by $3^y = x$ which can be expressed as $\log_3 x = y$. **63.** $16^{\frac{3}{4}} = 8$ **65.** 0.00765 N · m **67.** $\langle 13, 7 \rangle$ **69.** $y = \pm 70 \cos 4\theta$

Pages 744–748 Lesson 11-7

5. 8.66 yr **7.** 30.81 yr **9.** 9.73 yr **11.** logarithmic; the graph has a vertical asymptote **13.** exponential; the graph has a horizontal asymptote **15a.** $y = 1.0091(0.9805)^x$ **15b.** $y = 1.0091e^{-0.0197x}$ **15c.** 35.10 min **17.** $y = 40 + 14.4270 \ln x$ **19.** Take the square root of each side.

21a.

x	0	50	100	150	190
$\ln y$	1.81	2.07	3.24	3.75	4.25

21b. $\ln y = 0.0137x + 1.6833$ **21c.** $y = e^{0.0137x + 1.6833}$ **21d.** 117.4 persons per square mile **23a.** $\ln y$ is a linear function of $\ln x$. **23b.** The result of part **a** indicates that we should take the natural logarithms of both the x- and y-values.

$\ln x$	6.21	6.91	8.52	9.21	9.62
$\ln y$	4.49	4.84	5.65	5.99	6.19

23c. $\ln y = 0.4994 \ln x + 1.3901$ **23d.** $y = 4.0153x^{0.4994}$ **25.** 0.01 **27a.** $11.50 **27b.** $2645 **29.** about 109.6 ft **31.** 4 units left and 8 units down **33.** C

Pages 749–753 Chapter 11 Study Guide and Assessment

1. common logarithm **3.** logarithmic function **5.** mantissa **7.** linearizing data **9.** nonlinear regression **11.** 16 **13.** 81 **15.** $\frac{1}{3}$ **17.** $\frac{1}{8}x^{12}$ **19.** $2a^3b$

21.

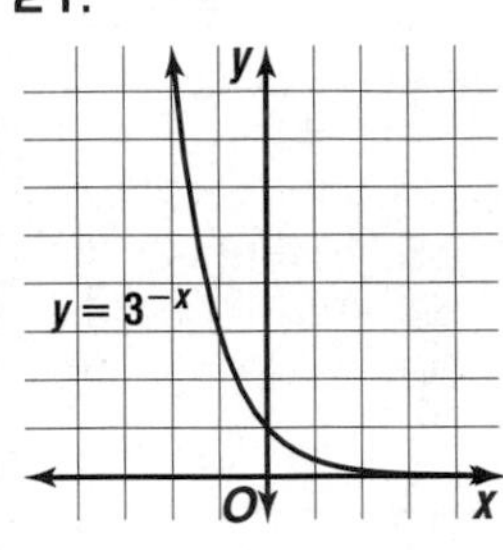

23.

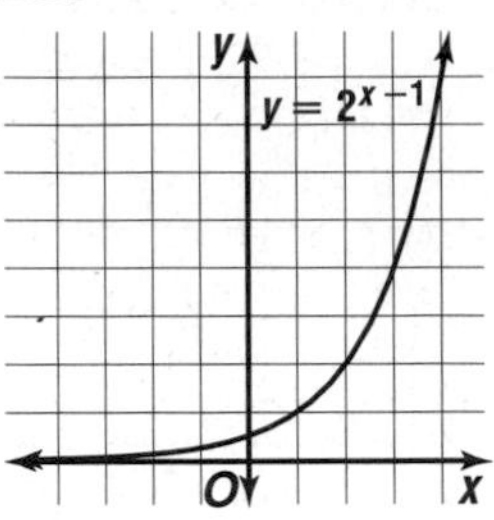

25.

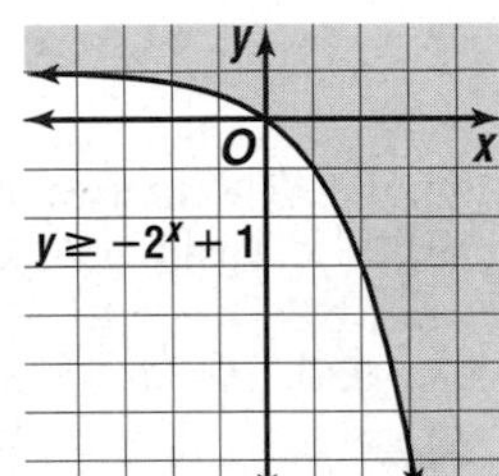

27. \$4788.85 **29.** \$21,647.86 **31.** $3^{-4} = \frac{1}{81}$ **33.** $\log_5 \frac{1}{25} = -2$ **35.** -3

37. -1 **39.** -1 **41.** 3 **43.** 16 **45.** 8

47.

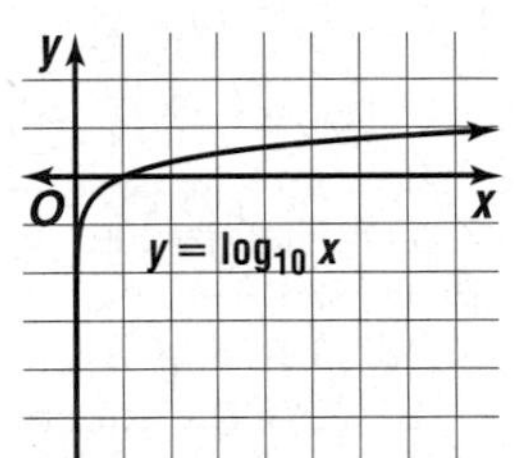

49. -3.5229 **51.** -1.8539 **53.** -8.04 **55.** $x \leq -4$

57. y = 3 log (x − 2)

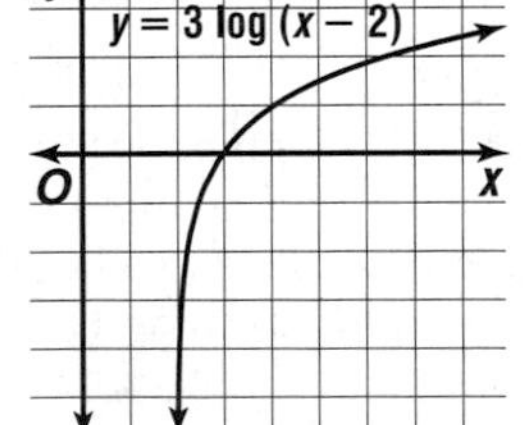

59. -3.42 **61.** 1.5283 **63.** 1.7829

65. 3.8982 **67.** -0.8967 **69.** $x \geq 2.5903$ **71.** $x < 2.20$ **73.** 13.52 **75.** 3561 yr **77.** 2014

Page 755 Chapter 11 SAT and ACT Practice

1. B **3.** E **5.** B **7.** D **9.** C

Chapter 12 Sequences and Series

Page 763–765 Lesson 12-1

7. 9, 17, 25, 33 **9.** -38 **11.** 15 **13.** 9, 14, 19, 24 **15.** 21 **17.** $-13, -19, -25, -31$ **19.** 7.5, 9, 10.5, 12 **21.** $b + 12, b + 16, b + 20, b + 24$ **23.** $-13n, -19n, -25n, -31n$ **25.** $2a + 16, 2a + 23, 2a + 30, 2a + 37$ **27.** 80 **29.** 13 **31.** 80 **33.** 4 **35.** $17 + \sqrt{5}$ **37.** -42.2 **39.** 12, 16.5, 21 **41.** $\sqrt{3}, \frac{12 + 2\sqrt{3}}{3}, \frac{24 + \sqrt{3}}{3}, 12$ **43.** -11 **45.** 1456 **47.** 7 **49.** $-8n + 14$ **51.** Let d be the common difference. Then, $y = x + d$, $z = x + 2d$, and $w = x + 3d$. Substitute these values into the expression $x + w - y$ and simplify. $x + (x + 3d) - (x + d) = x + 2d$ or z. **53.** 12 **55a.** 25 **55b.** 100 **55c.** Conjecture: The sum of the first n term of the sequence of natural numbers is n^2. Proof: Let $a_n = 2n - 1$. The first term of the sequence of natural numbers is 1, so $a_1 = 1$. Then, using the formula for the sum of an arithmetic series,

$$S_n = \frac{n}{2}(a_1 + a_n)$$
$$S_n = \frac{n}{2}[1 + (2n - 1)]$$
$$= \frac{n}{2}(2n) \text{ or } n^2$$

57. least: \$101, greatest: \$1001 **59.** \$285.77 **61.** $0.5\left(\cos \frac{\pi}{8} + \boldsymbol{i} \sin \frac{\pi}{8}\right)$, $0.46 + 0.19\boldsymbol{i}$ **63.** $\sqrt{3}x + y - 10 = 0$ **65.** $A = 70°28'$, $a = 4.2$, $b = 1.5$ **67.** $y = x - 1$ **69.** C

Pages 771–773 Lesson 12-2

7. 6; 144, 864, 5184 **9.** -4; $-115.2, 460.8, -1843.2$ **11.** $\frac{3}{2}$ **13.** 1, 3, 9, 27 **15.** \$28,211.98; \$39,795.78; \$79,185.19 **17.** -2.5; $-125, 312.5, -781.25$ **19.** $\frac{2}{5}$; $\frac{6}{125}, \frac{12}{625}, \frac{24}{3125}$ **21.** $\sqrt{2}$; $12, 12\sqrt{2}, 24$ **23.** $\boldsymbol{i}$; $1, \boldsymbol{i}, -1$ **25.** $\frac{1}{ab}, \frac{1}{a^3}, \frac{b}{a^5}, \frac{b^2}{a^7}, \frac{b^3}{a^9}$ **27.** $-\frac{243}{2048}$ **29.** $16\sqrt{5}$ **31.** $8\sqrt{2}$ **33.** 200, 40, 8 **35.** $-2, 6, -18, 54$ **37.** $\frac{605}{3}$ **39.** $-\frac{11{,}605}{512}$ **41a.** $b_t = b_0 \cdot 2^{2t}$ **41b.** 30,720 **41c.** Sample answer: It is assumed that favorable conditions are maintained for the growth of the bacteria, such as an adequate food and oxygen supply, appropriate surrounding temperature, and adequate room for growth. **43a.** \$11.79, \$30.58, \$205.72 **43b.** \$7052.15 **43c.** Each payment made is rounded to the nearest penny, so the sum of the payments will actually be more than the sum found in part **b**. **45.** $a_n = (-2)(-3)^{n-1}$ **47a.** \$25.05 **47b.** No. At the end of two years, she will have only \$615.23 in her account. **47c.** \$30.54 **49.** 13 weeks

51.

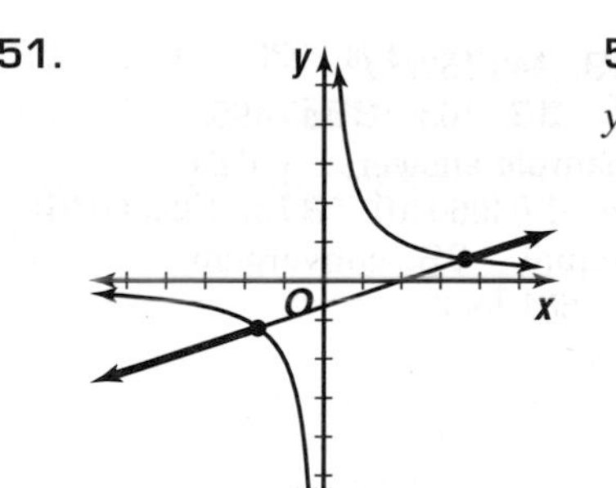

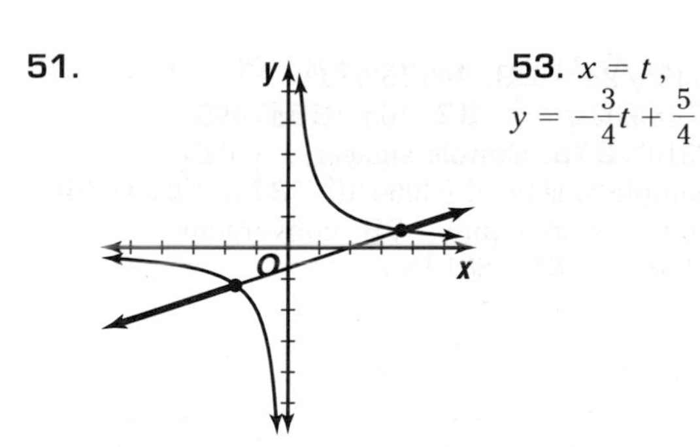

53. $x = t$, $y = -\frac{3}{4}t + \frac{5}{4}$

55. $y = 25 \sin\left(\frac{\pi}{2}t - 3.14\right) + 61$ **57.** 6

Pages 780–783 Lesson 12-3

5. 0; as $n \to \infty$, 5^n becomes increasingly large and thus the value $\frac{1}{5^n}$ becomes smaller and smaller, approaching zero. So the sequence has a limit of zero.

7. $\frac{3}{7}$; $\lim_{n\to\infty} \frac{3n-6}{7n} = \lim_{n\to\infty}\left(\frac{3}{7} - \frac{6}{7}\cdot\frac{1}{n}\right)$

$$= \lim_{n\to\infty}\frac{3}{7} - \lim_{n\to\infty}\frac{6}{7}\cdot\lim_{n\to\infty}\frac{1}{n}$$

$$= \frac{3}{7} - \frac{6}{7}\cdot 0 \text{ or } \frac{3}{7}$$

9. $5\frac{14}{111}$ **11.** $1\frac{1}{8}$ **13.** 125 m **15.** does not exist; simplifying the limit, we find that

$\lim_{n\to\infty}\frac{n^3-2}{n^2} = \lim_{n\to\infty}\left(n - \frac{2}{n}\right)$. $\lim_{n\to\infty}\frac{2}{n} = \lim_{n\to\infty} 2\cdot\frac{1}{n} = 2\cdot 0$ or 0, but as n approaches infinity, n becomes increasingly large, so the sequence has no limit.

17. $\frac{9}{2}$; $\lim_{n\to\infty}\frac{9n^3+5n-2}{2n^3} = \lim_{n\to\infty}\left(\frac{9}{2} + \frac{5}{2n^2} - \frac{1}{n^3}\right)$

$$= \lim_{n\to\infty}\frac{9}{2} + \lim_{n\to\infty}\frac{5}{2}\cdot\lim_{n\to\infty}\frac{1}{n^2} - \lim_{n\to\infty}\frac{1}{n^3}$$

$$= \frac{9}{2} + \frac{5}{2}\cdot 0 - 0 \text{ or } \frac{9}{2}$$

19. Does not exist; dividing by the highest powered term, n^2, we find $\lim_{n\to\infty}\frac{8 + \frac{5}{n} + \frac{2}{n^2}}{\frac{3}{n^2} + \frac{2}{n}}$ which as n approaches infinity simplifies to $\frac{8+0+0}{0+0} = \frac{8}{0}$. Since this fraction is undefined, the limit does not exist.

21. 0; as $n \to \infty$, 3^n becomes increasingly large and thus the value $\frac{1}{3^n}$ becomes smaller and smaller, approaching zero. So the sequence has a limit of zero.

23. 0,

$$\lim_{n\to\infty}\frac{5n+(-1)^n}{n^2} = \lim_{n\to\infty}\frac{5n}{n^2} + \lim_{n\to\infty}\frac{(-1)^n}{n^2}$$

$$= \lim_{n\to\infty}\frac{5}{n} + \lim_{n\to\infty}\frac{(-1)^n}{n^2}$$

$$= \lim_{n\to\infty}\frac{(-1)^n}{n^2}$$

As n increases, the value of the numerator alternates between -1 and 1. As n approaches infinity, the value of the denominator becomes increasingly large, causing the value of the fraction to become increasingly small. Thus the terms of the sequence alternate between smaller and smaller positive and negative values, approaching zero. So the sequence has a limit of zero.

25. $\frac{17}{33}$ **27.** $6\frac{7}{27}$ **29.** $\frac{29}{110}$ **31.** 64 **33.** 20

35. Does not exist; this series is geometric with a common ratio of 2. Since this ratio is greater than 1, the sum of the series does not exist. **37.** $3\frac{3}{5}$

39. $32 - 16\sqrt{3}$ **41a.** The limit of a difference equals the difference of the limits only if the two limits exist. Since neither $\lim_{n\to\infty}\frac{n^2}{2n+1}$ nor $\lim_{n\to\infty}\frac{n^2}{2n-1}$ exists, this property of limits does not apply.

41b. $-\frac{1}{2}$ **43.** No; if n is even, $\lim_{n\to\infty}\cos\frac{n\pi}{2} = \frac{1}{2}$, but if n is odd, $\lim_{n\to\infty}\cos\frac{n\pi}{2} = -\frac{1}{2}$. **45a.** $10\sqrt{2}$ ft

45b. $40 + 20\sqrt{2}$ ft or about 68 ft **47.** $-2, -1\frac{1}{3}, -\frac{8}{9}, -\frac{16}{27}$ **49.** $(6, -2)$; $(6 \pm \sqrt{5}, -2)$; $(8, -2)$, $(4, -2)$; $y = -\frac{1}{2}x + 5$, $y = \frac{1}{2}x - 1$ **51.** 42.75 miles, 117.46 miles

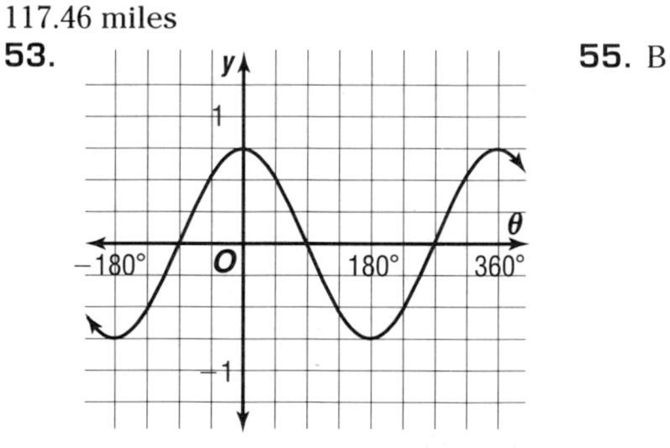

53. **55.** B

Pages 791–793 Lesson 12-4

5. convergent **7.** divergent **9.** convergent **11.** convergent **13.** convergent **15.** divergent **17.** convergent **19.** convergent **21.** convergent **23.** divergent **25.** convergent **27.** convergent **29.** divergent **31a.** No, MagicSoft let a_1 = 1,000,000 to arrive at their figure. The first term of this series is 1,000,000 · 0.70 or 700,000.

31b. \$2.3 million **33a.** Culture A: 1400 cells, Culture B: 713 cells **33b.** Culture B; at the end of one month, culture A will have produced 6000 cells while culture B will have produced 9062 cells.

$S_k \Rightarrow 1^2 + 3^2 + 5^2 + \cdots + (2k-1)^2 = \frac{k(2k-1)(2k+1)}{3}$

$S_{k+1} \Rightarrow 1^2 + 3^2 + 5^2 + \cdots + (2k-1)^2 + (2k+1)^2 = \frac{k(2k-1)(2k+1)}{3} + (2k+1)^2$

$$= \frac{k(2k-1)(2k+1) + 3(2k+1)^2}{3}$$

$$= \frac{[k(2k-1) + 3(2k+1)](2k+1)}{3}$$

$$= \frac{(2k^2+5k+3)(2k+1)}{3}$$

$$= \frac{(2k+3)(k+1)(2k+1)}{3}$$

Apply the original formula for $n = k + 1$.

$$\frac{(k+1)[2(k+1)-1][2(k+1)+1]}{3} = \frac{(k+1)(2k+1)(2k+3)}{3}$$

The formula gives the same result as adding the $(k + 1)$ term directly. Thus if the formula is valid for $n = k$, it is also valid for $n = k + 1$. Since the formula is valid for $n = 1$, it is also valid for $n = 2$, $n = 3$, and so on indefinitely. Thus, the formula is valid for all positive integral values of n.

17. $S_n \Rightarrow 7^n + 5 = 6r$ for some integer r. Step 1: Verify that S_n is valid for $n = 1$. $S_1 \Rightarrow 7^1 + 5$ or 12. Since $12 = 6 \cdot 2$, S_n is valid for $n = 1$. Step 2: Assume that S_n is valid for $n = k$ and show that it is also valid for $n = k + 1$.

$S_k \Rightarrow 7^k + 5 = 6r$ for some integer r

$S_{k+1} \Rightarrow 7^{k+1} + 5 = 6t$ for some integer t

$$7^k + 5 = 6r$$
$$7(7^k + 5) = 7 \cdot 6r$$
$$7^{k+1} + 35 = 42r$$
$$7^{k+1} + 5 = 42r - 30$$
$$7^{k+1} + 5 = 6(7r - 5)$$

Thus, $7^{k+1} + 5 = 6t$, where $t = 7r - 5$ is an integer, and we have shown that if S_n is valid, then S_{k+1} is also valid. Since S_n is valid for $n = 1$, it is also valid for $n = 2$, $n = 3$, and so on indefinitely. Hence, $7^n + 5$ is divisible by 6 for all integral values of n.

19. $S_n \Rightarrow 5^n - 2^n = 3r$ for some integer r. Step 1: Verify that S_n is valid for $n = 1$. $S_1 \Rightarrow 5^1 - 2^1$ or 3. Since $3 = 3 \cdot 1$, S_n is valid for $n = 1$. Step 2: Assume that S_n is valid for $n = k$ and show that it is also valid for $n = k + 1$.

$S_k \Rightarrow 5^k - 2^k = 3r$ for some integer r

$S_{k+1} \Rightarrow 5^{k+1} - 2^{k+1} = 3t$ for some integer t

$$5^k - 2^k = 3r$$
$$5^k = 2^k + 3r$$
$$5^k \cdot 5 = (2^k + 3r)(2 + 3)$$
$$5^{k+1} = 2^{k+1} + 3(2^k) + 6r + 9r$$
$$5^{k+1} - 2^{k+1} = 2^{k+1} + 3(2^k) + 6r + 9r - 2^{k+1}$$
$$= 3(2^k) + 15r$$
$$= 3(2^k + 5r)$$

Thus, $5^{k+1} - 2^{k+1} = 3t$, where $t = 2^k + 5r$ is an integer, and we have shown that if S_n is valid, then S_{k+1} is also valid. Since S_n is valid for $n = 1$, it is also valid for $n = 2$, $n = 3$, and so on indefinitely. Hence, $5^n - 2^n$ is divisible by 3 for all integral values of n.

21. Step 1: Verify that the formula is valid for $n = 1$. Since $\frac{1}{2}$ is the first term in the sequence and $\frac{1}{1+1} = \frac{1}{2}$, the formula is valid for $n = 1$. Step 2: Assume that the formula is valid for $n = k$ and derive a formula for $n = k + 1$.

$S_k \Rightarrow \frac{1}{1 \cdot 2} + \frac{1}{2 \cdot 3} + \frac{1}{3 \cdot 4} + \cdots + \frac{1}{k(k+1)} = \frac{k}{k+1}$

$S_{k+1} \Rightarrow \frac{1}{1 \cdot 2} = \frac{1}{2 \cdot 3} + \frac{1}{3 \cdot 4} + \cdots + \frac{1}{k(k+1)} + \frac{1}{(k+1)(k+2)} = \frac{k}{k+1} + \frac{1}{(k+1)(k+2)}$

$$= \frac{k(k+2)+1}{(k+1)(k+2)}$$
$$= \frac{k^2+2k+1}{(k+1)(k+2)}$$
$$= \frac{(k+1)^2}{(k+1)(k+2)}$$
$$= \frac{k+1}{k+2}$$

Apply the original formula for $n = k + 1$.

$$\frac{(k+1)}{(k+1)+1} = \frac{k+1}{k+2}$$

The formula gives the same result as adding the $(k + 1)$ term directly. Thus if the formula is valid for $n = k$, it is also valid for $n = k + 1$. Since the formula is valid for $n = 1$, it is also valid for $n = 2$, $n = 3$, and so on indefinitely. Thus, the formula is valid for all positive integral values of n.

23. Step 1: Verify that the formula is valid for $n = 1$. Since $S_1 \Rightarrow [r(\cos\theta + \boldsymbol{i}\sin\theta)]^1$ or $r(\cos\theta + \boldsymbol{i}\sin\theta)$ and $r^1[\cos(1)\theta + \boldsymbol{i}\sin(1)\theta] = r(\cos\theta + \boldsymbol{i}\sin\theta)$, the formula is valid for $n = 1$. Step 2: Assume that the formula is valid for $n = k$ and derive a formula for $n = k + 1$. That is, assume that $[r(\cos\theta + \boldsymbol{i}\sin\theta)]^k = r^k(\cos k\theta + \boldsymbol{i}\sin k\theta)$. Multiply each side of the equation by $r(\cos\theta + \boldsymbol{i}\sin\theta)$.

$[r(\cos\theta + \boldsymbol{i}\sin\theta)]^{k+1}$

$$= [r^k(\cos k\theta + \boldsymbol{i}\sin k\theta)] \cdot [r(\cos\theta + \boldsymbol{i}\sin\theta)]$$
$$= r^{k+1}[\cos k\theta\cos\theta + (\cos k\theta)(\boldsymbol{i}\sin\theta) + \boldsymbol{i}\sin k\theta\cos\theta + \boldsymbol{i}^2\sin k\theta\sin\theta]$$
$$= r^{k+1}[(\cos k\theta\cos\theta - \sin k\theta\sin\theta) + \boldsymbol{i}(\sin k\theta\cos\theta + \cos k\theta\sin\theta)]$$
$$= r^{k+1}[\cos(k+1)\theta + \boldsymbol{i}\sin(k+1)\theta]$$

When the original formula is applied for $n = k + 1$, the same result is obtained. Thus if the formula is valid for $n = k$, it is also valid for $n = k + 1$. Since the formula is valid for $n = 1$, it is also valid for

$n = 2$, $n = 3$ and so on indefinitely. Thus, the formula is valid for all positive integral values of n.

25. $S_1 \Rightarrow n^2 + 5n = 2r$ for some positive integer r. Step 1: Verify that S_1 is valid for $n = 1$. $S_1 \Rightarrow 1^2 + 5 \cdot 1$ or 6. Since $6 = 2 \cdot 3$, S_1 is valid for $n = 1$. Step 2: Assume that S_n is valid for $n = k$ and show that it is valid for $n = k + 1$.

$S_k \Rightarrow k^2 + 5k = 2r$ for some positive integer r

$S_{k+1} \Rightarrow (k + 1)^2 + 5(k + 1) = 2t$ for some positive integer t

$$\begin{aligned}(k + 1)^2 + 5(k + 1) &= k^2 + 2k + 1 + 5k + 5\\ &= (k^2 + 5k) + (2k + 6)\\ &= 2r + 2(k + 3)\\ &= 2(r + k + 3)\end{aligned}$$

Thus, if $k^2 + 5k = 2t$, where $t = r + k + 3$ is an integer, and we have shown that if S_n is valid, then S_{k+1} is also valid. Since S_n is valid for $n + 1$, it is also valid for $n = 2$, $n = 3$, and so on indefinitely. Hence, $n^2 + 5n$ is divisible by 2 for all positive integral values of n.

27. Step 1: Verify that $S_n \Rightarrow (x + y)^n = x^n + nx^{n-1}y + \frac{n(n-1)}{2!}x^{n-2}y^2 + \frac{n(n-1)(n-2)}{3!}x^{n-3}y^3 + \cdots + y^n$ is valid for n = 1. Since $S_1 \Rightarrow (x - y)^1 = x^1 = 1x^0y^1$ or $x + y$, S_n is valid for $n = 1$. Step 2: Assume that the formula is valid for $n = k$ and derive a formula for $n = k + 1$.

$$S_k \Rightarrow (x + y)^k = x^k + kx^{k-1}y + \frac{k(k-1)}{2!}x^{k-2}y^2 + \frac{k(k-1)(k-2)}{3!}x^{k-3}y^3 + \cdots + y^k$$

$$S_{k+1} \Rightarrow (x + y)^k(x + y) = (x + y)\left(x^k + kx^{k-1}y + \frac{k(k-1)}{2!}x^{k-2}y^2 + \frac{k(k-1)(k-2)}{3!}x^{k-3}y^3 + \cdots + y^k\right)$$

$$\begin{aligned}(x + y)^{k+1} &= x\left(x^k + kx^{k-1}y + \frac{k(k-1)}{2!}x^{k-2}y^2 + \frac{k(k-1)(k-2)}{3!}x^{k-3}y^3 + \cdots + y^k +\right.\\ &\qquad \left. y(x^k + kx^{k-1}y + \frac{k(k-1)}{2!}x^{k-2}y^2 + \frac{k(k-1)(k-2)}{3!}x^k - {}^3y^3 + \cdots + y^k\right)\\ &= x^{k+1} + kx^ky + \frac{k(k-1)}{2!}x^{k-1}y^2 + \cdots + xy^k + x^ky + kx^{k-1}y^2 + \frac{k(k-1)}{2!}x^{k-2}y^3 + \cdots + y^{k+1}\\ &= x^{k+1} + (k + 1)x^ky + kx^{k-1}y^2 + \frac{k(k-1)}{2!}x^{k-1}y^2 + \cdots + y^{k+1}\\ &= x^{k+1} + (k + 1)x^ky + \frac{k(k+1)}{2!}x^{k-1}y^2 + \cdots + y^{k+1}\end{aligned}$$

When the original formula is applied for $n = k + 1$, the same result is obtained. Thus if the formula is valid for $n = k$, it is also valid for $n = k + 1$. Since the formula is valid for $n = 1$, it is also valid for $n = 2$, $n = 3$, and so on indefinitely. Thus, the formula is valid for all positive integral values of n.

29. $8 - i$, $16 - i$, $32 - i$ **31.** $\frac{(x-2)^2}{4} + \frac{(y-5)^2}{25} = 1$; ellipse **33.** $y = \pm\frac{3}{4}\sin 2x$ **35.** B

Pages 829–833 Chapter 12 Study Guide and Assessment

1. d **3.** m **5.** k **7.** c **9.** b **11.** 6.9, 8.2, 9.5, 10.8 **13.** 6, 3.5, 1, −1.5, −4 **15.** 18 **17.** 36,044.8 **19.** 0.2, 1, 5, 25, 125 **21.** $62(1 + \sqrt{2})$ **23.** 6 **25.** 0 **27.** 2100 **29.** divergent **31.** $(3 \cdot 5 - 3) + (3 \cdot 6 - 3) + (3 \cdot 7 - 3) + (3 \cdot 8 - 3) + (3 \cdot 9 - 3)$ **33.** $\sum_{a=0}^{\infty}(2n - 1)$ **35.** $a^6 - 24a^5 + 240a^4 - 1280a^3 + 3840a^2 - 6144a + 4096$ **37.** $3360x^6$ **39.** $102{,}400m^6$ **41.** $2e^{i\frac{3\pi}{4}}$ **43.** $2\sqrt{2}e^{i\frac{7\pi}{4}}$ **45.** 0, 6, −12, 42 **47.** 4, $6 - 2i$, $7 - 3i$ **49.** $2 + i$, $5 - 1.5i$, $6.5 - 2.75i$

51. Step 1: Verify that the formula is valid for $n = 1$. Since the first term in the sequence is 1 and $\frac{1(1+1)}{2} = 1$, the formula is valid for $n = 1$. Step 2: Assume that the formula is valid for $n = k$ and derive a formula for $n = k + 1$.

$S_k \Rightarrow 1 + 2 + 3 + \cdots + k = \frac{k(k+1)}{2}$

$$\begin{aligned}S_{k+1} \Rightarrow 1 + 2 + 3 + \cdots + k + (k + 1) &= \frac{k(k+1)}{2} + \frac{2(k+1)}{2}\\ &= \frac{k^2 + k}{2} + \frac{2k + 2}{2}\\ &= \frac{k^2 + k + 2k + 2}{2}\\ &= \frac{k^2 + 3k + 2}{2}\\ &= \frac{(k+1)(k+2)}{2}\end{aligned}$$

Apply the original formula for $n = k + 1$.

$$\frac{(k+1)[(k+1)+1]}{2} = \frac{(k+1)(k+2}{2}$$

The formula gives the same result as adding the $(k + 1)$ term directly. Thus, if the formula is valid for $n = k$, it is also valid for $n = k + 1$. Since the formula is valid for $n = 1$, it is also valid for $n = 2$, $n = 3$, and so on indefinitely. Thus, the formula is valid for all positive integral values of n.

53. $S_n \Rightarrow 9^n - 4^n = 5r$ for some integer r. Step 1: Verify that S_n is valid for $n = 1$. $S_1 \Rightarrow 9^1 - 4^1$ or 5. Since $5 = 5 \cdot 1$, S_n is valid for $n = 1$. Step 2: Assume that S_n is valid for $n = k$ and show that it is also valid for $n = k + 1$.

$S_k \Rightarrow 9^k - 4^k = 5r$ for some integer r

$S_{k+1} \Rightarrow 9^{k+1} - 4^{k+1} = 5t$ for some integer t

$$
\begin{aligned}
9^k - 4^k &= 5r \\
9^k &= 4^k + 5r \\
9(9^k) &= (4^k + 5r)(4 + 5) \\
9^{k+1} &= 4^{k+1} + 5(4^k) + 20r + 25r \\
9^{k+1} - 4^{k+1} &= 4^{k+1} + 5(4^k) + 20r + 25r - 4^{k+1} \\
&= 5(4^k) + 45r \\
&= 5(4^k + 9r)
\end{aligned}
$$

Thus, $9^{k+1} - 4^{k+1} = 5t$, where $t = 4^k + 9r$ is an integer, and we have shown that if S_n is valid, then S_{k+1} is also valid. Since S_n is valid for $n = 1$, it is also valid for $n = 2$, $n = 3$, and so on indefinitely. Hence, $9^n - 4^n$ is divisible by 5 for all integral values of n.

55. \$117,987,860.30

Page 835 Chapter 12 SAT and ACT Practice

1. D **3.** A **5.** C **7.** D **9.** A

Chapter 13 Combinatorics and Probability

Pages 843–845 Lesson 13-1

5. 300 **7.** 720 **9.** 55,440 **11.** 15,504 **13.** 3,628,800 **15a.** 100,000 **15b.** 7290 **15c.** 999,900,000 **17.** 5040 **19.** dependent **21.** dependent **23.** 360 **25.** 840 **27.** 604,800 **29.** 6 **31.** 10 **33.** 6 **35.** 1 **37.** 168 **39.** 840 **41.** 2002 **43.** 420 **45a.** 22,308 **45b.** 144 **45c.** 792 **47.** 216,216 **49.** $P(n, n-1) \stackrel{?}{=} P(n, n)$

$$
\begin{aligned}
\frac{n!}{[n-(n-1)]!} &\stackrel{?}{=} \frac{n!}{(n-n)!} \\
\frac{n!}{1!} &\stackrel{?}{=} \frac{n!}{0!} \\
n! &= n!
\end{aligned}
$$

51a. 330 **52b.** 150 **53a.** 592

53b. Yes. Let h, t, and u be the digits.

$$
\begin{aligned}
&100h + 10t + u \\
&100h + 10u + t \\
&100t + 10h + u \\
&100t + 10u + h \\
&100u + 10t + h \\
+\ &100u + 10h + t
\end{aligned}
$$

$200(h + t + u) + 20(h + t + u) + 2(h + t + u) = 222(h + t + u)$

$\frac{222(h + t + u)}{6} = 37(h + t + u)$

55. 3025 **57.** 1.4 **59.** (2, 180°), (2, 0°) **61.** 0°, 180°, 360° **63.** $B = 63°$, $a = 7.7$, $c = 17.1$

Pages 849–851 Lesson 13-2

5. 22,680 **7.** circular; 3,628,800 **9.** circular; 39,916,800 **11.** 756 **13.** 907,200 **15.** 302,400 **17.** 3780 **19.** 126 **21.** circular; 39,916,800 **23.** circular; 40,320 **25.** circular; 5040 **27.** linear; 3,628,800 **29.** circular; 3,113,510,400 **31.** circular; $\approx 8.22 \times 10^{33}$ **33.** 2520 **35.** 46,200 **37a.** $\approx 7.85 \times 10^{17}$ **37b.** $\approx 1.41 \times 10^{51}$ **37c.** $\approx 6.04 \times 10^{52}$ **39.** 20 **41.** $x < 8.69$ **43.** $44 - 58\boldsymbol{i}$ **45.** about 3.31 inches

Pages 855–858 Lesson 13-3

5. $\frac{1}{3}$ **7.** 0 **9.** $\frac{4}{31}$ **11.** $\frac{18}{17}$ **13.** $\frac{3}{13}$ **15.** $\frac{5}{13}$ **17.** $\frac{1}{2}$ **19.** $\frac{7}{10}$ **21.** $\frac{1}{130}$ **23.** $\frac{3}{13}$ **25.** $\frac{1}{36}$ **27.** $\frac{2}{3}$ **29.** $\frac{11}{4}$ **31.** $\frac{22}{53}$ **33.** $\frac{92}{233}$ **35.** $\frac{4}{1}$ **37.** $\frac{1}{5}$ **39a.** $\frac{1}{720}$ **39b.** $\frac{999}{1}$ **41a.** $\frac{21}{1292}$ **41b.** $\frac{225}{421}$ **43.** $\frac{2}{9}$ **45.** 210 **47.** $2x$ **49.** $6\left(\cos\frac{5\pi}{4} + \boldsymbol{i}\sin\frac{5\pi}{4}\right)$, $-3\sqrt{2} - 3\sqrt{2}\boldsymbol{i}$ **51.** B

Pages 863–867 Lesson 13-4

5. dependent, $\frac{2}{15}$ **7.** exclusive, $\frac{10}{13}$ **9.** exclusive, $\frac{2}{13}$ **11.** ≈ 0.518 **13.** ≈ 0.032 **15.** $\frac{34}{39}$ **17.** independent, $\frac{25}{81}$ **19.** dependent, $\frac{2}{7}$ **21.** dependent, $\frac{19}{1{,}160{,}054}$ **23.** independent, $\frac{35}{1024}$ **25.** inclusive, $\frac{11}{36}$ **27.** inclusive, $\frac{8}{13}$ **29.** exclusive, $\frac{1}{4}$ **31.** exclusive, $\frac{11}{32}$ **33.** exclusive, $\frac{19}{33}$ **35.** $\frac{2}{221}$ **37.** $\frac{55}{221}$ **39.** $\frac{4}{7}$ **41.** $\frac{71}{210}$ **43.** $\frac{1}{21}$ **45.** $\frac{5}{7}$ **47.** $\frac{4}{5}$ **49.** $\frac{59}{143}$ **51.** $\frac{531}{1250}$ **53.** 0.93 **55a.** exclusive

55b.

Dead Battery

Flooded Engine

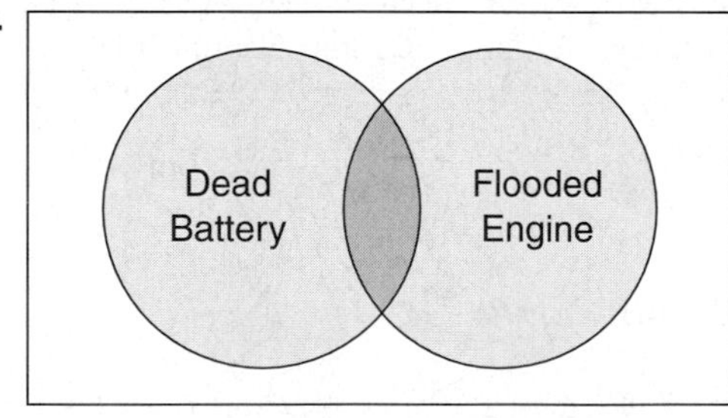

55c. $\frac{4}{5}$ **57.** 720 **59.** No, the spill will spread no more than 2000 meters away. **61.** \$11.50, \$2645 **63.** $\langle x - 1, y + 5\rangle = t\langle -2, -4\rangle$ **65.** B

Pages 871–874 Lesson 13-5

5. $\frac{1}{3}$ **7.** $\frac{1}{7}$ **9.** $\frac{1}{5}$ **11.** $\frac{2}{5}$ **13a.** $\frac{69}{70}$ **13b.** $\frac{2}{25}$ **13c.** $\frac{1}{25}$ **15.** $\frac{1}{2}$ **17.** $\frac{3}{5}$ **19.** $\frac{5}{8}$ **21.** $\frac{1}{13}$ **23.** 0 **25.** $\frac{2}{13}$ **27.** $\frac{5}{7}$ **29.** $\frac{2}{5}$ **31.** $\frac{3}{10}$ **33.** $\frac{1}{6}$ **35.** $\frac{1}{2}$ **37.** $\frac{19}{51}$

39. A = person buys something
B = person asks questions

$$P(A|B) = \frac{\frac{120}{500}}{\frac{150}{500}} \text{ or } \frac{4}{5}$$

Four out of five people who ask questions will make a purchase. Therefore, they are more likely to buy something if they ask questions. **41.** $\frac{5}{6}$

43. $P(A|B) = \frac{P(A \text{ and } B)}{P(B)}$ by definition. So, if $P(A) = P(A|B)$ then by substitution $P(A) = \frac{P(A \text{ and } B)}{P(B)}$ or $P(A \text{ and } B) = P(A) \cdot P(B)$. Therefore, the events are independent. **45.** 126 **47.** They are reflections of each other over the x-axis. **49.** $y = -5$ **51.** 54.7 ft^2 **53.** $\frac{1}{2}$ ft or 6 in. **55.** B

Pages 878–880 Lesson 13-6

5. $\frac{625}{648}$ **7.** $\frac{1}{7776}$ **9.** $\frac{1029}{2500}$ **11.** $\frac{768}{3125}$ **13.** $\frac{1}{81}$ **15.** $\frac{65}{81}$ **17.** $\frac{15}{128}$ **19.** $\frac{1}{1024}$ **21.** $\frac{1}{81}$ **23.** $\frac{11}{27}$ **25.** ≈ 0.058 **27.** $\approx 1.049 \times 10^{-4}$ **29.** ≈ 0.201 **31.** $\frac{1}{4}$ **33.** $\frac{3}{8}$ **35.** about 45% **37.** ≈ 0.0062 **39.** 0.807 **41a.** 0.246 **41b.** 0.246 **41c.** 0.41 **43.** $\frac{7}{26}$ **45.** 0.38 **47.** $0 - i\sqrt{2}$ **49.** about 101.1 cm and 76.9 cm **51.** 7/12

Pages 881–885 Chapter 13 Study Guide and Assessment

1. independent **3.** 1 **5.** permutation **7.** mutually exclusive **9.** conditional **11.** 6 **13.** 720 **15.** 20,160 **17.** 165 **19.** 3 **21.** 63 **23.** 50,400 **25.** 60 **27.** $\frac{1}{16}$ **29.** $\frac{1}{140}$ **31.** $\frac{1}{15}$ **33.** $\frac{1}{139}$ **35.** independent, $\frac{5}{1296}$ **37.** $\frac{9}{14}$ **39.** $\frac{1}{7}$ **41.** $\frac{2}{5}$ **43.** $\frac{2}{15}$ **45.** $\frac{1}{16}$ **47.** $\frac{5}{16}$ **49.** 2520 **51a.** $\frac{7}{15}$ **51b.** $\frac{1}{30}$

Page 887 Chapter 13 SAT and ACT Practice

1. D **3.** C **5.** E **7.** A **9.** B

Chapter 14 Statistics and Data Analysis

Pages 893–896 Lesson 14-1

5a.

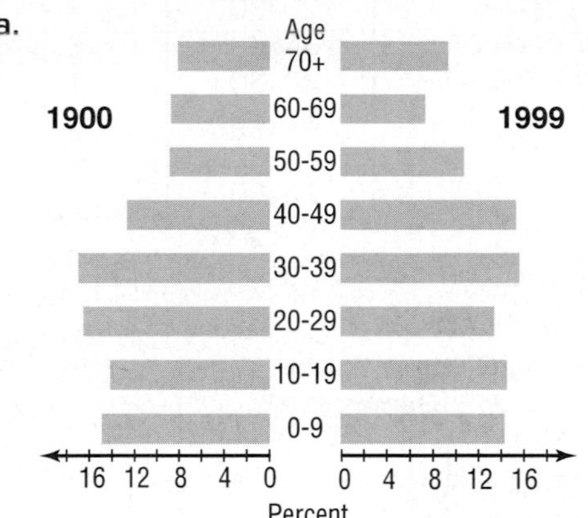

5b. In 1999, there are larger percents of older citizens than in 1990.

7a.

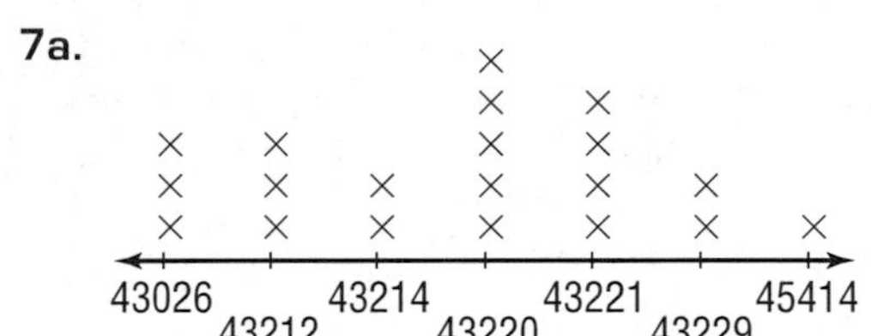

7b. 43220 **7c.** Sample answer: to determine where most of their customers live so they can target their advertising accordingly

9a.

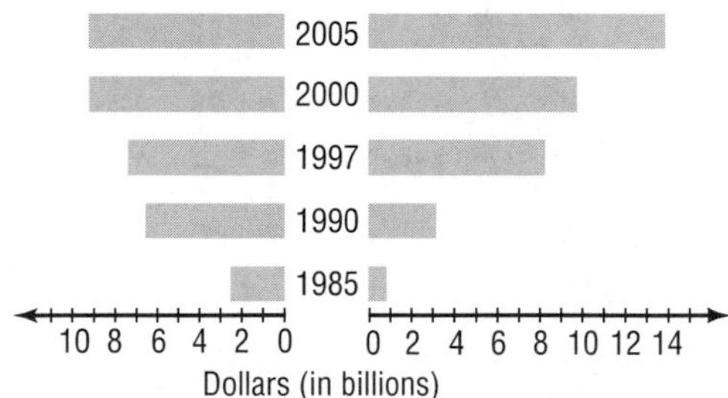

9b. Sales; the sales revenue is growing at a faster rate than the rental revenue. **11a.** 56 **11b.** Sample answer: 10 **11c.** Sample answer: 10, 20, 30, 40, 50, 60, 70, 80 **11d.** Sample answer: 15, 25, 35, 45, 55, 65, 75

11e. Sample answer:

Number of Nations	Frequency
10–20	2
20–30	3
30–40	8
40–50	1
50–60	1
60–70	2
70–80	1

11f. Sample answer:

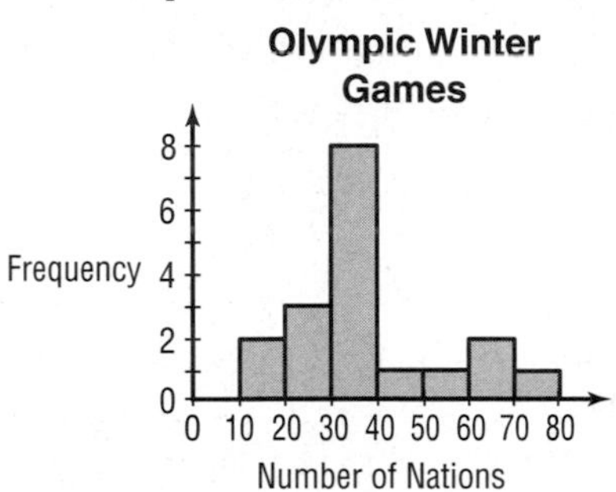

11g. Sample answer:

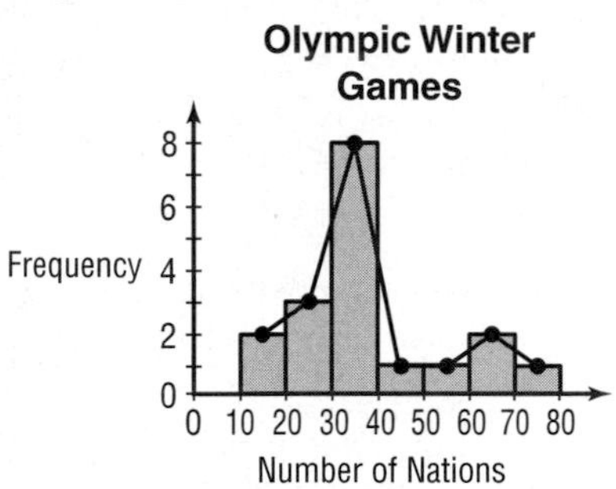

13a.

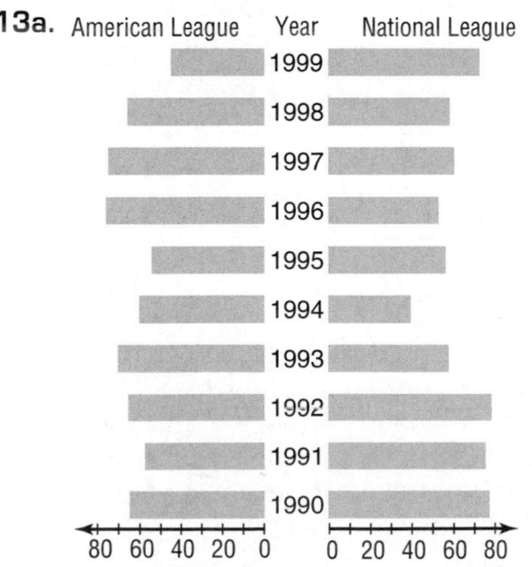

13b. Sample answer:

Stolen Bases	Frequency
30–40	1
40–50	1
50–60	6
60–70	5
70–80	7

13c. Sample answer:

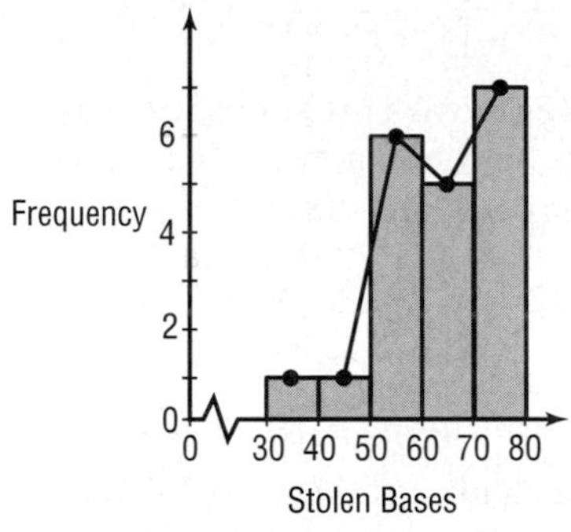

13d. 7 players **13e.** 2 players

15.

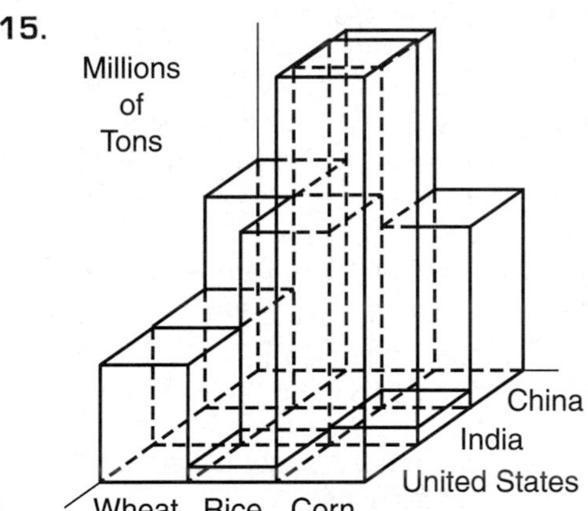

19. $-14c^6d$ **21.**

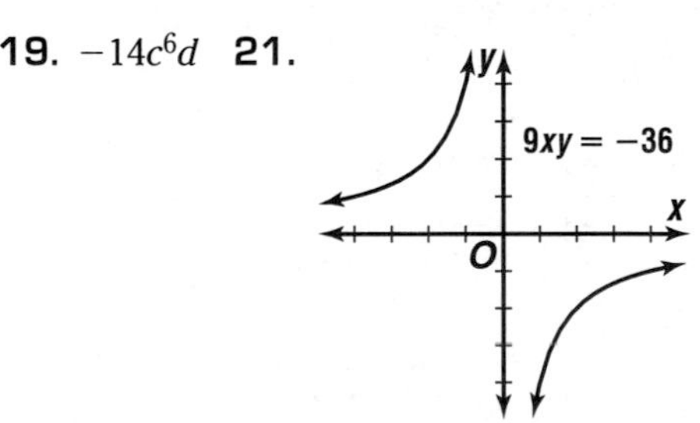

Pages 903–907 Lesson 14-2

5. 30.75; 27.5; 10 **7.** about 10,323; 10,500; 10,700

9a.

stem	leaf
0	6 7 7 7 9
1	3 3 4 4 5 6 7 8 9 9
2	0 0 0 1 1 1 1 1 3 3 8
3	0 1 1 1 2 4 4 6 8
4	1 1 1 2 7

1|3 = 13

9b. 23.55 **9c.** 21 **9d.** 21

9e. Since the mean 23.55, the median 21, and the mode 21 are all representative values, any of them could be used as an average. **11.** 5.4; 3; 3 **13.** 10.75; 11; 5 and 18 **15.** 8.5; 8.5; 6 and 11 **17.** about 45.8; 45; 45 **19.** 1088; 1090; 1180 **21a.** $1485, $3480, $4650, $1650, $2275, $1480, $780 **21b.** $15,800 **21c.** 100 employees **21d.** about $158 **21e.** $150–$160 **21f.** $155 **21g.** Both values represent central values of the data. **23.** 3 **25a.** about 425.6 **25b.** 400–450 **25c.** about 420.5 **27a.** Sample answer: {1, 2, 2, 2, 3} **27b.** Sample answer: {4, 5, 9} **27c.** Sample answer: {2, 10, 10, 12} **27d.** Sample answer: {3, 4, 5, 6, 9, 9} **29a.** about 215.2 **29b.** 200–220 **29c.** about 213 **29d.** about 215.9; 211 **29e.** The mean calculated using the frequency distribution is very close to the one calculated with the actual data. The median calculated with the actual data is less than the one calculated with the frequency distribution. **31a.** $87,800 **31b.** $61,500 **31c.** $59,000 **31d.** mean **31e.** mode **31f.** Median; the mean is affected by the extreme values of $162,000 and $226,000, and only two people make less than the mode. **31g.** Sample answer: I have been with the company for many years, and I am still making less than the mean salary. **33.** He is shorter than the mean (5'11.6") and the median (5'11.5"). **35.** dependent; $\frac{3}{55}$ **37.** $40,305.56 **39.** A

Pages 914–917 Lesson 14-3

5. 16; 8

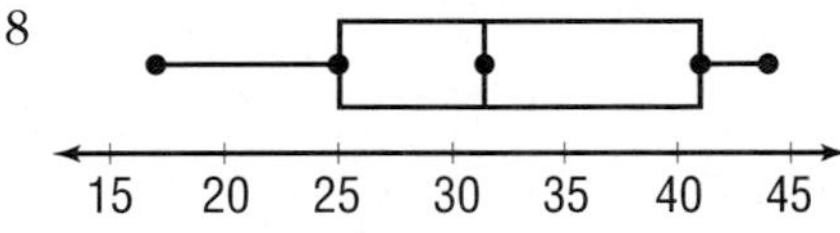

7. 30,250; about 13,226.39

9. 7; 3.5

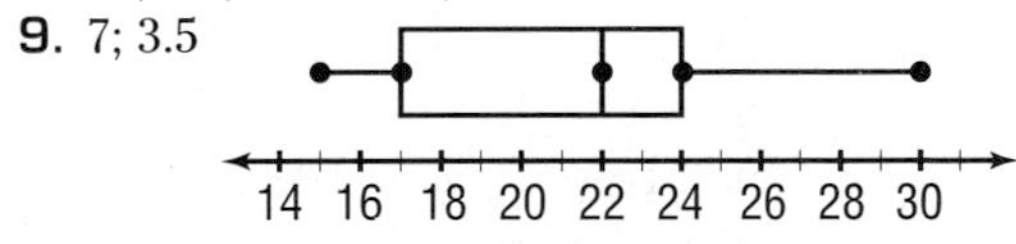

11. 2.9; 4.75

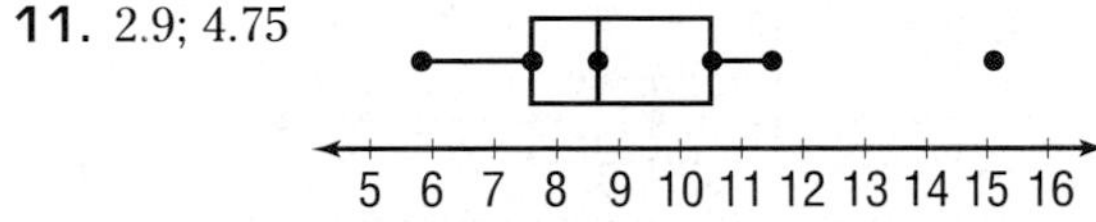

13. 211; about 223.14 **15.** 20.25; about 25.31 **17.** about 19.33; about 6.48 **19.** about 129.65; about 23.29 **21.** Sample answer: {15, 15, 15, 16, 17, 20, 24, 26, 30, 35, 45} **23a.** $2414, $2838, $4147 **23b.** 1733 **23c.** $20,480, $21,914

23d.

0 5000 10,000 15,000 20,000 25,000

23e. about 3507.18 **23f.** about 5643.35 **23g.** The data in the upper quartile is diverse. **25a.** 11 **25b.** about 2.94 **29a.** 45 **29b.** Sample answer: 10 **29c.** Sample answer: 20, 30, 40, 50, 60, 70

29d. Sample answer:

Programs Sold	Frequency
20–30	2
30–40	1
40–50	2
50–60	5
60–70	2

29e. Sample answer:

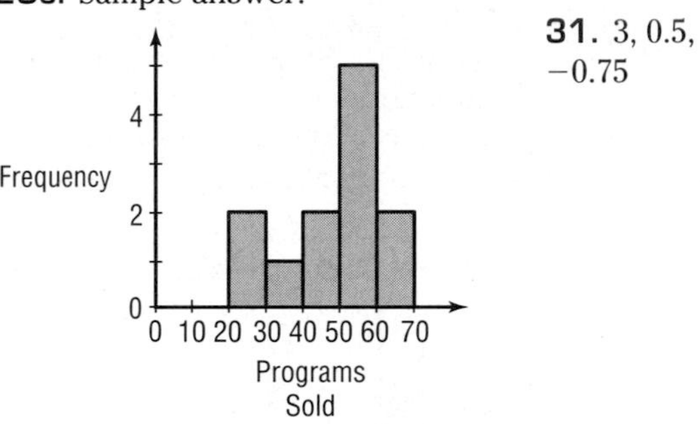

31. 3, 0.5, −0.75

Pages 923–925 Lesson 14-4

7a. 68.3% **7b.** 92.9% **7c.** 22.6–25.4 **7d.** 20.08–27.92

9a.

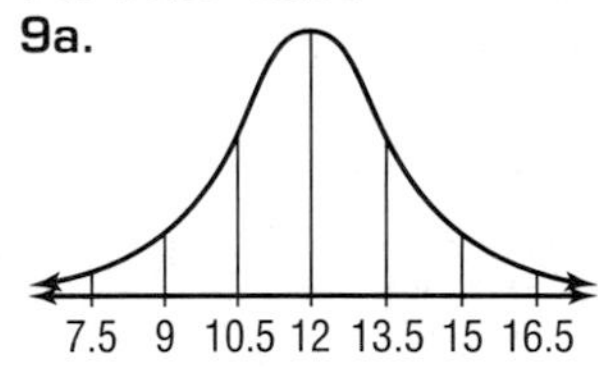

9b. 10.5–13.5 **9c.** 99.7% **9d.** 95.5% **11a.** 79.6–84.4 **11b.** 76.8–87.2 **11c.** 86.6% **11d.** 31.1% **13a.** 66.9% **13b.** 28.6% **13c.** 154 **15a.** $\frac{1}{64}, \frac{3}{32}, \frac{15}{64}, \frac{5}{16}, \frac{15}{64}, \frac{3}{32}, \frac{1}{64}$

15b.

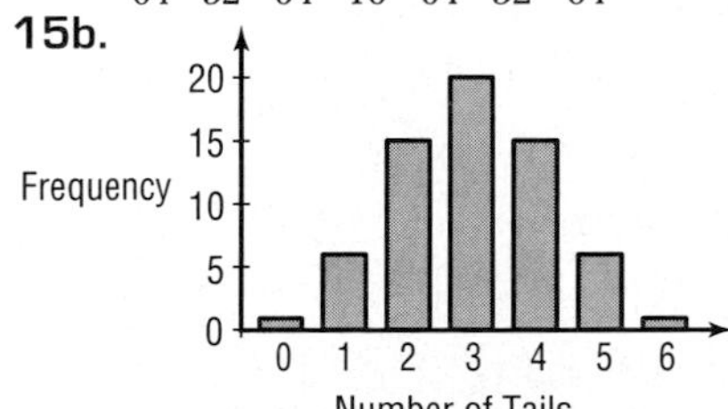

15c. 3 **15d.** about 1.2 **15e.** They are similar.

17a. 0.8% **17b.** 99.2% **19a.** 72 **19b.** 58 **19c.** 68–71 **21a.** about 2.55 mL **21b.** 57.6%

23. about 48.2; 45; 42 **25a.** Sample answer: $y = 0.05x^3 - 2.22x^2 + 29.72x + 366.92$ **25b.** Sample answer: 2553 students

Pages 930–932 Lesson 14-5

5. 7.3 **7.** 42.85–47.15 **9a.** about 0.29 **9b.** about 27.30–27.70 min **9c.** about 26.76–28.24 min **11.** about 0.37 **13.** about 0.53 **15.** about 0.70 **17.** about 333.07–336.93 **19.** about 77.81–82.19 **21.** 67.34–68.66 in. **23.** about 4524.21–4527.79 **25.** about 5.23–5.53 **27.** about 8.9% **29a.** 4.5 **29b.** 338.39–361.61 hours **29c.** Sample answer: 338 hours; there is only 0.5% chance the mean is less than this number. **31a.** about 0.57 **31b.** With a 5% level of confidence, the average family in the town will have their televisions on from 2.98 to 5.22 hours. **31c.** Sample answer: None; the sample is too small to generalize to the population of the city. **33a.** 750 h **33b.** 64 h **35.** 8.25; about 9.59 **37.** $\theta = 45°$ **39.** C

Pages 933–937 Chapter 14 Study Guide and Assessment

1. box-and-whisker plot **3.** standard error of the mean **5.** measure of central tendency **7.** bimodal **9.** histogram **11.** 5
13.

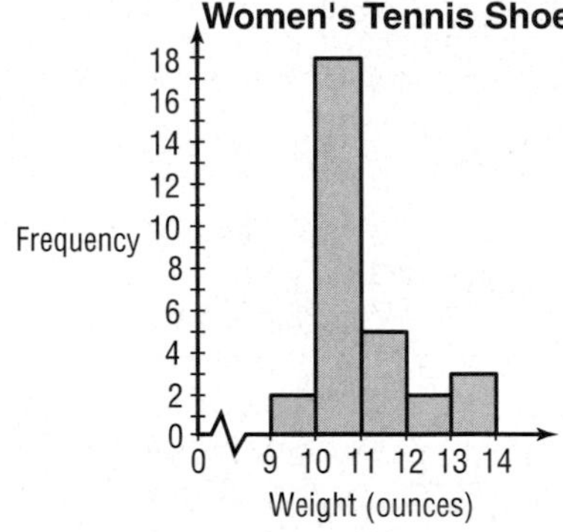

15. 210; 200; 200 **17.** 6.45; 6.5; 6.3 and 6.6 **19.** 3 **21.** 1.6 **23.** 95.5% **25.** 79.75–96.25 **27.** 143.25 **29.** about 0.16 **31.** 1.25 **33.** about 94.53–105.47 **35.** about 35.62–44.38 **37.** about 1.74–1.86 h **39.** about 1.71–1.89 h

41a.

stem	leaf
1	0 3 5 6 7 9
2	1 3 4 5
3	9 9

1|0 = 10

41b. 21.75 **41c.** 20 **41d.** 39

Page 939 Chapter 14 SAT and ACT Practice

1. D **3.** D **5.** E **7.** D **9.** C

Chapter 15 Introduction to Calculus

Pages 946–948 Lesson 15-1

5. −17 **7.** $-\frac{1}{4}$ **9.** $\frac{4}{15}$

11a.

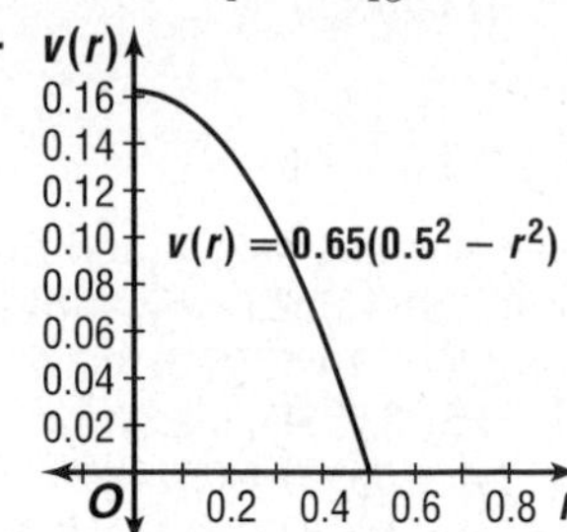

11b. 0 in./s **13.** 0; undefined **15.** −16 **17.** 0 **19.** 10 **21.** $\frac{3}{8}$ **23.** 0 **25.** −1 **27.** 4 **29.** −3 **31.** −1 **33.** 5 **35.** 2 **37.** −0.5 **39.** πa^2; letting c approach 0 moves the foci together, so the ellipse becomes a circle. πa^2 is the area of a circle of radius a. **41.** No; the graph of $f(x) = \sin\left(\frac{1}{x}\right)$ oscillates infinitely many times between −1 and 1 as x approaches 0, so the values of the function do not approach a unique number. **43.** 64 ft/s **45.** 15.684–16.716 mm **47.** $90x^3y^2$ **49.** $\frac{(x-5)^2}{16} + \frac{(y+2)^2}{9} = 1$ **51.** $\langle -7, -6 \rangle$; $\sqrt{85}$ **53.** −1 **55.** $y \to \infty$ as $x \to \infty$, $y \to -\infty$ as $x \to -\infty$ **57.** Yes; opposite sides have the same slope.

Pages 957–960 Lesson 15-2

5. $f'(x) = 2x + 1$ **7.** $f'(x) = -3x^2 - 4x + 3$ **9.** 4 **11.** $F(x) = \frac{1}{4}x^4 + \frac{4}{3}x^3 - \frac{1}{2}x^2 - 3x + C$ **13.** $8 **15.** $f'(x) = 7$ **17.** $f'(x) = -4$ **19.** $f'(x) = 3x^2 + 10x$ **21.** $f'(x) = 2$ **23.** $f'(x) = -6x + 2$ **25.** $f'(x) = 3x^2 - 4x + 5$ **27.** $f'(x) = 6x^2 - 14x + 6$ **29.** $f'(x) = 81x^2 - 216x + 144$ **31.** 3 **33.** 1 **35.** $F(x) = \frac{1}{7}x^7 + C$ **37.** $F(x) = \frac{4}{3}x^3 - 3x^2 + 7x + C$ **39.** $F(x) = 2x^4 + \frac{5}{3}x^3 - \frac{9}{2}x^2 + 3x + C$ **41.** $F(x) = 2x^3 - \frac{5}{2}x^2 - 21x + C$ **43.** $F(x) = \frac{1}{3}x^3 + 2x^2 + x + C$ **45.** Any function of the form $F(x) = \frac{1}{6}x^6 + \frac{1}{4}x^4 - \frac{1}{3}x^3 - x + C$, where C is a constant. **47.** $f'(x) = -\frac{1}{x^2}$ **49a.** $v(t) = 80 - 32t$ **49b.** 48 ft/s **49c.** $t = 2.5$ s **49d.** 103 ft **51a.** $r(p) = p(100 - 2p)$ **51b.** 25 cents
53a.

170 180 190 200 210 220 230 240 250 260

55. $-\frac{1}{27}$ **57.** $\left(x+\frac{1}{6}\right)^2+\left(y-\frac{7}{6}\right)^2=\frac{169}{18}$
59. $x = 8t - 3, y = 3t - 2$ **61.** about 214.9 m
63. D

Pages 966–968 Lesson 15-3
5. $\frac{26}{3}$ units^2 **7.** 72 **9a.** 576 ft **9b.** Yes; integration shows that the ball would fall 1600 ft in 10 seconds of free-fall. Since this exceeds the height of the building, the ball must hit the ground in less than 10 seconds. **11.** 9 units^2 **13.** 4 units^2
15. 312 units^2 **17.** $\frac{208}{3}$ units^2
19. $\lim_{n\to\infty}\sum_{i=1}^{n}\left(\sin i\frac{\pi}{n}\right)\cdot\frac{\pi}{n}$ **21.** $\frac{27}{2}$ **23.** 1088
25. $\frac{40}{3}$ **27.** 10.125 ft^2 **29.** 8000 ft^3
31a.

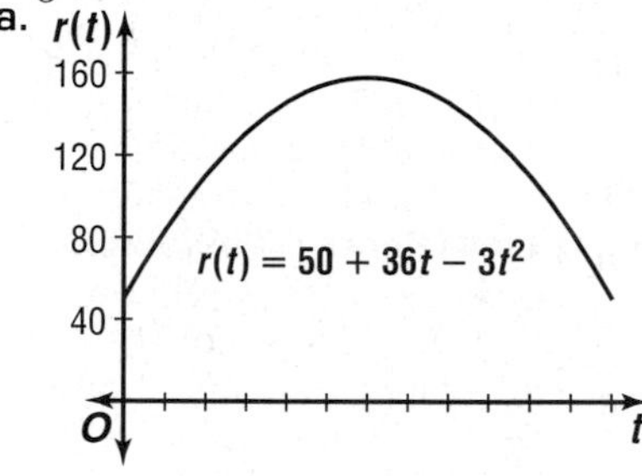

31b. \$1464 **31c.** \$122 **33.** $\frac{1}{2}\pi r^2$ **35.** 0
37. $\vec{\mathbf{u}} = \langle -1, -1, -10\rangle = -\vec{\mathbf{i}} - \vec{\mathbf{j}} - 10\vec{\mathbf{k}}$
39. $\frac{1}{2}, \frac{\pi}{5}$ **41.** C

Pages 973–976 Lesson 15-4
5. $\frac{2}{3}x^3 - 2x^2 + 3x + C$ **7.** $\frac{16}{3}$ units^2 **9.** $\frac{26}{3}$ units^2
11. $\frac{63}{2}$ **13.** 54 **15.** $\frac{1}{6}x^6 + C$ **17.** $\frac{1}{3}x^3 - x^2 + 4x + C$ **19.** $\frac{1}{5}x^5 + \frac{2}{3}x^3 - 3x + C$ **21.** $\frac{1}{3}x^3 - 3x^2 + 3x + C$ **23.** $\frac{13}{3}$ units^2 **25.** 64 units^2
27. $\frac{20}{3}$ units^2 **29.** $\frac{3}{4}$ unit^2 **31.** 686 **33.** 20
35. $\frac{34}{3}$ **37.** $\frac{9}{20}$ **39.** $\frac{413}{6}$ **41.** $\frac{15}{4}$ **43.** 18
45a. 44,152.52; 44,100 **45b.** 338,358.38; 338,350
47a. All are negative. **47b.** $-\frac{22}{3}$ **47c.** $\frac{22}{3}$

49a.

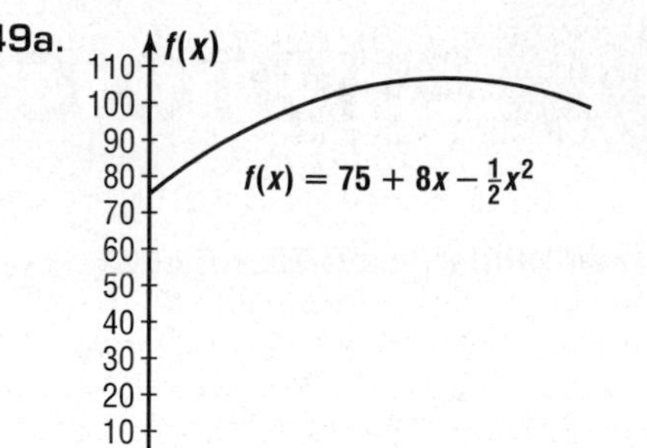

49b. \$93 **49c.** \$105 **51a.** 4.1×10^{16} Nm^2
51b. 6.3×10^9 J **53.** $f'(x) = 12x^5 - 6x$
55. $\frac{253}{4606}$ **57.** $(y + 1)^2 = -12(x - 6)$
59. 35.46 ft/s

Pages 977–981 Chapter 15 Study Guide and Assessment
1. false; sometimes **3.** false; indefinite **5.** false; secant **7.** false; derivative **9.** false; rate of change
11. $-1, -3$ **13.** -1 **15.** 0 **17.** 0 **19.** 4
21. $\frac{1}{5}$ **23.** $f'(x) = 8x + 3$ **25.** $f'(x) = 12x^5$
27. $f'(x) = 6x - 5$ **29.** $f'(x) = 2x^3 - 6x^2 + \frac{1}{3}$
31. $f'(x) = 35x^6 - 75x^4$ **33.** $F(x) = 4x^2 + C$
35. $F(x) = -\frac{1}{8}x^4 + \frac{2}{3}x^3 - \frac{3}{2}x^2 - 2x + C$
37. $F(x) = \frac{1}{3}x^3 - x^2 - 8x + C$ **39.** 4 units^2
41. $\frac{37}{3}$ units^2 **43.** 36 **45.** 28 **47.** $\frac{6}{5}x^5 + C$
49. $\frac{1}{3}x^3 + \frac{5}{2}x^2 - 2x + C$ **51.** $0.0000125m$
53a. 17.6 ft/s^2 **53b.** $v(t) = 17.6t$
53c. $d(t) = 8.8t^2$

Page 983 Chapter 15 SAT and ACT Practice
1. A **3.** D **5.** C **7.** C **9.** D

PHOTO CREDITS

Cover DigitalStock/CORBIS, (inset) Ellen Carey/ Photonica; **xi** Telegraph Colour Library/FPG; **xii** SuperStock; **1** Mark Burnett; **2–3** Yuji Tachibana/ Photonica; **4** A. & J. Verkaik/The Stock Market; **11** Mark Burnett; **12** Arthur Tilley/FPG; **13** William S. Helsel/Stone; **14** Barbara filet/Stone; **19** Bruce Byers/FPG; **20** Bruce Hands/Stone; **23** Marco Ugarte/AP/Wide World Photo; **27** Ryoichi Utsumi/Photonica; **30** Ken Biggs/Stone; **33** George B. Diebold/The Stock Market; **37** Photodisc; **43** SuperStock; **49** Andre Jenny; **50** Geoff Butler; **52** Gordon Wiltsie/Alpenimage, Ltd.; **55** Will & Deni McIntyre/Stone; **66** (t)Lester Letkowitz/The Stock Market, (b)Ron Kimball Photography; **71** John Todd Photography; **73** file photo; **75** NBA Baptist/DUOMO; **77** ©1999 The Cedar Rapids Gazette; **82** Rick Stewart/Allsport; **85** David Aubrey/The Stock Market; **88** The Everett Collection; **93** Photodisc; **95** Kean/ Archive Photos; **97** Courtesy Dr. Margaret H. Wright; **101** Gabe Palmer/The Stock Market; **103** Telegraph Colour Library/FPG; **104** Frederick C. Charles; **105** SuperStock; **107** Tom McCarthy/The Stock Market; **111** Gerald Zanetti/The Stock Market; **112** Michael Nelson/FPG; **114** Meinrad Faltner/The Stock Market; **116** Bud Fowle; **117** Telegraph Colour Library/FPG; **126** John Still/Photonica; **135** Y. Watabe/Photonica; **136** SIU/Peter Arnold, Inc.; **137** Addison Geary/Stock Boston; **142** Tony Freeman/PhotoEdit; **144** James Lemass/Liaison Agency; **145** Ariel Skelley/The Stock Market; **148** Jose L. Pelaez/The Stock Market; **149** Richard Laird/FPG; **154 157** SuperStock; **158** Aaron Haupt; **162** Michael Newman/PhotoEdit; **178** Jake Wyman/Photonica; **179** Spencer Grant/Liaison Agency; **183** SuperStock; **186** O'Brien Productions/CORBIS; **188** SuperStock; **190** Bilderberg/The Stock Market; **195** Jon Gipe/Photonica; **201** CORBIS/Bettmann; **204** Layne Kennedy/CORBIS; **209** David Brownell; **211** Ron Kimball Photography; **213** Elsa Hasch/ Allsport; **215** Vincent LaForet/Allsport; **221** William Taufic/The Stock Market; **222** Ben Radford/Allsport; **227** Aaron Strong/Liaison Agency; **229** Roger-Viollet; **231 234** SuperStock; **236** Mason Morfit/FPG; **242** Mark Burnett; **243** Telegraph Colour Library/FPG; **247** John M. Roberts/The Stock Market; **249** Chuck Pefley/ Stone/PNI; **251** Bill Robbins/Index Stock Imagery/PNI; **255** Porter Gifford/Liaison Agency; **256** Photodisc; **260** Horst Schafer/Peter Arnold, Inc.; **264** Laurence B. Aiuppy/FPG; **274–275** George B. Diebold/The Stock Market; **276** Bruce Carroll/Stone; **281** ESA/K. Horgan/ Stone; **284** Richard Megna/Fundamental Photographs; **289** Paul J. Sutton/DUOMO; **291** Mitchell B. Reibel/ NFL Photos; **292** Malcolm Emmons/NFL Photos; **297** SuperStock; **305** Bill Gallery/Stock Boston; **309** Don Pitcher/Stock Boston; **310** William R. Sallaz/DUOMO; **312** Photodisc; **313** Jonathan Daniel/Allsport; **317** Photo Researchers; **318** Alan Schein/The Stock Market; **319** Lynn Lockwood; **320** Chris Minerva/FPG; **325** Dave Jacobs/Stone; **326** SuperStock; **327** Donald Miralle/Allsport; **331** file photo; **332** Mark Wagner/Stone; **342** Telegraph Colour Library/FPG; **346** Jerry Driendl/FPG; **350** Alain Choisnet/The Image Bank; **354** courtesy The Children's Museum of Indianapolis; **356** SuperStock; **358** Chuck Fishman/ The Image Bank; **361** Doug Martin; **365** Mark Romine/ Liaison Agency; **367** Stock Montage/SuperStock; **371** Visuals Unlimited; **372 375** Richard Megna/ Fundamental Photographs; **378** James Blank/The Stock Market; **385** Daniel J. Cox/Stone; **389** Laurence Parent Photography; **392** Ralph H. Wetmore II/Stone; **402** Aaron Haupt; **403** Photodisc; **405** CORBIS/ Bettmann; **410** Elaine Shay; **417** Aaron Haupt; **420** Jess Stock/Stone; **426** Photodisc; **429** Dee Stuart/The Image Bank; **430** (t)Bob Daemmrich/Stock Boston, (b)Geographix; **434** VCG/FPG; **437** Photodisc, (inset)file photo; **442** Elena Rooraid/PhotoEdit; **443** David Parker/Science Photo Library/Photo Researchers; **445** The Lowe Art Museum, The U. of Miami/SuperStock; **448** Philippe Colombi/Photodisc; **455** Michael Maslan Historic Photographs/CORBIS; **456** Arthur Tilley/FPG; **461** William Whitehurst/The Stock Market; **462** CORBIS/Bettmann; **463** Gilbert Lundt/Temp Sport/CORBIS; **467** Joe Towers/The Stock Market; **474** Ed Wargin/CORBIS; **476** Tony Duffy/NBC/ Allsport; **484** Jim Sugar Photography/CORBIS; **489** Roger Ressmeyer/CORBIS; **493** Tony Freeman/ PhotoEdit; **495** Chris Cheadle/Stone; **498** Jeff Greenberg/PhotoEdit; **499** Ted Horowitz/The Stock Market; **501** David M. Dennis; **504** Michael R. Brown/Florida Today/Liaison Agency; **508** Aaron Rapoport/The Everett Collection; **510** Annie Griffiths Belt/CORBIS; **513** Steve Marcus/Las Vegas Sun; **515** Kunio Owaki/The Stock Market; **518** John Terence Turner/FPG; **522** Jim Sugar/CORBIS; **525** Wynn Miller/Stone; **527** Harry How/Allsport; **530** Beth Bergman Nakamura Photography; **532** D. Young Wolff/PhotoEdit; **533** VCG/FPG; **534** AKG London; **535** Ron Davis/Shooting Star International; **536** Peter Menzel/PNI; **541** Stuart Westmorland/Stone; **547** SuperStock; **550–055** Imtek Imagineering/Masterfile; **552** Philip Gould/CORBIS; **558** Bill Gallery/Stock Boston; **559** The Purcell Team/CORBIS; **560** Stephan Thompson/Pictor; **561** Charles Gupton/Stock Boston/PNI; **565** Julie Houck/Stock Boston; **566** Bobbi Lane/Stone; **572** Robin L. Sachs/PhotoEdit; **573** Ken Ross/FPG; **578** Darwin Dale/Photo Researchers; **584** Aaron Haupt; **585** David Noble/FPG; **593** Ross Harrison Koty/Stone; **598** Roger Ressmeyer/CORBIS; **599** Gregory Sams/Science Photo Library/Photo Researchers; **606** VCG/FPG; **614** Scott T. Smith/CORBIS, (inset)Absaroka Search Dog's Photo Archive, October 1998; **620** Joe Giblin/The Columbus Crew; **622** Bob Daemmrich/Stock Boston; **627** NASA; **629** Pete

Turner/The Image Bank; **630** The Image Bank; **639** Catherine Karnow/CORBIS; **647** CORBIS; **651** James Westwater; **652** Rob Matheson/The Stock Market; **657** NASA; **659** Bryan Yablonsky/DUOMO; **662** John Elk/Stock Boston; **666** World Perspectives/Stone; **668** Stephen Frisch/Stock Boston; **670** Rick Stewart/Allsport; **675** Telegraph Colour Library/FPG; **676** Louis Bencze/Stone; **680** Jon Riley/Stone; **683** David Young-Wolff/PhotoEdit; **684** Photofest; **694** Mike Brown/Liaison Agency, (inset)NASA; **700** SuperStock; **701** Photodisc; **703** Stephen Frisch/Stock Boston; **707** Klaus Hackenberg/Masterfile; **709** Bill Brooks/Masterfile; **712** CORBIS/ Bettmann; **715** Glen Allison/Stone; **720** A. Ramey/Stock Boston; **724** Tom Bean/PNI; **731** VCG/FPG; **732** Hugh Sitton/Stone; **736** Icon Images; **742** Karl Weatherly/CORBIS; **748** Joanna B. Pinneo/PNI; **756–757** Bruce Rowell/Masterfile; **759** Jeff Zaruba/Stone; **765** Nathan Bilow/Allsport; **768** Llewellyn/Pictor; **773** Jon Riley/Stone; **774** Michael Burchfield/AP/Wide World Photo; **779** Mark Burnett; **781** David Batterbury/CORBIS; **794** Telegraph Colour Library/FPG; **798** SuperStock; **800** Llewellyn/Pictor; **804** Laszlo Stern/Photex; **806** Kevin Schafer/CORBIS; **813** George Shelley/The Stock Market; **815** Daniel J. Cox/Stone; **822** Jonathan Meyers/FPG; **827** Tom Stewart/The Stock Market; **836** Kevin Horan/Stone; **839** Arthur Grace/Stock Boston; **841** Francis G. Mayer/CORBIS; **846** Icon Images; **851** Frank Siteman/Stone; **852** Jose L. Pelaez/The Stock Market; **854** Gamma Presse Images/Liaison Agency; **856** Mark Burnett; **859** file photo; **866** Buck Campbell/FPG; **869** Photodisc; **870** Icon Images; **872** Alvin E. Staffan; **875** Bill Horsman/Stock Boston; **878** Jeff Bates; **885** Icon Images; **888** Doug Wilson/CORBIS, (inset)John Biever/Sports Illustrated/TIME; **891** Joe Traver/Liaison Agency; **896** Elena Rooraid/PhotoEdit; **897** Archive Photos/The Image Bank; **900** Hiroyuki Matsumoto/Stone; **905** AFP/CORBIS; **907** Icon Images; **911** Bob Daemmrich/Stock Boston; **916** Rick Rickman/DUOMO; **917** Aaron Haupt; **920** Roy Ooms/Masterfile; **925** Rod Joslin photo; **927** R.W. Jones/CORBIS; **929** Imtek Imagineering/Masterfile; **937** David Madison/Stone; **940** Oliver Meckes/EOS/Gelderblom/Photo Researchers; **943** NASA; **946** G.W. Willis/BPS/Stone; **948** Lonny Kalfus/Stone; **954** Photodisc; **956** Bruce Forster/Stone; **961** Photodisc; **968** Andy Lyons/Allsport; **969** NASA; **970** Vladmir Pcholkin/FPG; **975** Jean Miele/The Stock Market; **976** Tom Stewart/The Stock Market; **981** Imtek Imagineering/Masterfile; **A1** (t)SuperStock, (b) The Lowe Art Museum, The U. of Miami/SuperStock.

INDEX

A

C

INDEX

F

G

H

I

N

Q

R

S

T

INDEX